REACTIONS AND SYNTHESIS IN SURFACTANT SYSTEMS

1. Nonionic Surfactants, *edited by Martin J. Schick* (see also Volumes 19, 23, and 60)
2. Solvent Properties of Surfactant Solutions, *edited by Kozo Shinoda* (see Volume 55)
3. Surfactant Biodegradation, *R. D. Swisher* (see Volume 18)
4. Cationic Surfactants, *edited by Eric Jungermann* (see also Volumes 34, 37, and 53)
5. Detergency: Theory and Test Methods (in three parts), *edited by W. G. Cutler and R. C. Davis* (see also Volume 20)
6. Emulsions and Emulsion Technology (in three parts), *edited by Kenneth J. Lissant*
7. Anionic Surfactants (in two parts), *edited by Warner M. Linfield* (see Volume 56)
8. Anionic Surfactants: Chemical Analysis, *edited by John Cross*
9. Stabilization of Colloidal Dispersions by Polymer Adsorption, *Tatsuo Sato and Richard Ruch*
10. Anionic Surfactants: Biochemistry, Toxicology, Dermatology, *edited by Christian Gloxhuber* (see Volume 43)
11. Anionic Surfactants: Physical Chemistry of Surfactant Action, *edited by E. H. Lucassen-Reynders*
12. Amphoteric Surfactants, *edited by B. R. Bluestein and Clifford L. Hilton* (see Volume 59)
13. Demulsification: Industrial Applications, *Kenneth J. Lissant*
14. Surfactants in Textile Processing, *Arved Datyner*
15. Electrical Phenomena at Interfaces: Fundamentals, Measurements, and Applications, *edited by Ayao Kitahara and Akira Watanabe*
16. Surfactants in Cosmetics, *edited by Martin M. Rieger* (see Volume 68)
17. Interfacial Phenomena: Equilibrium and Dynamic Effects, *Clarence A. Miller and P. Neogi*
18. Surfactant Biodegradation: Second Edition, Revised and Expanded, *R. D. Swisher*
19. Nonionic Surfactants: Chemical Analysis, *edited by John Cross*
20. Detergency: Theory and Technology, *edited by W. Gale Cutler and Erik Kissa*
21. Interfacial Phenomena in Apolar Media, *edited by Hans-Friedrich Eicke and Geoffrey D. Parfitt*
22. Surfactant Solutions: New Methods of Investigation, *edited by Raoul Zana*
23. Nonionic Surfactants: Physical Chemistry, *edited by Martin J. Schick*
24. Microemulsion Systems, *edited by Henri L. Rosano and Marc Clausse*
25. Biosurfactants and Biotechnology, *edited by Naim Kosaric, W. L. Cairns, and Neil C. C. Gray*
26. Surfactants in Emerging Technologies, *edited by Milton J. Rosen*
27. Reagents in Mineral Technology, *edited by P. Somasundaran and Brij M. Moudgil*
28. Surfactants in Chemical/Process Engineering, *edited by Darsh T. Wasan, Martin E. Ginn, and Dinesh O. Shah*
29. Thin Liquid Films, *edited by I. B. Ivanov*
30. Microemulsions and Related Systems: Formulation, Solvency, and Physical Properties, *edited by Maurice Bourrel and Robert S. Schechter*
31. Crystallization and Polymorphism of Fats and Fatty Acids, *edited by Nissim Garti and Kiyotaka Sato*
32. Interfacial Phenomena in Coal Technology, *edited by Gregory D. Botsaris and Yuli M. Glazman*
33. Surfactant-Based Separation Processes, *edited by John F. Scamehorn and Jeffrey H. Harwell*
34. Cationic Surfactants: Organic Chemistry, *edited by James M. Richmond*
35. Alkylene Oxides and Their Polymers, *F. E. Bailey, Jr., and Joseph V. Koleske*
36. Interfacial Phenomena in Petroleum Recovery, *edited by Norman R. Morrow*
37. Cationic Surfactants: Physical Chemistry, *edited by Donn N. Rubingh and Paul M. Holland*
38. Kinetics and Catalysis in Microheterogeneous Systems, *edited by M. Grätzel and K. Kalyanasundaram*
39. Interfacial Phenomena in Biological Systems, *edited by Max Bender*
40. Analysis of Surfactants, *Thomas M. Schmitt* (see Volume 96)

41. Light Scattering by Liquid Surfaces and Complementary Techniques, *edited by Dominique Langevin*
42. Polymeric Surfactants, *Irja Piirma*
43. Anionic Surfactants: Biochemistry, Toxicology, Dermatology. Second Edition, Revised and Expanded, *edited by Christian Gloxhuber and Klaus Künstler*
44. Organized Solutions: Surfactants in Science and Technology, *edited by Stig E. Friberg and Björn Lindman*
45. Defoaming: Theory and Industrial Applications, *edited by P. R. Garrett*
46. Mixed Surfactant Systems, *edited by Keizo Ogino and Masahiko Abe*
47. Coagulation and Flocculation: Theory and Applications, *edited by Bohuslav Dobiáš*
48. Biosurfactants: Production • Properties • Applications, *edited by Naim Kosaric*
49. Wettability, *edited by John C. Berg*
50. Fluorinated Surfactants: Synthesis • Properties • Applications, *Erik Kissa*
51. Surface and Colloid Chemistry in Advanced Ceramics Processing, *edited by Robert J. Pugh and Lennart Bergström*
52. Technological Applications of Dispersions, *edited by Robert B. McKay*
53. Cationic Surfactants: Analytical and Biological Evaluation, *edited by John Cross and Edward J. Singer*
54. Surfactants in Agrochemicals, *Tharwat F. Tadros*
55. Solubilization in Surfactant Aggregates, *edited by Sherril D. Christian and John F. Scamehorn*
56. Anionic Surfactants: Organic Chemistry, *edited by Helmut W. Stache*
57. Foams: Theory, Measurements, and Applications, *edited by Robert K. Prud'homme and Saad A. Khan*
58. The Preparation of Dispersions in Liquids, *H. N. Stein*
59. Amphoteric Surfactants: Second Edition, *edited by Eric G. Lomax*
60. Nonionic Surfactants: Polyoxyalkylene Block Copolymers, *edited by Vaughn M. Nace*
61. Emulsions and Emulsion Stability, *edited by Johan Sjöblom*
62. Vesicles, *edited by Morton Rosoff*
63. Applied Surface Thermodynamics, *edited by A. W. Neumann and Jan K. Spelt*
64. Surfactants in Solution, *edited by Arun K. Chattopadhyay and K. L. Mittal*
65. Detergents in the Environment, *edited by Milan Johann Schwuger*
66. Industrial Applications of Microemulsions, *edited by Conxita Solans and Hironobu Kunieda*
67. Liquid Detergents, *edited by Kuo-Yann Lai*
68. Surfactants in Cosmetics: Second Edition, Revised and Expanded, *edited by Martin M. Rieger and Linda D. Rhein*
69. Enzymes in Detergency, *edited by Jan H. van Ee, Onno Misset, and Erik J. Baas*
70. Structure–Performance Relationships in Surfactants, *edited by Kunio Esumi and Minoru Ueno*
71. Powdered Detergents, *edited by Michael S. Showell*
72. Nonionic Surfactants: Organic Chemistry, *edited by Nico M. van Os*
73. Anionic Surfactants: Analytical Chemistry, Second Edition, Revised and Expanded, *edited by John Cross*
74. Novel Surfactants: Preparation, Applications, and Biodegradability, *edited by Krister Holmberg*
75. Biopolymers at Interfaces, *edited by Martin Malmsten*
76. Electrical Phenomena at Interfaces: Fundamentals, Measurements, and Applications, Second Edition, Revised and Expanded, *edited by Hiroyuki Ohshima and Kunio Furusawa*
77. Polymer-Surfactant Systems, *edited by Jan C. T. Kwak*
78. Surfaces of Nanoparticles and Porous Materials, *edited by James A. Schwarz and Cristian I. Contescu*
79. Surface Chemistry and Electrochemistry of Membranes, *edited by Torben Smith Sørensen*
80. Interfacial Phenomena in Chromatography, *edited by Emile Pefferkorn*

81. Solid–Liquid Dispersions, *Bohuslav Dobiáš, Xueping Qiu, and Wolfgang von Rybinski*
82. Handbook of Detergents, *editor in chief: Uri Zoller*
    Part A: Properties, *edited by Guy Broze*
83. Modern Characterization Methods of Surfactant Systems, *edited by Bernard P. Binks*
84. Dispersions: Characterization, Testing, and Measurement, *Erik Kissa*
85. Interfacial Forces and Fields: Theory and Applications, *edited by Jyh-Ping Hsu*
86. Silicone Surfactants, *edited by Randal M. Hill*
87. Surface Characterization Methods: Principles, Techniques, and Applications, *edited by Andrew J. Milling*
88. Interfacial Dynamics, *edited by Nikola Kallay*
89. Computational Methods in Surface and Colloid Science, *edited by Małgorzata Borówko*
90. Adsorption on Silica Surfaces, *edited by Eugène Papirer*
91. Nonionic Surfactants: Alkyl Polyglucosides, *edited by Dieter Balzer and Harald Lüders*
92. Fine Particles: Synthesis, Characterization, and Mechanisms of Growth, *edited by Tadao Sugimoto*
93. Thermal Behavior of Dispersed Systems, *edited by Nissim Garti*
94. Surface Characteristics of Fibers and Textiles, *edited by Christopher M. Pastore and Paul Kiekens*
95. Liquid Interfaces in Chemical, Biological, and Pharmaceutical Applications, *edited by Alexander G. Volkov*
96. Analysis of Surfactants: Second Edition, Revised and Expanded, *Thomas M. Schmitt*
97. Fluorinated Surfactants and Repellents: Second Edition, Revised and Expanded, *Erik Kissa*
98. Detergency of Specialty Surfactants, *edited by Floyd E. Friedli*
99. Physical Chemistry of Polyelectrolytes, *edited by Tsetska Radeva*
100. Reactions and Synthesis in Surfactant Systems, *edited by John Texter*
101. Protein-Based Surfactants: Synthesis, Physicochemical Properties, and Applications, *edited by Ifendu A. Nnanna and Jiding Xia*

ADDITIONAL VOLUMES IN PREPARATION

Chemical Properties of Material Surfaces, *Marek Kosmulski*

Oxide Surfaces, *edited by James A. Wingrave*

Interfacial Electrokinetics and Electrophoresis, *edited by Ángel V. Delgado*

REACTIONS AND SYNTHESIS IN SURFACTANT SYSTEMS

edited by
John Texter
Strider Research Corporation
Rochester, New York

MARCEL DEKKER, INC.   NEW YORK • BASEL

Library of Congress Cataloging-in-Publication Data

Reactions and synthesis in surfactant systems/edited by John Texter.
p. cm.—(Surfactant science series ; 100)
Includes index.
ISBN 0-8247-0255-7 (acid-free paper)
1. Surface active agents. 2. Surface chemistry. I. Texter, J. (John) II. Surfactant science series ; v. 100.

TP994 .R42 2001
668′.1—dc21

2001028628

This book is printed on acid-free paper.

Headquarters
Marcel Dekker, Inc.
270 Madison Avenue, New York, NY 10016
tel: 212-696-9000; fax: 212-685-4540

Eastern Hemisphere Distribution
Marcel Dekker AG
Hutgasse 4, Postfach 812, CH-4001 Basel, Switzerland
tel: 41-61-261-8482; fax: 41-61-261-8896

World Wide Web
http://www.dekker.com

The publisher offers discounts on this book when ordered in bulk quantities. For more information, write to Special Sales/Professional Marketing at the headquarters address above.

Current printing (last digit):
10 9 8 7 6 5 4 3 2 1

PRINTED IN THE UNITED STATES OF AMERICA

Preface

This book presents a review of extant applications of surfactant technology in chemical synthesis as well as chemical reactions and catalysis. The applications and utilization of surfactants in diverse chemistries, including many areas of organic, inorganic, colloidal, surface, and materials chemistry, cover a very wide gamut. With the possible exception of the journal *Langmuir*, no single journal or professional publication ties together all these areas. It is therefore hoped that this volume will help practitioners and students increase the breadth of their appreciation of surfactant systems in various synthetic and practical applications. The burgeoning arena of template synthesis and nanotechnology indicates that major advances in nanoelectronics will not be forthcoming from further miniaturization programs, but must rely on dramatic improvements in the sophistication with which we wield molecular design and synthesis of surfactants and amphiphiles. Self-assembly principles have been clearly delineated, and the time has come to put this technology to work in chemical synthesis and processing technologies. The success of such endeavors will require closer collaboration among synthetic and physical scientists and engineers, and greater appreciation among practitioners in one discipline for the opportunities and limitations of key related disciplines.

The present volume is extremely wide in scope, covering a broad swath of organic, inorganic, surface, and colloidal chemistry and materials science united by the universal use of surfactant and amphiphile technology in each application. While the volume may seem too synthetic for some readers and too physical for others, it is intended to bring together many related areas and to facilitate closer collaboration between synthetic and physical practitioners in developing new applications and materials. The genesis of this broad arena is found in micellar catalysis, monolayer studies at the water–air interface, and inorganic particle precipitation. Although the general principles of monolayer formation at air–water interfaces and in self-assembled monolayers appear well defined, the controlled multilayer formation of composites, using surfactant templating and molecular recognition principles, is at an early stage in the development of practical applications and processes. Compartmentalization of reactants, as effected by self-organization in surfactant systems, is providing size and morphology control in synthesizing nanoparticulate inorganics and organics. These nanoparticulates are being incorporated into controlled arrays on mesoscales in furthering practical device development. While it was the editor's intention to cover all major application areas, some areas will regrettably have been overlooked, and for these omissions the reader is extended an apology. It is hoped that this book will stimulate readers to invent new applications areas, as naturally occurs in the bridging of disciplines.

The first of the five general parts of this volume comprises surfactant syntheses and electrochemical transformations. Part Two is basically physical–organic chemistry in surfactant systems. Included are micellar catalysis, reaction chemistry in microemulsions, electrocatalysis and electrosynthesis in various surfactant systems, and diverse applications involving emulsions, microemulsions, and vesicles. Parts Three and Four are focused on particle formation, organic and inorganic. Part Three addresses the role of surfactants in organic polymerizations and also provides a thorough review of polymerizable surfactants. Part Four examines particle formation and the role of surfactants in compartmentalizing precipitation chemistries, such as in reverse microemulsions and at surfactant interfaces. The precipitation of inorganic nanoparticles is not treated exhaustively, since an excellent recent volume edited by Tadao Sugimoto, *Fine Particles: Synthesis, Characterization, and Mechanisms of Growth* (Volume 92 of the Surfactant Science Series) treats this topic thoroughly. Part Five addresses syntheses and processing, via self-assembly, molecular recognition, and surfactant templating, on the supramolecular level. These categories cover a variety of themes, including:

Organic chemistry as influenced by surfactants and surfactant assemblies,
The production of organic particulates by chain polymerization in surfactant systems and by polymerizable surfactants,
The synthesis of inorganic nanoparticles using compartmentalized reaction chemistry,
The synthesis of supramolecular assemblies using surfactant assemblies as templates,
The formation of multilayer composites using surfactants and other growth-directing materials.

The interdisciplinary content presented bridges numerous areas of chemistry and materials science, including colloid and surface chemistry, organic synthesis and catalysis, inorganic synthesis and catalysis, electrochemical synthesis and electrocatalysis, inorganic–organic composites, and template-directed synthesis of mesoporous materials. This interconnected bridging makes this volume of interest to chemists and materials scientists of many persuasions. This volume is aimed at practicing industrial and academic scientists and engineers and at students involved in chemical and particle synthesis and processing.

Nearly 100 contributors collaborated in producing the 40 diverse chapters of this volume. I thank each of the authors for their contributions and patience as the volume progressed. I also thank my acquisitions editor, Anita Lekhwani, for her cheerful collaboration, and our production editor, Joseph Stubenrauch, for steadfastly coaxing and urging the volume forward and onward to completion.

John Texter

Contents

Contributors

Nicholas L. Abbott Department of Chemical Engineering, University of Wisconsin–Madison, Madison, Wisconsin

Joseph Akkara* U.S. Army Soldier Systems Center, Natick, Massachusetts

Nihal Aydogan Department of Chemical Engineering, University of Wisconsin–Madison, Madison, Wisconsin

Sukanta Banerjee Department of Chemical Engineering, Tulane University, New Orleans, Louisiana

Turgut Battal Department of Chemical Engineering, The Ohio State University, Columbus, Ohio

Ansgar Behler Cognis Deutschland GmbH, Düsseldorf, Germany

Manfred Biermann Cognis Corporation, Cincinnati, Ohio

Muriel Blanzat Laboratoire IMRCP, UMR CNRS 5623, Université Paul Sabatier, Toulouse, France

Janos B.Nagy Laboratoire de Résonance Magnétique Nucléaire, Universitaires Notre-Dame de la Paix, Namur, Belgium

Lucia Brinchi Department of Chemistry, University of Perugia, Perugia, Italy

Terri Carson[†] Department of Chemistry, University of North Carolina at Chapel Hill, Chapel Hill, North Carolina

Mario Castagnola Department of Chemistry, The Ohio State University, Columbus, Ohio

Carlos C. Co Department of Chemical Engineering, University of Delaware, Newark, Delaware

Nily Dan Department of Chemical Engineering, Drexel University, Philadelphia, Pennsylvania

Renko de Vries[‡] Department of Chemical Engineering, University of Delaware, Newark, Delaware

F. Debuigne Laboratoire de Résonance Magnétique Nucléaire, Universitaires Notre-Dame de la Paix, Namur, Belgium

Current affiliation:
*National Science Foundation, Arlington, Virginia.
[†]Polyurethanes Research and Development, Dow Chemical Company, Freeport, Texas.
[‡]Laboratory of Physical Chemistry and Colloid Science, Wageningen University, Wageningen, The Netherlands.

Joseph M. DeSimone Department of Chemistry, University of North Carolina at Chapel Hill, Chapel Hill, and Department of Chemical Engineering, North Carolina State University, Raleigh, North Carolina

Prabir K. Dutta Department of Chemistry, The Ohio State University, Columbus, Ohio

Jan B. F. N. Engberts Physical Organic Chemistry Unit, Stratingh Institute, University of Groningen, Groningen, The Netherlands

Janos H. Fendler Department of Chemistry, Clarkson University, Potsdam, New York

Jürgen-Hinrich Fuhrhop Institut für Organische Chemie (WE 2), Freien Universität Berlin, Berlin, Germany

D. Neil Furlong RMIT University, Bundoora, Victoria, Australia

Jane R. Galsworthy Infineum UK Ltd., Abingdon, Oxfordshire, England

Raimondo Germani Department of Chemistry, University of Perugia, Perugia, Italy

Roger W. Glyde Infineum UK Ltd., Abingdon, Oxfordshire, England

Christine Göltner* Max-Planck-Institute of Colloids and Interfaces, Golm, Germany

Ferdinand Gonzaga Laboratoire IMRCP, UMR CNRS 5623, Université Paul Sabatier, Toulouse, France

Brian Grady School of Chemical Engineering and Materials Science, University of Oklahoma, Norman, Oklahoma

Franz Grieser School of Chemistry, University of Melbourne, Parkville, Victoria, Australia

Karen Grieve School of Chemistry, University of Melbourne, Parkville, Victoria, Australia

Alain Guyot Laboratory for Polymerization Chemistry and Processes, CNRS, Villeurbanne, France

Jeffrey H. Harwell Institute of Applied Surfactant Research, and School of Chemical Engineering and Materials Science, University of Oklahoma, Norman, Oklahoma

Hans-Peter Hentze Department of Colloid Chemistry, Max-Planck-Institute of Colloids and Interfaces, Golm, Germany

Karlheinz Hill Cognis Deutschland GmbH, Düsseldorf, Germany

Krister Holmberg Department of Applied Surface Chemistry, Chalmers University of Technology, Göteborg, Sweden

J. D. Holmes Department of Chemical Engineering, University College, Cork, Ireland

Duncan C. Hone School of Chemical Sciences, University of East Anglia, Norwich, Norfolk, England

Jutta Hotz Department of Physical Chemistry, University of Basel, Basel, Switzerland

Martin In Complex Fluids Laboratory, CNRS-Rhodia, Cranbury, New Jersey

Glen Irvin Chemical Engineering Department, Tulane University, New Orleans, Louisiana

G. B. Jacobson Los Alamos National Laboratory, Los Alamos, New Mexico

L. Jeunieau Laboratoire de Résonance Magnétique Nucléaire, Universitaires Notre-Dame de la Paix, Namur, Belgium

Vijay T. John Department of Chemical Engineering, Tulane University, New Orleans, Louisiana

Keith P. Johnston Department of Chemical Engineering, University of Texas, Austin, Texas

**Current affiliation*: School of Chemistry, University of Bristol, Bristol, England.

Lana I. Jong* Department of Chemical Engineering, University of Wisconsin–Madison, Madison, Wisconsin

Eric W. Kaler Department of Chemical Engineering, University of Delaware, Newark, Delaware

David Kaplan Department of Chemical Engineering, Tufts University, Medford, Massachusetts

Jason S. Keiper Department of Chemistry, Emory University, Atlanta, Georgia

Jong-Mok Kim Department of Chemistry, Purdue University, West Lafayette, Indiana

Katharina Landfester Department of Colloid Chemistry, Max-Planck-Institute of Colloids and Interfaces, Golm, Germany

Armand Lattes Laboratoire IMRCP, UMR CNRS 5623, Université Paul Sabatier, Toulouse, France

C. T. Lee Department of Chemical Engineering, University of Texas, Austin, Texas

Stephen J. Lee Chemical Sciences Division, U.S. Army Research Office, Research Triangle Park, North Carolina

Yoon Seob Lee Department of Chemical Engineering, The Ohio State University, Columbus, Ohio

G. Li Department of Chemical Engineering, University of Texas, Austin, Texas

Sichu Li Department of Chemical Engineering, Tulane University, New Orleans, Louisiana

Raymond A. Mackay Department of Chemistry, Clarkson University, Potsdam, New York

Srinivas Manne Department of Physics, University of Arizona, Tucson, Arizona

Gary McPherson Department of Chemical Engineering, Tulane University, New Orleans, Louisiana

Wolfgang Meier Department of Physical Chemistry, University of Basel, Basel, Switzerland

Brian J. Melde Department of Chemistry, University of Minnesota, Minneapolis, Minnesota

Jean-Claude Micheau Laboratoire IMRCP, UMR CNRS 5623, Université Paul Sabatier, Toulouse, France

R. Nagarajan Department of Chemical Engineering, The Pennsylvania State University, University Park, Pennsylvania

John O'Haver Department of Chemical Engineering, University of Mississippi, University, Mississippi

Edgar A. O'Rear Institute of Applied Surfactant Research, and School of Chemical Engineering and Materials Science, University of Oklahoma, Norman, Oklahoma

Sijbren Otto† Physical Organic Chemistry Unit, Stratingh Institute, University of Groningen, Groningen, The Netherlands

Eric J. Paul Department of Chemical Engineering, Princeton University, Princeton, New Jersey

Emile Perez Laboratoire IMRCP, UMR CNRS 5623, Université Paul Sabatier, Toulouse, France

Ramannair Premachandran Department of Chemical Engineering, Tulane University, New Orleans, Louisiana

Robert K. Prud'homme Department of Chemical Engineering, Princeton University, Princeton, New Jersey

P. Psathas Department of Chemical Engineering, University of Texas, Austin, Texas

James F. Rathman Department of Chemical Engineering, The Ohio State University, Columbus, Ohio

Hans-Christian Raths Cognis Deutschland GmbH, Düsseldorf, Germany

Current affiliation:
*Department of Chemical Engineering, University of California, Davis, Davis, California.
†University Chemical Laboratory, University of Cambridge, Cambridge, England.

Isabelle Rico-Lattes Laboratoire IMRCP, UMR CNRS 5623, Université Paul Sabatier, Toulouse, France

Brian H. Robinson School of Chemical Sciences, University of East Anglia, Norwich, Norfolk, England

Laurence S. Romsted Department of Chemistry, Rutgers, The State University of New Jersey, New Brunswick, New Jersey

James F. Rusling Department of Chemistry, University of Connecticut, Storrs, Connecticut

Marie-Esther Saint Victor Cognis Corporation, Cincinnati, Ohio

Tetsuo Saji Department of Chemistry and Materials Science, Tokyo Institute of Technology, Tokyo, Japan

Gianfranco Savelli Department of Chemistry, University of Perugia, Perugia, Italy

Andreea Ruxandra Schmitzer Laboratoire IMRCP, UMR CNRS 5623, Université Paul Sabatier, Toulouse, France

Jason Y. Shin Department of Chemical Engineering, University of Wisconsin–Madison, Madison, Wisconsin

Blake A. Simmons Department of Chemical Engineering, Tulane University, New Orleans, Louisiana

Ramsharan Singh Department of Chemistry, The Ohio State University, Columbus, Ohio

Andreas Stein Department of Chemistry, University of Minnesota, Minneapolis, Minnesota

Sönke Svenson Corporate Research and Development–Chemical Sciences, The Dow Chemical Company, Midland, Michigan

Klaus Tauer Max-Planck-Institute of Colloids and Interfaces, Golm, Germany

John Texter Strider Research Corporation, Rochester, New York

Marc Thomalla Université Claude Bernard Lyon I, Villeurbanne, France

David H. Thompson Department of Chemistry, Purdue University, West Lafayette, Indiana

Günter Uphues Cognis Deutschland GmbH, Düsseldorf, Germany

Sharon L. Wells Department of Chemistry, University of North Carolina at Chapel Hill, Chapel Hill, North Carolina

J. L. Wolgemuth Department of Physics, University of Arizona, Tucson, Arizona

R. K. Workman Department of Materials Science and Engineering, University of Arizona, Tucson, Arizona

M. Z. Yates Los Alamos National Laboratory, Los Alamos, New Mexico

Weilie Zhou Advanced Materials Research Institute, University of New Orleans, New Orleans, Louisiana

1

Industrial Surfactant Syntheses

ANSGAR BEHLER Cognis Deutschland GmbH, Düsseldorf, Germany

MANFRED BIERMANN Cognis Corporation, Cincinnati, Ohio

KARLHEINZ HILL and HANS-CHRISTIAN RATHS Cognis Deutschland GmbH, Düsseldorf, Germany

MARIE-ESTHER SAINT VICTOR Cognis Corporation, Cincinnati, Ohio

GÜNTER UPHUES Cognis Deutschland GmbH, Düsseldorf, Germany

I. INTRODUCTION

For over 2000 years, humankind has used surfactants or surface-active ingredients in various aspects of daily life, for washing, laundry, cosmetics, and housecleaning. In the United States alone, over 10 billion pounds of detergents are used annually. Anionic surfactants represent 70–75% of the detergent market. Natural soaps are the oldest anionic surfactants and are used mainly in personal care and in the detergent industries. However, the development of more economical processes for the manufacture of surfactants has contributed to an increased consumption of synthetic detergents. Nonsoaps or synthetic detergents account for 84% of the total detergent market. In 1996, over 5 billion pounds of nonsoap surfactants were produced. In the Asia-Pacific region, the total surfactant consumption grows at an annual rate of 3.9% with a projection of 5.8 million tons in 2010. From a global perspective, the consumption and proportion of surfactants exhibit a different pattern for the North American and Western European regions compared with the Asia-Pacific region or Japan in particular. However, the major surfactants common (with respect to detergent) to all regions are linear alkylbenzene sulfonates (LASs), alcohol ether sulfates (AESs), aliphatic alcohols (AEs), alcohol sulfates (ASs), and soap.

In the past decades, new surfactants have proliferated mainly as nonionic or nonsoap surfactants offering unique properties and features to both industrial and household markets. Nonsoap surfactants are widely used in diverse applications such as detergents, paints, and dyestuffs; as specialty surfactants in home and personal care; and in the cosmetics and pharmaceutical industries. Since the 1960s, biodegradability and a growing environmental awareness have been the driving forces for the introduction of new surfactants. These forces continue to grow and influence the surfactant market and production. A new class of surfactants, carbohydrate-based surfactants, has gained significant interest and increased market share. Consequently, sugar-based surfactants, such as alkyl polyglycoside (APG*), are used as a replacement for polyoxyethylene alkylphenols (APEs) where biodegradability is a concern. They represent a new concept in compatibility and care.

*APG is a registered trademark of Cognis Deutschland GmbH.

Nonetheless, over 35 different types of surfactants are produced and used commercially in the formulation of home care, personal care, and industrial products. Contrary to many textbooks that elaborate on surfactant physical properties or formulation guidelines, this chapter approaches the surfactant topic from both synthesis and manufacturing perspectives. It offers a comprehensive overview of the most commonly used industrial surfactants with respect to their synthesis and manufacturing processes; their reactions and applications; and their physical, ecological, and toxicological properties. A concise and thorough description of the most pertinent synthesis routes is presented for the major types of surfactants predominantly used in the home and personal care industry. These surfactants are primarily anionic, nonionic, cationic, and amphoteric.

Also reviewed is the synthesis of surfactants derived from carboxylation, sulfation, and condensation of fatty acid and phosphoric acid derivatives. The most commonly used anionic surfactants are LASs, ASs, and AESs.

Nonionic surfactants are produced mainly by alkoxylation technology, although amine oxides under alkaline conditions are also classified as nonionic. Section III discusses the synthesis, production, and applications of the most commonly used ethoxylated surfactants such as alcohol ethoxylates, nonyl phenol ethoxylates and fatty acid ethoxylates, fatty amine oxides (FAOs), and fatty alkanolamides (FAAs).

Section IV is concerned with a class of biodegradable and highly compatible carbohydrate- or sugar-based surfactants such as sorbitan esters, sucrose esters, and glucose-derived esters. Their syntheses encompass a significant list of renewable raw materials, including sucrose from sugar beet or cane, glucose from starch, and sorbitol as the hydrogenated glucose derivative. The most commonly used sugar-based surfactants, such as APG and fatty acid glucamides (FAGs), are reviewed in depth.

The syntheses of cationic and amphoteric surfactants are reviewed in Sections V and VI, respectively. Cationic surfactants contain exclusively a quaternary tetracoordinated nitrogen atom (quaternary ammonium compounds). They are widely used as textile softeners in laundry formulations and in flotation. Amphoteric surfactants (including betaines) exhibit a zwitterionic character, i.e., they possess both anionic and cationic structures in one molecule.

Recent progress in the surfactant field focuses on polymeric, splittable, gemini, multifunctional, and biosurfactants.

II. ANIONIC SURFACTANTS

A. Carboxylates

1. Soaps

Soaps represent the oldest known class of surfactants. They have been known for at least 2300 years. In the period of the Roman Empire, the Celts produced soap from animal fats and plant ashes, which served as alkali. They gave this product the name "saipo" from which the word "soap" is derived [1]. The chemical nature of soaps, as alkali salts of long-chain fatty acids, was recognized many centuries later by Chevreul. He showed in 1823 that the process of saponification is a chemical process of splitting fat into the alkali salt of fatty acid and glycerine.

The term soap is mainly applied to the water-soluble alkali metal salts of fatty acids, although ammonia or triethanol amine salts are also used as technical soaps. Salts of fatty acids with heavy metals or with alkaline earth metals are water insoluble and are termed "metallic soaps." They possess no detergent or soaplike properties.

Generally, three different processes are suitable for the large-scale production of soaps:

1. The saponification of neutral oils (triglycerides)

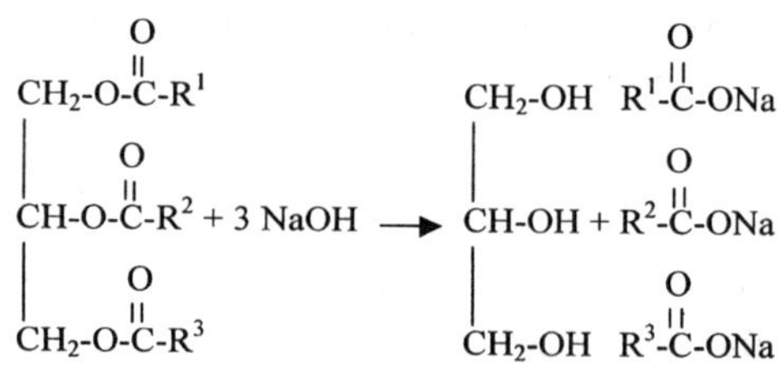

2. The saponification of the fatty acids obtained from fats and oils

$$R\text{-}\overset{O}{\overset{\|}{C}}\text{-}OH + NaOH \longrightarrow R\text{-}\overset{O}{\overset{\|}{C}}\text{-}ONa + H_2O$$

3. The saponification of the fatty acid methyl esters derived from fats and oils

$$R\text{-}\overset{O}{\overset{\|}{C}}\text{-}OCH_3 + NaOH \longrightarrow R\text{-}\overset{O}{\overset{\|}{C}}\text{-}ONa + CH_3OH$$

The most important industrial process is the saponification of the neutral oils and of the fatty acids. Both processes may be run in either batch or continuous mode. All types of fats and oils can be used in this process. The most important ones are tallow and coconut oil.

The main application of soap is in the personal care industry, followed by the detergent industry.

For the preparation of high-grade soaps, the basic soap must be very pure and free of unpleasant odors. The color quality and the odor of the basic soap are determined by the content of by-products. These impurities are of different origins:

1. Natural constituents of fats and oils (waxes, phosphatides, cerebrosides, sterols, fat-soluble vitamins, diol lipids, carotenoids, etc.)
2. Substances generated by oxidation processes during storage of the raw materials
3. Substances generated in the manufacturing process

By using special purification steps during the production process, these by-products are eliminated.

2. Ether Carboxylic Acids

The sensitivity of soaps to water hardness is a big disadvantage for many applications. In contrast, the alkyl polyoxyethylene carobxylic or alkyl (poly-1-oxapropen) oxaalkene carboxylic acids, or short ether carboxylic acids, exhibit an extreme water hardness resistance combined with good water solubility.

The starting material for ether carboxylic acids is fatty alcohol ethoxylates. Conversion to the ether carboxylic acid can be carried out by three different routes (Fig. 1).

The fatty alcohol ethoxylates can be carboxymethylated by reaction with monochloroacetic acid in the presence of sodium hydroxide [2] or through terminal oxidation of the fatty alcohol ethoxylate [3–5]. The ether carboxylic acid can also be synthesized by the addition of a vinylic system, i.e., acrylonitrile, to an oxyethylated fatty alcohol and subsequent hydrolysis. Ether carboxylic acids are temperature stable and resistant to alkali and hydrolysis, even under strong acidic or alkaline conditions.

Because of their advantageous ecological, toxicological, and physicochemical properties and good compatibility with representatives of all surfactant classes, ether carboxylic acids can be applied effectively in many fields. They are used in washing and cleaning agents as well as cosmetics. They are utilized as emulsifying and auxiliary agents in the textile, printing, paper, plastics, metalworking, and pharmaceutical industries [6].

The salts of ether carboxylic acids with a high degree of ethoxylation are considered to be very mild and skin-compatible surfactants. Therefore, they are particularly suitable for applications in cosmetics [7].

Ether carboxylic acids are also used for manual dishwashing detergents, carpet cleaners, and other household products [8].

In the plastics industry, ether carboxylic acids are employed as auxiliary agents for emulsion polymerization and as antistatic agents (or antistats).

They also exert a good corrosion-inhibiting effect and, therefore, ether carboxylic acids are also used as emulsifiers in drilling, rolling, and cutting oil emulsions and cooling lubricants [9].

B. Sulfonation Technology

The technology of sulfonation (C—S coupling reaction) and sulfation (C—O—S coupling reaction) can be realized by various processes. Only industrial processes that are of significant importance are discussed here. Those are sulfonation and sulfation or sulfoxidation and sulfochlorination (see Alkane Sulfonates).

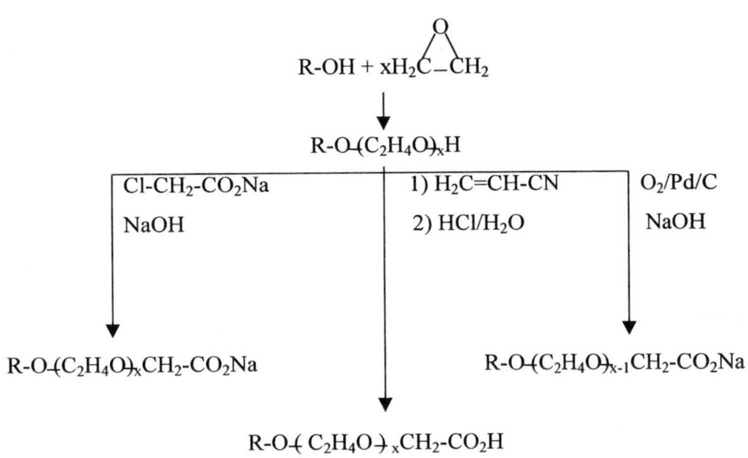

FIG. 1 Synthesis of ether carboxylic acids.

(*a*) *Sulfonation with Sulfur Trioxide.* Sulfonation with SO_3/air raised from sulfur has become the predominant technology for manufacturing sulfonation products [10–12]. The diluted SO_3 gas is generated by burning sulfur, followed by catalytic oxidation of SO_2 at a vanadium pentoxide contact (conversion). Alternative sources for gaseous SO_3 are liquid SO_3 and oleum (65%), which is not only hazardous in transport, handling, and storage but also more expensive.

The sulfonation is done mostly in falling-film reactor with 3–5% SO_3 in dry air (dew point < −60°C). A falling–film reactor, such as the Ballestra SULFUREX F system (Fig. 2), is a bundle of about 6-m-long reaction tubes in a shell in which heat exchange takes place with cooling water.

The organic raw material is fed to the top of the reactor and is distributed on the inner walls of the reaction tubes by identical annular slots. The contact time with SO_3 is relatively short to prevent undesired color-developing side reactions. After removal of the exhaust gas with a gas-liquid separator, the sulfonic acid is generally transferred to a neutralization loop. In some cases in which aging of the raw sulfonic acid is necessary to achieve a high degree of sulfonation (LAS, estersulfonates), a residence time is achieved by using an aging vessel or loop. Falling-film reactors of different designs are now available on the market.

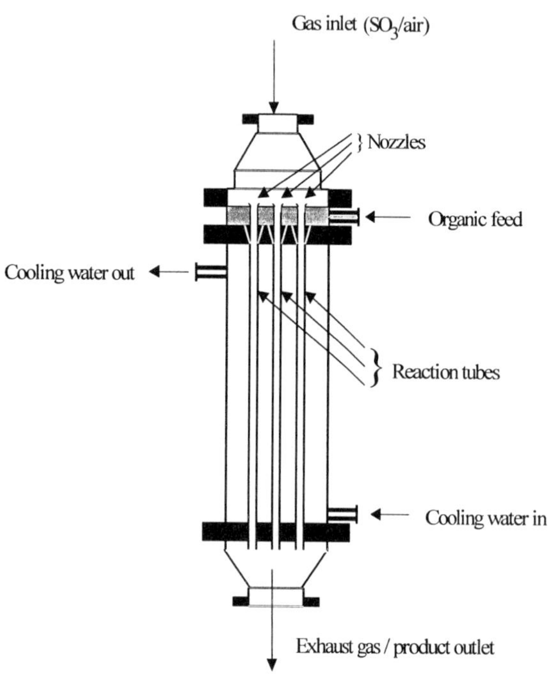

FIG. 2 Multitube sulfonation reactor.

(*b*) *Sulfonation with Chlorosulfonic Acid* [13]. Chlorosulfonic acid (CSA) is used in batch or continuous processes for the production of sulfates or ether sulfates on a relatively small scale:

$$ROH + ClSO_3H \rightarrow ROSO_3H + HCl$$

The HCl must be removed by degassing and absorbing; the sulfonic acid ester can be neutralized with the desired bases. This chemistry requires glass-lined steel or glass equipment. In contrast to falling-film reactors, the sulfation equipment takes less space and investment. The costs and handling of CSA are disadvantageous compared with those of sulfur trioxide.

(*c*) *Sulfonation with Amidosulfonic Acid ("Sulfamic Acid").* Amidosulfonic acid is a relatively seldom used sulfation agent. It is used, for example, to sulfate alkylphenol derivatives to avoid ring sulfonation by-products:

$$C_{12}H_{25}{-}C_6H_4{-}(O{-}CH_2{-}CH_2)_6{-}OH + H_2NSO_3H$$
$$\rightarrow C_{12}H_{25}{-}C_6H_4{-}(O{-}CH_2{-}CH_2)_6{-}OSO_3NH_4$$

Another example is the production of aliphatic ether sulfates [14].

1. Alkylarylsulfonates [10–12,15,16]

Linear alkylbenzene sulfonates (LABSs, LASs) or general alkylbenzene sulfonates (ABSs) have a long history, going back to the 1930s.

Using a Friedel-Crafts reaction of olefins with benzene in the presence of either aluminum chloride or hydrogen fluoride made alkylbenzene an economically attractive raw material for the synthesis of this class of anionic surfactant, which developed into the "workhorse" of detergents.

The first market product was tetrapropylenebenzenesulfonate (TPS) derived from α-dodecylene synthesized by tetramerization of propylene, giving a branched alkyl chain. Because of the insufficient biological degradability of the highly branched alkyl chain, which led to contamination of surface waters, TPS was replaced by the biologically more degradable LAS.

The linear alkylbenzene is structurally a nonuniform product. The most common product has a carbon number range of the alkyl chain from C_{10} to C_{13} (Scheme 1). The phenyl isomer distribution occurring therein is determined by the choice of catalyst. With use of $AlCl_3$, the content of 2-phenyl isomers is approximately 30% in mixture with 3-, 4-, 5-, and other phenyl isomers. In products of HF-catalyzed reaction, the content of 2-phenyl isomers is significantly lower at about 20%.

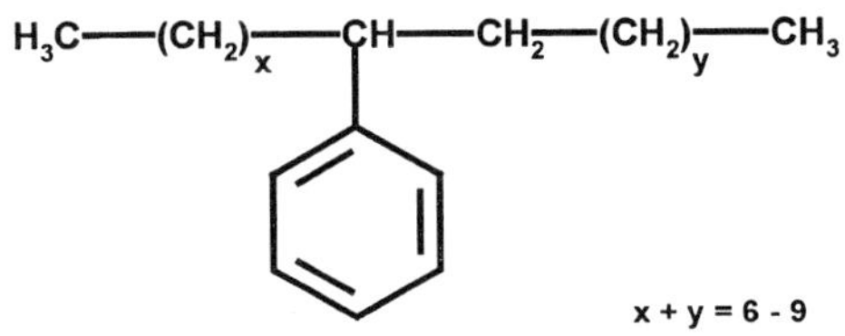

LAB

SCHEME 1

The sulfonation of alkylbenzenes [17–21] can be handled with oleum, sulfuric acid, or gaseous sulfur trioxide. The sulfonate group is introduced into the benzene ring primarily in the *p*-position. The process may be operated as either a batch or continuous process. The industrial sulfonation of LAB is accomplished today frequently with SO_3 in multitube falling-film reactors on a highly economical scale. The continuous sulfonation of alkylbenzene sulfonates is carried out at 40–50°C with a molar excess of 1–3% sulfur trioxide, diluted to 5–7 vol% in dry air.

During the sulfonation step, the desired sulfonic acids are not the only products. Anhydrides, called pyrosulfonic acids, are also formed as by-products (Scheme 2).

The content of alkylbenzenesulfonic acid can be increased with a postreaction (aging) step, which is necessary for a sufficient degree of sulfonation (Scheme 3). During aging, the pyrosulfonic acids can react with further alkylbenzene, sulfuric acid, or traces of water, increasing the content of alkylbenzenesulfonic acid.

Another undesirable side reaction is the formation of sulfones, which are part of the "free oil" content of LAS (Scheme 4). The reaction mixture is neutralized with sodium hydroxide solution. Aqueous pastes with up to 60% active substance content can be produced (Scheme 5). Other side reactions, for example, oxidation, whose chemistry is hard to state more precisely, give dark-colored by-products that can require bleaching of the aqueous LAS paste.

Unlike other sulfonation or sulfation products, the crude alkylbenzenesulfonic acid, although very corrosive, can be stored in the acid form. The anhydrides are converted to alkylbenzenesulfonic acid by addition of 1–2% water at 80°C in order to stabilize the product.

LAS is a good soluble anionic surfactant mainly for use in detergents [22]. It is moderately sensitive to water hardness. Most formulations contain surfactant mixtures in order to decrease sensitivity to water hardness and to enhance foam stability. The combinations are, for example, LAS with alkyl(ether) sulfates and/or noinionics.

LAS is completely biodegradable under aerobic conditions, resulting in high environmental safety. Degradation under anaerobic conditions (the relevance of which has been controversial [23–31]) is, as for sulfonate structures, poor. As LAS is and will continue to be the major component of detergent systems because of its good price/efficiency ratio, more environmental data are available for it than for any other surfactant (European Center for Ecotoxicology and Toxicology of Chemicals, ECETOC Technical Report No. 51, Brussels, 1992).

The processing of LAS toward compact detergent powders will have to be revised because of the sticky behavior of water-free products. Combinations of LAS

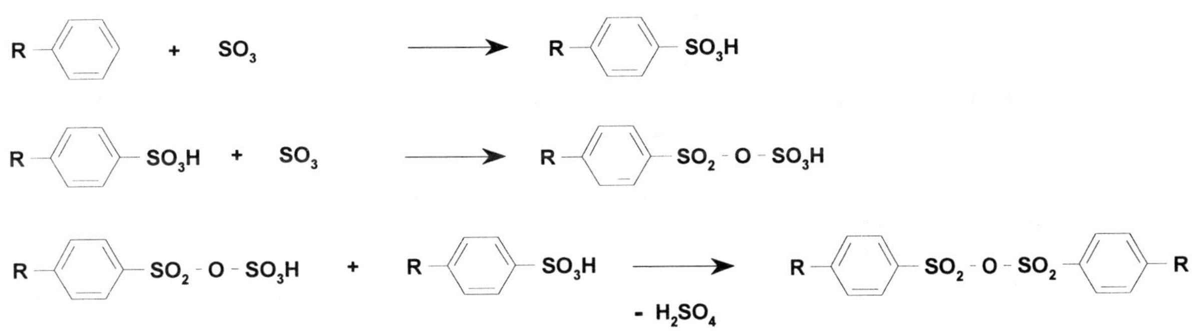

SCHEME 2

SCHEME 3

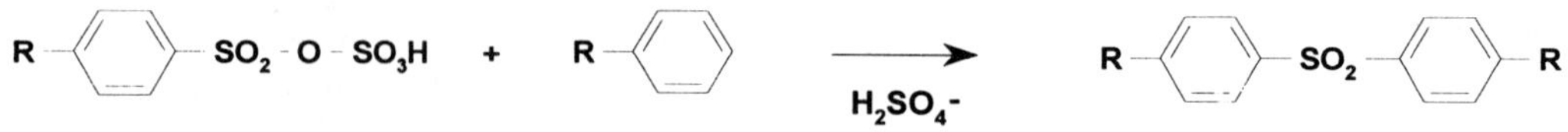

SCHEME 4

with alkyl sulfates are already employed because of the good crystallization of alkyl sulfates. Extension of the application of LAS to cosmetics was suggested by the use of the milder Mg salts [32].

2. Aliphatic Sulfonates

(*a*) *Alkane Sulfonates.* Sulfoxidation and sulfochlorination are the core technologies for the preparation of alkane sulfonates. Sulfoxidation, the older process, is more important than sulfochlorination.

Sulfoxidation. Sulfoxidation [33–37] is a photochemically induced process starting with sulfur dioxide, oxygen, and an *n*-alkane, normally in the range C_{12}–C_{18} or C_{14}–C_{17}. The radical chain reaction gives many isomers with mainly secondary sulfonate groups. The following sequence explains the reaction steps:

$$SO_2 + RH \xrightarrow[UV]{h\nu} R\bullet + HSO_2$$
$$R\bullet + SO_2 \rightarrow RSO_2\bullet$$
$$RSO_2\bullet + O_2 \rightarrow RSO_2OO\bullet$$
$$RSO_2OO\bullet + RH \rightarrow RSO_2OOH + R\bullet$$
$$RSO_2OOH + SO_2 + OH_2 \rightarrow RSO_3H + H_2SO_4$$

In practice, a paraffin-water mixture is contacted with SO_2 gas and oxygen at 30–40°C under irradiation with ultraviolet (UV) lamps. The process is run with an excess of paraffin in order to avoid the formation of multisubstituted products.

The excess of paraffin can be removed from the reaction mixture after cooling (with a separator) and can be recycled. Different work-up procedures have been established: the "Hoechst Light Water Technology" and the Hüls process. Both processes have in common separation and recirculation of the paraffin from the crude reaction product by extraction. Also, the sulfur dioxide can be removed by degassing and washing in order to be recycled. The sulfuric acid can be separated by phase separation or extraction.

The final product has to be bleached and neutralized, giving a yellowish paste with about 65% active matter.

Sulfochlorination. The sulfochlorination technology [37,38] is used for the conversion of paraffins or alkanes to alkane sulfonates. In a photochemically induced reaction, the paraffin is contacted by dry sulfur dioxide and chlorine:

$$RH + SO_2 + Cl_2 \xrightarrow[20-40°C]{h\nu\ (>400\ nm)} RSO_2Cl + HCl$$

The resulting sulfochloride is a mixture of approximately 94% mono- and 6% disulfochloride. In a subsequent hydrolysis step with NaOH solution at 80°C, the sulfonates are formed:

$$R{-}SO_2Cl + 2NaOH \rightarrow R{-}SO_3Na + NaCl$$

Alkane sulfonates are highly soluble surfactants and are preferably used in liquid products or concentrates. The trend to use renewable raw materials has reduced their use in household products to some extent. Typical applications are in detergents, personal care products, cleaners, and dishwashing detergents.

As is common to all sulfonates, alkane sulfonates are easily biodegradable under aerobic conditions [39] but fail under anaerobic conditions.

(*b*) *Olefin Sulfonates.* Alpha olefin sulfonates (AOSs) [40,41] are, in contrast to internal olefin sulfonates (IOSs), the most important products of this class. AOSs are mainly based on C_{12}–C_{18} alpha-olefins derived from ethylene oligomerization (Ziegler process). There is considerable interest in this class of surfactants today because they are derived from low-priced raw materials coupled with an inexpensive sulfonation process.

The most important sulfonation process works with SO_3 (Fig. 3), which adds in the primary step to the double bond of the olefin, giving a ring-structured sultone intermediate. Through different reaction steps of sultone formation, elimination, rearrangements, transi-

SCHEME 5

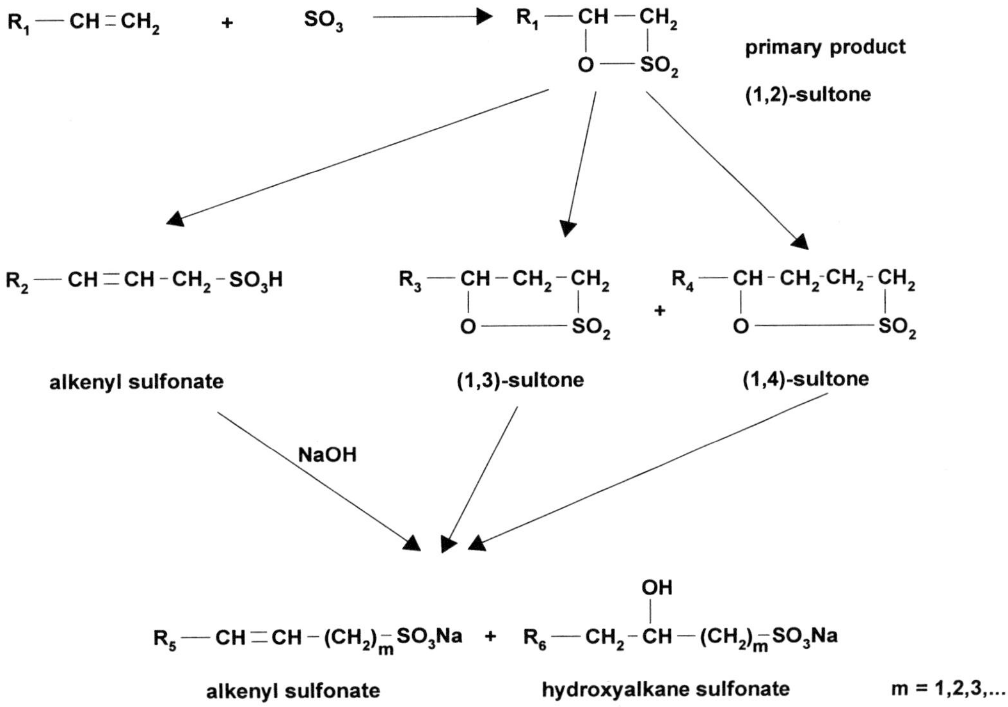

FIG. 3 Sulfonation of α-olefins with gaseous SO_3.

tions, and hydrolysis, a mixture of hydroxyalkane sulfonates and alkene sulfonates is obtained in a ratio of 30:70.

As far as surfactant properties are concerned, the alkenyl sulfonate is the more desirable structure. In any event, bleaching of the final product is necessary because of oxidation side reactions.

Because of the discussion of sultone intermediates [42], the use of AOS was limited. Through modern analytical methods, the sultones can be quantified, and the production process has been modified by adding a hydrolysis step, so that sultones need not be mentioned as a noteworthy component of AOS. The product can be regarded as safe for the consumer and the environment. AOS with a C_{14-16} alkyl chain is better foaming than C_{16-18} AOS. The sulfonate group gives high stability over a wide pH range. AOS is sensitive to water hardness. Typical applications are in detergents, shampoos, and cleansers [43–47].

α-Sulfo fatty acid methyl esters (MESs). Starting materials for α-sulfo fatty acid esters are fatty acid methyl esters, which are available from the transesterification of natural oils and fats. This low refined oleochemical raw material is sulfonated with SO_3/air. Ester sulfonates [48–59] are economically interesting surfactants, showing good detergency for the C_{16}–C_{18} MES event at low temperatures.

The sulfonation is quite a complex reaction (Scheme 6). Beside the desired ester sulfonate, MES contains methyl sulfate, α-sulfo fatty acids, and soap in amounts that depend on the manufacturing process. The first step is the insertion of SO_3 into the ester linkage (Fig. 4).

The primary reaction product, a mixed anhydride, can take up a second molecule of SO_3 via its enol form. The anhydride carrying two SO_3 units can lose one SO_3, which can react with another molecule of methyl ester. This "storage" of SO_3 is the reason for the necessary excess of SO_3 in this sulfonation reaction. The whole reaction sequence takes more time than is available with a falling-film reactor. Therefore, in order to achieve a high degree of sulfonation, aging is necessary. During the subsequent neutralization, the inter-

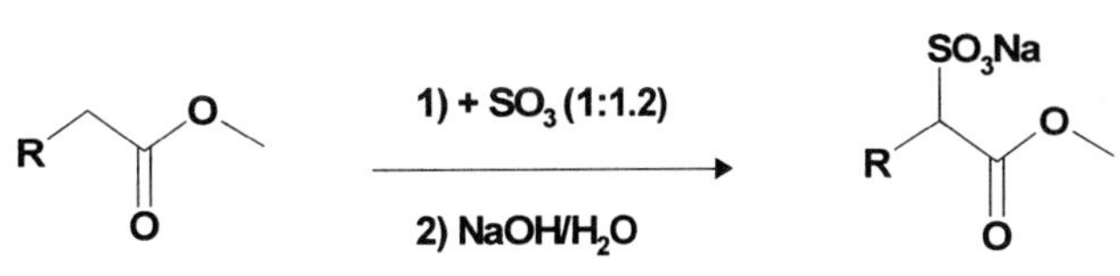

SCHEME 6

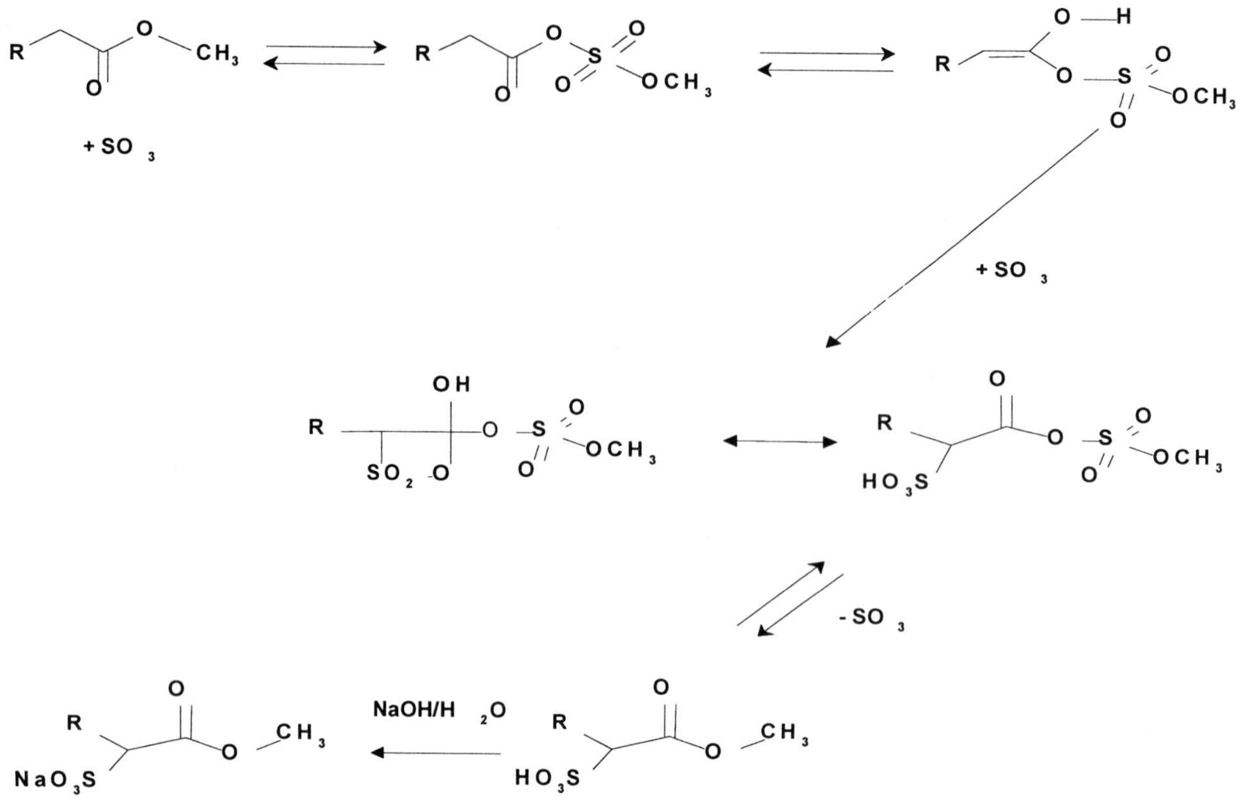

FIG. 4 Reaction mechanisms of the sulfonation of esters.

mediate anhydride of the α-sulfo acid is hydrolyzed to the disodium salt. To avoid this, hydrolysis of the α-sulfo acid anhydride with methanol is carried out. To achieve good color, bleaching of the sulfonic acid with hydrogen peroxide is necessary. The color of MES is dependent on the ester raw material. Raw materials with low iodine values (<0.1%) are suitable for the process.

The same reaction can be carried out with fatty acids, giving directly the disodium salt of the α-sulfo acid, which is of no industrial importance.

Sulfosuccinates. Sulfosuccinates (sulfosuccinic acid esters) are anionic surfactants based on maleic acid anhydride. One distinguishes mono- and dialkyl esters of the sulfosuccinic acid (Fig. 5).

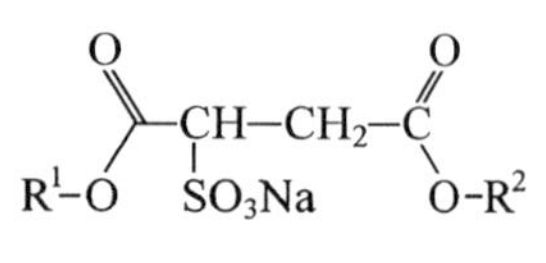

FIG. 5 Structure of sulfosuccinic acid esters. R^1, R^2 = H, alkyl, POE-alkyl.

Both mono- and diesters are obtained in a two-step process. In the first reaction step, maleic acid anhydride is esterified with compounds containing hydroxyl groups to the mono- or diester (Fig. 6). Diesters are mainly produced with alcohols. For monoesters, many different raw materials with hydroxyl groups are used, e.g., fatty alcohols and their ethoxylates and fatty acid alkanol amides and their ethoxylates [60]. Usual esterification catalysts such as *p*-toluenesulfonic acid are suitable as catalysts for diester production.

In the second reaction step, the maleic acid ester is sulfonated with an aqueous sodium hydrogen sulfite solution to obtain the corresponding sulfosuccinate (Fig. 7). In the case of the sulfosuccinic acid monoester, two regioisomeric sulfosuccinates are possible (Fig. 8). It was detected by ^{1}H nuclear magnetic resonance (NMR) analysis that the β position is preferred during sulfonation. The β/α ratio is approximately 4:1 (Henkel KGaA, unpublished results).

Sulfosuccinates are used in many different fields of application. Comprehensive overviews have been published [60,61–63]. Sulfosuccinic acid dialkyl esters are weakly foaming surfactants with good wetting power.

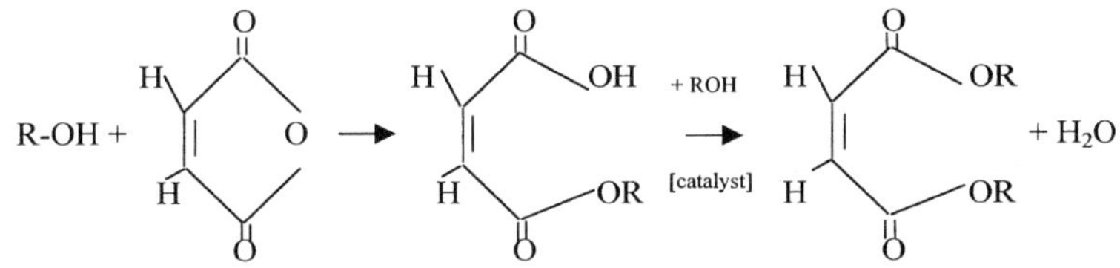

FIG. 6 Reaction scheme for the synthesis of maleic acid mono- and dialkyl esters. R = alkyl, POE-alkyl.

In particular, products based on *n*-octanol or 2-ethyl hexanol are distinguished by their outstanding wetting properties. Therefore, they are applied as "rapid wetting agents" in the textile industries [64]. In fiber technology, these products are used in spinning oils for nylon production. Furthermore, they are used in agriculture for pesticides as well as in paint formulations and in the leather industry. With regard to household products, the application of sulfosuccinic acid dialkyl esters is restricted to specific glass cleaners, e.g., for spectacle lenses or glass panes, as well as carpet shampoos.

In contrast to sulfosuccinic acid dialkyl esters, sulfosuccinic acid monoalkyl esters are good-foaming surfactants, especially products based on ethoxylated fatty alcohols, e.g., lauryl or myristyl polyoxyethylene (3) alcohol, which exhibit outstanding skin compatibility [65]. Because of their mildness to skin, large quantities of sulfosuccinic acid monoalkyl esters are used in personal care products such as shower gels, shampoos, and skin-cleaning agents. In particular, they are utilized in mild products such as baby shampoos or shampoos for sensitive skin. Their compatibility is very good, not only on sensitive skin but also on diseased skin [66]. Sulfosuccinic acid monoalkyl esters are very soluble in water and have good hard-water resistance with a low tendency to form calcium soaps. They exhibit high detergency that is synergistically enhanced in combinations with other surfactants [67]. Because of the hydrolysis-sensitive ester bond, their application is limited to a pH range of 6 to 8. In the industrial sector, sulfosuccinic acid monoalkyl esters are, for example, used as emulsifiers in emulsion polymerization. Both the mono- and dialkyl esters are readily biodegradable and have low toxicity [60].

3. Alcohol Sulfates (ASs)/Alcohol Ether Sulfates (AESs)

Through conversion of alcohol ethoxylates with sulfation agents, alcohol ether sulfates [68] are obtained. On a large technical scale, ether sulfates are produced mainly in a continuous process through mild conversion with sulfur trioxide in a "multitube falling-film" reactor. The importance of sulfation with chlorosulfonic acid has diminished considerably.

The reaction with SO_3 leads primarily to a pyrosulfate:

$$\begin{aligned} &R{-}(O{-}CH_2{-}CH_2)_n{-}OH + 2SO_3(g) \\ &\quad \rightarrow R{-}(O{-}CH_2{-}CH_2)_n{-}OSO_2OSO_3H \end{aligned}$$

The pyrosulfate primarily obtained is metastable and decomposes very rapidly in the presence of further alcohol into the desired sulfuric acid half-ester:

$$\begin{aligned} &R{-}(O{-}CH_2{-}CH_2)_n{-}OSO_2OSO_3H \\ &\quad + R{-}(O{-}CH_2{-}CH_2)_n{-}OH \\ &\quad \rightarrow 2R{-}(O{-}CH_2{-}CH_2)_n{-}OSO_3H \end{aligned}$$

The next step is neutralization with an aqueous solution of the respective base, e.g., NaOH, KOH, or NH_4OH,

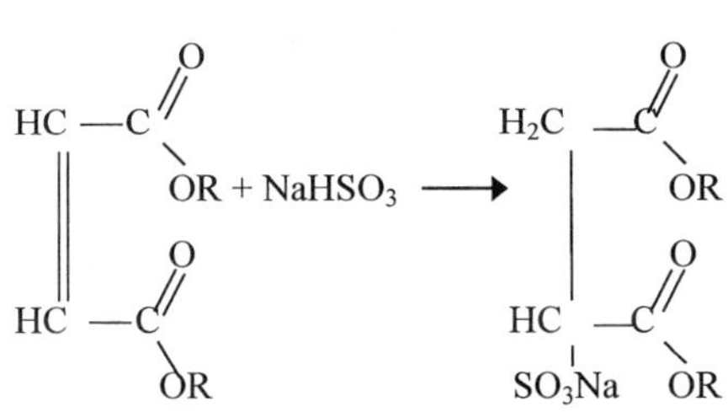

FIG. 7 Reaction scheme to sulfosuccinates. R^1, R^2 = H, alkyl, POE = alkyl.

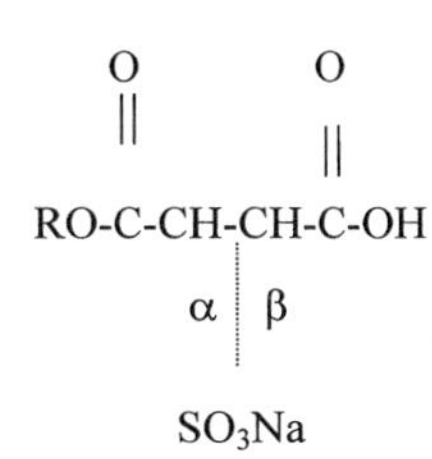

FIG. 8 Regiometric isomers of sulfosuccinic acid monoalkyl ester. R = alkyl.

or, in the case of organic amines without solvent, into the desired salt:

$$R—(O—CH_2—CH_2)_n—OSO_3H + NaOH$$
$$\rightarrow R—(O—CH_2—CH_2)_n—OSO_3Na + H_2O$$

Alcohol sulfates ($n = 0$) are interesting raw materials for detergents. Modern high-density, heavy-duty detergents require solid surfactants with excellent powder properties [69,70]. Water-free alkyl sulfates crystallize quite well. The production volume is steadily rising, whereas the capacity for LAS is decreasing. ASs can be classified as environmentally compatible, with complete biodegradation [71,72].

Ether sulfates are usually obtained as aqueous pastes with concentrations of approximately 30 or 70%. The process is continuous, and the short residence time between formation of the sulfuric acid half-ester and neutralization contributes to the high conversion rate of up to 98%. Because sulfur trioxide is oxidative, under unfavorable conditions and with qualitatively inadequate raw materials, discoloration (yellow and brown) of the product may occur. This can be eliminated in a bleaching step (preferably with H_2O_2) after neutralization. Ether sulfates are stable in the alkaline range but are easily hydrolyzed in an acid medium (autocatalytic reaction). The products are safe for the consumer and the environment A by-product that has been discussed is dioxane. The dioxane quantity can be reduced to <10 ppm, referring to 100% achievement, by technical measures such as conscientious process optimization and aftertreatment. These residual concentrations do not involve any health risk for the consumer.

Because of their good foaming power, alcohol ether sulfates are preferably used in foam baths, shampoos, and manual dishwashing detergents; in combination with sulfosuccinates, amphoteric surfactants, and amine oxides, they have synergistic effects with regard to skin compatibility.

4. Sulfated Oils and Glycerides

A sulfated oil is a reaction product of a sulfation agent, e.g., sulfuric acid, and a fatty oil. One can differentiate between natural fats and oil (triglycerides) and partially esterified glycerol (mono- or diglycerides) by the position of the sulfo group: an internal sulfo group exists in the case of sulfated fat or oil, and an external group at the end of the hydrophic chain exists in the case of partially esterified glycerol.

Sulfated oils were the first nonsoap organic surfactants. In 1834, F. F. Runge prepared a "sulfuric acid oil" from a mixture of olive oil and sulfuric acid and used it as a mordant. In 1875, sulfated castor oil, "Turkey-red oil," was introduced as the first commercial sulfated-type textile auxiliary [73].

The composition of sulfated oils is a complex mixture of sulfo esters, soap, water, fatty acids, and neutral oil. Most of the products are tailor made for special applications and end uses. Self-emulsifiable oils represent the largest group of applications. In addition, they are used as cutting oils for metalworking compositions, as oil sprays for insecticides, as spinning oils for textile processing, and as a so-called fat liquor in the leather industry.

In the group of sulfated mono- and diglycerides, sulfated fatty acid monoglycerides (Fig. 9) are the most important. As early as 1935, Colgate-Palmolive Peet Co. launched a soap-free shampoo formulation under the brand name "Halo" making use of monoglyceride sulfates as a surfactant component. Accordingly, monoglyceride sulfates are among the first synthetic surfactants used in cosmetics.

In 1954, Colgate-Palmolive claimed a continuous process for the preparation of monoglyceride sulfates [74]. In a first reaction step, glycerol is converted with oleum to the corresponding fatty acid glycerol sulfuric acid half-ester. This intermediate product is then transesterified with a triglyceride, usually coconut oil, thus leading to monoglyceride sulfate (the mechanism of this conversion is described in the following).

The resulting products were, for example, applied in the household cleaner "Vel," which was marketed by Colgate in the 1950s and 1960s. The industrial production process is a multistep process that uses 20% oleum as a sulfation agent. In the first reaction step, glycerol is reacted with oleum in such a way that all three hydroxyl groups of the glycerol are sulfated, thus forming a glycerol trisulfuric acid half-ester (Fig. 10). In a second reaction step, this glycerol trisulfuric acid half-ester is converted with a triglyceride, usually hardened coconut oil (molar ratio 2:1).

In a step similar to a transesterification reaction, sulfuric acid half-ester functions are now exchanged for a fatty acid residue, so that from two molecules of glycerol trisulfuric acid half-ester and one triglyceride mol-

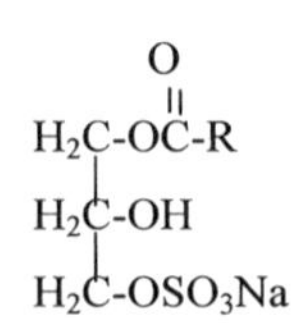

FIG. 9 Molecular formula of monoglyceride sulfate sodium salt. R = alkyl.

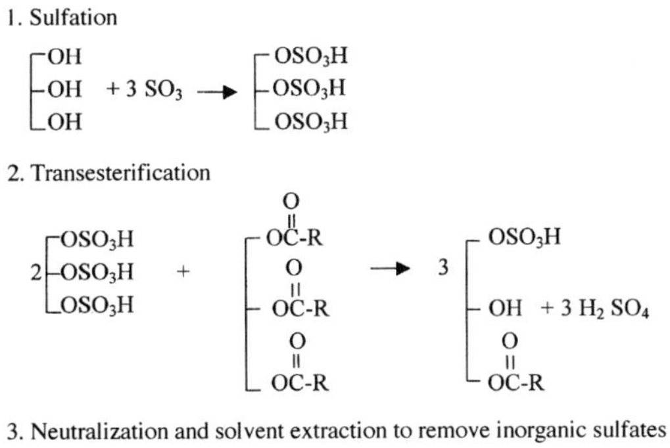

FIG. 10 Reaction pathway to fatty acid monoglyceride sulfates.

ecule, three molecules of monoglyceride disulfuric acid half-esters are obtained. According to Colgate, the sulfuric acid half-ester function in the *β*-position with the fatty acid ester function is unstable. Therefore, after neutralization, e.g., with ammonia or caustic soda solution, a 1,3-fatty acid monoglyceride sulfate is obtained in a highly selective process. Because of the high oleum excess during sulfation of the glycerol, large quantities of sodium sulfate are formed in the neutralization step. The unwanted salts are removed from the monoglyceride sulfate by means of an extraction process as follows: By adding alcohol with a low boiling point, e.g., ethanol, to the aqueous, neutralized surfactant solution, two liquid phases are formed—a heavier aqueous phase that is saturated with sodium sulfate and an alcohol phase that contains the monoglyceride sulfate. The salt-containing aqueous phase is separated, and the surfactant is obtained by evaporation of the ethanol. The resulting product yields the desired monoglyceride sulfate in a purity of approximately 80%; by-products are partial glycerides and fatty acid.

Patent literature describes further manufacturing processes for monoglyceride sulfates [75–81]. In the early 1990s, Henkel KGaA developed a continuous process for the preparation of monoglyceride sulfates [82]. In this process, technical grade monoglycerides are converted to the corresponding monoglyceride sulfates with gaseous SO_3 in a continuous falling-film reactor.

Monoglyceride sulfates, in particular those based on coconut oil, are soluble, high-foaming anionic surfactants. They are distinguished by excellent skin compatibility, which is comparable to that of mild anionic surfactants such as sulfosuccinate or ether sulfate [83]. Based on these properties, coconut monoglyceride sulfate was used in a hair shampoo as early as 1935, as mentioned earlier. In the 1950s and 1960s, the coconut-based monoglyceride sulfates, which Colgate produced on an industrial scale for sale under the brand names Arctic Syntex L and M and Monad G, were applied in many household products, e.g., a household cleaner with abrasive additives [84] and the household cleaner "Vel." In the United States, monoglyceride sulfates are still used as mild surfactants in syndet soaps [85]. In addition to lauryl sulfate, monoglyceride sulfates are described as surfactants for toothpastes and dental care products in combination with specific active substances [86–91]. The excellent skin compatibility of monoglyceride sulfates predestines these products for application in personal care products. A large variety of combinations of monoglyceride sulfates with other mild surfactants has been described for this field of application, e.g., combinations of monoglyceride sulfate with phosphoric acid esters [92], with succinic acid [93], with an aminophosphate surfactant [94] for skin-cleansing agents, and in combination with amino acids and amphoteric surfactants for hair shampoos [95].

Detergent mixtures of alkyl polyglycosides with monoglyceride sulfates show synergistic effects with regard to washing, rinsing, foaming, and cleansing power as well as its skin compatibility [96]. In this combination, high-performance and especially mild shaving preparations [97] or toothpastes [98] can be obtained.

5. Sulfated Alkanol Amides

The synthesis of sulfated alkanol amides has been reviewed in a previous volume of this series [99] comprising literature up to the early 1970s. That reference to basic preparation steps for sulfated alkanol amides is still up to date [100], so that recent developments focus on the fields of application and new starting materials.

To produce amide ether sulfates, alkanol amides may be sulfated directly or first ethoxylated and then sulfated, yielding amide sulfates (**1**) or amide polyoxyethylene sulfates (**2**) after subsequent neutralization with a base as shown in Fig. 11.

The most common alkanol amine basis for sulfated alkanol amides is definitely monoethanol amine [101], although *N*-alkyl-substituted as well as branched alkanol amines such as isopropanol amine have also been used [102]. Apart from these monoalkanol amides, polyhydroxy alkanol amides such as diethanol amides or 2,3-hydroxy propyl amides have also been prepared [103].

The corresponding alkanol amides are derived from (saturated or unsaturated) C_2- to C_{22}-carboxylic acids or hydroxy carboxylic acids [104], mainly from coco-

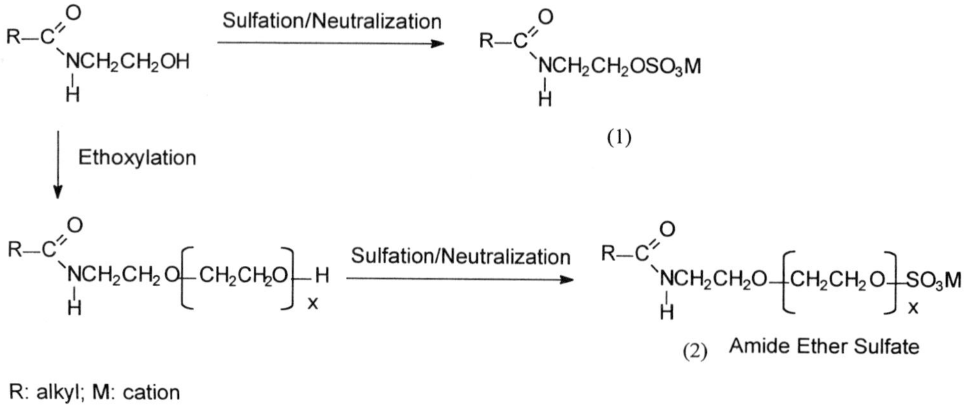

FIG. 11 Reaction paths to amide sulfates and amide polyoxyethylene sulfates.

nut or tallow-based feedstocks. A main drawback of the sulfation process for alkanol amides is the high viscosity of the sulfation mixture, which may be overcome by means of cosulfation with lower molecular weight alcohols [105], alkanol amines [106], fatty alcohols [107], or oxethylated fatty alcohols [108]. Because of a lower sulfation temperature, the products obtained by this route have an improved color. The choice of cations comprises ammonium (including alkyl and alkanol ammonium), alkali, and alkaline earth metals.

Sulfated alkanol amides are excellent foaming surfactants with good detergency [109]. Their hydrolytic stability as well as physicochemical data have been cited elsewhere [100]. Sulfated alkanol amides are used almost exclusively as cosurfactants together with anionic, nonionic, and sometimes cationic components.

Cosmetics (body, hair, and baby care) and manual dishwashing are the main fields of application because of the low skin irritancy of alkanol amide sulfates [110], which was observed during the late 1960s [111].

Alkanol amide sulfates are good lime soap dispersants and have thus been used in detergent compositions suitable for hard-water applications [112].

Technical applications concerning emulsion polymerization of ethylenically unsaturated monomers [113] or leather preparations [114] are related to the favorable emulsifying properties of sulfated alkanol amides. Sulfated alkanol amides have also been used as mold release [115] or antiadhesive reagents for rubber [116]. Together with cationic surfactants, alkanol amide sulfates may serve as dehydration promotion agents for the production of granular slag [117].

C. Fatty Acid Condensation Products

Fatty acid condensation products have a long history. Most of the products were developed during the early 1930s by the former IG-Farben and still have value as mild cosurfactants and as specialty primary surfactants.

1. Isethionates

Isethionates [118] are mild cosurfactants and are especially used in syndet bars. The largest market is the United States with about 20–30,000 tons per annum. The most common product has a C_{12}–C_{18} alkyl chain derived from coconut oil. The most recent development in isethionates was the ammonium cocoyl isethionate [119], which is an alternative to the poorly soluble sodium salt with regard to liquid formulations.

The condensation of fatty acid with sodium isethionate can be carried out in the presence of an esterification catalyst in a temperature range above 180–200°C.

$$\mathrm{HO{-}CH_2{-}CH_2{-}SO_3Na + R{-}CO_2H} \xrightarrow[-\mathrm{H_2O}]{\text{catalyst}} \mathrm{R{-}CO_2{-}CH_2{-}CH_2{-}SO_3Na}$$

An alternative route is the reaction corresponding to Schotten-Baumann with an acid chloride and sodium isethionate:

$$\mathrm{HO{-}CH_2{-}CH_2{-}SO_3Na + R{-}COCl} \rightarrow \mathrm{R{-}CO_2{-}CH_2{-}CH_2{-}SO_3Na + HCl}$$

The direct esterification route is more economical than the acid chloride route, but it yields a larger amount of unreacted fatty acid.

2. Taurates

Taurates [120,121] also belong to the class of fatty acid condensation products and can be prepared using the same reaction pathways as isethionates. The most common preparation starts with Na-*N*-methyltaurate:

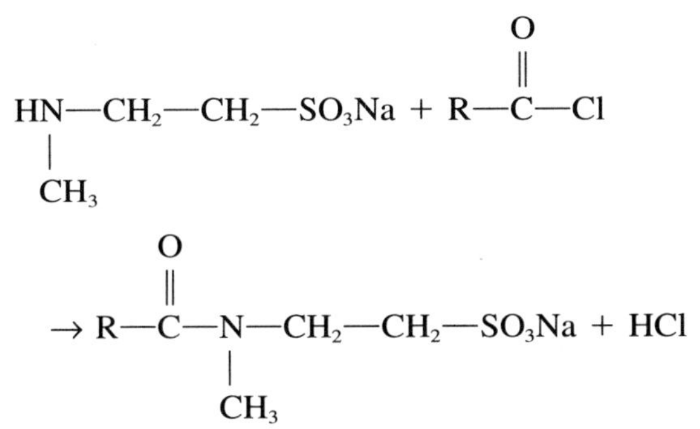

The alternative is the direct reaction of fatty acids with Na-*N*-methyltaurate at high temperatures:

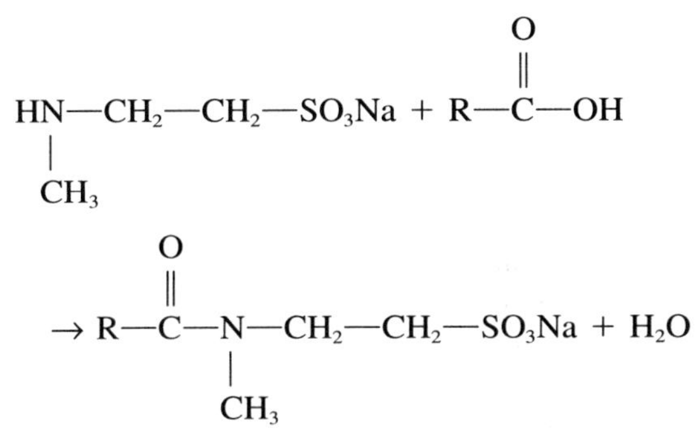

Taurates are more soluble in water than isethionates and are also very mild [121]. Taurates find applications in shampoos, toothpastes, and shaving foam and in various technical applications.

3. Sarcosinates

N-Acyl sarcosinates [122–124] are mild anionic surfactants derived from the amino acid sarcosine (*N*-methylglycine) and fatty acids. The reaction proceeds via the Schotten-Baumann route with an acid chloride and sodium sarcosinate:

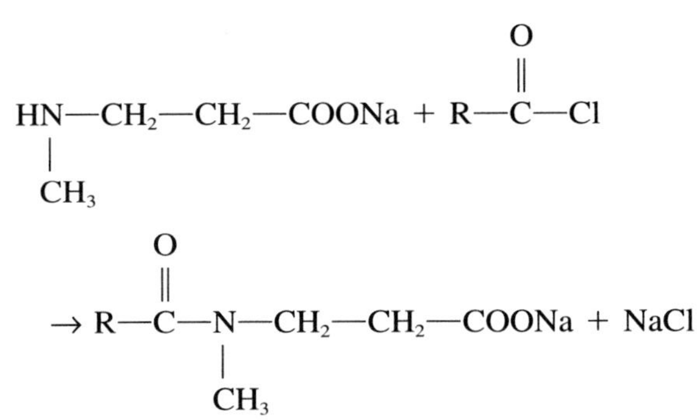

The *N*-acyl sarcosinic acid is insoluble in acid medium and can be isolated upon acidification. Aqueous solutions can be prepared with bases such as NaOH, KOH, NH_4OH, or triethanol-amine (TEA). Commercial products are available as 30% solutions in water or as spray-dried solids with different C chains from C_{12}, C_{14} to technical cocoyl sarcosinate (C_8–C_{18}) or oleyl sarcosinate.

Acyl sarcosinates are applied in numerous personal care products [125] such as shampoos, skin cleansers, bath additives, and toothpastes, where they show good foaming properties. Sarcosinates are, to some extent, compatible with cationic surfactants, which is an interesting point for formulations. Other applications are in corrosion inhibition, industrial emulsification, and detergents.

4. *N*-Acyl Amino Acids

The chemical preparation of *N*-acyl amino acids is basically the same as the manufacturing of sarcosinates. *N*-Acyl amino acids [126–130] are well known. With technical production capacities rising for the use of some products in nutrition, they have become attractive raw materials for the surfactant industry. Newer developments worth mentioning are *N*-acyl glutamate [131] or *N*-acyl aspartate. Sodium glutamate is used in food applications as a flavor enhancer, and L-aspartic acid is used as material for the popular artificial sweetener aspartame.

One example is the synthesis of *N*-lauroyl glutamate (Fig. 12). In contrast to the acylation of monocarboxylic amino acids, the reaction of sodium glutamate with acid chloride requires an additional water-miscible solvent such as acetone or isopropanol. The acylation product can be separated after acidification by filtration. It can be handled as dry powder in the acid form or neutralized with NaOH as a 25% solution.

Applications of *N*-acyl amino acid are mainly in the field of cosmetics because of its extreme mildness, good foaming, and cleaning properties.

5. Protein/Fatty Acid Condensates

Protein/fatty acid condensates [129,132–134] are produced on the basis of proteins as renewable raw materials. The hydrophilic part is a mixture of peptides, while the hydrophobic part is based on fatty acid. The peptide chain can vary in amino acid sequence and in molecular weight. Depending on the protein and the hydrolysis conditions, molecular weights of 600 to 5000 are achieved.

The acylation is carried out with a protein hydrolysate, a water-soluble peptide mixture obtained from the hydrolysis of an insoluble protein, and a fatty acid chloride (Fig. 13). The acyl group is normally formed by a C_8–C_{18} fatty acid. Proteins may come from collagen or vegetable sources such as soya or rice.

Protein/fatty acid condensates are processed as aqueous yellow to amber-colored solutions with 30–35% active matter or as powders.

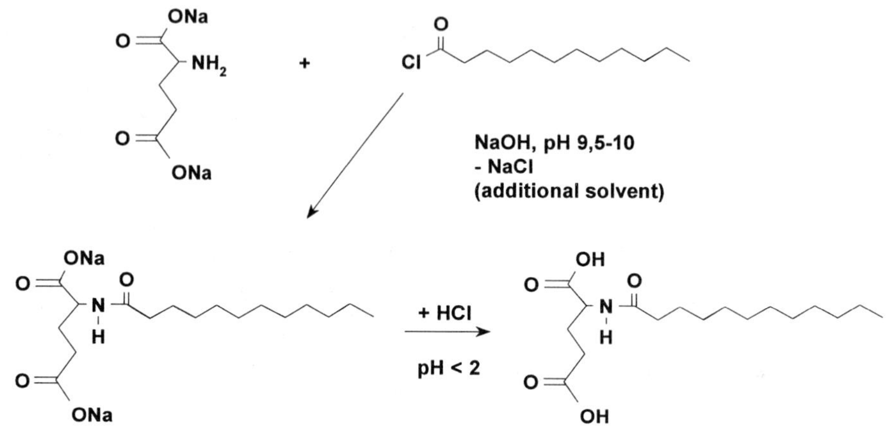

FIG. 12 Acylation of sodium glutamate.

Protein/fatty acid condensates are used mainly as mild surfactants in cosmetics, in the textile industry, and in laundry or dishwashing detergents.

D. Phosphoric Acid Derivatives

Another group of anionic surfactants with a significant market potential are phosphoric acid esters. They are based on fatty alcohols as well as on oxethylated alcohols [135] and may be used for special applications such as emulsifiers, wetting agents, antistats, lubricants, flotation auxiliaries, and corrosion inhibitors.

Regarding chemical properties, partial phosphoric acid esters show marked stability to hydrolysis except in strongly acidic conditions. The sensitivity to hard water is a real disadvantage for this surfactant class. Sometimes, compensation may be possible by application of phosphoric acid esters based on oxethylated alcohols.

Most of the practically used derivatives are mixtures of mono- and diesters; triesters are present only as minor components (Fig. 14).

Generally applied synthesis methods are "phosphations" by means of phosphorus pentoxide. The reaction is very complex and may occur as follows (Scheme 7).

The postulated molar ratio is obtained only if the phosphorus pentoxide used is of a highly pure grade. Because phosphorus pentoxide always contains polyphosphoric acid, at best a mole ratio of 1.2:1 mono- to diester is available. For the same reason, the final products often include pyrophosphates and *o*-phosphoric acid, which are formed in subsequent steps as follows, where polyphosphoric acid reacts with an alcohol (Scheme 8).

The pathway just described is also suitable for the preparation of monoalkyl phosphates, although they are always accompanied by *o*-phosphoric acid and, depending on the amount of alcohol, pyrophosphoric acid. The latter may be hydrolyzed by treatment with water at elevated temperatures.

Triesters based on oxethylated alcohols, which are used as cosmetic emulsifiers for ointments, creams, or lotions [136], can be produced by the reaction of al-

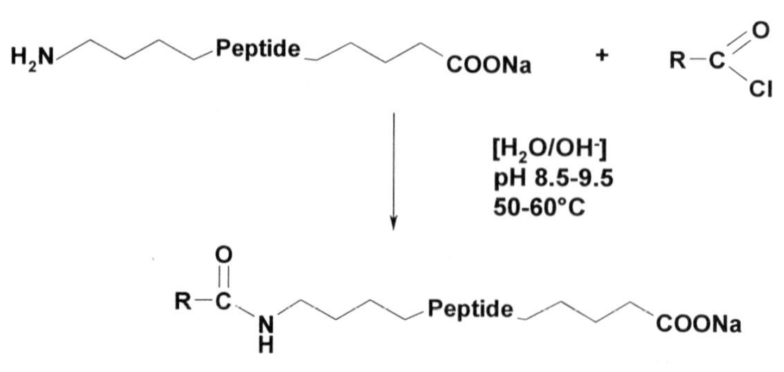

FIG. 13 Acylation of protein hydrolysate.

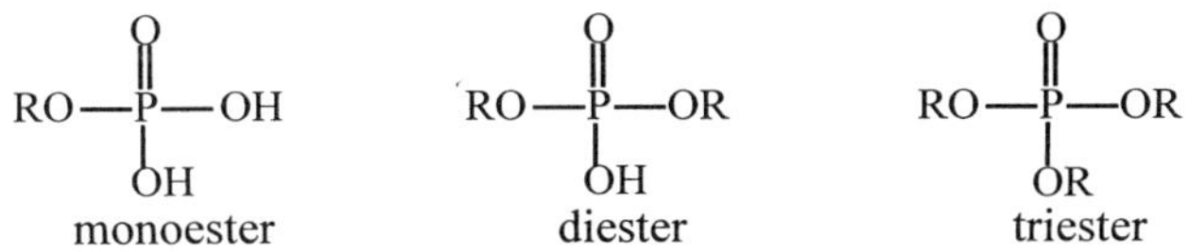

FIG. 14 Structures of phosphoric acid esters.

cohols with phosphorus oxychloride in the presence of tertiary amines as absorbents for the hydrochloride gas generated. This prevents, for the most part, the formation of alkyl chlorides as by-products [137] (Scheme 9). The relatively large quantity of amine salt must be removed by filtration.

Esterification directly with *o*-phosphoric acid is impossible. The reaction requires a temperature above 180°C, at which the ester formed will be destroyed, generating olefins [138]. In the presence of small amounts of basic components, however, the cleavage will be avoided [139]. But this technology has not found practical interest.

Mostly, the partial esters are sold in a neutralized form, usually as potassium or alkanolamine salts.

III. NONIONIC SURFACTANTS

A. Alkoxylation Technology

The most important technology in synthesizing nonionic surfactants is the reaction of alcohols, or other active hydrogen compounds, with alkylene oxides such as ethylene oxide (EO) and propylene oxide (PO) [140]. The reaction with ethylene oxide is used most frequently in order to increase hydrophilicity and thus the water solubility of alcohols and is commonly known as "ethoxylation" or, chemically more precisely, "oxethylation."

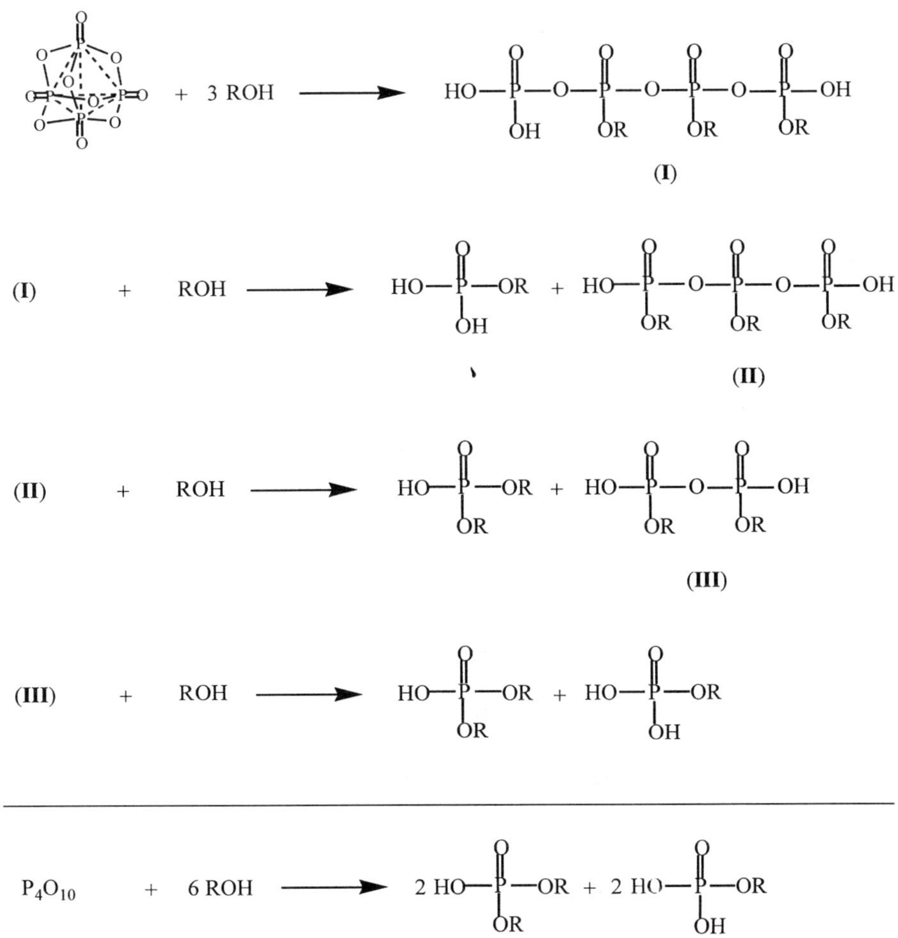

SCHEME 7

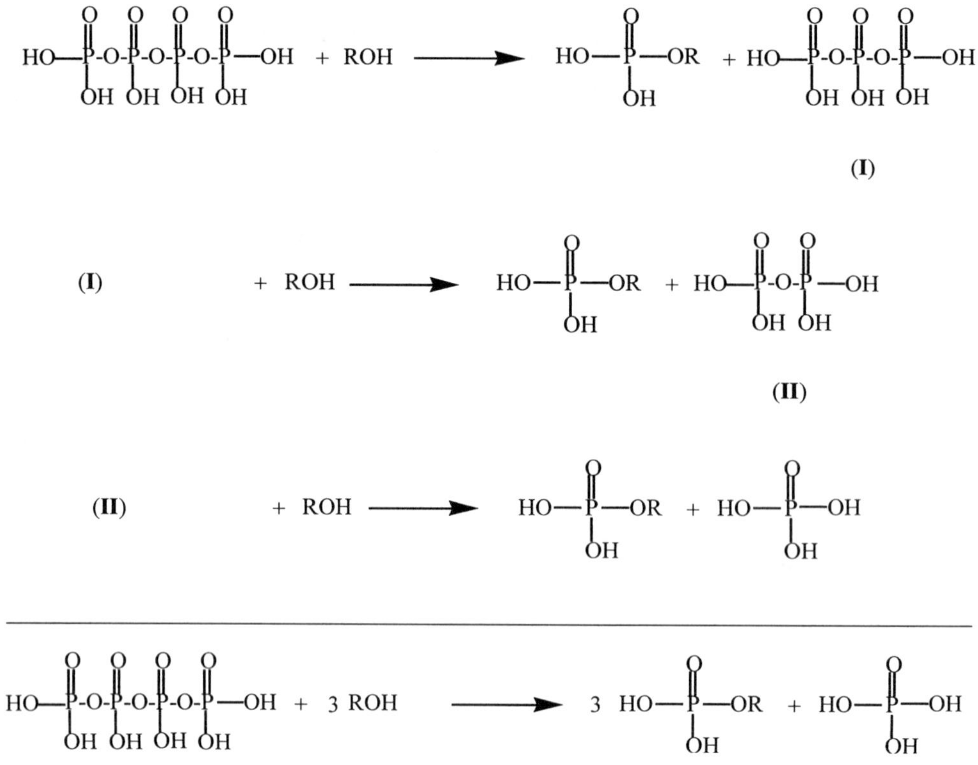

SCHEME 8

The reaction scheme is

$$R{-}OH + n\,\underset{\diagdown O \diagup}{H_2C{-}CH_2} \xrightarrow[\substack{120{-}180^{\circ}C \\ \text{max. } 5{-}7 \text{ bar}}]{\text{catalyst}} R{-}O{-}(CH_2{-}CH_2{-}O)_n{-}H$$

The ethoxylation reactions are normally carried out batchwise in a stainless steel reactor (Fig. 15) with a sparger of EO in the bottom, mixing of EO in an external circulation loop, or dosing with nozzles to the vapor phase. Temperatures range from 120 to 180°C at a pressure of 5–7 bar. The risk of spontaneous polymerization of EO with a high temperature jump has to be minimized by applying expensive computer control systems combining all safety features. The control system measures temperatures and pressure and automatically shuts the system down if critical limits are exceeded. There is also a safety lock to prevent catalyst mixtures from being back-mixed with ethylene oxide. Important criteria for safety are quick reaction with consistently low stationary concentrations of ethylene oxide and dilution with nitrogen.

Raw materials have to be dried before ethoxylation in order to reduce the undesired side reaction yielding polyethylene glycol. Traces of oxygen or air have to be removed before dosing EO to avoid the formation of explosive mixtures and to reduce discoloration. After the reaction, a posttreatment, applying vacuum to reduce traces of EO and sometimes 1,4-dioxane (stripping with steam), is necessary. The catalyst is normally neutralized with acetic acid, phosphoric acid, propionic acid, or lactic acid, resulting in soluble salts or salts that can be removed by filtration.

The reaction with propylene oxide gives the molecule a more hydrophobic character. Propylene oxide can be copolymerized with ethylene oxide in a random manner or blockwise in order to obtain certain product characteristics such as reduction of foaming, improvement of wetting, liquefaction, and others.

SCHEME 9

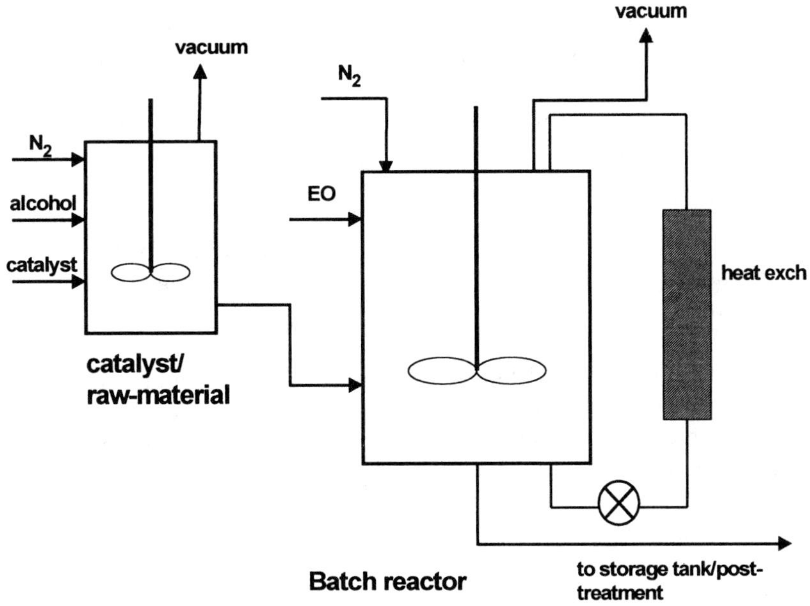

FIG. 15 Ethoxylation reactor.

Technical alkylene oxide derivatives always have a distribution of oligomers or polymers with a mean degree of polymerization, reflecting the mole ratio of the reaction of alcohol and EO. The product is not uniform in composition but has a distribution of homologues that strongly depends on the type of catalyst used [141]. Conventional alkaline catalysts such as KOH [142] or NaOMe give a relatively broad distribution (BRE = broad range ethoxylates, Fig. 16). However, it should be noted that special catalysts can be applied resulting in a narrow distribution (NRE = narrow range ethoxylates, Fig. 17).

The difference between BRE and NRE products from an application point of view is not as great as one might have predicted. The problem with NRE is the higher manufacturing expenditure and costs, making

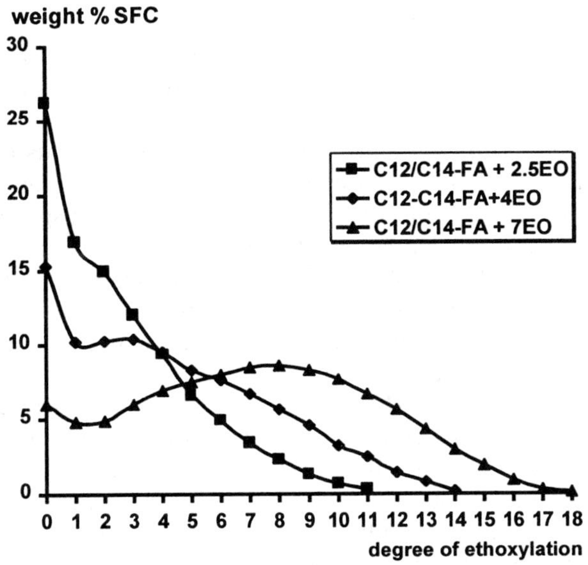

FIG. 16 Homologue distribution of broad-range ethoxylates.

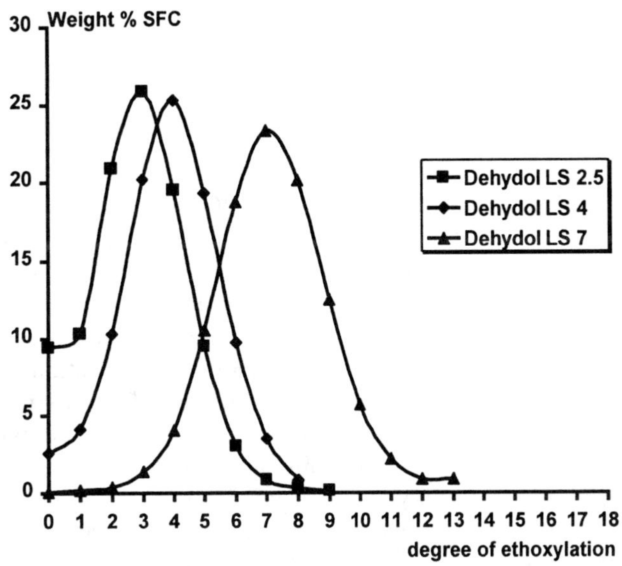

FIG. 17 Homologue distribution of narrow-range ethoxylates.

them unattractive for use as commodities. Areas of application can be found, however, where the improved physicochemical behavior gives enough benefits. Examples are thickeners for surfactant systems, cloud point extractions, lower odor, better solubility, etc.

1. Polyoxyethylene Alkyl Phenols (APEs) [143]

APEs have played an important role in the nonionic surfactant market. Most commonly these products are based on nonyl phenol or octyl phenol. Their application is almost universal because of their good performance characteristics. The alkyl phenols are normally *para*-substituted phenols with highly branched alkyl groups derived from the alkylation of phenol with olefins at an acidic catalytic contract such as boron trifluoride, acid montmorillonite-clay, or others. In contrast to nonyl phenol (nonyl is derived from propylene trimer), octyl phenol has a more defined structure with the alkyl chain coming from diisobutylene.

The process for the production of APEs is very similar to the ethoxylation of aliphatic alcohols. The homologue distribution is narrower than for alcohol ethoxylates because the first mole of EO reacts nearly quantitatively with the relative acidic alkyl phenol, giving a nearly statistical (Poisson) distribution for the rest of the chain.

APEs have a very broad application range, for example, in agriculture, detergents, cleaners, textile, paints, paper, and leather. The importance of APEs in European countries is decreasing because of environmental considerations [144,145]. There are environmentally persistent degradation products with possible estrogenic effects. The European detergent industry has agreed on self-limitation of the use of APEs. Numerous discussions and environmental risk assessments of APE are continuing.

2. Aliphatic Polyoxyethylene Alcohols (AEs)

Aliphatic alcohols are the most important source for nonionic surfactants made by ethoxylation or propoxylation [140]. The alcohols are derived either from natural fats and oils or from petrochemical raw materials.

Transesterification with methanol or esterification of fatty acids, resulting from the hydrolysis of natural triglycerides, followed by hydrogenation of the methyl ester results in straight-chain saturated or unsaturated alcohols in the alkyl range from C_8 to C_{18}.

Synthetic alcohols can be prepared from Ziegler or oxo processes [146–148]. The former results in even carbon numbers [149,150], which can compete with natural fatty alcohol; the latter results in branched-chain alcohols (hydroformylation of olefins). The oxo derivatives are more interesting for cold-water applications.

Polyethylene alcohols are mostly classified by their hydrophile-lipophile balance (HLB) [151–153], which is a measure of the balance between the hydrophilic EO headgroup and the lipophilic hydrocarbon tail.

The simplest way to calculate the HLB number is from the following equation:

$$\text{HLB} = E/5$$

where E is the amount of EO calculated in wt% of the molecule. Knowledge of the HLB value gives a rough guide to the application areas of AE:

HLB	Application
4–6	w/o—emulsifier
7–15	wetting agents
8–18	o/w—emulsifier
10–15	typical detergent
10–18	solubilizers

An important physical chemical property of AEs is the cloud point (ASTM D2024 test method). Whereas the solubility of ionic surfactants increases with temperature, polyoxyethylene alcohols become insoluble at high temperatures. The temperature at which the aqueous surfactant solution becomes cloudy is called the cloud point and is also a characteristic of the relation of the hydrophilic EO chain to the hydrophobic alkyl chain. This phenomenon can be explained by the breaking of hydrogen bonds that cause insolubility at high temperatures and is used as an important specification of AEs.

Interestingly, the detergency of AEs reaches its optimum near the cloud point, so the cloud point can be useful in the choice of the right surfactant for a specific application. Alternatively, the hydrophile of AE can be measured by clouding phenomena in mixed systems such as isopropanol/water or butyldiglycol/water.

Another systematic approach in choosing the right AE for application as an emulsifier is the phase inversion temperature (PIT) concept [154,155]. This refers to the phase inversion from oil/water (o/w) to w/o in a ternary system of oil, water, and surfactant.

The applications of AEs are widespread: detergents and cleansers, emulsifiers, textile, agriculture, intermediates for sulfonation, paper industry, and emulsion polymerization, to name a few. With regard to their importance, AEs have been studied very intensively

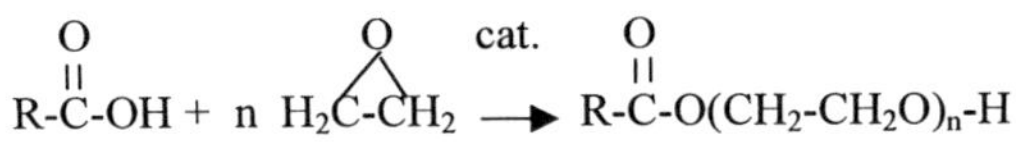

SCHEME 10

concerning their ecotoxicity and can be regarded as environmentally safe [156].

3. Ethoxylated Fatty Acids and Esters

(*a*) *Fatty Acid Ethoxylates.* Ethoxylated fatty acids can generally be obtained by two different methods: (1) esterification of fatty acids with polyethylene glycol and (2) Ethoxylation of fatty acids. In the esterification [157] mixtures of mono- and diacids, esters are formed because of the two hydroxyl groups in the polyethylene glycol that exhibit the same reactivity. By using an excess of fatty acid, the formation of diesters is favored [158]. Pure monoesters can be obtained by reaction of polyethylene glycol with basic acid, esterification of the borate obtained with fatty acid, and then selective splitting of the boric acid ester [159].

The ethoxylation of fatty acids is carried out in the presence of alkaline catalysts at temperatures between 120 and 200°C and a pressure of 1–5 bars (Scheme 10). Because monoesters of polyoxyethylene transesterify easily, polyoxyethylene diesters and free polyoxyethylene are also formed during the ethoxylation of fatty acids (Scheme 11). The molar ratios of monoesters to diesters to free polyoxyethylene (polyethylene glycol) are found to be approximately 2 : 1 : 1. For example, a 7 mole EO adduct of pentadecanoic acid contains monoester (48%), diester (40%) and polyethylene glycol (12%) [160]. A comprehensive overview of the chemistry and the properties of ethoxylated fatty acids has been published [161,162]. Their main areas of application are emulsifiers in metalworking, mold lubricants, textiles, and because of their toxicologic and dermatologic harmlessness, cosmetic and pharmaceutical formulations.

(*b*) *Ethoxylates of Fatty Acid Methyl Esters.* Raw materials that contain an active hydrogen atom in the molecule, e.g., fatty alcohol, fatty acids, anions, or partially esterified polyol esters, can easily be converted with ethylene oxide to the corresponding ethoxylates by using standard alkaline catalysts (e.g., $NaOCH_3$, KOH). Therefore, direct conversion of methyl esters is not possible if these conventional catalysts are used. However, new catalysts based on Ca/Al or Mg/Al compounds were developed that enable insertion of ethylene oxide into the ester bond and yield a monomethyl ether of an ethoxylated monoester or a fatty acid methyl ester ethoxylate (FMEO or MEE) [163–167] (Scheme 12).

It is assumed that the reaction takes place on the surface of the catalyst, where the bifunctional effect of acid-base active sites, caused by the different cations, results in a dissociative chemisorption of fatty acid methyl ester. This leads to a direct insertion of ethylene oxide, which involves coordinated anionic polymerization [164,168]. The products are obtained in high yields with a normal or narrow range homologue distribution. Because the product composition is very similar to that of alcohol ethoxylates, the general surfactant properties are comparable [163,165,167,169]. FMEOs can be used in dishwashing agents, household hard surface and all-purpose cleaning, as well as in industrial and institutional applications. Mid- to high-mole lauryl and tallow range FMEO exhibit high detergency. Compared with alcohol ethoxylates, FMEOs are less foaming and dissolve in water much faster without going through a gel phase [166]. FMEOs are readily biodegradable, exhibit low aquatic toxicity [166], and have outstanding dermatological compatibility [170,171]. Because of the ester bond, FMEOs are susceptible to hydrolysis. They are stable in aqueous formulations in a pH range from 3 to 9.

4. Ethoxylated Oils and Glycerides

Partially esterified glycerol with fatty acids, monoesters ("monoglycerides"), and diesters ("diglycerides") can be ethoxylated by using standard alkaline catalysts (e.g., $NaOCH_3$, NaOH, KOH) and standard reaction conditions. Purified triesters ("triglycerides") do not possess an active hydrogen atom in the molecule to offer ethylene oxide a reaction site (an exception is castor oil, which contains a secondary hydroxyl group). In this case the addition of water or glycerol, which causes the formation of partial glycerides, enables the reaction. In both cases using partial glycerides or triglycerides, a complex mixture of different products is obtained: free glycerol and glycerides (mono-, di-, tri-) and ethoxylated glycerol and glycerides as well as fatty acid ethoxylates and free polyethylene glycol. In

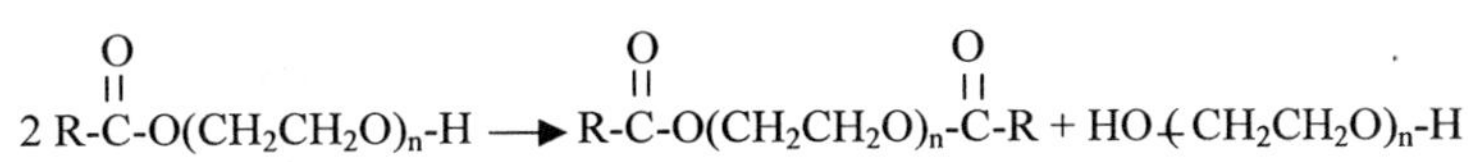

SCHEME 11

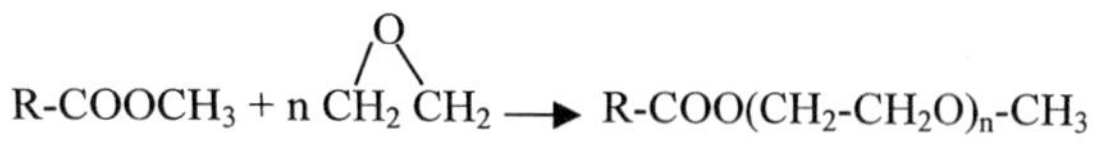

SCHEME 12

the case of castor oil ethoxylated with 40 moles of ethylene oxide, the product composition has been analyzed [172].

Besides the preceding ethoxylation products, higher esters were identified in which the secondary hydroxyl group is esterified with ricinoleic acid. Surprisingly, the degree of ethoxylation of the secondary hydroxyl group was rather low. Ethoxylated glycerol esters are used widely in the cosmetic industry, e.g., for emulsification, solubilization, refatting, thickening, improvement of the dermatological properties of basic surfactants, and improvement of skin feel [173]. Castor oil or hydrogenated castor oil ethoxylates play an important role in body care applications and as solubilizers or emulsifiers in pharmaceutical formulations.

5. Ethoxylated Amines and Alkanol Amides

The synthesis of ethoxylated amines can be separated into two reactions [174,175]. In the first step an amine is reacted with ethylene oxide to an amino alcohol in such a way that every H atom of the amine reacts with one ethylene oxide molecule (Scheme 13). The second step is the growth of the polyoxyethylene chain through reaction of more ethylene oxide with the hydroxyl groups of the amino alcohol (Scheme 14). Whereas in the first reaction step no catalyst is necessary, the second step requires a catalyst such a sodium or potassium hydroxide.

The most commercially available surfactants prepared by this method are the fatty amine ethoxylates. Another class of ethoxylated amines are the "Jeffamines" and the "Tetronics." Jeffamines are prepared by first ethoxylating a short-chain alcohol or glycol. This alcohol ethoxylate is then aminated to generate an amine [176,177]. Tetronics are made by ethoxylation of low-molecular-weight diamines, e.g., ethylenediamine. These products are tertiary amines and often consist of mixtures of ethylene oxide and propylene oxide adducts.

Ethoxylated amines have a wide range of applications [178]. They are used, for example, as emulsifiers,

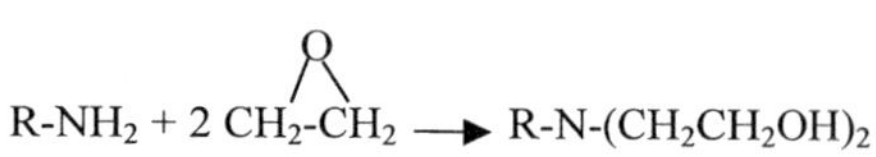

SCHEME 13

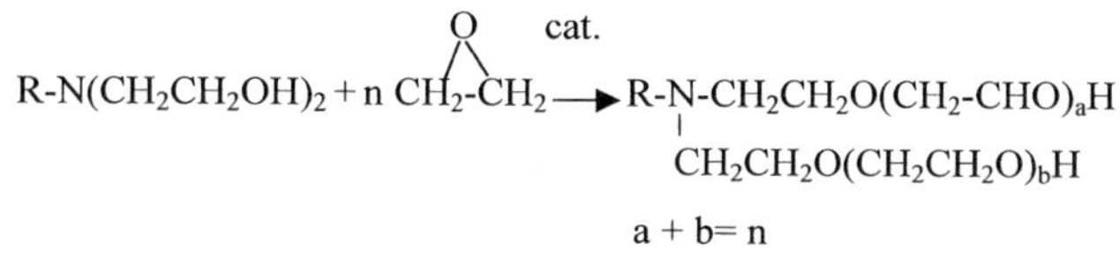

SCHEME 14

solubilizers, and antistat additives. This application range varies from cleaning and detergent formulations to additives in gasoline and drilling fluids.

B. Amine Oxides

A special class of surface-active substances are amine oxides, which belong in the class of nonionic components. This classification is true, however, only under alkaline and neutral conditions. In acid solutions they react weakly to form cationics.

The synthesis of amine oxides happens relatively simply by the reaction of tertiary amines with hydrogen peroxide in an aqueous medium according to Scheme 15. Because of the oxidizing environment, amine oxides sometimes contain nitroso amines, which are suspected of being carcinogenic. References [178–182] describe manufacturing methods to avoid these undesirable by-products.

The foaming, wetting, and cleaning properties as well as the ecotoxicological aspects of C_{12}–C_{18} alkyl dimethylamine oxides have been discussed previously [183]. Also, good thickening performance is mentioned.

Amine oxides are very important emulsifiers for many applications in which the reemulsification of absorbed components must be prevented. Such an application is possible because of the decomposition of amine oxides at temperatures above 100°C to yield olefins and derivatives of hydroylamine (Scheme 16). This reaction is taken advantage of, for example, in the application of waterproofing agents to textiles.

An interesting variety is the thermal decomposition of amine oxides that are based on ethers of dimethyl ethanolamine [184]. The vinyl ethers formed are known as reactive olefins [185]. Produced from a base of dimethyl ethanolamine esters, the preparation of vinyl esters becomes possible (Scheme 17).

IV. CARBOHYDRATE BASED SURFACTANTS

Carbohydrate-based surfactants are the final result of a product concept that is based on the greatest possible

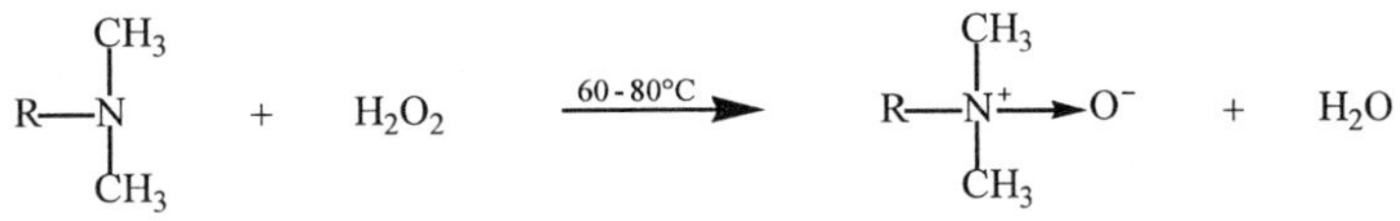

SCHEME 15

use of renewable resources. Whereas the derivatization of fats and oils to produce a variety of different surfactants for a broad range of applications has a long tradition and is well established [186], the production of surfactants based on fats, oils, and carbohydrates on a larger industrial scale is relatively new. The following will discuss the most important carbohydrate-based surfactants, such as sorbitan esters, sucrose esters, alkyl polyglucosides, and fatty acid glucamides, with primary focus on the glucose-derived products.

Considering the amphiphilic structure of a typical surfactant with a hydrophilic headgroup and a hydrophobic tail, it has always been a challenge to attach a carbohydrate molecule as the perfect hydrophilc group, due to the numerous hydroxyl groups, to a fat and oil derivative such as a fatty acid or a fatty alcohol [187]. Although scientists have reported numerous ways of making such linkages and have also described a large number of different carbohydrates used in such reactions, it is clear from an industrial perspective that only a few carbohydrates fulfill the criteria of price, quality, and availability to be an interesting raw material source. These include sucrose from sugar beet or sugarcane, glucose derived from starches, and sorbitol as the hydrogenated glucose derivative (Table 1). Most industrial developments in the field of sugar-based surfactants have concentrated, and still concentrate, on these carbohydrate feedstocks.

A. Sorbitan Esters

Sorbitan esters have been known for decades since the first industrial processes were established for their manufacture. One differentiates between a one-stage and a two-stage process (Fig. 18). In the first process, water is eliminated from sorbitol as a first step to form sorbitan, which is subsequently derivatized with fatty acid as a second step. In the second process, both reactions are carried out simultaneously [188]. Both methods have been developed for industrial scale production. Depending on the type and amount of fatty acid used, different types of sorbitan esters (e.g., laurates, oleates, or stearates) are produced with hydrophilic-lipophilic balance (HLB) values in a range of 1 to 8. To modify

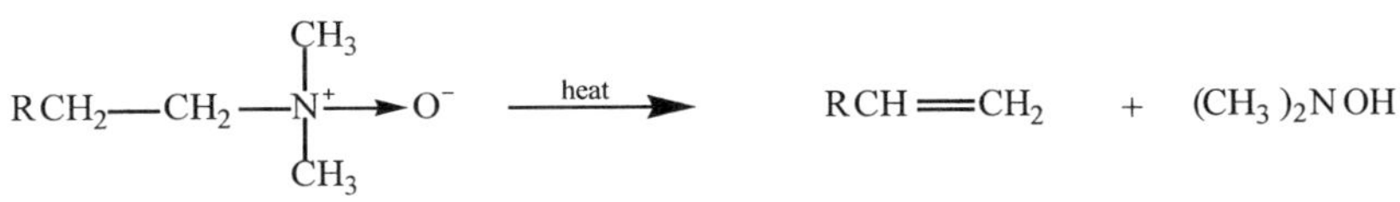

SCHEME 16

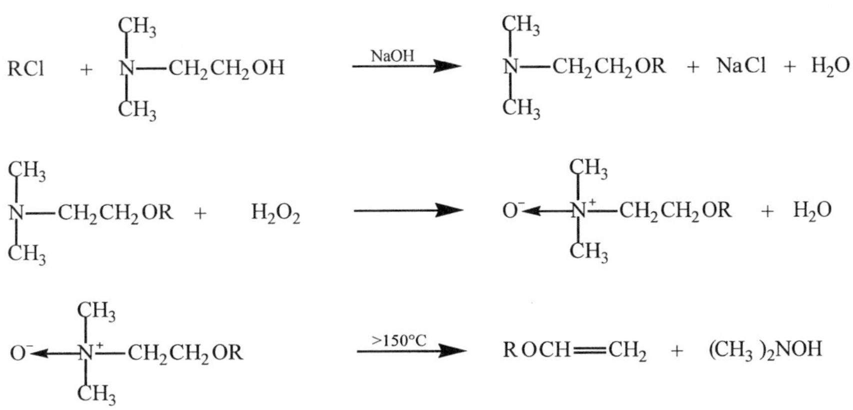

SCHEME 17

TABLE 1 Availability of Carbohydrate Raw Materials

Material	Production volume (t/a)[a]	Average price ($/kg)[b]
Sucrose	130,000,000	0.70–0.80
Glucose	16,000,000	0.55–1.12
Sorbitol	8,000,000	0.53–1.70

[a]Private communication (Cerestar, Henkel).
[b]According to Chem. Market. Rep. 3/97 and 7/97.

these relatively hydrophobic materials, it is common technology to derivatize the sorbitan esters further by reaction with ethylene oxide to produce sorbitan ester ethoxylates—or polysorbates for short—with HLB values of 10–17, depending on the number of ethylene oxide units attached (Fig. 19) [188a].

The main manufacturers for sorbitan ester products today are listed in Table 2. The total market size for sorbitan esters (including the ethoxylated products) is estimated to be approximately 25,000 tons per year. Mainly used as emulsifiers in pharmaceuticals, foods, cosmetic products, for emulsion polymerization and explosives, and for other technical applications, sorbitan ester products seem to have a relatively stable market size and there is obviously no attempt, and no need, for further development of this mature technology.

B. Sucrose Esters

The situation is different in the field of sucrose esters. Described as very mild with regard to their dermatological properties and approved as food additives in many countries, these products are perfect raw materials for personal care products, cosmetic applications, and food emulsifiers [189]. In Asia, one can find sucrose esters in special detergent products as well.

The problem in manufacturing sucrose esters is related to the high functionality of the sucrose molecule with eight hydroxyl groups, which compete during the derivatization step (Fig. 20). In a typical esterification reaction of sucrose with fatty acid methyl ester, a complex product mixture consisting of mono-, di-, tri-, tetra-, and pentaesters is formed (Fig. 21). These products are very hydrophobic and of limited application potential. Therefore, several methods have been developed to achieve higher selectivity in the reaction or provide economical purification procedures and, as a result, a high quantity of monoester. These include the use of solvents and fatty acid chlorides, special extraction and crystallization techniques, enzymatic catalyses, and equilibrium reactions. Figure 22 shows the increase of the monoester content in a sucrose laurate product achieved by alcoholysis with methanol [189c]. However, most of the methods remain limited to laboratory scale because of the process economics. Standard technology (conventional method, Fig. 23) is still transesterification combined with purification. Here, an optimized solvent-free process has been described (Fig. 23) [189b].

Today, the major producers of sucrose esters are Dia-Ichi Kogyo Seiyaku and Mitsubishi in Japan, Croda in the United States, Sisterna (a joint venture of Dai-Ichi with Suiker Unie from The Netherlands), and Goldschmidt in Germany (Table 2). It seems that the production capacities that exist today are much higher

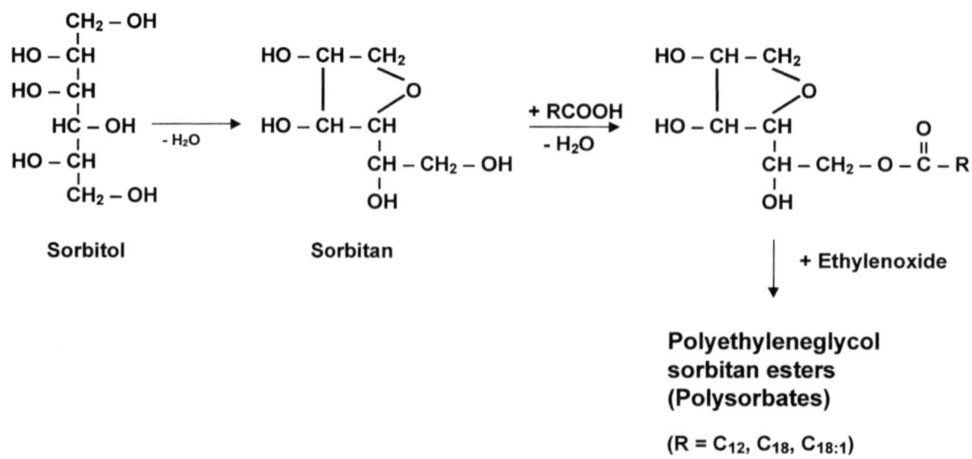

FIG. 18 Synthesis of sorbitan esters by intramolecular dehydration of sorbitol in the presence of acid (e.g., NaH_2PO_3) at about 150–200°C and subsequent base-catalyzed (e.g., Na_2CO_3) esterification with fatty acids (RCOOH) at 200–250°C.

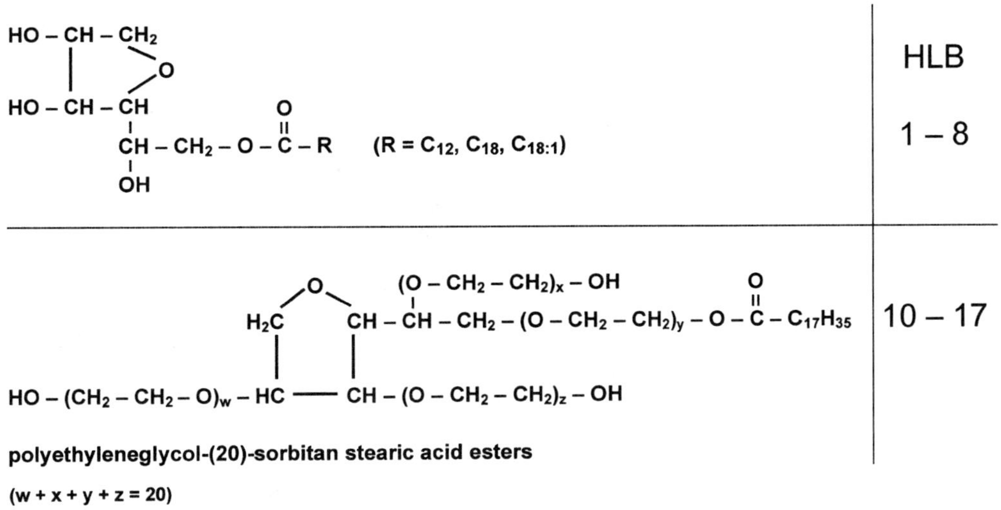

FIG. 19 Hydrophilicity of sorbitan esters.

TABLE 2 Fields of Application and Production Capacities for Sugar-Based Surfactants

Surfactants	Manufacturers	Fields of application	Production capacity, world (t/a)[a]
Sorbitan esters	Akcros, Dai-ichi Kogyo Seiyaku, Cognis, Kao, ICI, Montedison, PPG, Riken Vitamin, SEPPIC, Witco	Pharmaceuticals, personal care, food, fiber, agrochemicals, coatings, explosives	20,000
Sucrose esters	Croda, Dai-ichi Kogyo Seiyaku, Goldschmidt, Mitsubishi, Sisterna, Weixi Spark	Food, personal care, pharmaceuticals	<4,000
Alkyl polyglycosides	Akzo Nobel, BASF, Cognis, ICI, Kao, Nihon Seika, SEPPIC, Union Carbide	Personal care, detergents, agrochemicals, I + I	80,000
Fatty acid *N*-methyl glucamides	Pfizer/Hatco, Clariant	Detergents	40,000
Methylglucoside esters	Amerchol, Goldschmidt	Personal care, pharmaceuticals	10,000
Anionic alkyl polyglycoside derivatives	Pilot Chemical Co., Lamberti Spa.	Personal care	

[a]Estimated figures based on private communications and literature data, references given in the text.

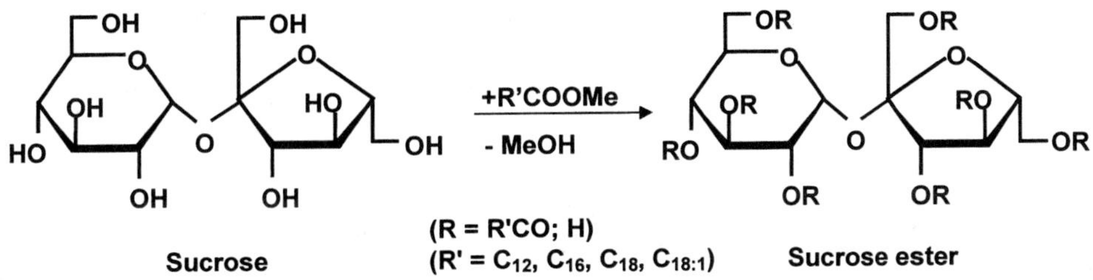

FIG. 20 Synthesis of sucrose esters by base-catalyzed (K_2CO_3) transesterification with fatty acid methyl esters (R′COOMe), usually carried out in solvents (e.g., dimethyl formamide at ~90°C) or microemulsions.

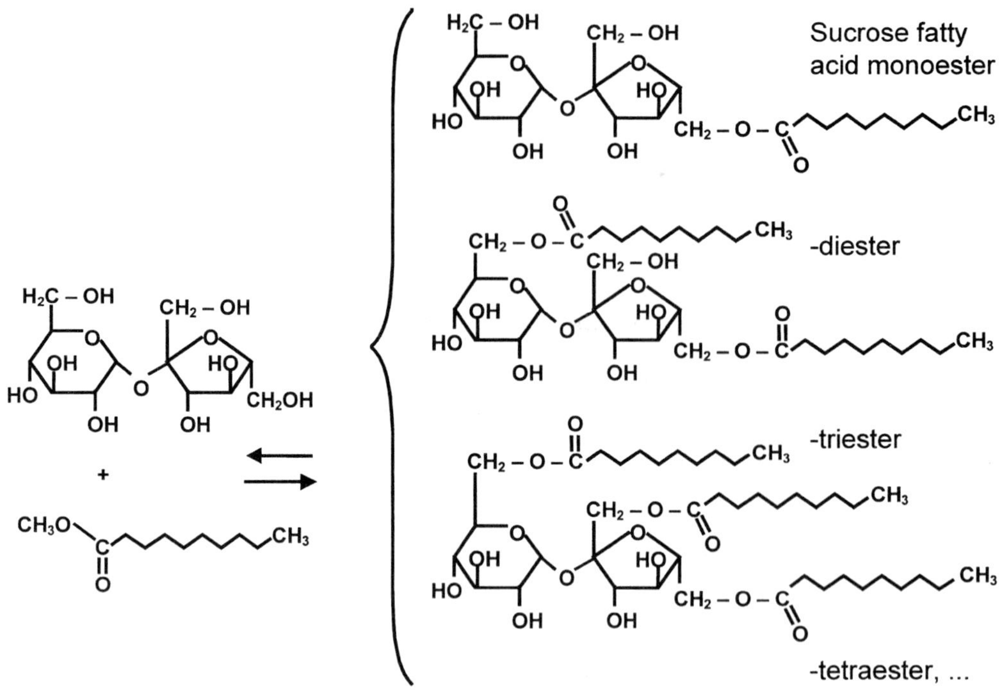

FIG. 21 Product composition and equilibrium in the synthesis of sucrose esters by transesterification with fatty acid methyl ester.

than the actual market potential, which is estimated to be less than 4000 tons per annum. However, demand and market volume could increase substantially if reaction processes, especially for the synthesis of high-mono products, can be further optimized.

C. Glucose-Derived Surfactants

The first step in overcoming the problem of nonselective derivatization of carbohydrates was achieved when Emil Fischer discovered the reaction of glucose with alcohol to form alkyl glucosides [190]. The glucosidation reaction is highly selective because of the hemiacetal function in the glucose molecule and the resulting high reactivity of the hydroxyl group at C-1. The same is true for the synthesis of fatty acid glucamides. Here the glucose molecule reacts initially with methylamine, which, after hydrogenation, selectively yields the glucamine as an intermediate [191]. Further derivatization with fatty acid methyl ester leads to the desired product.

1. Synthesis of Alkyl Polyglycosides

The first syntheses of alkyl polyglycosides were carried out more than 100 years ago. In the course of further developments, the reaction of glucose with alcohols was applied to long-chain alcohols with alkyl chains from C_8 to C_{16}. The result of the reaction is a complex mixture of alkyl mono-, di-, tri-, and oligoglycosides as a mixture of α- and β-anomers (Fig. 24). Therefore, the industrial products are called alkyl polyglycosides. The products are characterized by the length of the alkyl chain and the average number of glucose units linked to it—the degree of polymerization (DP) [192].

The crucial point with regard to the development of an industrial process was to establish reaction conditions that allowed the manufacturing of high-quality products under safe and economically acceptable conditions. This was achieved by optimizing the reaction parameters temperature, pressure, reaction time, and ratio of glucose to fatty alcohol. Of equal importance was the design of a special distillation technology to remove the excess fatty alcohol as smoothly as possible, as well as appropriate bleaching and stabilization in the final treatment step (Fig. 25). This so-called direct synthesis of alkyl polyglycosides is the currently preferred manufacturing mode. However, two-stage processes have been developed as well and are used, for example, by Hüls AG on a pilot plant scale. The breakthrough in the production of long-chain ($C_{12/14}$) alkyl polyglycoside occurred in 1992 with the inauguration of an approximately 25,000 tons per annum production plant for APG surfactants by Henkel Corporation in the

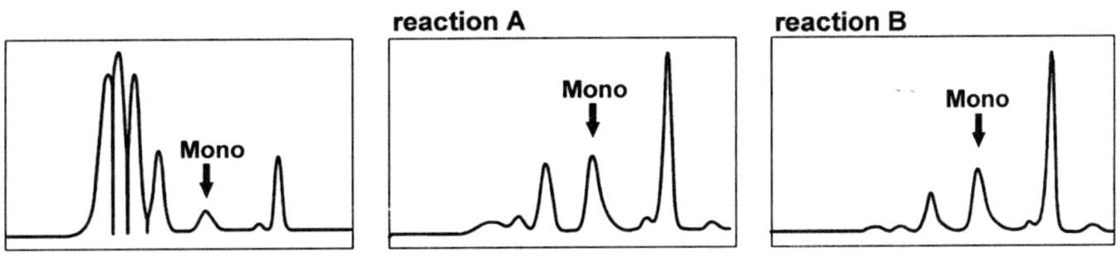

t	= 0 h	= 45 min	= 75 min
DE	= ca. 4	= ca. 1.8	= ca. 1.6
Mono	= 4%	= 48%	= 57%
Di	= 13%	= 35%	= 30%
Tri +	= >70%	= 17%	= 13%

(Mono/Di/Tri+ values: rel.)

FIG. 22 High mono sucrose esters by transesterification of sucrose in excess fatty acid methyl ester (without additional solvent) at about 140°C (~20 h) and simultaneous removal of methanol to form sucrose esters with a high degree of esterification (DE = ~4), subsequent alcoholysis of the sucrose ester by addition of methanol at 75°C (reaction A: 45 min, reaction B: 75 min) to form sucrose monoester and fatty acid methyl ester, and removal of residual methanol and fatty acid methyl ester using a thin-film evaporator (160°C, 0.2–0.3 mbar). Product analysis by high-performance liquid chromatography shows the increase of the sucrose monoester (Mono) content to 48% (DE = ~1.8, reaction A) and 57% (DE = ~1.6, reaction B) relative to sucrose diester (Di) and sucrose oligoesters (Tri+).

United States and in 1995 with the opening of a second plant of equal capacity by Henkel in Germany.

Today, the main producers of alkyl polyglycosides are Cognis, Seppic, ICI, Kao, Union Carbide, and BASF with an estimated total production capacity of approximately 80,000 tons per annum. The main applications for the $C_{12/14}$ alkyl polyglycosides are liquid dishwashing agents and detergents and personal care products. For the $C_{8/10}$ (or branched C_8) alkyl polyglycosides, there are hard surface cleaners, agrochemicals, and products for industrial, institutional, and personal care cleansing (Table 2).

2. Fatty Acid Glucamides

The synthesis to produce fatty acid glucamides involves the reaction of glucose with methylamine, under reductive conditions, to form the corresponding *N*-methylglucamine. In a subsequent reaction step, this intermediate is converted with fatty acid methyl ester to the corresponding fatty acid amide. Compared with

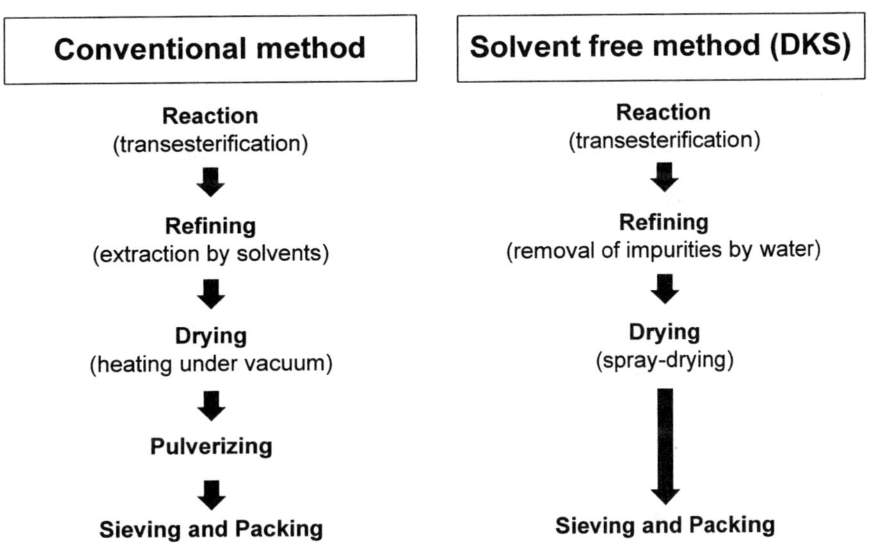

FIG. 23 Process schemes for the production of sucrose esters.

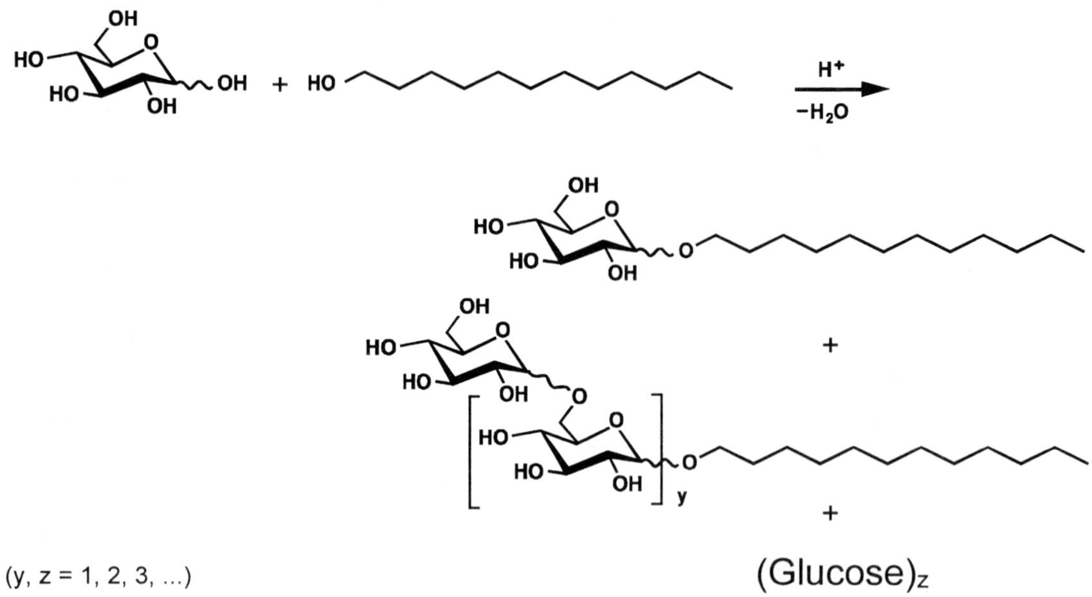

FIG. 24 Synthesis of alkyl polyglycosides by acid-catalyzed (*para*-toluenesulfonic acid) acetalization of glucose in molar excess of fatty alcohol (2- to 6-fold) and removal of water under vacuum at 100–120°C.

the alkyl polyglycosides, fatty acid glucamides are composed of only a single carbohydrate molecule attached to the fatty acid chain (Fig. 26). This is one reason why fatty acid glucamides are less soluble and tend to crystallize more easily from aqueous solutions.

Figure 26 shows the manufacturing scheme for the production of fatty acid glucamides. To avoid significant amounts of unreacted *N*-methylglucamine, which could be considered as potential precursors for nitrosamines, Procter & Gamble has developed an optional reaction with acetic anhydride in the finished product. Free secondary amines can be acetylated in this step, and the resulting acetates can remain in the final product [193].

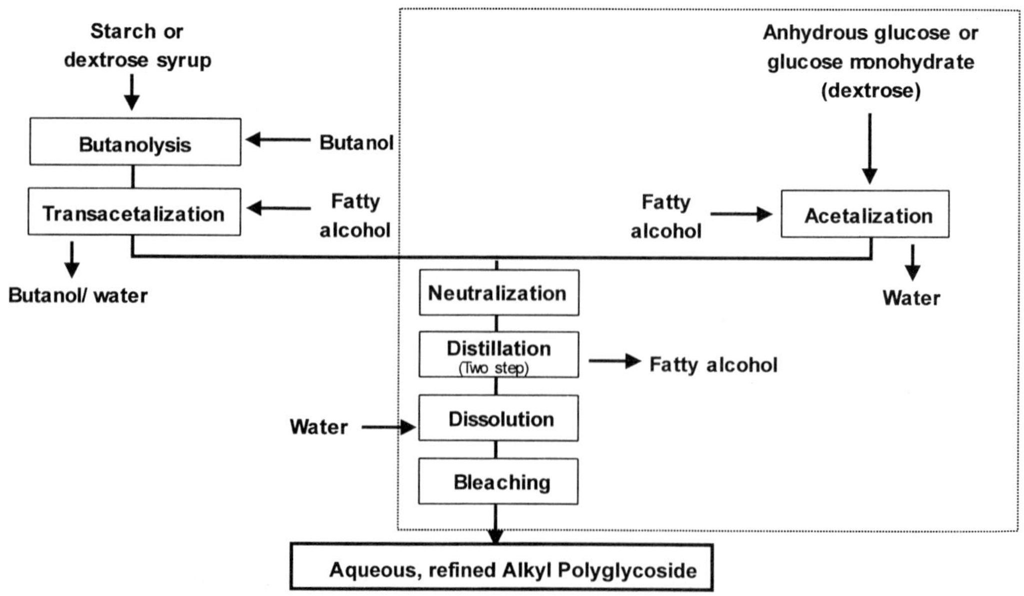

FIG. 25 Scheme for the production of alkyl polyglycosides.

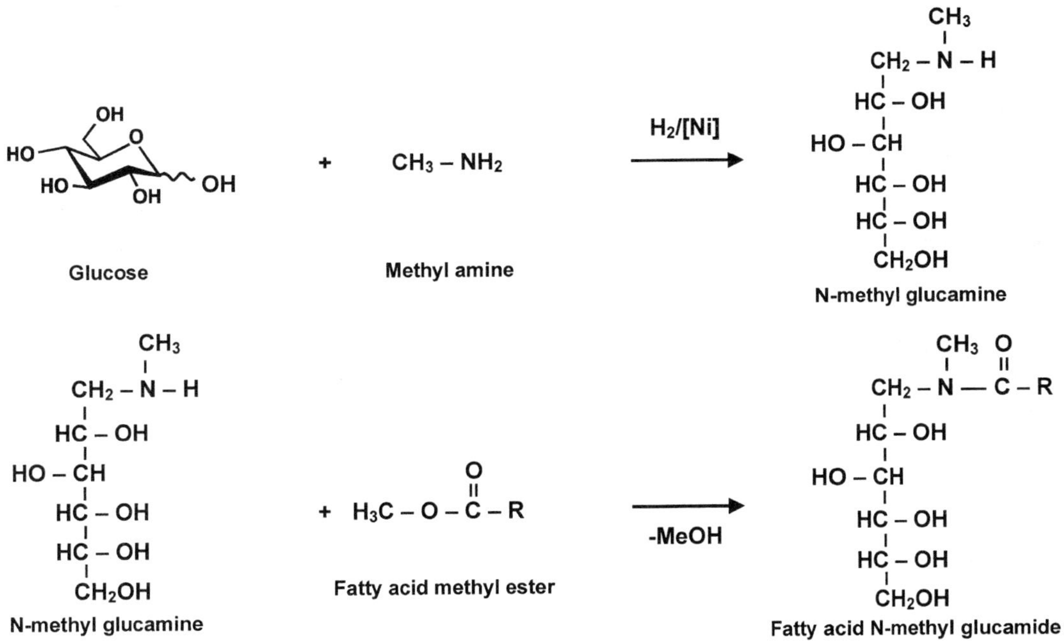

FIG. 26 Two-step synthesis of fatty acid glucamides by reductive alkylation of methylamine with glucose using Raney nickel as the hydrogenation catalyst to obtain *N*-methyl glucamine, which is acylated by base-catalyzed reaction with fatty acid methyl ester in a second step.

The existing production capacity is estimated to be 30,000 to 50,000 tons per annum active substance according to a study by Colin A. Houston & Associates [194]. Producers are Pfizer, Hatco in the United States, and Clariant (formerly Hoechst) in Germany (Table 2).

3. Properties of Alkyl Polyglycosides and Fatty Acid Glucamides

With regard to their basic physicochemical properties, such as surface and interfacial tension and critical micelle concentration, alkyl polyglycosides and fatty acid glucamides ($C_{12/14}$) are very comparable. There are slight differences in the basic foam behavior for the pure sugar-based surfactants as well as binary combinations. With regard to their ecological, toxicological, and dermatological properties, alkyl polyglycosides as well as fatty acid glucamides can be considered as surfactants with extraordinary product safety characteristics. This has been proved for both products in a series of detailed investigations. The results are published in several papers, mainly by Henkel and Procter & Gamble but also by independent research institutes [195].

Although it can be concluded that alkyl polyglycosides and fatty acid glucamides are more or less comparable with regard to their basic performance in detergents and dishwashing agents, there might be differences in specific product formulations. If, for example, the stability of concentrated manual dishwashing detergents is investigated, as in the case of a paste based on alkyl ether sulfate and alkyl polyethylene glycol ether, it is found that best results are obtained when alkyl polyglycosides are used as cosurfactants (Table 3) [196]. In general, glucose-derived surfactants have shown to be very efficient components in manual dishwashing detergents and liquid and powder detergents [196]. In contrast to alkyl polyglycosides, fatty acid glucamides are thus far not known in applications other than detergents.

In personal care products, alkyl polyglycosides represent a new concept in compatibility and care. They may be combined with conventional components and can even replace them in new types of formulations, leading to a broad spectrum of supplementary effects. With regard to foam, they are comparable to betaines and sulfosuccinates but do not match the foam volume of alkyl ether sulfates. On the other hand, alkyl polyglycosides can stabilize the foam of anionics in hard water and in the presence of sebum. The alkyl polyglycoside foam consists of finer bubbles and is more creamy than in the case of SLES (Fig. 27) [197a].

To demonstrate the large performance spectrum of alkyl polyglycosides, one more application should

TABLE 3 Stability of Concentrated Manual Dishwashing Detergents (Pastes)

Ingredients	Product 1 (wt%)	Product 2 (wt%)	Product 3 (wt%)	Product 4 (wt%)
Alkyl ether sulfate	10	10	10	10
Alkyl polyethylene glycol ether	15	15	15	15
Fatty acid alkanolamide	18	—	—	—
Alkylamidobetaine	—	18	—	—
Fatty acid glucamide	—	—	18	—
Alkyl polyglycoside	—	—	—	18
Appearance at 68°F (20°C)	Cloudy	Gel	Clear	Clear
Pour point	—	—	54°F (12°C)	32°F (0°C)
Storage test 3 weeks, 41°F (5°C)	Solid	Solid	Solid	Clear-liquid

be mentioned briefly. Alkyl polyglycosides ($C_{8/10}$ and $C_{12/14}$) have been shown to be substitutes for alkyl phenol ethoxylates in agrochemical formulations. They lead to higher salt tolerances and show good results as adjuvants in several post applied herbicides, such as control of giant foxtail in soybeans with Assure II (DuPont) and control of common lambsquarters in soybean with Pursuit (American Cyanamid). Currently, the short-chain products ($C_{8/10}$ and C_{9-11}) are approved as inert ingredients by the U.S. Environmental Protection Agency (USEPA) [198].

D. Derivatives of Alkyl Polyglycosides

In principle, two different approaches exist to combine a hydrophobic alkyl chain with the hydrophilic glucose molecule. These are glycosylation, the reaction of glucose with an alcohol as described earlier, and acylation, the esterification or amidation of a suitable glucose derivative, such as the alkyl polyglycosides.

Because alkyl polyglycosides are available in sufficient quantities and at competitive costs at present, their use as a raw material for the development of specialty surfactants has generated considerable interest. The derivatization of alkyl polyglycosides is currently being pursued with a goal to modify the surfactant properties of alkyl polyglycosides [192b,199]. A broad range of alkyl polyglycoside derivatives can be obtained by using relatively simple methods, for example, nucleophilic substitution. In addition to the reaction to esters or ethoxylates, ionic alkyl polyglycoside derivatives, such as sulfates and phosphates, can be synthesized. However, only a few products are established in the market: methylglucoside esters and a series of special esters based on alkyl polyglycosides.

1. Methyl Glucoside Esters

Esterification of methyl glucoside with methyl esters of stearic or oleic acid enhances the lipophilic character (Fig. 28). Methyl glucoside esters are, in contrast to

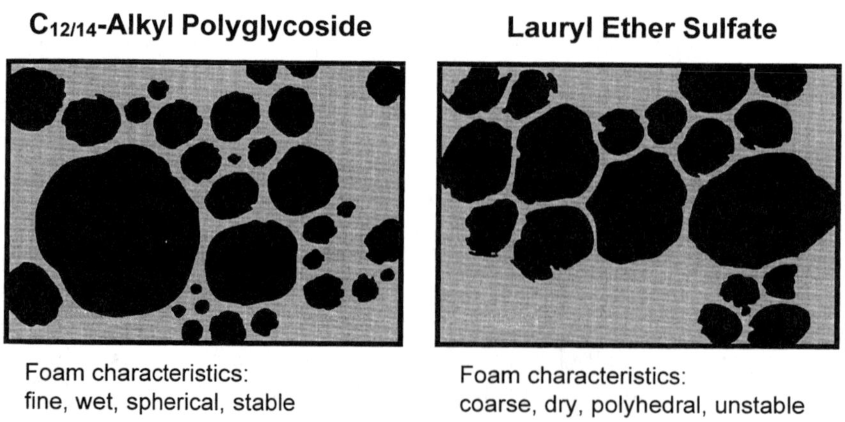

FIG. 27 Foam structure of surfactant solutions.

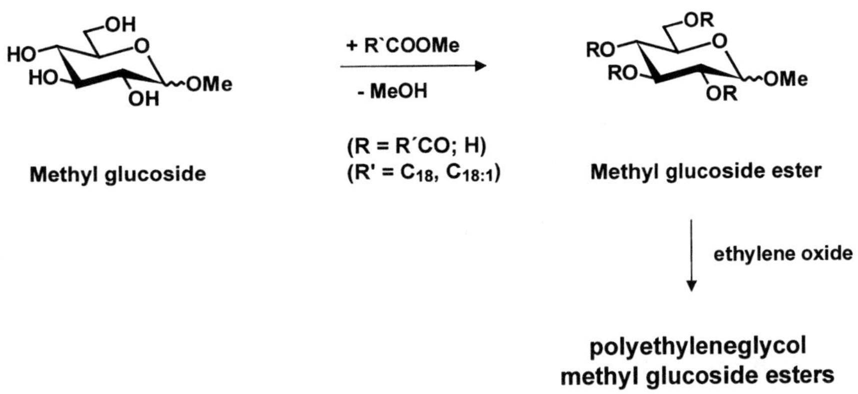

FIG. 28 Synthesis of methyl glucoside ester by base-catalyzed (K_2CO_3) transesterification of methyl glucoside with fatty acid methyl ester (R′COOMe) at 120–160°C.

alkyl polyglycosides with the same hydrophobic chain length, hardly soluble in water, but they exhibit excellent emulsification properties [189b,200]. They have found application as emollients, moisturizing and emulsifying agents, and thickeners for cosmetics. The hydrocarbon length and degree of substitution can be varied to obtain specific w/o emulsification behavior. These surfactants can be further ethoxylated to give rise to polyethylene glycol methyl glucoside esters.

Major manufacturers for methyl glucoside esters are Amerchol and Goldschmidt. The total market size, including the ethoxylated products, is estimated to be 10,000 tons per annum (Table 2).

2. Anionic Derivatives of Alkyl Polyglycosides

Cesalpina Chemicals, a subsidiary of Lamberti Spa., Italy, has introduced three nonionic alkyl polyglycoside esters (AGEs), namely citrates, sulfosuccinates, and tartrates, that can be used in personal care applications [201a]. The syntheses start with an alkyl polyglycoside, which is esterified with citric acid, maleic anhydride and subsequent sulfonation, and tartaric acid, respectively. Structures are shown in Fig. 29. The products will be marketed in the United States by Pilot Chemical Co. (Table 2) [201b].

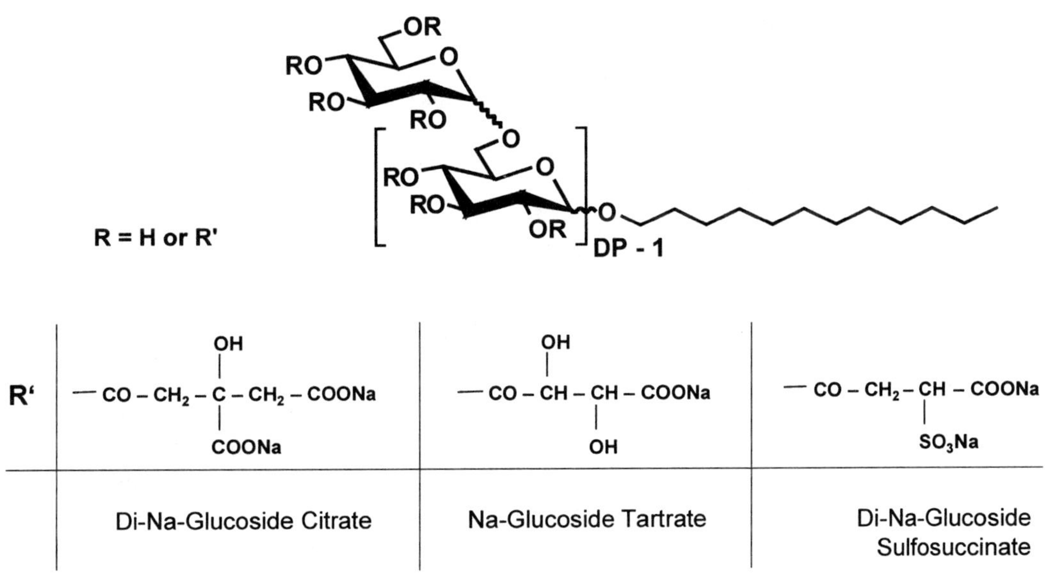

FIG. 29 Examples of anionic alkyl polyglycoside derivatives.

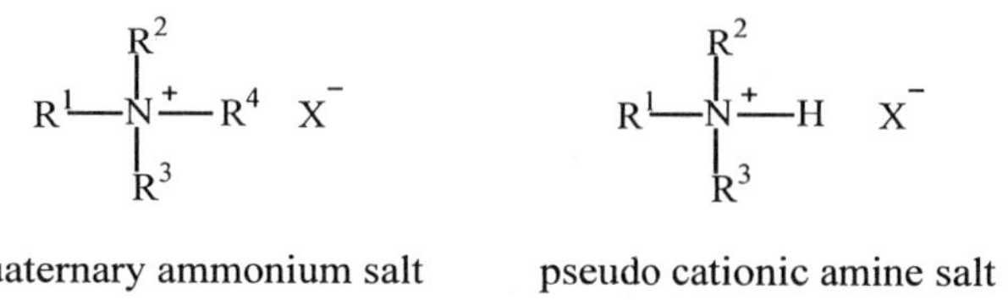

FIG. 30 Structures of tetra-coordinated ammonium salts.

V. CATIONIC SURFACTANTS

A. Syntheses of Cationic Surfactants

Commonly used cationic surfactants all contain a quaternary respective tetracoordinated nitrogen atom. To possess a cationic character, these tetraalkyl ammonium salts need at least one longer alkyl chain. The simple salts of long-chain amines are often erroneously described as quaternaries as well. However, contrary to true cationic surfactants, these so-called pseudocationics are formed by neutralization with acids and thus represent protonated amines (Fig. 30). The substantial difference between both types exists in the fact that the pseudocationics show surfactant (cationic) properties only at pH values significantly lower than 7. However, this section is devoted exclusively to true quaternary surfactants.

The product class of surfactants consists of a variety of types and preparation methods. Therefore, only some basic knowledge can be considered in the scope of this review. Quaternary ammonium compounds based on nonionic surfactants are described as a specialty in Ref. 202.

Quantitatively, the most important cationics are obtained by the reaction of tertiary amines with classical alkylating reagents such as methyl chloride, dimethyl sulfate, benzyl chloride, and, infrequently, trimethyl phosphate or methyl tosylate according to Scheme 18. The residues R^1–R^3 represent at least one longer alkyl chain, another longer alkyl chain, or a short alkyl group such as methyl. The R_4 stands for the alkyl or aryl part of the alkylating reagent, mostly methyl or benzyl, very rarely for a longer alkyl chain. Basically, the quaternization reaction is more quantitative if only one longer alkyl chain is present and the other chains are methyl groups. The quaternization reaction is carried out, normally, at a temperature between 80 and 100°C in substance or, depending on the viscosity and/or consistency, in a solvent.

SCHEME 18

Sometimes, primary or secondary amines may also serve as raw materials. First, however, they must be converted to tertiary amines by means of several alkylating methods, ultimately followed by the quaternization reaction (Scheme 19). The last reaction, Eq. (3), is of special interest because of the manufacturing of cationics on the basis of Guerbet alcohols [203].

Because of their easy biodegradability, so-called ester quats have become more important [204–206]. In principle, the synthesis is made by an esterification reaction of tertiary alkanolamines with 1.5–2 moles of fatty acids and the following quaternization (Scheme 20). Naturally, in practice, statistical mixtures of mono-, di-, and triester derivatives will be found.

Another type of ester quat is derived from the natural substance choline and can be understood as an ester thereof. The synthesis of such cholinesters is based on dimethyl ethanolamine as a raw material (Scheme 21).

Also remarkable are cationic protein derivatives (Fig. 31) that are used for cosmetic purposes. For preparation, a partially hydrolyzed protein is reacted with epichlorohydrin and then added to a tertiary amine [207].

Furthermore, polymeric cationics play an important role, not with regard to the quantity but with a view to the application. There are numerous possibilities for their synthesis. For example, acrylic or methacrylic esters of dimethyl ethanolamine are used as one part in copolymers. Afterward, they can be quanternized (Scheme 22). Special reasons can make desirable an exchange of the anion, which is normally prescribed by the alkylating agents. Reference 208 presents a solution for such demands (Scheme 23).

A completely different route for the synthesis of cationic surfactants is a quaternization reaction with alkylene oxide in the presence of water [209,210] as follows (Scheme 24). The free quaternary ammonium base produced may be neutralized with any acid. Primary and secondary amines can be used advantageously in this reaction. They form quaternary derivatives with hydroxyethyl or polyoxyalkyl ether groups if the amount of epoxide is appropriate [211].

During the reaction, the thermal instability of the free quaternary ammonium bases obtained in the aqueous environment is an unpleasant disadvantage of the procedure and results in a minor yield. But fortunately, if the amine is neutralized before the addition of alkylene oxide, high yields of cationics are available

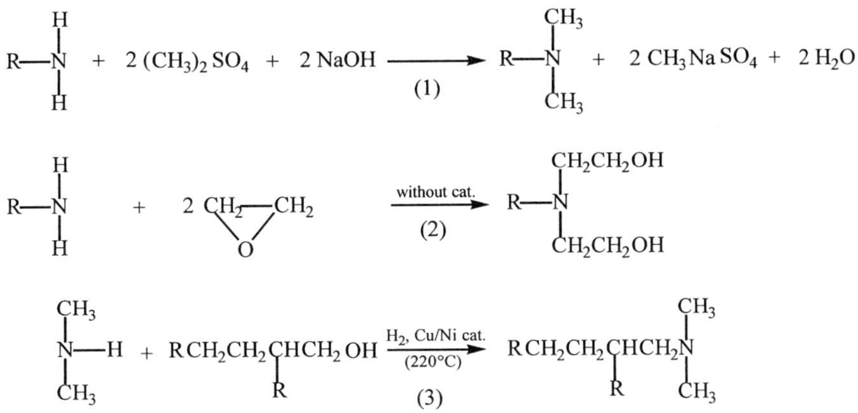

SCHEME 19

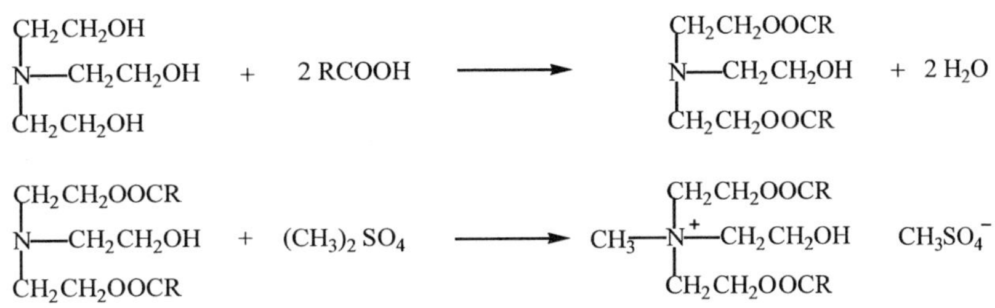

SCHEME 20

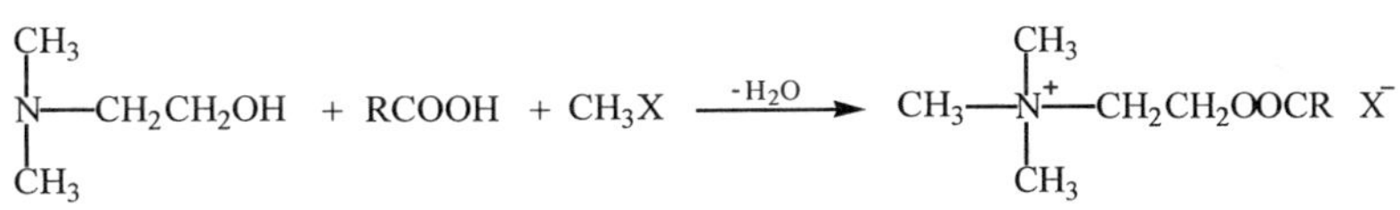

SCHEME 21

[212]. Inorganic as well as organic acids with a weak or strong character and even partial esters of polyvalent acids are suitable. The optimal reaction temperature is approximately 80°C. It should be mentioned that no strong acids force the formation of dioxane as a by-product if ethylene oxide was applied. Also, alkylene glycols are always present. Because of hydrolysis reactions, ester amines are unsuitable for this preparation method.

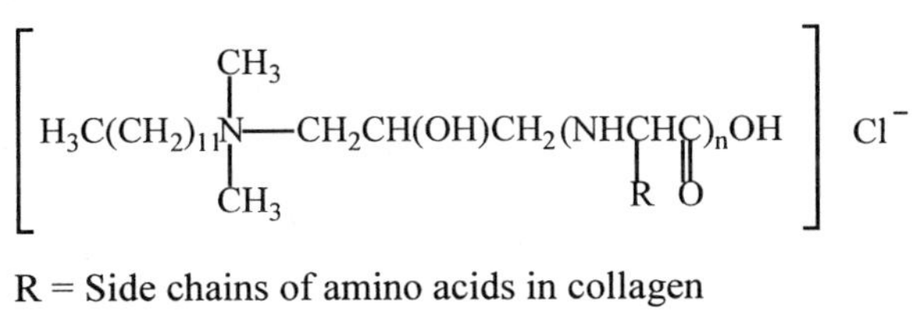

FIG. 31 Cationic protein derivative.

As per the following, a further interesting route for the preparation of polyquaternary compounds should be mentioned. Oxethylated fatty amines are polycondensed with dicarboxylic acids, i.e., adipic acid [213], and afterward quaternized with ethylene oxide [214] (Scheme 25).

Most of the quaternization reactions here are performed with ethylene oxide or propylene oxide as well. Analogous works with long-chain epoxides normally failed because of the heterogeneity of the aqueous mixtures with the tertiary amine salts. This problem, however, has been solved [215] by the use of a phase transfer catalyst (PTC) such as dimethyl distearyl ammonium chloride. Because of its versatility, many other reaction systems are conceivable (Scheme 26).

Another class with minor technological importance are the so-called betaine esters. The somewhat misleading name can be traced back to the property that

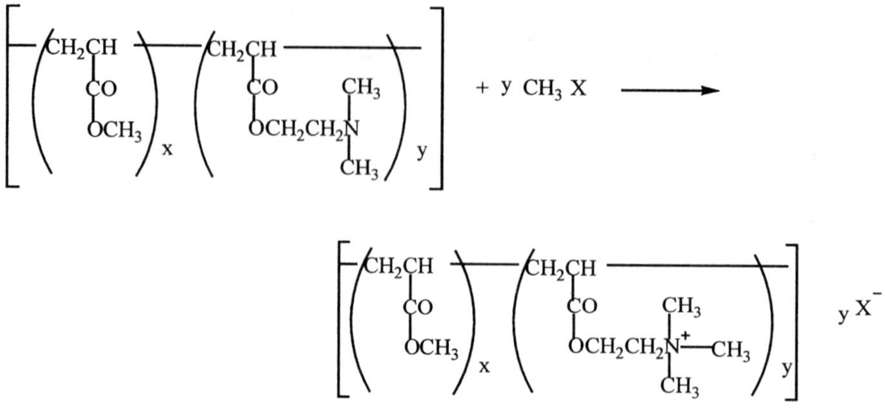

SCHEME 22

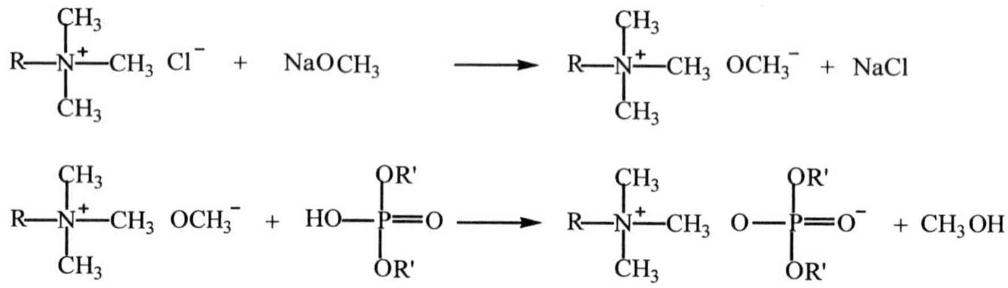

SCHEME 23

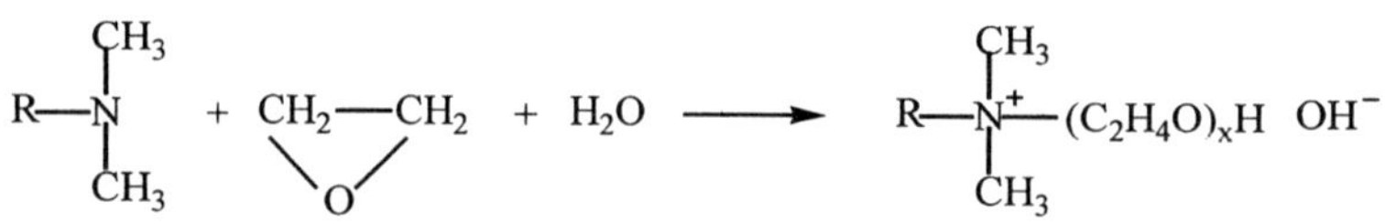

SCHEME 24

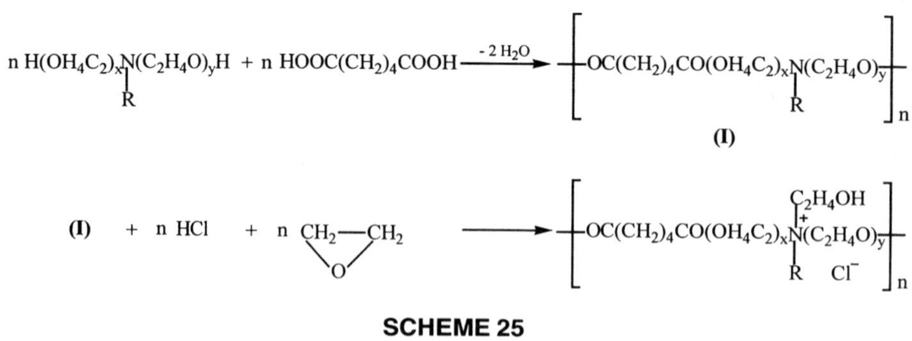

SCHEME 25

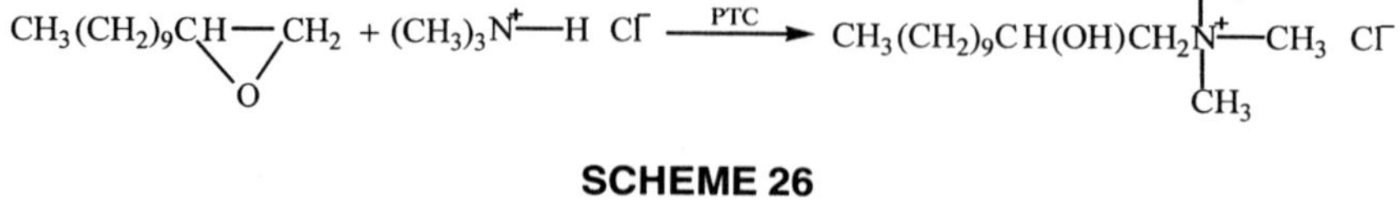

SCHEME 26

after hydrolysis of the ester group, a betaine results. The synthesis itself takes place by a quaternization of tertiary amines with, usually, chloroacetic acid esters. Basically, there are several possibilities:

1. The longer alkyl chain is part of the amine [216] [Eq. (4)].
2. The longer alkyl chain is part of the chloroacetic acid ester [217] [Eq. (5)].
3. Both reagents contain a longer alkyl chain [218] [Eq. (6)].

$$RN(CH_3)_2 + ClCH_2COOC_2H_5 \xrightarrow{(4)} R{-}N^+(CH_3)_2{-}CH_2COOC_2H_5\ Cl^-$$

$$CH_3N(CH_3)_2 + ClCH_2COOR' \xrightarrow{(5)} CH_3N^+(CH_3)_2{-}CH_2COOR'\ Cl^-$$

$$RCONH(CH_2)_3N(CH_3)_2 + ClCH_2COO(C_2H_4O)_x\ R' \xrightarrow{(6)} RCONH(CH_2)_3N^+(CH_3)_2{-}CH_2COO(C_2H_4O)_xR'\ Cl^-$$

Using the same principle, multiple betaine esters based on multifunctional tertiary amines or alcohols can be produced.

Also remarkable are cationic surfactants that are prepared by the addition of reactive substances, usually with an alcoholic group, with a preformed quaternary ammonium group. Specifically, glycidyl trimethylammonium chloride and 3-chloro-2-hydroxypropyl trimethylammonium chloride are suggested for such reactions (Scheme 27).

B. Properties of Cationic Surfactants

1. Physicochemical Behavior

Depending on the alkyl chain, the number of longer alkyl chains, the fundamental chemistry, and, sometimes, the kind of anion, the properties of cationic surfactants are altogether highly varied. The main feature for all quaternary ammonium compounds is the substantivity to almost any surfaces that are negatively charged. By choosing a suitable product, the characteristics of many substrates may be influenced. Examples are the softness of textiles, performance of laundry detergents, antistatic behavior, corrosion inhibition, flotation processes, hydrophobic finishes, microbial treatment of hard surfaces, fixing of dyes, etc. The field of applications is almost infinitely extensive.

Regarding ambient conditions, quaternary ammonium salts are considerably stable. Only at temperatures above 100°C do they decompose through a dealkylation reaction. The incompatibility with anionic surfactants is also disadvantageous. However, the incorporation of polyoxyethylene ether chains [202] prevents this "malfunction."

Ester quats as well as betaine esters are sensitive to hydrolysis, which makes them easily decomposable into the sometimes desired non-surface-active compounds. For special information, hydrolysis studies of betaine esters can be found in Ref. 219.

2. Ecological and Toxicological Behavior

Both properties depend, to a high degree, on the kind of molecule and the chain length of the alkyl group(s). The quaternary ammonium salts that contain ester functions are easily biodegradable, whereas the other types of cationic surfactants are not eliminated even after a longer adaptation time. A study of the biodegradation

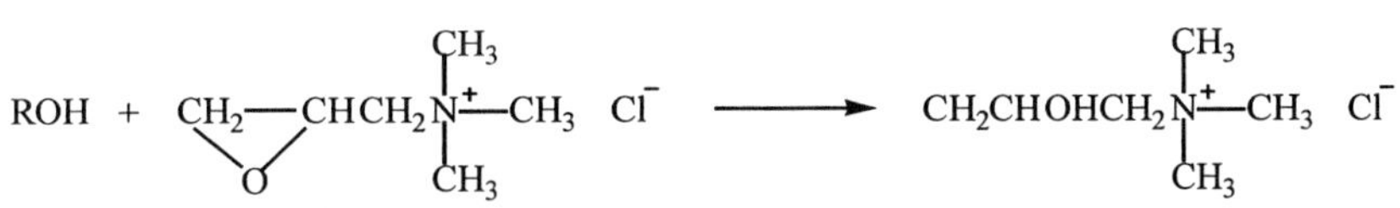

SCHEME 27

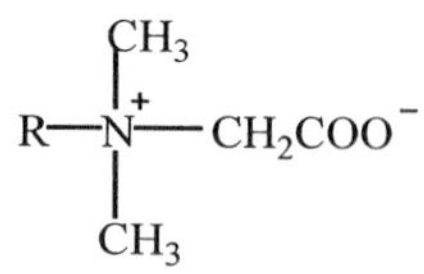

FIG. 32 Structure of betaines.

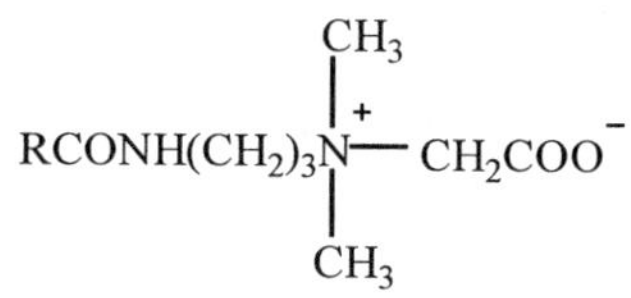

FIG. 33 Structure of an alkylamidopropyl betaine.

of the latter types [220] describes a pathway in which amine oxides are formed as intermediates.

Because of the variety of cationic surfactant types, it is impossible to make generally binding statements about toxicity within the scope of a short review. Only a few aspects should be highlighted.

Sometimes, the alkyl chain length may be responsible for toxicological effects. A comparison between behenyl and stearyl derivatives regarding eye and skin irritation showed that the behenyl chain dramatically lowers irritation potential [221].

Because of their excellent biodegradability, ester quats show moderate aquatoxicity (Biological Laboratories of Henkel KGaA, private communications, 1988). Also, human toxicity with regard to skin and mucous membrane irritation, acute toxicity (oral/dermal), mutagenicity, and sensitization has been evaluated as very low for ester quats with long alkyl chains. Medium-chain (C_8–C_{12}) ester quats, however, may act as biocides, as may other quaternary ammonium salts with a comparable alkyl chain length range.

VI. AMPHOTERIC SURFACTANTS

A. Betaines

It was approximately 1969 when betaines were proposed as cosurfactants for shampoo formulations [222]. The mildness to skin and eyes has been the decisive reason.

Regarding the chemistry, betaines are homologues of trimethyl glycinate, which was discovered more than a century ago in sugar beet (*Beta vulgaris*) juice. The general structure is depicted in Fig. 32. The residue R normally represents an alkyl or an alkylamido popyl group based on coconut or palm kernel oil. The main carbon chain distribution includes the C_8–C_{18} range. Sometimes products with a narrower chain distribution, i.e., C_{12}–C_{18}, or pure C_{12} are prominent. Because of difficult handling, long-chain betaines play only a subordinate role.

The synthesis proceeds relatively simply according to the following reaction (Scheme 28). In principle, the reaction must be understood as a quaternization reaction of a tertiary amine with monochloroacetate as an alkylating reagent. A high degree of conversion occurs only under the assumption that during the alkylation process, the salt form of chloroacetic acid is present because the free acid would block the amine function. Therefore, weakly alkaline conditions, analogous to the dissociation degree of chloroacetic acid, are recommended. Also, a slight excess of chloroacetate, usually the sodium salt, increases the yield. Depending on the type of tertiary amine used, a reaction temperature of 70–95°C is required. Normally, the reaction is carried out in an aqueous solution resulting in a final concentration of 30% betaine. Because of a tendency to gelatinize, slightly higher betaine concentrations may be possible only by adding special hydrotropes, such as polyols or fatty acids [223,224].

Besides the alkyl betaines already shown, alkylamidopropyl betaines (Fig. 33) are the predominant commercialized types. The preparation is a two-step process. First, fatty acids or their esters (glycerides or methyl esters) are condensed with, usually, dimethylaminopropyl amine, followed by reaction with sodium choroacetate [225] (Scheme 29). For a quantitative yield of intermediate, a redistillable excess of amine should be applied. The reaction temperature is limited by the amine boiling point of approximately 140°C. The following betainization reaction takes place under conditions similar to those described before.

Usually, the sodium chloride generated remains in the betaine solution. Sometimes, however, special applications require salt-free products. For this reason, the

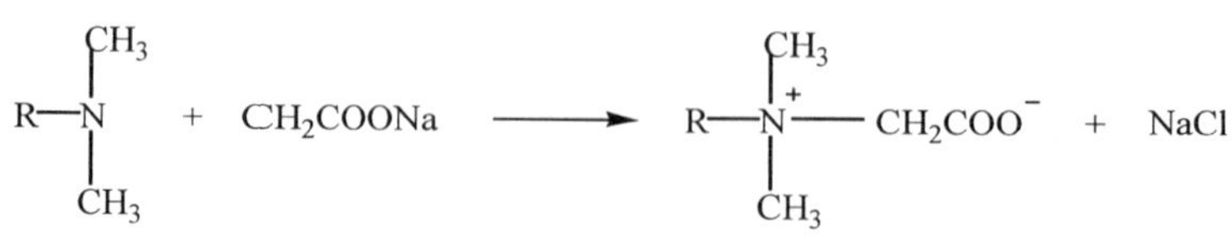

SCHEME 28

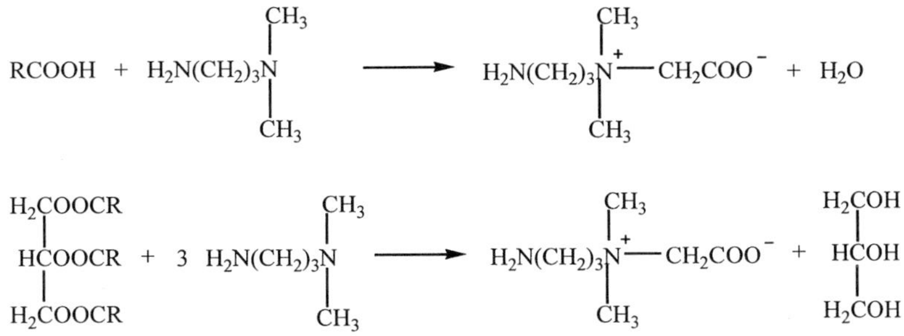

SCHEME 29

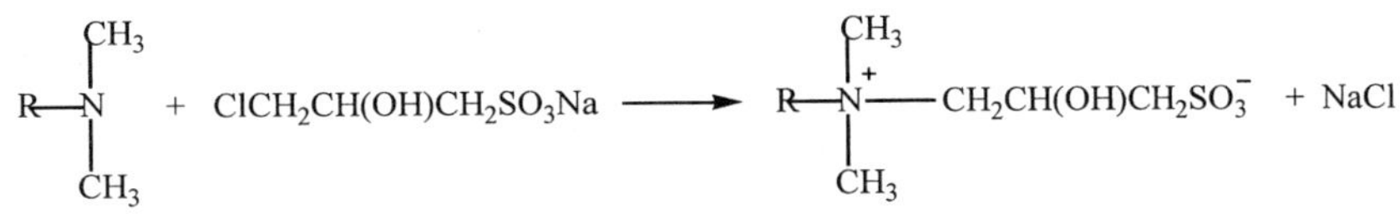

SCHEME 30

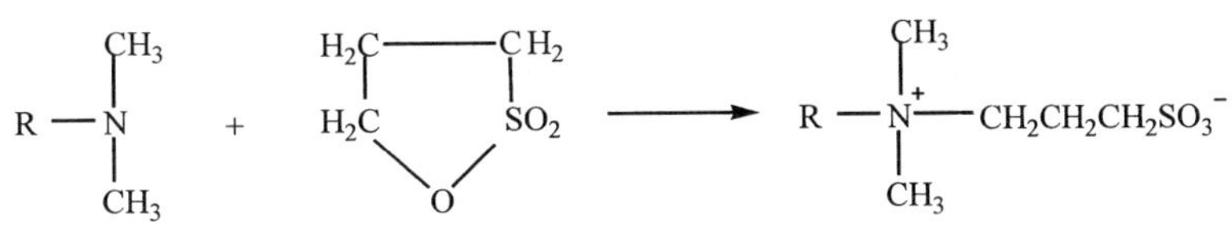

SCHEME 31

reaction is carried out in an alcoholic solution with subsequent filtration of the precipitated salt [226,227] or by means of ultrafiltration methods [228] (N. Kühne and G. Uphues, unpublished results, Henkel KGaA, 1991).

With regard to other by-products and trace impurities, environmental and product safety is becoming more and more relevant. A partial hydrolysis of monochloroacetate is responsible for a small content of harmless glycolic acid. Careful control of reaction conditions limits the amount. Concerning the removal of critical impurities, i.e., free amines and monochloro- and dichloroacetic acid, some hints are given in Ref. 229.

Another commercial betaine type is the sulfobetaines, also called sultaines. They are prepared similarly to the common types with chlorosulfonates instead of chloroacetate, usually 1-chloro-2-hydroxypropane sulfonate [230], as alkylating reagents (Scheme 30). In earlier times, propane sultone was the agent used for the manufacture of sulfobetaines. The high carcinogenic potential of sultones, however, prohibited the use of such substances more than 20 years ago (Scheme 31).

Of somewhat academic interest may be a synthesis route described for the preparation of sulfato betaines [231]. According to this method, a tertiary amine is reacted with sulfur trioxide followed by an insertion reaction of ethylene oxide. Ethylene or propylene carbonate is used as an inert solvent (Scheme 32).

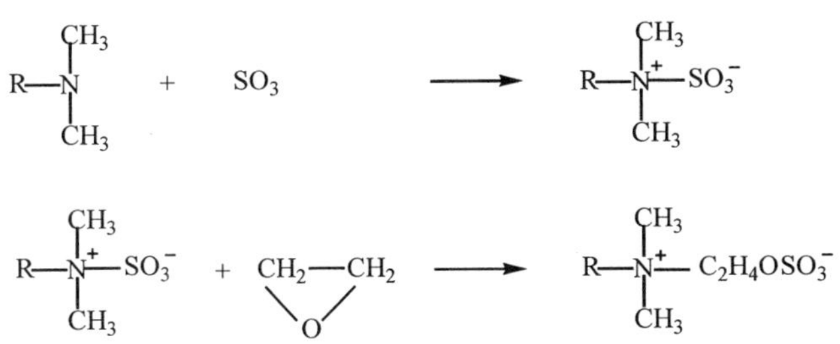

SCHEME 32

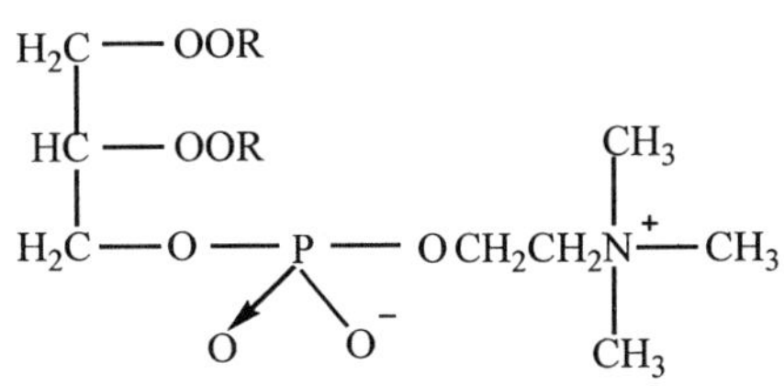

FIG. 34 Structure of α-lecithin.

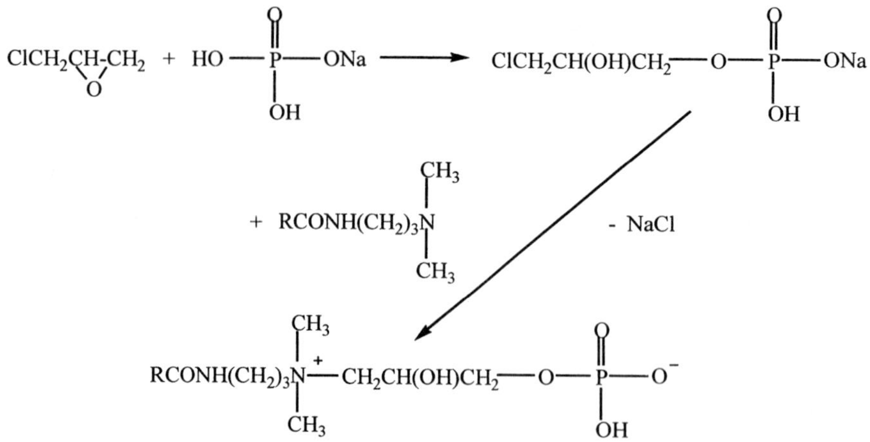

SCHEME 33

Lecithin (Fig. 34) is a further betaine type that is produced in a wide range in nature. Regarding the chemistry, lecithin can be designated as a phosphobetaine.

Hence, it is not surprising that synthesis routes for other phosphorus-containing betaines were developed. Admittedly, such substances have attained no great importance, but two interesting preparation methods should be mentioned.

On the basis of Ref. 232, one of many examples is presented. The resulting betaine could be interpreted as analogous to the betaine types already described (Scheme 33).

Another synthesis route is also remarkable [233]. By a kind of *Mannich reaction*, a phosphonate betaine has been obtained (Scheme 34).

B. True Amphoterics

Unlike betaines, true amphoterics do not contain a quaternary nitrogen atom. Simply speaking, the whole family of true amphoterics may be classified as amino acid derivatives. Depending on the strength of the ionic groups and the kind of alkyl residues present, they are capable of forming inner salts at different pH values, known as the isoelectric point or range.

Most types of true amphoterics marketed are derived from imidazolines, sometimes falsely designated as imidazolinium betaines (Fig. 35). But investigations [234,235] have proved that no ring structure exists in commercial products.

The synthesis seems to be relatively simple compared with that of betaines, although sodium mono-

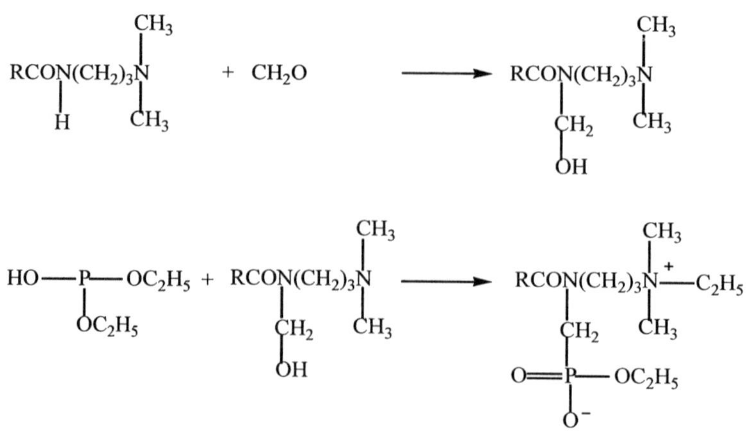

SCHEME 34

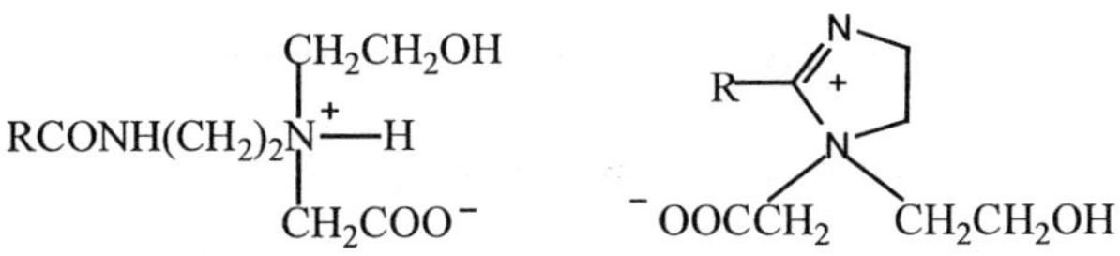

(a) True amphoteric (b) Real imidazolinium betaine

FIG. 35 Structures of imidazoline-based amphoterics.

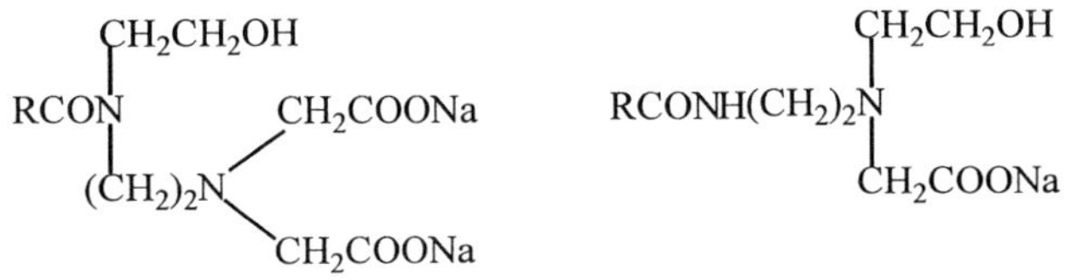

Diacetate Monoacetate

FIG. 36 Structures of imidazoline-based true amphoterics.

chloroacetate serves as the alkylating reagent. In practice, however, the chemistry is rather complicated. In relevant compendiums, monoacetates (Fig. 35a) as well as diacetates are mentioned. These differences have been traced back to the special chemistry of imidazolines.

The preparation itself takes place in a two-step reaction, at first forming an amino amide at a reaction temperature in the range of 150–180°C and ambient pressure, followed by the ring closure under additional vacuum conditions [236] (Scheme 35).

The most important property of imidazolines is instability in the presence of water at a pH value above 7. Even minor amounts of alkalinity suffice to open the ring system. Our own investigations (G. Uphues, unpublished results, Henkel KGaA, 1995) support the theory that the 2,3-double bond will be attacked, but depending on temperature, pH value, or amount of water present, the acyl group shifts more or less rapidly to the other nitrogen atom in the molecule, simulating a ring opening at the 1,2-position (Scheme 36). Both ring-opened substances are able to react with sodium monochloroacetate. Because of the primary amine function, the initial amido amine can add 2 moles of chloroacetate, whereas the other structure reacts with only 1 mole (Fig. 36).

Basically, the amount of alkali is equivalent to the amount of chloroacetate used. The "monoacetate/diacetate" ratio is influenced by the alkaline pH value during the reaction. The higher the pH value, the more monoacetate is formed. Because of the competitive situation with regard to the acyl shift and the alkylation reaction, in diacetates there are always monoacetates present.

For special applications, true amphoterics based on fatty amines are necessary. Preferably, they are synthesized by a *Michael* addition of methyl acrylate to fatty amines [237,239]. Depending on the amount of acrylate, mono and bis adducts are possible (Scheme 37).

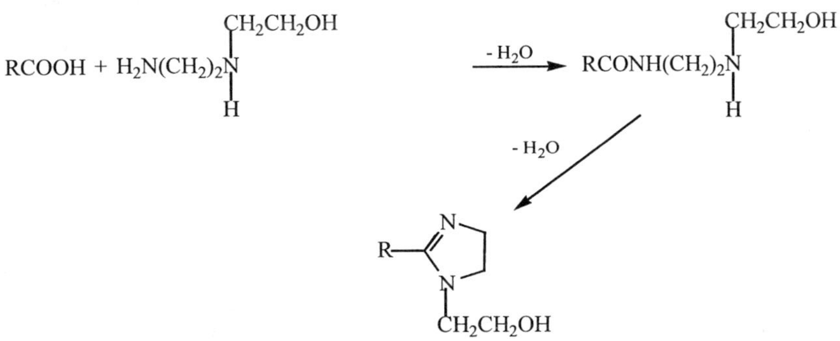

SCHEME 35

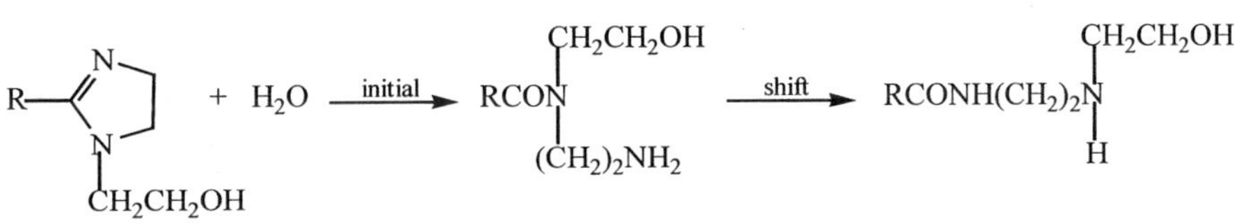

SCHEME 36

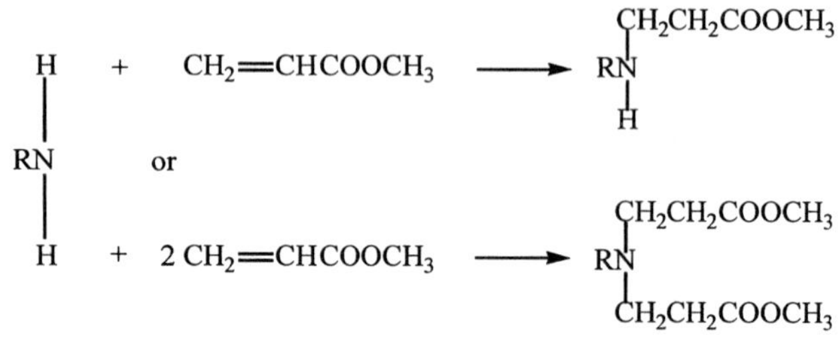

SCHEME 37

The methyl ester groups are hydrolyzed under pressure with various quantities of caustic in an autoclave, producing only sodium salts or mixtures of both salt and acid groups (Scheme 38).

Unfortunately, the methanol generated cannot be removed completely by less expensive methods. As modern cosmetic products require methanol-free ingredients, the so-called propionates are prepared by the addition of acrylic acid. This alternative procedure is restricted by the fact that only diadducts can be obtained. Usually, the reaction is carried out in a neutral aqueous solution forming the monosodium salt (Scheme 39).

Another route for the manufacture of salt-free true amphoterics is the addition of acrylic acid to ring-opened imidazolines. As the addition reaction runs slower than the shift of the acyl group just mentioned, essentially mono adducts are obtained (Scheme 40).

The specialized literature shows numerous other types and synthesis methods for amphoterics, but they have found only small or no commercial interest.

C. Properties of Amphoteric Surfactants

1. Physicochemical Behavior

The particular properties of amphoteric surfactants are related to their zwitterionic character. That means that both anionic and cationic structures are found in one molecule. Differences between betaines and true amphoterics are caused by changing behavior at several pH values.

Regardless of the pH value, betaines permanently represent a four-bonded nitrogen atom. Only at a very low pH value can the anionic group be protonated to take on a cationic character.

Unlike betaines, true amphoterics form salts at pH values higher than the isoelectric point. At lower pH values, the basic nitrogen is protonated and the molecule behaves like a cationic surfactant. So it is understandable that true amphoterics show the best application results outside the isoelectric range.

The amphoterics are mainly used as cosurfactants for cosmetic shampoo or dishwashing formulations, where they provide mildness to skin and hair, especially in blends with alkyl and alkyl ether sulfates. Another advantage is compatibility with most ionic surfactants. In addition, the general surfactant properties, i.e., wetting power, cleansing ability, foaming power, hard-water tolerance, and lime soap dispersibility, are excellent.

2. Ecological and Toxicological Behavior

A coco betaine, a cocoamidopropyl betaine, and a cocoamphoacetate were extensively tested with regard to

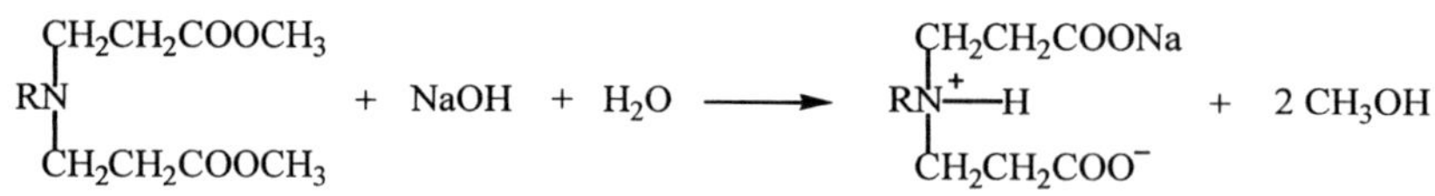

SCHEME 38

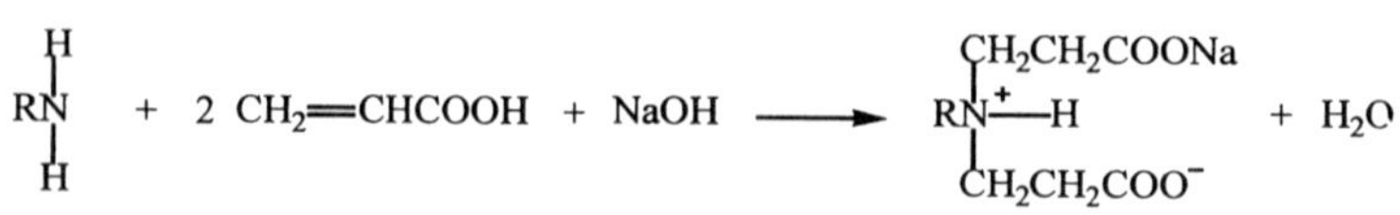

SCHEME 39

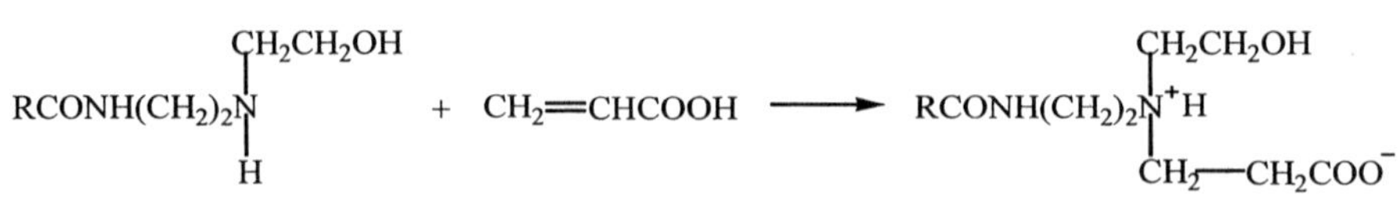

SCHEME 40

TABLE 4 Toxocological Behavior of Amphoteric Surfactants

Type of amphoteric surfactant	Acute toxicity (rat)	Irritation to skin[a] (rabbit)	Irritation to eye[a] (rabbit)	Sensitization (Magnusson-Kligman test)	Gene mutation (Ames test)	NOAEL[b] (mg/kg)
Coco betaine [35]	None	Yes	Yes	None	None	>250
Cocoamidopropyl betaine [36]	None	None	Yes	None	None	1000
Cocoamphoacetate [37]	None	Moderately	Slightly	None	None	>1000

[a]Concentration: 25% and 20%, respectively.
[b]Oral toxicity; NOAEL = no observed adverse effect level is the maximum dose tolerated in cumulative toxicity studies.

their environmental compatibility (Biological Laboratories of Henkel KGaA, private communications, 1996). They proved to be readily biodegradable in the stringent OECD (Organization for Economic Cooperation and Development) tests on ultimate biodegradation. As shown in the metabolite test, their degradation to CO_2, H_2O, inorganic salts, and biomass occurs quantitatively; i.e., no recalcitrant metabolites were formed. Using sewage treatment plant simulation tests, it was confirmed that they will be easily eliminated from wastewater.

The aquatic toxicity (toward algae, daphnia, and fish) of these substances is of the same order of magnitude as for other surface-active substances, ranging from toxic to moderately toxic (ratio of median effective concentration to median lethal concentration, EC_{50}/LC_{50}, >1–100 mg/L). For wastewater bacteria, these substances are minimally toxic.

According to their commercial importance, some toxicological data are presented for coco betaines, cocoamidopropyl betaines, and cocoamphoacetates [240–242]. The results are summarized in Table 4. More detailed toxicological information for cocoamidopropyl betaine is published in Ref. 243.

Whereas the ecological data indicate good environmental tolerance, the toxicological findings seem to reveal deficits with regard to skin and eye irritation values. These disadvantages, however, arise only at higher concentrations that do not conform to the practice. More important for a toxicological evaluation is the fact that amphoterics are usually combined with anionic surfactants, i.e., alkyl or alkyl ether sulfates. Besides other synergies, such blends have been found to be very mild to skin and mucous membranes [244–246].

REFERENCES

1. J. Davidsohn et al., *Soap Manufacture*, Vol. 1, Interscience Publishers, New York, 1953, p. 1.
2. K. Heyns and L. Basejewicz, Tetrahedron *9*:67 (1960); K. Heyns and H. Paulsen, Angew. Chem. *69*:600 (1957); H. Fiege and K. Wedemeyer, Angew. Chem. *93*:812 (1981).
3. H. A. Bruson, Org. Reactions *5*:79 (1949).
4. R. V. Christian and R. M. Hixon, J. Am. Chem. Soc. *70*:1333 (1948).
5. C. S. Hsia Chen, Org. Chem. *27*:1920 (1962).
6. G. Czichocki et al., Fat Sci. Technol. *94*:66 (1992).
7. A. Turowski, W. Skrypzak, A. K. Reng, and P. Jürges, Parfum. Kosmet. *76*:16 (1995).
8. L. Leusink and C. Broer, Seifen Ole Fette Wachse *119*: 810 (1993).
9. H. Meijer, Chem. Ind. *40*(4):58 (1988).
10. W. H. de Groot, Kluwer Academic Publishers, Dordrecht, 1991.
11. G. F. Moretti and I. Adami, Riv. Ital. Sostanze Grasse *75*(3):143–153 (1998).
12. D. W. Roberts, Org. Process Res. Dev. *2*(3):194–202 (1998).
13. F. J. Gohlke and H. Bergerhausen, Soap Chem. Spec. *43*(10):47–49, 62, 113, 186 (1967).
14. A. Zhu and W. Lin, Riyong Huaxue Gongye Issue *1*: 13–15 (1997).
15. K. S. Parasuram, Chem. Petrochem. J. *8*(11):9–13 (1977).
16. J. Falbe, *Surfactants in Consumer Products*, Springer-Verlag, Berlin, 1987.
17. D. W. Roberts, R. S. Ward, and P. J. Hughes, J. Surfactants Deterg. *1*(2):161–165 (1998).
18. J. Jensen, Sci. Total Environ. *226*(2,3):93–111 (1999).
19. L. Cohen, Tenside Surfactants Deterg. *34*:332–335 (1997).
20. D. W. Roberts and J. O. Morley, Comun. J. Com. Esp. Deterg. *27*:145–157 (1997).

21. N. M. Rockwell and Y. K. Rao, Proc. World Conf. Lauric Oils: Sources, Process. Appl., 1994, pp. 138–146.
22. D. W. Roberts, Org. Process Res. Dev. *2*(3):194–202 (1998).
23. E. Gonzalez-Mazo, J. M. F. Pajares, and A. Gomez-Parra, Environ. Sci. Technol. *32*:1636–1641 (1998).
24. F. Hirsinger, Chim. Oggi *16*(10):62–66 (1998).
25. U. Schoenkaes, Chim. Oggi *16*(9):9–13 (1998).
26. D. Prats, F. Ruiz, M. Rodriguez, P. Varo, and M. C. Garcia, Comun. J. Com. Esp. Deterg. *25*:171–185 (1994).
27. J. S. Leal, M. T. Garcia, R. Tomas, J. Ferrer, and C. Bengoechea, Tenside Surfactants Deterg. *31*:253–256 (1994).
28. P. Schoeberl, H. Klotz, R. Spiker, and L. Nitschke, Tenside Surfactants Deterg. *31*:243–252 (1994).
29. R. Fellows and F. W. Heywood, Spec. Publ. R. Soc. Chem. *107*(Ind. Appl. Surfactants III):63–81 (1992).
30. F. R. Schroeder, M. Schmitt, and U. Reichensperger, Waste Manage. (Oxf.) *19*(2):125–131 (1999).
31. D. C. McAvoy, S. D. Dyer, N. J. Fendinger, W. S. Eckhoff, D. L. Lawrence, and W. M. Begley, Environ. Toxicol. Chem. *17*:1705–1711 (1998).
32. L. Cohen, A. Moreno, and J. L. Berna, Tenside Surfactants Deterg. *35*:265–269 (1998).
33. I. Stanciu, L. Papahagi, R. Avram, C. Cristescu, I. Duta, and D. Savu, Rev. Chim. (Bucharest) *41*:885–888 (1990).
34. J. M. Quack and M. Trautmann, Tenside Deterg. *22*: 281–289 (1985).
35. H. Ramloch and G. Taeuber, Chem. Unserer Zeit *13*(5):157–162 (1979).
36. C. Beermann, Eur. Chem. News Norm. Paraffins Suppl. *36–7*:39–40 (1966).
37. J. Perkowski and J. S. Miller, Pol. Przem. Chem. *66*(1):9–13 (1987).
38. H. I. Joschek, Chem. Ztg. Chem. Appar. *93*:655–664 (1969).
39. H. Berenbold and K. Kosswig, Tenside Surfactants Deterg. *32*:152–154, 156 (1995).
40. R. Stephan, Tenside Surfactants Deterg. *35*:338–341 (1998).
41. N. M. Van Os, R. Van Ginkel, A. Van Zon, F. W. Heywood, E. L. Berryman, and J. K. Borchardt, Surfactant Sci. Ser. *56*(Anionic Surfactants):363–454 (1996).
42. D. W. Roberts, J. G. Lawrence, I. A. Fairweather, C. J. Clemett, and C. D. Saul, Tenside Surfactants Deterg. *27*:82–86 (1990).
43. E. Spiess, Pollena Tluszcze Srodki Piorace Kosmet. *34*(1–2):30–37 (1990).
44. S. K. Friedman, G. Barker, and M. Mausner, Soap Cosmet. Chem. Spec. *59*(9):35–36, 53 (1983).
45. G. Schneider, Seifen Ole Fette Wachse *95*:970–974 (1969).
46. S. K. Suri, M. S. Thakur, and S. Bhardwaj, J. Am. Oil Chem. Soc. *70*(1):59–64 (1993).
47. R. Stephan, Tenside Surfactants Deterg. *35*:338–341 (1998).
48. T. Ogushi, S. Matuura, S. Horie, and T. Maeda, Chuo Daigaku Rikogaku Kenkyusho Ronbunshu *3*:49–58 (1998).
49. M. Nakamura and K. Oba, Nihon Yukagakkaishi *46*: 957–967 (1997).
50. Y. K. Rao and B. Sajic, World Surfactants Congr. 4th *2*:383–391 (1996).
51. M. J. Schwuger and H. Lewandowski, Surfactant Sci. Ser. *74*(Novel Surfactants):139–177 (1998).
52. D. W. Roberts, C. J. Clemett, C. D. Saul, A. Allan, and R. A. Hodge, Comun. J. Com. Esp. Deterg. *26*: 27–33 (1995).
53. M. Fujiwara and T. Okano, Fragrance J. *23*(4):81–86 (1995).
54. T. Satsuki, K. Umehara, and Y. Yoneyama, J. Am. Oil Chem. Soc. *69*:672–677 (1992).
55. T. Inagaki, in *Proc. World Conf. Oleochem.* (T. H. Applewhite, ed.), AOCS, Champaign, IL, 1991, pp. 269–278.
56. M. Nakamura and K. Oba, Nihon Yukagakkaishi *46*: 957–967 (1997).
57. H. Baumann, Fett. Wiss. Technol. *92*(2):49–56 (1990).
58. T. Satsuki, Annu. Surfactants Rev. *1*(New Products and Applications in Surfactant Technology):132–155 (1998).
59. J. Drozd, in *Proc. World Conf. Oleochem.* (T. H. Applewhite, ed.), AOCS, Champaign, IL, 1991, pp. 256–268.
60. A. Domsch and B. Irrgang, in *Anionic Surfactants* (H. W. Stache, ed.), Surfactant Science Series, Vol. 56, Marcel Dekker, New York, 1995, p. 501.
61. J. A. Milm, R. Soc. Chem. (Ind. Appl. Surf.) *77*:76 (1990).
62. A. J. O'Lenick, Jr. and W. C. Smith, Soap Cosmet. Chem. Spec. *64*:36 (1988).
63. T. Schoenberg, Cosmet. Toiletries *105*:105 (1989).
64. K. Lindner, in *Tenside Textilhilfsmittel Waschrohstoffe*, Vol. 1, Wissenschaftliche Verlagsgesellschaft, Stuttgart, 1964, p. 747.
65. W. Kästner and P. J. Frosch, Fette Seifen Anstrichm. *83*:33 (1981).
66. J. Valeé, Parfum. Cosmet. Savons *4*:205 (1961).
67. G. Gawalek, in *Tenside*, Akademie Verlag, Berlin, 1975, p. 232.
68. X. Domingo, Surfactant Sci. Ser. *56*(Anionic Surfactants):223–312 (1996).
69. A. Syldath and K. H. Schmid, World Surfactants Congr. 4th 1:339–344 (1996).
70. S. Tagata, A. Ishikawa, and H. Hayashi, Chim. Oggi *14*(5):11–15 (1996).
71. J. Steber, P. Gode, and W. Guhl, Fett Wiss. Technol. *90*(1):32–38 (1988).

72. Cambridge, MA, USA. Report (1991), Issue ADL-REF-65913-VOL-1-PT-3; Order No. PB91-214015, 110 pp. Avail. NTIS. From: Gov. Rep. Announce. Index (U.S.) 1991, 91(20), Abstr. No. 155,700.
73. J. P. Sisley, Am. Dyestuff Rep. *43*:741 (1954).
74. J. Ross, U.S. Patent 2,693,479 to Colgate-Palmolive Company (1950).
75. J. K. Fincke, U.S. Patent 2,634,287 to Monsanto (1948).
76. F. W. Grag, U.S. Patent 2,979,521 to Colgate-Palmolive Company (1958).
77. (a) G. Gawalek, in *Tenside*, Akademie-Verlag, Berlin, 1975, p. 171; (b) M. Linfield, ed., *Anionic Surfactants*, Surfactant Science Series, Vol. 7, Marcel Dekker, New York, 1976, p. 219.
78. K. Yamashita et al., JP 50/70322 to Wako Pure Chemical Ind. (1975).
79. M. Shinohara, JP 78/77014 to Riken Vitamin Oil (1975).
80. H. Stühler, EP 267,518 to Hoechst AG (1987).
81. F. Ahmed et al., DE 3,831,446 to Colgate-Palmolive Company (1988).
82. A. Behler et al., DE 4,038,477 to Henkel KGaA (1990).
83. U. Zeidler, J. Soc. Cosmet. Chem. Jpn. *20*:17 (1986).
84. A. Straw et al., GB 2031455 to Colgate-Palmolive Company (1978).
85. F. Ahmed, EP 318729 to Colgate-Palmolive Company (1989).
86. H. Raaf et al., DE 2440802 to Blendax Werke Schneider (1974).
87. W. J. King et al., U.S. Patent 3956478 to Colgate-Palmolive Company (1975).
88. R. Zimmerer and R. Berg, EO 18019 to Procter & Gamble Company (1979).
89. A. Takahashi et al., JP 84/040940 to Lion Corp. (1984).
90. Y. Yamazaki, JP 85/161920 to Lion Corp. (1990).
91. I. D. Hill, U.S. Patent 4942034-A to Hill ID (1990).
92. S. Mitsyama, JP 03,153,796 to Tamanohada Soap Co., Ltd. (1991).
93. H. Takahashi et al., JP 85/269,578 to Sanyo Chem. Ind. Ltd. (1987).
94. K. Tsubone, JP 90/28099 to Kaneboo KKK (1990).
95. J. Shiomi, JP 93/95493 to Mandamu KK (1994).
96. A. Behler et al., WO 95/06702 to Henkel KGaA (1993).
97. W. Breitzke and A. Behler, DE 4328355 to Henkel KGaA (1993).
98. W. Breitzke and A. Behler, DE 4406748 to Henkel KGaA (1994).
99. J. K. Weil and A. J. Stirton, in *Anionic Surfactants* (W. M. Linfield, ed.), Surfactant Science Series, Vol. 7, Marcel Dekker, New York, 1976, pp. 224–230.
100. R. G. Bistline, Jr., in *Anionic Surfactants* (H. Stache, ed.), Surfactant Science Series, Vol. 56, Marcel Dekker, New York, 1995, pp. 633–634.
101. A. K. Reng, Parfuem. Kosmet. *61*:87 (1980).
102. (a) J. K. Weil, N. Parris, and A. J. Stirton, J. Am. Oil Chem. Soc. *47*:91 (1970); (b) M. Fukuda and T. Tamura, JP 62001448 to Lion Corp. (1987).
103. A. Fujio et al., JP 7070047 to Kao Corp. (1995).
104. A. El-Sawy, S. A. Essawy, M. M. El-Sukkary, and A. M. F. Eissa, Hung J. Ind. Chem. *20*:25 (1992).
105. R. G. Bistline, Jr., W. R. Noble, F. D. Smith, and W. M. Linfield, J. Am. Oil Chem. Soc. *54*:371 (1977).
106. M. Yoshio, S. Etsuo, and H. Yasushi, JP 55149243 to Nippon Oils and Fats (1980).
107. M. A. Podustov et al., SU 1051068 to Kharkov Polytechnic Institute (1983).
108. M. A. Podustov et al., SU 1544769 to Kharkov Polytechnic Institute (1990).
109. A. K. Reng and J. M. Quack, Seifen Ole Fette Wachse *102*:307 (1976).
110. (a) A. Turowski, W. Skrypzak, A. K. Reng, and P. Jürges, Parfuem. Kosmet. *76*:16 (1995); (b) J. Bohunek, G. Täuber, and H. Berghausen, DE 1815657 to Farbwerke Hoechst AG (1968).
111. G. Crass, Tenside Surfactants Deterg. *30*:408 (1993).
112. (a) T. Konekiyo et al., JP 1016898 to Mitsubishi Petrochemical Co. (1989); (b) P. Goffinet and R. Canaguier, EP 189687 to Union Generale de Savonnerie (1986).
113. F. Hironobu, K. Fumitaka, and M. Yoshio, JP 58015501 to Nippon Oils and Fats (1983).
114. H. H. Friese, U. Ploog, and F. Pieper, DE 3419405 to Henkel KGaA (1985).
115. R. Prather, U.S. Patent 4376088 to Upjohn Company (1983).
116. G. D. Kudinova et al., SU 1151470 to Belorussian Technological Institute (1985).
117. A. Yamamoto, Y. Arimoto, and M. Konno, BE 902736 to Chemical Works Company (1985).
118. P. J. Petter, Int. J. Cosmet. Sci. *6*(5):249–260 (1984).
119. W. C. Allison and J. E. Neibaur, Household Pers. Prod. Ind. *32*(5):92, 94, 96, 98, 100 (1995).
120. J. H. Kim, T.-Y. Kim, M.-J. Ju, and K.-D. Nam, Kongop Hwahak *7*(3):401–409 (1996).
121. K. Miyazawa and U. Tamura, Cosmet. Toiletries *108*(3):81–86 (1993).
122. H. Xu, F. Chen, and S. Wang, Huadong Ligong Daxue Xuebao *23*:720–724 (1997).
123. L. Daradics and J. Palinkas, Tenside Surfactants Deterg. *31*:308–313 (1994).
124. M. M. A. El-Sukkary, M. A. Salama, and D. A. Ismail, Al-Azhar Bull. Sci. *3*:619–624 (1992).
125. J. J. Crudden and J. V. Lazzaro, Cosmet. Exhib. Conf. Conf. Proc., 1994, pp. 351–391.
126. A. George, J. Modi, N. Jain, and P. Bahadur, Indian J. Chem. Sect. A Inorg., Bioinorg. Phys. Theor. Anal. Chem. *37A*:985–992 (1998).
127. B. Aha, M. Berger, B. Haase, J. Hermann, O. Keil, G. Machmüller, S. Müller, C. Waldinger, and M. Schnei-

der, Schriftenr. Bundesminist. Ernaehr. Landwirtsch. Forsten Reihe A: Angew. Wiss. (1997), Sonderh. (5. Symposium Nachwachsende Rohstoffe: Perspektiven fuer di Chemie, 1997), pp. 66–78.
128. S. Y. Mhaskar, R. B. N. Prasad, and G. Lakshminarayana, J. Am. Oil Chem. Soc. *67*:1015–1019 (1990).
129. K. Schäfer, J. Wirsching, and H. Hocker, Fett/Lipid *99*(6):217–222 (1997).
130. D. B. R. T. Braun, Cosmet. Toiletries *104*(9):87–89, 92–94, 96 (1989).
131. C.-C. Kwan and S.-T. Lin, Sci. Conf. Asian Soc. Cosmet. Sci. 3rd, 1997, pp. 266–270.
132. K. Berendes, S. Müller, and M. Schneider, Schriftenr. Nachwachsende Rohstoffe *10*(Biokonversion nachwachsender Rohstoffe):246–252 (1998).
133. H. Hensen and J. Kahre, Seifen Ole Fette Wachse J. *124*:806, 808–810, 812, 814–815 (1998).
134. A. Sander, E. Eilers, A. Heilemann, and E. Von Kries, Germany. Fett/Lipid *99*(4):115–120 (1997).
135. A. Behler, K.-H. Hill, A. Kusch, S. Podubrin, H.-C. Raths, and G. Uphues, in *Nonionic Surfactants* (N. M. van Os, ed.), Surfactant Science Series, Vol. 72, Marcel Dekker, New York, 1998, pp. 259 ff.
136. W. Skrypzak, A. K. Reng, and J. M. Quack, Parfuem. Kosmet. *60*:317 (1979).
137. T. Kurosaki and A. Manbu, DE 3,047,378 to Kao Soap Company (1979).
138. K. Sasse, in *Methoden der Organischen Chemie* (E. Müller, ed.), Thieme, Stuttgart, 1940, Vol. 12/2, p. 143.
139. G. Uphues and U. Ploog, EP 0,272,568 to Henkel KGaA (1987).
140. C. L. Edwards, Surfactant Sci. Ser. *72*(Nonionic Surfactants):87–121 (1998).
141. H.-C. Raths, Third CESIO International Surfactants Congress and Exhibition, London, June 1992.
142. E. Santacesaria, M. Di Serio, and R. Garaffa, Ind. Eng. Chem. Res. *31*:2413–2418 (1992).
143. D. Schmedding and V. L. Tatum, Int. Environ. Conf. Exhib. Bk. *1*:461–471 (1998).
144. D. T. Bennie, Water Qual. Res. J. Can. *34*(1):79–122 (1999).
145. C.-D. Hager, UK Annu. Surfactants Rev. *1*(New Products and Applications in Surfactant Technology):1–29 (1998).
146. B. Fell, Tenside Surfactants Deterg. *35*:326–337 (1998).
147. I. Kirshenbaum and E. J. Inchalik, in *Kirk-Othmer Encyclopedia of Chemical Technology*, 3rd ed., Vol. 16 (M. Grayson and D. Eckroth, eds.), John Wiley & Sons, New York, 1981, pp. 637–653.
148. P. I. Bhuva, Chem. Age India *38*(3):101–109 (1987).
149. P. H. Washecheck, ACS Symp. Ser. *159*(Monohydric Alcohols):87–100 (1981).
150. M. F. Gautreaux, W. T. Davis, and E. D. Travis, in *Kirk-Othmer Encyclopedia of Chemical Technology*, 3rd ed., Vol. 1 (M. Grayson and D. Eckroth, eds.), John Wiley & Sons, New York, 1978, pp. 740–754.
151. R. Aveyard, B. P. Binks, S. Clark, and P. D. Fletcher, J. Chem. Technol. Biotechnol. *48*(2):161–171 (1990).
152. J. Szymanowski, Tenside Surfactants Deterg. *26*:198–210 (1989).
153. L. Marszall, Surfactant Sci. Ser. *23*(Nonionic Surfactants):493–547 (1987).
154. T. Foerster, Surfactant Sci. Ser. *68*(Surfactants in Cosmetics (2nd Edition)):105–125 (1997).
155. A. Wadle, H. Tesmann, M. Leonard, and T. Förster, Surfactant Sci. Ser. *68*(Surfactants in Cosmetics (2nd Edition)):207–224 (1997).
156. T. Balson and M. S. B. Felix, Biodegrad. Surfactants 204–230 (1995).
157. M. Trautmann and P. Jürges, Tenside Deterg. *21*:57 (1984).
158. D. N. Bhattacharyya, S. Krishnan, R. Y. Kelkar, and S. V. Chikale, J. Am. Oil Chem. Soc. *61*:1925 (1984).
159. T. C. Rao, Y. Sita Rama Sastry, R. Subbarao, and S. Lakshminarayana, J. Am. Oil Chem. Soc. *54*:15 (1977).
160. H. Grossmann, Tenside Surfactants Deterg. *12*:16 (1975).
161. K. Kosswig, in *Nonionic Surfactants* (N. M. van Os, ed.), Surfactant Science Series, Vol. 72, Marcel Dekker, New York, 1998, p. 123.
162. G. Seipel, F. Wolf, and K. Löffler, Tenside Surfactants Deterg. *5*:132 (1968).
163. M. F. Cox and U. Weerasooriya, J. Surfactants Deterg. *1*:11 (1998).
164. I. Hama, H. Sasamoto, and T. Okamoto, J. Am. Oil Chem. Soc. *74*:817 (1997).
165. I. Hama, M. Sakaki, and H. Sasamoto, J. Am. Oil Chem. Soc. *74*:829 (1997).
166. A. Behler, B. Guckenbiehl, and H.-C. Raths, *Proceedings of the 87th AOCS Annual Meeting, Indianapolis*, AOCS, Champaign, IL, 1996, p. 79.
167. A. Behler, B. Guckenbiehl, and H.-C. Raths, Tenside Surfactants Deterg. *33*:64 (1996).
168. I. Hama, T. Okamoto, E. Hidai, and K. Yamada, J. Am. Oil Chem. Soc. *74*:19 (1997).
169. I. Hama, M. Sakaki, and H. Sasamoto, J. Am. Oil Chem. Soc. *74*:823 (1997).
170. M. F. Cox and U. Weerasoorya, J. Am. Oil Chem. Soc. *74*:847 (1997).
171. I. Hama, H. Sasamoto, T. Tamura, T. Nakamura, and K. Miura, J. Surfactants Deterg. *1*:93 (1998).
172. K. Müller, Tenside Surfactants Deterg. *3*:37 (1966).
173. A. Domsch and C. Möller, Cosmetics Conference Proceedings 1997, Düsseldorf, Verlag für Chemische Industrie, H. Ziolkowsky GmbH, pp. 258–273.
174. E. Santacesaria, M. Diserio, R. Garaffa, and G. Addino, Ind. Eng. Chem. Res. *31*:2413 (1992).
175. B. Bartha, L. Farkas, J. Morgos, P. Sallay, I. Rusznak, and G. Veress, J. Am. Oil Chem. Soc. *58*:650 (1981).

176. M. Cuscurida and E. L. Yeakey, U.S. Patent 5,103,062 to Texaco Chemical Company (1992).
177. J. F. Knifton and C. G. Naylor, U.S. Patent 5,344,984 to Texaco Chemical Company (1994).
178. M. D. Hoey and J. F. Gadberry, in Surfactant Science Series, Vol. 72 (N. M. van Os, ed.), Marcel Dekker, New York, 1998, pp. 163–175.
179. D. P. Bauer and T. K. Summerford, EP 0,320,694 to Ethyl Corp. (1989).
180. K. S. Laurenzo and D. P. Bauer, U.S. Patent 4,889,954 to Ethyl Corp. (1989).
181. K. S. Laurenzo, EP 0,356,918 to Ethyl Corp. (1990).
182. K. S. Laurenzo and J. D. Sauer, U.S. Patent 4,942,260 to Ethyl Corp. (1990).
183. H. Roerig and R. Stephan, Riv. Ital. Sostanze Grasse *68*(6):317–321 (1991).
184. Y. Hayashi, F. Shirai, T. Shimizu, Y. Nagano, and K. Teramura, J. Am. Oil Chem. Soc. *62*:555 (1985).
185. N. Baggett, in *Comprehensive Organic Chemistry* (J. F. Stoddart, ed.), Vol. 2, Pergamon Press, Oxford, 1979, p. 843.
186. (a) J. Falbe, ed., *Surfactants in Consumer Products—Theory, Technology and Application*, Springer, Berlin, 1986; (b) H. Baumann and M. Biermann, in *Nachwachsende Rohstoffe, Perspektiven für die Chemie* (M. Eggersdorfer, S. Warwel, and G. Wulff, eds.), VCH, Weinheim, 1993, pp. 33–55.
187. (a) P. Schulz, Chim. Oggi *33* (1992); (b) M. Biermann, K. Schmid, and P. Schulz, Starch/Starke *45*:281 (1993); (c) W. Ruback and S. Schmidt, in *Carbohydrates as Organic Raw Materials III* (H. van Bekkum, H. Röper, and A. G. J. Vorhagen, eds.), VCH, Weinheim, 1996, pp. 231–253; (d) B. Brancq, Seifen Ole Fette Wachse J. *118*:905 (1992).
188. (a) M. Biermann, F. Lange, R. Piorr, U. Ploog, H. Rutzen, J. Schindler, and R. Schmid, in Ref. 16, pp. 23–132; (b) S. Ropuszynski and E. Sczesna, Tenside Surfactants Deterg. *27*:350 (1990); (c) S. Ropuszynski and E. Sczesna, Tenside Surfactants Deterg. *22*:190 (1985).
189. (a) N. B. Desai, Cosmet. Toiletries *105*:99 (1990); (b) S. Nakamura, INFORM *8*:866 (1997); (c) K. Hill, J. Falkowski, S. Biermann, and C. Brand, DE 4212155 (1992), Henkel KGaA, Zucker-Aktiengesellschaft Uelzen-Braunschweig in Uelzen, Chem. Abstr. *120*: 194484 (1994).
190. E. Fischer, Ber. Dtsch. Chem. Ges. *26*:2400 (1893).
191. (a) P. Jürges and A. Turowski, in *Perspektiven Nachwachsender Rohstoffe in der Chemie* (H. Eierdanz, ed.), VCH, Weinheim, 1996, pp. 61–70; (b) H. Kelkenberg, Tenside Surfactants Deterg. *25*:8 (1988).
192. (a) K. Hill, W. von Rybinski, and G. Stoll, eds.), *Alkyl Polyglycosides—Technology, Properties and Applications*, VCH, Weinheim, 1997; (b) W. von Rybinski and K. Hill, in *Novel Surfactants, Preparations, Applications, and Biodegradability* (K. Holmberg, ed.), Marcel Dekker, New York, 1998, pp. 31–85; (c) D. E. Koeltzow and A. D. Urfer, J. Am. Oil Chem. Soc. *61*: 1651 (1984); (d) A. J. J. Straathof, H. van Bekkum, and A. P. G. Kieboom, Starch/Starke *40*:438 (1988); (e) Th. Böcker and J. Thiem, Tenside Surfactants Deterg. *26*:318 (1989).
193. (a) R. G. Laughlin, Y.-C. Fu, F. C. Wireko, J. J. Scheibel, and R. L. Munyon, in *Novel surfactants, Preparations, Applications, and Biodegradability* (K. Holmberg, ed.), Marcel Dekker, New York, 1998, pp. 1–30; (b) J. J. Scheibel, D. S. Connor, R. E. Shumate, and J. C. T. R. B. St. Laurent, EP-B 0558515, US-A 598462 (1990), Procter & Gamble, Chem. Abstr. *117*: 114045 (1992).
194. Colin A. Houston & Associates, Inc., Glucosamides: The challenge of a new sugar-based surfactant, 1993–1998, June 1996.
195. (a) W. Aulmann and W. Sterzel, in Ref. 192a, pp. 151–160; (b) W. Matthies, B. Jackwaerth, and H.-U. Kraechter, in Ref. 192a, pp. 169–176; (c) J. Steber, W. Guhl, N. Stelter, and F. R. Schröder, in Ref. 192a, pp. 177–188; (d) M. Stalmans, E. Matthies, E. Weeg, and S. Morris, Seifen Ole Fette Wachse J. *119*:794 (1993).
196. (a) K. Schmid, in *Perspektiven Nachwachsender Rohstoffe in der Chemie* (H. Eierdanz, ed.), VCH, Weinheim, 1996, pp. 41–60; (b) H. Andree, J. F. Hessel, P. Krings, G. Meine, B. Middelhauve, and K. Schmid, in Ref. 192a, pp. 99–130.
197. H. Tesmann, J. Kahre, H. Hensen, and B. A. Salka, in Ref. 192a, pp. 71–88; (a) H. Tesmann, J. Kahre, B. A. Salka, Cosmetics and Toiletries Manufacturing Worldwide (1996), pp. 21–33; (b) M. Weuthen, R. Kawa, K. Hill, and A. Ansmann, Fat. Sci. Technol. *97*: 209 (1995).
198. R. Garst, in Ref. 192a, pp. 131–138.
199. O. Rhode, M. Weuthen, and D. Nickel, in Ref. 192a, pp. 139–149.
200. Anonymous, Parfums Cosmet. Actual. *139*(Feb–Mar): 44 (1998).
201. (a) P. Bernardi, D. Fornara, L. Paglino, and T. Verzotti, *The 1st Concise Surfactants Directory*, Tekno Science, Milano, 1996, pp. 15–17; (b) Anonymous, Chem. Market. Rep. *17*(June):23 (1996).
202. A. Behler, K.-H. Hill, A. Kusch, S. Podubrin, H.-C. Raths, and G. Uphues, in *Nonionic Surfactants* (N. M. van Os, ed.), Surfactant Sciences Series, Vol. 72, Marcel Dekker, New York, 1998, pp. 273 ff.
203. K. Yahagi, N. Hoshino, and H. Hirota, JFSCC Paper No., 1988, pp. 71 ff.
204. M. Hofinger, H. Stühler, S. Billenstein, H. Berenbold, and J. M. Quack, EP 0,284,036 to Hoechst AG (1987).
205. W. Ruback and J. Schut, EP 0,295,385 to Hüls AG (1987).
206. S. Courdavault Duprat, L. Godefroy, P. Nivollet, D. Ray, Y. Storet, and J.-F. Vindret, EP 0,550,361 to Stepan Europe (1991).

207. A. Domsch, P. Busch, and H. Hensen, Seifen Ole Fette Wachse *114*:695 (1988).
208. M. Saiki, Y. Imai, and M. Takagi, EP 0,240,622 to Takemoto Yushi Kabashiki Kaisha (1986).
209. E. Ploetz and H. Ulrich, U.S. Patent 2,127,476 to I.G. Farbenindustrie AG (1938).
210. E. Ploetz and H. Ulrich, U.S. Patent 2,173,069 to I.G. Farbenindustrie AG (1939).
211. G. Demmering, DE 2,052,321 to Henkel KGaA (1970).
212. J. Kolbe, W. Kortmann, and J. Pfeiffer, EP 0,075,770 to Bayer AG (1982).
213. U. Ploog and M. Petzold, DE 3,032,216 to Henkel KGaA (1980).
214. M. Hofinger, A. K. Renk, and J. M. Quack, DE 3,345,156 to Hoechst AG (1983).
215. H. Rutzen, Fette Seifen Anstrichm. *84*:87 (1982).
216. O. Manzke, DE 1,618,026 to Wella AG (1967).
217. R. Bimczok, E. D. Racky, and G. Lang, EP 0,747,468 to Wella AG (1995).
218. J. R. Wechsler and M. Lane, U.S. Patent 4,370,272 to Stepan Chemical Company (1981).
219. L. Edebo, B. Ahlström, S. Allenmark, M. Bertilsson, E. Jennische, S. Lange, M. Lindstedt, and R. A. Thompson, in *Industrial Applications of Surfactants* (D. R. Karsa, ed.), Royal Society of Chemistry, Cambridge, 1992, pp. 184 ff.
220. S. F. Cancy, M. Thies, and H. Pardies, Chem. Commun. 2035 (1997).
221. K. F. Gallagher, Cosmet. Toiletries Mag. *109*(Dec):67 (1994).
222. A. Shansky, Am. Perfum. Cosmet. *84*:47 (1969).
223. A. Behler, G. Uphues, and P. Neumann, DE 4,340,423 to Henkel KGaA (1993).
224. C. Weitemeyer, W. Foitzik, H.-D. Käseborn, U. Begoihn, and B. Grüning, DE 4,207,386 to Th. Goldschmidt AG (1992).
225. R. Ernst and E. J. Miller, in *Amphoteric Surfactants* (B. R. Bluestein and C. L. Hilton, eds.), Surfactant Science Series, Vol. 12, Marcel Dekker, New York, 1982, pp. 71 ff.
226. V. Bade, EP 0,353,580 to Th. Goldschmidt AG (1989).
227. F. B. Downing and F. W. Johnson, U.S. Patent 2,129,264 to E.I. du Pont de Nemours & Company (1938).
228. P.-J. Derian, EP 0,736,521 to Rhone-Poulenc Chimie (1996).
229. G. Uphues, Fett/Lipid *100*:490 (1998).
230. N. Parris, C. Pierce, and W. M. Linfield, J. Am. Oil Chem. Soc. *54*:294 (1977).
231. G. Braun, C.-J. Tschang, C. Vamvakiris, and K. Glaser, EP 0,282,908 to BASF AG (1987).
232. M. Lindemann, R. Mayhene, A. O'Lenick, and R. Verdiccio, U.S. Patent 4,215,064 to Johnson & Johnson and Mona Industries Inc. (1978).
233. J. Giersberg and H.-J. Kollmeier, DE 3,826,805 to Th. Goldschmidt AG (1988).
234. H. Hein, H. J. Jaroschek, and W. Melloh, Fette Seifen Anstrichm. *80*:448 (1978).
235. G. Schwarz, P. Leenders, and U. Ploog, Fette Seifen Anstrichm. *81*:154 (1979).
236. F. D. Smith and W. M. Linfield, J. Am. Oil Chem. Soc. *55*:741 (1978).
237. A. F. Isbell, U.S. Patent 2,468,012 to General Mills Inc. (1945).
238. R. G. Freese, U.S. Patent 2,810,752 to General Mills Inc. (1957).
239. D. Aelony, U.S. Patent 2,814,643 to General Mills Inc. (1957).
240. W. Koehl and W. Sterzel, Toxicological Evaluation, Henkel-TTB Report 9700144 (1997).
241. W. Aulmann and W. Sterzel, Toxicological Evaluation, Henkel-TTB Report 9600144 (1996).
242. HELLID data set for 61791-32-0, dated January 19, 1998, Henkel KGaA.
243. Anonymous, Final report on the safety assessment of cocoamidopropyl betaine, J. Am. Coll. Toxicol. *10*:33 (1991).
244. K. Klein and O. Bator, Drug Cosmet. Ind. December: 38–42, 76–77 (1981).
245. P. Alexander, Manuf. Chem. August:54–57 (1985).
246. U. Zeidler and G. Reese, Arztliche Kosmetol. *13*:39 (1983).

2

Cleavable Surfactants

KRISTER HOLMBERG Chalmers University of Technology, Göteborg, Sweden

I. INTRODUCTION

By tradition, surfactants are stable species. Among the surfactant workhorses are: anionics such as alkylbenzene sulfonates and alkyl sulfates, nonionics such as alcohol ethoxylates and alkylphenol ethoxylates, and cationics such as alkyl quats and dialkyl quats; only alkyl sulfates are not chemically stable under normal conditions. Through the years, the susceptibility of alkyl sulfates to acid-catalyzed hydrolysis has been seen as a considerable problem, particularly well known for the most prominent member of the class, sodium dodecyl sulfate (SDS). The general attitude has been that weak bonds in a surfactant may cause handling and storage problems and should therefore be avoided.

More recently, the attitude toward easily cleavable surfactants has changed. Environmental concern has become one of the main driving forces for the development of new surfactants and rate of biodegradation has become a major issue. One of the main approaches taken to produce readily biodegradable surfactants is to build into the structure a bond with limited stability. For practical reasons the weak bond is usually the bridging unit between the polar headgroup and the hydrophobic tail of the surfactant, which means that degradation immediately leads to destruction of the surface activity of the molecule, an event usually referred to as the primary degradation of the surfactant. Biodegradation then proceeds along various routes depending on the type of primary degradation product. The ultimate decomposition of the surfactant, often expressed as amount of carbon dioxide evolved during 4 weeks exposure to appropriate microorganisms counted as a percentage of the amount of carbon dioxide that could theoretically be produced, is the most important measure of biodegradation. It seems that for most surfactants containing easily cleavable bonds, the value for ultimate decomposition is higher than for the corresponding surfactants lacking the weak bond. Thus, the strong tend toward more environmentally benign products favors the cleavable surfactant approach on two accounts.

A second incentive for the development of cleavable surfactants is to avoid complications such as foaming or formation of unwanted, stable emulsions after use of a surfactant formulation. Cleavable surfactants present the potential for elimination of some of these problems. If the weak bond is present between the polar and the nonpolar part of the molecule, cleavage will lead to one water-soluble and one water-insoluble product. Both moieties can usually be removed by standard work-up procedures. This approach has been of particular interest for surfactants used in preparative organic chemistry and in various biochemical applications.

A third use of surfactants with limited stability is to have the cleavage product impart a new function. For instance, a surfactant used in personal care formulations may decompose on application to form products beneficial to the skin. Surfactants that impart a new function after cleavage are sometimes referred to as functional surfactants.

Finally, surfactants that break down into nonsurfactant products in a controlled way may find use in specialized applications, e.g., in the biomedical field. For instance, cleavable surfactants that form vesicles or microemulsions can be of interest for drug delivery, provided the metabolites are nontoxic.

Most cleavable surfactants contain a hydrolyzable bond. Chemical hydrolysis is either acid or alkali catalyzed, and many papers discuss the surfactant breakdown in terms of either of these mechanisms. In the environment, bonds susceptible to hydrolysis are often degraded by enzymatic catalysis but few papers dealing with cleavable surfactants have dealt with such processes in vitro. Other approaches that have been taken include incorporation of a bond that can be destroyed by ultraviolet (UV) irradiation or use of an ozone-cleavable bond. This chapter is subdivided according to the type of weak linkage present in the surfactant. Emphasis is put on the development that has taken place in recent years.

II. ALKALI-LABILE SURFACTANTS

A. Normal Ester Quats

By the term ester quat one usually refers to surface-active quaternary ammonium compounds that have the general formula $R_4N^+X^-$ and in which the long-chain alkyl moieties, R, are linked to the charged headgroup by an ester bond and with X^- being a counterion. With normal ester quats one means surfactants based on esters between one or more fatty acids and a quaternized amino alcohol. Figure 1 shows examples of three different ester quats, all containing two long-chain and two short substituents on the nitrogen atom. The figure also shows the "parent," noncleavable quat. As can be seen, the ester-containing surfactants contain two carbon atoms between the ester bond and the nitrogen that carries the positive charge. Cleavage of the ester bonds of surfactants **II–IV** yields a fatty acid soap in addition to a highly water-soluble quaternary ammonium diol or triol. These degradation products exhibit low fish toxicity, and they are degraded further by established metabolic pathways. The overall ecological characteristics of ester quats are much superior to those of traditional quats as represented by compound **I** of Fig. 1.

During the past decade the dialkylester quats have to a large extent replaced the stable dialkyl quats as rinse cycle softeners, which is the single largest application for quaternary ammonium compounds. The switch from stable dialkyl quats to dialkylester quats may represent the most dramatic change of product type in the history of surfactants, and it is entirely environment driven. Unlike stable quats, ester quats show excellent values for biodegradability and aquatic toxicity [1,2]. Ester quats have also fully or partially re-

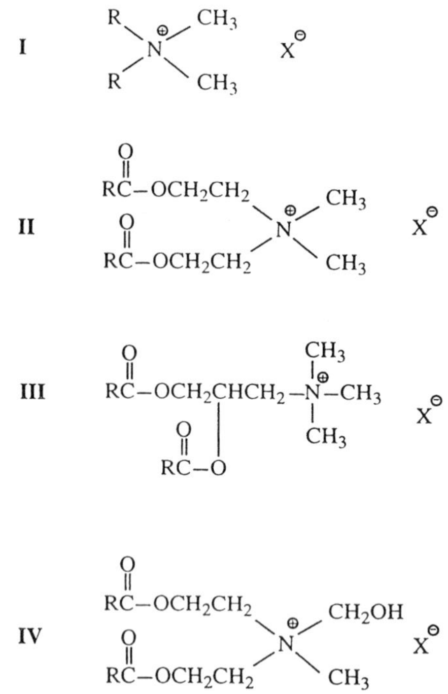

FIG. 1 Structures of one conventional quaternary ammonium surfactant (**I**) and three ester quats (**II–IV**). R is a long-chain alkyl, and X is Cl, Br, or CH_3SO_4.

placed traditional quats in other applications of cationics, such as hair care products and various industrial formulations [1].

The cationic charge close to the ester bond renders normal ester quats unusually stable to acid and labile to alkali. The strong pH dependence of the hydrolysis can be taken advantage of to induce rapid cleavage of the product. This phenomenon is even more pronounced for betaine esters, and the mechanism of hydrolysis is discussed in some detail in the following section. Figure 2 illustrates the pH dependence of hydrolysis of an ester quat. As can be seen, hydrolysis rate is at minimum at pH 3–4 and accelerates strongly above pH 5–6. Evidently, formulations containing ester quats must be maintained at low pH.

Esters of choline have attracted special attention because the primary degradation products, choline and a fatty acid, are both natural metabolites in the body. Thus choline esters should constitute a group of very nontoxic cationic surfactants. A series of choline esters were synthesized and evaluated as disinfectants with controlled half-lives [3,4] (Fig. 3). Compounds with an alkyl group, R, of 9–13 carbons showed an excellent antimicrobial effect. The in vivo hydrolysis was rapid, presumably due to catalysis by butyrylcholinesterase,

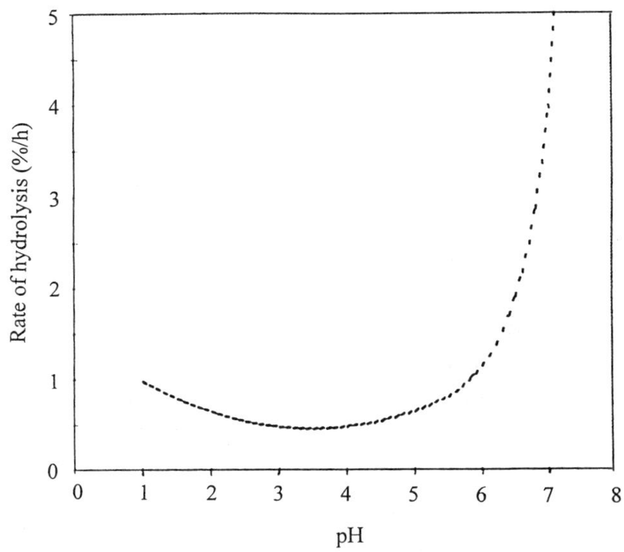

FIG. 2 Influence of pH on the hydrolytic stability of dicetylester of bis(2-hydroxyethyl)ammonium chloride at 25°C. (From Ref. 1.)

which is present in human serum and mucosal membranes.

B. Betaine Esters

The rate of alkali-catalyzed ester hydrolysis is influenced by adjacent electron-withdrawing or electron-donating groups. A quaternary ammonium group is strongly electron withdrawing. The inductive effect leads to decreased electron density at the ester bond; hence, alkaline hydrolysis, which starts by a nucleophilic attack by hydroxyl ions at the ester carbonyl carbon, is favored. Compounds II–IV of Fig. 1 all have two carbon atoms between the ammonium nitrogen and the —O— oxygen of the ester bond. Such esters undergo alkaline hydrolysis at a faster rate than esters lacking the adjacent charge, but the difference is not very large. If, on the other hand, the charge is at the other side of the ester bond, the rate enhancement is much more pronounced. Such esters are extremely labile on the alkaline side but very stable even under strongly acidic conditions [5]. The large effect of the quaternary ammonium group on the alkaline and acid rates of hydrolysis is due to a stabilization/destabilization of the ground state, as illustrated in Fig. 4. The charge repulsion, involving the carbonyl carbon atom and the positive charge at the nitrogen atom, is relieved by hydroxide ion attack but augmented by protonation. The net result is that, compared with an ester lacking the cationic charge, the rate of alkaline hydrolysis is increased 200-fold whereas the rate of acid hydrolysis is decreased 2000-fold [6]. For surface-active betaine esters based on long-chain fatty alcohols, the rate of alkaline hydrolysis is further accelerated by micellar catalysis [7]. Presence of large, polarizable counterions, such as bromide, can completely outweigh the micellar catalysis, however [8].

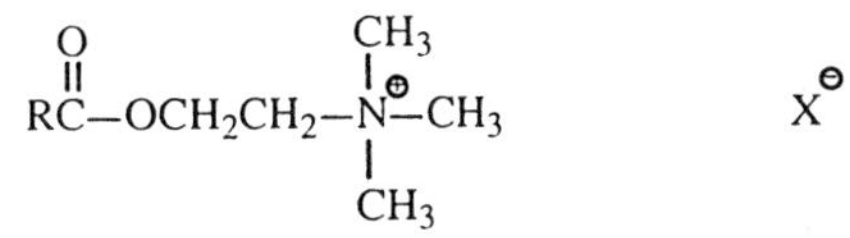

FIG. 3 Structure of a surface-active choline ester. R and X are the same as in Fig. 1.

The extreme pH dependence of surface-active betaine esters makes them interesting as cleavable cationic surfactants. Shelf life is long when they are stored under acidic conditions, and the hydrolysis rate will then depend on the pH at which they are used. Single-chain surfactants of this type have been suggested as "temporary bactericides" for use in hygiene products, for disinfection in the food industry, and in other instances where only a short-lived bactericidal action is wanted [7]. The patent literature also contains examples of betaine esters containing two long-chain alkyl groups [9–11]. Two examples are given in Fig. 5.

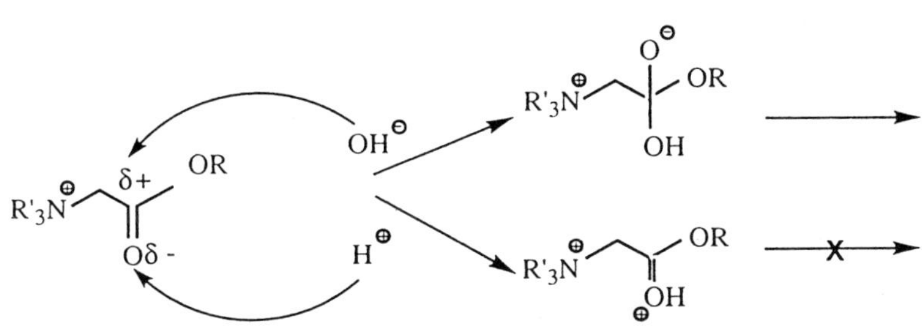

FIG. 4 Mechanism for the acid- and base-catalyzed hydrolysis of betaine ester.

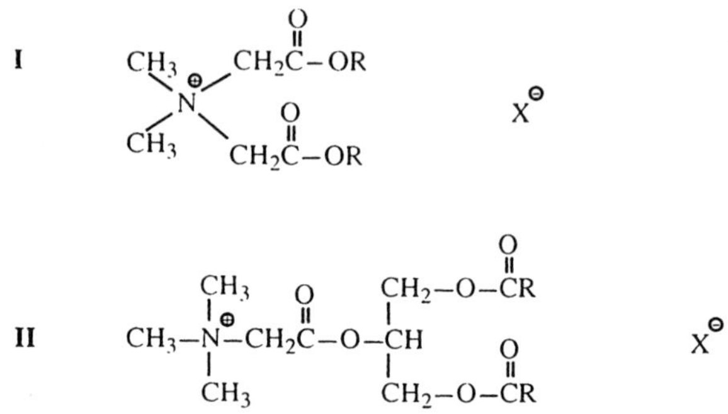

FIG. 5 Structures of two surface-active betaine esters. R and X are the same as in Fig. 1.

C. Monoalkyl Carbonates

Alcohol ethoxylates with short polyoxyethylene chains are viscous oils. Their incorporation into powder detergents is a well-known problem. Carbonate salts of such surfactants have been used as labile derivatives from which the surfactant can be readily regenerated. Such derivatives could be called "prosurfactants" by analogy with the term prodrug in medicine. Reaction of an alcohol ethoxylate with carbon dioxide gives a solid carbonate salt that decomposes under the alkaline washing conditions to give the starting nonionic surfactant and carbonate, as illustrated in Fig. 6 [12]. (Strictly speaking, the prosurfactant is also a surfactant although it is not meant to serve as such in the application step.) Conversion of an alcohol ethoxylate into a solid carbonate enables the incorporation of high levels of this surfactant into granular detergents of high bulk density.

D. Surfactants Containing the Si—O Bond

The silicon-oxygen bond is susceptible to both alkaline and acid hydrolysis. In addition, the bond is specifically cleaved by fluoride ions at relatively neutral pH. (In nonaqueous media, where the ions are not hydrated, the cleavage by F^- is extremely fast.) Single- and double-tailed cationic surfactants with the structures shown in Fig. 7 have been synthesized and tested with regard to degradation characteristics [13]. The route of preparation is relatively sophisticated, however, which means that such surfactants may be of limited practical value.

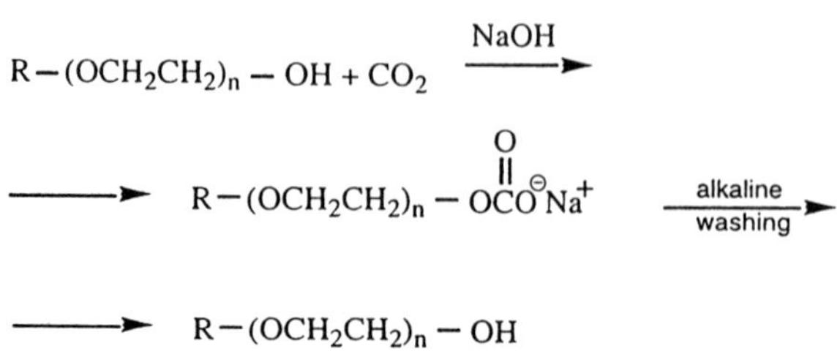

FIG. 6 Formation of a carbonate salt of a nonionic surfactant and subsequent regeneration of the starting surfactant during the washing step.

E. Surfactants Containing a Sulfone Group

An anionic and a cationic surfactant containing the ethylenesulfone moiety have been synthesized by oxidation of the corresponding sulfide [14]. These surfactants are stable in acid but break down to nonsurfactant products, a vinylsulfone and a phenol, in weak alkali, as shown in Fig. 8. The cleavage reaction is considerably faster for the cationic than for the anionic surfactant. This is mainly a micellar phenomenon: positively charged micelles are surrounded by a pseudophase of much higher hydroxyl ion activity than the bulk aqueous phase, and the reverse is true for negatively charged micelles. A comparative hydrolysis study with a nonsurfactant analogue of the anionic surfactant confirmed this view because the non-surface-active sulfone decomposed much faster than the surfactant.

F. Sugar Esters

Sugar esters have been receiving considerable attention, mainly because of developments in procedures for bio-organic synthesis. The main advantage of the biochemical route compared with conventional organic synthesis is the much higher regioselectivity obtained in the synthesis. A long reaction time is a typical disadvantage of the enzymatic process. Enzymatic synthesis of sugar esters has been thoroughly covered by Vulfson [15]. The topic will be briefly discussed in the following.

In a systematic investigation of the effect of the number of condensed hexose units on surfactant properties, monododecyl esters of glucose, sucrose (two sugar units), raffinose (three units), and stachyose (four units) were prepared by organic synthesis followed by careful chromatographic purification [16]. As can be seen from Fig. 9, all compounds had the acyl substituent at the 6-position of a glucose ring; i.e., the ester

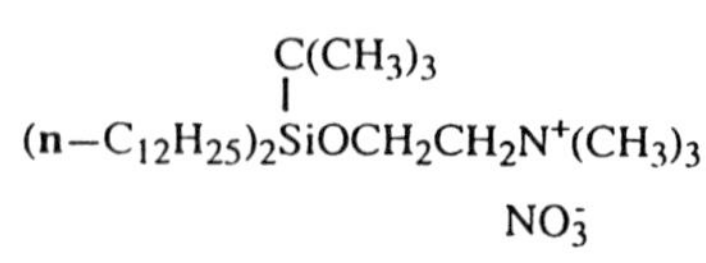

FIG. 7 Structure of a surfactant containing the Si—O bond.

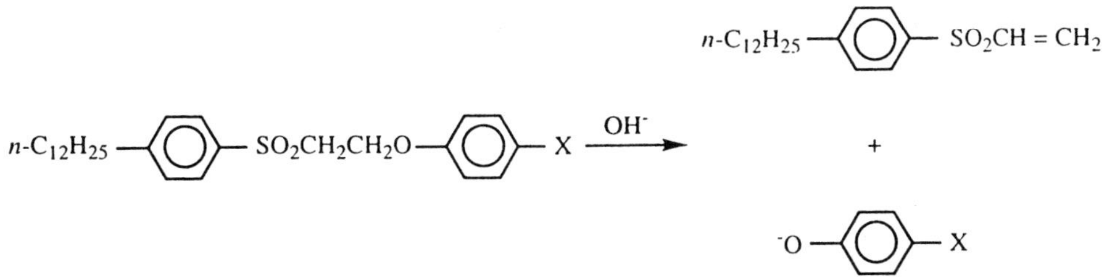

FIG. 8 Alkaline hydrolysis of a sulfone-containing surfactant. X may be $(CH_3)_3N^+$ or SO_3^-.

bond had the same environment in all four surfactants. The phase behavior and the surfactant properties of the compounds were studied. It was concluded that the self-assembly of the surfactants was governed primarily by geometric packing constraints, which, in turn, depended on the size of the polar headgroup. The phase behavior was practically independent of temperature and, as expected, none of the surfactants exhibited the clouding phenomenon characteristic of polyoxyethylene-based nonionic surfactants.

Enzymatic synthesis of sugar esters can be run either in an organic solvent [17,18] or under solvent-free conditions at reduced pressure [19,20]. In the latter process a relatively hydrophobic sugar derivative, e.g., a glucoside or an isopropylidene derivative, is employed. An interesting new development is the use of a microemulsion as the reaction medium [21]. In order to avoid difficult work-up problems, the reaction product, i.e., the ester surfactant, was used as microemulsion surfactant.

In a study aimed at optimizing the conditions of lipase-catalyzed sugar ester synthesis, several galactose and xylose esters were prepared by the solvent-free process starting from the isopropylidene derivative [22]. The monoester content was around 90% and the overall yield of the target ester ranged from 59 to 88%. Virtually no side products were formed, either in the course of the enzymatic reaction or in the subsequent removal of the isopropylidene group. This is very different from the complex product mixture obtained by organic synthesis, which is usually an acid- or base-catalyzed transesterification at elevated temperature.

Fatty acid esters of unmodified sugars (or sugar alcohols) were prepared in an organic solvent using immobilized lipase as the catalyst. Condensation water was continuously removed by refluxing through a desiccant under reduced pressure. Starting materials were glucose, fructose, sorbitol, xylitol, and the three fatty acids lauric, oleic, and erucic [23]. Physicochemical characterization of the sugar esters gave the expected result with efficiency and effectiveness of the surfactants mainly being dependent on the chain length of the fatty acid [24]. There was little difference in critical micelle concentration (cmc) between surfactants based on different sugars and the same fatty acid.

III. ACID-LABILE SURFACTANTS

A. Cyclic Acetals

Cyclic 1,3-dioxolane (five-membered ring) and 1,3-dioxane (six-membered ring) compounds, illustrated in Fig. 10, have been studied in depth by the groups of Burczyk, Takeda, and others as examples of acid-labile surfactants. They are typically synthesized from a long-chain aldehyde by reaction with a diol or a higher polyol. Reaction with a vicinal diol gives the dioxolane [25–27] and 1,3-diols yield dioxanes [28,29].

If the diol contains an extra hydroxyl group, such as in glycerol, a hydroxy acetal is formed and the remaining hydroxyl group can subsequently be derivatized to give anionic or cationic surfactants, as illustrated in Fig. 11. It is claimed that glycerol gives ring closure to dioxolane, yielding a free, primary hydroxyl group, but it is likely that some dioxane with a free, secondary hydroxyl group is formed as well. The free hydroxyl group can be treated with SO_3 and then neutralized to give the sulfate [30], it can be reacted with propane sultone to give the sulfonate [31], or it can be substituted by bromine or chloride and then reacted with dimethylamine to give a tertiary amine as polar group. Quaternization of the amine can be done in the usual manner, e.g., with methyl bromide [32]. An analogous reaction with pentaerythritol as diol yielded a 1,3-dioxane with two unreacted hydroxymethyl groups that can be reacted further, e.g., to give a dianionic surfactant [31]. The remaining hydroxyl group may also be ethoxylated, and such acetal surfactants have been commercialized [33]. The rate of decomposition in sewage plants of this class of nonionic surfactants is much higher than for normal ethoxylates [34].

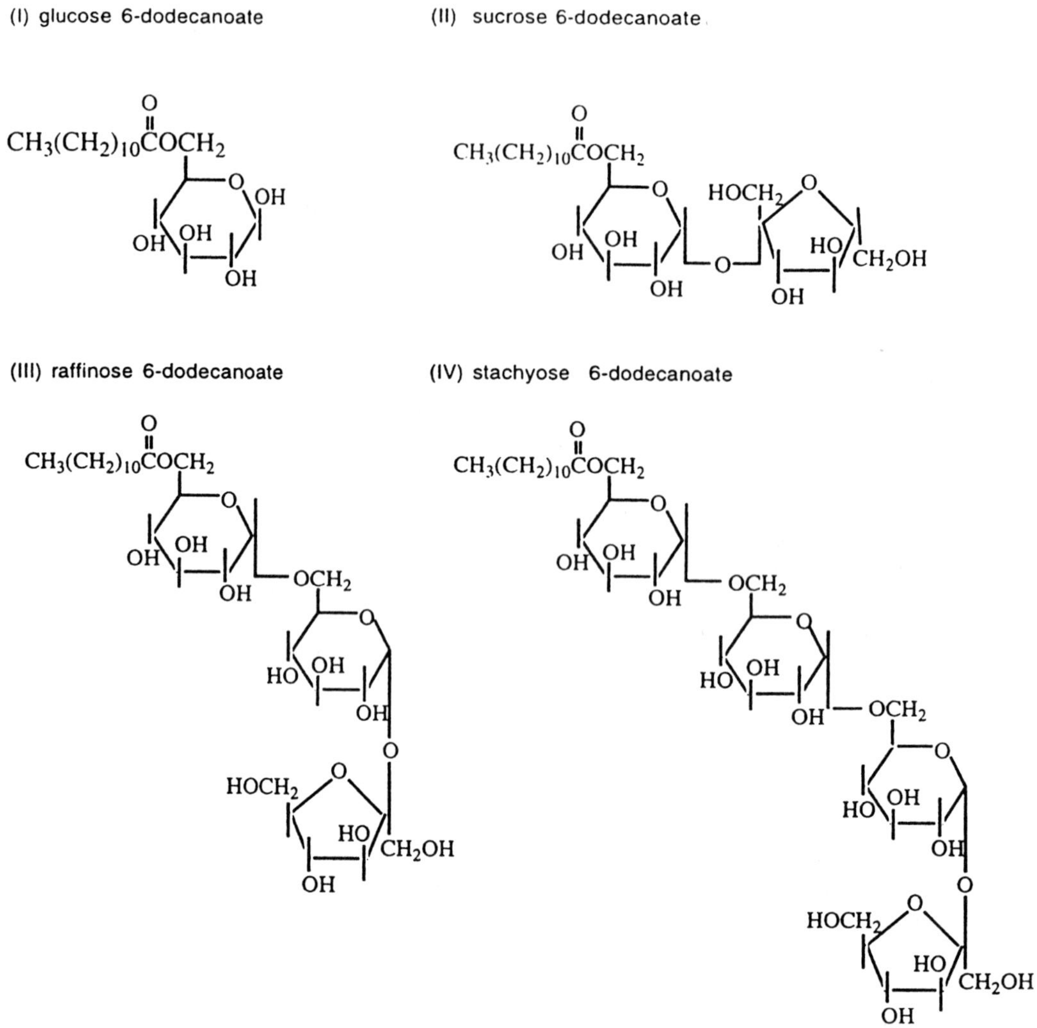

FIG. 9 Structures of surface-active sugar esters.

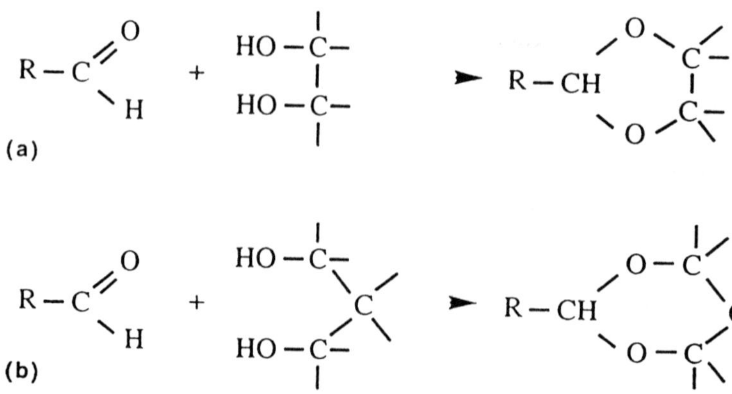

FIG. 10 Preparation of 1,3-dioxolane surfactant (a) and 1,3-dioxane surfactant (b) from a long-chain aldehyde and a 1,2- and a 1,3-diol, respectively.

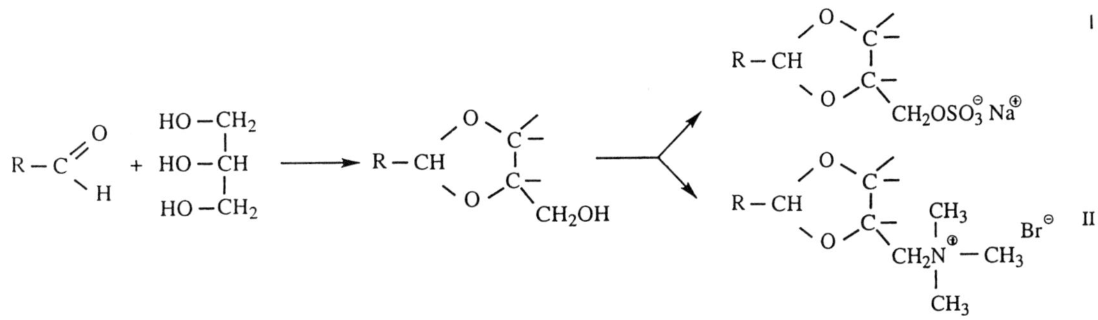

FIG. 11 Examples of anionic (**I**) and cationic (**II**) 1,3-dioxolane surfactants.

Hydrolysis splits acetals into aldehydes, which are intermediates in the biochemical *β*-oxidation of hydrocarbon chains. Acid-catalyzed hydrolysis of unsubstituted acetals is generally facile and occurs at a reasonable rate at pH 4–5 at room temperature. Electron-withdrawing substituents such as hydroxyl, ether oxygen and halogens reduce the hydrolysis rate, however [35]. Anionic acetal surfactants are more labile than cationic ones [25], a fact that can be ascribed to the locally high oxonium ion activity around such micelles. The same effect can also be seen for surfactants forming vesicular aggregates, again undoubtedly due to differences in the oxonium ion activity in the pseudophase surrounding the vesicle. Acetal surfactants are stable at neutral and high pH.

The advantage of using a cleavable acetal surfactant instead of a conventional amphiphile has been elegantly demonstrated in work by Bieniecki and Wilk [36]. A cationic 1,3-dioxolane derivative was used as surfactant in a microemulsion formulation that was employed as a reaction medium for an organic synthesis. When the reaction was complete, the surfactant was decomposed by addition of acid and the reaction product easily recovered from the resulting two-phase system. By this procedure, the problems of foaming and emulsion formation, frequently encountered with conventional surfactants, could be avoided.

The 1,3-dioxolane ring has been found to correspond to approximately two oxyethylene units with regard to effect on cmc and adsorption characteristics [27]. Thus, surfactant type I in Fig. 11 should resemble ether sulfates of the general formula R—$(OCH_2CH_2)_2OSO_3Na$. This is interesting because the commercial alkyl ether sulfates contain two to three oxyethylene units.

B. Acyclic Acetals

Alkyl glucosides, often somewhat erroneously referred to as alkyl polyglucosides or APGs, are cyclic compounds, but because the ring does not involve the two geminal hydroxyl groups of the aldehyde hydrate, they are included here in the category of acyclic acetals. Alkyl glucosides are by far the most important type of acetal surfactant. As this surfactant class has been the topic of several reviews [37–39], it will be only briefly outlined here.

Alkyl glucosides are made either by direct condensation of glucose and a long-chain alcohol or by transacetalization of a short-chain alkyl glucoside, such as ethyl glucoside, with a long-chain alcohol, in both cases using an acid catalyst (Fig. 12). The procedure leads to some degree of sugar ring condensation, the extent of which can be governed by various means, e.g., the ratio of long-chain alcohol to sugar.

The alkyl glucoside surfactants break down into glucose and long-chain alcohol under acidic conditions. On the alkaline side, even at very high pH, they are stable to hydrolysis. Their cleavage profile along with their relatively straightforward synthesis route makes these surfactants interesting candidates for various types of cleaning formulations.

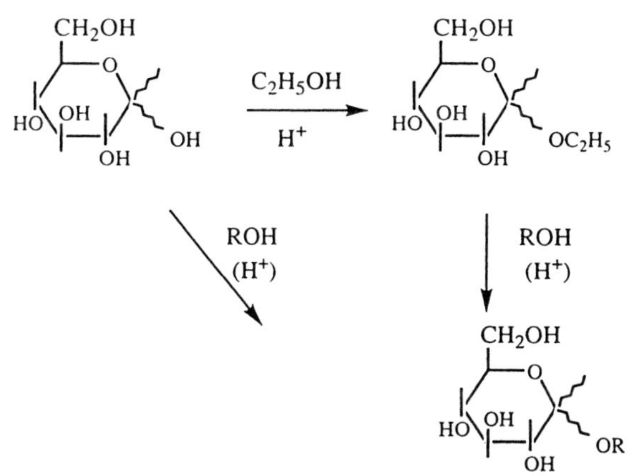

FIG. 12 Two routes of preparation of alkyl glucosides. R is a long-chain alkyl.

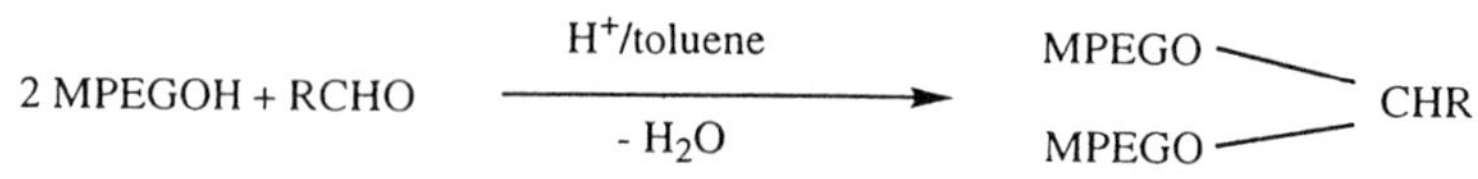

FIG. 13 Preparation of a cleavable surfactant containing two polyoxyethylene chains. R is a long-chain alkyl.

Polyoxyethylene-based cleavable surfactants have been synthesized by reacting end-capped poly(ethylene glycol) (PEG) with a long-chain aldehyde, as shown in Fig. 13 [40–42]. The physicochemical behavior of these surfactants resembles that of normal nonionics; for instance, they have the reverse solubility-temperature relationship and they exhibit a cloud point.

Acid hydrolysis of the labile polyoxyethylene-based surfactants yields PEG-monomethyl ether and long-chain aldehyde. It was found that the hydrolysis of these noncyclic acetal–linked surfactants was several orders of magnitude faster than that of cyclic acetal–linked surfactants [42]. This is important from a practical point of view because many applications of cleavable surfactants demand a rather high rate of breakdown. The hydrolytic reactivity increased as the hydrophobe chain length decreased if the hydrophiles were kept the same. This has been attributed to decreased hydrophobic shielding of the acetal linkage from oxonium ions. The structure of the hydrophobe, linear or branched, was not decisive of the hydrolysis rate, however, and neither was the size of the polar headgroup, i.e., the length of the PEG chains.

Ono et al. [43,44] have synthesized series of noncyclic acetal surfactants—anionics, nonionics, cationics and amphoterics—from a common allyl chloride intermediate (Fig. 14). It was found that the cmc values of these surfactants were lower than those of conventional surfactants of the same alkyl chain length. Furthermore, the efficiency of the surfactants, expressed as the concentration required to produce a 20 mN/m reduction in surface tension, was higher for the cleavable surfactants. Evidently, the connecting moiety, i.e., the group connecting the hydrophobic tail and the polar headgroup, gives a hydrophobic contribution to the amphiphilic properties.

A systematic study of hydrolysis rates was made with the four surfactant classes shown in Fig. 14. For a series of surfactants with the same hydrophobic tail and with the same connecting group, the time for complete decomposition was recorded. The results, shown in Table 1, constitute a nice illustration of the effect of

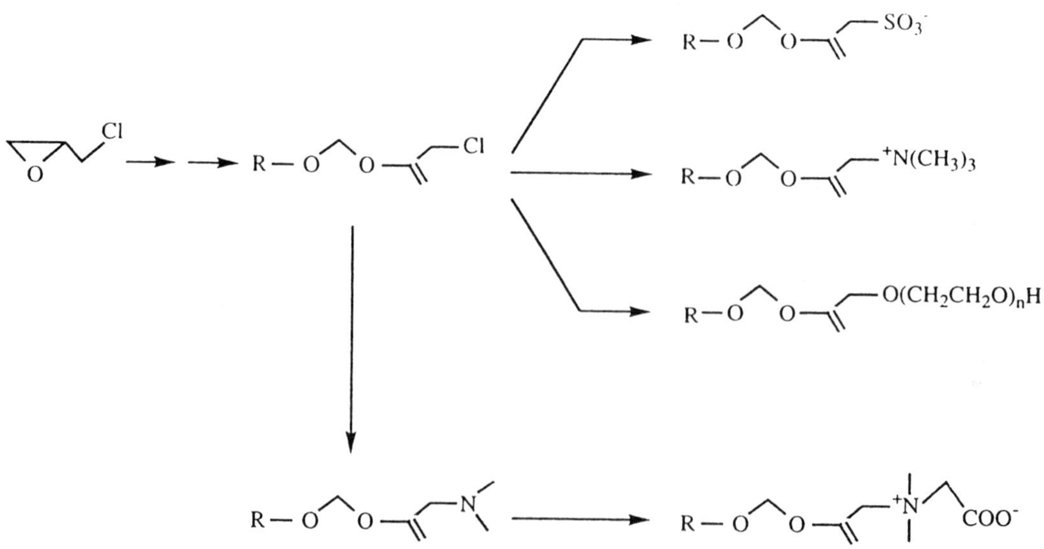

FIG. 14 Schematic synthesis routes of noncyclic acetal surfactants.

TABLE 1 Times for Complete Decomposition of Four Acetal-Based Surfactants at 25°C and at Varying Conditions[a]

Surfactant type	2% DCl	pD 1	pD 3
Anionic	Immediately	Immediately	30 min
Cationic	48 h	1 week	>2 weeks
Nonionic	Immediately	15 min	90 h
Amphoteric	3 h	24 h	>1 week

[a]Reactions were carried out in deuterated solvent to enable the hydrolysis reactions to be monitored by NMR.
Source: Ref. 61.

the micelle surface on the hydrolysis rate. With negatively charged micelles the reaction is very fast, with positively charged micelles the process is sluggish, and with the noncharged (or zero net charged) micelles the rate is intermediate.

C. Ketals

Surfactants containing ketal bonds can be prepared from a long-chain ketone and a diol in analogy with the reaction schemes given in Figs. 10 and 11 for the preparation of acetal surfactants [45]. Ketal-based surfactants have also been prepared in good yields from esters of keto acids by either of two routes, as shown in Fig. 15 [46–48].

The biodegradation profiles of the dioxolane surfactants of Fig. 15 are shown in Fig. 16 [47]. As expected, the degradation rate is very dependent on the alkyl chain length. The process is markedly faster for the labile surfactants (and particularly for structure I, which contains an extra ether oxygen) than for the conventional carboxylate surfactant of the same alkyl chain length used as reference. Ketal surfactants are in general more labile than the corresponding acetal surfactants [49]. As an example, a ketal surfactant kept at pH 3.5 was cleaved to the same extent as an acetal surfactant of similar structure kept at pH 3.0 [50]. The relative lability of the ketal linkage is due to the greater stability of the carbocation formed during ketal hydrolysis compared with the carbocation formed during acetal hydrolysis. (It is noteworthy that biodegradation of an acetal surfactant has been found to be faster than that of a ketal surfactant of very similar structure [47]. Evidently, there is no strict correlation between ease of biodegradation and rate of chemical hydrolysis.)

Jaeger has introduced the term "second-generation cleavable surfactant" for labile surfactants that on cleavage give another surfactant together with a small water-soluble species. The daughter surfactant generally has a higher cmc than the parent surfactant [51–54]. Figure 17 shows a typical example of a second-generation cleavable surfactant. The concept has been applied to a variety of structures, including phospholipid analogues [54] and several applications of this specific type of cleavable surfactants have been proposed in the papers by Jaeger et al.

Double-chain, double-headgroup second-generation surfactants have also been synthesized. The geometry of the molecules may be varied by the position of the link between the hydrocarbon tails. Both symmetrical and unsymmetrical cross-linkings with respect to the headgroups have been prepared [25,55,56]. These surfactants can be seen as examples of gemini surfactants, and in one approach labile gemini surfactants were synthesized that on acid treatment broke down into single-chain, single-headgroup surfactants [56]. They are of interest in model investigations, e.g., to study the morphology of aggregates. Their preparation is cumbersome, however, which means that their practical usefulness is limited.

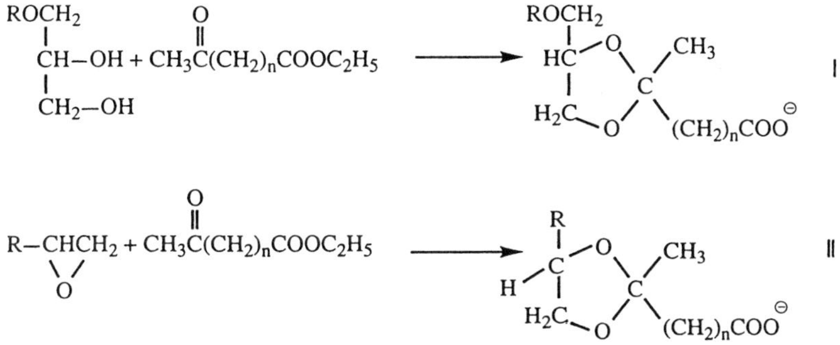

FIG. 15 Preparation of anionic 1,3-dioxolane surfactants from ethyl esters of keto acids.

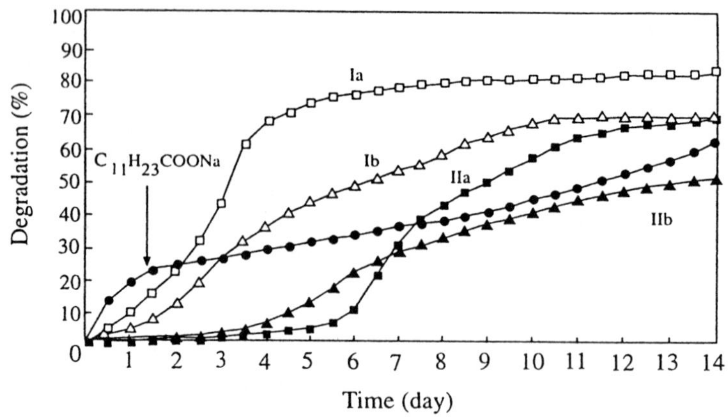

FIG. 16 Rate of biodegradation versus time for four ketal surfactants and for sodium decanoate as reference. I and II relate to the compounds of Fig. 6; (a) R = $C_{12}H_{25}$, n = 2; (b) R = $C_{16}H_{33}$, n = 2. (From Ref. 47.)

D. *Ortho* Esters

Ortho esters are interesting candidates for acid-labile surfactants. They are easily prepared from triethyl orthoformate (or a homologue thereof) and alcohols, as illustrated in Fig. 18; they are stable in alkali; and they decompose in acid by the same general mechanism as acetals and ketals [57]. Hydrolysis gives 1 mole of alkyl formate along with 2 moles of alcohol, as also shown in Fig. 18. One or more of the starting alcohols can be an end-capped PEG, in which case a nonionic polyoxyethylene surfactant is obtained [58].

An interesting feature of *ortho* esters is that they are much more labile in acid than both acetals and ketals. For instance, an *ortho* ester based on monomethyl-PEG decomposes to about 50% at pH 6 and to almost 100% at pH 5 after 1 h at room temperature [58].

The *ortho* ester concept gives molecules with three branches that may be the same or different. Figure 19 shows two examples: a block copolymer with two chains of polyoxypropylene and one chain of polyoxyethylene and a triple-tailed nonionic surfactant connected in the polar headgroup [59]. *Ortho* ester surfactants have recently been commercialized.

E. Surfactants Containing the N=C Bond

Jaeger et al. have synthesized surfactants consisting of two parts connected with a CONHN=C moiety. Each part is a surfactant of its own with a hydrophobic tail and a polar headgroup, and the two headgroups are of different sign [60]. The structure is shown in Fig. 20. As can be seen, the two charges are far apart in the molecule; thus, the type is conceptually different from double-chain zwitterionic surfactants such as phosphatidylcholine. Instead, they may be viewed as a kind of heterogemini surfactant.

Figure 20 also illustrates the acid-catalyzed breakdown of the surfactants. Hydrolysis into the cationic and the anionic surfactant parts occurs readily in weak

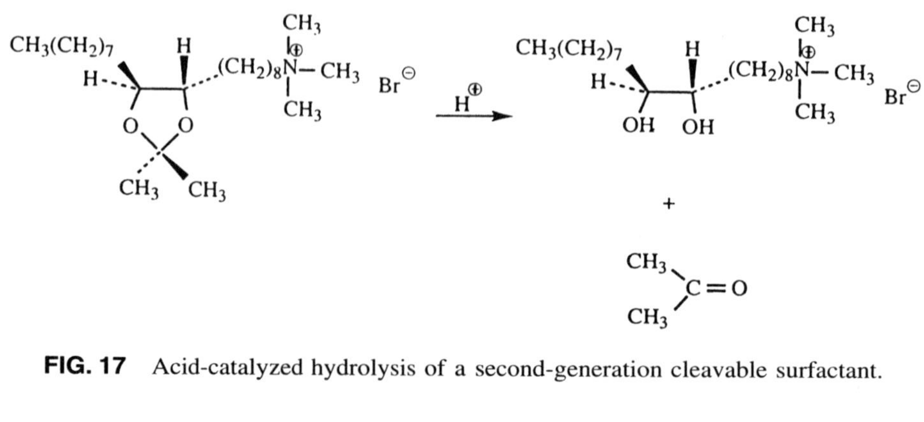

FIG. 17 Acid-catalyzed hydrolysis of a second-generation cleavable surfactant.

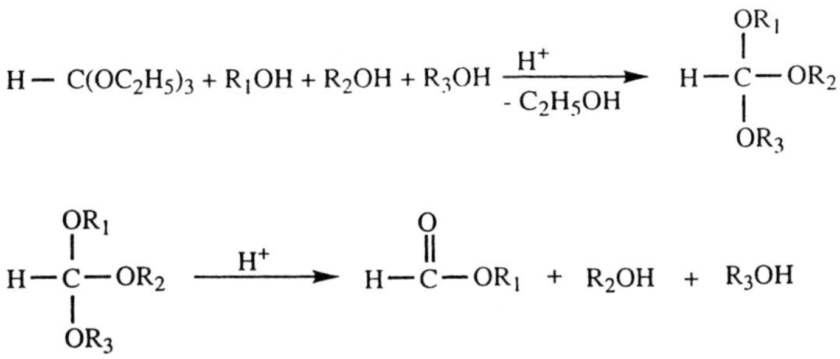

FIG. 18 Synthesis and hydrolysis of *ortho* esters. R_1, R_2, and R_3 are alkyl groups.

acid. The surfactant forms giant vesicles on sonication, and a suggested application is as entrapment and release devices that can be triggered by a change in pH from 7 to about 3.

IV. UV LABILE SURFACTANTS

The concept of triggering cleavage by UV light is attractive because it allows extremely fast breakdown of the surfactant to occur. An alkyl aryl ketone sulfonate, which bears some structural resemblance to alkylbenzene sulfonate surfactants, was synthesized [61]. This compound is photocleaved into a water-soluble aryl sulfonate and a mixture of two methyl-branched olefins, as shown in Fig. 21. The surfactant is of interest for solubilization of proteins because the work-up procedure is greatly facilitated by the instantaneous elimination of surfactant from the solution. The wavelength required for this type of photolysis, a so-called Norrish type II cleavage, is 300 nm and above. This low-energy radiation should be harmless to proteins.

Another approach has been to incorporate the light-sensitive diazosulfonate group between the polar headgroup and the tail of an anionic surfactant [62–64]. As can be seen from Fig. 22, these surfactants are also similar in structure to the commonly used alkylbenzene sulfonates. A comparison of cmc values for the diazosulfonate and the normal sulfonate surfactants with the same R substituent shows lower values for the former, indicating a contribution of hydrophobicity from the azo linkage. Photochemical cleavage yielded sulfate ion and the remaining diazonium compound, which was further photolyzed in a second step.

An interesting use of photolabile surfactants is as emulsifiers in emulsion polymerization [65,66]. The use of a photolabile emulsifier opens the possibility to control the latex coagulation process simply by exposing the dispersion to UV irradiation. The ionic headgroup of the surfactant will be split off by photolysis leading to aggregation of the latex particles. Such latexes could be of interest for coating applications.

A double-chain surfactant has been synthesized that contained Co(III) as complexing agent for two single-chain surfactants based on ethylenediamine in the polar headgroup. UV irradiation, or merely sunlight, causes reduction of Co(III) to Co(II). The latter gives a very

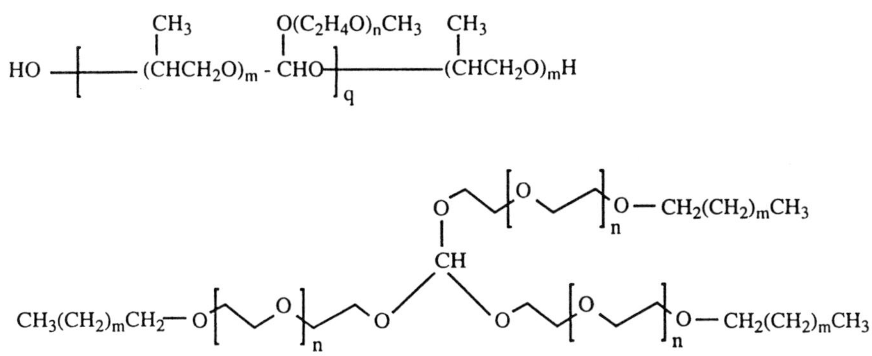

FIG. 19 Two examples of surface-active *ortho* esters.

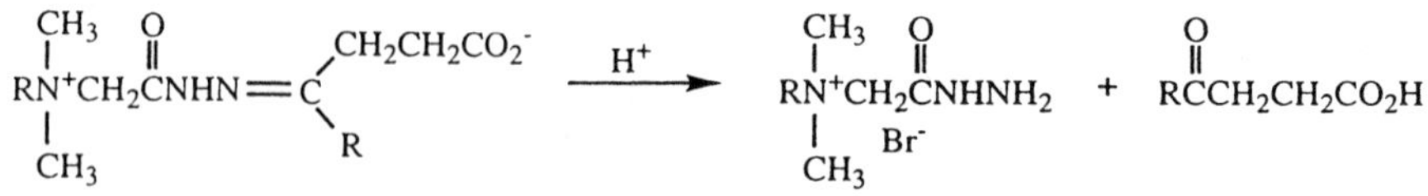

FIG. 20 Hydrolysis of a surfactant containing the N═C bond. R is a long-chain alkyl.

labile complex, and the double-chain surfactant immediately degrades into two single-chain moieties [67].

V. MISCELLANEOUS

Apart from the product classes already discussed, which include the most important types of cleavable surfactants, several more or less exotic examples of surfactants with limited half-lives have been reported. For instance, isethionate esters with a very high degree of alkali lability have been developed. These products, made by esterification of an alkyl polyoxyethylene carboxylic acid with the sodium salt of isethionic acid, have been claimed to be partially cleaved when applied to the skin [68]. Cleavable quaternary hydrazinium surfactants have been explored as amphiphiles containing a bond that splits very easily. The surfactants are cleaved by nitrous acid under extremely mild conditions [69]. Ozone-cleavable surfactants have been developed as examples of environmentally benign amphiphiles. These surfactants, which contain unsaturated bonds, break down easily during ozonization of water, which is a water purification process of growing importance [70].

Glucose-based surfactants having a disulfide linkage between the anomeric carbon of the sugar ring and the hydrophobic tail were synthesized and evaluated for use as solubilizing agents for membrane proteins [71]. Cleavage into nonsurfactant products was performed by addition of dithioerythritol, which is known to split disulfide linkages under physiological conditions.

Surfactants with thermolabile bonds have been synthesized and evaluated as short-lived surfactants. Amino oxide surfactants with an ether oxygen in the 2-position are examples of such structures. They decompose at elevated temperature to the corresponding vinyl ether [72].

VI. CONCLUDING REMARKS

Cleavable, or splittable, or chemodegradable surfactants are likely to become of increasing importance as the environmental concern with regard to surfactant formulations becomes even more widespread. The development that has occurred to this point has brought about a vitalization of the surfactants area in terms of new structures and synthesis strategies. The drive to make surfactants with bonds that break down in a controlled way to yield non-surface-active or less surface-active products has probably involved more creative thinking in terms of organic synthesis than any other area within the surfactant domain, possibly with the exception of the area of gemini surfactants. It will be interesting to monitor which of the many research av-

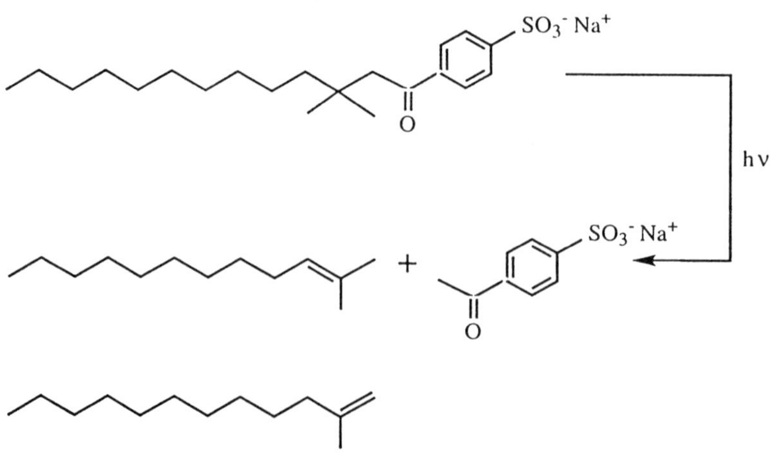

FIG. 21 Photocleavage of a surface-active alkylaryl ketone.

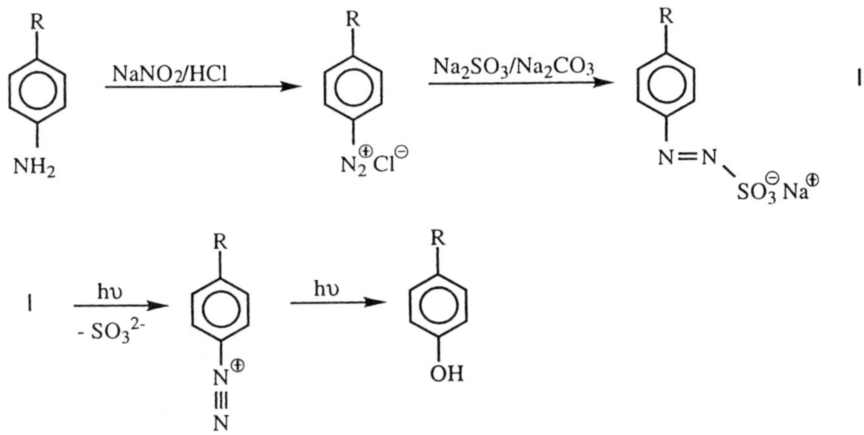

FIG. 22 Preparation and light-induced degradation of a diazosulfonate surfactant.

enues employed to prepare cleavable surfactants will turn out to become important commercial processes in the future.

REFERENCES

1. G. Krüger, D. Boltersdorf, and H. Lewandowski, in *Novel Surfactants* (K. Holmberg, ed.), Marcel Dekker, New York, 1998, pp. 115–138.
2. J. Waters, H. H. Kleiser, M. J. How, M. D. Barratt, R. R. Birch, R. J. Fletcher, S. D. Haigh, S. G. Hales, S. J. Marshall, and T. C. Pestell, Tenside Surfactants Deterg. *28*:460–468 (1991).
3. M. Chelminska-Bertilsson, A. Edebo, R. A. Thompson, and L. Edebo, Scand. J. Gastroenterol. *30*:670–674 (1995).
4. B. Ahlström, M. Chelminska-Bertilsson, R. A. Thompson, and L. Edebo, Antimicrob. Agents Chemother. *39*: 50–55 (1995).
5. M. Robson Wright, J. Chem. Soc. B 707–710 (1969).
6. R. P. Bell and F. J. Lindars, J. Chem. Soc. 1087–1098 (1954).
7. M. Lindstedt, S. Allenmark, R. A. Thompson, and L. Edebo, Antimicrob. Agents Chemother. *34*:1949–1954 (1990).
8. R. A. Thompson and S. Allenmark, J. Colloid Interface Sci. *148*:241–246 (1992).
9. G. D. Lichtenwalter, L. E. Miller, C. Siram, and E. H. Wahl, PCT Int. Patent Appl. WO 9325648 A1 931223 (1993).
10. L. M. Ilardi and S. A. Madison, U.S. Patent 5429755 A 950704 (1995).
11. H. J. Weissen and N. Porta, Eur. Patent Appl. EP 638639 A1 950215 (1995).
12. F. E. Hardy and A. D. Willey, PCT Patent Appl. WO 9525157 A1 950921 (1995).
13. D. A. Jaeger, M. Darlene Ward, and A. K. Dutta, J. Org. Chem. *53*:1577–1580 (1988).
14. D. A. Jaeger, C. T. Finley, M. R. Walter, and C. A. Martin, J. Org. Chem. *51*:3956–3959 (1986).
15. E. N. Vulfson, in *Novel Surfactants* (K. Holmberg, ed.), Marcel Dekker, New York, 1998, pp. 279–300.
16. I. Söderberg, C. J. Drummond, D. N. Furlong, S. Godkin, and B. Matthews, Colloids Surf. A *102*:91–97 (1995).
17. G. B. Oguntimein, H. Erdmann, and R. D. Schmid, Biotechnol. Lett. *15*:175–180 (1993).
18. A. Schlotterbeck, S. Lang, V. Wray, and F. Wagner, Biotechnol. Lett. *15*:61–64 (1993).
19. K. Adelhorst, F. Björkling, S. E. Godtfredsen, and O. Kirk, Synthesis *2*:112–115 (1990).
20. L. Guillardeau, D. Montet, N. Khaled, M. Pina, and J. Graille, Tenside Surfactants Deterg. *25*:342–344 (1992).
21. P. Skagerlind, K. Larsson, M. Barfoed, and K. Hult, J. Am. Oil Chem. Soc. *74*:39–45 (1997).
22. G. Fregapane, D. B. Sarney, S. G. Greenberg, D. J. Knight, and E. N. Vulfson, J. Am. Oil Chem. Soc. *71*: 87–91 (1994).
23. A. Ducret, A. Giroux, M. Trani, and R. Lortie, Biotechnol. Bioeng. *48*:214–221 (1995).
24. A. Ducret, A. Giroux, M. Trani, and R. Lortie, J. Am. Oil Chem. Soc. *73*:109–113 (1996).
25. D. A. Jaeger, Supramol. Chem. *5*:27–30 (1995).
26. A. Bieniecki, K. A. Wilk, and J. Gapinski, J. Phys. Chem. B *101*:871–875 (1997).
27. A. Sokolowski, A. Bieniecki, K. A. Wilk, and B. Burczyk, Colloids Surf. A *98*:73–82 (1995).
28. G.-W. Wang, X.-Y. Yuan, Y.-C. Liu, X.-G. Lei, and Q.-X. Guo, J. Am. Oil Chem. Soc. *72*:83–87 (1995).
29. G.-W. Wang, Y.-C. Liu, X.-Y. Yuan, X.-G. Lei, and Q.-X. Guo, J. Colloid Interface Sci. *173*:49–54 (1995).
30. A. Sokolowski, A. Piasecki, and B. Burczyk, J. Am. Oil Chem. Soc. *69*:633–638 (1992).
31. G.-W. Wang, X.-Y. Yuan, Y.-C. Liu, and X.-G. Lei, J. Am. Oil Chem. Soc. *71*:727–730 (1994).

32. K. A. Wilk, A. Bieniecki, B. Burczyk, and A. Sokolowski, J. Am. Oil Chem. Soc. *71*:81–85 (1994).
33. D. C. Galante and R. C. Hoy, Eur. Patent Appl. EP 0 742 177 A1 (1996).
34. R. C. Hoy and A. F. Joseph, INFORM *7*:428–429 (1996).
35. A. Piasecki, J. Prakt. Chem. *327*:731–738 (1985).
36. A. Bieniecki and K. A. Wilk, J. Phys. Org. Chem. *8*: 71–76 (1995).
37. W. von Rybinski, Curr. Opin. Colloid Interface Sci. *1*: 572–578 (1996).
38. W. von Rybinski, in *Alkyl Polyglycosides* (K. Holmberg, ed.), Marcel Dekker, New York, 1998, pp. 31–85.
39. D. Balzer, Tenside Surfactants Deterg. *33*:102–110 (1996).
40. B. Burczyk and A. Sokolowski, Tenside Surfactants Deterg. *15*:68–71 (1978).
41. A. Sokolowski and B. Burczyk, Tenside Surfactants Deterg. *18*:199–201 (1981).
42. C. Yue, J. M. Harris, P.-E. Hellberg, and K. Bergström, J. Am. Oil Chem. Soc. *73*:841–845 (1996).
43. D. Ono, A. Masuyama, and M. Okahara, J. Org. Chem. *55*:4461–4464 (1990).
44. D. Ono, A. Masuyama, T. Tanaka, and M. Okahara, Tenside Surfactants Deterg. *29*:412–417 (1992).
45. S. Yamamura, M. Nakamura, K. Kasai, H. Sato, and T. Takeda, J. Jpn. Oil Chem. Soc. *40*:1002–1006 (1991).
46. A. Masuyama, D. Ono, A. Yamamoto, T. Kida, Y. Nakatsuji, and T. Takeda, J. Jpn. Oil Chem. Soc. *44*:446–450 (1995).
47. D. Ono, S. Yamamura, M. Nakamura, T. Takeda, A. Masuyama, and Y. Nakatsuji, J. Am. Oil Chem. Soc. *72*:853–856 (1995).
48. D. Ono, A. Masuyama, Y. Nakatsuji, M. Okahara, S. Yamamura, and T. Takeda, J. Am. Oil Chem. Soc. *70*: 29–36 (1993).
49. T. Kida, N. Morishima, A. Masuyama, and Y. Nakatsuji, J. Am. Oil Chem. Soc. *71*:705–710 (1994).
50. B.-K. Song and K. Wolf, DWI Rep. *114*:549–554 (1995).
51. D. A. Jaeger, M. D. Ward, and A. K. Dutta, J. Org. Chem. *53*:1577–1580 (1988).
52. D. A. Jaeger, Y. M. Sayed, and A. K. Dutta, Tetrahedron Lett. *31*:449–450 (1990).
53. D. A. Jaeger and Y. M. Sayed, J. Org. Chem. *58*:2619–2627 (1993).
54. J. Y. Wang, R. A. Uphaus, J. Wang, and D. A. Jaeger, Thin Solid Films *242*:277–282 (1994).
55. D. A. Jaeger and S. G. G. Russell, Tetrahedron Lett. *34*:6985–6988 (1993).
56. D. A. Jaeger, S. G. G. Russell, and H. Shinozaki, J. Org. Chem. *59*:7544–7548 (1994).
57. R. Eliason and M. M. Kreevoy, J. Am. Chem. Soc. *100*: 7037–7041 (1978).
58. K. Bergström and P.-E. Hellberg, Swedish Patent Appl. 9602617-4 (1996).
59. K. Bergström and P.-E. Hellberg, Swedish Patent Appl. 9710241-3 (1997).
60. D. A. Jaeger, B. Li, and T. Clark, Jr., Langmuir *121*: 4314–4316 (1996).
61. W. W. Epstein, D. S. Jones, E. Bruenger, and H. C. Rilling, Anal. Biochem. *119*:304–312 (1982).
62. I. R. Dunkin, A. Gittinger, D. C. Sherrington, and P. Whittaker, J. Chem. Soc. Chem. Commun. 2245–2246 (1994).
63. I. R. Dunkin, A. Gittinger, D. C. Sherrington, and P. Whittaker, Synthesis J. Chem. Soc. Perkin Trans. 2 1837–1842 (1996).
64. O. Nuyken, K. Meindl, A. Wokaun, and T. Mezger, J. Photochem. Photobiol. A Chem. *85*:291–298 (1995).
65. O. Nuyken, K. Meindl, A. Wokaun, and T. Mezger, Macromol. Rep. *A32*:447–457 (1995).
66. T. Mezger, O. Nuyken, K. Meindl, and A. Wokaun, Polym. Mater. Sci. Eng. *73*:153–155 (1995).
67. D. A. Jaeger, V. B. Reddy, and D. S. Bohle, Tetrahedron Lett. *40*:649–652 (1999).
68. S. A. Madison, M. Massaro, G. B. Rattinger, and C. J. Wenzel, PCT Patent Appl. WO 9514661 A1 950601 (1995).
69. D. A. Jaeger, J. Wettstein, and A. Zafar, Langmuir *14*: 1940–1941 (1998).
70. A. Masayama, C. Endo, S. Takeda, and M. Nojima, Chem. Commun. *18*:2023–2024 (1998).
71. J. Cuomo, J. H. Merrifield, and J. F. W. Keana, J. Org. Chem. *45*:4216–4219 (1980).
72. Y. Hayashi, F. Shirai, T. Shimizu, Y. Nagano, and K. Teramura, J. Am. Oil Chem. Soc. *62*:555–557 (1985).

3

Gemini Surfactants and Surfactant Oligomers

MARTIN IN CNRS-Rhodia, Cranbury, New Jersey

I. INTRODUCTION

Covalent linking of several amphiphilic moieties at the headgroup level yields a "surfactant oligomer" (Fig. 1a). The surfactant oligomers we are going to deal with are higher homologues of the gemini surfactants (surfactant dimers) (Fig. 1c) [1]. They are distinguished from "oligomeric surfactants" (macrosurfactants), which consist of amphiphilic diblock copolymers (Fig. 1b), and are also currently the subject of active research [2]. In surfactant oligomers, the structural repeating unit is amphiphilic by itself. The chemical group that connects the amphiphilic moieties is of variable nature and length.

Gemini surfactants have been synthesized and patented for more than 50 years [3], especially cationic ones. They have become topical again in the last two decades and their properties have been the subject of several reviews [1,4–13]. They are of industrial and academic interest for diverse reasons. Cationic geminis were first described [14–16] and claimed to be good textile softeners whose action resists laundering and dry cleaning operations [17,18] or as efficient bactericidal and fungicidal agents with good diffusion properties and skin compatibility [19,20]. More recently Devínsky and colleagues, aiming to establish the relationship between biological activity and surfactant structure, synthesized and studied a large variety of cationic gemini surfactants [21]. Okahara and Ikeda have developed a new and easy synthetic route to oligo(ethylene glycol) diglycidyl ethers and described various types of derivatives including anionic gemini surfactants [6]. Zana and Talmon pointed out the importance of the length of the connecting chain between the headgroups, the "spacer," in controlling the microstructure of the self-assemblies [22]. Menger and Littau raised the question of the micellar structure of surfactants, which could not aggregate without exposing hydrocarbon moieties to water [23]. Finally, Rosen pointed out the unexpected effectiveness of these surfactants in lowering surface tension and their enhanced synergistic effect in mixtures. He analyzed the effects on performance properties [4], which could make gemini the "new generation of surfactants." The number of patents (which concern predominantly anionic and nonionic geminis) filed since then attests that this was rather convincing [24].

Gemini surfactants were considered the starting point for surfactant oligomers. Synthesis and physicochemical studies of surfactant trimers were reported [1,5]. Some surfactant oligomers have been synthesized and studied in other contexts. For instance, cationic lipids are mostly studied as carriers in the intracellular delivery of bioactive agents [25,26] and more specifically as nonviral transfecting agents [27]. Lipophilic di- and triamides have been used as ionophores for alkaline earth metal cations [28], and lipophilic cyclopolyamines are potential liquid membrane sensors for nucleosides [29]. The synthesis of higher homologues also provides an alternative path to the study of the transition from surfactant to polysoap behavior [30].

Gemini surfactants have shown many "unexpected properties," which are a posteriori rather well understood, and concepts long known [31,32] explain at least qualitatively these observations. Surfactant self-assembly results from two opposing forces. Attraction be-

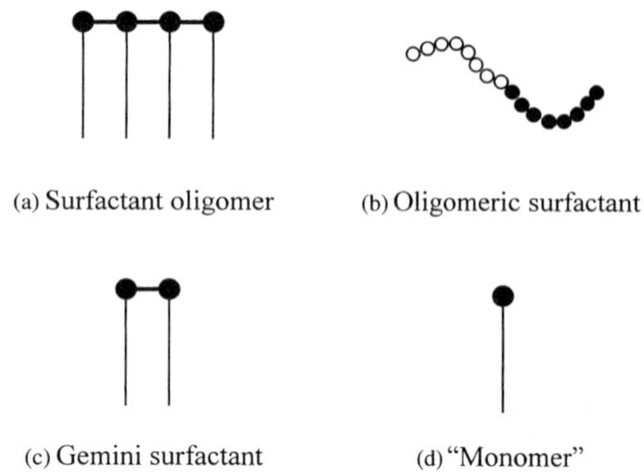

FIG. 1 Surfactant oligomers (a) are distinguished from oligomeric surfactants (b) by the fact that the structural repeating unit is amphiphilic by itself. Surfactant oligomers are higher homologues of gemini surfactants (c). The structural repeating unit (d) corresponds to a conventional surfactant, which will be referred to as the "monomer."

tween the hydrophobic tails induces the aggregation, and repulsion between the hydrophilic headgroups ensures the existence of a large interfacial area [31]. Classical ways to modify the micellization, the shape of the micelles, or the lyotropic behavior consist of tuning these opposing forces by varying the length of the hydrophobic tail and the nature of the headgroup. The concept of surfactant oligomers provides new parameters to tune this balance of opposing forces.

The degree of oligomerization *x*, i.e., the number of amphiphilic moieties in the surfactant, is a new variable parameter. The length of the spacer *s* can vary along with its hydrophobicity and rigidity. This makes it possible to achieve more direct and more efficient control of the optimal interfacial area per headgroup [32]. As in conventional surfactants, the length of the tail, *m*, and the chemical nature of the headgroup are possible chemical variables. These two variables are not specific to surfactant oligomers, but the study of their influence brings insight to the properties of surfactant oligomers. Gemini surfactants can also be nonsymmetric; i.e., both amphiphilic moieties can be different in terms of chain length and headgroup nature.

A large variety of surfactant oligomers will be discussed in the next section, where their synthesis is reviewed. Following Zana, the cationic surfactant dimers with simple structure are referred to as *m-s-m*, 2X when the spacer consists of a —$(CH_2)_s$— chain and *m-s/3-m*, 2X when the spacer consist of *s*/3 ethylene oxide units. X is the counter ion. Using the same logic, trimers are referred to as *m-s-m-s-m*, 3X. Some anionic surfactants with simple structure are designated in the same way but the headgroup nature is precise.

The new parameters *s* and *x* have a strong influence on the surface activity and the packing at the interface of surfactant oligomers as well as on their self-assembling properties in the bulk. This is described in Sections III and IV. Like conventional surfactants, oligomers have been used in analytical and synthesis chemistry, as selective receptors, as hosting or templating agents, and also as reactants. This will be the subject of the last section.

II. SYNTHESIS: STRUCTURE DIVERSITY

A. Surfactant Dimers (Gemini)

1. Cationic Geminis

Cationic surfactants are among the first gemini surfactants reported in the literature [3,14–20]. First claimed to be good fabric softeners or developed for their biological activity, they have also been studied for micellar catalysis. Easier to synthesize, they have been the materials of choice for fundamental studies. The influence of various parameters such as the length of the hydrophobic chains, their dyssymmetry, and the nature (hydrophilic or hydrophobic) and the length of the spacer has been studied. Some examples of cationic gemini surfactants are shown in Fig. 2.

Synthetic methods for preparing diquaternary ammonium (Fig. 2a, c, e, and g) [14,21,33–42] and dipyridinium (Fig. 2d, f, and h) [19,43] geminis rely on the same reactions (quaternization of a tertiary amine with bromoalkane) as those used for their corresponding monomer, except for the use of difunctional reagents.

Two routes can be distinguished. The first one proceeds by quaternization of a tertiary diamine (route 1) [14,21,33,38,42] as exemplified in Scheme 1 [14,20]. The second one couples two tertiary fatty amines with dibromoalkane (route 2) [36–39,41] as shown in Scheme 2 [38]. Two tertiary diamines can also be coupled with epichlorohydrin, giving a hydroxypropylene spacer [44,45].

Quaternary ammonium geminis with a hydrophilic spacer (Fig. 2b) have been synthesized using route 2 by reacting tertiary amines with α,ω-dibromo alcohols [46] or an α,ω-dibromo oligo(oxyethylene) [47–49] [the latter being synthesized via bromination of oligo(oxyethylene)glycol with phosphorus tribromide]. Other methods have been reported to produce diquaternary ammonium surfactants with oxyalkylene spacer [50].

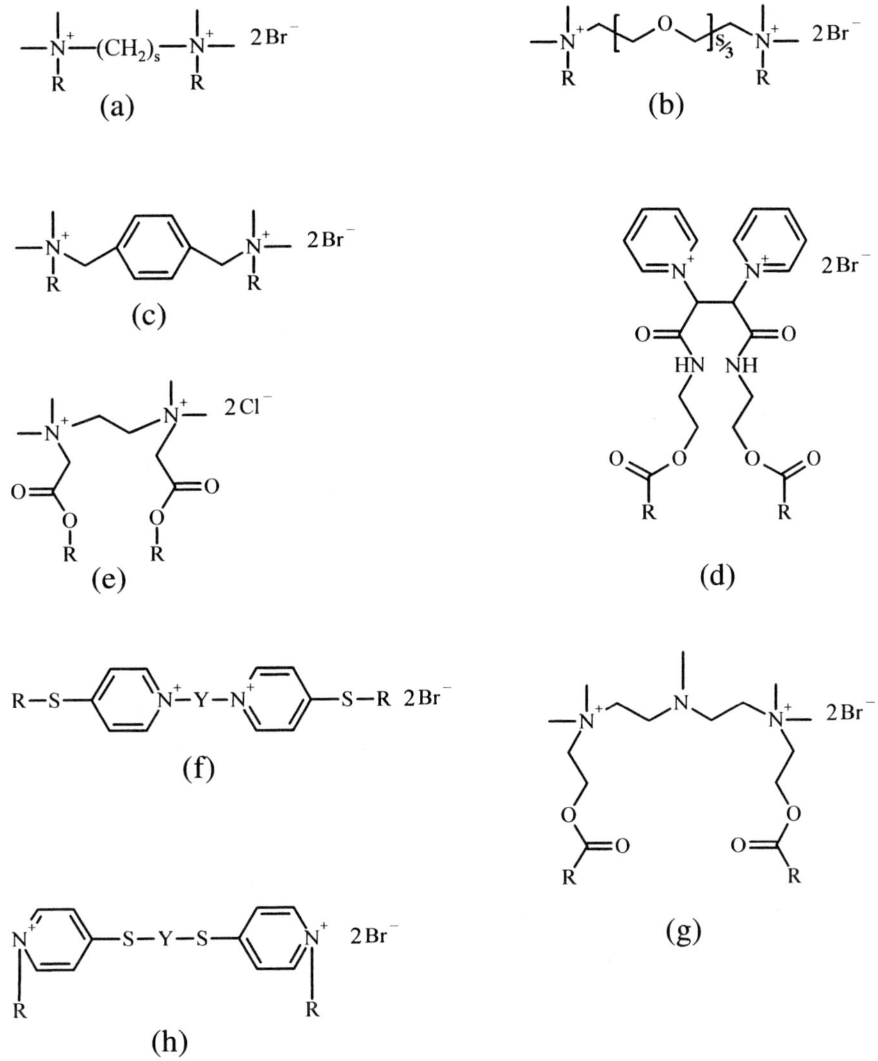

FIG. 2 Examples of cationic gemini surfactants.

2. Anionic Geminis

A large variety of anionic gemini surfactants (see Fig. 3: sulfonate 3a, sulfate 3b, phosphate 3c, carboxylate 3d) with hydrophilic spacers have been prepared from the corresponding fatty diols according to the conventional procedure used for classical surfactants [51–59] (Scheme 3, second step). Intermediate fatty diols have been prepared by the reaction of diglycidyl ethers (prepared according to Ref. 60) with the appropriate fatty alcohol, leading to gemini surfactants with a hydrophilic spacer (Scheme 3, first step). The intermediate fatty diols are nonionic gemini surfactants. They are not soluble in water, but their emulsifying properties

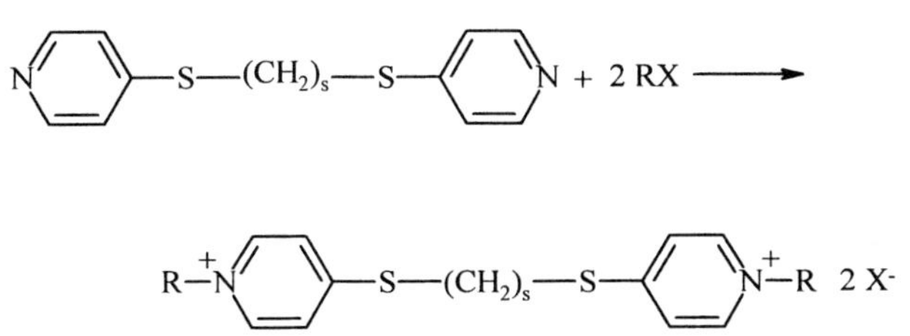

SCHEME 1

SCHEME 2

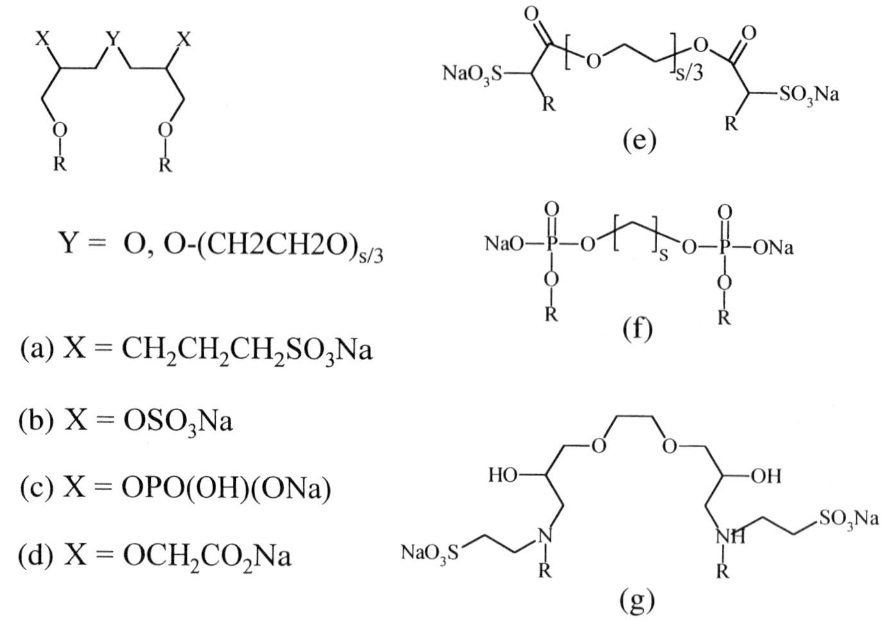

FIG. 3 Examples of anionic gemini surfactants.

[57] and insoluble monolayers at the air-water interface have been studied [61].

Sulfation of the diols proceeds with chlorosulfonic acid in the presence of glacial acetic acid in dichloromethane at room temperature [51] or chloroform at 0°C [52], followed by neutralization with aqueous sodium carbonate or alcoholic sodium hydroxide (see Scheme 3).

Phosphatation is carried out with polyphosphoric acid in dry benzene at 50°C. Use of phosphorus pentoxide leads to some undesirable dehydration of the secondary alcohol, and phosphorus oxychloride leads to a complex reaction mixture [53].

Disodium sulfonates (Fig. 3a) are prepared from 1,3-propanesultone in the presence of NaOH [51] or NaH in dry tetrahydrofuran (THF) at 60°C [51,54,55]. In all

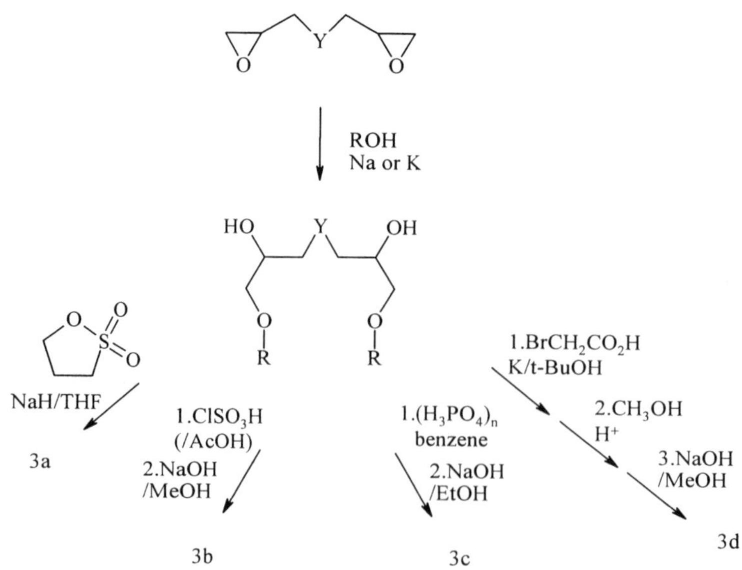

SCHEME 3

cases, pure products are obtained after extraction and separation by silica gel column chromatography or by high-performance liquid chromatography (HPLC).

The synthesis of dicarboxylate geminis (Fig. 3d) proceeds by reacting the diols with bromoacetic acid in *t*-BuOH under basic conditions, followed by esterification with methanol under acidic conditions. This ester is purified on a silica gel column and finally hydrolyzed by NaOH in methanol [59].

Taurine gemini surfactants (Fig. 3g) were synthesized by reaction of ethylene glycol diglycidyl ether with *N*-(alkyl)taurine in the presence of sodium carbonate in ethanol [56]. *N*-(Alkyl)taurines are prepared by reaction of the corresponding fatty amines with sodium 2-bromoethane-1-sulfonate. Anionic gemini surfactants with a hydrophilic spacer (Fig. 3e) were also synthesized by diesterification of polyethylene glycol with α-sulfonated fatty acids in carbon tetrachloride under reflux [62]. Esterification of polyethylene glycol with α-sulfonated acid (prepared as described in Ref. 63) yielded a monoester (about 50%), diester (about 25%), and unreacted PEG (25%); separation and purification of the components were carried out by reversed-phase chromatography [62]. The same types of sulfonate surfactants with a hydrophobic spacer has been synthesized by disulfonation of fatty diesters of adipic acid [64]. Another possible route to these compounds, coupling 2-hydroxy-1-alkanesulfonates with diacids, has proved unsuccessful [64]. The synthesis of other anionic gemini surfactants with a hydrophilic spacer of different lengths has been reported [65]. Dihydroxyl precursors such as tartaric acid have been used to prepare asymmetric anionic gemini surfactants [66].

Alkylphosphate geminis with a hydrophobic spacer (Fig. 3f) have been obtained by coupling alkyl phosphate tetramethylammonium salts with dibromoalkanes [67] or α,α'-dibromoparaxylene [39] according to Bauman's method [68]. This route relies on the fact that monoalkyl phosphates $[ROP(O)O_2]^{2-}$ are better nucleophiles than dialkyl phosphates $[(RO)_2P(O)O_2]^-$. Alternatively, Eibl's method [69] (phosphorylation of the diol using $POCl_3$ in the presence of triethylamine) has been used to synthesize phosphate geminis with a —$(CH_2)_s$— spacer (Fig. 3f) [70] or a rigid hydrophobic spacer [39]. Gemini glycerophosphates with a long flexible hydrophobic spacer have also been synthesized as models for archaebacterial membrane lipids [71].

Sarcosine-type surfactant dimers have been synthesized from fatty acid and ethylenediaminediacetic acid via the mixed anhydride method [72]. Their efficiency as flotation agents is improved as compared with the monomer, and strong synergistic effects have been observed on mixing with fatty amines.

3. Amino Acid Derivatives

The use of amino acids to prepare gemini surfactants offers a large variety of headgroup structures. Moreover, it facilitates the synthesis of enantiomerically pure surfactants. Amino acid-based gemini surfactants have improved biocompatibility. Some of them have been shown to be less hemolytic and less irritating. They have also proved to be good immunoadjuvants for the formulation of vaccines. Some examples are presented in Fig. 4.

Nonionic geminis (Fig. 4a) have been prepared by condensation of N^{α},N^{ε}-diacyl lysine with *N*,*N*-bis(methylpolyoxyethylene) amine [73]. Their structure is close to the structure of natural lecithin, and they do not rigorously correspond to surfactant dimers as defined in the introduction. The same type of compounds with alkyl chains of different lengths has been synthesized [74].

Gemini cationic surfactants with variable spacer lengths have been synthesized from arginine [75–78] (Fig. 4b). The synthesis proceeds in three steps: (1) protection of the guanido group of arginine with a nitro group, (2) coupling of two protected arginines via condensation of the α-carboxyl group to both primary amino groups of the α,ω-alkanediamine using benzotriole-1-yl-oxy-tris-(dimethylamino)-phosphonium hexafluorophosphate (BOP) in the presence of an activating base (DABCO), and (3) catalytic hydrogenation to unprotect the guanido group.

Anionic gemini surfactants have also been synthesized from L-cysteine (Fig. 4c) [79]. A chemoenzymatic route for the preparation of a variety of amino acid–based gemini surfactants has been developed [80]. Immobilized lipases efficiently catalyze the formation of diester (respectively diamide) from *N*-protected amino acids and α,ω-alkane-diols (respectively α,ω-alkanediamine) with hydrocarbon spacers of different lengths. Tyrosine- (Fig. 4d) and serine-based geminis were then obtained by acylation and acyl chloride followed by removal of the carbobenzyloxy-protecting group by catalytic hydrogenation under standard conditions. For glutamic acid-based geminis, the carboxyl group of the residue was esterified prior to lipase action. For lysine-based geminis, the *N*-protected acylated amino acid turned out to be a better substrate for the lipase than the *N*-protected amino acid itself [80].

4. Sugar Derivatives

Carbohydrate-based gemini surfactants present the advantage of being derived from a renewable source. The

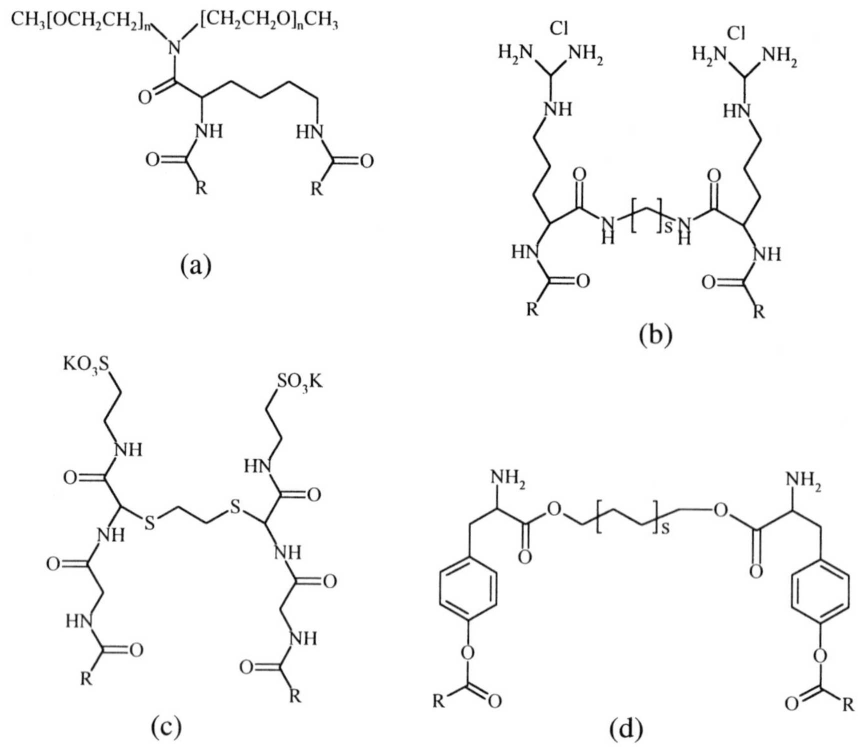

FIG. 4 Examples of amino acid-based gemini surfactants.

presence of two sugar headgroups is expected to enhance intra- and intermolecular hydrogen bonding, but it was also hoped that the gemini structure could lower the Kraft temperature (a point that often limits the practical use of polyalkylglucosides) [81]. Examples are given in Fig. 5.

The synthesis of the compound of Fig. 5b proceeds by reaction of carbohydrate lactone with a 2,2-dialkylpropane-1,3-diamine in methanol [82,83]. The diamine was obtained by dialkylation of malonitrile with bromoalkane in dimethyl sulfoxide (DMSO), followed by reduction of the nitrile groups into primary amine groups by $LiAlH_4$ in dry ether or with lithium in a liquid ammonia-ethanol mixture.

Sugar-based surfactants with variable spacer lengths (Fig. 5a) have been prepared by catalytic hydrogena-

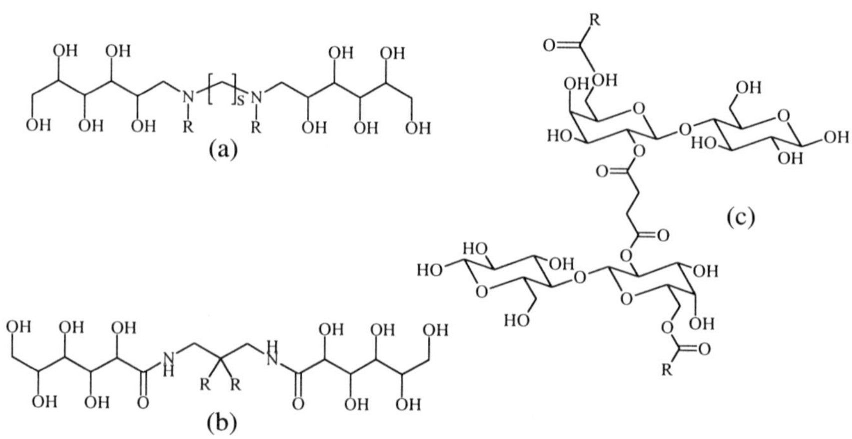

FIG. 5 Examples of sugar-based gemini surfactants.

tion of D-glucose and the appropriate α,ω-alkanediamine [84]. At that step, a bola-amphiphile is obtained; it is acylated with fatty acid anhydride to get a gemini surfactant.

Nonionic glucoside gemini surfactants with the two amphiphilic moieties linked at C-6 have been synthesized [85]. A large variety of xylose, glucose, galactose, and lactose (Fig. 5c) derived gemini surfactants, with different chain and spacer lengths, have been prepared from partially protected sugars (isopropylidene derivatives), using enzymes to introduce fatty acids regioselectively into carbohydrate moieties [86]. Both amphiphilic moieties were connected via different hydroxyl groups in the sugar molecule, and a heterodimer xylose-lactose (Fig. 6d) gemini was prepared [86].

5. Surfactant Heterodimer

Gemini surfactants can be nonsymmetric. This means that the two amphiphilic moieties can differ either in the length of the hydrophobic tail [87] or in the nature of the headgroup [88,89]. Some examples of surfactant heterodimers are presented in Fig. 6.

Cationic gemini surfactants with two hydrophobic chains of different lengths (Fig. 6a) were obtained in two steps from the permethylated diamine as already described [87]. The intermediate alkyldimethyl [1-(2-dimethylamino)ethyl] ammonium bromide is recrystallized in ether. Hybrid hydrocarbon-fluorocarbon cationic gemini surfactants have also been synthesized the same way [90].

The synthesis of gemini surfactants with two different hydrophilic headgroups (cationic-anionic, nonionic-nonionic, anionic-nonionic) (Fig. 6b–d) involves more steps. For example, the compound of Fig. 6c was synthesized in three steps: An alkyl dimethyl amine reacts first with ethylbromoacetate and then with hydrazine to give the surfactant $RMe_2N^+CH_2CONHNH_2Br^-$. The latter reacts with fatty keto acids, resulting in the surfactant of Fig. 6c. A ^{13}C nuclear magnetic resonance (NMR) study revealed that it was obtained as the *E* isomer with respect to the carbon-nitrogen double bond [88]. This study also provides an example of a cleavable spacer gemini surfactant (see Section II.A.7). The chemoenzymatic route to sugar-based surfactants described earlier allows the synthesis of nonionic surfactant heterodimers (Fig. 6d) [86].

6. Chiral Gemini Surfactants

Enantiomerically pure gemini surfactants derived from amino acids have already been mentioned. Other types of chiral surfactants have been synthesized from tartaric acid with anionic hydrophilic groups such as phosphates (Fig. 7a) [91], carboxylates, and sulfates [66] as well as cationic headgroups [66]. Chiral cationic gemini surfactants have been synthesized from chiral biphenyl (Fig. 7b) [92].

7. Functional Gemini Surfactants

(a) Cleavable Surfactants. Several types of cleavable gemini surfactants have been synthesized based on disulfide- [93], hydrazine- [88], acetal- [94–96], or ozone-cleavable double bonds [97].

The synthesis of cationic surfactants containing a disulfide bond in the spacer has been achieved by condensation of an alkyldimethylamino betaine with cystine or cystamine via the mixed anhydride method [93]. The betaine is first converted to a mixed anhydride by reaction with isobutyl chloroformate. The second step is the aminolysis of the anhydride by the amino group of cystine dimethyl ester or cystamine. In aqueous solutions the disulfide surfactants obtained decompose at room temperature at $pH > 8.0$ but are stable even at a higher temperature (50°C) at lower pH. These surfactants have potential applications in the textile and cosmetic fields because the disulfide bond can also react with thiol groups (of reduced keratin, for instance) (thiol-disulfide interchange) [93]. Disulfide-based phospholipid dimers have been synthesized from functionalized monomer surfactant in the micellar state. This was done to study nearest neighbor recognition in membranes [98].

The simplest synthesis of gemini surfactants with an acetal (1,3-dioxalane) based spacer proceeds by acid-catalyzed condensation of diethyl tartrate with fatty ketones followed by alkaline hydrolysis [94]. The synthesis presented in Ref. 95 is more involved and does not strictly give a gemini surfactant because the linkage between the two amphiphilic moieties is not located at the headgroup but in the middle of the fatty chains. In that respect, these surfactants provide interesting intermediate structures between gemini and bolaform surfactants [7].

Anionic gemini surfactants with an ozone-cleavable spacer (bearing a carbon-carbon double bond) [97] have been synthesized with unsaturated diglycidyl ether by the same method as in Ref. 54. Their ozonolysis has been studied by proton NMR [97].

(b) Miscellaneous. Ferrocenyl cationic geminis have been synthesized by the same procedure as the simple diquaternary ammonium gemini using bromoctylferrocene to quaternize permethylated propanediamine [99]. It was known that a variation of the oxidation state of the ferrocene moiety made it possible to control bulk

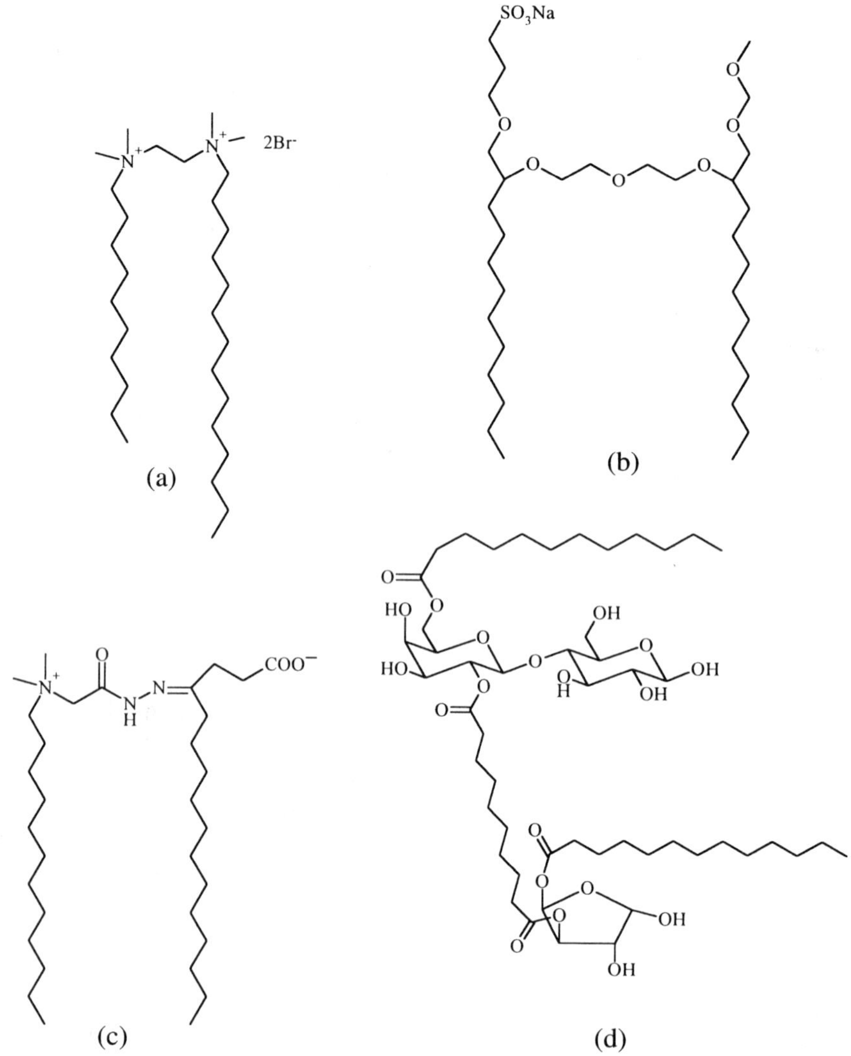

FIG. 6 Examples of surfactant heterodimers.

as well as surface properties of ferrocenyl surfactant solutions (see Table 26) [100]. With gemini ferrocenyl surfactants the range of concentration for which this control is efficient is greatly extended [99] (see Chapter 7 in this volume). Finally, anionic gemini surfactants with azo groups in the spacer have been synthesized and used as initiators for radical polymerization (inisurfs) [101]. As with any inisurfs, they suffer from a poor radical yield.

B. Oligomers

A large variety of surfactant oligomers, mostly trimers, has been reported in the literature. Their structures are presented in Fig. 8. Some of them are obtained pure; others consist of mixtures of surfactant oligomers with different oligomerization degrees.

Triazine-derived trialkyltriquaternary ammonium surfactants have been synthesized by quaternization of trialkyl triazine with dimethyl sulfate and claimed to be efficient antibacterial, antifungal, and antiviral agents with weaker toxicity [102].

A trimer of DTAB with $s = 3$, 12-3-12-3-12, 3Br, was first synthesized in a multistep procedure described in Scheme 4 [103]. Since then, its synthesis has been improved starting with bis(aminopropylamine) as shown in Scheme 5 [104]. The first step (permethylation) is carried out in acidic aqueous solution with formaldehyde and sodium borohydride as reducing agent, as described in Ref. 105. The second step consists of the

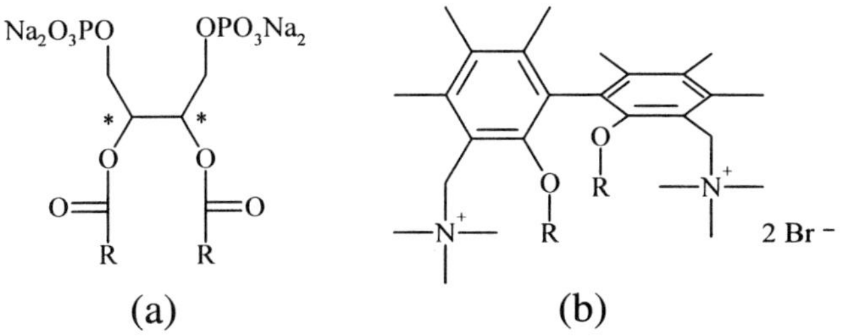

FIG. 7 Examples of asymmetric gemini surfactants.

quaternization with bromododecane in acetonitrile at 80°C. The purification proceeds by recrystallization [104].

8-3-8-3-8, 3Br, 16-3-16-3-16, 3Br (unpublished results), and the tetramer 12-3-12-4-12-3-12, 4Br (Fig. 8c) [104] have been synthesized in the same way. The preceding two-step procedure has been used to prepare *m*-6-*m*-6-*m*, 3Br (Fig. 8a) as well as triquaternary ammonium with a three-armed, star-shaped spacer (Fig. 8d) (unpublished results). For these surfactants with $s \neq 3$, permethylation could be performed via the Eschweiler-Clarke reaction (with a C_3 spacer, this reductive alkylation induces fragmentation of the triamine and yields the permethylated diamine [106]). The synthesis of 12-2-12-2-12, 3Br by the same route has been reported, but after the quaternization reaction, the trimer had to be purified from a mixture of dimer and trimer [107].

Another procedure to synthesize cationic surfactant trimers uses epichlorohydrin (Scheme 6) to produce cationic trimers with hydrophilic spacers (Fig. 8b) [108]. This route allows interesting variations in the structure. One can also obtain triquaternary ammonium surfactants with two fatty chains or diquaternary ammonium ones with three hydrophobic chains, depending on the nature of the amine that reacts with epichlorohydrin at the first step.

The synthesis of diglycidyl ether from diol has been generalized to polyglycidyl ether from polyols [109] and opened the path to the synthesis of sulfonate surfactant trimers with a hydrophilic star-shaped spacer (Fig. 8e) [110,111]. Other surfactants with unequal numbers of ionic groups and hydrophobic chains, triple-chain, double ionic groups, were synthesized from alkylglyceroldiglycidyl ether [112]. Triple-chain surfactants with hydrophobic chains of different lengths were also obtained [112].

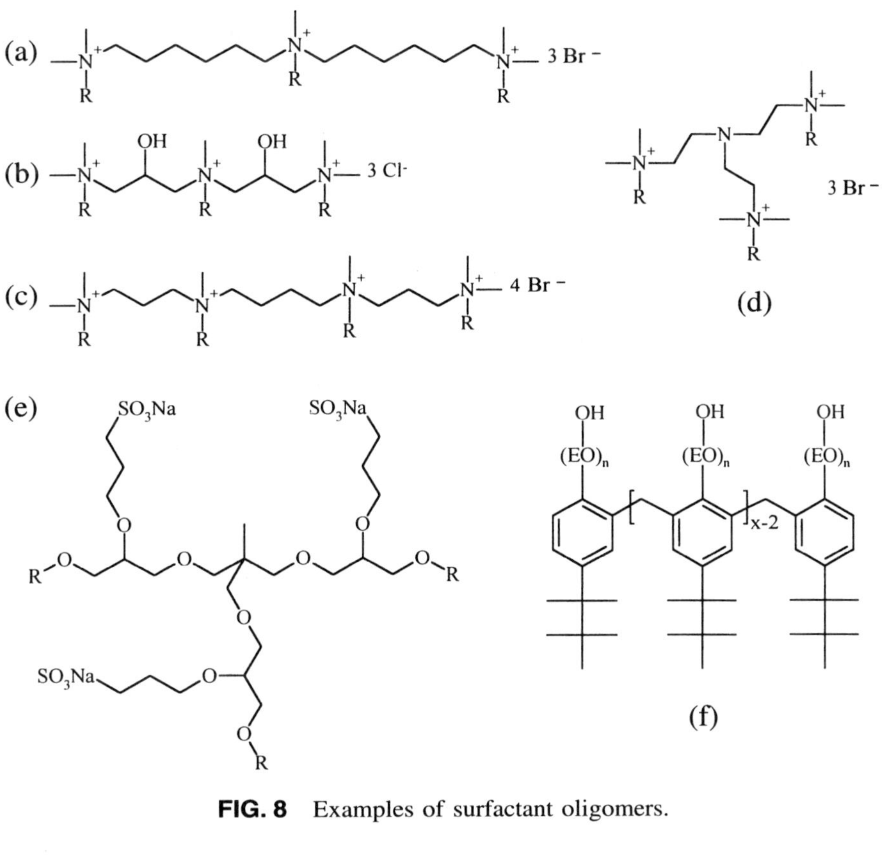

FIG. 8 Examples of surfactant oligomers.

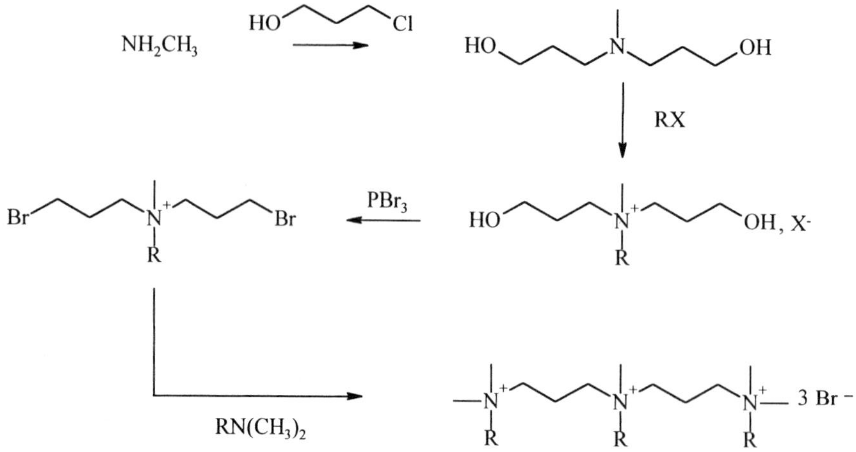

SCHEME 4

Nonionic surfactants with three hydrophilic heads and two lipophilic tails have been patented [113]. As Tiloxapol (Fig. 8g) [114], they probably consist of a mixture of different oligomerization degrees. The chemoenzymatic route to sugar-based gemini surfactant described earlier was also used to prepare nonionic surfactant trimers [86].

III. SURFACE ACTIVITY AND STRUCTURE AT INTERFACE

A. Air-Water Interface

The surface activity of soluble surfactant oligomers in aqueous solution has been extensively studied by tensiometry to determine their critical micelle concentration (cmc), to address their packing at the air-water interface, and to determine their performance properties (see Appendix, Section VII, Tables 5, 9, 10, 13, 15–22, 24–27, 29, 31–34, 36–46). The results of are often summarized by four parameters: C_{20}, γ_{cmc}, Γ_m, and the cmc. The C_{20}, which is the concentration needed to decrease the surface tension by 20 mN/m, characterizes the *efficiency* of a surfactant to lower surface tension. (The efficiency of the surfactant is actually often reported as $pC_{20} = -\log C_{20}$.) The C_{20} value reflects the partitioning of the unmicellized surfactant between the bulk and the interface and is related to the standard free energy of adsorption at the air-water interface [115]. The surface tension at the cmc, γ_{cmc}, characterizes the *effectiveness* of a surfactant in lowering surface tension. It is related to the maximum film pressure a surfactant can build up at the air-water interface Π_{cmc}, before self-assembling in the bulk is thermodynamically favorable. The Γ_m is the maximum excess surface concentration and is obtained from a γ-c plot through the Gibbs equation:

$$d\gamma = nRT\Gamma d \ln c \qquad (1)$$

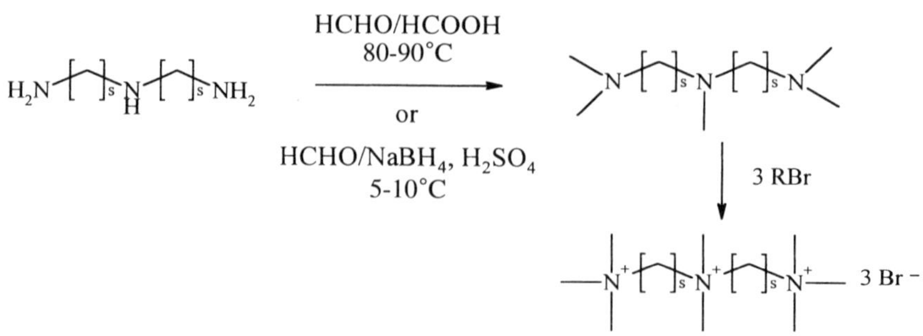

SCHEME 5

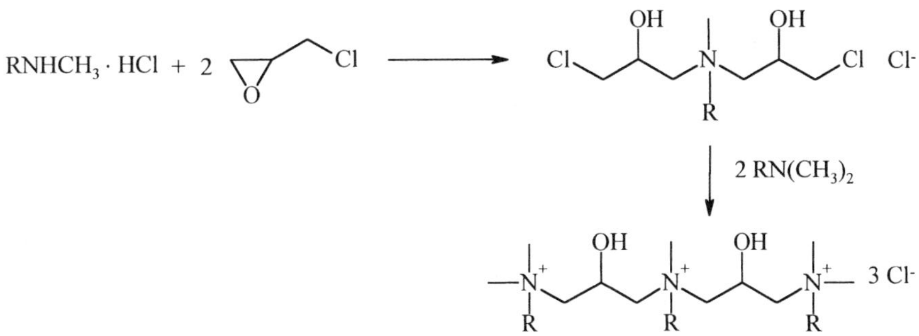

SCHEME 6

These four parameters are related by the following equation [115]:

$$\Pi_{cmc} = 20 + k\Gamma_m \log(\mathrm{cmc}/C_{20}) \quad (2)$$

From Γ_m, the minimum surface area per molecule of surfactant A_m, or per amphiphilic moiety a_m, has been determined.

Nonionic and long hydrophobic chain ionic surfactant oligomers are insoluble in water. The diagram of state of the insoluble monolayers they form at the air-water interface has been established with the Langmuir film balance. The Π-A isotherms obtained are characterized by the liftoff area A_L, the limiting surface area A_∞, and the collapse pressure Π_c. A_L corresponds to the highest surface area per molecule where a monolayer shows detectable resistance to compression. It is the inverse of the minimum surface concentration at which a surfactant builds up sensible pressure. A_∞ approximates the surface area per molecule at maximum compression and is obtained from the following relation:

$$A_\infty = A_c - \Pi_c(dA/d\Pi)_{\Pi c} \quad (3)$$

where A_c is the area per molecule at Π_c.

1. Efficiency in Lowering Surface Tension

The efficiency of surfactant oligomers in lowering surface tension is greater than that of conventional surfactants (see Appendix, Section VII, Tables 16, 17, 20, 21, 25, 32–34, 39, 40, 42–46). Typical C_{20} values for m = 12 conventional surfactants lie in the millimolar range, and for m = 12 gemini surfactant C_{20} values are currently close to 10^{-4} M. This is, of course, correlated with the lower cmc. The lower C_{20} of gemini surfactants as compared with conventional ones means that the standard free energy of adsorption is more negative. This can result a priori from either an increase in the standard chemical potential of the surfactant in the bulk or a decrease in the standard chemical potential at the interface. The observation that C_{20} decreases exponentially with the length of the alkyl chain in most of the gemini surfactants and with the same rate as it does for conventional surfactants suggests that the first hypothesis is the dominant factor. This is because the unfavorable contact between water and hydrocarbon for a gemini surfactant is twice that of the corresponding monomer.

When the length of the alkyl chains m exceeds a certain value, which is about 16 but varies with physicochemical conditions, the m dependence of C_{20} is weaker than expected and can reverse (see Tables 17, 21) [23,39,46,116,117]. In a few cases, C_{20} has been observed to increase with m. For anionic surfactant trimers (the spacer being star shaped), the C_{20} increases with m from m = 10 to 14 (see Table 44) [110,111]. These rather surprising results have been observed with surfactant oligomers whose spacer contains heteroatoms or aromatic rings but not with —$(CH_2)_s$— spacers. They have been interpreted in terms of premicellization [39,46,116,117]. An alternative explanation could be that intramolecular association occurs between the long alkyl chains. Such intramolecular interactions have been suggested on the basis of volumetric measurements [118]. If the surfactant molecule limits the contact between hydrocarbon chain and water by intramolecular hydrophobic association without losing too much conformational entropy, its chemical potential in water will be reduced and so will its tendency to adsorb at the air-water interface.

2. Effectiveness and Packing at the Air-Water Interface

The effectiveness of surfactant oligomers in lowering surface tension is not very different from that of conventional surfactants. The γ_{cmc} values for most surfactant oligomers lie between 30 and 40 mN/m. The dependence of γ_{cmc} on the alkyl chain length has been extensively studied for cationic geminis [39,45,46,119] and appears to vary with the composition of the spacer. For —$(CH_2)_s$— spacers (hydrophoblic and flexible), γ_{cmc} decreases slightly when m increases (see Table 13), as observed in conventional surfactants. However, when a heteroatom is present in the spacer (either S, O, or N), the m dependence of γ_{cmc} is nonmonotonic (see Tables 18–22), and long alkyl chain gemini surfactants are significantly less effective in reducing surface tension [119].

Short-spacer gemini surfactants ($s < 5$) are more effective than their corresponding conventional surfactants. However, increasing s decreases significantly the effectiveness (increase in γ_{cmc}) [40,46,52–56,119–123] as illustrated in Fig. 9. This s dependence of γ_{cmc} is confirmed with trimers. Short-spacer cationic trimers are more effective than short-spacer cationic dimers [104,107], and long-spacer cationic trimers are less effective than long-spacer cationic dimers (Table 43) [104].

The minimum surface area occupied by a surfactant molecule at the air-water interface A_m has been determined from the concentration dependence of the surface tension using the Gibbs equation. As already discussed [10], this is rigorously correct only when the surfactant is dissolved in brine because the prefactor n in the Gibbs equation is known ($n = 1$). In the absence of additional electrolyte, comparisons were made between analogous surfactants, taking $n = x + 1$. A neutron reflectivity study suggested that for the cationic gemini surfactant 12-s-12, 2Br, the correct value for n in the Gibbs equation is 2 instead of 3 [124]. Figure 10 shows that diquaternary ammonium geminis with a hydrophilic spacer —$(EO)_{s/3}$— are more densely packed at the air-water interface than their homologues with hydrophobic —$(CH_2)_s$— spacers. Note that Fig. 10 presents the surface area per amphiphilic moiety a_m, not per surfactant molecule. The a_m goes through a maximum between $s = 10$ and $s = 12$ in the case of hydrophobic spacers [121], whereas it increases monotonically for oxyethylene spacers [122]. The nonmonotonic behavior in the case of hydrophobic spacers was also observed with arginine-based cationic geminis (Fig. 4b) [77]. For the anionic gemini of Fig. 3d, the minimum surface area per molecule increases monotonically in the range of length studied (from one to four EO groups) (see Table 39) [59].

The nonmonotonic dependence of A_m and the position of the maximum have been accounted for theoretically [125,126] by considering the competition between the spacer geometrical characteristics (length and flexibility) and the interactions between the amphiphilic moieties. Monte Carlo simulations [127] have

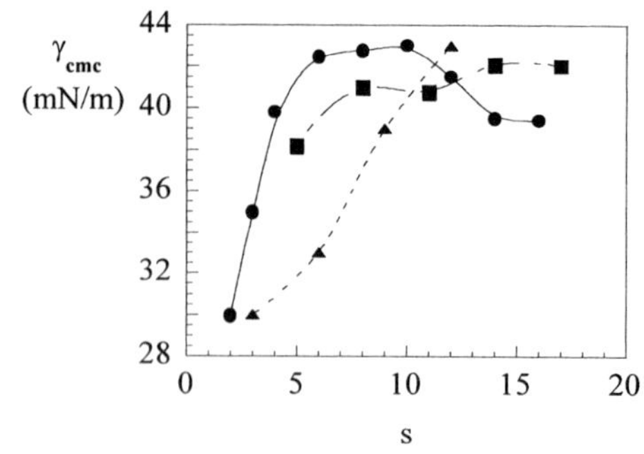

FIG. 9 γ_{cmc} versus spacer length, s, for 12—$(CH_2)_s$—12, 2Br (●) [121], 12—$CH_2(EO)_{s/3}CH_2$—12, 2Br (■) [122], and the compound of Fig. 3d, $m = 10$, Y = $O(EO)_{s/3-1}$ [59] (▲). For comparison, γ_{cmc} of DTAB is 39 mN/m.

reproduced the experimental observation that hydrophilic spacer geminis have a smaller specific surface area than hydrophobic spacer geminis. The possibility for hydrophilic spacers to buckle into water, where half of the space is forbidden for hydrophobic ones, explains these results. These simulations [127], however, did not reproduce the nonmonotonic dependence of A_m upon s for hydrophobic spacers.

The packing of an $m = 18$ cationic surfactant with a rigid phenyl spacer at the air-water interface has also been studied using the Langmuir film balance [39]. The high surface activity of geminis was readily observed in the pressure-area curve: $A_L = 2.40\ nm^2$/molecule, a value that is close to the square of the molecule dimension in its all-*anti* conformation. Upon compression the monolayer collapsed at about 0.76 nm^2/molecule. The same type of measurements done with succinimide surfactant monomers, dimers, and trimers

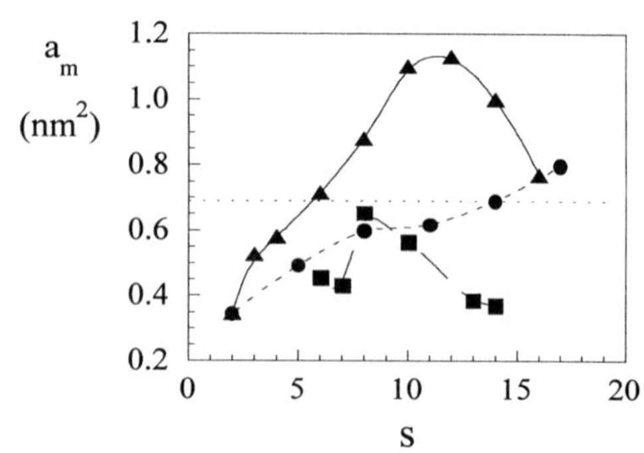

FIG. 10 Minimum surface area (at air/water interface) per amphiphilic moiety as a function of the spacer length for cationic dimers: 12—$(CH_2)_s$—12, 2Br—(▲) [121]; 12—$(EO)_{s/3}$—12, 2Br—(●) [122]; arginine-based surfactant (■) [77]. DTAB and the arginine monomer have the same specific surface area, which is represented by the dotted line.

[128] showed that A_L increases as the degree of oligomerization increases: 0.7, 1.3, and 1.7 nm^2/molecule for the C_{18} monomer, dimer, and trimer, respectively; A_∞ is 0.56, 0.96, and 1.22 nm^2/molecule, respectively. For the dimers and the trimers A_∞ was less sensitive to the alkyl chain length (comparison between $m = 8$, 12, and 18) than for the monomer and was determined by the structure of the headgroup. The Π-A curve has been established for glycerophosphate geminis with $m = s/2$ [71] and amphiphilic phtalocyanines, which can be considered as surfactant oligomers with a cyclic headgroup structure [129].

Neutron reflectivity studies of nonionic sugar derivative geminis [130] and cationic geminis [124] have been reported.

3. Foaming Ability and Foam Stability

Gemini surfactants have good foaming properties. Cationic gemini surfactants with short spacers have shown good foamability (with foam volume 10 times that obtained with DTAC) associated with good stability of the foam after 30 min for $m = 12$ and 14 [45]. The structure of the spacer does not influence the foamability to a large extent but seems to be an important parameter for the stability [45]. The same trends have been observed with anionic gemini surfactants studied by Okahara's group. With sulfate geminis (Fig. 3b), foamability and foam stability decrease as the spacer length increases [51,52]. Phosphate geminis (Fig. 3c) have shown very good foam stability for Y = $(EO)_1$ or $(EO)_2$ [53]. Sulfonate geminis (Fig. 3a) with $m = 12$ and short spacers produce about 30% more foam than the corresponding surfactant monomers. They still show good foamability for all spacer lengths, but foam stability is lost when the spacer contains more than two EO groups [54]. The foam stability can be improved by varying the composition of the spacer. For instance, sulfonate geminis with a sulfone group (see the structure in Table 32) in the spacer from very stable foams [57]. With carboxylate geminis (Fig. 3d), larger (35%) volumes of foam can be obtained as compared with the conventional carboxylate surfactant, but its stability is not greatly improved [59]. α-Sulfonated fatty acid oligoethylene glycol diester (Fig. 3e) showed some improvement in foam stability but not in foamability (Fig. 11) [62]. The foaming properties of surfactants containing an unequal number of hydrophobic chains and headgroups have also been studied [55,108,112].

The stability of soap films produced from dilute and semidilute cationic gemini surfactant 12-2-12, 2Br solutions have been studied with a thin-film balance [131]. Stable common black films can be produced

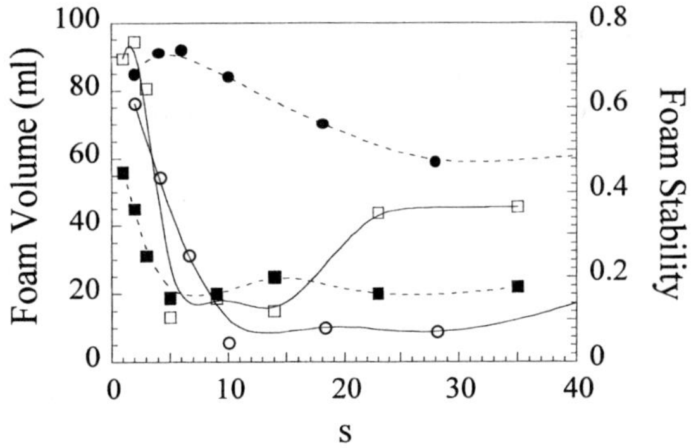

FIG. 11 Foaming properties of the gemini surfactants of Fig. 3e (squares) and of their corresponding monomers (circles). Filled symbols: volume of foam obtained right after shaking (foam ability). Empty symbols: the fraction of foam volume remaining after 30 min standing (foam stability) [62].

from 12-2-12, 2Br solutions at the cmc, but it is impossible to form stable films with the corresponding monomer (DTAB). Only longer chain cationic surfactants can produce stable films by themselves, and the formation of a stable film from DTAB solutions requires addition of a cosurfactant or salt [131,132]. Moreover, upon addition of salt, the 12-2-12, Br soap films undergo a sharp thickness transition from common black films to Newton black films 5–6 nm thick. DLVO theory accounts well for the thickness dependence of the disjoining pressure. It also suggests that the apparent charge density on 12-2-12 2Br films is one order of magnitude lower than on DTAB films (0.0047 C/m^2 instead of 0.046 C/m^2 for DTAB). This low charge density, in conjunction with a decrease in the hydration due to the spacer between the headgroups, explains the possible transition to Newton black films with 12-2-12. However, it does not explain the difference in film stability, which may be related to the high viscosity of 12-2-12 semidulute solutions. Correlation between foam stability and bulk viscosity has also been pointed out in Ref. [57].

B. Solid-Water Interface

1. Adsorption Isotherms

Multistep adsorption processes have been seen with several cationic gemini surfactants adsorbing onto silica (40 μm, washed several times with hydrochloric acid, specific surface area 29 m^2/g) [133]. The amount of surfactant adsorbed; the sodium, bromide, and proton concentration in the supernatant; and the electrophoretic mobility of the silica particles were measured along the binding isotherm. The first step consists of a rapid but small increase of the surfactant amount adsorbed, followed by a plateau that starts at the point of zero charge. It corresponds essentially to an exchange of the residual sodium ions bound to the silica. After the first adsorption plateau, whose broadness decreases as s increases, a second rapid increase in the amount adsorbed corresponds to the formation of surfactant aggregates at the interface. These aggregates (admicelles) bind bromide ions less than the corresponding bulk micelles (contrary to what has been observed with DTAB). Their positive charges induce a reduction of the pK_a of the silanol groups of the silica surface, as evidenced by the sharp drop of the pH (particularly with the short spacer gemini 12-2-12, 2Br) associated with the second step. At saturation, the amount of adsorbed surfactant is inversely proportional to the spacer length, s. It was suggested that for short-spacer gemini surfactants, the first step may involve charge redistribution at the silica interface. In a subsequent study, the same authors compared the adsorption isotherms of DTAB and 12-2-12, 2Br onto the same silica particles treated differently (with and without HCl wash) [134]. The adsorption mechanism of the monomer and the dimer on the unwashed and on the washed silica is qualitatively the same. However, the variation in the state of the surface induces quantitative differences: the amount of surfactant adsorbed at the point of zero charge and at saturation is larger with the unwashed silica. This makes the multistep mechanism difficult to observe with unwashed silica and may explain the results of other studies [104,135]. The adsorption of 12-2-12, 2Br starts at a much lower concentration than for the corresponding monomer DTAB, and the point of zero charge (PZC) of the particles is reached at a much lower concentration for 12-2-12, 2Br. However, the maximum amount differs only a little. Both surfactants keep on adsorbing at the silica surface even after their micellization in the supernatant, saturation being reached at an equilibrium concentration of about 1.5 times the cmc.

These results have been confirmed by force balance measurements and direct imaging with atomic free microscopy (AFM) [136]. The charges of the mica surface are neutralized (suppression of the repulsive double-layer force) at a bulk concentration of 1 to 5 μM of surfactant. As the concentration of surfactant increases, hydrophobicity of the surfaces increases (high pull-off force) and discrete surfactant monolayer patches grow and eventually merge. A further increase in concentration (5 μM to 0.1 mM) decreases the pull-off force, steadily increases the electrostatic repulsive force, and increases the compressed layer thickness, suggesting the formation of a bilayer. This step was observed directly by AFM to occur also by growth of patches. At a concentration of 2 mM, a full bilayer is formed. The force profile at 0.8 mM presents an extra repulsive force attributed to further adsorption on top of the bilayer. This interpretation is supported by AFM measurement of the surface roughness. The authors pointed out the time dependence of the force profiles obtained and concluded that there was a slow process of adsorption [136].

Adsorption isotherms of DTAB, 12-2-12, 2Br, and 12-2-12-2-12, 3Br on silica (0.3 μm and specific surface area 16.7 m^2/g) in 10^{-2} M NaBr have been established [135]. Electrophoretic mobility along the isotherm suggests that bilayers are formed with all surfactants, but a two-step adsorption process was observed only for the DTAB. The amount of surfactant adsorbed at saturation decreases from 57 to 48 μmol/g from the monomer to the dimer and down to 30 μmol/g for the trimer. A higher concentration of salt increases the amount of surfactant adsorbed at saturation, and no addition of salt reveals the two-step adsorption process for 12-2-12, 2Br has been observed only in the absence of salt [137]. The same studies have been carried out on laponite [138] and on titanium dioxide (bare or hydrophobically modified) [139]. The adsorption of cationic trimers of the same type with longer spacers (s = 3 and 6) onto silica has also been reported [104].

2. Interfacial Packing and Aggregate Geometry

From the amount of surfactant adsorbed at saturation, and knowing the specific surface area of the solid substrate, an average limiting surface area per surfactant molecule is readily obtained. For comparison, because the maximum amount of adsorbed surfactant depends on the state of the substrate surface, the surface area per amphiphilic moiety in gemini surfactants A_2 is normalized by the surface area of the corresponding monomer A_1 measured in the same series of experiments. For the cationic surfactant 12-s-12, 2Br, the normalized surface area increases linearly with the spacer length (Fig. 12) [133]. The average surface area occupied by an amphiphilic moiety is larger than that for DTAB except for $s = 2$. This spacer dependence is amplified with higher degrees of oligomerization (Fig. 13). For short spacers ($s = 2$ [135] and $s = 3$ [104]), the area per amphiphilic moiety slightly decreases with the degree of oligomerization. This means that each amphiphilic moiety is more densely packed in layers of surfactant oligomers with short spacers. However, for long spacers ($s = 6$), the surface area increases almost lin-

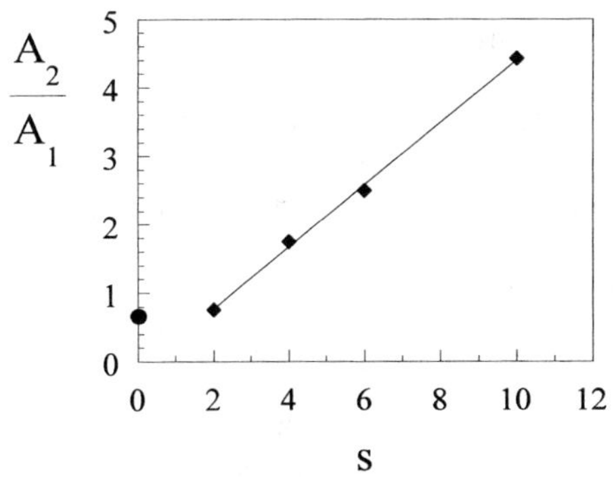

FIG. 12 Normalized limiting surface area (at silica/water interface) per amphiphilic moiety in 12-*s*-12, 2Br gemini surfactants as a function of the spacer chain length, *s* [133]. Normalization is done with respect to the limiting surface area of DTAB, A_1, measured in the same series of experiments; "*s* = 0" corresponds to DDAB [140].

early with the degree of oligomerization [104]. For comparison, results obtained for C_{12} multiple-chain surfactants with one cationic headgroup [140] have also been reported (Fig. 13). This illustrates the fine-tuning of the packing that can be achieved with surfactant oligomers by playing with the spacer length and the degree of oligomerization.

By AFM, using the precontact repulsive force (within the electrical double layer) [141,142], Manne et al. observed directly the aggregates formed by the cationic gemini surfactants 12-*s*-12, 2Br on the cleavage plane of mica [143]. The gemini surfactant with the shortest spacer, *s* = 2, which gives wormlike micelle in bulk solution, forms bilayers on mica surfaces. Bilayers were also observed with the double-chain surfactant DDAB, known to form vesicles in dilute solutions. Parallel cylinders are obtained when adsorbing the 12-4-12, 2Br surfactant and DTAB. These surfactants form spherical micelles in dilute solutions, which can slightly elongate at high enough concentration for the surfactant dimer. With the single-chain divalent surfactant, referred to as 12-2-1, 2Br, spherical admicelles form. From these observations, the authors concluded that the dimensionless packing parameter as defined in Ref. 32 to explain the morphology of micelles in the bulk determines the shape of the interfacial aggregates as well. However, the mica surface playing the role of a huge "counterion," the curvature of the aggregate at the interface can be (and most often is) lower than the curvature of the aggregate in the bulk.

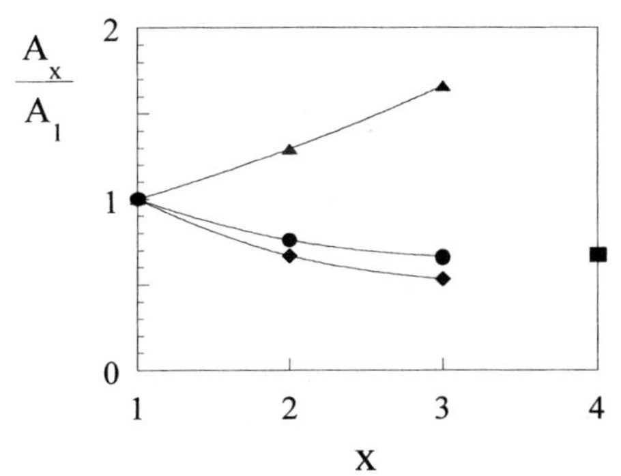

FIG. 13 Normalized limiting surface area (at silica/water interface) per amphiphilic moiety as a function of the degree of oligomerization *x* [104]; *s* = 3 (●); *s* = 6 (▲); surfactant tetramer of Fig. 8c (■); multiple chain surfactants, "*s* = 0" (◆) [140].

IV. STRUCTURE AND PROPERTIES OF SURFACTANT OLIGOMERS SELF-ASSEMBLIES

A. Critical Micelle Concentration

The cmc of surfactant oligomers has been measured by tensiometry, conductimetry, dye solubilization measurements. Gemini surfactants are characterized by a cmc that is 10 to 100 times lower than that of the corresponding conventional surfactant (monomer), the reduction factor being essentially determined by the cmc of the monomer.

Cmc values are reported in the tables of the Appendix. It can be seen that different methods can yield very different values. Some difficulties in determining the cmc have been reported rather often for geminis and surfactant oligomers. In conductivity measurements, ion pairing can sometimes interfere with micellization, especially with short-chain surfactants [144]. Slow adsorption at the interface may sometimes mask the cmc in surface tension measurements. This has already been discussed [1].

1. Alkyl Chain Length Dependence of the Cmc—Comparison with Monomers

The hydrophobic chain length *m* is not a variable specific to surfactant oligomers. However, the study of its influence on the cmc yields good insight into the micellization properties of surfactant oligomers. In most cases, the *m* dependence of the cmc is classical, meaning that the cmc decreases exponentially as the alkyl chain length increases (see Tables 14–24) [145,146]:

$$\ln \text{cmc} = A - B_m m \tag{4}$$

Figure 14 shows that the B_m factor is nearly inde-

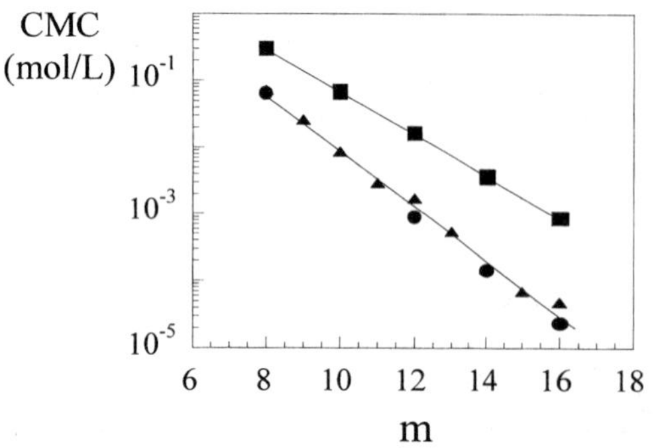

FIG. 14 Cmc versus alkyl chain length, *m*, in gemini and monomeric homologues: *m*-2-*m*, 2Br (●) [146]; *m*-6-*m*, 2Br (▲) [145]; monomer (■) [155].

pendent of the spacer length and rather close (but not equal) to that obtained for conventional surfactants. The same has been observed for the compounds of Fig. 2e when compared with their monomer (see Table 23) [42] and for the anionic gemini of Fig. 3e (see Table 30) [62]. This means that the free energy of transfer of one CH_2 from water into the micelle core, $\Delta G_{tr}(CH_2)$, is close in both types of surfactants [38,42,63]. This provides an important clue to understanding the low cmc values of surfactant oligomers. Gemini surfactants have lower cmc values than conventional ones because each molecule contains more methylene groups, which water does not like to solvate. The slope B_m is actually not equal for both types of surfactant, and the two straight lines in Fig. 14 become farther apart when m increases. (B_m is 1 for geminis and 0.7 for monomers, but for a correct comparison the B_m value of 1 should be divided by 2, because m corresponds to only half the number of methylene groups in geminis.) This means that the ratio cmc(monomer)/cmc(dimer) increases with m.

Thus, when going from the monomer to the dimer, the cmc is decreased by a factor that is related to the cmc of the monomer. This suggests that the standard free energy of micellization *per amphiphilic moiety* ΔG°_M is equal in both types of surfactants. This can be better understood by considering the micellization equilibrium, $NS_x \rightleftarrows [S_x]_N$, for an nonionic oligomer surfactant S_x which form micelles with aggregation number N (in number of surfactant molecules). The mass action model allows to relate the cmc (expressed in mole of amphiphilic moiety per liter) to the free energy of micellization per mole of amphiphilic moiety, as follows:

$$\Delta G^\circ_M = RT(1/x)\ln(\text{cmc}_x) - (1/x)RT\ln x \tag{5}$$

Assuming ΔG°_M to be independent of x leads to the following relation:

$$\text{cmc}_x = x.(\text{cmc}_1)^x \tag{6}$$

where cmc_x is the cmc of the oligomer and cmc_1 the cmc of the corresponding monomer. Equation (6) shows that the cmc of a surfactant dimer is twice the square of the cmc of the corresponding monomer. Thus, if we can assume that the free energy of micellization of an amphiphilic moiety does not depend on the structure of the surfactant it belongs to, then the reduction in cmc when going from the monomer to the dimer is half the cmc of the monomer.

ΔG°_M for cationic oligomers has been proved to be approximately independent of x and equal to -20 kJ per mol of amphiphilic moiety [104], using the following relation [147]:

$$\Delta G^\circ_M = RT(1/x + \beta)\ln \text{cmc} - (1/x)RT\ln x, \tag{7}$$

where $\beta = 1 - \alpha$ is the degree of association of the counterions to the micelles, and where the cmc is expressed in mole of amphiphilic moiety per liter.

Considering ΔG°_M independent of x is a very good first approximation. We can however notice small differences (see later Fig. 23) that are going to be discussed in Section II.A.3. Ion pairing and intramolecular interaction between the hydrophobic tails of the surfactant oligomer are among the factors that would decrease the absolute value of ΔG°_M and increase the cmc.

In establishing Eqs. (5) and (7), a mixing entropic term (common to any aggregation phenomenon) has been neglected. Indeed, Eqs. (5) and (7) are obtained by neglecting the molar micelle concentration $|[S_x]_N|$. This is reasonably close to the cmc and when N is large. However when x increases, we can expect N to tend to 1. (This supposes that the aggregation number n, expressed in number of alkyl chain, is independent of x and equal to Nx. This has been shown for cationic oligomers.) In other words, micellization of a surfactant oligomer is entropically more favored than micellization of the corresponding monomer, because part of the mixing entropy has already been lost at the synthesis step. The importance of this contribution largely depends on m and on the nature of the head group (charge and valence of the counter ions). The smaller m is, the more important is this contribution, because $|\Delta G^\circ_M|$ and $n = Nx$ decrease when m decreases. For the cationic oligomers with $m = 12$ and bromide counter-ion, this contribution is not significant when $x < 5$.

If x eventually increases enough to make the contribution of the mixing entropy of the amphiphlic moiety significant, a thermodynamic description of the micellization will become a polyelectrolytes problem where ion condensation and conformational entropy of the head group backbone have to be taken into account.

In Fig. 15, the cmc is plotted against $x \times m$, the total number of methylene groups belonging to the hydrophobic tails. It can be seen that, to get the same cmc with a dimer as with an $m = 16$ conventional surfactant, the surfactant dimer must contain 24 methylene groups in its hydrophobic chains. This may sound like a waste and is due to the fact that the additional methylene groups in gemini surfactants come with an additional headgroup. Thus, many of the methylene groups are close to a hydropholic charged group and do not trigger micellization. There is, however, an interesting benefit of this apparent waste. The vertical dotted lines in Fig. 15 correspond approximately to the maximum number of methylene groups a surfactant can contain to be soluble in water (meaning neither crystals nor mesophases) at room temperature. It is below 16 for monomers and above 32 for dimers and reaches at least 36 for trimers. Hence, the benefit of surfactant oligomers may not be as much to with reducing the cmc as compared with conventional surfactants. It is certainly to allow cmc values that could never be reached by increasing the length of a conventional surfactant because this would have induced a phase separation [108].

In cationic surfactant heterodimers, m-2-m', 2Br, the cmc depends not on the difference $(m - m')$ but on the total number of methylene groups in the surfactant [87].

The m dependence of the cmc values sometimes deviates from the exponential behavior described before when the alkyl chain becomes longer than a certain length (Fig. 16) and can sometime reverse (Table 16). This has been observed when the spacer contains either aromatic rings or oxygen atoms (oxyethylene or hydroxypropylene spacers) [23,39,46,116,117]. The onset

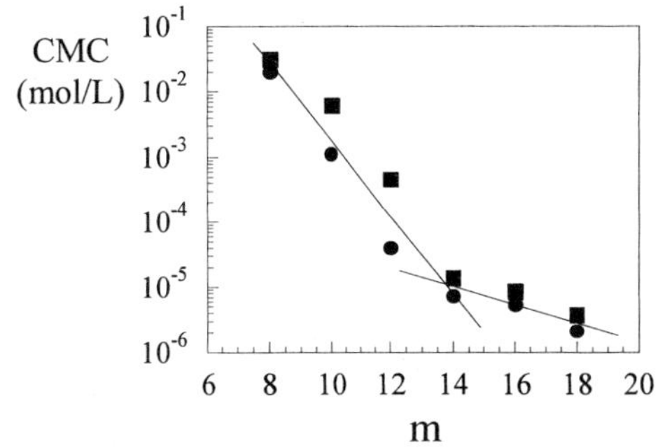

FIG. 16 Deviation from the classical exponential m dependence of the cmc of cationic gemini surfactant of Fig. 2c in 0.01 M NaCl (■), and 0.1 M NaCl (●) at 50°C [116].

length of this unusual behavior decreases with increasing ionic strength [116,117]. With the star-shaped anionic trimer (Fig. 8e), the cmc increases with m [110,111] (Fig. 17).

This unusual behavior is necessarily related to hydrophobic interactions. The formation of premicellar aggregates has been proposed to explain it [39]. Equilibrium constants for the formation of premicellar aggregates have been calculated from the difference between the experimental γ–log C plots and virtual plots established by extrapolating to high m values results obtained at low m values [46,116].

Alternative hypotheses should be considered with more attention. An increase of the cmc means a decrease in the absolute value of the free energy of micellization. This can result either from a decrease of the chemical potential of the free (unmicellized) surfactant μ_s° or from an increase of the chemical potential of the surfactant in the micelles μ_m°. In the same way, the unusual m dependence of the C_{20} mentioned earlier can result either from a decrease of μ_s° or from an increase of the chemical potential of the surfactant adsorbed at

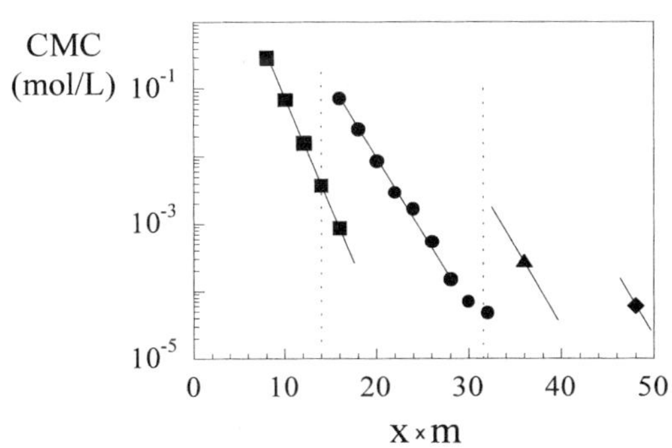

FIG. 15 Cmc versus number of methylene groups per surfactant molecule for cationic monomers (■), dimers (●), trimer (▲), and tetramer (♦).

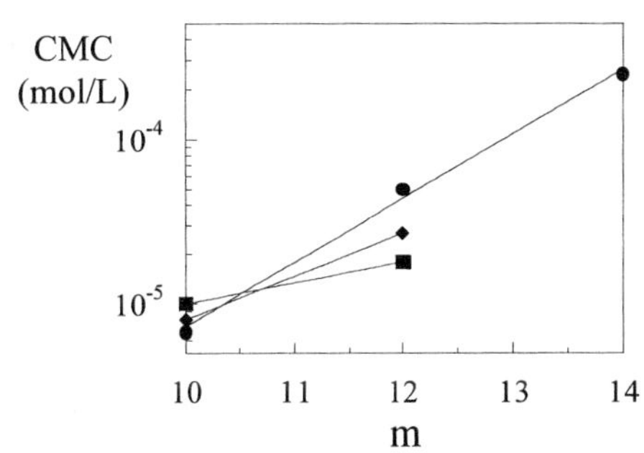

FIG. 17 Inverted m dependence of the cmc in anionic trimers with hydrophilic star-shaped spacers (see structure, Table 44) [110,111].

the air-water interface $\mu^\circ_{A/W}$. Premicellization indeed reduces the chemical potential in the bulk and explains lower surface activity and higher cmc. But an alternative, which would be less expensive in terms of mixing entropy, would involve intramolecular association between the alkyl chains of a surfactant oligomer. This hypothesis has been suggested by experimental observations of the change in enthalpy and in volume upon micellization [118]. Intramolecular association would decrease the hydrophobic hydration shell and thus decrease μ°_s and increase the cmc as well as C_{20}. The other alternative corresponds to an increase of μ°_m and could result, for instance, from the formation of micelles in which the hydrocarbon moieties would largely remain in contact with water. This would achieve Menger's initial goal [23] but would probably not lead to an increase in C_{20}.

A convincing explanation for the unusual *m* dependence of the cmc is still to be found. NMR conformational studies of the long-chain geminis such as the one carried out on $m = 8$ cationic surfactants with *o*-, *m*-, and *p*-phenylenedimethylene spacers [148] could be helpful. For this system, a neutron scattering study supports the formation of premicelles [149]. Also, an accurate determination of the Kraft temperature as done for 16-*s*-16, 2Br surfactants [150] would clarify the situation. Finally, Monte Carlo simulations, in which the cmc is defined as the surfactant concentration at which half of the surfactants have at least one surfactant as nearest neighbor [127,151], also yield an increase of the cmc as *m* increases.

2. Headgroup Nature Dependence of the Cmc

As mentioned before, the cmc of gemini surfactants is essentially determined by the cmc of the corresponding monomer. Hence the influence of the headgroup, and the dependence of the cmc on the nature of the headgroup, is the same as in the conventional surfactant. Figure 18 shows that for a given *s*/3, the cmc of anionic gemini surfactants varies with the headgroup nature as COONa > OPO(OH)(ONa) > $O(CH_2)_3SO_3Na$ > OSO_3Na [59].

3. Spacer Chain Length Dependence of the Cmc

The cmc of cationic surfactants *m*-*s*-*m*, 2Br varies nonmonotonically with *s* when the spacer is hydrophobic (Fig. 19) (Tables 3–7). It increases for short spacers up to four or five methylene groups, then decreases. This has been observed for $m = 8$ [148], 10 [152], 12 [38], and 16 [41]. Note that the comparison is done on a

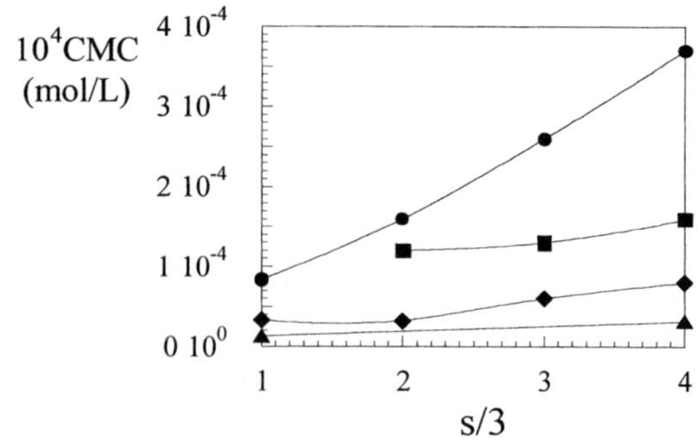

FIG. 18 Cmc versus spacer chain length for carboxylate (Fig. 3d) (●), phosphate (Fig. 3c) (■), sulfonate (Fig. 3a) (♦), and sulfate (Fig. 3b) (▲) disodium gemini with $m = 10$ [59].

linear scale. The same holds for anionic gemini surfactants (Table 35).

For the series $m = 12$, *s* has been varied up to high values [38,153]. The decreasing part of the curve has been explained by the increasing hydrophobicity of the surfactant. For *s* larger than or equal to 10, the cmc

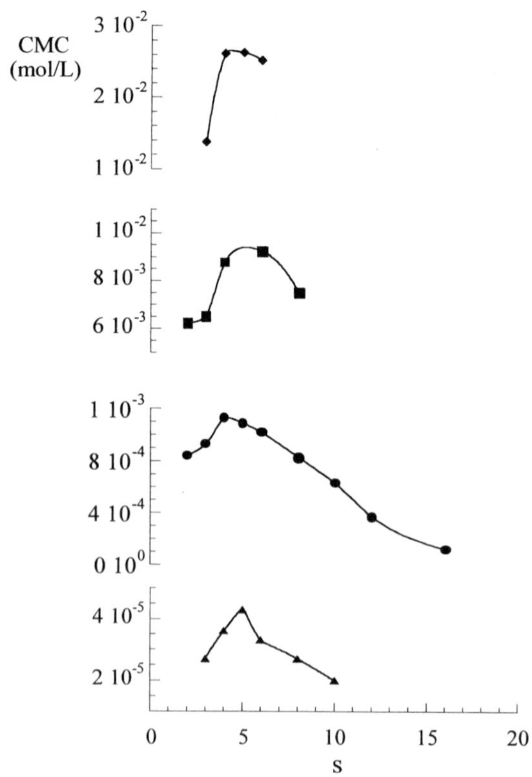

FIG. 19 Spacer length dependence of the cmc for cationic gemini *m*-*s*-*m*, 2Br: $m = 8$ (♦) [148]; $m = 10$ (■) [152]; $m = 12$ (●) [38]; $m = 16$ (▲) [41].

depends on s in the same way as it depends on m [38,153]:

$$\ln \text{cmc} = A - B_s s \quad (8)$$

The factor $B_s = 0.3$ is smaller than $B_{m/2} = 0.5$. When long enough, the hydrophobic spacer group can buckle into the micelle core. However, this is apparently not so easy considering the low value obtained for B_s, which is even smaller than that found for bolaform surfactants [154] (Fig. 20). The data of Fig. 20 suggest that the impediment to the spacer being inserted into the hydrophobic core of the micelle results from the combined effect of electrostatic interaction, loss of conformational entropy (as in a bolaform surfactant), and steric hindrance (as in double-chain surfactants).

The increase of the cmc with s when $s < 5$ (Fig. 19) is probably related to electrostatic interactions between the headgroups. When the spacer is short, part of the work against electrostatic repulsion necessary to bring the surfactants together upon micellization has already been done at the synthesis step as discussed in Ref. 99. This hypothesis is supported by several arguments. First, the cmc increases with s up to $s = 4–5$, which corresponds to the length of the spacer at which the interfacial surface area per amphiphilic moiety is equal in gemini and in conventional surfactants. The second argument relies upon Monte Carlo simulations. With ionic gemini surfactants, nonmonotonic variation of the cmc is observed, whereas the cmc of nonionic gemini surfactants increases monotonically with s [127].

This small electrostatic effect is clearly pointed out by considering the free energy of micellization per amphiphilic moiety. ΔG_M is slightly more negative for

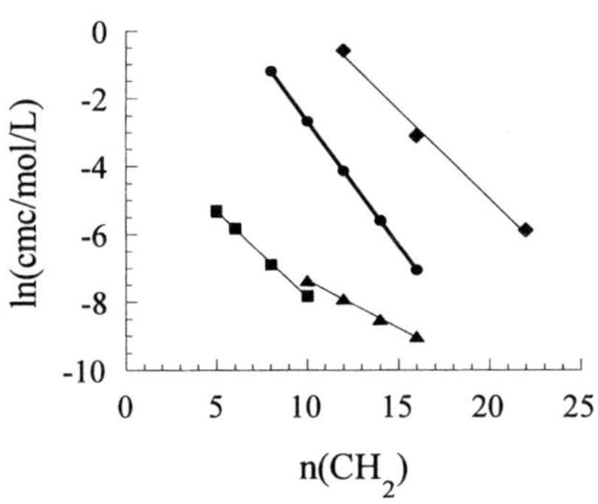

FIG. 20 Influence on the cmc of increasing the number of methylene groups n in the hydrophobic tail of a conventional surfactant [$C_nH_{2n+1}N(CH_3)_3$, Br] (●), in a bolaform surfactant [$(CH_3)_3NC_nH_{2n}N(CH_3)_3$, 2Br] (♦), in a double-chain surfactant [$C_{12}H_{25}C_nH_{2n+1}N(CH_3)_2$, Br] (■), and in the spacer of a gemini surfactant [$C_{12}H_{25}(CH_3)_2N(CH_2)_nN(CH_3)_2\ C_{12}H_{25}$, 2Br] (▲).

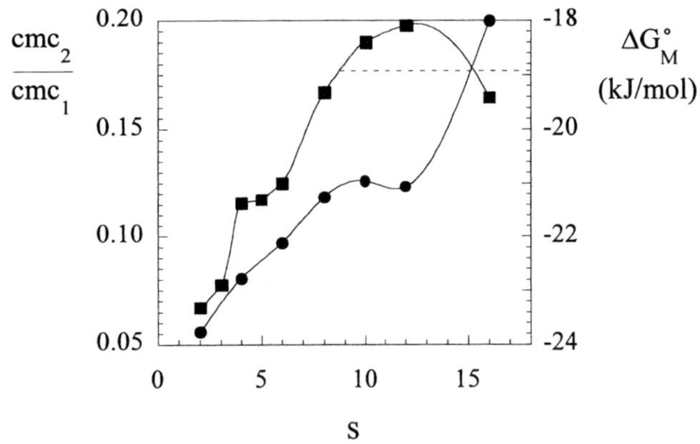

FIG. 21 Ratio of cmc of the cationic geminis 12-s-12, 2Br (cmc_2) over the cmc of the corresponding monomer $C_{12}H_{25}\cdot C_{s/2}H_{s+1}N(CH_3)_2$, Br ($cmc_2/cmc_1$) (●) and standard free energy of micellization per amphiphilic moiety ΔG_M (■) versus spacer length s.

short spacers than for long spacers (see Fig. 21) [104,152]. The differences in ΔG_M are not large but they are systematic, and the trend is confirmed with surfactant trimers and tetramers [104]. ΔG_M decreases with the degree of oligomerization for $s = 3$, whereas it is constant for $s = 6$ (see Fig. 23).

To understand the s dependence of the cmc, it is useful to compare the cmc of geminis 12-s-12, 2Br with the cmc of double-chain surfactants $C_{12}H_{25}C_{s/2}H_{s+1}N\cdot(CH_3)_2$, Br [155,156]. The latter correspond closely to the monomers in the sense that they have the same number of methylene groups per headgroup. Figure 21 shows that the reduction factor of the cmc, going from the monomer to the dimer, increases as the spacer increases. This increase is not regular and closely related to the variation of ΔG_M. Note that only for $s = 10$ and 12 is ΔG_M less negative than ΔG_M of DTAB.*

With hydrophilic oligo(oxyethylene) spacers, the cmc increases with s for cationic (see Tables 10 and 11) [47,48,49] as well as for anionic surfactants (see Table 30 and Fig. 18). A slightly lower cmc has been observed for cationic geminis of Fig. 2b with very large oxyethylene spacers (Table 11) [47]. When hydrophilic spacers become very large, gemini surfactants look like telechelic hydrophobicvally modified polymers, which are known to micellize at fairly low concentrations to

*This observation has to be related to the structure of the micelles, in which the spacer has been shown to remain largely in contact with water [167]. This might correspond to the situation Menger was expecting [23] and is actually unfavorable for micellization.

form flowerlike micelles [157]. This has, however, not been observed with anionic gemini surfactants of Fig. 3e with spacers consisting of up to 35 EO groups [62] (Table 30).

The chemical nature of the spacer influences the cmc value in a way that is sometimes difficult to rationalize. For intermediate spacer lengths, the cmc is lower when the spacer is hydrophilic than when the spacer is hydrophobic (see Tables 13 and 16). The cationic surfactant trimer with hydroxypropylene spacer 12-3*-12-3*-12, 3Cl (Fig. 8b) has a cmc of 9.6 mM, whereas 12-3-12-3-12, 3Cl has a cmc of 160 mM. Devínsky et al. observed slight changes in the cmc of cationic gemini surfactants with different spacers of the type —CH_2CH_2—X—CH_2CH_2—, with X = CH_2, NCH_3, O, or S [119]. In this study, the lipophilicity of the spacer seems to be the determining parameter.

The ionization degree of the micelles is independent of s for $m = 8$ (Table 3) but increases with s for $m = 10$ and 12 (Tables 4 and 5).

4. Oligomerization Degree

Figure 22 shows the decrease of the cmc as the degree of oligomerization x increases for an $m = 12$, $s = 3$ series of cationic surfactants [104]. The cmc appears to vary as a power law of the oligomerization degree for this series with $s = 3$, and the same is observed with $s = 2$ [107] and $s = 6$ [104] up to the trimer. The cmc of the cationic tetramer is 2.5 orders of magnitude lower than that of DTAB.

The ionization degree of the micelle at the cmc, α, is independent of x and is essentially determined by m and s (see Tables 3–5) [104]. With bromide counterions and for $m = 12$, $\alpha = 0.2$ for $s = 3$ and $\alpha = 0.3$ for $s = 6$, independent of x. The free energy of micellization decreases (becomes more negative) for short spacers but is constant for long spacers (Fig. 23).

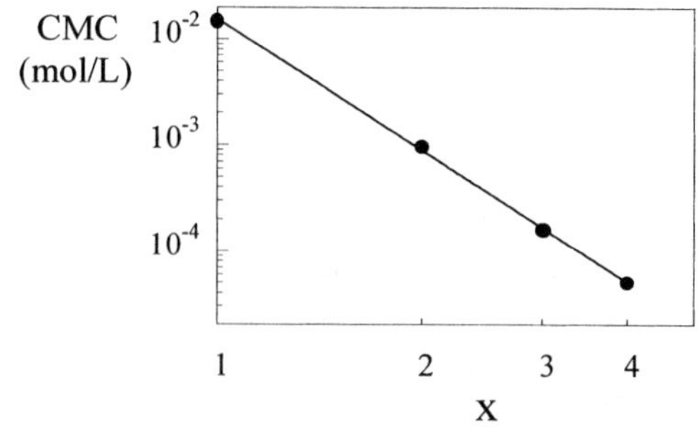

FIG. 22 Cmc versus oligomerization degree for cationic surfactant oligomers $m = 12$, $s = 3$ [104].

B. Micelle Properties

1. Micropolarity, Solubilization Capacity, and Emulsification

The micropolarity of cationic surfactant oligomers has been characterized by the value of the fluorescence intensity ratio I_1/I_3 of the first and third vibronic peaks in the emission spectra of micelle-solubilized pyrene [48,92,158]. It depends on the composition of the pyrene solubilization site, i.e., near the micelle-water interface. For cationic geminis with hydrophobic —$(CH_2)_s$— spacers, the micropolarity varies nonmonotonically with the spacer chain length, going through a maximum at about $s = 6$ [158]. Similar behavior has been obtained for the corresponding monomer, m-s/2 [158]. Micropolarity does not change with spacer length in cationic gemini 16-$(EO)_{s/3}$-16, 2Br [48]. The micropolarity of cationic surfactant oligomers does not depend on x [104,158] but slightly decreases with x in nonionic surfactant oligomers of Fig. 8g [114].

The solubilization capacity for transazobenzene, as expressed by the ratio of solubilizate concentration over surfactant concentration, increases monotonically with m in cationic geminis as with conventional surfactants (see Table 14) [159]. It varies nonmonotonically with s, going through a maximum at $s = 6$, where it is about twice that of $s = 2$ and five times that of $s = 12$ (Table 4).

The solubilization of toluene or n-hexane by cationic gemini surfactants is more efficient than by cationic surfactant monomers [40]. Cationic geminis have higher selectivity for toluene than for hexane [40]. Solubilization of β-naphthol in short spacer cationic surfactant dimers and trimers adsorbed to silica [135,137], titanium oxide [139], and laponite [138] increases with x. This is true when expressed as the molar ratio of β-naphthol adsolubilized to the adsorbed surfactants.

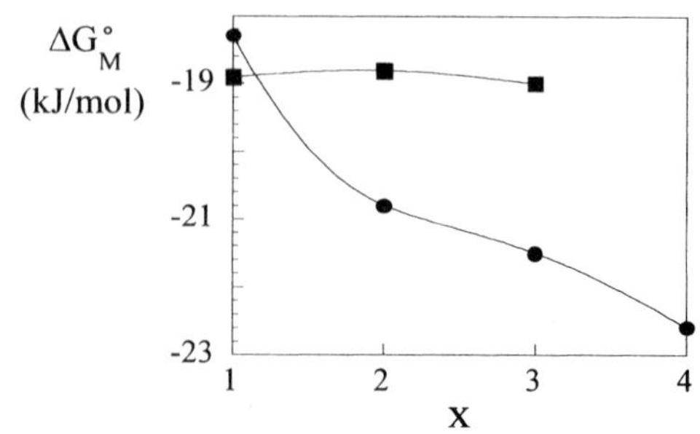

FIG. 23 Standard free energy of micellization per cationic amphiphilic moiety versus degree of oligomerization x for $s = 3$ (●) and $s = 6$, $m = 12$ (■) [104].

However, as pointed out in Ref. 135, the solubilization capacity decreases with x when expressed per dodecyl chain. This has been interpreted as a result of the increasing packing density as x increases.

The emulsification efficiency of cationic geminis was determined by comparing the rate of drop coalescence between heptane/water emulsions prepared from cationic geminis and their monomeric homologues. The drop coalescence was characterized by an exponential decay of the number of drops. The lifetime of the drops was found to be 1.7 times longer with the geminis [43]. The emulsification properties of nonionic diol geminis has also been studied [57].

The phase behavior of ternary water/styrene/C12 cationic gemini surfactants with hydrophobic [160] and hydrophilic [122] spacers of different lengths has been reported. The extension of the single-phase region that lies in the water corner of the phase diagram triangle depends strongly on the spacer chain length for hydrophobic —$(CH_2)_s$— spacers and more weakly for hydrophilic —$(EO)_{s/3}$— spacers. The extension of the single-phase region is strongly temperature dependent with —$(EO)_{s/3}$— spacer but weakly with —$(CH_2)_s$— spacers. At a fixed concentration of surfactant, between 5 and 20 wt%, only the 12-2-12, 2Br surfactant had poorer solubilization capacity than DTAB (expressed as the molar ratio of solubilized styrene over surfactant). With hydrophobic spacers, the solubilization capacity increases with s and is maximum for $s = 12$, at which it is six times higher than for DTAB. With hydrophilic spacers, it is maximum for spacers consisting of one or two EO groups. The influence of temperature, spacer rigidity, and oil size on the oil-water-geminis ternary phase diagram has also been studied by Monte Carlo simulation [161].

2. Microviscosity

The microviscosity of gemini surfactant micelles decreases when the spacer goes from 2 to 12 in a 16-s-16, 2Br series when the spacer is hydrophilic [41] but not much when it is hydrophobic [48,70,114,158].

The microviscosity depends, however, almost linearly on the degree of oligomerization x and is about 6 to 10 times (depending on the temperature) higher for the cationic tetramer than for the monomer [104,158]. This has also been observed for nonionic oligomer micelles, in which the dipyrenylpropane excimer lifetime is three to four times larger [114]. Microviscosity of admicelles on silica determined from the order parameter of a paramagnetic probe varies with the oligomerization degree in the order dimer > trimer > monomer [135] and increases with added salt [137].

C. Morphology of the Aggregates

The size and the shape of surfactant oligomer micelles have been studied by small-angle neutron scattering (SANS) [41,48,152,162–165] and directly observed by transmission electron microscopy at cryogenic temperature (cryo-TEM) [22,166,167]. Micellar shape can also be inferred from solubilization studies [159] or from the concentration dependence of the aggregation number determined by fluorescence quenching measurements [103,166].

The spacer length is a key parameter of the micelle morphology [22,166], as could have been deduced from the s dependence of the surface area. Hydrophobic short spacers ($4 \geq s$) reduce the preferred curvature of the micelles as compared with conventional surfactant micelles. With cationic geminis, wormlike micelles are obtained when $m = 12$. Spacers of intermediate length ($5 \geq s \geq 12$) favor the formation of spherical micelles. Interestingly, the spherical shape is preserved up to very high concentrations for $s = 10$ and 12 [168]. Thus, increasing s can be seen as releasing the spacer constraint for lower intermediate spacers but actually corresponds to a strong constraint for upper intermediate spacers. For long spacers ($s \geq 16$) vesicles are obtained as in the corresponding surfactant monomer (double chain) [22,166,167]. The trend is confirmed by SANS of 16-s-16, 2Br surfactants [164,165]. Because of electrostatic interactions, the scattered intensity presents a maximum at a finite wave vector q^*, which is inversely proportional to the distance between the micelles. At low enough concentration, the distance between the micelles reflects the size of the micelles. At a fixed low concentration (between 10 and 50 mM) of 16-s-16, 2Br surfactants, q^* increases as s increases from 5 to 12, suggesting that the aggregation number of the micelles decreases as s becomes larger. The scattered intensity 16-3-16, 2Br varies as q^{-2}. This has been interpreted as the signature of the formation of disks, but cryo-TEM micrographs showed vesicles and bilayer membrane fragments coexisting with wormlike micelles [166] (and this is consistent with $I \propto q^{-2}$).

The same results have been obtained with $m = 16$ phosphate geminis of Fig. 3f [70]. For $s = 2$, the scattered intensity varies as q^{-2}, which indicates zero curvature objects (disks or vesicles). For $s = 4$ wormlike micelles and for $s = 6$ or 10 prolate ellipsoids are formed.

With hydrophilic spacers (EO), the aggregation number of the micelles also depends on the length of the spacer. But even with the shortest spacer (one EO), the growth of the micelle is limited and the influence of the spacer length on the size of the micelles is less pronounced with hydrophilic spacers than with hydrophobic ones [41,48]. Nonionic sugar-based geminis of Fig. 5a, with m = 14, form cylindrical micelles when s = 6 and 8 and vesicles when s = 10 [84]. Nonionic sugar-based geminis of Fig. 5b form anisotropic micelles, cylindrical for $7 \geq m \geq 5$ and discoidal for m = 8 [163].

1. Wormlike Micelles of Cationic Surfactant Oligomers

(*a*) *Micellar Growth.* Cationic surfactant oligomers with short spacers are among the few systems in which the transition from spherical to very long wormlike micelles occurs at low concentration (below 10 wt%) and does not require addition of salt or cosurfactants or the presence of hydrophobic counterions. The formation of long wormlike micelles was directly observed by cryo-TEM in the 12-*s*-12, 2Br series [166,167]. However, the micellar growth can be described more quantitatively by looking at the concentration dependence of the aggregation number $N(c)$, which is obtained by SANS [152,162] or fluorescence quenching measurements [103,167]. The concentration dependence of the zero shear viscosity also gives information on micellar growth [169–171].

The $N(c)$ has been analyzed with the ladder model [172], which describes a prolate micelle as a cylinder capped at each end by hemispherical micelles. The concentration dependence of N is then expressed as

$$N = N_0 + 2K^{1/2}(c\text{-cmc})^{1/2} \tag{9}$$

where N_0 is the aggregation number at the cmc and c and cmc are mole fractions here.

$$K = \exp(E_c/kT) \tag{10}$$

where E_c is the end-cap energy, i.e., the excess chemical potential of a surfactant in the end caps compared with the chemical potential of a surfactant residing in the central cylindrical portion. This equation describes well the SANS results obtained for 10-2-10, 2Br geminis [152,162]. The parameters deduced from such an analysis are presented in Table 1. The end-cap energy decreases as the spacer length increases.

When applied to the data obtained by fluorescence quenching for the m = 12 series of cationic surfactant oligomers [103,167], the ladder model yields an end-cap energy of 11 kT for the 12-5-12, 2Br, 13 kT for the 12-3-12, 2Br, and 17 kT for the trimer 12-3-12-3-12, 3Br. Data in Fig. 24 show that when the tendency of the micelles to grow is strong, the ladder model applies only for the lowest concentrations. Moreover, the differences in end cap energy it yields are not pronounced in view of the large differences in viscosity observed at higher concentrations. Fluorescence quenching and SANS can provide information on the elongated micelle sizes only at low concentration. In this concentration range, the micelles are charged and not screened, and their growth is hindered by electrostatic repulsion [173]. Their growth rate results not only from the end-cap energy but also from an electrostatic contribution that favors the breaking of the micelles. This contribution decreases with increasing concentration. Three growth regimes have been distinguished [173]: at low enough concentrations, such that the Debye screening length is larger than the micelle size, the concentration dependence of the size is weak and the micelles are nearly monodisperse. As the concentration increases, a sharp crossover to a rapid

TABLE 1 Characteristic Parameters of the Micellar Growth of 10-*s*-10, 2Br Geminis [152,162]

s	N_0	E_c/kT
2	25	15
3	23	12
4	24	10
6	22	7
8	17	5
10	18	5
12	13	6

N_0 is the aggregation number at the cmc; E_c is the end-cap energy.

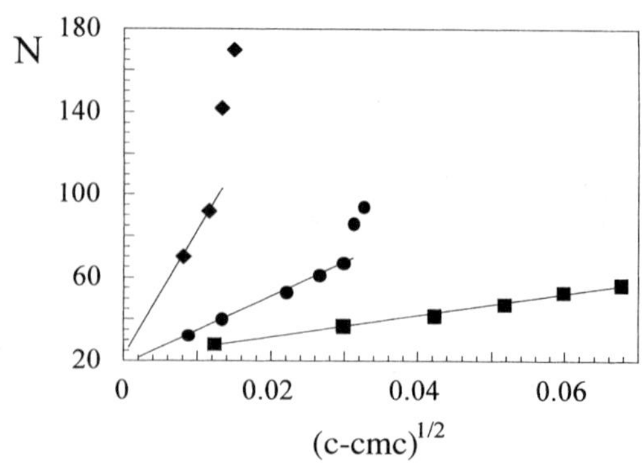

FIG. 24 Aggregation number of cationic surfactant oligomers: 12-5-12, 2Br (■); 12-3-12, 2Br (●); 12-3-12-3-12, 3Br (◆) [103,166] analyzed with the ladder model; c, the surfactant concentration, and the cmc are expressed in mole fraction.

growth regime occurs (when the Debye length equals the micelle size). The crossover concentration coincides with the concentration c^* at which the micelles begin to interact with each other (semidilute regime). As in the case of neutral micelles, the size distribution is large (exponential), but the characteristic size, N, grows faster than $c^{1/2}$ governing the growth of neutral micelles [Eq. (9)]. Its concentration dependence reads

$$N \approx 2c^{1/2} \exp[E/2k_BT - l_B a \nu^2/2c] \qquad (11)$$

where l_B is the Bjerrum length, a is the diameter of the micelle, c is the volume fraction and ν is an apparent charge density of the micelle. At higher concentration, the electrostatic contribution that tends to break the micelles results essentially from the entropy of the counterions near the end caps, where they are less tightly bound. The growth may be characterized by an effective power law:

$$N \propto c^{(\Lambda+1)/2} \qquad (12)$$

where Λ is related to the net charge on an end cap and depends only logarithmically on c [173].

The crossover concentration c^* from the slow to the fast growth regime is observed in the concentration dependence of zero shear viscosity [169,174]. For low concentration, the viscosity of surfactant oligomer solutions is not significantly higher than the viscosity of water. But at $c > c^*$, the viscosity increases very rapidly (Fig. 25).

According to Mackintosh et al. [173], $c^* \propto E_c^{-1/2}$ and this allows a comparison of the end-cap energy among various surfactant oligomers. In Table 2 we reported the values of c^* and $10/\sqrt{c^*}$, which is proportional to E_c, for various cationic surfactant oligomers. When m = 12, decreasing s from 3 to 2 doubles the end-cap energy. The same result is obtained by going from the dimer to the trimer, keeping s = 3 constant [170]. The rheology of heterodimers m-2-m', 2Br has been studied by Oda et al. [171]. Differences in alkyl chain length markedly affect the end-cap energy (see 12-2-12, 10-2-14, and 8-2-16 in Table 2).

End-cap energy values can be estimated rather indirectly from the dynamical properties of the systems (rheology or diffusion coefficient). For instance, in the plateau region of the relaxation spectrum, the storage modulus $G' = G_0$ is nearly independent of the frequency and the loss modulus G'' goes through a minimum. The ratio G_0/G''_{min} is related to the ratio of the length of the micelles to the mesh size of the entanglement network. The latter depends essentially on the concentration, whereas the temperature strongly determines the micelle length. The temperature dependence

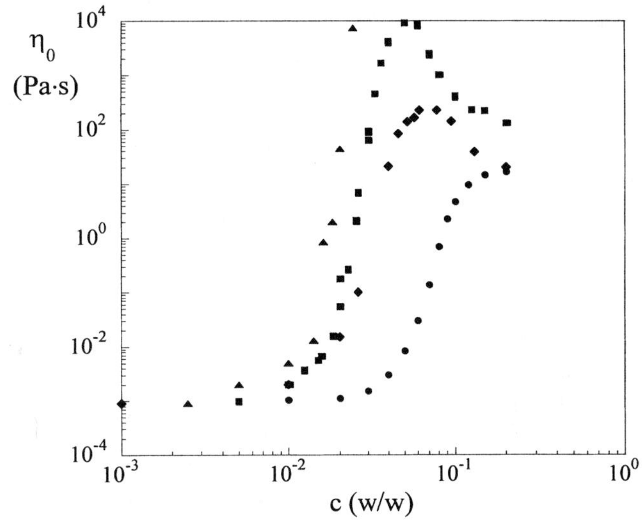

FIG. 25 Concentration dependence of the zero shear viscosity of aqueous solutions of 12-2-12, 2Br (♦), 12-3-12, 2Br (●), 12-3-12-3-12, 3Br (■), and 12-3-12-4-12-3-12, 4Br (▲). (From Ref. 104, with permission. Copyright (2000) American Chemical Society.)

of the ratio G_0/G''_{min} gives an end-cap energy of about 50 kT for 12-2-12, 2Br at 4% but lower values as the concentration increases (down to 30 kT at 10%) [169]. For the trimer 12-3-12-3-12, 3Br, the same analysis also leads to a decrease of the end-cap energy from 80 to 10 kT as the concentration increases from 4 to 10%

TABLE 2 Crossover Concentration from Slow to Fast Micellar Growth Regime c^* for Cationic Surfactant Oligomers

Surfactant	c^* (wt%)	$10/\sqrt{c^*}$
12-2-12	1	10
14-2-14	0.08	35
16-2-16	0.015	82
12-3-12	4	5
8-2-16	6	4
10-2-14	3	6
12-2-16	0.15	28
14-2-18	0.07	38
12-3-12-3-12	1	10
12-3-12-4-12-3-12	0.5	14

c^* is determined from viscosity measurements, as the concentration at which the viscosity is twice the viscosity of the water; $10/\sqrt{c^*}$ is directly proportional to the end-cap energy.

(unpublished results). The accuracy of such an analysis is, however, questionable and would require a better understanding of the relaxation mechanism to be proved.

Diffusion coefficient measurements by fluorescence recovery after photobleaching (FRAP) [175], interpreted assuming that the Cates model [176] holds in the concentration range studied, yields about 30 kT for 12-2-12, 2Br, 12-3-12, 2Br, and 16-4-16, 2Br, inconsistent with the differences observed in c^* [175].

The end-cap energy can also be estimated by fitting the concentration dependence of the zero shear viscosity in the fast growth regime [169]. This supposes assuming a relation between the viscosity, the length of the micelles L (proportional to N), and the concentration c and introducing it in the growth law for charged micelles [Eq. (11)]. This has been done for the 12-3-12, 2Br and 12-3-12-3-12, 3Br surfactants [170], assuming the Fuoss law:

$$\eta_0 = Lc^{1/2} \tag{13}$$

leading to E_c = 40 and 80 kT, respectively. Fuoss' law has been assumed, rather than the relations proposed in Ref. 169, because the micelles were probably not fully entangled in the concentration range in which the viscosity increases rapidly (see next section) [170].

According to the values of $10/\sqrt{c^*}$ and the estimated values of E_c for m = 12 surfactants, surfactants with a longer chain length (16-2-16, 2Br, for example) would have very high end-cap energies of several hundreds of kT. These are unreasonable values and show that large m surfactant oligomer solutions have been most often studied in a metastable state [171,177], and at equilibrium they exist as two-phase systems. (The Kraft temperature of 16-2-16, 2Br is 45°C [150].) In those conditions, the increase in viscosity probably does not reflect the growth of wormlike micelles.

Addition of DTAB to 12-2-12, 2Br decreases the size of the micelles, but DTAB does not concentrate in the end cap [178].

(*b*) *Rheology.* Cationic surfactant oligomers provide systems of charged and unscreened wormlike micelles, and many aspects remain to be understood in the dynamics of these polyelectrolyte type of living polymers [179].

For instance, as the concentration increases, η_0 goes through a maximum at a concentration c^{**} and then decreases (Fig. 25) [169,170,171]. The same is observed for relaxation time as well as for the diffusion coefficient [175]. This decrease has been attributed to a shortening of the micelles due to a theoretically predicted increase in ionization degree [169]. However, a decrease in the relaxation time and in the viscosity is also observed upon addition of salt [169,174], which is expected to increase the length of the micelles. [Upon further addition of salt, 12-2-12, 2Br solutions phase separate, leading to the coexistence of a dilute micellar phase and a lamellar phase [180]. Addition of salt to 12-3-12, 2Br solutions yields phase separation, but in this case a hexagonal phase is formed (unpublished results).]

It has been proposed that, in the concentration range where the viscosity increases rapidly, the systems were not fully entangled [170]. This is suggested by the strong concentration dependence of the elasticity at low concentration (Fig. 26). The crossover to a fully entangled regime seems to be reached at maximum viscosity concentration, c^{**}. For fully entangled systems, the elasticity is expected to vary as c^2, as observed for the dimer 12-3-12, 2Br system at c > 10 wt% (Fig. 26). The existence of a concentration regime of overlapping yet not fully entangled micelles may be a clue to understanding the nonmonotonic behavior of the viscosity. It also provides a simple explanation for the increase of the elastic plateau modulus upon addition of salt observed in 12-2-12, 2Br [181] and 12-3-12, 2Br solutions (unpublished results). It may also explain why the temperature dependence of viscosity for concentrations close the maximum in viscosity does not follow the Arrhenius law [169]. The extension of the concentration range where the micelles overlap and are yet not entangled (expressed as c^{**}/c^*) would reflect the rate at which they grow [170] and could provide a more reliable method to estimate E_c.

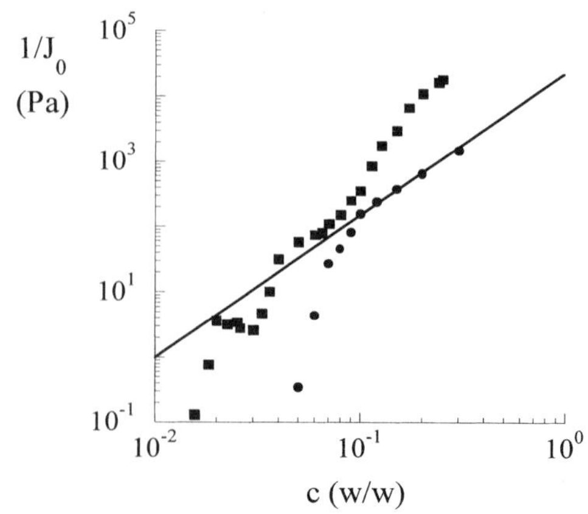

FIG. 26 Concentration dependence of the elasticity (inverse of the recoverable compliance) of 12-3-12, 2Br (●) and 12-3-12-3-12, 3Br (■) aqueous solutions [170].

The nonlinear rheology of 12-2-12, 2Br solutions has been subject to several studies. As in other wormlike micelle systems, shearing solutions of 12-2-12, 2Br close to c^* induces thickening [182–184]. The shear rate inducing thickening decreases with the concentration and increases with temperature [184]. Shear thickening is associated with anisotropy in neutron scattering [182] as well as in refractive index and conductivity [183]. Cryo-TEM micrographs of sheared samples [184] have shown that a phase separation occurs between surfactant-rich and surfactant-poor regions.

(*c*) *Branched and Closed-Looped Wormlike Micelles.* In trimer 12-3-12-3-12, 3Br solutions the extension of the overlapped yet unentangled regime is as broad as in the 12-3-12, 2Br system, although the c^* values suggest that the micelles are growing faster [170]. This could be good confirmation for the presence of branches that have been observed by cryo-TEM [104,185]. In the fully entangled regime, the elasticity of 12-3-12-3-12, 3Br solutions is enhanced as compared with 12-3-12, Br solutions (Fig. 26). This may also result from branching [170]. The formation of branched wormlike micelles has also been observed by cryo-TEM in 12-2-12, 2Br solutions [184] and has been obtained in molecular dynamics simulations [186].

Formation of a dominant population of closed loops in wormlike micellar systems, theoretically considered for a long time [32,187], has been achieved with the cationic surfactant tetramer 12-3-12-4-12-3-12, 4Br [188] (Fig. 27). The contour length distribution $N(L)$ of the closed-looped micelles has been determined from cryo-TEM micrographs (Fig. 28). At large contour lengths, the distribution observed scales as $N(L) \propto L^{-5/2}$, as expected from ring-chain equilibrium polymerization theory [187]. At small contour lengths, the ring closure probability depends on the rigidity of the micelles, and the maximum of the distribution at $L = 150$ nm corresponds to twice the persistence length [188].

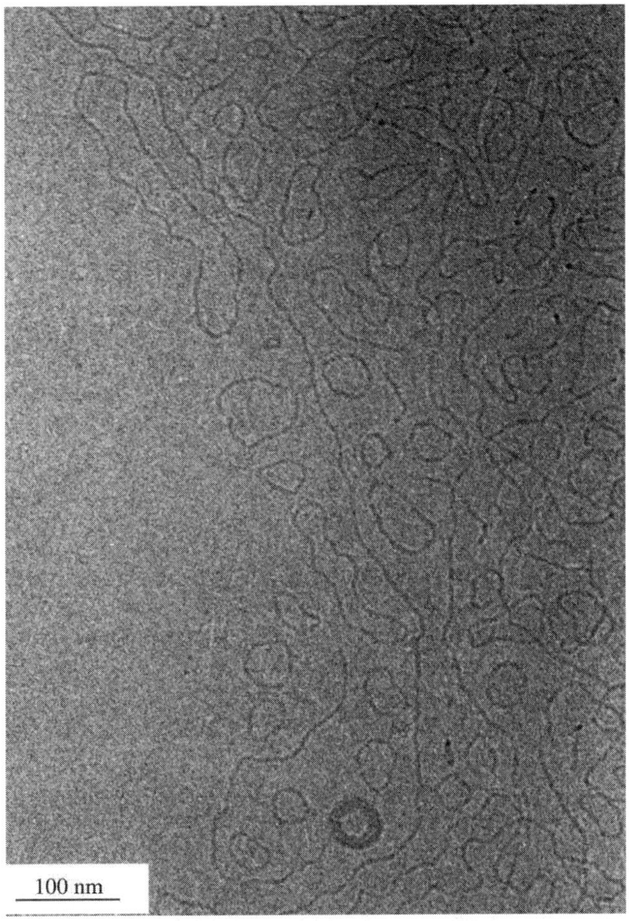

FIG. 27 Cryo-TEM micrographs of a vitrified 1% aqueous solution of 12-3-12-4-12-3-12 showing many closed loops coexisting with open wormlike micelles. (From Ref. 188, with permission. Copyright (1999) American Chemical Society.)

2. Vesicles and Other Low-Curvature Aggregates of Surfactant Oligomers

The formation of vesicles from gemini surfactants has been specifically reviewed [189]. Vesicles are obtained with $m = 12$ and $s > 16$ but also for $m = 16$ and $s < 4$ gemini surfactants. This is true for diquaternary ammonium [22,166,167] as well as diphosphates [67]. Vesicle formation has also been observed from nonionic sugar-based geminis such as in Fig. 5a (14-10-14) [84]. The cleavable heterodimer surfactant of Fig. 6c [88] forms vesicles that can be destroyed by acid catalysis under milder conditions (pH 3) than many other cleavable surfactants, a property that could find application in drug vectorization [88]. Other cleavable gemini surfactants have been shown to yield small unilamellar vesicles [96] having transition temperatures that are pH dependent.

An interesting result concerns the stereodependence of the fusion of vesicles [91]. Upon addition of Ca^{2+}, vesicles formed from (*S*, *S*) and (*R*, *R*) stereoisomers of surfactants in Fig. 7a undergo fusion whereas (*R*, *S*) isomer vesicles undergo fission. This observation has been correlated with the monolayer isotherms at the air-water interface. The surface area per headgroup is smaller for the meso compound than for the (*S*, *S*) isomer and decreases when Ca^{2+} ions are added whereas it increases for the (*S*, *S*) isomer [91].

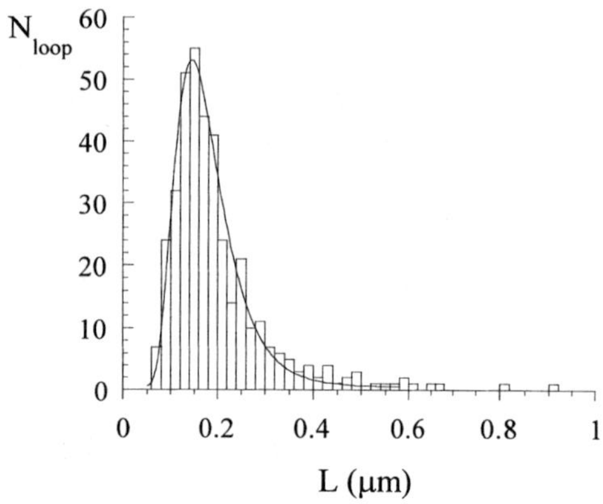

FIG. 28 Loop size distribution in 1% aqueous solution of 12-3-12-4-12-3-12, 4Br. The mode of the distribution corresponds to twice the persistence length (l_p = 75 nm). The decreasing part of the distribution scales as $L^{-5/2}$ as expected from the ring closure probability for Gaussian polymers. (From Ref. 188, with permission. Copyright (1999) American Chemical Society.)

Vesicles of diphosphate geminis 12-18-12, 2Na and 12-24-12, 2Na are characterized by a noncooperative phase transition observed by differential scanning calorimetry (DSC) and fluorescence depolarization, the midpoint of the transition range being about 45°C [67]. The transition is accompanied by a line broadening of the ^{1}H and ^{31}P NMR signals. X-ray diffraction suggests that the spacer hydrocarbon chain is membrane spanning in the 12-24-12, 2Na surfactant vesicles. This was previously observed with glycerophosphate gemini surfactants of the type *n*/2-*n*-*n*/2, 2Na [71]. The temperature transition is structure dependent and decreases as the pH increases. The authors pointed out that, for an equivalent thickness, the transition temperature was higher in membrane-spanning spacer gemini vesicles than in classical phosphoglyceride bilayer lipid membranes [71]. This could explain, from the evolutionary standpoint, why such lipids have been found in bacteria living under extreme thermal conditions [71].

As with conventional surfactants, mixtures of cationic and anionic surfactants can yield vesicles [190,191]. Bromide counterions of cationic geminis have been replaced by palmitate ions. The vesicles obtained have a higher transition temperature as expected with "catanionic" systems, but the interesting result is that this transition temperature decreases from 74 to 39°C when the spacer goes from 2 to 12. Large spacer surfactants also form vesicles that are more permeable to hydroxyl anions. Other catanionic systems involving geminis have been studied. For instance, addition of an anionic gemini to CTAB solutions induces a line broadening of the NMR signal of the methyl proton of CTAB, which has been interpreted as the formation of a network of cross-linked micelles [192]. This interpretation has, however, been contradicted by a cryo-TEM study that revealed the presence of vesicles and other large aggregates, suggesting that the system was close to precipitation [193].

Vesicles have also been obtained from asymmetric phosphate gemini surfactants mixed with L-histidine-based surfactants having two long alkyl chains [194]. The histidine surfactant is not soluble by itself. When mixed with (*R*, *S*) surfactant of Fig. 7a, at pH 6.5, large vesicles (150 to 750 nm in diameter) are obtained. With the (*R*, *R*) surfactant, ill-defined tubular structures are obtained. When added to the (*S*, *S*) isomer of the surfactant of Fig. 7a, 40-nm-wide right-handed helical ribbons are obtained with pitch of 90 nm rather independent of the composition. Similar helical ribbons are obtained with cationic 16-2-16 having L-tartrate or D-tartrate as counterion [195]. But in that case, the width and the pitch of the ribbons can be tuned by adjusting the enantiomeric excess, $(c_L - c_D)/(c_L + c_D)$ or by adding an excess of chiral counterion. It is worth mentioning that these geminis with a chiral counterion have the ability to gel halogenate organic solvents in the presence of a small amount of water [196]. Several factors seem necessary to gel the organic solvent: the short spacer, the chirality of the counterion, and its ability to hydrogen-bond. Other cationic geminis (m = 16, and 2,3-dimethoxybutane spacer) have shown the ability to self-assemble in chloroform in the presence of a small amount of water [197].

The vesicle-micelle transition has been studied by adding spherical micelle–forming surfactants such as DTAB and 12-10-12, 2Br to a vesicle-forming gemini surfactant 12-20-12, 2Br [198]. No wormlike micelle intermediate state has been observed. The vesicle-micelle transition can also be triggered by increasing the temperature of 16-3-16, 2Br solutions [199]. Addition of hexanol to solutions of 12-2-12, 2Br induces a transition from wormlike micelles to vesicles [177]. This transition has been studied by rheology and cryo-TEM observation. When the molar ratio, r, of hexanol to gemini increases, the viscosity and the size of the micelles first increase up to r = 1/8. Increasing r further (up to 1/3) leads to the formation of highly branched micelles coexisting with small vesicles of about 100 nm diameter. In this regime the viscosity decreases. Increasing r then induces the growth of the vesicles and eventually leads to phase separation [177].

D. Liquid Crystalline Phases

1. Lyotropic Behavior

Cationic gemini surfactants 12-*s*-12, 2Br (with $16 \geq s \geq 4$) form lyotropic mesophases stable from room temperature up to temperatures of 150–200°C [168]. The concentration range of stability of the micellar solutions broadens on increasing *s* and spans almost the entire phase diagram (up to 90%, where it starts to coexist with crystals) when $s = 10$ and 12. This has been interpreted as being due to a maximum mismatch between the length of the spacer and the length of the alkyl chains. For long spacers the range of the micellar phase narrows again, and with 12-16-12, 2Br surfactant, lyotropic mesophases are obtained at 30% [168]. The concentration range of stability of the lyotropic mesophases is widest for 12-8-12, 2Br. The mesomorphic behavior of 16-*s*-16, 2Br [200,201] has also been studied with emphasis on short spacers ($s = 1$, 2, 3, and 6). The Kraft boundary of the mesophases decreases with increasing *s* because of a disruption of the crystal packing. All these surfactants form a long rod nematic phase at the micellar (L_1)/hexagonal (H_1) boundary except for $s = 6$. For $s = 2$ and 3 the formation of an intermediate phase (Int) (noncubic liquid crystals with curvature intermediate between hexagonal and lamellar) is observed, but not for $s = 6$. Cubic bicontinuous (V_1) and lamellar (L_α) phases are obtained for all surfactants but require high temperatures (54°C and 76°C, respectively) for $s = 6$.

The succession of phases of cationic surfactant dimers and trimers with $m = 12$, $s = 3$ and 6 and various counterions has been surveyed using cross-polarized light microscopy and the water penetration technique (M. In and G. G. Warr, in preparation). Intermediate and bicontinuous cubic phases are confirmed for short spacers. More strongly binding counterions have a similar effect as a reduction in spacer length or an increase in oligomerization degree. This study again illustrates the fine tuning one can achieve by varying parameters specific to surfactant oligomers, namely the spacer length and the degree of oligomerization.

Lyotropic behavior of cationic gemini surfactants with hydrophilic (EO) spacers (Fig. 2b) has also been studied [49]. The concentration range of the L_1 phase broadens as *s*/3 increases as in hydrophobic spacer geminis.

The lyotropic behavior of the nonionic gemini of Fig. 5b with $m = 6$ was studied by polarized light microscopy and deuterium NMR spectroscopy [81]. At 27°C, as the concentration decreases, the succession of phases is S-L_α (89%), L_α-V_1 (75%), V_1-H_1 (71%), H_1-VB (62%), VB-L_1 (50%), where S stands for solid phase and VB for viscous birefringent phase. From ^{2}H NMR spectra it appears that L_α and H_1 phases always coexist with an isotropic phase (V_1 for L_α and L_1 for H_1). The VB phase was identified later as a biphasic region [163]. This sugar-based surfactant does not exhibit any cloud point in the range of temperature studied (0–100°C). The study has been completed for other chain lengths [163]. For $m = 5$, the viscous birefringent phase no longer appears. For $m = 7$ an upper critical solution temperature exists between 0.5 and 10%. The V_1 phase region is no longer seen but is supposed to exist in a very narrow range of concentration. For $m = 8$, the phase boundaries are less sensitive to temperature. The isotropic L_1 phase exists up to 0.8% and then coexists with an unidentified phase of optical texture resembling an L_α phase. A pure L_α phase is formed at 48%. The surfactant of Fig. 5b with $m = 9$ is insoluble in water up to 100°C [163].

2. Thermotropic Behavior

The thermotropic behavior of 16-*s*-16, 2Br surfactants has been reported [200,201]. DSC experiments show a main transition around 100°C corresponding to an S $\rightarrow$ VN (viscous neat phase) transition, with further small transitions at higher temperature. Optical microscopy shows that for short spacers ($s = 1$ and 2) the VN phase transforms into an isotropic phase at about 200°C (precise temperature transitions are difficult to obtain as a result of the progressive decomposition of the material). For longer spacers ($s = 3$ and 6), the system goes across to a smectic A phase at about 200–230°C. The transition at 100°C has an unexpectedly low enthalpy, which suggests that either the solid is not well ordered or the VN phase has significant conformational restrictions. X-ray diffraction in the VN phase indicates a tilted bilayer structure with disordered alkyl chains filling the space between ordered headgroup layers including the spacer. The degree of order in the headgroup layer depends on the spacer length (higher with a short spacer) and for longer spacers varies with the temperature. Thermotropic behavior was observed with cationic $m = 12$ gemini with $(EO)_{s/3}$ space [49] but not with their homologues with $(CH_2)_s$ spacer [168].

V. CHEMISTRY WITH GEMINI SURFACTANTS

A. Analytical Chemistry

1. Micellar Electrokinetic Chromatography

Micellar electrokinetic chromatography (MEKC) allows the separation of neutral compounds by electrophoresis on the basis of their differential partition between micelles and the solution. The cmc and the hydrophobicity of the micelles are critical parameters for the efficiency of this technique. Sulfonate gemini surfactants of Fig. 3a (Y = O) were able to separate various substituted naphthalene and benzene derivatives at concentrations as low as 7.5 mM [202]. Using sodium dodecyl sulfate (SDS) would have required 50 mM for the same resolution. The surfactant of Fig. 3a and SDS were also shown to have remarkably different selectivities. The separation of chlorophenols has also been carried out with the same type of surfactants [203]. Cationic surfactants have been used to separate ergot alkaloids by MEKC [204]. With all types of gemini surfactants, the retention factor increases linearly with surfactant concentration as expected [203,204]. Micellar-enhanced ultrafiltration, another micelle-based separation technique could also take advantage of the various morphologies observed in surfactant oligomer micelles, but as far as we know, no data have been reported in the literature.

2. Specific Electrodes

Di- and triamide surfactant dimers and trimers of the type shown in Fig. 29 have been synthesized in an attempt to prepare membrane systems having high selectivity for alkaline earth metal cations (Mg^{2+} and Ca^{2+}). Some of them have shown enough selectivity to allow measurements of Mg^{2+} activities in the millimolar range at physiological pH without interference by H_3O^+ ions [28]. The selectivity of these ionophores for Mg^{2+} over monovalent cations has been studied in detail [205,206] and the conditions for their use in membrane electrodes for application to human blood serum optimized [207].

The synthesis of lipophilic receptors has been reported [208] and is an active field of research. The synthesis of lipocyclopolyamines as potential adenosine mono-, di-, and triphosphates has already been mentioned [29]. More recently, the interfacial behavior of a new amphiphilic cage molecule [tetra-(*N*-2-tetra-decylcarboxamideoethyl)-tetraazacyclotetradecane] showing selective complexation of copper ions has been studied [209]. The use of this compound for liquid-liquid extraction has been demonstrated in the case of chlorinated solvents.

B. Micellar Catalysis

Bunton and coworkers, noticing the higher catalytic activity of polyelectrolytes compared with micelles, were interested in surfactants that would "combine some features of polyelectrolytes and detergents." They studied the catalytic activity of gemini surfactants 16-*s*-16, 2Br (s = 2, 4, and 6) for the reaction of hydroxide ion with chloro- and fluoro-2,4-dinitrobenzene and hydroxide and fluoride ion with *p*-nitrophenyl diphenyl phosphate [34]. Surfactants with s = 4 and 6 have shown much higher catalytic activity than CTAB for all the reactions studied. A short spacer (s = 2) or a rigid space (2,3-butinediyl) did not show any catalytic activity for reactions involving hydroxide ions. Cyclization of 2-(3-bromopropyloxy) and 2-(3-bromododecyloxy)phenoxide ion in micelles of cationic geminis (m = 16, spacer = butanediyl or 2,3-dimethoxybutanediyl) has been studied [197]. Intramolecular reactions are better suited to address the catalytic activity of micelles because they are independent of the reaction volume. The observed rate constant k_{obs} is three times higher with the butanediyl spacer gemini than with the 2,3-dimethoxybutanediyl spacer. The surfactant concentration dependence of k_{obs} showed a plateau, followed by an increase associated with the formation of larger aggregates. Under preparative conditions, i.e., with a large ratio of substrate to surfactant, only the product of intramolecular reaction was obtained [197].

C. Chemistry Hosted or Templated by Surfactant Oligomer Self-Assemblies

1. Synthesis of Colloidal Metallic Particles

A nonionic surfactant gemini known as Surfynol 465 has been used to synthesize gold [210,211] and silver [212] colloidal particles. The amphiphilic moieties in this surfactant consist of a branched alkyl chain with an oxyethylene headgroup. They are linked by an acetylenic spacer. The physicochemical properties of Surfynol 465 have been studied in detail by tensiometry, densitometry, osmometry, calorimetry, and UV and NMR spectroscopy [213–215]. The main properties are the following: cmc ranges from 10 to 16 mol/g, γ_{cmc} = 26 mN/m, $A_{a/w}$ = 0.64 nm^2 per molecule, cloud point at about 40°C, N_0 = 13. Mixing Surfynol 465 with chloroauric acid ($HAuCl_4$) or silver perchlorate ($AgClO_4$) leads to the formation of metallic particles in higher yield than with other synthesis methods. The mechanism has been studied by UV-visible and Raman scattering [212,216,217]. The particles form by successive first-order reduction reactions, where the acetylenic group of the surfactant acts as the reducing agent. The surfactant then adsorbs at the particle interface and stabilizes the colloidal solution. Anisotropic colloidal gold particles have been obtained by UV photolysis of $HAuCl_4$ in solutions of 12-2-12, 2Cl cationic gemini

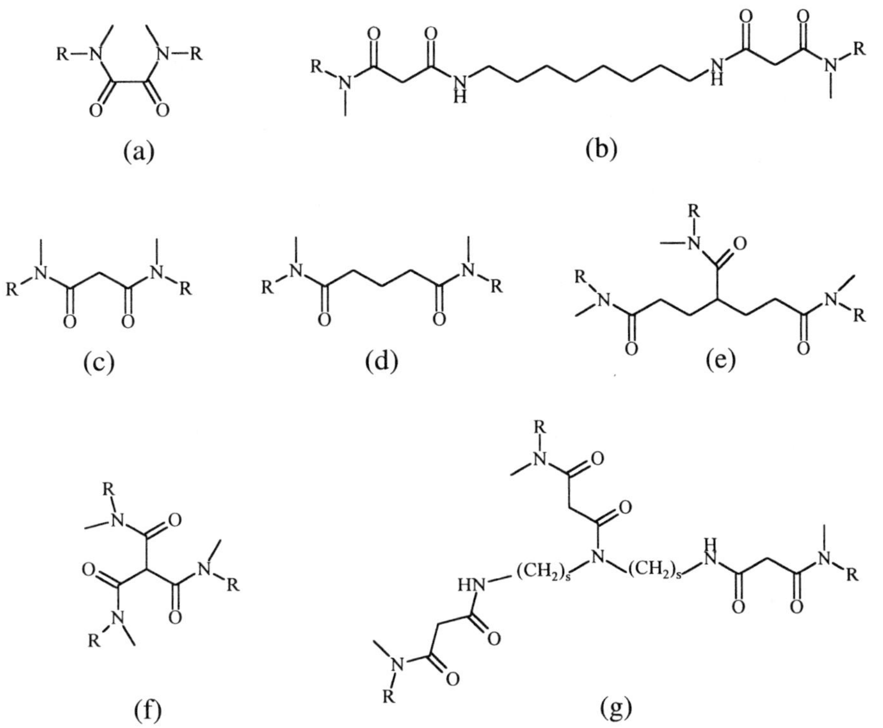

FIG. 29 Surfactant dimers and trimers used as alkaline earth metal cation ionophores [28].

surfactant [218]. The concentration necessary to produce these particles was 15 times lower than with CTAC. An excess of surfactant is necessary to disperse the insoluble complex 12-2-12, $2AuCl_4$ and allow efficient UV radiation, but if the surfactant/Au ratio is too high, polyhedral isotropic particles are obtained.

2. Emulsion Polymerization

The γ-radiation polymerization of styrene microemulsions described in Section IV.B.1 led to spherical latex particles whose size range could be controlled by the monomer/surfactant ratio but also depends on the spacer length of the gemini surfactant involved [122,160]. With —$(CH_2)_s$— spacers and in the absence of cross-linker, the latex particle size goes through a maximum at $s = 10$, but it goes through a minimum at $s = 6$ in the presence of cross-linker. The optimal microemulsion formulations leading to both small particles and high molecular weight are obtained with $s = 6$ geminis [160]. In such conditions, the molecular weight is four times that obtained in CTAB microemulsions. With —$(EO)_{s/3}$— hydrophilic spacers, the same type of behavior is observed. The spacer length leading to the maximum molecular weight (which in this case corresponds to the maximum particles size) is $s/3 = 3$ [122]. These observations led the authors to conclude that the polymerization behavior of ternary microemulsions with cationic geminis is rather independent of the chemical nature of the spacer provided that the flexibility of the interface is sufficient [122].

3. Gemini Surfactants as Templating Agents for Mesoporous Material

Tailoring the porosity of inorganic material is an issue for applications such as molecular sieving or selective catalysis. It has become a very active field of research since the pioneering work of scientists at Mobil Oil Research and Development, who used the cooperative self-assembling of silicate polyanions and cationic surfactant micelles to produce mesoporous silica [219] (see Chapter 40 in this volume). A large variety of surfactants have now been used, including cationic gemini surfactants [220–222]. The greatest advantage of using gemini surfactants for templating the formation of mesoporous material may be that they provide a new parameter, s, to control the structure, independently of m, which is the determining variable for the average pore size. However, the quality of the material obtained is also sensitive to *s* [220].

Typical syntheses proceed by mixing *m-s-m*, 2Br surfactants with tetraethyl orthosilicate (TEOS) in water (in molar proportions 0.06 : 1 : 150) under basic or acidic conditions at room temperature or/and under hydrothermal conditions for one or several days [221]. The crystallinity of the precipitate obtained is improved by subsequent hydrothermal treatment in fresh water. Most often, the structure of the ordered organic/inorganic composite material obtained is determined by the lyotropic behavior of the surfactant. Hence, short spacers favor lamellar phases (MCM-50) and intermediate spacers favor hexagonal phases (MCM-41) for $s < 10$. The 12-12-12, 2Br yields MCM-41 at room temperature [221], and a cubic structure (MCM-48) is obtained with 16-12-16, 2Br under hydrothermal conditions [220]. The use of gemini surfactants has been shown to reduce the synthesis time of high quality (in terms of crystallinity and pore size distribution) cubic mesoporous silica [220].

VI. CONCLUSIONS

The large variety of surfactant oligomer structures of interest today has been demonstrated. Their synthesis is sometimes involved but does not necessarily require new chemistry and can often rely upon well-known procedures. Surfactant oligomers have been shown to exhibit good surface activity. Their efficiency in lowering surface tension relies upon an adsorption triggered by several hydrophobic chains and interfacial interactions that take place at lower concentration because of their larger size. The cmc depends strongly on the degree of oligomerization and slightly on the spacer length. These two parameters, specific to surfactant oligomers, also determine the optimal area per headgroup and hence the morphology of the micelles as well as the mesoscopic behavior. Original structures, as compared with conventional monomeric surfactant, are obtained essentially for short spacers. Surfactant solutions behaving as polymers can be obtained and their viscosity controlled by the structure of the surfactant and not only through formulation. The stability of vesicles can also be controlled by varying the spacer characteristics.

The answer to the question "Is it worth going to higher oligomerization degree?" varies depending on the point of view. From the point of view of general applications, the answer is probably negative. The biggest gain is made from the monomer to the dimer, and the benefits of having higher degrees of oligomerization may not be worth the synthesis effort. For specialized surfactant applications and fundamental properties, it is the hope of the author that the reader's answer will be positive.

ACKNOWLEDGMENTS

I would like to take this opportunity to thank Vesna Vukov and Sharon Krauss, whose help in gathering the references, and sometime in translating them, has been invaluable. I am also thankful to Corine Gérardin and Amit Kulkarni for their comments on the manuscript. Last, many thanks are also extended to Dave Tracy and Raoul Zana for sharing their immense knowledge in the field of surfactant oligomers.

VII. APPENDIX

Notation:

Cmc: critical micelle concentration.
α: ionization degree of the micelles at the cmc.
γ_{cmc}: surface tension at the cmc.
pC_{20}: logarithm of the inverse of the surfactant concentration necessary to decrease the surface tension by 20 mN/m.
A: surface area per molecule at the air-water interface. In the absence of swamping electrolyte, the values reported in the following tables are obtained by taking $n = x + 1$ in the Gibbs equation [Eq. (9)].
k: number of solubilized dye molecules per surfactant molecule above the cmc.
KT: Kraft temperature.

A. Cationic Gemini Surfactants

1. Variable Spacer

(a) Hydrophobic Spacer

TABLE 3 8-*s*-8, 2Br

s	cmc (M)	α[a]
3	1.4×10^{-2}	0.70
4	2.6×10^{-2}	0.67
5	2.6×10^{-2}	0.69
6	2.5×10^{-2}	0.67
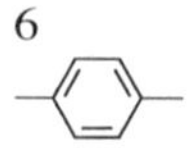	2.5×10^{-2}	0.54
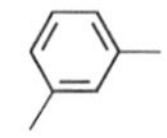	2.3×10^{-2}	0.62
	2.4×10^{-2}	0.62

[a]Conductimetry at 25°C [148].

TABLE 4 10-*s*-10, 2Br

s	cmc[a] (M)	k^a	cmc[b] (M)	α^b
2[c]	6.0×10^{-3}	99	6.2×10^{-3}	0.15
3	6.4×10^{-3}	106	6.5×10^{-3}	0.22
4	9.0×10^{-3}	121	8.7×10^{-3}	0.28
5	9.2×10^{-3}	147	—	—
6	8.7×10^{-3}	212	9.2×10^{-3}	0.29
8	6.8×10^{-3}	186	7.5×10^{-3}	0.32
10	4.7×10^{-3}	160	4.2×10^{-3}	0.37
12	3.7×10^{-3}	43	2.2×10^{-3}	0.39

[a]Dye solubilization (*trans*-azobenzene) at 20°C [159].
[b]Conductimetry at 25°C [152,162].
[c]Surface tension measurement by the maximum bubble pressure at 25°C gives cmc = 6.5×10^{-3} M and γ_{cmc} = 32 mN/m [40].

TABLE 5 12-*s*-12, 2Br

s	cmc[a] (M)	α^a	cmc[b] (M)	$\gamma_{cmc}{}^b$ (mN/m)	A[b] (Å^2)
2	8.4×10^{-4}	0.22	8.1×10^{-4}[c]	30[c]	69[c]
	(1.5×10^{-2})	(0.24)			
3	9.6×10^{-4}	0.22	9.1×10^{-4}	35.0	105
4	1.2×10^{-3}	0.28	1.0×10^{-3}	39.8	116
	(1.4×10^{-2})	(0.28)			
5	1.1×10^{-3}	0.29	—	—	—
6	1.0×10^{-3}	0.32	1.1×10^{-3}	42.5	143
	(1.1×10^{-2})	(0.32)			
8	8.3×10^{-4}	0.45	8.9×10^{-4}	42.8	176
	(7.0×10^{-3})	(0.44)			
10	6.3×10^{-4}	0.54	3.2×10^{-4}	43.0	220
12	3.7×10^{-4}	0.62	2.8×10^{-4}	41.5	226
14	2.0×10^{-4}	—	1.8×10^{-4}	39.5	200
16	1.2×10^{-4}	0.67	1.4×10^{-4}	39.4	154
	(1.0×10^{-3})	(0.62)			

[a]Conductimetry at 25°C [38]. Values in parentheses are for the corresponding monomer surfactant *m*-*s*/2, Br [155].
[b]Surface tension (ring method) at 25°C [121] unless otherwise specified.
[c]Surface tension (Wilhelmy plate open frame version) at 22°C [131].

TABLE 6 12-*s*-12, 2Cl

s	cmc (M)
2	1.3×10^{-3}
3	1.8×10^{-3}
4	1.3×10^{-3}
6	1.3×10^{-3}
10	6.3×10^{-4}
20	7.0×10^{-5}

Source: Ref. 153.

TABLE 7 16-*s*-16, 2Br

s	cmc[a] (M)	cmc[b] (M)	α[b]	cmc[c] (M)	KT[d] (°C)	cmc[e] (M)
2	1.4×10^{-5}	2.1×10^{-5}	0.60		45	3.4×10^{-5}
3	—	2.6×10^{-5}	0.35	2.5×10^{-5}		
4	3.2×10^{-5}	2.7×10^{-5}	0.56	2.7×10^{-5}	34	4.4×10^{-5}
5	—	—	—	3.6×10^{-5}		
6	6.5×10^{-5}	4.3×10^{-5}	0.43	4.3×10^{-5}	41	4.7×10^{-5}
8	—	3.3×10^{-5}	0.60	3.3×10^{-5}		
10	—	—		2.7×10^{-5}		
12	—			2.0×10^{-5}		

[a]Surface tension (ring method) at 25°C [34].
[b]Conductimetry at 25°C [38].
[c]Fluorescence spectroscopy at 30°C [41].
[d]Conductimetry [150].
[e]Conductimetry at 46.5°C [150].

TABLE 8

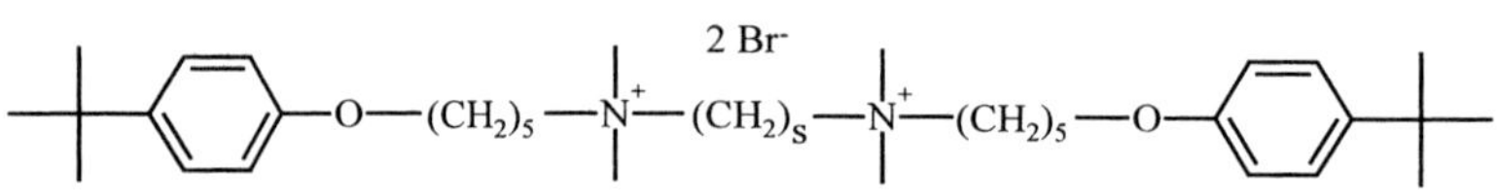

s	2	4	6	8	10	12
cmc (M)	3.2×10^{-3}	3.2×10^{-3}	3.1×10^{-3}	2.6×10^{-3}	2.3×10^{-3}	1.6×10^{-3}

Note: Dye solubilization [145].

TABLE 9

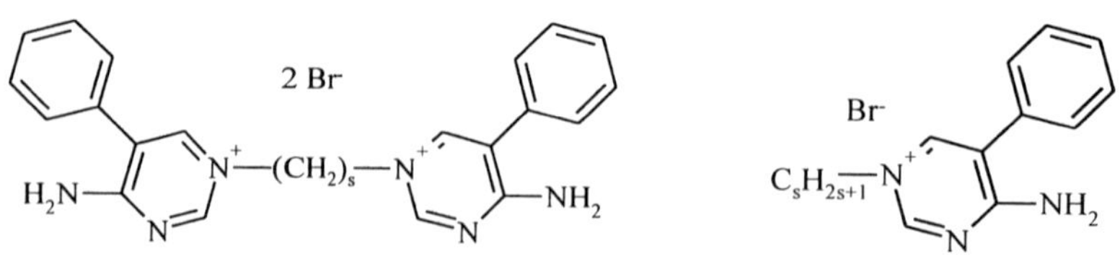

s	cmc (M)	γ_{cmc} (mN/m)	A (Å^2)	cmc (M)	γ_{cmc} (mN/m)	A (Å^2)
8	4.0×10^{-4}	45	314	1.0×10^{-2}	42	208
9	6.3×10^{-5}	43	444	8.3×10^{-4}	50	150
10	1.6×10^{-4}	44	381	9.1×10^{-3}	37	178
11	3.0×10^{-4}	50	335	5.0×10^{-4}	47	153
12	3.0×10^{-5}	34	189	7.9×10^{-3}	37	113

Note: Surface tension (du Nouy ring) at 25°C [43].

(b) Hydrophilic Spacer

TABLE 10

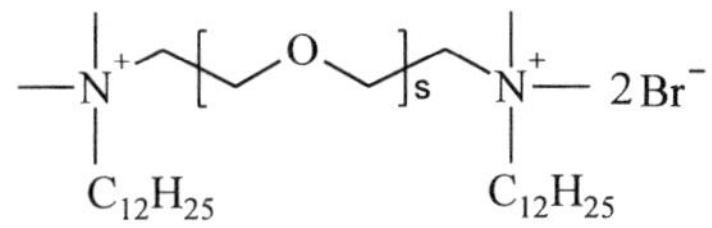

s	cmc (M)	γ_{cmc} (mN/m)
1	8.0×10^{-4}	38.2
2	9.2×10^{-4}	41.0
3	8.6×10^{-4}	40.8
4	9.0×10^{-4}	42.1
5	1.2×10^{-3}	42.0

Note: Surface tension (Wilhelmy plate) at 25°C [122].

TABLE 11

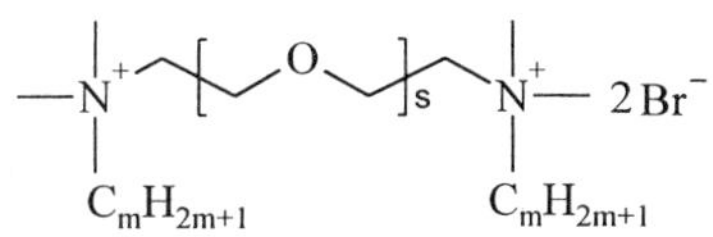

s-m	cmc1[a] (M)	cmc2[a] (M)	cmc[b] (M)	KT[c] (°C)	cmc[d] (M)
1-12	—	—	4.8×10^{-4}	—	
1-14	5.0×10^{-5}	2.5×10^{-4}	—	—	—
1-16	3.4×10^{-5}	1.5×10^{-4}	3.8×10^{-5}	28.8	1.9×10^{-5}
2-16					1.7×10^{-5}
3-16	6.0×10^{-5}	7.0×10^{-5}	6.5×10^{-5}	12.4	2.0×10^{-5}
7-16	1.6×10^{-4}	1.4×10^{-3}	1.6×10^{-4}	<0.0	—
1-18	1.2×10^{-5}	8.7×10^{-5}		40.5	—
3-18	3.8×10^{-5}	3.7×10^{-5}	3.3×10^{-4}	23.0	—
7-18	8.2×10^{-5}	6.6×10^{-4}	8.4×10^{-5}	4.0	—
1-22	7.2×10^{-6}	3.4×10^{-5}	—	53.2	—
3-22	1.4×10^{-5}	8.0×10^{-5}	—	40.6	—
7-22	2.4×10^{-5}	2.3×10^{-4}	2.3×10^{-4}	24.8	
20-22	1.5×10^{-5}	1.4×10^{-4}			

[a]Conductmetry at 25°C; cmc1 and cmc2 correspond respectively to the first and second breaks in the slope of the plot of the conductivity versus concentration [47].
[b]Surface tension (du Nouy ring method) 23°C [47].
[c]Conductimetry [47].
[d]Fluorescence spectroscopy (I_1/I_3 pyrene) at 30°C [48].

2. Variable Hydrophobic Chain

TABLE 12 *m*-2-*m*, 2Br

m	cmc[a] (mM)	KT[b]
12	1.4×10^{-3}	14
14	2.0×10^{-4}	33
16	3.4×10^{-5}	45

[a]Conductimetry 46.5°C [150].
[b]Conductimetry [150].

TABLE 13 *m*-5-*m*, 2Br

m	cmc[a] (M)	cmc[b] (M)	γ_{cmc}[b] (mN/m)	A^c (Å^2)
6	1.6×10^{-1}	4.2×10^{-1}	42.4	120
8	4.8×10^{-2}	4.9×10^{-2}	40.3	117
9	2.5×10^{-2}	1.5×10^{-2}	39.7	130
10	8.9×10^{-3}	7.7×10^{-3}	39.5	128
11	3.0×10^{-3}	2.3×10^{-3}	40.1	100
12	1.1×10^{-3}	1.0×10^{-3}	40.0	115
13	4.1×10^{-4}	4.9×10^{-4}	39.7	123
14	1.8×10^{-4}	2.0×10^{-4}	39.6	120
16	3.2×10^{-5}	9.3×10^{-6}	36.6	105

[a]Conductimetry at 20°C.
[b]Surface tension (ring method) at 20°C.
[c]Assuming $n = 2$ in the Gibbs equation [Equation (1)].
Source: Ref. 119.

TABLE 14 *m*-6-*m*, 2Br

m	cmc[a] (M)	cmc[b] (M)	k^b (μmol/mol)
8	2.5×10^{-2}	7.26×10^{-2}	69
9	2.2×10^{-2}	2.59×10^{-2}	123
10	8.2×10^{-3}	8.65×10^{-3}	212
11	2.7×10^{-3}	2.89×10^{-3}	229
12	1.2×10^{-3}	1.71×10^{-3}	228
13	4.6×10^{-4}	5.40×10^{-4}	282
14	1.50×10^{-4}	—	415
15	7.00×10^{-5}	—	432
16	4.80×10^{-5}	—	555

[a]Conductimetry at 20°C [145].
[b]Dye solubilization (*trans*-azobenzene) at 20°C [159].

TABLE 15

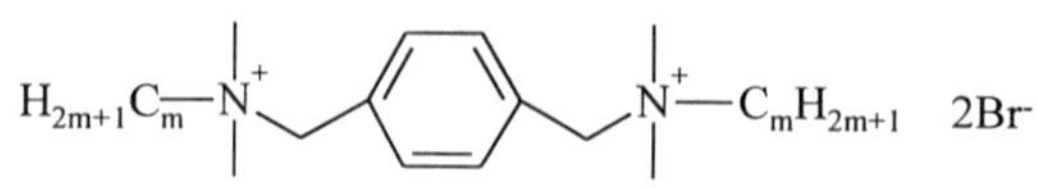

m	cmc (M)	γ_{cmc} (mN/m)
8	1.0×10^{-2}	38
12	1.0×10^{-3}	39
16	6.7×10^{-5}	41
18	3.7×10^{-4}	38

Note: Surface tension (ring method) at 50°C [39].

TABLE 16

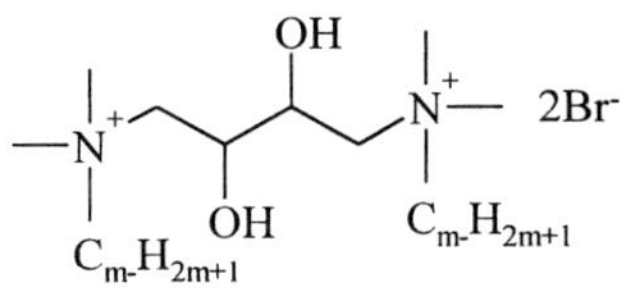

m	cmc[a] (M)	γ_{cmc}[a] (mN/m)	pC_{20}[a]	cmc[b] (M)	γ_{cmc}[b] (mN/m)	pC_{20}[b]	A[b] (Å$_2$)
8	—	—	—	1.1×10^{-2}	38.3	3.1	79.0
10	3.7×10^{-3}	35.5	3.24	3.3×10^{-4}	33.8	5.17	75.0
12	7.0×10^{-4}	35.4	3.89	6.0×10^{-6}	31.8	6.47	64.0
14	8.5×10^{-5}	36.0	5.50	1.0×10^{-6}	29.7	6.9	42.0
16	5.0×10^{-5}	41.4	5.50	—	—	—	—

[a](Wilhelmy plate) at 25°C [46].
[b]Surface tension (Wilhelmy plate) at 25°C, in 0.1 M NaBr [46].

TABLE 17

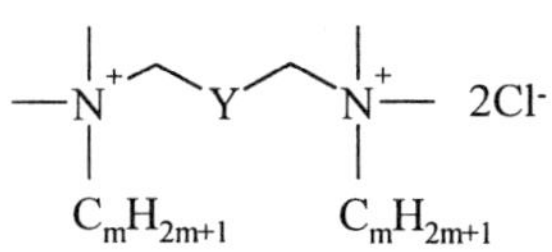

Y	m	cmc (M)	γ_{cmc} (mN/m)	pC_{20}
—CH_2—	12	9.8×10^{-4}	39.2	3.3
—CHOH—	12	7.8×10^{-4}	37	3.2
—$CHOHCH_2$—	12	6.5×10^{-4}	40.9	3.5
—CH_2—O—CH_2—	12	5.0×10^{-4}	39.2	3.6
—CHOH—	8			
	9			
	10	3.2×10^{-3}	36.5	2.9
	11			
	12	7.8×10^{-4}	37.0	3.2
	13			
	14	1.4×10^{-4}	39.0	4.4
	16	1.9×10^{-5}	42.2	5.3
—CH_2—	12	9.8×10^{-4}	39.2	3.3
	14	1.1×10^{-4}	41.8	4.5
	16	1.5×10^{-5}	42.0	5.4

Note: Wihelmy plate at 20°C. All surfactants have a Kraft point below 0°C [45].

TABLE 18

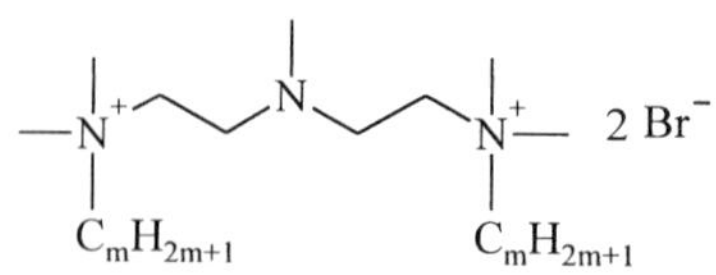

m	cmc[a] (M)	cmc[b] (M)	γ_{cmc}[b] (mN/m)	A[b] (Å^2)
8	3.1×10^{-2}	3.7×10^{-2}	37.1	98
9	1.8×10^{-2}	1.5×10^{-2}	38.0	100
10	8.4×10^{-3}	4.7×10^{-3}	35.5	98
11	3.0×10^{-3}	2.1×10^{-3}	37.5	103
12	1.1×10^{-3}	1.2×10^{-3}	36.1	108
13	3.3×10^{-4}	4.1×10^{-4}	38.0	111
14	1.2×10^{-4}	1.2×10^{-4}	38.0	103
15	4.3×10^{-5}	4.2×10^{-5}	40.1	113
16	1.2×10^{-5}	1.4×10^{-5}	42.9	119

[a]Conductimetry at 20°C.
[b]Surface tension (ring method) at 20°C.
Source: Ref. 119.

TABLE 19

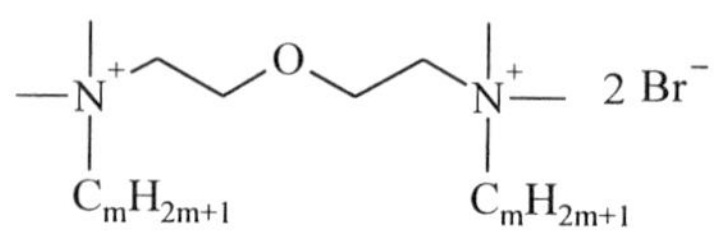

m	cmc[a] (M)	cmc[b] (M)	γ_{cmc}[b] (mN/m)	A[b] (Å^2)
8	4.4×10^{-2}	4.2×10^{-2}	41.7	128
9	2.2×10^{-2}	1.7×10^{-2}	40.4	133
10	7.7×10^{-3}	7.7×10^{-3}	40.8	135
11	2.6×10^{-3}	2.1×10^{-3}	39.1	127
12	1.0×10^{-3}	1.1×10^{-3}	38.2	129
13	3.8×10^{-4}	3.1×10^{-4}	39.7	126
14	1.9×10^{-4}	1.0×10^{-4}	40.1	138
16	6.4×10^{-6}	4.4×10^{-6}	47.2	141

[a]Conductimetry at 20°C.
[b]Surface tension (ring method) at 20°C.
Source: Ref. 119.

TABLE 20

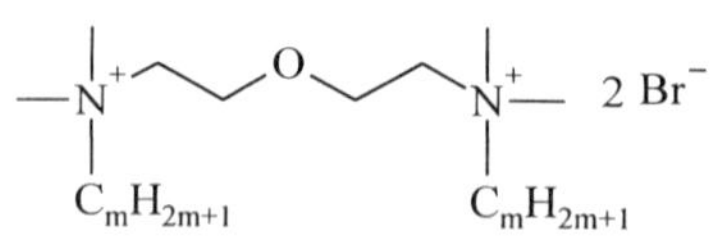

m	cmc[a] (M)	γ_{cmc}[a] (mN/m)	pC_{20}[a]	A[a] (Å^2)	cmc[b] (M)
10	5.6×10^{-3}				6×10^{-3}
12	5.2×10^{-4}	40	4.0	170	5×10^{-4}
14	5.8×10^{-5}	39	4.9	156	6×10^{-5}
16	6.6×10^{-6}	<40	5.7	102	6×10^{-6}
18	1.4×10^{-6}	40	6.3	84	6×10^{-6}
20	4.0×10^{-6}	55			2×10^{-5}

[a]Surface tension (Wilhelmy plate) at 25°C [117].
[b]Fluorescence spectroscopy of pyrenecarboxyaldehyde at 25°C [117].

TABLE 21

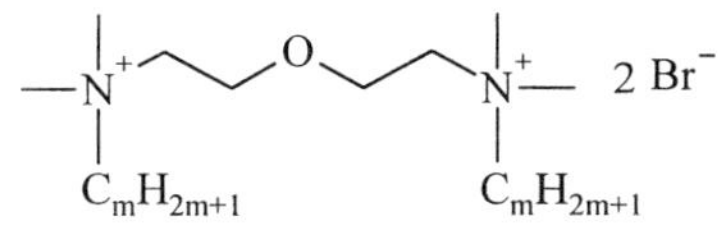

m	cmc[a] (M)	γ_{cmc}[a] (mN/m)	pC_{20}[a]	A[a] (Å^2)
12	3.0×10^{-5}	39	6.0	95
14	1.2×10^{-6}	36	7.3	79
16	9.5×10^{-7}	34	6.5	24
18	7.1×10^{-7}	30	6.4	14

[a]Surface tension (Wilhelmy plate) at 25°C, in 0.1 M NaCl [117].

TABLE 22

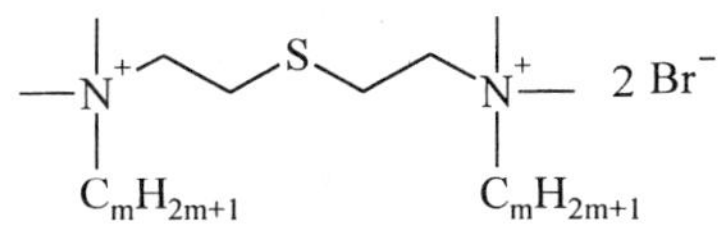

m	cmc[a] (M)	cmc[b] (M)	γ_{cmc}[b] (mN/m)	A[b] (Å^2)
6	1.3×10^{-1}	6.1×10^{-2}	28.7	79
8	3.7×10^{-2}	2.1×10^{-2}	33.1	88
10	6.2×10^{-3}	8.8×10^{-3}	35.0	85
12	9.6×10^{-4}	8.4×10^{-4}	34.5	84
14	1.0×10^{-4}	1.5×10^{-4}	33.9	114
16	9.7×10^{-6}	1.7×10^{-5}	41.9	121
18	1.0×10^{-6}	5.2×10^{-6}	40.9	120

[a]Conductimetry at 20°C.
[b]Surface tension (ring method) at 20°C.
Source: Ref. 119.

TABLE 23

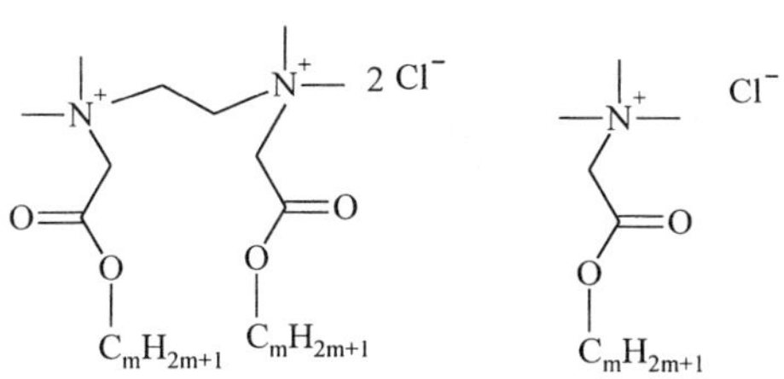

m	cmc_{dimer}[a] (M)	cmc_{dimer}[b] (M)	$cmc_{monomer}$[b] (M)
8	8.7×10^{-3}	8.1×10^{-3}	
9	5.2×10^{-3}		
10	4.0×10^{-3}	1.3×10^{-3}	
11	3.2×10^{-3}		
12	2.2×10^{-3}	3.0×10^{-4}	5.5×10^{-3}
14	1.2×10^{-3}	1.2×10^{-4}	1.9×10^{-3}
16	7.5×10^{-4}	4.0×10^{-5}	3.3×10^{-4}

[a]Conductimetry [123].
[b]Calorimetry at 25°C [42].

TABLE 24

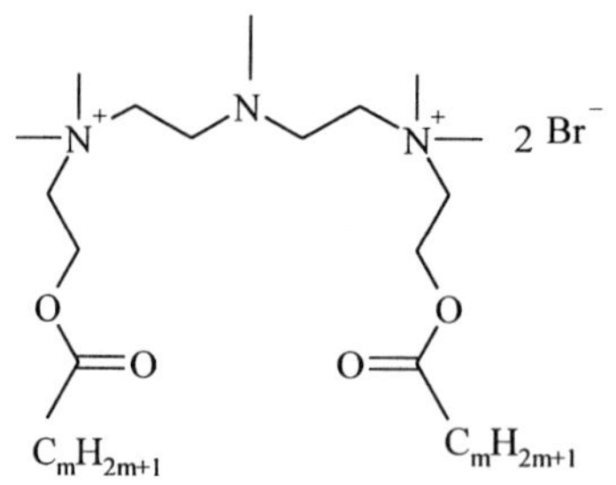

m	cmc[a] (M)	cmc[b] (M)	γ_{cmc}[b] (mN/m)	A[b] (Å^2)
6		4.8×10^{-2}		
7	4.7×10^{-2}	4.2×10^{-2}	38.8	101
8	9.1×10^{-3}	1.3×10^{-2}	31.1	119
9	6.2×10^{-3}	5.1×10^{-3}	36.3	109
10	2.2×10^{-3}	1.9×10^{-3}	34.2	110
11	7.8×10^{-4}	5.9×10^{-4}	37.1	118
11′	1.9×10^{-3}	2.0×10^{-3}	37.3	129
12	2.2×10^{-4}	3.0×10^{-4}	34.3	101
14	9.6×10^{-6}	1.7×10^{-5}	40.0	87
18″	9.9×10^{-6}	4.0×10^{-5}	52.6	155

[a]Conductimetry at 20°C.
[b]Surface tension (ring method) at 20°C.
11′: 10-Undecenoyl.
18″: Oleyl(cis-9-octadecenoyl)
Source: Ref. 120.

3. Miscellaneous

TABLE 25

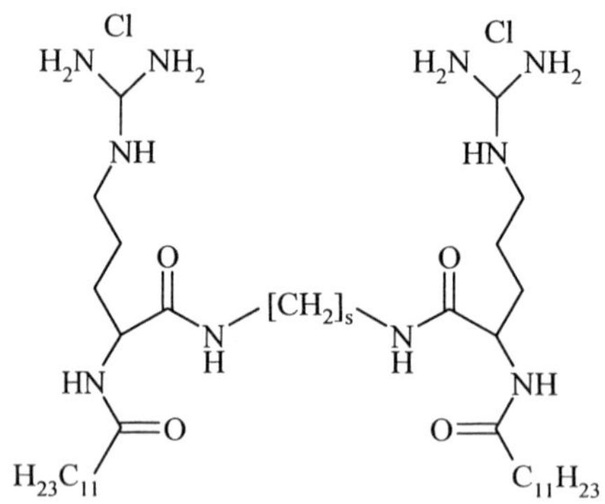

s	cmc[a] (M)	cmc[b] (M)	cmc[c] (M)	γ_{cmc}[c] (mN/m)	pC_{20}[c]	A[c] (Å^2)
2			9.5×10^{-6}	30	5.7	91
3	4×10^{-4}	4×10^{-4}	4.4×10^{-6}	35	5.9	86
4			2.8×10^{-6}	30	6.5	130
6	6×10^{-4}	4×10^{-4}	1.3×10^{-6}	30	6.7	113
9	3×10^{-4}	3×10^{-4}	2.8×10^{-6}	34	6.0	77
10			1.9×10^{-6}	34	6.2	74
Monomer	6×10^{-3}	5×10^{-3}	6.0×10^{-3}	33	3.2	67

[a]Conductimetry (chloride-selective electrode) at 25°C [78].
[b]Fluorescence spectroscopy of pyrene [78].
[c]Surface tension (Wilhelmy plate) at 25°C (aged 24–48 h) [77].

TABLE 26

Ferrocenyl

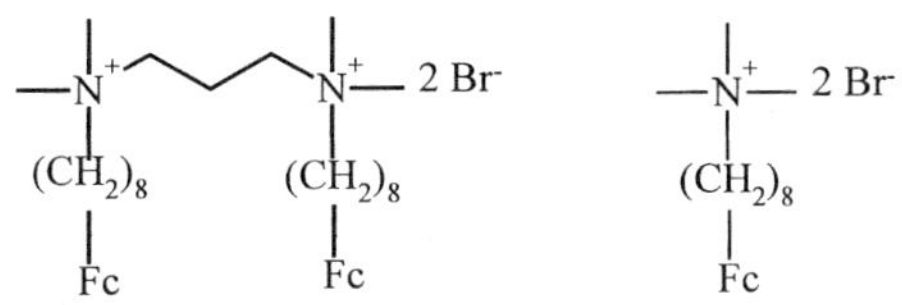

	cmc^a (M)	γ_{cmc} (mN/m)	A (Å^2)	cmc^a (M)	γ_{cmc} (mN/m)	A (Å^2)
Fc	$1–2 \times 10^{-5}$	50	39	2×10^{-3}	53	55
Fc^+	$>2 \times 10^{-3}$	<60	50	$>10^{-2}$	<60	65

[a]Surface tension (Wilhelmy plate) in 0.1 M Li_2SO_4 at pH 2 and $T = 25°C$ [99].

TABLE 27

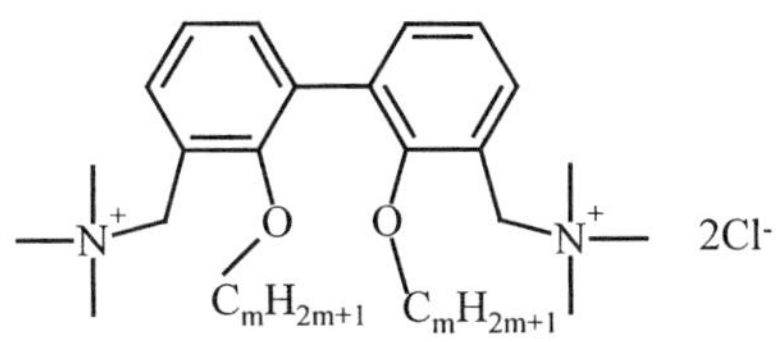

m	cmc^a (M)	$\gamma_{cmc}{}^a$ (mN/m)	A^a (Å^2)	α^b
1	7.23×10^{-2}	46.1	345	
6	9.4×10^{-3}	43.6	325	.73
12	1.9×10^{-4}	41.1	329	.56

[a]Surface tension (Wilhelmy plate) at 25°C.
[b]Conductimetry at 25°C.
Source: Ref. 92.

TABLE 28

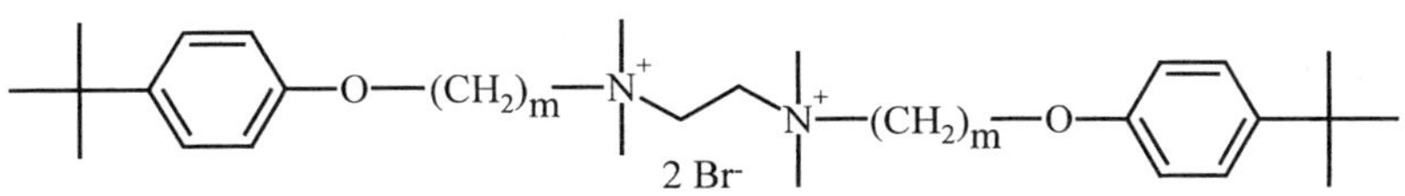

m	2	3	4	5	6	10
cmc (M)	7.0×10^{-3}	5.9×10^{-3}	4.2×10^{-3}	3.1×10^{-3}	2.1×10^{-3}	1.1×10^{-3}

Note: Conductimetry at 20°C [145].

B. Anionic Gemini Surfactants

1. Sulfates

TABLE 29

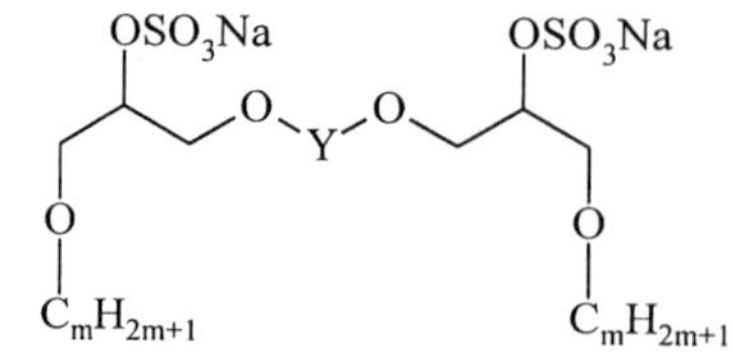

Y	$m = 8$ cmc (M)		$m = 10$ cmc (M)	
	cmc (M)	γ_{cmc} (mN/m)	cmc (M)	γ_{cmc} (mN/m)
CH_2CH_2	6.0×10^{-4}	29.2	1.3×10^{-5}	27.0
$CH_2CH_2CH_2CH_2$	1.5×10^{-3}	32.0	6.0×10^{-4}	32.5
$CH_2[CH_2OCH_2]CH_2$	1.8×10^{-4}	32.5	3.2×10^{-5}	32.0
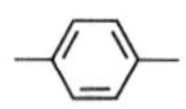	3.0×10^{-4}	41.8	3.5×10^{-5}	38.8
	1.7×10^{-4}	42.0	3.5×10^{-5}	39.5

Note: Surface tension (Wilhelmy plate) at 20°C [52].

TABLE 30

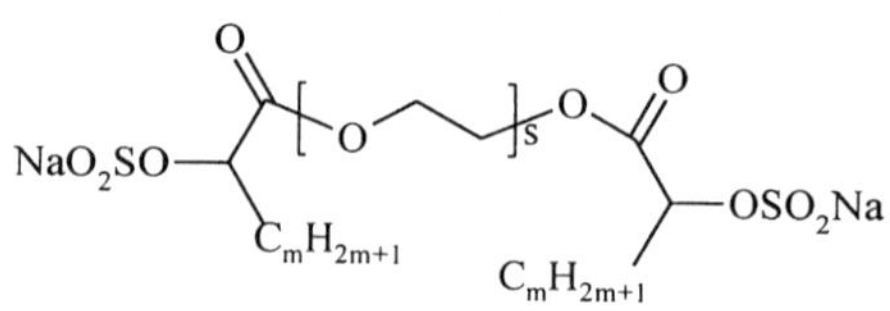

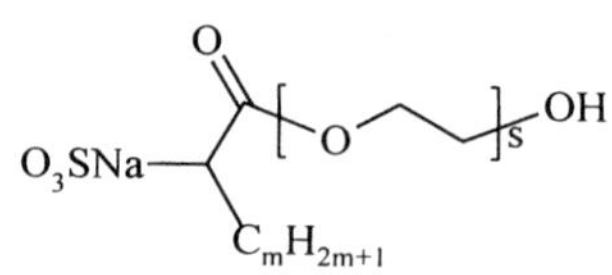

s	cmc (M) $m = 10$	$m = 12$	$m = 14$	$m = 16$	$m = 10$	$m = 12$	$m = 14$	$m = 16$
1	9.9×10^{-4}	2.5×10^{-4}	7.2×10^{-5}	1.7×10^{-5}	5.5×10^{-3}	1.0×10^{-3}	3.0×10^{-4}	7×10^{-5}
2	1.1×10^{-3}	2.7×10^{-4}	7.5×10^{-5}	1.8×10^{-5}	6.5×10^{-3}	1.3×10^{-3}	3.2×10^{-4}	8×10^{-5}
3	1.1×10^{-3}	2.9×10^{-4}	7.8×10^{-5}	1.9×10^{-5}	7.0×10^{-3}	1.7×10^{-3}	3.5×10^{-4}	9×10^{-5}
5	1.2×10^{-3}	3.2×10^{-4}	8.5×10^{-5}	2.1×10^{-5}	7.6×10^{-3}	1.9×10^{-3}	3.7×10^{-4}	1.0×10^{-4}
9	1.3×10^{-3}	3.4×10^{-4}	8.5×10^{-5}	2.2×10^{-5}	9.0×10^{-3}	2.1×10^{-3}	4.5×10^{-4}	1.1×10^{-4}
14	1.4×10^{-3}	3.6×10^{-4}	9.0×10^{-5}	2.3×10^{-5}	1.00×10^{-2}	2.2×10^{-3}	4.7×10^{-4}	1.2×10^{-4}
23	1.5×10^{-3}	3.7×10^{-4}	9.6×10^{-5}	2.5×10^{-5}	1.01×10^{-2}	2.4×10^{-3}	4.8×10^{-4}	1.3×10^{-4}
35	1.5×10^{-3}	3.8×10^{-4}	9.8×10^{-5}	2.6×10^{-5}	1.01×10^{-2}	2.4×10^{-3}	5.0×10^{-4}	1.3×10^{-4}

Note: Fluorescence spectroscopy of pinacyanol chloride at 25°C [62].

2. Sulfonates

TABLE 31

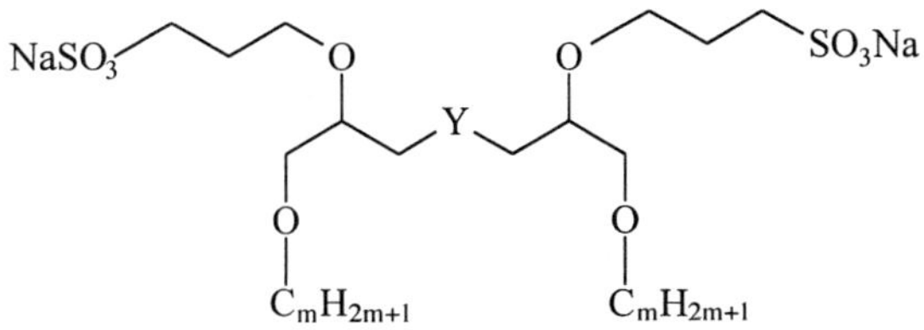

Y	m	cmc[a] (M)	γ_{cmc}[a] (mN/m)
	10	3.3×10^{-5}	28.0
—O—	12	1.4×10^{-5}	30.0
	14	2.5×10^{-5}	37.5
—OCH_2CH_2O—	8[b]	7.0×10^{-4}	29.2
	10	3.2×10^{-5}	30.0
—$O[CH_2CH_2O]_2$—	10	6.0×10^{-5}	36.0
—$O[CH_2CH_2O]_3$—	10	8.0×10^{-5}	35.0
—$O[CH_2]_4O$—	10	1.0×10^{-4}	36.0
—OCH2CH=CHCH2O—[c]	10	2.5×10^{-5}	33.0

[a]Surface tension (Wilhelmy plate) at 20°C from Ref. 54 unless otherwise specified.
[b]From Ref. 51.
[c]From Ref. 97.

TABLE 32

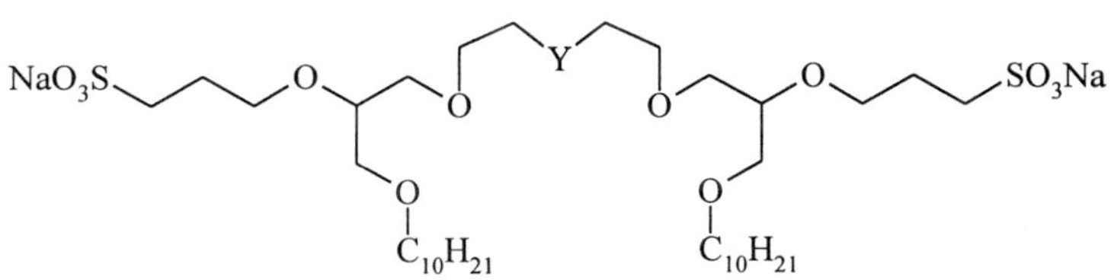

Y	In deionized water cmc (M)	γ_{cmc} (mN/m)	pC_{20}	In 0.1 M NaCl cmc (M)	γ_{cmc} (mN/m)	A (Å^2)
O	6.0×10^{-5}	36	4.3	2.4×10^{-5}	40	115
S	4.4×10^{-5}	34	5.7	2.0×10^{-5}	35	112
SO	1.1×10^{-4}	36	4.8	5.6×10^{-5}	41	116
SO_2	2.0×10^{-4}	39	4.4	6.0×10^{-5}	39	132

Note: Surface tension (Wilhelmy plate) at 20°C [57].

TABLE 33

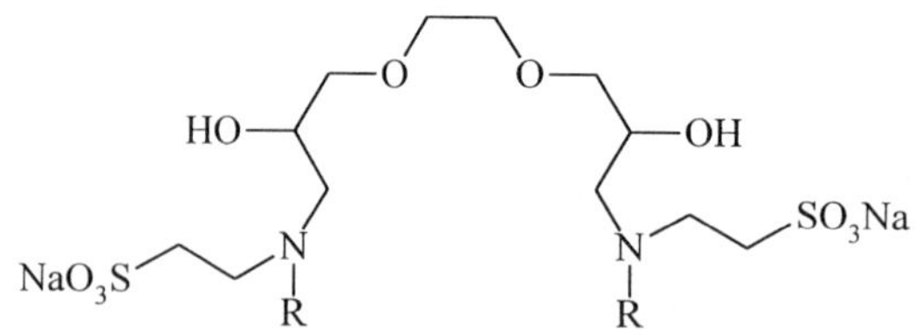

m	cmc (M)	γ_{cmc} (mN/m)	pC_{20}
8	1.9×10^{-3}	37.5	3.7
10	8.0×10^{-5}	37.0	5.2
12	6.1×10^{-6}	33.0	5.9
Monomer	1.0×10^{-2}	35.5	2.8

Note: Surface tension (Wilhelmy plate) at 20°C [56].

TABLE 34

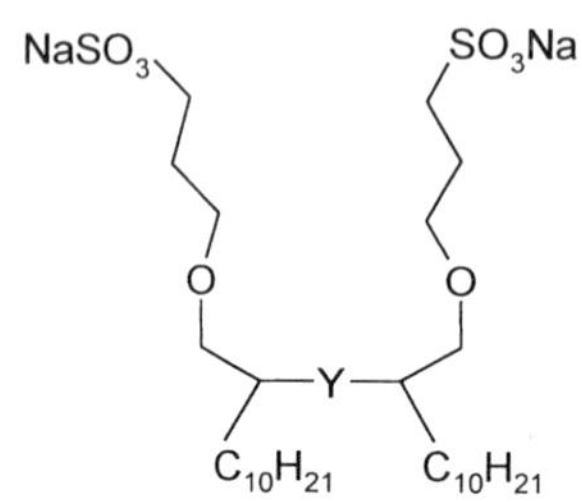

Y	cmc (M)	γ_{cmc} (mN/m)	pC_{20}
—O—	9.3×10^{-5}	31.4	8.6
—$(OCH_2CH_2)_3$O—	4.8×10^{-5}	35.5	6.5

Note: Surface tension (Wilhelmy plate) at 23°C [65].

3. Phosphates

TABLE 35

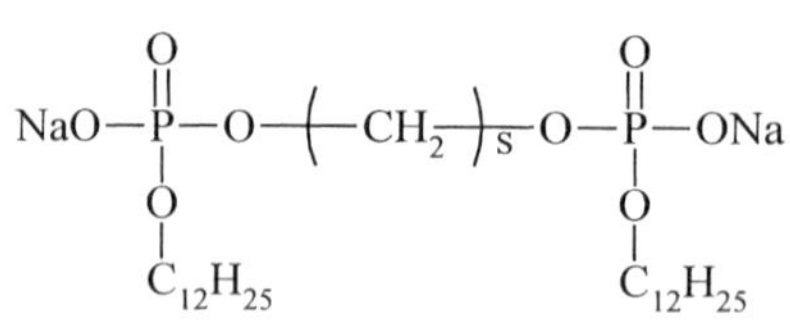

s	cmc[a] (M)	α[a]	cmc[b] (M)	cmc[c] (M)	cmc[d] (M)
6	3.5×10^{-4}	.84		1.9×10^{-4}	[2×10^{-4}, 5×10^{-4}]
8	4.8×10^{-4}	.62	1.5×10^{-4}	1.0×10^{-4}	[8×10^{-5}, 3×10^{-4}]
12	6.0×10^{-5}	.88	4.4×10^{-5}	3.5×10^{-5}	[2×10^{-5}, 5×10^{-5}]

[a]Conductimetry.
[b]Fluorescence pyrene.
[c]UV spectroscopy (pinacyanol chloride).
[d]Range of concentration from titration microcalorimetry at 30°C.
Source: Ref. 67.

TABLE 36

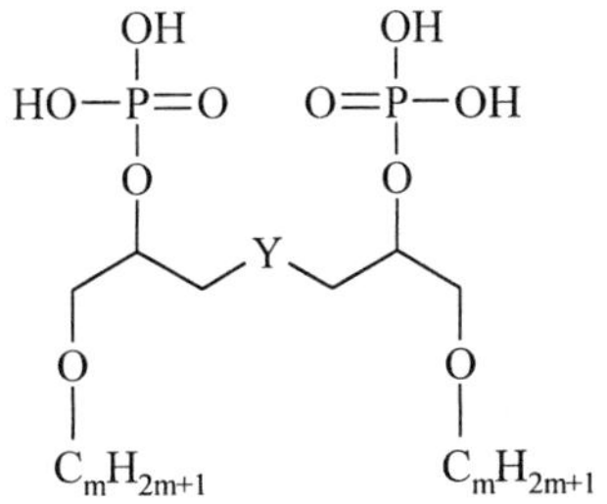

	Y	m = 8		m = 10	
		cmc (M)	γ_{cmc} (mN/m)	cmc (M)	γ_{cmc} (mN/m)
Disodium	$—OCH_2CH_2O—$	7.2×10^{-4}	32.0	1.2×10^{-4}	30.0
	$—O[CH_2CH_2O]_2—$	5.2×10^{-4}	30.5	1.3×10^{-4}	31.5
	$—O[CH_2CH_2O]_3—$	1.8×10^{-3}	32.0	1.6×10^{-4}	33.0
Tetrasodium	—O—	4.0×10^{-3}	36.0	8.5×10^{-4}	30.0
	$—OCH_2CH_2O—$	3.3×10^{-3}	41.0	3.6×10^{-4}	32.5
	$—O[CH_2CH_2O]_2—$	1.8×10^{-3}	42.0	3.1×10^{-4}	33.5
	$—O[CH_2CH_2O]_3—$	4.3×10^{-3}	38.5	3.6×10^{-4}	33.5

Note: Surface tension (Wilhelmy plate) at 20°C [53].

TABLE 37

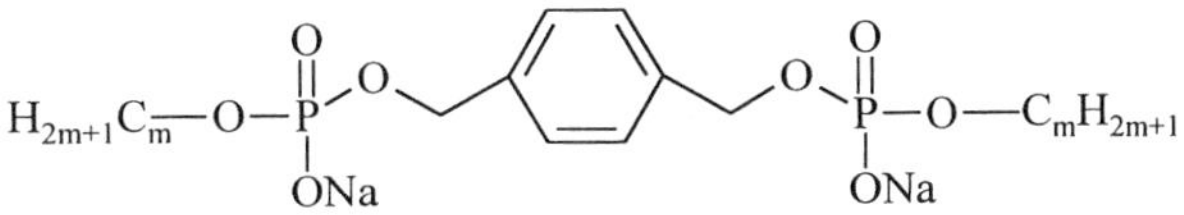

m	cmc[a] (M)	γ_{cmc}[a] (mN/m)	cmc[b] (M)	cmc[c] (M)	γ_{cmc}[c] (mN/m)
8	5.5×10^{-3}	42	1.6×10^{-2}	4.1×10^{-3}	42
12	1.7×10^{-4}	43	2.5×10^{-4}	1.1×10^{-4}	41
16	—	—	3.3×10^{-4}	2.6×10^{-3}	38

[a]Surface tension (ring method) at 23°C.
[b]^{23}Na NMR at 23°C.
[c]Surface tension (ring method) at 50°C.
Source: Ref. 39.

TABLE 38

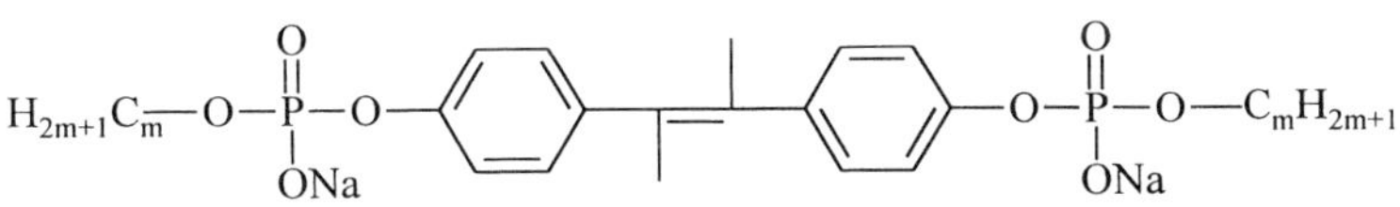

m	cmc[a] (M)	γ_{cmc}[a] (mN/m)	cmc[b] (M)	cmc[c] (M)	γ_{cmc}[c] (mN/m)
12	2.7×10^{-4}	46	4.8×10^{-4}	1.2×10^{-4}	45
16	1.6×10^{-3}	54	1.9×10^{-4}	7.1×10^{-4}	37
20	1.8×10^{-3}	60	2.5×10^{-4}	9.2×10^{-4}	40

[a]Surface tension (ring method) at 23°C.
[b]^{23}Na NMR at 23°C.
[c]Surface tension (ring method) at 50°C.
Source: Ref. 39.

4. Carboxylates

TABLE 39

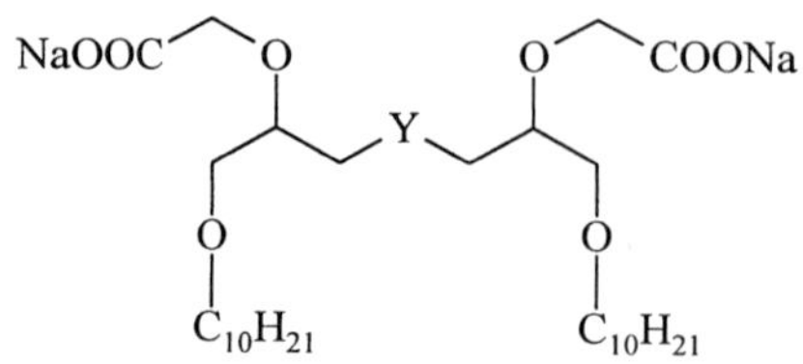

Y	In deionized water cmc (M)	γ_{cmc} (mN/m)	pC_{20}	In 0.1 M NaCl cmc (M)	γ_{cmc} (mN/m)	A (Å^2)
—O—	8.4×10^{-5}	30	5.4	9.8×10^{-6}	30	82
—OCH_2CH_2O—	1.6×10^{-4}	33	5.0	1.4×10^{-5}	35	94
—$O[CH_2CH_2O]_2$—	2.6×10^{-4}	39	4.4	2.5×10^{-5}	37	106
—$O[CH_2CH_2O]_3$—	3.7×10^{-4}	43	4.1	3.0×10^{-5}	39	110
—$O[CH_2]_4O$—	1.0×10^{-3}	36	4.1			

Note: Surface tension (Wilhelmy plate) at 20°C [59].

TABLE 40

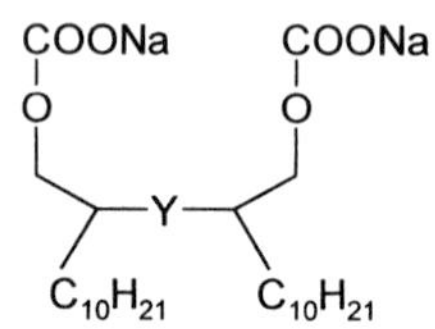

Y	cmc (M)	γ_{cmc} (mN/m)	pC_{20}
—O—	1.3×10^{-5}	26.4	5.4
—$(OCH_2CH_2)_3O$—	8.0×10^{-6}	38.7	4.4

Note: Surface tension (Wilhelmy plate) at 23°C [65].

C. Nonionic Gemini Surfactants

TABLE 41

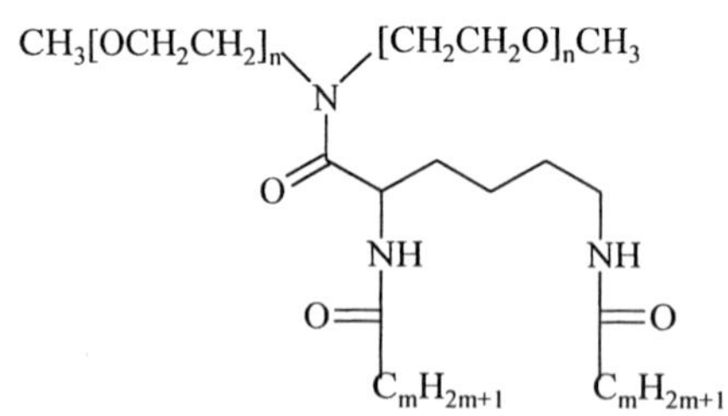

n-*m*	cmc (M)	γ_{cmc} (mN/m)	A (Å^2)
2-5	5.7×10^{-2a}	35.5	103
2-7	5.8×10^{-3a}	32.3	96
2-9	8.8×10^{-4b}	32.9	79
3-7	7.2×10^{-3b}	34.8	103
3-9	9.7×10^{-4b}	30.8	85
	6.6×10^{-5a}		

Note: Surface tension at [a] 25°C and [b] 6°C [73].

TABLE 42

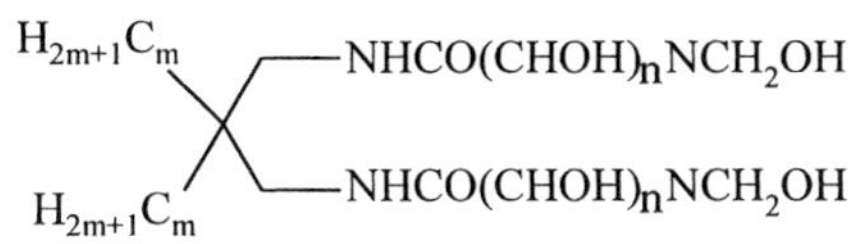

m-n	cmc (M)	cmc (wt%)	γ_{cmc} (mN/m)	pC_{20}	*A* (Å^2)	KT
5-3		0.29	—		66	
5-4	1.1×10^{-2}	0.709	36.8		81	
6-3		0.08	—	4.5		
6-4	1.4×10^{-3}	0.159	32.6	4.3	76	
7-4	1.6×10^{-4}	0.013	28.6		68	35
7-5		0.018	—	5.4	76	29
8-4	4.2×10^{-5}	0.0012	26.2		72	

Note: Surface tension (du Nouy ring) at 35°C [82,83,163].

D. Oligomer Surfactants

1. Cationic Trimers

TABLE 43

	cmc (M)	α	γ_{cmc} mN/m	*A* (Å^2)	pC_{20}
12-2-12-2-12, 3Br[a]	8.0×10^{-5}	0.93	25.2	128	
12-3-12-3-12, 3Br[b]	1.4×10^{-4}	0.24	32	148	
12-3-12-3-12, 3Cl[c]	3.3×10^{-4}	0.34			
12-6-12-6-12, 3Br[b]	2.8×10^{-4}	0.30	44	248	
12-3-12-4-12-3-12, 4Br[b]	6.0×10^{-5}	0.20			
12-3*-12-3*-12, 3Cl[d]	9.6×10^{-6}		32		5.36
12-3*-1-3*-12, 3Cl[d]	4.6×10^{-4}		39		3.68
12-3*-12°-3*-12, 2Cl[d]	6.2×10^{-6}		35		5.60
12-3*-1°-3*-12, 2Cl[d]	9.9×10^{-4}		42		3.34

[a]Surface tension (Wilhelmy plate) at 25°C [107].
[b]Cmc by conductimetry at 25°C; surface tension measured by pendant drop method at 25°C [104].
[c]Conductimetry at 25°C [103].
[d]Surface tension (Wilhelmy plate) at 20°C [108]; 3*, hydroxypropylene spacer; m°, tertiary amine headgroup.

2. Anionic Trimers

TABLE 44

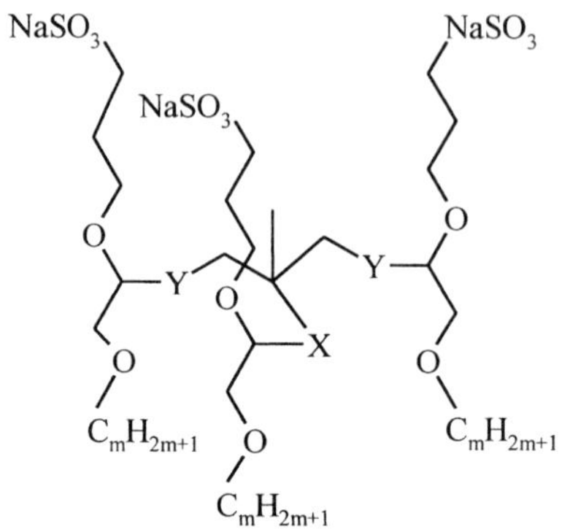

Y	X	m	cmc (M)	γ_{cmc} (mN/m)	pC_{20}
—O—	—CH_2O—[a]	10	6.8×10^{-6}	31.5	7.4
—O—	—CH_2O—[a]	12	5.0×10^{-5}	33.0	5.8
—O—	—CH_2O—[a]	14	2.5×10^{-4}	34.0	4.8
—OCH_2CH_2O—	—$CH_2OCH_2CH_2O$—[b]	10	8.0×10^{-6}	33.5	7.1
—OCH_2CH_2O—	—$CH_2OCH_2CH_2O$—[b]	12	2.7×10^{-5}	34.0	6.4
—OCH_2CH_2O—	—OCH_2CH_2O—[b]	10	1.0×10^{-5}	31.0	6.5
—OCH_2CH_2O—	—OCH_2CH_2O—[b]	12	1.8×10^{-5}	32.5	7.5

[a]Surface tension (Wilhelmy plate) at 20°C. m = 16 compound is insoluble even in hot water [110].
[b]Surface tension (Wilhelmy plate) at 20°C [111].

3. Unequal Number of Ionic Headgroups and Alkyl Chains

TABLE 45

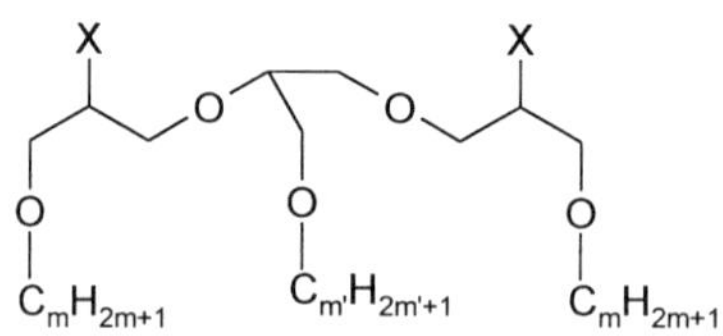

X	m	m'	cmc	γ_{cmc} mN/m	pC_{20}
CH2COONa	10	10	4.0×10^{-5}	29	5.7[a]
OSO3Na	10	10	9.0×10^{-6}	27	6.7[a]
	8	1	8.5×10^{-4}	36.5	3.9
	10	1	8.1×10^{-5}	36	5.0
$OCH_2CH_2CH_2SO_3Na$	8	8	4.6×10^{-5}	29	5.6
	10	8	1.6×10^{-5}	28	6.3[a]
	8	10	1.9×10^{-5}	28	6.4[a]
	10	10	1.4×10^{-5}	28	6.6[a]

[a]Extrapolated values.
Note: Surface tension (Wilhelmy plate) at 20°C [112].

TABLE 46

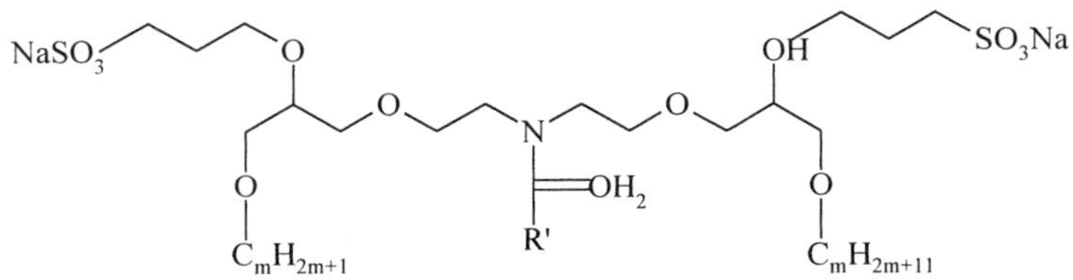

	m = 8			m = 10		
R′	Cmc (M)	γ_{cmc} (mN/m)	pC_{20}	CmC (M)	γ_{cmc} (mN/m)	pC_{20}
CH_3	5.2×10^{-4}	33.5	4.2	4.3×10^{-5}	32.5	5.3
C_9H_{19}	3.5×10^{-5}	31.0	5.6	8.0×10^{-6}	28.0	6.7[a]
$C_{11}H_{23}$	1.2×10^{-5}	30.5	6.5[a]	7.2×10^{-6}	27.5	6.9[a]

[a]Extrapolated values.
Note: Surface tension (Wilhelmy plate) at 20°C [55].

REFERENCES

1. R. Zana, in *Novel Surfactants. Preparation, Applications and Biodegradability* (K. Holmberg, ed.), Marcel Dekker, New York, 1998, pp. 241–277.
2. P. Alexandridis, Curr. Opin. Colloid Interface Sci. *1*: 490–501 (1996); Th. F. Tadros, Annu. Surfactants Rev. *1*:179–204 (1998).
3. Patents before 1990: Cationic: R. B. McConnell, U.S. Patent 3,855,235 (1974); R. B. McConnell, U.S. Patent 3,887,476 (1975); R. B. Login, U.S. Patent 4,764,306 (1988); R. B. Login, U.S. Patent 4,734,277 (1988); R. B. Login, U.S. Patent 4,812,263 (1989); R. K. Chauhuri, D. J. Tracy, and R. B. Login, U.S. Patent 4,886,890 (1989). Anionic: French Patent 1,055,420 (1954); French Patent 1,464,243 (1966); G. Reitz, and G. Boehmke, Great Britain Patent 1,503,280 (1978). Nonionic: F. Bersworth, U.S. Patent 2,524,218 (1950); F. Bersworth, U.S. Patent 2,530,147 (1950); F. Bersworth, U.S. Patent 2,532,391 (1950); Schmitz, G.B. Patent 1,149,140 (1967).
4. M. J. Rosen, Chemtech. *23*:30–33 (1993).
5. R. Zana, Curr. Opin. Colloid Interface Sci. *1*:566–571 (1996).
6. Y. Nakatsuji and I. Ikeda, Chim. Oggi/Chem. Today 40–43 (1997).
7. R. Zana, in *Specialist Surfactants* (I. D. Robb, ed.), Chapman & Hall, London, 1997, pp. 81–103.
8. E. Fisicaro, Cell. Mol. Biol. Lett. *2*(Suppl. 1):45–63 (1997).
9. E. Fisicaro, B. Róøycka-Roszack, G. Viscardi, and P. L. Quagliotto, Curr. Top. Colloid Interface Sci. 2:53–68 (1997).
10. R. Zana, in *Structure-Performance Relationships in Surfactants* (K. Esumi and M. Ueno, eds), Marcel Dekker, New York, 1997, pp. 255–283.
11. M. J. Rosen, Cosmetics Toiletries *113*:49–55 (1998).
12. M. J. Rosen and D. J. Tracy, J. Surfactants Deterg. *1*: 547–554 (1998).
13. E. N. Vulfson, Lipid Technol. *11*:31–36 (1999).
14. A. J. Wysocki and D. Taber, in *Cationic Surfactants* (E. Jungermann, ed.), Marcel Dekker, New York, 1970, pp. 71–146.
15. M. Katzman, U.S. Patent 2,217,683 (1940).
16. H.-J. Engelbert, U.S. Patent 2,445,319 (1948).
17. M. A. T. Rogers, U.S. Patent 2,386,141 (1945).
18. M. A. T. Rogers, U.S. Patent 2,386,143 (1945).
19. K. Thomae, German Patent 914387 (1963).
20. K. Thomae, U.S. Patent 3,121,088 (1964).
21. T. Imam, F. Devínsky, I. Lacko, D. Mlynarčík, and L. Krasnec, Pharmazie *38*:308–310 (1983).
22. R. Zana and Y. Talmon, Nature *362*:228–231 (1993).
23. F. M. Menger and C. A. Littau, J. Am. Chem. Soc. *113*:1451–1452 (1991).
24. Patents after 1989: Cationic: J. Li, M. Dahanayake, R. L. Reierson, and D. J. Tracy, U.S. Patent 5,643,498 (1997). Anionic: A. Behler, R. Piorr, and M. Schaefer, U.S. Patent 4,936,551 (1990); M. Okahara and A. Masuyama, U.S. Patent 5,160,450 (1992); B. Gruber, Ger. Offen. DE 4,232,414 A1 (1994); F. Wangemann, Ger. Offen. DE 4,321,022 A1 (1995); H. C. Raths and W. E. Noack, Ger. Offen. DE 4,401,565 (1995); R. J. Kaiser, U.S. Patent 5,487,778 (1996); R. J. Kaiser, U.S. Patent 5,507,863 (1996); R. J. Kaiser, U.S. Patent 5,599,933 (1997); R. Varadaraj and S. Zushma, U.S. Patent 5,493,050 (1996); R. Varadaraj and S. Zushma, U.S. Patent 5,585,516 (1996); T. Okano, M. Fukuda, J. Tanabe, M. Ono, Y. Akabane, H. Takahashi, N. Egawa, T. Sakatoni, H. Kanao, and Y. Yoneyanna,

U.S. Patent 5,681,803 (1997); H. C. Raths, Ger. Offen. DE 19,622,612 (1997); D. J. Tracy, J. Li, and J. M. Ricca, U.S. Patent 5,710,121 (1998); T. Kitsubi, M. Uno, K. Kita, Y. Fujikura, A. Nakano, M. Tosaka, K. Yahagi, S. Tamura, and K. Maruta, U.S. Patent 5,714,457 (1998). Nonionic: C. B. A. Briggs and A. R. Pitt, U.S. Patent 4,892,806 (1990); R. Corelli-Calvet, F. Brisset, J. Rivo, A. Lattes, and L. Godefroy, U.S. Patent 5,403,922 (1995); K. Adams, Eur. Patent Appl. EPO 688781 (1995); J. Scheibel, D. S. Connor, and E. Y. Fu, U.S. Patent 5,534,197 (1996); D. S. Connor, E. Y. Fu, and J. Scheibel, U.S. Patent 5,512,699 (1996); K. Tsubone, H. Nishio, and M. Kusumaru, Jpn. Kokai Tokkyo Koho JP 08,291,040 (1996); K. Tsubone, H. Nishio, and M. Kusumaru, Jpn. Kokai Tokkyo Koho JP 08,319,262 (1996); S. Wong, U.S. Patent 5,622,938 (1997). Zwitterionic: J. Li, M. Dahanayake, R. L. Reierson, and D. J. Tracy, U.S. Patent 5,643,498 (1997); A. Nakamo, T. Kitsuki, K. Kita, and M. Asuga, International Patent PCT WO 96/01800 (1996); K. Kwetkat, International Patent PCT WO 97/31890 (1997).

25. A. Haces and V. C. Ciccarone, International Patent PCT WO 95/17373 (1995).
26. E. Unger, D. Shen, and G. Wu, International Patent PCT WO 96/26179 (1996).
27. R. K. Scheule and S. H. Cheng, in *Artificial Self-Assembling Systems for Gene Delivery* (P. L. Felgner, M. J. Heller, P. Lehn, J. P. Behr, and F. C. Szocka, eds.), *ACS Conference Proceedings Series*, American Chemical Society, Washington DC, 1995, pp. 177–190.
28. D. Erne, N. Stojanac, D. Ammann, P. Hofstetter, E. Pretsch, and W. Simon, Helv. Chim. Acta *63*:2271–2279 (1980); F. Lanter, D. Erne, D. Ammann, and W. Simon, Anal. Chem. *52*:2400–2402 (1980).
29. G. Brand, M. W. Hosseini, and R. Ruppert, Helv. Chim. Acta *75*:721–728 (1992).
30. A. Laschewsky, Adv. Polym. Sci. *124*:1–86 (1995).
31. C. Tanford, *The Hydrophobic Effect: Formation of Micelles and Biological Membranes*, John Wiley & Sons, New York, 1973.
32. J. Israelachvili, D. J. Mitchell, and B. W. Ninham, J. Chem. Soc. Faraday Trans. 2 1525–1568 (1976).
33. J. C. H. Hwa, U.S. Patent 2,933,529 (1960).
34. C. A. Bunton, L. Robinson, J. Schaak, and M. F. Stam, J. Org. Chem. *36*:2346–2350 (1971).
35. I. Lacko, F. Devínsky, D. Mlynarčík, and Ľ. Krasnec, Acta Fac. Pharm. Univ. Comenianae *30*:109 (1977).
36. T. Imam, F. Devínsky, I. Lacko, D. Mlynarčík, and Ľ. Krasnec, Pharmazie 5 *38*:308–310 (1983).
37. F. Devínsky, L. Masárová, I. Lacko, and D. Mlynarčík, Collect. Czech. Chem. Commun. *49*:2819–2827 (1984).
38. R. Zana, M. Benrraou, and R. Rueff, Langmuir *7*: 1072–1075 (1991).
39. F. M. Menger and C. A. Littau, J. Am. Chem. Soc. *115*:10083–10090 (1993).
40. Th. Dam, J. B. F. N. Engberts, J. Karthäuser, S. Karaborni, and N. M. van Os, Colloid Surf. A *118*:41–49 (1996).
41. S. De, V. K. Aswal, P. S. Goyal, and S. Bhattacharya, J. Phys. Chem. *100*:11664–11671 (1996).
42. B. Różycka-Roszak, S. Witek, and S. Przestalski, J. Colloid Interface Sci. *131*:181–185 (1989).
43. S. Pegiadou-Koemtzopoulou, V. Papazoglou, and A. H. Kehayoglou, J. Surfactants Deterg. *1*:73–76 (2000).
44. T.-S. Kim, T. Hirao, and I. Ikeda, J. Am. Oil Chem. Soc. *73*:67–71 (1996).
45. T.-S. Kim, T. Kida, Y. Nakatsuji, T. Hirao, and I. Ikeda, J. Am. Oil Chem. Soc. *73*:907–911 (1996).
46. M. J. Rosen and L. Liu, J. Am. Oil Chem. Soc. *73*: 885–890 (1996).
47. H. C. Parreira, E. R. Lukenbach, and M. K. O. Lindemann, J. Am. Oil Chem. Soc. *56*:1015–1021 (1979).
48. S. De, V. K. Aswal, P. S. Goyal, and S. Bhattacharya, J. Phys. Chem B *102*:6152–6160 (1998).
49. M. Dreja, S. Gramberg, and B. Tieke, Chem. Commun. 1371–1372 (1998).
50. E. Girod, U.S. Patent 2,759,020 (1956).
51. M. Okahara, A. Masuyama, Y. Sumida, and Y.-P. Zhu, J. Jpn. Oil Chem. Soc. *37*:746–748 (1988).
52. Y.-P. Zhu, A. Masuyama, and M. Okahara, J. Am. Oil Chem. Soc. *67*:459–463 (1990).
53. Y.-P. Zhu, A. Masuyama, and M. Okahara, J. Am. Oil Chem. Soc. *68*:268–271 (1991).
54. Y.-P. Zhu, A. Masuyama, T. Nagata, and M. Okahara, J. Jpn. Oil Chem. Soc. *40*:473–477 (1991).
55. Y.-P. Zhu, A. Masuyama, Y.-I. Kirito, and M. Okahara, J. Am. Oil Chem. Soc. *68*:539–543 (1991).
56. A. Masuyama, T. Hirono, Y.-P. Zhu, M. Okahara, and M. J. Rosen, J. Jpn. Oil Chem. Soc. *41*:301–305 (1992).
57. Y.-P. Zhu, A. Masuyama, Y. Nakatsuji, and M. Okahara, J. Jpn. Oil Chem. Soc. *42*:86–94 (1993).
58. Y.-P. Zhu, K. Ishahara, A. Masuyama, Y. Nakatsuji, and M. Okahara, J. Jpn. Oil Chem. Soc. *42*:161–167 (1993).
59. Y.-P. Zhu, A. Masuyama, Y. Kobata, Y. Nakatsuji, M. Okahara, and M. J. Rosen, J. Colloid Interface Sci. *158*:40–45 (1993).
60. X.-P. Gu, I. Ikeda, and M. Okahara, Synthesis 649–651 (1985).
61. Y. Sumida, A. Masuyama, T. Oki, T. Kida, Y. Nakatsuji, I. Ikeda, and M. Nojima, Langmuir *12*:3986–3990 (1996).
62. T. Okano, N. Egawa, M. Fujiwara, and M. Fukuda, J. Am. Oil Chem. Soc. *73*:31–37 (1996).
63. T. Okano, J. Tanabe, M. Fukuda, and M. Tanaka, J. Am. Oil Chem. Soc. *69*:44–46 (1992).
64. A. van Zon, J. T. Bouman, H. H. Deuling, S. Karaborni, J. Karthaeuser, H. T. G. A. Mensen, and N. M. van Os, Tenside Surfactants Deterg. *36*:84–86 (1999).

65. P. Renouf, D. Hebrault, J.-R. Desmurs, J.-M. Mercier, C. Mioskowski, and L. Lebeau, Chem. Phys. Lipids *99*:21–32 (1999).
66. K. Aratani, T. Oida, T. Shimizu, and Y. Hayashi, Comun. J. Com. Esp. Deterg. *28*:45–56 (1998).
67. F. L. Duivenvoorde, M. C. Feiters, S. J. van der Gaast, and J. B. F. N. Engberts, Langmuir *13*:3737–3743 (1997).
68. R. A. Bauman, Synthesis 870–872 (1974).
69. H. Eibl, J. O. McIntyre, E. A. M. Fleer, and S. Fleischer, Methods Enzymol. *98*:623–632 (1983).
70. V. K. Aswal, S. De, P. S. Goyal, S. Bhattacharya, and R. K. Heenan, Phys. Rev. E. *59*:3116–3122 (1999).
71. J.-M. Kim and D. H. Thompson, Langmuir *8*:637–644 (1992).
72. C. Helbig, H. Baldauf, T. Lange, R. Neumann, R. Pollex, and E. Weber, Tenside Surfactants Deterg. *36*:58–62 (1999).
73. J. Seguer, C. Selve, M. Allouch, and M. R. Infante, J. Am. Oil Chem. Soc. *73*:79–86 (1996).
74. M. R. Infante, J. Seguer, A. Pinazo, and M. P. Vinardell, J. Dispersion Sci. Technol. *20*:621–642 (1999).
75. M. R. Infante, J. J. Garcia Dominguez, P. Erra, M. R. Juliá, and M. Prats, Int. J. Cosmet. Sci. *6*:275–282 (1984).
76. L. Pérez, J. L. Torres, A. Manresa, C. Solans, and M. R. Infante, Langmuir *12*:5296–5301 (1996).
77. L. Pérez, A. Pinazo, M. J. Rosen, and M. R. Infante, Langmuir *14*:2307–2315 (1998).
78. A. Pinazo, X. Wen, L. Pérez, M. R. Infante, and E. I. Franses, Langmuir *15*:3134–3142 (1999).
79. K. Jennings, I. Marshall, H. Birrell, A. Edwards, N. Haskins, O. Sodermann, A. J. Kirby, and P. Camilleri, J. Chem. Soc. Chem. Commun. 1951–1952 (1998).
80. R. Valivety, I. S. Gill, and E. N. Vulfson, J. Surfactants Deterg. *1*:177–185 (1998).
81. J. Eastoe, P. Rogueda, B. J. Harrison, A. M. Howe, and A. R. Pitt, Langmuir *10*:4429–4433 (1994).
82. C. B. A. Briggs and A. R. Pitt, U.S. Patent 4,892,806 (1990).
83. C. B. A. Briggs, I. M. Newington and A. R. Pitt, J. Chem. Soc. Chem. Commun. 379–380 (1995).
84. J. M. Pestman, K. R. Terpstra, M. C. A. Stuart, H. A. van Doren, A. Brisson, R. M. Kellogg, and J. B. F. N. Engberts, Langmuir *13*:6857–6860 (1997).
85. M. J. L. Castro, J. Kovensky, and A. Fernandez Cirelli, Tetrahedron Lett. *38*:3995–3998 (1997).
86. C. Gao, A. Millqvist-Fureby, M. J. Whitcombe, and E. N. Vulfson, J. Surfactants Deterg. *2*:293–302 (1999).
87. R. Oda, I. Huc, and S. J. Candau, J. Chem. Soc. Chem. Commun. 2105–2106 (1997).
88. D. A. Jaeger, B. Li, and T. Clark, Jr., Langmuir *12*: 4314–4316 (1996).
89. P. Renouf, C. Mioskowski, L. Lebeau, D. Hebrault, and J.-R. Desmurs, Tetrahedron Lett. *39*:1357–1360 (1998).
90. I. Huc and R. Oda, J. Chem. Soc. Chem. Commun. 2025–2026 (1999).
91. N. A. J. M. Sommerdijk, T. H. L. Hoeks, M. Synak, M. C. Feiters, R. J. M. Nolte, and B. Zwanenburg, J. Am. Chem. Soc. *119*:4338–4344 (1997).
92. T. Takemura, N. Shiina, M. Izumi, K. Nakamura, M. Miyazaki, K. Torigoe, and K. Esumi, Langmuir *15*: 646–648 (1999).
93. A. Pinazo, M. Diz, C. Solans, M. A. Pés, P. Era, and M. R. Infante, J. Am. Oil Chem. Soc. *70*:37–42 (1993).
94. D. Ono, T. Tanaka, A. Masuyama, Y. Nakatsuji, and M. Okahara, J. Jpn. Oil Chem. Soc. *42*:10–16 (1993).
95. D. A. Jaeger, S. G. G. Russell, and H. Shinozaki, J. Org. Chem. *59*:7544–7548 (1994).
96. D. A. Jaeger and E. L. G. Brown, Langmuir *12*:1976–1980 (1995).
97. A. Masuyama, C. Endo, S. Takeda, and M. Nojima, Chem. Commun. 2023–2024 (1998).
98. S. M. Krisovitch and S. L. Regen, J. Am. Chem. Soc. *114*:9828 (1992); S. M. K. Davidson and S. L. Regen, Chem. Rev. *97*:1269–1279 (1997).
99. B. S. Gallardo and N. L. Abbott, Langmuir *13*:203–208 (1997).
100. T. Saji, K. Hoshino, Y. Ishii, and M. Goto, J. Am. Chem. Soc. *113*:450–456 (1991).
101. A. Guyot, in *Novel Surfactants. Preparation, Applications and Biodegradability* (K. Holmberg, ed.), Marcel Dekker, New York, 1998, pp. 301–322.
102. K. Goldann and A. Kirstahler, German Patent 1 086 709 (1958).
103. R. Zana, H. Lévy, D. Papoutsi, and G. Beinert, Langmuir *11*:3694–3698 (1995).
104. M. In, V. Bec, O. Aguerre-Chariol, and R. Zana, Langmuir *16*:141–148 (2000).
105. A. G. Giumanini, G. Chiavari, and F. L. Scarponi, Z. Naturforsch. *30b*:820–821 (1975).
106. R. W. Alder, D. Colclough, and R. W. Mowlam, Tetrahedron Lett. *32*:7755–7758 (1991).
107. K. Esumi, K. Taguma, and Y. Koide, Langmuir *12*: 4039–4041 (1996).
108. T.-S. Kim, T. Kida, Y. Nakatsuji, and I. Ikeda, Langmuir *12*:6304–6308 (1996).
109. T. Kida, M. Yokota, A. Masuyama, Y. Nakatsuji, and M. Okahara, Synthesis 487–489 (1993).
110. A. Masuyama, M. Yokota, Y.-P. Zhu, T. Kida, and Y. Nakatsuji, J. Chem. Soc. Chem. Commun. 1435–1436 (1994).
111. Y. Sumida, T. Oki, A. Masuyama, H. Maekawa, M. Nishiura, T. Kida, Y. Nakatsuji, I. Ikeda, and M. Nojima, Langmuir *14*:7450–7455 (1998).
112. Y.-P. Zhu, A. Masuyama, Y.-I. Kirito, M. Okahara, and M. J. Rosen, J. Am. Oil Chem. Soc. *69*:626–632 (1993).

113. D. J. Tracy, R. Li, and J. Yiang, U.S. Patent 5,846,926 (1998).
114. O. Regev and R. Zana, J. Colloid Interface Sci. *210*: 8–17 (1999).
115. M. J. Rosen, *Surfactant and Interfacial Phenomena*, John Wiley & Sons, New York, 1989.
116. L. D. Song and M. J. Rosen, Langmuir *12*:1149–1153 (1996).
117. M. J. Rosen, J. H. Mathias, and L. Davenport, Langmuir *15*:7340–7346 (1999).
118. B. Róøycka-Roszack, E. Fisicaro, and A. Ghiozzi, J. Colloid interface Sci. *184*:209–215 (1996).
119. F. Devínsky, I. Lacko, F. Bittererová, and L. Tomečková, J. Colloid Interface Sci. *114*:314–322 (1986).
120. F. Devínsky, Ľ. Masárová, and I. Lacko, J. Colloid Interface Sci. *105*:235–239 (1985).
121. E. Alami, G. Beinert, P. Marie, and R. Zana, Langmuir *9*:1465–1467 (1993).
122. M. Dreja, W. Pyckhout-Hintzen, H. Mays, and B. Tieke, Langmuir *15*:391–399 (1999).
123. Y. F. Deinega, Z. R. Ul'berg, L. G. Marochko, V. P. Rudi, and V. P. Denisenko, *Kolloidn. Zh. 36*:649–653 (1974).
124. Z. X. Li, C. C. Dong, and R. K. Thomas, Langmuir *15*:4392–4396 (1999).
125. H. Diamant and D. Andelman, Langmuir *10*:2910–2916 (1994).
126. H. Diamant and D. Andelman, Langmuir *11*:3605–3606 (1995).
127. P. K. Maiti and D. Chowdhury, J. Chem. Phys. *109*: 5126–5133 (1998).
128. A. Tomlinson, T. N. Danks, D. M. Heyes, S. E. Taylor, and D. J. Moreton, Langmuir *13*:5881–5893 (1997).
129. E. J. Osburn, L.-K. Chau, S.-Y. Chen, N. Collins, D. F. O'Brien, and N. R. Armstrong, Langmuir *12*:4784–4796 (1996).
130. D. J. Cooke, J. R. Lu, E. M. Lee, R. K. Thomas, A. R. Pitt, E. A. Simister, and J. Penfold, J. Phys. Chem. *100*:10298–10303 (1996).
131. A. Espert, R. v. Klitzing, P. Poulin, A. Colin, R. Zana, and D. Langevin, Langmuir *14*:4251–4260 (1998).
132. V. Bergeron, Langmuir *13*:3474–3482 (1997).
133. C. Chorro, M. Chorro, O. Dolladille, S. Partyka, and R. Zana, J. Colloid Interface Sci. *199*:169–176 (1998).
134. M. Chorro, C. Chorro, O. Dolladille, S. Partyka, and R. Zana, J. Colloid Interface Sci. *210*:134–143 (1999).
135. K. Esumi, M. Goino, and Y. Koide, J. Colloid Interface Sci. *183*:539–545 (1996).
136. M. L. Fielden, P. M. Claesson, R. E. Verrall, Langmuir *15*:3924–3934 (1999).
137. K. Esumi, M. Goino, and Y. Koide, Colloids Surf. A *118*:161–166 (1996).
138. K. Esumi, Y. Takeda, M. Goino, K. Ishiduki, and Y. Koide, Langmuir *13*:2585–2587 (1997).
139. K. Esumi, S. Uda, M. Goino, K. Ishiduki, T. Suhara, H. Fukui, and Y. Koide, Langmuir *13*:2803–2807 (1997).
140. K. Esumi, M. Matoba, and Y. Yamanaka, Langmuir *12*:2130–2135 (1996).
141. S. Manne, J. P. Cleveland, H. E. Gaub, G. D. Stucky, and P. K. Hansma, Langmuir *10*:4409–4413 (1994).
142. S. Manne and H. E. Gaub, Science *270*:1480–1482 (1995).
143. S. Manne, T. E. Schäffer, Q. Huo, P. K. Hansma, D. E. Morse, G. D. Stucky, and I. A. Aksay, Langmuir *13*:6382–6387 (1997).
144. M. Frindi, B. Michels, H. Lévy, and R. Zana, Langmuir *10*:1140–1145 (1994).
145. F. Devínsky, I. Lacko, D. Mlynarčík, V. Račanský, and Ľ. Krasnec, Tenside Surfactants Deterg. *22*:10–15 (1985).
146. R. Zana and H. Lévy, Colloids Surf. A *127*:229–232 (1997).
147. R. Zana, Langmuir *12*:1208–1211 (1996).
148. N. Hattori, A. Yoshino, H. Okabayashi, and C. J. O'Connor, J. Phys. Chem. B *102*:8965–8973 (1998).
149. N. Hattori, H. Hirata, H. Okabayashi, and C. J. O'Connor, Colloid Polym. Sci. *277*:361–371 (1999).
150. J. Zhao, S. D. Christian, and B. M. Fung, J. Phys. Chem. B *102*:7613–7618 (1998).
151. P. K. Maiti and D. Chowdhury, Europhys. Lett. *41*: 183–188 (1998).
152. H. Hirata, N. Hattori, M. Ishida, H. Okabayashi, M. Frusaka, and R. Zana, J. Phys. Chem. *99*:17778–17784 (1995).
153. Z. R. Ul'berg and V. I. Podol'skaya, Kolloidn. Zh. *40*: 292–296 (1978).
154. R. Zana, S. Yiv, and K. M. Kale, J. Colloid Interface Sci. *77*:456–465 (1980).
155. R. Zana, J. Colloid Interface Sci. *78*:330–337 (1980).
156. R. Zana, in *Cationic Surfactants*, Marcel Dekker, New York, 1985, pp. 41–85.
157. M. A. Winnik and A. Yekta, Curr. Opin. Colloid Interface Sci. *2*:424–436 (1997) and references therein.
158. R. Zana, M. In, H. Lévy, and G. Duportail, Langmuir *13*:5552–5557 (1997).
159. F. Devínsky, I. Lacko, and T. Imam, J. Colloid Interface Sci. *143*:336–342 (1991).
160. M. Dreja and B. Tieke, Langmuir *14*:800–807 (1998).
161. K. M. Layn, P. G. Debenedetti, and R. K. Prud'homme, J. Chem. Phys. *109*:5651–5658 (1998).
162. N. Hattori, H. Hirata, H. Okabayashi, M. Furusaka, C. J. O'Connor, and R. Zana, Colloid Polym. Sci. *277*: 95–100 (1999).
163. J. Estaoe, P. Rogueda, A. M. Howe, A. R. Pitt, and R. K. Heenan, Langmuir *12*:2701–2705 (1996).
164. V. K. Aswal, S. De, P. S. Goyal, S. Bhattacharya, and R. K. Heenan, Phys. Rev. E *57*:776–783 (1998).
165. V. K. Aswal, S. De, P. S. Goyal, S. Bhattacharya, and R. K. Heenan, Phys. Rev. E *59*:3116–3122 (1999).
166. D. Danino, Y. Talmon, and R. Zana, Langmuir *11*: 1448–1456 (1995).
167. D. Danino, A. Kaplun, Y. Talmon, and R. Zana, in *Structure and Flow in Surfactant Solutions*, ACS

Symp. Ser. 578 (C. A. Herb and R. K. Prud'Homme, eds.), American Chemical Society, Washington, DC, 1994.
168. E. Alami, H. Lévy, R. Zana, and A. Skoulios, Langmuir *9*:940–944 (1993).
169. F. Kern, F. Lequeux, R. Zana, and S. J. Candau, Langmuir *10*:1714–1723 (1994).
170. M. In G. G. Warr, and R. Zana, Phys. Rev. Lett. *83*: 2278–2281 (1999).
171. R. Oda, I. Huc, J.-C. Homo, B. Heinrich, M. Schmutz, and S. Candau, Langmuir *15*:2384–2390 (1999).
172. P. J. Missel, N. A. Mazer, G. B. Benedek, C. Y. Young, and M. C. Carey, J. Phys. Chem. *84*:1044–1057 (1980).
173. F. C. Mackintosh, S. Safran, and P. Pincus, Europhys. Lett. *12*:697–702 (1990); S. Safran, P. Pincus, M. Cates, and F. C. Mackintosh, J. Phys. (Fr.) *51*:503–510 (1992).
174. S. J. Candau, P. Hebraud, V. Schmitt, F. Lequeux, F. Kern, and R. Zana, Nuovo Cimento *16D*:1401–1410 (1994).
175. J. Narayanan, W. Urbach, D. Langevin, C. Manohar, and R. Zana, Phys. Rev. Lett. *81*:228–231 (1998).
176. M. E. Cates, Macromolecules *20*:2289–2296 (1987); M. E. Cates, J. Phys. (Fr.) *49*:1593–1600 (1988); M. S. Turner and M. E. Cates, Langmuir *7*:1590–1594 (1991); R. Granek and M. E. Cates, J. Chem. Phys. *96*:4758–4767 (1992).
177. R. Oda, L. Bourdieu, and M. Schmutz, J. Phys. Chem. B *101*:5913–5916 (1997).
178. F. Schosseler, O. Anthony, G. Beinert, and R. Zana, Langmuir *11*:3347–3350 (1995).
179. L. J. Magid, J. Phys. Chem. B *102*:4064–4074 (1998).
180. E. Buhler, E. Mendes, P. Boltenhagen, J. P. Munch, R. Zana, and S. J. Candau, Langmuir *13*:3096–3102 (1997).
181. V. Schmitt and F. Lequeux, J. Phys. II (Fr.) *5*:193–197 (1995).
182. V. Schmitt, F. Schosseler, and F. Lequeux, Europhys. Lett. *30*:31–36 (1995).
183. R. Oda, F. Lequeux, and E. Mendes, J. Phys. (Fr.) *6*: 1–19 (1996).
184. R. Oda, P. Panizza, M. Schmutz, and F. Lequeux, Langmuir *13*:6407–6412 (1997).
185. D. Danino, Y. Talmon, H. Lévy, G. Beinert, and R. Zana, Science *269*:1420–1421 (1995).
186. S. Karaborni, K. Esselink, P. A. J. Hilbers, B. Smit, J. Karthäuser, N. M. van Os, and R. Zana, Nature *266*: 254–256 (1994).
187. G. Porte, J. Phys. Chem. *87*:3541–3550 (1983).
188. M. In, O. Aguerre-Chariol, and R. Zana, J. Phys. Chem. B *103*:7747–7750 (1999).
189. J. B. F. N. Engberts and J. Kevelam, Curr. Opin. Colloid Interface Sci. *1*:779–789 (1996).
190. S. Bhattacharya and S. De, Chem. Commun. 651–652 (1995).
191. S. Bhattacharya and S. De, Langmuir *15*:3400–3410 (1999).
192. F. M. Menger and A. V. Eliseev, Langmuir *11*:1855–1857 (1995).
193. R. Zana, H. Lévy, D. Danino, Y. Talmon, and K. Kwetkat, Langmuir *13*:402–408 (1997).
194. N. A. J. M. Sommerdijk, M. H. L. Lambermon, M. C. Feiters, R. J. M. Nolte, and B. Zwanenburg, Chem. Commun. 1423–1424 (1997).
195. R. Oda, I. Huc, and S. J. Candau, Angew. Chem. Int. Ed. *37*:2689–2691 (1998).
196. R. Oda, I. Huc, M. Schmutz, S. J. Candau, and F. C. MacKintosh, Nature *399*:566–569 (1999).
197. G. Cerichelli, L. Luchetti, G. Mancini, and G. Savelli, Langmuir *15*:2631–2634 (1999).
198. D. Danino, Y. Talmon, and R. Zana, J. Colloid Interface Sci. *185*:84–93 (1997).
199. V. K. Aswal, S. De, P. S. Goyal, S. Bhattacharya, and R. K. Heenan, J. Chem. Soc. Faraday Trans. *94*:2965–2967 (1998).
200. S. Fuller, N. N. Shinde, G. J. T. Tiddy, G. S. Attard, and O. Howell, Langmuir *12*:1117–1123 (1996).
201. S. Fuller, J. Hopwood, A. Rahman, N. N. Shinde, G. J. T. Tiddy, G. S. Attard, O. Howell, and S. Sproston, Liq. Cryst. *12*:521–529 (1992).
202. M. Tanaka, T. Ishida, T. Araki, A. Masuyama, Y. Nakatsuji, and M. Okahara, J. Chromatogr. *648*:469–473 (1993).
203. H. Harino, S. Tsunoi, J. Yoshioka, T. Araki, A. Masuyama, Y. Nakatsuji, I. Ikeda, and M. Tanaka, Anal. Sci. *14*:719–724 (1998).
204. K. Chen, D. C. Locke, T. Maldacker, J.-L. Lin, S. Aawassiripong, and U. Schurrath, J. Chromatogr. A *822*:281–290 (1998).
205. Z. Hu, T. Bührer, M. Müller, B. Rusterholz, M. Rouilly, and W. Simon, Anal. Chem. *61*:574–576 (1989).
206. M. Maj-Zurawska, M. Rouilly, W. Em Morse, and W. Simon, Anal. Chim. Acta *218*:47–59 (1989).
207. U. E. Spichiger, R. Eugster, E. Haase, G. Rumpf, P. Gehrig, B. Rusterholz, and W. Simon, Fresenius J. Anal. Chem. *314*:727–731 (1991).
208. A. E. Martell and S. Charberek, Jr., J. Am. Chem. Soc. *72*:5357–5361 (1950).
209. L. Galet, I. Pezron, W. Kunz, C. Larpent, J. Zhu, and C. Lheveder, Colloids Surf. A *151*:85–96 (1999).
210. S. Sato and H. Kishimoto, J. Surf. Sci. Technol. *8*: 209–216 (1992).
211. S. Sato, H. Sezaki, and H. Kishimoto, Prog. Colloid Polym. Sci. *93*:277–278 (1993).
212. S. Sato, N. Asai, and M. Yonese, Colloid Polym. Sci. *274*:889–893 (1996).
213. S. Sato and H. Kishimoto, J. Colloid Interface Sci. *123*:216–223 (1988).
214. S. Sato and H. Kishimoto, J. Colloid Interface Sci. *126*:108–113 (1988).

215. S. Sato and H. Kishimoto, in *Novel Surfactants in Solution 7* (K. L. Mittal, ed.), Plenum, New York, 1989, pp. 341–357.
216. S. Sato, Colloid Polym. Sci. *274*:1161–1169 (1996).
217. S. Sato, K. Toda, and S. Oniki, J. Colloid Interface Sci. *218*:504–510 (1999).
218. K. Esumi, J. Hara, N. Aihara, K. Usui, and K. Torigoe, J Colloid Interface Sci. *208*:578–581 (1998).
219. C. T. Kresge, M. E. Leonowicz, W. J. Roth, J. C. Vartuli, and J. S. Beck, Nature *359*:710–712 (1992); C. T. Kresge, M. E. Leonowicz, W. J. Roth, and J. C. Vartuli, U.S. Patent 5098684 (1992); J. S. Beck et al., J. Am. Chem. Soc. *114*:10834–10843 (1992).
220. P. Van der Voort, M. Mathieu, F. Mees, and E. F. Vansant, J. Phys. Chem. B *102*:8847–8851 (1998).
221. Q. Huo, R. Leon, P. M. Petroff, and G. D. Stucky, Nature *268*:1324–1327 (1995).
222. Q. Huo, D. I. Margolese, U. Ciesia, P. Feng, T. E. Gier, P. Sieger, R. Leon, P. M. Petroff, F. Schuth, and G. D. Stucky, Nature *368*:317–323 (1994).

4

New Glycolipids Having Biological Activities: Key Role of Their Organization

ARMAND LATTES, ISABELLE RICO-LATTES, EMILE PEREZ, and MURIEL BLANZAT Laboratoire IMRCP, UMR CNRS 5623, Université Paul Sabatier, Toulouse, France

I. INTRODUCTION

A. Overview

Surfactants are key components of the organized assemblies used in biological systems, and it is important to use an appropriate surfactant for the process under consideration. This requires synthesis of a range of suitable surfactants by modular methods leading to variations in the structures of the compounds. In this context, nonionic surfactants based on carbohydrates are very important in biology. They have potential pharmaceutical (biocompatible formulations), biochemical (extraction of membrane proteins), and medicinal applications. Carbohydrate-based surfactants have a water-soluble headgroup derived from a carbohydrate. This is linked by different functional groups to a hydrophobic part. Variations in the nature of the carbohydrate and in the hydrocarbon tail lengths determine the surface-active properties and applications of the resulting surfactant. Generally, these derivatives are not readily synthesized, as the starting carbohydrates require protection. Here we present routes, avoiding protection of the starting carbohydrates, to new series of surfactants derived from lactose and glucose reducing carbohydrates. Our synthetic methods for the preparation of such products enabled us to obtain molecules having one or two chains with different lengths or two carbohydrate heads bounded by hydrophobic spacers (bolaphilic compounds).

All of these compounds form aggregates such as direct micelles and/or vesicles in aqueous solutions. Some of them have interesting applications in the extraction of membrane proteins (surfactants with one tail). Bolaform-type surfactants leading to vesicles may find applications in drug formulation. Finally, soluble analogues of galactosyl ceramide (Galcer) possess activity against both human immunodeficiency virus (HIV) and *Aspergillus fumigatus*, a yeast responsible for opportunistic infections in patients with the acquired immunodeficiency virus (AIDS).

New double-chain and gemini catanionic analogues of Galcer were also easily prepared in two steps from lactose. They showed interesting anti–HIV-1 activities, acting as monomeric analogues of Galcer, and presented surface-active properties forming lamellar mesophases and vesicles.

B. Carbohydrates in Cellular Biochemistry

The role of carbohydrates in the functioning of living cells, and in biology in general, is more important than previously believed. Independently of their behavior as energy sources, carbohydrates are implicated in many biological events:

Cell surface oligosaccharides play crucial roles in pathogenic events: viral infection, bacterial adhesion, recognition associated with tumors, tumor metastasis, and other intercellular events (cell adhesion and cell-cell recognition, leukocyte adhesion in the immune response).

They are also antigenic determinants of the blood groups.

It is not surprising to find sugar-based compounds having biological applications in different domains, such as in:

Drug delivery systems targeting cells expressing "ose" receptors (glucose, galactose, mannose, etc.) with, for example, a preparation of sugar chain macromolecules as a drug carrier [1]

Antiviral drugs [2]

Anti-inflammatories (sulfated galactocerebrosides are potential anti-inflammatory agents [3]

Biocompatible formulations useful for solubilizing active principles and for realizing enzymatic reactions in dispersed media [4]

The extraction and purification of membrane proteins [5]

All of these applications are based on two fundamental properties:

The presence of recognition parts on the molecules, directed by the relative spatial orientation of the saccharide functions.

The total interaction energy with membrane constituents or receptors, strong enough to provide a chemical or biological response. In this case, the occurrence of multiple interactions (recognition) can be completed with the energy support of van der Waals and hydrophobic interactions. Surfactant molecules having sugar headgroups and hydrophobic tails would be one of the best families of compounds for exploring this field.

For this reason, we studied a new series of surfactants derived from reducing or oxidated sugar and polymers having pendant sugar functionalities. In order to have easy access to these molecules, we decided to develop syntheses without using protecting groups.

In this chapter, after a short description of the synthetic methods, we will attempt to establish a correlation between the structures and physicochemical properties of sugar-based surfactants and their different activities in extracting proteins, in acting as dispersed media for enzymatic reactions, and in medicinal applications.

II. PREPARATION OF NEOGLYCOLIPIDS

A. Use of the Properties of Reducing Sugars

Condensation of reducing sugars with a long-chain amine at room temperature gives *N*-alkylglycosylamines, but they cannot be used as surfactants because they spontaneously hydrolyze in aqueous solution to give the starting amines and ose [6]. These compounds can be stabilized in one of two ways:

1. Reduction of the $>$C=N— bond of the open form, which stabilizes the hydrogenated products and leads to *N*-alkylamino-1-deoxyglycitols [7]
2. Acylation of the glycosylamine to *N*-acetyl-*N*-alkylglycosylamine [8]

In this last case we avoid the secondary amine function and the eventuality of the formation of toxic *N*-nitrosamines when these surfactants are used as personal care products. Scheme 1 summarizes the reactions important in the preparation of these products. One obtains the *N*-alkylamino-1-deoxy lactitols with good yields and *N*-acetyl-*N*-alkyl-glucosyl and lactosyl amines with moderate yields [9].

Furthermore, structural analysis of these compounds by ^{1}H and ^{13}C nuclear magnetic resonance (NMR) demonstrate that they exist in a β anomeric conformation at the site of attachment of the acetyl group [10] (Table 1). This is in marked contrast to the results obtained with the *N*-alkyllactosylamines, which were observed in both α and β forms [11].

B. Double-Chain Surfactants

The ease of introduction of acetyl groups was the primary motivation to use a similar strategy to prepare double-chain surfactants. We have developed a simple route to a new family of compounds: the *N*-acyl-*N*-alkylamino-1-deoxyglycitols [11,12]. These compounds were synthesized in a three-stage process from nonprotected sugars (e.g., from lactose). The second chain was introduced using *N*-acylthiazolidine-2-thiones, which acylate the nitrogen atom without reacting with the hydroxyl groups of the sugar (Scheme 2).

C. Use of the Oxidized Form of Reducing Sugars

The oxidation of reducing sugars gives acids or lactones. By condensing such derivatives with amines or diamines [13,14], we get single-chain surfactants or bola-amphiphilic compounds (Scheme 3).

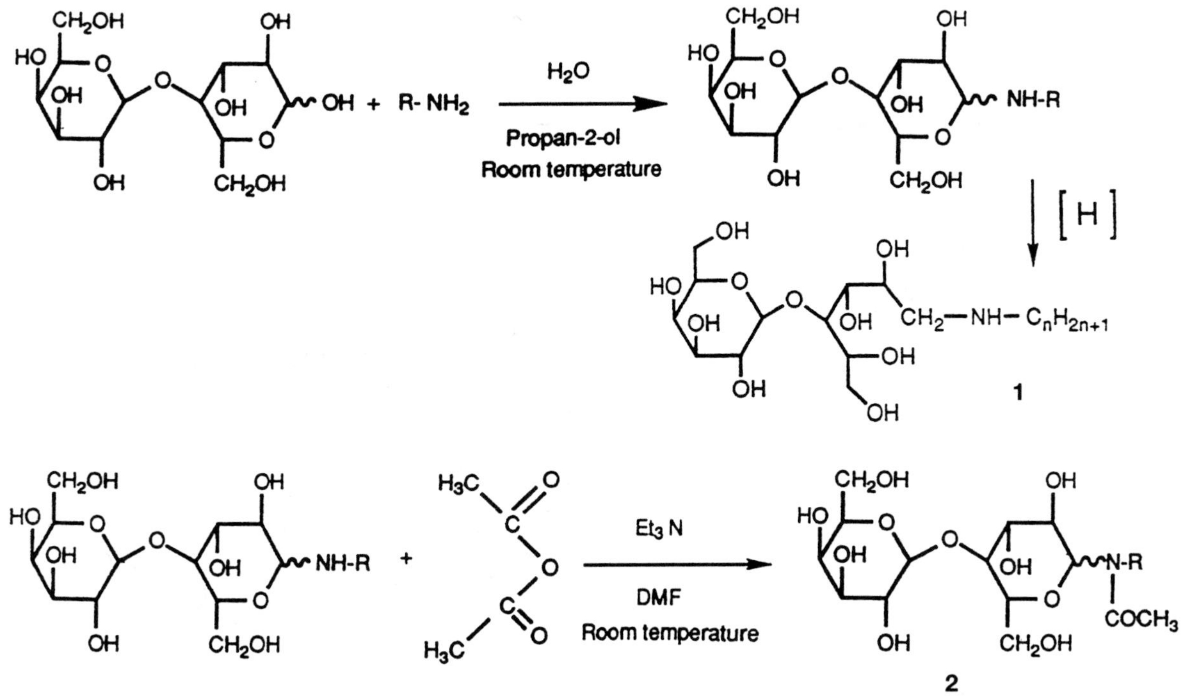

SCHEME 1 Synthetic scheme for preparation of the *N*-alkylamino-1-deoxylactitols (Form 1) and *N*-acetyl-*N*-alkyllactosylamines (Form 2).

Concerning the bola-amphiphiles, the first synthesized were symmetrical. Both heads had the same structure derived from glucose or lactose. By using amino acids in place of diamines, it was possible to synthesize a dyssymetric bola-amphiphile with one headgroup derived from lactose and the other (anionic) headgroup comprising a carboxylate group [15]. By acylation of the secondary amine, one also obtains a branched dyssymetric bola-amphiphile (Scheme 4).

TABLE 1 Characteristic ^{1}H and ^{13}C NMR Data for Two b Conformers of *N*-Acetyl-*N*-decyllactosylamine

Solvent	C_1 lactose		C_1N chain		C=O	CH_3—CO		C'_1 lactose	
D_2O C = 0.1 M	^{13}C δ (ppm)	^{1}H δ (ppm)	^{13}C δ (ppm)	^{1}H δ (ppm)	^{13}C δ (ppm)	^{13}C δ (ppm)	^{1}H δ (ppm)	^{13}C δ (ppm)	^{1}H δ (ppm)
Form 1	87.42	5.33	45.75	3.15 3.35	174.79	22.06	2.19	103.5	4.45
Form 2	83.93	4.95	41.90	3.15 3.35	174.57	22.06	2.15	103.5	4.43

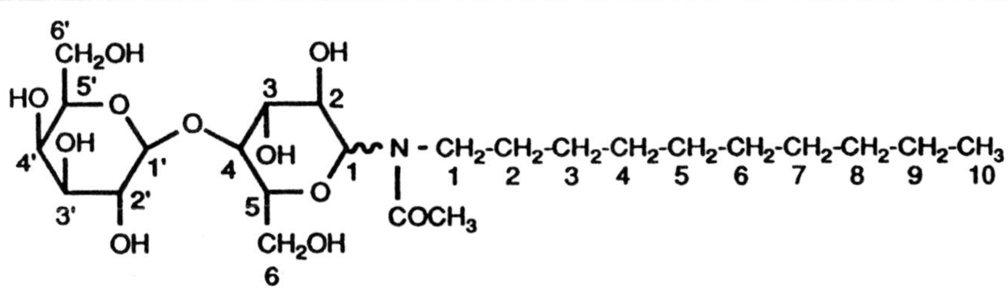

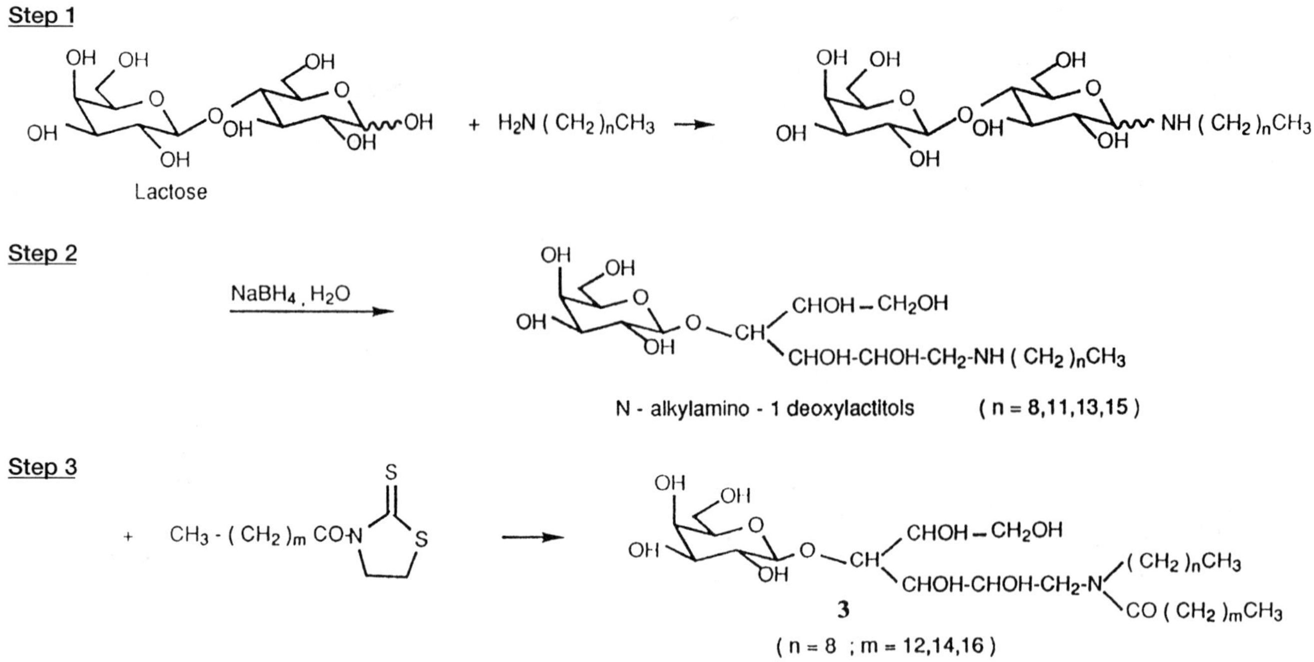

SCHEME 2 Synthetic scheme for preparation of the double-chain surfactants.

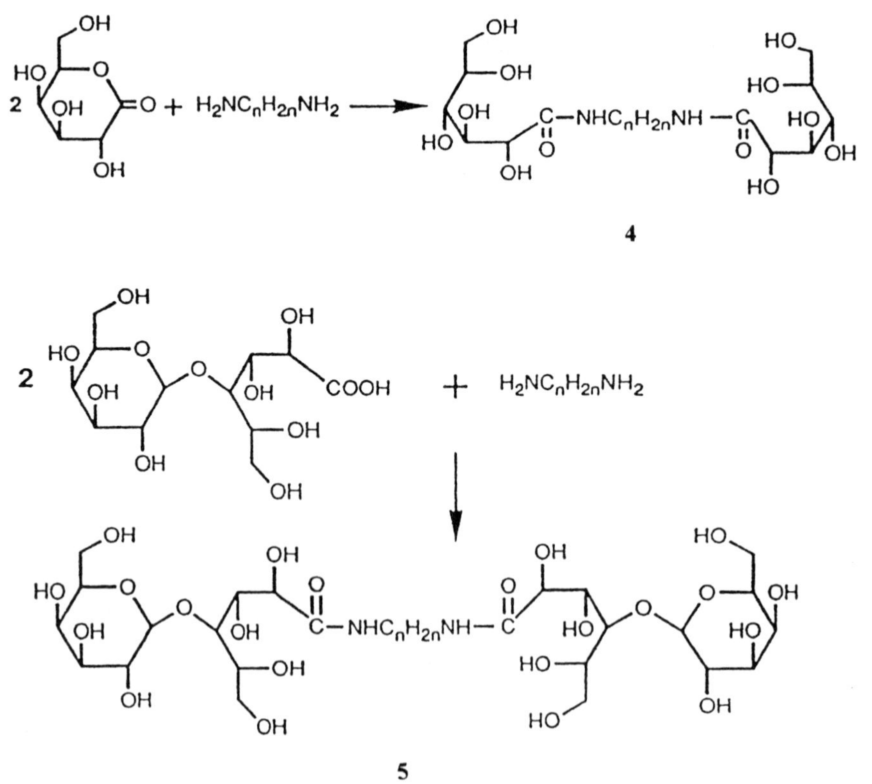

SCHEME 3 Synthesis of bis-gluconamido and bis-lactobionamidoalkanes.

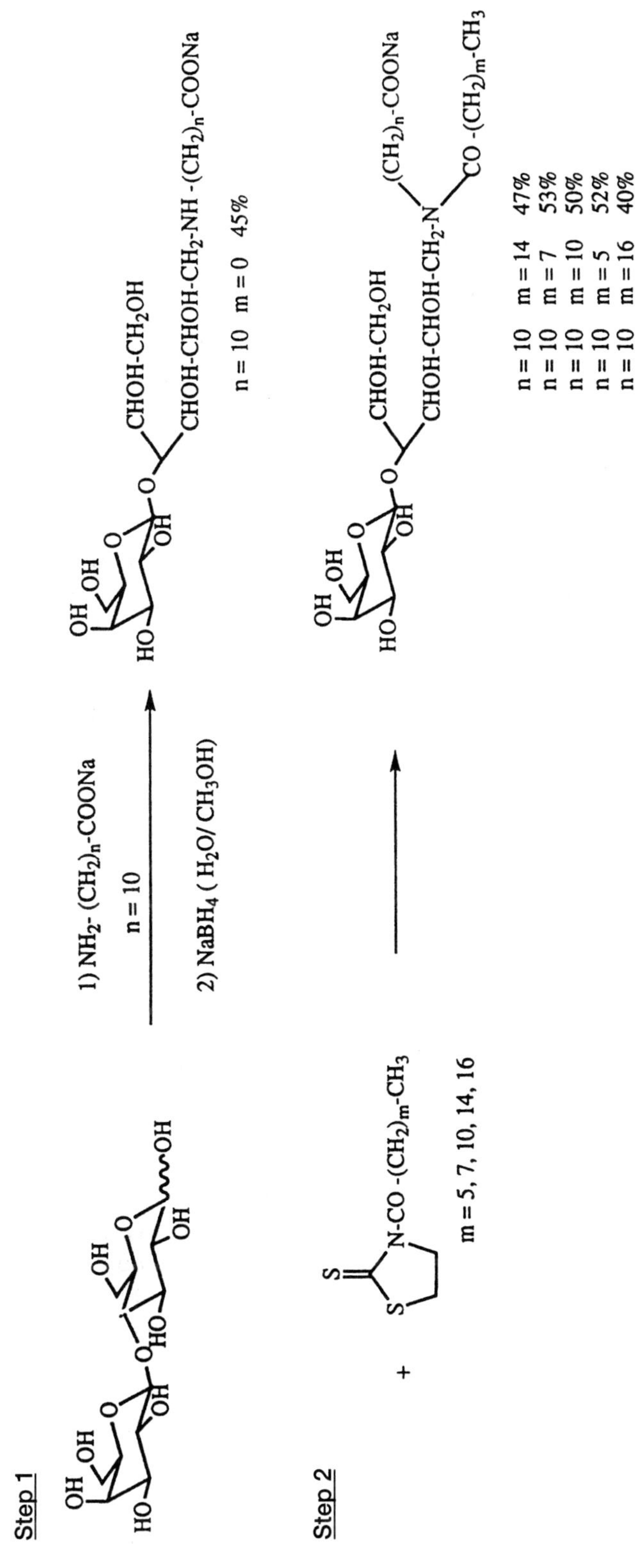

SCHEME 4 Synthesis of dyssymmetric bola-amphiphilic compounds.

TABLE 2 Critical Micelle Concentration and Aggregation Numbers of the *N*-Alkylaminolactitols

Surfactants	1b (C_8)	1c (C_9)	1d (C_{10})	1e (C_{12})
cmc (mol/L)	1.5×10^{-2}	5.6×10^{-3}	2.6×10^{-3}	6×10^{-4}
N	34	47	62	88

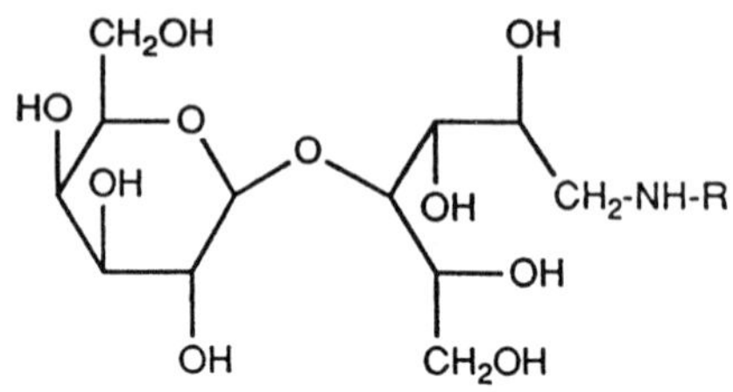

1

III. ASSOCIATIVE PROPERTIES OF SUGAR-BASED SURFACTANTS

A. Physical Properties

Table 2 [16] shows the physicochemical characteristics [critical micelle concentration (cmc) and aggregation numbers] of the *N*-alkylaminolactitols. Table 3 [17] shows the cmc of some *N*-acetyl-*N*-alkyllactosylamines.

For comparison with other surfactants, we can specify that the aggregation number of *N*-acetyl-*N*-dodecyllactosylamine is 64 [18]. *Double-chain surfactants* are sufficiently hydrophobic to be soluble in organic apolar solvents, where they form reverse micelles that are able to incorporate large quantities of water [12]. *Bola-amphiphilic molecules* do not give micelles if the spacer between both sugar headgroups is too short [19]. Nevertheless, Fig. 1 shows that they are good surfactants, without formation of micelles except when the spacer is long enough (C_{12}). Formation of vesicles has been observed under appropriate conditions.

TABLE 3 Critical Micelle Concentration of the *N*-Acetyl-*N*-alkyllactosylamines

R	Number	CMC (mol/L)
C_8H_{17}	2a	10^{-2}
C_9H_{19}	2b	5×10^{-3}
$C_{10}H_{21}$	2c	3×10^{-3}
$C_{12}H_{25}$	2d	0.5×10^{-3}
C_8H_{17}	β C_8 G[a]	1.9×10^{-2}
$C_{12}H_{25}$	β C_{12}M[b]	2.4×10^{-4}
$C_{10}H_{21}$	*N*-Decyl deoxy-1-lactitol amine	2.6×10^{-3}

[a] β-1-*n*-Octyl-D-glucopyranoside.
[b] β-1-*n*-Dodecyl-D-maltoside.

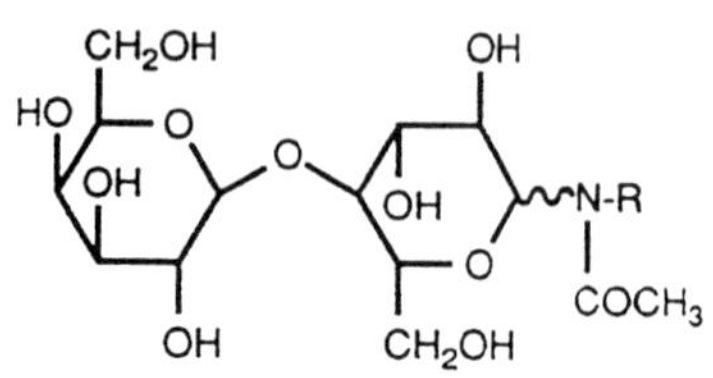

2

R = C_8H_{17}, 2a; C_9H_{19}, 2b; $C_{10}H_{21}$, 2c; $C_{12}H_{25}$, 2d

B. Models of Micelle Structure

Comparative analysis of the micellization of *N*-alkylaminolactitols by x-ray and neutron scattering indicated that the micelles consisted of an ellipsoidal aliphatic core surrounded by an ellipsoidal polar outer layer containing solvated polar heads. Figure 2 shows the scattering length density profiles [16].

These results show that the relatively high volume of the lactitol polar head of *N*-alkylaminolactitols did not prevent micellization. This model of micelle structure could account for the preferential packing of *N*-alkylaminolactitols into oblate ellipsoidal micelles rather than into prolate ellipsoids or spheres [16] (Fig. 3).

The same study with α- and β-1-*n*-dodecyl-maltosides shows that the structure formed by self-assembly depends on the configuration at the anomeric center:

The α-anomer forms quasi-spherical aggregates.
The β-maltoside forms large aggregates with an oblate ellipsoidal shape.

This observation throws more light on the influence of anomeric configurations during self-association of reducing sugar-based amphiphiles.

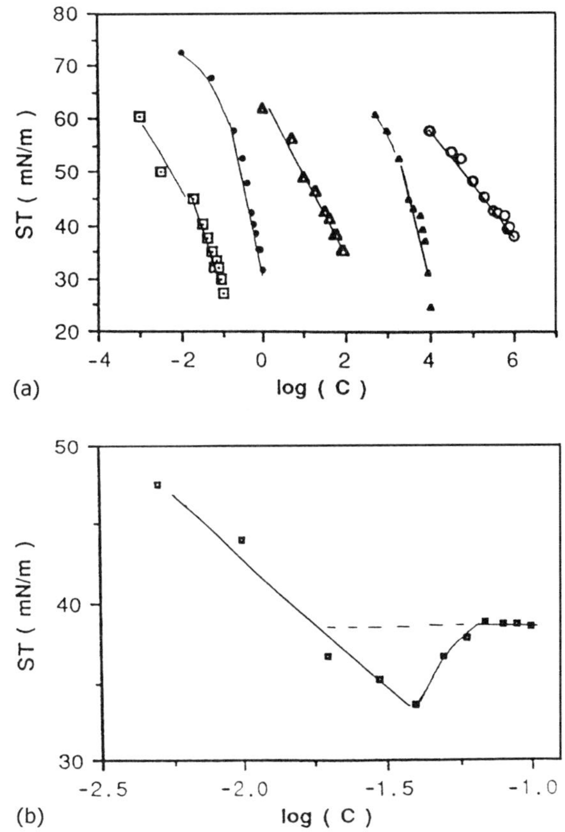

FIG. 1 (a) Surface tension of bis-lactobionamido alkanes at 40°C: (□) $n = 6$; (●) $n = 7$; (△) $n = 8$; (▲) $n = 9$; (○) $n = 10$. (b) Surface tension of bis-lactobionamido alkane (■, $n = 12$) at 40°C.

With an *N*-acetyl-*N*-dodecyllactosylamine mixture of β-conformers, the micelles are nearly spherical [20] (Fig. 4). The reduced aggregation number ($N = 65$) for this compound compared with that for *N*-dodecylaminolactitol ($N = 88$) can be explained by the increased steric hindrance at this site, which affects both the geometry of the micelles and their aggregation number (Table 4).

IV. EXTRACTION OF MEMBRANE PROTEINS

A. Extraction Efficiency

An important biological application of surfactants is the extraction of proteins from membranes. Extraction is a crucial step in the study of the mode of action of numerous membrane proteins, especially those of the receptor type.

In a first set of experiments we attempted to extract the ⟨⟨op⟩⟩ opiate receptors from the brain of the frog *Rana ridibunda*. A variety of agents have been used for the extraction of these fragile receptors. Digitonin has been found to be the least denaturing, solubilizing these receptors with the lowest loss of activity. Up to 50% solubilization has been obtained in the best cases, but the percentage depends on the purity of the commercially available product.

In the present study, we first tested all *N*-alkylaminolactitols and compared the results with those obtained with digitonin. The best result obtained was with a concentration of 1% of *N*-nonylaminolactitol (C_9). This surfactant gives a level of solubilization similar to those obtained with digitonin [21]. In addition, this derivative is advantageous compared with digitonin because it inhibits ligand binding to receptors, it is less expensive, and it has higher purity.

The second membrane protein we studied was the b6f complex of bacteriorhodopsin. In this case the best product was the aminolactitol having a C_{12} chain on the nitrogen atom [22].

Other experiments with different proteins showed that it is necessary, in a series of surfactants, to utilize amphiphilic molecules with different chain lengths. These results show the necessity of testing many molecules in the same family in order to obtain the best fit with a particular protein (kits of surfactants).

B. Relationship Between Micellar Shapes and the Solubilization of Proteins

When the solubilizing properties of surfactants are compared with their micellar shapes (Table 5), one can make an interesting observation [23]:

The best fit of the scattered intensity curves for all surfactants able to solubilize proteins was obtained with an oblate ellipsoidal model or, in the case of hecameg, a filamentous micelle.

All surfactants unable to extract proteins gave spherical micelles (for example, α-alkylmaltoside and *N*-acetyl-*N*-dodecyllactosylamine).

The differences in micellar geometries derive from the influence of the polar headgroup. The formation of oblate ellipsoidal micelles is compatible with low steric hindrance, especially at the junction between the polar headgroup and the hydrophobic part of the surfactant. The molecules of *N*-acetyl-*N*-dodecyllactosylamine, whose polar headgroup is bulky, form quasi-spherical micelles.

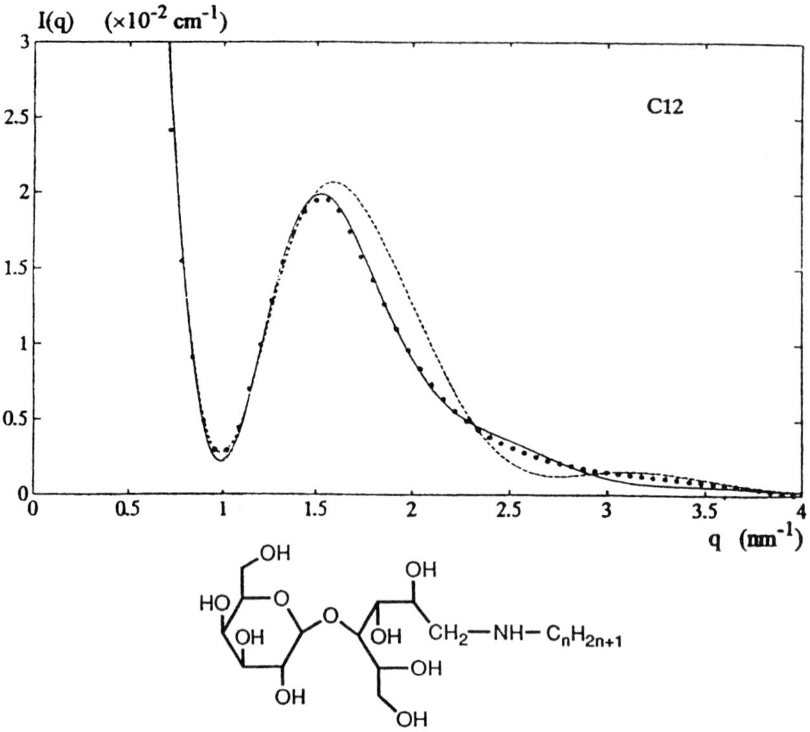

FIG. 2 Fit of experimental $I(q)$ curves for x-rays for the oblate ellipsoidal model (——) and the prolate ellipsoidal model (---) with a solution of C_{12} derivative.

In our opinion, surfactant molecules forming oblate ellipsoidal or filamentous micelles are able to penetrate into the membranes because they can adapt their conformation at the relatively flat surface of the membrane as at the elongated portion of the ellipse. Surfactant molecules forming spherical micelles cannot adapt as easily.

In the same field, bola-amphiphiles do not penetrate the membranes and are not able to extract proteins [24]. This behavior is interesting and may be useful in the design or formulation of organic components or drugs.

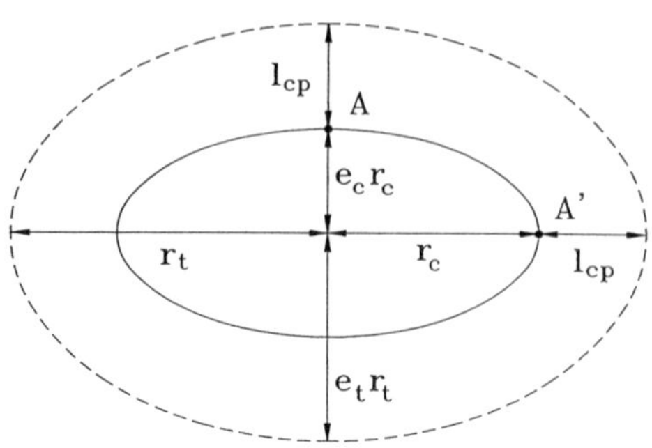

FIG. 3 Oblate ellipsoidal micelle.

V. APPLICATION OF SUGAR-BASED BOLAFORMS: EMULSIFICATION OF LIPOPHILIC COMPOUNDS

Linoleic acid can be oxidized by soybean lipoxygenase to form a conjugated diene hydroperoxide (Scheme 5).

If the enzyme is soluble in water, linoleic acid is not readily solubilized or dispersed in aqueous media. In fact, the activity of the enzyme toward this substrate is not fully realized without addition of a surfactant such as Tween 20 [22]. Unfortunately, this surfactant rapidly denatures the enzyme. We compared the behavior of sugar-based bolaforms with that of Tween 20 and found the following:

In many cases the specific activity of soybean lipoxygenase in the presence of bolaforms was higher than that of Tween 20 (Table 6).

In all cases there was little loss of activity in the presence of bola-amphiphilic surfactants (Table 7).

The sugar-based bolaforms formed globular systems with linoleic acid without denaturing the soybean lipoxygenase enzyme. They thus represent a good alternative to Tween 20 for this type of formulation in industrial applications [25].

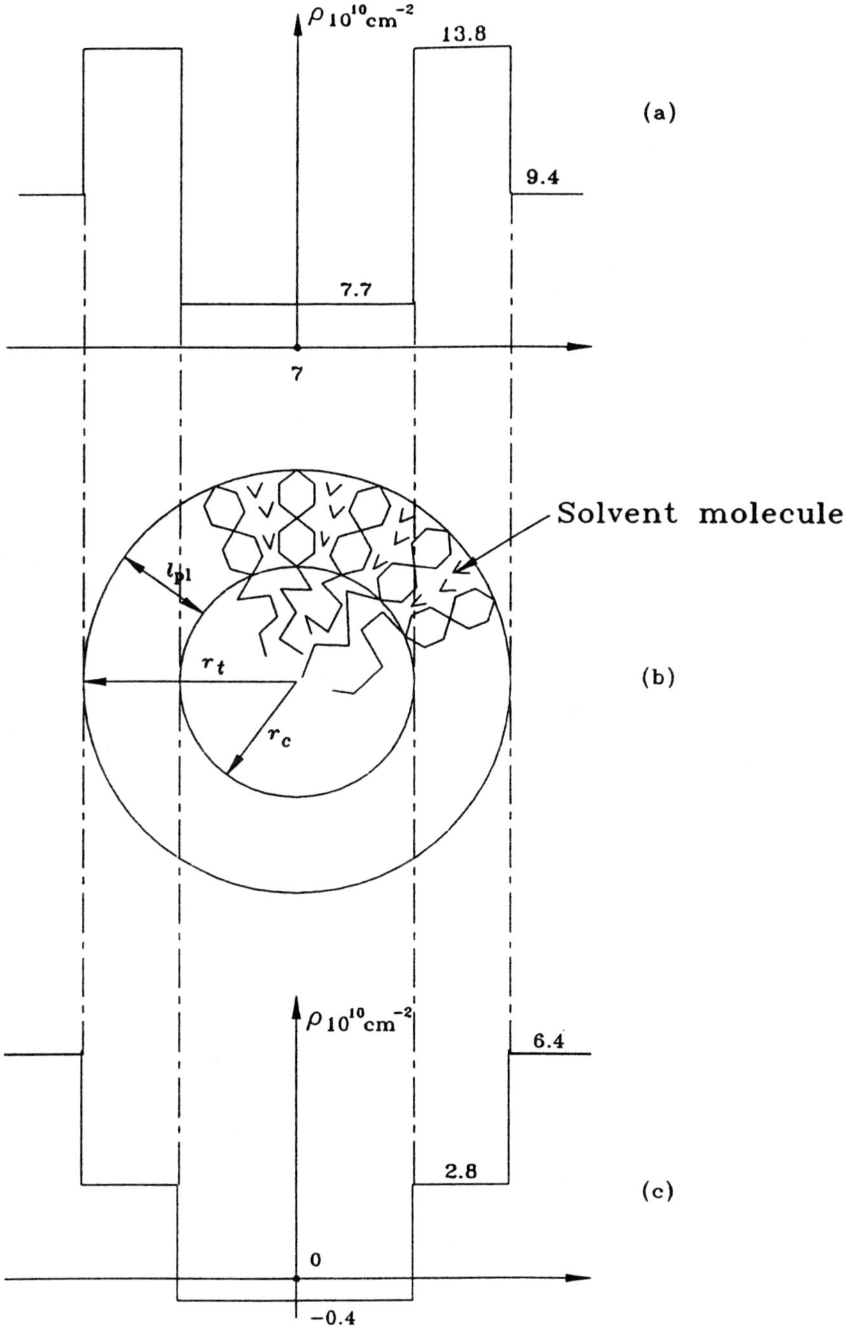

FIG. 4 *N*-Acetyl-*N*-dodecyllactosylamine spherical micelle.

TABLE 4 Theoretical Parameters Obtained from Modeling for Micelles of L (*N*-Dodecylaminolactitol), β (*N*-Dodecyllactobionamide), and Ac (*N*-acetyl-*N*-dodecyllactosylamine)

	e_t[a]	r_t (nm)	e_c	r_c (nm)	N	n_{sol}	l_{pl} (nm)	S(A) (nm^2)
L	0.71	3.67	0.55	2.38	88	33	1.29	0.37
B	0.67	3.39	0.55	2.46	95	15	0.93	0.37
Ac	1.0	2.74	1.0	1.75	64	20	0.99	0.60

[a]e_t, ellipticity of the external envelope of the micelle; e_c, ellipticity of the core; r_t and r_c, axial lengths; N, aggregation number; S, area of the polar head; l_{pl}, width of the polar outer layer; n_{sol}, number of molecules of water per polar head.

VI. SUGAR-BASED SURFACTANTS USEFUL IN MEDICINAL APPLICATIONS

A. Recognition Issues

Biological activity depends on different specific (recognition) or nonspecific (site passive occupation) factors. Possible interactions with membranes or their guests (proteins) essentially determine both activity and toxicity. We will provide two examples in which these interactions play a different role: possible integrations into phospholipid layers (antifungal activity) and better balancing of activity versus toxicity by inhibition of this penetration (antiviral activity).

B. Anti–*Aspergillus fumigatus* Activity

Patients with AIDS frequently develop infections with filamentous fungi (*Aspergillus* series) for which there are few effective treatments [26]. In view of the toxicity of the current antifungal agents (e.g., amphotericin B) and the emergence of resistance, especially to some azole compounds (such as itraconazole), new anti-*Aspergillus* agents are needed. Analysis of the fungal strains has shown the presence of membrane lectins bearing the galactose moiety [27]. It is recognized that these lectins play a fundamental role in the recognition and binding of infectious agents to carbohydrate receptors on target cells [28]. We have thus tested the prepared compounds bearing the galactose group for anti-*Aspergillus* activity.

TABLE 5 Comparison of the Solubilizing Properties of Surfactants with Their Micellar Shapes

Compounds	Micellar shapes	Protein solubilization
N-Octylaminolactitol	Ellipsoidal	+
N-Decylaminolactitol	Ellipsoidal	+
N-Dodecylaminolactitol	Ellipsoidal	+
N-Acetyl-*N*-dodecyllactosylamine	Spherical	−
N-Dodecyl-lactobionamide	Ellipsoidal	+
Hecameg	Filamentous	+
Dodecyl-β-maltoside	Ellipsoidal	+
Dodecyl-α-maltoside	Spherical	−

The results in Table 8 show that the tested compounds possessed antifungal activity, especially the derivative with $n = 8$ and $m = 16$, which had activity comparable to that of amphotericin B [29].

The mechanism of activity of amphotericin B was attributed, as far back as 1973, to the formation of pores in the plasma membrane, arising from the association of amphotericin B molecules with sterols, more specifically with ergosterol. The amphiphilic compounds studied have antifungal activity strongly associated with their lipophilicity. Based on this observation and the inactivity of bolaforms, we postulate the following hypothesis: *Amphiphilic compounds must interact with the membrane of the cell fungi*

Our experimental results, obtained by the study of product intercalation in a monolayer of dipalmitoylphosphatidylcholine (DPPC) and the observation of vesicles by electron microscopy, show that the active compounds studied here strongly interact with monolayers formed from DPPC and ergosterol. The compounds need to be present at levels 5 to 10 times more concentrated for interactions with monolayers containing cholesterol (as in human membranes) [30] to be

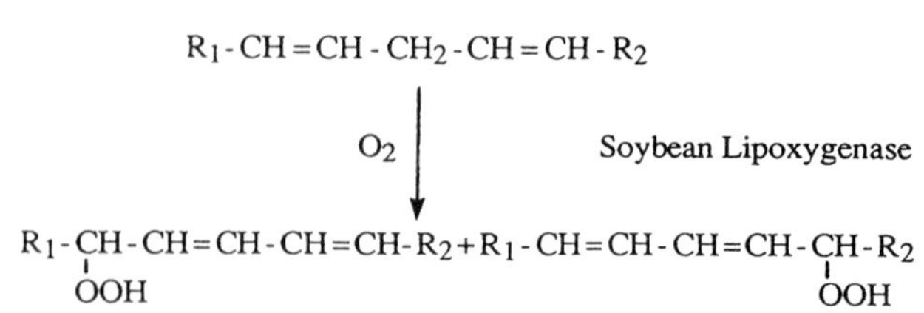

SCHEME 5 Oxidation of an unsaturated fatty acid by soybean lipoxygenase.

TABLE 6 Specific Activity (SA) of Soybean Lipoxygenase (LG) in the Presence of Bolaforms for Substrate Dispersal

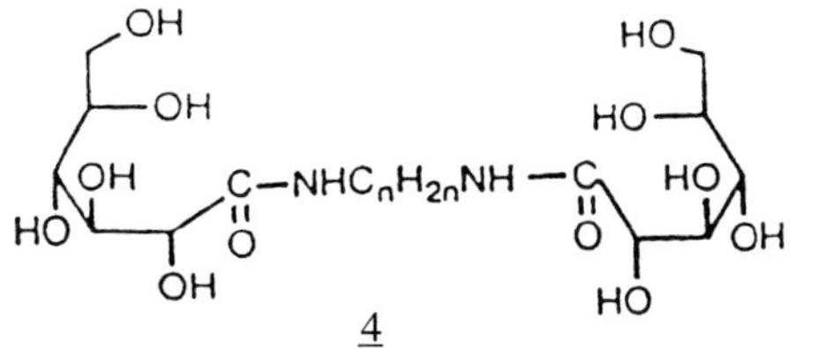

4

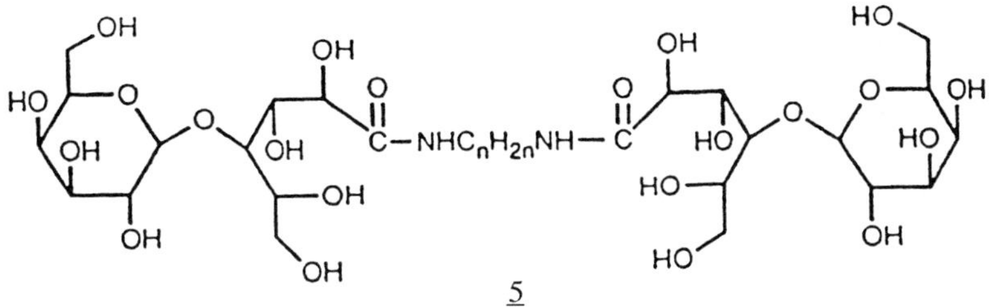

5

Surfactant		4a	4b	4c	4d	4e
n =	Tween 20	6	7	8	9	10
SA of soybean LG (IU/mg protein)	15,033	18,408	17,692	15,749	15,135	15,442
Surfactant	5a	5b	5c	5d	5e	5f
n =	6	7	8	9	10	12
SA of soybean LG (IU/mg protein)	15,033	18,467	21,200	21,000	19,600	17,800

observed. To optimize this family of compounds, we must improve their intercalation properties in ergosterol-containing membranes.

C. Soluble Sugar-Based Surfactants as Potential Inhibitors of HIV Infection

In a systematic study of CD4$^-$ nerve cells, Gonzalez-Scarano and coworkers characterized galactosyl ceramide (Galcer) (Fig. 5) as a potential receptor for HIV in the nervous system [31]. Fantini et al. also identified Galcer as the HIV receptor in colonic epithelial cells [32].

Soluble analogues of Galcer that mimic the Galcer receptor might be expected to possess anti-HIV activity. Bertozzi et al. were the first to describe the preparation of a Galcer analogue in a nine-stage process [33] (Fig. 6).

We have developed a simple route to two new families of amphiphilic analogues of Galcer:

The *N*-alkylamino-1-deoxylactitols
The *N*-acyl-*N*-alkylamino-1-deoxy lactitols

Several compounds were tested on CEM-SS cells infected with HIV-1:

N-Alkylamino-1-deoxylactitols have marked antiviral activity. The optimum was observed for *N*-dodecylamino-1-deoxylactitol.
N-Acyl-*N*-alkylamino-1-deoxylactitols were also found to possess anti-HIV activity.
N-Alkylamino-1-deoxylactitols were toxic above the cmc [34], having a detergent effect on cell membranes, but *N*-acyl compounds avoided this toxic detergent effect.

In order to lower the toxicity, we decided to develop a new class of anti–HIV-1 agents that interact with the HIV-1 surface envelope glycoprotein gp120 but also interact less strongly with the cell membranes. Therefore, we tried branched dyssymmetric bola-amphiphiles (e.g., CA 52) with one headgroup derived from lactose and the other comprising a carboxylate group (Fig. 7).

These soluble analogues of Galcer strongly interacted with the V3 loop of gp120 and showed good activity as anti-HIV agents [35]. The presence of a ga-

TABLE 7 Specific Activity (SA) and Relative Specific Activity (%) of Soybean Lipoxygenase (LG) in the Presence of Bolaforms after Incubation at 4°C for 60 h

Surfactant	Tween 20	4a	4b	4c	4d	4e	4f
SA of soybean LG (IU/mg protein)	573	11,863	13,294	11,863	12,067	10,840	12,067
Relative specific activity (%)	4	79	88	79	80	72	80

Surfactant	Tween 20	5a	5b	5c	5e	5f
SA of soybean LG (IU/mg protein)	573	10,523	15,032	13,529	13,530	15,033
Relative specific activity (%)	4	70	100	90	90	100

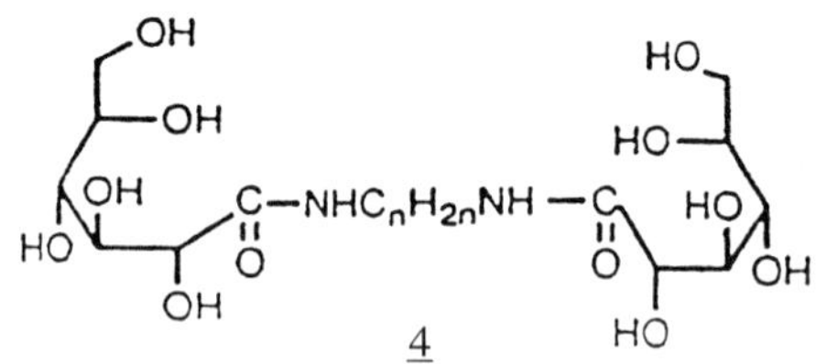

4

9

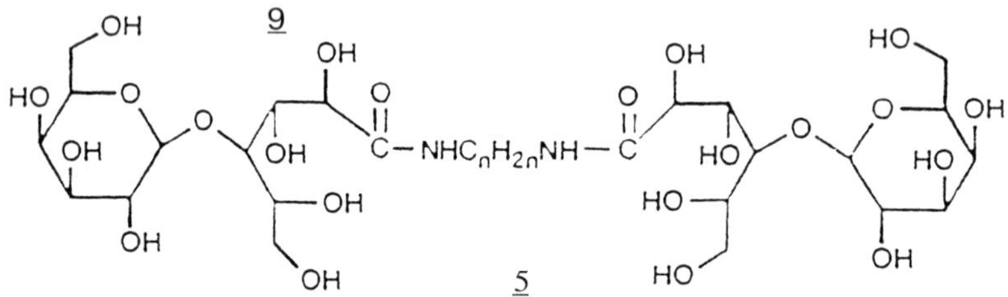

5

lactose unit is necessary for recognition. The hydrophobic long chain is also necessary to enhance binding. These compounds could be optimized as the base of potential low-cost anti–HIV-1 drugs that neutralize HIV infection through masking of the V3 loop.

In a second phase, we decided to improve the biological activity of Galcer analogues by replacing the carboxylate with another anionic motif. In previous work, Rideout and coworkers showed that sulfonated *N*-substituted naphthalimides were inhibitors of HIV in vitro [36]. Therefore we decided to use this motif to synthesize new bola-amphiphilic compounds with a sulfonaphthalimide group in place of carboxylate (Scheme 6). Preliminary results of biological studies

TABLE 8 Antifungal Activity of Glycolipids

Compounds	*n*	*m*	IC_{90}[a] (μg/mL)	IC_{90} (μmol/mL)	IC_{50} (μg/mL)	IC_{50} (μmol/mL)
Itraconazole	—	—	0.06	0.08	0.003	0.004
Amphotericin B	—	—	0.6	0.6	0.02	0.02
1c	9		>100	>212.9	>100	>212.9
1e	12		>100	>195.4	>100	>195.4
1f	14		28.7	53.1	15.2	28.1
1g	16		7.0	12.3	2.2	3.8
3d	8	12	18.0	26.5	5.0	7.4
3e	8	14	7.0	9.8	1.6	2.3
3f	8	16	1.8	2.4	0.5	0.6

[a] IC_n, Concentrations inhibiting the growth of *Aspergillus fumigatus* by 50% (n = 50) and 90% (n = 90).

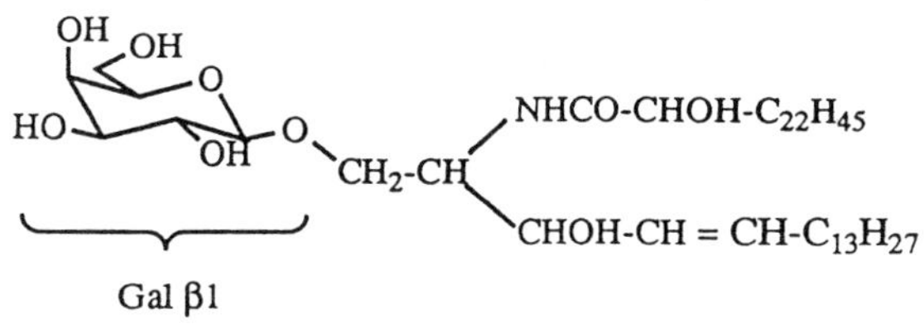

FIG. 5 Chemical structure of galactosyl ceramide (Galcer).

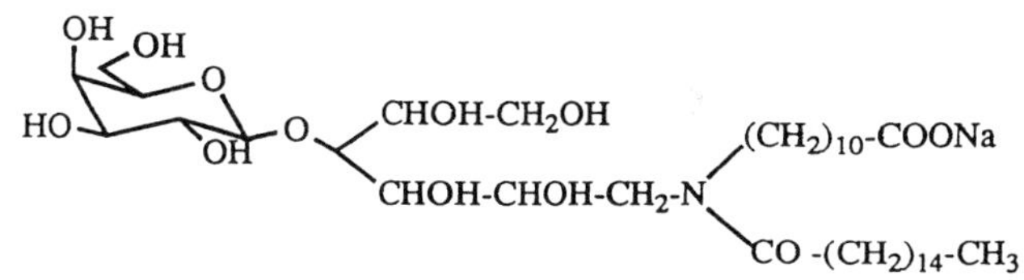

FIG. 7 Chemical structure of soluble analogue of Galcer (CA 52).

show very good activity of such bolaforms against HIV-1 gp120. The median inhibitory concentration (IC_{50}) for binding is, in some cases, less than half (0.4 μM) that of suramin (0.9 μM).

VII. NEW CATANIONIC GLYCOLIPIDS

A. Gemini Catanionics

Catanionic surfactant mixtures, such as two chains [37] or geminis, showing various aggregate microstructures (micelles, vesicles, and lamellar phases) have received increased attention [38]. Their mode of synthesis, achieved by mixing two oppositely charged surfactants, could be an alternative technique for easily preparing new two-chain or gemini glycolipids. Analogues of Galcer having these structures require multistep synthesis and purification; to overcome this drawback, we intended to synthesize catanionic analogues of galactosylceramide.

Although various two-chain glycolipids have been described in the literature [40], very few examples of gemini glycolipids have been reported [40,41].

B. Synthesis

We synthesized new analogues of Galcer, replacing the amide covalent bond by an amine-acid ionic bond (Scheme 7). *N*-Alkylamino-1-deoxylactitol and carboxylic acids stoichiometrically gave salts upon stirring in water, forming an ionic amine-acid bond. We thus obtained new catanionic glycolipids in high yields (96% in two steps from unprotected lactose) [42].

The catanionic monomeric species were characterized by mass spectrometry. The electrospray mass spectrometry technique is a useful method for characterizing molecules that form noncovalent complexes in solution [43] or in the gas phase [44]. Thus we have been able, for the first time, to observe directly monomeric species of catanionic surfactants.

C. Self-Association Properties

Surface tension measurements confirm the formation of micelles with bichain and gemini compounds when they are hydrophobic enough. Above the cmc, all compounds form mesophases in concentrated water solution at room temperature. With polarized light microscopy we observe lamellar phases, which indicate the possibility of vesicle formation.

Dynamic light scattering studies confirm the spontaneous formation of vesicles in a concentration range of 1 to 4 mM, and transmission electron microscopy (TEM) showed the aggregate morphology of vesicles.

D. Anti–HIV-1 Activities

Catanionic analogues showed interesting anti–HIV-1 activities (Table 9). The gemini compound with $n = 15$ and $m = 12$ is particularly interesting, providing both high anti-HIV activity (median effective concentration $EC_{50} = 0.5$ μM) and low toxicity ($CC_{50} > 100$ μM). In fact, it is more active and less toxic than previously studied Galcer analogues. Indeed, as previously ob-

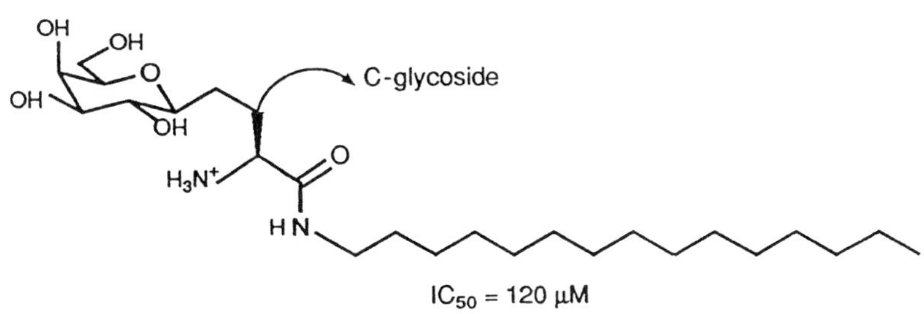

FIG. 6 Carbon-linked galactosphingolipid analogue (C. R. Bertozzi, Ref. 33).

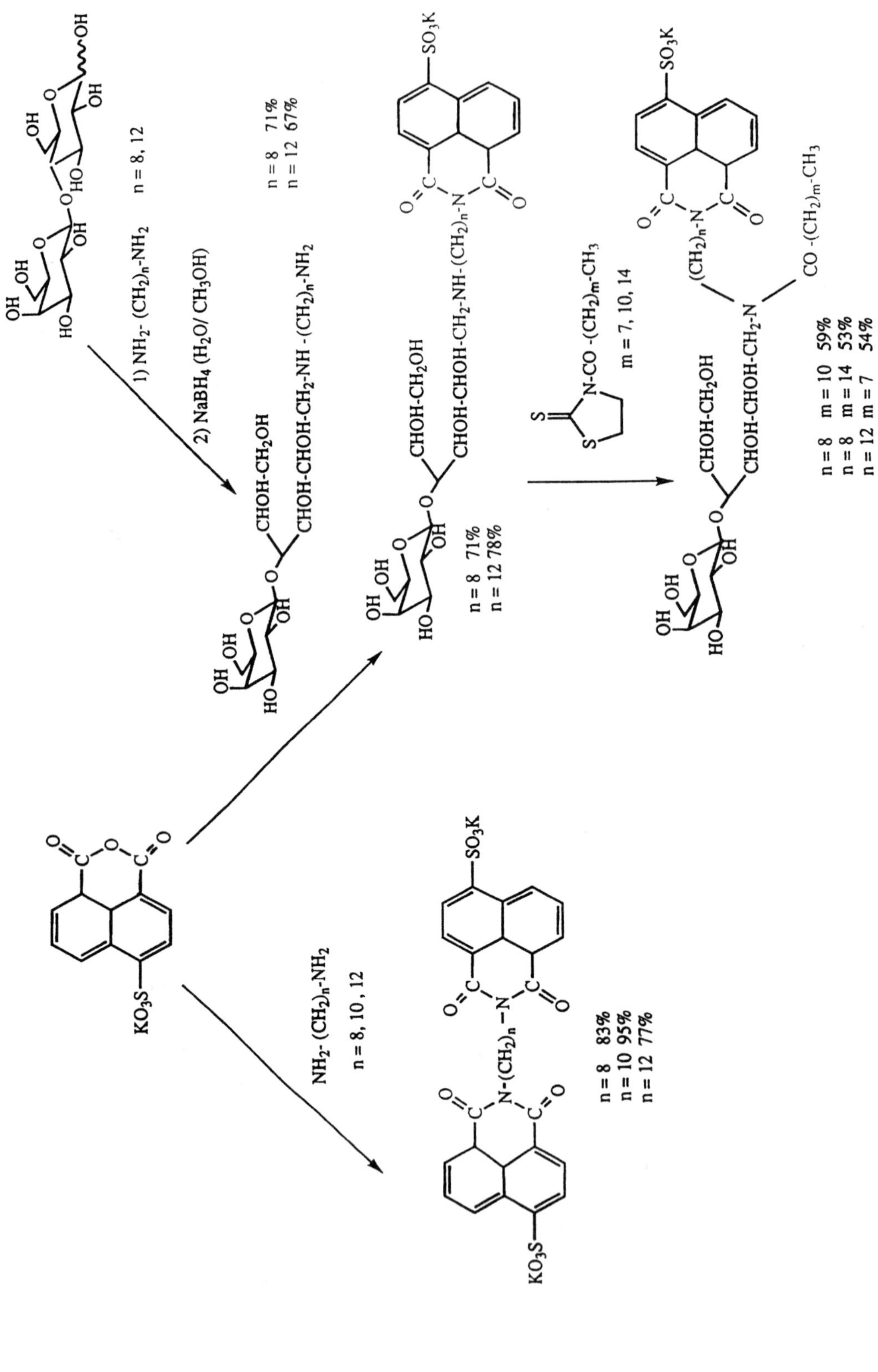

SCHEME 6 Synthetic scheme of *N*-substituted-1,8-naphthalimide derivatives.

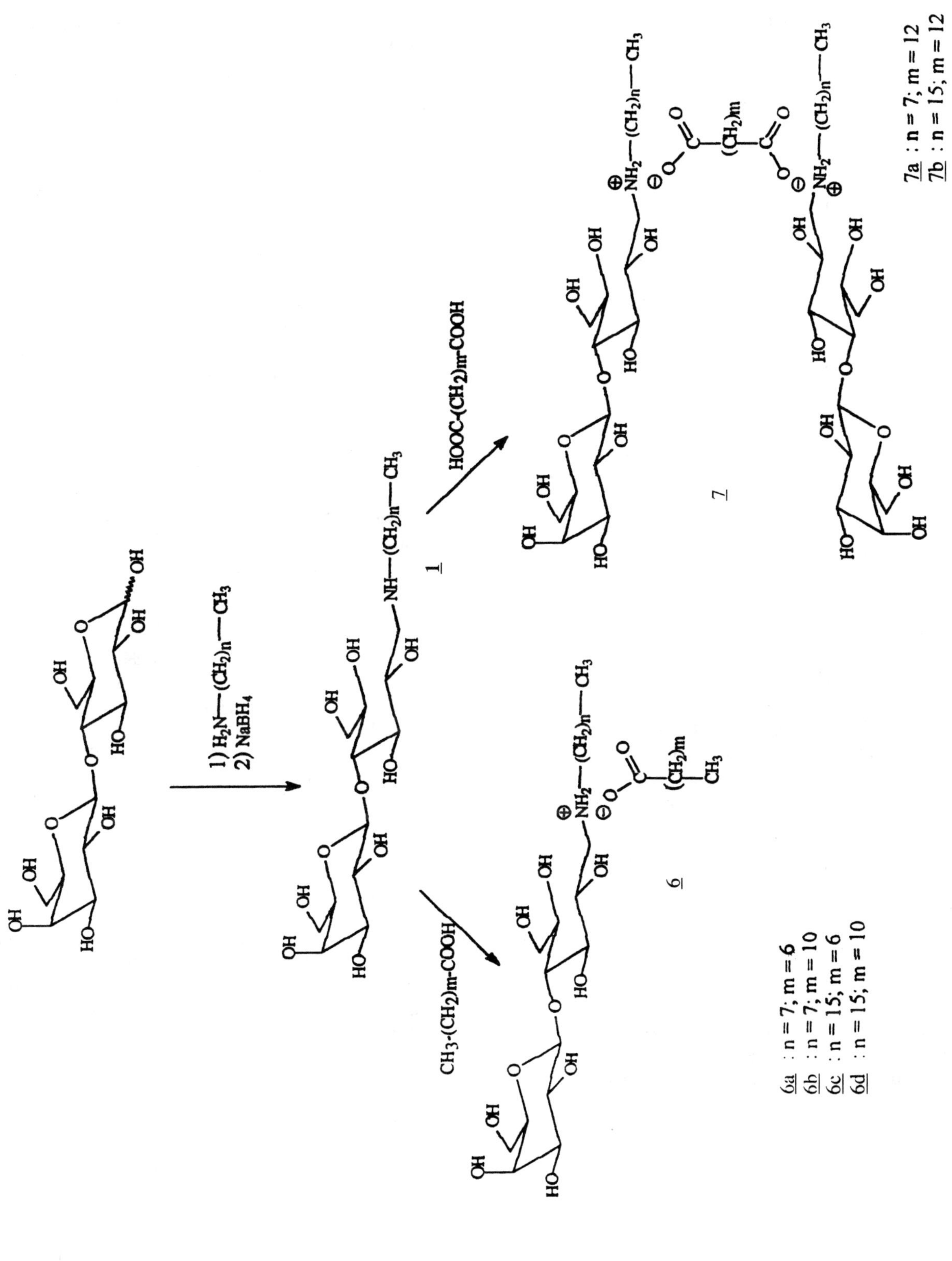

SCHEME 7 Synthetic scheme for preparation of catanionic compounds.

TABLE 9 Catanionic Anti-HIV Activities

Compound	EC_{50} (μM)	CC_{50} (μM)	IS	Log *P*
6a	>1000	>1000	—	1.7
6b	100	>100	>1	3.3
6c	16	38	2.3	4.9
6d	0.9	2.5	2.7	6.5
7a	500	600	1.1	2.1
7b	0.5	>100	>200	8.4
CA 52	50	220	4.5	4.5

served, the monomeric forms appear to be the most active, as the EC_{50} values are below the respective cmc.

The last results provide good reasons to use these easily synthesized catanionic compounds to prepare spontaneous surfactants with interesting properties, not only in therapeutic applications but also for encapsulation and vectorization.

VIII. CONCLUSIONS

In summary synthetic glycolipids are of interest for their biological applications. Therefore, depending on their structures, these new series of glycolipids have applications as mimics of natural ligands of proteins.

An interesting new strategy concerns the formation of gemini catanionic glycolipids. With this simplified synthesis we are now able to test many self-assembled entities offering considerable advantages over stepwise bond formation in the construction of active molecules.

REFERENCES

1. B. Brancq, L'Act Chim (AS-I) *2–3*:63–65 (1996).
2. M. Von Itzstein et al., Nature *363*:418–423 (1993).
3. L. A. Lasky, Science *258*:964–969 (1992).
4. J. S. Lazo and D. E. de Quinn, Anal. Biochem. *102*: 68–70 (1980).
5. G. W. Stubbs, H. G. Smith, and B. Litman, Biochim. Biophys. Acta *425*:46–50 (1976).
6. P. Latge, I. Rico, R. Garelli, and A. Lattes, J. Dispersion Sci. Technol. *12*:227–237 (1991).
7. P. Latge, I. Rico, A. Lattes, and L. Godefroy, French Patent 9005338 (1990).
8. F. Costes, M. El Ghoul, M. Bon, I. Rico-Lattes, and A. Lattes, Langmuir *11*:3644–3647 (1995).
9. M. El Ghoul, P. Latge, I. Rico, A. Lattes, and L. Godefroy, EP 924014038 (1992).
10. P. Latge, M. Bon, I. Rico, and A. Lattes, New J. Chem. *16*:387 (1992).
11. I. Rico-Lattes, J. C. Garrigues, E. Perez, C. André-Barrès, C. Madelaine-Dupuich, and A. Lattes, New J. Chem. *19*:341–344 (1995).
12. C. André-Barrès, C. Madelaine-Dupuich, and I. Rico-Lattes, New J. Chem. *19*:345–347 (1995).
13. R. Garelli-Calvet, F. Brisset, I. Rico, and A. Lattes, Synth. Commun. *23*:35–44 (1993).
14. R. Garelli-Calvet, F. Briset, I. Rico-Lattes, A. Lattes, and L. Godedroy, French Patent 91128511 (1991).
15. J. Fantini, D. Hammache, O. Delezay, N. Yahi, C. André-Barrès, I. Rico-Lattes, and A. Lattes, J. Biol. Chem. *11*:7245–7252 (1997).
16. C. Cupuy, X. Auvray, C. Petipas, R. Anthore, F. Costes, I. Rico-Lattes, and A. Lattes, Langmuir *12*:3162–3172 (1996).
17. I. Rico-Lattes and A. Lattes, Colloids Surf. A *123–124*: 37–48 (1997).
18. C. Dupuy, X. Auvray, C. Petipas, R. Anthore, I. Rico-Lattes, and A. Lattes, Langmuir *14*:91–98 (1998).
19. F. Brisset, R. Garelli-Calvet, C. Chebli, I. Rico-Lattes, and A. Lattes, New J. Chem. *20*:595–605 (1996).
20. C. Dupuy, X. Auvray, C. Petipas, I. Rico-Lattes, and A. Lattes, Langmuir *13*:3965–3967 (1997).
21. R. Garelli-Calvet, P. Latge, I. Rico, A. Lattes, and A. Puget, Biochim. Biophys. Acta *1109*:55–58 (1992).
22. (a) A. Dolev, W. W. K. Rohwedder, T. L. Munts, and H. J. Putton, Lipids *2*:33–38 (1967); (b) W. Grosh, F. Weber, and K. M. Fisher, Ann. Technol. Agric. *26*:133–137 (1977).
23. X. Auvray, C. Petipas, R. Anthore, I. Rico-Lattes, and A. Lattes, Langmuir *11*:433–439 (1995).
24. A. Lattes and I. Rico, Pour la Science *137*:44–50 (1992).
25. (a) M. J. Gibian and P. Vandenberg, Anal. Biochem. *163*:343–346 (1987); (b) B. Axelrod, T. M. Cheeselbrough, and S. Laakso, Methods Enzymol. *71*:441 (1981).
26. C. B. Moore, D. Law, and D. W. Denning, J. Antimicrob. Chemother. *322*:831–836 (1993).
27. M. H. Villas-Boas, H. Egge, G. Pohlentz, R. Hartmann, and E. Barreto-Bergter, Chem. Phys. Lipds *70*:11–19 (1994).
28. D. D. Manning, L. E. Strong, Xin Hu, P. J. Beck, and L. L. Kiessling, Tetrahedron *53*:11937–11952 (1997).
29. J. C. Garrigues, E. Perez, M. D. Linas, I. Rico-Lattes, J. P. Seguela, and A. Lattes, J. Mycol. Med. *6*:111–117 (1996).
30. J. C. Garrigues, I. Rico-Lattes, E. Perez, and A. Lattes, Langmuir *14*:5968–5971 (1998).
31. J. M. Harouse, S. Bhat, S. L. Spitalnik, M. A. Laughlin, K. Stefano, D. H. Silberg, and F. Gonzalez-Scarano, Science *253*:320 (1991).
32. J. Fantini, D. G. Cook, N. Nathanson, S. L. Spitalnik, and F. Gonzalez-Scarano, Proc. Natl. Acad. Sci. U. S. A. *90*:2700–2704 (1993).

33. C. R. Bertozzi, D. G. Cook, W. R. Kobertz, F. Gonzalez-Scarano, and M. D. Bernadski, J. Am. Chem. Soc. *14*:10639–10641 (1992).
34. B. Guidetti, C. André-Barres, I. Rico-Lattes, and A. Lattes, Perspect. Drug Discov. Des. *5*:2234–2422 (1996).
35. J. Fantini, D. Hammache, O. Delezay, N. Yahi, C. André-Barrès, I. Rico-Lattes, and A. Lattes, J. Biol. Chem. *272*:7245–7252 (1997).
36. D. Rideout, R. Schinazy, C. D. Pauza, K. Lovelace, L. C. Chiang, T. Calogeropoupole, M. MacCarthy, and J. M. Elder, J. Cell. Biochem. *51*:446 (1993).
37. (a) H. Fukuda, K. Kawata, and H. Okuda, J. Am. Chem. Soc. *112*:1635 (1990); (b) R. Zana and H. Levy, Langmuir *13*:402–408 (1997).
38. L. L. Brasher and E. W. Kaler, Langmuir *12*:6270–6276 (1996).
39. O. Lockhoff, Angew Chem. Int. Ed. Engl. *30*:1611 (1991).
40. M. J. L. Castro, K. Kovensky, and A. Fernandez-Cirelli, Tetrahedron Lett. *38*:3995–3997 (1997).
41. J. M. Pestman, K. R. Jerpstra, M. C. A. Stuart, H. A. Van Doren, A. Brisson, R. M. Kellog, and J. B. F. N. Engberts, Langmuir *13*:6857–6862 (1997).
42. M. Blanzat, E. Perez, I. Rico-Lattes, D. Promé, J. C. Promé, and A. Lattes, Langmuir *15*:6163–6169 (1999).
43. (a) B. L. Schwartz, K. J. Light, and R. D. Smith, J. Am. Soc. Mass Spectrom. *5*:201–204 (1994); (b) G. Pocsfawi, G. Di Landa, P. Ferranti, A. Ritieni, G. Randazzo, and A. Malomi, Rapid Commun. Mass Spectrom. *11*:265–272 (1997).
44. (a) J. A. Loo, Mass Spectrom. Rev. *16*:1–23 (1997); (b) R. L. Winston and M. C. Fitzgerald, Mass Spectrom. Rev. *16*:165–179 (1997).

5

Surfactants for Supercritical and Near-Critical Fluids

TERRI CARSON* and SHARON L. WELLS University of North Carolina at Chapel Hill, Chapel Hill, North Carolina

JOSEPH M. DESIMONE University of North Carolina at Chapel Hill, Chapel Hill, and North Carolina State University, Raleigh, North Carolina

I. INTRODUCTION

Over the past century, scientists have developed considerable interest in liquid and supercritical CO_2. In part, this interest has been sparked by environmental concerns involving the emission of volatile organic compounds (VOCs), the use of chlorofluorocarbons (CFCs) in polymer manufacturing and coating industries, and the enormous amounts of water used in numerous industrial processes. The use of liquid and supercritical CO_2 ($scCO_2$) as an alternative solvent choice is attractive because of the unique properties related to solvent strength. The solvent power of these fluids can often approach, and occasionally exceed, those of organic solvents without placing the environment at risk. An example where this idea has been employed is in the coffee industry. Supercritical CO_2 is used in the decaffeination process of coffee where dichloromethane was previously used [1,2]. In addition to the pollution prevention opportunities provided through the utilization of CO_2, increased energy efficiency is an important opportunity associated with CO_2 use. Because CO_2 has a very low heat of vaporization relative to water and organic solvents, CO_2-based processes offer the potential for significant energy savings. This would be especially true for water-intensive industries such as coatings, pulp and paper, textile dyeing, and polymer manufacturing. Although such uses of CO_2 are advantageous, their full development in other industrial processes, organic synthesis, and polymerizations has not been realized. However, with increased research efforts from both academic and industrial sectors, the use of dense CO_2 will show its value in the years to come.

The unique solvent characteristics of supercritical fluids were discovered over 100 years ago [3]. Common gases such as CO_2 and ethylene were pressurized and found to dissolve complex organic compounds. A fluid is defined as supercritical when its temperature and pressure are higher than their critical-point value (T_c, P_c). The solvent strength (dielectric constant) is highly tunable by changes in the temperature or pressure of the system. The fluid exists as a single phase possessing properties of both a liquid and a gas. Its density can be similar to that of liquids while it simultaneously has gaslike viscosities.

Among the research studies conducted using supercritical fluids, CO_2 has been the solvent of choice for many applications for a number of reasons. It is relatively nontoxic, nonflammable, easily recycled, inex-

**Current affiliation*: Dow Chemical Company, Freeport, Texas.

pensive, and readily available. CO_2 can be obtained from natural sources and isolated from exhaust streams of power plants and industrial plants that produce ethanol, ammonia, hydrogen, and ethylene oxide [4]. The one-component phase diagram of CO_2 is shown in Fig. 1. From this diagram it can be seen that the critical point of CO_2 is easily accessible (T_c = 31.1°C, P_c = 73.8 bar) in comparison with other supercritical solvents such as water (T_c = 374°C, P_c = 218 atm). Products formed in CO_2 are often isolated as dry powders upon CO_2 removal, thus eliminating energy-intensive drying processes or contamination of aqueous waste. Some companies have recognized these benefits and implemented CO_2-based processes. For example, Ford Motor Company uses Union Carbide's CO_2-based Unicarb process to deposit coatings onto bumpers as opposed to using more hazardous solvent-based paints and primers [5]. Dupont has begun work on a $40 million test plant where poly(tetrafluoroethylene) (Teflon) will be synthesized in CO_2, replacing chlorofluorocarbon solvents that are harmful to the earth's atmosphere [6].

However, despite the numerous attractive features and widespread applications, significant advances using CO_2 are hindered because of the low solubility of most hydrophilic compounds and polymers in CO_2. The solubility characteristics of CO_2 are very similar to those of perfluorohexane and represent a relatively low dielectric medium. The dielectric constant ranges from 1.01 to 1.45 for gaseous CO_2 and 1.60 to 1.67 for liquid CO_2 [7]. Some polar molecules such as methanol are able to dissolve in CO_2 because of its strong quadrupole moment, but others such as amides, ureas, urethanes, and azo dyes demonstrate poor solubility [8,9]. The solubility properties of a number of small molecules have been reviewed elsewhere [10,11]. High-molar-mass polymers generally have limited solubilities in CO_2 [12]. A number of researchers have focused their efforts on explaining this phenomenon, and some insight can be found in examining lattice fluid theory. According to this theory, three major factors govern the solubility of amorphous polymers in CO_2: (1) solute-solute interactions, (2) solute-solvent interactions, and (3) solvent-solvent interactions [13]. Johnston and coworkers conducted an extensive study quantitatively modeling experimental cloud point curves for various amorphous poly(alkylene oxide), poly(alkyl acrylate), and poly(dimethylsiloxane) homopolymers [14]. It was found that solubility is dependent upon polymer-polymer interactions and that polymer-CO_2 interactions play a secondary role. The only classes of polymers that have appreciable solubility in CO_2 at mild conditions (T < 100°C, P < 350 bar) are fluoropolymers and silicones [15–18]. We reported the synthesis of poly-

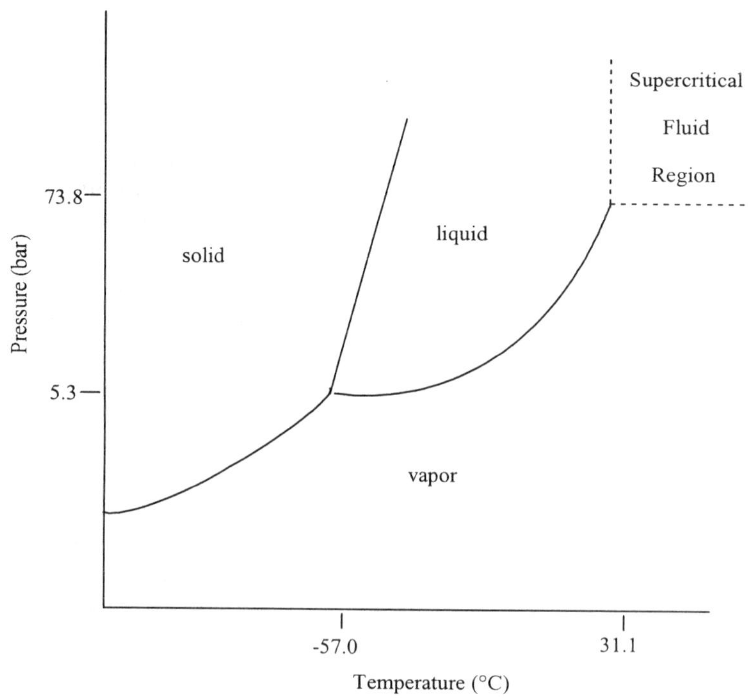

FIG. 1 One-component CO_2 phase diagram.

(1,1-dihydroperfluorooctyl acrylate) [poly(FOA)] using CO_2 as a continuous phase that yielded homogeneous solutions throughout the course of the reaction [18–21]. Krukonis and coworkers have demonstrated the solubility of poly(perfluoropropylene oxide) in CO_2 [22]. Barton and Kiran have reported on the high solubility of polydimethyl siloxane (PDMS) in CO_2 at approximately 450 bar [23–25].

Despite the inability of CO_2 to dissolve most polar or polymeric compounds, its solvency can be enhanced by the addition of surfactants that can drastically strengthen its effective solvating power for such substances. This chapter focuses on the efforts made toward the design and synthesis of surfactants for supercritical CO_2. These surfactants have been categorized into sections according to the "CO_2-philic" segment contained in the polymer (e.g., perfluoropolyethers, PDMS). A range of architectures have been explored, including homopolymers, block and graft copolymers, dendrimers, and small-molecule surfactants. Most often these materials were synthesized for the purpose of stabilizing hydrophilic or lipophilic compounds in CO_2 for such applications as cleaning, separations, or polymerizations.

II. FLUOROPOLYMER SURFACTANTS

A. Synthesis and Characterization in $scCO_2$

In 1992 our laboratory reported the homogeneous synthesis of a high molar mass amorphous fluoropolymer, poly(1,1′-dihydroperfluorooctyl acrylate) (PFOA) in CO_2 [18]. The reaction scheme is shown in Scheme 1. Polymerizations were extended to include the synthesis of statistical copolymers whereby FOA was copolymerized with monomers such as methyl methacrylate (MMA), styrene, ethylene, and butyl acrylate (BA). These copolymerizations proceeded homogeneously even with the addition of high concentrations of the comonomer (Table 1). Other fluorinated acrylate polymers have been synthesized in a similar manner, including poly[2-(*N*-methylperfluorooctane-sulfonamido)]ethyl acrylate, poly[2-(*N*-ethylperfluorooctane-sulfonamido)ethyl acrylate, and poly[2-(*N*-methylperfluorooctanesulfonamido)ethyl methacrylate [26]. Along with this investigation, decomposition rates and efficiency factors were measured for azobisisobutyronitrile (AIBN) in CO_2 and comparisons were drawn with conventional liquid solvents. It was found that AIBN decomposes at a rate 2.5 times slower in CO_2 than in benzene but with greater efficiency. This phenomenon

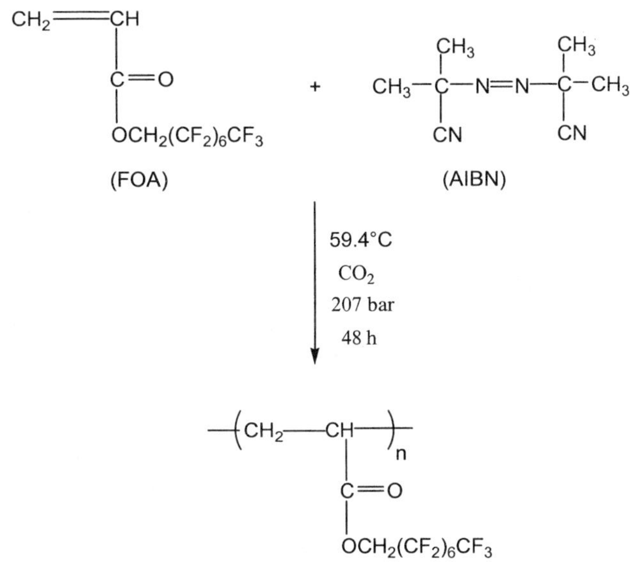

SCHEME 1 Synthesis of PFOA in $scCO_2$. The fluoroalkyl tail in PFOA contains about 25%—CF_3 branches.

can be explained in terms of a decreased solvent cage effect that exists in a low-viscosity supercritical medium. Hence, CO_2 can be considered an ideal solvent for free radical polymerizations with no chain transfer to solvent side reactions.

Solution properties of PFOA in CO_2 were studied using small-angle neutron scattering (SANS) [27]. McClain et al. demonstrated that SANS, after providing sufficient polymer-solvent contrast, was a viable technique to study homopolymer solutions in CO_2. SANS data were generated for a concentration series of both a high- and a low-molecular-weight PFOA in CO_2, $M_w = 1.4 \times 10^6$ g/mol and 1.1×10^5 g/mol, respectively. Samples were measured in dilute solution [$0.005 < C$ (g/mL) < 0.008] over a range of temperatures and pressures. The results of the SANS measurements are summarized in Table 2. The molecular weights measured by SANS were consistent with values expected from the synthetic conditions. The values for the radius of gyration (R_g) as a function of M_w were found to follow the function $R_g = (0.10 \pm 0.02)M^{1/2}$, where M is the molecular weight. This was the first study confirming that CO_2 is a thermodynamically good solvent for PFOA, as indicated by the measurement of positive values for the second virial coefficient A_2. In comparison, Martino et al. have shown that other CO_2-soluble polymers, poly(hexafluoropropylene oxide) (Krytox) and PDMS, have A_2 values that are zero and negative, respectively, at similar conditions [28]. Studies have also indicated that A_2 of PFOA not only is positive but

TABLE 1 Statistical Copolymers Containing FOA and Vinyl Monomers[a]

Copolymer	Feed ratio	Incorporated	Intrinsic viscosity (dL/g)
Poly(FOA-*co*-MMA)	0.47	0.57	0.10
Poly(FOA-*co*-styrene)	0.48	0.58	0.15
Poly(FOA-*co*-BA)	0.53	0.57	0.45
Poly(FOA-*co*-ethylene)	0.35	—	0.14

[a]Polymerizations were performed at 59.4 ± 0.1°C and 345 ± 0.5 bar for 48 h in CO_2. Intrinsic viscosity determinations were conducted in 1,1,2-trifluorotrichloroethane (Freon-113) at 30°C.

also is increased as the density of the CO_2 increases (S. Wells, M. Adam, M. Rubenstein, and J. M. DeSimone, in preparation).

DeSimone and coworkers also studied the synthesis of a CO_2-philic/hydrophilic amphiphilic graft copolymer [29]. Using the macromonomer technique, poly(FOA-*g*-PEO) copolymers were made and found to be completely soluble in CO_2 at 238 bar and 60°C (10 wt%). The microenvironment of these solutions was studied using solvatochromic characterization. Methyl orange (MO), a water-soluble dye that is insoluble in CO_2, was added to the CO_2 system in a solution with water. It was found that the PEO grafts enabled the solubility of the dye in CO_2, yielding bright orange solutions. This phenomenon was further confirmed by ultraviolet (UV) spectroscopy data that yielded a λ_{max} of 418 nm for the colored solution ($MO_{(aq)}$ has a λ_{max} of 464 nm). This blue shift is a result of decreasing solvent polarity, which has been observed by Zhu and Scheely [30].

The discovery of uptake of both methyl orange and water into the $scCO_2$ continuous phase by the PFOA-*g*-PEO graft copolymer led to studies using small-angle x-ray scattering (SAXS) [31]. Many copolymers molecularly designed for CO_2 applications are difficult to characterize because of their limited solubility in conventional solvents. Both SAXS and SANS have proved to be powerful techniques for obtaining structural information about aggregated systems in solution. The SAXS study by Fulton et al. was the first of many subsequent scattering experiments performed on aggregated amphiphilic polymeric systems in CO_2.

The PFOA-*g*-PEO polymer synthesized with 5 kg/mol PEO grafts was studied by SAXS. The experiments were performed at 60°C and three different pressures (470, 300, and 255 bar) in CO_2 in the presence of water. The water-to-surfactant ratio was 0.32. The data are shown in Fig. 2. As the pressure of the system was decreased, there was an increase in the scattered intensity. The increase is a result of not only the greater particle-solvent contrast at lower CO_2 densities but also the small increase in the size of the particles as pressure was decreased, as evidenced by the shift in the scattering peaks at lower scattering vectors (q). The oscillatory nature of the scattering curves is indicative of spherical core-shell structures. A depiction of the spherical micelle of PFOA-*g*-PEO with the collapsed PEO chains and water molecules within the core is shown in Fig. 3. A core-shell model was used to fit the SAXS data at low q. From the fits, the outer radii of the aggregates were found to be approximately 125 Å with relatively low polydispersities. The radii of the particles

TABLE 2 SANS Results for Concentration Series of Poly(FOA) in CO_2 at Various Conditions

Poly(FOA) sample	P (bar)	T (°C)	ρ^a (g cm^{-3})	R_g (Å)	A_2 ($\times 10^5$ cm^3 mol g^{-2})	M_w ($\times 10^{-6}$ g mol^{-1})
Low M_w	395	60	0.888	35 ± 0.15	9.5 ± 0.5	0.113 ± 0.006
High M_w	340	65	0.842	120 ± 13	1.9 ± 0.4	1.5 ± 0.4
High M_w	395	60	0.888	100 ± 6	4.1 ± 0.8	1.2 ± 0.3
High M_w	340	40	0.934	114 ± 9	2.5 ± 0.3	1.6 ± 0.3

[a]Density of pure CO_2 at these conditions.
Source: Adapted from Ref. 27.

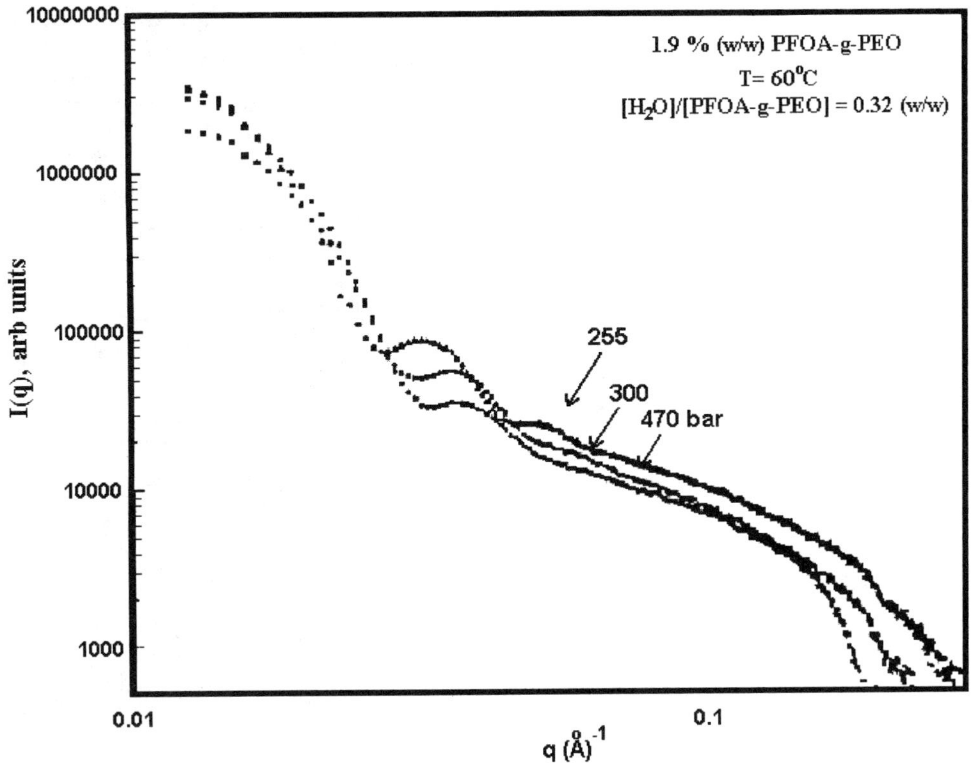

FIG. 2 Small-angle x-ray scattering spectra for 1.9% (w/w) PFOA-*g*-PEO in supercritical CO_2 at 60°C and three different pressures, 255, 300, and 470 bar. (Adapted from Ref. 30.)

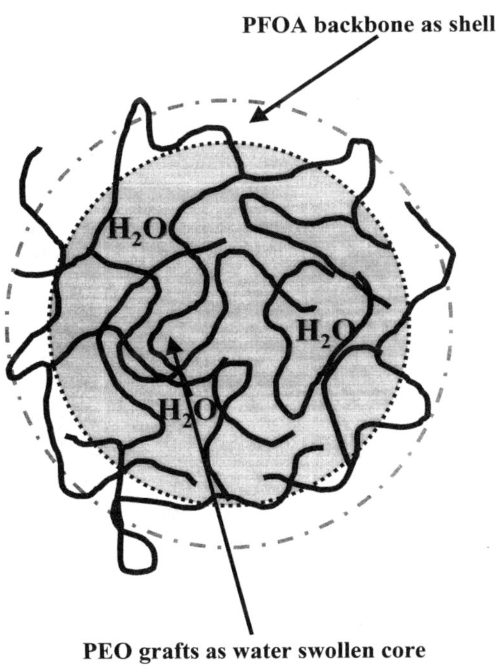

FIG. 3 Proposed structure of a PFOA-*g*-PEO graft copolymer micelle in supercritical CO_2. (Adapted from Ref. 30.)

increased when either the concentration was reduced or the pressure was decreased. The radius of the core was estimated to be 105 Å at the higher pressures and 125 Å at lower pressures. The number of PEO segments in the core was estimated to be about 600 based on the measured volume of the micelle core and PEO bulk density.

In attempts to mimic and further study the PFOA-*g*-PEO system, Chillura-Martino et al. performed SANS experiments on the same sample [28]. By taking advantage of the contrast differences between H_2O and D_2O, more conclusive data were obtained. The data showed not only that the micelles were swollen in the core but also that the shell was slightly swollen, suggesting that PEO segments also penetrate the shell.

Measurements were also performed on the polymer in the absence of water and showed that the micelle dimensions were smaller. Figure 4 shows the differences in the SANS data when H_2O and D_2O were present versus when there was no added water. The radius of gyration taken from a core-shell fit increased from ~56 Å (no added water) to 86 Å (H_2O swollen) and 136 Å (D_2O swollen).

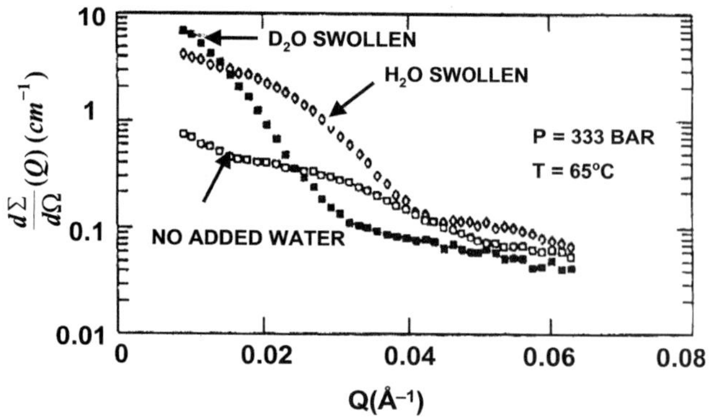

FIG. 4 The differential cross section per unit sample volume ($d\Sigma/d\Omega(Q)$) versus Q for PFOA-*g*-PEO graft copolymer in CO_2 before and after swelling with H_2O and D_2O. (Adapted from Ref. 28.)

B. Application in Heterogeneous Polymerizations

Materials useful as stabilizers in colloidal dispersions of lipophilic or hydrophilic polymers usually employ amphiphilic molecules. They contain anchoring segments that have an affinity for the polymer particles, most likely by physical adsorption, and a segment that is highly soluble in the continuous phase. DeSimone et al. demonstrated the amphiphilicity of PFOA and carried out the first successful dispersion polymerization in $scCO_2$ [32,33]. The polymerization of methyl methacrylate was carried out in a 10-mL high-pressure reaction view cell at 65°C and 204 bar using AIBN or a fluorinated derivative of AIBN as the initiator. Polymerizations conducted in the absence of stabilizer produced polymers in low yield. In remarkable contrast, the reactions with added stabilizer produced free-flowing powders in high yields upon removal of CO_2. Scanning electron micrographs displayed micrometer-sized particles with spherical morphologies and a relatively narrow size distribution (Fig. 5). Indeed, the amphiphilicity of PFOA contributed to the success of particle formation.

DeSimone et al. have also reported the successful dispersion polymerization of styrene in $scCO_2$ using amphiphilic diblock copolymers containing poly(styrene) (PS) and PFOA segments [34]. These materials were prepared via a controlled free radical method known as the iniferter technique. The detailed synthesis has been described by Guan and DeSimone and is illustrated in Scheme 2 [35]. With such block copolymers, polydisperse submicrometer-sized PS particles were produced by dispersion polymerization with spherical morphologies. It was also found that as the

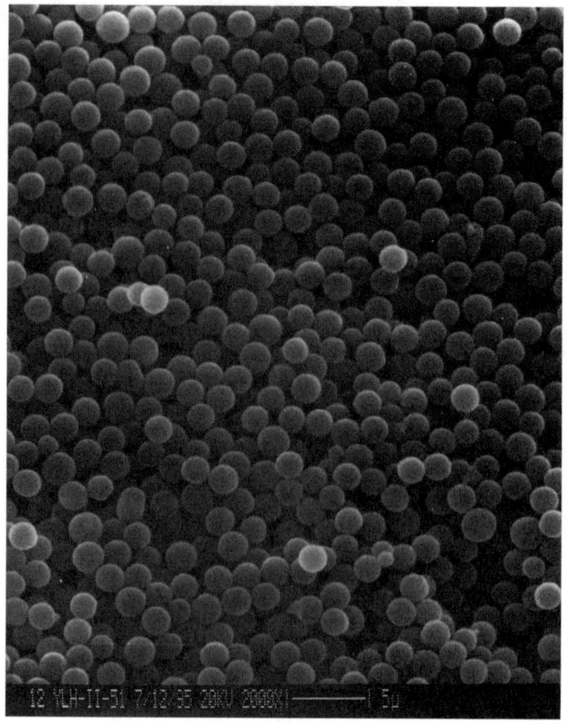

FIG. 5 Scanning electron micrograph of PMMA particles produced by a dispersion polymerization in $scCO_2$ using PFOA as stabilizer.

length of the stabilizing moiety increased, the particle size distribution decreased.

The PS-*b*-PFOA surfactants were further characterized by small-angle neutron scattering studies and found to self-assemble in solution to form multimolecular micelles [36]. Both SAXS and SANS measurements were performed on the samples at 65°C and 338 bar. The scattering curves confirm that spherical micelles are formed in solution. As with the PFOA-*g*-PEO samples, a core-shell model was used to fit the data (Fig. 6). Table 3 shows the polymer dimensions as well as the results from the fits. As the PFOA block length increases, as expected the thickness of the shell and size of the overall micelle increase. A core-shell model could not be applied to the 4K-*b*-245k copolymer because of the large asymmetry, which was expected to give rise to a different morphology. The form factor for an *f*-arm star polymer was applied to the data and gave results ($R_g \sim 200$ Å and $f \sim 7.7$ arms) consistent with those for other diblock copolymer samples.

After confirmation of micelle formation, the PS-*b*-PFOA surfactants were used to emulsify CO_2-insoluble PS oligomer [27]. SANS characterization of micelles of PS-*b*-PFOA surfactants in CO_2 (65°C, 340 bar) with

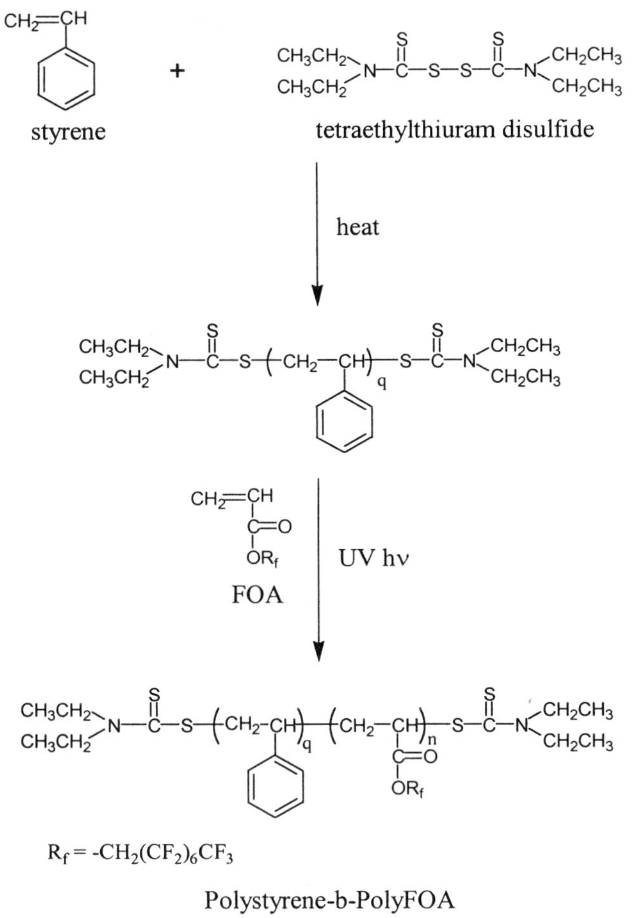

SCHEME 2 Synthesis of PS-*b*-PFOA via the iniferter technique.

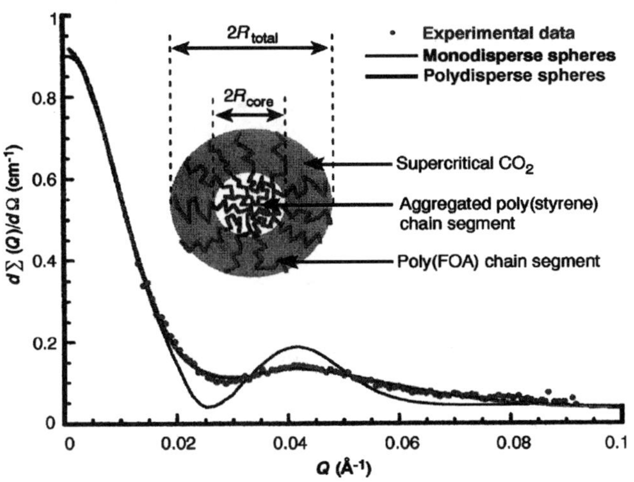

FIG. 6 Plot of $d\Sigma/d\Omega(Q)$ versus Q for 3.7^k-*b*-16.7^k PS-*b*-PFOA in $scCO_2$ (65°C, 340 bar) with fitting to a monodisperse and polydisperse core-shell model. (Adapted from Ref. 35.)

TABLE 3 Size of Core-Shell Micellar Aggregates for PS-*b*-PFOA in $scCO_2$ at 65°C and 338 bar as a Function of Shell (Corona) Block Length

Copolymer (PS-*b*-PFOA)	Aggregation no.	R_1 (Å)	R_2 (Å)
4k-*b*-17k	6.6	25.7	87.0
4k-*b*-40k	7.1	26.9	88.4
4k-*b*-61k	6.9	26.7	103.1
4k-*b*245k	<1 (core shell) 7.7 (*f*-arm)		

Source: Adapted from Ref. 35.

added hydrogenated and dueterated oligomer showed stabilization of >99% of the added oligomer into the core of the micelle. The micellar core volume increased with added oligomer as a function of oligomer concentration as can be seen in Fig. 7. There was an approximately eightfold increase in the volume of the micelle core with the addition of up to 20% (w/w) oligomer.

SAXS and SANS experiments provided conclusive data on the size and shape of micelles formed from PS-*b*-PFOA block copolymers in CO_2. However, to understand completely the CO_2-polymer interactions, complementary high-pressure, high-resolution (HPHR) nuclear magnetic resonance (NMR) spectroscopy experiments were conducted over a range of densities at a fixed temperature [37]. Figure 8 shows HPHR ^{1}H NMR spectra of the PFOA-*b*-PS recorded at 65°C and CO_2 densities between 0.26 and 0.85 g/cm^3. At low and intermediate CO_2 densities, the NMR detected only the

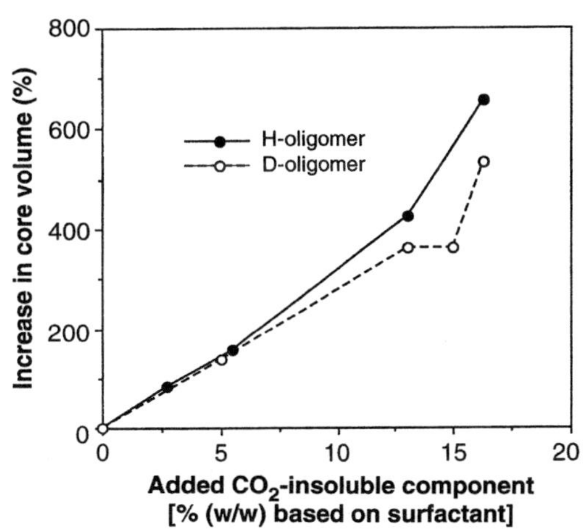

FIG. 7 Swelling of 3.7^k-*b*-39.8^k surfactant micelles in CO_2 (65°C, 340 bar) with PS oligomers. Surfactant concentration = 4% (w/v). (Adapted from Ref. 27.)

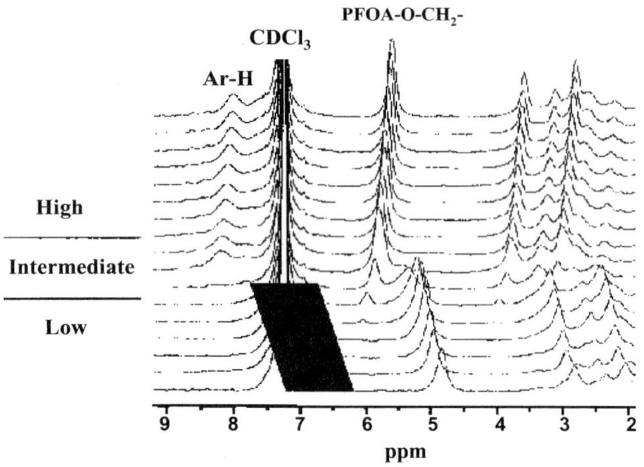

FIG. 8 High-pressure H NMR spectra of PS-*b*-PFOA in CO_2 at T = 66.6°C and CO_2 densities between 0.26 g/cm^3 (bottom trace) and 0.85 g/cm^3 (top trace). The shaded field indicates the region where signals from the PS block are expected for the spectra taken at low densities. (Adapted from Ref. 36.)

PFOA block. The aromatic signals for polystyrene remain undetected up to high CO_2 densities, indicating that the polystyrene was completely immobilized on the NMR time scale. This suggests that the PS is in the core and is neither solubilized nor plasticized by CO_2 in low and intermediate density ranges. In the high-density range, PS peaks appeared with intensities corresponding to about 35% of the total protons, indicating that a fraction of the micelle core became plasticized (Fig. 9).

In an effort to produce stable latexes of poly(vinyl acetate) (PVAc) and ethylene-vinyl acetate (EVA) in CO_2, DeSimone and coworkers have synthesized fluorinated and siloxane-based stabilizers including homopolymers, block copolymers, and reactive macromonomers [38]. The surfactants containing fluorinated acrylate segments were synthesized similarly to the PS-containing materials utilizing the iniferter technique. Polymerizations with these materials resulted in stable latexes at high conversion even in the absence of stirring. No apparent trend was observed in the results when the chemical composition of the stabilizers was varied. All reactions proceeded to high conversions and by visual observation produced white, opaque, and stable latexes. The solid latex particles could not be examined by electron microscopy because of the low T_g of PVAc, but the latex was characterized by turbidimetry. Polymerizations conducted in the presence of the fluorinated acrylate surfactants were found to be most turbid after 1 h and the surfactant with the longest blocks (PVAc: $M_n = 3.1 \times 10^4$ g/mol; PFOA: $M_n = 5.4 \times 10^4$ g/mol) produced the smallest diameter polymer particles.

The synthesis of highly cross-linked polymeric microspheres using $scCO_2$ as the solvent medium was reported [39]. These experiments were carried out with and without the addition of block copolymers of methyl methacrylate and perfluoroalkyl methacrylates that were prepared using screened anionic polymerizations. Polymerizations were conducted whereby divinylbenzene and ethylvinylbenzene were copolymerized at 65°C and 310 bar using AIBN as an initiator. In the absence of stabilizer and under certain specific conditions, relatively uniform, nonporous microspheres were produced with diameters ranging from 1 to 5 μm. Experimental variables such as the cross-linker ratio, monomer concentration, cross-linker structure, and mechanical agitation were examined. In an effort to produce particles under nonspecific conditions, the polymerizations were conducted in the presence of surfactant (0.25–3 wt%). At lower concentrations of surfactant, the microspheres showed a high degree of ag-

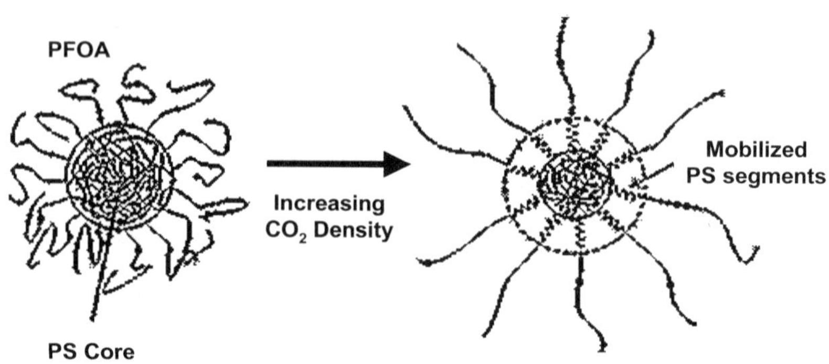

FIG. 9 Schematic representation of the solubilization of the PFOA-*b*-PS block copolymer in supercritical CO_2. After the CO_2 density was increased above the critical value, PS units at the core-shell interphase became mobilized. (Adapted from Ref. 36.)

gregation and a broad particle size distribution. Only when 3 wt% surfactant was used did the microspheres become more uniform and less aggregated. Furthermore, the particle sizes were much smaller (<0.5 μm). As a result of this smaller size and decreased agglomeration, the particles formed milky white suspensions in certain organic solvents in the presence of stabilizer that were stable for several weeks.

III. POLYSILOXANE SURFACTANTS

The use of silicone-based surfactants for applications in CO_2 is more attractive than the use of fluorinated materials for a number of reasons. Siloxane-based stabilizing systems are cheaper and more soluble in conventional solvents, which facilitates characterization. In addition, these surfactants can be synthesized via anionic polymerizations leading to narrow molecular weight distributions. Several groups have studied polysiloxanes in CO_2 and reported the high solubilities of these materials [25,40,41]. Hoefling and coworkers have designed and synthesized silicone-based amphiphiles and explored the relationship between structure and solubility via phase behavior experiments [42].

A. Developments in Dispersion Polymerizations

Yates and coworkers reported the design and synthesis of silicone-containing "ambidextrous" surfactants [43]. In this study surfactants were developed that are able to stabilize PMMA particles produced by dispersion polymerizations in CO_2, followed by their transfer to water, which results in an aqueous latex. Two surfactants were investigated, the first a block copolymer containing PDMS (M_w = 5500 g/mol) and methacrylic acid (M_w = 900 g/mol) and the second a graft copolymer of PDMS and pyrrolidonecarboxylic acid (M_w = 8500 g/mol). Although use of the graft copolymer produced smaller and more uniform PMMA particles ($\sim$3 μm), they rapidly flocculated in a dispersed aqueous buffer solution. However, the particles produced with the block copolymer resulted in an electrostatically stabilized latex at concentrations up to 10 wt% in buffered solutions of pH 8 and 11. The mechanism of stabilization is understood in terms of electrostatic repulsion whereby transfer of the particles from CO_2 to water results in collapsed PDMS chains on the surface of the particle while the acidic segment ionizes (Fig. 10). Characterization of surface charges was carried out using zeta potential measurements.

DeSimone et al. have reported the design and synthesis of siloxane-based block copolymers employed as stabilizers in the dispersion polymerizations of both styrene and vinyl acetate [38,44]. As discussed previously, the fluorinated counterparts were also prepared and employed in these systems. For the PS system, the siloxane-based polymers proved to be efficient stabilizers, resulting in high polymer yields and molecular weights. The best results were achieved with polymerizations in which shorter PDMS segments were employed (25K versus 65K). The surfactants used in the VAc polymerizations were not as effective as the fluoropolymer stabilizers. Collapsed latexes were observed at high conversion, which can be explained by observing the aggregation behavior of PDMS in pure CO_2 [28]. It was suggested that at low conversions, the VAc behaves as a cosolvent for the PDMS chains, allowing them to swell in the continuous phase, but as VAc is depleted the solvent polarity decreases and the PDMS chains collapse.

IV. PERFLUOROPOLYETHER SURFACTANTS

In 1985 Krukonis observed the solubility of perfluorinated polyethers in carbon dioxide while conducting fractionation studies on a commercially available high-molecular-weight perfluoropolyether oil [16]. This polymer was fractionated at 80°C over a pressure range from 82 to 275 bar. It was found that although this oil had been molecularly distilled prior to extraction, it could be separated into fractions having an even narrower molecular weight distribution using supercritical fluid extraction techniques with CO_2. Since this report, a considerable amount of research has focused on the design and synthesis of fluoroether functional surfactants. Hoefling and colleagues demonstrated the high CO_2 solubility of two nonionic perfluoropolyether surfactants, hydroxyaluminum bis[poly(hexafluoropropylene oxide) carboxylate] and poly(hexafluoropropylene oxide) carboxylic acid, and an anionic surfactant, sodium poly(hexafluoropropylene oxide) carboxylate [45,46]. The twin-tail surfactant was selected for its potential usefulness as a CO_2 thickener, but no appreciable increase in the solution viscosity was observed. Further work by Adamsky and Beckman employed an amide end-capped poly(hexafluoropropylene oxide) surfactant in the inverse emulsion polymerization of acrylamide in CO_2 [47]. These polymerizations were conducted in the presence of water, a cosolvent for the monomer, at a CO_2 pressure of 345 bar and 60°C using AIBN as the initiator. When surfactant was added to

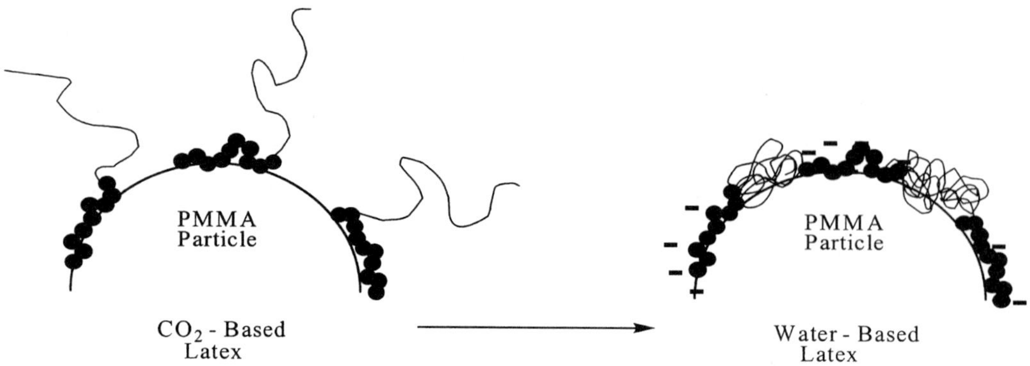

FIG. 10 Illustration of the mechanism of stabilization for the "ambidextrous" surfactant concept. (Adapted from Ref. 42.)

the polymerizations, the CO_2 solution had a milky white appearance similar to that of conventional inverse emulsion polymerizations. In addition, intrinsic viscosity measurements of the resultant polymer revealed significant molecular weights compared with conventional emulsion polyacrylamide. However, in the absence of any surfactant the solution appeared similar to that in a dispersion polymerization of an aqueous solution in CO_2. Phase separation occurred immediately when agitation ceased, and a large mass of precipitated polymer was collected.

The effect of structure on the CO_2 solubility of perfluoropolyether functional materials has been studied by Adamsky and Beckman who investigated perfluoropolyethers containing carboxylic acid, carboxylate, and hydroxy aluminum headgroups [48]. It was found that increasing the polarity of the polar headgroup induced the cloud point curve to move to higher pressures. The nature of the carboxylate counterion was also examined with respect to CO_2 solubility and the following trend was determined: $NH_4^+ > Na^+ > K^+ > Ca^+$. The fluoroether carboxylate salts also extracted thymol blue from an aqueous solution in CO_2.

Fluorinated polyether surfactants have been prepared for CO_2-based applications such as heavy metal extraction, dispersion polymerizations, and protein extraction [49]. In the case of heavy metal extraction, a perfluoropolyether containing a dithiol end group has been synthesized and found to extract as much as 98% of mercury from contaminated soil in laboratory-scale CO_2 extractions. It was determined that over 90% of the total extraction occurred during the first hour. The dispersion polymerization of MMA has also been carried out in CO_2 using a graft copolymer stabilizer, poly(MMA-*co*-hydroxyethyl methacrylate)-*g*-Krytox [50]. Factors such as the graft chain length and the graft density were varied in order to determine their effect on the PMMA colloid. It was found that increasing the graft density resulted in efficient stabilization, and a change in the graft length produced varied results depending on the backbone length.

Perfluoropolyethers derivatized with sorbitol ester, sulfate, and sulfonate groups have been synthesized to facilitate the extraction of proteins and polar molecules in CO_2 [49]. Emulsion studies were conducted with this surfactant system and the formation of three phase transitions was observed: emulsion in the continuous water phase (WI), emulsion in the continuous CO_2 (WII), and emulsion at the water-CO_2 interface (WIII). Furthermore, an investigation of the pressure effect on emulsion behavior revealed that the high compressibility of CO_2 allows WI $\rightarrow$ WIII $\rightarrow$ WII phase transitions that closely resemble those found with increased electrolyte concentrations.

Beckman et al. have reported the synthesis of biotin-functionalized fluoroether surfactants used in the extraction of avidin into CO_2 [51]. Extraction occurred by both inverse and three-phase emulsions but was twice as high for the three-phase system.

V. NONPOLYMERIC AMPHIPHILES IN CO_2

The formation of reverse micelles in supercritical alkanes was reported by researchers at the University of Texas at Austin and Battelle's Pacific Northwest Laboratories in the late 1980s [52–54]. The surfactants employed in this study were commercially available ionic amphiphiles with hydrocarbon tails, and they dramatically increased the solubility of amino acids, water-soluble polymers, proteins, and metal-containing compounds [53]. However, extension of the use of these surfactants in CO_2 had limited success because of the poor to negligible solubility of the surfactants.

To address this issue of solubility, new surfactants have been synthesized containing fluorocarbon tails. Fulton et al. studied reverse micelles formed from $F(CF_2)_{6-10}CH_2CH_2O(CH_2OCH_2O)_3$ or (FSO-100) in CO_2 using SAXS [31]. These studies indicated that the nonionic surfactant completely dissolved at high pressures and 65°C. The small PEO units and broad molecular weight distributions of the samples promoted the formation of polydisperse micelles of the order of 84 Å. Johnston and coworkers also demonstrated the use of a hybrid fluorocarbon-hydrocarbon surfactant $C_7F_{15}CH(OSO_3^-Na^+)C_7H_{15}$ for the purpose of creating water in carbon dioxide microemulsions [55]. The water-to-surfactant ratio in a single-phase microemulsion was found to be as high as 32 at 25°C and 231 bar. Over 150 surfactants were studied in $scCO_2$ microemulsion systems and none were shown to take up more than a few water molecules per surfactant molecule [56–58]. Subsequent work by this group employed an ammonium carboxylate perfluoropolyether surfactant, $[(OCF_2CF(CF_3))_n(OCF_2)_m]OCF_2COO^-NH_4^+$, in the formation of aqueous microemulsion droplets in CO_2. Indeed, the incorporation of CO_2 functional groups facilitates high solubility of water in these systems [59]. Further work concerning microemulsion formation in nonpolar supercritical fluids has been published in two reviews [60,61].

VI. DENDRITIC SURFACTANTS

Cooper et al. showed that dendrimers with fluorinated shells not only are soluble in CO_2 but also are powerful aids in transporting CO_2-insoluble molecules into the solvent [62]. Extending the methods developed by Meijer, a forth-generation hydrophilic dendrimer, DAB-denr-$(NH_2)_{32}$, was functionalized by a heptamer acid fluoride of hexafluoropropylene oxide, $CF_3CF_2\cdot CF_2(OCF(CF_3)CF_2)_5$-$OCF(CF_3C(O)F$ (Scheme 3) [63].

This generated a well-defined, unimolecular dendritic "micelle" with a CO_2-philic shell and a hydrophilic core. The functionalized dendrimer was insoluble in water and most common organic solvents but soluble in CO_2 at very moderate conditions (room temperature and 76 bar). However, the unfunctionalized dendrimer was insoluble in CO_2 under the same conditions. The radius of gyration and mass of the dendrimer measured in CO_2 at 25°C and 340 bar by SANS were 30.0 Å and 3.35×10^3 g/mol, respectively.

Spectroscopic analysis and extraction studies (Fig. 11) demonstrate that the dendritic micelle can transfer

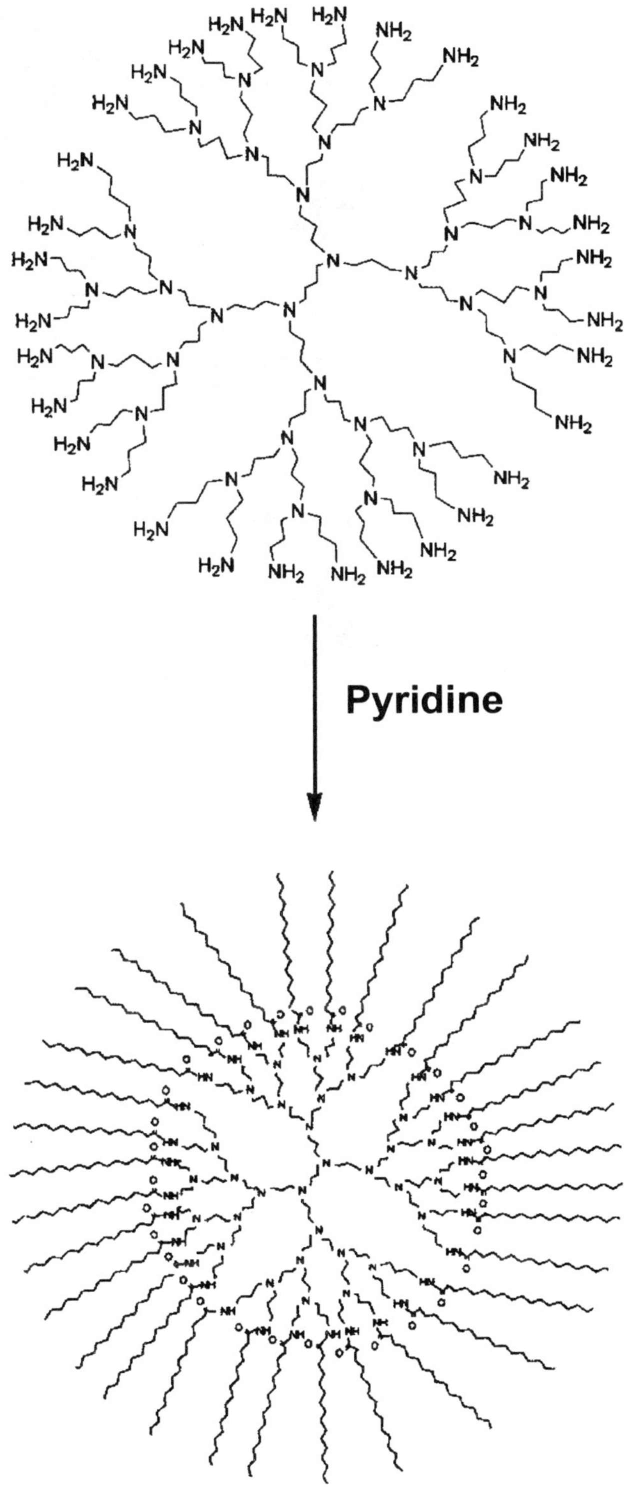

2=F(O) $CCF(CF_2)$-O-$(CF_2$-$CF(CF_3)$-$O)_3$-$CF_2CF_2CF_3$

~~~~ = -$CF(CF_3)$-O-$(CF_2$-$CF(CF_3)$-$O)_3$_$CF_2CF_2CF_3$

SCHEME 3 Synthesis of the unimolecular dendritic micelle.
~~~~

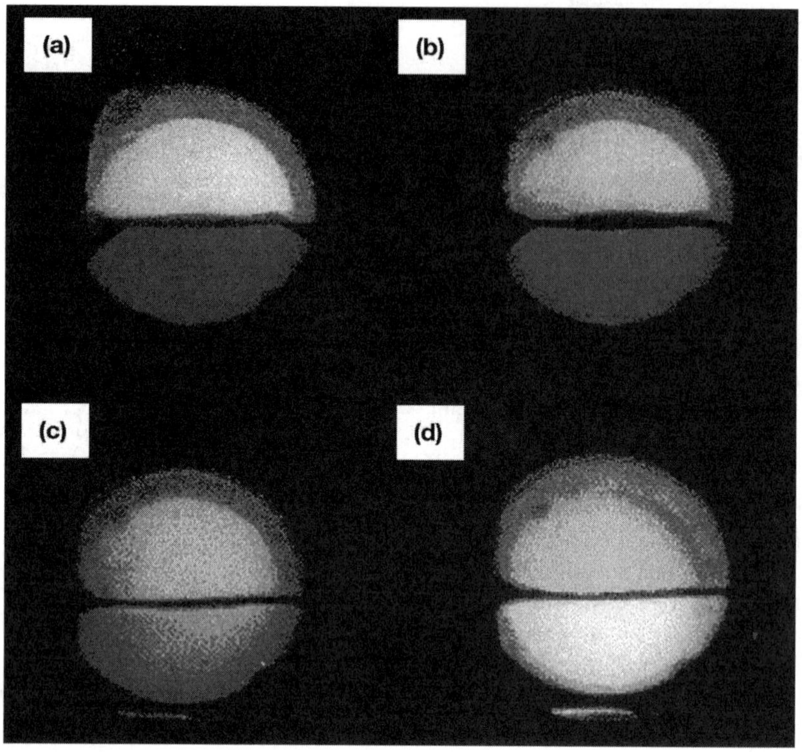

FIG. 11 Diffusion of methyl orange from an aqueous solution (upper phase in figure) into a unimolecular dendritic micelle in liquid CO_2 (lower phase) at 23.5°C, 340 atm. Video images taken 1 min (a), 6 min (b), 30 min (c), and 150 min (d) after the addition of CO_2. (Adapted from Ref. 61.)

methyl orange, a CO_2-insoluble ionic dye, from aqueous solution into carbon dioxide. When there was an excess of the dendrimer molecule, analysis showed that no residual dye was detected in the aqueous phase, indicating complete extraction from the aqueous layer. The maximum number of methyl orange molecules extracted per dendritic core was approximately 12, calculated from the ultraviolet spectroscopy. The maximum number of extracted molecules (seven) was lower when a larger dye molecule, rose bengal, was extracted.

VII. NEW FRONTIERS—REVERSIBLE CONTROL OF SELF-ASSEMBLY

McClain et al. were the first to postulate the existence of a critical micelle density (CMD), analogous to a critical micelle concentration, where polymeric surfactants undergo a unimer-to-aggregate transition in supercritical fluids on changing the solvent (quality) density [64]. In the PS-*b*-PFOA systems, when the solvent density was increased, smaller and more polydisperse micelles formed. The decrease in the size of the micelles was attributed to the increased solvation of both the PS and PFOA blocks, thereby decreasing the aggregation number and the size of the micelles. Although there was a decrease in aggregation number with increasing CO_2 density, the data did not indicate that the micelles actually broke apart into single chains in the range of densities studied.

A true CMD transition was realized when Triolo et al. studied the solution properties of PVAc-*b*-PFOA diblock copolymers in CO_2 by SANS and SAXS (R. Triolo et al., unpublished results). Figure 12 shows the SANS scattering curve for 6 w/v% of a PVAc-*b*-PFOA (PVAc: M_n = 10.3 kg/mol; PFOA: M_n = 63.1 kg/mol) sample in CO_2 at 45°C as a function of pressure. As the pressure (density) of the solution increased, there was a reduction in the size of the peak until the peak disappeared. The disappearance of this peak is indicative of the aggregate-to-unimer (single copolymer chains) transition, which occurred around 292 bar for this system. This first realization of a CMD has led to the synthesis, characterization, and study of a new series of molecularly engineered polymeric surfactants.

Buhler et al. were the first to use high-pressure light scattering methods to study the micellization and CMD

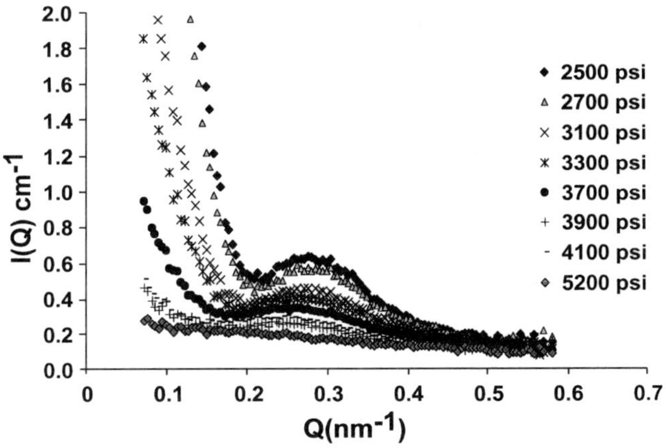

FIG. 12 SANS scattering curve for PVAc-*b*-PFOA in CO_2 as a function of pressure ($M_{n,\text{PVAc}}$ = 10.3 kg/mol; $M_{n,\text{PFOA}}$ = 63.1 kg/mol).

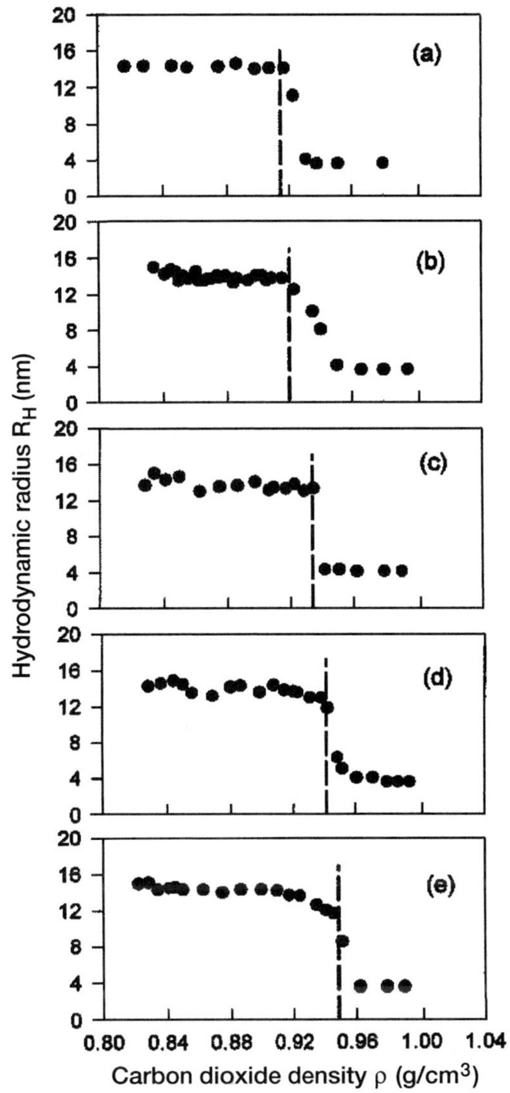

FIG. 13 Effect of the CO_2 density on the hydrodynamic radius for a copolymer concentration: (a) $c = 1.88 \times 10^{-3}$ g/cm^3, (b) $c = 3 \times 10^{-3}$ g/cm^3, (c) $c = 6 \times 10^{-3}$ g/cm^3, (d) $c = 1.125 \times 10^{-2}$ g/cm^3, and (e) $c = 2 \times 10^{-2}$ g/cm^3 at temperature T = 45°C. Dashed lines indicate the onset of the micelles-to-unimers transition. (Adapted from Ref. 64.)

behavior of polymeric surfactants in CO_2 [65]. Aided by high-power lasers, static and dynamic light scattering (DLS) methods have also proved to be powerful, reliable, and readily assessable techniques for studying the characteristics of polymers in CO_2. Very detailed studies of PVAc-*b*PFOA and PVAc-*b*-poly(1,1,2-tetrahydroperfluorooctyl acrylate) PTAN copolymers, both of which exhibited the CMD transition, have been reported [66]. Figure 13 shows the hydrodynamic size plotted against CO_2 density for a PVAc-*b*-PTAN (PVAc: M_n = 10.3 kg/mol/ PTAN: M_n = 60.4 kg/mol) solution. As predicted by SANS data, in the low-density range there are 15-nm spherical micelles. As the density was increased, the CMD transition region appeared leading into the 3- to 4-nm unimer region.

In addition to showing the CMD, DLS data were used to show the coexistence of unimers and micelles in the transition region. This coexistence prompted the construction of a copolymer surfactant–CO_2 binary phase diagram. The concentration (ranging from 0.0001 to 1 g/cm^3) versus carbon dioxide density phase diagram for the PVAc-*b*-PTAN surfactant at 45°C is shown in Fig. 14. Three regions on the phase diagram can be identified: a two-phase region at CO_2 densities below 0.82 g/cm^3 where the polymer was insoluble and phase separated into polymer-rich and solvent-rich phases, a region of spherical micelles at intermediate CO_2 densities, and a unimer region at high densities. The coexistence line connecting the micelle and unimer phases indicates not only the CMD at constant copolymer concentrations but also the critical micelle concentration at constant CO_2 density.

The CMD phenomenon implicates reversible control of self-assembly of polymeric surfactant in CO_2. To explore CMD possibilities further, a series of well-defined poly(*tert*-butylmethacrylate) (PTBM)-*b*-poly(fluorooctylmethacrylate) (PFOMA) samples with varying PTBM lengths have been synthesized (E. Yoshida et al., in preparation). These are the first series of diblock copolymers containing fluoropolymer segments made by anionic polymerization. The polymers have controlled molar masses and narrow molar mass and chemical composition distributions. With varied PTBM lengths, a series of phase diagrams have been generated

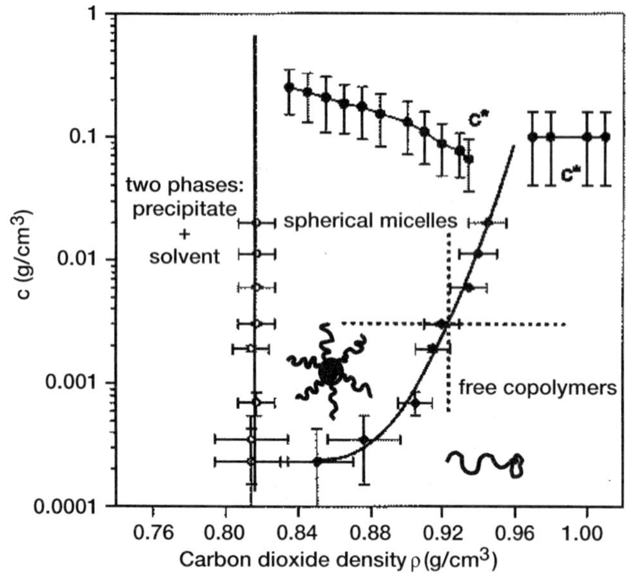

FIG. 14 Phase diagram in the copolymer–CO_2 density plane at a fixed temperature $T = 45°C$. Points (●) represent the cloud line (solubility line), points (♦) represent the spherical micelles-unimers transition, and points (○) represent the overlap concentration C^*. (Adapted from Ref. 65.)

for PTBM-*b*-PFOMA block copolymer solutions at 25°C. These phase diagrams illustrate how the CO_2-surfactant system can be manipulated and tailored to fit particular applications by small changes in both copolymer composition and CO_2 density. These innovative tools could have a major impact on CO_2 cleaning and extraction processes.

REFERENCES

1. K. Zosel, U.S. Patent 3,806,619 (1974).
2. R. Prasad, M. Gottesman, and R. Scarella, U.S. Patent 4,341,804 (1982).
3. P. Villard, J. Phys. *5*:455 (1896).
4. S. Barer and K. Stern in *Catalytic Activation of CO_2* (S. Barer and K. Stern, eds.), American Chemical Society, Washington, DC, 1988, p. 1.
5. D. Canelas and J. M. DeSimone, Chemistry in Britain August:38 (1998).
6. T. Romack, E. Maury, and J. DeSimone, Macromolecules *28*:912 (1995).
7. F. Keyes and J. Kirkwood, Phys. Rev. *36*:754 (1930).
8. K. O'Shea, K. Kirmse, M. Fox, and K. Johnston, J. Phys. Chem. *95*:7863 (1991).
9. J. Hyatt, J. Org. Chem. *49*:5097 (1984).
10. K. Bartle, A. Clifford, S. Jafar, and G. Shilstone, J. Phys. Chem. *20*:728 (1991).
11. M. King, A. Mubarak, J. Kim, and T. J. Bott, Supercrit Fluids *5*:296 (1992).
12. M. McHugh and V. Krukonis, *Supercritical Fluid Extraction: Principles and Practice*, 2nd ed., Butterworth-Heinemann, Stoneham, MA, 1994.
13. I. Sanchez and C. Panayiotou, in *Equation of State Thermodynamics of Polymer and Related Solutions* (I. Sanchez and C. Panayiotou, eds.), Marcel Dekker: New York, 1993, pp. 187–285.
14. M. O'Neill, Q. Cao, M. Fang, and K. Johnston, Ind. Eng. Chem. Res. *37*:3067–3079 (1998).
15. I. Yilgor, J. McGrath, and V. Krukonis, J. Polym. Bull. *12*:499 (1984).
16. V. Krukonis, Polym. News *11*:7 (1985).
17. Z. Guan, J. C. Y. Menceloglu, and J. DeSimone, Macromolecules *26*:2663 (1993).
18. J. DeSimone, Z. Guan, and C. Elsbernd, Science *257*: 945 (1992).
19. Z. Guan, J. Combes, C. Elsbernd, and J. DeSimone, Polym. Prepr. (Am. Chem. Soc. Div. Polym. Chem.) *34*:446 (1993).
20. P. Ehrlich, Chemtracts Org. Chem. *6*:92 (1993).
21. J. DeSimone, Int. Patent Appl. PCT/US93/01626 (1993).
22. V. Krukonis, Supercritical fluid fractionation: An alternative to molecular distillation, presented at AIChE Meeting, Houston, TX, March 1983.
23. G. Dris and S. Barton, J. Appl. Polym. Sci. *55*:773 (1995).
24. Y. Xiang and E. Kiran, Polymer *36*:4817 (1995).
25. X. Zhao, R. Watkins, and S. Barton, J. Appl. Polym. Sci. *55*:773 (1995).
26. Z. Guan, C. Elsbernd, and J. DeSimone, J. Polym. Prepr. (Am. Chem. Soc. Div. Polym. Chem.) *34*:329 (1992).
27. J. B. McClain, D. Londono, J. R. Combes, T. J. Romack, D. A. Canelas, D. E. Betts, G. D. Wignall, E. T. Samulski, and J. M. DeSimone, J. Am. Chem. Soc. *118*: 917–918 (1996).
28. D. Chillura-Martino, R. Triolo, J. B. McClain, J. R. Combes, D. E. Betts, D. A. Canelas, J. M. DeSimone, E. T. Samulski, H. D. Cochran, J. D. Londono, and G. D. Wignall, J. Mol. Struct. *383*:3 (1996).
29. E. Maury, H. Batten, S. Killian, Y. Menceloglu, J. Combes, and J. DeSimone, Polym. Prepr. (Am. Chem. Soc. Div. Polym. Chem.) *34*:664 (1993).
30. D. Zhu and Z. Schelly, Langmuir *8*:48 (1992).
31. J. L. Fulton, D. M. Pfund, J. B. McClain, T. J. Romack, E. E. Maury, J. R. Combes, E. T. Samulski, J. M. DeSimone, and M. Capel, Langmuir *11*:4241 (1995).
32. J. DeSimone, E. Maury, Y. Menceloglu, J. McClain, T. Romack, and J. Combes, Science *265*:356 (1994).
33. Y. Hsiao, E. Maury, and J. DeSimone, Macromolecules *28*:8159–8166 (1995).
34. D. Canelas, D. Betts, and J. DeSimone, Macromolecules *29*:2818 (1996).

35. Z. Guan and J. DeSimone, Macromolecules *27*:5527 (1994).
36. J. D. Londono, R. Dharmapurikar, H. D. Cachran, G. D. Wignall, J. B. McClain, D. E. Betts, D. A. Canelas, J. M. DeSimone, E. T. Samulski, D. Chillura-Martino, and R. Triolo, J. Appl. Crystallogr. *30*:690–695 (1997).
37. A. Dardin, J. B. Cain, J. M. DeSimone, C. S. Johnson, and E. T. Samulski, Macromolecules *30*:3593–3599 (1997).
38. D. Canelas, D. Betts, and J. DeSimone, Macromolecules *31*:6794 (1998).
39. A. Cooper, W. Hems, and A. Holmes, Macromolecules *32*:2156 (1999).
40. R. Mertsch and B. Wolf, Macromolecules *27*:3289 (1994).
41. C. Elsbernd, J. DeSimone, A. Hellstern, S. Smith, P. Gallagher, V. Krukonis, and J. McGrath, Polym. Prepr. (Am. Chem. Soc. Div. Polym. Chem.) *31*:673 (1990).
42. T. Hoefling, D. Newman, R. Enick, and E. J. Beckman, J. Supercrit. Fluids *6*:165 (1993).
43. M. Yates, K. Lim, G. Li, J. Shim, S. Maniar, K. Johnston, and S. Webber, Macromolecules *32*:1018–1026 (1999).
44. D. Canelas and J. DeSimone, Macromolecules *30*: 5673–5682 (1997).
45. T. Hoefling, D. Stofesky, M. Reid, E. Beckman, and R. Enick, J. Supercrit. Fluids *5*:237 (1992).
46. D. Newman, T. Hoefling, R. Beitle, E. Beckman, and R. Enick, J. Supercrit. Fluids *6*:205 (1993).
47. F. A. Adamsky and E. J. Beckman, Macromolecules *27*: 312–314 (1994).
48. A. Yazdi, C. Lepilleur, E. Singley, W. Liu, A. Adamsky, R. Enick, and E. Beckman, Fluid Phase Equilibria *117*: 297 (1996).
49. A. Yazdi and E. J. Beckman, Mater. Res. *10*:530 (1995).
50. C. Lepilleur and E. Beckman, Macromolecules *30*:745 (1997).
51. E. Ghenciu and E. Beckman, Ind. Eng. Chem. Res. *36*: 5366 (1997).
52. J. Fulton and R. Smith, J. Phys. Chem. *92*:2903 (1988).
53. R. Lemert, R. Fuller, and K. Johnston, J. Phys. Chem. *94*:6021 (1990).
54. K. Johnston, G. McFann, and R. Lemert, in ACS Symposium Series 406 (K. Johnston, G. McFann, and R. Lemert, eds.), American Chemical Society, Washington, DC, 1989.
55. K. Harrison, J. Goveas, K. Johnston, and I. O'Rear, Langmuir *10*:3536 (1994).
56. J. Oates, Ph.D. dissertation, University of Texas at Austin, 1989.
57. K. Consani and R. J. Smith, J. Supercrit. Fluids *3*:51 (1990).
58. G. McFann, K. Johnston, and S. Howdle, AIChE J. *40*: 543 (1994).
59. K. Johnston, K. Harrison, M. Clarke, S. Howdle, M. Heitz, F. Bright, C. Carlier, and T. Randolph, Science *271*:624 (1996).
60. K. Bartscherer, M. Minier, and H. Renon, Fluid Phase Equilibria *107*:93 (1995).
61. G. McFann and K. Johnston, *Microemulsions: Fundamental and Applied Aspects*, Marcel Dekker, New York, 1999.
62. A. I. Cooper, J. D. Londono, G. Wignall, J. B. McClain, E. T. Samulski, J. S. Lin, A. Dobrynin, M. Rubinstein, A. L. C. Burke, J. M. M. Frechet, and J. M. DeSimone, Nature *389*:368 (1997).
63. S. Stevelmans, J. Am. Chem. Soc. *118*:7398–7399 (1996).
64. J. B. McClain, D. E. Betts, D. A. Canelas, E. T. Samulski, J. M. DeSimone, J. D. Londono, H. D. Cochran, G. D. Wignall, D. Chillura-Martino, and R. Triolo, Science *274*:2049 (1996).
65. E. Buhler, A. V. Dobrynin, J. M. DeSimone, and M. Rubinstein, Macromolecules *31*:7347 (1998).
66. S. Zhou and B. Chu, Macromolecules *31*:7746–7755 (1998).

6

Acid- and Oxidatively Labile Vinyl Ether Surfactants: Synthesis and Drug Delivery Applications

JONG-MOK KIM and DAVID H. THOMPSON Purdue University, West Lafayette, Indiana

I. OVERVIEW OF LAMELLAR-HEXAGONAL PHASE TRANSITIONS TRIGGERED BY CLEAVABLE SURFACTANTS

Liposomes have evolved during the last 35 years from a laboratory curiosity to an effective delivery vehicle for several important drugs. This was made possible by overcoming initial problems with reproducible particle size, plasma stability, high encapsulation efficiency, long circulation times, and selective deposition in vivo that were encountered with first-generation carrier systems. It is now clear that the great potential of liposomes as drug delivery vehicles will not be fully realized until more effective targeting and membrane fusion mechanisms have been designed and incorporated in their formulations. This chapter summarizes our recent advances toward these goals and the membrane fusion aspect of this challenge.

An essential requirement of fusogenic liposomes is that they display compositional and mechanical stability during systemic circulation yet undergo sufficient structural transformation upon exposure to an applied stimulus that they promote rapid fusion with target cell membranes. Many candidate systems have been investigated for these properties [1–3]; however, there are no generally applicable systems that display *both* fusogenicity and plasma stability. Our work has focused on the development of new triggerable materials, based on the intrinsic lipid curvature model (Fig. 1) [4,5], that are designed to undergo L_α, H_I, or H_{II} phase transitions upon exposure to either acidic or oxidative environments. Plasmenylcholine-type phospholipids (Section II) have been utilized for their ability to promote the H_I phase, and 1,2-dioleoyl-*sn*-glycero-3-phosphoethanolamine (DOPE)—one of the most widely studied materials in membrane fusion research—has been employed in binary lipid mixtures to facilitate the formation of the H_{II} phase upon exposure to a triggering impulse (Section III).

II. PLASMENYLCHOLINE AND DISPLASMENYLCHOLINE LIPOSOME SYSTEMS

Plasmenylcholine (also known as plasmalogen) and diplasmenylcholine are naturally occurring Z-vinyl ether–linked phospholipids [6]. Although only one source of diplasmenylcholine has been reported, plasmenylcholine is widely distributed in mammals, particularly in electrically active tissues. Little is known about their role in biological systems. Plasmenylcholines and plasmenylethanolamines are highly enriched in arachidonate at the *sn-2* position. This observation has led to the proposal that their role in biology is as a nascent source of arachidonic acid for the initiation of signal transduction pathways. One structural study has even suggested that the *sn-1* Z-vinyl ether linkage enforces a unique glycerol backbone conformation at the membrane-water interface, thereby providing an interfacial signature for phospholipase recognition of the arachidonate-rich plasmenyl lipid pool [7]. An alternative proposal for their occurrence is to serve as a membrane-localized antioxidant [8]. The latter hypoth-

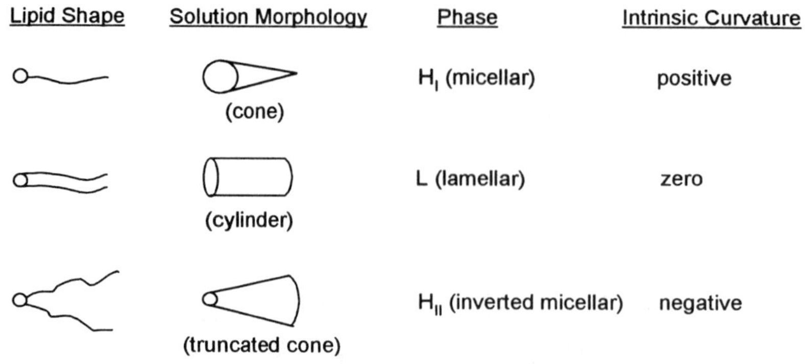

FIG. 1 Intrinsic curvature and lipid packing morphology.

esis is closely related to the strategy we have employed for photooxidative triggering of the lamellar-hexagonal I phase transition in plasmenylcholine-based liposome systems (Sections II.B and II.C). A second triggering pathway is also possible by taking advantage of the known sensitivity of vinyl ether bonds to acidic conditions (Scheme II.D) [9]. These reactions, both of which convert a double-chain lamellar phase forming lipid into single-chain H_I surfactants (Fig. 2), are the foundation for creating liposomes that can be triggered to undergo L_α-H_I phase transitions.

A. Lipid Synthesis

Early triggered-release experiments with plasmenylcholine liposomes utilized semisynthetic palmitoylplasmenylcholine (PPlsC) derived from bovine heart phosphatidylcholine [10–12]. Total syntheses have been developed in our laboratory for both plasmenylcholine (Fig. 3) [13] and diplasmenylcholine (Fig. 4) [14,15]. Bittman and coworkers have also recently reported a plasmenylcholine synthesis using alkynyl ethers as the key synthetic intermediate [16].

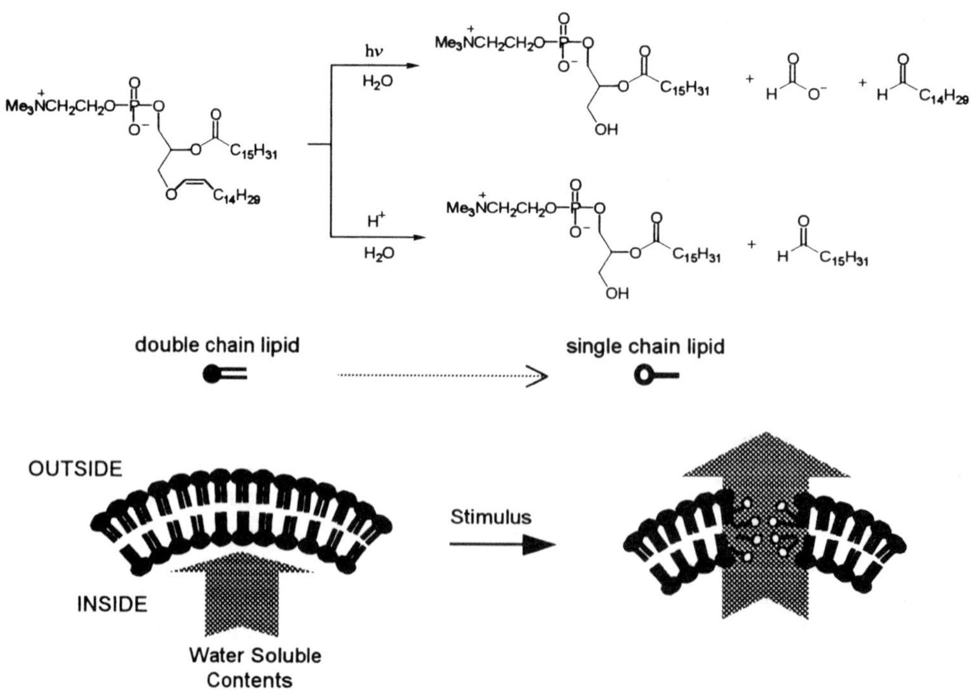

FIG. 2 Plasmenylcholine photooxidation and acid-catalyzed hydrolysis reactions (top) and the resulting L_α-H_I phase change that vinyl ether bond cleavage produces (bottom).

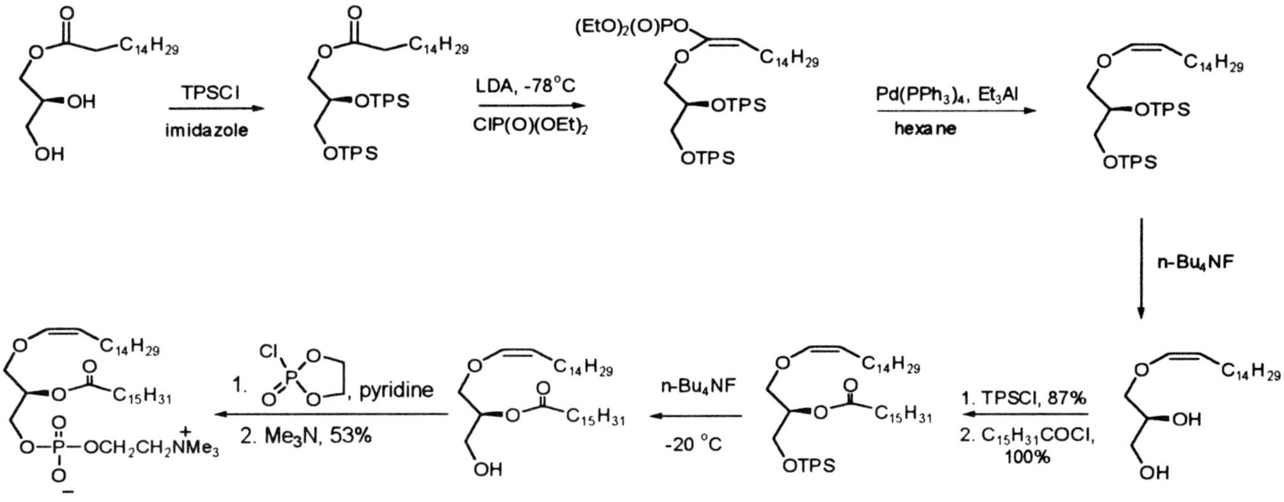

FIG. 3 Total synthesis of plasmenylcholine from monopalmitin [13].

B. Photooxidative Triggering of Plasmenylcholines [10,11]

Irradiation of PPlsC liposomes that contain bacteriochlorophyll (solubilized within the membrane bilayer) and calcein (entrapped in the inner aqueous compartment at self-quenching concentrations) produces release of contents as detected by calcein fluorescence dequenching (Fig. 5). Release rates were shown to be dependent on fluency, oxygen concentration, and plasmenylcholine concentration.

Electron microscopic (EM) evidence also indicated that the liposome morphology changed during the course of the photolysis, from lamellar to multilamellar with interlipidic particles, as a result of membrane-membrane fusion events. Time-resolved EM experiments also suggested that these changes occurred on the same kinetic time scale as contents release. A model involving membrane fusion at photooxidized lesions between adjacent liposomes was proposed for this system. Repetitive cycles of this process are thought to be responsible for the accumulation of multilamellar lipid.

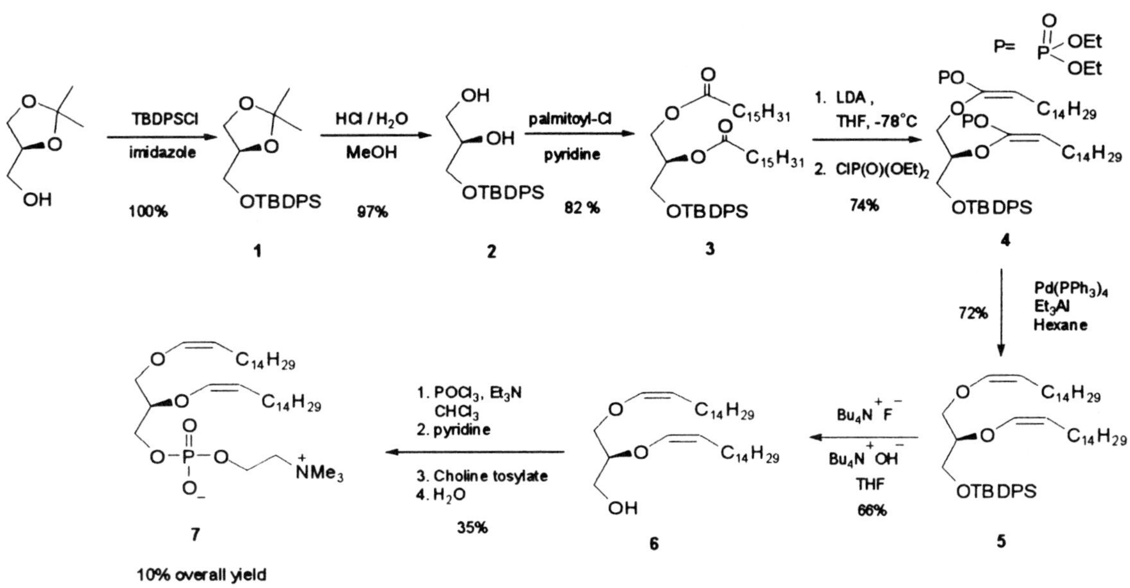

FIG. 4 Total synthesis of diplasmenylcholine from solketal [15,16].

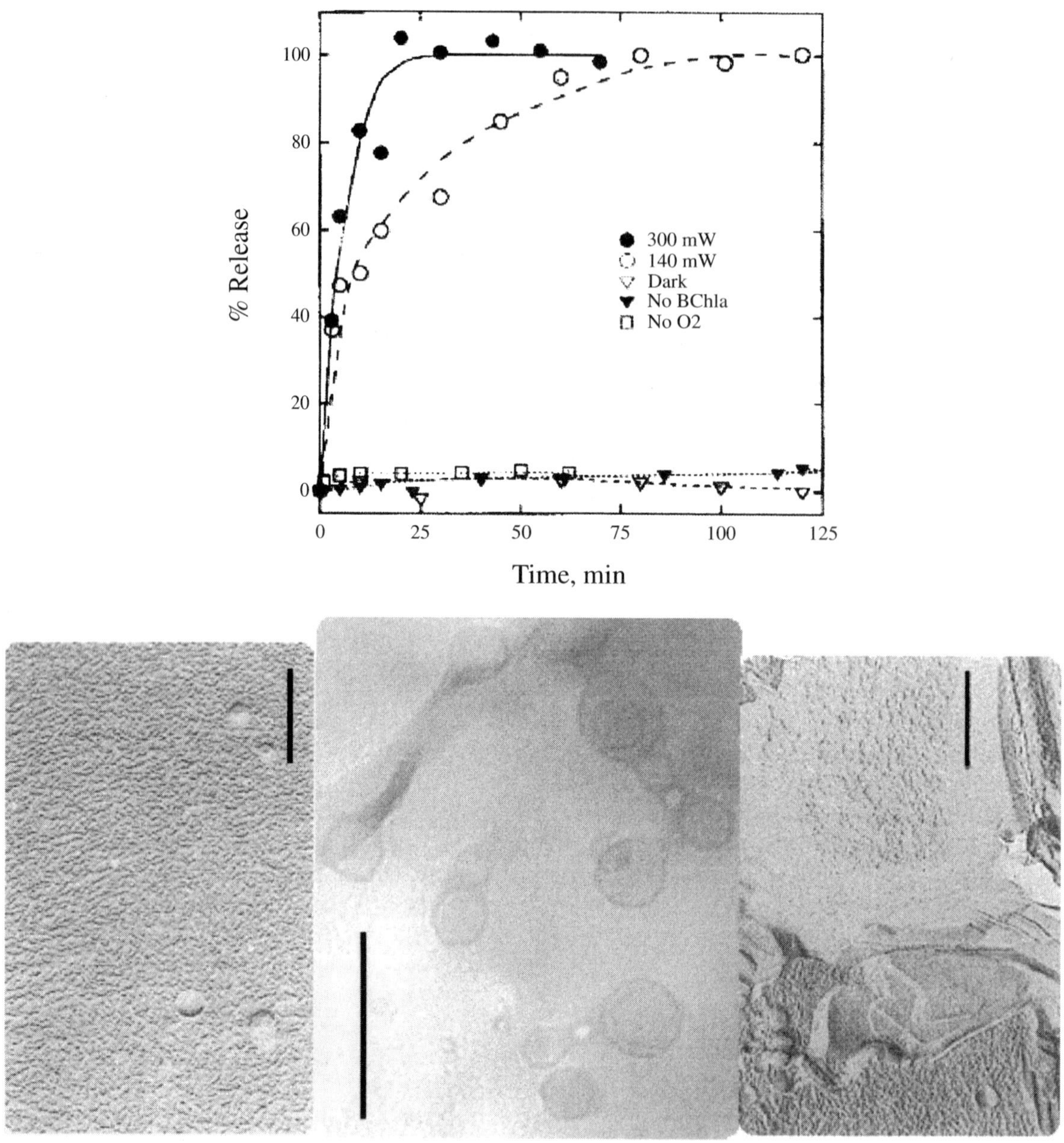

FIG. 5 (Top) Photorelease of calcein from Bchl:PPlsC liposomes using 800-nm excitation under various experimental conditions. (Bottom left) Freeze-fracture TEM micrograph before liposome irradiation. (Bottom center) Cryo-TEM micrograph taken after 5 min of irradiation. (Bottom right) Freeze-fracture TEM micrograph 48 h after liposome irradiation. (Data from Ref. 11.)

C. Cascade Triggering of Diplasmenylcholines

Photooxidation of diplasmenylcholine liposomes bearing internalized control elements such as catalysts or cofactors offers intriguing possibilities to either transduce or amplify the initial photochemical process. It can also reduce the amount of diplasmenylcholine required to elicit a release response. This principle was demonstrated using Bchl-containing dipalmitoylplasmenylcholine (DPPlsC) liposomes with internalized calcium ions. Because Ca^{2+} is a required cofactor for calcium-dependent phospholipase A_2 (CD-PLA_2) activity, release from a second population of CD-PLA_2–sensitive liposomes (e.g., dipalmitoylphosphatidylcholine, DPPC) can be controlled by phototriggering the release of Ca^{2+} from DPPlsC liposomes. This elicits a cascade response wherein phototriggered Ca^{2+} activates DPPC liposome contents release via CD-PLA_2 hydrolysis (Fig. 6). Bchl:DPPlsC liposomes irradiated at 800 nm in the presence of oxygen were observed to release their internalized calcium ions rapidly. After a brief lag period, presumably for CD-PLA_2 activation and hydrolysis of a critical amount of DPPC, contents release from the DPPC liposomes was also observed (Fig. 7). The cascade reaction is made possible in this case because diplasmenylcholine liposomes are insensitive to CD-PLA_2 hydrolysis, so they can retain their contents even in the presence of phospholipase. Because calcium ion concentrations are so tightly regulated in biology, there is a potentially wide array of Ca^{2+}-based cascade processes that could be addressed using this concept. Nonbiological reactions (e.g., calcium phosphate mineralization) could also be controlled in a spatiotemporal manner using this cascade reaction principle.

D. Drug Delivery via Acid-Catalyzed Triggering of Diplasmenylcholines [18]

Contents release from DPPlsC liposomes is also triggerable under acidic conditions, with release rates that vary as a function of solution pH (Fig. 8). Because the observed release rate at pH 4.5 gave a $t_{50\% \text{ release}}$ of approximately 90 min, these liposomes were tested for their ability to undergo intracellular triggering upon exposure to acidic endosomes that can achieve a pH of 5. A folate receptor–mediated uptake pathway was chosen as the route of entry into the endosomal compartment (Fig. 9). KB cells, cultured in folate-deficient media to stimulate the overexpression of folate receptors, were used as the test system for contents release from folate-targeted DPPlsC liposomes. After exposure to propidium iodide–loaded DPPlsC liposomes containing 0.5 mol% folate-conjugated poly(oxyethylene)-distearoylphosphatidylethanolamine (PEG-DSPE) for 4 h, the cells were washed and monitored for their distribution of propidium iodide fluorescence. Within 8 h of the washing step, the cells had become extensively stained with propidium iodide, suggesting that the liposomal contents had efficiently escaped both the liposomal and endosomal compartments. These observations were supported by a similar experiment with arabinofuranosylcytosine (AraC). As the data in Fig. 10

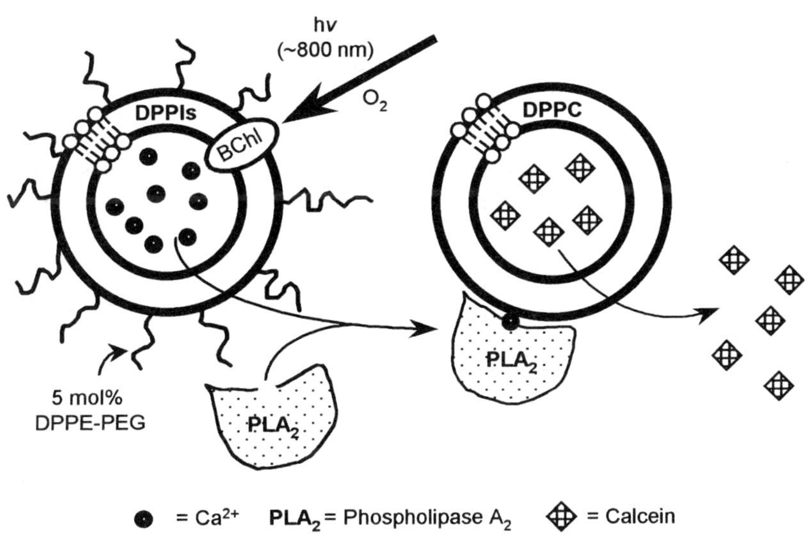

FIG. 6 Conceptual diagram of cascade triggering. (Adapted from Ref. 17.)

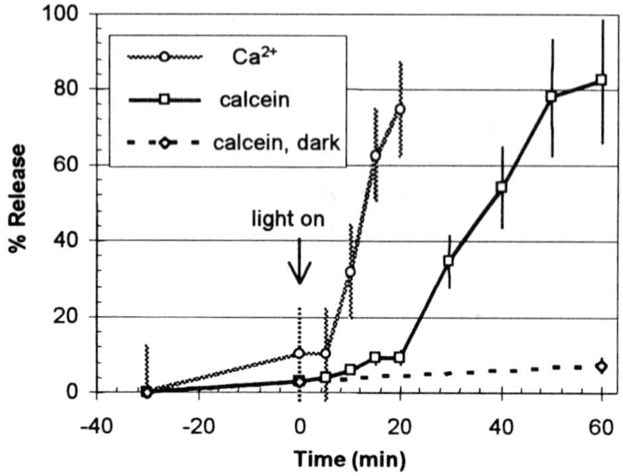

FIG. 7 Cascade release of Ca^{2+} and calcein from Bchl: PEG-DPPE:DPPlsC and DPPC liposomes, respectively, upon irradiation at 800 nm (37°C). Note that little appreciable release occurs from either liposome in the absence of light. (Data from Ref. 17.)

show, AraC delivered via folate-targeted DPPlsC liposomes is 6000-fold more cytotoxic than the free drug and 100-fold more toxic than the folate-targeted drug encapsulated within a pH-insensitive liposome. These results clearly demonstrate a synergistic enhancement of AraC activity when endosomal targeting and acid triggering mechanisms are combined within a DPPlsC liposomal delivery vehicle. They also suggest that substantially improved biological responses may be attainable with other water-soluble drugs if a similar delivery approach is used. This may be particularly important for peptides, plasmids, and oligonucleotides that have

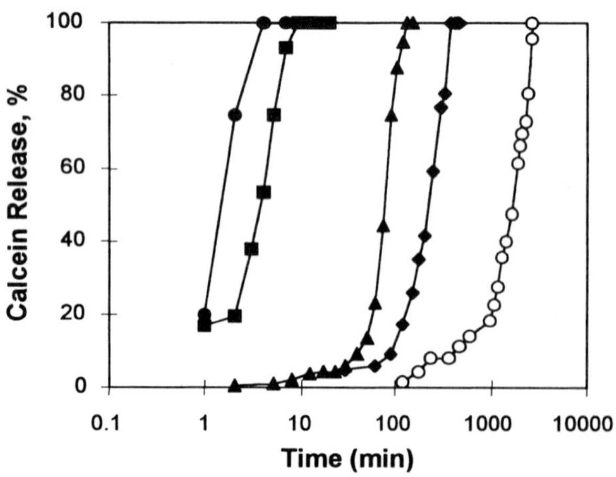

FIG. 8 Calcein leakage rates from DPPlsC liposomes at 37°C. (●) pH 2.3; (■) pH 3.2; (▲) pH 4.5; (♦) pH 5.3; (○) pH 6.3. (Data from Ref. 18.)

exceptionally low plasma stability and poor cell uptake characteristics because of their enzymatic lability, size, and hydrophilicity.

III. DEPEGYLATIVE TRIGGERING OF DOPE: VINYL ETHER–LINKED POLY(ETHYLENE GLYCOL) LIPOSOMES

Kirpotin et al. have reported a dePEGylative triggering approach using 3–6 mol% of a thiol-cleavable poly(ethylene glycol)-grafted distearoylphosphatidylethanolamine as guest lipid to stabilize the lamellar phase of DOPE [19]. Treatment of *N*-(2-(ω-methoxypoly(oxyethylene)-α-aminocarbonyl)ethyl)dithioproionyl-DSPE (mPEG-DTP-DSPE):DOPE liposomes with dithiothreitol (DTT) leads to cleavage of the liposomal steric stabilization layer and "unmasking" of the latently fusogenic, DOPE-rich liposomes. Clearance of PEG from the liposome surface destabilizes the lamellar phase, leading to membrane fusion and contents release promoted by the H_{II} phase. This strategy retains the appealing characteristics of plasma stable, long-circulating liposomes prior to thiol cleavage, but unfortunately, the high concentrations of thiolytic agent required (10 mM DTT) to effect liposome destabilization greatly limits the potential of this method in vivo. This problem has been addressed with a new thiol-cleavable PEG-lipid conjugate, but the concentrations of thiol required to activate the system are still quite high [20]. Lipid phase transitions induced by dePEGylation have also been described by Holland et al. [21]. This method is based on exchange of the PEG coating (2 mol% PE-PEG2000 in 1:1 POPS:DOPE liposomes) onto an acceptor population of "sink" 1-palmitoyl-2-oleylphosphatidylcholine (POPC) liposomes. DePEGylation of these liposomes produces membrane fusion in the presence of calcium ions with rates that are dependent on the molar ratio of PEG-lipid initially present as well as the chain length and degree of unsaturation of the lipid anchor.

A. Synthesis of a Vinyl Ether PEG Lipid Conjugate

We have also investigated dePEGylative triggering based on the same vinyl ether degradative reactions reported for plasmenylcholine and diplasmenylcholine liposomes (Fig. 11) [12,18]. DOPE liposomes were stabilized as a lamellar phase by incorporation of 1–5 mol% 1,2-di-*O*-(1*Z*′,9*Z*′-octadecadienyl)-*sn*-glycero-3-poly(ω-methoxyethylene[115]glycol)ate (BVEP). This

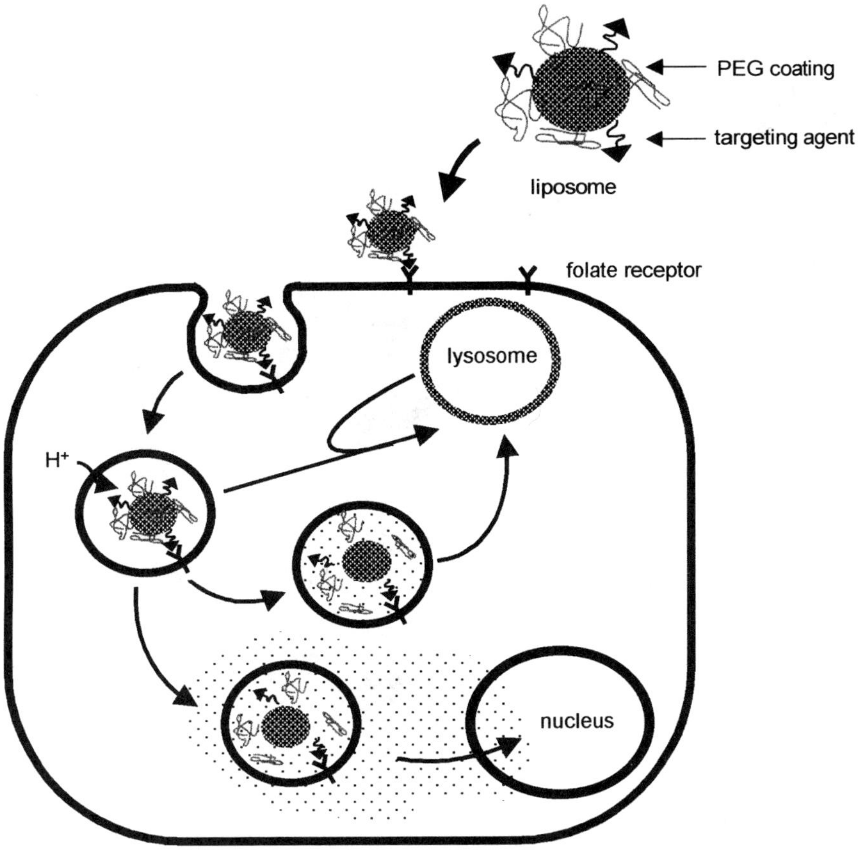

FIG. 9 Conceptual model of endosomal uptake and release pathway for targeted DPPlsC liposomes and their contents.

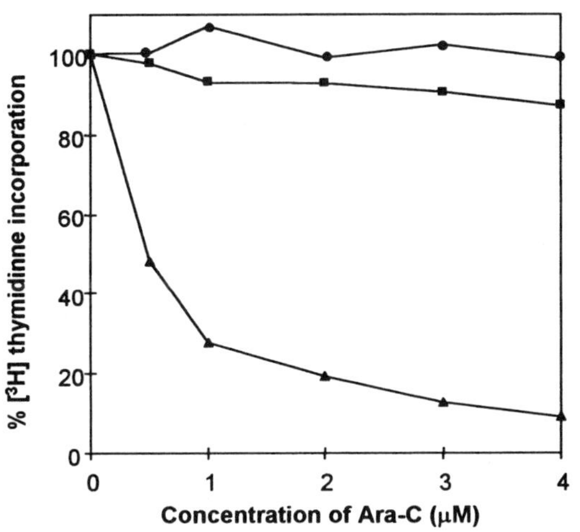

FIG. 10 Cytotoxicity of AraC to KB cells as a function of delivery vehicle. (●) Free AraC; (■) AraC encapsulated within folate-targeted egg phosphocholine liposomes (pH-insensitive control liposomes); (▲) AraC encapsulated within folate-targeted DPPlsC liposomes.

acid-labile PEG conjugate was synthesized by coupling the corresponding PEG acid with 1,2-di-*O*-(1*Z*′,9*Z*′-octadecadienyl)-*sn*-glycerol (compound **6** in Fig. 4) in the presence of dicyclohexylcarbodiimide.

B. Acid-Triggered Release Characteristics of DOPE:BVEP Liposomes

Acidification of binary DOPE:BVEP mixtures triggers vinyl ether hydrolysis and dePEGylation, with subsequent destabilization of the lamellar phase and contents release as the DOPE host lipid reverts back to its preferred hexagonal phase (Fig. 11). The onset of calcein leakage with time at pH 4.5 (Fig. 12) indicates that dePEGylative triggering can be achieved in acidic media at rates that are higher than in neutral solutions (pH 7.4). These acid-triggerable formulations are also relatively stable in 10% serum (data not shown). Collectively, these results suggest that the dePEGylative triggering approach may prove to be a useful technique

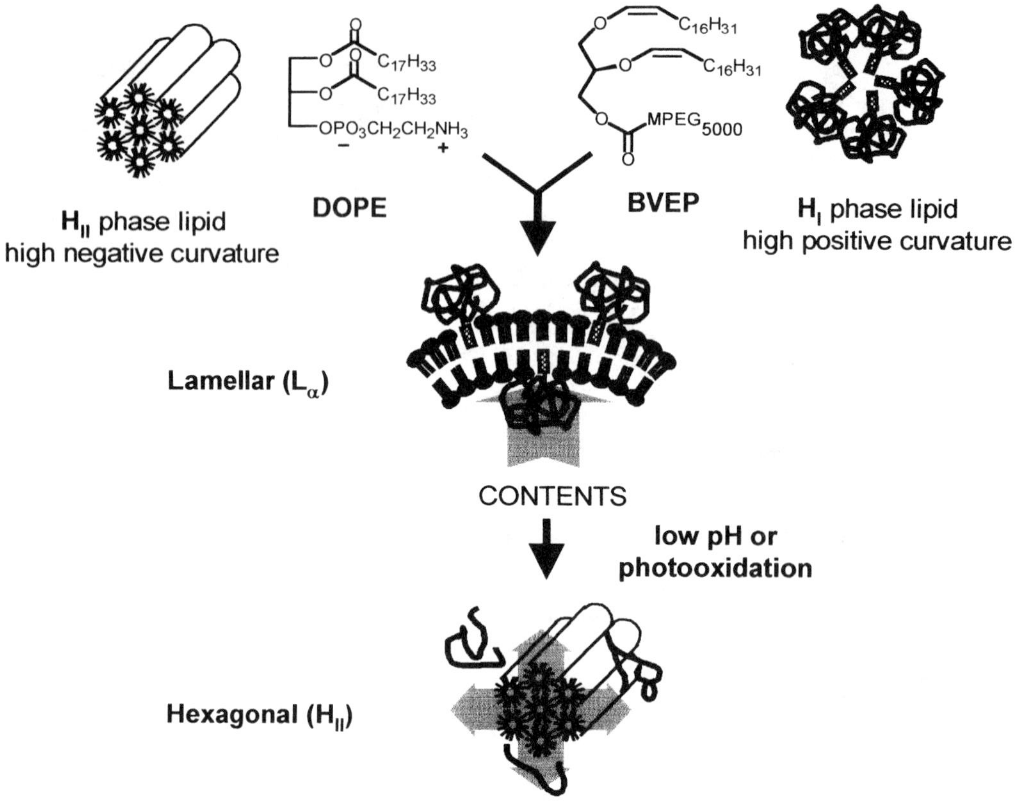

FIG. 11 Conceptual diagram for dePEGylative triggering to induce a lamellar-hexagonal II phase transition. DOPE, 1,2-dioleoyl-*sn*-glycero-3-phosphoethanolamine; BVEP, 1,2-di-*O*-(1′,9′-octadecadienyl)-*sn*-glycero-3-poly(ω-methoxyethylene-[115]glycol)ate or any other labile PEG-lipid.

for promoting endosomal escape of liposomal contents via liposome-endosome membrane fusion (Fig. 9).

IV. CATIONIC VINYL ETHER LIPIDS IN GENE DELIVERY

Many different cationic lipids have been synthesized for transfection applications. However, few of these compounds have been designed from a mechanistic standpoint to promote decomplexation and transgene expression in vivo. Because most of the effective DNA: cationic lipid complexes bear a net positive charge, their adsorption to the negatively charged cytoplasmic membrane surface leads, in most cases, to internalization via the endosomal uptake pathway. Gene expression can be limited at this stage if specific endosomal escape mechanisms have not been incorporated within the DNA delivery vector. It has become clear that the constitutive acidification that occurs within these transient compartments provides an intrinsic pathway for triggering intracellular contents release from appropriately designed vehicles. Our approach for improving gene expression, therefore, is to promote DNA:cationic lipid decomplexation and endosomal escape using the same acid-catalyzed vinyl ether hydrolysis chemistry described earlier that is known to promote cytoplasmic delivery of hydrophilic materials.

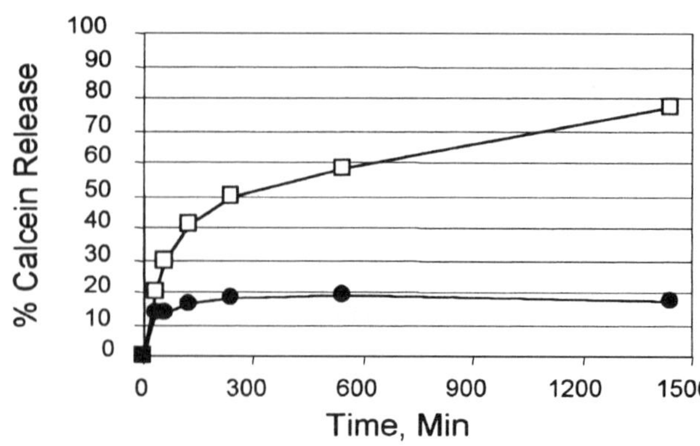

FIG. 12 Calcein release kinetics from 97:3 DOPE:BVEP liposomes at 37°C. (●) pH 7.4; (□) pH 4.5.

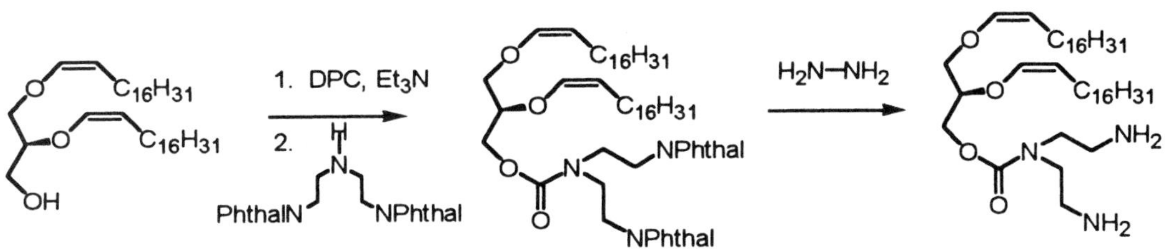

FIG. 13 Synthesis of BCAT from 1,2-di-*O*-(1′,9′-octadecadienyl)-*sn*-glycerol [22].

A. Synthesis of BCAT

An adaptation of the DPPlsC/DOPlsC synthesis (Fig. 4) was used to prepare 1,2-di-*O*-(1Z′,9Z′-octadecadienyl)-*sn*-glyceryl-3-*O*-carbamoyldiethylenetriamine (BCAT) (Fig. 13). The cationic headgroup was installed using dipyridyl carbonate (DPC) in a standard carbamate coupling methodology. Alcohol **6** (i.e., compound **6** in Fig. 4) was added to DPC, generating a mixed carbonate intermediate that was subsequently displaced with *N,N′*-diphthalamidyldiethylenetriamine, giving a protected carbamate product. This intermediate was then deprotected with hydrazine hydrate to give BCAT (free base) in 56% isolated yield from **6** (16% overall yield from (*S*)-(+)-2,2-dimethyl-1,3-dioxolane-4-methanol).

B. Uptake and Expression of DNA: Cationic Vinyl Ether Lipid Complexes

Flow cytometry was used to monitor cells treated with DNA complexes of BCAT, DCAT (a saturated analogue of BCAT), and DOTMA/Chol (a commercially available transfection agent). Psoralen-labeled DNA and green fluorescent protein (GFP) expression levels from a GFP vector were monitored as a function of lipid type (Fig. 14) (J. A. Boomer, D. H. Thompson, and S. Sullivan, submitted). After a 2.5-h exposure to DNA:cationic lipid complexes, the DOTMA/Chol formulation showed the highest levels of DNA uptake (two- to fourfold higher than BCAT or DCAT). GFP expression levels, however, were substantially better with the BCAT and DCAT formulations in spite of their lower uptake levels. These observations underscore the utility of decomplexation and endosomal escape from labile lipid carriers.

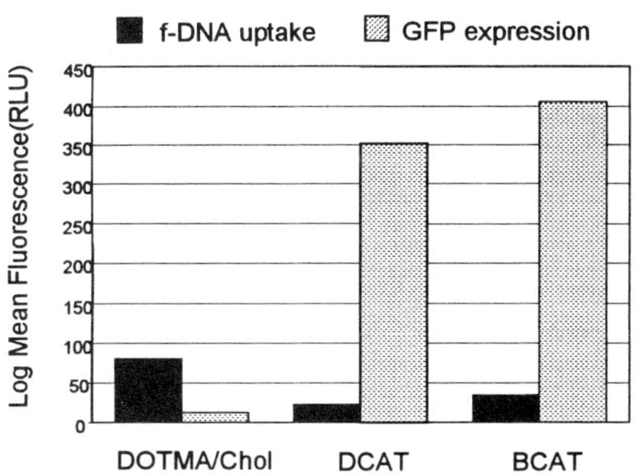

FIG. 14 Comparison of fluorescent DNA uptake with green fluorescent protein expression for various cationic lipid:DNA complexes in NIH 3T3 cells (J. A. Boomer, D. H. Thompson, and S. Sullivan, submitted).

V. CONCLUSIONS

These investigations indicate a promising future for drug and gene delivery applications using vinyl ether-based delivery systems. The most significant obstacles to their widespread use at this time are their lack of commercial availability and their relaively slow acid hydrolysis kinetics. Most of the desired applications of these materials will require activation of the lipid phase transition at pH values in the range of 5–6. This limitation can be addressed by varying the stereoelectronics of the vinyl ether or by increasing negative charge at the membrane interface. Once this problem has been surmounted, more flexible synthetic methods must be developed to simplify the preparation of many different classes of vinyl ether conjugate. Experiments in progress are aimed at addressing both of these impediments.

ACKNOWLEDGMENTS

The financial support of NIH Grant R01 55266 and the Purdue Research Foundation is greatly appreciated. The efforts of Mr. Junhwa Shin and Dr. Jeremy Boomer in assisting with the preparation of this manuscript are also gratefully acknowledged.

REFERENCES

1. O. V. Gerasimov, Y. Rui, and D. H. Thompson, in *Vesicles* (M. Rosoff, ed.), Marcel Dekker, New York, 1996, pp. 679–746.

2. O. V. Gerasimov, J. A. Boomer, M. M. Qualls, and D. H. Thompson, Adv. Drug Delivery Rev. *38*:317–338 (1999).
3. D. F. O'Brien and D. A. Tirrell, in *Bioorganic Photochemistry*, Vol. 2: *Biological Applications of Photochemical Switches* (H. Morrison, ed.), Wiley, New York, 1993, pp. 111–167.
4. P. R. Cullis and B. deKruijff, Biochim. Biophys. Acta *559*:399–420 (1979).
5. J. N. Israelachvili, S. Marcelja, and R. G. Horn, Q. Rev. Biophys. *13*:121–148 (1980).
6. F. Paltauf, Chem. Phys. Lipids *74*:101–139 (1994).
7. X. Han and R. W. Gross, Biochemistry *29*:4992–4996 (1990).
8. R. A. Zoeller, A. C. Lake, N. Nagan, D. P. Gaposchkin, M. A. Legner, and W. Lieberthal, Biochem. J. *338*:769–776 (1999).
9. J. R. Keeffe and A. J. Kresge, in *The Chemistry of Enols* (Z. Rappoport, ed.), Wiley, New York, 1990, pp. 399–480.
10. V. C. Anderson and D. H. Thompson, Biochim. Biophys. Acta *1109*:33–42 (1992).
11. D. H. Thompson, O. V. Gerasimov, J. J. Wheeler, Y. Rui, and V. C. Anderson, Biochim. Biophys. Acta *1279*: 25–34 (1996).
12. O. V. Gerasimov, A. Schwan, and D. H. Thompson, Biochim. Biophys. Acta *1324*:200–214 (1997).
13. Y. Rui and D. H. Thompson, Chem. Eur. J. *2*:1505–1508 (1996).
14. Y. Rui and D. H. Thompson, J. Org. Chem. *59*:5758–5762 (1994).
15. Y. Rui, Stereocontrolled total syntheses of plasmalogen and diplasmalogen lipids and their activities in an intracellular drug delivery system, Ph.D. dissertation, Purdue University, West Lafayett, IN, 1996.
16. D. Qin, H.-S. Byun, and R. Bittman, J. Am. Chem. Soc. *121*:662–668 (1999).
17. N. Wymer, O. V. Gerasimov, and D. H. Thompson, Bioconj. Chem. *9*:305–308 (1998).
18. Y. Rui, S. Wang, P. S. Low, and D. H. Thompson, J. Am. Chem. Soc. *120*:11213–11218 (1998).
19. D. Kirpotin, K. Hong, N. Mullah, D. Papahadjopoulos, and S. Zalipsky, FEBS Lett. *388*:115–118 (1996).
20. S. Zalipsky, M. Qazen, J. A. Walker, N. Mullah, Y. P. Quinn, and S. K. Huang, Bioconj. Chem. *10*:703–707 (1999).
21. J. W. Holland, C. Hui, P. R. Cullis, and T. D. Madden, Biochemistry *35*:2618–2624 (1996).
22. J. A. Boomer and D. H. Thompson, Chem. Phys. Lipids *99*:145–153 (1999).

7

Three Principles for Active Control of Interfacial Properties of Surfactant Solutions

JASON Y. SHIN, LANA I. JONG,* NIHAL AYDOGAN, and NICHOLAS L. ABBOTT
University of Wisconsin–Madison, Madison, Wisconsin

I. INTRODUCTION

A. Overview

Here we review three strategies that permit active control of interfacial properties of aqueous solutions of water-soluble surfactants. These strategies are based on the use of three water-soluble surfactants, each of which hosts either redox- or light-active groups, and make possible spatial and temporal control of the surface tension of aqueous solutions. The first strategy revolves around the oxidation and reduction of ferrocene-based surfactants such that changes in the oxidation state of the ferrocene substantially perturb the equilibrium partitioning of surfactant between an interface and bulk solution. The second strategy is based on reduction of a bolaform surfactant containing a disulfide bond. This transformation leads to the formation of monomeric, thiol-containing fragments. We demonstrate the use of this transformation to create nonequilibrium, interfacial states that possess surface tensions that are substantially lower than any equilibrium surface tensions measured in the same system. The third strategy revolves around a mixed surfactant system containing a water-soluble surfactant that hosts the photosensitive group azobenzene. We demonstrate that illumination of this surfactant system leads to substantial changes in both the states of aggregation of the surfactant in bulk solution and the dynamic surface tensions of freshly created surfaces of the solution. Finally, we illustrate how these three strategies for active control of surface tension can provide new principles for microfluidics where the behavior of liquids on millimeter and smaller scales are often dominated by interfacial stresses. We report the use of these surfactants to (1) achieve vectorial transport of droplets of liquid through a network of fluidic channels, (2) pattern the dewetting of liquids on energetically homogeneous surfaces, and (3) trigger, by using illumination, the release of pendant droplets of aqueous solutions from an array of capillaries.

**Current affiliation*: University of California, Davis, Davis, California.

B. Motivation

The three principles described in this chapter lead to spatial and temporal control of the interfacial properties of aqueous surfactant systems. The work described herein is motivated, in part, by the potential usefulness of these principles in microscale systems suitable for chemical synthesis and analysis. The development of simple and general principles for the pumping and positioning of liquids on submillimeter scales has the potential to enable fabrication of microanalytical systems that will make procedures such as blood chemistry analysis, flow cytometry and polymerase chain reactions, and DNA screening both rapid and inexpensive [1,2]. Current methods used to pump liquids within networks of channels, or to position liquids within arrays, generally rely on electrokinetic phenomena driven by high voltages (several kilovolts) [3,4], mechanical actuators that are complicated to fabricate and too expensive to be disposable [5], or passive fluid phenomena such as capillary wetting [6]. These methods are used to achieve largely serial manipulations of liquids within permanent channels that direct the liquid motion. Because the motions of liquids on submillimeter scales are generally dominated by the effects of interfacial stresses, we believe principles for spatial and temporal control of interfacial properties using redox- and light-active surfactants will provide useful ways to move, position, and mix liquids using simple but highly functional microfabricated structures.

In the following we review our recent efforts that have been directed toward this challenge. We first present three principles that permit active control of surface tension using redox- and light-active surfactants. Second, we illustrate the use of these principles for the control of liquids on millimeter and smaller scales.

II. PRINCIPLE 1. TRANSFORMATIONS BETWEEN EQUILIBRIUM STATES: FERROCENYL SURFACTANTS

A. Bulk Solution–Interface Equilibration

The first principle we describe for active control of interfacial properties of surfactant-based systems is based on use of redox-active surfactants and electrochemical methods [7–13]. The essential idea underlying the design of these surfactant systems is that a change in oxidation state of a surfactant can lead to a substantial change in the equilibrium partitioning of the surfactant between the bulk of the solution and the interface (Fig. 1a). The resulting transport of surfactant to or from the interface, as well as the change in state of surfactant at the interface, can lead to changes in interfacial properties.

Although the concept of "active control" of interfacial properties of surfactant systems necessarily implies the study of systems away from equilibrium states, the essential element of the strategy described here is that transformations of surfactants that are performed slowly relative to relaxation times of solutions will give rise to time-dependent properties of systems that can be qualitatively understood from knowledge of the equilibrium properties of the system. When dealing with a well-mixed solution of an ionic surfactant with millimolar activity, the characteristic time for equilibration of the surface of the solution with the bulk is likely to be only a few seconds [14–16]. For example, using a value of the surface area per surfactant molecule of 40 Å^2/molecule and a diffusion coefficient of

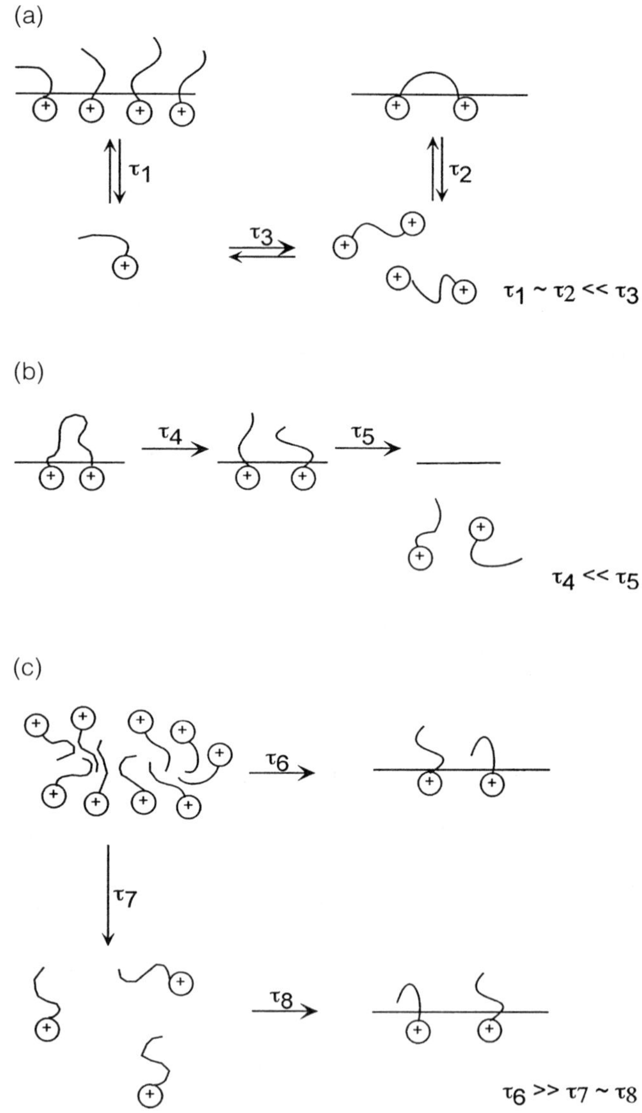

FIG. 1 Schematic illustration of three principles that permit active control of interfacial states of aqueous surfactant systems. (a) The first principle is based on the presence of an equilibrium between the interface and bulk solution. The transformation in the state of the surfactant (characterized by τ_3) takes place more slowly than processes that lead to the establishment of equilibrium between the interface and bulk solution (characterized by τ_1 and τ_2). (b) The second principle is based on a transformation in the state of surfactant hosted at an interface at a rate that is fast (characterized by τ_4) compared with the rate of desorption of the surfactant from the surface of the solution (characterized by τ_5). (c) The third principle is based on control of the extent of aggregation of the surfactant and thus its rate of transport to the surface of the solution. The time required to transport surfactant in aggregates to the surface (τ_6) is large compared with the time required to disrupt the aggregates (τ_7) and transport the monomeric species (τ_8) to the surface.

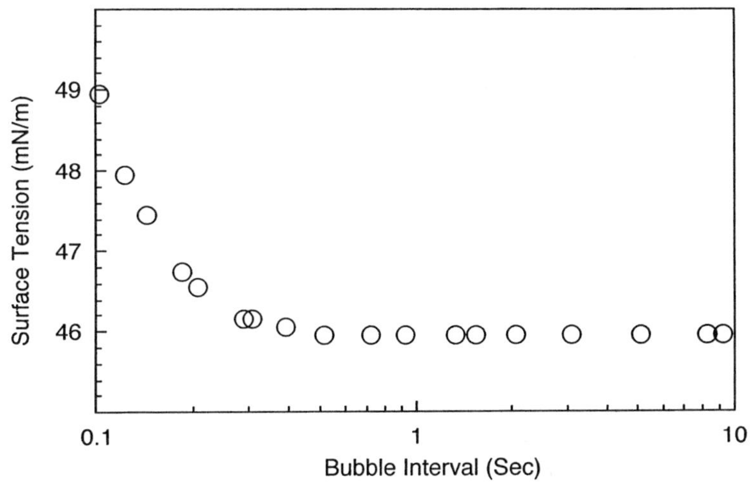

FIG. 2 Dynamic surface tension of an aqueous solution of 5 mM SDS measured using the maximum bubble pressure method. Surface tension is plotted as a function of the bubble interval. (Data from Ref. 16.)

10^{-6} cm^2/s, τ_d, the characteristic time for diffusive transport of surfactant from bulk solution to the surface, is only ~0.1 s (see Section IV). This estimate of the diffusion time is consistent with experimental measurements of dynamic surface tensions of aqueous solutions of ionic surfactants (Fig. 2). Thus, an electrochemical process that gives rise to a change in state of a solution of redox-active surfactant over a period of a few tens of seconds will probably result in a series of interfacial states that can be largely understood from knowledge of the composition of the bulk solution.

In summary, the strategy described in this section for active control of interfacial properties considers the system to follow a path of near-equilibrium states (i.e., the path is quasi-reversible). The design rule describing this pathway is, therefore, simply the equality of chemical potential of the surfactant species at the surface and in the bulk:

$$\mu_{\text{bulk}} = \mu_{\text{interface}} \tag{1}$$

B. Ferrocenyl Redox Chemistry

Here we illustrate this first principle for active control of interfacial properties using results from our studies of a series of ferrocenyl surfactants that have the structure $Fc(CH_2)_nN^+(CH_3)_3Br^-$, where Fc is the redox-active ferrocene group [7–13]. A surfactant with this structure ($n = 11$) was first synthesized by Saji and coworkers and subsequently used in studies of the deposition of thin films of organic dyes on electrodes [17–19]. The ferrocene group consists of an Fe^{2+} ion sandwiched between two cyclopentyldienyl anions, and it is therefore an electrically neutral complex. The complex has a volume of ~150 Å^3 (roughly three times the volume of a methyl group) and low solubility in water (5×10^{-5} M) [20]. Ferrocene, when hosted in surfactants having the preceding structure, can undergo a reversible one-electron oxidation in aqueous solution to form the ferrocenium cation. The redox potential for the oxidation is typically around 0.15 V (versus saturated calomel electrode, [SCE]). Oxidation of the ferrocene to the ferrocenium cation transforms the ferrocenyl surfactant with a single ionic headgroup to a surfactant with two ionic headgroups (one at each end of the molecule). We have synthesized ferrocenyl surfactants with $n = 8$ (**I**$^+$), 11 (**II**$^+$), and 15 (**III**$^+$) as well as a dimeric ferrocenyl surfactant (**DI**$^{2+}$) (Fig. 3).

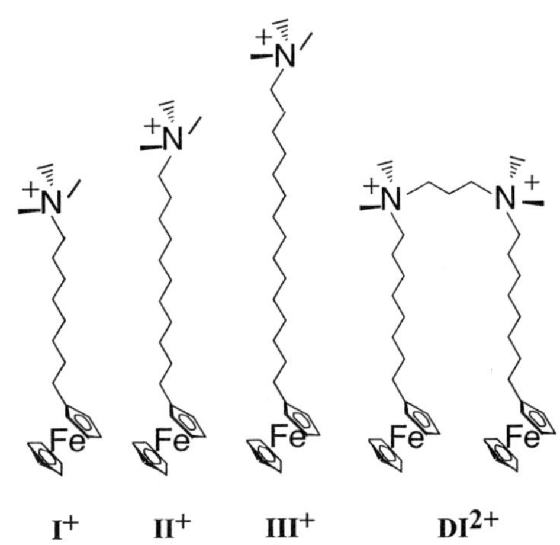

FIG. 3 Molecular structures of four ferrocenyl surfactants.

Figure 4 shows equilibrium surface tensions of aqueous solutions (0.1 M Li_2SO_4, pH 2) of II^+ and II^{2+} that were measured using the Wilhelmy plate method (and confirmed using the pendant drop and du Nouy methods). The reduced surfactant (II^+) has a critical micellar concentration (cmc) of 0.1 mM and a limiting surface tension of 49 mN/m (measured at concentrations above the cmc). The oxidized surfactant (II^{2+}) is not measurably surface active below concentrations of 0.1 mM. Oxidation of II^+ and II^{2+} at a concentration of 0.1 mM, therefore, changes the equilibrium surface tension of the aqueous solution from 49 to 72 mN/m. The oxidation of II^+ and II^{2+} is reversible over many cycles, and we have found that the reduction of II^{2+} to II^+ results in a return of the equilibrium surface tension to 49 mN/m (see later).

The results in Fig. 4 demonstrate that the equilibrium surface tension of an aqueous solution of 0.1 mM II^+ is substantially lower than the equilibrium surface tension of a 0.1 mM solution of II^{2+}. We have also measured dynamic surface tensions of solutions of these surfactants as the solutions were cycled in composition (> nine cycles) on time scales of minutes by repeatedly oxidizing and reducing II^+/II^{2+} (using electrochemical methods). The measurements of dynamic surface tension were performed with a maximum bubble pressure tensiometer (Fig. 5) with a bubble rate of 0.3 to 1.0 bubbles/s. Although the dynamic surface tensions of the aqueous solutions of 0.1 mM II^+ measured by using the maximum bubble pressure method are higher by ~8 mN/m than the equilibrium ones measured by using the Wilhelmy plate, the qualitative response of the system during the oxidation of II^+ to II^{2+} in Fig. 5 can be understood in terms of the changes in equilibrium surface tension shown in Fig. 4. That is, the process of oxidation of II^+ to II^{2+} leads to an increase in surface tension, whereas the reverse process leads to a decrease in surface tension. In the following, by using disulfide-based surfactants and transformations that are faster than those shown here using ferrocenyl surfactants, we demonstrate that the correspondence between dynamic surface tension during a change in state of a surfactant and the equilibrium properties of the system does not always hold.

The measurements shown in Fig. 4 reveal that the largest changes in equilibrium surface tension (~23 mN/m) upon oxidation of II^+ to II^{2+} occur at a concentration of 0.1 mM. At concentrations of surfactant above 0.1 mM, the change in surface tension upon oxidation of II^+ to II^{2+} is smaller than the maximum value (22 mN/m at 0.1 mM) because the surface tensions of the solutions of II^{2+} are lower than the surface tensions

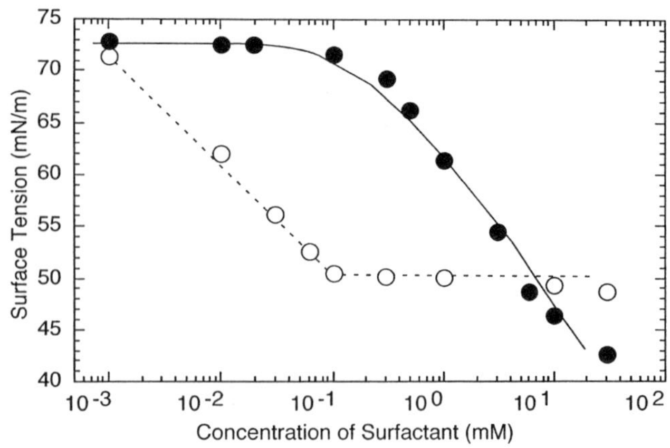

FIG. 4 Equilibrium surface tensions of aqueous solutions (0.1 M Li_2SO_4, pH 2, 25°C) of II^+ (open circles) and II^{2+} (filled circles) as measured by the Wilhelmy plate method. The lines show surface tensions predicted by a molecular thermodynamic model of Gibbs monolayers of these surfactants. See text for details.

of the aqueous solution of electrolyte (not containing surfactant). We found that the surface tension of aqueous solutions of II^{2+} decreases with increasing concentration of II^{2+} without any sign of a cmc. Surprisingly, at concentrations above 10 mM, oxidation of II^+ to II^{2+}

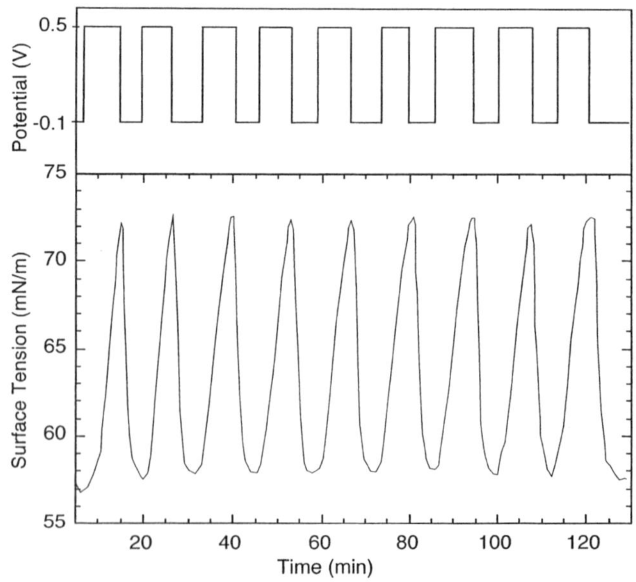

FIG. 5 Dynamic surface tensions of an aqueous solution (0.1 M Li_2SO_4, pH 2, 25°C) of 0.3 mM II^+/II^{2+} measured during repeated electrochemical cycling of the surfactant between oxidized (II^{2+}) and reduced (II^+) states. The dynamic surface tensions were measured by the maximum bubble pressure method at a bubble rate of 0.3 to 1 bubble/s.

no longer leads to an increase in surface tension but rather results in a *decrease* in the surface tension. This decrease in surface tension can be reversed by reduction of II^{2+} back to II^{+}. We also point out that the changes in surface tension shown in Fig. 4 have been observed when using both chemical [$Fe_2(SO_4)_3$ as oxidizer] and electrochemical oxidation of II^{+} to II^{2+}. By using the Gibbs adsorption equation [21], we estimate the limiting surface areas occupied by molecules of II^{+} and II^{2+} to be 85 ± 4 and 65 ± 4 Å^2, respectively. This result is also somewhat surprising because it suggests that oxidation of II^{+} to II^{2+} at concentrations greater than ~1 mM leads to an increase in the excess surface concentration of these surfactants even though the charge carried by each surfactant is increased upon oxidation.

In order to understand the origin of the interfacial behavior of II^{+} and II^{2+} just described, we developed a molecular thermodynamic model for Gibbs monolayers of these surfactants. The model, which is described in detail elsewhere [13], assumes that equilibrium exists between surfactants dissolved in the bulk solution and at the interface [Eq. (1)]. The results of the model are in good agreement with experimental measurements of surface tension (Fig. 4) and offer several insights into the balance of forces that controls the surface activity of ferrocenyl surfactants. First, the model reveals that the limiting surface area occupied by a molecule of II^{+} hosted within a Gibbs monolayer is large because II^{+} adopts a looped conformation at the surface of the aqueous solution (Fig. 6). This looped configuration is a result of an effective attraction between the ferrocene group of the surfactant and the aqueous subphase. Second, the model indicates that the desorption of surfactant from the surface of the solution, which accompanies oxidation at a 0.1 mM solution to II^{+} to II^{2+}, is caused by a reduction in the hydrophobic driving force for adsorption and a change in the electrostatic contribution to the standard free energy of formation of the Gibbs monolayer. In particular, the configurational (chain packing) contribution to the standard free energy of formation of the monolayer does not drive the oxidation-induced desorption of the surfactant from the surface of the solution because the surfactant adopts a looped conformation in both oxidized and reduced states. Third, the model also offers an explanation of the oxidation-induced *adsorption* of the ferrocenyl surfactant to the surface of a solution containing high (>1 mM) concentrations of the surfactant. As noted earlier, aqueous solutions of II^{+} have a cmc of 0.1 mM, and thus at concentrations greater than 0.1 mM II^{+}, the chemical potential of the surfactant changes little with concentration. In contrast to aqueous solutions of II^{+}, II^{2+} was found not to aggregate in solutions containing up to 30 mM of II^{2+} (the highest concentration investigated). That is, oxidation of II^{+} in aqueous solutions with concentrations of II^{+} between 0.1 and 30 mM leads to the dissolution of micelles of II^{+} to singly dispersed molecules of II^{2+}.

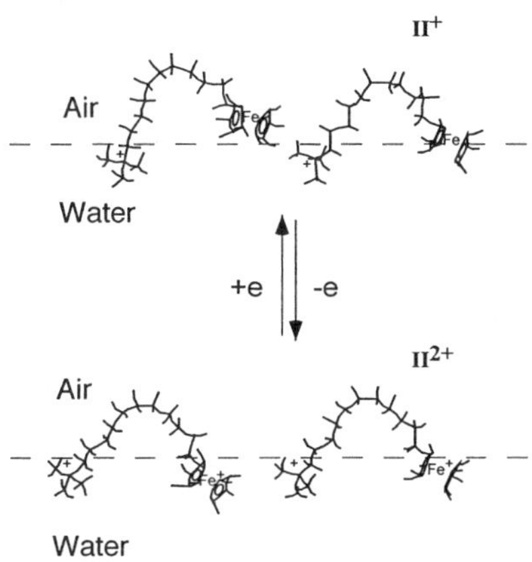

FIG. 6 Schematic illustrations of typical conformations assumed by II^{+} and II^{2+} within Gibbs monolayers at the surfaces of aqueous solutions.

Accompanying the disruption of the micelles is an increase in the cratic (concentration-dependent part of the chemical potential) contribution to the chemical potential of the surfactant in the aqueous solution (Fig. 7). For example, when the cratic contribution to the chemical potential is 0.58 (corresponding to the cmc of II^{+}), the surface area of the solution occupied by a molecule of II^{+} is 84 Å^2. With the same cratic contribution to the chemical potential, molecules of II^{2+} occupy a larger surface area. At higher surfactant concentrations, the area occupied by each molecule for II^{2+} continues to decrease with increasing concentration. When the cratic contribution to the chemical potential is 5.15 (7 mM of oxidized surfactant), the surface area occupied by each molecule of II^{2+} is calculated to be as small as 62 Å^2. In short, upon oxidation, the dissolution of micelles leads to an increase in the chemical potential of surfactant in bulk solution and thus a partitioning of surfactant toward the surface of the solution. The increase in the excess surface concentration of molecules as well as the increase in the charge carried by each surfactant molecule cause the decrease in surface tension of the solution that is measured upon oxidation at high concentrations of surfactant.

The experimental measurements of surface tensions of aqueous solutions of II^{+} show that oxidation and reduction can lead to large (up to 23 mN/m) and re-

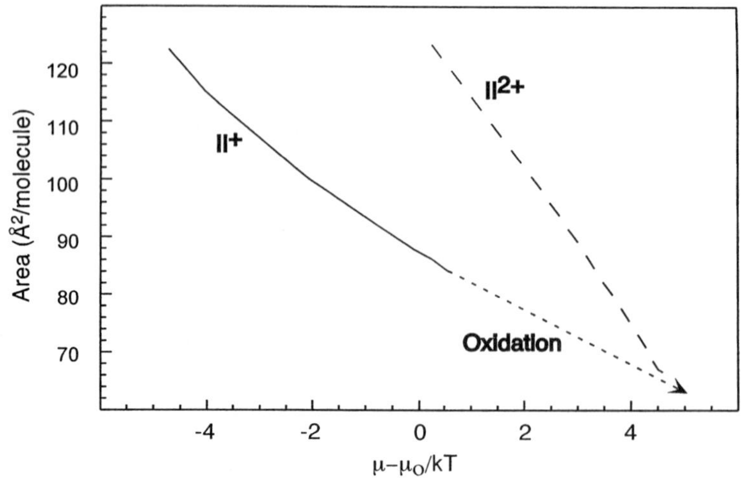

FIG. 7 Calculated surface areas occupied by molecules of II^+ and II^{2+} at surfaces of aqueous solutions as a function of the concentration-dependent part of the chemical potential. See text for details. The dashed arrow indicates the change in surface area per molecule that accompanies oxidation of II^+ to II^{2+} at the surface of the 7 mM solution of surfactant.

versible changes in surface tension. However, the range of concentrations over which these changes in surface tension take place is limited (concentrations close to the cmc, Fig. 4). In order to increase the window of concentrations over which large changes in surface tension can be driven by using ferrocenyl surfactants, we explored the interfacial properties of the dimeric surfactant DI^{2+} (Fig. 3) [10]. Surface tension measurements of the dimeric ferrocenyl surfactant show that the onset of surface tension reduction for the reduced surfactant occurs at a concentration that is almost three decades lower than that of the oxidized surfactant (Fig. 8a). The change in surface tension upon oxidation also occurs over a wider range of concentrations (Fig. 8b) for the dimeric surfactant as compared with I^+ (the corresponding monomeric surfactant). The maximal change in surface tension (~21 mN/m) occurs over almost two decades of concentration of the dimeric surfactant. We believe that the wide range of concentrations over which dimeric ferrocenyl surfactants can drive large changes in surface tension will make this class of surfactant useful for applications in microfluidics.

In summary, measurements of the surface tensions of aqueous solutions of ferrocene-based surfactants demonstrate that redox-active surfactants, when combined with use of electrochemical methods, permit large and reversible changes in the surface tensions of aqueous solutions. When the changes in oxidation state of the surfactant occur on time scales of tens of seconds, the surface tensions measured during the change in state of the system can be largely understood in terms of the equilibrium properties of the system. As described in Section V, this capability provides new ways to design electrochemical systems that can drive liquids into motion on millimeter and smaller scales.

III. PRINCIPLE 2. REACTION-INDUCED, EXCESS SURFACE CONCENTRATIONS: DISULFIDE-BASED SURFACTANTS

A. Interfacial States Far from Equilibrium

The second strategy we review for active control of interfacial properties of aqueous solutions of surfactants relies on the creation of interfacial states that are far from equilibrium. This principle involves transformations of the states of surfactants hosted at interfaces on time scales that are fast compared with their rates of relaxation (e.g., via desorption) (Fig. 1b). These "fast" transformations lead to transient, excess surface concentrations of the surfactants that are not in equilibrium with the bulk of the solution. Whereas the rate of accumulation of surfactant at a freshly formed interface is typically dominated by the rate of diffusion of surfactant molecules from the bulk of the solution to the interface, the kinetics of *desorption* of surfactant from a surfactant-laden interface is often substantially slower than that of adsorption because of the cohesive interactions between aliphatic chains of surfactants hosted at the interface. Past studies have reported that slow desorption kinetics give rise to transient states of surfactant-laden interfaces with properties that differ substantially from equilibrium ones. For example, Lin

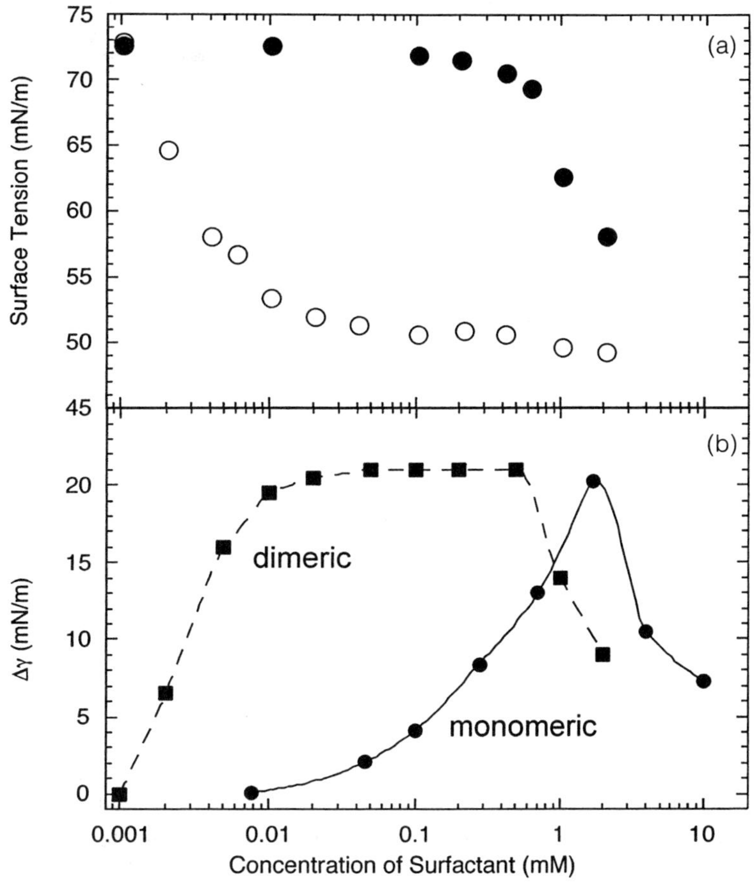

FIG. 8 (a) Equilibrium surface tensions of aqueous solutions (0.1 M Li_2SO_4, pH 2, 25°C) of DI^{2+} and DI^{4+}. (b) Comparison of the change in surface tension upon oxidation of DI^{2+} to DI^{4+} and I^+ to I^{2+}.

et al. [22,23] have reported measurements of the dynamic surface tension of aqueous solutions of a nonionic surfactant ($C_{12}E_8$) following reduction of the surface area of a pendant bubble. A 20% reduction in the surface area resulted in a transient lowering of surface tension of ~6 mN/m (Fig. 9). These authors interpreted the time-dependent evolution in surface tension to reflect the slow desorption of surfactant from the surface of the solution. The equilibrium surface tension was recovered in ~100 s.

Here we describe the generation of nonequilibrium states of an interface through transformations in the states of surfactants rather than changes in the surface area of the system. The properties of these nonequilibrium states can be strikingly different from equilibrium properties of the system. For example, whereas the transformation of state of the surfactant may ultimately lead to an increase in the equilibrium surface tension of a system, the transient state of the interface during the transformation can have a surface tension that is substantially lower than the initial surface tension of the system. Although the lifetimes of these transient interfacial states are strongly coupled to rates of desorption of surfactants from surfactant-laden interfaces (typically about 0.1–1 s^{-1} for ionic surfactants), we show in the following that it is possible to prolong these nonequilibrium interfacial states by replenishing the concentration of surfactant at the interface by transport to the surface. We report the creation of nonequilibrium states with lifetimes of minutes.

B. Bolaform Disulfide

We illustrate this second principle for active control of interfacial properties through the use of a bolaform surfactant that has a disulfide group in its aliphatic chain (**IV**) (Fig. 10). We report here the equilibrium and nonequilibrium properties that accompany the reduction of the disulfide group to thiols and the consequent fragmentation of the surfactant into monomeric products

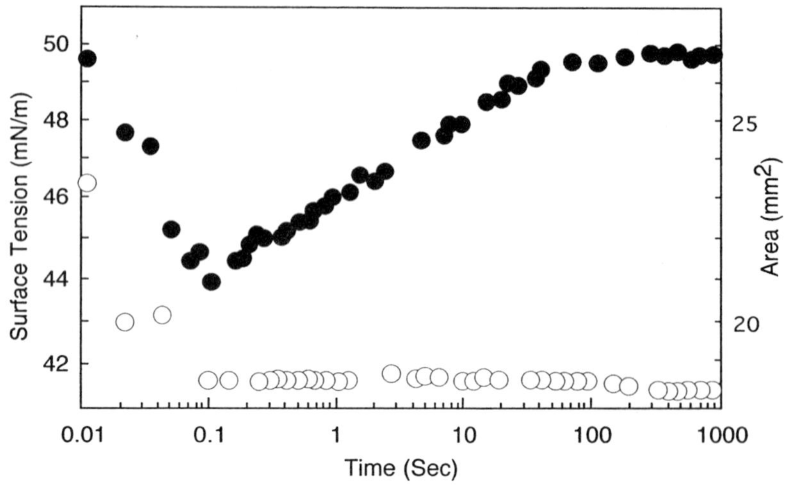

FIG. 9 Dynamic surface tension (filled circle) of an aqueous solution of $C_{12}E_8$ (7.32×10^{-6} M) upon reduction of the area (open circle) of the surface of the solution. (Data from Ref. 22.)

(**V**). We believe that this type of transformation is a particularly interesting one because the reduction of disulfide bonds to thiol groups can be accomplished by chemical [24], electrochemical [25], and photochemical [26] methods. Here we focus on the properties of this surfactant system during chemical reduction of the disulfide bond to thiol products by using dithiolthreitol (DTT) [27]. Next, we report the equilibrium interfacial properties of this system and then demonstrate the creation of nonequilibrium interfacial states that are not anticipated by consideration of the equilibrium properties of the system.

1. Equilibrium Surface Tension

Equilibrium measurements of the surface tensions (obtained using a pendant drop) of aqueous solutions of IV and V are reported in Fig. 11. Surfactant IV reduces the surface tension of its aqueous solution at concentrations that are substantially lower than the concentrations at which V (the thiol fragments) reduces the surface tension of its aqueous solution. We measured the equilibrium surface tension of a 1 mM solution of IV as 49 mN/m and estimated (using the Gibbs adsorption equation) the area occupied by each molecule of IV at the surface of the solution as 130 Å^2. In contrast, the equilibrium surface tension of an aqueous solution containing 2 mM of V is not measurably different from that of the aqueous solution of electrolyte that is free of V. The area occupied on average by a molecule of V is greater than 500 Å^2. This result is consistent with past studies of surface tensions of aqueous solutions of bolaform and monomeric surfactants in which the length of the aliphatic chain of the bolaform was twice that of the monomeric surfactant [28]. Unlike measurements of the ferrocene-based surfactants in Fig. 3, the measurements shown in Fig. 11 (and comparisons with other bolaform surfactants) demonstrate that the introduction of the disulfide bond into the aliphatic chains of IV, as well as the presence of the thiol group on V, has a relatively small influence on the equilibrium properties of these surfactant systems. This characteristic of disulfide/thiol groups ("invisibility") makes these surfactants especially interesting because structure-properties relationships established for classical surfactants can be readily used in the design of surfactants capable of being placed under active control.

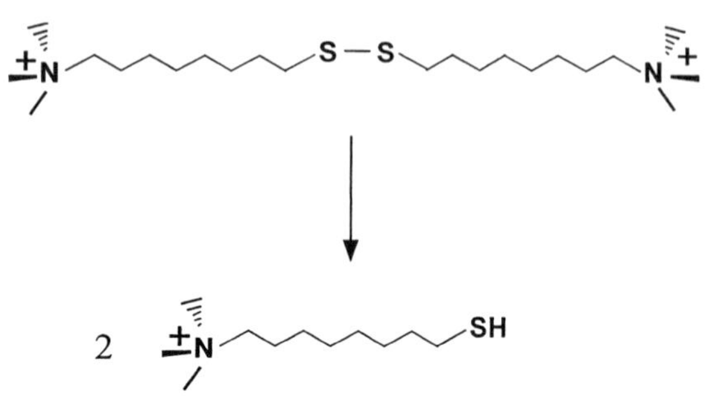

FIG. 10 Molecular structures of disulfide- and thiol-based surfactants.

2. Nonequilibrium Surface Tension

The equilibrium measurements of surface tension shown in Fig. 11 suggest that processes in which an aqueous solution of IV is slowly reduced to V will be

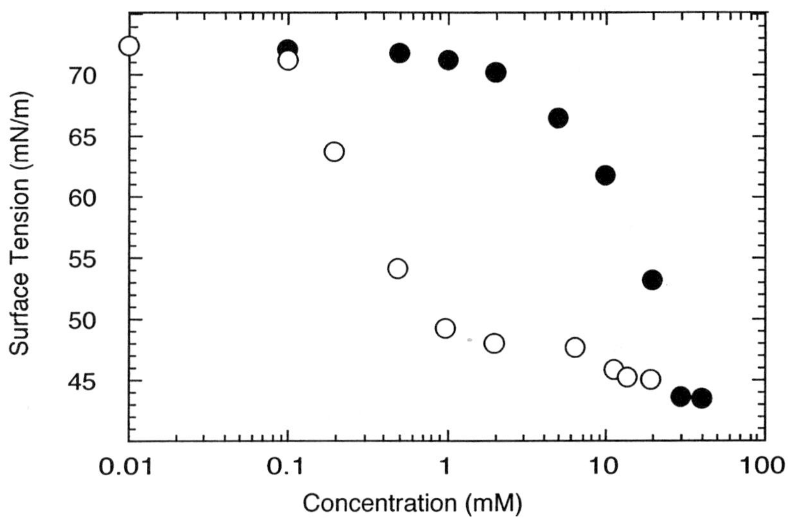

FIG. 11 Equilibrium surface tensions of aqueous solutions (50 mM phosphorus buffer) of IV (open circles) and V (filled circles) as a function of concentration as measured by the pendant drop method.

accompanied by an increase in surface tension. This increase in surface tension is largely the result of the partitioning of V away from the interface and into the bulk solution. Indeed, our experimental measurements do show that a slow reduction of IV to V is accompanied by a monotonic increase in surface tension (as described by principle 1 earlier). However, when the rate of reduction of IV to V is increased by addition of ~2 mM of reducing agent (DTT), we measure the surface tension *during* the reduction of IV to V and find that it assumes values that are far from any equilibrium ones measured in the system (Fig. 12). Indeed, the first 10 min of the transformation are accompanied by a *decrease* in surface tension by up to ~10 mN/m.

A number of experimental observations demonstrate that the minimum in surface tension observed in Fig. 12 can be understood in terms of a nonequilibrium, excess surface concentration of surfactant (IV and V)

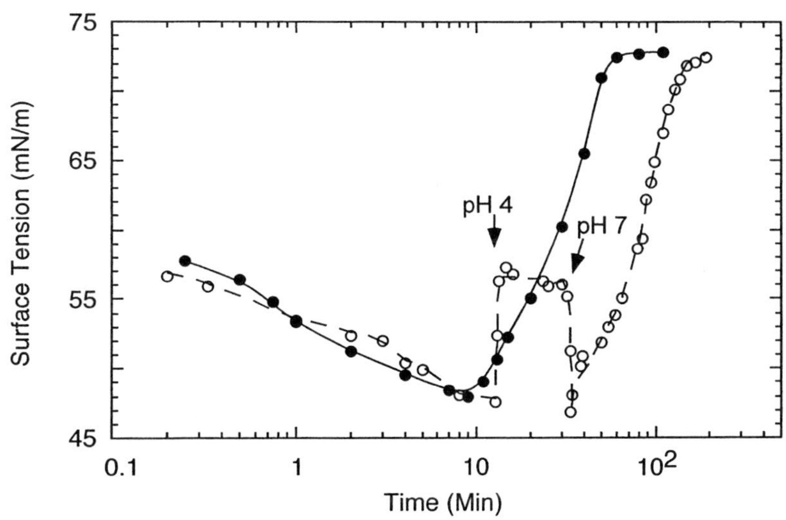

FIG. 12 Dynamic surface tensions measured during the reduction of a 1 mM solution of IV to a 2 mM solution of V (filled circles). The measurements were performed by the maximum bubble pressure method (bubble rate 3–10 bubbles/s) and 50 mM phosphate buffer at pH 7. Upon the addition of acid (and lowering of pH from 7 to 4) the reaction is quenched (open circles); readjustment of the pH to 7 results in resumption of the reaction.

that is generated through the reduction of IV to V by DTT at the surface of the solution. In particular, attempts to describe the presence of the minimum in terms of equilibrium properties of the system were unsuccessful. For example, we measured the equilibrium surface tensions of solutions containing mixtures of IV and V to determine whether the low surface tensions shown in Fig. 12 resulted from synergism [29]. The equilibrium surface tensions of the mixed surfactant systems were found to increase monotonically with composition of the system (Fig. 13), indicating that the minimum in Fig. 12 is not a result of synergism between IV and V. Further support for the reaction-dependent minimum is found through the influence of pH on the presence of the minimum. For example, upon addition of acid, the reaction leading to the reduction of IV to V is quenched and a rapid increase in surface tension is observed. When the system is returned to neutral pH, the reduction of IV to V resumes and the low surface tension is recovered. In addition, low surface tensions can be produced using a different reducing agent such as sodium sulfite.

These experimental observations lead us to conclude that the low surface tensions measured during the reduction of IV to V are the result of an excess surface concentration of surfactant that is formed by the reduction of IV to V at the surface of the solution. Support for this proposition has been derived from a simple diffusion-kinetic model that we have developed to describe the interfacial states (excess surface concentration and surface pressure) created by the transformation of IV to V. This model, which is described in detail elsewhere [30], combines a diffusion-kinetic model (to describe the adsorption, desorption, and transport of surfactant to and from the interface) and a surface equation of state to correlate the changes in excess surface concentration with changes in surface tension. The diffusion-kinetic model differs from past work [31–35] in that it accounts for the reaction leading to the conversion of IV to V within the Gibbs monolayer and bulk solution. We used the model to investigate the effect of the reaction on the surface excess concentration of IV and V. The predicted reduction of IV to V based on physically reasonable reaction rate constants, when combined with estimates of the rates of desorption of IV and V from the surface of the solution, was shown to lead to interfacial compositions (rich in V and lean in IV) with surface tensions that were smaller (by ~1–10 mN/m) than equilibrium ones corresponding to the same bulk composition of the solution. The model demonstrates that when the rate of formation of V is comparable to the rate of desorption of V from the surfactant-laden interface, an excess concentration of V builds up and thereby lowers the surface tension of the solution. Control of the rate of transformation of IV to V does, therefore, provide a means by which to exercise active control over the interfacial state of the system.

In summary, the results described here demonstrate that it is relatively straightforward to design a surfactant system such that nonequilibrium states of the interface can be accessed by transforming the surfactant between states. We believe these transient states may find use, for example, in processes that require control of interfacial states for short lengths of time (e.g., during emulsification).

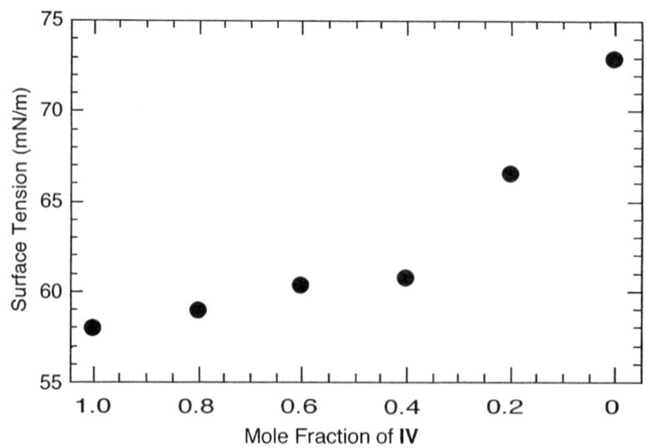

FIG. 13 Surface tensions of premixed solutions of IV and V (no reaction) as measured by the maximum bubble pressure method. The compositions of the mixtures correspond to compositions sampled during the reaction shown in Fig. 12.

IV. PRINCIPLE 3. DIFFUSION-LIMITED SURFACE EXCESS: AZOBENZENE-BASED SURFACTANTS

A. Effects of Aggregation on Surface Tension

The third principle we describe for active control of the interfacial properties of aqueous solutions of surfactant achieves control of *dynamic* surface tensions via manipulation of the state of aggregation of surfactant in bulk solution (Fig. 1C). Changes in the state of aggregation of surfactant in bulk solution can potentially give rise to changes in the dynamic surface tension of the solution through one of several mechanisms. For example, when dealing with diffusion-limited adsorption of surfactant to an interface, the characteristic time for surfactant molecules to diffuse to a freshly formed interface and form an interfacial assembly, t_D, is given by $t_D \sim \Gamma^2/C^2D$ [36], where Γ is the excess surface concentration of surfactant at equilibrium, C is the bulk concentration of surfactant, and D is the effective diffusion coefficient of surfactant in bulk solution. For aqueous solutions of ionic surfactants (e.g., a 10 mM solution of sodium dodecyl sulfate), the monomer activity is high (approximately equal to the cmc or 8 mM) and thus changes in the state of aggregation of surfactant will have relatively little effect on the effective diffusion coefficient (because transport of the surfactant will be dominated by the diffusion of monomer in solution). When dealing with mixed surfactant systems of anionic and cationic surfactant, however, the situation can be very different. First, the activity of the monomer in these solutions can be very low by virtue of the electrostatic interaction between surfactant molecules of opposite charge [37–41]. For example, Villeneuve et al. [39] have measured parts of the phase diagram of mixtures of sodium decyl sulfate and decyltrimethylammonium bromide. Whereas the cmc of decyltrimethylammonium bromide was ~100 mM, upon addition of sodium decylsulfate, the monomer concentration decreased by almost two orders of magnitude (Fig. 14). Under circumstances in which the activity of the surfactant is lowered to submillimolar concentrations, the diffusion of surfactant in solution is likely to be dominated by the transport of aggregates of surfactant and not monomeric species. Second, the size of the aggregates in these systems can be very large and thus they diffuse slowly [40,41]. At concentrations of mixed surfactants that are well below the cmc of the single-surfactant systems, these mixed solutions consist largely of vesicles in equilibrium with low concentrations of monomer [39]. By using the Stokes-Einstein relationship, the diffusion coefficient for aggregates with hydrodynamic diameters of ~150 nm (e.g., vesicles) is estimated to be ~10^{-8} cm^2/s. For systems of these types in which the extent of aggregation of surfactant can be manipulated (see later), a process of disassembly of large aggregates into monomers can plausibly increase the effective diffusion coefficient of surfactant in the system by roughly two orders of magnitude (from 10^{-8} to 10^{-6} cm^2/s). This change in diffusion coefficient can lead to similar changes in the time required for the surfactant to lower the surface tension of a freshly created interface [36].

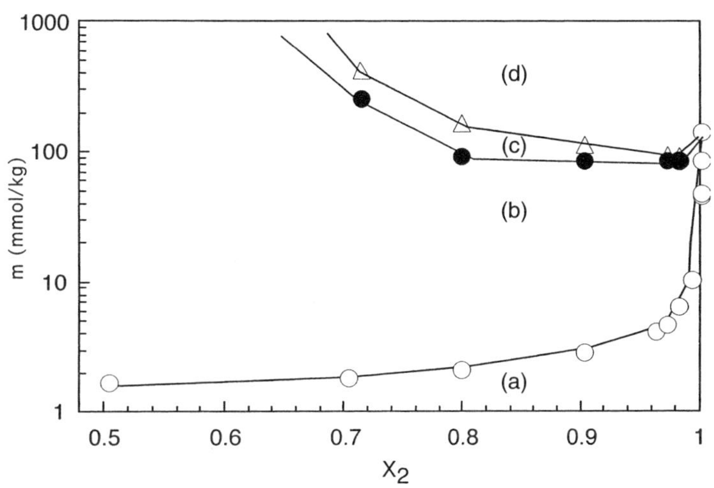

FIG. 14 Phase diagram for mixtures of decyl trimethylammonium bromide and sodium decylsulfate: (a) monomer, (b) monomer + vesicle, (c) monomer + vesicle + micelle, (d) monomer + micelle. *X*-axis is fraction of anionic surfactant in mixture. *Y*-axis is total surfactant concentration. (Data from Ref. 39.)

We also note a second mechanism by which changes in the extent of aggregation of a surfactant can influence the dynamic surface tension of a system. This mechanism is through changes in the lifetimes of aggregates of surfactants. For example, the lifetimes of foams have been correlated with the lifetimes of aggregates within solution [42]. The lifetimes of micelles in sodium dodecyl sulfate (SDS) solutions vary from milliseconds (near the cmc) to ~1 s at SDS concentrations well above the cmc (~150 mM). Several past studies have suggested that the lifetimes of micelles in solutions containing high concentrations of surfactant do influence dynamic surface tensions of these solutions [43–45].

B. Photoresponsive Surfactant

We illustrate this third approach to active control of interfacial properties by using a water-soluble surfactant that hosts the azobenzene moiety [46]. Azobenzene is a widely used photoactive group that can assume one of two isomeric states (*cis* or *trans*) depending on the wavelength of light used to illuminate the compound. The *trans* isomer absorbs ultraviolet (UV) light with a wavelength of 360 nm and is transformed upon illumination to the *cis* isomer. The *cis* isomer absorbs light with a wavelength of 460 nm, and illumination of the *cis* state with visible light causes the *cis* isomer to relax back to the *trans* isomer. Although a number of past studies have reported water-soluble and water-insoluble amphiphiles containing azobenzene [47–49], past studies based on water-soluble surfactants have been largely unsuccessful in driving substantial changes in surface tension upon illumination with light.

C. Catanionic System

Our approach is based on a mixed surfactant system containing a cationic, bolaform surfactant (VI) that hosts azobenzene (Fig. 15) and the anionic surfactant SDS. We designed and synthesized VI rather than a monomeric azobenzene surfactant because the conformations of bolaform surfactants within aggregates are constrained as compared with surfactants with a single polar headgroup. We hypothesized that the presence of constraints on the packing of bolaform surfactants would lead to larger changes in aggregation as compared with single-headed surfactants upon photoisomerization. In addition to the cationic bolaform surfactant, the system contained SDS, an anionic surfactant, so as to cause the monomer activity to be low. The influence of the attraction between cationic and anionic surfactants in solution (which leads to low monomer activity) is apparent in the small interfacial areas occupied by surfactants in aggregates of these mixtures of surfactants. For example, surfactant molecules hosted within vesicles formed from mixtures of cetyltrimethylammonium bromide and sodium octylsulfate occupy areas as small as 36 Å^2/molecule [40]. It is also plausible that the close packing of surfactant molecules will increase the impact of isomerization on the aggregation behavior through constraints imposed on the packing of the surfactant chains within the aggregates.

Dynamic and static light scattering measurements reveal that photoisomerization of aqueous solutions of 0.1 mM BTHA and 1.6 mM SDS leads to large and reversible changes in the extent of aggregation of surfactant in the system. Before illumination, the hydro-

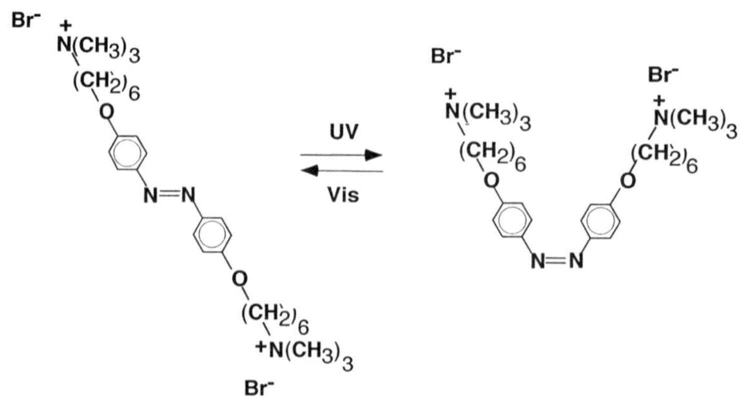

VI

FIG. 15 Photoisomers of azobenzene-based surfactant.

dynamic diameters of aggregates were measured as 150 nm ± 10 nm and the intensity of scattered light was 2500 kcnts/s. After 3 min of illumination with UV light, the measured hydrodynamic diameters were similar to that measured prior to illumination (140 nm ± 10 nm). However, the intensity of scattered light was lowered by almost a factor of 2 (1480 kcnts/s) after illumination with UV light. Because the intensity of light scattered from aggregates of similar size is proportional to the number density of aggregates in solution, these measurements indicate that illumination of the mixed surfactant system with UV light decreases the number density of aggregates in solution by a factor of 2. The dissolution of the aggregates will lead to a large change in the monomer (or small aggregate) concentration.

Measurements of the surface tension of the mixed VI and SDS surfactant system obtained using the du Nouy ring also reveal that isomerization is accompanied by large changes in the surface tension of the solution. The du Nouy ring method (Fig. 16a), which characterizes the surface of the solution with an age of ~10 s, shows that the dynamic surface tensions of the illuminated and nonilluminated solutions are quite different. The illuminated solution has a dynamic surface tension that is lower than that of the solution prior to illumination over a wide range of concentrations of VI. As shown in the following, these large, light-induced changes in surface tension can be exploited by control (by illumination) the release of liquids from capillaries.

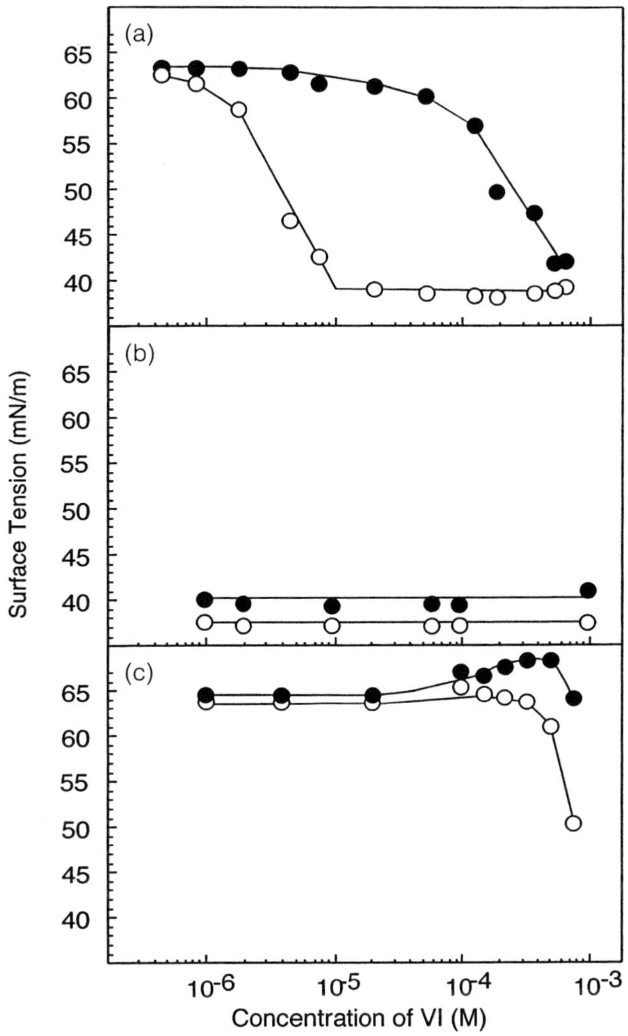

FIG. 16 Surface tensions of mixed solutions of VI and SDS measured before (filled circle) and after (open circle) illumination with UV light. (a) du Nouy ring method; (b) Wilhelmy plate method; (c) maximum bubble pressure method. All measurements were performed using aqueous solutions of 1.6 mM SDS and at 25°C.

D. Mass Transport

Several experimental observations support our view that changes in dynamic surface tension caused by illumination of mixed solutions of VI and SDS are largely controlled by the rate of delivery of surfactant onto the surface of the solution. First, by using a Wilhelmy plate (Fig. 16b), we have demonstrated that illumination of this surfactant system leads to relatively small changes in the equilibrium surface tension of the solution. That is, the difference in surface tension that is seen when the age of the surface of the solution is ~10 s largely vanishes when the system is provided sufficient time to deliver surfactant onto the surface of the solution. Second, we have measured the surface tensions of this system when the age of the surface of the solution is small compared with 10 s (by using the maximum bubble pressure method, Fig. 16c). Here, too, we observe that the influence of illumination on the dynamic surface tension of the system is small. In this case, however, the surface tension of the system is similar before and after illumination because the interface is starved of the mixture of surfactant. Figure 17 compares values of surface tension measured before and after illumination by UV light for solutions of 0.1 mM VI and SDS as measured with the maximum bubble pressure, du Nouy ring, and Wilhelmy plate methods.

Although our experimental investigation does not permit identification of one of the several possible mechanisms (see earlier) by which the change in state of aggregation of the surfactant in solution can change

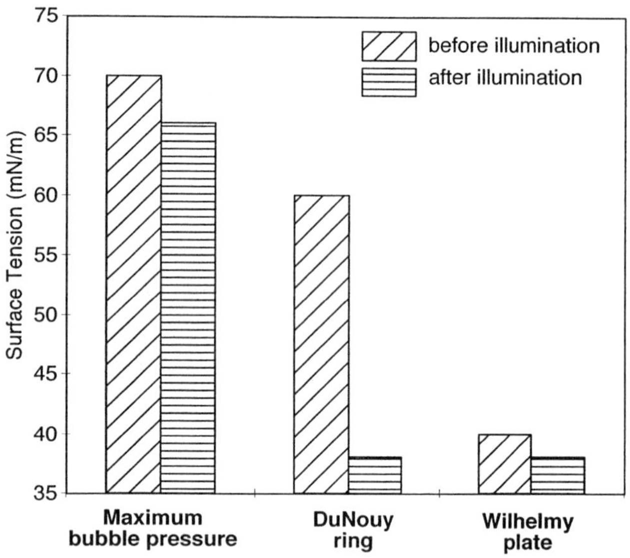

FIG. 17 Surface tensions of aqueous solutions containing VI (0.1 mM) and SDS (1.6 mM) before and after illumination with UV light.

the rate of delivery of surfactant to the interface, and thus the surface tension measured by the du Nouy ring, we point out that the time scales on which we observe the light to induce changes in surface tension are consistent with the influence of mass transport. As noted earlier, past studies have established that diffusion coefficients of aggregates of surfactants with hydrodynamic diameters of ~150 nm (e.g., vesicles) are $\sim 10^{-8}$ cm^2 s^{-1} whereas diffusion coefficients of monomeric surfactants are $\sim 10^{-6}$ cm^2 s^{-1}. A change in the effective diffusion coefficient from 10^{-8} to 10^{-6} cm^2/s will lead to changes in t_D by up to two orders of magnitude. For example, using values of $C = 0.3$ mM, $1/\Gamma = 50$ Å^2/molecule, and $D = 10^{-8}$ cm^2 s^{-1}, the characterisic diffusion time is ~100 s. That is, when transport of surfactant to the interface is controlled by the diffusion of aggregates, it takes ~100 s to transport a sufficient amount of surfactant to the interface to form a monolayer and thereby lower the surface tension. In contrast, if the transport of surfactant to the interface is dominated by diffusion of monomers ($D = 10^{-6}$ cm^2 s^{-1}), the characteristic diffusion time decreases to ~1 s. The change in characteristic diffusion times described earlier compares well with the dynamic surface tension behavior of the VI and SDS system before and after illumination (Figs. 16 and 17) and suggests that the dynamic surface tension of the mixed surfactant systems may well be controlled through the regulation of mass transport. In the next section, the capability to control dynamic surface tension by using light is used to demonstrate the controlled release of pendant drops from tips of capillaries.

V. ACTIVE CONTROL OF LIQUIDS ON MILLIMETER AND SMALLER SCALES

The final section of this chapter illustrates the manner in which the principles discussed earlier for active control of the interfacial properties of liquids can be exploited to form the basis of new ways to drive liquids into motion, release liquids from capillaries, and position liquids on surfaces in periodic arrays.

A. Electrochemical Control of Marangoni Effects

Electrochemical control of the oxidation state of the ferrocenyl surfactant II^+ can lead to large and reversible changes in the surface tension of its aqueous solutions. When dealing with solutions containing concentrations of II^+ of 0.1 mM, oxidation of II^+ to II^{2+} leads to a large increase in surface tension (Figs. 4 and 5). Here we illustrate that a simple arrangement of electrodes can be used to dispense or consume II^+ at localized regions of an aqueous system, thus creating gradients in the surface excess concentration of ferrocenyl surfactant between the electrodes. The resulting gradients in surface tension cause surface flows (Marangoni phenomena) directed away from the cathode and toward the anode. Figure 18 shows Marangoni flows created by reduction of II^{2+} to II^+ in a system with two working electrodes. Sulfur dust sprinkled on the surface is used to visualize the flow. Application of a reducing potential (−0.3 V vs. SCE) at the upper electrode and an oxidizing potential (0.3 V vs. SCE) at the lower electrode (Fig. 18b) causes the surface fluid to flow away from the cathode (Fig. 18c) and toward the anode. Reversal of the oxidizing and reducing potentials causes the surface to flow in the opposite direction (Fig. 18d).

We have also used electrochemical control of the surface activity of ferrocenyl surfactants to direct the flow of liquid through networks of channels. Figure 19 shows a top view of a network of four intersecting channels. Three of the four channels end at Pt electrodes, and the channel at the top ends at a reference electrode and a counterelectrode. The fluidic network was filled with 0.3 mM II^{2+} to a depth of ~1 mm. Small drops of a nematic liquid crystal (LC), 4-*n*-pentyl-4′-cyanobiphenyl, were placed on the surface of the aqueous solution within the channels so that the fluid

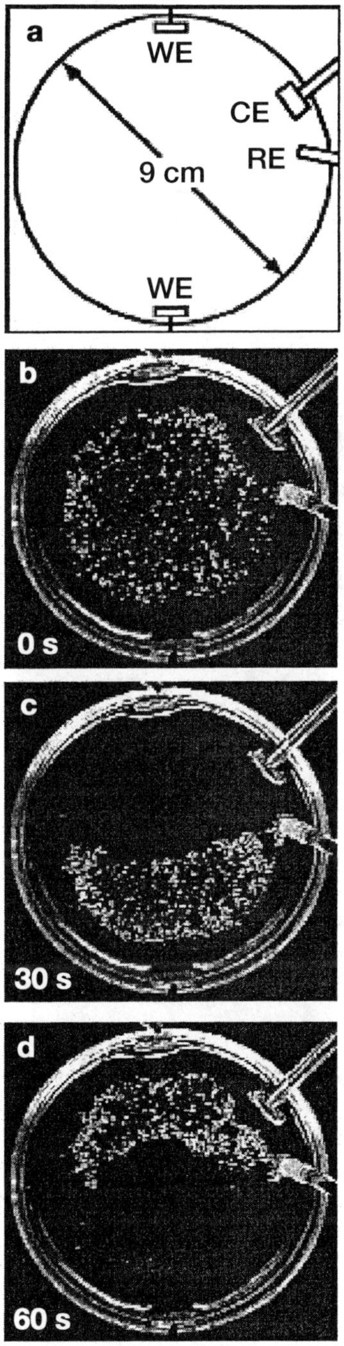

FIG. 18 Top view of petri disk filled with 0.1 mM II^+ (0.1 M Li_2SO_4, pH 2). Sulfur dust is sprinkled on top of the solution to visualize the displacement of fluid. (a) Electrode setup with two working electrodes, a counterelectrode, and a reference electrode. (b) Application of -0.3 V to the top working electrode and 0.3 V to the bottom electrode causes displacement of the surface (c). Reversal of working electrode potentials causes fluid displacement in the opposite direction (d).

flow could be visualized easily through crossed polarizers. Application of oxidizing (0.3 V vs. SCE) and reducing (-0.3 V vs. SCE) potentials to any pair of electrodes caused the LC droplets to be pumped between the two electrodes. In addition, the velocity of the droplets was controlled by the magnitude of the potential applied to the electrodes (Fig. 19d). The oxidation and reduction of II^+ were also used to demonstrate pumping of LC droplets across an unconfined surface (Fig. 19e). Whereas application of -0.3 V (vs. SCE) to either the top or left electrodes produced vertical or horizontal motion of the LC droplets across the image shown in Fig. 19e, respectively, simultaneous application of -0.3 V (vs. SCE) to the top and left electrodes produced a motion of the droplets in a direction given by the addition of the vectors of the single-electrode flows.

B. Addressable, Patterned Dewetting

Because localized changes in the interfacial tension of aqueous solutions (solid-liquid and liquid-vapor interfaces) can give rise to localized imbalances of force at the three-phase contact line of liquid supported on a surface and thereby place a liquid in motion, we have used ferrocenyl surfactants in strategies that permit the patterned wetting and dewetting of aqueous solutions on surfaces. We illustrate this capability here by using ferrocenyl surfactant to direct a thin film of liquid to dewet into a two-dimensional array of droplets supported on a surface (Fig. 20). The experiments are based on glass microscope slides that were patterned by evaporating 50 Å of titanium and 500 Å of gold through a micromachined aluminum mask. The patterned surface was made energetically homogeneous by immersing into a solution of 0.9 mM 11-mercaptoundecanoic acid and 0.1 mM hexadecyl mercaptan. The patterned electrode surface was inclined at ~20° from the horizontal with the lower edge touching a shallow pool of aqueous solution (0.01 M Li_2SO_4, pH 1.3) containing 0.3 mM II^+, an SCE, and a Pt counterelectrode. Surfactant solution pipetted onto the slide formed a continuous film (20 to 50 μm in thickness). When no external potential was applied to the electrodes, the film of liquid supported on the surface drained uniformly from the surface (that is, no fluid pattern was formed). However, when an oxidizing potential (0.5 V vs. SCE) was applied to the array of electrods, the aqueous solution dewet the gold electrodes while the glass surfaces remained covered by liquid (Fig. 20). Through this process, a continuous film of liquid was transformed into a periodic array of droplets.

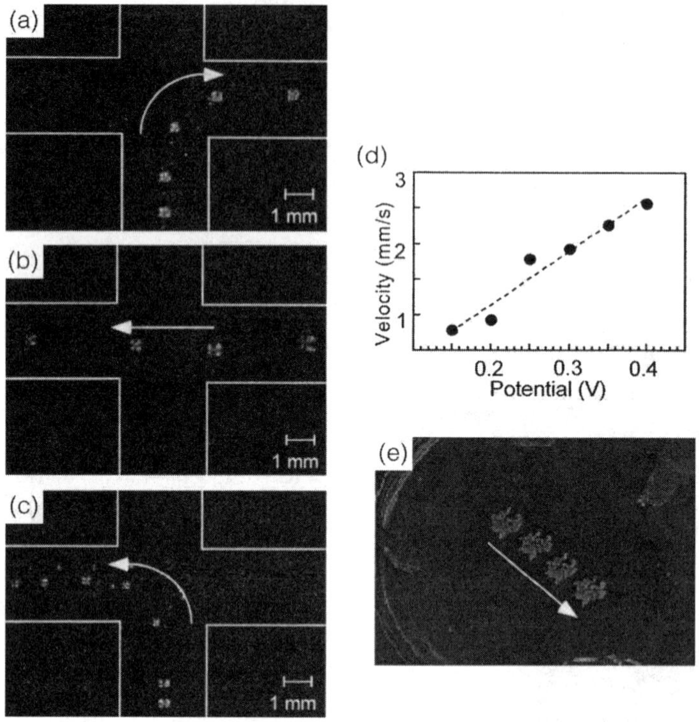

FIG. 19 Time lapse images of pumping of liquid crystal (LC) droplets across the surface of an aqueous solution of 0.3 mM II^{2+} (0.01 M Li_2SO_4) confined to four channels. Platinum electrodes protrude through the surface of the solution at the ends of the left, right, and lower channels. The end of the top channel contains an SCE and a counterelectrode. (a) Droplet of LC is dispensed at the bottom channel. The LC droplet is pumped by application of −0.3 V to the bottom electrode and 0.3 V to the right electrode. (b) Droplet of LC is pumped to the left electrode by application of −0.3 V to the right electrode and 0.3 V to the left electrode. (c) Droplet of LC dispensed at the bottom is pumped by application of −0.3 V to the bottom electrode and 0.3 V to the left electrode. (d) The velocity of the LC on the surface measured as a function of the potential used to pump fluid in a straight 4-mm-wide channel. The *x*-axis is the magnitude of the potential applied to the cathode and anode. (e) Sulfur particles supported on an unconfined surface pumped in the direction indicated by the arrow by application of −0.3 V to two electrodes at top and at left. All potentials are versus SCE.

Whereas gradients in surface tension caused by changes in the oxidation state of II^+ drive the phenomena shown in Figs. 18 and 19, the patterned dewetting of solutions of II^+ is driven by electrochemically induced changes in the contact angle of the solution on the surface. At equilibrium, no net force acts on the three-phase contact line of a liquid supported on a surface. An electrochemically induced increase in the local values of the interfacial free energies can, however, change the equilibrium contact angle from θ_i to θ_f and thus create a net force of magnitude $\gamma_{LV}(\cos\theta_i - \cos\theta_f)$ that leads to motion of the contact line. Figure 21a shows that static contact angles (measured to be constant for times greater than 10 to 20 s) of aqueous solutions of 0.3 mM II^+/II^{2+} placed on the surface of treated gold do increase with oxidation of II^+. A variety of experimental measurements (described elsewhere) demonstrate that II^{2+} does not measurably affect contact angles of these solutions. The increase in receding contact angle (0.3 mM II^+/II^{2+}) from <5° to 38° coincides with a decrease in the amount of surfactant adsorbed at the liquid-solid interface as measured by surface plasmon reflectometry (Fig. 21b). At 100% oxidation, no surfactant is adsorbed to the solid-liquid interface. This result strongly suggests that electrochemically induced desorption of surfactant from this interface plays a central role in causing the increase in receding contact angle and thus the dewetting phenomena in Fig. 20.

C. Light-Addressable Release of Liquid from Capillaries

Whereas the results described so far demonstrate the usefulness of ferrocenyl surfactants in manipulating liquids on millimeter and smaller scales, approaches to active control of surfactant solutions based on the use of light offer interesting possibilities for microfluidics

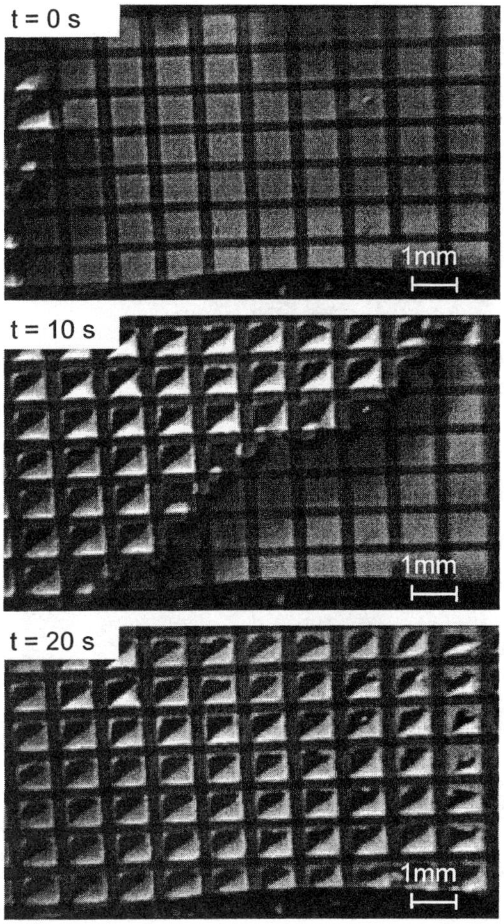

FIG. 20 Top view of dewetting of an aqueous film of 0.3 mM II^+ (0.01 M Li_2SO_4, pH 1.3) supported on a microscope slide patterned with a mesh of electrodes (dark region). Dewetting was induced at the top left corner of the microscope by application of 0.5 V (vs. SCE) to the electrode. The receding film leaves droplets on the regions of glass. The image was illuminated from the lower right side, causing shadows to form diagonal lines across the droplets.

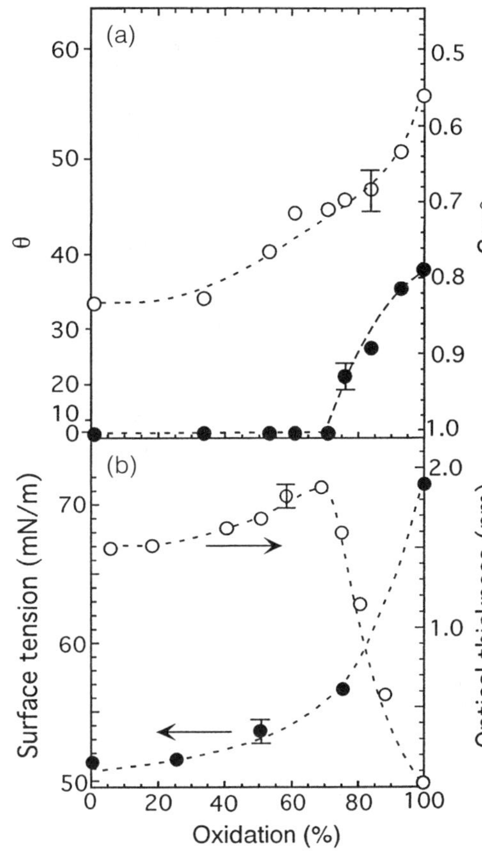

FIG. 21 (a) Static advancing (open circle) and receding contact angles (filled circle) of aqueous solutions of (0.01 M Li_2SO_4, pH 1.3) of 0.3 mM II^+ plotted as a function of the extent of oxidation of II^+ to II^{2+}. Contact angles were measured on the surface of treated films of gold (see text for details). (b) Surface tensions (filled circles) and optical thicknesses (open circle) of surfactant adsorbed to treated gold as a function of oxidation of 0.3 mM II^+ (0.01 M Li_2SO_4, pH 1.3). The optical thickness was measured using surface plasmon reflectometry.

because of the ease of addressing localized regions of a solution by patterned illumination. Here we illustrate the use of the light-sensitive surfactant VI in combination with SDS to trigger the release of pendant droplets of aqueous solution formed at the ends of capillaries. The experiment exploits the influence of light on the dynamic surface tensions of aqueous solutions of these surfactants (as described earlier). First, we formed a single pendant drop (9.5 μL) at the tip of a syringe (Fig. 22a) using 0.04 mM VI, 0.16 mM SDS, and 0.1 M NaBr. Illumination of the drop with UV light approximately 15, 30, or 45 s after formation of the drop caused the drop to release (Fig. 22b). We observed the shape of the droplet to evolve within 1 s of illumination and the droplet to release within 3–5 s of illumination. Without UV light, the decrease in surface tension is slow (see earlier discussion), causing the drops to hang from the capillary for times greater than 60 s.

We have also used the effect of light on the dynamic surface tension of mixed surfactant solutions of VI and SDS to trigger the release of selected droplets from an array of capillaries. Figure 23a and c show five pendant drops formed at the ends of capillaries. A simple system of shutters was positioned behind the droplets to permit illumination of selected droplets with UV light and thus promote their release. For example, illumination of the second and fourth drops of the array of five droplets caused their release (Fig. 23b) while the

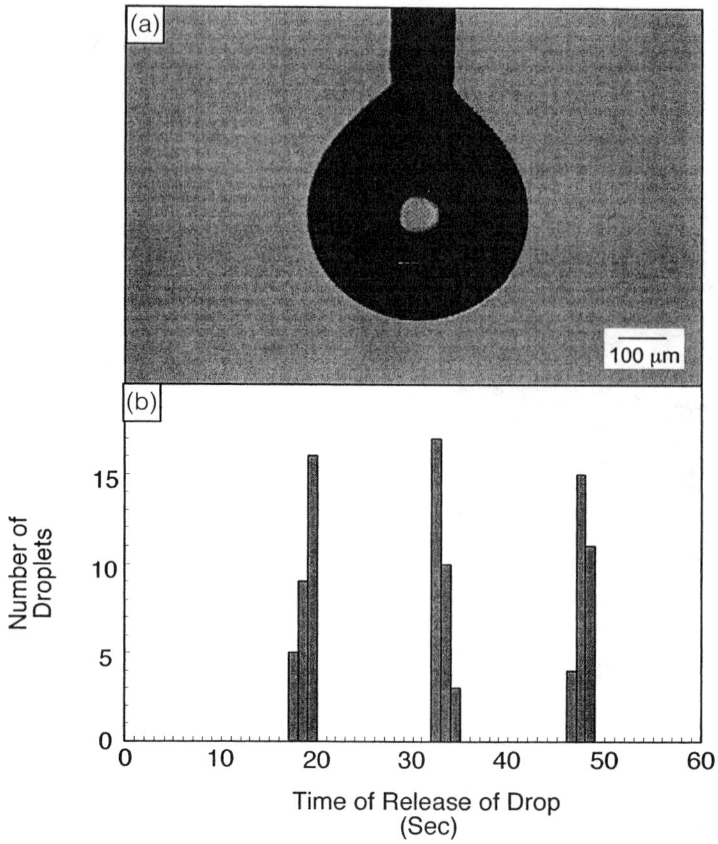

FIG. 22 Timed release of pendant droplets of an aqueous solution (0.04 mM VI, 0.16 mM SDS, 0.1 M NaCl) from the tip of a capillary using UV light. (a) Image of pendant drop. (b) Distribution of lifetimes of droplets following their illumination at 15, 30, and 45 s.

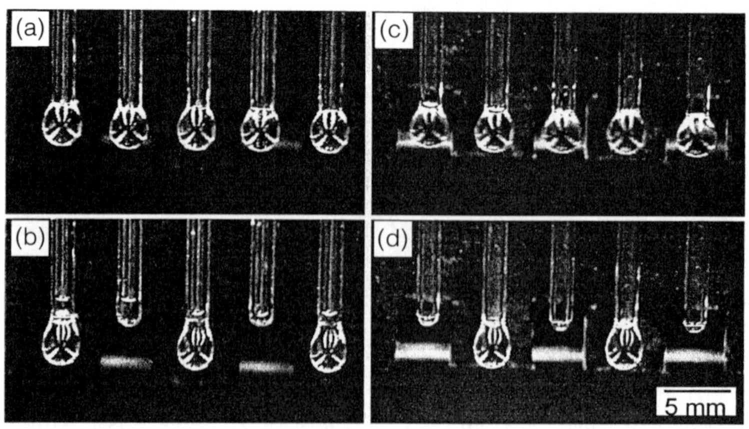

FIG. 23 Light-directed release of droplets of aqueous solution (0.04 mM VI, 0.16 mM SDS, 0.1 M NaCl) from an array of five capillaries. (a) Image of five droplets poised at the capillary tips. (b) Image of capillaries in part (a) after illumination of the second and fourth droplets with UV light. Illumination of the first, third, and fifth droplets of (c) causes their release (d).

first, third, and fifth drops remained attached. Alternatively, illumination of the first, third, and fifth drops caused their release (Fig. 23d), leaving the second and fourth drops attached.

ACKNOWLEDGMENTS

This work was supported in part by the Camille and Henry Dreyfus Foundation, the David and Lucile Packard Foundation, the donors of the Petroleum Research Fund, and the National Science Foundation (CTS-941014 and CTS-9502263).

REFERENCES

1. K. Peterson, Sens. Actuators A *56*:143 (1996).
2. J. L. DeRisi, V. R. Lyler, and P. O. Brown, Science *278*:680 (1997).
3. D. J. Harrison, K. Fluri, K. Seiler, Z. Fan, C. S. Effenhauser, and A. Manz, Science *261*:895 (1993).
4. H. Salimi-Mousavi, T. Tang, and D. J. Harrison, J. Am. Chem. Soc. *119*:8716 (1997).
5. G. T. A. Kovacs, *Micromachined Transducers Sourcebook*, McGraw-Hill, Boston, 1998.
6. E. Delemarche, A. Brenard, H. Schmid, B. Michel, and H. Biebuyck, Science *176*:779 (1997).
7. B. S. Gallardo, M. J. Hwa, and N. L. Abbott, Langmuir *11*:4209 (1995).
8. D. E. Bennett, B. S. Gallardo, and N. L. Abbott, J. Am. Chem. Soc. *118*:6499 (1996).
9. B. S. Gallardo, K. L. Metcalfe, and N. L. Abbott, Langmuir *12*:4116 (1996).
10. B. S. Gallardo and N. L. Abbott, Langmuir *13*:203 (1997).
11. B. S. Gallardo, V. K. Gupta, F. D. Eagerton, L. I. Jong, V. S. Craig, R. R. Shah, and N. L. Abbott, Science *283*: 57 (1999).
12. N. L. Abbott, Prog. Colloid Polym. Sci. *103*:300 (1997).
13. N. Aydogan, B. S. Gallardo, and N. L. Abbott, Langmuir *15*:722 (1999).
14. V. B. Fainerman, A. V. Makievski, and R. Miller, Colloids Surf. A *87*:61 (1994).
15. V. B. Fainerman, Colloids Surf. *62*:333 (1992).
16. K. J. Mysels, Langmuir *2*:423 (1986).
17. T. Saji, K. Hoshino, I. Yoshiyuki, and G. Masayuki, J. Am. Chem. Soc. *113*:450 (1991).
18. T. Saji, K. Hoshino, and S. Aoyagui, J. Am. Chem. Soc. *107*:6865 (1985).
19. R. A. Mackay and J. Texter, *Electrochemistry in Colloids and Dispersions*, VCH, New York, 1992.
20. G. Georges and S. Desmettre, Electrochim. Acta *29*:521 (1984).
21. D. K. Chattoraj and K. S. Birdi, *Adsorption and the Gibbs Surface Excess*, Plenum, New York, 1984.
22. S. Lin, R. Tsay, L. Lin, and S. Chen, Langmuir *12*:6530 (1996).
23. S. Lin, T. Lu, and W. Hwang, Langmuir *11*:555 (1995).
24. G. V. Lamourex and G. M. Whitesides, J. Org. Chem. *58*:633 (1993).
25. J. Zagal and P. Herrera, Electrochim. Acta *30*:449 (1985).
26. J. A. Barltrop, P. M. Hayes, and M. Calvin, J. Am. Chem. Soc. *76*:4348 (1954).
27. L. I. Jong and N. L. Abbott, Langmuir *14*:2235 (1998).
28. F. M. Merger and S. Wrenn, J. Phys. Chem. *78*:1387 (1974).
29. M. J. Rosen, *Surfactants and Interfacial Phenomena*, 2nd ed., Wiley, New York, 1989.
30. L. I. Jong, Ph.D. dissertation, University of California-Davis, 1999.
31. C. H. Chang and E. I. Frances, Colloids Surf. *100*:1 (1995).
32. R. P. Borwankar and D. T. Wasan, Chem. Eng. Sci. *38*: 1637 (1983).
33. R. P. Borwankar and D. T. Wasan, Chem. Eng. Sci. *41*: 199 (1986).
34. C. H. Chang and E. I. Frances, Colloids Surf. *69*:189 (1992).
35. B. Li, G. Geeraerts, and P. Joos, Colloids Surf. *88*:251 (1994).
36. J. K. Ferri and K. J. Stebe, J. Adv. Colloid. Interface Sci., in press.
37. J. Eastoe, J. Dalton, P. Rogueda, D. Sharpe, J. Dong, and J. Webster, Langmuir *12*:2706 (1996).
38. L. Zhang and G. Zhao, J. Colloid Interface Sci. *127*: 353 (1989).
39. M. Villeneuve, S. Kaneshina, T. Imae, and M. Aratone, Langmuir *15*:2029 (1999).
40. L. Brasher, Ph.D. Dissertation, University of Delaware, 1996.
41. M. T. Yatcilla, K. L. Herrington, L. L. Brasher, E. W. Kaler, S. Chiruvolu, and J. A. Zasadzinshi, J. Phys. Chem. *100*:5874 (1996).
42. B. K. Jha, A. Patist, and D. O. Shah, Langmuir *15*:3042 (1999).
43. J. Lucassen, Faraday Discuss. *59*:76 (1975).
44. R. Miller, Colloid Polym. Sci. *259*:1124 (1981).
45. E. Rillaerts and P. Joos, J. Phys. Chem. *86*:3471 (1982).
46. J. Y. Shin and N. L. Abbott, Langmuir *15*:4404 (1999).
47. T. Hayashita, T. Kurosawa, T. Miyata, K. Tanaka, and M. Igawa, Colloid Polym. Sci. *272*:1611 (1994).
48. L. Yang, N. Takisawa, T. Hayashita, and K. Shirahama, J. Phys. Chem. *99*:8799 (1995).
49. R. C. Ahuja, J. Maack, and H. Tachibana, J. Phys. Chem. *99*:9221 (1995).

8

Reactivity Control by Aqueous Amphiphilic Self-Assembling Systems

GIANFRANCO SAVELLI, RAIMONDO GERMANI, and LUCIA BRINCHI University of Perugia, Perugia, Italy

I. INTRODUCTION

Surfactants (*surf*ace-*act*ive *agents*) are compounds whose molecules are fitted with pronounced lipophilic and hydrophilic moieties; they are *amphiphilic* molecules. A process whereby dissolved surfactant molecules react to the repelling action of surrounding water is aggregation to form various kinds of supramolecular structures. Largely depending on the molecular architecture of the amphiphile, a wealth of three-dimensional structures can be formed ranging from spherical and rodlike micelles to multilayer structures and to complex biological membranes, whose matrix is a lipid bilayer composed of phospholipids and glycolipids, incorporating proteins [1–3]. Most studies of micellar systems have been carried out on synthetic surfactants in which the polar or ionic headgroup may be cationic (e.g., an ammonium ion), anionic (e.g., a carboxylate, sulfate, or sulfonate ion), nonionic (e.g., hydroxy compound), or zwitterionic (e.g., an amine oxide or a carboxy- or sulfobetaine). Surfactants are often given trivial or trade names, and abbreviations based on either trivial or systematic names are freely used (Fig. 1).

It has long been known that aqueous micelles can influence chemical reaction rates and equilibria [4–7]. Early studies of micellar effects on reaction rates and equilibria are described in extensive monographs [8–12]. Much of the impetus for the study of reactions in micelles is that they model, to a limited extent, reactions in biological assemblies, and the term "biomimetic chemistry" has been coined to describe this general area of study. Regarding the biological relevance of this area, it is important to mention the work of Luisi and colleagues on autopoietic self-reproduction [13–16] of micelles, seen as the simplest bounded structures (i.e., provided with a physically well-defined boundary). The term autopoietic indicates that the self-reproduction process is due to a reaction that takes place within a spherically closed boundary, a situation that, according to the proponents of the theory of autopoiesis, defines a prerequisite structural condition for minimal life. The biological relevance of this kind of work is greater in that vesicles instead of micelles were also used. In fact, vesicles (or liposomes) have the bilayer structure of biological membranes and they have been proposed as possible precellular systems [16]. For instance, vesicles prepared from oleic acid can catalyze the hydrolysis of surfactant precursor oleic anhydride: the first vesicle can solubilize the water-insoluble precursor, which is then efficiently hydrolyzed by the vesicles themselves, a process that is a typical autocatalysis [17,18]. It is also worth mentioning that, according to some authors, fatty acids might well have been the material for primordial cells walls [16].

A number of related self-assembling colloids can influence chemical rates and equilibria. Microemulsions generally contain water, an oil, a surfactant, and a co-

Dedicated to Clifford A. Bunton on the occasion of his 80th birthday.

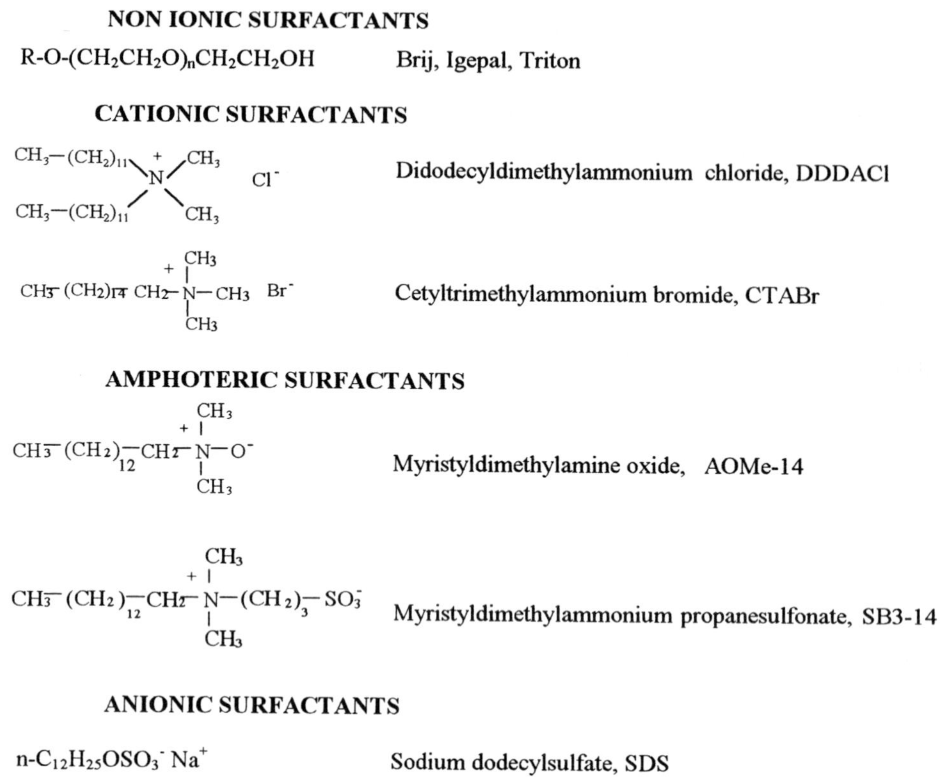

FIG. 1 Structure of representative surfactants.

surfactant, which is generally a medium-chain-length alcohol, amine, or similar polar organic molecule [3,9]. Oil-in-water (o/w) microemulsions are formed when water is the bulk solvent. Water-in-oil (w/o) microemulsions are formed when oil is the bulk solvent. They are akin to the reverse micelles that form when surfactants, usually with small amounts of water, are dissolved in an apolar organic solvent [8]. There have been reviews of the properties of microemulsions and effects upon reactions [19–22] and an interesting book about the structure and properties of reverse micelles [23]. Also, reverse vesicles, produced in a nonaqueous phase, with hydrophilic parts of the amphiphilic molecule oriented inside bimolecular layers have been investigated [24–26].

This chapter is restricted to discussion of thermal reactions in aqueous micellar systems. Our discussion is focused on organic reactions in aqueous micelles. The structure of micelles or other colloidal droplets is considered only to the extent needed to understand reactivity. Key features of micelles and similar colloidal aggregates are their interaction with water and their association with solutes. For example, an ionic micelle can bind a nonionic solute and also, by virtue of its charge, attract counterions. Therefore, it may affect reaction rates and equilibria by bringing reactants together or keeping them apart, but because the micelle can exert a medium effect it is necessary to separate the "medium" and "concentration" effects of the micelles.

Our choice of topics is dictated partly by our own interests but also by space requirements. The overall subject of reaction in submicroscopic aggregates has expanded so rapidly and in so many different directions that one has to focus one's interest on specific areas. The general principles that govern the effects of normal, aqueous micelles upon reaction rates and equilibria are considered first, and then we discuss some specific reactions and the relation of micellar effects to mechanism. We also briefly consider some nonmicellar species generated by amphiphiles that can also mediate reactivity, such as submicellar aggregates and vesicles. Special attention will be given to efforts to understand how the micellar influence on *chemical reactivity* of bound species is related to the structure of the monomers and of the microaggregates they form, with the aim of achieving a deeper understanding of the properties of noncovalent interactions. That promises to offer not only substantial understanding of the way nature "works" but also guidelines for the design and preparation of completely new chemical entities, of defined structure and function, useful for applications. A final section will briefly focus on applications of surfactant assemblies, particularly as related to environmentally benign preparative organic chemistry.

II. AQUEOUS MICELLAR AGGREGATES

Micelles (latin *mica(a)*, grain; + *ella*, diminutive suffix) are loose aggregates of amphiphiles in water or organic solvents (such as 1,2-diols, formamide) that form above a certain temperature (Krafft point) and concentration (critical micelle concentration, cmc) [3,8,9].

Below the Krafft temperature, clear micellar solutions become turbid and the amphiphile forms three-dimensional hydrated crystals. Below the cmc, micelles dissociate into monomers and small aggregates. Micellar solutions are stable and remain clear over years, although the individual micelle usually explodes and reforms within a few milliseconds [8]. The cmc decreases with increasing chain length of the apolar groups and is higher for ionic than for nonionic or zwitterionic micelles. For ionic micelles it is reduced by addition of electrolytes, especially those having low charge density counterions [27]. Onset of micellization is detected by sharp changes in such properties as surface tension, refractivity, or conductivity (of ionic micelles). To a first approximation, the solution is assumed to contain monomeric amphiphiles, whose concentration is given by the cmc, and fully formed micelles, with submicellar aggregates playing a minor role.

Micellization is a manifestation of the strong self-association of water and similar solvents and is an example of the hydrophobic or solvophobic effect that forces self-association of apolar materials [28,29]. The first reason for the formation of micellar aggregates is that the interactions between solvent molecules (usually water) are stronger than the interactions between the solvent and the solute [30]. This effect alone would lead to precipitation of the solute. In the case of amphiphiles that form micelles, however, the headgroups are strongly hydrated and repel each other. The hydration forces and steric forces [1] that are made responsible for this repulsion effect prevent crystallization above the Krafft point and also above the cmc. Where the formation of 3D crystals is impeded, the smallest possible droplet is formed, removing the alkyl chains from the solvent. The interactions between solvent molecules are therefore disturbed to a minimal extent, allowing the headgroup to be solvated with a minimal entropy loss. It is irrelevant whether the solvent contains clusters or not. Micelle formation occurs as a result of solvation of headgroups and nonsolvation of a solvophobic core [30]. Formation of micelles in water seems to require a minimum length of the hydrophobic alkyl group, which is 8 to 10 methylene (and methyl) groups; aggregation numbers are generally greater than 50 and increase with increasing hydrophobicity of the alkyl group [12]. Any factor that increases the balance of hydrophobicity over hydrophilicity stabilizes the micelle, as evidenced by decrease of the cmc and increase of the aggregation number. Normal aqueous micelles are generally formed from single-chain surfactants and chain branching inhibits micellization [12].

A. Structure of Micelles

Micelles have been investigated by an unusually wide variety of techniques including x-rays, nuclear magnetic resonance (NMR), electron spin resonance (ESR), fluorescence, static and dynamic light scattering, calorimetry, and kinetic probes [31]. Despite all this attention, micellar structure is still a controversial topic, although there seems to be a consensus on some points.

Ever since the discovery of micelles, theoretical models of micelle formation have in some way tried to account for the association-dissociation equilibrium that distinguishes micelles from other colloids. The earliest model is due to Hartley and regards the formation of a micelle as a chemical equilibrium between monomers, counterions, and micelles. This model is the so-called *Hartley micelle* [32] and involves a hydrocarbon-like interior surrounded by polar or ionic headgroups. The micelle is pictured as a roughly spherical aggregate containing 50–200 monomers with a radius approximately corresponding to the extended length of the hydrocarbon chain of the surfactant. Micellar headgroups and associated counterions are fully hydrated and are found in the *Stern layer*. Some of the counterions are bound within the shear surface, and many are located in the *Gouy-Chapman electrical double layer*, where they are dissociated from the micelle.

This model has quite appropriately been dubbed the *mass action (MA) approach* [33,34]. The MA approach was followed by the *phase separation (PS) model*, advanced by Stainsby and Alexander in 1950 [35], wherein the micelles are treated as a phase separate from that containing the monomeric species. The description of micelle formation has been presented by Evans et al. [36–38], who took the question of why counterions should bind to micelles as their point of departure. This has resulted in the *dressed micelles (DM) theory*. An extension of the DM theory that gives better agreement with experimental results at low salt concentrations was proposed by Hayter in 1992 [39]. It is, of course, of interest to compare theoretical values of the enthalpy of micelle formation with those obtained by a calorimetric experiment. A guide to the more current literature is given by van Os et al. [40–42], who have included a model for anionic, cationic, and nonionic surfactants in their reviews. A short review of models of micelle formation is given by van Os et al. [42], who tested the prediction of four different models with experimental enthalpies and concluded that a mass action model approach agrees best with experimental observations.

In this chapter, the structure of micelles or other colloidal droplets is considered only to the extent needed to understand reactivity. In Fig. 2 we show a simplified cartoon for the micellar structure.

A good model is that of Gruen, who has described a realistic model for a micelle [43,44] that involves a rather sharp interface between a dry hydrophobic hydrocarbon core and a region filled with surfactant headgroups, some of the counterions, and water, namely the Stern region. This model has been validated using molecular dynamics simulations [45,46] and is valid for both ionic and nonionic micelles.

Regarding computer simulations, there is significant interest in developing theoretical and simulation [47–49] models of the micellization process. Computer simulations are increasingly being used to study the structure and thermodynamics of micelles. Such simulations

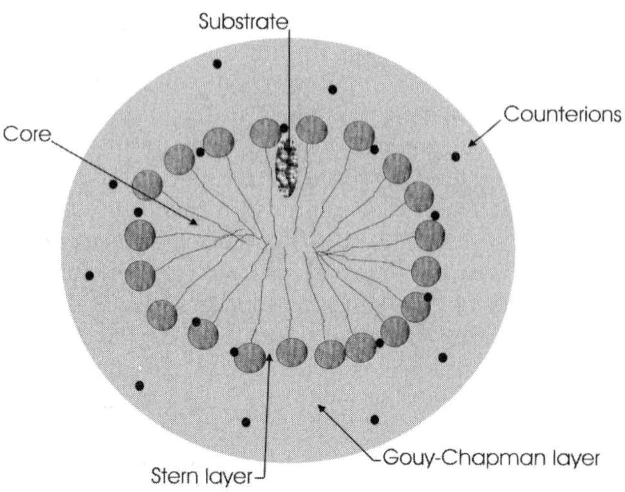

FIG. 2 Cartoon of a micelle.

can be classified into two broad categories. In the first, micelles of a certain size and shape are assumed to exist and the simulation probes their structure and short-time dynamics. Detailed atomistic models of the surfactant molecules can be studied by this method [47]. The other simulation approach is to study the spontaneous process of micelle formation. Floriano and Caponetti have developed another methodology based on grand canonical Monte Carlo simulation [50].

There are unanswered questions regarding micellar size and shape. The size of an approximately spherical micelle is geometrically constrained, but from a variety of measurements it appears that micelles grow with increasing concentration of amphiphile or added electrolyte and then become ellipsoidal [12]. Despite the obvious attraction of assuming a single micelle size, the presence of a distribution of aggregate sizes is well accepted. The polydispersity and growth of micelles in surfactant solution have been investigated using a wide variety of techniques over a range of different length scales, including viscosity measurements [51,52], NMR [53], and small-angle neutron scattering [54]. The equilibrium distribution of micelles may shift to larger or smaller average sizes, depending on the solution concentration, temperature, pressure, and other factors such as ionic strength. Experimental evidence for growth in surfactant solutions has shown increasing micellar aggregation numbers as a function of the total surfactant concentration [55], salt [51], and added organic solutes or cosurfactants [56]. In addition, the effect of aggregate interactions on the growth of micelles has been addressed in several studies, particularly for surfactants that form elongated or cylindrical micelles [54,57–60].

Micellar aggregates are assumed to be monodisperse for calculation purposes, but a model has been proposed that incorporates a distribution of micellar sizes self-consistently. The model is based on a mass action equilibrium approach that includes micelle-micelle interactions as a function of size for a multicomponent surfactant solution consisting of micellar aggregates, monomers, counterions and added electrolytes. The multicomponent model reflects experimental osmotic pressure data successfully [61]. However, fortunately for the kineticist, it seems that micellar size per se is not a dominant factor in determining chemical reactivity probably because it does not markedly affect the nature of the micelle-water interface.

B. Interfacial Water

Water penetration into micelles and water properties therein are other critically important points: knowledge of the properties of water at interfaces of organized assemblies at the microscopic level is a prerequisite for understanding the effect of these species on chemical reactivity and equilibria. Interfacial water is a reactant in spontaneous hydrolyses, but it is involved in, inter alia, solvation of reactants and transition states, solvation of surfactant headgroups and counterions, and proton transfer.

At one time or another, water was believed to penetrate the micelle completely, not to penetrate at all, or to reach to any intermediate depth. Neither of the two extremes, picturesquely described as the "fjord" and the "reef" models, is likely to represent the real situation. For example, Menger in 1979 [62] used space-filling models to demonstrate that there will be large voids in the micelle if eclipsing is to be avoided in the alkyl groups, and he concluded that these voids will accommodate water molecules, which penetrate deeply into the micelle. Right after him, Fromherz [63] postulated a different packing of monomers that also leads to partial exposure of alkyl groups to water, and Dill and Flory [64] showed how a micelle could be constructed with eclipsing of some of the alkyl groups and contact between methylene groups and water. A modified Dill-Flory model has been reported to be consistent with results of small-angle neutron scattering that exclude water penetration into the micellar interior [65]. However, the structure of the micelle is so dynamic that every segment of the hydrophobic alkyl group is exposed to water at some time, and in a small micelle over half the chain segments will be at the micellar surface at any given time. This theoretical prediction agrees with proton and fluorine NMR spectroscopy [66].

Properties of aqueous micelles are typically described by pseudophase models with a distinction being made between ionic and nonionic surfactant monomers in the bulk solvent and in the micellar pseudophase. Distinction between interfacial water at the micellar surface and that in the bulk has been made. A considerable amount of work on the properties of interfacial water has depended on probes [67] or the study of reactions whose rates are sensitive to the micellar microenvironment (see Section IV). A new method based on chemical trapping based on a dediazoniation reaction has been developed by Romsted and coworkers and used for determination of hydration numbers [68–70].

An interesting new method for investigating the properties of interfacial water is based on measurements of the deuterium isotope effect on the ^{1}H chemical shift of the solvent (H_2O plus D_2O), developed by El Seoud et al. [71,72]. It is evident that ionic and zwitterionic micelles generally increase structuring of the interfacial region, with values of the fractional factor, Φ, being lower for zwitterionic than for ionic micelles of otherwise similar structure.

C. Substrate Solubilization

One of the most important processes leading to micellar effects on reactions is the solubilization of organic compounds at the micellar surface. It is possible to solubilize water-insoluble substances or to increase the solubility of slightly soluble ones in aqueous micellar solutions. Solubilization has been applied empirically for centuries, but the fundamental bases of micellar solubilization were established only in the 1950s by McBain and Hutchinson [73] and by Elworthy [74].

The most realistic model for analyzing solute incorporation is undoubtedly that based on the mass action approach. In its general form [75,76], the solubilization process can be treated in terms of the stepwise addition of solute molecules, S, to aggregates containing -i-solute molecules. Making the assumptions that (1) the micellar aggregation number is independent of the presence of the solute, (2) the solute entry rate is independent of the number of solute molecules present in the micelle, (3) the solute exit rate is directly proportional to the number of solutes present in the micelle, and (4) the solubilization capacity of the micelle is "infinite," an interesting form of expressing the equilibrium constant for the incorporation of the first solute into an empty micelle, K_S, has been defined in terms of the concentration of micellized surfactant [Dn]:

$$K_S = \frac{[\mathrm{SDn}]}{[\mathrm{S_W}][\mathrm{Dn}]} \qquad (1)$$

An analogous solubilization constant can be written in terms of molarity of micelles, K_M; the two constants differ in magnitude by the aggregation number of the micelles. This definition of K_S, which in essence reduces the solubilization process to a pseudophase equilibrium of the solute between a micellized surfactant and the aqueous phase (Scheme 1), has played a central role in pseudophase models for analyzing micellar effects on reaction kinetics [11,12,77].

A wide range of experimental techniques and methods are available to measure the partitioning of nonionic solutes between water and micelles, such as cal-

$$\mathrm{S_W} + \mathrm{Dn} \overset{K_S}{\rightleftharpoons} \mathrm{SDn}$$

SCHEME 1

orimetry; ultraviolet-visible (UV-VIS), NMR, and ESR spectroscopies; laser light scattering; and luminescence probing [78,79]. Reliable values of binding constants are generally obtained provided that the solute is not very hydrophobic, that the amphiphile is in excess over substrate, and that the amphiphile concentration is well above the cmc, because solutes may induce micellization or bind to premicelles. Fluorescence quenching provides information on properties of association colloids and partitioning of solutes between the colloids and water, and this general method has been reviewed [80]. The binding of nonionic solutes has been treated by applying a linear solvation free energy relationship (LSER) [79,81–83]. The structural features that govern binding have been identified, with hydrophobic effects and solute basicity being the dominant terms. Distributions of solutes that cannot be examined experimentally, for example, because of their high reactivity in water, can be predicted on the basis of this approach.

Unlike homogeneous solvents, the inherently microheterogeneous micellar solutions provide a variety of solubilization environments, ranging (in principle) from the "hydrocarbon-like" core to the bulk water. Indeed, the distinctive feature of a micelle as a solvent is that it can provide not only different microenvironments for different molecules but also different microenvironments for different parts of the same molecule. Thus, polar molecules encounter a polar environment, hydrophobic ones have available a hydrocarbon-like medium, and amphipathic molecules should be able to orient themselves at the micellar surface with their hydrophobic portion extending into the hydrocarbon-like core.

The pathways and rates of reactions in micelles are affected by how deeply the solubilized species is located within the micelle. Both electrostatic and hydrophobic factors play a role in determining the binding site of a solute inside the micelle, and both the structure of the amphiphile and the solute are of great importance in determining the extent of solubilization and the penetration of solute into the micelle [31]. NMR is extensively used to elucidate locations of solubilized species. A large number of NMR parameters are significantly affected on adding a solubilizate to a micellar solution and are thus capable of giving some information on molecular aspects of solubilization [84]. Aromatic compounds are useful because of their sizable so-called ring current shifts in NMR, and such molecules as benzene and its derivatives were the first solutes studied by Eriksson and Gillberg [85]. Subsequent studies have provided more details on the interfacial location of aromatic solubilizates. An increase in the size of the surfactant headgroup also increases the depth of penetration of methyl naphthalene-2-sulfonate [86]. Kang et al., in their study of *n*-alkylphenothiazines, show that similarly sulfonated *N*-alkylphenothiazines reside near the micellar interface, unsulfonated ones penetrate more deeply into the micelle, and the degree of penetration depends on the length of the alkyl chain [87,88].

D. Ion Binding

A fundamental process in micellar catalysis or inhibition is ion binding to micelles. The major problem in the analysis of micellar rate effects is modeling the interfacial concentration and distribution of ionic reactants, although in favorable cases they can be directly measured [12,89].

Packing of headgroups and counterions at aggregate interfaces produces a high local counterion concentration; typical estimates are >1 M, much greater than counterion concentration in the surrounding aqueous pseudophase, which is usually in the range 0.001–0.01 M [1,11,12].

There are two models describing the ionic distribution at charged aqueous interfaces. In the pseudophase ion exchange (PIE) model, micellar surfaces are treated as selective ion exchangers saturated with counterions; using the Poisson-Boltzmann equation (PBE) modified for specific ion interactions, ion distributions are computed within a reaction region at the micelle surface [89] (see also Section III).

Distribution of counterions is usually characterized by the micellar degree of ionization, α, and for most ionic micelles α is in the range 0.1–0.3 and is seen to depend little on the overall concentration of counterions; in other words, the surface of an ionic micelle is treated as if it is saturated with counterions [11]. The remaining counterions are distributed in the diffuse Gouy-Chapman layer, and their distribution is governed by their nonspecific electrostatic interactions with the micelle, which can be regarded as a bulky macroion [12]. Values of α are determined for many ions, including different values for the same ion, by a variety of methods [31]. A new method is based on chemical trapping of "free" counterions in the aqueous pseudophase, based on the dediazoniation reaction developed by Romsted et al. [68,90]. Extensive kinetic data have been fitted by the PIE model with data on the ion-exchange parameters for pairs of ions [89], including estimates for Cl^- and Br^- and estimates based on the method of ion flotation of Warr et al. [91–93].

Counterions are bound primarily by the strong electrical field created by the headgroup, but also by specific interactions that are dependent on headgroup and counterion type [9,89,94]. The nature of the counterion has an important bearing on α for cationic surfactants: ions that have a lower hydration enthalpy, higher affinity for Dowex 2 (a tetralkylammonium ion resin), and a higher lyotropic number are more strongly bound to cationic micelles and highly effective in charge neutralization of the micellar headgroups. Generally speaking, we can say that micelles bind counterions selectively, and several properties such as size, shape, phase stability, binding of ions and neutral molecules, and their effects on the rates and equilibria of chemical reactions are sensitive to counterion concentration and type. Some lyotropic series or affinity orders have been established for the relative affinities of both anions and cations to micelles [31,94].

When salts are added to micellar solutions, the counterions of the salts compete for the ionic headgroup of micelles with the surfactant counterions that already exist in solution. Thus, displacement can occur, depending on the relative affinities of counterions for the headgroups. Interesting kinetic effects can arise. If the added salt is reactive, micellar rate enhancements are observed after displacement. If added ion is inert and the reactive ion is the surfactant counterion, the addition can cause inhibition. Kinetic salt effects are peculiar in micellar systems [31].

Ion selectivity has been treated thermodynamically, and it was concluded that partial ionic dehydration at the micellar interface is of crucial importance [95]. Evidence indicates that strongly hydrated ions, such as OH^- and F^- are readily displaced from cationic micelles by relatively weakly hydrated anions such as Br^- and Cl^- [89]. Results of a chemical trapping method are consistent with partial dehydration of counterions of CTABr and CTACl micelles at the sphere-to-rod transition [96].

Morini et al. [97] used ion-selective electrodes to demonstrate that in mixtures of DTABr and DTAOH, the breakpoints used to estimate the cmc differ for Br^-, OH^-, and the amphiphile cation; that the degree of ionization decreases continuously up to about 0.2 M amphiphile; and that the exchange constant for OH^- and Br^- depends on solution composition, consistent with theoretical models [98].

As for organic neutral substrates and also for organic anions, the structure of both the surfactant and the ion can affect location at the interface. Binding of salicylate and *o*-, *m*-, and *p*-nitrobenzoate anions to MTABr micelles has been examined by ^{1}H NMR spectroscopy [99]. Competition is treated by the PIE model, although these polarizable anions may perturb the micellar structure. With organic counterions there is often micellar growth, which is sensitive to counterion type. Two-dimensional NMR spectroscopy, capable of revealing spatial relationships among proximal protons (NOESY, ROESY), can be very useful, giving insight into structural details of micelle-solute interactions. Results with both 1D and 2D NMR show that the 3,5-dichlorobenzoate ion intercalates further into cationic rodlike micelles than does the 2,6-iosmeric ion into spherical micelles [100]. Other ^{1}H NMR data show that micellar growth in alkylpyridinium micelles to form entangled wormlike micelles depends on counterion structure; it was observed for *o*-hydroxybenzoate but not for the *p*-hydroxy isomer and was observed for *p*-chlorobenzoate but not for benzenesulfonates [101]. Growth induced by added organic anions is probably not due to charge reversal of the micelle surface, because the surface potential, estimated by using an indicator, approaches zero with added salicylate ion but does not change sign [102], consistent with Poisson-Boltzmann equation (PBE) treatments [77,89]. For the aromatic anion 2-naphthoate in cationic micelles, ^{1}H chemical shifts and NOESY spectra led to the conclusions that the aromatic section of the molecule is embedded in the palisade layer whereas the charged parts are located near the micellar interface, so they can still be solvated by water [103].

In spite of the neutral nature of zwitterionic surfactants and their well-documented insensitivity to ionic strength, electrolytes do bind to zwitterionic micelles. Binding of various ions to micelles was suggested in several papers [104–108], and it was unambiguously established from radioactive tracer, self-diffusion, and fluorescence quenching data [109,110]. Micellar aggregates of zwitterionic carboxybetaine, sulfobetaine, or phosphobetaine surfactant can be represented as a hydrophobic sphere surrounded by a spherical shell where positive charges are distributed on the inner layer and negative charges on the outer layer. In this purely electrostatic model, a positive electrostatic potential exists inside the shell as a function of the distance from the center of the sphere. The number of positive and negative charges is the same, but the surface where the positive charges lie is less extended. Positive charge density is consequently greater [110,111]. This model explains the strongest anion binding to zwitterionic micelles but does not explain the strongest binding of the softer anions [112]. Another failure of the model is the dependence of the electrostatic potential on the thickness of the double-charged shell, corresponding to the interchargе distance of the micellized monomers. The polymethylene arm is indeed flexible, and intercharge distance increases monotonically but not linearly with increasing number of methylenes [113–115]. Intercharge arms fold and loop inside the hydrophobic micellar core because of both electrostatic attraction between the two oppositely charged groups and hydrophobic interactions of the tether with the micellar interior. The surface areas of the monomers become larger as the intercharge arms become longer and as micelles become smaller [115–117]. As shown by self-diffusion and elastic and quasi-elastic light scattering studies, micellar dimensions increase on anion incorporation due to reduced electrostatic interactions between the two opposite charges and unfolding of the headgroup [117].

III. QUANTITATIVE TREATMENTS OF MICELLAR RATE EFFECTS

A. Pseudophase Model

Most kinetic treatments are based on the so-called pseudophase model. This model has been generally accepted on the reasonable assumption that for most activated thermal chemical reactions, transfer of material between water and micelle is so fast that reaction does not perturb the equilibrium distribution of reactants between the pseudophases. This generalization cannot be applied to photochemical reactions, where some steps of the reaction may be very rapid and therefore faster than solute transfer [9,118]. Provided that equilibrium is maintained between the aqueous and the micellar pseudophases, the overall reaction rate will be the sum of rates in water and in the micelles and will therefore depend on the distribution of reactants between each pseudophase and the appropriate rate constants in the two pseudophases.

Menger and Portnoy [119] developed a quantitative treatment that adequately described inhibition of ester saponification by anionic micelles. Micelles bound hydrophobic esters, and anionic micelles excluded hydroxide ions and so inhibited the reaction, whereas cationic micelles speeded saponification by attracting hydroxide ions [120]. Provided that only substrate distribution has to be considered, which is the situation for micelle-inhibited bimolecular or spontaneous unimolecular reactions, Scheme 2 shows the substrate distribution and reaction in each pseudophase [121].

In Scheme 2 Dn denotes micellized surfactant, S is substrate, subscripts W and M denote aqueous and micellar pseudophases respectively, and k'_W and k'_M are first-

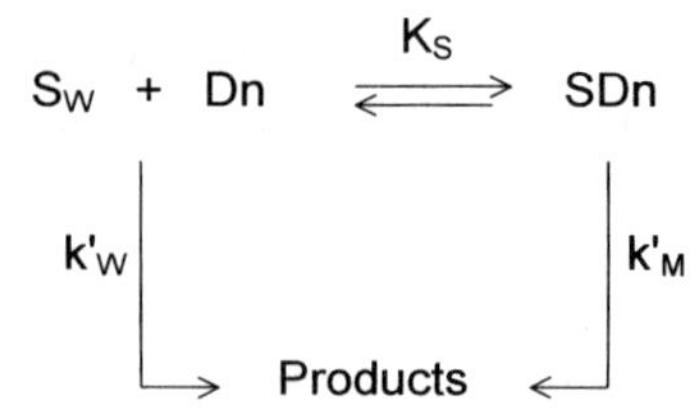

SCHEME 2

order rate constants. The binding constant, K_S, is written in terms of the molarity of micellized surfactant, but it could equally be written in terms of the molarity of micelles. The concentration of micellized surfactant is that of total surfactant less that of monomer, which is assumed to be given by the critical micelle concentration (cmc). The overall first-order rate constant k_{obs} is then given by Eq. (2).

$$k_{obs} = \frac{k'_W + k'_M K_S([D] - \text{cmc})}{1 + K_S([D] - \text{cmc})} \tag{2}$$

On the basis of Eq. (2), in the absence of micellized surfactant $k_{obs} = k'_W([Dn] = 0)$, and when the substrate is fully micelle bound ($K_S[Dn] \gg 1$ and $k'_M K_S[Dn] \gg k'_W$) we have $k_{obs} = k'_M$.

This equation is formally similar to the Michaelis-Menten equation of enzyme kinetics, although the analogy is limited because most enzymatic reactions are studied with substrate in large excess over enzyme. Equation (2) could be rearranged to give Eq. (3), which is formally similar to the Lineweaver-Burk equation and which permits calculation of k'_M and K_S provided that k'_W is known [119,120].

$$\frac{1}{k_{obs} - k'_W} = \frac{1}{k'_M - k'_W} + \frac{1}{(k'_M - k'_W)K_S[Dn]} \tag{3}$$

Variations of Eq. (2) have been developed to fit rate-[surfactant] profiles for spontaneous reactions that occur after a rapid preequilibrium. For instance, decomposition of aryl-2,2,2-trichloroethanols follows an (Elcb)R type of mechanism, i.e., rapid reversible deprotonation of the alcohol by OH^- followed by rate-determining loss of trichloro carbanion to give benzaldehyde product [122]. Micelle-bound substrates are completely deprotonated (because of the increased $[OH^-]$ at the micellar surface relative to bulk water) so that k_{obs} depends primarily on the distribution of the alkoxide ions between aqueous and micellar pseudophases and k_{obs}-[surfactant] profiles are fitted by Eq. (4), which is the form typically used for spontaneous reactions in micellar solutions (after correction for the extent of deprotonation of the trichloroethanols in the aqueous phase).

$$k_{obs} = \frac{K[OH_W^-](k'_M K_S[Dn] + k'_W)}{1 + K[OH_W^-](1 + K_S[Dn])} \quad (4)$$

In Eq. (4) K_S is the binding constant of an alkoxide ion to the micelle and K corresponds to the preequilibrium constant in water:

$$K = [RO_W^-]/[ROH_W][OH^-] \quad (5)$$

Equations (2) and (3) depend on some major assumptions, in particular that the cmc gives the concentration of monomeric surfactant, that rate and binding constants in the micellar pseudophase are unaffected by reactants and products, and that reagents do not react across the micelle-water interface. These equations have been used very extensively and have provided the basis for quantitative analysis of micellar rate effects. Much of this work has been reviewed in comprehensive monographs [8,12].

Equations (2) and (3) generally fail for bimolecular, micelle-assisted reactions. Equation (2) predicts that the first-order rate constants should reach a constant, limiting value at high surfactant concentration when the substrate is fully micelle bound, but rate maxima are observed for the corresponding nonsolvolytic bimolecular reactions when surfactant counterions are inert. The rate-surfactant concentration profiles can be treated quantitatively by taking into account the distribution of *both* reactants between water and micelles. This can be done by extending Eq. (2), and a simple formalism involves writing the first-order rate constants in terms of second-order rate constants in water and micelles and reactant concentrations in each pseudophase [10,123–126]. However, one immediately runs into the problem of defining concentration in the micellar pseudophase. One approach is to write concentration in terms of moles of reagent per liter of micelles or to assume some volume of the micellar pseudophase V_M in which reaction takes place. Another approach is to define concentration in the micellar pseudophase in terms of a mole ratio. Concentration is then defined unambiguously, and the equations take a simple form [11,126]. However, this approach does not allow direct comparison of second-order rate constants in aqueous and micellar pseudophases, and by evading one problem one faces another.

The first-order rate constants are written in Eqs. (6) and (7) as second-order rate constants, k_W and k_M, for reaction of a reactive anion, Y^-, where the mole ratio is $m_Y^S = [Y_M^-]/([D] - \text{cmc})$. Here and elsewhere, quantities in square brackets denote molarity in terms of total solution volume, which is approximately that of the aqueous pseudophase.

$$k'_W = k_W[Y_W^-] \quad (6)$$

$$k'_M = k_M m_Y^S = k_M \frac{[Y_M^-]}{[Dn]} \quad (7)$$

By combining Eqs. (2), (6), and (7), Eq. (8) is obtained for the observed pseudo-first-order kinetic constant,

$$k_{obs} = \frac{k_W[Y_W^-] + k_M K_S[Y_M^-]}{1 + K_S[Dn]} \quad (8)$$

where the dimensions of k_W are $M^{-1}\ s^{-1}$ and those of k_M are s^{-1}. This equation readily explains why first-order rate constants of micelle-assisted bimolecular reactions typically go through maxima with increasing surfactant concentration if the overall reactant concentration is kept constant and if surfactant counterion is inert. Addition of surfactant leads to binding of both reactants to micelles, and this increased concentration increases the reaction rate. Eventually, however, increase in surfactant concentration dilutes the reactants in the micellar pseudophase and the rate falls. This behavior supports the original assumption that substrate in one micelle does not react with reactant in another and that equilibrium is maintained between aqueous and micellar pseudophases.

Equation (8) and others that are essentially identical but are written in different ways can be applied to bimolecular micelle-assisted reactions provided that the distribution of both reactants can be determined.

In some cases, the problem is relatively simple. Ion binding can be estimated by conductivity. For example, Eq. (8) was applied to fit rate-[surfactant] profiles with the concentration of ions determined by conductivity for mixed systems, such as mixtures of CTABr and nonionic surfactant $C_{10}E_4$ [127]. Estimation of the extent of micellar binding is not a problem if the organic ion is very hydrophobic, because then it is completely micelle bound under essentially all conditions [123].

The problem is more difficult for bimolecular reactions of hydrophilic ions. The conductivity method is not useful for estimating β for CTAOH, because it seems that β increases with increasing surfactant concentration [128] and then β has to be calculated. A variety of theoretical treatments have been developed to estimate the ionic distribution between water and micelles [12,77] and some of them will be illustrated in this section.

The problem of quantifying the transfer equilibrium of a nucleophile between water and micelles is simplified by using functional amphiphiles in which the headgroup is the nucleophile. The mole fraction of nucleophile to micellized surfactant is therefore 1. In comicelles, allowance is made for dilution of functional headgroups by inert surfactants. This treatment has been applied to several reactions involving hydroxamate or oximate-functionalized surfactants [129,130] and reactions catalyzed by metallomicelles [131]. In this chapter we will not deal with functional micelles and comicelles. Anyway, it is worth mentioning that due to an additional catalyzing action, rate enhancements by functional micelles are considerably larger than those by nonfunctional micelles of comparable structure, and bifunctional micelles are better catalysts than monofunctional micelles [132–168]. The most widely reported functional groups are hydroxyl [166–168], imidazole [137,140,169,170], thiol [135,136, 163], and some —OH-bearing groups such as hydroxyalkyl [139,141,150–156,159–161,164,166,168,171], hydroximino alkyl [138,162,168], hydroxylamine [165], and hydroxamic acid [138].

Outstanding examples of functionalized micelles are the so-called metallomicelles, and they bind substrates with enzymelike efficiency. They are made of either transition metal complexes of surfactants containing imidazole or pyridine moieties or comicelles of ligand complexes with surfactants, and they are utilized in the hydrolysis of carboxylic and phosphoric esters and amides [145,172–180]. Aggregates of amphiphilic metal complexes (metallomicelles) [146,178–184] or metallovesicles [185,186] are attracting considerable attention as hydrolytic catalysts, potentially more powerful than their hydrophilic, monomeric, siblings. The impressive rate increases are a matter of fact, but their source is a much more delicate subject. Breslow and coworkers in 1986 [187] and later Kimura et al. [188] reported impressive increases in overall second-order rate constants in comicellar aggregates. Scrimin and coworkers [131], however, showed that reactivity of various metallomicelles toward phosphate triesters is accounted for by considering the concentrations of reactants in the reaction loci, i.e., in the micellar pseudophase. There was nothing unexpected in reactivities of these metallomicelles relative to those of the monomeric analogue complexes in water. The conclusion is that in hydrolyses catalyzed nucleophilically, reactivities of micellized metal complexes do not differ significantly from those of analogous, monomeric catalysts in water. This is in spite of the enhanced electrophilicity of the metal ion in micellized complexes indicated for Cu^{2+} from electrochemical evidence [146,189] and does not significantly depend on the particular structure and intrinsic reactivity of the substrate [190]. Therefore, the special micellar effects that are postulated to provide impressive accelerations of reactions are a chimera that disappears when transfer equilibria between the aqueous and the micellar pseudophases are taken into account [191].

1. Ion-Exchange Model

A first advance in the treatment of bimolecular reactions was made by Romsted [10,11]. The pseudophase ion-exchange (PIE) model of bimolecular ionic reactions was developed by Romsted to explain the rate maxima with increasing surfactant concentration, and inhibition by inert counterion competition between ions, e.g., Y^- and X^-, is written in terms of Eq. (9), which is similar to that used with ion-exchange resins.

$$K_X^Y = \frac{[Y_W^-][X_M^-]}{[Y_M^-][X_W^-]} \tag{9}$$

The micelle is assumed to be saturated with counterions, regardless of their nature or concentration; i.e., β is constant. The value of $[Y_M^-]$ can the be calculated from a mass balance equation in terms of K_X^Y and β. First-order rate constants are written as Eqs. (6) and (7); this last equation can be written in terms of a rate constant k_2^m with concentration as molarity at the micellar surface, as in Eq. (10), where V_M is the molar volume of the reactive region in the micellar pseudophase. This volume is probably approximately half that of the whole micelle [11], although it may depend on the nature of the reactants. Most estimates of V_M are in the range of 0.14–0.37 M^{-1} [11,12].

$$k_2^m = k_M V_M \tag{10}$$

Equations (6)–(9) can be combined, and the dependence of overall rate constants on concentrations of surfactant, ionic reagent, and added inert salt can be predicted in terms of β, k_M, k_W, and the ion exchange parameter [Eq. (9)], provided that the distribution of substrate between water and micelles is known [Eq. (1)].

This treatment fits a great deal of data, and it is still extensively used. It also fit kinetic micellar effects on inorganic reactions, such as the redox reaction between bromate and bromide ions in aqueous acidic medium in both cationic and anionic surfactants [192]. Values of the ion-exchange parameters [Eq. (9)] follow the prediction that very hydrophilic ions will be displaced by less hydrophilic ions; i.e, affinity of ions for micelles follows the Hofmeister series. The PIE model merely considers the nature of the surface of the micelle and not its size or overall structure. As noted later, models that involve solution of the Poisson-Boltzmann equation are sensitive to changes in micellar size and shape.

Development of the PIE model was a crucial breakthrough in the study of micellar rate effects, but it has been found to be a first approximation that breaks down under some conditions, especially when concentrations of the reactive ions are greater than about 0.1 M [193,194] and for competition involving very hydrophilic anions [195]. Furthermore, kinetic data are fitted with values of K_{Br}^{OH} generally in the range of 12–20, so that OH^- in large excess should displace Br^- from cationic micelles. However, fluorescence spectroscopy [196] shows that very hydrophilic anions, e.g., OH^- and F^-, are singularly ineffective in displacing Br^-, although competition between Br^- and other moderately hydrophilic anions follows the PIE model, based on both fluorescence and kinetic data. This is confirmed by evidence from the variation of the NMR line width of halide ions: the PIE does not fit NMR data for competition involving OH^- and F^- in CTABr or CTACl [98].

The PIE model is based on a postulated constant β, even though Eq. (9) predicts that ions have different affinities for micelles, and there is evidence that β is low for hydrophilic counterions. However, addition of a weakly bound dilute anion, e.g., OH^-, to CTABr probably does not markedly reduce β, so the assumption of a constant value is satisfactory in these cases.

Another problem with the PIE model in its original form is that the concept of ionic distribution between micelles following a step function and the concept of a constant β fail to explain the marked increases of rates of reactions of hydrophobic substrates such as elimination from DDT and related chlorides, in moderately concentrated OH^- [193,194]. Ionescu et al. postulated incursion of a reaction path across the Stern layer for the attack of OH^- in the aqueous pseudophase upon micelle-bound substrate [193,194]. Nome et al. [197] could fit dehydrochlorination of 1,1-diphenyl-2,2,2-trichloroethane (DTE) using Eq. (11):

$$[OH^-]_M = [OH^-]_M/[Dn]V + [OH^-]_W \qquad (11)$$

where V is the reactive volume in liters/mole of reactive region, i.e., using a modification of the PIE model, which includes the concept of variable α and the contribution of the counterion in the aqueous phase to the interfacial counterion concentration. The PIE treatment was also tested for reactions of 2,4-dinitro-1-chloronaphthalene (DNCN) and *p*-nitrophenyl diphenylphosphate (pNPDPP) in mixtures of OH^- and Cl^- or Br^-, with up to 0.5 M OH^- [198]. The fit with theory is satisfactory for dilute OH^-, but rates increase faster than predicted with increasing $[OH^-]$. Fits for reaction in 0.5 M OH^- would require a large increase in k_M relative to the value in dilute OH^- (Table 1). Alterna-

TABLE 1 Reactions with Concentrated Hydroxide Ion[a]

Substrate	$[OH^-]$, M	Surfactant	k_2^m, M^{-1} s^{-1} PIE	PBE	Mass action
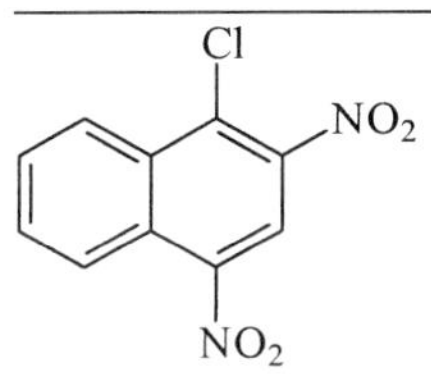	0.03	CTACl	0.014[a]	0.0115[b]	0.0168[a]
	0.5	CTACl	0.022[a]	0.0105[b]	0.0182[a]
	0.05	CTABr	0.013[b]	0.012[b]	0.0196[a]
	0.5	CTABr	0.017[a]	0.012[b]	0.0196[a]
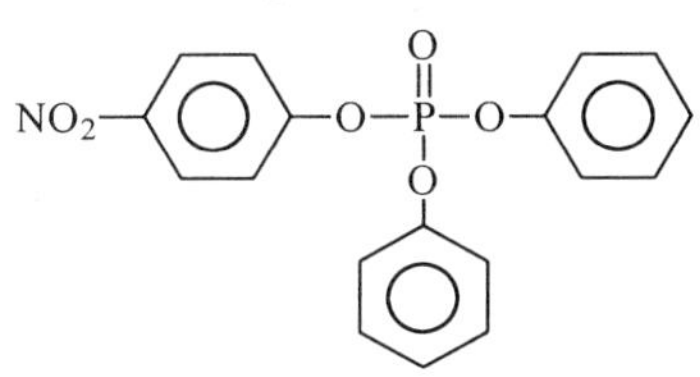	0.03	CTACl	0.076[a]	0.07[b]	0.0980[a]
	0.5	CTACl	0.161[a]	0.08[b]	0.133[a]
	0.05	CTABr	0.071[a]	0.075[b]	0.105[a]
	0.5	CTABr	0.143[a]	0.085[b]	0.140[a]
$(C_6H_5)_2CHCCl_3$	1	CTAOH	$6.50–6.80 \times 10^{-4}$[c]		

[a]For fitting parameters see Ref. 198.
[b]For fitting parameters see Ref. 217.
[c]For fitting parameters see Ref. 197; the range is for variations of k_2^m with various added salts.

tively, the two other models that follow have been used to investigate this problem (Table 1).

2. Mass Action Model

Values of β are not constant if counterions are very hydrophilic e.g., with CTAOH, CTAF, or CTAformate. Kinetic data for reactions of these anions in the absence of inert anions cannot be fitted with constant β because overall rate constants increase with increasing concentration of the reactive anion even though the substrate is fully micelle bound [199,200]. In these systems the micellar surface does not appear to be saturated with counterions. The kinetic data can be treated on the assumption that the distribution between water and micelles of reactive anion, e.g., Y^-, follows a Langmuir equation [200]:

$$K'_Y \frac{[Y_M^-]}{[Y_W^-]([Dn] - [Y_M^-])} \tag{12}$$

The Langmuir parameter K'_Y is low for hydrophilic ions and increases with decreasing hydrophilicity of the ion.

Although kinetic data can be fitted to the mass law model, its physical significance is uncertain. For example, ion binding is assumed to follow a site model, with the micelle having a number of binding sites whose occupancy depends on an affinity parameter, K'_Y, and on the concentration of ions in the aqueous pseudophase. In other words, α and β are not constant. But they will be approximately constant if K'_Y is large, as it should be for ions such as bromide. Alternatively, we could suppose that micelles of a hydroxide ion surfactant, for example, have a large size distribution and that micelles, regardless of their size, can bind nonionic substrates but that the small micelles are relatively ineffective at binding hydrophilic ions. On this basis, the ionic distribution represented by Eq. (12) may be due to an increase in the average size of the micelle upon addition of the counterion.

Another possibility is that the micellar reaction is not restricted to reactants in the Stern layer. There is not a great deal of physical evidence related to these questions. For example, CTAOH forms micelles that have small aggregation numbers and larger fractional charge, α, than micelles of CTACl or CTABr, but the evidence available at present is not sufficient to show that the properties of CTAOH micelles are constant over a range of conditions [200–202]. Despite these uncertainties, values of k_M for reactions of hydroxide ion in CTAOH and in mixtures of CTABr or CTACl with NaOH calculated using the ion-exchange or the mass action model agree reasonably well [12]. On extending the mass action model to systems containing both reactive and inert counterions, Eq. (12) has to be modified to include competition between Y^- (reactive) and X^- (inert) for the micellar surface, and two independent equations have to be considered, one for each ion, Eqs. (13) and (14).

$$K'_X = \frac{[X_M^-]}{[X_W^-]([Dn] - [X_M^-] - [Y_M^-])} \tag{13}$$

$$K'_Y = \frac{[Y_M^-]}{[Y_W^-]([Dn] - [Y_M^-] - [X_M^-])} \tag{14}$$

The ion-exchange constants, K_x^Y, in Eq. (9) should be related to the individual K'_X and K'_Y mass action constants, and this seems to be correct. A detailed data analysis shows that the value of the ion-exchange constant is consistent with values of the individual constants [203]. The advantage of using K'_X and K'_Y and not K_X^Y is that assumptions regarding values of β are eliminated. This model is also better than the ion-exchange model for reaction in moderately concentrated OH^-, where the ion exchange fails for $[OH^-] >$ 0.1 M but that based on K'_{Br} and K'_{OH} is satisfactory up to 0.5 M OH^- [198] (Table 1).

The concept of counterion binding to micelles in a well-defined Stern layer provides a very convenient means of describing micellar effects on reaction rates and equilibria, but there are serious questions about its physical significance. These equations fit a great amount of kinetic data in aqueous micelles, also with surfactants of different structures, and several of them will be discussed in Section V.

It has also been successfully used to fit data for reaction of Br^- with methyl naphthalene-2-sulfonate in micelles of CTABr and CTEABr modified by 1-butanol [204,205]. Rate constants at micellar surfaces are proportional to the mole ratio of bound Br^- and bound butanol, and their distribution (Br^- and butanol) between water and the micelles is described by the equation of the form of the Langmuir isotherm.

3. Poisson-Boltzmann Equation

The distribution of counterions around a colloidal macroion can be calculated in terms of Coulombic interactions between the macroion and the counterions, which are treated as point charges [206–209]. The cell model provides a straightforward method of estimating surface electrical potentials of spherical association colloids in terms of their radius and charge density and the ionic content and dielectric constant of the bulk solvent, which is typically water [209,210]. This treatment neglects specific interactions between a macroion, e.g., a micelle, and counterions and therefore does not explain the apparent differences in affinities of various counterions for micelles [10,11]. However, it gives values for α for SDS that are in reasonable agreement with experimental data, although specific binding of counterions in the micellar Stern layer is not taken into account [210]. The Poisson-Boltzmann equation (PBE) was used to fit, for instance, reactions of hydroperoxide ion (formed in situ from OH^- and hydrogen peroxide) with *p*-nitrophenyl diphenyl phosphate in CTACl and CTAOMs micelles [211] on the assumption that the hydroperoxide ion interacts only Coulombically with the micelle.

However, the high specific interactions between cationic micelles and anions such as Br^- or NO_3^- show that models based on purely Coulombic interactions of point-charge ions with charged surfaces do not explain the data. [The ion-exchange equation (9) explicitly considers the specificity of ion-micelle interactions.] It is necessary to include specific ion-micelle interactions that are largest for bulky ions that have low charge densities and are not very strongly hydrated [212–215]. One approach is to assume that counterions that bind specifically neutralize the charge of an equivalent number of micellar headgroups. These specific interactions are written in terms of Volmer or Langmuir isotherms, and the former is better for more concentrated ionic solutions [213–214]. The factor f, where f is the fractional coverage by counterions, describes neutralization of the micellar headgroups. In terms of the Volmer isotherms, f is given by Eq. (15):

$$f = \frac{\delta \exp(-f/(1 - f))[X_W^-]}{1 + \delta \exp(-f/1 - f)[X_W^-]} \tag{15}$$

where $[X_W^-]$ is the concentrations of X^- in moles per liter of the aqueous pseudophase and δ is a specificity parameter related to the non-Coulombic affinity of X^- for micellized cationic surfactant. Values of δ are low for hydrophilic ions and larger for polarizable ions.

Therefore, ions compete both Coulombically and specifically. On this hypothesis, an ion such as OH^-, which interacts only Coulombically ($\delta = 0$), should not completely expel Br^-, with its strong specific interaction (typically $\delta = 120\ M^{-1}$), from a cationic micelle, in agreement with spectroscopic evidence from fluorescence quenching [196] and from examination of the Br^- NMR line width [98].

This treatment, based on the solution of the PBE in spherical symmetry [209–215], involves a variety of assumptions. The solution is treated as an assembly of identical uniform spherical cells, each containing one micelle. The micellar surface is assumed to be smooth and of uniform charge density given by the micellar aggregation number and from the radius at the charged surface. Selected size parameters are consistent with physical measurement [77]. However, the treatment typically involves two disposable parameters. The first is an ion specificity coefficient, δ: OH^- was assumed to interact only Coulombically, which provides a reference, and allows specific interaction parameters to be assigned to other ions [212–214]. Micellar reactions are assumed to occur in a shell at the surface whose width, Δ, is the second disposable parameter. The first-order rate constant for reaction of a substrate, S, and an anionic reagent, Y^-, is given by Eq. (16) [212–214]:

$$k_{obs} = \frac{k_W[Y_M^-] + k_2^m K_S[Dn][Y_\Delta^-]}{1 + K_S[Dn]} \tag{16}$$

The rate constants in Eq. (16) have the same significance as in Eqs. (8) and (10), and $[Y_\Delta^-]$ is molarity integrated over the shell. (Reaction in the aqueous region can be neglected for hydrophobic substrates except with very dilute surfactant.)

In this model the substrate is assumed to be distributed uniformly over the micellar region of thickness Δ, as in the other pseudophase models, but counterion concentrations decrease in this region with increasing distance from the charged surface. The PIE model, on the other hand, assumes that concentrations of both nonionic substrate and ionic reagent are uniform within the reaction region, e.g., the Stern layer. The concept of ionic competition is common to both models, but in the PBE model counterions are assumed to compete in two different ways. Addition of counterions reduces the surface electrical potential of an ionic association colloid and therefore reduces the Coulombic attraction of counterions and repulsion of coions [209–215]. But counterions such as Br^- that interact specifically with the headgroups also neutralize their charge, and this reduction of the surface charge density further reduces the surface potential. Bivalent anions, such as SO_4^{2-}, sharply reduce the surface electrical potential [210] and the Coulombic attraction for other counterions, but, because they are strongly hydrated, they probably do not intercalate between micellar headgroups [214]. They are effective inhibitors of reactions of univalent anions, and this inhibition is fitted by the PBE model [213], whereas the PIE model in its simplest form does not fit the results. This and similar treatments include both nonspecific, Coulombic interactions and ion-specific interactions of ions that intercalate in the micellar surface.

PBE has been used to treat several rate effects associated with nucleophilic substitution [216] anions such as hydroxide, azide, bromide and chloride, and anionic concentration was treated by considering both specific and nonspecific interactions. It is important to recognize that the model applies to micelles whose charge is at the surface and not screened by bulky groups. It is not clear whether it can be applied usefully to micelles that have bulky headgroups, and here the mass action model is often successfully used [77]. As already noted, there is evidence for the failure of the PIE model for cationic micelles at high hydroxide concentrations, for anionic micelles at high hydrogen concentrations, and for both micelles at high salt concentrations. The PBE equation was used to fit data for reactions of OH^- with pNPDPP, DNCN, and 1,1,1-trichloro-2,2-bis(*p*-chlorophenyl)ethane (DDT): the treatment includes both specific and nonspecific ion-micelle interactions and considers a rough rather than a smooth micellar surface; i.e., a roughness parameter, σ, was introduced. It seems that it improves fits for reactions in concentrated OH^- [217] (Table 1).

There is also evidence that the PIE model is unable to explain results for positively [209,218–221] and negatively [203,221] charged substrates. Vera and Rodenas [222] and Chaimovich et al. [223] tried to explain results for negatively charged substrates by using them as micellar counterions, but in this way it is not possible to explain all the kinetic data. It was deduced that for charged substrates bound to charged micelles, it is necessary to consider electrostatic interactions depending on the micellar surface potential. A treatment developed by Rodenas and coworkers considers the ion distribution around micelles according to the Poisson-Boltzmann equation, by considering specific interaction between bromide ions of CTABr and the micellar surface; it was used and nicely explained results for hydrophilic charged substrates in CTABr micelles, such as basic hydrolysis of the positively charged crystal violet [224] and of negatively charged substrates such as acetylsalicylic acid and 3-acetoxy-2-naphthoic acid [225]. The treatment was further improved by considering specific interactions between all the ions in solution, micellar counterion and hydroxide reactive ions, and the micellar surface [226], and it has been applied to study the hydrolysis of crystal violet and of acetylsalicylic acid in cationic micelles of DTABr [227] (Table 2).

4. Treatment of Coion Reactions

Ionic micelles inhibit bimolecular reactions of coions by taking up the substrate and repelling the coion [8,9]. Menger and Portnoy showed that, to a first approximation, the extent of inhibition followed the extent of transfer of nonionic substrate into the micellar pseudophase where the reaction rate was small or zero [119]. However, Chaimovich et al. showed that there was a small but finite reaction of OH^- with very hydrophobic carboxylic esters in anionic micelles of SDS and that this reaction was accelerated by addition of NaCl [228]. They suggested that Na^+ competed with H_3O^+ at the micellar surface so that autoprotolysis of water leads to the presence of a finite concentration of OH^- at that surface.

Romsted and coworkers found corresponding results for the H_3O^+-catalyzed hydrolysis of hydrophobic acetals in cationic micelles of alkyltrimethylammonium chloride or bromide, although reactions were much slower than those of chemically similar hydrophilic acetals in aqueous strong acid [229,230]. The micellar reactions were accelerated by added salt, and NaBr was more effective than NaCl. The PIE model explains these results in terms of competition between cations and H_3O^+, or anions and OH^-, at the micellar surface with a consequent effect on the concentrations of lyate or lyonium ion (OH^- or H_3O^+, respectively).

There are finite rates of S_N2 reactions of thiocyanate and sulfite ions with methyl naphthalene-2-sulfonate in aqueous anionic micelles, but these reagents are conjugate bases of relatively strong acids, so any explanation based on autoprotolysis is inapplicable for these reactions and it is necessary to consider alternative models [231]. The dependence of concentrations of co- and counterions on distance from the surface of a spherical colloidal macroion can be predicted from electrostatic considerations, for example, by solution of the Poisson-Boltzmann equation [209,210,212–215]. The electrostatic potential decreases with increasing distance from the micellar surface. Added electrolyte reduces the surface potential and therefore also reduces the attraction of counterions. The net effect of the reduced potential and increased total counterion concen-

TABLE 2 Reactions of Charged Substrates[a]

Reaction	Surfactant	k_W, $M^{-1}\ s^{-1}$	k_2^m, $M^{-1}\ s^{-1}$
Crystal violet +	CTABr	0.201	0.260[b]
OH^-	DTABr	0.201	0.064[c]
Acetyl salicylate	CTABr	0.124	0.706×10^{-2}[b]
+ OH^-	DTABr	0.124	0.540×10^{-2}[c]

[a]From PBE treatment; for fitting parameters see references in b and c.
[b]From Ref. 226.
[c]From Ref. 227.

tration is that there is little change of counterion concentration at the surface [209,210,212–215]. But the decreased repulsion of coions and the increase of their total concentration lead to a sharp increase in their concentration adjacent to surface [231] and an increase of rate of reaction with a coion.

This treatment neglects specific ionic interaction with the micelle. If counterions intercalate the micellar surface and partially neutralize headgroup charges, the coion concentration at the surface will increase, as will the rate of reaction. Tetramethylammonium ion is more effective than Na^+ in speeding the reaction of OH^- with hydrophobic esters in solutions of SDS [231] and the H_3O^+-catalyzed hydrolysis of hydrophobic acetals in solutions of cationic micelles is faster when the counterion is Br^- rather than Cl^- [229,230]. These specific effects are consistent with the stronger binding of Me_4N^+ as compared with Na^+ for anionic micelles and of Br^- as compared with Cl^- for cationic micelles [11].

Rate-[surfactant] profiles for reaction of OH^- with fully bound hydrophobic esters in SDS and rate enhancements by added salts can be fitted by calculating the concentration of OH^- at the anionic micellar surface by solving the PBE in spherical symmetry [231]. The method has been applied only to the reaction of OH^- with *p*-nitrophenyl diphenyl phosphinate and alkanoate and not to the reaction with MeONs. This last substrate reaction with water makes a significant contribution to the overall rate in high [SDS] so that there is uncertainty about the contribution of reactions of the anions and they cannot be followed over the wide range of conditions needed for the simulation. Second-order rate constants are not very different from those in water or in cationic micelles (Table 3).

The effects of cationic surfactants and added salts on the H_3O^+-catalyzed hydrolyses of hydrophobic acetals [229,230] can also be explained in terms of an electrostatic model. Added salts increase the rate of re-

TABLE 3 Examples of Reactions of Anions in Anionic Micelles of SDS[a]

Substrate	$[X^-]$, M	Added salt, M	k_W, M^{-1} s^{-1}	k_2^m, M^{-1} s^{-1}
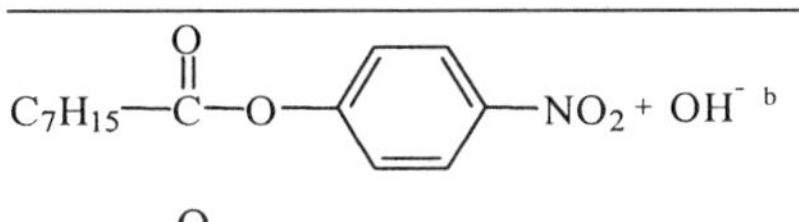 [b]	0.0193 0.0193	— NaCl, 0.20	8.8 6.1	0.7 0.7
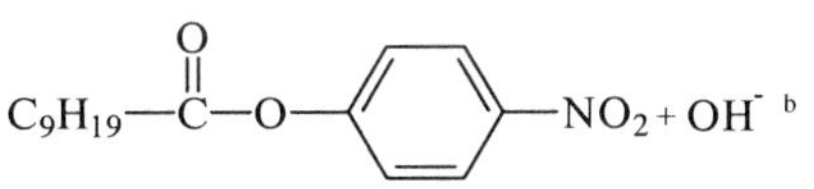 [b]	0.0193	— NaCl, 0.02–0.20	5.6 5.6	0.6 0.6
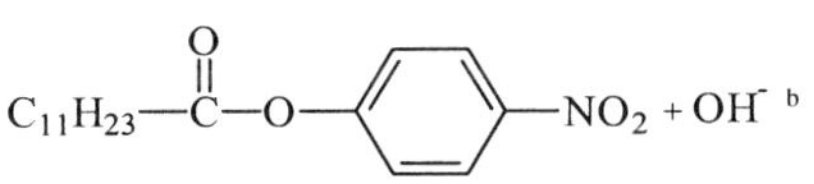 [b]	0.0193	—	4.0	0.65
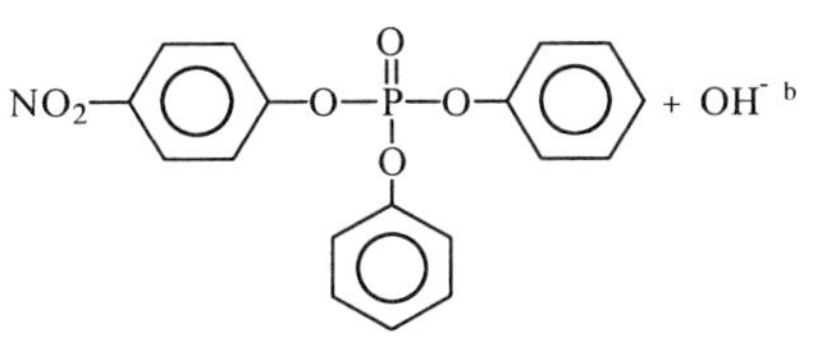 [b]	0.03–0.15 0.03–0.15	— NaCl, 0.30	8.7 3.65	3.75 3.75
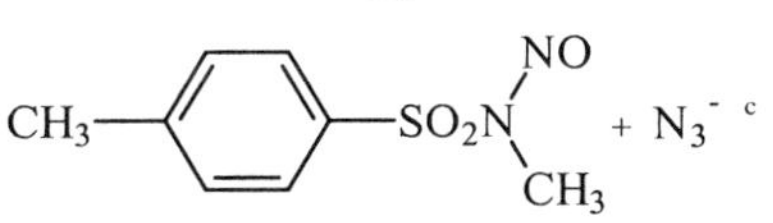 [c]	0.0029–0.0044	—	3.3×10^{-2}	$1.4–1.8 \times 10^{-3}$
$(CH_3CH_2CH_2)_2S + IO_4^-$ [d]	0.00203	Me_4NBr, 0.10–0.25	3.2	1.5
$(CH_3CH_2CH_2CH_2)_2S + IO_4^-$ [d]	0.00203	—		0.8
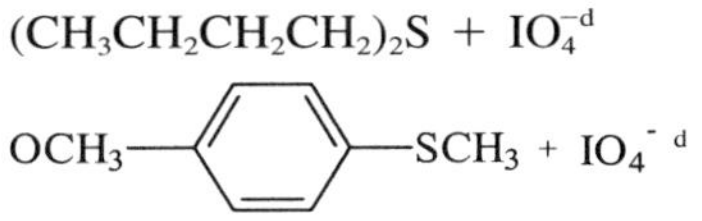 [d]	0.00203 0.00203 0.00203	— NaCl, 0.10–0.25 Me_4NBr, 0.10–0.25	1.55 1.55 1.55	0.6 0.5 0.4

[a]From PBE treatment; for fitting parameters see references in b, c, and d.
[b]From Ref. 231.
[c]From Ref. 234.
[d]From Ref. 233.

action of substrate fully bound to cationic micelles. The PIE model predicts that first-order rate constants, k_{obs}, will increase linearly with added salt concentration, but under some conditions increases are not linear [229, 230]. Plots of k_{obs} against [salt] have a break and slopes decrease at high [salt]. Application of the PBE model, as applied earlier to reactions of OH^- in SDS, predicts linear slopes at low [salt] and curvature at higher [salt] in agreement with experiment [232] (Table 3).

The same treatment to calculate the anion distribution in anionic micelles of SDS has been applied in kinetic studies of the oxidation of sulfides by periodate ion [233], the reactions of a series of nucleophiles (OH^-, SO_3^{2-}, N_3^-) with *N*-methyl-*N*-nitroso-*p*-toluensulfonamide (MNTS), and reactions of OH^- with 1,2-dichloroethyl-3-cyclohexyl-1-nitrourea in SDS micelles [234]. In the presence of SDS micelles the reaction of MNTS with azide ion appears to occur in both the aqueous and micellar pseudophases (Table 3), whereas its reaction with SO_3^{2-} and OH^- appears to occur exclusively in the aqueous phase, and it was explained on the basis of the location of the substrate and of the different anions (different polarizability and charge) in the micellar surface.

Khan has also studied effects of anionic micelles on rates of bimolecular reactions involving anionic reactants and observed increases in the rate of hydroxide ion–catalyzed hydrolysis of anionic acetyl salicylate [235] and of anionic *N*-phthaloylglycine [236]. He explained the increase in k_{obs} with an increase in [SDS] at constant [NaOH], thinking that the micelle-mediated reaction occurs in the Gouy-Chapman layer or at the junctural region of the Gouy-Chapman and the Stern layer, where the concentrations of water and hydroxide ions may be similar to their corresponding concentrations in the aqueous pseudophase. Such a reaction is regarded as a cross-border reaction. The mass action model has not been applied to coion reactions.

B. Other Treatments

Other general ways of treating micellar kinetic data should be noted. Piszkiewicz [237] used equations similar to the Hill equation of enzyme kinetics to fit variations of rate constants and surfactant concentration. This treatment differs from that of Menger and Portnoy [119] in that it emphasizes cooperative effects due to substrate-micelle interactions. These interactions are probably very important at surfactant concentrations close to the cmc because solutes may promote micellization or bind to submicellar aggregates. Thus, Eq. (2) and others like it do not fit the data for dilute surfactant, especially when reactants are hydrophobic and can promote micellization (see Section VI.B). Srivastava and Katiyar have also developed equations that attempt to take into account the way in which reactant interactions affect reactivity in micelles [238]. A refinement of Eq. (2) has been proposed including the possibility of different reaction domains within the micelle. The pseudophase model, distinguishing only between a micellar and an aqueous phase, turns into a model of three [239,240] or multiple [241,242] domains (multiple micellar pseudophase, MMPP), which is compatible with the transition state pseudoequilibrium constant approach of Kurz [243,244]. In the limit of an infinite number of domains, this model provides the exact rate constant as the integral over the domains with their local rate constant.

For a spontaneous reaction, for instance, there is a reaction in the first domain, i.e., the core of the micelle, with a rate constant k_C; a reaction in the Stern region, a second domain, with a rate constant k_S; and a reaction in bulk water, the third domain, with a rate constant k_W. Using the three-domain model and assuming that in the hydrophobic core no hydrolysis takes place, Engberts and colleagues [245] obtained the following equation, (17), for k_{obs} for a hydrolysis reaction:

$$k_{obs} = \frac{k'_W + k'_M K_M (V_M/V_W)}{1 + K_M (V_M/V_W)} \tag{17}$$

This relation resembles the ordinary Menger-Portnoy equation, but k'_M is now given by Eq: (18):

$$k'_M = k_S \left\{ \frac{V_M K_M - V_C K_{WS} K_{SC}}{V_M K_M} \right\} = k_Z \left\{ \frac{V_S K_{WS}}{V_M K_M} \right\} \tag{18}$$

where V_M and K_M are the micellar volume and the partition coefficient, respectively, V_C is the micellar core volume, K_{WS} is the water–Stern region partition coefficient, K_{SC} is the partition coefficient for the Stern region–core equilibrium, and V_S is the Stern region volume.

Alternatively, the application of the pseudophase model is considered as an adaptation of the Olson-Simonson model [246] to micellar systems. It can be demonstrated that the latter, frequently used in the interpretation of kinetic salt effects, is an alternative formulation of the Brønsted equation [247,248] that can be deduced from the transition state theory. For a reaction (A + B that yield products), this equation gives:

$$k_{obs} = k_0 \frac{\gamma_A \gamma_B}{\gamma_{\neq}} \tag{19}$$

where γ_A, γ_B, and $\gamma_{\neq}$ are the activity coefficients of the reactants and of the activated state, respectively, and k_0 is the reaction rate constant when this is carried out in the reference state. The possibility of applying the Brønsted equation to interpret the micellar kinetic effect was suggested several years ago by Bunton and Robinson [121], who later considered it again [249] but the model has not been frequently used, also because some authors have been against its use [250]. When ionic reactants are largely in the aqueous pseudophase, kinetic salt effects have been treated with a combination of the Brønsted equation and the extended Debye-Hückel approach for estimating activity coefficients. Sànchez and coworkers used this approach to fit rate constants of reactions of oppositely charged di- and trivalent ions in SDS [251], of like charged ions in SDS micelles [252], and the oxidation of $Fe(CN)_2(bpy)_2$ with $S_2O_8^{2-}$ in AOT micelles and microemulsions [253]. They also applied the equation to a system where both reactants are partitioned between the two pseudophases present in the system and studied a ligand substitution process in CTACl (Scheme 3) [254]. Interestingly, results from fits based on the Brønsted equation are equivalent to those based on the pseudophase model for this reaction, and this confirms the idea that both models are equally valid for interpretation of kinetic data in micellar systems.

$$Fe(CN)_5(4\text{-}CNpy)^{3-} + CN^- \rightarrow Fe(CN)_6^{4-} + 4\text{-}CNpy$$

(where 4-CNpy = 4-cyanopyridine)

SCHEME 3

IV. SPONTANEOUS, UNIMOLECULAR, AND WATER-CATALYZED REACTIONS

The quantitative treatment of micellar rate effects upon spontaneous reactions is simple in that the overall effect can be accounted for in terms of distribution of the substrate between water and the micelles and of the first-order rate constants in each pseudophase (Scheme 2). The micelles behave as a submicroscopic solvent and to a large extent their effects can be related to known kinetic solvent effects upon spontaneous reactions. It will be convenient first to consider unimolecular reactions and then water-catalyzed uni- and bimolecular reactions.

A. Unimolecular Reactions of Anionic Substrates

The solvent-dependent unimolecular decarboxylation of 6-nitrobenzisoxazole-3-carboxylate (6-NBIC) [255] and hydrolyses of 2,4-dinitrophenyl phosphate dianion ($DNPP^{2-}$) [256] have provided popular probes of micellar structures [12] because with fully bound substrate, rate effects are due wholly to changes in the relative free energies of the initial and transition states involving only one molecule or ion. Also, the cyclization of *o*-3-halopropoxyphenoxide ion (halo = Br, I) and the parameter $k_{M(I)}/k_{M(Br)}$ have been used as indicators of micellar properties because their values depend largely on interactions with phenoxide ions in the initial states and charge delocalized transition states, which differ for bromide and iodide substrates. These reactions have also been useful monomolecular models for S_N2 reactions; in fact, they can be regarded as intramolecular S_N2 reactions [257]. Kinetic analysis of these reactions can be carried out simply using the Menger-Portnoy equation [Eq. (2)].

The E1cb decompositions have been studied as a spontaneous decomposition of the anions formed in a rapid preequilibrium. Correia and coworkers [258] interpreted only qualitatively the inhibition of decomposition of *m*-nitrophenyl-9-fluorene carboxylate by CTABr by assuming that the ground state carbanion formed by deprotonation of the substrate at the micellar surface is more stabilized by trimethylammonium headgroups of surfactants than the transition state in which the charge is being transferred to the oxygen of the *m*-nitrophenyl group. This explanation is consistent with the rate enhancement they observed for the same reaction in anionic micelles and with the absence of any rate effect in nonionic micelles. Nome and coworkers [122] studied the decomposition of *p*-substituted aryl-2,2,2-trichloroethanols (Scheme 4) in aqueous NaOH and analyzed quantitatively the rate-surfactant profiles (Table 4) using Eq. (4) reported in Section III.

They analyzed a series of substrates. In the presence of surfactant there is a three- to fivefold decrease in

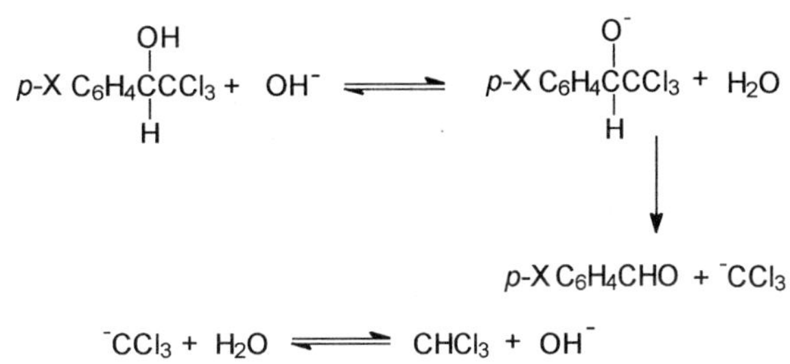

X= H, Me, OMe, Cl, NO_2

SCHEME 4

TABLE 4 Decomposition of *p*-Substituted Aryl-2,2,2,-trichloroethanols in Aqueous NaOH[a]

Substrate	Surfactant	$10^4\ k'_M$, s^{-1}
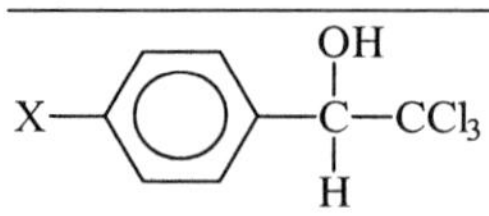		
X = H	CTABr	5.9
	CTACl	7.4
	CTAOH	8.9
X = CH_3	CTABr	6.9
	CTACl	8.3
	CTAOH	11.6
X = OCH_3	CTABr	9.7
	CTACl	10.0
	CTAOH	14.3
X = Cl	CTABr	5.2
	CTACl	5.8
	CTAOH	8.4

[a]From Ref. 122.

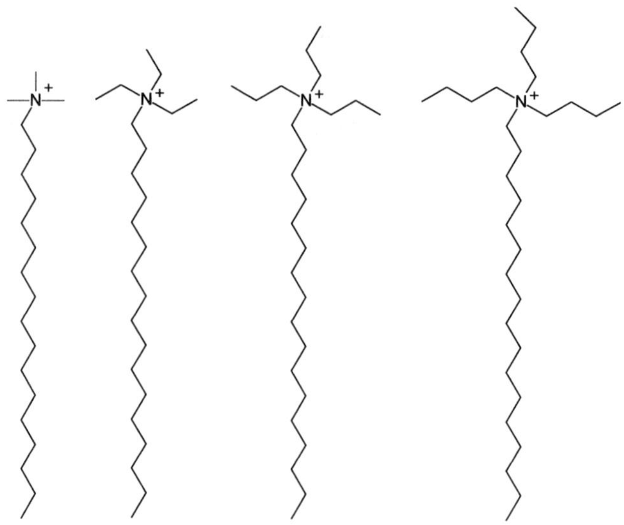

SCHEME 5

rate constant and the dependence of k'_M on substituent effect was also quantitatively analyzed in terms of the Hammett plot: ρ values are similar in water and in micelles of CTABr, CTACl, and CTAOH.

1. Variations in Surfactant Structure

A variety of new surfactants have been investigated using unimolecular spontaneous reactions as probes of micellar structure. Here we focus attention on studies of systematic variations in covalent surfactant structures. For instance, single-chain surfactants with increasing headgroup bulk (Scheme 5) have been investigated: CTABr, CTEABr, CTPABr, and CTBABr (Tables 5 and 6). For decarboxylation of 6-NBIC [259] and hydrolysis of $DNPP^{2-}$ [260] rate enhancements increase with increasing headgroup bulk (Table 5). Bulky groups should decrease the polarity at the micellar surface and, for dephosphorylation rate enhancements, have been associated with decreases in activation enthalpies [260].

Rate effects are not due solely to a change in the bulk of the *N*-alkyl group, because the morpholine (CMMABr) and the quinuclidine (MQBr) derivatives are little more effective than CTABr in accelerating decarboxylation [259]. The quinuclidinium and morpholinium moieties probably extend away from the micellar surface so that reaction takes place in a region exposed to water. Bulky alkyl groups in CTBABr do not extend into the water but are oriented along the micellar surface [86] to reduce water–alkyl group contact, so decarboxylation and dephosphorylation of bound substrates take place in a region of relatively low polarity. For decarboxylation, the effect of headgroup size is present to a limited extent in *p*-octyloxy surfactants (pOOTABr and pOOTBABr) [259]. For cyclization of *o*-3-halopropoxyphenoxide ion rate enhancement increases sharply in the sequence of surfactant headgroups (Table 6) [261] and increases are larger for reaction of the iodide than for the bromide. Micellar headgroups can control reactivity by excluding water from the surface, which increases the reactivity of the oxide ion but decreases the hydration of the leaving halide ion. But cationic headgroups interact readily with leaving halides, and more strongly with I^- than with Br^-, so they can provide electrophilic assistance to reactions.

A series of 1-alkyl-4-alkylpyridinium halide surfactants has been investigated by Engberts [262] (Table 5). Variation of the 1-alkyl substituent from methyl to ethyl and *n*-propyl and *n*-butyl resulted in increases in k'_M values, as for cetyltrialkylammonium surfactants. The same author also investigated other kinds of systematic variations (Table 5). Introduction of a rigid acetylenic segment in the center of the 4-dodecyl substituent and branching of the alkyl moiety hardly influenced the micellar catalytic effect, and the k'_M value still depended little on the alkyl chain length.

Replacement of a cationic by a zwitterionic headgroup does not inhibit decarboxylation of 6-NBIC [263], hydrolysis of $DNPP^{2-}$ [260], and cyclization of *o*-3-halopropoxyphenoxide ion [263] at micellar sur-

TABLE 5 Unimolecular Reactions of Anionic Substrates[a]

Substrate	Surfactant		$10^4\ k'_M$, s^{-1}	Ref.
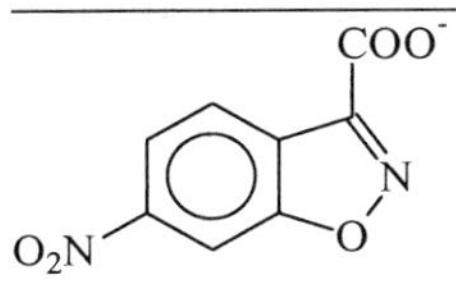	CTABr		≈3.3 (100)	259
	CTEABr		10 (330)	
	CTPABr		41 (1400)	
	CTBABr		85 (2800)	
	$(CDA)_2C_32Br$		7.7 (290)	
	CMMABr		4.8 (160)	
	MQBr		6.6 (220)	
	pOOTABr		>2.2 (>73)	
	pOOTBABr		3.7 (124)	
	CTACl		3.2 (110)	264
	$CTA(SO_4)_{0.5}$		3.9 (130)	
	CTAOTs		>6.6 (220)	
	CTAOH		1.8 (60)	
	DDDACl		>14 (500)[b]	
	SB3-14		6.6 (220)	263
	SB3-16		9.2 (307)	
	SBPr3-14		32 (1100)	
	AOMe-14		7.1 (240)	
	AOPr-14		>55 (1800)	
	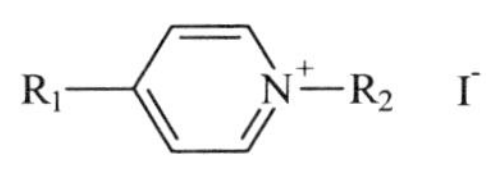			262[c]
	R_1	R_2		
	n-C_8H_{17}	CH_3	3.2 (42)	
	n-$C_{10}H_{21}$	CH_3	3.3 (43)	
	n-$C_{11}H_{23}$	CH_3	3.7 (49)	
	n-$C_{12}H_{25}$	CH_3	3.5 (46)	
	$—(CH_2)_4C{\equiv}C—C_6H_{13}—n$	CH_3	4.1 (54)	
	$—CH—(CH_3)(n\text{-}C_{10}H_{21})$	CH_3	3.9 (51)	
	$—(CH_2)_8C(CH_3)_3$	CH_3	4.2 (55)[d]	
			6.2 (82)[e]	
	$—(CH_2)_7CH(C_2H_5)_2$	CH_3	4.6 (61)	
	$—(CH_2)_5CH(n\text{-}C_3H_7)_2$	CH_3	4.0 (52)	
	n-$C_{12}H_{25}$	C_2H_5	5.2 (68)	
	n-$C_{12}H_{25}$	n-C_3H_7	6.2 (82)	
	n-$C_{12}H_{25}$	i-C_3H_7	8.6 (113)	
	n-$C_{12}H_{25}$	n-C_4H_9	9.8 (129)	
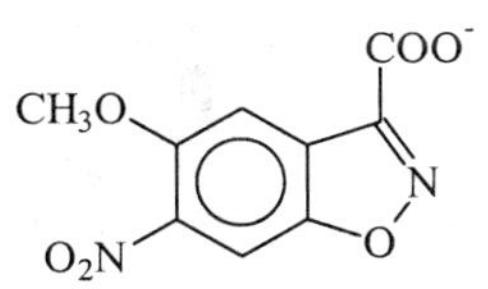	CTABr		0.048 (24)	280
	CTEABr		0.080 (40)	
	CTPABr		0.143 (71)	
	CTBABr		0.193 (96)	
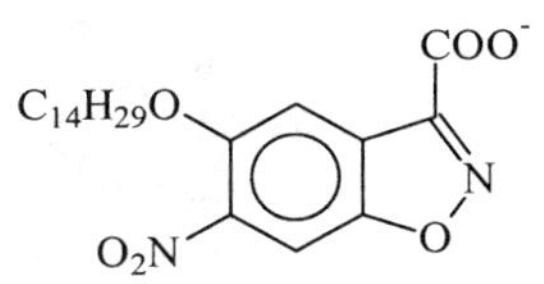	CTABr		0.068 (32)	280
	CTEABr		0.164 (78)	
	CTPABr		0.333 (159)	
	CTBABr		0.558 (266)	

TABLE 5 Continued

Substrate	Surfactant	$10^4\ k'_M$, s^{-1}	Ref.
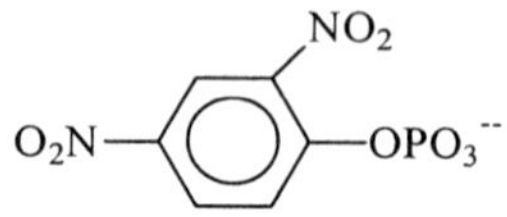	CTACl	2.4 (28)	265
	$CTA(SO_4)_{0.5}$	3.0 (35)	
	CPCl	3.8 (44)	
	CDMPCl	4.2 (49)	
	DDDACl	4.8 (56)	
	DDDABr	~7 (81)	
	$DDDA(SO_4)_{0.5}$	8.5 (99)	
	Bolaform (22) Br_2	1.6 (19)	
	Bolaform (22) SO_4	3.7 (43)	
	Bolaform (16) Br_2	>0.6 (7)	
	CTABr	2.2 (26)	260
	CTPABr	3.3 (38)	
	CTBABr	4.1 (48)	
	MQBr	3.1 (36)	
	SB3-14	2.8 (33)	
	SBEt3-14	2.8 (33)	
	SBPr3-14	3.2 (37)	
	SBBu3-14	3.6 (42)	
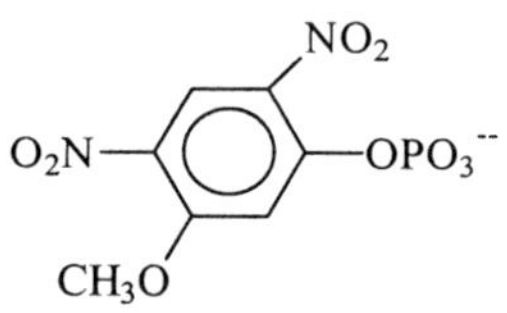	CTABr	1.7 (14)	f
	CTPABr	2.5 (21)	
	CTBABr	2.6 (22)	
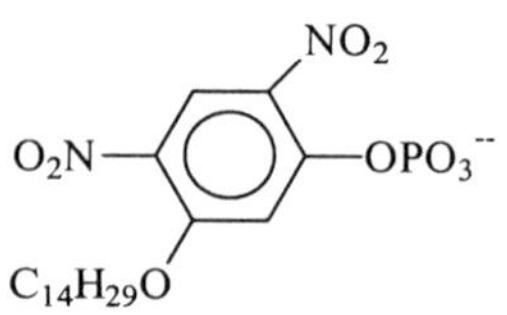	CTABr	1.1 (11)	f
	CTPABr	1.4 (14)	
	CTBABr	1.6 (16)	

[a]Values in parentheses are k'_M/k'_W; data are for 25°C if not otherwise specified.
[b]At [surfactant] 0.02 M.
[c]At 30°C.
[d]Spherical micelles.
[e]Rodlike micelles.
[f]M. Tugliani et al., Langmuir, in press, 2000.

faces (Tables 5 and 6), although under some conditions it may slow the overall reaction by decreasing micellar incorporation of anionic substrates. For hydrolysis of $DNPP^{2-}$, first-order rate constants continue to increase even up to 0.2 M sulfobetaine because the substrate is not strongly bound [260]. A carboxylate ion (such as 6-NBIC) will interact unfavorably with sulfonate or carboxylate centers in a betaine micelle, and the interaction will decrease as charge moves out of the carboxylate ion in the transition state. The situation is similar for spontaneous dephosphorylation [260] and for reactions in micellized amine oxides. Micellar catalysis is increased by an increase in headgroup bulk in cationic surfactants and by a change from cations to zwitterions, but the situation is more complicated when the two effects are combined. For example, decarboxylation is increased by a factor of about 2 to 3, dephosphorylation by a factor of about 1.3 [260], and cyclization of *o*-3-halopropoxyphenoxide ion by a factor of about 1.4 to 1.5 going from CTABr to SB3-14 [263], but with bulky headgroups, cationics are more effective than zwitterionic surfactants, especially for decarboxylation and dephosphorylation reactions.

Twin-tailed surfactants that spontaneously solubilize in water at relatively high concentration (for dilute surfactant see Section VI.A) are better catalysts than

TABLE 6 Cyclization Reactions[a]

Substrate		Surfactant	$10^4\ k'_M$, s^{-1}	Ref.
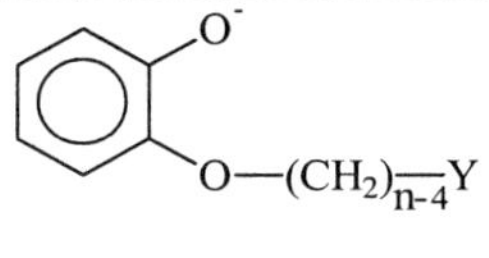		CTABr	5.5 (3.9)	261
		CTACl	5.6 (1.4)	
		$CTANO_3$	4.7 (3.4)	
		$CTA(SO_4)_{0.5}$	6.5 (4.6)	
Y = I	$n = 7$	CTEABr	10 (7.1)	
		CTPABr	26 (19)	
		CTBABr	41 (29)	
		SB3-14	8.4 (6.0)	263
		CB1-14	7.3 (5.2)	
		CB1-16	11 (7.9)	
		SB3-12	7.1 (5.1)	459
		$C_{12}E_{23}$	≈3.5 (2.5)	
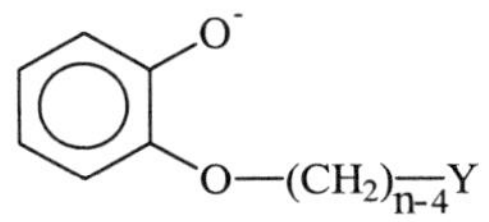		CTABr	4.1 (1.8)	261
		CTACl	4.4 (1.9)	
		$CTANO_3$	3.4 (1.5)	
		$CTA(SO_4)_{0.5}$	4.8 (2.1)	
Y = Br	$n = 7$	CTEABr	6.1 (2.6)	
		CTPABr	13.0 (5.6)	
		CTBABr	17.0 (7.3)	
		SB3-14	5.5 (2.4)	263
		SB3-16	6.2 (2.7)	
		SBPr3-14	13.0 (5.6)	
		CB1-14	4.6 (2.0)	
		CB1-16	6.5 (2.8)	
		SB3-12	4.9 (2.1)	459
		$C_{12}E_{23}$	>2.9 (1.2)	
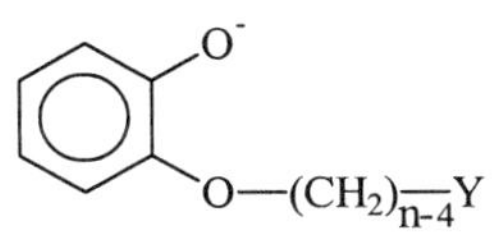				
Y = Br	$n = 10$	CTABr	0.059[b]	279
	$n = 12$	CTABr	0.036[b]	
	$n = 16$	CTABr	0.017[b]	
Y = Br	$n = 10$	CTBABr	0.130[b]	
	$n = 12$	CTBABr	0.094[b]	
	$n = 16$	CTBABr	0.037[b]	

[a]Values in parentheses are k'_M/k'_W.
[b]Values of k_W in water are not available, for the insolubility of the substrates.

CTABr for decarboxylation [264] and for hydrolysis of $DNPP^{2-}$ [265]. Packing of two hydrophobic alkyl chains could give lower surface charge density (Table 5).

As regards the counterion effect for decarboxylation of 6-NBIC, k'_M (Table 5 and Ref. 264) increases with decreasing fractional micellar charge, α, in the order CTAOTs > $CTA(SO_4)_{0.5}$ > CTABr > CTAOH. Micelles with high fractional charge, i.e., high surface charge, should interact most strongly with substrates. For hydrolysis of $DNPP^{2-}$, reactions are faster when SO_4^{2-} is the counterion than when Br^- is the counterion in solutions of single-chain cetyltrimethylammonium surfactants and also in solution of bolaform(22) and of twin-chain DDDAX surfactants [265]. For cyclization of *o*-3-halopropoxyphenoxide ion [261], rate constants follow the sequence $SO_4^{2-} > Cl^- > Br^- > NO_3^-$ (Table 6).

A considerable number of investigations have been reported on the surface and micellar properties of surfactants containing two hydrophilic or two hydrophobic groups in the molecule, called dicationic or gemini or dimeric surfactants. These include molecules that contain either a flexible hydrophilic [266–269], flexible hydrophobic [270,271], or rigid hydrophobic [272,273] linkage (spacer) between the two hydrophilic groups. The interest in these molecules appears to be due to their unusual surface and bulk properties. These include unusually high surface activity, low critical micelle concentration (cmc) values, and [274] unusual increases in cmc when the chain length of the alkyl group is increased beyond a critical length [272,273,275]. In decarboxylation of 6-NBIC [259] the dicationic surfactant $(CDA)_2C_22Br$ gives larger rate enhancement than CTABr, probably because the bridging methylene groups should force water molecules away from the micellar surface. An investigation of the cyclization of *o*-3-bromopropoxyphenoxide was carried out in aqueous micelles of 1,4-(*N*-hexadecyl-*N,N*-dimethylammonium)butane dibromide and (2*S*,3*S*)-2,3-dimethoxy-1,4-bis(*N*-hexadecyl-*N,N*-dimethylammonium)butane dibromide [276].

As for decarboxylation, dicationic surfactants are better catalysts than the corresponding monocationic ones, probably because the spacer decreases the extent of water penetration at the aggregate surface and the cyclization rate is increased in a less polar reaction site. But the rate-surfactant profile with dicationic surfactant is peculiar in that k_{obs} increases until a limiting value that is constant over a range of concentration but then increases again at high [surfactant] (only with the first of the two dicationic surfactants used).

Regarding decarboxylation, Engberts and colleagues [277] have reported that a plot of enthalpies versus entropies of activation show a linear relationship, giving an isokinetic temperature of 336 K for a series of different micellar structures, independent of the nature of the headgroup and counterion. The analysis gives different results in bilayer assemblies. Despite mechanistic differences between cyclization and decarboxylation, plots of log k'_M for cyclization versus log k'_M for decarboxylation are linear for a number of cationic and zwitterionic sulfobetaine micelles [278]. This result fits the widely used pseudophase model of rate effects.

2. Investigation of Variation in Substrate Structure

Systematic variations of the structure of the substrates that undergo unimolecular reactions have been investigated. Attention has been focused on substrate hydrophobicity.

Longer chain phenoxides, analogues of *o*-3-halopropoxyphenoxide bromide (Table 6), were investigated in CTABr and CTBABr [279] and values of k_{obs} at relatively high surfactant concentrations were found to increase with increasing headgroup bulk for all substrates, the effect being more important for the shorter chain and quite the same for all the longer chain ones (Scheme 7).

More recently, analogues of 6-NBIC [280] and of $DNPP^{2-}$ (M. Tugliani et al., Langmuir, in press, 2000), bearing a long alkyl chain of 14 carbon atoms, have been investigated (Schemes 8 and 9). Their reactivities were compared with those of their short-chain analogues (bearing an OCH_3 group in the same position) in order to separate accurately substituent electronic effects from substrate hydrophobicity effects. Reactivities have been studied in cationic surfactants with increasing headgroup bulk, and values of k'_M for fully bound substrates increase in the sequence CTABr < CTEABr < CTPABr < CTBABr (Table 5). For decarboxylation, micellar rate enhancements are decreased by introduction of an alkoxy group, based on k'_M for reaction of 6-NBIC, and surfactant headgroup effects are larger with the tetradecyloxy than with the methoxy derivative (Table 5). It is tempting to assume that the higher hydrophobicity of the former takes the reaction center deeper into the micellar interfacial region and away from water. For dephosphorylations, introduction of a long alkyl chain leads to a decrease in k'_M relative to that of the short-chain analogue. This decrease is insensitive to surfactant structure, its value being about 0.7.

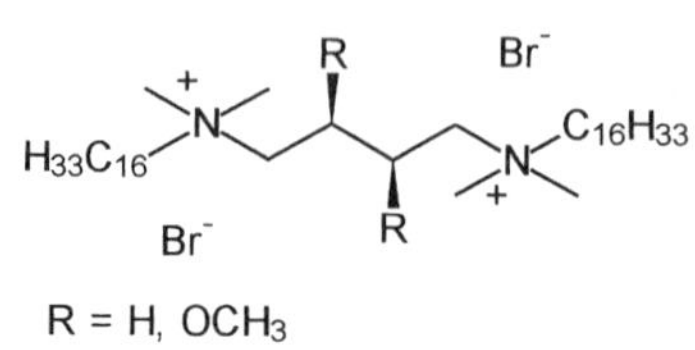

SCHEME 6

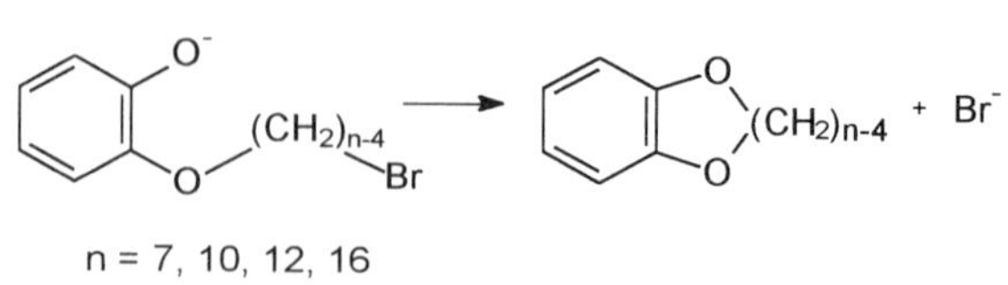

SCHEME 7

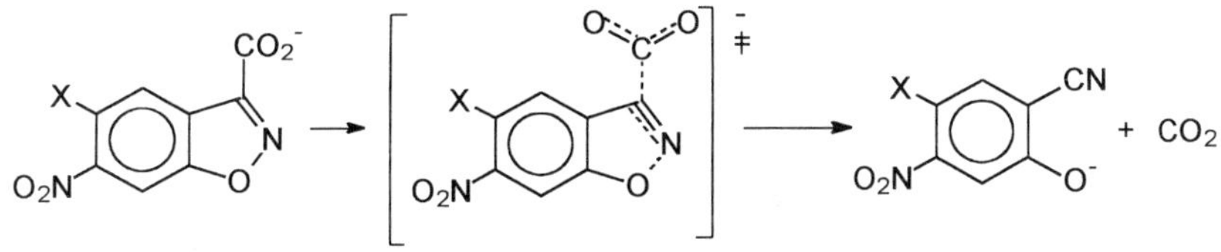

SCHEME 8

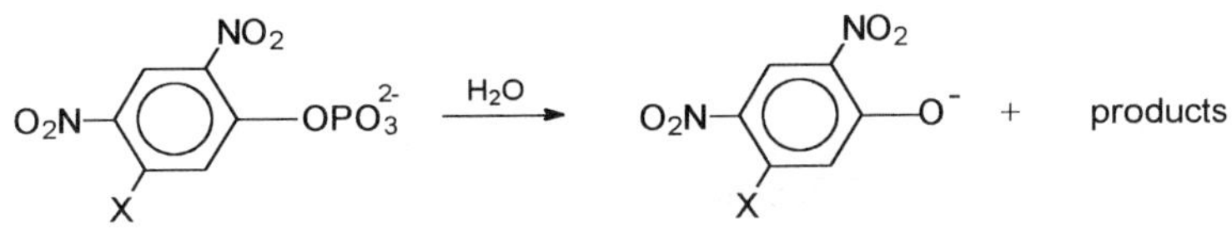

SCHEME 9

B. Water-Catalyzed, Uni- and Bimolecular Reactions

Spontaneous hydrolyses of alkyl halides, sulfonic esters, and acid chlorides and deacylations are typically micelle inhibited. The rate constant in the micellar pseudophase, k'_M, is generally simply estimated by analysis of rate-surfactant profiles, using the equation developed by Menger and Portnoy. These reactions are also inhibited by a decrease in the water content of aqueous-organic solvents [282], and qualitatively the micellar inhibitions are consistent with the interfacial region being less polar and aqueous than water. Comparisons of reaction rates in micelles and aqueous organic solvents have been used to estimate effective dielectric constant or polarities at micellar surfaces [281,282].

The high electrolyte content of ionic interfacial regions may also inhibit hydrolyses. Engberts and coworkers [283] (Table 7) have contended that observed micellar retardation of the hydrolysis of 1-benzoyl-3-phenyl-1,2,4-triazole and *p*-methoxyphenyl dichloroacetate is dominated by a salt effect. A comparison was made between medium effects in micellar solutions of cationic, anionic, and nonionic surfactants and in solutions of model compounds (tetramethylammonium bromide, TMABr, for cationic; sodium monomethylsulfate, NMS, for anionic; and tetra- and heptaethyleneglycol, TEG and HEG, for nonionic surfactants), and the micellar rate constants matched or nearly matched the rate constant for hydrolysis in the model compound solution at rather high concentrations. The matching occurs at 4.3 M aqueous solution of NMS for SDS micelles, at 4.2 M TEG for dodecylheptaoxyethylene glycol ether, and around 5 M TMABr for CTABr.

There is also a micellar charge effect that is related to reaction mechanism [281]. The charge effect of micelles on spontaneous hydrolyses appears to be related to charge asymmetry in the interfacial region and contrasting charge distributions in the transition states of bimolecular deacylations, S_N2 hydrolyses, or S_N1 hydrolyses. All the results to date are covered by a simple generalization: if bond making is dominant in the transition state $k^+/k^- > 1$, but if bond breaking is dominant $k^+/k^- < 1$, and compounds that react at the extremes of the S_N1-S_N2 spectrum fit the generalization (k^+ and k^- are values of k'_M in cationic and anionic micelles, respectively [281]). In the transition state for hydrolysis of sulfonate esters, or in deacylations, negative charge builds up on the organic moiety and interacts unfavorably with the anionic headgroup of an SDS micelle as compared with a cationic headgroup. Conversely, in an S_N1 reaction positive charge at the alkyl center in the transition state interacts unfavorably with a cationic headgroup [281]. Therefore, in analyzing kinetic micellar effects on spontaneous hydrolyses, not only polarity and water content but also surface charge distributions at micellar surfaces, which also depend on interactions between counterions and headgroups, have to be considered.

1. Investigation of Systematic Variations in Surfactant Structure

Systematic investigations of variations in surfactant charge and structure, including the effect of headgroup

TABLE 7 Water-Catalyzed Reactions[a]

Substrate	Surfactant	10^6 k'_M, s^{-1}	Ref.
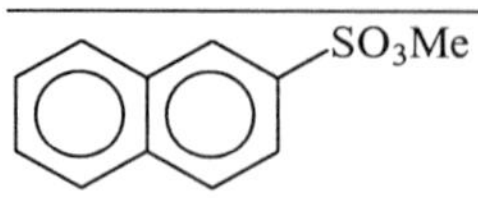	CTAOMs	7.33 (0.59)	284
	CTAOMs + $MeSO_3Na$	7.23 (0.58)	
	CTPAOMs	6.93 (0.55)	
	CTPAOMs + $MeSO_3Na$	7.01 (0.56)	
	SB3-14	7.03 (0.56)	
	SB3-14 + $MeSO_3Na$	7.94 (0.64)	
	SB3-14 + $MeSO_3H$	7.27 (0.58)	
	SB3-14 + $NaClO_4$	1.85 (0.15)	
	SBBu3-14	7.73 (0.62)	
	SBBu3-14 + $MeSO_3Na$	7.40 (0.59)	
	AOMe-14 + $MeSO_3H$	7.22 (0.58)	
	AOMe-14 + $MeSO_3H$	6.81 (0.54)	
	AOPr-14 + $MeSO_3H$	7.06 (0.56)	
	SDS	2.81 (0.22)	
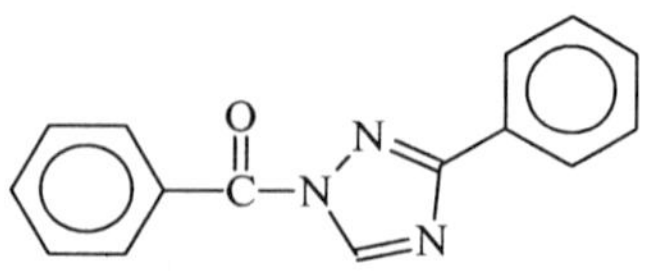	CTABr	67 (0.053)	283
	DTABr	126 (0.10)	
	CTACl	145 (0.115)	
	SDS	48 (0.038)	
	$C_{12}E_7$	58 (0.046)	
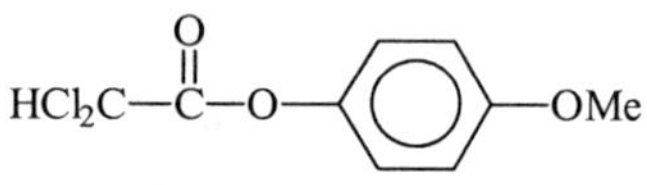	CTABr	16 (0.005)	283
	SDS	47 (0.015)	
	$C_{12}E_7$	6 (0.002)	
Substrate	**Surfactant**	**10^3 k'_M, s^{-1}**	**Ref.**
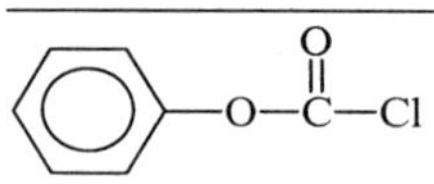	SDS	1.00 (0.07)	b
	CTACl	8.6 (0.64)	
	CTAOMs	8.3 (0.62)	
	CTABr	5.2 (0.39)	
	CTEABr	4.8 (0.36)	
	CTBABr	4.0 (0.30)	
	SB3-14	8.3 (0.62)	
	SBPr3-14	6.5 (0.49)	
	SB4-14	~7 (0.52)	
	SB5-14	<11 (0.82)	
	SB3-14 + $NaClO_4$	0.92 (0.07)	
	SBPr3-14 + $NaClO_4$	0.80 (0.06)	

[a]Values in parentheses are k'_M/k'_W.
[b]L. Brinchi et al., Eur. J. Org. Chem., in press, 2000.

salts and tether length in sulfobetaine surfactants, and the effect of specific salts have been carried out for hydrolysis of acyl derivatives and for spontaneous S_N2 hydrolysis of sulfonate ester [284] (L. Brinchi et al., Eur. J. Org. Chem., in press, 2000) (Table 7).

Cationic and sulfobetaine micelles have very similar effects on values of k'_M both for water-catalyzed deacylation of chloroformate (L. Brinchi et al., unpublished results) and for spontaneous S_N2 hydrolysis of methylnaphthalene-2-sulfonate [284] as well as for other spontaneous hydrolyses [281,285], indicating that their interfacial regions behave similarly as reaction media. An increase in the tether length of sulfobetaine surfactant seems to have little effect on inhibition of chloroformate hydrolysis. Anionic micelles of SDS strongly inhibit both hydrolyses in agreement with earlier evidence on other spontaneous hydrolyses [281].

Values of k'_M in cationic micelles decrease with increasing headgroup bulk for chloroformate hydrolysis as, to a lesser extent, in the S_N2 hydrolysis of methyl-

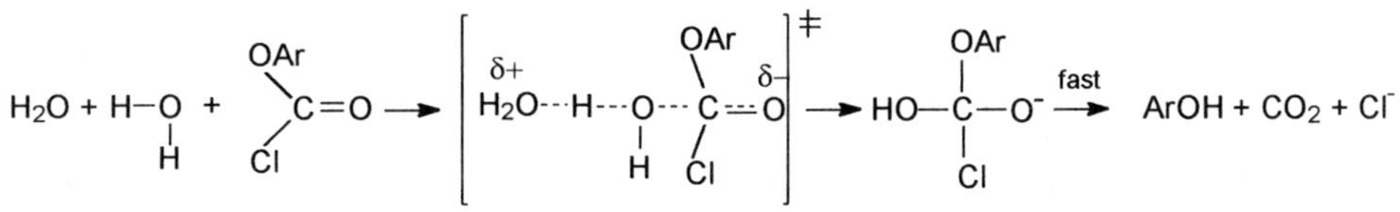

SCHEME 10

naphthalene-2-sulfonate. For chloroformate hydrolysis (Scheme 10), analysis could be made for cationic micelles with different counterions, and inhibition increases as the affinity of the counterion for CTA^+ increases; e.g., k_{obs} decreases in the sequence: $Cl^- > CH_3SO_3^- > Br^-$ for univalent ions. Hydrolysis in sulfobetaine micelles is inhibited by $NaClO_4$ (Table 7), which interacts strongly with micelles of SB3-14; solubilities of SB4-14 and SB5-14 are sharply increased by $NaClO_4$. Physical evidence (NMR spectroscopy and conductivity data) shows that ClO_4^- binds readily to micelles of SB3-14, which therefore become anionic [112], which by analogy with the behavior of SDS micelles should inhibit reaction. However, reaction in mixtures of sulfobetaines and $NaClO_4$ is, with high [$NaClO_4$], slightly slower than in high [SDS], where substrate is largely micelle bound. Hydrolysis of methylnaphthalene-2-sulfonate (Fig. 3) is slower in SB3-14 and $NaClO_4$ than in high [SDS], which shows that the rate decrease is due not only to the development of anionic character in micelles of SB3-14 but also to displacement of water from the interfacial region by ClO_4^-.

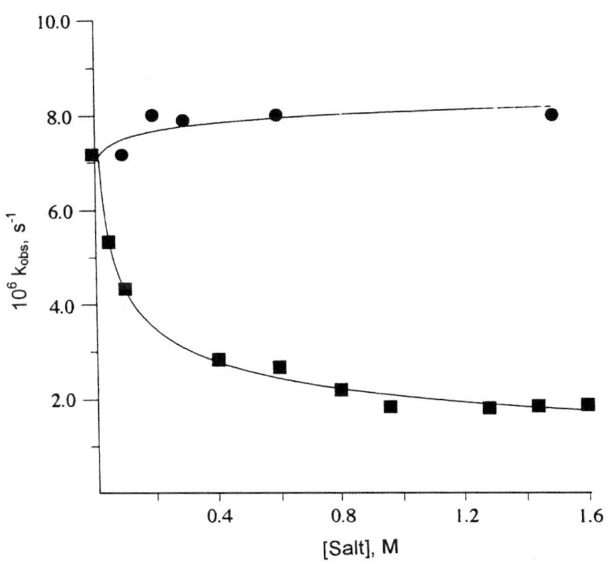

FIG. 3 Rate constants for the spontaneous hydrolysis of MeONs in 0.05 M SB3-14 with addition of $MeSO_3Na$ (●) or $NaClO_4$ (■).

Interfacial regions of betaine micelles are very open [114,117] and accessible to water, which will be displaced by ClO_4^- or other low-charged-density anions.

V. BIMOLECULAR REACTIONS

Micelles have effects on spontaneous reactions that can be related to the mechanism and to the properties of the micellar surface. Micellar inhibition of bimolecular reactions is also straightforward because micelles keep reactants apart. For micelle-assisted bimolecular nonsolvolytic reactions it is necessary to consider both medium and proximity effects. The term "micellar catalysis" is often used synonymously with "micellar rate enhancement," but a distinction is needed for bimolecular reactions in which changes in free energy of activation (effect on rate constant) should be separated from free energies of transfer from water to micelles (concentration effect).

Analysis of the variation of the overall rate constant of reaction with [surfactant] was discussed in Section III, and the treatment allows calculation of the second-order rate constants of reaction in the micellar pseudophase. These rate constants can be compared with second-order rate constants in water provided that both constants are expressed in the same dimensions, and typically the units are $M^{-1}\ s^{-1}$. The calculation of this rate constant, k_2^m [Eq. (10)], involves assumptions regarding the location of the substrate and the volume of reaction region at the micellar surface. In fits based on the PIE or mass action models, this volume has been approximated by the micellar molar volume, or that of the Stern layer, V_M. In the PBE model, reaction is assumed to take place in a shell whose thickness, Δ, has been approximated by the size of a cationic group. Inevitably, the comparison depends upon the assumed volume element of reaction; this may well differ from one type of micelle to another and probably also depends on the structure of the reactants and especially on the hydrophilicity of a reactive ion. Different investigators make different assumptions about values of this volume, but most estimates are within a factor of 2 [10,11]. In addition, some micelles grow with increasing concentration of surfactant and electrolyte, so these factors may influence the volume element of reaction. Fortunately, the uncertainties so introduced may be more apparent than real; for an approximately spherical micelle the volume of the Stern layer is approximately half the total volume of the micelle, and estimates of the volume element of reaction and of the second-order rate constants should therefore be within a factor of 2 provided that the reactant composition is uniform within the Stern layer. Estimates of the value of V_M, molar volume of reaction, vary from 0.14 to 0.3 M^{-1}, and a variety of values within this range is used [12]. Regarding the value of Δ, simulations of rate data for a variety of anionic nucleophiles have been made with Δ 2.4 Å [213,214]. Variations in this value for other kinds of reactions have been considered, and this variation affects values of k_2^m. For instance, values of 4 Å have also been used [233].

Most of our treatment will concern discussion of the source of micellar rate enhancement, as can be rationalized after a quantitative treatment to separate medium and concentration effects. Data for bimolecular reactions analyzed in terms of various models are given in Tables 1 to 3 and also Tables 8 to 12; values of the second-order rate constant in the micellar pseudophase and, when reported by the authors, values of k_2^m/k_W are given. For some reactions in Table 12, only overall rate effects have been reported, but in this section we will also consider some of them because they are of considerable chemical importance.

A. Relation to Mechanism

On the basis of the observation that most values of k_2^m/k_W are not very different from unity, especially in view of the approximations involved in their estimation, it is generally accepted that the major factor in micellar acceleration of bimolecular reactions of basic and nucleophilic anions is the increased ionic concentration in the interfacial region of cationic micelles. However, important relations of second-order rate constants to reaction mechanism and to charge dispersion are evident. It is interesting to mention the case of oxidation of sulfides by periodate ion. Despite extensive exclusion of the anion from an SDS micelle, the overall reaction is approximately twice as fast as in CTACl micelles, and the source of this rate enhancement is related to values of k_2^m/k_W being larger in SDS than in CTACl by two orders of magnitude [233]. This is rationalized in terms of an unfavorable interaction between the cationic headgroup of the cationic micelle and the developing positive charge on sulfur in transition state formation.

Regarding relations of values of k_2^m/k_W with the mechanism, it is useful to consider reactions of the same anion, for instance, hydroxide. Values of k_2^m/k_W are less than unity for carboxylic esters such as phenylbenzoates [286,287], *t*-butyl perbenzoate, and 2-naphthyl benzoate [288] and inorganic esters such as pNDPP [217]. Values of k_2^m/k_W are close to unity for S_N2 reaction with MeONs [289] but are greater than unity for aromatic nucleophilic substitution with DNCN [290] and E2 elimination from 2-phenethyl derivatives [291–293]. The variation in k_2^m/k_W is generally not large in view of differences in the parameters, especially V_M, used in the calculations, but they seem to be significant. The mechanisms of reactions of anionic nucleophiles with carboxyl derivatives and phosphate esters are similar in that negative charge tends to be localized on oxygen in the transition states, as shown for deacylation in Scheme 11.

The transition state here should be stabilized by hydrogen bonding to water molecules, which may be less available at the surface of an ionic micelle than in bulk water. The situation is different for aromatic nucleophilic substitution (where the transition state will be like a Meisenheimer complex with its negative charge delocalized over the nitroaromatic moiety) and for E2 elimination, where five centers are involved. These bulky, low-charge-density anions should interact favorably with cationic micellar headgroups and be stabilized by this interaction. We can also describe the differences between these reaction types in terms of Pearson's hard-soft description [294,295]. Cationic micellar headgroups interact best with soft bases, e.g., relatively large anions of low charge density such as bromide or arenesulfonate, or anionic transition states such as those for nucleophilic aromatic substitution or elim-

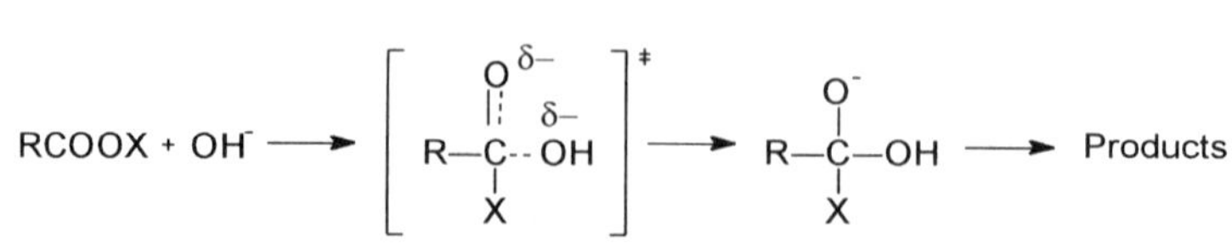

SCHEME 11

TABLE 8 Bimolecular Nucleophilic Substitutions

Substrate	Y	Surfactant	K_S, M^{-1}	K_X^Y	β	K'_Y, M^{-1}	K'_X, M^{-1}	$10^4\ k_M$, s^{-1}	k_2^m/k_W	Ref.
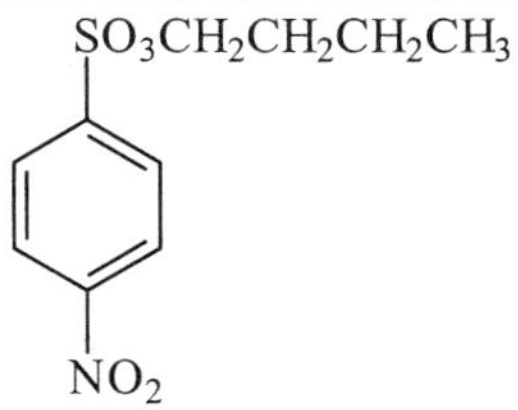	Br^-	CTABr	120	—	0.8			8	0.37	311
		CTABr	120			2000	—	8	0.37	
		CTAOMs	120	2.5				45	0.23	
		$CTA(SO_4)_{0.5}$	150	90				6	0.28	
		CTABr[a]	120					—	0.37	216
		CTAOMs[a]	120					—	0.23	
	Cl^-	CTACl	120	—	0.8			2.8	0.30	311
		CTACl	120	—		230	—	2.9	0.32	
		CTAOMs	115	2.0				1.6	0.18	
		$CTA(SO_4)_{0.5}$	150	65				1.75	0.20	
		CTACl[a]	120					—	0.3	216
		CTAOMs[a]	120					—	0.66	
	N_3^-	$CTAN_3$[a]	115					—	0.375	
		CTAOMs[a]	115					—	0.085	
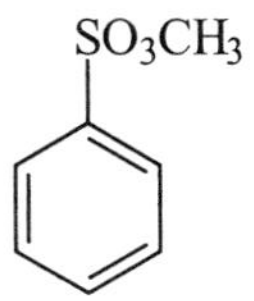	Br^-	CTABr	65	—	0.8			7.3	1.7	311
		CTABr	65			2200	—	6.1	1.4	
		CTAOMs	65	2.5				2.5	0.23	
		$CTA(SO_4)_{0.5}$	80	90				2.75	0.64	
		CTABr	65			2000	—	6.0	1.40	[f]
		CTEABr	65			1500	—	9.80	2.29	
		CTBABr	65			750	—	14.0	3.27	
	Cl^-	CTACl	50	—	0.8			1.6	0.8	311
		CTACl	70			220	—	1.5	0.8	
		CTAOMs	65	2				0.8	0.41	
		$CTA(SO_4)_{0.6}$	80	65				0.8	0.41	
	OH^-	CTAOH	65			55	—	18	0.24	293
		CTPAOH	65			25	—	14	0.19	
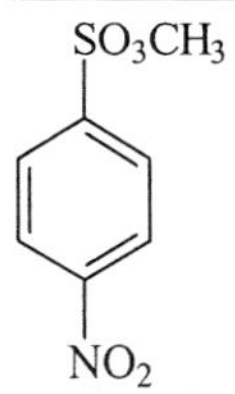	OH^-	CTAOH	58			55	—	140	0.50	293
		CTPAOH	58			25	—	165	0.58	
	Br^-	CTABr	50			2000	—	70.0	2.58	[f]
		CTEABr	58			1500	—	95.0	3.5	
		CTBABr	58			750	—	200	7.37	

TABLE 8 Continued

Substrate	Y	Surfactant	K_S, M^{-1}	K_X^Y	β	K'_Y, M^{-1}	K'_X, M^{-1}	10^4 k_M, s^{-1}	k_2^m/k_W	Ref.
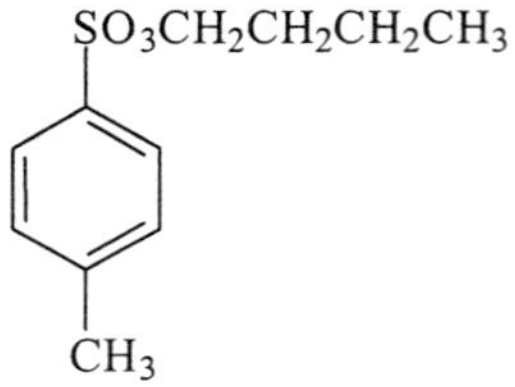	Br^-	CTABr	120	—	0.8			3.5	1.6	311
		CTABr	120			2200	—	3.4	1.6	
		CTAOMs	120	2.5				1.75	0.66	
		CTABr[a]	120					—	1.6	216
		CTAOMs[a]	120					—	0.66	
	Cl^-	CTACl	120	—	0.8			1.0	0.6	311
		CTACl	120			220	—	1.1	0.66	
		CTAOMs	120	2				0.6	0.37	
		CTACl[a]	120					—	0.6	216
		CTAOMs[a]	120					—	0.37	
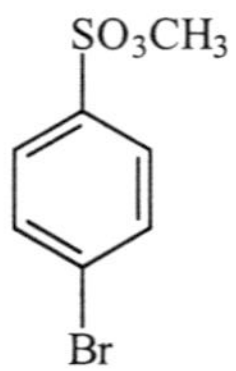	OH^-	CTAOH	130			55	—	50.0	0.44	[f]
		CTPAOH	130			25	—	48.0	0.43	
	Br^-	CTABr	130			2000	—	23.0	2.35	
		CTEABr	130			1500	—	34.5	3.53	
		CTBABr	130			750	—	60.0	6.13	
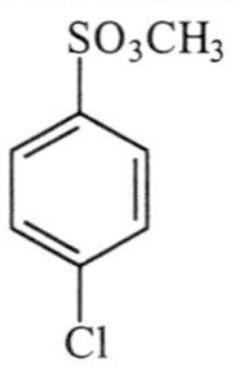	OH^-	CTAOH	120			55	—	47.0	0.44	[f]
		CTPAOH	120			25	—	44.0	0.42	
	Br^-	CTABr	120			2000	—	21.0	2.56	
		CTEABr	120			1500	—	21.0	2.56	
		CTBABr	120			750	—	54.0	6.75	
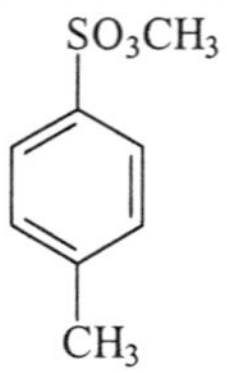	OH^-	CTAOH				55	—	12.0	0.26	[f]
		CTPAOH				25	—	8.20	0.17	
	Br^-	CTABr				2000	—	5.10	1.24	
		CTEABr				1500	—	7.60	1.85	
		CTBABr				750	—	10.0	2.43	
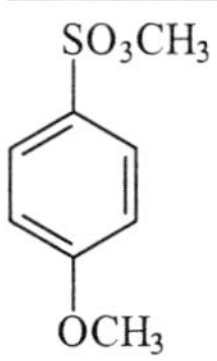	OH^-	CTAOH				55	—	7.10	0.18	[f]
		CTPAOH				25	—	4.75	0.12	
	Br^-	CTABr				2000	—	2.50	0.75	
		CTEABr				1500	—	3.90	1.16	
		CTBABr				750	—	4.80	1.43	

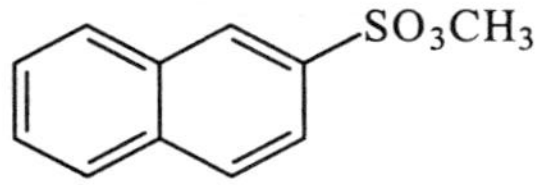	Br^-	CTABr	1000		475	—	1.34	1.76	296
		CTEABr	1000		350	—	2.00	2.63	
		CTPABr	1000		290	—	2.53	3.33	
		CTBABr	1000		155	—	3.51	4.62	
		CTABr	1000		2000	—	8.5	1.6	290
		CTEABr	1000		1500	—	11.0	2.1	
		CTPABr	1000		1100	—	15.0	2.9	
		CTBABr	1000		750	—	18.0	3.4	
		CTABr + BuOH		0.75–0.30		1.2[b]	9.1		204
		CTEABr + BuOH		0.56–0.18		1.4[b]	13.8		
		CTABr + BuOH ≤ 0.6 M	1500–530	0.75–0.46	400–53	1.4[b]	~9.2	~1.69	205
		CTABr + 0.8 M BuOH	510	0.32	23	1.4[b]	12	2.21	
		CTEABr + BuOH ≤ 0.6 M	1500–530	0.56–0.35	95–28	1.6[b]	16–15	~2.85	
		CTEABr + 0.8 M BuOH	510	0.23	13	1.6[b]	21	3.87	
		0.025 M CTABr + $C_{10}E_4$					10.0	1.84	127
		0.05 M CTABr + $C_{10}E_4$					11.5	2.12	
	$[Br^-]$ var.	[CTABr] + [$C_{10}E_4$] = 0.05 M					10.4	1.92	
	$[Br^-]$ const.	[CTABr] + [$C_{10}E_4$] = 0.05 M					13.0	2.39	
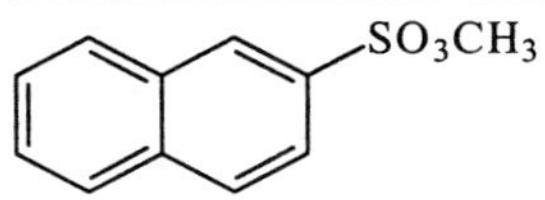	Br^-	SB3-13	1000		4.3	—	7.0		112
		SBEt3-14	1000		3.2	—	12		
		SBPr3-14	1000		2.6	—	18		
		SBBu3-14	1000		1.8	—	29		
		SB4-14	1000		4.3	—	9.5		
		SBPr4-14	1000		2.6	—	22		
		SB5-14	1000		4.3	—	10		
		SBPr5-14	1000		4.5	—	15		
		SB3-16[a]						1.82	315
		AOMe-14	500		1.8	—	12	2.1	317
		AOPr-14	500		1.5	—	18	3.1	
		AOMe-14 + H^+	500		20	—	12	2.1	
		AOPr-14 + H^+	500		15	—	18	3.1	
		MTABr	1000		475	—	9.5	1.7	
	Cl^-	CTACl	1000		80	—	0.23	1.53	296
		CTEACl	1000		70	—	0.39	2.60	
		CTPACl	1000		45	—	0.45	3.00	
		CTBACl	1000		45	—	0.48	3.20	
		SB3-16[a]						1.36	315

TABLE 8 Continued

Substrate	Y	Surfactant	K_S, M^{-1}	K_X^Y	β	K'_Y, M^{-1}	K'_X, M^{-1}	10^4 k_M, s^{-1}	k_2^m/k_W	Ref.
		OdTACl	1537			268	—	2.18	2.07	298
		CTACl	1093			74.2	—	1.82	1.73	
		MTACl	556			40.0	—	1.97	1.80	
		DTACl	310			42.0	—	1.67	1.53	
		DeTACl	154			22.9	—	1.61	1.53	
		OcTACl	0.65			0.66	—	0.64	1.0	
		OcTACl[c]	0.34			0.66	—	0.94	—	
		HeTACl[c]	0.005			0.0005	—	1.00	—	
		TMACl[c]	0.0044			0.0004	—	1.09	—	
		CTEACl	1000			70.5	—	2.50	2.92	297
		CTPACl	1000			50.0	—	2.51	2.92	
		CTBACl	1000			35.0	—	2.61	3.08	
		CTPeACl	213			18.8	—	9.01	10.5	
		CQCl	1000			165.0	—	3.20	3.75	
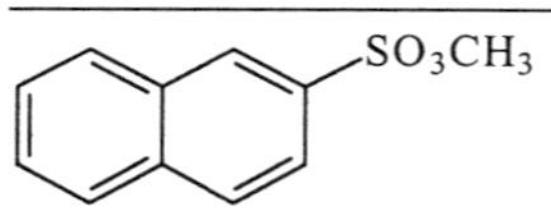	Cl^-	CDMHEACl	911			112.0	—	1.50	1.78	297
		CTHEACl	700			4300	—	0.91	1.08	
		SB3-16[a]					—		0.13	315
	OH^-	CTAOH	1090			55		19.5	0.34	289
		CTAOH + 0.05–0.5 M OH^-	1090			55		~22.7	~0.40	
		CTEAOH	1000			45	—	18.5	0.31	
		CTEAOH + 0.1–0.5 M OH^-	1000			45	—	~21.5	~0.37	
		CTPAOH	1000			25	—	15.0	0.25	
		CTPAOH + 0.1–1.0 M OH^-	1000			25	—	~22.3	~0.38	
		CTBAOH	1000			12	—	18.0	0.31	
		CTBAOH + 0.5–1.0 M OH^-	1000			12	—	~25.0	~0.42	
		CQOH	1000			50	—	23	0.40	
		CQOH + 0.5 M OH^-	1000			50	—	33	0.57	
	SO_3^{-2}	SB3-16[a]					—		0.34	315
	I^-	SB3-14				21		35		112
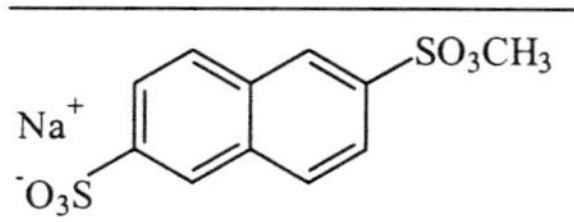	Br^-	CTABr	1000			2000	—	4.3	0.60	[g]
		CTPABr	1000			1100	—	14.9	2.1	
		CTBABr	1000			750	—	19.3	2.7	
	OH^-	CTAOH	1000			55	—	8.57	0.13	
		CTPAOH	1000			25	—	10.2	0.15	

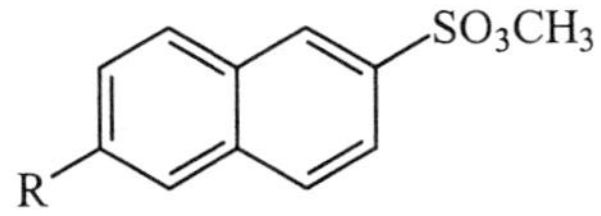

R = CH_3	Br^-	CTABr	2000	2000	—	7.90	1.4	h
		CTEABr	2000	1500	—	12.0	2.1	
		CTPABr	2000	1100	—	15.0	2.6	
		CTBABr	2000	750	—	18.0	3.2	
		CTAOMs	2000	2000	500	10.0	1.8	
		CTPAOMs	2000	1500	250	17.3	3.0	
	OH^-	CTAOH	2000	55	—	21.0	0.32	
		CTEAOH	2000	45	—	16.8	0.25	
		CTPAOH	2000	25	—	17.5	0.26	
		CTAOMs	2000	55	500	19.5	0.29	
		CTPAOMs	2000	25	250	14.5	0.22	
R = C_6H_{13}	Br^-	CTABr	3000	2000	—	9.80	1.7	h
		CTEABr	3000	1500	—	13.8	2.4	
		CTPABr	3000	1100	—	23.0	4.0	
		CTBABr	3000	750	—	27.5	4.8	
		CTAOMs	3000	2000	500	13.7	2.4	
		CTPAOMs	3000	1500	250	23.0	4.0	
	OH^-	CTAOMs	3000	55	500	24.5	0.37	
		CTPAOMs	3000	25	250	21.5	0.32	
R = $C_{12}H_{25}$	Br^-	CTABr	4000	2000	—	16.5	2.9	h
		CTEABr	4000	1500	—	20.5	3.6	
		CTPABr	4000	1100	—	29.0	5.1	
		CTBABr	4000	750	—	40.0	7.0	

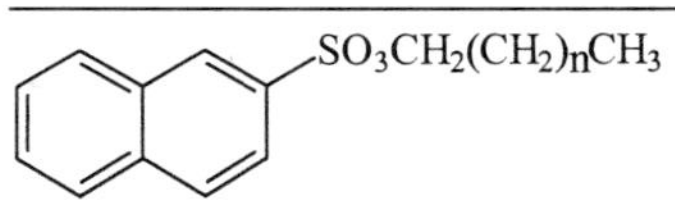

$n = 2$	Br^-	CTABr	2500	2000	—	0.670	1.8	h
		CTPABr	2500	1100	—	0.820	2.2	
		CTBABr	2500	750	—	0.910	2.5	
$n = 8$	Br^-	CTABr	3000	2000	—	1.15	3.1	h
		CTEABr	3000	1500	—	1.40	3.8	
		CTPABr	3000	1100	—	1.45	3.9	
		CTBABr	3000	750	—	1.56	4.2	

TABLE 8 Continued

Substrate	Y	Surfactant	K_S, M^{-1}	K_X^Y	β	K'_Y, M^{-1}	K'_X, M^{-1}	10^4 k_M, s^{-1}	k_2^m/k_W	Ref.
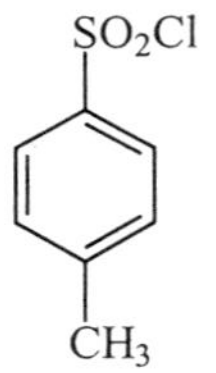	CH_3COO^-	CTAOAc						120	0.02	309
		CTACl						110	0.018	
	$C_5H_{11}COO^-$	CTAHexanoate						300	0.05	
		CTACl						270	0.045	
	$C_9H_{19}COO^-$	CTACl						400	0.093	
	$C_{13}H_{27}COO^-$	CTACl						600	—	
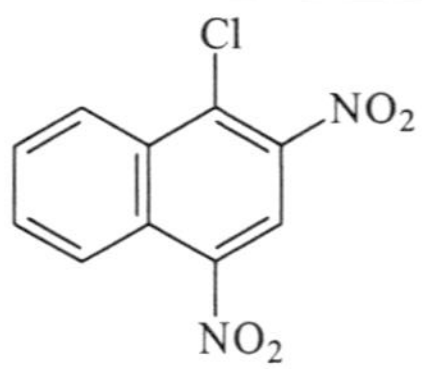	OH^-	CTABr	1600	14	0.80			~1000	2.7	290
		CTABr	1600			55	2000	1200	2.7	
		CTEABr	1600	14	0.75			~1200	2.8	
		CTEABr	1600			45	1500	~1275	2.8	
		CTPABr	1600	14	0.70			~1325	3.6	
		CTPABr	1600			25	1100	~1650	3.6	
		CTBABr	1600	14	0.60			~1975	4.7	
		CTBABr	1600			12	750	~2100	4.7	
		CCHDMABr	1600	14	0.75			~925		
		CCHDMABr	1600			50	1700	~1150		
		CMMBr	1600	14	0.75			~1250		
		CMMBr	1600			50	1700	~1200		
		MQBr	1600	14	0.75			~1000		
		MQBr	1600			50	1700	~1300		
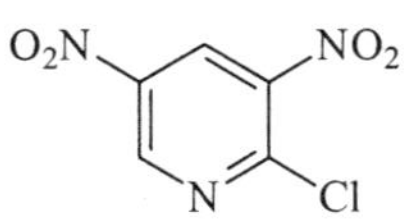	OH^-	CTACl[a]	600						0.5–4.2	216
	N_3^-	CTACl[a]	8.5						0.011	
		CTABr[a]	70						0.011	
		CTAN$_3$[a]	85						0.008	
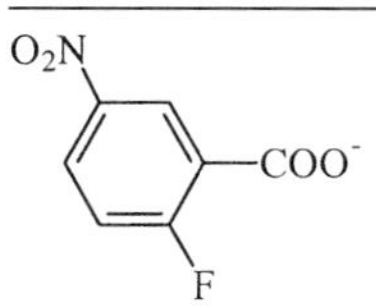	N_3^-	CTABr	34	1					1.1	301

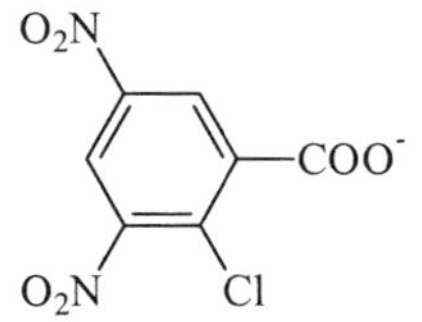	N_3^-	CTABr	3228	9	40	301
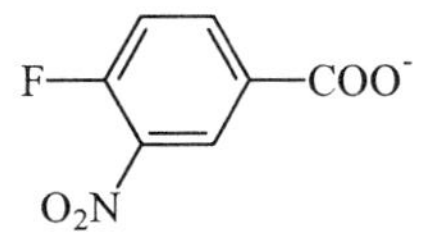	N_3^-	CTABr	196	2	0.5	301
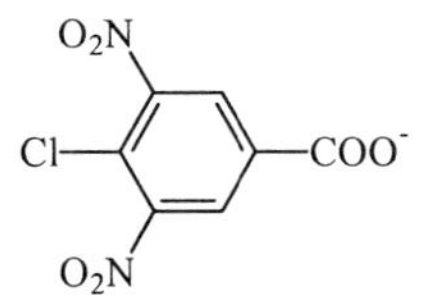	N_3^-	CTABr	1956	2	0.9	301
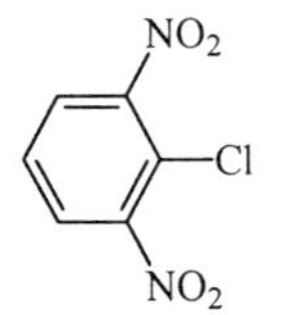	N_3^-	CTABr	240	1	10	301
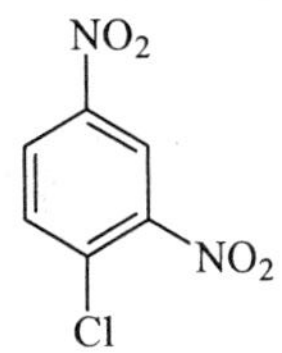	N_3^-	CTABr	67	2	52	301
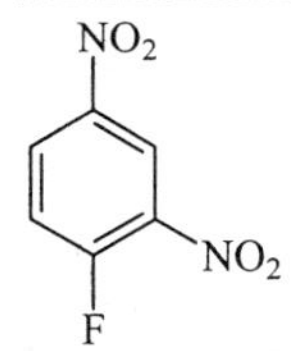	N_3^-	CTABr	17.5	2	48	301

TABLE 8 Continued

Substrate	Y	Surfactant	K_S, M^{-1}	K_X^Y	β	K'_Y, M^{-1}	K'_X, M^{-1}	10^4 k_M, s^{-1}	k_2^m/k_W	Ref.
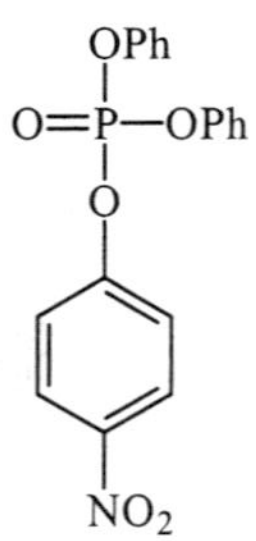	F^-	SB3-16							0.6	315
	OH^-	SB3-16[a]							1.25	
	IB^{-d}	SB3-16						59,000	0.83	
		CTAIB			~0.7			~60,000	0.84	
		CTAIB	10,000			1500		45,000	0.63	
	OIB^{-e}	SB3-16						59,000	0.83	
	HO_2^-	CTACl[a]	10,000						~0.18	211
		CTACl + KCl[a]	10,000						0.18	
		CTAOMs[a]	10,000						0.18	
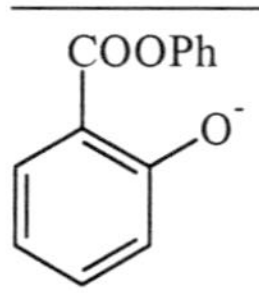	$C_4H_9NH_2$	CTABr	8500			~3		54.0	0.011	236
	Piperidine	CTABr	8400			~1		24.6	0.011	
	Pyrrolidine	CTABr	10,440			~0.7		64.7	0.011	

[a] PBE model.
[b] K_{BuOH}.
[c] Ion-exchange model in complexes of ion pairs and substrate.
[d] IB = *o*-iodose benzoate.
[e] OIB^- = 5-(octyloxy)-2-iodosobenzoate.
[f] L. Brinchi et al., Eur. J. Org. Chem., in press, 2000.
[g] L. Brinchi et al., unpublished results (or submitted).
[h] L. Brinchi et al., J. Colloids Interface Sci., in press, 2000.

TABLE 9 Bimolecular Eliminations by Hydroxide Ion

Substrate	Surfactant	K_S, M^{-1}	K_X^{OH}	β	K'_{OH}, M^{-1}	K'_X, M^{-1}	$10^4\ k_M$, s^{-1}	k_2^m/k_W	Ref.
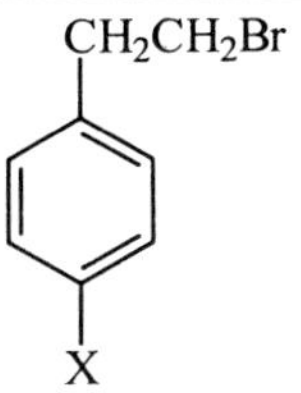									
X = Cl	CTABr	~420	27	0.76			~7.5	2.0	291
	CTAOH	230			30		~7.45	2.0	
X = H	$CTANO_3$	285	37	0.75			1.9		292
	CTABr	295	32	0.77			2.5		
	CTACl	305	7	0.75 + 0.5 [NaCl][a]			2.5		
	CTABr	~330	32	0.77	30		~2.5	1.1	291
	CTAOH	200					2.5	1.1	
	CTAOH	340			55		1.0	0.38	293
	CTPAOH	260			25		1.8	0.69	
	SB3-14	260			0.35		0.45	0.27	319
	SBBu3-13	260			0.25		1.63	0.63	
X = NO_2	$CTANO_3$	280	38	0.75			590		292
	CTABr	290	32	0.77			640		
	CTACl	310	9	0.72 + 0.5 [NaCl][a]			640		
	CTABr	~360	32	0.77			640	3.4	291
	CTAOH	250			30		640	3.4	
	CTAOH	420			55		400	2.3	293
	CTPAOH	420			25		1100	6.4	
	SB3-14	450			0.35		420	2.5	319
	SBBu3-14	500			0.25		1160	6.8	
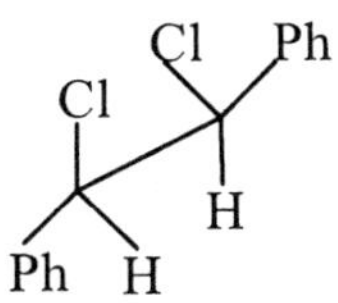	CTACl	3000			55	300	7.7	0.14	300
	CTAOH	3000			55	—	9.3	0.16	
	CTPAOH	3000			25	—	13.0	0.23	
	DDDACl	3000			55	300	14.0	0.25	
	DDDAOH	3000			55	—	10.0	0.17	
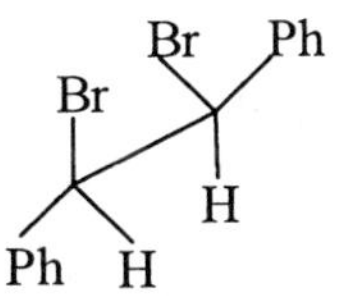	CTACl	3000			55	300	220	0.56	300
	CTAOH	3000			55	—	250	0.64	
	CTPAOH	3000			25	—	400	1.02	
	DDDACl	3000			55	300	435	1.11	
	DDDAOH	3000			55	—	270	0.69	

[a] β linearly increases with [NaCl].

ination. They interact less readily with hard bases, e.g., high-charge-density anions such as OH^-, or anionic transition states for deacylation. The relation between micellar kinetic effect and charge dispersion in the transition state seems to be a general phenomenon. In deacylation, electron-donating groups increase charge localization, whereas electron-withdrawing groups have the opposite effect, and there is a clear relation with values of k_2^m/k_W, which decrease by a factor of 8 as the substituent changes form CN to OMe [286]. Similar reasoning can be applied to the effect of the nitro group in elimination [291–293].

TABLE 10 Alkaline Hydrolyses

Substrate	Surfactant	K_S, M^{-1}	K_X^{OH}	β	K'_{OH}, M^{-1}	K'_X, M^{-1}	10^4 k_M, s^{-1}	k_2^m/k_W	Ref.
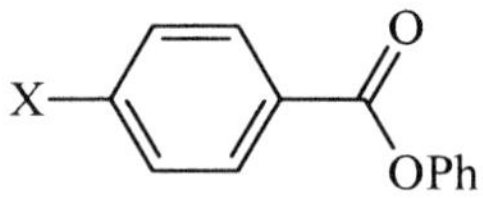									
X = H	CTABr	2000			55	2000	~8400	0.030	287
	CTEABr	2000			45	1500	~6400	0.023	
	CTPABr	2000			25	1100	5000	0.018	
	CTBABr	2000			12	750	~5900	0.021	
	MQBr	2000			50	1700	~7800	0.027	
	CTAOH	2000			55	—	1200	0.044	
X = OCH_3	CTABr	1500			55	2000	1400	0.018	286
	CTEABr	1500			45	1500	900	0.012	
	CTPABr	1500			25	1100	600	0.007	
	CTBABr	1500			12	750	500	0.006	
	MQBr	1500			50	1700	1100	0.014	
X = CH_3	CTABr	3000			55	2000	3000	0.021	286
	CTEABr	3000			45	1500	2200	0.016	
	CTPABr	3000			25	1100	1700	0.012	
	CTBABr	3000			12	750	1800	0.012	
	MQBr	3000			50	1700	2700	0.019	
X = Cl	CTABr	2000			55	2000	3900	0.057	286
	CTEABr	2000			45	1500	2800	0.041	
	CTPABr	2000			25	1100	2500	0.036	
	CTBABr	2000			12	750	3300	0.048	
	MQBr	2000			50	1700	3000	0.044	
X = CN	CTABr	2000			55	2000	55,000	0.12	286
	CTEABr	2000			45	1500	47,000	0.10	
	CTPABr	2000			25	1100	55,000	0.12	
	CTBABr	2000			12	750	72,000	0.16	
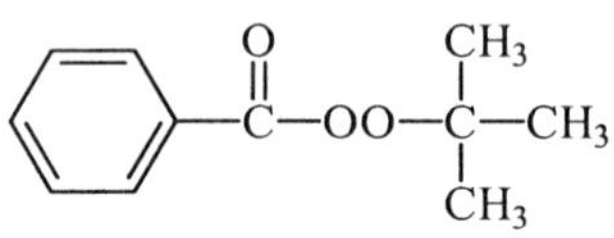	CTACl	~783	4	0.8			~967	0.042	288
	CTAOH	800			55		1100	0.042	
	CTACl[a]	~780						0.044	216
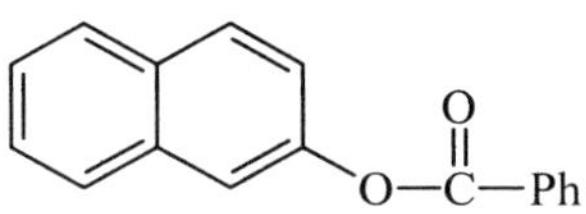	CTACl	2100	5	0.8			~1100	0.026	288
	CTAOH	2200			55		1100	0.026	
	CTACl[a]	2100						0.028	216
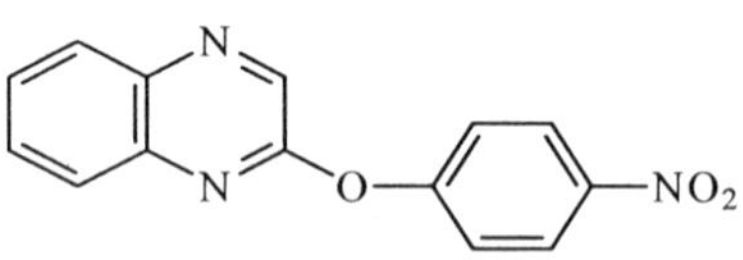	MTACl	6500	4	0.75			15.3	0.54	460
	MTABr	6500	8	0.75			12.7	0.64	
	CTACl	8000	4	0.8			18.1	0.62	
	CTABr	8000	8	0.8			17.0	0.60	
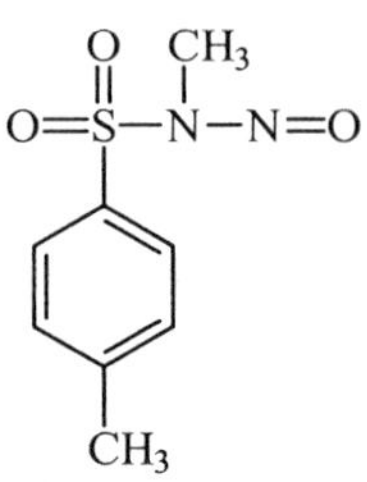	CTABr	184	23	0.7			1040	0.17	461

TABLE 10 Continued

Substrate	Surfactant	K_S, M^{-1}	K_X^{OH}	β	K'_{OH}, M^{-1}	K'_X, M^{-1}	$10^4\ k_M$, s^{-1}	k_2^m/k_W	Ref.
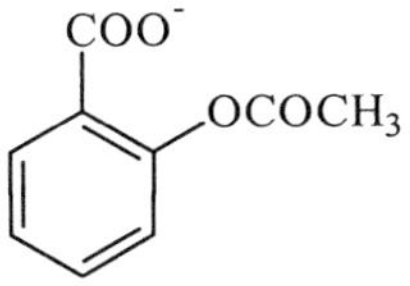	CTABr[a]							0.121	225
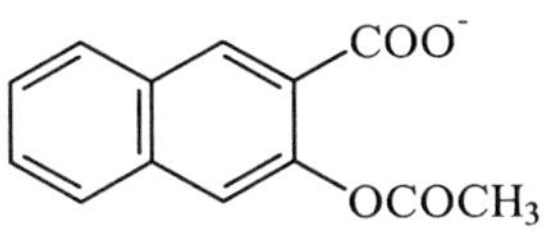	CTABr[a]							0.097	225

[a]PBE model.

B. Ion Specificity

The same kind of reasoning can explain the different behavior of different nucleophiles toward the same substrate. Values of k_2^m/k_W are less than unity for S_N2 reaction of MeONs with OH^- but are bigger for reaction of softer anions, such as chloride and bromide (Table 8) [289,290,296,297]. For S_N2 substitution with butyl 4-nitrobenzenesulfonate [216], values of k_2^m/k_W depend upon the hydrophilicity or polarizability of the nucleophile, i.e., follow the sequence $N_3^- > Br^- > Cl^-$ (Table 8). The intrinsic nucleophilicity of soft anions is different in water and in micelles, being higher in cationic micelles. This effect is related to partial disruption of soft ion hydration shells, as consistent with increases in NMR line width for bromide and chloride ions in cationic surfactants [86,297,298]. This effect on intrinsic reactivity of soft anions will be considered again later. Still specific interactions between the highly polarizable borohydride anion and cationic micelles have been invoked to rationalize the high inhibition in reductions of ketones observed by Cerichelli et al. (Table 11) [299]. Here, specific interactions with the reagent stabilize it, while the transition state with a partially negative oxygen atom is likely to be stabilized more effectively by water than by the micellar surface. Specific interactions between cationic headgroups and soft anions nicely explain the different electrophilic assistance to the leaving anion in intramolecular cyclization [261] (Table 6) of *o*-(3-halopropyloxy)phenoxide ion (halogen = I, Br), with values of $k_{W(I)}/k_{W(Br)}$ being 0.60, and values of $k_{M(I)}/k_{M(Br)}$ being 1.3 in cationic micelles of CTABr, and also in elimination from 1,2-dihalo-1,2-diphenylethanes (halogen = Br, Cl), with values of $k_{W(Br)}/k_{W(Cl)}$ approximately 7 in the absence of surfactant and values of $k_{M(Br)}/k_{M(Cl)}$ about 27–31 in cationic surfactants (Table 9) [300].

The behavior of azide ion in nucleophilic aromatic substitution is quite peculiar and leads to values of k_2^m/k_W much bigger than 1 [12]. Broxton et al. found values of k_2^m/k_W up to ~40–50 for the substrates in Scheme 12 [301]. Actually, these high values are observed only for some of the numerous substrates studied, with catalysis being smaller for other substrates (reported in Table 8) with fluoronitro compounds and for substrates that, on the basis of NMR studies, are more buried inside the micelle.

It is difficult to explain these results, although there seems to be a relation between the anomalous behavior of the azide ion in micellar reactions of aromatic substrates and its nucleophilicity in water and similar polar, hydroxylic solvents. Azide is a very powerful nucleophile toward carbocations, based on Ritchie and Sawada's N^+ scale, but in water it is much less reactive toward 2,4-dinitrohalobenzenes than predicted, whereas the reactivity of other nucleophiles fits the N^+ scale [302]. Therefore, the large values of k_2^m/k_W may reflect the fact that azide ion is unusually unreactive in

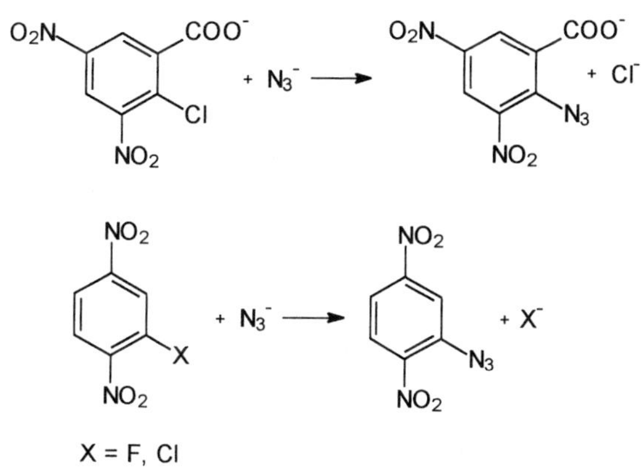

SCHEME 12

TABLE 11 Other Bimolecular Reactions

Reaction	Surfactant	K_S, M^{-1}	K_X^Y	β	$10^4\ k_M$, s^{-1}	k_2^m/k_W	Ref.
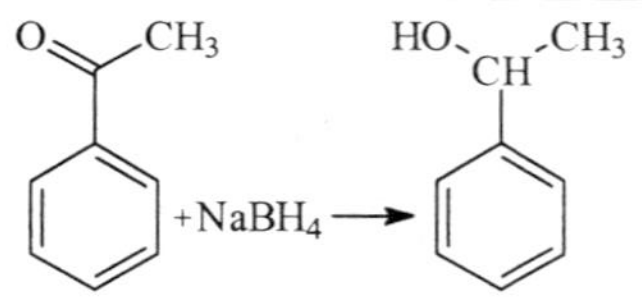	CTABr CTACl	31 27	2 4		44.3 60.0	0.036 0.048	299
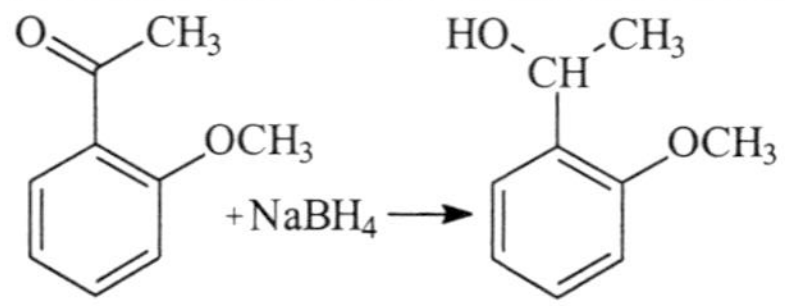	CTABr	52	2		31.0	0.029	299
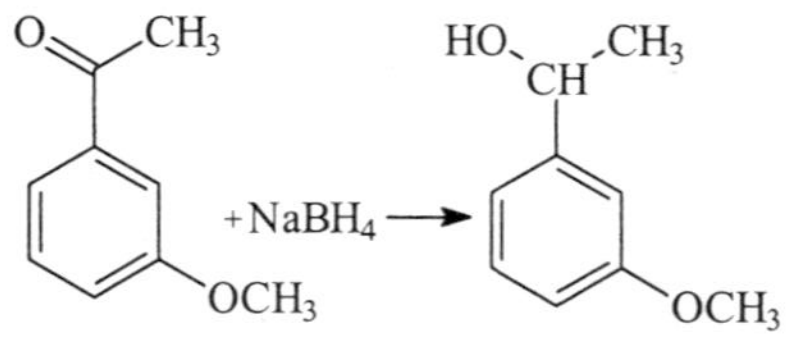	CTABr CTACl	85 60	2 4		57.5 75.0	0.028 0.036	299
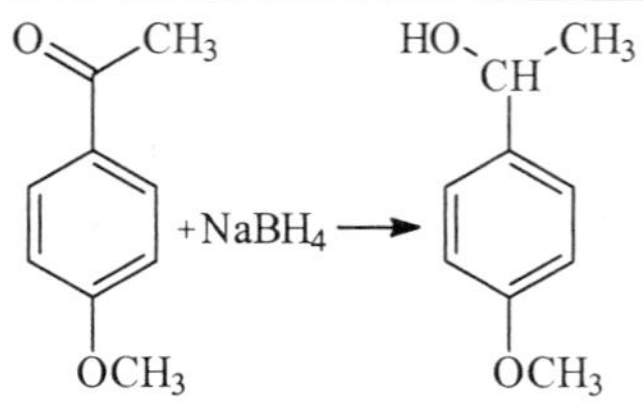	CTABr	70	2		3.80	0.018	299
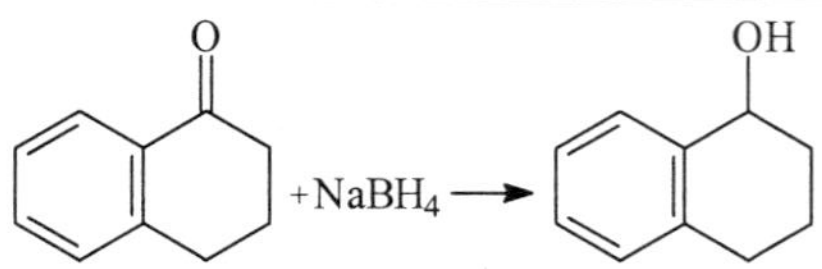	CTABr	100	2		39.0	0.022	299
$(CH_3CH_2CH_2)_2S + IO_4^- \rightarrow (CH_3CH_2CH_2)_2S{=}O$	CTACl[a]	140				0.004	233

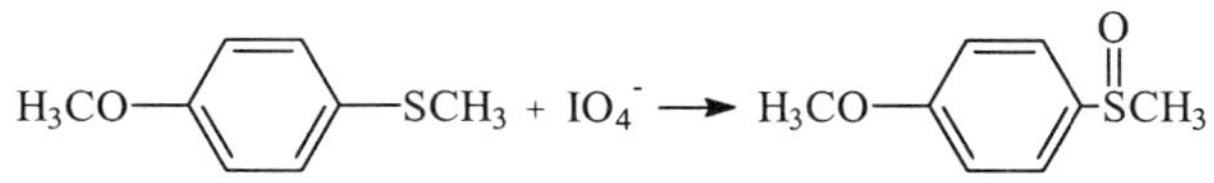	CTACl[a]	340		0.0008	233
$RCH{=}CH_2 + Br_2 \rightarrow$ Products					
$R = C_4H_9$	CTABr + NaBr ≤ 0.01 M			$\sim 3.6 \times 10^{-6}$	313
	CTABr + 0.1 M NaBr			5.9×10^{-7}	
$R = C_5H_{11}$	CTABr + NaBr ≤ 0.01 M			$\sim 2.2 \times 10^{-6}$	
	CTABr + 0.1 M NaBr			6.25×10^{-7}	
$R = C_6H_{13}$	CTABr + NaBr ≤ 0.01 M			$\sim 2.0 \times 10^{-6}$	
	CTABr + 0.1 M NaBr			5.9×10^{-7}	
$R = C_8H_{17}$	CTABr			1.1×10^{-6}	
	CTABr + 0.01–0.1 M NaBr			$\sim 4.4 \times 10^{-7}$	
$R = C_{10}H_{21}$	CTABr			1.9×10^{-6}	
	CTABr + 10^{-2} M NaBr			1.2×10^{-6}	
	CTABr + 0.1 M NaBr			5.7×10^{-7}	
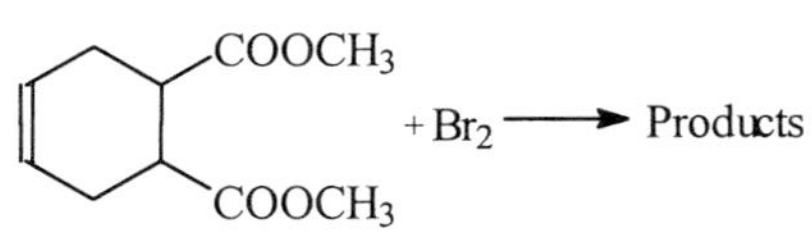	CTABr			3.9×10^{-6}	313
	CTABr + 10^{-2} M NaBr			2.9×10^{-6}	
	CTABr + 0.1 M NaBr			1.9×10^{-6}	
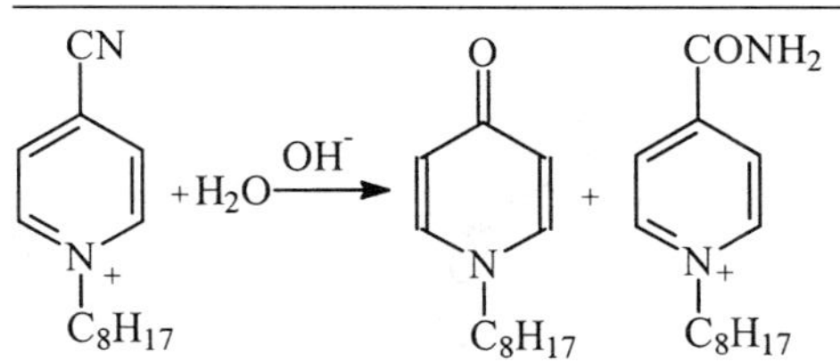	CTABr	0.8	0.08	0.071	307
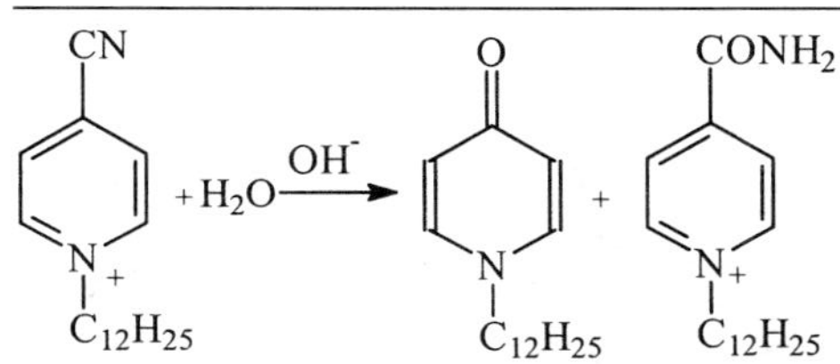	CTABr	35		0.071	307

[a] PBE model.

TABLE 12 Other Bimolecular Reactions with Other Treatments

Reaction	Surfactant	Remarks	Ref.
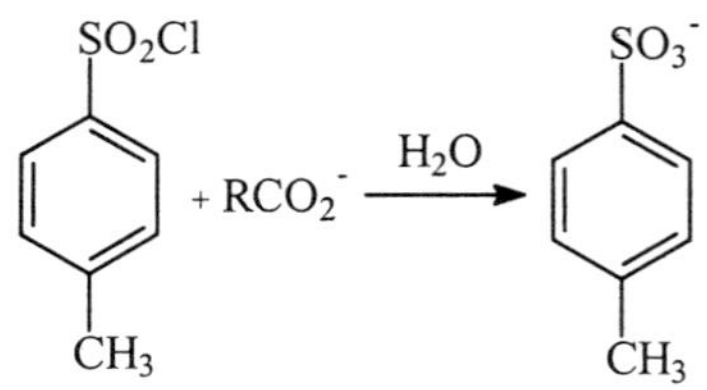 $R = H, CH_3, C_5H_{11}, C_9H_{19}, C_{13}H_{27}$	SDS	Anionic micelles of SDS inhibit the reaction and with hexanoate, decanoate, and tetradecanoate the observed first-order rate constants go through a minimum.	309
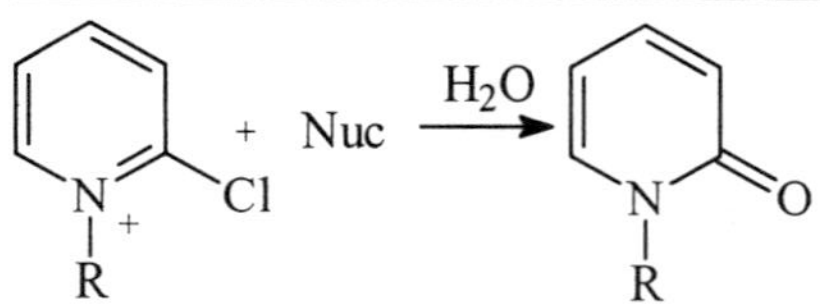 $R = CH_3, C_{10}H_{21}, C_{14}H_{29}$ $Nuc = CH_3NH_2, (CH_3)_2NH, (CH_3)_3N$	CTACl SDS	CTACl and SDS micelles catalyze the reactions with methyl- and dimethylamine. CTACl inhibits the reaction of methyl pyridinium with trimethylamine (this acts as a general base catalyst rather than as a nucleophile) and the reactivities of the other two substrates with methyl- and trimethylamine go through maxima. SDS inhibits the reactions of the three substrates with trimethylamine, with the inhibition increasing with increasing substrate hydrophobicity.	305
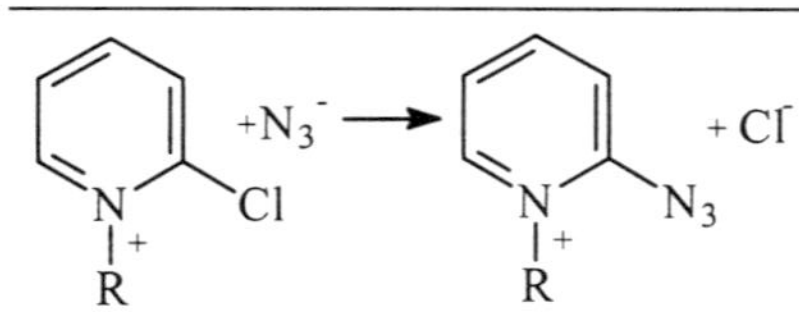 $R = CH_3, C_6H_{13}, C_{10}H_{21}, C_{12}H_{25}, C_{14}H_{29}, C_{16}H_{33}$	CTABr CTACl $CTAN_3$	Reactivity of the methyl derivative goes through a minimum with increasing [CTABr] due to weak substrate binding, whereas reactivity of the hexyl derivative increases as [CTABr] or [CTACl] increases due to cooperative binding of reactants. The most hydrophobic substrates' reactivities are strongly enhanced in CTABr and CTACl micelles (k_{rel} up to 1.5×10^4): the rate-surfactant profiles go through a maximum and rate enhancement was unexpectedly larger in CTABr than in CTACl. The profiles can't be fitted by an ion-exchange model, because the substrates can self-micellizate or induce micellization at low [surfactant], and micellar binding is cooperative. The minimum value of k_2^m/k_w for the hexadecyl derivative was estimated to be 440.	218
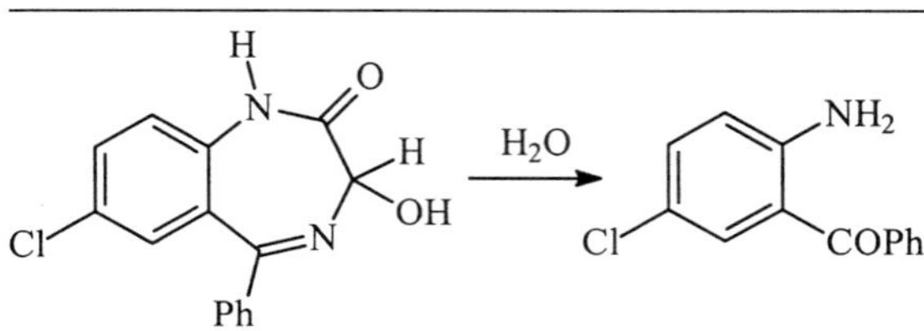	SDS CTABr	In the presence of SDS micelles the rate of the acid-catalyzed hydrolysis is faster than in water for all HCl concentrations and the pH-rate profile shows no plateau, indicating that there is a higher percentage of initial amide cleavage in SDS than in water. In the presence of CTABr micelles the rate of base-catalyzed hydrolysis is much slower than in water, probably because the NH group ($pK_a = 12.4$) is more ionized.	462

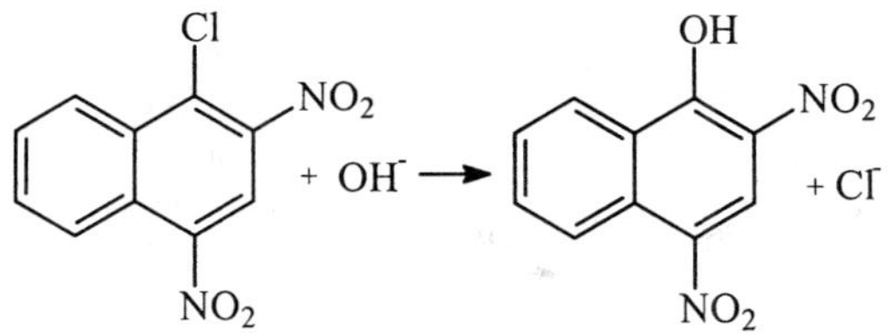	CB1-14 CB1-16	Zwitterionic micelles of CB1-14 and CB1-16 catalyze the hydroxy dehalogenation reaction, thanks to weak binding of hydroxide ions to the dipolar micelles. Indeed, catalysis is less effective in zwitterionic than in cationic micelles and increases linearly with increasing $[OH^-]$.	463
$[Fe(CN)_5L]^{-3} + L' \rightarrow [Fe(CN)_5L']^{-3} + L$ L, L′ = pyridine derivatives	Triton X-100 SDS CTABr	With uncharged L, the reaction is catalyzed, at about the surfactant cmc, especially when L is hydrophobic and the surfactant is neutral or anionic: $k_{rel} \cong 8$ in Triton X-100 with L = 4(1-butyl-pentyl)pyridine and L′ = 1,4 pyrazine. With cationic L the reaction rate is unaffected by Triton X-100 but is modestly speeded by SDS due to electrostatic and hydrophobic interactions. In reverse micelles of CTABr in *n*-hexanol, the most hydrophobic Ls are the best leaving groups. With L = 4(1-butyl-pentyl)pyridine, as the water content increases, k_{obs} tends to the value in water.	464
$[Fe(CN)_5(4\text{-CNPy})]^{-3} + CN^- \rightleftharpoons Fe(CN)_6^{-4} + 4\text{-CNPy}$ 4-CNPy = 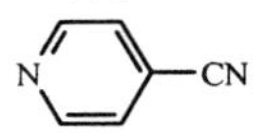	CTACl	The reaction follows a dissociative mechanism. Both pseudophase and Brønsted equation models fit kinetic data and calculate the same value of k_2^{rel} in water = 0.110 [rate constant for the reaction of CN^- with $Fe(CN)_5^{-3}$, relative to the rate constant of the reaction of 4-CNPy with the iron moiety].	254
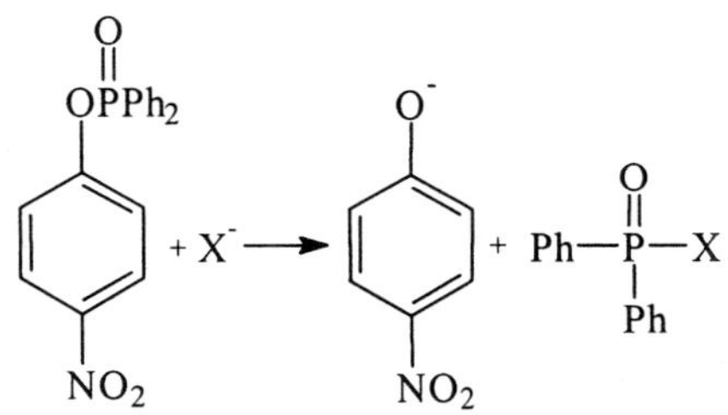 X = F, OH	$C_{12}E_{10}$ $C_{12}E_{23}$	In nonionic micelles of $C_{12}E_{10}$ and $C_{12}E_{23}$ the reaction is inhibited with both nucleophiles, due to incorporation of the substrate in the palisade layer, where ionic concentrations are lower than in water because of the large volume of the palisade layer and of the partial penetration of anions. OH^- is more reactive than F^- and it seems that the micellar environment has no effect on the second-order rate constants. Added hydrophobic anions, as *β*-naphthalene sulfonate, perchlorate, and *p*-toluene sulfonate, compete with the nucleophiles for the occupation of the micellar psuedophase and increase inhibition: k_{rel} = 0.30–0.60. Added $(C_7H_{15})_4N^+$ makes the micelles positively charged and speeds the reaction with F^- ($k_{rel} \cong 2$).	303

TABLE 12 Continued

Reaction	Surfactant	Remarks	Ref.
$[Fe(CN)_5H_2O]^{-3} + [\beta\text{-}Co(trien)(pzCO_2)]^{+2} \rightarrow [\beta\text{-}(trien)Co(\mu\text{-}pzCO_2)Fe(CN)_5]^{+2} + H_2O$ trien = triethylentetraammine $pzCO_2$ = pyrazinecarboxylate	SDS	In SDS micelles the reaction rate goes through a minimum in 0.03 M SDS due to complete Co(III) complex incorporation into micelles. A subsequent increase in [SDS] results in a decrease in the interfacial electrical potential, $\Delta\Psi$. It will favor the approaching process between the two oppositely charged reactants. This means that the electrostatic contribution to the activation free energy is the main factor that influences reactivity. Indeed, added alcohol and electrolytes decrease $\Delta\Psi$ and increase k_{obs} in the order $Li^+ < Na^+ < Cs^+$ and $Et_4N^+ < Pr_4N^+ < Bu_4N^+$.	465
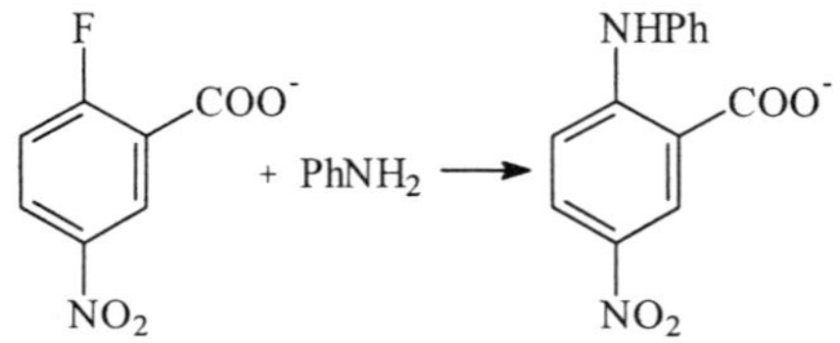	CTABr	The reaction is catalyzed in CTABr micelles, where both reaction centers are located at the interface: $k_{obs}^{min}/k_w = 2.33$ at 73°C.	314
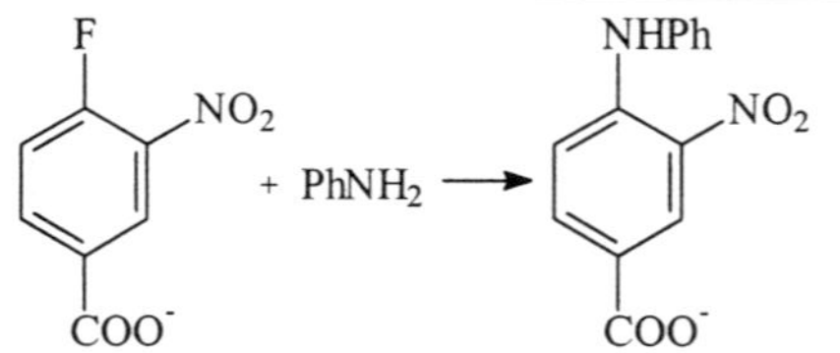	CTABr	The reaction is inhibited in CTABr micelles due to different locations of the two reactions centers: $k_{obs}^{min}/k_w = 0.30$ at 73°C.	314
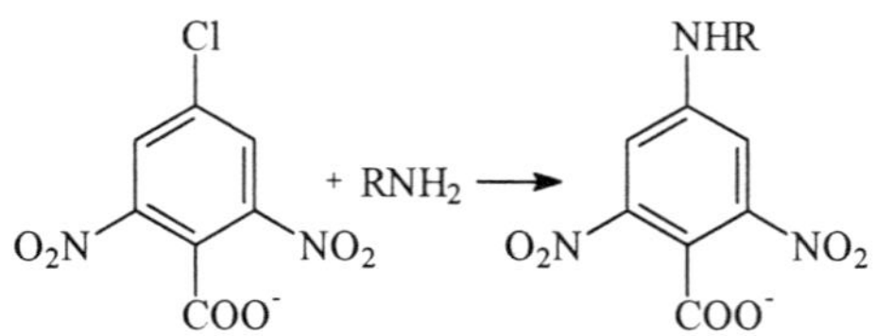 R = Ph, n-Pr	CTABr	The reaction with $PhNH_2$ is inhibited in CTABr micelles due to different locations of the two reaction centers: $k_{obs}^{min}/k_w = 0.14$ at 31°C. The reaction rate is instead unaffected in CTABr with n-$PrNH_2$: $k_{obs}^{min}/k_w = 1.06$ at 31°C.	314
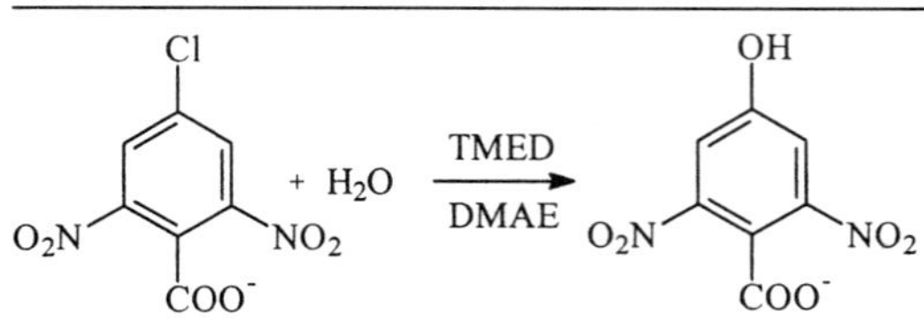 TMED = 1,1,2,2,-Tetramehylethylenediamine DMAE = Dimethylaminoethanol	CTABr	The two bulky tertiary amines act as specific-base catalysts and activate water as nucleophile. In CTABr the reaction is faster: $k_{obs}^{min}/k_w = 13.3$ at 73°C and 13.8 at 57°C for TMED and DMAE, respectively. As both substrate and amines are located in the micellar interior, water can penetrate inside the aggregates too.	314

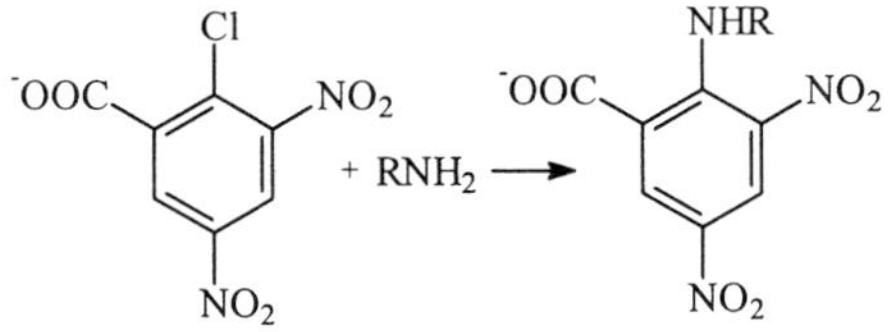 R = Ph, n-Pr	CTABr	Reactions with $PhNH_2$ and n-$PrNH_2$ are catalyzed in CTABr micelles because both reaction centers are located in the micellar interface: k_{obs}^{min}/k_w = 18.4 and 14.0 at 31°C for $PhNH_2$ and n-$PrNH_2$, respectively.	314
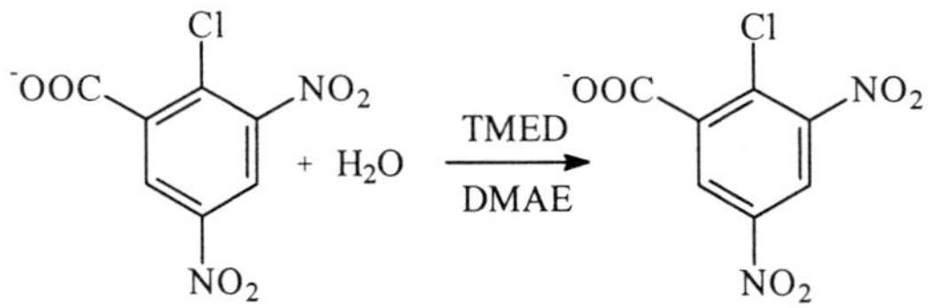 TMED = 1,1,2,2,-Tetramehylethylenediamine DMAE = Dimethylaminoethanol	CTABr	The two bulky tertiary amines act as specific-base catalysts and activate water as nucleophile. In CTABr the reaction is faster: k_{obs}^{max}/k_w = 6.4 at 73°C and 8.4 at 57°C for TMED and DMAE, respectively.	314
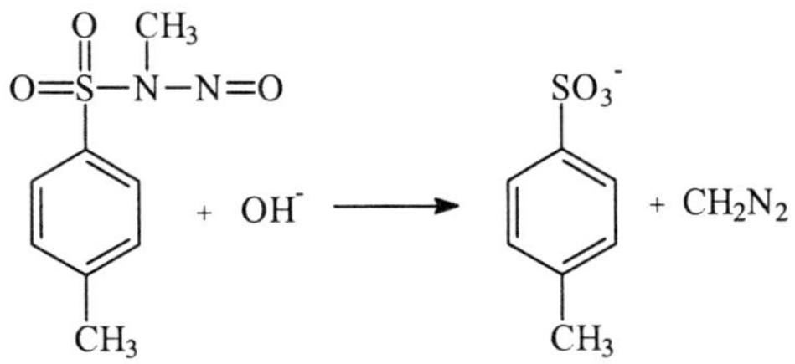	SDS CTABr MTABr $C_{16}E_{20}$	Alkaline hydrolysis is inhibited in SDS micelles due to exclusion of OH^- from the negatively charged micelle. The reaction in CTABr micelles was quantitatively treated by the psuedophase ion exchange model (see Table 10). In the presence of MTABr, k_{obs} values go through a maximum ($k_{rel} \cong 3$), in agreement with the model. In the nonionic surfactant $C_{16}E_{20}$ inhibition occurs, due to effective substrate association (K_s = 243 $cm^3\ g^{-1}$) and to the absence of significant amounts of OH^- in the micellar environment.	461
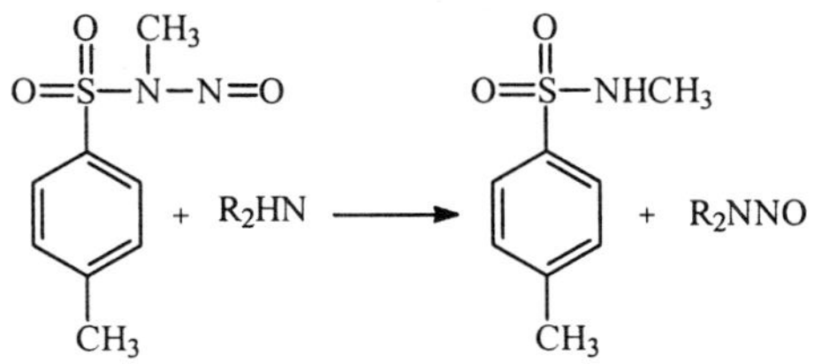 $R_2HN = (CH_2)_2NH$, CH_3NHCOO^- DMA SAR	SDS CTABr MTABr $C_{16}E_{20}$	In the presence of SDS the transnitrosation reaction with sarcosine (SAR) is inhibited by substrate micellar association (K_s = 75 M^{-1}) and anion repulsion from the negatively charged micellar surface. The same reaction is catalyzed by MTABr, where reaction rate goes through a maximum (k_{rel} = 1.7). The reaction with DMA is inhibited in SDS, CTABr, and $C_{16}E_{20}$.	461
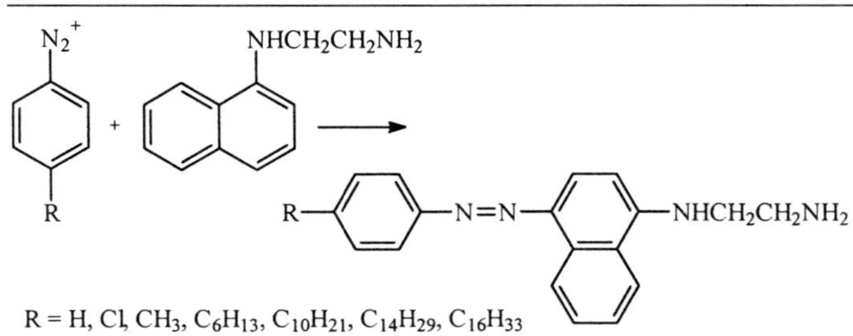 R = H, Cl, CH_3, C_6H_{13}, $C_{10}H_{21}$, $C_{14}H_{29}$, $C_{16}H_{33}$	CTACl SDS	The diazocoupling reaction is catalyzed by both CTACl and SDS micelles. Values of k_{rel} (= $k_{max}/k_{/PhN_2^+ in\ H_2O}$): R = Cl: 71.2 in SDS and 39 in CTACl; R = CH_3: 21 in SDS and 74 in CTACl; R = C_6H_{13}: 109 in SDS and 8.9 in CTACl; R = $C_{10}H_{21}$: 89.6 in SDS and 65.2 in CTACl; R = $C_{14}H_{29}$: 55 in SDS and 70 in CTACl; R = $C_{16}H_{33}$: 37.5 in SDS and 96.6 in CTACl.	308

TABLE 12 Continued

Reaction	Surfactant	Remarks	Ref.
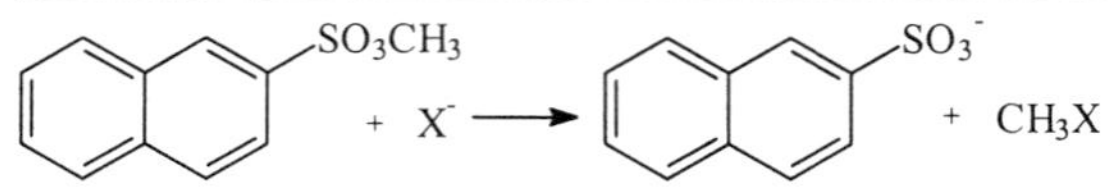 $X^- = OH^-, Br^-, SCN^-, SO_3^{-2}$	SDS	The reaction with Br^- is effectively suppressed by SDS, whereas the reactions with OH^-, SCN^-, and SO_3^{-2} are strongly but not completely inhibited by SDS. The Poisson-Boltzmann equation model was not applied because the reaction with water makes a significant contribution. Added inert salts such as NaCl and Me_4NCl speed micellar and nonmicellar reactions, the ammonium salt being more effective: in 0.18 M SDS, $10^5\ k_{obs}$ = 3.04 and 1.46 without salt, 4.55 and 3.85 in 0.4 M Me_4NCl for 0.1 M SO_3^{-2} and for 0.3 M SCN^-, respectively.	466
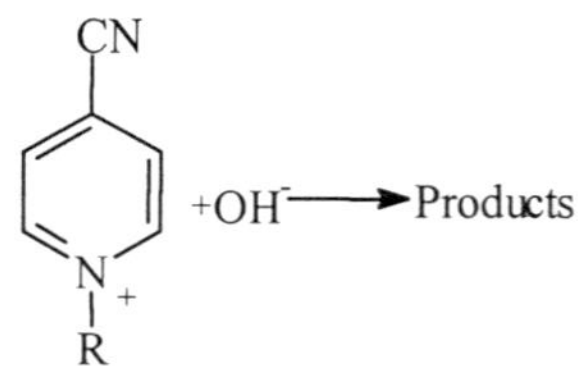 $R = C_{10}H_{21}$ (DeCP), $C_{12}H_{25}$ (DoCP)	SB3-16	Both substrates in alkaline hydrolyses are catalyzed by SB3-16 and kinetics are treated by means of an enzymatic model. For DeCP: $K_s = 120\ M^{-1}$, $k_M = 0.139\ s^{-1}$ at pH = 11.63; for DoCP: K_s = 1640 M^{-1}, $k_M = 0.125\ s^{-1}$ at pH = 11.63. Catalysis is reduced by adding Br^- or Cl^-, Br^- being more effective than Cl^-. The dissociation constants of the inhibitor-micelle-like complex, K_I, were calculated using a competitive inhibition model: K_I^{Cl}/K_I^{Br} = 4.5 for DoCP.	318
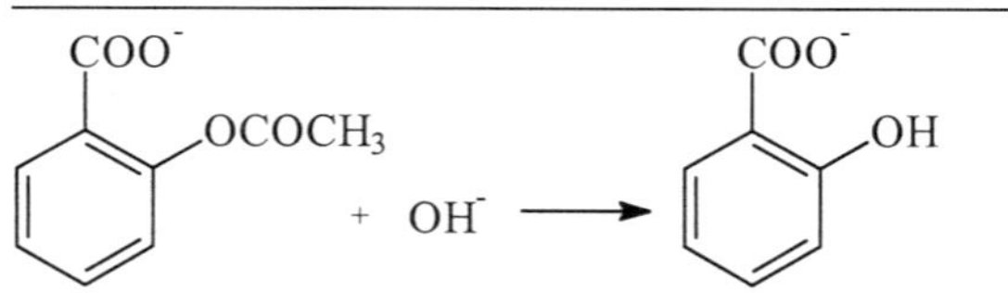	SDS	Alkaline hydrolysis of acetyl salycilate is speeded in anionic micelles of SDS, and the rate-surfactant profiles follow the empirical relationship: $k_{obs} = C + F$ [SDS]. Up to $[SDS]_T$ = 0.16 M, the observed data may be explained in terms of a dynamic pseudophase model, where $1 \gg K_{OH}[Dn]$ and $1 \gg K_s[Dn]$ (K_s = 0.13 M^{-1}). The fitting parameters show that the micelle-mediated reaction probably occurs in the interfacial region of the Stern and Gouy-Chapman layers, where counterions (nearly 30% of the total Na^+ ions) are loosely associated with SDS micelles.	235
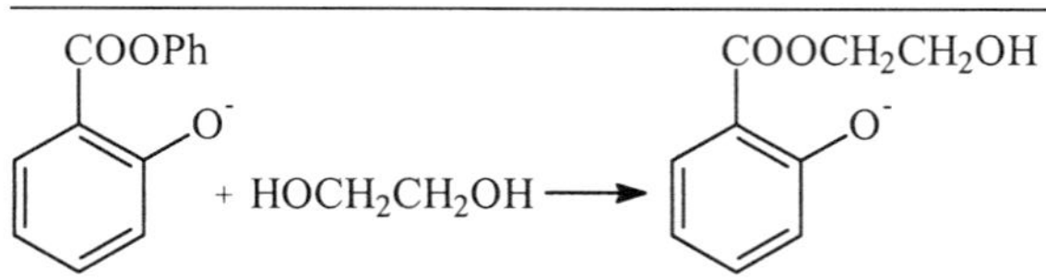	CTABr	At constant temperature and $[HOCH_2CH_2OH]_T$, k_{obs} values decrease as [CTABr] increases, and rate-surfactant profiles were quantitatively treated using a pseudophase model: K_s and micellar pseudo-first-order rate constant values (k_M^{ROH}) increase as $[HOCH_2CH_2OH]_T$ increases. Values of k_{NM}^{ROH}/k_M^{ROH} are >1 (k_{NM}^{ROH} = nonmicellar pseudo-first-order rate constant) due to loss of intramolecular general base catalysis and of $HOCH_2CH_2OH$ content in the micellar pseudophase. At 30°C, $10^4 k_M^{ROH}$ = 7.95, 12.6, and 20.6 s^{-1}; K_s = 6960, 2480, and 1740 M^{-1}; k_{NM}^{ROH}/k_M^{ROH} = 6.1, 5.6, and 3.9 in 20%, 30%, and 40% (v/v), respectively.	467

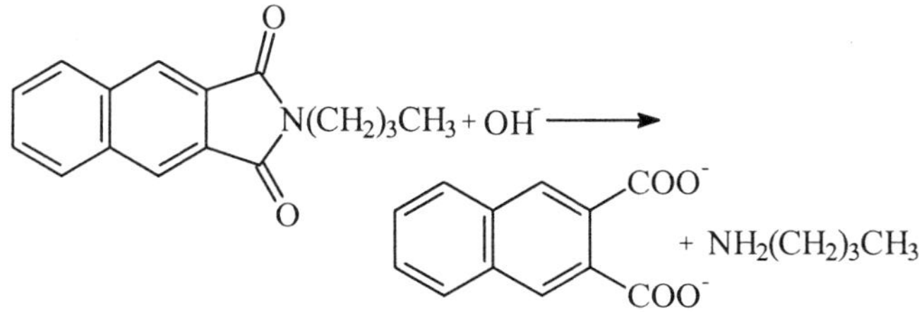	SB3-16	The alkaline hydrolysis is inhibited in SB3-16 aqueous micellar solution and kinetics are treated by means of an enzymatic model. At pH = 11.10, $K_s = 6.63 \times 10^3\ M^{-1}$, $k_M = 4.8 \times 10^{-4}\ s^{-1}$. Inhibition is stronger if Br^- or Cl^- is added, Br^- being more effective than Cl^-. The dissociation constants of the inhibitor-micelle-like complex, K_I, were calculated using a competitive inhibition model: $K_I^{Cl}/K_I^{Br} = 4.0$. Fluorescence quenching studies showed greater affinity of Br^- for the micellar surface: $K_{SV} = 1$ and $40\ M^{-1}$ for Cl^- and Br^-, respectively.	318
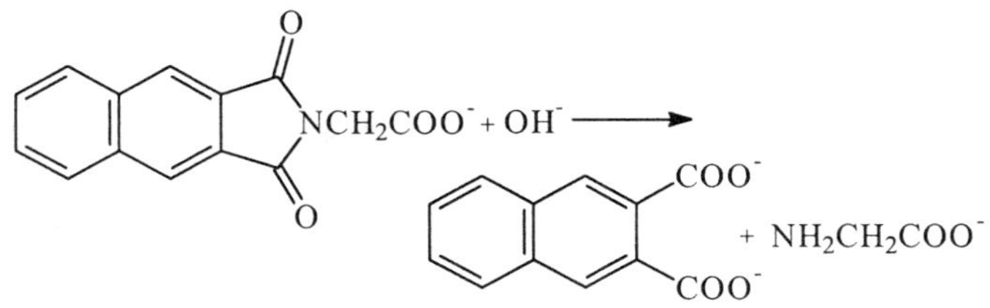	SDS	Psuedo-first-order rate constants for the alkaline hydrolysis of ionized *N*-phthalolglycine increase nearly 50% in 0.2 M SDS. The micellar reaction is supposed to occur in the Gouy-Chapman layer or in its junctural region with the Stern layer, where $[H_2O]_m \cong [H_2O]_w$ and $[OH^-]_m \cong [OH^-]_w$. The rate-determining step of the micelle-mediated reaction involves the collapse of the ion-pair complex formed between the ionized substrate and Na^+.	468
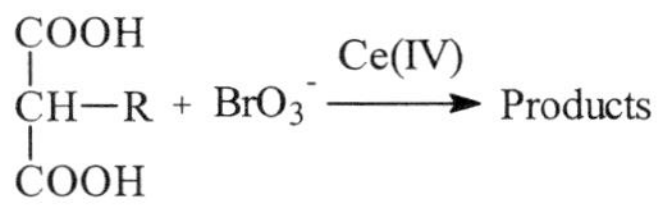 R = CH_3 (MeMA), Et (EtMA), $CH_2C_6H_5$ (BzMA)	$CTANO_3$ SDS	Addition of $CTANO_3$ or SDS has a little inhibitory effect on the redox process of Ce(IV) with MeMA and EtMA, whereas $CTANO_3$ significantly catalyzes and SDS retards to a larger extent the reaction of BzMA. Increasing amounts of $CTANO_3$ decrease the induction period of the oscilalting Ce(IV) Belousov-Zhabotinsky reactions. In 0.05 M SDS induction periods are shortened for MeMA and EtMA: from 410 to 341 min and from 361 to 344 min, respectively, and oscillation occurs for BzMA, too (IP = 763 min). $CTANO_3$ and SDS also affect oscillation period (τ) and duration of the raising portion of an oscillating circle (τ_{rais}): τ and τ_{rais} increase for MeMA and EtMA but decrease fo BzMA as $[CTANO_3]$ increases. In 0.05 M SDS τ and τ_{rais} are shorter for MeMA and EtMA and longer for BzMA with respect to the value obtained in 0.05 M $CTANO_3$.	310
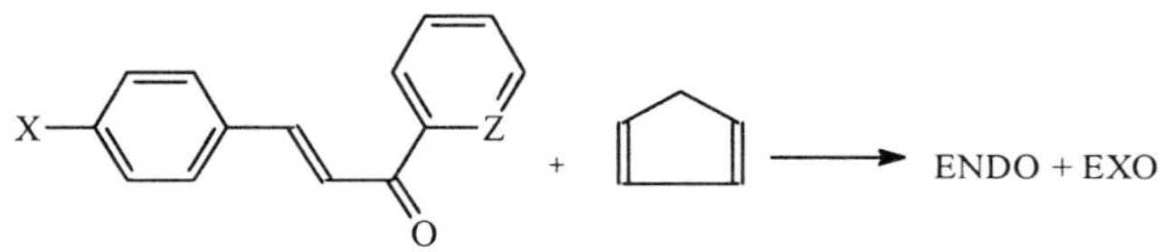 Z = N X = NO_2, Cl, H, CH_3, OCH_3, $CH_2SO_3^-\ Na^+$, $CH_2N^+(CH_3)_3\ Br^-$ Z = CH X = H	CTABr SDS $C_{12}E_7$ $Cu(DS)_2$ $Zn(DS)_2$	The Diels-Alder reaction is retarded by micelles of CTABr, SDS, and $C_{12}E_7$. Apparent micellar second-order rate constants are slightly smaller than those in water. Diene is located in the micellar interior and inhibition occurs both when dienophile is electrostatically repelled and attracted by the micellar surface. In the second case inhibition is even more pronounced because of different locations of the micelle-bound reactants, and the reaction senses a waterlike environment. Micelles of $Cu(DS)_2$ induce rate enhancements (up to 1.8×10^6 compared to the uncatalyzed reaction in acetonitrile), due to incorporation of the Lewis catalyst in the micellar environment and subsequent efficient complexation of the micelle-bound dienophile ($k_2^m = 5.9 \times 10^{-6}\ M^{-1}\ s^{-1}$ and $3.1 \times 10^{-5}\ M^{-1}\ s^{-1}$ for sulfonate and ammonium derivatives).	443

TABLE 12 Continued

Reaction	Surfactant	Remarks	Ref.
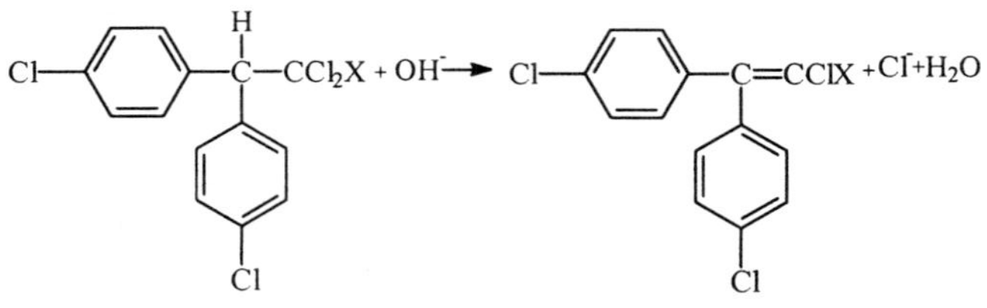 X = Cl (DDT), H (DDD)	MTABr	In the presence of hexanol, pentanol, butanol, and propanol both reactions are inhibited due to an increase of α. The ion exchange model fails in [NaOH] > 0.01 M. In these conditions the reaction is supposed to occur in the aqueous pseudophase.	469
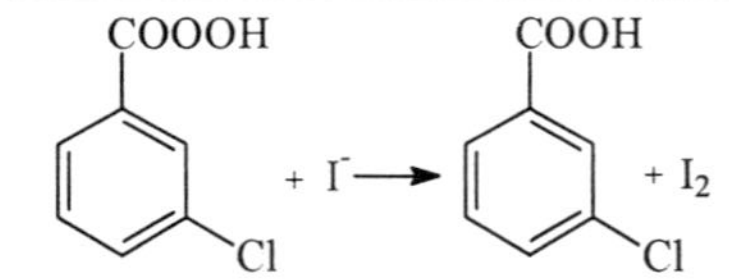 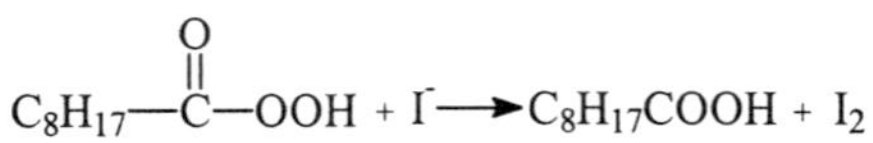	SDS $C_{12}E_{23}$ Triton X-100	Both reduction reactions are catalyzed in the presence of the nonionic surfactants $C_{12}E_{23}$ and Triton X-100, whereas they are inhibited in SDS. The kinetic data were fitted by a multiple micellar pseudophase model based on the transition state psuedoequilibrium constant approach. It models the micelle as *n* pseudophases consisting of adjacent layers extending from the middle and ascribes micellar catalysis or inhibition to differences between binding constants or reagents (K_{mic}^{PH}) and transition states (K_{mic}^{TS}). Best-fit parameters show that transition states are stabilized by nonionic micelles and destabilized by SDS, with respect to the peracids: $K_{mic}^{PH}/K_{mic}^{TS}$ = 0.40, 0.37, and 69.9 for pernonanoic acid; 0.30, 0.25, and 38.4 for 3-chloroperbenzoic acid in $C_{12}E_{23}$, Triton X-100, and SDS, respectively.	241
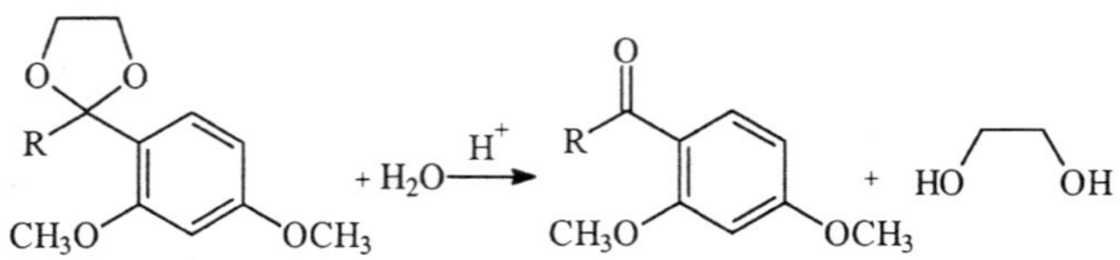 R = $C_{13}H_{27}$, $C_{15}H_{31}$	MTABr MTACl CTABr CTACl	Overall first-order rate constants for the acid hydrolysis of the dioxolanes decrease as [surfactant] increases, reaction never being completely suppressed. Added NaCl at constant [CTACl] increases residual activity, and a Poisson-Boltzmann equation model predicts the downward curvature of plots of k_{obs} against [NaCl]. The so calculated values of k_2^m are smaller than k_w: k_2^m = 0.23, ~0.20, 0.19, and 0.30 M^{-1} s^{-1} for the R = $C_{13}H_{27}$ derivative in MTACl, MTABr, CTACl, and in 0.02 M CTACl + NaCl, respectively; 0.16 and 0.24 M^{-1} s^{-1} for the R = $C_{15}H_{31}$ derivative in CTACl and in 0.02 M CTACl + NaCl, respectively. The PBE approach also predicts that with no added salt k_{obs} increases linearly with [surfactant] provided that reaction occurs wholly in the micellar psuedophase.	232

aromatic nucleophilic substitution in water rather than that it is abnormally reactive in micelles.

C. Kinetic Salt Effects, Effects of Other Additives

For several reactions, values of k_2^m/k_W are similar in the presence and absence of an inert counterion [12,77]. But if we look not at the value of k_2^m but at that of k_{obs}, addition of inert salts generally induces inhibition of reaction by changing the binding of reactive ions and the binding of the substrate (there is often linear experimental behavior for a K_S vs. [NaX] plot; for instance, see Ref. 292). The salt effects are specific, and both electrostatic and specific interactions play a significant role; effects are larger for low-charge-density and hydrophobic ions [89].

This ion specificity has been the subject of a deep investigation in nonionic and zwitterionic micelles, where the use of perchlorate ions is possible (in cationic micelles perchlorate ion associates with the cation and a precipitate forms). Reactions of OH^- and F^- with pNPDPP [303] (Table 12) in micelles of nonionic dodecyl polyoxyethylene glycol ethers $C_{12}E_{10}$ and $C_{12}E_{23}$ are inhibited by inert anions, but the effects are small except for bulkier inorganic anions such as ClO_4^- and also for organic ions such as naphthalene-2-sulfonate, ONs^-, and tosylate, $OTos^-$, that penetrate the palisade layer. The maximum value of the observed inhibition is a factor of ~0.3. The anion order for inhibition is $ONs^- > ClO_4^- > OTos^- > Br^- > SO_4^{2-}$, and this observed kinetic salt order is similar to that observed with ionic micelles [10,89]. On the other hand, addition of a cation, $(C_7H_{15})_4N^+$, accelerates reaction of F^- by attracting the anion into the micelle. Addition of ClO_4^- strongly inhibits reaction of Br^- with MeONs (Fig. 4), and this inhibition is effective even with NaBr in large excess over $NaClO_4$ and is evident for various sulfobetaines [112]. Changes in the cation from Na^+ to Cs^+, $(Me)_4N^+$, $(n\text{-}Bu)_4N^+$, have little effect on the binding of Br^- and on reactivity, as is consistent with the relatively weak, unspecific interaction of cations with sulfobetaine micelles [110,117].

Also, addition of nonionic surfactants $C_{10}E_4$ or 1-BuOH inhibits reaction of Br^- with MeONs in aqueous CTABr and CTEABr because they decrease the concentration of reactants at the micellar surface, but quantitative analysis shows that second-order rate constants at the micellar surface are very similar to those in the absence of additives [204,205,304] (Table 8).

However, values of k_2^m/k_W sometimes change if a counterion is present. This is consistent with there being a small, but significant, effect, possibly due to shrinkage of the volume element of reaction owing to the presence of additional electrolyte in the solution [12]. Alternatively, one could assume that added electrolyte increases the micellar radius, and provided that the headgroup area is constant the potential at the micellar surface will also increase [36]. Therefore, more ions, reactive and inert, will be attracted by the micelle. Evidence for changes in the structure of headgroups and possibly of the micelles of SB3-14 on addition of $NaClO_4$ has been given by an increase in ^{14}N NMR line width [112]. The increase is initially steep, probably because the location of ClO_4^- close to the quaternary ammonium center affects the environment and symmetry of the nitrogen, but there is thereafter a gradual increase in line width as the micelle become saturated with ClO_4^- due to a change in micellar structure, consistent with an increase in the aggregation number, N, at high [SB3-14] and high [$NaClO_4$] [112].

D. Substrate Structure

Micellar rate effects critically depend on the interaction of the micelle with the substrate. Two aspects of substrate structure play an important role: substrate hydrophobicity and the nature of its polar substituents.

1. Hydrophobicity

Association constants of nonionic solutes with aqueous micelles are sensitive to hydrophobicity (Section II.C), and there are a number of systems where apolar groups have been introduced with the aim of increasing association with micelles but also changing substrate locations and orientations in the interfacial region in order to induce chemo- or regioselectivity in product formation (as will be discussed further in Section VII).

There are several examples of the variation of reactivity with substrate hydrophobicity in terms of observed rate effects. *N*-Alkyl-2-chloropyridinium ions have been extensively investigated (Scheme 13).

Reactions with azide ion are enhanced by CTACl and CTABr, and rate enhancement increases markedly with substrate hydrophobicity, with values of k_{ref} (value of k_{obs} relative to reaction of methyl derivative in water) going from ~100 for the methyl derivative up to

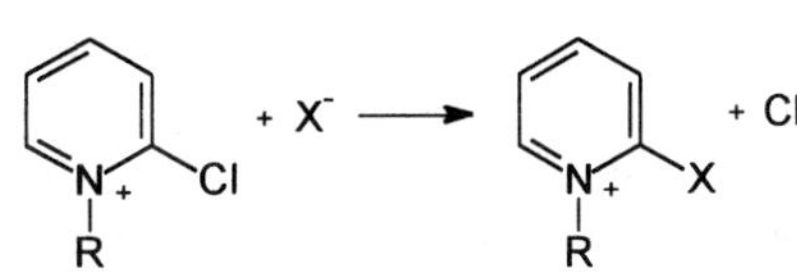

R = CH_3, C_6H_{13}, $C_{10}H_{21}$, $C_{12}H_{25}$, $C_{14}H_{29}$, $C_{16}H_{33}$

SCHEME 13

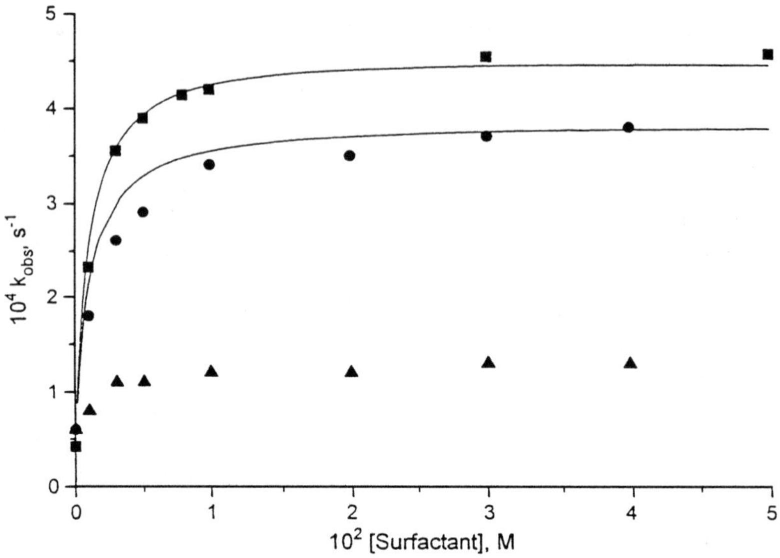

FIG. 4 First-order rate constants for reaction of MeONs with 0.5 M Br^- in SB3-14 with no $NaClO_4$ (■), 0.005 M $NaClO_4$ (●), 0.05 M $NaClO_4$ (▲). Solid lines are theoretical.

~15,000 for the *n*-hexadecyl derivative (Table 12) [218]. Reactions with alkylamines were also investigated, and CTACl, for example, slightly catalyzes (and for trimethylamine inhibits) reaction of the methyl derivative but accelerates greatly the reactions of the more hydrophobic derivatives (n-$C_{10}H_{21}$ and n-$C_{14}H_{29}$) [305]. Alkaline hydrolysis was accelerated by CTABr and SB3-16 for the long-chain derivative (n-C_8H_{17} and n-$C_{12}H_{25}$), and there was also an effect on the regioselectivity (see Section VII) [306,307]. Effects of substrate hydrophobicity on diazo coupling reactions have also been investigated (Table 12) [308], and the acceleration by CTACl increased greatly with increasing hydrophobicity of the electrophiles, whereas for reaction in SDS the rate increased with hydrophobicity from the 4-methyl to the 4-hexyl ion but then started to decline with further increase to 4-decyl, 4-tetradecyl, and 4-hexadecyl derivatives. Another kind of variation in hydrophobicity can be related to the second reagent. Al-Lohedan [309] examined reactions of a series of carboxylate ions, from formate to tetradecanoate, with 4-methylbenzenesulfonyl chloride in CTACl and in SDS (Scheme 14).

Cationic micelles give greater acceleration with increasing hydrophobicity of the carboxylate, whereas micelles of SDS inhibit reaction of the substrate with hydrophilic carboxylate (formate and acetate), but with hydrophobic carboxylate values of k_{obs} go to a minimum and then increase. Cavasino et al. [310] observed that different hydrophobicity of the substrates induces different behavior in the micellar effects on cerium(IV) oxidation of substituted malonic acid derivatives and on oscillatory parameters of the Belousov-Zhabotinsky system with these substrates. Benzyl-substituted malonic acid exhibits a peculiar kinetic and oscillatory behavior that has been related to its greater hydrophobic nature as compared with methyl and ethylmalonic acids.

Despite evident variations in k_{obs}, quantitative analysis indicates that changes in substrate hydrophobicity do not introduce high specificity in micelle-mediated reactions, and rate constants in the micellar pseudophases do not change significantly. For instance, k_M values for S_N2 reactions of various nucleophiles with MeONs and with benzenesulfonate in various cationic micelles are very similar (Table 8) [289,290,293,296, 311] as are values of k_M for deacylation of *p*-methyl and *p*-propyl- and *o*-propylbenzoates in various cationic micelles [286]. For bigger variations in substrate hydrophobicity, changes in k_M are often within a factor of about 2 [312], and results with a series of derivatives

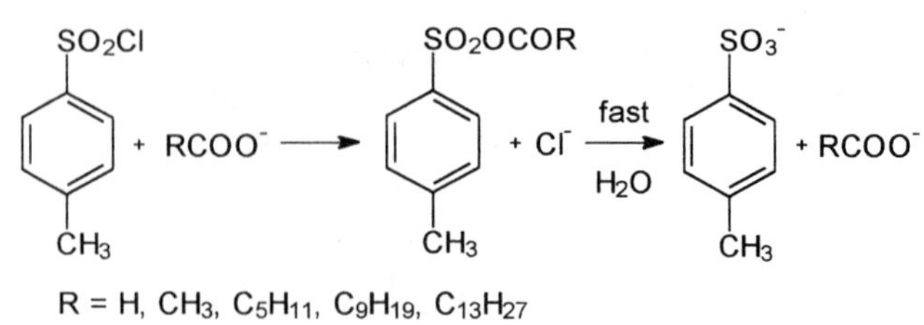

SCHEME 14

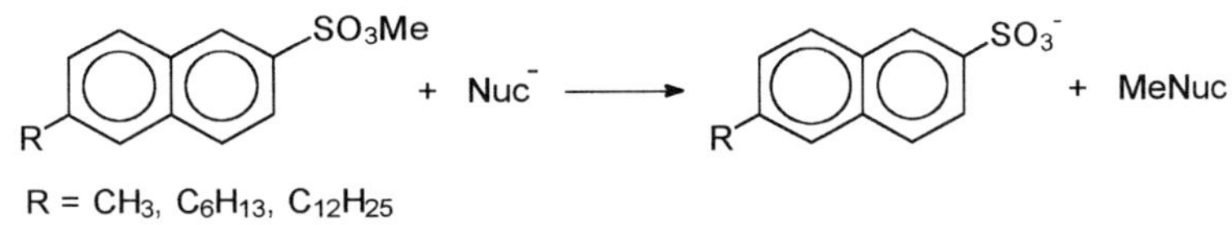

SCHEME 15

of MeONs with 6-alkyl substituents (alkyl = methyl, *n*-hexyl and *n*-dodecyl, Scheme 15) also show that values of second-order rate constants in the micellar pseudophase are affected little by substrate hydrophobicity (Table 8) (L. Brinchi et al., J. Colloids Interface Sci., in press, 2000).

A larger effect was observed in micellar effects on alkene brominations. Values of second-order rate constants of 1-alkenes depend on chain length, and the value of k_M for 1-decene is lower by a factor of about 7 than the value for 1-hexene (Table 11) [313]. It seems that there is an optimum chain length, because rate constants increase going from 1-decene to 1-dodecene. The different reactivities are related to different locations of the double bond of the alkenes in the micelles. The longer the alkene chain, the deeper the location of the double bond in the micelle, with a deeper location meaning a less polar medium and a slower reaction. Changes in the structure of the alkene also change the product distribution, as will be seen in Section VII.

2. Polar Substituents

Regarding polar substituents, we have already observed, when speaking about S_NAr reactions by azide ions, that reaction is highly micelle catalyzed only for some of the substrates used. This effect has been related to the location of the substrates, as suggested by NMR experiments (Table 12) [301]. Broxton and Marcou continued these kinds of studies, selecting some nitroactivated halobenzoates with charged carboxylate substituents in *ortho* and *para* positions to locate the substrates at the micellar interface and more deeply in the micellar core [314]. The carboxylate moiety plays a role in orienting the substrate in that it prefers to protrude from the micelle into the more polar aqueous pseudophase; on the other hand, nitro groups also affect location in that they prefer to be buried in the less polar region of micellar aggregates (Scheme 16).

The S_NAr reactions of aniline, which resides at the micellar surface, with compounds 1 and 2 (Scheme 16) are catalyzed by CTABr, whereas its reactions with compounds 3 and 4 are inhibited. Conversely, reactions of tertiary amines, which reside in the micellar interior, with more deeply buried substrates 3 and 4 were catalyzed more effectively than their reactions with substrates 1 and 2, residing at the micellar interface.

Another kind of variation associated with polar substituents is related to their electronic effects. Al-Lohedan studied S_N2 substitution of Cl^- and Br^- and with substituted alkylbenzenesulfonate [311] in CTACl, CTABr, CTAOMs, and $CTA(SO_4)_{0.5}$ (Table 8), and quantitative analysis showed a different effect of the nitro group in water and in micelles, with values of k_2^m/k_W being higher by a factor of ~2–3 for the methyl derivative with respect to the nitro derivative. Wilk and Burczyk analyzed elimination from 2-phenylethyl derivatives in CTAOH (Table 9) [291] and found values of k_2^m/k_W of 3.4, 2.0, and 1.1 for NO_2, Cl, and H substituents. A more direct comparison was carried out by comparing elimination from 2-phenylethyl derivatives and the S_N2 reaction of OH^- with methyl benzenesulfonates under the same set of experimental conditions and with the same quantitative treatments (Tables 8 and 9, Scheme 17) [293].

It is clear that introduction of a nitro group has a significant effect on relative reactivities in water and micelles, which is related to the extent of charge dispersion in formation of the transition state. For the S_N2 reactions of methyl benzenesulfonates the inductive effect of the nitro group increases reactivity by a factor

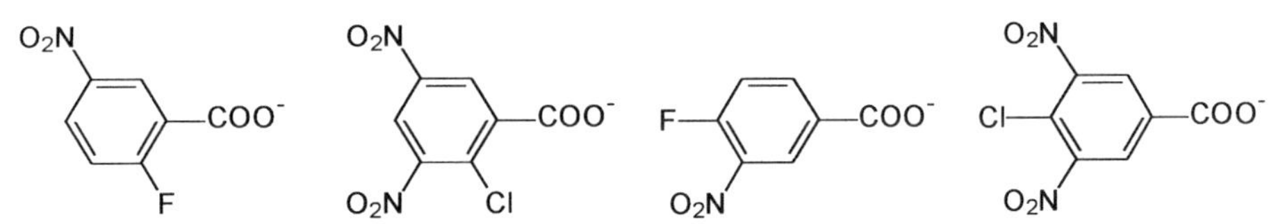

1 2 3 4

SCHEME 16

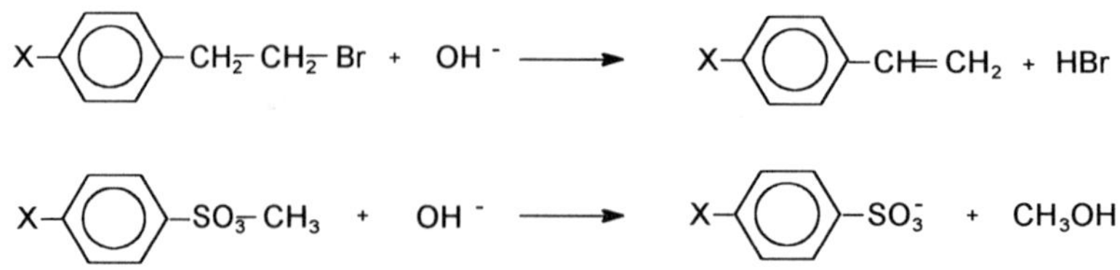

SCHEME 17

of ~4 in water and ~8 in CTAOH. However, there is a strong resonance interaction of the nitro group in the E2 reactions of phenylethyl derivatives, and rate increases by the nitro group are 67-fold in water and ~400-fold in CTAOH. In other words, micelles favor reactions with the most charge dispersion, even though they have the same molecularity and charge type.

More detailed analyses of substituent effects on reactions in micelles have been carried out by using the Hammett equation. An example of these analyses has already been mentioned in Section IV regarding E1cb reactions. For bimolecular reactions, values of ρ for reactions of aryl benzoates with OH^- [286] (Scheme 18), for example, are higher in cationic micelles of CTABr than in water, showing that the micellar interfacial region is less polar than water.

Reactions of methyl *p*-substituted benzenesulfonates with OH^-, Br^-, and H_2O in water and micelles, have also been studied (L. Brinchi et al., Eur. J. Org. Chem., in press, 2000). The analysis yielded values of ρ over a range of conditions and nucleophiles (Scheme 19).

Values of ρ show that S_N2 reactions of the methyl benzenesulfonates (for which $\rho = 1.13$, 0.94, and 1.00 in water for reaction with water, hydroxide, and bromide, respectively, and 1.41, 1.41, 1.58 in micelles of CTAOMs, CTAOH, CTABr) are less sensitive to electronic effects than saponification of phenyl benzoates, for which $\rho = 1.76$ in water and 2.6 in CTABr [286]. These differences are understandable in view of differences in relative locations of reaction centers and substituents in the two sets of reactions, which facilitate transmission of electronic effects in reactions of the phenyl benzoates. Attack of OH^- on phenyl benzoates involves rate-limiting addition to the acyl group, which is accelerated by electron-withdrawing substituents, but in S_N2 reactions we have to consider both nucleophilic attack and loss of the leaving sulfonate ion. Comparisons of second-order rate constants in water and micelles depend on parameters whose values are uncertain, e.g., those that describe interionic competition and the volume of the interfacial reaction region. We therefore have to be cautious in using kinetic data for the overall reaction to compare medium properties of water and micellar interfacial regions, because in some conditions kinetic fits are indeterminate for reactions of weakly interacting ions, e.g., OH^- in the presence of strongly interacting ions. However, comparison of values of ρ eliminates some of these uncertainties.

E. Surfactant Structure

Last but not least, an important factor that affects the micellar rate effect is the structure of the surfactant, as already discussed in Section IV. Regarding bimolecular reactions, few researchers studied systematic variations in surfactant structure if we exclude the change in counterion, already examined.

Concerning the systematic variation in the alkyl chain length of cationic surfactants Bacaloglu et al. examined the reaction of MeONs with Cl^- in a series of alkyltrimethylammonium chlorides (Table 8) [298]. Values of k_{obs} increase with the length of the alkyl chain, and the effect has been rationalized by quantitative analysis. In fact, the experimental association constant of Cl^- increases with alkyl chain length, but the value of k_2^m also increases in going from octyl surfactant (for which the value is approximately equal to that in water) to octadecyl surfactant, for which this

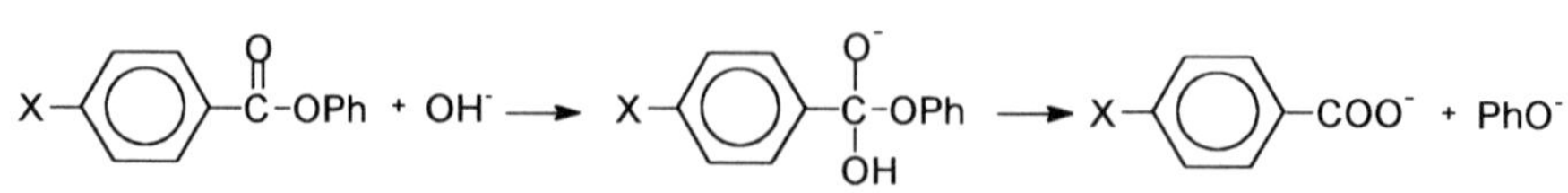

SCHEME 18

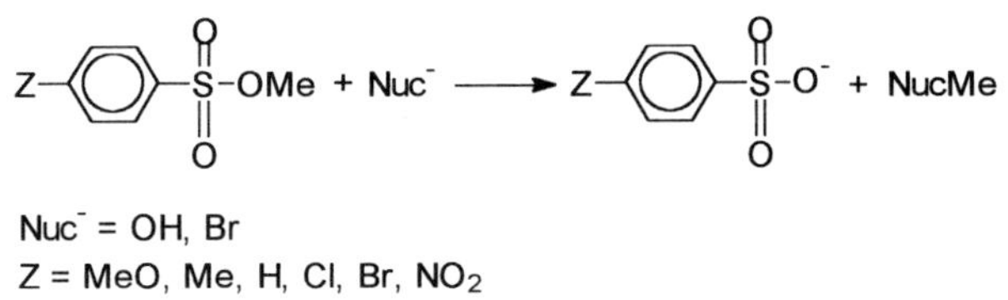

SCHEME 19

value is higher by a factor of ~3. The effect on second-order reaction rates in the micellar pseudophase has been related to partial disruption of the hydration shell of the nucleophile, as supported by a parallel increase in ^{35}Cl NMR line width.

Introduction of a covalently bound anionic charge to the cationic group, i.e., formation of a zwitterionic surfactant, does not suppress bimolecular reactions. Despite the absence of overall charge, micelles of zwitterionic surfactants can interact favorably with ions (see Section II), and this interaction induces high ionic discrimination. For instance, zwitterionic sulfobetaine micelles [315,316] and zwitterionic amine oxide surfactants (AOMe-14) [317] speed S_N2 reaction of soft ions with MeONs, and decrease, but do not suppress, reaction of even such hydrophilic and weakly associated anions as OH^- and SO_3^{2-}. In SB3-16 surfactants the relative reactivity of bromide and hydroxide is different from that in water, with $k_M(Br) > k_M(OH)$ [315] (Fig. 5).

Moreover, zwitterionic micelles were found to speed reaction of hydrophilic ions such as OH^- and also F^- in reactions with pNPDPP [315], to speed alkaline hydrolysis of *N*-decyl- and *N*-dodecyl-4-cyanopyridinium [318], and to speed E2 reaction from 4-nitrophenethyl bromide [319].

A particular effect on reactivity is related to headgroup bulk of cationic and also zwitterionic surfactants. For reactions of MeONs with Br^- and Cl^-, the observed constants increase with increasing headgroup bulk, whereas a decrease for reaction of OH^- with the same substrate is observed [289,296]. The binding constants of all counterions to micelles decrease with increasing headgroup bulk, as experimentally determined by various kinds of techniques [86,296–298], and that accounts for the decreased rate constants observed for reaction with OH^-. Quantitative analysis shows that values of k_M change little with surfactant for reaction of OH^- [289] but increase with increasing headgroup bulk size, modestly for reaction of Cl^- and more strongly for reaction of Br^- [296–298], so that $k_M(Br) > k_M(OH)$ for reactions in CTPAX and CTBAX, as in sulfobetaine SB3-16 and SBBu3-14 surfactants [315, 316]. Therefore, micelles generate an inversion of the reactivity sequence for anionic nucleophiles reacting

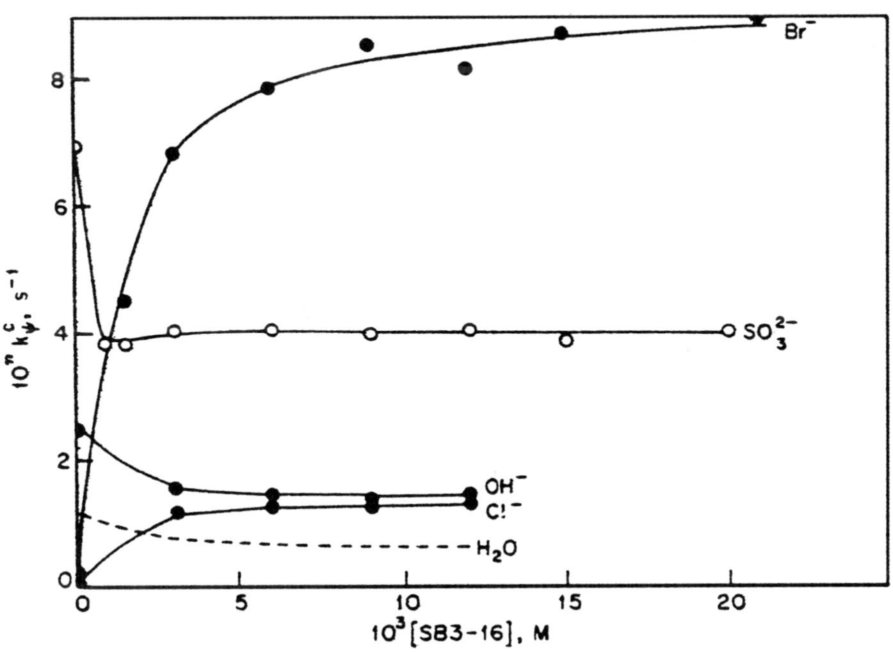

FIG. 5 Corrected rate constants for reactions of MeONs with anions in SB3-16; broken line represents reaction with water. For reaction with OH^-, Cl^-, Br^-, and H_2O, $n = 5$; for reaction of SO_3^{2-}, $n = 4$.

with a common substrate. There are clear differences between the behavior in cationic micelles of a very strongly hydrophilic anion, OH^-, a moderately hydrophilic ion, Cl^-, and a weakly hydrophilic ion, Br^-. The effects are probably due to differences in the hydration shell of the ions. Hydroxide ion interacts so strongly with water that its hydration should be little perturbed by cationic micelles, regardless of the headgroup, but hydration of Br^- and Cl^- is decreased when they interact with the micellar surface, with consequent increase in their nucleophilicity.

The increase in ion discrimination with increasing surfactant headgroup bulk is also observed in the increased electrophilic assistance to the leaving anion in intramolecular cyclization (Table 6) [261] of *o*-(3-halopropyloxy)phenoxide ion (halogen = I, Br), with values of $k_{M(I)}/k_{M(Br)}$ going from 1.3 in CTABr to 2.5 in CTBABr, and also in elimination from 1,2-dihalo-1,2-diphenylethanes (halogen = Br, Cl), with values of $k_{M(Br)}/k_{M(Cl)}$ going from ~27 in CTAOH to ~31 in CTPAOH (Table 9) [300].

The generalizations about the relation between reaction mechanism (in particular the softness of the transition relative to the initial state) and micellar kinetic effect can be extended to consideration of the role of headgroup bulk. For S_N2 reactions of OH^- with benzenesulfonate and MeONs, an increase in headgroup bulk slightly decreases k_M, but it slightly increases k_M for the reaction of methyl 4-nitrobenzenesulfonate. The headgroup effect is considerably larger for E2 than for S_N2 reactions of OH^-, where values of k_M in CTPAOH are larger than in CTAOH by factors >2. The situation is similar for increase in headgroup bulk in sulfobetaine surfactants; i.e., for eliminations from the same substrates, values of k_M in SBBu3-14 are larger than in SB3-14 by factors >2 [319] (Table 9). The conclusion that an increase in headgroup bulk favors reactions in which charge is dispersed in the transition state is consistent with observations on reactions of OH^- with phenyl *p*-substituted benzoates, where there is a clear relation between kinetic headgroup and electronic substituent effects (Table 10) as the headgroup is changed from NMe_3 to $N(n\text{-}Pr)_3$ and $N(n\text{-}Bu)_3$. Generally speaking, an increased dispersion of charge leads to higher values of $k_{M(CTPAOH)}/k_{M(CTAOH)}$.

Although second-order rate constants of reactions of ions with nonionic substrates are often similar in the aqueous and micellar pseudophases, relations between substrate structure and reaction mechanism and relative rate constants in aqueous and micellar pseudophases show that dispersion of negative charge in the transition state leads to higher rate constants in micelles, especially those with bulky headgroups. This relation between effects of bulk of the headgroup and charge dispersion, or "softness" of the transition state, seems to be general. It is evident in unimolecular decarboxylations and dephosphorylations and intramolecular cyclization, where kinetic analysis does not involve the transfer equilibrium of a second reactant.

VI. REACTIONS IN NONMICELLAR AGGREGATES

A. Vesicles

Closed bilayer vesicles formed by phospholipids were first described by Bangham and Horne [320] and are intensely explored structures. The work of Kunitake et al. showing that the synthetic surfactant dioctadecyldimethylammonium bromide, DODABr, forms vesicles marked the beginning of membrane mimetic chemistry [321]. This pioneering finding was soon followed by the description of vesicles prepared with a simple phosphate diester (sodium dihexadecyl phosphate, SDHP) [322]. Liposomes are vesicles made of phospholipids, and the assemblies formed from synthetic surfactants have been described as surfactant vesicles [9].

New vesicles have been prepared from a variety of surfactants, including single-chain surfactants [323, 324], synthetic glycolipids [325], and nonionic surfactants (referred to as niosomes) such as hexadecyldiglycerol ether, sorbitan monostearate [326], and sugar esters [327]. Mixtures of single-chain cationic and anionic micelles, referred to as catanionic systems, able to form vesicular systems are currently a subject of intensive investigation [328]. Surfactants used are often commercial ones, such as CTABr and SDS; vesicles are formed in coexistence with mixed micelles; and phase diagrams are currently under investigation [329,330], as are the morphologies of the aggregates [331]. A microscopic model for such mixed surfactant vesicles has been proposed [332].

Large vesicular systems, referred to as giant vesicles, have received great attention because of their large size. One of the simplest giant vesicles was produced simply by dispersing oleic acid in water at pH 8.5 [333]. This resulted in an oleic acid–oleate giant vesicle having a diameter of about 70 μm with a total of 10^{11} oleic acid–oleate molecules. They are particularly attractive to investigate because their formation and the behavior of one single supramolecular entity can be observed by light microscopy. The size of giant vesicles is typically in the range 10–100 μm, corresponding to an amphiphile aggregation number of typically 8×10^8–8×10^{10} per vesicle. The concept of self-reproduction has also been extented to giant vesicles [334]. An up-to-date review of this subject is the volume prepared as the proceedings of the Workshop "Giant Vesicles," which was held in Ascona, Switzerland, in 1998 [335]. Earlier work on giant vesicles by Ringsdorf et al. [336] and the current work of Menger and coworkers [337–339] on a variety of chemical and biochemical aspects of giant vesicles have been reviewed. Furthermore, much is known today about the physicochemical properties of giant vesicles and biomembranes thanks to the studies of Sackmann and coworkers [340].

The systems most investigated are vesicles formed by surfactants that have two *n*-alkyl groups. An excellent review on DODACl and SDHP vesicles by Carmona-Ribeiro [341] focuses attention on physicochemical characteristics of the systems, on analogies with and differences from phospholipid systems, and on their practical use. A wide range of applications take advantage of these systems, and some of them are briefly outlined in Section VII. Vesicles are single or multicompartment closed bilayer assemblies, and life without an inside-outside separation is unimaginable. The compartmentalization in vesicular media makes these systems quite interesting as membrane models. Here we focus attention on the properties of twin-chain surfactant vesicles as reaction media and on efforts to model the vesicular rate effects, taking into account the various reaction environments provided by vesicles.

Vesicles prepared with synthetic amphiphiles constitute useful microreactors where reaction rates can be finely controlled. DODABr and SDHP salts are sparingly soluble; vesicles are formed from dispersions of these salts by bath or tip sonication or by chloroform evaporation [342,343]. The structure of vesicles formed from a given surfactant depends on the preparation method. The first method yields multilamellar vesicles, the second gives small unilamellar vesicles (SUVs) and/or bilayer fragments, and the third gives large unilamellar vesicles (LUVs) [341]. Usually, vesicles obtained by sonication are not stable; they fuse and separation of phases occurs, and the ease of fusion depends on vesicular charge and the extent to which it is neutralized by added electrolyte. In contrast, vesicles prepared by chloroform evaporation are apparently stable for days [343].

Vesicles are multicompartmentalized microreactors capable of concentrating reactants or maintaining reactants separated in solution. They are permeable to apolar nonionic solutes and, if ionic, can bind counterions at the inner and outer surfaces; therefore, vesicles can catalyze reactions by significant factors. Counterion binding is also a crucial point in vesicles as in micelles. They play a major role in defining the permeability of vesicles formed by twin-chain cationic or anionic surfactants [344]. Carmona-Ribeiro and coworkers have investigated [345] the effect of the counterion in DODAX vesicles (X = Cl, Br, acetate). Vesicles with acetate as the counterion, i.e., the largest and the most hydrated, have the smallest size and the largest zeta potential. The phase equilibria of didodecyldimethylammonium surfactants have been shown to be sensitive to the counterion [346]. A multitechnique approach [347] used to study the aggregate structures showed that with OH^- and acetate as counterions, vesicles coexist with normal micelles within certain concentration ranges, above which micelles are the only stable aggregates.

Quantitative analysis of vesicle-modified reaction rates, made possible by using models initially developed for analyzing micellar rate effects, allows dissection of factors leading to catalysis or inhibition. It has been used extensively in analysis of the kinetic effect with surfactants such as didodecyldimethylammonium hydroxide and chloride in the region of their spontaneous solubility in water. Probably only micelles exist under these conditions, and they have been treated in Sections IV and V.

Mechanistic studies of organic reactivities in vesicles have focused on two questions: the application of the pseudophase model to reactions in vesicles and the reactions at the inner and outer vesicular surfaces. The complexity of vesicular systems offers multiple applications for reaction control.

The rates of vesicle-modified bimolecular reactions were first quantitatively analyzed using a pseudophase model with explicit consideration of ion exchange [348,349], with the models derived for micellar solutions. Quantitative analysis, using PIE, suggests that the rate enhancement is due primarily to reagent concentration in the dimensionally restricted environment provided by the vesicle, coupled with contributions from enhanced dissociation and reactivity of the nucleophile at the vesicle surface [350]. One of the conditions for the application of PIE models is the exchange of ions at the interface. Direct evidence for exchange and determination of selectivity constants for ion binding have been obtained directly by fluorescence quenching methods [351]. The fluorescence of 1-pyrenenonanoic acid (1-Py), incorporated in large vesicles of DODACl and DODABr, is quenched by iodide addition [352]. The selectivity constants for I^-/Br^- exchange ($K_{I/Br}$,) at the outer and inner surfaces are 8 and 7, respectively. The average value of $K_{I/Cl}$ at both interfaces is 31 ± 5. An exchange constant of ~4 for the Br^-/Cl^- exchange was calculated from the experimentally determined selectivity coefficients for exchange of halides with I^-. The value of $K_{Br/Cl}$ obtained for the exchange at the surface of DODACl vesicles is very similar to data obtained with positively charged micelles of comparable headgroup, i.e., cetyltrimethylammonium (CTA) halide [196].

Quantitative analysis of vesicle rate effects using models developed for micellar rate effects permits dissection of factors leading to catalysis or inhibition, but use of a single rate constant to represent reactivity in the vesicle ignores the fact that a solution containing vesicles has, in principle, several potential reaction sites, and no a priori theory predicts that reactivity at all the sites is equal or comparable. In a solution containing vesicles one can distinguish at least five reaction sites: (1) the inner compartment, (2) the internal interface, (3) the hydrophobic bilayer itself, (4) the external interface and (5) the aqueous phase [353] (Fig. 6).

With the purpose of site dissection, Chaimovich and Cuccovia prepared vesicles of various sizes, determined some of their physical properties, and developed theoretical and experimental tools for probing vesicular sites [354]. Small unilamellar vesicles (SUVs, hydrodynamic diameter, D_h < 50 nm) are not adequate for the purpose of site dissection because they do not permit the entrapment of analytically convenient amounts of substrate. Larger unimolecular vesicles (LUVs) of DODACl and SDHP (D_h > 300 nm) were obtained by chloroform vaporization at 70°C [355]. Several vesicular properties are size sensitive. In particular, the decrease in α with size is probably related to decreasing headgroup area and the increasing counterion association needed to relax the surface electrostatic potential [356]. The kinetic effects of vesicles in reaction rates are sensitive to the structural consequences of size variation. Even small differences in bilayer packing, with no changes in medium or aggregate composition, modulate the rates of chemical reactions occurring at vesicular interfaces. The rate-[surfactant] dependence obtained by studying the effect of small and large DODACl vesicles on the thiolysis (heptyl mercaptan) of esters (*p*-nitrophenyl octanoate) are quite different [357]. Smaller vesicles are two- to fivefold more efficient as reaction catalysts. The analysis with fit of the PIE model demonstrated that the size-dependent differences in kinetic efficiencies are attributable to differences in ion dissociation, substrate binding constants, and small changes in nucleophilic reactivity related to different packing of amphiphile in the bilayer of vesicles of different sizes [357].

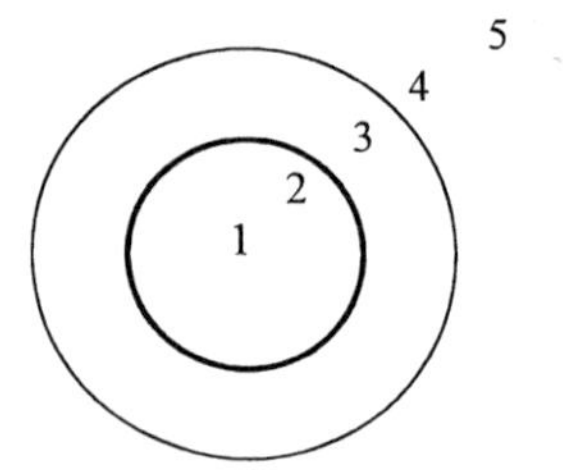

FIG. 6 Potential reaction sites in a vesicle: 1, inner compartment; 2, inner interface; 3, membrane; 4, outer interface; 5, intervesicle compartment.

Reactions at the inner aqueous compartment of unilamellar vesicles were investigated using water-soluble probes reacting with hydrophilic ions. The reaction probe chosen by Cuccovia and Chaimovich [358] to study the inner aqueous compartment of DODAX vesicles is the alkaline hydrolysis of *N*-methyl-4-cyanopyridinium ion (MCP). Reactivity of substrate in the large internal aqueous compartment is identical to that in free solution, and results indicate that slow OH^- diffusion through the membrane is responsible for the difference observed in the rate of MCP hydrolysis in the outer and inner aqueous compartments. Other experimental data demonstrating slow OH^- permeation have been reported [359].

In a solution containing vesicles, ions can reside in the continuous solution around the vesicles or in the internal aqueous compartment. The PBE equation was solved numerically assuming water- and ion-permeable hollow spheres and by treating specific ion adsorption using a Volmer isotherm [360]. The calculations suggest that the distribution of ions in the internal aqueous core of DODAX vesicles is measurable and that the value of the electrical potential at the vesicle center is not negligible at moderately low salt concentration. The PBE calculations showing appreciable concentrations of counterions in the inner aqueous compartment of the vesicle are consistent with results showing that the reactivity of OH^- in DODACl is comparable in the inner and outer compartments. Direct measurement of ion concentrations in the internal aqueous core of synthetic amphiphile vesicles has been performed using the dediazoniation method first described by Romsted (Section II). It can also be applied to the determination of Cl^- in the aqueous compartment of DODACl vesicles [354], and preliminary data suggest that the internal chloride ion concentration is consistent with calculations of ion distribution with the PBE equation.

Reactivity at the vesicular interfaces has been probed with substrates that bind and/or react preferentially in the inner or outer surfaces. The negatively charged 5,5′-dithiobis-(2-nitrobenzoic acid), DTNB, and OH^- bind to the positive surfaces of DODABr [361]. DTNB was selectively incorporated in the inner and/or outer surfaces of positively charged DODABr vesicles in order to probe reactivity with OH^- at both surfaces. Results show that the reactivity of OH^- at both surfaces is identical, and the reaction rate at the external surface can be modulated by changing the nature of the added salt. Addition of NaBr at the external aqueous compartment of DODABr vesicles in the presence of NaOH changes the ratio k_{out}/k_{in}, indicating that the externally bound OH^- and DTNB are exchanged by Br^-. These results, exemplifying selective reaction site control by surface composition, nicely demonstrate that differential ion binding at the surface can promote rate modulation of compartmentalized substrates. This in/out selectivity can be magnified in asymmetric vesicles, where the composition of the external leaflet is different from that in the internal interface [362–364]. In fact, natural selection has provided cells with a variety of bilayer bordered compartments with asymmetric distribution of lipids between inner and outer leaflets.

B. Premicelles

Very dilute amphiphiles do not significantly affect rates of many reactions, but with increasing concentration micelles form and rate or equilibrium constants change. Micellization occurs at the critical micelle concentration (cmc), which in some systems marks the onset of the rate increase, but rates often increase below the cmc, either because reactants induce micellization or because other species, so-called premicelles, are kinetically effective. As a result, a "kinetic" cmc is often used empirically in fitting rate-surfactant profiles, and Buckingham et al. noted physical evidence for the possible existence of premicelles [365]. The term "premicelles" seems to be applied to submicellar assemblies that form spontaneously or are generated by interactions with reactants. In the latter case it is not easy to distinguish between reactant-induced micellization and formation of premicelles [366]. Although the pseudophase model in its simplest form predicts that rate constants will increase only at the onset of micellization, they often increase monotonically well below the cmc, which could be ascribed to reactant-induced miscellization or to interactions with premicelles. For example, values of k_{obs} for reactions of hydrophobic substrates with small inorganic ions typically increase at [surfactant] < cmc and increase to maxima as predicted by pseudophase treatments. Similar behavior is often seen with micelle-mediated reactions of multivalent inorganic complexes and may be ascribed to interactions with premicelles [367]. Drennan et al. give a critical analysis of formation of complexes of Ni(II) in dilute sodium dodecyl sulfate where rate constants increase at surfactant concentrations below the cmc in water and show that it is not necessary to invoke the intervention of premicelles [366]. Micellization induced by interactions with multivalent inorganic ions is a common phenomenon in inorganic reactions.

Nonmicellar assemblies can influence reaction rates. Hydrophobic ammonium ions, which do not micellize, often increase reactivities, although generally concentrations are too low to allow physical identification of the assemblies [368–372]. In a few reactions in surfactants that generate micelles, there are extrema in plots of rate constants against [surfactant] at concentrations near or below the cmc [264,373–375]. These extrema cannot be ascribed to reactant-induced micellization, which gives monotonic changes in observed rate constants. It is easiest to identify this kinetic behavior in spontaneous, unimolecular reactions where only one species is partitioned between water and micelles or other assemblies. For example, in bimolecular ionic reactions one has to consider electrolyte effects on the cmc and competition between reactive and inert ions for the association colloids.

The expected kinetic form is observed for many spontaneous reactions, although k_{obs} often increases at surfactant concentrations below the cmc, and occasionally values of k_{obs} increase sharply with [surfactant] < cmc but then decrease and follow Eq. (2) at higher concentrations [264,373]. This anomalous behavior has been observed in two spontaneous reactions. The first is decarboxylation of 6-NBIC in a solution of DDDACl [264] and the second is cyclization of *o*-(ω-haloalkoxy)phenoxide with a long-chain alkyl tether in solutions of cationic surfactants CTABr and CTBABr and also of gemini surfactants $(CDA)_2C_42Br$ [373].

It has been speculated that these "premicellar" rate effects on cyclization require a hydrophobic interaction between substrate and, at most, a limited number of surfactant monomers. Alkoxy derivatives of 6-NBIC with groups of various lengths have been recently studied [280]; see also Section IV and Table 5.

Introduction of a methoxy substituent decreases rate constants of reactions of fully bound substrates in micelles, as predicted [255], and surfactant effects are similar to those for the unsubstituted derivative [264] except that micellar rate enhancements are lower. However, the behavior of the tetradecyloxy derivative is strikingly different. Values of k_{obs} are much higher than expected from the behavior of the methoxy derivative in dilute surfactant at concentrations well below the cmc and then decrease and become consistent with predictions. The surfactant concentrations at the rate maxima are in the sequence CTABr ≈ CTEABr > CTPABr > CTBABr. Values of k_{obs} increase so steeply in dilute CTPABr and CTBABr that only the descending part of the plots can be observed because of very low solubilities. The rate maxima in CTABr and CTEABr are at [surfactant] values that yield data for both the ascending and descending parts of the plots (Fig. 7).

Rate constants increase as substrate is transferred from water into surfactant-derived assemblies and then decrease to approximately constant values as substrate becomes micelle bound; i.e., the rate maxima demon-

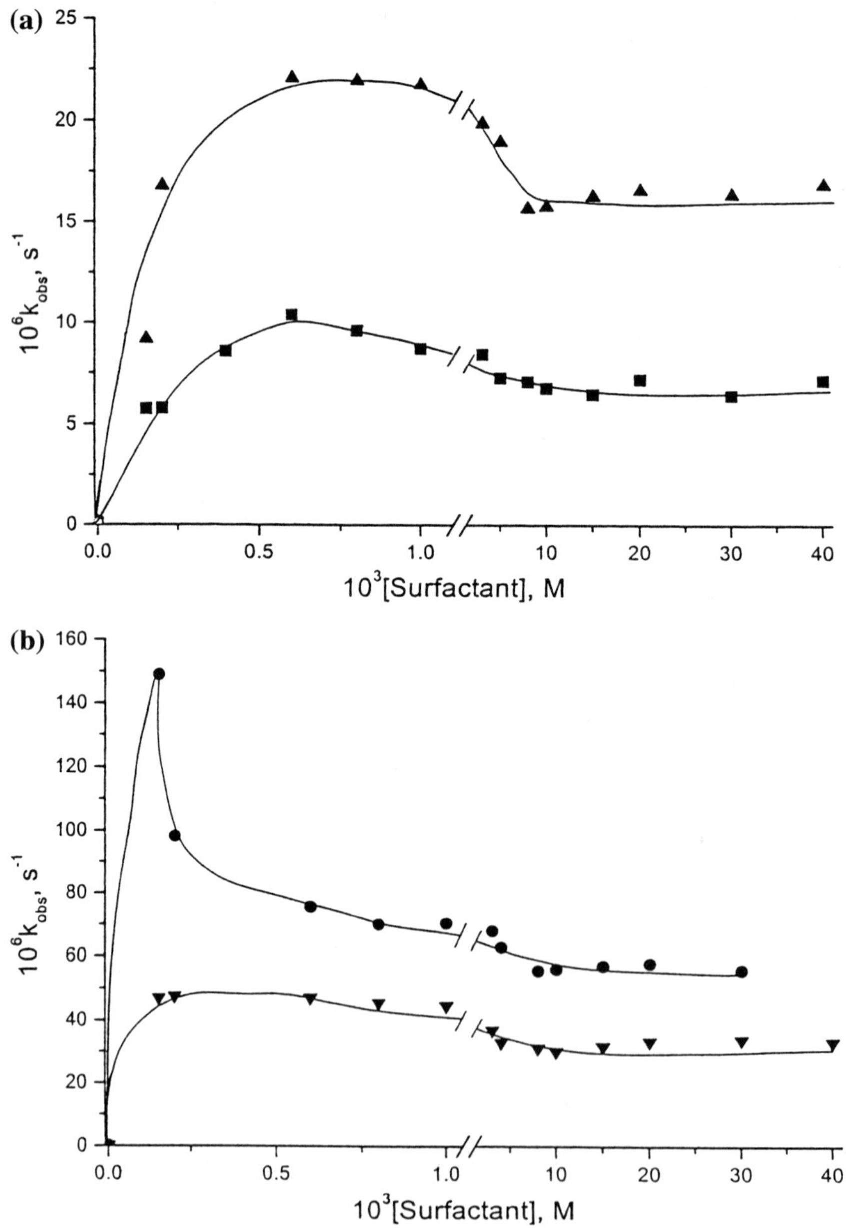

FIG. 7 (a) Decarboxylation of tetradecyloxy derivative of 6-NBIC in CTABr (■) and CTEABr (▲) and (b) in CTPABr (▼) and CTBABr (●). The lines are drawn to guide the eye, and rate constants in water are very close to zero.

strate the existence of three distinct reaction environments, depending on [surfactant]. The situation is similar to that for detection of premicelles by fluorescence spectra where the spectra are characteristic of aqueous, premicellar, and micellar environments [376]. The rate maxima cannot be ascribed to reactant-induced micellization. Once micelles form, they dissolve premicellar species, the substrate becomes micelle bound, and k_{obs} decreases toward the value of k'_M. The headgroup bulk influences the rate extrema: intervention of premicelles is most evident with CTBABr, and the initial increase of k_{obs} is seen in very dilute surfactant (1.5×10^{-4} M).

It is not possible to know the stoichiometry of the effective species in reactions in premicelles. The apparent maxima are obtained with approximately a 3-fold excess of CTABr and CTEABr, a 2-fold excess of CTPABr, and less than a 1.5-fold excess of CTBABr. The authors suggest a simple assumption—1:1 pairs are the active species—and in that event we could regard them as tight ion pairs held together by Coulombic forces and hydrophobic interactions of the long alkyl groups. Actually, small assemblies of substrate and hydrophobic quaternary ammonium ions can have a variety of conformations. For example, effects of submicellar cationic surfactants on the fluorescence spectrum of 2-*p*-toluidino-naphthalene-6-sulfonate were interpreted by assuming that long *n*-alkyl groups of the surfactant wrap around the organic moiety of the probe [376]. However, this conformation does not explain the premicellar rate increases, which require the hydrophobic quaternary ammonium ion to shield the carboxylate residue of long-chain substrate from water. Exclusion of water from the heterocyclic moiety would disfavor decarboxylation because, in the transition state, charge moves into this region with formation of a phenoxide ion [255]. Benzisoxazole carboxylate ions probably bind to micelles with the aryl residue inserting toward the apolar region, which leaves the carboxylate moiety adjacent to headgroups in the interfacial region. The rate increases in premicelles require association of the substrate with at most a few surfactant ions (Scheme 20). This association can be ascribed to the hydrophobic interactions of the long alkyl groups, but the rate increases require interactions to stabilize the transition relative to the initial state. If interactions in premicelles are similar to those in micelles, reactions should not be faster in the former. It is then necessary to assume that in premicelles the carboxylate ion is shielded from water, and in the transition state charge dispersion is favored by interaction with the quaternary ammonium ion (Scheme 20).

Derivatives of dianionic $DNPP^{2-}$ bearing a methoxy or the long-chain tetradecyloxy substituent in position 5 have also been investigated (M. Tugliani et al., Langmuir, in press, 2000); see also Section IV and Table 5.

As already observed for reactions of $DNPP^{2-}$, reaction rate constants for the methoxy derivative increase with [surfactant] and reach constant values characteristic of complete substrate association. When the headgroup bulk of the surfactant is increased, rate constants increase sharply on initial addition of cationic surfactant. Rate-surfactant profiles are characteristic of micelle-assisted reactions, and kinetic data fit the predicted behavior, based on Eq. (2). The introduction of a long alkyl chain in the substrate leads to complex variations of k_{obs} with [surfactant], as already observed in decarboxylations. Here values of rate constants go through a *maximum* at very dilute [surfactant] (in the range $8 \div 10 \times 10^{-5}$ M surfactant, well below the cmc), then through a *minimum* (at least when big head surfactants are present), and then they seem to reach a limiting constant value.

Small clusters form in dilute surfactants with an average of four surfactant molecules per substrate mole-

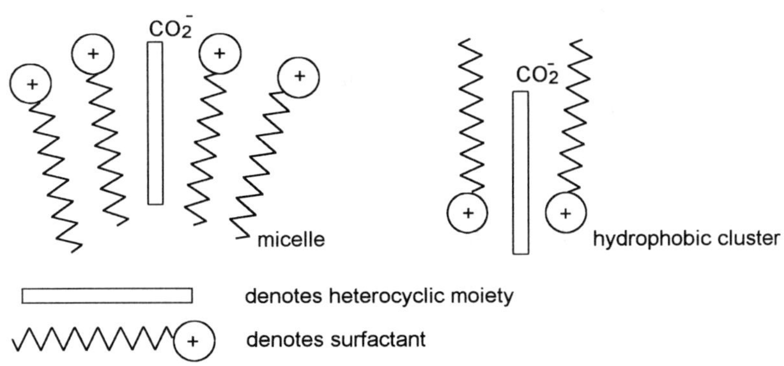

SCHEME 20

cule, and they are catalytically more effective than normal micelles because the relative free energy of activation is lower. Also for dephosphorylation nonmicellizing trioctylammonium salts speed hydrolysis [372,377]. Attempts to measure rate constants below 8 $\times 10^{-5}$ M failed because of solubility problems of substrate and/or products, probably related to ion pairing; therefore, even higher *maxima* could be possible at more dilute surfactant, especially with big-headgroup surfactants. Rate *maxima* seem not to be sensitive to headgroup bulk, and we have only a slight increase in k_{obs} in going from CTABr to CTBABr.

VII. APPLICATIONS OF SURFACTANTS

Self-organized systems composed of amphiphilic molecules have particular features that make them attractive, not only as relates to chemical reactivity aspects but also for a large variety of applications. For instance, surfactants have been used for extraction of metal ions [378]. Surfactant-based processes have attracted much attention in the past few years and it is now well established that metal ions can be removed by taking advantage of the association of the metal ion itself [379] or of the extractant/metal with the micellar entities. The removal of metal ions at low concentration can be performed in that way using ultrafiltration membranes with pore diameters smaller than the size of the particles. Such extraction processes operate in the absence of organic solvent and in 99% aqueous solution. Extraction processes using micellar particles as the extracting phase are currently under investigation in many places because of their potential in helping to solve specific environmental problems. These processes can completely avoid or at least considerably reduce the use of organic solvents. Not only undesirable metal ions but also, for instance, dyes [380] and organic pollutants can be removed [381].

Another field of possible development of the preceding processes is that of enantiomeric separations. There have been attempts to use chiral surfactants in order to perform selective transport of enantiomers [382–384]. Considering also chromatographic methods, for example, Camilleri et al. have investigated the use of anionic surfactants, which when used above their cmc in micellar electrokinetic capillary chromatography (MECC) allow the resolution of a number of structurally unrelated racemic mixtures [385–387]. The formation of noncovalent "diasteroisomers" between the micelles and the enantiomers leads to chiral discrimination if the energies of formation of these transient species differ sufficiently. Subsequently [388], they showed that the chiral separation efficiency of D-glucopyranoside anionic surfactants depends strongly on the orientation of the C_{14} hydrocarbon chain at the anomeric carbon center. Moreover, the possibility of using micellar particles for the transport of different solutes across liquid membranes has been investigated in the past 15 years [382,389–393].

The application of surfactant-based systems as drug delivery vehicles [394,395] is a growing research area that may develop further in the coming years. It is quite interesting that cationic amphiphiles are now widely used as an effective tool in delivering DNA into cells [396,397] even mammalian cells [398–402]. An investigation by Engberts and colleagues [403] yielded the conclusion that although membrane fusion may play a key role in DNA packaging, amphiphile/DNA complex formation [404] is probably a translocation via a "perturbed target membrane" mechanism, rather than by fusion, may be the mechanism by which nucleic acids are introduced into the cells. Mention should also be made of microemulsions incorporating fluorocarbons, which have specific potential for oxygen transport [405].

Bilayer-forming synthetic surfactants have been extensively used as membrane mimetic models, and some synthetic amphiphiles such as dihexadecyl phosphate or dioctadecyldimethylammonium salts have found many different uses in strategic applied areas [341]. In particular, synthetic cationic liposomes have been successfully employed to interact with negatively charged surfaces or biomolecules such as prokaryotic [406–408] or eukaryotic cells [409], antigenic proteins [410], nucleic acid [411,412], synthetic polymers and latex [413–416], and mineral surfaces [417–420]. In the following discussion, we focus our attention on chemical reactivity.

Aqueous association colloids as reaction media offer alternatives to the use of organic solvents, and there is considerable interest in their use in water as a reaction medium; they are attractive candidates in "green" chemistry [19,21,22,421,422]. For instance, Moss et al. [423,424] observed 1000- to 2000-fold rate enhancements in overall rate constants of hydrolyses of phosphate triesters catalyzed by different iodosocarboxylate ions of varying hydrophobicities in comicelles with CTACl and CTAOH. Here, product isolation is unimportant, but the solubilizing properties of micelles and their minimal toxicity compared with organic solvents make them useful as media for detoxification and organic synthesis. In the latter application, however, the amphiphile must be removed when the reaction is complete. In this case, as is often the case, studies of micellar effects on rates or products of organic reactions have been made with very low concentrations of reactants, and this small scale is not very useful to the synthetic organic chemist. An additional disadvantage is that surfactants complicate product separation by extraction or distillation, and to date most studies in this general area have been exploratory and aimed at solving these problems. Jaeger and coworkers have developed a more general approach based on the synthesis of chemically labile surfactants [425–427].

Surfactant-based reaction media have such kinds of features that make them useful in industrial-scale synthesis and interesting in developing "clean" processes. In fact, they are expected to be nontoxic and nonhazardous, they enhance reaction rates, reactions can usually be carried out under mild conditions, and in favorable cases surfactants can be separated and reused. Mashraqui et al. used micelles of CTABr as benign, economical, and mild reaction systems to afford chalcones [428] in high yield by an aldol reaction under weakly alkaline conditions and to prepare a variety of organic sulfides [429]. In both cases a preparative scale was used. Kobayashi et al. [430] demonstrated the utility of SDS in providing a hydrophobic micellar medium to perform $Ln(OTf)_3$- or $Cu(OTf)_2$-catalyzed Mannich-type reactions of aldehydes, amines, and silyl enolates to prepare β-amino ketones or esters in high yield without any side reaction adduct (Scheme 21).

They also carried out successfully three-component Mannich-type reactions of aldehydes, amines, and silyl enolates in water in the presence of dodecylbenzenesulfonic acid as a Brønsted acid–surfactant catalyst [431]. Other Mannich-type reactions in aqueous surfactant-rich media involved HBF_4-catalyzed condensation of aldehydes, amines, and silyl enolates for the synthesis of β-amino carbonyl compounds [432]. Kobayashi et al. also found that lanthanide triflates catalyze aldol reaction of silyl enolates with aldehydes with the aid of SDS [433,434] and also developed three-component coupling reactions of aldehydes, amines, and allyltributyltin in micellar systems [435]. The strategy of multiple assembly within micelles was even applied for the synthesis of porphyrins [436]. Micelles were viewed as potential wells able to bind products more tightly than reactants in catalyzing condensation reactions. Various functionalized porphyrins, difficult to make by direct aldehyde-pyrrole condensation in various organic solvents with micelle-like polarity, were conveniently prepared from the corresponding aldehydes in aqueous sodium dodecyl sulfate, with micelles directing the course of the synthesis. Rathman et al. [437,438] have developed so-called micellar phase transfer catalysis: addition of a phase transfer catalyst (tetrabutylammonium bromide) to surfactant systems results in a remarkable synergism, applied to the Williamson synthesis of phenyl butyl ether. The system provides higher reaction rates and conversions than observed in conventional micellar systems, and it is also successful at high reactant loading (>50 wt%) that exceeds the solubilization capacity of micellar solutions.

Moreover, reagent organization in and different microenvironments provided by aqueous micelles open the possibility of controlling product formation, providing chemo-, regio-, or stereoselectivity [9,31].

There are a number of examples of this type in the literature, and they are often easily explained in terms of the generally accepted model of kinetic micellar effects. Often the micelle-mediated specificity may be simply a polarity effect. Under given conditions, cationic micelles favor E2 over S_N2 (and S_N1) reactions. It has been shown that the reaction of phenethylnaphthalene-2-sulfonate in water gives 100% phenethylalcohol, whereas reaction in CTAOH gives 37% styrene and only 63% alcohol [439]. This preference is rationalized on the basis of previous kinetic studies [293], and it is related to a higher second-order rate constant at micellar surfaces than in water for elimination as compared with substitution. Politi and Chaimovich [307] studied the reaction of *N*-alkyl-4-cyanopyridinium ions (alkyl = Me, *n*-Bu, *n*-octyl, *n*-dodecyl) with hydroxide ion. Ratios of the reaction products, *N*-alkyl-4-pyridone and *N*-alkyl-4-carboxamidopyridinium, are shifted in favor of the pyridone in CTABr and SB3-12 for the long-chain substrates, whereas the product distribution is not sensitive to the presence of SDS. The regiochemical preference for pyridone is rationalized in terms of decreased polarity of the reaction microenvironment.

In other cases, regioselectivity induced by micelles has been rationalized in terms of a preorientational ef-

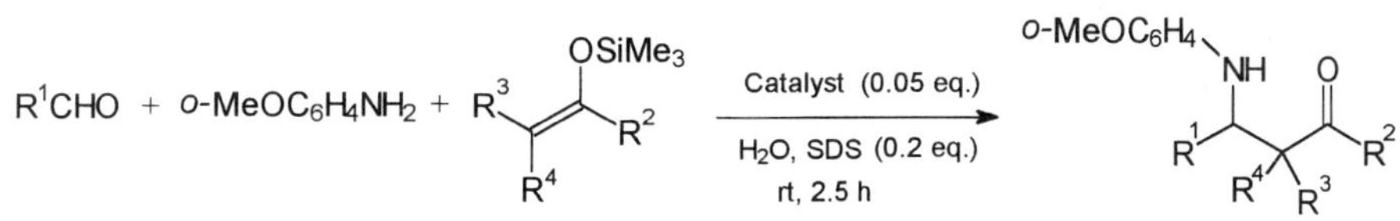

SCHEME 21

fect or anisotropy of the aggregate-water interface. For instance, surfactant control of the *ortho/para* ratio in the bromination of anilines was studied by Cerichelli et al. [440]. A range of different anilines were investigated in CTABr or $CTABr_3$, and the regioselectivity observed in water is opposite to that in micelles. High *ortho/para* ratios in micelles were observed and increased with greater steric hindrance in the *ortho* position. Further investigation showed a dependence not only on the substituents on the nitrogen of the aniline but also on the temperature, and that was explained by the authors as related to thermal shaking, which makes the interaction of substrate with the aggregate less specifically oriented [441].

Alignment of reactants at the micellar surface was at the basis of the observed regioselectivity control in Diels-Alder reactions in which both reactants were surface active. The Diels-Alder reaction is an important tool in organic synthesis, and there are a number of studies of this reaction in aqueous surfactant systems [442]. A deep kinetic investigation was carried out by Engberts and colleagues [443] (Table 12). Jaeger and Wang observed that when the reactions between surface-active diene (**1**, Scheme 22) and dienophile (**2**) were run in organic solvent, two regioisomers (**3** and **4**) were obtained in equal amounts, but when reactions were carried out in aqueous mixed micelles formed by the two surfactant reagents, one regioisomer (**3**) prevailed over the other (**4**) by a factor of 3 [442]. No effects have been observed when only diene was surface active [444], indicating that the regioselectivity is controlled by substrate orientation related to the long alkyl chain.

A considerable degree of chemoselectivity in S_N2 reactions of different sulfonate esters in solutions of OH^- and Br^- was obtained based on micelle-induced ion discrimination [439]. Cationic or zwitterionic surfactants have been shown to allow S_N2 reactions of Br^- to proceed quantitatively, even in alkaline solutions, and protect the alkyl bromide from subsequent reaction with OH^-. Affinities of these micelles for anions follow the Hofmeister series and are large for polarizable, "soft," low-charge-density ions such as Br^- and low for "hard" high-charge-density ions such as OH^-. The Br^- binds so much more strongly than OH^- to cationic and sulfobetaine micelles that nucelophilic attack of Br^- is strongly preferred over that of OH^- and the kinetically controlled product, the alkyl bromide, is protected from OH^-. Variations in surfactant structure with increasing headgroup bulk (use of CTBABr) further increase the ion discrimination and the chemoselectivity, and moreover it was possible to use a hexane-water system with extraction of the products and reuse of the aqueous surfactant. Chemoselectivity was also observed in the electrophilic bromination of olefins. A combination of kinetic study (reported in Section V) and product analysis for the bromination of a series of 1-alkenes in CTABr showed that decreased reactivities correspond to a decrease in bromohydrin product and correspond to different locations of the double bond of the alkenes in the micelle. The longer the alkene chain, the deeper the location in of the double bond: 1-decene gives the least bromohydrin [313].

The chemoselectivity in epoxidation of alkenes was studied. Olefins such as cyclooctene and cyclohexene could be oxidized with NaClO in homogeneous micellar media to give good yields of the epoxides in the presence of tailor-made micelle-bound metalloporphyrins [445].

The stereoselectivity of a reaction can also be altered

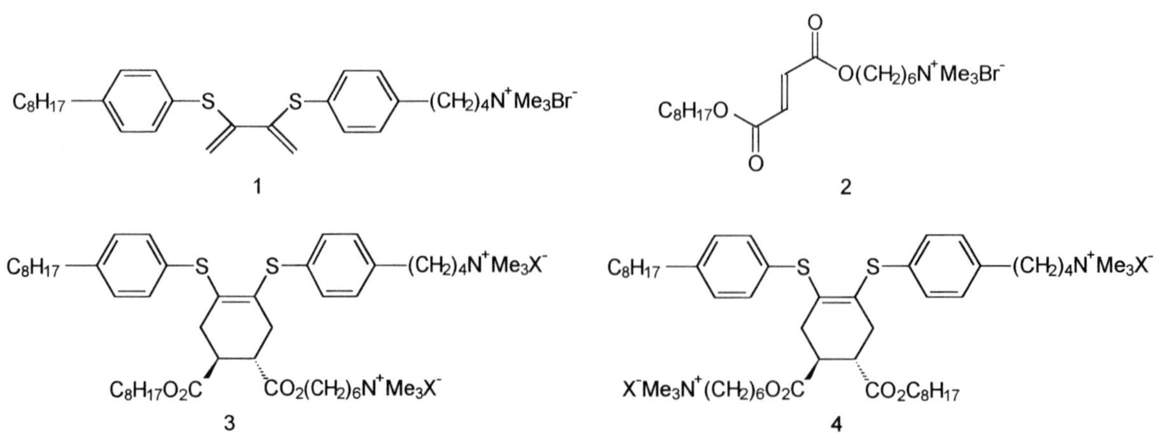

SCHEME 22 Diels-alder reaction between surface active diene (**1**) and surface active dienophile (**2**).

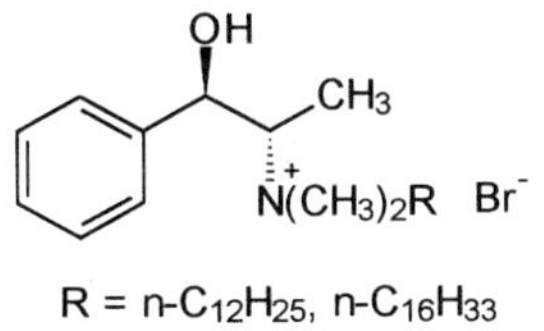

SCHEME 23 Homochiral quaternary ammonium salts.

in micellar systems. In this respect, the main factor to consider is the preorganizational effect of micelles. Denis et al. [446] obtained high regioselectivity and stereoselectivity in the reduction of α,β-unsaturated ketones to the corresponding allylic alcohol in the presence of glycosidic surfactants or amphiphilic carbohydrates. For instance, carvone was converted to (−)*cis*-carveol in 99% yield, 99% regioselectivity, and 93% *cis*-stereoselectivity.

The use of micelles to induce enantioselectivity is currently of high interest. The most widely studied stereochemical reactions are the hydrolyses of *p*-nitrophenyl esters of *N*-protected D or L amino acids in the presence of dipeptide or tripeptide catalysts [31]. Values of up to 131 for the k_L/k_D ratio in CTABr micelles have been reported [447]. However, asymmetric synthesis using chiral micelles as an asymmetric environment is a relatively new area. Goldberg et al. in 1978 first reported the reduction of prochiral ketones in an aqueous micellar solution, but with only 1.7% enantiomeric excess [448]. Zhang et al. have investigated the use of chiral cationic micelles [in several cases homochiral quaternary ammonium salts prepared from (−)-(1*S*,2*R*)-ephedrine, Scheme 23] in the reaction of many types of prochiral substates such as the epoxidation of chalcones, reduction of prochiral ketones, and oxidation of prochiral sulfides [449–451].

Zhang and Wu prepared β-hydroxy esters by the Reformatsky reaction [452] and oxiranes (Scheme 24) by the reaction of dimethylsulfonium methylide and aromatic aldehydes and ketones, with an enantiomeric excess up to 55% in the latter case [453].

Diego-Castro and Hailes [454] have used novel cationic chiral surfactants to study enantioselectivity in Diels-Alder reactions, and they had an enantiomeric

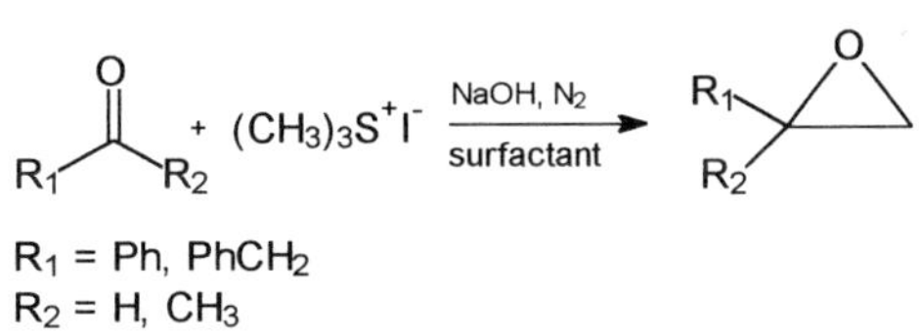

SCHEME 24 Enantioselective synthesis of oxiranes.

excess up to 15%, which compares with that cited for Diels-Alder in cyclodextrins.

High values of enantioselectivity were obtained by Brosch and Kirmse in the nitrous acid deamination of amines [455]. Deamination of 1-octamine affords mixtures of isomeric ocetenes, octanols, and octyl nitrites; the aggregation of the amine·$HClO_4$ in micelles induces the formation of dioctylether and 1-nitrooctane as additional products. The deamination of [1-^{2}H]-1-octamine in submicellar aqueous conditions (the cmc of the amine·$HClO_4$ is 0.105 M) gives [1-^{2}H]-1-octanol with ~95% inversion of configuration, and above the cmc the enantiomeric purity decreased, whereas [1-^{2}H]-1-nitrooctane was formed with ~90% retention of configuration. Selke et al. obtained what they called an impressive enhancement of the enantioselectivity for a hydroxy-containing rhodium(I) biphosphine catalyst in aqueous solutions by micelles [456]. The hydrogenation of some chelating olefinic substrates with a rhodium biphosphine catalyst is influenced by micelles of SDS, CTABr, and Triton X in water. They obtained up to ~77% enantiomeric excess, with a difference in enantiomeric excess with blanks of 70%.

VIII. CONCLUSIONS

Surfactants are now widely used in different fields, and they are being widely applied in developing the chemistry of the future [457]. Quite often, however, only a few marketed preparations are used, with almost an empirical approach.

This chapter illustrates how small modifications in the surfactant structure lead to changes in microaggregate structure, properties, and functions. The key point is that aggregates created by design contain what Lehn [458] would call "instructed molecular components."

In this chapter we have focused attention on new surfactants studied by or used for kinetic effects. In particular, regarding kinetics, we spoke about systematic variations in surfactant headgroup bulk, alkyl chain length, and the number of alkyl chains and about the use of gemini surfactants. Regarding applications, a greater range of surfactants have been encountered.

The ultimate goal is to develop the synthesis and applications of novel, more effective amphiphiles, capable of performing *selectively* the *function* that they were "engineered" for. Therefore, it seems important for physicochemical studies of micellar structure and properties to continue. It will become an important challenge to synthesize new amphiphiles with different headgroups, alkyl chains, and functional groups in order to achieve improved micellar effects.

ACKNOWLEDGMENTS

Our work in this area has depended on the help and enthusiasm of our colleagues and students: Giorgio Cerichelli, Giovanna Mancini, Luciana Lucchetti, Antonio Cipiciani, Nicoletta Spreti, Pietro Di Profio, Luisa Marte, Massimiliano Giovannini, Antonella Bartoletti, Simona Bartolini, Francesca Del Rosso, Vittoria Giacomini, Francesca Micheli, Monica Tugliani, and especially Clifford A. Bunton.

SYMBOLS

α	Degree of micellar ionization
β	Fraction of counterions bound to micelle, $\beta = 1 - \alpha$
cmc	Critical micelle concentration
[D]	Stoichiometric concentration of surfactant (detergent)
[Dn]	Concentration of micellized surfactant: generally [Dn] = [Dt] − cmc
K_S	Binding constant* of solute based on concentration of micellized surfactant (M^{-1})
K'_Y	Mass action binding constant* of ion Y (M^{-1})
k_{obs}	Observed first-order rate constant (s^{-1})
k'_W	First-order rate constant (s^{-1}) in the aqueous pseudophase
k_W	Second-order rate constant ($M^{-1}\ s^{-1}$) in the aqueous pseudophase
k'_M	First-order rate constant (s^{-1}) in the micellar pseudophase
k_M	Second-order rate constant in the micellar pseudophase (s^{-1}), with concentration expressed as mole ratio
V_M	Molar volume element of reaction in the micelle (M^{-1})
k_2^m	Second-order rate constant in the micellar pseudophase ($M^{-1}\ s^{-1}$): $k_2^m = k_M V_M$
K_Y^X	Ion-exchange constant* for ions of like charge
m_X^S	Mole ratio of X in the micelle
Δ	Width of shell of reactive region in the micelle
f	Fractional coverage by counterions
δ	Specificity parameter for ion association in the Volmer equation

*We use the word constants as usual in the literature, but we have to indicate that they are not thermodynamic constants.

SURFACTANTS

$(CDA)_2C_3 2Br$	1,3-Bis-(*N*-cetyl-*N*,*N*-dimethylammonium)-propane dibromide
$(CDA)_2C_4 2Br$	1,4-Bis-(*N*-cetyl-*N*,*N*-dimethylammonium)-butane dibromide
AOMe-14	Myristyldimethylamine *N*-oxide
AOPr-14	Myristyldipropylamine *N*-oxide
AOT	Bis(2-ethylhexyl) sodium sulfosuccinate
Bolaform (n) X_2	$Me_3N^+(CH_2)_nN^+Me_3 2X^-$
$C_{10}E_4$	Decyltetraoxyethylene glycol ether
$C_{12}E_7$	Dodecylheptaoxyethylene glycol ether
$C_{12}E_{10}$	Dodecyldecaoxyethylene glycol ether
$C_{12}E_{23}$	Dodecyltricosanoxyethylene glycol ether
$C_{16}E_{20}$	Cetyleicosanoxyethylene glycol ether
CB1-14	Myristyldimethylammonium carboxybetaine
CB1-16	Cetyldimethylammonium carboxybetaine
CCHDMABr	Cetylcyclohexyldimethylammonium bromide
CDMHEACl	Cetyldimethyl-2-(hydroxyethyl)ammonium chloride
CDMPCl	Cetyl-4-(dimethylamino)-pyridinium chloride
CMMBr	*N*-Cetyl-*N*-methylmorpholinium bromide
CPCl	Cetylpyridinium chloride
CQCl	Cetylquinuclidinium chloride
CQOH	Cetylquinuclidinium hydroxyde
$CTA(SO_4)_{0.5}$	Cetyltrimethylammonium sulfate
CTABr	Cetyltrimethylammonium bromide
$CTABr_3$	Cetyltrimethylammonium tribromide
CTACl	Cetyltrimethylammonium chloride
CTAF	Cetyltrimethylammonium fluoride
CTA formate	Cetyltrimethylammonium formate
CTAIB	Cetyltrimethylammonium *o*-iodosobenzoate
$CTAN_3$	Cetyltrimethylammonium azide
$CTANO_3$	Cetyltrimethylammonium nitrate
CTAOAc	Cetyltrimethylammonium acetate

CTAOH	Cetyltrimethylammonium hydroxyde
CTAOHexanoate	Cetyltrimethylammonium hexanoate
CTAOMs	Cetyltrimethylammonium methanesulfonate
CTAOTs	Cetyltrimethylammonium *p*-toluenesulfonate
CTBABr	Cetyltributylammonium bromide
CTBAOH	Cetyltributylammonium hydroxide
CTEABr	Cetyltriethylammonium bromide
CTEAOH	Cetyltriethylammonium hydroxide
CTHEACl	Cetyltrihydroxyethylammonium chloride
CTPABr	Cetyltripropylammonium bromide
CTPAOH	Cetyltripropylammonium hydroxide
CTPAOMs	Cetyltripropylammonium mesylate
CTPeACl	Cetyltripentylammonium chloride
$Cu(DS)_2$	Copper didodecylsulfate
DDDABr	Didodecyldimethylammonium bromide
DDDACl	Didodecyldimethylammonium chloride
DDDAOH	Didodecyldimethylammonium hydroxide
$DDDA(SO_4)_{0.5}$	Didodecyldimethylammonium sulfate
DeTACl	Decyltrimethylammonium chloride
DODABr	Dioctadecyldimetylammonium bromide
DODACl	Dioctadecyldimetylammonium chloride
DTABr	Dodecyltrimethylammonium bromide
DTACl	Dodecyltrimethylammonium chloride
DTAOH	Dodecyltrimethylammonium hydroxide
HDS	Hydrogen dodecyl sulfate
HeTACl	Hexyltrimethylammonium chloride
MQBr	*N*-Myristylquinuclidinium bromide
MTABr	Myristyltrimethylammonium bromide
MTACl	Myristyltrimethylammonium chloride
$MTANO_3$	Myristyltrimethylammonium nitrate
OcTACl	Octyltrimethylammonium chloride
OdTACl	Octadecyltrimethylammonium chloride
pOOTABr	Paraoctyloxybenzyltrimethylammonium bromide
pOOTBABr	Paraoctyloxybenzyltributylammonium bromide
SB3-14	Myristyldimethylammonium propanesulfonate
SB3-16	Cetyldimethylammonium propanesulfonate
SB4-14	Myristyldimethylammonium butanesulfonate
SB5-14	Myristyldimethylammonium pentanesulfonate
SBBu3-14	Myristyldibutylammonium propanesulfonate
SBEt3-14	Myristyldiethylammonium propanesulfonate
SBPr3-14	Myristyldipropylammonium propanesulfonate
SBPr4-14	Myristyldipropylammonium butanesolfonate
SBPr5-14	Myristyldipropylammonium pentanesolfonate
SDHP	Sodium dihexadecylphosphate
SDS	Sodium dodecyl sulfate
TMABr	Tetramethylammonium bromide
TMACl	Tetramethylammonium chloride
Triton X-100	Polyethylene glycol *tert*-octylphenyl ether
$Zn(DS)_2$	Zinc didodecylsulfate

SUBSTRATES

6-NBIC	6-Nitrobenzisoxazole-3-carboxylate
DDT	1,1,1-Trichloro-2,2-bis(*p*-chlorophenyl)-ethane
DNCN	2,4-Dinitro-1-chloro-naphthalene
$DNPP^{2-}$	2,4-Dinitrophenylphosphate dianion
DTE	1,1-Diphenyl-2,2,2-trichloroetane
DTNB	5,5′-Dithiobis-(2-nitrobenzoic acid)
MCP	*N*-Methyl-4-cyanopyridinium ion
MeONs	Methylnaphthalene-2-sulfonate
MNTS	*N*-Methyl-*N*-nitroso-*p*-toluenesulfonamide
pNPDPP	*p*-Nitrophenyl diphenylphosphate

REFERENCES

1. J. N. Israelachvili, *Intermolecular and Surface Forces*, Academic Press, London, 1985.
2. F. Vögtle, *Supramolecular Chemistry*, Wiley, Chichester, UK, 1991.
3. J. H. Fuhrhop and J. Koning, *Membranes and Molecular Assemblies: The Synkinetic Approach*, The Royal Society of Chemistry, 1994.
4. G. S. Hartley, Q. Rev. *2*:152–183 (1948).
5. E. F. J. Duynstee and E. Grunwald, J. Am. Chem. Soc. *81*:4540–4542 (1959).
6. J. L. Kurz, J. Phys. Chem. *66*:2239–2246 (1962).
7. V. A. Motsavage and H. B. Kostenbauder, J. Colloid Sci. *18*:603–615 (1963).
8. J. H. Fendler and E. J. Fendler, *Catalysis in Micellar and Macromolecular Systems*, Academic Press, New York, 1975.
9. J. H. Fendler, *Membrane Mimetic Chemistry*, Wiley, New York, 1982.
10. L. S. Romsted, in *Micellization, Solubilization and Microemulsion* (K. L. Mittal, ed.), vol. 2, Plenum, New York, 1977, pp. 489–530.
11. L. S. Romsted, in *Surfactants in Solution* (K. L. Mittal and B. Lindman, eds.), vol. 2, Plenum, New York, 1984, p. 1015.
12. C. A. Bunton and G. Savelli, Adv. Phys. Org. Chem. *22*:213–309 (1986).
13. P. L. Luisi and F. J. Varela, Origins Life Evol. Biosphere *19*:633–643 (1989).
14. P. A. Bachamann, P. Walde, P. L. Luisi, and J. Lang, J. Am. Chem. Soc. *113*:8204–8209 (1991).
15. P. A. Bachmann, P. L. Luisi, and J. Lang, Nature *357*: 57–59 (1992).
16. P. Walde, R. Wick, M. Fresta, A. Mangone, and P. L. Luisi, J. Am. Chem. Soc. *116*:11649–11654 (1994).
17. K. Morigaki, S. Dallavalle, P. Walde, S. Colonna, and P. L. Luisi, J. Am. Chem. Soc. *119*:292–301 (1997).
18. R. Wich, P. Walde, and P. L. Luisi, J. Am. Chem. Soc. *117*:1435–1436 (1995).
19. K. Holmberg, Adv. Colloid Interface Sci. *51*:137–174 (1994).
20. J. Sjoblom, R. Lindberg, and S. E. Friberg, Adv. Colloid Interface Sci. *95*:125–287 (1996).
21. R. Schomächer, J. Chem. Res. (S):92–93 (1991).
22. M. J. Schwuger, K. Stickdorn, and R. Schomacker, Chem. Rev. *95*:849–864 (1995).
23. M. P. Pileni, *Structure and Reactivity in Reverse Micelles*, Elsevier, Amsterdam, 1989.
24. H. Kunieda and K. Nakamura, Adv. Mater. *4*:291–293 (1992).
25. H. Kunieda, K. Nakamura, and D. F. Evans, J. Am. Chem. Soc. *113*:1051–1052 (1991).
26. H. Kunieda and M. Yamagata, J. Colloid Interface Sci. *150*:277–280 (1992).
27. P. Mukerjee and K. J. Mysels, *Critical Micelle Concentration of Acqeous Surfactants Systems*, National Bureau of Standards, Washington, DC, 1970.
28. C. Tanford, *The Hydrophobic Effect: Formation of Micelles and Biological Membranes*, 2nd ed., Wiley, New York, 1980.
29. A. Ben-Naim, *Hydrophobic Interactions*, Plenum, New York, 1980.
30. Y. Ishikawa, H. Kuwahara, and T. Kunitake, J. Am. Chem. Soc. *116*:5579–5591 (1994).
31. S. Tascioglu, Tetrahedron *52*:11113–11152 (1996).
32. G. S. Hartley and R. C. Murray, Trans. Faraday Soc. *31*:183–189 (1935).
33. P. H. Elworthy and K. J. Mysel, J. Colloid Interface Sci. *21*:331–347 (1966).
34. P. Mukerjee, Adv. Colloid Interface Sci. *1*:241–275 (1967).
35. G. Stainsby and A. E. Alexander, Trans. Faraday Soc. *46*:587–597 (1950).
36. D. F. Evans and B. W. Ninham, J. Phys. Chem. *87*: 5025–5032 (1983).
37. D. F. Evans, D. J. Mitchell, and B. W. Ninham, J. Phys. Chem. *88*:6344–6348 (1984).
38. D. F. Evans, M. Allen, B. W. Ninham, and A. Fouda, J. Solution Chem. *13*:87–191 (1984).
39. J. B. Hayter, Langmuir *8*:2873–2876 (1992).
40. N. M. van Os, G. J. Daana, and G. Haandrikman, J. Colloid Interface Sci. *141*:199–217 (1991).
41. N. M. van Os, J. R. Haak, and L. A. M. Rupert, *Physico-Chemical Properties of Selected Anionic, Cationic and Nonionic Surfactants*, Elsevier Science, Amsterdam, 1993.
42. N. M. van Os, B. Smit, and S. Karaborni, Rec. Trav. Chim. Pays-Bas *113*:181–185 (1994).
43. D. W. R. Gruen, J. Colloid Interface Sci. *84*:281–283 (1981).
44. D. W. Gruen and E. H. B. de Lacey, in *Surfactants in Solution*, vol. 1, New York, Plenum, 1984, pp. 279–306.
45. J. Bocher, J. Brickmann, and P. Bopp, J. Phys. Chem. *98*:712–717 (1994).
46. J. Shelley, K. Watanabe, and M. L. Klein, Int. J. Quantum Chem. Quantum Biol. Symp. *17*:103–117 (1990).
47. S. Karaborni and J. P. O'Connell, J. Phys. Chem. *94*: 2624–2631 (1990).
48. B. Smith, P. A. J. Hilbers, K. Esselink, L. A. M. Rupert, N. M. van Os, and A. G. Schlijper, Nature *348*: 624–625 (1990).
49. S. Karborni, K. Esselink, P. A. J. Hilbers, B. Smith, N. M. van Os, and R. Zana, Science *266*:254–256 (1994).
50. M. A. Floriano and E. Caponetti, Langmuir *15*:3143–3151 (1999).
51. A. Khatory, F. Kern, F. Lequeux, J. Appell, G. Porte, N. Morie, A. Ott, and W. Urbach, Langmuir *9*:933–939 (1993).

52. R. Mikulski, Pol. J. Chem. *67*:1623–1629 (1993).
53. M. Leaver, I. Furó, and U. Olsson. Langmuir *11*: 1524–1529 (1995).
54. E. Y. Sheu and S. H. Chen, J. Phys. Chem. *92*:4466–4474 (1988).
55. F. H. Quina, P. M. Nassar, J. B. S. Bonilha, and B. L. Bales, J. Phys. Chem. *99*:17028–17031 (1995).
56. S. Kumar, V. K. Aswal, H. N. Singh, P. S. Goyal, and K. ud-Din, Langmuir *10*:4069–4072 (1994).
57. V. K. Aswal, P. S. Goyal, S. V. G. Menon, and B. A. Dasannacharya, Physica B *213&214*:607–609 (1995).
58. M. Delsanti, A. Mussaid, and J. P. Munch, J. Colloid Interface Sci. *157*:285–290 (1993).
59. P. van der Schoot and M. E. Cates, Langmuir *10*:670–679 (1994).
60. R. Nagarajan, Langmuir *10*:2028–2034 (1994).
61. D. A. Amos, S. Lynn, and C. J. Radke, Lamgmuir *14*: 2297–2306 (1998).
62. F. M. Menger, Acc. Chem. Res. *12*:111–117 (1979).
63. P. Fromherz, Chem. Phys. Lett. *77*:460–466 (1981).
64. K. A. Dill and P. J. Flory, Proc. Natl. Acad. Sci. U.S.A. *78*:676–680 (1981).
65. K. A. Dill, D. E. Koppel, R. S. Cantor, J. D. Dill, D. Bendedouch, and S.-H. Cheng, Nature *309*:42–45 (1984).
66. B. Jönsson, P.-G. Nilsson, B. Lindman, L. Guldbrand, and H. Wennerström, in *Surfactant in Solution* (K. L. Mittal and B. Lindman, eds.), vol. 1, Plenum, New York, 1984, pp. 3–21.
67. F. M. Menger, and C. E. Mounier, J. Am. Chem. Soc. *115*:12222–12223 (1993).
68. A. Chauduri and L. S. Romsted, J. Am. Chem. Soc. *113*:5052–5053 (1991).
69. L. S. Romsted and J. Yao, Langmuir *15*:326–336 (1999).
70. L. S. Romsted and J. Yao, Langmuir *12*:2425–2432 (1996).
71. O. A. El Seoud, A. Blasko, and C. A. Bunton, Ber. Bunsenges. Phys. Chem. *99*:1214–1220 (1995).
72. O. A. El Seoud, A. Blasko, and C. A. Bunton, Langmuir *10*:653–657 (1994).
73. M. E. L. McBain and E. Hutchinson, *Solubilization and Related Phenomena*, Academic Press, New York, 1955.
74. P. H. Elworthy, A. T. Florence, and C. B. Macfarlane, *Solubilization by Surface/Active Agents and Its Application in Chemistry and the Biological Sciences*, Chapman & Hall, London, 1968.
75. M. Almgren, F. Grieser, and J. K. Thomas, J. Am. Chem. Soc. *101*:279–291 (1979).
76. Y. Moroi, H. Noma, and R. Matuura, J. Phys. Chem. *87*:872–876 (1983).
77. C. A. Bunton, in *Surfactants in Solution* (K. L. Mittal and D. O. Shah, eds.), vol. 11, Plenum, New York, 1991, p. 17.
78. L. Sepulveda, E. Lissi, and F. Quina, Adv. Colloid Interface Sci. *25*:1–57 (1986).
79. F. H. Quina, E. O. Alonso, and J. P. S. Farah, J. Phys. Chem. *99*:11708–11714 (1995).
80. E. Abuin and E. Lissi, Bol. Soc. Chil. Quim. *42*:113–134 (1997).
81. M. A. Rodriguez, E. O. Alonso, C. Yihwa, J. P. S. Farah, and F. H. Quina, Langmuir *15*:6770–6774 (1999).
82. H. Abraham, H. S. Chadha, J. P. Dixon, C. Rafols, and C. Treiner, J. Chem. Soc. Perkin Trans. *2*:887–894 (1995).
83. M. H. Abraham, H. S. Chadha, J. P. Dixon, C. Rafols, and C. Treiner, J. Chem. Soc. Perkin Trans. *2*:19–24 (1997).
84. B. Lindman, O. Soderman, and H. Wennerstrom, in *Surfactant Solutions: New Methods of Investigation* (R. Zana, ed.), Marcel Dekker, New York, 1987.
85. J. C. Eriksson and G. Gillberg, Acta Chem. Scand. *20*: 2019–2027 (1966).
86. R. Bacaloglu, C. A. Bunton, G. Cerichelli, and F. Ortega, J. Phys. Chem. *93*:1490–1497 (1989).
87. Y. S. Kang, P. Baglioni, and L. Kevan, J. Phys. Chem. *95*:7944–7947 (1991).
88. Y. S. Kang, H. J. D. Mamanus, and L. Kevan, J. Phys. Chem. *96*:7473–7478 (1992).
89. C. A. Bunton, F. Nome, F. H. Quina, and L. S. Romsted, Acc. Chem. Res. *24*:358–363 (1991).
90. I. M. Cuccovia, I. N. da Silva, H. Chaimovich, and L. S. Romsted, Langmuir *13*:647–652 (1996).
91. J. D. Morgan, D. H. Napper, G. G. Warr, and S. K. Nicol, Langmuir *10*:797–801 (1994).
92. G. G. Warr, Langmuir *13*:1451–1456 (1997).
93. N. Kellaway and G. G. Warr, J. Colloid Interfase Sci. *193*:312–314 (1997).
94. L. S. Romsted and C.-O. Yoon, J. Am. Chem. Soc. *115*:989–994 (1993).
95. J. D. Morgan, D. H. Napper, and G. G. Warr, J. Phys. Chem. *99*:9458–9465 (1995).
96. G. Mancini, C. Schiavo, and G. Cerichelli, Langmuir *12*:3567–3573 (1996).
97. M. A. Morini, P. C. Schulz, and J. E. Puig, Colloid Polym. Sci. *274*:662–668 (1996).
98. A. Blaskó, C. A. Bunton, G. Cerichelli, and D. C. McKenzie, J. Phys. Chem. *97*:11324–11331 (1993).
99. S. J. Bachofer and U. Simonis, Langmuir *12*:1744–1754 (1996).
100. P. J. Kreke, L. J. Magid, and J. C. Gee, Langmuir *12*: 699–705 (1996).
101. K. Bijma, E. Rank, and J. B. F. N. Engberts, J. Colloid Interface Sci. *205*:245–256 (1998).
102. M. A. Cassidy and G. G. Warr, J. Phys. Chem. *100*: 3237–3240 (1996).
103. S. J. Bachofer, U. Simonis, and T. A. Nowich, J. Phys. Chem. *95*:480–488 (1991).
104. C. A. Bunton, M. M. Mhala, and J. R. Moffat, J. Org. Chem. *52*:3832–3835 (1987).
105. C. A. Bunton, M. M. Mhala, and J. R. Moffat, J. Phys. Chem. *93*:854–858 (1989).

106. S. Brochsztain, P. Berci-Filho, V. G. Toscano, H. Chaimovich, and M. J. Politi, J. Phys. Chem. *94*:6781–6785 (1990).
107. M. S. Baptista and M. J. Politi, J. Phys. Chem. *95*: 5936–5942 (1991).
108. D. J. Hodge, R. G. Laughlin, R. H. Ottewill, and A. R. Rennie, Langmuir *7*:878–884 (1991).
109. N. Kamenka, Y. Chevalier, and R. Zana, Langmuir *11*: 3351–3355 (1995).
110. N. Kamenka, M. Chorro, Y. Chevalier, H. Levy, and R. Zana, Langmuir *11*:4234–4240 (1995).
111. M. da Silva Baptista, I. Cuccovia, H. Chaimovich, M. J. Politi, and W. F. Reed, J. Phys. Chem. *96*:6442–6449 (1992).
112. P. Di Profio, R. Germani, G. Savelli, G. Cerichelli, M. Chiarini, G. Mancini, C. A. Bunton, and N. D. Gillitt, Langmuir *14*:2662–2669 (1998).
113. G. Weers, J. F. Rathman, F. U. Axe, C. A. Crichlow, L. D. Foland, D. R. Scheuing, R. J. Wiersema, and A. G. Zielske, Langmuir *7*:854–867 (1991).
114. Y. Chevalier and P. Le Perchec, J. Phys. Chem. *94*: 1768–1774 (1990).
115. Y. Chevalier, Y. Storet, S. Pourchet, and P. Le Perchec, Langmuir *7*:848–853 (1991).
116. Y. Chevalier, F. Mélis, and J. P. Dalbiez, J. Phys. Chem. *96*:8614–8619 (1992).
117. Y. Chevalier, N. Kamenka, M. Chorro, and R. Zana, Langmuir *12*:3225–3232 (1996).
118. J. K. Thomas, *Chemistry of Excitation at Interface*, ACS Monograph, American Chemical Society, Washington, DC, 1984.
119. F. M. Menger and C. E. Portnoy, J. Am. Chem. Soc. *89*:4698–4703 (1967).
120. F. M. Menger, Pure Appl. Chem. *51*:999–1007 (1979).
121. C. A. Bunton and L. Robinson, J. Am. Chem. Soc. *90*: 5972–5979 (1968).
122. M. A. B. Marin, F. Nome, D. Zanette, C. Zucco, and L. S. Romsted, J. Phys. Chem. *99*:10879–10882 (1995).
123. K. Martinek, A. K. Yatsimirski, A. V. Levashov, and I. V. Berezin, in *Solubilization and Microemulsion* (K. L. Mittal, ed.), vol. 2, Plenum, New York, 1977, p. 489.
124. A. A. Balekar and J. B. F. N. Engberts, J. Am. Chem. Soc. *100*:5914–5920 (1978).
125. I. M. Cuccovia, E. M. Schroter, P. M. Montiero, and H. Chaimovich, J. Org. Chem. *43*:2248–2252 (1978).
126. C. A. Bunton, L. S. Romsted, and G. Savelli, J. Am. Chem. Soc. *101*:1253–1259 (1979).
127. C. A. Bunton, S. Wright, P. M. Holland, and F. Nome, Langmuir *9*:117–120 (1993).
128. M. F. Neves, D. Zanette, F. Quina, M. T. Moretti, and F. Nome, J. Phys. Chem. *93*:1502–1505 (1989).
129. C. A. Bunton, N. D. Gillitt, and H. J. Foroudian, Langmuir *14*:4415–4421 (1998).
130. C. A. Bunton, H. J. Foroudian, and N. D. Gillitt, Langmuir *15*:1067–1074 (1999).
131. C. A. Bunton, P. Scrimin, and P. Tecilla, J. Chem. Soc. Perkin Trans. *2*:419–425 (1996).
132. R. A. Moss, T. F. Hendrckson, and G. O. Bizzigotti, J. Am. Chem. Soc. *108*:5520–5527 (1986).
133. T. Kunitake, Y. Okahata, and T. Sakamoto, J. Am. Chem. Soc. *98*:7799–7806 (1976).
134. T. Kunitake, S. Shinkai, and Y. Okahata, Bull. Chem. Soc. Jpn. *49*:540–545 (1976).
135. R. A. Moss, R. C. Nahas, and T. J. Lukas, Tetrahedron Lett. 507–510 (1978).
136. R. A. Moss, G. O. Bizzigotti, T. J. Lukas, and W. J. Sanders, Tetrahedron Lett. 3661–3664 (1978).
137. W. Tagaki, D. Fukushima, T. Eiki, and Y. Yano, J. Org. Chem. *44*:555–563 (1979).
138. C. A. Bunton, F. H. Hamed, and L. S. Romsted, J. Phys. Chem. *86*:2103–2108 (1982).
139. C. A. Bunton, L.-H. Gan, and G. Savelli, J. Phys. Chem. *87*:5491–5498 (1983).
140. Y. Ihara, M. Nango, Y. Kimura, and N. Kuroki, J. Am. Chem. Soc. *105*:1252–1255 (1983).
141. C. A. Bunton and E. L. Dorwin, J. Org. Chem. *51*: 4093–4096 (1986).
142. T. J. Connolly and V. C. Reinsborough, Can. J. Chem. *70*:1581–1585 (1992).
143. Y. Ihara, M. Nango, and J. Koga, Bull. Chem. Soc. Jpn. *60*:815–817 (1987).
144. E. Dutkiewicz, A. Jakubowska, and W. Knoche, Z. Phys. Chem. (Munich) *171*:255–260 (1991).
145. K. Ogino, K. Nishi, H. Yamamato, T. Yoshida, and W. Tagaki, Tetrahedron Lett. *31*:7023–7026 (1990).
146. F. M. Menger, L. H. Gan, E. Johnson, and D. H. Durst, J. Am. Chem. Soc. *109*:2800–2803 (1987).
147. R. Fornasier, P. Scrimin, P. Tecilla, and U. Tonellato, J. Am. Chem. Soc. *111*:224–229 (1989).
148. G. De Santi, P. Scrimin, and U. Tonellato, Tetrahedron Lett. *31*:4791–4794 (1990).
149. K. Ogino, N. Kashihara, T. Ueda, T. Isaka, T. Yoshida, and W. Tagaki, Bull. Chem. Soc. Jpn. *65*:373–384 (1992).
150. K. A. Wilk, J. Phys. Chem. *95*:3405–3409 (1991).
151. W. B. Gruhn and M. L. Bender, J. Am. Chem. Soc. *91*:5883–5885 (1969).
152. C. A. Bunton, L. Robinson, and M. Stam, J. Am. Chem. Soc. *92*:7393–7400 (1970).
153. C. A. Bunton and L. G. Ionescu, J. Am. Chem. Soc. *95*:2912–2917 (1973).
154. C. A. Bunton, A. A. Kamego, and P. Ng, J. Org. Chem. *39*:3469–3471 (1974).
155. C. A. Bunton and S. Diaz, J. Org. Chem. *41*:33–36 (1976).
156. C. A. Bunton and S. Diaz, J. Am. Chem. Soc. *98*: 5663–5671 (1976).
157. C. A. Bunton and C. H. Paik, J. Org. Chem. *41*:40–44 (1976).
158. K. Martinek, A. V. Levashov, and I. V. Berenzin, Tetrahedron Lett. 1275–1278 (1975).

159. C. A. Bunton and M. McAneny, J. Org. Chem. *41*:36–39 (1976).
160. A. Pillersdorf and J. Katzhendler, Isr. J. Chem. *18*: 330–338 (1979).
161. J. Kulic, J. Sussenbek, and A. Ptacek, Collect. Czech. Chem. Commun. *58*:2891–2895 (1993).
162. J. Epstein, J. J. Kaminski, N. Bodor, R. Enever, J. Sowa, and T. Higuchi, J. Org. Chem. *43*:2816–2821 (1978).
163. R. A. Moss, G. O. Bizzigotti, and C.-W. Huang, J. Am. Chem. Soc. *102*:754–762 (1980).
164. R. A. Moss and K. Y. Kim, Isr. J. Chem. *25*:11–14 (1985).
165. C. A. Bunton, M. M. Mhala, and J. R. Moffatt, J. Org. Chem. *52*:3832–3835 (1987).
166. T. J. Broxton, R. J. Christie, and X. Sango, J. Org. Chem. *54*:1919–1922 (1989).
167. T. J. Broxton, J. Org. Chem. *56*:5958–5960 (1991).
168. K. A. Wilk, J. Phys. Chem. *96*:901–904 (1992).
169. J. M. Brown, C. A. Bunton, and S. Diaz, J. Chem. Soc. Chem. Commun. 971–972 (1974).
170. T. E. Wagner, C. Hsu, and C. S. Pratt, J. Am. Chem. Soc. *89*:6366–6368 (1967).
171. G. Biresaw and C. A. Bunton, J. Org. Chem. *51*:2771–2777 (1986).
172. K. Ogino, K. Inoue, and W. Tagaki, Tetrahedron Lett. *33*:4191–4194 (1992).
173. Y. Y. Lim, E. H. L. Tan, and L. H. Gan, J. Colloid Interface Sci. *157*:442–446 (1993).
174. T. J. Broxton and R. A. Cox, Can. J. Chem. *71*:670–673 (1993).
175. V. Faivre, A. Brembilla, and P. Lochon, J. Mol. Catal. *85*:45–56 (1993).
176. G. R. Newkome and C. N. Moorefield, Polym. Prep. (Am. Chem. Soc. Div. Polym. Chem.) *34*:75–76 (1993).
177. J. G. J. Weijnen, A. Koudjis, and J. F. J. Engbersen, J. Org. Chem. *57*:7258–7265 (1992).
178. P. Scrimin, P. Tecilla, and U. Tonellato, J. Org. Chem. *59*:4194–4201 (1994).
179. P. Scrimin, P. Tecilla, and U. Tonellato, J. Org. Chem. *56*:161–166 (1991).
180. P. Scrimin, P. Tecilla, and U. Tonellato, J. Org. Chem. *59*:18–24 (1994).
181. W. Tagaki and K. Ogino, Top. Curr. Chem. *128*:143–147 (1985).
182. R. Fornasier, P. Scrimin, P. Tecilla, and U. Tonellato, J. Am. Chem. Soc. *111*:224–229 (1989).
183. J. G. J. Weijnen, A. Koudijs, and J. F. J. Engbersen, J. Chem. Soc. Perkin Trans. 2:829–834 (1992).
184. J. G. J. Weijnen, A. Koudijs, and J. F. J. Engbersen, J. Org. Chem. *57*:7258–7265 (1992).
185. P. Scrimin, P. Tecilla, and U. Tonellato, J. Am. Chem. Soc. *114*:5086–5092 (1992).
186. M. C. Cleij, P. Scrimin, P. Tecilla, and U. Tonellato, Langmuir *12*:2956–2960 (1996).
187. S. H. Gellman, R. Petter, and R. Breslow, J. Am. Chem. Soc. *108*:2388–2394 (1986).
188. E. Kimura, H. Hashimoto, and T. Koioke, J. Am. Chem. Soc. *118*:10963–10970 (1996).
189. G. Ghirlanda, P. Scrimin, and L. A. Echegoyen, Langmuir *12*:5188–5194 (1996).
190. P. Scrimin, G. Ghirlanda, P. Tecilla, and R. A. Moss, Langmuir *12*:6235–6241 (1996).
191. P. Scrimin, P. Tecilla, U. Tonellato, and C. A. Bunton, Colloids Surf. A Physicochem. Eng. Aspects *144*:71–79 (1998).
192. A. Domínguez and E. Iglesias, Langmuir *14*:2677–2683 (1998).
193. F. Nome, A. F. Rubira, C. Franco, and L. G. Ionescu, J. Phys. Chem. *86*:1881–1885 (1982).
194. E. Stadler, D. Zanette, M. C. Rezende, and F. Nome, J. Phys. Chem. *88*:1892–1896 (1984).
195. T. J. Broxton and D. B. Sango, Aust. J. Chem. *36*: 711–717 (1983).
196. E. B. Abuin, E. Lissi, R. Araujo, R. M. V. Alexeio, H. Chaimovich, N. Bianchi, L. Miola, and F. Quina, J. Colloid Interface Sci. *96*:293–295 (1983).
197. L. C. Mello Ferreira, C. Zucco, D. Zanette, and F. Nome, J. Phys. Chem. *96*:9058–9061 (1992).
198. R. Germani, P. P. Ponti, G. Savelli, N. Spreti, C. A. Bunton, and J. R. Moffatt, J. Chem. Soc. Perkin Trans. 2:401–404 (1989).
199. C. A. Bunton, J. Frankson, and L. S. Romsted, J. Phys. Chem. *84*:2607–2611 (1980).
200. C. A. Bunton, L.-H. Gan, J. R. Moffatt, L. S. Romsted, and G. Savelli, J. Phys. Chem. *85*:4118–4125 (1981).
201. P. Lianos and R. Zana, J. Phys. Chem. *87*:1289–1291 (1983).
202. V. Athanassakis, J. R. Moffatt, C. A. Bunton, R. B. Dorshow, G. Savelli, and D. F. Nicoli, Chem. Phys. Lett. *115*:467–471 (1985).
203. E. Rodenas and S. Vera, J. Phys. Chem. *89*:513–516 (1985).
204. C. R. A. Bertoncini, F. Nome, G. Cerichelli, and C. A. Bunton, J. Phys. Chem. *94*:5875–5878 (1990).
205. C. R. A. Bertoncini, M. S. Neves, F. Nome, and C. A. Bunton, Langmuir *9*:1274–1279 (1993).
206. D. Stigter, J. Phys. Chem. *68*:3603–3611 (1964).
207. D. Stigter, Prog. Colloid Polym. Sci. *65*:45–52 (1978).
208. G. M. Bell and A. J. Dunning, Trans. Faraday Soc. *66*:500–508 (1970).
209. M. R. Mille and G. Vanderkooi, J. Colloid Interface Sci. *59*:211–224 (1977).
210. G. Gunnarsson, B. Jonsson, and H. Wennerstrom, J. Phys. Chem. *84*:3114–3121 (1980).
211. C. A. Bunton and H. J. Foroudain, Langmuir *9*:2832–2835 (1993).
212. C. A. Bunton and J. R. Moffatt, J. Phys. Chem. *89*: 4166–4169 (1985).
213. C. A. Bunton and J. R. Moffatt, J. Phys. Chem. *90*: 538–541 (1986).

214. C. A. Bunton and J. R. Moffatt, J. Phys. Chem. *92*: 2896–2902 (1988).
215. E. Rodenas and F. Ortega, J. Phys. Chem. *91*:837–840 (1987).
216. H. A. Al-Lohedan, J. Chem. Soc. Perkin Trans. *2*: 1707–1713 (1995).
217. C. A. Bunton and J. R. Moffatt, Langmuir *8*:2130–2134 (1992).
218. H. A. Al-Lohedan, J. Chem. Soc. Perkin Trans. *2*: 1401–1406 (1990).
219. F. Ortega and E. Rodenas, J. Phys. Chem. *90*:2408–2413 (1986).
220. F. Ortega and E. Rodenas, Can. J. Chem. *67*:305–309 (1989).
221. E. Rodenas and S. Vera, J. Phys. Chem. *90*:3414–3417 (1986).
222. S. Vera and E. Rodenas, An. Quim. Ser. A *84*:174–178 (1988).
223. H. Chaimovich, V. R. Correia, P. S. Araujo, R. M. V. Aleixo, and I. M. Cuccovia, J. Chem. Soc. Perkin Trans. *2*:925–928 (1985).
224. E. Rodenas, C. Dolcet, and M. Valiente, J. Phys. Chem. *94*:1472–1477 (1990).
225. C. Dolcet and E. Rodenas, Can. J. Chem. *68*:932–938 (1990).
226. C. Dolcet and E. Rodenas, Colloids Surf. A Physicochem. Eng. Aspects *75*:39–50 (1993).
227. E. Rodenas, C. Dolcet, M. Valiente, and E. C. Valeron, Langmuir *10*:2088–2094 (1994).
228. H. Chaimovich, R. M. V. Aleixo, I. M. Cuccovia, D. Zanette, and F. H. Quina, in *Solution Behavior of Surfactants* (K. L. Mittal and E. J. Fendler, eds.), vol. 2, Plenum, New York, 1982, pp. 949–973.
229. C. Armstrong, W. Gotham, P. Jennings, J. Nikles, L. S. Romsted, M. Versace, and J. Waidlich, in *Surfactants in Solution* (K. L. Mittal and E. J. Fendler, eds.), vol. 9, Plenum, New York, 1989, p. 197.
230. Z.-M. He, J. A. Loughlin, and L. S. Romsted, Bol. Soc. Chil. Quim. *35*:43–53 (1990).
231. C. A. Bunton, M. M. Mhala, and J. R. Moffatt, J. Phys. Chem. *93*:7851–7856 (1989).
232. A. Blaskó, C. A. Bunton, C. Armstrong, W. Gotham, Z. M. He, J. Nikles, and L. S. Romsted, J. Phys. Chem. *95*:6747–6750 (1991).
233. A. Blaskó, C. A. Bunton, and S. Wright, J. Phys. Chem. *97*:5435–5442 (1993).
234. S. Amado, L. G. Río, J. R. Leis, and A. Ríos, Langmuir *13*:687–692 (1997).
235. M. N. Khan, J. Colloid Interface Sci. *170*:598–601 (1995).
236. M. N. Khan, Z. Arifin, M. N. Lasidek, M. A. M. Hanifiah, and G. Alex, Langmuir *13*:3959–3964 (1997).
237. D. Piszkiewicz, J. Am. Chem. Soc. *99*:1550–1557, 7695–7697 (1977).
238. S. K. Srivastava and S. S. Katiyar, Ber. Bunsenges. Phys. Chem. *82*:1214–1219 (1980).
239. C. Minero, E. Premauro, and E. Pelizzetti, Langmuir *4*:101–105 (1988).
240. R. Da Rocha Pereira, D. Zanette, and F. Nome, J. Phys. Chem. *94*:356–361 (1990).
241. D. M. Davies, N. D. Gillitt, and P. M. Paradis, J. Chem. Soc. Perkin Trans. *2*:659–666 (1996).
242. D. M. Davies and S. J. Foggo, J. Chem. Soc. Perkin Trans. *2*:247–251 (1998).
243. J. L. Kurz, J. Am. Chem. Soc. *85*:987–991 (1963).
244. J. Kraut, Science *242*:533–540 (1988).
245. N. J. Buurma, A. M. Herranz, and J. B. F. N. Engberts, J. Chem. Soc. Perkin Trans. *2*:113–119 (1999).
246. A. R. Olson and J. R. J. Simonson, J. Chem. Phys. *17*:1167–1173 (1949).
247. J. N. Brønsted, Z. Phys. Chem. *102*:169–207 (1922).
248. J. N. Brønsted, Z. Phys. Chem. *115*:337–364 (1925).
249. C. A. Bunton, J. Mol. Liq. *72*:231–249 (1997).
250. E. H. Cordes, Pure Appl. Chem. *50*:617–625 (1978).
251. P. L. Lòpez-Cornejo, R. Jimènez, M. L. Moyà, F. Sànchez, and J. Burgess, Langmuir *12*:4981–4986 (1996).
252. R. Jimènez, M. M. Graciami, R. Rodriguez, M. L. Moyà, F. Sànchez, and F. Lòpez-Cornejo, Langmuir *13*:187–191 (1997).
253. F. Sanchez, M. L. Moyà, A. Rodriguez, R. Jemènez, C. Gòmez-Herrera, C. Yanès, and P. Lòpez-Cornejo, Langmuir *13*:3084–3089 (1997).
254. F. Muriel-Delgado, R. Jimènez, C. Gòmez-Herrera, and F. Sanchez, Langmuir *15*:4344–4350 (1999).
255. D. S. Paul and K. G. Kemp, J. Am. Chem. Soc. *97*: 7305–7312 (1975).
256. A. J. Kirby and A. G. Varvoglis, J. Am. Chem. Soc. *89*:415–423 (1967).
257. L. Mandolini, Adv. Phys. Org. Chem. *22*:1–111 (1986).
258. V. R. Correia, I. M. Cuccovia, M. Stelmo, and H. Chaimovich, J. Am. Chem. Soc. *114*:2144–2146 (1992).
259. R. Germani, P. P. Ponti, T. Romeo, G. Savelli, N. Spreti, G. Cerichelli, L. Luchetti, G. Mancini, and C. A. Bunton, J. Phys. Org. Chem. *2*:553–558 (1989).
260. F. Del Rosso, A. Bartoletti, P. Di Profio, R. Germani, G. Savelli, A. Blaskó, and C. A. Bunton, J. Chem. Soc. Perkin Trans. *2*:673–678 (1995).
261. G. Cerichelli, L. Luchetti, G. Mancini, M. N. Muzzioli, R. Germani, P. P. Ponti, N. Spreti, and G. Savelli, J. Chem. Soc. Perkin Trans. *2*:1081–1085 (1989).
262. J. B. F. N. Engberts, Pure Appl. Chem. *64*:1653–1660 (1992).
263. P. Di Profio, R. Germani, G. Savelli, G. Cerichelli, N. Spreti, and C. A. Bunton, J. Chem. Soc. Perkin Trans. *2*:1505–1509 (1996).
264. R. Germani, P. P. Ponti, G. Savelli, N. Spreti, A. Cipiciani, G. Cerichelli, and C. A. Bunton, J. Chem. Soc. Perkin Trans. *2*:1767–1771 (1989).
265. C. A. Bunton, E. L. Dorwin, G. Savelli, and V. C. Si, Rec. Trav. Pays-Bas *109*:64–69 (1990).

266. M. Okahara, A. Masuyama, Y. Sumida, and Y. P. Zhu, J. Jpn. Oil Chem. Soc. (Yukagaku) *37*:746–748 (1988).
267. Y. P. Zhu, A. Masuyama, and M. Okahara, J. Am. Oil. Chem. Soc. *68*:268–271 (1991).
268. M. J. Rosen, Z. H. Zhu, and X. Y. Hua, J. Am. Oil. Chem. Soc. *69*:30–33 (1992).
269. A. Masuyama, T. Hirono, Y. P. Zhu, M. Okahara, and M. J. Rosen, J. Jpn. Oil Chem. Soc. (Yukagaku) *41*: 301–305 (1992).
270. F. Devinsky, I. Lacko, F. Bittererova, and L. Tomeckova, J. Colloid Interface Sci. *114*:314–322 (1986).
271. R. Zana, M. Benrraou, and R. Rueff, Langmuir *7*: 1072–1075 (1991).
272. F. M. Menger and C. A. Littau, J. Am. Chem. Soc. *113*:1451–1452 (1991).
273. F. M. Menger and C. A. Littau, J. Am. Chem. Soc. *115*:10083–10090 (1993).
274. M. Rosen, J. Chem. Tech. *23*:30–33 (1993).
275. L. D. Song and M. J. Rosen, Langmuir *12*:1149–1153 (1996).
276. G. Cerichelli, L. Luchetti, G. Mancini, and G. Savelli, Langmuir *15*:2631–2634 (1999).
277. M. S. Patel, K. Bijma, and J. B. F. N. Engberts, Langmuir *10*:2491–2492 (1994).
278. G. Cerichelli, G. Mancini, L. Luchetti, G. Savelli, and C. A. Bunton, J. Phys. Org. Chem. *4*:71–76 (1991).
279. G. Cerichelli, G. Mancini, L. Luchetti, G. Savelli, and C. A. Bunton, Langmuir *10*:3982–3987 (1994).
280. L. Brinchi, P. Di Profio, R. Germani, V. Giacomini, G. Savelli, and C. A. Bunton, Langmuir *16*:222–226 (2000).
281. C. A. Bunton, in *Nucleophilicity* (J. M. Harris and S. P. McManus, eds.), Adv. Chem. Ser. No. 215, American Chemical Society, Washington, DC, 1987, pp. 425–441.
282. N. Fadnavis and J. B. F. N. Engberts, J. Org. Chem. *47*:152–154 (1982).
283. N. J. Buurma, A. M. Herranz, and J. B. F. N. Engberts, J. Chem. Soc. Perkin Trans. *2*:113–119 (1999).
284. L. Brinchi, P. Di Profio, R. Germani, G. Savelli, and C. A. Bunton, J. Chem. Soc. Perkin Trans. *2*:361–364 (1998).
285. C. A. Bunton, M. M. Mhala, and J. R. Moffatt, J. Org. Chem. *50*:4921–4924 (1985).
286. A. Bartoletti, S. Bartolini, R. Germani, G. Savelli, and C. A. Bunton, J. Chem. Soc. Perkin Trans. *2*:723–728 (1994).
287. R. Germani, G. Savelli, N. Spreti, G. Cerichelli, G. Mancini, and C. A. Bunton, Langmuir *9*:61–65 (1993).
288. H. A. Al-Lohedan, J. Chem. Soc. Perkin Trans. *2*: 1181–1185 (1989).
289. C. Bonan, R. Germani, P. P. Ponti, G. Savelli, G. Cerichelli, R. Bacaloglu, and C. A. Bunton, J. Phys. Chem. *94*:5331–5336 (1990).
290. R. Germani, G. Savelli, T. Romeo, N. Spreti, G. Cerichelli, and C. A. Bunton, Langmuir *9*:55–60 (1993).
291. K. A. Wilk and B. Burczyk, J. Phys. Chem. *93*:8219–8223 (1989).
292. K. A. Wilk, J. Phys. Chem. *93*:7432–7435 (1989).
293. L. Brinchi, P. Di Profio, R. Germani, G. Savelli, and C. A. Bunton, Langmuir *13*:4583–4587 (1997).
294. R. G. Pearson, Science *151*:1721–1727 (1966).
295. R. G. Pearson and J. Songstad, J. Am. Chem. Soc. *89*: 1827–1836 (1967).
296. R. Bacaloglu, C. A. Bunton, and F. Ortega, J. Phys. Chem. *93*:1497–1502 (1989).
297. R. Bacaloglu, C. A. Bunton, G. Cerichelli, and F. Ortega, J. Phys. Chem. *94*:5068–5073 (1990).
298. R. Bacaloglu, A. Blaskó, C. A. Bunton, G. Cerichelli, and F. Ortega, J. Phys. Chem. *94*:5062–5068 (1990).
299. G. Cerichelli, M. Coreno, and G. Mancini, J. Colloid Interface Sci. *158*:33–39 (1993).
300. L. Brinchi, R. Germani, G. Savelli, N. Spreti, R. Ruzziconi, and C. A. Bunton, Langmuir *14*:2656–2661 (1998).
301. T. J. Broxton, J. R. Christie, and R. P.-T. Chung, Aust. J. Chem. *42*:855–864 (1989).
302. C. D. Ritchie and M. Sawada, J. Am. Chem. Soc. *99*: 3754–3761 (1977).
303. C. A. Bunton, H. J. Foroudian, N. D. Gillitt, and C. R. Whiddon, J. Colloid Interface Sci. *215*:64–71 (1999).
304. C. A. Bunton, S. Wright, P. M. Holland, and F. Nome, Langmuir *9*:117–120 (1993).
305. H. A. Al-Lohedan, A. M. Al-Sulaim, A. S. Al-Ayed, and Z. A. Issa, J. Chem. Res. (S):470–471 (1993).
306. M. Politi, I. M. Cuccovia, H. Chaimovich, M. L. C. Almeida, J. B. S. Bonilha, and F. H. Quina, Tetrahedron Lett. 115–118 (1978).
307. M. J. Politi and H. Chaimovich, J. Phys. Org. Chem. *4*:207–216 (1991).
308. H. A. Al-Lohedan, A. M. Al-Sulaim, A. S. Al-Ayed, and Z. A. Issa, J. Chem. Res. (S) 100–101 (1993).
309. H. A. Al-Lohedan, J. Chem. Res (S) 200 (1991).
310. F. P. Cavasino, R. Cervellati, R. Lombardo, and M. L. Turco Livieri, J. Phys. Chem. *103*:4285–4291 (1999).
311. H. A. Al-Lohedan, J. Phys. Chem. *94*:8226–8231 (1990).
312. G. B. van de Langkruis and J. B. F. N. Engberts, J. Org. Chem. *49*:4152–4157 (1984).
313. G. Cerichelli, C. Grande, L. Luchetti, and G. Mancini, J. Org. Chem. *56*:3025–3030 (1991).
314. T. J. Broxton and V. Marcou, J. Org. Chem. *56*:1041–1044 (1991).
315. C. A. Bunton, M. M. Mhala, and J. R. Moffatt, J. Phys. Chem. *93*:854–858 (1989).
316. P. Di Profio, L. Brinchi, R. Germani, G. Savelli, G. Cerichelli, and C. A. Bunton, J. Chem. Soc. Perkin Trans. in press.
317. L. Brinchi, C. Dionigi, P. Di Profio, R. Germani, G. Savelli, and C. A. Bunton, J. Colloid Interface Sci. *211*:179–184 (1999).

318. S. Brochsztain, P. B. Filho, V. G. Toscano, H. Chaimovich, and M. J. Politi, J. Phys. Chem. *94*:6781–6785 (1990).
319. L. Brinchi, R. Germani, G. Savelli, and C. A. Bunton, J. Phys. Org. Chem. *12*:890–894 (1999).
320. A. D. Bangham and R. W. Horne, J. Mol. Biol. *8*: 660–668 (1964).
321. T. Kunitake, Y. Okahata, K. Tamaki, F. Kumamuro, and M. Yakayanagi, Chem. Lett. 387–390 (1977).
322. R. A. Mortara, F. H. Quina, and H. Chaimovich, Biochem. Biophys. Res. Commun. *81*:1080–1086 (1978).
323. Y. Q. Liang, L. X. Wu, Y. C. Tian, Z. Q. Zhang, and H. D. Chen, J. Colloid Interface Sci. *178*:703–713 (1996).
324. H. Hoffmann, D. Grabner, U. Hornfeck, and G. Platz, J. Phys. Chem. B *103*:611–614 (1999).
325. Y. Hisaeda, E. Ohshima, J. Kikuchi, and Y. Murakami, Tetrahedron Lett. *38*:6713–6716 (1997).
326. A. T. Florence, P. Arunothayanum, S. Kiri, M. S. Bernard, and I. F. Uchegbu, J. Phys. Chem. B *103*:1995–2000 (1999).
327. D. A. Vanhal, J. A. Bouwstra, A. Vanrensen, E. Jeremiasse, T. Devringer, and H. E. Junginger, J. Colloid Interface Sci. *178*:263–273 (1996).
328. M. Ambühl, F. Bangerter, P. L. Luisi, P. Skrabal, and H. J. Watzke, Langmuir *9*:36–38 (1993).
329. M. Villeneuve, S. Kaneshina, T. Imae, and M. Aratono, Langmuir *15*:2029–2036 (1999).
330. E. F. Marques, O. Regev, A. Khan, M. D. Miguel, and B. Lindman, J. Phys. Chem. B *102*:6746–6758 (1998).
331. M. Bergstrom and J. S. Pedersen, Langmuir *14*:3754–3761 (1998).
332. D. Duque, P. Tarazona, and E. Chacón, Langmuir *14*: 6827–6834 (1998).
333. R. Wick, P. Walde, and P. L. Luisi, J. Am. Chem. Soc. *117*:1435–1436 (1995).
334. A. Veronese and P. L. Luisi, J. Am. Chem. Soc. *120*: 2662–2663 (1998).
335. P. L. Luisi and P. Walde, *Giant Vesicles*, Wiley, Chichester, UK, 1999.
336. H. Ringsdorf, B. Schlarb, and I. Venzmer, Angew. Chem. Int. Ed. Engl. *27*:113–158 (1988).
337. F. M. Menger and K. D. Gabrielson, Angew. Chem. Int. Ed. Engl. *34*:2091–2106 (1995).
338. F. M. Menger and I. S. Keiper, Adv. Mater. *10*:888–890 (1998).
339. E. M. Menger and I. S. Keiper, Curr. Opin. Chem. Biol. *2*:726–732 (1998).
340. E. Sackmann, in *Structure and Dynamics of Membranes. Handbook of Biological Physics* (R. Lipowsky and E. Sackmann, eds.), vol. 1, Elsevier, Amsterdam, 1995, p. 213.
341. A. M. Carmona-Ribeiro, Synthetic amphiphile vescicles, Chem. Soc. Rev. 209–214 (1992).
342. A. M. Carmona-Ribeiro and H. Chaimovich, Biochim. Biophys. Acta *733*:172–179 (1983).
343. A. M. Carmona-Ribeiro, L. S. Yoshida, A. Sesso, and H. Chaimovich, J. Colloid Interface Sci. *100*:433–443 (1984).
344. Y.-C. Chung and S. L. Regen, Langmuir *9*:1937–1939 (1993).
345. D. B. Nascimento, R. Rapuano, M. M. Lessa, and A. M. Carmona-Ribeiro, Langmuir *14*:7387–7391 (1998).
346. C. Kang and A. Khan, J. Colloid Interface Sci. *156*: 218–228 (1993).
347. O. Regev, C. Kang, and A. Khan, J. Phys. Chem. *98*: 6619–6625 (1994).
348. J. H. Fendler and W. L. Hinze, J. Am. Chem. Soc. *103*:5439–5447 (1981).
349. I. M. Cuccovia, F. H. Quina, and H. Chaimovich, Tetrahedron *38*:917–920 (1982).
350. H. Chaimovich, J. B. S. Bonilha, D. Zanette, and I. M. Cuccovia, in *Surfactants in Solution* (K. L. Mittal and B. Lindmann, eds.), New York, Plenum, 1984, pp. 1121–1138.
351. E. A. Abuin, E. B. Lissi, E. Backer, A. Zanocco, and D. Whitten, J. Phys. Chem. *93*:4886–4890 (1989).
352. I. M. Cuccovia, H. Chaimovich, E. B. Lissi, and E. A. Abuin, Langmuir *6*:1601–1604 (1990).
353. J. Furhop and J. Mathieu, Angew. Chem. Int. Ed. Engl. *23*:100–113 (1984).
354. H. Chaimovich and I. M. Cucovia, Prog. Colloid Polym. Sci. *103*:67–77 (1997).
355. A. M. Carmona-Ribeiro, L. S. Yoshida, and H. Chaimovich, J. Phys. Chem. *89*:2928–2933 (1985).
356. I. M. Cuccovia, E. S. Feitosa, H. Chaimovich, and W. Reed, J. Phys. Chem. *94*:3722–3725 (1990).
357. M. K. Kawamuro, H. Chaimovich, E. B. Abuin, E. A. Lissi, and I. M. Cuccovia, J. Phys. Chem. *95*:1458–1463 (1991).
358. I. M. Cuccovia and H. Chaimovich, Bol. Soc. Chil. Quim. *35*:39–42 (1990).
359. R. A. Moss, S. Ganguli, Y. Okumura, and T. Fujita, J. Am. Chem. Soc. *112*:6391–6392 (1990).
360. E. S. Feitosa, A. Agostinho, and H. Chaimovich, Langmuir *9*:702–707 (1993).
361. I. M. Cuccovia, M. K. Kawamuro, M. A. K. Krutman, and H. Chaimovich, J. Am. Chem. Soc. *111*:365–366 (1989).
362. R. A. Moss, S. Bhattacharya, P. Scrimin, and S. Swarup, J. Am. Chem. Soc. *109*:5740–5744 (1987).
363. R. A. Moss, S. Bhattacharya, and S. Chatterjee, J. Am. Chem. Soc. *111*:3680–3687 (1989).
364. R. A. Moss, Pure Appl. Chem. *66*:851–858 (1994).
365. S. A. Buckingham, C. J. Garvey, and G. G. Warr, J. Phys. Chem. *97*:10236–10244 (1993).
366. C. E. Drennan, R. J. Hughes, V. C. Reinsborough, and O. O. Soriyan, Can. J. Chem. *76*:152–157 (1998).
367. M. del M. Graciani, A. Rodriguez, G. Fernandez, and M. L. Moya, Langmuir *13*:4239–4245 (1997).
368. T. Kunitake, Y. Okahata, R. Ando, S. Shinkai, and S. Hirakawa, J. Am. Chem. Soc. *102*:7877–7881 (1980).

369. C. A. Bunton, Y. S. Hong, L. S. Romsted, and C. Quan, J. Am. Chem. Soc. *103*:5784–5788 (1981).
370. G. Biresaw, C. A. Bunton, C. Quan, and Z.-Y. Yang, J. Am. Chem. Soc. *106*:7178–7184 (1984).
371. G. Biresaw and C. A. Bunton, J. Phys. Chem. *90*: 5849–5853 (1986).
372. G. Biresaw and C. A. Bunton, J. Phys. Chem. *90*: 5854–5858 (1986).
373. G. Gerichelli, G. Mancini, L. Luchetti, G. Savelli, and C. A. Bunton, Langmuir *10*:3982–3987 (1994).
374. A. Blaskó, C. A. Bunton, and H. Fourodian, J. Colloid Interface Sci. *163*:500–506 (1994).
375. R. Bacaloglu and C. A. Bunton, J. Colloid Interface Sci. *153*:140–151 (1992).
376. N. Shufang, K. R. Gopidas, N. J. Turro, and G. Gabor, Langmuir *8*:1271–1277 (1992).
377. C. A. Bunton and C. Quan, J. Org. Chem. *50*:3230–3234 (1985).
378. C. Tondre and M. Hebrant, J. Mol. Liq. *72*:279–294 (1997).
379. J. F. Scamehorn, S. D. Christian, and R. T. Ellington, in *Surfactant-Based Separation Processes* (J. F. Scamehorn and J. H. Harwell, eds.), Surfactant Science Series, Marcel Dekker, New York, 1989.
380. E. Pramauro, A. Bianco Prevot, P. Savarino, G. Viscardi, M. De la Guardia, and E. P. Cardells, Analyst *118*:23–27 (1993).
381. H. Uchiyama, S. D. Christian, E. E. Tucker, and J. F. Scamehorn, J. Colloid Interface Sci. *163*:493–499 (1994).
382. M. Hérbant, P. Mettelin, C. Tondre, J. P. Joly, C. Larpent, and X. Chasseray, Colloids Surf. A Physicochem. Eng. Aspects *75*:257–267 (1993).
383. W. L. Hinze, R. W. Williams, Z. S. Fu, Y. Suzuki, and F. H. Quina, Colloids Surf. *48*:79–94 (1990).
384. A. L. Creagh, B. B. Hasenack, A. Van der Padt, E. J. Sudhölter, and K. Van't Riet, Biotechnol. Bioeng. *44*: 690–698 (1994).
385. D. C. Tickle, G. N. Okafo, P. Camilleri, R. F. D. Jones, and A. J. Kirby, Anal. Chem. *66*:4121–4126 (1994).
386. V. de Biasi, J. Senior, J. A. Zukowski, R. C. Haltiwanger, D. E. Eggleston, and P. Camilleri, J. Chem. Soc. Chem. Commun. 1575–1576 (1995).
387. M. Bouzige, G. Okafo, D. Dhanak, and P. Camilleri, J. Chem. Soc. Chem. Commun. 671–672 (1996).
388. D. Tickle, A. George, K. Jennings, P. Camilleri, and A. J. Kirby, J. Chem. Soc. Perkin Trans. *2*:467–474 (1998).
389. A. Xenakis and C. Tondre, J. Phys. Chem. *87*:4737–4743 (1983).
390. C. Tondre and A. Xenakis, in *Surfactants in Solution* (K. L. Mittal and B. Lindman, eds.), vol. 3, Plenum, New York, 1984, pp. 1881–1896.
391. A. Derouiche and C. Tondre, J. Chem. Soc. Faraday Trans. 1 *85*:3301–3308 (1989).
392. A. Derouiche and C. Tondre, Colloids Surf. *48*:243–258 (1990).
393. C. Tondre and A. Derouiche, J. Phys. Chem. *94*:1624–1626 (1990).
394. D. W. Osborne, A. J. Ward, and K. J. O'Neill, J. Pharm. Pharmacol. *43*:451–454 (1991).
395. J. Kemken, A. Ziegler, and B. W. Muller, Pharm. Res. *9*:554–558 (1992).
396. T. Akao, T. Osaki, J. Mitoma, A. Ito, and T. Kunitake, Bull. Chem. Soc. Jap. *64*:3677–3681 (1991).
397. R. Levantis and J. R. Silvius, Biochim. Biophys. Acta *1023*:124–132 (1990).
398. P. Pinnaduwage, L. Schmitt, and L. Huang, Biochim. Biophys. Acta *985*:33–37 (1989).
399. P. L. Felgner, T. R. Gadek, M. Holm, R. Ronas, H. W. Chan, M. Wenz, J. P. Northrop, G. M. Ringold, and M. Danielsen, Proc. Natl. Acad. Sci. U.S.A. *84*:7413–7417 (1987).
400. B. J. Roessler and B. L. Davidson, Neurosci. Lett. *167*:5–10 (1994).
401. J. Y. Legendre and F. C. Szoka, Proc. Natl. Acad. Sci. U.S.A. *90*:893–897 (1993).
402. Y. Watanabe, H. Nomoto, R. Takezawa, N. Miyoshi, and T. Akaike, J. Biochem. *116*:1220–1226 (1994).
403. I. Van der Woude, H. W. Visser, M. B. A. Ter Beest, A. Wagenaar, M. H. J. Ruiters, J. B. F. N. Engberts, and D. Hoekstra, Biochim. Biophys. Acta *1240*:34–40 (1995).
404. H. Gherson, R. Ghirlando, S. B. Guttman, and A. Minsky, Biochemistry *32*:7143–7151 (1993).
405. G. Mathis, P. Leempoel, J. C. Ravey, C. Selve, and J. J. Delpuech, J. Am. Chem. Soc. *106*:6162–6171 (1984).
406. G. N. Tápias, S. M. Sicchierolli, E. M. Mamizuka, and A. M. Carmona-Ribeiro, Langmuir *10*:3461–3465 (1994).
407. S. M. Sicchierolli, E. M. Mamizuka, and A. M. Carmona-Ribeiro, Langmuir *11*:2991–2995 (1995).
408. L. M. S. Martins, E. M. Mamizuka, and A. M. Carmona-Ribeiro, Langmuir *13*:5583–5587 (1997).
409. A. M. Carmona-Ribeiro, F. Ortis, R. I. Schumacher, and M. C. S. Armelin, Langmuir *13*:2215–2218 (1997).
410. L. R. Tsuruta, W. Quintilio, M. H. B. Costa, and A. M. Carbona-Ribeiro, J. Lipids Res. *38*:2003–2011 (1997).
411. P. L. Felgner, Adv. Drug Delivery Rev. *5*:163–187 (1990).
412. J. P. Behr, Acc. Chem. Res. *26*:274–278 (1993).
413. A. M. Carmona-Ribeiro and B. R. Midmore, Langmuir *8*:801–806 (1992).
414. M. M. Lessa and A. M. Carmona-Ribeiro, J. Colloid Interface Sci. *182*:166–171 (1996).
415. L. R. Tsuruta and A. M. Carmona-Ribeiro, J. Phys. Chem. *100*:7130–7134 (1996).
416. L. R. Tsuruta, M. M. Lessa, and A. M. Carmona-Ribeiro, J. Colloid Interface Sci. *175*:470–475 (1995).
417. R. Rapuano and A. M. Carmona-Ribeiro, J. Colloid Interface Sci. *193*:104–111 (1997).

418. L. C. Salay and A. M. Carmona-Ribeiro, J. Phys. Chem. B *102*:4011–4015 (1998).
419. D. Katz, C. A. Kraaijeveld, and H. Snippe, in *The Theory and Practical Application of Adjuvants* (D. E. S. Stewart-Tull, ed.), Wiley, New York, 1994.
420. R. H. Smith and B. Ziola, Immunology *58*:245–250 (1986).
421. F. M. Menger and H. Park, Rec. Trav. Chim. Pays-Bas *113*:176–180 (1994).
422. S. G. Oh, J. Kizling, and K. Holmberg, Colloids Surf. A Physicochem. Eng. Aspects *97*:169–179 (1995).
423. R. A. Moss, A. T. Kotchevar, B. D. Park, and P. Scrimin, Langmuir *12*:2200–2206 (1996).
424. R. A. Moss and S. Bose, Tetrahedron Lett. *38*:965–968 (1997).
425. D. A. Jaeger and M. R. Frey, J. Org. Chem. *47*:311–315 (1982).
426. D. A. Jaeger, J. Jamrozik, T. G. Golich, M. W. Clennan, and J. Mohebalian, J. Am. Chem. Soc. *111*:3001–3006 (1989).
427. D. A. Jaeger, Y. M. Sayed, and A. K. Dutta, Tetrahedron Lett. *31*:449–450 (1990).
428. K. R. Nivalkar, C. D. Mudaliar, and S. H. Mashraqui, J. Chem. Res. (S):98–99 (1992).
429. C. D. Mudaliar and S. H. Mashraqui, J. Chem. Res. (S):174–175 (1994).
430. S. Kobayashi, T. Busujima, and S. Nagayama, Synlett 545–546 (1999).
431. K. Manabe, Y. Mori, and S. Kobayashi, Synlett 1401–1402 (1999).
432. T. Akiyama, J. Takaya, and H. Kagoshima, Synlett 1426–1428 (1999).
433. S. Kobayashi, T. Wakabayashi, and H. Oyamada, Chem. Lett. 831–832 (1997).
434. S. Kobayashi, T. Wakabayashi, S. Nagayama, and H. Oyamada, Tetrahedron Lett. *38*:4559–4562 (1997).
435. S. Kobayashi, T. Busujima, and S. Nagayama, J. Chem. Soc. Chem. Commun. 19–20 (1998).
436. R. P. Bonar-Law, J. Org. Chem. *61*:3623–3634 (1996).
437. C. Siswanto, T. Battal, O. E. Schuss, and J. F. Rathman, Langmuir *13*:6047–6052 (1997).
438. T. Battal, C. Siswanto, and J. F. Rathman, Langmuir *13*:6053–6057 (1997).
439. L. Brinchi, P. Di Profio, R. Germani, G. Savelli, and C. A. Bunton, Colloids Surf. A Physicochem. Eng. Aspects *132*:303–314 (1998).
440. G. Cerichelli, L. Luchetti, and G. Mancini, Tetrahedron Lett. *30*:6209–6210 (1989).
441. G. Cerichelli, G. Mancini, and L. Luchetti, Tetrahedron *50*:3797–3802 (1994).
442. D. A. Jaeger and J. Wang, Tetrahedron Lett. *33*:6415–6418 (1992).
443. S. Otto, B. F. N. Engberts, and J. C. T. Kwak, J. Am. Chem. Soc. *120*:9517–9525 (1998).
444. D. A. Jaeger, H. Shinozaki, and P. A. Goodson, J. Org. Chem. *56*:2482–2489 (1991).
445. D. Monti, P. Tagliatesta, G. Mancini, and T. Boschi, Angew. Chem. Int. Ed. Engl. *37*:1131–1133 (1998).
446. C. Denis, B. Laignel, D. Plusquellec, J. Y. Le Marouille, and A. Botrel, Tetrahedron Lett. *37*:53–56 (1996).
447. M. C. Cleij, W. Drenth, and R. J. M. Nolte, Rec. Trav. Chim. Pays-Bas *112*:1–6 (1993).
448. S. I. Goldberg, N. Baba, R. L. Green, R. Poudian, J. Stowers, and R. B. Dunlop, J. Am. Chem. Soc. *100*:6768–6769 (1978).
449. Y. M. Zhang and W. Li, Synth. Commun. *18*:1685–1689 (1988).
450. Y. M. Zhang, W. Q. Fan, P. Lu, and W. Wang, Synth. Commun. *18*:1495–1500 (1988).
451. Y. M. Zhang, C. L. Fu, and W. Q. Fan, Chin. J. Chem. *1*:89–96 (1990).
452. Y. Zhang and W. Wu, Tetrahedron Asymmetry *8*:3575–3578 (1997).
453. Y. Zhang and W. Wu, Tetrahedron Asymmetry *8*:2723–2725 (1997).
454. M. J. Diego-Castro and H. C. Hailes, J. Chem. Soc. Chem. Commun. 1549–1550 (1998).
455. D. Brosch and W. Kirmse, J. Org. Chem. *58*:1118–1121 (1993).
456. R. Selke, J. Holz, A. Riepe, and A. Börner, Chem. Eur. J. *4*:769–771 (1998).
457. G. M. Whitesides, Angew. Chem. Int. Ed. Engl. *29*:1209–1218 (1990).
458. J. M. Lehn, *Supramolecular Chemistry. Concepts and Perspectives. A Personal Account*, VCH, Weinheim, 1995.
459. G. Cerichelli, L. Luchetti, G. Mancini, G. Savelli, and C. A. Bunton, Langmuir *12*:2348–2352 (1996).
460. A. Cuenca, C. Bruno, and A. Taddei, Tetrahedron *50*:1927–1934 (1994).
461. A. Castro, J. R. Leis, and M. E. Peña, J. Chem. Soc. Perkin Trans. *2*:1221–1225 (1990).
462. T. J. Broxton and S. Wright, Aust. J. Chem. *44*:103–111 (1991).
463. A. Cipiciani and S. Primieri, Tetrahedron *47*(1):83–90 (1991).
464. J. Burgess and M. S. Patel, J. Chem. Soc. Faraday Trans. *89*:783–787 (1993).
465. G. Fernández, M. Graciani, A. Rodríguez, F. Ferranti, and M. L. Moyá, Langmuir *15*:4441–4446 (1999).
466. C. A. Bunton, M. M. Mhala, and J. R. Moffatt, J. Phys. Chem. *93*:7851–7856 (1989).
467. M. N. Khan and Z. Arifin, Langmuir *13*:6626–6632 (1997).
468. M. N. Khan, Langmuir *13*:2498–2503 (1997).
469. M. Muñoz, A. Rodríguez, M. Graciani, and M. L. Moyá, Langmuir *15*:1588–1590 (1999).

9

Diels-Alder Reactions in Micellar Media

SIJBREN OTTO* and JAN B. F. N. ENGBERTS University of Groningen, Groningen, The Netherlands

I. INTRODUCTION TO DIELS-ALDER REACTIONS

The Diels-Alder reaction is a [4+2]cycloaddition in which a diene (four-π component) reacts with a dienophile (two-π component) to provide a six-membered ring (Fig. 1). Six new stereocenters are formed in a single reaction step. Because the conformations of the double bonds are usually fully retained, the reaction is stereospecific and consequently the absolute configuration of the two newly formed asymmetric centers can be controlled efficiently. The Diels-Alder reaction is of great value in organic synthesis and is a key step in the construction of compounds containing six-membered rings [1]. A historic account of this important conversion has been published by Berson [2].

Homo Diels-Alder reactions involve only hydrocarbon fragments. If the diene or dienophile possesses heteroatoms in any of the positions a–f (Fig. 1), heterocyclic ring systems are formed (*hetero* Diels-Alder reactions).

Normal electron demand Diels-Alder reactions are promoted by electron-donating substituents in the diene and electron-withdrawing substituents in the dienophile. The opposite situation applies for *inverse electron demand* Diels-Alder reactions. *Neutral* Diels-Alder reactions are accelerated by both electron-withdrawing and electron-donating substituents.

**Current affiliation*: University of Cambridge, Cambridge, England.

Reactivity and selectivity in Diels-Alder reactions are often rationalized in terms of frontier molecular orbital (FMO) theory [3], emphasizing interactions between the highest occupied molecular orbital (HOMO) of one of the reaction partners and the lowest unoccupied molecular orbital (LUMO) of the other. During formation of the two new σ-bonds, orbital symmetry is conserved. Therefore no intermediate is involved and the pericyclic reaction is concerted. There is ample experimental and theoretical evidence for the concerted mechanism [4]. Only in relatively rare cases does the Diels-Alder reaction take place via a nonconcerted two-step mechanism, involving either a zwitterionic or a biradical mechanism and leading to modified stereochemistry. FMO theory has been useful in analyzing possible asynchronicity in the activation process and in predicting kinetically controlled regioselectivity for Diels-Alder processes involving asymmetric dienes in combination with asymmetric dienophiles [5].

Much attention has also been given to Diels-Alder reactions that provide *endo* and *exo* cycloadducts (Fig. 2). The *endo-exo* ratio is usually the result of relatively small differences in transition state energies which appear to be primarily determined by secondary orbital interactions [6,7]. The formation of the *endo* product is associated with the most compact activated complex and exhibits the most negative volume of activation. Apart from secondary orbital interactions, other factors have been proposed for explaining the *endo-exo* ratio, including steric effects, London-dispersion interactions, and solvent effects (e.g., [8]).

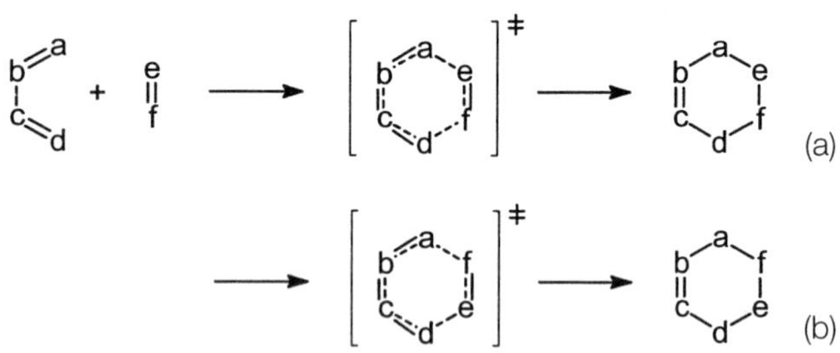

FIG. 1 Schematic representation of the Diels-Alder reaction. The versatility of the reaction is illustrated by the fact that heteroatoms are allowed at any of the positions a–f. Structures (a) and (b) indicate two regioisomeric products.

TABLE 1 Second-Order Rate Constants k_2 for the Dimerization of Cyclopentadiene in Solution and in the Gas Phase at 25°C

Solvent/state	k_2 (M^{-1} s^{-1})
Gas phase	6.9×10^{-7}
Neat	5.6×10^{-7}
Carbon tetrachloride	7.9×10^{-7}
Nitrobenzene	13×10^{-7}
Ethanol	19×10^{-7}

Source: Data from Ref. 9.

This chapter will describe micellar effects on Diels-Alder reactions with respect to both reaction rates and stereochemical aspects. For a proper understanding of the effects induced by micelles, we will first briefly review what is known about medium and catalytic effects on Diels-Alder cycloadditions.

A. Medium and Catalytic Effects on Diels-Alder Reactions

The Diels-Alder reaction is a textbook example of a reaction that is rather indifferent toward the choice of the solvent. An extreme example [9] is the dimerization of cyclopentadiene (Table 1), but for many other homo- and also for hetero Diels-Alder reactions, rate constants in a series of organic solvents vary only modestly. Nevertheless, attempts have been made to correlate kinetic data with solvent parameters, both for pure solvents and for binary mixtures [10,11]. Multiparameter analyses of solvent effects on Diels-Alder reactions have been carried out. For example, Gajewski [12] observed a dependence of rate constants for Diels-Alder reactions on the solvent α-parameter and the cohesive energy density. Intramolecular Diels-Alder reactions in highly viscous media have been related to the solvent density, which affects the translational motion of the reactants [13]. Rather unusual reaction media that have been employed for accelerating Diels-Alder reactions include solutions of lithium perchlorate in diethyl ether, dichloromethane, and nitromethane [14]. After considerable debate, it was argued that the substantial rate enhancements are largely due to Lewis acid catalysis by the lithium cation [15]. It is generally agreed that the small or modest solvent effects on the rates of Diels-Alder reactions are in accord with the concerted character of the cycloaddition that involves only a rather insignificant change of charge distribution during the activation process.

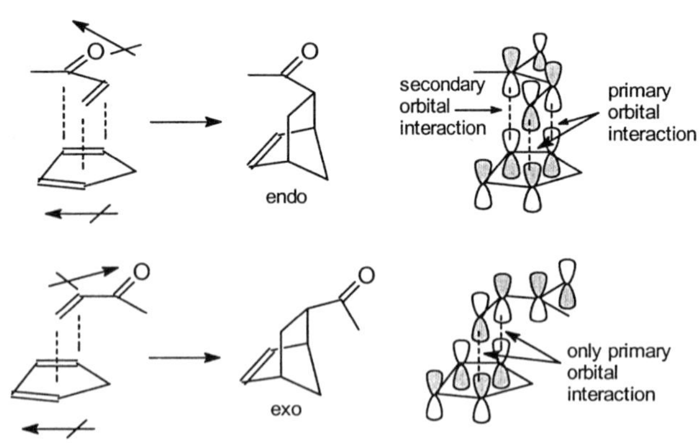

FIG. 2 *Endo* and *exo* pathway for the Diels-Alder reaction of cyclopentadiene with methyl vinyl ketone. As was first noticed by Berson, the polarity of the *endo* activated complex exceeds that of the *exo* counterpart due to alignment of the dipole moments of the diene and the dienophile [17]. The symmetry-allowed secondary orbital interaction that is possible only in the *endo* activated complex is usually invoked as an explanation for the preference for *endo* adduct exhibited by most Diels-Alder reactions.

The effect of the reaction medium on the regioselectivity of Diels-Alder reactions can be rationalized in terms of the FMO theory [16]. In particular, the hydrogen bond donating character of the solvent, as expressed in the α-parameter, affects the orbital coefficients on the terminal atoms of diene and dienophile.

Medium effects on the *endo-exo* ratios have received extensive attention, and Berson et al. have even based an empirical solvent polarity scale [$\Omega = \log(endo/exo)$] on the selectivity of the Diels-Alder reaction between methylacrylate and cyclopentadiene [17]. Solvent effects on the diastereofacial selectivity of the Diels-Alder process have also been examined and interpreted [18].

Other factors that have been studied with the aim of increasing the rate and stereoselectivity of Diels-Alder reactions include external pressure [19], ultrasound irradiation [20], and catalytic effects exercised by clays [21], alumina [22], silica gels [23], microporous organic crystals [24], antibodies [25], cyclodextrins [26,27], and supramolecular assemblies [28,29]. By far the most effective enhancements of rate and selectivity are induced by Lewis acids [30–33]. Studies of Lewis acid catalysis of Diels-Alder reactions have been almost completely restricted to organic solvents and binary mixtures of low water content. However, highly efficient Lewis acid catalysis has been observed for Diels-Alder reactions in purely aqueous media [34–38].

The remarkable effects of Lewis acids on the kinetics of Diels-Alder reactions have been known since the early 1960s [30] and may involve accelerations of 10^4 to 10^6 in organic solvents. Also the *endo/exo* selectivity may clearly respond to the presence of Lewis acids such as $AlCl_3 \cdot OEt_2$ [31]. Similarly, the regioselectivity [32] and diastereofacial selectivity [33] may be increased in the presence of Lewis acids.

The mechanism of Lewis acid catalysis can be understood with the aid of the FMO theory. Binding of the Lewis acid catalyst will lower the energy of the LUMO of the reactant to which it is coordinated. This binding will decrease the HOMO-LUMO energy difference, which will in turn increase the rate of the Diels-Alder cycloaddition. It has also been proposed that binding of the Lewis acid catalyst leads to increased secondary orbital interactions, thereby increasing the *endo/exo* ratio [39]. Other consequences of the coordination of a Lewis acid catalyst have also been considered, including an increase of asynchronicity in the formation of the activated complex [40]. Solvent effects on the efficiency of Lewis acid catalysis of Diels-Alder reactions have received relatively little attention [41]. At present, Lewis acid catalysis of Diels-Alder reactions is in everyday practice in synthetic organic chemistry.

B. Special Effects of Water on Diels-Alder Reactions

For a long time water was not a popular solvent for Diels-Alder reactions, although the pioneers Diels and Alder performed the reaction between furan and maleic acid in an aqueous medium in 1931 [42]. The latter experiment was repeated by Woodward and Baer in 1948, and a change in the *endo/exo* ratio was noted [43]. In 1973, Huisman et al. for the first time noticed a favorable aqueous effect on the rate of the same reaction, but the effect was not further explored [44]. Also, in two early patents the Diels-Alder reaction is mentioned in connection with water as the reaction medium [45].

A breakthrough came in 1980 in the work of Breslow and Rideout [26], who observed a substantial rate increase for simple Diels-Alder reactions in pure water. In subsequent extensive research it was shown that these remarkable kinetic aqueous medium effects are a general phenomenon [46–48]. Depending on the chemical structure of diene and dienophile, rate enhancements in water relative to organic solvents may amount to factors of more than 10^4. Rather soon after Breslow's pioneering work, synthetic applications of Diels-Alder reactions in aqueous media were explored in some detail, in particular by Grieco and his coworkers [48]. Of course, the often limited solubility of diene and dienophile is a major drawback. In elegant work, Lubineau et al. have tackled this problem by employing dienes that were rendered water soluble by the temporary introduction of a sugar moiety in the molecule [49]. The scaling up of aqueous Diels-Alder reactions has also been studied [50].

Ever since the early work of Breslow, many studies have been devoted to the identification of the special effects of the aqueous reaction medium that lead to the remarkable rate accelerations. These studies have been reviewed [46,47]. After considerable debate and controversy, it is now almost generally agreed that the enhanced reactivity in water is the result of two major effects: the *hydrogen bond donating capacity* of water and *enforced hydrophobic interactions* [51,52]. Previous suggestions that preassociation of the reactants in water played an important role were not substantiated. For example, vapor pressure measurements indicated that cyclopentadiene did not form aggregates at concentrations used in the kinetic measurements. Similar observations were made for methyl vinyl ketone, a popular dienophile in mechanistic studies of Diels-Alder reactions in water.

The peculiar nature of the Diels-Alder reaction in water was clearly revealed in a study in which Gibbs energies for the Diels-Alder reaction of cyclopentadiene with ethyl vinyl ketone over the whole mole fraction range in the mixture of 1-propanol with water were combined with Gibbs energies of transfer of the diene and dienophile from 1-propanol to the aqueous mixture and to pure water [51,53]. These data showed that the *initial state* (diene + dienophile) is significantly *destabilized* in water relative to 1-propanol (Fig. 3). By contrast, the *activated* complex has nearly equal chemical potentials in water and in 1-propanol. Consequently, in aqueous solution the hydrophobic parts of the activated complex have completely lost their nonpolar character

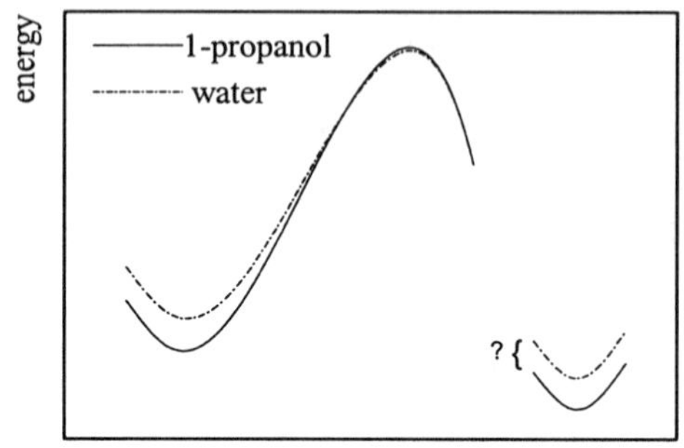

FIG. 3 Chemical potential of the initial state, the transition state, and the product of the Diels-Alder reaction between methyl vinyl ketone and cyclopentadiene in water as compared with 1-propanol. (Data from Ref. 53.)

as far as solvation is concerned. This conclusion has been confirmed in subsequent studies [52,54]. During the activation process of the Diels-Alder reaction, hydrophobic parts of the diene and the dienophile approach each other closely, a process that is particularly favorable in water ("enforced" hydrophobic interaction) compared with nonaqueous reaction media. The term "enforced" is used to stress the fact that the approximation of the nonpolar reagents is driven by the reaction and only enhanced by water. In addition, the electron redistribution that takes place during the activation process leads to an enhanced electron density at the carbonyl oxygen atom of ethyl vinyl ketone and a consequent enhanced propensity for hydrogen bond interaction with a hydrogen bond donating solvent. The small size of water molecules allows a particularly efficient interaction with hydrogen-bond acceptor sites. The medium effects on the chemical potentials, as shown in Fig. 3, are fully consistent with the operation of the hydrophobic and hydrogen-bonding effect. Beautifully detailed computational studies by Jorgensen et al. [55,56] led to similar conclusions and provided more quantitative insights into the relative importance of both solvation influences in water.

Attempts have been made to identify Diels-Alder reactions that are exclusively affected by either enforced hydrophobic interactions [57] or hydrogen-bonding effects [58]. The overall results confirmed the analysis and illustrated how the structures of diene and dienophile determine the magnitude of the aqueous rate acceleration. It appears well established now that the hydrophobicities of diene and dienophile as well as the polarizability of the activated complex play a key role in determining the acceleration of Diels-Alder reactions in water. These insights into the nature of the special effect of the aqueous medium are also of immediate relevance for understanding the effects of water on the *endo-exo* ratios and on the diastereofacial- and regioselectivity.

Finally, we briefly note that studies in the 1990s have shown that many other organic reactions benefit from the use of water as the reaction medium [48,59–62].

II. INTRODUCTION TO MICELLAR CATALYSIS

Micelles are highly dynamic, often rather polydisperse aggregates formed from single-chain surfactants [63,64] beyond the critical micelle concentration (cmc). Micellization is primarily driven by bulk hydrophobic interactions between the alkyl chains of the surfactant monomers and usually results from a favorable entropy change [65]. The overall Gibbs energy of the aggregate is a compromise of a complex set of interactions, with major contributions from headgroup repulsions and counterion binding (for ionic surfactants) [64].

The residence times of individual surfactant molecules in the micelle are typically of the order of 10^{-5}–10^{-6} s, whereas the lifetime of the micellar entity is about 10^{-3}–10^{-1} s. The size and shape of micelles are subject to appreciable structural variations. Average aggregation numbers are usually in the range of 40–150. For ionic micelles, a large fraction of the counterions are bound in the vicinity of the headgroups. The overall structure of the micelle is characterized by a situation in which the ionic or polar headgroups reside at the surface of the aggregates, where they are in contact with water, with the alkyl chains in the interior of the micelle forming a relatively dry hydrophobic core [66]. The alkyl chains of micellized surfactant molecules are not fully extended. Starting from the headgroup, the first two or three carbon-carbon bonds are usually *trans*, whereas *gauche* conformations are likely to be encountered near the center of the chain. As a result, the terminal methyl moieties of the chain can be located near the surface of the micelle and may even protrude into the aqueous medium [67]. Consequently, the micellar surface has a definite degree of hydrophobicity. Nuclear magnetic resonance (NMR) studies have shown that the hydrocarbon tails in a micelle are highly mobile and comparable in mobility to the chains in a liquid hydrocarbon [68]. The degree of water penetration into the micellar interior has long been a matter of debate. Small-angle neutron scattering studies have indicated that significant water penetration into the micellar core is unlikely [69].

Micellar catalysis of organic reactions has been extensively studied [70–76]. This type of catalysis is critically determined by the ability of micelles to take up all kinds of molecules. The binding is generally driven by hydrophobic and electrostatic interactions. The take-up of solutes from the aqueous medium into the micelle is close to diffusion controlled, whereas the residence time depends on the structure of the surfactant molecule and the solubilizate and is often of the order of 10^{-4}–10^{-6} s [77]. Hence, these processes are fast on the NMR time scale. Solubilization is usually treated in terms of a pseudophase model in which the bulk aqueous phase is regarded as one phase and the micellar pseudophase as another. This allows the affinity of the solubilizate for the micelle to be quantified by a partition coefficient P. Frequently P is expressed as the ratio of the mole fractions of solubilizate in the micellar pseudophase and in the aqueous pseudophase. However, for micelle-catalyzed reactions, it is more convenient to express P as a ratio of concentrations.

The time-averaged location of different solubilizates in or at a micelle has been a topic of contention [78]. Apart from saturated hydrocarbons, there is usually a preference for binding in the interfacial region, that is, at the surface of the micelle [79,80]. Such binding locations offer possibilities for hydrophobic interactions and avoid unfavorable disturbances of the interactions between the alkyl groups of the surfactant molecules in the core of the aggregate. The situation is, however, complicated, and the large volume of the interfacial region as compared with the core of the micelle should also be taken into account. The preferential binding of aromatic molecules at the micellar surface has been explained at least in part by the ability of the π-system of the molecule to form weak hydrogen bonds with water [81].

A. Kinetic Models

Kinetic studies of micellar catalysis and inhibition have been largely focused on organic reactions and the field has been reviewed extensively [70–76]. In these kinetic analyses the dependence of the rate constants on the surfactant concentration has usually been rationalized in terms of the pseudophase model assuming rapid exchange of the substrate(s) between the micellar and aqueous pseudophases. Different models have been developed for *uni*- and *bi*molecular reactions. For unimolecular reactions, the kinetic micellar effect depends on partitioning of the substrate between both pseudophases and on the rate constant in water (k_w) and in the micellar pseudophase (k_m). Menger and Portnoy [82] developed the classic model in 1967 and this model has been successfully employed ever since. The micellar rate effect k_m/k_w depends on the local medium at the substrate binding sites where the substrate experiences specific effects due to hydrophobic segments of the alkyl chains, the polar or ionic headgroups, and the counterions in case of ionic micelles.

For *bi*molecular reactions the analysis is much more complicated, and the overall kinetic effects are now also crucially affected by the local concentration of both reactants A and B in the micellar pseudophase. A classic approach has been advanced by Berezin et al. [71,83]. Again the pseudophase model is adopted, but now an independent assessment of at least one of the partition coefficients is required before the other relevant kinetic parameters can be obtained. The overall approach is illustrated in Fig. 4.

The apparent second-order rate constant (k_{app}), which is a weighed average of the second-order rate constants in the micellar pseudophase (k_m) and in water (k_w), is given by

$$k_{app} = \frac{k_m P_A P_B [S] V_{mol,S} + k_w (1 - [S] V_{mol,S})}{(1 + (P_A - 1)[S] V_{mol,S})(1 + (P_B - 1)[S] V_{mol,S})} \tag{1}$$

in which P_A and P_B are the micelle-water partition coefficients of A and B, respectively, defined as the ratios of the concentrations in the micellar and the aqueous phase, [S] is the concentration of surfactant, and $V_{mol,S}$ is the molar volume of the micellized surfactant. Accurate values of $V_{mol,S}$ are difficult to obtain, and the actual location of A and B in the aggregate may differ (see Section III.A). Usually, estimates of $V_{mol,S}$ are introduced into Eq. (1), leading to uncertainties in k_m. Despite these serious limitations, the kinetic analyses framed on the basis of Eq. (1) often produce reasonable results.

By far the most frequently analyzed types of bimolecular reactions are those involving an *ionic* reaction partner of the same charge type as the counterion

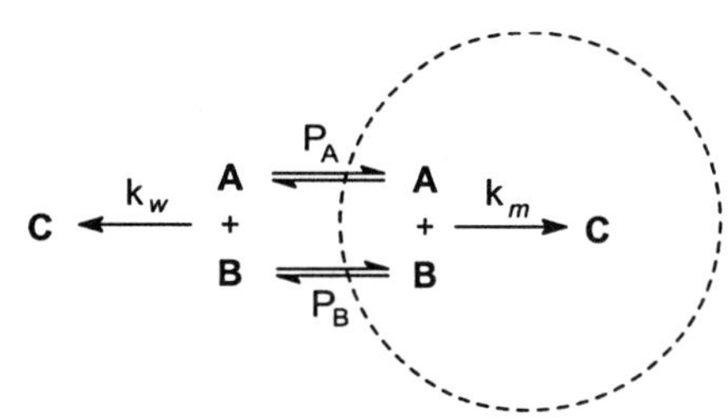

FIG. 4 Kinetic analysis of a bimolecular reaction A + B → C according to the pseudophase model.

of the ionic surfactant. Such processes are characterized by competition in binding between the reactive ion and the inert surfactant counterion. Pioneering work has been carried out by Romsted et al. [75], and the pseudophase ion-exchange model (PPIE) has been successfully applied in the micelle-catalyzed ionic bimolecular reactions. Again, it is often observed that the local microenvironment has only a modest influence on k_m/k_w and that the favorable entropic effect due to the increase of the local concentrations of both reactants in the micellar psuedophase is the dominant catalytic factor [84]. Over the years, the PPIE model has been severely tested; in particular, Romsted and his associates have advanced elegant methods for analyzing detailed aspects of counterion binding to micellar aggregates [85].

Studies of micellar catalysis of bimolecular reactions of uncharged substrates (such as most Diels-Alder reactions) have not been frequent. An example involves the reaction of 1-fluoro-2,4-dinitrobenzene with aniline in the presence of anionic and nonionic surfactants [86]. The apparent second-order rate constant (k_{app}) is increased relative to that in water as a result of compartmentalization of both reactants in the micelles. Interestingly, the second-order rate constant for reaction in the micellar pseudophase (k_m) was found to be roughly equal to or even lower than the rate constant in water. Similarly, the reaction of long-chain alkanethiols with *p*-nitrophenyl acetate [87] and the acylation of aryl oximes by *p*-nitrophenyl carboxylates [83] are catalyzed by micelles but, apart from local concentration effects, the influence of the micellar surface charge on the ionizaton constants of the SH and OH groups, respectively, must also be taken into account.

III. EFFECT OF MICELLES ON DIELS-ALDER REACTIONS

Because the diene and dienophile of the majority of intermolecular Diels-Alder reactions have a rather pronounced nonpolar character, efficient binding of both substrates to micelles is anticipated. This would imply that the effective reaction volume for the Diels-Alder reaction is significantly reduced, leading to micellar catalysis. Surprisingly, accounts of micelle-catalyzed Diels-Alder reactions are scarce. The first report of the influence of surfactants on Diels-Alder reactions stems from 1939, when the BASF company patented the use of detergents for promoting the yields of Diels-Alder processes in aqueous dispersions [45a]. Subsequently, more studies have appeared reporting beneficial effects of micellar systems on the yield of Diels-Alder reactions [88]. More mechanistically oriented studies have focused on the effect of micelles on the kinetics (Section III.A), the *endo-exo* selectivity (Section III.B), and the regioselectivity (Section III.C) of model Diels-Alder reactions. Also, the first example of modest enantioselectivity in a micelle-catalyzed Diels-Alder reactions has been reported (Section III.D). Finally, highly efficient micellar catalysis of a Diels-Alder reaction has been found for micelles with counterions that act as Lewis acid catalysts (Section III.E).

A. Effect of Micelles on the Rate of Diels-Alder Reactions

Studies of the kinetics of Diels-Alder reactions in the presence of micelles typically reveal only modest catalytic effects, and usually the apparent rate constants in micellar media are strikingly similar to the rate constants in water. Little effort was made to obtain second-order rate constants in the micellar pseudophases. We refer here to the work of Breslow et al. [89], who observed a small (15%) acceleration of the Diels-Alder reaction of cyclopentadiene with a number of dienophiles in the presence of sodium *n*-dodecyl sulfate (SDS) micelles as compared with water. Also, a modest micelle-induced decrease in the rate constant of a Diels-Alder reactions has been reported [90].

More detailed analyses have been performed by Hunt and Johnson [91], who studied the kinetics of the homo Diels-Alder reaction of 1,2-dicyanoethylene (**1**) with cyclopentadiene (**2**) as a function of the conentration of sodium dodecyl sulfate (SDS) surfactant. The presence of micelles induces a modest decrease of the rate of this reaction (Fig. 5). Enthalpies and entropies of activation of the reaction in micellar medium have been determined and compared with those in water, aqueous salt solutions, and organic solvents (Table 2). Gibbs energies, entropies, and enthalpies of activation for the reaction in micellar solutions resemble those in 0.5 M LiCl more than those in organic solvents or water. This seems to point toward the Stern region of the micelles as the prominent site for this Diels-Alder reaction.

Wijnen and Engberts [58] have studied the effect of SDS on another homo Diels-Alder reaction between 1,4-naphthoquinone (**4**) and cyclopentadiene (**2**). The results were compared with a structurally related retro Diels-Alder reaction (Fig. 6). Close to the cmc a modest acceleration of the former bimolecular Diels-Alder reaction was observed, whereas micelles induced a small inhibition of the retro Diels-Alder. However, this

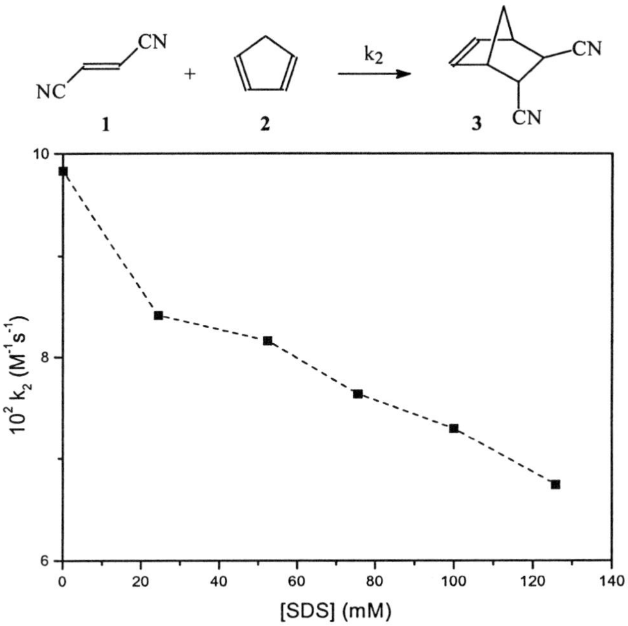

FIG. 5 Second-order rate constants for the Diels-Alder reaction of **1** with **2** at different concentrations of sodium dodecyl sulfate (SDS). (Data from Ref. 91.)

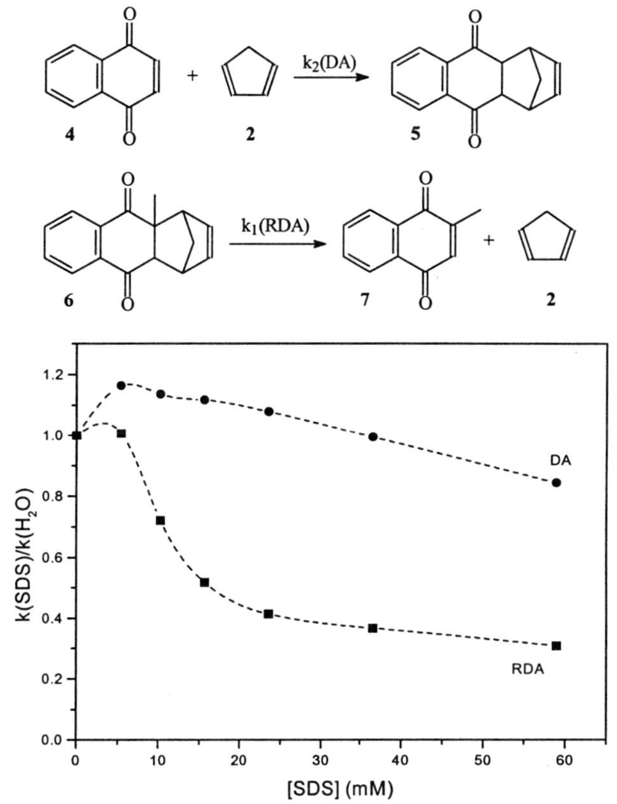

FIG. 6 Relative rate constants for the retro Diels-Alder reaction (■) of **6** and the bimolecular Diels-Alder reaction (●) of **4** with **2** at different concentrations of sodium dodecyl sulfate (SDS). (Data from Ref. 58.)

process is still considerably faster than that in organic solvents [58].

The same authors have studied a reversible hetero Diels-Alder reaction and compared it with an irreversible analogue (Fig. 7) [92]. This time the rates of both retro and bimolecular Diels-Alder reactions experienced a modest beneficial influence of the presence of SDS micelles. The equilibrum constant is somewhat displaced toward the adduct. This particular reaction is classified by Desimoni et al. [11] as a *type C* Diels-Alder reaction, signifying that it is almost insensitive to hydrogen bonding effects and that its rate is mainly governed by enforced hydrophobic interactions. This suggests that enforced hydrophobic interactions are slightly more efficient in the Stern region of the SDS micelles than in bulk water.

TABLE 2 Gibbs Energies, Enthalpies, and Entropies of Activation for the Diels-Alder Reaction of **1** with **2** in Different Media

Medium	$\Delta^{\theta}G^{\ddagger}$ (kJ/mol)	$\Delta^{\theta}H^{\ddagger}$ (kJ/mol)	$T\,\Delta^{\theta}S^{\ddagger}$ (kJ/mol)
0.05 M SDS	78.7	45.1	−33.6
Water	78.4	62.2	−16.2
0.5 M LiCl	77.8	41.9	−35.9
Ethanol	87.5	52.2	−35.3
Dioxane	89.1	48.5	−40.6

Source: Data from Ref. 91.

Van der Wel, Wijnen, and Engberts [57] have studied the influence of surfactants on the hetero Diels-Alder reaction of a cationic dienophile **12** with cyclopentadiene (Fig. 8). A 10-fold acceleration is induced by anionic SDS micelles, whereas nonionic Triton X-100 and cationic 1-*N*-dodecyl-4-methylpyridinium bromide have only modest effects on the rate of the reaction. The efficient catalysis by SDS most likely results from electrostatically enhanced binding of the dienophile to the micelles.

The presence of micelles does not lead to a significant alteration of the efficiency of an intramolecular Diels-Alder reaction [93] as compared with the process in pure water. The most detailed kinetic investigation of the effect of micelles on Diels-Alder cycloadditions has focused on the homo Diels-Alder reaction between 3-(*p*-substituted-phenyl)-1-(2-pyridyl-2-propen-1-one

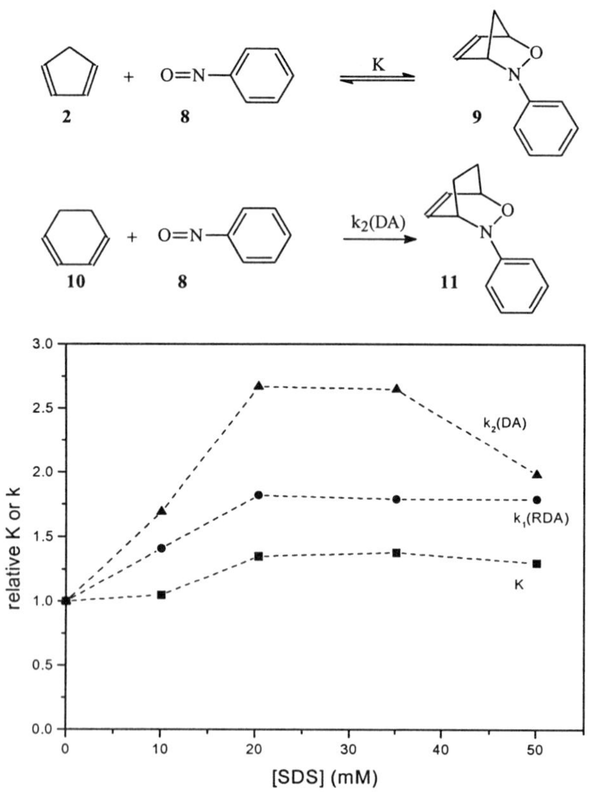

FIG. 7 Relative equilibrium constants for the reversible hetero Diels-Alder reaction of **8** with **2** (■), relative second-order rate constants of the addition of **8** to **10** (▲), and relative first-order rate constants for the retro Diels-Alder reaction of **9** (●) at different concentrations of sodium dodecyl sulfate (SDS). (Data from Ref. 92.)

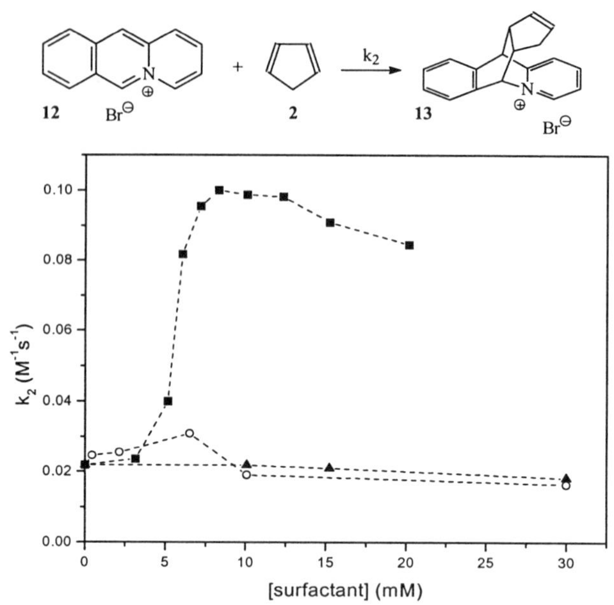

FIG. 8 Second-order rate constants for the reaction of **12** with **2** in aqueous solutions of sodium dodecyl sulfate (SDS) (■), Triton X-100 (○), and *N*-dodecyl-4-methylpyridinium bromide (▲). (Data from Ref. 57.)

dienophiles (**14a–g**) with cyclopentadiene (**2**) [94]. The influence of micelles of cetyltrimethylammonium bromide (CTAB), SDS, and dodecyl heptaoxyethylene ether ($C_{12}E_7$) on this process has been studied (Fig. 9). Note that the dienophiles can be divided into nonionic (**14a–e**), anionic (**14f**), and cationic (**14g**) species. A comparison of the effect of nonionic ($C_{12}E_7$), anionic (SDS), and cationic (CTAB) micelles on the rates of their reactions with **2** enabled assessment of the importance of electrostatic interactions in micellar catalysis or inhibition. The most important results of this study are summarized in Table 3.

Under the reaction conditions, the effect of micelles on the rate of the Diels-Alder reaction is obviously small and invariably results in a slight *inhibition* of the reaction. The most significant effect occurs for anionic **14f** in CTAB solution and for cationic **14g** in SDS solution. These are the two combinations for which one would expect essentially complete binding of the dienophile to the micelle as a result of favorable electrostatic interactions in addition to the hydrophobic interactions. Apparently, reaction in the micellar environment is slower than reaction in the bulk aqueous phase, despite the anticipated locally increased concentrations of the reactants in the micellar pseudophase. Also, in the case where electrostatic interactions inhibit binding of the dienophile to the micelle, i.e., **14f** in SDS and **14g** in CTAB solution, a retardation of the reaction is observed. In these cases the dienophile will most likely reside mainly in the aqueous phase. The retardation will result from a decrease in the concentration of **2** in this phase due to its partial solubilization by the micelles.

The kinetics of the aforementioned reactions have been analyzed in terms of the pseudophase model (Fig. 4). For the limiting cases of essentially complete binding of the dienophile to the micelle (**14f** in CTAB and **14g** in SDS solution) the following expression [95] was used:

$$\frac{1}{k_{\mathrm{app}}} = \frac{[\mathbf{2}]_{\mathrm{t}}}{k_{\mathrm{obs}}} = \frac{V_{\mathrm{mol,S}}}{k_{\mathrm{m}}}[\mathrm{S}] + \frac{V_{\mathrm{w}}}{P_2 \cdot V_{\mathrm{t}} \cdot k_{\mathrm{m}}} - \frac{\mathrm{cmc} \cdot V_{\mathrm{mol,S}}}{k_{\mathrm{m}}} \qquad (2)$$

Herein $[\mathbf{2}]_t$ is the total number of moles of **2** present in the reaction mixture, divided by the total reaction volume V_t; k_{obs} is the observed pseudo-first-order rate con-

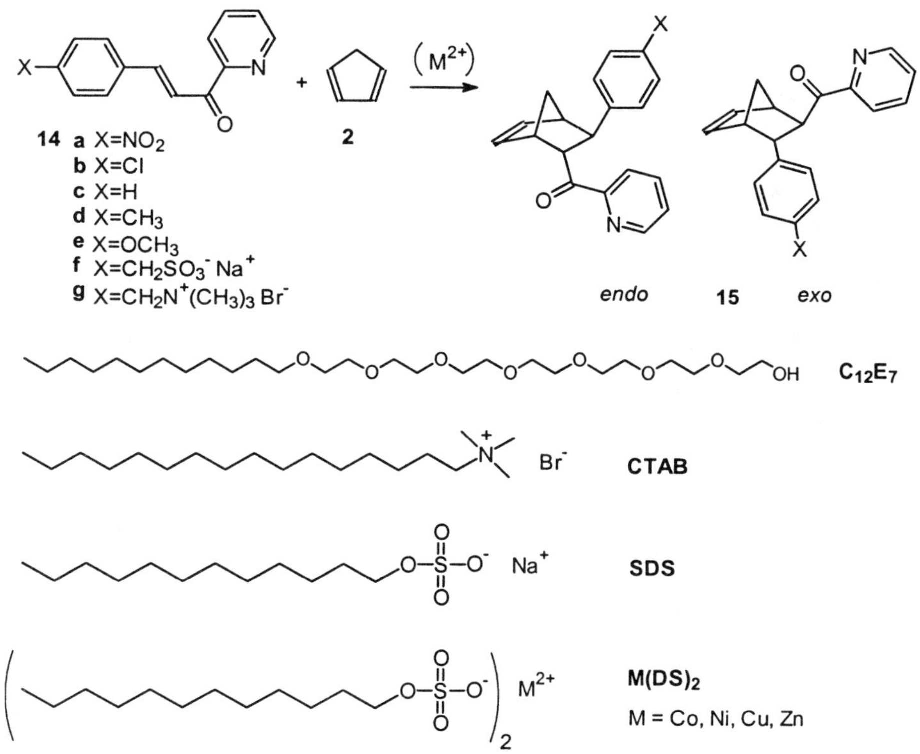

FIG. 9 A Diels-Alder reaction that is subject to Lewis acid catalysis in water.

stant; $V_{mol,S}$ is an estimate of the molar volume of micellized surfactant S; k_m and k_w are the second-order rate constants in the aqueous phase and in the micellar pseudophase, respectively; V_w is the volume of the aqueous phase; and P_2 is the partition coefficient of **2** over the micellar pseudophase and water expressed as a ratio of concentrations. From the dependence of $[\mathbf{2}]_t/k_{obs}$ on the concentration of surfactant, P_2 and k_m were obtained. Table 4 shows that the partition coefficients of **2** over SDS or CTAB micelles and water are similar. Comparison of the rate constants in the micellar pseudophase calculated using the pseudophase model with those in water (Tables 3 and 4) demonstrates a remarkable retardation induced by the micelles.

This retardation is unlikely to be a result of a micellar medium effect. Information concerning the mi-

TABLE 3 Influence of Micelles of CTAB, SDS, and $C_{12}E_7$ on the Apparent Second-Order Rate Constants ($M^{-1}\ s^{-1}$)[a] for the Diels-Alder Reaction of **14a**, **14f**, and **14g** with **2** at 25°C[b]

Medium[c]	14a	14f	14g
Water	4.02×10^{-3}	1.74×10^{-3}	2.45×10^{-3}
SDS	3.65×10^{-3}	1.44×10^{-3}	1.47×10^{-3}
CTAB	3.61×10^{-3}	0.283×10^{-3}	2.01×10^{-3}
$C_{12}E_7$	3.35×10^{-3}	1.62×10^{-3}	2.05×10^{-3}

[a]The apparent second-order rate constants are calculated from the observed pseudo-first-order rate constants by dividing the latter by the overall concentration of **2**.
[b][**14**] $\approx 2 \times 10^{-5}$ M; [**2**] = 2.0×10^{-3} M.
[c]All solutions contain 1.0×10^{-4} M EDTA in order to suppress catalysis by trace amounts of metal ions. The concentration of surfactant is 7.8 mM above the cmc of the particular amphiphile under reaction conditions.
Source: Data from Ref. 94.

TABLE 4 Analysis Using the Pseudophase Model: Partition Coefficients for **2** over CTAB of SDS Micelles and Water and Second-Order-Rate Constants for the Diels-Alder Reaction of **14f** and **14g** with **2** in CTAB and SDS Micelles at 25°C

Surfactant	Dienophile	P_2 (±10%)	k_m ($M^{-1}\ s^{-1}$) (±10%)
CTAB	**14f**	65[a]	5.9×10^{-6}
SDS	**14g**	49[a]	3.1×10^{-5}

[a]Corrected data; see Ref. 95.
Source: Data from Ref. 94.

croenvironment experienced by the Diels-Alder reactants was obtained from analysis of the *endo-exo* ratio of the reaction between **14c** and **2** in surfactant solution and in a number of different organic and aqueous media [94] (see also Section III.B). The results of the study clearly point toward a waterlike environment for the Diels-Alder reaction in the presence of micelles. The inhibitory effect of micelles is suggested to result from the fact that diene and dienophile are on average located in different parts of the micelles. The diene seems to prefer the hydrophobic center of the micelle, whereas the dienophile has a stronger affinity for the Stern region. Evidence comes from ^{1}H-NMR relaxation time studies in which paramagnetic ions are added to the micellar solutions [38,94]. Multivalent ions were used with a charge opposite to that of the surfactant headgroup, ensuring strong binding of these species to the Stern region of the micelles. As these paramagnetic ions enhance the relaxation of the protons in their vicinity, species bound to the Stern region will experience a more enhanced rate of relaxation from those residing in the core of the micelle. Comparison of Fig. 10 and Fig. 11 indeed demonstrates that the relaxation rate enhancement experienced by the diene is significantly smaller than that experienced by the dienophile.

In conclusion, the fact that micelles have a rather limited influence on the rate of bimolecular as well as retro and intramolecular Diels-Alder reactions suggests (1) that the micellar medium experienced by the reactants is not too different from water and (2) that concentration effect of the reactants in the micelles is not too efficient. The latter effect is probably a result of the fact that diene and dienophile prefer different binding sites in the micelle.

B. Effect of Micelles on the *Endo-Exo* Selectivity

Few detailed studies have been performed regarding micellar effects on *endo-exo* selectivities. Diego-Castro and Hailes [96] have studied the influence of micelles on the Diels-Alder reaction of cyclopentadiene with several alkyl acrylates of different chain lengths (methyl, ethyl, pentyl, heptyl, and nonyl). *Endo-exo* ratios in micellar media were strikingly similar to those in water irrespective of the length of the alkyl group in the dienophile. Unfortunately, the reactions were performed using a surfactant concentration close to the cmc, where solubilization of the reactants by the micelles is rather inefficient and the reaction is more likely to take place in bulk water than in the micelles.

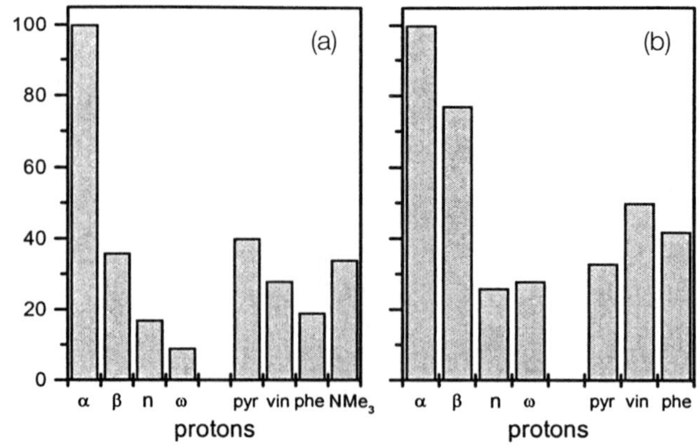

FIG. 10 Paramagnetic ion–induced spin-lattice relaxation rates (r_p) of the protons of **14c** (a) and **14g** (b) in SDS solution and of SDS in the presence of **14c** or **14g**, normalized to r_p for the surfactant α-CH_2. The solutions contained 50 mM SDS, 8 mM **14c** or **14g**, and 0 or 0.2 mM $DyCl_3$ and 0 or 0.6 mM cyclen. (Data from Ref. 94.)

Braun, Schuster, and Sauer [97] have studied the *endo-exo* ratio of the reaction of cyclopentadiene with acrylonitrile and butyl acrylate in micellar media. The *endo-exo* ratios were significantly larger than in organic solvents, which seems to point toward a highly polar micellar reaction medium. Unfortunately, no comparisons were made with the *endo-exo* selectivity in pure water. Otto et al. [94] have studied the effect of micelles of SDS, CTAB, and $C_{12}E_7$, on the *endo-exo* ratio of the Diels-Alder reaction of **14c** and **2** (Fig. 9). Com-

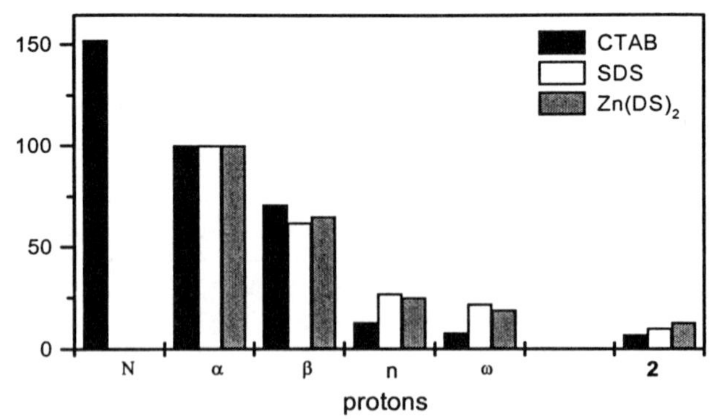

FIG. 11 Paramaganetic ion–induced spin-lattice relaxation rates (r_p) of the protons of **2** in CTAB, SDS, or $Zn(DS)_2$ solution and of these surfactants in the presence of **2**, normalized to r_p for the surfactant α-CH_2. The solutions contained 25 mM $Zn(DS)_2$, 50 mM CTAB or SDS, 3 mM **2**, and 0 or 0.4 mM $[Cu(EDTA)]^{2-}$ for CTAB solutions and 0 or 0.2 mM $Cu(NO_3)_2$ for SDS and $Zn(DS)_2$ solutions. (Data from Ref. 94.)

TABLE 5 *Endo-Exo* Product Ratios of the Diels-Alder Reaction of **14c** with **2** in Surfactant Solutions Compared with Water and Organic Solvents

Medium	%*Endo*-%*exo*
100 mM CTAB	86-14
100 mM SDS	88-12
100 mM $C_{12}E_7$	85-15
Water	84-16
Ethanol	77-23
Acetonitrile	67-33

parison of the results with those obtained for organic solvents and pure water (Table 5) demonstrates that the beneficial solvent effect of water is still present in the micelle-mediated reaction.

In summary, *endo-exo* selectivities in micellar media tend to be comparable to those in pure water [89] and significantly larger than those in organic solvents. Apparently, surfactants can be used in order to improve the solubility of the Diels-Alder reactants in water, without significant deterioration of the selectivity as compared with pure water. Interestingly, in microemulsions the *endo-exo* selectivity is reduced significantly [89,98].

C. Micellar Effects on the Regioselectivity

Significant work in this area has been carried out by Jaeger et al. An interesting issue that was addressed in the early days of micellar catalysis involves the question of how binding to specific sites in micelles could affect the stereochemistry of the reactions. For example, extensive structural changes in substrates were expected to influence the depth of penetration of the substrate into the micellar core with a concomitant change in the efficiency of the micellar catalysis. This expectation was not borne out in practice [99,100].

In fact, one could ask how "micellar binding sites" can be defined with sufficient precision to allow conclusions about the details of the relevant microenvironment and orientation of the substrate. In view of the micellar structure, it is more appropriate to consider a range of binding situations of small differences in Gibbs energy of binding and involving a range of substrate orientations. Most substrates in micelle-catalyzed reactions contain at least one polar substituent that prefers to bind at or close to the micellar surface and at least partly in direct contact with water. Solely apolar molecules, such as alkanes, will preferentially bind in the hydrophobic core of the micelle, assuming orientations that lead to a minimal disturbance of the chain packing of the surfactant molecules.

Jaeger et al. [101] examined how monohalogenation of alkyl phenyl ethers C_6H_5OR (R = n-C_5H_{11}, n-C_9H_{19}, and n-$C_{12}H_{25}$) by chlorine and bromine in micellar solutions of SDS and in vesicular solutions to give 4-XC_6H_4OR and 2-XC_6H_4OR exhibits *ortho*/*para* ratios and reaction rates different from those in aqueous buffer solutions in the absence of surfactants. Indeed, in the micelles the *o*/*p* ratio decreases with increasing length of R, whereas the second-order rate constant decreases in the series. These regioselectivity and kinetic data can be rationalized by assuming different solubilization sites for the aromatic ethers depending on the length of the R substituent. These differences lead to different reaction environments and concomitant kinetic differences. Lengthening of R is proposed to lead to solubilization "deeper" in the micelle and changes in the *o*/*p* preference.

In another series of studies, Jaeger et al. examined regioselectivity control of Diels-Alder reactions for cases in which the diene or both the diene and dienophile were amphiphilic molecules themselves. In a Diels-Alder process involving a cationic surfactant 1,3-diene with a neutral nonsurfactant dienophile, the orientational effects within the micellar aggregates were not sufficiently strong to overcome the intrinsically preferred regioselectivity of the reaction [102]. Modest regioselectivity was found for a Diels-Alder reaction of another cationic surfactant diene with cationic surfactant dienophiles [103,104]. The reactions were performed at 100°C, most likely decreasing the organizational abilities of the aqueous aggregate compared with those at lower temperatures.

A substantially larger regioselectivity [105] was found in a study employing amphiphilic diene **16** (cmc = 1.0×10^{-4} M) and amphiphilic dienophile **17** (cmc = 4.4×10^{-3}) (Fig. 12). The cycloadducts **18** and **19** were formed, which were separated by preparative reverse-phase HPLC and characterized by ^{1}H-NMR spectroscopy.

Since the substituents at carbons 1 and 2 in **17** are close to being electronically and sterically equivalent with respect to the dienophile reaction center, no regiochemical preference is anticipated in the absence of interfacial orientational effects in the mixed micelles formed from **16** and **17**. Evidence for this assumption was also obtained from an analysis of the regioselectivity of the Diels-Alder reaction of **20** and **21** in toluene. As expected, the two analogous cycloadducts were obtained in equal amounts. Interestingly, the re-

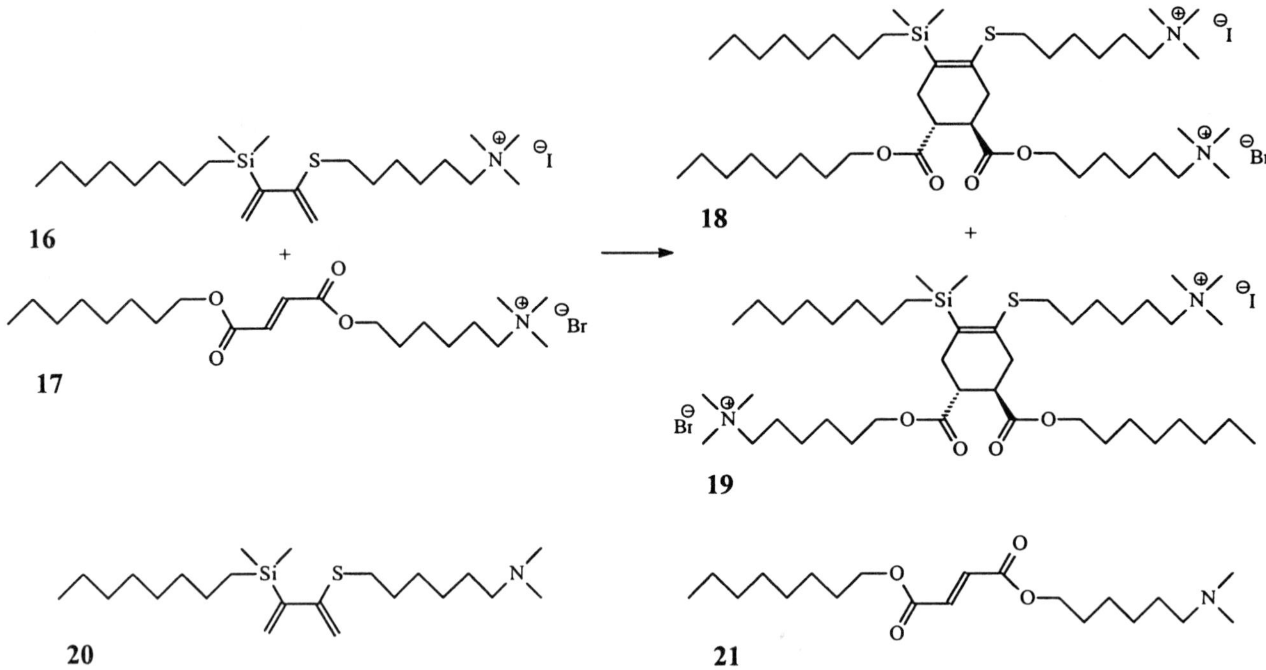

FIG. 12 A regioselective Diels-Alder reaction between a surfactant diene and a surfactant dienophile.

actions of **16** with **17** at concentrations above their cmc values gave an **18**:**19** ratio of 6.6:1. Therefore it is clear that interfacial and related orientational effects that result from surfactant aggregation can induce significant regioselectivity in a Diels-Alder reaction in aqueous solution.

D. Micellar Effects on the Enantioselectivity

Recently, a report appeared that described the first Diels-Alder reaction in aqueous chiral micellar media [106]. The novel (s)-leucine-derived chiral micellar amphiphile **22** was used as a catalyst for the Diels-Alder reaction of cyclopentadiene with *n*-nonly acrylate (**23**) (Fig. 13).

Preferential formation of the R-*endo* isomer was observed. Using a surfactant concentration of 11 mg L^{-1} and in the presence of 4.86 M LiCl, the yield was 75%, with an *endo*/*exo* ratio of 2.2 and an enantioselectivity of 15% (R). This result may be compared with the maximum enantioselectivity (21%) found for Diels-Alder reactions in the presence of cyclodextrins. In the absence of surfactant, the reaction in water gave a yield of 70% and an *endo*/*exo* ratio of 1.7. Further optimization of the structure of the chiral micellar catalyst might well lead to improved enantioselectivities. In this context it may be noticed that aqueous Diels-Alder reactions catalyzed by chiral Lewis acids may exhibit enantioselectivities up to 74% [36,37].

E. Effects of Micelles with Catalytically Active Counterions

The most efficient means of accelerating Diels-Alder reactions is catalysis by Lewis acids. In aqueous media this process is hampered by the strong interaction of the catalysts with water [62]. However, one example has been reported where this difficulty was overcome by modification of the dienophiles so that they can form a chelate with the catalyst ions (Fig. 9) [35–37].

The reaction of these dienophiles with cyclopentadiene in the *absence* of Lewis acid catalysts has been described in Section III.A. In that case introduction of micelles into the aqueous reaction mixture induced a modest retardation of the reaction.

Micellar catalysis of this reaction in combination with Lewis acid catalysis has been studied in detail [94]. The dodecyl sulfate surfactants $Co(DS)_2$, $Ni(DS)_2$, $Cu(DS)_2$, and $Zn(DS)_2$ containing catalytically active counterions are extremely potent catalysts for the Diels-Alder reaction between **14** and **2**. Figure 14 shows the dependence of the rates of the Diels-Alder reactions of **14c**, **14f**, and **14g** with **2** on the concentration of $Cu(DS)_2$. For all three dienophiles the apparent second-order rate constant for their reaction with **2**

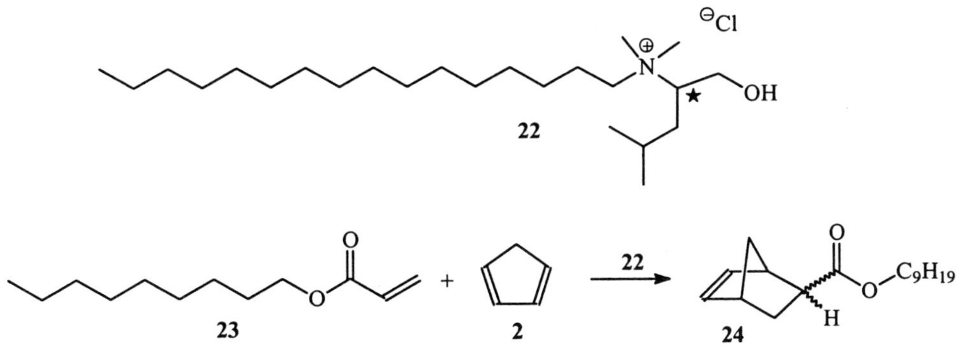

FIG. 13 The first example of enantioselectivity induced by a chiral surfactant in a micelle-catalyzed Diels-Alder reaction.

increases dramatically when the concentration of $Cu(DS)_2$ reaches the cmc (1.11 mM). Beyond the cmc, the dependence of the rate on the surfactant concentration is subject to two counteractive influences. At higher surfactant concentration, a larger fraction of dienophile will be bound to the micelle, where it reacts faster than in bulk water, resulting in an increase in the rate of the reaction. At the same time, the concentration of diene in the micellar pseudophase will drop with increasing surfactant concentration due to the increase in the volume of the micellar pseudophase. At higher surfactant concentrations the dienophile will be nearly completely bound to the micelles and the dilution effect will start to dominate the behavior. Together, these two effects result in the appearance of a rate maximum at a specific concentration of surfactant that is typical for micelle-catalyzed bimolecular reactions (see also Fig. 8). The position of the maximum depends primarily on the micelle-water partition coefficients of diene and dienophile.

Interestingly, the acceleration relative to the reaction in organic media in the absence of catalyst approaches enzymelike magnitudes: compared with the process in acetonitrile (second-order rate constant = 1.40×10^{-5} M^{-1} s^{-1}), $Cu(DS)_2$ micelles accelerate the Diels-Alder reaction between **14a** and **2** by a factor of 1.8×10^6.

Also the effects of cationic (CTAB) and nonionic ($C_{12}E_7$) surfactants on the Cu^{2+}-catalyzed reaction have been studied. However, these systems were much less efficient than $Cu(DS)_2$, suggesting that a local high concentration of catalyst ions in the Stern region of the micelles is a prerequisite for a highly efficient interaction with the dienophile.

The essentially complete binding of **14g** to the $Cu(DS)_2$ micelles allowed treatment of the kinetic data of Fig. 14 using the pseudophase model. Furthermore, complete binding of **14g** to the copper ions was assumed, which was supported by ultraviolet-visible analysis [94]. Using Eq. (2), a $Cu(DS)_2$-water distribution coefficient for **2** of 86 was obtained [95]. The second-order rate constant for reaction in the micellar pseudophase was calculated to be 0.21 M^{-1} s^{-1}. Comparison of this rate constant with those for the reaction in acetonitrile (0.472 M^{-1} s^{-1}) and ethanol (0.309 M^{-1} s^{-1}) seems to indicate a relatively apolar medium for the Diels-Alder reaction. This conclusion is hard to reconcile with the ionic character of two of the three reaction partners involved.

More insight into the local environment for the catalyzed reaction was obtained from the influence of substituents on the rate of this process in micellar and in different aqueous and organic solvents. The Hammett

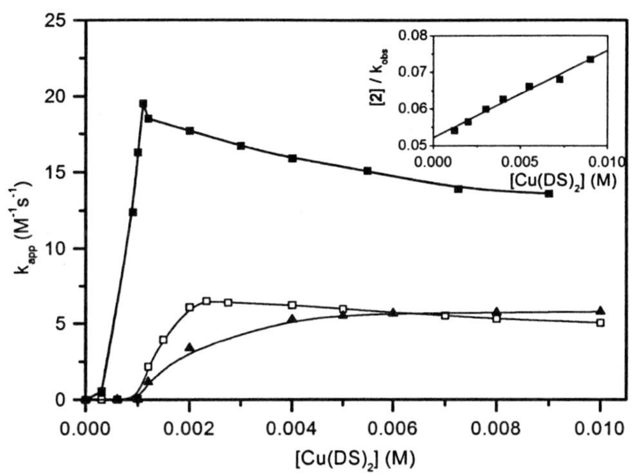

FIG. 14 Plots of the apparent second-order rate constant (k_{app}) versus the concentration of $Cu(DS)_2$ for the Diels-Alder reaction of **14c** (□), **14f** (▲), and **14g** (■) with **2** at 25°C. The inset shows the treatment of the data for the reaction of **14g** according to the pseudophase model. (Data from Ref. 94.)

ρ value in $Cu(DS)_2$ solution was found to resemble closely that in aqueous solution rather than those in organic solvents, suggesting an aqueous microenvironment for the reaction [94].

It appears that the outcome of the analysis using the pseudophase model (a rather apolar reaction environment) is not in agreement with experimental observations (an aqueous reaction environment). Apparently, the assumptions of the pseudophase model are not valid for the Diels-Alder reaction studied. In particular, the treatment of the micellar pseudophase as a homogeneous "solution" might not be warranted. As noted in Section III.A, there are strong indications that the diene and the dienophile reside on average in different parts of the micelle, the diene preferring the core and the dienophile the Stern region of the micelles. Additional paramagnetic ^{1}H-NMR relaxation rate studies of the binding location of the reactants in $Zn(DS)_2$ micelles further support this suggestion [38,94]. Surely, spatial separation of diene and dienophile will impede their reaction.

In summary, the use of anionic micelles with bivalent metal ions as catalytically active counterions can lead to accelerations of suitable Diels-Alder reactions of enzymelike magnitude. The high efficiency of these systems mainly results from the efficient interaction between dienophile and catalysts in the Stern region of the micelles, where both species are present in high local concentration. Even larger accelerations are anticipated upon modification of the diene so that this species also binds to the Stern region rather than in the core of the micelle. Examples of similar micellar systems have found application in synthetic organic chemistry [107].

IV. SUMMARY AND OUTLOOK

It is now well established that many Diels-Alder reactions, both of normal electron demand and of inverse electron demand, can be substantially accelerated by using water as the reaction medium. Also, *endo/exo* ratios are usually improved for aqueous media. These findings had important implications for further extending the versatility of Diels-Alder reactions in organic synthesis and for providing a stimulus for detailed studies of medium effects on pericyclic reactions. These interesting developments called for studies of Diels-Alder reactions in micellar solutions. By concentrating the diene and dienophile in the micellar reaction volume, further enhancements were anticipated. Furthermore, solubilization of the Diels-Alder reaction partners in the micelles could offer a solution for improving the otherwise limited solubility of diene and dienophile in water. Finally, effects on the Diels-Alder stereochemistry were expected. Specific binding could lead to regioselectivity, whereas the use of chiral micelle-forming surfactants would provide a possibility for obtaining enantioselectivity in appropriate Diels-Alder processes. Studies have illustrated the potential power to bring about these appealing results. Micellar catalysis of Diels-Alder reactions has been pursued and could indeed induce significant accelerations. Examples have been shown in this chapter. However, it is a requisite that diene and dienophile bind to rather similar binding sites in the micelle. In the case of, for example, an apolar diene and a moderately polar dienophile, the diene will preferentially reside in the core of the micelle and encounters with the dienophile, preferentially sitting at the micellar surface, will be hampered. The overall result will then be micellar inhibition rather than catalysis. Extreme rate enhancements can be obtained by combining micellar and Lewis acid catalysis. However, a specially designed dienophile is required for such a catalytic process.

Binding of dienes and dienophiles to micellar aggregates will certainly improve their solubilities in water and extend the potential for using aqueous reaction media for Diels-Alder reactions. The use of micelles with the aim of inducing favorable regioselectivity and enantioselectivity has had only modest success. However, it is anticipated that challenging developments in this area are possible through variation of the structural architectures of diene, dienophile, and micelle-forming amphiphile.

ACKNOWLEDGMENT

The authors gratefully acknowledge Miss H. E. Wolters for typing the manuscript.

REFERENCES

1. (a) G. Brieger and J. N. Bennett, Chem. Rev. *80*:63–97 (1980); (b) W. Oppolzer, Angew. Chem. Int. Ed. Engl. *16*:10–24 (1977); (c) J. Sauer and R. Sustmann, Angew. Chem. Int. Ed. Engl. *19*:779–807 (1980).
2. J. A. Berson, Tetrahedron Lett. *48*:3–17 (1992).
3. (a) K. Fukui, Acc. Chem. Res. *4*:57–64 (1971); (b) R. B. Woodward and R. Hoffman, Angew. Chem. Int. Ed. Engl. *8*:781–853, 1969.
4. K. N. Houk, Y. Li, and J. D. Evanseck, Angew. Chem. Int. Ed. Engl. *31*:682–708 (1992).
5. M. D. Rozeboom, I. Tegmo-Larsson, and K. N. Houk, J. Org. Chem. *46*:2338–2345 (1981).

6. R. Gleiter and M. C. Bohm, Pure Appl. Chem. *55*: 237–244 (1983).
7. R. B. Woodward and T. J. Katz, Tetrahedron *5*:70–89 (1959).
8. E. T., McBee, M. J. Keogh, R. P. Levek, and E. P. Wesseler, J. Org. Chem. *38*:632–636 (1973).
9. C. Reichardt, *Solvents and Solvent Effects in Organic Chemistry*, 2nd ed., VCH, Cambridge, 1990.
10. Examples: (a) B. Von Blankenburg, H. Fiedler, M. Hampel, H. G. Hauthal, G. Just, K. Kahlert, J. Korn, K.-H. Muller, W. Pritzkow, Y. Reinhold, M. Rollig, E. Sauer, D. Schnurpfeil, and G. Zimmermann, J. Prakt. Chem. *316*:804–816 (1974); (b) N. K. Sangwan and H. Schneider, J. Chem. Soc. Perkin Trans. 2 1223–1227 (1989).
11. G. Desimoni, G. Faita, P. Righetti, N. Tornaletti, and M. Visigalli, J. Chem. Soc. Perkin Trans. 2 437–441 (1989). (b) M. Burdisso, G. Desimoni, G. Faita, P. Righetti, and G. Tacconi, J. Chem. Soc. Perkin Trans. 2 845–850 (1989); (c) A. Corsico Coda, G. Desimoni, G. Faita, P. Righetti, and G. Tacconi, Tetrahedron *45*: 775–788 (1989); (d) G. Desimoni, G. Faita, and P. Righetti, Tetrahedron *46*:7951–7970 (1990); (e) G. Desimoni, G. Faita, D. Pasini, and P. Righetti, Tetrahedron *48*:1667–1674 (1992).
12. J. J. Gajewski, J. Org. Chem. *57*:5500–5506 (1992).
13. R. A. Firestone and S. G. Saffar, J. Org. Chem. *48*: 4783–4784 (1983).
14. P. A. Grieco, J. J. Nunes, and M. D. Gaul, J. Am. Chem. Soc. *112*:4595–4596 (1990).
15. (a) M. A. Forman and W. P. Dailey, J. Am. Chem. Soc. *113*:2761–2762 (1991). (b) G. Jenner and R. B. Salem, Tetrahedron *53*:4637–4648 (1997).
16. J. M. Mellor and C. F. Webb, J. Chem. Soc. Perkin Trans. 2 26–31 (1974).
17. J. A. Berson, Z. Hamlet, and W. A. Mueller, J. Am. Chem. Soc. *84*:297–304 (1962).
18. C. Cativiela, J. I. Garcia, J. Gil, R. M. Martinez, J. A. Mayoral, L. Salvatella, J. S. Urieta, A. M. Mainar, and M. H. Abraham, J. Chem. Soc. Perkin Trans. 2 653–660 (1997).
19. R. Van Eldik, T. Asano, and W. J. Le Noble, Chem. Rev. *89*:549–688 (1989).
20. T. P. Caulier and J. Reisse, J. Org. Chem. *61*:2547–2548 (1996).
21. C. Collet and P. Laszlo, J. Phys. Org. Chem. *8*:468–472 (1995).
22. R. M. Pagni, G. W. Kabalka, G. Hondrogiannis, S. Bains, and R. Kurt, Tetrahedron *49*:6743–6756 (1993).
23. C. Cativiela, J. I. Garcia, J. A. Mayoral, E. Pires, and A. J. Royo, Tetrahedron *51*:1295–1300 (1995).
24. K. Endo, T. Koike, T. Sawaki, O. Hayashida, H. Masuda, and Y. Aoyama, J. Am. Chem. Soc. *119*:4117–4122 (1997).
25. (a) A. A. P. Meekel, M. Resmini, and U. K. Pandit, J. Chem. Soc. Chem. Commun. 571–572 (1995); (b) M. Resmini, A. A. P. Meekel, and U. K. Pandit, Pure Appl. Chem. *68*:2025–2028 (1996).
26. R. Breslow and D. C. Rideout, J. Am. Chem. Soc. *102*: 7816–7817 (1980).
27. W.-S. Chung and J.-Y. Wang, J. Chem. Soc. Chem. Commun. 971–972 (1995).
28. (a) C. J. Walter, H. L. Anderson, and J. K. M. Sanders, J. Chem. Soc. Chem. Commun. 458–460 (1993); (b) C. J. Walter and J. K. M. Sanders, Angew. Chem. Int. Ed. Engl. *34*:217–219 (1995); (c) Z. Clyde-Watson, A. Vidal-Ferran, L. J. Twaman, C. J. Walter, D. W. J. McCallien, S. Fani, N. Bampos, S. R. Wylie, and J. K. M. Sanders, New J. Chem. *22*:493–502 (1998); (d) M. Maurus, Z. Clyde-Watson, L. J. Twyman, M. Nakash, and J. K. M. Sanders, Chem. Commun. 2265–2266 (1998).
29. (a) J. Kang and J. J. Rebek, Nature *385*:50–52 (1997); (b) J. Kang, G. Hilwersson, J. Santamaria, and J. Rebek Jr., J. Am. Chem. Soc. *120*:3650–3656 (1998).
30. P. Yates and P. Eaton, J. Am. Chem. Soc. *82*:4436–4437 (1960).
31. J. Sauer and J. Kredel, Tetrahedron Lett. *51*:6359–6364 (1966).
32. T. Inukai and T. Kojima, J. Org. Chem. *31*:1121–1123 (1966).
33. L. M. Tolbert and M. B. Ali, J. Am. Chem. Soc. *106*: 3806–3810 (1984).
34. S. Otto and J. B. F. N. Engberts, Tetrahedron Lett. *36*: 2645–2648 (1995).
35. S. Otto, F. Bertoncin, and J. B. F. N. Engberts, J. Am. Chem. Soc. *118*:7702–7707 (1996).
36. S. Otto, Boccalatti, and J. B. F. N. Engberts, J. Am. Chem. Soc. *120*:4238–4239 (1998).
37. S. Otto and J. B. F. N. Engberts, J. Am. Chem. Soc. *121*:6798–6806 (1999).
38. S. Otto, Catalysis of Diels-Alder reactions in water, Ph.D. dissertation, University of Groningen, 1998.
39. J. Motoyoshiya, T. Kameda, M. Asari, M. Miyamoto, S. Narita, H. Aoyama, and S. Hayashi, J. Chem. Soc. Perkin Trans. 2 1845–1850 (1997).
40. A. Venturini, J. Joglar, S. Fustero, and J. Gonzalez, J. Org. Chem. *62*:3919–3926 (1997).
41. V. D. Kiselev and A. I. Konovalov, Russ. Chem. Rev. *58*:230–249 (1989).
42. O. Diels and K. Alder, Ann *490*:243–294 (1931).
43. R. B. Woodward and H. Baer, J. Am. Chem. Soc. *70*: 1161–1166 (1948).
44. T. A. Eggelte, H. de Koning, and H. O. Huisman, Tetrahedron *29*:2491–2493 (1973).
45. (a) H. Hopff and C. W. Rautenstrauch, U. S. Patent 2,262,002 (1939); (b) L. C. Lane and C. H. J. Parker, U.S. Patent 2,444,263 (1948).
46. J. B. F. N. Engberts, Pure Appl. Chem. *67*:823–828 (1995).
47. W. Blokzijl and J. B. F. N. Engberts, in ACS Sym-

posium Series 568 (C. J. Cramer and D. G. Truhlar, eds.), American Chemical Society, Washington, DC, 1994, pp. 303–317.

48. P. A. Grieco, *Organic Synthesis in Water*, Blacky Academic and Professional, London, 1998.
49. A. Lubineau, E. Grand, and M.-C. Scherrman, Carbohydrate Res. *297*:169–174 (1997) and many previous papers.
50. G. Chen, Diss. Abstr. Int. *57*:6934-B (1997).
51. (a) W. Blokzijl, M. J. Blandamer, and J. B. F. N. Engberts, J. Am. Chem. Soc. *113*:4241–4246 (1991); (b)W. Blokzijl and J. B. F. N. Engberts, J. Am. Chem. Soc. *114*:5440–5442 (1992).
52. S. Otto, W. Blokzijl, and J. B. F. N. Engberts, J. Org. Chem. *59*:5372–5376 (1994).
53. W. Blokzijl, Ph.D. dissertation, University of Groningen, 1991.
54. A. Meijer, S. Otto, and J. B. F. N. Engberts, J. Org. Chem. *63*:8989–8994 (1998).
55. J. F. Blake and W. L. Jorgensen, J. Am. Chem. Soc. *113*:7430–7432 (1991).
56. J. F. Blake, L. Dongchul, and W. L. Jorgensen, J. Org. Chem. *59*:803–805 (1994).
57. G. K. van der Wel, J. W. Wijnen, and J. B. F. N. Engberts, J. Org. Chem. *61*:9001–9005 (1996).
58. J. W. Wijnen and J. B. F. N. Engberts, J. Org. Chem. *62*:2039–2044 (1997).
59. C. J. Li and T. H. Chan, *Organic Reactions in Aqueous Media*, Wiley, New York, 1997.
60. P. A. Grieco, Aldrichim. Acta *24*:59–66 (1991).
61. A. Lubineau, J. Auge, and Y. Queneau, Synthesis 741–760 (1994).
62. J. B. F. N. Engberts, B. F. L. Feringa, E. Keller, and S. Otto, Rec. Trav. Chim. Pays-Bas *115*:457–464 (1996).
63. B. Lindman and H. Wennerstrom, Top. Curr. Chem. *87*:1–83 (1980).
64. J. N. Israelachvili, *Intermolecular and Surface Forces*, Academic Press, London, 1992.
65. W. Blokzijl and J. B. F. N. Engberts, Angew. Chem. Int. Ed. Engl. *32*:1545–1579 (1993).
66. D. W. R. Gruen, Prog. Colloid Polymer. Sci. *70*:6–16 (1985).
67. J. Clifford, Trans. Faraday Soc. *61*:1276–1282 (1965).
68. H. Walderhaug, O. Soderman, and P. Stilbs, J. Phys. Chem. 88:1655–1662 (1984).
69. S. S. Berr, E. Caponetti, J. S. J. Johnson, R. R. M. Jones, and L. J. Magid, J. Phys. Chem. *90*:5766–5770 (1986).
70. E. H. Cordes and C. Gitler, Prog. Bioorg. Chem. *2*:1–53 (1972).
71. I. V. Berezin, K. Martinek, and A. K. Yatsimirskii, Russ. Chem. Rev. *42*:787–802 (1973).
72. C. A. Bunton, Pure Appl. Chem. *49*:969–979 (1977).
73. T. Kunitake and S. Shinkai, Adv. Phys. Org. Chem. *17*:435–487 (1980).
74. E. J. R. Sudholter, G. B. van de Langkruis, and J. B. F. N. Engberts, Rec. Trav. Chim. Pays-Bas *99*:73–82 (1980).
75. C. A. Bunton, F. Nome, F. H. Quina, and L. S. Romsted, Acc. Chem. Res. *24*:357–363 (1991).
76. S. Tascioglu, Tetrahedron *52*:11113–11152 (1996).
77. M. Almgren, F. Grieser, and J. K. Thomas, J. Am. Chem. Soc. *101*, 279–291 (1979).
78. F. Grieser and C. J. Drummond, J. Phys. Chem. *92*:5580–5593 (1988).
79. M. F. Emerson and A. Holtzer, J. Phys. Chem. *71*:3320–3330 (1967).
80. P. Stilbs, J. Colloid Interface Sci. *80*:608–610 (1981).
81. S. Suzuki, P. G. Green, R. E. Bumgarner, S. Dasgupta, W. A. Goddard III, and G. A. Blake, Science *257*:942–945 (1992).
82. F. M. Menger and C. E. Portnoy, J. Am. Chem. Soc. *89*:4698–4703 (1967).
83. A. K. Yatsimirski, K. Martinek, and I. V. Berezin, Tetrahedron *27*:2855–2868 (1971).
84. C. A. Bunton, Catal. Rev. Sci. Eng. *20*:1–59 (1979).
85. L. S. Romsted and J. Rao, Langmuir *15*:326–336 (1999).
86. S. J. Dougherty and J. C. Berg, J. Colloid Interface Sci. *49*:135–138 (1974).
87. W. Tagaki, T. Amada, Y. Yamashita, and Y. Yano, J. Chem. Soc. Chem. Commun. 1131–1132 (1972).
88. (a) V. K. Singh, B. N. S. Raju, and P. T. Deota, Synth. Commun. *17*:1103–1109 (1987); (b) V. K. Singh, B. N. S. Raju, and P. T. Deota, Synth. Commun. *18*:567–574 (1988); (c) M. S. Kumar and J. M. Rao, Tetrahedron *46*:5383–5388 (1990); (d) V. K. Singh and B. N. S. Raju, Indian J. Chem. *35B*:303–311 (1996); (d) K. Chiba, M. Jinno, A. Nozaki, and M. Tada, Chem. Commun. 1403–1404 (1997).
89. R. Breslow, U. Maitra, and D. Rideout, Tetrahedron Lett. *24*:1901–1904 (1983).
90. N. K. Sangwan and H.-J. Schneider, J. Chem. Soc. Perkin II 1223–1227 (1989).
91. I. Hunt and D. Johnson, J. Chem. Soc. Perkin Trans 2 1051–1056, 1991.
92. J. W. Wijnen and J. B. F. N. Engberts, Liebigs Ann. Recl. 1085–1088 (1997).
93. D. D. Sternbach and D. M. Rossana, J. Am. Chem. Soc. *104*:5853–5854 (1982).
94. S. Otto, J. B. F. N. Engberts, and J. C. T. Kwak, J. Am. Chem. Soc. *120*:9517–9525 (1998).
95. In the original paper [94] Eq. (2) and the data calculated with it contained a small error. This has now been corrected. We gratefully acknowledge Prof. P. Walde for pointing this out.
96. M. J. Diego-Castro and H. C. Hailes, Tetrahedron Lett. *39*:2211–2214 (1998).
97. R. Braun, F. Schuster, and J. Sauer, Tetrahedron Lett. *27*:1285–1288 (1986).
98. A. Gonzalez and S. L. Holt, J. Org. Chem. *47*:3186–3188 (1982).

99. G. A. van de Langkruis and J. B. F. N. Engberts, J. Org. Chem. *49*:4152–4157 (1984).
100. F. M. Witte and J. B. F. N. Engberts, J. Org. Chem. *50*:4130–4134 (1985).
101. D. A. Jaeger, M. W. Clennan, and J. Jamrozik, J. Am. Chem. Soc. *112*:1171–1176 (1990).
102. D. A. Jaeger, H. Shinozaki, and P. A. Goodson, J. Org. Chem. *56*, 2482–2489 (1991).
103. D. A. Jaeger and J. Wang, Tetrahedron Lett. *33*:6415–6416 (1992).
104. D. A. Jaeger and J. Wang, J. Org. Chem. *58*:6745–6755 (1993).
105. D. A. Jaeger and D. Su, Tetrahedron Lett. *40*:257–260 (1999).
106. M. J. Diego-Castro and H. C. Hailes, Chem. Commun. 1549–1550 (1998).
107. (a) S. Kobayashi, T. Busujima, and S. Nagayama, Synlett 545–546 (1999); (b) K. Manabe and S. Kobayashi, Synlett 547–548 (1999); (c) K. Manabe and S. Kobayashi, Tetrahedron Lett. 3773–3776 (1999).

10

Interfacial Compositions of Surfactant Assemblies by Chemical Trapping with Arenediazonium Ions: Method and Applications

LAURENCE S. ROMSTED Rutgers, The State University of New Jersey, New Brunswick, New Jersey

I. INTRODUCTION

A. Aims of the Chapter

The impetus for development of a method for determining interfacial compositions of surfactant assemblies based on the chemistry of arenediazonium ions came from several directions: (1) Most current methods for estimating the compositions of surfactant aggregates are based on physical methods that monitor only one component at a time [1]. (2) The understanding of chemical reactivity in surfactant assemblies in terms of pseudophase models developed to the point that they provide consistent qualitative and often quantitative interpretations of surfactant assembly effects on rate and equilibrium constants of chemical reactions [2–6]. (3) I sought an experimental method to answer a deceptively simple question: how can interfacial concentrations and distributions of two similar inorganic anions, e.g., Cl^- and Br^-, between cationic micelles and water be measured simultaneously? Ions bind selectively to interfacial regions of aggregates, but because of interferences, physical methods often cannot discriminate between two similar ions in the same solution. Chemical trapping discriminates between these and a variety of other anions. But it does much more. It also provides simultaneous estimates of their concentrations and that of water within the interfacial region over wide ranges of surfactant and counterion concentrations. Research over the past decade shows that the method has broad general utility because it works with both anionic and neutral nucleophiles and because it provides estimates of the distributions of nucleophiles between the oil-interfacial and water-interfacial regions of microemulsions. The method has the potential to provide information on interfacial concentrations of many of the important functional groups present in biological membranes and in commercial surfactant-based products.

The aim of this chapter is to describe the logic, practical aspects, and current and potential applications of the chemical trapping method. Section I characterizes the types of information that chemical trapping provides about properties of surfactant assemblies. Section II describes the basic assumption of the method and evidence for its validity. Section III focuses on practical aspects including optimal characteristics of a trapping reagent and its preparation and protocols for product analysis and measurement of dediazoniation rate constants. Sections IV and V show how the method can be used to obtain information on a variety of properties of surfactant assemblies: (1) hydration of the interfacial region; (2) interfacial counteranion concentrations in cationic and zwitterionic micelles and vesicles, including effects of sphere-to-rod transitions; (3) degrees of ionization, α, of cationic micelles; (4) coion concentrations around anionic micelles; (5) exchange constants for counteranion competition; (6) distribution constants of alcohols between aqueous-interfacial and oil-interfacial regions; (7) topologies of aggregate bound polypeptides; and (8) distributions of antioxidants in micro- and macroemulsions.

The experimental demands of the chemical trapping method are relatively modest. Syntheses of the arenediazonium salt probe, its water-soluble short-chain analogue, and its products are required, but they tend to be more tedious than complex. Most surfactants and other components are commercially available. A constant-temperature bath is needed for running reactions and an ultraviolet-visible (UV-Vis) spectrophotometer is required for measuring dediazoniation rate constants in previously untried systems. The workhorse tool is a high-performance liquid chromatograph (HPLC) fitted with a C-18 reverse-phase column and a UV detector for separation and quantitative analysis of reaction products. Adding an autosampler and computer for handling multiple samples and for data storage speeds collection of results. Unanticipated experimental problems sometimes appear, usually in the early stages of work with a new surfactant system. Once experimental protocols are established, results accumulate rapidly. Mathematical treatments are available for obtaining various types of information, but applications to multicomponent systems are still being developed.

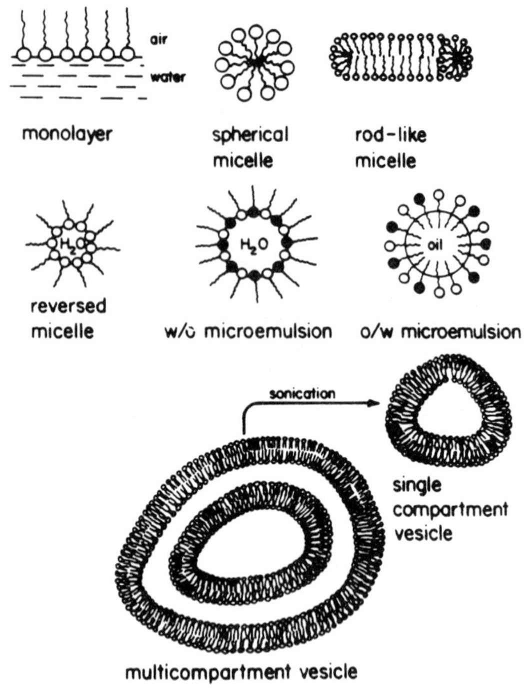

FIG. 1 Oversimplified images of some association colloid structures illustrating the organization of surfactant (open circle headgroups) and additives (filled circle headgroups). As in all surfactant assemblies, the headgroups (ionic surfactants also have counterions, not shown) are arrayed in a layer between aqueous and oil regions. (Adapted from Fig. 1 of Ref. 133 and used with permission of the author and the IUPAC.)

B. Background

Surfactants, or amphiphiles, are surface-active nonionic molecules or organic salts that, alone or in combination with a wide variety of other ionic and nonionic solutes, aggregate spontaneously and with a high degree of cooperativity in solution to form a variety of assemblies (or association colloids) whose structures depend both on solution composition and on the structures of the components, primarily the surfactant [7–10]. Figure 1 shows some typical association colloid structures. All surfactant assemblies in homogeneous solution share an underlying organizational structure: a fluid, hydrocarbon-like region separated from an aqueous region by an interfacial region with a thickness on the order of the diameter of the surfactant headgroup.

The balance of forces that drive aggregate formation and control aggregate structure and stability is determined in large part by the hydrophobic effect; that is, water molecules interact more strongly with themselves than with the nonpolar tails of the surfactant [7]. This escapist tendency is balanced by mutual interactions between the polar or ionic portions of the surfactant headgroups with water and ions in the interfacial region. Other amphiphilic or surface-active components, e.g., alcohols, benzene, polar solutes, and polypeptides, are generally assumed to associate with surfactant aggregates in a similar fashion, i.e., with their nonpolar portions buried in the aggregate core and their polar and charged portions maintaining contact with water [10,11]. Thus, the nonpolar regions of single and multicomponent aggregates tend to be hydrocarbon-like in minimal contact with water except at the interfacial region. However, their surfaces may contain a plethora of hydrated ions and polar molecules participating in a variety of water-mediated interactions, e.g., electrostatic, charge-dipole, induced dipole, and hydrogen bonding. The more components in a surfactant assembly, the more difficult it becomes to quantify the relationships between composition and aggregate structure and stability.

Changing the bulk solution composition and structures of the components alters interfacial composition and the balance of forces in this region that, because of the highly cooperative nature of surfactant aggregation, may change aggregate structure or induce macroscopic phase separation. For example, added salts generally lower the concentration at which surfactant monomers aggregate to form micelles and often increase the aggregation number, i.e., the number of monomers per micelle [11]. Micelle growth is often attributed to coulombic screening of headgroup repulsions by counterions permitting tighter headgroup packing [12]. But changes in micelle size and shape cannot be interpreted solely in terms of coulombic interactions because they also depend upon counterion type. For example, cetyltrimethylammonium bromide, CTABr, micelles undergo a sphere-to-rod transition when the aqueous Br^- concentration exceeds about 0.1 M, but over 1.0 M Cl^- is needed to induce this transition in CTACl micelles [13–16]. Chemical trapping provides new insight into the relationships between counterion type, interfacial counterion and water concentrations, and sphere-to-rod transitions (see Section IV.B.1).

Understanding relationships between solution composition and aggregate structure and stability in surfactant assemblies requires determining the compositions of their interfacial regions. This is a difficult task, especially in multicomponent systems. A variety of modern techniques, e.g., conductometry, potentiometry, and spectrophotometry [nuclear magnetic resonance (NMR), UV-Vis, fluorescence, electron spin resonance (ESR), infrared (IR), and circular dichroism (CD)], are currently used to examine compositions of these assemblies [1,10,11,17–21]. Some methods monitor only one component at a time, some are limited to narrow composition ranges, and others report on physical properties such as surface polarity but not composition. Moreover, these methods provide information only on the fraction bound of a given component. Only chemical trapping and molecular dynamics calculations [22,23] provide simultaneous estimates of concentrations (or number densities) of more than one component, including water, in the interfacial region.

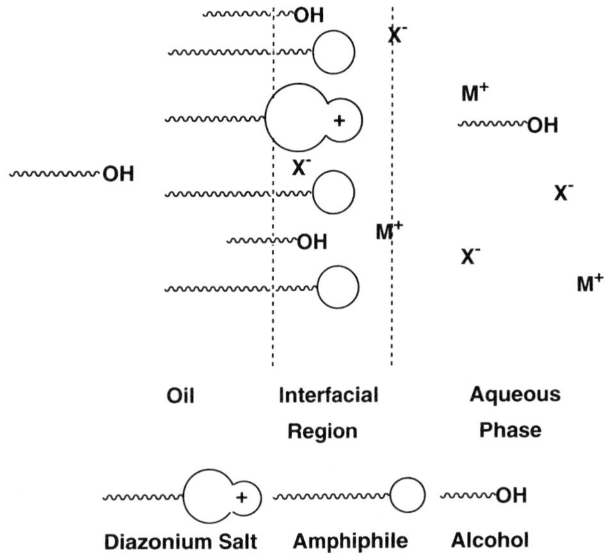

FIG. 2 Cartoon of a small section of a generic surfactant assembly interface illustrating the interfacial region flanked by the oil and aqueous regions, e.g., Fig. 1, and the location of a representative arenediazonium ion probe. The surfactant headgroups could be nonionic, zwitterionic, cationic, or anionic (respective counterions not shown). An alcohol solute may be in all three regions depending upon it hydrophobicity, interfacial structure, and the relative amounts of oil and water in the system. The distribution of added salt, M^+ and X^-, between the interfacial and aqueous regions depends on headgroup charge and anion and cation type. Water molecules (not shown) penetrate the interfacial region by hydrating the headgroups of the surfactants and polar solutes and merge into the oil region, surfactant tails, and any added oil (not shown). The boundaries of the interfacial region (dashed lines) are arbitrary, as are the sizes of the components.

C. What Does the Chemical Trapping Method "See" in the Interfacial Region of Surfactant Assemblies? What Information Does It Provide?

The chemical trapping method is a probe technique that reports on the concentrations of weakly basic nucleophiles in interfacial regions of surfactant assemblies [24]. The long-tailed arenediazonium ion, itself a surface active cation, is oriented with its reactive diazonio group in the interfacial region (Fig. 2). The hydrophobic tail drags the nonpolar portion of the probe into the oil region of the assemblies, but its hydrated cationic headgroup is anchored within the interfacial region. The ensemble of aggregate-bound arenediazonium ions react via spontaneous loss of dinitrogen to give highly reactive aryl cations that are trapped by *all* the weakly basic nucleophiles within the interfacial region but *not* by those in the nonpolar or aqueous regions (see Section II.A). Nucleophiles that will or should react via this mechanism include water, alcohols, and nonionic surfactants; the peptide bond; and many types of anionic concentrations, such as Cl^- and Br^-, and headgroups.

All the components in solution, including the probe, are assumed to be in dynamic equilibrium throughout the time course of the overall reaction and the totality of the aggregates in solution is described as a separate phase distributed throughout the surrounding bulk phase, either oil or water [11,17]. The aggregate-bound probe "sees" the interfacial regions of the totality of the aggregates and samples this region as a mixed solvent composed of surfactant headgroups, polar solutes, ions, and water. As with other reactions that generate reactive intermediates, the selectivity of dediazoniation reactions toward different nucleophiles is very small and product yields are proportional to interfacial nucleophile concentrations and the selectivity of the reaction toward them [25,26].

Chemical trapping differs from other interfacial probe methods in two important ways. (1) The dediazoniation reaction is unaffected by medium polarity and consequently provides no information on the medium properties of the interfacial region (Section II.B). (2) Concentrations are not expressed as fractions bound but as molarities of the interfacial volume sampled by the probe (Sections II.F and IV.A and B). Another unique feature of the method is that every experiment provides, in principle, an estimate of the concentration of water in the interfacial region from the yield of the phenolic product from reaction with interfacial water (Sections IV.A and IV.B.1). The compositional information obtained by chemical trapping reflects the specific interactions between components in the interfacial region that influence the size, shape, and stability of surfactant assemblies. This information is difficult to obtain by other methods. Finally, the method can be used to estimate distribution constants of components between aqueous-interfacial and oil-interfacial regions over wide ranges of solution compositions (Section IV.C).

II. THE LOGIC OF THE CHEMICAL TRAPPING METHOD

A. Arenediazonium Ion Chemistry

Arenediazonium ions have a rich, complex, and still not completely understood chemistry that depends upon substituent type, Y, the nature of the solvent, and reactants such as nucleophiles, X, and electron donors, D [25,27]. Three types of reactions are common (Scheme 1): (1) addition to the terminal nitrogen by strongly basic nucleophiles (see bottom pathway), (2) heterolytic loss of dinitrogen in the presence of weakly basic nuclephiles (see top pathway), and (3) homolytic displacement of dinitrogen by electron transfer from a donor molecule (see middle pathway). This chemical richness sets limits on the range of operational conditions under which arenediazonium ions can function as interfacial probes and is responsible for the formation

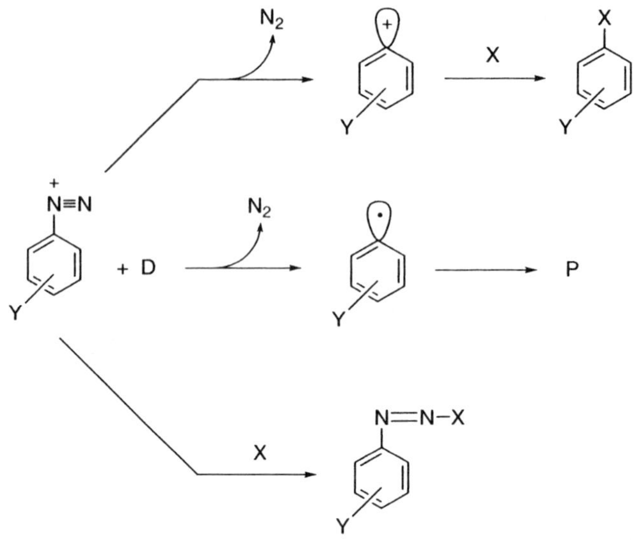

SCHEME 1

of certain side products (Sections II.G and III.B). However, in the absence of UV light and reducing and electron transfer reagents, and in neutral to acidic solutions, i.e., in water at pH < 7, arenediazonium ions generally react via rate-determining loss of N_2 to give aryl cations that are rapidly trapped by weakly basic nucleophiles (see top pathway) [25,28].

After some searching (Section III.A), 4-alkyl-2,6-dimethylbenzenediazonium ions z-ArN_2^+ were selected as the most versatile probes (Scheme 2). The long-chain derivative, 4-hexadecyl, $z = 16$, is used to probe interfacial regions and its short-chain analogue, 4-methyl, $z = 1$, provides the reference reaction in the absence of surfactant assemblies and probes the water pools of reverse assemblies (Section II.B.4). Scheme 2 illustrates the heterolytic pathway for competitive reaction with water and other weakly basic nucleophiles, X. The arenediazonium ions are prepared as their tetrafluoroborates (Section III.C). 16-ArN_2BF_4 is water insoluble and sufficiently hydrophobic that it binds strongly to surfactant assemblies except very near the critical micelle concentration (cmc) [24]. The reactive diazonio group of 16-ArN_2^+ is located in the interfacial region (Fig. 2). 1-ArN_2BF_4 is soluble in water and other polar solvents and is used to determine the selectivity of the dediazoniation reaction toward different nucleophiles, usually relative to water (Section II.D). We have shown by published and unpublished results that many weakly basic nucleophiles react with z-ArN_2^+ by the heterolytic mechanism, including the neutral nucleophiles H_2O, ROH (nonionic surfactants and alcohols), amides, and

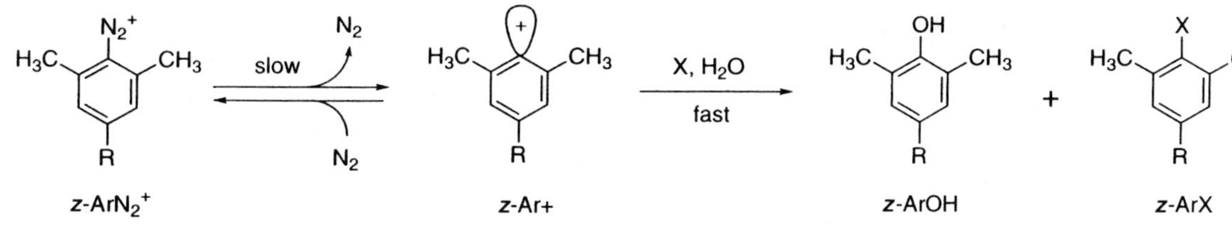

R = $C_{16}H_{33}$, z = 16; R = CH_3, z = 1; X = any weakly basic anionic or neutral nucleopihile

SCHEME 2

urea (at both N and O centers of the amide group) and the anionic nucleophiles X^- (Cl^-, Br^-, and I^-), RCO_2^-, RSO_4^-, and RSO_3^-. Scheme 3 summarizes reactions of z-ArN_2^+ with a variety of nucleophiles, both heterolytically and homolytically (phenols and antioxidants). The numbers in bold indicate the reference in which the preparation of the product is to be found and the bold asterisk (∗) indicates products obtained in as yet unpublished work. Note that Scheme 3 includes many of the weakly basic nucleophiles commonly found in both commercial surfactant systems and in biomembranes and proteins.

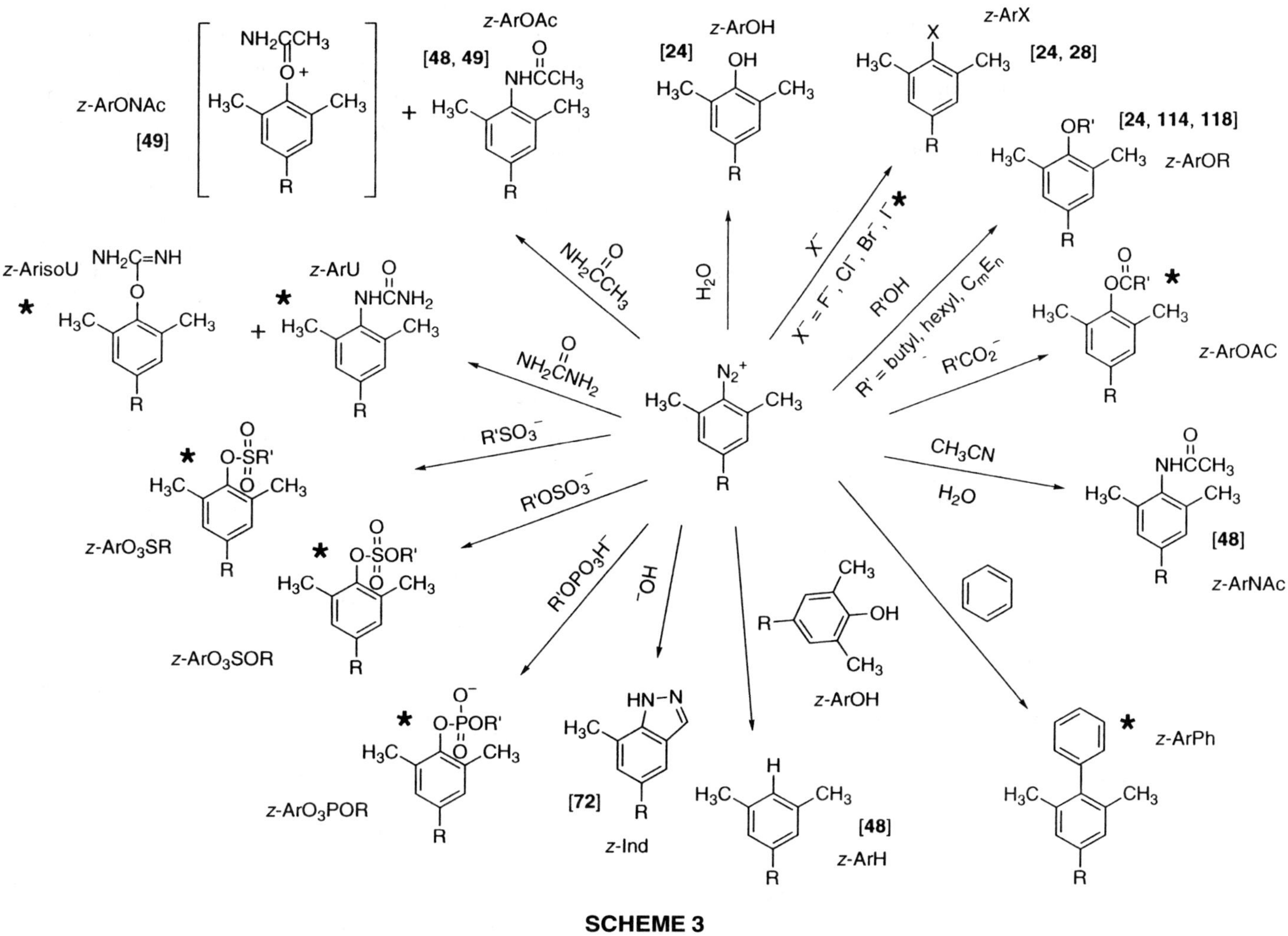

SCHEME 3

B. Important Characteristics of the Heterolytic Dediazoniation Reactions for Chemical Trapping

The fundamental assumption of the method is that the selectivity of the dediazoniation reaction of 16-ArN_2^+ toward two different nucleophiles within the interfacial region of a surfactant assembly is the same as that of 1-ArN_2^+ toward those nucleophiles in a reference bulk solution under comparable conditions [24]. This assumption is based on unique characteristics of heterolytic dediazoniations. Specific interpretations of the basic assumption are discussed in the section on the application of the method (Section IV).

Dediazoniation reactions are spontaneous, and rate-determining loss of N_2 precedes a very fast reaction with a nucleophile (Scheme 2). Observed dediazoniation rate constants, k_{obs}, are extraordinarily insensitive to solvent polarity and composition [25,28]. For example, the dediazoniation rate constant for the benzenediazonium ion is the same, within a factor of about 9, in concentrated sulfuric acid, methylene chloride, glacial acetic acid, methanol, and dilute aqueous acid [24,28]. In terms of transition state theory, this means that medium changes affect the ground and transition states to similar extents; i.e., the free energy of activation is essentially unchanged [29]. Ab initio calculations of charge distributions in the benzenediazonium ion and intermediate phenyl cation are very similar, consistent with the conclusion that the free energy of activation is approximately constant [30,31].

Chemical trapping results with z-ArN_2^+ and other arenediazonium ions are consistent with the spontaneous heterolytic mechanism. For example, k_{obs} values for dediazoniation of 1-ArN_2^+ in aqueous TMAX (X = Cl, Br) solutions (0.5–3 M) and 16-ArN_2^+ in aqueous micellar solutions of CTAX are essentially the same $\pm 20\%$ [24]. Rate constants for dediazoniation of 2-, 3-, and 4-methylbenzenediazonium ions are independent of HCl (0 to 1 M), NaCl (0 to 1 M), and $CuCl_2$ (0 to 0.05 M) concentrations [32,33]. Micelles of sodium dodecyl sulfate (SDS) and copper dodecyl sulfate have small effects on k_{obs} for 2-, 3-, and 4-methylbenzenediazonium ions [34].

The lifetime of the aryl cation has never been measured, but substantial evidence suggests that it must be less than 500 ps [35]. The lifetimes of more stable carbocations such as the diphenylmethyl in 1:1 MeCH:H_2O and in aliphatic alcohols [36] and isopropyl cation [37] in aqueous acetonitrile have been measured and are 750, about 70, and 50 ps, respectively. The hydride affinities of these carbocations are substantially less than that of the phenyl cation [38], suggesting that the lifetime of an aryl cation is governed by diffusion control. Lorand found that the selectives of benzene-, 2,4-dimethylbenzene-, and 2,4,6-trimethylbenzenediazonium ions toward Cl^- and Br^- are independent of solution viscosity, indicating that capture of aryl cations by halide ions is very rapid [39]. However, the lifetime of aryl cations is not zero. Zollinger and coworkers used ^{15}N labeling experiments under N_2 pressure in trifluoroethanol to show that the aryl cation lives long enough to permit scrambling of the two nitrogens and to exchange with dissolved dinitrogen during dediazoniation [40]. They interpreted these results in terms of formation of tight and solvent separate aryl cation-dinitrogen pairs [25].

The almost total insensitivity of dediazoniation rates to solvent effects and nucleophile concentrations means that the distribution of neutral and anionic nucleophiles in the immediate vicinity of the ensemble of ground state arenediazonium ions remains essentially unchanged through aryl cation and product formation. Thus, product yields reflect concentrations of nucleophiles within the immediate vicinity of the aryl cations and therefore within the immediate vicinity of the ground state arenediazonium ions. The mechanistic interpretation of dediazoniation selectivities (Section II.C) is similar to one used by Klumpp to describe ground state effects on relative reactivities of competing reactions that have the same or similar high-energy transition states [26]. Measured selectivities toward different nucleophiles are insensitive to properties of the reaction medium (Section II.D).

C. Relationship of the Dediazoniation Mechanism to Measured Selectivities

Scheme 4 shows a modified form of Zollinger's aryl cation · molecule pair mechanism, a preassociation route [41,42] that describes the heterolytic dediazoniation pathway in terms of ground state, arenediazonium · nucleophile, and intermediate aryl cation · nucleophile pairs for a competitive dediazoniation reaction with water and a second nucleophile, X, either neutral or anionic. The aryl cation is a transient intermediate and the steady-state assumption [43] for reactive intermediates is used to derive an expression for selectivities toward two different nucleophiles. Arenediazonium ions are assumed to exist as an ensemble of z-$ArN_2^+ \cdot H_2O$ and z-$ArN_2^+ \cdot X$ pairs and their concentrations depend on the value of the equilibrium constant, K_W^X, and the concentrations of H_2O and X. Product yields depend on the concentrations of these arenediazonium

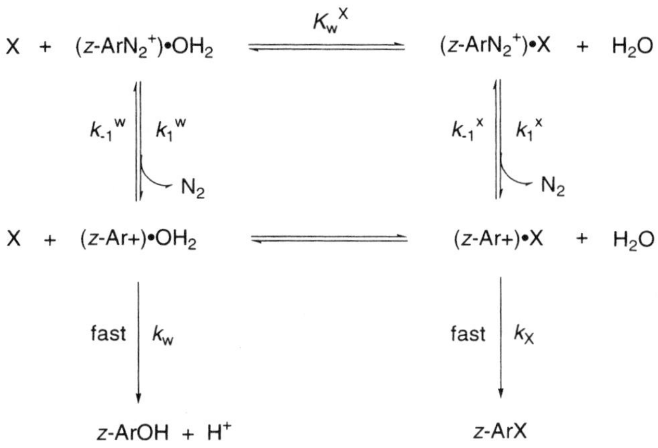

SCHEME 4

ion · nucleophile pairs and their rate constants for dediazoniation. As shown in the following, in this mechanism, selectivities are independent of rates of the fast product-forming steps from the aryl cation.

In bulk aqueous solutions K_W^X for the equilibrium between arenediazonium ion · nucleophile pairs is given by

$$K_W^X = \frac{[(z\text{-ArN}_2^+)\cdot X][H_2O]}{[(z\text{-ArN}_2^+)\cdot H_2O][X]} \tag{1}$$

where square brackets indicate concentration in moles per liter of solution volume here and throughout the text. Product yield ratios for these two nucleophiles are proportional to their rates of formation [26]:

$$\frac{\%(z\text{-ArX})}{\%(z\text{-ArOH})} = \frac{d[(z\text{-ArX})]/dt}{d[(z\text{-ArOH})]/dt} \tag{2}$$

Applying the steady-state approximation, we assume that:

$$\frac{d[(z\text{-Ar}^+)\cdot X]}{dt} = \frac{d[(z\text{-Ar}^+)\cdot H_2O]}{dt} = 0 \tag{3}$$

which leads to Eq. (4) for nucleophile X with the rate constants as defined in Scheme 4:

$$\frac{d[(z\text{-ArX})]}{dt} = \frac{k_X k_1^X[(z\text{-ArN}_2^+)\cdot X]}{k_X + k_{-1}^X[N_2]} \tag{4}$$

The rate constants for the forward, k_X, and back, k_{-1}^X, reactions should be near the diffusion-controlled limit and numerically similar, but $[N_2]$ is very small, about 10^{-3} M, the solubility of N_2 in water [44] and orders of magnitude smaller than the concentrations of H_2O, about 50 M, and other nucleophiles present. Note that Zollinger carried out dediazoniation reactions under N_2 pressure to observed exchange between dissolved N_2 (up to 3 M) and the diazonio group [40]. Thus, $k_X \gg k_{-1}^X[N_2]$ and Eq. (4) simplifies to

$$\frac{d[(z\text{-ArX})]}{dt} = k_1^X[(1\text{-ArN}_2^+)\cdot X] \tag{5}$$

A similar equation can be derived for the rate of formation of z-ArOH. Substituting the expressions for rates of formation of z-ArX and z-ArOH into Eq. (2) gives:

$$\frac{\%(z\text{-ArX})}{\%(z\text{-ArOH})} = \frac{k_1^X[(z\text{-ArN}_2^+)\cdot X]}{k_1^W[(z\text{-ArN}_2^+)\cdot H_2O]} \tag{6}$$

The ratio of arenediazonium ion · nucleophile pairs is given by Eq. (1). Substituting Eq. (1) into (6) gives

$$\frac{\%(z\text{-ArX})}{\%(z\text{-ArOH})} = \frac{K_W^X k_1^W[X]}{k_1^W[H_2O]} \tag{7}$$

Equation (7) is directly related to the definition for the selectivity in the trapping reaction, S_W^X:

$$S_W^X = \frac{\%(z\text{-ArX})}{\%(z\text{-ArOH})}\frac{[X]}{[H_2O]} = \frac{K_W^X k_1^W}{k_1^W} \tag{8}$$

Equation (8) shows that the selectivity of the reaction, S_W^X, equals the product of ratio of the rate constants for rate-determining loss of N_2 from the arenediazonuim ion · nucleophile pairs and the equilibrium constant for distributions of the two ion-nucleophile pairs in bulk solution and surfactant assemblies. Values of k_1^W have been estimated for a variety of arenediazonium ions in the absence of X [25,27,45], but independent estimates of K_W^X and k_1^X are not available. However, because heterolytic dediazoniation reactions are so insensitive to medium effects and nucleophile concentrations, the rate constant ratio should be approximately one, i.e., $k_1^W/k_1^X \approx 1$. This means that values of S_W^X should depend primarily on K_W^X, i.e., on the interactions of z-ArN$_2^+$ with nucleophilic and nonnucleophilic components.

D. Some Measured Selectivities of Various Nucleophiles

Selectivities have been measured in bulk solution for both nonionic and anionic nucleophiles and this section summarizes values obtained from various sources.

1. Nonionic Nucleophiles

Selectivities of these nucleophiles are essentially independent of nucleophile concentration. The selectivity of 1-ArN$_2^+$ toward BuOH compared with water, S_W^{BuOH}, is about 0.30 in BuOH-H_2O mixtures up to 1 M BuOH (saturation limit) and in 9:1 BuOH/H_2O solutions [24,46]. For the competitive reaction of MeOH and H_2O with the 2-methylbenzenediazonium ion in 0.01 M HCl at 35°C from 8 to 88% MeOH by weight, $S_{\mathrm{W}}^{\mathrm{MeOH}} = 0.41 \pm 0.03$ for 16 different MeOH concentrations [47]. For aqueous mixtures of tetraethylene and hexaethylene glycols, $S_{\mathrm{W}}^{\mathrm{ROH}} = 0.60$ [48]. This value is the same at 18 and 40°C, in the presence and absence of added HCl, and is independent of the molar ratio of water to oligooxyethylene glycol from 20:1 to 100:1. Both the amino N and acyl O of simple amides, acetamide, *N*-methylacetamide, *N,N*-dimethylacetamide [49,50], and urea (unpublished results) trap 1-Ar$^-$. The value of $S_{\mathrm{W}}^{\mathrm{O}}$ is approximately 1 for the three acetamides and urea, and $S_{\mathrm{W}}^{\mathrm{N}}$ is approximately 0.1 for acetamide and N-methylacetamide and about 0.4 for urea. No product was observed from trapping of the N on *N,N*-dimethylacetamide by 1-Ar$^+$ within our detection limits, about 1% [49].

2. Anionic Nucleophiles

Selectivities for competition between H_2O and anionic nucleophiles such as halogens range between approximately 2 to 15, depending upon the structure of the arenediazonium ion and the anion [24,32,33]. Selectivities toward CH_3CO_2H and $CH_3CO_2^-$ compared with water are essentially the same and about 2 between 0.5 and 5 M (unpublished results). The selectivities for Cl^- and Br^- decrease with added tetramethylammonium chloride, TMACl, and tetramethylammonium bromide, TMABr. As the anion concentration increases from 0.5 to 3.5 M, $K_{\mathrm{W}}^{\mathrm{Cl}}$ decreases from 7.8 to 4.6 and $K_{\mathrm{W}}^{\mathrm{Br}}$ from 14.5 to 8.5 [24]. However, the selectivity toward Br^- compared with Cl^-, $K_{\mathrm{Cl}}^{\mathrm{Br}}$, obtained from $K_{\mathrm{Cl}}^{\mathrm{Br}} = K_{\mathrm{W}}^{\mathrm{Br}}/K_{\mathrm{W}}^{\mathrm{Cl}}$ at equal salt concentrations is about 1.9 and independent of salt concentration.

What is most striking about these results is the almost total absence of selectivity, i.e., $S_{\mathrm{W}}^{\mathrm{X}}$ is near one (1). This result is particularly surprising for competition between anions and water for a cation. Such small selectivities are consistent with the heterolytic dediazoniation mechanism and the reactivity-selectivity principle; i.e., highly reactive intermediates show low selectivities [26]. The decrease in $K_{\mathrm{W}}^{\mathrm{Br}}$ and $K_{\mathrm{W}}^{\mathrm{Cl}}$ with increasing salt concentration is consistent with ionic strength effects on activity coefficient ratios of arenediazonium ion·halide ion complex to halide ion, Eq. (1) [24]. The equivalent expression for $K_{\mathrm{Cl}}^{\mathrm{Br}}$ has structurally similar complexes and anions in the numerator and denominator and salt effects on activity coefficients should cancel. As noted before, $K_{\mathrm{Cl}}^{\mathrm{Br}}$ is constant.

E. Application of the Pseudophase Model to Chemical Trapping

Scheme 5 illustrates application of the pseudophase model of aggregate effects on chemical reactivity to dediazoniations occurring in the bulk aqueous and interfacial regions [4–6]. A similar scheme may be written for reaction in bulk oil and interfacial regions. In the pseudophase model, the totality of the aggregates, micelles, microemulsions, vesicles, etc., is treated as a separate phase or, more correctly, a pseudophase distributed throughout the bulk aqueous or oil phase. The transfer rates of components, surfactant monomer, ions, water, and solutes, e.g., arenediazonium ions, are orders of magnitude higher than dediazoniation such that all components are at their equilibrium distributions throughout the time course of the reaction. In micelles and microemulsions, ions and molecules enter aggregates at rates near the diffusion-controlled limit, i.e., ≤1 ns, but exit rates depend upon hydrophobicity and, although slower, are still very fast, ≤1 μs [11,51,52]. The half-life for dediazoniation of *z*-ArN_2^+ is the order of 10^3 s [24], many orders of magnitude slower than component transfer rates.

This analysis shows that product yields from dediazoniations depend upon the fraction of arenediazonium ion located in the interfacial, oil, or aqueous regions of the surfactant assemblies; i.e., its location depends on the equilibrium constant or free energy of

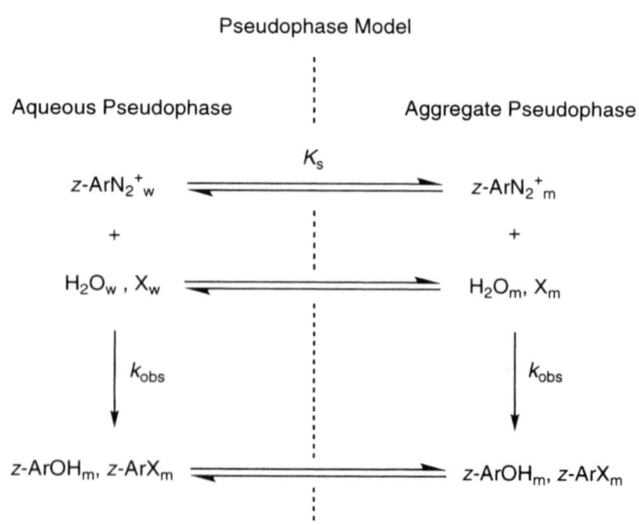

R = $C_{16}H_{33}$, *z* = 16; R = CH_3, *z* = 1

X = any weakly basic anionic or neutral nucleopihile

SCHEME 5

transfer between the three regions. Arenediazonium compounds are salts and their solubilities in the oil regions should be negligible. For 16-ArN_2^+, the hexadecyl chain ensures strong binding of the arenediazonium ion to the aggregates. The distribution (or binding) constant, K_S, for 16-ArN_2^+ (Scheme 5) cannot be measured directly because it decomposes spontaneously. However, it should be similar to that of *N*-1-hexadecyl-3-carbamoylpyridinium bromide, $K_S = 3.5 \times 10^3\ M^{-1}$ [53]. Assuming 16-ArN_2^+ has the same distribution constant, it would be >97% bound in 0.01 M CTABr [24]. Thus trapping results obtained at $\geq 10 \times$ cmc with 16-ArN_2^+ in cationic micelles should report only interfacial concentrations. Conversely, 1-ArN_2^+ is extremely water soluble and reaction occurs only in the aqueous pseudophase, $K_S \approx 0$ [54]. Because the structures of the 16-Ar+ and 1-Ar+ aryl cations are almost identical to those of their arenediazonium ion precursors, their distributions in solutions of surfactant assemblies should be the same as those of the ground state arenediazonium ions. These distribution assumptions will be violated only at the extremes, e.g., anionic micelles bind both short-chain arenediazonium ions [34] and 16-ArN_2^+ [55].

In summary, (1) all components are in dynamic equilibrium between aggregates and the surrounding bulk phase and (2) arenediazonium ions decompose spontaneously at essentially the same rate with weakly basic in any medium and therefore (3) product distributions reflect compositions of the phase in which the probe resides. In aqueous solutions, yields of 1-ArOH and 1-ArX from dediazoniation of 1-ArN_2^+ are proportional to total concentrations of the nucleophiles. Similarly, yields of 16-ArOH and 16-ArX from dediazoniation of 16-ArN_2^+ in surfactant assemblies are proportional to the interfacial molarities of the nucleophiles because its hydrophobicity ensures that the reactive diazonio group is located only in the interfacial region.

F. Basic Assumptions of the Pseudophase Model

To estimate interfacial concentrations of nucleophiles in surfactant assemblies from dediazoniation product yields, we must assume that S_W^X [Eq. (6)] for reaction in surfactant assemblies using the long-chain probe, $z = 16$, is the same as that for reaction in bulk solutions in the absence of surfactant assemblies using the short-chain probe, $z = 1$. That is, Eq. (9) holds *under comparable compositions in the interfacial and reference aqueous solutions*:

$$S_W^X = \frac{\%(1\text{-ArX})}{\%(1\text{-ArOH})}\frac{[X_W]}{[H_2O_W]} = \frac{\%(16\text{-ArX})}{\%(16\text{-ArOH})}\frac{X_m}{H_2O_m} \tag{9}$$

The validity of Eq. (9) is based on known low measured selectivities of weakly basic nucleophiles, their insensitivity to solution composition, and the insensitivity of k_{obs} to solvent polarity, which implies that the very rapid trapping of the aryl cation by nucleophiles will also be insensitive to changes in solvent polarity. Equation (9) states that when product yield ratios from reaction of 1-ArN_2^+ in an aqueous solution of known $[X_W]$ and $[H_2O_W]$ are the same as those from reaction of 16-ArN_2^+ in a solution containing surfactant assemblies, then the molar ratios of X_m and H_2O_m in the interfacial regions will be the same as that in water. This approach was used to estimate hydration numbers of nonionic micelles, where X is the terminal OH group of the surfactant (Section IV.A), and the K_{Cl}^{Br} exchange constant (Section IV.B.7).

Results in cationic micelles are treated differently because values of S_W^X for competition between anions, e.g., X = Cl^- or Br^-, and H_2O decreases gradually with added salt. To estimate interfacial molarities of anions and H_2O, we assume that when product yields in micelles and in the reference bulk aqueous solution are the same, their concentrations in micelles and the aqueous solution are the same. In brief, *when yields are the same, concentrations are the same*. This approach permits direct estimates of nucleophile concentrations in the interfacial region because the concentrations in the reference solution are known and because this approach automatically corrects for changes in S_W^X with salt concentration. This assumption was used to estimate interfacial molarities of counterions in Sections IV.B.1, 2, 4, and 6.

G. Limitations of the Chemical Trapping Method

Strongly basic nucleophiles such as OH^-, CN^-, SO_3^{2-}, and N_3^- react extremely rapidly with the terminal nitrogen (Scheme 1), typically within the mixing time of the reactants and orders of magnitude faster than dediazoniation [25,27]. These nucleophiles cannot be used in the chemical trapping method because there are no aryl cations that can be trapped by weakly basic nucleophiles and no information can be obtained on their interfacial concentrations. This limitation is unfortunate because basic nucleophiles, especially OH^-, have been used extensively in micellar catalysis studies [4,56,57].

The acid strength of a nucleophile's conjugate acid is an imperfect indicator of its nucleophilicity [58]. Weakly basic nucleophiles are generally conjugate bases of strong acids, e.g., Br^-, ROH, and H_2O are conjugate bases of strong acids [29]. However, pK_a values of HN_3 and CH_3CO_2H are 4.72 and 4.76, respectively [59], but N_3^- attacks the terminal nitrogen of the arenediazonium ion at diffusion-controlled rates and $CH_3CO_2^-$ does not and reacts via the heterolytic mechanism. If the measured rate constant, k_{obs}, significantly exceeds that for dediazoniation in water and is sensitive to nucleophile concentration, then the reaction may be occurring by attack at the terminal nitrogen or by electron transfer (Scheme 1). For example, in water in the absence of micelles, added $CuCl_2$ in aqueous NaCl speeds the breakdown of 4-nitrobenzenediazonium ion by a factor of 10 [60]. No reduced product, 4-nitrobenzene, is observed, suggesting that a chlorocuprate complex is speeding heterolysis. Added $CuCl_2$ has little effect on k_{obs} for the breakdown of 2-, 3-, and 4-methylbenzenediazoium ions in the presence [34] or absence [33] of SDS micelles. However, added SDS slightly speeds dediazoniation of 4-nitrobenzenediazonium ion and both SDS and $Cu(DS)_2$ induce the formation of product, 4-nitrobenzene, suggesting that micelles are changing the mechanism [61]. Electron withdrawing groups are known to promote the homolytic pathway (Scheme 1), and micelles may be enhancing this reaction [25,62]. Electron donors (Scheme 1) such as ascorbic acid speed the breakdown of 3-methylbenzenediazonium ions [63] and antioxidants speed the reduction of 16-ArN_2^+ to 16-ArH in nonionic micelles (unpublished results), probably by electron transfer [64].

High solution viscosity may also interfere with the chemical trapping reaction. If the surfactant solutions are too viscous, the arenediazonium ion and other components may not be in dynamic equilibrium [48]. The chemical trapping method has not been tried with surfactant assemblies in which the exchange of components is slow, e.g., vesicles. However, the pseudophase model has been used successfully to treat the effect of cationic vesicles on ester thiolysis [65] and the transfers of 16-ArN_2^+ may still be fast enough and vesicle structure sufficiently stable so that the requirement that the components be in their equilibrium distribution is satisfied.

III. PRACTICAL ASPECTS

A. Dediazoniations and Chemical Trapping in Aggregates: A Brief History

Selection of 16-ArN_2^+ as the most suitable probe of surfactant assembly interfaces was reached through trial and error, assessment of probe properties, and serendipity. Scheme 6 shows four different diazonuim ions that have been or are being used as probes of surfactant assemblies. In 1973 Moss and coworkers published the first dediazoniation reactions in micelles [66]. They demonstrated that micellization changes the stereochemistry of deamination of 2-aminooctane via in situ preparation of its alkane diazonium ion, **A**. The degree of inversion versus retention of configuration of the 2-octanol product from reaction with H_2O depends on counterion type. The stereochemical course changed significantly with increasing concentrations of "hydrophobic" counterions, e.g., ClO_4^- and tosylate, but was unaffected by concentrations of more hydrophilic counterions, e.g., Br^- and Cl^-. Singer and coworkers published the first dediazoniation chemical trapping experiment in 1982 [67]. They prepared holo-micelles, i.e., single-component micelles (in this case the substrate is also the surfactant) from hexyl and octyl alkanediazonium ions, **A**, from their alkylammonium precursors. Spontaneous decomposition of these micellized alkanediazonium ions in aqueous acid, HX (X = Cl, Br), gave mixtures of 1- and 2-substituted alcohols and haloalkanes, showing that micellization induces a large increase in the selectivity of the dediazoniation reaction toward Br^- and Cl^- compared with water. The same trends are observed with 16-ArN_2^+ [24]. The alkanediazonium ions have several disadvantages as probes.

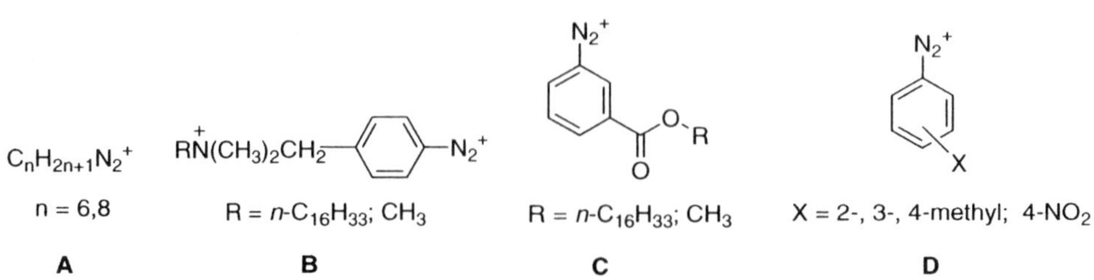

SCHEME 6

They decompose too rapidly to be isolated and purified. Acid and $NaNO_2$ are added to transform the amine precursor into the alkanediazonium ion in situ, which makes experiments difficult to run in the absence of salt. The products contain no chromophores and they must be isolated prior to GC analysis. Finally, product distributions may depend on the medium properties of the aggregate interface.

Our first attempt at chemical trapping was with probe **B**. The results demonstrated the viability of arenediazonium ions as probes, but **B** had some unanticipated weaknesses [68]. Comparison of product yields from dediazoniations of holomicelles of **B** (R = *n*-$C_{16}H_{33}$) and of its short-tail methyl analogue (R = CH_3) in aqueous solution showed an almost complete inversion of the bromo/phenolic product ratio as determined within the sensitivity of the NMR experiment, ~5%. At pH 4–6 in the absence of micelles, the short-chain arenediazonium ion gave only phenols and micelles of the long-chain arenediazonium ion gave only boromo product, but at pH 2 a significant amount of the phenol is formed. We now know that micelles accelerate the rates of electron transfer reactions of phenolic products with unreacted arenediazonium ion as the pH increases (see Section V.C), suggesting that in the pH range 4–6 the phenolic product is consumed by reaction with unreacted starting material. The utility of **B** is limited by its long dediazoniation half-life, about 10–20 h at ambient temperature, such that complete reaction takes 4–8 days—a serious test of patience and an analytical method.

Our second probe, **C**, was an arenediazonium ion with an ester group in the *meta* position. **C** was selected because syntheses of the long- and short-chain analogues are straightforward and the headgroup is a monocation [69,70]. The hexadecyl derivative was used in micelles and the methyl derivative in bulk aqueous solution. **C** traps both Br^- and Cl^- simultaneously at the micellar surface. From the ratio of the halo product yields we estimated ion-exchange constants between these two ions over a wide range of ionic strengths [69,70]. The protocol for estimating exchange constants and the results are discussed in Section IV.B.7. A major limitation of this probe is that the acid concentration in aqueous cationic micelles must be 0.1 M H^+ or higher with the long-chain probe in aqueous cationic micelles and 0.01 M or higher with the short-chain analogue in aqueous solutions to prevent formation of an unidentified yellow side product. The requirement of high acidity limits the use of this probe in membrane mimetic systems.

Probes **D** (Scheme 6) are in active use. Results with probes **D** have generated new analytical methods for measurement of dediazoniation rates (see Section III.E). Mancini et al. developed a chemical trapping approach based on the formation of bromonium ion from Br_2 and cyclohexene and their reaction with water to give bromohydrins and Cl^- to give chlorobromocyclohexane [16]. The halide ion/water product yield ratios showed marked increases at the sphere-to-rod transitions in CTABr and CTACl micelles, but interfacial counterion and water concentrations are difficult to estimate from the data because the results indicate micellar enhanced nucleophilicity of H_2O toward the bromonium ion. For results with z-ArN_2^+ in cationic surfactants, see Section IV.B.1.

B. Characteristics of z-ArN_2^+ as a Chemical Trapping Reagent

1. z-ArN_2BF_4 Salts Are Stable in the Solid State

Routine handling problems with arenediazonium salts are minimal, although repeated exposure to the atmosphere and prolonged storage promote reactions with water vapor to give z-ArOH and with BF_4^- to give z-ArF (the Schiemann reaction [27], Scheme 3). Care must be exerted because significant amounts of z-ArOH in the solid add to the z-ArOH yield from the trapping of H_2O by z-ArN_2^+ in solution. Formation of z-ArF reduces the quantity of arenediazonium ion in the sample but does not affect analyses of products as long as it is present in small amounts. Recrystallization removes both impurities (see Section III.C).

2. Competing Side Reactions

To date, the z-ArN_2^+ probes have shown few limitations on their utility. The alkyl groups minimize formation of side products as compared with earlier probes and competing reactions are minimal up to about pH 7. The primary competing reaction is formation of 16-ArH (Scheme 3), which comes from reaction of phenolic product with arenediazonium ion. This reaction is unimportant with 1-ArN_2^+ in aqueous solutions but sometimes becomes significant in surfactant assemblies that speed bimolecular reactions. This reaction competes with dediazoniation most effectively at surfactant concentrations just above the cmc, where the concentrations of aggregate-bound reactants are highest [4]. Note that the phenolic product is structurally similar to antioxidants, e.g., vitamin E, and we are using the reaction of 16-ArN_2^+ with antioxidants to estimate their distributions in microemulsions and emulsions (see Section V.C). Chaudhuri and coworkers discovered a second competing reaction, base-induced indazole formation, z-Ind (Scheme 3) [71]. This reaction consumes z-ArN_2^+ but not dediazoniation products, and as long as its yield is small, its formation does not affect quantitative analysis of interfacial concentrations from product yields formed by heterolysis.

3. z-ArN_2^+ Has Good Characteristics for Routine Quantitative Analysis

The half-life for dediazoniation is about 30 min at 40°C. This temperature has been used in many of our experiments because the reaction proceeds at a convenient rate and because much initial work was in aqueous CTABr, which has a Kraft point at about 25°C and tends to precipitate on standing. However, at 40°C, the CTABr and Br^- concentrations can be varied widely without problem. To date, we have used the probe at temperatures from 18 to 60°C [24,48]. Final z-ArN_2^+ concentrations are typically $\leq 1 \times 10^{-4}$ M. At this concentration, perturbation of surfactant assemblies by 16-ArN_2^+ should be minimal with surfactant concentrations $\geq$0.01 M, i.e., at surfactant/16-ArN_2^+ ratios of $\geq$100:1. In general, product mixtures, including the surfactants, are injected into the HPLC without work-up. The greatest HPLC analysis headache was with 1-ArN_2^+, which was eventually solved by a simple trick (see Section III.D).

C. Preparation of z-ArN_2BF_4 and Its Reaction Products

The arenediazonium ions are prepared as their tetrafluoroborates because these salts of arenediazonium ions have never been reported to explode, unlike those with other counterions [25], and they can be prepared from their distilled aniline precursors by a one-step nonaqueous procedure [72]. The 2,4,6-trimethylaniline precursor to 1-ArN_2BF_4 is available commercially. The biggest synthetic problem is preparing the precursor to 16-ArN_2BF_4, 4-hexadecyl-2,6-dimethylaniline. 2,6-Dimethylaniline is alkylated with 1-hexandecanol using anhydrous $ZnCl_2$ as Lewis acid catalyst at 260°C for about 1 day [73]. On cooling, the once violet product mixture turns into a hard black amorphous solid and laborious isolation and purification eventually give a white crystalline product with yield of ~10–20%. ^{1}H-NMR spectra from multiple syntheses show that alkylation occurs without competing 1,2 hydride shift and only at the 4- position of 2,6-dimethylaniline. The 3 and 5 aryl protons give a single signal in the aromatic region and only two ^{1}H signals are observed for —CH_3 groups, the ω methyl on the hexadecyl chain and the equivalent 2 and 6 methyls on the ring [24].

Dediazoniation product peaks in the HPLC chromatograms are identified by spiking experiments with independently prepared samples. These products are also used to prepare calibration curves for converting peak areas into product yields. Some 1-ArN_2^+ dediazoniation products are available commercially; others are prepared from the arenediazonium salts or by independent syntheses. Details of the preparation of products are in references numbered in bold adjacent to the products in Scheme 3. Products from work in progress are indicated with a bold asterisk.

Products from competing reactions are formed via the heterolytic pathway, except for the z-ArH and z-Ind products described in Section III.B. Formation of z-ArH and z-Ind can be minimized by making the solution more acidic or by increasing surfactant concentrations to inhibit these bimolecular reactions. Fluoro products, z-ArF, are probably formed via the Schiemann reaction with BF_4^- during storage [48] or in their nonaqueous stock solutions used to initiate dediazoniation; see Section III.D. Arenediazonium ion stock solutions are generally prepared in either MeCN or MeOH. Reaction with MeCN at the terminal nitrogen gives a reactive intermediate that is hydrolyzed by H_2O to an acetamide, z-ArOAc [48]. Reaction with MeOH gives an aryl methyl ether, z-ArOMe. All these products appear in HPLC chromatograms, but their yields are generally low, typically 1–2%, and because their formation only reduces the initial concentration of z-ArN_2^+. Their presence does not interfere with the analysis of the products from the trapping reaction.

D. Protocol for HPLC Analysis of Reactions and Products

The protocol for carrying out dediazoniation reactions in surfactant solutions is that commonly used for measuring rate constants in solution spectrophotometrically, i.e., injection of a concentrated substrate stock solution into a solution containing the other components [24]. Multiple solutions are prepared, typically in 5- to 10-mL volumetric flasks, with the needed concentrations of components, surfactant, acid, salt, and other additives (e.g., alcohols) and thermally equilibrated. A 5–100 μL concentrated stock solution, containing a weighed amount of z-ArN_2BF_4 is injected into each solution, which is rapidly but thoroughly mixed and returned to the temperature bath until reaction is complete, usually for $\geq$10 half-lives. The carrier solvent is on the order of 1–3% of the final solution volume. To minimize dediazoniation of z-ArN_2^+ in the stock solutions, they are prepared just before use and kept in an ice bath. Using a weighed amount of z-ArN_2BF_4 permits quantitative comparison of the total yield of products with the initial quantity of z-ArN_2BF_4. Conversion to products is generally quantitative, typically within ±10%. Yield variations are usually caused by combinations of weighing and measurement errors and uncertainties in calibrations of products by HPLC. Larger errors suggest unanticipated loss of products, e.g., by formation of products that are not detected by HPLC, such as ionic products that have similar retention times to ionic surfactants, or by formation of unknown products, e.g., "mystery peaks" in the HPLC chromatograms. Such products must be identified by the hard labor of isolation and identification, followed by confirmation by spiking experiments using independently synthesized compounds.

One initially refractory problem was eventually solved by a simple trick [24]. Reactions of 1-ArN_2^+ in aqueous TMABr and TMACl solutions gave good reproducible yields of 1-ArOH, but yields of 1-ArBr and 1-ArCl varied widely and sometimes decreased with increasing salt concentration—the opposite of what was expected. The problem was eventually attributed to the very low solubility of haloarenes in water, to their salting out by added salts, and to their vapor pressures. A simple calculation showed that their dediazoniation yields are so low that were they to vaporize completely they would occupy a volume smaller than the head space in the volumetric flask. Layering a small volume of cyclohexane, 50–100 μL, on the aqueous solution to dissolve phase-separated products solves the problem. The contents of the volumetric flask are then diluted with sufficient MeOH or *i*-PrOH to ensure that the solutions are homogenous prior to HPLC analysis. This procedure gives excellent reproducibility. This problem is not observed with 16-ArN_2^+ products, probably because of their higher molecular weights and their solubilization by surfactant assemblies.

Products from both 16-ArN_2^+ and 1-ArN_2^+ are separated on Microsorb-MV C-18 reverse-phase columns (4.6 mm inner diameter × 25 cm; 5-μm particle size). These inexpensive reverse-phase columns have proved sufficiently robust for many hours of use. Each analysis usually takes from 15 to 30 min. HPLC analysis with UV detection gives high reproducibility ±1–2% in peak area for duplicate to triplicate analyses with $\leq 1 \times 10^{-4}$ M z-ArN_2^+ in solution. Calibration curves for calculating product yields for each product are prepared from independently prepared and purified compounds at four to five concentrations and the correlation coefficient (cc.) are generally >0.99, spanning the range of interest and yields are obtained by interpolation. Once preliminary experiments establish that conversions to expected products are essentially quantitative and reproducible and that all significant peaks are accounted for, normalized product yields are calculated for the products from reactions with components in interfacial regions of surfactant assemblies. This procedure eliminates uncertainties caused by weighing and transfer errors and gives more reproducible results.

E. Measurement of Dediazoniation Rate Constants

The rate constant for dediazoniation should be determined each time the chemical trapping reaction is used with new components. As noted in Sections II.B and II.G, because of the extraordinary insensitivity of heterolytic dediazoniation rates to solution composition, a significant change in k_{obs} may mean a change in mechanism. Four methods are used for determining k_{obs} for dediazoniation that are applicable for different experimental conditions: following the loss of z-ArN_2^+ spectrophotometrically or electrochemically, trapping unreacted z-ArN_2^+ as an azo dye, and monitoring formation of dediazoniation products by HPLC. Details are in the indicated references.

1. Spectroscopy

In terms of ease and simplicity, UV spectrophotometry is the method of choice because, in general, z-ArN_2^+ absorbs much more strongly than the reaction products [24,32–34,47,60]. Typically, reaction is initiated by injection of a stock 30–50 μL of a 0.01–0.02 M freshly prepared stock solution of z-ArN_2^+ in MeCN or MeOH into an approximately 3-mL thermostated solution in a quartz cuvette containing the other components. The loss of z-ArN_2^+ is usually monitored at its λ_{max}. The change in absorbance, A, of the reaction is followed for about 10 half-lives. Values of k_{obs} are obtained from standard $\ln(A_\infty - A_t)$ versus t plots [74] and cc. ≥0.999. Other methods must be used when the spectrum of z-ArN_2^+ is completely masked.

2. Azo Dye Method

Because most dediazoniation reactions are relatively slow, their rate constants can be determined by sampling techniques. Bravo-Díaz and coworkers [32–34,47,60] took advantage of the very fast formation of azo dyes from arenediazonium ions and naphthoate anions (Scheme 7) to quench dediazoniation reactions. The quenching solutions are buffered at an optimal pH that maximizes deprotonation of the 2-naphthol to give its reactive anion and minimize side reactions such as z-Ind and z-ArH (Scheme 3) and diazotate formation

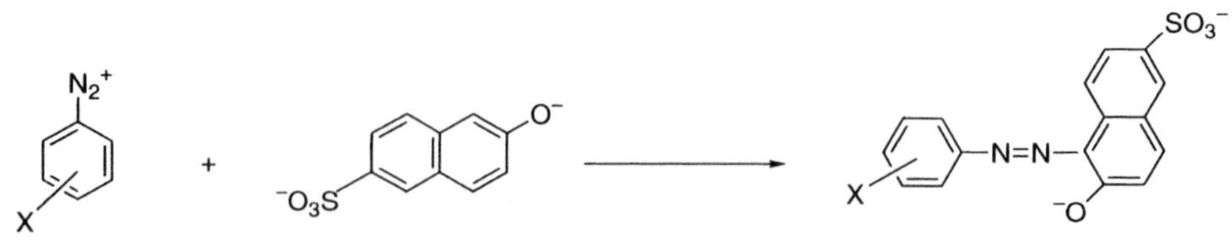

X = 2-, 3-, 4-methyl; 4-NO$_2$

SCHEME 7

(Section II.G). Values of k_{obs} are obtained from the increased yield of azo dye with time at its λ_{max} in the visible region of the spectrum. The yield of azo dye, and therefore unreacted z-ArN$_2^+$, is obtained from an absorbance versus azo dye calibration curve.

3. HPLC Method

The spectrophotometric and azo dye methods give k_{obs} in terms of the rate of disappearance of the arenediazonium ion. Because quenching halts the reaction, concentrations of dediazoniation products can be measured by HPLC and k_{obs} for their formation can be determined from their peak areas as a function of time [32–34,47,60].

4. Electrochemical Methods

Two methods have been developed by Bravo-Díaz and coworkers [64,75]. They monitored the loss of 3-methylbenzenediazonium ion by recording differential pulsed polarograms and by using differential pulsed voltammetry to monitor formation of phenolic product. They also showed that formation of azo dye produced by the quenching of 3-methylbenzenediazonium ion (Scheme 7) can be monitored by differential pulsed polarography.

Where comparable, the methods give similar values of k_{obs} for a particular arenediazonium ion. Results are generally consistent with the heterolytic mechanism of dediazoniation with the formation of a short-lived intermediate because k_{obs} values for loss of arenediazonium ion and product formation are the same. The primary exceptions are with 4-nitrobenzenediazonium ion (see Section II.G) and in the presence of antioxidants (see Section V.C).

IV. APPLICATIONS OF THE CHEMICAL TRAPPING METHOD

A. Hydration Numbers of and Terminal OH Distributions in Nonionic Micelles

Chemical trapping in nonionic micelles is a unique approach for estimating the hydration state of surfactant assemblies. Estimates of hydration numbers of nonionic micelles, e.g., the number of water molecules per ethylene oxide unit in the headgroup of a micellized oligooxyethylene monoalkyl ether, C$_n$E$_m$, by chemical trapping is conceptually straightforward and we have estimated them in mixed and holo-nonionic micelles [48,76]. Hydration numbers have been estimated by a variety of methods such as light scattering [77], sedimentation equilibrium [78], water (D$_2$O) self-diffusion by NMR [79–82], and ^{17}O magnetic relaxation [83]. All these methods are based on a measured change in a bulk property of the system and require information or assumptions about micellar size and shape to obtain the hydration number. Chemical trapping provides a fundamentally different approach because hydration numbers are obtained from aggregate-bound 16-ArN$_2^+$, which samples the composition of the interfacial region. Within the interfacial region, the probe reacts with H$_2$O to give 16-ArOH and the terminal OH groups of the surfactant, R′OH, to give 16-ArOR′ (Scheme 8). Hydration numbers are obtained from these data (see later) without assumptions about aggregate size and shape, but if the hydration numbers change with aggregate structure chemical trapping results will reflect that change. Chemical trapping requires reproducible product yield ratios and a value for the selectivity of the reaction toward terminal OH groups versus water. Product yield ratios are determined from HPLC peak areas by using calibration curves prepared with independently synthesized and purified products. Other minor products, as discussed in Section III.C, are also formed [48].

The average hydration number of a C$_m$E$_n$ micelle is defined as the number of water molecules, N_{MW}, per ethylene oxide unit containing n ethylene oxide units in its E$_n$ headgroup of the surfactant, R_{ROH}:

$$\text{Hydration number} = \frac{N_{MW}}{nN_{ROH}} \qquad (10)$$

The molar ratio of interfacial water to oligooxyethylene chains is given by the product of the selectivity of the

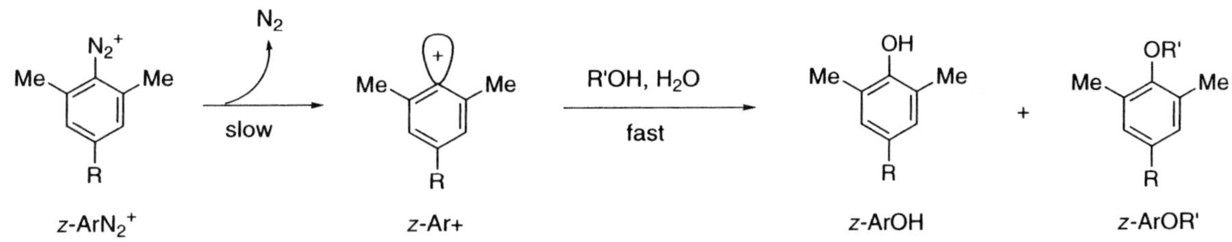

R' = $C_{10}H_{21}(OCH_2CH_2)_4$, $C_{10}H_{21}(OCH_2CH_2)_5$, $C_{12}H_{25}(OCH_2CH_2)_5$, $C_{12}H_{25}(OCH_2CH_2)_6$, $C_{16}H_{33}(OCH_2CH_2)_8$

$z = 16$, R = $C_{16}H_{33}$; $z = 1$, R = CH_3

SCHEME 8

reaction, S_W^{ROH}, and the yield ratio from trapping by interfacial water and terminal OH groups:

$$\frac{N_{MW}}{N_{ROH}} = S_W^{ROH} \frac{\%(16\text{-ArOH})}{\%(16\text{-ArOR}')} \tag{11}$$

The selectivity of the reaction within the interfacial region cannot be measured independently and is assumed to be equal to the selectivity, S_W^{EOH}, of the reaction of 1-ArN_2^+ toward oligoethylene glycols and water in their mixtures [48]:

$$S_W^{ROH} = S_W^{EOH} = \frac{\%(1\text{-ArOR}')}{\%(1\text{-ArOH})} \frac{N_w}{2N_E} = 0.6 \tag{12}$$

where %(1-ArOR'), %(1-ArOH), N_w, and N_E are, respectively, yields from reaction with a terminal OH group of the oligooxyethylene glycol and water and moles of water and oligooxyethylene glycol. The factor "2" corrects for the fact that oligoethylene glycols have two terminal OH groups and twice the probability of reacting with 1-ArN_2^+. As noted in Section II.D, S_W^{EOH} is independent of the ratio of oligoethylene glycol to water.

Our estimates of average hydration numbers in $C_{12}E_6$ micelles are consistent with current understanding of the properties of nonionic micelles. The interfacial regions are "wet" [82] and not almost "dry" [84]. At 40°C from 0.01 M to about 60 wt%, just below the liquid crystalline phase transition, the average hydration numbers decrease only slightly, from about 3.5 to 2.8. This gradual decline in hydration number and their numerical values are in good agreement with the results of water (D_2O) diffusion measurements, although at different temperatures [80]. Hydration numbers in 0.01 M $C_{12}E_6$ decrease linearly with temperature from 4.2 (20°C) to 2.9 (60°C). Above the cloud point at 50°C [81], solutions are phase separated. This steady decrease of hydration number through the cloud point shows that phase separation is not caused by a sudden dehydration of the interfacial region.

When the chemical trapping reaction is carried out in binary mixtures of nonionic micelles having different lengths of polyoxyethylene chains, e.g., $C_{10}E_4$ and $C_{16}E_8$, products are formed from trapping of OH groups of both surfactants [76]. The product yield ratios from reaction with terminal OH groups of the two surfactants are not proportional to their stoichiometric mole fraction ratios. The yield from the shorter oligooxyethylene chain is always in excess over its mole fraction in the micelles. Dr. Jihu Yao developed a novel interpretation of these results. He assumed that the motions of the oligooxyethylene chains obey a radial one-dimensional random walk. Figure 3a is a cartoon of the interfacial region of a mixed nonionic micelle illustrating the orientation of 16-ArN_2^+ within the interfacial region of a nonionic micelle composed of surfactants with different lengths of oligooxyethylene groups. The interfacial region is divided into layers one ethylene oxide unit thick and the chains are assumed to fold like a carpenter's ruler at the oxygens with the hydrocarbon core acting as an impenetrable wall. Figure 3b shows the six possible configurations of the E_4 oligooxyethylene chain, with each configuration being equally probable. Only the terminal OH groups within layer one (configurations 4 and 6) or layer two (configurations 3 and 5) are assumed to react with 16-ArN_2^+. The full treatment of the data provides two pieces of information (Fig. 4): (1) an excellent prediction, without disposable parameters, of the mole fraction yield excess from reaction with the short oligooxyethylene chain over that of the longer chain based on the probability that their terminal OH groups will be in layers 1 and 2, as illustrated for $C_{10}E_4/C_{16}E_8$ mixtures in Fig. 4a, and (2) estimates of the hydration numbers of layers 1 and 2 as illustrated for the same surfactants in Fig. 4b. Equally good results were obtained with five different binary mixtures of

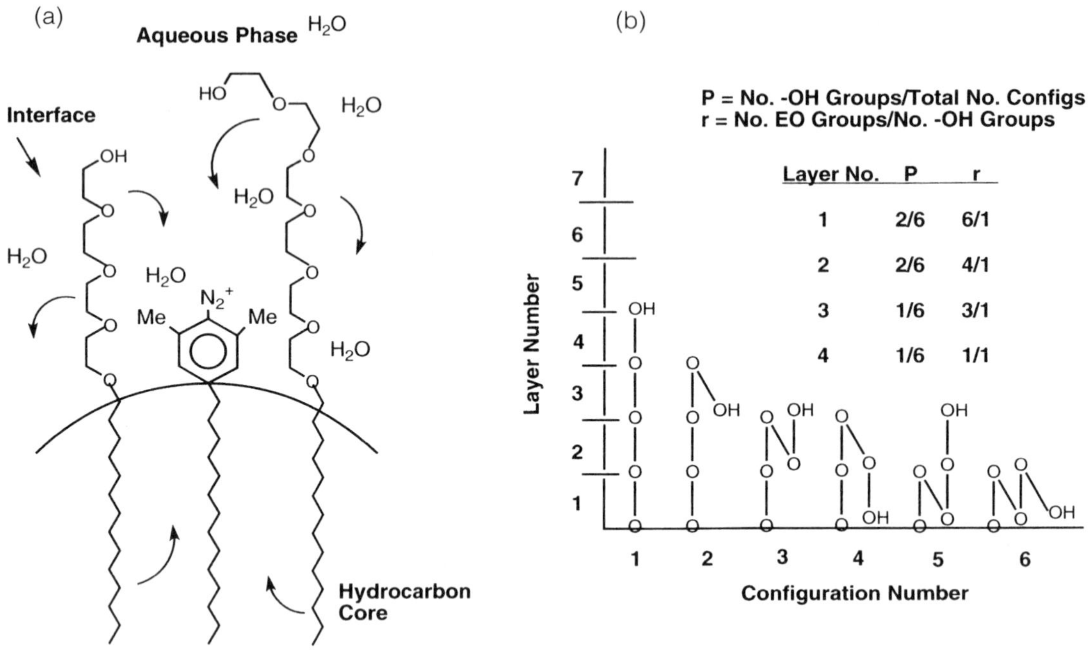

FIG. 3 (a) Cartoon of a small section of the core, interfacial, and aqueous regions of a mixed nonionic C_mE_4/C_mE_6 micelle showing the location of 16-ArN_2^+ with its reactive group adjacent to the micellar core and the flexibility of the alkyl and oligooxyethylene chains. (b) The six possible configurations of the tetraoxyethylene chain of C_mE_4 are based on the radial one-dimensional random walk model illustrating the locations of terminal OH groups and EO units; the sticks are —CH_2—CH_2— units. Layer 1 is adjacent to the hydrocarbon core of the micelles and each OH group is assigned to the same layer as that for the EO unit to which it is attached. The inset table shows the probability, P, of finding the OH group in each layer and the ratio, r, of EO to OH groups in each layer. (Adapted from Ref. 77. Reproduced with permission of the American Chemical Society.)

surfactants with different length tails and oligooxyethylene chains. In each case, the predicted excess mole fraction yield ratios are in excellent agreement with experimental results. The hydration numbers of layers 1 and 2 are essentially independent of the ratio of the two surfactants in the micelles and are somewhat lower than the average hydration numbers obtained for holo $C_{12}E_6$ micelles (see earlier).

These results show that chemical trapping provides a rapid, reliable method for estimating hydration numbers of holo and mixed nonionic micelles over wide ranges of solution compositions, aggregate structures, and temperatures. The binary mixed micelles studied behave "ideally"; i.e., there is random mixing of the surfactants in the aggregates. Chemical trapping may provide insight into the differences in properties of mixed nonionic micelles with very different oligooxyethylene chain lengths that are reported to behave "nonideally" [85,86]. The method may even be applicable to two-phase systems because 16-ArN_2^+ dissolves in the surfactant-rich phase.

B. Chemical Trapping in Ionic Surfactants

1. Interfacial Water and Anion Concentrations

In micellar solutions in which the interface region is composed of only two weakly basic nucleophilic components, their interfacial concentrations can be measured unambiguously by chemical trapping. For example, in CTACl micelles, 16-ArN_2^+ gives two products, 16-ArOH from reaction with H_2O and 16-ArCl from reaction with interfacial Cl^- (Scheme 4, X = Cl) [24]. The yields of these two products depend upon their concentrations in the interfacial region and the selectivity of the reaction toward these two nucleophiles, the first equality on the right-hand side of Eq. (13):

$$S_W^X = \frac{H_2O_m}{X_m}\,\frac{\%(16\text{-ArX})}{\%(16\text{-ArOH})} = \frac{[H_2O_w]}{[X_w]}\,\frac{\%(1\text{-ArX})}{\%(1\text{-ArOH})} \qquad (13)$$

As with chemical trapping in nonionic surfactants, selectivities of reaction at the micellar interface and in

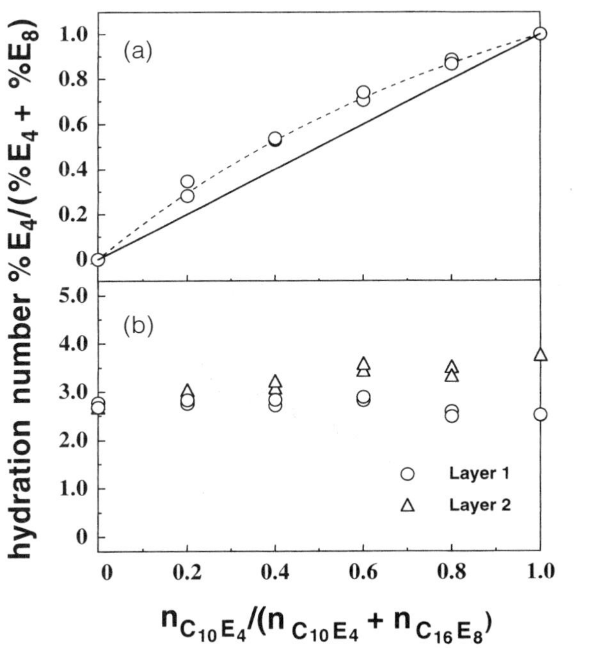

FIG. 4 Plots of product yield mole fraction (a) and hydration numbers (b) versus the mole fraction of $C_{10}E_4$ in mixed micelles of $C_{10}E_4$ and $C_{16}E_8$ from dediazoniation of 1×10^{-4} M 16-ArN_2^+ in 0.02 M total amphiphile at 18°C. In (A), the solid line has a slope of 1 and the dashed line is predicted by using the radial one-dimensional random walk model. In (B), hydration numbers are calculated from ratios of product yields from trapping of H_2O and terminal OH groups for layers 1 and 2. Details are given in Ref. 77.

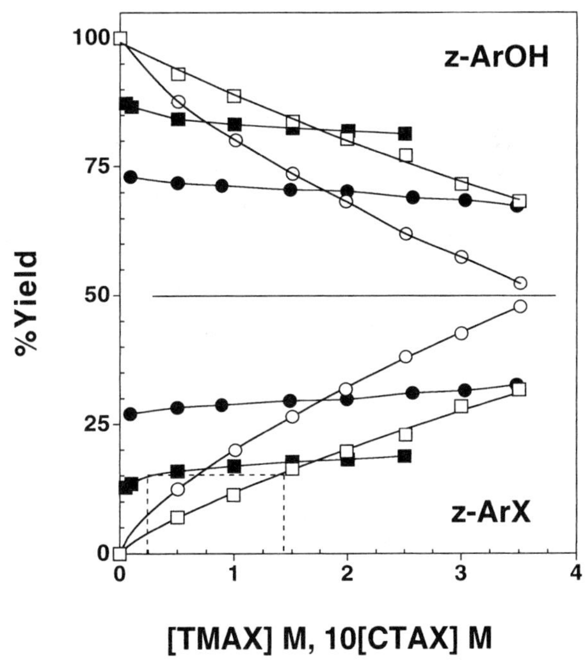

FIG. 5 Dediazoniation product yields at 40 ± 0.1°C from reaction with H_2O (upper curves) and halide ions, X (lower curves), with 1-ArN_2^+ in aqueous TMAX, 0.01 M HX (open symbols) and with 16-ArN_2^+ in aqueous CTAX, 0.01 M HX (closed symbols): (■, □) X = Cl; (●, ○) X = Br. Dashed lines show that when %(16-ArCl) = %(1-ArCl) = 15%, then Cl_m (in 0.025 M CTACl) = 1.4 M (in TMACl). Details are given in Ref. 24. Reproduced with permission of the American Chemical Society.

bulk aqueous solution are assumed to be the same; i.e., the equalities in Eq. (13) hold. However, as discussed in Section II.D, the measured selectivities toward Cl^- and Br^- versus H_2O in aqueous TMAX solutions decrease gradually with increasing salt concentration. To calculate X_m and H_2O_m based on the assumption discussed in Section II.F, we assume that when the yields of 16-ArX and 16-ArOH in a cationic micelle are the same as 1-ArX and 1-ArOH in an aqueous TMAX solution, then X_m and $(H_2O)_m$ in the micelles are the same as the $[X_w]$ and $[(H_2O)_w]$ in the aqueous TMAX solution [24].

Figure 5 illustrates graphically the calculation of Cl_m and $(H_2O)_m$ in CTACl micelles from their total concentrations in aqueous TMACl and CTACl solutions at 40°C in 0.01 M HCl. The dashed lines show that when %(16-ArCl) = %(1-ArCl) = 15%, Cl_m (in 0.025 M CTACl) = $[Cl_w]$ = 1.4 M. In practice, the product yields %(1-ArCl) and %(1-ArOH) versus [TMAX] M are fitted by equations that are used as standard curves for estimating interfacial concentrations. Details are given in Ref. 24.

Figure 6 shows the calculated values of Cl_m and H_2O_m in CTACl solutions with added TMACl, from 0 to 1.5 M at 40°C in 0.1 M HCl [87]. Several patterns are apparent in these data. X_m and H_2O_m lie on a series of almost parallel lines. Their slopes with increasing [CTACl] are gradual, increasing for X_m and decreasing for H_2O_m except in 1 and 1.5 M TMACl, for which they are distinctly curved. Added TMACl, at constant [CTACl], produces an approximately incremental increase in X_m and a proportional decrease in H_2O_m. The upward curvature at 1 and 1.5 M TMACl is in the vicinity of the salt concentrations required to induce the sphere-to-rod transition [13–16].

Figure 7 shows the results in Fig. 6 replotted against the aqueous $[Cl_w]$. $[Cl_w]$ is assumed to be equal to the sum of the concentrations of added TMACl and the concentration of counterions contributed by ionized CTACl micelles, α[CTACl], including a correction for

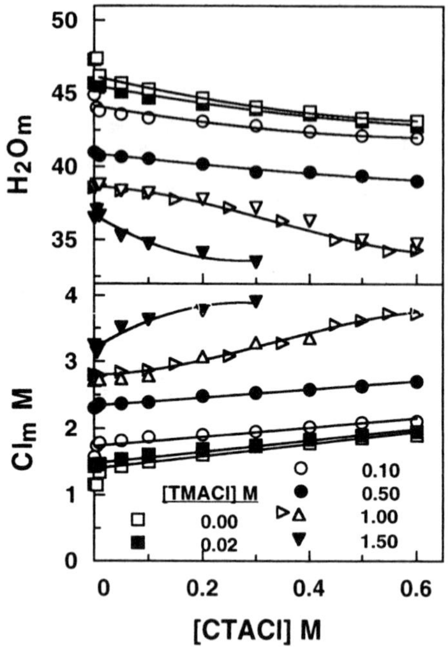

FIG. 6 Effect of increasing CTACl and increasing TMACl on Cl_m and H_2O_m in 0.1 M HCl at 40 ± 0.1°C. Cl_m values were calculated from an equation finding the %(1-ArCl) versus TMACl curve in Fig. 5. H_2O_m values were calculated from S_w^{Cl} values, normalized product yields, and Cl_m. Details are given in Ref. 88. Reproduced with permission of the American Chemical Society.

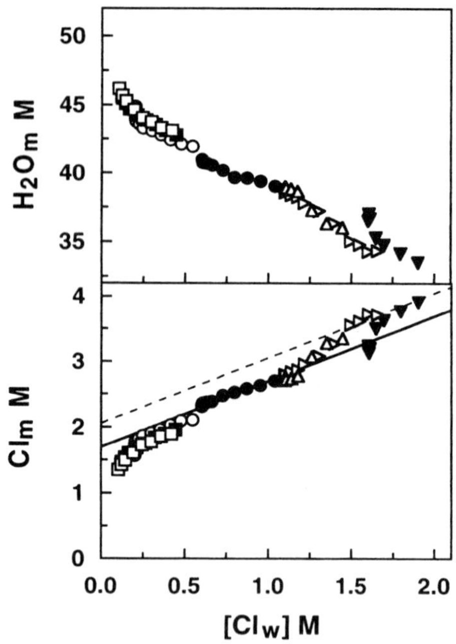

FIG. 7 Plots of Cl_m and H_2O_m versus $[Cl_w]$ at the optimal α value of 0.4 for CTACl/TMACl solutions. Data and symbols are as in Fig. 6. The straight line has a slope of 1, and the intercept was selected to give optimal contact with the linear portion of the curve. Note break from linearity at 1.2 M $[Cl_w]$. Details are given in Ref. 88. Reproduced with permission of the American Chemical Society.

the excluded volume of the micelles. Alpha is treated as a disposable parameter, and setting $\alpha = 0.4$ gives the smoothest profile for Cl_m values at all TMACl concentrations. Details are given in Ref. 87. The curves in Fig. 7 show that Cl_m and H_2O_m are continuous functions of $[Cl_w]$. The curves have three different regions: an initial relatively rapid rise in Cl_m, a linear section with a slope of 1, and a more rapid increase in Cl_m at about 1.2 M $[Cl_w]$ that appears again to approach a line with a slope of 1. Note that H_2O_m responds in a reciprocal fashion. The absence of a plateau contradicts common assumptions used to interpret catalysis in surfactant assemblies. The application of Langmuir isotherms to counterion binding requires that the interfacial counterion concentration saturates at high salt ([88,89] and references therein). In the pseudophase ion-exchange model, the interfacial counterion concentration is assumed to be constant and independent of the concentration of added counterion [4,56,90,91]. When two counterions are present, the sums of their interfacial concentrations are assumed to be constant. For a surfactant system with only one counterion, the interfacial concentration in the pseudophase model is defined as

$$X_m = \frac{\beta}{V_m} = \frac{[X_m]}{[D_n]V_m} \tag{14}$$

where the ratio of fraction of bound counterions, β ($\beta = 1 - \alpha$), divided by the molar volume available to the counterions in the interfacial region, V_m, is assumed to be constant. In two-site pseudophase models in which counterions are either bound or free in the aqueous pseudophase, β is defined as the molar concentration of bound counterions, $[X_m]$, divided by the concentration of micellized surfactant, $[D_n]$. In pseudophase models, V_m is generally assumed to be equal to the molar volume of the whole micelle or the molar volume of the interfacial region. In interpreting chemical trapping results, we assume that V_m equals the reaction volume sampled by the diazonio group of 16-ArN_2^+ and is the same as that available to the surfactant headgroups in the interfacial region.

The results in Fig. 7 suggest several new ways to view counterion binding in micelles. Both theoretical calculations [92–95] and experimental results [56] suggest that β is essentially independent of $[X_w]$. Thus the relatively rapid, approximately 30–40%, increase in Cl_m at low $[Cl_w]$ (<0.2 M) must be caused by a shrinkage of the interfacial volume available to counterions, i.e., V_m decreases. No theory is available to describe changes in the interfacial volume with solution composition. Above about 0.2 M Cl_w, the incremental increase in Cl_m is attributed to Cl^- entering the interfacial region at concentrations equal to $[Cl_w]$ and Eq. (14) becomes

$$X_m = \frac{\beta}{V_m} + [X_w] = \frac{[X_m]}{[D_n]V_m} + [X_w] \tag{15}$$

where the coefficient of $[X_w]$ is 1. Above about 1.2 M $[Cl_w]$, Cl_m increases significantly with a concomitant decrease in H_2O_m. Similar changes in Br_m are observed for CTABr micelles (not shown) [87], except that the marked increase in Br_m accompanying the sphere-to-rod transition occurs at about 0.1 M $[Br_w]$ [14]. Similar results are observed for surfactant micelles with triethyl, tri-*n*-propyl, and tri-*n*-butyl headgroups and Br^- counterions, except that no sphere-to-rod transition is observed [87]. The results were also used to estimate the ratio of water molecules to carbon atoms in the surfactant headgroups. For the bromide ion surfactants, this ratio decreases gradually with increasing headgroup size and may be connected to the fact that the tri-*n*-butyl surfactant phase separates at elevated Br^- concentrations [96].

These results provide an alternative to the "screening of headgroup repulsions by added counterion" interpretation of the sphere-to-rod transition [12]. The chemical trapping results are consistent with partial dehydration of headgroups and counterions at the transition that is governed by the free energy of hydration of the counterions. Interfacial Br^- should dehydrate more easily than Cl^- because it has a less negative free energy of hydration [97]. Obtaining chemical trapping results for a variety of surfactants over a range of counterion concentrations and types and temperatures may provide new insight into the factors responsible for micellar growth and other structural changes and bulk phase separation. Finally, chemical trapping estimates of interfacial concentration should be similar to those of counterion and water densities obtained by molecular dynamics simulations. For example, our estimates of Cl_m in the interfacial regions of CTACl micelles are in reasonable agreement with molecular dynamics estimates of Cl^- densities in the interfacial regions of decyltrimethylammonium ion micelles [22].

2. Counterion Effects on a Micellar Inhibited Reactions

CTACl and CTABr micelles strongly inhibit acid hydrolyses of micelle-bound ketals in aqueous solution, but at a constant surfactant concentration sufficient to completely bind the substrates, added NaCl (up to 5 M) and NaBr (up to 1 M) increase the rate constant significantly and this increase depends on anion type [98,99]. These results are consistent with the general observation that micelles inhibit reactions with coions, H^+ here, by excluding them from the interfacial region of the micelles. However, the increase in the rate constant with added salt violates the explicit assumption in the pseudophase ion-exchange model that interfacial counterion concentrations are constant because β and V_m are constant, Eq. (14), and the implicit assumption that the coion concentrations, i.e., H^+ and Na^+, are also constant. The Cl^-- and Br^--induced increases in ketal hydrolyses rates indicate that interfacial H^+ concentrations are increasing. The rate increases were successfully interpreted by assuming that (1) counterion effects of coion-catalyzed reactions are expressed by a Donnan equilibrium, (2) the interfacial counterion concentration is described by Eq. (15) or by using Cl_m and Br_m values obtained by chemical trapping, and (3) the stoichiometric H^+ concentration is corrected for the salt-induced change in its activity coefficient [99]. These results show that both the interfacial coion and counterion concentrations depend on counterion concentrations in the aqueous pseudophase.

3. Estimation of Degree of Ionization, α, of Cationic Micelles

Cuccovia and coworkers showed that product yields from chemical trapping of Cl^- and Br^- by 1-ArN_2^+ in the aqueous pseudophase around CTACl and CTABr micelles provide estimates of α that agree with literature values [54]. Good reproducibility was obtained despite the fact that the yields of 1-ArCl and 1-ArBr never exceed 4%. Alpha values were calculated from Eq. (16):

$$\alpha = \frac{[X_w] - \text{cmc}}{[D_t] - \text{cmc}} \tag{16}$$

where $[D_t]$ is the total surfactant concentration and $[X_w]$ is obtained from the product yields and standard curves. Fluorescence quenching experiments of $Ru(bpy)_3^{2+}$ by 1-ArN_2^+ support the basic assumption of this method, i.e., 1-ArN_2^+ does not bind to cationic micelles under the experimental conditions. Binding of 1-ArN_2^+ would increase measured 1-ArX yields significantly because of the much higher counterion concentrations in the interfacial region of the micelles. Alpha values in micelles with triethyl, tri-*n*-propyl, and tri-*n*-butyl headgroups obtained by trapping with 1-ArN_2^+ are also in agreement with literature values [87]. This method should work for all weakly basic anions trapped by *z*-ArN_2^+.

4. Counterion and Water Concentrations in Reverse Microemulsions

Chaudhuri and coworkers have developed two novel uses of chemical trapping by *z*-ArN_2^+ in reverse microemulsions of CTABr/isooctane/HexOH/H_2O [100, 101]. They showed that the method provides estimates of the water pool concentrations of water and counterions and, in combination with fluorescence quenching, estimates of the dimensions of the water pool and the interfacial region. Three products are obtained: 16-ArOH, 16-ArBr (X = Br), and 16-ArOR′ (X = R′OH = HexOH), Scheme 2. In the first paper, they estimated the [Br_w] and [H_2O_w] concentrations in the water pools as a function of w_0, the H_2O/CTABr molar ratio, and compared the results with rate constants for trypsin-catalyzed ester hydrolysis in reverse microemulsions [100]. As w_0 increases from 12 to 44, [Br_w] decreases from 1.91 to 0.29, [H_2O_w] increases from 4.38 to 55.4 M, and k_{cat} for trypsin-catalyzed ester hydrolysis increases from 3.91 to 18.1 s^{-1}, over fourfold. [Br_w] and [H_2O_w] were obtained from product yields from reaction of 1-ArN_2^+ with Br^- and H_2O and standard curves for 1-ArBr and 1-ArOH formation in aqueous TMABr solutions assuming that no 1-ArN_2^+ associates with the microemulsion interface (see evidence in the following). These results indicate that enzyme activity in water pools depends on water activity, ionic strength, or both and that chemical trapping can be used to probe the compositions of water pools in w/o droplets.

In the second paper [101], the thickness of the interfacial region and the radius of the water pool were estimated from interfacial and water pool Br^- concentrations and fluorescence quenching determinations of microemulsion droplet size in the same microemulsion system. Product yields from 16-ArN_2^+ were used to estimate interfacial Br_m and H_2O_m and product yields from 1-ArN_2^+ to estimate [Br_w] and [H_2O_w] in the water pools. Product from reaction with HexOH was observed only with 16-ArN_2^+. The 1-ArOHex from reaction of 1-ArN_2^+ and HexOH was below the limit of detection, 0.5%. This experiment shows that insignificant amounts of HexOH are dissolved in the water pools and insignificant amounts of 1-ArN_2^+ are associated with the aggregate interfaces. From these results, mass balance equations, and the assumption that the reverse microemulsions are spherical, they estimated the interfacial region thickness to be 6.3 Å and the radius of the water pool to be 29.1 Å for an aggregation number of 206. These values are in reasonable agreement with a compositionally similar system [102].

5. Interfacial Coion Concentrations

Cuccovia et al. used chemical trapping by 16-ArN_2^+ in micellar solutions of SDS (0.04 M) to obtain the first experimental estimates of coion concentrations, X_m, in the interfacial region of anionic micelles in added NaBr and NaCl up to 1.3 M [55]. The values of Cl_m and Br_m are the same at all salt concentrations, indicating that only coulombic interactions govern the association of these coions with SDS micelles. The interfacial coion concentrations were well fitted by the Poisson-Boltzmann equation taking into account the increase in the size of SDS micelles with added salt.

6. Interfacial Halide Ion Concentrations in Zwitterionic Micelles and Vesicles

The enzyme phospholipase A_2 (PLA_2) binds to anionic membranes and vesicles and catalyzes the hydrolysis of the *sn*-2-acyl chain of phospholipids. Jain and coworkers showed that PLA_2 binds poorly to zwitterionic surfaces, but added NaCl enhances PLA_2 binding and speeds hydrolysis of zwitterionic micelles and vesicles [103]. Chemical trapping experiments with 16-ArN_2^+ show that addition of 1 and 2 M NaCl to solutions of 1,2-diacyl zwitterionic phosphatidylcholines with dioctyl tails that make micelles and ditetradecyl tails and 1-hexadecyl and 2-oleoyl tails that make vesicles give Cl_m values that are 1 and 2 M in micelles and about 0.5 and 1 M in vesicles. The lower Cl_m values in vesicles are consistent with vesicles being more tightly packed than micelles. This evidence, combined with independent electrophoretic mobility results that show selective binding of anions over cations by other zwitterionic surfactant [104], supports the authors' conclusion that Cl^- binds to zwitterionic surfaces and more strongly than Na^+, giving them a net negative charge. This surface charge promotes binding of the cationic face of the enzyme to the surface and speeds hydrolysis of zwitterionic phospholipids. The authors also proposed a detailed general kinetic model for catalysis by interfacial enzymes.

Chemical trapping of Cl^- and Br^- in zwitterionic micelles of 3-(*N*-hexadecyl-*N*,*N*-dimethylammonio)-propane sulfonate (HPS) and in hexadecylphosphorylcholine (HDPC) show that binding of halide ions depends upon halide ion type, cation type and charge, and on the location of the surfactant headgroup charges in the interfacial region [105]. Several trends are observed. In HPS micelles, the positive charge is adjacent to the micellar core. The estimated interfacial halide ion concentrations are always in excess of their concentrations in the surrounding bulk phase up to 0.6 to 1.0 M added Na^+, TMA^+, and Ca^{2+}, and the Br_m/Cl_m ratio is always greater than one and decreases with increasing salt concentration. The effect of cations is more complex. Mg^{2+}, Ca^{2+}, and Li^+ enhance anion binding most, followed by Na^+, K^+, Cs^+, and Rb^+. TMA^+ has the smallest effect. In HDPC micelles, the negative charge is adjacent to the micellar core. The values of and trends in Br_m/Cl_m ratios and metal ion effects on halide ion binding are similar to those observed with HPS. However, except for Ca^{2+} and Mg^{2+}, the interfacial Br^- and Cl^- concentrations are always less than in the surrounding aqueous phase and Li^+ is less effective at increasing Cl_m than the divalent metal ions. The location of the reactive diazonio group within the interfacial region is not known, which makes comparisons of the two surfactant systems difficult. Nevertheless, these results show that anion binding depends on both cation type and the orientation of the surfactant headgroup charges and possibly the basicity of the headgroups; i.e., phosphate is a stronger base than sulfonate.

Both cations and anions show different affinity orders for micellar and many other types of surfactant assembly surfaces [56,105,106]. Eisenman's approach [107], which accounts for many of the metal ion affinity orders for different glasses used in glass electrodes, was based on the balance of interactions of hydrated metal ions with hydrated surfaces versus the strength of interactions between partially dehydrated surfaces and counterions. Diamond and Wright showed that the same affinity orders were observed with many biological surfaces [108,109]. A deeper understanding of ion binding in the interfacial regions of surfactant assemblies will require better characterization of the strengths of the interactions between headgroups, counterions, and water.

7. Counterion Exchange and Affinity

In the pseudophase ion-exchange model, ionic surfactant assemblies are treated as ion exchangers and competition between two ions is expressed by an ion-exchange constant [4–6,56,91]. For example, the exchange of Cl^- and Br^- at a cationic interface is given by

$$K_{Cl}^{Br} = \frac{Br_m[Cl_w]}{Cl_m[Br_w]} = \frac{[Br_m][Cl_w]}{[Cl_m][Br_w]} \tag{17}$$

Note that the interfacial ion concentration ratios can be expressed in moles per liter of either interfacial or solution volume. Equation (17) holds, provided the ions undergo 1:1 exchange; i.e., the total concentration of counterions in the interfacial region, X_m [Equation (14)], is constant and $X_m = Cl_m + Br_m$. Traditional experimental methods for estimating exchange constants are based on the displacement of one ion by another from the interfacial region and α must be assumed to be constant and the same for both ions to calculate an exchange constant from the data [56,110,111].

Chemical trapping offers another route to estimating exchange constants. The Br_m/Cl_m ratio can be estimated directly from the %16-ArBr/%16-ArCl product ratio multiplied by the selectivity constant. S_{Br}^{Cl}, of the arenediazonium ion toward Cl^- compared with Br^-.

$$\frac{Br_m}{Cl_m} = S_{Br}^{Cl}\,\frac{\%(16\text{-ArBr})}{\%(16\text{-ArCl})} \tag{18}$$

Arenediazonium ion **C**, Scheme 6, was used to estimate S_{Br}^{Cl} for CTAX micelles toward these two ions in aqueous NaBr and NaCl solutions [70]. Between 0.1 and 4 M NaX at 40°C, $S_{Br}^{Cl} = 0.75$. Unlike selectivities between neutral and anionic nucleophiles, S_{Br}^{Cl} is independent of ionic strength; see Section II.D. Values of K_{Cl}^{Br} were independent of total salt concentration but decreased from about 3.2 to 2.4 as the Br^-/Cl^- mole ratio increased from about 1:10 to about 10:1. The average value of K_{Cl}^{Br} is in reasonable agreement with those obtained by other methods [111]. The origin of the decrease in K_{Cl}^{Br} with increasing Br^-/Cl^- mole ratio is not understood. Unpublished experiments with 16-ArN_2^+ indicate that variations of K_{Cl}^{Br} disappear when the small, but real, effect of one added anion on the yield of the second in the determination of K_w^x values using 1-ArN_2^+ is taken into account.

In summary, chemical trapping gives reasonable estimates of the affinities of anions toward cationic surfaces and should be applicable to any anion that is trapped by the heterolytic mechanism, Scheme 1. In addition, Eq. (14) shows that K_{Cl}^{Br} or the exchange constant between any two ions can be estimated, possibly simultaneously, from ratios of the ion concentrations in the aqueous and micellar pseudophases obtained from product yield ratios from chemical trapping by 16-ArN_2^+ in cationic micelles with product yield ratios from chemical trapping by 1-ArN_2^+ in the aqueous phase. This approach would require no assumptions about α or V_m (see Section IV.C.1).

C. Alcohol Distribution Constants

The primary purpose of this section is to show how chemical trapping provides estimates of distributions of a cosurfactant, specifically alcohol, between the interfacial region and bulk solution, either water or oil. The major findings are that alcohol distribution constants in aqueous cationic microemulsions and reverse w/o microemulsions are independent of alcohol concentration and that these constants can be used to estimate water and alcohol mole ratios in bicontinuous microemulsions. The basic logic, assumptions, and important results are presented for alcohol distributions in cationic o/w microemulsions. Results for alcohol distributions in w/o and bicontinuous microemulsions, alcohol effects on alcohol and water concentrations, and the effect of oil type on alcohol distributions are briefly discussed. Full details are given in the indicated references.

1. Aqueous Cationic Microemulsions

Microemulsions [10,11,112] are thermodynamically stable homogeneous solutions of aggregates in dynamic equilibrium with their components including a surfactant (ionic, zwitterionic, or nonionic) and a cosurfactant, typically a medium chain length alcohol, that form three-component microemulsions in water or four-component microemulsions in water-oil mixtures (Fig. 1). In some systems, bicontinuous structures form at intermediate concentrations with oil and water regions in continuous contact with the surfactant and cosurfactant lining their interface. This organization and distribution of components is essentially the same in all microemulsions (Fig. 2). However, the relationships between interfacial concentrations of water and counterions and surfactant and microemulsion structure and solution composition are still not understood. Ionic surfactants are assumed to remain in the interfacial region, but alcohols and nonionic surfactants may distribute between aqueous interfacial and oil regions depending on their hydrophilic-lipophilic balance (HLB), i.e., the length of their alkyl tails versus the polarity of their headgroups. Alcohol distributions are often determined from the solubilities at the saturation limit of the alcohol in the microemulsion [18,113]. However, this method provides no information on the effect of alcohol concentration on its distribution. Other methods such as ultrafiltration [114], vapor pressure [115], and fluorescence quenching [116] provide information on alcohol distributions as a function of concentration. However, only chemical trapping gives information on alcohol distributions [113,117,118], interfacial water and alcohol concentrations, and, in principle, those of other weakly basic nucleophiles simultaneously.

As in experiments in reverse microemulsions (Section IV.B.4), trapping by 16-ArN$_2^+$ in microemulsions composed of CTABr/H_2O/R′OH gives three products. Two different alcohols were used, 1-butanol, BuOH, and 1-hexanol, HexOH. The primary assumption used in determining alcohol distributions from dediazoniation product yields is that in two or more solutions containing different R′OH and CTABr concentrations in which the 16-ArOR′ product yields are the same, the mole fraction of bound alcohol is the same; i.e., *when yields are the same, concentrations are the same.* A value for the selectivity of the reaction is not needed to estimate distribution constants. However, as noted in Section II.D, the selectivity of the chemical trapping reaction toward BuOH is independent of BuOH concentration in aqueous BuOH solutions and 16-ArOR′ product yields should be insensitive to the composition of the microemulsion interface.

2. Estimation of Distribution Constants, Mathematical Treatment, and Results

The mathematical formalism we use to determine alcohol distributions from 16-ArOR′ yields [113,117,118] is adapted from those used to estimate alcohol distributions by vapor pressure and fluorescence quenching methods [116,119]. Alcohol distributions are usually reported as mole fraction partition constants, K_A:

$$K_A = \frac{X_A}{Y_A} \tag{19}$$

where X_A and Y_A are the mole fractions of R′OH in the microemulsion and in the aqueous phases, respectively. X_A is given by

$$X_A = \frac{[\mathrm{R'OH_m}]}{[\mathrm{R'OH_m}] + [\mathrm{D_n}]} \tag{20}$$

X_A is defined the mole fraction of alcohol in the "dry" microemulsion in which bound H_2O is not included in the definition [120]. The mole fraction of alcohol in the aqueous pseudophase is given by

$$Y_A = \frac{[\mathrm{R'OH_W}]}{[\mathrm{R'OH_W}] + \mathrm{cmc} + [\mathrm{H_2O_W}]} \approx \frac{[\mathrm{R'OH_W}]}{[\mathrm{H_2O_W}]} \approx \frac{[\mathrm{R'OH_W}]}{55.5} \tag{21}$$

The first equality is the full definition of Y_A. We use the approximations given by the second and third equalities because (1) medium-chain-length alcohols lower the cmc significantly [18] and it is effectively zero under the experimental conditions [113], (2) the solubilities of BuOH and HexOH in water are low and their molarities are much lower than that of H_2O [121], and (3) the microemulsion aggregates never exceed more than 7% of the total solution volume and the excluded volume effect on $[R'OH_W]$ can be ignored [113].

Binding of R′OH, or any neutral solute, can also be described by a mass action distribution constant [120]. Concentrations are generally expressed in molarity but can be converted to mole fraction units [second right-hand-side equality, Eq. (22)] by using Eqs. (20) and (21):

$$K'_A = \frac{[R'OH_W]}{[R'OH_W][D_n]} = \frac{X_A}{(1 - X_A)(55.5Y_A)} \tag{22}$$

The following results show that values of partition constants, K_A, depend on [R′OH], but values of distributions constants, K'_A, are independent of [R′OH].

Figure 8 shows %16-ArOHex yields as a function of increasing total $[HexOH_t]$ at a series of $[CTAB_r]$. The curves were fitted with an exponential function: $\%16\text{-}ArOR' = A[R'OH_t]^B$ using an average value of B and separate values of A for each curve selected by iteration. The solid horizontal line intersects the curves at constant %16-ArOHex and the mole fraction of bound HexOH, X_A, is assumed to be the same at each

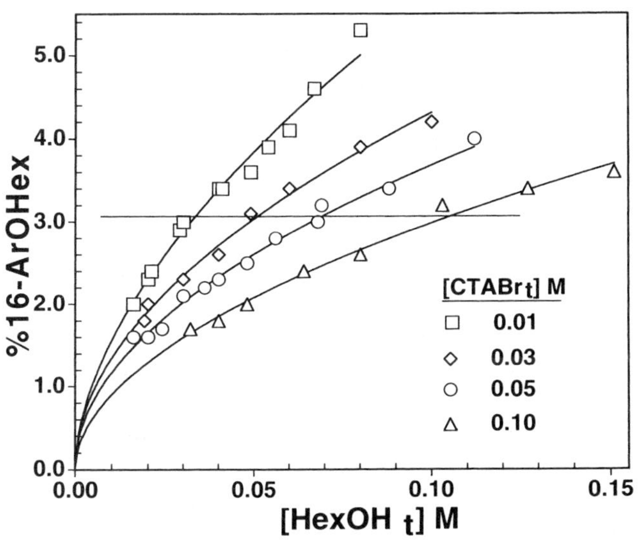

FIG. 8 Percent yields of ether product, 16-ArOHex, from reaction of $16\text{-}ArN_2^+$ with HexOH in CTABr microemulsions at four $[CTABr_t]$ values. Intersection points of fitted curves and the horizontal lines define $[R'OH_t]$, $[CTABr_t]$ data sets at constant %16-ArOHex, and X_A. Details are given in Ref. 114. Reproduced with permission of the American Chemical Society.

intersection point. To solve for X_A, an equation is needed that relates X_A to the total concentrations of HexOH and CTABr in the solutions. Substitution of the definitions of X_A, Eq. (20), and Y_A, Eq. (21), into the mass balance equation for the R′OH, Eq. (23), gives Eq. (24):

$$[R'OH_t] = [R'OH_m] + [R'OH_W] \tag{23}$$

$$[R'OH_t] = \frac{X_A[D_t]}{1 - X_A} + 55.5Y_A \tag{24}$$

Equation (24) expresses the basic assumption of the approach: any set of $[R'OH_t]$, $[CTABr_t]$ values that have the same interfacial or mole fraction concentration of micelle-bound R′OH, i.e., the same X_A, will give the same %16-ArOR′. Thus, the horizontal line drawn through the sets of data in Fig. 8 intersects each CTABr curve and the Y axis at constant X_A.

The data in Fig. 8 were used to calculate X_A, K_A, and K'_A at a series of alcohol and surfactant concentrations by a two-step process. First, a series of parallel lines at constant %16-ArOR′ are drawn through the data and the values of the $[R'OH_t]$, $[CTABr_t]$ pairs for each line are recorded. Two factors limit the %16-ArOR′ yield range over which X_A values can be estimated. (1) As %16-ArOR′ decreases, the CTABr curves approach each other, the differences in their values at the intersection points become less significant, and at some point errors are greater than differences in their values. Note that the method can be extended to lower alcohol and surfactant concentrations simply by increasing $[16\text{-}ArN_2^+]$, keeping in mind that if it is too high it may alter the micellar interface because $16\text{-}ArN_2^+$ is also a surfactant. (2) The upper limit at any one CTABr concentration is the solubility of R′OH. We restricted the analysis to interpolations within the experimental points. Second, values for each $[R'OH_t]$, $[CTABr_t]$ pair for each line give sets of parallel lines whose slopes and intercepts are used to calculate X_A, Y_A, K_A, and K'_A at each %16-ArOR′. Table 1 lists the results for HexOH and BuOH.

The most striking result in Table 1 is that the partition constant, K_A, depends on X_A, but the mass action binding constant, K'_A, does not. The variance in K'_A for BuOH is ±0.2% and in HexOH is ±2.5%. Note that K'_A is larger for HexOH than BuOH because HexOH is more hydrophobic. Our results are in good agreement with those obtained from solubility at the solubility limit [121] and are of the same order of magnitude as those obtained by ultrafiltration [114]. The constancy of K'_A indicates that BuOH and HexOH mix ideally with CTABr, that alcohol binding in these systems is

TABLE 1 Values of X_A and K_A and K'_A Obtained from the Slope/Intercept Ratios of Plots of $[BuOH_t]$ M and $[HexOH_t]$ M versus $[CTABr_t]$ M

%16-ArOBu	X_A	K_A	K'_A M^{-1}	%16-ArOHex	X_A	K_A	K'_A M^{-1}
7.0	0.835	54.8	5.98	3.5	0.510	859	31.6
6.5	0.818	60.6	5.98	3.0	0.435	967	30.8
6.0	0.797	67.2	5.97	2.5	0.356	1162	32.5
5.5	0.774	75.1	5.97	2.0	0.267	1348	33.1
5.0	0.745	84.5	5.97				
4.5	0.712	96.0	5.99				
4.0	0.671	109.6	5.99				
3.5	0.621	126.0	5.99				
3.0	0.561	146.4	6.01				
2.5	0.486	171.0	5.99				
2.0	0.398	201.2	6.02				

	Average[a]
BuOH	5.99 ± 0.01 ± 0.2%
HexOH	32.0 ± 0.8 ± 2.5%

[a]Average and percent average deviations.

driven by the hydrophobic effect, and that specific interactions in the interfacial region, e.g., differences in hydrogen binding or charge-dipole interactions between the alcohols and CTABr, are not significant. The constancy of K'_A also suggests that headgroups of BuOH and HexOH are located only in the interfacial region or that partitioning of these alcohols between the interfacial region and the microemulsion core does not depend on the fraction of bound alcohol.

3. Alcohol Distributions in Water-in-Oil and Interfacial Compositions of Bicontinuous Microemulsions

The approach to analyzing chemical trapping data in w/o microemulsions is conceptually that already outlined [117]. The equations usd to interpret 16-ArOR′ yields have a different form but the same basic meaning. Concentrations of CTABr, BuOH, and HexOH relative to oil are much higher in w/o microemulsions than in aqueous microemulsions. Consequently, the oil concentration has to be included explicitly in the definition of the mass action distribution constant because alcohols occupy a significant fraction of the total oil phase. Comparison of partition, K_A, and distribution, K'_A, constants for BuOH and HexOH in CTABr/hexadecane/R′OH/H_2O w/o microemulsions showed that, as in aqueous microemulsions, K_A values vary with mole fraction of the alcohol in the microemulsion interface but K'_A values do not; $K'_A = 23.1$ and 17.8 for BuOH and HexOH, respectively. The value for BuOH is probably larger because it is less soluble in hexadecane than HexOH.

Determination of the mass action distribution constants opens the possibility of estimating concentrations of alcohol and water, expressed as molar ratios of alcohol/surfactant, N_{ma}/N_s, and water/surfactant, N_{mw}/N_s, in the interfacial region of a four-component microemulsion in any single phase region of its phase diagram [117]. We obtained expressions for N_{ma}/N_s and N_{mw}/N_s for four-component BuOH microemulsions by combining equations for the distribution constants with those for the selectivity of the reaction toward BuOH compared with H_2O and the mass balance equations for BuOH and H_2O and with the 16-ArOH and 16-ArOBu product yields in the o/w, bicontinuous, and w/o regions of the microemulsions. The results suggest that transitions between approximately planar bicontinuous and spheroidal w/o and o/w droplet shapes are determined by the relative amounts of H_2O and BuOH in the interfacial region. The bicontinuous-w/o transition occurs when the molar concentration of interfacial H_2O drops below that of BuOH, indicating that there is insufficient H_2O available to hydrate the OH group of BuOH, the quaternary ammonium headgroup, and the Br^-. The bicontinuous-o/w transition occurs when the molar concentration of H_2O exceeds that of BuOH in the interfacial region and BuOH begins to dissociate from the microemulsions. Droplets are formed because the increasing hydration increases the fraction of interfacial volume occupied by the headgroups and because dissociation of BuOH changes aggregate packing.

Mass action binding constants for BuOH were also determined in CTABr/H_2O/BuOH/oil w/o microemulsions using the approach described before for five different oils: benzene, octane, hexadecane, tributyrin, and triolein [118]. The value of K'_A is not very sensitive to oil type, varying from 13.6 for triolein to 23.6 for benzene, despite the fact that the minimum amount of

BuOH required to make microemulsions varied significantly for each oil. The fraction of BuOH bound to the microemulsion interface is essentially constant when estimated in terms of the number of moles of each component but varies considerably when calculated in terms of the weight percent of the oil. We also found that the minimum amount of BuOH required to titrate mixtures of CTABr, H_2O, and oil to a clear homogeneous solution was considerably greater for the triglycerides than for the other oils. These observations suggest that microemulsions do not form as easily in triglycerides as in hydrocarbons because the solubility of an alcohol in triglycerides is considerably higher than in hydrocarbons, not because the interaction of the BuOH with the microemulsion interface is different.

V. CURRENT AND FUTURE APPLICATIONS

Heterolytic dediazoniation can provide estimates of concentrations of many different weakly basic nucleophiles in interfacial regions of surfactant assemblies and a number of new applications are suggested, including determining the topologies of aggregate-bound polypeptides. The homolytic reaction occurs with specific electron donors, for example, antioxidants, and in combination with the pseudophase kinetic model can be used to estimate distribution constants of antioxidants between the oil-interfacial and aqueous-interfacial regions of micelles, microemulsions, and perhaps emulsions.

A. Heterolytic Cleavage: Interfacial Hydration and Composition

One of the unique strengths of the chemical trapping reaction is that the yield of the phenolic product, 16-ArOH (Scheme 2), provides an immediate estimate of the amount of water in the interfacial region over wide ranges of composition. As shown in Section IV.A, product yields provide estimates of hydration numbers of nonionic micelles. The same approach should work with anionic micelles of surfactants with weakly basic headgroups such as alkyl sulfates, alkanesulfonates, alkanecarboxylates, alkylphosphates, and phospholipids (Scheme 3). Hydration numbers can be obtained as a function of surfactant concentration, surfactant chain length and headgroup structure, and counterion and additive type and their concentrations. The results should provide new information on the balance between the hydrophobic effect, driving aggregation, and interfacial composition, in particular hydration, which reflects the aggregation, size and shape, and phase state of the system. In cationic surfactants, chemical trapping provides information on hydration and interfacial counterion concentration. To date, we have focused primarily on nonionic surfactants and cationic surfactants with Cl^- and Br^- counterions. Many other counterions that have varying effects on aggregate size and shape should also be trapped, e.g., simple anions such as $CH_3O_2^-$, I^-, ClO_3^-, F^-, and NO_3^- and organic counterions such as RCO_2^-, RSO_3^-, where R may be alkyl or aryl (Scheme 2). Some organic counterions such as tosylate [122], salicylate [123], and certain substituted benzoates [124,125] induce formation of long flexible rodlike aggregates at low concentrations. The reasons for these special effects are not clear, and the answer may be related to the influence of aromatic counterions on the hydration of the interfacial region.

Certain mixtures of anionic and cationic surfactants form stable solutions that undergo spontaneous vesicle-micelle transitions with small changes in composition and salt concentration [126,127]. Here chemical trapping should provide information on interfacial concentrations of water, anionic headgroups, and anionic counterions on each composition side of the transition and the results should provide new information on the balance of forces controlling these transitions. Chemical trapping can also be used to estimate the interfacial concentrations of short-chain alcohols, e.g., MeOH and EtOH, and their distributions, which are currently unknown [18]. Preliminary results that show that 16-ArN_2^+ traps the amine N and acyl O of urea in aqueous cationic micelles to give stable *N*-arylurea and *O*-arylisourea products (unpublished results). Estimates of interfacial concentrations suggest that the interfacial urea concentration is the same as or slightly lower than its concentration in bulk solution. This surprising result indicates that urea is not bound selectively. This approach may provide important information on how urea and other small molecules affect micelle and perhaps protein stability. Finally, we have tagged anionic polyelectrolytes with 1-ArN_2^+ and obtained the first estimates of the counterion concentration in the condensed layer in the immediate vicinity of the polyelectrolyte. The results may eventually provide new insight into the interaction between like-charged polyelectrolytes in the semidilute concentration region [128].

B. Heterolytic Cleavage: Counting Peptide Bonds. Topologies and Orientations of Aggregate-Bound Polypeptides

We studied the mechanism of reactions of 1-ArN_2^+ with simple amides as models for the peptide bond. 16-Ar+

is trapped by both the acyl O and amine N of acetamide and *N*-methylacetamide and the acyl O of *N,N*-dimethylacetamide in aqueous solutions [49,50] (Scheme 8). Currently, this is the only reaction that tags an amide in water at neutral pH. Tagging of the acyl O by 1-Ar+ gives an imido ester, 1-ArOI (Scheme 9) that hydrolyzes rapidly to two sets of products: a 1-ArOH: amide pair and an ester (1-ArOAc):amine pair that comes from cleavage of the C—N bond. We also have preliminary evidence that 16-ArN_2^+ cleaves the amide bonds of micellized *N*-dodecylglycinate and *N*-dodecyl-*N*-methylglycinate and also tags their terminal carboxylate groups [129]. Chemical tagging and cleavage of amide bonds of polypeptides at aggregate interfaces should provide information on their topologies because 16-Ar+ will tag only the peptide bonds within the interfacial region. For example, the tagging and cleavage patterns of polypeptides with α-helical or β-sheet orientations in the interfacial region should differ. Similarly, the tagging and cleavage pattern of an α-helical peptide in the plane of the aggregate surface should differ from that of one oriented normal to the surface with only a section of the helix in the interfacial region. Tagging of amino acid side chains such as serine OH or glutamate CO_2^- should provide additional topological information. Liquid chromatographic–mass spectrometric (LC-MS) analysis of the tagging and fragmentation patterns will speed data accumulation.

C. Homolytic: Antioxidant Distributions

One unsolved problem in food chemistry is selecting the most efficient antioxidant for use in emulsified foods [130,131]. The problem is complex because antioxidant efficiency depends on both the reactivity of the antioxidant and its distribution within the emulsion. Information is becoming available on the reactivity of antioxidants [132], but their distributions are difficult to determine in opaque mixtures of emulsions. Antioxidants react with arenediazonium ions via the homolytic pathway [64] (Scheme 1). We have unpublished results showing that 16-ArN_2^+ reacts with antioxidants such as *t*-butylhydroquinone (TBHQ), α-tocopherol (vitamin E), propyl gallate, ascorbic acid (vitamin C), and BHA in nonionic micelles much faster than the heterolytic dediazoniation reaction. Under certain conditions (but not all), kinetics are clean, second order overall: first order in 16-ArN_2^+, first order in antioxidant, and the measured rate constant depends on surfactant concentration. We used the pseudophase kinetic model to determine the distribution constant of TBHQ in aqueous and reverse nonionic micelles (unpublished results). In aqueous micelles, binding constants of TBHQ obtained kinetically, by monitoring the loss of 16-ArN_2^+ spectrophotometrically, are in excellent agreement with those obtained by the spectral shift method and by the HPLC method based on monitoring the formation of 16-ArH (Section III.E). These results indicate that it may be possible to determine antioxidant distributions in emulsions and establish a scale of antioxidant efficiency by comparing rate constants for reaction with 16-ArN_2^+ in the interfacial region.

VI. CONCLUSIONS

The chemical trapping method provides a new approach to determining the compositions of the interfacial regions of surfactant assemblies. Product yields from the reaction via the heterolytic dediazoniation mechanism provide, in principle, simultaneous estimates of interfacial concentrations of all weakly basic nucleophiles associated with the aggregates. The method works for anionic counterions, hydration numbers of nonionic surfactants, and distributions of neutral solutes such as alcohols between the aqueous interfacial and oil regions of microemulsions. Chemical trapping with arenediazonium ions differs from spectrophotometric probe methods in that the target ions or molecules need not contain a chromophore. The reaction tags them with a chromophore to give stable products that are separated and analyzed by HPLC with UV detection. Changes in product yields provide information on the influence of one component on the interfacial concentration of another or its distribution between the oil, aqueous, and interfacial regions. Water-soluble arenediazonium ions provide similar information on concentrations of ions and molecules in the surrounding aqueous phase or inside water pools of reverse assemblies. Some components react with arenediazonium ions via different mechanisms. These reactions set limits on the chemical trapping method but also open up new possibilities. The work accomplished to date demonstrates the viability of the method. Much work remains to be done, especially in complex multicomponent systems.

ACKNOWLEDGMENTS

I am grateful for financial support from the following agencies: the Chemical Dynamics (CHE-9526206) and International Programs of the National Science Foundation (INT-97-22458) and the Center for Advanced

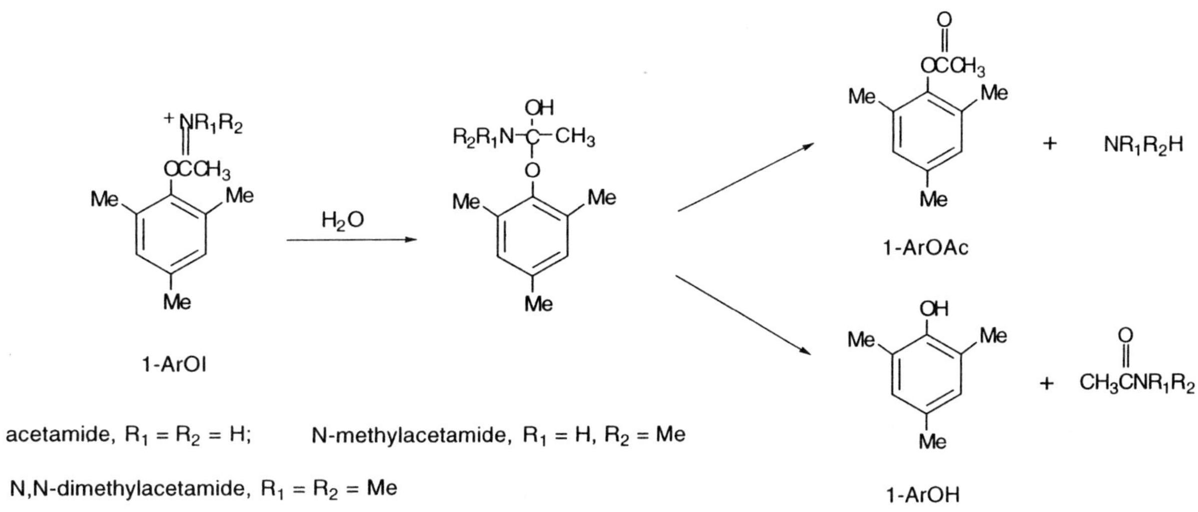

SCHEME 9

Food Technology at Rutgers University (Publication No. D105435-1-00). None of this would have been accomplished with out the contributions of many colleagues and students: Carlos Bravo-Díaz, Yu Cao, Arabinda Chaudhuri, Hernan Chaimovich, Iolanda Cuccovia, Sandro Froehner, Yan Geng, Matthew Harbowy, Zhen-Min He, Pat Jennings, Jason Keiper, John Loughlin, Barbara McKernan, Jackie Nikles, Faruk Nome, Frank Quina, Valdir Soldi, Jihu Yao, Dino Zanette, Jianbing Zhang, Lanzhen Zhuang, and especially Clifford A. Bunton. His unswerving support and encouragement has made this work a reality.

REFERENCES

1. R. Zana, ed., *Surfactant Solutions: New Methods of Investigation*, Marcel Dekker, New York, 1985.
2. S. Tascioglu, Tetrahedron *52*:11113–11152 (1996).
3. J. Sjoblom, R. Lindberg, and S. E. Friberg, Adv. Colloid Interface Sci. *65*:125–287 (1996).
4. C. A. Bunton, F. Nome, F. H. Quina, and L. S. Romsted, Acc. Chem. Res. *24*:357–364 (1991).
5. C. A. Bunton, J. Yao, and L. S. Romsted, Curr. Opin. Colloid Interface Sci. *2*:622–628 (1997).
6. C. A. Bunton and L. S. Romsted, in *Handbook of Microemulsion Science and Technology* (A. Kumar, ed.), Marcel Dekker, New York, 1999, pp. 457–482.
7. C. Tanford, *The Hydrophobic Effect: Formation of Micelles and Biological Membranes*, 2nd ed., Wiley, New York, 1980.
8. J. H. Fendler, *Membrane Mimetic Chemistry*, Wiley-Interscience, New York, 1982.
9. J. Israelachvili, *Intermolecular and Surface Forces*, 2nd ed., Academic Press, London, 1991.
10. K. S. Birdi, ed., *Handbook of Surface and Colloid Chemistry*, CRC Press, Boca Raton, FL, 1997, p. 763.
11. B. Jonsson, B. Lindman, K. Holmberg, and B. Kronberg, *Surfactants and Polymers in Aqueous Solution*, Wiley, Chichester, 1998.
12. A. Ben-Shaul and W. M. Gelbart, in *Micelles, Membranes, Microemulsions, and Monolayers* (W. M. Gelbart, A. Ben-Shaul, and D. Roux, eds.), Springer-Verlag, New York, 1994, p. 608.
13. H. Fabre, N. Kamenka, A. Khan, G. Lindblom, B. Lindman, and G. J. T. Tiddy, J. Phys. Chem. *84*:3428–3433 (1980).
14. T. Imae, A. Abe, and S. Ikeda, J. Phys. Chem. *92*: 1548–1553 (1988).
15. R. B. Dorshow, C. A. Bunton, and D. F. Nicoli, J. Phys. Chem. *87*:1409–1416 (1983).
16. G. Mancini, C. Schiavo, and G. Cerichelli, Langmuir *12*:3567–3573 (1996).
17. D. F. Evans and H. Wennerstrom, *The Colloidal Domain: Where Physics, Chemistry, Biology and Technology Meet*, VCH Publishers, New York, 1994.
18. R. Zana, Adv. Colloid Interface Sci. *57*:1–64 (1995).
19. B. Lindman and H. Wennerstrom, Top. Curr. Chem. *87*:32 (1980).
20. T. A. Cross and S. J. Opella, Curr. Opin. Struct. Biol. *4*:574–581 (1994).
21. L. K. Tamm and S. A. Tatulian, Q. Rev. Biophys. *30*: 365–429 (1997).
22. J. Bocker, J. Brickmann, and P. Bopp, J. Phys. Chem. *98*:712–717 (1994).
23. S. Bandyopadhyay, M. Tarek, and M. L. Klein, Curr. Opin. Colloid Interface Sci. *3*:242–246 (1998).

24. A. Chaudhuri, J. A. Loughlin, L. S. Romsted, and J. Yao, J. Am. Chem. Soc. *115*:8351–8361 (1993).
25. H. Zollinger, *Diazo Chemistry I: Aromatic and Heteroaromatic Compounds*, VCH Publishers, Weinheim, 1994.
26. G. W. Klumpp, *Reactivity in Organic Chemistry*, John Wiley & Sons, New York, 1982.
27. A. F. Hegarty, in *The Chemistry of the Diazonium and Diazo Groups*, Part 2 (S. Patai, ed.), John Wiley & Sons, New York, 1978.
28. C. G. Swain, J. E. Sheats, and K. G. Harbison, J. Am. Chem. Soc. *97*:783–790 (1975).
29. T. H. Lowry and K. S. Richardson, *Mechanism and Theory in Organic Chemistry*, 3rd ed., Harper & Row, New York, 1987.
30. R. Glaser and C. J. Horan, J. Org. Chem. *60*:7518–7528 (1995).
31. R. Glaser, C. J. Horan, and H. Zollinger, Angew. Chem. Intl. Ed. Engl. *36*:2210–2213 (1997).
32. M. C. Garcia-Meijide, C. Bravo-Diaz, and L. S. Romsted, Int. J. Chem. Kinet. *30*:31–39 (1998).
33. R. Pazo-Llorente, M. J. Sarabia-Rodiguez, C. Bravo-Diaz, and E. Gonzalez-Romero, Int. J. Chem. Kinet. *31*:73–82 (1999).
34. C. Bravo-Diaz, M. Soengas-Fernandez, M. J. Rodriguez-Sarabia, and E. Gonzalez-Romero, Langmuir *14*:5098–5105 (1998).
35. J. C. Scaiano and N. Kim-Thuan, J. Photochem. *23*:269–276 (1983).
36. J. E. Chateauneuf, J. Chem. Soc. Chem. Commun. 1437–1438 (1991).
37. J. P. Pezacki, D. Shukla, J. Lusztyk, and J. Warkentin, J. Am. Chem. Soc. *121*:6589–6598 (1999).
38. R. A. Carey and R. J. Sundberg, *Advanced Organic Chemistry*, Part A: *Structure and Mechanism*, 3rd ed., Plenum, New York, 1993.
39. J. Lorand, Tetrahedron Lett. *30*:7337–7340 (1989).
40. Y. Hashida, R. G. M. Landells, G. E. Lewis, I. Szele, and H. Zollinger, J. Am. Chem. Soc. *100*:2816–2823 (1978).
41. W. P. Jencks, Chem. Soc. Rev. *10*:345–375 (1981).
42. P. Rys, Acc. Chem. Res. *9*:345–351 (1976).
43. K. J. Laidler, *Chemical Kinetics*, Harper & Row, New York, 1987.
44. D. R. Lide, ed., *Handbook of Chemistry and Physics*, CRC Press, Boca Raton, FL, 1999.
45. M. L. Crossley, R. H. Kienle, and C. H. Benbrook, J. Am. Chem. Soc. *62*:1400–1404 (1940).
46. A. Chaudhuri and L. S. Romsted, J. Am. Chem. Soc. *113*:5052–5053 (1991).
47. R. Pazo-Llorente, M. J. Sarabia-Rodriguez, E. Gonzalez-Romero, and C. Bravo-Diaz, Int. J. Chem. Kinet. *31*:531–540 (1999).
48. L. S. Romsted and J. Yao, Langmuir *12*:2425–2432 (1996).
49. L. S. Romsted, J. Zhang, and L. Zhuang, J. Am. Chem. Soc. *120*:10046–10034 (1998).
50. L. R. Romsted, J. Zhang, and L. Zhuang, in *Proceedings of the 14th International Conference on Physical Organic Chemistry* (E. Humeres, ed.), Florianopolis, 1998, pp. 214–230.
51. E. A. G. Aniansson, S. N. Wall, M. Almgren, H. Hoffmann, I. Kielmann, W. Ulbricht, R. Zana, J. Lang, and C. Tondre, J. Phys. Chem. *80*:905–922 (1976).
52. E. A. G. Aniansson, J. Phys. Chem. *82*:2805–2808 (1978).
53. C. A. Bunton, L. S. Romsted, and C. Thamavit, J. Am. Chem. Soc. *102*:3900–3903 (1980).
54. I. M. Cuccovia, I. N. da Silva, H. Chaimovich, and L. S. Romsted, Langmuir *13*:647–652 (1997).
55. I. M. Cuccovia, A. Agostihno-Neto, C. M. A. Wendel, H. Chaimovich, and L. S. Romsted, Langmuir *13*:5302–5305 (1997).
56. L. S. Romsted, in *Surfactants in Solution* (K. L. Mittal and B. Lindman, eds.), Plenum, New York, 1984, Vol. 2, p. 1015.
57. C. A. Bunton and G. Savelli, Adv. Phys. Org. Chem. *22*:213–309 (1986).
58. W. P. Jencks, *Catalysis in Chemistry and Enzymology*, Dover, 1987.
59. W. P. Jencks and J. Regenstein, in *Handbook of Biochemistry* (H. A. Sober, ed.), Chemical Rubber Co., Cleveland, 1970, pp. J187–J226.
60. C. Bravo-Diaz, L. S. Romsted, M. Harbowy, E. Romero-Nieto, and E. Gonzalez-Romero, J. Phys. Org. Chem. *12*:130–140 (1999).
61. C. Bravo-Diaz, M. E. Romero-Nieto, and E. Gonzalez-Romero, Langmuir *16*:42–48 (2000).
62. I. Szele and H. Zollinger, Helv. Chim. Acta *61*:1721–1729 (1978).
63. C. Bravo-Diaz and E. Gonzalez-Romero, Anal. Chim. Acta *385*:373–386 (1999).
64. K. C. Brown and M. P. Doyle, J. Org. Chem. *53*:3255 (1988).
65. I. M. Cuccovia, F. H. Quina, and H. Chaimovich, Tetrahedron *38*:917–920 (1982).
66. R. A. Moss, C. J. Talkowski, D. W. Reger, and C. E. Powell, J. Am. Chem. Soc. *95*:5215–5224 (1973).
67. R. Singer, E. Eibler, and J. Sauer, Tetrahedron Lett. *23*:1135–1138 (1982).
68. R. A. Moss, F. M. Dix, and L. Romsted, J. Am. Chem. Soc. *104*:5048–5050 (1982).
69. J. A. Loughlin, Estimating the selectivity of cationic micelles toward bromide and chloride counterions using hydrophobic aryldiazonium salts, Ph.D., thesis, Rutgers, The State University of New Jersey, 1990.
70. J. A. Loughlin and L. S. Romsted, Colloids Surf. *48*:123–137 (1990).
71. R. Banerjee, P. K. Das, and A. Chaudhuri, Biochim. Biophys. Acta *1373*:299–308 (1998).
72. M. P. Doyle and W. J. Bryker, J. Org. Chem. *44*:1572–1574 (1979).
73. M. Wisniewski, Pol. J. Chem. *57*:593 (1983).

74. P. Zuman and R. Patel, *Techniques in Organic Reaction Kinetics*, John Wiley & Sons, New York, 1984, p. 340.
75. C. Bravo-Diaz, M. Soengas-Fernandez, M. J. Rodriguez-Sarabia, and E. Gonzalez-Romero, Int. J. Chem. Kinet. *32*:419–430 (2000).
76. L. S. Romsted and J. Yao, Langmuir *15*:326–336 (1999).
77. K. Streletzky and G. D. J. Phillies, Langmuir *11*:42–47 (1995).
78. C. Tanford, Y. Nozaki, and M. F. Rohde, J. Phys. Chem. *81*:1555–1560 (1977).
79. P. G. Nilsson, H. Wennerstrom, and B. Lindman, J. Phys. Chem. *87*:1377–1385 (1983).
80. P. G. Nilsson and B. Lindman, J. Phys. Chem. *87*: 4756–4761 (1983).
81. W. Brown, R. Johnsen, P. Stilbs, and B. Lindman, J. Phys. Chem. *87*:4548–4553 (1983).
82. P.-G. Nilsson, H. Wennerstrom, and B. Lindman, Chem. Scr. *25*:67–72 (1985).
83. G. Carlstrom and B. Halle, J. Chem. Soc. Faraday Trans. 1 *85*:1049–1063 (1989).
84. P. Baglioni, R. Bongiovanni, E. Rivara-Minten, and L. Kevan, J. Phys. Chem. *93*:5574–5578 (1989).
85. N. Nishikodo, Langmuir *8*:1718–1723 (1992).
86. M. Abe, H. Uchiyama, T. Yamaguchi, T. Suzuki, K. Ogino, J. F. Scamehorn, and S. D. Christian, Langmuir *8*:2147–2151 (1992).
87. V. Soldi, J. Keiper, L. S. Romsted, I. M. Cuccovia, and H. Chaimovich, Langmuir *16*:59–71 (2000).
88. C. Bonan, R. Germani, P. P. Ponti, G. Savelli, G. Cerichelli, R. Bacaloglu, and C. A. Bunton, J. Phys. Chem. *94*:5331–5336 (1990).
89. A. Blasko, C. A. Bunton, G. Cerichelli, and D. C. McKenzie, J. Phys. Chem. *97*:11324–11331 (1993).
90. L. S. Romsted, Rate enhancements in micellar systems, Ph.D. thesis, Indiana University, 1975.
91. L. S. Romsted, in *Micellization, Solubilization and Microemulsions* (K. L. Mittal, ed.), Plenum, New York, 1977, Vol. 2, pp. 489–530.
92. C. A. Bunton, M. M. Mhala, and J. R. Moffatt, J. Phys. Chem. *93*:7851–7856 (1989).
93. D. Stigter, Prog. Colloid Polym. Sci. *65*:45–52 (1978).
94. G. Gunnarsson and H. Gustavsson, J. Chem. Soc. Faraday Trans. 1 *78*:2901–2910 (1982).
95. P. Linse, G. Gunnarsson, and B. Jonsson, J. Phys. Chem. *86*:413–421 (1982).
96. S. A. Buckingham, C. J. Garvey, and G. G. Warr, J. Phys. Chem. *97*:10236–10244 (1993).
97. Y. Marcus, *Ion Solvation*, John Wiley & Sons, Chichester, UK, 1985.
98. C. Armstrong, W. Gotham, P. Jennings, J. Nikles, L. S. Romsted, M. Versace, and J. Waidlich, in *Surfactants in Solution* (K. L. Mittal, ed.), Plenum, New York, 1989, Vol. 9, pp. 197–209.
99. Z.-M. He, J. A. Loughlin, and L. S. Romsted, Bol. Soc. Chil. Quim. *35*:43–53 (1990).
100. P. K. Das, G. V. Srilakshmi, and A. Chaudhuri, Langmuir *15*:981–987 (1999).
101. P. K. Das, A. Chaudhuri, S. Saha, and A. Samata, Langmuir *15*:4765–4772 (1999).
102. M. Giustini, G. Palazzo, G. Colafemmina, D. M. Monica, M. Giomini, and A. Ceglie, J. Phys. Chem. *100*: 3190 (1996).
103. M. K. Jain, J. Rogers, B.-Z. Yu, J. Yao, L. S. Romsted, and O. G. Berg, Biochemistry *36*:14512–14530 (1997).
104. S. A. Tatulian, Biochim. Biophys. Acta *736*:189–195 (1983).
105. I. M. Cuccovia, L. S. Romsted, and H. Chaimovich, J. Colloid Interface Sci., in press.
106. Z.-M. He, P. J. O'Connor, L. S. Romsted, and D. Zanette, J. Phys. Chem. *93*:4219–4226 (1989).
107. G. Eisenman, *The Glass Electrode*, Interscience, New York, Interscience Reprint, 1965, p. 213.
108. J. M. Diamond and E. M. Wright, Annu. Rev. Physiol. *31*:581–646 (1969).
109. J. M. Diamond and E. M. Wright, J. Exp. Zool. *194*: 227 (1975).
110. C. Gamboa, L. Sepulveda, and R. Soto, J. Phys. Chem. *85*:1429–1434 (1981).
111. J. D. Morgan, D. H. Napper, and G. G. Warr, J. Phys. Chem. *99*:9458–9465 (1995).
112. S. E. Friberg and P. Bothorel, eds. *Microemulsions: Structure and Dynamics*, CRC Press, Boca Raton, FL, 1987.
113. A. Chaudhuri, L. S. Romsted, and J. Yao, J. Am. Chem. Soc. *115*:8362–8367 (1993).
114. C. Gamboa, A. Olea, H. Rios, and M. Henriquez, Langmuir *8*:23–26 (1992).
115. C. Treiner, J. Colloid Interface Sci. *90*:444–453 (1982).
116. E. B. Abuin and E. A. Lissi, J. Colloid Interface Sci. *95*:198–203 (1983).
117. J. Yao and L. S. Romsted, J. Am. Chem. Soc. *116*: 11779–11786 (1994).
118. J. Yao and L. S. Romsted, Colloids Surf. A Physicochem. Eng. Aspects *123–124*:89–105 (1997).
119. K. Hayase and S. Hayano, Bull. Chem. Soc. Jpn. *50*: 83–85 (1977).
120. L. Sepulveda, E. Lissi, and F. Quina, Adv. Colloid Interface Sci. *25*:1–57 (1986).
121. J. Gettins, D. Hall, P. L. Jobling, J. E. Rassing, and E. Wyn-Jones, J. Chem. Soc. Faraday Trans. 2 *74*:1957–1964 (1978).
122. P. A. Hassan, S. J. C. Andau, F. Kern, and C. Manohar, Langmuir *14*:6025–6029 (1998).
123. T. Shikata, S.-I. Imai, and Y. Morishima, Langmuir *14*: 2020–2026 (1998).
124. P. J. Kreke, L. J. Magid, and J. C. Gee, Langmuir *12*: 699–705 (1996).
125. L. J. Magid, Z. Han, G. G. Warr, M. A. Cassidy, P. D. Butler, and W. A. Hamilton, J. Phys. Chem. *101*: 7919–7927 (1997).

126. L. L. Brasher, K. L. Herrington, and E. W. Kaler, Langmuir *11*:4267–4277 (1995).
127. M. T. Yatcilla, K. L. Herrington, L. L. Brasher, E. W. Kaler, S. Chiruvolu, and J. A. Zasadzinski, J. Phys. Chem. *100*:5874–5879 (1996).
128. J. Ray and G. S. Manning, Langmuir *10*:2450–2461 (1994).
129. L. Zhuang, Mechanism of reaction of arenediazonium ions with the amide bond. Progress toward a chemical trapping method for determining topologies of peptides in biomimetic aggregates, Ph.D. thesis, Rutgers, The State University of New Jersey, 1998.
130. J. N. Coupland and D. J. McClements, Trends Food Sci. Technol. *7*:83–91 (1996).
131. W. L. Porter, Toxicol. Ind. Health *9*:93–122 (1993).
132. W. A. Pryor, J. A. Cornicelli, L. J. Devall, B. Tait, B. K. Trivedi, D. T. Witiak, and M. Wu, J. Org. Chem. *58*:3521–3532 (1993).
133. J. H. Fendler, Pure Appl. Chem. *54*:1809–1817 (1982).

11

Electro-Organic Synthesis in Macro- and Microheterogeneous Solutions: Emulsions, Micelles, and Related Systems

MARC THOMALLA Université Claude Bernard Lyon I, Villeurbanne, France

I. INTRODUCTION

A. Overview

The main features and the limitations of electrochemical conversions in emulsions are reported and the general features of electrochemical processes in microheterogeneous systems are briefly reviewed. The effects of heterogeneous suspensions and solutions on the selectivity and the efficiency of electro-organic synthesis are described. These effects are mainly governed by the "local" concentrations of the reactants, modifications made to the electrodes, and some specific interactions. All of these aspects are developed in this nonexhaustive review in order to emphasize the specificity and the range of applications available in different surfactant-based macro- and microheterogeneous systems.

B. Aqueous Electrolyte

Many organic substances are sparingly soluble in water [1] and their electrochemical conversion is very often performed in organic or hydroorganic solutions comprising acetonitrile, alcohols, etc. [2a]. However, for both technical (solvent cost and toxicity, low conductance of nonaqueous systems, etc.) and chemical reasons (reaction selectivities and yields, the need for inorganic species as reactants, etc.) it is often desirable to use aqueous electrolyte systems.

Practical electro-organic synthesis in aqueous solutions encounters a number of problems. These include the usable potential range of water, which is limited in the anodic direction, and the low solubility of organic depolarizers in water, which leads to poor space-time yields, as pointed out earlier by Fees and Wendt [3]. These problems may often be overcome by the use of emulsions, microemulsions, micelles, and related systems (vesicles, bilayers, etc.), all of which have specific influences on electrochemical reactions.

C. Emulsions

Emulsions are macroheterogeneous systems of two immiscible liquids (oil and water) that are thermodynamically unstable and require energy for their formation. There are two types of emulsions, oil in water (o/w) and water in oil (w/o), depending on which phase is dispersed in the other. The emulsification phenomenon and the role of surfactants in emulsion formation and stabilization have been extensively studied [4]. Emulsions are widely used in organic synthesis (phase transfer catalysis) [5] and in electrochemistry [3]. Electro-organic syntheses are often performed in o/w emulsions. The continuous aqueous phase contains the electrolyte and acts as a conducting medium for the transport of charge. The dispersed organic phase can be the substrate itself (when it is liquid at the working temperature) or a water-immiscible organic solvent in which the substrate is dissolved. In this last case, the partitioning of the substrate between the two phases is an important parameter that can be estimated by semiempirical methods [6,7].

D. Micelles, Microemulsions, and Mesophases

Micelles and microemulsions are "microheterogeneous" thermodynamically stable systems that are isotropic and optically clear solutions. On a microscopic scale, they present two pseudophases: a continuous one that can be aqueous (direct micelles, oil-in-water microemulsions) or organic (reverse micelles, water-in-oil microemulsions) and a dispersed one (micelles or microdroplets). In bicontinuous microemulsions both oil and water are continuous pseudophases. These structures are illustrated schematically in Fig. 1. Micelles and microemulsions need surfactants for their formation. The size and the shape of the micellar aggregates (spherical or cylindrical micelles, lamellar phases) depend on the concentration and nature of the surfactants. Insoluble surfactants with two or more hydrocarbon tails (for instance, didodecyldimethylammonium bromide, DDAB), which do not form micelles, can lead to the formation of vesicles (liposomes) or of lamellar liquid crystal phases. The formation and the features of such microheterogeneous organized media have been widely investigated, and for more details the reader is referred to other publications and reviews [8–13]. The solubilization of relatively water-insoluble organic substances in micellar solutions or microemulsions occurs by their incorporation into the micelles or the microdroplets [14]. The corresponding equilibria of the solutes between the two pseudophases are of primary importance in both kinetic (the pseudophase model) [15] and electrochemical studies [16–18]. The rate at which a substrate enters a micelle is essentially diffusion controlled and does not depend greatly on the molecular structure. However, the exit rate constant increases with the solubility of the solute in water and the values are in the range 10^3–10^7 s^{-1} [16–19a].

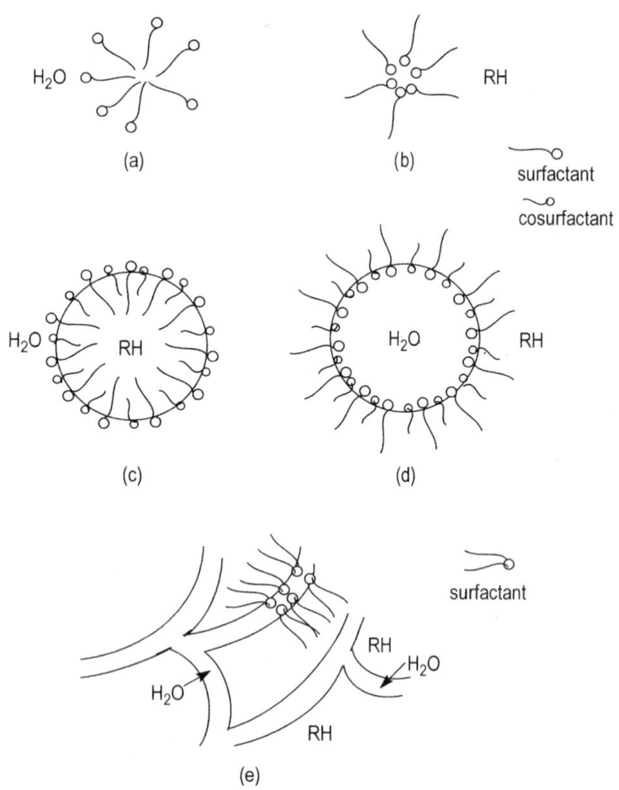

FIG. 1 Schematic representation of micelles [(a) direct, (b) reverse] and microemulsions [(c) o/w, (d) w/o, (e) bicontinuous].

Numerous studies dealing with thermochemical or photochemical organic reactions in micelles, microemulsions, and related systems address mechanistic and synthetic aspects [19b–25]. The effects of such media on electrochemical reactions have also been intensively studied [16–18]. These studies concern both the analytical electrochemistry or the mechanistic aspects and the synthetic applications that have been developed [26–28].

The aim of this chapter is to emphasize the specificity and the application in organic electrosynthesis of macro- and microheterogeneous systems by means of typical examples. However, the large number of publications in this field are not exhaustively reviewed.

II. GENERAL FEATURES OF ELECTROCHEMICAL PROCESSES IN O/W EMULSIONS

A. Mechanisms

Charge and mass transfer have been studied by Wendt and colleagues in conventional and trickle cells [3,29–32], by Alkire et al. in parallel and plate electrolyzers [33], and in flow-through porous electrodes [34]. In agreement with these studies and in particular with the mechanisms described by Wendt et al., two general cases must be considered. In the first one, all the processes take place in the aqueous phase (Fig. 2), and the organic dispersed phase serves only to saturate the aqueous phase continuously with the organic substrate (S). In the second case, the electrochemical or chemical conversion of the substrate occurs in the dispersed organic phase (Fig. 3). Three different mechanisms can

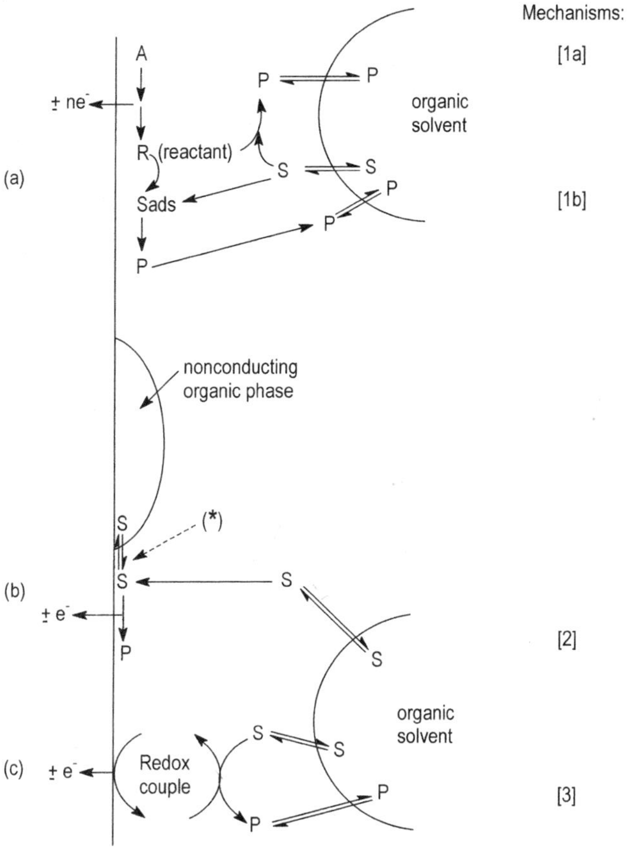

FIG. 2 Direct and indirect electrochemical processes in emulsions with the substrate conversion taking place in the aqueous phase. (*) Mass transfer enhancement due to the electrode wetting.

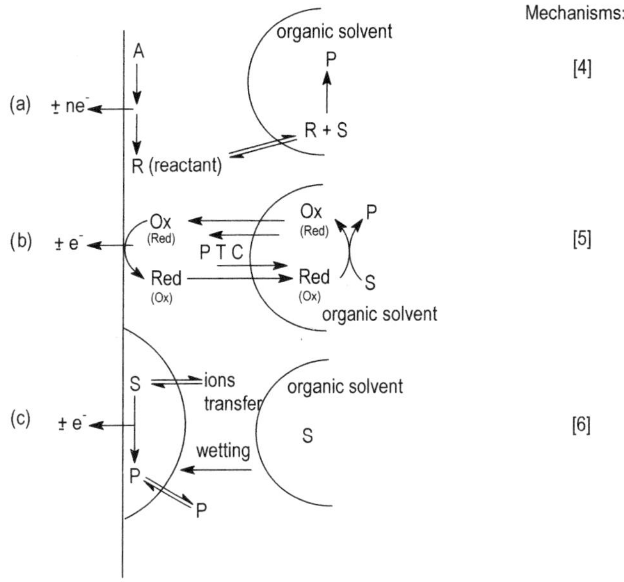

FIG. 3 Direct and indirect electrochemical processes in emulsions with the substrate conversion occurring in the organic phase.

be encountered in these two cases (i.e., six types of mechanism):

Mechanism 1. An electrochemically produced reactant on the electrode reacts chemically in the aqueous solution with the organic substrate (S).The reaction can occur either in the bulk aqueous electrolyte, *mechanism 1a*, or between adsorbed species, *mechanism 1b* (Fig. 2a).

Mechanisms 2 and 3. The electrochemical conversion of the organic depolarizer (S) takes place in the continuous aqueous phase either directly at the phase boundary electrode/electrolyte, *mechanism 2* (Fig. 2b), or indirectly in the bulk aqueous solution by a redox mediator system that is soluble only in the aqueous electrolyte, *mechanism 3* (Fig. 2c).

Mechanism 4. A reactant electrochemically formed in the aqueous solution is extracted by the organic phase, where it reacts chemically with the organic substrate (Fig. 3a).

Mechanisms 5 and 6. In these two last mechanisms, the electrochemical conversion of S takes place in the disperse organic phase either indirectly by the use of a phase transfer catalyst (PTC) that transfers a water-soluble mediator system into the organic solution, *mechanism 5* (Fig. 3b) or directly at the phase boundary electrode/organic phase, *mechanism 6* (Fig. 3c). This last case needs a contact between the electrode and the organic emulsified phase. This wetting phenomenon can be of primary importance in electro-organic synthesis (see the following).

B. Aqueous Solubility

The aqueous solubility of the substrate S is an important parameter whose influence depends on the reaction type. In mechanisms 1b and 2, low solubility in water can strongly reduce the process efficiency. In the other mechanisms, this parameter can be of less importance or on the contrary can be a favorable factor. These aspects are briefly discussed next.

As pointed out earlier by Feess and Wendt, *in the absence of electrode wetting by the organic phase*, the direct electrochemical conversion of a sparsely water-soluble organic substrate (mechanism 2) occurs with a poor space-time yield [3]. Indeed, the diffusion of the depolarizer S to the electrode is limited by its solubility in the aqueous electrolyte. As reported earlier [27a], this case is well illustrated by the anodic oxidation of

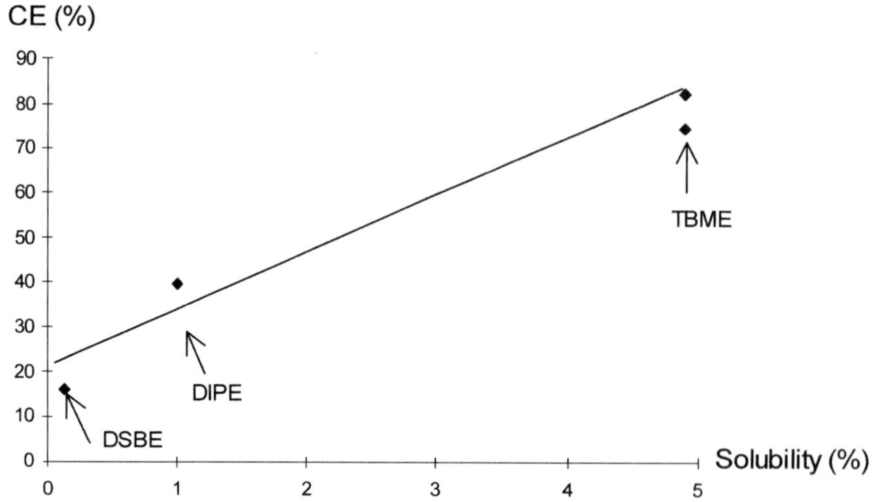

FIG. 4 Variation of current efficiencies with solubilities for the anodic oxidation (on Pt) of di-*sec*-butylether (DSBE), diisopropylether (DIPE) and *t*-butylmethylether (TBME) emulsified in 1 M H_2SO_4. CE values for the formation of R—CO—R′ + CH_3CO_2H (DSBE and DIPE) and $(CH_3)_3C$—OH (TBME). From Ref. 36. Data reprinted from Tetrahedron Vol. 47 (1991), pp 725–736, F. Beck, B. Wermeckes and W. Janssen, Anodic oxidation of isoalkylethers in aqueous electrolytes. Copyright 1991, with permission from Elsevier Science.

alipathic ethers emulsified in 1 M H_2SO_4 previously studied by Beck et al. [35,36]. Diisopropylether (DIPE) and di-*sec*-butylether (DSBE) give respectively acetone and a mixture of methylethylketone and acetic acid (as main products), and *tert*-butylmethylether (TBME) is cleaved into *tert*-butanol and methanol. The current efficiency of the process depends on the competition between the anodic oxidation of the water (oxygen evolution) and the electrochemical formation of alcohols and carbonyl compounds. As shown in Fig. 4, the current efficiency increases with the substrate solubility. The current efficiencies observed with the relatively more soluble TBME are much higher (about 80%) than those obtained with the poorly water-soluble DSBE and DIPE (about 16 and 40%, respectively) (Fig. 4). If the solubilities of the last two compounds are improved by a cosolvent (CH_3CN), the current efficiencies increase up to about 70–90% [36].

A similar limitation occurs with mechanism 1b. An illustrative example is given by the electrocatalytic hydrogenation (ECH) of unsaturated organic compounds emulsified in an aqueous solution. The ECH at Raney nickel cathodes constitutes a useful method for carrying out the hydrogenation of numerous unsaturated compounds under mild conditions [37]. The chemisorbed hydrogen generated in situ by hydronium ions or water reduction (reaction [1], Scheme 1) reacts with an unsaturated organic molecule (Y = Z) adsorbed on the catalyst (reaction [3]). The adsorption of the organic substrate (reaction [2]) is a key step in the overall process. Thus the ECH of limonene emulsified in an aqueous buffer (pH 2) does not occur, and the substrate is entirely recovered, whereas in an ethanol-water solution, *p*-menthene is obtained in good yield (62–88%) [38]. This striking difference between emulsions and

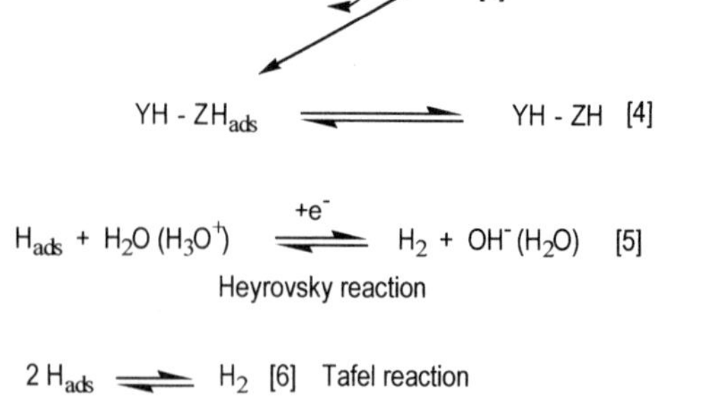

SCHEME 1 Mechanism for the electrocatalytic hydrogenation reaction (ECH) and the hydrogen evolution reaction (HER) [38].

homogeneous hydro-organic solutions has been attributed to the poor solubility of limonene in water, which prevents its adsorption on the cathode in an emulsified system [38].

C. Mass Transfer and Mediation Effects

Under suitable experimental conditions, the presence of the dispersed phase in o/w emulsions can lead to an increase of the mass transfer rate at the electrode [3,29–34,39). Thus the limitation due to the low water solubility of the organic depolarizer can be partially compensated. The enhancement of mass transfer results either from a "disturbance" of the mass transfer boundary layer on the electrode surface or from wetting of the electrode surface by the organic phase. When the organic phase is nonconducting, the reactant solubilized in the adsorbed organic drops is injected in the aqueous solution very near the electrode surface and mechanism 2 takes place (Fig. 2b). The parameters governing the overall mass transfer rate have been previously studied in flow cells with parallel plate electrodes [33] and with three-dimensional electrodes [3,30,31,34]. The latter ones (porous electrodes, packed or trickle bed electrodes) favor the mass transfer rate by providing a high surface density of the three-phase boundary (the electrode and the two liquid phases). These aspects are illustrated by two examples:

1. Anodic Oxidation of Butyltoluene

The anodic oxidation of *p-t*-butyltoluene emulsified in an aqueous electrolyte (1 M H_2SO_4) was carried out by Pangarkar and colleagues in a flow cell (filter-press cell). Under optimal conditions, *p-t*-butylbenzaldehyde was produced with a current efficiency of 73% and good selectivity [40].

2. Cathodic Reduction of Acetophenone

The cathodic reduction of acetophenone into alcohol and pinacol (Scheme 2) was studied by Cognet et al. both in a homogeneous aqueous solution saturated with acetophenone and in two-phase media [water/acetophenone and water/(acetophenone + toluene) emulsions] on a porous percolated pulsed electrode [41]. The pulsation of the electrolyte enhances the mass transfer and, in the case of emulsions, increases the three phase contacts between the electrode and the two phases. In all systems, the electroreduction occurs only at the aqueous phase/electrode interface. In the water-toluene emulsion, the ketone concentration in the aqueous electrolyte is nine times lower than that in the homogeneous aqueous solution under saturation conditions. However, in this emulsified medium, the detected intensity is only three times lower than in the homogeneous system. This results from a mass transfer enhancement. The latter has been attributed to a liquid/liquid transfer of the substrate between the two phases that compensates the loss of the substrate in the aqueous solution near the cathode [41].

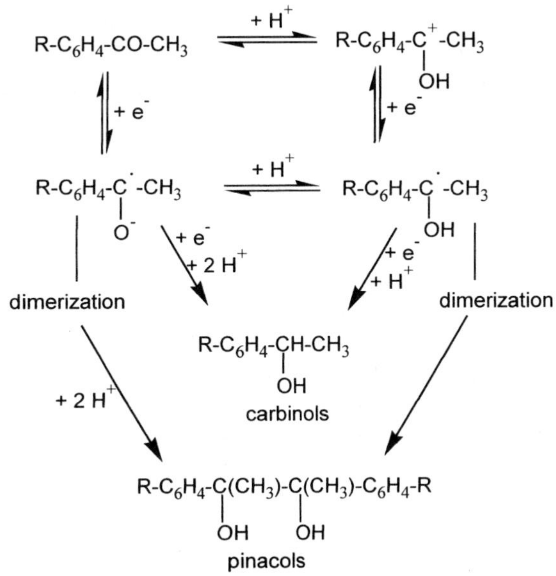

SCHEME 2 General simplified mechanism for the electroreduction of acetophenone (R = H) [2d,27g,41,150,151] and for *para*-substituted acetophenone (R = phenoxyalkyl ammonium or sulfate) [27g,175–177,181].

D. Emulsions in Flow Cells

With flow cells, stabilization of the emulsion could be necessary. This can be achieved with surfactants as in the preceding experiment dealing with the oxidation of *p-t*-butyltoluene [40]. In addition, surfactants enhance the mass transfer in different ways (smaller drops, better wetting, etc.) [31,34]. In the last example (the reduction of acetophenone) the emulsion was generated without surfactant by an external ultrasonic cell. Good current efficiency (60%) was obtained in the binary water/acetophenone system, but in the presence of toluene poorer results were observed [41]. Therefore, as suggested by the authors, it would be interesting to use a phase transfer agent or to activate the electrolyses by ultrasound [41]. This last aspect was studied (from an analytical point of view) by Marken and Compton with sonoelectrochemically modified electrodes in 1-octanol/water emulsions [42].

The limitation resulting from the low solubility of organic depolarizers in water can also be overcome by the use of redox mediator systems (mechanism 3). Organic, inorganic, and organometallic redox catalysts have been widely used in electro-organic synthesis. Basic concepts and numerous examples are collected in several reviews [2b,2c,3,43–45]. Some limitations can appear with inorganic catalysts in emulsified systems [3]. In particular, the rate of the indirect electrochemical conversion can be limited either by the transfer of a very poorly water-soluble compound from the organic phase to the aqueous phase or by a low rate for the redox reaction between the substrate and the mediator. Moreover, the electrochemical regeneration of the mediator is sometimes inhibited by the adsorption of the substrate or the products on the electrode [3]. These different difficulties may be solved in two ways:

1. Ex-Cell Methods

Ex-cell methods require two (or more) reactors. The difficulties arising from the regeneration of a mediator in the presence of organic substances and the interest in the ex-cell process are exemplified by the indirect oxidation of *p*-methoxytoluene (PMT) (4-methylanisol) to *p*-methoxybenzaldehyde (PMBA) (anisaldehyde) mediated by the redox couple Ce(IV)/Ce(III) (Scheme 3). Savall and Tzedakis have shown that PMT emulsified in an acidic aqueous solution containing ceric sulfate [formed by Ce(III) anodic oxidation] is oxidized in the aqueous phase (mechanism 3) into PMBA with good yields (from 65 to 86% at 50°C) and good selectivity [the overall yields of the by-products, *p*-methoxybenzoic acid (*p*-anisic acid) and *p*-methoxybenzyl alcohol (anise alcohol), vary between 9 and 20%] [46]. However, the adsorption and the polymerization of PMT and its derivatives on the platinized titanium electrode prevent Ce(IV) regeneration. Anode deactivation was avoided by recycling the aqueous phase in a separate electrolysis cell (ex-cell method) where Ce(IV) was regenerated with a high Faradic yield (90%) after careful elimination of the organic compounds from the aqueous electrolyte (two extractions with CH_2Cl_2 followed by flash distillation) [46]. Numerous indirect electrochemical oxidation or reduction syntheses have been published. Some examples are given in Refs. 3, 27a, 28, 47–50, and references cited therein.

2. PTC Use

The use of a PTC can transfer the redox mediator into the dispersed organic phase, where the chemical redox step occurs (mechanism 5). The advantages of this method were emphasized in 1982 by Feess and Wendt in their review [3]. They predicted "a promising scientific and technical development" for the application of phase transfer catalysis in electrochemistry. Indeed, since then, indirect electrochemical processes in the presence of PTC have been developed by several investigators (some nonexhaustive examples are given in Refs. 51 to 61). A typical example is given by the indirect oxidation of benzyl alcohol to benzaldehyde in the presence of both phase transfer catalysts and the redox mediator OCl^-/Cl^- (Scheme 4), which was extensively investigated [56–61].

Do and Chou carried out this reaction in a batch reactor (undivided cell) [56,57]. The reaction mechanism and kinetics and the factors that affect current efficiency (agitation, nature and concentration of PTC, pH, nature of the organic solvent, temperature, and current density) were studied in detail [56,57]. Only a small amount of alcohol is transferred into the aqueous phase and, therefore, the oxidation reaction occurs mainly in the organic phase. Depending on the experimental conditions, the rate-determining step becomes either the shuttling rate of the redox ions between the two phases or the rate of the mediator regeneration. Good current efficiency (80%) for the benzaldehyde formation was obtained with emulsions constituted by equal volumes of aqueous phase and dichloromethane

$$R\text{-}C_6H_4\text{-}CH_{3\,org} \rightleftharpoons R\text{-}C_6H_4\text{-}CH_{3\,aq}$$

$$[1]\quad R\text{-}C_6H_4\text{-}CH_{3\,aq} + 2\,Me^{n+} + H_2O \xrightarrow{k_1} R\text{-}C_6H_4\text{-}CH_2OH + 2\,Me^{(n-1)+} + 2\,H^+$$

$$[2]\quad R\text{-}C_6H_4\text{-}CH_2OH + 2\,Me^{n+} \xrightarrow{k_2} R\text{-}C_6H_4\text{-}CH{=}O + 2\,Me^{(n-1)+} + 2\,H^+$$

$$[3]\quad R\text{-}C_6H_4\text{-}CH{=}O + 2\,Me^{n+} + H_2O \xrightarrow{k_3} R\text{-}C_6H_4\text{-}CO_2H + 2\,Me^{(n-1)+} + 2\,H^+$$

$$Me^{n+} / Me^{(n-1)+} = Ce^{4+} / Ce^{3+} \text{ or } Mn^{3+} / Mn^{2+}$$

SCHEME 3 Indirect oxidation of toluene (R = H) and *p*-methoxytoluene (R = OCH_3) (3,46,85,86).

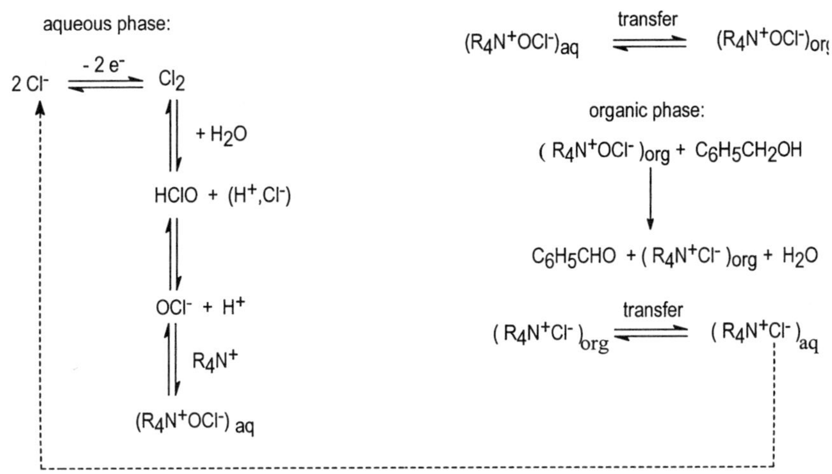

SCHEME 4 Indirect oxidation of benzyl alcohol with phase transfer catalysts. Simplified mechanism adapted from Ref. 57 (reactions and scheme I): J. Appl. Electrochem., Vol. 20 (1990), pp 979–980, J. S. Do and T. C. Chou, Kinetics of the anodic oxidation of benzyl alcohol in dichloromethane in the presence of redox mediator and phase transfer catalyst. Copyright 1990, with kind permission from Kluwer Academic Publishers.

in the presence of Bu_4NHSO_4 (PTC). Under these conditions, the wetting of the anode (graphite) by the organic phase is close to 40% [57]. Therefore direct oxidation of the alcohol can also occur in the organic phase (which presents electrical conductivity due to the presence of Bu_4N^+ OCl^- salt), but this direct process is of minor importance and the indirect oxidation represents the main part of the reaction [56,57].

More recently, Do and Do have performed the same indirect oxidation in a continuous-flow stirred-tank electrochemical reactor (CSTER) [58–60]. The kinetics of the different processes (i.e., the oxidation of benzyl alcohol into aldehyde and the oxidation of benzaldehyde into benzoic acid) and the numerous parameters controlling the yield and the selectivity of benzaldehyde formation were studied. The experimental results were correlated with extensively developed theoretical analysis [58–60].

The PTC can also be immobilized on a polymer and three phases are involved in the process: the aqueous continuous phase (solution) and two dispersed phases, the organic phase and the polymeric phase. Solid phase transfer catalysis has several advantages; in particular, the catalyst can easily be separated and reused. The indirect oxidation of benzyl alcohol in an emulsion (H_2O/CH_2Cl_2) with the redox mediator Cl^-/OCl^- and tri-*n*-butylamine or tri-*n*-butylphosphine (PTC) immobilized on a polymeric support has been demonstrated by Do and Chou [61]. The hypochlorite ion OCl^- formed by Cl^- oxidation (Scheme 4) is extracted from the aqueous solution to the polymer phase. The benzyl alcohol diffuses from the organic phase into the polymeric phase, where it reacts with the extracted OCl^- [61].

E. Organic Phase Effects

In mechanism 6, the depolarizer solubility in the aqueous electrolyte and its diffusion to the electrodes are not limiting factors. Among the different parameters reviewed by Feess and Wendt [3], three are primarily important:

1. Organic Phase Wetting of Electrodes

The wetting phenomenon, studied in detail by Wendt and colleagues [29–32], depends on several factors. These include the nature of the electrode material and the design, nature, and composition of the two phases (especially the density ratio between the aqueous and the nonaqueous solutions). Thus, the possibility of the electrode being wetted by an organic phase is reduced when its density is lower than that of the aqueous electrolyte. Addition of appropriate surfactants (to the aqueous phase) favors the wetting phenomenon by decreasing the surface tension differences at the three-phase boundary (organic solution/aqueous electrolyte/electrode). In some cases, a thin hydrophobic film is formed on the electrode (see the following).

2. Organic Phase Electrical Conductivity

Electrochemical conversion at the organic solution/electrode interface requires the organic phase to have a finite and not too low electrical conductivity. An ap-

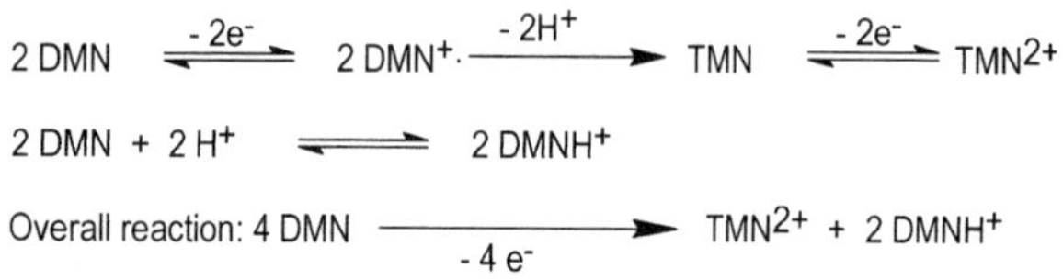

SCHEME 5 Electrodimerization of DMN. Adapted from Ref. 62: Electrochimica Acta, Vol. 35 (1990), p. 1081, N. Vettorazzi, H. Fernandez, J. J. Silber and L. Sereno, Electrodimerization of an aromatic amine in a two-phase solvent system. Copyright 1990, with permission from Elsevier Science.

propriate lipophilic electrolyte must therefore be present in the organic phase. Phase transfer catalysts can play this role, and typical examples are given by anodic nucleophilic substitutions under phase transfer catalysis conditions (see Section IV.B). Ionic reaction products can also supply the needed conductivity. This latter situation is well illustrated by the electrodimerization of *N,N*-dimethyl-1-napththylamine (DMN) into the *N,N,N′,N′*-tetramethyl-1,1′-naphthidine dication (TMN^{2+}) previously investigated by Sereno and colleagues [62] (Scheme 5). Controlled potential oxidation of DMN in a water-nitrobenzene emulsion was performed using 2 M $NaClO_4$ as supporting electrolyte in the aqueous phase. In this emulsion, there is wetting of the anode by the organic phase, whose density is higher than that of water. Direct electron transfer occurs at the electrode/nitrobenzene interface and the initially nonconducting organic phase becomes a conductor as the concentration of TMN^{2+} in nitrobenzene increases. This unusual situation requires the formation of a charged product, hydrophobic enough to remain in the organic phase.

3. The Ions Transfer at the Organic/Aqueous Phase Interface

In order to fulfill the electroneutrality of the organic solution, ions must migrate between the two phases. Thus, in the preceding example, the ClO_4^- anions transfer from the aqueous phase to the organic phase and the protons arising from the coupling reaction (Scheme 5) migrate in the reverse direction [62].

F. ITIES

Ion transfer at the aqueous-organic interface depends on its free enthalpy, and ionic surfactants can assist this transfer by changing the Galvani potential at the interface [3,31]. The thermodynamics and kinetics of ion transfer (with or without surfactants) at the interface of two immiscible electrolyte solutions (ITIES) have been very extensively studied; for more information the reader is referred to publications and reviews in this field [63–66].

G. Poorly Aqueous Soluble Organic Substrates

The poor water solubility of an organic substrate can become advantageous in the case of mechanisms 1a and 4. In mechanism 1a, with organic substrates sparsely soluble in water, the mass transport phenomenon and the rate of electrochemical formation of the reactant can be monitored in order to consume all the substrate solubilized in the aqueous solution (i.e., the transfer of the organic compound across the phase boundary becomes the limiting factor). Thus, a competitive anodic or cathodic reaction can be limited or suppressed.

An illustrative example is given by the indirect electrochemical epoxidation of 1-hexene described by Alkire and Köhler [67]. This compound, poorly soluble in water (6×10^{-4} M), is emulsified in an aqueous alkaline solution of sodium bromide. The molecular bromine produced by anodic oxidation of bromide ions react with 1-hexene. The nucleophilic addition of a hydroxide ion on the carbonium intremediate, followed by a base-catalyzed dehydrohalogenation, give 1,2-epoxyhexane, and Br^- is regenerated (Scheme 6). Under experimental conditions in which the rate of bromine formation is sufficiently high that the Br_2 concentration slightly exceeds the saturation concentration of hexene, epoxide is obtained with current efficiencies around 65%, and the total organic current efficiency (which includes the formation of by-products) reaches a maximum of about 98%. In those conditions, the reaction between hexene and Br_2 prevails over other possible competitive reactions [67].

In mechanism 4, the low water solubility of the substrate can prevent its direct electrochemical conversion at the aqueous electrolyte/electrode interface. The anodic thiocyanation of aromatic amines and phenols by two-phase electrolysis was performed and studied by Krishnan and Gurjar [68–70]. The electolyses were carried out in emulsions made up of an organic solvent (CH_2Cl_2 in most cases) and an acidic aqueous solution of ammonium thiocyanate. Anodic oxidation of SCN^- gave trithiocyanate $(SCN)_3^-$. This thiocyanating reagent (and/or the subsequently formed thiocyanogen), unstable in water, is extracted into the organic phase. This extraction prevents decomposition in the aqueous solution. The formation of thiocyano and isothiocyano derivatives occurs in the organic emulsified solution

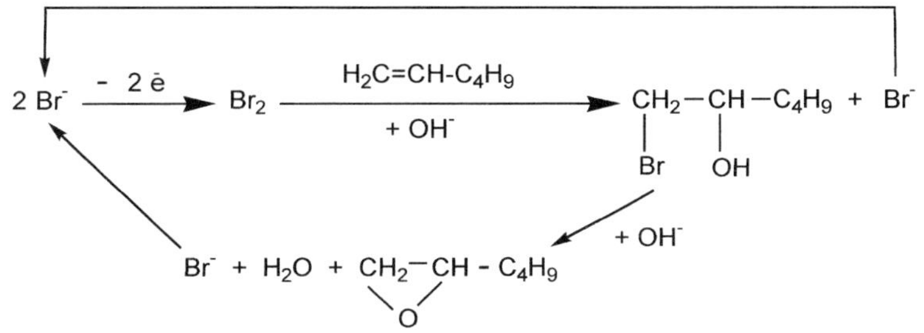

SCHEME 6 Electrochemical epoxidation of hexene. Adapted from Ref. 67: J. Appl. Electrochem., Vol. 18 (1988), p 406, R. Alkire and J. Köhler, Indirect electrochemical epoxidation of hexene in a liquid-liquid electrolyte. Copyright 1988, with kind permission from Kluwer Academic Publishers.

with medium to good yields (45 to 83%). With substrates sparingly soluble in water and having oxidation potentials close to that of SCN^-, in situ thiocyanation can be carried out because direct electro-oxidation of the substrate at the anode cannot take place in the aqueous solution (as a consequence of their low concentration in this phase) or in the organic solution, which is electrically nonconducting [68,69]. Although the electrode wetting by the organic phase was not clearly mentioned in the publication, the lower current intensity values, compared with those observed in the homogeneous aqueous electrolyte (Fig. 1 in Ref. 69), suggest some electrode wetting. Substrates soluble in the acidic electrolyte, such as amines, can be thiocyanated in two steps (i.e., the electrochemical formation of the reagent and its subsequent reaction with the substrate occur separately) [68].

III. GENERAL FEATURES OF ELECTROCHEMICAL PROCESSES IN MICELLES, MICROEMULSIONS, AND RELATED SYSTEMS

Electrochemical reactions have been performed mainly in micellar solutions, in oil-in-water (o/w) microemulsions, and in bicontinuous microemulsions, in which the aqueous pseudophase acts as a highly conducting continuous medium. Electrochemistry and electrochemical catalysis in these microheterogeneous systems have been the subjects of several reviews [16–18,27,71]. Therefore, only the main features related to solubilization equilibria and surfactant adsorption (the two most important parameters) will be briefly discussed. Detailed descriptions are given in Refs. 16–18 and 71 and in the references cited in these reviews.

A. Solubilization Equilibria

As pointed out in the introduction, the solubilization of an organic compound in a micellar solution corresponds to a fast equilibrium of the solute between the two pseudophases (the kinetics of such equilibria are in the microsecond time range). These solubilization equilibria influence both the diffusion of the organic depolarizer and its formal redox potential [16,18,27b,c]. Beyond the critical micelle concentration (cmc), in many cases, the diffusion of the organic solute to the electrode involves both the free molecules solubilized in the aqueous pseudophase and the molecules bound to micelles (Fig. 5). The apparent diffusion coefficients D_{app} of solutes in micellar systems have, therefore, lower values than those in homogeneous solutions (some examples are given in Tables 2, 3, and 4 of Ref. 18). As the D_{app} value decreases, the binding constant with the micelles increases. With a substrate totally bound, an approximate 10-fold decrease can be observed [18,72]. For a given redox couple, a shift in potential between that observed in an aqueous electrolyte and that obtained in a micellar solution can occur

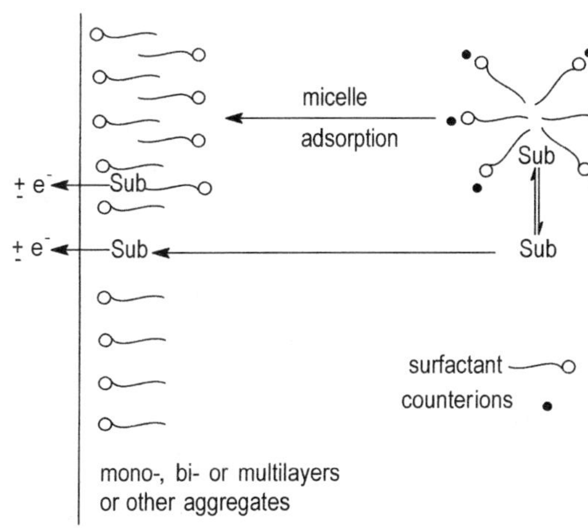

FIG. 5 Schematic representation of substrate adsorption and electron transfer in micellar systems.

when the interactions between the micellar aggregates and the two forms of the couple are different. This effect has been treated with the pseudophase partitioning model [16,18]. Similar general trends are encountered in o/w microemulsions (and in the few studies related to w/o microemulsions, lamellar dispersions, and vesicles), in particular concerning the diffusion [16–18,27d–f]. The electrochemical behaviors of water-soluble or oil-soluble substrates solubilized in bicontinuous microemulsions are very different from those observed in the other microheterogeneous systems. In bicontinuous microemulsions each pseudophase behaves as a continuous homogeneous medium, and the apparent electrochemical parameters are close to those measured in homogeneous solutions (some values are collected in Table 2 in Ref. 17 and in Table 6 of Ref. 18). Thus, the diffusion coefficients of oil-soluble depolarizers in bicontinuous microemulsions are similar to those obtained in homogeneous organic solvents (i.e., their values are an order of magnitude greater than those in o/w microemulsions) [16–18,27].

B. Surfactant Adsorption

The adsorption of surfactants onto electrodes is a fundamental parameter that can deeply influence electrochemical processes, the yields, and the selectivity of electroorganic syntheses.

Numerous studies are related to the adsorption of ionic and nonionic surfactants on different supports (e.g., metal oxides, graphite). Some examples are given in Refs. 73 to 75, and the main features are summarized in the aforementioned reviews [16–18,27,71]. The adsorption of surfactants from micellar solutions gives aggregates at solid-liquid interfaces whose shape, size, and organization (hemimicelles, spherical, cylindrical, or planar aggregates) are governed by the nature of the solid surface (hydrophilic or hydrophobic) [73] and the surfactant geometry [74]. The adsorption phenomenon also depends on the surfactant concentration. For instance, the adsorption isotherms of ionic surfactants onto hydrophilic solids (metal oxides) present four characteristic regions [75]. In very dilute solutions, surfactants adsorb as the individual monomers (region I). As the surfactant concentration increases (for concentrations of the order of 0.1 cmc), associative surfactant adsorption occurs (formation of aggregates such as hemimicelles) (region II). For higher concentrations the surface coverage grows until the formation of a bilayer on the whole surface (for concentrations close to the cmc) (region III). Above the cmc adsorbed amount remains constant (region IV) [16–18,71,75]. It must be noticed that the detailed nature of the adsorption phenomena in the different regions is not entirely clear and depends on the model used (see the introduction in Ref. 75).

The electrode potential represents an additional parameter that controls surfactant adsorption on electrodes in micellar [16,18,71] and microemulsion systems [17]. As pointed out by Rusling [71], a possible head-down orientation of ionic surfactants adsorbed on metals or carbone depends on the potential value. The potential needs to be positive of the point of zero charge (PZC) for anionic surfactants and cationic surfactants through anion counterion adsorption or negative of the PZC for cationic surfactants. For nonionic surfactants a head-down orientation may occur on both sides of the PZC. Moreover, aggregate structures similar to those formed on various supports, such as hemimicelles and mono-, bi-, or multilayers [16,18,71], may also be formed on electrodes, depending on several parameters (potential, nature of the surfactant and of the electrode surface, etc.).

Reactant associations with the adsorbed layers (or aggregates) of surfactant on the electrodes are similar to those with micelles and are governed by hydrophobic and coulombic interactions. The solubilization equilibria of a substrate (Sub) between the adsorbed aggregates of surfactant and the bulk micellar pseudophase can involve either the free molecules (Sub) solubilized in the aqueous electrolyte or the molecules associated with the micelles (the latter joining the adsorbed surfactant layer) (Fig. 5). The association phenomenon in the adsorbed layer can control the electron transfer kinetics according to the position of (Sub), namely specifically adsorbed to the electrode [which means the replacement of a surfactant molecule by (Sub) on the adsorption site] or solubilized in the surfactant layer (Fig. 5) [71,76,77]. The different models for electron transfer in micellar solutions may also be valid in microemulsions [17,71]. Besides the papers already mentioned, additional features and theoretical developments have been given by Lipkowski in a review that specifically deals with the effects of surfactant monolayers on the ion and electron transfer reactions [78]. In addition, the surfactant adsorbed layers can catalyze or inhibit the chemical steps (involved in the electrochemical conversion) by stabilizing or destabilizing the electrochemically produced intermediates [16–18], which means changes of the rate constants and/or of the local reactant concentrations as in the well-known micellar catalysis and inhibition phenomenon [15–25,79].

IV. EFFECTS OF HETEROGENEOUS SOLUTIONS ON THE EFFICIENCY AND THE SELECTIVITY OF ELECTRO-ORGANIC SYNTHESIS

From a preparative point of view, the effects of macro- and microheterogeneous systems lead to two main positive consequences in electrochemical synthesis: increased yields and improved process selectivities. These effects on the efficiency and the selectivity of electrochemical reactions can be schematically divided into three groups: effects resulting from the variation of the local concentrations of reactants and products, modifications of the electrodes, and solvent effects and other specific interactions.

A. Effects Resulting from Local Concentrations

In macro- and microheterogeneous systems, unwanted competitive reactions can be reduced or avoided by changing the "local" concentration of substrates, products, or intermediates. Therefore, both the selectivity and the efficiency (product and current yields) of the electrochemical synthesis may be improved.

1. Emulsified Systems

In emulsions, when the electrochemical conversion takes place in the aqueous electrolyte (mechanisms 1, 2, and 3), a product can be protected against further direct or indirect anodic or cathodic reactions by its extraction into the organic phase. This well-known protective effect increases with the product solubility in the organic solvent, and several examples, corresponding to the different mechanisms, can be found in the literature.

The anodic conversion of sodium dimethyldithiocarbamate (NaDTC) into tetramethylthiuram disulfide (TMT) published by Santic et al. [80] represents a classical example of product protection in the case of mechanisms 2 and 3. Indeed, in biphasic water-organic solvent emulsions, NaDTC direct and indirect (via hypochlorite) oxidation occurs in the aqueous phase (Fig. 6). In water-$CHCl_3$ emulsions, the product (and current) yield is higher (90%) than that obtained in homogeneous acetonitrile solution (68%). This effect increases as the TMT solubility in the organic solvent increases. Thus, yields and current efficiencies increase in the series benzene < CH_2Cl_2 < $CHCl_3$ (respectively 56, 85, and 91%). This yield improvement was ascribed to the extraction of TMT by the organic phase, which prevents further anodic or cathodic reactions, when the electrosyntheses were carried out in an undivided cell [80].

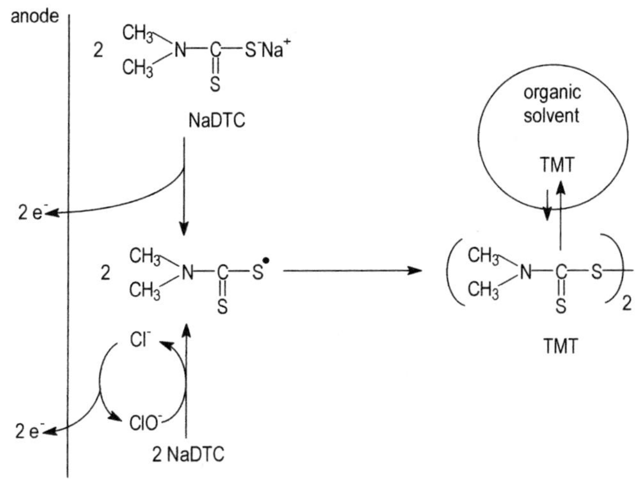

FIG. 6 Anodic direct and indirect oxidation of NaDTC in biphasic emulsions. Adapted from Scheme 1 in Ref. 80: J. Appl. Electrochem., Vol. 16 (1986), p 909, Z. Ibrisagic, V. Misovic, I. Tabakovic and I. Santic, Electrochemical synthesis of tetraalkylthiuram disulphides in emulsions. Copyright 1986, with kind permission from Kluwer Academic Publishers.

The electrochemical chlorination of various amine hydrochlorides performed in diaphragmless cell by Lyalin and Petrosyan [81] and the electro-oxidation on a nickel anode of benzyl alcohol in an undivided pulsed flow reactor by Cognet et al. [82] are two examples of product protection in the case of mechanisms 1a and 1b, respectively.

The main products of the electrochemical chlorination of primary and secondary amines are dichloroamines and monochloroamines, respectively (Scheme 7). In homogeneous aqueous solution, chlorination of $CH_3NH_3^+Cl^-$ gives the dichloro derivative (CH_3NCl_2) with low current efficiencies (14–16%) and the substrate conversions are below 20%. The authors ascribe the poor reaction efficiency to the easy direct cathodic reduction of the dichloroamine into amine, which can be rechlorinated at the anode. In aqueous NaCl/CCl_4 emulsions, the extraction of chloramine by the organic phase avoids its reduction. As a consequence, after optimization, the current yields and the conversions increase to 95% and 92%, respectively. Good current efficiencies (82–90%) and conversions (88–91%) are also observed in H_2O/CCl_4 emulsions for the chlorination of six other primary or secondary amine hydrochlorides [81].

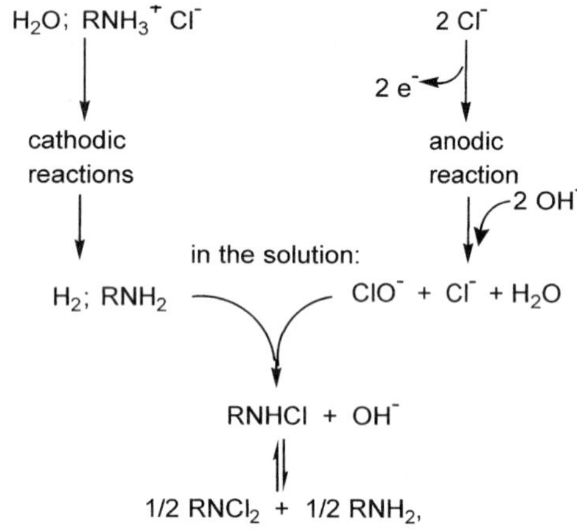

SCHEME 7 Oversimplified mechanism for the electrochemical chlorination of amine hydrochlorides. Adapted from Ref. 81: Russian J. Electrochem., Vol. 34 (1998), pp 1098–1102, B. V. Lyalin and V. A. Petrosyan, Electrosynthesis of alkylchloramines from hydrochlorides of the corresponding amines. Copyright 1998, with permission from International Academic Publishing Company MAIK "Nauka/Interperiodica."

The electro-oxidation of benzyl alcohol into benzaldehyde on Ni three-dimensional anodes in aqueous basic solutions (0.1 M KOH) corresponds to a heterogeneous chemical reaction (reaction 2, Scheme 8) between the electrogenerated (reaction 1) NiOOH and the adsorbed alcohol (mechanism 1b) [82–84]. In homogeneous aqueous solution, the aldehyde yields (11–32%) are strongly limited by further oxidation into acid. The competitive benzyl alcohol oxidation to benzoic acid (reaction 3) is the major process [82]. To prevent the unwanted aldehyde oxidation, electrolysis is performed in the presence of petroleum ether, which is dispersed in the aqueous phase using ultrasonication and maintained in emulsion by pulsation. The dispersed organic phase extracts the benzaldehyde, which strongly reduces further oxidation to acid. The best benzaldehyde yields (60–70%) obtained for a 16% organic solvent volume ratio are more than twice as high as that in homogeneous aqueous electrolyte [82]. Such a selective indirect oxidation of benzyl alcohol to aldehyde in emulsified solutions was also previously reported [83,84].

In principle, a dispersed immiscible liquid substrate could extract and protect a product, but in many cases this effect is limited. Thus, the direct anodic conversion of an emulsion of toluene (and substituted toluene) with a Mn^{3+}/Mn^{2+} redox mediator was studied extensively by Wendt and Schneider [85]. The reactions leading to alcohol, aldehyde, and acid (Scheme 3) take place in the aqueous phase. The oxidation of benzaldehyde cannot be entirely suppressed by its extraction into the organic toluene phase, and yield optimization requires the use of immiscible solvents and ex-cell procedures. Indeed, the rate of benzoic acid formation (reaction 3, Scheme 3) depends on the value of the pseudo-first-order constant k_3', which includes the concentration, C, of benzaldehyde in the aqueous solution ($k_3' = k_3C$). On decreasing the C value with continuous extraction (ex-cell process) with the use of an immiscible solvent, the k_3' value drops, and the benzaldehyde yield increases at the expense of benzoic acid formation [85,87]. In the same way, ex-cell processes and an aldehyde extraction by an immiscible organic phase are needed for the indirect electrochemical conversion of substituted toluenes to aldehydes [3,46,85–87].

A protective effect seems nonexistent during the oxidation at an $NiCo_2O_4$ anode for benzyl alcohol emulsified in aqueous KOH. Benzoic acid is specifically obtained with very good chemical and current yields (99% and 97%) [88]. If the mechanism taking place at this type of electrode is similar to that at Ni in basic medica (i.e., indirect oxidation involving NiOOH), as suggested by the authors [88], the aldehyde that could be formed is not protected at all after extraction by the emulsified starting material. However, the reactions involve adsorbed species. As a consequence, the competition between the adsorption and the extraction phenomena of substrate, products, and intermediates is a key parameter that depends on the nature of the electrode and the organic phase. Therefore, it could be different in the experiments performed on Ni and in those realized on an $NiCo_2O_4$ anode.

2. Substrate and Intermediate Solubilization

The solubilization of the starting substrate in a dispersed organic phase can also modify the selectivity of an electrochemical conversion occurring at the aqueous electrolyte/electrode interface. This point is well illustrated by the cathodic reduction of acetophenone reported by Cognet et al. (see Section II) in a pulsed flow reactor [41] (Scheme 2). In a homogeneous medium and in the heterogeneous one formed with only emulsified acetophenone, pinacol is the main product (molar yields of 60 and 70%, respectively) [41]. Under the same experimental conditions, the presence of emulsified toluene leads selectively to the alcohol (molar yield of 95%) as a consequence of the very small concentration of acetophenone in the aqueous phase (4.2×10^{-3} M, i.e., nine times lower than in saturation conditions), which disfavors the dimerization reaction [41].

$$Ni(OH)_2 + OH^- \underset{}{\overset{\bar{e}}{\rightleftharpoons}} NiOOH + H_2O \quad [1]$$

$$Ph\text{-}CH_2\text{-}OH_{ads} + 2\ NiOOH \longrightarrow Ph\text{-}CH{=}O + 2\ Ni(OH)_2 \quad [2]$$

$$Ph\text{-}CH_2\text{-}OH_{ads} + 4\ NiOOH + H_2O \longrightarrow Ph\text{-}CO_2H + 4\ Ni(OH)_2 \quad [3]$$

SCHEME 8 Oversimplified mechanism for the electro-oxidation of benzyl alcohol at Ni electrode in alkaline solution. Detailed processes are reported in Refs. 82–84.

When the substrate conversion takes place in the organic phase (mechanisms 4 to 6) an intermediate can be protected from undesirable reactions by its solubilization in the organic solvent. The anodic thiocyanation of aromatic amines and phenols in biphasic media (mechanism 4), reported in the first part of this chapter, is typical of such protective effects. The extraction of the thiocyanating reagent by the organic phase mitigates against its decomposition in the aqueous electrolyte and therefore allows the formation of thiocyanated derivatives with good yields [68–70]. Another example is given by the Ce(III)/Ce(IV) redox couple used to perform the indirect oxidation of several aromatic compounds (mechanism 5). Pletcher and Valdes have shown that the use of an emulsion (aqueous HNO_3/*n*-hexane) with a phase transfer reagent (PTC) allows the reactions to occur in an undivided cell [54,55]. With appropriate solvent and PTC, Ce(IV) is protected from a cathodic reduction by its extraction into the organic phase [54,55].

In the case of mechanism 6, the compartmentalization phenomenon can also strongly modify the efficiency of electrochemical synthesis, as shown by the electrodimerization of dimethylnaphthylamine (DMN) (Scheme 5) studied by Sereno et al. [62] (see Section II). In homogeneous nonaqueous solutions, the yield of the dimeric dication TMN^{2+} is at best 50% (overall reaction) because half of DMN is protonated (Scheme 5) and becomes electroinactive [89]. In emulsions, the DMN oxidation takes place at the electrode/nitrobenzene interface, and in order to maintain the electroneutrality of the organic phase, perchlorate anions migrate from the aqueous solution to the organic phase and the protons migrate in the reverse direction. The H^+ exclusion from the organic phase (which is enhanced by buffering the aqueous solution) reduces the extent of DMN protonation (Scheme 5) and therefore improves the TMN yield [62].

3. Micellar Systems

Reactant compartmentalization and variation of "local" concentrations are of primary importance in micellar catalysis and inhibition [19–25] and play an important role in electrochemical synthesis. Indeed, as pointed out in earlier reviews [18,27a,28], the influence of micelles on yields and selectivity of electrochemical processes results mainly from the variation of "local" reagent concentrations due to hydrophobic and coulombic interactions. In particular, the second-order rate constants in micellar pseudophases are often close to those in water [20]. Therefore, the modification of reaction rates in micellar systems results more often from changes of the reagent concentrations. Thus, the influence of micelles on yield and selectivity results essentially from concentration effects as illustrated by the following examples.

(*a*) *Anodic Nitration of Aromatics.* In previous work the electrochemical nitration of 1,4-dimethoxybenzene was performed in cationic (cetyltrimethylammonium bromide, CTAB, or benzethonium chloride), anionic (sodium dodecyl sulfate, SDS), and nonionic (Brij 35) aqueous micellar systems in the presence of $NaNO_2$ [90]. Nitration does not occur at the NO_2^- oxidation potential (0.8 V vs. SCE) (reaction 1, Scheme 9) but needs a potential that also allows the oxidation of the aromatic substrate ArH (1.5 V) (reaction 2). The best yields (70%) of 2,5-dimethoxynitrobenzene (the solely formed nitrated derivative) are obtained in nonionic micelles. Anionic and cationic micellar solutions lead, respectively, to medium (41%) and poor (0–28%) yields. These results rule out a nucleophilic addition of NO_2^- on the aromatic radical cation $ArH^{+\bullet}$ produced by the ArH anodic oxidation but are in agreement with a radical coupling between NO_2 and $ArH^{+\bullet}$ (reaction 3, Scheme 9) [90].

The same radical coupling was proposed by Sereno and coworkers [91] for the electrochemical nitration of

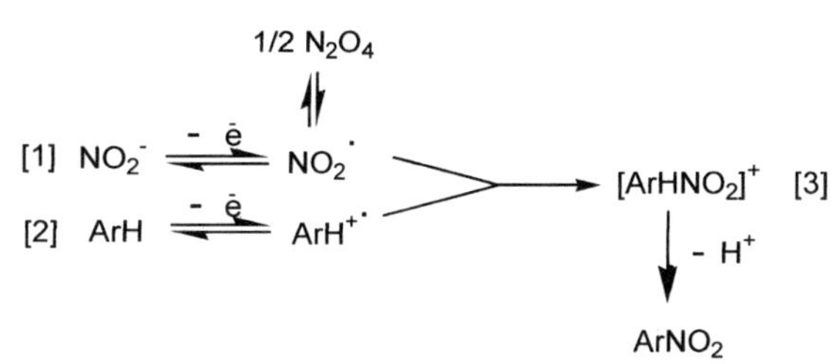

SCHEME 9 Anodic aromatic nitration [90].

naphthalene (NapH) in nonionic (Brij 35) micelles. The anodic oxidation of NapH solubilized in micelles gives the radical cation $NapH^{+\bullet}$, which participates in competitive reactions, including the following:

1. The radical coupling with the electrochemically formed NO_2 leading to 1- and 2-nitronaphthalene (NO_2NapH)
2. The formation of 1,2 and 1,4 naphthoquinones (NapHQ) by water addition
3. Dimerization leading to binaphthalene (BiNapH), which can be oxidized giving quinones or polymeric products

Without surfactant the nitration does not occur at all. Only the dimerization and the reaction with water take place. In nonionic micellar systems, 1- and 2-nitronaphthalenes are obtained with medium yield (34%) beside BiNapH and NapHQ. This result is interpreted in terms of an increase of "local" concentrations of NO_2 and $NapH^{+\bullet}$ in micelles, which favors the biomolecular coupling process. Moreover, the 1-NO_2NapH/2-NO_2NapH ratio is lower (16) than that obtained in acetonitrile (>50). This suggests a selectively induced by the micellar microenvironment that leads to preferred orientations of the substrate [91].

(*b*) *Anodic Cyanations.* Anodic cyanation of 1,3-dimethoxybenze in acetonitrile gives 2,4-dimethoxybenzonitrile (DMB) with a low yield (12%) [92]. This yield strongly increases (80%) if the reaction is carried out in aqueous cationic (CTAB) micellar solution [93,94]. The two reactants (DMB and CN^-) are associated with the cationic micelles (by electrostatic and hydrophobic interactions). Therefore, their local concentrations in the micellar pseudophase are higher than those in homogeneous organic solutions. This favors cyanation over other competitive reactions (i.e., polymerization). These preconcentration effects are inconsistent with anionic (SDS) micelles as a consequence of the electrostatic repulsion between the negatively charged micelles and CN^-. Thus, very low yields (4–8%) are obtained in anionic micellar solutions [93,94]. In addition, different mechanisms occur in cationic micelles and in other solutions (see following Section B).

(*c*) *Reduction of 4-Nitrosodiphenylamine.* The cathodic reduction of 4-nitrosodiphenylamine (NDPA) was studied in micellar solution by Tabakovic and coworkers [95]. The chemical dehydration step leading to quinonimine in this ECE electrochemical process (Scheme 10) is catalyzed by a base. The corresponding pseudo-first-order rate constant, k, exhibits a lower value (0.5 s^{-1}) in cationic CTAB micellar solution than in water (0.9 s^{-1}). As previously pointed out [27a], the k value could depend on the local OH^- concentration in the micellar pseudophase. In the vicinity of cationic micelles (the Stern layer), the exchange between the counterions OH^- and Br^- favors the Br^- more than the OH^- [96]. Therefore, the OH^- concentrations and, in consequence, the k value could be lowered in the micellar pseudophase. Thus the chemical step giving the quinonimine and the subsequent *p*-aminodiphenylamine (ADPA) formation are slowed down in micellar solutions. According to these k variations, preparative electrolysis with an Hg cathode give a mixture of *p*-hydroxyaminodiphenylamine and ADPA in micellar solution with high CTAB concentrations (2.7×10^{-2} M) whereas ADPA is the major product obtained in 94% chemical yield in water/toluene emulsions [97] and in 85% yield in CTAB micelles with low surfactant concentrations (2.7×10^{-3} M) [95].

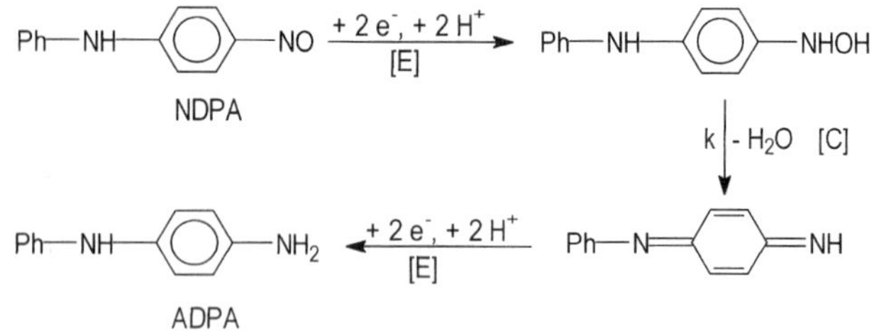

SCHEME 10 Electrochemical reduction of NDPA. Adapted from Ref. 95 (Scheme 1): J. Electroanal. Chem., Vol. 280 (1990), p 376, A. Davidovic, I. Tabakovic and D. Davidovic, L. Duic, Electrochemical reduction of p-nitrosodiphenylamine in a cationic micellar system. Copyright 1990, with permission from Elsevier Science.

(*d*) *Selectivity.* The reaction selectivity can also be controlled by the distribution of the substrate between the two pseudophases. The reduction of acetophenone at the Hg cathode in different micellar systems was studied in detail by the electrochemistry group at the University Blaise Pascal in France ([27g] and references cited therein). With CTAB in acidic aqueous buffer, the electrolysis performed at the potential of the radical reduction gave pinacol and carbinol (Scheme 2). At low acetophenone concentrations (3×10^{-2} and 8.5×10^{-2} M), the carbinol was mainly obtained (65–70%) and the pinacol was a minor product (30–35%). With increasing concentrations of ketone, this slightly water-soluble substrate (4.5×10^{-2} M [3]) became more and more solubilized in the micelles. The bimolecular dimerization process was favored by the high "local" concentrations of acetophenone in the micellar pseudophase and the percentage of pinacol grew to 83% [27g].

4. Lamellar and Vesicles Dispersions

Similar effects occur in these media, but such dispersions are seldom in organic electrosynthesis. The reader will find some examples of electrochemical processes in these dispersions in a review by Rusling [18].

5. Microemulsion Systems

Electro-organic conversions are often performed in o/w and bicontinuous microemulsions. W/O microemulsions are less often used. The properties of microheterogeneous systems (including microemulsions) and electrochemistry in these systems have been treated in various reviews [16–18,98]. Two of them deal specifically with microemulsions [17,98]. The effects of "local" reagent concentrations in microemulsions are often similar to those observed in micellar solutions. Thus, using microemulsion systems can be a way to improve yields, selectivities, and stereoselectivity. In addition, microemulsions allow the solubilization of greater amounts of substrate than micelles. For all of these reasons, the use of conductive o/w and bicontinuous microemulsions in organic electrosynthesis is a promising research field. From a preparative point of view, different kinds of electrosynthesis have been performed in conductive microemulsions:

(*a*) *Polymer Films.* The formation of polymeric films is generally realized in conductive o/w microemulsions [99–102], but some electropolymerizations have also been done in W/O inverse microemulsions [27h,103] (the influence of surfactants on electropolymerization is reported in the last section).

(*b*) *Cyanation of Aromatics.* The anodic cyanations of aromatic substrates in both cationic micelles (CTAB, 4×10^{-2} M) and microemulsions (CTAB/*n*-butanol/*n*-hexadecane/water, 33:33:17:17 wt%) have similar yields and selectivities [94].

(*c*) *Organohalide Dehalogenation.* The dechlorination of organic halides ([17] and references cited therein, [104–109]) and the detoxification of pesticides are important applications. Rusling and Zhang have shown that electrochemical catalytic dechlorination of polychlorinated biphenyls (PCBs) in a bicontinuous microemulsion is a promising technique for cleanup of soils and sediments contaminated with PCBs [108].

(*d*) *Indirect Oxidation of Organic Molecules.* Rusling et al. have reported the myoglobin-mediated electrochemical oxidation of styrene in cationic microemulsion [110]. Benzaldehyde and styrene oxide were obtained with 50-fold better yields than those observed with electrolysis in aqueous buffer. This yield improvement was attributed to better solubility of reactants in the microemulsion than in water [110].

(*e*) *Carbon Bond Formation.* Carbon bond formation (and cyclization) ([98] and references cited therein, [111–113]) represents one of the most exciting and promising aspects of the use of microemulsions in electro-organic synthesis, as illustrated by the following examples published by Rusling and coworkers [111,113,114]. Intermolecular and intramolecular (cyclization) carbon-carbon bond formation mediated by electrochemically generated $Co_{(I)}L$ from vitamin B_{12} was realized as shown in Scheme 11. In a first electrochemical step (reaction I) the cathodic reduction of vitamin B_{12} gives $Co_{(I)}L$, which reacts with alkyliodides R-I or with 2-bromoalkyl-2-cyclohexen-1-one (reaction II). This oxidative addition leads to the key intermediate $Co_{(III)}L$, which can be cleaved either photochemically (reaction III) or electrochemically (at -1.45 or -1.5 V vs. SCE) (reaction IV) giving respectively a radical $R^{\bullet}$ or a carbanion R^{-}. In the last step (reaction V) alkycyclohexanones or bicyclic compounds are obtained by inter- or intramolecular addition of $R^{\bullet}$ or R^{-} to the activated carbon-carbon double bond (cyclohexenone).

(*f*) *Photochemistry.* Using photolytic cleavage, the conjugated addition of primary alkyl iodides (*n*-butyl, *n*-octyl, *n*-dodecyl) to 2-cyclohexenone gives 3-alkyl cyclohexanones (Scheme 11) as the main product with good yields (68–81%) both in homogeneous DMF solution and in cationic (CTAB) and anionic (SDS) bicontinuous microemulsions [111]. In a similar way, the intramolecular cyclization of 2-(4-bromobutyl)-2-cyclohexenone to 1-decalone (Scheme 11, $n = 4$) occurs with high yields (75–90%) in the different homogeneous and microheterogeneous systems [111]. Photolytic and electrochemical alkyl-Co cleavages give comparable results for the intramolecular cyclization. However, smaller alkylcyclohexanone yields are observed when an electrochemical cleavage is used as a consequence of the competitive fast protonation of R^{-}. The yield of decalone may not be affected by this last process because the intramolecular cyclization is much faster. These results show that conductive microemulsions, less toxic and often less expensive than organic solvents, are useful systems for carbon-carbon bond formation.

(*g*) *Stereoselectivity.* The most striking effect of microemulsions is the strong increase of the stereoselectivity of the cyclization reaction. The relatively low ratio between the *trans* and *cis* isomers of 1-decalone

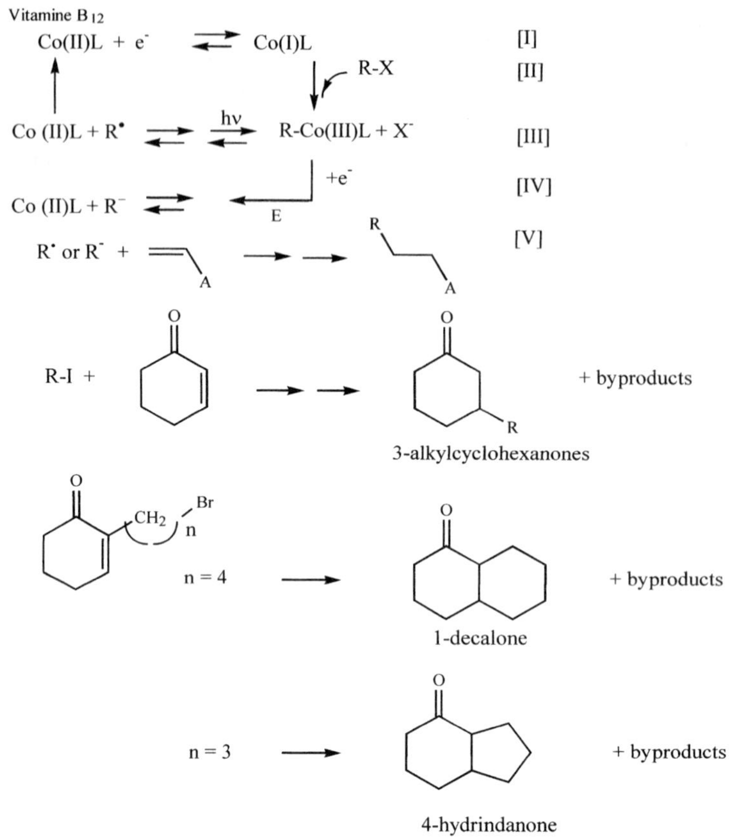

SCHEME 11 Mediated electrochemical carbon-carbon bond formation in microemulsions. Adapted from Ref. 111 and 113: J. Org. Chem. Vol. 61 (1996), pp 5972–5977, J. Gao, J. F. Rusling and D. L. Zhou, Carbon-carbon bond formation by electrochemical catalysis in conductive microemulsions, and Vol. 63 (1998), pp 218–219, J. Gao and J. F. Rusling, Electrochemical catalysis of a 5-endo-trig cyclisation in bicontinuous microemulsions. Copyright 1996 and 1998, with permission from The American Chemical Society.

(from 1 to 2.5) observed in organic solution (DMF) increases to 14 when the reaction is carried out in CTAB microemulsions. A detailed mechanistic study of this stereoselective formation of *trans*-1-decalone in microemulsions was published by Rusling et al. in 1999 [114]. In all 15 cationic and anionic microemulsions studied, 1-decalone is produced in good yield (61–91%) and with high values of the *trans/cis* ratio (77:23 to 95:5). The authors attributed the selective formation of the *trans* diastereoisomer to equilibration of isomers via a keto-enol tautomerization. The enol is catalyzed by the OH^- ions formed by water reduction, which occurs faster in microemulsions than in polar organic solvents as a consequence of the large amount of water in the former solutions [114]. The intramolecular cyclization of 2-(4-bromobutyl)-2-cyclohexenone was performed in microemulsion by Rusling et al. using vitamin B_{12} hexacarboxylate chemisorbed to nanocrystalline TiO_2 cathodes. The *trans/cis* ratios for product 1-decalone were similar to those obtained with vitamin B_{12} and carbon cathodes, but the turnover was better [115].

The 5-*endo*-trig cyclization of 2-(3-bromopropyl)-2-cyclohexenone (Scheme 11, $n = 3$) is disfavored and the photolytic cleavage of $Co_{(III)}L$ gives the cyclization product 4-hydrindanone with poor yields (21–24%) in both organic (DMF) solution and cationic (CTAB) microemulsions. The major produce is 2-allyl-2-cyclohexenone, whose formation is faster than the cyclization. Bad yields (7–19%) are also observed in organic and hydroorganic solutions when the cleavage is done electrochemically [113]. The carbanion R^- intermediary formed by the $Co_{(III)}L$ cleavage is quickly protonated, giving 2-propyl-2-cyclohexenone as the main product. Apparently, this protonation is strongly reduced in anionic and cationic microemulsions leading to 4-hydrindanone with good yields (62–70%). The authors suggest that the cyclization occurs at a site with low concentration of proton donor, i.e., in the oil phase or in the interfacial surfactant layer [113].

6. Summary

As evidenced by the preceding examples, the use of conductive bicontinuous microemulsions is a powerful method by which the control of electrochemical reactions can be realized. Thus the selectivity and even the diastereoselectivity of processes can be changed and oriented to the formation of the desired products. From this point of view, the work done by Rusling and coworkers represents a major contribution in this field.

In some cases, product extraction from micellar and microemulsion systems can be difficult, because of the great amount of surfactant that is often used. An original solution is given by Bunce et al. in two papers [116,117]. They report the electroreduction of hexachlorobenzene and DDT in an emulsion formed with 0.05 M sodium sulfate aqueous electrolyte containing 1% (v/v) heptane and 0.1% (v/v) Triton-SP175. Triton-SP175 is a nonionic surfactant with a hydrophobic tail attached to a polyethoxylate hydrophilic chain. The emulsion was indefinitively stable at neutral pH, but it was easily broken by a brief treatment with dilute acid. Below pH 3 the hydrophobic part is cleaved from the polyethoxylate chain and Triton-SP175 loses its surfactant properties. Thus the product extraction can be performed without difficulty [116,117].

In micelles, microemulsions, and related systems using surfactant, the solvent affects the specific interactions. Variation of the local concentrations of reactants and products can occur both in the bulk pseudophases and in the aggregates adsorbed on the electrodes. Therefore, electrode modification is an important parameter. This parameter can also play an important role in emulsions, as shown in the next section.

B. Effects Resulting from Electrode Modification

One of the most important phenomena in electrode modification is the formation of hydrophobic layers, either by wetting or by the dispersed organic phase (emulsions) or by adsorption of surfactant (micelles, microemulsions), with two main consequences:

The range of usable potential can become larger than that observed in a continuous conducting aqueous phase.

The kinetics of bimolecular reactions can be controlled by modification of reactant concentrations in the adsorbed layer.

1. Increase of the Potential Range

(a) Macroheterogeneous Systems (Emulsions). As pointed out in the first part of this chapter, in emulsions the electrode wetting by the organic phase can be important. This wetting can form a thin hydrophobic film on the electrode. This occurs during anodic nucleophilic substitutions in $H_2O—CH_2Cl_2$ emulsions under phase transfer catalysis conditions. A thin CH_2Cl_2 film is formed on the anode [118,119]. Using thin-film contactor cells, Fleischmann and coworkers have studied the optimal synthesis conditions and developed a simple theoretical model [120,121]. The film formation depends on the anode material, the nature of the phase transfer catalyst, and the electrolytes [118,122,123]. The aqueous phase oxidation cannot be entirely suppressed. Nevertheless, the presence of a CH_2Cl_2 film allows the oxidation of organic compounds in a large potential range (up to 1.8 V vs. SCE) [122–124]. Since the first study published by Eberson and Helgee in 1974 [125], two-phase electrosyntheses with phase transfer catalysts have been widely developed for cyanation [118,122,125–128], acyloxylation [120,121, 123,129–131], and chlorination [119,124] of several molecules.

A decrease of the aqueous electrolyte oxidation or reduction due to electrode wetting by a nonconducting organic phase is often encountered. This wetting phenomenon can be enhanced using hydrophobic electrodes prepared by composite plating nickel with hydrophobic oligomer particles. On this type of electrode the wetting by the organic droplets strongly reduces the water oxidation or reduction, and the electrochemical conversion of organic compounds emulsified in aqueous solutions has been achieved with medium to good yields [132].

(b) Microheterogeneous Systems (Micelles, Microemulsions). The adsorption of surfactants does not always enlarge the accessible potential range, and this range can be reduced in some cases. Thus, the adsorption of cationic surfactants can increase the local concentration (on the anode) of readily oxidizable counterions (e.g., CN^-, NO_2^-) whose enhanced oxidation reduces the anodic limit [94,133]. For instance, the anodic nitration of *N,N*-dimethylaniline occurs with good yields (60–85%) in cationic micellar solutions formed with cetyltrimethylammonium salts [27a,133]. But the medium current efficiencies (23–47%) observed with the counterion SO_4H^-, which is not oxidized at the working potential, drop to very low values (5–10%) with the easily oxidizable NO_2^- or Br^- counterions [133]. However, Franklin and coworkers have devel-

oped a series of applications of electrodes coated with hydrophobic Hyamine 2389 films (Hyamine 2389 is a cationic surfactant, mainly methyldodecylbenzyltrimethylammonium chloride) [134–144]. This film shifts the oxidation potential of water (alkaline solution) from 0.7 V to about 1.8 V (vs. SCE). It allows direct or indirect oxidation (involving the oxidized Hyamine as mediator) of numerous organic compounds in emulsions (CH_3CN–aqueous NaOH) or in Hyamine micelles [137,138,140]. Franklin and coworkers have also developed a Hyamine 2389–polystyrene filmed Pt anode stable in alkaline and acidic solution. The positive limit of the potential window is shifted up to about 2 V (vs. SCE) and numerous organic substrates give oxidation waves well defined on this modified electrode [142,144]. The reader can find a more detailed account of the electrodes coating in a review by Rusling [18].

2. Influence of Reactant Association with the Surfactant Adsorbed Layer

(*a*) *Emulsion Systems.* The presence of surfactants adsorbed on the electrode can strongly enhance the adsorption of an organic substrate emulsified in an aqueous electrolyte. The electrochemical conversion of the substrate is thus improved. This aspect is illustrated by the following examples.

The electrocatalytic hydrogenation (ECH) of unsaturated compounds (phenanthrene, limonene, and carvone) was performed on Raney Ni cathodes (Scheme 1) in hydro-organic solutions; in aqueous cationic (CTAB), anionic (SDS), and nonionic (Brij 35) micelles; and in emulsions with cationic surfactants (CTAB or DDAB) [27a,38,145,146]. In micellar solutions both current efficiency and the extent of hydrogenation increase compared with the homogeneous hydro-organic solutions. The highest effects are obtained with cationic micelles. *The improvement of the reaction efficiency results mainly from the cathode modification caused by the surfactant adsorption.* Indeed, the highest current efficiency (87%) for the ECH of carvone into four isomeric saturated alcohols was observed with this substrate emulsified in an aqueous electrolyte in the presence of very low surfactant concentrations (CTAB at 8.2×10^{-5} M) [38]. Moreover, the most important increase of current efficiency and extent of hydrogenation is obtained in emulsions in the presence of low amounts of didodeclydimethylammonium bromide (DDAB), a *cationic non-micelle-forming surfactant* [146].

This aspect is well demonstrated by comparing the ECH of limonene in different homogeneous and micro- or macroheterogeneous media (Table 1) [38,146]. The ECH of the sparsely water-soluble limonene emulsified in an aqueous buffer does not occur at all (entry 1, Table 1), and the hydrogen evolution reaction (HER) (reactions 5 and 6, Scheme 1) takes place alone. As already mentioned, this was attributed to the absence of substrate adsorption on the cathode. When limonene is entirely solubilized in a homogeneous hydro-organic solution, its ECH gives specifically *p*-menthene with good yields (62–66%) but with poor current efficiencies (18–19%) after the consumption of an excess of

TABLE 1 ECH (30°C, 5–7 F/mol) of Limonene in Homogeneous and Heterogeneous Solutions

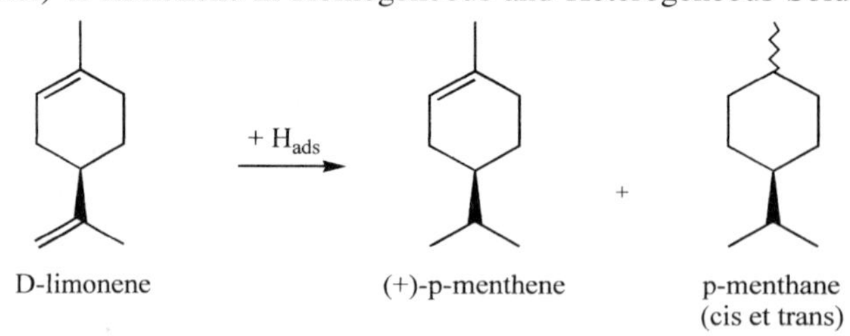

Entry	Solution	pH	Yield (%) *p*-Menthene	Yield (%) *p*-Menthane	Current efficiency (%)
1	Emulsified	2 (buffer)	0	0	0
2	Homogeneous (EtOH/H_2O)	4 (14)	62 (66)	<1	18 (19)
3	Micellar: CTAB 5.5×10^{-2} mol dm^{-3}	2 (12)	9 (44)	57 (25)	30 (27)
4	Emulsified + DDAB 2.7×10^{-4} mol dm^{-3}	10	9	75	70

Source: Data reprinted from Ref. 38 and 146: Thomalla et al. Can. J. Chem. Vol. 73 (1995), pp 804–815 (Table 1) and Vol. 75 (1997), pp 1529–1535 (Table 3).

electricity (i.e., 7 F/mol), in either acidic or alkaline (values in parentheses) media (entry 2, Table 1). Under the same conditions, *p*-menthene is not hydrogenated and is entirely recovered at the end of the electrolysis. In cationic (CTAB) micelles at pH, *p*-menthane becomes the major product (57% yield) and the current efficiency increases to 30% (entry 3). The improvement of the ECH efficiency resulting from the presence of CTAB is less pronounced in alkaline solution (values in brackets, entry 3) and *p*-menthene remains the major product. However, at pH 10 the addition of increasing amounts of DDAB leads to increases of the *p*-menthane yield and the current efficiency up to 75% and 70%, respectively (entry 4), where *p*-menthene becomes a minor product [146]. These maximum values are obtained for a low DDAB concentration (2.7×10^{-4} M). A study of adsorption phenomena shows strong adsorption of the organic compounds (substrate and product) in the presence of DDAB [146]. These results suggest that the improvement of the extent and the efficiency of the ECH of unsaturated substrates results essentially from the association of the substrates with the cationic surfactants adsorbed on the cathode [38,146]. Similar observations were made for the ECH of alkylphenols [147].

The ECH of 2-*t*-butylphenol emulsified in aqueous buffers (pH 2 and 9) is performed at 65°C on Raney nickel cathodes and leads to 2-*t*-butylcyclohexanone and 2-*t*-butylcyclohexanol [147a]. In the absence of surfactant, the alcohol is obtained with poor yields (<1 and 2%) (Table 2) and the ketone is the major product. In agreement with literature results (catalytic hydrogenation of phenol in water) [148] the yield of the ketone is better in alkaline buffer (37%) than in an acidic one (7%) (Table 2). In both media, after the consumption of the theoretical amount of electricity (6 F/mol), the reaction takes place with poor current efficiencies (5 and 27% at pH 2 and 9, respectively), and the HER is the main process. Addition of increasing amounts of DDAB enhances the ECH reaction at the expense of the HER. Moderate effects are observed at pH 2: the yield of 2-*t*-butylcyclohexanone increases from 7% to about 35%, the alcohol remains a minor product (yields 3–5%), and the current efficiency grows to a maximum value close to 30% (Table 2). On the contrary, a dramatic improvement of the ECH process occurs at pH 9 with low DDAB concentrations (within 5.4×10^{-5} and 2.2×10^{-4} M): the current efficiency and the overall product yield increase, respectively, from 27% to about 60% and from 39% to 67–69% (Table 2). Moreover, in the presence of 2.2×10^{-4} mol dm^{-3} DDAB, the alcohol/ketone ratio is 42 times as large as that observed without surfactant [147a]. This important increase of the ECH efficiency could result from a strong adsorption of phenate anions because of their association with the adsorbed surfactant layer through hydrophobic *and* electrostatic interactions. This hypothesis is supported by the ECH of 2,6-dimethylphenol.

TABLE 2 ECH (65°C, 6 F/mol) of 2-*t*-Butylphenol Emulsified in Aqueous Buffers

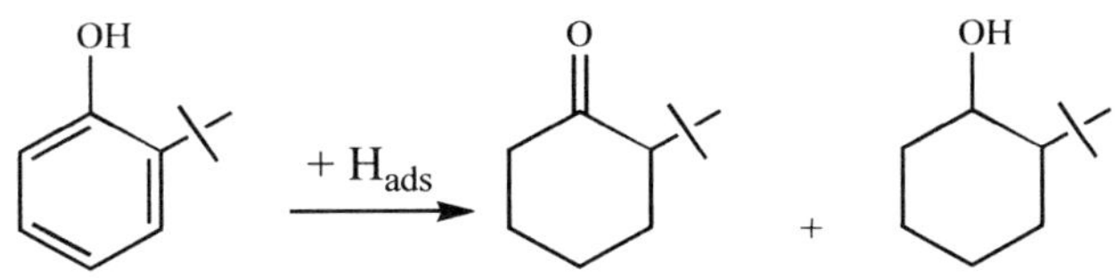

pH	[DDAB] (mol dm^{-3})	Recovered phenol (%)	Yield (%) 2-*t*-Butylcyclohexanone	Yield (%) 2-*t*-butylcyclohexanol	Current efficiency (%)
2	0	82	7	<1	5
	5.4×10^{-5}	69	20	3	16
	1.1×10^{-4}	58	34	5	28
	2.2×10^{-4}	63	20	3	17
9	0	48	37	2	27
	5.4×10^{-5}	20	31	37	57
	1.1×10^{-4}	23	24	43	59
	2.2×10^{-4}	18	22	47	62
	4.3×10^{-4}	34	18	37	49

Source: Ref. 147a.

At 65°C, 2,6-dimethylphenol (2×10^{-2} M) is entirely soluble in the aqueous buffers (pH 2 and 9), and thence its ECH takes place in a homogeneous phase [147b]. In all experiments the hydrogenation gives only 2,6-dimethylcyclohexanol. Small traces (0–2.5%) of this alcohol are formed in the acidic buffer. Even with increasing amounts of DDAB, the starting substrate is entirely recovered. On the contrary, at pH 9, the presence of DDAB has a strong effect on the alcohol yield, which grows from 8 to 70% when the surfactant concentration increases from 0 to 2.2×10^{-4} M. These results can be correlated with the adsorption of 2,6-dimethylphenol. The substrate adsorption at the interfaces is deduced from the decrease of the UV absorption of the dimethylphenol solubilized in the aqueous buffers. At pH 2, whatever the DDAB concentration is, dimethylphenol remains in solution, and its adsorption at the interfaces is very weak (close to zero). At pH 9 the weak adsorption (5–7%) observed without DDAB increases with the surfactant concentration and reaches values close to 40% for the highest concentrations (2×10^{-3}–4.5×10^{-3} M). The results as a whole suggest that the lack of hydrogenation at pH 2 is mainly due to weak substrate adsorption. The latter increases at pH 9 with the DDAB concentration and leads to a strong enhancement of the efficiency of the ECH reaction [147b].

This last example clearly illustrates the importance of surfactant adsorption in electrochemical processes in homogeneous aqueous electrolyte. This can also occur in homogeneous hydroorganic solutions. Thus, the adsorption of cationic and nonionic surfactants can modify the mechanisms of the electrochemical reduction of 4,4′-dinitrodibenzyle solubilized in ethanol-water (50:50) solutions [149].

(*b*) *Micelles, Microemulsions, and Related Systems.* As pointed out in the first part of this chapter, the interactions between the reactants and the surfactant aggregates adsorbed on the electrodes can lead to effects similar to those occurring in the bulk microheterogeneous media. Therefore, the selectivity and, in some cases, the diastereoselectivity of electrochemical processes can be modified. For instance, the repartition between 1-phenylethanol and 2,3-diphenyl-2,3-butanediol formed by the acetophenone reduction on a Pb cathode (Scheme 2) was studied by Ito et al. in different aqueous micellar solutions [150]. Without surfactant, in aqueous Na_2SO_4 electrolyte, the carbinol/pinacol molar ratio is close to one. This ratio increases only slightly (3.1–3.4) in the presence of nonionic and anionic surfactants. A much higher value (17) is obtained in cationic (CTAB) micellar solution. The study of the influence of cathode potential and of CTAB concentration on product distribution suggests that *the selectivity depends on the surfactant concentration on the cathode*. A strong enhancement of carbinol formation is also observed for the benzaldehyde electroreduction. It can be examined by stabilization of the anionic intermediates in the cationic surfactant adsorbed layer (and/or micelles and formation of ion pairs with the ammonium headgroups of the surfactant), which favors the second electron transfer (Scheme 2) [150]. However, Mousty and Mousset have shown that this stabilization is not the sole key parameter. These authors performed the acetophenone reduction on a mercury cathode in alkaline solution [27g,151]. In aqueous emulsions without surfactant, electropinacolization is the main process, and the carbinol represents only 30% of the two products. This percentage increases slightly (37–53%) in the presence of cetyltrimethylammonium bromide, chloride, or sulfate (CTAX, 10^{-2} M). On the contrary, it drops to low values (12–16%) in the presence of cetyltripropylammonium salts (CTPAX, 10^{-3} M). The two cationic surfactants stabilize anionic intermediates, and the observed difference has been interpreted in terms of hydrophobicity of the surfactant adsorbed layers. The CTAX adsorbed layer is not hydrophobic enough to lower the protonation reaction leading to carbinol formation. On the contrary, with cetyltripropylammonium salts a more compact hydrophobic adsorbed layer is formed, which reduces the protonation reaction. Consequently, the dimerization process is favored at the expense of the carbinol [151]. Moreover, the electroreduction of acetophenone on Hg cathodes in aqueous buffers with and without surfactants, previously realized by the Blaise Pascal group, gave different product distributions [27g]. These results show that the repartition between pinacol and carbinol is controlled by several parameters: the pH, the concentration and nature of the surfactants, the ratio between the substrate and the surfactant concentration, and adsorption phenomena [27g]. Therefore, an extensive interpretation must take into account concentration effects and stabilization phenomena.

In many cases, a clear distinction between the effects resulting from microstructures in the bulk solution and those related to the organized adsorbed layers of surfactant is not obvious. However, most of the "micellar" effects result from these adsorbed aggregates. The rate enhancement of bimolecular reactions in adsorbed surfactant layers (resulting from high reactant concentrations in a restricted reaction volume) can be much higher than that occurring in diffusing micelles. In this last case, the statistical distribution of reactants among the micelles lowers the number of reactive aggregates containing the two reactants. Therefore, the control of bimolecular reactions by coassociation with adsorbed surfactant layers is an important aspect of electrochemical synthesis in microheterogeneous systems. As a consequence, different kinds of modified electrodes with thick surfactant coatings containing organic or organometallic mediators have been developed and used to perform electrochemical catalytic conversions (mainly catalytic reduction of organohalides) in microheterogeneous media [27i,152–154]. Actually, this field is increasing in development, and numerous publications deal with the study of the catalytic properties of electrodes modified by lamellar liquid crystal surfactant films incorporating redox proteins [153–164]. Rusling and coworkers have developed catalytic electrodes using polyion adsorption [165] or polyester sulfonic acid ("polymeric surfactant") films ([166] and references cited therein). The development and use of such electrodes represent a borderline subject for this chapter, and for more details the reader is referred to other publications and reviews [17,18,27i,98,152, 159,164].

3. Passivation

Passivation phenomena are another important aspect of electrode modification. Electrode passivation strongly reduces the efficiency of the electrochemical conversion. It results from either electroinactive polymeric film formation, adsorption of products, or adsorption of reactants (autoinhibition). Electrode passivation can be diminished or suppressed by the use of surfactants (micelles, microemulsions) or disperse immiscible organic phases (emulsions).

An example of the efficiency of emulsified organic phases in suppressing passivation resulting from product adsorption is given by the electrochemical conversion of 4-nitrosodiphenylamine (NDPA) to *p*-aminodiphenylamine (ADPA) (Scheme 10) studied by Fioshin et al. [97,167,168]. In hydro-organic alkaline solutions the ADPA adsorption inhibits the NDPA electroreduction and prevents the use of graphite or mercury cathodes. Thus low current efficiencies (<35%, depending on the nature of the cathode) are obtained (Fig. 7) [167,168]. In the presence of benzene, *p*-xylene, or toluene emulsified in an aqueous methanolic phase, ADPA extraction by the organic disperse phase leads to high increases in current efficiencies (up to about 70%) (Fig. 7). In addition, graphite or mercury cathodes can be used with medium current efficiencies, 32 and 50%, respectively [97].

In many cases the reduction of autoinhibition by an emulsion is not so efficient. For instance, in 1999 Novikov and coworkers studied the feasibility of the electrochemical synthesis of *t*-butyl-2-benzothiazolesulfenamide (TBBS), an efficient rubber accelerator [169]. The anodic oxidation of 2-mercaptobenzothiazole (MBT) in alkaline media gives di-2-benzothiazole disulfide (DBDS), which reacts with *t*-butylamine (TBA) giving TBBS. With increasing MBT and current density, the electrochemical conversion is strongly inhibited by the adsorption of DBDS, which is poorly soluble in the reaction medium. Even in the presence of an emulsified organic solvent the reaction goes very slowly. Therefore, a more efficient alternative pathway was studied. The electrolysis was conducted in chloride solutions and the *t*-butylmonochloramine (obtained in the reaction of TBA with electrochemically formed ClO^-) reacts with MBT giving TBBS [169]. Such a limitation is often encountered in the case of indirect electrochemical reactions. Indeed, as indicated in the first part of this chapter, during indirect electrochemical conversion, the mediator regeneration can be more or less inhibited by substrate or product adsorption. When the solubilization of substrate and product by an organic dispersed phase does not reduce the inhibition enough, the ex-cell method may be an answer to this problem. The indirect *p*-methoxytoluene oxidation by the Ce(IV)/Ce(III) redox couple reported earlier (Scheme 3) illustrates this aspect. In this particular example, the inhibition results from adsorption and polymerization reactions of the aromatic substrate [46].

Passivation by polymer formation often occurs with aromatic compounds. Thus the anode fouling that hampers the electrochemical oxidation of chlorinated phenols (depollution aim) results from the deposition of oligomers on the anode as showed by Bunce et al. [170]. The use of micellar or microemulsion systems can reduce the unwanted competitive polymerization as shown by the anodic nitration of *N*,*N*-dimethylaniline (NDMA). This reaction occurs in acetonitrile with low yields (7%) [171] as a consequence of competitive oligomer formation. Although anodic NDMA polymerization takes place in aqueous solution [172], in cationic aqueous micellar solutions, the polymerization is reduced and high yields (depending on NDMA concentration) are obtained (55–85%) [27a,133]. This effect corresponds partially to a solvent effect (see the following).

Solubilization of starting substrate or of products in micellar or microemulsified pseudophases can also prevent electrode passivation. For example, the indirect reduction, involving the Ti^{4+}/Ti^{3+} (TiO_2/TiOOH) redox

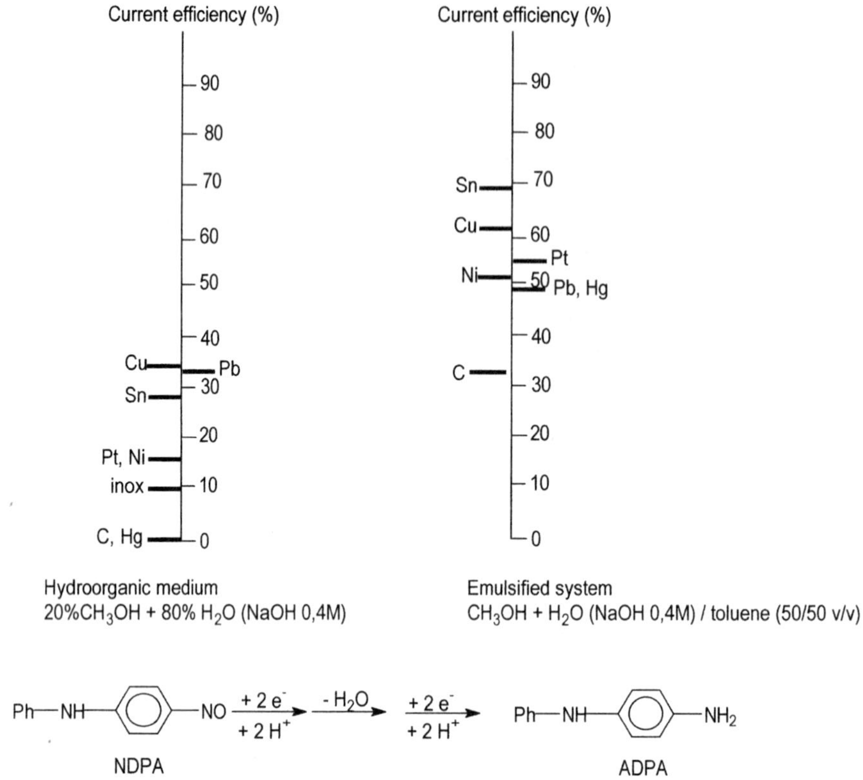

FIG. 7 Influence of the medium (homogeneous hydro-organic and biphasic emulsion) and the cathode nature on the current efficiency of the electrochemical NDPA conversion. From Ref. 97, 168. Data reprinted from Soviet Electrochem., Vol. 13 (1977), p. 498, G. M. Sarsenbaeva, O. N. Novikova, I. A. Avrutskaya and M. Ya. Fioshin, Electrosynthesis of 4-aminodiphenylamine, and Vol. 14 (1978), p 406, I. A. Avrutskaya, G. M. Sarsenbaeva, P. D. Tsar'kov and M. Ya. Fioshin, 4-aminodiphenylamine electrosynthesis with organic solvent extraction. Copyright 1977, 1978, with permission from Kluwer Academic/Plenum Publishers.

couple, of aromatic nitrocompounds was performed on Ti/TiO_2 electrodes by Anantharaman and colleagues [173]. Most of the 18 nitro compounds studied exhibited good solubility in 1 M H_2SO_4 aqueous solution, with the exception of 2-chloronitrobenzene. Because of its low solubility, this last compound becomes strongly adsorbed on the electrode and therefore inhibits the redox reaction. This inhibition is suppressed by a small amount (0.01%) of CTAB, which improves the solubility of the nitrocompound [173]. In some cases the presence of surfactants can induce an unwanted decrease of the reaction efficiency. Thus, the electroreduction at a lead cathode of 4-nitrosoantipyrine (NAP) giving 4-aminoantipyrine was realized by Nankov et al. in aqueous acidic solutions [174]. Because of the poor solubility of NAP in both aqueous and organic media, suspensions of NAP were used. The electrochemical conversion is limited by the rate of dissolution of the solid phase and inhibited by an excess of NAP. In order to increase the product and current yields, anionic and nonionic surfactants were added. The presence of surfactant led to effects opposite to those expected. In all cases, both chemical yields and current efficiencies were lower than without additives. The decrease of current efficiencies was attributed to hydrogen evolution promoted by the surfactants [174].

Besides the solubilization of substrate and/or products, surfactants are often strongly adsorbed on electrodes, and the suppression of inhibition phenomena could result mainly from their preferential adsorption. Mousset and coworkers have studied the electrochemical reduction at a mercury cathode of 4-acetyl-1-phenoxyalkyl ammonium or sulfate salts of various chain lengths [175–177]:

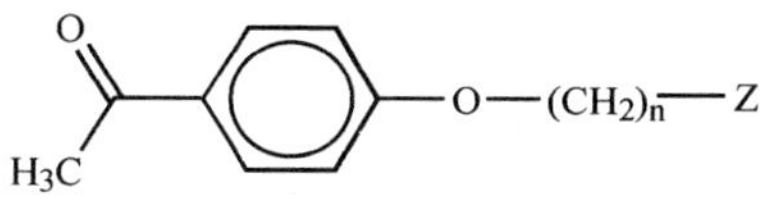

$Z = N^+(CH_3)_3\ Br^-$ or $OSO_3^-\ Na^+$ and n = 3 to 10

With all compounds, autoinhibition by reduction products was observed in aqueous acidic or alkaline media. In the case of acetylphenoxyalkylammonium substrates, the inhibition was suppressed by addition of cationic (CTAB) or nonionic (Brij 35) surfactants. This results only from surfactant adsorption and occurs with all substrates including those bearing the shorter alkyl chains ($n = 3, 4$) which are not associated with micelles [175,176].

C. Solvent Effects and Specific Interactions

The importance of solvent effects is well known in organic synthesis [178]. Changing an organic aprotic solvent to an aqueous one can modify the rate constants of competitive processes. The nature of the solvent can also influence the kinetics of electron transfer and electrode reactions [179,180]. Thus solvent effects can be important for both chemical and electrochemical steps of electrochemical conversions. These effects must be operative both in emulsions and in micellar (or microemulsified) systems. In emulsions the electrochemical conversion can be performed either in the aqueous phase or in the organic phase, thus changing the reaction rates. In micelles the electrochemical conversion often takes place in an aqueous environment for the following reasons:

Organic compounds are often localized in the outer layer of ionic micelles (the palisade layer), a region widely open to the aqueous pseudophase [14,19a].

The charged intermediate species (ions, ion radicals) are often more soluble in water than their parent molecules. In the absence of specific interactions with the ionic headgroups of surfactants, these intermediates migrate in the aqueous pseudophase, where subsequent reactions take place.

As mentioned earlier, the anodic nitration of *N,N*-dimethylaniline (NDMA) gives nitrated compounds with yields higher (55–85%) in cationic micelles than in homogeneous acetonitrile solution (7%). The suppression of competitive polymerization (responsible for the low yields) by cationic micelles is mainly due to an aqueous environment. Indeed, in acetonitrile-water solutions (30–80% H_2O in CH_3CN), the yields (60–70%) are close to those observed in cationic micelles [27a,133].

The anodic cyanation of 1,2-dimethoxybenzene (1,2-DMB) has been studied in different homogeneous and heterogeneous media by several authors [92–94,122,126]. The partitioning between the two products, 2-methoxybenzonitrile (2-MBN) and 3,4-dimethoxybenzonitrile (3,4-DMBN), was strongly dependent on the nature of the medium. In homogeneous organic solutions (CH_3CN, CH_2Cl_2) 2-MBN was mainly obtained (78–100%) and 3,4-DMBN was only a minor product (0–22%) [92,122]. In emulsions, micelles, and microemulsions, higher amounts (35–50%) of 3,4-DMBN were observed [93,94,122,126]. This increase of 3,4-DMBN formation can result from the aqueous environment, especially with nonionic and anionic micelles. In anionic micelles the reaction between CN^- and the cation radical $DMB^{+\bullet}$ anodically formed occurs in the aqueous pseudophase. Moreover, in cationic micellar systems, cyanation involves radical coupling between $CN^\bullet$ and $DMB^{+\bullet}$ (both electrochemically formed) as opposed to the "classical" CN^- addition on $DMB^{+\bullet}$ that occurs in the other homogeneous or heterogeneous systems [94]. This change of mechanism results specifically from an increase of CN^- oxidation in the cationic surfactant adsorbed layer [94]. Specific effects play a primary role both in micelles in the bulk solution and in adsorbed aggregates (or layers) on the electrodes.

As pointed out in the previous section with the reduction of acetophenone, specific electrostatic interactions can modify the selectivity [27g,151]. They can also change the reaction diastereoselectivity. At pH 10.6, the *dl/meso* pinacol ratio grows from 1.1 in emulsions without surfactant to 3 (or 2.6) in the presence of $(CTA)_2SO_4$ (or CTAPX) adsorbed layers. Mousty and Mousset have interpreted this striking enhancement of the *dl/meso* ratio by the formation of the less hindered ammonium-cetyl ion pair, which favors the formation of the *dl* form [27g,151].

Similar general trends, with some additional effects due to the micellization of substrates bearing long alkyl chains, are observed for the electroreduction at the Hg cathode of 4-acetyl-1-phenoxyalkyl ammonium or sulfate salts [27g,177,181] performed by Mousset et al. in aqueous buffers at pH values from 2.5 to 10.8 (Scheme 2).

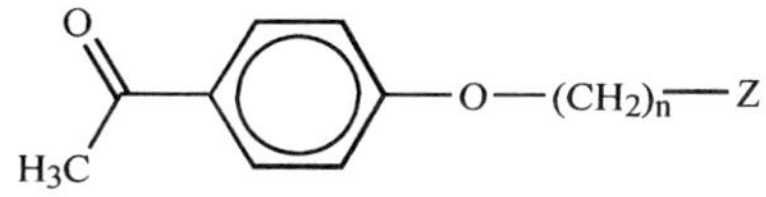

$Z = N^+(CH_3)_3\ Br^-$ substrate **1** (n = 3) **2** (n = 10)

or $OSO_3^-\ Na^+$ substrate **3** (n = 3) **4** (n = 8) **5** (n = 10)

Thus, the interaction between the cetyl anion radical (or anion) and the ammonium headgroups of a cationic surfactant (CTAB) or of the substrate itself (reductions of **1** and **2** without surfactant) leading to intimate ion-pair intermediate formation seems to be responsible for the following salient results:

In alkaline media, the major product of the reduction of **1** performed at the potential of the second electron transfer (see Scheme 2) is the pinacol (75%) in the absence of surfactant and the carbinol (74%) in the presence of CTAB. As suggested by the authors, this drastic change could result from the stabilization of the cetyl anion by the cationic headgroups [181].

In the same way, the formation of this less hindered ion-pair intermediate was invoked to explain the exclusive formation of the *dl* pinacol isomer by reduction of **1** at pH 10.3 [181].

In neutral and basic media, the reduction of **2** gives mainly the carbinol (74–77%) [181] as a consequence of the interaction of the cetyl anion radical and the ammonium headgroup of **2**. This interaction requires bending of the alkyl chain [177].

In the absence of such specific interactions, the reduction at pH 10.5 of **3**, **4**, and **5** leads to the preferential formation of pinacol (40–61%) [177]. At pH 8, the alcohol is the major product (57 and 62%) of **3** and **4** reduction, whereas pinacol is preferentially formed (58%) with compound **5** as a consequence of the micellization of this substrate. The formation of micelles also leads to preferential orientation of the alkyl chains inducing some modifications of the *dl/meso* pinacol ratio in the cases of substrates **2** and **5** [177,181].

Interactions between surfactants and substrates or intermediates very often modify the apparent redox potential. The potential shift can influence the selectivity of the process. For instance, in acidic solutions, the reduction of **1** performed at the potential of the first electrochemical step (Scheme 2) gives 27% carbinol and 73% pinacol. In the presence of CTAB (10^{-2} M), the pinacol is exclusively obtained. This results from the better separation between the two electrochemical steps due to potential shifts [27g]. On the other hand, at pH 8 the interaction of the CTAB ammonium headgroups with the electrochemically formed intermediates favors the second electron transfer, and the amount of carbinol produced [27g] in the presence of CTAB is higher (93%) than in its absence (80%).

Effects due to the orientation of the substrate in cationic surfactant layers or micelles were reported by Franklin and coworkers for the electrochemical oxidation of organic compounds catalyzed by high-valent oxides [27j]. These electrochemical conversions occurring in aqueous suspensions represent a particular application of surfactants in electrosynthesis ([182] and references cited therein). Likewise, the presence of dodecylbenzenesulfonate (DBS) allows the indirect anodic oxidation (by means of Mn^{3+}/Mn^{2+}redox mediator) of anthracene to anthraquinone in a slurry electrolyte (5.5 M H_2SO_4) system. The main factors affecting the current efficiency were studied by Chou and Chen [183]. In particular, without DBS the yield of anthraquinone was zero. Increasing the DBS concentration from 3.56×10^{-5} to 3.28×10^{-4} increased the yield from 0.93 to 17.5%. Thus DBS allows the reaction between the lipophilic anthracene particles and the hydrophilic Mn^{3+} ions [183].

At last, the electrosynthesis of conducting polymers has been performed in aqueous anionic and nonionic micellar solutions. The properties (stability, conductivity, etc.) and structure of the polymeric films are more or less modified by the presence of surfactants. The electropolymerization of pyrrole in water with various concentrations of sodium dodecyl sulfate or dodecylbenzene sulfonate provides typical examples of the effect of surfactant on polymeric structure [184]. The films formed with surfactant concentrations below the cmc present a smooth and compact thin film structure. Above the cmc, the adsorbed micelles lead to the formation of perpendicularly oriented and columnar structures [184]. This use of surfactants, leading to modified electrodes, is a borderline field for the subject of this chapter, and for more information the reader is referred to the publication (and references cited therein) by Lacaze and coworkers [185].

V. CONCLUSIONS

Macro- and microheterogeneous systems present some common features (i.e., reactant distribution between phases or pseudophases, formation of hydrophobic layers, solvent effects) that influence the selectivity and the efficiency of electrochemical synthesis. In many cases the adsorption of surfactants on the electrodes plays a primary role both in emulsions and in micellar and microemulsion systems. The development of catalytic electrodes with adsorbed surfactant aggregates or layers and the use of ordered systems (in particular bicontinuous microemulsions) show great promise for electro-organic synthesis in the future.

REFERENCES

1. E. J. Baum, *Chemical Property Estimation. Theory and Application*, Lewis, New York, 1998, pp. 73–121.

2. H. Lund and M. M. Baizer, eds., *Organic Electrochemistry*, Marcel Dekker, New York, 1991, (a) H. Lund, pp. 253–327, (b) H. Lund, pp. 1199–1215, (c) J. Simonet, pp. 1217–1263, (d) M. M. Baizer, pp. 433–436.
3. H. Feess and H. Wendt, in *Technique of Electroorganic Synthesis*, Part III (N. L. Weinberg and B. V. Tilak, eds.), John Wiley & sons, New York, 1982, pp. 81–177.
4. B. P. Binks, ed., *Modern Aspects of Emulsion Science*, The Royal Society of Chemistry, Cambridge, 1998.
5. M. E. Halpern, ed., *Phase-Transfer Catalysis, Mechanisms and Syntheses*, American Chemical Society, Washington, DC, 1997.
6. A. Leo, C. Hansch, and D. Elkins, Chem. Rev. *71*: 525–616 (1971).
7. D. J. W. Grant and T. Higuchi, *Solubility Behavior of Organic Compounds*, John Wiley & Sons, New York, 1990, pp. 58–61, 113–115, 307–398.
8. G. D. Rees and B. H. Robinson, Adv. Mater. 5:608–619 (1993).
9. W. M. Gelbart, A. Ben-Shaul, and D. Roux, eds., *Micelles, Membranes, Microemulsions and Monolayers*, Springer-Verlag, New York, 1994.
10. R. G. Laughlin, *The Aqueous Phase Behavior of Surfactants*, Academic Press, London, 1994.
11. B. Kronberg, M. Costa, and R. Silveston, in *New Horizons* (R. T. Coffey, ed.), An AOCS/CSMA Detergent Industry Conference. AOCS Press, Champaign, IL, 1996, pp. 63–69.
12. D. O. Shah, ed., *Micelles, Microemulsions and Monolayers*, Marcel Dekker, New York, 1998.
13. R. D. Neuman and T. H. Ibrahim, Langmuir 15:10–12 (1999) (and references cited).
14. S. D. Christian and J. F. Scamehorn, eds., *Solubilization in Surfactant Aggregates*, Marcel Dekker, New York, 1995.
15. C. A. Bunton, J. Mol. Liquids *72*:231–249 (1997).
16. R. A. Mackay, Colloids Surf. A Physicochem. Eng. Aspects *82*:1–28 (1994).
17. J. F. Rusling, in *Modern Aspects of Electrochemistry* No. 26 (B.E. Conway, J. O'M. Bockris, and R. E. White, eds.), Plenum, New York, 1994, pp. 49–104.
18. J. F. Rusling, in *Electroanalytical Chemistry*, Vol. 18 (A. J. Bard, ed.), Marcel Dekker, New York, 1994, pp. 1–88.
19. J. H. Fendler, *Membrane Mimetic Chemistry*, John Wiley & Sons, New York, 1982, (a) pp. 27–35, 227–229, (b) pp. 293–491.
20. C. A. Bunton and G. Savelli, Adv. Phys. Org. Chem. 22:213–309 (1986).
21. L. S. Romsted, C. A. Bunton, and J. Yao, Curr. Opin. Colloid Interface Sci. *2*:622–628 (1997).
22. M. A. Fox, in *Topics in Current Chemistry*, Vol. 159 (J. Mattay, ed.), Springer-Verlag, Heidelberg, 1991, pp. 67–101.
23. V. Ramamurthy, R. G. Weiss, and G. S. Hammond, in *Advances in Photochemistry*, Vol. 18 (D. H. Volman, G. S. Hammond, and D. C. Necker, eds.), John Wiley & Sons, New York, 1993, pp. 67–234.
24. T. Wolff, Adv. Colloid Interface Sci. *59*:31 (1995).
25. I. Rico-Lattes and A. Lattes, in *Homogeneous Photocatalysis*, (M. Chanon, ed.), John Wiley & Sons, New York, 1997, pp. 309–353.
26. G. L. McIntire, Cri. Rev. Anal. Chem. *21*:257–278 (1990).
27. R. A. Mackay and J. Texter, eds., *Electrochemistry in Colloids and Dispersions*, VCH, New York, 1992, (a) M. Thomalla, pp. 273–289, (b) E. Fisicaro, C. Minero, and E. Pelizetti, pp. 25–32, (c) J. R. Kirchoff, J. D. Skelton, and K. T. Brooks, pp. 43–53, (d) M. Han and E. Kaifer, pp. 105–117, (e) E. Dayalan, S. Qutubuddin, and J. Texter, pp. 119–135, (f) A. Berthod, pp. 137–146, (g) C. Mousty and G. Mousset, pp. 291–301, (h) E. Garcia, L. E. Oppenheimer, and J. Texter, pp. 257–270, (i) J. F. Rusling, N. Hu, H. Zhang, D. J. Howe, C. L. Miaw, and E. C. Couture, pp. 303–318, (j) T. C. Frankin, J. Darlington, R. Nnodimele, and R. C. Duty, pp. 319–329.
28. M. Thomalla, Actualite Chim. *10*:61–65 (1998).
29. R. Dworark and H. Wendt, Ber. Bunnsenges. Phys. Chem. *81*:728–732 (1977).
30. R. Dworak, H. Feess, and H. Wendt, AICHE Symp. Ser. Vol. 75 *185*:38–43 (1979).
31. H. Feess and H. Wendt, J. Chem. Technol. Biotechnol. *30*:297–312 (1980).
32. H. Feess and H. Wendt, Ber. Bunsenges. Phys. Chem. *85*:914–919 (1981).
33. P. Y. Lu and R. C. Alkire, J. Electrochem. Soc. *131*: 1059–1067 (1984).
34. J. M. Fenton and R. C. Alkire, J. Electrochem. Soc. *135*:2200–2209 (1988).
35. B. Wermeckes and F. Beck, Electrochim. Acta *30*: 1491–1500 (1985).
36. F. Beck, B. Wermeckes, and W. Janssen, Tetrahedron *47*:725–736 (1991).
37. J. M. Chapuzet, A. Lasia, and J. Lessard, in *Electrocatalysis* (J. Lipkowski and P. N. Ross, eds.), Wiley-VCH, New York, 1998, pp. 155–196.
38. P. Chambrion, L. Roger, J. Lessard, V. Beraud, J. Mailhot, and M. Thomalla, Can. J. Chem. *73*:804–815 (1995).
39. M. M. Letord-Quemere, F. Coeuret, and J. L. Legrand, Electrochim. Acta *33*:881–890 (1988).
40. A. S. Vaze, S. B. Sawant, and V. G. Pangarkar, J. Appl. Electrochem. *28*:623–626 (1998).
41. C. Belmant, P. Cognet, J. Berlan, G. Lacoste, P. L. Fabre, and J. M. Jud, J. Appl. Electrochem. *28*:185–191 (1998).
42. F. Marken and R. G. Compton, Electrochim. Acta *43*: 2157–2165 (1998) (and references cited).
43. S. Torii, *Electroorganic Synthesis, Methods and Ap-*

plication, Part I: *Oxidations*, Kodansha VCH, Tokyo, 1985, pp. 279–306.
44. E. Steckhan, in *Topics in Current Chemistry*, Vol. 142 (E. Steckhan, ed.), Springer-Verlag, Berlin, 1987, pp. 1–69.
45. D. Kyriacou, *Modern Electroorganic Chemistry*, Springer-Verlag, Berlin, 1994, pp. 166–191.
46. T. Tzedakis and A. Savall, J. Appl. Electrochem. *27*: 589–597 (1997).
47. D. Floner, L. Laglaine, and C. Moinet, Electrochim. Acta *42*:525–529 (1997).
48. T. Jan, D. Floner, and C. Moinet, Electrochim. Acta *42*:2075–2079 (1997).
49. M. Chollet, J. L. Burgot, and C. Moinet, Electrochim. Acta *44*:201–206 (1998).
50. J. J. Jow and T. C. Chou., J. Appl. Electrochem. *18*: 298–303 (1988).
51. D. Pletcher and N. Tomov, J. Appl. Electrochem. *7*: 501–504 (1977).
52. R. E. W. Jansson and N. R. Tomov, J. Appl. Electrochem. *10*:583–586 (1980).
53. J. Lin-Cai and D. Pletcher, J. Electroanal. Chem. *152*: 157–162 (1983).
54. D. Pletcher and E. M. Valdes, Electrochim. Acta *33*: 499–507 (1988).
55. D. Pletcher and E. M. Valdes, Electrochim. Acta *33*: 509–515 (1988).
56. J. S. Do and T. C. Chou, J. Appl. Electrochem. *19*: 922–927 (1989).
57. J. S. Do and T. C. Chou, J. Appl. Electrochem. *20*: 978–985 (1990).
58. J. S. Do and Y. L. Do, Electrochim. Acta *39*:2037–2044 (1994).
59. J. S. Do and Y. L. Do, Electrochim. Acta *39*:2299–2309 (1994).
60. J. S. Do and Y. L. Do, Electrochim. Acta *39*:2311–2319 (1994).
61. J. S. Do and T. C. Chou, J. Appl. Electrochem. *22*: 966–972 (1992).
62. N. Vettorazzi, H. Fernandez, J. J. Silber, and L. Sereno, Electrochim. Acta *35*:1081–1088 (1990).
63. J. Koryta, Electrochim. Acta *33*:189–197 (1988).
64. Z. Samec and Kakiuchi, in *Advance in Electrochemical Science and Engineering*, Vol. 4 (H. Gerischer and C. W. Tobias, eds.), VCH, Weinheim, 1995, pp. 297–361.
65. W. Schmickler, J. Electrochem. Chem. *460*:144–148 (1999).
66. L. Murtomäki, K. Kontturi, and D. J. Schiffrin, J. Electroanal. Chem. *474*:89–93 (1999).
67. R. Alkire and J. Köhler, J. Appl. Electrochem. *18*: 405–409 (1988).
68. P. Krishnan and V. G. Gurjar, Synth. Commun. *22*: 2741–2744 (1992).
69. P. Krishnan and V. G. Gurjar, J. Appl. Electrochem. *23*:268–270 (1993).
70. P. Krishnan and V. G. Gurjar, J. Appl. Electrochem. *25*:792–796 (1995).
71. J. F. Rusling, Colloids Surf. A Physicochem. Eng. Aspects *123–124*:81–88 (1997).
72. J. Georges and S. Desmettre, Electrochim. Acta *29*: 521–525 (1984).
73. S. Manne, Prog. Colloid Polym. Sci. *103*:226–233 (1997).
74. S. Manne, T. E. Schäffer, Q. Huo, P. K. Hansma, D. E. Morse, G. D. Stucky, and I. A. Aksay, Langmuir *13*:6382–6387 (1997).
75. K. Nagashima and F. D. Blum, J. Colloid Interface Sci. *214*:8–15 (1999).
76. J. F. Rusling, Interface (Electrochem. Soc.) *6*:26–31 (1997).
77. P. Joo, Colloids Surf. A Physicochem. Eng. Aspects *141*:337–346 (1998).
78. J. Lipkowski, in *Modern Aspects of Electrochemistry*, No. 23 (B. E. Conway, J. O'M. Bockris, and R. E. White, eds.), Plenum, New York, 1992, pp. 1–99.
79. X. Chen, X. Zeng, Y. Chen, and S. Chen, J. Colloid Interface Sci. *214*:447–449 (1999).
80. Z. Ibrisagic, V. Misovic, I. Tabakovic, and I. Santic, J. Appl. Electrochem. *16*:907–912 (1986).
81. B. V. Lyalin and V. A. Petrosyan, Russ. J. Electrochem. *34*:1098–1102 (1998).
82. P. Cognet, J. Berlan, G. Lacoste, P. L. Fabre, and J. M. Jud, J. Appl. Electrochem. *26*:631–637 (1996).
83. H. J. Schäfer, Top. Curr. Chem. *142*:101–117 (1987) (and references cited).
84. T. C. Wen, S. M. Lin, and J. M. Tsai, J. Appl. Electrochem. *24*:449–454 (1994).
85. H. Wendt and H. Schneider, J. Appl. Electrochem. *16*: 134–146 (1986).
86. G. Kreysa and H. Medin, J. Appl. Electrochem. *16*: 757–767 (1986).
87. H. Wendt, in *Recent Advances in Electroorganic Synthesis* (S. Torii, ed.), Elsevier, Amsterdam, 1987, pp. 453–456.
88. P. Cox and D. Pletcher, J. Appl. Electrochem. *21*:11–13 (1991).
89. M. C. Miras, J. J. Silber, and L. Sereno, Electrochim. Acta *33*:851–858 (1988).
90. E. Laurent, G. Rauniyar, and M. Thomalla, Bull. Soc. Chim. Fr. *I*:78–79 (1984).
91. M. N. Cortona, N. R. Vettorazzi, J. J. Silber, and L. E. Sereno, J. Electroanal. Chem. *470*:157–165 (1999).
92. S. Andreades and E. W. Zahnov, J. Am. Chem. Soc. *91*:4181–4190 (1969).
93. E. Laurent, G. Rauniyar, and M. Thomalla, Nouv. J. Chim. *6*:515–517 (1982).
94. G. Rauniyar and M. Thomalla, Bull. Soc. Chim. Fr. *II*:156–167 (1989).
95. A. Davidovic', I. Tabakovic, D. Davidovic', and L. Duic', J. Electroanal. Chem. *280*:371–379 (1990).
96. P. Caubère, *Le transfert de phase et son utilisation en chimie organique*, Masson, Paris, 1982, pp. 30–39.

97. I. A. Avrutskaya, G.M. Sarsenbaeva, P. D. Tsar'kov, and M. Ya Fioshin, Sov. Electrochem. (translated from Electrokhimiya) *14*:405–408 (1978)
98. J. F. Rusling and D.-L. Zhou, J. Electroanal. Chem. *439*:89–96 (1997).
99. K. L. N. Phani, S. Pitchumani, S. Ravichandran, S. Tamil Selvan, and S. Barathey, J. Chem. Soc. Chem. Commun. 179–181 (1993).
100. D. A. Kaplin and S. Qutubuddin, Synth. Met. *63*:187–194 (1994).
101. S. Tamil Selvan, A. Mani, S. Pitchumani, and K. L. N. Phani, Electroanal. Chem. *384*:183–186 (1995).
102. A. Manna, K. Bandyopadhyay, K. Vijayamohana, P. R. Rajamohana, S. Sainkar, and B. D. Kulkarni, Langmuir *14*:84–90 (1998).
103. H. Sze On Chan, L. Ming Gan, C. Har Chew, L. Ma, and S. How Seow, J. Mater. Chem. *3*:1109–1115 (1993).
104. M. O. Iwunze, N. Hu, and J. F. Rusling, J. Electroanal. Chem. *333*:353–361 (1992).
105. S. Zhang and J. F. Rusling, Environ. Sci. Technol. *27*: 1375–1380 (1993).
106. S. Schweizer, J. F. Rusling, and Q. Huang, Chemosphere *28*:961–970 (1994).
107. Q. Huang and J. F. Rusling, Environ. Sci. Technol. *29*: 98–103 (1995).
108. S. Zhang and J. F. Rusling, Environ. Sci. Technol. *29*: 1195–1199 (1995).
109. D. L. Zhou, H. Carrero, and J. F. Rusling, Langmuir *12*:3067–3074 (1996).
110. A. C. Onuoha, X. Zu, and J. F. Rusling, J. Am. Chem. Soc. *119*:3979–3986 (1997).
111. J. Gao, J. F. Rusling, and D.-L. Zhou, J. Org. Chem. *61*:5972–5977 (1996).
112. J. F. Rusling, D.-L. Zhou, and J. Gao, in (R. D. Weaver, F. Fischer, F. R. Kalhammer, and D. Mazur, eds.), *Fundamentals and Potential Applications of Electrochemical Synthesis* (Proceedings Vol. 97–6), The Electrochemical Society, Pennington, NJ, 1997, pp. 137–149.
113. J. Gao and J. F. Rusling, J. Org. Chem. *63*:218–219 (1998).
114. J. Gao, K. N. Njue, J. K. N. Mbindyo, and J. F. Rusling, J. Electroanal. Chem. *464*:31–38 (1999).
115. J. K. N. Mbindyo and J. F. Rusling, Langmuir *14*: 7027–7033 (1998).
116. S. G. Merica, C. E. Banceu, W. Jedral, J. Lipkowski, and N. J. Bunce, Environ. Sci. Technol. *32*:1509–1514 (1998).
117. S. G. Merica, W. Jedral, S. Lait, P. Keech, and N. J. Bunce, Can. J. Chem. *77*:1281–1287 (1999).
118. S. R. Ellis, D. Pletcher, P. Gough, and S. R. Korn, J. Appl. Electrochem. *12*:687–691 (1982).
119. S. R. Forsyth and D. Pletcher, in *Recent Advances in Electroorganic Synthesis* (S. Torii, ed.), Elsevier, Amsterdam, 1987, pp. 61–66.
120. M. Fleischmann, C. L. K. Tennakoon, H. A. Bampfield, and P. J. Williams, J. Appl. Electrochem. *13*: 593–602 (1983).
121. M. Fleischmann, C. L. K. Tennakoon, P. Gough, J. H. Steven, and S. R. Korn, J. Appl. Electrochem. *13*:603–610 (1983).
122. E. Laurent, G. Rauniyar, and M. Thomalla, J. Appl. Electrochem. *14*:741–748 (1984).
123. E. Laurent, G. Rauniyar, and M. Thomalla, J. Appl. Electrochem. *15*:121–127 (1985).
124. S. R. Forsyth, D. Pletcher, and K. P. J. Healy, J. Appl. Electrochem. *17*:905–913 (1987).
125. L. Eberson and B. Helgee, Chem. Scripta *5*:47–48 (1974).
126. L. Eberson and B. Helgee, Acta Chem. Scand. *B29*: 451–456 (1975).
127. L. Eberson and B. Helgee, Acta Chem. Scand. *B31*: 813–817 (1977).
128. L. Eberson and B. Helgee, Acta Chem. Scand. *B32*: 313–316 (1978).
129. L. Eberson and B. Helgee, Acta Chem. Scand. *B32*: 157–161 (1978).
130. S. R. Ellis, D. Pletcher, P. H. Gamlen, and K. P. Healy, J. Appl. Electrochem. *12*:693–699 (1982).
131. E. Laurent, G. Rauniyar, and M. Thomalla, C. R. Acad. Sci. Paris II *295*:339–342 (1982).
132. Y. Kunugi, P. C. Chen, T. Nonaka, Y. B. Chong, and N. Watanabe, J. Electrochem. Soc. *140*:2833–2836 (1993).
133. V. Beraud, Thesis. Lyon, France, 1993, pp. 20–57.
134. T. C. Franklin and T. Honda, in *Micellization, Solubilization and Microemulsions*, Vol. 2 (K. L. Mittal, ed.), Plenum, New York, 1977, pp. 617–626.
135. T. C. Franklin and L. Sidarous, J. Electrochem. Soc. *124*:65–69 (1977).
136. T. C. Franklin and T. Honda, Electrochim. Acta. *23*: 439–444 (1978).
137. T. C. Franklin and M. Iwunze, J. Electroanal. Chem. *108*:97–106 (1980).
138. T. C. Franklin and M. Iwunze, Anal. Chem. *52*:973–976 (1980).
139. T. C. Franklin and M. Iwunze, J. Am. Chem. Soc. *103*: 5937–5938 (1981).
140. T. C. Franklin, M. Iwunze, and S. Gipson, in *Inorganic Reactions in Organized Media* (S. L. Holt, ed.), ACS Symp. Ser. 177, American Chemical Society, Washington, DC, 1982, pp. 139–55.
141. T. C. Franklin and S. Gipson, Surf. Technol. *15*:345–355 (1982).
142. T. C. Franklin and M. Ohta, Surf. Technol. *18*:63–76 (1983).
143. T. C. Franklin and T. Jimbo, Surf. Technol. *24*:143–155 (1985).
144. T. C. Franklin and S. Mathew, in *Surfactants in Solution*, Vol. 10 (K. L. Mittal, ed.), Plenum, New York, 1989, pp. 267–286.

145. P. Chambrion, J. Lessard, J. Mailhot, and M. Thomalla, Extended Abstracts 179th Meeting of Electrochem Soc., 1991, p. 856.
146. V. Beraud, J. Lessard, and M. Thomalla, Can. J. Chem. *75*:1529–1535 (1997).
147. N. Rekik, Thesis, Lyon, France 1999, (a) pp. 94–110, (b) pp. 86–94.
148. A. A. Wismeijer, A. P. G. Kieboom, and H. van Bekkum, Rec. Tray. Chim. Pays-Bas *105*:129–136 (1986).
149. V. Danciu, A. M. Martre, P. Pouillen, and G. Mousset, Electrochim. Acta *37*:2001–2008 (1992).
150. S. Ito, Y. Kodama, A. Kitani, and K. Sasaki, in *Novel Trends in Electroorganic Synthesis* (S. Torii, ed.), Springer Verlag, Berlin, 1998, pp. 201–204.
151. C. Mousty and G. Mousset, New J. Chim. *16*:1063–1070 (1992).
152. J. F. Rusling, Acc. Chem. Res. *24*:75–81 (1991).
153. J. F. Rusling and A. E. F. Nassar, J. Am. Chem. Soc. *115*:11891–11897 (1993).
154. A. E. F. Nassar, J. M. Bobbitt, J. D. Stuart, and J. F. Rusling, J. Am. Chem. Soc. *117*:10986–10993 (1995).
155. J. F. Rusling and D. J. Howe, Inorg. Chim. Acta *226*: 159–169 (1994).
156. J. F. Rusling and A. E. F. Nassar, Langmuir *10*:2800–2806 (1994).
157. A. C. Onuoha and J. F. Rusling, Langmuir *11*:3296–3301 (1995).
158. A. E. F. Nassar, Y. Narikiyo, T. Sagara N. Nakashima, and J. F. Rusling, J. Chem. Soc. Faraday Trans. *91*: 1775–1782 (1995).
159. A. E. F. Nassar, W. S. Willis, and J. F. Rusling, Anal. Chem. *67*:2386–2392 (1995).
160. Q. Huang, Z. Lu, and J. F. Rusling, Langmuir *12*: 5472–5480 (1996).
161. Z. Lu, Q. Huang, and J. F. Rusling, J. Electroanal. Chem. *423*: 59–66 (1997).
162. A. E. F. Nassard, Z. Zhang, N. Hu, J. F. Rusling, and T. F. Kumosinski, J. Phys. Chem. B *101*:2224–2231 (1997).
163. M. Ciureanu, S. Goldstein, and M. A. Mateescu, J. Electrochem. Soc. *145*:533–541 (1998).
164. J. F. Rusling, Acc. Chem. Res. *31*:363–369 (1998).
165. Y. M. Lvov, Z. Lu, J. B. Schenkman, X. Zu, and J. F. Rusling, J. Am. Chem. Soc. *120*:4073–4080 (1998).
166. J. Yang, N. Hu, and J. F. Rusling, J. Electroanal. Chem. *463*:53–62 (1999).
167. G. M. Sarsenbaeva, I. A. Avrutskaya, and M. Ya Fioshin, Sov. Electrochem. (translated from Electrokhimiya) *13*:245–247 (1977).
168. G. M. Sarsenbaeva, O. N. Novikova, I. A. Avrutskaya, and M. Ya Fioshin, Sov. Electrochem. (translated from Electrokhimiya) *13*:497–500 (1977).
169. G. R. Mukminova, I. A. Avrutskaya, and V. T. Novikov, Russ. J. Electrochem. (translated from Electrokhimiya) *35*:841–845 (1999).
170. J. D. Rodgers, W. Jedral, and N. Bunce, Environ. Sci. Technol. *33*:1453–1457 (1999).
171. A. Laurent, E. Laurent, and P. Locher, Electrochim. Acta *20*:857–862 (1975).
172. R. Holze and C. H. Hamann, Tetrahedron *47*:737–746 (1991).
173. M. Noel, C. Ravichandran, and P. N. Anantharaman, J. Appl. Electrochem. *25*:690–698 (1995).
174. T. A. Arkhipova, I. A. Avrutskaya, and N. A. Nankov, Sov. Electrochem. (translated from Electrokhimiya) *28*:862–866 (1992).
175. C. Mousty, P. Pouillen, and G. Mousset, J. Electroanal. Chem. *236*:253–265 (1987).
176. C. Mousty and G. Mousset, J. Colloid Interface Sci. *128*:427–436 (1989).
177. L. Dolenski, G. Mousset, and C. Mousty, Bull. Soc. Chim. Fr. *133*:235–242 (1996).
178. C. Reichardt, *Solvents and Solvent Effects in Organic Chemistry*, VCH, New York, 1990.
179. A. P. Abbott and J. F. Rusling, J. Phys. Chem. *94*: 8910–8912 (1990).
180. Z. Galus, in *Advances in Electrochemical Science and Engineering* (H. Gerischer and C. W. Tobias, eds.), VCH, Weinheim, 1995, pp. 217–295.
181. C. Mousty, B. Cheminat, and G. Mousset, J. Org. Chem. *54*:5377–5380 (1989).
182. T. C. Franklin, R. Nnodimele, and J. Kerimo, J. Electrochem. Soc. *140*:2145–2150 (1993).
183. T. C. Chou and C. H. Cheng, J. Appl. Electrochem. *22*:743–748 (1992).
184. K. Naoi, Y. Oura, M. Maeda, and S. Nakamura, J. Electrochem. Soc. *142*:417–422 (1995).
185. N. Sakmeche, S. Aeiyach, J. J. Aaron, M. Juini, J. C. Lacroix, and P. C. Lacaze, Langmuir *15*:2566–2574 (1999).

12

Mediated Electro-Organic Synthesis in Microemulsions

JAMES F. RUSLING University of Connecticut, Storrs, Connecticut

I. INTRODUCTION

Microemulsions are clear fluids made from water, oil, and surfactant. They are heterogeneous on the nanometer scale and provide cheaper and less toxic alternatives to organic solvents for mediated (catalytic) electrochemical syntheses. This chapter reviews current understanding of the reaction dynamics and design of mediated electrochemical syntheses in microemulsions. Aspects of mass transport and interfacial dynamics in microemulsions are discussed with respect to their influence in these synthetic processes. Examples of synthetic pathway control using microemulsions are presented. Strategies for design of catalytic films on electrodes for synthesis in microemulsions are discussed. Electrochemical synthesis in microemulsions may be a viable future approach to environmentally friendly, "green," methods of making fine chemicals.

Electrolytic methods are well suited to relatively clean organic chemical syntheses. These processes employ reducing or oxidizing power controlled by externally applied voltages on electrodes, thereby avoiding chemical reducing or oxidizing agents. Degradation products of chemical redox agents do not cause disposal problems because no chemical oxidants or reductants are needed. Electrochemical methods can be applied to a wide range of organic syntheses [1–3].

Although many elegant electro-organic syntheses have been developed using aprotic organic solvents, water is preferable for industrial processes because of its low cost and ease of handling. Further, keeping the environment and ourselves healthy requires that modern industrial chemical processes strive for cleanliness and low toxicity, and water is ideal from this point of view.

Perhaps the most successful industrial organic electrosynthesis, the Monsanto process for production of adiponitrile, uses a water-organic emulsion. This process [2] employs a saturated solution of reactant acrylonitrile in aqueous buffer containing a tetraethylammonium salt. There is no solvent other than water. Product is removed and new reactant is added in a continuous flow reactor. A layer of tetraalkylammonuim ions on the electrode facilitates dimerization of acrylonitrile and inhibits formation of the unproductive two-electron reduction product.

Water is a poor solvent for many desired reactants in electro-organic syntheses. It is also a good proton donor and a source of hydroxide ion from electrolysis of water, properties that may be undesirable for some reactions. *Microemulsions* made from water, oil, and surfactants represent alternatives that can retain a high water content but have much wider-ranging solubilizing ability than water alone. Microemulsions are optically clear, stable, microheterogeneous fluids. They have excellent solubilizing power for ionic, polar, and nonpolar solutes [4]. Although they can contain considerable water, proton donor properties at the site of reaction can often be controlled. They are less toxic and less expensive than pure aprotic organic solvents [5].

Thus, the combination of electrolysis and microemulsions may have an important role in clean, or

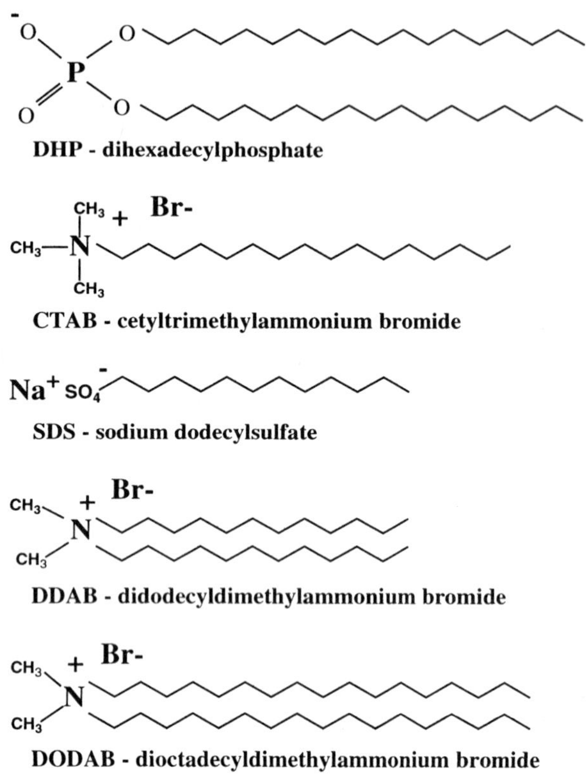

FIG. 1 Some surfactants commonly used in conductive microemulsions.

green, synthetic processes of the future. This chapter presents the current state of knowledge regarding electrochemical synthesis in microemulsions. In the next section, we briefly review the properties of microemulsions and microemulsion structures with an emphasis on requirements for use in electrochemical synthesis. Next, the important topics of mass transport of reactants and dynamics and kinetics at electrode surfaces in microemulsions are discussed. Following this, results for a series of synthetic applications including dehalogenations, carbon-carbon bond formation, and cyclizations are presented. The last section of this chapter covers the development of catalytic films designed specifically for electrochemical synthesis in microemulsions.

II. CONDUCTIVE MICROEMULSIONS

As mentioned earlier, microemulsions are optically clear, thermodynamically stable fluids made from water, oil, and surfactant (Fig. 1). A cosurfactant such as a long-chain alcohol may be required, and salt can be added to improve conductivity. These fluids are less toxic and often less expensive than alternative organic solvents [5]. Their properties can be tuned for specific applications by adjusting their compositions. Although microemulsions appear homogeneous to a visual observer, they are actually microheterogeneous. Various nanometer-scale structures in these fluids are stabilized by surfactant at the oil-water interfaces. This leads to a vanishingly small interfacial tension, which stabilizes the fluid structures.

Conductive microemulsions (μEs) are most suitable for electrochemical synthesis. These include oil-in-water (o/w) and bicontinuous microemulsions [6–8]. Conductivity is imparted by ionic surfactants or by added inert electrolyte. These fluids hold significant promise for controlling rates of mediated electrochemical syntheses.

Oil-in-water microemulsions feature a continuous water phase with dispersed surfactant-coated oil droplets (Fig. 2). The continuous phase is not packaged in

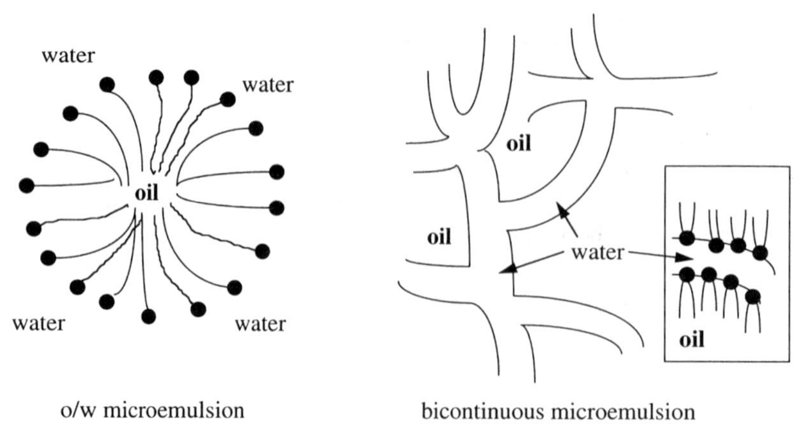

FIG. 2 Oversimplified representations of microemulsion structures.

discrete structures but is continuous throughout the fluid. The oil droplets in o/w microemulsions are about 5–50 nm in diameter.

Water-in-oil (w/o) microemulsions have surfactant-coated water droplets dispersed in a continuous oil phase and are usually poorly conductive. Bicontinuous microemulsions have *both water and oil* as continuous phases. They contain dynamic intertwined networks of oil and water, with surfactant monolayers at the interfaces (Fig. 2) [8,9]. Three-component bicontinuous microemulsions made with didodecydimethylammonium bromide feature dynamic water tubules with diameters of 2 to 10 nm [10]. Bicontinuous microemulsions can have sufficient conductivity for electrolytic applications when ionic surfactants or inorganic salts are used as components. Some surfactants used in microemulsions are shown in Fig. 1.

The conductivity of o/w and bicontinuous microemulsions containing ionic species results from the continuity of the water phases. Conductivities of these fluids can approach those of homogeneous aqueous electrolyte solutions, and they are suitable for electrolysis with conventional electrodes made from carbon or metals. When the surfactant is ionic, a charged double layer exists at the oil-water interface, leading to interfacial potentials up to 100 mV [8]. Interfacial potential may be particularly important for chemical steps that occur in interfacial regions during electrochemical synthesis.

Bicontinuous microemulsions with similar amounts of water and oil may offer unique advantages for *mediated* electrochemical synthesis. In mediated synthesis, a chemical mediator (or catalyst, often a metal complex) effectively delivers electrons between the electrode and an organic reactant. In such cases, bicontinuous microemulsions can solubilize an ionic metal complex in the water phase and the organic reactant in the oil phase. If the phases are intimately mixed in a fluid with high interfacial area, the rate of the key reaction between the electrochemically activated mediator and the organic reactant will not be inhibited and may even be enhanced [11]. Mediators are usually chosen for electrochemical reversibility and unique reactivity properties of the active form.

Efficient mass transport is of great practical importance to a heterogeneous reaction process such as electrochemical synthesis [1–3]. In bicontinuous microemulsions, ionic and polar species diffuse in the water phase and nonpolar species diffuse in the oil phase. Thus, mass transport tends to be faster than in o/w microemulsions. In the latter fluids, nonpolar molecules will be transported along with the larger oil droplets, and their mass transport will be slowed down [7,8].

Understanding the dynamics of electrochemical processes in microemulsions rests on an understanding of the dynamic nature of microemulsion structures. The structures represented in Fig. 2 are continually rearranging on the submillisecond time scale. Studies have revealed two relaxation processes in microemulsions containing alcohol cosurfactants [12,13]. One process occurs in 10 ns or less and is associated with alcohol exchange between the oil-water interface and the continuous phase. A slower process occurring in 10 ns or more involves surfactant exchange between the interface and the continuous phase.

Thus, there is often rapid molecular exchange of surfactant and cosurfactant between the interface and continuous phases. Systematic solute exchange studies in microemulsions are rare. However, solute dissociation from micelles has rate constants in the range 10^3–10^7 s^{-1}, and recapture rates are close to diffusion limited [13]. Thus, solutes residing at o/w interfaces of microemulsions may have similarly rapid "on-off" dynamics.

III. DYNAMICS IN MICROEMULSIONS

A. Mass Transport of Electroactive Solutes

Mass transport is of key importance in electrochemical synthesis because reactants or mediators must travel to the electrode and accept or donate electrons for the reaction to proceed. Diffusion of electroactive probes in unstirred microemulsions is reasonably well understood [7,8]. Understanding has been facilitated by the fact that the peak current (I_p) for a reversibly reduced or oxidized probe in electrochemical experiments such as cyclic voltammetry (CV) depends on the diffusion coefficient D. The relevant relationship for CV is called the Randles-Sevcik equation [14]:

$$I_p = 0.4463nFAC^*(nF/RT)^{1/2}\nu^{1/2}D'^{1/2} \qquad (1)$$

where n is the number of electrons transferred per molecule, F is Faraday's constant, R is the gas constant, T is temperature in Kelvins, and ν is scan rate. Equation (1) predicts a linear plot of I_p versus $\nu^{1/2}$ for a reversible reduction or oxidation of an electroactive species. The slope of this plot can be used to estimate D', the *apparent* diffusion coefficient of the electroactive species in the microemulsion.

There are a number of ways that mass transport can be influenced by microheterogeneous structures of microemulsions. This is why electrochemically measured D values must be expressed as *apparent* or *conditional* values, D'. Mackay and coworkers found that D' for water-soluble, unbound electroactive ions in microemulsions is correlated with the volume fraction of oil (ϕ_c) [7]:

$$D' = D_0(1 - \phi_c)^{m-1} \qquad (2)$$

where D_0 is the diffusion coefficient of the free ion in water and m is a fitted coefficient. The D' values of unbound water-soluble ions in microemulsions can be interpreted in terms of obstruction by oil droplets in o/w microemulsions and are typically 40–80% of those in water. We can also expect D' values of electroactive species residing in a single microphase of bicontinuous microemulsions to be mildly obstructed by the bicontinuous network.

Ions bound to droplets or to interfacial regions and molecules dissolved in oil droplets in microemulsions can have D' values about 10-fold smaller than the actual D value of that species in the corresponding homogeneous solvent. For o/w or w/o microemulsions with a single distribution of droplet sizes, D' is given by [8]:

$$D' = f_a D_0 + f_b D_1 \qquad (3)$$

where f_a is mole fraction of unbound probe, f_b is the fraction of bound probe, and D_0 and D_1 are the diffusion coefficients of the free probe and the droplet in the medium. Equilibrium expansions of Eq. (3) that allow estimation of binding parameters from D' versus concentration data are available [6–8]. Amphiphilic reactants with both ionic and hydrophobic character can also bind to interfaces and oil droplets, resulting in retardation of mass transport.

While ions bound to microemulsion droplets can have approximately 10-fold smaller D' values compared with D_0 [7,8], the situation in bicontinuous microemulsions is more complex. Here, ions bound to interfacial surfactant may diffuse at rates similar to those of the lateral diffusion of interfacially bound surfactant molecules. Free molecules and ions in individual phases of the microemulsion may have somewhat obstructed diffusion. A few examples are illustrative:

1. The uncharged molecule ferrocene had D' values of 0.3–0.7 $\times$ 10^{-6} cm^2 s^{-1} in o/w microemulsions of viscosity 1.7 cP made with tetralkylammonium surfactants and about 3 $\times$ 10^{-6} cm^2 s^{-1} in bicontinuous microemulsions with viscosities of 10–12 cP [15]. The lower values in the o/w microemulsions reflect binding to oil droplets, whereas water-insoluble ferrocene can diffuse more freely in the oil phases of the bicontinuous systems.
2. An amphiphilic ion strongly bound to o/w interfaces in bicontinuous microemulsions made with didodecyldimethylammonium bromide (DDAB) had a D' of 0.14 $\times$ 10^{-6} cm^2 s^{-1} [16], whereas ferrocene had D' values of 3–6 $\times$ 10^{-6} cm^2 s^{-1} in similar microemulsions [17]. Here the ampiphilic ion most likely diffuses along o/w interfaces, while ferrocene travels in the continuous oil phase. The DDAB microemulsions had bulk viscosities of 19–38 cP, and these D' data show that unlike that in homogeneous solvents, diffusion of species in microemulsions does not necessarily follow the Stokes-Einstein equation [8]. In other words, D' need not be inversely related to bulk viscosity.

We conclude from the preceding discussion that respectable mass transport rates can be obtained even in microemulsions with relatively large bulk viscosity. In terms of ease of handling, there is a practical upper limit to viscosity of microemulsions used for electrochemical synthesis, which relies on convective mass transport as well as diffusion. High conductivity and reasonable viscosity (e.g., below 12 cP) appear to be desirable properties for electrochemical synthesis.

B. Dynamics and Interfacial Structures at Electrodes

Key electron transfer and chemical events in electrochemical synthesis occur at the interface of the electrode with the fluid medium. Electron transfer between a reactant or catalyst and an electrode is often the initial event in a synthetic pathway. In a microemulsion, the surface of the electrode may be coated with adsorbed surfactant and possibly other molecular components of the system [11]. This adsorbed layer may have a significant effect on rates of electron exchange between electrodes and reactants [6,8]. Some knowledge of the supramolecular structure and dynamics of electrode-fluid interfaces is necessary for a complete understanding of interfacial factors that influence electrochemical reactions in microemulsions. In this section, we discuss studies of the electrochemical influence of adsorbed films on electrodes in microemulsions. As this topic has been much less studied than electrode interfaces in micellar solutions, we will use the latter as a comparative benchmark.

In micellar solutions, entry of a reactant into a dynamic adsorbed film of surfactant on the electrode probably precedes electron transfer [18]. This results in a mixed adsorbate layer of reactant and nonelectroactive surfactant on the electrode prior to electron transfer. Depending on the reactant's ability to compete with nonelectroactive surfactant for surface sites on the electrode, electron transfer can involve adsorbed reactant or reactant that approaches the electrode to within a distance roughly equivalent to the diameter of the adsorbed surfactant's headgroup.

We employed ferrocene surfactant probes **1–4** (Fig. 3) to study adsorbate structure, interactions, and dynamics at electrodes in dilute micellar solutions. Interesting details in general accord with reactant entry into a surfactant film on the electrode were obtained. Relatively stable, ordered films were formed on glassy carbon electrodes from very dilute solutions of **3**. Residence times on the electrode in micromolar solutions were 4.5 s for Fc-C_8, 14 s for Fc-C_{12}, and 66 s for Fc-C_{16} [19], which can be taken as estimates of surfactant residence times on carbon electrodes in dilute surfactant solutions.

Voltammetric studies in a microemulsion made from the electroactive surfactant Fc-C_{16} (Fig. 3), *n*-octane, 1-butanol, 0.1 M NaBr (2.3:2.3:4.6:90.8) suggested that roughly a bilayer of Fc-C_{16} was adsorbed onto glassy carbon electrodes [18]. When an electrode equilibrated with this microemulsion was transferred into water, the coverage of Fc-C_{16} decreased to a monolayer in about an hour. This result suggested surfactant coverage on electrodes in microemulsions qualitatively similar to that in micellar solutions.

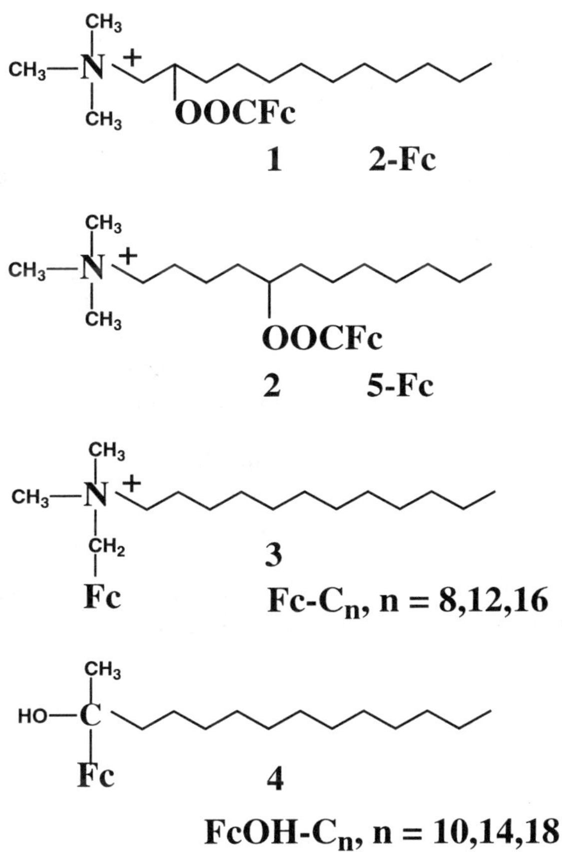

FIG. 3 Amphiphilic ferrocene (Fc) derivatives used to probe electrode interfaces.

Comparisons of electron transfer (ET) rate constants of ferrocene surfactants **1** and **2** in microemulsions and micellar solutions reflected different dynamics at these two electrode-fluid interfaces. Apparent standard heterogeneous electron transfer rate constants ($k^{\circ\prime}$) for oxidations of 2-Fc (**1**) and 5-Fc (**2**), respectively, at glassy carbon electrodes were similar in homogeneous dimethylformamide (DMF) and dimethylsulfoxide (DMSO) on Pt and glassy carbon electrodes [20,21]. However, in aqueous micellar CTAB, electron transfer rate constants were in the order Fc > 2-Fc > 5-Fc, with 10-fold differences in successive values. This can be rationalized by assuming that 2-Fc and 5-Fc are oriented headgroup down on the electrode prior to electron transfer. Adsorbed CTA^+ on the electrode apparently assists in ordering **1** and **2** at the electrode surface prior to electron transfer.

In a bicontinuous CTAC microemulsion, $k^{\circ\prime}$-values for Fc, 2-Fc, and 5-Fc on glassy carbon were smaller than in acetonitrile. Ferrocene had a $k^{\circ\prime}$ twice as large as that of 2-Fc and 5-Fc in the microemulsion [21], but values for 2-Fc and 5-Fc were the same within experimental error. These results suggest a more random orientation of these reactants at electrodes in the CTAC microemulsion and more disorder and mobility in the electrode-fluid interface in the CTAC microemulsion than in micellar CTAB solutions.

Amphiphilic ferrocene alcohols (**4**) $FcOHC_{10}$, $FcOHC_{14}$, and $FcOHC_{18}$ gave one-electron oxidations that were nearly reversible and diffusion controlled in microemulsions of DDAB, CTAC, or sodium dodecyl sulfate (SDS). Apparent diffusion coefficients in the microemulsions suggested that the alcohols are distributed between the oil-water interfaces and the oil phase of the microemulsions. Increasing the chain length favored binding at the oil-water interface [22]. These results are in agreement with studies of water-in-oil microemulsions suggesting that the presence of sufficient cosurfactant can improve interfacial fluidity and facilitate electron transfer at electrodes [23–25]. Little evidence was found for strong adsorption of any ferrocene alcohol at electrode-microemulsion interfaces.

Although a detailed dynamic molecular view of the microemulsion-electrode interface is not yet available, a qualitative picture emerges via comparison with the better understood micellar solution-electrode interface. Surfactant seems to be adsorbed onto charged electrodes from both types of fluids. However, the kinetic studies are consistent with adsorbate layers that are more disordered and more fluid in microemulsions than in micellar solutions [11]. On the other hand, the rate of surfactant ejection from these adsorbate layers is probably slow enough, i.e., in the range of seconds, that the surface composition in microemulsions can be considered static during fast electrochemical processes. Also, entry of reactants into the surface adsorbate layers does not seem to be detected by conventional voltammetry. Although details may differ, electrochemistry in most microemulsions appears similar to that in homogeneous electrolyte solutions.

IV. KINETICS OF MEDIATED REACTIONS

A. Mediating Electrochemical Reactions

Electrochemical synthesis can be mediated by a reversibly reduced or oxidized species (called the mediator or catalyst) that can shuttle electrons between electrodes and reactants [26]. This approach can be used for reactants that do not directly exchange electrons with electrodes or to decrease the activation energy (i.e., the applied potential) required for a specific oxidation or reduction. Alternatively, a mediator activated at an electrode can link to a reactant, followed by bond cleavage and trapping of a cleavage fragment by another reagent to yield the desired product.

Reductions of organic compounds mediated by transition metal complexes have been developed to a high degree of synthetic versatility [27–32]. Carbon-carbon bonds between an alkyl halide RX and an activated olefin can be formed using Co(II) complexes as mediators (Scheme 1):

$$\text{Co(II)L} + e^- \rightleftharpoons \text{Co(I)L} \qquad \text{(at electrode)} \tag{4}$$

$$\text{RX} + \text{Co(I)L} \xrightarrow{k_1} \text{R—Co(III)L} + \text{X}^- \tag{5}$$

$$\text{R—Co(III)L} + (\text{H}^+, e^-, \text{or } h\nu) + \text{CH}_2\text{=CHZ} \rightarrow \text{R—CH}_2\text{CH}_2\text{Z} + \text{Co(II)L} \tag{6}$$

SCHEME 1 Mediated electrochemical bond formation.

where X = Cl, Br, or I and Z = electron-withdrawing group. These reactions begin with donation of an electron from an electrode to mediator Co(II)L (L = macrocyclic ligand) to form Co(I)L [Eq. (4)]. The Co(I)L reacts with alkyl halide RX [Eq. (5)] in what is often the rate-determining step (rds) to yield intermediate —Co(III)L. When R—Co(III)L is cleaved at an electrode or with light in the presence of an activated olefin [Eq. (6)], a carbon-carbon bond is formed. In principle, these reactions have good atom economy [33] in that all reactant atoms end up in the product, leaving no undesirable side products.

Vitamin B_{12}, cobaloximes, Co(salen), and other cobalt complexes react with alkyl halides by pathways similar to Scheme 1 to give alkyl-cobalt complexes. The carbon-cobalt bonds can be cleaved by visible light, electrolysis, or reducing agents to give carbon-centered radicals or anions that can add in situ to activated olefins to form carbon-carbon bonds [Eq. (6)]. Such methods have been used to synthesize lactones, pheromones, prostaglandins, C-glycosides, optically active olefins and alcohols, and cyclic molecules [27–32]. In this section we discuss factors that control the kinetics of key steps in mediated electrochemical synthesis in microemulsions.

B. Control by Mediator Formal Potential

Mediated electrochemical reactions often have bimolecular rate-determining steps. When the rds occurs in the bulk microemulsion and reactant partition is not rate limiting, the reaction rate can be controlled primarily by the intrinsic activation free energy of the bimolecular rds. For a single reactant with a series of mediators operating in reductive mode in a homogeneous solvent, the activation free energy is controlled by the difference in reduction potential between reactant and mediator [34]. Thus, a plot of the log of the reaction rate constant (k_1) versus mediator formal potential ($E^{\circ\prime}$) should be linear. Bimolecular reactions controlled by intrinsic activation free energy in microemulsions should display a similar linear plot of log k_1 versus $E^{\circ\prime}$.

Rates of catalytic reductions of *trans*-1,2-dibromocyclohexane (DBCH) [Eq. (7)] [35] were found to give such a linear plot of log k_1 versus $E^{\circ\prime}$ (Fig. 4) in microemulsions. Here, k_1 is the apparent or conditional second-order rate constant estimated by voltammetry. The mediator complexes resided in the water phase of bicontinuous microemulsions of oil, water, and DDAB,

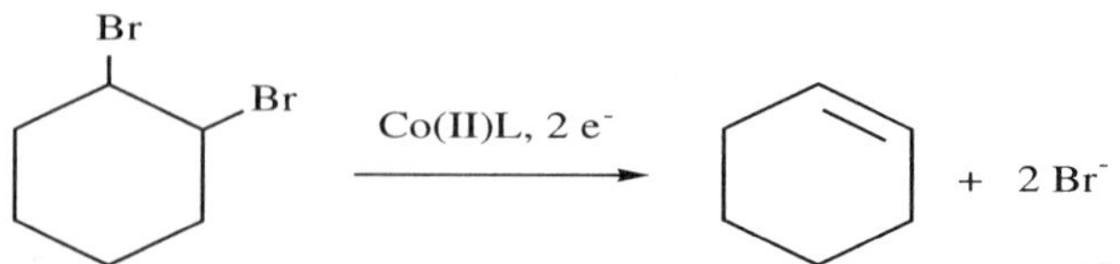

(7)

while DBCH was in the oil phase. For homogeneous DMF and the DDAB microemulsion, data for a series

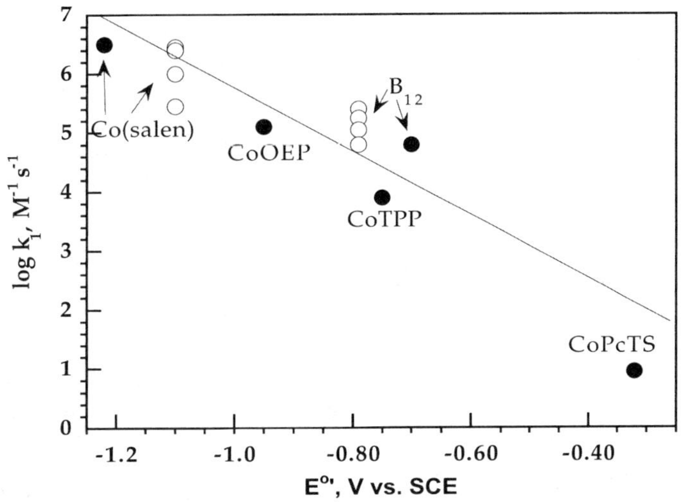

FIG. 4 Influence of catalyst formal potential ($E^{\circ\prime}$) on log k_1 for reaction of Co(I)L with DBCH in a bicontinuous microemulsion (○) of DDAB/water/dodecane (21 : 39 : 40) and in DMF (●) for dissolved catalysts vitamin B_{12}, Co(salen), cobalt phthalocyanine-tetrasulfonate (CoPCTS), cobalt tetraphenylporphyrin (CoTPP), and cobalt octaethylporphyrin (CoOEP). Points from reactions in the microemulsion (○) represent apparent k_1 values for 0.4, 0.5, 1.0, and 2.0 mM catalyst concentrations in order of decreasing log k_1.

of Co(I) complex mediators were fit by the same log k_1 versus $E^{\circ\prime}$ Co(II)/Co(I) regression line (Fig. 4). Thus, kinetics were controlled mainly by intrinsic activation free energies dependent on the reversible reduction potential of Co(II)/Co(I). Even though reactants reside in different phases in the microemulsion, their distribution between phases was less important than the mediator's formal potential. A similar linear plot was found for the reduction of benzyl bromide in microemulsions [36].

Formal potentials of mediators in the microemulsion depend on specific interactions, such as those with cationic surfactant headgroups or the influence of water phase pH [35]. Such interactions offer a mode of kinetic control via tailoring microemulsion composition to control the $E^{\circ\prime}$ of the mediator.

We also found that apparent k_1 values in the DDAB microemulsion depended slightly on mediator concentration but not on DBCH concentration. The dependence of rate on mediator concentration is illustrated by the open circles in Fig. 4, showing decreases in *apparent* log k_1 as the mediator concentration was increased. This effect is less important than the mediator formal potential, the major factor controlling k_1.

Dynamic scanning electrochemical microscopy (SECM) was applied in an attempt to understand further interfacial effects in these reactions in microemulsions [37]. We used SECM to interrogate directly the kinetics of catalytic reduction of DBCH at an interface between water and benzonitrile with and without adsorbed surfactant.

The reaction rate between the Co(I)L form of vitamin B_{12} generated electrochemically in the water phase and DBCH in benzonitrile was probed by an ultramicroelectrode (UME) tip in SECM. Whereas only an overall *apparent* reaction rate constant can be obtained by voltammetry in a microemulsion, SECM measured the relevant interfacial kinetics directly. The Co(I)L generated at the UME tip in the water phase diffuses to the interface and reacts with DBCH at the liquid-liquid interface. The kinetic influence of reactant concentration, potential drop across the liquid-liquid interface, and interfacial surfactants was investigated. As in the microemulsion, the apparent pseudo-first-order rate constant for DBCH reduction was independent of DBCH, but it depended somewhat on the concentration of vitamin B_{12}.

Results of interfacial SECM and voltammetry in microemulsions suggest that the kinetics of reduction of DBCH by B_{12}Co(I) are more complex at liquid-liquid interfaces than the second-order rate-limiting process in homogeneous organic solvents. Reasons for the mediator concentration dependence of k_1 are unclear. Possible explanations include a concentration-dependent shift to mixed kinetic control at the interface [11].

The study of DBCH reduction just described uncovered interfacial effects in a complex two-electron catalytic reduction process but did not provide direct kinetic information on carbon-carbon bond-forming reactions. To provide more insight into interfacial influences on these reactions, we investigated the S_N2 processes in Eq. (5) in the absence of the follow-up reduction process [Eq. (6)], i.e., under conditions where R—Co(III)L was stable [35]. Rate constants for reactions of Co(I)L with butyl- and dodecyliodide in microemulsions and homogeneous DMF and DDAB microemulsions also gave linear plots of log k_1 versus $E^{\circ\prime}$ (Fig. 5), with points for both fluids falling on the same straight line. In this case, the apparent k_1 in the microemulsion depended on neither reactant nor mediator concentration, reflecting true second-order kinetics at the microemulsion interface. Thus, interfacial kinetic complications found for mediated reduction of DBCH would seem to be a special case.

C. Control by Reactant Preconcentration on the Electrode

The rate of a bimolecular rds can be enhanced by inclusion of reactants in an adsorbed surfactant layer on an electrode [26]. High reactant concentrations in this restricted reaction volume for a bimolecular rds such as Eq. (5) can lead to practical increases in reaction rate. This type of rate enhancement was identified for reductions of alkyl vicinal dibromides to olefins using adsorbed metal phthalocyanines (MPCs) as mediators in DDAB microemulsions [38]. Apparent rate constants for reduction of DBCH and 1,2-dibromobutane estimated by voltammetry were up to 400-fold faster in the microemulsion than in homogeneous organic solvent, supporting the reactant surface preconcentration view. Enhancing rates in this way requires finding the right combination of reactants, microemulsion components, and electrode to facilitate such coadsorption and is largely a trial-and-error endeavor.

Apparent log k_1 values for reduction of DBCH with adsorbed metallophthalocyanine catalysts in a microemulsion were not correlated with $E^{\circ\prime}$ values of mediators [38]. Here, intrinsic activation free energy for the reaction does not control the rates, which are enhanced by the preconcentration effect.

D. Control by Reactant Partitioning

In this situation, reactants in a bimolecular rds reside separately in oil and water phases and the reaction occurs at the o/w interface. For interfaces with insufficient surface area or slow dynamics, the reaction rate may be slowed down compared with a homogeneous solution. Obviously, this is a mode of kinetic control of mediated electrochemical reactions in microemulsions that should usually be avoided.

Reductions of oil-soluble vicinal dibromides mediated by water-soluble vitamin B_{12} in a water-in-oil microemulsion made with the surfactant Aerosol OT (sodium bis-2-ethylhexylsulfosuccinate) had 40-fold smaller rates [39] than in bicontinuous DDAB microemulsions. Decreased rates in the w/o microemulsion

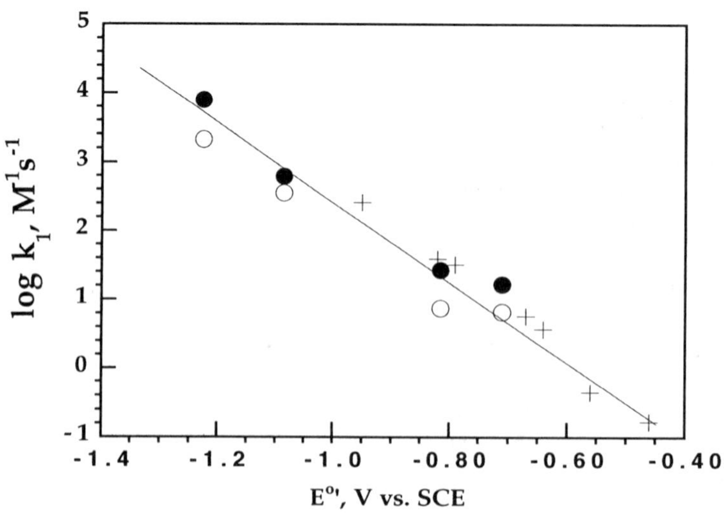

FIG. 5 Influence of catalyst formal potential ($E^{\circ\prime}$) on log k_1 for S_N2 reactions of Co(I)L with butyl bromide (●, +) and (○) dodecyl bromide using various cobalt complexes (see Ref. 35). Open and closed circles at −1.09 V and −0.82 V represent experiments in bicontinuous microemulsion of DDAB/water/dodecane (21 : 39 : 40). All other data obtained in homogeneous DMF solvent.

were attributed to the physical separation of vitamin B_{12}, which resided in water droplets, from the nonpolar vicinal dibromides in the continuous oil phase. The Aerosol OT w/o microemulsion used had an interfacial area about fourfold smaller than the bicontinuous DDAB microemulsion [35]. The smaller interfacial area of the w/o system can decrease the frequency of reactant encounters. Interfacial dynamics may also play a role in controlling reaction rates in such systems.

V. MAKING CARBON-CARBON BONDS

This section describes successful mediated syntheses in microemulsions employing Scheme 1 with cobalt complexes as mediators. All the synthetic pathways begin with formation of the active Co(I)L species at the electrode [Eq. (4)] followed by reaction with an organic halide to yield R—Co(III)L [Eq. (5)].

A. Converting Bibenzyl to Benzyl Bromide

Reduction of benzyl bromide can proceed by two possible pathways. Formation of benzyl radical results in dimerization to bibenzyl. Reduction to benzyl anion yields toluene. By controlling the reduction potential by the correct choice of mediator, complete conversion of benzyl bromide to bibenzyl or toluene was obtained in a DDAB microemulsion [36]. Although good yields of bibenzyl were also obtained by this reaction in DMF, selective production of toluene was not achieved.

The Co(I)L form of vitamin B_{12} formed at about -0.8 V versus SCE gave a benzyl-Co(III)L intermediate, which was reduced at -1.1 V versus SCE and yielded bibenzyl (**1**, Scheme 2) as the sole product of a radial pathway in the microemulsion. Co(salen) was also used to mediate this reaction. In this case, the reduction potential of benzyl-Co(III)(salen) was about -1.4 V, negative enough that benzyl radicals are reduced to anions to give toluene (**2**) as sole product (Scheme 2). A radical or anionic pathway can be chosen by controlling the reduction potential of benzyl-Co(III)L via controlling the nature of the cobalt complex.

B. Unsymmetrical Carbon-Carbon Bonds and Cyclizations

Reactions following the general pathway of Scheme 1 [Eqs. (4)–(6)] were mediated by vitamin B_{12} at an electrode potential of -0.85 V versus SCE. Conjugated additions of primary alkyl iodides of different chain lengths to 2-cyclohexen-1-one yield 3-alkyl cyclohexanones **3** (Scheme 3). Reaction of Co(I)L with the alkyl

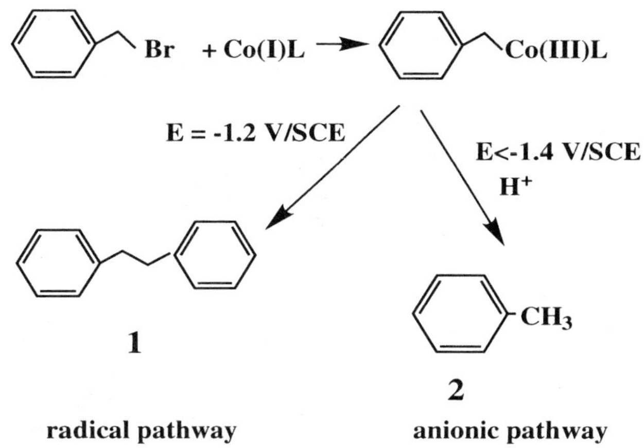

SCHEME 2

iodides gave stable Co-alkyl complexes. Cleavage of these complexes by visible light or a potential of -1.45 V in the dark were compared. Addition of the alkyl radicals to the activated double bond of 2-cyclohexen-1-one gave 70–80% yields of products **3** using -0.85 V + light [40]. Much poorer yields of **3** were obtained at -1.45 V without light because of competing reduction of the alkyl iodides to olefins and alkanes.

If alkyl halide and activated olefin moieties are present in the same molecule, Scheme 1 can be used for cyclization reactions. For example, (4-bromobutyl)-2-cyclohexene-1-one (**4**) was converted to *cis*- and *trans*-1-decalone **5** in microemulsions (Scheme 4) [40]. Yields of 90% *cis*- and *trans*-1-decalone were obtained with either light or electrochemical cleavage of the alkyl-cobalt bond in microemulsions and in DMF.

The cyclization in Scheme 4 gave remarkable stereoselectivity in microemulsions, with about 93:7 *trans*/*cis* ratios, but poor stereoselectivity in homogeneous DMF. This reaction was investigated in 15 o/w and bicontinuous microemulsions. Neither microemulsion composition, surfactant type (cationic or anionic), nor chirality of the mediator significantly influenced the *trans*/*cis* ratio of 1-decalone [41]. Selective formation of *trans*-1-decalone in microemulsions was attributed to equilibration of isomers via keto-enol tautomeriza-

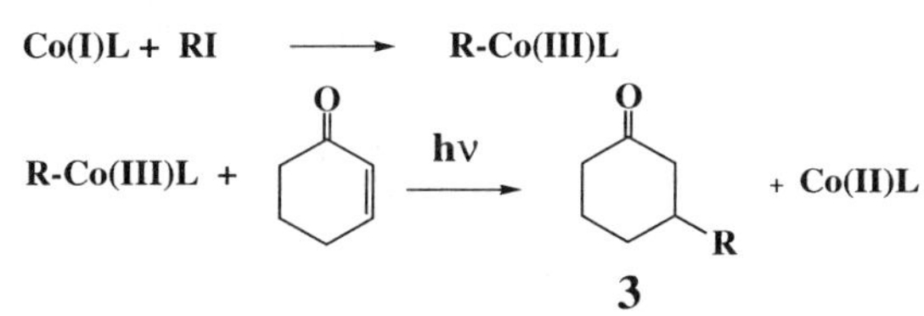

SCHEME 3

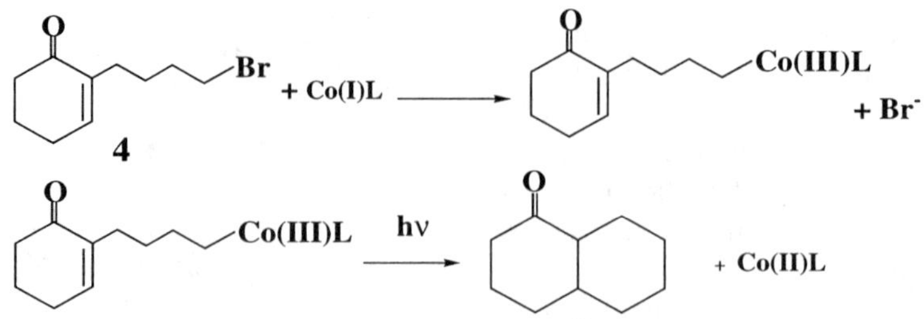

4

5

SCHEME 4

tion catalyzed by hydroxide ions formed by coelectrolysis of water during the reaction. *trans*-1-Decalone was most efficiently produced in 2-h electrolyses in microemulsions with large volume fractions of water. Organic solvents with trace amounts of water gave little stereoselectivity in similar periods, but longer equilibration times produced larger proportions of *trans*-1-decalone.

Results show that microemulsions are synthetically useful media for carbon-carbon bond formation mediated by electrochemically generated Co(I)L from vitamin B_{12}. Photolytic cleavage of C—Co bonds gave higher yields of intermolecular addition products than electrochemical bond cleavage.

Microemulsions were shown to provide advantages for other cyclization reactions as well. Electrochemical catalysis using vitamin B_{12} at −1.5 V versus SCE at carbon electrodes in bicontinuous microemulsions made with CTAB and SDS was used to convert 2-(3-bromopropyl)-2-cyclohexene-1-one **6** to the 5-endo-trig cyclization product 4-hydrindanone **7** in 62–70% yield (Scheme 5) [42]. The same reaction conditions gave only 19% of **7** in DMF, 7% in MeOH, and 10% in MeOH/water (4:1). About 20% yields of **7** in the CTAB microemulsion and in DMF were obtained using visible light during electrolyses. The alkyl radical formed at −0.9 V by visible-light cleavage of the alkylcobalt intermediate led to poor yields of **7**. In electrolyses at −1.5 V without light, the microemulsions seem to inhibit protonation of a key carbanion intermediate that would otherwise lead to nonbicyclic products [42].

Radical cyclization of bromoacetal **8** was previously reported via catalysis by low-valent cobaloxime species generated by electrochemical or chemical reduction in MeOH or MeOH/H_2O to make bicyclic heterocycle **9** [43]. Selective formation of saturated or unsaturated oxygen heterocycles from bromoacetals was catalyzed by an iodocobaloxime in acetonitrile and DMF [44]. A high concentration of the cobalt catalyst and mild reducing conditions led to unsaturated products. Low concentrations of cobalt complex and negative reduction potentials afforded saturated products (Scheme 6).

Similar selectivity for reduction of **8** was achieved in a microemulsion. Electroreduction of **8** was catalyzed by vitamin B_{12} at −0.85 V on a carbon cloth cathode irradiated with visible light. The unsaturated product **9** was produced in 61–70% yield in a CTAB microemulsion or MeOH [45]. Alternatively, electrochemical cleavage of the organocobalt adduct was effected by electrolysis at −1.5 V in the dark. This reaction afforded **9** when 0.2 equivalents of catalyst were used, but 0.01 equivalents of catalyst gave saturated product **10** at 60% yield in the microemulsion, with <1% **9**. The use of 0.01 equivalents of catalyst in MeOH gave a 50% overall yield of a mixture of **9** and **10** in the ratio 21:79.

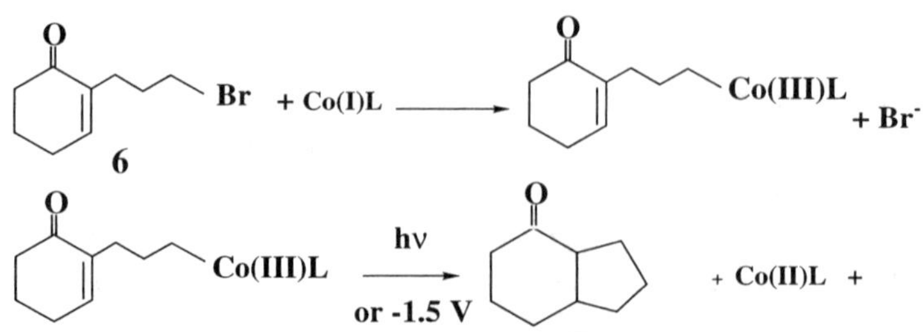

6

7

SCHEME 5

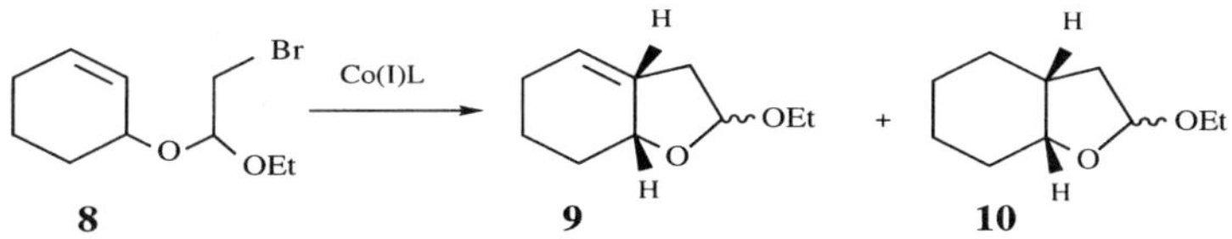

SCHEME 6

VI. CATALYTIC ELECTRODES IN MICROEMULSIONS

Considerable effort has been directed toward chemical modification of electrodes with catalytic films [46,47]. The technological utility of such electrodes for synthetic purposes would be enormous, holding the exciting promise of reusable electrode surfaces specifically designed for production of fine chemicals in cheap, low-toxicity media using only electricity and reactants. The catalytic electrode could lower process costs by decreasing the power requirements of the reactions. Practical applications of such electrodes for synthesis in microemulsions place stringent requirements on stability of the catalytic film because of the excellent solubilizing power of the microemulsions for polymers and other film materials.

An initial study of a catalytic electrode in microemulsions involved vitamin B_{12} hexacarboxylate [$B_{12}(COOH)_6$] (see below) chemisorbed onto a high-surface-area nanocrystalline titanium dioxide cathode [48]. Strong chemisorption of this cobalt complex occurs via interactions of the carboxylate groups with the TiO_2 surface. Compared with dissolved vitamin B_{12}, $B_{12}(COOH)_6$-titanium dioxide cathodes were catalytically more efficient in a CTAB microemulsion for reduction of dibromocyclohexane to cyclohexene [Eq. (7)]. 1,2-Dibromocyclohexane was converted to cyclohexene with 100 h^{-1} turnover using $B_{12}(COOH)_6$-TiO_2 cathodes, representing a fivefold increase in turnover number compared with the same reaction catalyzed by vitamin B_{12} in solution on a bare carbon electrode.

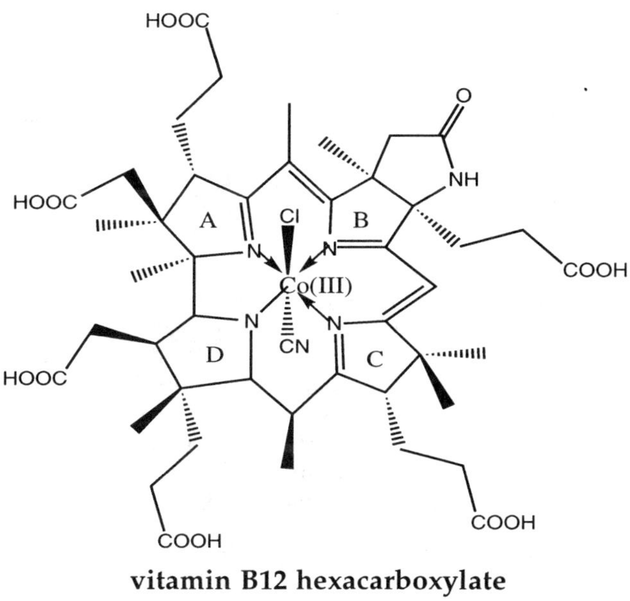

vitamin B12 hexacarboxylate

Cyclization of **4** in the microemulsion (Scheme 4) gave similar *trans/cis* ratios of **5** averaging 95:5 for both soluble and adsorbed catalyst systems. However, $B_{12}(COOH)_6$-titanium dioxide cathodes gave more than 100-fold larger turnover numbers in microemulsions [48]. In DMF, coated TiO_2 gave poor stereoselectivity for this reaction and poorer turnover than in the microemulsion.

Binding vitamin B_{12} hexacarboxylate onto TiO_2 electrodes improved electron transfer rates and catalytic efficiency while retaining essential features of cobalt complex–mediated catalysis in microemulsions. Unfortunately, a large fraction of the cobalt complex was lost from the cathode surface during synthetic electrolyses. Improved operational stability of a catalytic film would seem to require covalent linkage from electrode to film, with all film components also covalently linked.

A recent approach to stable catalytic films in our laboratory featured a polymeric scaffold attached to carbon electrode surfaces by covalent bonds. Multiple functional groups on the polymer were used to make bonds to the oxidized graphite surfaces and also to link to catalyst molecules. Specifically, catalytic films were constructed by covalently binding poly-L-lysine (PLL) onto oxidized pyrolytic graphite and carbon cloth [48]. Covalent amide linkages were then made between PLL and the cobalt corrin $B_{12}(COOH)_6$ (Fig. 6). These films could be used for 4–5 days in microemulsions.

The PLL-$B_{12}(COOH)_6$ films gave reversible electron transfer for the key catalytic Co(II)/Co(I) redox couple and exhibited characteristic voltammetric features of surface-confined electrochemistry. Formal potentials of the Co(II)/Co(I) couple in the films were controlled by the concentration of electrolyte in the fluid and by coulombic interactions of the films with surfactant. The

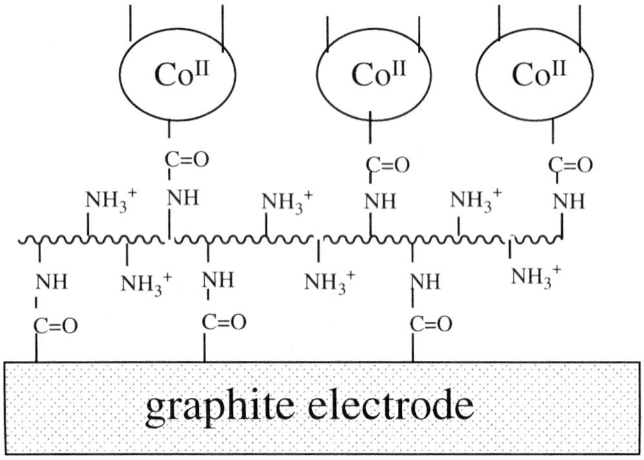

FIG. 6 Conceptual representation of fully covalently bound PLL-$B_{12}(COOH)_6$ films on graphite electrodes.

PLL-$B_{12}(COOH)_6$ films demonstrated good catalytic activity in microemulsions for reduction of vicinal dibromides to olefins, for dechlorination of trichloroacetic acid, and for alkylation of an activated olefin. Turnover numbers for conversion of dibromocyclohexane to cyclohexene in microemulsions were threefold or more larger for PLL-$B_{12}(COOH)_6$ on carbon cloth cathodes than for the $B_{12}(COOH)_6$-titanium dioxide cathodes [49].

VII. PROSPECTS FOR THE FUTURE

Development of mediated electrochemical synthesis in microemulsions is still in the early stages. Research using dissolved mediators has so far uncovered ways in which microemulsions can enhance and control kinetics of mediated electrochemical reactions, and some general kinetic principles have been proposed. We also have a qualitative understanding of surface films on electrodes in microemulsions and how they can influence synthetic reactions. Examples have been presented in which reaction pathways are favorably controlled in microemulsions. However, a detailed understanding of reaction pathway control awaits fundamental studies on a wider variety of systems. Nevertheless, microemulsions with dissolved electrochemical mediators presently appear quite attractive for specific applications to "green" synthetic processes.

An exciting and challenging new frontier involves design and control of catalytic electrodes in microemulsions. The technological promise of reusable catalytic electrodes designed for a range of mediated synthetic reactions in inexpensive, low-toxicity fluids is enormous. Preliminary work suggests new forms of control of mediated reactions in microemulsions via fine tuning of catalytic film construction and interactions of films with microemulsion components [49]. For this potential to be realized, a fundamental understanding of how film construction and molecular interactions can be used to influence reaction efficiency and pathways needs to be developed.

ACKNOWLEDGMENTS

The author's work described herein was supported by grants CTS-9306961, CTS-9632391 and CTS-9982854 from the National Science Foundation. The author thanks students and colleagues named in joint publications for their fine contributions to this work and especially Professors James M. Bobbitt and Albert J. Fry for helpful discussions, experiments, and suggestions.

REFERENCES

1. M. M. Baizer and H. Lund, eds., *Organic Electrochemistry*, 2nd ed., Marcel Dekker: New York, 1983.
2. D. Pletcher and F. C. Walsh, *Industrial Electrochemistry*, 2nd ed., Blackie Academic, London, 1993.
3. A. J. Fry, *Synthetic Organic Electrochemistry*, 2nd ed., Wiley-Interscience, New York, 1989.
4. M. Bourrel and R. S. Schechter, *Microemulsions and Related Systems*, Marcel Dekker, New York, 1988.
5. S. Friberg, Adv. Colloid Interface Sci. *32*:167–182 (1990).
6. J. F. Rusling, in *Electroanalytical Chemistry*, Vol. 18 (A. J. Bard, ed.), Marcel Dekker, New York, 1994, pp. 1–88.
7. R. A. Mackay, Colloids Surf. *82*:1–23 (1994).
8. J. F. Rusling, in *Modern Aspects of Electrochemistry*, No. 26 (B. E. Conway and J. O'M. Bockris, eds.), Plenum, New York, 1994, pp. 49–104.
9. D. F. Evans, D. J. Mitchell, and B. W. Ninham, J. Phys. Chem. *90*:2817–2825 (1986).
10. D. F. Evans and B. W. Ninham, J. Phys. Chem. *90*: 226–234 (1986).
11. J. F. Rusling and D.-L. Zhou, J. Electroanal. Chem. *439*:89–96 (1997).
12. R. Zana and J. Lang, in *Solution Behavior of Surfactants*, Vol. 2 (K. J. Mitall and E. J. Fendler, eds.), Plenum, New York, 1980, pp. 1195–1206.
13. R. Zana, in *Surfactants in Solution*, Vol. 4 (K. J. Mitall and P. Botherel, eds.), Plenum, New York, 1986, pp. 115–130.
14. A. J. Bard and L. R. Faulkner, *Electrochemical Methods*, Wiley, New York, 1980.

15. J. Gao, C. K. Njue, J. K. N. Mbindyo, and J. F. Rusling, J. Electroanal. Chem. *464/1*:31–38, 1999.
16. G. Gounili, C. L. Miaw, J. M. Bobbitt, and J. F. Rusling, J. Colloid Interface Sci. *153*:446–456 (1992).
17. M. O. Iwunze, A. Sucheta, and J. F. Rusling, Anal. Chem. *62*:644–649 (1990).
18. J. F. Rusling, Colloids Surf. *123–124*:81–88 (1997).
19. W. Peng, D.-L. Zhou, and J. F. Rusling, J. Phys. Chem. *99*:6986–6993 (1995).
20. A. P. Abbott, G. Gounili, J. M. Bobbitt, J. F. Rusling, and T. F. Kumosinski, J. Phys. Chem. *96*:11091–11095 (1992).
21. G. Gounili, J. M. Bobbitt, and J. F. Rusling, Langmuir *11*:2800–2805 (1995).
22. X. Zu and J. F. Rusling, Langmuir *13*:3693–3699 (1997).
23. E. Garcia, S. Song, L. E. Oppenheimer, B. Antalek, A. J. Williams, and J. Texter, Langmuir *9*:2782–2785 (1993).
24. F. Garcia and J. Texter, J. Colloid Interface Sci. *162*: 262–264 (1994).
25. B. Antalek, A. J. Williams, E. Garcia, and J. Texter, Langmuir *10*:4459–4467 (1994).
26. J. F. Rusling, Acc. Chem. Res. *24*:75–81 (1991).
27. R. Scheffold, in *Modern Synthetic Methods*, Vol. 3 (R. Scheffold, ed.), Wiley, New York, 1983, pp. 355–439.
28. R. Scheffold, S. Abrecht, R. Orlinski, H.-R. Ruf, P. Stamouli, O. Tinembart, L. Walder, and C. Weymuth, Pure Appl. Chem. *59*:363–372 (1987).
29. R. Scheffold, Chimia *39*:203–211 (1985).
30. S. Torii, Synthesis 873–886 (1986).
31. G. Pattenden, Chem. Soc. Rev. *17*:361–382 (1988).
32. R. Scheffold, in *Electroorganic Synthesis* (R. D. Little and N. L. Weinberg, eds.), Marcel Dekker, New York, 1991, pp. 317–322.
33. B. M. Trost, Science *254*:1471–1477 (1991).
34. J.-M. Savéant, Adv. Phys. Org. Chem. *26*:1–130 (1990).
35. D.-L. Zhou, J. Gao, and J. F. Rusling, J. Am. Chem. Soc. *117*:1127–1134 (1995).
36. D.-L. Zhou, H. Carrero, and J. F. Rusling, Langmuir *12*:3067–3074 (1996).
37. Y. Shao, M. V. Mirkin, and J. F. Rusling, J. Phys. Chem. *101*:3202–3208 (1997).
38. G. N. Kamau and J. F. Rusling, Langmuir *12*:2645–2649 (1996).
39. A. Owlia, Z. Wang, and J. F. Rusling, J. Am. Chem. Soc. *111*:5091–5908 (1989).
40. J. Gao, J. F. Rusling, and D.-L. Zhou, J. Org. Chem. *61*:5972–5977 (1996).
41. J. Gao, C. K. Njue, J. K. N. Mbindyo, and J. F. Rusling, J. Electroanal. Chem. *464/1*:31–38 (1999).
42. J. Gao and J. F. Rusling, J. Org. Chem. *63*:218–219 (1998).
43. (a) H. Bhandal, G. Pattenden, and J. J. Russell, Tetrahedron Lett. *27*:2299–2302 (1986); (b) M. Ladlow and G. Pattenden, Tetrahedron Lett. *25*:4317–4320 (1984); (c) H. Bhandal, V. F. Patel, G. Pattenden, and J. J. Russell, J. Chem. Soc. Perkin Trans. I 2691–2701 (1990).
44. B. Giese, P. Erdmann, T. Gobel, and R. Springer, Tetrahedron Lett. *33*:4545–4548 (1992).
45. H. Carrero, J. Gao, J. F. Rusling, C.-W. Lee, and A. J. Fry, Electrochemical Acta *45*:503–512 (1999).
46. R. W. Murray, in *Electroanalytical Chemistry*, Vol. 13 (A. J. Bard, ed.), Marcel Dekker, New York, 1984, pp. 191–368.
47. R. W. Murray, ed., *Molecular Design of Electrode Surfaces*, Vol. 22, Techniques of Chemistry Series, Wiley-Interscience, New York, 1992.
48. J. K. N. Mbindyo and J. F. Rusling, Langmuir *14*: 7027–7033 (1998).
49. (a) D.-L. Zhou, C. K. Njue, and J. F. Rusling, J. Am. Chem. Soc. *121*:2909–2914 (1999); (b) C. K. Njue and J. F. Rusling, J. Am. Chem. Soc. *122*:6459–6463 (2000).

13

Chemical Activation in Micelles, Pseudomicelles, and Microemulsions

ISABELLE RICO-LATTES, ARMAND LATTES, EMILE PEREZ, FERDINAND GONZAGA, and ANDREEA RUXANDRA SCHMITZER Laboratoire IMRCP, UMR CNRS 5623, Université Paul Sabatier, Toulouse, France

I. INTRODUCTION

A. Overview

A variety of novel synthetic reactions have been performed in organized molecular systems (OMSs). These organized media possess a number of advantages, including solubilization of substances that are not soluble in the continuous phase of the OMS, localization of reactants and products, selective orientation, and stabilization of various entities in the various stages of the reaction.

The reactivity of substrates can be directed in these systems so that competing reactions may be selectively oriented to either one of the reaction pathways (cyclization or polymerization, for example). We illustrate this strategy in two model processes: (1) intramolecular cyclization and formation of lactams and (2) polymerization and formation of latexes. In both of these cases it is possible to perform reactions with high yields and selectivity.

To increase performance, OMS could be extended to chiral synthesis. Development of chiral surfactants, such as those with sugar or amino acid headgroups, has so far led to only slight enantiomeric excess. The reason for these failures is the dynamics of micelles. By using amphiphilic dendrimers, similar to rigid unimolecular micelles, we have, for example, been able to reduce prochiral ketones with highly asymmetric induction.

B. Organized Molecular Systems

Organized media have the combined advantages of homogeneous (which they are macroscopically) and heterogeneous (which they are microscopically) media. They can thus be defined as microscopically heterogeneous solvent media. Such systems have emerged in the natural evolution of chemistry and are not encompassed in the field of supramolecular chemistry. The implementation of complex systems, which may comprise a large number of components, places the emphasis on systems rather than species.

1. Expectations from Organized Molecular Systems

Liquid organized media can both solubilize and localize reactants. Substances dissolved in these solutions will distribute throughout the continuous phase as well as in objects such as micelles and mesophase surfactant aggregates. This compartmentalization of reactants and products can be exploited to modify concentrations locally, prevent approach of unwanted reactants, or orient molecules so that they react selectively [1]. For example, it is thus possible to perform reactions that could be carried out only in the solid phase [2].

2. Choice of Organized Molecular Systems

Surfactants are key components of such organized systems used in studies of chemical reactivity, and it is important to use an appropriate surfactant for the pro-

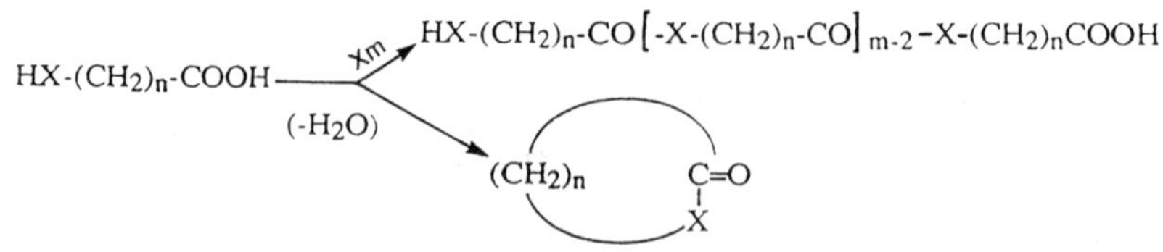

SCHEME 1 Competition between polymerization and cyclization of bifunctional compounds.

cess under consideration. This may require synthesis of a range of suitable surfactants. The overall objective largely dictates the choice of the organized molecular systems (direct or reverse micelles, microemulsions, etc).

To illustrate these considerations some concrete examples will be discussed. We have chosen these examples in order to emphasize the peculiarities and the advantages of such media:

- Demonstration of the occurrence of some reactions at an interface
- Localization of reactants and products
- Development of formulations by simplifying the number of components in the medium, referred to as the "principle of molecular economy"
- Generalization of the amphiphile concept
- Dynamic behavior of micellar media preventing the occurrence of stereoselective reactions and compelling the interface to become rigid in order to provide asymmetric reactions

II. INTRAMOLECULAR CYCLIZATION AND FORMATION OF LACTAMS

A. Intramolecular Cyclization

The presence of two complementary functional groups in the same molecule can lead to competition between two reactions, polymerization and cyclization (Scheme 1).

The preparation of compounds with medium and large rings is of considerable interest in view of the number of naturally occurring compounds with 7 to 20-membered rings, specially lactones and lactams [3]. Polymerization can be avoided by carrying out the reaction at high dilution. Nevertheless, the formation of these rings is entropically unfavorable and intermolecular processes cannot always be circumvented, even with high dilution techniques. To favor cyclization, Mukaiyama has described the use of a carboxyl activating agent: *N*-methyl-2-chloropyridinium iodide (C_1ClPyI) [4].

Previous studies of cyclization reactions in micellar environments include lactonization in reverse micelles [5] and, more recently, in aqueous micelles, where Lennox et al. observed enhanced cyclization rates due to the looping of the substrate at the interface [6].

We utilized this effect for both lactonization and lactamization by using *N*-hexadecyl-2-chloropyridinium iodide (C_{16}ClPyI) instead of the Mukaiyama reagent [7]. To prepare C_{16}ClPyI, in view of the difficulty of direct alkylation, we developed a new two-step procedure illustrated in Scheme 2.

The first stage is a phase transfer reaction giving a 94% yield of *N*-hexadecyl-2-pyridone. The second stage is a one-pot process that affords *N*-hexadecyl-2-chloropyridinium iodide in 80% yield. Then we designed experiments to determine the nature of aggregates formed in the solvent used for the cyclization reaction with the classical Mukaiyama reagent, 1,2-dichloroethane. Owing to the low cohesion energy density and polarity of this solvent, inverse micelles would be expected. The results obtained at 25°C are illustrated in Fig. 1.

In 1,2-dichloroethane, the change in slope was observed around 4.5×10^{-5} M. Therefore, micellar effects could be produced in a very large range of concentrations during cyclization processes.

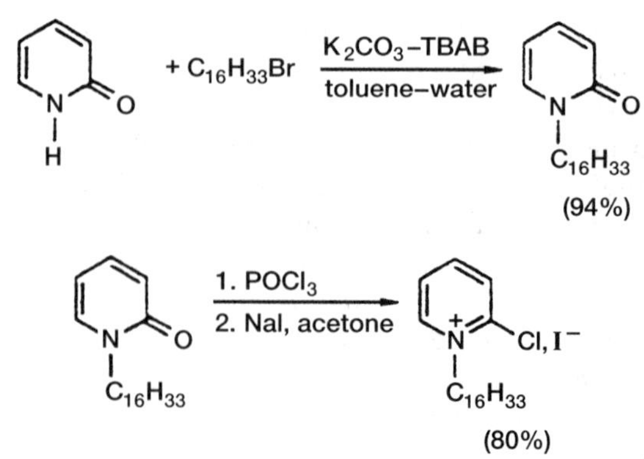

SCHEME 2 Synthesis of *N*-hexadecyl-2-chloropyridinium iodide (C_{16}PyCl, I).

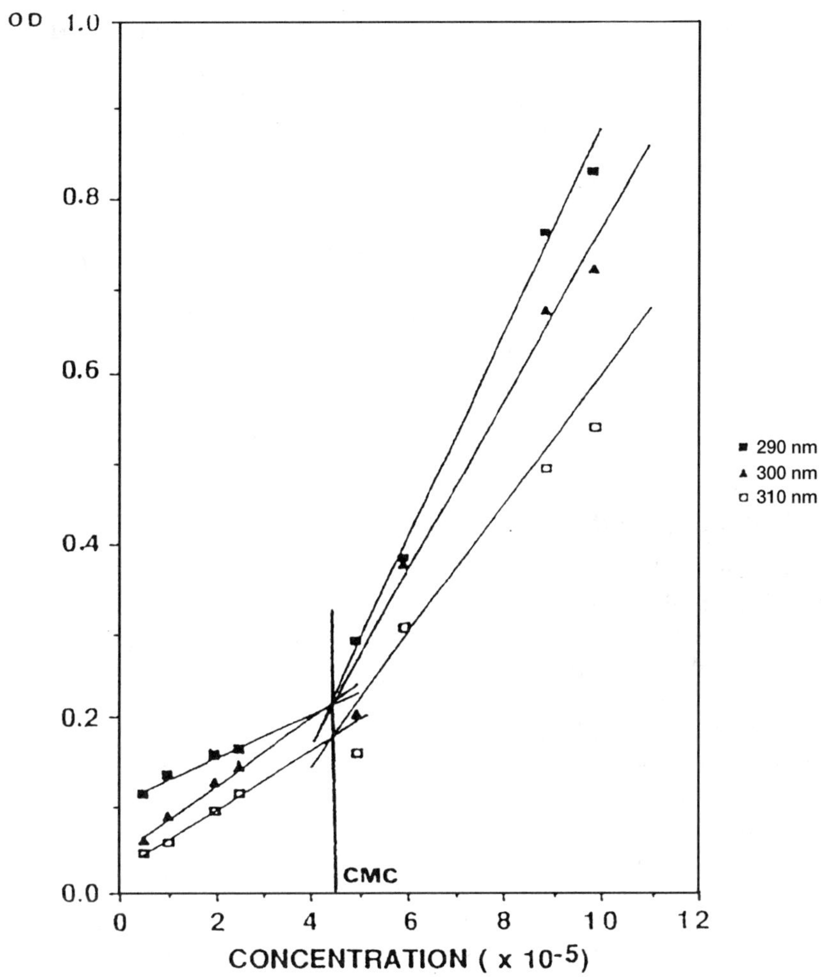

FIG. 1 Plot of absorbance OD (optical density) against concentration (M) of C_{16}ClPyI in 1,2-dichloroethane at 25°C.

To confirm the nature of the micelles, we examined the solubilization of water in micelles of C_{16}ClPyI in 1,2-dichloroethane. We found that water could not be solubilized, either with or without triethylamine (in the same proportions as those used in the cyclization reaction).

Without addition of water, the water content (Karl Fischer method) was 0.02% in 1,2-dichloroethane. Under the reflux conditions, the activating agents (C_1 or C_{16}) were hydrolyzed, but with dry argon as an inert atmosphere, this reaction was reduced. The fact that C_{16}ClPyI was more readily hydrolyzed than C_1ClPyI can be attributed to a micellar catalytic effect for the long-chain derivative. The OH^- ions have greater affinity for the polar micellar core than for the apolar continuous phase. Figure 2 shows the high concentration of OH^- ions within the micelle, enhancing hydrolysis.

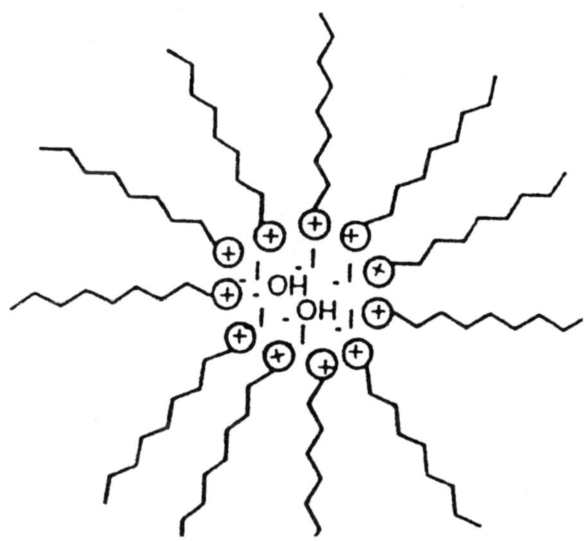

FIG. 2 Hydrolysis of C_{16}ClPyI in 1,2-dichloroethane.

TABLE 1 Lactonization of $HOC_{15}H_{31}CO_2H$ (S) Activated by C_1PyCl, I or $C_{16}PyCl$, I at 80°C; (S) = 0.028 M in 1,2-Dichloroethane

Entry	Activating agent (A)	Cyclization conditions: Argon	Cyclization conditions: Ratio (A/S)	Yield of lactone (%)
1	C_1PyCl, I	−	2:1	37
2	$C_{16}PyCl$, I	−	2:1	23
3	C_1PyCl, I	+	2:1	41
4	$C_{16}PyCl$, I	+	2:1	36

TABLE 2 Lactamization of $H_2NC_{11}H_{23}CO_2H$ (S) Activated by C_1PyCl, I or $C_{16}PyCl$, I at 80°C; S = 0.028 M in 1,2-Dichloroethane

Entry	Activating agent (A)	Cyclization conditions: Argon	Cyclization conditions: Ratio (A/S)	Yield of lactam (%)
1	C_1PyCl, I	−	1:1	20
2	$C_{16}PyCl$, I	−	1:1	35
3	C_1PyCl, I	−	2:1	24
4	$C_{16}PyCl$, I	−	2:1	32
5	C_1PyCl, I	+	1:1	24
6	$C_{16}PyCl$, I	+	1:1	54

B. Lactonization and Lactamization Activated by C_1ClPyI or $C_{16}ClPyI$

The preceding results show that both reagents are sensitive to traces of water. This is the reason we compared cyclization reactions in the presence of an atmosphere of dry argon. The results of the *lactonization* of 16-hydroxydecanoic acid carried out in 1,2-dichloroethane according to the method of Mukaiyama, but with a five-fold higher substrate concentration (0.028 M), are listed in Table 1.

Lactamization of 12-aminododecanoic acid was also carried out under the same conditions. Although the starting ω-amino acid was only weakly soluble in 1,2-dichloroethane, the lactam was quite soluble in this solvent. Yields of macrocyclic derivatives, determined by GPC (gas phase chromatography), are listed in Table 2.

The following comments can be made about the results in Tables 1 and 2:

1. Yields were higher in the presence of an atmosphere of dry argon, in agreement with the results on hydrolysis.
2. There was a micellar effect for lactamization but not for lactonization.

To understand this effect we have to consider the mechanism shown in Scheme 3.

The observed difference in yield between lactonization and lactamization with $C_{16}ClPyI$ as activating agent was accounted for by differences in solubility of the substrate in the continuous medium. *The efficiency of cyclization appeared to be inversely related to the solubility of the substrate in the solvent.*

If the substrate is completely soluble in the organic solvent, polymerization tends to occur before the substrate has had time to react at the interface.

If the substrate is poorly soluble in the organic solvent, it will be solubilized at the interface where cyclization will take place.

The differences in yield between these two substrates can be accounted for by the high solubility of

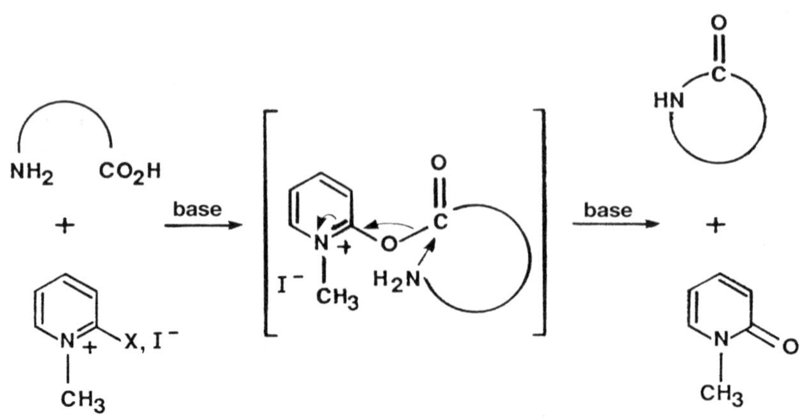

SCHEME 3 Mechanism of the reactions observed with bifunctional compounds and activation reagents.

the ω-hydroxy acids and the poor solubility of the ω-amino acids in 1,2-dichloroethane. These two situations are illustrated in Figs. 3 and 4.

3. The value of a microstructured medium for preparation of lactams was demonstrated. By using *N*-hexadecyl-2-chloropyridinium iodide as activating agent and surfactant and by curbing the hydrolysis reaction, we doubled the yield of lactam normally obtained with the Mukaiyama reagent.

The reaction could also be carried out at high substrate concentration, avoiding both high dilution conditions and modifications of the starting ω-amino acid.

III. POLYMERIZATION OF NORBORNENE AND NORBORNENE DERIVATIVES

Norbornene and its derivatives can be polymerized via two different routes: vinyl polymerization and ring opening (Fig. 5) The route chosen depends on the catalyst employed:

Catalysts based on transition metals such as W, Ru, Mo, Os, or Ir lead to ring opening (Fig. 5, structure III).

Catalysts based on palladium, such as those of the Ziegler-Natta type, lead to vinyl polymerization (Fig. 5, structure I).

In previous studies we examined the polymerization of norbornene by $RuCl_3 \cdot 3H_2O$ in micellar solutions and microemulsions.

A. Polymerizing Metathesis of Norbornene

It is known that norbornene can polymerize by metathesis. In an alcoholic medium (1-butanol) the catalyst $RuCl_3 \cdot 3H_2O$ leads to a stereoselective polymerization (95% *trans*) [8].

The addition of formamide, after a threshold concentration has been reached, gives rise to a polynorbornene with 45% *cis* and 55% *trans* isomers [9]. At this minimal concentration of formamide, we observed an abrupt alteration in the viscosity of the polymer (Table 3). In the absence of surfactant, this was attributed to the formation of systems resembling microemulsions favored by the different natures of formamide and norbornene (immiscible) [9]:

Low amounts of formamide: droplets of formamide are dispersed in a continuous phase of norbornene and alcohol where the reaction takes place.

Larger amounts of formamide: droplets of norbornene are dispersed in a continuous phase of formamide and alcohol. The reaction takes place at the interfaces of the norbornene droplets.

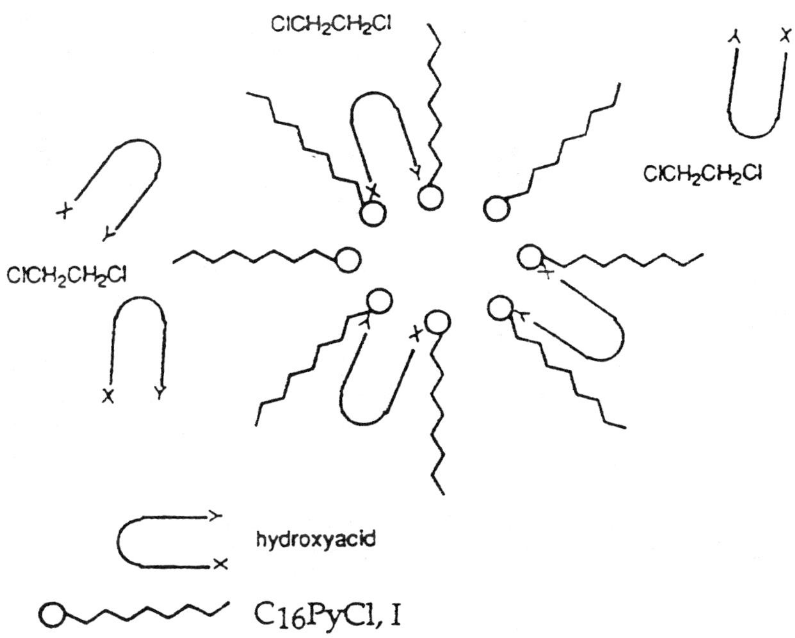

FIG. 3 Cyclization of an ω-hydroxy acid that is completely soluble in the continuous phase.

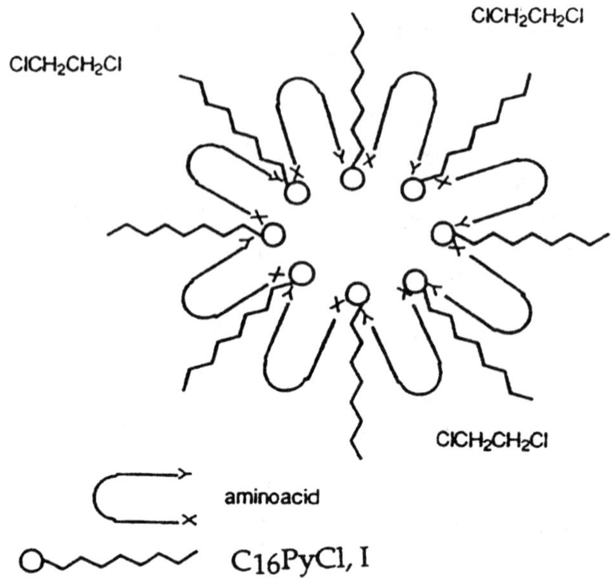

FIG. 4 Cyclization of an ω-amino acid that is poorly soluble in the continuous phase.

It can be seen that these two situations are akin to the extreme cases of H_2O/oil or oil/H_2O microemulsions. This hypothesis was verified by carrying out the polymerization in a true microemulsion whose ternary diagram is reproduced in Fig. 6.

1. In microemulsion 2, where the dispersed phase is composed of droplets of formamide and the continuous phase is norbornene, *trans* stereoselectivity is observed.
2. In microemulsion 1, where droplets of norbornene form in the formamide continuous phase, the stereoselectivity is 40% *cis*.

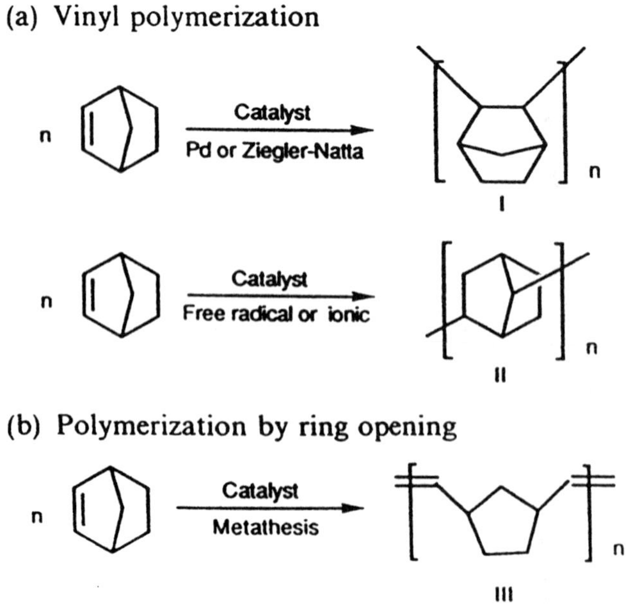

FIG. 5 Different routes of norbornene polymerization.

These examples of spontaneous organization can thus be seen to be quite general. Such organization processes may help account for the unexpected influence of certain reaction conditions, where minimal alterations in reaction conditions can produce marked changes in the products obtained.

B. Vinyl Polymerization of Norbornene and Formation of Latex

Norbornene can undergo a polymerization of the vinyl type using catalysts based on palladium. Among the palladium-based catalysts, $PdCl_2$ has been relatively little used [10] because it gives low yields. It is generally replaced by more active catalysts, which have the drawback of requiring lengthy preparation [11,12]. $PdCl_2$ also has the disadvantage of being insoluble in most organic solvents, although it is slightly soluble in water (0.7 g L^{-1}). We investigated the polymerisation of norbornene catalyzed by $PdCl_2$ dispersed in water and in emulsions [13,14].

1. Polymerization in Water Dispersion

Norbornene is polymerized at 70°C. At this temperature it is a liquid compound. The reaction is carried out in

TABLE 3 Norbornene Polymerization in Butanol Medium with $RuCl_3$, $3H_2O$ in the Presence of Formamide

Ru/butanol-1/norbornene (moles)	$HCONH_2$/butanol-1	$\sigma_c \pm 0.05$[a]	Yields (%)[b]	(η) $(m^3\ kg^{-1})$[c] $\times 10^{-2} \pm 0.05 \times 10^{-2}$
1/370/100	0	0.05	85	1.05
1/370/100	0.2	0.05	82	1.05
1/370/100	1.2	0.05	74	1.15
1/370/100	1.5	0.45	63	0.65
1/555/100	2.3	0.4	—	—
1/370/100	3.4	0.4	40	0.60
1/227/100	5.5	0.4	—	—
1/74/100	29.8	0.35	—	—

[a]Proportions of *cis* double bonds.
[b]Yields after purification.
[c]Intrinsic viscosity (T = 30°C, benzene, Ubbelohde viscosimeter).

water under vigorous stirring, which leads to the formation of norbornene droplets dispersed in the aqueous medium. Because palladium chloride is slightly soluble in water, the reaction starts at the droplet interface and then progresses inward.

Polymerization was rapid, giving a crude yield of 70% within 24 h. The polymer obtained was in the form of a powder insoluble in solvents normally used in such studies, indicating that it had high molecular weight or a very cross-linked structure. The infrared (IR) spectroscopic data showed that vinyl-type polymerization (structure I) had taken place.

2. Polymerization in SDS/Water Emulsion

The preceding results show that water does not inhibit the reaction, so it was suggested that reaction in emulsion might lead to novel polymer-type latex structures. We selected an anionic surfactant, sodium dodecyl sulfate (SDS), to favor the presence of Pd^{2+} at the interface. At 1, 3, and 6% SDS, 20% norbornene, the solutions consist of an oil-in-water emulsion. In the presence of a low quantity of $PdCl_2$ (0.37%), the polymerization reaction occurs. We investigated the effect

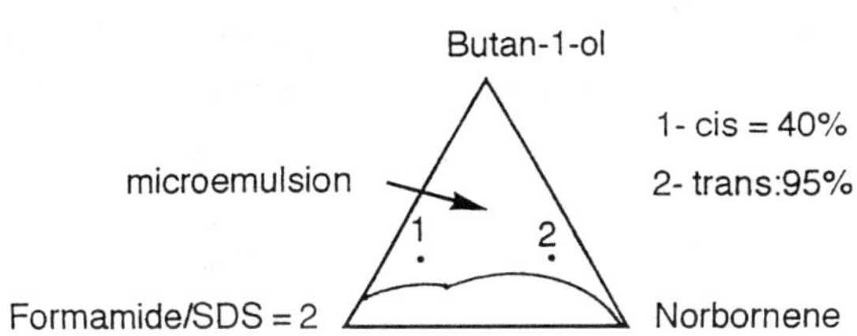

FIG. 6 Pseudoternary phase diagram for the system formamide/SDS (= 2), norbornene, and 1-butanol at 58°C and methatetic reaction compositions of norbornene in two mixtures (1 and 2).

of the concentration of SDS on the size of latex particles (Table 4). At 3 and 6% SDS, latex was produced in the form of nanoparticles with a size range (9.5–16.9 nm) generally observed only in microemulsion polymerization.

Tacticity studies were difficult to carry out on such polymers because of their high molecular weight. Stereochemistry and tacticity of linkages were deduced from the mechanism and the nuclear magnetic resonance (NMR) data. An atactic structure for the polymers was proposed. The possible linkages illustrated in Fig. 7 were suggested.

The results showed for the first time that it is possible to use $PdCl_2$ as a catalyst to polymerize norbornene in water dispersions and in aqueous emulsions. In this last case, the nanoparticles obtained were found to have the smallest size obtained to date (<10 nm in diameter) and were highly monodisperse.

C. Vinyl Polymerization of Norbornene Derivatives

The preparation of small latexes has been extended to latexes of water-soluble polymers bearing functional groups such as alcohol and sugar residues [15]. Polymerization of 5-hydroxymethyl bicyclo [2.2.1]-2-heptene is illustrated in Scheme 4.

With this functional monomer the polymerization was carried out in suspension in water without surfactant. Under these conditions we needed to use a hydrosoluble catalyst, $PdCl_2$ (TPP TS)$_2$ [16], to get a latex (Fig. 8).

Under the same conditions as in the previous reaction with unsubstituted norbornene, we obtained the polymer with 80% yield. The latex obtained was highly

TABLE 4 Influence of SDS Concentration on Polymer Properties and Yield after Reaction for 24 h at 70°C

SDS % (weight)	Size (nm)[a]	$[\eta]$ (dL/g)[b]	Mn[c]	Yield (%)[d]
1 (3.5×10^{-2} M)	—	—	630	80
3 (0.1 M)	9.5–16.9	1.6×10^{-2}	840	90
6 (0.2 M)	9.5–16.9	3.3×10^{-2}	1010	90

[a]Diameters determined by light scattering at 25°C.
[b]Intrinsic viscosity in cyclohexane at 25°C.
[c]Molecular weights determined by cryometry.
[d]Yield of polymer after purification.

monodisperse. The particle diameters were narrowly distributed around 316 nm.

Polymerization of sugar-substituted norbornenes has also been achieved. We have studied two sugar-substituted norbornenes: *N*-methylnorbornylgluconamide and *N*-methyllactobionamide. The monomers were prepared as indicated in Scheme 5.

The polymerization in water catalyzed by $PdCl_2$ $(TPTS)_2$ gave very good yields of the corresponding polymers. Soluble in water, they formed stable aggregates, which were characterized by transmission electron microscopy. The gluconamide polymers yielded very small (9.5 nm) nanoparticles. The lactobionamide polymers yielded polydisperse (10–350 nm) particles. Comparable size results were obtained by light scattering (9.5 and 440 nm for the two polymers, respectively). This approach appears to be an original method for preparing latexes by addition polymerization of norbornene and functionalized norbornenes in water emulsions or solutions.

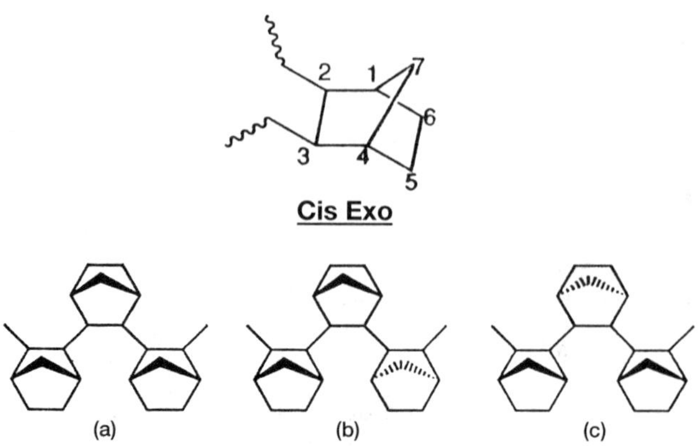

FIG. 7 Possible stereochemistry of norbornene polymers.

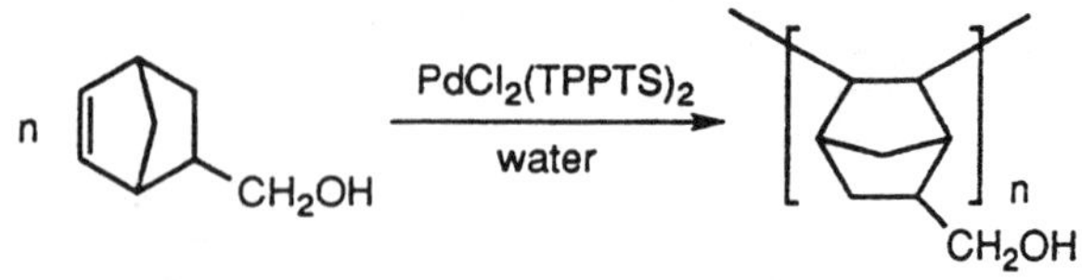

SCHEME 4 Polymerization of 5-hydroxymethyl bicyclo [2.2.1]-2-heptene catalyzed by $PdCl_2$ $(TPPST)_2$.

IV. ASYMMETRIC INDUCTION AT THE "PSEUDOMICELLAR" INTERFACE OF A CHIRAL AMPHIPHILIC DENDRIMER

Such novel latexes bearing sugar residues gave us the idea of evaluating their behavior in mimicking the stereochemistry of enzymatic reactions. Chiral micelles have been used to attempt asymmetric induction, but only a few positive examples are described in the literature (the yield varying from 1.7 to 27% for reduction of aromatic ketones) [17,18]. The essential reason for all of these failures is the dynamics of the micelles. The chiral interface is not rigid enough to provide stable asymmetric complexes.

In organic liquids or in water, latexes from colloidal objects more stable in shape and more rigid than micelles, even if their conformation depends on their heterogeneous of homogeneous state in solution. Reactions with the newly prepared glycopolymer latexes are currently under investigation, but in order to optimize the environmental effect of chiral residues, we decided to prepare dendrimers having polyhydroxylated groups derived from glucose at their periphery.

We described the synthesis of new amphiphilic dendrimers derived from PAMAM (Poly(amidoamine)) involving this multiple attachment of chiral groups [19] (Fig. 9). The procedure for preparing the dendrimers G(*n*)G is similar to procedures previously described for synthesizing glycopolymers derived from norbornene.

These new polymeric compounds could be considered rigid unimolecular micelles. We have chosen to study, as a model reaction, the reduction of prochiral ketones by sodium borohydride at the chiral interface of these dendrimers. The third-generation dendrimer G(3)G (Fig. 9) has been utilized because it is the first

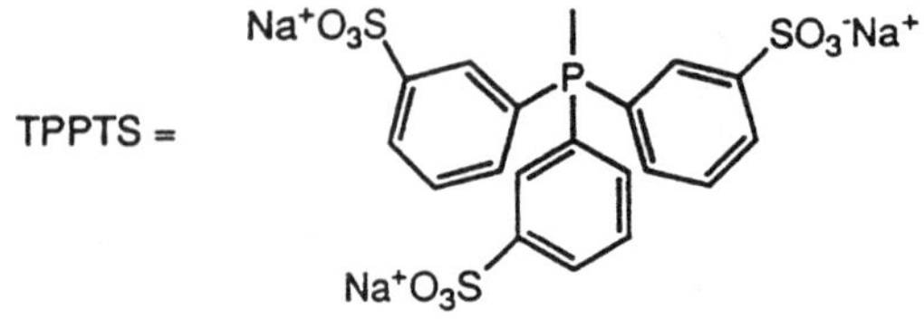

FIG. 8 Structure of $PdCl_2$ (TPP TS)$_2$.

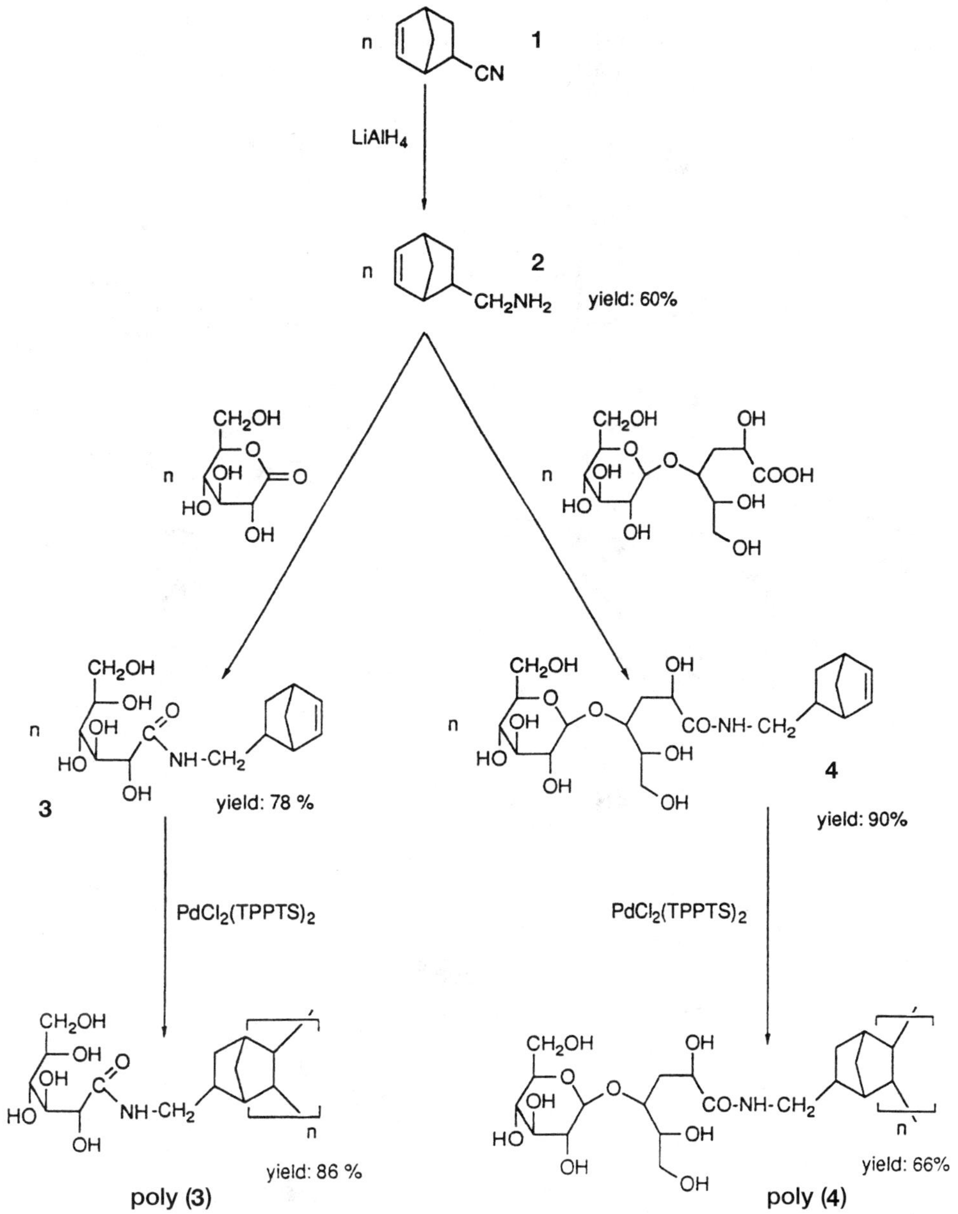

SCHEME 5 Synthesis and polymerization of *N*-methylnorbornylgluconamide and *N*-methylnorbornyllactobionamide.

generation leading to a nearly spherical shape similar to a micelle and it possesses a closed and densely packed structure. The reduction was performed in two solvents, water and THF [20]. Results are summarized in Table 5.

One can observe in Table 5 that highly asymmetric induction is obtained in the presence of a chiral dendrimer. The best result at this time is with acetophenone (yield = 99.4%). The induction is highest when the reduction is performed under heterogeneous conditions.

V. CONCLUSIONS

In the first part of this chapter we described the important effect of a novel carboxyl-activating agent. The organization of this agent in micelles in the reaction medium facilitates cyclization giving rise to a micellar effect. Because cyclization takes place at the interface and polymerization occurs essentially in the continuous phase, the solubility of the substrate in the solvent (continuous phase) is of particular importance. This

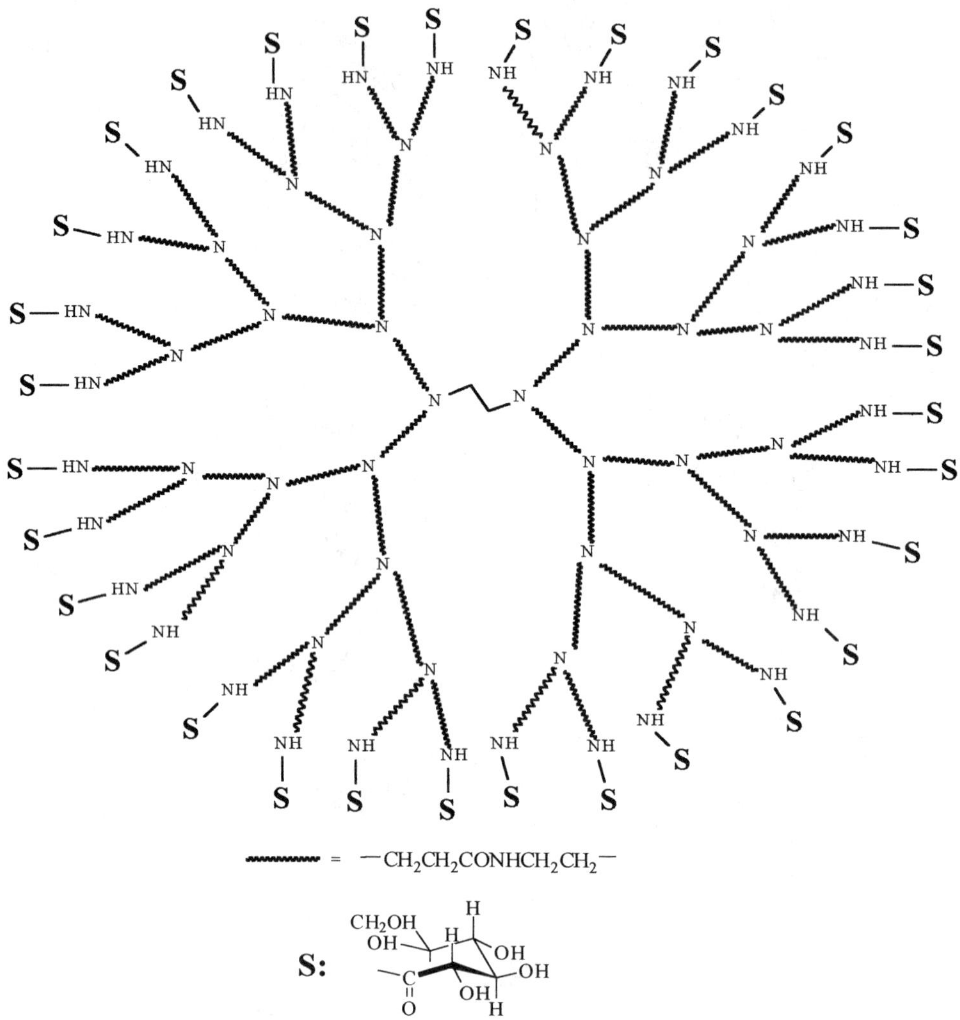

FIG. 9 Structure of amphiphilic dendrimer G(3)G.

TABLE 5 Enantioselective Reduction of Prochiral Ketones by $NaBH_4$ in the Presence of G(3)G

Ketones	Solvent	T (°C)	Chemical yield (%)	Ee (%)	Abs. conf. of alcohol
Cyclohexyl phenyl	Water	25	95	50	*S*
Cyclohexyl phenyl	THF	25	97	97	*S*
Butyl phenyl	THF	25	96	100	*S*
Propyl phenyl	THF	25	90	99.9	*S*
Methyl phenyl	THF	25	92	82	*S*
Methyl phenyl	THF	0	94	99.4	*S*

Ee = Enantiomeric excess.
Abs. conf. = absolute configuration.

method would be especially suited to cyclization of peptides, which are poorly soluble in organic solvents.

Polymerizing metathesis of norbornene is a good model for demonstrating that organic reactions may be influenced by the *spontaneous formation of aggregates*. Also with norbornene, the first example of latex synthesis by a nonradical organometallic olefin polymerization was described as emanating from reaction in aqueous emulsions with SDS.

The preparation of small latexes has also been extended to latex bearing functional groups such as alcohol or sugar residues. The new glycopolymers seem interesting for their potential biological applications.

Finally, an amphiphilic chiral dendrimer has been prepared in the same way. It acts as a highly pseudomicellar enantioselective ligand for the reduction of prochiral ketones.

REFERENCES

1. A. Lattes and I. Rico-Lattes, C. R. Acad. Sci. Paris *324* (Ser. IIb):575–587 (1997).
2. H. Amarouche, C. de Bourayne, M. Rivière, and A. Lattes, C. R. Acad. Sci. Paris *298* (Ser. II):121–124 (1984).
3. I Paterson and M. M. Mansouri, Tetrahedron *41*:3569–3624 (1985).
4. T. Mukaiyama, Angew. Chem. Int. Ed. Engl. *18*:707–721 (1979).
5. D. A. Jayer and J. T. Ippoliti, J. Org. Chem. *46*:4964–4968 (1981).
6. I. Wei, A. Lucas, J. Yue, and R. B. Lennox, Langmuir *7*:1336–1341 (1991).
7. I Rico, K. Halvorsen, C. Dubrule, and A. Lattes, J. Org. Chem. *59*:415–420 (1994).
8. E. A. Ofstead, J. P. Lawrence, M. L. Senyek, and N. Calderon, J. Mol. Catal. *8*:227–237 (1980).
9. E. Perez, N. Alandis, J. P. Laval, I. Rico, and A. Lattes, Tetrahedron Lett. *28*:2343–2346 (1987).
10. C. Tanielan, A. Kiennemann, and T. Osparpucu, Can. J. Chem. *57*:2022–2027 (1979).
11. A. Sen, Wang Lai Ta, and R. R. Thomas, J. Organometal. Chem. *358*:567–575 (1988).
12. A. Sen and Wang Lai Ta, J. Am. Chem. Soc. *104*:3520–3522 (1982).
13. P. Eychenne, E. Perez, I. Rico, M. Bon, A. Lattes, and A. Moisand, Colloid Polym. Sci. *271*:1049–1054 (1993).
14. L. Puech, E. Perez, I. Rico-Lattes, M. Bon, and A. Lattes, New J. Chem. *21*:1235–1242 (1997).
15. L. Puech, E. Perez, I. Rico-Lattes, M. Bon, A. Lattes, and A. Moisand, New J. Chem. *21*:1229–1234 (1997).
16. C. Larpent and H. Patin, Appl. Organomet. Chem. *24*, 43–46 (1987).
17. S. I. Goldberg, N. Baba, R. L. Green, R. Pandiar, and J. Stowers, J. Am. Chem. Soc. *100*:6768–6769 (1978).
18. Y. Zhang and P. Sun, Tetrahedron Asymmetry *7*(1a): 3055–3058 (1996).
19. A. Schmitzer, E. Perez, I. Rico-Lattes, A. Lattes, and S. Rosca, Langmuir *8*:4397–4403 (1999).
20. A. Schmitzer, E. Perez, I. Rico-Lattes, and A. Lattes, Tetrahedron Lett. *40*:2947–2950 (1999).

14

Reactions and Synthesis in Microemulsions and Emulsions in Carbon Dioxide

KEITH P. JOHNSTON, C. T. LEE, G. LI, and P. PSATHAS University of Texas, Austin, Texas

J. D. HOLMES University College, Cork, Ireland

G. B. JACOBSON and M. Z. YATES Los Alamos National Laboratory, Los Alamos, New Mexico

I. INTRODUCTION

The environmentally benign, nontoxic, and nonflammable fluids water and carbon dioxide (CO_2) are the two most abundant and inexpensive solvents on earth. Recently discovered water/CO_2 (W/C) and CO_2/water (C/W) microemulsions [1,2] and emulsions [3,4] offer new possibilities in waste minimization by replacing organic solvents in separations, reactions, and materials formation processes. Microemulsion droplets 2 to 10 nm in diameter are optically transparent and thermodynamically stable, whereas kinetically stable emulsions and latexes in the range of 200 nm to 10 μm are opaque and thermodynamically unstable. These studies in CO_2 result from a foundation that was built for water/oil microemulsions [5,6] and polymer latexes [7] in ethane and propane as reviewed [8–10]. Because CO_2 is nonpolar (unlike water) and has weak van der Waals forces (unlike lipophilic phases), it may be considered a third type of condensed phase. Consequently, polymers with low cohesive energy densities and thus low surface tensions are the most soluble in CO_2 (e.g., fluoroacrylates [11], fluorocarbons, fluoroethers [12], siloxanes, and to a lesser extent propylene oxide).

Microemulsions and emulsions of water and CO_2 have the ability to function as a somewhat "universal" solvent medium by solubilizing high concentrations of polar, ionic, and nonpolar molecules. A brief overview is given of the formation and stability of these dispersions with an emphasis on surfactant structure. We then describe the use of these dispersions as novel media for phase transfer reactions without requiring toxic organic solvents or phase transfer catalysts. The high interfacial area is shown to facilitate phase transfer reactions between hydrophilic nucleophiles and hydrophobic, CO_2-soluble, inorganic or organic substrates. Also, W/C microemulsions have been used to host enzymes for catalytic transformations of CO_2-soluble substrates. Examples are discussed in which W/ethane and W/C microemulsions serve as templates to control the size of copper, silver, and CdS nanoparticles. The next section provides a mechanistic description of steric stabilization of organic polymer emulsions and latexes in CO_2 that complements Chapter 5 in this book on polymerization in CO_2. Finally, novel bifunctional and trifunctional ambidextrous surfactants have been developed for steric stabilization of latexes in CO_2, which can be transferred to water to form electrostatically stabilized latexes.

II. WATER-IN-CO_2 MICROEMULSIONS

A fundamental understanding of colloid and interface science for surfactant design is emerging on the basis of studies of interfacial tension, γ, and surfactant adsorption at CO_2/water and CO_2/organic interfaces [3,13]. The effect of surfactants on the interfacial tension between water and supercritical fluids is a key property for describing emulsions and microemulsions

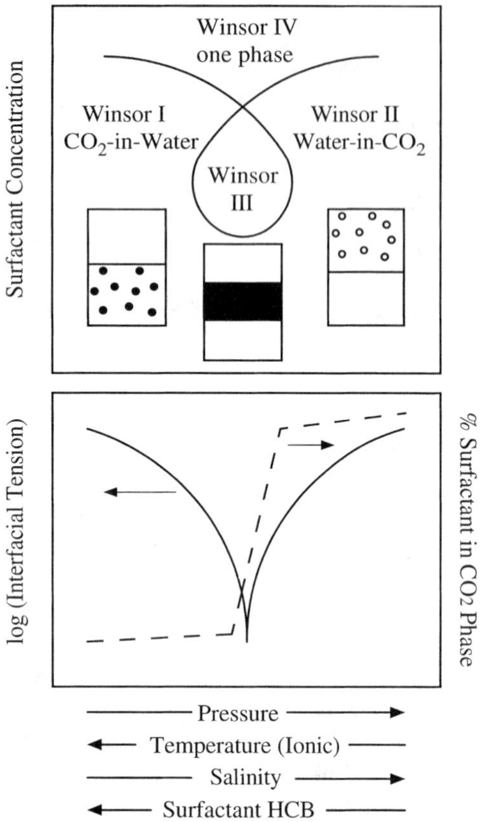

FIG. 1 Schematic representation of the relationship between phase behavior and interfacial tension for water/CO_2/ionic surfactant mixtures as a function of formulation variables.

[3], as shown in Fig. 1. A minimum in γ is observed at the phase inversion point, where the system is balanced with respect to the partitioning of the surfactant between the phases [14]. Upon changing any of the formulation variables away from this point, for example, the hydrophilic/CO_2-philic balance (HCB) in the surfactant structure, the surfactant will migrate toward one of the phases. This phase usually becomes the external phase, according to the Bancroft rule. For example, a perfluoropolyether (PFPE) surfactant with a low HCB such as PFPE $COO^-NH_4^+$ [molecular weight (MW)2500] forms upper phase W/C microemulsions with an excess water phase [15]. Studies of γ versus HCB have been reported for block copolymers of propylene oxide and ethylene oxide and of poly(dimethylsiloxane) (PDMS) and ethylene oxide [3]. The low interfacial tensions and high surfactant adsorption for PFPE $COO^-NH_4^+$ surfactants explain in part their ability to stabilize W/C microemulsions.

For PFPE $COO^-NH_4^+$ W/C microemulsions Fourier transform infrared (FTIR), ultraviolet-visible, fluorescence, and electron paramagnetic reconance experiments have demonstrated the existence of an aqueous microdomain in CO_2 with a polarity approaching that of bulk-water [2]. From small-angle neutron scattering (SANS), the water droplet radius is 3.5 nm for a molar water-to-surfactant ratio of 30 [16]. Surfactants that have been reported to stabilize W/C microemulsions include PFPE $COO^-NH_4^+$ [2,16,17], the hybrid hydrocarbon-fluorocarbon surfactant, $C_7F_{15}CH(OSO_3^-\ Na^+)C_7H_{15}$ [1], and the surfactant di(1*H*,1*H*,5*H*-octafluoro-*n*-pentyl) sodium sulfosuccinate (di-HCF_4) [18]. Organic-in-CO_2 microemulsions have also been formed for MW 600 poly(ethylene glycol) (PEG600) and for polystyrene oligomers [19,20]. For microemulsions with low water volume fractions (typically $\phi < 0.02$) the absolute amount of dissolved hydrophiles can be somewhat small, and this can limit reaction rates for phase transfer catalysis [21]. Highly concentrated W/C microemulsions ($\phi \sim 0.5$) have been formed (C. T. Lee et al., in press) providing much higher solubilization.

The pH inside microemulsion droplets is typically 3 as determined with fluorescence [17] and absorbance [18,22] probes. Inorganic and organic bases and buffers, such as NaOH, can be used to control the aqueous pH in PFPE $COO^-NH_4^+$-stabilized microemulsions from 3 to values from 5 to 7.

III. WATER-IN-CO_2 EMULSIONS

Stable W/C emulsions for either liquid or supercritical CO_2 have been formed with the surfactants (PFPE $COO^-)_2\ Mn^{+2}$ [23], PFPE $COO^-NH_4^+$ (672 to 7500 MW) [15], and block copolymer surfactants composed of PDMS and poly(acrylic acid) (PAA) or poly(methacrylic acid). The emulsions were formed by shear through a 130-μm capillary or a small orifice. The ratio of water to CO_2 has been varied from 9:1 to 1:9 with less than 1 wt% surfactant. In contrast, W/C microemulsions contain less than 5 wt% water for this level of surfactant due to the much higher interfacial area.

Emulsion stability contours are shown for the surfactant PDMS-*b*-PAA (20K-0.7K) as a function of pH of the aqueous phase in Fig. 2. Equal volumes of water and surfactant were emulsified across a 130-μm capillary tube with 1 wt% surfactant. The stability is defined by the time required for the emulsion to settle 20%. At a low pH of 3, the COOH groups do not ionize and observations of phase behavior indicated that the surfactant greatly prefers CO_2 to water, consistent with the extremely hydrophobic PDMS block. As the density of CO_2 is lowered, the degree of flocculation of

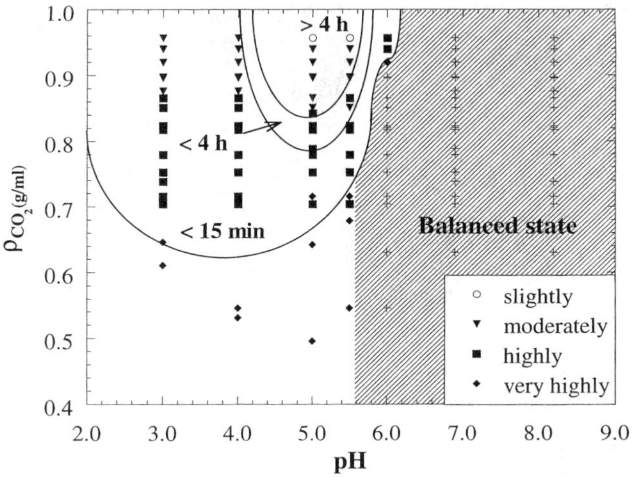

FIG. 2 Stability contours for W/C emulsions stabilized by PDMS-*b*-PAA (20k-0.7K) in terms of the time for 20% settling.

the emulsion increases from moderately to very highly flocculated as shown in Fig. 3. With an increase in pH to 4 and then 5 by adding concentrated NaOH, ionization of the COOH group strengthens the surfactant-water interaction on the hydrophilic side of the interface, raising the surfactant adsorption and emulsion stability to greater than 4 h. Even when the density was lowered to 0.6 g/mL, it took 4 h for the emulsion to phase separate completely, illustrating the resistance to coalescence. At a pH of 6, the surfactant reaches the balanced state, as defined in Fig. 1, and favors the two phases equally. Here the emulsion is highly flocculated at all densities. Emulsions are destabilized by flocculation because of attraction between droplets and coalescence of the flocs due to thinning of the continuous phase between droplets. Emulsions become more stable with a change in any of the formulation variables away from the balanced state (Fig. 3). The resulting increase in interfacial tensions and thus interfacial tension gradients resists film drainage (Marangoni-Gibbs stabilization). For surfactants that are even more hydrophilic, such as PDMS-*b*-PMA (5K/2K), an increase in pH above the pK_a of PAA resulted in inversion to a C/W emulsion. This type of inversion was also observed for PFPE-$COO^-NH_4^+$ surfactants on raising the temperature or lowering the pressure [15].

IV. ORGANIC REACTIONS IN W/C MICROEMULSIONS AND EMULSIONS

The W/C microemulsions and emulsions have been utilized in heterogeneous reactions between CO_2-soluble organic compounds and hydrophiles including nucleophilic ions. They have also been used to disperse hydrophilic hydrogenation catalysts, which may then be recovered by reducing the pressure to separate the aqueous and CO_2 phases.

The reduction of the electron paramagnetic resonance probe 4-hydroxy-2,2,6,6-tetramethyl piperidino-1-oxy (4-hydroxy-TEMPO) was studied in PFPE $COO^-NH_4^+$ W/C microemulsions [2]. The spectra indicated a rapid reduction of the nitroxide by reaction with NH_4^+ within the polar microemulsion core. The rate constant was much larger than in bulk aqueous solution, in part because of a high degree of nitroxide orientation in the restricted environment of the microemulsion core.

The reaction of a CO_2-soluble organic substrate, benzyl chloride, with a hydrophilic nucleophile, KBr, has been accomplished in a PFPE $COO^-NH_4^+$ W/C microemulsion [21]. The microemulsion containing KBr in the water droplets was formed in a variable-volume view cell. Benzyl chloride was injected into the microemulsion solution with a 100-μL sample loop. Periodic samples were obtained with the sample loop, expanded and recovered with ethanol, and analyzed by gas chromatography (GC). Figure 4 shows the conversion of benzyl chloride to benzyl bromide in a W/C microemulsion with a water-to-surfactant ratio $w_0 = 9.7$ (corrected for water in bulk CO_2). The yield was an order of magnitude higher in W/C microemulsions than in conventional water-in-oil microemulsions at similar conditions, probably because of the low interfacial viscosity. Furthermore, benzoyl chloride and *p*-nitrophenyl chloroformate were hydrolyzed in W/C microemulsions with rate constants an order or magnitude faster than those in W/O microemulsions. It was found that changing the size and thereby the polarity and nucleophilicity of the water droplet could affect the reaction rates because of the different transition states of these substrates.

Relatively few reactions have been reported in water-in-oil emulsions with the exception of heterogeneous polymerizations, in part due to the difficulty in breaking the emulsion to isolate the reaction products. A W/C emulsion may be broken simply by reducing the pressure. It is likely that the pressure may be manipulated to fractionate products and to recover CO_2 and surfactant

In W/C macroemulsions, much larger amounts of water can be dispersed than in W/C microemulsions. The relatively large interfacial area in W/C emulsions, due to the large amounts of both phases, can be advantageous for phase transfer reactions. The hydrolysis

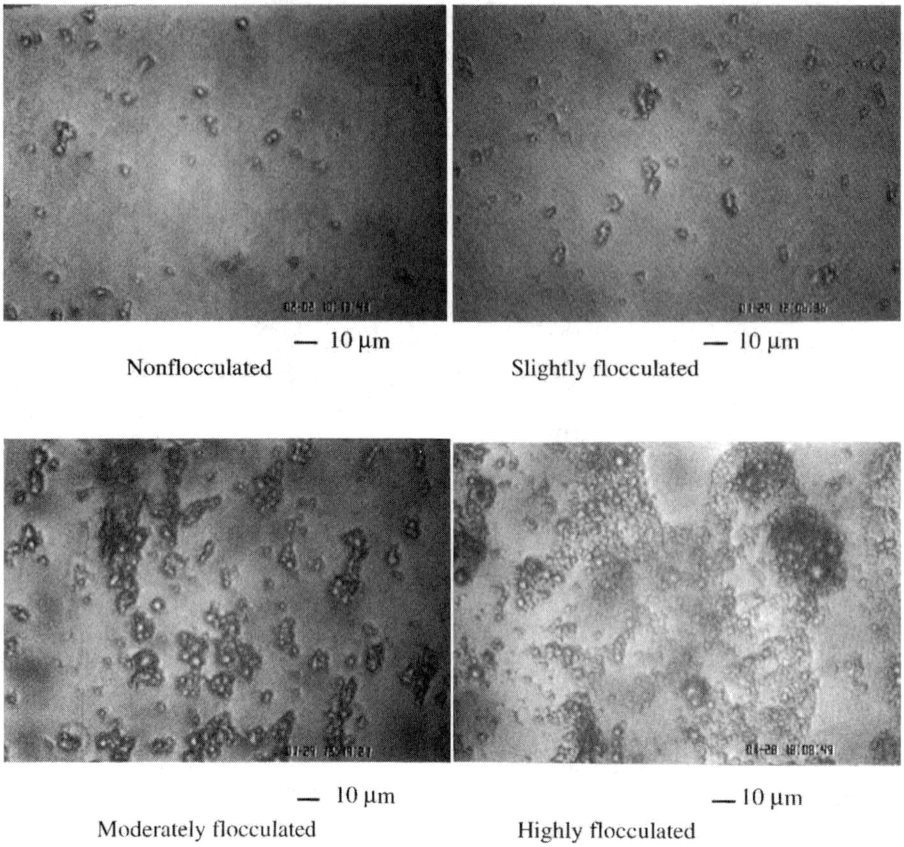

FIG. 3 Photomicrographs of degree of flocculation for W/C emulsions stabilized by PDMS-*b*-PMA and PDMS-*b*-PAA.

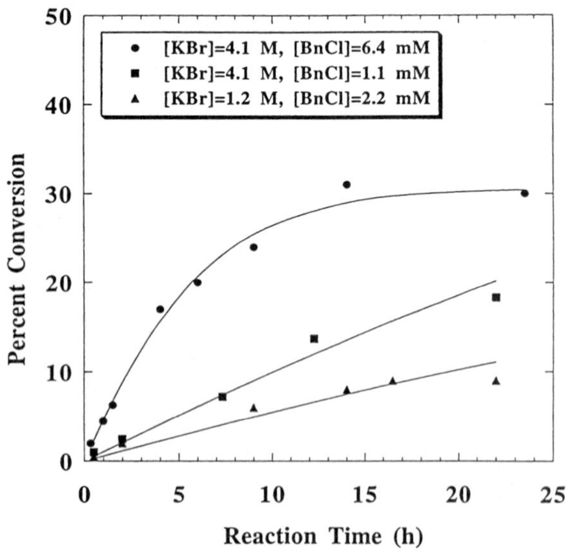

FIG. 4 Conversion of benzyl chloride to benzyl bromide at 65°C and 276 bar in a W/C microemulsion formed with 1.4 wt% PFPE $COO^-NH_4^+$ at various initial concentrations of KBr in water and benzyl chloride in CO_2.

of benzoyl chloride has been studied in a W/C emulsion formed with PFPE-Mn surfactant and equal weights of water and CO_2. Figure 5 shows the conversion of benzoyl chloride to benzoic acid versus time, indicating that nearly complete conversion was achieved after 1 h (T. W. Randolph et al., in press). The reaction was also performed in a two-phase water-CO_2 system without surfactant, which was stirred and sheared in the same manner as the emulsion. As shown in Fig. 5, the presence of surfactant increased the reaction rate markedly, illustrating the effect of the high interfacial area for an emulsion.

A promising approach for phase transfer reactions between water and CO_2 or solids and CO_2 is to use a phase transfer catalyst to shuttle ions between phases. The kinetics of the nucleophilic displacement of benzyl chloride with a bromide ion (solid phase) in CO_2, with acetone as a cosolvent were studied [24]. The phase transfer catalyst tetraheptylammonium bromide transported Cl^- and Br^- ions between the solid and CO_2 phases. Equilibrium conversions were obtained within 5% in 48 h or less.

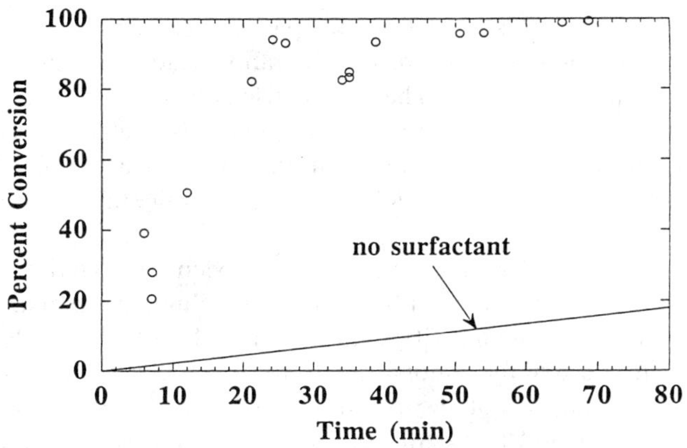

FIG. 5 Hydrolysis of benzoyl chloride to benzoic acid in a W/C emulsion formed with 7 mM PFPE-Mn surfactant and equal amounts of water and CO_2 at 25°C and 270 bar. The line indicates the conversion without surfactant.

The difficulty of catalyst separation and recovery continues to create economic and environmental barriers to the broader industrial application of homogeneous catalysts, despite the remarkable activity and selectivity attainable through sophisticated ligand design in these systems. Recently, W/C emulsions have been utilized to disperse water-soluble rhodium-phosphine complexes for the hydrogenation of alkenes [25]. After reaction, the emulsion can be broken by simply decreasing the pressure to achieve product separation and catalyst recycle. Significant increases in catalyst activity, relative to conventional biphasic water/organic systems, resulted from a combination of higher hydrogen concentration, due to miscibility in the CO_2 phase, and increased interfacial surface area by emulsion formation.

V. ENZYMATIC REACTIONS IN MICROEMULSIONS

The discovery that enzymes remain active in supercritical CO_2, in the presence of small amounts of water has led to the development of novel applications of such "low-water" systems in chemical synthesis [26]. In these examples the enzyme was immobilized or suspended in CO_2.

The formation of water-in-CO_2 (W/C) microemulsions that have the ability to act as "universal solvents" for solubilizing both polar and apolar molecules including enzymes, provides a unique medium for enzyme-catalyzed biotransformations, which have been studied in water/oil microemulsions. Proteins and amino acids have been extracted from water into water/oil microemulsions formed in supercritical ethane and liquid propane [27–29]. The solubility and selectivity have been tuned with pressure by forming and breaking the microemulsions. The group of Bright [2] initially demonstrated that the protein bovine serum albumin (BSA) (MW 67,000) could be solubilized in a PFPE $COO^-NH_4^+$-stabilized W/C microemulsion. BSA compartmentalized within the water droplets was shown to experience an aqueous environment similar to that of native BSA in buffered solution. Fluoroether functional surfactants have also been used to extract subtilisin Carlsberg from aqueous buffer and cell culture media into CO_2, with recovery accomplished by depressurization. In this case water and CO_2 formed a three-phase emulsion with the surfactant in the middle phase or a C/W emulsion. Some of the protein was sacrificed to aid the formation of the emulsion. Protein concentration in the emulsion was found to be as high as 19.5 mg/mL.

Holmes et al. [18] performed two enzymatic reactions, the lipase-catalyzed hydrolysis of *p*-nitrophenol butyrate and lipoxygenase-catalyzed peroxidation of linoleic acid, in W/C microemulsions stabilized by a fluorinated dichain sulfosuccinate surfactant (di-HCF_4). The activities of both enzymes in the W/C microemulsion environment were found to be essentially equivalent to those in a water/heptane microemulsion stabilized by Aerosol OT, a surfactant with the same headgroup as di-HCF_4. The inherent condition of low pH of the water droplets in the microemulsion, ~3 [17,18,22], was improved by using the buffer 2-(*N*-morpholino)ethanesulfonic acid (MES) to fix the pH in the range 5–6. The resulting initial rate profile for the "bioconversion" of linoleic acid by soybean lipoxygenase, which requires gaseous O_2 as a substrate, is shown in Fig. 6. This graph clearly shows a progressive increase in the rate of reaction, in this case oxidation, with increasing enzyme concentration and demonstrates the viability of enzyme-catalyzed reactions within a pH-controlled W/C microemulsion.

VI. INORGANIC REACTIONS IN MICROEMULSIONS

A. Spectroscopy

Clarke et al. [30] showed that W/C microemulsions may be used as a new environment for inorganic chemistry by studying well-known inorganic reactions. An advantage of this approach is that gaseous reactants may be mixed with CO_2 at much larger concentrations than in the case of organic solvents. Aqueous acidified potassium dichromate ($K_2Cr_2O_7$) in the microemulsion

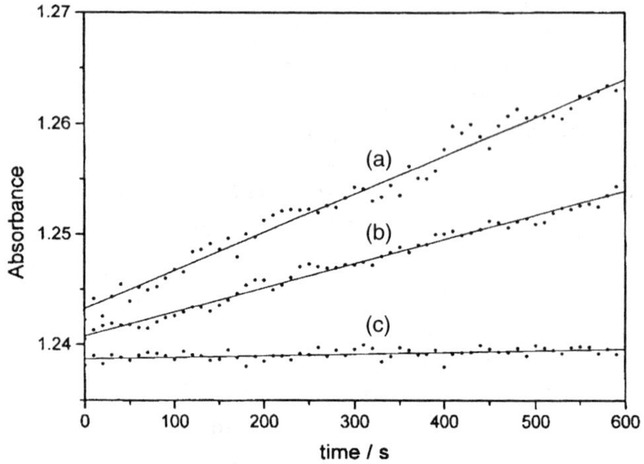

FIG. 6 Reaction profiles (monitored at $\lambda = 234$ nm) for the peroxidation of linoleic acid by soybean lipoxygenase in a di-HCF_4–stabilized W/C microemulsion containing dissolved O_2 gas with [linoleic acid] = 2 mM: (a) [enzyme] = 0.54 μg cm^{-3}; (b) [enzyme] = 0.27 μg cm^{-3}; (c) control (no lipoxygenase added). $[MES]_{aq}$ = 100 mM, [di-HCF4] = 30 mM, $w_0 = 10$, $T = 20$°C, $P = 450$ bar.

droplets was reacted with SO_2, which is highly miscible with CO_2. The orange $K_2Cr_2O_7$ solution disappeared and was replaced by a clear blue-green solution with spectral features indicative of Cr(III) ions (at 430 and 600 nm) in aqueous solution. In a second example, sodium nitroprusside $Na_2[Fe(CN)_5(NO)]\cdot 2H_2O$ was reacted with H_2S in CO_2, a common reaction for the assay of low levels of H_2S in sour water. Both of these reactions demonstrate that "bulk" water is sufficiently well formed to support solubility of ionic species in a CO_2-continuous microemulsion and that high levels of dissolved gases in CO_2 can be reacted with ions in the droplet core.

B. Synthesis of Metal and Metal Oxide Nanoparticles

Cason and Roberts [31] have demonstrated the production of nano-sized metallic copper particles in AOT reverse micelles in compressed propane and supercritical ethane through the reduction of $Cu(AOT)_2$. The particles were formed in compressed propane at 184 bar and 37°C at concentrations of 7×10^{-2} M AOT, 7×10^{-3} M $Cu(AOT)_2$, and 2.1×10^{-2} M hydrazine reducing agent; a water-to-surfactant ratio of 10; and 10% (v/v) isooctane as a cosolvent. With transmission electron microscopy (TEM), it was seen that ~14-nm particles were produced in propane. The particle size of copper is correlated with the ratio of the absorbance at 566 nm to that at 500 nm ($\varepsilon_{566}/\varepsilon_{500}$). With this benchmark, Fig. 7 shows that the particle growth rates in supercritical ethane are considerably faster than those in liquid isooctane. The lower viscosity increases collision frequencies between droplets. Also, ethane exhibits weaker interactions with the AOT surfactant tails than isooctane, which increases micelle-micelle interactions.

The W/C microemulsions have been exploited for the synthesis of metallic and semiconductor nanoparticles. By reducing silver nitrate, Ji et al. [32] were able to harvest silver nanoparticles from a W/C microemulsion. Analysis of the plasmon resonance peak at 400 nm indicated that samples collected at intervals of 20 and 10 min were ~4 nm in diameter. A subsequent decrease in the intensity of the plasmon band, over a period of 1 h, was attributed to the slow flocculation of nanoparticles.

Holmes et al. [33] synthesized semiconductor nanoparticles of cadmium sulfide, in PFPE $COO^-NH_4^+$ stabilized W/C microemulsions. The exciton peaks are shown in Fig. 8. At $w_0 = 5$ and 10 the exciton energies are 3.86 and 3.09 eV, corresponding to mean particle radii of approximately 0.9 and 1.8 nm, respectively. These radii are consistent with measurements by TEM, shown in Fig. 9, as well as the microemulsion droplet diameters from SANS [16]. Despite the low viscosity of CO_2, resulting in higher droplet collision frequencies, agglomeration is minimized and the particles remain restricted by the water core radius (r_w) after a reaction period of 10 min. The templating effect possibly arises from the difficulty in containing two particles in one micelle at a particle diameter approaching r_w. Also, particles of a size equivalent to r_w presumably become surrounded by surfactant molecules, which act as a protective agent. W/C microemulsions, therefore, serve as an effective medium for compartmentalized growth of nanoparticles with controllable size and relatively narrow size distributions.

VII. STABILITY OF ORGANIC POLYMER EMULSIONS AND LATEXES

Copolymers containing CO_2-philic segments such as poly(1,1-dihydroperfluorooctyl acrylate) [34], poly(perfluoropropylene oxide) [35], and poly(dimethylsiloxane) are CO_2-philic and have been used to stabilize sterically latexes that form during dispersion polymerization. Steric stabilization may prevent flocculation if (1) the adsorbed or grafted surfactant provides high surface coverage, and (2) the length of the solvated "CO_2-philic segment" is sufficient to overcome van der Waals attraction between the particle cores, and (3) the sol-

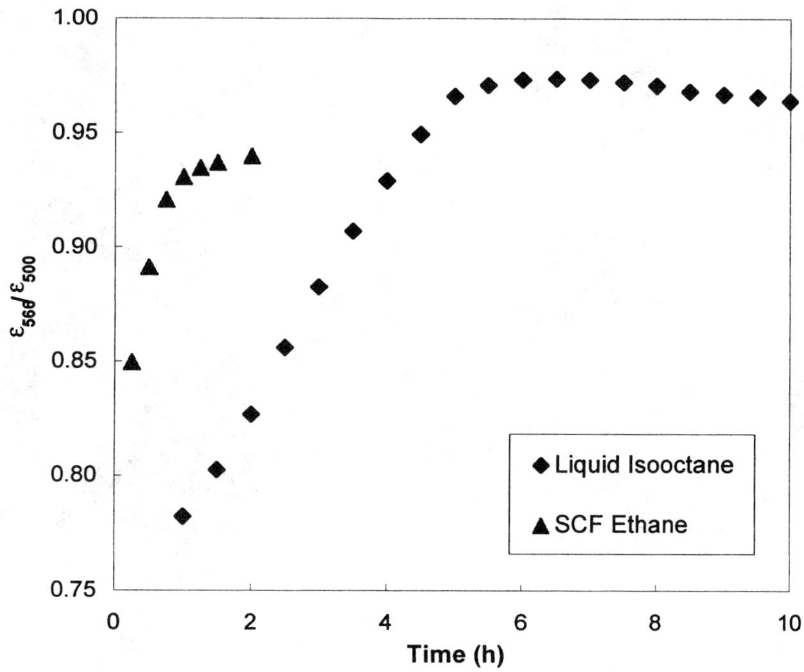

FIG. 7 Change in extinction coefficient ratio, $\varepsilon_{566}/\varepsilon_{500}$, with time for copper particles synthesized in liquid isooctane and SCF ethane at $w_0 = 8$ [31].

vation of the CO_2-philic segment is sufficient to prevent interparticle attractive interactions between these segments. With theory [36] and computer simulation [37], an analogy has been observed between polymer-supercritical fluid phase separation in bulk and the flocculation of surfaces with grafted polymer (see Fig. 10). As the density is lowered, the polymer and supercritical fluid first phase separate at the upper critical solution density (UCSD). At this density, the force between surfaces with grafted chains becomes attractive, indicating flocculation. This point is called the critical flocculation density (CFD). In each case the solvent expands away from the polymer chains, resulting in either precipitation of a polymer-rich phase (bulk phase separation) or flocculation between surfaces. This expansion is entropically driven. To date, the simulations have been performed only for symmetric systems where the segment-segment interactions on the polymer match the segment-solvent interactions.

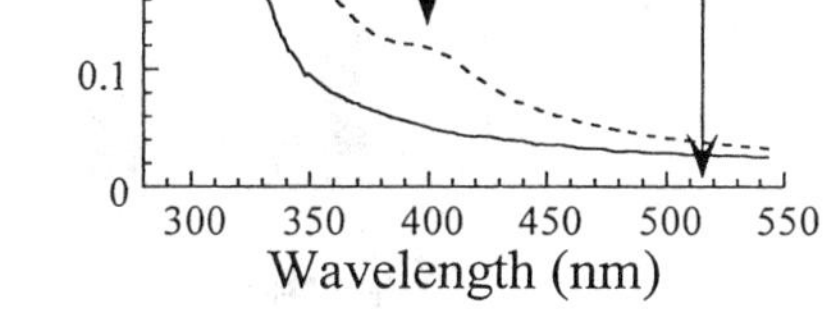

FIG. 8 UV-visible absorbance spectra of CdS particles prepared in PFPE $COO^-NH_4^+$ (3 wt%) stabilized W/C microemulsions: (a) $w_0 = 10$ and (b) $w_0 = 5$.

Steric stabilization of poly(2-ethyl hexyl acrylate) (PEHA) emulsions in CO_2 was studied with static and dynamic light scattering [38,39]. These emulsions were stabilized with poly(1,1-dihydroperfluorooctyl acrylate) (PFOA)–based surfactants. The CFD agreed very closely with the UCSD of PFOA in CO_2 at the same temperature, in agreement with theory and simulation. For submicrometer silica particles with grafted PDMS (MW up to 22,000), a different result was obtained (M. Z. Yates and K. P. Johnston, in press). The particles were unstable and flocculated well above the UCSD of the PDMS-CO_2 binary system. Further work is needed to understand why these colloids were not stable.

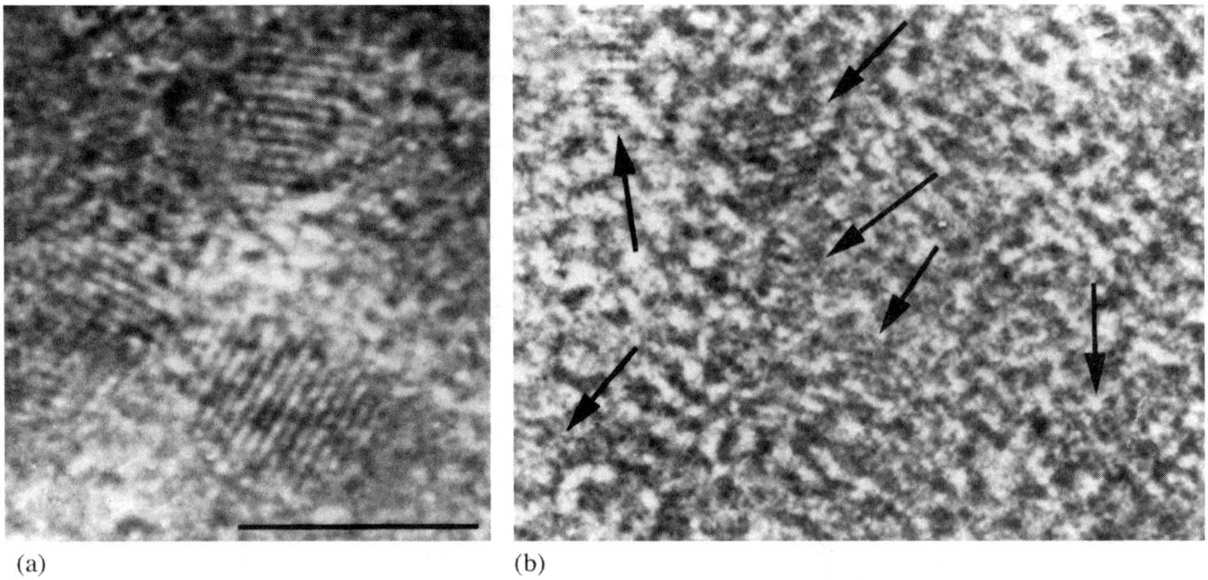

FIG. 9 TEM images confirming the particle size for CdS produced in PFPE $COO^-NH_4^+$ W/C microemulsions with $w_0 = 5$ (a) and $w_0 = 10$ (b). The lattice fringes are clearly seen for (a). Scale = 5 nm in both images.

VIII. IN SITU STUDIES OF THE MECHANISM OF DISPERSION POLYMERIZATION

Dispersion polymerization is discussed in detail in this book by DeSimone et al. (Chapter 5). Here we describe studies of the mechanism that complement the preceding scattering and theoretical studies of steric stabilization of organic emulsions. Turbidimetry has been used to study dispersion polymerization of PMMA [40] and poly(vinyl acetate) [41]. The average particle size, particle number density, and overall surface area were measured versus time during particle formation. Coagulative nucleation and controlled coagulation regions were governed by the amount of stabilizer available relative to the total surface area of the dispersion. At the end of the controlled coagulation region, which can

FIG. 10 Schematic illustrating the analogy between colloid flocculation behavior and phase behavior of the stabilizer in bulk solution. As density is lowered, separation of solvent from chains in bulk solution resembles separation of solvent from chains on surfaces, which produces flocculation.

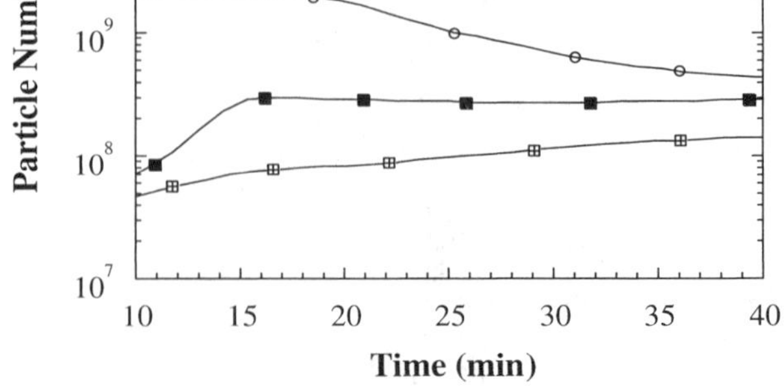

FIG. 11 Particle number density measured by turbidimetry for dispersion polymerization of poly(vinyl acetate) stabilized with homopolymers and block copolymers [41].

last tens of minutes, the particle number density approaches the final value. For the dispersion polymerization of vinyl acetate, the particle number density was obtained for the stabilizers: PDMS homopolymer, PDMS-*b*-PVAc, a poly(1,1-dihydroperfluorooctyl acrylate) homopolymer (PFOA), and PFOA-*b*-PVAc as shown in Fig. 11. The measured particle number densities after 40 min agreed with those for diluted samples at the completion of the reaction, again suggesting that the number density is fixed early in the reaction. The PDMS-based surfactants produce larger particles than the PFOA-based surfactants, suggesting that they provide less resistance to flocculation and coagulation. In addition, for good stabilizers, PVAc particles are several times larger than PMMA particles due to slower nucleation rates for PVAc because it is more miscible with CO_2. With fewer nucleated particles, larger particles are produced for the same overall volume fraction.

IX. AMBIDEXTROUS SURFACTANTS

The concept of an "ambidextrous" surfactant that is active at an organic-CO_2 and organic-water interface has been demonstrated as shown in Fig. 12. The surfactants included a poly(dimethylsiloxane) block, a poly(methacrylic acid) (PMA) or poly(acrylic acid) block, and in some cases a third component, methyl methacrylate units in a random or block architecture [42]. In CO_2, PMMA particles produced by dispersion polymerization are sterically stabilized by the PDMS segments with particle diameters given in Table 1. Upon transforming the surfactant-coated particles to water (or buffered water) at up to 40% by weight, the PDMS block collapses onto the latex surface and some of the PMA groups are ionized, producing electrostatic stabilization. The surfactant is ambidextrous in that it stabilizes two different types of interfaces, each by a different stabilization mechanism. The size of the latex particles does not change upon transfer to water on the basis of measurements by dynamic light scattering, as shown in Table 1. In water, only about 1% of the AA or MA groups ionize, as determined from the zeta potential, yet this provides sufficient electrostatic stabilization. The change in surface charge from positive to negative with an increase in pH is consistent with the pK_a of the AA and MA groups. The particles produced with trifunctional surfactants composed of PDMS, methyl methacrylate (anchor groups), and PMA or PAA were much smaller and less agglomerated than those produced with bifunctional PDMS-*b*-PMA. The third functionality, MMA, enhances adsorption at the PMMA surface, leading to smaller particles. Consequently, much more stable latexes were formed upon transfer to water. With this concept, it is possible to produce latexes in CO_2 with minimal waste, which can be vented, shipped as dry powder, and resuspended in water to form an aqueous latex for coatings applications.

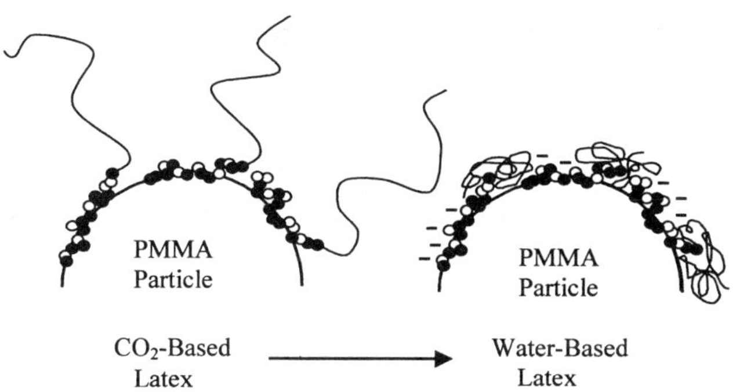

FIG. 12 Trifunctional "ambidextrous" surfactant for steric stabilization of organic/CO_2 latexes and electrostatic stabilization of organic/water latexes [42].

TABLE 1 Average Particle Diameter (μm) for PMMA Latex in CO_2 (Scanning Electron Microscopy, SEM) and Water (Dynamic Light Scattering, DLS)

Surfactant	(SEM)	(DLS)
PDMS-*b*-P(tBA-*co*-AA) (10,000/800–500)	1.31	1.04
PDMS-*b*-PMMA-*b*-PAA (10,000/1500/700)	1.18	0.89
PDMS-*b*-P(MMA-*co*-MA) (5000/1100–500)	0.78	0.68

REFERENCES

1. K. L. Harrison, J. Goveas, K. P. Johnston, and E. A. O'Rear III, Langmuir *10*:3536–3541 (1994).
2. K. P. Johnston, K. L. Harrison, M. J. Clarke, S. M. Howdle, M. P. Heitz, F. V. Bright, C. Carlier, and T. W. Randolph, Science *271*:624–626 (1996).
3. S. R. P. da Rocha, K. L. Harrison, and K. P. Johnston, Langmuir *15*:419–428 (1999).
4. G. B. Jacobson, C. T. Lee, S. R. P. da Rocha, and K. P. Johnston, J. Org. Chem. *64*:1207–1210 (1999).
5. J. L. Fulton and R. D. Smith, J. Phys. Chem. *92*:2903–2907, (1988.
6. K. P. Johnston, G. McFann, and R. M. Lemert, Am. Chem. Soc. Symp. Ser. *406*:140–164 (1989).
7. D. H. Everett and J. F. Stageman, Faraday Disc. Chem. Soc. *65*:230–241 (1978).

8. K. A. Bartscherer, M. Minier, and H. Renon, Fluid Phase Equilibria. *107*:93–150 (1995).
9. J. L. Fulton, in *Microemulsions: Fundamental and Applied Aspects* (P. Kumar, ed.), Marcel Dekker, New York, 1999, pp. 629–650.
10. G. J. McFann and K. P. Johnston, in *Microemulsions: Fundamental and Applied Aspects* (P. Kumar, ed.), Marcel Dekker, New York, 1999, pp. 281–307.
11. J. M. DeSimone, Z. Guan, and C. S. Elsbernd, Science *257*:945–947 (1992).
12. E. J. Singley, W. Liu, and E. J. Beckman, Fluid Phase Equilibria. *128*:199–219 (1997).
13. K. Harrison, Interfacial tension measurements of CO_2-water systems and formation of water-in-CO_2 microemulsions, dissertation, University of Texas at Austin, 1996.
14. R. Aveyard, B. P. Binks, S. Clark, and P. D. I. Fletcher, J. Chem. Tech. Biotechnol. *48*:161–171 (1990).
15. T. C. Lee, P. A. Psathas, K. P. Johnston, J. deGrazia, and T. W. Randolph, Langmuir *15*:6781–6791 (1999).
16. R. G. Zielinsky, S. R. Kline, E. W. Kaler, and N. Rosov, Langmuir *13*:3934–3937 (1997).
17. E. D. Niemeyer and F. V. Bright, J. Phys. Chem. *102*: 1474 (1998).
18. J. D. Holmes, D. C. Steytler, G. D. Rees, and B. H. Robinson, Langmuir *14*:6371–6376 (1998).
19. K. L. Harrison, K. P. Johnston, and I. C. Sanchez, Langmuir *12*:2637–2644 (1996).
20. J. B. McClain, D. E. Betts, D. A. Canelas, E. T. Samulski, J. M. DeSimone, J. D. Londono, H. D. Cochran, G. D. Wignall, D. Chillura-Martino, and R. Triolo, Science *274*:2049 (1996).
21. G. B. Jacobson, C. T. Lee, and K. P. Johnston, J. Org. Chem. *64*:1201–1206 (1999).
22. J. D. Holmes, K. J. Ziegler, M. Audriani, C. T. Lee, P. A. Bhargava, D. C. Steytler, and K. P. Johnston, J. Phys. Chem. B. *103*:5703–5711 (1999).
23. K. P. Johnston, G. B. Jacobson, C. T. Lee, C. Meredith, S. R. P. DaRocha, M. Z. Yates, J. DeGrazia, and T. W. Randolph, in *Microemulsions, Emulsions, and Latexes in Supercritical Fluids* (P. Jessop and W. Leitner, eds.), Wiley-VCH, Weinheim, 1999.
24. A. K. Dillow, S. L. Yun, D. S. Suleiman, D. L. Boatright, C. L. Liotta, and C. A. Eckert, Ind. Eng. Chem. Res. *35*:1801–1806 (1996).
25. G. B. Jacobson, C. T. Lee, K. P. Johnston, and W. Tumas, J. Am. Chem. Soc. *121*:11902 (1999).
26. A. J. Mesiano, E. J. Beckman, and A. J. Russell, Chem. Rev. *99*:623–633 (1999).
27. R. D. Smith, J. L. Fulton, J. P. Blitz, and J. M. Tingey, J. Phys. Chem. *94*:781–787 (1990).
28. R. M. Lemert, R. A. Fuller, and K. P. Johnston, J. Phys. Chem. *94*:6021 (1990).
29. G. A. Ayala, S. Kamat, E. J. Beckman, and A. J. Russell, Biotech. Bioeng. *39*:806 (1992).
30. M. J. Clarke, K. L. Harrison, K. P. Johnston, and S. M. Howdle, J. Am. Chem. Soc. *119*:6399–6406 (1997).
31. J. B. Cason and C. B. Roberts, J. Phys. Chem. B, in press.
32. M. Ji, X. Chen, C. M. Wai, and J. L. Fulton, J. Am. Chem. Soc. *121*:2631–2632 (1999).
33. J. D. Holmes, P. A. Bhargava, B. A. Korgel, and K. P. Johnston, Langmuir *15*:6613–6615 (1999).
34. J. M. DeSimone, E. E. Maury, Y. Z. Menceloglu, J. B. McClain, T. J. Romack, and J. R. Combes, Science *265*: 356 (1994).
35. C. Lepilleur and E. J. Beckman, Macromolecules *30*: 745–756 (1997).
36. J. C. Meredith and K. P. Johnston, Macromolecules *31*: 5518–5528 (1998).
37. J. C. Meredith, I. C. Sanchez, K. P. Johnston, and J. J. D. Pablo, J. Chem. Phys. *109*:6424–6434 (1998).
38. M. O'Neill, M. Z. Yates, K. L. Harrison, K. J. Johnston, D. A. Canelas, D. E. Betts, J. M. DeSimone, and S. P. Wilkinson, Macromolecules *30*:5050–5059 (1997).
39. M. Z. Yates, M. L. O'Neill, K. P. Johnston, S. Webber, D. A. Canelas, D. E. Betts, and J. M. DeSimone, Macromolecules *30*:5060–5067 (1997).
40. M. L. O'Neill, M. Z. Yates, K. P. Johnston, C. D. Smith, and S. P. Wilkinson, Macromolecules *31*:2848–2856 (1998).
41. D. A. Canelas, D. E. Betts, J. M. DeSimone, M. Z. Yates, and K. P. Johnston, Macromolecules *31*:6794–6805 (1998).
42. M. Z. Yates, G. Li, J. J. Shim, S. Maniar, and K. P. Johnston, Macromolecules *32*:1018–1026 (1999).

15

Reactions in Multiphase (Liquid/Liquid) Micellar Systems

TURGUT BATTAL and JAMES F. RATHMAN The Ohio State University, Columbus, Ohio

I. INTRODUCTION

Although multiphase reactions are central to many industrial processes, they are not yet understood nearly as well as their single-phase counterparts. Studies of homogeneous single-phase reaction systems have provided a wealth of information for understanding how variables such as pressure, temperature, and composition affect rate constants, selectivity, reaction mechanism, and activation energy. Single-phase systems are also of significant practical importance; however, the use of multiphase systems is expected to become much more prevalent because of their potential for reducing the environmental impact of chemical manufacturing processes. The fundamental and applied knowledge for various gas-solid, liquid-solid, and gas-liquid two-phase systems is substantial, but the design of reaction processes with multiple liquid phases remains problematic. Perhaps one reason that studies of liquid-liquid reaction systems are far less common is that it is often possible to use an organic solvent or solvent mixture to achieve single-phase conditions. From an environmental standpoint, this approach is often no longer acceptable.

Multiphase reactions are especially complex in systems that are also microheterogeneous, comprising bulk phases containing surfactant aggregates that themselves influence the reaction kinetics. This chapter aims to explore potential advantages of multiphase microheterogeneous systems over their single-phase counterparts. Although such systems are more complex, their multiphase nature can be exploited to achieve a number of desirable outcomes: the need for organic solvents can be eliminated, and postreaction product separation can be accomplished with greater ease. Multiphase systems also provide a unique reaction environment that can be designed to enhance reaction rates, to improve selectivity, and even to synthesize novel materials.

II. CURRENT UNDERSTANDING OF MULTIPHASE LIQUID REACTION SYSTEMS

The objective of this research addresses reactions in which a water-soluble reactant present in an aqueous solution must be combined with a hydrophobic liquid reactant. If neat hydrophobic reactant is added, the reaction rate is limited by the contact surface area, mass transfer rate, and equilibrium distribution between the bulk phases. Various strategies have been devised for obtaining high rates of reaction in such systems. These are briefly summarized next.

A. Addition of an Organic Solvent to Solubilize Both Reactants

The most common method for combining immiscible reactants is to add a protic or aprotic solvent to improve the mutual solubilities of both reactants in at least one of the bulk liquid phases [1–3]. Although it is effective, the observed increase in reaction rate is generally modest and maximizing conversion often requires increased temperature, which in turn may lead to the generation of undesired by-products. The primary problem with this approach is that, for economic and environmental reasons, the solvent must be recovered.

B. Phase Transfer Catalysis

In 1951, Jarrouse [4] reported that two-phase liquid-liquid reaction rates were enhanced by adding small catalytic quantities of a quaternary salt. The term phase transfer catalysis (PTC) was coined by Starks [5] and describes reactions in which the reactants are initially present in immiscible liquid phases [1,3,6–9]. The role of the phase transfer catalyst is to form an ion pair or complex with the hydrophilic reactant in the aqueous phase and shuttle it into the bulk organic phase. The catalysts include onium salts, crown ethers, and alkali metal salts. PTC has been demonstrated to be effective for many general classes of reaction, including alkylation, arylation, condensation, elimination, and polymerization. Extension of the basic PTC technique to organometallic PTC applications, e.g., carbonylation of halides to carboxylic acids using CO_2-$(CO)_8$, has also been reported [10]. Although less common, another useful approach is inverse phase transfer catalysis, in which the catalyst shuttles the hydrophobic reactant into the bulk aqueous phase. Examples include reactions involving partially water-soluble pyridines or derivatives that form complexes with an organic substrate [11]. Aryl anhydrides can be produced with good selectivity from acid chlorides by this process. Industrial applications of PTC reactions are increasing, especially in the fine chemical area. Several examples of industrial-scale processes using phase transfer catalysis have been described [12]. Among the most important applications are the processes for making benzylpenicillins and other drugs derived by *O*- and *N*-alkylation of various heterocycles, processes for making synthetic pyrethroids such as cypermethrin and other analogues, and reactions of alkyl chlorides with nitriles.

Various detailed reaction mechanisms have been proposed for PTC; a general scheme is illustrated in Fig. 1. The quaternary cation (Q^+) is the phase transfer catalyst in this case, Y^- is the reactive anion, Z^- is the chemically inert counterion of the phase transfer catalyst salt, and RX the hydrophobic organic reactant. The desired reaction occurs primarily in the bulk organic phase. The overall rate of product formation is a function of mass transport rate of the QY ion pair across the liquid-liquid interface and reaction rate constant. For a system subjected to continuous mechanical agitation, diffusion effects within the continuous liquid phase can usually be ignored; however, diffusion rates of components in the dispersed liquid phase may be important. Evans and Palmer [13] quantitatively modeled a PTC process in which both interfacial mass transfer and chemical reaction kinetics were included. Melville and Goddard [14] presented an analysis of mass transfer in solid-liquid phase transfer catalysis, and a model describing the liquid-liquid phase transfer–catalyzed reaction by application of two-film theory was analyzed by Wang and Wu [15].

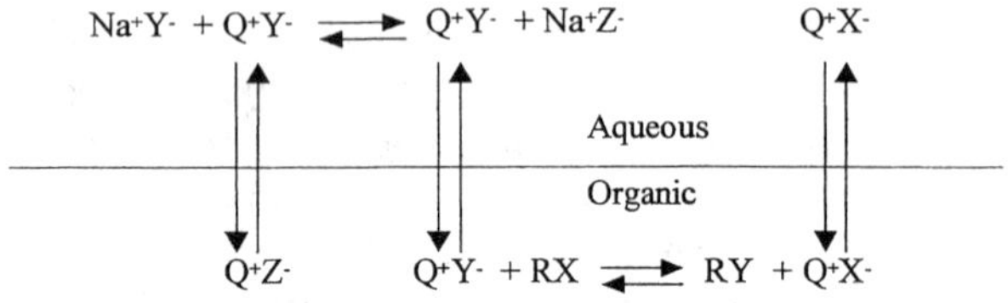

FIG. 1 Schematic representation of steps of typical phase transfer catalysis reaction system.

C. Single-Phase Micellar Catalysis

The potential of aqueous surfactant solutions as replacements for organic solvents is great because micellar catalysis has already been shown to be applicable to reactions that are important in many chemical industries [7,16–31]. Micelles are colloidal aggregates of surfactant molecules that form spontaneously in solution. Micelle formation depends upon numerous factors, including surfactant structure and concentration, solvent properties, temperature, and the presence of other components. In aqueous media, normal micelles form with the hydrophobic surfactant "tail" groups in the micellar core and with the hydrophilic surfactant "head" groups exposed to the continuous water phase. Substances that are virtually insoluble in water are often much more soluble in aqueous surfactant solutions because of solubilization in micelles. A lipophilic solute may be concentrated either in the micelle core or in the palisade layer, the more polar outer region of the micelle near the headgroups.

In the early 1960s, the discovery that solutions containing surfactant micelles could affect bimolecular reaction rates led to a surge of research activity in this area. Given an appropriate surfactant for a particular reaction, micellar solubilization and electrostatic forces act to concentrate both lipophilic and hydrophilic reactants locally near micelles, resulting in dramatic increases in reaction rates (Fig. 2). The surfactant itself is chemically inert, and so this phenomenon is commonly referred to as micellar catalysis. Although most of the research focused primarily on measuring and modeling the reaction kinetics, a few studies demon-

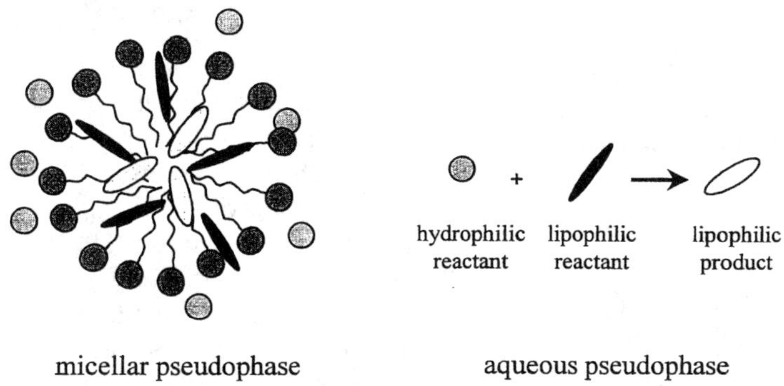

FIG. 2 Schematic representation of one-phase micellar catalysis containing normal micelles in the aqueous phase.

strated that micellar solutions could be employed not only to increase reaction rates but also in some cases to improve product selectivity and even to perform stereospecific catalysis [28]. Reaction conditions used in most prior studies are much different from those typically encountered in industrial processes, so difficulties in extrapolating results from fundamental research to an industrial scale are not trivial. As a result, despite the vigorous research activity in the area of reaction kinetics in surfactant solutions, at present very few industrial processes are based on this technology.

Single-phase micellar catalysis is often modeled quantitatively in terms of the pseudophase separation model. This approach treats micelles collectively as a separate thermodynamic phase from the surrounding solvent in which they are present. As it is commonly applied to micellar catalysis, the usual assumption is that the equilibrium distribution of reactants between the phases is not displaced by the occurrence of the chemical reaction. The justification for this is that the micellization and solute exchange processes between water and micelles are generally faster than the actual rate of reaction, so the overall rate is the sum of rates in water and micelles. Another common assumption is that micellar properties are unaffected by incorporation of the reactants. This assumption is valid only at high surfactant/solute ratios, so that a single micelle will on average contain at most only one or two reactant molecules. The pseudophase model has been applied to a wide variety of reactions in the presence of micelles and with added electrolyte [17]. Modifications have been proposed to account for the competitive counterion binding of reactive and inert counterions and the effects of pH [23,32]. The efficiency of micellar catalysis in single-phase systems relies on two principal factors: the localized concentration of reactants by the micelles and differences in the reactivity of the components in the bulk aqueous and micellar environments that result from differences in the transition state. The contribution of the concentrating effect to the overall acceleration of a reaction is essentially determined by the strength of ionic and hydrophobic interactions between the reactant molecules and the micelles as well as by the reaction order, and transition state changes could result from, for example, specific orientations adopted by solutes solubilized in a micelle. An important conclusion from prior studies is that rate enhancements of bimolecular and higher order reactions are not due to large changes in the kinetic rate constants but instead are a consequence of the localized concentration of reactants at the micelle-solution interface.

Even in dilute reactant concentration regimes, the development of more descriptive models of micellar catalysis is complicated by the microheterogeneity of the micellar environment. The location and orientation of a solute molecule in a micelle are not random but rather determined by interactions between the solute and the surfactant tails in the core and heads at the micelle-solution interface. Even if orientational effects are ignored, determining the correct volume to use when estimating the local concentrations of species solubilized in a micelle, or bound electrostatically at the micelle surface, is not obvious but is necessary for the development of accurate models. Current models, for example, use counterion concentrations based on an estimated volume of the Stern layer.

From an applications standpoint, potential problems with single-phase micellar catalysis include the competitive binding between reactive and nonreactive counterions, which can greatly reduce the rate of reaction, and the fact that the surfactant must be ionic and opposite in charge compared with the water-soluble reactant. Some of the most extensively studied cationic surfactants in the micellar catalysis literature either are known to be toxic or have not been approved for use or release into biological systems.

D. Micellar Phase Transfer Catalysis

Micellar phase transfer catalysis (MPTC) [27,28,30] is a process that combines the best aspects of conventional phase transfer catalysis and micellar catalysis while at the same time avoiding some of the associated problems of these methods. MPTC reaction systems consist of reactants, water, surfactant, and a phase transfer catalyst—no organic solvent is used. The surfactant acts to solubilize and emulsify the lipophilic reactant, while the phase transfer catalyst shuttles the hydrophilic reactant from the aqueous phase into the micellar environment where the reaction takes place. Because electrostatic binding of one of the reactants is not required, competitive counterion binding is not an issue, and so biodegradable nontoxic surfactants can be used. Cationic, nonionic, and even anionic surfactants may be considered. MPTC can therefore be employed using a much wider range of surfactants than conventional micellar catalysis.

E. Reactions in Two-Phase Micellar Systems

There is an upper limit to the amount of material that can be solubilized in a given surfactant solution; beyond this limit, liquid-liquid phase separation occurs. A schematic of a two-phase liquid-liquid system with surfactant micelles present is shown in Fig. 3. Two-phase reactions are of particular interest in many fields of chemistry such as industrial and chemical engineering [33], organic synthesis [30,34–38], molecular and chiral recognition [39], membrane chemistry [40], and prebiotic chemistry [41]. Despite this broad interest, fundamental kinetic data are scarce. Qualitatively, work by the author [30] and others [37,38,42,43] has shown that the heterogeneous nature of two-phase systems can be exploited to obtain some distinct advantages over single-phase systems. These include localized concentration effects at interfaces, improved reaction rates, and ease of product separation. With the exception of emulsion polymerization, fundamental studies and industrial applications of micellar catalysis in multiphase liquid systems are virtually nonexistent. The complexity of multiphase liquid reaction systems containing surfactants makes their quantitative description especially challenging. In addition to all the factors included in conventional models for single-phase micellar catalysis, additional phenomena that must be considered are interface mass transport, transport phenomena within the disperse phase, and physical and chemical properties of the dispersed phase and liquid-liquid interface. Commercial adoption of this technology presents additional processing concerns, but real-

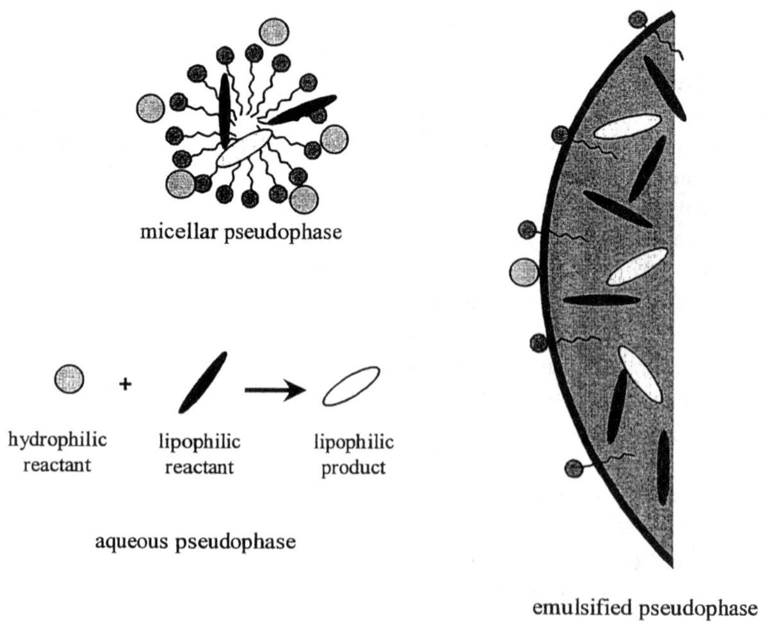

FIG. 3 Simplified schematic of the various reaction environments present in an oil-in-water emulsion containing normal micelles in the aqueous phase.

istically these have for the most part already been addressed and solved, as emulsion technology (creating, stabilizing, handling, and breaking emulsions) is advanced and well understood. In the area of modeling reactions in liquid-liquid emulsion systems, the literature is also quite extensive, although almost entirely for systems containing either no surfactant (poor emulsions in which continuous mechanical agitation is required to disperse the droplet phase) or surfactant at concentrations below the critical micelle concentration (cmc) so that micelles are not present.

Liquid-liquid two-phase reactions in which products have an influence on the interfacial properties are a novel class of nonlinear chemical systems; kinetic studies of such systems have been presented only recently [42,43]. Highly nonlinear behavior giving rise to kinetic bistability in a continuously stirred tank reactor has been reported in a biphasic (liquid-liquid) system [42]. The kinetics of the alkaline hydrolysis of ethyl alkanoates has been studied [43] and a kinetic model for this particular system has been developed. The model takes into account the results of thermodynamic calculations that have been performed to estimate the average size and stoichiometry of aggregates containing sodium alkanoate esters; these values were subsequently used as fixed parameters in the kinetic model. Johnston et al. [37,38] demonstrated that organic synthesis between hydrophilic nucleophiles and CO_2-soluble reactants can be performed in a water-in-carbon dioxide (w/c) microemulsion containing an environmentally benign surfactant. They also reported the synthetic reaction between a hydrophobe, benzyl chloride, and a hydrophilic nucleophile, KBr, in water-in-carbon dioxide and carbon dioxide-in-water emulsions. Higher yields of benzyl bromide were obtained in w/c and c/w emulsions as compared with water-in-octane emulsions. Yields were found to be much higher in two-phase emulsion systems than in one-phase microemulsion systems.

Emulsion polymerization continues to be widely used for preparing polymers, agricultural products, novel materials, and pharmaceuticals. Its widespread use results from superior heat transfer and rheological properties that make temperature control and product handling much easier. Polymer synthesis performed in a dispersing medium allows better control of particle size and heat dissipation. Colloidal polymer particles in water, also known as latexes, are commonly manufactured by emulsion polymerization. Their typical diameters are 50–500 nm. They have great potential value for the study of fundamental heterogeneous reactions because of their high surface area, ease of synthesis, and ease of synthetic control of properties via hydrophile and lipophile content [44]. Prerequisite components for emulsion polymerization include the reactive monomer (30–60% by volume), a dispersing medium (usually water), an initiator (generally soluble in dispersing medium), and an emulsifying agent. The effects of emulsion polymerization conditions, initiator concentration, kind and amount of emulsifier, temperature, ratio of water to monomer, ionic strength, and stirring speed on polymerization rate, particle size, and particle size distribution have been studied extensively [45]. The general theory of the emulsion polymerization process was first presented by Harkins [46] with the mathematical description of the theory being subsequently given by Smith and Ewart [47]. Harkins stated that the main loci of the emulsion polymerization were the surfactant micelles and polymer particles. Three possible nucleation loci have been widely discussed in the literature: (1) micelles, (2) aqueous phase via homogeneous (or coagulate) nucleation, and (3) monomer droplets. Smith and Ewart developed the rate of polymerization equations for three limiting cases: case 1, where the average number of free radicals per particle is much less than one; case 2, where the average number of free radicals per particle equals 0.5; and case 3, where the average number of free radicals per particle is greater than one. Three intervals have been described from these qualitative and quantitative analyses of the emulsion polymerization process whose main characteristics are as follows:

Interval I: The number of polymer particles increases with time; the rate of polymerization increases; micelles and monomer droplets are present

Interval II: The number of polymer particles and rate of polymerization are constant; micelles have disappeared at the beginning of this interval but monomer droplets are still present

Interval III: The number of polymer particles is constant; the rate of polymerization decreases as the concentration of monomer, present only in the polymer particles, decreases.

III. EXPERIMENTAL MATERIALS AND METHODS

A. Materials

Dimethyldodecylamine oxide (DDAO, 30% in water), tetrabutylammonium bromide (TBAB, 99%), and 1-bromobutane (98%) were from Fluka Chemical Corporation. Butylphenyl ether (99%), epichlorohydrin (99%), sodium dodecyl sulfate (SDS, 97%), and cetyl-

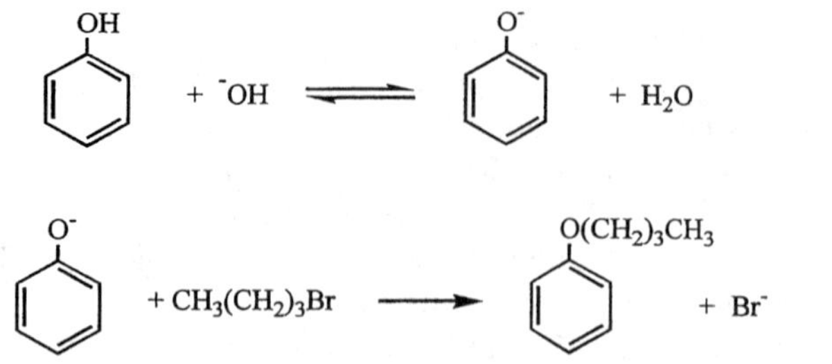

FIG. 4 Synthesis of butylphenyl ether in aqueous sodium hydroxide solution.

trimethylammonium chloride (CTAC, 25% in water) were from Aldrich Chemical Company. Phenol (89% aqueous solution) was from Mallinckrodt. Sodium hydroxide pellets (98.7%) were from Jenneile Enterprises. Distilled deionized water from a Barnstead MP-1 still was used in all experiments.

B. Reaction Systems

The first reaction system investigated in this work is the preparation of unsymmetrical phenolic ethers via alkylation with alkyl halides [48], the Williamson reaction. The specific reaction of interest is the synthesis of butylphenyl ether in aqueous sodium hydroxide solution. Phenol is depronated in alkaline solution to form the hydrophilic phenolate ion, which is then reacted with a hydrophobic reactant, bromobutane, to form butylphenyl ether (Fig. 4). The desired reaction may proceed at four different localities in the two-phase MPTC reaction system: (1) reaction of phenolate ion with 1-bromobutane at the liquid-liquid interface; (2) PTC, reaction of TBA^+:phenolate complex with 1-bromobutane in the dispersed droplet phase; (3) MC, reaction of phenolate ion with 1-bromobutane solubilized in normal micelles in the aqueous phase; and (4) MPTC, reaction of TBA^+:phenolate complex with 1-bromobutane solubilized in micelles. The formation of symmetrical ethers may possibly be significant; e.g., dibutyl ether may be produced by

$$CH_3(CH_2)_3Br + {}^-OH \rightleftharpoons CH_3(CH_2)_3OH + Br^-$$
$$CH_3(CH_2)_3OH + {}^-OH \rightleftharpoons CH_3(CH_2)_3O^- + H_2O$$
$$CH_3(CH_2)_3O^- + CH_3(CH_2)_3Br \rightleftharpoons CH_3(CH_2)_3O(CH_2)_3CH_3 + Br^-$$

Because primary alcohols are weaker acids in aqueous solution than water itself, formation of the alkoxide from an alcohol does not occur in aqueous sodium hydroxide solutions, even at high NaOH concentrations. For this reason, formation of symmetrical ethers is generally assumed to be negligible [49]; however, the apparent pK_a of an alcohol solubilized in a micelle may be different from that in aqueous solution, so this assumption is questionable in aggregated systems. Furthermore, even if dialkyl ether formation is indeed negligible, conversion of the alkyl halide to alcohol may occur at high NaOH concentrations and represents an undesired side reaction in processes intended to synthesize the alkylphenyl ether.

The second reaction system investigated in this work is the etherification of epichlorohydrin to form glycidyl ethers (Fig. 5). The specific reaction of interest is the synthesis of phenolic glycidyl ether in aqueous sodium hydroxide solution. Phenol is depronated at high pH to form hydrophilic phenolate ion, which then reacts with epichlorohydrin to form the hydrophobic glycidyl ether. At high initial alkalinity, epichlorohydrin may react with OH^- ion in the aqueous phase to form glycerol, an undesired product in this case.

An important difference between the two reaction systems studied here is the solubility characteristics of lipophilic reactants, bromobutane and epichlorohydrin. Unlike that of bromobutane, the solubility of epichlorohydrin in water is relatively high and certainly not negligible. Depending on the nature of reaction solutions, it is possible to form a single-phase solution with epichlorohydrin concentrations up to 0.5 M without ad-

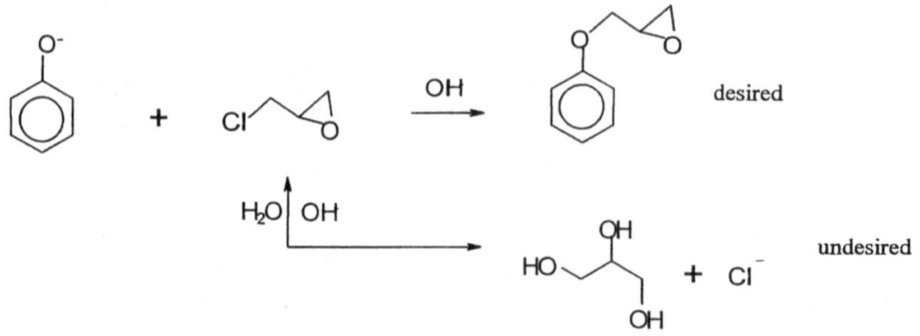

FIG. 5 Synthesis of glycidyl ether in aqueous sodium hydroxide solution.

dition of surfactant to the system. Therefore, reaction in aqueous phase should be considered and cannot be neglected in interpreting kinetic data. Glycidyl ethers, the products in the second reaction system, are compounds that have a chemically stable ether bond and a highly reactive epoxy bond separated by a methylene unit. They are used in the synthesis of drugs for the treatment of cardiovascular diseases [50] and in cosmetic products [51]. Glycidyl ethers with more than eight carbon atoms are also used in the synthesis of some plastics [52]. These glycidyl ethers can also be alkylated or sulfated to manufacture gelling agents, emulsifiers, solubilizers, or surfactants [53].

C. Methods

Reactions were performed by placing 75 to 150 g of the initial solution in a 200-mL round-bottom flask and stirring continuously using a mechanical mixer equipped with a polytetrafluoroethylene (PTFE) stirring blade rotating at 350 rpm. Temperature was maintained at (55 ± 1)°C using a heating mantle and jacketed reactor vessel, closed to minimize loss by evaporation. Samples taken from the reaction vessel were diluted with NaOH solution to a final phenol concentration in the range 0.00010 to 0.00020 mol/kg and analyzed by ultraviolet (UV) spectroscopy. Progress of each reaction was followed for 7–8 h. A quartz sample cell with a 1-cm path length was used, and UV spectra were collected using a Hewlett Packard model 8453 diode array spectrophotometer with 1-nm resolution. Appropriate blank and calibration measurements were performed. Concentrations of phenol and the ether product were calculated simultaneously using standard peak fitting algorithms.

IV. RESULTS AND DISCUSSION

A. Alkylation of Phenol

Figure 6 presents the catalytic effects observed for the alkylation of phenol with 1-bromobutane in various reaction systems. The systems compared are no surfactant and no phase transfer catalyst, phase transfer catalyst alone, surfactant alone, and surfactant plus phase transfer catalyst. All systems were two-phase emulsions and initially contained 2.0 mol/kg of both phenol and 1-bromobutane, representing a total reactant loading of 50% (w/w). In the absence of both surfactant and phase transfer catalyst, the reaction proceeded very slowly, with phenol conversion less than 20% observed after 400 min. Reactions performed with TBAB but no surfactant proceed by conventional phase transfer catalysis; as shown, the phenol conversion reached 30% over the same time interval, indicating that the phase transfer catalyst alone significantly increased the reaction rate, as expected.

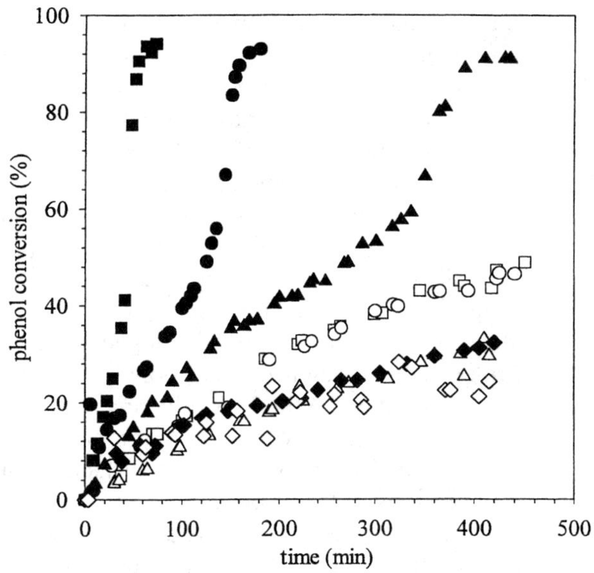

FIG. 6 Effect of MPTC on synthesis of butylphenyl ether at 55°C. All solutions initially contained 2.0 mol/kg phenol, 2.0 mol/kg 1-bromobutane, 6.0 mol/kg NaOH, and water. Surfactant concentrations were (■, □) 0.20 mol/kg CTAC or (●, ○) DDAO or (▲, △) SDS or (♦, ◇) no surfactant. Empty symbols denote runs without TBAB and filled symbols denote runs with 0.10 mol/kg TBAB.

Reactions with surfactant but no TBAB proceed by conventional micellar catalysis. For the CTAC system in Fig. 6, the conversion was only slightly higher at any given time than in the TBAB-only system. The surfactant plays multiple roles in these reaction systems: emulsifier, solubilizing agent, and possibly even phase transfer catalyst. The cmc of CTAC in water at 60°C is 0.0015 M; added NaOH and bromobutane both act to decrease the cmc, so at the high concentrations used here only a small fraction of surfactant is present in monomer form. For reactions performed using surfactant concentrations below 0.0010 mol/kg, reaction rates and conversions were not significantly different from the results shown in Fig. 6 for no CTAC and no TBAB, despite the fact that sufficient surfactant was present to form a good emulsion of 1-bromobutane droplets dispersed in the aqueous medium. The fact that no catalytic effect was observed for CTAC at submicellar concentrations indicates that (1) reaction at the liquid-liquid interface contributes very little to the overall conversion, so that emulsification by a surfactant is, in and of itself, not sufficient to increase the net rate of production, and (2) CTA^+ monomers are not efficient phase transfer catalysts. The profile for the DDAO-only system was similar to the CTAC-only data in Fig. 6, a somewhat surprising result because in single-phase systems (at much lower reactant loading) cationic surfactant systems exhibit better catalysis than nonionic surfactants due to counterion binding of phenolate. One possible explanation is that, at the high phenol/surfactant ratio used here, solubilization of 1-bromobutane in micellar DDAO is higher than in CTAC; higher solubilization of the lipophilic reactant in DDAO compensates for the lack of counterion binding in DDAO as compared with CTAC systems, so that the observed reaction rates for these two surfactants are similar. Conversions in the SDS-only system were significantly lower, as expected because anionic SDS micelles electrostatically repel the reactive phenolate anions, inhibiting the reaction.

The most significant result in Fig. 6 is that the phenol conversion is much higher in the system containing both CTAC and TBAB than either of these components by themselves. This strong synergistic effect clearly illustrates the advantage of micellar phase transfer catalysis over conventional PTC or MC. For solutions containing both CTAC and TBAB, the observed enhancements in reaction rate and conversion are due primarily to the formation of the TBA^+:phenolate ion pair. This complex is only sparingly water soluble and so is solubilized into the micelle pseudophase, thereby facilitating the desired reaction; i.e., the phase transfer catalyst shuttles the hydrophilic reactant into the micellar "pseudophase" for reaction with the solubilized lipophile.

Another important observation in Fig. 6 is the conversions obtained with nonionic (DDAO) and anionic (SDS) surfactants. As expected, the reaction is fastest in the presence of cationic surfactant (CTAC), but the advantage of MPTC is clearly seen in that high conversions are also obtained using nonionic (DDAO) and even anionic (SDS) surfactants. Addition of the phase transfer catalyst TBAB to these surfactant systems resulted in phenol conversions of $\sim$95% after 200 min in DDAO and after 400 min in SDS. Although the apparent reaction rates decrease in the order CTAC > DDAO > SDS, essentially the same high conversion can be achieved using any of these surfactants. The catalysis observed for SDS is especially significant, because addition of TBAB effectively overcomes the electrostatic repulsion between phenolate and the micelle surface, resulting in a much greater reaction rate. The fact that high conversion can be attained with SDS/TBAB emphasizes a key advantage of micellar phase transfer catalysis, because the charge of the surfactant is less important than in conventional micellar catalysis. This provides greater flexibility in designing a process for a particular reaction and permits other factors such as surfactant toxicity, cost, solubilization capacity, and ease of separation to be considered when selecting the surfactant.

An interesting feature of the surfactant/TBAB micellar phase transfer catalysis data shown in Fig. 6 is that reaction profiles in two-phase systems often exhibit what appears to be an autocatalytic effect, as evidenced by a sharp increase in the slope of the conversion versus time data, generally at some point when the conversion was in the range 20–60%, depending on the surfactant. The reason for this phenomenon is not clearly understood, but it has been observed in many systems under two-phase conditions. The sharp increase in reaction rate could indicate an abrupt change in micelle shape and/or orientation of 1-bromobutane solubilized in micelles. Conductivity measurements, commonly used to characterize the continuous phase in emulsions [26], were made at various times during the reaction. These experiments showed that the reaction systems were water continuous at all times, so the abrupt change in reaction rate is not due to an inversion from an o/w emulsion to a w/o emulsion.

Figure 7 represents the difference in reaction profiles observed between one- and two-phase emulsion MPTC systems. As shown in Fig. 7, the reaction rate of alkylation of phenol with 1-bromobutane in a single-phase MPTC system is much slower and exhibits a plateau at much lower conversion compared with a two-phase MPTC system. The higher conversions attained in two-phase systems can be attributed to a number of factors. First, the undesired side reaction in which 1-bromobutane is converted to butanol was not observed in the two-phase reactions. In single-phase systems, this reaction resulted in maximum phenol conversions less than 50%; however, in two-phase systems, phenol conversions of 95% or greater were routinely obtained. Second, the emulsified phase acts both as a continuous source of the lipophilic reactant and as a sink into which the butylphenyl ether product preferentially partitions. In one-phase systems, the micellar solubilization capacity is not exceeded and so the lipophilic ether product remains solubilized in the micelles. This could affect the location and orientation of the bromobutane reactant in the micelle, which in turn could decrease the rate of reaction. In two-phase systems the emulsified phase acts as a continuous source of the hydrophobic reactant and as a sink into which the product preferentially partitions. Because the product does not accumulate in the micelles, more reactant can be solubilized and the reaction proceeds to a much higher conversion. Reaction rates and conversions for MPTC systems are affected by concentrations of surfactant, type of phase transfer catalyst, and NaOH concentration. Results have been summarized elsewhere [30].

B. Etherification of Epichlorohydrin

Figure 8 presents the catalytic effects observed for the synthesis of glycidyl ether in surfactant solutions. Reaction solutions initially contained 0.2 mol/kg of both phenol and epichlorohydrin, representing a total reactant loading of 4 wt%. Because the solubility of lipophilic reactant (epichlorohydrin) in the aqueous phase is relatively high (~0.5 mol/kg), all systems were initially single phase. The glycidyl ether product is sparingly soluble so that reaction solution remains single phase only until the product concentration exceeds the solubility capacity of the surfactant solution. In the absence of surfactant, the reaction proceeded at high rate, with a phenol conversion of 80% observed after 200 min indicating that the reaction rate in aqueous pseudophase is moderately fast because of the high solubility of epichlorohydrin in water. The addition of surfactant DDAO at a concentration well above its cmc increases the reaction rate because of the contribution of conventional micellar catalysis to the overall rate; however, this effect is slight (90% conversion, as compared with 80% without surfactant over the same time interval). At the high surfactant concentrations used here only a small fraction of surfactant is present in monomer form. The micelles effectively solubilize the product of this reaction, delaying the eventual phase separation at higher product concentrations.

Figure 9 presents the effect of surfactant concentration observed for the synthesis of glycidyl ether. Reaction solutions initially contained 1.5 mol/kg of both phenol and epichlorohydrin, representing a total reactant loading of 30 wt%. Because the solubility of epichlorohydrin in water is ~0.5 mol/kg, all systems but 0.6 mol/kg DDAO were initially two phase despite the fact that surfactant was present in the reaction solutions. In the absence of surfactant, the reaction proceeded at a high rate with phenol conversion of 87% observed after 200 min, indicating that the reaction rate in aqueous pseudophase is fast and high conversions can be achieved even in two-phase systems where reaction rates can be limited by mass transfer of species between different phases. Reaction with 0.1 mol/kg

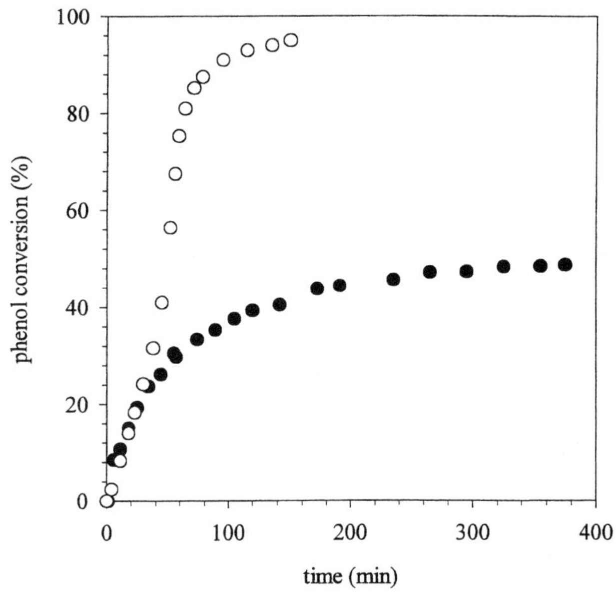

FIG. 7 Comparison of single-phase (●) and two-phase (○) MPTC in the synthesis of butylphenyl ether at 55°C in micellar CTAC solutions. Solutions initially contained (●) 0.2 mol/kg phenol, 0.2 mol/kg 1-bromobutane, 0.6 mol/kg NaOH, 0.01 mol/kg TBAB, 0.05 mol/kg CTAC, and water and (○) 2.0 mol/kg phenol, 2.0 mol/kg 1-bromobutane, 6.0 mol/kg NaOH, 0.1 mol/kg TBAB, 0.5 mol/kg CTAC, and water.

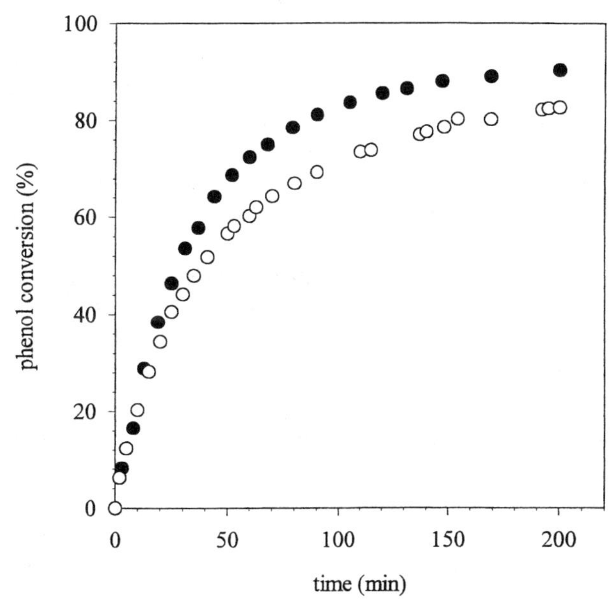

FIG. 8 Effect of MC on synthesis of glycidyl ether at 55°C. All solutions initially contained 0.2 mol/kg phenol, 0.2 mol/kg epichlorohydrin, 0.25 mol/kg NaOH, and water. Concentrations (mol/kg) of DDAO: (○) 0.0; (●) 0.3.

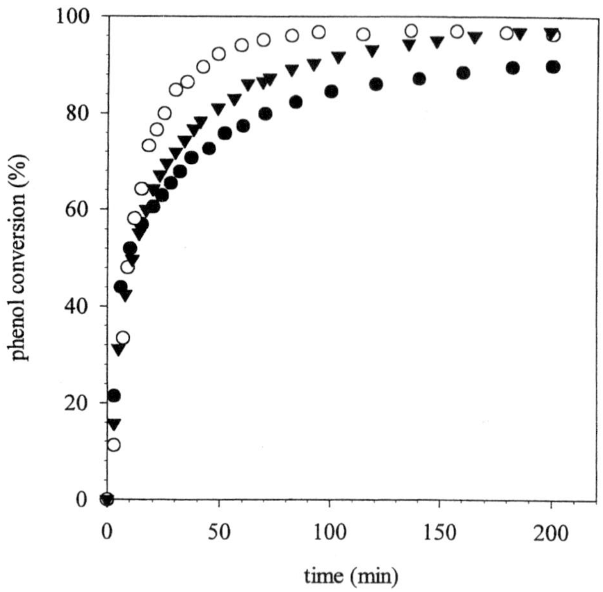

FIG. 9 Effect of surfactant concentration on synthesis of glycidyl ether at 55°C. All solutions contained 1.5 mol/kg phenol, 1.5 mol/kg epichlorohydrin, 1.5 mol/kg NaOH, and water. Concentrations (mol/kg) of DDAO: (●) 0.0; (▼) 0.1; (○) 0.6.

DDAO showed improvement in conversion, reaching 96% over the same time interval.

Figure 10 presents the effects of the various surfactants on the synthesis of glycidyl ether. Results were obtained for the same reaction system as Fig. 9 with nonionic (DDAO), cationic (CTAC), and anionic (SDS) surfactants. The reaction profiles for each system were similar and the conversions were comparable. These results illustrate an important difference between single- and two-phase reaction systems. In single-phase systems (low reactant loading), cationic micelles catalyze this reaction much better than nonionic surfactants due to counterion binding of phenolate, and anionic surfactants actually inhibit the reaction because of the electrostatic repulsion between the reactive phenolate anions and the micellar surface. In two-phase systems, the type of surfactant headgroup has much less effect on the reaction profile. As in alkylation of phenol, this provides much greater flexibility in designing a process for a particular reaction and permits other factors such as surfactant toxicity, cost, solubilization capacity, and ease of separation to be considered when selecting the surfactant. Moreover, for reactions performed using surfactant concentrations below the cmc, reaction rates and conversions were not significantly different than the results for no surfactant and no TBAB, despite the fact that sufficient surfactant was present to form a good emulsion of epichlorohydrin droplets in the aqueous medium. Therefore, emulsification of the reaction system (increasing the surface area of epichlorohydrin droplets) does not seem to play a major role in improvement of reaction rate and conversion profiles.

As shown in Fig. 11, the solution alkalinity also strongly influences this reaction. The minimum required amount of NaOH is determined by the phenol concentration because the solution must remain sufficiently alkaline during the reaction to convert phenol fully to phenolate. Figure 11 presents results for initial NaOH/phenol molar ratios of 0.5, 1.0, 1.25, and 3.0. Unlike the effect on the alkylation of phenol [30], increasing NaOH concentration did not exert a strongly proportional effect on the observed rates of reaction. This is mostly due to increasing excess hydroxide ions, which promote the side reaction and consume epichlorohydrin. High NaOH concentrations would also be expected to influence mass transfer rates of ions and lipophilic molecules between bulk phases and between the aqueous and micellar pseudophase.

C. Factors to Be Considered in Modeling Multiphase Micellar Reaction Systems

Developing a quantitative model for reaction systems is necessary in order to determine concentration profiles for components that cannot be easily measured experimentally, to predict behavior under other conditions, and to gain insight into the various transport, thermodynamic, and kinetic processes that affect the overall system. The strategy for the systems of interest here is to combine models for reactions in two-phase emulsions (no micelles) and one-phase micellar solutions, as suggested in Fig. 12. Experiments on these systems are performed to determine rate constants, which are then assumed to be valid in two-phase micellar systems. For both alkylation of phenol and etherification of epichlorohydrin in multiphase surfactant systems, three separate reaction locales must be considered: micelle, aqueous solution, and liquid-liquid interface as illustrated in Fig. 3. Applying the pseudophase approach, the overall reaction rate for any given species can be expressed by summing the rate equations, including interphase transport of reactants and products, for the various reaction sites.

Mass transfer rates in emulsions obviously depend on the liquid-liquid surface area, and so droplet size distribution and coalescence or breakup of the dispersed phase are important parameters. Surfactant adsorption at the liquid-liquid interface stabilizes the

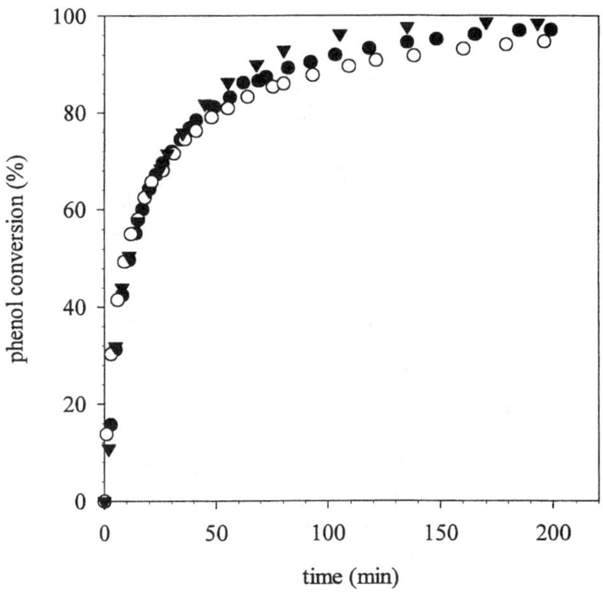

FIG. 10 Effect of type of surfactant on synthesis of glycidyl ether at 55°C. All solutions contained 1.5 mol/kg phenol, 1.5 mol/kg epichlorohydrin, 1.5 mol/kg NaOH, 0.1 mol/kg surfactant, and water. Types of surfactants: (○) SDS; (●) DDAO; (▼) CTAC.

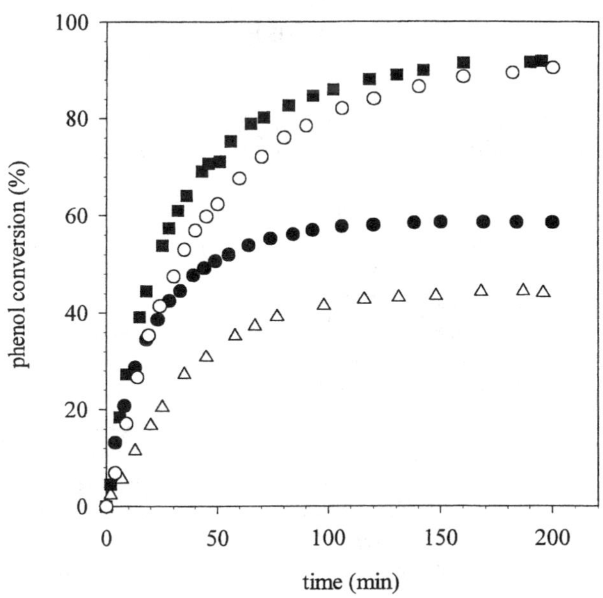

FIG. 11 Effect of initial alkalinity on synthesis of glycidyl ether at 55°C. All solutions contained 0.2 mol/kg phenol, 0.2 mol/kg epichlorohydrin, 0.2 mol/kg DDAO, NaOH, and water. Concentrations (mol/kg) of NaOH: (△) 0.1; (●) 0.6; (○) 0.2; (■) 0.25.

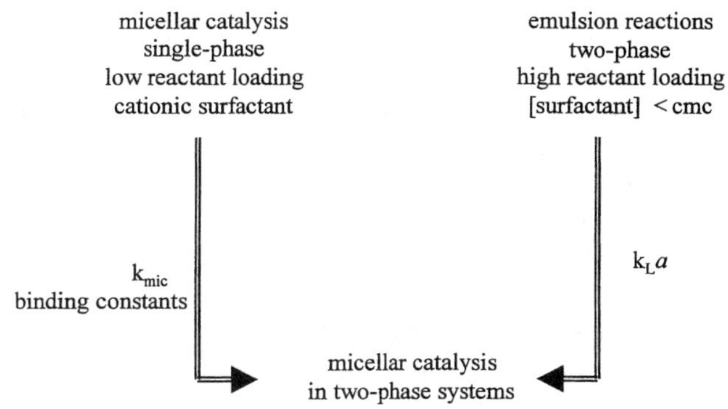

FIG. 12 Strategy for modeling two-phase micellar catalysis.

emulsion, helping to maintain a high contact area, but this surfactant monolayer may also provide additional resistance to mass transport. Alternatively, the adsorbed surfactant may concentrate hydrophilic reactants by electrostatic attraction, facilitating the reaction at the interface. A quantitative description of mass transfer in these systems can be developed using the local rate modeling approach, exemplified by the film, penetration, and surface renewal theories of mass transfer [54]. The description of mass transfer in these models is based on a conceptual parameter (the film thickness in the film theory, surface renewal rate in the surface renewal theory, etc.) that can be determined only by experimental measurements. The effect of chemical reaction on mass transfer rates is then described in terms of an enhancement factor. These models describe the transport reaction processes in a microscopic fluid or flow element with the implicit assumption that such phenomena are not influenced by the neighboring fluid or flow elements. Thus, phenomena such as interference between neighboring concentration fields are not accounted for. Nevertheless, these approaches have proved to be of immense value in understanding and quantifying the effects of chemical reaction on mass transfer rates and have come to occupy a prominent place in industrial practice.

Appropriate assumptions must be made to avoid having to deal with the large number of variables that arise in a completely general analysis of multiphase micellar reactions. The assumption of equal rate constants in the micellar and aqueous pseudophases has been shown to be acceptable for many single-phase micellar reactions; however, in more concentrated systems this assumption is less likely to be valid. For example, in the systems studied here the local concentrations of reactants solubilized and/or bound to the micelle can be as high as 5 M; the observed increased rate of reaction is probably due to both high concentration of reactants at the micelle surface and the fact that the chemical environment near the micelle surface is much different from that in the bulk. Product effects must also be considered for these systems: product formed in the micellar phase may remain solubilized in the micelle, partition into the aqueous phase, or partition into the dispersed droplet phase. For a reaction in which the product is lipophilic, the emulsion droplets initially contain pure lipophilic reactant, but as the reaction progresses, the droplet becomes more dilute with respect to reactant because of partitioning of the product into the dispersed phase. The model must therefore take into account the effects of the changing composition of the droplets with time. As discussed in earlier sections, variation in micellar properties (shape, size, etc.) during the course of reaction and difficulty in interpreting localized concentrations and orientations of reactants or products solubilized in the micelle are other factors that need to be considered to develop useful descriptive models for two-phase micellar catalysis.

REFERENCES

1. C. M. Starks, C. L. Liotta, and M. Halpern, *Phase-Transfer Catalysis*, Chapman & Hall, New York, 1994.
2. E. V. Dehmlow and S. S. Dehmlow, *Phase-Transfer Catalysis*, VCH Weinheim, New York, 1993.
3. W. P. Weber and G. W. Gokel, *Phase-Transfer Catalysis on Organic Synthesis*, Springer Verlag, New York, 1977.
4. J. Jarrouse, C. R. Hebd. Seances Acad. Sci. *252*:1424 (1951).
5. C. M. Starks, J. Am. Chem. Soc. *93*:195 (1971).
6. A. Loupy, J. Sansoulet, E. Diez-Barra, and J. B. Carriollo, Synth. Commun. *21*:1465 (1991).
7. F. S. Sirovsky, Russ. Chem. Bull. (Izv. Akad. Nauk. Ser. Khim.) *11*:2122 (1995).
8. S. N. Tan, R. A. Dryfe, and H. H. Girault, Helv. Chim. Acta *77*:231 (1994).
9. M. L. Wang and H. M. Yang, Ind. Eng. Chem. Res. *29*: 522 (1990).
10. H. Alper, Adv. Organomet. Chem. *19*:183 (1981).
11. L. J. Mathias and R. A. Vaidya, J. Am. Chem. Soc. *108*: 1093 (1986).
12. H. H. Freedman, Pure Appl. Chem. *58*:857 (1986).
13. K. J. Evans and H. J. Palmer, AIChE Symp. Ser. *77*: 104 (1981).
14. J. B. Melville and J. D. Goddard, Chem. Eng. Sci. *40*: 2207 (1985).
15. M. L. Wang and H. S. Wu, Chem. Eng. Sci. *46*:509 (1991).
16. T. J. Broxton and S. Wright, Aust. J. Chem. *44*:103 (1991).
17. C. A. Bunton, Catal. Rev. Sci. Eng. *20*:1 (1979).
18. C. A. Bunton and G. Savelli, Adv. Phys. Org. Chem. *22*:213 (1986).
19. C. A. Bunton, Micellar rate effects upon organic reactions, in *Kinetics and Catalysis in Microheterogeneous Systems* (M. Gratzel and K. Kalyanasundaram, eds.), Vol. 38, Marcel Dekker, New York, 1991, pp. 13–47.
20. E. H. Cordes, Pure Appl. Chem. *50*:617 (1978).
21. C. A. Bunton and J. R. Moffatt, Langmuir *8*:2130 (1992).
22. J. Fendler and E. Fendler, *Catalysis in Micellar and Macromolecular Systems*, Academic Press, New York, 1975.
23. L. S. Romsted, A General kinetic theory of rate enhancements for reactions between organic substrates and hydrophilic ions in micellar systems, in *Micellization Solubilization and Microemulsions* (Proc. Int. Symp.), New York, 1976.
24. L. S. Romsted, Micellar Effects on Reaction Rates and Equilibria, In: K. L. Mittal, B. Lindman, eds., Surfactants in Solution, New York: Plenum Press, 1984, pp. 1015–68.
25. M. J. Rosen, *Surfactants and Interfacial Phenomena*, 2nd ed., John Wiley & Sons, New York, 1989, pp. 196–202.
26. S. Ross and I Morrison, *Colloidal Systems and Interfaces*, John Wiley & Sons, New York, 1988.
27. C. Siswanto, T. Battal, O. E. Schuss, and J. F. Rathman, Langmuir *13*:6047 (1997).
28. F. A. L. Van der Horst, M. H. Post, J. J. M. Holthuis, and U. A. T. Brinkman, Chromatographia *28*:267 (1989).
29. R. Zana, J. Colloid Interface Sci. *78*:330 (1980).
30. T. Battal, C. Siswanto, and J. F. Rathman, Langmuir *13*:6053 (1997).
31. D. Perez-Bendito and S. Rubio, Trends Anal. Chem. *12*: 9 (1993).
32. H. Chaimovich, R. M. V. Aleixo, I. M. Cuccovia, D. Zanette, and F. H. Quina, The quantitative analysis of micellar effects on chemical reactivity and equilibra: An evolutionary overview, in *Solution Behavior of Surfactants: Theoretical and Applied Aspects* (E. J. Fendler and K. L. Mittal, eds.), Plenum, New York, 1982, pp. 949–973.
33. M. M. Sharma and A. K. Nanda, Trans. Inst. Chem. Eng. *46*:T44–T52 (1968).
34. F. M. Menger, Chem. Soc. Rev. *1*:229 (1972).
35. P. R. Kust and J. F. Rathman, Langmuir *11*:3007 (1995).
36. A. Lubineau, Chem. Ind. (Lond.) *4*:123 (1996).
37. G. B. Jacobson, C. T. Lee, S. R. P. daRocha, and K. P. Johnston, J. Org. Chem. *64*:1207 (1999).
38. G. B. Jacobson, C. T. Lee, and K. P. Johnston, J. Org. Chem. *64*:1201 (1999).

39. M. Shinitzky and R. Haimovitz, J. Am. Chem. Soc. *115*: 12545 (1993).
40. M. Seno, Y. Shiraishi, S. Takeuchi, and J. Otsuki, J. Phys. Chem. *94*:3776 (1990).
41. D. W. Deamer and A. G. Volkov, Oil/water interfaces and the origin of life, in (A. G. Volkov and D. W. Deamer, eds.), *Liquid-Liquid Interfaces* CRC Press, Boca Raton, FL, 1996, pp. 363–374.
42. T. Buhse, V. Pimienta, D. Lavabre, and J. C. Micheau, J. Phys. Chem. *101*:5215 (1997).
43. T. Buhse, R. Lavabre, R. Nagarajan, and J. C. Micheau, J. Phys. Chem. *102*:10552 (1998).
44. J. J. Lee and W. T. Ford, J. Org. Chem. *58*:4070 (1993).
45. Z.-Z. Yu, B.-G. Li, M.-J. Cai, B.-F. Li, and K. Cao, J. Appl. Polym. Sci. *55*:1209 (1995).
46. W. D. Harkins, J. Am. Chem. Soc. *69*:1428 (1947).
47. W. V. Smith and R. H. Ewart, J. Chem. Phys. *16*:592 (1948).
48. B. Jursic, Tetrahedron *44*:6677 (1988).
49. H. H. Freedman and R. A. Dubois, Tetrahedron Lett. *38*:3251 (1975).
50. G. Mouzin, H. Cousse, and A. Stenger, European Patent Applications EP 37777, EP 37778, EP 37781 (1980).
51. H. Sebag and G. Vanlerberghe, French Patent FR 2 574 786 (1984).
52. R. Pozniak and J. Chlebicki, Polym. J. Chem. *52*:1283 (1978).
53. A. Messert and H. Baumann, German Patent DE 2 943 353 C2 (1988).
54. L. K. Doraiswamy and M. M. Sharma, *Heterogeneous Reactions: Analysis, Examples and Reactor Design*, Vol. 2, Wiley-Interscience, New York, 1984.

16

Chemical Detoxification in Amphiphilic Systems

RAYMOND A. MACKAY Clarkson University, Potsdam, New York

I. INTRODUCTION

This chapter deals with the use of amphiphilic systems for the detoxification of hazardous or unwanted chemicals. Whereas the principles will be seen to be of rather broad applicability, the focus on specific systems will be narrower and will be discussed in this section.

A. Amphiphilic Systems

These systems are restricted here to "organized" fluids based on surfactants, such as micellar solutions, microemulsions, emulsions, and lyotropic mesophases. The surfactants considered, for reasons to be discussed later, are principally cationic, although nonionic surfactants may be useful for separation applications. The former are generally quaternary ammonium salts with one or two long alkyl chains, and the latter can range from polyoxyethylene (PEO) alkyl ethers to PEO/polypropylene oxide (PPO) block copolymers. All of the amphiphilic systems considered here contain water in some amount as one of the components. It may be noted at the outset, however, that nonaqueous amphiphilic systems have been developed that contain a polar organic solvent (e.g., formamide) in place of water. In all cases, the chemical(s) to be detoxified will be introduced into the condensed-phase amphiphilic system, along with any necessary reagents. In this regard, some clarification of terms will be useful. Detoxification is normally taken to mean that the material under consideration is chemically transformed into a nonhazardous (or much less hazardous) substance or substances; decontamination is more generic and may refer to either chemical destruction (detoxification) or simply physical removal from the area of concern. Section IV.B deals with the use of surfactant phase equilibria for physical separation. Phase separation may be employed either for physical decontamination or for further treatment and removal of reaction products following chemical reaction. The specific amphiphilic systems considered in this chapter are almost exclusively microemulsions and are described in Section II. This is because the basic principles for reactions in all types of amphiphilic systems are essentially the same, and microemulsions are generally the most practical class of these fluid media for physical and/or chemical decontamination. However, a few other types of amphiphilic systems will also be briefly considered.

B. Chemical Systems

A large range of environmentally and industrially hazardous chemicals are potential candidates for detoxification, ranging from pesticides to polychlorinated biphenyls (PCBs). However, much of the research involving amphiphilic systems has been directed toward the destruction of chemical warfare agents, and the types of structures, reagents, and systems employed have been reviewed by Yang et al. [1]. In these cases, and indeed all cases in which amphiphilic systems may be of use as dissolution and reaction media, the chemical to be decontaminated generally has one major characteristic: limited solubility in water. However, the chemical reactions employed to destroy or neutralize the agents are normally either hydrolysis or oxidation using aqueous-based reagents such as hydroxide or hypochlorite salts. The surfactant-based systems therefore

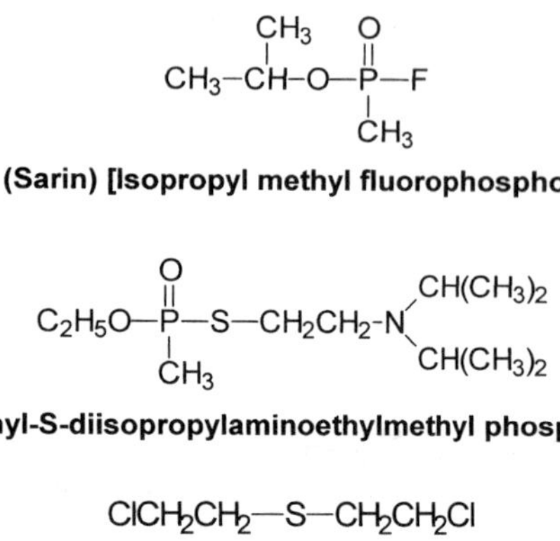

GB (Sarin) [Isopropyl methyl fluorophosphonate]

VX [O-Ethyl-S-diisopropylaminoethylmethyl phosphothiolate]

HD (Mustard) [Bis (2-chloroethyl) sulfide]

FIG. 1 Structures of chemical agents.

contain the essential ingredients to be able to solubilize sufficient quantities of both oil- and water-soluble substances and also function as media in which reactions between them can occur at reasonable rates. The structures of the three prototypical chemical agents are shown in Fig. 1. Two of the three, the nerve agents, are organophosphorus compounds, as are many insecticides such as paraoxon and malthion. For this reason, many of the chemical reactions studied involve hydrolysis, particularly of organophosphate esters. The physical properties and many chemical reactions of these and other hazardous substances, as well as many alternative detoxification strategies, have been documented by Greenpeace International [2].

II. MICROEMULSIONS

A. Nature and Types

A great deal has been published on the preparation and characterization of microemulsions and their use in technical processes [3]. Some applications are discussed in other chapters of this book. In general, microemulsions are transparent, isotropic, fluid dispersions of comparable amounts of two immiscible liquids stabilized by one or more surfactants or a surfactant-cosurfactant mixture. The two liquids are normally water (W) and an oil (O). A wide variety of ionic and nonionic surfactants have been employed, frequently in combination with a cosurfactant. Although the most commonly used cosurfactants are medium-chain-length alcohols, a number of other relatively low-molecular-weight polar fluids have been shown to be effective [6]. The term "microemulsion" that has been applied to these systems is in most cases a misnomer, because thermodynamic stability appears to be the rule rather than the exception. Nonetheless, the name has stuck, and it remains convenient to describe these single-phase fluids using two-phase emulsion terminology. Thus, the surfactant/cosurfactant mixture will be referred to as the "emulsifier" (E). Because the microemulsion literature is now fairly voluminous, only a brief description of the salient features of the microstructure of these fluids will be given here for the reader who is unfamiliar with them and to provide the context for the presentation of microemulsion decontamination reactions.

There are three generally recognized microemulsion phases: upper, lower, and middle. The upper phase is of the water-in-oil (W/O) type, consisting of nanosized water droplets dispersed in the (continuous) oil. The lower phase is of the oil-in-water (O/W) type. The middle phase is "bicontinuous," with no extended aggregates of either type and with a rapidly fluctuating oil-water interface. However, because the rates of the chemical reactions that will be considered here are all slow compared with the time frame for exchange of surfactant in these systems (on the order of 10^{-5} s), most of these micro- or nanostructural features become unimportant. The single most important feature is the existence of both oil and water domains and a very large O-W interfacial area on the order of 10^9 cm^2/dm^3 (100 m^2/g). The bulk of the surfactant and much of the cosurfactant, when present, is located at the O-W interface, being of course responsible for creating this interface. The pseudophase model of microemulsions considers these three domains to be equivalent to three phases in equilibrium: an aqueous phase consisting mainly of water containing a small amount of cosurfactant, a similar oil phase, and an interphase consisting of the surfactant, the remainder of the cosurfactant, and a small amount of oil and water.

It is also important to keep in mind that microemulsions are not equivalent to true solutions (molecular dispersions) of the components. Each domain essentially retains its characteristic solubility for dissolved substances. The extent to which a solute dissolves in a given microemulsion, and also its distribution among the three general pseudophase locations, will be essentially controlled by its solubility in the oil, water, and interphase and the volume fractions of each. A final consideration in the case of ionic surfactants is the charge at the oil-water interface. In an O/W system, the oil nanodroplets carry a charge due to unbound surfactant counterions, whereas this charge is effectively neutralized in a W/O system. In both cases, however, the effective ionic strength on the aqueous side of the O-W interface is very high, on the order of a few moles

dm^{-3}. A general schematic of typical pseudo-three-component phase maps showing one-phase microemulsion regions in four-component systems is given in Fig. 2. It should be emphasized that actual systems can contain separated one-phase regions as well as liquid crystalline, sponge, and cubic phases.

B. Considerations for Use

In this section the practical requirements for the use of amphiphilic systems in general and microemulsions in particular will be examined. First, the nature of what is to be decontaminated must be considered. In the most general sense, this can be grouped into two major categories, materiel and bulk chemical. By materiel is meant equipment, vehicles, soil, buildings, and by extension even personnel. Bulk chemical means that the agent has either been collected by some means or is already stored in containers of some type. Most of the materiel categories will require some type of spray application of the decontaminant, whereas the bulk chemical and some categories of materiel (e.g., small equipment, soil) may be amendable to a batch process. In general, application to either category at a fixed installation will have requirements that differ in some respects from those of field application. At a fixed site, it is possible to mix the reagents at or just prior to use, use a reasonable volume of solution, and have a rate of reaction that is not exceptionally fast as long as the reaction goes to completion. In the field, a smaller volume of preferably premixed reagent must be able to react rapidly, also to completion. At a fixed site, advantage can be taken of the phase behavior, as discussed later. In the field, the system must be stable under ambient conditions, which in some circumstances can be quite extreme.

Whether the decontamination solution is a single system or two or more components that must be mixed on use, all most be stable in storage under variable temperature conditions for long periods of time. As microemulsions are in general (thermodynamically) one-phase systems, they will spontaneously reform in their stable temperature range even if they have phase separated by a thermal excursion outside this range. Microemulsions composed of ionic surfactants generally have higher thermal stability than nonionics, although this does depend on the hydrophile-lipophile balance (HLB) of the nonionic. Mixed surfactants can also be used, as shown in Fig. 3. A change in ionic strength can also affect the HLB of the surfactant and cause either a phase change or phase separation. Because a variety of salts, including reagents, buffers, and environmental compounds, will be present, the stability of the microemulsions with respect to these materials is important. The microemulsions or other amphiphilic systems must be able to function in the presence of oil, dirt, etc. and must be nontoxic, noncorrosive, and nonflammable. They must be able to remove (dissolve) and neutralize the agent from various surfaces without damaging the surfaces (e.g., paint). The resulting residue that must be disposed of should be of as small a volume as possible because even decontaminated agents must generally be disposed of as a category of hazardous waste. The cost of use must also be taken into account with respect to the nature of the application.

A final note regarding the compositional range of existence of the microemulsion itself: The one-phase region should be large so that significant quantities of the agent can be accommodated and the system will not separate ("break") if diluted by reasonable amounts of either oil or water. One example of a four-component system with a wide compositional range is shown in Fig. 4.

III. REACTIONS

A. Hydrolysis

This is perhaps the most extensively studied class of reactions for the decontamination of toxic substances such as organophosphorus compounds. Because basic hydrolysis is the preferred general method, the reaction is promoted by cationic surfactants and retarded by anionic surfactants. One of the most widely employed model substrates in both micellar and microemulsion media is PNDP, *p*-nitrophenyl diphenyl phosphate [7]. There are several reasons why this substrate is quite convenient. It is reasonably stable and readily prepared in sufficient quantities for kinetic studies. It is a solid, which is conveniently weighed, and the reaction can be followed spectrophotometrically as a result of release of the *p*-nitrophenoxide ion, which absorbs at 400 nm [Eq. (1)]. The very low solubility of PNDP in water ensures complete solubilization in the oil pseudophase of the microemulsion or micelle. In fact, even more water-soluble esters such as the diethyl or dihexyl, which are partitioned in aqueous micellar solution, are effectively all in the oil pseudophase in microemulsions [8].

$$(C_6H_5)_2P(O)OC_6H_5NO_2 + OH^- \rightarrow (C_6H_5)_2P(O)(OH) + O_2NC_6H_5O^- \quad (1)$$

The effect of surface charge on the rate of reaction is best seen by comparing the rate in systems stabilized

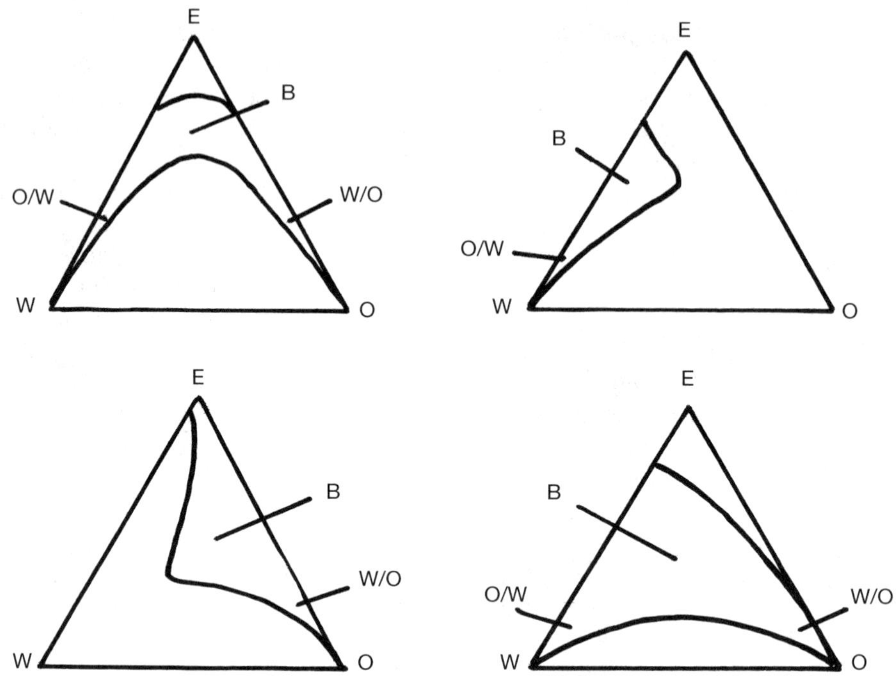

FIG. 2 Schematic microemulsion phase maps, where E is a surfactant, mixed surfactant, or surfactant/cosurfactant mixture, W is water, and O is the oil. The symbols O/W, W/O, and B refer to oil-in-water, water-in-oil, and bicontinuous (middle) microemulsion phases (see text).

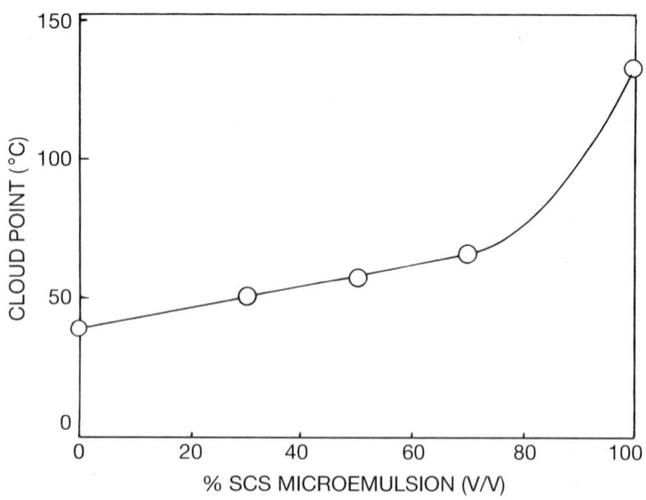

FIG. 3 Thermal stability of a mixed ionic/nonionic microemulsion. Compositions (wt%) are as follows. Ionic: sodium cetyl sulfate (SCS), 12.4; 1-pentanol, 19.4; heavy mineral oil, 8.7; water, 59.5. Nonionic: Tween 40, ICI trade name for polyoxyethylene (20) sorbitan monopalmitate, 56.0; 1-pentanol, 16.2; heavy mineral oil, 7.8; water, 20.0.

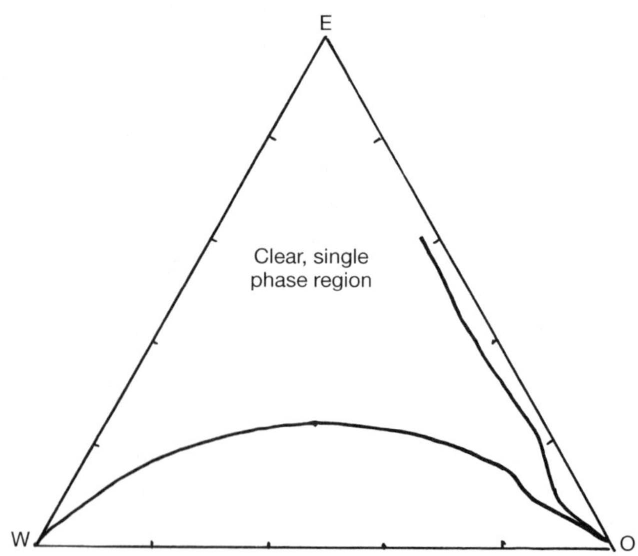

FIG. 4 Pseudo-three-component phase maps of the cetyltrimethylammonium bromide (CTAB)/1-butanol/tributyl phosphate (TBP) water system at 25°C. Axes, by weight percent, are W = water, O = TBP, E = CTAB/1-butanol (1:1, w/w).

by cationic surfactants with that in a comparable nonionic microemulsion. There is a great deal of evidence that in most cases the principal effect of a reaction occurring at a microemulsion oil-water interface is the same as that responsible for micellar "catalysis," namely the concentration of reagents. In other words, there is little effect of the medium per se (i.e., solvent effect), and the rate depends mainly upon the local concentration of substrate (phosphate ester) and nucleophile (hydroxide ion) in the reaction (interphase) region. This effect is shown in Fig. 5 for a hexadecane-in-water microemulsion stabilized by the nonionic surfactant polyoxyethylene (10) akyl ether (Brij 96) or the cationic surfactant cetyltrimethylammonium bromide (CTAB) and various cosurfactants. In this case the nucleophile is iodosobenzoate (IBA^-), to be discussed in more detail later. Similar curves are obtained for all anionic nucleophiles, including hydroxide. The rate constants are corrected for the compartmentalization of nucelophile in the aqueous pseudophase by using a phase volume corrected rate constant $k_{2\phi}$ given by Eq. (2):

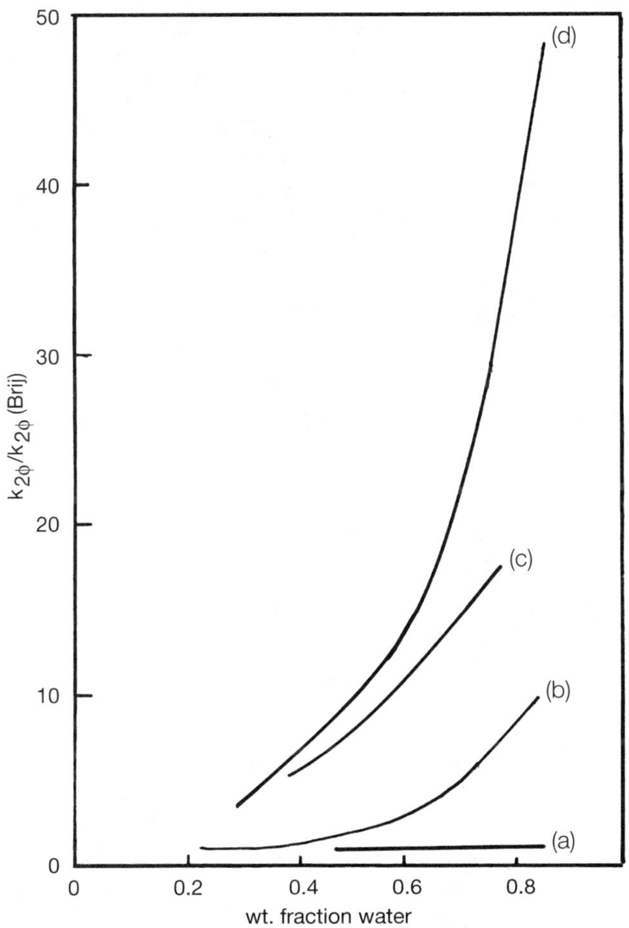

FIG. 5 Relative phase volume–corrected rate constants ($k_{2\phi}$) versus weight fraction water for the IBA-catalyzed hydrolysis of PNDP in various water/hexadecane microemulsions stabilized by various surfactant/cosurfactant mixtures. (a) Brij 96/1-butanol; (b) CTAB/1-butanol; (c) CTAC/dibutylformamide; (d) CTAB/2-methyl pyrrolidinone and Adogen 464.

$$k_{2\phi} = k_2/(1 - \phi) \qquad (2)$$

Here, k_2 is the observed second-order rate constant for the reaction [Eq. (1)] and ϕ is the volume fraction of the "oil" phase. Because essentially none of the surfactant and little of the cosurfactant is in the aqueous portion, the phase volume is approximated by Eq. (3):

$$\phi = 1 - wg \qquad (3)$$

where w is the weight fraction of water and g is the specific gravity of the microemulsion.

Three features in Fig. 5 illustrate the points just mentioned. First, $k_{2\phi}$ for the nonionic system is independent of ϕ (i.e., water content), consistent with the dependence of the rate constant on local concentration. Second, for the CTAB microemulsions, $k_{2\phi}$ increases with increasing water content, consistent with an increasing degree of bromide counterion dissociation (α). An increase in α should lead to a larger surface potential (ψ) and consequently a higher local (interface) concentration of hydroxide ion in the reaction region, roughly equivalent to the Stern layer of a micelle. Based on the evidence that the intrinsic rate constants in comparable ionic and nonionic microemulsions differ only by reason of the surface potential, an effective value of ψ can be calculated from Eq. (4):

$$k_{ionic}/k_{nonionic} = \exp(\psi/kT) \qquad (4)$$

Values of ψ are given in Table 1 for IBA^- as the nucleophile and are in good agreement with values obtained from other measurements.

As might be expected, the use of a medium-chain-length quaternary ammonium salt as a cosurfactant leads to an increased reaction rate compared with alcohol or formamide cosurfactants. Interestingly, the use of a pyrrolidinone as cosurfactant gives comparable results. It might be expected that *N*-butyl formamide (NBF), which has a much higher dielectric constant than *n*-butanol, would produce a greater degree of counterion dissociation. However, this cannot be the sole explanation for the effect of the pyrrolidinones, because they have a dielectric constant intermediate between those of the alcohol and the formamide. The direct participation of the zwitterionic form of the

TABLE 1 Effective Surface Potentials (ψ)

	Surface potential (mV)[c]	
Cosurfactant[a,b]	0.40	0.75
1-butanol	33	52
DBF[c]	72	83
MP[c]	73	91
Adogen 464	87	87

[a]The surfactant in all cases is CTAB.
[b]Dibutyl formamide (DBF) and 1-methyl-3-pyrrolidinone (MP).
[c]Value of surface potential at volume fractions of 0.40 and 0.75, corresponding approximately to 60% and 25% water by weight, respectively. Similar values were obtained using hydroxide as the nucleophile.
Source: data from Ref. 6.

N—O bond has been ruled out [9]. However, more recent results indicate that cyclohexyl pyrrolidinone forms aggregates in water similar in size to surfactant micelles (F.R. Longo, personal communication), and it may be that the nanostructural features of these microemulsions differ in some respect.

Third, the phosphate ester hydrolysis is essentially unaffected by ionic strength because it is a reaction between a neutral molecule and an ion. Therefore, even though the effective ionic strength in the microemulsion "Stern" layer is very high, there is no effect of added salt on the rate constants. Reactions between ions such as the reaction of 1-decyl-3-amido pyridine with cyanide are greatly affected by ionic strength. Interestingly, when this reaction is carried out in microemulsion, the rate constant is significantly decreased by an increase in ionic strength with nonionic surfactant but unaffected with cationic (CTAB) surfactant [10]. Also, at first unexpectedly, the rate with nonionic surfactant and no added salt is higher than with CTAB. This is again, however, a manifestation of the very high (on the order of 3M) ionic strength in the Stern layer. Therefore, except for possible effects on phase stability, added salts should have little or no effect on the interfacial kinetics of hydrolysis reactions in ionic microemulsions.

B. Catalytic Hydrolysis

In an attempt to increase the rate of hydrolysis of phosphate esters and related compounds, a variety of additional catalysts have been employed in micellar and microemulsion media, including metal ions [11,12] and iodosobenzoates (IBAs) mentioned in the preceding section [9,13]. The IBA reagents deserve special note because they are both potent nucleophiles and true catalysts for the hydrolysis of reactive esters or phosphates, particularly when employed as *N*-alkyl-*N*-methyl-*N,N*-bis(3-carboxy-4-iodoso)benzylammonium bromides [14]. The basic reaction scheme is shown in Fig. 6. The IBA is fully ionized at pH 8 in a CTAB/1-butano/hexadecane/water microemulsion, and the rate is given by Eq. (5) [15]:

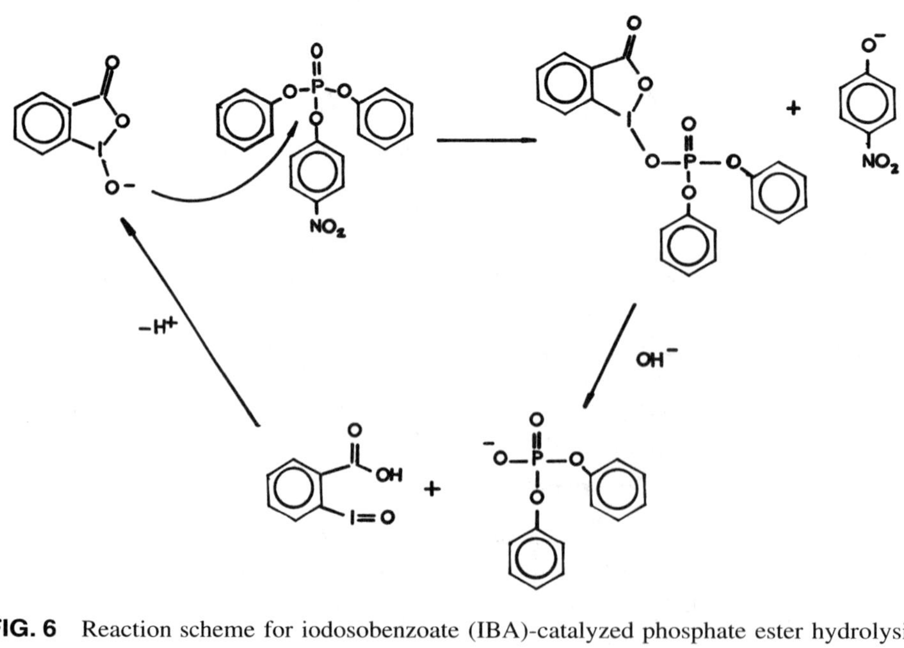

FIG. 6 Reaction scheme for iodosobenzoate (IBA)-catalyzed phosphate ester hydrolysis.

$$\text{Rate} = (0.14[\text{OH}^-] + 1.2[(\text{IBA}^-])[\text{ester}] \quad (5)$$

From this result it is seen that the same rate is obtained at pH 8 using 0.1 mM IBA as at pH 11 in the absence of IBA. The principal advantage of the use of this catalyst is that it permits the use of mildly alkaline media for the decontamination of pesticides, G-type agents, and related materials. It should be stressed again that these are true turnover catalysts, and there is no net consumption of IBA as indicated in Fig. 6. The actual stoichiometric reagent is the buffer itself, which can be bicarbonate, borate, etc. As might be expected, the rate is increased as the water-soluble IBA anion is derivatized with longer alkyl chains, presumably by increasing the local concentration in the interfacial reaction zone. Similar effects are observed in both micelles and microemulsions, although the kinetics in the latter are slower than in the former.

C. Oxidation and Reduction

1. Oxidation

Although phosphate esters related to the G-type agents can be effectively detoxified by hydrolysis, the V-type phosphorothiolates and sulfides such as mustard (HD) cannot [1]. A variety of oxidants have been examined for this purpose, and the commercial oxidant oxone was shown to be quite effective for both VX and HD [16]. The active compound in oxone is $KHSO_5$, and it was necessary to use a mixture of water and organic solvent (*N*-methyl-2-pyrrolidinione) to dissolve the agents. The sulfide is converted to the sulfoxide and then quantitatively to the sulfone, while three equivalents of oxidant are required for VX. The G-type compounds are hydrolyzed under these conditions, but very slowly. A microemulsion system was developed to embody all of these concepts: dissolution, hydrolysis, and oxidation. The system (MCBD) consists of water, tetrachlorethylene, cetyltrimethylammonium chloride (CTAC), and a small amount of tetra(*t*-butyl)ammonium hydroxide as a cosurfactant. The active reagents were Fichlor (oxidant) and borate buffer (hydrolyzer) with sodium-2-nitro-4-iodoxy benorate (IBX) as the hydrolysis catalyst. This is essentially an oxidized form of IBA, discussed earlier [17]. However, the rate enhancement for the more polar (water-soluble) phosphates was slight, and the use of chlorinated solvents is now precluded by environmental considerations.

Strong oxidants such as potassium permanganate and sodium hypochlorite were shown to react immediately with 2-chloroethylethyl sulfide (CEES, or half-mustard) (R. A. Mackay, unpublished results). As expected, the reaction was very fast in microemulsions stabilized by cationic surfactant (CTAB) but slower in systems containing anionic sodium dodecyl sulfate (SDS). With potassium dichromate, no reaction occurred. These microemulsions contained 1-butanol as cosurfactant, and the permanganate and hypochlorite also reacted with the alcohol. Menger and Elrington [18,19] examined the reaction of hypochlorite with CEES in hydrocarbon (cyclohexane)/water microemulsions stabilized by SDS with 1-butyl alcohol as cosurfactant and showed that the sulfide was converted rapidly and completely to the sulfoxide. In this case it was proposed that the alcohol acts as an intermediary between the oil-soluble half-mustard and the water-soluble hypochlorite by conversion to a butyl hypochlorite at the microemulsion oil-water interface. Subsequently, a hexadecane/CTAC/*t*-butanol system containing hypochlorite was shown to be effective for both oxidative and hydrolytic reactions [20]. Menger and Rourk [21] also have developed a microemulsion formulation containing propylene glycol, a nonionic surfactant, and 1-hexanol as cosurfactant that can resist freezing and phase separation at −18°C. This complex system, at pH 10–12 with peroxide as the oxidant, was shown to be effective for carrying out both oxidative and hydrolytic decontamination reactions.

2. Reduction

Much less attention has been paid to the use of reducing agents than to oxidants, both in the basic exploration of chemistry in micelles and microemulsions and as detoxification reagents. Jaeger et al. [22] examined the borohydride reduction of mono- and dicarbonyls using sodium borohydride in a CTAB/1-butanol/hexadecane microemulsion containing aqueous potassium hydroxide. The reductions were only slightly faster in microemulsion than in aqueous isopropanol. However, for reduction of α,β-unsaturated ketones, some amount of 1,4-reduction was observed in microemulsion but only 1,2-reduction was observed in homogeneous solution. The reduction of 0.2 M CEES by 0.3 M sodium borohydride at pH 11 in a similar microemulsion containing about 21% water by weight has been examined (R. A. Mackay, unpublished results). No reaction occurs in the absence of the sulfide, and there is about a 5-minute incubation period before the reaction begins. Although slow (about 1½ h), the reaction goes to completion and the microemulsion does not phase separate. Based on mass spectral analysis, the gaseous product is dimethyl sulfide with no trace of hydrogen chloride or sulfide or of chlorine. The likely reaction is given in Eq. [6]:

$$NaBH_4 + ClCH_2CH_2—S—S—CH_2CH_3 \rightarrow NaCl + (C_2H_5)_2S + [BH_3] \quad (6)$$

Even at pH 11, borohydride is not stable in solution for long periods. The more stable cyanoborohydride was used at pH 6. Unfortunately, it is also less reactive. Although this reagent could slowly reduce ketones in microemulsion, no sulfide reduction was observed.

D. Other Systems

1. Electrochemical and Enzymatic Reactions

Although the principal focus of detoxification studies in amphiphilic systems has been on the use of chemical reagents and catalysts, other systems have been examined, including electrochemical, photochemical, and enzymatic. Photochemical reactions can be effective, particularly in suspensions of semiconductor sols such as TiO_2, but little practical use for detoxification in amphiphilic media has been demonstrated. There has also not been widespread use of electrochemistry for this purpose. However, one notable example is the catalytic dechlorination of 4,4′-dichlorobiphenyl- and Aroclor-contaminated clays in a bicontinuous microemulsion at a lead cathode [23]. Soils were contaminated with these polychlorinated biphenyls (PCBs) and then electrochemically dechlorinated in a didodecyldimethylammonium bromide/dodecane/water system with a current efficiency of 50% over a 12-h period.

The use of enzymes as catalysts for reactions in microemulsions has been examined (e.g., [24]). Many investigations have shown that enzymes can function effectively in surfactant media. Because biological redox systems are generally complex, the focus has been on hydrolytic enzymes such as the organophosphorus acid (OPA) anhydrases [1]. However, only enzymes effective for G-type agents were available, and these agents are readily hydrolyzed by nonenzymatic catalysts (see Section III.B). One report indicated, however, that when organophosphorus hydroxylase (OHP) from *Pseudomonas diminuta* was covalently linked to cotton, pesticides and G- and V-type agents were readily hydrolyzed [25]. The availability of stabilized enzymes may in the future lead to greater use of them in surfactant-based media because they can operate under mild conditions. This would be particularly useful on easily damaged surfaces, including human skin.

2. Unconventional Cosurfactants

As already noted, cosurfactants other than medium to long-alkyl-chain alcohols can be used as cosurfactants. Perhaps the earliest application of these polar substances to decontamination was by Seiders [26], who employed sulfones to form microemulsions with CTAB, tributyl phosphate, and CEES as the oils. Sulfolane (tetramethylene sulfone) is a high-boiling (130°C/6.5 mm Hg), nontoxic, and inexpensive industrial solvent and was able to solubilize very significant quantities of both oils. Sulfolane has a high freezing point (20°C), and 3-methyl sulfolane, with a considerably lower freezing point, is equally effective as a cosurfactant. A major potential advantage of this type of cosurfactant is its resistance to oxidation (see Section III.C.1). Cosurfactants such as the pyrrolidinones (Section III.A) also exhibit such resistance and were used because of their ability to dissolve and penetrate polymeric materials. Therefore, they could also replace the use of chlorinated oils such as tetrachloroethylene, which were employed for their solvent-penetrating power (Section III.C.1). Other cosurfactants examined have ranged from *N,N*-dibutyl formamide to Adogen 464, a commercially available tri(octyl-decyl)methyl ammonium chloride. The latter has the advantage of producing high hydrolytic reaction rates but the disadvantage of providing more viscous fluids.

E. Comparison of Micelles and Microemulsions

It was stated earlier that all reactions occurring at a surfactant-stabilized oil-water interface are governed by the same principles irrespective of the nature of the particular amphiphile system. This is true as long as the reaction kinetics are slow compared with the rate of surfactant exchange, which is certainly the case here. In other words, micellar "catalysis" is really due to an increase in concentration of the reagents in the interfacial zone or Stern layer in which reaction is taking place and not to an effect on the intrinsic rate constant [12]. This can be clearly demonstrated by data for the reaction of hydroxide with PNDP [6]. As per Eq. (2), $k_{2\phi}$ is the observed phase volume–corrected second-order rate constant, and $k^{\circ}_{2\phi}$ is the "intrinsic" value in the reaction zone, which would be obtained if the actual interfacial concentration had been used. Then, following Eq. (4):

$$k^{\circ}_{2\phi}(\mathrm{m})/k^{\circ}_{2\phi}(\mathrm{ME}) = k^{\circ}_{2\phi}(\mathrm{m})/k^{\circ}_{2\phi}(\mathrm{ME}) \exp\left(\frac{\psi_{\mathrm{m}} - \psi_{\mathrm{ME}}}{26}\right) = k^{\circ}_{2\phi}(\mathrm{m})(1 - \phi_{\mathrm{m}})/[k^{\circ}_{2\phi}(\mathrm{ME})(1 - \phi_{\mathrm{m}})] \quad (7)$$

Here, m = micelle, ME = microemulsion, and ψ is the surface potential in mV obtained by the application of Eq. (4) [13,27]. For CTAB micelles and microemulsion

ψ_m = 130 mV, ψ_{ME} = 36 mV at ϕ_{ME} = 0.4, and k_2(m)/k_2(ME) = 30. The value of $\phi_m \ll 1$ and can be neglected. Then,

$$k^{\circ}_{2\phi}(\mathrm{m})/k^{\circ}_{2\phi}(\mathrm{ME}) = 1.3 \approx 1$$

and the intrinsic rate constant for this reaction in both micelles and microemulsions is essentially the same.

IV. UTILIZATION OF PHASE BEHAVIOR

A. Foams and Lyotropic Liquid Crystals

Use may also be made of amphiphilic mesophases for detoxification, and two such applications are briefly considered here. Foams have already been examined for purposes of decontamination. For example, the U.S. Navy has explored the utilization of a standard shipboard aqueous fire fighting foam with added reagents. One report [28] describes a cationic surfactant–stabilized foam containing a hydroxyl peroxide carbone and hydrotropes that reportedly react with both chemical (G and V nerve agents and mustard) and biological (spores) agents. For any application that involves spraying a liquid decontamination fluid, whether at an installation or in the field, foam application can be a great advantage in terms of conserving the amount of material used. For example, in the application of a microemulsion an ideal scenario is the formation of a foam sufficiently stable to be applied but then collapsing to a liquid film for maximum agent dissolution and reaction effectiveness. Conditions for foam formation in both three-phase emulsion and microemulsion systems have been examined [29,30]. In the latter, the lamellar liquid crystalline (LLC) phase plays a key role in stabilizing the foam. Depending upon the specific conditions under which a liquid decontaminant is employed, evaporation of volatile components (e.g., water, cosurfactant, oil) may be important [31].

The use of lyotrophic liquid crystalline phases, such as the lamellar phase already referred to, also has potential for decontamination. Refer to a (partial) schematic phase diagram as shown in Fig. 7. The is meant only to illustrate the point, and many such systems with a variety of components are known ranging from three to five basic materials, not including reagents, buffers, etc. Composition X in the LLC phase may be used as a "gel" consistency concentrate that can be diluted with water in the application device to composition Z, which is then sprayed or foamed on. Alternatively, dilution to composition Y gives an emulsion that may adhere better to the surfaces on which it is applied. If water is the most volatile component, evaporation results in the formation of a surface LLC film, which can be removed by a water rinse.

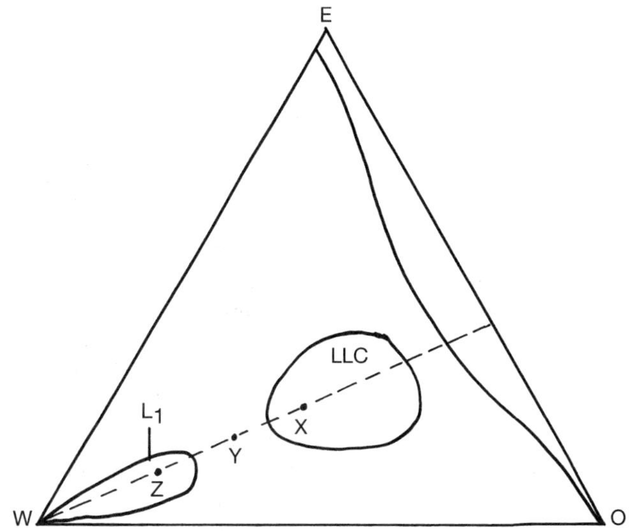

FIG. 7 Partial schematic phase diagram showing an isotropic aqueous (L_1) phase and a lamellar liquid crystalline (LLC) phase. The apexes are water (W), an oil (O), and emulsifier (E), which may be a single or mixed surfactant or a surfactant/cosurfactant blend.

B. Separations

There will be circumstances under which the hazardous material or its detoxified products must be collected for further treatment or disposal, respectively. In this regard, as noted in Section II.B, the final volume must be as small as possible. In effect, this is a separations problem that can be attacked in amphiphilic systems by taking advantage of their phase behavior [32]. This will again be illustrated by using microemulsions as the model. As discussed in Section II.A, microemulsions are of three "types": O/W, W/O, or bicontinuous. Under appropriate conditions of composition and temperature, the O/W and W/O phases can exist in equilibrium with oil and aqueous phases, respectively. In these two-phase equilibria the O/W microemulsion phase is normally denser than the oil phase and is thus the "lower" phase. Similarly, the W/O is the "upper" phase microemulsion. The bicontinuous phase will be in equilibrium with both oil and aqueous phases and thus in the "middle." It may be noted in passing that the middle phase at some optimal composition will have a minimum interfacial tension with regard to both oil and water and hence good penetrating power into pores and cracks. These equilibria from one to two or three phases can be controlled by varying an appropriate field parameter (i.e., composition, salinity, temperature). One schematic but concrete example is given in the following to illustrate compositional extraction. Use can be made of temperature changes in a similar fashion.

Suppose, as is frequently the case (see Section I.C), that an oil-soluble toxic substance is being reacted to form water-soluble products that must be removed and disposed of. Consider a water (W), surfactant (S), oil (O) system as shown in Fig. 8. In this example, the agent is a component of the oil in some proportion. Naturally, as the agent is reacted, the phase boundaries change somewhat. This facet, although important, is being ignored here for simplicity. In Fig. 8, composition B is (by weight) 22% W, 8% S, and 70% O and is the reaction phase. After the reaction is complete, composition A (94% W, 3% S, 3% O) is added to B in the weight ratio 3:17 to give an overall composition C (33% W, 7% S, 60% O). After stirring to ensure equilibration of all components, the phases are left to separate, which is generally quite rapid. Phase A, which now contains the bulk of the water-soluble reaction products, can be separated off and phase B "recycled" for reaction of more agent. Clearly, as implied before, there is some loss of material in each cycle so that small amounts of "makeup" surfactant would have to be added from time to time. Also, the reaction composition would not actually be composition B at the phase boundary but would be somewhat inside the single-phase region (L_2) and similarly for the aqueous phase (L_1). However, after mixing and separation the phases would have the compositions A and B, and some makeup water and oil would also be required.

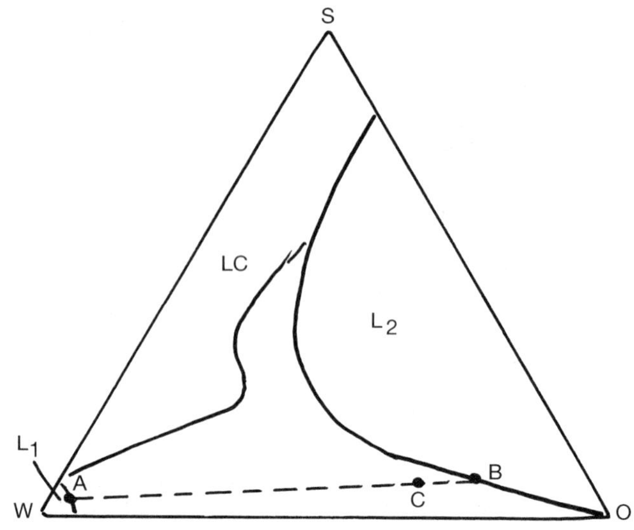

FIG. 8 Phase diagram of a surfactant (S)/water (W)/hydrocarbon (O) system. L_1 and L_2 are isotropic aqueous and oil phases, respectively.

A final note related to separations should include mention of cleavable surfactants [33]. The concept is that, after reaction is complete, the surfactant would be cleaved into nonamphiphilic components by addition of another reagent or change in pH and/or temperature. With the surfactant destroyed, phase separation would occur. Although not of general utility, this may find application in a limited set of circumstances, for example, where it is impractical to achieve phase separation by physical means or where necessary facilities are not available.

V. FUTURE DIRECTIONS

There is a great deal of potential for applications of amphiphilic systems to detoxification of hazardous materials, including chemical warfare agents. However, much of this potential is as yet unrealized and current practice is still primarily "soap and water." The reasons have to do with the rather stringent conditions, discussed or alluded to in the preceding sections, necessary for either commercial or military success. Some of these conditions have to do with the development of suitable reagents or catalysts. The availability of more generally useful enzymes, both hydrolytic and redox, should drive further studies in microemulsion and related systems. In situations in which evaporative water loss is a problem, totally nonaqueous systems may be of value. For purposes of recyclability (Section IV.B) or more environmentally friendly processes, use of surfactant-stabilized water in liquid or supercritical CO_2 microemulsions should be explored [34,35]. In this regard, the use of ionic fluid/surfactant systems remains virtually unexplored [36]. Although there has been work on the combination of hydrotropes with surfactant-based systems, their application to detoxification also remains largely unexplored. Another area that has not been well investigated with regard to decontamination applications in amphiphilic systems is the combination of chemical and physical properties. For example, under some circumstances it may be desirable to react and/or encapsulate the toxic material by applying a fluid medium that can then be polymerized [37]. Therefore, the use of polymerizable surfactants could also be of considerable interest [38].

REFERENCES

1. Yu-Chu Yang, J. A. Baker, and R. Ward, Chem. Rev. *92*:1729–1743 (1992).
2. A. Picardi, P. Johnston, and R. Stringer, *Alternative Technologies for the Detoxification of Chemical Weapons: An Information Document*, Greenpeace International, Washington, DC, 1991.
3. M.-J. Schwuger, K. Stickdorn, and R. Schomaecker, Chem. Rev. *95*:849–864 (1995).
4. K. L. Mittal and P. Kumar, *Handbook of Microemulsion Science and Technology*, Marcel Dekker, New York, 1999.
5. C. Solans and H. Kunieda, *Industrial Applications of Microemulsions*, Marcel Dekker, New York, 1996.
6. R. A. Mackay, L'Actualitie Chimique May–June:161–167 (1991).
7. V. Athanassakis, C. A. Bunton, and D. C. McKenzie, J. Phys. Chem. *90*:5855–5862 (1986).
8. R. A. Mackay and C. Hermansky, J. Phys. Chem. *85*: 739–744 (1981).
9. S. M. Garlick, H. D. Durst, R. A. Mackay, K. G. Haddaway, and F. R. Longo, J. Colloid Interface Sci. *135*: 508–519 (1990).
10. L. Damaszewski and R. A. Mackay, in *Structure-Performance Relationships in Surfactants* (M. J. Rosen, ed.), American Chemical Society, Washington, D.C. 1984, pp. 175–186.
11. F. Tafesse, Inorg. Chim. Acta *269*:287–291 (1998).
12. P. Scrimin, P. Tecilla, U. Tonellato, and C. A. Bunton, Colloids Surf. A Physicochem. Eng. Aspects *144*:71–79 (1998).
13. B. A. Burnside, B. L. Knier, R. A. Mackay, H. Dupont Durst, and F. R. Longo, J. Phys. Chem. *92*:4505–4510 (1988).
14. R. A. Moss, R. Fujiyama, H. Zhang, Y.-C. Chung, and K. McSorley, Langmuir *9*:2902–2906 (1993).
15. R. A. Mackay, F. R. Longo, B. L. Knier, and H. D. Durst, J. Phys. Chem. *91*:861–864 (1987).
16. Y.-C. Yang, L. L. Szafraniec, W. T. Beaudry, and D. K. Rohrbaugh, J. Am. Chem. Soc. *112*:6621–6627 (1990).
17. A. R. Katritzky, B. L. Duell, H. D. Durst, and B. L. Knier, J. Org. Chem. *53*:3972–3978 (1988).
18. F. M. Menger and A. R. Elrington, J. Am. Chem. Soc. *112*:8201–8203 (1990).
19. F. M. Menger and A. R. Elrington, J. Am. Chem. Soc. *113*:9621–9624 (1991).
20. F. M. Menger and H. Park, Rec. Trans. Chim. Pays-Bas *113*:176 (1994).
21. F. M. Menger and M. J. Rourk, Langmuir *15*:309–313 (1999).
22. D. A. Jaeger, M. D. Ward, and C. A. Martin, Tetrahedron *40*:2691–2698 (1984).
23. S. Zhang and J. F. Rusling, Environ. Sci. Technol. *29*: 1195–1196 (1995).
24. C. P. Singh, Enzymatic reactions in microemulsions and monolayers, University of Florida, 1996.
25. Chem. Eng. News, September 13, 1999, p. 40.
26. R. P. Seiders, in *Macro and Microemulsions: Theory and Applications* (D. O. Shah, ed.), ACS Symposium Series 272, American Chemical Society, Washington, DC, 1985, pp. 265–273.
27. R. A. Mackay, Adv. Colloid Interface Sci. *15*:131–156 (1981).
28. Chem. Eng. News, March 8, 1999, p. 10.
29. H. Kuneida and S. E. Friberg, Colloids Surf. *21*:17–26 (1986).
30. S. E. Friberg and C. Solans, Langmuir *2*:121–126 (1986).
31. S. E. Friberg, T. Young, R. A. Mackay, J. Oliver, and M. Breton, Colloids Surf. A Physicochem. Eng. Aspects *100*:83–92 (1995).
32. J. Scamehorn and J. Marwell, eds., *Surfactant-Based Separation Process*, Surfactant Science Series 33, Marcel Dekker, New York, 1989.
33. D. A. Jaeger, C. A. Martin, and T. G. Golick, J. Org. Chem. *49*:4545–4547 (1984).
34. G. B. Jacobson, C. T. Lee, Jr., and K. P. Johnston, J. Org. Chem. *64*:1201–1206 (1999).
35. J. D. Holms, D. C. Steytler, C. D. Rees, and B. H. Robinson, Langmuir *14*:6371–6376 (1998).
36. S. E. Friberg, Q. Yin, F. Pavel, R. A. Mackay, J. Holbrey, K. R. Seddon, and P. A. Aikens, submitted.
37. C. H. Chew, T. D. Li, L. H. Gan, C. H. Quek, and L. M. Gan, Langmuir *14*:6068–6076 (1998).
38. M. Dreja, W. Pyckhout-Hintzen, and B. Tieke, Macromolecules *31*:272–280 (1998).

17

Colloidal Chemistry of Lubricating Oils

DUNCAN C. HONE and BRIAN H. ROBINSON University of East Anglia, Norwich, Norfolk, England

JANE R. GALSWORTHY and ROGER W. GLYDE Infineum UK Ltd., Abingdon, Oxfordshire, England

I. INTRODUCTION

The main functions [1] of the lubricant used in internal combustion (IC) engines are (1) to create a film between moving surfaces in order to reduce friction and wear, (2) to act as a coolant by heat removal, and (3) to provide suspension of contaminants. However, it should be noted that oil formulations used in IC engines are subject to extremely adverse operating conditions. These include high temperatures (200–300°C) and pressures and the potential for contamination by water and unburned fuel through a process known as "blow-by" [2]. For oils to function effectively, it is necessary to introduce a range of *additives* to the lubricant formulation. Additives are chemical substances that impart new or unique properties or enhance existing properties of the base oil. There are many types of lubricant additive [3–5], each with a different function, and typically about 10 different additives are blended with the base (hydrocarbon) oil to generate a fully formulated lubricant. The specific type and amount of each additive depend on the specific application. With such a complex mixture, questions of synergy, or compatibility, of the various additives can be very important. Correctly formulated, these additives provide optimal engine performance over a wide range of operating conditions. The two main colloidal additives present are the *detergent* additive for the neutralization of acidic species and the *dispersant* additive for dispersing carbon and other colloidal particulate contaminants.

A range of carbon-based (organic) acids, e.g., RCOOH, where R is usually an alkyl group, can be produced as a result of oxidative degradation of the hydrocarbon base oil. In addition, gases, e.g., SO_2, that are formed in the combustion process when sulfur-containing fuels are used can react with water to produce inorganic acidic species such as H_2SO_4, together with HCl and HNO_3. Unless neutralized, these acids attack metal surfaces, which leads subsequently to corrosion of engine components and to polymerization of organic species. This latter process results in resin formation and the buildup of tarry deposits in critical regions of the engine, e.g., piston rings. Consequently, there is a requirement to continuously neutralize these acids that form during the operation of an engine. Because large amounts of acid must be neutralized, it is convenient to supply "basicity" to the oil in a concentrated form by means of an overbased additive (so-called detergent) formulation. This class of additive [6,7] enables large amounts of sterically stabilized colloidally dispersed base to be present in the oil. The base is essentially a metal carbonate, e.g., $CaCO_3$ or $MgCO_3$. The stabilizing surfactant is generally an alkylbenzene sulfonate, phenate, or salicylate.

The synthesis and production of detergent additives are covered extensively in the patent literature and the essential procedures have been summarized elsewhere [7,8]. The reaction system typically consists of a metal oxide or hydroxide, surfactant, and aromatic solvent.

Carbon dioxide is passed through the system to form the metal carbonate. Typically, calcium, magnesium, and sodium metals are employed, and the surfactant alkyl chains are highly branched (C_{12} to C_{60}) with a high molecular weight. The mechanisms involved in these detergent particle synthesis reactions are complex and not yet fully understood [9], but the process can be precisely controlled to produce reproducible materials. It has been suggested that the small particle size and monodispersity obtained are due to the initial presence of water-in-oil microemulsions within the system [10,11]. The formation of colloidal calcium hydroxide particles has also been reported [12], again via formation of water-in-oil microemulsions.

II. STRUCTURAL CHARACTERIZATION TECHNIQUES

The techniques of small angle x-ray (SAXS) and neutron (SANS) scattering can provide a detailed structural characterization of overbased detergent additive systems, although traditionally ultracentrifugation has been the method of choice. The overbased sulfonates (in particular calcium and magnesium based) have received the greatest attention, no doubt a consequence of their prevalence in the automotive industry. Markovic, Ottewill, and coworkers [13–16] carried out a detailed study of both calcium and magnesium carbonate detergents stabilized by sulfonate surfactants. They showed that the particle core diameter and the stabilizing adsorbed layer thickness can be independently determined by contrast matching techniques, and they proposed a concentric shell model consisting of a spherical core of metal carbonate surrounded by a curved monolayer (shell) of surfactant. This model has also been suggested by several other authors [9,17–21] and is widely accepted as giving an accurate description of the colloidal particle. However, there are many questions still not resolved, such as the extent of surface coverage by the surfactant molecules, nature of the surface region (e.g., roughness on the molecular scale and composition), composition and structure within the core, extent of oil solvent penetration into the surfactant layer, nature of the interaction between the surfactant headgroup and the carbonate layer or core, and the lateral mobility of surfactant on the particle surface. A schematic picture of the overbased detergent additive is shown in Fig. 1.

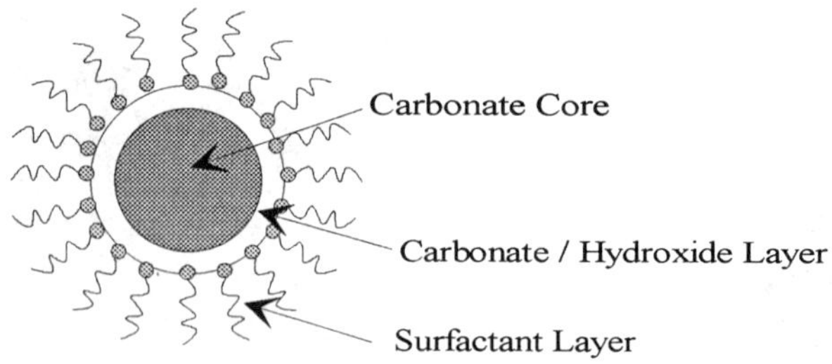

FIG. 1 Schematic representation of an overbased detergent additive.

Typically, the carbonate core radius is in the range 1–10 nm and the surfactant layer thickness 1–5 nm. These values are dependent on the individual chemical constituents used and the experimental conditions that are followed in the synthesis procedure. For a given detergent, the degree of polydispersity is small. Both salicylate [22] and phenate [23,24] stabilized systems have been analyzed and found to have a structure similar to that of the sulfonate species. To date, relatively little theoretical work has been carried out on these systems. Molecular dynamics simulations [25] have been carried out in an attempt to determine the extent of coverage of the particle surface. O'Sullivan et al. [26] used a combination of SANS and and SAXS in a structural study but complemented this with wide-angle x-ray scattering (WAXS) to show that in their system the particle core consisted of amorphous calcium carbonate and not a crystalline form (e.g., calcite, vaterite). This is thought to be necessary in order to stabilize the metal carbonate as larger core structures (crystalline) cannot be so readily stabilized. Extended x-ray absorption fine structure spectroscopy (EXAFS) [27] confirmed this view and showed that calcium is surrounded by nearly six oxygen neighbors at a distance of 0.24 nm.

The particle interactions do not change significantly when their concentration is changed. In concentrated solutions they behave essentially like classical hard spheres as measured through the structure factor in SANS [15]. Several authors [13,27] have reported that the detergent particles remain stable when extracted from their original hydrocarbon medium. This is often a necessary step in sample preparation prior to inspection by the experimental technique of choice. These observations indicate that the detergent particles also remain stable when diluted in pure solvent, suggesting that the particle remains coated with surfactant. The role of the stabilizing surfactant has been investigated in terms of its dynamic properties. Mansot et al. [27] suggested that the surfactant is tightly bound to the particle surface and does not take part in any exchange process with surfactant in the bulk solvent, as is the case in classical micellar systems. Miller et al. [28], however, reported that there is a dynamic equilibrium between surfactant molecules at the particle surface and in the solvent. Fluorescence studies [29,30] have shown that the rigidity of the surfactant layer is reduced on addition of small amounts of low-molecular-weight alcohols (methanol, ethanol). The aromatic moiety of the stabilizing sulfonate surfactant was used as an intrinsic fluorescent probe in these studies.

Energy-filtered electron microscopy (EFEM) [31,32] showed the presence of both carbon and calcium within the particle core, which is surrounded by an organic shell of surfactant molecules. Quasi-elastic light scattering (QELS) or photon correlation spectroscopy allows simple measurements to be made in which sample preparation is straightforward. The use of QELS has been reported [33,34] for the determination of the hydrodynamic radius of the detergent particle.

Electrophoretic mobilities of overbased detergent additives have also been determined [28] using phase-analysis light scattering (PALS). Perhaps surprisingly, a significant charge was found to be present at the particle-liquid interface. The surface charge was found to be dependent upon the solvent and chain length of the stabilizing surfactant, particles stabilized by shorter chain length surfactants exhibiting a more negative surface potential,

Steytler et al. [35] described a novel approach to monitoring the structural changes of the particle (+ surfactant layer) that occur in the liquid-to-solid transition of the solvent induced by change in temperature and/or pressure around 6°C, which was the freezing point of the solvent (cyclohexane) used. It was shown that it was possible to induce interdigitation of the stabilizing surfactant layers and particle-particle interactions were shown to be weakly repulsive, consistent with the operation of steric interactions. A quantitative analysis of interparticle forces associated with interpenetration of the layers was possible using SANS to determine the separation of the particles as a function of the applied pressure. Kurilo and Glavati [36] commented that during such an interaction the hydrocarbon chain of the surfactant does not undergo any appreciable deformation or compression. Interparticle forces have also been measured using an alternative method, the Langmuir film balance [37], and the particle sizes measured were in good agreement with SANS data.

III. PERFORMANCE EVALUATION TECHNIQUES AND TESTS

Detergent additives are classified by their total base number (TBN) [38]. This is defined as the equivalent milligrams of KOH per gram of sample, which is usually in the form of a liquid concentrate. This value is 200 to 400 for a typical detergent additive. The TBN value alone does not give any indication of the neutralization *efficiency* of the detergent additive. Generally, the higher the value of the TBN the more effective it is likely to be as more basic material is present. However, for a range of detergents (various metal carbonates, stabilizing surfactant, etc.) with the same TBN, there will undoubtedly exist a large spread in neutralization efficiencies, reflecting the difference between the additives at the molecular and colloidal level. The efficiency of a detergent can be measured in a number of ways. The most realistic test of a detergent is to actually run it in an engine under operational conditions, with appropriate protocols for assessing performance. This procedure is obviously expensive, inconvenient, and time consuming, especially in a research and development area where there would be many candidate samples to be screened.

Therefore, a range of laboratory-based bench tests have been developed in order to evaluate detergents. These fall into two main classes, direct and indirect. Direct measurement of acid neutralization involves the use of a procedure that monitors a change in a characteristic physicochemical property of either the acid or base in the system. Indirect methods are closer to the engine analysis, and the effect of acid on a metal part is analyzed in terms of corrosive wear, which is indicative of how effectively the acid has been neutralized. Several direct measurement techniques rely on the simple reaction between acid and the metal carbonate core, with the evolution of CO_2 being monitored.

A stirred emulsion system of detergent in a base oil with aqueous sulfuric acid can be studied using a Warburg rig to monitor the pressure of CO_2 as neutralization proceeds. The time to half the total pressure change is generally recorded as an indicator of neutralization efficiency. However, this technique does not give any insight into the mechanistic pathway of acid neutralization, and there is the difficulty that the reaction rate depends on stirring speed because the contact interfacial area between oil and water depends on this.

Roman [39] developed an apparatus for studying acid neutralization in a thin film of oil by monitoring the evolution of CO_2, again as a change in the measured pressure. This method is known as NAMO, neutralization ability of marine oils. Sulfuric acid is used to mimic the predominant acid found in such systems (because diesel fuels used in marine engines contain high levels of sulfur). The test was able to discriminate between different detergents, and mechanistic information was also claimed. It was proposed that the reaction occurs in three stages: (1) acidic gas dissolves in the oil film and forms droplets stabilized by surfactant present, (2) acid droplets collide with overbased particles, and (3) the reaction slows as acid and base are consumed. These studies indicate that acid neutralization occurs during encounters between the overbased particle and the solubilized aqueous acid; however, the acid droplets are not clearly defined in this study.

The use of a solid-state microsensor consisting of a microchip pressure transducer to measure CO_2 evolution was reported by Wohltjen et al. [40]. The sensor was designed for TBN field measurements (on board ships) and was not intended to give mechanistic information. The neutralization of organic acids (namely carboxylic acids) by detergent additives stabilized by an alkyl benzene sulfonate can be monitored by infrared spectroscopy [41]. The carbonate/sulfonate ratio can be obtained through integration of 865 and 1065 cm^{-1} infrared bands. Initial neutralization was not accompanied by a change in the carbonate/sulfonate ratio, and thus Papke [41] concluded that noncarbonated base (presumably hydroxide) is located around the surface of the particle (as shown in Fig. 1). When one considers that approximately 10% of the base in such systems is likely to be noncarbonated and given the large ratio of surface area to volume of such colloidal particles, this observation is perhaps to be expected. It was found that not all acid in the system is neutralized even under conditions in which an excess of base is present. This finding is likely to be associated with the fact that the particle surface becomes saturated with acid (anion). This may or may not exchange with the solvent, but in any case base located within the solid core is immobile and cannot reach the surface. It is even possible that only OH^- reacts with the weak acid.

Techniques for the indirect measurement of acid neutralization include the sequence IID engine test developed in 1978. This is a 32-h engine test followed by inspection and visual rating of 65 metallic engine parts for rust. A new test method, known as the ball rust test (BRT), has been developed [42] that is designed as a simpler bench test replacement for the sequence IID engine test [43]. This laboratory-based test allows the screening of multiple samples in a relatively short time. A steel ball bearing is immersed in a sample of oil and an acidified aqueous phase is introduced under controlled conditions over an 18-h period. The aqueous phase is a mixture of HCl, HBr, and acetic acid. The ball bearing is then analyzed using an optical imaging system and is rated in terms of surface reflectance.

The Plint TE/77 rig [44] was designed as a screening tool prior to engine tests. Corrosive wear is continuously monitored throughout the test and is used to evaluate the performance of the oil.

The rotating diffusion cell (RDC) was adapted by Albery et al. [45] and Lewis [46] to monitor the kinetics and mechanism of acid neutralization occurring at a planar interface under controlled hydrodynamic conditions such that, by extrapolation to infinite rotation speed, the influence of diffusion to and from the interface (mass transfer) on the kinetics is eliminated. The uptake of aqueous acid (as monitored by a pH-stat apparatus) into the oil phase was postulated to take place via formation of w/o microemulsion droplets stabilized by excess free surfactant present in the detergent formulation. Neutralization of the solubilized acid would then take place during an effective collision with the detergent particle.

Attempts have also been made [47] to model TBN depletion as a function of system variables (e.g., engine design, operational conditions, oil type).

IV. STOPPED-FLOW TECHNIQUE APPLIED TO MEASURING ACID NEUTRALIZATION

In order to obtain both kinetic and mechanistic information concerning the acid neutralization process in these systems, we have developed over a number of years at the University of East Anglia [46,48–50] a spectrophotometric assay based on the solubilization of water-soluble pH indicator molecules in a hydrocarbon medium. It is perhaps surprising that very few studies have been carried out on the interaction of nanodroplets (microemulsions) and nanoparticles, especially from a dynamic viewpoint. Such studies will therefore provide fundamental insights into the interactions in such systems.

An acidified aqueous phase is dispersed in a hydrocarbon solvent (*n*-heptane, *n*-decane) in the form of w/o microemulsion droplets stabilized by sodium bis(2-ethylhexyl) sulfosuccinate (Aerosol OT or AOT). (It is believed [39] that excess free surfactant present will solubilize acid in similar microemulsion droplets in a real engine scenario.) Droplets stabilized by AOT are thermodynamically stable and are found in the one-phase isotropic L_2 region of the ternary water/oil/surfactant phase diagram. In these studies the conditions are such that the composition of the system is located far from the upper and lower phase boundaries so that the droplet phase separation does not occur either on standing or during reaction. The droplet size is determined by the water/surfactant molar ratio, w [51–53], and in systems stabilized by AOT the extent of polydispersity of the droplets is small [51,52]. The presence of strong acid added to the droplets attenuates the phase diagram but measurements are still made within the one-phase region. A variety of mineral acids have been studied and the solubilized acid can be readily changed in order to simulate the required conditions. For instance, sulfuric acid may be used when assessing the performance of marine diesel additives because of the high level of sulfur found in such fuels. Mixtures of acids can also be studied because they are also readily solubilized. With such a droplet dispersion, the structure of the acid droplet is very well defined and understood. Acid neutralization is monitored spectrophotometrically by means of a stopped-flow instrument by recording the change in absorbance of a water-soluble pH indicator (e.g., methyl orange) that is located in the water core of the microemulsion droplet.

Using the stopped-flow method, reactions can be monitored in times as short as 10 ms. The reaction is initiated on mixing a droplet microemulsion with a diluted solution of detergent, the diluent being the same hydrocarbon solvent as used for the continuous phase of the w/o microemulsion. This is usually *n*-decane, but in principle a base oil (e.g., Stanco 150) could also be used if the temperature is increased to lower the viscosity. Prior to reaction with detergent ($t = 0$), the water pool is acidified (pH 1) by injecting a small volume of acid into an AOT/oil reversed micelle ($w = 0$) system. Hence the pH indicator will be in its acidic form. The acid droplet dispersion is then mixed with the detergent dispersion. As the reaction proceeds, there is an observed color change when the acid is neutralized and the indicator is converted to its basic form. By using pH indicators with different pK_a values, the amount of acid that has been neutralized can be readily measured and so the reaction can be monitored at different stages. The absorbance is conveniently recorded at λ_{max} for either the acid or base form of the pH indicator, and as neutralization progresses there is a corresponding decrease or increase, respectively, in the measured absorbance. Most experiments, e.g., those represented in Fig. 2, were performed using excess base. In contrast to the situation with the weak acid (carboxylic acid), we find that 100% of the acid is neutralized by the same stoichiometric amount of base, suggesting a very different mechanism of acid neutralization for the two cases. Figure 2 shows stopped-flow traces obtained using three different pH indicators, methyl orange, 4-nitrophenyl-2-sulfonate, and Nile blue, which correspond to different pH ranges. (It should be noted that the pK_a measured in water is shifted to some extent in the microemulsion.)

The time for the absorbance to reach $\Delta A/2$ is recorded as $t_{1/2}$ (s), and this value is subsequently used as a direct measure of the ability of the detergent to neutralize the acid. The shorter the time $t_{1/2}$, the faster the acid neutralization reaction and thus the "better" the detergent. The reactions are generally fast, in the range 100 ms–10 s, so that conventional spectrophotometry is not possible. The exact pH within the water

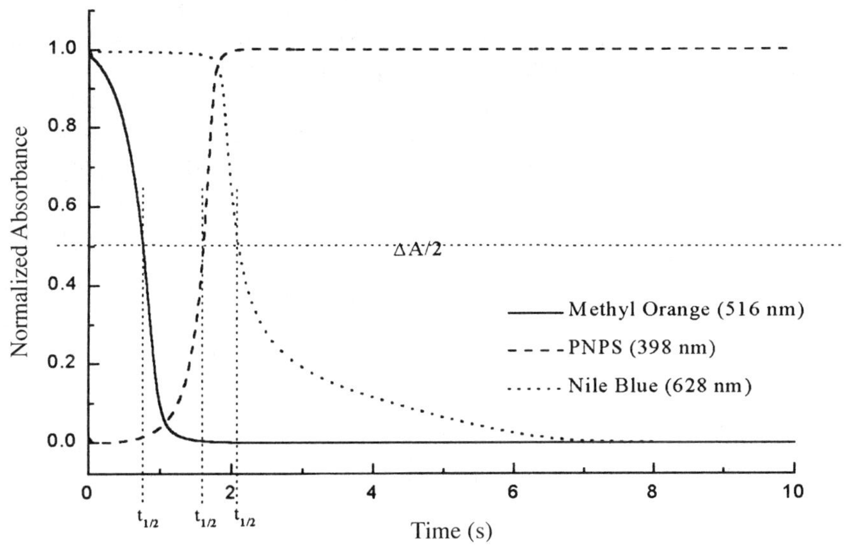

FIG. 2 Stopped-flow traces for acid neutralization.

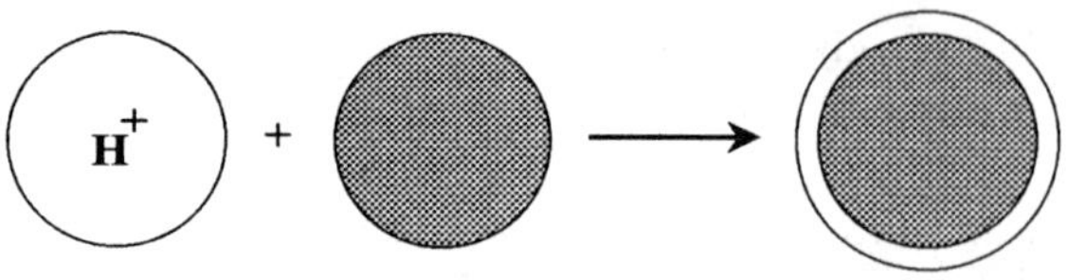

MECHANISM I

pools at $t_{1/2}$ is a complicated value to determine and is discussed more fully in papers on related systems [54–59]. The pH at $t_{1/2}$ is, however, independent of the detergent samples evaluated. It has been confirmed by small-angle neutron scattering studies that the droplets retain their integrity during the neutralization process. This observation has implications for the mechanism because the particles clearly do not interact with the droplets in such a way that the contents of the droplet become incorporated in the particle. For the screening analysis of detergent samples, 4-nitrophenyl-2-sulfonate was chosen as the pH indicator as it monitors (changes color) around neutral pH.

The amount of acid and base in the system at the start of the reaction is easily calculated, so neutralization can readily be carried out under a variety of conditions (e.g., large excess of base, similar amounts of acid and base). The technique has the advantage that many experimental parameters can be systematically investigated. Studies on the w/o microemulsion droplets have shown that they remain stable on variation of a range of experimental parameters, which can include temperature [60–62], pressure [61–63], alkane chain length of the solvent [52,64], and droplet volume fraction [51,65]. This means that we have a system in which we can change many experimental parameters (temperature, solvent, acid concentration, etc.) while maintaining the integrity of the test solution (the w/o microemulsion droplets). These are all extremely important parameters when considering the implication of the test results for the real engine situation. The technique is currently being developed in order to investigate the effect of pressure.

The stopped-flow methodology not only allows the determination of the neutralization ability of detergent additives but also provides kinetic information from which mechanistic information can be elucidated [66,67]. Four broad mechanistic schemes can be visualized for the acid neutralization process, three of which are based on an interaction of the droplet directly with the particle. The four schemes (I–IV) are shown schematically next. The surfactant monolayers stabilizing both the particle and droplet have been omitted for clarity (with the part exception of mechanism II).

Mechanism I involves fusion of the droplet and particle leading to neutralization of the acid. This mechanism is not thought to make a significant contribution as it is known that the droplets retain their integrity over time (i.e., stay as separate entities) when mixed with the detergent particles.

Mechanism II involves transfer of H^+ as an ion pair through the oil medium (i.e., surfactant-assisted transfer). This also involves surfactant transfer from the droplet to the particle. However, because it can be shown that neutralization is equally facile when cationic surfactants (e.g., dimethydidodecyl ammonium bromide, DDAB) are used, this mechanism is also ruled out [67].

The transfer of H^+ *from* the droplet *to* the particle, as shown in mechanism III, occurs during an encounter between the droplet and particle. The two colloidal species then separate, the droplet having been depleted of acid. However, this mechanism, although it is superficially attractive, does not give any explanation of how empty (neutral water pool) AOT droplets become basic when mixed with the detergent additive and so cannot be the main mechanism involved.

Mechanism IV indicates that a process of base abstraction from the particle to the droplets occurs. There is the possibility of two different processes occurring: either (a) direct transfer of base to an acid droplet or (b) indirect transfer via "empty" droplets, with acid neutralization occurring during subsequent droplet-droplet interactions. Both processes (a) and (b) must

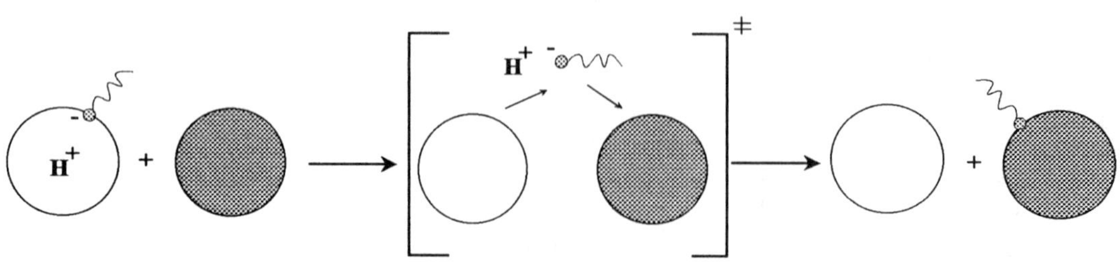

MECHANISM II

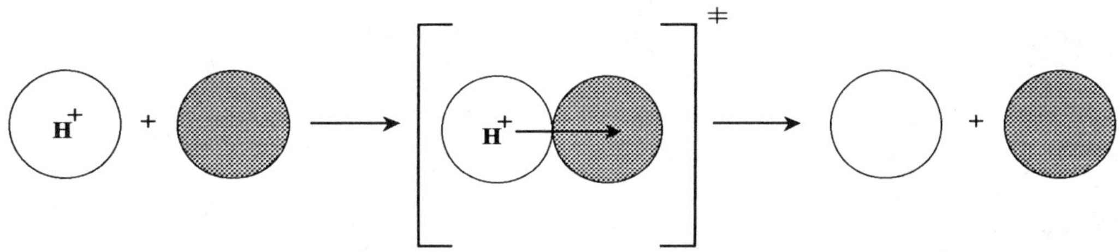

MECHANISM III

operate and this is the preferred mechanism (shown in more detail in Fig. 3). It is consistent with the important observation that the base is readily extracted from particles to neutral water droplets, increasing their pH to values considerably >7.

Considering mechanism IV in more detail, the acid neutralization reaction process involves three main steps. First, the detergent particle and w/o microemulsion droplet undergo encounters during translational diffusion (Brownian motion) through the bulk hydrocarbon solvent. Second, during an effective (sticky) encounter (which will happen in only a small fraction of collisions) a channel is formed between the solid core of the particle and the water pool of the droplets. A dissolution process then proceeds in which base is transferred from the particle into the water droplet. This will happen at a rate that is independent of the acid occupancy of the droplet. The average occupancy is about 1 H^+ per droplet at the start of our experiment but decreases to << 1 as the reaction proceeds. However, transfer into "empty" AOT-stabilized droplets is expected to take place at the same rate. The reaction between two droplets containing acid and base is facile.

Finally, the two colloidal species separate (the droplets maintaining their integrity) and a fast acid-base proton transfer reaction occurs in the water pool of the droplet. Previously [52], the homogeneous reaction between H^+- containing droplets and OH^-- containing droplets has been studied kinetically. These reactions are very fast with rate constants approaching 10^7 dm^3 mol^{-1} s^{-1}. It is interesting then that many parallels exist between the two types of systems. The investigation of additive synergy is also possible by this new technique. The neutralization ability of a detergent can be dra-

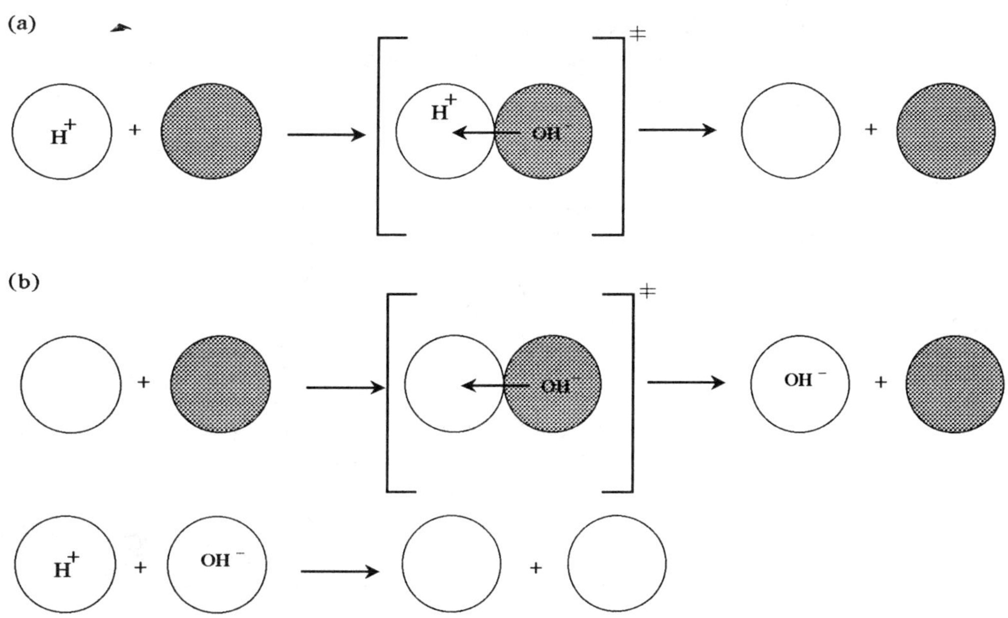

MECHANISM IV

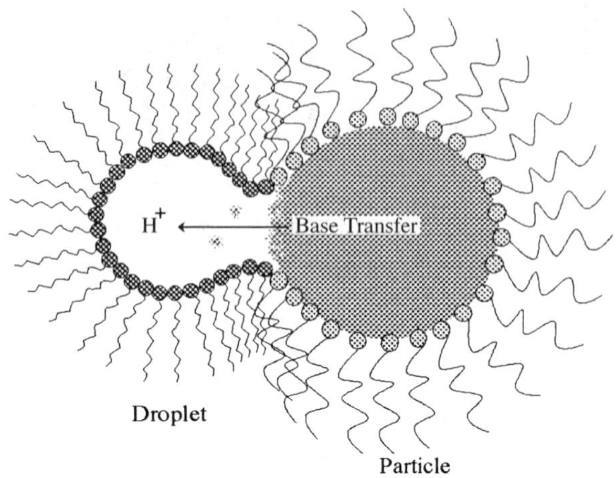

FIG. 3 Schematic diagram of additive-droplet interaction.

matically affected by the inclusion of further additives (e.g., dispersants, antiwear additives) in the oil. Figures 4 and 5 show the effect of adding dispersant to a detergent. Both sets of data refer to the same droplet system and the same additive concentration (hence TBN).

The detergents are both stabilized by the same surfactant, a high-molecular-weight (~1000) alkyl benzene sulfonate, and are at the same TBN. The essential difference is that detergent 1 contains magnesium and detergent 2 contains calcium. The first observation is that the magnesium-based detergent is more efficient at neutralizing the inorganic acid than the calcium-based detergent (by a factor of 10). This might well reflect the difference in dissolution rates between the magnesium and calcium bases. On addition of dispersant (a high-molecular-weight polyisobutenyl succinimide), the values of $t_{1/2}$ change dramatically.

One can see clearly that the dispersant has an advantageous effect on detergent 2; however, the value of $t_{1/2}$ has increased for detergent 1, and so it appears that the dispersant inhibits the neutralizing action of this detergent. An explanation for these observations is beyond the scope of this chapter, but the essential message is that the technique not only gives information on detergent-only-based samples but also readily yields vital information concerning additive interactions. It is found that a good correlation exists between detergents tested with this technique and by both the ball rust test (J. R. Galsworthy, personal communication) and Fourier transform infrared [68]. Similar trends have also been observed for the effects of temperature [67] and metal detergent type [47]. However, to compare results from different methods directly could be misleading. The chemical reactions involved are dependent on many experimental parameters, with each test having its own unique set of protocols (temperature, flow rate, etc.). No one technique can be used to investigate fully the in situ performance of detergents, and an engine field test will continue to be the definitive test of overall lubricant performance under operational conditions of an engine.

V. SUMMARY

This chapter starts with a discussion of the main colloidal components of a modern engine oil formulation, including a review of the principal methods used to investigate the properties of these systems. One particular methodology is discussed in detail; this is a new method for studying the neutralization behaviour of model overbased detergent additive systems. This class

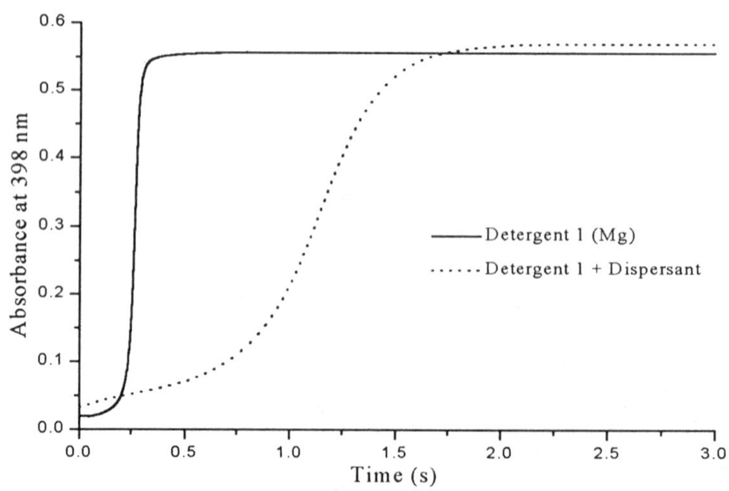

FIG. 4 Magnesium-based detergent 1.

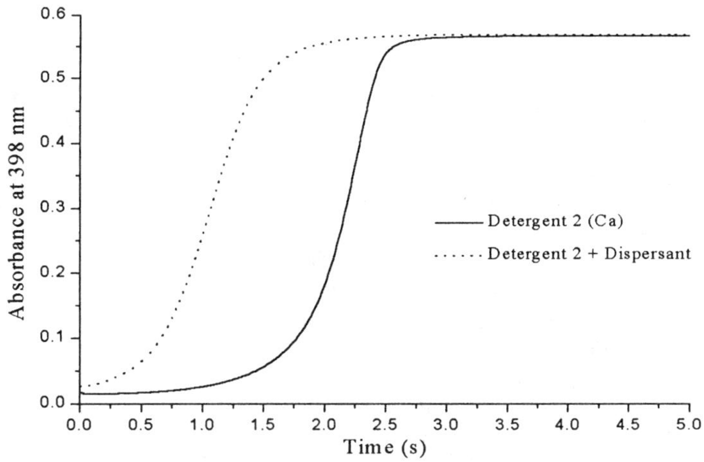

FIG. 5 Calcium-based detergent 2.

of additive is an extremely important component of modern engine oil formulations and is employed for the neutralization of acidic species that would otherwise build up in the engine and cause corrosive wear. It is thought that the presence of water and excess free surfactant in engine oil formulations will disperse acid in the form of water droplets similar to those of the model w/o microemulsion droplets. The use of water-soluble pH indicators allows the change in the mean pH of the dispersed water to be monitored for these reactions in a relatively facile way using conventional stopped-flow instrumentation. More complex additive systems of commercial importance have also been investigated. These systems are those in which a number of other additives (e.g., dispersant) are blended with the overbased detergent. Using the stopped-flow technique described, it has been possible to investigate the effect of these additional additives on the acid neutralization properties of the overbased detergent, and it is shown that the approach can provide a new simple and effective diagnostic test for monitoring the performance efficiency of lubricant detergent additives.

ACKNOWLEDGMENTS

BHR and DCH would like to dedicate this chapter to John Marsh (of Exxon Chemical Ltd) who has always been very supportive of our work in this area.

REFERENCES

1. C. C. Colyer and W. C. Gergel, *Chemistry and Technology of Lubricants*, 2nd ed., Blackie, London, 1997.
2. J. B. Heywood, *Internal Combustion Engine Fundamentals*, McGraw-Hill, New York, 1988.
3. G. P. Wood, J. Inst. Pet. *55*:194–204 (1969).
4. T. V. Liston, J. Jpn. Soc. Tribol. *40*:280–285 (1995).
5. R. C. Watkins, Phys. Technol. *15*:321–328 (1984).
6. P. Salino and P. Volpi, Ann. Chim. *77*:145–165 (1987).
7. J. F. Marsh, Chem. Ind. 470–473 (1987).
8. C. Belle, C. Beraud, D. Faure, R. Gallo, P. Hoornaert, J. M. Martin, and C. Rey, J. Chim. Phys. Phys. Chim. Biol. *87*:93–104 (1990).
9. J. P. Roman, P. Hoornaert, D. Faure, C. Biver, F. Jacquet, and J. M. Martin, J. Colloid Interface Sci. *144*: 324–339 (1991).
10. K. Kandori, N. Shizuka, K. Konno, and A. Kitahara, J. Dispers. Sci. Technol. *8*:477–491 (1987).
11. K. Kandori, K. Konno, and A. Kitahara, J. Colloid Interface Sci. *122*:78–82 (1988).
12. B. Delfort, M. Born, A. Chive, and L. Barre, J. Colloid Interface Sci. *189*:151–157 (1997).
13. I. Markovic, R. H. Ottewill, D. J. Cebula, I. Field, and J. F. Marsh, Colloid Polym. Sci. *262*:648–656 (1984).
14. I. Markovic and R. H. Ottewill, Colloid Polym. Sci. *264*:65–76 (1986).
15. I. Markovic and R. H. Ottewill, Colloid Polym. Sci. *264*:454–462 (1986).
16. R. H. Ottewill, E. Sinagra, I. P. Macdonald, J. F. Marsh, and R. K. Heenan, Colloid Polym. Sci. *270*:602–608 (1992).
17. S. M. Kurilo, O. L. Glavati, V. V. Tsukruk, and V. V. Shilov, Chem. Technol. Fuels Oils *26*:311–315 (1990).
18. S. Giasson, D. Espinat, T. Palermo, R. Ober, M. Pessah, and M. F. Morizur, J. Colloid Interface Sci. *153*:355–367 (1992).
19. A. N. Agaev, S. M. Kurilo, K. I. Sadykhov, N. I. Mishchenko, and S. A. Bagirova, Chem. Technol. Fuels Oils *29*:141–144 (1993).
20. S. M. Kurilo, O. L. Glavati, T. P. Tantsyura, A. I. Marchenko, and G. G. Kravchuk, Colloid J. USSR *48*:309–312 (1986).
21. B. Delfort, B. Daoudal, and L. Barre, Tribol. Trans. *42*: 296–302 (1999).

22. O. L. Glavati, S. M. Kurilo, G. G. Kravchuk, Y. T. Gordash, V. V. Shilov, V. V. Tsukruk, and O. A. Lokhonya, Chem. Technol. Fuels Oils, *25*:273–275 (1989).
23. J. A. Griffiths, R. Bolton, D. M. Heyes, J. H. Clint, and S. E. Taylor, J. Chem. Soc. Faraday Trans. *91*:687–696 (1995).
24. J. A. Griffiths and D. M. Heyes, Langmuir *12*:2418–2424 (1996).
25. D. J. Tobias and M. L. Klein, J. Phys. Chem. *100*: 6637–6648 (1996).
26. T. P. O'Sullivan, M. E. Vickers, and R. K. Heenan, J. Appl. Crystallog. *24*:732–739 (1991).
27. J. L. Mansot, M. Hallouis, and J. M. Martin, Colloids Surf. A Physicochem. Eng. Aspects *71*:123–134 (1993).
28. J. F. Miller, B. J. Clifton, P. R. Benneyworth, B. Vincent, I. P. Macdonald, and J. F. Marsh, Colloids Surf. *66*:197–202 (1992).
29. T. C. Jao and K. L. Kreuz, J. Colloid Interface Sci. *126*: 622–628 (1988).
30. T. C. Jao and W. S. Joyce, Langmuir *6*:944–948 (1990).
31. J. M. Martin, J. L. Mansot, and M. Hallouis, Ultramicroscopy *30*:321–327 (1989).
32. J. M. Martin, J. L. Mansot, M. Hallouis, and H. Tenailleau, Microsc. Microanal. Microstruct. *1*:93–102 (1990).
33. E. R. Arndt and K. L. Kreuz, J. Colloid Interface Sci. *123*:230–237 (1988).
34. R. Makhloufi, J. L. Mansot, E. Hirsch, J. Wery, S. J. Candau, D. Thomas, and J. P. Rolland, Colloid Polym. Sci. *273*:242–251 (1995).
35. D. C. Steytler, B. H. Robinson, J. Eastoe, K. Ibel, J. C. Dore, and I. Macdonald, Langmuir *9*:903–911 (1993).
36. S. M. Kurilo and O. L. Glavati, Colloid J. USSR *51*: 38–42 (1989).
37. J. H. Clint and S. E. Taylor, Colloids Surf. *65*:61–67 (1992).
38. American Standards and Test Methods, Total Base Number of Petroleum Products by Potentiometric Perchloric Acid Titration, D2896-IP 276, 1983.
39. J.-P. Roman, Proceedings of the International Congress on Combustion Engines (CIMAC), Copenhagen, 1998, pp. 913–926.
40. H. Wohltjen, N. L. Jarvis, M. Klusty, N. Gorin, C. Fleck, G. Shay, and A. Smith, Lubr. Eng. *50*:861–866 (1994).
41. B. L. Papke, Tribol. Trans. *31*:420–426 (1988).
42. C. Kim, C. C. Kuo, D. M. Marchand, C. Kim, C. C. Kuo, and D. M. Marchand, *Development of the Ball Rust Test—A Laboratory Test Replacing the Sequence IID Engine Test*, SAE, Tulsa, OK, 1997, Vol. 972961, pp. 49–59.
43. C. C. Kuo, D. M. Marchand, W. Y. Lam, C. C. Kuo, D. M. Marchand, and W. Y. Lam, *The Effects of Oil Additives in the Ball Rust Test*, SAE, Tulsa, OK, 1997, Vol. 972923, pp. 27–34.
44. A. Woord and T. Garner, Measuring neutralisation performance and corrosive wear of cylinder oils on the Plint TE/77 reciprocating rig, 1996.
45. W. J. Albery, R. A. Choudhery, N. Z. Atay, and B. H. Robinson, J. Chem. Soc. Faraday Trans. I *83*:2407–2419 (1987).
46. J. Lewis, Mechanism of action of overbased oil additives, Ph.D. thesis, University of East Anglia, Norwich, 1987.
47. G. K. Sharma and O. P. Chawla, Tribol. Int. *21*:269–274 (1988).
48. D. C. Steytler, A study of reactions occurring in water-containing reversed-micelle systems, Ph.D. thesis, University of Kent, Canterbury, 1981.
49. P. D. I. Fletcher, Ph.D. thesis, University of Kent, Canterbury, 1982.
50. N. M. Perrins, Interactions in surfactant stabilised colloidal micro-systems, Ph.D. thesis, University of Kent, Canterbury, 1985.
51. S. H. Chen, Annu. Rev. Phys. Chem. *37*:351–399 (1986).
52. P. D. I. Fletcher, A. M. Howe, and B. H. Robinson, J. Chem. Soc. Faraday Trans. I *83*:985–1006 (1987).
53. M. Kotlarchyk, J. S. Huang, and S. H. Chen, J. Phys. Chem. *89*:4382–4386 (1985).
54. O. A. El Seoud and A. M. Chinelatto, J. Colloid Interface Sci. *95*:163–171 (1983).
55. O. El Seoud and R. C. Vieira, J. Colloid Interface Sci. *93*:289–292 (1983).
56. O. A. El Seoud, A. M. Chinelatto, and M. R. Shimizu, J. Colloid Interface Sci. *88*:420–427 (1982).
57. C. J. Drummond, F. Grieser, and T. W. Healy, J. Chem. Soc. Faraday Trans. I *85*:521–535 (1989).
58. C. J. Drummond, F. Grieser, and T. W. Healy, J. Chem. Soc. Faraday Trans. I *85*:537–550 (1989).
59. C. J. Drummond, F. Grieser, and T. W. Healy, J. Chem. Soc. Faraday Trans. I *85*:561–578 (1989).
60. C. Oldfield, B. H. Robinson, and R. B. Freedman, J. Chem. Soc. Faraday Trans. *86*:833–841 (1990).
61. S. Clark, P. D. I. Fletcher, and X. L. Ye, Langmuir *6*: 1301–1309 (1990).
62. J. Eastoe, W. K. Young, B. H. Robinson, and D. C. Steytler, J. Chem. Soc. Faraday Trans. *86*:2883–2889 (1990).
63. J. Eastoe, B. H. Robinson, and D. C. Steytler, J. Chem. Soc. Faraday Trans. *86*:511–517 (1990).
64. J. L. Fulton and R. D. Smith, J. Phys. Chem. *92*:2903–2907 (1988).
65. J. Eastoe, B. H. Robinson, D. C. Steytler, and D. Thornleeson, Adv. Colloid Interface Sci. *36*:1–31 (1991).
66. D. C. Hone, B. H. Robinson, D. C. Steytler, J. A. Cleverley, and R. W. Glyde, Can. J. Chem. *77*:852–857 (1999).
67. D. C. Hone, B. H. Robinson, D. C. Steytler, J. R. Galsworthy, and R. W. Glyde, Langmuir, *16*(2):340–346 (2000).
68. J. R. Galsworthy, R. W. Glyde, N. Z. Diggs, K. R. Gorda, D. C. Hone, and B. H. Robinson, Abstr. Pap. Am. Chem. Soc. *218*:U249–U249 (1999).

18

Giant Vesicles as Microchemical Vessels

STEPHEN J. LEE U.S. Army Research Office, Research Triangle Park, North Carolina

JASON S. KEIPER Emory University, Atlanta, Georgia

I. INTRODUCTION

A colloid scientist may spend a lifetime working with surfactants, only coming to understand their behavior separated by the cold glass of a flask or cuvette. Indeed, tensiometry, spectroscopy, and other techniques paint portraits of structure and function, but ordinarily we are not privy to the lilliputian universe of the association colloid. Over the past few years, however, a remarkable system has come to prominence that allows intimate study of individual colloidal particles. It is the giant vesicle—a curved microscopic structure 1–500 μm in diameter. In the simplest form, giant vesicles can be thought of as "stripped down" versions of biological cells—vessels composed of membrane-forming amphiphiles surrounded by and encapsulating water.

The giant vesicle's physical features allow for achievement of experiments that yield information distinct yet complementary to that obtained from smaller vesicle systems—direct and dynamic visualization of physical, chemical, and biochemical membrane phenomena. The primary basis for this is that the giant vesicle's size permits experimentation using an optical microscope; thus single vesicles may be isolated, observed, manipulated, and probed employing well-established cell biology methods. Figure 1 depicts four common modes of study. Figure 1A shows a phase-contrast micrograph of a population of giant phospholipid vesicles, spherical in shape with sizes up to approximately 60 μm in diameter. Typically, a vesicle from among such a bevy can be chosen for analysis of membrane fluctuations or manipulation with a glass micropipette, for example. Micropipettes, as will be discussed more in depth later, have proved vital for the aspiration-based measurements of physical properties of lipid membranes. They also serve as tools for physically maneuvering giant vesicles, as demonstrated in Fig. 1B, where a "held" vesicle doped with a small percentage of anionic lipid adheres to another vesicle of cationic charge. Another oft-used microtool is the glass injection pipette, for delivery of solutions to the exterior or interior of giant vesicles. Entry of a glass micropipette into a giant vesicle is demonstrated in Fig. 1C with successful injection of a dye solution into the inner compartment of a vesicle. Finally, Fig. 1D shows an image of giant vesicles marked with a fluorescent lipid. Fluorescence microscopy allows giant vesicle researchers an alternative way to study membrane form and activity.

Several reviews have appeared highlighting important aspects of giant vesicle science [1–8] and provide excellent bases for understanding the cytomimetic behavior and membrane biophysics of giant vesicles. The present chapter begins with a brief chronicle of key contributors and works on giant vesicles (GVs) that established much of the foundation for the current interest in GVs. The remainder surveys developments of chemical, biochemical, and materials aspects of giant vesicles in the 1990s. We will discuss vesicle formation and manipulation and newly emerging ways to study giant vesicles with instruments such as the confocal microscope and optical trap. Finally, we offer specu-

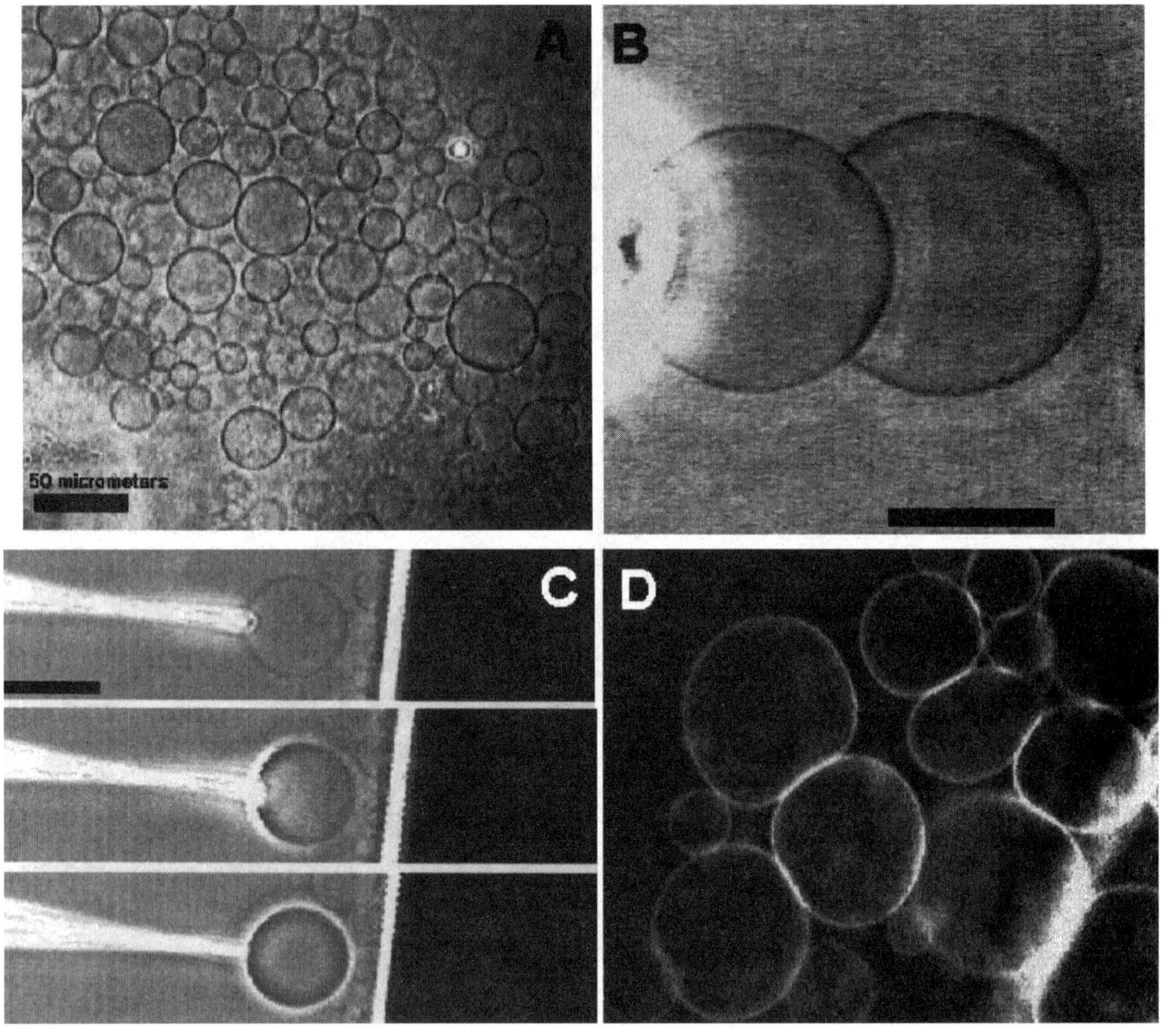

FIG. 1 (A) Phase-contrast micrograph of a population of phospholipid giant vesicles. (B) Two adhered vesicles manipulated with a glass micropipette (left). (C) Demonstration of microinjection of a dye solution into a phospholipid giant vesicle. (D) Fluorescence micrograph of giant vesicles containing a small percentage of a membrane-bound fluorescent marker. Bar = 50 μm.

lation on the future of giant vesicles, especially in regard to their potential utility beyond being a simple model system.

II. KEY GIANT VESICLE FOUNDATIONS

Reeves and Dowben first described the preparation of "giant" vesicles, up to 10 μm in diameter, in 1969 [9]. The study of lipid vesicles or liposomes was not even a decade old at that point [10], but the dimensions of the thin-walled giant vesicles were significant, opening the possibility for microscopic investigation. Formation of microscopic giant vesicles (diameters > 1 μm) requires lower energy input than formation of the submicroscopic smaller vesicle systems (diameters < 1 μm), which are regularly prepared by breaking down larger lipid structures via extrusion or sonication. Generally, hydration of a suitable lipid film with purified water or buffer under the influence of ambient thermal and vibrational energy input has been used in most giant vesicle investigations. This simple method and variants thereof, however, can lead to a variety of vesicle structures having irregular nonspherical morphologies and multiple bilayers ("lamellarity") [3,6]. A revolutionary advance in the reproducible preparation of giant, unilamellar vesicles came about with the "electroformation" method developed by Angelova and coworkers [6,11], which involves subjecting hydrating lipid films to an alternating current. Although the exact mechanism of electroformation is still not well understood, it is believed that the alternating current induces small dipole shifts in the water, allowing water to enter lipid films subtly and progressively push out spherical giant unilamellar (single bilayer) vesicles.

In the 1970s and 1980s, Evans, Waugh, and Needham advanced elegant micropipette aspiration techniques for giant vesicles to ascertain fundamental physical parameters of membranes. In essence, a giant vesicle ($\geq 10\ \mu m$) held by a micropipette can be subjected to a defined inward suction that can be correlated with the geometry of the aspirated structure to derive a number of physical properties of lipid membranes (e.g., compressibility, thermoelasticity, lipid molecular areas, free energies of adhesion). The study of rugged and deformable red blood cells using micropipette techniques [12–15] paved the way for extension to the compositionally simpler giant vesicles [16–20]. Thanks to the aspiration method, there is a standard, straightforward method for direct evaluation of lipid membrane properties, allowing great strides in the understanding in the areas of lipid phase transitions, membrane-membrane adhesion energies, and the effect of cholesterol on membranes, to name a few.

Much of our current understanding of membrane energy, morphology, curvature, and bending concepts arose from the theoretical and experimental work of the German biophysicists Helfrich and Sackmann and the Slovenian biophysicists Svetina and Zeks. Helfrich developed theoretical analyses explaining the elastic behavior of lipid bilayer membranes that have had a profound impact on the understanding of living and nonliving membrane systems [21]. Membrane expansivity and elasticity concepts were further extended by Svetina and Zeks [22,23] into what has now evolved into the "area-difference-elasticity" (ADE) model of bilayer membranes [24]. The key aspect of this model is the distinction in area and bending energy between inner and outer monolayers of a bilayer membrane and how subsequent differences can lead to morphological changes. Sackmann and coworkers, in particular, have made great strides in using the giant vesicle to bridge the gap between experiment and theory in regard to such morphological transformations and thermal effects on lipid membranes [25–28].

It was discovered by Kunitake and Okahata in 1977 that nonphospholipid synthetic surfactants, including ammonium salts such as didodecyldimethylammonium bromide (DDAB), are capable of forming small and giant vesicles in aqueous media [29]. This was an important finding because it opened a new forum for synthetic chemists in membrane science. Perhaps the greatest contributions in this area came from Ringsdorf and coworkers, who described a fascinating and creative career in the area in a seminal 1988 review, including aspects of photochemistry, lipid domains, and recognition processes in membranes [30]. Their work opened the door for the use of giant vesicles not only in biophysics but also in the chemical, biochemical, and material sciences.

III. CHEMISTRY

Studying the chemistry of giant vesicles can be approached in two ways: (1) through control of vesicle chemical composition before preparation (lipids, fluorescent probes, polymers, etc.) and (2) adding chemicals or externally changing the chemical nature of the systems after the vesicles are formed. It will be shown here that both paths are used in different circumstances.

An important means to evaluate the effect of externally added chemicals on giant vesicles is through use of glass microinjection pipettes. Menger and coworkers have examined both natural and synthetic lipid giant vesicles extensively using manipulation and microinjection techniques [3,31]. Injections of various chemicals can be performed on the outer surface of a giant vesicle, allowing researchers to evaluate solution gradients of various chemicals on lipid membranes. For example, addition of a KI/I_2 solution to the surface of a DDAB giant vesicle led to a stable rupture of the spherical structure, causing the formation of a "nanocup" (Fig. 2). The fact that the reagents were introduced only on a limited area of the giant vesicle, rather than a bathing solution around it, made this intermediate form available for observation. This type of gradient is essentially impossible when working with smaller vesicle systems. Alternatively, as shown in Fig. 1c, internal microinjections can be performed, as first demonstrated by Menger and Lee [31]. Luisi and coworkers provided an in-depth analysis of internal microinjection of solutions of chemical agents using electroformed giant vesicles [32].

Microinjection techniques again played a key role in Menger's examination of the effect of the saponin digitonin on giant vesicles [33]. Digitonin, a steroid glycoside conjugate, is well known to form supramolecular complexes with sterols within lipid membranes (Fig. 3). Although most studies of this membrane interaction have used small vesicles or cells, giant vesicles allowed dynamic observation of the dramatic effect of digitonin on cholesterol-containing vesicles, including rupture and tubule formation. The work sug-

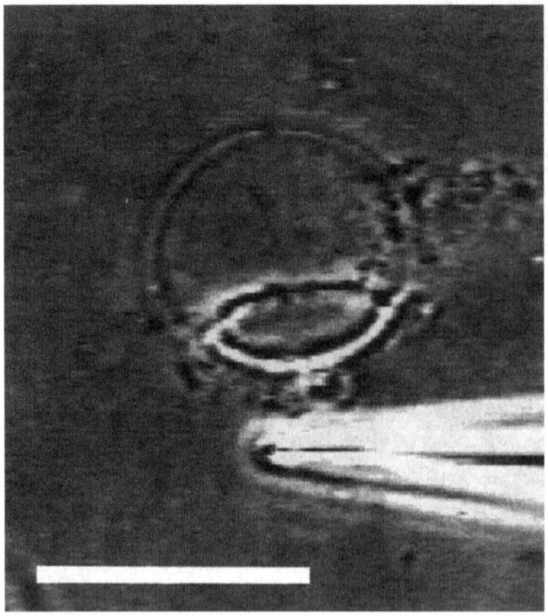

FIG. 2 A large defect (a "nanocup") formed on the surface of a DDAB giant vesicle by microinjection of KI/I_2 solution. Bar = 100 μm.

gests that giant vesicles are an accommodating and informative medium for the study of the membrane activity of natural products.

Electrostatic interactions of lipid membranes are of high interest, especially in relation to liposomal gene therapy (i.e., the interactions of negatively charged DNA with positively charged carrier vesicles; see Chapter 35). By control of the chemical composition of vesicle populations, it is possible to examine the influence of ionic charge on giant vesicles, such as the interaction of oppositely charged vesicles. Lehn and coworkers were the first to examine vesicle-vesicle adhesion with both oppositely charged small vesicles and mixed solutions of giant vesicles [34]. This was soon followed by Menger et al., who used micropipette manipulation techniques to study isolated vesicles (as seen in Fig. 1b) [35,36]. Although those studies found only evidence for adhesion and not fusion, using only up to 20 mol% charged lipid components in the respective cationic and anionic vesicles, a later report demonstrated that oppositely charged vesicles with higher amounts of charged lipids were capable of fusion or hemifusion events [37].

Fluorescently labeled molecules are vital tools for the study of cells and membrane processes, as synthetic probes have found use in both spectroscopy and microscopy [38]. It has followed that fluorescence microscopy has been widely employed in the study of giant vesicles in many contexts. A few examples are cited here. Upon microinjection of a micellar solution of quinolinium surfactant fluorophore (Fig. 4a) on the surfaces of immobilized giant vesicles, the surfactant inserted into the membranes [39]. The surfactant was capable of "flip-flopping" the bilayer membranes; accordingly, such a probe provided a way to differentiate between unilamellar, multilamellar, and oligovesicular giant vesicles. Giant vesicles can also be prepared containing fluorescently labeled agents, including a modified phospholipid (Fig. 4b) used by Devaux and coworkers in photobleaching experiments [40] as well as modified polymers, such as the fluorescein-labeled polysaccharide pullulan that incorporates into POPC giant vesicles [41]. The labeled pullulan in that example was modified with cholesteryl groups to ensure binding to the lipid membranes. This work also featured the use of a confocal microscope, a tool capable of high-resolution, three-dimensional fluorescence imaging. The incredible potential of this instrument in the study of giant vesicles was further realized by Korlach et al., who were able to image discrete phase separations in giant vesicles composed of dilauroylphosphocholine (DLPC) and dipalmitoylphosphocholine (DPPC) [42]. The probes depicted in Fig. 4c and d possessed preferential affinities for discrete, coexisting phases of DLPC and DPPC, respectively. Multiple scans of a vesicle hemisphere afforded extraordinary images of phase separation, a feat made possible with the proper choice of fluorescent probes (Fig. 5). Finally, another innovative microscopic technique, two-photon fluorescence imaging, has also been used with giant vesicle systems to monitor phase transition behavior [43].

Chemo-osmotic effects appear to be responsible for expulsion or "birthing" of daughter vesicles from within a larger giant vesicle. Menger and coworkers have twice reported expulsion processes with DDAB vesicles induced by either external addition of octyl glucoside [44] or simple dilution or heating processes (Fig. 6) [45]. Phosphocholine giant vesicles were also subject to spontaneous expulsion of inner vesicles by use of optical tweezers [46,47]. It was postulated that laser-induced osmotic forces were responsible for expulsion events.

Beyond micropipettes and laser traps, other means of manipulating giant vesicles have been reported, with the researchers relying on the design of appropriate chemical systems. "Magnetic liposomes" were prepared by Menager and Cabuil by entrapping maghemite particles within phospholipid giant vesicles by

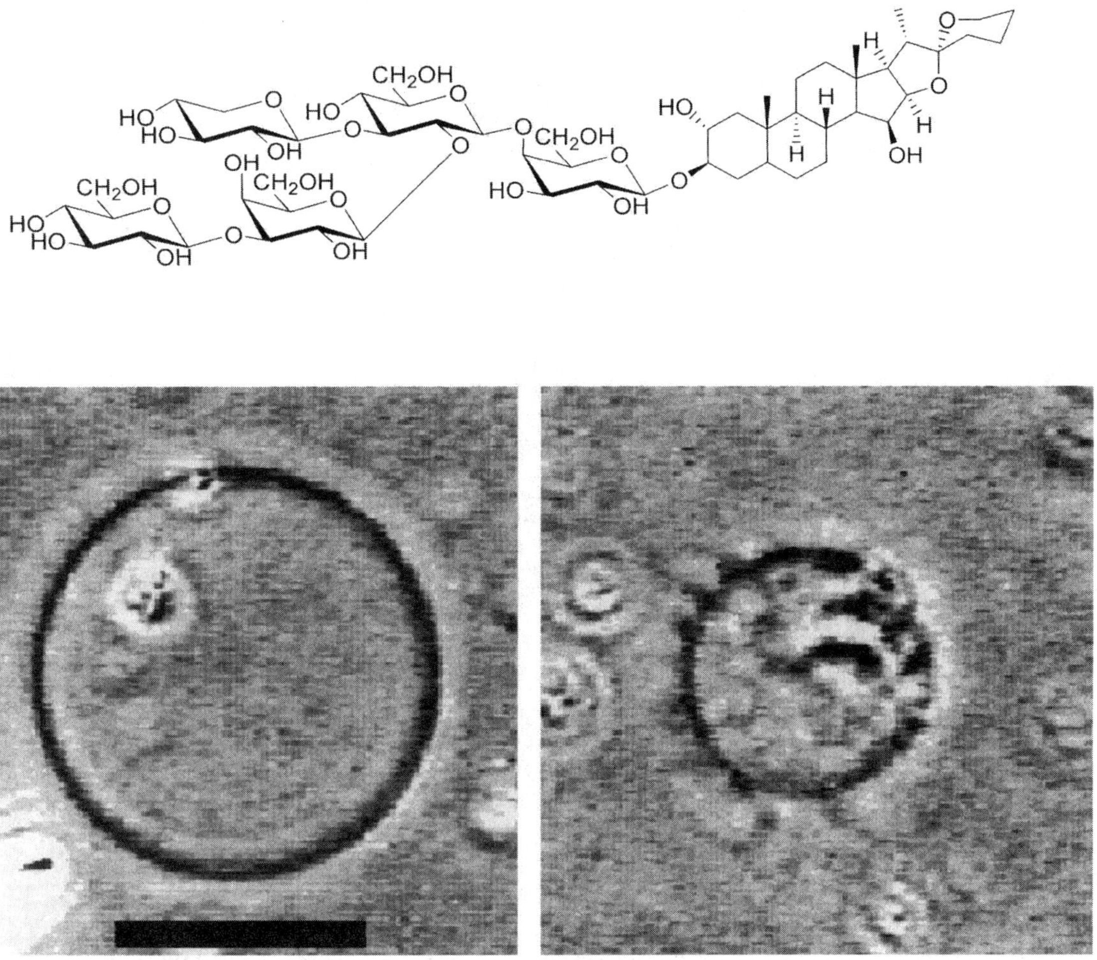

FIG. 3 Digitonin (top structure) can induce rupture of cholesterol-containing giant vesicles. Bar = 25 μm.

an emulsion process [48]. The iron particles provided a pliable inner fluid subject to the control of a magnetic field, leading to controlled deformations of the lipid bilayer. Magnetic vesicles well demonstrate the capability vesicles have to sequester one fluid phase within a different bulk phase. Irradiating polymerizable lipids within giant vesicles has also been accomplished [49] and even demonstrated in tandem with perforation of "polymerized" vesicles with applied electrical pulses [50].

We close this section by mentioning studies on "autocatalytic" vesicle formation performed by Luisi and coworkers [51,52]. The basic concept was that under appropriate conditions and with necessary chemical "ingredients," vesicles composed of a specific lipid could "self-catalyze" formation of their own kind by reaction with a water-insoluble lipid precursor. For example, oleic acid vesicles would catalyze the hydrolysis of water-immiscible oleic anhydride to form more oleic acid and thus more vesicles observable with light microscopy. They have thus far demonstrated this concept with giant vesicles using the oleic acid/oleic anhydride tandem [51] as well as with novel phosphatidyl nucleosides (Fig. 7) [52].

IV. BIOCHEMISTRY

In studies to understand better the interactions of enzymes, proteins, and other biomolecules with membranes, giant vesicles are versatile systems that can be modified to suit experimental design. With the vesicle's lipid skeleton and capacity for compartmentalization, it is not difficult to imagine building up increasingly complex systems that mimic biochemical cellular phenomena. For example, gathering information on the behavior of enzymes in GVs may eventually lead to the

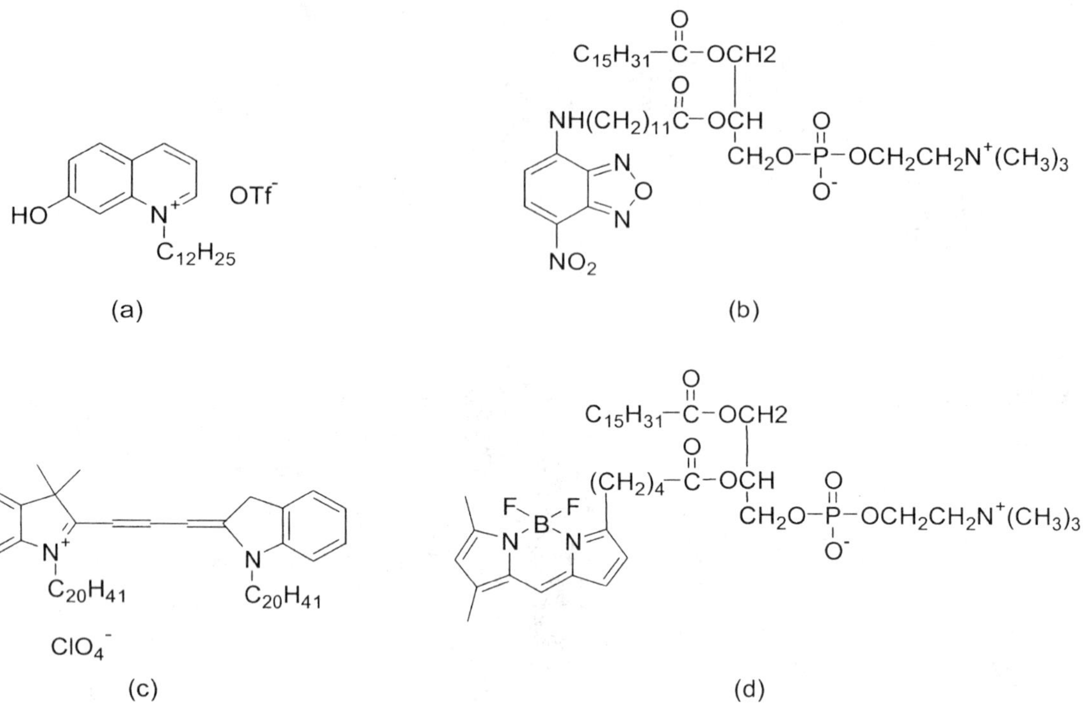

FIG. 4 Structures of membrane-soluble fluorescent probes.

realization of elaborate metabolic pathways from a "lifeless" system [53]. Some significant advances in this area are described here.

One of the important features of giant vesicles is that they provide a cell-size bilayer structure of well-defined composition, allowing quantitative evaluation of biochemical phenomena. This is well demonstrated in their use in high-resolution "patch clamp" experiments that provide electrophysiological analysis of ion flow across lipid membranes. Gonzalez-Ros et al. were among the first to utilize giant vesicles in this manner with reconstituted lipids and ion channels [54]. Others have used similar methods to assay for the presence of ion channels in proteins [55] and examine thylakoid membrane systems [56].

Luisi, Walde, and coworkers have evaluated a variety of enzymes for their interactions with giant vesicles. Using microinjection techniques, snake venom phospholipase PL-A2 was introduced to the exteriors and interior compartments of POPC giant vesicles [57]. Hydrolysis of the constituent phospholipids disrupted vesicle structure rapidly (<1 min) with external addition of PL-A2, whereas internal injection led to slower disintegration. The discrepancy was attributed to a higher local concentration gradient of enzyme with external microinjections and a more homogenous distribution to the inner monolayer with internal microinjections. Luisi and Wick followed up this work by demonstrating the first step of the "salvage pathway" in the interior of giant vesicles, preparing lysolipids from the enzyme *sn*-glycerol-3-phosphate-acyltransferase [53]. The lysolipid caused morphological changes in the vesicles observable by videomicroscopy.

Fluorescence microscopy was critical in experiments by Walde et al. showing the action of phospholipase D (PL-D) on POPC giant vesicles [58]. Microinjection of PL-D at the surfaces of giant vesicles caused no apparent changes in the structures. On incorporation of a Texas Red–DOPE probe in POPC vesicle membranes, dramatically reduced fluorescence intensity was observed after exposure to PL-D, indicating that the enzyme was acting on the phospholipids but did not disturb the vesicle integrity.

DNA-lipid interactions are of growing interest in transfection systems. Giant vesicles have begun to serve as models for these interactions. Yoshikawa and coworkers demonstrated entrapment of T4 DNA in phosphocholine giant vesicles [59]. Microinjection of DNA (oligonucleotides of 250 bp) caused endocytotic events within PC/sphingosphine (cationic lipid) vesicles [60]. A threshold concentration of sphingosine proved necessary for "endocytosis," and a partial

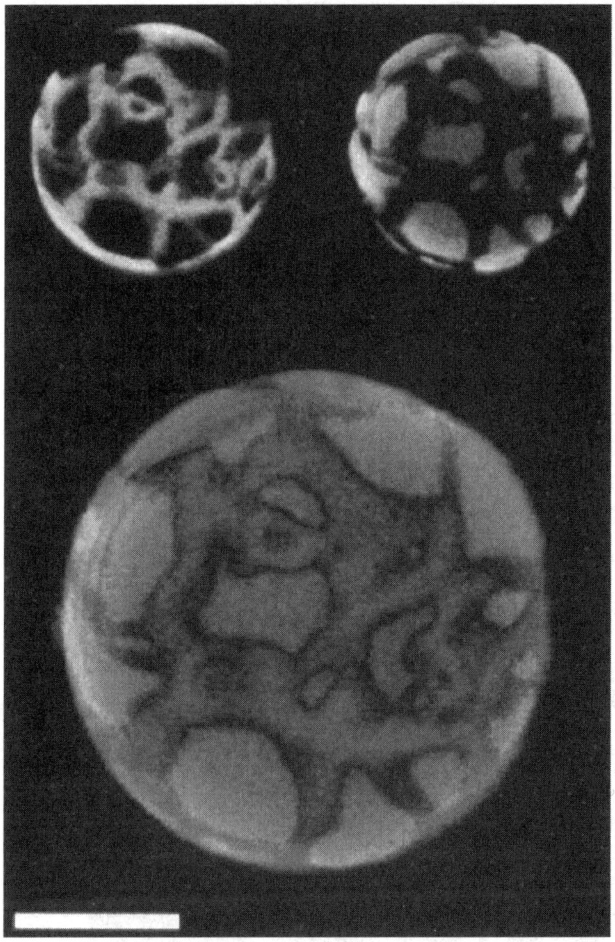

FIG. 5 Confocal micrograph of a mixed phospholipid giant vesicle containing dyes C and D (Fig. 4) preferentially partitioned in domains. Bar = 10 μm. (From Ref. 42. Copyright 1999 National Academy of Sciences.)

model for the phenomenon was proposed, involving inverse micelle intermediates.

Other biopolymers have also proved amenable to study with giant vesicles. DMPC giant vesicles doped with 1 mol% of an ATP-functionalized lipid resulted in specific binding of the protein actin at the surfaces [61]. It was possible, with the addition of EDTA, to remove actin only from the outer monolayer of a vesicle (Fig. 8). Consequently, thermal transitions were evaluated for vesicles with and without actin coating the outer vesicle surface. The transitions were found to be lower for the outer "actin-free" vesicles, as determined by budding processes (30°C with outer actin, 26°C without outer actin). Only outward budding process were observed for the actin-removed vesicles, as the bound inner actin prevented inner bending deformations.

Hammer et al. have published a thorough series of works evaluating the effect of influenza virus fusion peptides on membrane bilayers [62–64]. In their experiments, Hammer's group studied the insertion of fusion peptides into phospholipid vesicles via micropipette aspiration techniques, including how specific changes in peptide sequences alter their membrane behavior [64]. Understanding such peptides is important for the understanding of how viruses fuse with cell membranes—such fundamental information can be used to combat viruses or better use them as vectors.

V. MATERIALS AND APPLICATIONS

Although submicroscopic vesicle and liposome systems are constantly targeted for applications, the practical utility of giant vesicles for pharmaceuticals or cosmetics is limited by relative fragility under shear flow [65] and relative polydispersity in their formation. Nevertheless, a number of promising giant vesicle systems have opened possibilities for their use as novel materials and other applications.

Jaeger and coworkers, who have examined giant vesicles formed by potentially useful acid-labile "cleavable" lipids [66], developed a novel vesicle system for the detection and decontamination of mustard-like compounds (Fig. 9) [67]. Lipid a is capable of forming bilayer vesicles and entrapping water-soluble agents, such as a fluorescent dye. It is also capable of reacting with mustard simulant b, forming c, a non-amphiphilic sulfur compound of presumably lower toxicity than b. As a vesicle composed of a reacts with b in aqueous buffer, the vesicles are morphologically "wounded," allowing release of an entrapped fluorescent marker that serves to signal the presence of b. These experiments, carried out with small and giant vesicles, represent a "dual purpose" (detection and decontamination) approach to designing systems to combat chemical warfare agents.

Polymerization within fluid bilayers was exploited by Evans et al. to form nanoscale "conduits and networks" from giant vesicles [68]. Elegant vesicle manipulation and aspiration techniques allowed the extraction of tubules from excess bilayer area. The doped lipid tubules were then polymerized into solid structures of controlled length and thickness. Beyond the biological relevance of tubules [39], the authors describe the utility of such networks in conduction, enzyme immobilization, and sensor devices.

Menger et al. have described the sequential layering of lipid bilayers by micromanipulation of giant vesicles [35]. Individual vesicles of opposite charge adhere, and

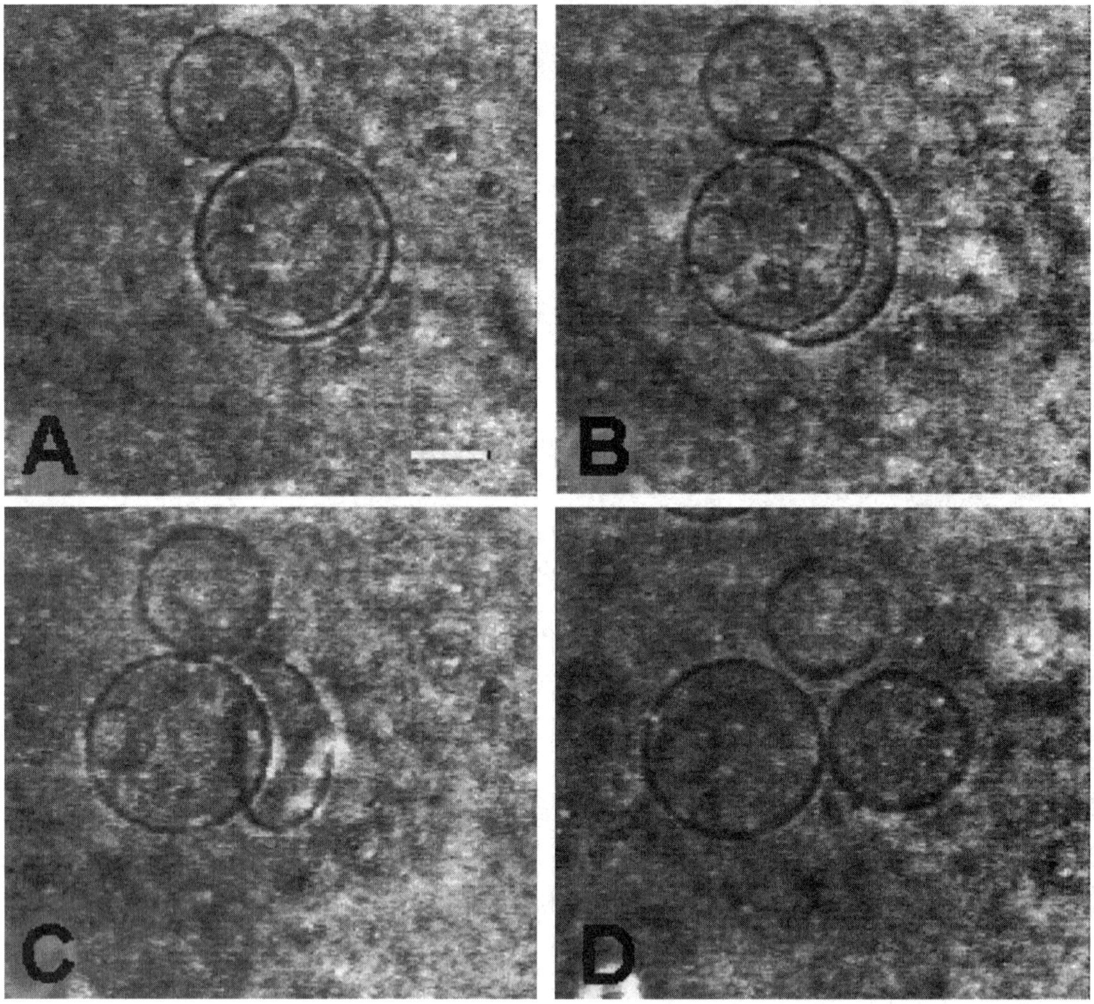

FIG. 6 Expulsion of a "daughter" DDAB vesicle from a larger giant vesicle shell, induced by heating. Bar = 12 μm.

if the electrostatic attraction is great enough to overcome membrane bending repulsion, one adhered vesicle will collapse onto its partner, thus either fully or partially encapsulating the intact vesicle. This process can be repeated, providing a vesicle structure analogous to layered Langmuir-Blodgett films. Whereas L-B films possess an air-film interface, however, layered vesicles surround and are surrounded by liquid. Progress has been made in chemically triggering the layering process in a selective manner [33]. (Controlled encapsulation of small vesicle systems to produce micrometer-scale "vesosomes" has been demonstrated, although with an estimated efficiency of only 5–15% [69].) Needham and coworkers have also utilized coating processes to encapsulate synthetic secretory granule mimics with lipid bilayers [70], and Fromherz et al. have adsorbed giant phospholipid vesicles onto silicon chips in the hope of producing a hybrid semiconductor device [71].

So-called steric stabilization of vesicle systems for drug delivery garnered a great deal of attention in the 1990s [72]. The basic premise of such biomaterials is to include poly(ethylene glycol) grafts in small percentages with conventional lipid vesicles, providing an inert surrounding that sterically blocks recognition of the vesicles by the body's defenses. Giant vesicles have served as extremely useful models for determining the in situ effectiveness of PEG grafts for blocking molecular and biomacromolecular access to bilayers. Evans et al. were able to evaluate the adhesion properties between GVs containing PEG grafts in PEG aqueous solutions [73]. Needham and coworkers, in two notable examples, evaluated vesicles with PEG grafts (molecular weight 750) using micropipette procedures. Their

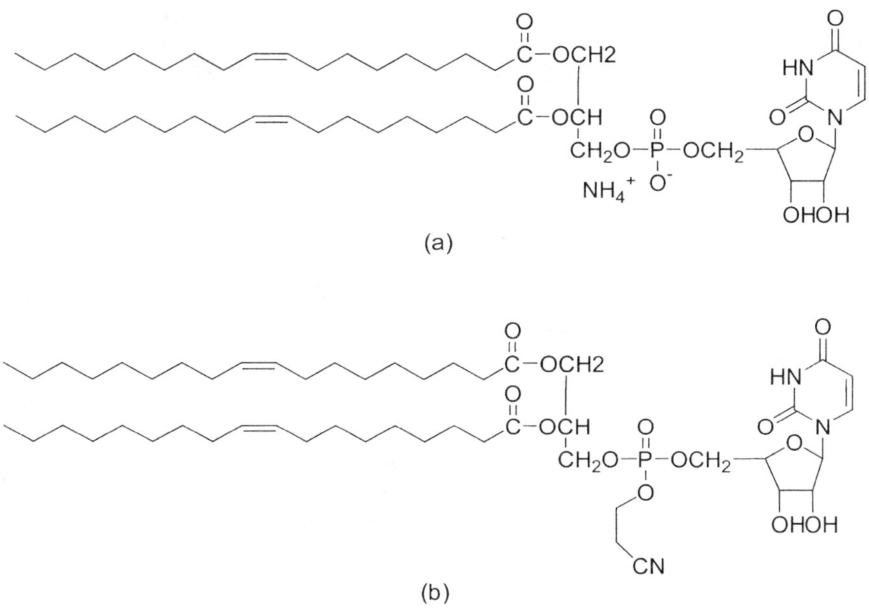

FIG. 7 Lipid a and lipid precursor b in Luisi and coworkers' autocatalysis experiments.

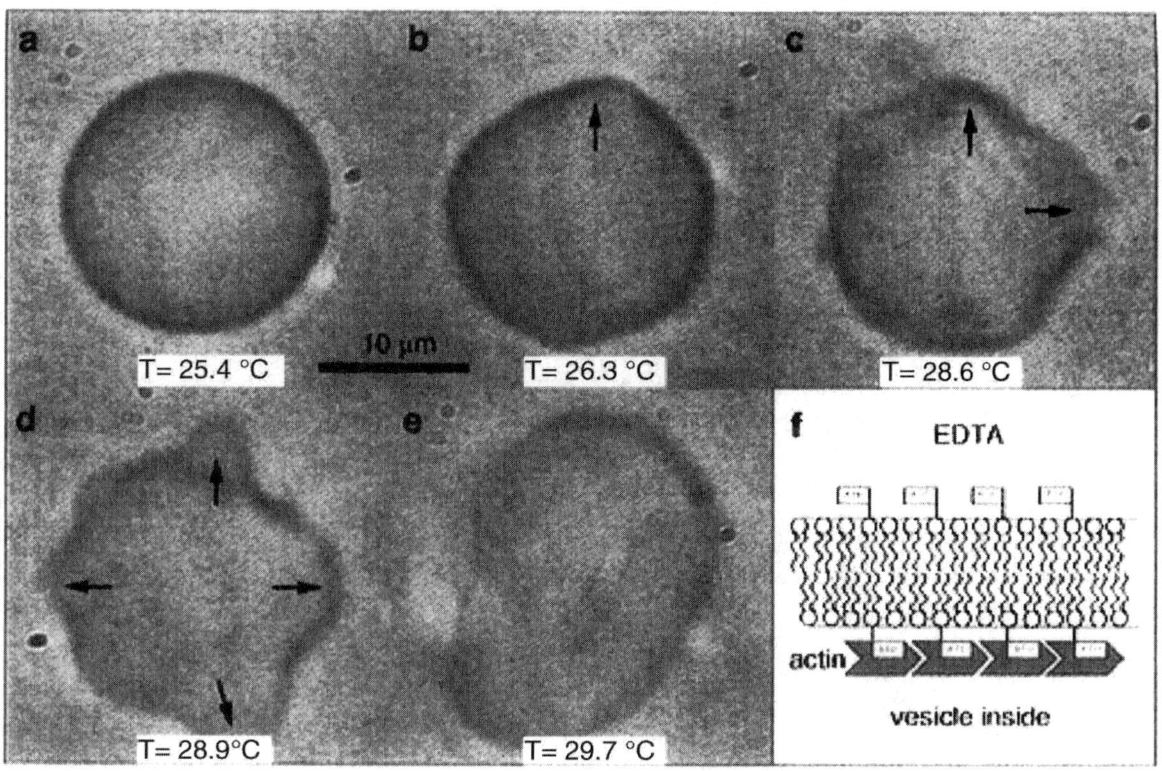

FIG. 8 Influence of temperature on the morphology of vesicles with actin bound only to the interior monolayer. (From Ref. 61. Copyright 1996 American Chemical Society.)

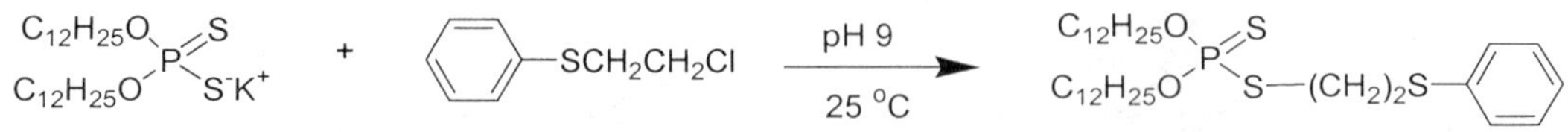

FIG. 9 Decontamination reaction with giant vesicles developed by Jaeger and coworkers.

extremely sensitive methods allowed them to demonstrate PEG graft–induced inhibition of monooleoylphosphocholine (MOPC) micelle fusion with the membrane, whereas surfactant monomers could more readily penetrate into the bilayers [74]. Their techniques allowed discrimination between MOPC monomers, oligomers, and micelles. (The groups of Sackmann [75] and Zhelev [76] have analyzed the effect of "exchangeable" surfactants on giant vesicle membranes not containing PEG lipid.) In another set of experiments, variable concentrations (0–10 mol%) of PEG-grafted lipids were determined for the inhibition of avidin-biotin interactions between two apposed vesicle surfaces [77]. With 2 mol% PEG lipid, avidin-biotin interaction was four times less, and 10 mol% PEG lipid fully inhibited interaction. Needham's experiments show how giant vesicles can be used to evaluate fusion and recognition processes that strike to the core of the steric stabilization principle.

Amphiphilic diblock copolymers capable of forming giant vesicle structures have emerged as promising materials. Polyethyleneoxide-polyethylethylene polymers of number average molecular weights of ~ 3900 g/mol were used to form "polymersomes" by Hammer and coworkers [78]. Pipette aspiration techniques demonstrated that the polymersomes had bending and deformabilities similar to those of phospholipid bilayers but proved to be much stronger and less permeable. Such properties immediately bring to mind drug delivery possibilities, and it is exciting to consider that such polymers, like lipids, ideally can be structurally modified and tailored for desired vesicular properties. Jenekhe and Chen reported another example of polymeric giant vesicles [79]. Instead of forming aqueous vesicular dispersions, however, these poly(phenylquinoline)-block-polystyrene amphiphiles (Fig. 10) aggregated into stable micrometer-sized vesicles, lamellae, and cylinders in mixed organic solvents trifluoroacetic acid and dichloromethane at different ratios. Control of aggregate structure was available by varying block sizes, solvent ratios, and solvent drying rates. In addition, the researchers demonstrated encapsulation of up to 10^{10} fullerene molecules in the interior of their assemblies.

Finally, optical trapping techniques provided Pouligny and coworkers with a means of manipulating both giant lipid vesicles and microscopic latex spheres [80]. This enabled them to study the adhesion of the spheres to the phosphocholine vesicle surfaces, "drag" vesicles in solution by manipulating the spheres, create complexes of multiple vesicles, insert the spheres within the vesicle aqueous compartments, and subsequently expel them (Fig. 11). With the precise control of the laser, vesicles can potentially be used to create defined networks or clusters. Similar experiments using glass and latex microspheres permitted the determination of the shear viscosity of phosphocholine vesicle membranes [81], allowing novel characterization of these potentially useful "soft materials" [8].

VI. SUMMARY

We hope that we have demonstrated the utility and potential of giant vesicles, microscopic association colloids that have increased in interest since 1990. If it appears that giant vesicle research is disparate and without a unified focus, it should not be misconstrued as a weakness of the field but rather as an indication of its outward growth and expanding relevance to many areas of science. The coming years should find the giant vesicle an increasingly popular system among those studying membrane interactions with polymers, viruses, and enzymes. It may also be possible that the giant vesicle will provide the perfect chassis for an "artificial" functioning metabolic system from a collection of chemicals and biochemicals.

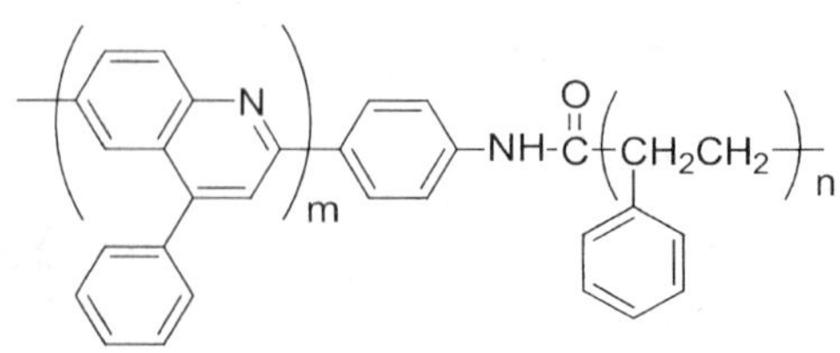

FIG. 10 Poly(phenylquinoline)-block-polystyrene amphiphiles prepared by Jenekhe and Chen.

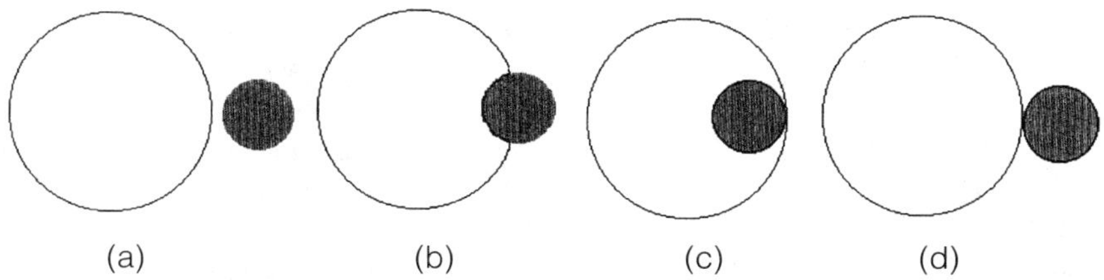

FIG. 11 Schematic sequence of optical trap manipulations of a latex sphere (gray circle) and phospholipid giant vesicle (open circle) by Pouligny and coworkers. (a) Introduction of sphere and vesicle. (b) Adsorption of latex sphere to vesicle surface. (c) Internalization of latex sphere, accompanied by lipid coating of the latex sphere's surface. (d) Expulsion of the latex sphere.

REFERENCES

1. R. Lipowsky, Nature *349*:475–481 (1991).
2. D. Needham, in *Permeability and Stability in Lipid Bilayers* (E. DiSalvo and S. Simon, eds.), CRC Press, Boca Raton, FL, 1995, pp. 49–76.
3. F. M. Menger and K. D. Gabrielson, Angew. Chem. Int. Ed. Engl. *34*:2091–2106 (1995).
4. U. Seifert, Adv. Phys. *46*:13–137 (1997).
5. F. M. Menger and J. S. Keiper, Adv. Mater. *10*:888–890 (1998).
6. F. M. Menger and M. I. Angelova, Acc. Chem. Res. *31*:789–797 (1998).
7. F. M. Menger and J. S. Keiper, Curr. Opin. Chem. Biol. *2*:726–732 (1998).
8. D. Needham, MRS Bull. *24*:32–40 (1999).
9. J. P. Reeves and R. M. Dowben, J. Cell. Physiol. *73*: 49–60 (1969).
10. A. D. Bangham, M. M. Standish, and J. C. Watkins, J. Mol. Biol. *13*:238–245 (1965).
11. M. I. Angelova and D. S. Dimitrov, Faraday Discuss. Chem. Soc. *81*:303–311 (1986).
12. R. P. Rand, Biophys. J. *4*:303–316 (1964).
13. E. A. Evans, R. Waugh, and L. Melnik, Biophys. J. *16*: 585–595 (1976).
14. R. Waugh and E. A. Evans, Biophys. J. *26*:115–131 (1979).
15. E. A. Evans, Methods Enzymol. *173*:3–35 (1989).
16. E. A. Evans and R. Waugh, J. Colloid Interface Sci. *60*: 286–298 (1977).
17. E. Evans and R. Kwok, Biochemistry *21*:4874–4879 (1982).
18. E. Evans and D. Needham, Faraday Discuss. Chem. Soc. *81*:267–280 (1986).
19. E. Evans and D. Needham, J. Phys. Chem. *91*:4219–4228 (1987).
20. D. Needham and E. Evans, Biochemistry *27*:8261–8269 (1988).
21. W. Helfrich, Z. Naturforsch. Teil C *28*:693–703 (1973).
22. S. Svetina and B. Zekš, Biomed. Biochim. Acta *42*:S86 (1983).
23. S. Svetina and B. Zekš, Eur. Biophys. J. *17*:101–111 (1989).
24. H.-G. Döbereiner, E. Evans, M. Kraus, U. Seifert, and M. Wortis, Phys. Rev. E *55*:4458–4474 (1997).
25. E. Sackmann, H.-P. Duwe, and H. Engelhardt, Faraday Discuss. Chem. Soc. *81*:281–290 (1986).
26. K. Berndl, J. Käs, R. Lipowsky, E. Sackmann, and U. Seifert, Europhys. Lett. *13*:659–664 (1990).
27. J. Käs and E. Sackmann, Biophys. J. *60*:825–844 (1991).
28. H.-G. Döbereiner, J. Käs, D. Noppl, I. Sprenger, and E. Sackmann, Biophys. J. *65*:1396–1403 (1993).
29. T. Kunitake and Y. Okahata, J. Am. Chem. Soc. *99*: 3860–3861 (1977).
30. H. Ringsdorf, B. Schlarb, and J. Venzmer, Angew. Chem. Int. Ed. Engl. *27*:113–158 (1988).
31. F. M. Menger and S. J. Lee, Langmuir *11*:3685–3689 (1995).
32. P. Bucher, A. Fischer, P.-L. Luisi, T. Oberholzer, and P. Walde, Langmuir *14*:2712–2721 (1998).
33. F. M. Menger and J. S. Keiper, Angew. Chem. Int. Ed. Engl. 37:3433–3435 (1998).
34. V. Marchi-Artzner, L. Jullien, L. Belloni, D. Raison, L. Lacombe, and J.-M. Lehn, J. Phys. Chem. *100*:13844–13856 (1996).
35. F. M. Menger, J. S. Keiper, and S. J. Lee, Langmuir *13*:4614–4620 (1997).
36. F. M. Menger and J. S. Keiper, Angew. Chem. Int. Ed. Engl. *36*:248–250 (1997).
37. D. P. Pantazatos and R. C. MacDonald, J. Membrane Biol. *170*:27–38 (1999).
38. R. P. Haugland, *Handbook of Fluorescent Probes and Research Chemicals*, 6th ed, Molecular Probes, 1996.
39. F. M. Menger, S. J. Lee, and J. S. Keiper, Langmuir *12*:4479–4480 (1996).
40. L. Mathivet, S. Cribier, and P. F. Devaux, Biophys. J. *70*:1112–1121 (1996).
41. T. Ueda, S. J. Lee, Y. Nakatani, G. Ourisson, and J. Sunamoto, Chem. Lett. 417–418 (1998).
42. J. Korlach, P. Schwille, W. W. Webb, and G. W. Feigenson, Proc. Natl. Acad. Sci. USA *96*:8461–8466 (1999).
43. L. A. Bagatolli and E. Gratton, Biophys. J. *77*:2090–2101 (1999).

44. F. M. Menger and K. Grabrielson, J. Am. Chem. Soc. 1567–1568 (1994).
45. F. M. Menger, S. J. Lee, and J. S. Keiper, Chem. Commun. 957–958 (1998).
46. J. D. Moroz, P. Nelson, R. Bar-Ziv, and E. Moses, Phys. Rev. Lett. *78*:386–389 (1997).
47. R. Bar-Ziv, E. Moses, and P. Nelson, Biophys. J. *75*: 294–320 (1998).
48. C. Menager and V. Cabuil, J. Colloid Interface Sci. *169*: 251–253 (1995).
49. M. Dvolaitzky, M. A. Guedeau-Boudeville, and L. Léger, Langmuir *8*:2595–2597 (1992).
50. J.-C. Bradley, M. A. Guedeau-Boudeville, G. Jandeau, and J.-M. Lehn, Langmuir *13*:2457–2462 (1997).
51. R. Wick, P. Walde, and P. L. Luisi, J. Am. Chem. Soc. *117*:1435–1436 (1995).
52. A. Veronese and P. L. Luisi, J. Am. Chem. Soc. *20*: 2662–2663 (1998).
53. R. Wick and P. L. Luisi, Chem. Biol. *3*:277–285 (1996).
54. G. Riquelme, E. Lopez, L. M. Garcia-Segura, J. A. Ferragut, and J. M. Gonzalez-Ros, Biochemistry *29*: 11216–11222 (1990).
55. P. Regueiro, J. Monreal, R. S. Diaz, and F. Sierra, J. Neurochem. *67*:2146–2154 (1996).
56. C. Enz, T. Steinkamp, and R. Wagner, Biochim. Biophys. Acta *1143*:67–76 (1993).
57. R. Wick, M. I. Angelova, P. Walde, and P. L. Luisi, Chem. Biol. *3*:105–111 (1996).
58. V. Dorovska-Taran, R. Wick, and P. Walde. Anal. Biochem. *240*:37–47 (1996).
59. N. Kumazawa, Y. S. Mel'nikova, and K. Yoshikawa, Macromol. Symp. *106*:219–223 (1996).
60. M. I. Angelova, N. Hristova, and I. Tsoneva, Eur. Biophys. J. *28*:142–150 (1999).
61. L. Schmitt and R. Tampé, J. Am. Chem. Soc. *118*: 5532–5543 (1996).
62. S. A. Soltesz and D. A. Hammer, Biophys. J. *68*:315–325 (1995).
63. M. L. Longo, A. J. Waring, and D. A. Hammer, Biophys. J. *73*:1430–1439 (1997).
64. M. L. Longo, A. J. Waring, L. M. Gordon, and D. A. Hammer, Langmuir *14*:2385–2395 (1998).
65. N. Shahidzadeh, D. Bonn, O. Agurre-Chariol, and J. Meunier, Phys. Rev. Lett. *81*:4268–4271 (1998).
66. D. A. Jaeger, B. Li, and T. Clark Jr., Langmuir *12*: 4314–4316 (1996).
67. D. A. Jaeger, C. L. Schilling III, A. K. Zelenin, B. Li, and E. Kubicz-Loring, Langmuir *15*:7180–7185 (1999).
68. E. Evans, H. Bowman, A. Leung, D. Needham, and D. Tirrell, Science *273*:933–935 (1996).
69. S. A. Walker, M. T. Kennedy, and J. A. Zasadzinski, Nature *387*:61–64 (1997).
70. P. F. Kiser, G. Wilson, and D. Needham, Nature *394*: 459–462 (1998).
71. P. Fromherz, V. Kiessling, K. Kottig, and G. Zeck, Appl. Phys. A *69*:571–576 (1999).
72. D. D. Lasic and D. Needham, Chem. Rev. *95*:2601–2628 (1995).
73. E. Evans, D. J. Klingenberg, W. Rawicz, and F. Szoka, Langmuir *12*:3031–3037 (1996).
74. D. Needham, N. Stoicheva, and D. V. Zhelev, Biophys. J. *73*:2615–2629 (1997).
75. W. Häckl, U. Seifert, and E. Sackmann, J. Phys. II Fr. *7*:1141–1157 (1997).
76. D. V. Zhelev, Biophys. J. *75*:321–330 (1998).
77. D. A. Noppl-Simson and D. Needham, Biophys. J. *70*: 1391–1401 (1996).
78. B. M. Discher, Y.-Y. Won, D. S. Ege, J. C.-M. Lee, F. S. Bates, D. E. Discher, and D. A. Hammer, Science *284*:1143–1146 (1999).
79. S. A. Jenekhe and X. L. Chen, Science *279*:1903–1907 (1998).
80. C. Dietrich, M. Angelova, and B. Pouligny, J. Phys. II Fr. *7*:1651–1682 (1997).
81. K. Velikov, C. Dietrich, A. Hadjiisky, K. Danov, and B. Pouligny, Europhys. Lett. *40*:405–410 (1997).

19

Electroless Plating of Organic Pigment Thin Films Using Surfactants with an Azobenzene Group

TETSUO SAJI Tokyo Institute of Technology, Tokyo, Japan

I. INTRODUCTION

In 1985 we demonstrated that micelles formed by cationic surfactants with a ferrocenyl group can be broken into monomers when the surfactants are oxidized chemically or electrochemically and that a solubilizate is released from the micelles as the micelles dissociate [1,2]. This phenomenon was then applied to the electrochemical formation of organic thin films using the same surfactants [3,4]. Later, we illustrated the preparation of organic pigment thin films by this method using nonionic ferrocenyl surfactants (FPEG) [5–7]. These later studies showed that the organic pigment particles are not actually dissolved in the micelles formed by the surfactants but are dispersed by having the surfactants adsorbed onto the particle surface. The pigment particles are destabilized when the surfactants adsorbed on the particles are oxidized, and finally the particles deposit on the electrode (Fig. 1). This method has the following advantages:

1. Thin films of a wide variety of organic compounds that are soluble in micellar solutions and dispersible in surfactant solutions may be prepared.
2. The film-forming compounds are not electrolyzed and are the same as the starting compounds.
3. No organic solvent is needed.
4. Film thickness is easy to control.
5. Preparation of large-area films is feasible.

However, we could not prepare a film on a base metal using this surfactant because the metal oxidizes at the potential of electrolysis for the film formation [+0.5 V vs. saturated calomel electrode (SCE)]. This is ascribable to the relatively high ionization potential of the ferrocenyl group. We may overcome this deficiency by using a surfactant that loses its amphiphilic function by reduction. Subsequently, we discovered a surfactant with an azobenzene group (AZPEG, Fig. 2) that loses its amphiphilic function when reduced. This loss of surfactancy enables formation of a thin film on a base metal, e.g., copper, nickel, and aluminum, by immersing in the dispersion without electrolysis [8,9]. In this chapter, we present the details of our studies on preparations of organic pigment thin films using this azobenzene surfactant.

II. EXPERIMENTAL

A. Preparation of the Surfactant and Dispersions

AZPEG was prepared from 4-[(4-hexylphenyl)azo]-phenol (AZOH) in two steps (Fig. 3). Dispersions containing 1–2 mM (1 M = 1 mol dm^{-3}) AZPEG, 0.1 M HCl, and 10–30 mM pigment were prepared by sonicating the mixture for 10 min twice using strong sonication (Ultrasonic Disruptor, Tomy Seiko UR-200P) and stirred for 2 days. Usually, a film was prepared successfully only by using such sonication. When a milder sonicator used for washing was substituted, sonication and stirring for longer times were necessary for film formation.

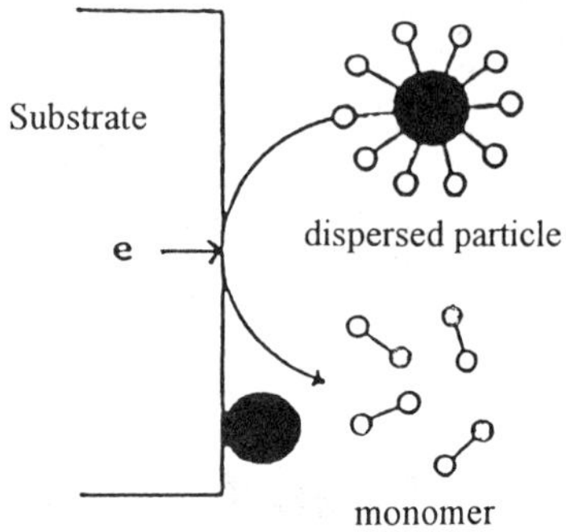

FIG. 1 Scheme of film formation.

B. Materials

Metal plates were cleaned by polishing with the metal polishing reagent PIKAL (Nihon Maryo-Kogyo Co. Ltd.) followed by sonication with acetone for 5 min. Indium tin oxide (ITO) was cleaned by sonication with acetone for 5 min. Most of the pigments in Table 1 were donated from producing companies and were used as received. Some of the pigments were washed with distilled water in order to remove unidentified impurities, which influence cohesive forces among these pigments and the substrate.

C. Film Formation

Electroless platings on base metals were made by immersing the base metal plates in dispersions formulated as already described for fixed times and then by washing with distilled water. Contact platings on ITO and noble metals were made by immersing both the substrate and aluminum plates in the pigment dispersion, where these two plates were short-circuited with a metal clip. The amount of the pigment deposited on the substrate was estimated by absorption of the aqueous dispersion that was obtained by redispersing the film into an aqueous solution of 5 mM Brij 35.

III. RESULTS AND DISCUSSION

A. Properties of Surfactants

The critical micelle concentration (cmc) of AZPEG in a 0.1 M HCl aqueous solution was determined by the

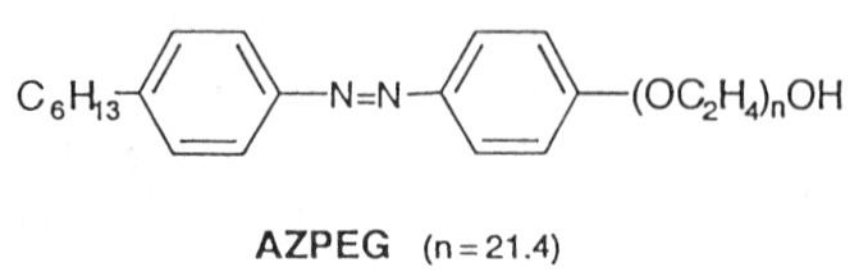

FIG. 2 Molecular structure of a surfactant with an azobenzene group (AZPEG).

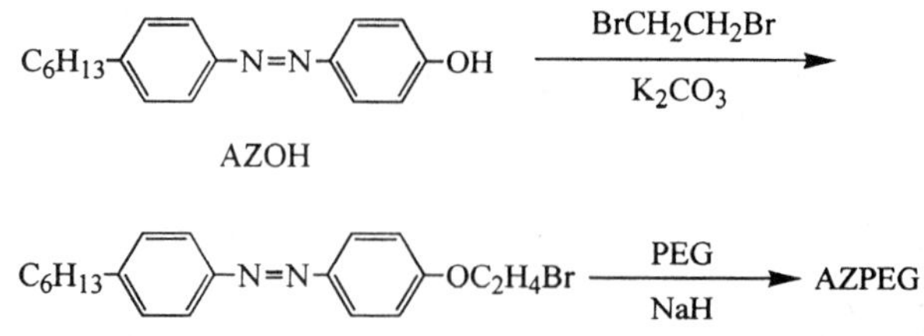

FIG. 3 Route of preparation of AZPEG.

concentration dependence of the molar absorption coefficient to be 7 μM [8]. The cyclic voltammogram of AZPEG in a 0.1 M HCl aqueous solution shows an irreversible reduction peak at -0.2 V versus SCE. Disruption of AZPEG micelles was proved by monitoring the uptake of (water-insoluble) aniline blue (AB). AZPEG (50 μM) in 0.1 M aqueous HCl solubilized 92 μM AB. In contrast, solutions of 50 μM reduced AZPEG solubilized less than 2 μM AB. This difference suggests that the micelles of AZPEG dissociate into the monomers formed by reduction owing to the enhancement of their hydrophilic character [8].

B. Electroless Plating

In the following discussion, as a typical pigment, β-type copper phthalocyanine (CuPc) was used. A blue film with reflectance from the metal surface was formed on nickel plate by immersion in a dispersion containing 2.0 mM AZPEG, 0.1 M HCl, and 28 mM CuPc for a few minutes without electrolysis. Similar CuPc blue films were also obtained on copper, iron, lead, tin, zinc, and aluminum (and their alloys, e.g., brass, phosphor bronze, nickel silver, and duralumin) plates by similar methods. Aluminum plates need longer times (40 min) for the same film thickness as that on the other base metal plates owing to the existence of an oxide film. In the cases of zinc and chromium plates, uniform films were not formed owing to evolution of hydrogen gas. Such electroless plating may be explained by the chemical reduction of AZPEG with these base metals because the standard potentials of these base metals are more negative than the reduction potential of AZPEG (-0.2 V). On the other hand, blue films from CuPc were not formed on the noble metals, on ITO, or on stainless steel plates.

C. Contact Plating

In the field of metal plating, the term "contact plating" has been used to denote the deposition of a metal without the use of an outside source of current by immersion of substrate in a solution in contact with another

TABLE 1 Results of Film Formation of Pigments Using Surfactants with an Azobenzene Group

Pigment	Product name	Appearance of film
Yellows		
Disazo Yellow AAMX	Seikafast Yellow 2600	Yellow
Disazo Yellow HR	Seikafast Yellow 2700	Reddish yellow
Cromophtal Yellow GR	Cromofine Yellow 5910	Reddish yellow
Reds		
Perylene Vermillion	Indofast Brilliant Scarlet (R-6335)	Red
	Aliogen Red L 3870HD	Red
Dianthraquinoyl Red	Cromophtal Red A3B	Red
Perylene Red	Paliogen Red L 3910HD	Red
Perylene Maroon	Perindo Maroon R-6424	Dark red
Perylene Scarlet	Indofast Brilliant Scarlet (R-6500)	Red
Pyranthrone Red	Paliogen Red L 3530HD	Red
Purples		
Rhodamin B Lake	Fast Rose Lake B	Reddish purple
Methyl Violet Lake	Bronze Violet GI	Bluish purple
Dioxazine Violet	Cromofine Red 6820	Bluish purple
Blues		
Phthalocyanine Blue R	Cromofine Blue B145	Reddish blue
Phthalocyanine Blue G	Cromofine Blue 4920	Greenish blue
Phthalocyanine Blue E	Heliogen Blue L6700F	Reddish blue
Heliogen Blue G	Heliogen Blue L6700F	Greenish blue
Indanthron Blue	Cromophtal Blue A3R	Reddish blue
Greens		
Phthalocyanine Green	(Tokyo Kasei)	Bluish green
Phthalocyanine Green 6Y	Heliogen Green D9360	Yellowish green
Black		
Carbon black	Denka Black	Black
	Toka Black 7550F	Black

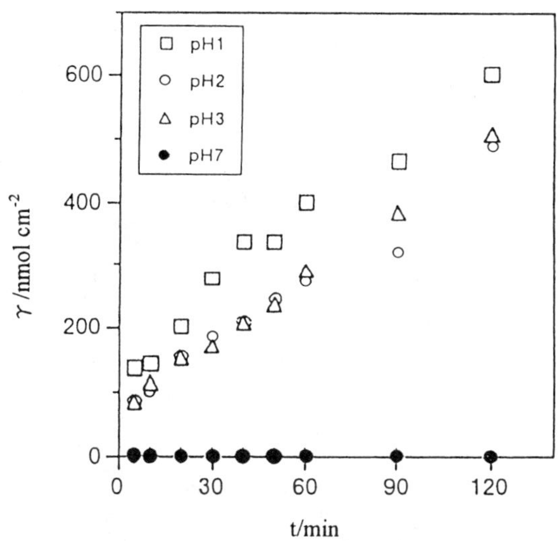

FIG. 4 CuPc film (thickness) coverage (γ) versus immersion time (t) of the substrate in the aqueous dispersion containing 1 mM AZPEG, 17.5 mM CuPc, and 0.1 M HCl.

metal (e.g., aluminum or zinc) [10]. We can apply this technique to the formation of pigment films on noble metals and on ITO plates. A transparent blue film of CuPc was formed on an ITO plate by immersing both ITO and aluminum plates in a CuPc dispersion, where these two plates were short-circuited with a metal clip. A blue film was also formed on silver, palladium, gold, platinum, and stainless steel plates by the same method [9].

D. Rate of Film Growth

Formation of the film depends on the pH of the dispersion. Under acidic conditions (pH $\leq$ 3), a film was formed (Fig. 4) [11]. This may be explained by the following mechanism: AZPEG is reduced to hydrazobenzene, followed by acid-catalyzed formation of the aniline derivative, and loses its function as a surfactant because of enhancement of the tail group's hydrophilicity [12]. The film thickness of the CuPc film increases with immersion time. The thickness of the film

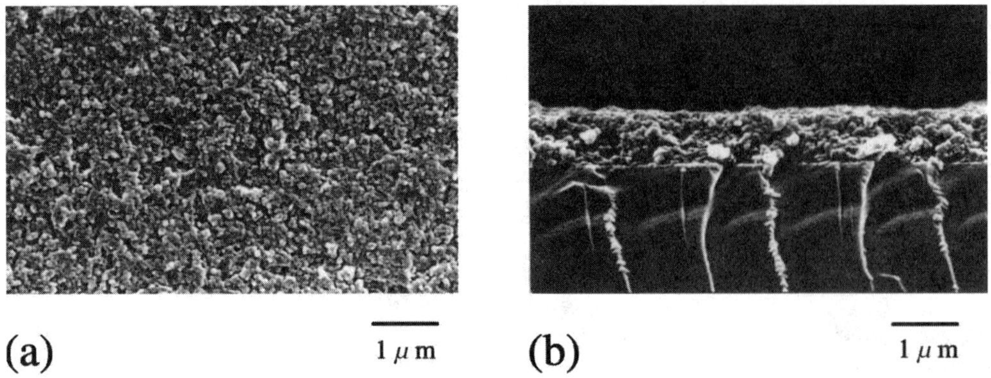

FIG. 5 Scanning electron micrographs of a surface (a) and a cross section (b) of a perylene vermillion film on ITO.

increased to more than 10 μm during overnight immersion of the substrate. This may be explained by the fact that the surfactant can penetrate into the film owing to the existence of small spaces and interstices between pigment particles in the film and eventually reaches the substrate surface. The rate of the film growth also depends on the concentration of AZPEG. The rate is fastest when the pigment surface is covered with AZPEG and the concentration of free AZPEG is lowest [12] because the free AZPEG disturbs the pigment deposition.

E. Scope of Pigments

We have prepared films of a wide variety of pigments. Table 1 lists the pigments that we have successfully plated to date. These pigments satisfy following conditions:

1. They are not soluble in water and are hydrophobic.
2. The pigment particles are dispersible. Particle sizes must be submicrometer. Particles larger than 1 μm lead to precipitation and sedimentation because Brownian motion no longer dominates particle motion.

These films exhibit the intrinsic color of the pigment used for preparation of the respective dispersions. The films of carbon black (CB) break when washed with water owing to weak cohesive forces among the carbon black (pigment) particles and the substrate. Composite films of carbon black/metal phthalocyanine (MPc, M = copper or iron) stable against washing were obtained using a dispersion containing less than 70 wt% of CB [13].

F. Film Properties

Typical representative scanning electron micrographs of a perylene vermillion film prepared by this contact plating method, shown in Fig. 5, indicate that this film has a uniform thickness and is composed of particles. These particles have the same size and shape as those used for the preparation of the plating dispersion. Most of these films are transparent. Visible light does not scatter very much because the size of these particles in the films is less than the half of wavelength of the incident light. The absorption spectrum of a film of CuPc prepared by contact plating is very similar to those of β-type CuPc films prepared by vacuum sublimation (Fig. 6) [9]. The absorption spectrum of an

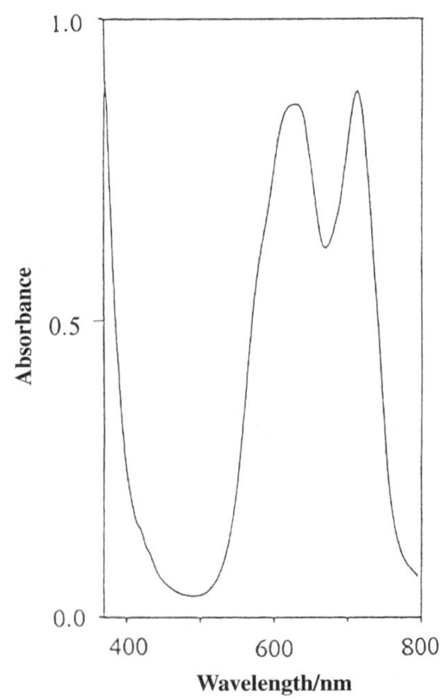

FIG. 6 Absorption spectrum of a film of CuPc film on ITO prepared by the contact plating method using an aqueous dispersion containing 1 mM AZPEG, 17.5 mM CuPc, and 0.1 M HCl and an immersion time of 1 min.

aqueous dispersion prepared by washing this film with a 5 mM Brij 35 aqueous solution consisted of broad peaks and was also very similar to that for a 5 mM Brij 35 aqueous dispersion of CuPc. This similarity of absorption spectra indicates that the crystalline form of the pigment is maintained throughout the film preparation processes, which means that we can control the crystalline structure of the film. The AZPEG and CuPc content in the film were 3–4 and 96–97 wt%, respectively, which means that the film is mainly composed of pigment [11].

G. Mechanism of Film Formation

Such film formation may be explained by a mechanism similar to that obtained for ferrocenyl surfactants [7]: AZPEG is reduced to the hydrazo derivative and its efficacy as a surfactant dramatically decreases owing to enhancement of the hydrophilicity of the tail group. The concentration of the surfactant in the vicinity of a substrate decreases to less than the cmc. The surfactant adsorbed on the particles is desorbed, and the resulting particle destabilization leads to the deposition of the particles on the substrate (Fig. 1).

H. Reinforcement of Film

The mechanical strength of the films is not great because no binder has been incorporated. The following methods have been proposed for film reinforcement:

1. Overcoating pigment films with a polymer film using an organic solution of polymer, e.g., 5% poly(methyl methacrylate) (PMMA) ethylacetate solution.
2. Composite plating of polymer/pigment [14]. Thermal treatment of this film melts polymer particles, which then function as a binder.
3. Composite plating of Ni/pigment film [15]. Electrochemical deposition of Ni on the substrate covered with a pigment film gives a stable film. The deposited Ni serves as a binder and provides high adhesion to the substrate.

I. Preparation of Color Filters

Films prepared on ITO prepared from only one pigment do not satisfy the spectral requirements of color filters for liquid crystalline display (LCD) devices. Such color filters meeting the spectral requirements for LCD applications can be prepared, however, by using dispersions containing two or three pigments that are appropriately spectrally balanced. Figure 7 shows the transmittance spectra of three primary color filters prepared by this method, which transmit more than 80% of visible light at peak wavelengths.

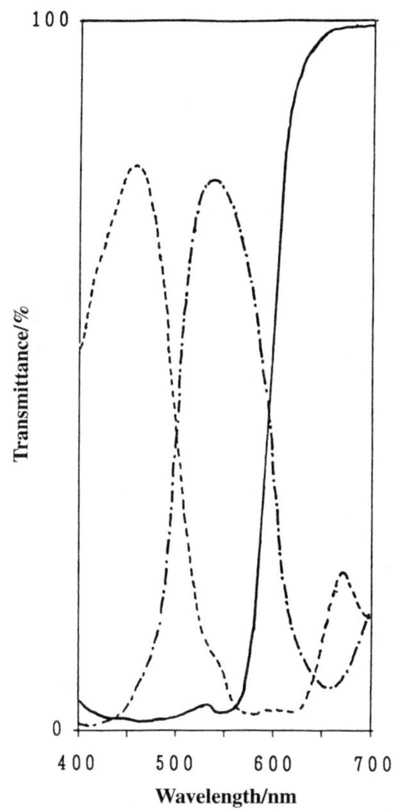

FIG. 7 Transmittance spectra of the three primary color filters prepared on ITO by this contact plating method.

IV. CONCLUSIONS

Experimental results, properties, and the scope of pigmented films that can be formed by electroless plating methods using electroactive azobenzene surfactants have been presented. In addition of the advantages of using FPEG, advances of using AZPEG include the following:

1. Films can be prepared simply by immersion of the substrate into an appropriately formulated dispersion.
2. Base metals can be used as substrates.
3. Pigment crystallinity can easily be maintained in the film.
4. The preparation of AZPEG surfactants is easier than that of ferrocenyl surfactants.

Such an electroless plating technique still has some limitations:

1. The substrate has to be an electrical conductor or have an intrinsic ability to reduce the surfactants (AZPEG).
2. The film-forming material has to be dispersible by the electroactive surfactants.

Such limitations will probably decrease after further research and development. We expect that the electroless plating method will be more widely applied in the future.

ACKNOWLEDGMENTS

We thank R. Ohki for the electron micrograph data. This work was partially supported by a Grant-in-Aid for Scientific Research (C) (No. 11650840) and that on the Priority Area of "Electrochemistry of Ordered Interfaces" (No. 11118223) from the Ministry of Education, Science, Sports and Culture, Japan.

REFERENCES

1. T. Saji, K. Hoshino, and S. Aoyagui, J. Am. Chem. Soc. *107*:6865–6868 (1985).
2. T. Saji, K. Hoshino, and S. Aoyagui, J. Chem. Soc. Chem. Commun 1985:865 (1985).
3. K. Hoshino and T. Saji, J. Am. Chem. Soc. *109*:5881–5883 (1987).
4. K. Hoshino, M. Goto, and T. Saji, Chem. Lett. 1988: 547–550 (1988).
5. T. Saji, Chem. Lett. 1988:693–696 (1988).
6. T. Saji and Y. Ishii, J. Electrochem. Soc. *136*:2953–2956 (1989).
7. T. Saji, K. Hoshino, Y. Ishii, and M. Goto, J. Am. Chem. Soc. *113*:450–456 (1991).
8. T. Saji, K. Ebata, K. Sugawara, S. Liu, and K. Kobayashi, J. Am. Chem. Soc. *116*:6053–6054 (1994).
9. T. Saji, Y. Igusa, K. Kobayashi, and S. Liu, Chem. Lett. 1995:401–402 (1995).
10. F. A. Lowenhein, in *Electroplating Engineering Handbook* (A. K. Graham, ed.), Reinhold, New York, 1955, p. xiii.
11. Y. Igusa, Ms dissertation, Tokyo Institute of Technology, 1995.
12. Y. Ito, Ms dissertation, Tokyo Institute of Technology, 1999.
13. H. Kondo, Y. Ohsawa, and T. Saji, Jpn. Soc. Col. Mat. *71*:541–547 (1998).
14. T. Saji and K. Ebata, Bull. Chem. Soc. Jpn. *66*:3091–3093 (1993).
15. N. K. Shrestha and T. Saji, J. Surf. Fin. Soc. Jpn. *46*: 1066–1067 (1995).

20

Kinetic and Thermodynamic Modeling of Micellar Autocatalysis

JEAN-CLAUDE MICHEAU Laboratoire IMRCP, UMR CNRS 5623, Université Paul Sabatier, Toulouse, France

R. NAGARAJAN The Pennsylvania State University, University Park, Pennsylvania

I. INTRODUCTION

A. Overview

Chemical systems in which molecular aggregates such as micelles are reported to catalyze their own formation have attracted considerable scientific interest. An example is the alkaline hydrolysis of ethyl esters, which displays autocatalytic kinetics. This reaction has been discussed in the literature from the viewpoint of self-replication of molecules in organized compartments such as micelles. The reaction products from the hydrolysis of ethyl alkanoates are ethanol and the corresponding sodium alkanoates. Because the sodium alkanoates are capable of self-assembling into micellar structures, the observed autocatalysis has been attributed to micellar catalysis. However, in spite of their common autocatalytic behaviors, the hydrolysis products from different esters have significantly differing properties. Specifically, the product from the C-4 ester does not form micelles under the reaction conditions, whereas that from the C-8 ester does form micelles. Therefore, autocatalysis cannot be attributed solely to the formation of micelles. More importantly, anionic sodium alkanoate micelles cannot possibly catalyze (in the conventional sense in which catalysis is understood) the hydrolysis reaction involving anionic hydroxyl ions. Therefore, a general kinetic model that can describe the observed autocatalysis is needed in all cases.

Such a model is presented in this chapter. The model is built on a sequence of six events during the hydrolysis of ethyl alkanoates: formation of an ethyl ester–aqueous interface, dissolution of ester in the aqueous phase, hydrolysis of the ester in the aqueous phase, micellization of product sodium alkanoate, solubilization of ester into micelles, and micelle-mediated transport of ester from the organic to the aqueous phase. The dissolution of ester into the aqueous phase is promoted by the salting-in effect of product sodium alkanoate and by the solvent effect of product ethanol. The kinetic model demonstrates that for C-4, taking into account salting-in and solvent effects is sufficient to reproduce the autocatalytic behavior. For C-8, the main step is the micelle-mediated phase transfer of ethyl esters into the aqueous phase. For C-6, in addition to the salting-in and solvent effects and the micelle-mediated ester transport, one has to invoke the possibility that the micelles play yet another role of trapping ester molecules during the initial phase of the reaction without allowing their rapid release into the aqueous phase. Comparison with kinetic experiments performed on different chain lengths of ethyl esters, under different mixing conditions, and for different phase volume ratios has been used to validate the proposed kinetic model of autocatalysis.

B. BLL Reaction

Chemical systems in which molecular aggregates such as micelles have been reported to catalyze their own formation have attracted considerable scientific interest [1]. A prominent example of such processes is the basic hydrolysis of ethyl octanoate in a biphasic system, which was described originally by Bachmann, Luisi, and Lang [2] and is referred to in this chapter as the BLL reaction. The experimental system they employed is very simple and consists of an organic phase of neat ethyl ester placed above an aqueous alkaline solution. The reacting medium is maintained at 90°C under reflux for many hours, keeping mild stirring conditions so as not to perturb the macroscopic interface between the two immiscible liquids. It is found that there exists a well-defined time period after the start of the reaction during which very little activity takes place. At the end of this quiescent period, the reaction suddenly takes off at a dramatic rate and the overlaying ester is rapidly consumed (Fig. 1).

A qualitative interpretation of the observed kinetic behavior was proposed by the authors (BLL) assuming a micellar catalytic process. The hydrolysis of ester yields ethanol and amphiphilic sodium octanoate,

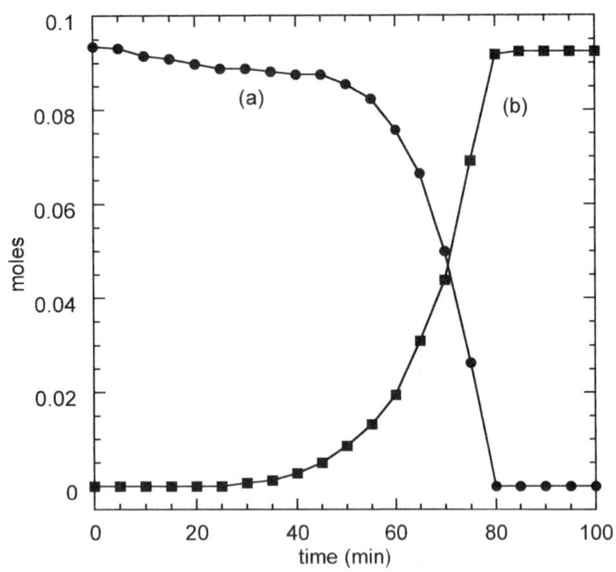

FIG. 1 Kinetics of the biphasic basic hydrolysis of ethyl octanoate performed in a 250-mL round-bottom flask at 90°C by vigorous mixing of 70 mL aqueous 3 M NaOH and 19 mL neat ethyl octanoate with a 25 × 6 mm magnetic bar at 800 rpm. (a) Evolution of ethyl octanoate (total volume of the supernatant organic phase). (b) Evolution of the concentration of the sodium octanoate in the aqueous phase. As in the original BLL reaction, we have observed a quiescent period followed by strong acceleration. Note that organic volume (a) and sodium octanoate concentration (b) exhibit symmetrical kinetic behavior.

which is known to form anionic micelles in aqueous medium. At the beginning of the reaction, only the ester molecules that are molecularly dissolved in water are available for hydrolysis. Therefore, the amount of product formed is small, reflecting little activity. Correspondingly, the concentration of the product (surfactant) sodium octanoate in the water phase is below the critical value necessary for micellization to start. This initial time period of low activity is observed as the lag time in the kinetic experiments. The surfactant concentration progressively increases with time and eventually reaches the critical micelle concentration (cmc) at which micelles come into existence. Once formed, the micelles can capture large numbers of ester molecules at the interface and carry them into the bulk aqueous phase where hydrolysis occurs, leading to autocatalytic production of the surfactant.

Starting from this qualitative interpretation, several independent authors have proposed more quantitative approaches. First, Billingham and Coveney [3], then Chimadzhew et al. [4], and later Maestro [5] investigated several macroscopic kinetic models by using exclusively the original BLL experimental results. In these models, the complete autocatalytic process has been artificially divided into two different steps depending on the surfactant concentration being below or above the cmc. None of these models was able to reproduce accurately the effect of added surfactant on the observed kinetics. To address this deficiency, an alternative approach has been proposed by Coveney et al. [6,7]. The principal idea behind this approach is to formulate a global nonequilibrium model wherein all the steps are coupled and without postulating an a priori set cmc. The numerical simulations of this nonequilibrium model show the highly dynamic nature of the entire process, including the reversible self-assembly and breakup of the micelles. Although these models bring new concepts to the literature on reaction kinetics, several basic questions remain unresolved. For example, the possibility of anionic micellar catalysis is questionable because one of the reactants is the negative hydroxide ion OH^-. In these conditions, micellar catalysis is not expected [8]. Another serious limitation of these modeling efforts is the failure to obtain a more accurate quantitative description of the original BLL experiment in order to exclude or validate the proposed mechanism by its ability to fail or fit quantitatively the experimental kinetic curves.

In order to develop a comprehensive and sufficiently realistic kinetic model for the observed autocatalytic behavior, we experimentally revisited [9] the BLL reaction via systematic measurements of the kinetics of the biphasic hydrolysis of C-4 to C-8 ethyl alkanoates. In all cases, the autocatalytic behavior was observed; that is, there was acceleration of the reaction rate during the course of the reaction. The development of the kinetic model describing such autocatalytic behavior is outlined in Section II. The model has been validated by comparison with experiments in Section III. To reduce the number of free parameters present in the kinetic model, all thermodynamic variables appearing in the model have been theoretically predicted. For this purpose, thermodynamic calculations at 80°C and 3 M ionic strength have been performed to estimate the solubility of ethyl alkanoates in the aqueous medium, the dependence of this solubility on the concentration of reaction products ethanol and sodium octanoate, the cmc of sodium alkanoates, the average size and stoichiometry of aggregates in the presence of ester, and the localization of the ester within the aggregates. The prediction of all these thermodynamic variables is described in Section IV. The last section briefly summarizes the main conclusions. In the following text, the reactions are numbered with the notation re., the physicochemical processes with the notation pr., and the equations with the notation eq.

II. MODELING OF AUTOCATALYTIC BIPHASIC ALKALINE HYDROLYSIS OF ETHYL ALKANOATES

In order to construct a kinetic model, we take into consideration all of the main reactive species and the various macroscopic physicochemical processes that are expected to occur in the oil/water biphasic hydrolysis of long-chain ethyl esters. These liquid-liquid reaction systems can be visualized as consisting of three phases or pseudophases. The organic phase, which is considered to contain only the neat ester, will be denoted as E_{org}.* The aqueous phase initially contains hydroxide and sodium ions, but during the reaction several other species accumulate: dissolved ester (E_{aq}), free surfactant molecules (S^-, denoted as S), ester-containing aggregates (ECAs) that consist of an average of g' surfactants and p ester molecules, empty micelles (M) containing an average of g surfactants, and ethanol (EtOH). In between, we have considered an interfacial pseudophase (E_{int}), which corresponds to a small volume in which the density profiles of ester and water are functions of the distance from their respective bulk phases. E_{int} corresponds to the part of the bulk organic phase that is directly in contact with the aqueous phase. The sequence of events in the biphasic alkaline hydrolysis of C-4 to C-8 ethyl alkanoates in a stirred tank reactor can be summarized by the following six main steps: formation of an ethyl ester–aqueous interface, dissolution of ester in the aqueous phase, hydrolysis of the ester, micellization of product sodium alkanoate, solubilization of ester into micelles, and micelle-mediated transport of ester into the aqueous phase. We will now develop the details of each step.

A. Formation of an Ethyl Ester-Aqueous Interface

For an undisturbed oil (ethyl ester)-water interface in a reactor, the number of ester molecules located at the interface (n_{Eint}) can be assumed proportional to the reactor cross section. On applying mechanical stirring, emulsification of oil occurs. This effect increases with increasing agitator speed. Kinetic studies of the emulsification of oil in a continuous water phase have been performed by Polat et al. [10] in a laboratory reactor relatively similar to ours. It was found that dispersion proceeds quite rapidly (with a half-time of the whole process around 1 min) and reaches a steady value asymptotically. In these conditions, n_{Eint} depends on the relative rate of stirring and coalescence, the oil/water volume ratio, and the presence of surfactants. Surfactants are known to decrease the interfacial tension by adsorption at the interface, thereby decreasing the energy needed to break a droplet and creating new interfacial areas. Taking into account all of these parameters that influence the size of the interface, we have suggested an empirical relation for n_{Eint} on phenomenological grounds. See Section III.D.2 [Eq (1)]. We note that a more detailed fundamental study of the interface is beyond the scope of this chapter.

B. Dissolution of Ester in the Aqueous Phase

The saturation solubility $[\text{Solub}]_0$ of long-chain alkyl esters in aqueous solution is generally low and decreases appreciably with increasing chain length, similar to the behavior of hydrocarbons in water. Section IV.A provides a brief summary of the thermodynamic relations governing the aqueous solubility of ethyl esters. The solubility of ethyl esters in water is increased by the presence of various additives such as ethanol

*Tests using sodium alkanoates or water have shown that their dissolution in the ester phase is negligible. Moreover, 200 Mhz NMR measurements using $CDCl_3$ as solvent have shown that during the reaction the volume ratio of ethanol within this phase always remained less than 8%.

and sodium alkanoates that enter the system as the products of the hydrolysis reaction. The increase in the solubility of the ester is due to the so-called solvent or salting-in effects [11]. Thermodynamic treatments of these solubility enhancements due to the solvent and salting-in effects are described in Section IV.B.

Table 1 lists the saturation solubility of the ethyl esters in the aqueous phase $[Solub]_0$, the correcting coefficient α that accounts for the salting-in effect, and the molar volumes of C-4 to C-8 ethyl alkanoates. For simplicity, we have assumed that the solubility enhancement due to ethanol is similar to the salting-in effect and describable by the coefficient α.

From the kinetic modeling point of view, dissolution can be seen as an equilibrium between ester in the bulk organic phase (E_{org}) and ester in the aqueous solution (E_{aq}).

$$E_{org} \rightleftharpoons E_{aq} \qquad \text{(re. 1)}$$

See Sections III.D.3 and III.D.8 (pr. 1) for more details related to this equilibrium.

C. Hydrolysis of Ethyl Ester

We have assumed that the hydrolysis reaction occurs only in the bulk water phase and can be represented as

$$E_{aq} + OH^- \rightarrow S^- + EtOH \qquad \text{(re. 2)}$$

The hydrolysis at the interface region has been neglected because of the expected lack of hydroxide ions in this pseudophase. These simplifying assumptions are supported by the studies of Sharma and Nanda [12] and Engel and Hougen [13], who used ester hydrolysis under biphasic liquid-liquid conditions in order to compare the rate of bulk hydrolysis with the rate of mass transfer across the oil-water interface.

The kinetics of alkaline hydrolysis of ethyl alkanoates have been studied by Evans et al. [14] under monophasic conditions (85% ethanol or 70% acetone in aqueous-organic solvent mixture). It has been found that the second-order rate constants increase with the water content in the reacting medium and that they are almost independent of the carbon chain length. From the reported activation energy and after extrapolation to 100% water solvent, the hydrolysis second-order rate constant in water has been estimated to be around 60 $M^{-1}\ s^{-1}$ at 80°C and 17 $M^{-1}\ s^{-1}$ at 60°C. These rate constants would be significantly modified if the hydrolysis is catalyzed. However, anionic micelles are not known to catalyze basic hydrolysis of long-chain esters because of mutual electrostatic repulsion [15] between the hydroxyl ions and the surfactant anions. Therefore, the rate constants just estimated are applicable to re. (2) in the present model.

D. Micellization of Product Sodium Alkanoates

Following Becker-Döring equations [16] or Aniansson and Wall theory [17], one can represent the micellization process by the stepwise association and dissociation of the micelles (i.e., only one surfactant monomer at a time enters and leaves the micelles):

$$S_{j-1} + S \rightleftharpoons S_j \qquad \text{(re. 3)}$$

where S_j denotes a micellar aggregate containing j surfactant monomers ($j = 2, 3, \ldots$). Assuming monodisperse micelles, the whole set of reversible addition and fragmentation processes can be contracted to a one-step micellization equilibrium. The gth order kinetics reflects the fact that micellization requires (on average) the assembly of g surfactant monomers:

$$gS \rightleftharpoons M \qquad \text{(re. 4)}$$

We have confirmed by numerical simulation that the simplified representation of a multistep aggregation equilibrium by a one-step monomer-micelle equilibrium is satisfactory for our macroscopic modeling of biphasic hydrolysis. This one-step model (re. 4) has the capacity to generate a cmc in a self-consistent manner, i.e., by its own structure.

The equilibrium parameters associated with this one-step model, namely the average micelle aggregation number g and the critical micelle concentration cmc,

TABLE 1 Saturation Solubility $[Solub]_0$, Solubility Enhancement Coefficient α, and Molar Volume of Ester V_m Obtained from Thermodynamic Calculations at 80°C and 3 M Ionic Strength

	C-4	C-5	C-6	C-7	C-8
$[Solub]_0$ (M)	5.49×10^{-3}	1.70×10^{-3}	5.30×10^{-4}	1.60×10^{-4}	5.11×10^{-5}
α (M^{-1})	0.56	0.64	0.71	0.78	0.85
V_m (M^{-1})	0.132	0.149	0.166	0.182	0.199

have been determined on the basis of predictive molecular thermodynamic calculations of aggregate size distributions carried out in nonreactive solutions. These calculations are briefly described in Section IV.C. The predicted size distributions of the sodium alkanoate aggregates are shown in Fig. 2 for chain lengths of C-4 to C-8.

The aggregate size distributions have been calculated for various values of total surfactant concentration. From the size distribution, the weight average aggregation numbers are determined and are plotted as a function of the total surfactant concentration in Fig. 3. This plot can be used to determine the cmc by identifying it as the concentration corresponding to a sharp change in the aggregation number. The cmc values are determined for various surfactant tail lengths to be approximately as follows: C-4, cmc > 3 M; C-5, cmc = 1 M; C-6, cmc = 0.3 M; C-7, cmc = 8.3×10^{-2} M; and C-8, cmc = 2.5×10^{-2} M. The weight average aggregation number (g) has been estimated at a surfactant concentration higher than the cmc for each ethyl ester: C-4, $g < 6$; C-5, $g = 12$; C-6, $g = 24$; C-7, $g = 36$; and C-8, $g = 47$. These results are in accordance with the experiments of Danielsson et al. [18], who demonstrated that association occurs in solutions of alkanoates but only for tails bearing more than four carbon atoms. See Section III.D (pr. 5).

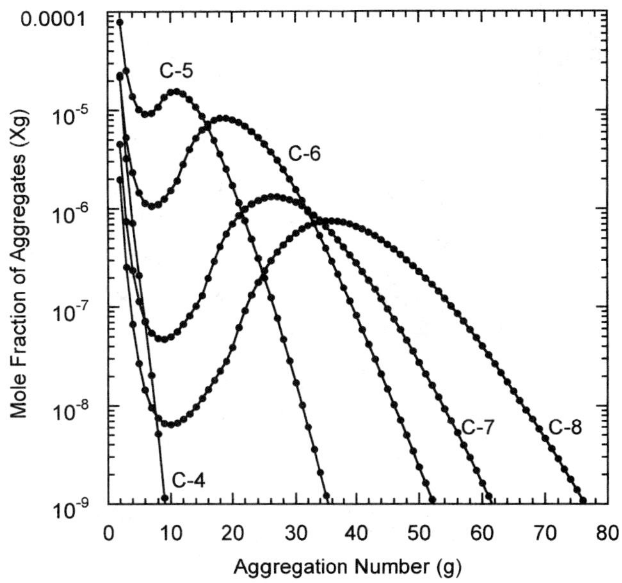

FIG. 2 Mole fraction of aggregates (X_g) versus aggregation number (g) for C-4 to C-8 alkanoates at 80°C and in the presence of 3 M NaCl salt, estimated from thermodynamic calculations. Surfactant concentrations are C-4, 1.15 M; C-5, 0.96 M; C-6, 0.39 M; C-7, 0.112 M; and C-8, 0.058 M. Note that C-4 does not micellize under these conditions.

E. Solubilization of Ester into Micelles

Solubilization refers to the large increase in the amount of solubilizate molecules in aqueous surfactant solutions (far beyond their solubility limit in water) caused by the incorporation of hydrophobic solubilizates into the micelles formed in surfactant solutions. Thermodynamic calculations discussed in Section IV.D show that ester-containing aggregates (ECAs) are formed even at low surfactant concentrations. This is indeed the case at the beginning of reaction, before the critical micelle concentration (corresponding to solubilizate-free conditions) has been reached. These calculations also suggest that the size of ECAs increases as a function of the total surfactant concentration. The increase in aggregate size results from an increase in both the average number of surfactant molecules (g') and the average number of solubilized ester molecules (p) in the micelle. Figure 4 shows the predicted weight average aggregation number of the micelles saturated with the solubilizate (ethyl ester), and Fig. 5 provides the predicted average number of ester molecules solubilized in a micelle.

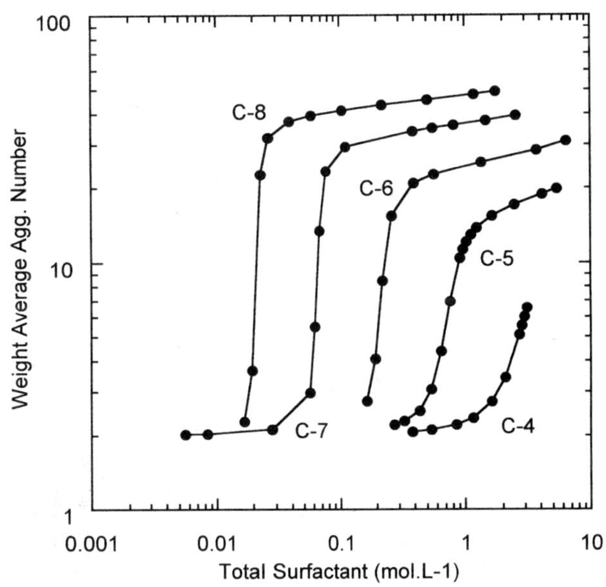

FIG. 3 Weight average aggregation number versus total surfactant concentration of various sodium alkanoates at 80°C and in the presence of 3 M ionic strength, estimated from thermodynamic calculations.

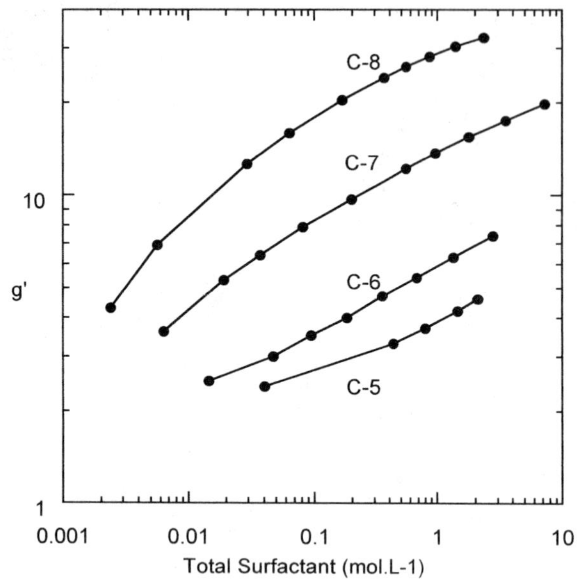

FIG. 4 Weight average aggregation number of micelles (g') in the presence of ester as solubilizate versus total concentration of surfactant at 80°C and in the presence of 3 M ionic strength, estimated from thermodynamic calculations.

The thermodynamic calculations distinguish between solubilized ester molecules located in the core (where they are protected against hydrolysis) and those in the periphery (where they are more accessible). The predicted molar ratio of the number of solubilized ester molecules in the periphery (namely the surfactant tail region of the aggregate) to that in the core is plotted in Fig. 6 as a function of the average aggregation number of the micelle for various surfactant chain lengths.

For smaller micelles, the ester molecules have a greater tendency to be located in the core. As a result, at the beginning of the reaction, smaller aggregates store ester molecules in the core, where they are protected against hydrolysis. Later, larger aggregates with more ester located in the periphery participate in the phase transfer process. The identification of different regions of micelle where solubilizate molecules can be incorporated has been discussed in the literature. For example, Mukerjee et al. [19] suggested that the interior of a micelle is divided into two regions: the surface region (palisade layer) and the core region. Depending on its polarity, the solubilizate is preferentially solubilized in either of the two regions. Nonpolar molecules such as aliphatic hydrocarbons are solubilized mainly in the core region, and more polar molecules such as

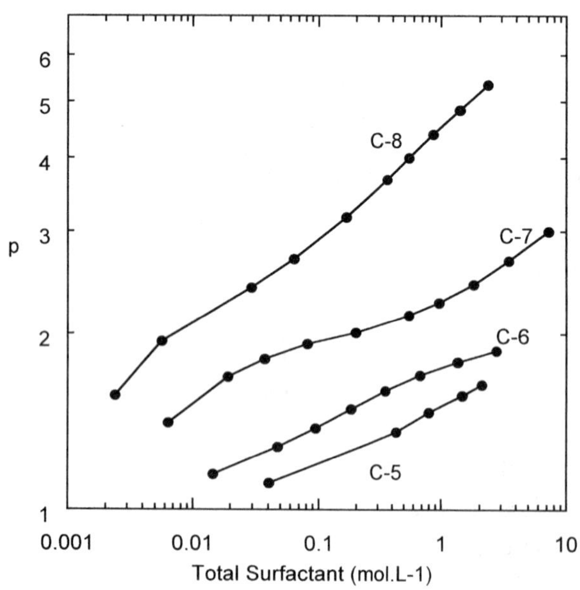

FIG. 5 Average number of ester molecules (p) solubilized in each micelle versus total concentration of surfactant at 80°C and in the presence of 3 M ionic strength, estimated from thermodynamic calculations.

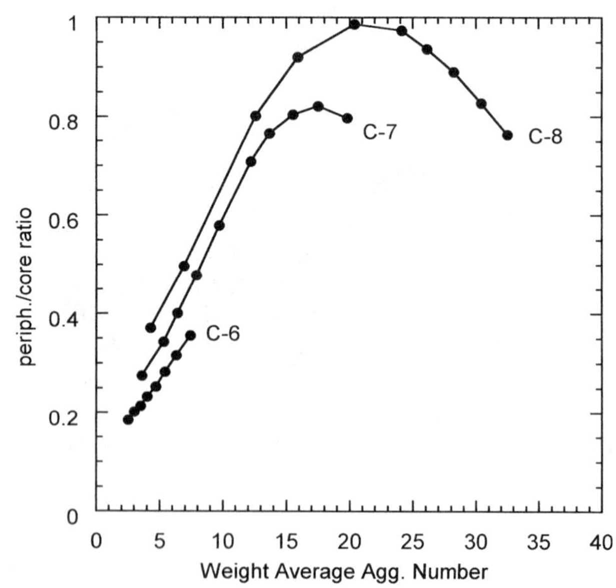

FIG. 6 Distribution of solubilizate ester between the periphery and the core (plotted as the ratio of moles in the periphery to the moles in the core) in sodium alkanoate micelles as a function of the weight average aggregation number. Note that except for C-6 (which generates only small aggregates), the periphery-to-core ratio increases and then decreases for larger micelles.

alcohols are dissolved mainly in the surface layer. The concept of two regions of solubilization has been incorporated in the thermodynamic theory of solubilization formulated by Nagarajan et al. [20] and by Jönsson et al. [21]. It has also been shown [20] that the same solubilizate molecule (for example, an aromatic hydrocarbon) can be located in both the core and the periphery of the micelle depending upon its relative amount in the micelle.

The kinetic mechanism of solubilization of oil into micelles in aqueous medium has been investigated by Karaborni et al. [22] using molecular dynamics simulations. They concluded that (1) there is dissolution of oil in the water phase before oil is captured by micelles, (2) there is exchange of oil between the oil droplet and the micelle during a soft collision, and (3) there is adsorption of surfactants on oil droplets followed by the collective desorption of surfactants and oil from the adsorbed interface. According to Plucinski and Nitsch [23], the solubilization kinetics are controlled by exchange of oil between the oil droplet (E_{org}) and the micelle (ECA) during a soft collision. The rate of solubilization is proportional to the oil solubility in water and increases when the concentration of micelles increases.

For modeling purposes, a one-step solubilization process (i.e., with a high-order kinetic rate) has been used instead of a multistep description, considering, however, two sizes of ECA. Smaller ECAs are characterized as having ester molecules stored in the core, where they are temporarily protected against hydrolysis. Owing to their small size, they show lower cooperativity, i.e., lower kinetic order. On the contrary, larger aggregates exhibit higher cooperativity and more phase transfer possibilities because in this case ester is closer to the periphery. Two parallel solubilization processes with two sizes of ECA and two cooperativities have been taken into account.

$$p_i E_{org} + g_i' S \rightleftharpoons ECA_i \qquad \text{(re. 5)}$$

where p_i and g_i' (i = 1 or 2) are, respectively, the number of ester and surfactant molecules in smaller (i = 1) or larger (i = 2) ECA_i. See Section III.D (pr 3.1 and pr 3.2).

F. Micelle-Mediated Transport of Ester into the Aqueous Phase

Hebrant and Tondre [24] have studied the transport of amino acids through liquid membranes mediated by reverse micelles. They have shown that transport dynamics seem to be governed mostly by the rate of release rather than by the rate of uptake. Kinetics of liquid-liquid extraction by micelles have been investigated by Otsuki and Seno [25]. They concluded that oil molecules transfer across the oil-water interface, diffuse for a while, and then are taken up by micelles and diffuse again. The authors found that the rate of transfer shows a strong correlation with the water solubility of oil. From this literature survey, it appears that ECAs play the role of mass transfer carriers. In these conditions, a process symmetrical with solubilization has been considered in our kinetic modeling approach. ECAs are dispersed in the bulk aqueous phase, where they dissociate releasing g' surfactant and p ester molecules. See section III.D (pr 4.1 and pr 4.2).

$$ECA_i \rightleftharpoons p_i E_{aq} + g_i' S \qquad \text{(re. 6)}$$

III. VALIDATION OF THE MODEL

Several independent experiments have been performed in order to validate the main features of our model described in the previous section.

A. Short-Chain-Length Ester: Ethyl Butanoate (C-4)

Autocatalytic kinetic behavior has been observed experimentally in the hydrolysis of ethyl butanoate, the concentration range used for this experiment being 0 to 1.8 M. Because sodium butanoate is known not to form aggregates in this concentration range, the three steps of micellization, solubilization, and micelle-mediated transport of ethyl ester are not relevant to understanding the observed kinetics. Under these conditions, the observed autocatalysis can be attributed only to salting-in and solvent effects.

At the beginning of the reaction, the rate-determining step is the dissolution process. The rate does not depend on the available quantity of oil. Hydrolysis thus takes place as soon as some ester is present in the aqueous phase. As the reaction proceeds, the products sodium butanoate and ethanol accumulate in the water phase. Consequently, the concentration of the ethyl ester in the aqueous phase increases because of the salting-in effect due to sodium butanoate and the solvent effect due to ethanol. The rate of dissolution of the ester increases continuously until all the organic phase has disappeared. One can observe from Fig. 7 that model simulations performed taking only the first three kinetic steps postulated in Section III describe quantitatively the experimental kinetic results.

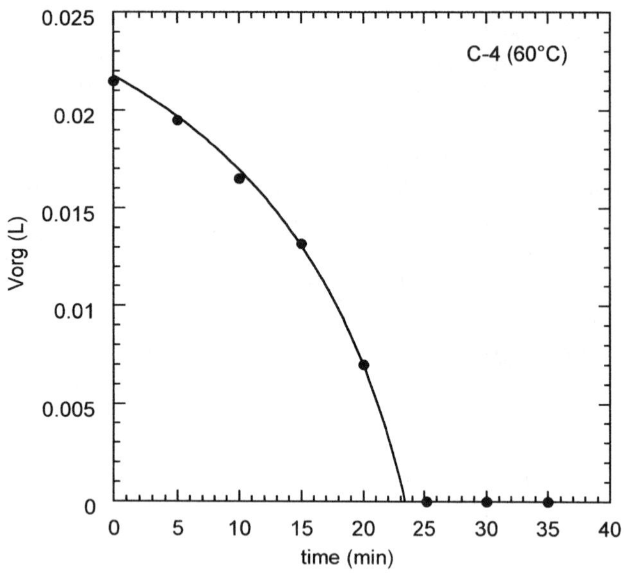

FIG. 7 Evolution of the volume of the organic phase in the biphasic hydrolysis of ethyl butanoate using 3 M NaOH aqueous solution. $T = 60°C$, $V_{org0}/V_{tot} = 22\%$, stirring rate 800 rpm in a round-bottom flask. (●) Experimental points; (——) simulation by the model; only one parameter has been fitted: $k_{-1} = 5.15 \times 10^{-1}\ s^{-1}$.

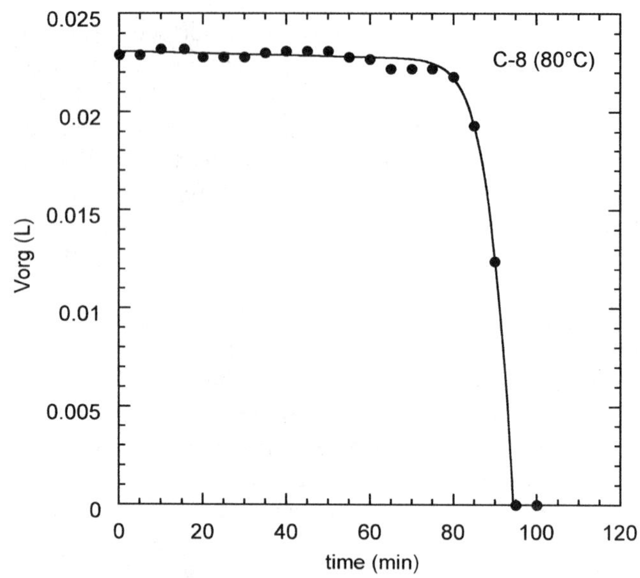

FIG. 8 Evolution of the volume of the organic phase in the biphasic hydrolysis of ethyl octanoate in a 250-mL round-bottom flask using 3 M NaOH aqueous solution. $T = 80°C$, $V_{org0}/V_{tot} = 22\%$, stirring rate 800 rpm. (●) Experimental points; (——) simulation by the model. Fitted parameters are $k_{32} = 2.87 \times 10^{41}$; $k_{-32} = 2.3 \times 10^{-2}$; $k_{42} = 1.56$; $k_{-42} = 10^{-10}$; $k_{-5} = 10^{-13}$; k_{-1} has the same value as in C-4.

B. Long-Chain-Length Ester: Ethyl Octanoate (C-8)

The reaction with ethyl octanoate shows a much more pronounced nonlinear behavior than that observed in the case of ethyl butanoate. In particular, the acceleration of the reaction rate that follows a long induction time is much stronger and abrupt. Solvent and salting-in effects are far from being sufficient to explain this strong acceleration behavior. Indeed, the contributions to autocatalysis from solvent and salting-in effects are considerably smaller (compared with the case of ethyl butanoate) because of the much lower solubility of ethyl octanoate in water, as noted previously (see Table 1). The role of ECAs must be taken into account to explain the observed autocatalysis. The ECAs ensure solubilization of ethyl octanoate in the micelles and thereby facilitate the phase transfer of ethyl octanoate from the organic to the aqueous phase, where it is rapidly hydrolyzed, releasing more and more surfactant molecules in the water phase. Figure 8 shows the result of simulations based on the model in which all the kinetic steps postulated in Section II have been included.

The model shows good quantitative agreement with the experimental data. Considering two sizes for ECAs is not necessary for a satisfactory fitting. After an induction period, during which surfactant molecules accumulate to reach a critical aggregation concentration, phase transfer mediated by ester-containing aggregates takes place and strong acceleration is observed. Finally, when all the ester has been consumed, empty micelles are formed.

C. Medium-Chain-Length Ester: Ethyl Hexanoate (C-6)

In this case, the kinetic data shown in Fig. 9 are characterized by a significant initial slope and by a large reaction extent (about 30% of ester is consumed) before any marked acceleration starts; autocatalysis seems to be inhibited for a while.

Our first attempts to fit the experimental kinetic curves using salting-in and solvent effects and only one size for the ECAs (as was done for C-8 ester hydrolysis) met with various difficulties. Specifically, if we try to reproduce the initial slope, then the overall reaction time is found to be too short, phase transfer then taking place too early in the reaction. On the other hand, if we attempt to obtain the correct overall reaction, then the initial rate becomes too small (shown as dashed line in Fig. 9). Considering only one size for

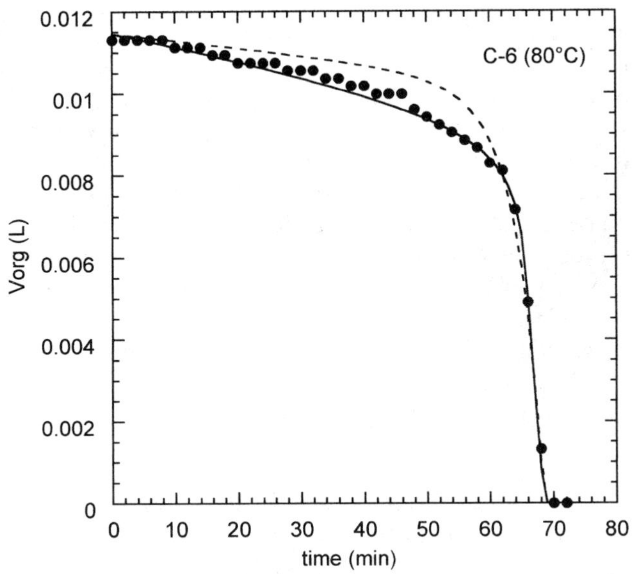

FIG. 9 Kinetics of ethyl hexanoate biphasic hydrolysis performed in a 18.9 cm^2 section cylindrical flask at 600 rpm and T = 80°C, V_{org} = 11.5 mL; V_{tot} = 50 mL. (●) Experimental points; (dashed line) best fit assuming only one size of aggregates; (continuous line) best fit assuming two sizes of aggregates. Fitted parameters are $k_1 = 116$; $k_{3.1} = 2.07 \times 10^6$; $k_{-3.1} = 64.3$; $k_{4.1} = 0.06$; $k_{-4.1} = 9.82$; $k_{3.2} = 8.19 \times 10^7$; $k_{-3.2} = 9.83 \times 10^3$; $k_{4.2} = 137$; $k_5 = 1.52 \times 10^{-4}$; Stir = 183; β = 128.

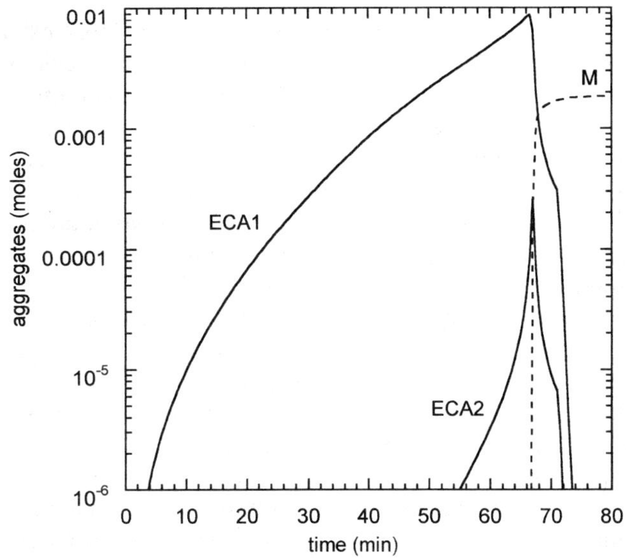

FIG. 10 Numerical simulation of the evolution of aggregates ECA_1, ECA_2, and M during biphasic hydrolysis of ethyl hexanoate. Fitted parameters are the same as in Fig. 9. Note that although the acceleration period occurs around 30% extent (a value close to the cmc of sodium hexanoate in our experimental conditions), empty micelles (M) appear only at the end of reaction.

the ECAs is therefore clearly not sufficient to describe the observed C-6 kinetics accurately. It is necessary to assume two sizes for the aggregates: ECA_1 (the smaller) are formed early in the reaction with low cooperativity and accumulate ester molecules protected from hydrolysis; ECA_2 (the larger) are formed later with higher cooperativity and contribute significantly to phase transfer. Figure 10 shows that smaller aggregates play the role of a storage reservoir.

To get further insight into this phenomenon, several complementary experiments were carried out. Experiments without emulsifying the mixture (i.e., with a well-defined interface) show that the extent of reaction before acceleration is not affected by the size of the interface (Fig. 11).

These experiments show that the size of the interface does have an effect on the overall reaction time, but they allow us to exclude any "storage" effect due to large adsorption at the interface. To check the initial volume and stirring effects, several other experiments were carried out with decreasing initial volume of ester, from 11.4 mL (0.069 mol) to 2.6 mL (0.016 mol). Because 30% extent of reaction (observed before any ac-

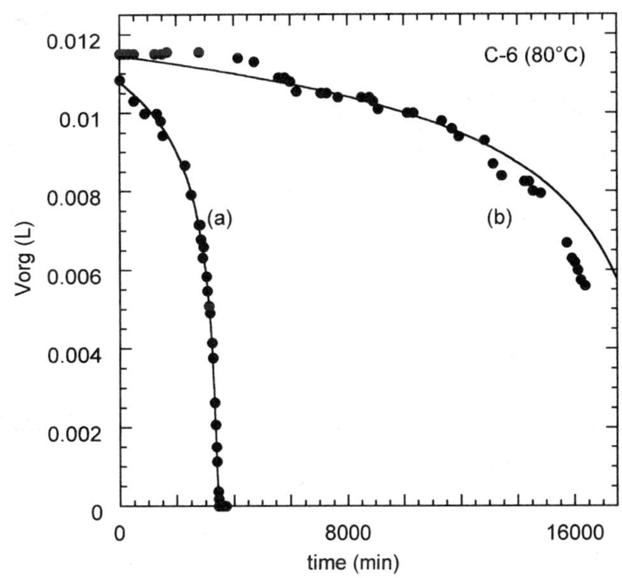

FIG. 11 Kinetics of ethyl hexanoate biphasic hydrolysis in nonemulsified conditions (Stir = 0). (a) Int_0 = cross section, Int_0 = 18.9 cm^2; (b): Int_0 = 3.3 cm^2. Fitted parameters are the same as in Fig. 9 except $k_{3.2} = 8.19 \times 10^7$; $k_{4.2} = 137$.

celeration effects) corresponds to 0.021 mole of ester consumed, one would expect no significant acceleration to be observed for experiments in which the initial number of moles of ester is smaller than this value. As shown in Fig. 12 (kinetic curves a, d, e, and f), this is indeed what is observed experimentally: the experiments (a, d, and e) show a marked acceleration around 30% extent, but the experiment (f) involving less than 0.021 mole of ester does not. Moreover, the overall reaction time increases as the initial volume of the organic phase decreases. It is suggested that this phenomenon is related to the dependence of reaction time on the size of the interface. Assuming that when the stirring rate is maintained constant (600 rpm), the size of the droplets formed by emulsification remains the same, decreasing the organic phase volume decreases the number of ester droplets and hence the interface size and therefore the overall reaction time increase. Finally, we carried out two more experiments using faster stirring rates. The kinetic curves obtained support the preceding interpretation. As expected, as the stirring rate is increased, the reaction time decreases (Fig. 12a, b, and c).

Kinetic analysis of the biphasic hydrolysis of C-5 and C-7 ethyl alkanoates has not yet been performed using the model with two aggregate sizes. But preliminary studies show that the interpretation of the kinetics of the C-5 experiments requires consideration of both salting-in and solvent effects or low cooperative phase transfer by only one size of ECA. In this particular case it can be shown that salting-in and low cooperative phase transfer have similar consequences for the observed kinetics. For C-7 experiments, it has been shown that as in the C-8 case, high-order cooperative formation of one size of ECA is the predominant process. Improvement of the fit is, however, obtained by considering a few percent of stored ester (for instance, in the smaller ECA_1 aggregates).

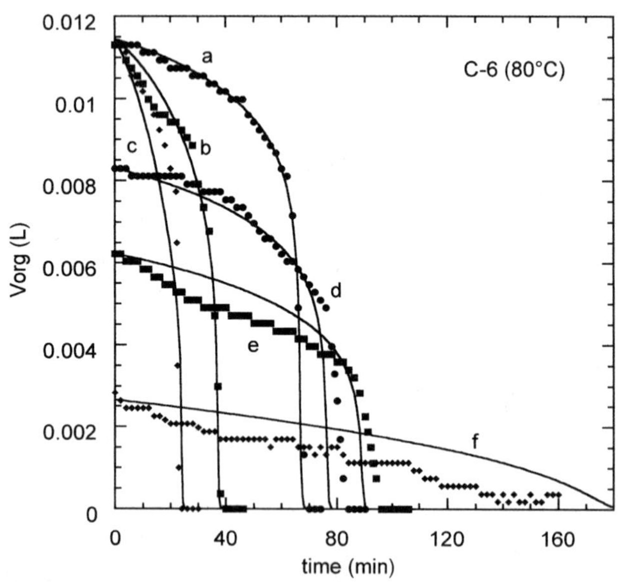

FIG. 12 Effect of initial organic volume (a) = 11.4 mL, (d) = 8.3 mL, (e) = 6.2 mL, (f) = 2.6 mL and stirring rate (a), (d), (e), and (f) = 600 rpm, (b) = 800 rpm, (c) = 1050 rpm on the kinetics of the biphasic hydrolysis of ethyl hexanoate at 80°C. Fitted parameters are the same as in Fig. 9 except for (a), (d), (e), and (f) Stir = 183; (b): Stir = 497; (c): Stir = 1013.

D. Skeleton Mechanism, Rate Laws, and Kinetic Parameters

More details of the kinetic model just discussed are given in this section. The details include notations, reaction schemes, reaction rate expressions, and assumptions made to simplify the kinetic model.

1. List of Species

E_{org}: ethyl alkanoate in the bulk organic phase
E_{aq}: ethyl alkanoate dissolved in the bulk aqueous phase
OH^-: hydroxide ion
S: free sodium alkanoates in the bulk aqueous phase
EtOH: ethanol
ECA_1: smaller ester-containing aggregates made up of g'_1 surfactants and incorporating p_1 ester molecules in the core
ECA_2: larger ester-containing aggregates made up of g'_2 surfactants and incorporating p_2 ester molecules in the periphery
M: empty micelles made up of g surfactants

2. List of Variables

$$n_{\mathrm{Eint}} = (\mathrm{Int}_0/s_0)(1 + \mathrm{Stir}(V_{\mathrm{org}}V_{\mathrm{aq}}/V^2_{\mathrm{tot}}))(1 + \beta(n_{\mathrm{S}}/V_{\mathrm{aq}})) \quad \text{(eq. 1)}$$

n_{X} = number of moles of species x
n_{Eint} = number of ethyl ester molecules at the aqueous-organic interface.

3. Processes and Rate Laws

$$\mathrm{E_{org}} \rightleftharpoons \mathrm{E_{aq}} \qquad r_1 = k_1 n_{\mathrm{Eint}} \exp(\alpha(n_{\mathrm{EtOH}} + n_{\mathrm{S}})/V_{\mathrm{aq}}) \qquad \text{(pr. 1)}$$

$$r_{-1} = k_{-1} n_{\mathrm{Eint}} n_{\mathrm{Eaq}}/V_{\mathrm{aq}}$$

$$\mathrm{E_{aq}} + \mathrm{OH^-} \rightarrow \mathrm{S} + \mathrm{EtOH} \qquad r_2 = k_2 n_{\mathrm{Eaq}}(n_{\mathrm{OH^-}})/V_{\mathrm{aq}} \qquad \text{(pr. 2)}$$

$$p_1\mathrm{E_{org}} + g'_1\mathrm{S} \rightleftharpoons \mathrm{EAC_1} \qquad r_{3.1} = k_{3.1} n_{\mathrm{Eint}}(n_{\mathrm{S}}/V_{\mathrm{aq}})^{g'_1} \qquad \text{(pr. 3.1)}$$

$$r_{-3.1} = k_{-3.1} n_{\mathrm{ECA_1}}$$

$$\mathrm{ECA}_1 \rightleftharpoons p_1\mathrm{E}_{aq} + g_1'\mathrm{S} \quad r_{4.1} = k_{4.1} n_{\mathrm{ECA}_1} \qquad \text{(pr. 4.1)}$$

$$r_{-4.1} = k_{-4.1} n_{\mathrm{Eaq}} (n_\mathrm{S}/V_{aq})^{g_1'}$$

$$p_2\mathrm{E}_{org} + g_2'\mathrm{S} \rightleftharpoons \mathrm{ECA}_2 \quad r_{3.2} = k_{3.2} n_{\mathrm{Eint}} (n_\mathrm{S}/V_{aq})^{g_2'} \qquad \text{(pr. 3.2)}$$

$$r_{-3.2} = k_{-3.2} n_{\mathrm{ECA}_2}$$

$$\mathrm{ECA}_2 \rightleftharpoons p_2\mathrm{E}_{aq} + g_2'\mathrm{S} \quad r_{4.2} = k_{4.2} n_{\mathrm{ECA}_2} \qquad \text{(pr. 4.2)}$$

$$r_{-4.2} = k_{-4.2} n_{\mathrm{Eaq}} (n_\mathrm{S}/V_{aq})^{g_2'}$$

$$g\mathrm{S} \rightleftharpoons \mathrm{M} \quad r_5 = k_5 n_\mathrm{S}^g V_{aq}^{1-g} \qquad \text{(pr. 5)}$$

$$r_{-5} = k_{-5} n_\mathrm{M}$$

where r_j is the rate of process j (in mol min^{-1}), rate constant k_j is expressed in the usual units (i.e., in L^{g-1} mol^{1-g} min^{-1} for a gth order reaction), index -j refers to the reverse process j.

4. Kinetic Parameters

Rate constants: first order: k_1, $k_{-3.1}$, $k_{4.1}$, $k_{-3.2}$, $k_{4.2}$, k_{-5}
Second order: k_{-1}, k_2
Higher order: $k_{3.1}$ and $k_{-4.1}$ (g_1 + 1), $k_{3.2}$ and $k_{-4.2}$ (g_2 + 1), k_5 (g)
Relationships between kinetic parameters:

$$k_1 = k_{-1}[\mathrm{Solub}]_0$$

$$k_5 = k_{-5}\,\mathrm{cmc}^{1-g}/g^2 \qquad \text{(Benjamin's formula [26])}$$

5. Thermodynamic Parameters

S_0: molecular area of ethyl alkanoates (around 10^4–10^5 m^2 mol^{-1})
V_m: molecular volume of ethyl alkanoates
[Solub]$_0$: saturation solubility of ethyl alkanoates in aqueous phase
cmc: critical micelle concentration of sodium alkanoates
α: correction factor for salting-in or solvent effects
β: interfacial tension correction factor
g: average aggregation number of empty micelles
g_i': average number of surfactant molecules in ester-containing aggregates ECA$_i$
p_i: average number of ester molecules in an ester-containing aggregate ECA$_i$

6. Experimental Parameters

Int$_0$: cross section of the undisturbed interface (no emulsification in experiments)
Stir: rate of stirring (emulsification in experiments)
$V_{org} = V_m n_{\mathrm{Eorg}}$ (volume of the organic phase)
$V_{aq} = V_{tot} - V_{org}$ (volume of the aqueous phase)
V_{tot} = total volume (oil + water)

7. Differential Equations

The whole set of differential equations has been built according to the matrix of the stoichiometric coefficients listed in Table 2.

8. Comments on Eq. (1) and Rate Laws

(a) *n_{Eint} and Interface Area.* n_{Eint} is given by the empirical Eq. (1). The nondimensional term $V_{org}V_{aq}/(V_{tot})^2$ accounts for the dependence of the interface area on the volume fractions of both organic (V_{org}/V_{tot}) and aqueous phases (V_{aq}/V_{tot}). We have assumed a quasi-steady-state value for the size of the interface and we have also used a linear dependence on the Stir parameter representing the intensity of mixing and the β parameter representing the influence of interfacial tension on the interfacial area generated.

(b) *Dissolution Rate of Ethyl Ester in Aqueous Phase: r_1 and r_{-1}.* The overall rate (i.e., the time constant to reach the saturation solubility) has been taken to be proportional to n_{Eint}. As the saturation solubility does not depend on the size of the interface, this factor has been put in both dissolution and release rate. The dissolution rate also depends on the concentration of the reaction products, both surfactant (salting-in effect) and ethanol (solvent effect), and follows an exponential law. Because solubility is independent of the available

TABLE 2 Matrix of the Stoichiometric Coefficients of the Whole Mechanism

d/dt	r_1	r_{-1}	r_2	r_{31}	r_{-31}	r_{32}	r_{-32}	r_{41}	r_{-41}	r_{42}	r_{-42}	r_5	r_{-5}
n_{Eorg}	−1	1	0	$-p_1$	p_1	$-p_2$	p_2	0	0	0	0	0	0
n_{Eaq}	1	−1	−1	0	0	0	0	p_1	$-p_1$	p_2	$-p_2$	0	0
$n_{\mathrm{OH^-}}$	0	0	−1	0	0	0	0	0	0	0	0	0	0
n_S	0	0	1	$-g_1'$	g_1'	$-g_2'$	g_2'	g_1'	$-g_1'$	g_2'	$-g_2'$	$-g$	g
n_{EtOH}	0	0	1	0	0	0	0	0	0	0	0	0	0
n_{ECA1}	0	0	0	1	−1	0	0	−1	1	0	0	0	0
n_{ECA2}	0	0	0	0	0	1	−1	0	0	−1	1	0	0
n_M	0	0	0	0	0	0	0	0	0	0	0	1	−1

First column corresponds to the rate of evolution dn_X/dt of the species X in mol min^{-1} units. See text for values of p_i, g_i', and g. For example, the differential equation for the first species in the table will be $dn_{\mathrm{Eorg}}/dt = -r_1 + r_{-1} - p_1 r_{31} + p_1 r_{-31} - p_2 r_{32} + p_2 r_{-32}$.

quantity of solute, the rate law r_1 is zeroth order with respect to n_{Eorg}.

(*c*) *Rate of Aggregation:* $r_{3.1}$ *and* $r_{3.2}$. In order to account for the different properties of small and large aggregates during the C-6 reaction course where surfactant concentration varies from 0 to 1.38 M, we have taken two extreme values of p and g' from Figs. 4 and 5. Noting that at low surfactant concentrations only molecular clusters such as dimers and trimers are likely, we have chosen $p_1 = 1$ and $g_1' = 2$ for ECA_1 (this corresponds to a surfactant concentration of 0.01 M at the beginning of the reaction) and $p_2 = 2$ and $g_2 = 6$ for ECA_2 (corresponding roughly to a surfactant concentration of 1.0 M at the end of reaction). Aggregation processes are assumed to be gth order (g, g_1', or g_2') in surfactant concentration. Rate laws $r_{3.1}$ and $r_{3.2}$ are zeroth order in n_{Eorg} and first order in n_{Eint} because the more the organic phase is in contact with the water bulk, the faster the exchange occurs.

(*d*) *Micelle-Mediated Transport of Ethyl Ester:* $r_{4.1}$ *and* $r_{4.2}$. The dissociation rate of ECA has been taken to be first order. Rate laws $r_{-4.1}$ and $r_{-4.2}$ have been taken first order in n_{Eaq}. Stoichiometric coefficients p have been used in the mass balance equations.

IV. THEORETICAL ESTIMATION OF THERMODYNAMIC PARAMETERS

In formulating the model for autocatalysis, a number of thermodynamic variables appear, including the solubility $[\mathrm{Solub}]_0$ of ethyl alkanoates in aqueous solution, the constant α accounting for the influence of ethyl alcohol and sodium alkanoate on the aqueous solubility of ethyl alkanoate, the average aggregation number g of a solubilizate-free micelle M, the critical micelle concentration (cmc), and the number g' of surfactant molecules and p' of ethyl alkanoate molecules in the ECA. These thermodynamic variables (listed in Table 2 or shown in Figs. 3 to 6) have not been treated in our work as free fitting parameters of the kinetic model but instead have been calculated a priori from molecular properties using molecular thermodynamic approaches. Methods of estimation of these thermodynamic variables are briefly described next.

A. Solubility of Ethyl Alkanoates in Water

The experimental solubility data of ethyl alkanoates in water at 298 K have been correlated [27] as a function of the number N of carbon atoms in the alkanoate ($N = 8$ for octanoate). One obtains $\Delta G_S^\circ = 1.317 + 0.688$ N, expressed in units of kcal/mol K, where ΔG_S° is the free energy change associated with dissolution and is equal to $RT \ln X$, where X is the mole fraction solubility of the ester in water. Experimental data on the dependence of solubility on temperature and salt concentration for ethyl alkanoates are not available. Therefore, the corrections for temperature and salt effects have been made using information available for alkanes. The corrections are made using a group contribution procedure and account for all the CH_2 and CH_3 groups in the ethyl alkanoate but ignore any correction for the COO group. In the presence of NaCl, the free energy $\Delta G_S^\circ/RT$ is estimated [28] to change by $0.384C$ for the CH_3 group and by $0.064C$ for the CH_2 group, where C is the molar concentration of the added salt. We do not have information about this correction term at other temperatures, and therefore this correction is taken as temperature independent. From solubility data for alkanes [27] we know that $\Delta G_S^\circ/RT$ for the CH_2 group is 1.496 at 298 K and 1.311 at 353 K. For the CH_3 group, $\Delta G_S^\circ/RT$ is 3.536 at 298 K and 3.548 at 353 K. Therefore the change in temperature from 298 to 353 K will cause a change in solubility given by $\Delta G_S^\circ/RT$ of -0.185 for CH_2 and 0.012 for CH_3. Taking into account the salt and temperature effects on solubility, we can calculate the solubility at 353 K and 3 M NaOH using the solubility information at 298 K and 0 M NaOH as follows:

$$
\begin{aligned}
\ln X(T = 353\ \mathrm{K}, C = 3\ \mathrm{M}) = \ln X(T = 298\ \mathrm{K}, C = 0\ \mathrm{M}) \\
+ [0.185 n_{\mathrm{CH_2}} - 0.012 n_{\mathrm{CH_3}}] \\
- C[0.064 n_{\mathrm{CH_2}} + 0.384 n_{\mathrm{CH_3}}]
\end{aligned}
$$

The first correction term is for the temperature dependence and the second is for the salt dependence. The numbers of methylene and methyl groups in ethyl alkanoate are denoted by $n_{\mathrm{CH_2}}$ and $n_{\mathrm{CH_3}}$, respectively. The mole fraction solubility data X are converted to molar concentration $[\mathrm{Solub}]_0$ by multiplying by 55.55.

B. Solubility Enhancement due to Ethyl Alcohol and Sodium Alkanoate

The solubility of ethyl alkanoate in water is affected by the presence of ethyl alcohol and sodium alkanoate, both of which are products of the hydrolysis reaction. Both contribute to an increase in the aqueous solubility of ethyl alkanoates by modifying the structure of water. To describe the influence of ethyl alcohol, we can view the problem as that of the solubility of ethyl alkanoate in a mixed solvent consisting of ethyl alcohol and water. The solubility in the mixed solvent (X_{mix}) can be represented in terms of the solubility in the pure solvents water and ethanol (X_W and X_E) by applying the framework of any suitable solution theory. For example, application of the Flory-Huggins solution model yields (eq. 3) [29]

$$\ln X_{mix} = \phi_W \ln X_W + \phi_E \ln X_E + \chi_{WE}\phi_E\phi_W \qquad \text{(eq. 3)}$$

where ϕ_W and ϕ_E denote the volume fractions of water and ethanol in the mixed solvent, and χ_{WE} is the interaction parameter between water and ethyl alcohol. Because the volume fraction of ethyl alcohol that would be present under the reaction conditions is small (for example, 1 M ethyl alcohol would represent a volume fraction of about 0.06) and the interaction parameter term is smaller than the other terms, the solubility X_{mix} can be approximately represented by the expression (eq. 4)

$$X_{mix} \approx X_W \exp\left(\phi_E \ln \frac{X_E}{X_W}\right) =$$

$$X_W \exp\left(0.0585 C_E \ln \frac{X_E}{X_W}\right) = X_W \exp(\alpha C_E) \qquad \text{(eq. 4)}$$

In obtaining the preceding expression, we have replaced ϕ_E by $0.0585C_E$, where C_E is the molar concentration of ethanol and α denotes the coefficient of C_E appearing within the exponent. The solubilities X_E and X_W are calculated using known group contributions at 25°C, namely -0.178 kT and -0.935 kT for CH_2 and CH_3 groups when ethanol is the solvent and -1.425 kT and -3.875 kT when water is the solvent. The group contribution for the COO group can be estimated for water using available solubility data, but such information is unknown in the case of ethanol as the solvent. One may anticipate that the contributions for both water and ethanol would be comparable given their affinity for the polar COO group. Consequently, the ratio X_W/X_E would not be affected significantly by the COO group contribution. Also, the temperature dependences of X_E and X_W would approximately cancel each other and thus to a first approximation, α is temperature independent.

The influence of sodium alkanoates on the solubility of ethyl alkanoates can be described by the concepts of salting in and salting out applied to solutions containing electrolytes. One can write the solubility in the presence of an electrolyte as a function of the electrolyte concentration using the relation of the form (eq. 5)

$$\ln\left(\frac{X(C)}{X(C=0)}\right) = kC \qquad \text{(eq. 5)}$$

where k is the salting-in or salting-out equilibrium constant and C is the molar concentration of the added salt. For inorganic ions, k in the preceding expression is a negative constant and the solute is salted out, implying that its solubility is lowered by the addition of salt. For an organic electrolyte (sodium alkanoate in the present case), depending on the importance of the organic part, the constant k can be positive and the solute is thus salted in, implying that its solubility is enhanced by the addition of the organic salt.

Quantitative methods for determining the salting-in constant k are not sufficiently well developed. Experimental estimates for k based on measured solubility data are thus more commonly used. In principle, k will depend on the organic ion as well as the solute molecule. The equilibrium constant k has been found to increase linearly with the alkyl chain length for large organic ions. For benzoic acid as the solute with long-chain quaternary ammonium ions, it has been found that k has a methylene group contribution of 0.07 M^{-1}. The absolute values of k lie in the range of 0.35 to 0.91 M^{-1} for total carbon numbers of 4 to 12 in the ammonium ions. We observe that the incremental variation in the parameter α that accounts for the influence of ethanol on the solubility of ethyl alkanoates is also 0.07 per methylene group (see Table 2). Because no direct measurements of k relevant to our system are presently available and the incremental variation in α compares with that in k, we assume that α can be equated to k for simplifying our calculations. Therefore, the solubility X of ethyl alkanoates in the presence of ethanol and sodium alkanoate can be written as (eq. 6)

$$X = X(\text{EtOH} = 0, \text{S} = 0)\exp \alpha([\text{EtOH}] + [\text{S}]) \qquad \text{(eq. 6)}$$

where [EtOH] and [S] are the molar concentrations of ethanol and sodium alkanoate in the aqueous solution.

C. Micellization Variables g, cmc, and K_m for Sodium Alkanoates

The aggregation characteristics of sodium alkanoates ($C_{n-1}H_{2n-1}COONa$) such as the cmc, the average aggregation number of micelles, the variance of the micelle size distribution, and the micellization equilibrium constant (step 8 in the reaction scheme) can all be predicted a priori using the molecular thermodynamic theory formulated by Nagarajan and Ruckenstein [30]. For a surfactant solution containing micelles of various aggregation numbers g, the equilibrium condition of a minimum in the Gibbs free energy stipulates (eq. 7)

$$\mu_g^\circ + kT \ln X_g = g(\mu_1^\circ + kT \ln X_1) \qquad \text{(eq. 7)}$$

where X_1 and X_g are the mole fractions of the singly dispersed molecules and aggregates of size g, respectively and μ_1° and μ_g° are their respective standard chemical potentials, defined as those corresponding to infinitely dilute solution conditions. In order to calculate X_g, we need an explicit expression for

$$\Delta\mu_g^\circ = \frac{\mu_g^\circ}{g} - \mu_1^\circ \qquad \text{(eq. 8)}$$

which is the difference in the standard chemical potential between a surfactant molecule in an aggregate of size g and a singly dispersed surfactant molecule in the solvent as a function of the size and shape of the micelles.

From a geometrical point of view, micelles of small aggregation numbers pack as spheres and larger micelles pack into globular or ellipsoidal shapes. The geometrical properties of the spherical and ellipsoidal micelles are dependent only on the aggregation number g and have been described before [30]. The standard state free energy difference term $\Delta\mu_g^\circ$ has a number of contributions that arise from the changes accompanying the aggregation process. These contributions account for the following factors: (1) the surfactant tail is removed from contact with the solvent and is transferred to the hydrophobic core of the micelle $(\Delta\mu_g^\circ)_{tr}$, (2) the surfactant tail inside the micelle has a conformation different from that in a pure hydrocarbon liquid because of packing constraints imposed inside the micelle $(\Delta\mu_g^\circ)_{def}$, (3) the formation of the micelle creates an interface between the hydrophobic micellar core and the solvent $(\Delta\mu_g^\circ)_{int}$, (4) the polar headgroups of the surfactants at the micelle surface exhibit steric repulsions $(\Delta\mu_g^\circ)_{ste}$, and (5) the polar headgroups also exhibit at the micelle surface mutual electrostatic repulsions $(\Delta\mu_g^\circ)_{ionic}$. These expressions are functions of temperature T, the molar concentration of added electrolyte C_{add}, and the micellar size (represented by the aggregation number g).

The molecular constants necessary for the predictive calculations are estimated from the molecular structure of sodium alkanoates. The molecular volume v_s and the extended length l_s of the surfactant tail is calculated [30] from the group contributions of methylene groups and the terminal methyl group. Only two molecular constants specific to a headgroup are needed. One is the cross-sectional area a_p of the headgroup, which is estimated to be 0.11 nm^2 for sodium alkanoates. The other is the distance from the hydrophobic core surface to the position where the counterion is located, which is estimated as $\delta = 0.555$ nm.

All the predicted results correspond to the experimental conditions of 80°C and the presence of 3 M NaCl in the surfactant solution. The predicted weight-average aggregation number g as a function of the total concentration X_{tot} $(=X_1 + \Sigma\, gX_g)$ of sodium alkanoate in solution is plotted in Fig. 3. The mole fractions X are converted to molar concentrations C by multiplying by 55.5. One may notice that for C-4 alkanoate, no aggregate formation occurs up to a surfactant concentration of 2 M. For other tail lengths, one can observe that the aggregation number is increasing with increasing surfactant concentration. This is a typical behavior anticipated [30] when the aggregation numbers are small. Indeed, this behavior corresponds to the presence of a somewhat polydispersed distribution of aggregates in solution. The aggregation numbers listed in Table 2 are the values predicted corresponding to a concentration of 1 M sodium alkanoate. A sharp transition in the plot of X_1 against the total concentration $X_{tot} = X_1 + \Sigma\, gX_g$ is used to predict the cmc values listed in Table 2.

D. Solubilization Variables *g′* and *p*

When sodium alkanoates and ethyl alkanoates are both present, solubilizate (ethyl alkanoate) containing aggregates (designated as ECA in the kinetic model) can form at surfactant concentrations that are lower than the cmc calculated for solubilizate-free surfactant solutions. To predict the aggregation number g' and the number of solubilizate molecules p present in an ECA aggregate, one can adopt [30] exactly the same approach as that used for micelle formation. The concentration of aggregates made up of g surfactant molecules and j solubilizate molecules can be written by analogy with the micelle size distribution equation

$$X_g = X_1^g f^{j/g} \exp - \left(\frac{g\Delta\mu_g^\circ}{kT}\right), \qquad \text{(eq. 9)}$$
$$\Delta\mu_g^\circ = \frac{\mu_{gj}^\circ}{g} - \mu_1^\circ - \frac{j}{g}\mu_o^H, \qquad f = \frac{X_1^\circ}{X_{1o}^S}$$

where μ_{gj}° is the standard chemical potential of an aggregate containing g surfactant and j solubilizate molecules, X_g is a function of both g and j, the chemical potential of the solubilizate in the pure solubilizate phase is μ_o^H, X_{1o}^S is the saturation solubility of the solubilizate in water, and f is the fractional saturation of the solubilizate in water. The molar solubilization ratio corresponding to the maximum solubilization possible is obtained by taking $f = 1$ in the preceding equation. This condition occurs when there is an excess phase of solubilizate coexisting with the aqueous solution. This is the case for our experiment, and the predicted results thus correspond to the condition $f = 1$. Knowing the distribution X_g as a function of g and j, the average aggregation number g' and the average number of solubilizate molecules p in an aggregate can be calculated.

The factor $\Delta\mu_g^\circ$ is the difference in the standard chemical potential for a surfactant molecule and (j/g) solubilizate molecules present in an aggregate with respect to a singly dispersed surfactant molecule in water and (j/g) solubilizate molecules in the bulk solubilizate phase. All the free energy contributions considered for micelle formation are relevant for the micelles containing the solubilizate, with the understanding that the geometrical properties of the aggregates are now influenced by the number of solubilizate molecules present. Further, the presence of the solubilizate modifies the interfacial tension between the micelle core and water. Also, one has to consider the entropy and the enthalpy of mixing of solubilizate and surfactant molecules in the micelle. Detailed quantitative expressions for such free energy contributions have been formulated before [30] along with geometrical relations for aggregates containing the solubilizates.

The calculated results for g' and p are plotted in Figs. 4 and 5 as functions of the total concentration of sodium alkanoate in solution. One can notice that both g' and p are dependent on the total concentration of the surfactant and increase with increasing surfactant concentration. As mentioned before, this is a feature characteristic of systems in which the aggregation numbers are small. The values listed in Table 2 that are used in the kinetic model are those predicted at a total surfactant concentration of 1 M. The solubilization model considers that some solubilizate molecules are present in the inner core of aggregates, constituting a pure pool of solubilizates, and the remaining solubilizate molecules are distributed in the region of the surfactant tail where they mix and interact with the tails (periphery). The calculated distribution of solubilizate molecules between the periphery and core regions of the aggregates is plotted in Fig. 6.

V. CONCLUSIONS

The mechanism of the biphasic hydrolysis of ethyl alkanoates has been established using thermodynamic calculations and kinetic modeling. The origin of the autocatalytic effect depends on the chain length of the ethyl ester. In the case of ethyl butanoate, the autocatalysis results from the enhancement of ester solubility caused by the reaction products sodium butanoate and ethanol, which are responsible for salting-in and solvent effects. For ethyl octanoate, phase transfer mediated by ester-containing aggregates is the main reason for the occurrence of autocatalytic behavior. In the case of esters of intermediate chain length (ethyl hexanoate), phase transfer is also the main autocatalytic process, but two sizes of ester-containing aggregates need to be invoked to explain phase transfer inhibition in the beginning of the reaction. Small aggregates temporarily store ester, and larger ones transport it into the aqueous phase, where rapid hydrolysis takes place. Pure micelles are formed as the final product when ester has been totally consumed; they have to be considered as an inactive end product. Although we are aware that drastic simplifications have been made, we think that the main reacting species, paths, couplings, and features of the mechanism have been correctly identified.

As a general property, the rate of oil-water biphasic reaction is independent of the remaining amount of supernatant organic phase. As a consequence, biphasic liquid-liquid reactions display intrinsically zero-order kinetics. If the reaction products are able to change the physicochemical properties of the medium, for instance, by increasing the saturation solubility of the organic solute into the aqueous phase, autocatalytic behavior is expected. In this respect, it is interesting to note that nonlinear kinetics in the course of phase transfer catalytic reaction were observed in 1973 by Starks and Owens [31] during the biphasic cyanide displacement on 1-halooctanes. Other liquid-liquid biphasic experiments exhibiting autocatalytic behavior have been described by Rathman et al. [32]—for instance, the synthesis of *N,N*-dimethyldodecylamine *N*-oxide from *N,N*-dimethyldodecylamine and hydrogen peroxide or the synthesis of alkylphenyl ethers from alkyl halides and phenates in two-phase systems—and by Walde et al. [33] during the formation of fatty acid vesicles from alkaline hydrolysis of octanoic and oleic anhydrides. The behavior of this last autocatalytic reaction is so striking that several authors have attempted a kinetic modeling approach. First, Mavelli and Luisi [34] assumed two interfacial reactions: a slow one at the macroscopic interface and a rapid one at the vesicle surface. Then Coveney and Wattis [35] gave a nonequilibrium macroscopic description of vesicles formation and self-replication. From a more general point of view, it can be concluded that the dynamics of all these reactions show induction periods very similar to those we have found for the biphasic alkaline hydrolysis of ethyl alkanoates. This may point out that a nonlinear phase transfer takes place in these systems as well.

Liquid-liquid biphasic reactions in which reaction products have an influence on the interfacial properties appear to be a new class of nonlinear chemical systems. Kinetic bistability in a continuous stirred tank reactor (CSTR) during the biphasic alkaline hydrolysis of ethyl octanoate [36] appears to be the first experimental example of such highly nonlinear behavior. Further examples of systems in which autocatalytic behavior is expected can be found in classical organic chemistry, such as the sulfonation of aromatic compounds or acetalization of sugars.

REFERENCES

1. P. R. Kust and J. F. Rathman, Langmuir *11*:3007–3012 (1995).
2. P. A. Bachmann, P. L. Luisi, and J. Lang, Nature *357*: 57–59 (1992).
3. J. Billingham and P. V. Coveney, J. Chem. Soc. Faraday Trans. *90*:1953–1959 (1994).
4. Y. A. Chimadzhew, M. Maestro, and F. Mavelli, Chem. Phys. Lett. *226*:56–62 (1994).
5. M. Maestro, Mol. Eng. *6*:391–403 (1996).
6. P. V. Coveney, A. N. Emerton, and B. M. Boghosian, J. Am. Chem. Soc. *118*:10719–10724 (1996).
7. P. V. Coveney and J. A. D. Wattis, Proc. R. Soc. Lond. A *452*:2079–2102 (1996).
8. P. S. Raghavan and V. S. Srinivasan, Proc. Indian Acad. Sci. (Chem. Sci.) *98*:199–206 (1987).
9. (a) T. Buhse, R. Nagarajan, D. Lavabre, and J. C. Micheau, J. Phys. Chem. *101*:3910–3917 (1997); (b) T. Buhse, D. Lavabre, R. Nagarajan, and J. C. Micheau, J. Phys. Chem. *102*:10552–10559 (1998). (c) J. Tixier, V. Pimienta, T. Buhse, D. Lavabre, R. Nagarajan, and J.-C. Micheau, Colloids and Surfaces A: Physicochem. Eng. Aspects *167*:131–142 (2000).
10. H. Polat, M. Polat, and S. Chander, AIChE J. *45*:1866–1874 (1999).
11. (a) V. Srinivas, G. A. Rodley, K. Ravikumar, W. T. Robinson, M. M. Turnbull, and D. Balasubramanian, Langmuir *13*:3235–3239 (1997); (b) V. Srinivas and D. Balasubramanian, Langmuir *14*:6658–6661 (1998).
12. M. M. Sharma and A. K. Nanda, Trans. Inst. Chem. Eng. *46*:T44–T52 (1968).
13. A. J. Engel and O. A. Hougen, AIChE J. *9*:724–729 (1963).
14. (a) D. P. Evans, J. J. Gordon, and H. B. Watson, J. Chem. Soc. 1439–1444 (1938); (b) G. Davies and D. P. Evans, J. Chem. Soc. 339–345 (1940).
15. (a) F. M. Menger and C. E. Portnoy. J. Am. Chem. Soc. *89*:4698 (1967); J. Am. Chem. Soc. *90*:1875 (1968); (b) J. H. Fendler and E. J. Fendler, *Catalysis in Micellar and Macromolecular Systems*, Academic Press, New York, 1975.
16. R. Becker and W. Döring, Ann. Phys. *24*:719 (1935).
17. E. A. G. Aniansson, S. N. Wall, M. Almgren, H. Hoffmann, I. Kielmann, W. Ulbricht, R. Zana, J. Lang, and C. Tondre, J. Phys. Chem. *80*:905–922 (1976).
18. I. Danielsson and P. Stenius, J. Colloid Interface Sci. *37*:264–280 (1971).
19. P. Mukerjee, J. R. Cardinal, and N. R. Desai, in *Micellization, Solubilization, and Microemulsions* (K. L. Mittal, ed.), Plenum, New York, 1977, p. 241.
20. R. Nagarajan, M. A. Chaiko, and E. Ruckenstein, J. Phys. Chem. *88*:2916–2922 (1984).
21. (a) M. Aamodt, M. Landgren, and B. Jönsson, J. Phys. Chem. *96*:945–950 (1992); (b) M. Landgren, M. Aamodt, and B. Jönsson, J. Phys. Chem. *96*:950–961 (1992).
22. S. Karaborni, N. M. van Os, K. Esselink, and P. A. J. Hilbers, Langmuir *9*:1175–1178 (1993).
23. P. Plucinski and W. Nitsch, J. Phys. Chem. *97*:8983–8988 (1993).
24. M. Hebrant and C. Tondre, Anal. Sci. *14*:109–115 (1998).
25. J. Otsuki and M. Seno, J. Phys. Chem. *95*:5234–5238 (1991).
26. L. Benjamin, J. Phys. Chem. *68*:3575–3581 (1964).
27. M. H. Abraham, J. Chem. Soc. Faraday Trans. 1 *80*: 153–181 (1984).
28. C. Tanford, *The Hydrophobic Effect*, Wiley, New York, 1973.
29. R. Nagarajan and E. Ruckenstein, Self-assembled systems, in *Equation of State for Fluids and Fluid Mixtures* (J. V. Sengers, ed.), Elsevier, Amsterdam, in press.
30. R. Nagarajan and E. Ruckenstein, Langmuir *7*:2934–2969 (1991).
31. C. M. Starks and R. M. Owens, J. Am. Chem. Soc. *95*: 3613–3617 (1973).
32. C. Siswanto, T. Battal, O. E. Schuss, and J. F. Rathman, Langmuir *13*:6047–6052 (1997).
33. P. Walde, J. Wick, M. Fresta, A. R. Mangone, and P. L. Luisi, J. Am. Chem. Soc. *116*:11649–11654 (1994).
34. F. Mavelli and P. L. Luisi, J. Phys. Chem. *100*:16600–16607 (1996).
35. P. V. Coveney and J. A. D. Wattis, J. Chem. Soc. Faraday Trans. *94*:233–246 (1998).
36. T. Buhse, V. Pimienta, D. Lavabre, and J. C. Micheau, J. Phys. Chem. A *101*:5215–5217 (1997).

21

Emulsion Polymerization

KLAUS TAUER Max-Planck-Institute of Colloids and Interfaces, Golm, Germany

I. INTRODUCTION

Emulsion polymerization belongs to the class of heterophase polymerizations that comprises a wide variety of different techniques generally characterized by their heterogeneous nature. This can mean liquid in gas, solid in gas, liquid in liquid, and solid in liquid heterogeneous systems. The participation of stabilizers is important in the cases in which at least one component stays liquid throughout the whole process. Considering only liquid dispersion media (or continuous phases), then suspension, microsuspension, miniemulsion, microemulsion, and dispersion polymerization belong with emulsion polymerization to the polymerization techniques leading to polymer dispersions. A polymer dispersion is considered to consist of a polymer that is finely distributed in a liquid dispersion medium in the form of stable individual particles. It does not matter whether the polymer is a solid or a highly viscous fluid at a given temperature. In principle, the liquid forming the continuous phase can be any liquid in which the polymer is insoluble. In the following, water is considered exclusively as the dispersion medium because these systems are the most technically important; furthermore, the significance of water as a dispersion medium will increase considerably in the future for environmental reasons.

It is useful to define emulsion polymerization in a general way as polymerization or copolymerization in aqueous systems of any combinations of monomers that lead to water-insoluble polymers or copolymers in the form of individual polymer particles with a size distribution of diameters in a range typically lower than 10 μm. The polymer or copolymer particles swell after nucleation with the monomers. These swollen particles represent the reaction loci where most of the monomer is polymerized. In many cases (especially in industrial systems), emulsifiers are present during the polymerization to stabilize the large interfacial area. Any kind of polymerization mechanism can be employed provided the initiation system is stable in water.

A clear demarcation from other aqueous heterophase polymerization techniques is necessary and possible. In the cases of suspension, microsuspension, miniemulsion, and microemulsion polymerization, the monomer must be only slightly water soluble as it has to form a separate phase mainly in the shape of spherical droplets whose size is controlled by proper choice of the dispersing technique (stirring, ultrasonic treatment, homogenization) in combination with the stabilizing system. The droplet size decreases in the order suspension > microsuspension > miniemulsion > microemulsion polymerization. The polymerization recipes are designed in such a way (for instance, oil-soluble instead of water-soluble initiators) that the polymerization takes place mainly inside the preformed monomer droplets. In these techniques the stabilizers have to support the emulsification process and the stabilization of the monomer droplets, whereas in case of emulsion polymerization a separate free monomer phase must not necessarily be present. Moreover, the monomer can be fed continuously into the reactor either as neat monomer or as emulsion with the additional advantage of being able to polymerize most of the time at a conversion corresponding to the polymerization rate maximum.

It is obvious that emulsion polymerization is a large topic in its own right with nearly 90 years of history and an extensive literature including monographs, textbooks, conference proceedings, and almost 800 original papers and patent applications per year. Emulsion polymerization is commercially the most important process for effecting the preparation of polymer dispersions. Almost 7% of the world polymer production is produced as a polymer dispersion, which corresponds to 10^7 tons calculated as dispersion with 50% solids content [1]. Concerning the variety of applications of polymer dispersions, the reader is referred to excellent overviews [2,3].

It is possible to accomplish an emulsion polymerization in very different ways, as illustrated by the following examples. The simplest emulsion polymerization recipe comprises only two components: a monomer that can undergo thermal polymerization (e.g., neat styrene) and water. Stirring at elevated temperatures (about 90°C) leads to polymerization mainly in the bulk monomer phase but also results in a turbid aqueous phase. The solid content of the water phase is less than 1%, and its examination by electron microscopy reveals the existence of polystyrene particles with an average size of about 100 nm.

Contrary to this example, polymerization recipes for industrially important polymer dispersions comprise up to six monomers, frequently more than two emulsifiers, more than one initiating system, and a few other aids such as biocides, defoaming agents, and plasticizers for supporting film formation [4]. The monomer-to-water ratio is adjusted in such a way that a solids content typically between 40 and 60% or even higher is obtained. The amounts of surfactants and initiator (mainly persulfate) are typically between 0.5 and 2% (w/w) relative to the monomers and 0.5% (w/w) relative to water, respectively.

According to the definition of emulsion polymerization given earlier, another limiting case for a recipe consists of an aqueous monomer solution (i.e., no free monomer phase) in the presence or absence of surfactants. The radicals can be produced either by light or by thermal decomposition of water-soluble initiators (azo- or peroxo- compounds). Indeed, such systems have been attracting researchers continuously for more than 50 years [5–12]. Depending on the particular conditions (type and concentration of components such as monomer, initiator, emulsifier, and reactor material), particles will be nucleated or not [11,12].

The words stabilizer and emulsifier and surfactant will be used interchangeably in the text to refer to a substance that is amphiphilic in nature, consisting of hydrophobic and hydrophilic parts, and is able to adsorb at any interfaces present in the aqueous phase. Stabilizers can be either polymeric or monomeric and their hydrophilic groups can be either charged or uncharged. It is not an exaggeration to say that all kinds of stabilizers have been applied in emulsion polymerizations.

This chapter is organized in such a way that after briefly considering some historical aspects, an attempt is made to outline a more general mechanistic picture of emulsion polymerization. This mechanism is discussed especially with respect to nucleation, particle swelling, and particle growth, always emphasizing the role of surfactants and emulsifiers. Then follow some general remarks regarding the role of stabilizers during both the polymerization and the application of polymer dispersions. Finally, technical realizations of emulsion polymerization are briefly described before the chapter is finished with remarks on unresolved problems and possible future developments. The peculiarities of different monomers or monomer combinations in emulsion polymerization are not subject matter of this chapter but are more general aspects of emulsion polymerization.

II. HISTORICAL DEVELOPMENT

For comprehensive and detailed information concerning the early days of emulsion polymerization, the reader is referred to excellent reviews by Whitbey and Katz [13,14] and Hohenstein and Mark [15,16]. The development of heterophase polymerization techniques is closely connected to the history of synthetic rubber. The birth of emulsion polymerization can be tracked to 1909, when attempts were made to reproduce the mild conditions during the latex synthesis employed by nature in order to improve the properties of synthetic rubbers prepared by bulk polymerization initiated with metallic sodium [17]. Consequently, the first materials used as stabilizers were naturally occurring biopolymers such as gelatin, egg albumin, starch, milk, and blood serum. These substances are rather hydrophilic protective colloids and are not typical surfactants. No catalysts were used, and the polymerization was started by keeping the reaction mixtures in an autoclave stirred for several weeks at elevated temperatures in the presence of oxygen (air). In the second half of the 1920s, work on the heterophase polymerization of dienes made significant progress when typical surface-active molecules (ammonium, sodium, and potassium oleates as well as alkyl aryl sulfonates) were applied [18–20] for the first time. Furthermore, these disclosures specify the simultaneous use of surfactants and initiators (water as well as monomer soluble peroxides) and hence can be considered as the start of catalyzed emulsion polymerization. These particular experimental results from 70 years ago that use a single surfactant were sufficient to enable the manufacture of polymeric dispersions with relatively high solids at considerably increased polymerization rates compared with surfactant-free polymerizations. This advance was an essential breakthrough in the development of emulsion polymerization techniques.

Emulsion polymerization grew rapidly in importance in the following 20 years mainly due to activities in German and U.S. companies. Once the strategic and economic importance of this polymerization technique was recognized, its further development in both countries was supported and sponsored by government funding. In Germany, the "Kunststoffkommission" (Plastics Committee) [21] was established, and in the United States the "Rubber Reserve Company; Synthetic Rubber Program of the United States Government" [22] was formed.

A special advantage of the emulsion polymerization technique was soon recognized. It is possible to obtain simultaneously both high polymerization rates and high molecular weights in such a radical polymerization process. Other technical advantages include ease of maintaining almost isothermal conditions during the polymerization because of good heat transfer through the aqueous phase and the possibility of easily removing unreacted monomers by steam stripping. Therefore this polymerization technique not only was used for dienes but also was applied to a variety of monomers and monomer mixtures including styrene [23,24], acrylic and methacrylic acid esters [25], vinyl esters [26], vinyl chloride [27], and ethylene [21]. It was also found possible to prepare a variety of polymer dispersions by homo- as well as copolymerization that were either used as dispersions or the solid polymer was further processed after coagulation. The first such example for a direct application was a polyacrylate dispersion for leather finishing [28].

During these early days of emulsion polymerization, the pace was mainly determined by industrial researchers and hence the span of time from first laboratory results to technical application was relatively short. Unfortunately, most of the early results are summarized in nonpublic research reports or published in the form of patents. There is only one report on emulsion polymerization in the open literature before 1939, and it is in the form of an abstract. Fikentscher reported on the occasion of the annual meeting of the Plastics Division of the German Chemical Society in 1938 on "Emulsion polymerization and technical exploitation" and gave interesting information concerning the knowledge at this time in Ludwigshafen [29]. First, he pointed out that the polymerization does not take place inside the monomer droplets, rather it is the monomer that is dissolved in the aqueous emulsifier solution that is polymerized. Through diffusion from the reservoir droplets the monomer concentration is maintained constant in the aqueous phase as long as droplets are present. Second, the polymerization recipe has to be fitted for each monomer or monomer combination to find optimal polymerization conditions; furthermore, if the polymer dispersion is directly applied, it is necessary to consider by the design of the polymerization recipe requirements arising from the particular application. Finally, Fikentscher emphasized the economic importance of emulsion polymerization at this time for Germany, where the amounts of directly applied polymer dispersions based on acrylic acid ester, vinylic ester, and vinyl ether monomers were increasing rapidly. However, the amount of polymer produced by emulsion polymerization but processed as powder was much greater, with Buna S (synthetic rubber) ranking first, followed by poly(vinyl chloride).

After World War II, more and more research activities to investigate the polymerization mechanism were started at universities and in independent research institutes worldwide. It is worth mentioning a few landmarks in the historic development of emulsion polymerization regarding mechanism and theory. Harkins [30,31] developed a general mechanistic picture of emulsion polymerization with two essential features. First, he considered two loci of particle formation, the monomer swollen micelles and the aqueous phase. The latter becomes more and more important with decreasing emulsifier concentration. Second, he identified the monomer swollen polymer particles as the locus in which nearly all of the polymer is formed.

The most important contribution to emulsion polymerization theory was published in 1948 by Smith and Ewart [32]. These authors developed a quantitative theory of the kinetics of radical polymerization in isolated loci (monomer swollen polymer particles) where the free radicals are supplied to the loci from an external source (aqueous phase). The centerpiece of the Smith-Ewart theory is the famous and generally valid recursion formula for the number of particles containing a given number of growing radicals and therefrom derived an average number of radicals per particle, $\bar{n}$. With respect to $\bar{n}$, three cases are of interest: (1) where $\bar{n} \ll 1$, (2) where $\bar{n} \approx 0{,}5$, and (3) where $\bar{n} \gg 1$. An approximation for the number of particles (N) formed assuming case 2 (e.g., $\bar{n} \approx 0{,}5$) suggests that N depends on the 3/5 power on the surfactant concentration (C_E), to the 2/5 power on the rate of formation of free radicals, and should decrease to the $-2/5$ power on the average growth rate of a particle. The assumptions made to derive these relations are that particle formation stops when the emulsifier micelles disappear and the rate of polymerization per particle is constant, independent of the particle size or of the rate of entrance of free radicals. Note that the constant rate of polymerization per particle means that both the monomer and the radical concentration in the particle are constant. All these assumptions are quite restrictive, and hence one would expect that these relations are fulfilled only under very special circumstances. Indeed, even for styrene, which is believed to be a very good monomer to meet the assumptions, experimental results are known confirming [33] as well as contradicting [34] the $N \propto S^{3/5}$ relation.

Gerrens and his colleagues at BASF in Ludwigshafen have contributed greatly to the knowledge of emulsion polymerization kinetics of technically important monomers [33,35–37]. Fascinating work consists of the laboratory-scale experiments on the continuous emulsion polymerization of styrene, where the oscillation phenomena in latex surface tension and particle size distribution have been carefully investigated [38,39].

The development of a quantitative theory of nonmicellar particle formation in emulsion polymerization is entirely owing to Fitch and Tsai [40–42]. In 1965, Fitch proposed that if a growing macroradical in solution becomes insoluble, it precipitates and forms a particle [43]. Initiation, capture of growing radicals by existing particles, and coagulation of the single-chain particles are considered as individual steps. A few years later, Hansen and Ugelstad contributed considerably to homogeneous particle formation in emulsion polymerization both theoretically [44] and experimentally [34,45]. This kind of homogeneous nucleation theory in emulsion polymerization is today called HUFT theory for Hansen, Ugelstad, Fitch, and Tsai. Hansen and Ugelstad also contributed to the general kinetics of emulsion polymerization with a landmark paper published in 1976 [46]. They developed a procedure to calculate $\bar{n}$ and its dependence on radical entry, exit, initiation, and termination in the aqueous phase.

Napper, Gilbert, and coworkers made many important contributions to our understanding of emulsion polymerization. They pointed out the role of coagulation of primary particles during the nucleation period [47]. Furthermore, they did a great deal on developing methods to estimate rate constants for entry, exit, termination, and propagation [48].

The Emulsion Polymers Institute at Lehigh University headed by Vanderhoff and El-Aasser over the past decades has contributed enormously to the whole field of polymer dispersions, covering nearly all topics from the kinetics of different heterophase polymerization techniques to polymer particle characterization methods and particle morphology control. The development of miniemulsion polymerization to a topic in its own right over the past 25 years is strongly connected with El-Aasser and his coworkers [49]. It is interesting to note that this development—the polymerization inside preformed monomer droplets with diameters between 100 and 300 nm—started together with Ugelstad [50].

Emulsion polymerization research has grown enormously worldwide with strong research groups working in nearly all industrialized and developing countries. Techniques have been developed to prepare extremely monodisperse particles with special functionalities for medical applications [51] and to design the particle morphology as well as the properties of the polymer dispersions in a desired way [52]. Reviews of emulsion polymerization have been published almost regularly, and it is possible to refer the reader to a short list of excellent monographs and overviews published during the past 45 years [53–59].

III. KINETICS AND MECHANISM

A. Rate of Polymerization

The basic equation of emulsion polymerization kinetics with respect to monomer conversion is Eq. (1):

$$r_P = -\frac{dM_T}{dt} = k_P C_M N_P \bar{n} + r_{P,W} \tag{1}$$

where r_P is the rate of polymerization M_T is the overall monomer concentration in M, t is the time in s, k_P is the propagation rate constant in $M^{-1}\ s^{-1}$, C_M is the monomer concentration within the latex particles in M, N_P is the particle concentration in M, and $\bar{n}$ is the dimensionless average number of radicals per particle. Note that the product $\bar{n}N_P$ corresponds to the overall radical concentration in radical solution or bulk kinetics if the polymerization in the aqueous phase is neglected. The monomer conversion rate in the aqueous phase is $r_{P,W}$ ($r_{P,W} = k_{P,W} M_W R_W$), where $k_{P,W}$ denotes the propagation rate constant in $M^{-1}\ s^{-1}$, M_W the monomer concentration in M, and R_W the radical concentration in the aqueous phase in M.

An important assumption leading to Eq. (1) is that there are only two principal reaction loci in an emulsion polymerization: the aqueous phase and the monomer swollen polymer particles. Thus, initiation of the polymerization inside the monomer droplets (if present) takes place only to a very minor extent and can be neglected at least if water-soluble initiators are used. Again, this is a completely different situation compared with suspension, microsuspension, miniemulsion, and microemulsion polymerization, where the major part of the polymerization takes place inside the monomer phase.

Equation (1) looks simple but the solution is complicated as C_M, N_P, and $\bar{n}$ are complex functions of time. Furthermore, $r_{P,W}$ depends mainly on the monomer solubility in water as well as on the partition of radicals and monomer between particles and water, respectively. A detailed discussion is, however, outside the scope of this contribution. The reader is referred to a summary of a NATO Advanced Study Institute [58] and textbooks [55–57].

Emulsion polymerizations can be carried out as discontinuous (batch), semicontinuous (feed), or continuous processes. In the case of semicontinuous processes, either the neat monomer or a monomer emulsion containing water, stabilizer, monomer, and initiator is fed into the reactor over a certain time period. Feed procedures are frequently applied in industrial polymerizations. A nice prescription for a laboratory scale feed polymerization can be found in Ref. 4. Furthermore, emulsion polymerizations can be subdivided in ab initio and seed processes. Ab initio means that particle nucleation and particle growth take place consecutively in the same reactor. A seed process means that particle nucleation and particle growth are spatially separated and the growth process is usually controlled in such a way that nucleation of new particles does not occur unless the preparation of latexes with multimodal particle size distributions is desired. There are three distinct stages during an ab initio emulsion polymerization that are generally characterized as follows:

Stage I: particle formation where $dN_P/dt > 0$
Stage II: particle growth where $dv/dt \geq 0$ (v, volume of a swollen particle)
Stage III: monomer starvation where $dC_M/dt < 0$ and $dv/dt < 0$ (shrinkage of the monomer swollen particles)

Equation (1) contains all essential quantities that are necessary to describe even quantitatively the conversion time curve of an emulsion polymerization provided k_P and the functions $C_M(t)$, $N_P(t)$, and $\bar{n}(t)$ are known. However, determining these quantities is not that easy. The minor part of the problems arises from k_P, as today, with pulsed polymerization techniques, exact determinations of k_P values are possible even under the conditions of an emulsion polymerization [60]. It is a good approximation to assume that $r_{P,W}$ is constant during stage I and stage II. With respect to $\bar{n}(t)$, the situation is easy for so-called zero-one systems in which a radical inside a particle will immediately terminate when a second radical enters. In that case $\bar{n}(t)$ = constant = 0.5 (Smith-Ewart case 2) and we have to determine only $C_M(t)$ and $N_P(t)$. It is easy to keep $N_P(t)$ constant during the polymerization (after particle formation is finished or during seed polymerizations when a seed is added and particle formation is avoided) by a proper choice of the stabilizing conditions. Concerning $C_M(t)$, the assumption is made that it is constant as long as a free monomer phase is present or the monomer feed is constant, e.g., during stage II [56,57]. However, the situation regarding $C_M(t)$ will be considered later in more detail.

These assumptions result in an $r_P(t)$ curve for an ab initio emulsion polymerization with an emulsifier concentration above the critical micelle concentration (cmc) as shown schematically in Fig. 1. As long as N_P increases, r_P increases as well (stage I). When dN_P/dt = 0, stage II is reached and the rate of polymerization is constant until the monomer concentration declines accompanied by a decrease in r_P during stage III (curve a in Fig. 1). However, it may also happen that the situation for the growing radicals inside the particles during stage III changes in such a way that r_P increases for a short period of time (curve b in Fig. 1) due to the

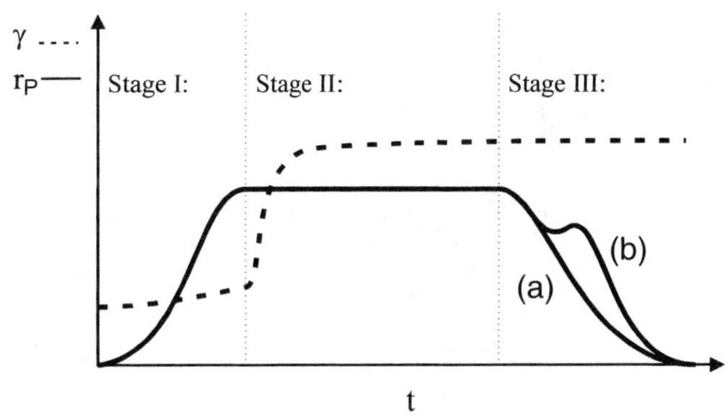

FIG. 1 Schematic drawing of the change of the rate of polymerization (r_P) and the latex surface tension (γ) with polymerization time (t) during an ab initio emulsion polymerization obeying Smith-Ewart case 2 kinetics: (a) without Norrish-Trommsdorff effect and (b) with Norrish-Trommsdorff effect.

Norrish-Trommsdorff effect or gel effect [61,62]. Reaction calorimetry allows on-line monitoring of the rate of polymerization during emulsion polymerizations. Systematic investigations of the emulsion polymerization of styrene in the presence of sodium dodecyl sulfate (SDS) at a concentration higher than the critical micelle concentration and potassium persulfate (KPS) as initiator (a classical Smith-Ewart case 2 system) revealed that the constant rate period during stage II does not necessarily occur [63]. There is experimental evidence that the end of the nucleation period (stage I) and the start of stage III (monomer starvation) take place at the same conversion range between 36 and 40%.

Similar results have been observed for butyl methacrylate emulsion polymerization started with SDS and KPS. The results depicted in Fig. 2 show no period during the polymerization with a constant polymerization rate [the heat flow (HF) corresponds directly to the rate of polymerization] and also a decrease in the rate at a conversion of about 40%. Note that the water solubilities of *n*-butyl methacrylate and styrene differ only slightly [64,65]. One characteristic feature of these reaction rate profiles is the maximum in the heat flow at the beginning of the polymerization, which is independent of the KPS concentration in the same conversion range of a few percent and which seems to be a

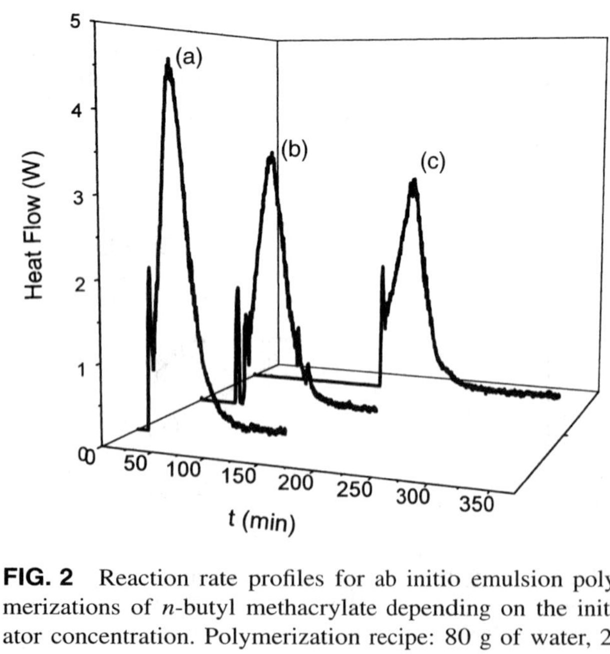

FIG. 2 Reaction rate profiles for ab initio emulsion polymerizations of *n*-butyl methacrylate depending on the initiator concentration. Polymerization recipe: 80 g of water, 20 g of *n*-butyl methacrylate, 0.2 g of SDS, and (a) 0.2 g of KPS, (b) 0.1 g of KPS, and (c) 0.05 g of KPS; calorimeter RM2-S (ChemiSens, Lund, Sweden).

special feature of this monomer in emulsion polymerization [66]. Meanwhile, many examples are known which confirm that the reaction rate profiles measured with calorimetric techniques deviate more or less from the Smith-Ewart case 2 predictions [63,67–71].

These results clearly show that the schematic drawing in Fig. 1 is not a general situation in emulsion polymerization kinetics but a special case that is only rarely fulfilled in ab initio polymerizations. The schematic drawing with respect to changes of the latex surface tension is verified by experimental results, as discussed later.

It is interestingly to note that in the case of the much more water-soluble methyl methacrylate, almost ideal behavior was observed with respect to r_P. The polymerization was started with the redox system ceric ion and poly(ethylene glycol) with a molecular weight of 10^4 g mol^{-1}. In that case a poly(ethylene glycol) radical starts the polymerization and stabilizes the resulting block copolymer particles. The reaction rate profile shows an increase in the heat flow at the beginning followed by a long time period in which the heat flow stays constant and finally an extremely high gel peak before r_P almost linearly decreases [72]. This behavior is qualitatively comparable to results of electron spin resonance investigation of methyl methacrylate emulsion polymerizations [73].

The reason for the observed deviations from the ideal Smith-Ewart case 2 behavior is the invalidity of any of the preceding assumptions. If either $N_P(t)$, $C_M(t)$, or $\bar{n}(t)$ is not constant, r_P is also not constant during stage II if it is assumed that the case in which the changes cancel one another out is rather unlikely. Consequently, the development of N_P, C_M, and $\bar{n}$ during an emulsion polymerization requires additional attention and will be discussed in the following paragraphs.

B. Particle Nucleation

Before discussing particle nucleation in detail, it seems useful to try a definition. Looking at the aqueous phase polymerizations in the absence of emulsifier and a free monomer phase, it is straightforward to identify the nucleation process as formation of the second, the polymer phase (first-order transition). Consequently, all events after this phase formation such as coagulation of the particles do not belong to particle nucleation.

Three basically different nucleation mechanisms are discussed in scientific papers:

Micellar nucleation, according to which a particle is formed per se when a radical enters a micelle [56].

Precipitation mechanism (frequently called homogeneous nucleation, HUFT theory), according to which a particle is formed when a single growing chain becomes water insoluble and precipitates [56,57].

Aggregation nucleation, according to which nucleation occurs when a critical supersaturation of growing or dead oligomers in the aqueous phase is reached and this solution becomes unstable and separates in to a polymer phase and a less concentrated aqueous phase [74]. (In contrast to the precipitation mechanism, this is a multichain process.)

The phenomenon of particle formation is probably the subject of most controversy in the scientific discussion of ab initio emulsion polymerization. There are at least two reasons for this. The first is that nucleation in emulsion polymerization is considered by most of the heterophase polymerization community to be very special and was not considered in connection with other nucleation phenomena such as bubble and droplet nucleation or crystallization processes. But especially in these fields, very well developed theoretical approaches have existed for more than 70 years [75–77] but have remained unnoticed for a long period of time. In 1975, Barrett used the classical nucleation theory (CNT) and the Flory-Huggins theory of polymer solutions to consider nucleation in dispersion polymerization without directly connecting it with polymerization kinetics [78]. This is the more surprising as the CNT was applied in 1950 by LaMer and Dinegar to a chemical reaction also leading to a colloidal system: the hydrochloric acid–catalyzed formation of sulfur hydrosols starting with thiosulfate [79].

The second reason is connected to the experimental problems of resolving particle nucleation in emulsion polymerization. Particle nucleation occurs at an extremely low conversion or solids content and there is no direct single method that allows the detection of the onset of particle nucleation. For instance, in the case of a surfactant-free styrene emulsion polymerization started with KPS, the onset of nucleation was clearly detected by a combination of on-line turbidity with on-line conductivity measurements [74,80]. In that particular case, nucleation occurred after a prenucleation period—in which aqueous phase polymerization takes place—of 431 s. At the moment of nucleation, 1.76×10^{13} particles are formed per cm^3 of water with an average particle size of 13 nm. The amount of polymer or better oligomer formed up to this moment is 2.13×10^{-5} g per cm^3 of water [81]. Most of the conclusions with respect to nucleation in emulsion polymerization have been based on experimental results obtained far away from the nucleation point. Hence, almost no hard experimental facts exist supporting or rejecting one or the other idea on nucleation. Harkins pointed out that the particles formed in the absence of micelles exhibit no essential difference in behavior from particles initiated in the presence of surfactant micelles [31].

Nucleation processes are an important part of our world. They concern researchers in meteorology, geology, biology, chemistry, and medicine, in food and beverage companies, and in the production of silicon single crystals and other crystallization processes dealing with nucleation phenomena under the umbrella of general thermodynamic considerations fitted to the particular problem. Emulsion polymerization research should make a step forward in this direction.

It is also necessary to address a problem that is a direct consequence of the lack of conversation between the nucleation community and the heterophase polymerization community. Both communities use the same technical terms but mean different things. For instance, in the heterophase polymerization community, heterogeneous and homogeneous nucleation mean nonmicellar and micellar nucleation mechanisms, respectively. In the more general point of view of the nucleation community, homogeneous nucleation occurs when interfaces have no influence, whereas in the case of heterogeneous nucleation the process is influenced by any interfaces present in the reaction system. Consequently, both "micellar" and "homogeneous" nucleation in emulsion polymerization can either occur homogeneously or heterogeneously.

It is frequently believed and stated that the Smith-Ewart theory and the relation $N_P \propto S^{3/5}$ are based on micellar nucleation. But this not true. It is entirely owing to Roe, who showed that the same scaling laws result if the assumption is made that the particles are generated by a homogeneous reaction in the water phase [82]. The emulsifier micelles serve as a reservoir from which the single emulsifier molecules diffuse to the newly created particle-water interface and impart stability to the particles. Newly formed particles can be stabilized as long as a certain amount of free (not adsorbed) surfactant is present.

There are several arguments against a micellar nucleation mechanism even if micelles are present. It was stated by Roe [82] "that some factor other than the presence of micelles exerts a strong effect on particle generation." He came to this conclusion from a simple but impressive experimental fact, namely that it is possible to generate equal particle numbers in styrene emulsion polymerizations even if the micelle concentrations in the case of chemically different emulsifiers (potassium laurate and a nonionic emulsifier at concentrations below and well above the cmc, respectively) are widely different. Twenty years earlier, Staudinger [83] carried out emulsion polymerizations of styrene and butadiene at constant initiator and emulsifier concentrations. He used potassium dodecanoate and octadecanoate, respectively, and observed a lower initial rate of polymerization for the runs with the dodecanoate soap although there was more potassium dodecanoate in the micellar form than potassium octadecanoate. Also, Robb [84] obtained evidence for nonmicellar particle nucleation in styrene emulsion polymerization with persulfate as initiator and sodium decyl sulfate and sodium dodecyl sulfate as emulsifier.

A last example is the famous paper of Priest [85] concerning the emulsion polymerization of vinyl acetate, a much more water-soluble monomer than styrene. He introduced the idea of a single-chain precipitation as nucleation mechanism followed by an interparticle combination (coalescence or coagulation) in the case of stabilizer-free polymerizations.

So, the question arises, how do surfactants contribute specifically to nucleation or to the development of the particle concentration?

Figure 3 shows a plot of ln N vs. ln C_E for styrene emulsion polymerization initiated with persulfate and varying emulsifier concentration for potassium octadecyl sulfonate (Amphoseife, curve a) and pure SDS (dodecanol free, curve b). Although the actual conditions are not identical with respect to temperature and KPS concentration, the extreme difference between both types of surfactants is surprising. Curve a proves the Smith-Ewart prediction (exponent 0.61), whereas curve b has a much higher exponent (2.67). The higher exponent indicates a higher emulsifier efficiency regarding particle stabilization of the SDS compared with the C_{18} alkyl sulfonate. These results are understandable only if the Krafft temperatures of both surfactants are considered. The Krafft temperature (T_K) is defined as the temperature at which the solubility equals the cmc and hence reflects an equilibrium between surfactants in solution and hydrated crystals. The experiments with the Amphoseife as surfactant were carried out at 45°C, but the T_K of sodium octadecyl sulfonate is between 57 and 70°C [86]. Taking into account that T_K can be shifted a little to lower temperatures (no pure water as dispersion medium and counterion influence), the polymerizations were carried out very close to T_K. Under these conditions it is very likely that not all of the surfactant was available for stabilization. In the case of SDS with a T_K between 8 and 16°C [86] the difference between T_K and the polymerization temperature (60°C) is so large that all of the surfactant is available for stabilization.

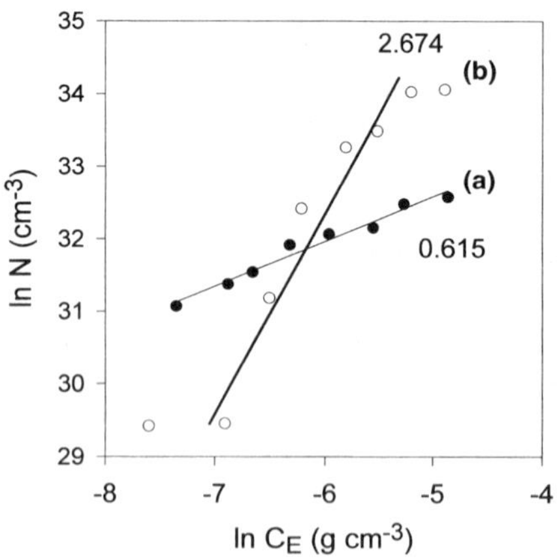

FIG. 3 Double logarithmic plot of the particle numbers at the end of a styrene ab initio emulsion polymerization (N per cm^3 of water) depending on the emulsifier concentration (C_E in grams per cm^3 of water). Curve a: emulsifier, Amphoseife C18; initiator, 0.361 g potassium peroxodisulfate per liter of water, 45°C (data from Ref. 33). Curve b: emulsifier, sodium dodecyl sulfate; initiator, 0.6 g potassium peroxodisulfate per liter of water, 60°C (data from Ref. 34).

In a series of very well designed experiments Dunn and Al-Shahib investigated the influence of the emulsifier chain length on the emulsion polymerization of styrene [87–89]. They used C_8, C_{10}, C_{12}, C_{14}, C_{16}, and C_{18} sodium alkyl sulfates at equal concentrations of 60 mM and came to the conclusion that the number of micelles initially present does not determine the particle concentration but rather the total surface area of the micelles governs the particle number. Figure 4, which combines the particle numbers of Al-Shahib and Dunn [87] with the number of micelles according to Aniansson et al. [90], shows that there is a negative or almost no correlation between both quantities. On the other hand, a strong correlation exist between the particle numbers and the cmc of the surfactants (Fig. 5a) and the adsorption energies (Fig. 5b), respectively. The adsorption energies were calculated according to Lunkenheimer et al. [91].

How surfactants assist with the time development of the particle concentration can now be answered. The cmc of a given surfactant and the adsorption energy are measures of the surface activity or of the stabilizing power for a given surfactant. The lower the cmc and the higher the adsorption energy, the higher is the surface activity. The same concentration provided the total micellar surface as discussed by Dunn [89] is also a measure of the surface activity. Consequently, a more

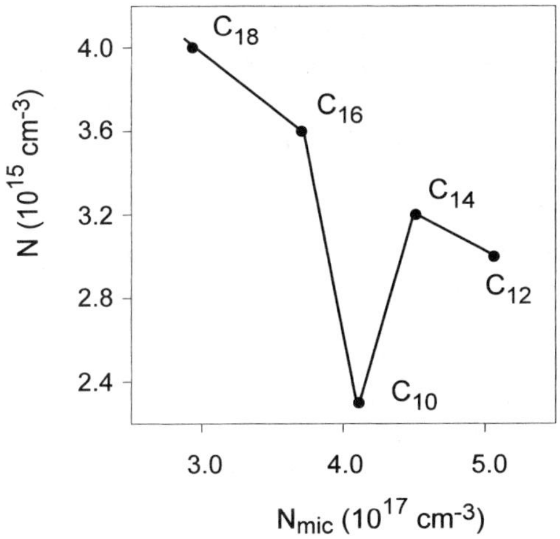

FIG. 4 Plot of the particle numbers (N per cm^3 of water) at the end of an ab initio emulsion polymerization of styrene as a function of the number of micelles (N_{mic} per cm^3 of water) for sodium alkyl sulfonate surfactants with varying alkyl chain lengths between C_{10} and C_{18T} and a concentration of 60 mM. Particle numbers are from Ref. 87 and micelle number was calculated with data from Ref. 90. Polymerization recipe: 80 mL of styrene, 200 mL of water, 12 mL of 0.1 M sodium hydroxide solution, 0.424 g of potassium peroxodisulfate, 60°C.

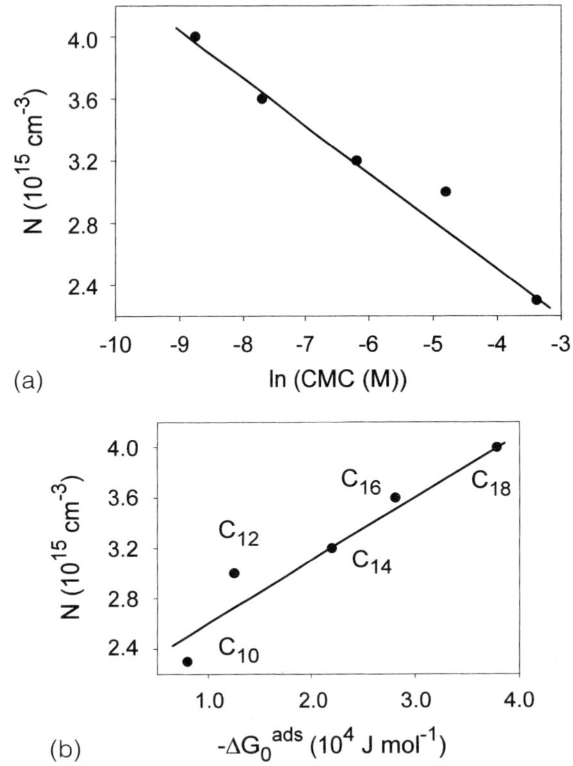

FIG. 5 (a) Dependence of the particle numbers (N per cm^3 of water) at the end of an ab initio emulsion polymerization of styrene on the logarithm of the critical micelle concentration for sodium alkyl sulfonate surfactants with varying alkyl chain lengths between C_{10} and C_{18}. Particle numbers are from Ref. 87 and the critical micelle concentrations are from Ref. 90. (b) Dependence of the particle numbers (N per cm^3 of water) at the end of an ab initio emulsion polymerization of styrene on the absorption energy (ΔG_0^{ads}) of sodium alkyl sulfonate surfactants with varying alkyl chain lengths between C_{10} and C_{18}. Particle numbers are from Ref. 87 and the adsorption energies are from Ref. 91.

plausible explanation of these results is that after particles have been formed the stabilizing power of the surfactant is responsible for the observed effects in the sense that the higher the chain length the stronger the adsorption and consequently the better the stability and the larger the surface area or the higher the number of particles that can be stabilized. Finally, a higher number of particles leads to a higher rate of polymerization.

With the CNT it is possible to answer the question of how a surfactant assists with nucleation. A centerpiece of a model for particle nucleation based on the CNT is a free energy equation for the nucleus formation (ΔG_N) [Eq. (2)].

$$\Delta G_N = (-c_1 m \ln S + c_2(mj)^{2/3}\gamma_{21} + g_{edl})g_{het} \qquad (2)$$

where m is the number of chains forming a nucleus and j is their chain length, S is the supersaturation, c_1 and c_2 are constants [92], γ_{21} is the interfacial tension between the nucleus and the dispersion medium, g_{edl} is the contribution of the electrical double layer [93], and g_{het} is the heterogeneous contribution due to the presence of substrates with interfaces [94]. As the dispersion medium in water, it is always necessary to consider the influence of charges. The g_{edl} can be expressed as $g_{edl} = -c_3\sigma\Psi_0(mj)^{2/3}$ where c_3 is a constant, σ is the surface charge density, and Ψ_0 is the surface potential. Furthermore, g_{edl} depends on the ionic strength, the dielectric constant of the dispersion medium, the size of the nucleus, and the temperature. The g_{het} can be expressed in a simple way as $g_{het} = (2 + \alpha_c)(1 - \alpha_c)^2/4$, where α_c is obtained with Young's equation as function of the contact angle (Θ) at which the nucleus makes contact with the substrate $\alpha_c = \cos\Theta = (\gamma_{31} - \gamma_{32})/\gamma_{21}$. The interfacial tensions γ_{31} and γ_{32} denote the tensions between the substrate and the dispersion medium and between the nucleus and the substrate, respectively. With the approximations $g_{edl} = 0$ and $g_{het} = 1$, Eq. (2) can be used easily together with the radical polymerization kinetics and Flory-Huggins solution theory to model particle nucleation as described [74,92].

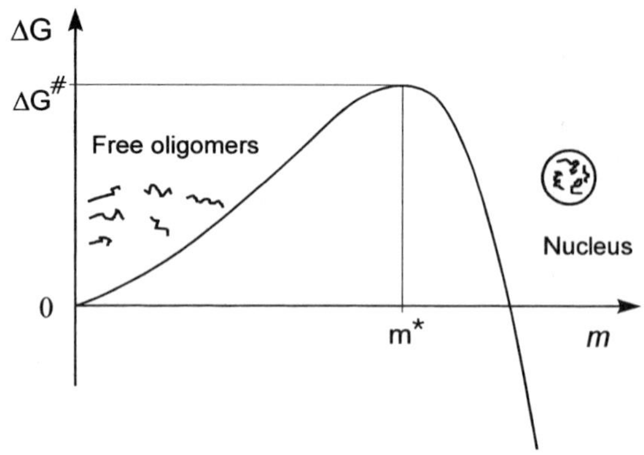

FIG. 6 Schematic drawing of the course of the free energy (ΔG) during nucleation and its dependence on the number of molecules (m) forming a nucleus. $\Delta G^{\#}$, free energy of activation of nucleation; m^{*}, number of molecules forming the critical nucleus.

An important difference between the aggregation nucleation model and the micellar and precipitation nucleation models is the occurrence of an activation energy. Figure 6 shows schematically the change of ΔG with dependence on m. The oligomers have to form a nucleus of a critical size (m^{*}) to surmount the energy barrier ($\Delta G^{\#}$). Nuclei with a size smaller than m^{*} are unstable and dissolve, whereas nuclei with a size larger than m^{*} survive and continue to grow. As the rate of nucleation is an exponential function of $\Delta G^{\#}$, the nucleation process is very sensitive to small changes in the conditions and to impurities.

Essential features of the aggregation nucleation model based on the CNT have been verified experimentally in a qualitative way in the case of styrene emulsion polymerization [74,80,95]. Consequently, the sketch in Fig. 7 reflects the course of a surfactant-free styrene emulsion polymerization. After the initiation, the reaction starts in the aqueous phase with the formation of water-soluble oligomers (prenucleation period). When a critical supersaturation is reached, nucleation occurs. The real nucleation process is very fast: in a time period of less than 1 s a huge number of particles is formed. The fate of the particles after their nucleation depends on the particular conditions. A decrease in the particle number as shown in Fig. 7 due to a coalescence process was observed in the case of a surfactant-free polymerization. Note that this is qualitatively the same development of N with the polymerization time as predicted by Priest [85] for the much more water soluble vinyl acetate. At an SDS concentration higher than the cmc, after the jump during the nucleation step a further increase in N was observed [96]. In the case of the nonionic surfactant Antarox CO880 the particle number remained unchanged after the nucleation. In that case it was possible to use conductivity measurements to detect the onset of nucleation. An almost linear decrease in the duration of the prenucleation period was measured with increasing Antarox CO880 concentration, spanning a range from below to above the cmc [74]. This is exactly the way a surfactant contributes specifically to nucleation. A surfactant can lower γ_{21} and in this way influence the free energy of nucleation. Ionic surfactants have an additional influence on nucleation as they may also change g_{edl}.

It was shown that the nucleation process in emulsion polymerization is influenced by the reactor material and hence the nucleation is heterogeneous [11]. Consequently, g_{het} is of some importance. If Θ is zero degrees, then g_{het} is zero and hence nucleation takes place

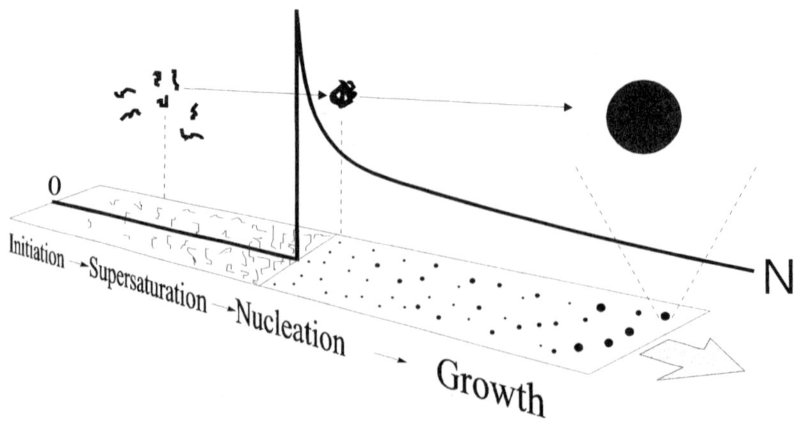

FIG. 7 Schematic drawing of the development of the particle number during a surfactant-free ab initio emulsion polymerization.

as soon as supersaturation is reached ($S \geq 1$). On the other hand, if Θ is 180 degrees g_{het} is 1 and the nucleation is homogeneous, which means that interfaces or substrates have no influence. If $0 < g_{het} < 1$, nucleation is to a certain extent heterogeneous and $\Delta G^{\#}$ is lower than in the homogeneous case.

An important quantity within the framework of the CNT is the supersaturation. S is defined as the ratio of concentration to solubility. Nucleation occurs at a critical supersaturation, which can be related to a critical nucleation concentration of oligomers in the aqueous phase (CNC). Concerning the nucleation in emulsion polymerization, these considerations lead to two important conclusions. First, for a given monomer, not only the chain length of the oligomers (j) but also the chemical nature of the end groups is important. Consequently, chemically different initiators lead to different supersaturation necessary for nucleation even for the same monomer. Furthermore, for persulfate it has been known for almost 50 years that due to its oxidizing capability, sulfate ion radicals and hydroxyl radicals are formed [97–99]. In an aqueous dispersion medium the situation is still more complicated as radicals can take part in variety of additional reactions [100]. This also seems to be true for carbon radicals [101]. It was possible to show with matrix-assisted laser desorption ionization time-of-flight mass spectroscopy that the polymer inside the particles of a styrene emulsion polymerization initiated with KPS and with 2,2′-azobis(2-amidinopropane)dihydrochloride, respectively, has besides the expected end groups a variety of other end groups such as hydroxyl, hydrogen, and carboxylic groups [81,102,103]. These reactions are the reason that in the aforementioned case of a thermally initiated surfactant-free emulsion polymerization of styrene, the particles possess some stability. Obviously, if two oligomers have the same j but one has a hydroxyl and the other a sulfate end group, their solubilities will be different. The oligomers with the lower solubility will nucleate earlier provided the critical S is reached.

Second, the surfactants and the monomers that are present in the aqueous phase also have an influence on the solubility of the oligomers. It is not only solubilization by micelles, but increased solubility is also due to the freely dissolved molecules. Before starting and during an emulsion polymerization the dispersion medium is generally not a pure aqueous phase. Under these conditions it is important that dissolved organic matter leads to an increase in the solubility of other organic compounds [104]. That this really may have an influence on nucleation was shown in the case of an aqueous phase styrene polymerization. In some cases, in the presence of surfactants, no particles have been detected, whereas in the corresponding surfactant-free polymerizations particles have been detected [11,12]. The conclusion is that surfactants may cause an increase in the CNC and under certain conditions, for instance monomer starvation, their presence can even prevent nucleation.

Nucleation means the formation of the second reaction locus—the polymer particles. This has strong consequences with respect to the distribution of all participants in the reaction system. Surfactants, monomers, and radicals have to redistribute, and according to Eq. (1) the kinetics will change as well.

C. Swelling of Polymer Particles

Just after nucleation the polymer particles start to interact with their environment. This corresponds first and foremost to the uptake of monomer. Most monomers are good solvents for their own polymer but the polymer particles in an aqueous dispersion are not dissolved by the monomer. This phenomenon resembles the behavior of macroscopic gels, which also do not dissolve upon interacting even with a good solvent. Hence, the technical term swelling is used in both cases. Note that swelling of macroscopic gels is an original topic of colloid science [105,106]. It is necessary to come back to this analogy later.

The understanding of the swelling behavior of latex particles is important for several reasons:

1. To verify the mechanism of emulsion polymerization with the monomer swollen polymer particles as the main reaction locus [30–32,107–110]
2. To control the monomer concentration at the main reaction locus [111–113]
3. To control the copolymer composition in the case of more than one monomer [114–116]
4. Because it determines the viscosity inside the particles and hence influences the kinetics [48]
5. For the preparation of large monodisperse particles [117–123]
6. For controlling particle morphology and structure [124–136]
7. For controlling and reducing the residual monomer concentration during the high conversion period of the polymerization [48,137]

Different methods have been applied to investigate the swelling of latex particles. Most popular are methods based on vapor pressure measurements to investigate both the swelling kinetics and equilibrium swelling [113,138–141]. But other methods have also been employed such as conductivity measurements [142]; scanning angle reflectometry, which allows direct observation of the uptake of the swelling agent by the particles [143]; and density measurements for investigating the equilibrium swelling [144]. Note that swelling experiments require extremely stable latexes as swelling leads to conditions favoring latex instability (increase in surface area and decrease in internal viscosity). Furthermore, as pointed out by LaMer and Gruen, investigating the swelling by means of size measurements leads to unambiguous and useful results only if the initial system is monodisperse. For polydisperse systems, the Laplace or Kelvin effect causes an increase in the polydispersity [145].

An important topic that is discussed from time to time is whether swelling leads to a homogeneous or heterogeneous distribution of the swelling agent inside the latex particles. A heterogeneous morphology is discussed for the particles in a styrene emulsion polymerization that consist of a polymer-rich core and a monomer-rich shell. The controversial discussion at the beginning of the 1970s can be followed in Refs. 146 and 147 with arguments supporting a core-shell structure and in Refs. 148 and 149 with arguments favoring a homogeneous morphology. In the middle of the 1990s, spatial inhomogeneities in poly(methyl methacrylate) latexes swollen with methyl methacrylate under equilibrium conditions were deduced from small-angle x-ray scattering data [150]. The authors concluded that there was a depletion of the polymer chains near the particle surface due to the wall repulsion effect and estimated a thickness of approximately 2 nm for the outer shell where the monomer is enriched. The answer to whether swelling leads to homogeneous particles is not easy to find. First of all it is necessary to distinguish between dynamic swelling during the polymerization and equilibrium swelling in nonpolymerizing systems. In the first case, monomer diffusion from the droplets or the gas phase has to compete with the monomer consumption due to polymerization. The rate of monomer diffusion through the aqueous phase is normally higher than the polymerization rate [111]. But there is some evidence that diffusion through the particle-water interface is hindered [111,113]. This is probably due to condensed surfactant layers. If mass transfer resistance is high, it can, of course, influence a homogeneous monomer distribution throughout the particle during polymerization. An inhomogeneous monomer distribution throughout a polymerizing particle may also occur for larger particles with higher $\bar{n}$ values. In the case of equilibrium swelling, however, an inhomogeneous monomer distribution is not easy to understand. A possible explanation will be given in the following.

Morton et al. [109] calculated the change in the chemical potential during swelling of latex particles [Eq. (3)]. Here the Flory-Rehner term [151] (last term in the brackets on the right-hand side) for cross-linking is added in order to be more general.

$$\left(\frac{2\gamma}{r}\right)\frac{V_1}{RT} = -\left[\ln(1-\phi_2) + (1-1/j)\phi_2 + \chi\phi_2^2 + \frac{V_1\rho_2}{\bar{M}_C}(\phi_2^{1/3} - \phi_2/2)\right] \qquad (3)$$

In Eq. (3) γ is the interfacial tension between particle and dispersion medium, r is the swollen particle radius, V_1 is the molar volume of the swelling agent, RT is the thermal energy, ϕ_2 is the polymer volume fraction at equilibrium, χ is the Flory-Huggins interaction parameter, ρ_2 is the polymer density, $\bar{M}_C$ is the average molecular weight between two cross-links in the network. If the latex particle is not cross-linked ($\bar{M}_C \Rightarrow \infty$), Eq. (3) is identical to the Morton-Kaizerman-Altier (MKA) equation [109].

Figure 8 compares measured monomer volume fraction (ϕ_1)–particle size (D) curves with predictions using Eq. (3), where $\phi_1 = 1 - \phi_2$ is valid. Between ϕ_1 and C_M the relation $\phi_1 = C_M V_{mon}$ holds, where V_{mon} is the molecular volume of the monomer. Note that D is

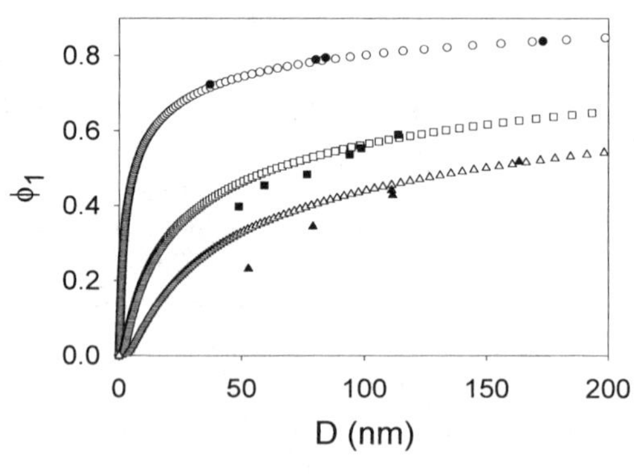

FIG. 8 Change of the monomer volume fraction inside the latex particles (ϕ_1) with the swollen particle size (D) at equilibrium swelling for polystyrene particles swollen with styrene ($j = \infty$). The open symbols are calculated data according to Eq. (3) with $\chi = 0.45$, 25°C and $\gamma = 3$ mN m^{-1} for (○), $\gamma = 30$ mN m^{-1} for (□), and $\gamma = 70$ mN m^{-1} for (△). The experimental data are from Ref. 107 for (●) and from Ref. 142 for (■) and (▲). (●) Polystyrene latex in the presence of free potassium laurate; (■) cleansed polystyrene latex with covalently bound sulfonate groups; (▲) cleansed polystyrene latex with covalently bound gluconosiloxane groups.

the swollen particle diameter, which is related to the unswollen size (D_0) according to $(D/D_0)^3 = 1/\phi_2$. The common feature of the experimental data is the increase in ϕ_1 with D. As this increase is perceptible up to particle sizes over 200 nm, some doubts arise regarding the assumption of a constant monomer concentration in the particles during interval II. A quantitative comparison of the data presented in Fig. 8 shows that at the same particle size the experimental monomer volume fraction differs up to a factor of 3. This illustrates the additional importance of parameters other than particle size and chemical nature of the polymer. The samples for series 2 and 3 were prepared with reactive surfactants and extensively cleaned to remove all residual recipe components as well as oligomers formed during the polymerization [144]. These latexes represent extremely stable neat polystyrene spheres with covalently bound stabilizing groups. In contrast, the latexes of series 1 contain such amounts of free potassium laurate emulsifier that the surface is still completely saturated at the swelling equilibrium [109]. This means that swollen micelles as well as adsolubilized swelling agent in the surfactant layer at the particle water interface lead to that high ϕ_1 values [109,152].

It was shown that the apparently increased swelling of a poly(vinyl acetate) latex with increasing emulsifier concentration was due to a specific interaction of vinyl acetate monomer with the ethoxylated nonyl phenol surfactant [153]. The vinyl acetate monomer was enriched in the surfactant layer. The swelling agent should also be enriched at the particle-water interface for entropic reasons in the presence of adsorbed surfactant. Hence, the adsorbed surfactant is probably responsible for the observed inhomogeneous particle structure at equilibrium swelling [150]. Figure 8 shows that γ is an important parameter to shift the calculated curves in the vicinity of the experimental data. This is almost perfectly possible for series 1 but not so good for series 2 and 3. Furthermore, a curve through the experimental data of series 2 and 3 will reach the origin only if the dependence of ϕ_1 on D is changed for smaller diameters. Note that the γ values needed for series 2 and 3 are much too high. For neat polystyrene surfaces the interfacial tension with respect to water is 32.7 mN m^{-1} at 20°C [154]. Even a value of 30 mN m^{-1} is much too high as the latexes of series 2 are stabilized with sulfonate groups. According to Eq. (3), it is possible to estimate γ and χ values by plotting $[-\ln(1 - \phi_2) + \phi_2]/\phi_2^2$ against $1/(r\phi_2^2)$. For series 1 this leads to reasonable values ($\gamma = 2.2$ mN m^{-1} and $\chi = 0.474$), whereas for series 2 and 3 much higher γ values and unrealistic χ parameters [96] are obtained.

In conclusion, the MKA equation with reasonable values for γ and χ in the case of neat polystyrene latexes stabilized with covalently bound sulfonate and gluconosiloxane surfactants predicts much higher monomer volume fractions than are measured. In contrast, in the case of polystyrene latexes in the presence of excess potassium laurate, where the micelles and the surfactant layer at the particle-water interface mainly contribute to swelling, the agreement between theory and experiment is excellent. A possible solution of this problem requires a modification of the MKA equation in such a way that the contribution of the polymer core to swelling is considered in greater detail.

The basic assumption of the MKA equation that only the change in the interfacial free energy counterbalances the swelling force is questionable. The crucial point is that a polymer latex particle behaves differently compared with a solvent droplet or a gas bubble or a micelle of the same size. The difference is an attractive force between the polymer chains that, in addition to the interfacial free energy, counteracts swelling. This attractive force between polymeric chain segments has been proved for a variety of polymers in semidilute solutions by osmotic compressibility measurements [155]. It is worth noting that upon dilution, the attractive force is reversed in a repulsion between the chains. A swollen latex particle represents a highly concentrated polymer solution (at $\phi_1 = 0.5$ the concentration is higher than 1 g per cm^3 swelling agent) and thus attractive forces exist between the polymer segments. Consequently, calculating the change in the chemical potential during swelling of latex particles needs an additional term to resist swelling. The simplest possibility is the introduction of $P\,dV$, the volume work, where P is a pressure and V the particle volume. The same procedure as used by Morton et al. [109] leads to Eq. (4).

$$\left(\frac{2\gamma}{r} + P\right)\frac{V_1}{RT} = -\left[\ln(1 - \phi_2) + (1 - 1/j)\phi_2 + \chi\phi_2^2 + \frac{V_1\rho_2}{\bar{M}_C}(\phi_2^{1/3} - \phi_2/2)\right] \tag{4}$$

where P can be regarded as swelling pressure similar to that for the swelling of macroscopic gels. In order to reach an equilibrium with the swelling agent, if the gel is not maximally swollen, a pressure must act on the gel, otherwise it would try to reach the maximum swelling. This pressure is called swelling pressure and is identical to the osmotic pressure of a solution in considering the gel as membrane [106]. In the case of latex particle swelling, the surface tension pressure

(γ/r) and the swelling pressure together balance the chemical potential. The limiting cases of Eq. (4) regarding $r \rightarrow \infty$ and $\phi_2 \rightarrow 0$ are the osmotic pressure or swelling pressure equation and the Young-Laplace equation, respectively. This seems to be very reasonable as it is an expression for the position of polymer colloids between colloid and polymer chemistry.

The results depicted in Fig. 9 prove that the consideration of an additional resistance to swelling is very satisfying. The theoretical curves are shifted with surface tensions of 7 and 14 mN m^{-1} in the range of the experimental data for series 2 and 3, respectively. The grading in the γ values between all series seems to be reasonable. Compared with the experiments of series 2 and 3, where the aqueous phase is pure, the experiments of series 1 contain several organic solutes leading to a decrease in γ. The difference in γ between series 2 and 3 is due to the fact that the latexes of series 2 have sulfonate groups whereas those of series 3 are sterically stabilized by gluconosiloxane groups. A first way to solve the problem regarding the change in the size dependence of ϕ_1 that still exists has been proposed in Refs. 144 and 156. Concerning the influence of surfactants, the swelling experiments described in these papers are unique, as for the first time latexes in the absence of free or adsorbed surfactants have been investigated. If surfactants are present, their specific contribution to swelling has to be taken into account.

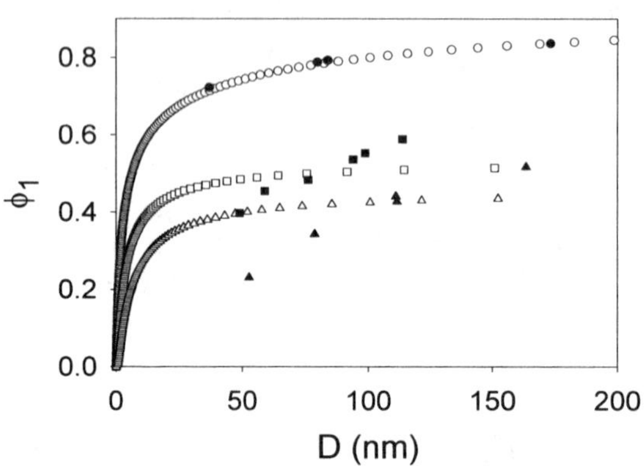

FIG. 9 Change of the monomer volume fraction inside the latex particles (ϕ_1) with the swollen particle size (D) at equilibrium swelling for polystyrene particles swollen with styrene taking into account a swelling pressure ($j = \infty$). The open symbols are calculated data according to Eq. (4) with χ = 0.45, 25°C and γ = 3 mN m^{-1} and P = 0 for (○), γ = 7 mN m^{-1} and $P = 1.5 \times 10^6$ N m^{-2} for (□), γ = 14 mN m^{-1} and $P = 2.5 \times 10^6$ N m^{-2} for (△). The experimental data are from Ref. 107 for (●) and from Ref. 142 for (■) and (▲). (●) Polystyrene latex in the presence of free potassium laurate; (■) cleansed polystyrene latex with covalently bound sulfonate groups; (▲) cleansed polystyrene latex with covalently bound gluconosiloxane groups.

Besides the particle size, the interfacial tension, and surfactants, the Flory-Huggins interaction parameter, the cross-linking of the core, and the degree of polymerization also influence the swelling of latex particles. The influence of both cross-linking and χ is straightforward as the better the solubility of the polymer, the better the swelling. Hence, cross-linked particles swell less than un-cross-linked particles and monomer-polymer combinations with high χ values swell less [vinyl chloride and poly(vinyl chloride) with $\chi \approx 0.9$, for example] than systems with χ values ≤ 0.5, which is the case for many monomer-polymer combinations.

The Flory-Huggins theory predicts that the solubility of a polymer depends on its chain length. Consequently, this is also the case for the swelling of polymer latex particles as illustrated in Fig. 10. The effect that the swelling is better the shorter the chains has an important influence during emulsion polymerization as it counteracts the low swelling tendency of the small particles just after nucleation. Furthermore, the uptake of monomer leads to an increase in the mobility of the chains and hence is a prerequisite for the formation of spherical particles if the glass transition temperature of the polymer is higher than the polymerization temperature. Figure 11 shows that well-defined polystyrene oligomers prepared by anionic polymerization terminated with sulfonate groups form nonspherical particles

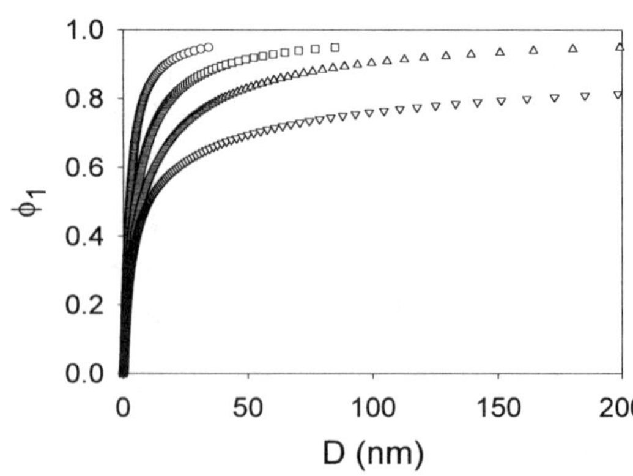

FIG. 10 Change of the monomer volume fraction inside the latex particles (ϕ_1) with the swollen particle size (D) at equilibrium swelling for polystyrene particles swollen with styrene for variation in the polymer chain length j. The open symbols are calculated data according to Eq. (4) with P = 0; χ = 0.45; 25°C; γ = 5 mN m^{-1}; and j = 2 for (○), j = 5 for (□), j = 12 for (△), and $j = 10^3$ for (▽).

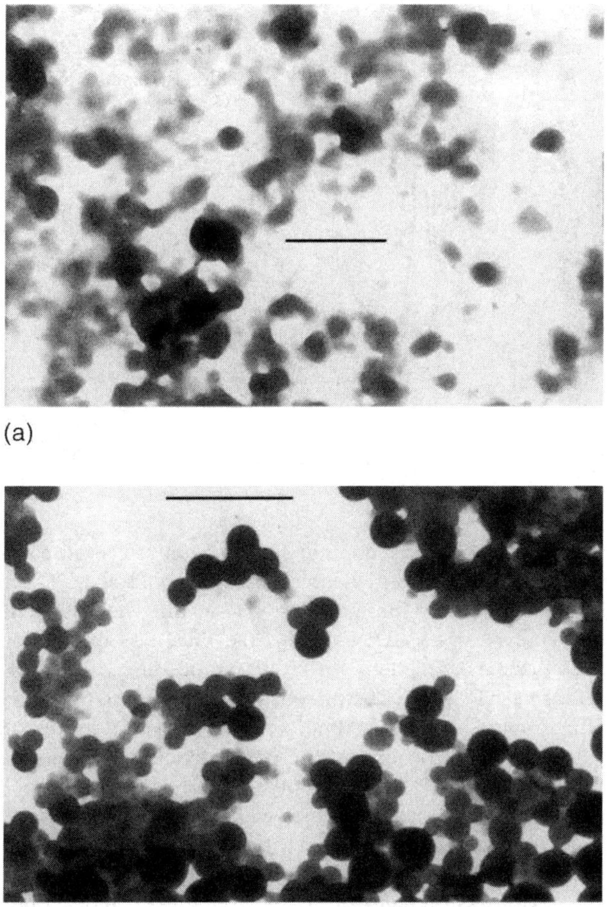

(a)

(b)

FIG. 11 Transmission electron microscopy (TEM) pictures of aqueous dispersions of chemically well-defined polystyrene oligomers with sulfonate end groups (average degree of polymerization 12). (a) Original dispersion; (b) dispersion swollen with toluene. Bars indicate 1 μm.

with a crystalline-like appearance (see Fig. 11a) upon dispersion in water [81]. Upon swelling with tetrahydrofuran, the swollen particles have a spherical shape that corresponds to the minimum free interfacial energy (see Fig. 11b).

Furthermore, the j dependence of particle swelling was successfully employed to develop methods for the preparation of large monodisperse particles by Ugelstad and coworkers [117–120]. They used a low-molecular-weight compound inside the particles (an agent promoting swelling) that is insoluble in water so that it cannot diffuse through the aqueous phase into the monomer droplets. Another way to enhance the swelling capacity of latex particle in the absence of a swelling promoter is to use a monomer emulsion of a slightly water-soluble monomer with a droplet size smaller than the size of the seed particles [121]. These monomer droplets are diffusion unstable in the presence of polymer particles and dissolve by diffusion into the polymeric seed particles. This process is similar to Ostwald ripening, where smaller particles dissolve due to the higher Laplace pressure ($2\gamma/r$) and the larger particles grow. A very interesting modification that avoids the emulsification process is the dynamic swelling method developed by Okubo and coworkers [123]. The principle of this method is that a monomer solution is treated in the presence of polymer particles in such a way that the conditions for the monomer to stay in solution become worse and tiny droplets are formed that are diffusion unstable according the principles mentioned earlier.

Swelling of colloid polymer particles is a fascinating field of colloid chemistry. Although many achievements have been made, a general description of latex particle swelling during emulsion polymerization is lacking. For instance, problems with the application of Eqs. (3) and (4) to model the swelling behavior may arise because of the assumption that γ remains unchanged upon swelling and the limitations for most of the polymer solvent systems regarding the validity of the Flory-Huggins theory. Furthermore, the quantification of the emulsifier influence that is present during the polymerization is still an open question.

D. Particle Growth

After discussing particle formation and particle swelling, the parameter that remains for further discussion according to Eq. (1) is $\bar{n}$. There is no doubt that particle growth takes place via polymerization of the monomer inside the monomer swollen particles. Hence, besides swelling, particle growth requires the presence of growing radicals inside the particles.

As mentioned earlier, the famous Smith-Ewart recursion formula describes in a generally valid way the population of radicals among the particles during an emulsion polymerization [Eq. (5)]. According to Smith and Ewart [32], the population of radicals inside the particles (n) is influenced by three reactions: entry [$dn/dt = \rho'/N$], exit [$dn/dt = -k_0 a(n/v)$], and termination inside the particles [$dn/dt = -2k_t n((n - 1)/v)$], where v and a are the volume and the surface of the particles, respectively, ρ' is the overall entry rate into all particles, and k_0 and k_t are the exit and termination rate constants, respectively. Note that the reaction rate constants are defined per volume of the average particle.

For these three reactions, the steady-state condition for the particle concentration containing n radicals is given by

$$N_{n-1}(\rho'/N) + N_{n+1}k_0a((n+1)/v) + N_{n+2}k_t((n+2)\cdot(n+1)/v) = N_n\{(\rho'/N) + k_0a(n/v) + k_tn((n-1)/v)\} \quad (5)$$

The left-hand side of Eq. (5) describes the formation of particles with n radicals and the right-hand side describes their disappearance. As $\Sigma_{n=1}^{\infty} nN$ corresponds to the total radical concentration inside particles, the average number of radicals per particle is

$$\bar{n} = \frac{1}{N}\sum_{n=1}^{\infty} nN$$

Besides the mathematical problems associated with finding an analytical solution to the recursion formula, the individual entry, exit, and termination reactions represent very complex problems. For an overview regarding the state of the art, the reader is referred to Refs. 56 and 57. Equation (5) contains more problems than mentioned explicitly. For instance, radical exit has to be treated self-consistently with chain transfer reactions occurring inside the particles, as the small radicals formed diffuse much faster through the particle and their exit is more probable compared with growing radicals. If these radicals react with monomer in the aqueous phase, the oligomers should have different properties than oligomers formed from primary initiator radicals. Consequently, one may speculate whether both types of oligomers enter the particles at the same rate. If a radical enters a growing particle, instantaneous termination takes place provided the particle is not too large. In larger particles more radicals can grow independently of each other. So the question arises whether the termination depends on both particle size and number of growing radicals per particle.

The influence of recipe components and latex properties on the average number of radicals per particle can be summarized as $\bar{n} \propto N^{-x_1}$, CTA^{-x_2}, D^{x_3}, I^{x_4} where CTA means chain transfer agent and I is the initiator concentration. The exponents x_1, x_2, x_3, and x_4 mean a certain power of the particular dependence, which is very likely not unity. The dependence of $\bar{n}$ on the emulsifier concentration is not so easy to predict as it depends on whether the particle number or the particle diameter is more affected by changing the emulsifier concentration. Stabilizers may also influence $\bar{n}$ by obstructing entry of radicals by forming a condensed or a hairy layer surrounding the particle core. The reaction rate profiles depicted in Fig. 12 give an impression of the influence of the hydrodynamic layer thickness formed by different stabilizers on the kinetics of an emulsion polymerization. Although there is a distinct difference between curves c and d, the general kinetics seem to be almost independent of whether the primary radicals have the same charge sign as the stabilizer layer or are uncharged. The thickness of the stabilizer layer has a dominating influence on the entry and exit processes and thus on the reaction rate profile.

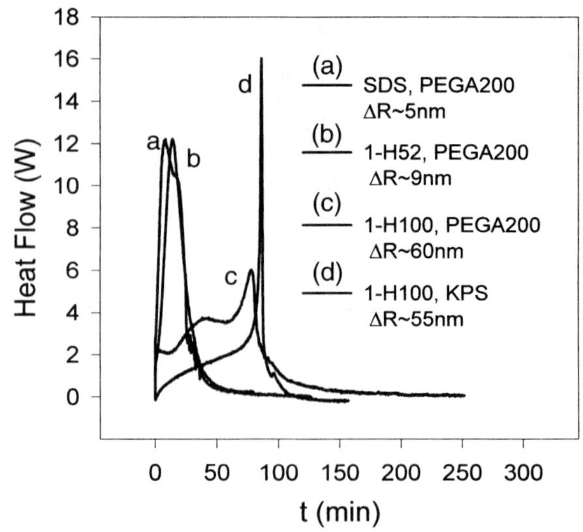

FIG. 12 Heat flow–time curves (reaction rate profiles) for ab initio emulsion polymerizations of styrene at 82°C with different stabilizer-initiator systems leading to differently decorated particles. ΔR indicates the thickness of the hairy layer. For the detailed polymerization procedure see Ref. 183. Calorimeter RM2-S (ChemiSens, Lund, Sweden). (a) SDS as initiator and a symmetrical poly(ethylene glycol) azo initiator with a molecular weight of the poly(ethylene glycol) of 200 g mol^{-1} (PEGA200). (b) Poly(styrene sulfonate)-*b*-poly-(ethylethylene) blockcopolymer with a degree of sulfonation of almost 52% and 233 styrene sulfonate units, 215 styrene units, and 42 ethyl ethylene units as stabilizer (1-H52). (c) Poly(styrene sulfonate)-*b*-poly(ethylethylene) blockcopolymer with a degree of sulfonation of almost 100% and 448 styrene sulfonate untis and 42 ethyl ethylene units as stabilizer (1-H100) and PEGA200. (d) Poly(styrene sulfonate)-*b*-poly(ethylethylene) blockcopolymer with a degree of sulfonation of almost 100% and 448 styrene sulfonate units and 42 ethyl ethylene units as stabilizer and KPS as initiator.

Monomer consumption inside the particles leads to volume growth of the individual latex particles. However, with beginning of the monomer starvation—when the continuous flow of monomer into the particles decreases—the swollen particles in the reactor start to shrink. The shrinkage is stronger the larger the density difference between monomer and polymer.

Growth by monomer consumption is in most cases the desired growth mechanism as the aim of emulsion polymerization is the conversion of monomer into polymer. But note that a particle can also grow in volume by coagulation or coalescence processes with other particles. This may happen during the polymerization as the system tries to reduce its free interfacial energy or may be done intentionally to achieve some special application properties [157]. In both cases it is important to control stability and to avoid a catastrophic coagulation. The former case frequently occurs during emulsifier-free polymerizations after nucleation as mentioned before but was also observed in apparently well-stabilized systems with emulsifier concentrations well above the cmc. For instance, a continuous decrease in the particle concentration was observed during the whole course of an ab initio vinyl chloride emulsion polymerization [158,159].

Figure 13 illustrates particle coalescence by means of an N-time curve and by a plot of the particle size versus the cubic root of the conversion (X). The decrease in N is almost over one order of magnitude, and the D-$X^{1/3}$ plot deviates from the straight line that it should follow if N is constant during the particle growth. This deviation occurs in a direction indicating stronger particle growth than by monomer consumption alone. The decrease in N occurred only in the polymerizing system. As soon as the polymerization was stopped either by the addition of radical scavengers or by turning off the monomer feed, the decrease in N stopped, too [158]. After restarting the polymerization, the particle concentration continued to decrease. This effect was called reaction-induced particle coalescence and could be modeled with the assumption that the exit of a radical is the coalescence rate-determining step [160]. It turned out that the reaction-induced particle coalescence is a special feature of vinyl chloride emulsion polymerization. A vinyl chloride polymerization is characterized by a high chain transfer constant to the monomer, and hence radical exit frequently occurs leading to local destruction of the stabilizer layer, thus lightening particle coalescence.

Well-balanced stabilization is also required during the particle growth period, just as in the case of particle swelling, as the total interfacial area increases. An increasing particle surface results in a decreasing degree of coverage of the particle interface with emulsifier (E_S). Figure 14 shows such a development in the case of an ab initio vinyl chloride polymerization in which E_S decreases to less than 10%. This result proves the schematic drawing in Fig. 1 regarding the change of γ in the course of an ab initio emulsion polymerization as a decrease in E_S also means a decrease in the concentration of the free surfactant in the aqueous phase due to the adsorption equilibrium and hence an increase in γ.

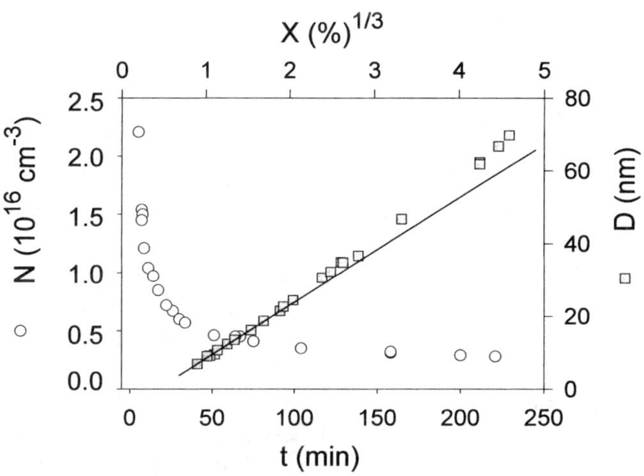

FIG. 13 Plots of the particle number (N) versus time (t) and the particle diameter (D) versus the cubic root of the conversion ($X^{1/3}$) for an ab initio emulsion polymerization of vinyl chloride at 50°C. For a description of the polymerization procedure see Ref. 156. Recipe: 4 g SDS and 1.6 g KPS per liter of water, 500 g of monomer, 690 g of water.

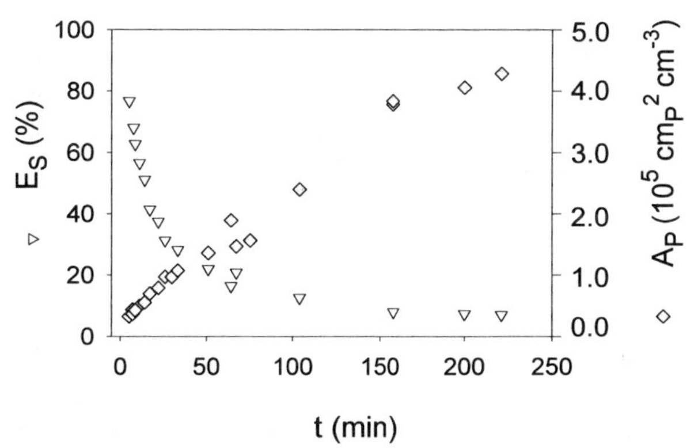

FIG. 14 Change of the total particle water interfacial area (A_P) measured as cm^2 of particle surface per cm^3 of water and of the degree of coverage of the particle surface with emulsifier during an ab initio emulsion polymerization of vinyl chloride at 50°C. For a description of the polymerization procedure see Ref. 156. Recipe: 4 g SDS and 1.6 g KPS per liter of water, 500 g of monomer, 690 g of water.

IV. TECHNICAL REALIZATIONS

Compared with condensation polymerizations or bulk polymerizations, the technical realization of emulsion polymerizations on both laboratory and industrial scales is easy. The minimum equipment needed is a container that can be closed and that is resistant to water and the monomers. In the case of gaseous monomers, the container must withstand the corresponding vapor pressure. This type of reactor, for instance, beverage bottles of different styles with crown caps or caps with a septum to allow sampling, was employed for many investigations during the early days of emulsion polymerization [161]. These so-called bottle reactors were fixed to a rotating shaft and immersed in a thermostated bath. With more than 100 bottles per shaft, it was possible to investigate many recipe variations simultaneously or to acquire conversion time data with sampling volumes large enough for the necessary latex and polymer characterizations. Nowadays for laboratory investigations stirred vessels with heating and cooling jackets are widely used, although, especially if larger sampling volumes are desired, the bottle reactors still have some advantages.

In industrial processes, stirred reactors with numerous technical variations are mainly used. Besides jacket cooling, the heat of polymerization can be removed by various techniques such as hot cooling or internal fittings inside the reactor. Reactors for commercial processes have a size between 10 and 200 m^3 and are pressure proof up to 2000 atm if necessary [4]. The most important procedures for effecting polymer dispersions on a technical scale are semibatch or feed processes, which are very flexible regarding product properties. Depending on the required properties with respect to particle size distribution, molecular weight distribution, chemical composition in the case of copolymerization, and particle morphology, numerous feeding policies have been developed. A variety of emulsion polymerizations [for example poly(vinyl chloride) and rubber production] are carried out continuously mainly in continuous stirred tank reactors (CSTRs) or in a series consisting of up to eight CSTRs. In the latter case, intermediate feed streams are frequently applied to control copolymer composition and particle morphology. A specific and very exciting feature of continuous processes in stirred tank reactors is the occurrence of damped or sustained oscillations even at the steady state. Besides CSTRs, continuous loop reactors are used because of greater reactor productivity. For instance, a 5 m^3 batch CSTR can be replaced by a 0.05 m^3 continuous loop reactor [162]. For more detailed information concerning emulsion polymerization in continuous reactors see reviews in Refs. 163–166. Today, from a commercial point of view, pure batch processes play only a minor role.

In large-scale technical reactors it is much more necessary to attach importance to the hydrodynamic conditions than in laboratory-scale reactors. If the stirring conditions are not optimized, problems can arise regarding the homogeneity of the reaction mixture as well as the heat removal in the case of insufficient mixing. In contrast, if the shear forces are too high the latex may become unstable with respect to coagulation [4]. Is there an influence of the hydrodynamic conditions or of the stirrer speed on the kinetics besides the effect that any dispersion has only a limited shear stability. It is generally assumed that agitation has no effect and this topic is ignored in almost all textbooks. However a few published results indicate an influence if the inert gas for purging contains traces of oxygen as well as if a chain transfer agent has to diffuse out of the monomer droplets into the particles [35], if the monomer diffusion to the reaction loci is influenced [167], or if the emulsifier distribution between the interfaces is changed considerably [168]. All these investigations have been carried out by increasing the stirrer speed starting from well-mixed conditions. Recently, drastic changes in both the kinetics and the polymer and particle properties have been reported when the stirrer speed is increased starting from values so low that a bulky free monomer phase still exists on top of the reactor [11,12]. Increasing the stirrer speed leads to an increase in the polymerization rate, the particle size, and the molecular weight. The particle diameters and the number average molecular weights increased from about 10 to 20 nm and from about 8×10^3 to 4×10^4 g mol^{-1}, respectively, when the stirrer speed was increased by a factor of 8. Under the same conditions, the polymerization time was reduced by a factor of 6. This was observed for a styrene emulsion polymerization (recipe: 125 g of H_2O, 1.5 g of styrene, 0.5 g of SDS, 0.2 g of KPS, 70°C) in a reaction calorimeter RM200S (ChemiSens, Lund, Sweden).

In conclusion, many different technical realizations exist to prepare polymer dispersions ranging from biosynthesis in many plants over bottle reactors up to high-tech and computer-controlled production systems. For completeness, it is necessary to mention that emulsion polymerizations have also been carried out in space under conditions of zero gravity during two *Columbia* missions in 1982 [169] under the supervision of the Lehigh Emulsion Polymers Institute.

V. ROLE OF SURFACTANTS DURING EMULSION POLYMERIZATION

There is a very special connection between emulsion polymerization and emulsifiers originating from the fact that emulsifiers have a direct influence on the course of the polymerization, the properties of the dispersions, and the properties during the final applications. Consequently, one of the most common problems in the industrial world of emulsion polymerizations is that of matching a polymer dispersion formulation by making a proper choice of surfactant or surfactant combination to obtain a required end effect. This has been known in industry since the 1930s [29], and dealing with surfactants plays a major role in the industrial preparation of polymer dispersions. The application of surfactants in polymer dispersions has a bivalent character as the demands upon the surfactants during the polymerization and during the application are contradictory.

All advantages of the emulsion polymerization techniques are directly connected with the dispersed state and hence with the colloidal stability and the stabilizing system. Some important advantages are [170]:

The existence of a thermodynamically stable large interfacial area in an aqueous environment allows easy removal of the polymerization heat.

The viscosity of the dispersion medium is low and independent of the degree of polymerization of the polymer formed.

The monomer consumption takes place outside the monomer phase, thus allowing feed procedures and hot cooling as result of which control of the polymerization is possible.

Control of the monomer concentration at the main reaction locus is possible in such a way that the reactor can operate constantly at the maximum rate of polymerization.

It is possible to carry out with comparably little technical effort batch, semibatch, and continuous polymerizations in which the feed of any additional components is possible.

The particle size and the size distribution can be tailored in a desired range between 50 nm and 10 μm.

The use of emulsifiers, water-soluble polymers, or other auxiliary materials offers further possibilities to modify the product properties.

Surfactants distribute in all phases and hence they act in all phases. In the aqueous phase they form micelles and lead to increasing solubility of organic compounds in water. As neat colloidally dispersed systems with hydrophobic interfaces attracting each other unscreened with a rate that is given by the fast coagulation kinetics first considered by von Smoluchowski [171], the most important aspect of surfactants is, however, their ability to adsorb at interfaces. Thus, they lower the interfacial tension and impart stability to particles, droplets, and bubbles. The latter may lead to foaming, which must be suppressed by the addition of defoamers. In technical polymerizations the stability is mainly endangered by electrolytes [172]. Furthermore, surfactants may interact in very specific ways with all kinds of water-soluble polymers [173].

The way stabilizers impart stability depends on the particular stabilizing system. Ionic stabilizers lead to an electrostatic repulsion between the particles, nonionic emulsifiers stabilize via a steric mechanism, and polyelectrolyte block copolymers act, for instance, as electrosteric stabilizers. Naturally occurring polymers were the first substances used to impart stability in emulsion polymerization. Subsequently, a variety of synthetic polymers were developed for applications in emulsion polymerization [174,175]. Stabilizers based on poly(ethylene glycol) (PEG) as hydrophilic blocks are widely used [175–178]. In this case the distribution of the emulsifiers between all phases may considerably influence the kinetics. Poly(ethylene glycol) is soluble in water and in many organic solvents, including many monomers, and excludes other polymers from its aqueous solutions [179]. At the elevated temperatures at which the polymerizations are carried out, it is more soluble in the monomer phase than in the aqueous phase. Consequently, it redistributes with increasing conversion more and more into the aqueous phase and hence may lead to the stabilization of a second particle generation [180,181] or to a special particle morphology as shown for PEG-*b*-polystyrene and PEG-*b*-poly(methyl methacrylate) block copolymer particles [72].

Stabilization of colloids is a large area in its own right, and for detailed and comprehensive information the reader is referred to Refs. 157 and 174. As a rule, stability against electrolytes and freeze-thaw stability are increased by nonionic stabilizers and/or polymeric stabilizers. In industrial emulsion polymerizations a few percent of an ionic comonomer is frequently employed in order to form a polymeric stabilizer during the polymerization, thus imparting additional stability. As a new class of polymeric stabilizers, polyelectrolyte block copolymers are worthy of brief mention. If the polyelectrolyte block is long and stretched into the aqueous phase, as in the case of curves c and d of Fig.

11, where the layer thickness is about 60 nm, the particles possess extraordinary salt stability [182]. For instance, these particles are stable over months in the presence of 3 M sodium chloride.

In contrast to the preparation of polymer dispersions, for most of their applications, colloidal instability is required at a certain moment during the application as in the case of film formation. In a coating the surfactants can migrate through the polymer layer and assemble in hydrophilic domains, leading to enhanced water uptake and subsequently to blooming or blushing [183,184]. For most applications it would be very useful if the stabilizers, which have to contain hydrophilic parts as long as the polymer is dispersed in the aqueous environment, would lose these parts and hence their hydrophilicity together with the evaporating water during the application. A possible way to avoid this drawback of surfactants during the application is to fix them in place. This may be done by using either reactive surfactants that are covalently bound to the polymer chains (see Chapter 28) or high-molecular-weight surfactants that are unable to migrate considerably. For instance, a polystyrene with a molecular weight of 4×10^4 g mol^{-1} and subsequently sulfonated to a degree of about 50% acts as a good stabilizer in emulsion polymerization [185]. It forms no extended hydrodynamic layer because of multiple adsorption points along the chain and results in a reaction rate profile similar to that in the case of SDS as emulsifier (see Fig. 11, curves a and b).

A problem that is inherent in commercial surfactants is the extreme difficulty of preparing these surfactants in a chemically pure form. Some of the main sources of the imperfections are homologues as well as isomers of the main components, isomers with respect to the placement of the hydrophilic groups, oversubstitution of hydrophilic groups, inorganic electrolytes as impurities, and hydroperoxidation of alkyl or alkyl ether chains. Fluctuations of the composition are sometimes observed between batches made by the same producer and certainly between samples supplied by different producers. These fluctuations are a serious source of problems regarding reproducibility but also regarding the comparability of nominal identical polymerizations in different laboratories.

In conclusion, any new surfactant for an emulsion polymerization is faced with the problem that it is only an auxiliary agent that has to meet strict cost requirements and has to improve existing solutions in two contradictory fields (polymerization and application), but this is also a challenge for further research.

VI. OPEN PROBLEMS AND FUTURE DEVELOPMENTS

Over the past 90 years, enormous knowledge has been accumulated on emulsion polymerization resulting in an extremely wide variety of products for very different applications. Sophisticated models have been developed that require knowledge of almost 50 basic kinetic and thermodynamic constants [160,186–188] because of the complex heterogeneous nature of the process, of which only a few are known precisely. Some problems in the case of modeling co- and terpolymerizations have been discussed [189].

A summary of on-line techniques suited for emulsion polymerizations can be found in Refs. 190–193. Procedures are also available for the control of copolymer composition and to avoid a composition drift [194–196]. Improvement with regard to process analysis and control can be expected with the development and application as much as possible of on-line sensors in combination with reaction calorimeters.

It would be advantageous to be able to control online the level of free surfactant during an emulsion polymerization. Surface tension measurements have been carried out quasi–on line by bypassing the reaction mixture through a bubble tensiometer [197,198]. On-line measurements of surface tension directly inside stirred reactors have been reported at elevated temperatures [199]. This now allows control of the level of free surfactant as well as determination of surfactant properties directly under reaction conditions. Almost all the quantitative data characterizing surfactant properties used today for modeling have not been determined by polymerization temperatures. As the surface tension shows a distinct temperature dependence, improvements can be expected [200,201] if surfactant data obtained at the polymerization temperature become available.

A great deal of work is going on regarding the adaptation of controlled radical polymerization processes to emulsion polymerization. All modes of controlled radical polymerization techniques are investigated, as it is promising to combine the advantages of emulsion polymerization with the possibilities of controlling the molecular weight and preparing well-defined block copolymers [202–213].

Besides the almost continuous development of new products and the improvement of existing products, several attempts have been made to use other polymerization mechanisms in aqueous emulsion polymerization. The goal is to develop initiator systems allowing stereospecific emulsion polymerization and hence control of the molecular structure (modes of addition of the monomer molecules, branching, and isomerism). This has been an ongoing research topic since the 1960s, when dienes were polymerized with rhodium and palladium salts [214–216]. Recently, the emulsion polymerization of ethylene with organonickel metal complexes was reported for the first time [217]. Even more exciting is the cationic polymerization of 1,3,5,7-tetramethylcyclotetrasiloxane [218] and *p*-methoxystyrene (controlled cationic polymerization) in aqueous emulsion [219].

Without any doubt, the meaning of emulsion polymerization as an environmentally friendly aqueous reaction system will continue to increase in the future. It is a field with growing attraction for scientists from different disciplines, reflecting its enormous potential as well as its multidisciplinary character.

REFERENCES

1. D. Distler, in *Wäßrige Polymerdispersionen: Synthese, Eigenschaften, Anwendungen* (D. Distler, ed.), Wiley-VCH, Weinheim, 1999, pp. 1–12.
2. D. Distler, *Wäßrige Polymerdispersionen: Synthese, Eigenschaften, Anwendungen*, Wiley-VCH, Weinheim, 1999.
3. A. J. DeFusco, K. C. Sehgal, and D. R. Bassett, in *Polymeric Dispersions: Principles and Applications* (J. M. Asua, ed.), Kluwer Academic Publishers, Dordrecht, 1997, pp. 379–396.
4. D. Distler, in *Wäßrige Polymerdispersionen: Synthese, Eigenschaften, Anwendungen* (D. Distler, ed.), Wiley-VCH, Weinheim, 1999, pp. 13–21.
5. J. H. Baxendale, M. G. Evams, and G. S. Park, Trans. Faraday Soc. *42*:155–169 (1946).
6. J. H. Baxendale, M. G. Evams, and G. S. Park, Trans. Faraday Soc. *42*:668–684 (1946).
7. V. J. Frilette and W. P. Hohenstein, J. Polym. Sci. *III*: 22–31 (1948).
8. F. A. Bovey and I. M. Kolthoff, J. Polym. Sci. *V*:487–504 (1950).
9. R. M. Fitch, M. B. Prenosil, and K. J. Sprick, J. Polym. Sci. Part C *27*:95–118 (1969).
10. R. M. Fitch and C. Tsai, Polym. Lett. *8*:703–710 (1970).
11. K. Tauer and C. Schellenberg, Nucleation in heterogeneous polymerization, Proceedings of International Symposium on Polymers in Dispersed Media, Lyon, 1999, pp. 6–10.
12. K. Tauer, C. Schellenberg, and A. Zimmermann, Macromol. Symp. *150*:1–12 (2000).
13. G. S. Whitbey and M. Katz, Ind. Eng. Chem. *25*: 1204–1211 (1933).
14. G. S. Whitbey and M. Katz, Ind. Eng. Chem. *25*: 1338–1348 (1933).
15. W. P. Hohenstein and H. Mark, J. Polym. Sci. *1*:127–145 (1946).
16. W. P. Hohenstein and H. Mark, J. Polym. Sci. *1*:549–580 (1946).
17. F. Hofmann and K. Delbrück, German Patent 250.690 to Farbenfabriken Bayer (1909).
18. H. Fikentscher, H. Gerrens, and H. Schuller, Angew Chem. *72*:856–867 (1960).
19. M. Luther and C. Heuck, German Patent 558.890 to IG Farbenindustrie (1927).
20. R. P. Dinsmore, U.S. Patent 1.732.795 to Goodyear Company (1927).
21. Mod. Plast. *23*:153–220 (1946).
22. F. A. Bovey, I. M. Kolthoff, A. I. Medalia, and E. J. Meehan, *Emulsion Polymerization*, Interscience, New York, 1955, pp. 1–25.
23. K. Meisenburg, DRP 568.907 to IG Farbenindustrie (1930).
24. K. Meisenburg, DRP 573.648 to IG Farbenindustrie (1930).
25. H. Fikentscher and C. Heuk, DRP 654.989 to IG Farbenindustrie (1930).
26. W. Starck and H. Freudenberger, DRP 727.995 to IG Farbenindustrie (1934).
27. H. Mark, H. Fikentscher, J. Hengstenberg, and G. von Susich, U.S. Patent 2.068,424 (1932).
28. E. Scharf and H. Fikentscher, DRP 615.219 to IG Farbenindustrie (1931).
29. H. Fikentscher, Angew. Chem. *51*:433 (1938).
30. W. D. Harkins, J. Am. Chem. Soc. *69*:1428–1444 (1947).
31. W. D. Harkins, J. Polym. Sci. *V*:217–251 (1950).
32. W. V. Smith and R. H. Ewart, J. Chem. Phys. *16*:592–599 (1948).
33. E. Bartholomé, H. Gerrens, R. Herbeck, and H. M. Weitz, Z. Elektrochem. *60*:334–348 (1956).
34. F. K. Hansen and J. Ugelstad, J. Polym. Sci. Polym. Chem. Ed. *17*:3047–3067 (1979).
35. H. Gerrens, Fortschr. Hochpol. Forsch. *1*:234–328 (1959).
36. H. Fikentscher, H. Gerrens, and H. Schuller, Angew. Chem. *72*:856–864 (1960).
37. H. Gerrens, Dechema Monogr. *49*:53–73 (1964).
38. H. Gerrens, K. Kuchner, and G. Ley, Chem. Ing. Tech. *43*:693–698 (1971).
39. G. Ley and H. Gerrens, Makromol. Chem. *175*:563–581 (1974).
40. R. M. Fitch and C. H. Tsai, Polym. Lett. *8*:703–710 (1970).
41. R. M. Fitch and C. H. Tsai, in *Polymer Colloids* (R. M. Fitch, ed.), Plenum, New York, 1971, pp. 73–102.
42. R. M. Fitch, Br. Polym. J. *5*:467–483 (1973).
43. R. M. Fitch, Off. Dig. J. Paint Technol. Eng. *37*:32–48 (1965).
44. F. K. Hansen and J. Ugelstad, J. Polym. Sci. Polym. Chem. Ed. *16*:1953–1979 (1978).

45. F. K. Hansen and J. Ugelstad, J. Polym. Sci. Polym. Chem. Ed. *17*:3033–3045 (1979).
46. J. Ugelstad and F. K. Hansen, Rubber Chem. Technol. *49*:536–556 (1976).
47. G. Lichti, R. G. Gilbert, and D. H. Napper, J. Polym. Sci. Polym. Chem. Ed. *21*:269–291 (1983).
48. R. G. Gilbert, *Emulsion Polymerization: A Mechanistic Approach*, Academic Press, New York, 1995, pp. 78–140.
49. M. S. El-Aasser and C. M. Miller, in *Polymeric Dispersions: Principles and Applications* (J. M. Asua, ed.), Kluwer Academic Publishers, Dordrecht, 1997, pp. 109–126.
50. J. Ugelstad, M. S. El-Aasser, and J. W. Vanderhoff, J. Polym. Sci. Polym. Lett. Ed. *11*:503–513 (1973).
51. J. Ugelstad, P. Stenstad, L. Kilaas, W. S. Prestvik, A. Rian, K. Nustad, R. Herje, and A. Berge, Macromol. Symp. *101*:491–500 (1996).
52. V. L. Dimonie, E. S. Daniels, O. L. Shaffer, and M. S. El-Aasser, in *Emulsion Polymerization and Emulsion Polymers* (P. A. Lovell and M. S. El-Aasser, eds.), Wiley, New York, 1997, pp. 293–326.
53. F. A. Bovey, I. M. Kolthoff, A. I. Medalia, and E. J. Meehan, *Emulsion Polymerization*, Interscience, New York, 1955.
54. D. C. Blackley, *Emulsion Polymerisation Theory and Practice*, Applied Science Publishers, London, 1975.
55. V. I. Eliseeva, *Polimernije Dispersii (Russ.)*, Chimija, Moscow, 1980.
56. R. G. Gilbert, *Emulsion Polymerization A Mechanistic Approach*, Academic Press, New York, 1995.
57. R. M. Fitch, *Polymer Colloids: A Comprehensive Introduction*, Academic Press, New York, 1997.
58. J. M. Asua, ed., *Polymeric Dispersions: Principles and Applications*, Kluwer Academic Publishers, Dordrecht, 1997.
59. A. M. van Herk, in *Polymeric Dispersions: Principles and Applications* (J. M. Asua, ed.), Kluwer Academic Publishers, Dordrecht, 1997, pp. 17–30.
60. P. A. Lovell and M. S. El-Aasser, eds., *Emulsion Polymerization and Emulsion Polymers*, John Wiley, New York, 1997.
61. R. G. W. Norrish and R. R. Smith, Nature *150*:336–337 (1942).
62. E. Trommsdorff, H. Köhle, and P. Lagally, Makromol. Chem. *1*:169–198 (1947).
63. L. V. De La Rosa, E. D. Sudol, and M. S. EL-Aasser, J. Polym. Sci. Part A Polym. Chem. *34*:461–473 (1996).
64. R. S. Corley, in *Monomers* (E. R. Blout and H. Mark, eds.), Interscience, New York, 1951.
65. A. L. Ward and W. J. Roberts, in *Monomers* (E. R. Blout and H. Mark, eds.), Interscience, New York, 1951.
66. K. Tauer, H. Müller, C. Schellenberg, and L. Rosengarten, Colloids Surf. A Physicochem. Eng. Asp. *153*: 137–142 (1999).
67. G. Fevotte, T. F. McKenna, S. Othman, and A. M. Santos, Comput. Chem. Eng. *22*(Suppl):S443–S449 (1998).
68. L. Jansson, H. Nilsson, C. Silvergren, and B. Törnell, Thermochim. Acta *118*:97–105 (1987).
69. E. Özdeger, E. D. Sudol, M. S. El-Aasser, and A. Klein, J. Polym. Sci. Part A Polym. Chem. *35*:3813–3825 (1997).
70. E. Özdeger, E. D. Sudol, M. S. El-Aasser, and A. Klein, J. Polym. Sci. Part A Polym. Chem. *35*:3827–3835 (1997).
71. E. Özdeger, E. D. Sudol, M. S. El-Aasser, and A. Klein, J. Polym. Sci. Part A Polym. Chem. *35*:3837–3846 (1997).
72. K. Tauer, M. Antonietti, L. Rosengarten, and H. Müller, Macromol. Chem. Phys. *199*:897–908 (1998).
73. H.-Y. Parker, D. G. Westmoreland, and H.-R. Chang, Macromolecules *29*:5119–5127 (1996).
74. K. Tauer and I. Kühn, in *Polymeric Dispersions: Principles and Applications* (J. M. Asua, ed.), Kluwer Academic Publishers, Dordrecht, 1997, pp. 49–65.
75. M. Vollmer and A. Weber, Z. Phys. Chem. *119*:277–301 (1925).
76. R. Becker and W. Döring, Ann. Phys. *24*:719–752 (1935).
77. J. Frenkel, J. Chem. Phys. *7*:538–547 (1939).
78. K. E. J. Barrett, *Dispersion Polymerization in Organic Media*, John Wiley & Sons, New York, 1975.
79. V. K. LaMer and R. H. Dinegar, J. Am. Chem. Soc. *72*:4847–4854 (1950).
80. I. Kühn and K. Tauer, Macromolecules *28*:8122–8128 (1995).
81. K. Tauer, R. Deckwer, I. Kühn, and C. Schellenberg, Colloid Polym. Sci. *277*:607–626 (1999).
82. C. P. Roe, Ind. Eng. Chem. *60*:20–33 (1968).
83. J. J. P. Staudinger, Chem. Ind. (Lond.) *67*:563–568 (1948).
84. I. D. Robb, J. Polym. Sci. Part A-1 *7*:417–427 (1969).
85. W. J. Priest, J. Phys. Chem. *26*:1077–1082 (1952).
86. N. M. van Os, J. R. Haak, and L. A. M. Rupert, *Physico-chemical Properties of Selected Anionic, Cationic and Nonionic Surfactants*, Elsevier, Amsterdam, 1983, pp. 1–107.
87. W.-A. G. R. Al-Shahib and A. S. Dunn, J. Polym. Sci. Polym. Chem. Ed. *16*:677–684 (1978).
88. A. S. Dunn and W. A. Al-Shahib, in *Polymer Colloids II* (R. M. Fitch, ed.), Plenum, New York, 1980, pp. 619–628.
89. A. S. Dunn, in *Emulsion Polymerization* (I. Piirma, ed.), Academic Press, New York, 1982, pp. 221–245.
90. E. A. G. Amiansson, S. N. Wall, M. Almgren, H. Hoffmann, I. Kielmann, W. Ulbricht, R. Zana, J. Lang, and C. Tonore, J. Phys. Chem. *80*:905–922 (1976).
91. K. Lunkenheimer, G. Czichocki, R. Hirte, and W. Barzyk, Colloids Surf. A Physicochem. Eng. Asp. *101*: 187–197 (1995).

92. K. Tauer and I. Kühn, Macromolecules *28*:2236–2239 (1995).
93. W. Wu and G. H. Nancollas, J. Colloid Interface Sci. *182*:365–373 (1996).
94. A. Laaksonen, V. Talanquer, and D. W. Oxtoby, Annu. Rev. Phys. Chem. *46*:489–524 (1995).
95. I. Kühn, Untersuchungen zum Mechanismus der Teilchenbildung während der Emulsionspolymerisation, University Potsdam, Potsdam, Germany, 1996.
96. K. Tauer, I. Kühn, and H. Kaspar, Prog. Colloid Polym. Sci. *101*:30–37 (1996).
97. I. M. Kolthoff and I. K. Miller, J. Am. Chem. Soc. *73*: 3055–3059 (1951).
98. D. A. House, Chem. Rev. *62*:185–203 (1962).
99. E. J. Behrman and J. O. Edwards, Rev. Inorg. Chem. *2*:179–206 (1980).
100. G. V. Buxton, C. L. Greenstock, W. P. Helman, and A. B. Ross, J. Phys. Chem. Ref. Data *17*:513–886 (1988).
101. K. Tauer, Polym. Adv. Technol. *6*:435–440 (1995).
102. K. Tauer and R. Deckwer, Acta Polym. *49*:411–416 (1998).
103. R. Deckwer, Oligomere in der Emulsionspolymerisation von Styrol—Eigenschaften und Charakterisierung, Ph.D. dissertation, University Potsdam, Potsdam, Germany, 1999.
104. W. J. Lyman, in *Handbook of Chemical Property Estimation Methods—Environmental Behavior of Organic Compounds* (W. J. Lyman, W. F. Reehl, and D. H. Rosenblatt, eds.), American Chemical Society, Washington, DC, 1996, pp. 2-1–2-52.
105. A. Kuhn, *Wörterbuch der Kolloidchemie*, Theodor Steinkopff, Dresden, 1932, pp. 124–128.
106. J. Stauff, *Kolloidchemie*, Springer Verlag, Berlin, 1960, pp. 682–690.
107. W. Smith, J. Am. Chem. Soc. *70*:3695–3702 (1948).
108. S. H. Herzfeld, A. Roginsky, and M. L. Corrin, J. Polym. Sci. *V*:207–215 (1950).
109. M. Morton, S. Kaizerman, and M. W. Altier, J. Colloid Sci. *9*:300–312 (1954).
110. J. L. Gardon, J. Polym. Sci. Part A-1 *6*:2859–2879 (1968).
111. B. W. Brooks, Br. Polym. J. *3*:269–373 (1971).
112. I. A. Maxwell, J. Kurja, G. H. J. Van Doremaele, A. L. German, and B. R. Morrison, Makromol. Chem. *193*:2049–2063 (1992).
113. J.-I. Kim, J. W. Vanderhoff, and A. Klein, J. Appl. Polym. Sci. *58*:279–290 (1995).
114. I. A. Maxwell, J. Kurja, G. H. J. Van Doremaele, and A. L. German, Makromol. Chem. *193*:2065–2080 (1992).
115. P. D. Armitage, J. C. De La Cal, and J. M. Asua, J. Appl. Polym. Sci. *51*:1985–1990 (1994).
116. Y. G. Duront and D. C. Sundberg, J. Appl. Polym. Sci. *58*:1607–1618 (1995).
117. J. Ugelstad, Makromol. Chem. *179*:815–817 (1978).
118. J. Ugelstad, K. H. Kaggerud, F. K. Hansen, and A. Berge, Makromol. Chem. *180*:737–744 (1979).
119. J. Ugelstad, K. H. Kaggerud, and R. M. Fitch, in *Polymer Colloids II* (R. M. Fitch, ed.), Plenum, New York, 1980, pp. 83–93.
120. J. Ugelstad, P. C. Mork, A. Berge, T. Ellingsen, and A. A. Khan, in *Emulsion Polymerization* (I. Piirma, ed.), Academic Press, New York, 1982, pp. 383–413.
121. L. H. Jansson, M. C. Wellons, and G. W. Poehlein, J. Polym. Sci. Polym. Lett. Ed. *21*:937–943 (1983).
122. J. Ugelstad, H. R. Mftuakamba, P. C. Mork, T. Ellingsen, A. Berge, R. Schmid, L. Holm, A. Jorgendal, F. K. Hansen, and K. Nustad, J. Polym. Sci. Polym. Symp. *72*:225–240 (1985).
123. M. Okubo, T. Yamashita, and M. Shiozaki, J. Appl. Polym. Sci. *60*:1025–1031 (1996).
124. D. I. Lee, Makromol. Chem. Macromol. Symp. *33*: 117–131 (1990).
125. D. C. Sundberg, A. P. Casassa, J. Pantazopoulos, M. R. Muscato, B. Kronberg, and J. Berg, J. Appl. Polym. Sci. *41*:1425–1442 (1990).
126. M. Okubo, Makromol. Chem. Macromol. Symp. *35/36*:307–325 (1990).
127. C. L. Winzor and D. C. Sundberg, Polymer *33*:4269–4279 (1992).
128. Y. C. Chen, V. L. Dimonie, O. L. Shaffer, and M. S. El-Aasser, Polym. Int. *30*:185–194 (1993).
129. E. J. Sundberg and D. C. Sundberg, J. Appl. Polym. Sci. *47*:1277–1294 (1993).
130. S. Omi, T. Kohmoto, and M. Iso, Polym. Int. *30*:499–504 (1993).
131. Y. C. Chen, V. L. Dimonie, and M. S. El-Aasser, J. Appl. Polym. Sci. *45*:487–499 (1992).
132. L. J. Gonzales-Ortiz and J. M. Asua, Macromolecules *28*:3135–3145 (1995).
133. L. J. Gonzales-Ortiz and J. M. Asua, Macromolecules *29*:383–389 (1996).
134. L. J. Gonzales-Ortiz and J. M. Asua, Macromolecules *29*:4520–4527 (1996).
135. M. Okaniwa, J. Appl. Polym. Sci. *68*:185–190 (1998).
136. X. Z. Kong and E. Ruckenstein, J. Appl. Polym. Sci. *71*:1455–1460 (1999).
137. A. R. Berens, Pure Appl. Chem. *53*:365–375 (1981).
138. E. J. Meehan, J. Am. Chem. Soc. *71*:628–633 (1949).
139. E. Vanzo, R. H. Marchessault, and V. Stannett, J. Colloid Sci. *20*:62–71 (1965).
140. H. Nilsson, C. Silvergren, and B. Törnell, Eur. Polym. J. *14*:737–741 (1978).
141. G. Löhr, in *Polymer Colloids II* (R. M. Fitch, ed.), Plenum, New York, 1980, pp. 71–81.
142. L. F. J. Noel, R. Q. F. Janssen, W. J. M. van Well, A. M. van Herk, and A. L. German, J. Colloid Interface Sci. *175*:461–469 (1995).
143. E. A. van der Zeeuw, L. M. C. Sagis, and G. J. M. Koper, Macromolecules *29*:801–803 (1996).
144. M. Antonietti, H. Kaspar, and K. Tauer, Langmuir *12*: 6211–6217 (1996).

145. V. K. LaMer and R. Gruen, Trans. Faraday Soc. *48*: 410–415 (1952).
146. M. R. Grancio and D. J. Williams, J. Polym. Sci. Part A-1 *8*:2617–2629 (1970).
147. P. Keusch and D. J. Williams, J. Polym. Sci. Polym. Chem. Ed. *11*:143–162 (1973).
148. D. H. Napper, J. Polym. Sci. A-1 *9*:2089–2091 (1971).
149. J. L. Gardon, J. Polym. Sci. Polym. Chem. Ed. *11*: 241–251 (1973).
150. J. Bolze and M. Ballauff, Macromolecules *28*:7429–7433 (1995).
151. P. J. Flory and J. Rehner, J. Chem. Phys. *11*:521 (1943).
152. B. Kitiyanan, J. H. Ohaver, J. H. Harwell, and S. Osuwan, Langmuir *12*:2162–2168 (1996).
153. K.-L. Gosa, J. Dispersion Sci. Technol. *17*:577–589 (1996).
154. B. R. Vijayendran, J. Appl. Polym. Sci. *23*:733–742 (1979).
155. W. Burchard, Prog. Colloid Polym. Sci. *78*:63–67 (1988).
156. H. Kaspar, Untersuchungen zur Gleichgewichtsquellung polymer Nanopartikel, Ph.D. dissertation, University Potsdam, Potsdam, Germany, 1996.
157. R. H. Ottewill, in *Polymeric Dispersions: Principles and Applications* (J. M. Asua, ed.), Kluwer Academic Publishers, Dordrecht, 1997, pp. 31–48.
158. K.-H. Goebel, H.-J. Schneider, W. Jaeger, G. Reinisch, Acta Polymerica 33:49–62, (1982).
159. K. Tauer, B. R. Paulke, I. Müller, W. Jaeger, and G. Reinisch, Acta Polym. *33*:287–298 (1982).
160. K. Tauer, G. Reinisch, H. Gajewski, and I. Müller, J. Macromol. Chem. *A28*:431–460 (1991).
161. F. A. Bovey, I. M. Kolthoff, A. I. Medalia, and E. J. Meehan, *Emulsion Polymerization*, Interscience, New York, 1955, pp. 271–324.
162. C. Abad, J. C. de la Cal, and J. M. Asua, in *Polymeric Dispersions: Principles and Applications* (J. M. Asua, ed.), Kluwer Academic Publishers, Dordrecht, 1997, pp. 333–347.
163. B. Jacobi, Angew. Chem. *64*:539–543 (1952).
164. G. W. Poehlein and D. J. Dougherty, Rubber Chem. Technol. *50*:601–638 (1977).
165. G. W. Poehlein, Polym. Int. *30*:243–251 (1993).
166. F. J. Schork, in *Polymer Reactor Engineering* (C. McGreavy, ed.), Chapman & Hall, London, 1994, pp. 148–202.
167. K. Arai, M. Arai, S. Iwasaki, and S. Saito, J. Polym. Sci. Polym. Chem. Ed. *19*:1203–1215 (1981).
168. M. Nomura, M. Harada, W. Eguchi, and S. Nagata, J. Appl. Polym. Sci. *16*:835–847 (1972).
169. J. W. Vanderhoff, M. S. El-Aasser, F. J. Micale, E. D. Sudol, C. M. Tseng, A. Silwanowizc, D. M. Kornfeld, and F. A. Vicente, J. Dispersion Sci. Technol. *5*:231–246 (1984).
170. G. Markert, Angew. Makromol. Chem. *123/124*:285–306 (1984).
171. M. von Smoluchowski, Z. Phys. Chem. *92*:129 (1917).
172. M. F. Kemmere, J. Meuldijk, A. A. H. Drinkenburg, and A. L. German, J. Appl. Polym. Sci. 69:2409–2421 (1998).
173. B. Lindman and K. Thalberg, in *Interaction of Surfactants with Polymers and Proteins* (E. D. Goddard and K. P. Ananthapadmanabhan, eds.), CRC Press, Boca Raton, FL, 1993, pp. 203–276.
174. D. H. Napper, *Polymeric Stabilization of Colloidal Dispersions*, Academic Press, New York, 1983.
175. I. Piirma, *Polymeric Surfactants*, Marcel Dekker, New York, 1992.
176. R. Brown, B. Stützel, and T. Sauer, Macromol. Chem. Phys. *196*:2047–2064 (1995).
177. J.-L. Mura and G. Riess, Polym. Adv. Technol. *6*:497–508 (1995).
178. C.-S. Chern, Y.-C. Liou, and W.-Y. Tsai, J. Macromol. Sci. *A33*:1063–1075 (1996).
179. J. M. Harris, in *Poly(ethylene glycol) Chemistry Biotechnical and Biomedical Applications* (J. M. Harris, ed.), Plenum, New York, 1992, pp. 1–14.
180. I. Piirma and M. Chang, J. Polym. Sci. Polym. Chem. Ed. *20*:489–498 (1982).
181. I. Piirma and T. S. Maw, Polym. Bull. *11*:497–504 (1984).
182. K. Tauer, H. Müller, L. Rosengarten, and K. Riedelsberger, Colloids Surf. A Physicochem. Eng. Asp. *153*: 61–74 (1999).
183. J. Dickstein, Polym. Prepr. *27*:427–428 (1986).
184. S. Lam, A. C. Hellgren, M. Sjöberg, K. Holmberg, H. A. S. Schoonbrood, J. M. Asua, K. Tauer, D. C. Sherrington, and A. Montoya Goni, J. Appl. Polym. Sci. *66*:187–198 (1997).
185. H. Müller, W. Leube, K. Tauer, S. Förster, and M. Antonietti, Macromolecules *30*:2288–2293 (1997).
186. K. W. Min and W. H. Ray, J. Macromol. Sci. Rev. Macromol. Chem. *C11*: 177–255 (1974).
187. C. Kiparissides, J. F. MacGregor, and A. E. Hamielec, J. Appl. Polym. Sci. *23*:401–418 (1979).
188. E. M. Coen, R. G. Gilbert, B. R. Morrison, H. Leube, and S. Peach, Polymer *39*:7099–7112 (1998).
189. A. M. van Herk and A. L. German, Macromol. Theory Simul. *7*:557–565 (1998).
190. W.-D. Hergeth, in *Polymeric Dispersions: Principles and Applications* (J. M. Asua, ed.), Kluwer Academic Publishers, Dordrecht, 1997, pp. 243–256.
191. M. Morbidelli, G. Storti, and S. Siani, in *Polymeric Dispersions: Principles and Applications* (J. M. Asua, ed.), Kluwer Academic Publishers, Boston, 1997, pp. 257–266.
192. W.-D. Hergeth, in *Polymeric Dispersions: Principles and Applications* (J. M. Asua, ed.), Kluwer, Dordrecht, 1997, pp. 267–288.
193. E. G. Chatzi, O. Kammona, and C. Kiparissides, J. Appl. Polym. Sci. *63*:799–809 (1997).
194. A. Urretabizkaia, E. D. Sudol, M. S. El-Aasser, and J.

M. Asua, J. Polym. Sci. Part A Polym. Chem. *31*: 2907–2913 (1993).

195. M. Morbidelli and G. Storti, in *Polymeric Dispersions: Principles and Applications* (J. M. Asua, ed.), Kluwer Academic Publishers, Dordrecht, 1997, pp. 349–361.
196. J. R. Leiza and J. M. Asua, in *Polymeric Dispersions: Principles and Applications* (J. M. Asua, ed.), Kluwer Academic Publishers, Dordrecht, 1997, pp. 363–378.
197. F. J. Schork and W. H. Ray, J. Appl. Polym. Sci. *28*: 407–430 (1983).
198. D. C. H. Chien and A. Penlidis, J. Rev. Macromol. Chem. Phys. *C30*:1–42 (1990).
199. K. Tauer, C. Dessy, S. Corkery, and K.-D. Bures, Colloid Polym Sci. *277*:805–811 (1999).
200. N. B. Vargaftik, B. N. Volkov, and L. D. Voljak, J. Phys. Chem. Ref. Data *12*:817–820 (1983).
201. K. S. Birdi, in *Handbook of Surface and Colloid Chemistry* (K. S. Birdi, ed.), CRC Press, Boca Raton, FL, 1997, pp. 71–118.
202. J. Krstina, C. L. Moad, G. Moad, E. Rizzardo, and C. T. Berge, Macromol. Symp. *111*:13–23 (1996).
203. M. Bouix, J. Gouzi, B. Charleux, J.-P. Vairon, and P. Guinot, Macromol. Chem. Rapid Commun. *19*:209–213 (1998).
204. C. Marestin, C. Noel, A. Guyot, and J. Claverie, Macromolecules 31:4041–4044 (1998).
205. T. Nishikawa, M. Kamigaito, and M. Sawamoto, Macromolecules *32*:2204–2209 (1999).
206. J. Qui, S. G. Gaynor, and K. Matyjaszewski, Macromolecules *32*:2872–2875 (1999).
207. T. Prodpran, V. L. Dimonie, E. D. Sudol, and M. S. El-Aasser, Nitroxide-mediated living free radical miniemulsion polymerization, Proceedings of the ACS PMSE (Vol. 80) Spring Meeting, Anaheim, CA, 1999, pp. 534–535.
208. S. G. Gaynor, J. Qui, and K. Matyjaszewski, Controlled/"living" radical polymerization applied to water-borne system, Proceedings of the ACS PMSE (Vol. 80) Spring Meeting, Anaheim, CA 1999, pp. 536–537.
209. Y. G. Duront, Miniemulsion polymerization applications and continuous processes, Proceedings of the ACS PMSE (Vol. 80) Spring Meeting, Anaheim, CA, 1999, pp. 538–540.
210. D. Charmot, P. Corpart, H. Adam, S. Zard, and T. Biadatti, Controlled radical polymerization in dispersed media, Proceedings of International Symposium on Polymers in Dispersed Media, Lyon, 1999, pp. 15–16.
211. J. Claverie, S. Kanagasabapathy, and D. Christie, Living radical polymerization in emulsion, Proceedings of International Symposium on Polymers in Dispersed Media, Lyon, 1999, pp. 17–18.
212. M. Lansalot, B. Charleux, J.-P. Vairon, R. Pirri, and P. Tordo, Nitroxide-mediated controlled radical polymerization of styrene in aqueous dispersed media, Proceedings of International Symposium on Polymers in Dispersed Media, Lyon, 1999, pp. 19–20.
213. B. Charleux, Controlled radical polymerization in dispersed media, Proceedings of EPF Workshop on Controlled Radical Polymerization, Paris, 1999, p. L5.
214. R. E. Rinehart, H. P. Smith, H. S. Witt, and H. Romeyn, Jr., J. Am. Chem. Soc. *83*:4864–4865 (1961).
215. R. E. Rinehart, H. P. Smith, H. S. Witt, and H. Romeyn, Jr., J. Am. Chem. Soc. *84*:4145–4147 (1962).
216. A. J. Canale and W. A. Hewett, Polym. Lett. *2*:1041–1045 (1964).
217. A. Tomov, J. P. Broyer, and R. Spitz, Emulsion polymerization of ethylene in water medium catalyzed by organotransition metal complexes, Proceedings of International Symposium on Polymers in Dispersed Media, Lyon, 1999, p. 36.
218. S. Maisonnier, J. C. Favier, M. Masure, and P. Hemery, Polym. Int. *48*:159–164 (1999).
219. K. Satoh, M. Kamingaito, and M. Sawamoto, Macromolecules *32*:3827–3832 (1999).

22

Free Radical Polymerization in Microemulsions

CARLOS C. CO, RENKO DE VRIES,* and ERIC W. KALER University of Delaware, Newark, Delaware

I. INTRODUCTION

Microemulsions are thermodynamically stable one-phase solutions of immiscible substances such as water (or another polar component, e.g., formamide) and oil (hydrocarbon, fluorocarbon, or silicone) that self-assemble in the presence of one or more surfactants into a wide range of microstructures. Depending on the temperature, composition, molecular architecture, and hydrophobicity or hydrophilicity of these components, various microstructures can exist. As the oil-to-water ratio is increased, a progression of microstructures, ranging from oil-swollen micelles dispersed in water to bicontinuous structures and finally to water-swollen micelles dispersed in oil, is typically observed. Furthermore, the oil- or water-swollen micelles can adopt spherical, ellipsoidal, or rodlike geometries. With some exceptions, microemulsions are optically transparent because the length scales of the oil and water domains are usually less than 10 nm. The rich variety of microstructures possible within microemulsions have made them the subject of great scientific and practical interest.

The substitution of the oil by monomers or the addition of water-soluble monomers followed by subsequent polymerization has been attempted for essentially all possible types of microstructures. However, in many cases, the original microstructure is not preserved because of the rapid diffusion of the monomer to the localized regions where polymerization is taking place. In addition, the forces that govern self-assembly are usually too weak to suppress microstructural changes caused by the differences in physical properties between the monomer and polymer. Numerous review articles [1–3] (see the companion chapters in Part 3 of this volume) discussing the polymerization in self-assembled structures are available, and no attempt is made here to review comprehensively polymerizations in all types of microstructures. Instead, this chapter focuses on the latest developments in the quantitative modeling of free radical polymerizations of oil-in-water microemulsions containing droplet microstructures. A detailed understanding of polymerizations in this simpler geometry should be useful in developing quantitative models for polymerizations in bicontinuous phases, vesicles, and liquid crystalline phases.

A. Why Study Microemulsion Polymerization?

The free radical polymerization of microemulsions containing droplet microstructures provides a route for the synthesis of latex particles as small as 5 nm in radius that are difficult to prepare via emulsion polymerization [4]. These small latex nanoparticles are useful for coating the internal surfaces of microporous materials and can also be used as seed particles for emulsion polymerizations. For larger latex particles (~15 nm in radius), the molecular weights of the polymers formed are extraordinarily high (>20 million daltons) because a particle typically contains only one polymer chain. The synthesis of such high-molecular-weight polymers within these nanoparticles leads to polymers with increased stereoregularity [5] and, possibly, chain knotting.

**Current affiliation*: Wageningen University, Wageningen, The Netherlands.

B. Differences Between Microemulsion and Emulsion Polymerizations

The preparation of latexes via emulsion polymerization is a large-scale commercial process that is very well understood. Gilbert [6] provides a comprehensive experimental and theoretical review of emulsion polymerization. The following brief and qualitative summary of emulsion polymerization serves only to highlight the differences between microemulsion and emulsion polymerization processes.

In typical nonseeded emulsion polymerizations, monomer is dispersed by surfactant stabilization and vigorous stirring to form large monomer droplets (1 to 10 μm) and smaller monomer-swollen micelles ($\sim$3 nm). During the first interval of emulsion polymerization, surface-active oligomeric free radicals, generated by the reaction between initiator-derived radicals and monomer partitioned in the aqueous phase, nucleate the swollen micelles to form polymer particles. Typically, the concentration of surfactant is such that when 0 to 10% of the monomer has been converted, all the surfactant becomes adsorbed on the polymer particles. At this point, monomer-swollen micelles disappear and particle nucleation ceases.

In contrast, as monomer is added to a surfactant solution to form a microemulsion, equilibrium microstructure forms without stirring. All the monomer in a microemulsion exists within monomer-swollen micelles and no large droplets are present. Microemulsions typically have higher surfactant concentrations than used to form emulsions, so usually the amount of surfactant that adsorbs on the polymer particles is negligible. Consequently, monomer-swollen micelles are present throughout the polymerization and particle nucleation ceases only when all the monomer has been consumed. The rate at which nucleated particles grow is directly proportional to the concentration of monomer at the locus of polymerization within the polymer particles. The concentrations of monomer with the polymer particles during microemulsion and emulsion polymerizations are controlled by different mechanisms.

The large monomer droplets in emulsions constitute a separate thermodynamic phase wherein the chemical potential of monomer is essentially constant. Thus, for emulsion polymerizations, the concentration of monomer within the polymer particles is set by a balance between surface and monomer-polymer interaction free energies [7,8]. Typical monomers are good solvents for their polymers. Hence, the interaction between monomer and polymer would result in polymer particles that are highly swollen with monomer. In this case, the extent to which the polymer particles are swollen by monomer is limited only by interfacial tension and the availability of monomer.

A key feature that differentiates microemulsion from emulsion polymerizations is the absence of large monomer droplets. In the absence of a separate monomer phase, the chemical potential of the monomer within the swollen micelles is set primarily by the curvature elastic properties of the surfactant monolayers and can change as the polymerization proceeds (C. C. Co et al., submitted). Thus, in contrast to emulsion polymerizations, the concentration of monomer within the polymer particles is governed by a balance between the curvature energy of the micelles and monomer-polymer interaction free energy. This theory qualitatively implies that the monomer-swollen micelles have an optimal radius of curvature that corresponds to the radius of the micelles at the solubilization limit. As the amount of monomer is progressively reduced, the micelles compete more effectively with the polymer particles for monomer. This effect underscores the essential interplay of phase behavior and microstructure in quantitatively modeling microemulsion polymerizations.

II. POLYMERIZABLE MICROEMULSIONS

A. Phase Behavior

The literature abounds with papers on microemulsion polymerization. However, the majority of these studies tend to focus on the polymerization aspects and the preparation of materials with unique morphologies. This approach results in rather limited phase behavior studies for a large variety of polymerizable microemulsions, so it is difficult to construct a unified view of the situation.

A powerful general approach for studying the phase behavior of oil/water/surfactant systems has been presented by Kahlweit et al. [9,10]. This approach entails mapping out the phase behavior on vertical sections of the Gibbs phase prism. This systematic approach has been applied by O. Lade (submitted) in an exhaustive phase behavior study of methacrylates with nonionic ethoxylated alkyl surfactants in water.

The phase behavior of ionic surfactants is usually insensitive to changes in temperature. Hence, mixtures of ionic surfactants are often used to improve their usefulness and flexibility because mixing provides additional degrees of freedom in phase space. A useful example of this approach toward using ionic surfactants was described by Lusvardi et al. [11,12], who studied the phase behavior of methacrylates using a mixture of dodecyltrimethylammonium bromide (DTAB) and didodecyldimethylammonium bromide (DDAB) surfactants. The molecular architecture of DTAB favors the formation of oil-in-water (o/w) microemulsions, whereas the twin hydrocarbon tails of DDAB favor inverse w/o microemulsions. On mixing these two surfactants, a one-phase channel forms in the appropriate phase prism and it becomes possible to form microemulsions with an arbitrary water-to-oil ratio.

Systematic phase behavior studies make it possible to choose the most efficient surfactants and compositions for a microemulsion polymerizing a given monomer. In addition, mapping out the phase behavior thoroughly makes it possible to choose polymerization conditions that permit a systematic study. For example, by first mapping out the phase behavior of a series of methacrylates in mixtures of DTAB, DDAB, and water, one can find a common set of conditions under which each of the methacrylates forms a one-phase microemulsion [12]. By polymerizing these methacrylate microemulsions under similar conditions, the effect of differences in monomer properties on the kinetics and polymer properties can be elucidated.

The primary information from these phase behavior studies that is needed to model monomer partitioning is the location of the phase boundary, i.e., the maximum amount of monomer that can be microemulsified (α_{max}) under the polymerization conditions. In Section IV.D we discuss how the monomer concentration at the locus of polymerization depends on the difference between the amount of monomer in the microemulsion (α) and α_{max}. The primary experimental technique for measuring monomer partitioning is small-angle neutron scattering (SANS), which is a quantitative probe of the microstructure of these microemulsions.

B. Microstructure and Characterization

The microstructural length scales present within microemulsions range from 1 to more than 100 nm, so small-angle x-ray scattering (SAXS) and SANS are the methods of choice for quantitative structural investigations. Analysis of small-angle scattering (SAS) spectra using Glatter's generalized indirect Fourier transform (GIFT) technique permits almost unmistakable assignment of spherical, cylindrical, or planar geometries as well as accurate estimates of their dimensions [13]. Direct imaging using electron microscopy on vitrified samples provides further verification of these results. Nevertheless, SAS is not without limitations. For example, it is not possibly to establish bicontinuity using SAS; instead self-diffusion nuclear magnetic resonance (NMR) is the most reliable technique. Similarly, the GIFT method for analyzing SAS data is not applicable to all systems, such as polymerized microemulsions containing both latex particles and micelles. Consequently, analysis of SAS spectra of microemulsions containing monomer-swollen micelles and their polymerized counterparts is best done by fitting SAS spectra calculated by assuming certain geometric models for the micelles and polymer particles. Excellent reviews of these SAS analysis techniques are available [14–17]. Therefore, the following discussion presents only a brief introduction to some specific approaches used to analyze the SAS spectra to measure structure and even to record the evolution of the microstructure during microemulsion polymerization.

Figure 1 shows the calculated scattering spectra for a dilute solution of monodisperse latex particles. The scattering intensity [$I(q)$] can be interpreted in terms of the scattering contribution from each particle [$P(q)$],

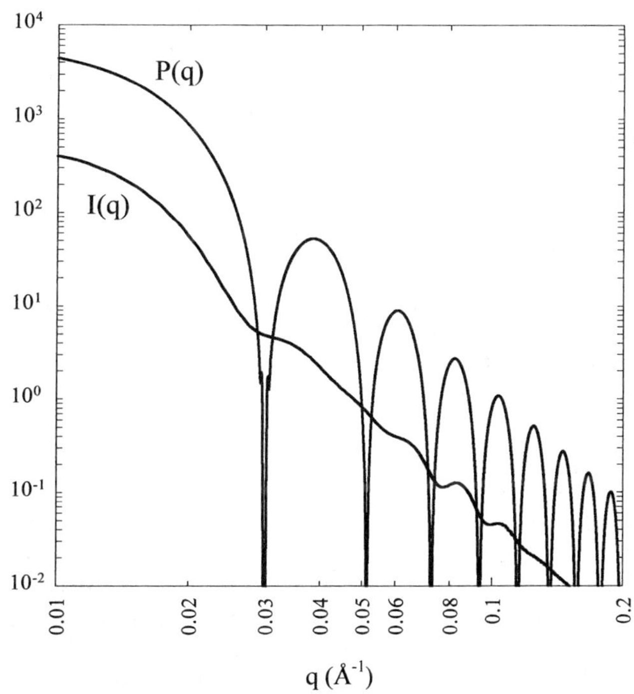

FIG. 1 Variation of the form factor [$P(q)$] for monodisperse hard spheres with a radius of 150 Å. Sharp minima in $P(q)$ are smeared out by polydispersity in the particle size distribution. For a dilute solution of such particles (1 vol%), the structure factor [$S(q)$] is approximately unity and the smooth $I(q)$ curve results if the latex particles have a 10% size polydispersity.

scaled for the number density (n) of particles, and corrected for their nonrandom positioning due to interparticle interactions [$S(q)$].

$$I(q) = nP(q)S(q) \tag{1}$$

The form factor [$P(q)$] depends on the size, shape, and composition of the individual latex particles. In the simplest case where the particles are modeled as monodisperse spheres of radius, R, and volume, V, with a uniform scattering contrast relative to the solvent ($\Delta\rho$),

$$P(q) = \left(3(\Delta\rho)V\,\frac{\sin(qR) - qR\cos(qR)}{(qR)^3}\right)^2 \tag{2}$$

For $qR < 1$, $P(q)$ varies little and is approximately equal to $(V\Delta\rho)^2$. At $qR = 4.5$, $P(q)$ goes through the first of an infinite number of sharp minima. However, these sharp minima are typically smeared or smoothed out by polydispersity in the particle size distribution as shown in Fig. 1. In general, form factors can be calculated for any particular structure.

The structure factor [$S(q)$] accounts for the nonrandom positioning of the scattering centers due to interactions between the scattering objects. The arrangement of particles or micelles depends on their potential of mean force, and reliable descriptions of this potential exist for colloidal structures. The potential of mean force yields $S(q)$ via application of one of a family of well-studied liquid-state models, and computationally convenient methods exist to aid in data fitting [18–20]. For dilute solutions of weakly interacting particles, such as a dilute solution of latex particles, the structure factor correction is approximately unity and can often be neglected. However, for microemulsions consisting of micelles made up of ionic surfactants, the structure factor contribution is a critical component of the scattering intensity.

Figure 2 shows the calculated scattering spectra for a hypothetical microemulsion consisting of monodisperse micelles made up of ionic surfactants. The peak in the scattering intensity is due to the periodic variation of the structure factor, and its position is an indication of the relative distance between the micelles. The sharp minima in the form factor and scattering intensity are seldom observed because of polydispersity in the size and shape of the micelles as well as background scattering.

Polydispersity in the size or shape of the micelles and polymer particles can be taken into account approximately by decoupling the correlation between position and size or orientation [21]. However, for highly polydisperse or bidisperse samples, e.g., polymerized microemulsions containing both micelles and particles, more rigorous approaches are necessary. In the case of hard-sphere systems, Vrij [22] has derived an analytical result for the scattering intensity that is particularly convenient for model fitting SAS spectra. For more complex pair interaction potentials, the partial structure factors between individual components of the mixture must be numerically calculated to obtain the total scattering intensity [18].

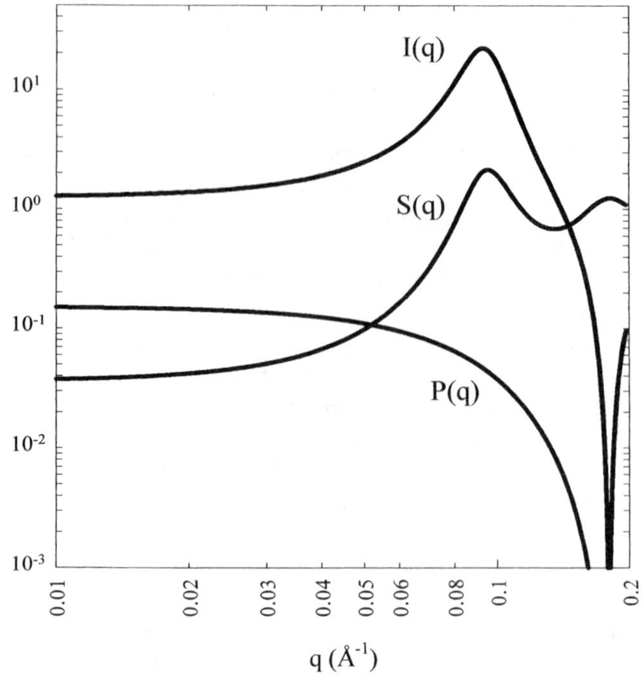

FIG. 2 Variation of the form factor [$P(q)$], structure factor [$S(q)$], and scattering intensity [$I(q)$] for a hypothetical microemulsion consisting of monodisperse spherical micelles with a radius of 25 Å and a volume fraction of 0.15. The structure factor is calculated using the Percus-Yevick closure [19,20] and assuming that the micelles interact via an effective hard-sphere radius of 35 Å.

III. EXPERIMENTAL TECHNIQUES

A. Kinetics

For the most accurate kinetic studies, inhibitors must be completely removed from the monomer and this typically requires vacuum distillation. Oxygen must be completely removed from the water and the reactor must be kept at a positive pressure of nitrogen or argon. Higher purity grades of nitrogen or argon are recommended, especially for kinetic studies at low initiator concentrations. However, the sensitivity of these reactions to oxygen can be used advantageously as a means of quenching the reaction rapidly as samples are taken from the reactor.

The monomer content in polymerizable microemulsions with droplet-type microstructures is typically only a few percent and rarely exceeds 10%. Accuratcly measuring the polymerization kinetics in a mixture of monomer, polymer, surfactant, and water presents significant analytical challenges. The accuracy of most spectroscopic techniques is significantly reduced by the need to correct for the increasing turbidity of the samples as the concentration of latex particles increases. Chromatographic techniques are complicated by the presence of polymer, surfactant, and water. Gas chromatography may be feasible for volatile and less reactive monomers with thermally stable surfactants but is not generally applicable. Because of the low concentration of analyte, bulk techniques to reduce interferences due to either polymer, surfactant, or water often lead to large errors in conversion measurement (>5%). Titration of the monomer double bonds by bromination, hydrogenation, ozonolysis, and other rapid chemical reactions may be possible, but side reactions involving the surfactant and water are often problematic. The most reliable and general techniques for measuring the polymerization kinetics are densimetry, calorimetry, and a gravimetric technique we use in our laboratory.

It is routinely possible to measure the density of liquids to five significant digits by measuring the frequency of vibration of a U-tube filled with the sample; such is the principle of operation of Anton Paar densimeters. Although the relationship between measured density and conversion is often highly linear, it is prudent to confirm this by gravimetry. The accuracy of these U-tube densimeters can be seriously compromised by gradual phase separation of the sample or changes in the viscosity of the samples. Under optimal conditions, conversion measurements with less than ±1% error can be achieved.

Reaction calorimeters, e.g., the Mettler RC1, typically have adequate sensitivity to monitor polymerization kinetics provided that the reactor head is heated to prevent condensation. The accuracy of this technique is relatively poor (±5%) if it is necessary to relate conversion to the heat generated on an absolute scale. The accuracy of the technique is significantly improved if the final conversion can be determined independently using a different technique such as gravimetry.

Accurate gravimetric measurement of the polymer formed is possible if the monomer is adequately volatile. This technique is typically feasible only for monomers with volatilities at least equal to that of *n*-hexyl methacrylate (boiling point 90°C at 14 mm Hg). Surfactants with very low but nonnegligible volatility such as ethoxylated alkyl surfactants are incompatible with this technique. In this procedure, approximately 5-mL samples are extracted from the reactor directly into preweighed 20-mL ampules, sealed with a septum, then frozen in liquid nitrogen. Condensate from the ampules is washed off with acetone, and then the amount of sample in the ampules is measured by difference. The water and monomer are then evaporated by positioning the ampules horizontally overnight under a gently blowing air stream. A volatile solvent (e.g., acetone) in which the surfactant is insoluble is used to extract residual volatiles from the cake, and the ampules are air dried again after rotating them 180° in the rack. After thorough air drying, the samples are dried further in a vacuum oven initially kept at 30°C for 12 h and then ramped up to 60°C for 6 h. The samples are cooled to ambient temperature and then reweighed. The precision of the 30 to 40 conversion measurements obtained for each reaction is routinely better than ±0.5% provided the weighings are performed carefully on an analytical balance. Note that the buildup of static electricity on the thoroughly dried glass ampules can seriously affect the accuracy of digital balances and care should be taken to minimize frictional contact. The accuracy of these measurements is within ±1% and is limited by the accuracy in preparing the original microemulsion.

The most reliable way to obtain rates of polymerization is by differentiating accurate conversion versus time data using the cross-validation smoothing spline algorithm developed by Craven and Wahba [23]. Statistical validity is then examined by using a bootstrapping algorithm [24] to account for errors in measuring the conversion and time.

B. Molecular Weight Distribution

Accurate measurement of the molecular weight distribution of polymers prepared by microemulsion polymerization has been hampered by their very high molecular weights. Because of the unique morphology of polymers prepared by microemulsion polymerization and the relatively low quality of high-molecular-weight commercial standards, absolute molecular weight determination using light scattering detectors is required. However, columns capable of separating high-molecular-weight polymers typically shed and interfere with light scattering detectors. Only with the advent of nonshedding columns for high-molecular-weight separations from Polymer Laboratories has it been possible to perform these measurements properly. Even then, the high molecular weight of these polymers frequently results in overloaded columns, and it is necessary to work at very low concentrations such that the concentration of polymer is less than 10 ppm at the peak. Careful static light scattering measurements on the original unfiltered samples give molecular weights that are consistent with those obtained chromatographically, so shear degradation during the chromatographic separation is minimal.

C. Particle Size Distribution

Numerous experimental techniques exist for measuring particle size. For polydisperse samples, fractionation techniques, e.g., field flow fractionation (FFF), capillary hydrodynamic chromatography, or centrifugation, are typically necessary for accurate estimation of the particle size distribution. However, the very small particles prepared by microemulsion polymerization may cause difficulties in applying these fractionation techniques. Direct imaging by transmission electron microscopy (TEM) gives an excellent indication of the particle size distribution, but care must be taken to avoid artifacts related to sample preparation and the sizing of only a small number of particles.

Quasi-elastic light scattering (QLS) is a popular and relatively easy to use technique for estimating the particle size distribution of latexes prepared by microemulsion polymerization. However, particle size measurements obtained using QLS can be highly misleading and these results must be interpreted very carefully. Therefore, we discuss some of the caveats here in detail.

QLS measurement must be performed in the absence of interactions between the particles. This can be ascertained by checking that no variation in the measured particle size occurs upon further dilution. A volume fraction of approximately 0.1% is an adequate starting point for these dilution studies. Destabilization of some latexes may occur if they become highly diluted with pure water. In these cases, a solution of the same surfactant used to prepare the microemulsion at a concentration slightly below the critical micelle concentration (cmc) should be used. For typical latexes, multiple scattering is no longer a concern at the dilute concentrations necessary to minimize the interactions.

The light scattering intensity is, to a first approximation, proportional to the sixth power of the particle radius. Deconvoluting QLS data to yield an accurate size distribution in terms of normalized number density versus radius is often impossible. Experimental errors and the strong dependence of the scattering intensity on radius typically result in numerical instabilities that preclude this type of analysis. Therefore, the particle size distribution measured using QLS is typically plotted as intensity versus radius. For polydisperse samples, this may result in a gross overestimation of the average particle size and underestimation of the polydispersity. Modern techniques that take advantage of the weaker scattering of large particles at larger scattering angles and simultaneously analyze QLS data obtained at multiple angles can provide better estimates [25]. Nevertheless, the use of complementary techniques such as SAS and electron microscopy is required to confirm QLS results.

D. Monomer Partitioning

The rate of polymerization is directly proportional to the monomer concentration at the locus of polymerization. Clearly, any progress in understanding the polymerization kinetics as well as the molecular weight and particle size distributions requires measurement of the monomer concentration. Traditional techniques for determining latex swelling by solvents are not applicable because it is impossible to separate the monomer-swollen polymer particles and monomer-swollen micelles without affecting the partitioning of monomer. Until recently, the only independent measurements of the monomer concentration in polymer particles have been based on pulsed laser polymerization (PLP) of styrene microemulsion [26]. Unfortunately, the PLP technique is restricted to low conversions (<10%). QLS has been used by some researchers to monitor the swelling of polymer particles. However, as discussed in Section III.C, accurate particle size measurements are possible only when multiple scattering effects and interparticle interactions can be ignored. Dilution of the samples to avoid these complications strongly affects the swelling equilibrium and severely limits the accuracy of monomer partitioning results acquired using QLS.

On-line SANS investigations of *n*-hexyl methacrylate (C6MA) microemulsion polymerizations using mixed DTAB and DDAB surfactants indicate that the polymer particles are not significantly swollen with monomer [27]. However, the monomer concentration in the polymer particles could not be reliably extracted from these SANS spectra because many fitting parameters are required and the fits may not be unique. To overcome this, it is necessary to simplify the experiment and models by developing assumptions that can be validated post hoc.

This alternative approach takes advantage of the observation that the size of the unswollen polymer particles varies little with conversion [28–30]. In addition, because of the small size of the polymer particles, no diffusion limitations are expected and monomer partitioning is governed by equilibrium thermodynamics. Thus, it is possible to use equilibrium mixtures of the fully polymerized microemulsions with the original unpolymerized microemulsion to mimic a polymerizing microemulsion (C. C. Co et al., submitted). Figure 3 shows model calculations for such an experiment in which a fully polymerized microemulsion is mixed with the original microemulsion in a 1:1 ratio to mimic a 50% conversion sample. Initially, the scattering intensity is reduced by a factor of approximately 2 due to the reduction in number density of the polymer particles. However, as the particles are swollen with monomer, a recovery in the scattering intensity is observed that gives a quantitative measure of the monomer volume fraction or concentration within the polymer particles.

The entire monomer concentration profile can be obtained by simultaneously fitting the SANS spectra of mixtures containing varying ratios of the fully polymerized and unpolymerized microemulsions to a model. This model assumes that the unswollen particle size distribution can be characterized in terms of the average diameter and standard deviation of a discretized Schultz distribution. The number density of each discrete population then scales directly with conversion. Accordingly, the diameter of each discrete population scales volumetrically with the monomer volume fraction in the polymer particles that is used as a single fitting parameter for each SANS spectrum. Figure 4 shows that the unswollen particle size distribution, obtained by simultaneously fitting the SANS spectra at all conversions, is in excellent agreement with independent cryo-TEM measurements (C. C. Co et al., submitted). The robustness of this technique stems from the very small number of adjustable parameters used to fit each SANS spectrum.

Note that at the low q values over which the SANS spectra are fitted, the monomer-swollen micelles and D_2O act together as an effective solvent whose scattering is negligible compared with that of the polymer particles. Furthermore, because of the low volume frac-

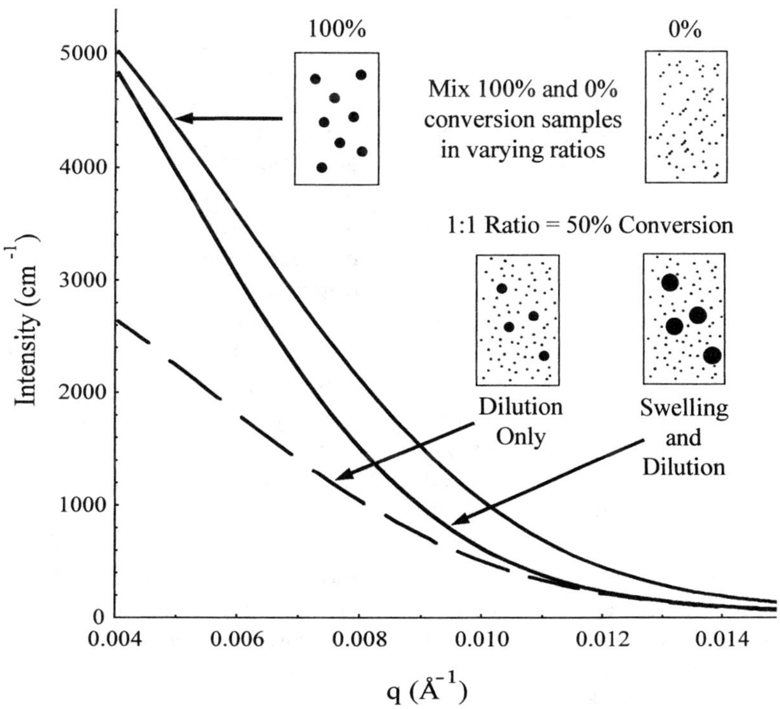

FIG. 3 Schematic of equilibrium SANS swelling experiment in which fully polymerized microemulsions are mixed with the original unpolymerized microemulsions at varying ratios to simulate different levels of conversion. For the case of 1:1 mixing, the number density of the polymer particles is diluted by a factor of 2, which leads to a commensurate initial decrease in the scattering intensity. Equilibrium partitioning is then rapidly achieved and the polymer particles are swollen with monomer, resulting in recovery of the scattering intensity.

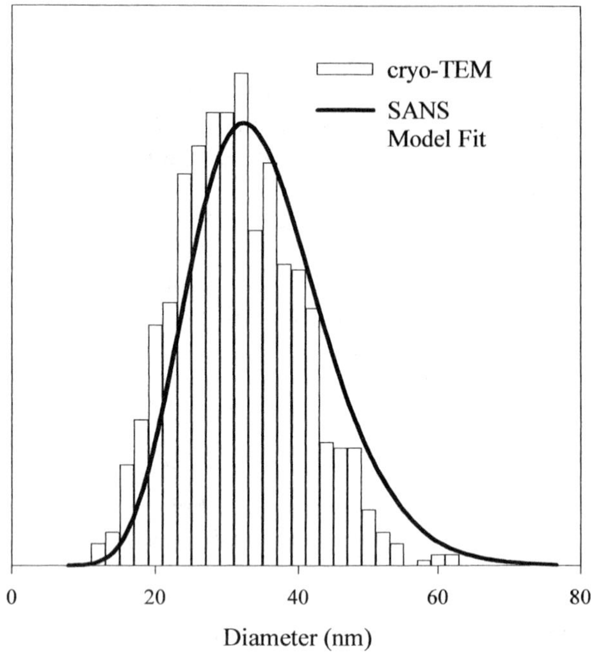

FIG. 4 Comparison of the particle size distribution of poly(nC4MA) obtained by fitting the equilibrium SANS spectra and cryo-TEM imaging (courtesy of Stefan Burauer, Institute for Physical Chemistry, University of Cologne). Details of the experiment are given by C. C. Co et al. (submitted).

tion of particles, interactions between the polymer particles are almost negligible and can be approximated roughly as that between hard spheres. Consequently, absolute scattering intensities can be calculated readily using a method due to Vrij [22].

Finally, it is possible to confirm that these equilibrated mixtures of fully polymerized and unpolymerized microemulsions are indeed representative of the microemulsion as it polymerizes. This is done by comparing the full SANS spectra of these equilibrated mixtures with those obtained from on-line SANS experiments as shown in Fig. 5. The remarkably good agreement between the on-line and equilibrium SANS spectra validates this approach and attests that the monomer partitioning results measured using equilibrated samples accurately represent the partitioning of monomer during actual microemulsion polymerizations. With the exception of very low conversion samples (<5%), the agreement between the on-line and equilibrium SANS spectra generally extends all the way to low q values. This result confirms the previous assumption of a relatively constant unswollen particle size as predicted by the Morgan model for the particle size distribution described in the next section.

IV. MECHANISM AND THEORY

A. Morgan Model for Kinetics and Size Distributions

The rate of polymerization within each particle containing a single growing polymer chain ($k_p c_m$) is directly proportional to the propagation rate constant (k_p) and the local concentration of monomer (c_m). The overall rate at which monomer is converted to polymer is then

$$\frac{\partial f}{\partial t} = \frac{k_p c_m N^*}{M_0} \tag{3}$$

where f is the extent of conversion, N^* is the number of growing radical chains, and M_0 is the total concentration of monomer in the entire microemulsion [31]. To apply this equation to describe the microemulsion polymerization kinetics, it is necessary to know how c_m and N^* vary as the polymerization proceeds.

For both of the two limiting cases in which the monomer either partitions mainly in the polymer particles or into the micelles, the monomer concentration would decrease approximately linearly with increasing conversion, i.e., $c_m = c_0(1 - f)$. For the case in which the monomer does not swell the polymer and partitions predominantly in the micelles, the initial monomer concentration (c_0) can be calculated if one assumes that the surfactant hydrocarbon tails act merely as diluents. The partitioning of monomer between the polymer particles and micelles has been directly measured using the SANS technique described in Section III.D. For all systems investigated, an approximately linear relationship between monomer concentration and conversion was observed (C. C. Co et al., submitted). However, neither limit was approached and monomer partitions more or less equally between the polymer particles and micelles. A new model for monomer partitioning has been developed to describe these results and is summarized in Section IV.D.

Several factors influence the number of growing radicals as a function of time. Experimentally, the initiator half-life is long relative to the time of reaction, so new radicals are produced at a constant rate. In that case, and in the absence of biradical termination, the number of growing radical chains would simply increase linearly with time.

$$N^* = 2k_d[I]t \tag{4}$$

Here, k_d and $[I]$ are the decomposition rate constant and concentration of the initiator, respectively. For monomers with very low water solubility, biradical termination in both the polymer particles and the aqueous

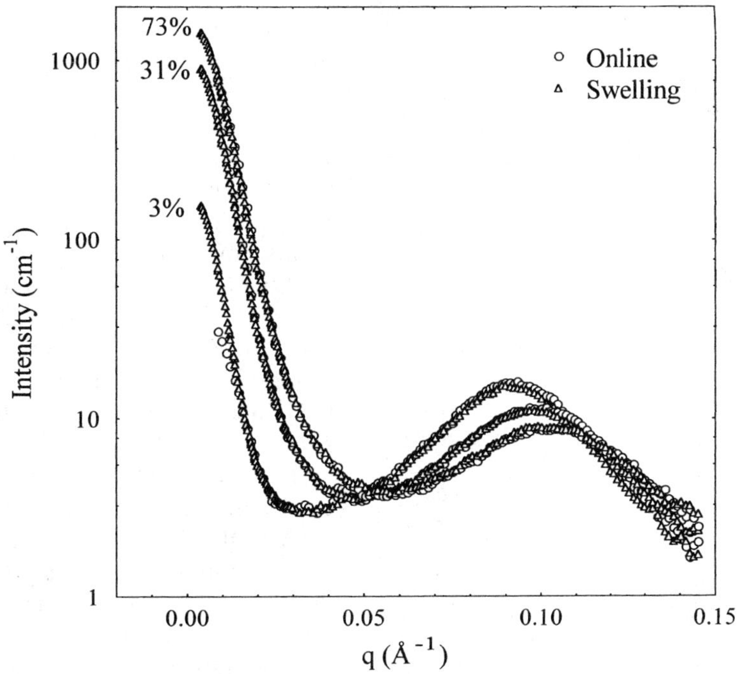

FIG. 5 Comparison of the equilibrium (C. C. Co et al., submitted) and on-line SANS spectra.

phase can be neglected. This assumption was carefully justified by Morgan et al. [31], who also assumed a linear monomer concentration profile to arrive at the following expression for the conversion as a function of time.

$$f = 1 - \exp\left(-\frac{k_p c_0 k_d [I]}{M_0} t^2\right) \quad (5)$$

Figure 6 shows a comparison of this kinetic model with experimentally measured kinetics for C6MA microemulsion polymerizations.

Remarkably, this kinetic model also predicts that the maximum rate of polymerization will always occur at

$$f_{max} = 1 - \frac{1}{\sqrt{e}} = 0.39 \quad (6)$$

for the polymerization of any microemulsion, subject of course to the two assumptions outlined previously. For the microemulsion polymerization of *n*-hexyl methacrylate (C6MA) with mixed DTAB-DDAB surfactants, the maximum rate of polymerization indeed occurs at 39% conversion (Fig. 7). Moreover, when literature values for k_p and k_d are used together with a value of c_0 consistent with the situation in which monomer partitions predominantly in the micelles, the model quantitatively predicts the experimental kinetics.

Using the same assumptions used to derive this kinetic model, Morgan and Kaler [30] have put forward an analytical equation describing the evolution of the molecular weight distribution (MWD) during microemulsion polymerization. The basic idea is to use the kinetic model to calculate (1) the previous time (t_1) when a chain must be initiated to grow to a length of $\mathcal{L}$ monomer units at time t and (2) the time period necessary for one propagation event at time t_1. The number of chains initiated in this time period at time t_1 is then equal to the number of chains (n) with length $\mathcal{L}$ at time t. In typical microemulsion polymerization systems, chain transfer to monomer followed by radical exit from the growing particle appears to be the dominant mechanism for chain transfer. Accounting for chain transfer probabilities results in a complex but analytical expression for the chain length or molecular weight distribution. Because the model assumptions imply that each particle contains a single polymer chain, the MWD can be readily converted to a particle size distribution (PSD).

The validity of this single-chain assumption can be tested by comparing the MWD measured using GPC/MALLS/RI with the MWD calculated from the PSD measured using cryo-TEM assuming that the particles consist of single polymer chains. Such a comparison is shown in Fig. 8, where the peaks of the MWD ex-

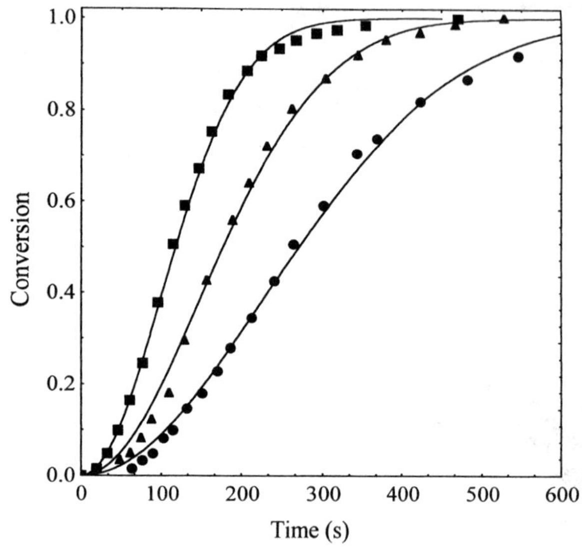

FIG. 6 Model and experimental kinetics of C6MA microemulsion polymerizations with DTAB/DDAB surfactants for varying V50 initiator/C6MA monomer weight ratios (■, 0.25 wt%; ▲, 0.10 wt%; ●, 0.05 wt%). Lines are calculated from the Morgan kinetic model [Eq. (5)] treating the group of constants ($k_p c_0 k_d / M_0$) as a single adjustable parameter.

tracted from cryo-TEM imaging match nearly exactly the MWD obtained using GPC/MALLS/RI (C. C. Co et al., submitted). The discrepancy in the width of the distributions is an artifact of the difficult chromatographic separation. A typical plot of the theoretical PSD at different levels of monomer conversion is shown in Fig. 9. Note that the most probable particle size remains constant at ~150 Å and is a consequence of continuous formation of new particles and termination of growing particles by chain transfer. This "steady-state" behavior in the MWD and PSD has been verified by particle size measurements using QLS and SANS [27] in conjunction with molecular weight measurements (C. C. Co et al., submitted).

The success of the kinetic and size distribution models for the microemulsion polymerization of C6MA in DTAB-DDAB microemulsions gives the impression that everything is understood. However, examination of the polymerization kinetics of C6MA in DTAB-only microemulsions (Fig. 10) reveals that the maximum rate of polymerization occurs at ~30%. For styrene microemulsion polymerizations, the maximum rate of polymerization shifts further to ~20% conversion. In-

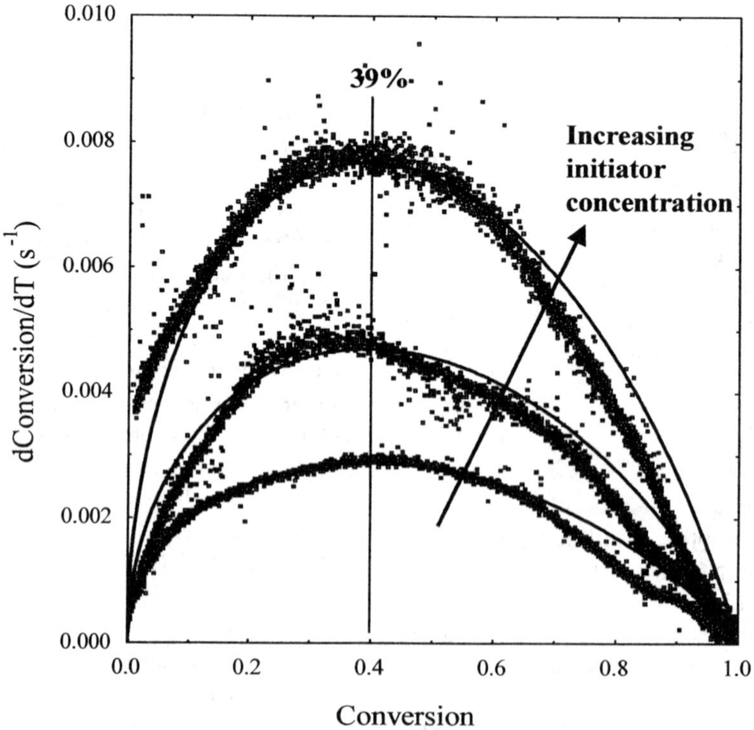

FIG. 7 Model and experimental rates of polymerization as a function of conversion for C6MA microemulsion polymerizations with DTAB/DDAB surfactants.

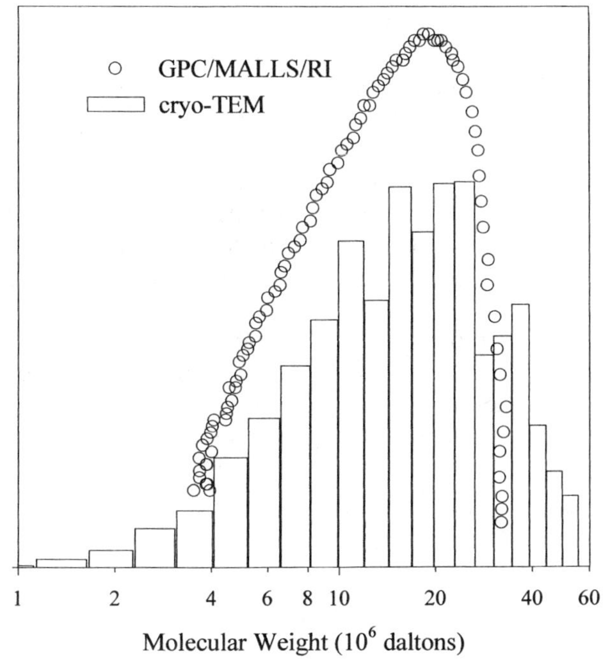

FIG. 8 Comparison of the molecular weight distribution of poly(nC4MA) calculated from the cryo-TEM particle size distribution (Fig. 4) assuming that the particles consist of single polymer chains with GPC/MALLS/RI results (courtesy of Patricia Cotts, Dupont Central Research & Development, Wilmington, DE). Details of the experiment are given by C. C. Co et al. (submitted).

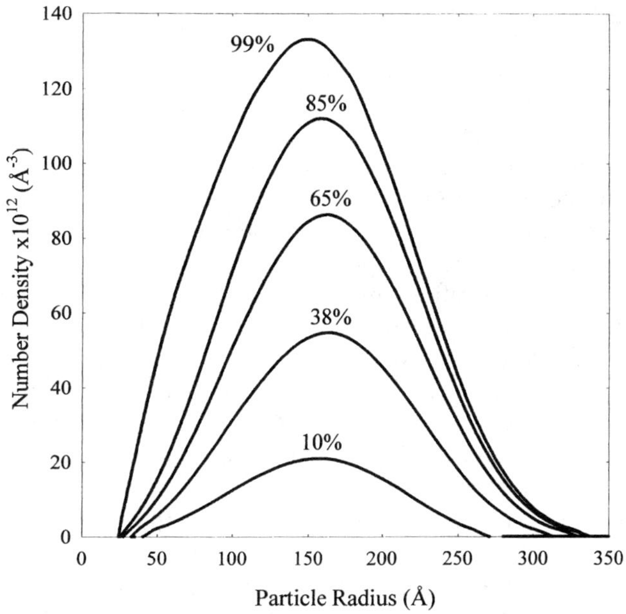

FIG. 9 Particle size distribution of C6MA as calculated using Morgan's model [27,30].

deed, Nomura and Suzuki [32] rederived the Morgan kinetic model and concluded that it is valid for styrene microemulsion polymerizations only at very low conversions. However, these observations should come as no surprise given the assumptions of a monomer concentration profile that decreases linearly with conversion coupled with a radical concentration profile that increases linearly with time. Indeed, the shift in the maximum rate of polymerization to ~30% for C6MA polymerized in DTAB-only microemulsions can be explained by the slight nonlinearity in the monomer concentration profile (C. C. Co et al., submitted). However, the further shift of the maximum rate to even lower conversions in the case of styrene is probably due to a combination of nonlinearities in the monomer concentration profile with biradical termination and glass transition effects that were not considered in the original model. These additional complications are discussed in the following sections.

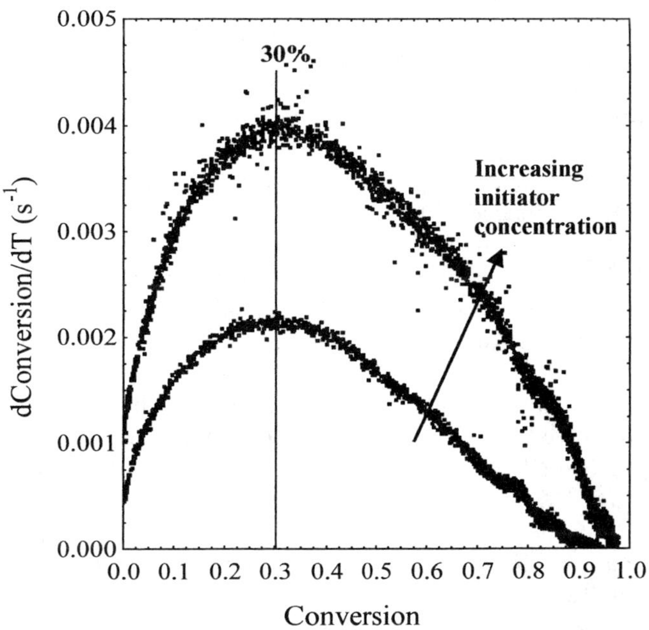

FIG. 10 Experimental C6MA polymerization kinetics in DTAB-only microemulsions (R. de Vries et al., submitted).

B. Initiator Artifacts

A secondary but nevertheless important result discussed by Morgan et al. [31] is related to artifacts that may arise due to initiator side reactions. For example, because of side reactions with water, potassium persulfate (KPS) solutions become increasingly acidic and, at pH values below 3, KPS decomposes predominantly through a nonradical pathway [33]. Consequently, free radical generation in KPS-initiated microemulsion polymerizations ceases prematurely, and this is manifested by a shift in the maximum rate of polymerization to ~20% conversion. However, on buffering the microemulsion at neutral pH through the addition of 2 mM potassium bicarbonate, the maximum rate of polymerization shifts back to ~39% conversion and the overall rate of polymerization is increased sixfold. Nevertheless, the mechanism of KPS free radical initiation is very complex and this strong pH dependence is not generally observed. Given the complex nature of KPS initiation, microemulsion polymerization studies are best performed using azo-based initiators such as V50 that exhibit simpler initiation pathways. Note that accurate kinetic studies involving ultraviolet (UV) initiation are impossible because of the increasing turbidity of the microemulsions as they polymerize.

C. Biradical Termination Effects

The ratio of the number of micelles to the number of polymer particles during and after the polymerization is ~1000 [27,34]. In addition, only a small fraction of the polymer particles actually contain a growing radical chain. Consequently, it seems perfectly logical to ignore the possibility of biradical termination that occurs when a free radical species enters a growing particle. Morgan et al. [31] carefully justified that this assumption is indeed valid for free radical species derived from highly water-insoluble monomers such as C6MA.

In other cases, this situation may not hold, and a more detailed accounting of termination effects is warranted. In general, water-soluble initiators homolytically dissociate to form highly reactive free radicals. These initiator-derived radicals react with monomer dissolved in the aqueous phase to form the primary free radical (IM·) so rapidly that this reaction becomes diffusion limited [35]. The second propagation step (IM· + M) is much slower and has a rate constant comparable to the polymerization rate constant (k_p). The second source of free radicals in the aqueous phase comes from monomeric radicals (M·) that are generated by chain transfer to monomer and exit the particle, which now contains a dead polymer. Upon entering a monomer-swollen micelle, either radical species may initiate polymerization or exit without initiating polymerization. Which event occurs depends on the relative residence and propagation time scales. For typical monomeric radicals, successful initiation occurs only after ~10^2 entry and exit events (R. de Vries et al., submitted). Consequently, the probability that these aqueous phase radicals enter a growing polymer particle and thus result in biradical termination may not be negligible.

Based on these ideas, a more quantitative analysis yields the following equation for the number of growing particles (N^*) in the presence of biradical termination (R. de Vries et al., submitted).

$$\frac{\partial N^*}{\partial t} = \rho \frac{1 - p_{\text{term}}/p_{\text{prop}}}{1 + p_{\text{term}}/p_{\text{prop}}} - 2k_{\text{tr}}C_{\text{mon}}^{(\text{part})}N^* \frac{p_{\text{term}}/p_{\text{prop}}}{1 + p_{\text{term}}/p_{\text{prop}}} \tag{7}$$

The first term represents termination effects due to primary free radicals that are generated at a rate ρ. The second term represents the rate of biradical termination due to transfer-generated monomeric radicals. These radicals are generated at a rate that depends on the chain transfer constant (k_{tr}) and the concentration of monomer in polymer particles ($C_{\text{mon}}^{(\text{part})}$). The ratio $p_{\text{term}}/p_{\text{prop}}$ simply represents the relative probabilities for termination and propagation for each entry event and can be estimated by:

$$\frac{p_{\text{term}}}{p_{\text{prop}}} \approx \frac{1}{\tau_{\text{res}}k_p C_{\text{mon}}^{(\text{mic})}} \frac{N^*}{N} \tag{8}$$

Here, τ_{res} is the residence time of the free radical species within the micelle and $C_{\text{mon}}^{(\text{mic})}$ represents the concentration of monomer in the micelles. Accurate estimates of the residence times are generally unavailable. Nevertheless, the residence times are expected to scale with the water solubilities of the free radical species, which in turn depend on the aqueous solubility of the monomer. As demonstrated by de Vries et al. (submitted), this theory, in conjunction with other experimental observations, can partly explain why styrene microemulsion polymerizations exhibit a maximum rate of polymerization at about 20% conversion instead of 39% as predicted by the original kinetic model.

D. Monomer Partitioning

As mentioned previously, the assumption of a linear monomer concentration profile at the locus of polymerization was based on analyzing the limiting cases in which (1) monomer partitions predominantly in the polymer particles and (2) monomer partitions mainly in the micelles. The first limiting case is the result of an approximate thermodynamic analysis that utilizes a Flory-Huggins approach to estimate the chemical potential of monomer in both the polymer particles and micelles [29]. This model predicted that at low conversions (<5%), essentially all the monomer partitions in the polymer particles and the monomer concentration decreases linearly afterward. However, on-line SANS results clearly showed that this is not true for C6MA microemulsion polymerizations with DTAB-DDAB surfactants [27]. Instead, the on-line SANS results and kinetic data indicated that the opposite limit where the monomer does not swell the polymer particles was the case. The on-line/equilibrium SANS experiments described previously have permitted the direct measurement of the monomer concentration within the polymer particles.

Monomer partitioning measurements for styrene/DTAB/D_2O microemulsions have been performed at monomer loadings of $\alpha = 3$, 5, and 7.5% and a constant surfactant concentration of $\gamma = 12\%$. The phase boundary for this microemulsion lies at $\alpha = 8.2\%$. As the phase boundary is approached, the volume fraction of monomer in the polymer particles increases and the overall profile becomes increasingly nonlinear (Fig. 11). These two key features are captured by the theoretical curves as calculated using the following model for monomer partitioning (C. C. Co et al., submitted).

Assuming rapid diffusion, and hence thermodynamic equilibrium, the chemical potential of monomer in the micelles ($\mu_{\text{mon}}^{(\text{mic})}$) and that in the polymer particles ($\mu_{\text{mon}}^{(\text{part})}$) must be equal. Again, the Flory-Huggins equation is used to estimate the chemical potential of monomer in the polymer particles ($\mu_{\text{mon}}^{(\text{part})}$). However, the calculation of $\mu_{\text{mon}}^{(\text{mic})}$ is based on the curvature elastic energy and translational entropy of the micelles that can be expressed in terms of the actual (R_{mic}) and maximum possible ($R_{\text{mic,max}}$) micelle radii.

$$\mu_{\text{mon}}^{(\text{mic})} = -\frac{6V_{\text{mon}}}{R_{\text{mic}}^3}\left\{k_c c_0(R_{\text{mic,max}} - R_{\text{mic}}) + \frac{k_B T}{4\pi}\ln\left(R_{\text{mic}}/R_{\text{mic,max}}\right)\right\} \quad (9)$$

Here, V_{mon} is the molecular volume of the monomer. $R_{\text{mic,max}}$ corresponds to the radii of the monomer swollen micelles at the phase boundary ($\alpha = 8.2\%$) and can be calculated by mass and surface area balances. Likewise, these balances are used to calculate the volume fraction of monomer in the polymer particles in terms of R_{mic}, which is the radius adjusted to equate $\mu_{\text{mon}}^{(\text{part})}$ and $\mu_{\text{mon}}^{(\text{mic})}$. The product of the bending modulus and spontaneous curvature $k_c c_0$ remains as a single adjustable parameter in fitting all three sets of data simultaneously, and the fitted value of $k_c c_0$ (2.4 kT/nm) is within the range of reported values (C. C. Co et al., submitted).

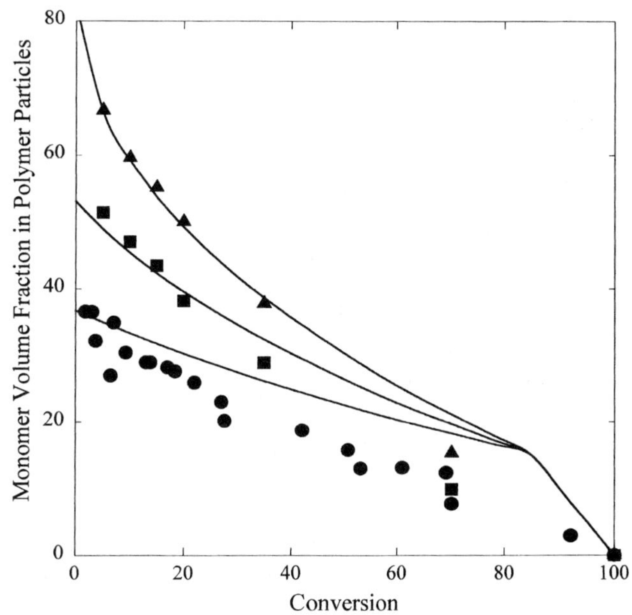

FIG. 11 Styrene monomer concentration profile as measured using the equilibrium SANS technique at varying monomer loadings (●, $\alpha = 3$ wt%; ■, $\alpha = 5$ wt%; ▲, $\alpha = 7.5\%$ wt%). The phase boundary lies at $\alpha = 8.2$ wt%. The solid lines were calculated using the monomer partitioning model described in the text.

E. Glass Transition Effects

Glass transition effects can be observed even at low conversions during microemulsion polymerizations. A study of the kinetics of *n*-butyl methacrylate (nC4MA) and *t*-butyl methacrylate (tC4MA) polymerizations with DTAB surfactant in water demonstrates this clearly. nC4MA and tC4MA have almost identical aqueous solubilities (3.4 and 4.3 mM at 60°C, respectively) and propagation rate constants (1015 and 1040 L mol^{-1} s^{-1}, respectively), which rules out any significant differences in the polymerization kinetics related to biradical termination effects. Monomer partitioning experiments similar to that shown previously for styrene also show that the concentration profiles are essentially identical (C. C. Co et al., submitted). However, Fig. 12 shows that under the same polymerization conditions, the kinetic profile for tC4MA tracks that of nC4MA until ~15% conversion, after which the tC4MA rate starts dropping off (R. de Vries et al., submitted). Polymer glass transition temperature (20°C and 105°C, respectively) seems to be the only property that distinguishes these two monomers adequately to explain the variance in their kinetic profiles.

The kinetic profile of tC4MA is reminiscent of that for styrene, which also has a maximum rate of polymerization at ~20% conversion. This observation makes it tempting to consider glass transition effects as well for styrene microemulsion polymerizations. After all, polystyrene has a glass transition temperature of 100°C and an aqueous solubility of 4.6 mM that is very comparable to that of nC4MA and tC4MA. However, kinetic profiles obtained at higher temperatures (80°C) and monomer loadings ($\alpha = 7.5\%$), where glass transition effects are not expected, still manifest a maximum rate of polymerization at ~20% conversion.

The relative importance of biradical termination and glass transition effects in styrene microemulsion polymerization is still not clear. The lower propagation rate constant of styrene (340 L mol^{-1} s^{-1}) compared with butyl methacrylates leads to higher biradical termination rates [Eq. (8)] that can quantitatively explain the observed kinetics. However, glass transition effects cannot be ruled out, especially for reactions at lower temperatures and monomer loadings. There also remains the remote possibility that the growing polystyrene particles become glassy at the earliest stages of particle growth. In this situation, monomer partitioning may be diffusion controlled. Unfortunately, both online and equilibrium SANS monomer partitioning experiments offer no insight into this issue because the number of growing polymer particles constitutes only a very small fraction of the total number of polymer particles.

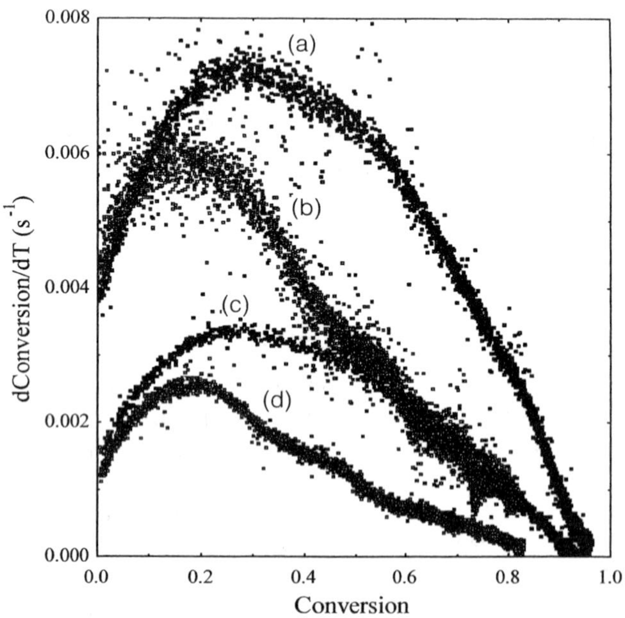

FIG. 12 Kinetic profiles for nC4MA and tC4MA microemulsion polymerizations initiated with 0.063 and 0.25 wt% V50 initiator relative to monomer. (a), nC4MA 0.25 wt%; (b), tC4MA 0.25 wt%; (c), nC4MA 0.063 wt%; (d), tC4MA 0.063 wt%.

V. CONCLUSIONS/DIRECTIONS FOR FUTURE RESEARCH

Experimental and theoretical studies have greatly advanced our mechanistic understanding of microemulsion polymerization. Nevertheless, several important issues have yet to be resolved. For example, the relative importance of biradical termination and glass transition effects has yet to be understood. Experimentally, the extent of biradical termination during microemulsion polymerization may be probed via relaxation studies on microemulsions initiated using γ-radiation. Systematic kinetic studies of monomers with relatively high water solubility and low polymer glass transition temperatures may also yield insight into this question. Alternatively, the same information may be obtained by systematically studying the kinetics of nC6MA or nC4MA polymerizations at higher temperatures (>>60°C) using more thermally stable azo initiators. Understanding these mechanistic details will be the key to developing better models for the molecular weight and particle size distribution.

Typical microemulsion polymerization recipes call for the use of at least an equal amount of surfactant and monomer, which makes the polymerization and surfactant removal processes prohibitively expensive. Intermittent monomer addition during microemulsion polymerization has been reported to reduce the surfactant requirements by a factor of 10 or more without increasing the average particle size [36,37]. It is still unclear why the existing polymer particles do not act as seed particles and grow continuously as more monomer is added. Thoroughly understanding the mechanism of this process may result in an economically feasible microemulsion polymerization process.

REFERENCES

1. F. Candau, *Polymerization in Organized Media* (C. M. Paleos, ed.), Gordon & Breach, Philadelphia, 1992, pp. 215–282.
2. S. D. Desai, R. D. Gordon, A. M. Gronda, and E. L. Cussler, Curr. Opin. Colloid Interface Sci. *1*:519–522 (1996).
3. W. Meier, Curr. Opin. Colloid Interface Sci. *4*:6–14 (1999).
4. M. R. Ferrick, J. Murtagh, and J. K. Thomas, Macromolecules *22*:1515–1517 (1989).
5. S. C. Pilcher and W. T. Ford, Macromolecules *31*: 3454–3460 (1998).
6. R. G. Gilbert, *Emulsion Polymerization: A Mechanistic Approach*, Academic Press, San Diego, 1995.
7. M. Morton, S. Kaizerman, and M. W. Altier, J. Colloid Sci. *9*:301–312 (1954).
8. E. Vanzo, R. H. Marchessault, and V. Stannett, J. Colloid Sci. *20*:62–71 (1965).
9. M. Kahlweit, R. Strey, D. Haase, H. Kunieda, T. Schmeling, B. Faulhaber, M. Borkovec, H. F. Eicke, G. Busse, F. Eggers, T. H. Funk, H. Richmann, L. Magid, O. Soderman, P. Stilbs, J. Winkler, A. Dittrich, and W. Jahn, J. Colloid Interface Sci. *118*:436–453 (1987).
10. M. Kahlweit and R. Strey, Angew. Chem. Int. Ed. Engl. *24*:654–668 (1985).
11. K. M. Lusvardi, A. P. Full, and E. W. Kaler, Langmuir *11*:487–492 (1995).
12. K. M. Lusvardi, K. V. Schubert, and E. W. Kaler, Langmuir *11*:4728–4734 (1995).
13. B. Weyerich, J. Brunner-Popela, and O. Glatter, J. Appl. Crystallogr. *32*:197–209 (1999).
14. E. W. Kaler, in *Modern Aspects of Small-Angle Scattering* (H. Brumberger, ed.), Kluwer Academic Publishers, Dordrecht, 1995, pp. 329–354.
15. S. H. Chen, Annu. Rev. Phys. Chem. *37*:351–399 (1986).
16. J. S. Pedersen, Curr. Opin. Colloid Interface Sci. *4*: 190–196 (1999).
17. J. B. Hayter, in *Physics of Amphiphiles* (V. Degiorgio and M. Corti, eds.), North Holland, Amsterdam, 1985, pp. 59–93.
18. S. R. Kline and E. W. Kaler, J. Appl. Crystallogr. *29*: 427–434 (1996).
19. J. P. Hansen and I. R. McDonald, *Theory of Simple Liquids*, 2nd ed, Academic Press, London, 1986.
20. K. S. Schmitz, *Macroions in Solution and Colloidal Suspension*, VCH, New York, 1993.
21. M. Kotlarchyk and S. H. Chen, J. Chem. Phys. *79*: 2461–2469 (1983).
22. A. Vrij, J. Chem. Phys. *71*:3267–3270 (1979).
23. P. Craven and G. Wahba, Numer. Math. *31*:377–403 (1979).
24. B. Efron and R. Tibshirani, Stat. Sci. *1*:54–74 (1986).
25. G. C. Bryant, C. Abeynayake, and J. C. Thomas, Langmuir *12*:6224–6228 (1996).
26. B. G. Manders, A. M. van Herk, and A. L. German, Makromol. Chem. Rapid Commun. *14*:693–701 (1993).
27. C. C. Co and E. W. Kaler, Macromolecules *31*:3203–3210 (1998).
28. F. Bleger, A. K. Murthy, F. Pla, and E. W. Kaler, Macromolecules *27*:2559–2565 (1994).
29. J. S. Guo, E. D. Sudol, J. W. Vanderhoff, and M. S. El-Aaser, J. Polym. Sci. Polym. Chem. Ed. *30*:691–702 (1992).
30. J. D. Morgan and E. W. Kaler, Macromolecules *31*: 3197–3202 (1998).
31. J. D. Morgan, K. M. Lusvardi, and E. W. Kaler, Macromolecules *30*:1897–1905 (1997).
32. M. Nomura and K. Suzuki, Macromol. Chem. Phys. *198*:3025–3039 (1997).
33. E. J. Behrman and J. O. Edwards, Rev. Inorg. Chem. *2*:179–206 (1980).
34. A. P. Full, E. W. Kaler, J. Arellano, and J. E; Puig, Macromolecules *29*:2764–2775 (1996).
35. I. A. Maxwell, B. R. Morrison, D. H. Napper, and R. G. Gilbert, Macromolecules *24*:1629–1640 (1991).
36. W. H. Ming, F. N. Jones, and S. K. Fu, Polym. Bull. *40*:749–756 (1998).
37. M. Rabelero, M. Zacarias, E. Mendizabal, J. E. Puig, J. M. Dominguez, and I. Katime, Polym. Bull. *38*:695–700 (1997).

23

Heterophase Polymerization in Inverse Systems

KATHARINA LANDFESTER and HANS-PETER HENTZE Max-Planck-Institute of Colloids and Interfaces, Golm, Germany

I. INTRODUCTION

Inverse heterophase polymerizations are essential techniques for the industrial production of water-soluble high-molecular-weight polymers and are modern means of polymer synthesis in fundamental research. The most important contributions for technical applications are improved process monitoring (e.g., temperature control) and the low viscosity of latexes (polymer dispersions). For example, a 3 wt% aqueous solution of polyacrylamide is very viscous, whereas a 50 wt% dispersion of acrylamide is still liquid and therefore easier to process. Modern approaches not only aim at optimized formulations for the synthesis of popular polymers but also contribute fascinating examples of material design on a nanometer scale. Some examples of the synthesis of new materials with outstanding and fascinating properties are presented here to demonstrate the versatility and the potential of inverse heterophase polymerizations.

After a brief classification of inverse heterophase polymerizations, an introduction to heterophase formulation is given. The following sections discuss the state of the art of free radical polymerizations in inverse emulsions, including current and potential future applications.

II. CLASSIFICATION OF INVERSE HETEROPHASE POLYMERIZATIONS

Historically, emulsions were classified as oil-in-water (direct or conventional emulsion) or water-in-oil (inverse emulsion) systems. One primary criterion for categorization is the surface tension driving force, $\Delta\gamma = \gamma_{A\text{-}H} - \gamma_{O\text{-}L}$, where $\gamma_{A\text{-}H}$ is the surface tension between the aqueous phase and the hydrophilic moiety and $\gamma_{O\text{-}L}$ the surface tension between the organic phase and the lipophile. This surface tension driving force delineates the dispersion structure. If $\Delta\gamma$ is less than zero, an oil-in-water dispersion forms, whereas if $\Delta\gamma$ is larger than zero, a water-in-oil dispersion is favored.

Because of the different emulsion types, heterophase polymerizations can also be subdivided into microemulsion, (macro)emulsion, miniemulsion, suspension, and dispersion polymerization. This categorization follows a suggestion of an international group of researchers [1]. The main characteristics are the following:

A. Microemulsion Polymerization

In this case, the criterion is that a threshold emulsifier concentration exists above which, for a given organic and aqueous phase, a thermodynamically stable microemulsion is spontaneously formed. During microemulsion polymerization, the polymerization starts from this thermodynamically stable state. Because the initiation cannot be obtained in all microdroplets simultaneously, polymer chains are formed only in some droplets. A thermodynamically nonequilibrium state usually leads to an increase of the particle size.

B. (Macro)emulsion Polymerization

Below a threshold of surfactant amount, which also depends on the temperature, monomer concentration, and the chemistry of the emulsifier, no microemulsion but kinetically stabilized macroemulsions are formed. Therefore, large monomer droplets of 1–10 μm diameter stabilized by surfactant and empty or monomer-swollen surfactant micelles coexist in the initial state. The water-soluble initiator forms oligoradicals from slightly water-soluble monomer units. These oligoradicals then enter the micelles and start to form particles. During polymerization, the monomer diffuses through the water phase to the micelles in order to sustain polymer particle growth. In comparison with suspension polymerization, polymerization is considered (macro)-emulsion polymerization if either one of the following criteria is satisfied: (1) the kinetics, as defined by the average number of macroradicals per polymer particle ($\bar{n}$), are not significantly larger than 1, or (2) the mechanism of particle nucleation occurs outside the monomer droplet.

The term "emulsion polymerization" is used in the literature for the process we defined for clarity as macroemulsion polymerization (in comparison with micro- and miniemulsion polymerization). In the following, we use emulsion polymerization synonymously with macroemulsion polymerization.

C. Miniemulsion Polymerization

In terms of the stability of the emulsion and in terms of the size of the resulting particles, miniemulsions are in between macroemulsions and microemulsions. Miniemulsions are classically defined as aqueous dispersions of relatively stable oil droplets within a size range of 50 to 500 nm prepared by shearing a system containing oil, water, a surfactant, and a hydrophobe [2]. The principle can be extended to systems containing a monomer, a continuous phase that is immiscible with the monomer phase, a surfactant, and an osmotic agent that shows extremely low solubility in the continuous phase. The addition of an osmotic agent suppresses Ostwald ripening. In the case of low solubility of the monomer in the continuous phase, monomer molecules in the continuous phase and in the droplets can exchange to a low extent, but a pseudoequilibrium is established [3]. In miniemulsion polymerization, each droplet can be seen as a minireactor because in the ideal case no effective transport of monomer takes place during the reaction and each droplet becomes a particle.

D. Suspension Polymerization

In the case of large monomer droplets in the continuous phase, nucleation occurs predominantly in the droplets, and each polymerizing droplet behaves as an isolated batch polymerization reactor. The notation "suspension polymerization" is reserved for systems where nucleation occurs in the monomer droplets and the average number of radicals per particle $\bar{n}$ is very high (10^{2-6}). This is usually obtained if the droplets are larger than 1 μm. Therefore, emulsions and suspensions should be distinguished by the particle nucleation mechanism and the kinetics [1].

E. Dispersion Polymerization

In dispersion polymerization, the monomer is soluble in the continuous phase, but the polymer precipitates and is stabilized by emulsifier molecules in order to form particles.

F. Nomenclature Difficulties

However, in the literature there is some confusion about the different types of inverse polymerizations because either important parameters were overlooked or the distinction between inverse microemulsion, (macro)emulsion, miniemulsion, suspension, and dispersion polymerization is not always strict and, therefore, sometimes difficult to categorize.

Hunkeler et al. [1,4] introduced the term "microsuspension" by subcategorizing inverse macroemulsions according to a second criterion: the level of surfactant with respect to the critical micelle concentration (cmc). At emulsifier levels above the cmc where micelles can play a significant role in the polymerization process it was called emulsion or inverse emulsion polymerization, whereas at emulsifier levels below the cmc the term suspension polymerization was suggested. This is true as long as the droplets are large enough. But in the case of suspension polymerization at high emulsifier concentrations, the particle sizes are reduced to such an extent that the total interfacial area is large relative to the droplet volume.

Under such conditions, radical reactions within the interfacial layer become competitive with the propagation, termination, and transfer reactions occurring within the monomer droplet. The onset of such phenomena is presumed for a diameter of 1–10 μm. To distinguish these processes from suspensions, which are characterized by radically inert interfaces, the nomenclature microsuspension has been proposed [5]. However, in later publications the authors also use the term "inverse emulsion polymerization" for this process [6].

III. CONCEPTS FOR HETEROPHASE FORMULATIONS

Historically, the first concept for the formulation of heterophases was developed in 1913. The so called Bancroft rule implies that more hydrophilic surfactants are suitable stabilizers for oil-in-water (o/w) emulsions and oil-soluble surfactants are more likely to form stable w/o emulsions [7].

Modern approaches tried to find a way to quantify the ability of a surfactant to stabilize a certain heterophase. Since Griffin [8] first proposed the hydrophilic-lipophilic balance (HLB) concept, it has been extensively applied in industrial and fundamental studies. Griffin's work has led to the concept of the required HLB, which states that for each nonaqueous phase in an emulsion, an HLB exists at which the emulsion stability is optimal. In principle, the HLB parameter correlates the structure and effectiveness of ethylene oxide–based surfactants. The relationship between emulsifier HLB and emulsion structure is given in Table 1. Unfortunately, it has proved difficult to arrive at a theoretical definition of this quantity.

Davies [9] attempted to put HLB on a thermodynamic basis and expressed it in terms of structural factors. Based in part on Davies' theory, Lin and coworkers [10,11] related the HLB of a surface-active agent to the cmc. Several analytical procedures have been proposed [12] for determining the HLB of emulsifiers from structural factors. However, they all suffer from the disadvantage that the emulsifiers are examined in isolation without a detailed consideration of the microscopic or surface environment in the emulsion [13] or emulsifier-emulsifier interactions. Therefore, the data are unreliable for a quantitative evaluation of stability. Shinoda [14] pointed out that the HLB depends on the balance of the emulsifier at the oil-water interface, the nature of the oil phase, and the additives in the aqueous and oil phases. Furthermore, Ford et al. [15] studied the stabilization of water-in-oil emulsions using oil-soluble emulsifiers with xylene as the organic phase. They found that the HLB values did not show any correlation with emulsion behavior. The presence of one or several comonomers and the addition of electrolytes may also drastically affect the aqueous phase composition as well as the interactions at the interface [16].

TABLE 1 Relation Between HLB and Microemulsion Structure

HLB range	Structure
4–8	Water in oil (inverse)
8–10	Bicontinuous
12–15	Oil in water (direct)

Another approach for the quantitative prediction of the stability of inverse emulsions was chosen by Ni and Hunkeler [17] by utilizing an artificial neural network. They used the percentage of organic phase separation after a 2-week period as the standard metric.

The cohesive energy ratio (CER) concept developed by Berrbower and Hill [18] for the stability of classical emulsions was the basis for an optimized process for the effective formulation of microemulsions [19]. The basic assumption of the CER concept is that the partial solubility parameters of the oil (δ_o^2) and the surfactant lipophilic tail (δ_L^2), as well as those of water (δ_w^2) and the hydrophilic head (δ_H^2) are perfectly matched. When these conditions are satisfied, one gets a relationship for the surfactant hydrophilic-lipophilic balance that correlates the structure with the effectiveness of the surfactants. One obtains the ratio of the cohesive energies of the lipophile for the oil and the hydrophile for water. This ratio determines the structure and phase relation of the dispersion. Therefore, it is possible to calculate the required HLB for a given oil. In addition to the chemical structure and HLB matches, Beerbower and Hill [18] showed that appropriate properties of molar volumes between oil and lipophile greatly enhance the formation of a stable emulsion.

This concept shows the advantage of paraffinic continuous phases and allows the optimization of the emulsifier level and the determination of the HLB of the surfactant blend. By appropriately matching the molar volume and the solubility parameters of the continuous phase with the lipophilic portion of the emulsifier, inverse microemulsions can be formulated at low stabilizer levels.

IV. INVERSE HETEROPHASE POLYMERIZATION

In the following, the different types of inverse heterophase polymerizations are discussed in detail. Table 2 gives an overview of different surfactant systems and monomers used for the polymerizations in inverse heterophases. This list is not exhaustive but is intended to illustrate the breadth of systems examined.

A. Inverse Microemulsion Polymerization

In 1943, Hoar and Schulman [72] reported for the first time transparent systems that were formed by oil, water, and surfactants. In 1959, Schulman, Stoeckenhuis,

TABLE 2 Overview of Inverse Heterophase Polymerization

Emulsifier	Chemical structure	Monomer	Continuous phase	Initiator	Type of polymerization	References
Span 60	Sorbitan monostearate	Sodium *p*-vinylbenzene	*o*-Xylene	BPO, $K_2S_2O_8$	E	Vanderhoff [20] (1962).
Span 80	Sorbitan monooleate	Acrylic acid	Kerosene	$K_2S_2O_8$	E	Jiang [21] (1996)
Span 80	Sorbitan monooleate	Acrylic acid	Kerosene	^{60}Co γ ray	E	Jiang [22] (1996)
SMO	Sorbitan monooleate	AM/DMAEA AM/DMAEM	Isopar-K	AIBN, $K_2S_2O_8$	E	Hunkeler [23] (1991)
SMO	Sorbitan monooleate	Acrylic acid	Isopar M	AIBN, ADVN, AIBA, AIHEA	E	Reichert [24] (1984)
SMO	Sorbitan monooleate	Acrylamide	Toluene	α-/β-KGA	E	Ghosh [25] (1992)
Span 80	Sorbitan monooleate	Acrylic acid, acrylic acid/AM	Toluene	$K_2S_2O_8$	S	Omidian [26] (1994)
Span 80	Sorbitan monooleate	Acrylic acid	Toluene	$K_2S_2O_8$	S	Askari [27] (1993)
Span 80/ethyl cellulose	Sorbitan monooleate (degree of substitution: 2.42–2.53, N-14 type of Hercules)	Acrylic acid	Toluene	$K_2S_2O_8$	S	Askari [27] (1993)
Span 80/Tween 80	Sorbitan monooleate/polyethylene glycol sorbitan monooleate	Acrylic acid	Aliphatic hydrocarbons	AIBN $(NH_4)_2S_2O_8$	E	Kriwet [28] (1998)
Span 80/Tween 80	Sorbitan monooleate/polyethylene glycol sorbitan monooleate	Acrylamide		BPO	E	McKechnie [29] (1982)
Span 80/OP10	Sorbitan monooleate	AM/MADQUAT	Kerosene	$K_2S_2O_8$	E	Ge [30] (1997)
Span 80/OP10	Sorbitan monooleate	AM/MADQUAT	Kerosene	^{60}Co γ ray	E	Ge [31] (1997)
Span 80/Tween 85	Sorbitan monooleate/polyethylene sorbitan trioleate	Vinylpyrrolidone	Isopar M	^{60}Co γ ray	E	Taylor [32] (1994)
G946/Tween 85	Sorbitan monooleate/polyethylene sorbitan trioleate	Acrylamide	Isopar M	ADVN	μE/E	Hernandez-Barajas [33] (1997)
Arlacel 83/ G1086	Sorbitan sesquiolate/polyethylene $(EO)_{40}$ sorbitol hexaoleate	AM/NaAA	Isoparaffinic mixture	AIBN UV light	μE	Candau [34] (1986)
Arlacel 83/ G1086	Sorbitan sesquiolate/polyethylene $(EO)_{40}$ sorbitol hexaoleate	MADQUAT			μE	Buchert [35] (1990)
Arlacel 83/ G1086	Sorbitan sesquiolate/polyethylene $(EO)_{40}$ sorbitol hexaoleate	Acrylamide	Isopar M	ADVN	μE/E	Hernandez-Barajas [33] (1997)
Arlacel 83/ G1086	Sorbitan sesquiolate/polyethylene $(EO)_{40}$ sorbitaol hexaoleate	NaAMPS/MADQUAT/ AM	Isopar M	AIBN, UV	μE	Ohlemacher [36] (1996)
Arlacel 83/ G1086	Sorbitan sesquiolate/polyethylene $(EO)_{40}$ sorbitol hexaoleate	MADQUAT/NaAMPS/ AM/BAM	Isopar M	AIBN, UV	μE	Neyret [37] (1997)
Arlacel 83/ G1086	Sorbitan sesquiolate/polyethylene $(EO)_{40}$ sorbitol hexaoleate	AM/NaAA	Isopar M	AIBN (+UV)	μE	Candau [38] (1986)
Arlacel 83/ G1086	Sorbitan sesquiolate/polyethylene $(EO)_{40}$ sorbitol hexaoleate	Acrylamide	Isopar M	AIBN UV	μE	Holtzscherer [39] (1986)

Arlacel 83/ Tween 80	Sorbitan sesquioleate polyethylene $(PEO)_{20}$ sorbitan monooleate	NaAMPS/MADQUAT	Isopar M	AIBN	μE	Corpart [19] (1993)
Montane 83/ Montanox 80	Sorbitan sesquiolate polyoxyethylene sorbitan trioleate	Acrylamide	Toluene	AIBN	E	Graillat [40] (1986)
	Sorbitan sesquioleate C_{18} terminated acrylamide polymer	AM/acrylic acid	Toluene	AIBN	E	Glukhikh [41] (1987)
T10/Arlacel 83/ G1096	*t*-Octyl-phenoxy-PEO_{10} methacrylate/sorbitan sesquiolate/polyethylene sorbitan hexaoleate	T10/AM/NaAA	Isopar M	AIBN, UV	bic μE	Candau [42] (1999)
Arlacel 83, G1096, G1086, Arlacel 85, Tween 80	Sorbitan sesquiolate/polyethylene sorbitan hexaoleate/sorbitan trioleate/polyethylene(PEO_{20})-sorbitan monooleate	AM/NaAA	Isopar M	AIBN, UV	bic μE	Candau [42] (1999)
AOT	Sodium bis(2-ethylhexyl sulfosuccinate)	Acrylamide	Toluene		μE	Carver [43] (1989)
AOT	Sodium bis(2-ethylhexyl sulfosuccinate)	Acrylamide	Toluene	AIBN, $K_2S_2O_8$	μE	Candau [44] (1985)
AOT	Sodium bis(2-ethylhexyl sulfosuccinate)	AM/St	Toluene	BPO, $K_2S_2O_8$	μE	Barton [45] (1991)
AOT	Sodium bis(2-ethylhexyl sulfosuccinate)	Acrylamide Acrylic acid Salt of acrylic acid	Paraffinic hydrocarbon C_{14-18}	$(NH_4)_2S_2O_8$	E	Benda [46] (1997)
AOT	Sodium bis(2-ethylhexyl sulfosuccinate)	MMA/HEMA	Toluene	AIBN (60C)	μE	Dresco [47] (1999)
AOT	Sodium bis(2-ethylhexyl) sulfosuccinate	Acrylamide	Toluene	AIBN	μE	Leong [48] (1982)
AOT	Sodium bis(2-ethylhexyl) sulfosuccinate	AM/MMA AM/St	Toluene	AIBN, BPO, $(NH_4)_2S_2O_8$	μE	Vaskova [49] (1990)
AOT/Brij 30	Sodium bis(2-ethylhexyl sulfosuccinate)/polyethylene glycol dodecylether	AM/BAM/AA, AM/ BAM/ADAH	Hexane	TMEDA/ $(NH_4)_2S_2O_8$	μE	Daubresse [50] (1996)
AOT/SDS	Sodium bis(2-ethylhexyl sulfosuccinate)	Acrylamide	Toluene	BPO, $K_2S_2O_8$	μE	Barton [51] (1996)
AOT/ cyclohexanol	Sodium bis(2-ethylhexyl) sulfosuccinate	Acrylamide	Decane	Benzophenone derivatives UV	μE	Fouassier [52] (1986)
CTAB/hexanol	Hexadecyltrimethylammonium bromide	Acrylamide	Kerosene	KPS	μE	Yan [53] (1998)
PEM	Pentaerythriolmyristate	Acrylamide	Isooctane	ADVN	E	Pross [54] (1998)
Igepal CO-430, MYRJ45	Polyoxyethylene(4) nonylphenol Polyoxyethylene(8) steracid acid	Acrylamide	White spirit	$NaHSO_3$-$K_2S_2O_8$	S	Dimonie [55] (1982)
EO(4)NP	Polyoxyethylene(4) nonylphenol	Acrylamide	Cyclohexane	$NaHSO_3$-$K_2S_2O_8$	S	Dimonie [56] (1992)
Tetronic 1102	Polyoxyethylene adduct of polyoxypropylene diamine adduct	Acrylamide	*o*-Xylene	BPO	E	Vanderhoff [57] (1984)

TABLE 2 Continued

Emulsifier	Chemical structure	Monomer	Continuous phase	Initiator	Type of polymerization	References
Sentamid-5	Amide of stearic acid polyoxyethylated	Acrylamide	Toluene	$K_2S_2O_8$	E	Kurenkov [58] (1978)
	N,N-bishydroxyethyl tall oil amide	Acrylic acid	Isopar M	$Na_2S_2O_8/KBrO_3$	E	Liu [59] (1997)
HB239	Polyester-polyethylene oxide	Acrylamide	Isopar M	AIBN	E	Hernandez-Barajas [60] (1995)
HB239/Arlacel 83/Tween 85	Polyester-polyethylene oxide Sorbitan sesquiolate Polyethylene sorbitan trioleate	AM/DMAEA AM/DMAEM	Isopar M	AIBN	E	Ni [61] (1996)
HB246/G946	Polyester-polyethylene oxide sorbitan monooleate	AM/DMAEA AM/DMAEM	Isopar M	AIBN	E	Hernandez-Barajas [62] (1997)
PS-co-PEO	Amphiphilic copolymer	Acrylamide	Toluene	AIBN, UV	μE	Leong [63] (1981)
SE copolymer	Polystyrene-polyethylene $M_w = 1000–6000$ g mol^{-1}	Acrylic acid	Toluene	ADVN	D	Fengler [64] (1994)
SE3030	Polystyrene-polyethylene $M_w = 6000$ g mol^{-1}	Acrylic acid/dicarboxylic dichloride Acrylic acid/diglycidyl ether	Benzene	ADVN	D	Dauben [65] (1996)
PVME	Poly(vinyl methyl ether)	Acrylamide	*t*-Butylalcohol	$(NH_4)_2S_2O_8$	D	Ray [66] (1997)
KLE3729	Poly[(ethylene-butylene)-co-ethyleneoxide]	Acrylamide HEMA	Cyclohexane Hexadecane	AIBN	ME	Landfester [66a] (2000)
EmpiLan NP5	(Polyoxyethylene)$_{5–9}$ nonyl phenol	Aniline	Petroleum ether	$Na_2S_2O_8$	μE	Gan [67] (1993)
Polymerizable stabilizer C_{10}MA	Didecyldimethylammonium methacrylate	C_{10}MA	Toluene	AIBN	Reverse micelles	Hammouda [68] (1995)
C_{10}MA	Didecyldimethylammonium methacrylate	NaMA/C_{10}MA	Toluene	AIBN, UV	μE	Moumen [69] (1999)
Polymerizable stabilizer DUSS	Di(10-undecenyl)sulfosuccinate	DUSS or DUSS/diethylfumarate	*n*-Hexane	AIBN	μE	Nagai [70] (1993)
—	—	AM/BMA AM/methacrylic acid	Methanol Ethanol *n*-Propanol	AIBN, V-501	D	Kawaguchi [71] (1991)

Polymerization type: E, (macro)emulsion polymerization; μE, microemulsion polymerization; bic μE, bicontinuous microemulsion; ME, miniemulsion polymerization; S, suspension polymerization; D, dispersion polymerization.
Continuous media: Isopar M/K, isoparaffinic mixtures.
Initiators: ADVN, 2,2′-axo-bis(2,4-dimethyl valeronitrile); AIBA, azoisobutyroamidin; AIBN, azoisobutyronitrile; AIHEA, azoisobutyro-*N*-hydroxyethyl-2-amidin; BPO, benzoyl peroxide; TMEDA, *N,N,N′,N′*-tetramethylenediamine; V-501, azobis(4-cyanopentanoic acid).
Monomers: AM, acrylamide; BAM, *N,N′*-methylene bisacrylamide; DMAEA, dimethylaminoethylacrylate; DMAEM, dimethylaminoethylmethacrylate; HEMA, hydroxyethylmethacrylate; MADQUAT, 2-methyloyloxy-ethyltrimethylammonium chloride; MMA, methyl methacrylate; NaAA, sodium acrylate; NaMA, sodium methacrylate; Na-AMPS, sodium-2-acrylamide-2-methylpropane sulfonate; St, styrene; T10, *t*-octylphenoxy-poly(oxyethylene) methacrylate, with 10 ethylene oxide units.
Other abbreviations: PEO, poly(ethylene oxide).

and Prince [73] called these systems "microemulsions." Microemulsions are defined as thermodynamically stable, at least ternary mixtures of two immiscible liquids, stabilized by surfactant or a mixture of surface-active agents [74]. They are isotropic, transparent, or opaque and form spontaneously on mixing or stirring. Most common are microemulsions that consist of oil, water, surfactant, and often also cosurfactant. The surface-active agents reduce the interfacial energy close to zero, which enables the formation of well-defined aggregation structures on a length scale of several nanometers. Depending on the ratio of the two liquid phases and the structure of the surfactant, normal (or direct) globular (o/w, oil-in-water), inverse (or reverse) globular (w/o, water-in-oil), or bicontinuous microemulsions are formed.

Even though microemulsions have been known for a long time, it took nearly 40 years before the first free radical polymerizations in microemulsions were performed. The first microemulsion polymerizations were investigated by Stoffer and Bone [75,76], who polymerized methyl methacrylate (MMA) and methyl acrylate (MA) in normal microemulsions. In the same period, Leong and Candau [48,77] performed the first successful polymerizations of acrylamide (AM) in inverse microemulsions. As current and comprehensive reviews about microemulsion polymerization are available [74,78,79], including the companion Chapter 22 in this volume, the intent of this chapter is to give a brief summary of the main features of this modern polymer synthesis technique and to highlight the developments that give the directions for future fundamental research and applied polymer chemistry on this exciting field. The main characteristics of direct and inverse microemulsion polymerizations are:

- Synthesis of nanosized particles with a relatively narrow size distribution that contain polymers with high molecular weight ($>10^6$ g/mol)
- High rate of polymerization (complete conversion often within 5–40 min compared with several hours in emulsion polymerization)
- Low viscosity of microlatexes
- High stability against coagulation (on the time scale of months to years)

Beside the many possibilities and advantages of microemulsion polymerization, some disadvantages also exist, which so far prevent the use of this technique in technical applications on a broader scale. The most challenging factor is the expensive formulation (relatively low monomer and high surfactant contents compared with conventional emulsion polymerization). New approaches have been developed to overcome these problems through improving our understanding of polymerization mechanisms. The current literature is providing fascinating examples of the versatility of inverse microemulsion polymerization in the synthesis of new materials, nanocomposites, and catalysts.

1. Surfactants for Inverse Microemulsion Polymerization

In contrast to normal microemulsions, which contain the monomer as dispersed oil nanodroplets in a continuous polar phase (e.g., water or formamide [80]), in inverse microemulsions nanodroplets of aqueous monomer solution are dispersed in a continuous oil phase. The function of the surfactant is the formation of the microemulsion by decreasing the interfacial tension and the solubilization of the monomer solution. Furthermore, the surfactant has to stabilize the system throughout the polymerization process to prevent phase separation, gel formation, and coagulation. Often cosurfactants are used, such as aliphatic alcohols with short hydrocarbon chains (e.g., 2-propanol or octanol), to achieve the formation of a single-phase microemulsion. The cosurfactant interacts with the surfactant at the interface, penetrates the surfactant monolayer [81], and affects the partitioning of the monomer in the continuous phase [82]. Also, the water-soluble monomers themselves usually act as cosurfactants, as they consist of a polymerizable hydrophobic vinyl group and a polar group.

^{1}H and ^{13}C spectra were used to study environmental changes and surfactant transitions in acrylamide-water/bis(2-ethylhexyl)sulfosuccinate (AOT)/toluene microemulsions [83] due to monomer addition. The monomers increase the flexibility and fluidity of the interface [83]. This leads to an extension of the single-phase microemulsion domain in the phase diagram. Carver et al. [43] hypothesized that added acrylamide facilitates attractive interactions between particles, increases the contact time of such particles during collisions, and increases interfacial flexibility with concomitant transitory pore opening. The existence of an acrylamide-rich region in the shell of the water pool was documented by Candau et al. [77].

The following are general guidelines for the formulation of inverse microemulsions for free radical polymerizations:

- Chemical compatibility (in terms of solubility parameters and molar volume) between oils and the hydrophobic moieties of the surfactant improves microemulsion formation.

Good solubility of the polymer and the cosurfactant leads to higher microlatex stability.

Addition of cosolvent to the polymer enhances microlatex stability (water usually acts as cosolvent in inverse systems of liquid monomers, e.g., acrylic acid).

In cases of thermal initiation of nonionically stabilized microemulsions, stable systems exist only well below the cloud point of the surfactant.

The influence of electrolytes can play an important role in microemulsion formulation [84–86]. The phase behavior of nonionic surfactants is generally more temperature dependent and less dependent on the electrolyte concentration than that of ionic surfactants. But even for nonionically stabilized microemulsions, addition of electrolyte causes salting-in and salting-out effects, which reduce or enhance the stability of the microemulsion [39]. Addition of electrolyte also affects the structure of the water. Two types of water, namely interfacial and bulklike, were detected by Fourier transform infrared (FTIR) spectroscopy, nuclear magnetic resonance (NMR) spectroscopy, electron spin resonance (ESR), and near infrared spectroscopy [87]. The relative amounts were found to depend on the water-to-surfactant ratio and also on the nature of the organic solvent. The interactions between the water and polar groups of the surfactant (AOT) change with the addition of electrolyte [88].

Nowadays, inverse microemulsions typically contain about 10–15 wt% of their total mass as surfactant (occasionally up to 30 wt%), compared with 2–4 wt% surfactant in inverse emulsion formulations. However, one has to consider that the monomer content in inverse microemulsions is quite low, typically about 15 wt% compared with 30–60 wt% in conventional emulsions. Therefore, a systematical approach to optimization of microemulsion formulations is necessary to improve the efficiency, especially for industrial applications on a broader scale. The most systematic approach to optimize the formulation of inverse microemulsions for polymerizations was achieved by Candau, Pabon, and Antquetil [42]. By blending surfactants with different HLB values, the HLB value of the surfactant mixture can be adjusted. The modified cohesive energy ratio (CER) model was used for the calculation of the required HLB value of the surfactant mixture for a given continuous oil phase to find the optimal surfactant mixture. The calculation of the HLB value neglected the effect of monomer on the HLB and the interfacial properties, which was thoroughly investigated for acrylamide [16,89] and MADQUAT [35,90]. Nevertheless, the surfactant content could be drastically reduced from about 10 to 5.5 wt%. Increasing the monomer content to 22 wt% leads automatically to a transition to a bicontinuous microemulsion, as the monomer amount increases the flexibility and decreases the curvature of the interface. It can be anticipated that future studies will also improve the efficiency of globular microemulsion polymerization formulations by using this concept.

The first inverse microemulsion systems for polymerizations contained up to 10 times more surfactant and cosurfactant than monomer [52]. The cosurfactants were alcohols (2-propanol, cyclohexanol, and octanol) [48], which are also radical transfer agents. They decrease the molecular weight and thereby often decrease the end-use efficiency of the resulting polymer. These first formulations were improved by the use of mixtures of AOT and nonionic surfactants or nonionic surfactant blends (e.g., of sorbitan ester). The efficiency increased, less surfactant was necessary for stabilization, and radical transfer by the cosurfactant was avoided by banishing alcohols as cosurfactants. In the following, a variety of examples of formulations for inverse microemulsion polymerization are given.

The first polymeric stabilizer used for inverse microemulsion polymerization was a polystyrene-co-poly(ethylene oxide) copolymer [63]. The steric stabilization by polymeric surfactants was investigated by Hernandez-Barajaz and Hunkeler [91]. Cationic surfactants have also been used for inverse microemulsion polymerization: Yan et al. [92] reported the polymerization of acrylamide in hexadecyltrimethylammonium bromide (CTAB)/hexanol-stabilized inverse microemulsions. Barton and Stillhammerova [51] reported the use of a mixture of two anionic surfactants. The more hydrophilic sodium dodecyl sulfate (SDS) and the more hydrophobic AOT were utilized as surfactants for the polymerization of acrylamide initiated by dibenzoyl peroxide. The presence of SDS increases the one-phase microemulsion phase region and decreases the acrylamide polymerization rate and the resulting particle size. In the case of the polymerization of the cationic monomer MADQUAT, it was shown [35,90] that the application of a blend of sorbitan sesquiolate (Arlacel 83, HLB = 3.7) and a sorbitan monooleate with 20 ethylene oxide units (Tween 80, HLB = 15) was well adapted to the microemulsification of the aqueous monomer solution in cyclohexane. For the copolymerization of particles of acrylamide with acrylates in microemulsions, an emulsifier blend (HLB = 9.30) of sesquiolate sorbitan (Arlacel 83, HLB = 3.7) and a polyoxyethylene sorbitol hexaoleate with 40 ethylene oxide residues (GG 1086, HLB = 10.2) was successfully employed [34]. For the use of microlatexes in biomedical applications, lecithin-stabilized biocompatible, nontoxic microlatexes were prepared in normal microemulsions [93]. Inverse microemulsions can also be formed with lecithin [94–96]. However, to our knowledge these systems have so far not been applied to inverse microemulsion polymerization.

The use of polymerizable surfactants is a special case of polymerizations in inverse systems. Using surfactants that form inverse micelles and that have a polymerizable double bond in the polar headgroup [97–99] results in particles about 60 nm in diameter. Nagai et al. [100] reported the polymerization of sodium(10-undecenyl)sulfosuccinate (DUSS) with the electron-accepting monomer diethyl fumarate (EF). The copolymerization in the reverse micellar system was greatly accelerated by solubilization of water. The use of a surfactant with a polymerizable counterion (didecyldimethylammonium methacrylate, DMAMA) resulted in extremely small 2- to 5-nm particles consisting of polyelectrolyte-surfactant complexes [68]. By using this surfactant for copolymerization with sodium methacrylate (NaMA), latex particles 12 nm in size were obtained [69].

2. Thermodynamics, Kinetics, and Mechanism

Because of their thermodynamic stability, microemulsions form spontaneously. Before polymerization the systems are clear and transparent. During the course of reaction they usually turn into translucent to opaque dispersions. The observed polymer particle size is typically bigger (20–120 nm) than the primary microdroplet size (<10 nm). At the same time the number of particles decreases compared with the number of oil droplets that existed in the microemulsion before polymerization. In other cases the microemulsion turns into a lamellar system [101] or undergoes other phase transitions [102]. Attempts to generate an identical dispersion compared with the microlatex by redispersing polymers synthesized in microemulsions by using otherwise unchanged formulations failed [44]. The particle size and the size distribution are usually much larger. All of these observations and results show that during polymerization a change of the thermodynamic stability of the system occurs. This is because the phase behavior of the polymers generated is usually expected to be quite different from the phase behavior of the monomer solutions. Therefore one limiting aim of microemulsion polymerization is to keep as much as possible of the preorganized microemulsion structure during the course of polymerization by choosing appropriate formulations and reaction parameters.

Models provide a better understanding of the elementary steps of the microemulsion polymerization mechanism. Even though there are contradictions in the literature that show no general scheme for the polymerization kinetics in inverse microemulsions, which is valid for all systems, there are some controlling features:

- Continuous particle nucleation [44] causes an uni-oligo molecular polymerization, resulting in particles with one up to a few polymer chains [44] (compared with several thousand in conventional emulsion polymerization). This suggests kinetics that do not follow the Smith and Ewart theory.
- Nucleation can occur in the continuous nonpolar phase, in the monomer phase, or in both phases depending on the solubility of the initiator in these phases. The use of water-soluble or oil-soluble initiators usually does not significantly affect the characteristics of the resulting microlatex [45].
- The polymerization rate curves show two stages, an increase with a maximum typically about 20–40% of conversion followed by a monotonic decrease. The constant-rate interval, which is characteristic of emulsion polymerizations, is missing.
- High conversions (>90%) are reached in a short period of time, often within 5–40 min.
- The number of particles and the particle size increase during the polymerization process.

A qualitative model that describes the scenario of inverse microemulsion polymerization schematically is the CLF model (Fig. 1), established by Candau, Leong, and Fitch [44]. At the beginning of the polymerization the aqueous monomer solution is homogeneously dispersed within the nonpolar continuous phase (Fig. 1a). The kinetics of the microemulsion formation cause fluctuations of the nanodroplets around their equilibrium size. The initiation occurs by entry of radicals inside the primary microemulsion droplets or by homonucleation. In a second step, particle growth takes place by diffusion of monomer through the continuous phase (primary growth) and by particle collision (secondary growth) during the polymerization process (Fig. 1b). Besides polymer particles, empty micelles are formed that can serve as reservoirs for the continuous nucleation of new latex particles. At the end of the polymerization, micelles and latex particles coexist (Fig. 1c). The size of the particles is typically much larger and their number much lower than those of the primary microdroplets.

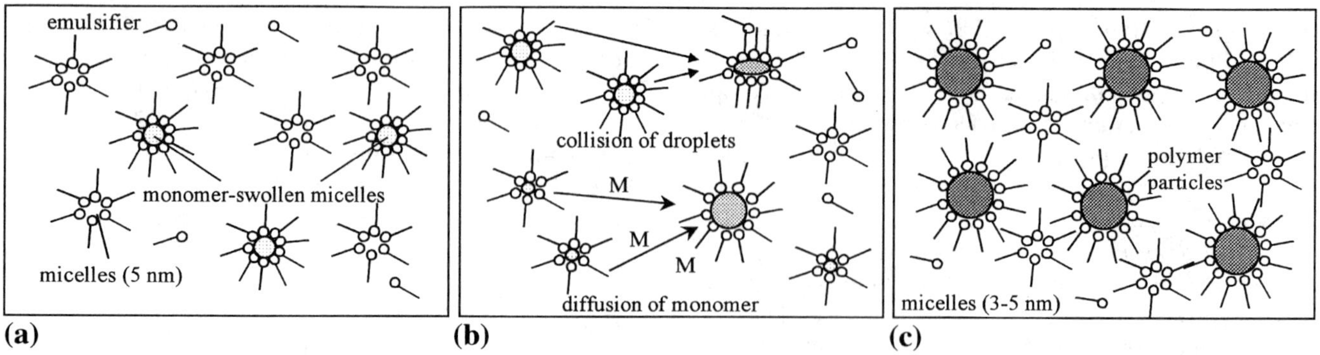

FIG. 1 Schematic representation of inverse microemulsion polymerizaion. (a) At the beginning of the polymerization, the aqueous monomer solution is homogeneously dispersed within the unpolar continuous phase. (b) In a second step, particle growth takes place by diffusion of monomer through the continuous phase and by particle collision during the polymerization process. (c) At the end of the polymerization, micelles and latex particles coexist. The size of the particles is typically much larger and their number much lower than those of the primary microdroplets.

Most quantitative models for microemulsion polymerization were developed for normal o/w microemulsions (see Chapter 22) [103–106]. A mathematical model for inverse microemulsion polymerization was presented [107]. Assuming pseudo-steady-state conditions, the parameters taken into account were the balance of monomer concentration and initiator concentration; the balance of the particle population and the radicals; the partitioning of monomer, oil, and surfactant; the number of micelles; and the balance of the growing polymer chains' molecular weight. The predictions of the resulting particle size and the time dependence of the conversion were compared with experimental data for the photoinitiated polymerization of 2-methacryloyl oxyethyl trimethylammonium chloride (MADQUAT) stabilized by a blend of nonionic emulsifiers [108]. On the whole, a good fit of predicted and experimental data was achieved. Applying this model to other systems will broaden the data basis and provide deeper insights.

The use of fluorescence spectroscopy facilitates the investigation of molecular changes in the interior of particles during polymerization [109,110]. It was shown that the decrease of acrylamide during polymerization is monotonic regardless of the fluorescent probe localization. Future studies will give a better understanding of the data and allow further interpretations. Other important parameters that determine mechanism and kinetics are the locus of initiation and the dynamics of the microemulsion, e.g., in terms of monomer exchange between microdroplets by percolation.

(*a*) *Locus of Initiation.* The nature of the initiator determines the locus of initiation. The locus is the water pool of inverse micelles, e.g., in the case of ammonium peroxodisulfate (APS), or the continuous oil phase, e.g., in the case of dibenzoyl peroxide (BPO). The partitioning of the monomer and/or free radical initiator between water and oil phases of an inverse microemulsion is very important for determining the formation and growth mechanism of polymer particles. If radical reactions are limited to only one phase in the inverse system, e.g., to the water phase, the formulation of a mechanism for the polymer particle formation is straightforward [44]. In the case of partitioning the (co)monomer and the initiator in both phases of the dispersion system, polymerization of the monomer in the oil phase contributes to the overall polymer particle formation.

The decrease of the acrylamide polymerization rate observed for the APS initiator in percolating inverse microemulsion reflects the lowering of the acrylamide concentration in the main locus of propagation (alteration of the acrylamide-rich regions by the formation of water channels in percolating inverse microemulsions) [111]. On the other hand, the increased concentration of acrylamide in the water channels augments the probability of successful competition of acrylamide in the reaction with free radicals. Thus the retardation of acrylamide polymerization in the presence of Fremy's salt is lower in percolating systems than in nonpercolating systems [112].

In the case of a continuous inverse microemulsion polymerization, it is important to work at low temperatures. Therefore, the redox system sodium metabisulfite/ammonium persulfate in a 1 : 1 mole ratio was used for initiation [113]. It was shown that the particle size is not affected by the amount of initiator. Removal of the oxygen allowed high conversion, and the molecular weight decreased with increasing amount of initiator.

Another study showed that the polymerization rate of acrylamide was independent of the type of initiator [azobisisobutyronitrile (AIBN) or APS], but the initiation by an oil-soluble initiator led to a more complex mechanism [45]. The oligoacrylamide radicals and acrylamide oligomers, after reaching their limit of solubility in toluene, precipitate and are captured by AOT micelles or form aggregates that are finally also captured by inverse AOT micelles. Oligoacrylamide radicals penetrate through the interphase of the inverse micelle into the acrylamide-rich region in the surface layer of the water pool of the inverse micelle [77]. In the acrylamide-rich region, the growth of oligoacrylamide radicals continues. The locus of propagation for the polymerization of acrylamide in inverse microemulsions is thus initially the oil phase but later mainly the acrylamide-rich phase in the surface layer of the water pools. The acrylamide-rich phase as a locus of propagation is responsible for the high polymerization rates of acrylamide beyond a conversion of 10%.

Fouassier et al. [52] studied the photopolymerization of acrylamide as an alternative means of initiation in inverse microemulsions.

(b) Kinetics in Nonpercolating Systems. In the case of the acrylamide (AM) polymerization within an inverse microemulsion, it has been shown that the polymerization rate R_p and the number-average molecular mass M_n depend on the acrylamide concentration in the dispersed phase. For a nonpercolating inverse microemulsion the dependences are described by the equation

$$R_p(\text{or } \bar{M}_n) \propto [\text{AM}]^x \tag{1}$$

where x has a value of 1.8 for R_p and 1.4 for M_n [112]. It was found that at low emulsifier concentrations the polymerization rate depends strongly on the concentration of emulsifier, but the effect is negligible for high emulsifier concentrations. Similar behavior was observed for the particle size, namely that R_p decreased sharply at low emulsifier concentration and much more smoothly later [114]. Comparable observations have been reported by Zekhnini [115]. The dependence of the number of particles N_p on the amount of emulsifier changed from $N_p \propto [\text{emulsifier}]^{3.2}$ at low emulsifier concentrations to $N_p \propto [\text{emulsifier}]^{0.6}$ at high emulsifier concentrations. The effect of the emulsifier is stronger than in normal microemulsion polymerization [116]. This suggests that the lowest emulsifier concentration was too close to the stability limit for microemulsions and partial coagulation occurred.

For the kinetics of the inverse microemulsion polymerization of MADQUAT stabilized with a blend of nonionic emulsifiers and initiated by UV irradiation, it was found that during polymerization R_p first increased, reached a maximum, and then decreased. This effect was due to the combined effect of continuous nucleation and the decrease of the concentration of the monomer in the polymer particles as the polymerization proceeded. The correlation $R_p \propto [\text{AIBN}]^{0.54}$ was found, suggesting that a second-order process for radical loss was operative [114]. These determinations facilitated the development of this quantitative model [117].

(c) Kinetics in Percolating Systems. In percolating inverse microemulsions, inverse micelles form semipermanent aggregates. The impetus for the formation of these aggregates might be provided by the strings of connected micelles in a percolating system [43] suggested by quasi-elastic light scattering (QELS) and viscosimetry. Acrylamide at the interface induces strong interparticle attractions because the addition of acrylamide is sufficient to convert a nonpercolating (only water) to a percolating system [43]. The formation of water channels between water pools of inverse micelles in a percolating inverse microemulsion leads to an average acrylamide concentration in the dispersed water phase [118]. This process inevitably influences the location of acrylamide at the water-oil interface. As a consequence, for a given acrylamide/water mass ratio, the concentration of acrylamide at the locus of the polymerization reaction of the percolating inverse microemulsion is different from that of a nonpercolating inverse microemulsion. This situation influences both the kinetics and the molecular mass [112].

Percolation decreases the rate of polymerization of acrylamide initiated by APS [119]:

$$R_p(\text{or } \bar{M}_n) \propto [\text{AM}]^x \tag{2}$$

where for percolating inverse microemulsions the respective values of x are 1.1 and 0.4 for R_p and M_n, respectively. For the polymerization of acrylamide initiated by BPO, percolation decreases the time interval of slow oil phase polymerization. This can be ascribed to the increased rate of capturing the acrylamide oligomer radicals by inverse micelles and to efficient mass transfer between oil and water phases in percolating inverse microemulsions.

3. Characteristics of Microlatexes

The properties of the resulting polymer dispersions (microlatexes) and the polymers are directly correlated with the mechanism of the polymerization. For polymer characterization, the latex particles are usually precipitated in a large excess of nonsolvent, purified by extraction or washing, and redissolved in a suitable solvent (e.g., water or methanol). Afterward, standard techniques of polymer characterization can be used for analysis, such as dynamic light scattering (DLS), gel permeation chromatography (GPC), viscometry, or differential scanning calorimetry (DSC). Especially in the case of cross-linking polymerizations, where the spherical structure of the polymer particle is kept by covalent bonding, transmission electron microscopy (TEM) and scanning electron microscopy (SEM) are also useful analytical tools. A brief summary of the results of polymer characterization is as follows:

Decreasing particle size is achieved by increasing the surfactant/monomer ratio. The higher surfactant content enables the stabilization of the larger total interfacial area provided by smaller particles (as described in a simple geometrical model for normal microemulsion polymerization [74]).

Increasing the monomer content [39] or the temperature [34] increases the resulting particle size (among other factors) by increasing the number of collisions and the diffusion rate of the monomer.

Increasing the initiator concentration usually results in smaller particles [74], probably because of reduced primary particle growth in the presence of many simultaneously growing polymer chains.

Comparison of the latex particle size with the gyration radius of the dispersed polymers shows that the polyacrylamide chains in microlatex particles are in a highly collapsed state [44].

The molecular weight of the polymers is typically $>10^6$ g mol^{-1} and increases with higher monomer contents, lower surfactant/monomer ratios, lower initiator concentrations, and lower temperatures (DLS, GPC, viscometry) [78].

The use of alcohols as cosurfactants and other chain transfer agents lowers the molecular weight of the polymer obtained.

DSC measurements show no differences in the glass temperatures of polymers synthesized in solution, inverse emulsion, or inverse microemulsion polymerization [120].

Beside the parameters already mentioned that control particle size and molecular weight, the choice of the monomer has the greatest influence on the polymer properties. An overview of different monomers and surfactants used in inverse microemulsion polymerization is given in Table 2. Most common is the use of acrylamide, acrylates, and MADQUAT. The polymers obtained are not only interesting model systems for the study of inverse microemulsion polymerization but also interesting candidates for applications of high-molecular-weight, water-soluble monomers of Section V. In addition to the polymerization parameters discussed in this chapter, copolymerization and functionalization can be used to control the polymer and particle characteristics of microlatexes and polymers with outstanding and fascinating properties.

4. Copolymerization and Functionalization

In general, there are two different ways to achieve the functionalization of polymer latexes [121]. One is the functionalization by polymer analogue reactions; the other, more convenient and more common way, is the addition of a functionalized compound (e.g., a comonomer) in a one-step-procedure. According to the latter strategy, amine [122], carboxyl [123], hydroxyl [124], mercapto [125], and sulfonate [126] functionalized latexes have been synthesized in heterophases. Depending on the method of monomer addition (batch or feed polymerization) and the physiochemical properties of the monomers, different latex morphologies can be obtained [127].

The main advantage of copolymerization in microemulsions, compared with copolymerization in homogeneous solutions or emulsions, is the possibility of achieving very homogeneous products, even from monomers that are not suitable for copolymerization in solution or emulsion. The copolymerization parameters of monomer pairs are strongly affected by the microstructure of the reaction medium, as a comparison of reactivity ratios has shown [78]. In microemulsions the comonomers are preferably oriented toward the water-oil interface and charge effects are shielded that way. But as the kinetics of microlatex formation are also strongly affected by monomer transport through the interface, this process also has some influence on the copolymerization behavior. Principally, two different cases of copolymerizations can be distinguished, as follows.

(a) Copolymerization of Hydrophilic and Hydrophilic Monomers. The most remarkable effect of copolymerizations in microemulsions is the improvement of structural homogeneity as the copolymerization parameters tend toward unit. In the case of the copolymerization of acrylamide with sodium acrylate, the acrylate content of the copolymer was varied between 10 and

55 mol% [34]. By ^{13}C NMR studies it was shown that the monomer sequence distribution is perfectly random and obeys Bernoullian statistics [128]. The reactivity ratios are close to unity.

By using biunsaturated vinyl monomers, cross-linked microlatexes or so-called microgels are obtained. One study investigated the effect of copolymerization of acrylamide with the cross-linker *N,N′*-methylenebisacrylamide (BAM) on kinetics, polymer particle size, and the degree of swellability. Among other effects, variations of conversion curves and particle sizes were found [129].

Inverse microemulsion polymerization has also been applied to the synthesis of copolymers containing both positive and negative charges along the polymer chain, i.e., polyampholytes [19,130]. The importance of structure homogeneity for the net charge distribution of linear ampholyte terpolymers based on sodium-2-acrylamido-2-methylpropanesulfonate (NaAMPS), 2-(methacryloyloxy)-ethyltrimethylammonium chloride (MADQUAT), and acrylamide (AM) has been demonstrated [36]. By viscometry and dynamic light scattering it was shown that quasi-neutral chains exhibited an extended conformation when they were solubilized in low-salt solutions, which hints at a low charge excess. Increasing the salt content resulted in more compact conformations.

Compared with their macroscopic counterparts (polyampholyte macrogels), little attention has been paid so far to their colloidal counterparts. Cross-linking of NaAMPS, MADQUAT, and AM using *N,N′*-methylenebisacrylamide (BAM) leads to polyampholyte microgels with diameters in the range 85–116 nm. The flocculation behavior in aqueous solutions has been investigated with respect to effects of charge densities and electrolyte addition [37]. Cross-linked polyelectrolyte microgels are interesting model systems for the systematic investigation of the origin of polyelectrolyte effects due to conformational transitions [131].

(*b*) *Copolymerization of Hydrophilic and Hydrophobic Monomers.* The copolymerization of hydrophilic and hydrophobic monomers results in amphiphilic polymers. In an early study, acrylamide was copolymerized with the oil-soluble monomer methyl methacrylate (MMA) using the oil-soluble initiator AIBN. In addition to the copolymer, homopolymers of polyacrylamide (PAM) and poly(methylmethacrylate) (PMMA) were found. The total conversion of MMA was lower than 10%. The composition of the copolymer was almost independent of the comonomer ratio, probably because of a constant molar ratio of the monomers at the interface as the locus of the copolymerization [49,111].

The locus of copolymerization and the nature of the w/o interlayer are also important factors for the copolymerization of acrylamide with styrene in the system toluene/styrene/AOT/water/acrylamide. Also in this case, copolymer and homopolymers were formed. The molar fraction of acrylamide in homopolymer compared with copolymer is about 45–55 [119].

Addition of small amounts of the polymerizable surfactant T10 [*t*-octylphenoxy-poly(oxyethylene) methacrylate, with 10 ethylene oxide units] to the copolymerization of AM and sodium acrylate (NaAA) resulted in hydrophobically modified, water-soluble copolymers [42]. The polymers show interesting rheological properties, e.g., shear thickening, shear thinning, and viscoelasticity. Very low amounts of hydrophobic entities in the copolymer structure (0.5 mol% T10) increase the low-shear viscosity of a 0.2 wt% solution by a factor of 1000 compared with the nonmodified copolymer.

B. Inverse (Macro)emulsion Polymerization

Vanderhoff et al. [20] developed a heterophase water-in-oil polymerization process that they termed "inverse emulsion polymerization," analogous to the direct oil-in-water process. Inverse emulsion polymerization comprises emulsification of a water-miscible monomer, usually in aqueous solution, in a continuous oil medium using a water-in-oil emulsifier. Stabilization was achieved sterically, and the initiating species could reside either in the dispersed water phase (in analogy to suspension polymerization) or in the continuous phase (in analogy to the classical emulsion polymerization) to give a colloidal dispersion of water-swollen polymer particles in oil. The final product is usually a colloidal dispersion of hydrophilic polymer particles in a continuous oil phase. The average particle size of inverse latexes is usually 100–1000 nm, compared with the original droplet size of 100 nm to several micrometers.

In contrast to microemulsions, macroemulsions are only kinetically stable. This means that besides the formulation, other parameters such as reactor geometry and stirring speed can have a strong influence on the course of reaction and the product properties. Many of the different ways of performing polymerizations in direct emulsions have also been established for inverse emulsion polymerization, such as batch, feed, and seeded polymerization. A list of different systems including the surfactant, the monomer, the initiator, and the continuous phase as reported in the literature is given in Table 2.

1. Surfactants for Inverse (Macro)emulsion Polymerization

In principle, for the formulation of inverse emulsions, surfactants very similar to those known for inverse microemulsions are used, but the required amount is usually much smaller and is in the range 2–4% by weight. There are only a few papers on the use of AOT as surfactant in such systems [46]. Sorbitan esters of fatty acids such as sorbitan monooleate are common surfactants for the formulation of the inverse emulsions [24,132], even if the resulting polymerized dispersion usually shows limited stability. Dimonie et al. [55] used polyoxyethylene(8) stearic acid or polyoxyethylene(4) nonylphenol as emulsifier and potassium persulfate and sodium bisulphite as the redox initiator in an acrylamide system. Kurenkov et al. [58] used Sentamid-5 (amide of polyoxyethylated stearic acid) as emulsifier and potassium persulfate as initiator. Usually, nonionic stabilizers are blended to achieve an overall HLB value of 4–6 to better prevent particle coalescence. McKechnie [29] used a blend of Span 80 and Tween 80 as emulsifier and benzoylperoxide as initiator. He illustrated the relation between stabilization and the amount of emulsifier. The Lehigh group [57] used Tetronic 1102 (polyoxyethylene adduct of polyoxypropylene ethylene diamine) as emulsifier for the formulation of an acrylamide emulsion in *o*-xylene.

For the inverse emulsion polymerization of aqueous acrylamide solution in toluene, a blend of a low-HLB Montane 83 and a high-HLB Montanox 85 was used in order to determine the influence of the high-HLB emulsifier on the initial polymerization rate [40]. Despite the small particles, the final dispersions with sorbitan monooleate and sorbitan sesquiolate were not very stable for long storage times. Dispersions with higher stability were obtained with the use of triblock polymeric surfactants such as polyester–polyethylene oxide–polyester (HB 239) [60,133] prepared by reacting condensed 12-hydroxystearic acid with polyethylene oxide [33,134]. Steric stabilization using these polymers is achieved when the polyethylene oxide chain is anchored to the interface and the poly(12-hydroxystearic acid) end is free to move in the continuous phase. The stabilization is enhanced relative to the low molecular fatty acid ester derivatives because of the inherent advantages of triblock surfactants as well as the larger extended length of the poly(12-hydroxystearic acid) (115 Å compared with 20–22 Å for C_{18} sorbitan esters) [135]. In another study in which small amounts of sorbitan fatty acid esters and polyethoxylated sorbitans were blended with the block copolymer, nonsettling polymerizable inverse emulsions at low surfactant and high monomer concentrations could also be obtained [61].

Kurenkov et al. [136] investigated the emulsifier-free polymerization of acrylamide in a water-toluene medium in the absence and in the presence of ionogenic monomers (sodium or potassium styrene sulfonate).

2. Kinetics and Mechanism

The level of understanding of inverse emulsion polymerization has significantly improved over the past 30 years. During the 1980s efforts were dedicated to the elucidation of the reaction mechanism, kinetic measurements, and reactor modeling of the inverse emulsion polymerization processes. Even so, the published data are often inconsistent with each other because the kinetic behavior of inverse emulsion polymerization is complex and is specific to a given monomer-emulsifier-initiator-continuous phase set.

The features of the proposed mechanism are mainly based on the nature of the initiator, which controls the initiation process as well as the colloidal stability behavior of the growing polymer particles during the polymerization process. Apparently, there is a difference in the initiation process with water-soluble and oil-soluble initiators, as already seen for microemulsions. Whereas in the case of inverse microemulsions, an oil-soluble initiator leads to more complex kinetics, the water-soluble initiator does so in case of inverse emulsions. When an oil-soluble initiator is used, the kinetics of inverse emulsion polymerization seem to resemble those of a conventional emulsion polymerization process. The micellar model can be put into practice for the explanation of the polymerization mechanism only in some cases [132]. The inverse emulsion polymerization initiated by a water-soluble initiator probably takes place as a solution polymerization in a microparticle [5,137]. Other differences can cause reaction among the components of the polymerization system [44]. For example, the redox reaction between the emulsifier AOT and initiator APS strongly influences the dependence of the polymerization rate on the concentration of the emulsifier. Because of the great importance of the different methods of initiation, we will focus on this point in detail.

(*a*) *Initiation by Oil-Soluble Initiators.* The kinetics and mechanism of an inverse emulsion polymerization are expected to be similar to those observed in the conventional emulsion polymerization because the initiator is dissolved in the continuous phase. There it can form radicals, followed by reaction with monomer units that are dissolved to a low extent in the continuous phase. The macroradicals formed can enter into the micelles swollen with monomer or into the particles.

One important feature is the droplet and particle size throughout the polymerization. This seems to follow a different mechanism than that in conventional emulsion polymerization. In the latter case, monomer diffuses from large monomer droplets to the particles through the water phase during the polymerization process. Contrary to that, in the case of inverse emulsion polymerization with oil-soluble initiators, the evolution of the droplet is reported to follow a balance between dispersion and coalescence of the emulsion droplets. Graillat et al. [40] showed for the system acrylamide in toluene initiated by the oil-soluble AIBN that two populations of droplets or particles existed in both the initial emulsion and the final latex. The large droplets underwent a sharp decrease in size at a certain percentage of conversion depending on the agitation. It was suggested that this results from the balance between dispersion and coalescence of the emulsion droplets.

The determination of the initiation locus and the rate of polymerization is of great interest. Systems have been reported with kinetics that are close to those for conventional emulsion polymerization. Deviation from the conventional kinetics is mostly a result of the nature of the emulsifier.

Kinetics similar to conventional emulsion polymerization. When an oil-soluble initiator is used, the kinetics of inverse emulsion polymerization seem to resemble those of a conventional emulsion polymerization process. This is schematically presented in Fig. 2. The initiation step (Fig. 2a) occurs in the continuous phase, and radicals or macroradicals have to reach the monomer phase where the particles are nucleated. Large monomer droplets act as a monomer reservoir and monomer diffuses through the continuous phase toward the reaction locus. As the particles grow, they are stabilized by absorbing surfactant from surrounding micelles. After micelles disappear, no new polymer particles are generated and the number of particles remains constant throughout the polymerization (Fig. 2b). At the end, the monomer in the monomer droplets has been consumed and only the monomer in the monomer-swollen particles has to be polymerized to obtain the final latex (Fig. 2c). Pross et al. [138] reported that in the case of acrylamide in isooctane stabilized by a surfactant of high purity, pentaerythriolmyristate, and initiated by 2,2′-azobis(2,4-dimethylvaleronitril) (ADVN, Wako V-65), kinetics of the Smith-Ewart (case 2) type with an average radical number of 0.5 are found. The inverse emulsion polymerization shows the following features of a conventional emulsion polymerization:

Particle nucleation and monomer consumption take place in the monomer droplets.
Mass transfer of monomer and radical entry are not rate limiting [139].
The partial order of the emulsifier concentration is positive.
Chain termination is predominately second order with respect to the polymer radical concentration at initiator concentrations above 0.09 mmol L^{-1}.

The inverse emulsion polymerization of aqueous acrylamide solution in toluene using a selected blend of emulsifiers (Montane 83 and Montanox 85) and oil-

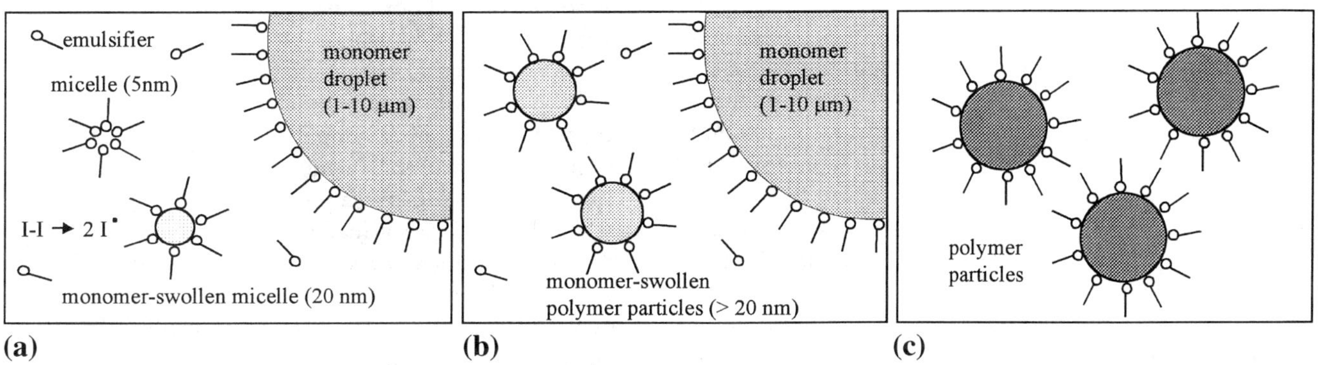

FIG. 2 Schematic representation of inverse emulsion polymerization. (a) One starts from large monomer droplets and surfactant micelles in the water phase. The initiation step occurs in the continuous phase and radicals or macroradicals have to reach the monomer phase, where the particles are nucleated. (b) During the polymerization, the monomer diffuses through the water phase. (c) At the end, particles with a diameter usually larger than 100 nm are obtained.

soluble azo compounds as initiators is also found to behave, for the most part, like the conventional emulsion polymerization process.

Deviation from the kinetics of conventional emulsion polymerization. Numerous papers report kinetics that are different from those of conventional emulsion polymerization. It is confusing that all three Smith-Ewart cases are reported. In one of the first papers about the kinetics of inverse emulsion polymerization, Vanderhoff et al. [20] assumed a Smith-Ewart case 1 model with an oil-soluble initiator. The average number of radicals per particle was found to be incredibly small (0.008 to 0.2). At that time it was proposed that the initiation step is due to the slightly dissolved initiator, benzoyl peroxide, inside the aqueous phase. Other authors found that the kinetics resemble a Smith-Ewart case 3 model [40,140], and Platkowski et al. [138] proposed a Smith-Ewart case 2 model. In an effort to explain this discrepancy, it was found [5] that the kinetics are strongly influenced by the nature of the surfactant. This can lead to the following unusual steps during inverse emulsion polymerization:

A reaction between the macromonomers with the hydrophilic moiety of an interfacial surfactant molecule can take place.
A long-chain reaction with radically active functional groups on the emulsifier has to be considered.
A chain length–dependent mass transfer of primary radicals from the continuous to the disperse phase might be possible.

In most of the papers in which different kinetics compared with conventional emulsion polymerization kinetics were reported, the inverse emulsion polymerization was carried out using a fatty ester of sorbitan and an aliphatic continuous phase. For example, McKechnie [29] used different blends of Span 80 and Tween 80 for the emulsion polymerization of acrylamide and reported kinetic and molecular weight data for a wide range of benzoyl concentrations. It was found that there are unusual dependences of the polymerization rate on emulsifier concentration ($R_p \propto [E]^{0.2}$) and on the initiator concentration ($R_p \propto [I]^{1.0}$). The rate dependence on initiator indicates that a unimolecular termination competes with bimolecular termination, and the relationship between rate and emulsifier concentration is an indirect indication of the radical activity of the surfactant molecules. Many of the studies have also shown that the use of sorbitan esters of fatty alcohols, such as sorbitan monooleate, leads to a degradative chain transfer reaction resulting in branched homopolymers and lower polymerization rates. Therefore, the low rates of polymerization of acrylamide in inverse emulsion polymerization with sorbitan monooleate can be explained in terms of the emulsifier's chain transfer activity [5,23]. Sorbitan monooleate has an unsaturated carbon in the middle of its hydrophilic tail and five labile hydroxy functional groups on its surfactant head. These radically active functional groups can react with primary radicals in the continuous phase, lowering the polymerization rate and increasing the molecular weight.

A systematic study has been reported by the Lehigh group [57] on an acrylamide system using a surfactant of different chemical nature, Tetronic 1102 (polyoxyethylene adduct of polyoxypropylene ethylenediamine adduct), and benzoyl peroxide as initiator. However, because of the nature of the emulsifier, the formation of multicelled emulsion droplets consisting of Tetronic emulsifier molecules surrounding aqueous acrylamide subdroplets was found to affect the polymerization kinetics.

In the inverse emulsion polymerization of aqueous acrylamide solution in toluene using a selected blend of emulsifiers (Montane 83 and Montanox 85) and oil-soluble azo compounds as initiators, it was found that not only the nature of the surfactant but also the composition of different surfactants in a surfactant blend is of interest. There is a minor effect of the high-HLB emulsifier concentration on the initial polymerization rate that is reflected in the molecular weight values. The use of mercapto-acrylamide oligomers results in a lower molecular weight. It is suggested that the initiation process follows a mechanism in which the primary radicals are mainly produced in the oil phase or possibly in the interfacial layer. In the former situation, the radical or oligoradical may be captured by the monomer droplets or cause a homogeneous nucleation in the oil phase by reacting with the dissolved acrylamide molecule; in the latter case, the radicals diffuse into the interior of the monomer droplets. Other experimental data also support [141] an initiation step from radicals formed in the interfacial layer of emulsifier and captured by the droplets.

The kinetics are affected not only by the chemical role of the surfactant but also by its physical role. This could be shown by inverse emulsion polymerization involving the block copolymer surfactant HB239. In this case no chemical reaction with the surfactant is expected to take place. Indeed, it was found that the rate of polymerization of acrylamide in inverse emulsion polymerization is higher when the block copolymeric surfactant is utilized in comparison with sorbitan monooleate, and the rate of polymerization does not depend on the emulsifier concentration. The initial rate of polymerization of an inverse emulsion of acrylamide using HLB239 is found to be:

$$R_p \propto [M]^{1.0}[I]^{1.0}[E]^0 \quad (3)$$

The first-order dependence on the initiator concentration suggests that unimolecular termination dominates. The first-order dependence with respect to the monomer implies standard free radical initiation and propagation steps. The zeroth-order dependence with respect to the emulsifier suggests a purely physical role of the emulsifier [6,60]. However, the weight average molecular weight of the resulting polymer was found to be lower because of a transfer reaction to the hydrophilic part of the emulsifier [56].

The influence of the emulsifier on the polymerization rate can be studied by comparing the systems with a system synthesized in absence of an emulsifier. In the case of emulsifier-free polymerization [136] of acrylamide in a water-toluene medium, in the absence and in the presence of ionogenic monomers (styrene sulfonate of sodium and potassium), the kinetics were found to follow a 0.46 order dependence on the initiator concentration when AIBN was used as initiator. Factors that influence the particle diameter, such as the rate of agitation and the impeller size and type, also have a minor influence on the rate of polymerization.

(*b*) *Initiation by Water-Soluble Initiators.* Apparently, the inverse emulsion polymerization initiated by a water-soluble initiator is expected to take place in a microparticle and the kinetics should resemble those for a solution polymerization [5,137]. In their pioneering work on inverse emulsion polymerization of sodium *p*-vinylbenzene sulfonate, Vanderhoff et al. [20] investigated the effect of temperature, emulsifier concentration, type, and concentration of the emulsions. Based on TEM micrographs showing very small droplets (in the range of 20 nm), it was postulated that particle nucleation occurs in emulsion droplets as well as in monomer-swollen micelles if present.

From the kinetic and molecular weight results, Vanderhoff et al. [20] assumed a Smith-Ewart case 2 model in the case of a water-soluble initiator. But the Smith-Ewart micellar theory cannot be used to explain the polymerization mechanism because the initiator is dissolved not in the continuous phase but in the dispersed phase. The initiation reaction starts in finely dispersed droplets of the aqueous solution of the monomer.

The evolution of particle size during polymerization is of great interest for the interpretation of the mechanism. Many papers do not mention the droplet size versus the particle size or the dependence of particle size on conversion. In some papers the evolution of the particle size distribution versus conversion is observed [40,54]. This is mostly in the case of oil-soluble initiators, and it was discussed with the balance between dispersion and coalescence of the emulsion droplets. Contrary to these findings, in the case of water-soluble initiation, most of the published data show that the droplet and particle size seems to remain constant throughout the conversion [28,30,33,46]. For this behavior two explanations may be discussed:

The mechanism of coalescence and breakup of droplets takes place.
The droplets are stable throughout the polymerization.

The question of whether the identity of the droplets during polymerization is maintained also leads directly to the discussion of miniemulsion polymerization (see Section IV.C). However, carefully designed experiments in which monomer and initiator were in different droplets and therefore different droplets had to collide for polymerization to occur supported the mechanism of coalescence and breakup of droplets [59]. The largest droplets act as monomer reservoirs and their size is found to be affected by the stirring rate. The dispersion under high shear seems to be more efficient at relatively high conversions. The smaller droplets, which are insensitive to the stirring rate, are sensitive to Brownian motions and they coagulate with other polymer particles. Coalescence and breakup of aqueous droplets (see Fig. 3) take place simultaneously under continuous agitation and have a significant effect on polymerization, drop/particle size, and distribution [59]. The particle size distribution is broad for the inverse latexes. This is expected for particles produced by a breakup coalescence mechanism. However, it is interesting to note that Kriwet et al. [28] have observed that the use of water-soluble initiators leads to particles of 1–10 μm, whereas in the case of oil initiators, smaller particles of about 80–150 nm were formed.

There are numerous studies of inverse emulsion polymerizations, some of which are discussed here. Studies of Kurenkov et al. [142] using KPS initiator and Sentamid-5 emulsifier show that the polymerization rate increased with increasing emulsifier concentration (up to 20% conversion) and then became independent of emulsifier concentration. Molecular weight was found to decrease with increasing initiator, emulsifier, and toluene concentrations. For the kinetics of acrylamide stabilized by Tetronic 1102 and initiated by KPS [141] the rate of polymerization was found to be:

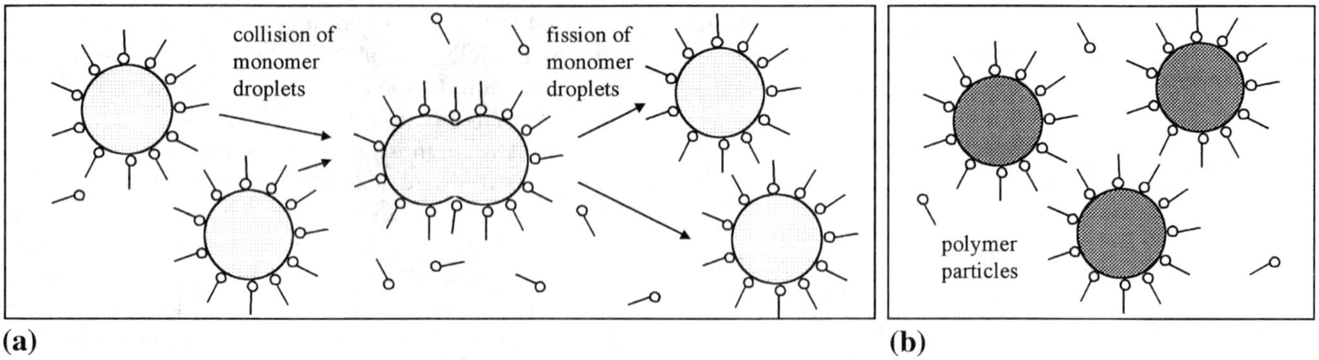

FIG. 3 (a) Mechanism of coalescence and breakup of aqueous droplets. (b) This mechanism can lead to the same particle size as it is observed for the droplet size in the starting emulsion.

$$R_p \propto [I]^{0.4}[M]^{1.0}[E]^{0.6} \quad (4)$$

Benda et al. [46] found that the kinetics of the persulfate-initiated emulsion polymerization are independent of the type of monomer: acrylamide and acrylic acid (sodium or ammonium salt) show similar kinetics and the polymerization can be described by the following equation:

$$R_p \propto [I]^{0.5}[M]^{1.5}[E]^{0.1} \quad (5)$$

The kinetics of the KPS-initiated inverse emulsion polymerization of aqueous sodium acrylate solutions in kerosene with Span 80 as emulsifier were studied [21]. The conversion-time curves are S shaped. The following expressions have been obtained for the maximum rate of polymerization and the molecular weight of the polymers under the experimental conditions investigated:

$$R_{max} \propto [KPS]^{0.78}[M]^{1.5}[Span\ 80]^{0.1} \quad (6)$$

$$\bar{M}_v \propto [KPS]^{-0.37}[M]^{2.9}[Span\ 80]^{-0.2} \quad (7)$$

Other studies show that the polymerization rate can have up to a 1.7 order dependence on the monomer concentration [137]. The rate of polymerization for an inverse emulsion photopolymerization of acrylamide using the water-soluble initiator α-ketoglutaric acid (α-KGA) and the emulsifier Span 80 was found to be [25]:

$$R_{max} \propto [Intensity\ of\ light]^{0.5}[I]^{0.5}[M]^{1.28}[Span\ 80]^{-0.42} \quad (8)$$

The resulting molecular weights are higher than in the case of the water-soluble 4,4′-azobis-4-cyanopentanoic acid (ACPD). This is due to the fact that ACPA generates single radical pairs on dissociation, whereas α-KGA (and β-KGA) gives rise to triplet pairs [143].

In conclusion, the kinetics of the inverse emulsion polymerization initiated by water-soluble initiators depends mainly on:

The type and amount of the surfactant ($R_p \propto [E]^{0.1-0.6}$)
The amount of initiator ($R_p \propto [I]^{0.4-0.8}$)
The type and amount of monomer ($R_p \propto [M]^{1.0-1.7}$)

A dependence of the polymerization rate on the surfactant and initiator concentration was observed in the case of oil-soluble initiators, and it was also obtained for the inverse emulsion polymerization initiated by water-soluble initiators. This behavior was explained by reactions between the radicals and the emulsifier. However, in the case of water-soluble initiators, the dependence on the amount of initiator follows a lower order (0.4 to 0.8) than for the oil-initiated inverse emulsion polymerization. This low order is consistent with the order in emulsifier-free polymerization [136] of acrylamide in a water-toluene medium, for which in both the absence and the presence of ionogenic monomers (styrene sulfonate of sodium and potassium) a 0.5 order dependence related to the initiator is observed in the case of KPS. This deviation from the first-order rate with respect to the monomer is observed only in the case of water-initiated inverse emulsion polymerization.

The deviation from the first-order rate with respect to the monomer is between 1 and 1.7 and usually has been explained by complex or cage effect theory or a hybrid of them. Hunkeler and Hamielee [137] assumed the occurrence of persulfate in three forms: dissolved, compact cage fragments, and diffuse cage fragments. The explanation of the polymerization mechanism is also connected to unusually low polymerization temperatures, and Benda et al. [46] discussed a redox reaction of AOT and APS. The initiation reaction takes place in the water phase and was found to go over the complex stage. The ammonium persulfate and acrylamide complex decomposes into two unpaired radicals capable of propagation. This suggestion is supported by the almost sesquimolecular rate order with respect to the monomer for acrylamide and acrylic acid. Formation of the complex accounts for the enhanced decay of ammonium persulfate at low temperatures.

The initiation can also be performed using water-soluble redox initiators [56,144,145]. In inverse emulsion polymerization using water-soluble initiators, the two components of a redox pair have to be introduced separately. Usually, the oxidizing part is dissolved in the aqueous monomer, which is then dispersed in an oil stabilized with surfactant, and the reducing part is then added to start polymerization. An alternative is that both oxidant and reductant are added separately to the emulsion of aqueous monomer in an agitated oil phase. Therefore, the polymerization system may consist of either two or three different droplets and the distribution of initiator(s) is heterogeneous in nature. This also supports the mechanism of coalescence and breakup of aqueous droplets [59,146,147]. The initial polymerization rate in an inverse emulsion polymerization of acrylic acid in Isopar-M and *N,N*-bishydroxyethyl tall oil amide as surfactant was found to be:

$$R_p \propto [AA]^{2.01}[Na_2S_2O_5]^{0.70}[KBrO_3]^{0.76}[E]^{-0.47} \quad (9)$$

A combination of bimolecular and monomolecular termination modes, a chain transfer of the surfactant, and an oxidizing role for the oxygen molecules were suggested [147]. This complicated mechanism has to be considered because a coalesced aqueous drop might undergo further coalescence and breakup.

Modeling of inverse emulsion polymerization applying Monte Carlo methods has also been done [148]. This permits the calculation of kinetics as well as different product distributions. This method is not restricted to a steady-state assumption, and chain length–dependent reactions, such as cross-linking and long chain branching, can easily be included.

(*c*) *Initiation by Radiation.* The initiation of polymerization in inverse emulsions can also be achieved by radiation using a ^{60}Co source [32]. In principle, in the case of radiation-initiated polymerization, kinetics close to those with the water-soluble initiator are expected because the nucleation and polymerization are limited to the monomer phase. The polymerization of vinylpyrrolidone in an isoparaffinic hydrocarbon with an emulsifier blend of Span 80 and Tween 85 results in high-molecular-weight polyvinylpyrrolidone obtained at high polymerization rates ($R_p \approx$ 10–35 mol L^{-1} s^{-1}) to high conversions (90–95%). The particle sizes are in the range 200 nm to 2 μm.

In the case of ^{60}Co γ-ray–initiated inverse emulsion polymerization of aqueous sodium acrylate solutions in kerosene with Span 80 as emulsifier, the polymerization rate was found to be:

$$R_p \propto D^{0.9}[M]^{1.5}[E]^{0.4} \quad (10)$$

This kinetic analysis suggests a dose rate–independent polymerization process for the system such that there was only one active passage of radiation through a droplet; e.g., all polymer radicals resulting from a radiation passage were terminated before another passage [22]. In another system, the copolymerization of (2-methacryloyloxyethyl) trimethyl ammonium chloride and acrylamide was performed in an inverse emulsion polymerization using gamma rays for initiation. The system was emulsified in kerosene with a blend of Span 80 and OP10. The rate of polymerization can be presented by:

$$R_p \propto D^{0.87}[M]^{1.37}[E]^{0.53} \quad (11)$$

where D is the dose rate [30,31].

3. Inverse Emulsion Copolymerization

Copolymerization reactions allow the combination of nonionic, anionic, and cationic monomers in order to obtain polymers with new properties. Inverse emulsion copolymerization studies of acrylamide and methacrylic acid with AIBN as initiator were reported by Glukhikh et al. [41]. The kinetic behavior of the copolymerization is strongly pH dependent. This can be partly explained by the wide differences in the reactivities of MAA monomer molecules and MAA-ended macroradicals, but it is also due to the partition of the ionic comonomer between the organic and the aqueous phase of the emulsion under acidic conditions. The latex stability under acidic conditions is poor. An experimental investigation of the inverse emulsion copolymerization of acrylamide and quaternary ammonium cationic monomers, dimethylaminoethylacrylate (DMAEA), and dimethylaminoethylmethacrylate (DMAEM) has been carried out using sorbitan monooleate [23] or a block copolymeric surfactant (HB246) whose hydrophilic moiety is polyethylene oxide and whose hydrophobic moiety is poly(12-hydroxy stearic acid) [62]. The reaction was started by AIBN or KPS and the following observations were made:

Nucleation and polymerization occur within the monomer droplets.

Heterophase diffusion-limited oligoradical precipitation is the predominant initiation reaction.
Unimolecular termination with interfacial species is competitive with the bimolecular process.
Propagation and termination were not found to be influenced by the nature of the polymerization system, proceeding at equal rates in solution and inverse emulsion.

A kinetic expression suggested for the inverse emulsion copolymerization of (2-methacryoyloxyethyl) trimethyl ammonium chloride with acrylamide using KPS as initiator is [30]:

$$R_p \propto I^{0.52}[M]^{1.50}[E]^{0.38} \tag{12}$$

NMR measurements have shown that the choice of the monomer pairs leads to different compositions. The microstructure of acryloylethyltrimethylammonium chloride turned out to be more homogeneous than the copolymer (2-methacryoyloxyethyl) trimethyl ammonium chloride with acrylamide [149].

It was also found that not only does the choice of the monomer set influence on the composition but also the surfactant has a strong influence on the quality of the copolymer produced [62]. For example, more uniform copolymers of acrylamide and DMAEA can be synthesized using the block copolymer as emulsifier at faster production rates in comparison with sorbitan monooleate when using batch reactors. But even in the first case a composition drift is observed because DMAEA reacts faster than acrylamide. By feeding the monomers, a more homogeneous composition can be obtained. In addition, artificial neural networks have been used to predict the copolymer composition as a function of reaction conditions and conversion [150].

4. Inverse Macroemulsion Polymerization in Supercritical Carbon Dioxide as an Alternative Medium

Supercritical fluids (materials at temperatures and pressures above their critical values) possess intriguing physical properties that make them interesting media in which to conduct polymerizations. For polymerization in CO_2, fluorinated surfactants are required [151]. Yates et al. [152] investigated two steric stabilizers, poly(1,1-dihydroperfluorooctyl acrylate (PFOA) and a block copolymer, PS-*b*-PFOA. For emulsions stabilized with these polymers, the critical flocculation density (CFD) is very near the Θ point of the stabilizing moiety in CO_2. Just below the CFD, emulsions stabilized with PFOA exhibit a sharp increase in average droplet size, followed by a sharp decrease, indicating flocculation and subsequent sedimentation [153]. Emulsions stabilized by PS-*b*-PFOA exhibit a lower flocculation rate below the CFD because of greater surfactant adsorption and absence of bridging flocculation.

C. Inverse Miniemulsion Polymerization

In the case of inverse miniemulsion systems, hydrophilic monomers were miniemulsified by high shear in a nonpolar medium, e.g., cyclohexane or hexadecane containing a surfactant suitable for inverse emulsions (see Fig. 4a). In order to provide osmotically stabilized droplets, water or a salt was added as a "lipophobe" to the monomer phase. Polymerization in carefully prepared miniemulsions should result in latex particles of about the same size as the initial droplets, and a virtually 1:1 copying of the droplets to the particles with respect to their sizes can be obtained (see Fig. 4b and c) [154], as shown for direct systems by a combination

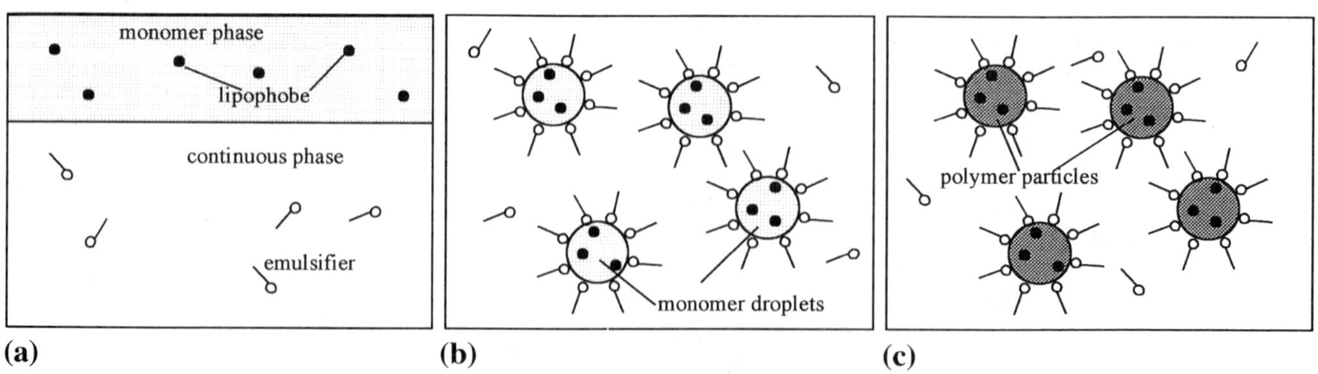

FIG. 4 Schematic representation of inverse miniemulsion polymerization. (a and b) In a first step, relatively stable oil droplets with interfacial tensions larger than zero and droplet sizes within the range 50 to 500 nm are prepared by shearing a system containing oil, water, surfactant, and a water-insoluble hydrophobe. (c) These minidroplets can be polymerized to polymer latex particles, ideally in a 1:1 copying process.

of SANS, surface tension measurements, and conductometry [155]. It was found that inverse systems exhibit characteristics similar to those of direct (oil-in-water) miniemulsions (K. Landfester et al., Ref. 66a):

The formation of a miniemulsion requires high mechanical agitation to reach a steady state given by a rate equilibrium of droplet fission and fusion.

The dispersed miniemulsions are osmotically stable but critically stabilized with respect to colloidal stability.

The interface energy between the oil and water phase is greater than zero. The surface coverage of the miniemulsion phases by surfactant molecules is not complete.

The osmotic stability of miniemulsion droplets results from an osmotic pressure in the droplets that controls the solvent or monomer evaporation. The osmotic pressure results from the addition of a lipophobe, which has extremely low solubility in the continuous phase.

During the polymerization in miniemulsions the growth of particles can be suppressed. The monomer diffusion is balanced by a high osmotic background of the lipophile, which makes the influence of the polymer less serious.

The amount of surfactant required to form a polymerizable miniemulsion with surfactant was $0.015 < S < 0.25$ (where S is the mass ratio of surfactant to monomer) and covers the whole range from inverse suspension to inverse microemulsion polymerization.

Inverse miniemulsions were generated with the polar monomers acrylic acid, hydroxyethyl methacrylate, and acrylamide in cyclohexane or hexadecane as an unpolar continuous phase, and the miniemulsions were polymerized to latexes (K. Landfester et al., Ref. 66a). Rather small and narrowly distributed latexes in a size range 50 nm $< d <$ 200 nm were made of acrylic acid, acrylamide, and hydroxyethylacrylate. Nonionic amphiphilic block copolymers with poly(ethylene-co-butylene) tails turned out to be very efficient stabilizers. Depending on the system, the surfactant loads can be as low as 1.5 wt% per monomer, which is very low for an inverse polymerization reaction and clearly underlines the applicability. It was found that with increasing amount of emulsifier, the particle size decreases as expected. But as already seen for direct miniemulsions, the demand of surface per surfactant molecule increases with decreasing particle size. This means that the smaller the particles are, the higher the coverage of the particles with surfactants is in order to obtain stable latex particles.

It might be possible that some of the inverse emulsion polymerization processes with water-soluble initiators are very close to miniemulsions because in some cases the droplet and particle size seems to be constant throughout the polymerization. Especially in the case of copolymeric emulsifier [33], high stability of the droplets is expected. The ionic initiator can overtake the function of the osmotic agent because it is not soluble in the continuous phase. However, the use of high shear to obtain defined starting conditions would be important to ensure a well-defined droplet size with a small distribution.

D. Inverse Suspension Polymerization

Inverse suspension polymerization involves the dispersion of a water-soluble monomer in a continuous organic phase. The rather low consumption of surfactant, the very low amount of coagulum, and the ease of recovery of the polymer from the reaction medium are important advantages of this type of polymerization. Emulsifier levels are typically very low with 2–5% of the organic phase and are below the critical micelle concentration. Inverse micelles have not been detected during nucleation, and the polymerization proceeds in the monomer droplets. The dispersion is thermodynamically unstable and requires both continuous vigorous agitation and the addition of a low-HLB steric stabilizer. This forms a condensed electrically neutral interfacial layer and prevents coalescence. The monomer droplets are typically between 1 and 200 μm in diameter and are controlled by the Weber number of the mixture. A scheme for inverse suspension polymerization is presented in Fig. 5.

1. Surfactants for Suspension Polymerization

Sorbitan monooleate, Span 80, has also been reported as a suitable surfactant for polymerization in inverse suspension polymerization [27]. If the dispersion is stabilized by a mixture of the low-molecular-weight surfactant Span 80 and macromolecular emulsifier ethyl cellulose (degree of substitution 2.42 to 2.53, N type of Hercules), better control of the size and the morphology is obtained. Polyoxyethylene(4) nonylphenol has also been used as a effective surfactant for inverse suspension polymerization [56].

2. Kinetics and Mechanism

The initiation and polymerization of an inverse suspension should take place only in the droplets. Because of the large size of the droplets and therefore contrary to inverse emulsion polymerization initiated by water-sol-

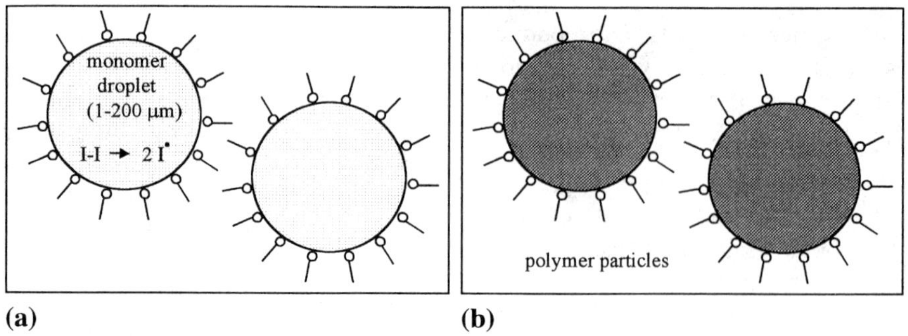

FIG. 5 Schematic representation of inverse suspension polymerization. (a) It involves the dispersion of a water-soluble monomer in a continuous organic phase; the initiator is dissolved in the monomer droplets. (b) After polymerization, large polymer particles (1–200 μm) are obtained.

uble initiator, the polymerization is not affected by the interfacial layer formed by emulsifier molecules. Omidian et al. [156] reported the preparation of superabsorbent polymers by inverse suspension polymerization using different acrylic acid/sodium acrylate compositions and different amounts of cross-linking agents. The particle size distribution was narrow (300–400 μm). Partially substituting the ionic monomers with nonionic acrylamide broadened the distribution considerably. Large superabsorbent polymers based on acrylic acid with sizes of about 200 μm and with undefined morphology have been synthesized by using Span 80 and toluene [27]. Toluene as the continuous phase was heated with sorbitan monooleate. The monomer blend was then added dropwise under constant agitation. The monomer droplet size was inversely related to the emulsifier level. The suspension polymerization took place with a high polymerization rate, reached the maximum conversion, and resulted in a linear (gel-free) polyacrylamide characterized by a high molecular weight of 10^6 to 10^7 [27]. If the dispersion was stabilized by a mixture of low-molecular-weight surfactant Span 80 and macromolecular ethyl cellulose emulsifiers (degree of substitution 2.42 to 2.53, N type of Hercules), better control of the size and the morphology was obtained.

Dimonie et al. [55] investigated the persulfate-initiated polymerization of acrylamide in the presence of a low emulsifier concentration. They proposed an inverse suspension process with the following reaction steps: starting with a water-in oil dispersion of aqueous monomer droplets, which are very instable, the addition of $K_2S_2O_8$ rapidly causes an increase in viscosity, followed by gel formation and subsequently phase inversion. As the polymerization proceeds and enough polymer is formed, the gel is broken into smaller particles by stirring. This mechanism emphasizes the important roles of the emulsifier type and concentration, salt concentration, reaction temperature, and stirring intensity in the physicochemical interfacial behavior during polymerization.

It is reported that acrylamide polymerization by inverse suspension differs from the widely accepted model for suspension polymerization of vinyl monomers [56]. The similarities between these two processes are conventional ones, and they concern mainly the shape and size of particles obtained at the end of the polymerization. The changes of topochemistry were shown by an inverse suspension polymerization carried out by dispersing the concentrated solution of acrylamide in cyclohexane containing ethoxylated nonylphenol ethers [EO(4)NP] and using a redox initiator ($NaHSO_3$-$K_2S_2O_8$). Several distinct stages were found: coarse water-in-oil dispersion, concentrated oil-in-water emulsion, bicontinuous system, and coarse polymer-water dispersion in oil. The main reaction, which determines the limitation of the molecular weight, was the chain transfer with the emulsifier EO(4)NP.

E. Inverse Dispersion Polymerization

Water-soluble polymers also include the preparation of a nonaqueous dispersions in which a water-immiscible monomer in organic solvent solution is polymerized using an oil-soluble initiator, so that the polymer precipitates as it is formed. In the presence of a suitable oil-in-oil emulsifier, spherical particles 100 nm to 10 μm in diameter are formed from the precipitating polymer. Thus, these polymerizations begin as precipitation polymerizations but become emulsion polymerizations upon stabilization of the polymer particles. A scheme for inverse dispersion polymerization is presented in Fig. 6.

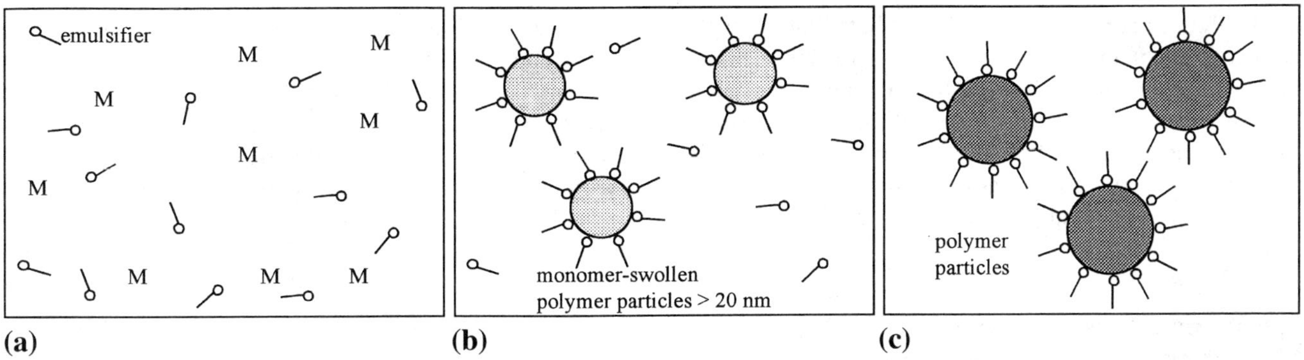

FIG. 6 Schematic representation of inverse dispersion polymerization. (a) Water-immiscible monomer is dissolved in an organic solvent in the presence of a water-in-oil emulsifier; the polymer precipitates as it is formed. (b) Spherical particles 100 nm to 10 μm in diameter are formed from the precipitating polymer. The particles still grow by diffusion of monomer, and (c) after consumption of the monomer, polymer particles are obtained.

1. Surfactants for Dispersion Polymerization

For the process of dispersion polymerization, polyacrylic particles were stabilized by block copolymers (polystyrene-*b*-polyethylenoxide). The use of this stabilizing system enables the production of polyacrylate particles with particle size ranging from 50 to 300 nm depending on the composition of the surfactant (block length) and its concentration [64]. Ray and Mandal [66] carried out dispersion polymerization of acrylamide using poly(vinyl methyl ether) (PVME) as the polymeric stabilizer.

2. Kinetics and Mechanism

Baade and Reichert [132] suggested a dispersion polymerization model for the acrylamide system using different nonionic emulsifiers (sorbitan derivatives) in two oil phases (isoparaffinic mixture and toluene). One starts from a solution of acrylic acid in toluene, and during the course of polymerization, the polymer precipitates and is stabilized by emulsifiers. Their model revealed that radicals from the oil phase diffused into the hydrophilic phase in which polymerization took place when the oil-soluble initiator (azodimethylvaleronitrile or AIBN) was used. In this case, they reported a kinetic expression for the polymerization rate (R_p) that is initiator concentration dependent (1.0 order) and emulsifier and emulsifier concentration dependent (-0.2 order), instead of the 0.5 order dependence related to the initiator concentration in the case of a water-soluble initiator.

The mass transfer of primary radicals from the oil into the water phase is considered to be the rate-determining step of the reaction [157]. The rate of polymerization in an acrylic system showed strong autocatalytic behavior. The maximum of the polymerization rate and the corresponding time of appearance depend strongly on the water content. Because the polymerization started in the continuous phase and no micelles of the block copolymer were detected, a homogeneous nucleation process was proposed [158].

Using dicarboxylic dichloride or diglycidyl ether as cross-linking agent for polyacrylic acid particles synthesized by inverse dispersion polymerization, the particles can be linked with each other and a gel with three-dimensional network structure can be formed [159,160]. Ray and Mandal [66] carried out dispersion polymerization of acrylamide using poly(vinyl methyl ether) (PVME) as the polymeric stabilizer, KPS as the initiator, and various alcohols or their mixtures with water as the polymerization medium. Polydisperse spherical as well as oval particles are formed. The simultaneous presence of both species suggests the coalescence of particles of similar size leading to polydispersity. The increase of stabilizer concentration leads to a decrease in the particle size and a decrease of the molecular weight. The particle size also decreases with increasing *t*-butyl alcohol (TBA) concentration in the TBA-water mixture. TBA-water mixtures exhibit cosolvency toward PVME, the solvency becoming highest at about 70 vol% TBA.

Very similar to dispersion polymerization is precipitation polymerization. In this case, the polymer formed is usually not stabilized by a surfactant. Precipitation polymerization of acrylamide and *N,N'*-methylbisacrylamide (MBA) was carried out in alcohols [71]. Large, bulky, and porous particles were formed by the polymerization. On the contrary, copolymerization of acrylamide and MBA with a certain amount of methacrylic acid resulted in the formation of fine monodisperse hydrogel microspheres. This resembles a dispersion polymerization process. This result was attributed to the contribution of methacrylic acid units to the stabilization of particles formed at the initial stage of polymerization and the enhancement of swelling of the particles by monomer and alcohol. It could be shown that the size of the microspheres varied from 0.2 to 1.3 μm as a function of the solubility parameter of the dispersant [161].

F. Mixed Cases of Inverse Heterophase Polymerization

Since their respective inventions by Vanderhoff et al. [20] and Leong and Candau [48], inverse emulsion and inverse microemulsion polymerizations have shared the common objectives of synthesizing high-molecular-weight homo- and copolymers with controlled properties. However, both have unique advantages and disadvantages. For example, the lower emulsifier level required for inverse emulsions led to earlier and more extensive commercialization. However, these latexes are thermodynamically unstable. Furthermore, process improvements that have been implemented to enhance the colloidal stability, such as the addition of cosurfactants, have generally decreased the molecular weight of the resulting product and the end-use efficiency. One approach has been discussed in which a thermodynamically unstable inverse macroemulsion is transformed into a thermodynamically stable inverse microemulsion [33].

Heterophase water-in-oil polymerizations of acrylamide were performed in the presence of blends of nonionic stabilizers at 20% monomer concentration. The initial monomeric system is located outside the inverse microemulsion domain, yet close to the inverse macroemulsion/inverse microemulsion phase boundary. A turbid, viscous, and unstable dispersion is produced at the outset and during the conversions. This evolves to an inviscid and nonsettling system at high conversions. Transparent inverse latexes can also be produced provided that the polymerizations are conducted semiadiabatically. Independent of the conversion, the particles were found to be 150 nm. The system follows an inverse macroemulsion–like mechanism. The hybrid inverse microemulsion/inverse macroemulsion polyacrylamides produced herein have a smaller diameter in the aqueous phase than those produced by either solution polymerization or true inverse emulsion polymerization. This is probably due to a large number of intermolecular interactions, such as hydrogen bonds, which are induced by the collapsed nature of the polymer chains in the inverse microemulsion droplets. The molecular weight and the diameter of the final latex are independent of the polymerization conditions such as initiator level, the hydrophilic-lipophilic balance, the temperature, and physical changes occurring during the polymerization. From the kinetic point of view, the molecular weights of these systems are controlled by the transfer to monomer, and transfer to interfacial emulsifier is the polymerization rate-controlling step. The polymerization process produces final latexes that are transparent and nonsettling with particle sizes smaller than 150 nm.

V. APPLICATIONS

The production of water-soluble polymers in the United States is a multibillion dollar industry, and polyacrylamide and its copolymers constitute the majority of this market. Typical applications are flocculants for wastewater treatment, paper manufacturing, superadsorber, coatings, flotation aids, and aqueous viscosity modifiers in enhanced oil recovery and latex paint systems [33,162–164]. Polyacrylamide and its anionic and cationic copolymers are also applied as coagulants and flocculants in waste and potable wastewater treatment applications, as pushing fluids in enhanced oil recovery, as drag reduction agents and drilling fluids, and for process water clarification in industries such as mining and papermaking [165].

A. Latexes Obtained by Inverse Microemulsion Polymerization

Polymers synthesized in inverse microemulsions are interesting candidates for all applications of water-soluble polymers. As microlatexes combine high molecular weights with small particle sizes, they show some advantages compared with conventional polymers, such as low viscosities and higher surface areas. Some copolymers can be obtained only by microemulsion polymerization, especially amphiphilic copolymers of water- and oil-soluble monomers. Because of the huge surface area of microlatex particles in the dispersion ($\sim$100–300 $m^2\ g^{-1}$), they are outstanding adsorbents, e.g., for proteins, enzymes, and for the immobilization of antibodies [166]. Therefore one important field of application is targeted drug delivery systems [167,168]. The functionalized small microlatexes can pass the blood-brain barrier and have longer circulation times in the body [169] compared with functionalized emulsion latexes (typically 100 nm to 10 μm in diameter).

Gan et al. [67] reported the polymerization of aniline in inverse emulsion polymerization. Because polyaniline is a conducting material, there might be a potential for this kind of particle.

Besides the listed applications, many new applications will arise from the use of functionalization techniques, e.g., applications of nanohybrids for materials science, reactive particles, and catalysis. This will enable the synthesis of new materials with outstanding properties. One example is the synthesis of superparamagnetic hydrogel particles [170]. In a single microemulsion, magnetite nanoparticles were prepared and then coated by copolymerization of methacrylic acid (MAA) and hydroxyethyl methacrylate (HEMA). The size of the hybrid particles varied between 160 and 320 nm depending on the surfactant/monomer ratio. In a similar approach, superparamagnetic colloids were generated within a continuous hydrogel, which was synthesized in a bicontinuous microemulsion [171]. Silica particles prepared in an inverse microemulsion can be encapsulated, e.g., by a semicontinuous emulsion polymerization of ethyl acrylate (EA) [172]. All these examples show how different properties of inorganic and organic compounds can be combined within nanostructured hybrid materials.

The synthesis of reactive, functionalized microgels by immobilization of enzymes can be achieved in a two-step procedure [50]. The copolymerization of acrylamide (AM) and *N,N′*-methylene bisacrylamide (BAM) with *N*-acryloyl-1,6-diaminohexane or acrylic acid resulted in amine- or carboxylic acid–functionalized microgels. In a second step, alkaline phosphatase was physically entrapped by adsorption at the particle surface. The functionalized particles show enzymatic activity toward the hydrolysis of *p*-nitrophenylphosphate.

Using cross-linked microgels of poly-MADQUAT or sulfonated microgels as exotemplates for the controlled growth of noble metal colloids results in metal-polymer modules that can act as endotemplates for the synthesis of functionalized mesoporous silica supports [173]. The metal-polymer nanoparticles generate the porosity of the silica matrix (by calcination) in which the noble metal colloids (e.g., rhodium, platinum, palladium) are entrapped. The functionalized silica shows catalytic activity in the hydrogenation of dehydrolinalol.

These are only a few examples, which were chosen to demonstrate the versatility and the huge impact for materials science provided by inverse microemulsion polymerization.

B. Latexes Obtained by (Macro)emulsion Polymerization

The inverse emulsion polymerization process can also be used for the encapsulation of polypyrrole particles of about 100 nm by acrylamide. The encapsulation increases the processability of the conducting particles by preventing reagglomeration of the particles. It also improves the electrical stability of the particles because the shell of insulating polymer can act as a physical barrier against the dedoping effect [174].

Xie et al. [175] reported the synthesis in inverse emulsion polymerization of poly(lithium acrylate), which can be used as an electrorheological fluid. Such a dispersion can undergo changes in rheological properties, such as development of a yield stress and increased viscosity upon application of kV mm^{-1} electric fields [176].

Because poly(acrylic acid) micro- and nanoparticles yielded excellent bioadhesive properties in an in vitro assay, they may be suitable for the encapsulation of peptides and other hydrophilic drugs [28].

C. Latexes Obtained by Suspension Polymerization

By utilizing the suspension polymerization technique on water-soluble monomers such as acrylic acid, hydrogels with a high capacity for retaining water can be obtained. For some fields of application, such as agricultural and horticultural use, the exact particle size distribution is of minor importance because superadsorbent unmodified systems can be used. But for medical and hygienic uses such as sanitary napkins and baby diapers, some modification of size is necessary. Such particles can be achieved by the use of macromolecular stabilizers [27]. To obtain superadsorbers with a high capacity for adsorption and desirable kinetics, some optimization (chiefly in terms of neutralization degree, type and amount of the cross-linking agent, and monomer concentration) must be made [26].

VI. CONCLUSION AND OUTLOOK

The main aim of this chapter was to review critically the current literature in the fascinating field of inverse heterophase polymerization. The main interest is in carrying out the process in order to obtain stable latexes of water-soluble monomers with high molecular weight at low surfactant content. To understand the underlying kinetics thoroughly, the mutual interplay of monomer, surfactant, initiator, and the continuous phase is of high interest.

The challenge for the future is to combine the advantages of the different types of inverse heterophase polymerization. In particular, high stabilities such as those of inverse microemulsions are desired for systems with low surfactant content as is usual for the other types of inverse heterophase polymerization. Homogeneous functionalization of the particles can potentially be used to create complex polymer superstructures for different applications such as catalysis, selective ionic binding, and detoxification. Incorporation of inorganic material within water-soluble latex particles with well-defined structures will be the focus of future research. Polymerization in inverse heterophases is a rapidly developing field in which major scientific breakthroughs are to be expected. The numerous possible applications testify that the tool of modern heterophase polymerization is very powerful. We are standing at the beginning of fascinating developments for industry and fundamental research.

ACKNOWLEDGMENTS

We especially thank Markus Antonietti for his great support, his advice, and his help. Financial support by the Fonds der Chemischen Industrie, the DAAD, and the Max Planck Society is gratefully acknowledged.

REFERENCES

1. D. Hunkeler, F. Candau, C. Pichot, A. E. Hamielec, T. Y. Xie, J. Barton, V. Vaskova, J. Guillot, M. V. Dimonie, and K. H. Reichert, Adv. Polym. Sci. *123*:115–133 (1994).
2. E. D. Sudol and M. S. El-Aasser, in *Emulsion Polymerization and Emulsion Polymers* (P. A. Lovell and M. S. El-Aasser, eds.), Wiley, Chichester, 1997, p. 699.
3. K. Landfester, N. Bechthold, F. Tiarks, and M. Antonietti, Macromolecules *32*:5222–5228 (1999).
4. D. Hunkeler and A. E. Hamielec, ACS Symp. Ser. *480*:24–41 (1992).
5. D. Hunkeler, A. E. Hamielec, and W. Baade, Polymer *30*:127–142 (1989).
6. J. Hernandez-Barajas and D. Hunkeler, Polymer *38*: 437–447 (1997).
7. W. D. Bancroft, J. Phys. Chem. *17*:501–519 (1913).
8. W. C. Griffin, Soc. Cosmet. Chem. *1*:311 (1949); W. C. Griffin, Soc. Cosmet. Chem. *5*:249 (1954).
9. J. T. Davies, Int. Congr. Surface Activity 2nd, London, 1957, *1*:426.
10. I. J. Lin and P. Somsundaran, J. Colloid Interface Sci. *37*:731–743 (1971).
11. I. J. Lin, J. P. Friend, and Y. Zimmeles, J. Colloid Interface Sci. *45*:378–385 (1973).
12. P. Sherman, in *Emulsion Science* (P. Sherman, ed.), Academic Press, New York, 1968, chapter 3.
13. J. Boyd, G. Parldnson, and P. Sherman, J. Colloid Interface Sci. *41*:359–370 (1972).
14. K. Shinoda, Proc. Int. Congr. Surface Activity 5th, 1969, *2*:275.
15. R. E. Ford, L. Ford, and C. G. Furmidge, J. Colloid Interface Sci. *22*:331–341 (1966).
16. C. Holtzscherer and F. Candau, Colloids Surf. *29*:411–423 (1988).
17. H. Ni and D. Hunkeler, J. Dispersion Sci. Technol. *19*: 551–569 (1998).
18. A. Beerbower and H. W. Hill, *McCutcheon's Detergents and Emulsifiers Annual*, Allured Publishing Co. Ridgewood, NJ, 1971, p. 223.
19. J. M. Corpart and F. Candau, Colloid Polym. Sci. *271*: 1055–1067 (1993).
20. J. W. Vanderhoff, E. B. Bradford, M. L. Tarkowski, J. B. Shaffer, and R. M. Wiley, Adv. Chem. Ser. *34*:32 (1962).
21. P.-Y. Jiang, Z.-C. Zhang, and M.-W. Zhang, J. Polym. Sci. Polymn. Chem. *34*:695–699 (1996).
22. P.-Y. Jiang, Z.-C. Zhang, and M.-W. Zhang, Macromolecules *29*:6710–6716 (1996).
23. D. J. Hunkeler and A. E. Hamielec, Polymer *32*:2626–2640 (1991).
24. K. H. Reichert and W. Baade, Angew. Makromol. Chem. *124*/*125*:361–381 (1984).
25. S. K. Ghosh, M. Nazimuddin, and B. M. Mandal, Makromol. Chem. Rapid Commun. *13*:583–586 (1992).
26. H. Omidian, S. A. Hashemi, F. Askari, and S. Nafisi, J. Appl. Polym. Sci. *54*:241–249 (1994).
27. F. Askari, S. Nafisi, H. Omidian, and S. A. Hashemi, J. Appl. Polym. Sci. *50*:1851–1855 (1993).
28. B. Kriwet, E. Walter, and T. Kissel, J. Controlled Release *56*:149–158 (1998).
29. M. T. McKechnie, Proc. Conf. Emulsion Polym., London, June 1982, paper 3.
30. X. Ge, Q. Ye, Z. Zhang, and G. Chu, J. Appl. Polym. Sci. *67*:1005–1010 (1998).
31. X. Ge, Q. Ye, X. Xu, Z. Zhang, and Q. Sun, Radiat. Phys. Chem. *50*:253–258 (1997).
32. M. B. Taylor, R. D. Gilbert, V. T. Stannett, and J. L. Williams, J. Appl. Polym. Sci. *53*:1385–1390 (1994).
33. J. Hernandez-Barajas and D. Hunkeler, Polymer *38*: 5623–5641 (1997).
34. F. Candau, Z. Zekhnini, F. Heatley, and E. Franta, Colloid Polym. Sci. *264*:676–682 (1986).
35. P. Buchert and F. Candau, J. Colloid Interface Sci. *136*:527–540 (1990).
36. A. Ohlemacher, F. Candau, J. P. Munch, and S. J. Candau, J. Polym. Sci. Polym. Phys. *34*:2747–2757 (1996).
37. S. Neyret and B. Vincent, Polymer *38*:6129–6134 (1997).
38. F. Candau, Z. Zekhnini, and J. Durand, Colloid Polym. Sci. *264*:676–682 (1986).
39. C. Holtzscherer, J. P. Durand, and F. Candau, Colloid Polym. Sci. *265*:1067–1074 (1987).

40. C. Graillat, C. Pichot, A. Guyot, and M. S. El-Aasser, J. Polym. Sci. Polym. Chem. *24*:427–449 (1986).
41. V. Glukhikh, C. Graillat, and C. Pichot, J. Polym. Sci. Polym. Chem. *25*:1127–1161 (1987).
42. F. Candau, M. Pabon, and J. Y. Anquetil, Colloid Surf. A Physicochem. Eng. Asp. *153*:47–59 (1999).
43. M. T. Carver, E. Hirsch, J. C. Wittmann, R. M. Fitch, and F. Candau, J. Phys. Chem. *93*:4867–4873 (1989).
44. F. Candau, Y. S. Leong, and R. M. Fitch, J. Polym. Sci. Polym. Chem. Ed. *23*:193–214 (1985).
45. J. Barton, Makromol. Chem. Rapid Commun. *12*:675–679 (1991).
46. D. Benda, J. Snuparek, and V. Cermak, Eur. Polym. J. *33*:1345–1352 (1997).
47. P. A. Dresco, V. S. Zaitsev, R. J. Gambino, and B. Chu, Langmuir *15*:1945–1951 (1999).
48. Y. S. Leong and F. Candau, J. Phys. Chem. *86*:2269–2271 (1982).
49. V. Vaskova, V. Juranicova, and J. Barton, Makromol. Chem. *191*:717–723 (1990).
50. C. Daubresse, C. Grandfils, R. Jerome, and P. Teyssie, Colloid Polym. Sci. *274*:482–489 (1996).
51. J. Barton and M. Stillhammerova, Angew. Chem. *237*: 113–122 (1996).
52. J. P. Fouassier, D. J. Lougnot, and I. Zuchowicz, Eur. Polym. J. *22*:933–938 (1986).
53. F. Yan, C. Zheng, X. Zhai, and D. Zhao, J. Appl. Polym. Sci. *63*:747–754 (1998).
54. A. Pross, K. Platkowski, and K.-H. Reichert, Polym. Int. *45*:22–26 (1998).
55. M. V. Dimonie, C. M. Boghina, N. N. Marinescu, M. M. Marinescu, C. I. Cincu, and C. E. Oprescu, Eur. Polym. J. *18*:639–645 (1982).
56. M. Dimonie, N. Marinescu, D. S. Vasilescu, O. Shaffer, H. Caldarar, A. Caragheorgheopol, and J. Languri, J. Macromol. Sci. Pure Appl. Chem. *A29*:277–296 (1992).
57. J. W. Vanderhoff, F. V. Ditefano, M. S. El-Aasser, R. O'Leary, O. M. Shaffer, and D. G. Visioli, J. Dispersion Sci. Technol. *5*:323–363 (1984).
58. V. F. Kurenkov, T. M. Osipova, E. V. Kuznetsov, and V. A. Mtagchenkov, Vysokomol Soedin Ser. B *20*: 647–650 (1978).
59. Z. Liu and B. W. Brooks, J. Appl. Polym. Sci. *66*: 2191–2197 (1997).
60. J. Hernandez-Barajas and D. Hunkeler, Polym. Adv. Technol. *6*:509–517 (1995).
61. H. Ni and D. Hunkeler, J. Dispersion Sci. Techn. *18*: 123–160 (1996).
62. J. Hernandez-Barajas and D. J. Hunkeler, Polymer *38*: 449–458 (1997).
63. Y. S. Leong, G. Riess, and F. Candau, J. Chim. Phys. *78*:279–284 (1981).
64. S. Fengler and K.-H. Reichert, Angew. Makromol. Chem. *225*:139–152 (1995).
65. M. Dauben, K. H. Reichert, P. Huang, and J. Fock, Polymer *37*:2827–2830 (1996).
66. B. Ray and B. M. Mandal, Langmuir *13*:2191–2196 (1997).
66a. K. Landfester, M. Willert, M. Antonietti, Macromolecules *33*:2370–2376 (2000).
67. L. M. Gan, C. H. Chew, H. S. O. Chan, and L. Ma, Polym. Bull. *31*:347–350 (1993).
68. A. Hammouda, T. Gulik, and M. P. Pileni, Langmuir *11*:3656–3659 (1995).
69. N. Moumen, M. P. Pileni, and R. A. Mackay, Colloids Surf. A Physicochem. Eng. Asp. *151*:409–417 (1999).
70. K. Nagai, H. Satoh, and N. Kramoto, Polymer *34*: 4969–4973 (1993).
71. H. Kawaguchi, Y. Yamada, S. Kataoka, Y. Morita, and Y. Oktsuka, Polym. J. *23*:955–962 (1991).
72. J. P. Hoar and J. H. Schulman, Nature *152*:102–104 (1943).
73. J. H. Schulman, W. Stoeckenhuis, and L. Prince, J. Phys. Chem. *63*:1677–1680 (1959).
74. M. Antonietti, R. Basten, and S. Lohmann, Macromol. Chem. *196*:441–466 (1995).
75. J. O. Stoffer and T. Bone, J. Dispersion Sci. Technol. *1*:37–54 (1980).
76. J. O. Stoffer and T. Bone, J. Polym. Sci. Polym. Chem. Ed. *18*:2641–2648 (1980).
77. F. Candau, Y. S. Leong, G. Pouyet, and S. Candau, J. Colloid Interface Sci. *101*:167–183 (1984).
78. F. Candau, in *Polymerization in Organized Media* (C. Paleos, ed.), Gordon & Breach, Philadelphia, 1992, pp. 215–282.
79. J. Sjoblom, R. Lindberg, and S. E. Friberg, Adv. Colloid Polym. Sci. *65*:125–287 (1996).
80. K. M. Lusvardi, K. V. Schubert, and E. W. Kaler, Ber. Bunsenges. Phys. Chem. *100*:373–379 (1996).
81. R. Strey and M. Jonstromer, J. Phys. Chem. *96*:4537–4542 (1992).
82. J. S. Guo, E. D. Sudol, J. W. Vanderhoff, H. J. Yue, and M. S. El-Aasser, J. Colloid Interface Sci. *149*: 184–196 (1992).
83. B. Antalek, A. J. Williams, E. Garcia, and J. Texter, Langmuir *10*:4459–4467 (1994).
84. P. Becher, J. Colloid Interface Sci. *17*:325–333 (1962).
85. M. Schick, J. Colloid Interface Sci. *17*:801–813 (1962).
86. H. Schott, J. Colloid Interface Sci. *43*:150–155 (1973).
87. T. Kawai, K. Hamada, N. Shindo, and K. Kon-No, Bull. Chem. Soc. Jpn. *65*:2715–2719 (1992).
88. H. MacDonald, B. Bedwell, and E. Gulari, Langmuir *2*:704–708 (1986).
89. C. Holtzscherer and F. Candau, J. Colloid Interface Sci. *125*:97–110 (1988).
90. F. Candau and P. Buchert, Colloids Surf. *48*:107–122 (1990).
91. J. Hernandez-Barajaz and D. J. Hunkeler, Abstracts of Papers of the American Chemical Society 1993, 206: 127.

92. F. Yan, C. Zheng, X. Zhai, and D. Zhao, J. Appl. Polym. Sci. *67*:747–754 (1998).
93. M. Antonietti, R. Basten, and F. Grohn, Langmuir *10*: 2498–2500 (1994).
94. S. Avramoitis, A. Xenakis, and P. Lianos, Prog. Colloid Polym. Sci. *100*:286–289 (1996).
95. S. Avramiotis, V. Bekiari, P. Lianos, and A. Xenakis, J. Colloid Interface Sci. *194*:326–331 (1997).
96. S. Backlund, M. Rantala, and O. Molander, Colloid Polym. Sci. *273*:293–297 (1995).
97. G. Voortmans, C. Jackers, and F. De Schryver, Macromolecules *21*:1977–1980 (1988).
98. C. A. Parker, Proc. R. Soc. *A220*:104 (1953).
99. C. G. Hatchard and C. Parker, Proc. R. Soc. *A235*:516 (1956).
100. K. Nagai, H. Satoh, and N. Kramoto, Polymer *34*: 4969–4973 (1993).
101. A. P. Full, J. E. Puig, L. U. Gron, E. W. Kaler, J. R. Minter, T. H. Mourey, and J. Texter, Macromolecules *25*:5157–5164 (1992).
102. L. M. Gan, C. H. Chew, S. C. Ng, and S. E. Loh, Langmuir *9*:2799–2803 (1993).
103. J. S. Guo, E. D. Sudol, J. W. Vanderhoff, and M. S. El-Aasser, J. Polym. Sci. Polym. Chem. *30*:691–702 (1992).
104. J. S. Guo, E. D. Sudol, J. W. Vanderhoff, and M. S. El-Aasser, J. Polym. Sci. Polym. Chem. *30*:703–712 (1992).
105. J. D. Morgan and E. W. Kaler, Macromolecules *31*: 3197–3202 (1998).
106. J. D. Morgan, K. M. Lusvardi, and E. W. Kaler, Macromolecules *30*:1897–1905 (1997).
107. A. Saenz de Buruaga, J. C. de la Cal, and J. M. Asua, J. Polym. Sci. Polym. Chem. *37*:2167–2178 (1999).
108. A. Saenz de Buruga, I. Capek, J. C. de la Cal, and J. M. Asua, J. Polym. Sci. Polym. Chem. *36*:737–748 (1998).
109. M. A. Chung, J. Adams, and J. Fuhrmann, Polym. Bull. *40*:195–202 (1998).
110. I. Lacik, J. Barton, and G. G. Warr, Macromol. Chem. Phys. *196*:2223–2236 (1995).
111. V. Vaskova, Z. Hlouskova, V. Juranikova, and J. Barton, Makromol. Chem. Macromol. Chem. Phys. *193*: 1445–1449 (1992).
112. J. Barton, J. Tino, Z. Hlouskova, and M. Stillhammerova, Polym. Int. *34*:89–96 (1994).
113. A. Saenz de Buruaga, J. de la Cal, and J. M. Asua, J. Appl. Polym. Sci. *72*:1341–1348 (1999).
114. A. Saenz de Buruga, I. Capek, J. C. de la Cal, and J. M. Asua, J. Polym. Sci. Polym. Chem. *36*:737–748 (1998).
115. Z. Zekhnini, Ph.D. thesis, Strasbourg University, 1986.
116. I. Capek and P. Potisk, J. Polym. Sci. Polym. Chem. *33*:1675–1683 (1995).
117. A. Saenz de Buruaga, J. C. de la Cal, and J. M. Asua, J. Polym. Sci. Polym. Chem. *37*:2167–2178 (1999).
118. J. Barton, Polym. Int. *30*:151–156 (1993).
119. J. Barton, Makromol. Chem. Macromol. Symp. *53*: 289–306 (1992).
120. I. Janigova, K. Csomorova, M. Stillhammerova, and J. Barton, Macromol. Chem. Phys. *195*:3609–3614 (1994).
121. C. Pichot, Makromol. Chem. Macromol. Symp. *35/36*: 327–347 (1990).
122. K. Sakota and T. Okaya, J. Appl. Polym. Sci. *20*: 1725–1733 (1976).
123. R. S. Molday, W. J. Dreyer, A. Rembaum, and S. P. S. Yen, J. Cell Biol. *64*:75–88 (1975).
124. H. Tamai, M. Hasegawa, and T. Suzawa, J. Appl. Polym. Sci. *38*:403–412 (1989).
125. K. G. I. Nilsson, J. Immunol. Methods *122*:273–277 (1989).
126. J. H. Kim, M. Chainey, M. S. El-Aasser, and J. W. Vanderhoff, J. Polym. Sci. Polym. Chem. Ed. *27*: 3187–3199 (1989).
127. S. Kirsch, A. Doerk, E. Bartsch, H. Sillescu, K. Landfester, H. W. Spiess, and W. Maechtle, Macromolecules *32*:4508–4518 (1999).
128. F. Candau, Z. Zekhnini, and F. Heatley, Macromolecules *19*:1895–1902 (1986).
129. V. Vaskova, M. Stillhammerova, and J. Barton, Chem. Pap. Chem. Zvesti *48*:355–361 (1994).
130. J. M. Corpart, J. Selb, and F. Candau, Makromol. Chem. Macromol. Symp. *53*:253–265 (1992).
131. M. Antonietti, A. Briel, and S. Förster, J. Chem. Phys. *105*:7795–7807 (1996).
132. W. Baade and K. H. Reichert, Eur. Polym. J. *20*:505–512 (1984).
133. J. Hernandez-Barajas and D. J. Hunkeler, Polymer *38*: 437–447 (1997).
134. N. O'Connor, L. J. Barker, and G. Ryles, U.S. Patent 5,200,448 (1993).
135. G. Bognolo, in *Industrial Applications of Surfactants* (R. Karsa, ed.), Royal Society of Chemistry, Cambridge, 1990.
136. V. F. Kurenkov, A. S. Verizhnikova, E. V. Kuznetsov, and V. A. Myagchenkkov, Izv. Vuz. Khim. Kh. Tekh. *25*:221–226 (1982).
137. D. Hunkeler and A. E. Hamielec, ACS Symp. Ser. *467*:82–104 (1991).
138. K. Platkowski, A. Pross, and K.-H. Reichert, Polym. Int. *45*:229–238 (1998).
139. W. Huebinger and K. H. Reichert, *Polymer Reaction Engineering* (K. H. Reichert and W. Geiseler, eds.), VCH, Berlin, 1989, p. 207.
140. W. Baade and K. H. Reichert, Makromol. Chem. Rapid. Commun. *7*:235–241 (1986).
141. G. Visioli, Ph.D. thesis, Lehigh University, Bethlehem, PA, 1984.
142. V. F. Kurenkov, T. M. Osipova, E. V. Kuznetsov, and V. A. Myagchenkow, Vysokomol. Soedin Ser. *B20*: 647–650 (1978).

143. S. K. Gosh and B. M. Mandal, Polymer *34*:4287–4290 (1993).
144. V. F. Kurenkov, Izv. Vuz. Khim. Technol. *29*:92–95 (1986).
145. J. Texter, L. E. Oppenheimer, and J. R. Minter, Polym. Bull. *27*:487–494 (1992).
146. Z. Liu and B. W. Brooks, Polymer *40*:2181–2188 (1999).
147. Z. Liu and B. W. Brooks, J. Polym. Sci. Polym. Chem. *37*:313–324 (1999).
148. K. Platkowski and K.-H. Reichert, Polymer *40*:1057–1066 (1999).
149. C. Vu and J. Cabestany, J. Appl. Polym. Sci. *42*:2857–2869 (1991).
150. H. Ni and D. Hunkeler, Polymer *38*:667–675 (1997).
151. F. A. Adamsky and E. J. Beckmann, Macromolecules *27*:312–314 (1994).
152. M. Z. Yates, M. L. O'Neill, K. P. Johnston, S. Webber, D. A. Canelas, D. E. Betts, and J. M. DeSimone, Macromolecules *30*:5060–5067 (1997).
153. S. R. Reddy, D. H. Melik, and H. S. Fogler, J. Colloid Interface Sci. *82*:116–127 (1981).
154. C. M. Miller, E. D. Sudol, C. A. Silebi, and M. S. El-Aasser, J. Polym. Sci. Polym. Chem. Ed. *33*:1391–1408 (1995).
155. K. Landfester, N. Bechthold, S. Förster, and M. Antonietti, Macromol. Rapid Commun. *20*:81–84 (1999).
156. H. Omidian, S. A. Hashemi, P. G. Sammes, and I. Meldrum, Polymer *40*:1753–1761 (1999).
157. K. H. Reichert and W. Baade, Angew. Macromol. Chem. *123*/*124*:361–380 (1984).
158. S. Fengler and K.-H. Reichert, Angew. Makromol. Chem. *225*:139–152 (1995).
159. M. Dauben, K. Patkowski, and K.-H. Reichert, Angew. Makromol. Chem. *250*:67–83 (1997).
160. M. Dauben, K. H. Reichert, P. Huang, and J. Fock, Polymer *37*:2827–2830 (1996).
161. H. Kawaguchi, K. Fujimoto, M. Saito, T. Kawasaki, and Y. Urakami, Polym. Int. *30*:225–231 (1993).
162. M. J. Snowden, P. Thomas, and B. Vincent, Analyst *118*:1367 (1993).
163. M. S. Wolfe, Polym. Mater. Sci. Eng. *61*:398 (1989).
164. S. Evani and G. D. Rose, Polym. Mater. Sci. Eng. *57*:477 (1987).
165. D. Hunkeler and J. Hernandez-Barajas, in *Polymeric Materials Encyclopedia* (J. C. Salamone, ed.), Vol. 9, CRC Press, New York, 1996, pp. 3322–3333.
166. S. Slomkowski, Prog. Polym. Sci. *23*:815–874 (1998).
167. C. Larpent and T. F. Tadros, Colloid Polym. Sci. *269*:1171–1183 (1991).
168. C. Vauthier-Holtzscherer, S. Benabbou, G. Spenlehauer, M. Veillard, and P. Couvreur, S.T.P. Pharma. Sci. *1*:109 (1991).
169. E. Allemann, G. Gurney, and E. Doelker, Eur. J. Pharm. Biopharm. *39*:173–191 (1993).
170. P. A. Dresco, V. S. Zaitsev, R. J. Gambino, and B. Chu, Langmuir *15*:1945–1951 (1999).
171. M. Breulmann, H. Cölfen, H. P. Hentze, M. Antonietti, D. Walsh, and S. Mann, Adv. Mater. *10*:237–241 (1998).
172. P. Espiard, A. Revillon, A. Guyot, and J. E. Mark, ACS Symp. Ser. *492*:387–404 (1992).
173. N. T. Whilton, B. Berton, L. Bronstein, H. P. Hentze, and M. Antonietti, Adv. Mater. *11*:1014–1018 (1999).
174. P. Degobert and F. Henry, Polym. Adv. Technol. *6*:362–370 (1994).
175. H.-Q. Xie, D. Tian, P. He, and J. Guo, J. Appl. Polym. Sci. *68*:2169–2174 (1998).
176. W. M. Winslow, J. Appl. Phys. *20*:1137–1140 (1949).

24

Vesicular Polymerization

JUTTA HOTZ and WOLFGANG MEIER University of Basel, Basel, Switzerland

I. INTRODUCTION

Vesicles or liposomes are spherically closed lipid bilayers that enclose an aqueous core and are dispersed in water or an aqueous buffer (see Fig. 1 for a schematic representation). The bilayer is composed of individual lipid molecules, which typically consist of a water-soluble headgroup covalently attached to hydrophobic hydrocarbon chains. Because of this amphiphilic nature, the molecules form aggregates in water in which their hydrophobic tails are shielded as far as possible by the hydrophilic headgroups from the surrounding aqueous environment. Typically, lipid molecules show very low solubility in aqueous media, which—together with their molecular geometry—forces them into bilayer aggregates [1].

Usually phospholipids are the most common constituent of vesicles or liposomes. However, Kunitake and Okahata discovered in 1977 [2] that synthetic amphiphiles are also able to form bilayer membrane structures, and synthetic compounds, such as long-chain quaternary ammonium salts or even block copolymers, are finding increasing interest [3–5].

There exist many different methods for the preparation of vesicles and liposomes. The interested reader may find more details of the preparation of vesicles, for example, in Ref. 6, where this topic has been extensively reviewed. Vesicles can be formed, for example, by hydration of thin lipid films, extrusion of extended bilayer structures through pores of defined width, ultrasonification of lipid dispersions, detergent dialysis of lipid-detergent mixed micelles, injection of organic lipid solutions into water, reverse phase evaporation, or spontaneously, just by mixing micellar solutions of oppositely charged surfactants [7], to mention only the most common methods [6].

The amphiphilic molecules of the lipid bilayer of the vesicles or liposomes can be in a gel (or solidlike) phase or in a liquid crystalline (or fluid) state. In the low-temperature gel phase the individual molecules usually show orientational and positional ordering with, for example, good packing of the hydrocarbon chains, typically in an all-*trans* conformation. This phase melts at the phase transition temperature T_c into the less ordered liquid crystalline phase, with disordered hydrocarbon chains. This different degree of order has direct consequences for the mobility of the lipid molecules within the bilayer: it typically differs by about two orders of magnitude for the two different phases, e.g., the later diffusion coefficient being about 10^{-10} cm^2 s^{-1} in the gel phase and about 10^{-8} cm^2 s^{-1} in the liquid crystalline phase [8]. The dynamics of the lipid molecules in the bilayer and their phase behavior have, for example, a direct influence on the stability of the whole vesicle or the permeability of the lipid membrane [6].

The self-closed bilayer structure of vesicles can be regarded as a simple model for biological membranes that, similarly to the natural prototype, can serve as a permeability barrier to shield the aqueous core from the surrounding medium or can be used as a support matrix to embed membrane proteins. Moreover, vesicles or liposomes can also be loaded with various polar and nonpolar substances and used as a vehicle able to transport the entrapped substances across membranes and other hydrophobic barriers or even to perform targeted delivery. These unique properties have led to a multi-

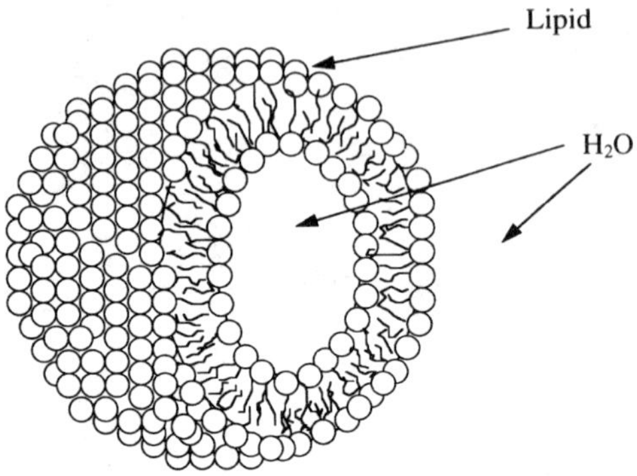

FIG. 1 Schematic presentation of an unilamellar vesicle.

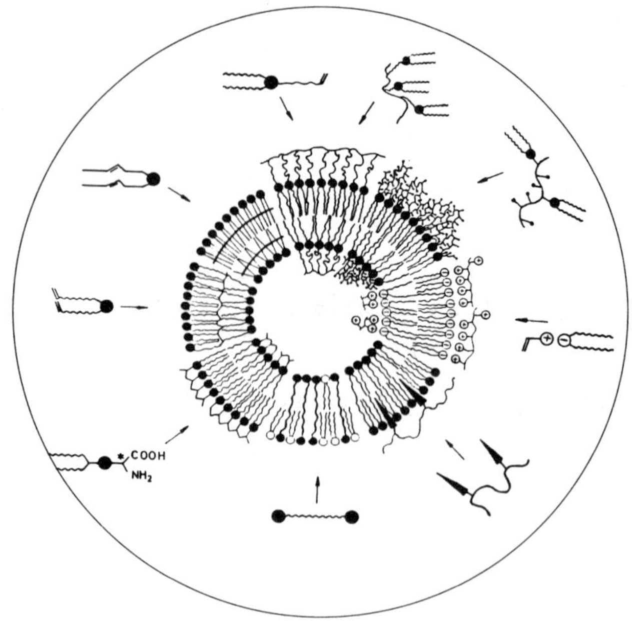

FIG. 2 Schematic presentation of several methods for stabilizing lipid vesicles. (From Ref. 12.)

tude of applications of vesicles and liposomes in various fields of science and technology, reaching from their use in basic studies on the shape of biological cells, membrane and membrane protein function [7], and chemical catalysis [9] or even as templates in biomimetic materials chemistry [10] to applications as drug delivery systems, medical diagnostics, transfection vectors, new hydrogels in cosmetics [11], or self-healing paints [6].

However, the usually only very limited stability of the vesicles frequently represents a serious problem, especially for applications. It is well known that their stability can be significantly enhanced using their interactions with polymers [12] (see Fig. 2). In this context, so-called hydrophobically modified water-soluble polymers are of special interest. These polymers carry a low fraction of hydrophobic anchor groups along their hydrophilic polymer backbone. These anchor groups can, for example, be inserted into the lipid membrane of vesicles, thus immobilizing the polymer chains at the surface of the vesicle and leading to steric stabilization [13]. The so-called "Stealth liposomes" are perhaps the most famous example and have found applications as drug carriers in pharmacy [6,13].

Another possibility to stabilize lipid membranes and, in addition, to control the physical properties or the morphology of the vesicular system is their polymerization. This can be realized by modifying the membrane-forming lipids with polymerizable groups [12] or by dissolving conventional hydrophobic monomers in the lipid bilayer of the vesicle [14–19]. Beyond this, such polymerization in a two-dimensional fluid is also interesting from a theoretical point of view, e.g., the scaling behavior of the newly formed polymer chains within the restricted environment in the two-dimensional membrane [20,21], and can be used to prepare polymer particles with unusual structures and morphologies, such as polymer hollow spheres or quasi-two-dimensional polymers [19].

Studies of polymerized vesicles derived from polymerizable lipids were pioneered in the late 1970s and early 1980s [22–25], and this is still an active field of research. Polymerization of conventional hydrophilic or hydrophobic monomers in vesicle dispersions has attracted less attention, although their polymerization leads to interesting new materials with exceptional structures [19]. However, it is well known that hydrophobic substances can be solubilized to a certain degree in the hydrophobic part of lipid bilayers [26]. It is therefore probable that hydrophobic monomers can be polymerized in the interior of the lipid bilayer. In this case, the vesicle serves only as a template, determining the structure of the newly formed polymer, e.g., eventually a two-dimensional polymer which provides an internal scaffold of the bilayer membrane without being covalently attached to the lipid molecules.

In this chapter we review the developments in the area of polymerization in vesicular structures. Because several reviews of the polymerization of reactive lipids in vesicles and liposomes have already appeared [12,27–32], we will focus mainly on some materials and polymer chemical aspects. We summarize the polymerization of polymerizable lipids and that of conventional monomers in vesicles separately to put some emphasis on the recent revival of the latter method.

II. POLYMERIZATION OF REACTIVE LIPIDS IN VESICLES

A. General Aspects

Since the end of the 1970s and beginning of the 1980s [22–25] it has been well known that polymerizable groups can be incorporated in bilayer-forming lipids. Up to now, the polymerization of such reactive amphiphiles has been the most common method in the field of polymerization of vesicular structures. Synthetic procedures have been developed to prepare such amphiphilic monomers derived from nearly every naturally occurring or synthetic lipid. The polymerizable moiety can be localized anywhere along the molecular unit of the lipid, i.e., along the tail groups or attached to the hydrophilic headgroup [12] (see Fig. 3).

For the headgroup-functionalized lipids the polymerizable group can be bound covalently or electrostatically as a counterion. The latter, i.e., the polymerization of reactive counterions, can be regarded as an interfacial polymerization whereby the vesicle surface serves as a template for the newly formed polyelectrolyte [33]. Some aspects of such systems (which can be regarded as intermediate between the vesicular polymerization of reactive lipids and that of conventional monomers) are discussed at the end of this section.

Derivatization of lipids has been reported with a great variety of polymerizable groups, such as butadiene, styryl, diacetylene, vinyl, acryloyl, methacryloyl, or sorbyl, just to mention those most commonly used [12] (see Fig. 4).

After formation of the vesicles, free radical polymerization is usually used to convert the reactive lipids into the corresponding polymers. The polymer chain reaction in the lipid bilayer of vesicles can be initiated using thermal, UV irradiation–mediated, or redox chemical generation of radical species [32]. Although the free radical polymerization is constrained to two dimensions because of the organized structure of the bilayer, the rather high mobility of the individual lipid molecules, especially in the liquid crystalline phase, permits the monomers to diffuse to the growing chain ends [8].

However, other polymerization reactions in vesicles have been performed. This is especially interesting in the context of the preparation of biodegradable polymerized vesicles that could be useful as drug carrier systems. Typical examples are the pH-triggered ring opening polymerization of lipoyl-functionalized lipids, which proceeds via S—S linkages [34] and the polycondensation of long-chain α-amino acid esters to polypeptide vesicles [35,36]. Interestingly, the bulk polymerization of α-amino acid esters is not possible. Obviously, the organized structure of the lipid bilayer leads to a rather high density of well-aligned reactive groups, which seems to be a basic requirement for polypeptide formation.

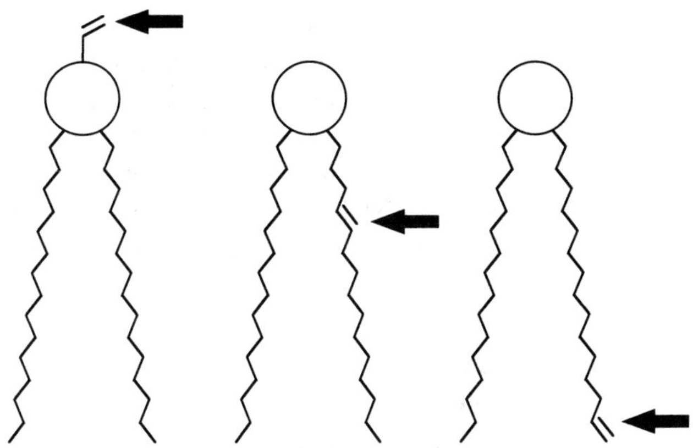

FIG. 3 Schematic presentation of the different locations of polymerizable groups in reactive lipids.

Another case where the alignment of the polymerizable groups is also of crucial importance is the poly-

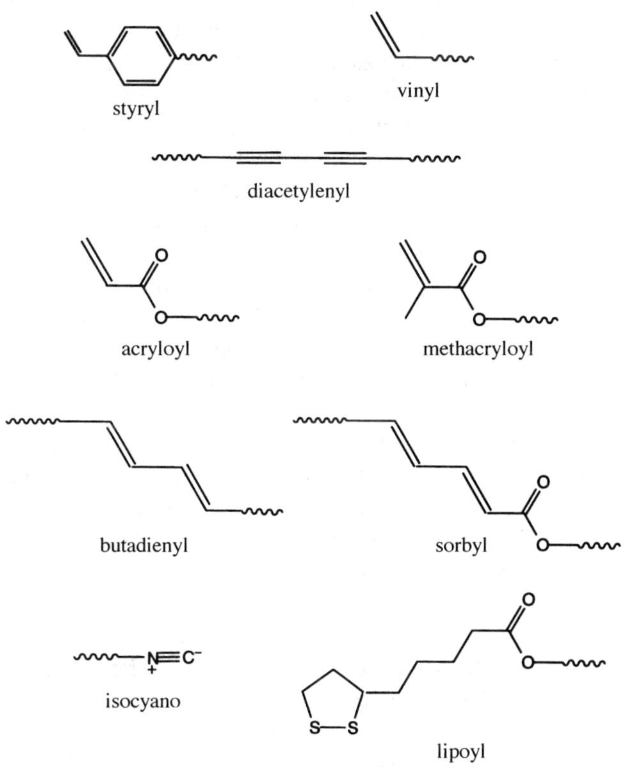

FIG. 4 Some representative examples of polymerizable groups that have been introduced in reactive lipids.

merization of diacetylene lipids [25]. Whereas the free radical polymerization of the most reactive groups is possible in both the fluid, liquid crystalline phase and the solid-analogous gel phase of the bilayer structure, diacetylene lipids can be polymerized efficiently only in the higher ordered gel phase because of the well-known topochemically controlled nature of the polymerization [37].

B. Free Radical Polymerization of Reactive Lipids

To learn more about the mechanism of the polymerization in vesicles, it is necessary to perform kinetic studies and, hence, to investigate the influence of the monomer and the initiator concentration on the reaction kinetics. However, one essential point of polymerization in vesicular structures is that the monomers and the initiator (e.g., in the case of the hydrophobic AIBN) are constrained within the lipid bilayer and are not homogeneously distributed in the bulk, as in conventional solution polymerization. The polymerization occurs within the individual vesicles, which are essentially isolated from each other on the time scale of the experiment; i.e., molecular exchange does not play a role [38]. Variation of the monomer concentration within the vesicles can, therefore, be achieved only by incorporating nonpolymerizable, conventional lipids in the bilayers. The relative fraction of polymerizable lipid in the resulting two-dimensional solution can be used as a measure for the monomer concentration, which enters into the kinetics of the polymerization [38,39].

It has been shown that for low conversions the rate of polymerization shows the same concentration dependence as a conventional solution radical chain reaction, i.e., $r_p \propto [M][I]^{0.5}$. However, at higher conversion ratios, this kinetic behavior changes, leading to an $r_p \propto [M]^2$ relation and becoming independent of [I] [38].

This change has been attributed to a reduced termination rate relative to the propagation rate because of the reduced mobility of the growing polymer chains within the constrained environment of the bilayer. This can be expected to increase the lifetime of the reactive chain ends of the growing polymers, thereby lowering the probability for termination by bimolecular recombination or disproportionation of chains. The diffusion of the small initiator radical fragments within the bilayer is less hindered and, therefore, primary termination (i.e., the reaction of initiator fragments with the active chain ends) becomes more likely. In conventional solution polymerization in isotropic media, primary termination usually occurs at high initiator concentrations and/or in highly viscous solvents.

This termination mechanism in the bilayers is also reflected in the observed dependence of the degree of polymerization on the monomer and initiator concentration. Whereas at low conversions the kinetic chain length depends on [M] and $[I]^{-0.5}$ as in common solution polymerization, at high conversion ratios a molecular weight dependence scaling with $[M]^2$ and $[I]^{-1}$ has been reported [39].

The average degree of polymerization depends, as in conventional solution polymerization, on the relative stability of the propagating species. Degrees of polymerization ranging from 2 up to about 10^4 have been reported, depending on the mode of initiation, the phase structure of the bilayer, the type and location of the reactive group, and others [27–32].

In the case of photoinduced polymerization using short-wavelength UV light, a significant dependence of the polymer chain length on the irradiation time has been observed: polymer chain degradation is in competition with chain growth, leading to lower degrees of polymerization for longer irradiation times [40].

The overall rate of a thermally initiated free radical polymerization of the monomeric lipids has been shown to be quite similar to that of solution polymerization of the respective underlying conventional monomers [38]. However, the difference in the reactivity of different polymerziable groups, such as acryloyl and methacryloyl groups, seems to be "squeezed out" [38]. The highly ordered state in the bilayer has a major influence and minimizes the reactivity difference between the monomers.

The influence of the state of order in the bilayer on the rate of polymerization is also reflected in investigations of the temperature dependence of the kinetics of a redox-initiated free radical polymerization [41]. Interestingly, despite the different mobility of the monomeric lipid molecules in the solidlike gel phase and the liquid crystalline phase of the lipid bilayer, only a very moderate difference in both the overall rate of polymerization and the degree of polymerization has been found for the respective phase structures [41]. Obviously, the lower mobility of the monomeric lipids in the gel phase compared with the liquid crystalline phase can be partly compensated. The reason may be that the all-*trans* conformation of the lipid tails in the gel phase leads to a more favorable orientation of the reactive groups with respect to the propagation reaction compared with the liquid crystalline phase.

The polymerization of monomeric lipids in vesicles can also be performed in a cross-linking manner, depending on the number of polymerziable groups per monomeric lipid [42]. Whereas the polymerization of lipids containing only one single reactive moiety typically leads to the formation of linear polymer chains, lipids bearing more than one reactive group generally yield covalently cross-linked polymer network structures. As a result of such a cross-linking polymerization, the whole vesicle or at least each half of the lipid bilayer of the vesicle is one single molecule, thus locking in the structure and the shape of the aggregate existing during the cross-linking reaction [30]. This leads to significantly different physical properties of the whole vesicle compared with linearly polymerized lipids. Such covalently cross-linked structures show stability of shape, solid-state properties such as elasticity, increased chemical stability, or drastically lowered solubility [30].

In isotropic cross-linking polymerizations, usually mixtures of mono- and bi- or higher functional monomers are used to prepare the three-dimensional polymer network structures. Usually only a very small fraction (often below 1 mol% in the monomer mixture) of the cross-linking agent is necessary to exceed the gel point, i.e., to form a system (vesicle)-spanning covalently cross-linked polymer network structure. In contrast to such monomers in isotropic media, the lipid monomers in a vesicle are confined to the quasi-two-dimensional fluid of the lipid bilayer. As a consequence, the cross-linking polymerization leads to a two-dimensional network structure. However, the geometric restrictions within the bilayer can be expected to have a significant influence on the cross-linking reaction. Indeed, investigations with mixtures of polymerizable lipids bearing one and two reactive groups at the end of their hydrophobic tails revealed that a substantially higher mole fraction of bifunctional monomer, i.e., about 30 mol%, is necessary to reach the gel point within the bilayer of the vesicles [43]. This inefficiency of the cross-linking reaction could be proved using different methods that are sensitive to the changes in the physical properties associated with the onset of gelation, e.g., the lateral diffusion behavior of a small molecule probe in the bilayer or the detergent solubilization of the vesicles [43] (see Fig. 5).

The low effectiveness of the cross-linking reaction has been attributed to the preferred conformation of the lipid chains in the bilayer structure, which may favor side reactions such as intramolecular cyclization leading to network defects. Indeed, this seems to be supported by the fact that locating the reactive groups closer to the hydrophilic headgroups (which reduces the probability for such side reactions) increases significantly the efficiency of the cross-linking reaction; i.e., only about 10 mol% of the cross-linking agent is required for gelation [32].

C. Properties of Polymerized Vesicles

An interesting aspect of the polymerization of reactive lipids in the bilayers of vesicles is that this method generally offers the possibility to engineer various physical properties of the whole vesicular systems. One example, is the aforementioned stabilization effect. The stability and flexibility of biological cell membranes usually arise from the coupling between the lipid bilayer and biopolymer networks, for example, the cytoskeleton [44], the spectrin network of erythrocytes [45,46], or the murein network of gram-negative bacteria [47,48]. In some sense, the polymerized lipids can be considered to mimic the role of such polymeric scaffolds, thereby stabilizing the synthetic membrane structure.

Indeed, polymerized vesicles remain stable for months and are usually not destroyed by the addition of detergents or a few percent of an organic solvent (e.g., alcohol). In this context, vesicles stabilized by a cross-linking polymerization are an especially telling example: the cross-linked polymer network structure allows the vesicles to preserve their spherical shape, even in the absence of water [30].

In contrast to that, the lipid polymerization can also be used to produce controlled labile domains in the lipid membrane. This can be realized, for example, by polymerization of vesicle bilayers constituted from lipid mixtures consisting of reactive lipids and nonpolymerizable lipids. It is well known that such polymerizations may cause phase separation phenomena in the membrane; i.e., separation of the polymerized and the nonpolymerized lipids into enriched domains [30]. The more labile, nonpolymerized domains of such partially polymerized vesicles can be "uncorked" using detergents, solvent, or even enzymatic lysis to yield skeletonized vesicles consisting of a spherically closed polymerized bilayer with multiple holes [12] (Fig. 6). If such vesicular skeletons are charged and the holes are small enough, they can be plugged or unplugged with suitable ions upon variation of the pH. Such vesicles with a switchable permeability should be valuable systems for controlled release of trapped molecules [12,30].

Typically, polymerization of the lipids in the bilayer causes changes in both the bilayer fluidity and permeability and the phase behavior of the whole lipid membrane [12]. Whereas linear polymerization usually

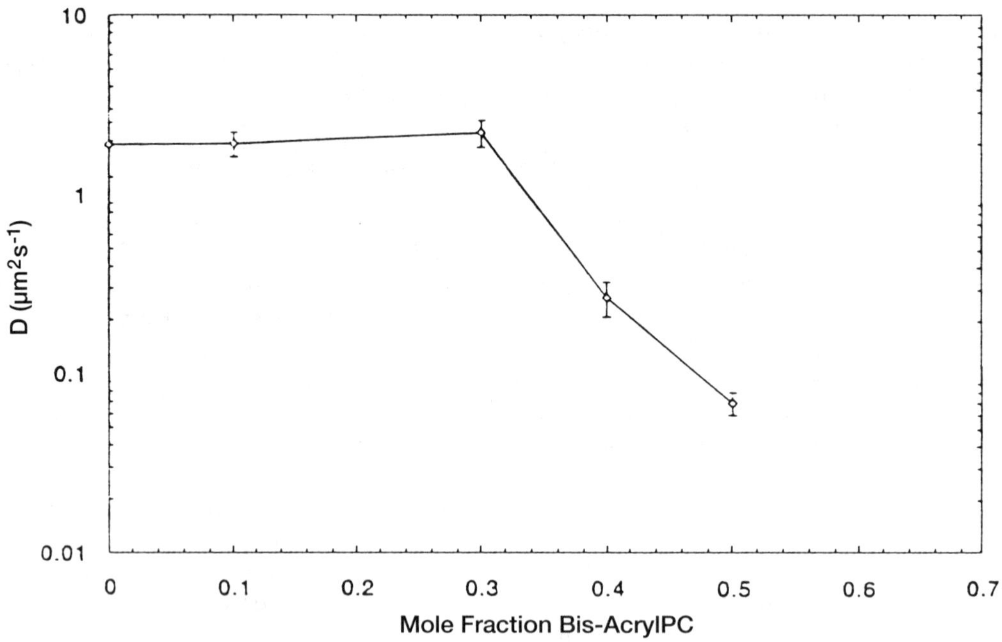

FIG. 5 Lateral diffusion coefficient in polymerized mixed mono- and bis-substituted lipid bilayers as a function of the mole fraction of bis-substituted lipid. (From Ref. 43.)

causes only quite moderate changes, the formation of a cross-linked polymer network structure leads to dramatic changes. For example, the lateral diffusion of low molecular bilayer components is substantially retarded by the cross-linking polymerization of reactive lipids. The bilayer permeability decreases upon formation of linear polymers in the membrane by a factor of about 2–5, and the formation of a two-dimensional polymer network structure decreases the permeability by at least two orders of magnitude [49,50].

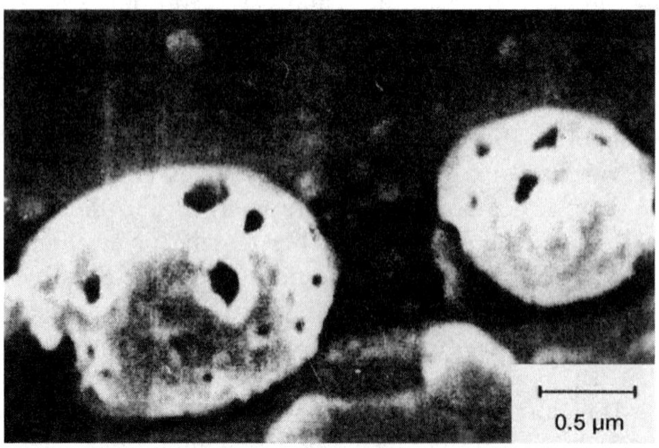

FIG. 6 Raster electron micrograph of a skeletonized vesicle with multiple holes. (From Ref. 12.)

This allows, for example, maintaining a pH gradient of several units across completely polymerized vesicles [51], although protons or hydroxide ions are usually able to permeate almost instantaneously through conventional (unpolymerized) bilayer membranes [7]. Similarly, polymerization via the hydrophobic tails of the lipids normally changes the thermotropic phase behavior of the membranes. The bilayers often lose their fluid phase behavior upon polymerization because of the covalent linkage of the lipid tails. Polymerization via the hydrophilic headgroups generally leads to lower conformative restrictions for the lipid tails; consequently, in this case the gel-to-liquid crystalline phase transition of the bilayer is preserved [12].

An extremely interesting example of polymerization via the hydrophilic headgroups is the polymerization of reactive counterions. This polymerization can also be regarded as a "template polymerization" occurring at the interface of the vesicles, whereby the shape and dimensions of the vesicles are imprinted on the newly formed polymer [33]. The resulting polyelectrolytes are attached via electrostatic interactions to the vesicle surface, which led to the term "liposomes in a net" [12]. The polyelectrolyte chains (or the covalently cross-linked polyelectrolyte network structure) can be attached to the inner and outer surface of the vesicles or to one of these surfaces. The latter can be achieved, for example, by ion exchanging the outer counterions with charged, polymerizable counterions and subsequently polymerizing them [12]. A similar polymerization of only the inner leaflet of the vesicle bilayer, but using covalently attached polymerizable moieties, has also been performed using β-nitrostyrene as the reactive group [52].

Interestingly, the permeability of polyelectrolyte-stabilized vesicles has been shown to be identical to those of the monomeric vesicles [12]. Consequently, the polymer provides in this case—similarly to the spectrin network in erythrocytes—just the physical stability of the membrane while the permeability behavior is determined by the lipid bilayer.

As with the cross-linking polymerization of the reactive lipids themselves, it is possible to produce a two-dimensional polymeric network at the surface of the vesicles using mixtures of reactive counterions bearing one or more than one polymerizable groups. Because of their cross-linked polymer network structure, these polymer particles are also able to retain their hollow sphere morphology even after removal of the vesicle-forming lipid molecules [53]. The resulting so-called ghost vesicles are promising systems that could, eventually, serve as responsive nanocapsules for the encapsulation of pharmaceutically active substances. One interesting point in this context is that the mesh size of the polymer network structure controls the porosity of the polymeric membrane. Molecules having dimensions smaller than this mesh size should be able to pass the shell of the polymer particles without hindrance; molecules bigger than this mesh size should not be able to permeate through the polymer network.

III. POLYMERIZATION OF CONVENTIONAL MONOMERS IN VESICLES

A. General Aspects

In the first part, we have summarized some aspects of the polymerization of reactive lipids in vesicular structures. Characteristic of these systems is that the newly formed polymer chains or polymer network structures are usually covalently attached to the lipid bilayer. It is obvious that the covalent coupling must dramatically influence the structure and dynamics of the whole membrane. As already mentioned, this influence can be reduced to some extent if the polymer is attached via electrostatic interactions to the bilayer structure [12]. However, still another possibility is to produce conventional hydrophilic or hydrophobic polymers in the respective domains of the vesicles, which are bound to the vesicular structure simply by steric constraints and/or hydrophobic interactions. Up to now, there exist only a few studies dealing with such polymerization of hydrophilic or hydrophobic monomers to the respective polymer structures within the hydrophilic or hydrophobic parts of lipid vesicles [14–19,54].

As mentioned before, vesicles or liposomes are able to solubilize hydrophobic substances to a certain degree [26]. Such compounds are usually dissolved in the hydrophobic part of the lipid bilayer. If such substances also carry polymerizable groups, their subsequent polymerization should lead to the formation of polymer chains entrapped in the interior of the membrane. In contrast to polymerizable lipids, the polymer chains are now simply dissolved within the alkane part of the bilayer forming lipids, and, hence, they have a minor influence on the overall physical properties of the membranes.

Similarly, hydrophilic substances can be encapsulated in the aqueous core of vesicles [54]. Subsequent polymerization leads to the formation of hydrophilic polymer chains whose dimensions are directly determined by the monomer concentration and the dimensions of the enclosed water pool of the vesicles. This geometrical restriction is a direct consequence of the very low permeability of the lipid bilayer against the hydrophilic monomers, which prevents an intervesicular exchange of monomers on the time scale of the experiment.

A cross-linking polymerization of, for example, a mixture of acrylamide and *N,N′*-methylenediacrylamide can be used to produce a hydrophilic polymer gel in the interior of the vesicle, similar to the cytoskeleton of biological cells [44]. The resulting polymer particles are a direct cast of the aqueous core of the vesicle and are able to preserve their shape and dimensions even after their isolation from the vesicles [54].

One special feature of vesicular polymerization of conventional hydrophilic or hydrophobic monomers is that the different compartments provided by the self-assembly of the lipid molecules generally serve only as a template that determine size and shape of the resulting polymers. Hence, it is possible to use nearly every natural or synthetic lipid without any modification. In most reported studies of vesicular polymerization of conventional monomers, usually synthetic lipids such as dioctadecyl dimethylammonium chloride (DODAC) or bromide (DODAB), sodium di-2-ethylhexyl phosphate (SEHP), or even spontaneously formed vesicles, prepared from mixtures of cationic and anionic surfactants [14–19], have been used (perhaps primarily for budget reasons). Also, any natural lipid would provide a suitable matrix. Moreover, a combination of polymerziable lipids and conventional monomers could be incorporated in the templating vesicles to yield hybrids with interesting new polymer structures (J. Hotz and W. Meier, to be published).

B. Swelling of Lipid Bilayers with Hydrophobic Monomers

In the following we will focus mainly on the polymerization of hydrophobic monomers such as styrene, divinyl benzene, or long-chain alkyl acrylates or alkyl methacrylates in vesicles. Such hydrophobic substances are usually dissolved within the hydrophobic part of the lipid bilayers. The usual polymerization procedure therefore requires, as a first step, the swelling of the vesicles with the hydrophobic monomers. This is usually performed by simply mixing an adequate amount of the water-insoluble monomer with a suspension of preformed vesicles and stirring this mixture until the monomer has dissolved [14–19].

In this context, it is, therefore, of crucial importance to know more about this swelling process. Of course, the incorporation of hydrophobic substances into lipid bilayers should not exceed a certain saturation concentration. Above this concentration the monomer is no longer homogeneously distributed within the bilayers, as has been shown for toluene in phospholipid vesicles at concentrations above the saturation value [55]. Moreover, exceeding this saturation concentration may also disrupt the whole bilayer structure, thus converting the system to a conventional emulsion or even leading to the formation of a separate monomer phase in the presence of intact vesicles.

Generally, the degree of solvent uptake by the vesicles is restricted by the interfacial free energy of the bilayers. This has been shown by using a thermodynamic model based on classical Flory-Huggins theory, which allows the description of the equilibrium partitioning of hydrophobic solvents between water and phospholipid vesicles [56]. Although these findings provide a theoretical basis for the understanding of bilayer swelling and may even allow certain predictions concerning the saturation concentration of individual substances, experimental investigations are also required to provide information about the local distribution of monomers within the bilayers and the resulting physical properties of swollen lipid membranes. In this context, investigations of not only vesicles or free-standing lipid membranes (so-called black lipid membranes) [57] but also lyotropic liquid crystalline phases [58] or even microemulsions [59,60] may yield valuable information.

The swelling of the lipid bilayers can be expected to influence the phase behavior and the ultrastructure of the lipid membrane. It has been shown that, for example, long-chain alkanes are preferentially oriented parallel to the palisade layer of the lipid alkyl chains within the bilayer, thereby increasing the gel-to-liquid crystalline phase transition temperature [58]. In contrast, short-chain alkanes are accumulated in the central part, in the region between the two underlying lipid monolayers, thereby disturbing the alkyl chain orientation of the lipid molecules. This leads to a decrease of the phase transition temperature of the membrane [58]. Moreover, this difference in the locus of solubilization for long- and short-chain alkanes also provides an explanation for the decreasing saturation concentration of alkanes with increasing chain length [58].

It is obvious that the overall thickness of the lipid membranes has to increase upon solubilization of hydrophobic substances. However, the maximum swelling of the membrane typically leads to an increase from about 3 nm to about 5 nm [57,58]. This is, however, a negligible effect compared with the overall diameter of typical small unilamellar vesicles, which is about 100 nm. Usually, therefore, no dimensional changes of the underlying vesicles can, as long as the monomer concentration stays below the saturation concentration in the membranes, be detected upon swelling of such vesicles with monomer [14–19].

C. Polymerization of Hydrophobic Monomers in Lipid Bilayers

The free radical polymerization of the hydrophobic monomers incorporated in the lipid membranes of vesicles can be initiated, similarly to the polymerization of reactive lipids, by ultraviolet (UV) irradiation or thermal or redox chemical radical generation. Like emulsion or microemulsion polymerization systems, these systems consist of at least three components, water, monomer, and surfactant or lipid. However, in vesicular systems, in contrast to the emulsions and microemulsions, no influence of the polymerization rate on the nature of the initiating species, e.g, hydrophilic or hydrophobic, has been detected [16].

The local monomer concentration is always very high in the interior of the vesicles because of the restricted volume within the hydrophobic part of the lipid bilayers. Moreover, the mobility of the individual monomer molecules is rather high compared with the alkyl chains of the lipid molecules, and obviously the anisotropic environment within the bilayer induces a preferred molecular orientation of the monomers (and simultaneously of the reactive groups) that favors the polymerization reaction. As a consequence, the overall rate of polymerization has been shown always to be very high in vesicular polymerization [14,16].

The conversion from monomer to polymer can be followed, for example, by ^{1}H nuclear magnetic resonance (NMR) or UV spectroscopy, depending on the monomer. The kinetics have been found to be similar to those of a homogeneous phase of polymerization or the polymerization of reactive lipids in vesicles with a rate dependence of $r_p \propto [M][I]^{0.5}$, whatever the initiating species is [16]. Obviously, in contrast to emulsion and microemulsion polymerization, in vesicular polymerization the intervesicular exchange (which would lead to a different dependence of the polymerization rate) plays only a minor role on the time scale of the reaction. This is also reflected in the observation that each vesicle contains only one polymer particle, whose size is directly related to that of vesicle [18,19].

Because of the rather low reaction volume in the compartmentalized structure of the system, the number of radical species entering the bilayer per time unit is quite low. As a result, the time for chain growth is rather high until termination occurs; i.e., polymers with a degree of polymerization of about 10^4 are typically found and the conversion of the chain reaction is always relatively high, always above 90% [14] (Fig. 7).

Like the polymerization of reactive lipids carrying more than one polymerizable group, the polymerization of hydrophobic monomers in the lipid membrane can also be performed in a cross-linking manner. In contrast to the very inefficient cross-linking reaction to polymerizable lipids (see earlier), for the hydrophobic monomers in vesicle bilayers about 0.1 mol% of a bifunctional monomer is enough to exceed the gel point in the interior of the bilayer [19]. This can be shown, for example, by isolation of the resulting polymer particles from the lipid bilayer. Hence, the fraction of cross-linking agent necessary for gelation is quite similar to that of an isotropic cross-linking reaction [19].

In contrast to linear polymerization, which simply results in a hydrophobic polymer chain dissolved in a quasi-two-dimensional liquid, the cross-linking polymerization can be expected to lead to a polymer particle consisting of one single polymer molecule with the shape of a hollow sphere whose the inner and outer shape is directly controlled by the underlying vesicle (see Fig. 8 for a schematic representation).

D. Properties of Lipid Membranes with a Hydrophobic Polymer Scaffold

Because of the nearly two-dimensional geometry of the whole lipid bilayer, it is extremely difficult to visualize directly such a homogeneous polymer shell enclosed in the interior of the membrane and, hence, to get information about the localization of the newly formed polymer particles. However, electron microscopy and light scattering experiments on polymerized vesicle dispersions clearly show that the size and shape of the vesicles are not altered upon polymerization. Moreover, there are no separate polymer particles present in the

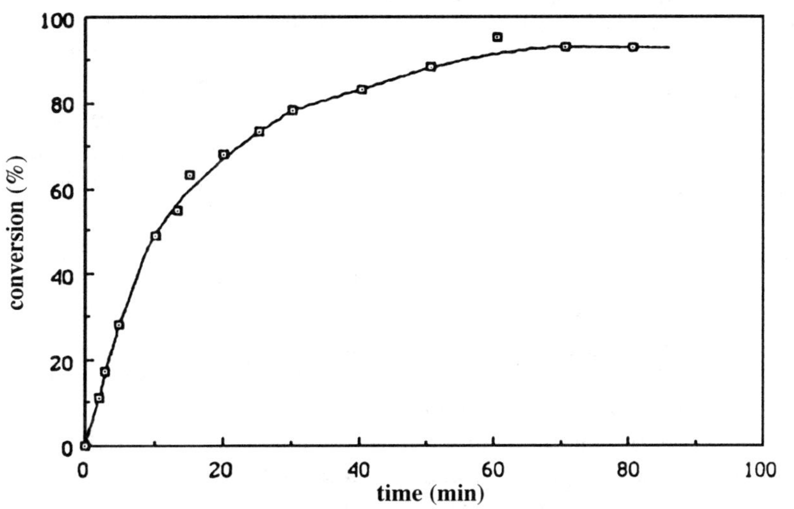

FIG. 7 Conversion of isodecyl acrylate during vesicular polymerization at 60°C. (From Ref. 16, reprinted by permission of John Wiley & Sons, Inc.)

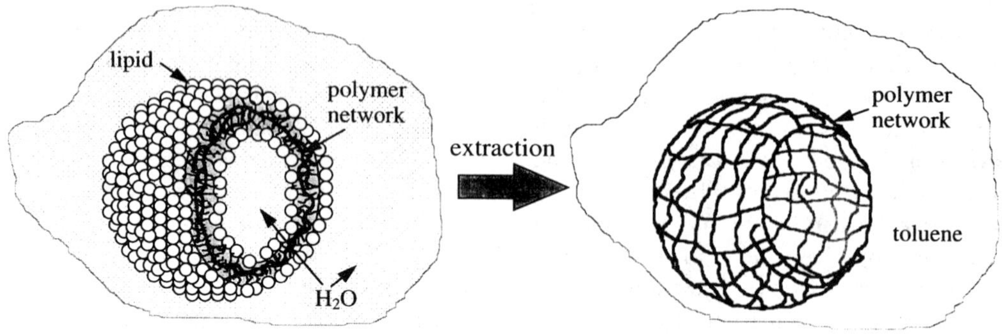

FIG. 8 Schematic presentation of the polymer hollow sphere formed upon cross-linking vesicular polymerization of hydrophobic monomers. The hollow sphere morphology is preserved even after extraction of the lipid.

system and the polymerized vesicles show a significantly increased lifetime and have been found to be stable against detergent lysis [14–19]. Taking into account all of these findings, they can be explained only by the fact that the newly formed hydrophobic polymer is indeed incorporated in the interior of the lipid bilayer.

As already mentioned, one essential point of such systems is that the lipid molecules are not covalently attached to this polymer scaffold. As a result, their lateral mobility [14], the phase behavior [14], and even the permeability of the polymer-containing membranes [14,15] have been shown to undergo no significant changes compared with the polymer-free reference systems. This holds for both linear polymer chains and cross-linked polymer network structures [14], at least as long as the cross-linking density of the network structure is not too high, i.e., the mesh size of the network is large enough.

The mechanical stabilization of the polymer scaffold containing bilayers has been demonstrated using planar lipid membranes as model systems [33]. Rupture of the membranes can be induced by carefully applying single electric field pulses across the membrane. The critical voltage causing the breakdown of the membrane provides direct information about the membrane's mechanical stability [61]. This voltage has been shown to increase by a factor of about 5 in the presence of the hydrophobic polymer scaffold, thus reflecting a considerable mechanical stabilization [33]. This stabilization is also reflected by the fact that polymer-containing vesicles could be directly visualized using atomic force microscopy in the liquid tapping mode [17]. This method can usually not be used with unpolymerized vesicles because of their high fragility.

E. Properties of Vesicle-Templated Polymer Particles

In the context of membrane stabilization as well as with respect to possible applications of the resulting hollow polymer particles, it is essential to know more about the structure and behavior of the polymer scaffold in the membrane. Only if the quasi-two-dimensional polymer network is distributed homogeneously within the membrane of a vesicle does the cross-linking polymerization lead to a spherically closed shell. Otherwise, only fragments should be formed.

However, it was shown that the structure of the polymer itself could be successfully visualized [18,19]. Whereas cryogenic transmission electron microscopy on polystyrene in DODAB vesicles clearly shows a phase separation within the lipid bilayer leading to a so-called parachute-like morphology (i.e., small spherical polymer particles linked to the vesicle bilayers [18] in the respective DODAC-alkylmethacrylate system), confocal laser scanning microscopy (CLSM) and scanning electron microscopy (SEM) investigations provide evidence for a homogeneously closed polymer hollow sphere morphology [19] (Fig. 9). The differences between these two systems probably arise from the fact that the alkyl chain milieu of the lipid bilayer represents a poor solvent for polystyrene and poly(alkyl methacrylates) show better compatibility. Moreover, styrene is a good solvent for poly(styrene), and hence the monomer can be expected to be taken up by the nascent polymer particles. In contrast, alkyl methacrylates are a poor solvent for their corresponding polymers and consequently the growing polymer will preferentially form a compact polymer core in the center of the monomer-swollen hydrophobic part of the lipid

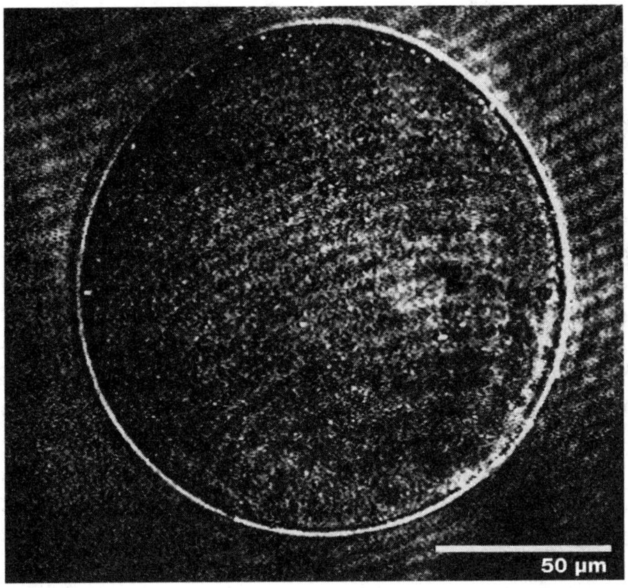

FIG. 9 Confocal laser scanning micrograph of a polymer hollow sphere prepared in a giant vesicle. (From Ref. 19.)

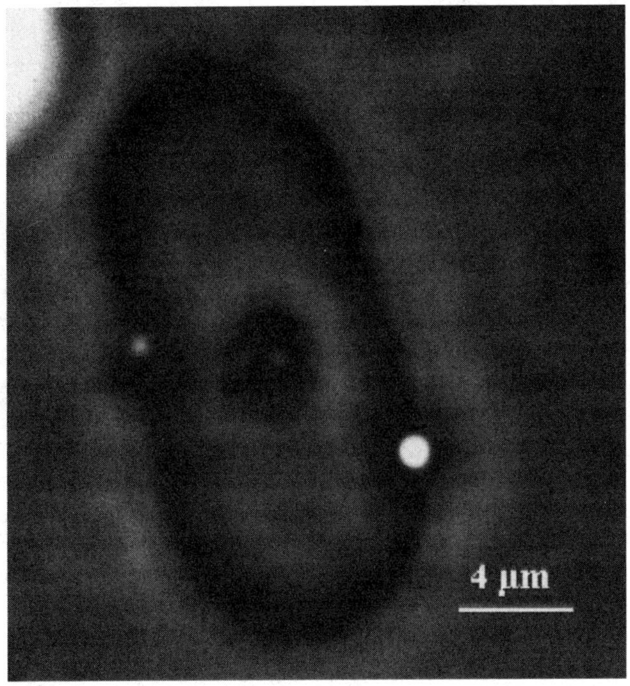

FIG. 10 Micrograph (obtained by combined phase contrast and fluorescence microscopy) of the phase separation induced by polymerization of styrene/divinyl benzene (molar ratio 1:1) in a giant dimyristoyl phosphatidylcholine (DMPC) vesicle. The bright spots in the membrane region are the phase-separated polymer particles. (Courtesy of E. Brückner and H. Rehage.)

bilayer. Similar arguments have been successfully applied to model the chain length and particle size distribution of styrene and alkyl methacrylate microemulsion polymerization [59,60].

Another difference between the two systems arises from the relative monomer concentrations. Whereas a polymer hollow sphere morphology of poly(alkyl methacrylates) was reported for molar ratios of monomer to lipid smaller than 1 [19], in the styrene-DODAB system this ratio was 2 [18]. This is also supported by the observation of phase separation within the lipid bilayers upon solubilization of high concentrations of toluene in dipalmitoylphosphatidylcholine vesicles, which led to morphologies very similar to that of the vesicular parachutes [55] (Fig. 10).

An interesting aspect of the formation of especially cross-linked polymer particles in vesicular dispersions arises from the fact that—in contrast to linear polymers—they should be able to retain their structure even after their isolation from the lipid matrix. This has been shown for cross-linked poly(alkyl methacrylates) by CLSM, SEM, and light scattering investigations [19]. Although the particles contract considerably after their isolation from the lipid membrane, they preserve their spherical shape. Their dimensions remain, however, always directly proportional to those of the underlying vesicles [19] (Fig. 11). This is not surprising because the polymer chains can be expected to be forced into a nearly two-dimensional conformation in

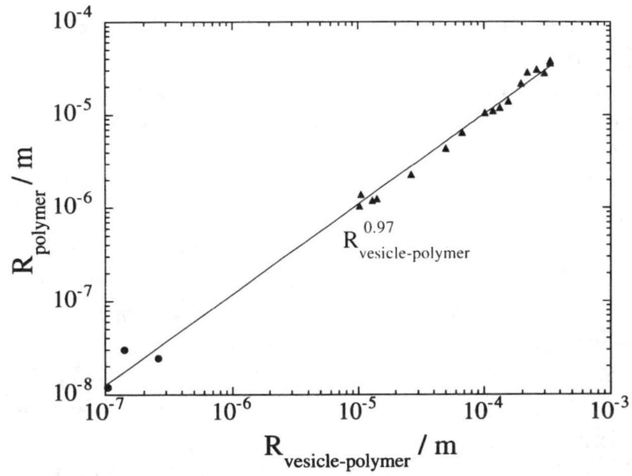

FIG. 11 Radius of polymer particles $R_{polymer}$ as a function of the radius of the templating vesicles $R_{vesicle\text{-}polymer}$. (From Ref. 19.)

the interior of the lipid membrane. After their liberation from the membrane, the polymer chains can gain entrophy by adopting a three-dimensional conformation. To do this, such spherically closed polymer shells have to shrink and the thickness of their shells increases. Up to now it has, however, not been fully clarified why the polymers retain their spherical shape (without collapsing) even in the dry state. However, similar observations with polymerized liposomes [62] and the results of computer simulations on two-dimensional polymers [21] (where it has been shown that the so-called flat phase is the stable state of such polymers) seem to support these observations.

Polarizing microscopy investigations of polymer particles obtained by polymerization in giant vesicles show that the shells of these polymer hollow spheres are birefringent in the dry state. This birefringence vanishes upon swelling the polymer with organic solvents such as toluene. Obviously, the polymer backbone of the pure polymer adopts an ordered conformation. One possibility would be an accordion-like folding of the polymer chains normal to the surface of the sphere (J. Hotz and W. Meier, to be published).

The extent of the observed contraction of the particles depends sensitively on the cross-linking density of the polymer network structure. The contraction increases with increasing cross-linking density, showing the same scaling behavior as branched polymers upon variation of their number of branches [19]. For the highest cross-linking densities the particles contract to about 1/10 of the original size of the templating vesicles. Consequently, the overall shell thickness of the pure polymer hollow spheres increases by a factor of up to about 100. Detailed light scattering investigations of the behavior of such polymer hollow spheres in solution seem to reflect a transition from the hollow spheres to the behavior of branched polymers with decreasing cross-linking density (J. Hotz and W. Meier, to be published).

Although size and shape of the resulting polymer particles are directly determined by the templating vesicles, the polymer scaffold can be modified rather easily using conventional chemical reactions. It has been shown, for example, that polymer hollow spheres made from poly(styrene) can be converted into water-soluble polyelectrolyte hollow spheres by sulfonation (polyanionic hollow spheres) or by reaction with chloro-dimethyl ether and subsequent quaternization with trimethyl amine (polycationic hollow spheres). Similarly, the saponification of poly-*tert*-butylacrylate leads to (also polyanionic) poly(acrylic acid) hollow spheres or polymerization of *N*-isopropylacrylamide in vesicles to poly(*N*-isopropylacrylamide) (PNIPAM) hollow spheres [63].

An extremely interesting aspect that all of the preceding examples of such water-soluble hollow spheres have in common arises from their reversible pH-, electrolyte concentration–, or temperature-dependent swelling. The dimensions of poly(acrylic acid) hollow spheres, for example, change considerably with variation of the pH of the solution, analogous to the coil extension observed for linear poly(acrylic acid) chains or poly(acrylic acid)-based gels [63,64].

At low pH (<5) the particles are in a compact, contracted state. With increasing pH, the acrylic acid groups increasingly dissociate. This leads to increasing repulsive electrostatic interactions between the identically charged acrylate groups along the polymer backbone, which finally results in a strong expansion of the hollow spheres. The extent of this expansion also depends, at a given pH and ionic strength, on the cross-linking density of the polymer network structure of the spherical shell and on the presence of hydrophobic comonomers [63].

This pH-dependent transition influences considerably the permeability of such polyelectrolyte shells. As with the analogous pH-controlled structural changes observed in the protein coats of certain viruses [e.g., the CCM ("cowpea chlorotic mottle") virus] [65,66], gated pores are opened (closed) during this transition in the polyelectrolyte shells. These pores enable free molecular exchange between the interior of the hollow sphere and the bulk medium. Consequently, this process allows a pH-switchable control of the permeability of such polyelectrolyte envelopes. This structural transition can be used to selectively entrap and release water-soluble substances and even water-soluble polymers (see Fig. 12 for a schematic representation).

Furthermore, such polyelectrolyte hollow spheres can also serve as "nanoreactors," for example, for the preparation of inorganic nanoparticles with controlled size and morphology. This could be demonstrated with the help of several model systems [e.g., magnetite, $CaCO_3$ or $BaSO_4$ in poly(acrylic acid) hollow spheres] (W. Meier, to be published). In this context, the pH- or electrolyte concentration–dependent gating mechanism can be used for the controlled loading of the envelopes with the precursor substances for the crystallization. The dimensions of the newly formed particles are controlled by the crystallization in this spatially restricted environment in the interior of the hollow spheres, and their morphology can be influenced by specific interactions of the polyelectrolyte shells with certain crystallographic planes of the growing nuclei [67].

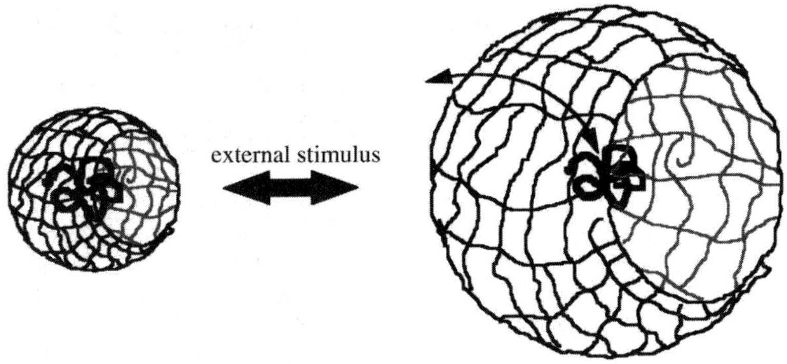

FIG. 12 Schematic presentation of the external stimulus-controlled entrapment or release of a polymer chain in or from a polymer hollow sphere.

IV. SUMMARY AND OUTLOOK

It is well known that polymerization processes can be used to enhance the stability of existing materials and structures. This concept has already been applied for about 20 years to vesicles or liposomes to lock in the fragile structure of lipid bilayers. In the meantime, the resulting polymerized vesicles increasingly find their way into industrial, pharmaceutical, and diagnostic applications.

Regarding this polymerization process in vesicles from a more materials chemistry point of view, it can be regarded as a template polymerization during which the lipid vesicle imprints its morphology on the newly formed polymer, thus allowing structure control at the nanometer level. This becomes even more obvious if one considers the vesicular polymerization of conventional monomers. Here the formation of the polymer is more or less decoupled from the chemical constitution of the individual lipid molecules, which now provide only the structure guiding medium for the polymerization process. Whereas the shapes and sizes of the resulting polymer particles are directly determined by the underlying vesicles, the polymer backbone can simply be modified by conventional chemical reactions. This makes it possible to adapt the chemical constitution of such polymer hollow spheres to the given application. Such nanocapsules are promising materials not only for drug delivery and gene therapy but also as nanoreactors for the controlled crystallization of inorganic nanoparticles in their interior and, hence, the formation of organic-inorganic composite materials.

ACKNOWLEDGMENTS

We thank E. Brückner and Prof. Dr. H. Rehage for generously providing nonpublished results. Financial support from the Swiss National Science Foundation is gratefully acknowledged.

REFERENCES

1. J. N. Israelachvili, D. J. Mitchell, and R. Horn, J. Chem. Soc. Faraday Trans. II *72*:1525–1568 (1976).
2. T. Kunitake and Y. Okahata, J. Am. Chem. Soc. *99*: 3860–3861 (1977).
3. T. Kunitake, Angew. Chem. Int. Ed. Engl. *31*:709–726 (1992).
4. L. Zhang and A. Eisenberg, Science *268*:1728–1731 (1995).
5. S. A. Jenekhe and X. L. Chen, Science *279*:1903–1906 (1998).
6. D. D. Lasic, *Liposomes: From Physics to Applications*, Elsevier Science Publishers, Amsterdam, 1993.
7. E. W. Kaler, A. K. Murthy, B. E. Rodriguez, and J. A. Zasadzinski, Science *245*:1371–1374 (1989).
8. P. F. Fahey and W. W. Webb, Biochemistry *17*:3046–3053 (1978).
9. J. M. Lehn, Angew. Chem. Int. Ed. Engl. *29*:1304–1319 (1991).
10. J. H. Fendler, *Membrane Mimetic Approach to Advanced Materials*, Springer Verlag, Berlin, 1994.
11. W. Meier and J. Schreiber, Patent Application WO 98/13025.
12. H. Ringsdorf, B. Schlarb, and J. Venzmer, Angew. Chem. Int. Ed. Engl. *27*:113–158 (1988).
13. D. D. Lasic, Angew. Chem. Int. Ed. Engl. *33*:1685–1698 (1994).
14. J. Murtagh and J. K. Thomas, Faraday Discuss. Chem. Soc. *81*:127–136 (1986).
15. J. Kurja, R. J. M. Nolte, I. A. Maxwell, and A. L. German, Polymer *34*:2045–2049 (1993).
16. N. Poulain, E. Nakache, A. Pina, and G. Levesque, J. Polym. Sci. Part A Polym. Chem. *34*:729–737 (1996).
17. J. D. Morgan, C. A. Johnson, and E. W. Kaler, Langmuir *13*:6447–6451 (1997).

18. M. Jung, D. H. W. Hubert, P. H. H. Bomans, P. M. Frederik, J. Meuldijk, A. M. van Herk, H. Fischer, and A. L. German, Langmuir *13*:6877–6880 (1997).
19. J. Hotz and W. Meier, Langmuir *14*:1031–1036 (1998).
20. P. Eggl, D. Pink, B. Quinn, H. Ringsdorf, and E. Sackmann, Macromolecules *23*:3472–3480 (1990).
21. G. Gompper and D. A. Kroll, Curr. Opin. Colloid Interface Sci. *2*:373–381 (1997) and references cited therein.
22. D. Day, H.-H. Hub, and H. Ringsdorf, Isr. J. Chem. *18*: 325–329 (1979).
23. S. L. Regen, B. Czech, and A. Sing, J. Am. Chem. Soc. *102*:6638–6640 (1980).
24. D. S. Johnston, S. Sanghera, M. Pons, and D. Chapman, Biochim. Biophys. Acta *602*:57–69. (1980).
25. D. F. O'Brien, T. Whitesides, and R. Klingbiel, J. Polym. Sci. Polym. Lett. Ed. *19*:95–101 (1981).
26. M. Abe, K. Ogino, and H. Yamauchi, in *Solubilization in Surfactant Aggregates* (S. D. Christian and J. F. Scamehorn, eds.), Marcel Dekker, New York, 1995.
27. D. F. O'Brien and V. Ramaswami, Encyc. Polym. Sci. Eng. *17*:108–135 (1989).
28. A. Singh and J. M. Schnur, in *Phospholipids Handbook* (G. Cevc, ed.), Marcel Dekker, New York, 1993, pp. 233–291.
29. D. F. O'Brien, Trends Polym. Sci. *2*:183–188 (1994).
30. B. A. Armitage, D. F. Bennett, H. G. Lamparski, and D. F. O'Brien, Adv. Polym. Sci. *126*:54–84 (1996).
31. S. Okada, S. Peng, W. Spevak, and D. Charych, Acc. Chem. Res. *31*:229–239 (1998).
32. D. F. O'Brien, B. Armitage, A. Benedicto, D. E. Bennett, H. G. Lamparski, Y.-S. Lee, W. Srisiri, and T. H. Sisson, Acc. Chem. Res. *31*:861–868 (1998).
33. J. Hotz and W. Meier, Adv. Mater. *16*:1387–1390 (1998).
34. A. Sadownik, J. Stefely, and S. L. Regen, J. Am. Chem. Soc. *108*:7789–7791 (1986).
35. R. Neumann and H. Ringsdorf, J. Am. Chem. Soc. *108*: 487–490 (1986).
36. R. Neumann, H. Ringsdorf, E. V. Patton, and D. F. O'Brien, Biochim. Biophys. Acta *898*:338–348 (1987).
37. G. Wegner, Makromol. Chem. *154*:35–48 (1972).
38. J. Lei and D. F. O'Brien, Macromolecules *27*:1381–1388 (1994).
39. T. D. Sells and D. F. O'Brien, Macromolecules *27*:226–233 (1994).
40. H. Meier, I. Sprenger, M. Bärmann, and E. Sackmann, Macromolecules *27*:7581–7588 (1994).
41. J. Lei, T. M. Sisson, H. G. Lamparski, and D. F. O'Brien, Macromolecules *32*:73–78 (1999).
42. S. Kölchens, H. G. Lamparski, and D. F. O'Brien, Macromolecules *26*:398–400 (1993).
43. T. M. Sisson, H. G. Lamparski, S. Kölchens, A. Elayadi, and D. F. O'Brien, Macromolecules *29*:8321–8329 (1996).
44. K. R. Porter, Mod. Cell Biol. *2*:259–302 (1983).
45. J. D. A. Fishman and T. L. Steck, J. Supramol. Struct. *1*:233–248 (1973).
46. A. Elgsaater, B. I. Stokke, A. Michaelson, and D. Branton, Science *234*:1217–1223 (1986).
47. A. Clark and S. B. Ross-Murphy, Adv. Polym. Sci. *83*: 57–192 (1987).
48. J. M. Guysen and G. D. Schockman, Microbiol. Ser. *1*: 37 (1973).
49. K. Dorn, R. T. Klingbiel, D. P. Specht, P. M. Tyminski, H. Ringsdorf, and D. F. O'Brien, J. Am. Chem. Soc. *106*:1627–1633 (1984).
50. J. Stefely, M. A. Markowitz, and S. L. Regen, J. Am. Chem. Soc. *110*:7463–7469 (1988).
51. J. H. Fendler and P. Tundo, Acc. Chem. Res. *17*:3–8 (1984).
52. B. J. Ravoo, W. D. Weringa, and J. B. F. N. Engberts, Langmuir *12*:5773–5780 (1996).
53. S. L. Regen, J.-S. Shin, J. F. Hainfeld, and J. S. Wall, J. Am. Chem. Soc. *106*:5756–5757 (1984).
54. V. P. Torchilin, A. L. Klibanow, N. N. Ivanov, H. Ringsdorf, and B. Schlarb, Makromol. Chem. Rapid Commun. *8*:457–460 (1987).
55. E. Brückner and H. Rehage, Prog. Colloid Polym. Sci. *109*:21–28 (1998).
56. I. A. Maxwell and J. Kurja, Langmuir *11*:1987–1993 (1995).
57. D. W. R. Gruen and D. A. Haydon, Pure Appl. Chem. *52*:1229–1240 (1980).
58. T. J. McIntosh, S. A. Simon, and R. C. MacDonald, Biochim. Biophys. Acta *597*:445–463 (1980).
59. J. D. Morgan and E. W. Kaler, Macromolecules *31*: 3197–3202 (1998).
60. C. C. Co and E. W. Kaler, Macromolecules *31*:3203–3210 (1998).
61. M. Winterhalter, K.-H. Klotz, and R. Benz, in *Electromanipulation of Cells* (U. Zimmerman and G. A. Neil, eds.), CRC Press, Boca Raton, FL, 1996, pp. 137–171.
62. M. Dvolaitzky, M. A. Guedeau-Boudeville, and L. Léger, Langmuir *8*:2595–2597 (1992).
63. J. Hotz and W. Meier, *Functionalized Polymer Hollow Spheres by Template Polymerization in Vesicles*, Bunsentagung, Dortmund 1999.
64. K. Dusek, *Responsive Gels: Volume Phase Transitions I*, Adv. Polym. Sci. 109, Springer, Berlin, 1992; K. Dusek, *Responsive Gels: Volume Phase Transitions II*, Adv. Polym. Sci. 110, Springer, Berlin, 1993.
65. J. A. Speir, S. Munshi, G. Wang, T. S. Baker, and J. E. Johnson, Structure *3*:63–78 (1995).
66. T. Douglas and M. Young, Nature *393*:152–155 (1998).
67. S. Mann, *Biomimetic Materials Chemistry*, VCH, New York, 1996.

25

The Use of Surfactant Self-Assembly in the Enzymatic Synthesis of Novel Polymers

GLEN IRVIN, SUKANTA BANERJEE, RAMANNAIR PREMACHANDRAN, BLAKE A. SIMMONS, SICHU LI, VIJAY T. JOHN, and GARY MCPHERSON
Tulane University, New Orleans, Louisiana

JOSEPH AKKARA* U.S. Army Soldier Systems Center, Natick, Massachusetts

DAVID KAPLAN Tufts University, Medford, Massachusetts

WEILIE ZHOU University of New Orleans, New Orleans, Louisiana

I. OVERVIEW

Surfactant microstructures provide a unique environment for the enzymatic synthesis of polymers. The ability of surfactants to create large oil-water interfaces can be exploited for such biocatalysis, where the enzyme is resident in the aqueous phase and the monomer is resident in the oil phase or at the oil-water interface. The polymerization of substituted phenols using an oxidative enzyme, horseradish peroxidase (HRP), is described in this chapter. Two surfactant microstructures are used in the synthesis: (1) the system of AOT water-in-oil microemulsions and (2) a novel gel system formed by the addition of lecithin and water to AOT water-in-oil microemulsions. The monomers partition to the oil-water interface, and polymerization is extremely feasible. The synthesized polymer also acquires the morphology of interconnected microspheres with controllable internal densities. In the case of the gel system, it is possible to form spherical superclusters of interconnected microspheres. The morphology evolution is perhaps a consequence of monomer prealignment at the oil-water interface. Functional properties of the polymer based on chemical structure and new applications based on morphology are described.

**Current affiliation*: National Science Foundation, Arlington, Virginia.

II. INTRODUCTION

A. Background

The use of enzymes in synthesis represents an inherently environmentally benign approach to chemical processing. In addition to catalytic function at or near ambient conditions, the exquisite specificity of enzyme biocatalysis can be potentially exploited to develop processes with minimal side products. Advances in large-scale enzyme production and purification and synthetic enzyme mimetic chemistry are eventually expected to lead to enzyme-based processes that are economically viable [1]. The use of enzymes to synthesize polymers is one such area of research undergoing rapid development [2]. There is a wonderful aspect of biomimetics here, as enzymes function in the biological world in the synthesis and degradation of polysaccharides, proteins, polyphenols, and polynucleic acids. If such enzyme action can be exploited in the in vitro synthesis of polymers with commercial significance, it may be possible to obtain polymers with unique materials properties that are inherently biodegradable and, in specific cases, biocompatible. Examples of enzyme biocatalysis for polymer synthesis include the use of lipases for polyester synthesis [3], peroxidases for polyphenol synthesis [4], and cellulases for polysaccharide synthesis [5].

In developing enzymes for synthesis, the reaction environment plays a vital role in determining the activity and catalytic efficiency of the enzyme. The ability of enzymes to function in nonaqueous environments when properly conditioned has brought about the development of an entire field of biocatalysis in organic solvents [6]. A parallel development is the field of enzyme chemistry in surfactant systems, in particular the system of water-in-oil microemulsions, conventionally referred to as reversed micelles [7].

B. Polyphenols and Enzymatic Synthesis

This chapter describes concepts of enzymatic polymer synthesis in surfactant systems, focusing on the synthesis of polyphenols and polyaromatic amines. Phenolic polymers have a variety of conventional applications in making resins for coatings, laminates, etc. [8]. The traditional technology for making these polymers involves a formaldehyde-based high-temperature process in which undesirable side reactions lead to poor control of polymer structure and molecular weight. In addition, concern over the toxicity of formaldehyde necessitates the study of alternative technologies to produce such polymers. The enzymatic approach using an oxidative enzyme such as horseradish peroxidase obviates the need for formaldehyde. The reaction has a mechanistic analogy to the synthesis of lignin [9] and is illustrated through the simplified mechanism shown in Fig. 1, where reaction is initiated by the addition of H_2O_2. Phenoxy radical centers initially formed on the monomer or growing chain migrate to the *ortho* positions (the *para* position, although mechanistically allowed, is less favored), after which coupling through condensation occurs.

In principle, the enzymatically synthesized polyphenolics have applications similar to those of the phenol-formaldehyde resins that are chemically synthesized [10]. However, there are unique aspects related to the enzymatically synthesized material. In contrast to phenol-formaldehyde polymers, these polyphenolics lack the intervening methylene bridge between the aromatic groups as illustrated in the simplified mechanistic scheme of Fig. 1. The polymer is therefore conjugated and thus has a variety of potential applications in electro-optics [11]. A specific property that is being investigated is the use of these materials in nonlinear optics (NLO), in particular for applications based on the optical third-order nonlinear susceptibilities ($\chi^{(3)}$) [4,11]. In such materials, higher order terms become significant in the expansion of the material polarization (P) in terms of an applied electric field (E).

$$P = \chi^{(1)}E + \chi^{(2)}E^2 + \chi^{(3)}E^3 + \cdots$$

Details of the physics behind NLO materials can be found in several excellent sources (e.g., Prasad and Williams [12]). The relevance of NLO polymers to applications in optical switching, waveguide technology, laser protection, etc. is well recognized. Thus, the enzymatically synthesized polymers have some useful applications in electro-optics [13]. The polyaromatic amines synthesized through the enzymatic process are especially promising as materials with high $\chi^{(3)}$ values of up to 10^{-7} e.s.u., among the highest reported for organic materials [13].

The synthesis of polyphenols and polyaromatic amines in surfactant microstructures has some interesting aspects. The monomers are typically water insoluble, and there is therefore an intrinsic incompatibility with the enzyme, which is water soluble. Therefore, there is a need to devise systems to promote efficient enzyme-monomer contact. If two-phase oil-water sys-

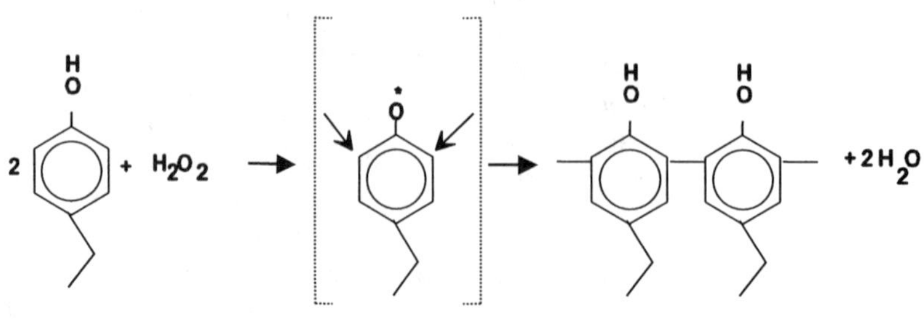

FIG. 1 Simplified schematic of the polymerization of alkyl-substituted phenols. The arrows (in the phenoxy radical) indicate coupling in the *ortho* position.

tems are used, it is intuitive that the agitation needed to enhance enzyme-monomer contact may shear deactivate the enzyme. A second approach is to activate the enzyme in organic solvents, and this has been shown to be feasible in polyphenol synthesis [4,6]. A third approach is the use of surfactants to incorporate water into an organic phase (or vice versa) so that the enzyme and the monomer can exist in close proximity with minimal agitation. This is the subject of the present chapter. We describe two systems: the system of water-in-oil microemulsions (reverse micelles), where the enzyme is encapsulated in microaqueous water pools, and the system of a novel, bicontinuous gel phase, where the enzyme is immobilized in percolating water channels. These systems lead to some novel aspects of morphology development in the synthesized polymer, and thus to some new applications.

III. POLYMER SYNTHESIS IN REVERSE MICELLES

The AOT (bis 2-ethylhexyl sodium sulfosuccinate) water-in-oil microemulsion system is an effective system for synthesizing polyphenolics and polyaromatic amines. Figure 2 illustrates the system and the rationale

(a)

(b)

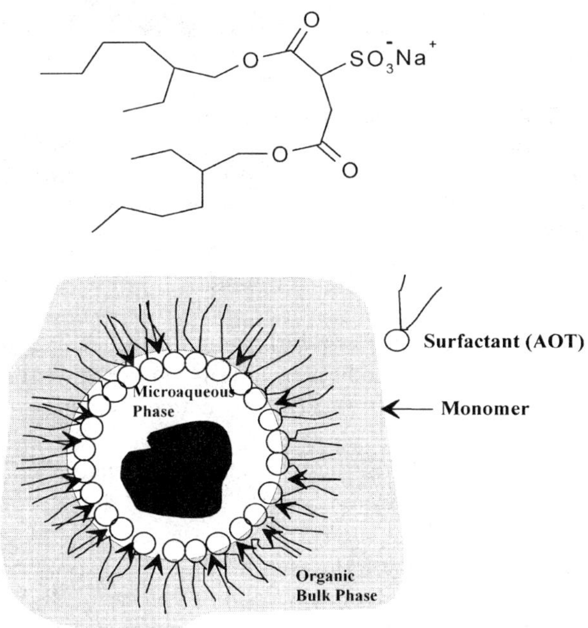

FIG. 2 (a) Chemical structure of the anionic surfactant sodium bis(2-ethylhexyl) sulfosuccinate. (b) Schematic of enzyme solubilized in the micelle and monomer partitioning to the micelle interface. The arrow refers to the monomer.

for synthesis in this medium. The enzyme is catalytically active when solubilized in the water core, and the organic bulk phase helps support monomer and chain solubilization. An interesting aspect of synthesis in reversed micelles is the partitioning of the polar monomer to the oil-water interface (depicted by the arrow in Fig. 2b with the head of the arrow representing the hydroxyl moieties of the monomer). In addition, hydrogen bonding between the surfactant headgroup and the monomer influences monomer orientation and partitioning at the water-oil interface. A strong orientation of such hydrogen bonding is the perturbation of the vibrational frequencies of the surfactant C═O groups upon addition of the monomers [14]. Such partitioning may result in prealignment of the monomers before synthesis and also provides a means of monomer replenishment to the vicinity of the enzyme. In addition, such surfactant-monomer interactions significantly enhance monomer solubility in the reaction medium. If this is indeed true, the implication is that the hydroxyl groups are on the same side of the polymer backbone as shown in Fig. 1. But we must note that Fig. 1 is not really representative of the polymer structure because the ring-to-ring C—C bond depicted does not have the correct bond angle to the ring (120°). In simple molecular models, it can be seen that *ortho*-coupling with proximal hydroxyls on the same side of the backbone does induce a natural curvature of the polymer. Indeed, intramolecular hydrogen bonding leads to a folding of the polymer if all the hydroxyls lie on the same side of the backbone [15]. We will return to the concept of polymer curvature in the interpretation of the observations.

Polymerization in reverse micelles is very easy to carry out [16]. In a typical experiment, the enzyme (HRP) dissolved in 0.01 M HEPES buffer (pH 7.5) is added to a dry reversed micellar solution of AOT in isooctane, followed by the addition of the monomer, 4-ethylphenol (EP). The enzyme concentration in the buffer is adjusted so that the final enzyme concentration in the micellar solution is 0.5 mg/mL. The reaction mixture typically has the overall composition 0.5 M AOT, 0.15 M EP, 0.5 mg/mL HRP, and a w_0 of 15 (w_0 is the water-to-surfactant molar ratio). The reaction is initiated by the addition of H_2O_2 in aliquots to minimize enzyme deactivation. The reaction is rapid, and within 5–10 min of reaction initiation, over 80% of monomer conversion occurs, with the polymer precipitating out of solution [16].

Figure 3 illustrates the fascinating microsphere morphology of polymers synthesized and precipitated from reaction in reversed micellar solutions. We have found

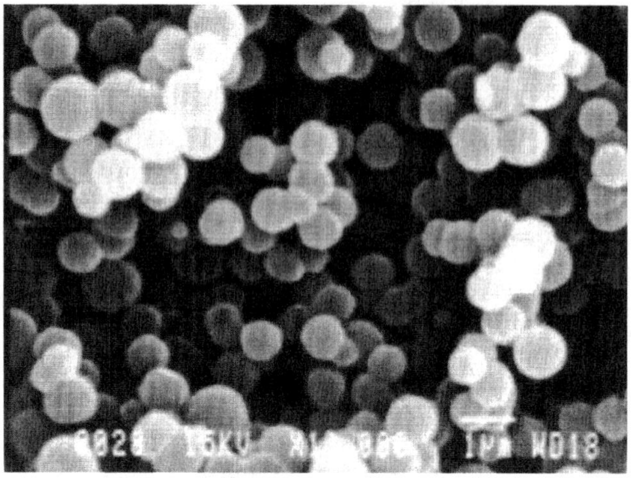

FIG. 3 Scanning electron micrograph of polymer microspheres prepared through synthesis in reverse micelles.

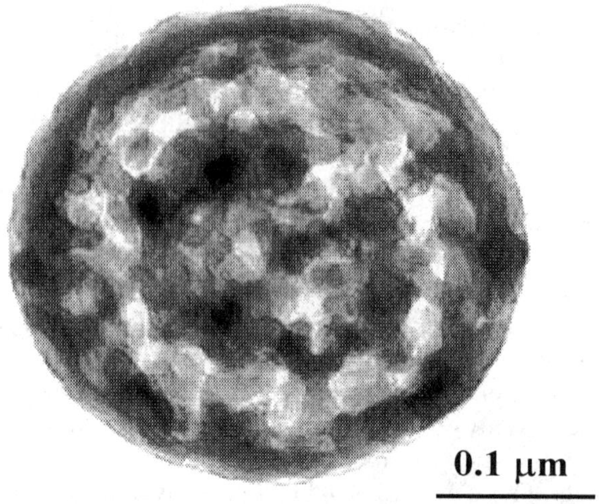

(a)

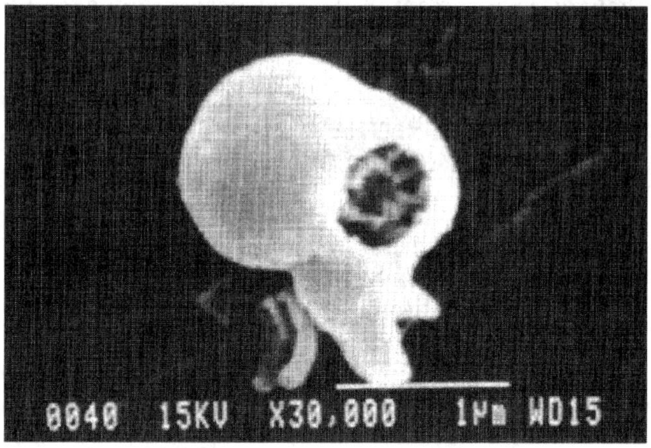

(b)

FIG. 4 (a) Transmission electron micrograph of a polymer microsphere at an initial stage (5 min reaction time) illustrating the low internal density. (b) Scanning electron micrograph illustrating the internal structure of a polymer microsphere.

that as long as the initial surfactant-to-monomer ratio is 2:1 or higher (preferably 3:1) the polymer is always precipitated in the morphology of microspheres [16]. Many of the microspheres are not independent but are connected to each other. Although it appears that the microspheres are a consequence of polymerization around micelle interfaces, the size of these microspheres (0.5–1 μm) is significantly larger than the micelle size (0.5–5 nm depending on the water level in the micelle).

The large sizes of the microspheres in comparison with micelle dimensions indicate the lack of a direct templating effect produced by synthesis at the micelle oil-water interface. However, in studying the growth characteristics of these microspheres, we have found that they initially have very low internal density. The microspheres appear to be made up of minute interconnected spherical patches [17,18] that may originate from reaction on the micelle periphery. This can be seen from the transmission electron micrograph of Fig. 4a. The scanning electron micrograph of Fig. 4b, where the shell of a microsphere is broken by gentle sonication, appears to correlate well with the fact that the internal density of the microsphere is initially low. In fact, the internal density of the microspheres can be controlled by adjustment of reaction time and H_2O_2 addition. This is because the microspheres contain enzyme and unreacted monomer, and polymerization continues even after the overall morphology has been achieved, leading to densification. Figure 5 illustrates partial densification of the microspheres as observed by transmission electron microscopy. Over a period of about 24 h, all microspheres fully densify [17]. At the other extreme, at very early reaction times (<5 min), sampling of the polymer simply indicates interconnected patches with no overall defined spherical morphologies.

The internal morphology and densification characteristics of the microspheres suggest the following mechanism. The initial patches are a consequence of reaction at the oil-water interface. The polymer is "sticky" because of the large number of hydroxyl groups, which are capable of hydrogen bonding both within the chain and between chains. The intrachain hydrogen bonding may lead to the curvature of the chain, and interchain hydrogen bonding can create the connections to polymer growing on separate micelles

FIG. 5 Transmission electron micrograph illustrating microspheres at various stages of densification.

during micelle collisions. The interplay between these effects leads to the interconnected patches of Fig. 4a.

The second interesting observation is that during precipitation, the polymer also encapsulates solutes located in the water core of the micelle. The densification observation is clearly a consequence of peroxidase becoming encapsulated in the microspheres. Realizing this, we have attempted to exploit the phenomenon to synthesize polymer-nanoparticle composites. The hypothesis is as follows. Reversed micelles have been proposed as a microenvironment within which to synthesize inorganic materials that do not grow beyond the nanometer size range [19]. Enhanced band gap semiconductor materials (CdS, TiO_2), magnetic particles, etc. are examples of novel nanoparticles that have been synthesized in reversed micelles. We have attempted to make polymer-nanoparticle composites by first synthesizing the nanoparticles in the micelles and then adding monomer, enzyme, and H_2O_2 to initiate polymer synthesis. When the polymer precipitates out (in spherical morphologies) it incorporates a significant amount of the nanoparticles. For example, we have synthesized superparamagnetic iron oxide in reversed micelles [20,21]. In such synthesis, the particle size approaches magnetic domain size, leading to thermally induced randomization of magnetic dipole orientations. The particles therefore do not exhibit magnetic characteristics in the absence of a field (the term superparamagnetism is usually applied to particles that are nonmagnetic in the absence of a field, as paramagnetic materials are, but upon application of an external field exhibit magnetization far in excess of paramagnetic materials). We then incorporated these particles into the polymer as already described, allowing the particles to densify. Figure 6 is a cut section transmission electron micrography of a polymer particle; the dark specks are the ferrite particles uniformly distributed across the section. The cut section also indicates a significant amount of ferrite entrapment (up to 6% of the polymer weight). We also note that the particle size increases (the particle fluffs up) upon incorporation of the nanoparticles. The particle deformation in the micrograph is a consequence of the cut section procedure.

In addition to applications evolving from the morphological characteristics of the polymer particles synthesized enzymatically in surfactant microstructures, it is possible to derive some novel functional materials by changing the monomer structure. Figure 7 illustrates chemical structures of some of the monomers that can be used in the synthesis of modified phenolic polymers and copolymers with novel functional characteristics. For example, the multiring naphthol- and hydroxypyrene-based polymers are intrinsically photoluminescent [17,22]. Because these polymers are conjugated, there may be opportunities to develop new classes of electroluminescent and photoluminescent polymers. The

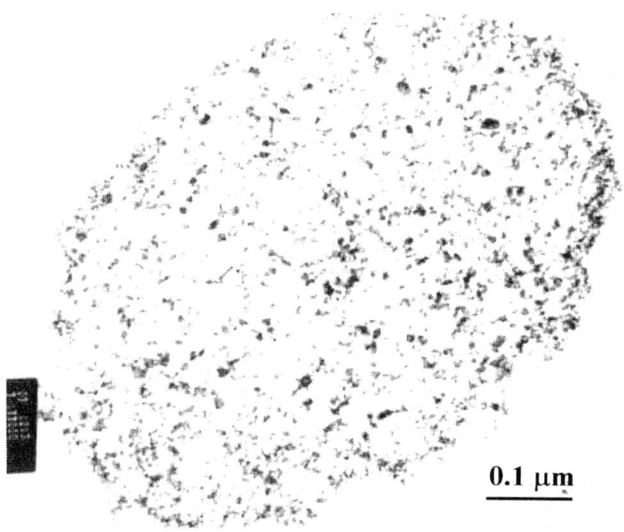

FIG. 6 Cut section transmission electron micrograph of ferrite nanoparticles within a polymer microsphere. The ellipsoidal cross section is induced by the cut section procedure.

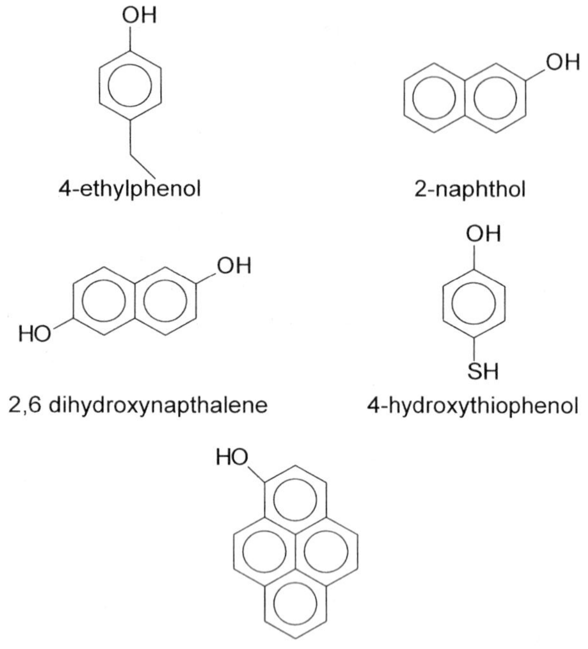

FIG. 7 Some of the monomers that can be used for the synthesis of functional polymers.

use of monomers such as hydroxythiophenol allows the synthesis of polymers that are capable of binding to inorganic sulfide nanoclusters (e.g., CdS, PbS) with semiconductor properties [23]. This may again open up new possibilities in the preparation of polymer-nanoparticle films and composites with electro-optical properties. Polymers made with dihydroxynaphthalene may be redox active and may have electrochemical applications in battery and sensor development [24].

IV. SYNTHESIS IN A NOVEL SURFACTANT-BASED GEL SYSTEM

As described in the previous section, synthesis in reversed micelles leads to microspherical polymer particles that are precipitated out of solution. In an effort to maintain polymer solubility in the reaction medium, we have exploited a surfactant-based gel system to sustain the polymer during synthesis. The gel is formed by a novel transformation from low-viscosity AOT water-in-soil microemulsions to a highly viscous, rigid state. Specifically, we have found that the addition of lecithin (phosphatidylcholine) to these water-in-oil microemulsions can result in the formation of a rigid gel when additional water is added to the system. These gel systems are a significant variant of the fascinating organogels discovered by Luisi and coworkers [25] based on the system lecithin/water/cyclohexane and extensively characterized through various spectroscopic and scattering techniques [26–28]. The earlier case essentially is one in which the water content of the gel system is less than 10 wt%. In the present system, the amount of water can be greater than 50 wt% with retention of gel stability. In other words, the gel phase can be sustained with equal volume fractions of water and the organic phase, implying the presence of spatially immobilized extended hydrophilic and hydrophobic microstructures. The gel is neither an organogel nor a hydrogel but could, for example, be an immobilized bicontinuous structure of water and hydrocarbon networks. A typical gel has the composition AOT/lecithin/isooctane/water 15:14:29:42 (wt%).

Figure 8 illustrates the zero shear viscosity and conductivity trends as the water content of the system is varied. The measurements were made by adding water to a system containing 0.8 M AOT and 0.4 M lecithin in isooctane. The quantity w_0 is the molar ratio of water to AOT and is the quantity typically used to characterize the water content in AOT water-in-oil microemulsions. As Fig. 8 indicates, gelation states at a w_0 of about 50 with a rigid gel formed at a w_0 value of 65. The zero shear viscosity increase of six to seven decades indicates the magnitude of rigidification from a low-viscosity solution. It is also interesting to note that the gelation point correlates rather well with a significant increase in electrical conductivity. The sharp rise in electrical conductivity is perhaps indicative of the formation of percolating water channels in the gel. The ^{31}P nuclear magnetic resonance (NMR) data of Fig. 9 provide a correlation with molecular properties. The large increase in line width at gelation indicates rigidification of the phosphate headgroups of lecithin. Sharp line widths are typically obtained when high molecular rotational mobility allows resonance at exactly the same frequency for all headgroup ^{31}P nuclei. When molecular mobility on the NMR time scale is reduced, the line width is broadened as there is a spread of frequencies at which the headgroup ^{31}P nuclei come into resonance with the applied magnetic field. It is also interesting to note that as the water content is further increased, the phosphate headgroups regain some mobility, although the system is still in the gel state.

It is therefore hypothesized that this is a medium in which aqueous synthesis can be combined with organic synthesis, leading to structured composite materials with novel application possibilities. For example, the organic phase could be used in the synthesis of hydrophobic polymers and the aqueous phase can be used in

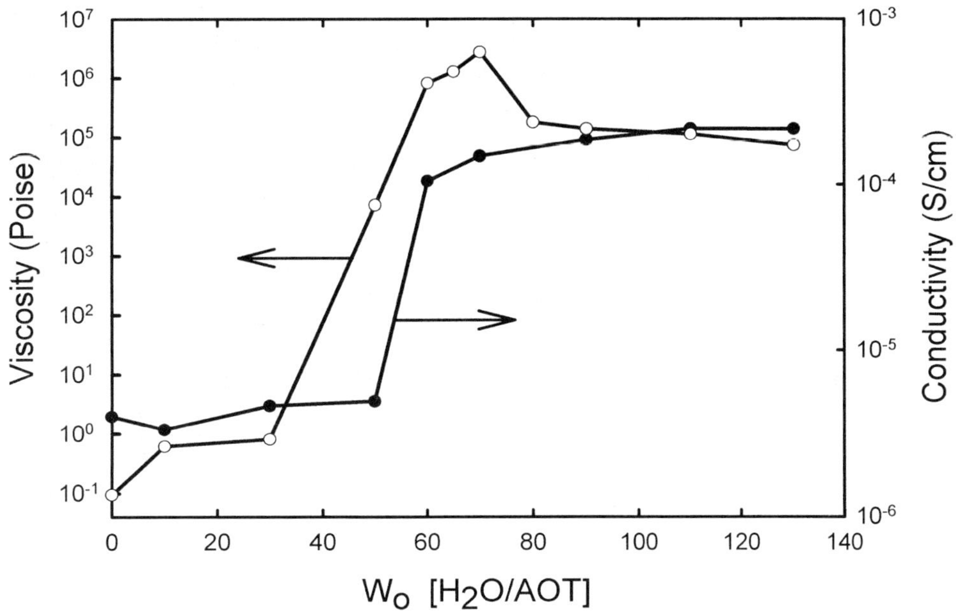

FIG. 8 Viscosity and electrical conductivity data illustrating the conversion from liquid to gel systems upon adding water to AOT-lecithin systems.

the synthesis of hydrophilic polymers or inorganic materials. Thus, there is the possibility of obtaining structured polymer-polymer nanocomposites or polymer-ceramic nanocomposites. The oil-water interface in these systems may also be exploited in interfacial polymerization. The microaqueous phase may also be used to sustain biomolecules, leading to novel systems for enzyme biocatalysis or drug delivery. And if synthesis is carried out in one of the microphases (organic or aqueous), there is the possibility of generating materials with structured porosities, leading to membrane and separations applications.

Our objective in conducting enzymatic polymerization in these systems has been to sustain polymer solubility in the reaction phase. Monomer and enzyme are loaded into the gel, the monomer being soluble in the organic phase (isooctane) and the enzyme being encapsulated in the percolating microaqueous phase. The reaction is initiated by injecting H_2O_2 into the system. The results of these early experiments are striking, and we are still in the early stages of understanding this system. The polymer still has the microsphere morphology but the microspheres are extensively connected. But if the surfactants are washed off in bulk and the polymer particles collected and imaged, we see the formation of superclusters as shown in Fig. 10. We still have not imaged the interior of these particles but presume they are also made up of the patches of Fig. 4a.

The retention of microsphere morphology in the AOT-lecithin gel system where there is no agitation during synthesis further points to the concept of monomer prealignment leading to curvature of polymer chains. The molecular model of Fig. 11 illustrates the curvature of the chains obtained when all hydroxyl groups lie on the same side of the polymer backbone. When the polymer is unfolded, for example, by dissolving the polymer in a polar solvent with hydrogen bonding acceptor groups (e.g., acetone), the open structure shown in Fig. 11 is realized. Our observation is that enzyme-catalyzed polyphenol synthesis does not lead to any specific morphology when the reaction is carried out in monophasic organic solvents. On the other hand, synthesis in self-assembled surfactant systems leads to the microsphere morphology, revealing that monomer prealignment may lead to oriented chains that fold as a consequence of bond geometry. These end up being the initial patches in the system. Interactions between the patches and the continued tendency of the chains to fold may lead to the microspheres. Finally, if the microspheres are extensively connected as in the case of synthesis in the gel system, the entire assembly could fold to form the superclusters.

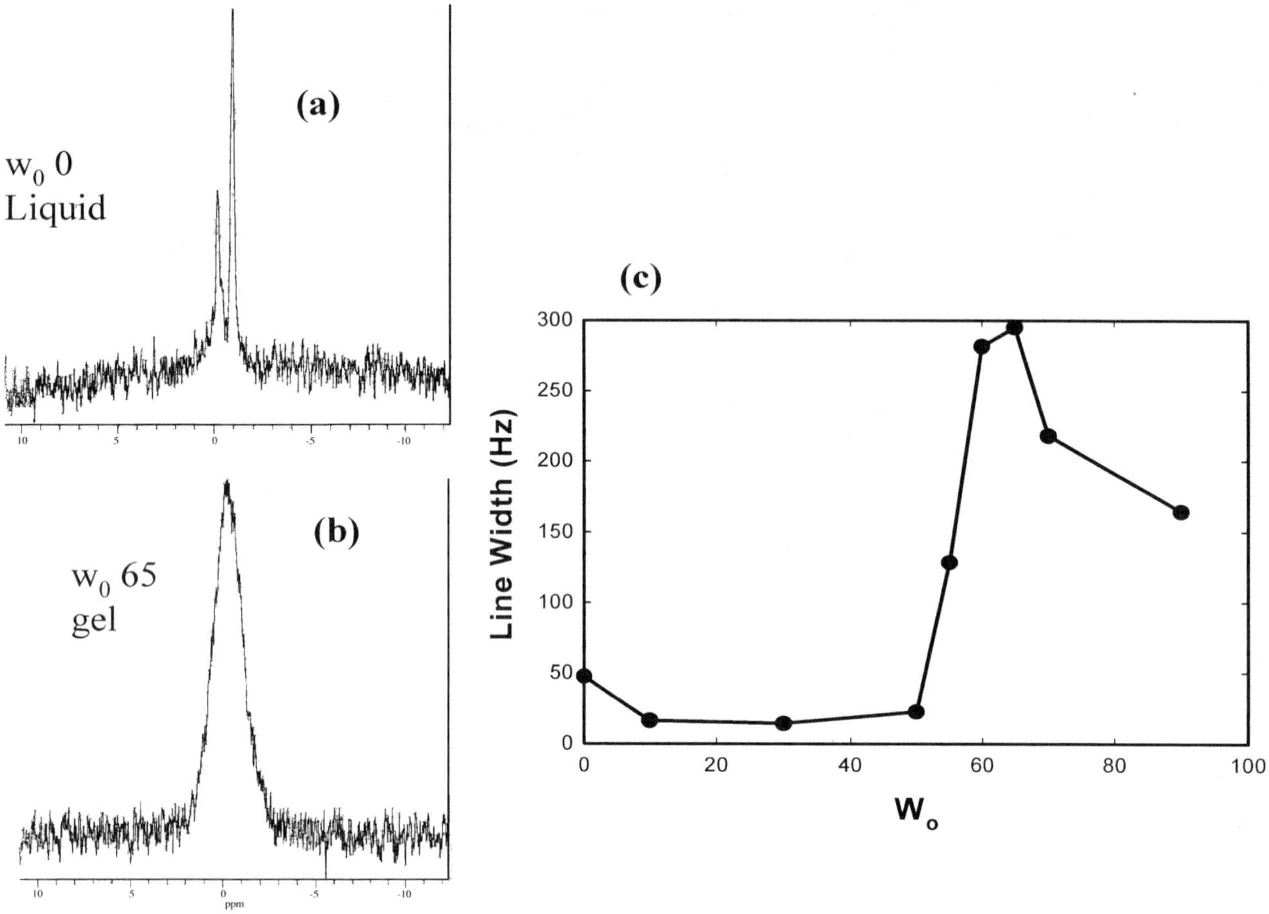

FIG. 9 ^{31}P NMR image of the lecithin headgroup indicating rigidity of phosphate groups as the liquid system evolves to a gel system upon addition of lecithin.

(a) (b)

FIG. 10 Scanning electron micrographs of supercluster morphology upon recovery of the polymer from the gel phase.

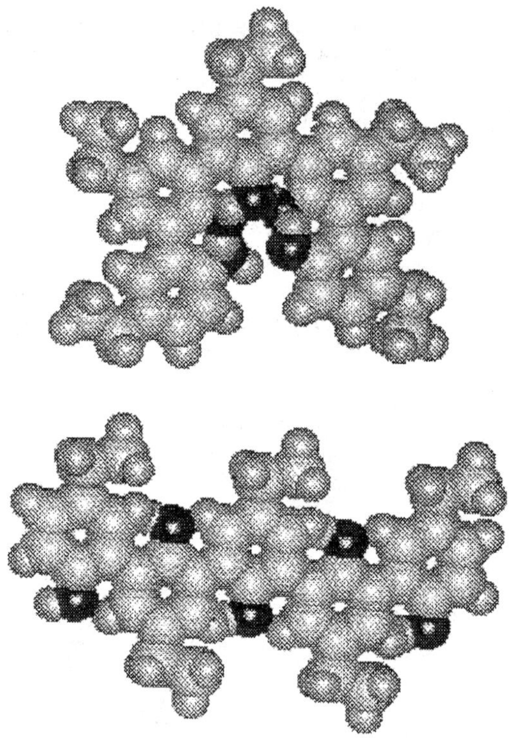

FIG. 11 Molecular model of polyphenol folding as a consequence of monomer alignment at an oil-water interface.

V. SUMMARY

Self-assembled surfactant systems can be exploited in the biocatalytic synthesis of structured polymers with unique properties that are a consequence of both chemical structure and morphology. Although a precise templating effect from the surfactant microstructure may not be immediately evident, it is clear that the surfactant self-assembly plays an important role in the development of polymer morphology. These systems may be exploited in the development of a variety of polymer-polymer and polymer-ceramic composites. For example, enzymatic polymer synthesis could be coupled with traditional free radical polymer synthesis with vinyl monomers to make novel structured composites. It is also possible to combine enzymatic polymer synthesis with inorganic cluster synthesis to make structured polymer-ceramic nanocomposites. Continuing work seeks to systematize these studies to develop clear correlations between surfactant microstructure and material morphology.

ACKNOWLEDGMENT

Support from the U.S. Army and DARPA (Grant MDA972-97-1-0003) is gratefully acknowledged.

REFERENCES

1. J. Dordick, *Biocatalysts for Industry*, Plenum, New York, 1991.
2. R. A. Gross, D. Kaplan, and G. Swift, *Enzymes in Polymer Synthesis*, ACS Symposium Series 684, American Chemical Society, Washington, DC, 1998.
3. A. K. Chaudhary, E. J. Beckman, and A. J. Russell, in *Enzymes in Polymer Synthesis* (R. A. Gross, D. Kaplan, and G. Swift, eds.), ACS Symposium Series 684, Washington, DC, 1998.
4. J. A. Akkara, K. J. Senecal, and D. L. Kaplan, J. Polym. Sci. *29*:1561 (1991).
5. S. Kobayashi, K. Kasiwa, T. Kawaskai, and S. Shoda, J. Am. Chem. Soc. *113*:3079 (1991).
6. J. Dordick, M. A. Marletta, and A. M. Klibanov, Biotechnol. Bioeng. *30*:31 (1987).
7. S. Barbaric and P. L. Luisi, J. Am. Chem. Soc. *103*: 4239 (1981).
8. P. W. Kopf, in *Encyclopedia of Polymer Science and Engineering*, Vol. 11, John Wiley & Sons, New York, 1985.
9. B. Halliwell and J. M. C. Gutteridge, *Free Radicals in Biology and Medicine*, Clarendon Press, Oxford, 1989.
10. G. A. Millis, in *Kirk-Othmer Encyclopedia of Chemical Technology* (H. M. Mark and D. F. Othmer, eds.), Vol. 5, John Wiley & Sons, New York, 1979.
11. M. Ayyagari, F. Bruno, S. Tripathy, K. Marx, D. Kaplan, J. Akkara, and D. Rao, in *Polymers and Other Advanced Materials: Emerging Technologies and Business Opportunities* (P. N. Prasad, J. E. Mark, and T. J. Fai, eds.), New York, Plenum, 1996.
12. P. N. Prasad and D. J. Williams, *Introduction to Nonlinear Optical Effects in Molecules and Polymers*, John Wiley & Sons, New York, 1991.
13. J. A. Akkara, F. A. Aranda, D. V. G. L. N. Rao, and V. T. John, in *Electrical and Optical Polymer Systems* (D. L. Wise, ed.), Marcel Dekker, New York, 1998.
14. A. M. Rao, V. T. John, R. D. Gonzalez, J. A. Akkara, and D. L. Kaplan, Biotechnol. Bioeng. *41*:531 (1993).
15. S. Banerjee, R. Premachandran, M. Tata, V. John, G. McPherson, J. Akkara, and D. Kaplan, Ind. Eng. Chem. Res. *35*:3100 (1996).
16. C. Karayigitoglu, N. Kommareddi, V. John, G. McPherson, J. Akkara, and D. Kaplan, Mater. Sci. Eng. C *2*:165 (1995).
17. S. Banerjee, P. Rammannair, K. Wu, V. T. John, G. McPherson, J. A. Akkara, and D. L. Kaplan, in *Enzymes in Polymer Synthesis* (R. A. Gross, D. Kaplan, and G. Swift, eds.), ACS Symposium Series 684, Washington, DC, 1998, p. 125.
18. X. Xu, N. Kommareddi, M. McCormick, V. T. John, G. L. McPherson, J. A. Akkara, and D. L. Kaplan, Mater. Sci. Eng. C *4*:161 (1996).
19. M. P. Pileni, J. Phys. Chem. *97*:6961 (1993).
20. N. Kommareddi, M. Tata, C. Karayigitoglu, V. T. John, G. L. McPherson, M. F. Herman, C. J. O'Connor, Y. S.

Lee, J. Akkara, and D. L. Kaplan, Appl. Biochem. Biotechnol. *51/52:*241 (1995).

21. N. Kommareddi, M. Tata, C. Karayigitoglu, V. T. John, G. L. McPherson, M. F. Herman, C. J. O'Connor, Y. S. Lee, J. Akkara, and D. L. Kaplan, Chem. Mater. *8*:801 (1996).
22. R. Premachandran, S. Banerjee, V. T. John, G. L. McPherson, J. A. Akkara, D. L. Kaplan, and M. Ayyagari, Macromolecules *29*:6452 (1996).
23. R. Premachandran, S. Banerjee, V. T. John, G. L. McPherson, J. Akkara, and D. L. Kaplan, Chem. Mater. *9*: 1342 (1998).
24. N. Wang, B. D. Martin, S. Parida, D. G. Rethwisch, and J. S. Dordick, J. Am. Chem. Soc. *117*:12885 (1995).
25. P. L. Luisi, R. Scartazzini, G. Haering, and P. Schurtenberger, Colloid Polym. Sci. *268*:356 (1990).
26. P. Schurtenberger, L. J. Majid, P. Lindner, and P. L. Luisi, J. Phys. Chem. *95*:4173 (1991).
27. D. Capitani, E. Rossi, A. L. Segre, M. Guistini, and P. L. Luisi, Langmuir *9*:685 (1993).
28. D. Capitani, A. L. Segre, F. Dreher, P. Walde, and P. L. Luisi, J. Phys. Chem. *100*:15211 (1996).

26

Material Synthesis by Polymerization in Surfactant Mesophases

ERIC J. PAUL and ROBERT K. PRUD'HOMME Princeton University, Princeton, New Jersey

I. INTRODUCTION

Polymerizations in surfactant solutions are used to produce polymers or materials that have geometries or functionalities that cannot be achieved using traditional bulk or solution polymerizations. There are two main research areas: polymerizable surfactants and monomer polymerizations templated by surfactant self-assembly. Polymerizable surfactants (i.e., amphiphilic monomers) are used to form structured materials on a nanometer scale. The structure is stabilized by the subsequent polymerization of the surfactant. This area is covered extensively in a chapter by Alain Guyot and Klaus Tauer in this volume. We will be focusing on the second area of polymerization in surfactant systems: monomer polymerization where the surfactant provides a template for the subsequent polymerization. In most cases the surfactant can be removed and the templated material remains. This type of surfactant-templated polymerization can be further divided into inorganic or organic monomer systems.

Inorganic polymers are formed from ceramic precursors such as tetraethyl orthosilicate, which can undergo hydrolysis to form a polymeric backbone (in this case amorphous SiO_2). The early fundamental work was done on the MCM-41 and similar systems by a group at Mobil Research and Development Corporation's Central Research Laboratory [1–6]. The Mobil group was able to template hexagonal and cubic phase surfactant systems to produce porous ceramics with uniform pores between 15 and 100 Å. McGrath and others at Princeton University were able to template with the L_3 phase (also called the sponge phase) to form a porous silica glass [7].

Although there is much work in the area of inorganic templating, this chapter focuses on polymerizing organic monomers in surfactant systems. In these types of systems there are three different reasons to polymerize in surfactant systems. The first involves the polymerization of incompatible monomers. The surfactant is used to mix the monomers on the molecular level. Polymerization of the mixture gives a copolymer of monomers that could not be in close enough contact to form a polymer without the help of surfactants. This method has been used to produce copolymers between water-soluble monomers and hydrophobic monomers to make hydrophobically modified polymers [8,9]. The second method is polymerization in constrained spaces. Surfactant systems can form small domains that can confine monomer movements such that the monomer behaves as if it is in a lower dimensional system. Polymerization in lamellar systems allows the polymerization in a two-dimensional framework. The kinetics of the polymerization inside the lamella can be drastically different from that in the bulk. This can lead to polymers with different microstructures than those polymerized in bulk [10].

This chapter focuses on a third objective for polymerizing monomer-surfactant systems: producing a polymeric gel that is a direct copy of the original surfactant system. The goal of this method is to use the surfactant as a template for polymerization. With the structure frozen in by the polymer, the surfactant and other nonpolymerized materials can be extracted to form a pore network. The ideal polymerization has the polymer directly following every detail of the surfactant structure, giving a *one-to-one* template. In many cases one-to-one templating is not achieved but a material is produced that is dependent on the initial structure of the system. In these *directed polymerizations*, as we will call them, the surfactant is used to form a template that is important to the final structure but without being directly related to the final structure. The change in structure could be as simple as a size change or as complex as phase separation of the surfactant to form a porous network.

Polymerizing monomer in a surfactant template has some advantages over polymerization of amphiphilic monomers. First is that using standard monomers means that there is no need to synthesize surfactant monomers with the correct properties. Normal surfactants can be used instead. Because the phase diagrams of many surfactants are known, it is easy to modify a known system for use as a template. Adding monomer to these systems does affect the phase diagram, but many of these effects can be predicted, thus making it easier to find the target surfactant structure than with a newly synthesized monomeric surfactant.

II. SURFACTANT SYSTEMS

A. Structure

Because of the interactions between the hydrophilic headgroups and the hydrophobic tail of the surfactants, they tend to form aggregates spontaneously when placed into water, oil, or a mixture of the two. The aggregates are thermodynamically stable and therefore long lived. This interaction can be represented by a geometric packing parameter, P [11]. Because the headgroups of surfactants are either hydrated or charged, they prefer to maintain a certain distance from their nearest neighbors, which is represented by the area, a_0. Tail interactions, from energy and enthalpic differences between the solvent and the environment of hydrophobic tails, lead to another area based on the length and volume of the tail, v/l_c. The packing parameter is the ratio of these two areas ($P = v/a_0 l_c$). For certain values of P, the surfactant packs into different preferred geometries, which are shown in Fig. 1.

Normal phases (Fig. 1a–c), where water is continuous, have packing parameters less than one, and reverse phases (Fig. 1d and e), where oil is continuous, have packing parameters greater than one. When spheres (Fig. 1a) are in random order, the system is called a micellar solution. If the spheres close pack into a lattice, a discrete cubic liquid crystal can be formed [12]. As the packing parameter is increased, cylinders (Fig. 1b) are formed. In dilute solution when the orientation of the cylinders is random (i.e., uncorrelated), they are called rodlike micelles, but when close packed into a lattice, they form a hexagonal liquid crystal [11].

Bilayer sheets (Fig. 1c) can give many structures. When P is close to one, the bilayers are parallel to each other and form the lamellar liquid crystal [11]. If the packing parameter is less than one, these sheets can no longer remain parallel to each other and may fold back on themselves, trapping solvent in a bilayer container called a vesicle [11]. The bilayers may also maintain a curved shape and fill space by following an infinite periodic minimal surface, IPMS [12]. An IPMS is a geometry that is periodic in three directions and has

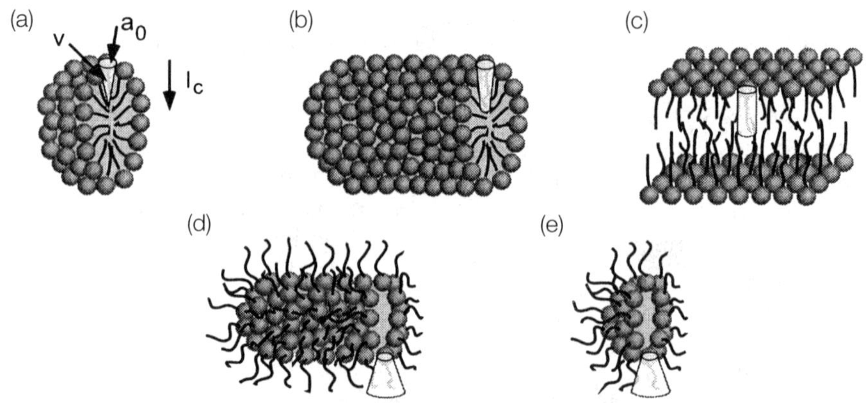

FIG. 1 Packing parameter $P = v/a_0 l_c$. (a) A spherical micelle ($P < 1/3$) with a geographical representation of the three values that make up P. (b) A cylindrical micelle ($1/3 < P < 1/2$). (c) A bilayer sheet ($1/2 < P < 2$). (d) A reverse cylindrical micelle ($2 < P < 3$). (e) A reverse spherical micelle ($P > 3$).

constant mean curvature. When the bilayer conforms to this geometry, both the water and oil fractions are continuous in three directions, and because the IPMS gives a cubic crystal pattern, these liquid crystals are known as bicontinuous cubic liquid crystals [12]. The packing of bicontinuous cubics as well as all liquid crystals is thermodynamically controlled, so the structures conform to a regular array of fixed size [13]. A surfactant phase diagram containing a bicontinuous cubic phase, as well as other phases, can be seen in Fig. 2. Once past the value $P = 1$, the geometries and ordered phase repeat in reverse order but with the oil and solvated surfactant tails becoming the continuous phase, i.e., reverse phases.

Along with the ordered liquid crystal structures there are random geometries that are also thermodynamically stable. These systems are called microemulsions. They are related to regular emulsions in that they are dispersions of two immiscible liquids stabilized by surfactants. Unlike emulsions, microemulsions are thermodynamically stable and optically transparent. The basic structure of microemulsions is a swollen micellar solution. When water is the solvent, the systems are known as oil-in-water (O/W) microemulsions. Similarly, if the oil is the continuous phase and the water is within a reverse micelle, the system is a water-in-oil (W/O) microemulsion. When the amounts of oil and water are nearly equal, a bicontinuous structure can be formed. These structures are the disordered analogues of the cubic and lamellar phases.

B. Properties

Some easy-to-measure properties can be used to identify the different structures surfactant systems form. The first of these is birefringence. If a system is anisotropic, it can display birefringence. Micellar solutions, including W/O and O/W microemulsions, have random orientations of the micelles in solution. Even if the micelles are rodlike, the overall effect is an isotropic solution with no birefringence. This also applies to vesicular solutions, which are isotropic. Cubic systems are also isotropic because the ordering is equivalent along all three axes. The last systems, hexagonal and lamellar, both show birefringence. This is due to the structure of these phases derived from the packing of rods and sheets. These structural units both have two axes with the same type of structure and a third axis with a different one. This leads to a different refractive index along this third axis, which leads to the birefringence of these samples.

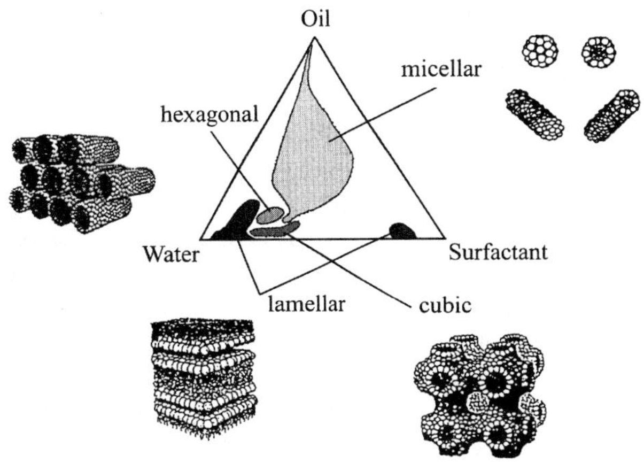

FIG. 2 Typical surfactant phase diagram containing micellar, hexagonal, lamellar, and bicontinuous cubic phases. (Adapted from Refs. 14 and 15.)

A second property of these systems that is useful for identifying the phase of the system is viscosity. Spherical micelle solutions are often not very viscous. When the micelles become less spherical, there is an increase in intermicelle interactions in the flow, which increases the viscosity. When the rods form a hexagonal phase, there is a dramatic increase in the viscosity. The highest viscosity is normally obtained with cubic phases. In terms of viscosity, bicontinuous cubics can range from hard brittle gels to flowable viscous liquids [16]. Although cubics have high viscosity, they are still liquid-like at the molecular level, giving rise to diffusion coefficients for water and oil species that are of the same order of magnitude as that of the bulk [17]. On the other hand, lamellar phases show lower viscosity than either cubic or hexagonal phases. This is due to the layer like order of the system, which has very little resistance to shear [11].

Along with these simple techniques, there are other more advanced methods to study surfactant systems. Small-angle x-ray scattering (SAXS) is used predominately to identify the liquid crystal structures. SAXS can give both the space group (i.e., the crystal structure) and the lattice parameter (i.e., the domain size) of a given liquid crystal [18]. The bicontinuous nature of cubics and microemulsions can be probed with a variety of techniques. The first is pulsed field gradient spin echo nuclear magnetic resonance (NMR), which can measure the diffusion coefficient of each component in the system [15]. If both the oil and water have diffusion coefficients similar to that of the bulk, the system is most likely bicontinuous, whereas a discrete cubic or microemulsion has one of the components confined, giving low diffusion coefficients for this species.

A simpler but less definitive method than the NMR technique to determine bicontinuity is conductivity measurements. A conductivity meter can monitor the mobility of the ions in the water-rich region of a system. At high water concentrations, an O/W microemulsion shows conductivity that is comparable to the conductivity of an electrolyte solution, indicating a bicontinuous structure [19]. The conductivity of the low-water-content W/O microemulsion, in which the water domains are discrete, is about 5 orders of magnitude lower than that of the bicontinuous system and about 6 to 10 orders of magnitude greater than that of the apolar solvent alone [19]. The limitation of the conductivity measurements is that they can not confirm whether an oil continuous system is present.

III. POLYMERIZATIONS

A. Micellar and Globular Microemulsions

First we will look at polymerizations in micelles and globular microemulsions. These two types of polymerization differ only on the size of the "reactor" volume. There is a continuous change from micelles to globular microemulsions as micelles increase in size due to solubilized monomer. Microemulsions allow polymers to be formed with high molecular weights at rapid reaction rates because the free radicals grow in isolation and therefore termination by radical combination is minimized, unlike the situation in solution polymerization. The main product of globular microemulsion polymerization is thermodynamically stable latex particles of 50 nm or smaller [20]. Novel products include copolymers of normally immiscible monomers [8,9]. Hydrophobic monomers are contained in the interior of the swollen micelle, and hydrophilic monomers are in the continuous phase. Because the two immiscible monomers are in close proximity, polymerization yields copolymers that range from random to block depending on the conditions.

There are many sources that give more information about polymerizations in globular microemulsions and micelles, including an excellent review by Candau [20]. Two key pieces of information are used to determine the mechanism of polymerization [20]. The first is that the final latex spheres are larger than the initial swollen micelles. Unswollen micelles are about 3 nm in size and become swollen to between 5 and 10 nm. Upon polymerization, the final particle size is between 25 and 50 nm. Along with this information is the fact that there are few polymer chains per latex particle, sometimes as few as one. There is continuous nucleation of particles, giving more particles as time increases [20]. This is different from emulsion polymerization, in which the number of particles does not increase from the initial value but the size and number of chains per particle do. From this information, a basic mechanism for the polymerization in globular microemulsions can be constructed.

The first stage of the mechanism is initiation from the continuous phase [20,21]. Because there are many more uninitiated monomer-swollen micelles in a reaction mixture than growing polymer particles, a radical has the greatest probability of being captured by an uninitiated monomer-swollen micelle. Stage II involves particle growth, which can occur by two methods. In the first, growing polymer receives more monomer by coalescence with neighboring micelles. The second involves the diffusion of monomer from uninitiated micelles through the continuous phase to the growing polymer chain. Because the instantaneous monomer concentration of a growing polymer particle is below the equilibrium concentration, the monomer diffuses from uninitiated micelles to the growing chains. Finally, at the end of the polymerization large polymer particles, stabilized by surfactant, and excess surfactant, in the form of unswollen micelles, remain. Figure 3 illustrates the polymerization mechanism.

It is easy to see that the surfactant plays an important role in the polymerization process. The first job of the surfactant is to isolate the monomers so that small particles can be formed. As the reaction occurs, the surfactant stabilizes the growing polymer chain but does little to keep the particles at their initial size. Although the particles do grow, the surfactant structure limits growth by creating a boundary the monomers must cross. These facts show that the template acts as a *directed* template because the final products do not follow the original surfactant structure on a one-to-one basis.

B. Vesicles

Vesicles are an important surfactant structure that is being investigated for use in drug delivery applications. Polymerization of vesicles is used to increase the life span of the vesicle by preventing spontaneous rupture and rearrangement. This increased lifetime allows the vesicle to deliver a drug to the desired area more effectively [22]. Most researchers attempt to polymerize the amphiphile [23], but polymerizing a normal mono-

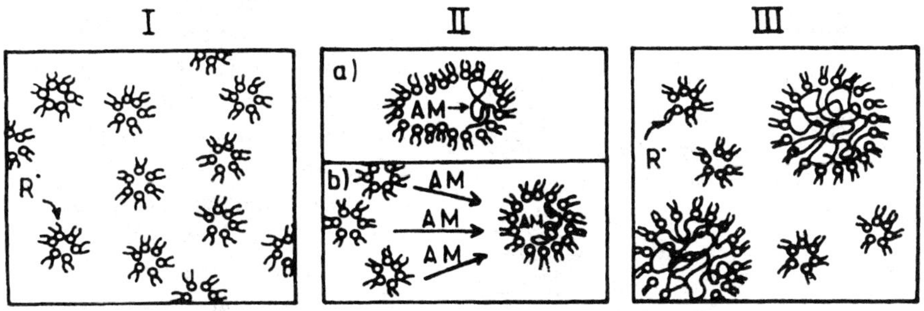

FIG. 3 Polymerization mechanism in reverse AOT globular microemulsions. (I) Before polymerization: AOT micelles ($d \approx$ 6 nm). (II) Polymer particle growth (a) by collisions between particles; (b) by monomer diffusion through the toluene phase. (III) End of polymerization: polymer particles ($d \approx$ 40 nm) plus small micelles ($d \approx$ 3 nm). (Reproduced from Ref. 21, Fig. 6, p. 84, with kind permission from Kluwer Academic Publishers.)

mer within the bilayer can also lead to the desired material. Also, by cross-linking the polymer in the bilayer, it is possible to create structures that are stable without the surfactant present [24,25].

Polymerization in vesicle bilayers should be straightforward, but there are many problems. For instance, polymerization within the bilayer can cause rearrangement of the surfactant structure. This is caused by either loss of entropy due to the collapse of polymer chains or incompatibility of the polymer with the surfactant [22]. The monomer may be soluble, but the decreased entropy of the polymer chain may cause phase separation. The desired final structure is a hollow polymer particle, but this is not the only possible structure. One possible geometry is a latex particle [26]. This geometry occurs when monomer from surrounding vesicles diffuses to an initiated vesicle, adding to its mass and causing the formation of a latex particle. This is similar to the mechanism by which latex particles are produced in microemulsions. To combat the formation of latex particles, one lowers the monomer concentration below saturation levels in both the continuous and vesicle phases, which reduces the driving force for diffusion and thus breaks the mechanism for particle growth.

Another geometry seen in vesicle polymerization is the parachute-like geometry, a vesicle-polymer hybrid seen by Jung and coworkers [22]. This geometry consists of a small polymer particle attached to a vesicle bilayer. There are two reasons for the origin of this geometry. One is that the polymerization is localized to a single spot on the vesicle surface. The monomer diffuses through the bilayer to the growing chain to form a particle within the bilayer. The second reason involves migration and aggregation of incompatible polymers inside the bilayer. As the polymer chain grows, it becomes incompatible with the surfactant membrane. The chain phase separates but is contained in the bilayer because it is hydrophobic. It continues to diffuse in the bilayer until it aggregates with other insoluble chains. If this second method is the mechanism for the formation of the parachute-like geometry, it may be possible to immobilize the polymers in the bilayer structure by cross-linking the chains to prevent migration. Holz and Meier [24] were able to use cross-linking to stabilize the polymerization to produce templated hollow spheres. If the polymerization is faster than diffusion, the network will form a continuous quasi-two-dimensional structure in the bilayer that cannot phase separate into a single particle because of the presence of the internal surfactant layer. Once the network is formed, the surfactant can be removed to give a hollow polymer particle, but one with smaller dimensions than the vesicle. The size change is due to the fact that the quasi-two-dimensional network is constrained by the internal surfactant layer. When this layer is removed, the network can contract into a fully dense shell structure.

In each of the three geometries, the surfactant template plays an important role. In the case of the latex particle formation, the vesicle acts similarly to a globular microemulsion, giving the same directed polymerization. With the parachute geometry, the vesicle acts as a container for the monomer, limiting the size of the final polymer particle. The hollow sphere geometry is the closest to a one-to-one template of a surfactant system. The only change upon removal of the surfactant is a size change. This could be eliminated by creating a more rigid polymer or by reducing tension in the polymer produced, which could be accomplished by changing monomers or by adjusting the cross-link density.

C. Hexagonal and Lamellar

Although templating the lamellar or hexagonal phase would produce novel materials, there are few examples in the literature [27]. Antonietti and coworkers [28,29] have done work in this area, but they had little success in templating these materials one-to-one. Because they also investigated bicontinuous cubics and microemulsions, this work is discussed in the next section. Laversanne [30] claimed to have templated both lamellar and hexagonal systems, but some researchers are not sure the data are sufficient to support the claim [28]. Because Lavarsanne investigated a bicontinuous cubic phase in the same work, this is also discussed in the next section.

One of the reasons that there is little information in this area is the low viscosity of these phases, which allows rearrangement of the structure due to thermal disruptions during polymerization [31]. Therefore it is more difficult to polymerize a one-to-one template of the structure and freeze the novel structure of the lamellar or hexagonal phases.

Another problem with these systems is that the interfaces are flexible and fluctuating [27]. The surfaces of these systems are flat only at sizes smaller than the persistence length of the film and can easily undergo deformations. As many monomers are surface active, polymerization causes curvature changes at the interface that can cause the system to phase separate into a new geometry.

A final problem that is important for lamellar phases is the crumpling of polymer sheets. It has been shown that a polymerized non-self-avoiding membrane undergoes a crumpling transition [32–34]. At low temperatures the membrane remains flat and its radius of gyration, R_g, is directly proportional to the membrane size, L. At higher temperatures, the membrane crumples with R_g proportional to $(\ln L)^{1/2}$. This theoretical membrane may be analogous to a polymer sheet in θ solvent because there is a balance of steric repulsion and van der Waals attraction that leads to zero excluded volume [35]. If a successful templating of a lamellar phase leads to polymer sheets in a θ solvent, the polymer sheets may not be observed because of the crumpling transition.

D. Bicontinuous Cubics and Microemulsions

Surfactant templates can also be used to produce porous gels by utilizing bicontinuous cubics and microemulsions. Because bicontinuous surfactant systems are continuous in both the surfactant and solvent phases, the structure can be frozen using monomers added to the solvent and removing the surfactant can form a pore network. The main benefit of using these bicontinuous structures is that they remove a problem inherent in typical porous gels: the coupling of the porosity to the mechanical properties.

In a nontemplated gel, the distance between cross-links in the polymer network determines the pore size of the gel. To make the pores larger, the distance must be increased either by adding more solvent or decreasing the amount of cross-linker. As the shear modulus is proportional to the cross-link density, there is an inverse relationship between pore size and mechanical strength. In a bicontinuous templated gel, the porosity and pore size are determined by the kind and amount of surfactant, while the cross-linker and monomer concentrations in the solvent determine the mechanical properties. This allows the cross-link density to be varied without changing the porosity, thus reducing the coupling between the pore size and the mechanical properties. Of course, increasing the porosity decreases the strength of the gel because of the absolute decrease in the amount of the gel, but this can be offset by increasing the modulus of the solidified continuous phase with increased cross-linker levels.

In a one-to-one template, the pore network would result from the three-dimensional continuous network that the surfactant formed. For a bicontinuous cubic system this gives a structure with periodic ordering that has monodisperse pore sizes, which are desirable in many applications. In the case of a bicontinuous microemulsion, the network would have less structure but still would be continuous. However, if the surfactant structure only "directs" the templating, then the pore network does not necessary follow the original structure. The pore network may still be continuous but it may have an altered structure: for a cubic phase polymerization, the periodic order and monodisperse pore sizes may be lost, and for a disordered microemulsion the size scale may be altered. It is therefore important to understand the mechanism behind each type of templating in bicontinuous systems.

Another difference between bicontinuous cubics and microemulsions is the viscosity of the systems. Bicontinuous cubic surfactant systems have a high viscosity because of their structure, which leads to long rearrangement times. Bicontinuous microemulsions have faster rearrangement kinetics and thus have a lower viscosity. This difference can be important in successful templating because long rearrangement times allow the system to polymerize before the template has a chance to change its arrangement; therefore, the type of bicontinuous system can affect the outcome of the templating.

Work on using bicontinuous cubic liquid crystals as templates for polymerization was begun by Anderson, who used the method to produce monodisperse microporous materials [31,36,37]. The cubic template was used because it was thought that it could produce materials that have three important features: monodisperse pores on the micrometer scale, highly interconnected pores, and high porosity. Also, the high viscosity of the cubic system was believed to hinder rearrangement of the structure during polymerization. Anderson reported producing cubic templated materials using a variety of monomers (styrene, methyl methacrylate, and acrylamide). The structures were verified via SAXS and transmission electron microscopy (TEM) and showed structures that were one-to-one templates of the cubic system, but later researchers have not successfully repeated these experiments.

Subsequent work in this area did not give as clear results, giving materials that did not follow the cubic template directly. In our own laboratory, cubic systems formed using Brij 78 as surfactant, toluene, and a monomer solution containing acrylamide were polymerized to give a porous polymer gel [38]. The starting materials were transparent but became opaque upon polymerization. After removing the surfactant and toluene and allowing the gel to swell to equilibrium in deionized water, the gel was again transparent. When the size of the pores was checked by determining whether a large macromolecule (blue dextran 2000, Pharmacia Fine Chemicals, $R_g \approx 40$ nm) would diffuse into the pore structure, it was found that the pores were larger than expected. This, combined with the gel becoming opaque during polymerization, led us to believe that the structure was disrupted during polymerization. By making a small change in the system by adding a second monomer (2-acrylamido-2-methyl-1-propanesulfonic acid) (AMPS) to the mixture, we were able to produce gels that did not become opaque during polymerization and had the correct pore size range as determined by the diffusion experiment [38]. This indicates the sensitivity of the interaction between the polymer gel chemistry and the surfactant mesophase chemistry in which the gel is confined.

Antonietti and coworkers noted the same problem when templating not only bicontinuous cubics and microemulsions but also hexagonal and lamellar systems [28,29,39–41]. They observed that the pore sizes of the templated materials were orders of magnitude larger than expected from the original template. These systems also became opaque during polymerization, suggesting that there was phase separation. Using SAXS and TEM, they were able to propose a mechanism for the changes in pore size (Fig. 4). The system begins as a homogeneous surfactant liquid crystal. As polymerization begins, polymer demixes from the liquid crystal because of the incompatibility between the growing polymer and the surfactant. The polymer begins to phase separate, with its domain growth controlled by the anisotropic transport properties in the liquid crystal phase. Finally, the polymer domains grow to form an interconnected gel structure. The surfactant-rich region is embedded in the polymer gel structure and can be continuous. Antonietti showed that the surfactant-rich phase can still maintain a liquid crystal ordering and will give comparable SAXS powder patterns, although the *d* spacing in the liquid crystal decreases upon polymerization. As the polymer-rich phase and the surfactant-rich phase both compete for solvent, the surfactant-rich phase becomes more concentrated, decreasing the *d* spacing of the liquid crystal. Therefore, the ex-

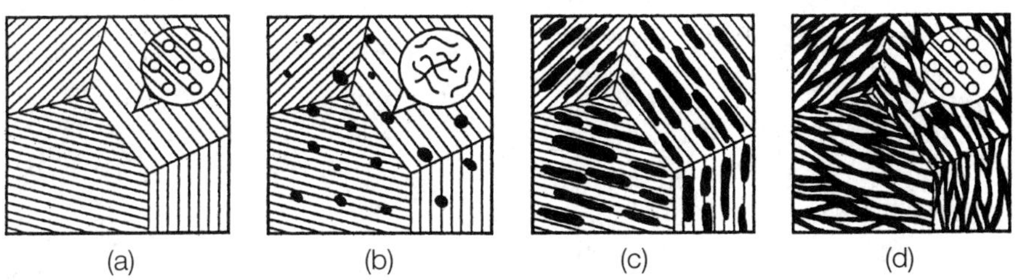

FIG. 4 Scheme of the proposed reaction scenario. The reaction starts from a homogeneous lyotropic liquid crystalline phase (a). The polymer, once formed, demixes from the continuous lyotropic phase (b, polymer phase drawn in black). The growth of the polymer domains is controlled by the anisotropic transport properties of the liquid crystalline phase (c). An increase in the amount of polymer leads to the interconnected gel structure (d), embedding the liquid crystalline phase with the majority of the surfactant. This subphase can be continuous, too. (Reproduced with permission from *Langmuir* 1998, *14*, 2670–2676. Copyright 1998 American Chemical Society.)

istence of a liquid crystal after polymerization, verified by SAXS, cannot be the lone determinant of whether templating is one-to-one.

Work by Laversanne on bicontinuous cubic, lamellar, and hexagonal phases indicates that it is possible to produce templated materials from these phases [30]. Using acrylamide as the monomer, he was able to form gels in each phase that were slightly turbid upon polymerization. Laversanne's explanation was that the turbidity was due to scattering at grain boundaries, because data from SAXS and optical microscopy indicated that there was one-to-one templating and no phase separation. Antonietti et al. [28] believed that there was not enough evidence to determine whether there was one-to-one templating in this case. The SAXS data show that there were changes in the *d* spacing after polymerization in all three templated systems. This indicates that phase separation could be occurring. Coupling this with the fact that the samples became slightly turbid upon polymerization gives a stronger case for phase separation. Although optical microscopy did not show any apparent phase separation, the possibility cannot be ruled out without the use of other measurement techniques.

Buban and coworkers [42] also had problems with rearrangement and phase separation when trying to template bicontinuous microemulsions. They wanted materials with high surface areas and thus used surface area measurements to characterize their materials. Systems with hydrophobic monomers (methyl methacrylate, styrene, and butyl methacrylate) became opaque upon polymerization and gave surface areas that were as high as 70 m^2/g. They expected values closer to 750 m^2/g based on the microemulsion composition and the surfactant interfacial area. Also, measurements of the pore size distribution using nitrogen sorption and mercury porosimetry gave pore radii of 250 Å, which were an order of magnitude larger than the size based on the surfactant geometry. These two measurements indicate some type of rearrangement, which Burban tried to define by using SAXS measurements. The SAXS measurements showed that there was a peak corresponding to the microemulsoin characteristic size both before and after the polymerization. During polymerizaion, a peak begins to form that corresponds to the 250 Å peak found in the nitrogen sorption experiments. On the basis of these data, Burban believed that the structure was similar to that of kernels of popcorn struck together. Each kernel kept the microemulsion structure, but the larger pores came from the packing of the kernels.

One problem with Burban's explanation is that it does not account for the loss of surface area seen in the templated microemulsion. One possible explanation is that the structure was being changed after the extraction of the surfactant, as there was no longer a microemulsion peak in the SAXS patterns after extraction. A better explanation comes from Antonietti et al. [28], who claim that this is an instance of phase separation. During polymerization the system separates into microemulsion-rich regions in a polymer matrix. The postpolymerization samples still show the microemulsion SAXS peak because the microemulsion is still there, but because there is no polymer templated there the structure disappears upon extraction, giving a low-surface-area material.

In the hydrophilic monomer (acrylic acid and methacrylic acid) systems that Burban studied, the systems did not become opaque during the polymerization step [42]. However, these materials all gave low surface areas compared with theoretical values, which they believed was due to the material collapsing upon drying.

Although Antonietti and Burban did not produce materials that were one-to-one templates of the original surfactant mesophases, they were able to produce porous materials. These materials do not have the morphology of the original templates, but their structures may be useful in some application. Anderson's work and work done in our own laboratory indicate that one-to-one templating of the bicontinuous cubic phase is possible. One question is, how does one design a system in order to preserve the one-to-one templating? This question is answered in the next section.

IV. CONTROLLING TEMPLATE BREAKTHROUGH

Template breakthrough is due to the thermodynamic influence of the growing polymer chain on the surfactant template [28], so understanding the thermodynamics of confined polymer chains should help in finding solutions for breakthrough. First calculated by Casassa and coworkers [43–45] and later presented by de Gennes [35], the energy, ΔG, to confine a polymer to a given tube of diameter D is

$$\Delta G = kT \frac{R_0^2}{D^2} \tag{1}$$

where R_0 is the average square end-to-end distance of a polymer chain. The equation is valid for values of R_0 greater than D. De Gennes [35] also showed that

$$R_0 \propto M^{3/5} \tag{2}$$

where M is molecular weight. Therefore, a polymerization proceeds (i.e., R_0 becomes larger), the energy needed to confine the chain in the liquid crystal geometry increases.

This leads to the first adjustment that can be made to prevent breakthrough, changing the size of the polymer produced. Decreasing the molecular weight of the polymer decreases R_0 and thus the energy needed to confine the polymer. Molecular weight can be represented by $\bar{x}_n$, the number average degree of polymerization, which is a measurement of a polymer chain based on length rather than weight of the polymer. In free radical polymerization, $\bar{x}_n$ is given by

$$\frac{1}{\bar{x}_n} = \frac{k_t}{k_p^2} \frac{v_p}{[M]^2} + C_M + C_S \frac{[S]}{[M]} + C_I \frac{[I]}{[M]} \quad (3)$$

where k_t and k_p are rate constants; C_M, C_S, and C_I are ratios of chain transfer rate to propagation rate for monomer, chain transfer agent, and initiator, respectively; v_p is the rate of polymer growth; and [M], [S], and [I] are concentrations of monomer, chain transfer agent, and initiator, respectively [46]. Many surfactant systems are sensitive to temperature, so the rate constants cannot be adjusted easily. Therefore the only way to change $\bar{x}_n$ is to change the concentrations of the reacting species.

Because the concentration of monomer is set by the desired mechanical properties of the final product, either increasing the initiator or increasing the amount of chain transfer agents decreases $\bar{x}_n$. One caveat is that decreasing the molecular weight will affect the gelation of the polymer in systems in which a polymer gel is produced. Odian [47] gives the critical extent of reaction, p_c, for the onset of gelation as

$$p_c = \frac{[A] + [B]}{[B]\bar{x}_w} \quad (4)$$

where [A] is the concentration of monomer, [B] is the concentration of cross-linker, and $\bar{x}_w$ is the weight average degree of polymerization. On decreasing $\bar{x}_w$, p_c becomes larger, so a higher conversion is needed before the gel is formed. The value of p_c should be kept low so that the gel can form before rearrangement can occur. This means that the concentration of cross-linker must be increased.

The next aspect to look at is the surfactant system, because the confinement energy is due to the surfactant structure. Therefore, breakthrough can be controlled through surfactant system changes. The first surfactant system adjustment to keep breakthrough from occurring is to design the surfactant mesophase to have very slow rearrangement dynamics. This will be evidenced as a high solution viscosity. Low-viscosity systems can rearrange on the time scale of polymerization [31]; therefore stress due to growing polymer chains will easily cause breakthrough before the polymer reaches the gel point. Anderson and Ström [31] believed that using bicontinuous cubic phases, which have the highest viscosity of any surfactant mesophases, was key to producing a one-to-one template. For cubic surfactant solutions equilibration times are long, sometimes on the order of weeks [31].

Another way to combat breakthrough is to change surfactants. If surfactants are used that have slower exchange dynamics, they may be able to resist breakthrough. One type of surfactant is the counterion-coupled gemini surfactant or "cocogem." Cocogems are normal ionic surfactants that are coupled to a polyfunctional organic counterion to form a multitailed surfactant. Although these systems produce more stable surfactant mesophases, they were not able to produce one-to-one templates for Antonietti et al. [28,39,41]. The use of longer surfactants also kinetically stabilizes the surfactant systems. This can also lead to longer characteristic dimensions in the surfactant structure, which leads to larger values of D and thus a lowering of the energy needed for confinement.

A final method proposed by Laversanne [30] deals with preventing phase changes. The phase diagram for surfactant mesophases is generally quite sensitive to composition. Adding monomer to either the aqueous or the hydrophobic phase changes the polarity of the fluid phase and packing parameter of the surfactant and, therefore, changes the phase boundary, as can be seen in Fig. 5. As the polymerization of a system proceeds,

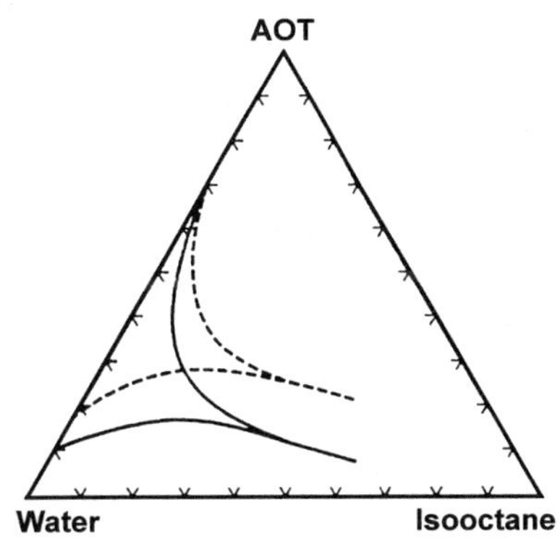

FIG. 5 Schematic phase diagram of the ternary system water/AOT/dodecane (solid line) showing the existence domain of the lamellar phase, superposed on the diagram for water/acrylamide/AOT/isooctane. The existence domain of the lamellar phase of the quaternary system water/acrylamide/AOT/isooctane (with an acrylamide/water ratio of 0.1) is indicated by a dotted line. (Reproduced with permission from *Macromolecules* 1992, *25*, 489–491. Copyright 1992 American Chemical Society.)

the loss of monomer can cause changes in the phase of the surfactant system, thus redefining the template before the system has been completely templated. To prevent this, choices of monomer and surfactant should be made so that the phase boundaries do not change upon the addition of monomer. Areas of overlap between the phase boundaries with and without monomer are compositions where changes in polarity during polymerization will not be the cause of template breakthrough.

It may be possible to extend this method to include stability toward the polymer as well, such that addition of polymer to the system does not change the phase of the system. By finding out how the surfactant mesophase changes with the addition of small and large polymers at various concentrations, the stability of the surfactant system during polymerization could be predicted.

V. CONCLUSIONS

Polymerizations templated by surfactant mesophases can produce interesting mesoporous materials for various applications. Both inorganic and organic monomers can be polymerized using this method, although the inorganic field is more advanced than the organic field because of its earlier start. Our emphasis has been on *directed* versus *one-to-one* templating. Although both types of templating can produce useful materials, only one-to-one templating accurately reproduces the exact structure of the original surfactant mesophase. Successful templating (i.e., one-to-one templating) depends on both thermodynamics and kinetics. Thermodynamic issues include the energy needed to confine the polymer in a confined space. This energy, generated from the entropy of the polymer, requires cross-linking of the polymer in order to trap the structure before the confinement of the surfactant template is lost. Another important thermodynamic issue is the solvent polarity change when monomer is added to the system, which affects the phase boundaries of the surfactant mesophase. The kinetics of rearrangement of the surfactant structure versus the polymerization time are another important issue that must be considered. If the polymerization is slow compared with the surfactant rearrangement, the growing polymer will disrupt the template before a gel is formed. This condition favors the use of bicontinuous cubic phases because of their long rearrangement times compared with other surfactant mesophases. As researchers understand these systems and their thermodynamic and kinetic issues better, we should see an increase in successful examples of templating organic materials in surfactant mesophases.

REFERENCES

1. J. S. Beck, J. C. Vartuli, W. J. Roth, M. E. Leonowicz, C. T. Kresge, K. D. Schmitt, C. T. W. Chu, D. H. Olson, E. W. Sheppard, S. B. McCullen, J. B. Higgins, and J. L. Schlenker, J. Am. Chem. Soc. *114*:10834–10843 (1992).
2. J. S. Beck, J. C. Vartuli, G. J. Kennedy, C. T. Kresge, W. J. Roth, and S. E. Schramm, Chem. Mater. *6*:1816–1821 (1994).
3. J. S. Beck and J. C. Vartuli, Curr. Opin. Solid State Mater. Sci. *1*:76–87 (1996).
4. C. T. Kresge, M. E. Leonowicz, W. J. Roth, J. C. Vartuli, and J. S. Beck, Nature *359*:710–712 (1992).
5. J. C. Vartuli, K. D. Schmitt, C. T. Kresge, W. J. Roth, M. E. Leonowicz, S. B. McCullen, S. D. Hellring, J. S. Beck, J. L. Schlenker, D. H. Olson, and E. W. Sheppard, Chem. Mater. *6*:2317–2326 (1994).
6. J. C. Vartuli, C. T. Kresge, W. J. Roth, K. D. Schmitt, S. B. McCullen, J. S. Beck, M. E. Leonowicz, J. D. Lutner, and E. W. Sheppard, Abstr. Pap. Am. Chem. Soc. *209*:27-PETR (1995).
7. K. M. McGrath, D. M. Dabbs, N. Yao, I. A. Aksay, and S. M. Gruner, Science *277*:552–556 (1997).
8. I. Lacik, J. Selb, and F. Candau, Polymer *36*:3197–3211 (1995).
9. H. Yu and D. W. Grainger, Macromolecules *27*:4554–4560 (1994).
10. M. Miyata, in *Polymerization in Organized Media* (C. M. Paleos, ed.), Gordon & Breach, Philadelphia, 1992, pp. 327–367.
11. M. R. Porter. *Handbook of Surfactants*, 2nd ed., Blackie Academic & Professional, New York, 1994, p. 324.
12. K. Fontell, Adv. Colloid Interface Sci. *41*:127–147 (1992).
13. V. Percec, H. Jonsson, and D. Tomazos, in *Polymerization in Organized Media* (C. M. Paleos, ed.), Gordon & Breach, New York, 1992, pp. 1–104.
14. H. T. Davis, J. F. Bodet, L. E. Scriven, and W. G. Miller, in *Physics of Amphiphilic Layers* (J. Meunier, D. Langevin, and N. Boccara, eds.), Springer-Verlag, Berlin, 1987, pp. 310–327.
15. K. Fontell and M. Jansson, Prog. Colloid Polym. Sci. *76*:169–175 (1988).
16. K. Fontell, Colloid Polym. Sci. *268*:264–285 (1990).
17. P. Ström and D. M. Anderson, Langmuir *8*:691–709 (1992).
18. R. J. Young and P. A. Lovell, *Introduction to Polymers*, 2nd ed., Chapman & Hall, New York, 1991, p. 443.
19. M. Giustini, G. Palazzo, G. Colafemmina, M. D. Monica, and A. Ceglie, J. Phys. Chem. *100*:3190–3198 (1996).

20. F. Candau, in *Polymerization in Organized Media* (C. M. Paleos, ed.), Gordon & Breach, Philadelphia, 1992, pp. 215–282.
21. F. Candau, in NATO ASI Series C No. 303 (F. Candau and R. H. Ottwill, eds.), Kluwer Academic Publishers, Boston, 1990, pp. 73–96.
22. M. Jung, D. H. W. Hubert, P. H. H. Bomans, P. M. Frederik, J. Meuldijk, A. M. van Herk, H. Fischer, and A. L. German, Langmuir *13*:6877–6880 (1997).
23. C. M. Paleos, in *Polymerization in Organized Media* (C. M. Paleos, ed.), Gordon & Breach, Philadelphia, 1992, pp. 283–326.
24. J. Hotz and W. Meier, Langmuir *14*:1031–1036 (1998).
25. J. D. Morgan, C. A. Johnson, and E. W. Kaler, Langmuir *13*:6447–6451 (1997).
26. N. Poulain, E. Nakache, A. Pina, and G. Levesque, J. Polym. Sci. Polym. Chem. *34*:729–737 (1996).
27. C. Holtzscherer, J. C. Wittmann, D. Guillon, and F. Candau, Polymer *31*:1978–1985 (1990).
28. M. Antonietti, C. Goltner, and H. P. Hentze, Langmuir *14*:2670–2676 (1998).
29. M. Antonietti, R. A. Caruso, C. G. Goltner, and M. C. Weissenberger, Macromolecules *32*:1383–1389 (1999).
30. R. Laversanne, Macromolecules *25*:489–491 (1992).
31. D. M. Anderson and P. Ström, ACS Symp. Ser. *384*: 204–224 (1989).
32. L. Peliti and S. Leibler, Phys. Rev. Lett. *54*:1690–1693 (1985).
33. Y. Kantor and D. R. Nelson, Phys. Rev. Lett. *58*:2774–2777 (1987).
34. E. Guitter, F. David, S. Leibler, and L. Peliti, Phys. Rev. Lett. *61*:2949–2952 (1988).
35. P.-G. de Gennes, *Scaling Concepts in Polymer Physics*, Cornell University Press, Ithaca, NY, 1979, p. 325.
36. D. M. Anderson, U.S. Patent 5238613 (1993).
37. D. M. Anderson, U.S. Patent 5244799 (1993).
38. J. Hittle, Synthesis and characterization of porous gels via polymerization of the surfactant cubic phase, Senior thesis, Princeton University, Princeton, NJ, 1996.
39. M. Antonietti and H. P. Hentz, Adv. Mater. *8*:840–844 (1996).
40. M. Antonietti and H. P. Hentze, Colloid Polym. Sci. *274*:696–702 (1996).
41. H. P. Hentze, C. G. Goltner, and M. Antonietti, Ber. Bunsenges. Phys. Chem. Chem. Phys. *101*:1699–1702 (1997).
42. J. H. Burban, M. T. He, and E. L. Cussler, AIChE J *41*: 907–914 (1995).
43. E. F. Casassa, J. Polym. Sci. B *5*:773–778 (1967).
44. E. F. Casassa and Y. Tagami, Macromolecules *2*:14–26 (1969).
45. E. F. Casassa, Macromolecules *9*:182–185 (1976).
46. F. W. J. Billmeyer, *Textbook of Polymer Science*, 3rd ed., John Wiley & Sons, New York, 1984, p. 578.
47. G. Odian, *Principles of Polymerization*, McGraw-Hill, New York, 1970, p. 652.

27

Admicellar Polymerization

JOHN O'HAVER University of Mississippi, University, Mississippi

BRIAN GRADY University of Oklahoma, Norman, Oklahoma

JEFFREY H. HARWELL and EDGAR A. O'REAR Institute of Applied Surfactant Research, and University of Oklahoma, Norman, Oklahoma

I. INTRODUCTION

The phenomena of surfactant micelles, solubilization of molecules in surfactant micelles, and subsequent reactions in these surfactant micelles have been known and studied for more than 50 years. The fact that surfactants adsorb to solid surfaces has also been known for a long time. These separate concepts came together in the mid-1980s when the nature of adsorbed surfactant aggregates (admicelles) became better understood, particularly the similarities between admicelles and micelles [1]. Several studies clearly indicated that admicelles had the ability to solubilize other compounds (adsolubilization), just as solubilization occurs in micelles. This background led to examining the possibility of polymerizations occurring in these adsorbed aggregates, just as polymerizations take place in micelles. Admicellar polymerization was the term assigned to describe this new process.

Research concerning surfactant adsorption, admicelles, adsolubilization, and reactions within admicelles has progressed steadily since that point. The work has expanded to include a wide variety of substrates, surfactants, adsolubilizates, reactions, and reagents. Admicellar polymerization fundamentals and possible applications of the technology have been and continue to be examined. Admicellar polymerization has opened up an entirely new area of chemistry, with much yet to be learned.

II. ADMICELLAR POLYMERIZATION: THE PROCESS

As shown in Fig. 1, admicellar polymerization is typically a four-step process, with some of the steps being performed concurrently for certain applications. The solvent used for this process is water, although other solvents could possibly be used. Step 1 consists of surfactant adsorption to form a bilayer on a solid surface. Several factors must be considered in this step, including the nature of the substrate, the polymerization scheme to be utilized, the monomers used, and the type of surfactant. Most substrates can acquire a surface charge in aqueous solution, a charge that is typically pH dependent. The pH value at which the surface charge is completely neutralized is called the point of zero charge, or PZC. At a pH above the PZC, the surface of the substrate is negatively charged; below the PZC, the surface is positively charged. Silicon dioxide has a PZC of approximately 2.9, titanium dioxide a PZC of 5.8, and aluminum oxide a PZC of 9.1 [2]. Thus, at a pH of 5.8, the SiO_2 surface would have a negative charge, the TiO_2 surface would have no net charge, and the Al_2O_3 surface would have a positive charge. Adsorption by ionic surfactants is achieved by adjusting the surface charge on the substrate to be opposite that of the surfactant headgroup.

Many other factors must be considered in choosing a particular surfactant for this process. The substrate

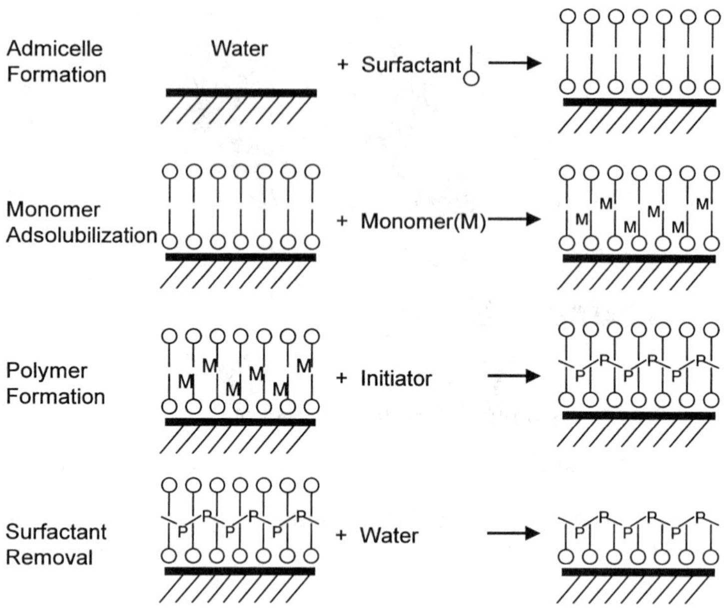

FIG. 1 Schematic of the steps in the admicellar polymerization process.

cannot be soluble in water at the pH necessary for surfactant adsorption. For example, amorphous silica is fairly insoluble in water at pH values ranging from 5 to 8, but as the pH increases above 8, the solubility increases significantly and may interfere with the process. Another factor is that the reagents necessary for the polymerization must not react with or precipitate the surfactant. In addition, the surfactant must remain adsorbed under reaction conditions, i.e., increased temperature. The type of monomer or monomers chosen depends on the final application and may also influence the choice of surfactant. Surfactants that are chemically similar to the monomer to be used typically increase the amount of adsolubilization. The final consideration is the type of surfactant to be used. Nonionic surfactants rather than ionic surfactants may be chosen to prevent interference with the components of the polymerization reaction and/or to adsorb to very nonpolar surfaces. Because of the wide number of surfactants currently available and the number of criteria that must be fulfilled, choosing an appropriate surfactant for an admicellar polymerization can be a lengthy process.

Figure 2 represents a schematic of an adsorption isotherm, which is a plot of surfactant adsorbed versus equilibrium surfactant concentration on a log-log scale. When surfactant is added to a water solution containing a solid surface, the surfactant can migrate to one of four environments: to the bulk solution, to the solid surface, to a micelle, or to the vapor-liquid interface. In systems under consideration in this chapter, the amount of surfactant at the vapor-liquid interface is exceedingly small and is neglected. In a typical admicellar polymerization, the bulk concentration must be below the critical micelle concentration (cmc) because emulsion polymerization would significantly degrade performance and introduce significant complications. The plateau at high surfactant adsorption levels can indicate the onset of the cmc, but this plateau can also indicate that maximum bilayer coverage was reached. Differentiating between reaching the cmc and reaching maximum bilayer coverage is straightforward because one typically assumes that the cmc changes very little with the addition of the solid substrate.

Surfactant adsorption can be a fairly rapid process, with 90% occurring within 2 h for some surfaces. How-

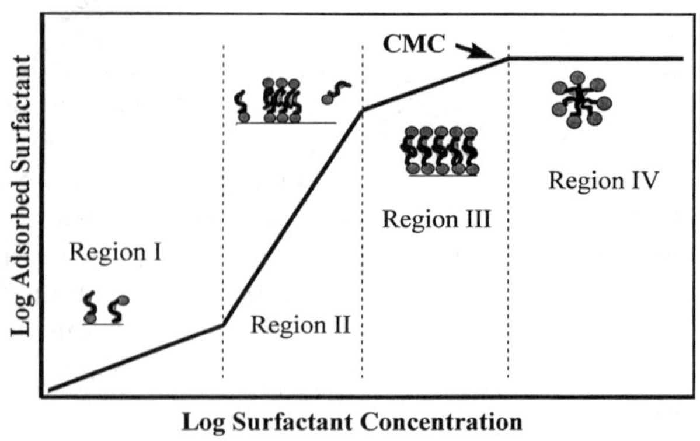

FIG. 2 Schematic of an adsorption isotherm.

ever, on very porous substrates plateau adsorption may take as long as 12 to 24 h. The amount of adsorbed surfactant corresponding to the maximum amount of bilayer coverage scales roughly with the surface area of the substrate; typical values are on the order of 50 μmol/g for nonporous glass fibers to 10 times this value for porous silica or alumina. One significant unanswered question is the effect of incomplete surfactant adsorption on the admicellar polymerization process. Even in the case where the cmc does not interfere with surfactant adsorption, maximum bilayer coverage does not necessarily imply that the entire surface is covered by a bilayer. However, maximum bilayer coverage is not needed to have significant polymerization and, in fact, is probably not desirable for some applications.

Step 2 in the process is adsolubilization or the solubilization of the monomer(s) into the adsorbed surfactant aggregates. This step may be done after the first step or simultaneously with it. Small molecules, such as monomer molecules, can shift the adsorption isotherm [3], but the effect is typically small and is usually neglected. Compatibility between the monomer(s) and the surfactant is necessary in order to maximize the amount of adsolubilization. Another possibly important factor is the solubility of the monomer in water; however, monomer concentrations are typically so low that solution polymerization does not seem to occur even when non-water-soluble initiators are used. The molar concentration ratio of monomer to surfactant is of the order of 1:1 in typical admicellar polymerizations, which contrast significantly with the 100:1 or higher ratio in typical emulsion polymerizations.

Step 3 is the polymerization of the monomer. Very little is known about the way in which the reaction is initiated or the properties of the formed polymer. What is known is discussed in the next section.

Step 4 involves surfactant removal, usually by repeated or continuous contact of the modified solids with an excess of water. As Fig. 1 implies, removing all of the surfactant is usually difficult, especially on porous substrates. Mass balances on porous systems suggest that even after vigorous washing, fully half of the surfactant remains on the modified particle.

III. FUNDAMENTALS OF ADMICELLAR POLYMERIZATION

Three patents were filed on the admicellar polymerization process in the mid to late 1980s and early 1990s [4–6]. As these dates imply, this process was invented in the mid-1980s and hence was only approximately 15 years old at the time this chapter was written. This chapter describes almost completely what is known about this process, and the reader should realize that the unknown far outweighs the known. This lack of knowledge is driven by two facts: (1) it is difficult to produce enough polymer for detailed studies and (2) it is difficult (and in the case of porous substrates, practically impossible) to extract the polymer entirely without degrading the polymer. This latter characteristic is partly responsible for the good adhesion that these modified substrates can have to a polymer in a polymer-matrix composite.

Wu et al. published the first two papers [3,7] examining the admicellar process in detail. Styrene was polymerized in admicelles of sodium dodecyl sulfate (SDS) on Al_2O_3 using a water-soluble persulfate initiator. Wu observed that SDS adsorbs well onto a 100 m^2/g Al_2O_3, with a maximum adsorption of almost 800 μmol/g in the presence of ethanol and styrene. The SDS bilayer was able to adsolubilize approximately one styrene molecule per two surfactant molecules. After polymerization, the modified alumina was observed to be very hydrophobic, indicating a fundamental change in the nature of the surface of the alumina. This behavior was strikingly exhibited in a series of photographs [8] showing that the unmodified powder wet easily in water whereas the modified powder did not, even under vigorous stirring. This change in wetting behavior is regarded as one of the fundamental characteristics of admicellar polymerization.

The kinetics of admicellar polymerization were the subject of the second of these two papers. The authors measured the styrene concentration of the supernatant and the polymer concentration on the surface. To determine the latter, the alumina powder was ground and then extracted with tetrahydrofuran. It is not clear how the authors verified that the polymer was completely removed because no details are given. No attempt was made to measure the molecular weight of the polymer, which did not allow a determination of detailed termination kinetics. The authors determined that mass transfer of the supernatant monomer into the admicelle was rate limiting, and hence an adjustable parameter was added in fits of a Smith-Ewart type of model to experimental data. This effort also indicated that termination of radicals inside the admicelles was dominating. Finally, the propagation rate constant calculated via model fits to the data agreed reasonably well with rate constants calculated from emulsion polymerization.

Other specific aspects of admicellar polymerization have been studied over the past 10 years. One area is the relationship between the surfactant, substrate, and monomer. The simplified schematic presented in Fig. 1 is misleading because monomer is drawn from the solution after polymerization begins. Simple calculations show that in many cases, the majority of repeat units in the polymer structure are from monomers not initially adsolubilized in the admicelle. A thorough study of how much of the monomer comes into the admicelle after the initiator has been added has not been made.

Because of the importance of this issue to other processes, considerably more is known about how the initial amount of adsolubilized monomer changes with surface identity and surfactant type. The amount of monomer adsorbed per surfactant molecule is not just a function of the surfactant-monomer pair. Chen [9] utilized SDS and styrene to modify the surface of TiO_2 and showed that the maximum adsorption amount was approximately one styrene molecule per surfactant molecule rather than the 0.5:1 ratio for alumina. Lai et al. [10] were the first to examine a nonhydrocarbon system as well as among the first to examine adsolubilization with a gaseous monomer. They examined the adsorption of sodium perfluoroheptanoate surfactants onto alumina and the adsolubilization of tetrafluoroethylene (TFE) monomer into the admicelles. Monomer adsolubilization depended on pressure, but the maximum was nearly five monomers per surfactant molecules. As Lai et al. noted, these results are questionable because gaseous monomers can condense in the pores of the alumina. However, capillary condensation probably did not occur in this system because of the strong repulsive interactions between the alumina and the TFE monomer and the fact that such behavior tends to increase the amount of monomer apparently adsorbed by orders of magnitude. Table 1 shows that the maximum monomer/surfactant ratio varies significantly for different monomers in cetyltrimethlyammonium bromide (CTAB) admicelles. The effect of partial surfactant adsorption on admicellar polymerization has not been studied to our knowledge; the studies just discussed had adsorbed surfactant levels at or near plateau adsorption.

Table 1 shows another interesting effect, enhancement of adsolubilization of one monomer through the addition of a second monomer. Little enhancement occurs in styrene adsorption through the addition of either butadiene or isoprene, but adsorption of the latter monomers increases dramatically with styrene addition. This type of enhancement could be due to some specific order of the adsolubilized monomers within the bilayer, or the monomers could induce some fundamental change in the bilayer. In either case, the reactivity ratios of the comonomers might be significantly different from those found for other types of polymerizations. A study of copolymer architecture is definitely needed.

A unique characteristic of this technique is the wide variety of substrates that can be used for admicellar polymerization. O'Haver and coworkers [11] performed admicellar polymerization on the surfaces of precipitated and fumed silicas as well as unpublished work on several other substrates including clays and powdered metals. Other types of surfaces include nickel flake [12] and glass fibers [13]. Apparently, the only significant issue to be resolved for a particular surface is the question of what surfactant to use. The surfactant type does not seem to affect the reaction, although obviously the amount of surfactant adsorbed does, and the amount of surfactant adsorption can be

TABLE 1 Maximum Adsolubilization of Various Monomers on Two Silica Substrates Showing Enhanced Absorption of One Component after Addition of a Second Component

Monomer	Surfactant	Substrate BET surface area (m^2/g silica)	Maximum adsolubilization (μmol/g)	Monomer-to-surfactant ratio
Styrene	CTAB	141	500	1.85:1
Styrene	CTAB	170	850	1.6:1
Isoprene	CTAB	170	550	1.0:1
Styrene with isoprene	CTAB	170	850	1.6:1
Isoprene with styrene	CTAB	170	1750	3.2:1
Styrene with 1,3-butadiene	CTAB	141	500	1.9:1
1,3-Butadiene	CTAB	141	425	1.6:1
1,3-Butadiene with styrene	CTAB	141	1575	5.8:1

The surfactant concentration was held constant for a given silica. The constant concentration was approximately 20% removed from the plateau concentration in both cases.

significantly different even within a homologous series. Polymerizable surfactants can even be incorporated in the polymer [14].

The most obvious effect of admicellar polymerization on the properties of a surface is the change in wetting properties mentioned previously. A highly hydrophilic surface can become highly hydrophobic after admicellar polymerization if a hydrophobic monomer is used. The effect of admicellar polymerization on the surface properties of porous precipitated silicas with surface areas between 100 and 200 m^2/g has been quantitatively established by O'Haver et al. [11,15]. Pore properties were measured before and after modification, and nearly all samples exhibited the following trends. First, a decrease in the BET N_2 and mercury porosimetry surface areas was found, probably due to the polymer and remaining surfactant blocking the entrance to smaller pores. Second, there was a slight increase in mean aggregate diameter. Whether this was due to polymer chains forming bridges between aggregate particles or just changes related to processing is unclear. Third, there was a slight increase in the average pore diameter, probably caused by the blockage of the smallest pores. However, similar measurements by Wu [8] on modified alumina powder indicated a decrease rather than an increase in pore size.

The continuity, thickness, and morphology of the film have been studied as well. In the study by Wu et al. [3], the thickness of a polystyrene film on glass slides was measured as approximately 3.5 nm at short reaction times and 13 nm at long reaction times. The authors noted that the films seemed more or less continuous, although this claim is difficult to justify because ellipsometry is not an entirely appropriate technique to evaluate this property. The contact angle of poly(tetrafluoroethylene) (PTFE) admicellar films was almost that of a pure PTFE sheet, indicating that the coverage was continuous, at least over the length scale that wetting studies probe. However, scanning tunneling microscopy studies of nickel flakes coated with polypyrrole clearly indicate the formation of patchy polypyrrole surfaces, which is confirmed by the observation that the bulk conductivity of the packed pure flake does not change [12]. The continuity of the film is almost certainly a function of both surface chemistry and topology.

The most extensive studies carried out to date on admicellar film properties have been performed in the laboratories of two of the coauthors using atomic force microscopy (AFM). Mica was used for this study because the effects of surface roughness could be minimized. In addition, the film was lifted off the surface and folded back upon itself so that film thickness and nucleation morphology could be studied. A schematic of the observed morphology at low film thicknesses is shown in Fig. 3. The films seem to begin growing at a point, because AFM images indicate ellipsoidally shaped particles. These studies indicate that the individual elliposids impinge on each other very quickly when the film thickness reaches only a few nanometers, so an AFM image at this point has almost flat regions with periodic valleys. At much higher film thicknesses, this morphology is not evident and the film is smooth.

The properties of the admicellar polymer, such as its molecular weight, branching, defects (i.e., head-to-head vs. head-to-tail), and tacticity, are essentially unknown. The molecular weight of the polystyrene that could be extracted from silica via hot THF has been measured [11]. Generally, the molecular weights were quite low, with M_n and M_w approximately 3000 g/mol and 30,000 g/mol, respectively. The authors clearly noted that the extraction was not complete and almost certainly significantly underrepresented the actual molecular weight of the polymer.

All monomers tested have been able to undergo reaction using admicellar polymerization. Even polypropylene, which cannot be formed in an emulsion, can undergo reaction in an admicelle [16]. One unpublished study suggested that condensation polymers can be synthesized in this manner, but the inability to extract polymer in a quantitative fashion makes this assertion questionable. Finally, monomers that react via a charge transfer mechanism, e.g., pyrrole, can be polymerized using this technique [17]. Unlike hydrophobic adsolubilizates, pyrrole shifts the adsorption isotherm toward higher bulk concentrations (a higher bulk surfactant concentration is required for a given amount of surfactant adsorption on the surface), indicating that for this monomer, adsolubilization and polymerization probably occur in the headgroup region.

Initiator concentrations are extremely high, on the order of a 10:1 monomer/initiator concentration ratio. This ratio is typically on the order of 1000:1 for emul-

FIG. 3 Schematic showing the morphology of an admicellar film (gray) at a film thickness of a few nanometers on a flat mica surface (black).

sion polymerization. The source of this discrepancy is not initiator per se because the initiator concentration is of the same order as in an emulsion polymerization. However, the amount of monomer added in a typical admicellar polymerization is on the order of 100 times less than that in a typical emulsion polymerization. Initiator type and amount were studied in detail in the kinetic study described earlier, but the monomer/initiator ratio was in the range 1:1. Decreasing initiator concentration increased the induction time, which is typical of emulsion polymerization systems. The lowest initiator concentration, which corresponded to a monomer/initiator ratio of 1:1.25, had an induction time of approximately 45 min. Such long initiation times are typically not encountered in most admicellar polymerizations; which suggests that at least some of this effect was due to radical quenching impurities. Still, the extremely large initiator concentrations are troubling. Water-insoluble initiators are also able to initiate free radical polymerization inside admicelles; the confounding effect of solution polymerization seems to be avoided because monomer concentrations are so low.

IV. APPLICATIONS OF ADMICELLAR POLYMERIZATION

The inventors of this process recognized almost immediately that admicellar polymerization might have significant commercial application. At present, however, no commercial processes have been developed using this technique, although one specific application was patented [18]. The remainder of this section describes three major areas of commercial exploration for admicellar polymerization.

A. Interfacial Adhesion Improvement in Polymer-Matrix Composites

Mechanical properties of polymer-matrix composites are strongly influenced by the interface between the two components; in fact, virtually all commercially important composites have strong adhesion between the polymer and the filler. In some cases, strong adhesion between the filler and polymer occurs naturally because of the intrinsic chemical identities of the two species; however, in other cases compatibilizating agents are required. Admicellar polymerization is typically less expensive than conventional compatibilization techniques, i.e., silane coupling. Further, admicellar polymerization might be able to raise interfacial adhesion versus conventional techniques, although the following studies indicate comparable behavior rather than an improvement. Finally, admicellar polymerization could be particularly useful in systems where no other modification technique is available.

A series of papers explored the effect of silica modification using admicellar polymerization on the properties of a rubber formulation developed to mimic automobile tires [15,19,20]. Table 2 presents data for a multitude of industrially important properties before and after modification. Some formulations provide significant improvements in important physical properties of the rubber compounds, including maximum torque, elongation to break, cut growth resistance, and tear energy, without compromising other properties. Perhaps the most important result was the notable decrease in cure times. Silica normally significantly increases curing time because surface silanol groups inhibit curing agents. This decrease seems to occur for almost any particular type of admicellar polymer. The use of modified silica should reduce the amount of, or even possibly eliminate, additional promoting agents utilized to restore the cure rate of silica-filled tires back to that of carbon black–filled tires.

Commercial silicas are modified by having surface silanol groups react with coupling agents before incorporation in tires. The performance of admicellar-modified silicas and silane-modified silicas was compared in rubber compound physical testing [20]. Results showed comparable performance for the two types of silicas, although with one anomaly. Cut growth resistance, which had been improved in all previous formulations, was significantly decreased with the particular admicellar polymerization–modified silicas tested.

TABLE 2 Selected Butadiene Homo- and Copolymer Rubber Compound Physical Properties

Property	Control	PB	SBR	4MS-B
T_{90} cure time (min)	4.4	2.0	2.1	3.8
Maximum torque (dnM)	23.0	22.1	23.4	27.0
Break strength (MPa)	20.6	21.9	21.4	21.9
Elongation to break	657	622	723	671
20% modulus (MPa)	0.63	0.67	0.64	0.70
100% modulus (MPa)	1.41	1.48	1.39	1.43
300% modulus (MPa)	3.85	3.95	4.17	3.61
Ratio, M300/M100	2.7	2.5	3.0	2.5
Tear energy (N/mm)	11.5	19.1	15.4	17.1
Cut growth (mm)	17.0	15.1	10.3	15.4
G' at 2% strain (MPa)	3.66	3.16	3.14	3.66
G'' at 2% strain (MPa)	0.382	0.327	0.344	0.289

PB-3, polybutadiene homopolymer; SBR-1, styrene/butadiene copolymer; 4MS-B, 4-methoxystyrene/butadiene copolymer [19].

However, the compounds were mixed for the length of time recommended for the addition of carbon black. A significant increase in the amount and energy of mixing is needed to incorporate silica well into rubber. The poorly mixed silica particles would provide weakened areas in the rubber matrix, promoting cut growth. It is thought that better mixing would prevent this from occurring.

The amount of admicellar polymer at surfactant levels near the plateau adsorption level also affects performance. An optimal level of admicellar polymer exists for the greatest improvement in mechanical properties. In other words, adding too much polymer to the surface can degrade mechanical properties. N. Chinpan et al. (in preparation) examined the impact of the amount of polymer deposited on the surface on rubber physical testing results. In testing modified silicas with 5 to 40 g of monomer per kg of silica, the results indicated that intermediate loading levels afforded the best overall improvement of rubber physical properties. Another study (R. Kudisri et al., in preparation) examined whether the admicellar polymerization process could be utilized to modify inexpensive, nonreinforcing clay fillers in order to improve the physical properties of rubber compounds made with these fillers. Rubber compounds made with the modified clays showed an increased cure rate and improvements in tensile strength, elongation at break, tear strength, and hardness with decreases in the abrasion loss, flex cracking resistance, fatigue to failure property, and compression set properties.

Admicellar polymerization can also be used to improve interfacial adhesion in rigid as well as elastomeric composites. Epoxy-matrix composites made from admicellar-treated cloth gave significantly higher strengths in three-point bend tests than composites made from untreated cloth as shown in Table 3 [21]. In fact, the strength of a composite made with admicellar-modified fibers was comparable to that of a composite made with the commercial fiber that had been treated with coupling agent. Single-fiber pullout tests, although quantitatively not conclusive, qualitatively indicated behavior consistent with better adhesion in the silane- and admicellar-treated fibers than in the untreated fibers. The admicellar polymer was a styrene-isoprene copolymer. Surprisingly, composites made with a polyester resin showed no significant improvement in properties. This result showed that obtaining good adhesion in admicellar polymerization depends on the chemical identities of the polymer and admicellar polymer; similar results were also found for elastomeric composites as well.

TABLE 3 Flexural Test Results of Epoxy-Glass Mat Composites

Condition	Percent elongation at break	Flexural strength (psi)
Silane treated	16.5 ± 3.0	44,600 ± 16,800
Untreated	11.1 ± 3.6	24,100 ± 6,800
Admicellar treated	17.0 ± 4.4	38,200 ± 4,100

The epoxy used was EPON 828 and the error represents 1 standard deviation from duplicate measurements of different samples.

B. Modification of Surface Wetting Behavior and Friction Coefficient

Poly(tetrafluoroethylene) (PTFE, DuPont trademark is Teflon®) is well known for its extremely low surface energy, which has led to its use in no-stick cookware, for example. Admicellar polymerization could be used to coat almost any surface with a thin layer of PTFE to reduce the friction coefficient. Similarly, the dramatic changes in wetting can also be useful for many applications.

High-speed electronic storage devices were the focus of a paper by Lai et al. [10]. Alumina plates were coated with a thin layer of PTFE, and contact angles and friction coefficients were measured. The contact angle jumped from 52° for alumina to an average of 112° for the coated alumina, and the contact angle for pure PTFE was measured as 116°. The results for friction coefficient were not as encouraging. The friction coefficient dropped approximately 20% upon coating with PTFE; however, the friction coefficient of pure PTFE is approximately three times lower than that of alumina.

C. Conductivity Enhancement in Conductive Composites

Electrically conductive composites are mixtures of an insulating polymer and a conducting filler. Charge is carried in these systems by the conductive filler, and the composite is conductive if a continuous path of conductive material exists in the material. A schematic representing the relationship between conductivity and volume fraction of filler is shown in Fig. 4. The steep rise in conductivity occurs over a fairly narrow volume fraction region, and this region is termed the critical region. The plateau region signals the end of the critical region, and the conductivity changes very little with increasing volume fraction of filler.

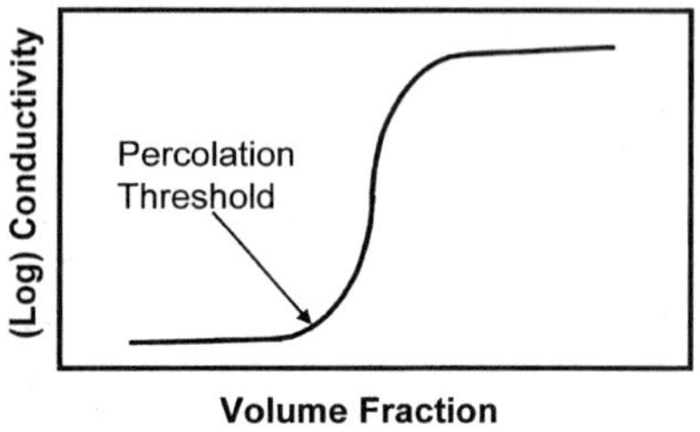

FIG. 4 Representative plot of electrical conductivity of a polymer-matrix composite where the polymer is insulating and the filler is conductive. The jump in conductivity is more than 15 orders of magnitude for many metal fillers.

Once the plateau region is reached, the amount of charge that the conductive phase can carry depends on three factors. Two of these resistances are not relevant to this discussion, but the third, the interfacial resistance where two conductive particles meet, is apparently altered with admicellar polymerization. Nickel flake having an electrically conductive polymer, polypyrrole, formed on its surface via admicellar polymerization produced a composite with two to three orders of magnitude higher conductivity than the composite made with the unmodified flake [12]. This modification affects only the plateau conductivity; the rest of the percolation diagram, including the volume fractions corresponding to the critical regions, are unchanged. Composites made from modified alumina, which has a conductivity value between the values for a truly conductive and a truly insulating filler, have a conductivity four orders of magnitude higher than that of composites made from the unmodified alumina.

V. POTENTIAL APPLICATIONS

The three examples discussed in the previous sections are certainly not the only possible applications of admicellar polymerization. Several other potentially important commercial applications are possible. A few of these are discussed next.

Admicellar polymerization can be used to develop novel and inexpensive high-performance liquid chromatography (HPLC) packings. The method may be used with such a variety of monomers and monomer combinations that the surface of the silicas may be quickly customized for a variety of applications [22].

Hydrophobic silicas are widely utilized in the personal care and petroleum industry as viscosity modifiers. This process could be used to produce these silicas cheaply and potentially to custom modify them to produce silicas with a given hydrophobicity or range of hydrophobicities. The hydrophobic nature of the surface can be changed by forming differing amounts of polymer on the surface or by changing the types of polymer formed.

VI. CONCLUSIONS

The development of the admicellar polymerization process has opened up a new field of chemistry. Its study has resulted in a significant increase in our understanding of interfacial phenomena and confined reactions. In addition, admicellar polymerization has potential applications in polymer production, separation processes, composite materials, and surface engineering. As previously stated, although much has been learned about the phenomenon, much more remains unknown.

REFERENCES

1. J. H. O'Haver, J. H. Harwell, L. L. Lobban, and E. A. O'Rear, in *Solubilization in Surfactant Aggregates* (S. D. Christian and J. F. Scamehorn, eds.), Surfactant Science Series *55*:277–296, 1995.
2. P. V. Brady, *Physics and Chemistry of Mineral Surfaces*, CRC Press, Boca Raton, FL, 1996.
3. J. Wu, J. H. Harwell, and E. A. O'Rear, Langmuir *3*: 531–537 (1987).
4. J. H. Harwell and E. A. O'Rear, U.S. Patent 4,770,906 (1988).
5. J. H. Harwell and E. A. O'Rear, U.S. Patent 4,900,627 (1990).
6. J. H. Harwell and E. A. O'Rear, U.S. Patent 5,106,691 (1992).
7. J. Wu, J. H. Harwell, and E. A. O'Rear, J. Phys. Chem. *91*:623–634 (1987).
8. J. Wu, J. H. Harwell, E. A. O'Rear, and S. D. Christian, AIChE J. *34*:1511–1518 (1988).
9. H. Chen, Master's thesis, University of Oklahoma, 1992.
10. C. Lai, J. H. Harwell, E. A. O'Rear, S. Komatsuzaki, J. Arai, T. Nakakawaji, and Y. Ito, Langmuir *11*:905–911 (1995).
11. J. H. O'Haver, J. H. Harwell, E. A. O'Rear, L. J. Snodgrass, and W. H. Waddell, Langmuir *10*:2588–2593 (1994).
12. W. B. Genetti, W. L. Yuan, B. P. Grady, E. A. O'Rear, C. L. Lai, and D. T. Glatzhofer, J. Mater. Sci. *33*:3085–3093 (1998).
13. S. Sakhalkar and D. E. Hirt, Langmuir *11*:3369–3373 (1995).
14. D. T. Glatzhofer, G. Cho, C. L. Lai, E. A. O'Rear, and B. M. Fung, Langmuir *9*:2949–2954 (1993).
15. W. H. Waddell, J. H. O'Haver, L. R. Evans, and J. H. Harwell, J. Appl. Polym. Sci. *55*:1627–1641 (1995).

16. N. Duman, Master's thesis, University of Oklahoma, 1994.
17. G. P. Funkhouser, M. P. Arevalo, D. T. Glatzhofer, and E. A. O'Rear, Langmuir *11*:1443–1447 (1995).
18. W. H. Waddell, L. R. Evans, J. H. Harwell, and J. H. O'Haver, U.S. Patent 5,426,136 (1995).
19. J. H. O'Haver, J. H. Harwell, L. R. Evans, and W. H. Waddell, J. Appl. Polym. Sci. *59*:1427–1435 (1996).
20. V. Thammathadanukul, J. H. O'Haver, J. H. Harwell, S. Osuwan, N. Na-Ranong, and W. H. Waddell, J. Appl. Polym. Sci. *59*:1741–1750 (1996).
21. B. P. Grady, E. A. O'Rear, L. S. Penn, and A. Pedicini, Polym. Composites *19*:579–587 (1998).
22. A. E. Riviello, J. F. Scamehorn, S. D. Christian, and J. H. O'Haver, J. Surfactants Deterg. *2*:69–78 (1999).

28

Polymerizable and Polymeric Surfactants

ALAIN GUYOT CNRS, Villeurbanne, France

KLAUS TAUER Max-Planck-Institute of Colloids and Interfaces, Golm, Germany

I. INTRODUCTION

A. Overview

In emulsion polymerization, surfactants are needed to control particle size and particle size distribution as well as to ensure the stability of the dispersion during the polymerization process. They may, however, cause a number of practical difficulties when the dispersions are to be used subsequently. These drawbacks are due to the fact that the surfactants can be desorbed from the particle surface. In order to get nondesorbable surfactants, there are two possibilities. One can create covalent bonds between the surfactant and the material of the particles, i.e., use polymerizable surfactants, or use surfactants so strongly adsorbed onto the particle surface that they cannot be desorbed. This chapter is, consequently, divided in two parts. First, the state of the art is described, and outstanding problems are discussed. Second, recent developments are critically described.

B. Benefits of Strong Attachment

The use of polymerizable and polymeric surfactants in emulsion polymerization and in dispersion polymerization is expected to result in certain benefits. The first benefit that can be expected is an improvement in the stability of the particle dispersion (latex). Under certain constraints, such as high shear, classical surfactants such as sodium dodecyl sulfate (SDS) tend to desorb and cease to protect the latex against flocculation. High-shear conditions can occur even in the polymerization reactor if a high solids content is to be produced (this is a general trend in industrial practice). High shear may also occur during transport through pipes and tubing. Stability problems also occur under freezing conditions, when conventional surfactants often desorb. However, reactive surfactants are able to maintain stability under such conditions, as compared with nonreactive surfactants having very similar structures [1].

After preparation, latexes can be flocculated, as done in the rubber industry for natural rubber and synthetic rubbers such as polychloroprene, styrene-butadiene copolymers (SBR), nitrile rubber (NBR), and some plastics such as acrylonitile-butadiene-styrene (ABS) copolymers. In such cases, the water used in the flocculation treatment can be polluted by all the components adsorbed onto the surface of the particles and released during the treatment or the washing process. Among these components are surfactants if they are not strongly attached to the particles. Some experiments carried out with simple polymerizable surfactants (surfmers) bearing maleic functionality and used in polystyrene latexes demonstrated that the amount of surfactant left in the water upon flocculation by addition of calcium salts can be reduced by a factor of 4 as compared with conventional surfactants [2]. This result was confirmed in the rubber industry for SBR after measurement of the oxygen demand of the water used in the flocculation treatment, which was strongly reduced.

Polymeric latexes are also heavily used in the coatings industry, e.g., in decorative and industrial paints, in textile sizers, in paper coating, in nonwoven textiles, and in adhesive formulations. In such cases, a film-forming material is used with a rather low glass transition temperature, T_g. The film is formed upon coalescence of the latex particles. The coalescence process involves first the evaporation of water, so that the particles become close packed. Then they are deformed because of the capillary forces in a kind of hexagonal network. Finally, interdiffusion of macromolecules take place across the boundaries of the deformed particles. This process of interdiffusion is known as film maturation and generally takes several weeks for completion. The final state is a clear, continuous thin film.

During the first step of evaporation, a phase separation takes place between the organic and water phases. The surfactants that are water soluble tend to migrate in the water phase after desorption from the surface of the particles. The volume of the water phase rapidly decreases and tends to be divided into small domains surrounded by the organic phase, in which these domains are almost trapped. Part of the surfactant is able to escape and to migrate toward the surfaces of the film (both the interface with air and that with the substrate supporting the film). This migration process leads to a loss of adhesion on the substrate, which is strongly detrimental for many applications.

This trend toward the formation of a large number of small hydrophilic domains is also detrimental for many applications, even after the film has become completely dry. These hydrophilic domains make the material sensitive to water when it is exposed later to humidity. Then the diffusion of water molecules through the organic material of low T_g cause the swelling of such domains. These domains behave as mechanical defects and tend to reduce the dimensional stability of the material. It was recognized early on that the use of reactive surfactants in the synthesis of latex particles can strongly reduce the sensitivity of the coatings to exposure to moisture [3–10]. For instance, Biale [8] reported that latexes made using poly(oxyethylene)-based polymerizable surfactants provide binders having markedly improved flow and leveling properties without exhibiting any significant decrease of gloss, film build, or wet adhesion characteristics. Later, the use of surfactants with about the same structure was shown to improve greatly the dimensional stability of the films dipped in water for rather long periods of time [1].

There are two main ways to obtain particles from which the surfactant cannot desorb. The first is to use reactive surfactants, i.e., surfactants that participate in one of the reactions involved in the polymerization process. Most often such a process is a radical polymerization process, so the reactivity may involve either an initiation reaction, a chain transfer reaction, or a propagation reaction. The corresponding products are referred to, respectively, as an *inisurf* (combining the character of an initiator and that of a surfactant), a *transurf* (which is both a transfer agent and a surfactant), and a *surfmer* (i.e., a monomer and also a surfactant). In all these cases a covalent bond is formed between the reactive surfactant and the polymer latex, provided the two functions are actually working.

When the surfactant is an inisurf (both surfactant and initiator), it is expected that its incorporation will be definite as soon as it is efficient. Indeed, the efficiency of an initiator is defined as the proportion of the radicals produced that are actually initiating a polymer chain. Then the efficiency of the initiator is a number between 1 and 0. In most cases that number is around 0.6. However, when the initiator is an inisurf, whether it is a peroxide or an azo compound, that number is very small. For instance, Kusters et al. [11] used as an inisurf a symmetrical azo compound prepared by reacting an ethoxylated nonylphenol with an azocarboxy derivative in the presence of a cyclohexylcarbodiimine. Using sophisticated techniques, such as gamma radiolysis, they found a value as low as 4×10^{-4}. Attempts to improve the results by choosing an asymmetrical azo with a small hydrophilic part, expected to escape easily from the cage to the water phase, were no more successful. However, inisurfs seem to be able to initiate emulsion polymerizations rather well as compared with some commercial recipes [12].

In the cases of transurfs (surfactants reacting as transfer agents) the situation is quite different, although the number of studies is much more limited. Pioneering work carried out by a student of Fitch [13] involves a thiol-ended alkylsulfonate, $HSC_{11}H_{22}SO_3Na$. The product is able to stabilize a polystyrene latex from a batch emulsion polymerization, but no report has been published in the open literature and there are no data about the incorporation in the thesis work. Later, Vidal et al. [14] reported working with a nonionic thiol-terminated alkyl ethoxylated product, $HSC_{11}H_{22}O(CH_2CH_2O)_nH$ with n = 17 to 90. The transfer constant of these compounds is rather high: 17 and 14 for n = 17 and 90, respectively [14]. Stable polystyrene latexes were obtained, and the incorporation yield, as measured from nuclear magnetic resonance (NMR) analysis of the latex previously washed by serum exchange using ultrafiltration techniques, was around 33% in most cases. There are two reasons for the limitation of the incorporation yield with transurf containing thiol groups. The first is the high reactivity of the thiol, so that it works mainly in the water phase and causes the formation of water-soluble oligomers. This is especially the case if the length of the hydrophilic sequence of ethylene oxide units is long [15]. The second reason is also related to the high reactivity of the thiol group. When the hydrophobicity of the transurf is high enough, it may be rather strongly adsorbed and then does react at the surface of the particles. However, it produces oligomers that are easily extracted with alcohol.

For these two reasons, it is better to use a less reactive transfer agent with a transfer constant close to 1. This has been done using a transfer agent working with an addition fragmentation type of mechanism (F. Vidal et al., submitted). the structure of the transurf is the following:

$$CH_2{=}CH(C_6H_5)CH_2SCOO(CHCH_3CH_2O)_8 \cdot (CH_2CH_2O)_{45}CH_3$$

It was prepared by condensation of the corresponding block copolymer of ethylene oxide and butylene oxide with, respectively, 45 and 8 units, with 2-(2-phenyl-2-propenylthio)propanoic acid in the presence of dicyclohexylcarbodiimide. The acid was prepared according to the procedure of the Australian team who discovered the technique of addition fragmentation [16]. When a radical $R\cdot$ is attacking the styrenic structure of the transurf, a fragmentation process takes place with the formation of an unsaturated monomer $RCH_2C(C_6H_5){=}CH_2$ and a mercaptopropionic ester derivative of the surfactant, which initiate the polymerization and thereby attach the amphiphilic block copolymer to the polymer. NMR analysis of dialyzed samples of a batch polystyrene latex indicates an incorporation yield between 28 and 75%. As shown by gradient polymer elution chromatography (GPEC), the remaining part of the transurf-produced oligomers stay in the serum. The transfer constant of the product was measured as 0.3 at 50°C in bulk. So it was demonstrated that the reduction of that transfer constant has been a benefit in terms of the incorporation yield.

The case of the surfmers is much more interesting because it is concerned with a much broader variety of structural features and has been the subject of a large number of studies. As already stated, polymerizable surfactants are expected to be reactive in the polymerization process. They produce copolymers with the main monomers. Therefore, the compounds responsible for the stabilization of the latex particles are actually polymeric in nature. The surfmers are soluble in the water phase and, at least below the critical micelle concentration (cmc), they react molecularly in the water phase with the initiator. The composition of the copolymers should be dependent on the solubility of the monomer(s) in the water phase and also on the cmc of the surfmer.

The second way to produce polymer latexes with very little chance of the stabilizing surfactant desorbing from the particle surface is to use polymeric surfactants. The surfmers produce such polymeric surfactants. Homopolymers of surfmers are called "polysoaps." They are expected to be water soluble, like their monomers, and more so than their copolymers with the main hydrophobic monomers. On the other hand, they also contain hydrophobic segments, which may cause them to be strongly adsorbed onto hydrophobic surfaces, such as the surfaces of most latex particles. A comparative study of normal (nonreactive) surfactants, polymerizable surfactants, and their polysoap homopolymers was published by Lawschewsky et al. [17]. Both the surfmers and the polysoaps were shown to give stable latexes with high polymerization rates and yields. It was observed in several cases that the use of polysoaps gave better results in terms of stability (long-term storage stability versus strong electrolytes).

There have been many studies of the stabilization of latex particles [chiefly polystyrene or poly(methyl methacrylate)] using nonionic polymeric surfactants. These stabilizers are most often block copolymers that work through a steric stabilization mechanism. The hydrophobic sequence is strongly adsorbed to the particle surface. If the hydrophobic part is compatible with the latex, it can be linked to the particle through entanglements. The hydrophilic sequence, generally poly(ethylene oxide), protrudes out from the particle surface and extends into the contiuous phase, forming a polymer shell and protecting the latex against flocculation. For instance, Piirma et al. [18] have published an important study of stabilization of polystyrene latexes using block copolymers of polystyrene-*b*-poly(ethylene oxide). They determined that the best stabilizing properties are obtained with a 10-unit length of the polystyrene sequence and a hydrophilic sequence of 30 to 50 units. Such a finding points out the importance of the hydrophilic-lipophilic balance (HLB). The stabilization of latexes using either block or graft copolymers has been reviewed by Piirma [19].

It can be theorized that electrosteric stabilization should be preferred to steric stabilization because the advantages of electrostatic stabilization can be combined with those of steric stabilization. This approach has been followed by Tauer [20] using block copolymers of hydrogenated polybutadiene and polystyrene more or less sulfonated. Excellent stability can be obtained when the particles become core-shell with the shell made of the quite fully sulfonated polystyrene sequence. Different situations can be observed, depending on the structure of the block copolymer and its degree of functionalization. For instance, partial sulfonation can lead to a structure with polystyrene trains remaining adsorbed on the particles, and hydrophilic sulfonated loops ensure good steric and electrostatic stabilization.

This chapter is divided in two main parts. The first deals with the use of polymerizable surfactants, but owing to many published reviews [21–26], only the latest and most pertinent progress will be discussed. Obviously, the most important problem is the actual incorporation of the surfmer on the surface of the latex particles. The second part is devoted to the use of polymeric surfactants. Emphasis wil be put on polymers bearing charges. These have not been studied much, until recently, and give excellent latex stabilization.

II. POLYMERIZABLE SURFACTANTS IN EMULSION POLYMERIZATION AND RELATED PROCESSES

A. State of the Art

The pioneering work on the use of reactive surfactants in emulsion polymerization was published by Greene et al. [27–28] of Dow. In that study, the authors started from a styrene-butadiene copolymer latex of 134 nm prepared in the presence of sodium 9- (and 10-) acrylamidostearate (NaAAS) as surfmer in batch at 70°C. This base latex was covered with various amounts of NaAAS, calculated so that the surface coverage should be between 20 and 100% of the full coverage (at saturation of the surface after cleaning the latex). Then in situ polymerization of the adsorbed surfmer was carried out in the presence of potassium peroxodisulfate (KPS). Then the surfmer was shown to be at least partially immobilized, and the yield of immobilization was observed to be 100% at low coverage and to decrease to about 70% at full coverage. Part of the surfmer can be extracted upon treatment over ion-exchange resins as either nonreacted monomer or water-soluble oligomers. Another series of latexes was covered with similar amounts of surfmer, but then no polymerization was effected. Then the two series of latexes, one just covered by the surfmer and the other with immobilized surfmer, were submitted to a mechanical stability test by subjecting these latexes to high-shear conditions (2700 rpm, 30 min, 30°C). It was observed that the immobilized series was much more stable [27]. These researchers also tested the electrolyte stability of the two series of latexes and found that the immobilized series provides the best stability. These latexes were also used to check certain features of the DLVO theory [28].

A styrenic surfactant, styrene sodium dodecyl sulfonate ether (SSDSE), with a cmc of 2×10^{-3} M, has been prepared by Tsaur and Fitch [29] and used for styrene polymerization, either with a photoinitiator at 20°C or with KPS at 70°C. In the first case SDS must be added to keep the latex stable, in concentrations larger than that of the SSDSE, to obtain monodisperse particles in the range 140–260 nm with a surface charge density between 1.9 and 6.5 $\mu C/m^2$. With KPS, it is possible to get good control of the particle size, and a logarithmic plot of the number frequency diameter, D_n, versus SSDSE concentration was linear in the range 150–400 nm (the surface charge density being between 4.0 and 14.0 $\mu C/m^2$). Much larger sizes can be obtained from a seed polystyrene latex covered with SSDSE and further polymerized. If the amount of the surfmer is kept below the cmc, no water-soluble polymers are produced and the immobilization yield is from 10 to 60% with a corresponding charge density from 10 to 100 $\mu C/m^2$. The distribution of charge density seems narrow, as judged from the data from electrophoresis experiments. However, if the concentration of SSDSE is larger than the cmc, polyelectrolyte is formed and several cycles of centrifugation, washing, and redispersion are necessary to separate these water-soluble polymers from the latex.

$$CH_2{=}CH{-}C_6H_4{-}O{-}(CH_2)_{12}SO_3Na$$

SSDSE

A latex with only carboxylate charges has been produced by Pichot et al. [30] using an azo-carboxy initiator and sodium acrylamido undecanoate (SAU) as surfmer with a cmc at 25°C of 5×10^{-3} M. Copolymerization of styrene (S) and butyl acrylate (B) was carried out. The reactivity ratios of copolymers of the surfmer were measured or estimated from the *Q*, *e* scheme and using the partition coefficients of SAU between the water and the organic phase (also measured). A simulation of the copolymerization process showed an S shape of the conversion curve of the SAU, which indicates that most of the surfmer is polymerized at the end of the polymerization process. Polyelectrolytes of rather high molecular weight were formed, causing flocculation. Before the flocculation, the particle number goes through a maximum. The largest part of the carboxylic groups (10 to 50%) remain buried inside the particles, the immobilization yield being limited to 20–30%. The remaining SAU was unreacted or produced polyelectrolytes. If, instead of ad initio polymerization, one uses SAU to cover a seed latex, the immobilization yield reaches only a slightly higher value (25–35%).

$$CH_2{=}CH{-}CONH{-}(CH_2)_{10}COONa$$
SAU

Urquiola et al. [31,32] carried out a rather important study of a commercially available surfactant, TREM-LF 40 (sodium alkyl allyl sulfonate, SAAS), in comparison with its nonreactive hydrogenated equivalent, HSAAS. In the polymerization of vinyl acetate, the latter behaves as a normal surfactant in terms of polymerization kinetics and control of the particle size. On the other hand, the reactivity ratio of SAAS in its copolymerization with vinyl acetate shows a small retardation effect, but the main effect is a large transfer effect, the transfer constant being 0.011. In emulsion copolymerization, oligocopolymers are found in the water phase in amounts increasing with the amount of surfmer. The polymerization rate decreases upon increasing the concentration of surfmer, the particle size also decreases, and the initiator concentration has no effect on the particle size. Competitive growth experiments have been carried out with a mixture of two families of particles using either SAAS or HSAAS. Normal growth behavior with decreasing size differences was observed in the case of HSAAS. In contrast, using SAAS, the big particles undergo more rapid growth than the smaller ones. Such behavior is due to more important adsorption of SAAS on the higher surface area of the small particles, which causes a stronger retardation effect, probably caused by the transfer activity of that allylic compound and, consequently, a high exit rate and a high termination rate in the water phase. Modeling showed that it is the transfer ability of the surfmer, rather than the reactivity in copolymerization, that causes the retardation of the polymerization rate [33].

The pioneering work of Ottewill et al. [34,35], using a methacrylic macromonomer of polyethylene oxide, inspired two new industrial processes at ICI. The first one, called "Aquersymer," is actually a dispersion polymerization process; it uses a mixture of water and alcohol as the serum. The second, called NIAD (nonionic aqueous dispersion), is a real emulsion process. These processes, in addition to the macromonomer, can use allylic surfmers. Improved steric stabilization is claimed [36].

The field was reviewed in 1994 [22], when a European network was started to study the topic further. An important set of reactive surfactants (Tables 1 and 2), covering the three kinds of reactants (inisurfs, transurfs, and surfmers), was prepared, characterized, and applied in emulsion polymerization of either styrene or copolymers of styrene, butyl acrylate, and acrylic acid (up to more than 50% solid). Some of these surfactants were also applied in dispersion polymerization. The results obtained by this network are discussed later in this chapter together with other recent developments. However, it might be mentioned here that some of these data have already been discussed, at least partly, in some reviews [24–26].

B. Problems to Be Solved

When using polymerizable surfactants, a few questions should be addressed. If the structure of the surfactant is fixed, because the surfactant is commercially available, for instance, the questions that need to be addressed are, in what amount and with what procedure should the surfactant be employed to obtain the desired latex (particle size, solids content, etc.)? However, if the structure and the reactivity of the surfmer are not well suited for the expected purpose, the problem becomes one of determining a suitable structure (ionic or nonionic, reactivity of the polymerizable group, hydrophilic-lipophilic balance).

The main problem, related to the expected property of nondesorable surfactant, is to know to what extent the polymerizable surfactant is actually incorporated in the polymeric material and, more precisely, on the particle surface. Surprisingly, there are few studies in which a precise answer to this incorporation question has been obtained. Most often, only the conversion of the surfmer has been determined and a convincing analysis of the location of incorporation in or on the particle is missing. Progress on this issue is described in the following.

C. Latest Developments

1. European Network

As already mentioned, this network has been able to prepare a rather large set of reactive surfactants, including inisurfs, transurfs, and surfmers, either nonionic (Table 1) or ionic (Table 2), with both anionic and cationic compounds). All of these products were tested in styrene emulsion polymerizations, but only a few of them were used for core-shell latexes with film-forming properties. This was the case for a set of anionic surfmers including either a methacrylic group

TABLE 1 Ionic Surfactants

Anionic inisurfs	
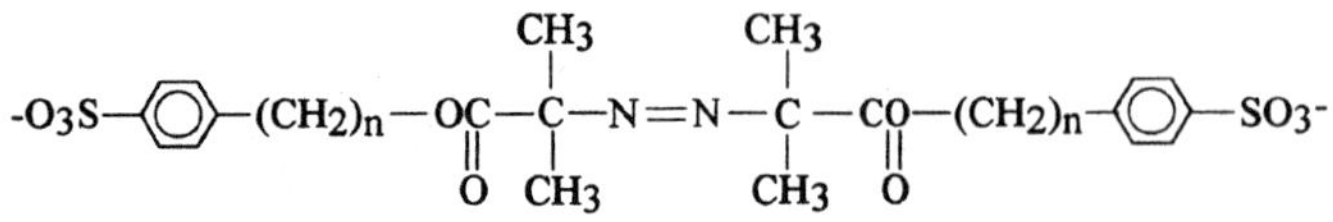	
Bis [2-(4′-sulfophenyl)alkyl]-2,2′-azodiisobutyrate ammonium salts	$n = 1, 2, 3$
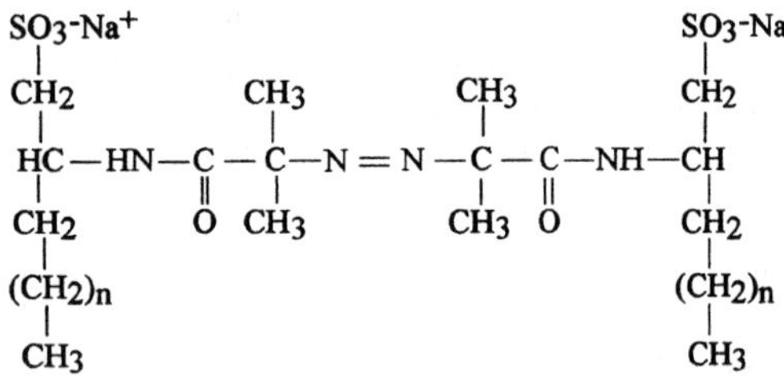	
2,2′-azobis(*N*-2′-methylpropanoyl-2-amino-alkyl-1-sulfonate)s	$n = 4, 12, 14$
Anionic surfmers	
A3 Hemiester of maleic anhydride (Na salt)	
$CnH_{2n+1}OCOCH{=}CH{-}COO{-}COONa$	$n = 8, 10, 12, 14, 16, 18$
A4 Alkyl maleate propyl sulfonate	
$CnH_{2n+1}OCOCH{=}CH{-}COO{-}(CH_2)_3SO_3Na$	$n = 8–14$
A5 Sodium 11-methacryloyl-undecan-1-oyl sulfate	
$CH_2{=}CH{-}COOC_{11}H_{22}SO_4Na$ CH_3	
A6 Sodium 11-crotonoyl undecan-1-oyl sulfate	
$CH_3{-}CH{=}CH{-}COOC_{11}H$	
Cationic surfmers	
C1 Alkyl maleate trimethylamino ethyl bromide	
$C_nH_{2n+1}OCOCH{=}CH{-}COO{-}CH_2{-}CH_2N(CH_3)_3^{+-}Br$	$n = 12$
C2 Alkylmaleate alhyl pyridinium bromide	
$C_nH_{2n+1}OCOCH{=}CH{-}COOC_mH_{2m}NC_5H_5Br$	$n = 1, 6, 10, 12$ $m = 2, 6, 11$

Source: Adapted from Ref. 12.

(A5), a crotonic group (A6), or a maleic group (A4) [36]. From styrene-butylacrylate copolymer seeds (<40 nm) prepared in the presence of SDS, the core-shell copolymer comprised a mole ratio of 49.5:49.5:1 styrene/butyl acrylate/acrylic acid and was grown up to 50% solids content. The conversion of all the monomers, including the surfmer, was determined. Using the methacrylic surfmer A5, flocculation took place before the end of the process. It was concluded that the surfmer was too reactive and was consumed before the end of the process. There are two possible mechanisms by which a very reactive surfmer can cause early flocculation. The first is the production of water-soluble high polymers, causing bridging flocculation. The second is burying of the surfmer by polymer converted later in the process. This leaves no surfactant to protect the particle surface. In the case of the crotonic A6, strong retardation of the polymerization is observed due to the allylic reactivity of the surfmer. Finally, the best of the three was shown to be the maleic A4. Its reactivity was moderate at the beginning of the process, except versus the styrene, but it hardly copolymerizes with the acrylics, and it remains only partly converted. Its conversion remains limited unless the particle surface area is large enough. The copolymerization of the surfmer A4 with styrene was studied further [37], and it was shown that

TABLE 2 Nonionic Surfactants

N1	Maleic diester $CH_3O(CH_2CH_2—O)_{45}COCH=CH—COOC_{12}H_{25}$	surfmer
	Block copolymers EO-BO	
N2	$CH_3O(CH_2CH_2O)_{45}—[CH_2CH(C_2H_5)O]_mH$	precursors m = 8, 9, 12
N3	Me $EO_{45}—BO_9—CO—(CH_3)_2C(CN)—N=N—R$ R = *tert*-butyl or $—C(NCH_2CH_2OH)_3$	inisurf
N4	Me $EO_{45}BO_{12}—OCOCH_2SH$	transurf
N5	Me $EO_{45}BO_9—S—C(C_6H_5)—CH=CH_2$	transurf
N6	Me $EO_{45}BO_{15}—CH_2—C(C_6H_5)—CH=CH_2$	surfmer (styrenic)
N7	Me $EO_{45}BO_9CONH—C(CH_3)_2—C_6H_4—C(CH_3)=CH_2$ α Methyl styrenic isocyanato derivative (TMI)	surfmer
N8	Me $(EO)_{45}BO_8OCO—CH=CH_2$	surfmer (acrylic)
N9	Me $(EO)_{45}(BO)_{12}OCNH—CO—CH(CH_3)=CH_2$	surfmer (methacyloyl isocyanate)
	Block copolymers EO-CL	
N10	Me $(EO)_{45}—[(CH_2)_4—O]_{10}H$	precursor polycaprolactone
N11	Me $(EO)_{45}—(CL)_{10}—OCOCH=CH—COOH$	maleic surfmer
N12	Me $(EO)_{45}(CL_{10}—CONH—C(CH_3)_2—C_6H_4C(CH_3)=CH_2$	

Source: Ref. 12.

the initial polymerization rate was higher for the surfmer than for the styrene and resulted in the formation of a rather large amount of oligomers. It may be suspected that such oligomers are possibly alternated co-oligomers because of the presence of a charge transfer complex. That result is at variance with other published data in which these maleic surfmers were compared with their fumarate, more reactive, isomers [38]. However, sulfur titration did show that the actual incorporation yield of A4 on the surface was limited at 45% [37].

A comparison of the three anionic surfmers A4, A5, and A6 led the authors to postulate what the optimal behavior of a reactive surfactant should be [39]. It was stated that the surfmer should be moderately reactive at the beginning of the process, so that it would not be buried in the interior of the polymer particle, and also it should not cause the formation of water-soluble high polymers able to cause bridging flocculation. On the other hand, it should possess high reactivity versus the main monomers at the end of the polymerization so that it would be located on the surface of the particles. A set of strategies have been defined in order to reach that second objective. The first is to offer a high surface area near the end of the polymerization in order to favor the adsorption of the residual surfmer. In an experiment in which the final conversion of the surfmer was limited to 82%, a new charge of the seed was introduced with a small amount of monomer so that the specific surface area offered was doubled. Upon continuing the polymerization, the surfmer conversion reached 100%. The second strategy was to adapt the reactivity of the mixture of the monomers to fit that of the surfmer, at least at the end of the process. It was observed that the reactivity of A4 was better with vinyl monomers (vinyl acetate) than with acrylates. In an initial experiment to check that strategy, vinyl acetate was introduced at the end of the polymerization, and again, upon continuing the polymerization, the conversion of the surfmer went to completion. The same result was obtained when the initial mixture of the feed contained some vinyl acetate. Indeed, the conversion of the styrene and of the acrylic monomers is much more rapid than that of vinyl acetate, so the vinyl acetate tends to accumulate during the feed process. For that reason, at the end of the process, it becomes the main residual monomer and copolymerizes easily with the surfmer, causing full conversion of the surfmer. An obvious drawback of that strategy is that one produces a copolymer somewhat different from the targeted composition.

Finally, two other strategies were suggested but have not been demonstrated. One of these was to use miniemulsions instead of emulsions, and the second was to use a programmed feeding procedure in order to fit the addition of the surfmer to the conversion of the main monomers. Up to now, there has been nothing published about the use of reactive surfactants in miniemulsion polymerization. The second strategy has also not really been followed. Ottewill et al. [40] introduced a nonionic macromonomer at various steps of styrene conversion in a seeded emulsion polymerization. They did not observe any renucleation of the macromonomer. Addition was carried out either at the beginning or near the end of the feed procedure; on the other hand, the latexes were stable only if the additions were done near the end. In our laboratory, we followed the same procedure, using a nonionic styrenic surfmer [41]. A stable latex at 40% solids was used as a seed. The addition of the nonionic styrenic surfmer to the feed resulted in flocculation unless the addition was carried out after 75% conversion of the feed. Then improvement of the stability versus electrolyte addition was observed. It seems that the trend for flocculation, which was observed upon premature addition of the nonionic surfmer, was caused by a nonsuitable HLB balance, as shown later on [1].

The very simple surfmer A3 was used in a seeded copolymerization of a mixture of styrene and butyl acrylate monomers with continuous feeding of both monomers and surfmer. The incorporation yield of the surfmer was estimated as follows: The latex was washed first and then covered with SDS up to saturation of the surface. The amount of SDS used to achieve that coverage was subtracted from the amount needed to cover the same area of a similar latex (same particle size) produced with the same initiator but with a conventional and nonreactive surfactant. From the data obtained, the incorporation yield was estimated as 75%. The nonionic surfmer N1 was also prepared from A3 upon condensation (esterification) with the monomethyl ether of polyethylene glycol (with 45 monomer units [42]. In the same kind of core-shell copolymer synthesis, the incorporation yield, estimated from NMR measurements, can reach 70%. The corresponding latex was observed to resist to more than 10 freeze-thawing cycles [43]. Films have been obtained from these latexes prepared with both A3 and N1 as surfactants. When the latexes were previously washed, so that they were protected with only the grafted surfmers, the films were first dried (30 days) and then dipped into water. Their water uptake was very limited (3 and 8%, respectively). The same kind of latex prepared using SDS showed a water uptake of 140% under similar conditions.

As shown in Fig. 1, the deformation energy, which is built in two steps before reaching a plateau value (full maturation), displays very different behavior for the two kinds of films. The first step occurs early when a surfmer has been used (Fig. 1b), but in the case of SDS that step is much longer (Fig. 1a). Such a result probably means that the healing process between the

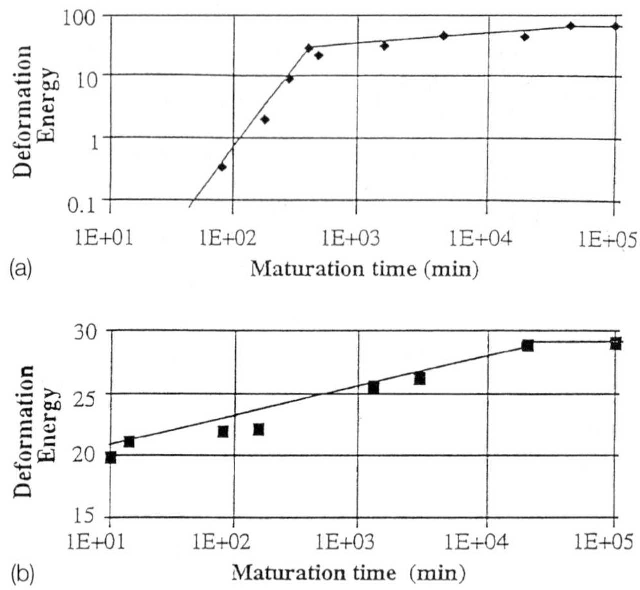

FIG. 1 Comparative development of deformation energy for films produced in the presence of SDS (a) or the A3 reactive surfactant (b). (From Ref. 12.)

packed particles, after they have been deformed, is much more rapid when using the surfmer. Important consequences should result in term of blocking properties. In addition, the stress-strain properties are much better in the case of a dry film than when the water uptake of the film has reached even a limited value (Fig. 2).

2. Further Studies

After the end of the European Network program, further studies were continued in various laboratories. Abele et al. [44] prepared a set of maleic reactive surfmers, including hemiamides and hemiesters as well as the corresponding succinic nonreactive surfactants, and

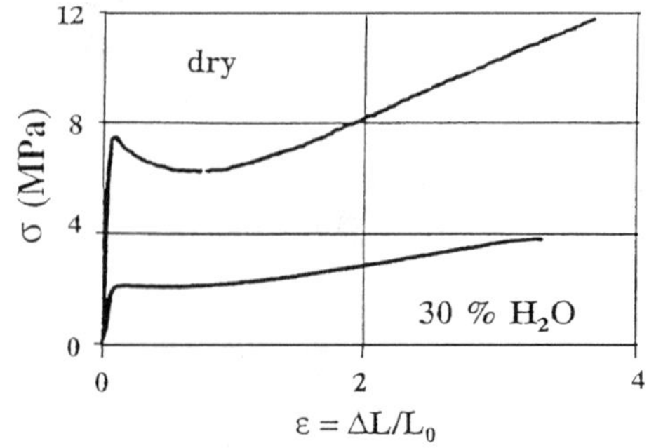

FIG. 2 Stress-strain curve of films either dry or containing 30% water and prepared with SDS as surfactant. (From Ref. 12.)

used them in either batch emulsion polymerization of styrene or seeded core-shell copolymerization of styrene and butyl acrylate. All of these surfactants facilitated the preparation of stable latexes. Conductimetric titration of the carboxylic groups indicated that between 33 and 68% of the surfactant was grafted or strongly adsorbed to the particle surface. Very little difference was observed between the reactive and the nonreactive surfactants. The same group also prepared maleic and succinic cationic and zwitterionic surfactants [45]. These surfactants were again used in both kinds of emulsion polymerization (batch and seeded core-shell), where they provided good stability. Films were formed with the core-shell latexes of the two previous latex series.

The mechanical properties (stress-strain curves) show systematically stronger initial modulus and ultimate break points in the cases in which the nonreactive surfactants were used (see an example in Fig. 3). However, the reverse was true for the water uptake. Then the behavior was better in the case of the surfmers. An example of the water uptake kinetics, which follow Fick's law, is shown in Fig. 4. A diagram comparing the behavior of three series of maleic surfactants is shown in Fig. 5. It turns out that the best products are the hemiamides.

In the same laboratory, series of styrenic nonionic and anionic surfmers and nonreactive surfactants with similar structures were prepared using living anionic ring-opening block copolymerization of successively butylene oxide and ethylene oxide, initiated with a potassium vinylbenzyl alcoolate. The living polymer was then killed in the same pot with water (to produce the nonionic surfactant, or it was used to open the ring of the propane sultone further to give the anionic sulfonate) [47].

These surfmers, as well as the corresponding nonreactive surfactants, were used in emulsion polymerization of core-shell all-acrylic monomers (core of polymethyl methacrylate, shell of copolymer methyl methacrylate–butyl acrylate). Depending on the HLB balance, stable latexes were obtained in most cases with nonreactive or reactive products. The best results in terms of stability versus the various tests (addition of electrolytes, ethanol test, resistance to shear, freeze-thawing) were obtained, as shown in Table 3, with hydrophilic compounds having a high cmc of 1.8×10^{-4} M (0.4 g/L). The latexes in Table 3 were produced from a seed of PMMA of 100 nm diameter and fed with a mixture of MMa and BuA to get particles 240 nm in diameter at 25% solids. In most cases, the targeted diameter was obtained and the latexes were monodisperse with only a small amount of coagulum. Stability versus the freeze-thawing test was obtained only with a few surfmers (VB-7, 34-S, and H).

The incorporation yield was estimated using ^{1}H NMR analysis in normal water with about 10% deuterated water added to provide the necessary lock signal. Higher D_2O contents sometimes cause exchangeable (NH or others) protons to disappear [48]. A set of irradiation techniques, now commonly available in most spectrometers [49], have been proposed that allow suppression of the huge normal water proton signal. It was shown that the surfmer causes the formation of copolymers containing a large majority of monomer units with a few surfmer units. These copolymers can be almost fully extracted upon extensive washing through ultrafiltration. However, normally they remain strongly adsorbed and, in most cases, do resist the ethanol extraction test. The NMR analysis of the latex, according to this approach, showed that 9% of the surfmer is located on the surface of the latex with only about 5% remaining in the serum. An example spectrum is shown in Fig. 6.

Only a part of the hydrophilic sequence of poly(ethylene oxide) participates in the steric stabilization, and about a third to a half of that sequence remains strongly attached to the particle. The corresponding data are shown in Table 4. The NMR analysis was limited to anionic surfmers, obtained through the opening of the ring of propane sultone by the living chain-end anion, which, being strongly hydrophilic, is supposed to be only in the water phase, where it can be used as an internal reference. Attempts to extend this quantitative study to the nonionic surfmer failed because of some adsorption on the particle surface of other calibration standards tried.

3. Surfmers in Latexes for Biotechnologies

Pichot's group carried out a set of studies to prepare and characterize surfmers having hydrophilic arms terminated with reactive groups for grafting to polystyrene or other conventional latexes. The general purpose of these preparations was to attach antibodies to the latexes so that they could be used safely in diagnostic assays. The surfmers were actually bifunctional, with a polymerizable end and at the other end of the hydrophilic arm a reactive coupling group.

They started with a styrenic macromonomer of polyvinyl alcohol prepared using the aldol group transfer polymerization technique [50] with an aldehyde group at the end. Later, polystyrene latexes were functionalized with it in a step-growth procedure [51]. The incorporation yield of the surfmer was higher if polar

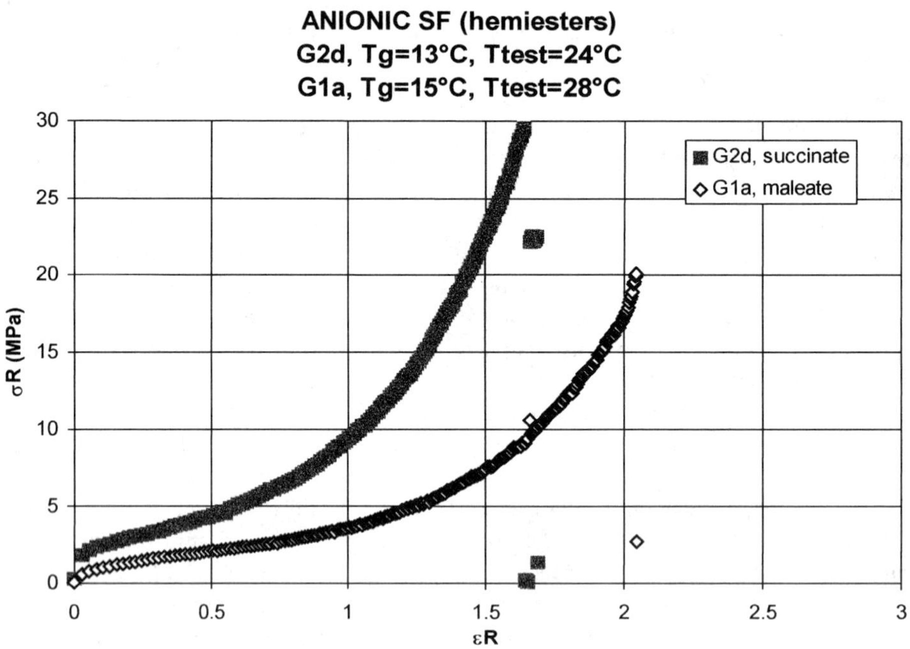

FIG. 3 Stress-strain curves of the films obtained from latexes prepared using hemiesters maleic (G1) or succinic (G2). (From Ref. 46.)

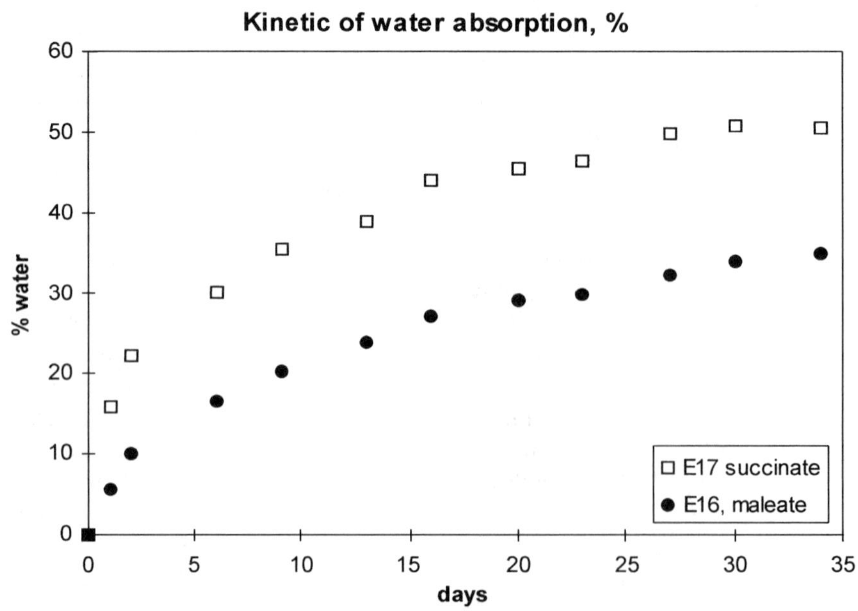

FIG. 4 Kinetics of water absorption of films from latexes prepared using zwitterionic surfactants: succinic (squares) and maleic (black rings). (From Ref. 46.)

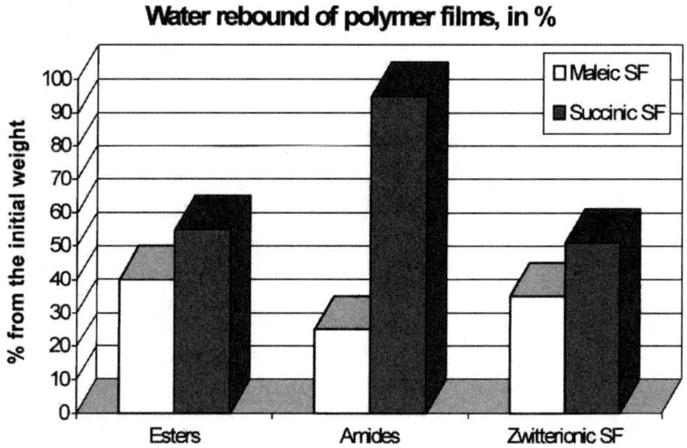

FIG. 5 Comparison of the water uptake of films prepared using the three series of surfactants, (hemiesters, hemiamides, zwitterionics). (From Ref. 46.)

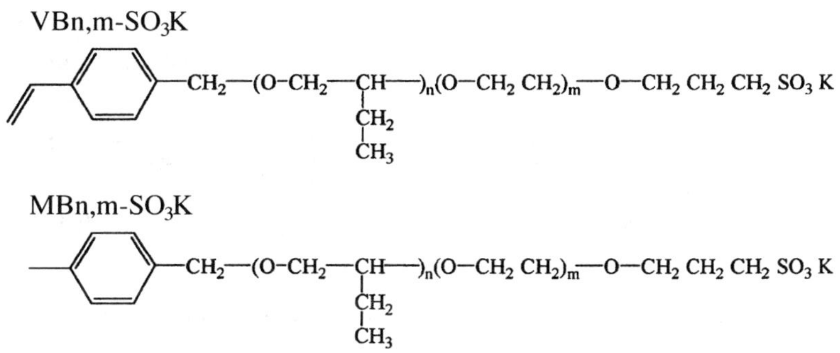

SCHEME 1 Stabilizer structures.

TABLE 3 Results with Styrenic Latexes

Latex[a]	Surf.[b]	Coag. (%)	D_n (nm)	Ip	Stability test: $MgSO_4$	Ethanol	Freeze	Shear
E47	SDS	2	249	0.05	Floc	Floc	Floc	ng
E48	NP 30	1.6	237	0.04	Stable	Floc	Floc	nd
E49	NOS 25	0.4						
E50	VB-10, 36-S	1.3	230	0.07	Floc	Stable	Floc	nd
E52	VB-6, 16-S	0.7	274	0.04	Stable	Floc	Floc	nd
E56	MB-6, 33-S	0.6	245	0.07	Stable	floc	floc	Coag
E57	VB-12, 45-S	4.1	234	0.04	Floc	Stable	Floc	nd
E62	VB-7, 34-H	2.3	277	0.05	Stable	Stable	Stable	Stab
E60	VB-7, 34-S	0.6	258	0.05	Stable	Stable	Stable	nd
E61	VB-12, 45-S	0.2	218	0.08	Stable	Stable	Floc	nd
E63	MB-6, 33-H	1.4	238	0.05	Stable	Floc	Floc	Stab

[a]E47, E48, and E49 are from commercial nonreactive surfactants.
[b]VB, styrenic (vinylbenzyl); MB, methylbenzyl (nonreactive); 12, 45, 12 PO units and 45 eo units; S, sulfonated (anionic); H, OH ended (nonionic)
Source: Data from Ref. 1.

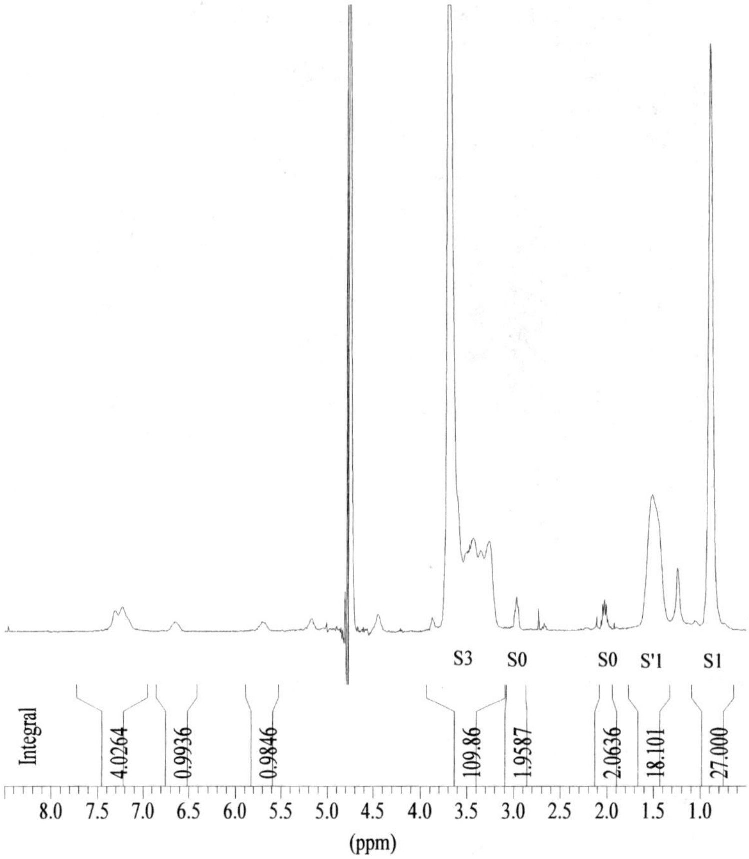

FIG. 6 Example of NMR spectrum of a styrenic anionic block-copolymer VB-9, 20-S. (From Ref. 48.)

TABLE 4 Analysis by ^{1}H NMR in Water (with Water Suppression) of Latexes Produced Using Etyrenic Anionic Block Copolymers PO-EO

Latex	Surfactant 1–3 phm*	Mobile EO units	% free in water	% on the particles
E50	VB10, 36-SO_3K 1–3	25	4.2	95.8
E52	VB6, 17-SO_3K 1–3	8	12.6	87.4
E57	VB12, 45-SO_3K 1–3	11	5.2	94.8
E61	VB6, 17-SO_3K 1.3–3.9	14	4.8	95.2
E59	MB10, 34-SO_3K 1–3	12	9.4	90.6
E60	VB7, 34-SO_3K 1–3	12	5.7	94.3
E64[a]	VB7, 34-SO_3K 1–3	12	4.4	95.6
E65	VB7, 34-SO_3K 0.5–1.5	10	4.6	95.4
E56	MB6, 33-SO_3K 1–3	8	14.8	85.2
E66[b]	*VB7, 34-SO_3K 1–3*	*15*	*6.6*	*93.4*
E68[b]	*VB7, 34-SO_3K 1–3*	*13*	*4.8*	*95.2*

[a]40% solid contents instead of 25%.
[b]Emulsion copolymerization composed of MMA/BA/AA 50/49/1 (AA, acrylic acid).
*phm = per hundred of expected mass.
Source: Ref. 49.

monomers were used in the growth process (acrylonitrile or MMA). Then the group turned to a sugar derivative obtained by condensation of hexyl methacrylate on a disaccharide; this derivative was also grafted to a polystyrene latex [52,53]. Finally, they used a liposaccharide for the same purpose [54]. In all of these studies, they took great care to characterize the reactivity of their surfmer (reactivity ratio measurements) and also their incorporation yield from NMR analysis of the latexes in deuterated water, where only the hydrophilic parts are visible.

More recently, Nagai et al. [55] functionalized polystyrene microspheres with a series of activated ester surfmers. These esters were obtained by condensation of unsaturated carboxylic acids (methacrylate, vinylbenzoate, acrylamidoundecanoate, decene-1-yl) onto hydroxyphenyldimethylsulfonium methyl sulfate. That technique leads to a high density of reactive groups at the surface of the latexes.

4. Polymerizable Surfactants in Other Emulsion-Like Processes

Emulsion-like processes are processes involving, for example, microemulsions, miniemulsions, dispersions, or micellar polymerizations.

(*a*) *Microemulsions.* The main interest in using microemulsion polymerization is the possibility of producing very small particles, in the range of 10–30 nm in diameter. Such small particles would be expected to result in improvements of latex properties, such as better gloss in paint films. One of the drawbacks is the need for large amounts of surfactants, even if the amount can be somewhat reduced by the use of cosurfactants, which are often low-molecular-weight alcohols (typically pentanol). Of course, a possible solution for that problem is the use of polymerizable surfactants or cosurfactants.

Not much has been published about the use of polymerizable surfactants to produce nanolatexes. All are quite recent papers. The first one to appear [56] was based on a polymerizable product, similar to CTAB, namely 11-(acryloyloxy) undecyl trimethylammonium bromide (AUTMAB).

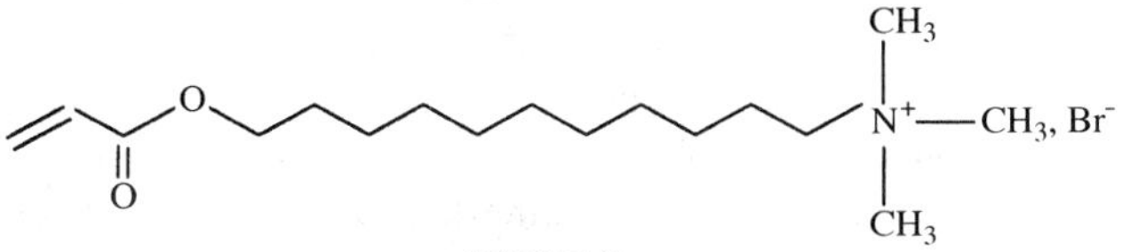

AUTMAB

The polymerization was carried out with γ-irradiation in transparent mixtures of styrene, water, and surfactant. Using a weight ratio between surfactant and monomer of 4, polymer particles of 20.7 nm diameter were obtained. However, the duration of the polymerization was much longer than in the case of CTAB. In addition to the copolymer particles, a large amount of AUTMAB was just homopolymerized. The particle size distribution was rather broad.

Subsequently, the same authors [57] compared the behavior of AUTMAB with another surfmer in which the polymerizable group is near the hydrophilic part of the surfmer. In that case the particle size remains that of the initial microemulsion, but they were linked together by a few polymer chains so that the final product was similar to a gel.

A more recent paper [58] suggested the use of hydroxypropyl methacrylate (HMPA) and similar polymerizable products as cosurfactant. Polymerization of styrene with 100% conversion was achieved at room temperature using either an oil-soluble photoinitiator or a water-soluble redox system (H_2O_2-ascorbic acid). The resulting latexes have sizes around 20 nm. The composition of the copolymer is similar to that of the monomer mixture when using a redox system, where the radicals are produced in the water phase, whereas a two-step polymerization giving rise to a block copolymer is produced when using the oil-soluble initiator. In the latter case, styrene (inside the core of the particles) is polymerized first. The ratio between the monomer and surfactants remains low, but the cosurfactant become incorporated in the latex so that the cleaning process of the latex, needed because of the use of a large amount of surfactant, can be simplified. No data were given about the particle size distribution.

Favero et al. [59] replaced SDS with the simple maleic surfmer A3 (Table 1) and used the reactive cosurfactant HMPA, as well as an initiator redox system, to polymerize styrene microemulsions. They obtained latexes with small particles of 20–30 nm containing all the monomer and cosurfactant, but only a part of the surfmer; which was not fully converted.

(*b*) *Miniemulsions.* Miniemulsion polymerization is a process in which very small submicrometer monomer droplets are produced and directly polymerized. The stabilizer system involves the use of a mixture of surfactant and hydrophobic additive and allows a shelf life of the miniemulsion of several days to weeks. However, at variance from the case of microemulsions, the so-called cosurfactant, actually a hydrophobic additive, works through a thermodynamic anti-Ostwald ripening mechanism. The hydrophobic additive typically has essentially vanishing solubility in the continuous phase and appreciable solubility in the miniemulsion droplet phase. The mole ratio of monomer to hydrophobic additive can be denoted *n*. When monomer dissolves in the continuous phase (to a very low extent) so as to be (diffusion) transported to larger monomer droplets and yield larger droplets with lower radii of curvature (Ostwald ripening), a thwarting osmotic presure gradient is set up because the mole ratio *n* in the continuous phase cannot be maintained due to the vanishing solubility of the hydrophobic additive. Hence, dissolution of monomer in the continuous phase is greatly retarded, and Ostwald ripening of miniemulsion droplets is blocked. Typical hydrophobic additives are hexadecane and cetyl alcohol.

Another report concerns the use of reactive cosurfactants in miniemulsion polymerization [60]. In that study dodecyl methacrylate (DMA) and stearyl methacrylate (SMA) were used as cosurfactants with SDS and compared with cetyl alcohol (CA) and hexadecane (HD). It has been shown that DMA behaves like CA and SMA displays a behavior similar to that of hexadecane in terms of droplet size stability as well as the particle size distribution of latexes. However, the distribution obtained using these reactive hydrophobic additives are in both cases somewhat narrower than for the model additives.

(*c*) *Dispersion Polymerizations.* In dispersion polymerization, the monomer is initially soluble in the surrounding medium, but the polymer is not soluble and precipitates as soon as it is formed. In the presence of a suitable surfactant, the precipitate particles are stabilized, generally as small spheres of submicrometer or micrometer size. The initial work, described in a book by Barrett [61], used organic solvents (typically polymerization of styrene or methyl methacrylate in heptane), but more recently polar mixtures of water and light alcohols have been used to produce monodisperse micrometer-sized particles in the presence of hydrophilic polymers such as cellulosic compounds, polyvinyl alcohol, or polyvinylpyrollidone (PVP) [62–64].

The Japanese group of Kobayashi [65,66] discovered that using macromonomers of polyoxazoline allows control of the particle size of polymethyl methacrylate with much less stabilizer than when using a high-molecular-weight polymer of the same structure. Typically, 0.2 to 0.8% of macromonomer can be used instead of 5 to 30% of high polymer to give monodisperse ($D_w/D_n < 1.03$) particles 1 to 4 μm in diameter. Styrenic macromonomers of polyethylene oxide have been studied by Kawaguchi et al. [67] and showed good stabilization efficiency. The authors developed a mechanistic model based upon an adaptation of the model of Paine [68] allowing prediction of the final particle size.

A more systematic study has been carried out by Lacroix-Desmazes and Guyot [69] to compare a set of hydrophilic macromonomers of polyoxyethylene with a variety of polymerizable groups with the corresponding amphiphilic products with a hydrophobic chain end. Whatever the nature of the macromonomer (styrenic, methacrylic, maleic), better performances (in terms of particle size monodispersity, coagulum formation, and incorporation yield) were obtained with macromonomers than with the corresponding amphiphilic products with a hydrophobic chain end. The incorporation yield, however, was rather limited. The macromonomer stabilizers led to a rather good fit with the Kawaguchi model involving very rapid particle nucleation, which may be checked but does not account for the change in solvency of the medium owing to the conversion of the monomer [70]. The reason for the better performances of the macromonomers as compared with true surfactants is not clear. It may be related to the reactivity of the stabilizer in the solution. The large particles (>2 μm) produced are stable, depending chiefly on the grafting density of the macromonomer. When the length of the chain is about 50 monomer units, the latexes are resistant to freeze-thawing (provided the area covered by each grafted macromonomer is not greater than 6 nm^2).

It has been shown that it was possible to stabilize vinyl acetate dispersion polymerization in supercritical CO_2 using perfluoroalkyl polymerizable surfactants [71].

(*d*) *Micellar-Emulsion Polymerization.* In micellar polymerization, all the monomer that swells a micelle is polymerized from the same radical. In the case of emulsion polymerization, the radical is expected to come from the water phase, where it should polymerize the water-soluble monomers. In micellar-emulsion polymerization, block copolymers, with segments of water-soluble monomer units, followed by a hydrophobic segment of the monomer swelling the micelle can be produced if the radical commences polymerization in the continuous phase and then transfers to the swollen micelle. If the radical then returns to the water phase, it will continue to polymerize the water-soluble monomer, and then it may enter another micelle and so on. A multiblock copolymer will then result with alternating hydrophilic and hydrophobic blocks. The hydrophilic segments present a random distribution, but the hydrophobic segments should have a quasi-monodisperse distribution because the swollen micelles are all expected to be about the same size.

This topic has been studied thoroughly by Candau's group in order to produce associative thickeners. They prepared and used styrenic cationic surfmers [72] and extended the concept to other monomers that can be polymerized photochemically [73]. They further carried out studies to investigate the mechanism of micellar-emulsion polymerization [74]. To that purpose, they used a variety of initiators and inhibitors having different water solubilities. It was concluded that most often the radicals were produced in the water phase and had to enter the micelles before the polymerization could take place. A clear picture of the system appeared more recently [75]. The authors studied three systems. In the first one, a hydrophobic acrylamido monomer-swollen SDS micelle was copolymerized with acrylamide. In the second system, they used the styrenic cationic surfmer N16. In the last system, they used mixed micelles of N16 and a similar but nonpolymerizable surfactant, B16. Figures 7, 8, and 9 show schematically the structure of the resulting polymers.

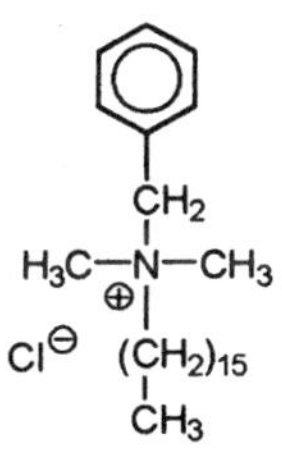

B16

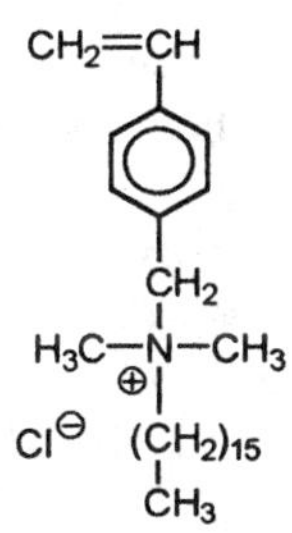

N16

In the first system, the hydrophobic monomer swelling the SDS micelle is polymerized in the same polymer chain, but most of the SDS is found in the serum (Fig. 7). In the second system, the hydrophilic part of the surfmer remains associated with the main polymer chain (Fig. 8). At variance with these results, it was found that if the surfmer is mixed with a nonpolymerizable surfactant, the surfmer units are randomly distributed in the polyacrylamide polymer (Fig. 9). Surprisingly, the thickening efficiency of the last system is a thousandfold higher than that of the former. It is suggested that the difference is due to the intramolecular nature of the hydrophobic interactions when the micelles contain only the surfmer.

The same group of authors reported a very interesting surfmer with a fluorocarbon hydrophobic chain [76]. The solution, micellar polymerization, and associative properties of these surfmers have also been reported [77]. They also display strong intramolecular in-

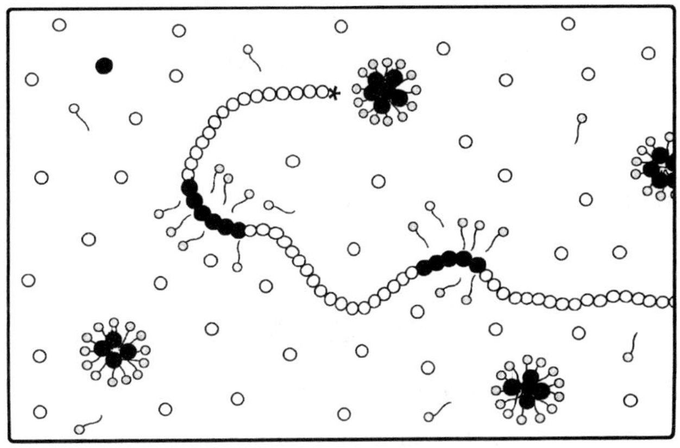

FIG. 7 Schematic representation of the process I acrylamide (○), hydrophobic monomer (◎), and SDS. (From Ref. 75.)

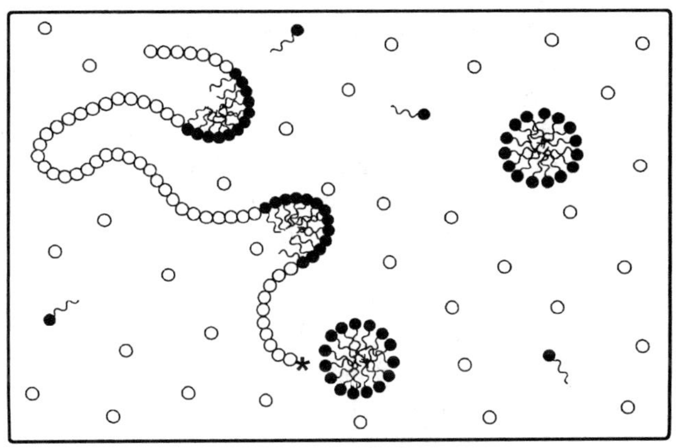

FIG. 8 Schematic representation of the copolymerization process II: acrylamide (○) and polymerizable surfactant. (From Ref. 75.)

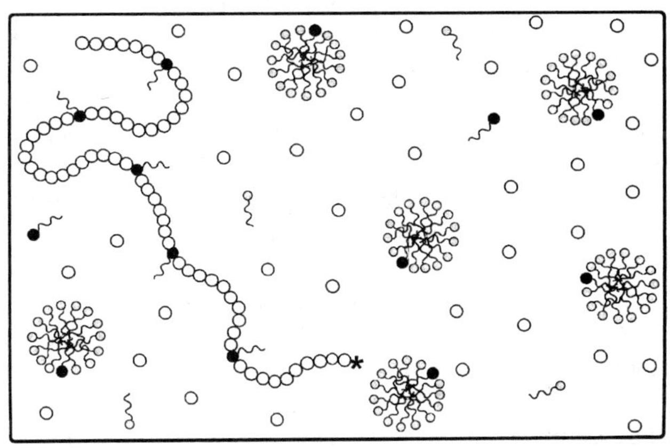

FIG. 9 Schematic representation of the copolymerization process III: acrylamide (○) and a mixture of polymerizable (surfmer) and nonpolymerizable surfactants (black points for the polymerizable group of the surfmer). (From Ref. 75.)

teraction and low viscosity. The most interesting behavior of such fluorocarbon acrylamido surfmers is that they are able to produce multicompartment polymeric micelles when they are copolymerized with their hydrocarbon analogous derivatives because of their strong incompatibility [78].

III. POLYMERIC STABILIZERS IN HETEROPHASE POLYMERIZATION

A. The Meaning of Polymeric Stabilizers

Another way to fix stabilizers in place or to reduce the migration thereof in the final application of a polymer dispersion is to use polymeric stabilizers that alone, because of their larger mass, diffuse much slower than conventional low-molecular-weight surfactants. The use of water-soluble polymeric surfactants has a long history in heterophase polymerization. For the first attempts to reproduce the mild conditions of the formation of natural rubber latexes in the laboratory at the beginning of the 20th century, naturally occurring biopolymers such as gelatin, egg albumin, starch, milk, and blood serum were used as colloidal stabilizers (see Chapter 21 in this volume). Today, a huge variety of different polymeric stabilizers are known, although only a few classes have found technical application in heterophase polymerizations. These have been mainly nonionic products based on poly(ethylene glycol) (PEG) and in situ formed copolymers with methacrylic or acrylic acid that are frequently applied as comonomers. Nonionic surfactants are mainly polymeric surfactants as the hydrophilic group in these surfactants is almost exclusively a PEG chain.

The interaction between polymer molecules and polymeric colloidal particles can lead per se to fascinating phenomena, as the two different polymers are incompatible provided they do not strongly interact and their concentrations are not too low. Consequently, mixing (stabilization) and demixing (flocculation) may occur. Polymers as stabilizers or flocculants are of great practical interest in a broad range of industrial processes [79].

Compared with low-molecular-weight surfactants, polymeric stabilizers possess more degrees of freedom regarding the fine tuning of their properties. These are the molecular weight or molecular weight distribution and, as most of the polymeric stabilizers are copolymers, the distribution of the chemically different units along the chain. Altering these properties may lead to drastic changes in the stabilizing behavior. Another interesting difference compared with the behavior of low-molecular-weight surfactants is that polymers (in these particular cases mainly homopolymers) can stabilize a dispersion without being adsorbed but simply by their presence in the dispersion medium. This effect is called depletion stabilization, but a flocculation of a dispersion by depletion effects is also possible. Whether depletion stabilization or flocculation occurs depends on the temperature and concentrations of both the polymer and the particles and strongly on the particle-to-polymer coil size ratio [80]. These facts illustrate that the design of an optimal polymeric stabilizer is not an easy task and that polymeric stabilizers offer more possibilities than low-molecular-weight surfactants to tailor properties. The latter is presumably the reason why nature employs polymeric stabilizers in various applications, for instance, to stabilize milk or the natural rubber latex *Hevea brasiliensis*. Furthermore, even at very low concentrations polymeric stabilizers affect in a very specific way the properties of a colloidal system, for instance, the aggregation behavior, the colloidal stability, and the rheology [79,81,82].

In emulsion polymerization, transport of matter (monomers, radicals) through the particle interface is essential and hence a rigid polymeric stabilizer layer may influence the polymerization kinetics considerably (see Chapter 21). As this section deals mainly with the application of polymers as stabilizers in heterophase polymerizations, some other very interesting subjects of current interest in the greater scientific field are only briefly mentioned. The first topic is the formation of polymeric or polymer-coated nanoparticles by consecutive adsorption of oppositely charged polyelectrolytes. The nanometer-sized templates can be either polymeric latex particles [83–85] or biomedical objects such as enzyme crystals (F. Caruso et al., submitted). This layer-by-layer adsorption also allows the preparation of composites consisting of polystyrene cores and, for instance, multilayered shells of poly(diallyldimethylammonium chloride) and silica particles [86] or metallosupramolecular coordination polyelectrolytes [87], respectively. Note that instead of synthetic polyelectrolytes, oppositely charged proteins can be employed to form such multilayered shells [88]. A review article regarding this topic has appeared [89]. The second example is the use of polymeric stabilizers [in that particular case poly(2-hydroxyethylmethacrylate-co-sytrene-4-sulfonic acid sodium salt] for the preparation of colloidal dispersions of conducting polymers [90]. For example, such dispersions are very promising in applications as waterborne conductive coatings for corrosion protection. A third example is the stabilization of inorganic dispersions such as titanium dioxide by block copolymers. Diblock copolymers of 4-vinyl pyridine and sodium methacrylate prepared by anionic polymerization of 4-vinyl pyridine followed by *tert*-butyl methacrylate and subsequent hydrolysis of the *tert*-butyl methacrylate block to the sodium methacrylate are effective stabilizers for aqueous dispersions of alumina-coated titanium dioxide with a solids content up to 80 wt% [91].

This part of the chapter begins with some general remarks on polymers as stabilizers for polymer dispersions followed by special developments in the field for various kinds of heterophase polymerization techniques, particularly suspension, dispersion, and emulsion polymerization.

B. Stabilizing Polymer Dispersions by Polymers—Basic Principles

Polymeric stabilizers may be either nonionic or ionic or a combination of both. Consequently, polymers may contribute to the stability of a colloidal system via all known mechanisms, i.e., electrostatic, steric, and electrosteric stabilization or any mixture of these. However, compared to low-molecular-weight surfactants, polymeric stabilizers behave distinctly differently. For instance, selecting a polymeric surfactant for a particular application is not an easy undertaking as classical concepts such as the HLB value or the reduction in the interfacial tension are not applicable alone. Other effects have to be taken into consideration, such as osmotic contributions, the conformation of both adsorbed and dissolved polymer chains, and entropic and elastic forces. Hence, a general expression for the total free energy of interaction between colloidal particles stabilized with polymeric surfactants (ΔG_{toti}) should be written as

$$\Delta G_{\text{toti}} = \Delta G_{\text{at}} + \Delta G_{\text{es}} + \Delta G_{\text{st}} + \Delta G_{\text{est}}$$

where ΔG_{at} is the attractive energy, ΔG_{es} and ΔG_{st} are the electrostatic and steric repulsive energies, respectively, and ΔG_{est} is the electrosteric repulsive energy. For a detailed discussion of steric stabilization the reader is referred to excellent monographs by Napper [79] and Piirma [92]. The benefit of steric stabilization with nonionic polymers is that such particles are much less sensitive to process fluctuations and they exhibit, compared with purely electrostatically stabilized particles, increased stability against electrolytes. Furthermore, these stabilizers are equally effective at low and high solids contents and they are able to stabilize particles in both aqueous and nonaqueous dispersions.

Regarding electrosteric stabilization, a few more words are necessary as this is much more a generic term than the simple combination of electrostatic and steric stabilization. In principle, there two possibilities for electrosteric stabilization. Either the charges are on the particle surface and the steric contribution results from an uncharged macromolecule additionally adsorbed or the charges are directly on the polymer chain attached to the particles. The latter case has been treated theoretically by Pincus [93]. The important conclusion of his treatment is that polyelectrolyte chains attached to particles should give this system extraordinary stability against added salt if the thickness of the polyelectrolyte layer is much larger than the particle radius.

C. Polymeric Stabilizers for Heterophase Polymerization Techniques

The number of polymeric structures suitable to act as a colloidal stabilizer is almost infinite. Under particular conditions, every kind of polymeric architecture including homopolymers and copolymers (e.g., random, comblike, star, or multiarm) and block copolymers can act as a stabilizer of colloidal systems. On the other hand, each heterophase polymerization technique requires stabilizers or stabilizer mixtures with specific properties. Note that the stabilizer properties have to meet the demands of both the polymerization process and the final application of the polymer.

1. Suspension Polymerization

In suspension polymerization, so-called colloidal stabilizers are frequently applied that are basically water-soluble polymers instead of low-molecular-weight surfactants. Typical stabilizers are poly(vinyl alcohol) (PVAL) or hydroxypropyl celluloses and polycarboxylic acids for vinyl chloride and acrylonitrile-butadiene-styrene suspension polymerizations, respectively [94]. There is evidence that these stabilizers are adsorbed at the interface between the organic phase and water [94]. Furthermore, the stabilizers in suspension polymerization lower the interfacial tension of the aqueous phase and in this way support the formation of the starting monomer emulsion. An oil-soluble initiator starts the polymerization inside the monomer droplets, whose size almost corresponds to the size of the final polymer beads or grains (corrected by the shrinkage due to the density difference between monomer and polymer). The stabilizer has to support the emulsification of the organic phase (droplet formation) by shear forces. The droplets move within the shear field inside the reactor and may dissolve again if they are in regions of lower forces and will be formed again in a region of higher shear forces as long as the conversion is below a certain value called the point of particle identity. Because of the much larger particles (much lower interfacial area) the amount of stabilizer required in suspension polymerization is 0.06–0.1 wt% based on the monomer weight, which is much lower than for emulsion polymerization (usually 1–5 wt%). For a more detailed description of the suspension polymerization process see Refs. 94 and 95.

The influence of the properties of poly(vinyl alcohol) as stabilizer in vinyl chloride suspension polymerizations is an important research topic. For instance, the influence of an acetalization of PVAL [96] and a thermo-oxidative treatment [97] on the grain morphology of poly(vinyl chloride) (PVC) was investigated. It was found that the modified stabilizers led to less porous PVC particles with substantially higher bulk densities and lower plasticizer absorption.

The size of the vinyl chloride droplets formed during the emulsification process depends on the stirrer speed and on the PVAL concentration [98]. The extent of coalescence rises slowly with mixing time, is almost proportional to the stirrer speed, and decreases when the PVAL concentration is increased. Furthermore, the coalescence rate depends on the degree of hydrolysis of the stabilizer. The degree of hydrolysis of the poly(vinyl acetate) (PVAC), the precursor of PVAL, also influences the morphology, the porosity, and the rigidity of the PVC particles [99]. To fulfill the need for higher grain porosity facilitating faster removal of residual vinyl chloride monomer and faster, more uniform plasticizer uptake, so-called secondary suspending agents are now applied. These are PVAC polymers with a degree of hydrolysis less than 55% that are also employed as PVAL stabilizers (highly hydrolyzed PVAC polymers) [100].

It was found that the mean particle diameter of the PVC beads decreases with increasing solution viscosity of the PVAL stabilizer but increases with increasing interfacial tension between the aqueous and the vinyl chloride monomer phase [101]. Furthermore, during the initial state of a vinyl chloride suspension polymerization, PVC chains are grafted onto the PVAL molecules, forming a membrane around the droplets. The particular polymerization conditions (type of PVAL, temperature, hydrodynamic conditions, etc.) govern the properties of the membrane, which influence the morphology of the growing PVC beads and the properties of the final resin [102].

A review of the styrene-divinylbenzene copolymerization system covering literature up to 1995 has been published [103]. In the following some more recent developments related to styrene suspension polymerization are reported. The interfacial tension between an aqueous PVAL solution and styrene monomer was investigated as a function of the PVAL concentration and the temperature [104]. Based on these investigations, a configuration model was developed to explain the results obtained both in unstirred and stirred reactions.

PVAL was also used as stabilizer in the preparation of microcapsules containing pigments and polymers [105]. In a first step, a mixture of pigments, core monomers, free radical initiator, and an oil-soluble shell monomer was dispersed in an aqueous PVAL solution, followed by an interfacial polycondensation and a thermal polymerization of the core monomer. It was found that the PVAL takes part in the interfacial condensation reaction.

It is interesting to know the development of the particle size distribution during polymerization for controlling the process and polymer properties. A technique based on light scattering measurements has been developed for the aqueous suspension polymerization of methyl methacrylate (MMA) with benzoyl peroxide as initiator and agarose as stabilizer [106].

An extension of the scope of the suspension polymerization technique to monomers and initiators that cannot be used because of their solubility or reactivity in conventional systems (oil-in-water or water-in-oil systems) is possible if perfluorocarbon (PFC) fluids are used as dispersion medium. Because of their unique properties (low surface tension, bad solvency for nonfluorinated compounds, and chemical inertness), perfluorocarbon fluids represent quasi-universal dispersion media [107]. From the colloid chemical viewpoint, the choice of a stabilizer for a polymerization in PFC fluids poses a very interesting problem. In general, no stabilizer is needed to obtain a monomer emulsion under normal dispersing conditions as the surface tension of the PFC fluids is in the range 12–20 mN m^{-1} and is much lower than the corresponding values of the monomers. In contrast to suspension polymerizations in other dispersion media (aqueous as well as nonaqueous), the stabilizer must not be soluble in the dispersion medium. Of all systems investigated [stabilizer-free or poly(methacrylic acid), sodium dodecyl sulfate, sodium perfluorooctanoate, and a copolymer containing perfluorocarbon and poly(ethylene glycol) tails as stabilizers, respectively], sodium dodecylbenzenesulfonate led to the smallest particles in the case of a styrene divinylbenzene polymerization. This is explained by an interaction of the aromatic group with the droplets/particles, thus effectively preventing coalescence or coagulation processes.

Finally, attempts have been made to carry out controlled radical polymerizations in suspension polymerization mediated by both 2,2,6,6-tetramethylpiperidine-*N*-oxyl (TEMPO) [108] and a ruthenium complex in water or alcohol systems [109]. In these cases the stabilizers also play an important role, as any kind of transfer strongly influences the kinetics and may lead to loss of control [110].

2. Dispersion Polymerization

As in suspension polymerization but not emulsion polymerization, the stabilizer in dispersion polymerization is usually a polymer. Typically amphiphatic polymers are used that are able to adsorb at the interface between the particles and the dispersion medium. Good reviews covering all aspects of the dispersion polymerization technique have been published and are highly recommended to the more interested reader [111,112]. A characteristic feature of dispersion polymerization is that the dispersion medium is a solvent for the monomer(s) and the stabilizer(s) but not for the final high-molecular-weight polymer(s). This makes dispersion polymerization unique among all heterophase techniques as during a batch polymerization the solvency of the dispersion medium changes drastically in proportion to the conversion. The stabilizer has to face these changes and must impart colloidal stability to the system under changing conditions regarding the state of solution.

Table 5 gives a summary of recent papers devoted to dispersion polymerization, placing special emphasis on the different stabilizers put to use. Furthermore, this table clarifies the many different possibilities to vary the particular conditions in dispersion polymerization compared with the other heterophase techniques. The main reason for this is the possibility to tune the properties of the continuous phase to the special needs for radical, anionic, or oxidative polymerization mechanisms and also for a variety of different monomer-polymer combinations.

The easily accessible particle size range of a few micrometers makes dispersion polymerization a promising tool for the preparation of particles for various kinds of separation techniques. These applications often require cross-linked particles. But the use of a cross-linker sometimes causes difficulties. A detailed discussion of the problems encountered with the use of the cross-linker divinylbenzene in the preparation of polystyrene particles can be found in Ref. 175.

Another point is worth mentioning. In different papers it has been pointed out that a certain degree of grafting of the polymeric stabilizers to the particles is necessary to impart sufficient colloidal stability [122,136,139]. In this context macromonomers (see Table 5), polymeric initiators [174], and polymers with chain-transfer active groups [136,139] play a special role. A review concerning the radical polymerization of PEG macromonomers in disperse systems reports applications in both dispersion and emulsion polymerization [176].

3. Emulsion Polymerization

The variety of polymeric stabilizers used for purely water-based emulsion polymerizations is limited compared with the apparently infinite number of structures available for dispersion polymerization. This is due to the fact that the stabilizers must be soluble in the continuous phase. From a technical point of view, poly(ethylene glycol) is the most important polymer or building block of polymeric stabilizers for emulsion polymerization. PEG takes an outstanding position as technical products are available in a wide variety and its solutions possess unique properties [177]. The properties of PEG-based polymeric stabilizers in aqueous emulsion polymerizations have been reviewed extensively [178,179] and will not be considered here in detail. For the application of macromonomers in emulsion polymerization (also mainly based on PEG) the reader is referred to Ref. 176. Different types of macromonomers with respect to the reactive group (maleate, methacryl, vinylbenzyl) based on PEG as the hydrophilic part have been investigated in a comprehensive study as stabilizers in the batch emulsion polymerization of styrene [180]. In contrast to nonionic, water-soluble polymers (PVAL, PEG, hydroxyethylcellulose, potato starch dextrin) that lead after nucleation to secondary agglomeration during the polymerization, the macromonomers are able to stabilize the particles very effectively. In principle, similar macromonomers can be utilized either in aqueous dispersion or in emulsion polymerization. Examples can be found in Ref. 146. This paper also describes the first examples of sulfobetaine-based macromonomers. These macromonomers were synthesized by derivatizing the poly(2-(dimethylamino)ethyl methacrylate) macromonomers with 1,3-propane sultone. These macromonomers have proved to be effective electrosteric stabilizers for styrene emulsion polymerization at high electrolyte concentrations (1 M NaCl).

TABLE 5 Overall View of Stabilizer Used in Dispersion Polymerization

Stabilizer/continuous phase	Monomer/initiator	Ref.
Oxidative polymerization		
Ethylhydroxyethyl cellulose/H_2O-EtOH	Pyrrole/APS or $FeCl_3$	113, 114
Poly(vinyl methyl ether)/1.25 M HCl in H_2O	Aniline/APS	115
Water-soluble polymer/1,4 dioxane-H_2O	Phenol, cresols/HRP	116–118
Methylcellulose/H_2O-EtOH	Aniline/APS	119
Poly(vinyl alcohol), poly(*N*-vinyl pyrrolidone)/HCl-H_2O	Aniline/APS	120
Poly(vinyl alcohol)/H_2O	3,5-Xylidine/APS	121
Ethylhydroxyethyl cellulose/H_2O or H_2O-EtOH	Aniline/APS	122
PEG with a single pendant aniline per chain/DC1-D_2O	Aniline/SPS or KIO_3	123
Carboxymethyl cellulose/EtOH-H_2O (1:1)	Aniline/APS	124
Supercritical CO_2 ($scCO_2$) as continuous phase		
Poly(1,1-dihydroperfluorooctylacrylate) (PFOA)	MMA/AIBN	125, 126
Poly(dimethylsiloxane)-methyl methacrylate	MMA/AIBN	127, 128
Poly(1*H*,1*H*,2*H*,2*H*-perfluorooctyl methacrylate-*b*-PMMA)	Divinylbenzene/AIBN	129
Poly(1,1-dihydroperfluorooctyl methacrylate) or PFOA	Styrene/AIBN	130
Poly(1,1-dihydroperfluorooctyl methacrylate) or PFOA	Fluorinated (meth)acrylates, MMA, DMAEMA/Cu(0)/CuCl/bipyridine derivatives	131
Encapsulation of inorganic materials		
PEG/H_2O, Fe_3O_4 particles	Styrene-co-HEMA	132
Poly(*N*-vinyl pyrrolidone)/H_2O-EtOH, silica particles	Styrene/AIBN	133, 134
PEG/H_2O-EtOH, Fe_3O_4 particles	Styrene/AIBN	135
Reactive stabilizers (macromonomers and others)		
PEG with terminal thiol group or poly(*N*-vinyl pyrrolidone)/H_2O-EtOH	Styrene/AIBN	136, 137
PVAL- and poly(methacrylic acid) with terminal vinylbenzyl groups/H_2O-EtOH	MMA/AIBN	138
Poly(vinyl alcohol) with terminal thiol group/EtOH	Styrene/AIBN	139
PEG-maleate or PEG-maleate-alkyl/H_2O-EtOH	Styrene/AIBN	140, 141
Vinylbenzyl-terminated poly(acrylic acid)/H_2O-EtOH	MMA/AIBN	142
ω-Methoxy-PEG-undecyl-α-methacrylate/H_2O-EtOH	Styrene/AIBN	143, 144
Poly(dimethylsiloxane)-methyl methacrylate/$scCO_2$	MMA/AIBN	127, 128
Poly(*N*-isopropylacrylamide)-macromonomer/EtOH or H_2O-EtOH	Styrene/AIBN	145
Macromonomers based on 2-(diethylamino)ethyl methacrylate or other tertiary amine methacrylates/EtOH	Styrene/AIBN	146
Nonylphenol-PEG-monomaleate/different alcohols	Styrene/AIBN	147
Methacrylate-terminated phthalate glycol polyester/$CHCl_3$-EtOH	MAA-*co*-glycidyl-methacrylate/BPO	148
Anionic polymerization		
PS-*b*-polybutadiene/*n*-hexane	Styrene/*sec*-butyl lithium	149, 150
Poly(*t*-butylstyrene)/*n*-hexane	Styrene/*sec*-butyl lithium	151
General papers (no specific feature)		
PVAL (35% hydrolyzed)/CH_3OH	Styrene/AIBN	152
Poly(*N*-vinylpyrrolidone) and Na-dioctyl sulfosuccinate as costabilizer/EtOH	Butadiene co styrene/AIBN	153
PEO-PPO and dextrane/H_2O-HCl	Ethyl cyanaoacrylate/phosphoric acid	154
Poly(*N*-vinylpyrrolidone)/isopropanol-H_2O	Styrene and DVB/AIBN	155
2-Ethylhexyl methacrylate copolymer/isooctane	VAC/AIBN	156
Poly(*N*-vinylpyrrolidone)/H_2O-EtOH	MMA/AIBN and ceric ammonium nitrate	157
Poly(methacrylic acid)-*co*-ethylacrylate/H_2O-EtOH	Styrene	158, 159
Poly(2-(dimethylamino)ethylmethacrylate)-*b*-alkylmethacrylate/alcohols (C_1OH-C_8OH)	Styrene/AIBN	160
Poly(*N*-vinylpyrrolidone) or poly(*N*-vinylpyrrolidone)-*co*-(vinyl acetate) or PVAC/EtOH-2-methoxy ethanol	Styrene/AIBN	161
Polyepichlorhydrine/alcoholic media	Styrene	162
PVAL/EtOH-H_2O	MMA/AIBN	163
Polyacrylic acid/EtOH-H_2O	Styrene/AIBN	164
PVAL/H_2O-EtOH	*n*-Butyl methacrylate onto PS-seed/BPO	164
Poly(vinyl methyl ether)/H_2O-*tert*-butyl alcohol	Acrylamide/APS	165
PS-*b*-polybutadiene/*N*,*N*-dimethylforamide-toluene	4-Vinylpyridine co MAA/AIBN	166

TABLE 5 Continued

Stabilizer/continuous phase	Monomer/initiator	Ref.
Na-alginate-$CaCl_2/H_2O$	*N*-Isopropylavrylamide/KPS-teramethylenediamine	167
Poly(methacrylic acid)/MeOH-H_2O	MMA or styrene onto PMMA seed/AIBN	168
Poly(*N*-vinylpyrrolidone) and Aerosol OT/EtOH	Styrene-*co*-urethane acrylates/AIBN	169
Poly(vinyl methyl ether)/H_2O-*tert*-butyl alcohol	Acrylamide/AIBN	170
Cellulose acetate butyrate/toluene-2-methylpropan-1-ol	2-Hydroxyethylmethacrylate-*co*-ethylene dimethacrylate/BPO	171, 172
Poly(*N*-vinylpyrrolidone)/EtOH-H_2O	Styrene co glycidylmethacrylate/AIBN	173
Polydimethylsiloxane/different solvency for polydimethylsiloxane (heptane, toluene, 1,4-dioxane, etc.)	MMA/Polydimethylsiloxane-azo-initiator	174

Abbreviations: APS, ammonium peroxodisulfate; HRP, horseradish peroxidase; SDS, sodium dodecyl sulfate; AIBN, 2,2′-azobisisobutyronitrile; DMAEMA, *N,N*-dimethylaminoethyl methacrylate; HEMA, hydroxyethyl methacrylate; MAA, methacrylic acid; BPO, benzoyl peroxide; DVB, divinyl benzene; PS, polystyrene; KPS, potassium peroxodisulfate; VAC, vinyl acetate; EtOH, ethanol; MeOH, methanol; C_1OH–C_8OH, methanol to octanol.

Another interesting class of substances for emulsion polymerizations are macroinitiators, which are considered here to be exclusively polymers with one or more thermolabile groups inside the polymeric backbone [181–189]. Polymers with pendant thermolabile groups along the chain are outside the scope of this contribution as these form one polymeric and one low-molecular-weight radical. A proper choice of the polymerization recipe will reduce the components to water, monomer(s), and the macroinitiator [190–192]. The resulting particles consisting of block copolymers and the particular structure thereof (di-, tri-, or multiblock copolymers) are determined by the mode of the termination reaction [190–192].

Neat PEG chains can also be used as reactive steric stabilizers in emulsion polymerization either as a reducing agent in the ceric ion redox system [190,192] or with a terminal thiol group as chain transfer agent [14]. Many water-soluble polymers show a strong tendency to grafting if they are present during a radical polymerization. This can be utilized to prepare polymer dispersions with covalently bound polymers. A systematic investigation concerning the grafting efficiency of potassium alginate, nitrolignin, methyl cellulose, hydroxyethyl cellulose, and sodium carboxymethyl cellulose during a methyl acrylate polymerization started with APS is described in Ref. 193. The graft copolymers formed during the first stage of the polymerizations act in the further course as stabilizers for the particles.

One family of polymers produced by emulsion polymerization is strongly connected with a polymeric stabilizer, namely polymer dispersions based on vinyl acetate and prepared with PVAL. Furthermore, the history of technical emulsion polymerizations is strongly connected with vinyl acetate using PVAL as emulsifier [194]. PVAL is prepared by hydrolysis of poly(vinyl acetate) and it is likely that hydrolysis of vinyl acetate also takes place during the polymerization in the aqueous phase. The resulting alcohol units (1) influence the particle nucleation and (2) contribute to latex stability. Alcoholic groups are also incorporated in the final polymer by reaction between a growing poly(vinyl acetate) chain with the PVAL forming a graft copolymer [195]. Hence, poly(vinyl acetate)-PVAL can be considered a unique example in heterophase polymerization of a "symbiosis-like" polymer-stabilizer system.

PVAL is also frequently used as a stabilizer for monomers other than vinyl acetate. An interesting example is the stabilization of poly(DL-lactide-*co*-glycolide) nanoparticles prepared by the spontaneous emulsification solvent diffusion (SESD) process [196]. The SESD process means, as the first step, the preparation of a poly(DL-lactide-*co*-glycolide) solution in organic solvents followed by emulsification in an aqueous PVAL solution and subsequent evaporation of the solvents. The solid polymer phase is separated by preparative ultracentrifugation and after that redispersed in pure water. Particles with diameters as small as 100 nm can be obtained. It was found that the degree of hydrolysis of the PVAL is more important than the degree of polymerization in influencing the particle size and the redispersibility. In general, PVAL is the key stabilizer in redispersible powders that are frequently applied as binders in various branches of the construction chemicals industry [197].

It is necessary, however, to mention that other emulsifiers can be used in vinyl acetate emulsion polymerization. Earlier, gum arabic (polyarabic acid) was successfully used as stabilizer. But a drawback—not only of this natural stabilizer—is that from time to time it behaved unpredictably, leading to polymerization irregularities and heavy precipitation [198]. Systematic investigations with other polymeric stabilizers such as hydroxyethyl cellulose [199] and poly(methacrylic acid) [200] for poly(vinyl acetate) have also been made.

In many technical emulsion polymerization recipes (meth)acrylic acid is part of the monomer mixture contributing to the stability by electrosteric stabilization due to in situ formation of charged copolymers. It is suspected that a hairy layer around the particles, besides contributing to colloidal stability, also influences all entry and exit processes. Indeed, in the case of polymerizing styrene onto polystyrene seed particles decorated with a poly(styrene-co-acrylic acid) layer (formed during and before seeded copolymerization with styrene and acrylic acid onto polystyrene seed particles), it was shown that radical entry and exit were greatly reduced compared with the same particles with only electrostatic stabilization [201]. See Chapter 21 to see the influence of polymeric layers of different thickness but the same chemical composition on the polymerization kinetics determined with reaction calorimetry.

Various approaches use separately prepared organic acids containing copolymers as electrosteric stabilizers. An interesting example of such copolymers if poly (2-vinylnaphthalene-alt-maleic acid)-graft-polystyrene, which is a fluorescent polymer with significant surface activity [202]. The naphthalene unit acts as a fluorescence probe and has been used to carry out investigations concerning the orientation of the adsorbed stabilizer.

Amphiphilic derivatives of poly(acrylic acid) and their utilization as stabilizers in styrene emulsion polymerization are described in Ref. 203. The polymers were synthesized either by reaction of poly(acrylic acid) with tetradecylamine or by copolymerization of acrylic acid with styrene. The emulsion polymerization results reveal that a molecular weight of 5000 g mol^{-1} and a hydrophobic modification of 5 mol% (tetradecylamine) are an optimal composition within the investigated composition range.

To improve the properties of epoxy acrylate coatings, emulsion polymerization can be conducted in the presence of the epoxy resin with poly(acrylic acid) stabilizers. It has been shown that the type of amine used to neutralize the stabilizer significantly influences the particle size and the conversion [204].

Progress in the field of ionic polymerization techniques now allows the preparation of well-defined block copolymers and tailor made as stabilizers for emulsion polymerization. Concerning amphiphilic, nonionic block copolymers (see the monographs of Refs. 92 and 179, an overview of the application of block copolymers as polymeric surfactants in latex technology is recommended [205]. This paper places special emphasis on the synthesis of small latexes with polystyrene-*b*-PEG di- and triblock and PMMA-*b*-PEG diblock copolymers and hairy latexes with PMMA-*b*-poly(acrylic acid) copolymers. The swelling of the poly(acrylic acid) hairy layer upon neutralization is demonstrated. Furthermore, the benefit of such latexes in a controlled agglomeration process with latex particles stabilized with sodium dodecyl sulfate is mentioned.

Polyelectrolyte block copolymers with strong ionic blocks such as quaternary ammonium or sulfonate groups are of special interest as such stabilizers should be active over a broader pH range than carboxylic acid polymers. A few examples of the application of such polymers in emulsion polymerization are referred to in the following.

The first paper was published at the beginning of the 1990s and deals with the application of poly(alkyl methacrylate-*b*-sulfonated glycidyl methacrylate) as stabilizer in acrylate emulsion polymerization [206]. These are very interesting stabilizers as the sulfonation of the glycidyl methacrylate leads to a sulfonate and a hydroxyl group. Thus, these systems are a very special class of electrosteric stabilizers. Unfortunately, the length of the glycidyl methacrylate block in the polymers investigated was not that high, so only hydrodynamic layers with a maximum thickness of 10 nm have been obtained. The critical coagulation concentration of a poly(ethyl acrylate) latex has been determined to be on the order of 1.2 M KCl [206]. However, according to a theoretical treatment mentioned earlier, the stability of colloids against added salt stabilized with polyelectrolyte chains should be even higher for a still higher hydrodynamic layer thickness [93].

The application of another type of polyelectrolyte block copolymer, namely poly(ethyl ethylene)-*b*-poly(styrene sulfonate), with different degrees of sulfonation and block lengths is described in Refs. 20, 208, and 209. The polyelectrolyte block of one of these stabilizers had a length of almost 113 nm assuming a fully extended zigzag shape, which is much longer than in the previous study mentioned. The behavior of this class of stabilizers in emulsion polymerization and its influence thereof on the latex properties is rather complex. For instance, the adsorption pattern strongly depends on the degree of sulfonation. Complete sulfonation leads to "porcupine" particles [93] where the polyelectrolyte block is stretched to almost 70 nm into the aqueous phase. These particles possess, as theoretically predicted [93], an extraordinary stability against added electrolytes (>3 M NaCl). The hydrodynamic layer thickness (δ) scales with the electrolyte concentration (c_s): $\delta \propto c_s^{-\alpha}$. According to a theory of Pincus [93], the exponent α is 1/5 provided that δ is much larger than the particle radius (R). In a series of experiments with polystyrene model latexes it was found [20,208,209] that α depends on the ratio δ/R. For $\delta/R > 0.7$ the theoretically predicted value 1/5 for α was determined. For larger particles ($\delta/R < 0.7$), α values of 1/10 were obtained [209].

If the degree of sulfonation of the polyelectrolyte block is about 50%, the adsorption pattern changes completely. The polymer is still water soluble and a very efficient stabilizer because of multiple adsorption points leading to so-called ringlet adsorption patterns [208]. However, if the stabilizer concentration is higher than a critical value the hydrophobic points along one chain can adsorb on different particles, leading to aggregate formation [20] and bridging flocculation. In contrast to the porcupine particles, the ringlet particles are very sensitive to added electrolyte. The coagulation concentration with NaCl is in the range of some 10 mM, and consequently it is advisable to use nonionic initiators during the polymerization. It is noteworthy that it is possible to achieve a ringlet adsorption behavior with any kind of a statistical copolymer consisting of hydrophilic and hydrophobic segments. Thus, a partly sulfonated polystyrene [20], a partly quaternized poly(4-vinyl pyridine) [209], and a partly sulfonated poly(butadiene) or poly(isoprene) (K. Tauer et al., submitted) are very efficient stabilizers in emulsion polymerization as long as nonionic initiators are used. Partly sulfonated poly(butadiene) and poly(isoprene) represent reactive polymeric surfactants as they will be grafted to the particles via the remaining double bonds. Furthermore, the hydrophobic parts of these reactive polymeric surfactants can easily find the equilibrium configuration in the adsorbed state because of the low glass transition temperature. Note that the stabilizing efficiency of the ringlet-type polymeric surfactants per surfactant weight is as high as for conventional low-molecular-weight, ionic surfactants. Of course, the efficiency per stabilizer molecule is much higher. A single molecule of a 50% sulfonated polystyrene with a number average molecular weight of 40,000 g mol^{-1} (before the sulfonation) is able to stabilize a polystyrene particle about 15 nm in diameter [209].

The synthesis and application as stabilizer in emulsion polymerization of styrene initiated with KPS of a poly(styrene sulfonate)-*b*-polystyrene are described in Ref. 210. The block copolymer was prepared by nitroxide-mediated controlled radical polymerization and had a composition of 11–12 styrene units and 30–31 styrene sulfonate units.

The same controlled radical polymerization method was used to prepare cationic amphiphilic block copolymers starting from poly(vinyl benzyl chloride)-*b*-polystyrene and subsequent quaternization with trimethyl alkyl amine [211]. Latexes with high surface charge densities have been prepared by batch emulsion polymerization of styrene initiated with H_2O_2. Depending on the amount of block copolymer stabilizer, a surface charge density of up to 161 $\mu C\ cm^{-2}$ has been obtained for 16.6% stabilizer relative to the monomer weight. The particle size is smaller the higher the stabilizer concentration and the smaller the number of styrene units in the stabilizer for a given cationic block length. The lowest stabilizer concentration reported is 5% relative to the monomer weight and hence much higher than in the case of the poly(ethyl ethylene)-*b*-poly(styrene sulfonate) stabilizers [20,208].

By copolymerization of vinyl acetate and *N*-vinyl formamide and subsequent two-step hydrolysis, a PVAL-*co*-poly(vinyl amine) was prepared and used as stabilizer for styrene emulsion polymerization [212]. The resulting latexes have a cationic net charge although the polymerization was started with persulfate initiator.

A further class of polymeric stabilizers comprises polymers prepared from surface-active monomers by either homopolymerization or copolymerization leading to so-called polysoaps. With the keyword "polysoap" numerous papers can be found in the scientific literature over the last 5 years [77,213–235], but only two (to the best of our knowledge) deal with the utilization of polysoaps as stabilizers in emulsion polymerization [17,235]. In a comprehensive study, different cationic polysoaps have been employed in a semibatch styrene emulsion polymerization started either with a cationic initiator (2,2′-dimethyl-2,2′-axo-*N*-benzylpropionamidine) hydrochloride or with KPS [17]. Compared with the cationic initiator, KPS can lead to some trouble during the polymerization due to electrostatic interactions. The following polysoaps have been investigated: a homopolymer of poly(((methacryloyloxy)undecyl)-*N*,*N*′-dimethylamino-*N*-2-hydroxyethyl ammonium bromide), a copolymer of ((methacryloyloxy)ethyl)-*N*,*N*′-dimethyl-*N*-decyl ammonium bromide, and ((methacryloyloxy)ethyl)trimethyl ammonium bromide, a copolymer of ((methacryloyloxy)ethyl)-*N*,*N'*-dimethyl-*N*-undecyl ammonium bromide and ((methacryloyloxy)ethyl)trimethyl ammonium bromide. The copolymers investigated were different with respect to the hydrophobic-hydrophilic composition. If the hydrophilic-hydrophobic balance in the polysoap molecule is in a proper range (i.e., the content of the hydrophobic component is higher than a certain critical value), emulsion polymerization results in stable latexes with a much higher surface tension compared with the case with classical low-molecular-weight surfactants.

IV. CONCLUSIONS

Many studies have been made of synthesis of polymerizable surfactants as well as their use in various processes of heterophase polymerization (chiefly emulsion polymerization). Recent studies have stressed the associated benefits. The latexes produced are more stable with electrolyte addition, they are more stable under shear, and they are more robust after being subjected to freezing. Further, when used in the preparation of film-forming latexes, polymerizable surfactants yield various improvements in the film properties (mainly the blocking properties) and in the behavior of the films in the presence of water. This last benefit seems to be a very constant property of the introduction of reactive surfactants in the polymerization recipes. A further example of this property is shown in Fig. 6 of Ref. 1, where film samples, aged for 30 days, were immersed in water for 50 days. Films produced using styrenic polymerizable block copolymer surfactants maintain their dimension (length) much better than those derived from similar but nonreactive dispersants. In addition, the former films remain more cohesive than the latter, which are very fragile after the immersion treatment.

However, an explanation of these improved properties is far from complete, and much more work is needed. A study of the concentration profile of surfactant within such films should be a great help in explaining the beneficial behavior in the presence of water. It will be useful to determine whether there are hydrophilic domains within such films and whether there is a trend for migration of the surfactant toward the film surface. Such migration might interfere with the adhesive properties of the surface.

It is known that the latex stability is very dependent on the structure of the polymerizable surfactant (HLB balance, reactivity of the polymerizable group), but systematic analysis of these structural features has not yet been carried out. Finally, the optimization of a recipe using surfmers for a particular application remains a very empirical task, and that may explain why this class of surfactants has not yet been used for any important market, even if a few products can be obtained on a commercial basis.

Polymeric surfactants are a fascinating class of stabilizers, as they combine different stabilization mechanisms in one molecule, such as electrostatic and steric or electrostatic, steric, and depletion. Besides variations in the chemistry of the hydrophobic and hydrophilic parts, the arrangement of these segments along the polymeric chain, the degree of polymerization, and the molecular weight distribution offer additional possibilities for fine tuning various properties.

The different examples illustrate the enormous versatility of applications of polymeric surfactants and illustrate the potential of this class of stabilizers. Nevertheless, one may get the impression that especially for emulsion polymerization, the application of polymeric stabilizers, chiefly in the case of polymers with charged blocks, is just in an initial stage of basic research. The authors believe that the potential of polymeric stabilizers is today far from being completely explored or understood, as almost each day modern chemical methods allow access to new structures with improved performance and the possibilities to create polymers with new structures are almost unlimited.

REFERENCES

1. Q. Soula, A. Guyot, N. Williams, J. Grade, and T. Blease, J. Polym. Sci. A Polym. Chem. *37*:4205 (1999).
2. A. Guyot and A. Goux, J. Appl. Polym. Sci. *65*:2289 (1997).
3. S. Onodera, S. Yamamoto, T. Tamai, and H. Takahashi, Jpn. Patent 06239908 (1994).
4. R. H. Tang and P. M. Chakrabarti, U.S. Patent 5296627 (1994).
5. K. Yokota, A. Ishihara, and H. Shinike, U.S. Patent 5324862 (1994).
6. K. Kinoshita, Jpn. Patent 94/49108 (1994).
7. S. Usai, Jpn. Patent 94/65551 (1994).
8. J. Biale, U.S. Patent 5326814 to Rhom & Haas (1994).
9. D.E. Patent 3239527 to Rhom GmBh (1984).
10. G.B. Patent 1427789 to Bayer AG (1982).
11. I. M. Kusters, R. H. Napper, R. G. Gilbert, and A. L. German, Macromolecules *25*:7043 (1992).
12. A. Guyot, K. Tauer, J. M. Asua, S. Van Es, C. Gauthier, A. C. Helgren, D. C. Sherrington, A. Montoya-Goni, M. Sjöberg, O. Sindt, F. Vidal, M. Unzué, H. A. S. Schoonbrood, E. Schipper, and P. Lacroix-Desmazes, Acta Polym. *50*:57 (1999).

13. C. C. Fifield, Ph.D. thesis, University of Connecticut, Univ. Microfilms Int. 5869, Ann Arbor MI, 1992.
14. F. Vidal, J. Guillot, and A. Guyot, Colloid Polym. Sci. *273*:999 (1995).
15. F. Vidal and A. Guyot, New J. Chem. *19*:1081 (1995).
16. E. Rizzardo, G. F. Meijs, and S. H. Thang, U.S. Patent 5.385.996 (1995).
17. D. Cochin, A. Lawschewski, and F. Nallet, Macromolecules *30*:2278 (1997).
18. I. Pirma and J. R. Lenzotti, British Polymer J. *21*:45 (1989).
19. I. Pirma, Polymer surfactants and polymer colloid stabilization, in *The Polymeric Materials Encyclopedia*, CRC, New York, cd rom, 1996.
20. K. Tauer, M. Antonietti, L Rosengarten, and H. Muller, Macromol. Chem. Phys. *199*:897–908 (1998).
21. A. Guyot and K. Tauer, Adv. Polym. Sci. *111*:45 (1994).
22. K. Tauer, Polym. News *20*:342 (1995).
23. A. Guyot, Curr. Opin. Colloid Interface Sci. *1*:580 (1996).
24. K. Tauer, in *Polymeric dispersions: Principles and Applications* (J. M. Asua, ed.), Kluwer Academic Publishers, Dordrecht, 1997.
25. J. M. Asua and H. A. S. Schoonbrood, Acta Polym. *49*:671 (1998).
26. A. Guyot, in *Novel Surfactants* (K. Holmberg, ed.), Surface Science Series, Vol. 74, Marcel Dekker, New York, 1998.
27. B. W. Greene and D. P. Sheetz, J. Colloid Interface Sci. *32*:90, 96 (1970).
28. B. W. Greene and F. L. Saunders, J. Colloid Interface Sci. *33*:393 (1970).
29. S. L. Tsaur and R. M. Fitch, J. Colloid Interface Sci. *115*:450 (1987).
30. J. L. Guillaume, C. Pichot, and J. Guillot, J. Polym. Sci. A Polym. Chem. *28*:137 (1990).
31. M. B. Urquiola, V. L. Dimonie, E. D. Sudol, and M. S. El Aasser, J. Polym. Sci. A Polym. Chem. *30*:2619, 2631 (1992).
32. M. B. Urquiola, E. D. Sudol, V. L. Dimonie, and M. S. El Aasser, J. Polym. Sci. A Polym. Chem. *31*:1403 (1993).
33. R. H. Ottewill and R. Satguranathan, Colloid Polym. Sci. *265*:845 (1987).
34. R. H. Ottewill and R. Satguranathan, Colloid Polym. Sci. *266*:547 (1988).
35. A. L. Palluel, M. J. Westby, C. W. Bromley, S. P. Davies, and A. J. Backhouse, Makromol. Chem. Macromol. Symp. *35/36*:509 (1990).
36. H. A. S. Schoonbrood, M. J. Unzué, O. J. Beck, J. M. Asua, A. Montoya-Goni, and D. C. Sherrington, Macromolecules *30*:6024 (1997).
37. H. A. S. Schoonbrood, M. J. Unzué, J. I. Amalvy, and J. M. Asua, J. Polym. Sci. A Polym. Chem. *35*:2561 (1997).
38. K. H. Goebel and K. Stälher, Polym. Adv. Technol. *6*:452 (1994).
39. H. A. S. Schoonbrood and J. M. Asua, Macromolecules *30*:6034 (1997).
40. R. H. Ottewill and R. Satguranathan, Colloid Polym. Sci. *273*:379 (1995).
41. F. Thenoz, O. Soula, and A. Guyot, J. Polym. Sci. A Polym. Chem. *37*:2251 (1999).
42. T. Hamaide, A. Zicmanis, C. Monnet, and A. Guyot, Polym. Bull. *33*:133 (1994).
43. O. Sindt, C. Gauthier, T. Hamaida, and A. Guyot, J. Applied Polymer Sci. *77*:2768 (2000).
44. S. Abele, C. Graillat, A. Zicmanis, and A. Guyot, Polym. Adv. Technol. *10*:301 (1999).
45. S. Abele, A. Zicmanis, C. Graillat, C. Monnet, and A. Guyot, Langmuir *15*:1033, 1045 (1999).
46. S. Abele, C. Gauthier, C. Graillat, and A. Guyot, Polymer. *41*:1147 (2000).
47. O. Soula and A. Guyot, Langmuir *15*:7956 (1999).
48. O. Soula, R. Petiaud, M. F. Llauro, and A. Guyot, Macromolecules *32*:6938 (1999).
49. S. Braun, H. O. Kalinowski, and S. Berger, *100 and More Basic Experiments*, VCH, New York, 1996, pp. 170, 173, 390.
50. B. Charleux and C. Pichot, Polymer *34*:195 (1993).
51. B. Charleux, C. Pichot, and M. F. Llauro, Polymer *34*:4353 (1993).
52. M. T. Charreyre, P. Boulanger, C. Pichot, B. Mandrand, and M. F. Llauro, Makromol. Chem. *194*:117 (1993).
53. M. T. Charreyre, P. Boulanger, T. Delair, B. Mandrand, and C. Pichot, Colloid Polym. Sci. *271*:668 (1993).
54. J. Revilla, A. ElAïssari, C. Pichot, and B. Gallot, Polym. Adv. Technol. *6*:455 (1994).
55. K. Nagai, T. Ohashi, R. Kaneko, and T. Tanigushi, Colloid Surf. Phys. Chem. Asp. *153*:133 (1999).
56. M. Dreja and B. Tieke, Macromol. Rapid Commun. *17*:825 (1996).
57. M. Dreja, W. Pickout-Hintzen, and B. Tieke, Macromolecules *31*:272 (1998).
58. C. Larpent, E. Bernard, J. Richard, and S. Vaslin, Macromolecules *30*:354 (1997).
59. C. Favero, C. Graillat, and A. Guyot, Macromol. Symp. *150*:235 (2000).
60. C. S. Chem and T. J. Chen, Colloid Polym. Sci. *275*:546, 1060 (1997).
61. K. E. J. Barrett (ed.), *Dispersion Polymerization in Organic Media*, Wiley, New York, 1975.
62. C. K. Ober, Makromol. Chem. Macromol. Symp. *35/36*:87 (1990).
63. Y. Almog, S. Reich, and M. Levy, Br. Polym. J. 131 (1982).
64. S. Chen, E. D. Sudol, and M. S. ElAasser, J. Polym. Sci. Polym. Chem. *32*:1087 (1994).
65. S. Kobayashi, H. Uyama, J. H. Choi, and Y. Matsumoto, Polym. Int. *30*:265 (1993).

66. S. Kobayashi, H. Uyama, S. W. Lee, and Y. Matsumoto, J. Polym. Sci. A Polym. Chem. *31*:3133 (1993).
67. S. Kawaguchi, M. A. Winnik, and K. Ito, Macromolecules *28*:1159 (1995).
68. A. J. Paine, Macromolecules *23*:3109 (1990).
69. P. Lacroix-Desmazes and A. Guyot, Polm. Bull. *37*: 183 (1996).
70. P. Lacroix-Desmazes and A. Guyot, Colloid Polym. Sci. *274*:1129 (1996).
71. D. A. Canelas, D. E. Betts, and J. M. De Simone, ACS Polym. Preprints *35*:628 (1997).
72. D. Cochin, R. Zana, and F. Candau, Macromolecules *26*:5755, 5765 (1993).
73. F. Candau and R. Zana, Encycl. Polym. Mater. *6*:4287 (1996).
74. V. Vaskova, D. Renoux, M. Bernard, J. Selb, and F. Candau, Polym. Adv. Technol. *6*:441 (1995).
75. J. Selb and F. Candau, *Association Polymers in Aqueous Solution*, ACS Symp. Ser. (J. E. Glass, ed.), 765, 2000, p. 95–108.
76. K. Stähler, J. Selb, P. Barthelemy, B. Pucci, and F. Candau, Langmuir *14*:4765 (1998).
77. K. Stähler, J. Selb, and F. Candau, Colloid Polym. Sci. *276*:860 (1998).
78. K. Stähler, J. Selb, and F. Candau, Langmuir, *15*:7565 (1999).
79. D. H. Napper, *Polymeric Stabilization of Colloidal Dispersions*, Academic Press, London, 1983.
80. A. Bringer, E. Eisenriegler, F. Schlesener, and A. Hanke, Eur. Phys. J. B *11*:101–109 (1999).
81. J. W. Goodwin, in *Colloid Science* (D. H. Everett, ed.), Vol. 2, The Chemical Society, London, 1976, pp. 246–293.
82. R. H. Ottewill, in *Emulsion Polymerization and Emulsion Polymers* (P. A. Lovell and M. S. El-Aasser, eds.), Wiley, New York, 1976, pp. 59–121.
83. F. Caruso, H. Lichtenfeld, E. Donath, and H. Möhwald, Macromolecules *32*:2317–2328 (1999).
84. F. Caruso, E. Donath, and H. Möhwald, J. Phys. Chem. B *102*:2011–2016 (1998).
85. E. Donath, G. B. Sukhorukov, F. Caruso, S. A. Davis, and H. Möhwald, Angew. Chem. Int. Ed. *37*:2201–2205 (1998).
86. F. Caruso, R. A. Caruso, and H. Möhwald, Science *282*:1111–1114 (1998).
87. D. G. Kurth, F. Caruso, and C. Schüler, Chem. Commun. 1579–1580 (1999).
88. F. Caruso and H. Möhwald. J. Am. Chem. Soc. *121*: 6039–6046 (1999).
89. F. Caruso, Chem. A Eur. J. *6*:413–416 (2000).
90. F. Lux, Synth. Met. *102*:1190–1193 (1999).
91. S. Creutz and R. Jerome, Langmuir *15*:7145–7156 (1999).
92. I. Piirma, *Polymeric Surfactants*, Marcel Dekker, New York, 1992.
93. P. Pincus, Macromolecules *24*:2912–2919 (1991).
94. E. A. Grulke, in *Encyclopedia of Polymer Science and Engineering* (J. I. Kroschwitz, exec. ed.) Vol. 16, New York, 1989, pp. 443–473.
95. H. G. Yuan, G. Kalfas, and W. H. Ray, J. Macromol. Sci. Rev. Macromol. Chem. Phys. *C31*:215–299 (1991).
96. V. Macho, M. Fabini, M. Rusina, S. Bobula, and M. Harustiak, Polymer *35*:5773–5777 (1994).
97. V. Macho, M. Fabini, M. Rusina, S. Bobula, and M. Harustiak, Chem. Pap. Chem. Zv. *49*:102–105 (1995).
98. M. Zerfa and B. W. Brooks, J. Appl. Polym. Sci. *60*: 2077–2086 (1996).
99. M. Zerfa and B. W. Brooks, J. Appl. Polym. Sci. *65*: 127–134 (1997).
100. K. D. Prince, Plast. Rubber Compos. *28*:105–108 (1999).
101. D. H. Lee, J. Chem. Eng. Jpn. *32*:97–103 (1999).
102. F. Lerner and S. Nemet, Plast. Rubber Compos. *28*: 100–104 (1999).
103. E. Vivaldolima, P. E. Wood, A. E. Hamielec, and A. Penlidis, Ind. Eng. Chem. Res. *36*:939–965 (1997).
104. R. Olayo, E. Garcia, B. Garciacorichi, L. Sanchezvazquez, and J. Alvarez, J. Appl. Polym. Sci. *67*:71–77 (1998).
105. H. K. Mahabadi, T. H. Ng, and H. S. Tan, J. Microencap. *13*:559–573 (1996).
106. G. Polacco, M. Palla, and D. Semino, Polym. Int. *48*: 392–397 (1999).
107. D. W. Zhu, Macromolecules *29*:2813–2817 (1996).
108. G. Scmidt-Naake, M. Drache, and C. Taube, Angew. Makromol. Chem. *265*:62–68 (1999).
109. T. Nishikawa, M. Kamigaito, and M. Sawamoto, Macromolecules *32*:2204–2209 (1999).
110. H. Fischer, J. Polym. Sci. A Polym. Chem. *37*:1885–1901 (1999).
111. E. D. Sudol, in *Polymeric Dispersions: Principles and Applications* (J. M. Asua, ed.), Kluwer Academic Publishers, Boston, 1997, pp. 109–126.
112. J. L. Cawse, in *Emulsion Polymerization and Emulsion Polymers* (P. A. Lovell and M. S. El-Aasser, eds.), Wiley, New York, 1976, pp. 743–761.
113. T. K. Mandal and B. M. Mandal, Polymer *36*:1911–1913 (1995).
114. T. K. Mandal and B. M. Mandal, J. Polym. Sci. Polym. Chem. *37*:3723–3729 (1999).
115. P. Banerjee and B. M. Mandal, Macromolecules *28*: 3940–3943 (1995).
116. H. Uyama, H. Kurioka, and S. Kobayashi, Chem. Lett. 795–796 (1995).
117. H. Kurioka, H. Uyama, and S. Kobayashi, Polym. J. *30*:526–529 (1998).
118. H. Uyama, H. Kurioka, and S. Kobayashi, Colloids Surf. A Physicochem. Eng. Asp. *153*:189–194 (1999).
119. D. Chattopadhyay and B. M. Mandal, Langmuir *12*: 1585–1588 (1996).
120. J. Stejskal, P. Kratochvil, and M. Helmstedt, Langmuir *12*:3389–3392 (1996).

121. M. Okubo, T. Masuda, and T. Mukai, Colloid Polym. Sci. *276*:96–99 (1998).
122. D. Chattopadhyay, S. Banerjee, D. Chakravorty, and B. M. Mandal, Langmuir 1544–1547 (1998).
123. M. T. Gill, S. E. Chapman, C. L. Dearmitt, F. L. Baines, C. M. Dadswell, J. G. Stamper, G. A. Lawless, N. C. Billingham, and S. P. Armes, Synth. Met. *93*: 227–233 (1998).
124. P. Banerjee, Eur. Polym. J. *34*:841–847 (1998).
125. Y. L. Hsiao, E. E. Maury, S. Mawson, and K. P. Johnston, Macromolecules *28*:8159–8166 (1995).
126. Y. L. Hsiao and J. M. DeSimone. J. Polym. Sci. A Polym. Chem. *35*:2009–2013 (1997).
127. M. L. O'Neill, M. Z. Yates, K. P. Johnston, C. D. Smith, and S. P. Wilkinson, Macromolecules *31*:2848–2856 (1998).
128. M. L. O'Neill, M. Z. Yates, K. P. Johnston, C. D. Smith, and S. P. Wilkinson, Macromolecules *31*:2838–2847 (1998).
129. A. I. Cooper, W. P. Hems, and A. B. Holmes, Macromolecules *32*:2156–2166 (1999).
130. H. Shiho and J. M. DeSimone, J. Polym. Chem. A Polym. Chem. *37*:2429–2437 (1999).
131. J. H. Xia, T. Johnson, S. G. Gaynor, K. Matyjaszewski, and J. M. DeSimone, Macromolecules *32*: 4802–4805 (1999).
132. X. H. Li and Z. G. Sun, J. Appl. Polym. Sci. *58*:1991–1997 (1995).
133. E. Bourgeat-Lami and J. Lang, J. Colloid Interface Sci. *197*:293–308 (1998).
134. E. Bourgeat-Lami and J. Lang. J. Colloid Interface Sci. *210*:281–289 (1999).
135. X. B. Ding, Z. H. Sun, G. X. Wan, and Y. Y. Jiang, Chin. J. Polym. Sci. *17*:129–134 (1999).
136. E. Bourgeat-Lami and A. Guyot, Polym Bull *35*:691–696 (1995).
137. E. Bourgeat-Lami and A. Guyot, Colloid Polym. Sci. *275*:716–729 (1997).
138. K. Ishizu and N. Tahara, Polymer *37*:1729–1734 (1996).
139. H. Miyazaki, K. Terada, T. Sato, H. Maruyama, and T. Okaya, J. Appl. Polym. Sci. *60*:2149–2157 (1996).
140. P. Lacroix-Desmazes and A. Guyot, Macromolecules *29*:4508–4515 (1996).
141. P. Lacroix-Desmazes and A. Guyot, Polym. Adv. Technol. *8*:608–615 (1997).
142. K. Ishizu, M. Yamashita, and A. Ichimura, Polymer *38*:5471–5474 (1997).
143. J. Liu, L. M. Gan, C. H. Chew, C. H. Queck, H. Gong, and L. H. Gan, J. Polym. Sci. A Polym. Chem. *35*: 3575–3583 (1997).
144. J. Liu, C. H. Chew, S. Y. Wong, L. M. Gan, J. Lin, and K. L. Tan, Polymer *39*:283–289 (1998).
145. M. Q. Chen, T. Serizawa, and M. Akashi, Polym. Adv. Technol. *10*:120–126 (1999).
146. S. F. Lascelles, F. Malet, R. Mayada, N. C. Billingham, and S. P. Armes, Macromolecules *32*:2462–2471 (1999).
147. D. Donescu, M. Teodorescu, L. Fusulan, and C. Petcu, J. Dispersion Sci. Technol. *20*:1085–1096 (1999).
148. B. Moustafa and A. Faizalla, J. Appl. Polym. Sci. *73*: 1793–1798 (1999).
149. M. A. Awan, V. L. Dimonie, and M. S. El-Aasser, J. Polym. Sci. A Polym. Chem. *34*:2633–2649 (1996).
150. M. A. Awan, V. L. Dimonie, and M. S. El-Aasser, J. Polym. Sci. A Polym. Chem. *34*:2651–2664 (1996).
151. J. Kim, S. Y. Jeong, K. U. Kim, Y. H. Ahn, and R. P. Quirk, J. Polym. Sci. A Polym. Chem. *34*:3277–3288 (1996).
152. J. V. Dawkins, D. J. Neep, and P. L. Shaw, Polymer *35*:5366–5368 (1994).
153. R. Hu, V. L. Dimonie, E. D. Sudol, and M. S. El-Aasser, J. Appl. Polym. Sci. *55*:1411–1415 (1995).
154. A. Tuncel, H. Cicek, M. Hayran, and E. Piskin, J. Biomed. Mater. Res. *29*:721–728 (1995).
155. E. Piskin, K. Kesenci, N. Satiroglu, and O. Genc, J. Appl. Polym. Sci. *59*:109–117 (1996).
156. O. Pekcan, Eur. Polym. J. *32*:117–124 (1996).
157. V. Bulmus, A. Tuncel, and E. Piskin, J. Appl. Polym. Sci. *60*:697–704 (1996).
158. M. Laus, C. Dinnella, G. Lanzarini, and A. Casagrande, Polymer *37*:343–347 (1996).
159. C. Dinnella, M. Doria, M. Laus, and G. Lanzarini, Biotechnol. Appl. Biochem. *23*:133–140 (1996).
160. F. L. Baines, S. Dionisio, N. C. Billingham, and S. P. Armes, Macromolecules *29*:3096–3102 (1996).
161. H. Bamnolker and S. Margel, J. Polym. Sci. A Polym. Chem. *34*:1857–1871 (1996).
162. M. Laus, M. Lelli, and A. Casagrande, J. Polym. Sci. A Polym. Chem. *35*:681–688 (1997).
163. V. Bulmus, H. Ayhan, and E. Piskin, Chem. Eng. J. *65*:71–76 (1997).
164. T. Yamashita and M. Okubo, Colloid Polym. Sci. *275*: 214–219 (1997).
165. B. Ray and B. M. Mandal, Langmuir *13*:2191–2196 (1997).
166. K. Takahashi, S. Miyamori, H. Uyama, and S. Kobayashi, Macromol. Rapid. Commun. *18*:471–475 (1997).
167. M. Bayham and A. Tuncel, J. Appl. Polym. Sci. *67*: 1127–1139 (1998).
168. K. Takahashi, H. Uyama, and S. Kobayashi, J. Macromol. Sci. Pure Appl. Chem. *A35*:1473–1487 (1998).
169. J. W. Kim and K. D. Suh, Colloid Polym. Sci. *276*: 870–878 (1998).
170. B. Ray and B. M. Mandal, J. Polym. Sci. A Polym. Chem. *37*:493–499 (1999).
171. D. Horak, M. Karpisek, J. Turkova, and M. Benes, Biotechnol. Prog. *15*:208–215 (1999).
172. D. Horak, J. Polym. Sci. A Polym. Chem. *15*:3785–3792 (1999).
173. W. Yang, J. Hu, Z. Tao, L. Li, C. Wang, and S. Fu, Colloid Polym. Sci. *277*:446–451 (1999).

174. K. Nakamura, K. Fujimoto, and H. Kawaguchi, Colloid Surf. A Physicochem. Eng. Asp. *153*:195–201 (1999).
175. B. Thomson, A. Rudin, and G. Lajoie, J. Polym. Sci. A Polym. Chem. *33*:345–357 (1995).
176. I. Capek, Radical Polymerization Polyelectrolytes 145 PG: 1–55, 1999.
177. J. M. Harris, in *Poly(ethylene glycol) Chemistry* (J. M. Harris, ed.), Plenum, New York, 1992, pp. 1–14.
178. I. Piirma, *Polymeric Surfactants*, Marcel Dekker, New York, 1992, pp. 127–164.
179. V. M. Nace, ed., *Nonionic Surfactants—Polyxyalkylene Block Copolymers*, Marcel Dekker, New York, 1996.
180. R. Brown, B. Stutzel, and T. Sauer, Macromol. Chem. Phys. *196*:2047–2064 (1995).
181. R. Walz, B. Bömer, and W. Heitz, Makromol. Chem. *178*:2527–2534 (1977).
182. O. Nuyken and R. Weidener, Adv. Polym. Sci. *73/74*: 145–199 (1986).
183. R. G. Santos, J. G. Zilliox, J. E. Herz, and G. J. Beinert, Eur. Polym. J. *28*:233–242 (1992).
184. I. Cakmak, Macrmol. Rep. *A31*(Suppl. 3 and 4):333–337 (1994).
185. Y. Haneda, H. Terada, M. Yoshida, A. Ueda, and S. Nagai, J. Polym. Sci. A Polym. Chem. *32*:2641–2652 (1994).
186. H. Terada, Y. Haneda, A. Ueda, and S. Nagai, Macromol. Rep. (Suppl. 1 and 2):173–178 (1994).
187. C. I. Simionescu, G. David, A. Ioanid, V. Paraschiv, G. Riess, and B. C. Simionescu, J. Polym. Sci. A Polym. Chem. *32*:3123–3132 (1994).
188. H. Kitano and J. Kawabata, Macromol. Chem. Phys. *197*:1721–1729 (1996).
189. H. Kitano and J. Kawabata, Macromol. Chem. Phys. *197*:3657–3666 (1996).
190. K. Tauer, Polym. Adv. Technol. *6*:435–440 (1995).
191. T. Nagamune, A. Ueda, and S. Nagai, J. Appl. Polym. Sci. *62*:359–365 (1996).
192. K. Tauer, M. Antonietti, L. Rosengarten, and H. Müller, Macromol. Chem. Phys. *199*:897–908 (1998).
193. V. N. Kislenko, Colloid Surf. A Physicochem. Eng. Asp. *152*:199–203 (1999).
194. M. S. El-Aasser and J. W. Vanderhoff, eds., *Emulsion Polymerization of Vinyl Acetate*, Applied Science Publishers, London, 1981.
195. G. S. M. Gonzales, V. L. Dimonie, E. D. Sudol, H. J. Yue, A. Klein, and M. S. El-Aasser, J. Polym. Sci. A Polym. Chem. *34*:849–862 (1996).
196. H. Murakami, Y. Kawashima, T. Niwa, T. Hino, H. Takeuchi, and M. Kobayashi, Int. J. Pharm. *149*:43–49 (1997).
197. H. Zeh and H. Baumgartl, JOCCA Surf. Coat. Int. *78*: 132–138 (1995).
198. H. Warson, in *Emulsion Polymerization of Vinyl Acetate* (M. S. El-Aasser and J. W. Vanderhoff, eds.), Applied Science Publishers, London, 1981, pp. 1–10.
199. D. Donescu, K. Gosa, I. Diaconescu, M. Mazare, and N. Carp, in *Emulsion Polymerization of Vinyl Acetate* (M. S. El-Aasser and J. W. Vanderhoff, eds.), Applied Science Publishers, London, 1981, pp. 203–214.
200. W. E. Daniels, J. W. Vanderhoff, C. T. Enos, J. G. Iacoviello, S. M. Ahmed, and J. Frost, in *Emulsion Polymerization of Vinyl Acetate* (M. S. El-Aasser and J. W. Vanderhoff, eds.), Applied Science, London, 1981, pp. 191–202.
201. E. M. Coen, R. A. Lyons, and R. G. Gilbert, Macromolecules *29*:5128–5135 (1996).
202. T. Cao, W. P. Yin, and S. E. Webber, Macromolecules *27*:7459–7464 (1994).
203. M. Antonietti and M. C. Weißenberger, Macromol. Rapid Commun. *18*:295–302 (1997).
204. D. de Wet-Roos, J. H. Knoetze, B. Cooray, and R. D. Sanderson, J. Appl. Polym. Sci. *71*:1347–1360 (1999).
205. G. Riess, Colloid Surf. A Physicochem. Eng. Asp. *153*:99–110 (1999).
206. L. Leemans, R. Fayt, Ph. Teyssié, and N. C. de Jaeger, Macromolecules *24*:5922–5925 (1991).
207. K. Tauer, H. Müller, L. Rosengarten, and K. Riedelsberger, Colloid Surf. A Physicochem. Eng. Asp. *153*: 75–88 (1999).
208. K. Tauer, M. Antonietti, H. Müller, and L. Vorwerg, Pitture Vrnici—Paints Varnishes *73*:31–46 (1997).
209. H. Müller, Polymere Stabilisatoren in der Emulsionspolymerization, Ph.D. dissertation, Universität Potsdam, Germany, 1997.
210. M. Bouix, J. Gouzi, B. Charleux, J. P. Vairon, and P. Guinot, Macromol. Rapid Commun. *19*:209–213 (1998).
211. W. Jaeger, U. Wendler, A. Lieske, and J. Bohrisch, Langmuir *15*:4026–4032 (1999).
212. K. D. Campbell, D. J. Sagl, and J. W. Vanderhoff, J. Dispersion Sci. Technol. *19*:785–804 (1998).
213. S. E. Friberg, T. Moaddel, and E. D. Sprague, J. Phys. Chem. *98*:13414–13417 (1994).
214. J. Kevelam and J. B. F. N. Engberts, Langmuir *11*: 793–797 (1995).
215. G. J. Wang and J. B. F. N. Engberts, Eur. Polym. J. *31*:409–417 (1995).
216. G. J. Wang and J. B. F. N. Engberts, J. Org. Chem. *60*:4030–4038 (1995).
217. A. Laschewsky, Adv. Polym. Sci. *124*:1–86 (1995).
218. Y. Hu, M. C. Kramer, C. J. Bourdeaux, and C. L. McCormick, Macromolecules *28*:7100–7106 (1995).
219. O. Anthony and R. Zana, Langmuir, *12*:3590–3597 (1996).
220. Y. X. Hu, R. S. Armentraut, and C. L. McCormick, Macromolecules *30*:3538–3546 (1997).
221. Y. X. Hu, G. L. Smith, M. F. Richardson, and C. L. McCormick, Macromolecules *30*:3526–3527 (1997).
222. Y. Rouault and C. Marques, J. Phys. II *7*:903–911 (1997).

223. V. G. Babak, M. Rinaudo, J. Desbrieres, G. A. Vikhoreva, and M. C. Michalski, Mendeleev Commun. *4*:149–151 (1997).
224. V. S. Zdanowicz and U. P. Strauss, J. Phys. Chem. B *102*:40–43 (1998).
225. T. Sata, K. Mine, Y. Tagami, M. Higa, and K. Matsusaki, J. Chem. Soc. Faraday Trans. *94*:147–153 (1998).
226. Y. Yang, R. Prudhomme, K. M. McGrath, P. Richetti, and C. M. Marques, Phys. Rev. Lett. *80*:2729–2732 (1998).
227. I. Iliopoulos, Curr. Opin. Colloid Interface Sci. *3*:493–498 (1998).
228. T. Sata, Y. Tagami, and K. Matsusaki, J. Phys. Chem. B *102*:8473–8479 (1998).
229. V. Babak, I. Lukina, G. Vikhoreva, J. Desbrieres, and M. Rinaudo, Colloids Surf. A Physicochem. Eng. Asp. *147*:139–148 (1999).
230. D. Cochin and A. Laschewsky, Macromol. Chem. Phys. *200*:609–615 (1999).
231. A. Benjelloun, P. Lochon, and A. Brembilla, J. Mol. Catal. A Chem. *140*:215–223 (1999).
232. S. Bayoudh, A. Laschewsky, and E. Wischerhoff, Colloid Polym. Sci. *277*:519–527 (1999).
233. A. Lieske and W. Jaeger, Tenside Surfactants Deterg. *36*:155–161 (1999).
234. M. R. Anklam, D. A. Saville, and R. K. Prud'homme, Langmuir *15*:7299–7307 (1999).
235. W. Jaeger, U. Wendler, A. Lieske, J. Bohrisch, and C. Wandrey, Macromol. Symp. in press.

29

Organic Particle Precipitation

JOHN TEXTER Strider Research Corporation, Rochester, New York

I. INTRODUCTION

This chapter addresses precipitation and condensation processes for forming organic particulates in homogeneous solutions and in multiphase fluids. Both soft and hard particles are discussed. Soft particles such as micelles and microemulsion droplets sometimes have only fleeting dynamical lifetimes, but they provide nanoscale compartmentalization that is useful, sometimes, for templating other particulates. Other soft particles, such as emulsion droplets of oil in water, can be used to template the growth of larger droplets [1], they can provide micrometer-scale compartmentalization for forming relatively large droplets and beads [2], and they can be used to compartmentalize condensation of solid amorphous and crystalline organic materials (see Section II).

Comminution and related grinding and fluid-energy milling processes are widely used industrially for preparing micrometer-scale and submicrometer-sized powders and dispersions of organic materials [3]. However, such methods are not very applicable to organic solids that are soft or waxy or solids that are explosive. Also, such processes often do not yield particle sizes appropriate for a particular application. Dissolution or vaporization followed by condensation, precipitation, or recrystallization represents a diverse class of processes for producing small-particle organic formulations with sizes ranging from a few nanometers to tens or hundreds of micrometers and with physical states ranging from liquid to amorphous solid to crystalline.

A. Physical State

The production of well-characterized and monodisperse organic particles by precipitation is much less developed than inorganic particle precipitation. This is mainly due to the high number of internal degrees of freedom in typical organic molecules of interest in comparison with inorganic solids often derived from relatively simple salts or oxides. Large numbers of internal degrees of freedom such as chiral centers and rotational degrees of freedom make for the occurrence of a plethora of polymorphs, when crystallization can be achieved, and make amorphous states of condensation the most easily obtained in rapid precipitation processes.

Organic precipitation of dyes and pigments to produce press cakes suitable for purification and storage but requiring further treatment (typically comminution) for particle size reduction is well-established art [4]. Emphasis in such processes is often on how to promote particle growth so that crystals can be washed and filtered expeditiously and without clogging.

Both amorphous and crystalline physical states can be advantageous under certain conditions. One or the other type of physical state may have a preferred property, especially with respect to color or ultraviolet (UV)-visible spectrum. Amorphous physical states can lead to untoward behavior, such as unwanted crystallization and concomitant changes in other physical properties. Transmission electron micrographs (TEMs) of various crystalline *Color Index* pigments [5] are il-

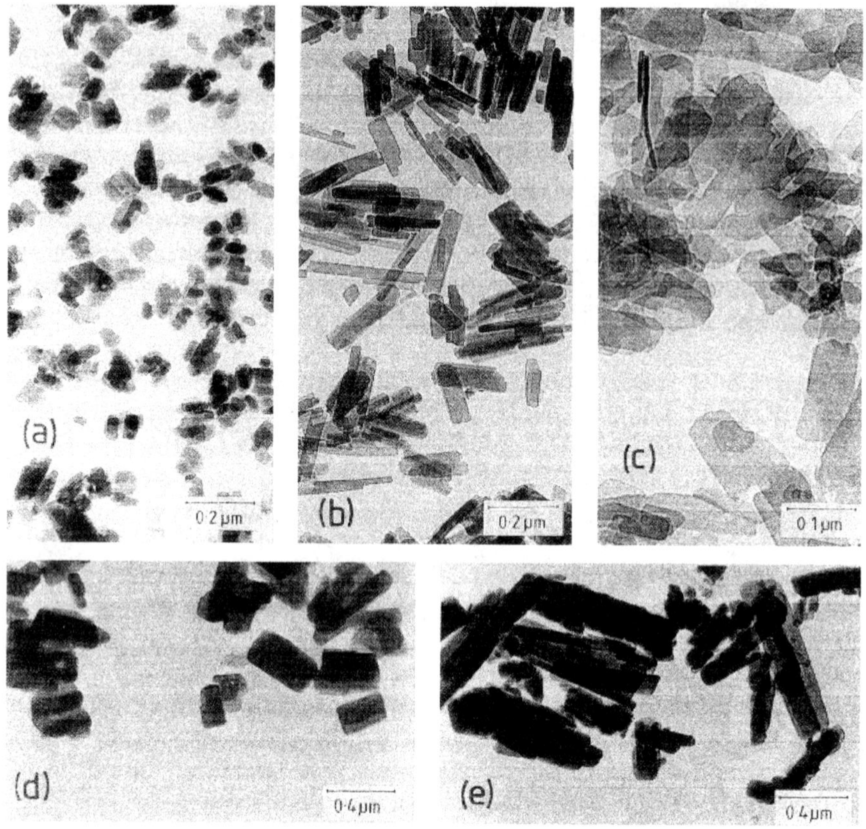

FIG. 1 Transmission electron micrographs of crystalline organic pigments: (a,b) Blue 15:3; (c) Red 57:1; (d) Violet 19; (e) Yellow 110. (From Ref. 6.)

lustrated in Fig. 1 [6]. Dispersions of the pigments illustrated in Fig. 1 are routinely achieved by comminution processing to create a stable suspension. A challenge before us is to learn how to precipitate various organic pigments and materials as small particles of controlled size in a crystalline state with well-defined habit. A further challenge is to learn to do so while obtaining a colloidally stable dispersion, where viscosity and particle-particle interactions are minimized.

B. Compartmentalization

Particle size during precipitation can usually be controlled by some sort of compartmentalization. In cases relating to condensation from supersaturated solutions, particle size is essentially a kinetic or diffusion-controlled process wherein a de facto reaction field or region of significant concentration gradient directs the flux of condensing species at a growing particle. Such compartments may be thought of as reaction or diffusion-limited condensation zones. Alternatively, physical or interfacial boundaries may serve to define nanoreactors or microreactors that limit the size of particles formed therein. Such cases are easier to visualize conceptually and include precipitation within emulsion droplets and aerosol droplets as well as within hollow capsules (see Section IX.B).

C. Driving Forces

1. Free Energy Within Single Phase Domains

An important class of organic particles are those that form spontaneously because of chemical free energy driving forces. This class comprises micelles, microemulsions, reverse micelles, reverse microemulsions, and some vesicles [7]. In such systems one needs only to bring the various chemical components into physical contact and thermodynamic evolution will drive codissolution and mixing until a thermodynamically isotropic solution is formed. The supramolecular particles formed therein from amphiphilic molecules are in a dynamic equilibrium with the rest of the solution phase. The use of such particles may be quite limited, as changes in intensive variables such as temperature and concentration often cause transformation or loss of the particles of interest or the formation of multiple-phase domains. However, such thermodynamically stable particles can serve as useful templates from which kinetically stabilized (thermodynamically unstable) particles can be formed.

2. Supersaturation and Nucleation

If S_∞ is the solubility of a material having an infinite radius of curvature, or in simpler terms the macroscopic equilibrium solubility, the supersaturation is defined as $[S/S_\infty - 1]$, where S is the local solubility. The greater the degree to which a condensation volume element is devoid of heterogeneous nucleation material (seed nuclei, dust, etc.), the higher the supersaturation has to be in order to induce nucleation. An in-depth discussion of nucleation theory is beyond the scope of this chapter; see Refs. 8 and 9 for reviews. However, several qualitatively useful generalizations should be noted.

As more nuclei are formed, smaller particle sizes usually obtain because a given amount of mass is distributed over a greater number of centers. This effect also often leads to slightly to dramatically increased viscosity because as particles decrease in size the average separation distance decreases and interparticle interactions increase. The faster a given supersaturation is reached, the greater will be the number of nuclei formed.

Nuclei tend to dissolve at a given supersaturation, and this instability becomes more severe as the size of nuclei decreases. This effect is described by the well-known Gibbs-Thompson (Ostwald-Freundlich) equation [10]. This equation,

$$kT \ln \frac{S_r}{S_\infty} = \frac{2\gamma \bar{u}}{r}$$

relates the solubility S_r of a particle of radius r to that of a macroscopic particle, S_∞, the particle-continuous phase interfacial free energy, γ, and the molecular volume, $\bar{u}$, of the organic species. This equation shows that the propensity for dissolution (or increase in solubility) is inversely proportional to the particle's radius. This effect gives rise to the ever present driving force for particle *growth* by Ostwald ripening, where small particles (nuclei) dissolve and reprecipitate or condense on larger particles.

3. Metastable Two-Phase Boundary Formation

Most organic particles when condensed quickly from a multicomponent liquid adopt a spherical morphology. This result is a consequence of surface free energy minimization, where the interfacial free energy per unit volume is a minimum, because spheres yield the lowest surface-to-volume ratio. This driving force is also prevalent even in the precipitation of nanoparticulate inorganics, including metals and diverse oxides. Subsequently, however, further free energy may be given up by crystallization into one or more polymorphs accessible to the material. These transformations lead to the development of particular facets at the particle surface, characteristic of crystalline habits.

II. DISPERSIONS FROM EMULSIONS

A. General Preparative Methods

1. Emulsification of Organics

A widely applicable technique for preparing amorphous organic nanoparticulates is to dissolve the water-insoluble organic substrate in a water-immiscible solvent having a relatively high vapor pressure (e.g., ethyl acetate) in order to form an "oil phase" containing the substrate. Such water-immiscible solvents are often known as "auxiliary" solvents. This dissolution step is often aided by heating. The amount of heating (temperature of dissolution) is determined by the thermal stability of the organic substrate, the viscosity required during emulsification, and the amount of auxiliary solvent utilized. It is sometimes possible to omit the auxiliary solvent, or nearly so, when the melting point of the organic substrate is not too high. It is generally desired to obtain dissolution of the organic substrate using the least amount of auxiliary solvent necessary so as to minimize the amount of auxiliary solvent (VOCs, volatile organic components) that must subsequently be removed.

This organic solution is then emulsified with an aqueous solution containing stabilizers such as surfactant and polymeric dispersants (e.g., gelatin, polyvinyl alcohol, polyvinylpyrrolidone). Generally, increasing surfactant levels lead to decreasing particle sizes, as the specific surface area stabilized increases in direct proportion to the surfactant added. Charged surfactants provide charge stabilization. Nonionic surfactants and polymeric stabilizers provide steric stabilization. All of the surfactants when adsorbed to the aqueous-oil interface lower the interfacial tension and facilitate smaller droplet production with a given amount of input shear energy. Many high-shear methods for achieving emulsification are available. The available shear methods in decreasing order of effectiveness are high-pressure homogenization, colloid milling with rotor-stator devices, ultrasonic mixing, and very high speed stirring.

2. Transformation of Emulsions to Dispersions

The final steps center around removal of the auxiliary solvent. Two methods are used in practice, evaporation and washing.

(*a*) *Evaporation.* On the laboratory scale, evaporation is easily done using a rotovap apparatus. The emulsion is circulated over a large surface area under reduced pressure and the water-immiscible solvent gradually diffuses out of the emulsion into the vapor phase, where it is vented or condensed for recycling. Larger scale industrially practical flow systems generally condense the auxiliary solvent for recycling. These techniques are practical only for certain classes of solvents, as the auxiliary solvents must exhibit a sufficiently high vapor pressure. Ethyl acetate and cyclohexanone are two examples of auxiliary solvents well suited for evaporative removal. There are, however, several limitations of evaporative solvent removal. One limitation is that evaporation may be too slow at the desired processing temperature. Another is that the amorphous physical state of the organic substrate may be too unstable at the processing temperature and untoward physical ripening, particle growth, or crystallization may occur.

(*b*) *Washing.* Although the auxiliary solvents are water immiscible, they usually have finite water solubility. It is sometimes preferred to use countercurrent washing techniques to remove unwanted auxiliary solvent. If the emulsion is not gelled, then constant-volume dialysis or ultrafiltration can be used effectively to wash out the auxiliary solvent. If the emulsion gels, as is easily obtained when using gelatin or some other gelling polymeric stabilizer, one can chill (causing gelation) and then chunk the emulsion to produce a high surface area. This gelled emulsion is then repeatedly washed with a stream of wash water (containing an appropriate electrolyte to control swelling) to remove the auxiliary solvent.

(*c*) *Condensation.* As the auxiliary solvent is removed and the temperature of the emulsion cools, the solubility of the organic substrate decreases. After the auxiliary solvent is gone, one usually obtains an amorphous (solid) state for the organic substrate. In some cases, particularly when the substrate has few degrees of internal freedom, intraemulsion particle crystallization may occur. However, for most organics having molecular weights >300 or so, condensation into an amorphous solid state is the rule rather than the exception. Low-vapor-pressure organic solvents or plasticizers can be included in such dispersion formulations for various chemical and physical reasons. Such solvents, plasticizers, and additives typically do not very effectively solubilize the substrate. Sufficient water-insoluble solvent, dramatically less water miscible than auxiliary solvents, can be included to prevent condensation of the organic substrate and to keep the substrate in a solution state.

B. Applications

Such methods have been in use on an industrial scale for many decades in the manufacture of photographic papers and films. Applications to pharmaceutical preparation have begun and are currently a focus area where current research is being done. Other diverse applications of this condensation approach include non-cross-linked polymer bead production.

1. Photographic Applications

Image dye formation in most color films and papers occurs via coupling with oxidized color developers [11]. The couplers that form dye upon reaction with oxidized developer are generally only sparingly soluble in water and are incorporated into particular layers of photographic elements as submicrometer particles prepared by emulsification techniques [12]. Typical examples of such couplers are illustrated in Table 1.

The coupler dispersion particles are usually in the size range 50–300 nm. Particle size is an important physical parameter, particularly with respect to its influence on the kinetic reactivity and dye-forming efficiency of such dispersions [13,14]. Such couplers are often formulated with a low-vapor-pressure organic solvent in order to modify coupling reactivity or dye hue. High-vapor-pressure auxiliary solvents, such as ethyl acetate and cyclohexanone, are often used temporarily to facilitate dissolution of coupler at a given processing temperature.

In formulations using auxiliary solvent, 1 part coupler, $\frac{1}{4}$ to 2 parts low-vapor-pressure plasticizer or coupler solvent, and 4–6 parts auxiliary solvent are combined with mild heating and low-shear mixing in order to effect dissolution. Such a solution is typically termed the "oil" phase. In parallel, an aqueous solution containing dispersing aids such as surfactant and gelatin is prepared, and this solution is typically referred to as the "aqueous" phase. These oil and aqueous phases are then mixed under mild shear to effect crude emulsification, and then this emulsion is passed through high

TABLE 1 Amphiphilic Color-Forming Photographic Couplers

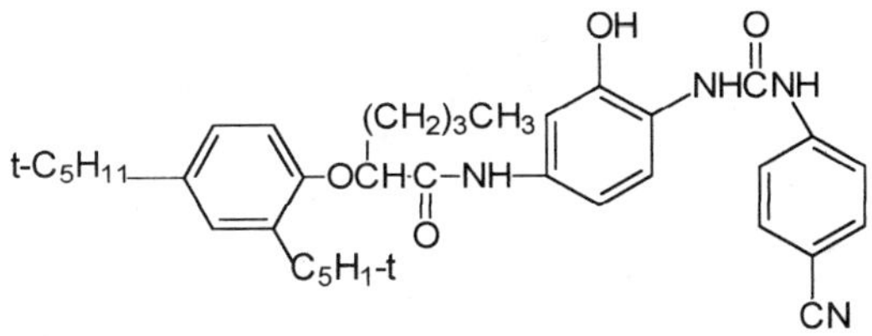

I

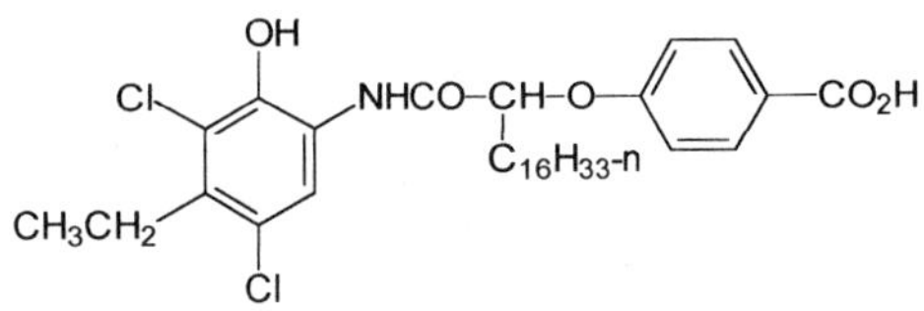

II

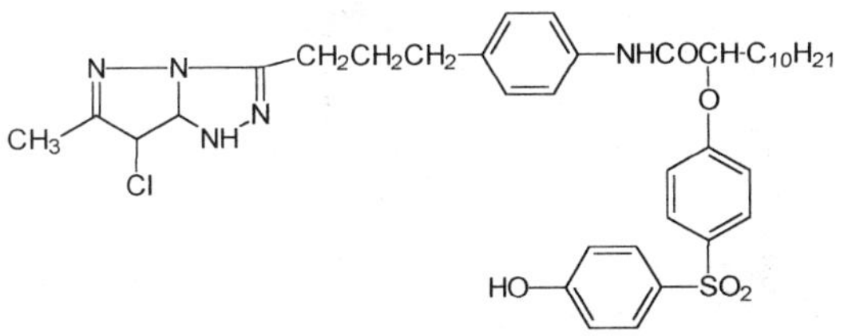

III

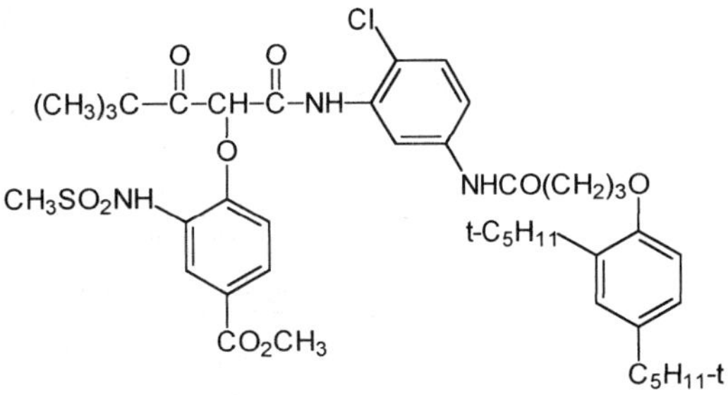

IV

shear to produce a submicrometer-sized emulsion. Colloid mills and diverse types of homogenizers are widely used in the art to effect such emulsification. When gelatin has been included as a dispersing aid (it is a good steric stabilizer and binder), it is often convenient to chill the emulsion to effect gelation, then chunk it to produce a high surface area, and then subject the emulsion in this state to washing processes to remove the auxiliary solvent. As an alternative to chilling and gelling, the auxiliary solvent may be removed by evaporation under reduced pressure. After this stage of auxiliary solvent removal, the oil-in-water emulsion has been thereby transformed into a dispersion comprising amorphous coupler-coupler solvent particles. Only rarely are such couplers significantly soluble in their respective coupler solvents. Two general modifications of these processing steps are encountered in industrial applications.

(*a*) *NS Dispersions.* One modification simply omits the use of coupler solvents (low-vapor-pressure organic solvents or plasticizers). Such omission is most often made in connection with couplers that have sufficient reactivity without added solvent. Auxiliary solvents are still often used, but after their removal, such dispersions are often termed "NS" dispersions (no solvent).

(*b*) *Direct Dispersions.* Another important modification in such formulations is the omission of auxiliary solvents. The driving force for formulating without auxiliary solvents is that one avoids the expense of the auxiliary solvent *removal* steps. However, the price that must be paid, typically, is that the oil phase must be heated to a much higher temperature than is necessary when formulating without auxiliary solvents. The organic composition to be dispersed is simply heated to a suitable liquidus temperature and then emulsified or homogenized to form an emulsion. The primary concern is the thermal stability of the substance being liquefied and dispersed. With couplers that are sensitive to aerial oxidation, such elevated heating is often done under a nitrogen blanket.

(*c*) *Examples.* Examples of such coupler dispersions are illustrated in the TEMs of Fig. 2. All of the formulations illustrated used aqueous phases that comprised gelatin and Alkanol-XC (Du Pont) as dispersing aids (Structure 1). The dispersion of coupler **I** illustrated in Fig. 2a was formulated 1 : 1 (w/w) with di-*n*-butylphthalate and utilized ethyl acetate as auxiliary solvent, and this auxiliary solvent was removed by evaporation. The amorphous particles obtained are nearly spherical, and the particle size distribution is polydisperse, ranging from 50 to 400 nm in diameter. An NS dispersion of coupler **II** is illustrated in Fig. 2b. The carboxy functionalty of this coupler makes it highly surface active, so that no added coupler solvent is necessary to achieve efficient coupling. Again, ethyl acetate was used as auxiliary solvent to facilitate emulsification. A very fine particle size dispersion was obtained after removal of the auxiliary solvent. Particles

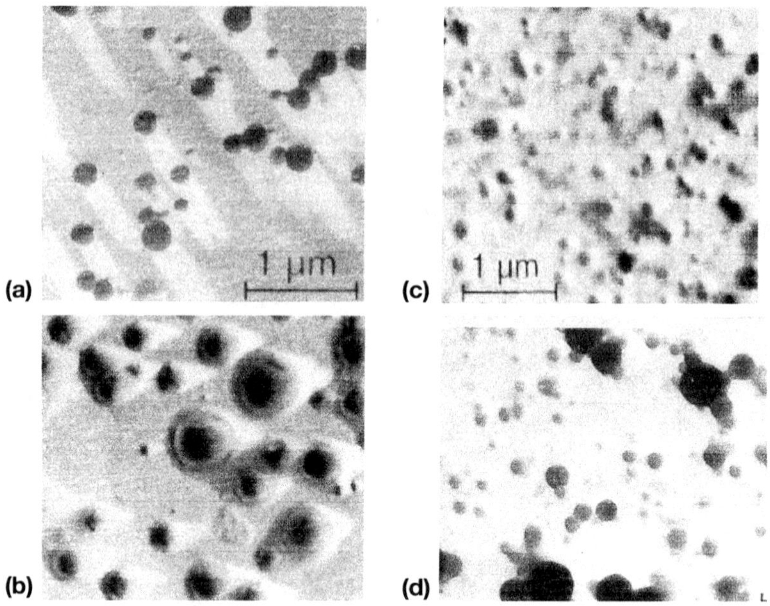

FIG. 2 Transmission electron micrographs of coupler dispersions prepared by emulsification (homogenization): (a) coupler **I** formulated 1 : 1 (w/w) with di-*n*-butylphthalate; (b) coupler **II** formulated without added plasticizer; (c) coupler **III** formulated 1 : ½ with tricresylphosphate; (d) coupler **IV** formulated 1 : ¼ with di-*n*-butylphthalate. (Adapted from Ref. 12.)

in the range 50–300 nm are evident in Fig. 2b. Coupler **III** was formulated 1 : ½ with tricresyl phosphate as coupler solvent and emulsified by colloid milling, and the ethyl acetate auxiliary solvent was removed by evaporation. The resulting dispersion particles are illustrated in Fig. 2c, where it is seen clearly that the dispersion particles *are not homogeneous*. The shadowing shows that while these dispersion particles are probably spherical in the bulk, they are sufficiently soft to deform on the TEM grids. Moreover, the deformations appear to be conelike, as is evident from the shadows. Figure 2d shows a dispersion of yellow dye forming coupler **IV** formulated 1 : ¼ with di-*n*-butylphthalate. This dispersion was prepared without the use of auxiliary solvent, and emulsification was done using high-pressure homogenization. Such oil phases are typically fairly viscous during homogenization, and a consequence of this viscosity is a broader particle size distribution. The particles illustrated in Fig. 2d span a range of 50–500 nm in diameter.

2. Pharmaceutical Applications

Homogenization is becoming more prevalent in the pharmaceutical literature as a means for preparing amorphous and crystalline dispersions of active drug substances and of excipients. Nearly all of the problems currently being encountered in formulation work have been encountered (and overcome) much earlier in the photographic technology literature. A prime example is illustrated by the so-called hot homogenization technique.

(*a*) *Hot Homogenization.* A schematic of the hot homogenization process is illustrated in Fig. 3. This method is essentially isomorphic to the direct dispersion process described above for dispersing couplers without auxiliary solvent. The overarching goal is to produce fine-particle crystallites of pharmaceutical agent, encapsulated and stabilized by an immiscible coating of lecithin. The lecithin and agent are heated to dissolution and then emulsified, typically by homogenization, with a hot aqueous phase to produce an emulsion [15]. This emulsion is then cooled, during which process (it is hoped that) the liquid disperse phase solidifies and crystallizes to produce a fine particle dispersion. During this crystallization, the pharmaceutical agent and stabilizer (lecithin) phase sepa-

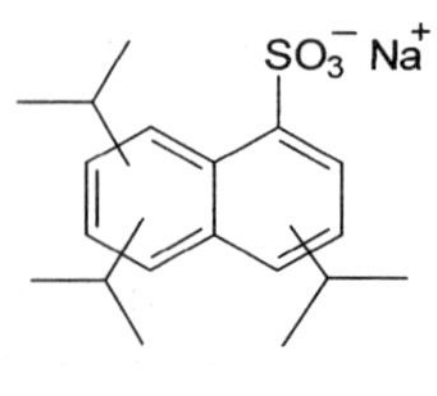

STRUCTURE 1 XC (Alkanol-XC).

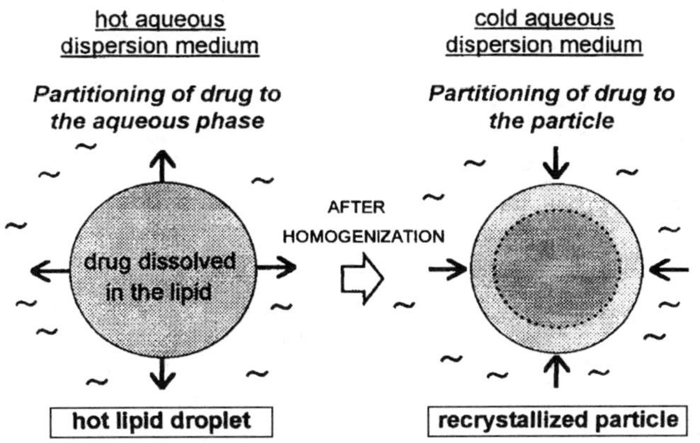

FIG. 3 Hot homogenization process for dispersing pharmaceutical agents. Liquification of drug compound and stabilizer, followed by (hot) homogenization, yields emulsion droplets pictured on left. Subsequent cooling leads to solidification of core drug compound surrounded by stabilizer (right). (Adapted from Ref. 15.)

rate, with the latter forming a stabilizing coating around the former.

Severe crystallization problems can be encountered in such processes, especially since lecithins are not particularly well suited as stabilizers and, furthermore, particularly because it is difficult to control when and how the liquid agent phase will crystallize after cooling. As an example, consider the hot homogenization of a model pharmaceutical agent where lecithin is used as the stabilizer. A composition of model compound and lecithin was heated to effect liquefaction of the organic phase, and then this phase was emulsified in water using homogenization. Particle size analysis soon after emulsification showed that a submicrometer dispersion was obtained. However, after 5 days of storage, particle size analysis revealed that in addition to the submicrometer mode corresponding to the original dispersion, a second significant mode in the 10–20 μm range had appeared, suggesting that a major portion of the mass had been transformed into large microcrystallites.

This untoward crystallization was the natural consequence of several features. First, there was no matrix stabilization, so dispersion particles were free to undergo binary collisions during storage. Such collisions can activate and nucleate crystallization processes. Furthermore, lecithin itself is not an ideal stabilizer, in general. Because it was believed that the lecithin and model compound were immiscible at storage temperature, it is presumed that phase separation drives these two components apart. It is unknown whether lecithin spontaneously wets the amorphous or crystalline surfaces and facets of the model compound once the compound solidifies upon cooling. In addition, it is difficult to control exactly when a liquid disperse phase will crystallize. Although cooling is sufficient to produce an amorphous solid phase, the free energy of such phases is generally higher than that of a crystalline phase. Such particles are in a supercooled physical state and have an effectively higher solubility than lower free energy crystallites have. Such dispersions are therefore primed for untoward crystal growth by Ostwald ripening, in which large (lower free energy) crystallites are formed from metastable, amorphous solid dispersion particles. In such cases, monomer transport is by diffusion, and binary particle collisions are not required for transport, although such collisions can provide the necessary activation for nucleation of a crystalline phase.

(*b*) *Cholesteryl Acetate.* A general flowchart illustrating dissolution, homogenization, precipitation, auxiliary solvent extraction, and recycling is shown in Fig. 4. Application of this process to produce cholesteryl acetate (CholA) (Structure 2) particles under various auxiliary solvent evaporation regimes consistently resulted in amorphous CholA particles [16]. Particle sizes obtained were insensitive to the conditions of auxiliary solvent removal and were mainly controlled by the emulsion droplet sizes following homogenization. When toluene was used as the auxiliary solvent and removed by evaporation and POE-(20)-sorbitan monooleate was used as emulsifier, particle sizes in the 300–500 nm range were obtained using 3% (w/w) emulsifier and in the 100–150 nm range at 10% (w/w) emulsifier, as a function of the disperse phase of the emulsion. Particle sizes correlated very strongly with starting emulsion droplet sizes, even though the auxiliary solvent was removed by dramatically different processes. A summary of these results is given in Table 2. The final dispersion sizes are fairly uniform no matter how the auxiliary solvent was removed, whether by evaporation or by washing (dialysis). Perhaps more surprising is the close correspondence between the dispersion sizes and the starting emulsion droplet sizes, even though the toluene has been removed. At the highest CholA loading, 50% (w/w), there was essentially no difference in size upon solvent removal. At only 5% loading, the final dispersion sizes were still 75% of the starting emulsion sizes. A detailed analysis of this surprising result is not yet available, but several factors probably contribute. First, there is probably some incorporation of water during the evaporation and wash-

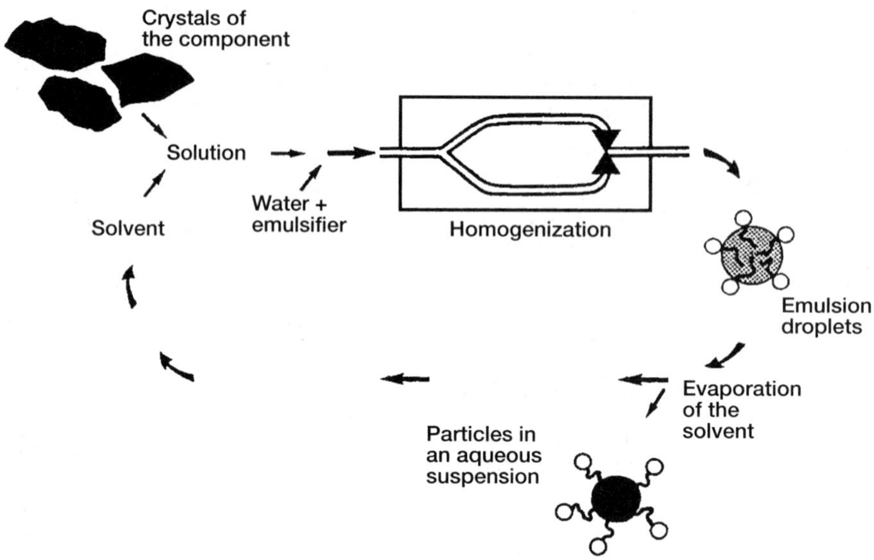

FIG. 4 Cyclic process for producing organic dispersions by emulsification. The active component is dissolved in an auxiliary solvent and combined with an aqueous emulsifier solution. This two-phase mixture is then homogenized to produce an emulsion. The auxiliary solvent is then removed by evaporation (or other means) and recycled. (From Ref. 16.)

ing steps to remove the auxiliary solvent. Cholesteryl derivatives are known to form hydrates. Second, the direct evidence indicated that the resulting dispersion particles were amorphous and therefore solidified in a sort of diffusion-limited aggregation sense that preserved volume and retarded compaction and densification during condensation. Third, the emulsifier most likely intercalated to some degree in the periphery of the dispersion particles and provides a significant shell volume, especially due to the 20-unit poly(oxyethylene) hydrophilic part of the headgroup, which amounts to 20–30% of the particle diameters.

3. Polymeric Bead Formation

A particularly straightforward path for producing polymeric beads from polymer produced in melt or solution polymerization is to dissolve the polymer in an auxiliary solvent, emulsify this polymer solution, and then remove the auxiliary solvent to leave polymer beads in suspension. As the auxiliary solvent is removed, the supersaturated polymer condenses to form a bead. Of course, beads may be produced using many different polymerization pathways, as discussed in several other chapters in this volume (see Part 3).

This approach to bead formation has several advantages. First, all of the technologies available for controlling particle size in emulsification are applicable. For example, the controlled production of large (1 μm diameter and larger) beads by using limited coalescence [17] is available. The production of beads having mixtures of various polymers and polymer fractions essentially requires only that all of the ultimate components be soluble in the same auxiliary solvent. Whether microscopic phase segregation occurs during condensation or subsequently is another matter, however.

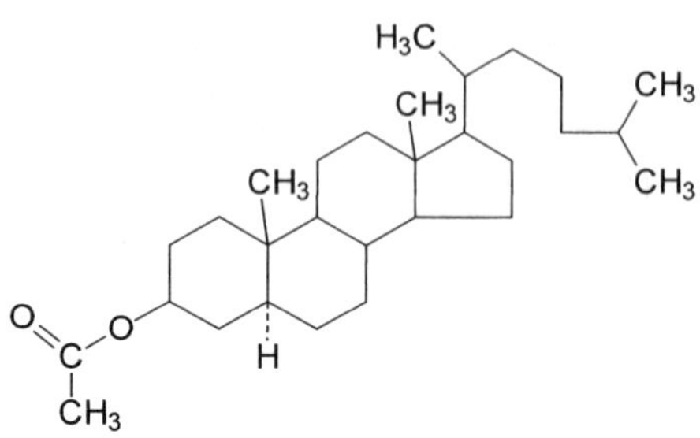

STRUCTURE 2 CholA.

III. PRECIPITATION BY SOLVENT SHIFTING

Changes in solubility resulting from physical modification of a solvent's ability to dissolve an organic substrate are central to nearly all of the precipitation processes discussed in this chapter, as well as the condensation processes discussed in the previous section. Solvation power changes dramatically with most intensive variables, including volume fraction, mole fraction, temperature, and concentrations of additives,

TABLE 2 Correlation of Particle Sizes of CholA Dispersions, Prepared According to Different Regimes of Auxiliary Solvent (Toluene) Removal, with Starting Emulsion Droplet Particle Sizes

	Mean particle size (nm)				
CholA [% (w/w)]	Emulsion droplet	Evaporation at 100 mbar	Evaporation at 1 bar	Washing by dialyis	Dialysis with toluene-saturated water
5	157	108	108	117	111
10	149	127	133	140	
50	193	193	195	197	197

salts, and binding and nucleation sites. In this section we focus upon the controlled dissolution of substrate in a "good" solvent (sufficient to produce a single-phase solution), followed by mixing with a *miscible* but poor solvent (typically water). A poor solvent can also be constructed by having sufficient components that diminish solvent quality toward certain substrates, such as certain salts.

A. Flow Systems

Solvent shifting processes are often done in batch mode, and such methods are standard in bulk precipitation and crystallization. However, batch solvent shifting processes generally do not yield very uniform particle size distributions; continuous flow systems are better suited for producing good particle uniformity. Tee-mixing comprising two axially opposing flow streams, such as pictured in Fig. 5, is very effective and various sizes are available to accommodate flows of varying volume. Subsequent mixing downstream may or may not be utilized and can consist of flow ultrasonication, high-energy milling, and static mixing. Turbulent flow usually suffices.

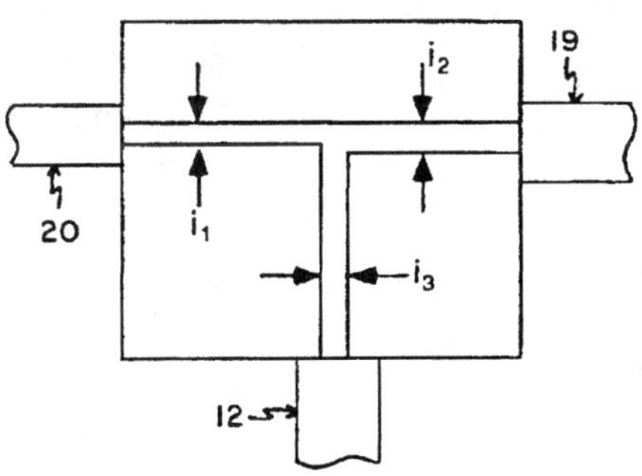

FIG. 5 Simple tee-mixing head fashioned for small-scale manufacturing by solvent shifting. Such a mixing head is easily machined from plastics, and even simpler mixing heads can be fashioned from tee-pipe and tubing fittings. Opposing reactant flows enter through 19 and 20; the mixed flow exits through 12.

B. Salting Out

Some organic pigments may be precipitated from solution by so-called salting out procedures. A solution of the pigment is prepared and then salt, often containing a metal ion that interacts (binds) specifically with the dye anion, is added. When the solubility product of the metal ion complex of the dye is low, the metallized dye or pigment precipitates. Such metallized pigments and dyes are known as "lakes." Production of lakes by encapsulating inorganic pigment particles with organic dyes and pigments is an important process for producing specialized colorful and pearlescent pigments, and these processes are discussed further in Section IX.

1. Organic-Inorganic Composites

Wu and Matijević [18] demonstrated the importance of agitation and surfactant level in controlling particle size and particle aggregation in $BaCl_2$-induced precipitations of Red Dye No. 6 (**V**) (Structure 3) to produce the so-called barium lake of **V**. Precipitation of dye solutions (0.024 M) with aqueous $BaCl_2$ (0.02 M) produced thin needles typically of the order of 1.5 μm in length and 100 nm in diameter. These needles were not primary single crystals but comprised many smaller nanosize subunits. When precipitation was done in the presence of a commercial surfactant, Daxad 11G, primary subunits about 30–40 nm in largest dimension

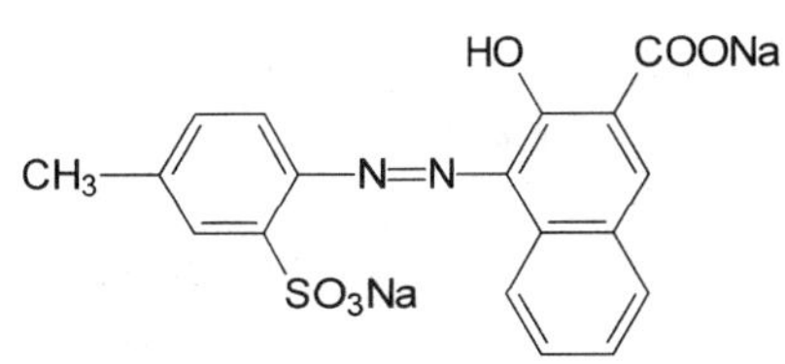

STRUCTURE 3 V (Red dye No. 6).

were formed. These nanoparticles were amorphous by x-ray diffraction and were negatively charged, due to the adsorbed anionic surfactant, with an electrophoretic mobility of about -0.8 $\mu m\ s^{-1}\ V^{-1}$. Such surfactant-aided precipitations produced stabilized nanodispersions with much improved optical covering power and dramatically decreased scattering in the long-wavelength region of the visible spectrum.

Another example consists of coprecipitation of the anionic dye 3 Mordent Blue (3MB, Table 3) with aluminum sulfate to produce fairly monodisperse particles in the 200–300 nm diameter range comprising Al/dye ratios of 1:0.001 to 1:0.1 [19]. Such composite precipitated particles are produced by a hydrothermal process, and good size monodispersity arises as a result of kinetically controlled chemistry to produce hydrous oxides. The aluminum sulfate and dye solutions are aged at elevated temperatures. The aluminum hydrolyzes water to produce aluminum hydroxide and coprecipitation occurs with the anionic dye.

An alternative form of composite pigment has been described by Wu et al. [20]. Nanoparticulate silica (Ludox CL), 5–12 nm in diameter, was surface treated to produce an alumina-type surface with resultant particles about 13–20 nm in diameter. These particles were then reacted with various organic dyes, including Red Dye No. 6 (**V**), Acid Yellow 1, Acid Blue 25, and Guinea Green B (Table 3) to form inorganic (core)-organic (shell) composite pigment particles. The amount of **V** incorporated ranged from 12 to 30% (w/w) relative to the inorganic weight and increased with specific surface area of the inorganic core. Inorganic cores 15 nm in diameter supported saturation organic shells of 30, 19, 70, and 60% (w/w), for Red Dye No. 6, Acid Yellow 1, Acid Blue, and Guinea Green B, respectively. Such coating is analogous to the formation of so-called aluminum lakes.

2. Protein Coacervation

Submicrometer-sized gliaden particles, derived from wheat gluten, have been produced [21] by precipitation from nonsolvent mixtures and are potential excipients for delivery of pharmaceutically active compounds such as *trans*-retinoic acid [22]. An aqueous ethanol solution of gliadin is infused with an aqueous saline solution that induces, by salting out, protein coacervation and particle formation. Ethanol evaporation (rotary evaporator) expedites the coacervation. Particles are isolated by centrifugation, resuspended, and optionally cross-linked with glutaraldehyde. Various aqueous ethanol, aqueous ethylene glycol, and aqueous propylene glycol mixtures were examined from a solubility

TABLE 3 Dye Structures for Forming Inorganic-Organic Core-Shell Composite Pigments

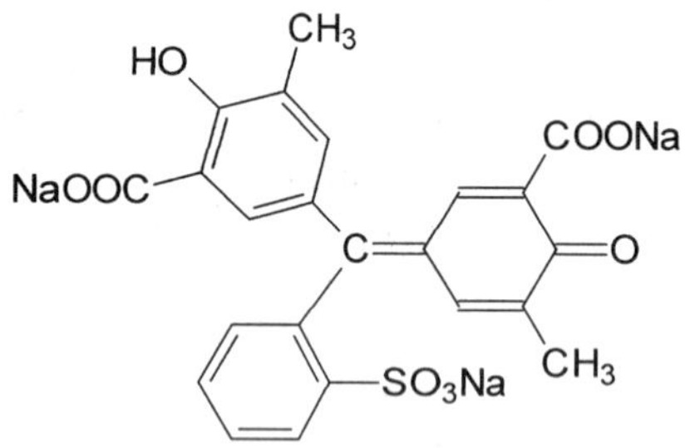

3MB

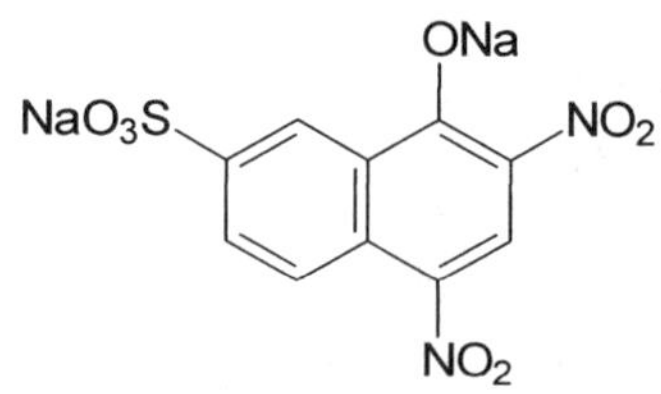

Acid Yellow 1

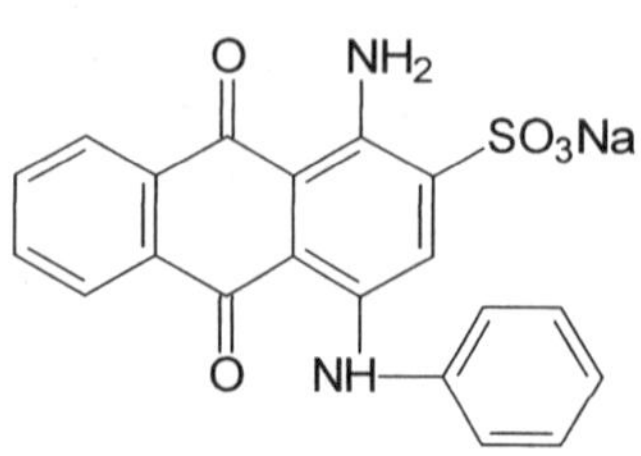

Acid Blue 25

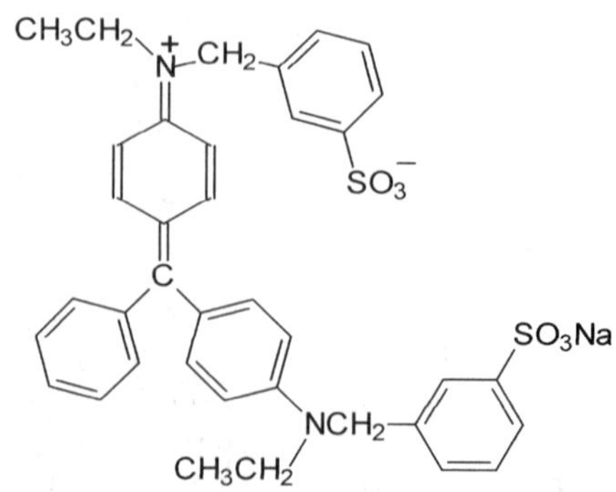

Guinea Green B

parameter perspective in examining solvent and processing effects on particle size.

C. Miscible Solvent Mixing

A practical development scale flow system is illustrated in Fig. 6 for the case of dispersing organic pigments as submicrometer dispersion particles. The pigment is dissolved in a suitable water-miscible solvent and fed into the reactor in a shear field. As droplets of the pigment solution are formed in the nonsolvent, water, counterdiffusion flows of water into the particles and solvent out of the droplets occur, leading to supersaturation of the pigment, followed by nucleation and growth.

1. Carotenoids

The precipitation of β-carotenoids in aqueous gelatin solutions by solvent shifting has been studied by Horn and coworkers [23,24]. The carotenoid is first dissolved in a water-miscible solvent such as an alcohol or ketone at an elevated temperature. This solution is then pumped into a mixing chamber, where it is mixed with an aqueous gelatin solution. The temperature drop and lower solubility in water induce precipitation to produce nanosize particles sterically stabilized by the gelatin. The presence of a surfactant, ascorbylpalmitate (AscP) (Structure 4), in the aqueous gelatin solution yields smaller particles than obtained otherwise. Results are summarized in Table 4.

Gelatin is well known as an excellent steric stabilizer. It is also surface active and has been used for decades in photographic dispersion technology in the preparation of emulsions and dispersions as described in Section II. The results in Table 4 were obtained using two separate types of gelatins. Type A had a molecular weight of 240,000 and type B had a molecular weight of 300,000. The isoelectric point (IEP) of type A gelatin was determined to be 9.5, while that of type B gelatin was 5. The isoelectric points of the precipitated particles depend on the IEP of the type of gelatin used but do not depend on the surfactant, AscP. The surfactant will be either nonionic (in ring-closed form) or anionic at pH above 5 or so if the carboxylate group is activated by ring opening. The surfactant has a significant effect on particle size for both types of gelatins.

2. Color Instant Image Dyes

The production of 100–200 nm diameter dispersions of organic pigments for color instant photographic applications by solvent shifting was well demonstrated by Gutoff and Swank [25]. Two pigments, one yellow (Yel) (Structure 5) and one cyan (Cyan) (Structure 6),

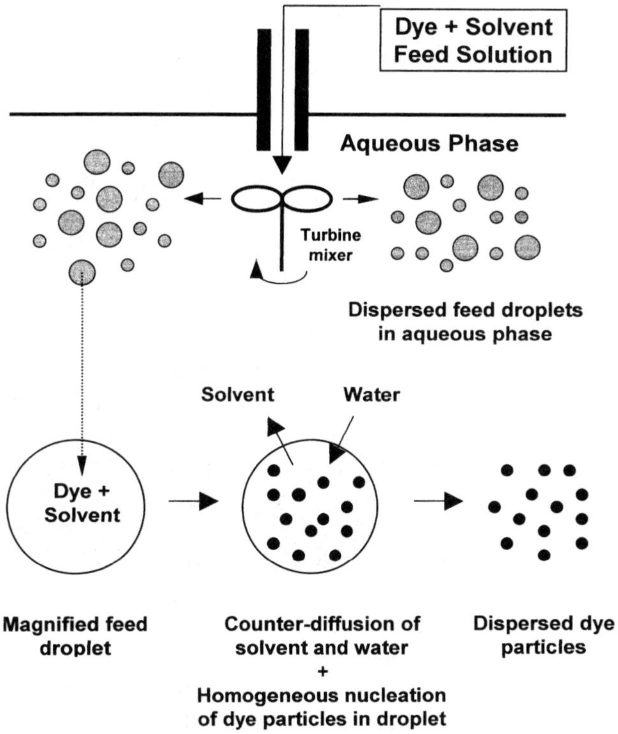

FIG. 6 Miscible solvent shifting reactor illustrating organic pigment solution fed into reactor containing aqueous stabilizers. The shear field in the reactor produces droplets of feed solution that immediately begin to undergo countercurrent solvent exchange, with water diffusing into the droplets and organic solvent diffusing out of the droplets. All of these processes lead to supersaturation with respect to the organic pigment and subsequently to homogeneous nucleation of pigment particles. (Courtesy of M. C. Brick, H. J. Palmer, and T. J. Whitesides, to be published.)

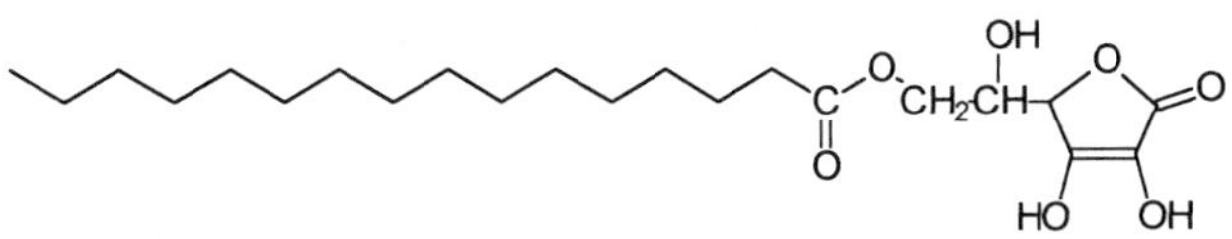

STRUCTURE 4 AscP.

TABLE 4 Precipitated β-Carotenoid Nanoparticle Properties

Gelatin type	Weight % surfactant	Particle diameter (nm)	Isoelectric point (pH)
A	0	346	6.0
A	8	262	6.8
B	0	327	4.8
B	8	157	4.7

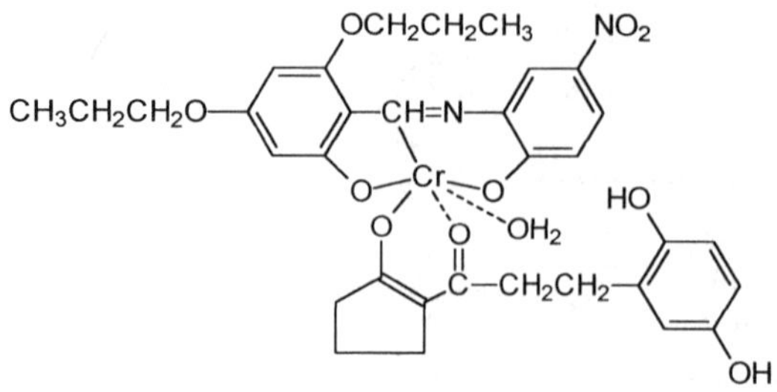

STRUCTURE 5 Yel.

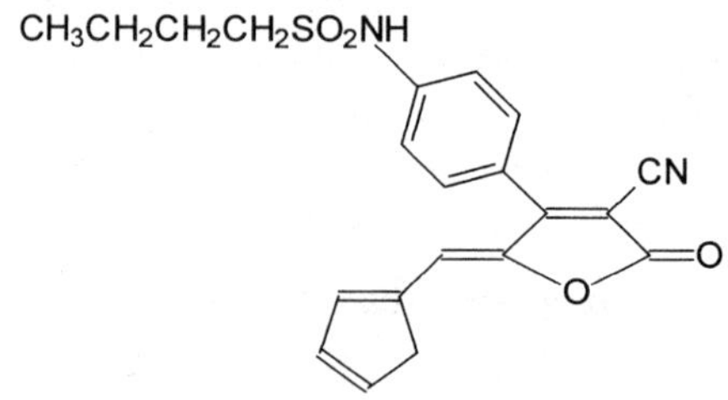

STRUCTURE 7 YPG.

were dissolved in water-miscible solvent (acetone/isopropanol and acetone, respectively) and fed into an aqueous stabilizer solution in a tee-mixer to produce submicrometer amorphous pigment particles. Both surfactants (Tamol 731) and polymeric dispersants (polyvinylpyrrolidone, PVP) were used to provide stabilization of the resulting dispersions. After precipitation, the auxiliary solvent was removed in a sieve tray column. The essentially auxiliary solvent–free dispersion was cycled in a reboiler at the bottom of the column, and the residence time in this reboiler was correlated with final average particle size and was shown to be an effective means for growing particles to an aim size.

3. Photographic Couplers and Pigments

Solvent shifting to prepare a very small particle size dispersion of amphiphilic couplers is a well-established photographic art. Godowsky and Duane [26] disclosed water-miscible solvent shifting to disperse molecular couplers and polymeric couplers. The couplers were typically fairly amphiphilic, owing to carboxy,-sulfonic acid, and methyl ester substituents.

Recent work has demonstrated homogeneous nucleation in miscible solvent shifting for the yellow organic pigment YPG (Stucture 7) precipitated out of various solvents (dimethylacetamide, methyl pyrrolidone, dimethyl sulfoxide, and acetonitrile). Using an aqueous phase containing both a surfactant, sodium dodecyl sulfate (SDS), and a polymeric stabilizer, PVP, Brick et al. (M. C. Brick et al., to be published) have shown that submicrometer particles such as those illustrated in Fig. 7a can be generated. Nucleation rates as a function of supersaturation ratio are illustrated in Fig. 7b. The four different water-miscible solvents all yield about the same power law correlation between nucleation rate and the supersaturation ratio over a range in ratio from 500 to 10^5.

4. Pharmaceutical Dispersions

Specialized batch precipitations for pharmaceutical preparations have been demonstrated by Violante and coworkers [27–29] for a variety of organic materials. A general flow diagram is illustrated in Fig. 8. A basic limitation of this process is the need to use preparative centrifugation to isolate suitable size fractions of the drug dispersions. The water-insoluble pharmaceutical is dissolved in a water-miscible solvent and is precipitated by infusion of an aqueous or nonaqueous precipitating liquid (miscible with the original solvent). Or-

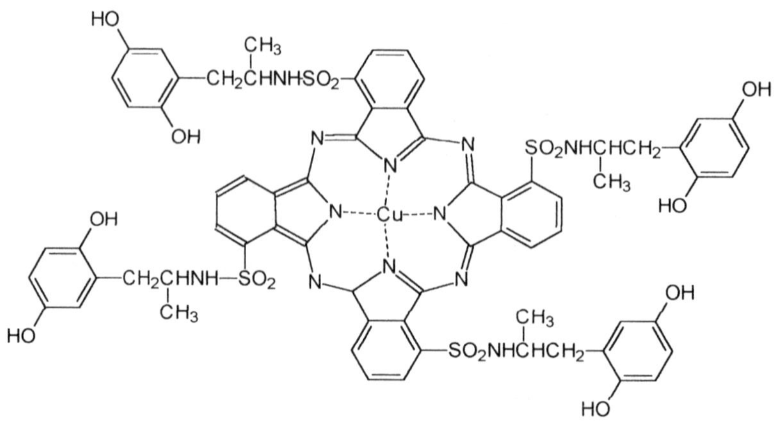

STRUCTURE 6 Cyan.

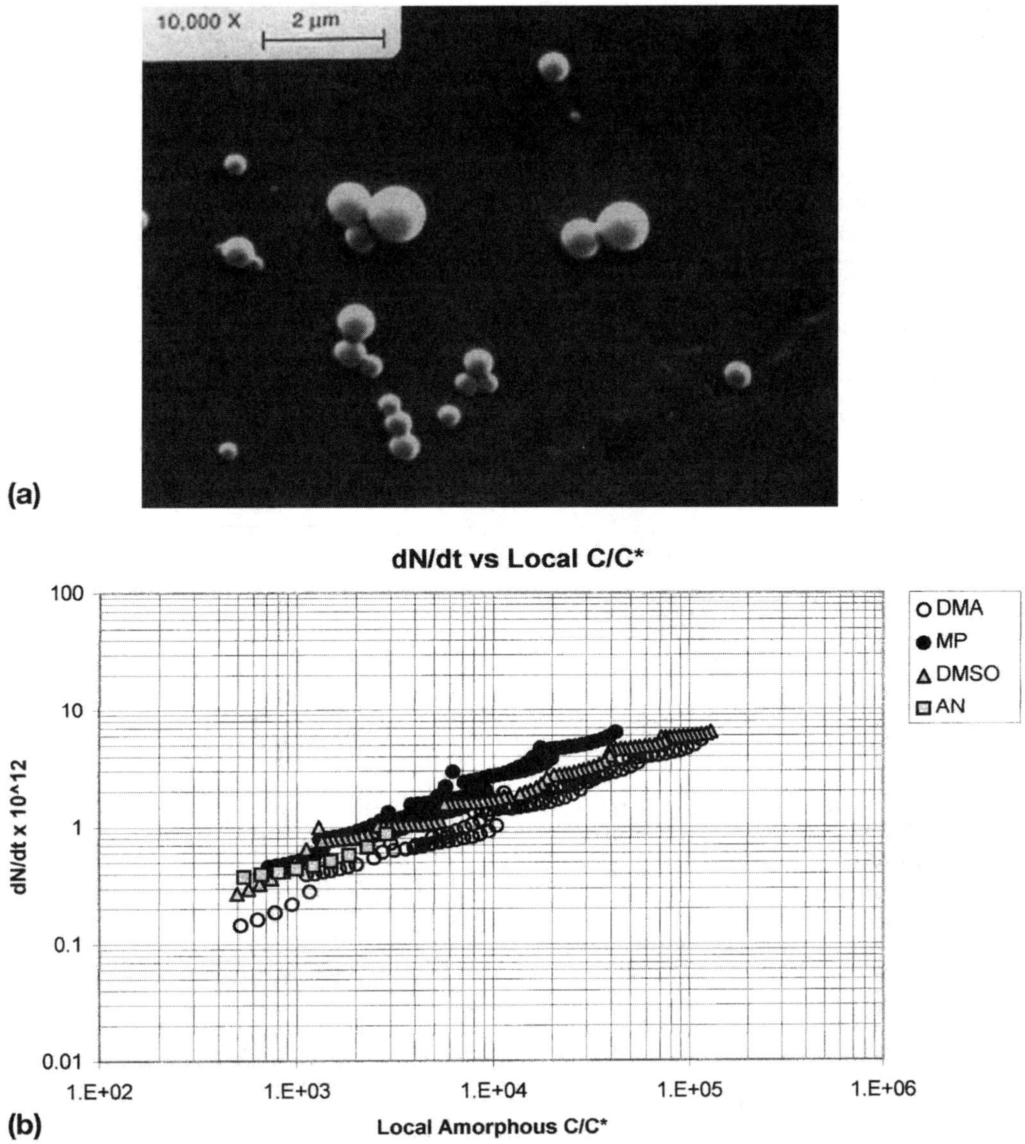

FIG. 7 (a) Scanning electron micrograph of organic pigment YPG particles made by miscible solvent shifting; (b) correlation of nucleation rate with supersaturation ratio for YPG precipitation out of various organic solvents using SDS and PVP as particle stabilizers; units of *dN/dt* are particle per second. DMA, dimethylacetamide; MP, methylpyrrolidone; DMSO, dimethyl sulfoxide; AN, acetonitrile. (Courtesy of M. C. Brick, H. J. Palmer, and T. J. Whitesides, to be published.)

ganic solvent removal is done by washing and centrifugation, and particles ranging in size from 10 nm to 5 μm can be obtained depending on formulation and process parameters. For example, the image contrast agent iodipamide ethyl ester (IEE) (Structure 8) can be precipitated as 100 nm to 2 μm particles from 1:2 dimethyl sulfoxide (DMSO)/ethanol by 5% (w/w) aqueous PVP, where average size can be controlled by the aqueous dispersant infusion rate. Increasing infusion rate of nonsolvent produces smaller particles. Increasing temperature produces larger particles. Another example is provided by 2,2′,4,4′-tetrahydroxybenzophenone (THBP) (Structure 9) precipitated from DMSO by an aqueous serum albumin solution to produce 500-nm-diameter particles.

Frank et al. [30] have reported a particularly simple solvent process to produce 2-μm particles of probucol (PBCOL) (Structure 10) stabilized by SDS and PVP. An ethanol solution of probucol is dispersed into aqueous stabilizer solution with moderate shear. Counter-

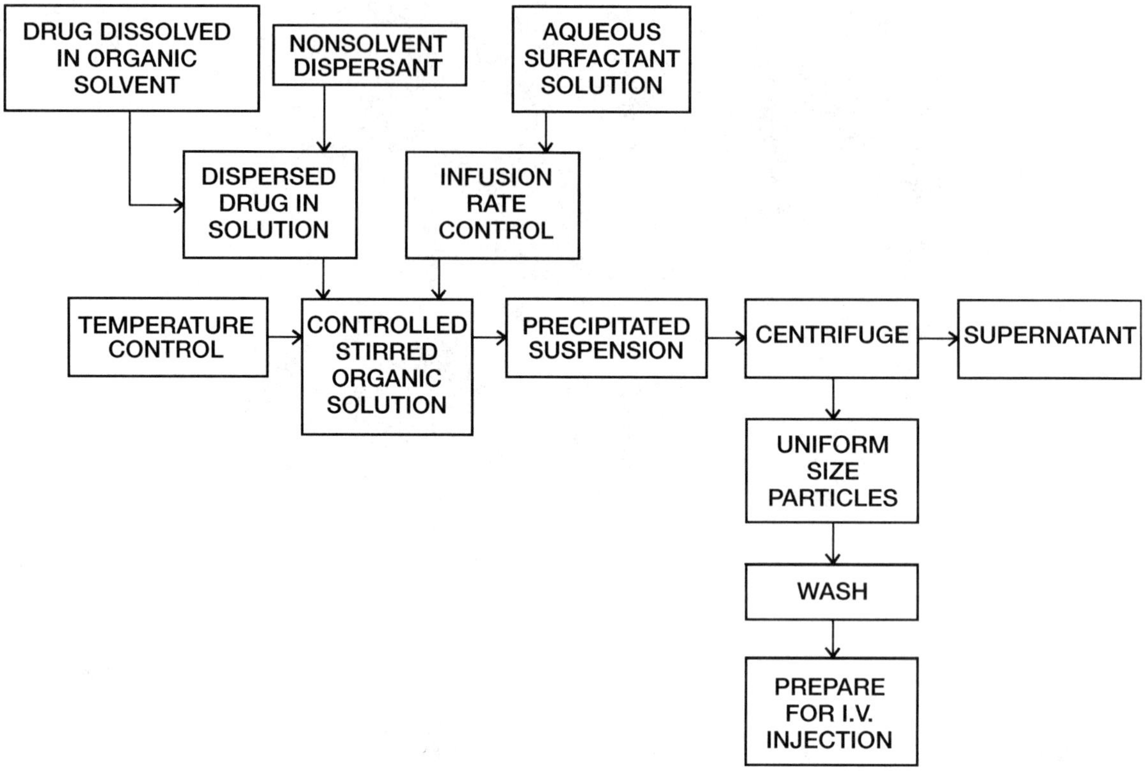

FIG. 8 Batch solvent shifting precipitation process for producing pharmaceutical dispersions suitable for intravenous injection. Nonsolvents bring the organic solution close to saturation or place the solution into a supersaturated state and often include solvents such as lower alcohols, e.g., ethanol. (From Ref. 27.)

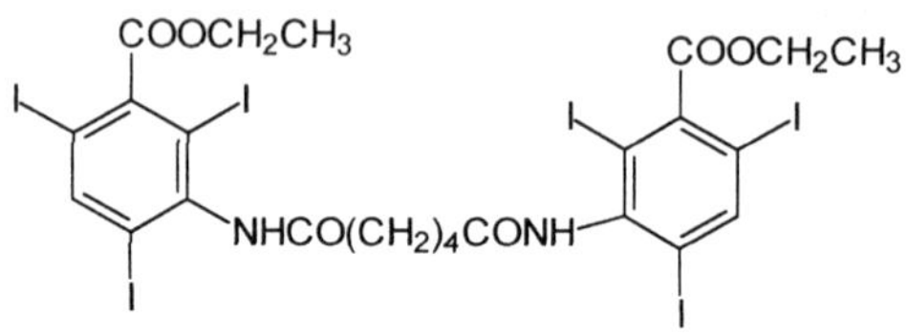

STRUCTURE 8 IEE.

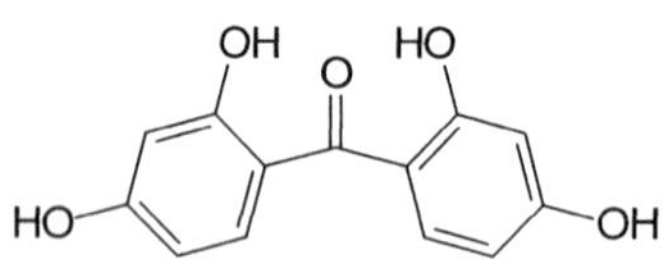

STRUCTURE 9 THBP.

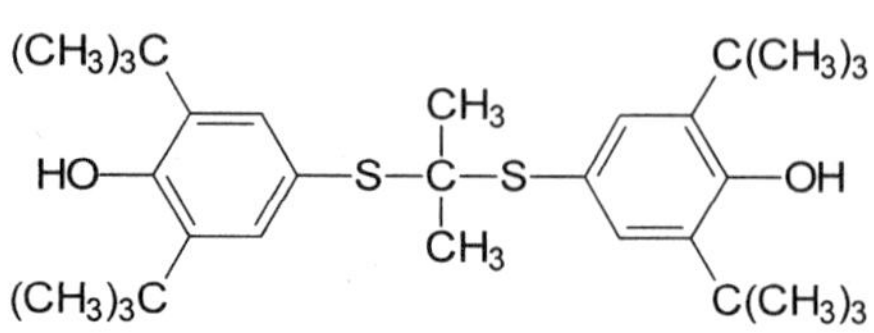

STRUCTURE 10 PBCOL.

diffusion leads to supersaturation and precipitation of the PBCOL particles.

5. Block Copolymer Self-Assembled Particles

A very interesting class of spherical and vesicular particles in the micrometer size range has been discovered by Jenekhe and Chen [31,32] based on block copolymer self-assembly. The block copolymers are derived from rigid rods of poly(phenylquinoline) and random coils of polystyrene (PPQ-b-PS) (Structure 11). Morphology control is achieved by using different solvent combinations that are good solvents or nonsolvents for

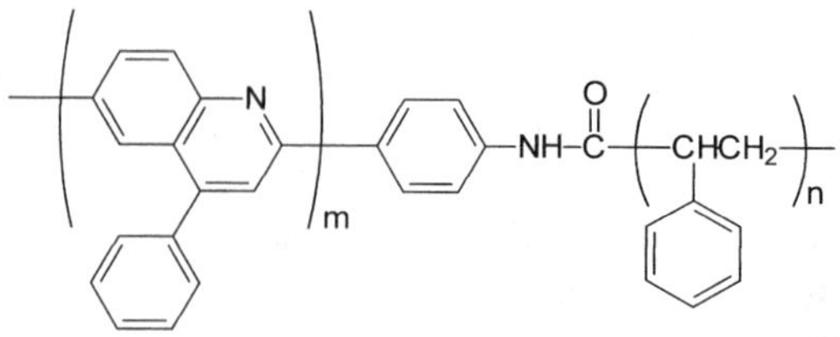

STRUCTURE 11 PPQ-b-PS.

one or the other of these two block materials. For example, trifluoroacetic acid and dichloromethane and toluene are good solvents for the PPQ rod blocks, but the PS coils are insoluble in these solvents. Such diblocks then tend to form bilayer structures and upon solvent evaporation yield various vesicular morphologies (spherical, ellipsoidal). Particles having largest dimensions on the order of μm were obtained, depending on evaporation rate and solvent mixture proportions. These are remarkably large size dimensions to obtain by self-assembly. Such particles have also been prepared in fullerene suspensions in such a way that the interior of the particles was filled with fullerenes [32].

IV. SUPERCRITICAL FLUID METHODS

A variety of supercritical and near-critical fluid processes have been introduced in the past 20 years for small particle precipitation. Most of these techniques combine aspects of aerosol particle formation with solvent shifting aspects, where the critical or near-critical fluid may function as the primary solvent for the organic principal compound or as an antisolvent.

Such fluids are of interest for several reasons. A primary motivation is that a variety of useful solvents can be made critical at rather moderate temperatures. A summary of critical temperatures and pressures for some useful fluids is given in Table 5. Carbon dioxide is the most favored supercritical fluid because it is inexpensive, has low critical conditions (pressure and temperature), is not flammable, and is environmentally "green" and nontoxic. Carbon dioxide is particularly favored for pharmaceutical formulations. Processes using supercritical fluids usually can be very easily "dried" because such fluids quickly vaporize upon venting at atmospheric pressure.

A. Rapid Expansions of Supercritical Fluids

The most broadly studied process termed rapid expansion of supercritical solutions (RESS) involves the dissolution of the organic principal compound in near-critical or supercritical solvent followed by controlled expansion of the solution at reduced pressure. This expansion is very analogous to aerosol processes for producing droplets. In RESS processes, however, there is no significant heat transfer problem to be overcome during the expansion and "evaporation" stage. As the supercritical solvent expands, its solubilizing power decreases, and the organic solute becomes supersaturated and condenses into either an amorphous solid state or a crystalline morphology. This kind of solvent shifting may be thought of as one-dimensional solvent shifting, as there is typically only one solvent component, the supercritical (or near-critical fluid).

TABLE 5 Critical Values for Supercritical Fluids

Solvent	P_c (atm)	T_c (°C)
Carbon dioxide	74	31
Nitrous oxide	72	36
Ethylene	51	10
Trifluoromethane	47	26
Chlorotrifluoromethane	39	29

A laboratory-scale RESS system is illustrated in Fig. 9 [33]. The process is direct and simple. Dissolution of the organic in the near-critical or supercritical fluid occurs in the extraction chamber. Expansion with concomitant precipitation occurs in the next stage, the U-tubes in this instance. Early applications of this process [33] to dodecanolactam (DL), β-estradiol, lecithin (Lec), and a blue azo-type pigment (Azo) demonstrated some of the potential of this technology. See Table 6 for structures. All of these organics were extracted (dissolved) at 55°C and 5000 psi in supercritical carbon dioxide, where the flow rate was maintained at 5–10 standard liters per minute, and were collected in the sequential U-tubes as illustrated in Fig. 6.

Dodecanolactam polymerizes to form Nylon 12 and was available as a powder of irregular particles 5–10 μm in largest dimension. Under the operating conditions just described it is soluble at about 15% (w/w). Precipitation in this RESS system produced primary particles that were needles in the range of 10–30 μm in length and less than 1 μm in diameter. β-Estradiol was initially available as a polydisperse powder with largest dimensions in the range of micrometers to hundreds of micrometers. Reprecipitation in this RESS process produced particles uniformly less than 1 μm in largest dimension. Lecithin (Lec) is an important stabilizer and excipient in pharmaceutical formulations, and it often comes with many impurities such as phosphatidylinositol and phosphatidylenthanolamine, so that it is often soft and gummy and difficult to disperse by comminution processing. The Lec examined in this process consisted of large particles hundreds of micrometers in diameter. Application of the RESS process produced a very fine powder with primary particle sizes ranging up to several micrometers. A blue pigment, Navy Blue (Azo), was obtained as particulates in the range of 50–150 μm in largest dimension. This material was transformed into a fine powder having particle

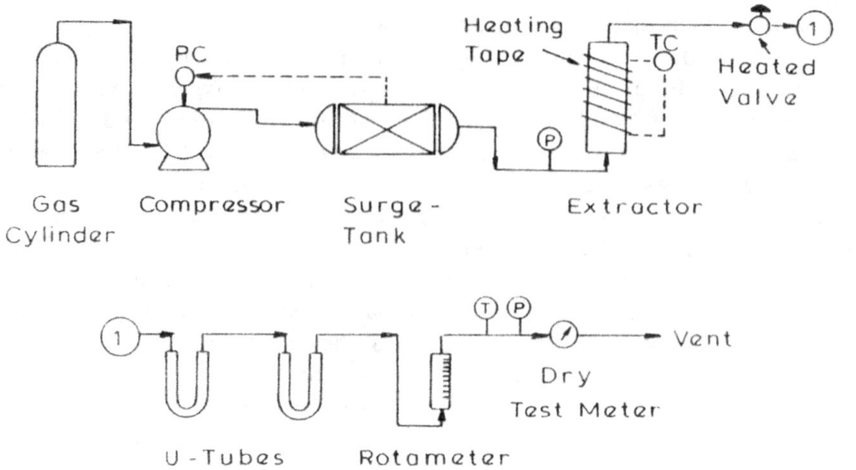

FIG. 9 RESS flow system for precipitation of organic particulates from supercritical solutions. PC, pressure controller; P, pressure guage; TC, temperature controller; T, thermocouple. (From Ref. 33.)

sizes in the range of a few micrometers. The RESS process was also applied to polypropylene (PP) (Structure 12), although instead of carbon dioxide, supercritical propylene (140°C, 3500 psi) was used. The as supplied material consisted of large clumps scores of micrometers in diameter. The supercritical reprecipitated powder comprised very small particles (a few micrometers and smaller) that formed linear strings or linear clusters up to 50 μm in length. The morphology and size yielded by the RESS process vary dramatically with the target organic species. Solubility in the supercritical fluid, and how this solubility changes (decreases) as fluid density decreases, is a key parameter in modifying resulting particle size (growth) and how primary particles might tend to congeal during the process. Application of additional chemical components as stabilizers is an area that has not yet been very broadly investigated.

B. Gas Antisolvent Process

Another important supercritical process is the gas antisolvent (GAS) process. This process requires that the solute be essentially insoluble in the supercritical fluid and soluble in an auxiliary solvent. The auxiliary solvent will generally have at least limited miscibility with the supercritical fluid. The organic solute is dissolved in the auxiliary solvent to produce a solution. The supercritical fluid is then introduced into this solution, causing it to expand and causing, by "two-dimensional" solvent shifting (shifting the balance from predominantly one solvent component to another), the organic solute to precipitate as small particles. It is important that the organic substrate be soluble in the auxiliary solvent and that the gas antisolvent (supercritical fluid) have finite solubility in the auxiliary solvent as well. This last requirement is necessary in order to be certain that the supercritical fluid will effectively expand the auxiliary solvent. In the absence of such mutual solubility, the injected supercritical fluid will tend not to "take the auxiliary solvent with it" while it expands, and two-phase separation at the outset will render the process relatively ineffective. It is also important that the organic substrate not be too soluble in the supercritical fluid, as the goal is to create a supersaturated solution during expansion from which the substrate will condense or precipitate. If the substrate has appreciable solubility in the gas antisolvent, then a much narrower window of pressure and temperature will be available for creating supersaturation.

Gas antisolvent injection rates and extents of expansion are two control variables that can effectively be used to control particle size and polydispersity. A range of expansion "paths" is illustrated in Fig. 10. Curve A represents a long-time and low-expansion extreme case. The degree of expansion corresponds to the threshold pressure (THP), where weak turbidity corresponding to the first visual detection of precipitating particles is visually evident. Curve B represents a path that yields a continuously varying particle size distribution. The step-shaped path C yields a polydisperse distribution with discrete particle size modes generated during each plateau period. Stepping to a higher plateau induces secondary (and tertiary) nucleation and the growth of smaller particles. Path D yields small monodisperse particles.

TABLE 6 Structures of Organics Reprecipitated by Supercritical Processing

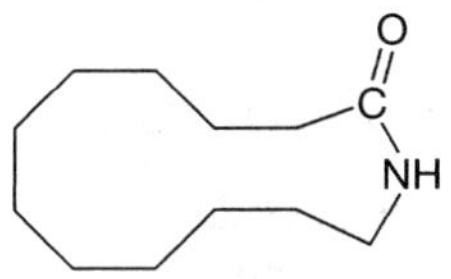

DL

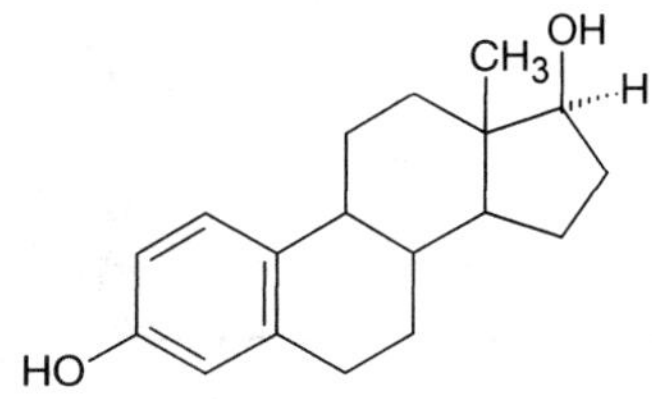

β-Estradiol

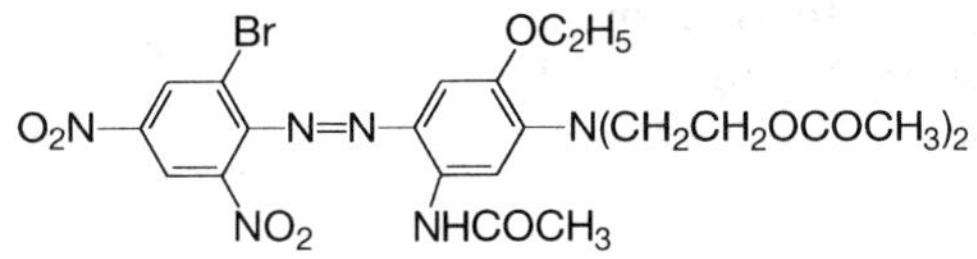

Azo

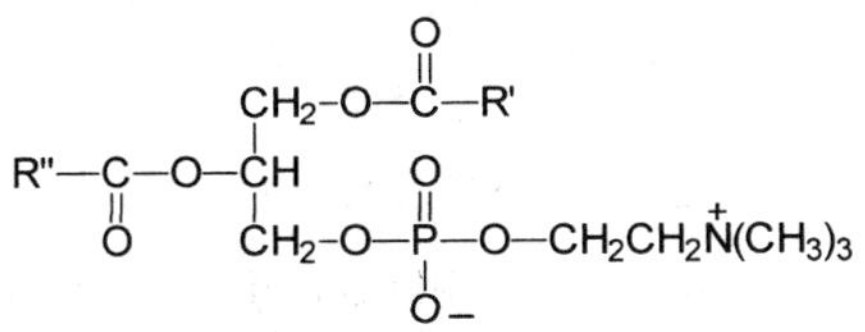

phosphatidyl choline

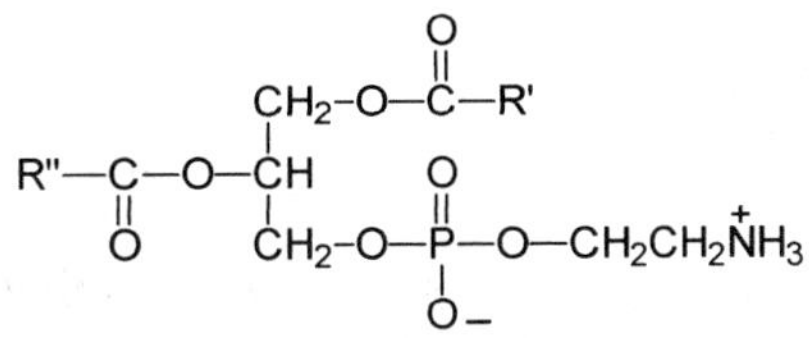

phosphatidyl ethanolamine

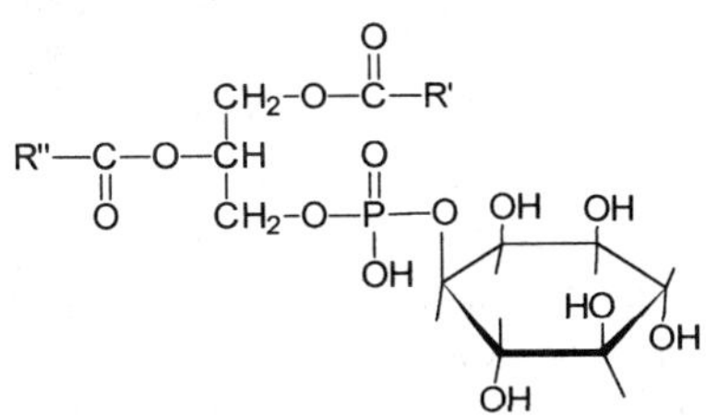

phosphatidyl inositol

Lec

This GAS process has been applied [34] to nitroguanidine (NG) (Structure 13) dissolved in dimethylformamide (DMF), an explosive that is not a prime candidate for comminution processing. The initial material consisted of crystalline needles in the range of 100 μm in length and about 5 μm in diameter. Using very rapid expansion paths, fine NG particles in the 3 μm range were obtained. Similar results were obtained using carbon dioxide and expanding rapidly (5 s) a 12% (w/w) solution of NG in DMF to 750 psi at 20°C or by rapidly expanding a 10% solution in critical chlorodifluoromethane to 75 psi at 20°C. When very slow (30 min) expansion was done at the threshold pressure (80 psi at 22°C) of a 5% NG solution in DMF using chlorodifluoromethane injection, quite large spheres in the range of 30–60 μm in diameter were obtained. When intermediate rates of expansion or paths were used, intermediate particle sizes were obtained.

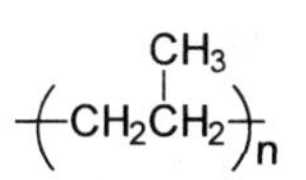

STRUCTURE 12 PP.

C. Solution Enhanced Dispersion by Supercritical Fluids

An alternative process is the so-called solution enhanced dispersion by supercritical fluids, SEDS, pro-

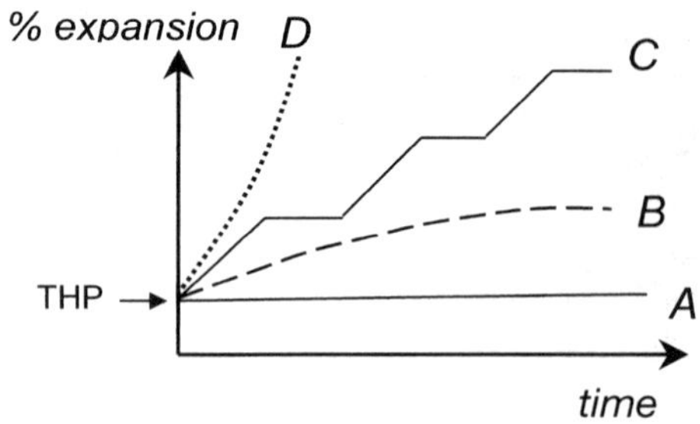

FIG. 10 Expansion paths for antisolvent (supercritical fluid) addition. THP denotes the threshold pressure for the onset of particle precipitation. Path A yields large, monodisperse particles. Path B yields polydisperse particles. Path C yields polydisperse particles, with discrete modes evident. Path D yields small monodisperse particles.

$H_2NC(NH)NHNO_2$

STRUCTURE 13 NG.

cess. It is a solvent shifting process somewhat similar to the GAS process. However, in the SEDS process, the supercritical fluid and a solution of the organic principal compound in an auxiliary solvent are introduced into the expansion chamber simultaneously through a coaxial nozzle assembly [35,36]. This coaxial flow produces a dispersion of the two fluid phases. The auxiliary solvent is extracted into the supercritical fluid. This extraction leads to supersaturation of the organic principal and subsequent precipitation as small particles. Application of this SEDS process to salmeterol xinafoate (SX) (Structure 14) using acetone as auxiliary solvent (0.5% w/v) and supercritical carbon dioxide (300 bar, 35°C) yielded crystalline particles of 7.5 μm median spherical diameter. Flow rates and working pressures could be used as control variables to modify particle sizes.

V. PRECIPITATION BY pH SHIFTING

Weak acids such as pharmaceuticals, organic pigments, and dyes are often precipitated by acidifying a concentrated solution of a soluble anionic form. The in situ precipitation of such species to produce controlled submicrometer particle sizes while maintaining viscosity at a sufficiently low level to facilitate convective mixing and transport is often a challenge. This is because condensing organics readily form amorphous states and aggregate in almost a diffusion-limited fashion, leading to network and gel structures. We stress that pH shifting is a specialized example of solvent shifting.

A. Organic Pigments

The nanoprecipitation of methine oxonol pigments of the general structure MO5 (Structure 15) by pH shifting has been described for mono-, tri-, and pentamethine (n = 1, 2, and 3, respectively) oxonols [37–39]. Such pigments have a variety of useful functions in various kinds of conventional photographic products, including microfilm, movie film, and reversal films. Such pigments are easily dissolved in dilute alkali because of the weakly acidic carboxylic acid groups and because of the fairly acidic oxonol proton. Note that two resonance forms for delocalization along the oxonol backbone can be written; this contributes to the acidity of these hydroxyl protons. Subsequent precipitation by

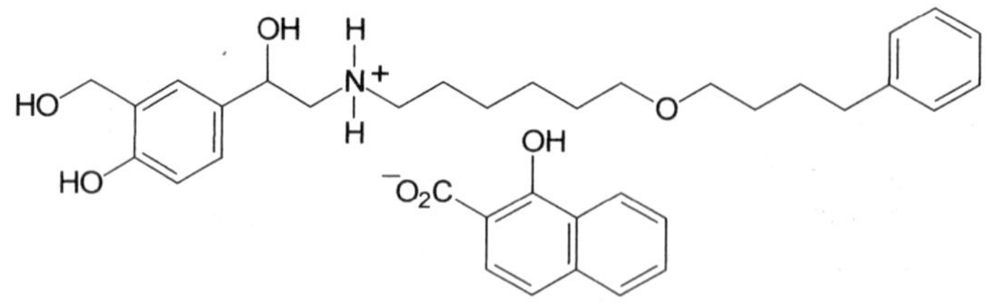

STRUCTURE 14 SX.

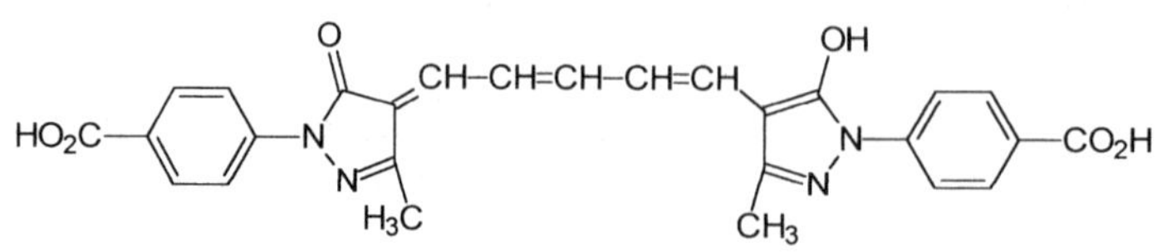

STRUCTURE 15 MO5.

acidification in the presence of various dispersing aids can produce very fine particle sizes and domain lengths at weight concentrations of up to 1%. Because of the carboxyl bifunctionality, such species tend locally to form hydrogen bonds that can yield lengthy nematic strings in one dimension but quite nanoscale cross sections normal to the nematic vector.

An ultrafine dispersion of MO5 was prepared using a double-jet acid precipitation procedure [37,38] in which the pH was kept constant at 5.2 during precipitation. The pigment feed solution at 0.073 M was prepared by dissolution at pH 9 with dilute 2 N NaOH. The reactor was initially charged with water, gelatin, and Alkanol-XC (sodium di- and triisopropyl naphthalene sulfonate in about equal molar proportions). The pigment solution was fed into the reactor at a constant rate, and 2 N aqueous H_2SO_4 was added under feedback control in order to maintain a set pH (5.2). The amorphous precipitated particulates exhibited a nematic-type structure as revealed by cryo-TEM. The proton uptake at completion corresponded to only approximately 64%, so the balance of the sites was neutralized with sodium ions. The peculiar physical state obtained had serendipitous optical effects, as illustrated in Fig. 11. There one can see that an optical absorption envelope is obtained that is dramatically different from that obtained by comminution of the fully protonated pigment. Thus, we see that such a precipitation regimen yields a different physical state than obtained by presscake precipitation. Similar long-wavelength absorption envelopes have been produced for trimethine and monomethine analogues of MO5.

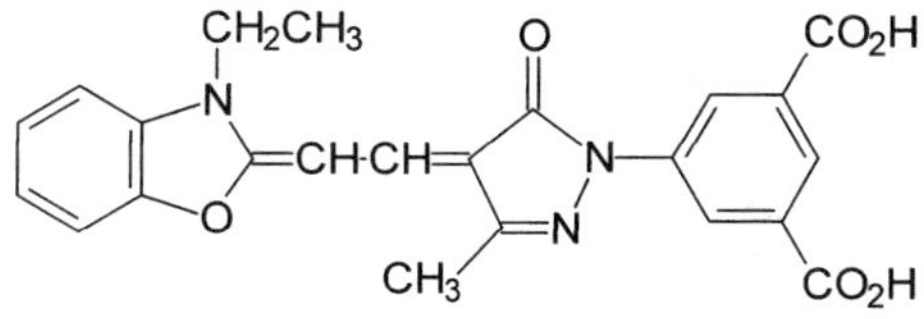

STRUCTURE 16 **VI**.

Some organic precipitations can yield crystalline habits. An example is given by pigment **VI** (Structure 16) precipitated similarly to the process just described for MO5 [39]. A slightly alkaline feedstock solution of **VI**, 1.9% by weight, was prepared by dissolving the pigment with 2 N NaOH at pH 7.5. Double-jet precipitation at pH 5 was done at 24°C using polyvinyl alcohol (molecular weight 15,000 to 30,000) in this reactor as a stabilizer. The precipitation yielded moderate-sized (up to several micrometers in largest dimension) microcrystallites. A comparison microcrystalline dispersion of **VI** was prepared by comminution and yielded comparably sized microcrystallites as determined by optical microscopic analysis. Thin-film coatings of these dispersions yielded the absorption spectra illustrated in Fig. 12. These spectra are essentially indistinguishable, as would be expected for comparably sized pigment particles in the same crystalline physical state. Thus, there are some organics that do crystallize relatively rapidly.

B. Pharmaceuticals

Smaller molecular weight weak acids are conventionally precipitated by acid-base techniques. Precipitation

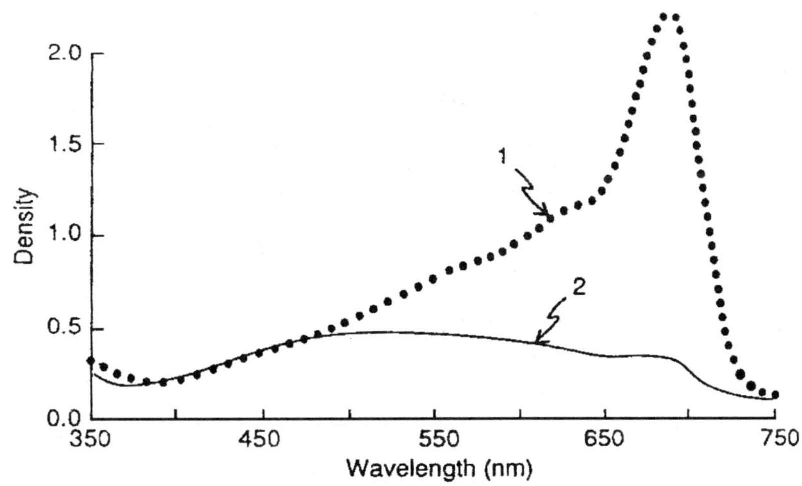

FIG. 11 Visible absorption spectrum (curve 1) of MO5 dispersion coated at 108 mg/m^2 in a thin film using gelatin (1.61 g/m^2) as a binder. Curve 2 is for a fully protonated counterpart of MO5 prepared by comminution processing and coated at 69 mg/m^2 similarly with gelatin.

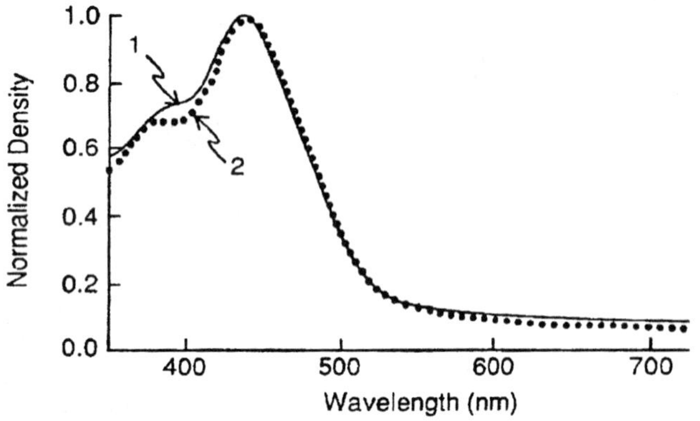

FIG. 12 Visible absorption spectrum (curve 1) of precipitated **VI** dispersion coated at 107 mg/m^2 in a thin film using gelatin (1.61 g/m^2) as a binder. Curve 2 is a comparison dispersion prepared by comminution processing and coated at 320 mg/m^2 similarly with gelatin.

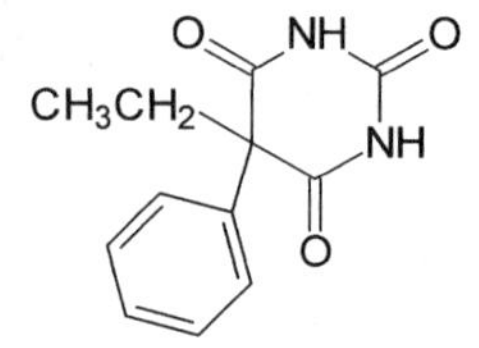

STRUCTURE 17 PB.

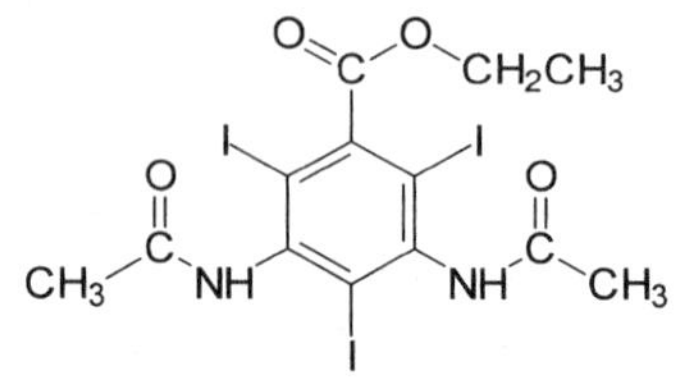

STRUCTURE 18 IA.

conditions affect particle size and physical state; often various polymorphs are accessible, depending on the precipitation process. A good example is the precipitation of phenobarbitone (PB, phenobarbital) (Structure 17). Thirteen polymorphs have been reported under various conditions. Generally increasing particle size is obtained [40] as the precipitating acid solution is made more dilute. This trend is consistent with elementary nucleation concepts in that the more slowly nucleation is induced, the fewer nuclei are created. This leads to larger crystals. Precipitation temperature was also observed to affect size, as well as polymorph form. Over the 10–50°C interval, form VIII of PB was produced and particle size increased with increasing temperature, from about 3 to 9 μm in largest dimension. As the temperature was increased further, form II of PB was produced and the size dropped below 3 μm. Form VIII is a monohydrate [40].

Direct precipitation of the triiodobenzoate ester IA (Structure 18), a radiographic contrast agent, was shown to be a reasonable direct alternative to comminution techniques for preparing aqueous dispersions [41]. An aqueous solution was prepared using 5 g of IA, 5 g of water, and 11.15 g of 20% aqueous NaOH. This mixture was heated to 56°C and then cooled to room temperature to yield a transparent solution. An aqueous stabilizer solution was prepared using 2.1 g of 60% Polystep B23 (Stephan) and 0.1 g of Aerosol OT in 125 g of water. These two solutions were then combined and neutralized using 75 mL of 15% propionic acid. The resulting dispersion was dialyzed to remove salts. The average particle size was about 250 nm in largest dimension. Upon autoclaving, the average size increased to about 350 nm.

A similar process was disclosed, but with the substrate molecule used to derive an aqueous stabilizer [42]. The benzoate derivative, DA (sodium diatrazoate) (Structure 19), was prepared to serve as a surface-active species with high affinity for IA surfaces. A dispersion was formulated using 20 g IA, 20 g of water, and 45 g of 20% aqueous NaOH. This mixture was dissolved by heating to 75°C and cooling to room temperature. The aqueous stabilizer solution was prepared using 0.4 g DA, 6.5 g Tetronic T-908, 500 g water, and 1 g 20% aqueous NaOH. These two solutions were combined and then neutralized with 300 g of 15% propionic acid to form IA particles about 230 nm in size.

A further important variation of such precipitation processes is to use structurally similar materials to restrain particle growth. Such materials are known as crystal growth modifiers. When effective, they serve to restrain growth, thereby producing smaller particles (and a higher number density of particles per unit volume). One approach to designing such growth modifiers is to model them upon the substrate molecular structure but to alter the structure sufficiently that appreciable solid-state mutual solubility is avoided, while

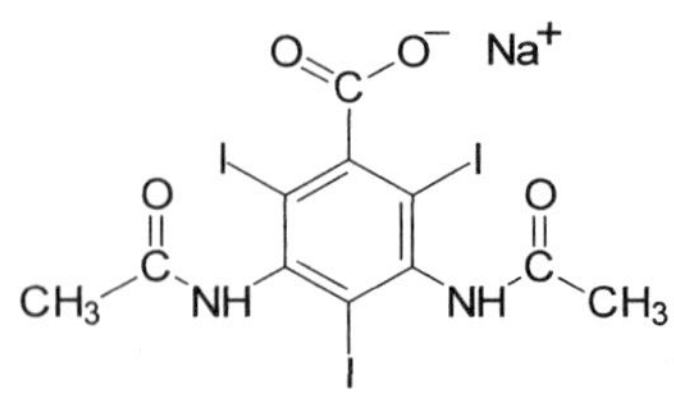

STRUCTURE 19 DA.

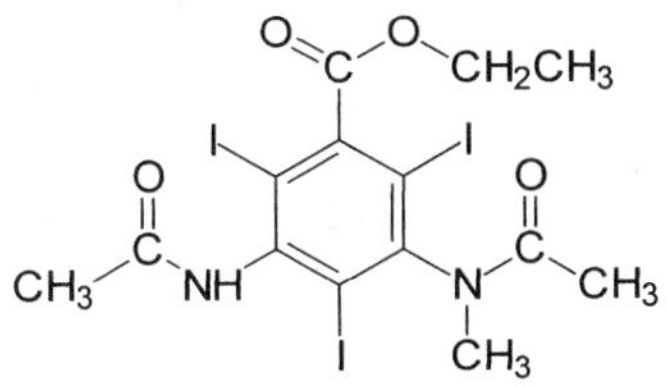

STRUCTURE 20 CGM.

intense surface activity, through structural correspondence, is achieved. An example applied to the precipitation of IA [43] is the modifier CGM (Structure 20). A mixture of 4 g IA, 1 g CGM, and 5 g water was dissolved using 11.5 g 20% aqueous NaOH and by heating to 55°C. After cooling, this solution was combined with aqueous stabilizer comprising 125 g water, 1.165 g Tetronic T-908, and 0.1 g Aerosol OT. This alkaline solution was then neutralized with 75 mL of 15% propionic acid to produce a dispersion, and this dispersion was then dialyzed to remove salt. The resulting particle size was about 194 nm. The size was dramatically smaller than the 274 nm obtained when another gram of IA was substituted for the CGM in this formulation.

VI. MICELLES AND RELATED SYSTEMS

Micelles are thermodynamically stable nanoparticles that provide important templates for various kinds of materials, including thermodynamically metastable particles as discussed in Section B. Excellent discussions of the thermodynamic driving forces for forming micelles [44] and for forming mixed micelles [45] are available, so these topics will not be elaborated here.

A. Effects of Solvent and pH Shifting on Solubility

Solvent shifting is a process by which the solubility of a component is altered by changing the composition of the solvent. High solubility in one solvent can be mitigated by addition of another solvent in which the soluble component is much less or only sparingly soluble. Similarly, high solubility can often be obtained by ionizing an organic species, such as a weak carboxylic acid, in dilute alkali solution. Subsequent neutralization by pH shifting, or dropping the pH through the p*K* of the acid, generally yields a significant drop in solubility and is often accompanied by precipitation of the weak acid component.

B. pH Shifting Method of Priest

1. Dissolution and Micellization

The use of pH shifting to utilize mixed micelle formation of alkali-soluble hydrophobic organic compounds and to stabilize such mixed micelles kinetically appears to have been disclosed first by Priest [46]. Such organic compounds have some weak acid functionality, such as —COOH, —OH, —SO_2NH—R, and the like, that can be ionized at sufficiently high pH. Technologically useful candidates for such processes are color-forming couplers used to form indoaniline image dyes in photographic films and papers. Typical examples include structures **VII** to **XI** in Table 7. Ionization generally increases solubility in aqueous or an aqueous water-miscible solvent system to about 0.5% (w/w) or more. When ionized, these compounds are surface active and form micelles and mixed micelles with other surfactant or amphiphilic compounds. Upon lowering the pH by addition of mineral or organic acid, the solubility of these compounds decreases dramatically, and if the micelles in which these compounds reside are not kinetically stabilized, macroscopic crystals of the protonated species can readily form or other untoward effects may occur.

2. Stabilization

Once the pH has been lowered and the amphiphilic ionized molecules have been reprotonated, one can no longer rely upon chemical free energy forces for stabilization because the most stable states have shifted to multiphase crystalline states. The kinetic stabilization of such reprotonated species can be achieved by more or less standard methods, such as charge, steric, and matrix stabilization techniques.

(*a*) *Charge Stabilization.* The use of ionic surfactants, such as anionic and cationic surfactants having charged groups that remain charged at pH >1, is advantageous in producing mixed micelles with the weakly acidic organic compounds. After acidification and reprotonation of the weakly acidic organic compounds, the ionic surfactants remain charged and provide charge stabilization as long as the ionic strength of the suspension is suitably low and the charged groups are not excessively shielded by counterions.

(*b*) *Steric Stabilization.* The use of nonionic surfactants and polymeric stabilizers (nonionic and polyelectrolytes) can provide effective stabilization of the reprotonated particles. Such use of polymeric stabilizers in conjunction with surfactants is illustrated in Fig. 13, where interfacial tension at the air-water or hydrocarbon-water interface is illustrated as a function of sur-

TABLE 7 Amphiphilic Color-Forming Photographic Couplers

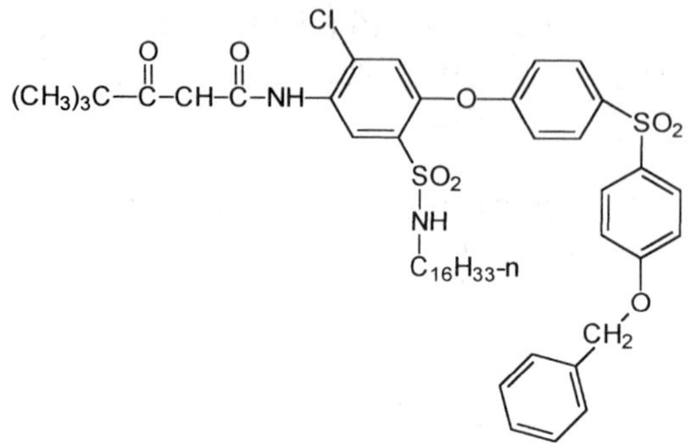

VII

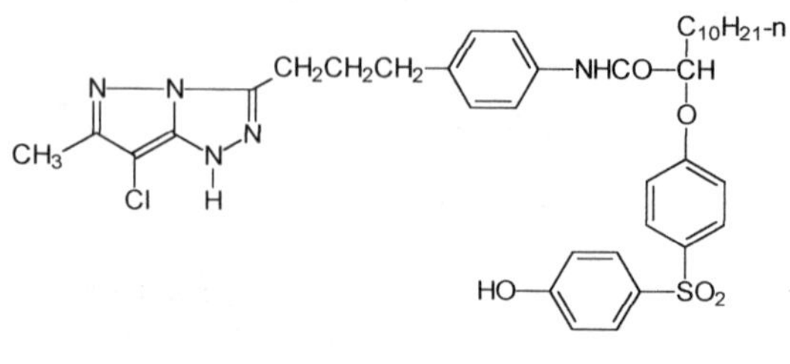

VIII

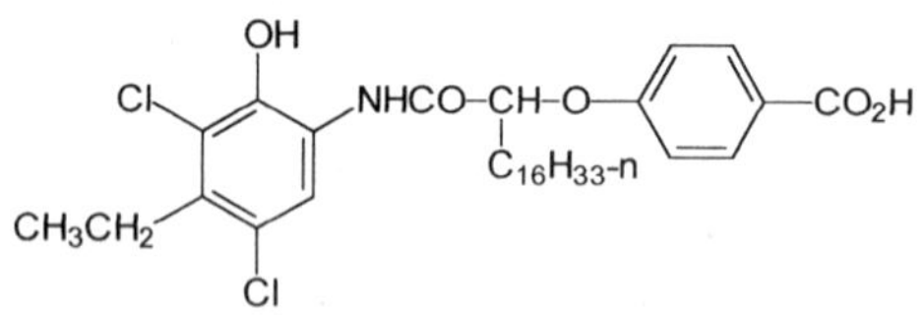

IX

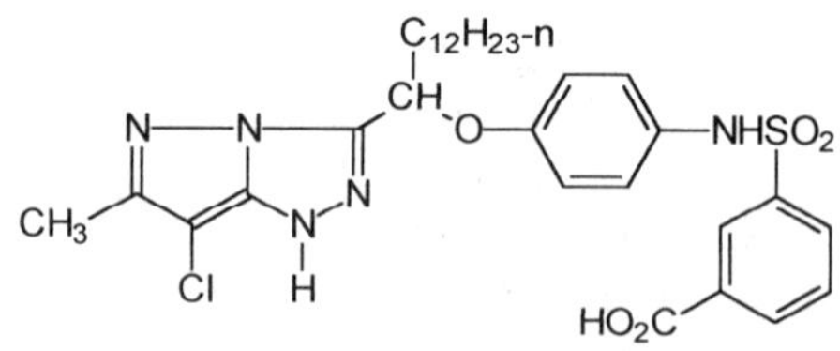

X

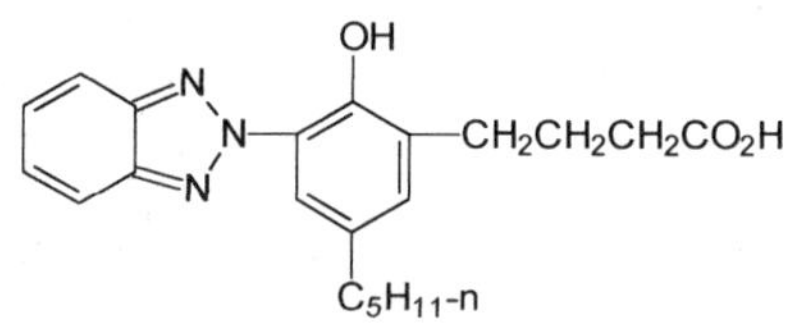

XI

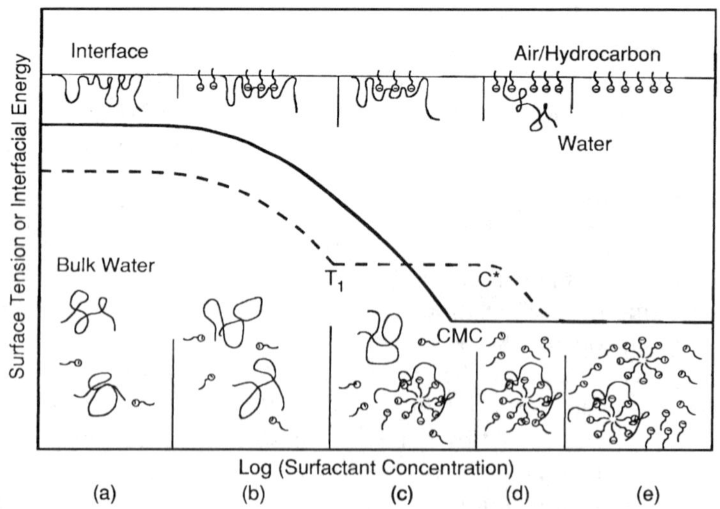

FIG. 13 Schematic for surface tension lowering as a function of surfactant concentration with and without added polymeric stabilizer that interacts with the surfactant. The full line corresponds to the case of no polymer, and the dashed curve indicates aggregation processes occurring in the presence of polymer. (Reproduced by permission from Ref. 47; copyright American Chemical Society.)

factant concentration [47]. The self-assembly of surfactant and polymeric stabilizer in the bulk and at the interface is illustrated by cartoons in five different regimes of surfactant concentration (a–e). In the low surfactant concentration limit (region a) the polymeric stabilizer adsorbs at the interface and little complexation between surfactant and polymer occurs in the bulk. At higher concentration (region b) the situation in bulk solution has not changed much, but surfactant adsorbs to the interface. As the concentration rises further (region c), the polymer nucleates surfactant aggregation in the bulk. The surfactant concentration at point T_1 is the so-called critical aggregation concentration. At higher concentration (region d), polymer desorbs (at concentration C*) from the interface in order to bind to aggregates in the bulk. In the highest concentration region (e) all of the polymer is involved in aggregates and then conventional surfactant micelles form. In the absence of polymer the surfactant critical micelle concentration (cmc) is found as illustrated by the solid line in Fig. 13.

(*c*) *Matrix Stabilization.* Organic particulates are stabilized against collision-based aggregation and growth processes when the viscosity of the continuous phase is increased to the point where particle diffusion is shut down or dramatically retarded. Such "matrix" stabilization can be achieved by using thickeners or associative thickeners to increase viscosity. Another very effective approach is to use a polymeric solution in the continuous phase that forms a gel network. Upon gelation, the mobility of particles therein is effectively quenched.

(*d*) *Chemical Stabilization.* Most complex organic molecules that are greater than a few hundred daltons in molecular mass and that are amphiphilic and have hydrolyzable linkages are susceptible to a variety of hydrolysis reactions and are pH sensitive. The rigors of dissolution through ionization can also lead to untoward chemical decomposition, as increasing pH leads to increasing rates of nucleophilic attack by hydroxyl ion and concomitant hydrolysis chemistry. The susceptibility of alkali-soluble hydrophobic organic compounds to such hydrolysis chemistry varies widely with the particular structure and species. The pH range and electrolyte composition during the dissolution and micellization processes must therefore be controlled in order not to induce significant decomposition during the initial stages of particle preparation. Often water-miscible solvents must be chosen to aid dissolution while permitting lower alkali levels to be used [48,49]. Phase diagrams can be particularly useful in such cases in identifying stability domains for the organic amphiphiles of interest.

3. Examples

Coupler **VII** (200 mg) and 100 mg of *N*,*N*,*N'*,*N'*-tetraethylphthalamide dissolved in diethoxymethane were mixed with 1.5 mL of 0.4 N KOH. The dimethoxymethane was removed in a nitrogen stream to yield a clear micellar solution. About 3 mL of 10% (w/w) type V gelatin was added and the pH was adjusted to 10. Butyrolactone (0.12 mL) was added to protonate the coupler and to lower the pH. The dispersion that formed also gelled and imparted matrix stabilization. The dispersion was stable for more than a month at room temperature and could be melted easily by moderate heating.

A flowchart of this assisted micellization approach is detailed in Fig. 14. Surfactant and substrate (coupler) solutions are combined under appropriate pH and sol-

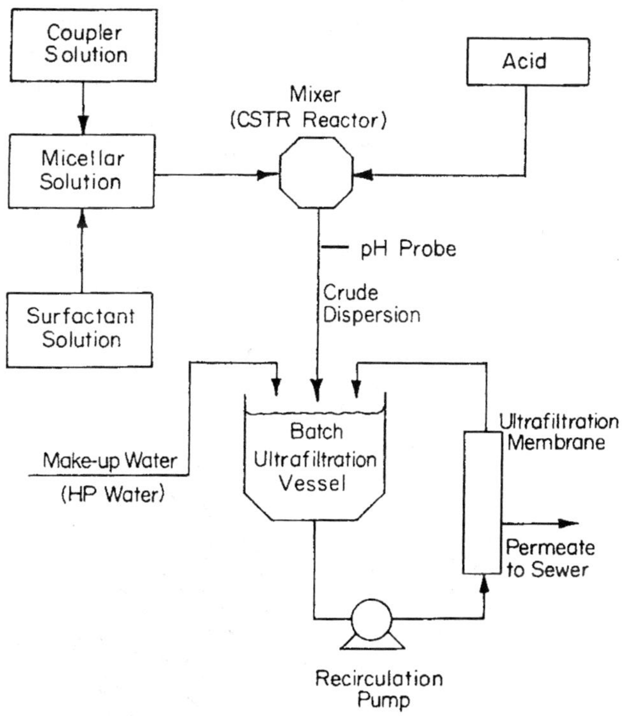

FIG. 14 Schematic block diagram of nanoparticle precipitation using pH shifting in mixed micellar systems. Mixed micelles of surfactant and amphiphilic coupler (or other amphiphilic organic compound containing weak acid groups) are formed using chemical free energy. The mixed micelles are reacted with acid in a controlled manner in order to reprotonate the amphiphilic coupler and to lock in the nanoscale structure. Excess salt is then removed by ultrafiltration. (From Ref. 48.)

vent conditions to form a mixed micellar solution. This solution is fed into a reactor, where the substrate species are reprotonated under controlled conditions, thereby creating metastable nanoparticles of substrate and surfactant. The dispersion so produced is then subjected to washing by ultrafiltration or other dialysis means in order to remove unwanted salts and/or solvents.

A major limitation that applies to such alkali-assisted dissolution techniques, especially for multifunctional organic molecules, is untoward decomposition and hydrolysis. For example, the coupler **X** (Table 7) is not stable in aqueous alkali when dissolved. However, mixtures with *n*-propanol are stable for appreciable times. A ternary stability diagram is illustrated in Fig. 15. This diagram illustrates a composition regime in which **X** is stable for longer than it takes to undergo the dissolution-reprotonation process illustrated in Fig. 14. The dispersion process involved preparing a Dupanol ME surfactant solution comprising 900 g surfactant, 4.1 kg PVP, and 329 kg water, and a solution of **X** (8.2 kg) also comprising 18.1 kg *n*-propanol and 13.7 kg of 1 M aqueous NaOH, both at 25°C. In less than 20 min these two solutions were mixed as shown in Fig. 14, metered into a continuous stirred tank reactor (CSTR) at 18.4 kg/min, and mixed with a 15% (w/w)

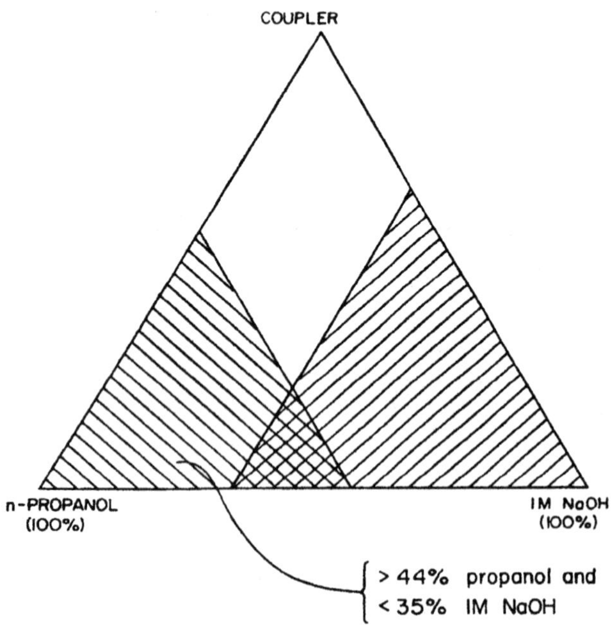

FIG. 15 Stability diagram for coupler **X** in mixtures of *n*-propanol and aqueous 1 M NaOH. The region of stability is given on the left by the noncrosshatched region. (From Ref. 48.)

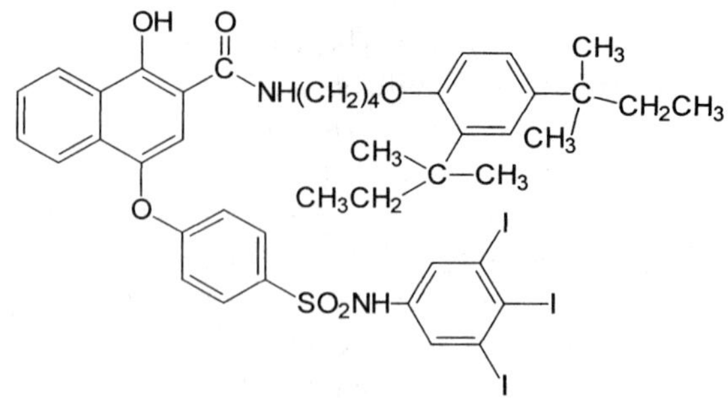

STRUCTURE 21 PL-IA.

acetic acid solution metered at 300 g/min. The resulting dispersion was washed for three turnovers at constant volume, pH adjusted to 5.2, and concentrated by ultrafiltration to 5.5% **X**. This dispersion was optically transparent and was stable for more than 3 weeks [49].

A newer approach to drug delivery is the combination of linking chemistry with amphiphilic water-insoluble carriers developed for photographic applications. The various technologies useful for dispersing photographically useful compounds can then be brought to bear on preparing nanosize dispersions of pharmaceutically useful compounds that are linked to photographically useful carrier species. For example [50], consider PL-IA (Structure 21), which contains a triiodo group, thus qualifying as an image contrast agent (for radiography). A substrate solution of 5 g PL-IA, 5 g *n*-propanol, and 20% aqueous NaOH was prepared, heated to 55°C, and cooled to room temperature. This solution was combined with 125 g water and 45 g of 33% aerosol A102 and then neutralized with 15 g of 15% propionic acid. The dispersion was then dialyzed to remove salts. The average particle size obtained was only 12 nm. The bioavailability of such formulations is extremely high.

C. Stabilization by Polymerization

Kline [51] has illustrated the stabilization of micelles by polymerizing polymerizable counterions in the headgroup region. The system studied comprised rodlike micelles of cetyltrimethylammonium 4-vinylbenzoate (CTVB) (Structure 22). Highly viscous solutions of CTVB are formed at millimolar concentration and comprise highly entangled rodlike micelles about 40 Å in diameter. After free radical polymerization of the 4-vinylbenzoate counterion, cylindrical rods about 400 Å in length were formed from micellar solutions 0.1% and 1.0% in CTVB. The initial viscosities of these solutions were quite different, but essentially identical reaction products were obtained after polymerization.

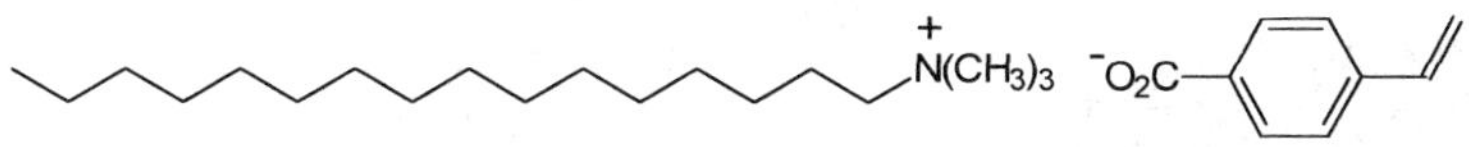

STRUCTURE 22 CTVB.

While the entangled micelles in the highly viscous CTVB solutions may be branched, the rodlike micelles stabilized by the polymerization appear shorter (400 Å) and behave in some respects as uncharged rods. These micelles with their linear polymer coating appear stable for times in excess of a year. These suspensions may be dried and then redispersed in water; the rods retain their dimensionality. These suspensions can also be put through freeze-thaw cycles without including surfactant crystallization. On the whole, this approach appears an effective exoskeletal way to stabilize soft micellar structures.

D. Vesicles

Vesicles and liposomes are important particulate objects for many reasons, particularly with respect to drug delivery, and are extensively reviewed [52,53]. The production of vesicles is usually by some shear-assisted means, and such approaches fall outside the scope of our present chapter. However, an important exception is a class of thermodynamically stable micelles discovered by the Kaler-Zasadzinski collaboration [54,55]. Isotropic micellar solution domains in ternary water, cationic surfactant, anionic surfactant systems were discovered that contained thermodynamically stable vesicles. The first report [54] used cetyltrimethylammonium tosylate (CTAT) and sodium dodecylbenzene sulfonate (SDBS). One domain in which vesicles formed spontaneously was characterized as being primarily cationic surfactant, with less than a stoichiometric ratio of anionic, and another domain comprised mainly anionic surfactant, with a smaller amount of cationic. At stoichiometric equivalence, the surface-active ions precipitate to form an insoluble double tail salt, so excess population by one of the species is necessary to obtain stability and curvature. It is believed that the ratios of different surfactants vary among the inner and outer halves of the bilayer membrane. Recent work has shown that unilamellar and double lamellar vesicles may be spontaneously formed (E. W. Kaler and J. A. Zasadzinski, private communication, 1999).

E. Nanodiscs

As a logical extension of the formation of thermodynamically stable vesicles, the controlled use of mixtures of cationic and anionic surfactants (catanionic solutions) has produced thermodynamically stable nanodiscs of essentially bilayer thickness and diameters ranging from tens of nanometers to several micrometers [56]. A key limitation in the stabilization of thermodynamically stable structures deriving from catanionic solutions is the electrostatic screening length dictated by the associated ionic strength. A 5% (w/w) catanionic solution may have an electrolyte concentration of about 0.1 M from the counterion salts formed. Such electrostatic screening promotes the formation of closed vesicles over discs.

In this preparation of nanodiscs the catanionic solutions were prepared from myristic acid (MyrA) (Structure 23) and cetyltrimethylammonium hydroxide (CTAOH) (Structure 24). Solutions were purged with nitrogen to eliminate carbonate. The protons and hydroxyl ions combine to form water, so the net ionic contributions to ionic strength emanate from the hydroxyl ions associated with the excess cationic surfactant level. Such steps helped maintain the ionic strength close to 10 μM and the conductivity at about 10 μS cm^{-1}. A screening length of about 100 nm was obtained, rather than a legnth of about 10 nm if alkali and halide counterions had been used. The cationic excess results in excess hydroxyl species, and pH varied between 7 and 13, depending on the degree of excess.

The physical properties of the nanodiscs obtained vary with the compositional variables in this pseudoternary system (water, myristic acid, and CTAOH). The total amount of surfactant is denoted c (wt%) and the mole fraction of myristic acid to total surfactant is denoted r. A single-phase region (U+) in the water-rich corner yielded stable nanodiscs. In the presence of atmospheric CO_2, this region U+ is bounded from above by $c = 0.1\%$. In the presence of atmospheric CO_2, nanodiscs are found in a two-phase region with a lamellar phase. Herein disc diameters vary from 30 nm to sev-

STRUCTURE 23 MyrA.

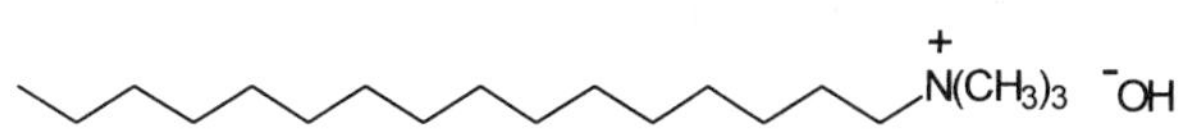

STRUCTURE 24 CTAOH.

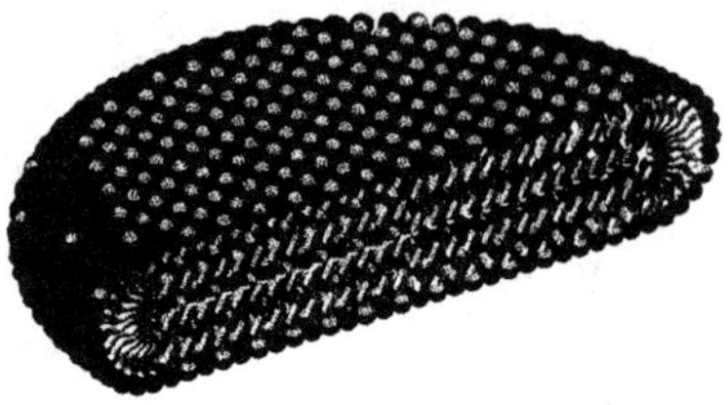

FIG. 16 Perspective cross-section view of nanodisc particles synthesized from mixtures of anionic and cationic surfactants. The particle surfaces comprise ion pairs (de facto double-tail zwitterionic surfactants) and the rounded edges comprise an excess of cationic surfactant. The central layer of the nanodisc comprises hydrocarbon chains that appear to adopt a near-crystalline packing arrangement. (Courtesy of Professor Thomas Zemb.)

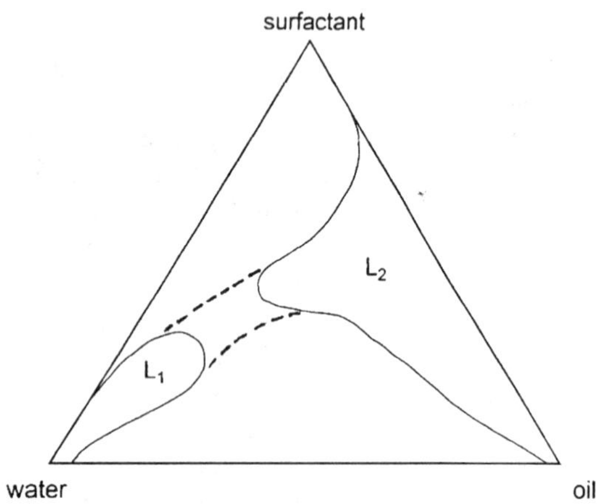

FIG. 17 Ternary phase diagram of surfactant, oil, water microemulsion. Isotropic single-phase oil-in-water (L_1) and water-in-oil (L_2) microemulsion domains are illustrated. In some systems, or with variation in some field variable (temperature, fourth chemical component concentration), these L_1 and L_2 domains may become simply connected, as illustrated by the dotted lines. In such circumstances an irregular bicontinuous microemulsion comprising interdigitated oil and water domains connects these L_1 and L_2 domains.

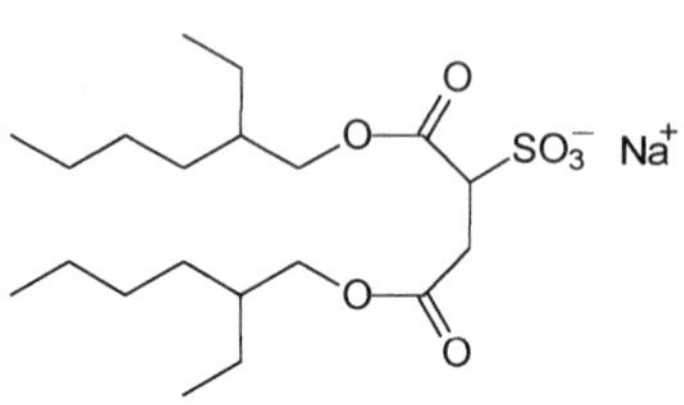

STRUCTURE 25 AOT.

eral micrometers. A schematic cross-sectional view of such discs is shown in Fig. 16.

VII. MICROEMULSIONS

A. Summary of Phase Diagrams

Microemulsions are thermodynamically stable micellar emulsions of two immiscible liquids stabilized by a third chemical component, a surfactant. A typical ternary phase diagram is illustrated in Fig. 17. There are usually two isotropic microemulsion domains in such systems [57]: an L_1 domain, corresponding to a nominally oil-in-water droplet structure, and an L_2 domain, corresponding to a water-in-oil droplet structure. Reverse microemulsions comprising water cores surrounded by a surfactant layer have become increasingly used as nanoreactors for exploring various kinds of chemistries, including inorganic precipitations [59], inorganic polymerizations [60], organic polymerizations [61], and enzymatic catalysis [62]. Until recently, there had been no reports of organic precipitations of nanoparticles in such microemulsion media.

B. Solvent Shifting

The first example of precipitating organics in a reverse microemulsion has been described [63], and these processes are described more fully in Chapter 30 and in some recent manuscripts (F. Debuigne, L. Jeunieau, M. Wiame, and J. B.Nagy, submitted). The reverse microemulsion system was a water-in-heptane microemulsion prepared with Aerosol OT (AOT) (Structure 25). The organic substrates, rhodiarome (RhoD) (Structure 26) and rhovanil (Rhov) (Structure 27), were dis-

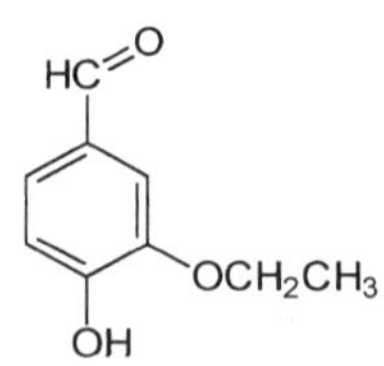

STRUCTURE 26 RhoD.

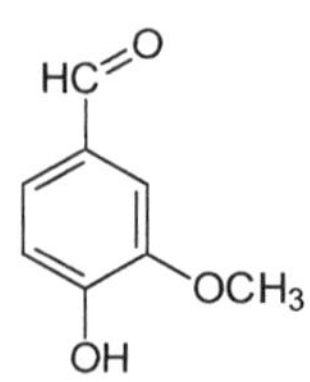

STRUCTURE 27 RhoV.

solved in water-miscible solvents such as acetone, ether, and ethanol and then added dropwise to the reverse microemulsion system. The weight of added active solution was generally equal to the weight of water solubilized in the microemulsion. Substrate concentrations were varied over the 50 to 400 g/L range, but this concentration did not affect particle size much. Average particle diameters in the range of 5–7 nm were obtained. These examples illustrate compartmentalization by the water pools in the microemulsions. The auxiliary solvent solution of the organic principal is directed to the microemulsion droplets where precipitation occurs. The same process has been used to produce nanocrystallites of cholesterol (F. Debuigne et al., submitted).

VIII. GAS CONDENSATION

The formation of inorganic ultrafine particles by gas evaporation, gas condensation, and related vacuum–gas phase processing methods has been very well developed for producing nanoparticulate metals, oxides, and ceramic precursors [64–66]. Such methods have been successfully extended to certain organic particulates. Such techniques use one method or another to vaporize the organic compounds, often into a low pressure or partially evacuated chamber. In the gas phase, collisions among molecules lead to condensation of clusters and particles, which may aggregate further with one another before adhering to a solid collection device or filter. Such techniques are limited to organic materials that will vaporize with heating without decomposing. A purported advantage of this gas condensation method is that the resulting particles have a more narrow particle size distribution than obtained, for example, by standard comminution methods. Another is that such fine particulates can be readily dispersed in water, even though the underlying organic material may be hydrophobic.

This gas condensation method has been successfully applied [67] to relatively low molecular weight compounds such as anthrahacene, carbazole, β-carotene, chloramphenicol, cortisone acetate, phthalocyanine, and pyrene. It has also been applied to some polymers, including polyethylene, polyvinyl alcohol, polyvinyl chloride, and polystyrene. Particle size can be controlled by temperature and pressure, particularly the partial pressure of added inert noble gases. Figure 18 shows a fluorescence micrograph of pyrene particles produced by vapor phase condensation. The nominally 100-nm particles in Fig. 18a were produced at 150°C in a 0.1 torr helium background. The pyrene was heated in a crucible about 20 cm away from the target substrate (glass). The 3-μm-diameter particles of Fig. 18b were obtained similarly, except a helium gas pressure of 5 torr was used. Identical conditions except for a lower partial pressure of helium (0.1 torr) yield pyrene particles about 100 nm in diameter. An important difference for most organics precipitated in this way is that they do not tend to form fractal aggregates as readily as metals and ceramic precursors produced by such methods.

IX. ENCAPSULATION METHODS

A particularly effective way to stabilize organic particles is to precipitate them inside a matrix (matrix stabilization) or inside a hollow particle or cavity. The particle or cavity wall (shell) effectively prevents organic nuclei from coming in contact with other cores by virtue of a (steric) "cell wall." An alternative approach would be to precipitate an organic particle by means such as detailed earlier and then encapsulate by one or more mechanisms of microencapsulation, such as coacervate formation, or by some condensation type of polymerization [68]. The use of preformed hollow capsules or hollow particles provides very effective limitations on the size of the resulting precipitated particles because of the compartmentalization of the precipitation chemistry.

An alternative encapsulation approach to having the organic precipitated particle in the *core* region of an encapsulated composite particle is to layer the precipitated organic material on the *exterior* of a templating core material. Yet another approach is to produce composite particles by precipitating organic materials in a matrix that forms disperse particles.

A. Precipitation in Reverse Emulsions

Precipitation in so-called liquid membranes or multiple reverse emulsions offers a convenient approach to selective extractions for analytical and preparative purposes. For example [69], the selective removal of cupric ions from commercial acid leach solutions containing copper sulfate and contaminating ions such as ferric, ferrous, and nickel ions can be accomplished by combining ion-specific ligands with an organic precipitating species, such as oxalate. The acid oxalate precipitating aqueous solution is dispersed as a water-in-oil emulsion, such as kerosene, using an oil-soluble surfactant, such as Span 20 (sorbitan laurate). A ligand specific for cupric ion, such as 2-hydroxy-5-nonylacetophenone, is included in the oil phase. This water-in-oil emulsion is then crudely emulsified with the

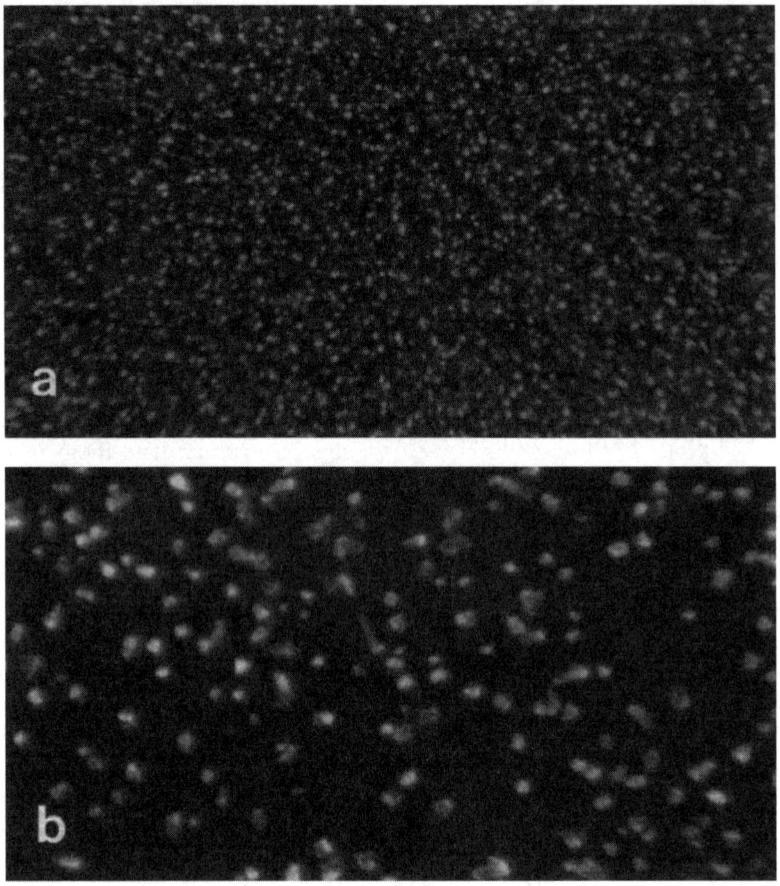

FIG. 18 Fluorescence micrographs of pyrene produced by a gas condensation process. (a) 100-nm-diameter particles produced at 0.1 torr helium background; (b) nominally 3-μm-diameter particles produced at 5 torr helium background pressure. (Adapted from Ref. 68.)

aqueous metal ion solution to produce a water-in-oil-in-water double emulsion. The ligand selectively transports copper across the kerosene membrane, wherein copper precipitates in the interior aqueous droplets as the oxalate. The size of such precipitates is compartmentalized by the inner water droplets and is significantly less than obtained in batch precipitations of copper and oxalate.

B. Precipitation in Encapsulating Spheres

Recently, the layer-by-layer polyelectrolyte-nanoparticle self-assembly technique has been applied to polymer particle templates as a means for producing hollow silica spheres. Such spheres may be produced over a wide range of micrometer and submicrometer inner diameters, and the thickness and permeability of the walls may be varied by proven formulation variations. Such spheres have been shown to be venues for organic particle precipitation [70]. The spheres are suspended in a solvent for the organic solid to be precipitated, with subsequent addition of a solution of the principal compound, or alternatively they are suspended in a solution of the principal compound. The compound then permeates the sphere wall, so that the sphere is filled with a solution of the principal agent. A nonsolvent, miscible with the original solvent, is then added slowly to the suspension. The nonsolvent also permeates the sphere walls and then initiates precipitation of the principal compound inside the hollow silica spheres by a solvent shifting process. Alternatively, pH shifting may be used. Nucleation sites or other as yet not well understood factors typically induce precipitation inside the spheres prior to precipitation in the continuous phase. Thus, precipitation internally sets up radical diffusion gradients that feed more active compound into the spheres. This modified solvent shifting method produces encapsulated particulates for possible applications in drug delivery and in organic pigments.

A variety of colloidal particles may be used for templating. For example, human erythrocytes have been templated using nine alternating layers of 4-poly (styrene sulfonate, sodium salt) and poly(allylamine hydrochloride) [70]. After wall formation, the templating erythrocyte contents were removed by oxidation with sodium hypochlorite. The weak acid 6-carboxyfluorescein (6CF; see Table 8 for structure) was solubilized at 10^{-3} M at pH 10.5 and imbibed into empty capsules. The 6CF was then precipitated by lowering the pH to 6 with addition of dilute aqueous HCl. Scanning electron microscopy images showed the nominally 5 μm diameter particles to make up a distinguishable core region about half the diameter of the particles. The lack of precipitate in the continuous phase points to the probable preponderance of heterogeneous nucleation sites inside the hollow particles and possibly in the alternating shell layers. Weak bases such as rhodamine 6G (6G) can also be precipitated in such hollow shells by pH shifting. Increasing pH leads to precipitation of 6G from a saturated solution in the presence of erythrocyte-templated hollow capsules. A rapid pH increase to 12 results in rodlike 6G needles in the continuous phase; a slower pH increase to 10 leads to solely intracapsule 6G rod precipitation with no "extracellular" material; milder and slower pH change to 8 leads to amorphous 6G precipitation in the cores. Salting out and solvent shifting have also been demonstrated in such hollow capsules for dyes such as pseudoisocyanine hexafluorophosphate (PIC) and bis(dimethylamino)heptamethine (DMHM) [70].

C. Nanocolorant Precipitation

An interesting process for dispersing organic dyes has been disclosed by BASF [71]. This process combines aspects of emulsification, precipitation, and miniemulsion polymerization in achieving small particle dispersions of organic colorants. The primary step involves being able to solubilize colorant (dye) at high load (10–20% w/w) in a polymerizable vinyl monomer, such as acrylates, styrenes, and methacrylates. In addition to the polymerizable monomer, a relatively high fraction of cross-linking monomer (bifunctional monomer) is added. A hydrophobic additive, such as hexadecane, is then added so as to osmotically stabilize the resulting miniemulsion from untoward Ostwald ripening and growth. This colorant-monomer solution is then homogenized under pressure to produce a submicrometer-sized miniemulsion, with particle sizes less than 200 nm. Suspension polymerization is then initiated using a thermal initiator or by using UV irradiation. When cross-linking is sufficiently fast and dense, the colorant is kinetically trapped within the cross-linked matrix. Full polymerization typically produces a polymeric matrix in which the colorant has low solubility, and so it condenses locally to the extent possible before further transport is impeded by the cross-linking of the polymer network. Thus nanodomains of organic colorant are encapsulated as metastable embryos within the highly cross-linked polymeric matrix. Phase separation on the mesoscale of the particle size is prevented. The phase separation that occurs on the nanoscale corresponds to domains that are nearly molecular in their electronic absorption properties, and extremely high coloring power is obtained.

TABLE 8 Organic Dyes Precipitated in Hollow Capsules

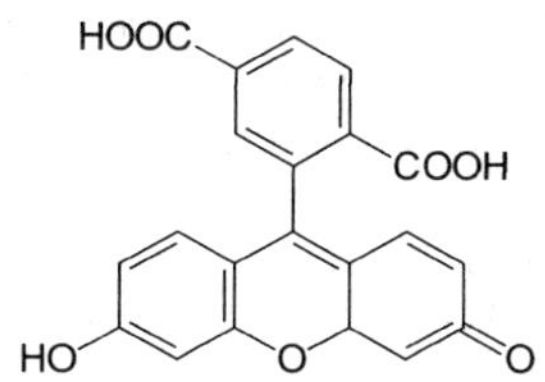

6CF

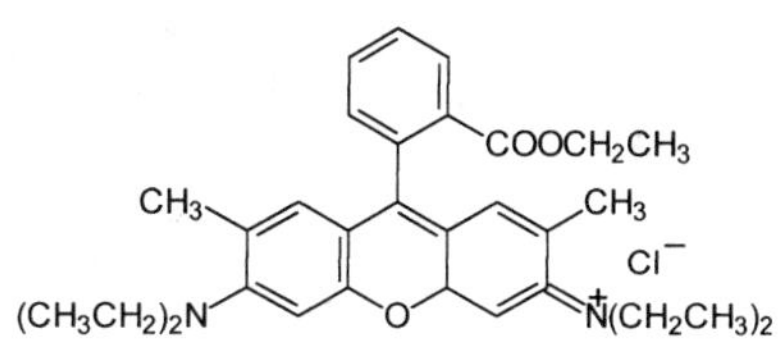

6G

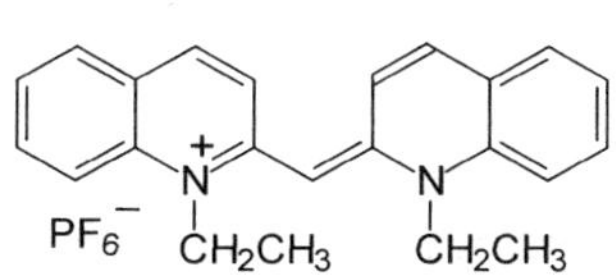

PIC

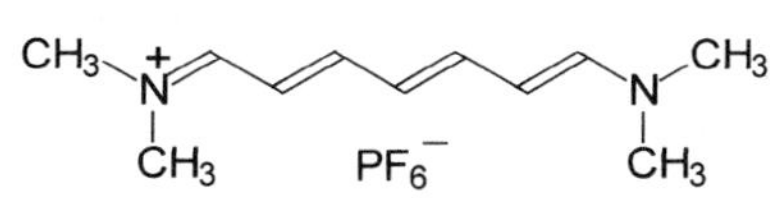

DMHM

ACKNOWLEDGMENTS

Thanks are extended to M. Christine Brick for permission to present some of her preliminary Ph.D. thesis results, including Figs. 6 and 7, on dispersion of organic pigments by miscible solvent shifting. We are also indebted to Professor B.Nagy of Namurs and to F. Debuigne for permission to cite some of her Ph.D. research on precipitation of organics in microemulsions prior to publication.

REFERENCES

1. B. P. Binks, J. H. Clint, P. D. I. Fletcher, and S. Rippon, Langmuir *14*:5402–5411 (1998).
2. K. Naito, K. Ohe, and N. Tsuji, U.S. Patent 4,21,862 (1980).
3. T. C. Patton, *Paint Flow and Pigment Dispersion*, 2nd ed., Wiley, New York, 1979, pp. 376–512.
4. R. B. McKay and F. M. Smith, in *Dispersion of Powders in Liquids* (G. D. Parfitt, ed.), 3rd ed., Applied Science Publishers, London, 1981, pp. 471–505.
5. Society of Dyers and Colourists, *Color Index*, 3rd ed., 1982.
6. R. B. McKay, in *Technological Applications of Dispersions* (R. B. McKay, ed.), Marcel Dekker, New York, 1994, pp. 143–176.
7. E. W. Kaler, K. L. Herrington, A. K. Murthy, and J. N. Zasadzinski, J. Phys. Chem. *96*:6698–6707 (1992).
8. J. L. Katz and M. D. Donohue, Adv. Chem. Phys. *40*: 137–155 (1979).
9. I. Leubner, Curr. Opin. Colloid Interface Sci. *5*:151–159 (2000).
10. J. W. Mullin, *Crystallization*, 3rd ed., Butterworth-Heinemann, Oxford, 1993.
11. J. Kapecki and J. Rodgers, in *Kirk-Othmer Encyclopedia of Chemical Technology*, 4th ed., Vol. 6, Wiley, New York, 1993, pp. 965–1002
12. J. Texter, J. Imag. Sci. Technol. *39*:340–354 (1995).
13. J. Texter, D. S. Ross, and T. Matsubara, J. Imag. Sci. *34*:123–131 (1990).
14. J. Texter, J. Photog. Sci. *36*:14–17 (1988).
15. C. Schwarz, W. Mehnert, J. S. Lucks, and R. H. Muller, J. Contr. Rel. *30*:83–96, 1994.
16. B. Sjöström, B. Bergenståhl, M. Lindberg, and A. C. Rasmuson, J. Dispersion Sci. Technol. *15*:9–117 (1994).
17. T. H. Whitesides and D. S. Ross, J. Colloid Interface Sci. *169*:48–59 (1995).
18. G. Wu and E. Matijević, J. Dispersion Sci. Technol. *19*: 903–913 (1998).
19. A. Tentoria, E. Matijević, and J. P. Kratohvil, J. Colloid Interface Sci. *77*:418–426 (1980).
20. G. Wu, A. Koliadima, Y.-S. Her, and E. Matijević, J. Colloid Interface Sci. *195*:222–228 (1997).
21. C. Duclairoir, E. Nakache, H. Marchais, and A.-M. Orecchioni, Colloid Polym. Sci. *276*:321–327 (1998).
22. I. Expeleta, J. M. Irache, S. Stainmesse, C. Chabenat, J. Gueguen, Y. Popineau, and A.-M. Oricchioni, Int. J. Pharm. *131*:191–200 (1996).
23. H. Auweter, V. André, D. Horn, and E. Lüddecke, J. Dispersion Sci. Technol. *19*:163–184 (1998).
24. D. Horn and E. Lüddecke, in *Fine Particles Science and Technology—From Micro to Nanoparticles* (E. Pelizetti, ed.), NATO Advanced Science Institutes Series 3, Vol. 12, Kluwer Academic Publishers, Dordrecht, 1996, pp. 761–775.
25. E. B. Gutoff and T. F. Swank, AIChE Symp. Ser. (No. 193) *76*:43–51 (1980).
26. L. Godowsky and J. J. Duane, U.S. Patent 2,870,012 (1959).
27. M. R. Violante and R. T. Steigbigel, U.S. Patent 4,783,484 (1988).
28. M. R. Violante and H. W. Fischer, U.S. Patent 4,826,689 (1989).
29. M. R. Violante and H. W. Fischer, U.S. Patent 4,997,454 (1991).
30. S. Frank, J.-E. Löfroth, and L. Bostanian, U.S. Patent 5,780,062 (1998).
31. S. A. Jenekhe and X. L. Chen, Science *279*:1903–1907 (1998).
32. S. A. Jenekhe and X. L. Chen, Science *283*:372–375 (1999).
33. V. Krukonis, Supercritical fluid nucleation of difficult-to-comminute solids, Proceedings 1984 Annual Meeting, AIChE, San Francisco, November 25–30, 1984, paper 140f.
34. P. M. Gallagher, M. P. Coffey, V. J. Krukonis, and N. Klasutis, in *Supercritical Fluid Science and Technology* (K. P. Johnston and J. M. L. Penninger, eds.), American Chemical Society, Washington, DC, 1989, pp. 334–354.
35. M. H. Hanna, P. York, D. Rudd, and S. Beach, Pharm. Res. *12*:S141 (1995).
36. P. York and M. Hanna, in *Respiratory Drug Delivery V* (R. N. Dalby, P. R. Byron, and S. J. Farr, eds.) Interpharm Press, Buffalo Grove, IL, 1996, pp. 231–239.
37. J. Texter, U.S. Patent 5,274,109 (1993).
38. J. Texter, U.S. Patent 5,326,687 (1994).
39. J. Texter, W. B. Travis, and V. J. Flow, U.S. Patent 5,624,467 (1997).
40. G. S. Riley, SCI Monogr. *38*:267–281 (1973).
41. P. Bagchi, P. H. Karpinski, and G. L. McIntire, U.S. Patent 5,560,932 (1996).
42. P. Bagchi, G. L. McIntire, and J. R. Minter, U.S. Patent 5,716,642 (1998).
43. P. Bagchi, R. P. Scaringe, and H. W. Bosch, U.S. Patent 5,665,331 (1997).
44. J. Israelachvili, *Intermolecular & Surface Forces*, 2nd ed., Academic Press, London, 1992, pp. 341–365.
45. M. J. Rosen, *Surfactants and Interfacial Phenomena*, 2nd ed., Wiley, New York, 1989, pp. 393–418.

46. W. J. Priest, Res. Disc. (No. 16468), December 1977, pp. 75–80.
47. K. Chari, J. Minter, N. Ferris, and P. Amato, J. Phys. Chem. *97*:2640–2645 (1993).
48. K. Chari and J. T. Beck, U.S. Patent 5,256,527 (1993).
49. K. Chari, Chem. Technol. *29*:49–55 (1999).
50. P. Bagchi, R. C. Stewart, G. L. McIntire, and J. R. Minter, U.S. Patent 5,662,883 (1997).
51. S. P. Kline, Langmuir *15*:2726–2732 (1999).
52. M. Rosoff, ed., *Vesicles*, Marcel Dekker, New York, 1996.
53. R. Arshady, ed., *Microspheres, Microcapsules and Liposomes*, Vol. 1, *Preparation & Chemical Applications*, Citus Books, London, 1999.
54. E. W. Kaler, A. K. Murthy, B. E. Rodriquez, and J. A. Zasadzinski, Science *245*:1371–1374 (1989).
55. L. L. Brasher and E. W. Kaler, Langmuir *12*:6270–6276 (1996).
56. T. Zemb, M. Dubois, B. Deme, and T. Gulik-Krzywicki, Science *283*:816–819 (1999).
57. S. E. Friberg and P. Bothorel, eds., *Microemulsions: Structure and Dynamics*, CRC Press, Boca Raton, FL, 1987.
58. J. Eastoe and B. Wayne, Curr. Opin. Colloid Interface Sci. *1*:800–805 (1996).
59. K. Kon-No, in *Structure-Performance Relationships in Surfactants* (K. Esumi and M. Ueno, eds.), Marcel Dekker, New York, 1997, pp. 551–582.
60. T. Li, J. Moon, A. A. Morrone, J. J. Mecholsky, D. R. Tallham, and J. H. Adair, Langmuir *15*:4328–4334 (1999).
61. N. Moumen, M.-P. Pileni, and R. A. Mackay, Colloids Surf. A *151*:409–417 (1999).
62. M.-P. Pileni, ed., *Structure and Reactivity in Reverse Micelles*, Elsevier, Amsterdam, 1989.
63. F. Debuigne, L. Jeunieau, M. Wiame, and J. B. Nagy, The synthesis of rhodiarome nanoparticles by direct precipitation in a microemulsion, in *Proceedings of the 73rd ACS Colloid and Surface Science Symposium* (P. E. Laibinis and T. A. Hatton, eds.), MIT, Cambridge, 1999, paper 90.
64. R. W. Siegel, Mater. Sci. Eng. B *19*:37–43 (1993).
65. R. W. Siegel, Sci. Am. *275*:74–79 (1996).
66. C. Hayashi, R. Uyeda, and A. Tasaki, eds., *Ultra-Fine Particles*, Noyes Publications, Westwood, NJ, 1997.
67. H. Toyotama, in *Ultra-Fine Particles* (C. Hayashi, R. Uyeda, and A. Tasaki, eds.), Noyes Publications, Westwood, NJ, 1997, pp. 286–292.
68. F. Lim, ed., *Biomedical Applications of Microencapsulation*, CRC Press, Boca Raton, FL, 1984.
69. M. Yang, G. A. Davies, and J. Garside, Powder Technol. *65*:235–242 (1991).
70. G. Sukhorukov, L. Dähne, J. Hartmann, E. Donath, and H. Möhwald, Adv. Mater. *12*:112–115 (2000).
71. T. Clemens, A. J. Boehm, and S. Kielhorn-Bayer, Polym. Prepr. *41*:24–25 (2000).

30

Synthesis of Inorganic and Organic Nanoparticles in Microemulsions

L. JEUNIEAU, F. DEBUIGNE, and JANOS B.NAGY Universitaires Notre-Dame de la Paix, Namur, Belgium

I. INTRODUCTION

The synthesis of monodisperse nanometer-size nanoparticles is of great technological and scientific interest. Quantum size effects are particularly of interest because they lead to interesting mechanical, chemical electrical, optical, magnetic, electro-optical, and magneto-optical properties that are quite different from those reported for bulk materials [1–4]. Nanoparticles not only are of basic physical interest but also have resulted in important technological applications, such as in catalysts, high-performance ceramic materials, microelectronic devices, and high-density magnetic recordings [5–7].

The synthesis of nanoparticles in microemulsion allows one to obtain monodisperse particle sizes. In some cases it is possible to control the size of the particles by variation of the microemulsion droplet radius and of precursor concentrations.

Although the synthesis of inorganic particles in microemulsion is already widespread, only polymer nanoparticles have been synthesized in microemulsion media, as far as organic particles are concerned [8–10]. In this chapter, it will be shown that it is also possible to synthesize organic particles by precipitation in microemulsions.

We emphasize some of the fundamental aspects of monodisperse nanoparticles formation. Two models are proposed for the formation of the particles: the first is based on the LaMer diagram; the second is based on the thermodynamic stabilization of the particles. In the first case the particle size varies as a function of either the size of the inner water cores or the precursor concentration in the microemulsions; in the second case the particle size is independent of these parameters.

A. Preparation of Nanoparticles Using Microemulsions

1. Description of the Microemulsions

A water-in-oil microemulsion is a thermodynamically stable, optically transparent dispersion of two immiscible liquids stabilized by a surfactant. The important properties are governed mainly by the water-surfactant molar ratio (R = [H_2O]/[surfactant]). This factor has often been linearly correlated with the size of the water droplets.

Nanoparticles have been synthesized in different reverse microemulsion systems. Ternary phase diagrams for some of these are shown in Fig. 1, where the isotropic reverse microemulsion (L_2) domain boundaries are illustrated.

The anionic AOT (bis(2-ethylhexyl)sodium sulfosuccinate)/heptane/water system is one of the best characterized microemulsion systems [11,12]. The cationic cetyltrimethylammonium bromide (CTAB)/hexanol/water system contains hexanol as the organic phase; in many other formulations hexanol plays the role of cosurfactant [13]. The nonionic penta(ethylene glycol) dodecyl ether (PEGDE)/hexane/water system was studied by Friberg and Lapczynska [14]. The reverse micellar droplets have a cylindrical shape in which the

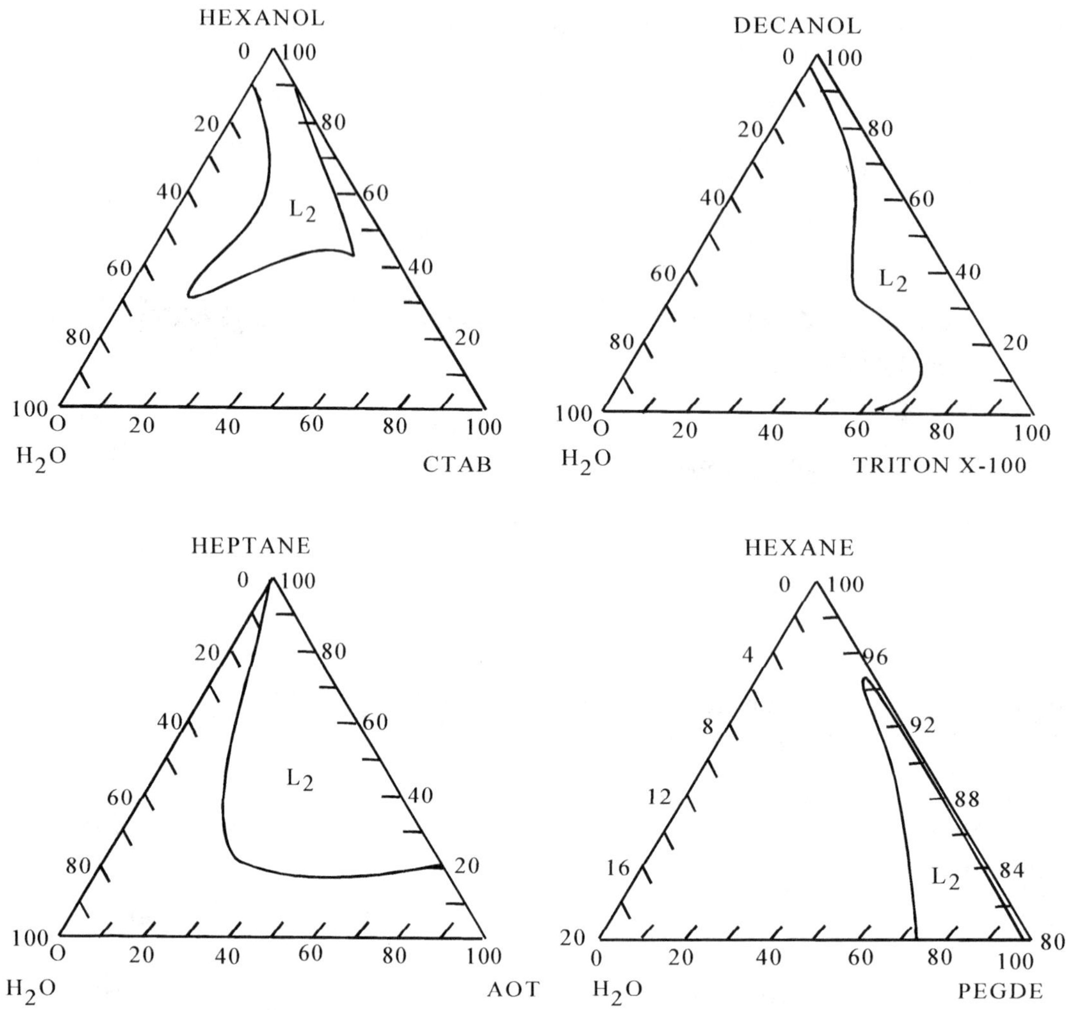

FIG. 1 Ternary phase diagrams for various microemulsion systems. Reverse microemulsion L_2 domains are illustrated.

surfactant molecules stay parallel to each other, forming a bilayer impregnated with water. The Triton X-100 (*p*-(1,1,3,3-tetramethylbutyl)phenyl-polyethoxy-ethanol)/decanol/water system has been characterized by Ekwall et al. [15,16].

2. Mechanism of Synthesis of Nanoparticles in Microemulsion

The aqueous droplets in reverse microemulsions continuously collide, coalesce, and break apart, resulting in a continuous exchange of solution content between nanodroplets. In fact, the half-life of the exchange reaction between the droplets is of the order of 10^{-3}–10^{-2} s [17,18].

Two models have been proposed to explain the variation of the size of the particles with the precursor concentration and with the size of the aqueous droplets. The first is based on the LaMer diagram [19,20]. This diagram (Fig. 2) has been proposed to explain the precipitation from solution in terms of supersaturation, nucleation, and growth. It illustrates the variation of the concentration with time during a reaction of precipitation and is based on the principle that nucleation is the limiting step in the reaction of precipitation. In the first step, the concentration increases continuously with increasing time. As the concentration reaches the critical supersaturation value, nucleation occurs. This leads to a decrease of the concentration. Nucleation continues as the concentration C^*_{max} falls to C^*_{min}, at which point particle growth is presumed to replace particle nucleation. Later, the decrease of the concentration is due to the growth of the particles by diffusion. This growth occurs until the concentration reaches the equilibrium solubility value.

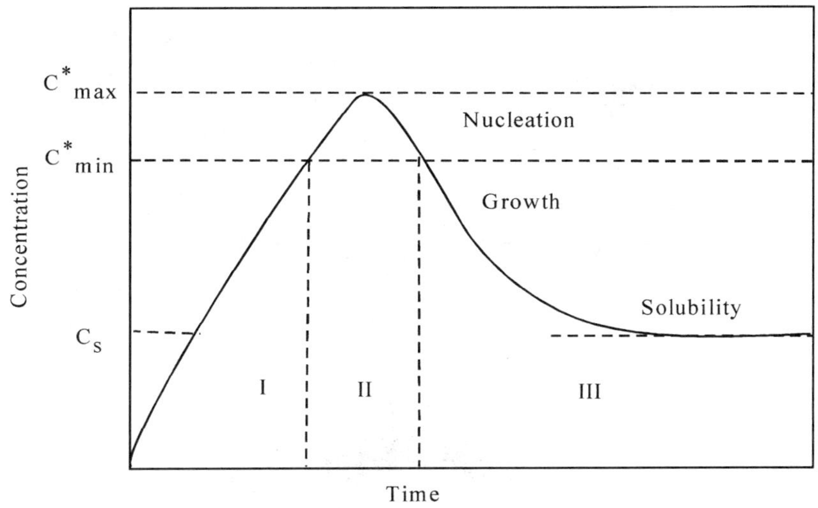

FIG. 2 LaMer diagram.

This model should be applicable to microemulsion media because microemulsions are optically isotropic solutions (thermodynamically) even though they have complex microheterogeneous structure. We assume that nucleation occurs in the first part of the reaction, and later on only the growth of the particles occurs. If this model is followed, the size of the particles will increase continuously with the concentration of the precursor or a minimum in the variation of the size with the concentration can also be observed. This is because the number of nuclei at the end of the nucleation phase is assumed to remain constant and the increase of concentration (at levels less than C^*_{min}) leads to an increase in the growth of the particles.

The second model we evaluate is the thermodynamic stabilization of the particles. In this model the particles are thermodynamically stabilized by the surfactant. The size of the particles stays constant when the precursor concentration and the size of the aqueous droplets vary.

These two models are limiting and approximate. The diagram of LaMer does not take into account the stabilization of the particles by the surfactant, and the thermodynamic stabilization does not take into account that the nucleation of the particles is more difficult than the growth by diffusion.

II. SYNTHESIS OF INORGANIC PARTICLES

A. Synthesis of Silver Halide Particles

Particles of silver bromide, silver chloride, and silver bromochloride have been synthesized. The method of preparation is the following: Two microemulsions are prepared, one containing silver nitrate, the other containing potassium bromide or chloride. The two microemulsions are mixed under red light or in the dark. The principle of the formation of such AgBr particles is illustrated in Fig. 3. Dynamical droplet fusion and fission mix the reagents. Many different kinds of inorganic precipitations have been executed using this reaction approach. In addition, many reductions to produce nanoparticle metals have been done, where a reducing agent is included in one of the microemulsions being mixed. Inorganic oxides have also been synthesized using hydrothermal syntheses in reverse microemulsions.

1. AgBr Particles in AOT/Heptane/Water Microemulsions

Silver bromide particles have been synthesized [21] in the AOT/heptane/water microemulsion system. Figure 4 shows the variation of the particle size as a function of $AgNO_3$ concentration. The particle size increases monotonously with increasing $AgNO_3$ concentration and approaches a plateau at high concentrations. These results have been successfully interpreted by using the LaMer diagram. In fact, the nucleation occurs only at the beginning of the reaction and the other reactants are used for the growth of the particles [22–24].

These results can be rationalized by computing the number of nuclei (N_n) formed by N_M inner water cores, the Poisson distribution of Ag^+ ions in the water cores, and the sum of probabilities $\Sigma_{k=i}^{\infty} p_k$ of containing i ions or more per water core.

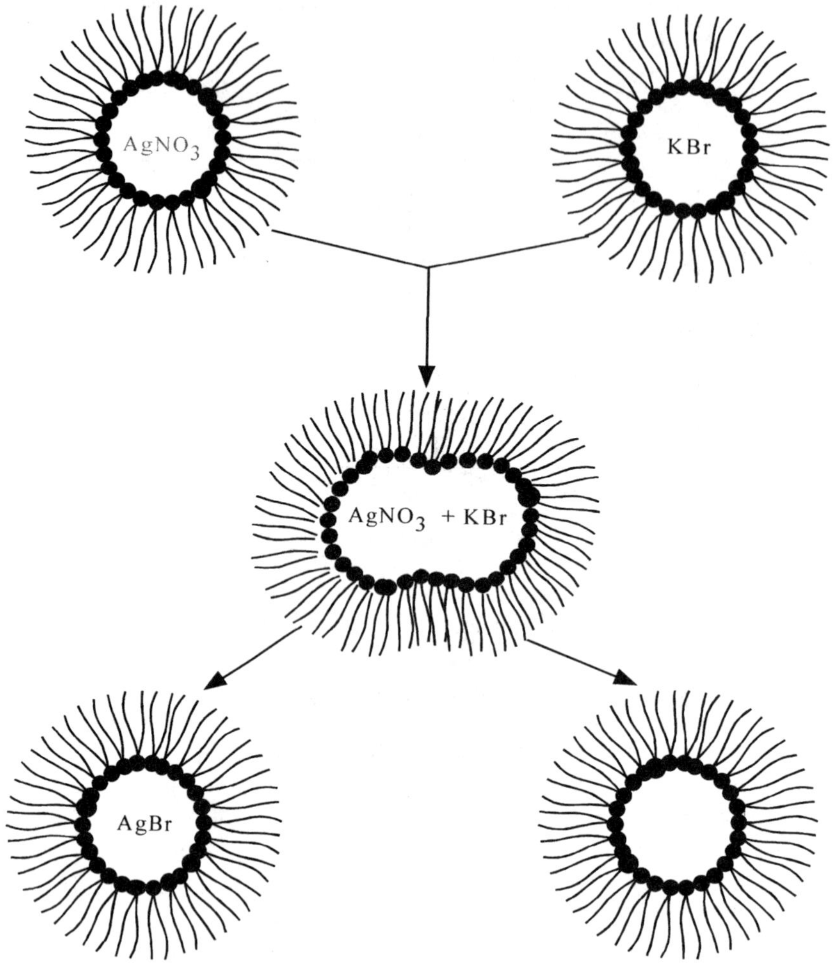

FIG. 3 Silver bromide precipitation in reverse microemulsions.

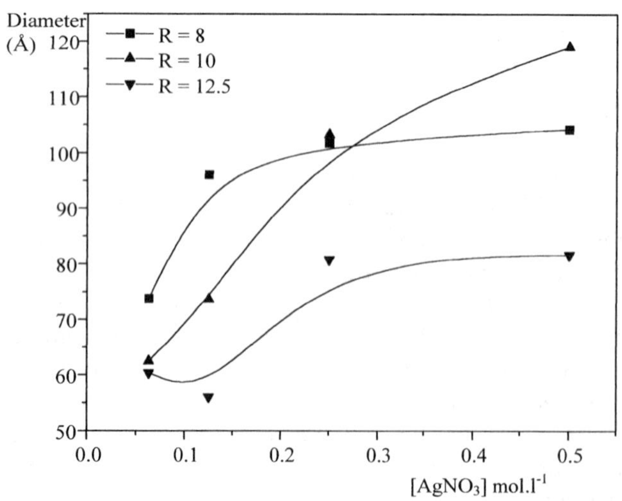

FIG. 4 Diameter of AgBr particles as a function of $AgNO_3$ concentration (M).

The number of nuclei (N_n) is computed from a simple formula taking into account the size distribution of the particles:

$$N_n = \frac{N_{op}}{\sum_i m_i/M_t}$$

where N_{op} is the number of observed particles on a photomicrograph, $\Sigma_i m_i$ is the mass of the particles ($m_i = dV_i$ with $d_{AgBr} = 6.473$ g cm^{-3}), and M_t is the total mass of AgBr formed. The corresponding data are reported in Table 1.

The average number of inner water cores is computed from the volume of $AgNO_3$ solution in the microemulsion divided by the volume of one inner water core of radius r_M:

TABLE 1 Important Parameters for the Formation of AgBr Colloidal Particles of Size d[a]

$[AgNO_3]$ (mol L^{-1})	d (Å)	N_n/N_M	N_{Ag^+}/N_M	$\left(\sum_{k=i}^{100} p_k\right)^2$	F
$R = 8$					
0.0063	73.8	2×10^{-4}	1.17	0.4756	4.2×10^{-4}
0.125	96.1	3×10^{-4}	2.25	0.8003	3.7×10^{-4}
0.250	101.8	4×10^{-4}	3.57	0.9445	4.2×10^{-4}
0.500	104.3	8×10^{-4}	6.96	0.9980	8.0×10^{-4}
$R = 10$					
0.063	62.5	7×10^{-4}	1.52	0.6104	1.1×10^{-3}
0.125	73.7	1×10^{-3}	4.23	0.9711	1.0×10^{-3}
0.250	103.3	8×10^{-4}	5.89	0.9945	8.0×10^{-4}
0.500	119.1	1×10^{-3}	10.06	0.9999	1.0×10^{-3}
$R = 12.4$					
0.063	60.4	1×10^{-2}	2.88	0.8909	1.1×10^{-2}
0.125	56.1	8×10^{-3}	5.96	0.9948	8.0×10^{-3}
0.250	80.8	2×10^{-3}	10.09	0.9999	2.0×10^{-3}
0.500	81.7	6×10^{-3}	23.85	1.0000	6.0×10^{-3}
$R = 20$					
0.063	65.3	5×10^{-3}	11.28	1.0000	5.0×10^{-3}
0.125	61.6	1×10^{-2}	17.53	1.0000	1.0×10^{-2}

[a] N_n/N_M, number of nuclei per water core; N_{Ag^+}/N_M, mean number of silver ions per water cor; p_k, Poisson distribution of silver ions; F, proportionality factor (see text).

$$N_M = \frac{M_{sol.AgNO_3}/d_{sol.AgNO_3}}{\frac{4}{3}\pi r_M^3}$$

The r_M values are computed from the relation between r_M and R [25]. The densities of the aqueous solutions of $AgNO_3$ are taken from Ref. 26.

The mean number of $AgNO_3$ units per water core (N_{Ag^+}/N_M or λ) is then computed (Table 1). The probability of having $k\,Ag^+$ ions per water core is given by Poisson statistics:

$$p_k = \frac{\lambda^k e^{-\lambda}}{k!}$$

where k is an integer. Table 1 also shows the square sum of probabilities $(\sum_{k=1}^{100} p_k)^2$ of containing one Ag^+ ion or more per inner water core. Indeed, it can be assumed that one Ag^+ ion can lead to the formation of one surfactant-stabilized AgBr entity, which can be considered the initialization of an embryonic AgBr particle, although such a minimal embryo is unstable. The limit of 100 was chosen instead of ∞ because, for practical reasons, the probability of having more than 100 ions in a water core is negligible.

It was proposed previously that

$$N_n = FN_M \sum_{k=1}^{\infty} p_k$$

where F is a scaling factor. This formula is valid when an aqueous solution of reactant is added to the microemulsion containing another reactant. As in the present case when two microemulsions are mixed, the correct formula is

$$N_n = FN_M \left(\sum_{k=1}^{\infty} p_k\right)^2$$

A constant value of F for all the experimental concentrations corresponds to a good value of i. If i is put equal to 1 or 2, the value of F is reasonably constant for $R = 8$ and 10 microemulsions. The F values are somewhat less scattered for $i = 1$ ($F \approx 4 \times 10^{-4}$–$1 \times 10^{-3}$), and it is this value which is taken as the minimum number of AgBr units to form an (unstable) embryonic nucleus (Table 1). For microemulsions with $R = 12.4$ and 20, a value of 7×10^{-3} is more suitable for F. This rather low value of F shows that only approximately one out of every 1000 water droplets leads to the formation of a nucleus for $R = 8$ and 10 microemulsions and approximately one out of every 100 water droplets for the $R = 12.4$ microemulsion. This small value emphasizes the fact that nucleation occurs very early after mixing, and once the nuclei are formed they grow to yield monodisperse silver bromide particles.

(*a*) *Characterization of the Particles* [27]. In the numerous studies concerning the synthesis of nanoparticles in microemulsion medium, the localization of water after the nanoparticle synthesis has never been determined. Two models can be proposed (Fig. 5). In the first one the particles are surrounded by a layer of water; in the second the surfactant molecules (the AOT are directly adsorbed on the particles and only a small amount of water is present.

In order to discriminate between these two models, ^{2}H nuclear magnetic resonance (NMR) measurements of deuterated water in the microemulsion have been carried out. Two NMR lines were observed in the ^{2}H NMR spectra (Fig. 6) for the various microemulsions without particles of silver bromide.

If the spectrum is taken for a very low R value, such as $R = 0.5$ (Fig. 7), three NMR lines are observed. These lines are not due to the presence of impurities. In fact, their intensity does not decrease as the amount of water decreases, and these lines stem from different types of water molecules, as indicated by their different relaxation times T_1. In fact, for $R = 1$ the following relaxation times T_1 were obtained at 273 K: 321 ms for the broader line, 804 ms for the line situated at -3.50 ppm and 1087 ms for the line situated at -3.95 ppm. As the variation of relaxation time with temperature indicates that we are in a region where the relaxation time increases with decreasing temperature, these two lines correspond to water molecules less mobile and therefore more in contact with the surfactant molecules.

Generally, three kinds of water may exist in a microemulsion medium: "bulk" water in the center of the water core; "bound" water, which interacts with the

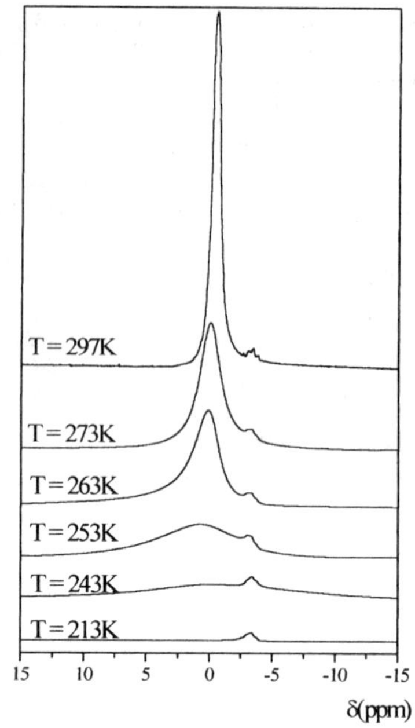

FIG. 6 NMR spectra of the deuterated water in the microemulsion for $R = 3.1$.

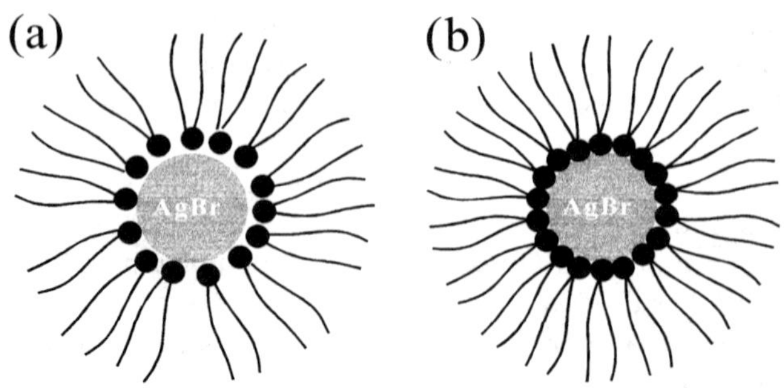

FIG. 5 Two models of the nanoparticles stabilized in the microemulsion media: (a) the particle is surrounded by a layer of water; (b) AOT is directly adsorbed on the particle.

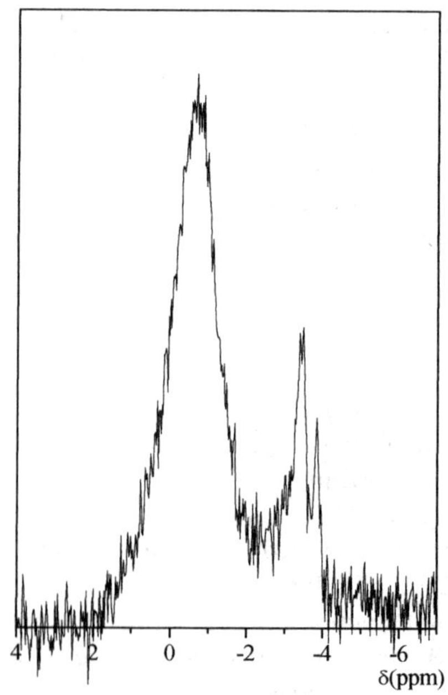

FIG. 7 NMR spectrum of deuterated water in AOT/heptane/water microemulsion for $R = 0.5$ at $T = 297$ K.

hydrophilic part of the surfactant molecule; and "trapped" water, which is trapped in the interface in the form of monomers or dimers [28]. Bulk water molecules are normally not present for *R* values below 6–10, where all the water molecules are structured because of their interaction with Na^+ counterions and the strong dipole of the AOT polar group [29]. In this case, where the ratio $R = [H_2O]/[AOT]$ is 3.1, only two kinds of water molecules should be expected. Therefore, it is assumed that the two NMR lines observed here correspond to bound water and to trapped water. In order to check this assumption, the same experiment was done for higher *R* values. The chemical shift increases with the *R* value until reaching approximately that of the pure deuterated water (used as reference) while the line width at half-height decreases with *R* (Fig. 8).

This variation has already been observed [29] and is the result of a fast exchange (faster than 2×10^{10} s^{-1}) between the bulk water and the bound water. At low *R* values, the observed chemical shift comes from the variation of the number of hydrogen bonds in which the water molecules are involved. In fact, the water molecules adsorbed at the interface (or solvating the Na^+ ions) form fewer hydrogen bonds, provoking a high-field chemical shift. This decreasing number of hydrogen bonds has previously been shown by Wong et al using ^1H NMR experiments [30].

Furthermore, if the NMR spectra are recorded at lower temperatures, the NMR line corresponding to the bound water decreases because of the freezing of this kind of water (the bandwidth becomes too large to be detectable) (Fig. 6). In fact, the freezing point of bound water seems to be about 243 K inside these reverse micelles. This corresponds to a decrease of the freezing point of water with decreasing droplet size. For example, the freezing point of water in a droplet corresponding to $R = 4.5$ in AOT/water/2,2,4-trimethylpentane is at around 241 K [31]. On the other hand, the line corresponding to the trapped water shows no freezing and its intensity remains quasi-constant.

In order to distinguish between these two models of AgBr stabilization (see earlier), the NMR experiments already mentioned have also been carried out in the presence of silver bromide nanoparticles. As the only difference between the two experiments is the presence of silver bromide particles, all observed differences must be due to the particles. In the presence of these particles, the quantity of trapped water is larger, as shown by comparison of the spectra in the presence and absence of nanoparticles (Fig. 9). It could be hypothesized that the particles repel the bound water into the interface and, as a consequence, the amount of trapped water increases. The total intensity is also greater in the presence of silver bromide particles,

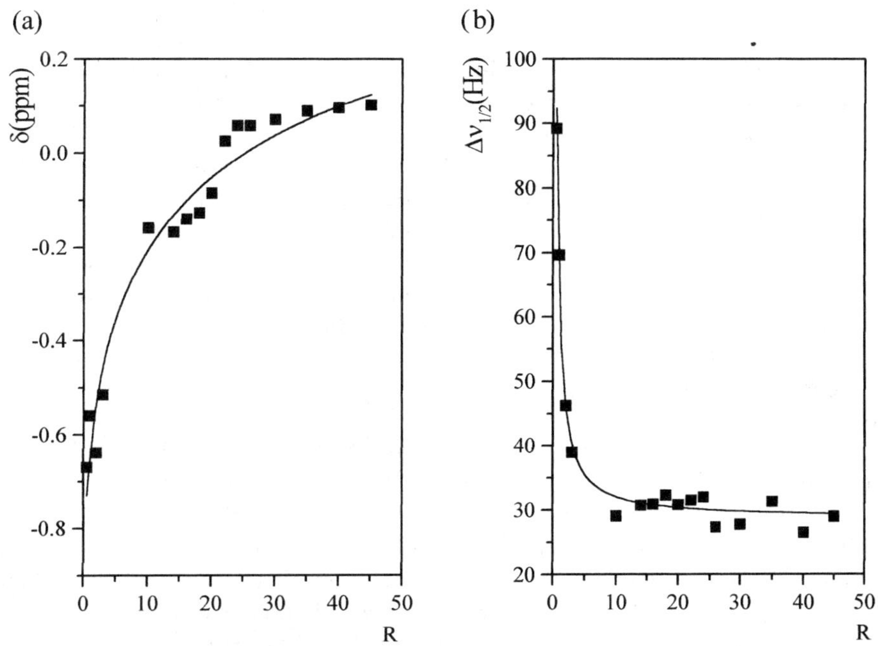

FIG. 8 (a) Variation of the ^2H chemical shift as a function of the *R* factor. (b) Variation of the line width as a function of the *R* factor.

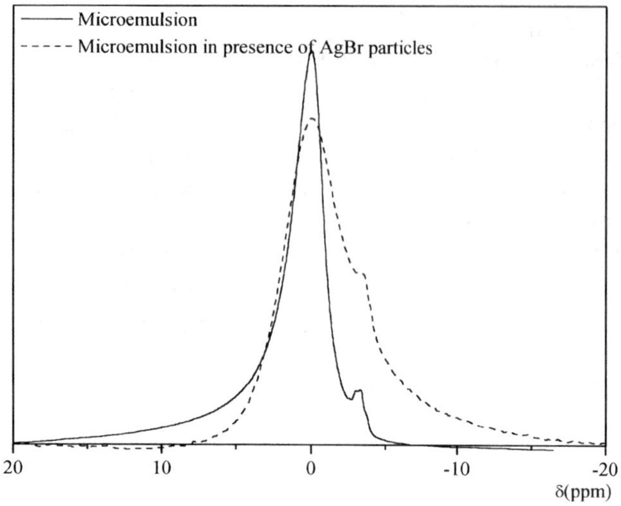

FIG. 9 NMR spectrum of the deuterated water in the microemulsion (full line) and in presence of AgBr particles (dotted line) at 263 K.

stemming also from the greater importance of the trapped water. In fact, this water freezes at a lower temperature. Furthermore, not all the water cores of the microemulsion are occupied by a particle. Only 1 water core out of 1.3×10^4 is occupied by a particle. Hence, if the microemulsion structure stayed constant, with the same number of water molecules in each water core, no influence on the NMR spectra could be observed upon addition of AgBr.

A greater amount of trapped water favors the stabilization model of Fig. 5b, where the particles are in closer contact with the interfacial layer. However, the NMR line of the adsorbed water could overlap that of the trapped water. In order to check this hypothesis, the number of water molecules per AOT has been calculated. The spectrum in Fig. 9 has been decomposed into two bands corresponding, respectively, to the bound water and to the trapped water. The difference in intensity of the two NMR lines corresponding to the trapped water in the spectra without and with AgBr particles gives the amount of water trapped or adsorbed on the particles. The number of AOT molecules per particle has been calculated by using a spherical surface corresponding to a 46 Å diameter and by using a surface area of 41 $Å^2$ for the polar part of the AOT molecule [32]. It has been computed that if all the line intensity corresponded to the trapped water, there could be 2000 water molecules per AOT molecule. As the trapped water is considered to be in the form of monomers or dimers, this value is too high to correspond only to water molecules trapped in the interface. Hence, it has to be supposed that the additional water molecules so computed are adsorbed on the AgBr particles and the NMR lines of the trapped water and of the adsorbed water overlap.

If it is assumed that all these additional water molecules are adsorbed on the particles, the number of water monolayers can be calculated by using the van der Waals radius of a water molecule. Approximately 1000 monolayers of water are estimated, and this number is much too high because it does not take into account that the number of trapped molecules increases by repelling the water molecules in the interface. These two arguments, the observation of an NMR line corresponding to the adsorbed water molecules and the estimation of the number of water monolayers, favor the structural model of Fig. 5a. Hence, we adopt this model.

In order to quantify the amount of water adsorbed on the nanoparticles by another method, a microemulsion in which the particles sedimented has also been examined. This microemulsion was obtained by adsorption of pseudoisocyanine on the particles. This dye provokes rapid sedimentation of the particles [33], and a ^2H NMR spectrum was taken after sedimentation of all the particles. From this spectrum, it has been established that 68% of the water is adsorbed on the particles. In order to visualize this result, the number of water monolayers has been estimated to correspond to about 4600 monolayers. This value is unrealistically high and would correspond to a water core diameter of 2.6 μm. Such cores should be highly light scattering, and as the colloidal suspension is transparent, the number of water molecules bound to the silver halide particles must be highly overestimated by this approach. Such a large amount of water in the precipitate can be explained only if the sedimented particles form a sort of gel where a large amount of water is required. This gelation was previously shown in the case of Co_2B nanoparticles prepared from a microemulsion of CTAB/*n*-hexanol/water [34]. This great amount of adsorbed water molecules also favors the structural model of Fig. 5a.

2. AgBr Particles in AOT/*p*-Xylene/Water Microemulsions

Astonishingly, the average diameters of the AgBr nanoparticles prepared in microemulsions of AOT/*p*-xylene/water remain quite constant, whatever the concentrations of precursor salts or the size of the water nanodroplets. Figure 10 shows the average diameters of the nanoparticles as a function of the salt concentration and the *R* values. All the average diameters seem

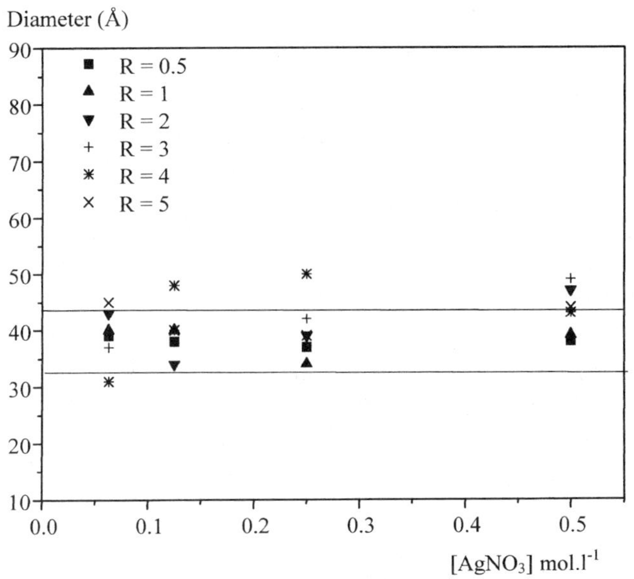

FIG. 10 Average diameters of the AgBr nanoparticles prepared in the AOT/*p*-xylene/water microemulsion.

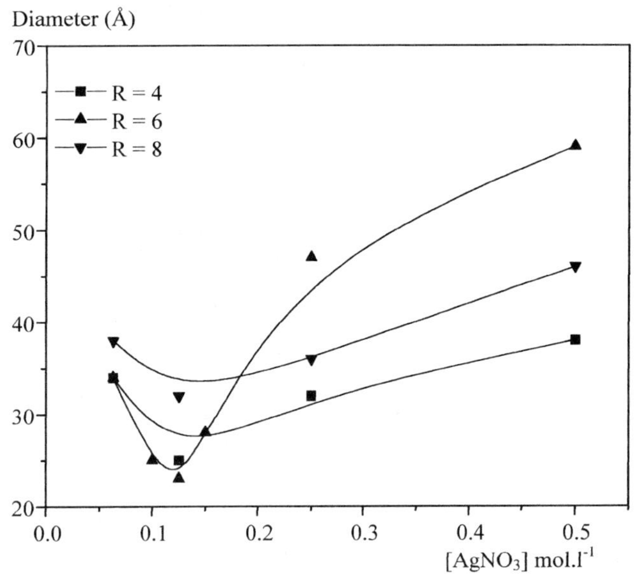

FIG. 11 Average diameter of the AgCl nanoparticles prepared in the system AOT/*n*-heptane/water as a function of the concentration of precursor salts and the value of *R*.

to lie between 33 and 43 Å. The reproducibility of the size measurements has been estimated to be approximately 9 Å. As we have not observed any correlation between these synthesis parameters and the average diameter of the nanoparticles, we conclude that in this microemulsion system, the nucleation and growth of the nanoparticles do not follow the LaMer diagram. The size of the particles seems to be thermodynamically stabilized by the adsorbed surfactant.

3. Synthesis of AgCl

The preparation of silver chloride nanoparticles has also been studied in the AOT/*n*-heptane/water microemulsion. A size dependence on the synthesis parameters has been observed (Fig. 11). The diameter seems to pass through a minimum value at around 0.125 M $AgNO_3$.

The synthesis of AgCl particles seems to follow a model between that represented by the LaMer diagram and that which we term thermodynamic stabilization. The variation of the particles size seems to correspond to the LaMer diagram, but other factors appear to favor the thermodynamic stabilization of the particles. In fact, the size of the particles (and the number of nuclei) does not vary with the contact surface between the two microemulsions during the precipitation. If the LaMer diagram were followed, the number of nuclei should increase with the contact surface between the microemulsions and the size of the particles should decrease. It is not astonishing that the synthesis of AgCl particles is situated between the LaMer diagram and the thermodynamic stabilization of the particles because these two models are only limiting cases.

4. Ag(Cl,Br) in AOT/*n*-Heptane/Water Microemulsions

Nanoparticles of mixed silver halides as silver chlorobromide [Ag(Cl,Br)] have been prepared in the system AOT/*n*-heptane/water. Figure 12 shows an interesting behavior of the size variation as a function of the percentage of chloride in the silver chlorobromide. One may notice that from 0 to 20% chloride, the diameter of the nanoparticle goes from that of pure AgBr to that of pure AgCl. This suggests that the particles are not homogeneous but that the chloride is mainly located at the surface of the particles. It may de due to faster nucleation of the silver bromide particles. In fact, the K_{sp} values of AgBr and AgCl are, respectively, 7.7×10^{-13} mol^2/L^2 and 1.56×10^{-10} mol^2/L^2. The nuclei of AgBr are the first ones formed and following the diagram of LaMer, the growth of the particles is made by the AgCl. The size of the particles is governed by the interaction between the AgCl and the AOT, which explains the constant size of the particles for the greater percentages of chloride. This shows that thermodynamic stabilization is involved in the synthesis of AgCl particles. If only the LaMer diagram were involved, the size of the particles should be determined by the nu-

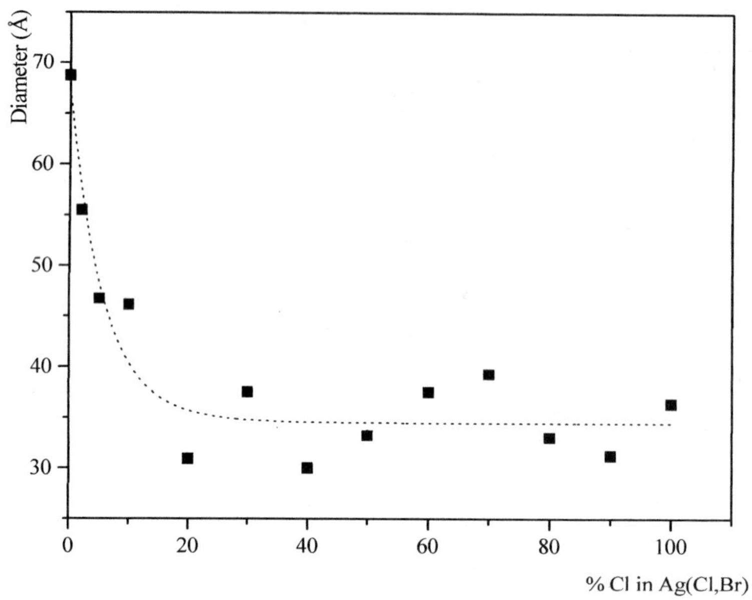

FIG. 12 Average diameter as a function of the percentage of chloride in the silver chlorobromide nanocrystals.

cleation and would therefore be similar to the size of the AgBr particles as the nucleation is easier for AgBr than for AgCl. One would expect to get two different populations of AgBr and AgCl, but the size distributions of these mixed silver halide particles were monodisperse. This supports the homogeneity of the Ag(Cl,Br) particle populations.

B. Metal Boride and Metal

Two different methods were used for preparing metal boride and metal nanoparticles. The two reactants were dissolved in the same microemulsion system and then mixed with vigorous stirring (scheme II in Fig. 13). This method led to the smallest size and was used whenever the aqueous solutions of the reactants were stable enough. The nanoparticles of Pt, ReO_2, and Pt-ReO_2 were all prepared following scheme II.

The Ni_2B, Co_2B, and Ni-Co-B particles were prepared by adding an aqueous solution of $NaBH_4$ to a microemulsion containing the metal salt (scheme I in Fig. 13). Moreover, these particles were prepared in a glove box under an argon atmosphere by adding dropwise a threefold excess of aqueous $NaBH_4$ solution at 0°C with vigorous stirring. The expected microemulsion composition was achieved after complete mixing of the reactants. At the end of the reaction, the temperature was raised to room temperature until complete hydrolysis of excess $NaBH_4$ occurred.

1. Synthesis of Nickel Boride Particles

Monodisperse colloidal nickel boride and cobalt boride particles are synthesized by reducing, with $NaBH_4$, the metallic ions solubilized in the water cores of the microemulsions. The $NaBH_4/MCl_2$ ratio was held equal to 3 because larger particles were obtained for a lower value; the particle size remained constant above that ratio [22,23,35].

The composition of the particles was determined by XPS (x-ray photoelectron spectroscopy) to be, respectively, Ni_2B and Co_2B. In every case, the size of particles (2.5–7.0 nm) is much smaller than that obtained by reduction of Ni(II) or Co(II) in water (300–400 nm) or in ethanol (250–300 nm), and the size distribution is quite narrow (±0.5 nm).

Figure 14 shows the dependence of the nickel boride particle size on the water content in the microemulsion and on the Ni(II) ion concentration. The average size of the particles decreases with decreasing size of the inner water core (decreasing water content), and a complex behavior is observed as a function of the Ni(II) ion concentration. A minimum is detected at a concentration of approximately 5×10^{-2} M. These observations can be understood if one analyzes the nucleation and the growth processes of the particles following the LaMer diagram.

(*a*) *Quantitative Aspects of Particle Formation.* The same model can be used as already described; the following equation has to be used as only the precursor salt initially stays in the microemulsion medium.

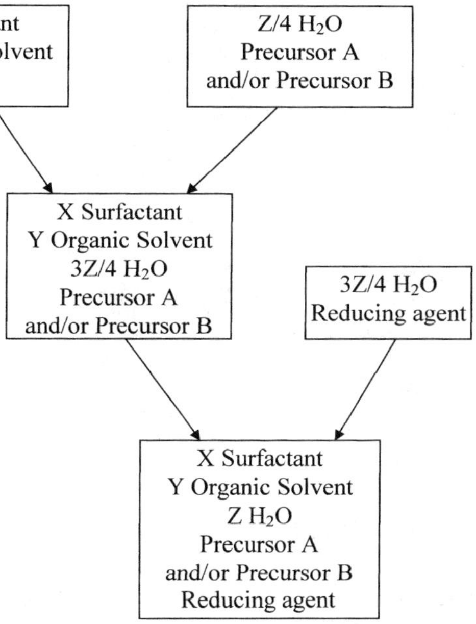

Scheme I

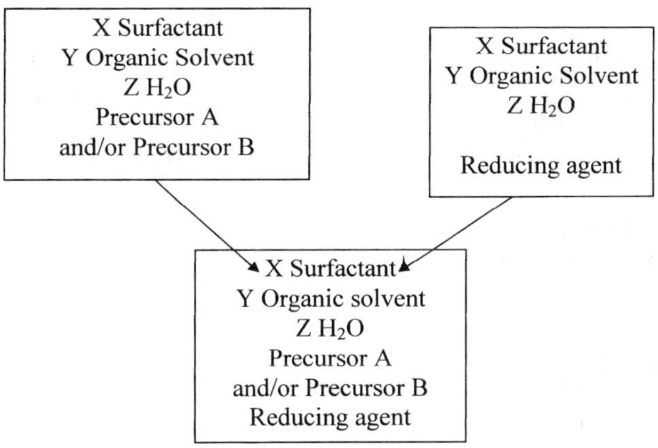

Scheme II

FIG. 13 Method of preparation of monodisperse particles.

$$N_n = FN_M \sum_{k=i}^{\infty} p_k$$

The diameter of the particles is systemically higher than the diameter of the inner water cores. For all the particles synthesized, we calculated the proportionality factor F by systemically varying the value of the minimum number of ions required to form a nucleus (i). Only if i takes the value 2 is the factor F reasonably constant. This value of 2 seems logical as two atoms of nickel are needed to form a nickel boride particle. The order of magnitude of the factor F is always 10^{-3}. This means that at the very beginning of the reduction, i.e., when the nuclei are formed, only one aggregate per thousand leads to the formation of metal boride particles.

There is another indication that the nucleation occurs at the very beginning of the reduction. Indeed, the average radius of the water cores used for the calculation of the formation parameters is measured for the system containing only three quarters of the total amount of water, which is the composition of the solution before the addition of the reducing agent. If the final composition is used, however, no coherent results based on the preceding analysis can be obtained.

The order of magnitude of the factor F is constant, but its value decreases with increasing water content in the microemulsion. This phenomenon can be easily understood because the rearrangement rate of the microemulsion decreases with the amount of water and hence the number of aggregates reached by the reducing agent before rearrangement decreases. As the number of nuclei formed decreases for a constant concentration of precursor ions, the particle size increases with increasing water content.

The results of Table 2 also allow us to explain the minimum in the particle size as a function of the concentration of precursor ions (see Fig. 14). For a constant microemulsion composition, at low ion concentration, only a few water cores contain the minimum number of ions (two) required to form a nucleus; hence, a few nuclei are formed at the very beginning of the reduction, and the metal boride particles are relatively large. When the ion concentration increases, the number of ions per water core increases and the number of nuclei obtained by reduction increases faster than the total number of ions (Fig. 15). This results in a decrease in the particle size. When more than 80% of the water cores contain two or more ions, the number of nuclei formed remains quasi-constant with increasing ion concentration. Hence, the size of the particles increases again.

Figure 14 also shows the particle size as a function of water content in the microemulsions for different Ni(II) concentrations. An increase in the average diameter is observed with increasing proportion of water. The decrease in the number of micellar aggregates (N_M) with water (Table 2) is accompanied by an increase in their size. For the same Ni(II) concentration with respect to water (i.e., for the same probability of collision between the ions in the same water core), the total number of nuclei formed in the early stage of the reduction decreases with increasing water concentration, and more ions can participate in the growth process. This results in an increase in the particle size.

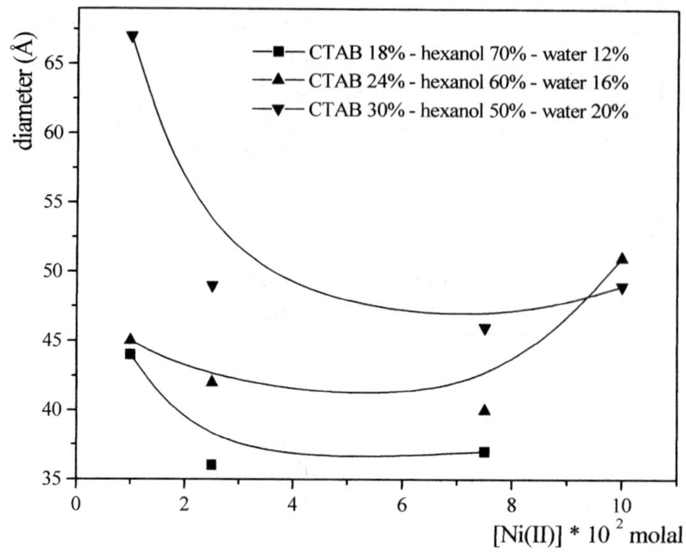

FIG. 14 Variation of the average diameter (in Å) of the nickel boride particles as a function of Ni(II) ion molal concentrations for different percentages in water (the indicated percentage corresponding to the composition after adding $NaBH_4$).

(*b*) *$(Ni,Co)_2B$ Nanoparticles.* Particles of Co_2B have been synthesized under the same conditions. Furthermore, mixed particles of $(Ni,Co)_2B$ have also been synthesized by using a microemulsion containing $NiCl_2$ and $CoCl_2$ in different proportions. Figure 16 shows the variation of the size of the particles as a function of the molar fraction in Co(II).

The F values for nickel boride and cobalt boride particles are quite different. For the former the value obtained for F is 3.2×10^{-3} and for the latter 17.4×10^{-3}. As for these experiments the rearrangement rate of the microemulsion system is constant in the first approximation, the difference between F values is probably due to different solvation of the two types of

TABLE 2 Important Parameters for the Formation of Ni_2B Colloidal Particles

$[Ni(II)] \times 10^{-2}$ m	r_M[a]	d (Å)	N_n/N_M[b]	$\Sigma_{k=2}^{\infty} p_k$	$F \times 10^3$[c]
CTAB 18%/hexanol 70%/water 12%					
1.00	1.02	44	8.63×10^{-5}	0.0347	2.5
2.60	1.22	36	7.09×10^{-4}	0.3661	1.9
5.10	1.37	32	2.82×10^{-3}	0.8713	3.2
7.70	1.47	37	3.38×10^{-3}	0.9899	3.4
CTAB 24%/hexanol 60%/water 16%					
1.00	1.17	45	9.14×10^{-5}	0.0415	2.2
2.50	1.32	42	4.05×10^{-4}	0.3265	1.2
7.50	1.54	40	2.24×10^{-3}	0.9752	2.3
10.00	1.57	51	1.52×10^{-3}	0.9964	1.8
CTAB 30%/hexanol 50%/water 20%					
1.00	1.34	67	3.31×10^{-5}	0.0589	0.6
2.50	1.48	49	2.83×10^{-4}	0.3732	0.8
7.50	1.68	46	1.51×10^{-3}	0.9780	1.5
10.00	1.72	49	1.78×10^{-3}	0.9973	1.8

[a]Values given for the system containing three quarters of the total amount of water.
[b]Values given for 1 kg of solution.
[c]Corrections factor from $N_n = FN_M \Sigma_{k=2}^{\infty} p_k$.

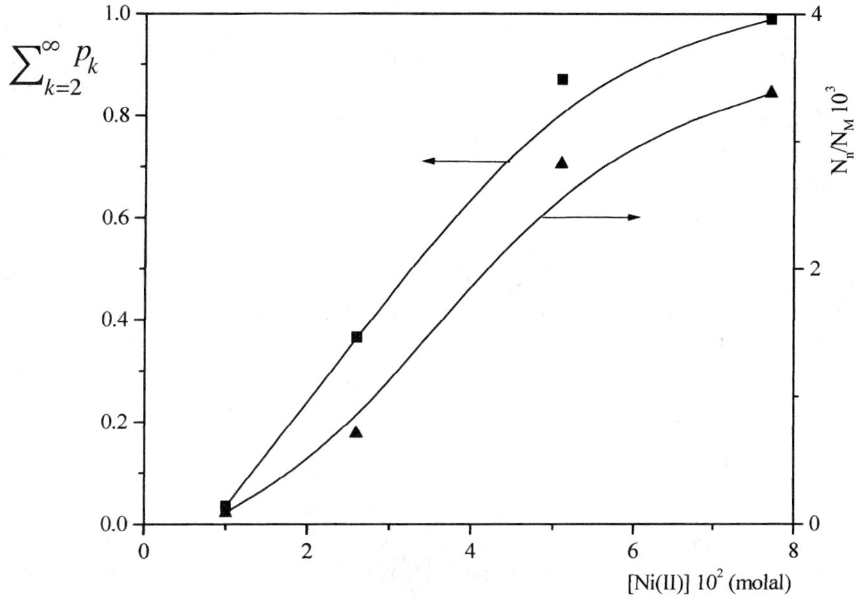

FIG. 15 Variation in the number of nuclei formed per aggregate and the probability of having two or more ions per aggregate as a function of Ni(II) concentration in the microemulsion CTAB 18%/hexanol 70%/water 12%.

ions at the interface. The Co(II) ions contain, on average, one hexanol molecule in their first coordination shell, while the Ni(II) ions are multiply-coordinated with hexanol at the interface [22]. Hence, the mobility of the latter is lower, and the probability of collision between the two reduced Ni atoms required to form a nucleus is also lower. In other words, the rate of nucleation is higher for cobalt boride than for nickel boride particles.

As it has been previously shown that the formation of Ni_2B and Co_2B follows the LaMer diagram, it is the nucleation that plays a predominant role in the deter-

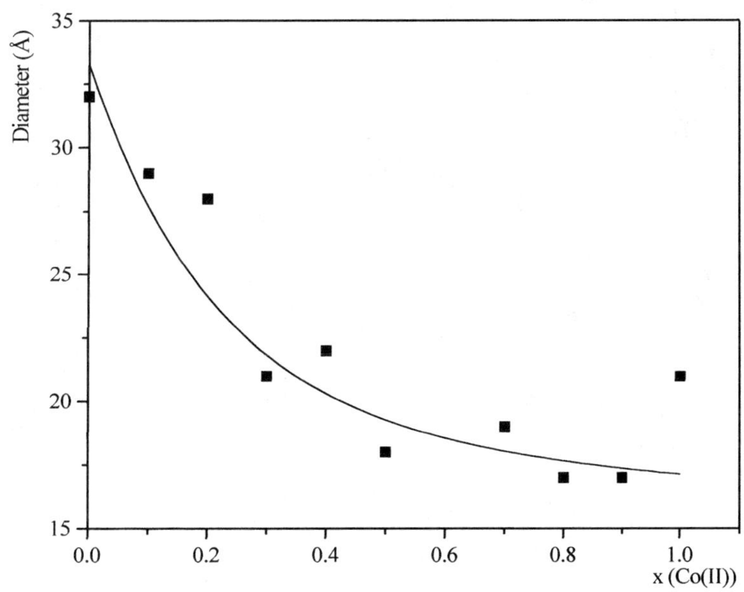

FIG. 16 Variation of the size of $(Co,Ni)_2B$ as a function of the molar fraction in Co(II). The composition of the microemulsion system is CTAB 18.0%/hexanol 70.0%/water 12.0%. The total salt concentration is 5.0×10^{-2} molal.

mination of the particle size. As nucleation is easier for the Co_2B particles, the size of the mixed particles is determined more by the cobalt ions than by the nickel ions. This is shown in Fig. 16, where, in fact, the size of the particles did not vary linearly with the molar fraction in cobalt. The particle size approaches the size of the Co_2B particles more rapidly than linearly. As nucleation is easier for the Co_2B particles, the number of nuclei (and consequently the size of the particles) is determined by the cobalt concentration. The nickel ions are used for the growth of the particles. This leads to the formation of inhomogeneous particles (more nickel ions are situated at the surface of the particles).

2. Size of Platinum and Rhenium Dioxide Particles

(*a*) *Synthesis of Platinum Particles.* Platinum particles have been synthesized in different microemulsion systems. The monodisperse Pt particles prepared from H_2PtCl_6 dissolved in the CTAB/hexanol/water microemulsion had an average diameter of 4.0 ± 0.5 nm, and their size was not dependent on the H_2PtCl_6 concentration (5×10^{-3}–2×10^{-2} M with respect to water) [36]. An aqueous solution of hydrazine containing a 10-fold molar excess of hydrazine with respect to H_2PtCl_6 had an initial pH of 10. It should be noted that the metal particle precursor is soluble in both the dispersed inner water core and the continuous (or hexanol) phases. The size of the particles is thus determined by the thermodynamic stabilization of the particles as no variation of the size of the particles was observed.

It is interesting to note that particles of a similar size were obtained, independently of water and H_2PtCl_6 concentration, from the AOT/heptane/water microemulsions [37]. Colloidal Pt particles were prepared following schemes I and II of Fig. 13 in PEGDE/hexane/water microemulsions. A nonionic surfactant, PEGDE, was used to form a microemulsion of composition PEGDE 9.5%/hexane 90%/water 0.5%. Only K_2PtCl_4 was tested as precursor salt, however, because it is insoluble in the organic medium.

Figure 17 shows the variation in the size of the Pt particles obtained following scheme I as a function of initial K_2PtCl_4 concentration. The standard deviation was small in all cases studied. The particle diameter increases monotonically with increasing K_2PtCl_4 concentration and approaches a plateau at high concentration. This variation shows that the synthesis of the Pt particles follows the LaMer diagram in this microemulsion system.

If the particles are prepared following scheme II, where the two microemulsions containing the precursor K_2PtCl_4 and the reducing agent N_2H_4, respectively, are mixed together, smaller sizes are obtained. Indeed, the

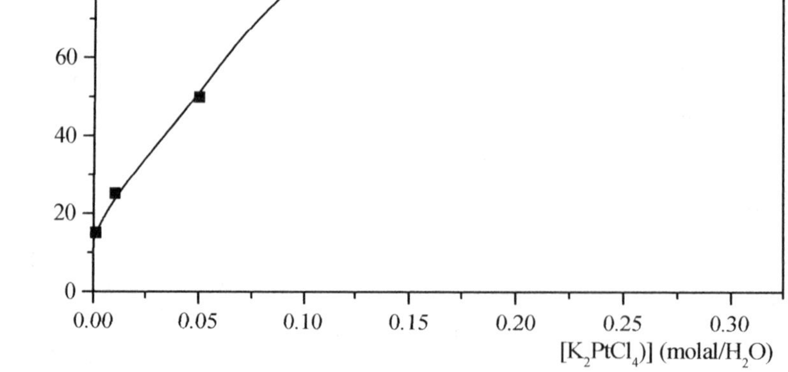

FIG. 17 Variation of the Pt average diameter as a function of K_2PtCl_4 concentration with respect to water prepared according to scheme I of Fig. 13.

Pt particles prepared from the microemulsion with [K_2PtCl_4] of 0.1 M with respect to water have a diameter of 3.5 ± 0.5 nm, whereas the diameter is much greater (9.0 ± 1.0 nm) if scheme I is used. Figure 17 illustrates the variation of the average diameter of the Pt particles as a function of the concentration of K_2PtCl_4 prepared by scheme I.

The average size of the Pt particles obtained by the method of scheme I can be explained in a first approximation by the diffusion of the aqueous solution through the organic phase being slower than the exchange between the water cores. Although in the PEGDE/hexane/water microemulsion no separate spherical droplets are present, the water is probably the dispersed phase in the microemulsion. The structure of the microemulsion is better represented as a lamellar aggregate where the surfactant molecules are associated head to head along a cylinder.

(*b*) *ReO_2 Particles.* Monodisperse ReO_2 particles were obtained by reducing $NaReO_4$ with hydrazine in the PEGDE 9.5%/hexane 90%/water 0.5% microemulsion system following scheme I of Fig. 13. The presence of ReO_2 was confirmed by XPS experiments. However, the $NaReO_4$ is only partially reduced under these conditions.

Figure 18 illustrates the variation in particle size as a function of $NaReO_4$ concentration. Once again the size of monodisperse particles approaches a plateau for high $NaReO_4$ concentrations, and this behavior is quite similar to that of the Pt particles. This indicates that the synthesis of the ReO_2 particles follows the LaMer diagram.

(*c*) *Pt-ReO_2 Particles.* Monodisperse Pt-ReO_2 particles were prepared following scheme I from the PEGDE/hexane/water microemulsion using a total ion concentration [K_2PtCl_4] + [$NaReO_4$] = 0.1 molal with respect to water. Figure 19 shows the variation in the particle size as a function of the mole fraction x of K_2PtCl_4.

It is surprising that up to $x = 0.7$ the diameter of the particles remains quasi-constant and is close to that of the pure ReO_2 particles. For higher initial [K_2PtCl_4], the diameter of the particles increases monotonically to reach that of the pure Pt particles. The electrochemical potential of $PtCl_4^{2-}$ and ReO_4^- are, respectively 0.73 and 0.51 V. Furthermore, two electrons are needed for the reduction of K_2PtCl_4 and three electrons for the reduction for $NaReO_4$. The nucleation should thus be easier for the platinum particles than for the ReO_2 particles. It can thus be concluded that the ReO_2 is dispersed on the Pt particles. As the particles size is constant for low values of the K_2PtCl_4 molar fraction, it can be concluded that the size of the particles is not determined by the nucleation, as for the $(Ni,Co)_2B$ particles, but by the interaction between the ReO_2 and the surfactant. This shows the importance of the thermodynamic sta-

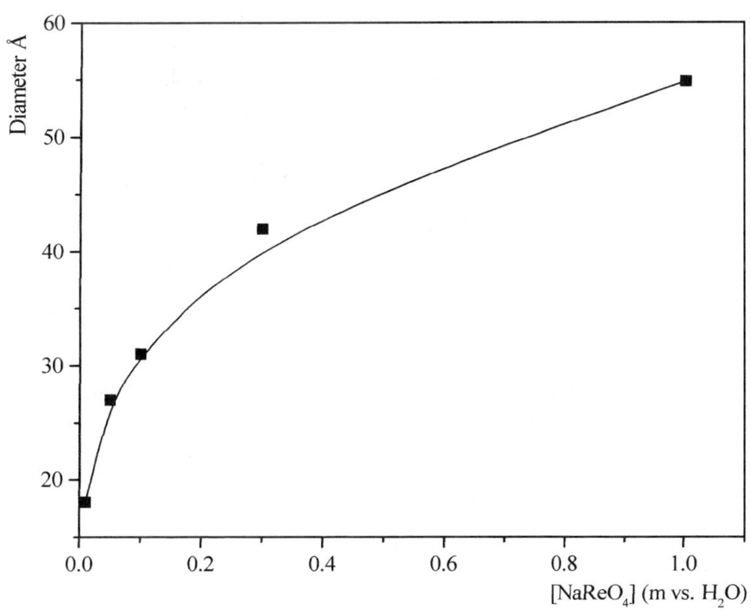

FIG. 18 Variation of the ReO_2 average diameter as a function of $NaReO_4$ concentration with respect to water prepared according to scheme I of Fig. 13.

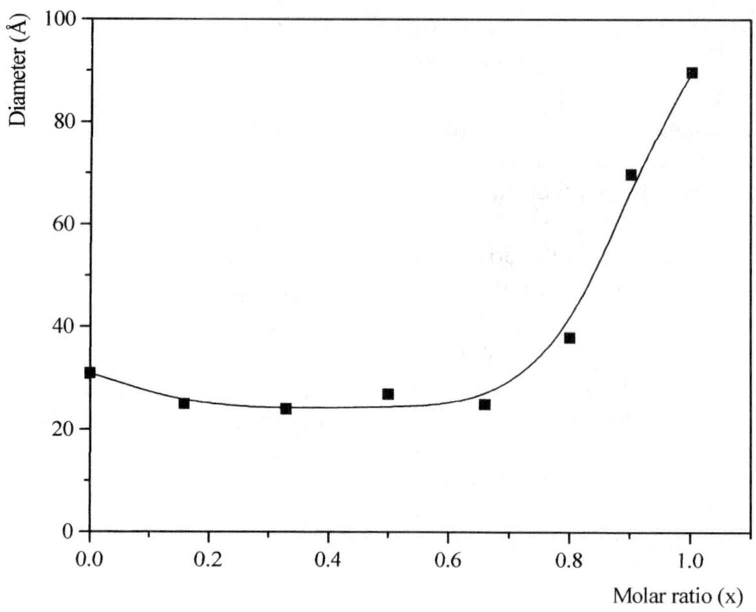

FIG. 19 Variation in Pt-ReO_2 particle size as a function of ratio x of K_2PtCl_4 ($[K_2PtCl_4] + [NaReO_4] = 0.10$ molal with respect to water).

bilization in a case where the particle size seems to be determined by the LaMer diagram. A similar variation has been observed in the case of Ag(Cl,Br) particles.

All these results are different from those one might expect on the basis of a mechanical mixture. Indeed, in that case a bimodal distribution would be expected, at least for $x \geq 0.5$, based on the different sizes of the separate Pt and ReO_2 particles.

III. SYNTHESIS OF ORGANIC PARTICLES

A. General Considerations

Several types of organic nanoparticles have been synthesized recently in some microemulsions. The active compounds are cholesterol, rhodiarome, and rhovanil (aroms) (Fig. 20). The microemulsions used are AOT/heptane/water, Triton/decanol/water, and CTAB/hexanol/water.

The general preparation of these organic nanoparticles has been described [38,39]. It consists of direct precipitation of the active compound in the aqueous cores of the microemulsion. After their preparation, nanoparticles are stained with iodine vapor and observed by transmission electron microscopy (Philips EM301) [40,41]. A transmission electron micrograph of rhodiarome nanoparticles is presented in Fig. 21.

A hypothesis for such nanoparticle formation has been proposed [38,39,41]. This hypothesis consists of several stages. A solution of active compound in an appropriate solvent is added to the microemulsion. The active compound goes to the aqueous cores (by diffusion) and partitions inside by crossing the interfacial film. The solvent plays a role in this transport to the aqueous cores. The active compound precipitates in the aqueous cores because of its insolubility in water, and nuclei are thus formed. These nuclei can grow because of the exchange of active compound between the aqueous cores. At the end, nanoparticles are stabilized by the surfactants.

B. Nanoparticles of Cholesterol Prepared in Different Microemulsions

Figure 22 represents the evolution of nanoparticle size as a function of R at a fixed concentration of cholesterol solution in chloroform. The cholesterol is precipitated in an AOT/heptane/water microemulsion. It has to be noted that the total amount of cholesterol added increased with increasing R, as the volume of chloroform solution was equal to that of the water in each microemulsion. The mean particle size was in the range 30–60 Å and a minimum was observed for a certain R value. Although it is tempting to put forward a hypothesis for this local minimum, the overall variability does not justify a detailed discussion at this time. In these precipitations the amount of chloroform increases with R and the relative amount of chloroform in the

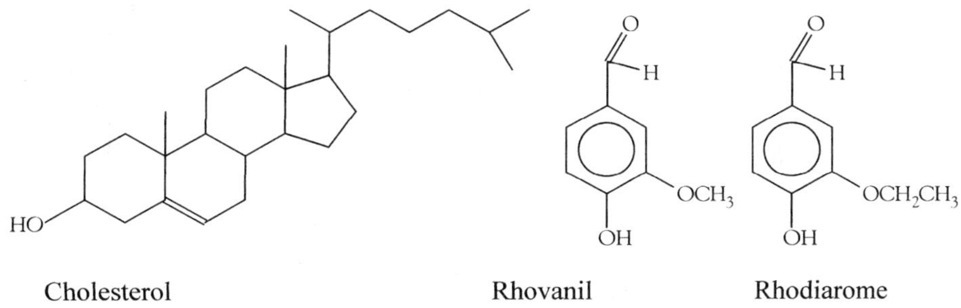

FIG. 20 Structures of the active compounds.

solvation sphere could depend on the size of the particles. In order to check the veracity of this hypothesis, another series of precipitations was done in which the same amount of cholesterol solution in chloroform (0.3 mL) was added to the various microemulsions with different *R* values (Fig. 23).

In this case, the particle size is constant as a function of *R*. The size of the particles is hence controlled by the thermodynamic stabilization. But in this experiment, the amount of cholesterol was also constant. If the concentration of chloroform remains constant, the corresponding quantity of cholesterol may be sufficient to allow the thermodynamic stabilization of the particle.

Figure 24 shows the variation of nanoparticle size as a function of the concentration of cholesterol in the same microemulsion system. No local minimum appears. The size of the particles is thus controlled by thermodynamic stabilization by the surfactant. A certain size is favored. In fact, in this experiment the amount of water stays constant, and it is perhaps the difference in the number of water molecules per water core that produces a difference in the nucleation.

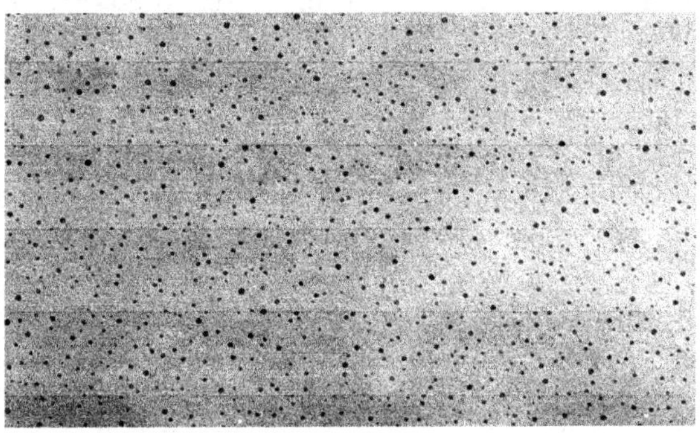

FIG. 21 Photograph of rhodiarome nanoparticles synthesized with a solution of rhodiarome in acetone (50 g/L) in AOT/heptane/water microemulsion (scale 96,000×).

Nanoparticles of cholesterol have also been synthesized in two other microemulsion systems: Triton/decanol/water and CTAB/hexanol/water. Similar experiments have been carried out. In the two cases, the nanoparticle size is independent of the factor *R* and also of the concentration of the cholesterol solution. The particles are thus thermodynamically stabilized by the surfactants at a certain favored size. The nanoparticles are stable for months, no precipitate appears, and the final solutions are still limpid.

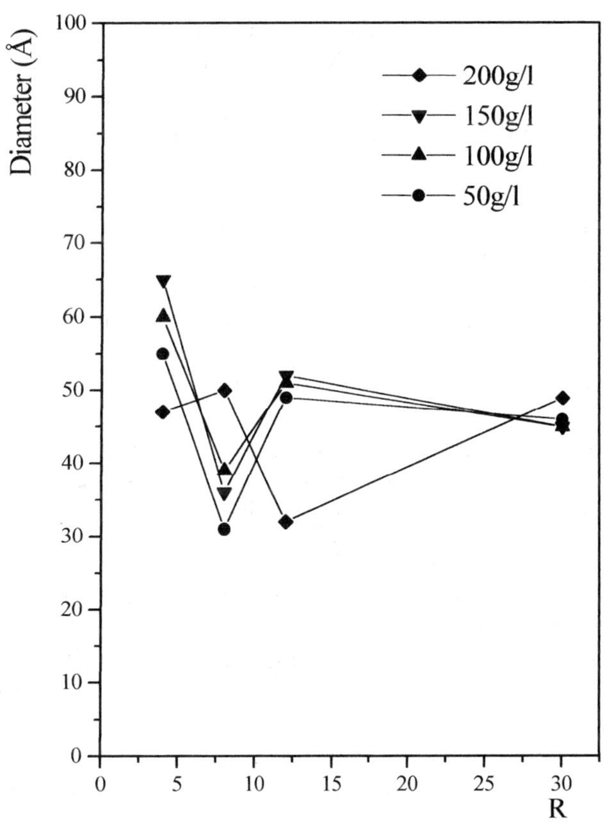

FIG. 22 Variation of the nanoparticle size of cholesterol as a function of *R* at a fixed concentration.

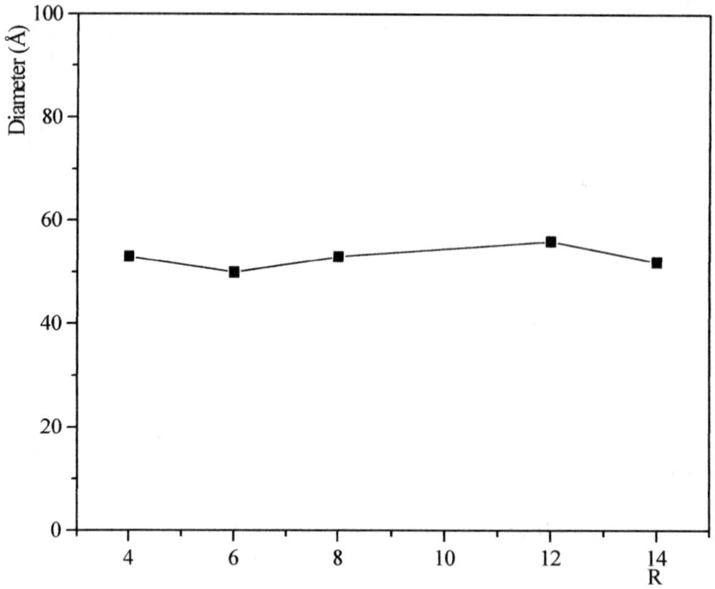

FIG. 23 Variation of the nanoparticle size of cholesterol as a function of *R* at a fixed concentration (50 g/L).

C. Rhodiarome (or Rhovanil) in AOT/Heptane/Water Microemulsion

1. Influence of the Factor *R* and of the Concentration of Active Compound

An example is presented for the formation of nanoparticles of rhovanil. A solution of rhovanil in acetone (50 g/L) is used. Figure 25 shows the variation of the mean diameter as a function of *R*. The nanoparticle size is relatively constant as a function of *R* and ranges between 45 and 62 Å for the four concentrations. It is the same in the other cases. The nanoparticle size is independent of the factor *R*. The second parameter studied is the concentration of the active compound in the solvent. Figure 26 shows a certain constancy of the size, which ranges between 45 and 70 Å. In these two cases, it appears the nanoparticle size is essentially determined by thermodynamic stabilization by the surfactant molecules at a certain size.

2. Influence of Stirring

Figure 27 shows that the diameter of the particles synthesized using a magnetic stirrer for mixing the solutions is higher than the diameter of those synthesized in the presence of ultrasonification (especially for smaller *R*).

The following explanation is proposed. The available energy that is required to favor the mixing of the active compound in acetone solution in the microemulsion is more important for the ultrasound bath. The active molecule is dispersed better in the microemulsion in the case of ultrasound. A greater number of nuclei are formed in contact with the aqueous cores, and the size of the nanoparticles is smaller than in the case of magnetic stirring. This difference between the two methods indicates a contribution of the LaMer diagram. In fact, this indicates the importance of the nucleation and that the growth is easier than the nucleation of the particles.

3. Solubility Limitations

A microemulsion formulation can accommodate only a limited quantity of active compound (solution) to form nanoparticles without macroscopic phase separation. In Table 3, for a given factor *R*, a certain volume of rhodiarome solution, with a constant concentration (400 g/L), is tolerated. As the number of active compound molecules per aqueous core increases, the corresponding interaction with surfactant molecules increases at the interface. The optimal radius of curvature is perturbed and a phase separation appears (emulsion failure). Further, as the factor *R* increases, the tolerable quantity of rhodiarome in acetone decreases because the microemulsion with more water is a poorer solvent for the added rhodiarome-acetone solution.

4. Effect of Principal Compound Solution Volume

Figure 28 shows the variation of particle size as a function of *R* for two volumes (5 and 50 mL) of rhodiarome (50 g/L) in acetone. No significant difference in the diameter of monodisperse nanoparticles is observed.

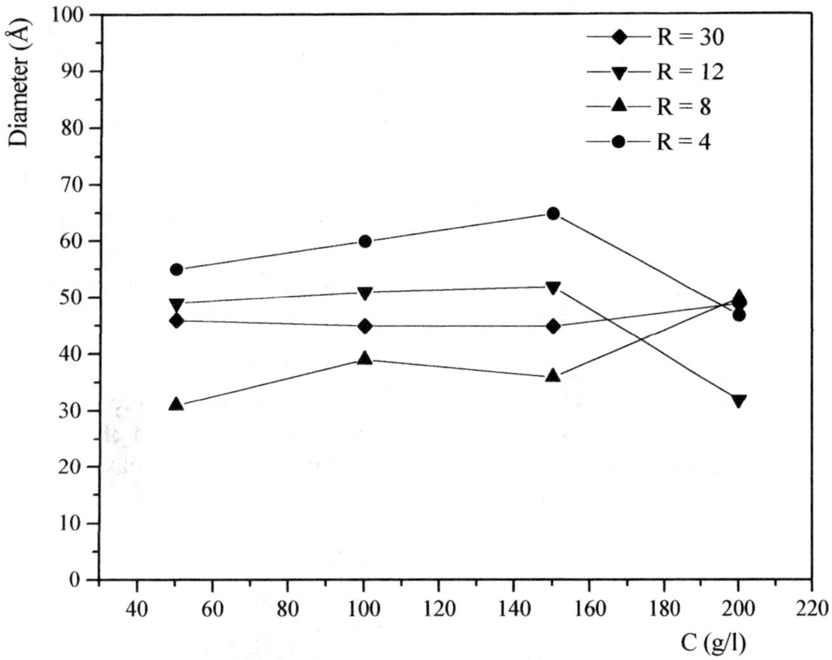

FIG. 24 Variation of the nanoparticle size of cholesterol as a function of the concentration.

Thermodynamic stabilization of the particles is suggested with a small LaMer contribution. Indeed, in a small volume, the active compound may not be uniformly dispersed in the microemulsion prior to nucleation and precipitation. Fewer nuclei are formed at the beginning of the reaction, but the greater size obtained is not very significant.

5. Influence of Auxiliary Solvents

When the active compound is added as a solid phase into the microemulsion, no particles are observed because of insufficient dispersion, slow dissolution, and lack of a thermodynamic driving force for dissolution. Solvents such as acetone probably play the role of a vector because the active compound must be carried to

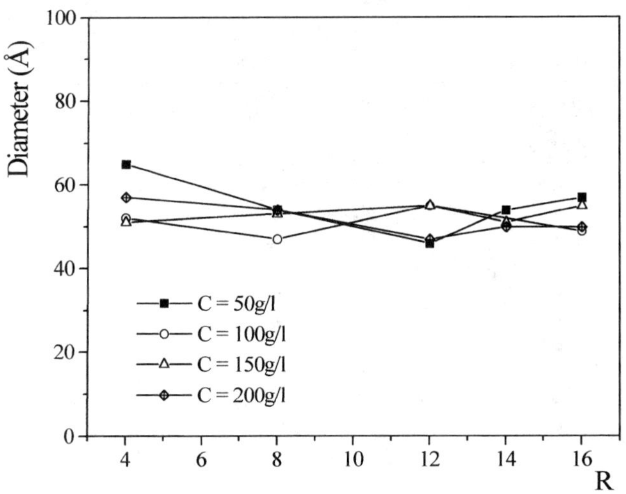

FIG. 25 Variation of the nanoparticle size of rhovanil as a function of *R*.

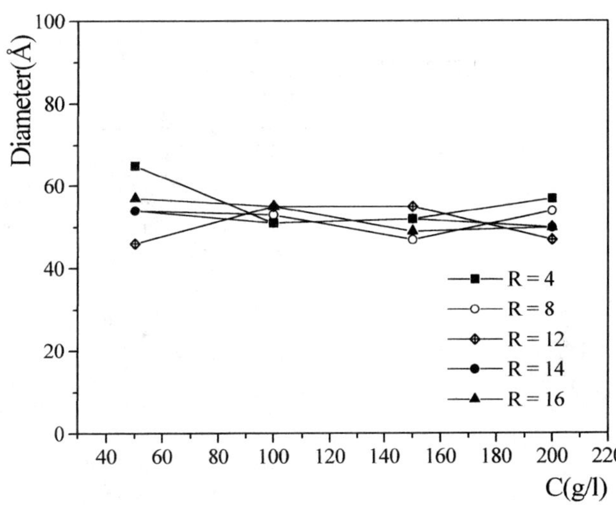

FIG. 26 Variation of the nanoparticle size as a function of the concentration of rhovanil in acetone (50 g/L).

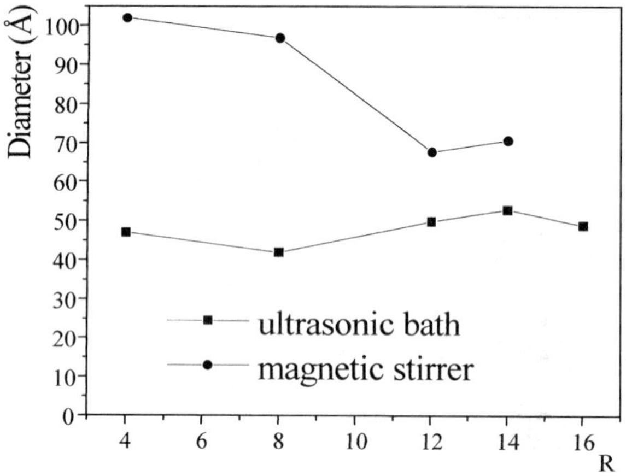

FIG. 27 Variation of the nanoparticle size of rhovanil in the case of the use of a magnetic stirrer or ultrasound bath.

the aqueous cores in order to induce nanoparticulate precipitation.

Two microemulsions are used in a precipitation. The first contains the active compound in solid form and the second contains the auxiliary solvent (acetone). The final solution (obtained by mixing the two solutions) is stable after treatment for 15 min with ultrasound. In this experiment, the volume is 5 mL of a solution of rhodiarome in acetone (50 g/L) in a vessel of 25 mL and the results are compared with those in the case of the use of one microemulsion (Fig. 29).

The diameters obtained are between 45 and 65 Å, and no obvious minimum or maximum is observed. Thermodynamic stabilization of the nanoparticles appears to control in this system. Several solvents such as acetone, ethanol, or ether are used in order to dissolve the active compound. There is no apparent influence of these solvents on the nanoparticle size. These auxiliary solvents can thus be vectors for the transport of the active compound toward the aqueous cores. Avoidance of such auxiliary solvents leads to poor dispersion and aggregates are observed. These auxiliary solvents do not appear to play a significant role in the precipitation, other than their role in uniformly distributing the active compound throughout the microemulsion.

6. Effects of Time

Figure 30 shows no preponderant changes of nanoparticle size as a function of time. The size ranges between 50 and 80 Å for the case of a rhodiarome (in acetone) precipitation. The final solutions are still stable and transparent after a long period of time. No sedimenting agglomerates appeared after more than a year, so these nanoparticles are very effectively thermodynamically stabilized.

A solution of rhodiarome in acetone (100 g/L) added drop by drop into a microemulsion yielded the results shown in Fig. 31. Different samples were taken at different times after injection.

The size did not significantly change as a function of time. The ultrasound treatment can thus be reduced. The particles are first formed by direct precipitation in the aqueous cores and are simultaneously thermodynamically stabilized by the surfactants.

7. Recovery of Nanoparticles

Some potential pharmaceutical applications can be considered if less toxic solvents are used. Thus the residual solvents (heptane, for example) are evaporated under

TABLE 3 The Amount of Active Compound Tolerated in Microemulsion

R	0.25 mL (0.1 g)	0.3 mL (0.12)	0.35 mL (0.14 g)	0.4 mL (0.16 g)	0.45 mL (0.18 g)	0.5 mL (0.20 g)	0.6 mL (0.24 g)	0.7 mL (0.28 g)
48	●	●	●	●	●	●	●	●
46	◇	●	●	●	●	●	●	●
44	◇	●	●	●	●	●	●	●
42	◇	●	●	●	●	●	●	●
40	◇	●	●	●	●	●	●	●
38	◇	●	●	●	●	●	●	●
36	◇	●	●	●	●	●	●	●
34	◇	◇	◇	●	●	●	●	●
32	◇	◇	◇	◇	●	●	●	●
30	◇	◇	◇	◇	●	●	●	●
28	◇	◇	◇	◇	◇	◇	◇	●
26	◇	◇	◇	◇	◇	◇	◇	◇

◇, Limpid and stable solutions.
●, Solutions with two phases.

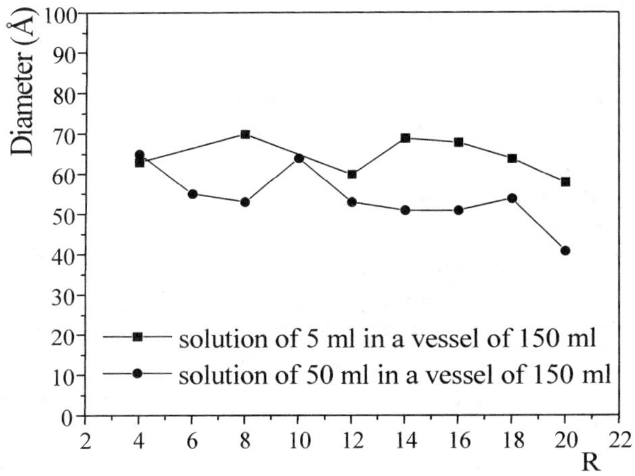

FIG. 28 Variation of rhodiarome nanoparticle size as a function of *R*: influence of the volume of microemulsions.

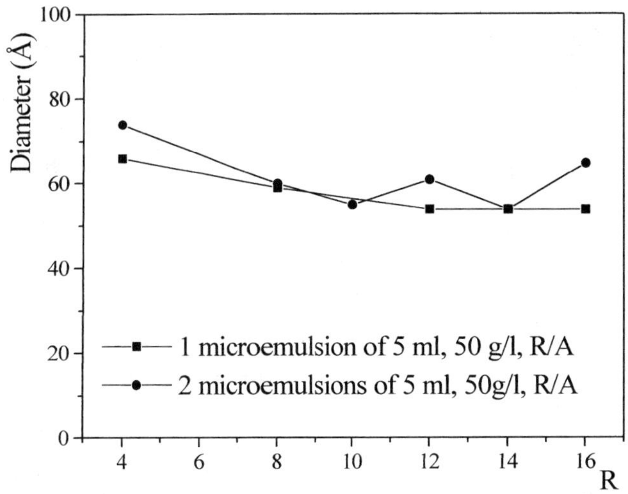

FIG. 29 Variation of rhodiarome nanoparticle size as a function of *R*: particles synthesized in one or two microemulsions.

vacuum and the nanoparticles stabilized by surfactants are recovered. These particles are reintroduced in distilled water under ultrasound in order to obtain a limpid and stable solution. Figure 32 shows the variation of the nanoparticle size as a function of *R*.

The sizes range between 57 and 63 Å and do not change after particle recovery. The nanoparticles are thus thermodynamically stabilized by the surfactants. The change of the medium does not influence the nanoparticle size. More biocompatible microemulsions have also been used in order to allow their use in drug delivery [39,41].

IV. CONCLUSIONS

This chapter has emphasized the mechanism of formation of particles in microemulsions. Two models have been proposed: the LaMer diagram and the thermodynamic particle stabilization model. These two models are relatively simple. The LaMer diagram model is based on the separation between the nucleation and the growth stages. It is consistent with the mechanism proposed by López-Quintela and Rivas [42] for Fe nanoparticles obtained in AOT microemul-

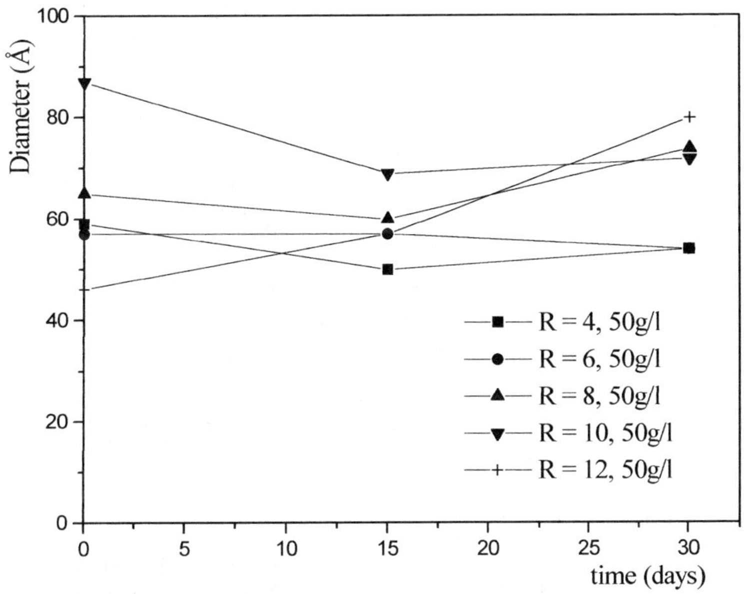

FIG. 30 Evolution of particle size of rhodiarome nanoparticles as a function of time.

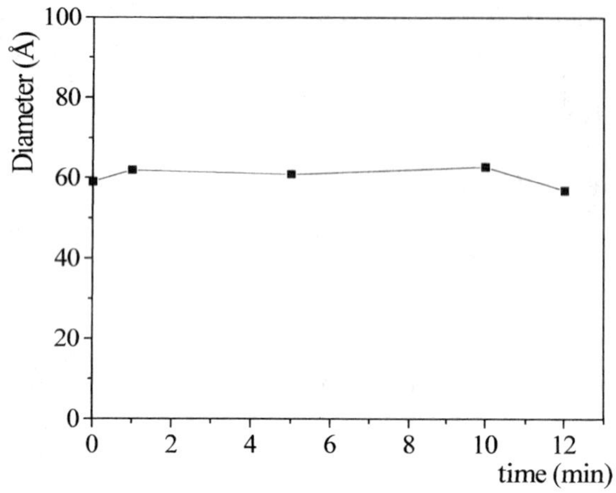

FIG. 31 Variation of particle size of rhodiarome nanoparticles as a function of time after the injection of the active compound in microemulsion.

sions using a stopped-flow technique. Nucleation implies an increase in the number of scattering centers (number of particles) for a given observation window, and, therefore, it gives an increase in the scattered intensity. On the contrary, the growth of particles is associated with a decrease of the scattered intensity because the observation window corresponds to the diffraction of smaller particles that are disappearing during the growth process. The presence of this maximum (although not well defined) has also been spectrophotometrically detected by Towey et al. [43] for the formation of CdS in AOT microemulsions. This is an illustration of the LaMer diagram, as following this diagram the nucleation occurs only at the beginning of the reaction. Theoretical calculations have been done

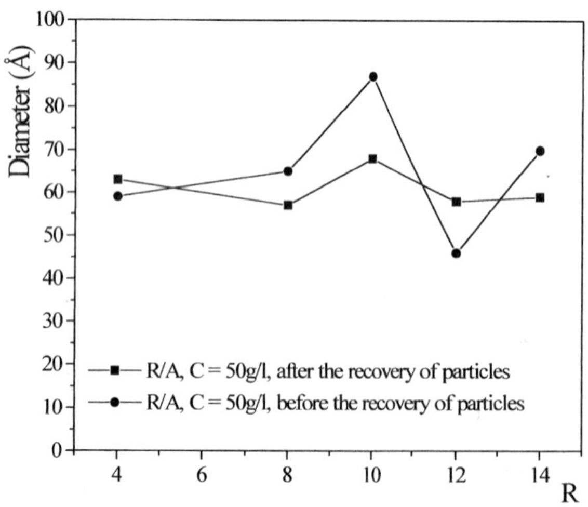

FIG. 32 Variation of the nanoparticle size of rhodiarome as a function of *R* before and after nanoparticle recovery.

by Tojo et al. [44] involving the study of the influence of the concentration and of the film flexibility and of the kinetic exchange constant between the droplets using the difference between the nucleation and the growth of the particles. Thermodynamic stabilization is less documented in the literature, but an example shows the formation of secondary monodisperse spherical particles by coagulation of the primary particles [45].

Whether the precipitation reaction follows the LaMer diagram or thermodynamic stabilization depends on the microemulsion phase diagram and on the nature of the particles synthesized. As an example, the synthesis of AgBr particles follows the LaMer diagram in the AOT/heptane/water microemulsion system, but it follows the thermodynamic stabilization model in the AOT/*p*-xylene/water system. The difference between the two systems can come from the adsorption of the *p*-xylene molecule on the particles of AgBr. In fact, the adsorption of *p*-xylene on the AgBr particles has been shown in the study of the adsorption of pseudoisocyanine on these particles [46]. In CTAB/hexanol/water microemulsions, the formation of Ni_2B particles follows the LaMer diagram model, but that of Pt follow the thermodynamic stabilization of the particles. The difference could stem from a difference in the adsorption of the surfactant on the particles. Mixed particles have also been synthesized. Not all the particles are homogeneous and the size of these particles does not vary linearly with their composition.

In the case of organic nanoparticles, all of the particle preparations seem to follow thermodynamic stabilization. However, the cholesterol synthesized in the AOT/heptane microemulsion may be an exception to this generalization. This may be due to a specific interaction of the surfactant with the particles.

ACKNOWLEDGMENT

L. J. thanks F.R.I.A. for financial help.

REFERENCES

1. J. H. Fendler and F. C. Meldrum, Adv. Mater. *7*:607 (1995).
2. G. A. Ozin, A. Kuperman, and A. Stein, Angew. Chem. Int. Ed. Engl. *28*:359 (1989).
3. J. Belloni M. Mostafavi, J.-L. Marignier, and J. Amblrad, J. Imaging Sci. *35*:68 (1991).
4. A. Henglein, J. Phys. Chem. *97*:5457 (1993).
5. R. P. Andres, R. S. Averback, W. L. Brown, L. E. Brus, W. A. Goddard III, A. Kaldor, G. Louie, M. Moscovits, P. S. Peercy, S. J. Riley, R. W. Siegel, F. Spaepen, and Y. Wang, J. Mater. Res. *4*:704 (1989).

6. R. W. Siegel, MRS Bull. *15*:60 (1990).
7. Y. T. Han, MRS bull. *14*:13 (1989).
8. M. J. Lawrence, Eur. J. Drug Metab. and Pharmacokinet. *3*:257 (1994).
9. B. Sjöström, B. Bergenstahl, and B. Kronberg, J. Pharm. Sci. *82*:584 (1993).
10. E. Allémann, R. Gurny, and E. Doelker, Eur. J. Pharm. Biopharm. *39*:173 (1993).
11. J. Rouvière, J.-M. Couret, M. Lindheimer, J.-L. Dejardin, and R. Marrony, J. Chem Phys. *76*:289 (1979).
12. C. Cabos and P. Delord, J. Appl. Crystallogr. *12*:502 (1979).
13. S. I. Ahmad and S. Friberg, J. Am. Chem. Soc. *94*:5196 (1972).
14. S. Friberg and I. Lapczynska, Prog. Colloid Polym. Sci. *56*:16 (1975).
15. P. Ekwall, L. Mandell, and K. Fontell, J. Colloid Interface Sci. *33*:215 (1970).
16. P. Ekwall, L. Mandell, and K. Fontell, Mol. Cryst. Liq. Cryst. *8*:157 (1969).
17. P. D. I. Fletcher, A. M. Howe, and B. H. Robinson, J. Chem. Soc. Faraday Trans. 1 *83*:985 (1987).
18. S. S. Atik and J. K. Thomas, Chem. Phys. Lett. *79*:351 (1981).
19. V. K. LaMer and R. H. Dinegarn, J. Am. Chem. Soc. *72*:4847 (1950).
20. T. Sugimoto, Adv. Colloid Interface Sci. *28*:65 (1987).
21. Ph. Monnoyer, A. Fonseca, and J. B.Nagy, Colloids Surf. *100*:233 (1995).
22. J. B.Nagy, E. G. Derouane, A. Gourgue, N. Lufimpadio, I. Ravet, and J. P. Verfaillie, in *Surfactants in Solution* (K. L. Mittal, ed.), Vol. 10, Plenum, New York, 1989, p. 1.
23. J. B.Nagy and A. Claerbout, in *Surfactants in Solution* (K. L. Mittal and D. O. Shah, eds.), Vol. 11, Plenum, New York, 1991, p. 363.
24. A. Claerbout and J. B.Nagy, Stud. Surf. Sci. Catal. *63*: 705 (1991).
25. J. Eastoe, B. H. Robinson, A. J. W. G. Visser, and D. C. Steytler, J. Chem. Soc. Faraday Trans. *87*:1899 (1991).
26. *Handbook of Chemistry and Physics*, 67th ed., CRC Press, Boca Raton, FL, 1986.
27. L. Jeunieau, W. Verbouwe, E. Rousseau, M. Van der Auweraer, and J. B.Nagy, Langmuir, *16*:1602 (2000).
28. T. K. Jain, M. Varshney, and A. Maitra, J. Phys. Chem. *93*:7409 (1989).
29. H. Hauser, G. Haering, A. Pande, and P. L. Luisi, J. Phys. Chem. *93*:7869 (1989).
30. M. Wong, J. K. Thomas, and T. Nowak, J. Am. Chem. Soc. *99*:4730 (1977).
31. P.-O. Quist and B. Halle, J. Chem. Soc. Faraday Trans. I *84*:1033 (1988).
32. M. A. J. Rodgers and M. Da Silva, Chem. Phys. Lett. *78*:256 (1974).
33. L. Jeunieau and J. B.Nagy, Colloids Surf. *151*:419 (1999).
34. I. Ravet, Ph.D. thesis, FUNDP, 1988.
35. J. B.Nagy, A. Gourgue, and E. G. Derouane, Stud. Surf. Sci. Catal. *16*:193 (1983).
36. A. Wathelet, Bachelor's thesis, FUNDP, Belgium, 1984.
37. A. Khan-Lodhi, B. H. Robinson, T. Towey, C. Herrmann, W. Knoche, and U. Thesing, in *The Structure, Dynamics and Equilibrium Properties of Colloidal Systems*, NATO ASI Ser C 324 (D. M. Bloor and E. Wyn-Jones, eds.), Kluwer, Dordrecht, 1990, p. 373.
38. F. Debuigne, L. Jeunieau, M. Wiame, and J. B.Nagy, Langmuir, in press.
39. F. Debuigne, Bachelor's thesis, DEA, FUNDP, 1999.
40. *Electron Microscopy and Photography*, A. Kodak databook.
41. F. Debuigne, Bachelor's thesis, FUNDP, 1997.
42. M. A. López-Quintela and J. Rivas, J. Colloid Interface Sci. *158*:466 (1993).
43. T. F. Towey, A. Khan-Lodhi, and B. H. Robinson, J. Chem. Soc. Faraday Trans. *86*:3757 (1990).
44. C. Tojo, M. C. Blanco, F. Ricadulla, and M. A. López-Quintela, Langmuir *13*:1970 (1997).
45. L. Lerot, F. Lefrand, and P. De Bruycker, J. Mater. Sci. *26*:2353 (1991).
46. L. Jeunieau and J. B.Nagy, Appl. Organomet. Chem. *12*:341 (1998).

31

Colloidal Nanoparticles and Nanoparticulate Films Grown at the Air-Water Interface

JANOS H. FENDLER Clarkson University, Potsdam, New York

I. INTRODUCTION

The preparation and characterization of size-quantized nanoparticles are receiving ever increasing attention by material scientists, physicists, chemists, and biologists. Mechanical manipulation is the predominant strategy physicists have employed in their preparations of nanoparticles and nanostructured materials. At the low end of the scale, this involves the exhaustive grinding or milling of bulk materials [1]. Examples at the high technological end include single-atom transfer from one site to another, relocation of small molecular clusters from surfaces, and etching or deposition of materials in subnanometer regions [2]. Generation of atomic and molecular clusters via gas condensation in an ultrahigh vacuum [3] falls between these two extremes. In general, any nanoparticle can be fabricated by physical methods and band gap engineering permits the construction of semiconductor superlattices with any desired nanoscale architecture. The high cost involved in these methods does not, however, conveniently lend itself to the large-scale production of advanced nanoparticles and nanostructured materials.

Chemists, by vocation and definition, are makers of molecules. Rather than "breaking down" materials, they "build them up" from their elements, often by innovative routes. Increasingly, chemists are turning their attention to the synthesis of molecular clusters and to the formation and stabilization of colloidal nanoparticles. Versatility and the relative ease of scale-up are the advantages of the chemical approach to nanoparticle preparations. Polydispersity, the presence of impurities, and the lack of an established protocol for layering uniform two- and three-dimensional particulate networks are the disadvantages of this approach.

Nature routinely performs elegant and efficient nanoparticle preparations in a more advanced manner than either physicists or chemists. Not only are the appropriate and monodisperse nanoparticles synthesized, but they are processed into higher level organizations [4]. Accomplishments of the humble magnetotactic bacterium serve to illustrate the point. The bacterium is capable of producing 20 to 25, 45 $\pm$ 8 nm diameter, spherical, single-domain Fe_3O_4 (magnetite) particles in the cytoplasmic membrane that are nicely aligned along its body [5]. The bacterium uses these monodisperse magnetites, in connection with the magnetic field of the earth, for navigation toward warmer waters.

The mimicking of biology in general, that of biomineralization and the functioning of the biological membrane in particular, has led to the development and burgeoning of bio-organic (and bioinorganic) chemistry, of biomimetic materials chemistry [4], and of the membrane mimetic approach to advanced materials preparations [6]. The membrane mimetic approach relies on the construction of templates and/or compartments in which nanoparticles are generated in situ or into which they are incorporated. The templates and compartments are designed to imitate such essential functions of the biological membrane as organization and compartmentalization in distinct microenvironments. Zeolites and related molecular sieves, pillared clays and clay organocomplexes, porous glasses, graphite and metallic tubes, and polymeric membranes have been used as templates [7]. Monolayers, Langmuir-Blodgett (LB) films, self-assembled monolayers and multilayers, micelles, surfactant vesicles (liposomes), bilayer lipd membranes (BLMs), cast multibilayers, and polymers have been used as compartments [6]. The terms templates and compartments are somewhat arbitrary and often used interchangeably. Similarly, membrane mimetic chemistry and biomimetic materials chemistry are closely related. The latter has been used to describe biologically inspired advanced materials synthesis by molecular tectonics. The phrase molecular tectonics (Greek *tekton* = builder) has been coined by chemists to describe the construction of supramolecules that integrated molecular synthesis and self-assembly into larger structures.

The preparation of nanoparticles and the construction of nanostructured materials by membrane mimetic approaches are the long-term research objectives in our laboratories. Advantage has been taken of membrane mimetic systems to provide chemical, spatial, and dimensionality control for the in situ generation and stabilization of ultrasmall metallic, semiconducting, and magnetic particles and particulate films [6].

In this chapter, our work on the membrane mimetic preparation of nanoparticles under monolayers and their subsequent transfer to solid substrates are surveyed.

II. EXPERIMENTAL METHODOLOGIES

Both chemical and electrochemical routes have been developed for the in situ generation of nanoparticles under monolayers [6].

The experimental setup used for the chemical generation and in situ monitoring of nanocrystalline particulate films is illustrated in Fig. 1 [8]. Typically, a surfactant monolayer is spread on an aqueous solution of the metal salt precursor of the nanoparticles and crystallization is induced by injection of the reactant gas into the closed system. Facilities are provided for determining surface pressure versus surface area and surface potential versus surface area isotherms in the film balance placed under the glass cover. Reflectivities, angle-dependent reflectivities, Brewster angle and fluorescence microscopies, and nonlinear optical parameters can also be monitored during the nanoparticle formation under the monolayer.

The experimental setup used for the electrochemical generation of nanocrystalline silver particulate films is illustrated in Fig. 2 [9]. A 1.0-mm-diameter, 3-cm-long silver electrode is immersed in the subphase (aqueous silver nitrate solution) and electrical connection is made through a 20-μm-diameter platinum electrode floated (subsequent to monolayer formation) on the water surface at the middle of the trough. Ten to 20 min after monolayer formation, a potential of 1.8–1.9 V is applied across the electrodes (keeping Pt negative) by means of a DC power supply. With time, silver particles grow concentrically, forming larger and larger circles at the monolayer-water interface. The rate of this two-dimensional growth is typically 1–2 cm^2/h. No silver particles are observed upon the application of the same potential to the water surface in the absence of

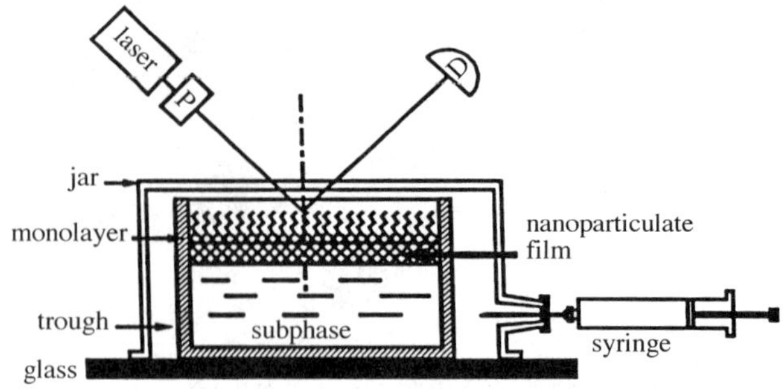

FIG. 1 The experimental arrangement used for the in situ generation of nanoparticulate films from their precursors (one of them is in the subphase under the monolayer, and the other as a gas, infused through the monolayer after its injection by the syringe). The arrangement used for the in situ reflectivity measurements (P = polarizer, D = detector) is also illustrated.

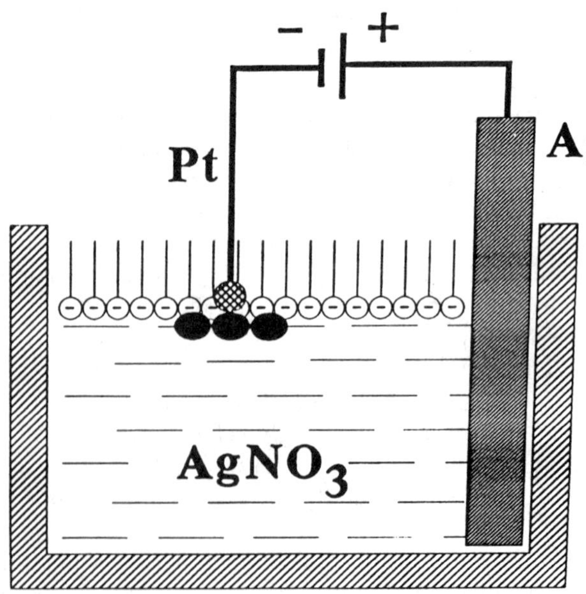

FIG. 2 Setup used for the electrochemical generation of nanocrystalline silver particulate films.

surfactants or to monolayers prepared from positively charged surfactants. Negatively charged monolayers are essential to the electrochemically generation of silver particles; they provide binding sites for silver ions that are reduced at the cathodic surface.

To date, cadmium sulfide, zinc sulfide, lead sulfide, cadmium selenide, and lead selenide semiconductor particulate and silver and gold metallic nanoparticulate films have been chemically grown, in situ, under monolayers in our laboratories [6,10]. The formation of nanoparticulate films under monolayers by similar methodologies has been reported by other research groups [11].

III. MECHANISM OF PARTICLE GROWTH

Evolution of a nanocrystalline particulate film, as illustrated by the formation of sulfide semiconductor particulate films (Fig. 3), has been discussed in terms of the following steps [8]:

1. Formation of metal-sulfide bonds at a large number of sites at the monolayer-aqueous interface

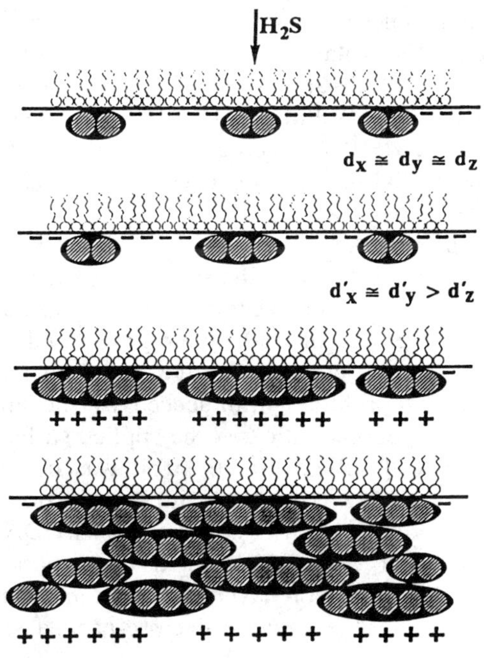

FIG. 3 Proposed schematics of the initial and subsequent growth of a monolayer-supported, porous, size-quantized semiconductor particulate film (SQSPF). The *dx* and *dy* dimensions are in the plane and the *dz* dimension is normal to the plane; they refer to the earliest observable particles. *d'x*, *d'y*, and *d'z* are dimensions in the plane and are normal to the plane; they refer to particles that were observed at later stages of their growth.

2. Downward growth of well-separated nanocrystalline metal sulfide particles
3. Coalescence of clusters into interconnected arrays of semiconductor particles
4. Formation of the "first layer" of a porous sulfide semiconductor particulate film composed of 20- to 40-Å-thick, 30- to 80-Å-diameter particles
5. Diffusion of fresh metal ions to the monolayer headgroup area
6. Formation of a "second layer" of the porous sulfide semiconductor particulate film (by using steps 1, 2, and 3)
7. Buildup of "subsequent layers" of the sulfide semiconductor particulate film (by using steps 1, 2, and 3) up to a plateau thickness (~300 Å for CdS and ~3500 Å for ZnS) beyond which the film cannot grow

The presence of a monolayer with an appropriate surface charge is essential to sulfide semiconductor particulate film formation. In the absence of a monolayer, infusion of H_2S over an aqueous metal ion solution results in the formation of large, irregular, and polydispersed metal sulfide particles that precipitate in the bulk solution before settling to the bottom of the trough.

An important aspect of generating nanoparticles and nanoparticulate films under monolayers is that they can be transferred to substrates at any stage of their growth for ex situ characterizations and then used as devices and sensors.

IV. EPITAXIAL GROWTH OF SEMICONDUCTOR NANOCRYSTALLITES

Oriented growth requires matching the crystal lattice of the surfactants, constituting the monolayers, with that of the incipient nanocrystallites. Such epitaxial matching has been achieved by growing lead sulfide [12], lead selenide [13], and cadmium sulfide [14] under monolayers prepared from arachidic acid and from mixtures of arachidic acid and octadecylamine. This approach is illustrated here by the description of the epitaxial growth of lead sulfide nanocrystallites.

Exposure of an aqueous lead ion solution to hydrogen sulfide in the absence of monolayers results in the formation of large (several millimeters long) irregular cubic crystalline lead sulfide crystals. Conversely, exposing an arachidic acid monolayer–coated aqueous lead nitrate solution to hydrogen sulfide gas in the system illustrated in Fig. 1 results in the formation of well-oriented, relatively monodisperse equilateral triangular lead sulfide nanocrystallites (Fig. 4). The size of the crystals is dependent on the rate of crystal growth. Infusion of H_2S for only 5 min yielded crystals with sides of a mean length of 297 Å, and a reaction time of 30 min produced significantly larger crystals of a mean side length of 607 Å [L]. Selected area electron diffraction of the crystalline films showed "single crystal" patterns, indicative of an epitaxial relationship between the lead sulfide particles and the crystalline monolayer. Reciprocal lattice spots corresponding to {220}, {422}, {440}, etc. forms of planes of the cubic lead sulfide structure were identified and demonstrated that all of the crystals nucleated and grew from {111} basal planes. Lead sulfide crystals were also grown under arachidic acid monolayers that were maintained at lower surface pressures. Even gaseous state monolayers provided a substrate for the epitaxial growth of lead sulfide. Circular domains of epitaxially oriented lead sulfide particles were located, presumably having grown from crystalline domains of arachidic acid that were surrounded by disordered molecules in the gas phase [12].

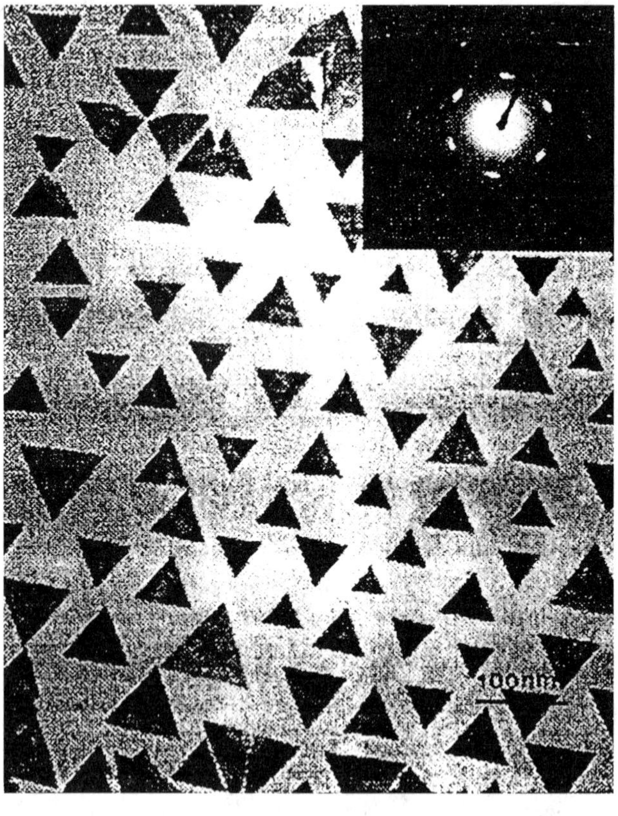

FIG. 4 Transmission electron micrograph of a PbS particulate film. The film was formed by the infusion of H_2S to an AA monolayer, floating on an aqueous 5.0×10^{-4} M $Pb(NO_3)_2$ solution in a circular trough, for 45 min. The PbS particulate film was deposited on an amorphous carbon–coated, 200-mesh copper grid. The bar represents 100 nm. Inset: Electron diffraction of a PbS particulate film domain. Limiting aperture was applied to cover an area 2 μm in diameter.

The mechanism of oriented crystal growth has been rationalized by comparison of the structures of the arachidic acid monolayer and the lead sulfide crystals. Synchrotron x-ray studies of arachidic acid monolayers in their solid states showed that they comprise fully extended molecules with a planar zigzag conformation. The arachidic acid molecules are oriented approximately normal to the liquid surface in a hexagonal close-packed array and exhibit a lattice constant of a = 4.85 Å. An experimentally obtained lattice constant of arachidic acid monolayers on lead nitrate of a = 4.81 Å, as derived from surface pressure versus surface area isotherms, was considered to be in good agreement with the published data and was utilized in the analysis. Lead sulfide possesses an NaCl-type cubic structure with a lattice constant of a = 5.9458 Å. Epitaxial growth of lead sulfide from the {111} face resulted from the geometrical complementarity between the arachidic acid monolayer and the {111} lead sulfide face (Fig. 5). The Pb-Pb and S-S interionic distances of 4.20 Å in the lead sulfide {111} plane geometrically matched the d{100} spacing of 4.16 Å for arachidic acid; the spatial mismatch between the crystals is only of the order of 1%.

The investigations of epitaxial lead sulfide growth were extended by doping the supporting arachidic acid monolayer with octadecylamine [15]. The size and orientation preference of lead sulfide grown under mixed arachidic acid–octadecyl amine monolayers were shown to be profoundly influenced by the arachidic acid/octadecylamine ratio and the applied surface pressure. The lead sulfide growth habit was observed to change from [111] to [001] with a reduction in the arachidic acid/octadecylamine ratio (AA/ODA) from 1:0 and 5:1 to 2:1. The Π versus A isotherms were identical for these monolayer compositions, indicating maintenance of the hexagonal close-packed structure [15].

The differences in morphology (Fig. 6) between equilateral-triangular PbS-I, right-angle-triangular PbS-II (epitaxially grown under monolayers, prepared from AA/ODA = 1:0 and AA/ODA = 1:1), and disk-shaped PbS-III (nonepitaxially grown under monolayers, prepared from hexadecylphosphate) manifested themselves in different spectroelectrochemical behavior

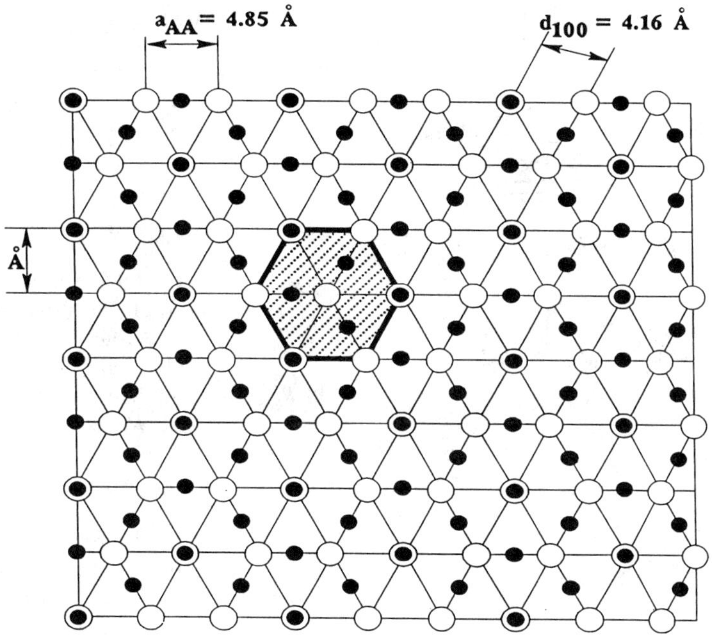

FIG. 5 Schematic two-dimensional representation of the proposed overlap between Pb^{2+} ions and AA headgroups. (○) AA headgroup; (●) Pb^{2+}; (⊙) Pb^{2+} and AA headgroups. A unit cell is highlighted by the dotted area enclosed by heavy lines.

[16]. Specifically, marked differences were observed in the potential-dependent absorption spectra of PbS-I, PbS-II, and PbS-III. Biasing the epitaxially grown PbS nanoparticulate films to negative potentials (from −0.5 to −1.1 V) increased the intensity of absorption in the ultraviolet region. In contrast, no change in the absorption at wavelengths longer than 700 nm was observed in the nonepitaxially grown PbS-III nanoparticulate film on changing the potential from 0 V to −1.5 V. Absorption spectra of the optically transparent conductive glass (i.e., the control) remained unaltered upon biasing the potential between +0.5 and −1.5 V. The near-infrared absorption is likely to correspond to the spectrum of trapped charge carriers. Increase of this absorption results from the accumulation of trapped conduction-band electrons at negative bias potentials in PbS-I and PbS-II. Indeed, absorbances for PbS-II at 750 nm were found to decrease with increasing applied positive potential linearly to −0.6 V, after which they remained unaltered. The point of inflection, −0.50 ± 0.05 V, may be taken to correspond to the flatband potential, V_{fb}, of the PbS-II nanoparticulate film.

Marked differences between PS-I, PS-II, and PS-III also manifested themselves in capacitance versus potential and photocurrent curves. The rise of the photocurrents at negative potentials is characteristic for *p*-type semiconductors. The onset of photocurrent is considered to correspond to the flatband potential. Although for PS-II it coincides with the flatband potential determined from the potential-dependent long-wavelength absorption spectra, in the absence of other evidence the onset of photocurrent cannot be meaningfully attributed to flatband potentials in the present system.

Dependence of the absorbance on the applied potential, as well as the observed photocurrent and voltage-dependent capacitances, reflects a complex interplay between the electron population in the electronic bands, in the traps (whose levels correspond to bulk imperfections), and in the available surface states, in addition to the ongoing interfacial electrochemical and photoelectrochemical processes.

V. CONCLUSION

Generation of nanoparticles and nanoparticulate films under monolayers has demonstrated the viability of this colloid chemical approach to advanced materials synthesis. The versatility of the approach permitted the fabrication of simple and composite nanoparticles, nanoplatelets and nanostructured films; two- and three-

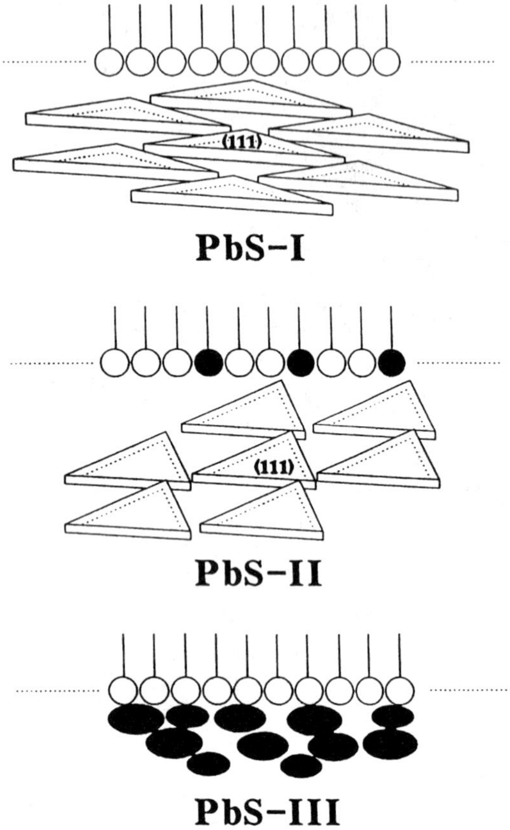

FIG. 6 Schematics of equilateral-triangular PbS-I epitaxially grown under monolayers prepared from AA, PbS-II epitaxially grown under monolayers prepared from mixtures of AA and ODA, and PbS-III nonepitaxially grown under monolayers prepared from hexadecylphosphate.

dimensionally size-quantized nanoparticles; and epitaxially grown nanocrystallites. The information obtained has considerably aided the design of self-assembled films potentially usable in optical, electro-optical, and electronic devices.

ACKNOWLEDGMENT

Support of this work by the New York State Science and Technology Foundation and Clarkson University's Center for Advanced Materials Processing (CAMP) is gratefully acknowledged.

REFERENCES

1. E. Hellstern, H. J. Fecht, Z. Fu, and W. L. Johnson, J. Appl. Phys. *65*:305–312 (1998).
2. P. Avouris, Acc. Chem. Res. *27*:159–165 (1994); E. Hartmann, P. Marquardt, J. Ditterich, and H. Steinberger, Adv. Colloid Interface Sci. *46*:221–262 (1993).
3. R. W. Siegel, in *Materials Science and Technology: A Comprehensive Treatment, Cluster Assembly of Nanophase Materials*, VCH Publishers, Weinheim, 1991, pp. 583–614.
4. S. Mann, *Biomimetic Materials Chemistry*, VCH Publishers, New York, 1996.
5. D. A. Bazylinski, A. J. Garratt-Reed, and R. B. Frankel, Microsc. Res. Tech. *27*:389–401 (1994).
6. J. H. Fendler, *Membrane-Mimetic Approach to Advanced Materials*, Springer-Verlag, Berlin, 1994, pp. 1–235.
7. G. A. Ozin, Adv. Mater. *4*:612–649 (1992); G. A. Ozin, Adv. Mater. *6*:71–76 (1994).
8. J. H. Fendler, Isr. J. Chem. *33*:41–46 (1993).
9. N. A. Kotov, E. D. Zaniquelli, F. C. Meldrum, and J. H. Fendler, Langmuir *9*:3710–3716 (1993).
10. J. H. Fendler and F. C. Meldrum, Adv. Mater. *7*:607–632 (1995).
11. S. X. Ji, C. Y. Fan, F. Y. Ma, X. C. Chen, and L. Jiang, Thin Solid Films *242*:16–20 (1994).
12. X. K. Zhao, J. Yang, L. D. McCormick, and J. H. Fendler, J. Phys. Chem. *96*:9933–9939 (1992).
13. J. H. Fendler, Supramol. Chem. *6*:209–216 (1995).
14. J. P. Yang, F. C. Meldrum, and J. H. Fendler, J. Phys. Chem. *99*:5500–5504 (1995).
15. J. P. Yang and J. H. Fendler, J. Phys. Chem. *99*:5505–5511 (1995).
16. X. K. Zhao, L. D. McCormick, and J. H. Fendler, Adv. Mater. *4*:93–97 (1992).

32

Formation of Nanoparticles in Organized Amphiphilic Films

KAREN GRIEVE and FRANZ GRIESER University of Melbourne, Parkville, Victoria, Australia

D. NEIL FURLONG RMIT University, Bundoora, Victoria, Australia

I. INTRODUCTION

Thin organic films containing colloidal particles have been of interest for some time because of their possibly synergistic properties. By surrounding nanoparticles with organic/amphiphilic groups, the useful photo, electrical, or reactive properties of the particles may be enhanced, and at the same time, the organic film provides a robust, readily manipulable and transferable medium. These particle-films have been prepared in a number of ways and can be loosely classified into three groups:

1. Amphiphilic or polyelectrolyte films in which metal or semiconductor particles have been prepared in situ (in the films). These usually involve the reaction of an incorporated metal ion with a selected reactant, which may be a gaseous or solution species.
2. Films of amphiphiles/polyelectrolytes/polymers assembled with colloidal particles previously formed. The particles may be
   (a) Formed using the amphiphile used in film assembly as a template, as in the epitaxial formation of particles at a Langmuir monolayer, or in vesicles, or
   (b) Previously prepared in an unrelated solution.
3. Films of amphiphiles into which the preprepared particles are allowed to percolate and be adsorbed.

These methods are schematically illustrated in Fig. 1.

The in situ production of particles, method 1, has most frequently been undertaken using Langmuir-Blodgett (LB) films containing metal or semiconductor particles. As the formation of particles in situ offers considerable possibilities of particle-film synergy, these films will be considered in some detail in this chapter. LB films have also been made in which particles have been introduced at the precasting (2) or postcasting (3) stage. LB films produced from Langmuir monolayers and particles formed epitaxially under them, or from expanded vesicles of nanoparticles (method 2a), also offer possible synergistic properties with the surfactant headgroup intimately associated with the particles. A major reason for the use of LB techniques to produce particle-films is the fine, molecular-level control over film thickness and particle distribution that can be achieved, especially if the particles are synthesized in situ.

The technique of layer-by-layer (LbL) assembly, in which particle films are built up through electrostatic interactions, offers less precision at the assembly stage but produces more robust films with wider possible applications and will also be discussed in this chapter. The conditions required for the manufacture of these films are considerably less rigorous than those for LB films, which require absolute cleanliness and special apparatus. It is only recently that particles have been prepared in situ in these films (method 1). Before this, particles were always prepared in colloidal solution and

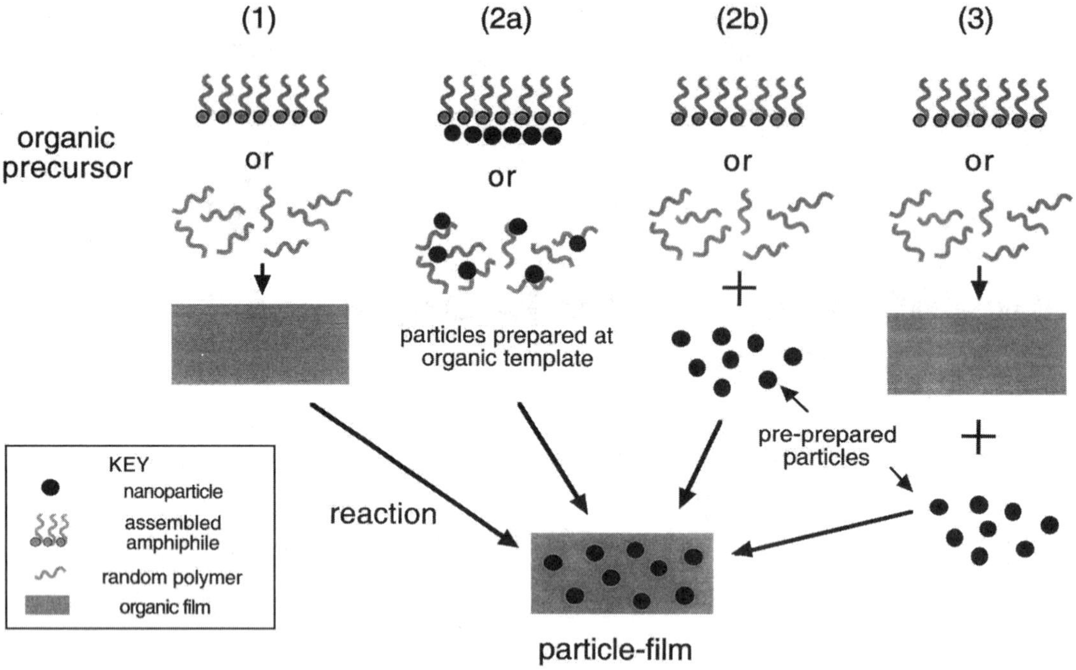

FIG. 1 Scheme of the methods for preparing amphiphilic particle-films. For an outline of the methods, refer to the text in the Introduction. Note that two types of organic precursor have been illustrated to include ordered precursors such as Langmuir monolayers and disordered precursors such as polyelectrolytes.

then assembled as part of the film. By prepreparing particles, although the possible synergic reactions between the particle surface and the amphiphile may be lost, it is possible to incorporate a larger variety of particles into films and be sure of their properties independent of the film.

Both the techniques of LB and LbL films can be used to prepare extremely thin, molecularly controlled particle-films that are useful for electro-(optical) applications such as diodes and photocells. On the other hand, to test properties such as photobleaching, films with very high optical absorbance are required and it is far more convenient to use films of self-assembled polyelectrolytes, such as Nafion, in which particles have been synthesized in situ (method 1). These films can be compared with particle-films prepared by sol-gel technologies. Other, thicker organic particle-films usually involve the incorporation of prepared colloidal particles into the film at the film building stage (method 2). Methods include electropolymerization, optical polymerization, and spin-casting.

As stated, this chapter will concentrate on the thinner sequentially assembled films produced by the LB and LbL techniques. Nanoparticles incorporated into these films include both semiconductors and metals; however, there has been considerably more research into semiconductor particles, and this bias will be reflected here. Following a discussion of the preparation and physical characteristics of these films, an outline of the results of photoelectrochemical investigations will be presented. To emphasize the use of particle-films as devices, photobleaching results for self-assembled films will also be described.

II. TYPES OF FILMS

A. Langmuir-Blodgett Films

Langmuir-Blodgett (LB) films have been prepared and studied for over 70 years [1,2], but it is only in the last 25 that they have been used for the in situ synthesis of nanoparticles [3,4] and even more recently for the incorporation of premade particles [5]. They are prepared from compressed amphiphilic monolayers (Langmuir monolayers) formed at the air-water interface. The amphiphile is transferred to the substrate a monolayer at a time, in a highly controlled manner, with hydrophilic or hydrophobic attraction binding the layers.

1. Film Preparation

The method of preparation is illustrated in Fig. 2, which shows the transfer of a surfactant monolayer onto a hydrophobic substrate as it passes through the

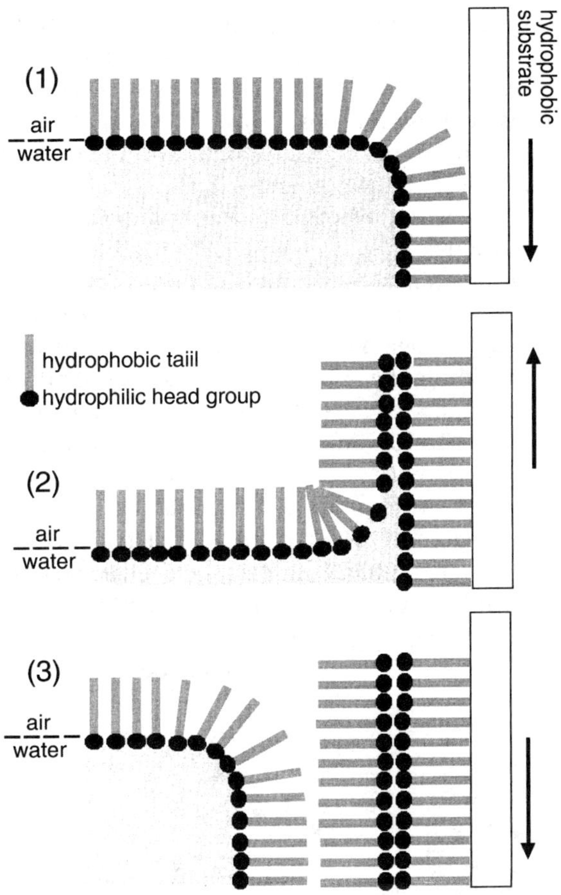

FIG. 2 The film transfer process for a hydrophobic substrate passing through a Langmuir monolayer. The formation of the first three layers of a Y-type LB film is illustrated.

air-water interface (1). The surface of the substrate becomes hydrophilic because the headgroups face outward, and so when it is passed back up through the monolayer, it may be coated with another layer of the amphiphile, this time bonding through its hydrophilic portion (2). If the substrate is hydrophilic, the first layer is transferred when the substrate leaves the subphase. By repeating this "dipping" process, films of a known number of bilayers can be prepared. The structure of the film may be lamellar (formed of bilayers; Y-type) or identically oriented monolayers (X- and Z-type) depending on whether transference occurs each time the substrate passes through the monolayer or only when the substrate enters or leaves the subphase, respectively. The surface pressure of the monolayer is maintained during transfer by continual reduction in the monolayer area, effectively compensating for the area of amphiphile transferred. By comparing the area of the substrate to the reduction in area necessary to maintain a constant surface pressure, the efficiency of the transfer process can be measured.

Properties of LB films are affected by factors such as the stability of the monolayer, subphase conditions including pH and the presence of electrolytes, temperature, the surface pressure of the monolayer when it is transferred (transfer pressure), and the speed at which the substrate passes through the monolayer (transfer speed). Other factors that can affect the films are the time allowed for drying between dips and the orientation of the substrate with respect to the surface. The nature of the substrate, in particular its roughness, hydrophobicity, or hydrophilicity, is also very important. Detailed information about the manufacture and properties of LB films can be found elsewhere [6–8].

(*a*) *Nanoparticles in Langmuir-Blodgett Films.* Semiconductors and metals can be produced by the reaction of a metal ion with a reacting gas such as hydrogen sulfide or a reducing gas such as hydrazine, respectively. If the reaction is carried out in the presence of a stabilizing medium, the product can be in the form of nanosized particles [9]. Equation (1a) is an example of the reaction between a divalent metal ion and a sulfide ion to produce II-VI semiconductor nanoparticles, Equation (1b) of a monovalent metal ion and hydrazine to produce metal.

$$M^{2+} + S^{2-} \xrightarrow{\text{stabilizer}} \text{size-quantized MS} \qquad (1a)$$

$$M^{+} \xrightarrow{\text{stabilizer, } N_2H_4} \text{size-quantized } M^0 \qquad (1b)$$

(*b*) *Premade Particles.* Nanoparticles can be incorporated into LB films by a variety of mechanisms. Particles can be preprepared in colloidal solutions of micelles or vesicles in which a surfactant(s) or polyelectrolyte acts as the stabilizer. These can be introduced into the subphase of a Langmuir monolayer prior to film transfer. The particles become associated with the monolayer amphiphile and are transferred with it [5,10–27]. This process is an example of synthesis type 2b by the classification used in the Introduction. A further way of preparing nanoparticles is to use vesicles as the stabilizing/capping medium. Such vesicles have been shown to open up in aqueous solution to form monolayer films at the subphase surface, with the particles attached to the monolayer (type 2a) [28,29]. Stabilized particles can also be spread directly as the monolayer, either with a supporting amphiphile [30–32] or cross-linking/binding agent [33] or, if the colloidal stabilizer has particular properties, as the sole component of the Langmuir-type monolayer (type 2b) [34–46]. These films can then be transferred using the Langmuir-Blodgett technique. A further method for introducing semiconductor nanoparticles to LB films is one developed by Fendler and coworkers whereby particles are formed epitaxially at a Langmuir monolayer prior to film transfer (type 2a) [10,28,47–52]. In all these methods the prepared nanoparticles are incorporated in the LB film as it is formed; however, it is possible that particles can be introduced after the film is formed. Both metal [53–56] and semiconductor [57] particles were found to diffuse into similar surfactant bilayer films formed by thermal evaporation (type 3). The scheme in Fig. 3 shows the various methods for the incorporation of nanoparticles in LB films. Of the methods for incorporating premade particles, those using opening vesicles or Langmuir monolayer epitaxy are most likely to involve significant nanoparticle-amphiphile interaction. One way to ensure this interaction is to form the particles in situ.

(*c*) *In Situ Formation of Nanoparticles.* A method for preparing nanoparticles in situ within fatty acid LB films (type 1 in the Introduction) was developed by Barraud and coworkers in their efforts to build conducting layers within amphiphilic films [3,4]. The basic procedure involves the reaction of a metal ion–fatty acid LB film with a reactive species such as H_2S or N_2H_4 as in Eqs. (1a) and (1b). As many metal ions undergo a pH-dependent exchange with the carboxylic acid proton of the fatty acid [58,59], the film is prepared by either transferring a surfactant monolayer over a subphase that contains the metal ion at an appropriate pH [60] or immersing the fatty acid film in a solution containing the metal ion (again at an appropriate pH) [3,4]. The former method is used more frequently as the metal ion in the subphase often acts to stabilize the Langmuir monolayer and facilitates high-quality film transfer [8]. To form metal nanoparticles, the metal salt fatty-acid LB film is exposed to a reducing gas such as N_2H_4 [3,4,61,62] or CO [63] or is photoreduced [64]. Another reductant, H_2, has been used to reduce CuS particles in LB films [65]. To form chalcogenide or halide semiconductor nanoparticles, the film is exposed to an appropriate gas [H_2E (E = S, Se, Te for chalcogenides) or HX (X = Cl, Br, I for halides)]. The general

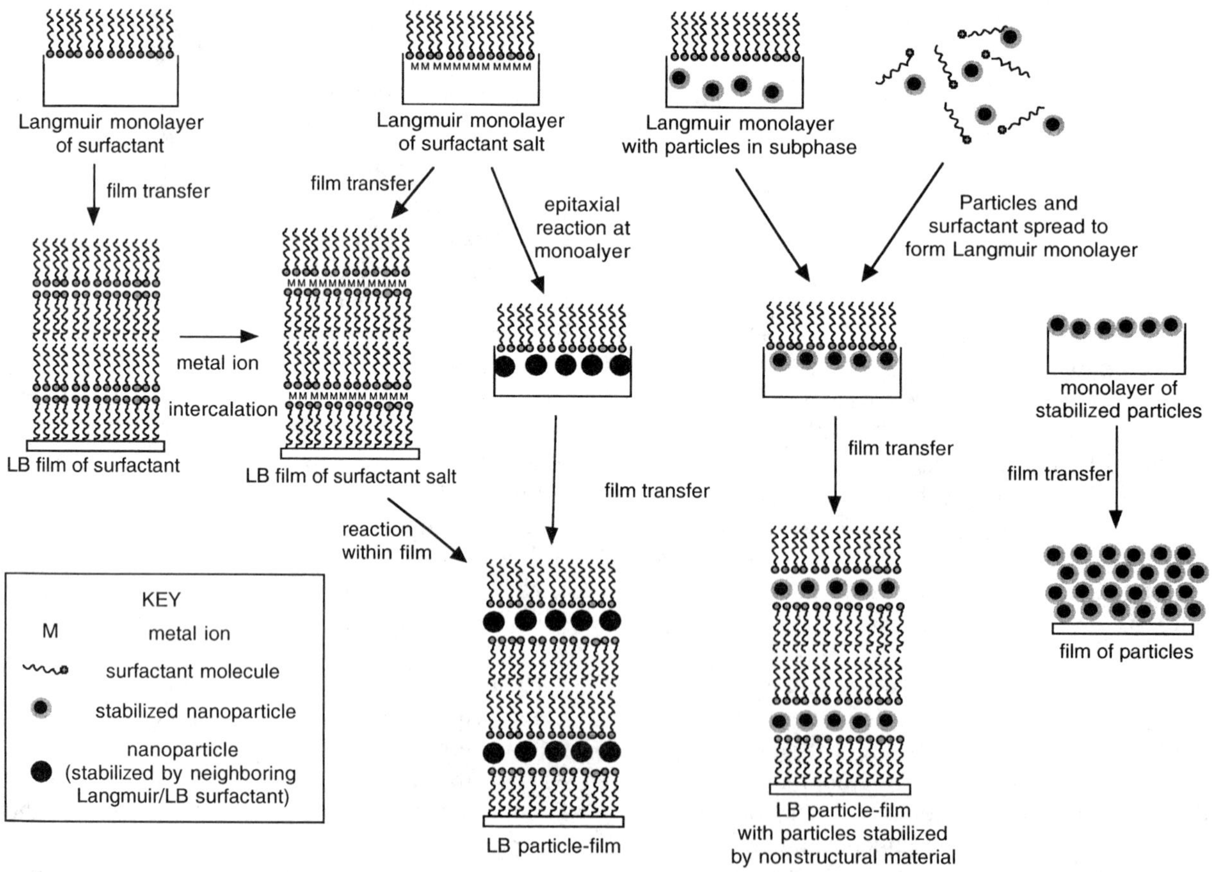

FIG. 3 A scheme of the methods that have been used to incorporate nanoparticles into LB films.

reaction mechanism is illustrated by the reaction of a cadmium arachidate film and H_2S as shown in Eq. (2a).

$$Cd^{2+}\{^{-}OOC(CH_2)_{18}CH_3\}_2 + H_2S_{(g)} \rightarrow CdS + 2HOOC(CH_2)_{18}CH_3 \quad (2a)$$

For the purposes of this chapter, the surfactant name will be abbreviated, so arachidic acid is designated ArH and cadmium arachidate $CdAr_2$. Similar abbreviations for behenic acid (BeH), steric acid (StH), and dioctadecyl ammonium bromide/chloride (DDAB/C) and their corresponding salts will also be used. Using this nomenclature, Eq. (2a) becomes Eq. (2b):

$$CdAr_2 + H_2S_{(g)} \rightarrow CdS + 2ArH \quad (2b)$$

The reaction to form particles is usually initiated in a multilayered LB film. It is possible, however, to allow the reaction to take place after each bilayer is transferred [66]. This method has been used to form different kinds of semiconductor particles in sequential layers of the film [67].

Amphiphile-stabilized, nanosized TiO_2 can be made by the hydrolysis of films formed from ArH monolayers spread on subphases containing $TiCl_4$ [68]. The pyrolysis of similar films leads to the formation of quantum-sized TiO_2 particles but destroys the amphiphilic part of the film [69,70]. Palladium likewise can be formed by the thermal decomposition of palladium-containing films [71]. In an analogous fashion, the formation of CdO particles by the exposure of cadmium-containing films to ultraviolet (UV) light and ozone is accompanied by loss of the amphiphile [72,73]. The latter methods are useful for making particles but, as the amphiphilic part of the film is destroyed in the process, will not be considered further.

The easiest method for establishing the size-quantized nature of the particles formed is to analyze the absorption spectra. Blue-shifted absorption spectra, such as those shown in Fig. 4, are a clear indication of size quantization [9,74,75]. The absorption edges of the CdS, CdSe, and CdTe particles in this figure are all shifted by 0.4–0.3 eV from the band gap of the bulk material. The types of nanoparticles that have been formed in LB films are listed in Table 1. It can be seen that in addition to simple metals and semiconductors, more complex particles can be made. For example, CdS-CdSe core-shell particles have been made by exposing a CdS/BeH film to H_2Se [113,114,119] and Te^{2-} can replace some Se^{2-} to form CdSeTe core-shell particles [114]. For the former reaction, it has been calculated that a monolayer of CdSe is formed around the CdS core. When the subphase of the Langmuir layer

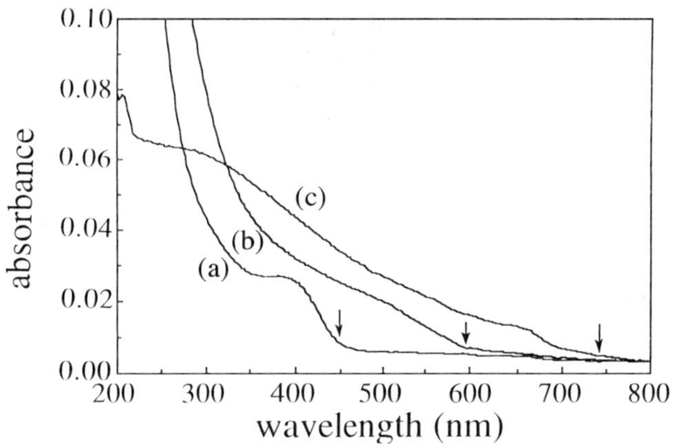

FIG. 4 Absorbance spectra of 19-layer cadmium nonacosa-10,12-diyonate LB films after exposure to (a) H_2S; (b) H_2Se; (c) H_2Te. The spectra are corrected for the absorbance of the plate and the amphiphile, and the arrows indicate the absorption onset of the spectra [114].

contains a mixture of metal ions, the effect on the formation of particles within the LB film depends on the nature of the metal ions. If the subphase contained Zn^{2+} and Cd^{2+}, or Cd^{2+} and Mn^{2+}, mixed semiconductors $Zn_{1-y}Cd_yS$ [98] and $Cd_{1-x}Mn_xS$ [120,121] were formed. The metal ratio in the particles reflected that of the subphase. However, when both Hg^{2+} and Cd^{2+} are in the subphase, it has been shown by analysis of UV-visible spectra that discrete HgS and CdS particles are formed [107]. The different reaction rates of Cd^{2+} and Hg^{2+} with S^{2-} have been used to explain the result. Mixed subphases of Cd^{2+} and Pb^{2+} produced only $PbSt_2$ films and PbS particles [98]. This may be due to the higher stability expected of $PbSt_2$ over $CdSt_2$.

Nanoparticles are formed throughout the whole LB film and not just at the surface. This is shown by the linear relationship commonly observed between the number of layers in a film and the UV-visible (UV-vis) absorbance of the film at a particular wavelength [116,117]. An example of this relationship is shown in Fig. 5 for an ArH film containing HgS nanoparticles [107].

2. Control of Particle Size

Particle dimensions in size-quantized particles determine their optical absorption and redox characteristics, and the ability to tune these properties is fundamental to nanoparticle synthesis and application. Particle size is thought to be restricted by either or both of two processes. One process is known as chemical "capping" and results from interaction between a stabilizing surfactant and the surface of a nanoparticle. Examples

TABLE 1 Semiconductor and Metal Particles That Have Been Made In Situ in LB Films[a]

Particle	Amphiphile	References
CdS	ArH	[60,76–93]
	StH	[94–103]
	BeH	[104–107]
	other	[94,108–117]
CdSe	ArH	[118]
	NDA	[113,114]
CdTe	NDA	[113,114]
CdS-CdSe	ArH	[119]
	NDA	[113]
CdS/PbS	StH	[96–98]
Cd-ZnS	StH	[98]
Cd-MnS	ArH	[120,121]
CdS-PbS & MgS	ArH	[67]
CdS-HgS	BeH	[107]
HgS	ArH	[122]
	BeH	[106,122–124]
CuS	ArH	[61]
	StH	[125]
	BeH	[105,126]
	other	[65,96,112]
PbS	ArH	[127]
	StH	[96,97,101,128–133]
	other	[110,112,134]
CoS	StH	[135]
ZnS	StH	[96,97,136]
	BeH	[105]
$PtS_{(2)}$	ArH	[137]
	DDAB	[62]
PdS	ArH	[137]
	DDAB	[62]
CdI_2	ArH	[138–140]
$CdCl_2$	ArH	[140]
$CdBr_2$	ArH	[140]
PbI_2	StH	[97]
AgI	ArH	[66]
TiO_2	ArH	[68]
Ag	ArH	[66]
	BeH	[3,4]
Au	ArH, DDAB	[63]
	DDAC	[64]
Cu	ArH	[61]
	BHDB	[65]
Pd	DDAB	[62]

[a]The amphiphiles used have been listed, particularly straight-chain fatty acids, for comparative purposes. As there is insufficient space to list all amphiphiles, those which have been used less frequently are usually classified under "other."

Abbreviations: ArH, arachidic acid; StH, stearic acid; BeH, behenic acid; NDA, nanocosa-10,12-diyonic acid; DDAB, dimethyldioctadecylammonium bromide; DDAC, dimethyldioctadecylammonium chloride; BHDB, 2,4-dihydroxybenzilidine-4′-(hexadecylamino)-benzylamine.

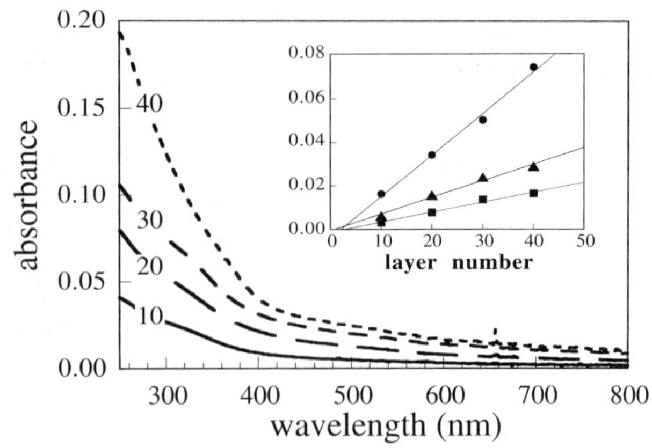

FIG. 5 UV-vis absorbance spectra (corrected for the absorbance of the fatty acid and the quartz substrate) of 10-, 20-, 30-, and 40-layer $HgAr_2$ films exposed to H_2S for 15 min. The number of layers is indicated on each spectrum. Inset: Absorbance as a function of layer number at 350 nm (●), 400 nm (▲), and 600 nm (■) [33].

are the use of thiols [141] or phosphates [142] as stabilizers in the preparation of colloidal solutions of CdS. The other process is the purely physical confinement presented by host media such as zeolites [143] and glasses [144]. In LB films, it is likely that the growth of the particles is restricted by both the physical confines of the layers and the capping action of the surfactant molecules. The methods are effective as particles produced in LB films are usually of the order of 1–4 nm in diameter.

(*a*) *Particle Size Determination.* The method most frequently used to determine the dimensions of particles in LB films involves the comparison of a feature of the UV-vis absorbance spectrum with a calibration curve relating the diameter of the particles to the chosen feature. Figure 6 is an example a calibration curve for CdS particles. The blue shift (increased energy) in the absorbance of smaller particles due to size quantization effects [9,75,145] is clearly demonstrated by this curve. The data have been taken from numerous literature sources in which the average sizes of particles in solution have been resolved by methods such as transmission electron microscopy (TEM) and x-ray diffraction (XRD), and the inflection point (minimum of the first derivative) of the absorption spectrum has also been determined. The inflection point does not represent the band gap of the average-sized particle; however, the minimum of the second derivative, which more accurately describes this band gap [145,146], can be difficult to find precisely in LB films, which often display low absorbance and significant amounts of scat-

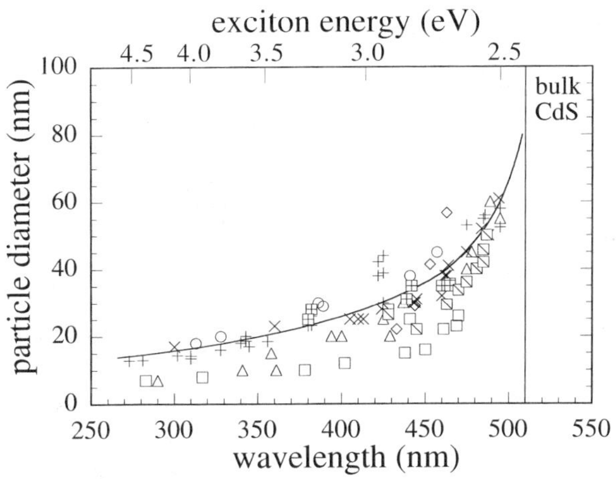

FIG. 6 A calibration curve of inflection point wavelengths in absorbance spectra of colloidal CdS nanoparticles as a function of experimentally determined (TEM and XRD) particle sizes. The data have been collected from a large number of literature sources [249]. The solid line is a theoretical curve often used in this method of calibration [142].

ter. The curve (solid line) in Fig. 6 [147] has frequently been used to estimate particle diameters from spectra inflection points, and, as can be seen from the figure, it should provide a reasonable result.

Statistical methods that have been used to determine the sizes of particles produced in LB films include TEM and atomic force microscopy (AFM). TEM images are most useful for highly contrasting metal particles such as those in Fig. 7 [62–64], but the low contrast between surfactant and particle makes analysis of semiconductor particle-films difficult. It is possible to obtain indirect estimates of the sizes of particles in films from TEM images taken when the particles have been washed from a film onto a grid [114] or when the surfactant is removed by heating under vacuum [77]. AFM measurements have also been used [61,62,79,122,137]; however, for these it is necessary to remove the surfactant in a process that tends to result in growth of the particles. The effects of removing the amphiphile will be discussed further in the next section.

(*b*) *Factors Affecting Particle Size.* The size of particles formed in LB films is influenced by conditions at all stages of preparation, i.e., when the film is made, during the reaction to form the nanoparticles, and after the particles have been formed.

During film preparation there are many factors that influence the size of the particles which will form, those that have received the most attention being the structure of the film and the concentration of metal ions

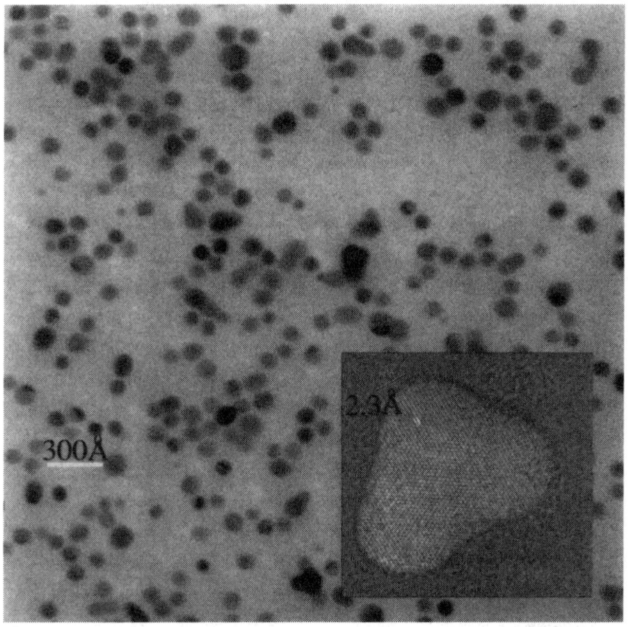

FIG. 7 TEM image of gold nanoparticles in an ArH film. ArH was compressed over an $[Au(en)(en\text{-}H)]^{2+}$ subphase and the grid was dipped through the surface three times before exposure to CO. The insert shows an HRTEM image of a gold particle with a lattice spacing of 2.3 Å [63].

in the film. Film structure can be varied simply by changing the nature of the amphiphile. More structured films, such as those prepared using calixarenes [103,117], tend to produce smaller particles than films of straight-chain amphiphiles. Straight-chain amphiphiles in turn produce smaller particles than disordered bilayer films [94]. A systmatic investigation in which hydrocarbon interaction and film stability were increased by decreasing the ratio between carboxylic acid groups and hydrocarbon chains in poly maleic acid octadecanol ester (PMAO) amphiphiles showed a similar inverse relationship between structure and size [111,134]. Another method for increasing film structure is to increase transfer pressure. When higher transfer pressures were used in the formation of Cd, Cu, and Zn behenate films, smaller particles resulted [105]. It is possible that the higher degree of structure brought about by the various means discussed above creates films with less flexibility. Thermodynamic considerations would favor the creation of more, smaller particles with less disruption of the film structure. The effect of film structure is not simple, however, as altering the amphiphile or transfer pressure may also alter the particle concentration in the film, and as will be discussed next, this too is an important factor in determining the size of particles formed.

The metal ion/amphiphile ratio in an LB film can be altered by varying the subphase pH and therefore the proportion of metal ions exchanged [103]. The metal ion concentration/area ratio can readily be altered by changing the molecular area of the surfactant [94] and the concentration of carboxylate functions in the amphiphile [134]. In both these instances, an increased (reactive) ion concentration led to an increase in particle size. Figure 8 illustrates the relation between the concentration of Cd^{2+} ($[Cd^{2+}]$) in the film and the size of CdS particles created in the form of the band gap (E_g) of the resulting films. $[Cd^{2+}]$ was varied in this instance by the use of amphiphiles with different molecular areas and also by the intercalcation of more Cd^{2+} into the films after particle formation, by a method discussed later in Section II.A.2.c. The metal ion/amphiphile ratio can also be lowered by introducing into the subphase an inert metal ion such as Ca^{2+}, which dilutes the amount of reactive Cd^{2+} in the films [113,114]. However, this change was insufficient to decrease the size of the particles produced unless dihexadecyl phosphate was mixed with the 16-8 diyonate amphiphile. The role of the phosphate is unclear, but its influence on the size of the particles may be related to amphiphile-metal interaction. To date there has been no systematic study published on the effect different surfactant headgroups exert during particle formation. Such an investigation into the phenomenon of capping would significantly enhance our understanding of particle-film synthesis.

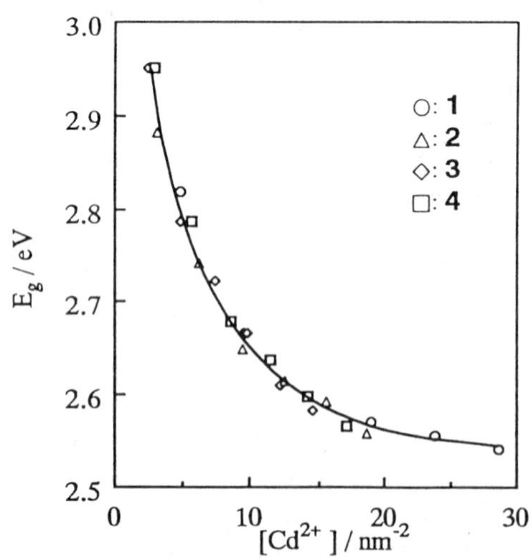

FIG. 8 The band gap energy (E_g) of CdS as a function of the concentration of Cd^{2+} per unit area ($[Cd^{2+}]$). The concentration was altered by varying the headgroup of the amphiphile (the legend refers to the amphiphiles described in the original text) and by a series of intercalation-sulfidation cycles [94].

A further, final method of controlling the metal ion/area ratio, namely that of controlling the surface pressure, was introduced earlier. In this case, an increased concentration led to smaller particles [105]. The smaller size of the particles may be caused by the increased structure of the film. This apparent anomaly to the generally observed size-concentration dependence serves to illustrate the complicated relationship between film structure and ion-amphiphile interaction during the synthesis of particles within LB films.

As another contributing factor, the thickness of the film might be expected to influence the size of particles produced, but results are inconclusive with evidence both supporting [87] and opposing [60,103,113] an impact. The contrasting results may be explained by differences in conditions when the films were transferred.

At the reaction stage, two factors that have been observed to affect the size of particles produced are temperature and the presence of water. Higher temperatures predictably led to the formation of larger particles when $CdAr_2$ was exposed to H_2S [82,83]. In other investigations, smaller particles were formed when care was taken to exclude water from the reacting H_2S than when water-saturated H_2S was used (R. S. Urquhart et al., submitted). Both these results can be explained by kinetic considerations, with water and heat facilitating the coalescence of CdS molecules within the film and assisting amphiphile rearrangement.

Removal of the amphiphile after particle formation can lead to growth of the particles. For example, washing films with a solvent known to dissolve the amphiphile leads to an increase in the size of the particles remaining on the substrate. The spectra in Fig. 9 were taken before and after a CdS/ArH film was washed with chloroform. The red shift in the spectrum corresponds to an increase in particle size from 2.2 to 2.9 nm, and this accords well with other published observations [87]. Heating the film to remove some of the surfactant also led to a red shift in the absorbance spectra of CdS particles. By altering the annealing conditions, the particles could be grown in a controlled fashion [90]. The introduction of wet N_2 gas to CdS/ArH films, in which CdS was formed by the exposure of $CdAr_2$ films to dry H_2S, also caused a red shift in the UV-vis absorbance spectra (R. S. Urquhart et al., submitted). This too may indicate that the size restriction of nanoparticles in these films is to some extent a kinetic phenomenon, similar to that observed for solution colloids. The processes of annealing and of dissolving

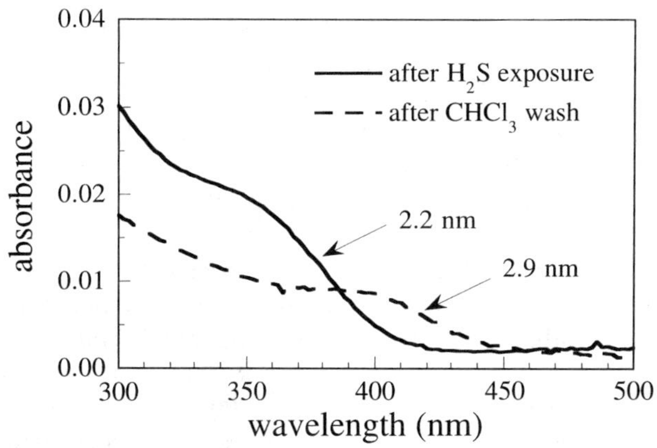

FIG. 9 UV-vis absorbance spectra (corrected for absorbance due to the fatty acid and the substrate) of a 40-layer ArH film containing CdS before and after immersion in a stirred solution of chloroform. The inflection point of the spectra, which was used to determine the average particle size given, is indicated by the arrows [107].

surfactant would open and/or destroy the structure of the films and, like the addition of water, allow or promote coalescence of the molecules/particles. There is further discussion of the conditions affecting the particle growth in the section on mechanism. It is important to note that in spite of all the coalescence that may occur, either when the particles are being made or afterward, bulk material is not formed and the dimensions of the particles remain nanoscale.

(*c*) *Stepwise Growth of Particles.* It is also possible to grow particles in support films by the addition of more reactants in a two-step process. Semiconductor particles such as CdS, PbS, and HgS have been grown in LB films by the relatively simple procedure shown schematically in Fig. 10. After particle formation more metal ions are introduced into the film by immersing it in a solution containing the metal ion at an appropriate pH. The carboxylic acid proton on the fatty acid, which has been restored during semiconductor synthesis, is again exchanged for the metal ion. Reexposure of the film to the chalcogenide gas results in growth of the particles. By repeating these intercalation and gassing steps in a cyclical fashion, the particles can be grown stepwise [77,122,126,136]. (This cycle will be referred to as intercalation-sulfidation as H_2S is the most commonly used reacting gas.) Figure 11 shows the UV-vis absorption spectra for ArH films containing CdS that has been subjected to three intercalation-sulfidation cycles after the initial particle formation (four exposures to H_2S in total). The spectrum is red shifted with each

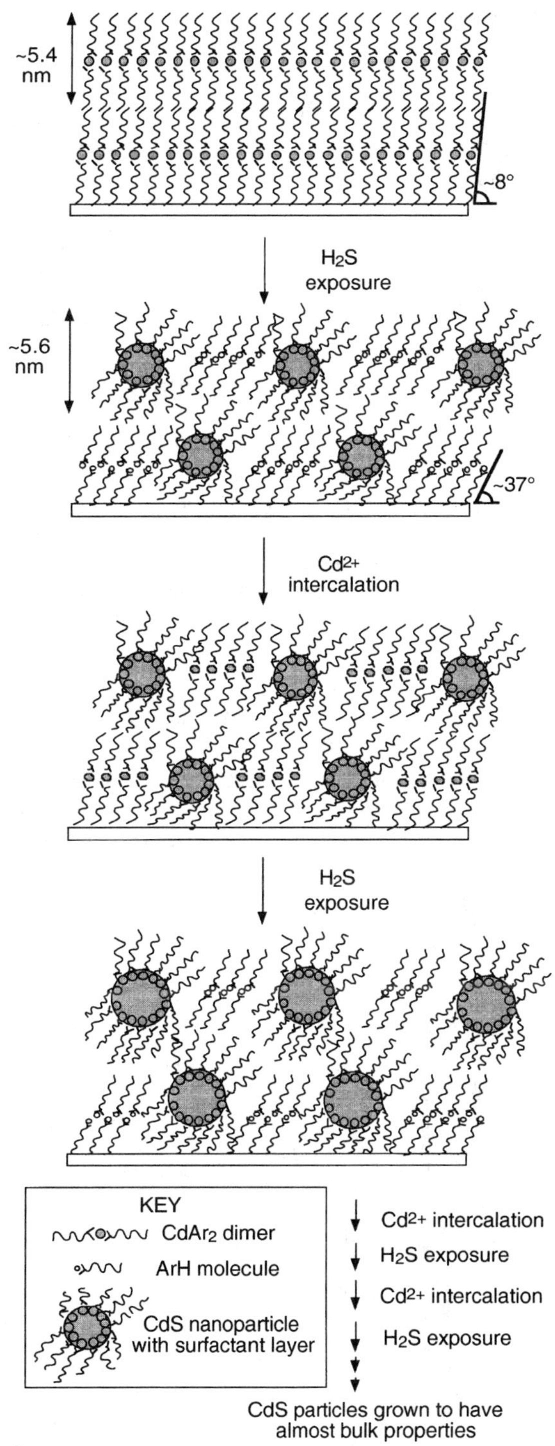

FIG. 10 A schematic diagram showing the effect of exposure of a $CdAr_2$ film to H_2S. As CdS nanoparticles form, the film is disrupted, becomes thicker, and develops a greater degree of tilt. The CdS particle is drawn to show the arachidate molecules associated with surface cadmium ions in a "capping" effect. The growth of the particles by intercalation in Cd^{2+} solution and further sulfidation is also illustrated.

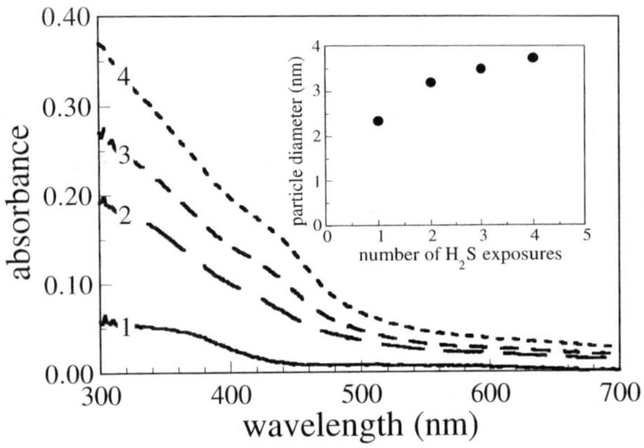

FIG. 11 UV-vis absorbance (corrected for the absorbance of the amphiphile and the substrate) of CdS in 19-layer ArH films grown by repeating the intercalation-sulfidation cycles. (a), (b), (c), and (d) refer to films that have undergone 1, 2, 3, and 4 sulfidation steps, respectively. Inset: Particle size as determined from the absorption spectrum as a function of sulfidation steps.

cycle, indicating particle growth, and the scatter in the film increases as the structure of the film is disrupted. The growth in particle size can also be seen in AFM measurements taken after one, three, and five sulfidation steps (Fig. 12). As the particles grow during the washing step prior to AFM measurements, the sizes cited are not those of the particles in the film, but the growth of the particles can be seen clearly and is in

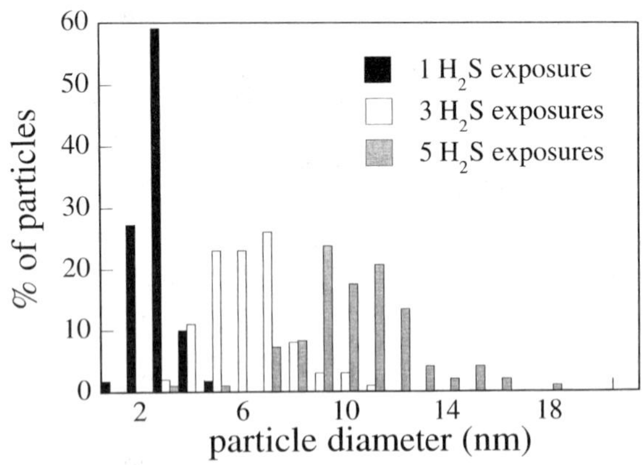

FIG. 12 Particle size distributions of CdS formed by exposing bilayer films of ArH to H_2S determined by AFM measurements. The films were prepared on mica and subjected to washing in chloroform before analysis. The particles were grown by repetitions of the intercalation and sulfidation cycle and the histograms show the distributions determined for particles formed after 1 (solid), 3 (unfilled), and 5 (shaded) exposures to H_2S, respectively [79].

accord with the sequential red shift in the spectra. Semiconductor nanoparticles treated by this stepwise process include CdS [77–79,94,95,97,98,136], CuS [126], HgS [122], PbS [96,97,128], ZnS [96,97], and AgI [66]. Metal nanoparticles of Ag [63,64,66] have also been grown on or in LB films by analogous cycles of Ag^+ intercalation and reduction.

By taking advantage of the intercalation step to introduce a second metal to the film, mixed semiconductor (possibly core-shell) particles have been produced. The introduction of Cd^{2+} to PbS/StH and ZnS/StH films caused the displacement of Pb and Zn ions and mixed particles formed on exposure to H_2S [98]. Dipping CdS/StH films into Pb^{2+} and continuing with repeated intercalation/sulfidation cycles seemed, however, to produce discrete PbS particles in the film [96,98], similar to the binary system of HgS and CdS observed in ArH from mixed subphases as discussed earlier.

3. Characterization of Films

(*a*) *Particle Shape.* It has been claimed by many that planes rather than particles are formed within the films. This view usually results from an interpretation of XRD and Fourier transform infrared (FTIR) measurements that show maintained crystallinity and small changes in interlayer spacings within the films during the exposure of fatty acid salt films to H_2S [95,99,135,136]. There is considerable evidence, however, that in almost all instances particles are formed. That spheroidal particles are produced in the films is supported by TEM images [60] such as those of Ag particles in ArH [63] shown in Fig. 7. AFM images of ArH films containing CdS showed protrusions in an otherwise smooth surface that were attributed to particle formation [102]. Further evidence for the formation of particles is the production of features of the FTIR spectrum consistent with RCOOH facing dimers when a CdAr film is exposed to H_2S [86]. The formation of CdS in planes would prevent observation of such dimers. Rutherford backscattering (RBS) analysis has indicated particle formation [84], whereas x-ray diffraction/reflection studies have been interpreted to suggest the existence of disklike structures of CdS in ArH films [92]. It has been claimed that the shape of the particles formed is determined by the ability of the molecules formed to coalesce into particles [see Eqs. (3) and (4)], which is more difficult in more highly structured films. The theory has been extended to claim that in perfectly formed Y-type films with high concentrations of highly interactive surfactant, layers of semiconductor would be formed. This has been used to explain the very different XRD spectra for CdS/StH films of very high transfer pressure (37.5 mN m^{-1}) compared with those of similar films prepared at lower pressures [99,100].

(*b*) *Mechanism of Particle Formation.* Semiconductor nanoparticles appear to be formed in a two-step process similar to that of colloids in solution. To give an example of a well-studied system, the formation of metal chalcogenide particles by reaction of a divalent metal ion with H_2E requires the following steps:

$$M^{2+} + H_2E \rightarrow MeE + 2H^+ \quad (3)$$

$$nME \rightarrow (ME)_n \quad (4)$$

The first equation (3) produces metal sulfide molecules that then diffuse and coalesce together to form particles according to Eq. (4). If a fatty acid is the matrix, the protons released from Eq. (3) form carboxylic acids and the overall equation becomes

$$M^{2+}\{^{-}OOC(CH_2)_nCH_3\}_2 + H_2E_{(g)} \rightarrow ME + 2HOOC(CH_2)_nCH_3 \quad (5)$$

It is difficult to differentiate the two mechanisms as they occur to some extent simultaneously.

FTIR spectroscopy and quartz crystal microbalance (QCM) gravimetry have both been used to monitor the extent of the reactions taking place within LB films. Reactions of the type depicted in Eq. (5) have been shown to reach less than 100% completion according to grazing angle FTIR spectra taken before and after H_2S exposure, such as that in Fig. 13. The peak due to the symmetric stretching mode of unprotonated $RCOO^-$ (1400–1500 cm^{-1}) is significantly reduced but does not disappear on H_2S exposure, even though the stretching modes associated with the protonated RCOOH (C=O, 1700 cm^{-1}; O—H, 3000–3200 cm^{-1}) become apparent [78]. A similar incomplete reaction of all $RCOO^-$ molecules has been observed by FTIR analysis of other fatty acid systems [91,99,100,111,131,137]. Other studies, however, have determined that the band caused by metal-associated carbonyl groups (1700 cm^{-1}) does disappear entirely on H_2S exposure [87,96,105,123,124,136]. If not all the carboxylate groups are reprotonated by the reaction with H_2S, then it is thought that some surfactant molecules remain associated with the metal ion and it is presumed that these form a capping layer around the semiconductor particle, as represented diagrammatically in Fig. 10. The less than 100% reprotonation of the carboxylic acid salts is supported by evidence of FTIR spectra taken before and after sublimation of ArH from a CdS/ArH film [78]. Bands due to RCOOH were observed

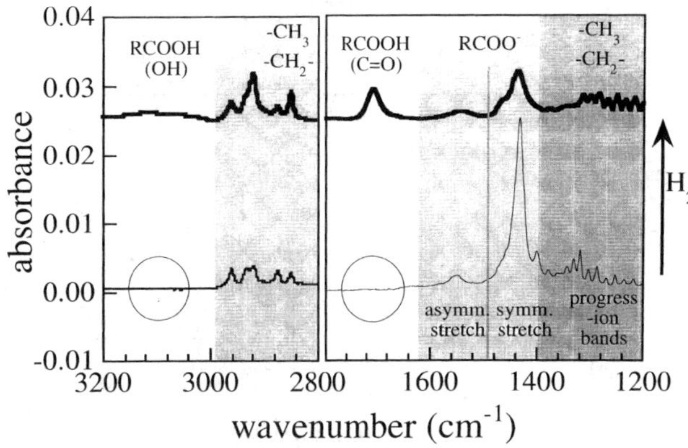

FIG. 13 Grazing angle FTIR spectra of a 20-layer CdAr film (lower) and an identically prepared film exposed to H_2S gas for 2.5 h (upper). The spectrum of the film after sulfidation has been displaced vertically for clarity. The structures responsible for the major bands are shown at the top of the plot. The circles on the lower spectrum indicate the regions that would show bands if protonated carbonyl groups were present [78].

to disappear after heating, whereas those due to $RCOO^-$ changed little, indicating that they were associated with the CdS remaining on the substrate.

The association of the surfactant with the particles is supported by evidence from QCM measurements. QCM microgravimetry can be used to determine the mass changes in a film caused by a reaction and, by comparing the change in mass determined with that predicted from the stoichiometry of Eq. (2a), the extent of reaction is ascertained. Figure 14 shows an example of H_2S uptake with time of a $CdAr_2$ film exposed to H_2S, with the percentage uptake calculated from the stoichiometry of Eq. (2a) and the mass of fatty acid salt film when initially transferred and left to dry. It can be seen that the uptake of H_2S is about 80% of the stoichiometric amount. Similar results have been found elsewhere [87]. In other instances, more than 100% reaction has been recorded, although these did not take into account the effect of H_2S on the QCM electrode [76,79]. A larger mass may also be explained by the formation of elemental chalcogenide occasionally observed by x-ray plasmon spectroscopy (XPS) [62,118,125]. The existence of a capping layer around the particles helps explain the maintenance of the quantum-sized nature of the particles when most of the amphiphile has been washed or evaporated away. It also explains the greater than expected mass remaining on a QCM electrode after thermal evaporation of the amphiphile [115].

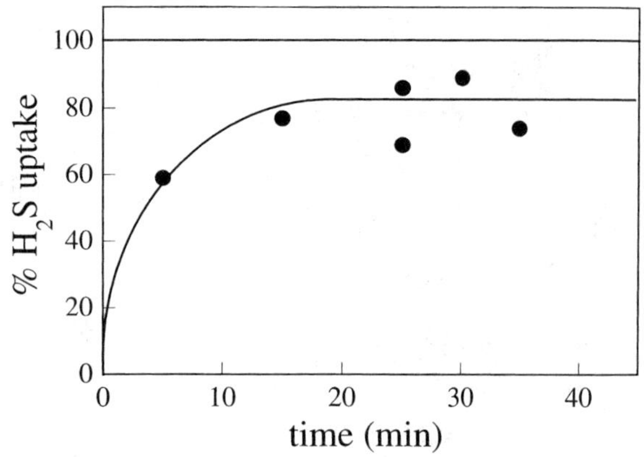

FIG. 14 Percentage conversion of $CdAr_2$ to CdS on exposure of a 19-layer film to H_2S as a function of time as determined by QCM measurements. The percentage is calculated according to the stoichiometry of Eq. (5) and the mass change during film transfer. The films were placed under vacuum before each measurement, and the frequency change of blank QCM electrodes on exposure to H_2S has also been taken into account.

The reaction taking place during ion intercalation is the same as that taking place at a monolayer in the formation of a fatty acid salt. It can be represented by Eq. (6).

$$M^{2+} + 2HOOC(CH_2)_nCH_3 \rightarrow M^{2+}\{^-OOC(CH_2)_nCH_3\}_2 + 2H^+ \quad (6)$$

This reaction has also been followed by FTIR and QCM measurements. FTIR measurements have shown that all the bands due to RCOOH virtually disappear after prolonged immersion in M^{2+} solution, indicating a complete reaction with the available surfactant molecules [78,98,128,136]. QCM measurements show [77,79] that the mass change due to Cd^{2+} adsorption into a CdAr film is consistent with the stoichiometry in Eq. (6).

(*c*) *Effects on Film Structure.* The CdS/ArH particle-film has been extensively studied. A representation of the changes in film structure that accompany the formation of CdS particles in ArH LB films can be seen in Fig. 10. Deposited films of $CdAr_2$ are well-ordered bilayer structures with a long orientation of 8 ± 5° to the normal axis [148] and a bilayer thickness measured as 53.4 Å [76], 54 ± 1 Å [87], 55 Å [93], or 55.2 Å [91] as measured by surface plasmon resonance (SPR), XRD, x-ray scattering (XRS), or ellipsometry, respectively. On exposure to H_2S, CdS particles form in layers determined by the hydrophilic headgroups of the surfactant molecules [91]. Both the tilt and the thickness of the film have been observed to increase, accommodating the formation of particles. The new tilt was measured by XRD to be 37° to the surface normal [91], in accordance with the 25–40° commonly observed for protonated fatty acid films. The tilt can also be discerned from grazing angle FTIR (GA-FTIR) spectra (Fig. 13), where the bands due to methyl groups (2800–3000 cm^{-1}) increase in intensity due to increased colinearity with the incident beam. The increase in bilayer thickness has usually been measured to be slight, ranging from 1 to 6 Å [60,76,93], although increases as high as 12 Å have been found [87].

Evidence for the maintenance of a layered structure, although somewhat disrupted, on H_2S exposure can be seen in the broadened but still discernible peaks in bilayer spacing in XRD spectra [91] and the maintenance of the progression bands in FTIR spectra (the bands at less than 1400 cm^{-1} in the spectra in Fig. 13) [78]. Moderate film disruption has been confirmed by AFM [102] and ellipsometric measurements. The observed changes can all be explained by the creation of well-structured microdomains within the films. Further intercalation and sulfidation steps resulting in the growth of the particles lead to further disruption of the film, and this can be seen readily in the increased scatter in the UV-vis adsorption spectra such as that shown in Fig. 11 and somewhat less clearly in changes in the progression bands in GA-FTIR experiments (Fig. 13) [78]. Although the data discussed pertain to the formation of CdS in ArH films, similar increases in tilting and bilayer thickness and disorder during particle formation have been observed for most LB films, and the majority of references in Table 1 contain some information about the effects of particle formation. Although a decrease in film thickness has been observed for the formation of CdS in StH [101], this has been explained by the high pressure of H_2S used [87].

Films have sometimes been observed to change spontaneously after particle formation, with coagulation seen in AFM measurements [102] and migration toward the substrate caused by sulfide reaction indicated by RBS analysis [84]. However, in general the effect of particle formation introduces a degree of film disruption while maintaining overall short-range structure throughout the whole of the film.

(*d*) *Rate of Reaction.* It seems that there are both chemical and physical factors that govern the rate of reaction of M^{2+} with H_2E. The formation of ME molecules [Eq. (3)] seems to be dependent on the immediate chemical environment of M^{2+}. High degrees of rigidity and structure in the films, and the absence of water, restrict the rate at which the ME molecules coalesce [Eq. (4)].

In general, it is difficult to distinguish the reactions corresponding to Eqs. (3) and (4) as they usually occur simultaneously. Structure, however, is known to play a role in the kinetics of particle formation, as it does in limiting the size of particles formed. Investigations by Peng and coworkers into the effect of film preparation on reaction rate illustrate this very well. For example, reactions in films with greater structure caused by a greater dipping speed [131], or higher transfer pressure accompanied by an increased dipping speed [99], were significantly slower as shown by FTIR spectra. The choice of metal ion, too, can influence the structure of the film and hence the rate of reaction. Long-range order, which is greater in $PbSt_2$ than $CdSt_2$ [149], and tight packing, greater in $ZnBe_2$ and $CuBe_2$ than $CdBe_2$ [105], has been shown to slow the rate of reaction. The properties of the surfactant can also affect the structure of the film and the rate of reaction as shown by the rates of HgS formation in ArH and BeH [137]. The pressure of gas used in the reaction may also affect the rate of particle formation [99,100,125], and the diffusion of gas through the film has been shown to be governed by the film structure, illustrated by a dependence of reaction rate on the thickness (number of layers) of the film [99,100]. The low pressure of H_2S used in these experiments may partially explain this result.

Experiments have shown the chemical rate-determining step to be the deprotonation of the sulfide. The protons released by the sulfidation of the metal ion need a "sink," usually provided by the carboxylate group of the fatty acid. By removing possible sources of sinks, the whole reaction is slowed or stopped. This is illustrated by the reaction of H_2S with films formed from diallyl dimethyl ammonium bromide (DDAB) and $[PdCl_4]^{2-}$ [62]. $[PdCl_4]^{2-}$ readily undergoes hydrolysis in the subphase conditions to form $[PdCl_n(OH)_m(H_2O)_l]^{2-}$, but this can be suppressed by the addition of HCl to the subphase. Films formed in the presence of HCl, therefore, contained no base (i.e., proton sink) in the coordination sphere of the ion. In these films there was no noticeable reaction on exposure to H_2S (similar to the nonreaction observed by Ichinose et al. for Cd^{2+} in already protonated amino amphiphilic films [150]), and these may possibly be compared to similar films containing $[PtCl_4]^{2-}$ and $[PtCl_6]^{2-}$ (complex ions that do not hydrolyze), which took days to react completely [62]. Spectra taken of $(PdCl_4)(DDA)_2$ films made with and without HCl before and after exposure to H_2S are shown in Fig. 15. It can be seen that for the film where the $[PdCl_4]^{2-}$ is

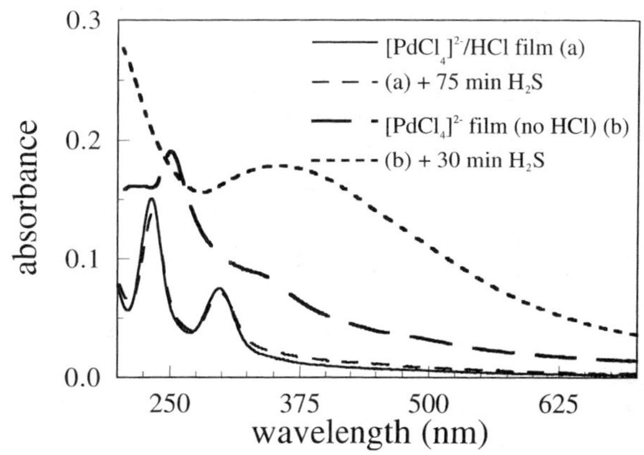

FIG. 15 UV-vis absorbance (corrected for the absorbance of the amphiphile and the substrate) of LB films formed from DDAB on $[PdCl_4]^{2-}$ and $[PdCl_n(OH)_m(H_2O)_l]^{2-}$. The latter film was made by the inclusion of HCl in the subphase. The spectra are taken before and after H_2S exposure for the times indicated [62].

hydrolyzed, the spectrum is changed due to PdS formation after only 30 min. On the other hand, 75 min of exposure to H_2S was insufficient to change the spectrum of the film containing the unhydrolyzed complex. The times required for complete particle forming reaction have varied from minutes to 368 h for H_2S with $(PdCl_4)(DDA)_2$ [62] but are difficult to compare given the large range of conditions and factors that must be taken into account.

In essence, nanoparticles can be readily formed in LB films. The time necessary for particles to form, the size they reach, and the degree to which the films are altered by their formation are all factors that depend on the individual constituents of the film and the conditions of its preparation and treatment. These considerations, if understood sufficiently, could provide means to tune the properties of the particle-films to some extent and optimize them for specific applications.

B. Self-Assembling Films

LB films are ideal for manufacturing particle-films with molecular-layer precision and close interaction between the film medium (amphiphile) and the particle surface. Their fastidious manufacture is, however, a disadvantage when relatively thick films are required. Self-assembled films can also be used for the in situ formation of nanoparticles. As the particle surface interacts with charged groups in the film, the possibly synergistic properties of the matrix and nanoparticles are retained, but films up to many micrometers in thickness can be prepared.

1. Nafion

An example of a self-assembling material used in the preparation of nanoparticle-films is the cation-exchange polymer Nafion. As we have first-hand experience of its use, it will be discussed in some detail, although similar particle-films can be formed using other polymers such as polyethylene–polymethacrylic acid copolymer [151, 152].

(*a*) *Film Preparation.* Nafion has the "intelligence" to form a regular matrix of cavities connected by narrow tunnels. The cavities and tunnels are lined by sulphonate (SO_3^-) groups, which undergo exchange with M^{2+} in a process similar to the proton exchange in Langmuir monolayers and LB films outlined earlier. Exposure of the membrane to gaseous or aqueous S^{2-} results in the formation of semiconductor nanoparticles. Countercurrent techniques where H_2S and Cd^{2+} have been introduced simultaneously from opposite directions [153] have been claimed to lead to more uniform films [154] than those produced by sequential cation exchange and chalcogenide exposure. The types of particles prepared in Nafion membranes are listed in Table 2.

(*b*) *Control of Particle Size.* In the first experiments of this type [153,160,161,170,171,173,174], the particles formed tended to be relatively large particles (e.g., up to 1 μm). However, particles in the tight confinement region of size quantization have been formed since and considerable control over particle size has been developed. Methods for controlling particle size include the use of dry chalcogenide gas [162,165–167,176] (wet gas was used in the initial experiments). The slow rate of particle growth in films exposed to dry H_2S allows the growth to be arrested by instant degassing when the particles reach the desired size. It is then a relatively simple procedure to grow the particles to different sizes by Ostwald ripening by exposing the films to water at various temperatures [167,178] with the possible addition of more H_2S [165]. The use of Na_2S instead of H_2S provides another method for producing larger particles [165,167] and sonication has been used to grow the particles [165]. Controlling the concentration of M^{2+} in the films limits the size of particles that can be formed. Immersion in a dilute solution of M^{2+} [176] and dilution of the concentration of active ions by the introduction of inert ions such as Ca^{2+} [165] are two methods which have been used. The synthesis conditions can also influence the crystal structure of the particles prepared [174]. When the films are constructed to probe the physicochemical properties of nanoparticles, ammonia is often introduced to the Nafion film prior to M^{2+} exchange, and this serves to passivate the film [164,167,178]. Examples of absorbance spectra showing the effect of film treatment on particle size can be seen in Fig. 16. Platinum metal has been prepared at and under the surface of Nafion films by reacting exchanged Pt^{2+} with $NaBH_4$ [155]. In a method corresponding to 2a in the Introduction, Fe_2O_3 has been made by the hydrolysis of Fe^{3+} in a Nafion solution prior to the film being cast [179]. Particle size was controlled by the metal ion/sulfonate ratio. Particles of TiO_2 have also been incorporated as Nafion films were cast (method 2b in the Introduction) and grown through an aging process, but it was not clear whether the particles were on the surface or within the

TABLE 2 Nanoparticles That Have Been Formed Within Nafion Films

Particle	References
Pt	[155]
TiO_2	[156–158]
SiO_2	[159]
CdS	[145,154,156,160–169]
CdS.ZnS	[170–173]
CdS/Pt	[174]
CdSe	[165,175]
PbS	[176]
FeS_2	[177]

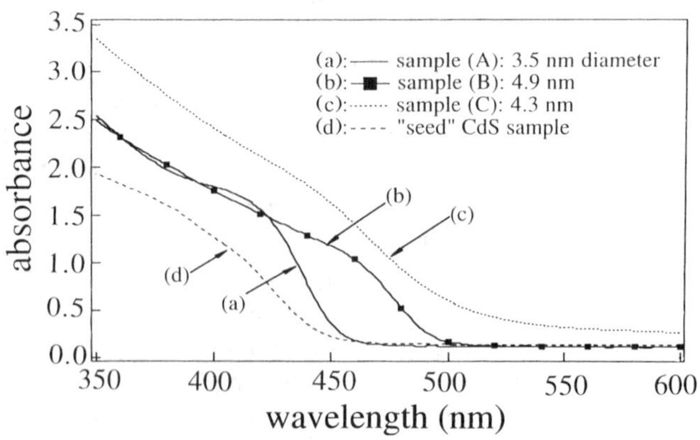

FIG. 16 Absorbance spectra of CdS nanoparticles in Nafion films. The diameter of the particles estimated from the inflection points (indicated) is also shown. Sample (d) was prepared by exposing a dried Cd^{2+} exchanged film to ammonia gas and then to H_2S. Samples (a) and (b) underwent similar treatment to (d) but were subsequently immersed in water at room temperature (a) or boiling water (b). Sample (c) was prepared by immersing the dried Cd^{2+} exchanged film in Na_2S solution [167].

cavities of the film [158,180]. When nanoparticles form within Nafion films, the molecules coalesce inside already existing cavities, so considerably less research has been conducted on structural and mechanistic effects than has been undertaken for LB films, although the structure of the particles formed was found to be dependent on the method of preparation [153,174].

2. Cast Bilayer Films

Another example of a self-assembling amphiphile that has been used to form nanoparticles is tetradecyl-*N*-[[4-[6-(*N*,*N*′,*N*′-ethylenediamino)-hexyl]oxy]benxoyl]-L-glutamate (DTG). The cadmium salt of this surfactant ($Cd(DTG)_2$) was exposed to dry H_2S resulting in the formation of CdS particles about 4 nm in diameter [181,182]. If the film was cast as a multibilayer structure prior to Cd^{2+} intercalation, the product of H_2S exposure depended on the nature of the Cd^{2+} salt used ($CdCl_2$ or $CdBr_2$) [150]. CdS has also been formed in amphiphilic cast films containing cyclams as size-restricting units. The particles did not interact strongly with the amphiphile once formed, and this hindered their potential use [183]. Lead halide nanoparticles have been prepared by similar procedures in other bilayer cast films [184]. The diffusion of nanoparticles into similar multibilayer films created by thermal evaporation was discussed earlier.

C. Layer-by-Layer Assembly

The layer-by-layer (LbL) assembly technique has been known variously as the alternate or layer-by-layer (self) assembly, the ionic self-assembled monolayer, or the molecular deposition method. The films are similar to LB films in the sequential nature of their assembly, but the degree of control in LbL films is not as fine. It has also been necessary for the components, usually stabilized nanoparticles and polyelectrolytes for particle-film formation, to be formed prior to film assembly (method 2b in the Introduction). Nanoparticle-films made by this method have been widely studied as they are very simple to prepare and a wide range of materials can be incorporated [185–187]. As the preprepared particles are usually surrounded by a stabilizer that interacts with the other material in the film, the surface of the semiconductor or metal is not in contact with the film. This makes these films different from the LB and self-assembled films discussed earlier.

1. Film Preparation

This method of assembling thin films has been developed as a technique for forming polyelectrolyte films [185,188]. The method is simple and involves the immersion of a charged substrate alternately in solutions of oppositely charged polyelectrolyte. (The substrate is usually rinsed between steps.) With each immersion, polyelectrolyte is adsorbed, overcompensating the existing surface charge and thereby reversing it. In this way, a film of oppositely charged layers is built up. LbL films can be prepared on any charged surface and substrates are not restricted to flat surfaces. Colloidal silica particles are an example of an unusual substrate that has been used [189–191]. Charged particles (it is usually the stabilizer that bears the charge) can take the place of polyelectrolytes, and the assembly process for the case of a cationic polyelectrolyte and anionic particles is illustrated in Fig. 17.

Charged particles that have been incorporated in polyelectrolyte films include biological entities such as proteins [192–195] and virus particles [196], dyes [197,198], latex particles [199], inorganic charged particles such as clay platelets [200–204], and oxides such as oxidized graphite [205], silica [190,191,206,207], and zirconia [208]. Ultrasmall inorganic particles have also been incorporated into films using layer-by-layer assembly. These include semiconductor nanoparticles such as CdS [200,201,209], CeO_2 [210], TiO_2 [200,201, 210,212], CdSe [212–214], Fe_3O_4 [215,216] and PbS [200,201], and large metal complexes [189,217,218]. Gold particles have also been incorporated in this way [219–221]. The films containing the nanoparticles just listed have consisted of alternate layers of polyelectrolyte and particles, often including varieties of both in the films. Polyelectrolytes commonly used have included the anion polystyrene sulfonate (PSS and cations poly(allyl hydrochloric acid) (PAH) and poly(diallyldimethyl ammonium chloride) (PDADMAC). Semiconducting polyelectrolyte precursors such as polyphenylenevinylene (PPV), which are polymerized after the films are formed, have also been used to prepare LbL light-emitting diodes (LEDs) [213,214].

Bipolar molecules have also been used to form films, such as the assemblies of bipolar pyridinium and PbI_2 nanoparticles prepared by Gao and coworkers [222,223]. The interaction between the particles and the polymer may be more than simple electrostatic attraction, with displacement of existing surface molecules [217], coordination complexes [224], and specific (biochemical) interactions [225] all being used to bind the constituents. Films have also been prepared consisting purely of oppositely charged particles [226,227]. The preemptive work of Iler, in fact, concerned the buildup of oppositely charged alumina fibrils and silica [228]. A similar technique of layer-by-layer deposition using bridging molecules such as dithiols to replace existing

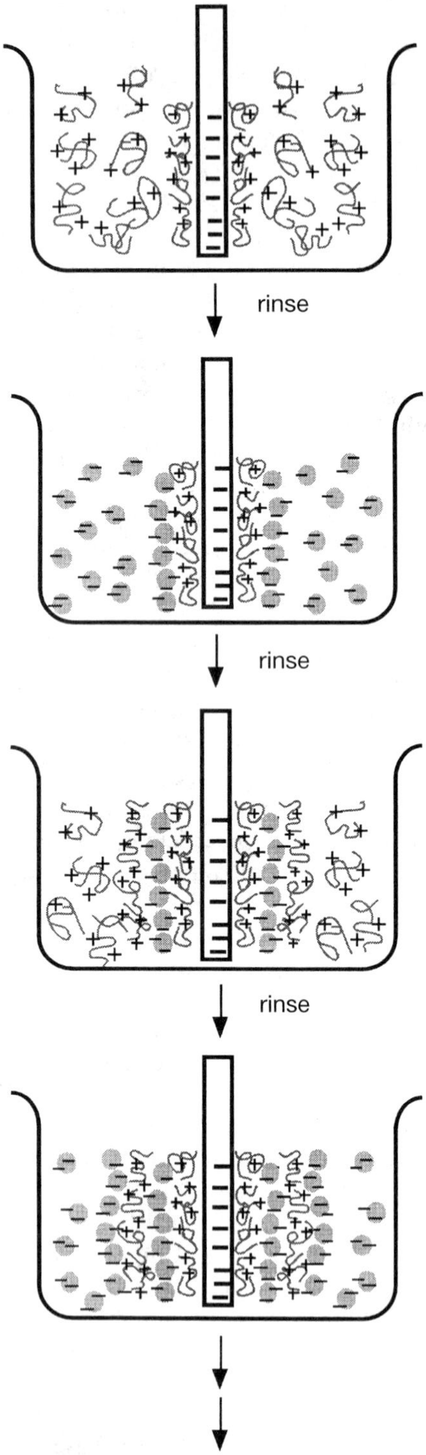

FIG. 17 The layer-by-layer assembly process for the production of a film on a negatively charged substrate, incorporating a cationic surfactant and anionic particles. The substrate is alternately immersed in the solutions and rinsed in between until a film of the desired numer of layers is achieved.

stabilizers, and which form covalent rather than electrostatic bonds, has been used to construct particle-films [212,229–234]. Also, in an innovation that increases the possible interaction of particle and film, CdS [235] and Cu_2S [236] nanoparticles have been prepared in situ by exposing M^{2+} loaded PSS/poly 4-vinylpyridine (PVP) films to H_2S. In situ synthesis of particles in a manner similar to that in LB films may also be possible in films such as those prepared by Decher and Hong [236] and Gao and coworkers [223] in which bipolar amphiphiles have been incorporated; however, this is an area that is yet to be investigated.

Given the simple nature of the assembly process, it is straightforward to build films with multiple components for particular purposes. Films containing different semiconductor nanoparticles, deposited either in sequence [200,201,205] or premixed [210] before absorption, are examples of slightly more complicated systems. As another example, polyelectrolyte films containing ordered layers of metal particles and insulating clay have been prepared for use as small-scale capacitors [220]. Multicomponent protein films able to undergo sequential enzymatic reactions have also been manufactured [191,225].

2. Characterization of Films

The underlying mechanism of alternate adsorption has been verified by a variety of techniques. For polyelectrolyte films, zeta potential [237] measurements have revealed definite reversals of charge with the successive adsorption of layers. More general techniques have been used for films containing particles. Cationic particles (TiO_2) that adsorb onto anionic polyelectrolytes (PSS) were found not to adsorb onto a cationic silanated surface, verification of the significance of electrostatic attraction [211]. A similar effect was noted as a result of pH. SiO_2 is negatively charged at pH 10 and would not adsorb onto an anionic poly(styrenesulfonate) surface [210]. Contact angle studies have supported the notion of alternately layered films with the contact angle dependent on the nature of the most recent component added [211,215].

(*a*) *Film Structure.* It has been observed by UV-vis spectroscopy measurements that the amount absorbed with successive layers is consistent for up to 60 layers for TiO_2/PSS films [211] and hectorite/PDADMAC [238]. An example of the type of spectra that can be obtained is shown in Fig. 18. The inserted plot shows the consistent nature of film buildup. Similar linear plots can be obtained from QCM gravimetry measurements when the total mass added is plotted as a function of the number of bilayers deposited [209]; how-

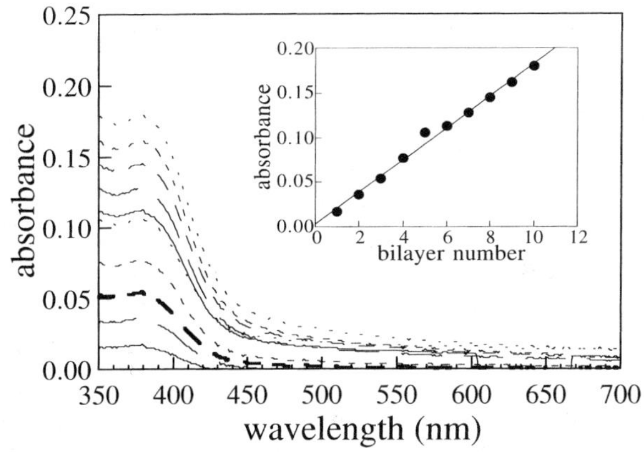

FIG. 18 UV-vis absorbance spectra for the alternate absorption of CdS [mercaptoethanol stabilized; [Cd^{2+}] (0.02 M)]; PDADMAC (1% w/v); all solutions pH 11. Spectra were taken after each CdS layer was absorbed. Inset: Absorbance at 380 nm as a function of the number of bilayers [1 bilayer = 1 layer (PDADMAC) + 1 layer (CdS)].

ever, the data are most frequently presented as a staircase showing the total mass as it changes with each deposition [207,210] such as the example for CdS and PDADMAC in Fig. 19. This technique has the advantage of showing changes with each step, even for non-absorbing materials. Analyses by ellipsometry also reveal a linear relationship, in this case between film thickness and the number of immersion cycles [203,208,217,219]. Surface roughness was found to be independent of the number of layers, indicating a consistent buildup of layers [213].

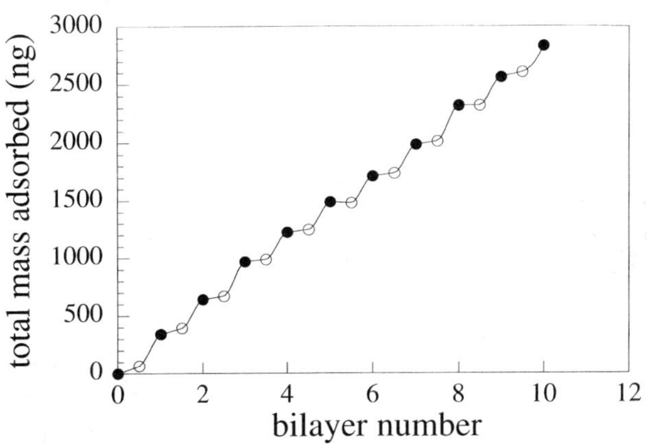

FIG. 19 The total mass adsorbed on the QCM (both sides) as function of the layer number on the surface of a 9-MHz QCM electrode as alternate layers of PDADMAC (●) and CdS (○) are deposited by the LbL technique. Solution conditions: CdS [mercaptoethanol stabilized; [Cd^{2+}] (0.02 M)]; PDADMAC (1% w/v); all solutions pH 11.

Sometimes it has been observed that linearity is not achieved in the first (up to five) pairs of immersions, and many groups have found it necessary to pretreat a substrate surface before adsorbing the first significant layer. Examples of pretreatment include the use of polyelectrolytes such as PSS and PEI [poly(ethyleneimine)] [210] and silanization [222]. The application of an LB monolayer has also been shown to be a suitable foundation for regular polyelectrolyte film buildup [239]. Others, however, have claimed that pretreatment is unnecessary for the assembly of high-quality films [201,202]. For polyelectrolyte films of PTAA/PAH, UV-vis measurements for up to 10 layers showed very little dependence on the pretreatment of the glass substrate [240].

Films formed by the LbL technique are relatively imprecise in structure and there is thought to be considerable leakage between the layers, in particular, in films containing particles where it is impossible to form a film of 100% surface coverage. Studies have shown that if polyoxometalate particles are absorbed onto multilayers of polyelectrolytes, then the thickness of the preceding layers affects the number of particles absorbed, indicating considerable leakage into the polyelectrolyte layer [218]. There is evidence that particles can cross intervening polyelectrolyte layers. The UV-vis absorbance spectra and XRD reflectivity measurements of gold nanoparticles in LbL films showed diminishing interaction as the thickness of the polyelectrolyte multilayers between the nanoparticles increased [221]. Other gold-containing films with single polyelectrolyte layers between the particles demonstrated resistivities comparable to those of bulk gold [219] (others introduced clay particles to increase resistivity and formed a capacitor [220]). The same group found that the layer spacing was less than the diameter of the gold particles [219], indicating interpenetration of particles between layers, a conclusion consistent with the three-dimensional tight packing of particles observed in scanning electron microscopy (SEM) images of SiO_2-containing films [207]. Many groups have found peaks in x-ray reflection and diffraction analyses indicating a periodic, layered structure, even for films in which the particles were adsorbed with only a single deposition of polyelectrolyte between them [200,203, 209,213], whereas others have been unable to resolve layer spacings [218]. The range of contrasting observations may be caused by the large number of factors that can influence the film during its formation.

(b) Particle Concentration and Film Thickness. The most significant variables in LbL particle-films are the

thickness of the films and the concentration of the particles. These factors are very closely linked and can be controlled to a certain extent by the solutions used to form the films. Film formation is an adsorption process, so constituent solution concentration and the amount of time for which the substrate is immersed in solution both play a role. The plot in Fig. 20 illustrates the effects of exposure time and the concentration of particles in solution for films made from PDADMAC and mercaptoethanol-stabilized CdS nanoparticles (CdS/ME). The UV-vis absorbance of the film is proportional to the concentration of particles in the film and can be seen to depend on time of exposure only when more dilute solutions of CdS are used. In this example the equilibrium number of particles in each layer was independent of particle concentration. Others have found that the saturation amount depends on the concentration of the particle solution [199,207,210].

As film formation is electrostatically driven, the charge density on the constituents and the presence of screening electrolytes will also play a role. The lower the charge density of the polyelectrolyte, the larger the loops formed when it absorbs onto the surface. This in turn allows the adsorption of greater amounts of nanoparticle into the succeeding layer and thicker films are formed [199]. Apart from the resulting dependence on the identity of the polyelectrolyte, the number of particles in the film will be affected by the pH dependence of charge on weak electrolytes and on the particle surface. This is especially noticeable in bioactive particles [193].

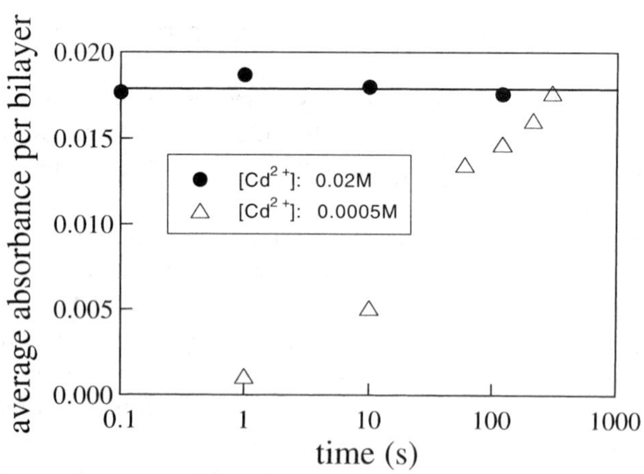

FIG. 20 The average UV-vis absorbance (at 380 nm) per bilayer for the assembly of CdS (mercaptoethanol stabilized) and PDADMAC (1% w/v) as a function of immersion time in the CdS solution. The concentration of the Cd^{2+} in the colloidal solution prior to H_2S exposure ([Cd^{2+}]) was 0.02 M. Films were made from this solution (●) and from the same solution diluted by a factor of 40 (△).

The presence of salt further complicates matters as increased ionic strength of the colloidal solution acts to screen charges and allows tighter packing of particles and a higher concentration of particles within the film [199,206,207,210]. SEM images show close-packed particles (not layers) with a very smooth surface at NaCl concentrations of 0.1 M. The use of higher salt concentrations led to increased surface roughness, similar to that seen in films containing only polyelectrolytes. Increased ionic strength of the polyelectrolyte solution has a similar effect on charge and allows much denser layers to form. More tightly packed particle layers have also been prepared by compacting each PDADMAC layer by microwave treatment [216]. This treatment reduces surface roughness and produces more ordered nanoparticle layers on TEM analysis. The degree of packing, and therefore the amount of particles absorbed in each layer, is also influenced significantly [199,210,221] or weakly [206] by the size of the particles.

Many factors have been shown to influence the nature of LbL particle-films. It has been postulated that the enormous influence of variables such as particle and salt concentration is due in part to the mismatch between the rigidity and charge density of the particle surface and the polymer [210]. A further way to increase the concentration of particles in the films and its thickness is to apply a potential to the substrate as the adsorption process is under way. The application of a positive potential to a substrate increased the amount of anionic clay adsorbed at each step [200].

It is common practice to dry the film between immersions, especially when QCM or UV-vis measurements are to be made, and this may also affect the properties of the film. Drying was shown to alter film surface structure reversibly when it allowed the adsorption of a PSS layer onto a previously deposited PSS layer, a situation that did not occur if the film remained undried [241]. Any effect drying may have on the long-term properties of a film has yet to be established. Films prepared by LbL assembly have been found to be substantially harder and more stable than those prepared by sputter coating, especially when subjected to polymer burning and annealing treatment [208].

The application of LbL films to practical devices depends to some extent on the ability to control film thickness and particle concentration. This can be done by varying the conditions of film preparation as already outlined. Given the enormous variety of constituents that can be incorporated in prescribed orders into these films, LbL assembly has become another tool in the particle-film toolbox.

III. PHYSICOCHEMICAL PROPERTIES OF FILMS

The possible uses for semiconductor nanoparticles in photo-optic, optoelectronic, and photochemical applications is a major driving force behind the research being carried out in this area. The electronic and optical properties of nanoparticles can be tuned over a wide range by varying the size of the particles and their immediate environment and by combining them with other photo/opto/electroactive materials including other semiconductors and metals. To utilize the properties of nanoparticles it is often necessary for a significant area of nanoparticles to be exposed to a light source and/or an electrically active environment. Amphiphilic films provide a means of creating these domains while maintaining a fine degree of control over the scale of the film in the third dimension. By building extremely thin films, one can utilize the size-quantized properties of the semiconductors in the nanoscopic dimensions necessary for modern devices.

A. Photoelectrochemistry

Examples of uses of particle-films are LEDs [213,214, 242,243] made from semiconductor nanoparticles and semiconducting polymers in LbL films. Similar films using nonconducting polyelectrolytes have been used in photocells [201] and, with the addition of clay particles, in capacitors [220]. Photoelectrochemical [154, 158,176] and photocatalytic [153,156–158,172,174] cells have been constructed using various semiconductors in Nafion films. Nanoparticulate CdS grown in situ in LB films [79,119], PbS incorporated in LB-like films [33], and TiO_2 electrophoretically deposited in films [244] are examples of photoactive semiconductor particles in other types of films. Single-electron tunneling in CdS particles formed in LB films in a phenomenon of much interest to the electronics industry [245]. For optimal use of the size quantization effects of semiconductor nanoparticles, the luminescence or redox potentials reflect the increasing band gap of smaller particles. An example is the size-dependent photovoltages seen in CdS/Nafion films [158,166,176,179] and CdS/glutaraldehyde films [33]. Following is a discussion of some investigations carried out by our group into thin-film photoelectrodes made from nanoparticulate CdS and certain LbL and LB films in PEC cells.

1. Langmuir-Blodgett Films

Particle films can be applied to almost any substrate surface as long as the surface has been rendered sufficiently charged or hydrophilic/phobic, depending on the type of film being constructed. If the film is assembled on a conducting substrate, it can serve as an electrode of a photoelectrochemical cell. Of course, it is necessary that the film is stable under the cell conditions and that the ions of the electrolyte can permeate through the film.

LB films of CdS grown in situ in ArH were prepared on indium tin oxide (ITO)-coated quartz, with the size of the CdS nanoparticles increased by the intercalation-sulfidation method described earlier. Figure 21 shows typical *I-V* curves of a cell with this electrode in dark and light conditions. The fill factor of the light curve is 0.25, indicating significant inefficiencies, which are probably due to surface defects. When films containing differently sized particles were used as the photoelectrode, size-dependent photovoltages were observed. Figure 22 is a plot of photovoltage as a function of particle size for 19-layer CdS-ArH films. If size quantization was being observed, an increased photovoltage due to an increased band gap would be seen in smaller particles. Figure 22 shows the opposite effect, with a decrease in photovoltage accompanying a decrease in particle size. One possible explanation is the insulating effect of the surfactant in the LB film. Long-chain fatty acids are known for their insulating properties and have been used in some electronic devices to prevent elec-

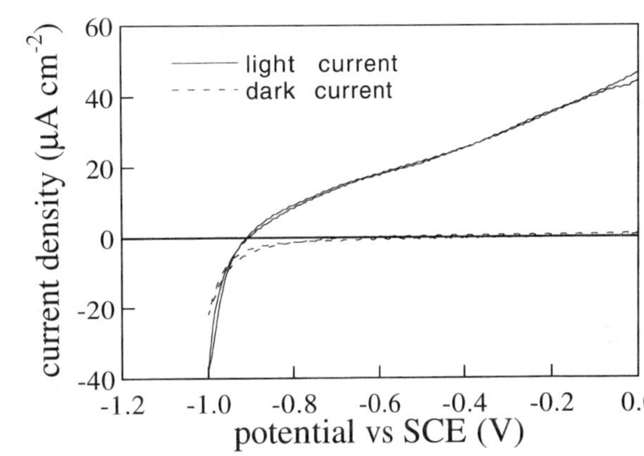

FIG. 21 Typical dark and light cycle voltammograms for CdS (4 nm) in a 19-layer ArH LB film. The electrolyte is Na_2SO_3 (1 M, pH 7.25) and the potential was cycled from 0 to −1.0 V and back at 10 mV min^{-1}.

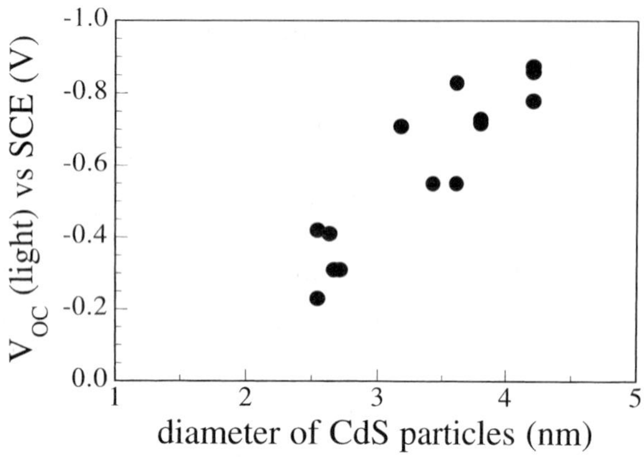

FIG. 22 Photovoltage as a function of particle size for CdS/ArH particle films. The electrolyte was Na_2SO_3 (1.0 M, pH 7.25).

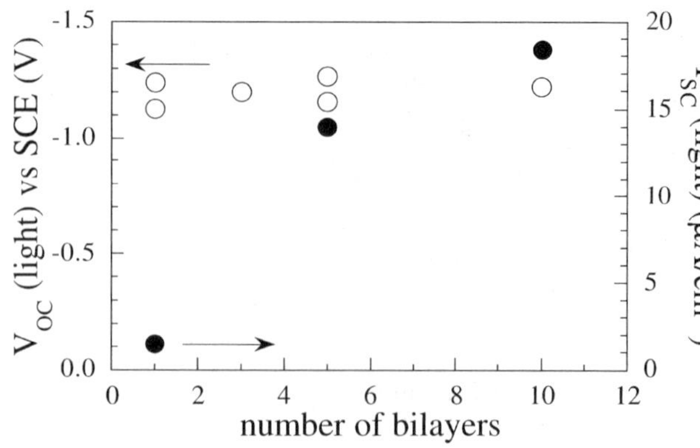

FIG. 23 V_{OC} (○) and I_{SC} (●) as a function of the number of bilayers in CdS (mercaptoethanol stabilized; [Cd^{2+}]: 0.02 M, 2.5 nm) and PDADMAC (1% w/v) LbL film electrodes in Na_2SO_4 (0.5 M) and TEA (20 mM) electrolyte at pH 9.65.

tron transport. In LB particle-films, smaller particles (2–3 nm) are separated by relatively large volumes of surfactant when compared with films containing the same concentration of larger particles (4–6 nm) that are thought to be close enough to allow interparticle electron transfer. The large particles would provide electron-conducting pathways through the film, overcoming the photovoltage-diminishing retardation due to the surfactant.

2. Layer-by-Layer Films

Thin-film photoelectrodes have also been constructed using the LbL technique to build alternate layers of PDADMAC and colloidal stabilized CdS on ITO plates. The preprepared nanoparticles were stabilized in solution by mercaptoethanol (CdS/ME) and hexametaphosphate (CdS/HMP). The size of the particles was controlled by adjusting the pH of the solution prior to the addition of H_2S in a method similar to that used by Henglein and coworkers [142]. As the pH was increased from 7 to 11, particle sizes increased from 3.5 to 7 nm for CdS/HMP and from 2.5 to 3.5 nm for CdS/ME.

In PEC cells with working electrodes of LbL particle-films on ITO-coated glass, the photocurrent was found to be dependent on the number of layers, whereas the photovoltage was found to be independent on film thickness (Fig. 23), results similar to those found for LB films. The films produced by LbL assembly were more stable than LB films and have been illuminated for over 15 h with no reduction in photovoltage magnitude. Results for photovoltage as a function of particle diameter (determined from Henglein's calibration curve) are shown in Fig. 24. It seems that, in contrast to the results for the LB film electrodes, a size quantization effect is seen, with the photovoltages of smaller particles significantly greater than that of bulk CdS. The photovoltages observed were lower than expected from the band gaps seen in the absorbance spectra. This is clearly seen for the larger particles in Fig. 24, which showed photopotentials lower than those of the bulk. The explanation may lie in either the resistance of the polyelectrolyte surrounding the particles or the presence of surface states that would lower the effective conduction band of the semiconductor particles. To understand the significance of these factors, experiments need to be undertaken to determine the

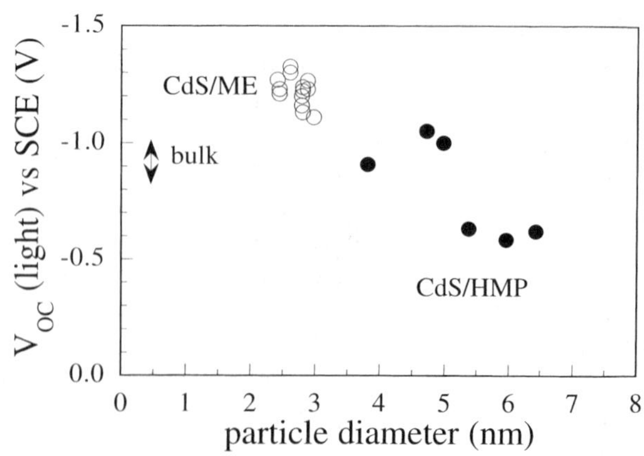

FIG. 24 Photovoltage produced by LbL particle-films of CdS [stabilized by mercaptoethanol (○) and HMP (●)] and PDADMAC as a function of particle size (estimated from UV-vis absorbance spectra). The electrolyte was Na_2SO_4 (0.5 M) and TEA (20 mM) at pH 9.65.

effects of capping agents and particle density on the cell characteristics.

3. Electrolyte Effects

The results outlined for LB films and LbL films are themselves not directly comparable as different electrolytes have been used in the cells. Electrolyte conditions are very important to quantitative analysis, as will be illustrated for the CdS/ME/PDADMAC system in Na_2SO_4/triethanolamine (TEA) electrolyte. When the concentration of TEA is increased, the magnitude of the photovoltage is also observed to increase (Fig. 25). A similar, although less significant, tendency is observed for an increase in pH. Electron donors or acceptors must be adsorbed onto the semiconductor surface to enable electron transfer in <0.3 ns after excitation, to prevent electron-hole recombination and facilitate current flow. Increasing the concentration of the donor in solution can increase the amount adsorbed on the particle surface, decreasing the number of deeper surface traps and increasing the magnitude of the photovoltage possible. The effect of pH cannot be so simply explained as adsorption and/or redox characteristics may be responsible. TEA (pK_a 7.8) has fewer charged species at higher pH; therefore adsorption will be greater than at low pH, where repulsion between amine groups and positive cadmium sites on the particle surface may occur. At lower pH, TEA, being more likely to be positively charged, is less likely to act as a hole scavenger. Its increased oxidation rate and potential at higher pH may be sufficient to explain the effects observed.

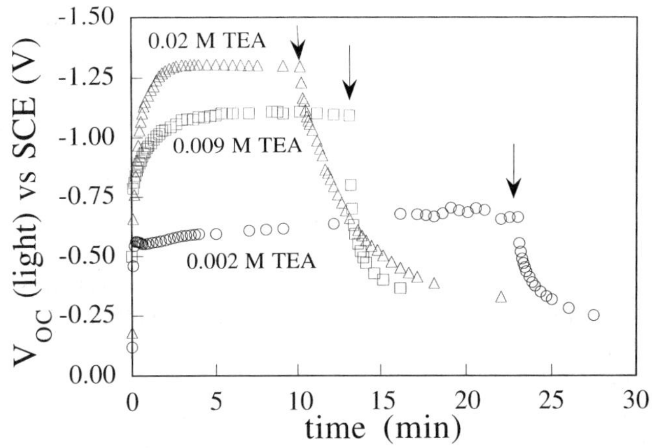

FIG. 25 Photovoltage as a function of illumination time for CdS (mercaptoethanol stabilized, 2.5 nm) and PDADMAC (1% w/v) LbL film electrodes. The electrolyte was Na_2SO_4 (0.5 M) and TEA [0.02 M (△); 0.009 M (□); 0.002 M (○)] at pH 9.65. The arrows indicate when the lamp was turned off.

B. Photobleaching

In addition to their unusual electronic/redox properties, size-quantized semiconductor particles exhibit nonlinear optical (NLO) properties, an example of which can be seen in transient absorbance bleaching and recovery [163,178]. The recovery time for size-quantized semiconductor bleaching depends on electron-hole recombination kinetics and is of the order of femtoseconds. As silicon electronic switches operate on a picosecond time frame, there is an order of magnitude advantage in using optical switches incorporating nanoparticles. Optical switches, like photoelectrodes, require a thin, optically transparent film, and LB films are ideal. To investigate photobleaching properties, however, it is often necessary that the absorbance of the films is high (e.g., >2 at 430 nm) and very thick LB films (e.g., over 5000 layers) would have to be prepared. As the assembly of such thick LB films is impractical, CdS was produced in situ in self-assembled films of Nafion [167,182] and ditetradecyl-*N*-[4-[[6-(*N*,*N*′,*N*′-trimethylethylenediamino) - hexyl]oxy]benxoyl] - L - glutamate (DTG) [167,181,182]. The sizes of the particles formed varied depending on the reaction conditions. The samples were irradiated with 400-nm laser pulses from a femtosecond laser and transient photobleaching spectra were measured.

An example of a time profile for photobleaching is shown in Fig. 26. The maximum bleaching for this sample of 4.9-nm CdS in Nafion was about 1 ps. By fitting the data with bioexponential kinetics, a fast com-

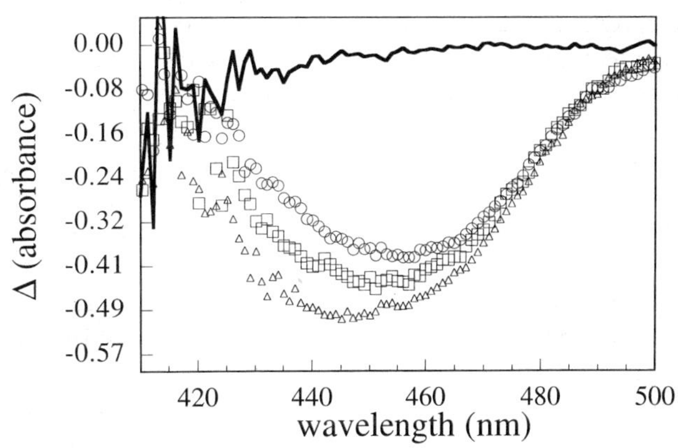

FIG. 26 Time-resolved transient bleaching spectra of 4.9-nm CdS particles in Nafion (pump energy = 15 μJ per pulse) after 0 (——), 2.6 (△), 3.9 (□), and 6.5 (○) ps. The films were excited at 400 nm with femtosecond pulses [167].

ponent of lifetime 8 ps (45%) and a slow component were revealed. Different-sized particles gave different results but of similar orders of magnitude. It is thought that two types of electron-hole recombination are responsible for the two observed components of photobleaching recovery: direct recombination and recombination from surface states. Electrons trapped in surface states or defects have longer recombination times than those undergoing direct recombination. It is, however, possible to modify the surface of CdS to eliminate the surface traps by the formation of a $Cd(OH)_2$ layer on the surface [141]. Transient photobleaching of aqueous colloidal CdS modified in this way has shown faster recovery times than for untreated samples [141].

IV. OVERVIEW

The use of nanoparticles in electronic and optical devices is an area of increasing interest that has stimulated an expanding field of research. Nanoscale particle-films often display optical transparency and size quantization effects, the combination of which makes them particularly attractive for use in photo-optic, -voltaic, and -electronic applications. Amphiphilic and polyelectrolytic films are very attractive media for the formation, manipulation, and preservation of nanoparticles, and the incorporation of nanoparticles in several types of films has been discussed.

Thin organic films can incorporate nanoparticles in a variety of ways. The particles can be prepared separately and then incorporated in the organic film or can be synthesized using the functionality of the organic material to "cap" the particles. The layer-by-layer film assembly technique is best suited for the incorporation of premade particles that have a surface charge. The charge is usually associated with the stabilizing agent on the surface and facilitates the alternate assembly with oppositely charged polyelectrolytes. Characteristics of LbL films, such as thickness and particle density, are influenced by factors that affect the charge and charge interactions of the species involved in the film structure. Solution pH, background salt, and surface charge density all play a role. As the films are constructed one layer at a time, considerable control can be exerted over the placement of the constituents of the films and their thickness can be limited to the nanometer scale. Thicker films, with much higher optical densities, are sometimes required for testing phenomena such a photobleaching. In these cases, it is often convenient to form the particles in situ, in self-assembling films.

The growth of the particles is limited by the spaces within the film structure, which may be cavities or bilayer spacings, and often there is significant interaction between the functional groups of the self-assembling polymer and the nanoparticle surface. This internal particle-matrix interaction is also present when particles are formed using self-assembling amphiphile structure to assist in growth restriction, such as when particles are synthesized under Langmuir monolayers and in vesicles. It is particularly evident when particles are grown within the confines of a Langmuir-Blodgett film. The film structure restricts the growth of the particles and evidence suggests that the particles are capped by the surrounding amphiphile, in a manner similar to that found for solution colloids. The particle size is influenced by the preparation conditions, which affect parameters such as reactant ion concentration and film structure, but once formed they can be grown within the films by a number of techniques. The film tends to be disrupted by the formation of particles but maintains an overall bilayer structure and a degree of crystallinity. Similarly to LbL films, LB films are assembled in a sequential manner, but with a greater degree of control on a molecular level. They are, however, less robust than LbL films and limited to the incorporation of a smaller range of materials.

The possibilities for manipulating the properties of LB particle films lie mostly in altering the nature of the amphiphile. As an example, more durable films have been prepared by polymerizing the surfactant after particle formation [109,113]. As well as limiting the size, the amphiphile might be used to control the shape of the particles. Triangular and/or rodlike particles are often formed at Langmuir monolayers [10] and in vesicles [246]. It is conceivable that a similar variety of particle shapes could result from in situ synthesis in LB films, widening the potential applications of these films. Another way to modify properties by altering constituents is to use a functionalized surfactant with photophysical-chemical properties. This could act as a sensitizer and undergo photoexcited electron transfer with a film-incorporated semiconductor particle, in a similar manner to the dye-semiconductor system outlined by Kamat [247,248]. Such a system would fully realize the potentially powerful synergistic properties of amphiphile particle-films, which as yet remain unexplored.

ACKNOWLEDGMENTS

This work was supported by the Australian Research Council. KG acknowledges the receipt of an Australian

Postgraduate Award and support from the Advanced Mineral Products Research Centre. We are also grateful to our many collaborators for their contributions in studying the particle-films reported here.

REFERENCES

1. A. Pockels, Nature *43*:437 (1891).
2. I. Langmuir, J. Am. Chem. Soc. *39*:1848–1906 (1917).
3. J. Leloup, P. Maire, A. Barraud, and A. Ruaudel-Texier, J. Chim. Phys. Phys. Chim. Biol. *82*:695–700 (1985).
4. A. Ruaudel-Texier, J. Leloup, and A. Barraud, Mol. Cryst. Liq. Cryst. *134*:347–354 (1986).
5. S. Xu, X. K. Zhao, and J. H. Fendler, Adv. Mater. *2*: 183–185 (1990).
6. G. L. J. Gaines, *Isoluble Monolayers at Liquid-Gas Interfaces*, Interscience, New York, 1966.
7. G. Roberts, ed., *Langmuir-Blodgett Films*, Plenum, New York, 1990.
8. A. Ulman, *An Introduction to Ultrathin Organic Films from Langmuir-Blodgett to Self-Assembly*, Academic Press, San Diego, 1991, p. 121 and references therein.
9. A. Henglein, Chem. Rev. *89*:1861–1873 (1989).
10. J. H. Fendler and F. C. Meldrum, Adv. Mater. *7*:607–632 (1995).
11. Y. Tian, C. Wu, and J. H. Fendler, J. Phys. Chem. *98*: 4913–4918 (1994).
12. X. Peng, Y. Zhang, J. Y. Ang, B. Zou, L. Xiao, and T. Li, J. Phys. Chem. *96*:3412–3415 (1992).
13. X.-G. Peng, M.-L. Gao, Y.-Y. Zhao, S.-H. Kang, Y.-H. Zhang, Y. Zhang, D.-J. Wang, L.-Z. Xiao, T.-J. Li, and H.-Y. Chen, Chem. Phys. Lett. *209*:233–237 (1993).
14. J. Yang, X. G. Peng, T. J. Li, and S. F. Pan, Thin Solid Films *243*:643–646 (1994).
15. Y. S. Kang, D. K. Lee, and P. Stroeve, Thin Solid Films *327–329*:541–544 (1998).
16. Y. S. Kang, S. Rishbud, J. Rabolt, and P. Stroeve, Langmuir *12*:4345–4349 (1996).
17. X. Yuan, L. Cao, H. Wan, G. Zeng, and S. Xi, Thin Solid Films *327–329*:33–36 (1998).
18. L. S. Li, J. Zhang, L. J. Wang, Y. Chen, Z. Hui, T. J. Li, L. F. Chi, and H. Fuchs, J. Vac. Sci. Technol. B *15*:1618–1622 (1997).
19. L. S. Li, Z. Hui, Y. Chen, X. T. Zhang, X. Peng, Z. Liu, and T. J. Li, J. Colloid Interface Sci. *192*:275–280 (1997).
20. K. S. Mayya, V. Patil, P. M. Kumar, and M. Sastry, Thin Solid Films *312*:300–305 (1998).
21. K. S. Mayya, V. Patil, and M. Sastry, J. Chem. Soc. Faraday Trans. *93*:3377–3381 (1997).
22. K. S. Mayya, V. Patil, and M. Sastry, Langmuir *13*: 2575–2577 (1997).
23. M. Sastry, K. S. Mayya, V. Patil, D. V. Paranjape, and S. G. Hegde, J. Phys. Chem. B *10*:4954–4958 (1997).
24. K. S. Mayya and M. Sastry, J. Phys. Chem. B *101*: 9790–9793 (1997).
25. K. S. Mayya and M. Sastry, Langmuir *14*:74–78 (1998).
26. M. Sastry, K. S. Mayya, and V. Patil, Langmuir *12*: 5921–5928 (1998).
27. K. S. Mayya and M. Sastry, Langmuir *14*:6344–6346 (1998).
28. X. K. Zhao, S. Xu, and J. H. Fendler, Langmuir *7*: 520–524 (1991).
29. J. Jin, L. S. Li, Y. Q. Tian, Y. J. Zhang, Y. Liu, Y. Y. Zhao, T. S. Shi, and T. J. Li, Thin Solid Films *327–329*:559–562 (1998).
30. R. E. Schwerzel, Polym. Prepr. *35*:116–117 (1994).
31. R. E. Schwerzel, Naval Res. Rev. *4*:24–29 (1994).
32. R. T. Schwerzel, K. B. Spahr, J. P. Kurmer, V. E. Wood, and J. A. Jenkins, J. Phys. Chem. A *102*:5622–5626 (1998).
33. T. Torimoto, N. Tsumura, M. Miyake, M. Nishizawa, T. Sakata, H. Mori, and H. Yoneyama, Langmuir *15*: 1853–1858 (1999).
34. F. C. Meldrum, N. A. Kotov, and J. H. Fendler, J. Phys. Chem. *98*:4506–4510 (1994).
35. F. C. Meldrum, N. A. Kotov, and J. A. Fendler, Langmuir *10*:2035–2040 (1994).
36. F. C. Meldrum, K. A. Kotov, and J. H. Fendler, Chem. Mater. *7*:1112–1116 (1995).
37. N. A. Kotov, F. C. Meldrum, C. Wu, and J. H. Fendler, J. Phys. Chem. *98*:2735–2738 (1994).
38. N. A. Kotov, F. C. Meldrum, and J. H. Fendler, J. Phys. Chem. *98*:8827–8830 (1994).
39. N. A. Kotov, G. Zavala, and J. H. Fendler, J. Phys. Chem. *99*:12375–12378 (1995).
40. J. H. Fendler, Curr. Opin. Colloid Interface Sci. *1*: 202–207 (1996).
41. B. O. Dabbousi, C. B. Murray, M. F. Rubner, and M. G. Bawendi, Chem. Mater. *6*:216–219 (1994).
42. Y. Tian and J. H. Fendler, Chem. Mater. *8*:969–974 (1996).
43. S. Yae, I. Nakanishi, Y. Nakato, N. Toishima, and H. Mori, J. Electrochem. Soc. *141*:3077–3081 (1994).
44. J. R. Heath, C. M. Knobler, and D. V. Leff, J. Phys. Chem. B *101*:189–197 (1997).
45. T. Nakaya, Y.-J. Li, and K. Shibata, J. Mater. Chem. *6*:691–697 (1996).
46. S. Doherty and D. Fitzmaurice, J. Phys. Chem. *100*: 10732–10738 (1996).
47. J. H. Fendler, Isr. J. Chem. *33*:41–46 (1993).
48. X. K. Zhao, Y. Yuan, and J. H. Fendler, J. Chem. Soc. Chem. Commun. 1248–1252 (1990).
49. X. K. Zhao and J. H. Fendler, J. Phys. Chem. *95*: 3716–3723 (1991).
50. X. K. Zhao, L. McCormick, and J. H. Fendler, Chem. Mater. *3*:922–935 (1991).

51. K. C. Yi and J. H. Fendler, Synth. Met. *71*:2109–2110 (1995).
52. Z. Y. Pan, X. J. Liu, S. Y. Zhang, G. J. Shen, L. G. Zhang, Z. H. Lu, and J. Z. Liu, J. Phys. Chem. *101*: 9703–9709 (1997).
53. M. Sastry, V. Patil, and K. S. Mayya, Langmuir *13*: 4490–4492 (1997).
54. V. Patil and M. Sastry, Langmuir *13*:5511–5513 (1997).
55. V. Patil and M. Sastry, Langmuir *14*:2707–2711 (1998).
56. M. Sastry, V. Patil, and Sainkar, J. Phys. Chem. B *102*: 1404–1410 (1998).
57. V. Patil and M. Sastry, J. Chem. Soc. Faraday Trans. *93*:4347–4353 (1997).
58. J. A. Spink and J. V. Sanders, Trans. Faraday Soc. *51*: 1154–1165 (1955).
59. E. D. Goddard and J. A. Ackilli, J. Colloid Sci. *18*: 585–595 (1963).
60. E. S. Smotkin, C. Lee, A. J. Bard, A. Campion, M. A. Fox, T. E. Mallouk, S. E. Webber, and J. M. White, Chem. Phys. Lett. *152*:265–268 (1988).
61. D. J. Elliott, D. N. Furlong, and F. Grieser, Colloids Surf. A *141*:9–17 (1998).
62. D. J. Elliot, D. N. Furlong, T. R. Gegenbach, F. Grieser, R. S. Urquhart, C. L. Hoffman, and J. F. Rabolt, Colloids Surf. A *103*:207–219 (1995).
63. D. J. Elliott, D. N. Furlong, F. Grieser, P. Mulvaney, and M. Giersig, Colloids Surf. A *129–130*:141–150 (1997).
64. I. Moriguchi, N. Fujiyoshi, R. Sakamoto, Y. Teraoka, and S. Kagawa, Colloids Surf. A *126*:159–166 (1997).
65. G. Hemakanthi and A. Dhathathreyan, Langmuir *15*: 3317–3320 (1999).
66. C. Fan, T. Lu, B. Li, and L. Jiang, Thin Solid Films *286*:37–39 (1996).
67. V. Erokhin, P. Facci, L. Gobbi, S. Dante, F. Rustichelli, and C. Nicolini, Thin Solid Films *329*:503–505 (1998).
68. D. N. Furlong, R. S. Urquhart, F. Grieser, K. Matsuura, and Y. Okahata, J. Chem. Soc. Chem. Commun. 1329–1330 (1995).
69. D. V. Paranjape, M. Sastry, and P. Ganguly, Appl. Phys. Lett. *63*:18–20 (1993).
70. M. Sastry, S. Pal, D. V. Parajape, and P. Gaguly, J. Electron Spectrusc. Relat. Phenomena *67*:163–172 (1994).
71. E. Maassen, B. Tieke, G. Jordan, and W. Rammensee, Langmuir *12*:5595–5600 (1996).
72. C. L. Mirley and J. T. Koberstein, Langmuir *11*:2837–2839 (1995).
73. C. L. Mirley and J. T. Koberstein, Langmuir *11*:1049–1052 (1995).
74. Y. Wang and N. Herron, J. Phys. Chem. *95*:525–532 (1991).
75. L. Brus, J. Phys. Chem. *90*:2555–2560 (1986).
76. N. J. Geddes, R. S. Urquhart, D. N. Furlong, C. R. Lawrence, K. Tanaka, and Y. Okahata, J. Phys. Chem. *97*:13767–13772 (1993).
77. R. S. Urquhart, D. N. Furlong, H. Mansur, F. Grieser, K. Tanaka, and Y. Okahata, Langmuir *10*:899–904 (1994).
78. R. S. Urquhart, C. L. Hoffman, D. N. Furlong, N. J. Geddes, J. F. Rabolt, and F. Grieser, J. Phys. Chem. *99*:15987–15992, 1995.
79. H. S. Mansur, F. Grieser, M. S. Marychurch, S. Biggs, R. S. Urquhart, and D. N. Furlong, J. Chem. Soc. Faraday Trans. *91*:665–672 (1995).
80. C.-Y. Liu and A. J. Bard, J. Phys. Chem. *93*:3232–3237 (1989).
81. K. Asai, K. Ishigure, and H. Shibata, J. Lumin. *63*: 215–222 (1995).
82. K. Asai, T. Yamaki, K. Ishigure, and H. Shibata, Thin Solid Films *277*:169–174 (1996).
83. K. Asai, T. Yamaki, S. Seki, K. Ishigure, and H. Shibata, Thin Solid Films *284–285*:541–544 (1996).
84. T. Yamaki, K. Asai, and K. Ichigure, Chem. Phys. Lett. *273*:376–380 (1997).
85. T. Yamaki K. Asai, K. Ichigure, and H. Shibata, Radiat. Phys. Chem. *50*:199–205 (1997).
86. C. T. Ewins and B. Stewart, Thin Solid Films *284–285*:49–52 (1996).
87. P. Facci, V. Erokhin, A. Tronin, and C. Nicolini, J. Phys. Chem. *98*:13323–13327 (1994).
88. V. Erokhin, P. Facci, S. Carrara, and C. Nicolini, J. Phys. D Appl. Phys. *28*:2534–2538 (1995).
89. P. Facci, A. Diaspro, and R. Rolandi, Thin Solid Films *327–329*:532–535 (1998).
90. P. Facci and M. P. Fontana, Solid State Commun. *108*: 5–9 (1998).
91. A. Dhanabalan, H. Kudrolli, S. S. Major, and S. S. Talwar, Solid State Commun. *99*:859–862 (1996).
92. S. Vitta, T. H. Metzger, S. S. Major, A. Dhanabalan, and S. S. Talwar, Langmuir *14*:1799–1803 (1998).
93. J. K. Basu and M. K. Sanyal, Phys. Rev. Lett. *79*: 4617–4620 (1997).
94. I. Moriguchi, F. Shibata, Y. Teraoka, and S. Kagawa, Chem. Lett. 761–762 (1995).
95. I. Moriguchi, K. Hosoi, H. Nagaoka, I. Tanaka, Y. Teraoka, and S. Kagawa, J. Chem. Soc. Faraday Trans. *90*:349–354 (1994).
96. I. Moriguchi, H. Nii, K. Hanai, H. Nagaoka, Y. Teraoka, and S. Kagawa, Colloids Surf. A *103*:173–181 (1995).
97. I. Moriguchi, Y. Teraoka, and S. Kagawa, Kobunshi Ronbunshu *52*:765–772 (1995).
98. I. Moriguchi, K. Matsuo, M. Sakai, K. Hanai, Y. Teraoka, and S. Kagawa, J. Chem. Soc. Faraday Trans. *94*:2199–2204 (1998).
99. Z. Pan, J. Liu, X. Peng, T. Li, Z. Wu, and M. Zhu, Langmuir *12*:851–853 (1996).

100. Z. Pan, X. Peng, T. Li, and J. Liu, Appl. Surf. Sci. *108*:439–447 (1997).
101. V. Erokhin, L. Feigin, G. Ivakin, V. Klechkovskaya, Y. Lvov, and N. Stiopina, Makromol. Chem. Macromol. Symp. *46*:359–363 (1991).
102. X. Yang, G. Wang, and Z. Lu, Supramol. Sci. *5*:549–552 (1998).
103. A. V. Nabok, T. Richardson, F. Davis, and C. J. M. Stirling, Langmuir *13*:3198–3201 (1997).
104. F. N. Dulstev and L. L. Sveshnikova, J. Struct. Chem. *38*:673–677 (1997).
105. F. N. Dulstev and L. L. Sveshnikova, Thin Solid Films *288*:103–107 (1996).
106. C. Zylberajch, A. Ruaudel-Texier, and A. Barraud, Synth. Met. *27*:B609–B614 (1988).
107. D. J. Elliot, D. N. Furlong, and F. Grieser, Colloids Surf. A, *15*:101–110 (1999).
108. K. C. Yi and J. H. Fendler, Langmuir *6*:1519–1521 (1990).
109. R. Zhu, Y. Wei, C. Yuan, S. Xiao, Z. Lu, and H. J. Schmitt, Solid State Commun. *84*:449–451 (1992).
110. X. Peng, R. Lu, Y. Zhao, L. Qu, H. Chen, and T. Li, J. Phys. Chem. *98*:7052–7055 (1994).
111. L. S. Li, L. Qu, R. Lu, X. Peng, Y. Zhao, and T. J. Li, Thin Solid Films *327–329*:408–411. (1998).
112. L. Wang, L. Li, L. Qu, R. Lu, X. Peng, Y. Zhao, and T. Li, Mol. Cryst. Liq. Cryst. Sci. Tech. Sec. A Mol. Cryst. Liq. Cryst. *294*:473–478 (1997).
113. D. J. Scoberg, F. Grieser, and D. N. Furlong, J. Chem. Soc. Chem. Commun. 515–517 (1991).
114. F. Grieser, D. N. Furlong, D. J. Scoberg, I. Ichinose, N. Kimizuka, and T. Kunitake, J. Chem. Soc. Faraday Trans. *88*:2207–2214 (1992).
115. D. N. Furlong, R. Urquhart, F. Grieser, K. Tanaka, and Y. Okahata, J. Chem. Soc. Faraday Trans. *89*:2031–2035 (1993).
116. H. Xiong, H. Li, Z. Wang, X. Zhang, J. Shen, M. Gleiche, L. Chi, and H. Fuchs, J. Colloid Interface Sci. *211*:238–242 (1999).
117. A. V. Nabok, T. Richardson, C. McCartney, N. Cowlam, F. Davis, C. J. M. Stirling, A. K. Ray, V. Gacem, and A. Gibraud, Thin Solid Films *329*:510–514 (1998).
118. R. S. Urquhart, D. N. Furlong, T. Gengenbach, N. J. Geddes, and F. Grieser, Langmuir *11*:1127–1133 (1995).
119. H. S. Mansur, F. Grieser, R. S. Urquhart, and D. N. Furlong, J. Chem. Soc. Faraday Trans. *91*:3399–3404 (1995).
120. T. Yamaki, T. Yamada, K. Asai, and K. Ishigure, Thin Solid Films *327–329*:586–590 (1998).
121. T. Yamaki, T. Yamada, K. Asai, K. Ishigure, and H. Shibata, Thin Solid Films *329*:581–585 (1998).
122. D. J. Elliot, D. N. Furlong, T. R. Gengenbach, and F. Grieser, Colloids Surf. A *102*:45–46 (1995).
123. C. Zylberajch, A. Ruadel-Texier, and A. Barraud, Thin Solid Films *179*:9–14 (1989).
124. A. C. Zylberajch, A. Barraud, H. Roulet, and G. Dufour, Appl. Surf. Sci. *52*:323–327 (1991).
125. H. Chen, X. Chai, Q. Wei, Y. Jiang, and T. Li, Thin Solid Films *178*:535–540 (1989).
126. J. Leloup, A. Ruaudel-Texier, and A. Barraud, Thin Solid Films *210/211*:407–409 (1992).
127. V. Erokhin, P. Facci, S. Carrara, and C. Nicolini, Thin Solid Films *284–285*:891–893 (1996).
128. I. Moriguchi, K. Hanai, Y. Teraoka, S. Kagawa, S. Yamada, and T. Matsu, Jpn. J. Appl. Phys. *34*:L323–L325 (1995).
129. X. Peng, Q. Wei, Y. Jiang, X. Chai, T. Li, and J. Shen, Thin Solid Films *210/211*:401–403 (1992).
130. X. Peng, S. Guan, X. Chai, Y. Jiang, and T. Li, J. Phys. Chem. *96*:3170–3174 (1992).
131. X. Peng, H. Chen, S. Kan, Y. Bai, and T. Li, Thin Solid Films *242*:118–121 (1994).
132. G. W. Min, X. M. Yang, Z. H. Lu, and W. Yu, J. Vac. Sci. Technol. B *12*:1984–1987 (1994).
133. R. Zhu, G. Min, Y. Wei, and H. J. Schmitt, J. Phys. Chem. *96*:8210–8211 (1992).
134. L. S. Li, L. Qu, L. Wang, R. Lu, X. Peng, Y. Zhao, and T. J. Li, Langmuir *13*:6183–6187 (1997).
135. X. Luo, Z. Zhang, and Y. Liang, Langmuir *10*:3213–3216 (1994).
136. I. Moriguchi, F. Shibata, Y. Teraoka, and S. Kagawa, J. Chem. Soc. Chem. Commun. 1401–1402 (1991).
137. D. J. Elliot, D. N. Furlong, T. R. Gengenbach, F. Grieser, R. S. Urquhart, C. L. Hoffman, and J. F. Rabolt, Langmuir *11*:4773–4778 (1995).
138. J. K. Pike, H. Byrd, A. A. Morrone, and D. R. Talham, J. Am. Chem. Soc. *115*:8497–8498 (1993).
139. J. K. Pike, H. Byrd, D. R. Talham, and A. A. Morrone, Thin Solid Films *243*:510–514 (1994).
140. J. K. Pike, H. Byrd, A. A. Morrone, and D. R. Talham, Chem. Mater. *6*:1757–1765 (1994).
141. N. Herron, Y. Wang, and H. Ecker, J. Am. Chem. Soc. *112*:1322–1326 (1990).
142. L. Spanhel, M. Haase, H. Weller, and A. Henglein, J. Am. Chem. Soc. *109*:5649–5655 (1987).
143. Y. Wang and N. Herron, J. Phys. Chem. *91*:257–260 (1987).
144. A. Ekimov, J. Lumin. *70*:1–20 (1996).
145. Y. Wang and N. Herron, Phys. Rev. B *42*:7253–7255 (1990).
146. L. Katsikas, A. Eychmüller, M. Giersig, and H. Weller, Chem. Phys. Lett. *172*:201–204 (1990).
147. H. Weller, H. M. Schmidt, U. Koch, A. Fojtik, S. Baral, A. Henglein, W. Kunath, K. Weiss, and E. Dieman, Chem. Phys. Lett. *124*:557–560 (1986).
148. J. F. Rabolt, F. C. Burns, N. E. Schlotter, and J. D. Swalen, J. Chem. Phys. *78*:946–952 (1983).
149. A. Malik, M. K. Durbin, A. G. Richter, K. G. Huang, and P. Dutta, Thin Solid Films *284–285*:144–146 (1996).
150. I. Ichinose, N. Kimizuka, and T. Kunitake, J. Phys. Chem. *99*:3736–3742 (1995).

151. Y. Wang, A. Suna, W. Mahler, and R. Kasowski, J. Chem. Phys. *87*:7315–7322 (1987).
152. W. Mahler, Inorg. Chem. *27*:435–436 (1988).
153. M. Krishnan, J. R. White, M. A. Fox, and A. J. Bard, J. Am. Chem. Soc. *105*:7002–7003 (1983).
154. K. Honda, A. Kuwano, K. Chiba, A. Ishikawa, and H. Miyama, Chem. Lett. 195–198 (1988).
155. Y.-J. Hong and S. M. Oh, Sens. Actuator B Chem. *32*: 7–13 (1996).
156. J. Prekumar and R. Ramaraj, J. Mol. Catal. A Chem. *132*:21–32 (1998).
157. F.-R. F. Fan, H.-Y. Liu, and A. J. Bard, J. Phys. Chem. *89*:4418–4420 (1985).
158. H. Miyoshi, S. Nippa, H. Uchida, H. Mori, and H. Yoneyama, Bull. Chem. Soc. Jpn. *63*:3380–3384 (1990).
159. Q. Deng, K. M. Cable, R. B. Moore, and K. A. Mauritz, J. Polym. Sci. B Polym. Phys. *34*:1917–1923 (1996).
160. J. P. Kuczynski, B. H. Miloslavljevic, and J. K. Thomas, J. Phys. Chem. *88*:980–984 (1984).
161. N. Kakuta, J. M. White, A. Campion, A. J. Bard, M. A. Fox, and S. E. Webber, J. Phys. Chem. *89*:48–52 (1985).
162. Y. Wang and W. Mahler, Opt. Commun. *61*:233–236 (1987).
163. Y. Wang, W. Mahler, and A. Suna, J. Opt. Soc. Am. B *6*:808–813 (1989).
164. E. F. Hillinski, P. A. Lucas, and Y. Wang, J. Chem. Phys. *89*:3435–3441 (1988).
165. E. S. Smotkin, R. M. Brown Jr., L. K. Rasenberg, K. Salomon, A. J. Bard, A. Campion, M. A. Fox, T. E. Mallouk, S. E. Webber, and J. M. White, J. Phys. Chem. *94*:7543–7549 (1990).
166. K. R. Gopidas and P. V. Kamat, Proc. Ind. Acad. Sci. *105*:505–512 (1993).
167. H. Inoue, R. S. Urquhart, T. Nagamura, F. Grieser, and H. Sakaguchi, Colloids Surf. A *126*:197–208 (1997).
168. P. Nandakumar, C. Vijayan, Y. V. G. S. Murti, K. Dhanalakshmi, and G. Sundararajan, Bull. Mater. Sci. *20*:549–582 (1997).
169. P. Nandakumar, A. R. Dhobale, Y. Babu, M. D. Sastry, C. Vijayan, Y. V. G. S. Murti, K. Dhanalakshmi, and G. Sundararajan, Solid State Commun. *106*:193–196 (1998).
170. M. F. Finlayson, K.-H. Park, N. Kakuta, A. J. Bard, A. Campion, M. A. Fox, S. E. Webber, and J. M. White, J. Lumin. *39*:205–214 (1988).
171. N. Kakuta, K.-H. Park, M. F. Finlayson, A. J. Bard, A. Campion, M. A. Fox, S. E. Webber, and J. M. White, J. Phys. Chem. *89*:5028–5031 (1985).
172. A. Ueno, N. Kakuta, K. H. Park, M. F. Finlayson, A. J. Bard, A. Campion, M. A. Fox, S. E. Webber, and J. M. White, J. Phys. Chem. *89*:3828–3833 (1985).
173. N. Kakuta, K. H. Park, M. F. Finlayson, A. Ueno, A. J. Bard, A. Campion, M. A. Fox, S. E. Webber, and J. M. White, J. Phys. Chem. *89*:732–734 (1985).
174. A. W.-H. Mau, C.-B. Huang, N. Kakuta, A. J. Bard, A. Campion, M. A. Fox, M. White, and S. E. Webber, J. Am. Chem. Soc. *106*:6537–6542 (1984).
175. K. R. Gopidas and P. V. Kamat, Mater. Lett. *9*:372–378 (1990).
176. H. Miyoshi, M. Yamachika, H. Yoneyama, and H. Mori, J. Chem. Soc. Faraday Trans. *86*:815–818 (1990).
177. J. M. Zen, G. Chen, F. R. Fan, and A. J. Bard, Chem. Phys. Lett. *169*:23–26 (1990).
178. Y. Wang, A. Suna, J. McHugh, E. F. Hilinski, P. A. Lucas, and R. D. Johnson, J. Chem. Phys. *92*:6927–6939 (1990).
179. H. Miyoshi, K. Tanaka, H. Uchida, H. Yoneyama, H. Mori, and T. Sakata, J. Electroanal. Chem. *295*:71–78 (1990).
180. H. Yoneyama, Res. Chem. Intermed. *15*:101–111 (1991).
181. H. Inoue, R. S. Urquhart, T. Nagamura, F. Grieser, H. Sakaguchi, and D. N. Furlong, J. Polym. Sci. Tech. *11*:73–76 (1998).
182. T. Nagamura, H. Inoue, H. Sakaguchi, F. Grieser, R. Urquhart, D. J. Elliot, and D. N. Furlong, Proc. SPIE-Organic-Inorganic Hybrid Materials for Photonics, *3469*:174–181 (1998).
183. N. Kimizuka, T. Miyoshi, I. Ichinose, and T. Kunitake, Chem. Lett. 2039–2042 (1991).
184. N. Kimizuka, T. Maeda, I. Ichinose, and T. Kunitake, Chem. Lett. 941–944 (1993).
185. G. Decher, Science *277*:1232–1237 (1997).
186. G. Decher, M. Eckle, J. Schmitt, and B. Struth, Curr. Opin. Colloid Interface Sci. *3*:32–39 (1998).
187. J. H. Fendler, Chem. Mater. *8*:1616–1624 (1996).
188. G. Decher and J. Schmitt, Prog. Colloid Polym. Sci. *89*:160–164 (1992).
189. S. W. Keller, S. A. Johnson, E. Brigham, E. H. Yonemoto, and T. E. Mallouk, J. Am. Chem. Soc. *117*: 12879–12880 (1995).
190. F. Caruso, H. Lichtenfeld, M. Giersig, and H. Möhwald, J. Am. Chem. Soc. *120*:8523–8524 (1998).
191. F. Caruso, R. A. Caruso, and H. Möhwald, Science *282*:1111–1114 (1998).
192. Y. Lvov, K. Ariga, and T. Kunitake, Chem. Lett. 2323–2326 (1994).
193. Y. Lvov, K. Ariga, I. Ichinose, and T. Kunitake, J. Am. Chem. Soc. *117*:6117–6123 (1995).
194. M. Onda, Y. Lvov, K. Ariga, and T. Kunitake, J. Ferment. Bioeng. *82*:502–506 (1996).
195. F. Caruso, D. N. Furlong, K. Ariga, I. Ichinose, and T. Kunitake, Langmuir *14*:4559–4565 (1998).
196. Y. Lvov, H. Haas, G. Decher, H. Möhwald, A. Mikhailov, B. Mtchedlishvily, E. Morgunova, and B. Vainshtein, Langmuir *10*:4232–4236 (1994).
197. H. Fukumoto and Y. Yonezawa, Thin Solid Films *327–329*:748–751 (1998).
198. D. Yoo, A. Wu, J. Lee, and M. F. Rubner, Synth. Met. *85*:1425–1426 (1997).

199. T. Serizawa, H. Takeshita, and M. Akashi, Langmuir *14*:4088–4094 (1998).
200. J. H. Fendler, in *Semiconductor Nanoclusters* (P. V. Kamat and D. Meisel, eds.), Elsevier Science, Amsterdam, 1997, pp. 261–276.
201. N. A. Kotov, I. Dékány, and J. H. Fendler, J. Phys. Chem. *99*:13065–13069 (1995).
202. N. A. Kotov, T. Haraszti, L. Turi, G. Zavala, R. E. Geer, I. Dékány, and J. H. Fendler, J. Am. Chem. Soc. *119*:6821–6831 (1997).
203. G. S. Ferguson and E. R. Kleinfeld, Adv. Mater. *7*: 414–416 (1995).
204. Y. Lvov, K. Ariga, I. Ichinose, and T. Kunitake, Thin Solid Films *284–285*:797–801 (1996).
205. N. A. Kotov, I. Dékány, and J. H. Fendler, Adv. Mater. *8*:637–641 (1996).
206. A. Krozer, S.-A. Nordin, and B. Kasemo, J. Colloid Interface Sci. *176*:479–484 (1995).
207. K. Ariga, Y. Lvov, M. Onda, I. Ichinose, and T. Kunitake, Chem. Let. 125–126 (1997).
208. A. Rosidian, Y. Liu, and R. O. Claus, Adv. Mater. *10*: 1087–1091 (1998).
209. M. Gao, X. Zhang, B. Yang, F. Li, and J. Shen, Thin Solid Films *284–285*:242–245 (1996).
210. Y. Lvov, K. Ariga, M. Onda, I. Ichinose, and T. Kunitake, Langmuir *13*:6195–6203 (1997).
211. Y. Liu, A. Wang, and R. Claus, J. Phys. Chem. B. *101*: 1385–1388 (1997).
212. T. Cassagneau, T. E. Mallouk, and J. H. Fendler, J. Am. Chem. Soc. *120*:7848–7859 (1998).
213. M. Gao, B. Richter, S. Kirstein, and H. Möhwald, J. Phys. Chem. B *102*:4096–4103 (1998).
214. M. Gao, B. Richter, and S. Kirstein, Adv. Mater. *9*: 802–805 (1997).
215. Y. Liu, A. Wang, and R. O. Claus, Appl. Phys. Lett. *71*:2265–2267 (1997).
216. M. A. Correa-Duarte, M. Giersig, N. A. Kotov, and L. M. Liz-Marzan, Langmuir *14*:6430–6435 (1998).
217. S. W. Keller, H.-N. Kim, and T. E. Mallouk, J. Am. Chem. Soc. *116*:8817–8818 (1994).
218. F. Caruso, D. G. Kurth, D. Volkmer, M. J. Koop, and A. Müller, Langmuir *14*:3462–3465 (1998).
219. Y. Liu, Y. Wang, and R. O. Claus, Chem. Phys. Lett. *298*:315–319 (1998).
220. D. L. Feldheim, K. C. Grabar, M. J. Natan, and T. E. Mallouk, J. Am. Chem. Soc. *118*:7640–7641 (1996).
221. J. Schmitt, G. Decher, W. J. Dressick, S. L. Brandow, R. E. Geer, R. Shashidar, and J. M. Calvert, Adv. Mater. *9*:61–65 (1997).
222. M. Gao, M. Gao, X. Zhang, Y. Yang, B. Yang, and J. Shen, J. Chem. Soc. Chem. Commun. 2777–2778 (1994).
223. M. Gao, X. Zhang, B. Yang, and J. Shen, J. Chem. Soc. Chem. Commun. 2229–2230 (1994).
224. E. Hao, L. Wang, J. Zhang, B. Yang, X. Zhang, and J. Shen, Chem. Lett. *1*:5–6 (1999).
225. Y. Lvov, K. Ariga, I. Ichinose, and T. Kunitake, J. Chem. Soc. Chem. Commun. 2313–2314 (1995).
226. E. Hao, B. Yang, J. Zhang, X. Zhang, J. Sun, and J. Shen, J. Mater. Chem. *8*:1327–1328 (1998).
227. F.-U. Fulda, A. Kampes, L. Krasemann, and B. Tieke, Thin Solid Films *327–329*:752–757 (1998).
228. R. K. Iler, J. Colloid Interface Sci. *21*:569–594 (1966).
229. T. Nakanishi, B. Ohtani, K. Shimazu, and K. Uosaki, Chem. Phys. Lett. *278*:233–237 (1997).
230. T. Nakanishi, B. Ohtani, and K. Uosaki, Jpn. J. Appl. Phys. *36*:4053–4056 (1997).
231. T. Nakanishi, B. Ohtani, and K. Uosaki, J. Phys. Chem. *102*:1571–1577 (1998).
232. T. Nakanishi, B. Ohtani, and K. Uosaki, Jpn. J. Appl. Phys. Part 1 Regul. Pap. Short Notes Rev. Pap. *38*: 518–521 (1999).
233. M. Brust, R. Etchenique, E. J. Calvo, and G. J. Gordillo, J. Chem. Soc. Chem. Commun. 1949–1950 (1996).
234. M. Brust, D. Bethall, C. J. Kiely, and D. J. Schriffin, Langmuir *14*:5425–5429 (1998).
235. H. Xiong, Z. Zhou, Z. Wang, and X. Zhang, J. Chem. Supramol. Sci. *5*:623–626 (1998).
236. G. Decher and J. D. Hong, Ber. Bunsenges. Phys. Chem. *95*:1430–1434 (1991).
237. N. B. Hoogeveen, M. A. Cohen Stuart, G. J. Fleer, and M. R. Böhmer, Langmuir *12*:3675–3681 (1996).
238. E. R. Kleinfeld and G. S. Ferguson, Science *265*:370–373 (1994).
239. Y. Lvov, F. Essler, and G. Decher, J. Phys. Chem. *97*: 13773–13777 (1993).
240. M. Ferriera and M. F. Rubner, Macromolecules *28*: 7107–7114 (1995).
241. G. Decher, Y. Lvov, and J. Schmitt, Thin Solid Films *244*:772–777 (1994).
242. V. L. Colvin, M. C. Schlamp, and A. P. Alivisatos, Nature *370*:354–357 (1994).
243. B. O. Dabbousi, M. G. Bawendi, O. Onitsuka, and F. Rubner, Appl. Phys. Lett. *66*:1316–1318 (1995).
244. H. Uchida, S. Hirai, T. Torimoto, S. Kuwabata, T. Sakata, H. Mori, and H. Yoneyama, Langmuir *11*:3725–3729 (1995).
245. V. Erokhin, S. Carrara, H. Amenitch, S. Bernstorff, and C. Nicolini, Nanotech *9*:158–161 (1998).
246. M. P. Pileni, T. Gulik-Krzywicki, J. Tanori, A. Filankembo, and J. C. Dedieu, Langmuir *14*:7359–7363 (1998).
247. P. V. Kamat, Chem. Rev. *93*:267–300 (1993).
248. P. V. Kamat, in *Nanoparticles and Nanostructured Films* (J. H. Fendler, ed.), Wiley-VCH, Weinheim, 1998, pp. 207–233.
249. K. Grieve, The physicochemical properties of nanocrystalline semiconductor particles in thin organic films. Ph.D thesis, University of Melbourne, Parkville, VIC, in preparation.

33

The Role of Steric Constraints and Intermolecular Interactions in the Formation of Surfactant Phases

SÖNKE SVENSON The Dow Chemical Company, Midland, Michigan

I. INTRODUCTION

The self-aggregation and self-organization of surfactants and related amphiphilic compounds are of essential importance in numerous areas of industrial applications and scientific research. Industrial applications include the use of surfactants in cleaning and personal care products, as emulsifiers for paints and dyestuffs, and as formulating agents in pharmaceuticals. Surfactants are used as templates for the production of molecular sieves and related inorganic materials and the formation of metal and mineral nanocrystals of specific sizes and shapes. Surfactants find application in the development of organic-inorganic composite materials, biomaterials, and as surface coatings for medical devices. As an integral part of complex fluids, surfactants are used in the oil field industry and as drag reducers in heating fluids. In scientific research, self-organizing amphiphiles are models for biological membranes. They find use in studies of membrane processes and as carriers for drugs and other biologically relevant agents. Micellar solutions are important mediators of chemical reactions between otherwise immiscible compounds. These are just a few examples of many applications detailed in this volume that are based on the self-aggregation of surfactants. The understanding of the driving forces behind these processes and the development of reliable models that predict the size and shape of the aggregates and allow tailoring molecules for specific purposes are therefore of essential importance in all of these areas.

This chapter will focus on the interplay between the geometrical shape of the molecules and their intermolecular interactions based on isotropic forces and how this interplay influences the size and shape of surfactant mesophases and self-assembled structures. The question of whether there are general rules or models that predict the experimental observations will be addressed. The phase behavior of single-chain surfactants with mainly isotropic interactions will be reviewed, followed by the phase behavior of branched, double-chain, and multiple-chain surfactants. The review will continue with examples of the phase behavior of mixed surfactants with opposite charges. Finally, the use of the *monomer solubility* as a hands-on measure to tailor the aggregation behavior will be proposed. The emphasis of this review will always be on the chemical structure of the monomers. Other means to influence the aggregation and phase behavior—for example, the addition of salt, change of pH or temperature, and the presence of cosurfactants—will not be discussed in detail. The phase behavior and self-aggregation of surfactants have been described and reviewed in greater detail in numerous publications [1–23].

II. SELF-AGGREGATION AND PHASE BEHAVIOR OF SINGLE-CHAIN SURFACTANTS

Surfactant aggregation in water above the critical micelle concentration (cmc) produces a wide variety of structures held together by physical interaction forces.

These structures include micelles of various shapes such as spheres, disks, and cylinders; essentially spherical vesicles and liposomes; and planar bilayers. At high enough concentrations, most surfactants eventually form mesophases or liquid crystals. In these homogeneous phases, the surfactant monomers are assembled into several possible geometries, the most common ones being lamellar (parallel stacks of surfactant bilayers), hexagonal (spherical or cylindrical aggregates are hexagonally close packed), and cubic. A schematic phase diagram for a single-chain surfactant is shown in Fig. 1a. All of these structures have two general features in common. First, the structures are

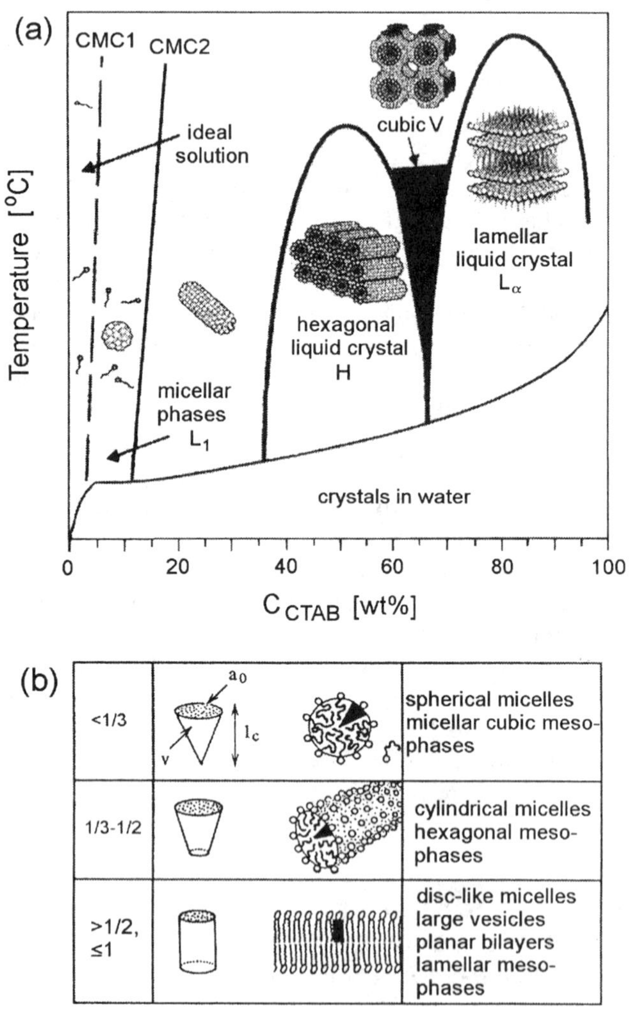

FIG. 1 (a) Schematic phase diagram of the cationic surfactant cetyltrimethylammonium bromide (CTAB) in water. (Adapted from Ref. 24.) (b) Mean (dynamic) packing shapes of surfactants and the structures they form. (Adapted from Ref. 27.)

dynamic in nature. Surfactant monomers are constantly joining and leaving an aggregate on a time scale that can be as rapid as microseconds. This limits the lifetime of any one aggregate, which can be on the order of milliseconds for a small species such as a spherical micelle. Second, the differences in energy between these various structures are quite small. The consequence is that surfactants can often be transformed readily between the various types of aggregate simply by small changes in solution conditions such as concentration, temperature, pH, or electrolyte concentration.

The unifying principle behind all the aggregation phenomena described here is the hydrophobic effect, i.e., the tendency of water molecules (or other solvent molecules connected by hydrogen bonds) to maintain their internal structure. The standard free energy of transfer of a single hydrocarbon molecule from an oil phase into water is therefore large and positive. For the same reason, water molecules are trying to "concentrate" randomly distributed hydrophobic tails of surfactants onto the water surface or, with increasing concentration, into micelles and other aggregates [16–19,21,23,25,26]. The intermolecular interactions between surfactant monomers are attractive van der Waals forces between the hydrocarbon chains and repulsive forces between charged and hydrated headgroups. Both forces are isotropic. Their interplay with each other and the geometry of the monomers will determine the size and shape of the surfactant aggregates.

The desire to *predict* size and shape of these aggregates has led to the "packing parameter" concept, which is mainly based on geometrical considerations. The packing properties of surfactants depend on the optimal headgroup area a_0 (defined by the equilibrium between hydrocarbon tail attraction and headgroup repulsion); the volume v of the hydrocarbon chain(s), which is (are) assumed to be fluid and incompressible; and the critical chain length l_c (a semiempirical parameter, because it represents a somewhat vague cutoff distance beyond which hydrocarbon chains can no longer be considered as fluid) [27–30]. A few examples of packing parameters $v/a_0 l_c$ of surfactants and the structures they form are shown in Fig. 1b. This concept is apparently too simple to predict the aggregate structures of the CTAB phase diagram displayed in Fig. 1a. Even though the curvatures of cylindrical micelles and single rods and spheres of the hexagonal and cubic mesophases are quite similar, the size and shape of the whole aggregates are very different. The transformation from one structure to the next is caused solely by a change in surfactant concentration and should not affect the shape of the monomers. Following are examples of some of the shortcomings of the packing parameter concept. The hydrocarbon chains are considered entirely fluid and not opposing any distortion until they are extended beyond l_c. Any effects that may exist as a result of curvature or other distortions of the molecular packing are not considered. However, there is evidence that the chains in micelles are more fluid than in bilayers [27,31]. The charge and degree of hydration of the polar group, and the choice of counterion affect the headgroup area a_0 [16,25]. Despite these shortcomings, the concept may be helpful in some areas of practical application when applied carefully.*

The phase behavior of surfactants mixed with water shows both universal and particular features. Many of the same phases are formed in widely different systems, including lamellar, hexagonal cylindrical, disordered lamellar, and intermediate mesophases. Other aspects of the phase behavior are sensitive to the amphiphile type. Figure 2 illustrates some universal as well as nonuniversal aspects of the phase behavior in water of four different single-chain surfactants with nonionic, anionic, cationic, and zwitterionic headgroups. With increasing surfactant concentration, there are transitions from an isotropic micellar solution (L_1) to a hexagonal packing of cylinders (H), then to a bicontinuous cubic phase (V), and finally to a lamellar phase (L_α) [33–37]. The order-order transitions in each case are mainly lyotropic, that is, they are driven by changes in concentration, not temperature. These similarities suggest that universal features of the surfactants, such as volume-filling constraints, entropies and energies of mixing, and entropies of surfactant chain conformation, are causing the phase transitions. However, there are some differences among the phase diagrams. For the anionic surfactant, there is a small intermediate mesophase (X) located between the hexagonal and cubic phases. This phase seems to consist of rodlike aggregates in some sort of a bicontinuous network with lower symmetry than the cubic phase [36]. For the cationic surfactant, a micellar cubic phase (I) consisting of cubic packing of spheres is wedged between the spherical micelles and hexagonal cylinders [32]. The lamellar phase (L_α) shifts to higher surfactant concentration in the order nonionic, zwitterionic, cationic, and anionic. The hexagonal phase (H) is enlarged for the zwitterionic surfactant compared with the others.

Although the different headgroup characters (EO_6, COO^-, Me_3N^+, and PC; see Fig. 2) clearly influence the phase behavior, the enlargement of the anionic headgroup area by insertion of a methyl group in the α-position to it is tolerated. Sodium (*R*)-2-methyldecanoate and the racemic mixture show essentially the same phase behavior as sodium decanoate [36]. The packing parameter should change in favor of a more conical shape of the monomers, resulting in the preferred formation of aggregates of higher curvature. This has not been confirmed experimentally. The situation is different in cases of cationic trialkylammonium surfactants with triethyl, tripropyl, tributyl, and tripentyl headgroups instead of trimethyl. As expected, the cmc is decreasing with increasing hydrophobicity of the headgroups. But at the same time the counterion association values are decreasing as well, which results in almost identical values of the free energy of micellization for the trimethyl, triethyl, and tripropyl headgroups. Tributylammonium surfactants with C_{12} and C_{14} alkyl chains, however, spontaneously demix into dilute and concentrated conjugate phases on warming [37]. This lower consolute behavior in surfactant solutions has been found before for nonionic surfactants such as poly(oxyethylene)alkyl ethers and cationic surfactants in concentrated electrolyte solutions and has been explained by a change in the dipole moment due to a change in the headgroup conformation (nonionics) and a structural change from spherical to rodlike or wormlike micelles (cationics) [38–42]. The tributylammonium surfactants form small spherical micelles at all concentrations because of the somewhat bulky headgroup. In contrast to the spontaneous micelle formation usually observed for surfactants, there is evidence that tributylammonium monomers form premicellar aggregates that continuously grow to form spheres. This leaves a relatively high concentration of free monomers in solution, which are believed to screen the repulsive interactions between the micelles and allow micelles to form clusters and eventually separate to form the concentrated phase. The even higher tendency of the tripentylammonium surfactant to separate from water re-

*When applied less than carefully, the concept unfortunately can result in some irritating statements, e.g., when the molecular packing of amphiphiles within fiber structures below the gel-to-liquid crystalline phase transition temperature (=*the hydrocarbon chains are not fluid*) is "explained" by the packing parameter. Obvious mismatches between the fiber curvature and the molecular shape are then taken as evidence for the presence or absence of hydration shells around the headgroups. A list of examples given here certainly would be incomplete and therefore unfair. Readers with experience in this area will have their own examples, and newcomers will soon find some when they are keeping this comment in mind.

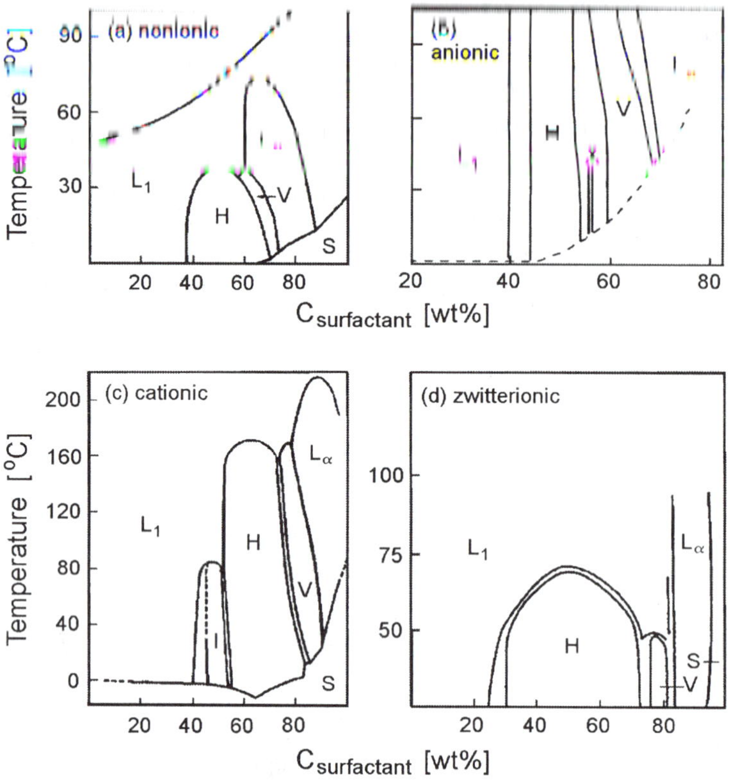

FIG. 2 Phase diagrams of four surfactants in water: (a) nonionic hexaethyleneglycol mono *n*-dodecylether ($C_{12}E_6$), (b) anionic sodium 2-methyldecanoate, (c) cationic dodecyltrimethylammonium chloride (DTAC), and (d) zwitterionic 1-oleoyl-*sn*-glycero-3-phosphocholine (1-OPC). L_1 stands for isotropic micellar solution; H, L_α, I, and V, respectively, stand for hexagonal, lamellar, micellar cubic, and bicontinuous cubic mesophases; X denotes a mesophase of rodlike aggregates in some sort of a bicontinuous network; and S stands for crystalline solid. (Adapted from Refs. 32 and 36.)

sults in the formation of metastable solutions and spontaneous crystallization [37]. The effect bulkier cationic headgroups have on the phase diagram and the alkyl chain length dependence of the phase separation are displayed in Fig. 3.

The phase diagram changes again when the anionic surfactant sodium sulfopropyl octadecyl maleate (SSPOM) is studied, whose sulfopropyl headgroup is connected via a maleic acid spacer to a C_{18} hydrocarbon chain. SSPOM behaves like other anionic surfactants at temperatures above the Kraft boundary ($T_K \sim 37°C$), displaying an isotropic micellar phase (L_1) followed by a hexagonal phase (H). Uncommon properties appear in the temperature range below T_K where crystals and a gel-like phase (G_β) with lamellar structure exist (Fig. 4a). This gel phase is metastable but long lived. Wide-angle x-ray scattering data suggest that the molecules within the bilayers of the gel phase are densely packed and interdigitated with the hydrocarbon chains in the crystalline state [43]. The maleic acid spacer has two effects. First, it introduces a fixed *gauche*-bend into the molecule that facilitates the interdigitated packing. Second, it provides attractive anisotropic intermolecular interactions in the form of hydrogen bonds between ester carbonyls and α-methylene hydrogens of adjacent hydrocarbon chains, which stabilize the lamellar molecular packing within the G_β phase. Although the molecules are more cone shaped than the common single-chain surfactants because of the *gauche*-bend, any resulting curvature of the aggregates is compensated by the formation of linear hydrogen bridges.

The introduction of a second charge within the headgroup region should strongly favor aggregates of higher curvature because of the increasing headgroup repulsion. The composition phase diagrams shown in Fig. 4b for the divalent dodecylpentamethyl-1,3-propylenebis-(ammonium chloride) (DoPPDAC) and dipotassium dodecylmalonate (K_2DoM) surfactants display the same succession of phases known from monovalent

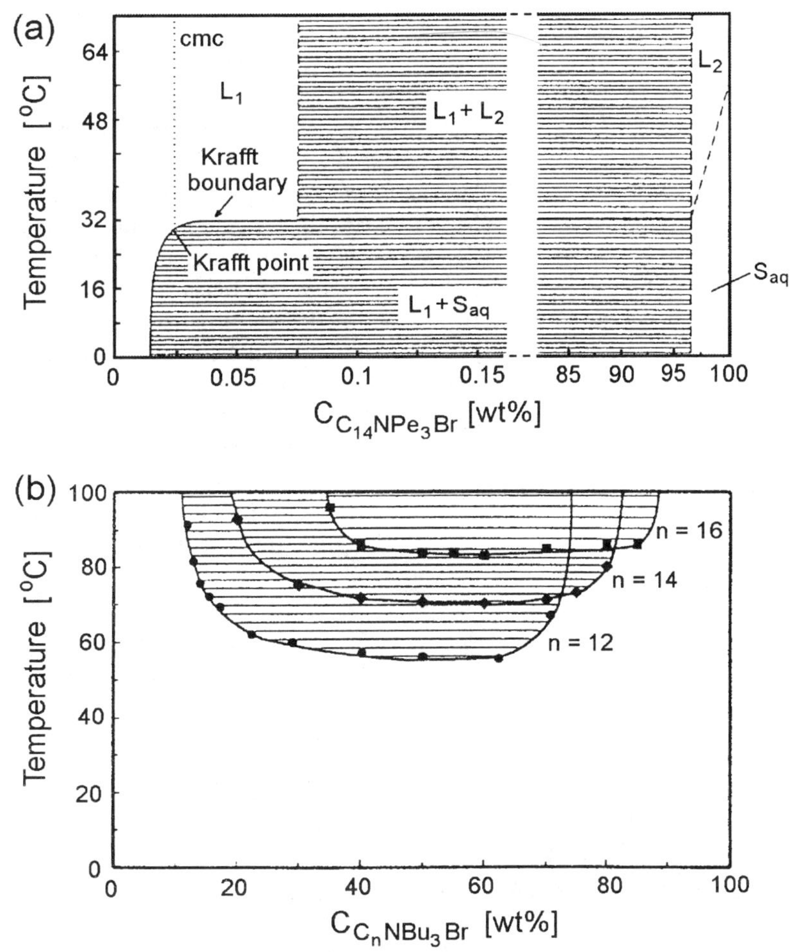

FIG. 3 (a) Partial phase diagram of cationic tetradecyl tripentylammonium bromide ($C_{14}NPe_3Br$) in water. A straight line denotes a measured phase boundary, a broken line an inferred boundary, and the dotted line indicates the locus of the cmc. (b) Partial binary phase diagrams of alkyl tributylammonium bromides (C_nNBu_3Br) in water showing a two-phase region for dodecyl (n = 12), tetradecyl (n = 14), and hexadecyl (n = 16) alkyl chains. Horizontal hatching represents isothermal tie lines through a two-phase region. L denotes isotropic liquid phases and S_{aq} denotes hydrated cystalline solids. (Adapted from Ref. 37.)

surfactants, but they differ in that the phase transitions are shifted toward higher surfactant concentrations. A discontinuous cubic phase, a hexagonal phase, and a mixture of liquid crystalline phases and hydrated crystals follow an isotropic micellar solution. The discontinuous cubic phase consists most likely of prolate-shaped micelles and might contain two different phases in case of the surfactant K_2DoM. A lamellar phase, however, has been observed for divalent surfactants only after mixing with monovalent surfactants to dilute the total charge density [44].

Another approach to increasing the headgroup area a_0 and force curvature into the aggregates is realized in the "dicephalic" surfactant 1,3-bis(1-imidazolyl)-2-propyl octadecanoate (BIPO), which has two imidazole headgroups linked to a single C_{18} hydrocarbon chain. The headgroup charge (= repulsive interaction) can be controlled by the pH of the solution (pK_{a1} 3.6 and pK_{a2} ~7). Instead of the expected spherical micelles, BIPO forms approximately 1000-nm large multilamellar vesicles at pH 5.5 (monoprotonated cationic headgroups) and extended bilayer structures at pH 9.5 (deprotonated neutral headgroups), which transform into polycrystalline platelets upon aging as revealed in freeze-fracture electron micrographys (Fig. 4c) [45]. BIPO is a second example for a situation in which linear hydrogen bridges between protonated imidazole headgroups and/or ester carbonyls and α-methylene protons interfere with a higher curvature of the aggregates. Replacement of the imidazole moieties by disodium phosphate

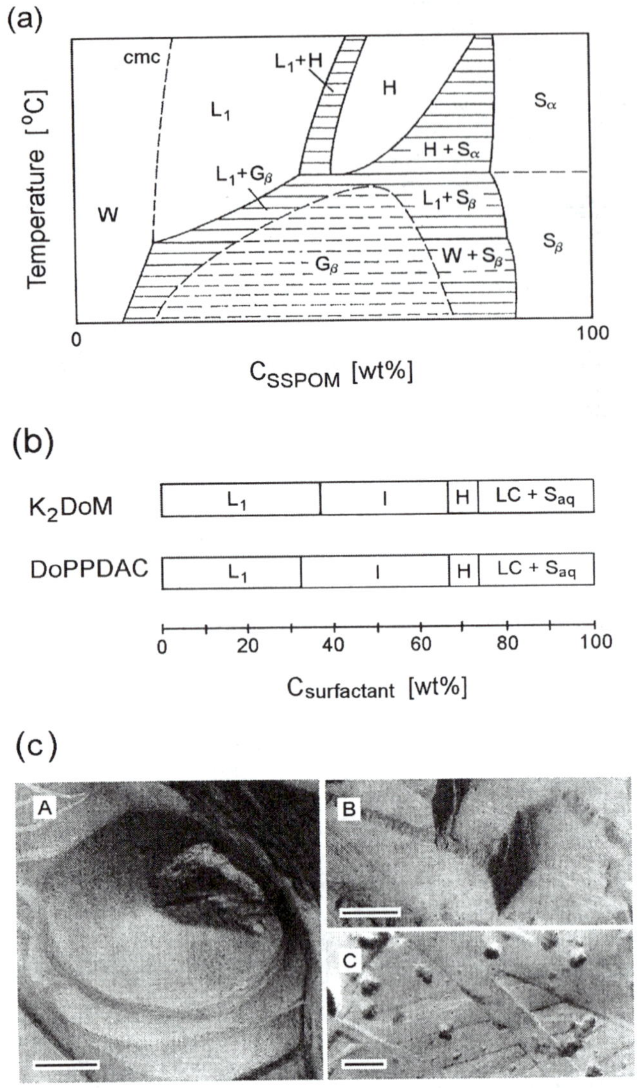

FIG. 4 (a) Schematic phase diagram of the anionic surfactant sodium sulfopropyloctadecyl maleate (SSPOM) in water. Note the uncommon metastable G_β phase at temperatures below the Kraft boundary. (Adapted from Ref. 43.) (b) Composition phase diagram of dipotassium dodecylmalonate (K_2DoM) and dodecylpentamethyl-1,3-propylenebis(ammonium chloride) in D_2O at 25°C. (Adapted from Ref. 44.) W denotes the aqueous phase below the cmc; L_1, I, and H stand for the isotropic micellar, discontinuous cubic, and hexagonal phases, respectively; LC denotes unspecified liquid crystalline phases; S_{aq}, S_α, and S_β denote hydrated surfactant crystals and crystal modifications above and below the chain melting temperature, respectively. (c) Freeze fracture electron micrographs of dispersions of the "dicephalic" surfactant 1,3-bis(1-imidazolyl)-2-propyl octadecanoate (BIPO) in water prepared at (A) pH 5.5, and (B,C) pH 9.5 after aging for 16 h (B) and 1 week (C). Bars represent 250 nm. (Reprinted in part with permission from Ref. 45. Copyright 1999 American Chemical Society.)

groups results in the formation of a dicephalic surfactant that aggregates in water to form bundles of fibers. At pH 7.0 (monoprotonated cationic headgroups), the fibers have uniform diameters of 6.5 nm (approximately twice the molecular length) and lengthens up to 15 μm (aspect ratio > 2000). As expected, the curvature of the aggregates is high, but again a network of linear hydrogen bridges prevents the formation of spherical micelles in favor of extended rods [46]. Both dicephalic surfactants give clear evidence that the aggregation behavior can be changed dramatically by screening the headgroup repulsion and turning it into an attractive force.

A related single-chain surfactant with an imidazole headgroup is based on urocanic acid (4-imidazole-acrylic acid), a metabolite of histidine that is of biological interest. Urocanic acid can be alkylated at both nitrogen atoms of the imidazole moiety and through esterification of the carboxylic acid group. *N*-Methyl-*N*-alkyl-urocanic esters with a methyl group and a dodecyl (C_{12}) hydrocarbon chain connected to the second nitrogen and the ester carbonyl or vice versa, that is, N^1,N^3-dimethyl urocanic dodecyl ester and N^1-dodecyl-N^3-methyl urocanic methyl ester, are cationic surfactants. Regardless of which position the C_{12} chain is connected to, both surfactants form micelles in water at a very similar critical micelle concentration (e.g., 5.4×10^{-4} and 6.5×10^{-4} with chloride as a counterion) [47].

The influence of the headgroup size on the phase behavior of nonionic oxyethylene (C_iEO_j) surfactants [also called poly(oxyethylene) and poly(ethyleneoxide) surfactants, although the typical number of EO units is well below 20 and thus hardly "poly"] is displayed in Fig. 5. The increase of the length of the linear EO_j headgroup from 7 to 20 units at a constant hydrocarbon chain length results in a significant change of the phase behavior. When the headgroup contains 7 EO units, hexagonal (H) and lamellar (L_α) phases are produced above certain surfactant concentrations (Fig. 5a). An increase of the headgroup size to 11 EO units produces a bicontinuous cubic phase (V) wedged between the hexagonal and lamellar mesophases. The lamellar phase is much smaller as before (Fig. 5b). An additional increase of the headgroup to 20 EO units eliminates the lamellar phase and a large micellar cubic phase (I) appears before the hexagonal mesophase (Fig. 5c) [48]. More detailed studies of the micellar cubic phase (I) revealed a more complicated phase behavior within this concentration range. The surfactant dodecaoxyethylene mono-*n*-dodecyl ether ($C_{12}EO_{12}$), for example, forms three different cubic phases (space

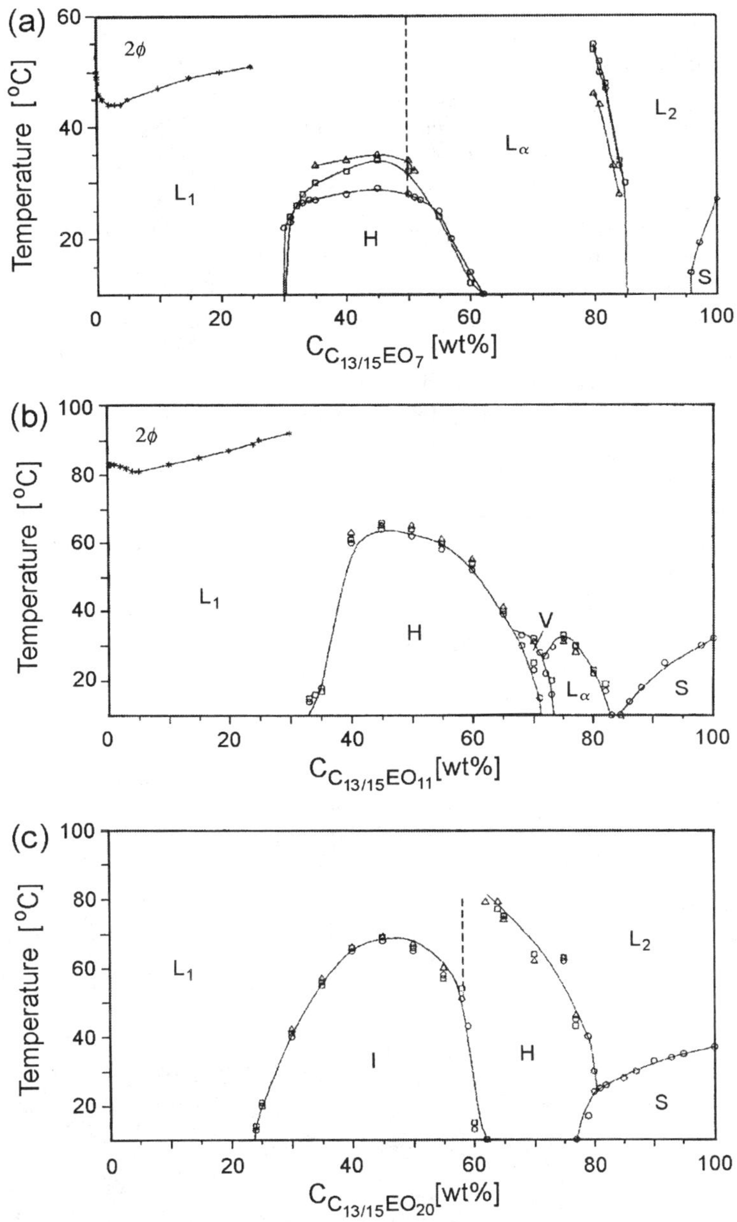

FIG. 5 Phase diagrams of nonionic tridecyl/pentadecyl poly(oxyethylene) surfactants ($C_{13}/C_{15}EO_x$ with $x = 7$, 11, 20 and $C_{13}/C_{15} = 66\%:34\%$ mixture) in water. The data points were measured by rheology, differential scanning calorimetry, microscopy, and visual studies. L denotes isotropic liquid phases; H, L_α, I, and V stand for hexagonal, lamellar, micellar cubic, and bicontinuous cubic mesophases, respectively; S stands for crystalline solid; and 2ϕ denotes a two-phase region (lower consolute region). The broken lines represent an equilibrium between phases. (Adapted from Ref. 48.)

groups; P*m*3*n*, I*m*3*m*, and F*m*3*m*) at concentrations between 30 and 60 wt% in water [49]. A large increase of the headgroup size while keeping the hydrocarbon chain at about the same length as before ($C_{17}EO_{84}$) again changes the phase behavior to some extent. The micellar cubic phase consists of only one phase over the concentration range 20–60 wt% (space group: I*m*3*m*). This phase is composed of discrete micellar building blocks, which appear to be nonspherical, e.g., the aggregation number increases with concentration from 52 ± 3 monomers at 10 wt% to 107 ± 9 monomers at 40 wt% [50]. Similar results have been reported for $C_{18:1}EO_j$ oxyethylene surfactants whose extremely pure $C_{18:1}$ oleyl alcohol hydrocarbon chains are connected to EO_j headgroups of varying sizes between 0 (oleyl alcohol) and 20 EO units. The increase of the oxyethylene chain, that is, the increase of the hydrophile-lipophile balance (HLB) of the monomers, corresponds to an increasing curvature of the surfactant layers toward water.

Various self-aggregating structures have been found: hexagonal and lamellar liquid crystals, four kinds of isotropic liquid crystals, a sponge phase, and a reverse hexagonal liquid crystal (Fig. 6a) [51]. An interesting observation has been made on the $C_{18:1}EO_{50.8}$-water system. This binary mixture forms an aqueous micellar (L_1) and a discontinuous cubic (I_1) phase. Upon addition of *m*-xylene, the I_1 phase changes to a hexagonal phase (H_1), a bicontinuous cubic phase (V_1), and finally a lamellar phase (L_α) with increasing oil content. Although the mean curvature of the surfactant aggregates changes from positive (curved toward oil) to zero, a negative (curved toward water) curvature is not achieved because of the steric hindrance caused by the long $EO_{50.8}$ chains. The L_α phase solubilizes a large amount of oil, and eventually very stable reverse vesicles form in the oil-rich region [52]. These phase transformations are caused by penetration of *m*-xylene molecules into the palisade layer of the surfactant molecules, thus reducing the water content within the layer. A phase diagram displaying the position of the reverse vesicle phase and a micrograph of this phase are shown in Fig. 6b. It should be noted that the geometry of the molecules is not affected by the observed change of the mean curvature. The area per molecule is almost independent of the geometry of the interface [53–55].

A related example is the effect a change in temperature has on the phase behavior of ternary oxyethylene mixtures. At constant temperature, the structural sequence of normal spheres to planar bilayers via cylinders is observed as a function of decreasing water concentration in the phase diagram of the ternary $C_{12}EO_5$-water-decane mixture (Fig. 7) [53]. The phase diagram is strikingly similar to that observed for the binary $C_{12}EO_5$-water mixture despite the presence of

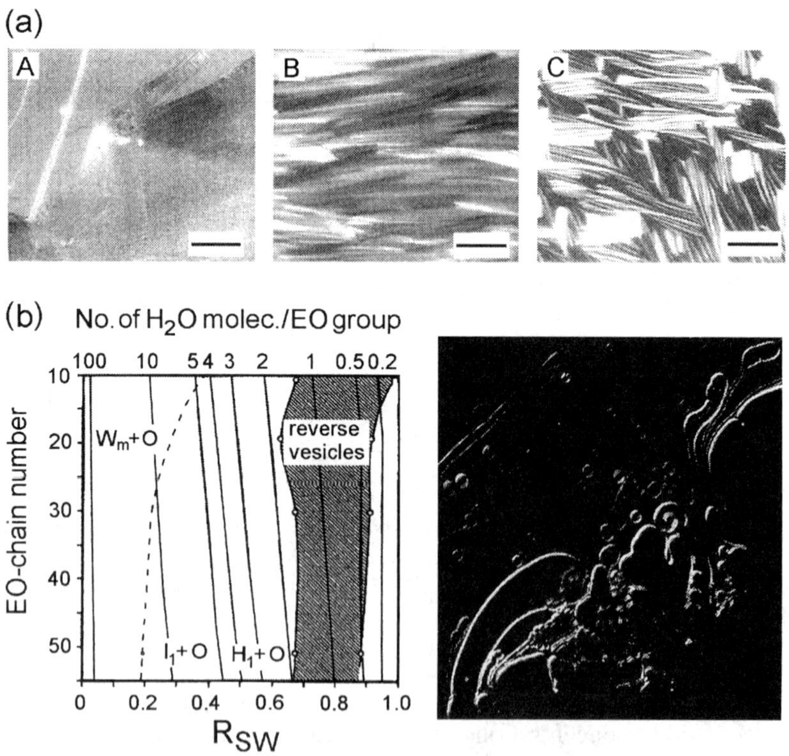

FIG. 6 (a) Typical birefringent textures of the liquid crystal phases found in the $C_{18:1}EO_j$-water system: (A) lamellar phase; (B) hexagonal phase; and (C) reverse hexagonal phase. Bars = 50 μm. (Reprinted with permission from Ref. 51. Copyright 1997 American Chemical Society.) (b) Phase diagram of $C_{18:1}EO_j$-water-*m*-xylene at 25°C as a function of the surfactant-to-water ratio R_{SW} and the number of oxyethylene groups (note the decreasing number of water molecules per EO group that leads to the formation of reverse vesicles); micrograph of reverse vesicles and myelin strands formed in the oil-rich region. (Reprinted with permission from Ref. 52. Copyright 1999 American Chemical Society.)

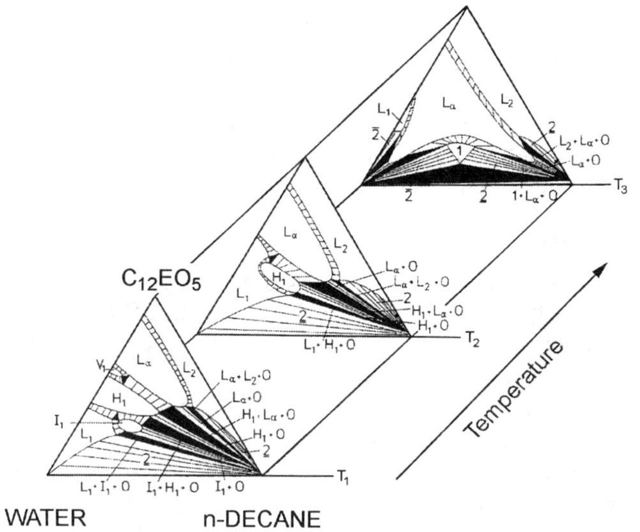

FIG. 7 Phase prism of temperature-dependent changes in the behavior of a ternary $C_{12}EO_5$-water-decane mixture showing some individual isothermal cuts. T_3 corresponds to the balanced temperature where the spontaneous curvature is zero (38°C in the present system). At T_1 and T_2, the spontaneous curvature is toward oil (positive). The L_3 phase is not included because it is formed at temperatures above T_3. L denotes isotropic liquid phases; L_α, H_1, I_1, and V_1 stand for lamellar, hexagonal, micellar cubic, and bicontinuous cubic phases, respectively; O denotes oil, and $\underline{2}$ and $\overline{2}$ denote oil-rich and water-rich two-phase regions. (Adapted from Ref. 53.)

the oil [56]. With increasing temperature, the phase behavior can be understood in terms of a monotonic decrease of the spontaneous or preferred curvature. Oil-swollen micelles are formed at low temperature, hence their curvature is toward the oil (positive). Upon heating, the micelles grow from spheres along the lower phase boundary to elongated micelles, consistent with a decrease in the mean curvature. At an intermediate temperature (38°C in the present system), the spontaneous curvature is zero and only structures with this mean curvature are stable, that is, either a balanced bicontinuous microemulsion containing equal volumes of water and oil or a lamellar phase. Upon further heating, the lamellar phase transforms to the liquid L_3 phase because its monolayer mean curvature is toward the water (negative). The strong temperature dependence of the spontaneous curvature originates from the fact that oxyethylene chains are increasingly less water soluble at higher temperatures; that means the equilibrium concentration of water in the palisade layer of the nonionic $C_{12}EO_5$ film decreases [53,57]. Because the area per surfactant molecule remains constant, as mentioned before, the mean curvature of the aggregates has to decrease from sphere to cylinder to lamellae with decreasing water concentration. A more detailed discussion of the phase behavior of nonionic C_iEO_j surfactants and the effect of additives is given in a review [58]. Although the relationship between a change in headgroup size and the resulting change in phase behavior should be more obvious for nonionic surfactants, which lack the additional counterion interactions mentioned earlier for charged surfactants, the experimental data for oxyethylene surfactants indicate that a prediction of the phase behavior solely based on packing parameters is insufficient.

Nevertheless, there are striking similarities many oxyethylene systems share and which may have some predictive power. It appears that very few empirical parameters are needed to reduce the scales of composition, temperature, length, and interfacial tension at ambient pressure. Relevant parameters are (1) the mean temperature T_m of the three-phase body (roughly the HLB temperature); (2) the temperature sensitivity of the phase behavior (equivalent to the temperature dependence of the spontaneous curvature), characterized by the extent ΔT of the three-phase body; (3) the fraction of surfactant γ_i necessary to solubilize equal volumes of water and oil at T_m, which is inseparably linked to the maximum length scale in the system; and (4) the minimum interfacial tension at T_m [59,60]. A very detailed study of the phase behavior of the system *n*-butyl monoglycol ether (C_4EO_1)-water-dodecane between 22 and 82°C and a comparison with 23 other systems of the type C_iEO_j-water-*n*-alkane are given in Ref. 59. It should be noted that the geometrical structure of the surfactant monomers is not one of the relevant parameters necessary to define a microemulsion. In fact, studies have shown that the behavior of long- and short-chain C_iEO_j surfactants is similar [61,62]. The headgroup area at the water-oil interface was found to be independent of the number of carbon atoms (C_i) of the alkyl chain but to depend strongly, and nearly linearly, on the headgroup size (EO_j) of the surfactant [63]. Most interestingly, the results reported in Ref. 59 suggest that the *monomer solubility of the surfactant in water and oil* might be the key to describing the main features of these microemulsion systems. This point will be discussed in more detail later in this chapter.

The addition of a sulfate group to the oxyethylene head leads to the commercially very important family of alkyl ethoxy sulfate surfactants [C_nH_{2n+1}—$(OCH_2CH_2)_m$—OSO_3Na] [64–66]. These surfactants exhibit an aggregation behavior intermediate between that of nonionic and anionic surfactants, determined by the interplay between steric interaction of hydrated EO_j groups and electrostatic repulsion of the sulfates. The relative strength of both forces depends on the number of EO groups, and the length of the hydrocarbon chain influences the hydrophile-lipophile balance. Dodecyl ethoxy sulfate surfactants with two EO groups form spherical micelles in water, which undergo a sphere-to-rod transition in the presence of multivalent counterions such as Ca^{2+} and Al^{3+} [65]. A more thorough study involving dodecyl ethoxy sulfates with one, two, four, and six EO groups revealed that micelles formed by surfactants with one and two EO groups undergo a one-dimensional growth beyond a threshold NaCl concentration due to the salt screening of the sulfate repulsion. In contrast, micelles formed by surfactants with four and six EO groups remain spherical over the entire salt concentration and temperature range used in the study because of stronger steric hindrance between these larger headgroups. A phase separation at high temperatures analogous to the behavior of nonionic C_iEO_j surfactants has not been observed so far [66].

Replacement of the linear hydrocarbon chain C_iEO_j surfactants by a trisiloxane group leads to another family of industrial important nonionic surfactants [67–69]. Trisiloxane or $M(D'E_n)M$ surfactants [M denotes the trimethylsiloxy group $(CH_3)_3—SiO_{1/2}—$; D′ stands for $—O_{1/2}Si(CH_3)(R)O_{1/2}—$, where R is a oxyethylene group (E_n) attached to the silicon atom through a propyl spacer] behave differently from conventional hydrocarbon surfactants in being surface active in nonaqueous media. Although the hydrophobicity of the MD′M group (which includes the propyl spacer) in water is similar to that of a linear $C_{12}H_{25}$ hydrocarbon chain, the shape of this hydrophobic group is quite different. Its "umbrella" structure is shorter and wider than $C_{12}H_{25}$ and has a larger volume (e.g., MD′M: 9.7 Å length and 530 $Å^3$ volume; $C_{12}H_{25}$: 15 Å length and 350 $Å^3$ volume). Solubility and self-aggregation of trisiloxane surfactants derive from the oxyethylene headgroup; that is they become less water soluble with increasing temperature, analogous to the behavior of C_iEO_j surfactants. Trisiloxane surfactants differ from these linear analogues in that their phase behavior resembles that of C_iEO_j surfactants having longer hydrophilic EO_j chains because of the umbrella structure of the MD′M group [e.g., $M(D'E_8)M$ behaves more like $C_{12}EO_5$ than $C_{12}EO_8$] [68]. The binary phase diagram of $M(D'E_{12})M$ is dominated by the hydrophilic character of the EO_{12} headgroup. An isotropic phase is formed at all concentrations in a wide temperature range between 12 and 43°C, the lower consolute temperature (Fig. 8a). In this one-phase region, $M(D'E_{12})M$ forms spherical micelles, cylindrical micelles, random interconnected bilayers, and inverted microstructures at high surfactant concentrations. A hexagonal (H_1) and a lamellar (L_α) phase have been found below 12°C [67]. Reduction of the hydrophilic character of $M(D'E_{12})M$ by reducing the length of the EO_{12} headgroup results in a richer phase behavior [68]. In the phase diagram of $M(D'E_6)M$, a water-rich isotropic phase (W) is formed at surfactant concentrations below 0.1 wt%, followed by a phase ($W + L_\alpha$) consisting of unilamellar vesicles in a wide size distribution (50–500 nm). Most of the phase diagram is dominated by a large lamellar phase. A 3–5°C wide, continuous band of a transparent and isotropic phase is present above L_α at all concentrations between 1 and 80 wt%. The water-rich end of this band has been labeled L_3 and the surfactant-rich end L_2. The L_3 phase exhibits strong flow birefringence, and the L_2 phase is isotropic with no birefringence. A continuous evolution between two phases with no phase boundary is not atypical but has been reported for other systems [70]. In general, trisiloxane $M(D'E_n)M$ surfactants follow the same self-aggregation principles as linear C_iEO_j surfactants. They favor microstructures of higher positive curvature with increasing size of the EO_j group [68,69].

The introduction of a trimethylsilyl-terminated alkyl chain $[(CH_3)_3—Si—(CH_2)_n—]$ as hydrophobe instead of the bulky trisiloxane group causes another peculiarity of the phase behavior. Trimethylsilane surfactants display solution properties similar to those of the trisiloxane analogues but have the advantage of higher hydrolytic stability. Their phase behavior has been studied intensely [71–75]. Trimethylsilane surfactants behave similar to nonionic C_iEO_j surfactants; however, they differ in that there are two disconnected lamellar mesophases in the phase diagram, a large concentrated one and a small diluted one. As an example, the phase diagram of $CH_3Si(CH_2)_6(OCH_2CH_2)_5OCH_3$ ($Me_3SiC_6E_5$) in water is shown in Fig. 8b [76].

Probably the most important family of single-chain amphiphiles with zwitterionic headgroups consists of 1-acyl-*sn*-glycerophosphocholines or lysophosphocholines (lyso-C_nPC). Lysophosphocholines are constituents of many biological membranes. They derive from natural phosphocholines by enzymatic hydrolysis of the acyl chain in the 2-position of the glycero backbone [77–79]. The phase behavior of lyso-C_nPC molecules strongly depends on the length of the hydrocarbon chain. In general, they behave similarly to other surfactants as they show the same sequence of isotropic micellar solution (L_1), hexagonal phase (H_1), and lamellar phase (L_α) before the region of hydrated crystals

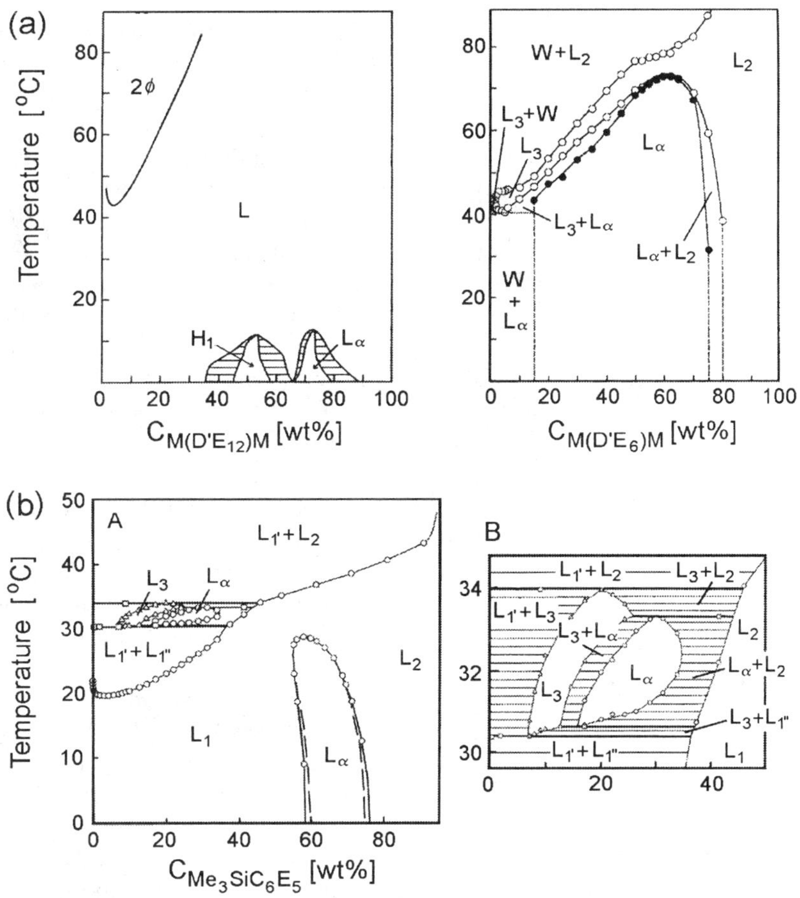

FIG. 8 (a) Binary phase diagram of the trisiloxane surfactants M(D′E_{12})M and M(D′E_6)M in water. For details see text. (Adapted from Ref. 67 and 68.) (b) Binary phase diagram of (A) the trimethylsilane surfactant $CH_3Si(CH_2)_6(OCH_2CH_2)_5OCH_3$ ($Me_3SiC_6E_5$) in water showing two separate lamellar L_α phases and (B) an enlarged portion of the diagram. (Adapted from Ref. 76.) L and W denote isotropic liquid phases; L_α and H, respectively, stand for lamellar and hexagonal phases; and 2ϕ denotes a two-phase region.

(Fig. 9). Nevertheless, there are some peculiarities in the diagrams. Lauroyl (lyso-C_{12}PC), myristoyl (lyso-C_{14}PC), and palmitoyl (lyso-C_{16}PC) lysophosphocholines have a cubic phase (I_1) wedged between the L_1 and H_1 mesophases, which is absent in the diagram of the homologous stearoyl (lyso-C_{18}PC) lysophosphocholine (Fig. 9a and b). Lyso-C_{18}PC, on the other hand, has a cubic mesophase (V_1) located between the H_1 and L_α phases. The presence of two additional methylene groups (e.g., palmitoyl C_{16} compared with stearoyl C_{18}) should not affect the geometry of the lysophosphocholines but clearly affects their aggregation behavior. Insertion of a *cis* C=C double bond as present in the hydrocarbon chain of oleoyl (lyso-$C_{18:1}$PC) lysophosphocholine does not change the sequence of mesophases but changes the phase transition temperatures compared with the saturated lyso-C_{18}PC. The transition from crystalline to liquid crystalline phases takes place above room temperature in case of lyso-C_{18}PC (Fig. 9b and c) [77,78]. The micelles within the L_1 phase of saturated lysophosphocholines remain spherical over the whole concentration range of the phase, whereas the micelles of the unsaturated lyso-$C_{18:1}$PC seem to be elongated and polydispersed. The cubic phase (I_1) consists of disklike micelles with an axial ratio of about two, arranged in a cubic lattice as found for other surfactants [78].

In general, zwitterionic surfactants are of interest because of their insensitivity to changes of the ionic strength and their compatibility with living matter; e.g., they are used in protein solubilization without denaturation and in low-irritant shampoos. Single-chain zwitterionic surfactants form spherical micelles in water because of their usually bulky polar headgroups (>50–60

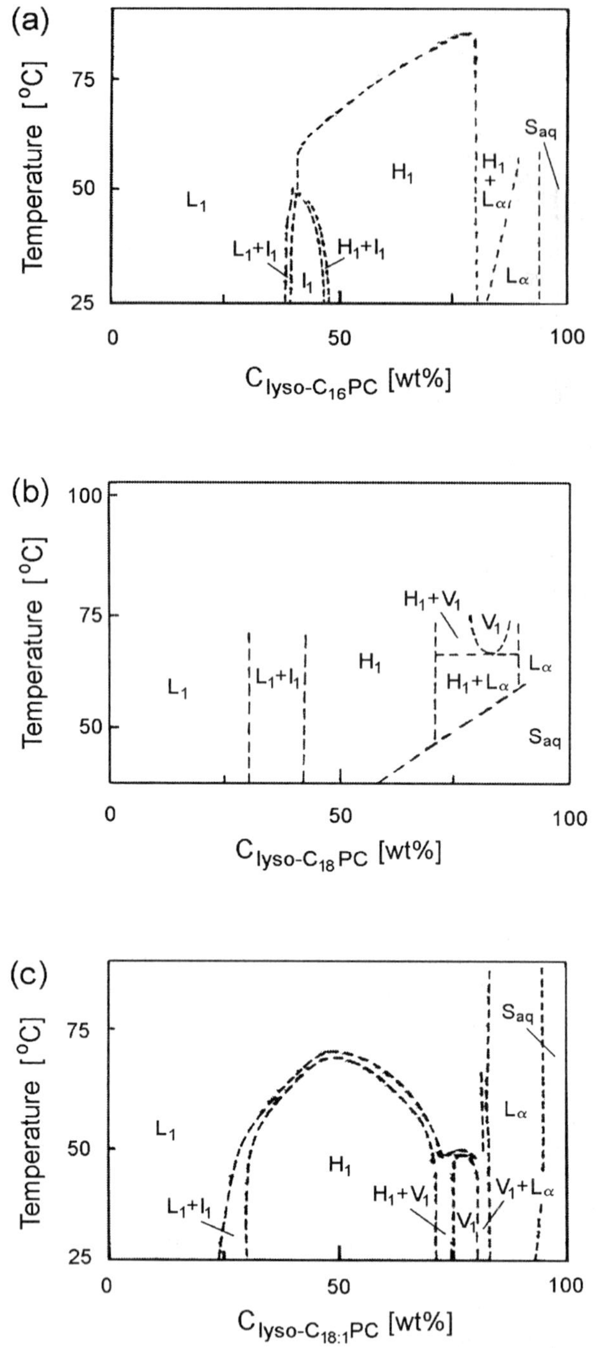

FIG. 9 Phase diagrams of (a) palmitoyl lysophosphocholine (lyso-C_{16}PC), (b) stearoyl lysophosphocholine (lyso-C_{18}PC), and oleoyl lysophosphocholine (lyso-$C_{18:1}$PC) in water. Note the presence of cubic I_1 and V_1 phases dependent on the acyl chain length, and the change in phase transition temperatures dependent on the degree of saturation of the acyl chains. L_1 denotes an isotropic liquid phase; L_α, H_1, I_1, and V_1 stand for lamellar, hexagonal, micellar cubic, and bicontinuous cubic phases, respectively; and S_{aq} denotes hydrated crystals. (Adapted from Ref. 77.)

Å^2). At concentrations above 60–70 wt% they aggregate to form mesophases of high curvature, such as micellar cubic (I_1) and hexagonal (H_1) phases. The phase boundaries are often parallel to the temperature axis, resulting in weak temperature sensitivity of the phases [80–82]. The aggregation behavior of zwitterionic surfactants depends on the flexibility of the spacer between the two charges. Long hydrophobic alkyl spacers ($(-CH_2-)_n$ with $n > 7$, for example, fold back into the hydrophobic region of the aggregates, thus allowing the charges to get closer to each other [83,84]. Zwitterionic *N*-alkylpyridiniocarboxylate surfactants with the COO^- group in either the 3- or 4-position at the pyridinium ring (e.g., C_{12}-Py-*m*-CO_2 with $m = 3$, 4), on the contrary, have a fixed distance between the charges. They display a rather unusual phase behavior as shown in Fig. 10a. The sequence of mesophases with increasing concentration is as expected: L_1, I_1, and H_1. A temperature increase at a constant surfactant concentration between 60 and 68 wt%, however, also results in phase transitions form L_1 to H_1 via I_1. The position of the COO^- group does not affect the sequence of mesophases but shifts the transition temperatures. The I_1 phase consists of a periodic arrangement of globular micelles, which can be slightly anisotropic (space group *Pm3n*), while the H_1 phase contains micellar rods with the alkyl chains in the liquid state (space group *p6m*). The different dipole moments of the headgroups due to the 3- versus 4-position of the carboxylate COO^- to the ammonium N^+ group affects the behavior in dilute solution and thus the cmc. However, this dependence vanishes with increasing concentration of the zwitterionic surfactants [85].

Another approach to changing the hydrophile-lipophile balance and to increasing attractive intermolecular interactions without major changes of the geometrical structure of the surfactants involves the use of rigid aromatic segments within the hydrocarbon chain. These molecules combine the structural element of a liquid crystal–forming compound [e.g., *N*-(4-methoxybenzylidene)-4-butylaniline, MBBA] with the structural element of a micelle-forming surfactant (e.g., cetyltrimethylammonium bromide, CTAB). For this purpose, azobenzene ($C_n-C_6H_4-N{=}N-C_6H_4-C_m$), benzylideneaniline ($C_n-C_6H_4-N{=}CH-C_6H_4-C_m$), and salicylideneaniline [$C_n-C_6H_4-N{=}CH-$(2-OH) $C_6H_3-C_m$] containing hydrocarbon chains have been connected to cationic trialkylammonium headgroups [86–89]. The aromatic segments reduce the flexibility of the alkyl chains (thus interfering with one of the requirements of the packing parameter concept) and improve intermolecular interactions due to the stacking of the aromatic rings. Bilayer formation with and without interdigitation of opposing half-layers has been observed [86,87]. Depending on the position of the aromatic group within the hydrocarbon chains, that is, the length of the C_n and C_m segments, the azobenzene derivative forms stable bilayer membranes for the following ratios: n:m = 12:2–12; n:m = 10:5–10; and n:m = 8:10. These results clearly indicate the higher importance of the C_n length for the bilayer formation [86]. The benzylideneaniline derivative forms unilamellar vesicles (n:m = 12:5), short fibers with diameters of 60–70 Å (n:m = 12:10), and large unstructured aggregates (n:m = 8.10). The related salicylideneaniline derivative, which has a hydroxy group in the α-position to the CH= group, forms not very well-defined fibrous structures with diameters around 150 Å regardless of the C_n:C_m ratio (Fig. 10b) [88]. Several other rigid segments have been used, linear and bent ones with two and three aromatic rings. Whereas molecules containing a linear segment tend to assemble in a neat, two-dimensional arrangement (membrane), molecules carrying a bent segment favor more radial structures such as rods and tubules. For more detailed information, see the reviews in Refs. 86 and 89.

The concept of a rigid segment is driven even farther by the use of a three- and four-ring spiro system that has no rotational freedom as a hydrocarbon chain, connected to a trimethylammonium headgroup. Both compounds behave like surfactants and form micelles. The cmc of the three-ring spiro surfactant of 0.14 M turned out to be identical to the cmc of the related octyl trimethylammonium bromide with a similar chain length. On the other hand, the cmc is 3.3 times larger than that of the undecyl trimethylammonium bromide, a conventional surfactant with the same number of carbon atoms within the hydrocarbon chain. This observation and the

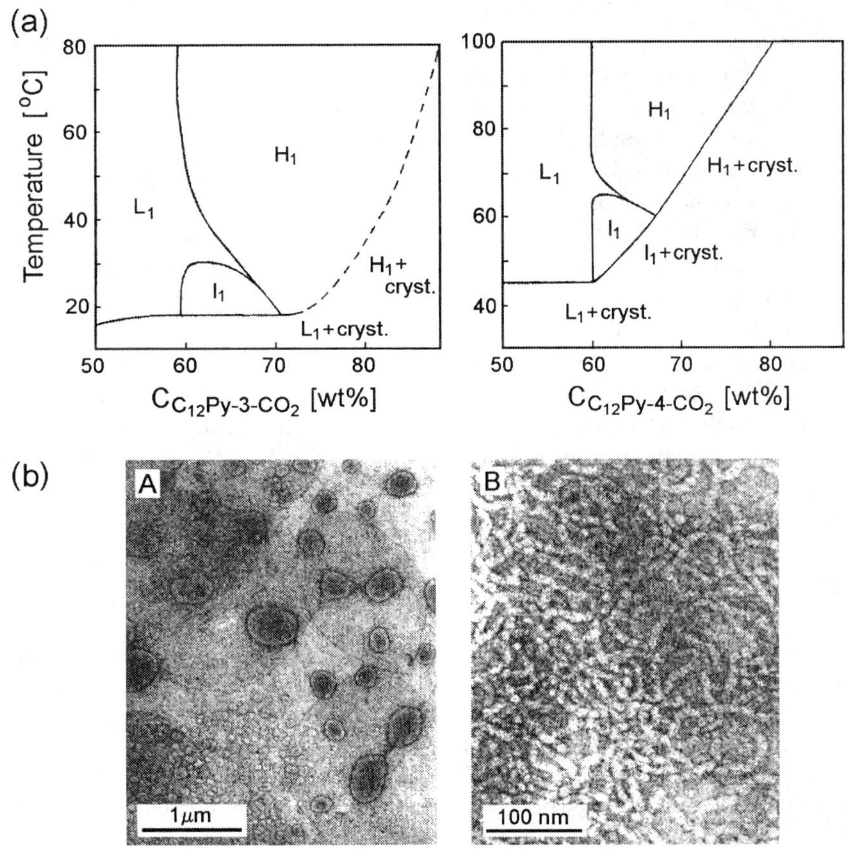

FIG. 10 (a) Binary phase diagrams of zwitterionic N-dodecylpyridinio-3-carboxylate (C_{12}-Py-3-CO_2) and N-dodecylpyridinio-4-carboxylate (C_{12}-Py-4-CO_2) in water. L_1 denotes an isotropic liquid phase; I_1 and H_1, respectively, stand for micellar cubic and hexagonal mesophases. (Adapted from Ref. 85.) (b) Transmission electron micrographs of trimethylammonium surfactants containing (A) benzylideneaniline and (B) salicylideneaniline as rigid segments in their hydrocarbon chain. (The C_n/C_m ratios are 12:5, and 8:10; for details see text). (From Ref. 88.)

fact that micelles could not be detected by light scattering (which had a lower size limit of 40 Å) indicate the formation of very small aggregates of perhaps only 10–20 molecules by the spiro surfactant. The spiro surfactants are another example of molecules that are violating the packing concept based on the molecular geometry [90].

One final way, discussed here, to change the hydrophile-lipophile balance of a single-chain surfactant without causing major changes in the molecule's geometry involves the use of fluorocarbon chains [91–97]. Fluorocarbon chains are more hydrophobic and stiffer than hydrocarbon chains (note the similarity to the rigid segment discussed earlier). They possess less conformational freedom because of the larger energy difference between *gauche* and *trans* conformations. As a consequence, fluorosurfactants have a much stronger capacity to self-aggregate in water into discrete molecular assemblies: (1) their cmc corresponds to a hydrocarbon analogue with 1.5–1.7 as many carbon atoms in the tail; (2) even very short, single-chain fluorinated amphiphiles form stable vesicles without the need for supplementary attractive interactions; and (3) sturdy microtubules were obtained from nonchiral, non-hydrogen-bonding single-chain fluorosurfactants. Like hydrocarbon surfactants, fluorosurfactants form micelles and various mesophases in the same sequence as found for hydrocarbons. However, because of the chain stiffness, they often form structures with less curvature, and they have a higher tendency to form intermediate phases between the H_1 and L_α phases. The formation of intermediate phases in surfactant-water mixtures and the self-assembly of fluorocarbon surfactants have been reviewed [91–93]. Therefore, only a few examples will be discussed here. The cationic fluorosurfactant 1,1,2,2-tetrahydroperfluorodecyl-pyridinium chloride (HFDePC) aggregates in water to form micellar (L_1), hexagonal (H), centered rectangular (R), and centered trigonal (T) mesophases with increasing concentration.

Evidence for a random mesh phase and a lamellar phase was found at even higher concentrations (Fig. 11a) [94]. One specific feature of fluorosurfactants is displayed in Fig. 11b. Partial screening of the headgroup repulsion by addition of a small amount of salt is sufficient to transform spherical micelles of HFDePC into threadlike micelles. The related fluorocarbon surfactant 2-hydroxy-1,1,2,3,3-pentahydroperfluoroundecyldiethylammonium chloride (I-C11) forms threadlike micelles even in the absence of salt. Spherical and threadlike micelles coexist in salt-free aqueous solution at a concentration of 50 mM [95]. Perfluoroalkylated single-chain surfactants with a neutral dimorpholinophosphoramidate (CF-MPA) or a zwitterionic phosphocholine (CF-PC) headgroup aggregate in water to form rigid tubules and flexible fibers. CF-MPA tubules form readily from lipid films by hydration at 50°C with gentle swirling by hand within hours after preparation of the suspension. They have a length distribution of 10 to 50 μm and diameters between 100 and 500 nm. The tubules are fragile and transform into large vesicles when heated above 40°C. CF-PC fibers, on the other hand, grow within 1 to 4 months from their suspensions. They are flexible, have a length distribution of 50 to 500 μm with diameters between 1 and 5 μm, and are sometimes branched. The reversible fiber-to-large vesicle transformation occurs at 50°C (Fig. 11c). The hydrocarbon analogues form only micelles, even when the hydrocarbon chain is extended to adjust the cmc to those of the fluorosurfactants [96]. The packing parameter concept is useful only in predicting the formation of vesicles versus micelles when adjusted to the specifics of the CF groups. CF_2 and CF_3 groups have higher volumes than the corresponding CH_2 and CH_3 groups (e.g., CF_2 41 Å^3 and CH_2 27 Å^3), and fluorocarbon chains are slightly longer than hydrocarbons. The contribution per CX_2 group to the length of the fully extended fluorocarbon chain is 1.30 Å (hydrocarbon: 1.25 Å), and the cross-sectional area is 31.5 Å^2 (hydrocarbon: 21.4 Å^2) [95,97]. The surface area per headgroup a_0 remains the same for the fluorinated and hydrogenated analogues. The significantly larger volume of the fluorocarbon chains results in a truncated-cone geometry of the fluorosurfactants that favors vesicles, whereas the cone geometry of the analogous hydrocarbon surfactants favors micelles [97]. Besides these adjustments, the packing parameter concept has the same shortcomings as mentioned earlier for hydrocarbon surfactants when used to explain the transformation of vesicles into either rigid tubules or flexible fibers below the gel-to-liquid crystal phase transition temperature T_c.

III. PHASE BEHAVIOR OF BRANCHED, DOUBLE-CHAIN, AND MULTIPLE-CHAIN SURFACTANTS

The phase behavior of branched, double-chain, and multiple-chain surfactants is discussed in this section. The packing parameter $v/a_0 l_c$, (the volume v) changes because of the presence of a branch or additional hydrocarbon chains. The most thorough study of the effect the shape of the hydrophobe has on the aggregation behavior of surfactants has been carried out on asymmetrical and symmetrical nonionic oxyethylene surfac-

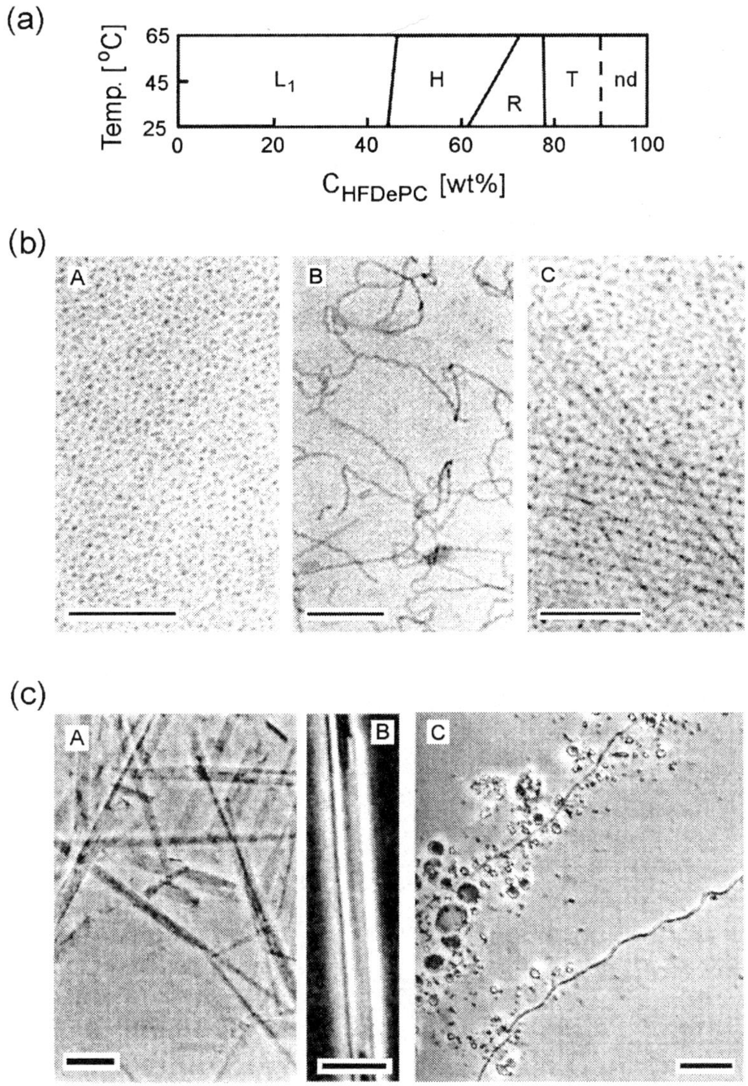

FIG. 11 (a) Partial phase diagram of the cationic fluorocarbon surfactant 1,1,2,2-tetrahydroperfluorodecylpyridinium chloride (HFDePC) in D_2O. L denotes isotropic liquid phases; L_α, H, R, and T stand for lamellar, hexagonal, centered rectangular, and centered trigonal (rhombohedral) mesophases, respectively; nd denotes a not safely determined portion of the diagram that might consist of a random mesh phase and a lamellar phase. (Adapted from Ref. 94.) (b) Cryo-transmission electron micrographs of (A) monodispersed spherical micelles of 50 mM HFDePC in water; (B) long, flexible, and entangled threadlike micelles of 25 mM HFDePC in 150 mM aqueous NaCl; and (C) spherical micelles in coexistence with rather stiff threadlike micelles of 50 mM 2-hydroxy-1,1,2,3,3-pentahydroperfluoroundecyldiethylammonium chloride (I-C11) in water. Bars = 100 nm. (Reprinted in part with permission from Ref. 95. Copyright 1999 American Chemical Society.) (c) Phase-contrast optical micrographs of (A) tubules obtained from CF-MPA in water; (B) a magnified single tubule, note the aqueous core visible in the upper part of the micrograph; and (C) the reversible transformation at 50°C of flexible fibers into large vesicles. Bars_{ABC} = 8, 4, and 8 μm. (From Ref. 96.)

tants of the series $C_kC_nGE_8M$ [C_k symbolizes an *n*-butyl (C_4) or *tert*-butyl (C_4^t) chain; C_n stands for a decyl (C_{10}) or dodecyl (C_{12}) chain; G denotes a triglyceryl unit (—OCH_2—CHO—CH_2O—); and E_8M stands for octaoxyethylene monomethyl ether] [98–100]. The double-chain surfactants $C_4^tC_{10}GE_8M$, $C_4C_{10}GE_8M$, and $(C_7)_2GE_8M$ have the same hydrophile-lipophile balance and volume ratio of hydrocarbon chains to headgroup, V_l/V_h. According to the packing parameter, this should result in identical phase behavior of all three surfactants in water. This is clearly not the case, as one can see by comparison of the phase diagrams displayed in Fig.

12a. At temperatures below the lower critical consolute temperature T_c of the miscibility gap $L + L_1$, the phase behavior is very similar with respect to the crystalline coexistence region and the sequence of mesophases. The isotropic micellar solution (L_1) is followed by the normal hexagonal (H_1), bicontinuous cubic (V_1), and lamellar (L_α) phases with increasing surfactant concentration. At highest concentrations, an isotropic reverse solution (L_2) is formed.

However, there are some differences within the phase diagrams. The L_α phase of $C^t_4C_{10}GE_8M$ is smaller than those of $C_4C_{10}GE_8M$, and $(C_7)_2GE_8M$ and surrounded by the isotropic L_1 solution, whereas L_α of the other surfactants has a boundary with the liquid-liquid phase $L_1 + L_2$. In addition, the phase behavior of $C^t_4C_{10}GE_8M$ at temperatures above T_c is quite different from that of $C_4C_{10}GE_8M$ and $(C_7)_2GE_8M$ (Fig. 12a) [98]. Different geometrical structures are possible with the same volume of the hydrophobic part, as shown in Fig. 12b. The packing parameter concept based on the V_l/V_h ratio therefore has to be extended by a structure parameter S that gives information about

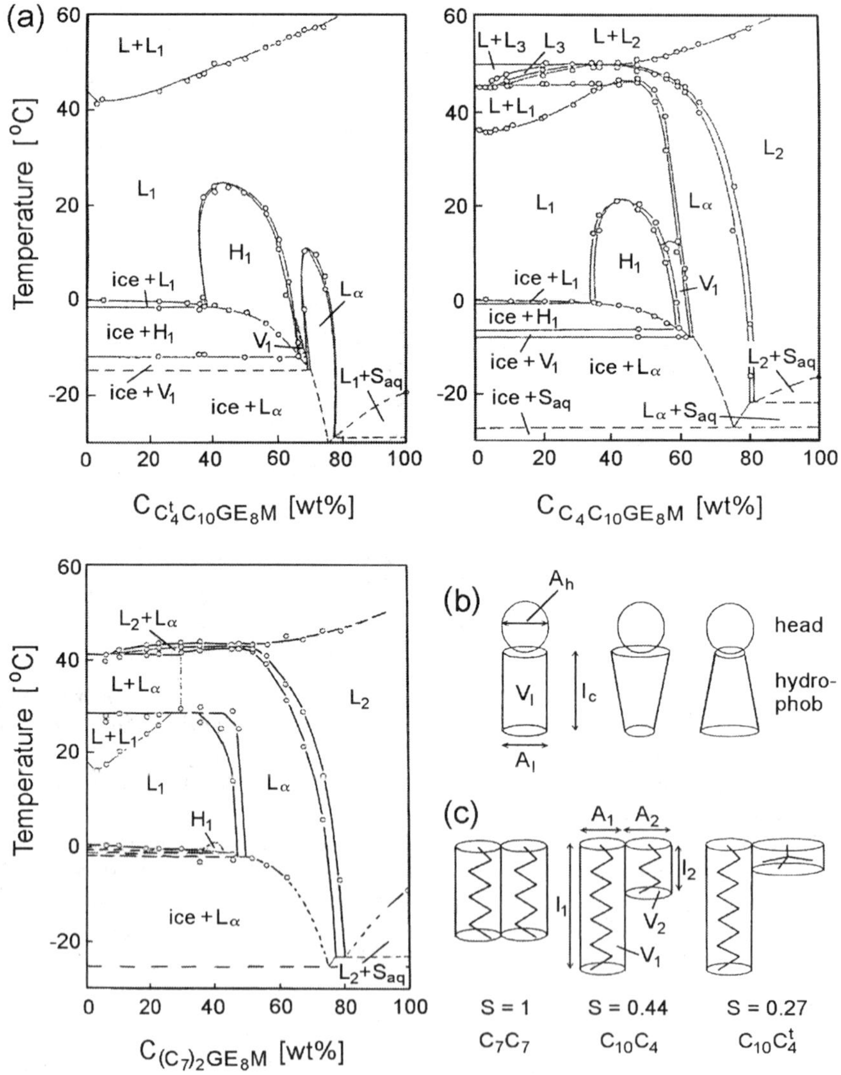

FIG. 12 (a) Phase diagrams of the double-chain surfactants $C^t_4C_{10}GE_8M$, $C_4C_{10}GE_8M$, and $(C_7)_2GE_8M$, which have the same HLB and ratio of the hydrophobic and hydrophilic volumes. L denotes isotropic liquid phases; L_α, H_1, and V_1 stand for lamellar, hexagonal, and bicontinuous cubic phases, respectively; and S_{aq} denotes hydrated crystals. (b) Hypothetical structures of surfactants with the same parameters V_l, l_c, and A_h. (c) Hydrocarbon part of symmetrical and asymmetrical double-chain surfactants with their structure parameter S. (Adapted from Ref. 98.)

the degree of asymmetry of the hydrocarbon part. Examples of structure parameters S derived from the molecular geometry of the surfactants $C_4^tC_{10}GE_8M$, $C_4C_{10}GE_8M$, and $(C_7)_2GE_8M$ are shown in Fig. 12c. Although the modified packing parameter is more suitable to describe the phase behavior of these surfactants, packing differences such as interdigitation of opposing half-layers are still not considered. For example, the symmetrical surfactant $(C_7)_2GE_8M$ forms an L_α phase without interdigitation, whereas the asymmetrical analogue $C_4C_{10}GE_8M$ interdigitates. This results in a different thermal stability of the L_α phases as observable in the respective phase diagrams. The series of C_kC_n GE_8M surfactants allows some additional structural considerations. For example, elongation of the decyl chain by two methylene units (e.g., $C_4C_{10}GE_8M$ versus $C_4C_{12}GE_8M$) does not significantly affect the phase behavior [98].

In a related study, the effect a branched headgroup [$C_{14}G(E_4M)_2$ = Y-shape] and a branched hydrophobe [$(C_7)_2GE_8M$ = V-shape] have on the binary phase behavior and on the behavior of ternary mixtures of both surfactants in water has been investigated. In accordance with the packing parameter, $C_{14}G(E_4M)_2$ with its large headgroup area prefers aggregates of high curvature such as spherical micelles, micellar cubic (I_1) phase, and hexagonal (H_1) phase. The V-shaped $(C_7)_2GE_8M$, on the other hand, aggregates to form structures of low curvature such as cylindrical or disklike micelles and an extended lamellar (L_α) phase. The phase behavior of the mixture depends on the mixing ratio α of the surfactants. The packing parameter concept would therefore predict the phases only in a qualitatively correct way when extended by the mixing parameter α [99]. Systematic changes of the hydrophile-lipophile balance of $(C_n)_2GE_mM$ surfactants by varying n and m result in a phase behavior that is in accordance with the packing parameter concept. Increasing the hydrophobic volume by increasing n stabilizes phases with low curvature, and increasing the hydrophilic area by increasing m stabilizes phases of high curvature [100]. A related study investigated the aggregation behavior of diamines into bilayers with respect to the packing parameter. The compounds employed were 1,2-ethanediamine and 1,4-butanediamine connected to one or several hexadecyl (C_{16}) hydrocarbon chains. The substitution pattern included *N*-alkyl, *N*-dialkyl, *N,N′*-dialkyl, *N*-alkyl-*N′*-dialkyl, and *N*-dialkyl-*N′*-dialkyl derivatives of both diamines [101].

Although the most compact systematic study of the phase behavior of surfactants with regard to changes in their geometrical structure has been conducted on nonionic oxyethylenes (see earlier), the most often studied double-chain compounds are cationic surfactants. The binary phase diagram of didodecyldimethylammonium bromide (DDAB or $diC_{12}NBr$) in water displays an isotropic solution (L_1), a fluid phase (L_3) that is optically isotropic at rest but shows flow birefringence, and a viscous permanently birefringent L_α phase. A two-phase region ($L_1 + L_\alpha$) is located between L_3 and L_α, containing L_α aggregates dispersed in the L_1 phase. Below the melting temperature of the hydrocarbon chains are two phases, denoted as $L_1 + L_\beta$ and L_β. These two phases cannot be distinguished from $L_1 + L_\alpha$ and L_α by visual inspection, but differential scanning calorimetry (DSC) studies show the chains in the frozen state (Fig. 13a). The L_3 phase is metastable and vanishes after 90 days of equilibration and slow cooling to am-

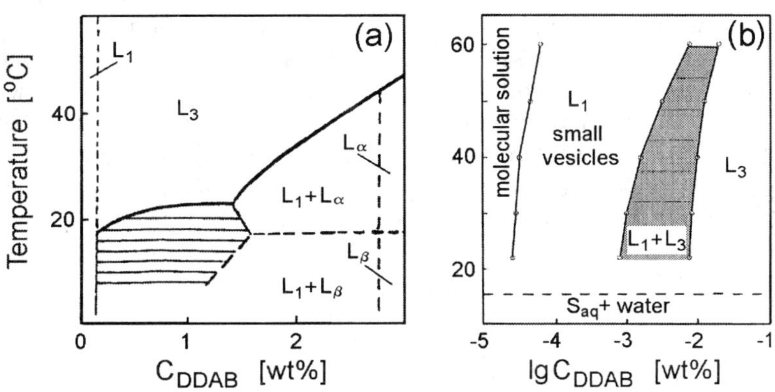

FIG. 13 Binary phase diagram of the double-chain surfactant didodecyldimethylammonum bromide (DDAB) in water (a) at concentrations below 3 wt% and (b) in the very dilute region between 10^{-5} and 10^{-1} wt%. Note the spontaneous formation of small vesicles upon dilution of the L_3 phase. L denotes isotropic liquid phases; L_α and L_β stand for lamellar phases above and below the melting temperature of the hydrocarbon chains; and S_{aq} denotes hydrated crystals. (Adapted from Ref. 102 and 105.)

bient temperature. The phase diagram of the homologous dihexadecyldimethylammonium bromide (DHAB or diC_{16}NBr) is virtually the same as that of DDAB [102].

The structures of the L_α and L_3 phases of DDAB have been studied by freeze-fracture electron microscopy. The micrographs clearly reveal the lamellar structure of the L_α phase and the bicontinuous network of two aqueous subvolumes separated by a random bilayer network that builds the structure of the L_3 phase [70]. A more recent study of DDAB reports the existence of two thermodynamically stable lamellar phases. The dilute L_α phase is stabilized mainly by electrostatic forces, whereas hydration forces stabilize the condensed $L_{\alpha'}$ phase [103]. The condensed $L_{\alpha'}$ phase seems to be stable up to very high concentrations of the surfactant. In the phase diagrams of the homologous diC_{10}NBr and the asymmetrical double-chain surfactant C_8C_{16}NBr, and the L_α phase is followed by a second liquid crystalline phase, which is believed to be hexagonal. The second mesophase vanishes in all cases when the samples are heated to temperatures above 80–90°C and isotropic solutions form [104]. Dilution of the L_3 phase of DDAB results in the spontaneous formation of unilamellar vesicles with a mean diameter of 33 nm. The vesicles are a part of four distinct regions that are present at low concentrations: the L_3 phase, a two-phase region consisting of small vesicles and large multilayer aggregates, the vesicular solution (L_1 phase), and the molecular solution (Fig. 13a). The phase boundaries in the diagrams of homologous dialkydimethylammonium bromides with decyl (C_{10}) and octyl (C_8) chains are shifted to higher surfactant concentrations. For example, diC_{10}NBr forms only three phases (molecular and vesicular solutions and two-phase region), and diC_8NBr forms two phases (molecular and vesicular solutions) within the concentration range used in the study. The unilamellar vesicles of both surfactants have slightly smaller mean diameters of 28 and 30 nm. The reported phase behavior and the spontaneous formation of vesicles are considered to be in accordance with the packing parameter concept. The packing parameters of diC_nNBr surfactants ($P = 0.58$–0.62) are within the range of $\frac{1}{2} < P < 1$ where the formation of vesicles or lamellae is expected [105]. However, whereas DDAB forms small vesicles with a diameter of 33 nm upon dilution from the L_3 phase, it will form giant unilamellar vesicles with diameters of 10–200 μm when 0.1–0.2 mg DDAB is immersed in 450 μL of water without any energy input. These giant vesicles have been intensely studied as mimics of living cells, and chemically induced birthing and foraging as well as aggregation, budding, and fusion have been observed [106,107].

Dodecyldimethylammmonium bromide has also been one of the first synthetic compounds successfully used to form artificial bilayer membranes (BLMs). Upon ultrasonification of an aqueous DDAB dispersion, single- and multiwalled vesicles with a layer thickness of 30–50 Å have been found [108]. It is important to make a distinction between liquid crystalline mesophases and bilayer membranes. The viscosity and stability of mesophases originate from intermicellar forces, that is, lattice forces. In contrast, a bilayer membrane should be able to maintain its structural integrity without relying on lattice forces. The presence of lamellar multilayer mesophases therefore does not warrant the formation of isolated bilayer membranes. Their formation requires a self-assembling ability greater than that of liquid crystalline dispersions.

The structural elements used in a systematic survey of the aggregation behavior of surfactants or amphiphiles are displayed in Fig. 14a. These elements include the hydrophobic tail and hydrophilic headgroup as well as a spacer moiety and a connector that allows connecting two or more hydrocarbon chains to one headgroup. In Fig. 14b and c, examples of double-chain surfactants having cationic ammonium heads; anionic sulfate, phosphate, or carboxylate heads; and nonionic oxyethylene headgroups are shown together with examples of zwitterionic headgroups and some of the connectors used in the survey. Despite their different compositions, all of these surfactants have the ability to aggregate in water to bilayer membranes, either open BLMs in the form of lamellae or closed structures in the form of vesicles and liposomes, when both alkyl chains are at least 10 carbon atoms long [86]. To probe the packing parameter concept and its usefulness in predicting BLM formation, surfactants with three, four, and seven hydrocarbon chains have been prepared [109–111]. The large volume of the hydrocarbon part should force the molecules into a lamellar or even inverted bilayer structure. However, the critical micelle concentrations (cmc) of the triple-chain surfactants shown in Fig. 15a are close to those found for double-chain compounds, suggesting that the third chain has only a little effect on the aggregation behavior in water.

More important is the influence of spacer and connector structures on the aggregate morphology. All triple-chain compounds form bilayer membranes except for the tridodecylmethylammonium bromide, which has no structural element that *orients* the three alkyl chains. The space-filling model for this surfactant

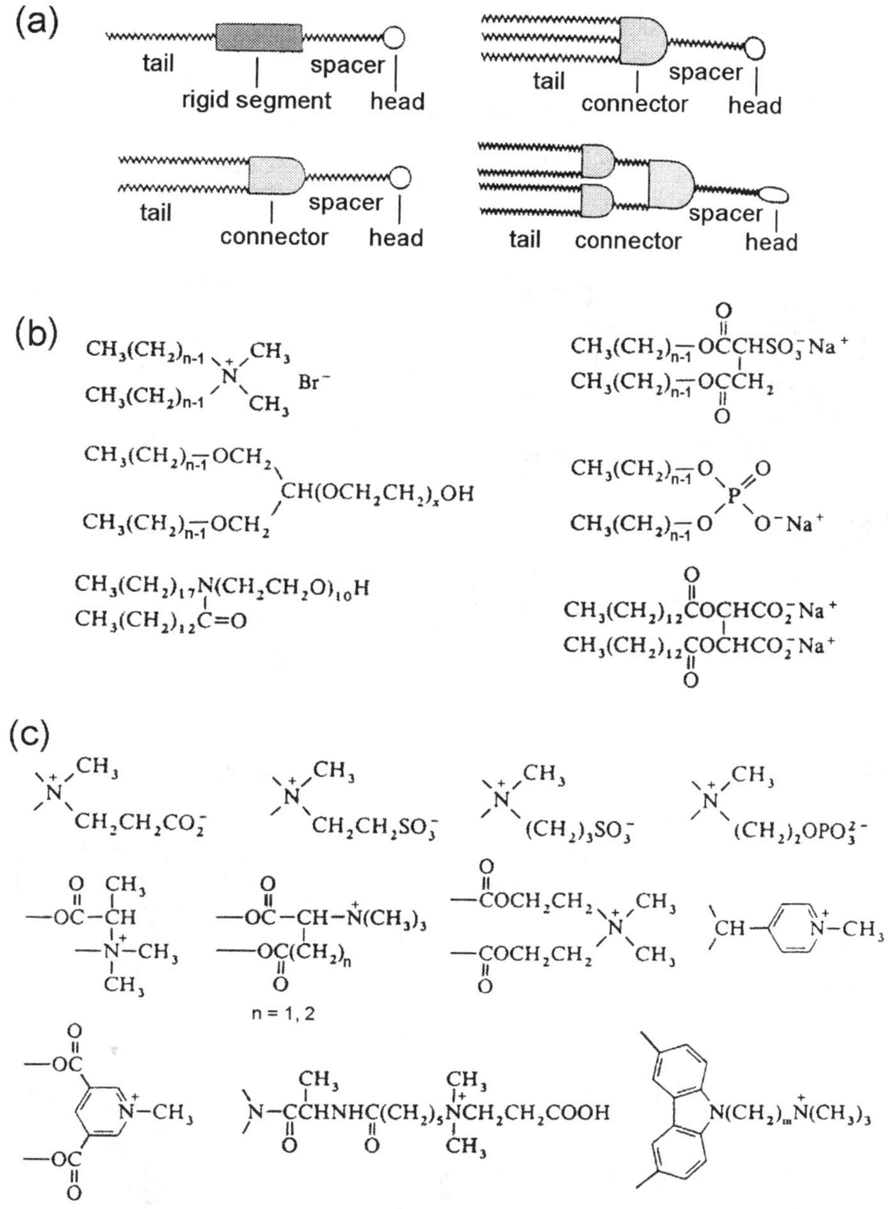

FIG. 14 (a) Structural elements of bilayer-forming amphiphiles. (b) Examples of double-chain surfactants with cationic, anionic, and nonionic headgroups that form bilayer membranes (lamellae and/or vesicles). Note the surfactant with two acyl chains and two carboxylate heads, connected by a single C—C bond, which carries the structural element of a gemini surfactant. (c) Examples of zwitterionic headgroups and connectors used in the synthesis of double-chain surfactants. (Adapted from Ref. 86.)

clearly shows rather disordered hydrocarbon chains close to the ammonium atom because of its tetrahedral configuration (Fig. 15a). The triple chains of the other surfactants are well aligned because of the presence of ester connectors, thus allowing the ammonium head to protrude into the surrounding water layer without conformational constraints. This order requirement and the finding that these triple-chain surfactants show endothermic order-disorder transitions in their DSC thermograms upon heating indicate that the hydrocarbon chains are not in the fluid state as required for the packing parameter concept [109]. The same observations hold true for the four-chain ammonium bromides used in the formation of bilayer membranes. The four hydrocarbon chains are connected via ester functions to glutamic acid and lysine residues. The presence of sev-

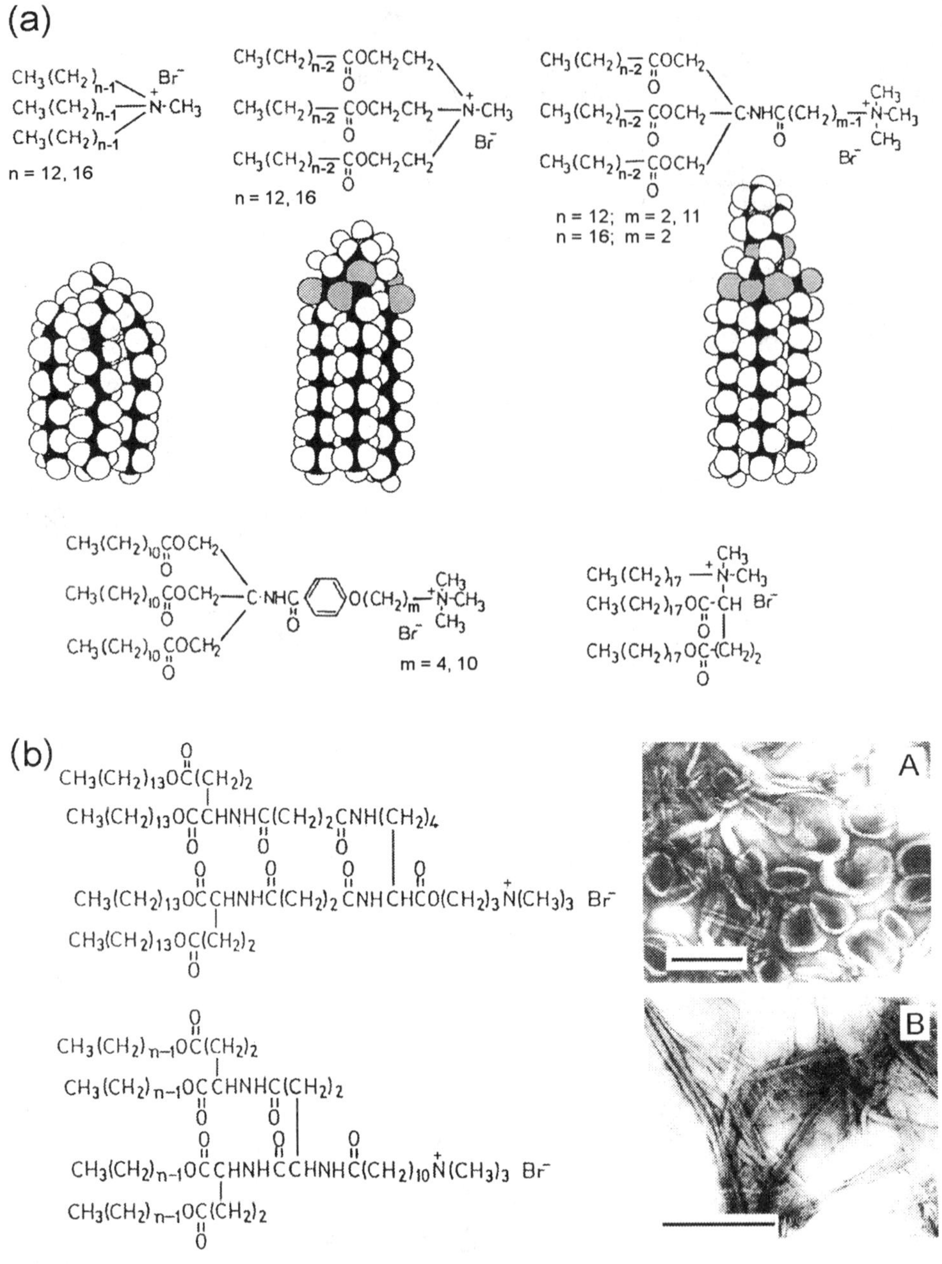

FIG. 15 (a) Chemical structures and space-filling models of triple-chain surfactants used to form bilayer membranes. Note the increasing packing order of hydrocarbon chains in the models going from left to right. (Adapted from Ref. 108). (b) Chemical structures of four-chain surfactants and electron micrographs of vesicles and tubules formed by them in water. Bars = 200 nm. (From Ref. 110.)

eral amide groups allows the formation of stabilizing hydrogen bridges as mentioned earlier for single-chain surfactants containing a maleic acid moiety. The aggregation behavior of the four-chain compounds is thus based not solely on isotropic repulsive and attractive forces but to some extent also on anisotropic attractive interactions. Both compounds aggregate in water to form vesicular and tubular structures (Fig. 15b) [110]. The seven-chain "heptopus" surfactant lacks any ability to form curved structures but is used to form ordered monolayer arrays at the air-water interface [111].

An interesting family of surfactants for which the

packing parameter concept is hardly suitable to describe their aggregation behavior consists of "gemini" surfactants. Geminis are dimeric single-chain surfactants usually connected by an alkyl spacer below the headgroup region; that is, they consist of two hydrocarbon chains and two headgroups. Most gemini surfactants aggregate in water to form micelles. For a review of geminis and related surfactant oligomers refer to the contribution by Martin In in Part 1 of this book. Only two examples of bilayer membrane–forming gemini surfactants are presented here. The first example describes the aggregation behavior of a dianionic gemini. The connection of two octadecyl (C_{18}) chains and two phosphate headgroups via tetrol connectors derived from erythritol and tartaric acid results in the formation of optically active and *meso* gemini surfactants. These diphosphate surfactants aggregate in water at pH 7 to form planar bilayer membranes, which rearrange upon ultrasonication into unilamellar vesicles. The vesicles have diameters of 15–25 nm (*S,S* and *R,R*) and 50–100 nm (*R,S*) and consist of intercalated bilayers with a wall thickness of 38 ± 2 Å. The addition of Ca^{2+} ions triggers fusion of the (*R,R*) and (*S,S*) vesicles but fission and eventually tubule formation of the (*R,S*) vesicles. The wall thickness of the (*R,S*) aggregates increases to 66 ± 1 Å, indicating a nonintercalated bilayer membrane (Fig. 16a). This surprisingly different behavior of the optically active and *meso* surfactants is explained by intermolecular versus intramolecular interactions with the Ca^{2+} ions. Calcium-mediated intermolecular interactions between phosphate groups of adjacent molecules stabilize large surfaces of low curvature, thus favoring large vesicles. Intramolecular interaction, that is, complexation of Ca^{2+} by the two phosphates of one surfactant molecule, screens the charge and reduces the headgroup area at the interface to water, thus triggering the rearrangement of the molecules into nonintercalated bilayers and subsequently destabilizing the vesicles in favor of smaller aggregates. [112].

The second example employs a didodecyl (diC_{12}) gemini surfactant with one cationic ammonium head

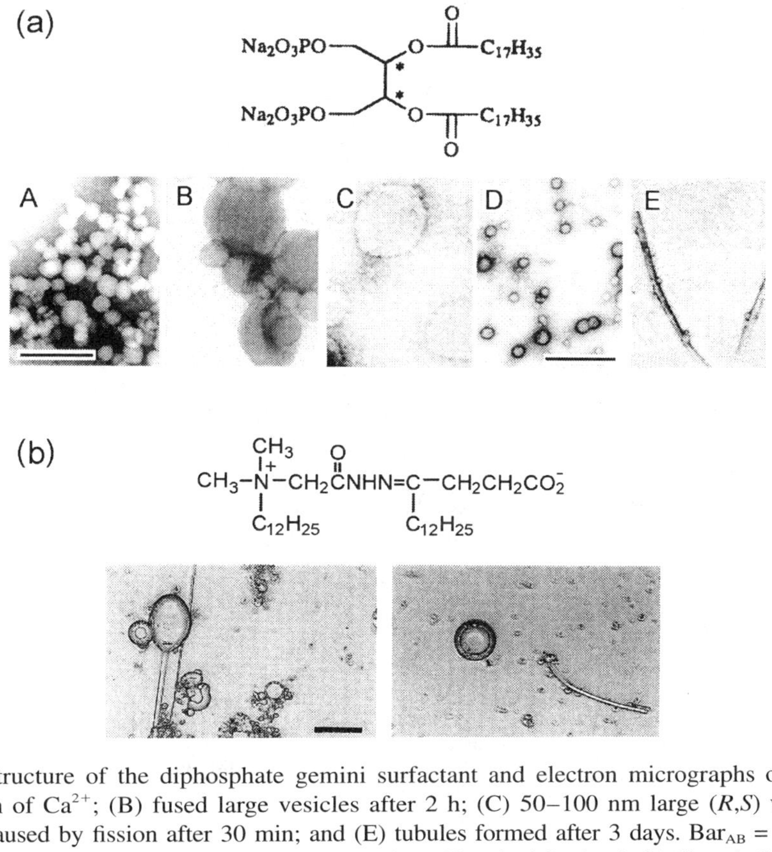

FIG. 16 (a) Chemical structure of the diphosphate gemini surfactant and electron micrographs of (A) small (*S,S*) vesicles immediately after addition of Ca^{2+}; (B) fused large vesicles after 2 h; (C) 50–100 nm large (*R,S*) vesicles before addition of Ca^{2+}; (D) small vesicles caused by fission after 30 min; and (E) tubules formed after 3 days. Bar_{AB} = 100 nm; Bar_{C-E} = 250 nm. (Reprinted with permission from Ref. 112. Copyright 1997 American Chemical Society.) (b) Chemical structure of the ammonium and carboxylate double-headed gemini surfactant and phase-contrast optical micrographs of giant vesicles, tubules, and filaments formed in water. Bar = 50 μm. (Reprinted with permission from Ref. 113. Copyright 1996 American Chemical Society.)

and one anionic carboxylate head. Upon hydration of the dry surfactant film, the gemini monomers aggregate in water to form giant vesicles, tubules, and filaments that can be observed by phase-contrast optical microscopy (Fig. 16b). The giant vesicles are stable over weeks at room temperature but can easily be destroyed at pH 3 by hydrolysis of the C=N double bond within the —$CH_2CONHN{=}CCH_2CH_2$— spacer unit. The formation of stable bilayers can be explained by back folding of the spacer into the hydrophobic membrane region, thus bringing both headgroups close together. The molecules then assume a near-cylindrical shape wherein the cross-sectional area of the double headgroup is slightly larger than that of the alkyl chains. Such a shape is appropriate for the formation of bilayer membranes [113]. Both examples indicate the difficulties that arise from using the packing parameter concept. In the first example, the interaction with Ca^{2+} ions is more influential in the packing of molecules within the bilayer than the geometry of the single molecules. In the second example, the not so obvious (and not so *predictable*) back folding of the spacer unit determines the molecular shape and thus the curvature of the aggregates, which actually covers a wide range as one can see from Fig. 16b.

The biologically most important and most often studied double-chain molecules with charged, neutral, or zwitterionic headgroups are phospholipids. Several reviews are available; therefore, the discussion of their phase behavior presented here will be short [114–127]. Phospholipids aggregate at temperatures below the melting temperature T_c of the hydrocarbon chains into lamellar gel phases consisting of hexagonally ordered chains with disordered headgroups. Depending on the tilt the chains have with respect to the layer normal, these phases are denoted L_β (no tilt) or $L_{\beta'}$ (tilted). The L_β phase is actually a sequence of three separate phases, differing in the azimuthal angle ϕ of the tilt direction with respect to the underlying hexagonal chain lattice. In the low-hydration $L_{\beta F}$ phase, the tilt is directed toward a face of the hexagonal net ($\phi = 0°$), whereas the high-hydration $L_{\beta I}$ phase the tilt lies toward the nearest neighbor chain ($\phi = 30°$). The $L_{\beta L}$ phase in between both has azimuthal angles ϕ that continuously vary between 0° and 30°. Phospholipids with a large mismatch between the cross-sectional areas of the hydrated headgroups and the hydrocarbon chains often form an interdigitated $L_{\beta I}$ phase. The formation of a rippled lamellar phase ($P_{\beta'}$) is observed in the phase diagram of saturated phospholipids at temperatures close to the $L_{\beta'}$-to-L_α phase transition. The ripples are asymmetric, have a periodicity of the order of 15–20 nm, and there is a phase shift in the ripple pattern from layer to layer producing a two-dimensional oblique lattice. Ripples usually form in a cooperative manner at a pretransition temperature T_p.

The main reason for the rippling of lipid bilayers is the tendency of the polar headgroups to achieve a certain degree of fluidity and hydration while the hydrocarbon chains remain ordered. Phospholipids, consequently, undergo a pretransition only if the polar headgroups are sufficiently hydrophilic and hydrated and if the interchain packing is sufficiently weak. At least 12 ± 2 water molecules must be associated with each headgroup for a bilayer undulation to be feasible [122,123]. Another two-dimensional modulated phase, the "hats and saddle" or "egg carton" structure, has attracted theoretical and experimental attention [128–131]. Above the chain melting temperature T_c, the chains form the normal lamellar L_α phase (Fig. 17a). Diacyl phospholipids tend not to form the hexagonal H_I phase, whereas the reverse hexagonal H_{II} phase is very common in phospholipids having small, weakly hydrated headgroups and attractive head-head interactions. As an example, the partial phase diagrams of dipalmitoylphosphatidylcholine (DPPC) and dipalmitoylphosphatidylethanolamine (DPPE) in water are displayed in Fig. 17b. DPPE has a less hydrophilic headgroup than DPPC, causing a smaller L_α phase for this lipid. For a similar reason, the gel phase is metastable, the ripple phase $P_{\beta'}$ is not observed, and the reverse hexagonal H_{II} phase is very prominent in the phase diagram of DPPE [114].

The main determinant of the phospholipid phase behavior is not so much the geometrical shape of the molecules but the type of hydrocarbon chains because of the strong chain dependence of the melting entropy. Increasing the chain length thus causes all transition temperatures at which the interfacial packing density decreases to become higher. The transition enthalpy and entropy increase with the chain length in an approximately linear manner. The more polar a given lipid, the stronger is this length dependence. Chain unsaturation and branching effectively decouple the parts of the hydrocarbon chain on either side of the double bond or the branching point, and the longest parallel segment of the chain determines the chain melting temperature T_c. These chain modifications thus reduce the length of parallel, strongly interacting chain segments and mimic the consequences of chain shortening. This effect is approximately 50% weaker in a *trans* than in a *cis* configuration. Increasing the number of double bonds per chain causes only a small further decrease of T_c. Asymmetry between the hydrocarbon chains has

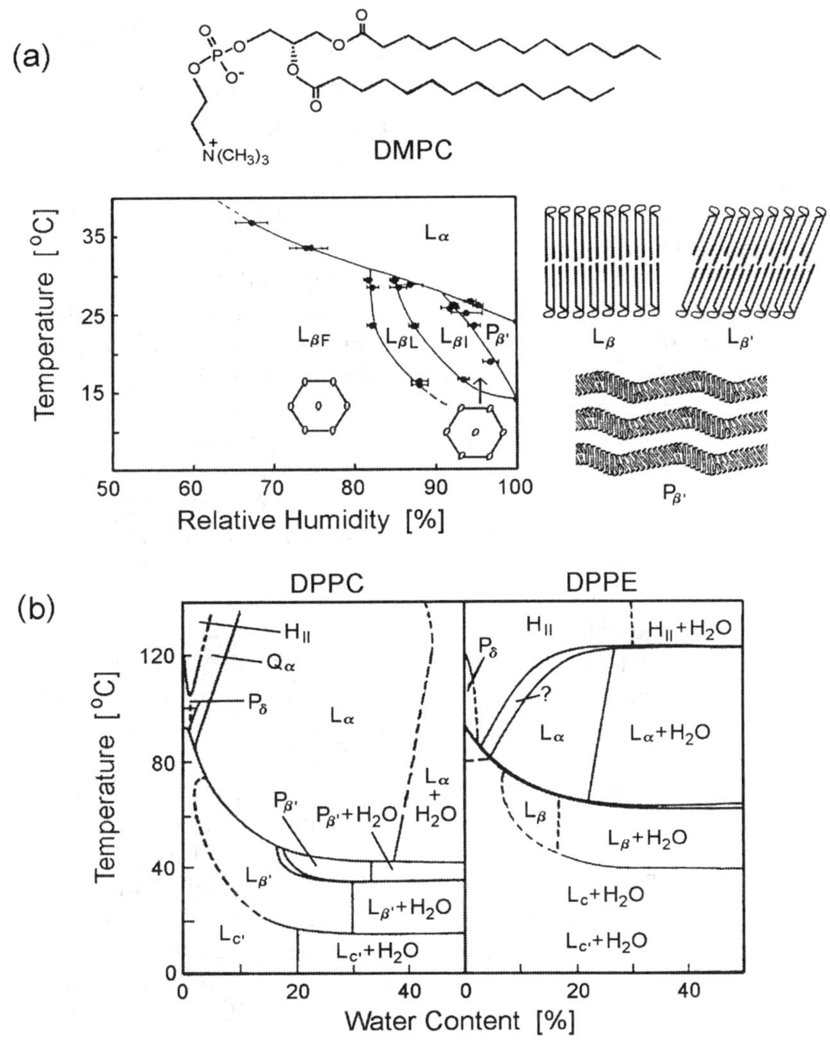

FIG. 17 (a) Chemical structure of dimyristoylphosphatidylcholine (DMPC), the temperature-humidity phase diagram of DMPC showing the gel phases and the L_α phase above the chain melting temperature T_c, and cartoons of the L_β, $L_{\beta'}$, and $P_{\beta'}$ phases. For details see text. (Adapted from Ref. 132 and 133.) (b) Partial phase diagram of dipalmitoyophosphatidylcholine (DPPC) and -ethanolamine (DPPE). For details see text. L_c and $L_{c'}$ denote crystalline phases; L_β and $L_{\beta'}$ stand for lamellar gel phases with untilted and tilted hydrocarbon chains, respectively; $P_{\beta'}$, P_δ, and Q_α denote rippled, ordered ribbon, and cubic phases, respectively; L_α and H_{II} stand for lamellar fluid and reverse hexagonal mesophases, respectively. (Adapted from Ref. 114.)

mainly the same consequence. T_c decreases with increasing asymmetry unless the hydrocarbon chains interdigitate in the gel phase, which results in an increase of the chain melting phase transition enthalpy by approximately a factor of 2. Finally, the chain attachment to the glycerol backbone also affects the phase behavior. The melting temperatures of dialkylphospholipids (ether bond) compared with those of diacylphospholipids (ester bond) at 1–5°C higher [122].

Apart from the hydrocarbon chains, the most decisive factor for the phase behavior of phospholipids is the lipid polarity. Phospholipids at high pH, which as a rule are at their maximum polarity, respond more strongly to headgroup variations than do lipids at low pH. Other factors such as the ionic or zwitterionic character, or the *size of the headgroups*, play a much smaller role. In general, the thermodynamic significance of each headgroup and its interactions with neighboring headgroups and water molecules increases with the relative hydrophilicity of the polar residues and with the degree of unsaturation of the hydrocarbon chains. For example, unsaturated or short-chain phosphatidylcholine and phosphatidylglycerol, which belong to the most polar phospholipids, are more sensitive to headgroup effects than saturated or long-chain analogues or the less polar phosphatidylserine, phosphatidylethanolamine, or phosphatidic acid. Deprotonation of phospholipids always increases the lipid's polarity and thus its sensitivity to headgroup effects. The method used to achieve this is unimportant, whether it is done by increasing the pH, by replacement of lipid-bound protons by other ions or chemically bound methyl groups, or by breaking interlipid hydrogen bonds. All methods cause qualitatively similar thermodynamic effects, e.g., they lower the chain melting temperature. The T_c of nearly all common phospholipids steadily decreases with increasing pH of the suspending solution and increases during acidification. The influence the headgroup size has on the stability of the bilayer membranes is less important than is often believed. This factor is of significance only when the size variations are crucial for the interlipid bonding patterns. DPPC analogues, for example, that have minimal direct interactions between the PC headgroups all melt between 40 and 44°C regardless of the alkyl chain lengths between phosphate and ammonium groups. On the contrary, the melting temperature of various phosphatidylethanolamines decreases considerably with increasing headgroup length. However, the decrease is highly nonlinear, indicating the significance of direct head-head interactions [122,134,135]. Although the phase behavior of phospholipids clearly follows certain rules, the observations described so far give strong evidence that geometrical considerations based on the molecular shape are insufficient to explain these observations and *predict* the phase behavior.

The following examples describe phospholipids whose hydrocarbon chains have been modified, thus effecting the geometrical shape of the molecules and their hydrophile-lipophile balance. In general, it has been found that a small nonpolar substituent located at the beginning or the end of one or both acyl chains of a saturated phosphocholine has only a modest effect on the phase transition temperature T_c but has a substantial effect on T_c when placed in the middle of the chain. In contrast, a polar substituent (e.g., a keto group) has an only modest effect on T_c regardless of where it is positioned along the acyl chains [135]. To prove these observations further and to test the effect substituents have on the aggregation behavior of phosphocholines, distearoylphosphocholine (DSPC) has been modified by placing a dimethylketal group at the beginning, middle, and end of one octadecanoyl chain and by placing a *vic*-diol group in the middle of a C_{18} chain (Fig. 18a). The ketal groups of adjacent chains can participate in dipole-dipole interactions, while the diol groups can interact via stronger hydrogen bonds. Nevertheless, all four compounds form vesicles of size distributions comparable to that found for the parental DSPC (27 ± 3 nm). The phase transition temperature, however, changes from 55.1°C for DSPC to 48.5, −15.8, and 53.2°C for compounds with the ketal group at the beginning, middle, and end of the C_{18} chain.

These changes are similar to those found after incorporation of a methyl group into the C_{18} chain, although the ketal group is larger than methyl and contains C—O bonds. The presence of the *vic*-diol lowers the phase transition temperature moderately to 46–49°C as compared with DSPC and thus behaves similarly to a midchain keto group but enhances the transition enthalpy ΔH considerably from 42.3 to 67 kJ/mol. This effect has been attributed to the stabilization of the bilayer membranes by interchain hydrogen bonds, which are absent in keto groups. The results obtained with the dimethylketal and *vic*-diol compounds are in accord with the "statistical bend" model for phosphocholine bilayers that assumes the development of a critical bend near the center of the chains on going from the gel to the liquid crystalline state of the bilayer. Substituents near the center of one or both acyl chains can stabilize this bend, whereas those near the headgroup or the end of the chains have little effect. Stabilization of the bend results in lower T_c and ΔH values [137]. However, it depends on the individual structure of a midchain polar substituent to which extent it stabilizes the bilayer; this is how much it lowers ΔH.

In the second example, the two hexadecanoyl chains of DPPC have been exchanged by either one or two phytanyl chains or by one phytanyl and one eicosyl (C_{20}) chain. Phytanyl chains, having four methyl groups in the 3, 7, 11, 15-positions distributed along the C_{16} chain, are bulkier than their unsubstituted analogues. All chains are connected to the glycerol backbone via ether bonds instead of ester bonds to improve the molecules' temperature stability (Fig. 18b). Despite the presence of bulky phytanyl chains and a longer C_{20} chain, all three molecules aggregate in water to form stable unilamellar vesicles with diameters of 20–100 nm, which is a size distribution close to that found for DPPC vesicles in a control experiment. What has changed, however, are the phase transition temperature T_c and the stability of the bilayer membranes against leakage of 5(6)-carboxyfluorescein and sodium chloride. The chain melting temperature T_c in the case of DPPC is 41°C, whereas the corresponding phase transition for diPhyPC, C_{16}PhyPC, and C_{20}PhyPC occurs

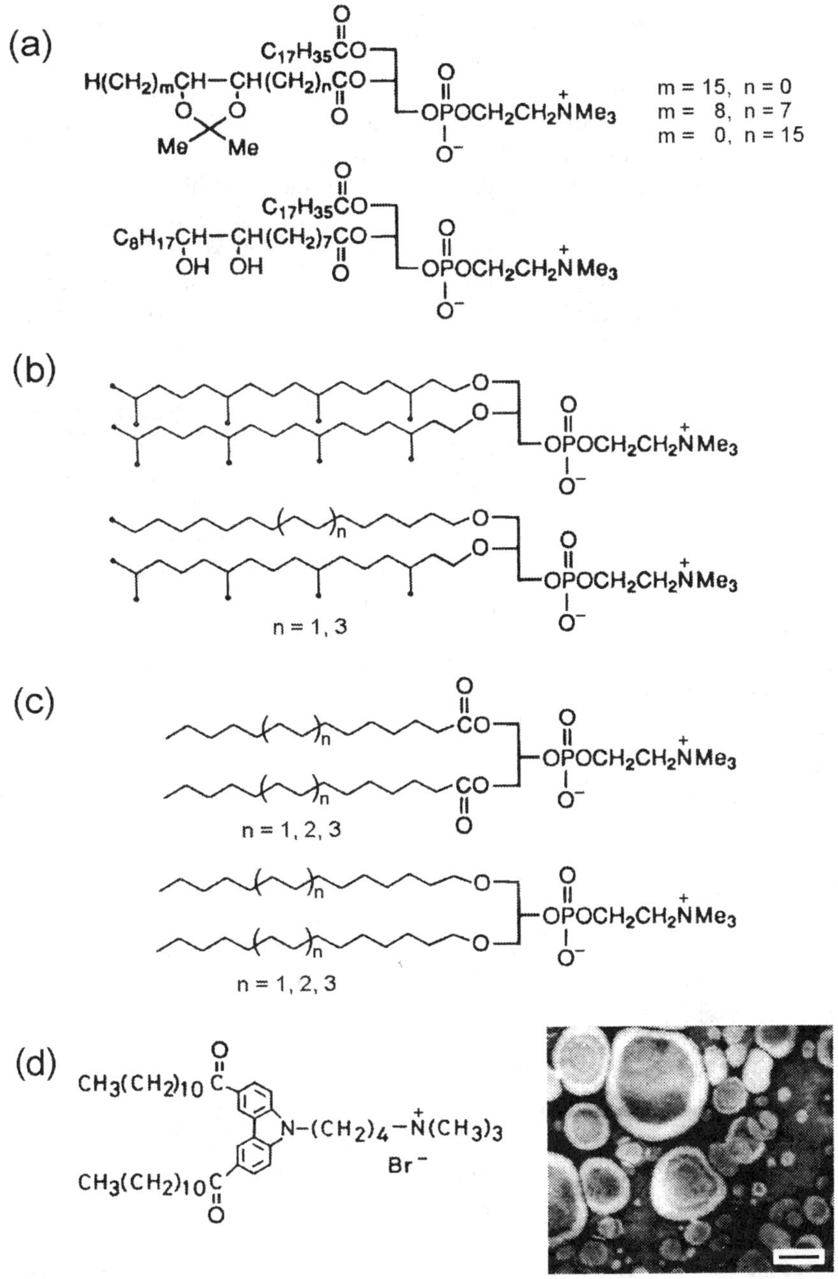

FIG. 18 Chemical structures of (a) phosphocholines carrying a dimethylketal in the beginning, middle, and end of one C_{18} chain and a *vic*-diol in the middle of one C_{18} chain (adapted from Ref. 136); (b) phosphocholines with one and two 3,7,11,15-tetramethyl C_{16} (phytanyl) chains and cone C_{20} chain replacing linear C_{16} chains (adapted from Ref. 138); (c) unnatural phosphocholines with acyl and alkyl chains in 1,3-position along the glyceryl backbone (adapted from Ref. 139) and trimethylammonium bromide with acyl chains in 3,6-position at a carbozole connector and a transmission electron micrograph of the vesicular bilayers formed in water; bar = 100 nm (from Ref. 140).

below $-30°C$ and around $-11°C$ for the two chimeric lipids. DPPC vesicles having their bilayer membranes in the gel state are stable against leakage of dyes and ions but become leaky at higher temperatures. DiPhyPC vesicles, on the other hand, are stable up to 70°C but have slight leakage even at room temperature. The chimeric lipids C_{16}PhyPC and C_{20}PhyPC combine the positive features of both compounds; they are stable upon heating to 70°C and they do not have a slight persistent leakage at room temperature [138].

Another way to change the cross-sectional area of the hydrocarbon chains involves the use of 1,3-diacyl-*rac*-glycero-2-phosphocholines and their 1,3-dialkyl analogues instead of natural 1,2-disubstituted phosphocholines (Fig. 18c). Nature's preference for 1,2-disubstituted phospholipids could suggest a superior packing behavior of these compounds to form stable bilayer membranes compared with the unnatural 1,3-derivatives. The bilayer properties of 1,3-phospholipids have been examined thoroughly using electron microscopy, differential scanning calorimetry, substrate entrapment, ion permeation across the bilayer, and phase separation. All 1,3-derivatives aggregate in water to form stable bilayer membranes in essentially the same way as natural phospholipids. The 1,3-diacyl phosphocholines form preferably multiwalled vesicles and to a small extent (ribbonlike) lamellae, whereas the 1,3-dialkyl compounds form flattened vesicles and/or the ribbon structure [140].

It has also been reported that some 1,3-phospholipids tend to form a viscous solution after ultrasonication above T_c and cooling to ambient temperature. X-ray diffraction studies of 1,3-ester and 1,3-ether phosphocholines give evidence that the bilayer membranes consist of at least partly interdigited molecules ($L_{\beta I}$ phase) [141,142]. The hydrocarbon chain length does not change the aggregate structure significantly. The phase transition temperature and enthalpy of 1,3-phospholipids are in close range with those of the 1,2-analogues. Similar observations have been made for the remaining comparisons. The somewhat surprising results of the study establish that 1,3-phospholipids form bilayers whose physicochemical characteristics are essentially the same as those of the natural 1,2-phospholipids, regardless of the hydrocarbon cross-sectional area [139]. The two acyl chains are spread even further apart using a carbazole connector in a related study of the aggregation behavior of a trimethylammonium surfactant. Nevertheless, the double-chain surfactant aggregates to closed bilayer membranes in form of single- and multiwalled vesicles with a size distribution of 20–200 nm (Fig. 18d) [140].

Fluorination as a way to change the hydrophile-lipophile balance has been applied to phospholipids as well. Phosphocholine and ethanolamine lipids having one or two perfluoro or partly fluorinated acyl and alkyl chains in 1,2- and 1,3-positions along the glyceryl backbone generally display a behavior similar to that of the hydrocarbon analogues. They form unilamellar vesicles with diameters between 50 and 100 nm regardless of the number of fluorinated chains or the 1,2- versus 1,3-position or glycerol. The slow conversion of vesicles into ribbonlike structures at room temperature is observed only for 1,3-lipids with a fluorinated tail on both hydrocarbon chains and a chain length of at least 17 carbon atoms. No interdigitation has been found in the bilayers of fluorinated lipids. A marked contrast in the phase behavior of fluorolipids and their hydrocarbon analogues is the large half-width of the chain melting transition peak in the DSC thermograms. The longer the CF_2 tail, the lower the cooperativity of the phase transition and the wider its temperature range. The impact of an ester or ether connection is insignificant in most cases. Ether phospholipids whose hydrophobic chains are ended by a CF_6 or CF_8 tail, however, display T_c values 6–9°C lower than those of their corresponding ester analogues, the opposite behavior to that found for hydrocarbon esters and ethers.

In the case of the hydrocarbon lipids, the different behavior is explained by closer packing of the ether chains owing to the absence of the carbonyl groups. The higher attractive interaction between the alkyl chains thus outweighs the loss of intralayer stabilization by hydrogen bonds involving the carbonyl groups. The larger cross-sectional area of the CF_6 and CF_8 tails prevents such closer chain contact, and the presence or absence of carbonyl hydrogen bonds becomes decisive. The thermodynamic parameters of 1,2- and 1,3-disubstituted phospholipids are essentially the same, paralleling the trend found for the hydrocarbon analogues. A more detailed evaluation of the thermodynamic parameters reveals their dependence on the number of fluoro chains, their length, and the degree of fluorination within the chains. Replacement of one hydrocarbon chain by its perfluorinated equivalent results in a decrease of T_c, whereas T_c increases with the length of the CF_2 segment for a given CH_2 spacer. In addition, the degree of fluorination has a larger impact on T_c than the overall length of the chain. The phase behavior of phospholipids with fluorinated segments in both chains follows less straightforward trends [92,143]. Besides these rather minor differences, the aggregation and phase behavior of double-chain fluoro compounds seem to be much less affected by a hydrocarbon-fluo-

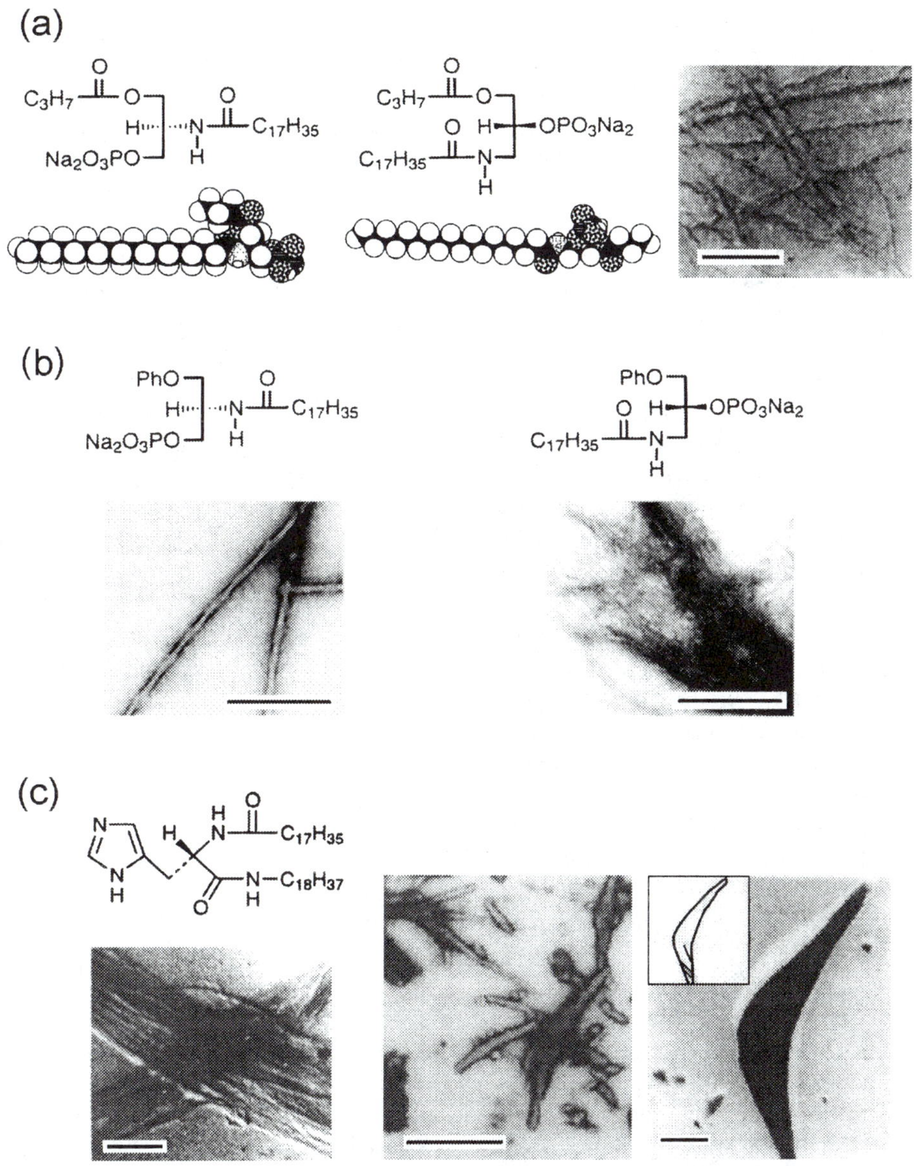

FIG. 19 (a) Chemical structures, space-filling models, and electron micrograph of 1,2- and 1,3-disubstituted double-chain diester phosphate surfactants. The micrograph shows left-handed helices of the 1,3-isomer. Bar = 500 nm. (From Ref. 144.) (b) Chemical structures and electron micrographs of 1,2- and 1,3-disubstituted phosphate surfactants with ester and ether connections. Note the helices of the 1,2-isomer and the fiber bundles of the 1,3-isomer. Bars = 1000 nm and 500 nm. (From Ref. 144.) (c) Chemical structure of a double-chain histidine surfactant and electron micrographs showing long fibers of the protonated surfactant at pH 2.5 and boomerang structures formed in the presence of copper triflate (ratio 1:4). Bars = 250 nm, 1 μm, and 250 nm. (From Ref. 145.)

rocarbon replacement as observed for single-chain fluoro surfactants.

This section ends with the presentation of three examples of structures generated by self-aggregation of double-chain surfactants based on directed anisotropic interactions in the form of amide hydrogen bonds and metal coordination. Contrary to the almost identical aggregation behavior of 1,2- and 1,3-disubstituted hydrocarbon and fluorocarbon phospholipids, the first two examples demonstrate remarkable differences between both substitution patterns when the aggregate structure is determined by amide hydrogen bonds. The two double-chain phosphates displayed in Fig. 19a are obtained after reaction of the appropriate *N*-acylated aziridine with dibenzyl phosphate and subsequent debenzylation. Both compounds have an (*R*) configuration but differ in the position of the phosphate and amide groups along the backbone. The 1,2-disubstituted compound aggregates in water to form planar structures, whereas the 1,3-diacyl isomer produces left-handed helical strands coagulated to ropelike structures. The single strands of the rope appear to be tubules with a diameter of 22 nm. X-ray powder diffraction experiments revealed a repetitive distance of 40 and 46 Å for both compounds, indicating interdigitated molecular packing within the bilayers. Replacement of the ester bound C_4 chains by ether bound phenyl rings results in a pair of 1,2- and 1,3-isomers, which again show quite different aggregation behavior. The 1,2-derivative forms vesicles with diameters of 500–1000 nm when dispersed in water at pH 6.5. These vesicles slowly rearrange to form ribbons containing interdigited bilayer membranes. After protonation of the phosphate groups at pH 2.5, however, the ribbons start to twist and form left-handed helices, which ultimately transform into tubular structures. The 1,3-derivative, on the other hand, aggregates directly into coagulated fiber structures with interdigitated membranes that are stable for at least a week (Fig. 19b) [144].

The long-chain histidine derivative presented in the third example is insoluble in water. After protonation of the headgroup at pH 2.5, the double-chain histidine aggregates to form very long thin fibers, some showing a right-handed twist, which assemble side by side to yield bundles. In a 1:4 mixture with copper triflate, boomerang-like scrolls were generated (Fig. 19c). The boomerangs are of different thickness but all show left-handed twists. Boomerang formation has been suggested to start by stretching and twisting at opposite sides of vesicle membranes, resulting in a thickened center and thin twisted ends of the boomerang. Eventually the boomerangs become thinner and also show the helical twist at the center region. This model was derived from the spiral growth of protoplasts induced by the action of fusicoccin. In case of the histidine derivative, the coordination of the copper ions is the driving force behind the formation of this interesting structure [145]. Packing parameter concepts based on the molecular geometry will fail to predict any of these aggregation behaviors in a reliable way.

IV. PHASE BEHAVIOR OF MIXTURES OF SURFACTANTS

The phase behavior of mixed surfactant systems has been reviewed [14,15,115,119,121,146]. Some examples will be presented that allow drawing some conclusions about the importance of the molecular geometry in describing and modeling mixed systems. Equimolar mixtures of oppositely charged single-chain surfactants aggregate in water to planar layers because of electrostatic attraction between headgroups in addition to van der Waals attraction between hydrocarbon chains. The critical aggregation concentration (cac) of mixed cationic dodecyltrimethylammonium bromide (DTAB) and anionic sodium dodecyl sulfate (SDS), for example, is about two decades in concentration lower than the cmc of the pure components for a wide range of mixing ratios [147]. It has been generally observed that for a hydrocarbon chain length longer than C_8 for both cationic and anionic surfactants, the equimolar catanionic mixture is practically insoluble in water, and the phase diagram is dominated by the formation of a lamellar liquid crystalline phase in the water-poor region of the surfactant-water system. Equimolar mixtures of surfactants having C_8 hydrocarbon chains form precipitates at very high water contents, which redissolve to form stable vesicle and micellar solution phases upon addition of an excess amount of one of the parent compounds. A more interesting phase behavior appears at nonequimolar mixtures of oppositely charged surfactants.

The binary phase diagrams of cationic dodecyltrimethylammonium chloride (DTAC), anionic sodium nonanoate (SN), and the catanionic mixture in water at 40°C as well as the pseudoternary phase diagram of this system are displayed in Fig. 20. It can be seen from Fig. 20a that the DTAC surfactant forms the well-known sequence of mesophases with increasing surfactant concentration: micellar solution (L_1), micellar cubic phase (I), hexagonal phase (H), bicontinuous cubic phase (V), and lamellar phase (L_α). On the contrary, the phase behavior of SN is rather simple. The system forms a micellar solution followed by a hexagonal phase with a limited stability range. There are two important differences between the phase behavior of the catanionic mixture and that of the parent surfactants. First, the only mesophase found is the lamellar L_α phase; phases of higher curvature are absent. Second, the extension of the solution phase (L_1) is larger for the catanionic system (total surfactant concentration 47 wt%) than for the cationic (43 wt%) and anionic (37 wt%) surfactant systems under identical conditions [148].

The isothermal pseudoternary phase diagram DTAC-SN-D_2O at 40°C is shown in Fig. 20b. The triangle sides correspond to the different binary axes; thus the bottom side corresponds to the SN-water system,

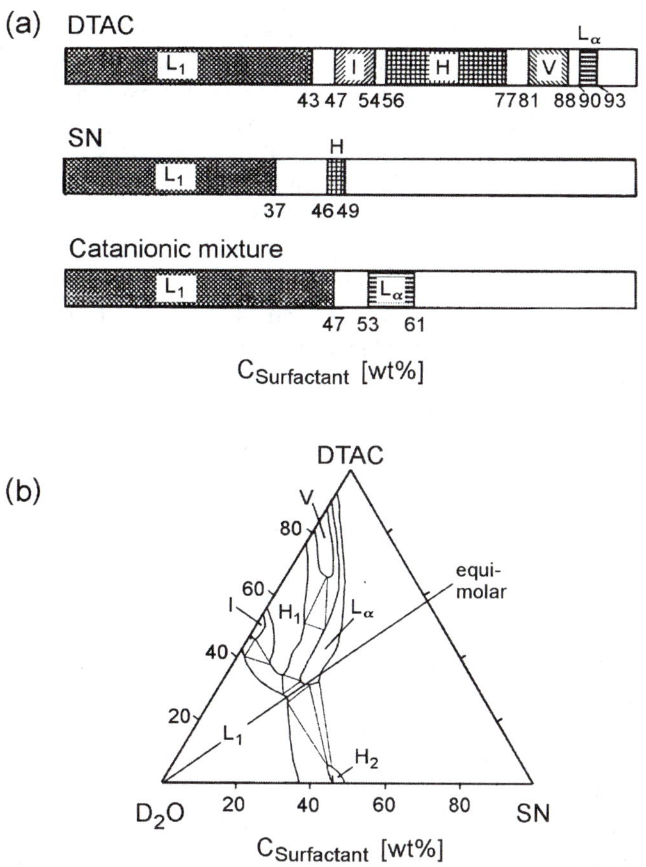

FIG. 20 (a) Binary phase diagrams of DTAC-water, SN-water, and the catanionic mixture–water systems at 40°C. Blank areas correspond to two-phase regions. (b) Pseudoternary DTAB-SN-^2H_2O system at 40°C. Notations are as follows: L_1 isotropic solution; H_1 DTAC-rich hexagonal phase; H_2 SN-rich hexagonal phase; I micellar cubic phase; V bicontinuous cubic phase; and L_α lamellar phase. (Adapted from Ref. 148.)

and the left and right ones correspond, respectively, to the DTAC-water and DTAC-SN systems. Mixtures above the equimolar line are rich in DTAC, whereas samples below the line are rich in SN. The phase diagram contains an extended water-rich isotropic solution and five single liquid crystalline phases (two normal hexagonal phases, one lamellar phase, and two cubic phases) and one large region of hydrated surfactant crystals. It should be noted that the mixed system does not form any new phases, but the phase formed by one parent surfactant-water system can solubilize the other surfactant. For example, adding anionic SN to the micellar solution of DTAC results in micellar growth, and large asymmetric micelles (e.g., wormlike or rodlike) are found at the 1:1 molar ratio. Upon further addition of SN, a second aggregate species appears in addition to the asymmetrical micelles that is identical to the spherical micelles of pure SN [148].

In dilute aqueous mixtures of dodecyltrimethylammonium bromide (DTAB) and sodium dodecyl sulfate (SDS) at total surfactant concentrations below 5 wt%, the water-rich corner of the pseudoternary phase diagram displays an even more diverse phase behavior (Fig. 21a). A clear and colorless two-phase region consisting of an isotropic liquid and a crystalline precipitate (I + S) dominates the DTAB-rich side of the phase diagram. This two-phase region gives way to a micellar solution when the total surfactant concentration is greater than the cmc for the majority component (e.g., cmc_{DTAB} = 0.46 wt%). The SDS-rich side of the phase diagram shows a narrow one-phase lobe at a mixing ratio of 35:65 CTAB to SDS, which appears bluish to the eye and contains large (>200 nm) and polydisperse vesicles. A two-phase region containing vesicles and precipitate (V + S) is located between the equimolar mixture and the vesicle phase. Samples prepared at mixing ratios between the vesicle and micellar phases are viscoelastic at lower concentrations and turbid and birefringent at higher concentrations. This might be a multiphase region or a metastable dispersion. When the total surfactant concentration is greater than the cmc for SDS (cmc_{SDS} = 0.23 wt%), the SDS-rich micellar phase appears. Because of the formation of the 1:1 precipitate, the DTAB-SDS-water mixture yields five components: DTAB, SDS, NaBr, DTA^+DS^-, and water. A consistent picture of the micelle-to-vesicle transition emerges from combined time-resolved fluorescence quenching (TRFQ), video-enhanced differential interference contrast microscopy (VEM), and cryo-transmission electron microscopy (cryo-TEM). SDS-rich micelles grow to an aggregation number ratio of 20:80 DTAB to SDS with increasing DTAB concentration. Further increase of the DTAB concentration results in the formation of long rodlike micelles, which abruptly transform into large and polydisperse vesicles above a certain DTAB limit [149].

The very similar geometrical shape of both single-chain surfactants, having linear hydrocarbon chains of the same length (C_{12}), fosters the molecular packing into lamellar aggregates. The phase behavior is therefore dominated by the formation of crystalline precipitate and polydisperse vesicles with low curvature. The tendency to form precipitates increases with the length of the hydrocarbon chains, as has been found studying the aggregation behavior of mixed alkylammonium chlorides and sodium alkyl sulfates with decyl, dodecyl, and tetradecyl chains [150]. A mismatch between

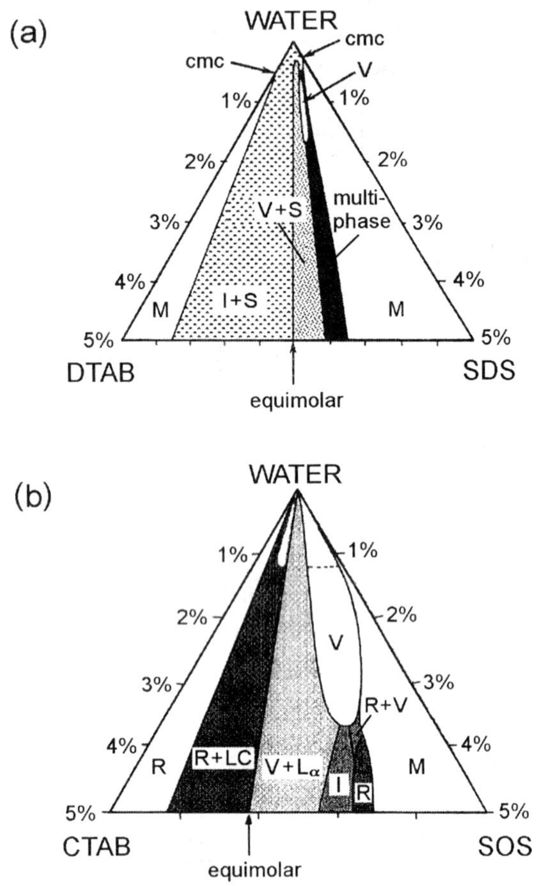

FIG. 21 (a) Water-rich corner of the pseudoternary DTAB-SDS-water at 25°C. One-phase regions contain micelles (M) or vesicles (V), two-phase regions (shaded) consist of either clear liquid and preciptate (I and S) or vesicles and precipitate (V and S); the dark shaded region is viscoelastic and of unknown composition. (Adapted from Ref. 149.) (b) Pseudoternary phase diagram of CTAB-SOS-water at 25°C. The one-phase regions represent SOS-rich vesicles (V) and micelles (M), and CTAB-rich rodlike micelles (R); shaded two-phase regions consist of CTAB-rich rodlike micelles and vesicles (R + V), SOS-rich vesicles and lamellar phases (V + L_α), and an isotropic liquid with precipitate along the equimolar line. Very small amounts of turbid clouds form in the SOS-rich vesicle lobe above the dashed line. (I) represents an unresolved multiphase region. (Adapted from Ref. 151.)

the hydrocarbon moieties of both surfactants should thus reduce the stability of the precipitates and improve vesicle formation. This assumption has been confirmed by replacing DTAB by cetyltrimethylammonium bromide (CTAB) and SDS by sodium octyl sulfate (SOS). Although the molecular geometry of CTAB (C_{16}) and SOS (C_8) remains the same as that of DTAB (C_{12}) and SDS (C_{12}), the interchain interactions and packing properties between CTAB and SOS have changed. As expected, vesicles form spontaneously over a wide range of compositions of the pseudoternary CTAB-SOS-water mixture in both CTAB-rich and SOS-rich solutions, creating a vesicle phase much larger than the one observed in the DTAB-SDS-water mixture (Fig. 21b). Moreover, the vesicles are unilamellar and less polydisperse. The bilayer properties of the vesicles depend on the ratio of CTAB and SOS, with CTAB-rich bilayers stiffer than SOS-rich ones. Whenever the hydrocarbon chain length is unequal, the extent of the vesicle lobe is largest for mixtures rich in the shorter tailed surfactant, as apparent from Fig. 21. The close packing of vesicles sets the limit of the vesicle phase. Close packing will occur at lower concentrations when the vesicles are large. Above this concentration, the surfactants form a multilamellar phase. Two modes of transition between micelles and vesicles have been identified. The first one in CTAB-rich solutions and SOS-rich solutions at higher SOS concentrations occurs between rodlike micelles and vesicles. It is first order and results in macroscopic phase separation. The second mode of transition occurs in SOS-rich mixtures at low SOS concentration and exhibits no phase separation. Instead, small micelles abruptly transform into vesicles over a narrow range of surfactant concentration. It should be noted that these vesicles represent a thermodynamic stable phase because they form under thermodynamic control [151]. Other mixtures of surfactants with mismatching hydrocarbon parts, for example, cetyltrimethylammonium tosylate (CTAT) and sodium dodecylbenzenesulfonate (SDBS), show a similar low tendency to form crystalline precipitates in favor of stable vesicles with high curvature [152,153].

Replacement of bromide by chloride in the system dodecyltrimethylammonium chloride (DTAC) and sodium dodecylbenzenesulfonate (SDBS) results in two remarkable observations. First, the transition from micelles to vesicles happens continuously in mixtures of DTAC and SDBS. This is in contrast to the abrupt first-order transition observed in the other mixtures discussed earlier. There is no observable two-phase region separating the phases containing micelles and vesicles. Second, the vesicle phase is significantly smaller as in the presence of bromide. As mentioned earlier, the extent of the vesicle lobe appears to be set by the condition of close packing of vesicles. In the presence of strong intervesicle interactions, the limiting packing density of vesicles is reached at a comparatively low surfactant concentration, thus limiting the size of the vesicle phase. Chloride ions are more hydrated than bromide ions and therefore less effective in shielding the charge of the vesicles. It is likely that the lower shielding leads to the observed size reduction of the vesicle lobe [154]. The prominent role electrostatic interactions play in the phase behavior of oppositely charged surfactants becomes evident when the charge distribution is changed by addition of a monovalent salt. For example, the addition of 5 wt% sodium bromide to a sample containing 2 wt% of a 3:7 weight ratio mixture of CTAB and SOS destabilizes the vesicle phase and leads to micelle formation. The addition of salt results in an increased insertion of SOS monomers into the vesicle surface, thereby increasing the surface charge and triggering the reorganization of the mixed surfactants to form smaller, more curved micelles. This assumption has been verified by contrast variation in small-angle neutron scattering (SANS) experiments [155,156].

Not only vesicle formation but also the formation and composition of mixed micelles strongly depend on composition and electrostatic condition of the surfactant mixture. For example, cationic surfactants have a greater tendency to be incorporated into mixed micelles than anionic ones. This has been attributed to either differences in micellar size or differences in the interaction between headgroups and counterions. Consequently, one can find rodlike micelles on one side of the phase diagram but spherical ones on the other side [150,151]. The formation of unusual disklike micelles in equilibrium with a lamellar L_β phase has been observed. The crucial requirement for obtaining highly stable colloidal solutions and nanodisk self-assembly is a high osmotic pressure induced by unscreened electrostatic repulsion. This condition is met in pure catanionic surfactant solutions that contain only recombining H^+ and OH^- counterions plus counterions of the component in excess. Mixing anionic myristic acid ($C_{13}H_{27}COOH$ = MA) and an excess of cationic cetyltrimethylammonium hydroxide [$C_{16}H_{33}N^+(CH_3)_3$ OH^- = CTAOH] in carbonate-free water results in the formation of a lamellar phase with the hydrocarbon chains in the frozen state (L_β) and the formation of large disklike aggregates with diameters of 2–3 μm. The size of the dispersed disks decreases from 3 mm to 30 nm when the positive surface charge increases by decreasing the molar ratio of MA to CTAOH from 0.45 to 0.39. The nanodisk formation requires ion pairing on planar faces coexisting with highly curved interfaces forming the edges. Therefore, part of the excess cationic surfactants form the edges of the disk. The final size is a balance between the entropy of mixing and electrostatic coupling between disks. Increasing the excess charge favors small disks until the phase transition toward wormlike micelles occurs [157].

Contrary to the strong effects the hydrocarbon chain length and symmetry have on the phase behavior of catanionic mixtures, the chemical composition of the surfactant headgroups has been found to have little effect [148]. Furthermore, the same sequence of phases is formed in mixtures of divalent and nonionic surfactants as in mixtures of oppositely charged monovalent surfactants. The general trend is a shift of the individual phase transitions toward higher surfactant concentrations when the total headgroup charge z increases, but the succession of phases with increasing surfactant concentration is not affected to any considerable extent. However, the formation of the micellar cubic phase (I) is strongly favored by a high average surfactant charge [158].

It should be noted that the choice of solvent, H_2O versus D_2O, seems to have an impact on the phase behavior of (catanionic) surfactants. This observation is important because often the phase behavior is studied using different techniques. Some of them, the most obvious ones being nuclear magnetic resonance (NMR) and SANS-based studies, require the use of D_2O instead of H_2O. In Fig. 22, pseudoternary phase diagrams of cetyltrimethylammonium tosylate (CTAT) and sodium dodecylbenzenesulfonate (SDBS) in H_2O and D_2O are shown. The main features of the phase diagrams are generally the same with H_2O and D_2O, but there are several exceptions. First, the boundary between vesicle and lamellar phases occurs at higher water contents in D_2O, resulting in smaller lobes of the vesicle phases. Second, the transition from a single lamellar phase into the multiphase region occurs at lower concentrations of added SDBS in D_2O. This observation and the fact that solutions of CTAT in D_2O are significantly more viscous than those in H_2O suggest more elongated rodlike micelles in D_2O. Finally, there is no inversion in the density of the lamellar phase L_α-because D_2O has a higher density than H_2O [153]. One should thus keep in mind that the determination of surfactant phase diagrams based on techniques that use H_2O and D_2O in the same study will result in some inaccuracy of the phase boundaries.

All of the preceding indicates that the phase behavior of catanionic mixtures is only indirectly related to the geometrical shape of the monomers. The main concern in developing pseudoternary systems with rich phase behavior is the prevention of crystallization. Small changes in the chain length hardly affect the individual geometrical shape but cause large changes in the phase behavior. The mismatch between hydrocar-

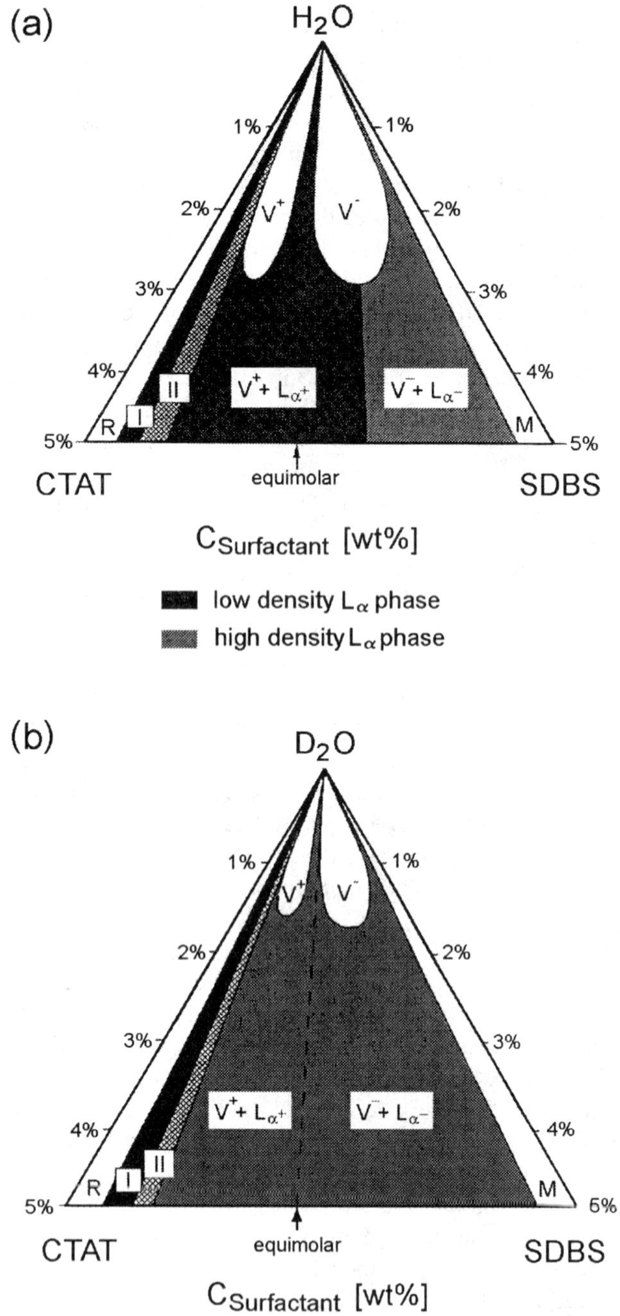

FIG. 22 Pseudoternary phase diagrams for CTAT-SDBS mixtures in (a) H_2O at 25°C and (b) D_2O at 28°C. Differences in the phase behavior are not caused by the small temperature difference but by the solvents. Notations are as follows: One-phase regions are CTAT-rich vesicles (V^+) and rodlike micelles (R), SDBS-rich vesicles (V^-) and micelles (M); two-phase regions are shaded and are V^+ and CTAT-rich lamellar phase ($L_{\alpha+}$), V^- and SDBS-rich lamellar phase ($L_{\alpha-}$), R and hexagonal phase (zone I), and an isotropic liquid with precipitate along the equimolar line. There is also an unresolved multiphase region (zone II). (Adapted from Ref. 153.)

bon tails of cationic and anionic surfactants is thus not as much triggering the formation of aggregates with high curvature as it is allowing these aggregates to form by preventing the competing formation of more stable lamellar aggregates. Chemical structure and charge of the headgroups, on the other hand, may affect the geometrical shape but have almost no input on the phase behavior. Finally, the addition of salt and the choice of solvent clearly change the phase behavior but have no direct impact on the geometrical shape of the molecules. Instead, they change the intermolecular interactions based on hydration and electrostatic forces. Consequently, it is often possible to relate an observed experimental fact to the rather fuzzy packing parameter concept, but it is not possible to *predict* the aggregation behavior (at best) on a more than qualitative level.

The packing behavior is certainly affected by the presence of double-chain surfactants. Pseudoternary mixtures of cationic double-chain surfactant didodecyldimethylammonium bromide (DDAB) and anionic single-chain surfactant sodium dodecyl sulfate (SDS) differ from the catanionic mixtures discussed so far in the tendency to form aggregates of higher curvature because of the larger space requirement of the double-chain moiety. For example, a micellar cubic phase (I) is observed in the DDAB-SDS-water mixture at surfactant ratios at which the single-chain catanionic DTAB-SDS-water system forms a lamellar phase (Figs. 21a and 23b). The binary phase behavior of DDAB and SDS in water is displayed in Fig. 23a. Single-walled vesicles are formed at very low DDAB concentrations and begin to coexist with double-walled vesicles and tubular structures upon increasing the concentration to 0.5 wt%. This region is followed by two lamellar phases, one with high (D_I) and the other with low (D_{II}) water content, which coexist within a wide concentration range (D_I and D_{II} are also denoted L_α and $L_{\alpha'}$ by other authors). The binary phase diagram of SDS is richer, with an isotropic phase (L_1) containing spherical micelles with a mean aggregation number of 60–70 at low concentration. With increasing SDS concentrations rodlike micelles appear, followed by a hexagonal phase with long rodlike micelles packed in a hexagonal array. The phase diagram continues with a lamellar phase and crystal formation at high SDS concentrations.

The pseudoternary DDAB-SDS-D_2O phase diagram displays several interesting features (Fig. 23b). The different regions of the lamellar phases have very different stability ranges. Phase D_I, which has a large stability range in the binary DDAB-water mixture, can incorporate only small amounts of SDS before a transition to the L_1 or some other phase is induced. On the con-

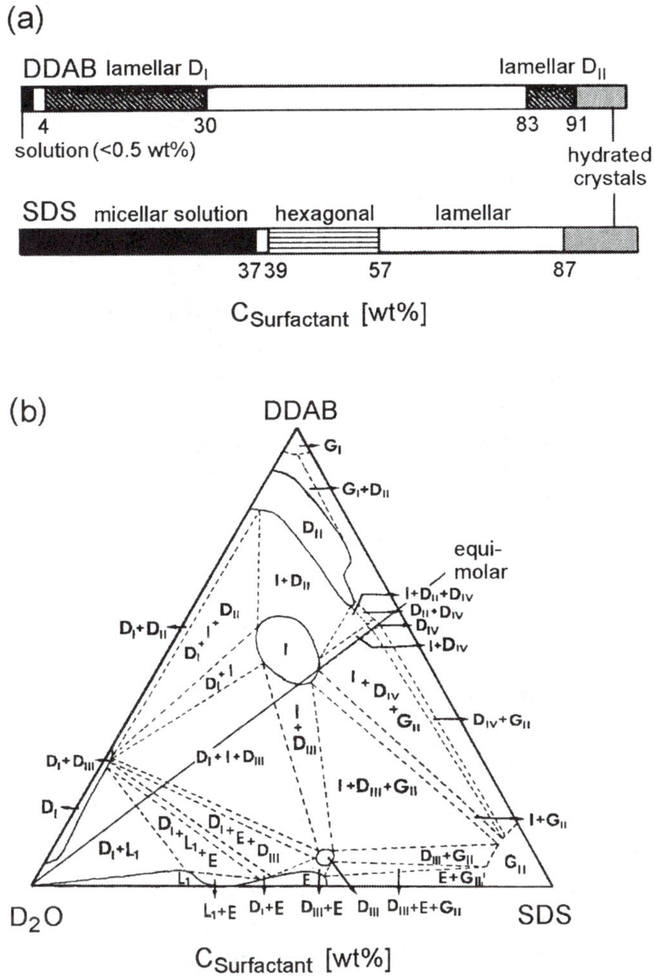

FIG. 23 (a) Binary phase diagrams of DDAB-water and SDS-water at 40°C with blank areas corresponding to two-phase regions. (b) Pseudoternary DDAB-SDS-water phase diagram at 40°C, where two- and three-phase regions are approximately as shown. The notations are: L_I isotropic solution; E hexagonal phase; I cubic phase; D_I, D_{II}, D_{III}, D_{IV} lamellar liquid crystalline phases; and G_I, G_{II} surfactant crystals. (Adapted from Ref. 159.)

trary, the D_{II} lamellar phase can accommodate very large amounts of SDS. The isotropic solution and hexagonal mesophase found in the binary SDS-water system are subject to phase transitions at rather low concentrations of DDAB in the ternary mixture. Furthermore, the pseudoternary mixture displays a number of phase regions that are not observed in the two binary surfactant-water systems, at least not at the same temperature. This is another difference in comparison with catanionic mixtures based on single-chain surfactants such as the DTAC-SN-water system. These new phase regions include two lamellar phases of limited areas of existence, the D_{III} phase close to the hexagonal phase and D_{IV} phase. Furthermore, there is a wide region of a cubic phase (I) at the center of the phase diagram. Cubic phase samples are clear, optically isotropic, and extremely stiff. Finally, in the water-rich corner of the diagram are two very small separate regions of spontaneous vesicle formation, with an excess of either surfactant (not visible in Fig. 23b). Vesicles formed at the DDAB-rich side of the equimolar line have a mean diameter of 500 nm, which increases with dilution, whereas vesicles on the SDS-rich side have a mean diameter of 260 nm, which seems to be invariant with dilution. At mixtures along the equimolar line, however, precipitation occurs even at very high dilution [159].

The pseudo-triple-chain system DDAB-SDS-water forms only normal and bicontinuous-type aggregates. Replacement of SDS by the anionic double-chain surfactant sodium bis(2-ethylhexyl)sulfosuccinate (AOT) enlarges the hydrophobic volume of the mixture. AOT consists of two branched acyl chains connected via the relatively bulky C_4 succinate connector to the small SO_3^- headgroup. This molecular geometry should favor lamellar or even reverse-type mesophases, and, as expected, the binary AOT-water system is dominated by a wide lamellar phase (10–70 wt%) followed by a small bicontinuous cubic phase (73–80 wt%) and a reverse hexagonal phase at concentrations above 82 wt%. The tendency of AOT to favor reverse structures is even preserved in the phase behavior of the catanionic mixture. Equimolar amounts of positively charged DDAB and negatively charged AOT, forming the catanionic surfactant didocecyldimethylammonium bis(2-ethylhexyl)sulfosuccinate (DDAOT), aggregate in very dilute aqueous solution of less than 1 wt% to multiwalled polydisperse vesicles. A stable precipitate follows the vesicle solution at concentrations up to 10 wt%, which redissolves upon further addition of DDAOT, giving way to a hexagonal phase in equilibrium with an isotropic solution. A single hexagonal phase exists at surfactant concentrations between 90 and 95 wt%, and above 95 wt% an equilibrium with hydrated DDAOT crystals is formed. From the position in the phase diagram it was concluded that the hexagonal phase consists of reverse-type hexagonal rods. Although the complete pseudoternary DDAB-AOT-water phase diagram is as complex as the DDAB-SDS-water system displayed in Fig. 24b, its main feature is the coexistence of two reverse hexagonal phases in the AOT-rich region. The reverse hexagonal phase of the catanionic mixture is in equilibrium with the reverse hexagonal phase that originates from the binary AOT-water system [160].

The main aggregate structure found for mixtures of oppositely charged single-chain surfactants such as DTAB and SDS is a lamellar phase. Bicontinuous cubic and lamellar phases dominate the mixtures of double-chain DDAB and single-chain SDS. Finally, the mixtures of double-chain surfactants DDAB and AOT preferably form reverse hexagonal phases. The increasing tendency of these surfactant mixtures to form structures of reverse curvature is caused by the geometry of the molecules involved, e.g., the increasing space requirement of the hydrophobic part. But this observation is obvious and does not require a packing parameter concept to understand. However, some kind of theoretical model is necessary to understand and eventually predict the complex phase behavior of pseudoternary surfactant mixtures as shown in Fig. 23b, but simple geometrical considerations, even when extended by electrostatic assumptions, are insufficient to serve this purpose.

A very interesting study in this context is the experimental and theoretical evaluation of the phase behavior of the cationic double-chain surfactant DDAB mixed with the nonionic double-chain glycolipid *N*-dodecanoyl-*N*-nonyl lactitol ($LC_{11}C_9$). The ternary phase diagram of the DDAB-$LC_{11}C_9$-water systems at 25°C is very similar to the diagram of the binary DDAB-water mixture, containing two lamellar phases (L_α and $L_{\alpha'}$) together with two distinct critical points that close two lamellar two-phase regions (Fig. 24a) [161]. This phase behavior has been calculated using osmotic pressure measurements based on essentially three intermolecular interactions, hydration forces, electrostatic forces, and van der Waals attraction, and an adhesion energy related to the sugar headgroups. Adhesion has been designated as any lowering of the osmotic pressure from the value imposed by hydration and electrostatic interactions. The variation of the osmotic pressure was determined as a function of the sample composition. Two parameters have been used to describe the sample composition in the phase triangle, x_{LC11C9}, the molar fraction of $LC_{11}C_9$ compared with DDAB, and D, the periodicity of the lamellar phase ($D = \delta/1 - \Phi_{water}$, where δ is the bilayer thickness and Φ_{water} is the water volume fraction.

The theoretical phase diagram for the DDAB-$LC_{11}C_9$-water mixture, displayed in Fig. 24a, has been evaluated based on the force balance in the ternary system. Several hypothetical diagrams shown in Ref. 162 demonstrate how the force balance affects size and position of the lamellar two-phase regions. The diagrams reveal that the hydration force increasing with x_{LC11C9} and the induction of a steric stabilization explain the appearance of the first critical point in the DDAB-rich part of the diagram. The second critical point comes from the lowering of electrostatic repulsion due to the low cationic headgroup density and from the strong adhesion between sugar headgroups. Differences between the experimental and calculated phase diagrams may result from mixing entropies and possible cluster formation in the bilayer, which are neglected in the model, and the curvature energy that may be significant when the lamellar phase has a multilayered onion structure [163]. It should be noted that the aggregation behavior of double-chain DDAB and $LC_{11}C_9$ can be described very well by a model based on variable intermolecular interactions instead of geometrical dimensions based on the molecular shape.

Studies of the phase behavior of phospholipids in the presence of surfactants are strongly stimulated by their applications in biotechnology, where solubilization of cell membranes by addition of a micelle-forming surfactant into the extracellular medium is the most commonly used method of extraction and concentration of membrane proteins. Following is an example of such a phospholipid-surfactant mixture, which has been used to shed more light on the underlying theoretical model. The phase behavior of nonionic double-chain egg phosphatidylcholine (EPC) and the single-chain surfactant octylglucoside (OG) is determined by the tendency of EPC to aggregate into extended lamellar bilayers forming closed vesicles and the opposite tendency of OG to form small strongly curved micelles. The phase behavior of the mixture can be described in several steps. Addition of OG to a solution of EPC vesicles results in partitioning of the surfactant between vesicle bilayers and the bulk solution until the bilayers become saturated with OG. Further addition of surfactant results in the formation of mixed threadlike micelles, which grow in numbers at the expense of the mixed vesicles. Finally, above a certain OG concentration only elongated micelles are left in solution and continuous addition of surfactant starts to reduce the micellar length (Fig. 24b) [163].

The composition-induced transition between micelles and vesicles is generally assumed to depend solely on the surfactant-to-lipid ratio of the aggregates and to be independent of the total concentration of the compounds in water. The vesicles are regarded as one extended closed bilayer and the micelles are treated as just one threadlike micelle, which is long enough that the inhomogeneity in its structure related to the two end caps can be neglected. The experimental data

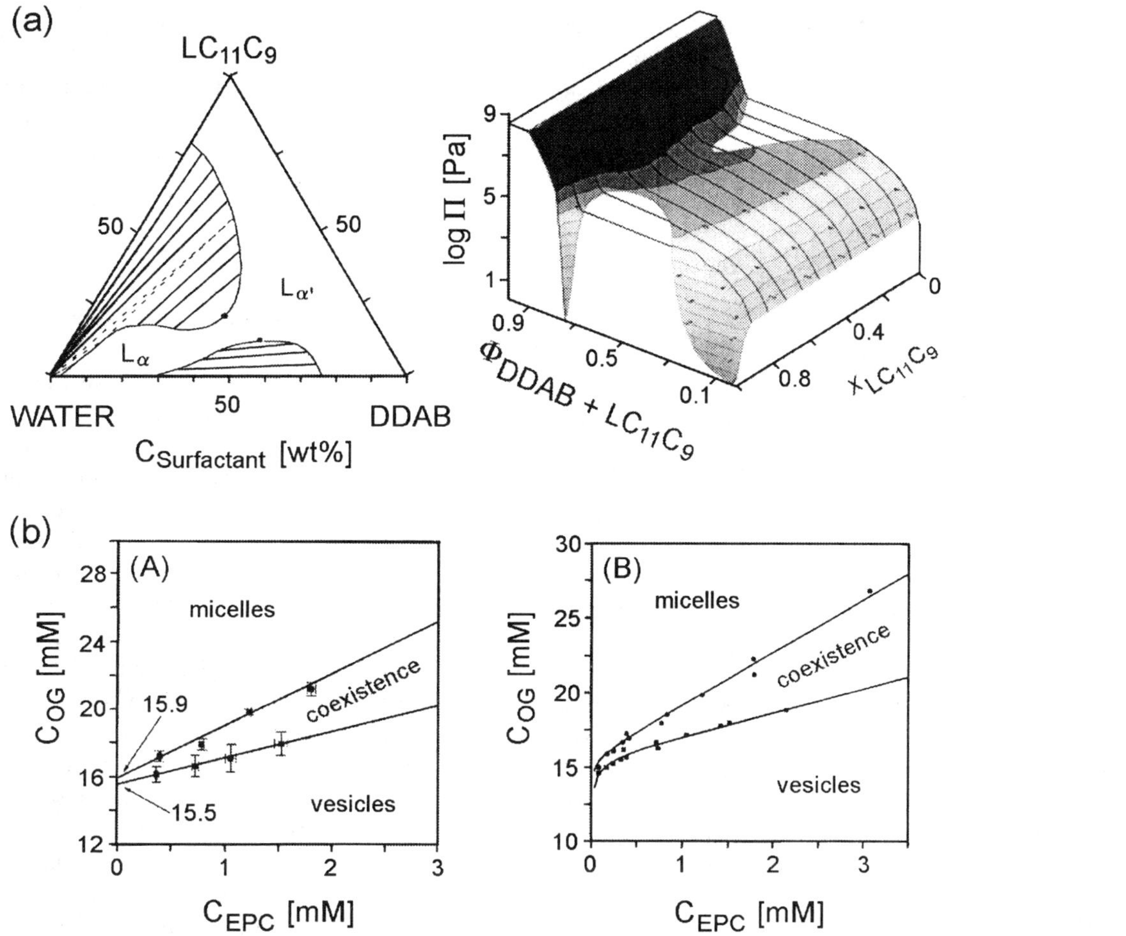

FIG. 24 (a) Ternary phase diagram of the DDAB-$LC_{11}C_9$-water mixture at 25°C and the three-dimensional representation of the equation of state II used to describe the phase behavior. Regions with tie lines in the phase diagram represent lamellar two-phase regions, and the two dots represent the two resulting critical points. The thick black lines in the 3D representation correspond to the pressure-versus-distance curves calculated for a given X_{LC11C9}. The two white areas are the calculated lamellar two-phase regions. (Adapted from Ref. 162.) (b) Phase diagrams of egg phosphatidylcholine (EPC)-octylglucoside (OG)-water (A) based on a model that considers only the EPC/OG ratio within the aggregates and (B) experimental data points and fitting curve based on a revised model considering the finite size of threadlike micelles and the EPC and OG concentration in solution. Note the different behavior at low EPC concentration. (Adapted from Ref. 164.)

points representing the phase boundaries seems to form a straight line. Extrapolation of both phase boundaries to lipid concentration $L = 0$ should give identical values for the surfactant concentration in the lipid-free solution. This is not the case, as indicated by the phase diagram in Fig. 24b. The intercepts are at 15.5 and 15.9 mM, a small but reproducible difference. A revised model based on thermodynamic considerations has thus been developed, which accounts for the finite length of the the threadlike micelles, their repartitioning in the water volume, and the effects of the end caps. As a consequence, the chemical potentials in the micellar phase no longer depend only on the surfactant-to-lipid ratio but also depend on the absolute aqueous concentrations of the lipid and surfactant in water. It should be noted that the purely thermodynamic approach does not use any model assumptions about the micellar structure (and thus the geometrical shape of the molecules generating this structure!). One result of the revised thermodynamic model is the prediction that the phase boundaries in the range of relatively low lipid concentrations have to deviate from straight lines and adopt convex shapes. This prediction has been verified by measurements of the phase boundaries in the EPC-OG-water system at lower concentrations than those studied before. The theoretically derived phase boundaries are in good agreement with the experimental results, as indicated by the phase diagram in Fig. 24b [164]. This is another example in which the phase behavior is not explained by individual geometrical factors. Instead, interactions between molecules and the composition of the solution are taken into account.

V. MODELS DESCRIBING THE PHASE BEHAVIOR OF SURFACTANTS

The examples discussed so far clearly indicate the shortcomings of a packing parameter concept that treats molecules like bricks, with the geometrical shape of the single brick determining the structure to which they can assemble, but ignoring the fact that bricks held together by a glue can form a vast variety of structures not related at all to the shape of the single unit. In case of surfactant molecules as well as other amphiphiles, the glue is present in the form of intermolecular interactions. In general, any concept or model that ignores the interactions of molecules with each other and with the chemical and physical conditions of the bulk solution will fail to explain the aggregation behavior. The simplicity of the packing parameter concept has not only lured many researchers into its use but also triggered some rigorous statements against it. To clear the ground, two of them will be cited here.

> The most intuitively appealing characterization is to visualize the shape of the volume occupied by a molecule in the H_{II} phase as "tapered," i.e. smaller in cross-sectional area at the head group than at the tail. This may be quantitatively specified by a dimensionless shape parameter given by v/al, where v is the molecular volume, a is the area at the lipid-water interface, and l is the length of the tail. Such a "shape concept" characterization is often misunderstood and leaves much to be desired. Rigorously speaking, one must refer to the shape which minimizes the overall free energy of a given molecular volume under a given set of conditions. This bears only a weak resemblance to the steric shape of the molecule because lipids are highly flexible and because factors such as charge, hydrogen bonds, etc. strongly affect the free energy. If v/al is taken to be a characterization of the shape of the mean volume actually assumed in a given phase, then v/al is simply a tautological description of the phase and has no predictive value, because the shape changes sharply at the phase transition. Therefore, v, a, and l should not be taken to refer to the actual molecular dimensions, but rather to the dimensions which the system would prefer in the absence of other constraints; this poses serious problems of definition of v, a, and l and of measuring the values for real systems [119].

The second statement is even more outspoken.

> Structural modeling can be a very formal affair! Take, for example, an ideal sphere with a bimolecular diameter in the shape of a micelle and divide it by the aggregation number. A cone with a surface a_0 of the circle on the top, a length l_c and a volume v is obtained. So far, so good. But now, to make sense out of the cone, a_0 is called the head group area, l_c a critical length "specified for a given lipid," and v becomes the hydrocarbon chain volume. As a result, one now has defined nicely an "average molecule," using the shape of an aggregate which is less than ill-defined. This does not sound very promising and true enough, things get worse! A "critical packing parameter" is defined via the cone's measurements and conclusions such as "SDS in low salt is a cone" and "SDS in high salt is a truncated cone" are drawn. Such pseudo-structural, mean-dynamic-packing models do not rationalize the appearance of molecular assemblies, but mystify them. Not a single crystal structure, NMR spectrum or molecular model gives evidence of a cone-like shape of any known amphiphile and one cannot derive it from a molecular assembly, from which the models are as far apart as spherical droplets, cubic blocks and irregular reefs [165].

After all, do any models exist that allow a reliable prediction of the phase behavior of surfactants? Following are some examples of models that are more suitable for achieving this goal than the packing parameter concept. The phase transitions observed for some phospholipids, changing from lamellar phase (L_α) to inverse hexagonal phase (H_{II}) via bicontinuous cubic phase (V) with increasing temperature, are explained by the interplay of the spontaneous radius of curvature R_0 of the bilayer membrane and the hydrocarbon chain packing energy E_{hc}. The in-plane forces at the hydrophilic surface of a monolayer include electrical charge, hydrogen bond, and other interactions that are generally different from those of the hydrophobic surface. The resultant forces are usually functions of the molecular area at the depth of the monolayer in question. If the sum of the in-plane forces is balanced for an area larger at the tail end than at the headgroup end, then the monolayer may have an effective moment that yields a minimum-energy configuration in which the monolayer is bent with a concave headgroup surface. In this case, the monolayer is said to have a spontaneous curvature. Spontaneous curvature is a thermodynamic property of the monolayer that has the dimensions of a curvature; it is therefore measurable and usefully describes many lipid monolayers.

The hydrocarbon chain packing energy E_{hc} results from the conformation of the hydrocarbon chains. The number of *gauche* rotamers is determined primarily by a competition between the energy of introducing *gauche* rotamers and the resultant increase in entropy of the chains. In the lamellar geometry, there is no geometrical constraint which dictates that the mean chain length for any molecule needs to be different from that for any other. In the H_{II} phase, however, given uniformly curved (e.g., cylindrical) water cores, some of the tails have to reach further than others to fill the hydrocarbon lattice at near uniform density. Therefore, not all molecules are at the minimum of the free energy with respect to the chain extension. A lamellar phase thus has a low energy with respect to the hydrocarbon chain packing but a high energy with respect to bending of the monolayer. The situation is reversed in case of the H_{II} phase, with low energy due to the bend but high energy due to the chain packing. The L_α-to-H_{II} phase transition occurs at a temperature at which the sum of the two energies in a curved geometry falls below the sum in a lamellar geometry [119].

The stability of vesicles made of mixed single-chain catanionic surfactants has been modeled based on a similar thermodynamic approach. In an earlier model, the vesicle stability has been connected to the tendency of "1-2" surfactant pairs to have a bond distance different from the average of "1-1" and "2-2" pairs, resulting in a release of curvature frustration upon forming vesicles [166]. There are two objections to this model. First, the molecular interactions are specific, whereas vesicle formation has been observed for a wide range of surfactants. Second, it is well known that molecules exchange ("flip-flop") between the inner and outer layers of a vesicle membrane as well as between membrane and bulk solution, especially in the case of thermodynamically stable vesicles. This exchange eventually results in an average distribution of the surfactants in both half-layers and a spontaneous curvature equal to zero.

In a more recent thermodynamic approach, neither intervesicular interactions nor specific headgroup interactions are invoked. The free energy of a geometrically closed bilayer is modeled with consideration of (1) the curvature dependence of the bilayer tension; (2) the gain in free energy upon bringing a hydrocarbon chain from water into an alkane bulk phase; (3) the contribution that arises when the electrical charges from headgroups, counterions, and coions are concentrated in a restricted volume near the hydrocarbon-water interface; (4) the loss of conformational flexibility of hydrocarbon chains whose headgroups are restricted at the interface compared with the freedom of the chains in bulk alkanes; (5) a number of less known effects such as shielding of the hydrocarbon-water contact by the headgroups, different hydration of headgroups, counterions, and coions near the hydrocarbon bilayer, and repulsive correlation interactions between headgroups, counterions, and coions that have not been accounted for in the Poisson-Boltzmann approximation; and finally (6) the free energy of mixing the aggregated monomers in the vesicle. Summing up the different contributions then derives the total excess energy of forming a single monolayer out of monomers in solution. The detailed model calculation for the dodecylammonium chloride–sodium dodecyl sulfate mixture reveals a steep rise of the work of bending a planar bilayer into a closed vesicle at an equimolar mixture of the surfactants. Likewise, the bending work increases as the mole fraction of either of the surfactants approaches unity, resulting in a minimum of the free bending energy on each side of the equimolar composition [167]. This model calculation is in good agreement with experimental observations of vesicle lobes in phase diagrams (e.g., see Fig. 22). It is noteworthy that the molecular geometry is only indirectly present in the calculation; most parameters reflect intermolecular interactions and the conditions of the bulk solution.

The Poisson-Boltzmann cell model is an approach that has been successful in describing several features of binary monovalent surfactant–water systems and observations made for the isotropic phase of a divalent surfactant–water mixture [168]. Consequently, this model has also been applied to describe the ternary phase diagram of a mixture of divalent and monovalent surfactants having a charge of the same sign in water. The transitions from the micellar solution to the hexagonal phase and the lamellar phase are modeled on the basis of spherical micelles, circular cylinders, and flat bilayers of infinite extension in two dimensions. Micelles and cylinders are considered to be stiff and of monodisperse size distribution. The interior of the aggregates is modeled as a pure liquid hydrocarbon, and the headgroups are confined to the surface of the aggregates. The thermodynamic model includes the following important contributions to the free energy: (1) electrostatic interactions between the surfactants, (2) the entropy of mixing the surfactants in the aggregates, (3) the interfacial energy contribution that is assumed to be proportional to the area each surfactant molecule exposes to the water, and (4) the entropy of mixing the micelles. Repulsive forces ("hydration forces") are not included in the model calculations. These forces contribute significantly to the free energy at surfactant concentrations above 50 wt%. No fitting parameters have been applied to the model.

Examples of an experimentally determined and a calculated phase diagram are shown in Fig. 25. The question of whether spherical micelles form a micellar solution or a discontinuous cubic phase has not been addressed in the model. Therefore the spherical region of the model diagram would correspond to the region with both cubic and micellar solution phases of the experimental diagram. Similarly, the circular cylindrical and bilayer regions of the model diagram correspond to the hexagonal and lamellar phase regions of the experimentally determined diagram. Keeping this in mind, both diagrams are in good agreement. The inclusion of repulsive hydration forces would not change the sequence of phases but shift the phase boundaries toward higher water contents. The effect would be more pronounced the lower the water content of the system [44].

Very good agreement between experimental data and a theoretical fitting using the UNIQUAC model has been observed for binary nonionic single-chain poly(oxyethylene) surfactant–water mixtures at weight fractions of water between 0.2 and 1.0. The liquid-liquid equilibrium phase diagrams of C_6EO_2, C_6EO_3, and C_7EO_3, ranging from the lower critical consolute solution temperature to 70°C, have been measured [169].

The techniques most often employed to model the phase behavior of surfactants are molecular dynamics (MD) and Monte Carlo (MC) simulations. The application of computer simulation to the field of self-assembly has been the subject of two reviews [170,171]. Two MD simulations describe the phase behavior of nonionic single-chain poly(oxyethylenes). In the first example, a coarse-grained model, parametrized to yield the phase behavior of the corresponding physical system, is used instead of a realistic model with its severe spatial and temporal limitations. The surfactants within the model are composed of chains of Lennard-Jones sites connected by harmonic springs. The solvent is modeled as a Lennard-Jones fluid. Different solution conditions (temperature and concentration) as well as

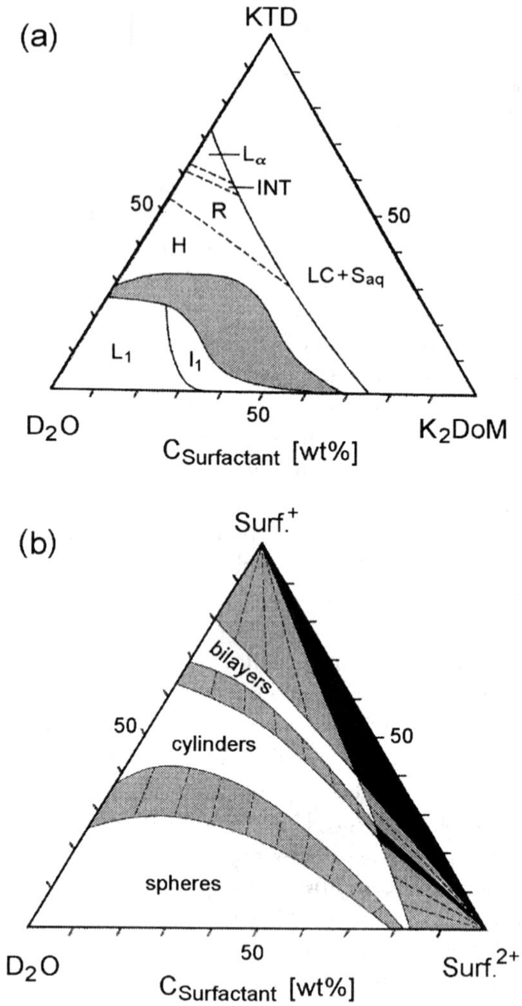

FIG. 25 (a) Composition phase diagram of the dipotassium dodecylmalonate (K_2DoM)-potassium tetradecanoate (KTD)-D_2O system at 50°C. L_1, I_1, H, R, L_α, and INT denote micellar solution, discontinuous cubic, hexagonal, ribbon, lamellar, and intermediate phases, respectively; LC + S_{aq} stands for liquid crystalline phases plus hydrated surfactant crystals. Shaded regions are multiphase regions, and hatched lines are approximate tie lines. (b) Theoretically calculated phase diagram for a ternary divalent surfactant-monovalent surfactant-D_2O system. Precipitated surfactant crystals and surfactants in spherical, circular cylindrical, and planar aggregates have been considered in the calculations. Single-phase, two-phase, and three-phase regions are white, shaded, and black, respectively. (Adapted from Ref. 44.)

different model parameters (size of headgroup and tail and their ratio and strength of interaction between the different sites) have been studied with the aim of gaining insight into how far this simplified MD technique can be used to observe self-organization of amphiphilic systems at higher concentrations. Despite the simplified approach, it has been found that the surfactants aggregate to form very different micellar phases. Lamellar and bicontinuous phases have been found as well, but no hexagonal ordering [172]. In the second example, the properties of the lamellar L_α phase of the binary $C_{12}EO_2$-water mixture have been modeled using constant pressure and temperature (NPT) MD simulations. The calculated interlamellar spacing and the area per surfactant are in reasonable agreement with x-ray diffraction results. The water molecules form hydrogen-bonded bridged structures linking the oxygen atoms of the same surfactant chain. This interaction stabilizes a *gauche* conformation of the headgroups [173].

A very thorough study of the phase behavior of nonionic, cationic, anionic, and zwitterionic single-tail surfactants based on Monte Carlo simulations has been carried out. The phase diagrams are determined by MC lattice simulations for idealized symmetric and asymmetric molecules mixed with single-site "oil" and "water" molecules. At surfactant concentrations above 20 wt%, the simulations show the formation of liquid crystalline phases, including smectic, hexagonal, and gyroid cubic mesophases, and body-centered cubic (BCC) packing of spherical micelles. The locations of the phases in the diagrams of asymmetric surfactants in "water" are shifted relative to those for symmetric molecules in a way that favors phases whose interfaces curve in a way that leaves bulkier groups on the convex side of the interface. Many aspects of the predicted phase behavior, including the compositions at which transitions among ordered phases occur, compare favorably with experimental observations [32]. MC simulations have also been employed to study the phase behavior of ternary surfactant-water-oil mixtures. Several short-chain surfactants with varying tail and headgroup size have been studied, and quantitative phase diagrams for both symmetric and asymmetric molecules have been determined. Two- and three-phase coexistence regions as well as the formation of microemulsions have been observed [174].

The solubility of a compound in a surfactant micelle has been successfully modeled using lattice-based MC simulations. Points of interest were the phase behavior of the surfactant-solute-solvent system and the examination of the locus and extent of the solubilization of the solute in micelles as a function of the solute hydrophobicity and chain length. A novel method based on the distribution of the solute in clusters of different sizes has been developed to study the solute phase behavior in the absence and presence of the surfactant [175]. Generally, the thermodynamic models of mixed micelle formation can be divided into two categories.

In the first category, the process is treated as a reversible chemical reaction and the mass action approach is used. The reactants are the surfactants, the bound counterions, and solubilizates when appropriate. The standard free energy associated with this reaction is in reference to a standard state in which the reactants are dispersed at infinite dilution. This type of model is appropriate because micelle formation is reversible. Micelles have a finite lifetime, as do the individual components that form a micelle. The aggregation number is a dynamic variable and, at any instant, various individual micelles will be composed of different numbers of surfactant molecules. In the case of mixed micelles, the number of surfactant molecules of a given structure may also vary from micelle to micelle.

The second approach is to consider the micelle as a separate bulk phase having an infinite lifetime. The chemical potentials of the components in the micellar phase (although not a true thermodynamically different phase) can then be equated to those of the molecules that are dispersed in the aqueous phase. Various models for the chemical potential can be used to account for the nonideal behavior in both the micellar and solution phases. One of the most popular models of this type of treatment is the regular solution theory [176]. The phase separation model defines the proportions of various components that exist within the micellar phase, and it provides equations for the cmc of mixtures. However, it does not address factors related to the size of the micelles or their aggregation number.

A new model for micellization behavior of binary surfactant mixtures in solution has thus been developed. The significant features of the model are the consideration of asymmetric behavior of micellization in binary surfactant mixtures and the prediction of changes in the structure of mixed micelles with concentration that are dependent on the overall ratio of the surfactant components and their chemical characteristics. Based on the results obtained, a schematic model for the structural changes of mixed micelles has been proposed [177]. The model introduces a packing parameter P^* to describe the random mixing in mixed micelles. This packing parameter, however, changes as a function of the overall mixing ratio of the surfactants, similarly to the relationship of packing density of two different particles as a function of mixing ratio, instead of being a rather static geometrical factor of the single molecules.

All model calculations give reasonable to good results in modeling the phase behavior of binary and ternary surfactant-water and surfactant-water-oil mixtures. In addition, the solubility of a solute in micelles and the formation of mixed micelles can be modeled successfully. The common features of these approaches are the importance of intermolecular forces and the equilibrium between these forces; the geometry of the molecules is only indirectly involved in the calculations. Even in the one approach that uses a packing parameter term, this packing parameter is a variable that changes with the composition of the system rather than a constant number for an individual molecule.

Whereas the model calculations mentioned so far depend heavily on a physical or physicochemical background, the following example is based on an organic chemical approach. The synkinetic approach treats the formation of mesophases and higher organized supramolecular structures like a synthesis. In a synthesis, appropriate precursors ("synthons") form a desired chemical molecule in a controlled way. In exactly the same way, surfactants and other amphiphiles ("synkinons") form a desired mesophase or supramolecular structure when equipped with the right geometry, hydrophile-lipophile balance, and, most important, the appropriate functional groups to engage in certain intermolecular interactions. Synkinesis is thus the target-oriented synthesis of noncovalent molecular aggregates. Figure 26 displays some examples of typical synkinons and their target structures [165].

VI. MONOMER SOLUBILITY—A HANDS-ON MEASURE TO TAILOR THE AGGREGATION BEHAVIOR

Is there any measure between the trial-and-error approach (whether or not based on geometrical shapes) and rather time-consuming model simulations that would help in designing surfactants and other amphiphiles for a certain purpose, e.g., forming mesophases or supramolecular structures of a desired shape? What common macroscopic measure is affected by the molecular shape, intermolecular interactions, the concentration, the presence of cosolutes, the amount of salt, the pH, the temperature, the temperature gradient of heating or cooling, and the characteristics of the solvent? All these factors affect the *solubility of the monomers*. The hydrophile-lipophile balance of monomers, the character and strength of interactions between them, and the conditions of the bulk solution determine their solubility.

Molecules will obviously not form aggregates in water when they are too soluble, and they will form only amorphous lamellar structures when they are too insoluble. For example, finding the right cooling rate that allows specific molecules to organize perfectly into single crystals is the permanent challenge in crystallography. Of course, a crystallographer is restricted in the choices to influence the aggregation behavior of a specific molecule; e.g., changing the molecular structure or adding cosolutes or even a salt is not very helpful in obtaining the structure of the true low-energy crystal of this molecule. However, if the goal is creating a certain mesophase or fiber structure or tubule with a specific inner diameter, then all of the variables mentioned can be employed to realize this goal. The monomer solubility as a measure was mentioned earlier in the context of the phase behavior of ternary C_iEO_j-water-*n*-alkane mixtures and the factors influencing this behavior [59]. Some examples based on single-chain *n*-alkylhexonamides that will support the use of the monomeric solubility as a guide in optimizing the aggregation behavior of these compounds to form well-defined structures follow.

Galactonic, mannonic, gluconic, talonic, and gulonic headgroups each connected to an *n*-octylamide tail aggregate on cooling from hot aqueous solution to form ribbons, rolled up planar sheets, well-defined helices, ill-defined twisted fibers, and lamellar crystals, respectively (Fig. 27) [178,179]. In addition, planar hexonamide crystals are formed by bent or linear monomers, although one linear conformation contains a geometrically unfavorable pair of 1,3-*syndiaxial* oxygen atoms. No close interaction between the geometrical shape of the monomers and the curvature of the aggregates has been found [180–186]. This diverse aggregation behavior can therefore not be explained by geometrical constraints but results from strong and oriented attractive forces in the form of hydrogen bonds between the molecules. Depending on the individual stereochemistry, these hydrogen bonds provide intermolecular interactions of different strengths, thus affecting the solubility of the hexonamides. The solubility is therefore an excellent measure to explain the different aggregate structures.

The extremely low solubility of galactonamide in boiling water (<0.5 wt%) allows only one-dimensional growth of the aggregates, resulting in the formation of ribbons. The ribbons form at relatively high temperatures at which hydrogen bonds are still weak. Consequently, no specific curvature is present in the ribbons, except for a physical twisting of some ribbons because of the edge tension. The approximately sixfold higher solubility of mannonamide (3.0 wt%) results in a higher local concentration of the monomers and a two-dimensional growth of the aggregates to form planar

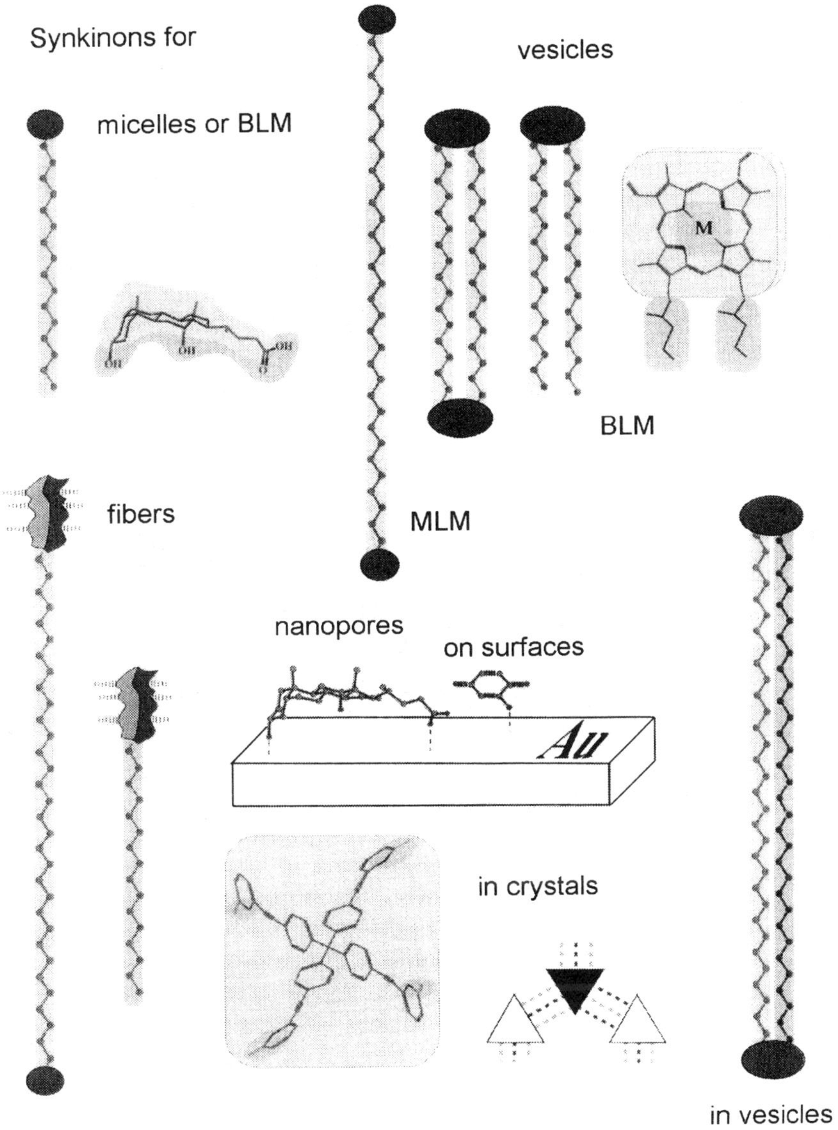

FIG. 26 Schematic models of a few typical synkinons for some membranes and molecular assemblies. The light gray and dark gray shaded areas represent hydrophobic and hydrophilic regions; BLM and MLM denote bilayer lipid membrane and monolayer lipid membrane. (Adapted from Ref. 165.)

bilayer sheets, which eventually roll up into cigarlike structures. Again, the aggregation occurs at temperatures too high to allow strong interactions based on oriented hydrogen bonds. The almost 20-fold higher solubility of gluconamide (50 wt%) and the much lower temperatures during aggregate formation enable the molecules to organize slowly into well-defined helices. The handedness of the respective helices reflects the D- or L-configuration of the parent sugar headgroups. Another increase of the solubility as measured for talonamide (70 wt%) indicates weakened intermolecular forces. Talonamide molecules still aggregate at low temperatures to form twisted fibers, but the fibers are ill defined because of the weak intermolecular interactions. Finally, the even weaker intermolecular forces between gulonamide molecules (solubility >100 wt%) hinder any fiber formation. Instead, lamellar crystals form spontaneously at room temperature from often supersaturated solutions. The observations suggest the existence of an "optimal" solubility (=speed of aggregate formation), which allows the formation of stable aggregates that reflect the specific interactions

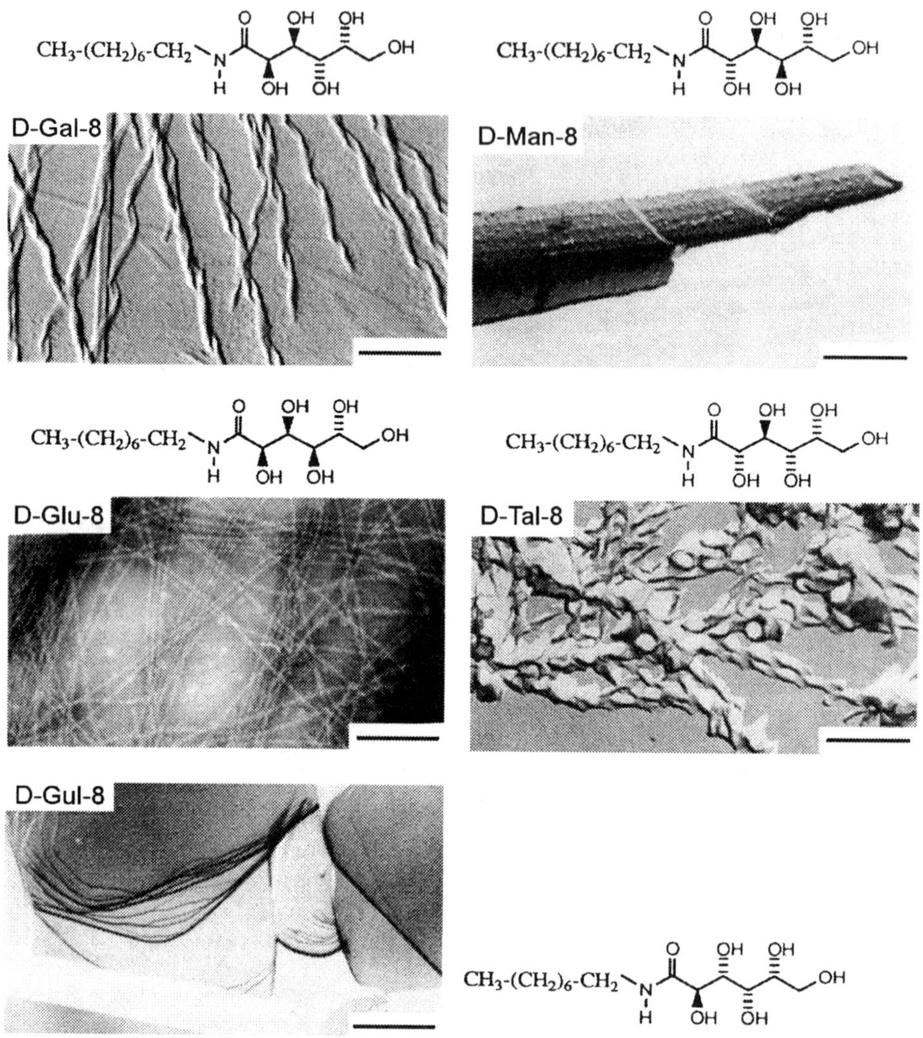

FIG. 27 Chemical structures and transmission electron micrographs of aqueous solutions of *n*-octyl galactonamide (D-Gal-8); mannonamide (D-Man-8); gluconamide (D-Glu-8); talonamide (D-Tal-8); and gulonamide (D-Gul-8). Bars = 500 nm. (From Ref. 187.)

between the monomers. This situation is realized in the case of gluconamide. Lower solubility (=faster aggregation) results in stable aggregates that do not reflect specific interactions between the monomers (galactonamide and mannonamide), whereas higher solubility (=slower aggregation) would allow the formation of aggregates reflecting specific monomer interactions, but the low aggregate stability limits their lifetime (talonamide).

Increasing the length of the alkyl chain and thus the hydrophobic interaction while keeping the headgroup unaltered should lower the solubility and therefore change the aggregation behavior. The hydrophobic interaction increases by 1.1 kT per CH_2 group [26]. For example, gluconamide headgroups connected to decyl, dodecyl, and tetradecyl chains form virtually the same helices as the octyl homologue. The pentadecyl gluconamide, however, aggregates in water to twisted ribbons. The lower solubility of the pentadecyl gluconamide prevents the organization of the monomers into well-defined helices. Lowering the solubility of the more soluble octyl gulonamide by replacing the octyl chain by a hexadecyl chain, on the other hand, initiates the formation of fiber aggregates. Hexadecyl gulonamide first forms twisted ribbons, which eventually grow into flexible tubules (Fig. 28). Finally, the pres-

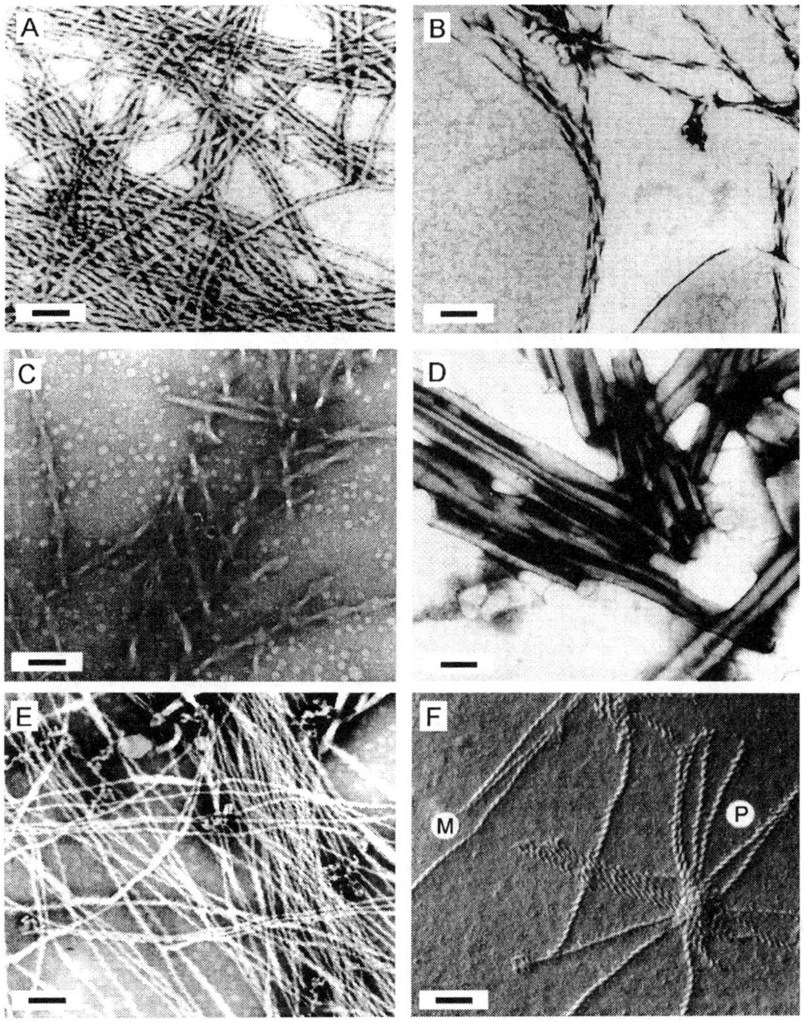

FIG. 28 Transmission electron micrographs of aqueous solutions of (A) helices of tetradecyl gluconamide; (B) twisted ribbons of pentadecyl gluconamide; (C) twisted ribbons of hexadecyl gluconamide, which (D) grow into flexible tubules; (E) helices of pentadecyl gluconamide formed in the presence of SDS micelles; and (F) helices from octadecyl mannonamide formed in the presence of SDS micelles. Bars A–E = 200 nm and Bar F = 150 nm. (From Ref. 187.)

ence of surfactant micelles as cosolute improves the solubility of the hexonamides through the formation of mixed micelles. The incorporation of the alkyl chains into the micellar core reduces the local concentration of the hexonamide molecules in the bulk solution and leads to a slower and more organized aggregation. Pentadecyl gluconamide in the presence of 15% SDS (with regard to the gluconamide) forms well-defined helices instead of the twisted ribbons observed before. The almost water-insoluble octadecyl mannonamide forms right-handed (P) and left-handed (M) helices from hot water–SDS solution instead of lamellar sheets (Fig. 28) [187].

These examples clearly indicate the close relationship between solubility and aggregate formation. It would be foolish, of course, to assume this is a linear relationship in such a way that one can conclude that 0.5 wt% solubility results in ribbons and 50 wt% solubility leads to the formation of helices. This is not another packing parameter concept! However, educated guesses about factors that influence the solubility of the molecules under investigation will provide more reliable guidance on how to affect their aggregation behavior than pure trial-and-error approaches. This statement is true for studies not only of a family of homologous compounds but also (at least) of a group of similar molecules.

VII. CONCLUSIONS

The binary phase behavior of surfactants in water based on mainly isotropic interactions (e.g., headgroup repulsion and hydrophobic attraction) as well as the phase behavior of catanionic mixtures has been reviewed with the emphasis on whether geometrical considerations in the form of a packing parameter concept will allow prediction of the observed behavior. It has been found that intermolecular forces between surfactant monomers and the condition of the bulk solution are of more importance. Several model simulations have been presented that provide reasonable to good agreement between the calculated phase behavior and experimental results. Finally, the solubility of the surfactant monomers as a hands-on measure to tailor molecules for a certain aggregation behavior has been proposed and its feasibility has been demonstrated with several examples.

REFERENCES

1. R. G. Laughlin, *The Aqueous Phase Behavior of Surfactants*, Academic Press, New York, 1994.
2. J. M. Richmond, *Cationic Surfactants—Organic Chemistry*, Surfactant Science Series Vol. 34, Marcel Dekker, New York, 1990.
3. D. N. Rubingh and P. M. Holland, *Cationic Surfactants—Physical Chemistry*, Surfactant Science Series Vol. 37, Marcel Dekker, New York, 1990.
4. C. Gloxhuber and K. Künstler, *Anionic Surfactants—Biochemistry, Toxicology, Dermatology*, 2nd ed., Surfactant Science Series Vol. 43, Marcel Dekker, New York, 1992.
5. H. Stache, *Anionic Surfactants—Organic Chemistry*, Surfactant Science Series Vol. 56, Marcel Dekker, New York, 1995.
6. M. J. Schick, *Nonionic Surfactants—Physical Chemistry*, Surfactant Science Series Vol. 23, Marcel Dekker, New York, 1987.
7. N. M. Van Os, *Nonionic Surfactants—Organic Chemistry*, Surfactant Science Series Vol. 72, Marcel Dekker, New York, 1997.
8. D. Balzer and H. Lüders, *Nonionic Surfactants—Alkylpolyglucosides*, Surfactant Science Series, Marcel Dekker, New York, 2000.
9. E. Lomax, *Amphoteric Surfactants*, 2nd ed., Surfactant Science Series Vol. 59, Marcel Dekker, New York, 1996.
10. K. Holmberg, *Novel Surfactants—Preparation, Applications, and Biodegradability*, Surfactant Science Series Vol. 74, Marcel Dekker, New York, 1998.
11. S. E. Friberg and B. Lindman, *Organized Solutions—Surfactants in Science and Technology*, Surfactant Science Series Vol. 44, Marcel Dekker, New York, 1992.
12. A. K. Chattopadhyay and K. L. Mittal, *Surfactants in Solution*, Surfactant Science Series, Vol. 64, Marcel Dekker, New York, 1996.
13. N. M. Van Os, J. R. Haak, and L. A. M. Rupert, *Physico-Chemical Properties of Selected Anionic, Cationic, and Nonionic Surfactants*, Elsevier Science, Amsterdam, 1993.
14. K. Ogiono and M. Abe, *Mixed Surfactant Systems*, Surfactant Science Series Vol. 46, Marcel Dekker, New York, 1992.
15. P. M. Holland and D. N. Rubingh, *Mixed Surfactant Systems*, ACS Symposium Series, Oxford University Press, New York, 1999.
16. D. Myers, *Surfaces, Interfaces, and Colloids: Principles and Applications*, 2nd ed., Wiley-VCH Publishers, New York, 1999.
17. D. F. Evans and H. Wennerström, *The Colloidal Domain: Where Physics, Chemistry, Biology, and Technology Meet*, 2nd ed., Wiley-VCH Publishers, New York, 1999.
18. D. O. Shah, *Micelles, Microemulsions, and Monolayers*, Marcel Dekker, New York, 1998.
19. M. N. Jones and D. Chapman, *Micelles, Monolayers, and Biomembranes*, Wiley-VCH Publishers, New York, 1994.
20. P. Kumar and K. L. Mittal, *Handbook of Microemulsion Science and Technology*, Marcel Dekker, New York, 1999.
21. K. Esumi and M. Ueno, *Structure-Performance Relationships in Surfactants*, Surfactant Science Series Vol. 70, Marcel Dekker, New York, 1997.
22. C. A. Herb and R. K. Prud'homme, *Structure and Flow in Surfactant Solutions*, ACS Symposium Series, Oxford University Press, New York, 1994.
23. M. J. Rosen, *Surfactants and Interfacial Phenomena*, 2nd ed., Wiley-VCH Publishers, New York, 1989.
24. N. K. Raman, M. T. Anderson, and C. J. Brinker, Chem. Mater. *8*:1682–1701 (1996).
25. H. H. Clint, *Surfactant Aggregation*, Blackie and Son, London, 1992.
26. C. Tanford, *The Hydrophobic Effect: Formation of Micelles and Biological Membranes*, 2nd ed., John Wiley & Sons, New York, 1980.
27. J. N. Israelachvili, *Intermolecular and Surface Forces*, 2nd ed., Academic Press, San Diego, 1991.
28. D. J. Mitchell and B. W. Ninham, J. Chem. Soc. Faraday Trans. II *77*:601–629 (1981).
29. H. Hoffmann and G. Ebert, Angew. Chem. Int. Ed. *27*:902–912 (1988).
30. C. Hall, G. J. T. Tiddy, and B. Pfannemüller, Liq. Cryst. *9*:527–537 (1991).
31. D. W. R. Gruen and E. H. B. de Lacey, in *Surfactants in Solution*, Vol. 1 (K. L. Mittal and B. Lindman, eds.), Plenum, New York, 1984, pp. 279–306.
32. R. G. Larson, J. Phys. II Fr. *6*:1441–1463 (1996) and references cited therein.

33. P. Mariani, L. Q. Amaral, L. Saturni, and H. Delacroix, J. Phys. II Fr. *4*:1393–1416 (1994).
34. J. C. Blackburn and P. K. Kilpatrick, J. Colloid Interface Sci. *149*:450–471 (1992).
35. P. O. Eriksson, G. Lindblom, and G. Arvidson, J. Phys. Chem. *91*:846–853 (1987).
36. M. B. Sjöbom, H. Edlund, and B. Lindström, Langmuir *15*:2654–2660 (1999).
37. S. A. Buckingham, C. J. Garvey, and G. G. Warr, J. Phys. Chem. *97*:10236–10244 (1993).
38. T. Imae and S. Ikeda, J. Colloid Interface Sci. *113*: 449–455 (1986).
39. T. Imae, M. Sasaki, A. Abe, and S. Ikeda, Langmuir *4*:414–418 (1988).
40. B. Lindman, A. Carlsson, G. Karlström, and M. Malmsten, Adv. Colloid Interface Sci. *32*:183–203 (1990).
41. G. Porte, J. Phys. Chem. *87*:3541–3550 (1983).
42. T. Imae and S. Ikeda, J. Phys. Chem. *90*:5216–5223 (1986).
43. H. von Berlepsch, Langmuir *11*:3667–3675 (1995).
44. H. Hagslätt, O. Söderman, and B. Jönsson, Langmuir *10*:2177–2187 (1994).
45. N. A. J. M. Sommerdijk, K. J. Body, A. M. A. Pistorius, M. C. Feiters, R. J. M. Nolte, and B. Zwanenburg, Langmuir *15*:7008–7013 (1999).
46. N. A. J. M. Sommerdijk, T. L. Hoeks, K. J. Booy, M. C. Feiters, R. J. M. Nolte, and B. Zwanenburg, Chem. Commun. 743–744 (1998).
47. M. C. Monje, H. Mawlawi, A. Lattes, and M. Rivière, New J. Chem. *17*:231–235 (1993).
48. G. T. Dimitrova, T. F. Tadros, P. F. Luckham, and M. R. Kipps, Langmuir *12*:315–318 (1996).
49. P. Sakya, J. M. Seddon, R. H. Templer, R. J. Mirkin, and G. J. T. Tiddy, Langmuir *13*:3706–3714 (1997).
50. B. Håkansson, P. Hansson, O. Regev, and O. Söderman, Langmuir *14*:5730–5739 (1998).
51. H. Kunieda, K. Shigeta, and K. Ozawa, J. Phys. Chem. B *101*:7952–7957 (1997).
52. H. Kunieda, K. Shigeta, and M. Suzuki, Langmuir *15*: 3118–3122 (1999).
53. M. S. Leaver, U. Olsson, H. Wennerström, R. Strey, and U. Würz, J. Chem. Soc. Faraday Trans. *91*:4269–4274 (1995).
54. U. Olsson and P. Schurtenberger, Langmuir *9*:3389–3394 (1993).
55. K. Fukuda, U. Olsson, and U. Würz, Langmuir *10*: 3222–3229 (1994).
56. R. Strey, R. Schomäcker, D. Roux, F. Nallet, and U. Olsson, J. Chem. Soc. Faraday Trans. *86*:2253–2261 (1990).
57. G. Karlström, J. Phys. Chem. *89*:4962–4964 (1985).
58. R. Strey, Ber. Bunsenges. Phys. Chem. *100*:182–189 (1996).
59. S. Burauer, T. Sachert, T. Sottmann, and R. Strey, Phys. Chem. Chem. Phys. *1*:4299–4306 (1999).
60. T. Scottmann and R. Strey, J. Chem. Phys. *106*:8606–8615 (1997).
61. M. Kahlweit and R. Strey, Angew. Chem. Int. Ed. Engl. *24*:654–668 (1985).
62. B. M. Knickerbocker, C. V. Pesheck, L. E. Scriven, and H. T. Davis, J. Phys. Chem. *83*:1984–1990 (1979).
63. T. Scottmann, R. Strey, and S. H. Chen, J. Chem. Phys. *106*:6483–6491 (1997).
64. M. J. Schwuger, in *Structure/Performance Relationships in Surfactants* (M. J. Rosen, ed.), ACS Symposium Series 253, American Chemical Society, Washington, DC, 1984, chapter 1.
65. R. G. Alargova, K. D. Danov, J. T. Petkov, P. A. Kralchevsky, G. Broze, and A. Mehreteab, Langmuir *13*: 5544–5551 (1997).
66. N. Zoeller and D. Blankschtein, Langmuir *14*:7155–7165 (1998).
67. X. Li, R. M. Washenberger, L. E. Scriven, and H. T. Davis, Langmuir *15*:2267–2277 (1999).
68. X. Li, R. M. Washenberger, L. E. Scriven, and H. T. Davis, Langmuir *15*:2278–2289 (1999).
69. H. Kunieda, H. Taoka, T. Iwanaga, and A. Harashima, Langmuir *14*:5113–5120 (1998).
70. R. Strey, W. Jahn, G. Porte, and P. Bassereau, Langmuir *6*:1635–1639 (1990).
71. R. M. Hill, *Silicone Surfactants*, Surfactant Science Series Vol. 86, Marcel Dekker, New York, 1999.
72. X. Zhu, W. G. Miller, L. E. Scriven, and H. T. Davis, Colloids Surf. A *90*:63–78 (1994).
73. T. Stoebe, R. M. Hill, M. D. Ward, and H. T. Davis, Langmuir *13*:7276–7281 (1997).
74. T. Svitova, H. Hoffmann, and R. M. Hill, Langmuir *12*:1712–1721 (1996).
75. M. He, R. M. Hill, Z. Lin, L. E. Scriven, and H. T. Davis, J. Phys. Chem. *97*:8820–8834 (1993).
76. R. Wagner and R. Strey, Langmuir *15*:902–905 (1999).
77. G. Arvidson, I. Brentel, A. Khan, G. Lindblom, and K. Fontell, Eur. J. Biochem. *152*:753–759 (1985).
78. P. O. Eriksson, G. Lindblom, and G. Arvidson, J. Phys. Chem. *91*:846–853 (1987).
79. N. Maurer, E. Prenner, F. Paltauf, and O. Glatter, Biochim. Biophys. Acta. *1192*:167–176 (1994).
80. O. Söderman, G. Carlström, M. Monduzzi, and U. Olsson, Langmuir *4*:1039–1044 (1988).
81. P. G. Faulkner, A. J. I. Ward, and D. W. Osborne, Langmuir *5*:924–926 (1989).
82. C. LaMesa, B. Sesta, M. G. Bonicelli, and G. F. Ceccaroni, Langmuir *6*:728–731 (1990).
83. Y. Chevalier, F. Mélis, and J. P. Dalbiez, J. Phys. Chem. *96*:8614–8619 (1992).
84. Y. Chevalier, Y. Storet, S. Pourchet, and P. Le Perchec, Langmuir 7:848–853 (1991).
85. J. Amrhar, Y. Chevalier, B. Gallot, P. Le Perchec, X. Auvray, and C. Petipas, Langmuir *10*:3435–3441 (1994).

86. T. Kunitake, Angew. Chem. Int. Ed. Engl. *31*:706–726 (1992).
87. Y. Ishikawa, T. Nishimi, and T. Kunitake, Chem. Lett. 165–168 (1990).
88. T. Nishimi, Y. Ishikawa, R. Ando, and T. Kunitake, Rec. Trav. Chim. Pays-Bas *113*:201–208 (1994).
89. T. Kunitake, Y. Okahata, M. Shimomura, S. Yasunami, and K. Takarabe, J. Am. Chem. Soc. *103*:5401–5413 (1981).
90. F. M. Menger and J. Ding, Angew. Chem. Int. Ed. Engl. *35*:2137–2139 (1996).
91. M. C. Holmes, Curr. Opin. Colloid Interface Sci. *3*: 485–492 (1998).
92. M. P. Krafft and J. G. Riess, Biochimie *80*:489–514 (1998).
93. M. Monduzzi, Curr. Opin. Colloid Interface Sci. *3*: 467–477 (1998).
94. K. Wang, G. Orädd, M. Almgren, T. Asakawa, and B. Bergenståhl, Langmuir *16*:1042–1049 (2000).
95. K. Wang, G. Karlsson, M. Almgren, and T. Asakawa, J. Phys. Chem. B *103*:9237–9246 (1999).
96. F. Giulieri, M. P. Krafft, and J. G. Riess, Angew. Chem. Int. Ed. Engl. *33*:1514–1515 (1994).
97. F. Giulieri and M. P. Krafft, Colloids Surf. A *84*:121–127 (1994).
98. K. Kratzat and H. Finkelmann, J. Colloid Interface Sci. *181*:542–550 (1996).
99. K. Kratzak, C. Stubenrauch, and H. Finkelmann, Colloid Polym. Sci. *273*:257–262 (1995).
100. K. Kratzak and H. Finkelmann, Colloid Polym. Sci. *272*:400–408 (1994).
101. E. V. Shumilina and Y. A. Shchipunov, Colloids Surf. A *121*:23–26 (1997).
102. M. Dubois and T. Zemb, Langmuir *7*:1352–1360 (1991).
103. M. Dubois, T. Zemb, N. Fuller, R. P. Rand, and V. A. Parsegian, J. Chem. Phys. *108*:7855–7869 (1998).
104. G. G. Warr, R. Sen, D. F. Evans, and J. E. Trend, J. Phys. Chem. *92*:774–783 (1988).
105. T. F. Svitova, Y. P. Smirnova, S. A. Pisarev, and N. A. Berezina, Colloids Surf. A *98*:107–115 (1995).
106. F. M. Menger and K. Gabrielson, J. Am. Chem. Soc. *116*:1567–1568 (1994).
107. F. M. Menger and N. Balachander, J. Am. Chem. Soc. *114*:5862–5863 (1992).
108. T. Kunitake and Y. Okahata, J. Am. Chem. Soc. *99*: 3860–3861 (1997).
109. T. Kunitake, N. Kimizuka, N. Higashi, and N. Nakashima, J. Am. Chem. Soc. *106*:1978–1983 (1984).
110. N. Kimizuka, H. Ohira, M. Tanaka, and T. Kunitake, Chem. Lett. 29–32 (1990).
111. P. Bissel, M. Onda, K. Yoshihara, H. Koyano, K. Ariga, T. Kunitake, Y. Oishi, and K. Suehiro, Langmuir *15*:1791–1795 (1999).
112. N. A. J. M. Sommerdijk, T. H. L. Hoeks, M. Synak, M. C. Feiters, R. J. M. Nolte, and B. Zwanenburg, J. Am. Chem. Soc. *119*:4338–4344 (1997).
113. D. A. Jaeger, B. Li, and T. Clark Jr, Langmuir *12*: 4314–4316 (1996).
114. G. Cevc and D. Marsh, *Phospholipid Bilayers: Physical Principles and Models*, Wiley, New York, 1987.
115. G. Cevc, *Phospholipids Handbook*, Marcel Dekker, New York, 1993.
116. D. Marsh, *Handbook of Lipid Bilayers*, CRC Press, Boca Raton, FL, 1987.
117. D. M. Small, *Handbook of Lipid Research*, Vol. 4, Plenum, New York, 1989.
118. R. Lipowsky and E. Sackmann, *Structure and Dynamics of Membranes*, Vol. 1A, *Handbook of Biological Physics*, Elsevier, Amsterdam, 1995.
119. S. M. Gruner, J. Phys. Chem. *93*:7562–7570 (1989).
120. K. Larsson, J. Phys. Chem. *93*:7304–7314 (1989).
121. J. M. Seddon, Biochim. Biophys. Acta *1031*:1–69 (1990).
122. J. M. Seddon, Ber. Bunsenges. Phys. Chem. *100*:380–393 (1996).
123. F. C. MacKintosh, Curr. Opin. Colloid Interface Sci. *2*:382–387 (1997).
124. P. Ekwall, Adv. Liq. Cryst. *1*:1–142 (1975).
125. G. Lindblom and L. Rilfors, Biochim. Biophys. Acta *988*:221–256 (1989).
126. D. Marsh, Chem. Phys. Lipids *57*:109–120 (1991).
127. P. R. Cullis, M. J. Hope, and C. P. S. Tilcock, Chem. Phys. Lipids *40*:127–144 (1986).
128. W. Helfrich, Liq. Cryst. *5*:1647–1658 (1989).
129. R. Goetz and W. Helfrich, J. Phys. II Fr. *6*:215–223 (1996).
130. B. Klösgen and W. Helfrich, Biophys. J. *73*:3016–3029 (1997).
131. B. Klösgen and W. Helfrich, Eur. Biophys. J. *22*:329–340 (1993).
132. J. M. Seddon and G. Cevc, in *Phospholipids Handbook* (G. Cevc, ed.), Marcel Dekker, New York, 1993, pp. 403–454.
133. G. S. Smith, E. B. Sirota, C. R. Safinya, R. L. Plano, and N. A. Clark, J. Chem. Phys. *92*:4519–4529 (1990).
134. D. Bach, I. Bursuker, H. Eibl, and I. R. Miller, Biochim. Biophys. Acta *514*:310–319 (1978).
135. G. Cevc, J. Phys. *50*:1117–1134 (1989).
136. D. A. Jaeger, W. Subotkowski, J. Wang, E. Kubicz-Loring, and H. Shinozaki, Langmuir *13*:1844–1849 (1997) and references therein.
137. F. M. Menger, M. G. Wood, Jr., Q. Z. Zhou, H. P. Hopkins, and J. Fumero, J. Am. Chem. Soc. *110*: 6804–6810 (1988).
138. K. Yamauchi, K. Doi, and M. Kinoshita, Biochim. Biophys. Acta *1283*:163–169 (1996).
139. T. Kunitake, Y. Okahata, and S.-I. Tawaki, J. Colloid Interface Sci. *103*:190–201 (1985).
140. Y. Ishikawa, T. Nishimi, and T. Kunitake, Chem. Lett. 25–28 (1990).
141. J. Y. Kim, J. Mattai and G. G. Shipley, Biochemistry *26*:6592–6598 (1987).

142. E. N. Serrallach, R. Dijkman, G. H. De Haas, and G. G. Shipley, J. Mol. Biol. *170*:155–174 (1983).
143. V. Ravily, C. Santaella, P. Vierling, and A. Gulik, Biochim. Biophys. Acta *1324*:1–17 (1997).
144. N. A. J. M. Sommerdijk, P. J. J. A. Buynsters, H. Akdemir, D. G. Geurts, A. M. A. Pistorius, M. C. Feiters, R. J. M. Nolte, and B. Zwanenburg, Chem. Eur. J. *4*: 127–136 (1998).
145. N. A. J. M. Sommerdijk, M. H. L. Lambermon, M. C. Feiters, R. J. M. Nolte, and B. Zwanenburg, Chem. Commun. 455–456 (1997).
146. B. Kronberg, Curr. Opin. Colloid Interface Sci. *2*:456–463 (1997).
147. E. H. Lucassen-Reynders, J. Lucassen, and D. Giles, J. Colloid Interface Sci. *81*:150–157 (1981).
148. H. Edlund, A. Sadaghiani, and A. Khan, Langmuir *13*: 4953–4963 (1997) and references therein.
149. K. L. Herrington, E. W. Kaler, D. D. Miller, J. A. Zasadzinski, and S. Chiruvolu, J. Phys. Chem. *97*: 13792–13802 (1993).
150. N. Filipovic-Vincekovic, M. Bujan, D. Dragcevic, and N. Nekic, Colloid Polym. Sci. *273*:182–188 (1995).
151. M. T. Yatcilla, K. L. Herrington, L. L. Brasher, E. W. Kaler, S. Chiruvolu, and J. A. Zasadzinski, J. Phys. Chem. *100*:5874–5879 (1996).
152. E. W. Kaler, A. K. Murthy, B. Rodriguez, and J. A. Zasadzinski, Science *245*:1371–1374 (1989).
153. E. W. Kaler, K. L. Herrington, A. K. Murthy, and J. A. Zasadzinski, J. Phys. Chem. *96*:6698–6707 (1992).
154. O. Söderman, K. L. Herrington, E. W. Kaler, and D. D. Miller, Langmuir *13*:5531–5538 (1997).
155. L. L. Brasher, K. L. Herrington, and E. W. Kaler, Langmuir *11*:4267–4277 (1995).
156. L. L. Brasher and E. W. Kaler, Langmuir *12*:6270–6276 (1996).
157. T. Zemb, M. Dubois, B. Demé, and T. Gulik-Krzywicki, Science *283*:816–819 (1999).
158. H. Hagslätt and K. Fontell, J. Colloid Interface Sci. *165*:431–444 (1994).
159. E. Marques, A. Khan, M. da Graca Miguel, and B. Lindman, J. Phys. Chem. *97*:4729–4736 (1993).
160. A. Caria and A. Kahn, Langmuir *12*:6282–6290 (1996).
161. F. Ricoul, M. Dubois, T. Zemb, M. P. Heck, A. Vandais, D. Plusquellec, I. Rico-Lattes, and O. Diat, J. Phys. Chem. B *102*:2769–2775 (1998).
162. F. Ricoul, M. Dubois, L. Belloni, T. Zemb, C. André-Barrès, and I. Rico-Lattes, Langmuir *14*:2645–2655 (1998).
163. E. Opatowski, M. M. Kozlov, and D. Lichtenberg, Biophys. J. *73*:1448–1457 (1997).
164. Y. Roth, E. Opatowski, D. Lichtenberg, and M. M. Kozlov, Langmuir *16*:2052–2061 (2000).
165. J.-H. Fuhrhop and J. Köning, *Membranes and Molecular Assemblies: The Synkinetic Approach*, The Royal Society of Chemistry, Cambridge, 1994.
166. S. A. Safran, P. A. Pincus, D. Andelman, and F. C. MacKintosh, Phys. Rev. A *43*:1071–1078 (1991).
167. M. Bergström, Langmuir *12*:2454–2463 (1996).
168. B. Jönsson and H. Wennerström, J. Phys. Chem. *91*: 338–352 (1987).
169. H.-H. Lai and L.-J. Chen, J. Chem. Eng. Data *44*:251–153 (1999).
170. R. G. Larson, Curr. Opin. Colloid Interface Sci. *2*: 361–364 (1997).
171. S. Karaborni and B. Smit, Curr. Opin. Colloid Interface Sci. *1*:411–415 (1996).
172. B. Fodi and R. Hentschke, Langmuir *16*:1626–1633 (2000).
173. S. Bandyopadhyay, M. Tarek, M. L. Lynch, and M. L. Klein, Langmuir *16*:942–946 (2000).
174. A. D. Mackie, K. Onur, and A. Z. Panagiotopoulos, J. Chem. Phys. *104*:3718–3725 (1996).
175. S. K. Talsania, L. A. Rodríguez-Guadarrama, K. K. Mohanty, and R. Rajagopalan, Langmuir *14*:2684–2692 (1998).
176. D. N. Rubingh, in *Solution Chemistry of Surfactants*, Vol. 1 (K. L. Mitall, ed.), Plenum, New York, 1979, pp. 337–348.
177. L. Huang and P. Somasundaran, Langmuir *13*:6683–6688 (1997).
178. J.-H. Fuhrhop, P. Schnieder, J. Rosenberg, and E. Boekema, J. Am. Chem. Soc. *109*:3387–3390 (1987).
179. J.-H. Fuhrhop, P. Schnieder, E. Boekema, and W. Helfrich, J. Am. Chem. Soc. *110*:2861–2867 (1988).
180. S. Svenson, J. Köning, and J.-H. Fuhrhop, J. Phys. Chem. *98*:1022 (1994).
181. S. Svenson, B. Kirste, and J.-H. Fuhrhop, J. Am. Chem. Soc. *116*:11969 (1994).
182. C. André, P. Luger, S. Svenson, and J.-H. Fuhrhop, Carbohydr. Res. *230*:31–40 (1992).
183. C. André, P. Luger, S. Svenson, and J.-H. Fuhrhop, Carbohydr. Res. *240*:47–56 (1993).
184. J.-H. Fuhrhop, S. Svenson, P. Luger, and C. André, Supramol. Chem. *2*:157–171 (1993).
185. J.-H. Fuhrhop and S. Svenson, in *Kinetics and Catalysis in Microheterogeneous Systems* (M. Grätzel and K. Kalyanasundaram, eds.), Surfactant Science Series, Vol. 38, Marcel Dekker, New York, 1991, pp. 273–301.
186. V. Zabel, A. Müller-Fahrnow, R. Hilgenfeld, W. Saenger, B. Pfannemüller, V. Enkelmann, and W. Welte, Chem. Phys. Lipids *39*:313 (1986).
187. S. Svenson, Ph.D. thesis, Free University, Berlin, 1993.

34

Stereochemistry of Lipid Micelles and Vesicles That Survive Drying

JÜRGEN-HINRICH FUHRHOP Freien Universität Berlin, Berlin, Germany

I. INTRODUCTION

Spherical micelles in water are dynamic species that break up and reform. On a microsecond time scale, single surfactant molecules retreat from and travel toward micelles, and within milliseconds the whole micelle, typically made of less than 100 amphiphilic molecules, disintegrates [1,2]. The reason for the short lifetime of micelles lies in the large distance and repulsive interaction between hydrated headgroups and the relatively high critical concentration of monomers that is needed to form micelles [critical micelle concentration (cmc) $\cong 10^{-2}$ M]. Detergent micelles in water can only be observed by cryo-transmission electron microscopy (TEM) [3,4] (Fig. 1) because they disintegrate immediately on solid surfaces. Their diameter corresponds directly to the length of two detergent molecules, typically 4–5 nm. Micelles dissolve compounds that are not soluble in the bulk solvent. Detergent micelles usually take up just one water-insoluble dye [5] molecule or a hydrogen-bonded dimer [6].

Vesicles have a much lower critical concentration of monomers in water, namely 10^{-5} M or lower. They are made of mono- or bilayer membranes containing tens of thousands of water-insoluble amphiphilic molecules. They are stable for months in aqueous solutions and survive ultrafiltration as well as gel chromatography. On drying they collapse first to flat disks with bulgy edges and then dissipate [7] (C. Böttcher, unpublished) (Fig. 2). Aqueous phase vesicles do not survive on solid-air or liquid-liquid interfaces. If one dries vesicles on carbon grids, gold surfaces, or mica, one invariably observes collapse to circular structures with a flat bilayer in the center. Vesicles are also disrupted at water-solvent interfaces. When the vesicle bilayer in water is in the fluid state and is brought in contact with immiscible solvents, monomers are deposited at the interface [8] and monolayers are formed if the bulk concentration is above 10^{-5} M. The monolayer at a water-CCl_4 interface is much less ordered than at the water-air interface. Long alkyl chains stand perpendicular in the organic phase, but infrared spectra show disordering.

Fluid micellar rods (hexagonal phase) are obtained in concentrated aqueous solutions of amphiphiles that form spherical micelles in dilute solution. They may be isolated in dry form under favorable circumstances (cast film, liquid crystals) [9]. Fluid vesicular fibers, on the other hand, are formed upon swelling of the soft crystals of insoluble lipids in water [10] (Fig. 3). They fade away upon drying. No binding forces between the headgroups are observed in these fibrous molecular assemblies, and crystallization does not occur. Another important tubule system is made of water-soluble urea and long-chain alkanes in water. These inclusion compounds form upon crystallization and decompose again in water [11]. Here one has a case where isolation of noncovalent fibers in the solid state is easy, but the fibers do not survive in water.

Stability rises drastically if the monomeric amphiphiles are replaced by amphiphilic diblock copolymers, e.g., of polystyrene (PS) or polyethylethylene (PEE), and smaller blocks of polyacrylic acid (PAA), polyethylene glycol (PEG), or polyvinyl pyridine (PVP). Such

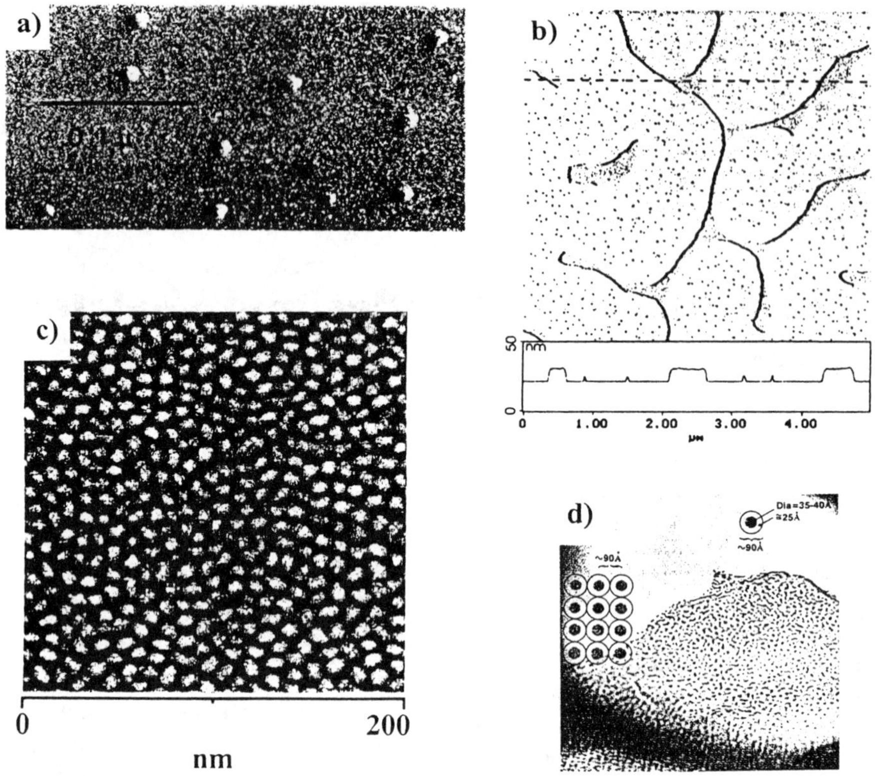

FIG. 1 Micrographs of spherical micelles. (a) Cryo-TEM of an eicosane sulfate micelle fixated by rapid freezing (from Ref. 3). (b) AFM of a polystyrene-thiophene$_{11}$-polystyrene triblock copolymer on mica from a toluene solution. (c) TEM of a closed film of the same polymer (from Ref. 14). (d) TEM of dendrimer micelles (from Ref. 18).

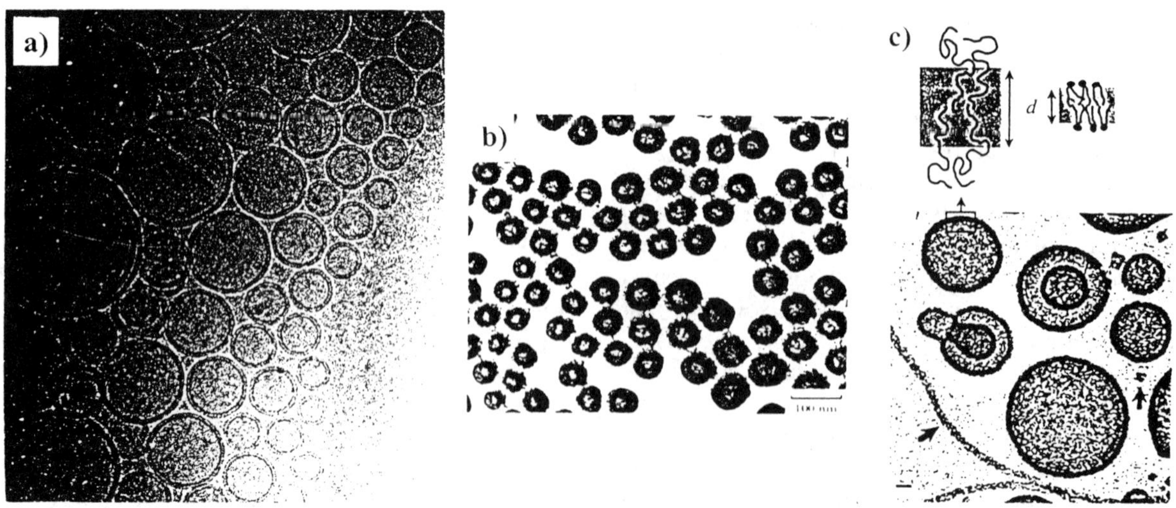

FIG. 2 Micrographs of spherical vesicles. (a) Typical cryo-TEM of lecithin-cholesterol vesicles. (b) TEM of (styrene)$_{200}$ (acrylate)$_8$ vesicle in water/DMF 4:1. The mean lamellar thickness is 30 nm (from Ref. 15). (c) Aqueous suspension of *t*-butyl-[CH_2-$CH(C_2H_5)$]$_{37}$-$(CH_2CH_2O)_4$-H copolymer. The mean lamellar thickness is 8 nm (from Ref. 19).

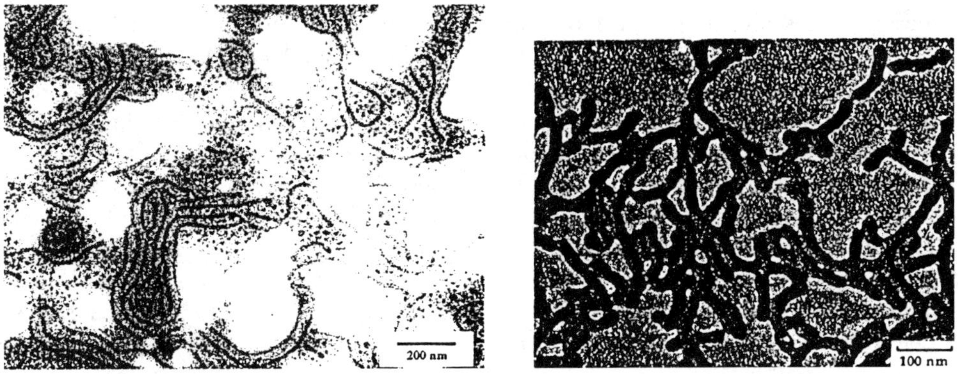

FIG. 3 TEMs of micellar rods made of cross-linked diblock polymers. (Left) From cyclopentane solutions of polystyrene-poly(2 cinnamoyl methacrylate) (from Ref. 13). (Right) From polystyrene-polyacrylate in water (from Ref. 15).

amphiphiles with very large immiscible blocks of solvophilic and solvophobic monomers tend to form spherical or ellipsoidal micelles. Because covalent bonds do not allow separation into two bulk phases, one observes the aggregation of soluble blocks into domains within the matrix of insoluble blocks. The resulting micellar domains can then be observed by TEM. Diameters of 10–200 nm are typical. These polymeric micelles are usually formed in organic solvents, e.g., methanol/acetone or toluene (Fig. 1) and dissolve inorganic salts or colloids [12–16] in their cores. Heck reactions, for example, could be carried out in toluene instead of dimethylformamide (DMF) and similarly unpleasant polar solvents. Palladium acetate in copolymer-stabilized toluene solutions was exceptionally stable and remained active after 50,000 turnover cycles, which involve the same number of Pd^{2+}/Pd^0 conversions [17]. Another type of stable micelle is made of just one single spherical dendrimer molecule. They are usually based on three or four building blocks of first-generation monodendrons, which are often connected in the center. The maximal number of such monodendrons in a spherical micelle is, depending on size, between 8 and 18. Dendrimers can also easily be visualized by TEM without freezing in the solvent [18] (Fig. 1).

Amphiphilic diblock copolymers of long PS and short PAA blocks give micellar spheres, cylinders, and vesicles under different conditions. Polymers consisting of cores of insoluble blocks surrounded by a thin shell of soluble blocks are called crew-cut. The morphology of their aggregates is, in addition to the usual repulsive and attractive interactions in micelles, controlled by the surface torsion at the core-corona interface at the onset of micellization. This interaction can be modified by the addition of ions and change of solvents. Hard spherical micelles are, for example, formed in DMF, and vesicles appear in tetrahydrofuran (THF) because the central core swells. In TEM such polymer vesicles appear as three-dimensional (3D) spheres with a dent in the center. They contain so much material that flattening does not occur upon drying. Copolymers of PEG and EE (PEG_{40}-PEE_{37}), however, produce perfectly unilamellar vesicles, which are osmotically active (Fig. 2) [19]. In contrast to lipid vesicles in water, the polymer membranes in organic solvents have not been shown to build up osmotic pressure. The physical properties of the thick polymer walls in fluid droplets are hardly differentiated in micelles and vesicles. The strict separation of two water volumes by spherical membranes remains a unique property of lipidic vesicles. Reverse micelles made of block polymers dissolve inorganic salts (e.g., a palladium catalyst for reactions between organic substrates).

Cylindrical micelles made of amphiphilic copolymers appear in a variety of morphologies in the solid state. Most of them are made of covalent block copolymers, where the solvent preferentially solvates only one of the blocks. TEM usually shows striations close to the size of the corresponding spherical micelles. The coagulation to cylinders is reversible on raising the temperature or changing the solvent. Cylinder formation is controlled by a balance between separation of the lyophobic tails and the respulsion between the solvated headgroups. The relatively large surface energy at the hemispherical ends of the molecular cylinders then provides a driving force for the formation of long cylinders. Most of the polymer micelles are, however, reverse micelles in organic solvents. The polymers are quite insoluble and growth of the individual noncovalent fibers is limited. Typically one obtains average lengths around 500 nm and diameters of 20 nm (Fig. 3, left side).

Covalent polymeric tubules are, to the best of our knowledge, not obtained from diblock polymers. They are, however, ubiquitous in nature. Typical examples are rolled-up β-sheets or interwound helices of proteins [20], the helices of starch, which entrap I_3^- as a blue polymer with an inner diameter of about 1 nm [21], and cellulose tubules of 1 μm diameter. Only the latter are useful for entrapment in the form of hollow rayon fibers [22]. The walls of these biopolymers are rigid and do not dissolve anything, in contrast to the artificial micellar polymers described so far. Water-soluble compounds may be entrapped within the central, water-filled cavity. In synthetic polymers a hollow center is obtained only by mechanical means (Fig. 4).

The following section deals with solid noncovalent micellar and vesicular spheres, rods, and tubules. It will be shown that strong binding interactions between the headgroups do not necessarily lead to 3D crystals but that highly curved structures can be maintained without solvation spheres.

II. AN ISOLABLE, NONCOVALENT SPHERICAL MICELLE

There are, to the best of our knowledge, no reports on spherical micelles that survive isolation from water or, in the case of reverse micelles, from organic solvents. Drying leads to powders or oil droplets. The reason for this instability of nanometer spheres lies in the necessity to maintain the solvation sphere around the headgroups. If headgroup repulsion vanishes, curvature fades away and bulk materials are formed. This assumption may, however, not be true if one connects the headgroups by strong amide hydrogen bonds in three directions on the surface of a sphere. We have, for example, played with long-chain amide derivatives, e.g., of tris-(aminoethyl)-amine but obtained vesicles only upon sonication (M. Skupin, J.-H. Fuhrhop, unpublished). Dissolution in hot water and cooling never produced the expected micelles with a hydrogen-bonded network on the surface, but ill-defined precipitates were found.

The first stable micelles were formed when we precipitated amphiphile **1** with a ruthenium tris-(bipyridine) dichloride headgroup by addition of hexafluorophosphate anions. One of the bipyridine units carried a long-chain malonic diester link, which aggregated in water to form bilayer membranes. We expected formation of a planar bilayer or, after sonication, vesicles. Cryoelectron microscopy as well as TEM of negatively stained material at low-dose electron irradiation showed, however, multilayered micelles instead of vesicles (Fig. 5). A 45-Å-wide bilayer of a spherical micelle could be reproducibly observed in the center of the smaller micelles. The sequence of events in the formation of the onion-type micelles was then elucidated by cryo-TEM a few seconds after the addition of PF_6^-. At first, large vesicles filled with microcrystalline debris of the ruthenium complex amphiphile appeared. After continued sonication or longer aging, the crystallites disappeared and the vesicles filled up to the center with alternating ruthenium and alkyl layers. No planar multilayers were ever observed in aqueous media. The micellar solution was then placed on solid mica or gold surfaces and atomic force microscopy (AFM) pictures were taken in the trapping mode. Perfect spheres with about equal width and height were observable for several hours. After days, some flattening occurred [23].

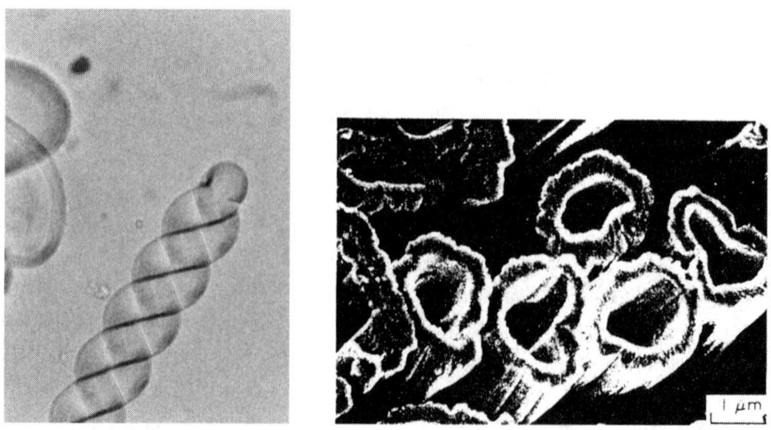

FIG. 4 Light micrographs of vesicular tubules. (Left) Protrusions from lecithin crystals in water. The central hole has a width of about 50 nm (Prof. I. Sakurai, personal communication). (Right) Spun fiber of acetylated cellulose. The central hole has a width of several micrometers (from Bever, 1985, Ref. 22).

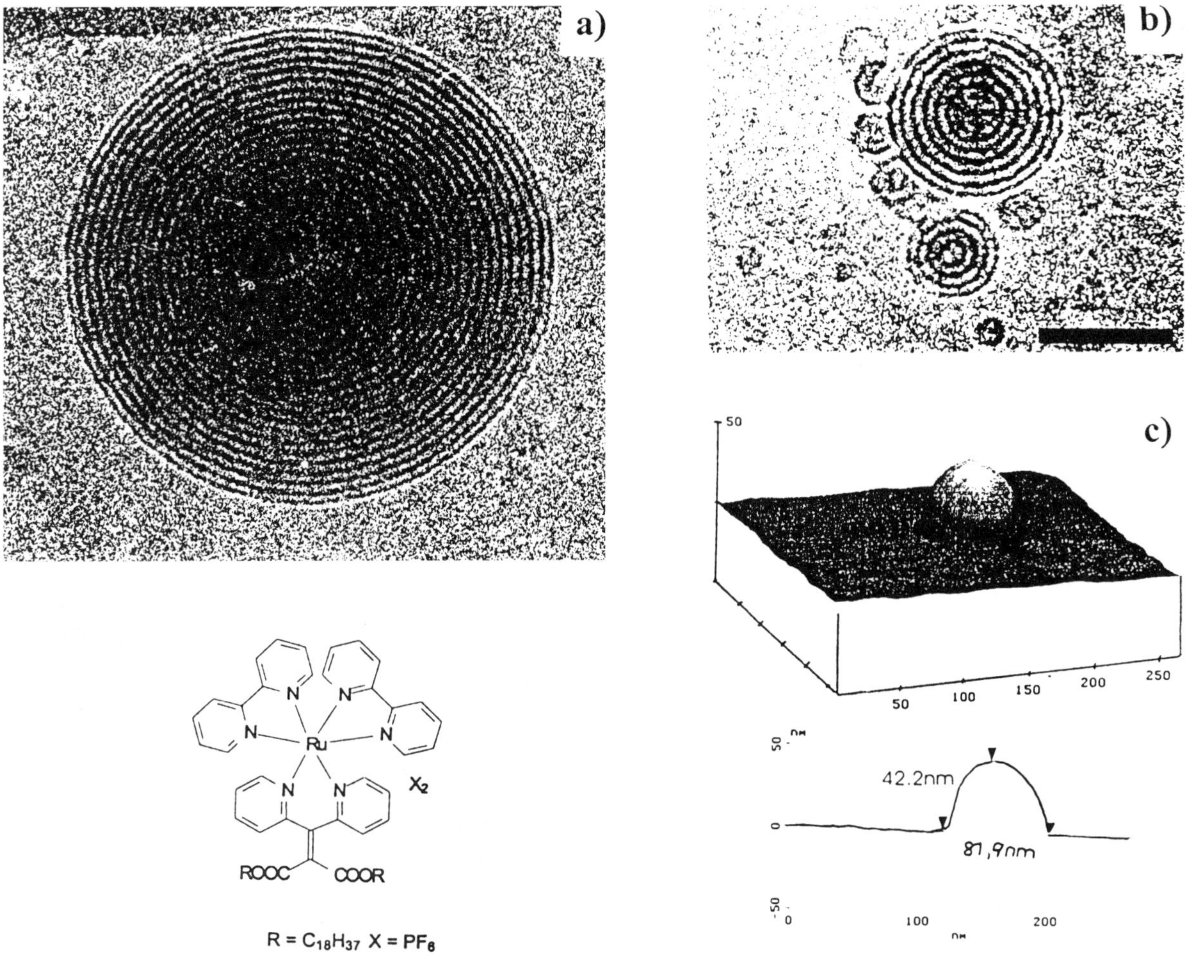

FIG. 5 Two TEMs of spherical micelles of **1**. (a) Dried samples on a carbon grid. (b) Cryo-TEM in water. (c) AFM of the micelles on mica. The black and white stripes have a thickness of 2–3 nm each. They correspond to ruthenium complex and interdigitated alkyl bilayers (see Fig. 6) (from C. Draeger et al., 1999, from Ref. 23).

The cross-section anion of the headgroup (100 $Å^2$) is so much larger than that of the two alkyl chains (40 $Å^2$) that interdigitation takes place. Furthermore, the ruthenium hexafluorophosphate headgroups show a unique tendency to dimerize directly on the side of the cube opposite to the position of the alkyl chains. The hydrophobic alkyl chains therefore stretch to both sides of the headgroup region and interdigitate on both sides. The growth of this interdigitated and double-sided network occurs in three dimensions and produces the spherical shape of multilamellar vesicles at first and after extended sonication multilayered micelles with a CH_2 bilayer in the center (Fig. 6).

We think that the combination of large headgroups with two long alkyl chains will in general produce stable spheres. Currently, we are synthesizing osmium and platinum analogues, but all kinds of water-stable divalent metal complexes of appropriate size and solubility should also work. Such particles may then be applied as light-collecting entities (Ru, Os) or catalytic particles (Pd, Pt) as well as water-soluble dyes (multivalent Fe, W, Mo compounds). The advantage of these metal complex micelles to metal/salt-loaded polymers as described in the introduction is the well-defined environment of and the distance between metal ions. There should also be no environmental problem with these assemblies. The micelles decompose simply on heating; the components are hydrolyzable to fatty alcohols, 2,2′-bipyridine, and malonate. The perfect organization of the pseudocrystalline micelle also causes the main disadvantage of the material as compared with block polymers. Swelling of the latter in organic solvent–water mixtures produces loose aggregates that readily dissolve substrates and reagents. The large solid

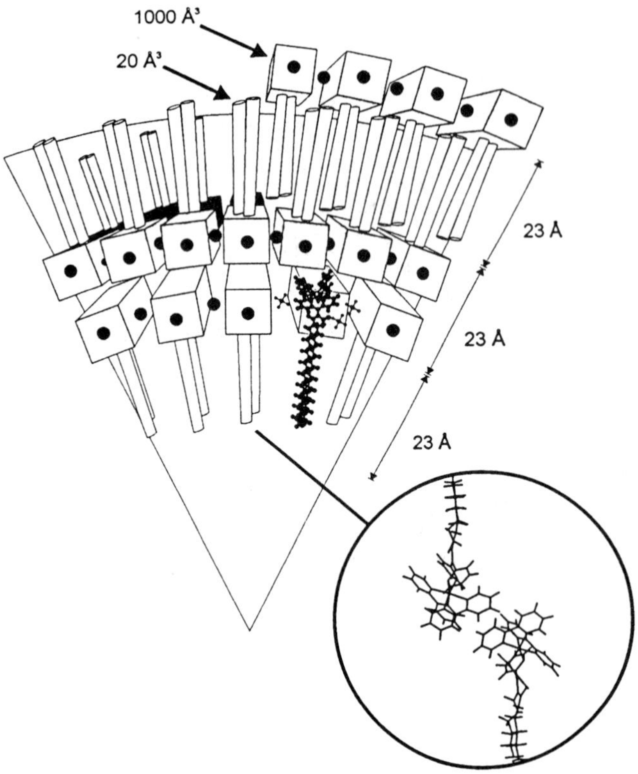

FIG. 6 Molecular model of the multilayered, isolable micelles made of an amphiphilic ruthenium complex (from Ref. 23).

micelles can react only on the surface. It therefore seems mandatory to produce much smaller and uniform micelles before application in catalysis becomes feasible. Application as light-collection cells, on the other hand, is also straightforward with the large micelles and is currently under investigation.

The question remains: Why are the ruthenium complex micelles stable without hydration spheres? Why do they survive drying and do not rearrange to form 3D crystallites when all repulsive hydration forces disappear? A possible answer is that the micelles are only kinetically stable. We propose that at first vesicles are formed upon ultrasonication by disruption of crystallites with interdigitated multilayers. These vesicles have a very low critical concentration (cmc) and upon further sonication fill up their interior with concentric rings in water. Interdigitation of the bilayers stabilizes the densely packed alkyl bilayer. The headgroup bilayer is less rigid and highly hydrated. Upon drying, the curvature does not disappear because the interdigitated alkyl bilayers keep the headgroups in place and thus allow 30–40% of empty space in the dried ruthenium layers. This then opens the possibility of introducing water-soluble reagents into the gaps by hydration-dehydration cycles using aqueous solutions.

III. A DRY, NONCOVALENT SPHERICAL VESICLE

Vesicles survive in contrast to micelles in gel chromatography. Their cmc is usually so low ($<10^{-5}$ M) that loss by dissociation is negligible under appropriate conditions. If, however, the water content of the vesicle is removed, the vesicle collapses. TEM after evaporation shows flat disks with bulges around them. Polymerization of the monomers within a preformed vesicle hardly changes the situation. Although the vesicles then partly survive replacement of water by ethanol, the rigidity of the membrane is not enhanced by the formation of polymer blocks in the order of 10-mers [24,25].

The situation changes if the flexible oligomethylene chains of vesicle bilayers are replaced by rigid polyenes and if four of these linear chains are fixed on an equally rigid macrocycle. Porphyrin **2** with four bixin side chains forms vesicles in water upon sonication. The TEMs show perfect spheres, which do not flatten in contact regions [26]. This is very unusual. In general, one observes more or less complete flattening of both vesicles in contact areas such as in soap bubble foams. The appearance of contact points between perfect spheres is already strong evidence for the presence of rigid membranes (Fig. 7). Irradiation of the orange-red vesicles with visible light leads to colorless, red-fluorescent polymerized vesicles. These polymers with a white cross-linked membrane also produce the same glasslike appearance of a shift membrane (not shown).

The same porphyrin-carotene amphiphile **2** also forms stable monolayers on water. Irradiation again produces a white polymer cloth of 40 Å thickness and a surface of about 10^{3} cm^{2}. It has, so far, not been possible to transfer this monomolecular cloth to solid surfaces. AFM shows an interrupted double layer of micellar structures with a thickness of about 10 nm. The reasons for and mechanisms of the rearrangement of the polymeric monolayer are not understood. It was also found that the polyene bola-amphiphiles alone or a soft covalent arrangement of several bixin molecules in a single, bola-amphiphilic molecule did not automatically lead to stiff membranes. Bixin itself, for example, readily formed polymerizable vesicles upon sonication, but they were soft and unstable.

When an aqueous suspension of the stiff vesicles made of **2** was transferred to gold or mica surfaces, the vesicles survived drying. AFM indicated spheres of

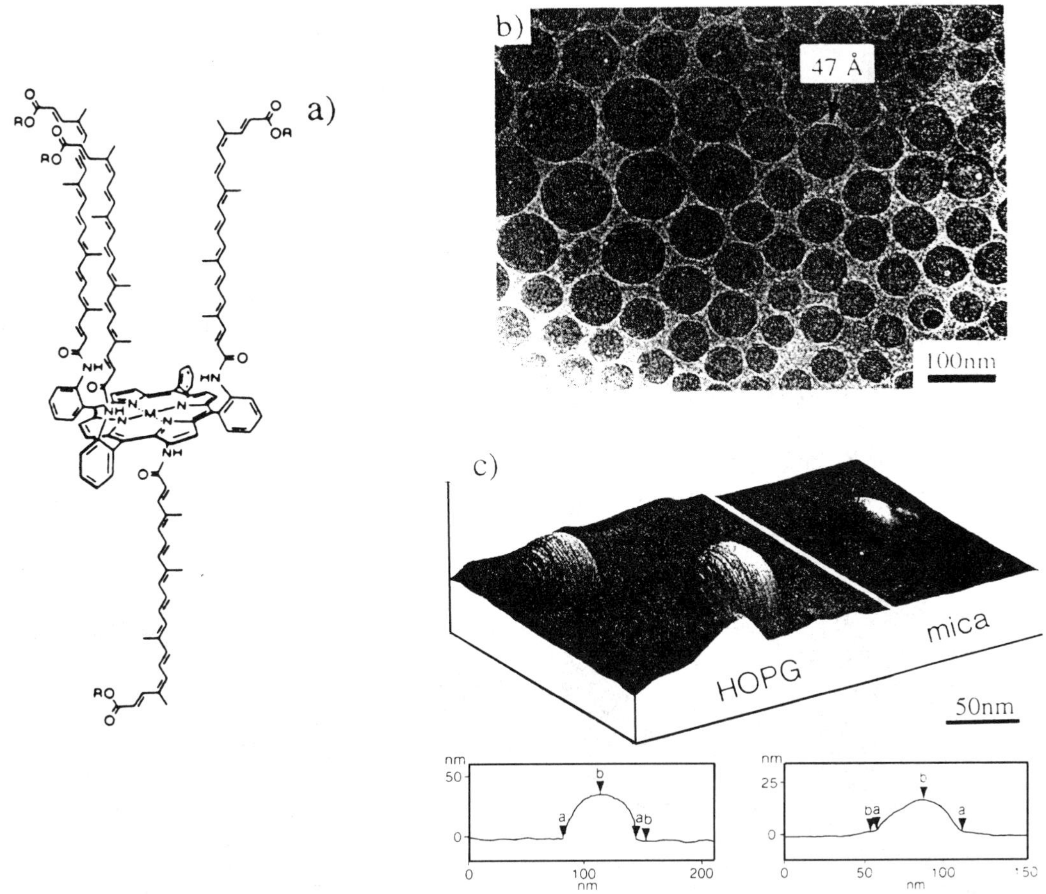

FIG. 7 Micelle made of the carotenoid porphyrin. (a) Cryo-TEM and (b) AFM on graphite and mica (from Ref. 26).

equal width and height (Fig. 7). Slow flattening occurred only after several days.

The glasslike behavior of the porphyrin-bixin monolayer refers only to the overall appearance of the spheres. The membrane is by no means smooth or electrolyte impermeable on a molecular level. The vesicles shown in Figs. 5 and 6, to the contrary, are osmotically totally inactive. Addition of 1 M NaCl causes neither precipitation nor shrinkage. This means that the membrane is perforated, which also becomes evident from model building. There are, on average, two polyene chains ending with a carboxylate group on each side of the porphyrin plane. The cross-sectional area of the polyene chains is about 60 Å^2, and it is more than 200 Å^2 for the porphyrin plane.

Stiffness of the membrane in aqueous solution (cryo-TEM) or the dry state (AFM) does, therefore, not come from a crystal-like assembly as in the ruthenium-micelle case (Fig. 5) but from a regular porphyrin layer in the center of the membrane and an irregular crisscross of stiff polyene chains above and below it. The stable micelle made of the ruthenium complex **1** survives drying because the dense interdigitated monolayer cannot rearrange to form 3D crystals. The vesicle made of the tetrabixinato-porphyrin does not flatten upon drying because the irregular and stiff porphyrin-carotenoid abatis cannot become disentangled.

A comparison of the solid, isolable bixinatoporphyrin vesicle with the vesicles obtained from polymers (see the introduction) shows that both are very different from the fluid lecithin-type vesicles of nature. The polymer vesicles are multilayered fat droplets that entrap some solvent upon swelling with organic solvents. The stiff porphyrin-bixin membrane has the 5 Å thickness of a lipid membrane but no barrier properties. Both types of artificial materials can, nevertheless, be applied as carriers for reactive materials in colloidal solutions.

IV. DRY, NONCOVALENT MICELLAR FIBERS

Solid micelles and vesicles do not exist in nature, but solid amphiphilic fibers are ubiquitous in the form of covalent helical proteins, polysaccharides and nucleic acids. Introduction of one or two secondary amide groups, the structure forming motiff of proteins, converts non-covalent spherical micelles to linear fibers. The conformation of the monomers and the corresponding fibrous aggregates may then be modified by interactions between chiral centers in the headgroup region. The beginning of this development has been summarized already in an earlier article in this series [27]. It was shown there, that left- and right-handed lipid bilayer helices combined to form racemic sheet structures, that twisted ribbons may be filled up by more lipid molecules to give vesicular tubules and that alloys of two different carbohydrate amphiphiles could be formed.

Later, it was shown that *N*-dodecyl-D-gluconamide (see Fig. 9) formed extended quadruple helices (Fig. 8a–c) upon cooling aqueous solutions from a temperature >80°C to room temperature [28,29]. The yield of these fibers was close to quantitative when rearrangement to crystals and thicker assemblies was prevented by addition of sodium dodecyl sulfate (SDS) micelles. These micelles obviously dissolve crystallites and lead the dissolved molecules back to the fiber. These fibers can be lyophilized and stored in solid form for years. After resuspension in water at room temperature, un-

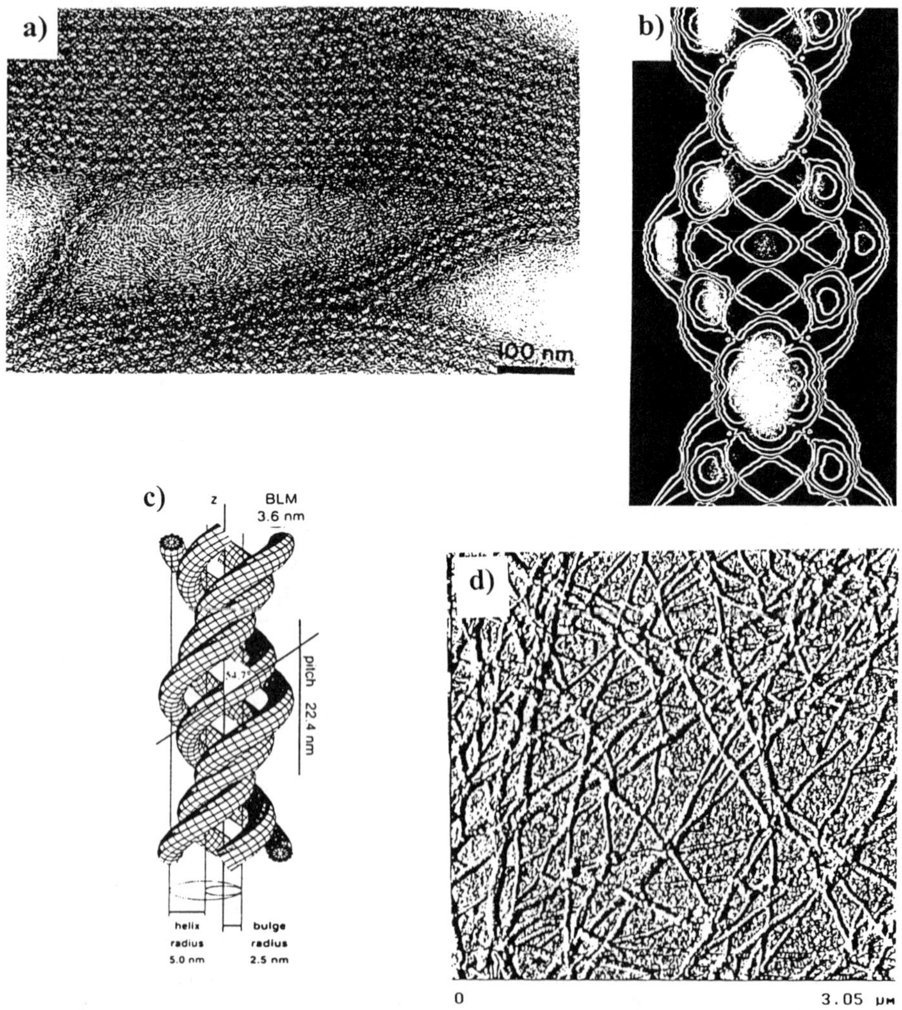

FIG. 8 (a–c) TEM, image analysis and model of the quadruple helix *N*-octyl-D-gluconamide (uranyl acetate negatively stained) (from Ref. 28). (d) AFM of the nonstained fiber on mica (from Ch. Messerschmidt, unpublished).

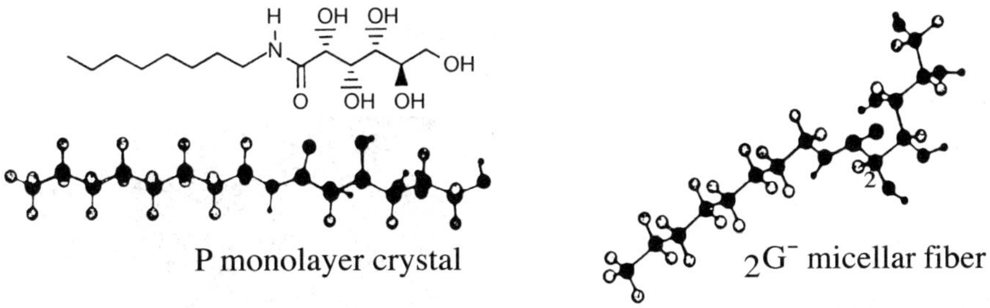

FIG. 9 Molecular conformation of *N*-octyl-D-gluconamide **3** in the fiber shown in Fig. 8. A *gauche* bend in the headgroup allows strong hydration of the micellar fiber without disrupting the connecting hydrogen bond chain (from Ref. 30).

changed fibers appeared. Transfer to solid surfaces was also possible, and a net of quadruple helices was detected by AFM (C. Messerschmidt and J.-H. Fuhrhop, unpublished results) (Fig. 8d).

Diastereomeric glyconamides do not form tightly wound helical micelles in water. They appear as rolled-up sheets (mannon) or ill-defined twisted ribbons (galacton). The beautifully uniform and well-defined structure of the gluconamide fibers allowed a detailed analysis by solid-state ^{13}C nuclear magnetic resonance (NMR) spectroscopy combined with x-ray analysis, solution NMR, and infrared spectroscopy. As a result, the curved molecular conformation shown in Fig. 10 was established. In water, the empty room left within the bend is filled with water molecules. Upon drying, intra- and intermolecular hydrogen bridges between the headgroup hydroxyls stabilize the structure and help to keep the curvature. Kinetic stability is provided by strong and well-defined hydrogen bonds between individual OH groups of the carbohydrate chains [30,31].

The hydrocarbon skeleton of micelle-forming amphiphiles and bola-amphiphiles has been replaced by porphyrins bearing the same gluconamide or amino and amino acid headgroups [32–34]. These dyes of essentially square shape (7 × 7 Å) then stack to form hydrogen-bonded micelles. The thickness of these fibers corresponds essentially to the sum of the length of two side chains plus the 7 Å of the porphyrin core. These dye fibers can also be isolated in solid form, stored for months, and then resuspended in water. Another example is the ultrathin fiber of a tin(IV)-porphyrin bisgluconamide (Fig. 10) [33,34], which is essentially held together by stacking and hydrogen bonding between chloride axial ligands and H_3O^+. Porphyrin fibers usually do not fluoresce, whereas the monomers do.

Similar fibers are also formed between cyanine dyes in water [35] (Jelly or Scheibe aggregates). These fibers are *not* of micellar character but are held together only by dipole and van der Waals forces. In this case, a reverse fluorescence behavior was observed: only the high molecular aggregate fluoresces, not the monomers.

The micellar fibers described here are very easy to prepare on a large scale. The monomers can be synthesized on the kg scale. Fiber formation in water occurs spontaneously and quantitatively upon heating-cooling cycles or pH change. They can also be deposited on solid surfaces without any change of structure. So far, however, there is no obvious application for this kind of new material.

V. DRY, NONCOVALENT VESICULAR FIBERS

Several amphiphiles and bola-amphiphiles with one or two secondary amide links form vesicular tubules in water. Long-chain diamides with amino and L-lysine headgroups dissolve in water at 85°C and pH 4 to a maximum concentration of 2×10^{-4} M. Upon raising the pH to 10.5, long uniform tubules appear with an inner diameter of 50 nm and a membrane thickness of 4.4 nm. The inside of the tubules is filled with water and can be stained with heavy metal salts by imbibement [36,37] (Fig. 11).

The tubules were isolated in dry form, stored for 2 years, and resuspended in water. No degradation of the tubes and length/diameter ratios of up to 10^3 were observed. Similar tubules have also been obtained from gluconamide/galactonamide bilayers [37,38] and from amphiphilic porphyrins [39]. These fibers look similar to asbestos fibers and can be made on a large scale. They decompose upon addition of ethanol or heating above 75°C in water or to 150°C in air. Because these fibers are made of fatty acids or amines and glucose or amino acids, they are readily biodegradable.

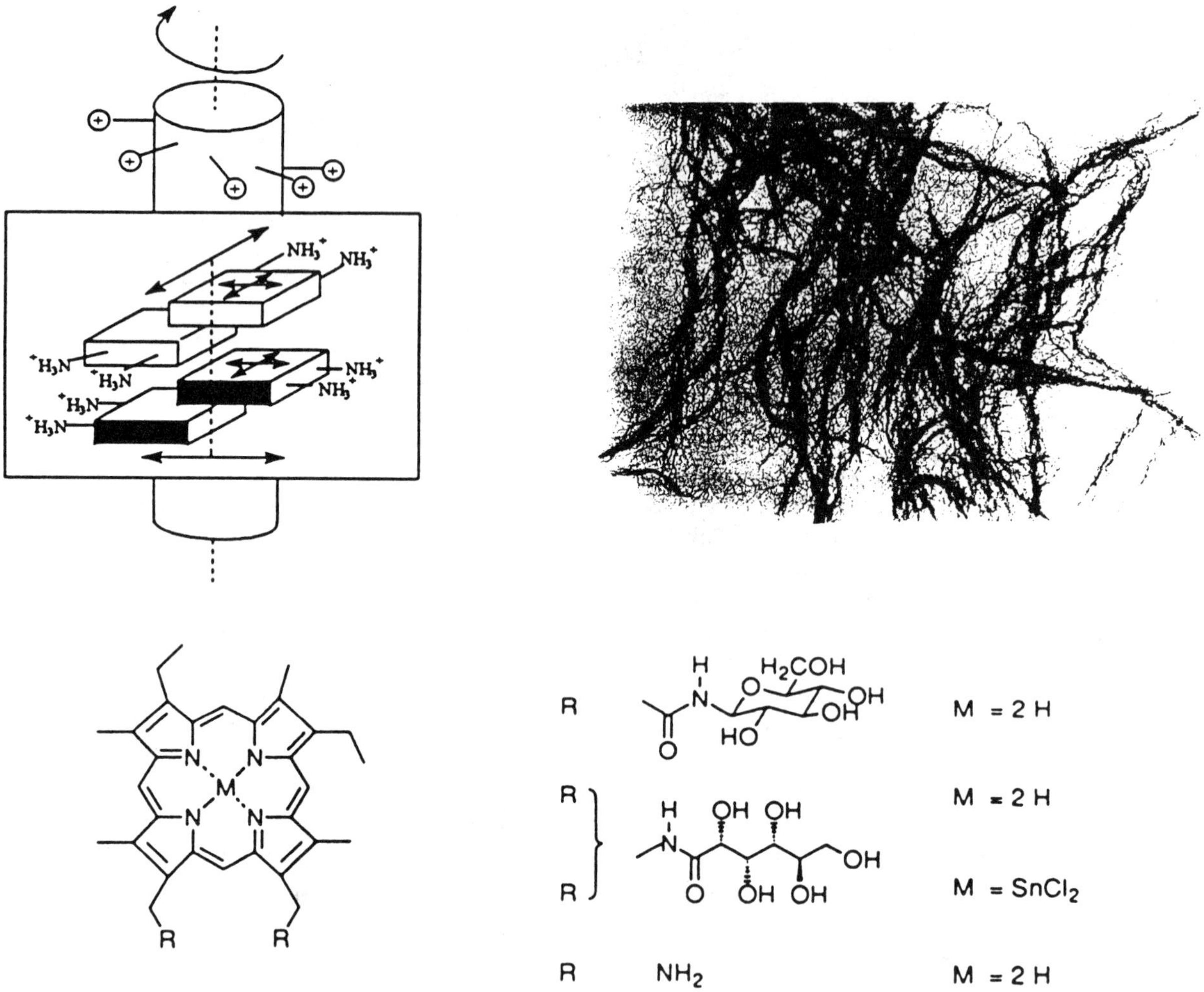

FIG. 10 TEM and model of a noncovalent, micellar porphyrin fiber typical for protoporphyrin-type amphiphiles (from Ref. 33).

VI. CONCLUSION

Spherical and fibrous micelles and vesicles have been obtained for the first time in solid form. They survive drying, which removes the repulsive hydration forces. They are kinetically stable for different reasons:

1. The spherical micelle made of the ruthenium complex **1** cannot crystallize because the interdigitating bilayers do not allow flattening of curvature upon drying. The headgroup regions are hydrated in aqueous solution, but removal of the hydration wafer headgroup does not disturb the bilayer of the ruthenium complex.
2. The spherical vesicle made of porphyrin-bixin **2** does not collapse upon removal of the entrapped water because the stiff amphiphiles cannot disentangle. A mixture of porphyrin atropisomers also prevents formation of 3D crystals.
3. The rigid gluconamide fiber keeps its shape in the dry state because the bent conformation of the headgroup is stabilized by strong intra- and intermolecular hydrogen bonds. Dehydration does not destroy the curvature.
4. The most stable of all amphiphilic assemblies are vesicular tubules, which are stabilized by van der Waals interactions between the alkyl chains, hydrogen bond chains between the headgroups, and very low cmc values below the melting point of the amide hydrogen bond chains.

The only common structural motif in all of these solid spheres, rods, and tubules is an ultrathin ($\leq$5 nm) membrane. The rigidity of the different membrane sys-

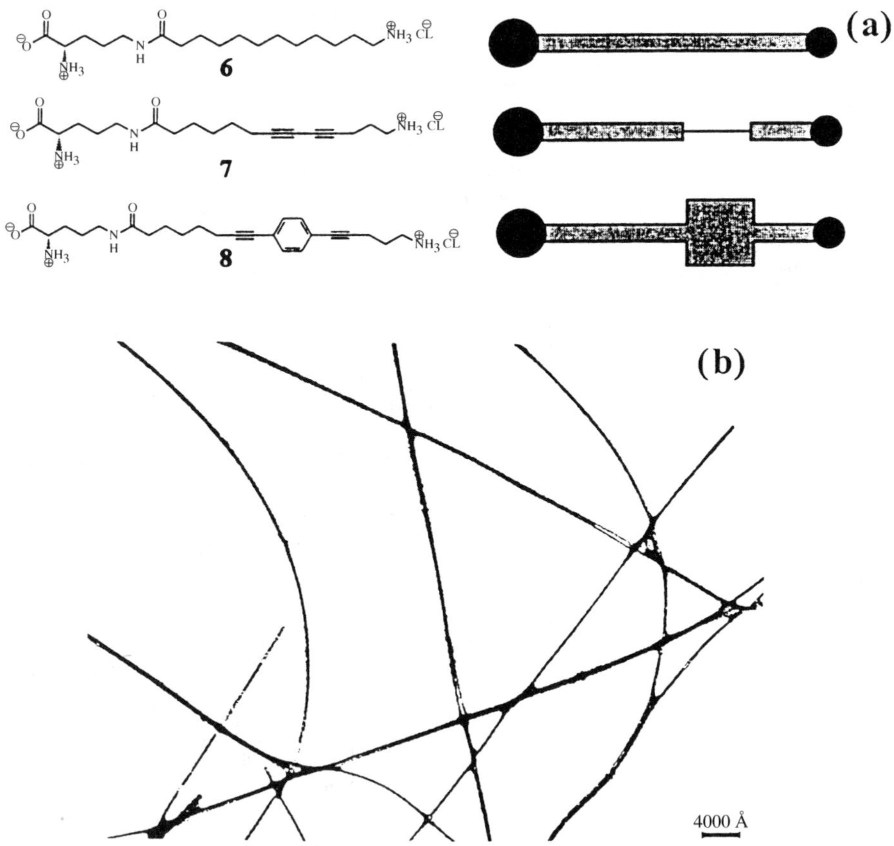

FIG. 11 TEM and model of the monolayered tubules made of bola-amphiphiles with two different headgroups, e.g., amino and amino acid groups, and secondary amide links in the hydrophobic chain. The diacetylene units lead to polmeric fibers upon UV irradiation (from Ref. 36).

tems is, however, not coupled with a uniform, crystal-like ordering. The polyene carboxylates lie in a chaotic crisscross ordering. The ruthenium complex bilayer forms concentric, interdigitated bilayers of extremely different curvature. The gluconamide quadruple helices produce exquisite regularity in pitch, fiber width multiplicity, and molecular conformation. The vesicular tubules, made of amino acid bola-amphiphiles, all have the same inner diameter of 50 ± 10 nm and the membrane thickness is exactly 4.1 nm. The mesoscopic measures of the assemblies in the given examples are usually uniform, stable, and reproducible. They compare favorably to covalent biopolymers and the best unidisperse polymers.

The dry lipid assemblies are, on the other hand, so hard that they do not dissolve anything. The selective dissolution of soft metal colloids and hard polyols, which works perfectly with fluid polymer and lipid membrane systems, is out of reach. The colloids can be used only as reactive particles in the same way as functional proteins. They need to be loaded with fitting "coenzymes" and substrates will react only on the surfaces.

Defined surface clefts have not been realized with the solid membraneous particles described here. The ruthenium complex micelle is currently mixed with corresponding osmium, platinum, and palladium analogues. The micellar fibers have been decorated on the surface with colloidal metal spheres, but no interparticle connection could be achieved. The vesicular tubules can be filled with all kinds of salts, e.g., silver nitrate. Attempts to produce continuous silver wires by reduction so far have failed.

A new approach, which combines the motifs of rigid amide hydrogen bond chains and the self-assembly of membranes in order to form enzyme-like surface gaps, is based on successive binding of flat and upright standing amphiphiles on the surface of gold electrodes or colloids. At first, a porphyrin is bound flatly on the gold surface, and this is then embedded in a rigid monolayer

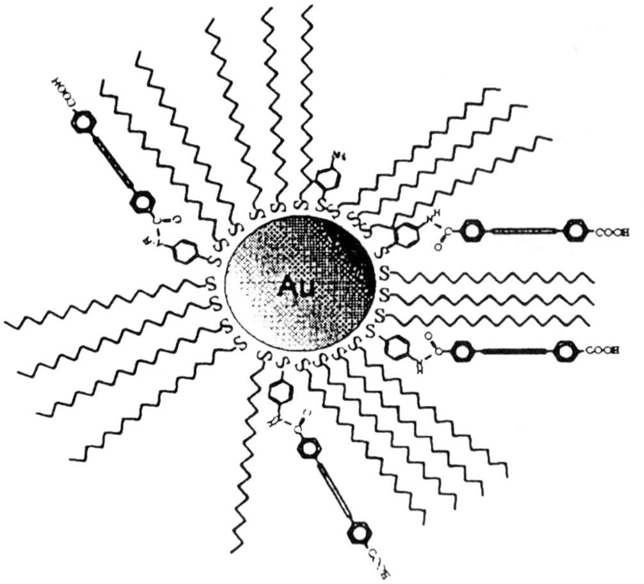

FIG. 12 Self-assembled membranes containing reactive angstrom gaps on colloidal particles of micellar size ($d = 10$–20 nm). Water- or solvent-soluble particles are obtained depending on the headgroups of the amphiphiles (from J.-H. Fuhrhop, unpublished).

containing two secondary amide group hydrogen bond chains. The resulting hydrophobic porphyrin-sized gaps take up water and bind fitting solutes in the entrapped "hydrophobic water" volume [40,41]. These reactive membrane systems can then be transferred to colloidal gold particles and other smooth colloids (Fig. 12).

REFERENCES

1. J. H. Fendler, *Membrane Mimetic Chemistry*, Wiley, New York, 1982.
2. J.-H. Fuhrhop and J. Köning, *Molecular Assemblies and Membranes*, Monographs in Supramolecular Chemistry (J. F. Stoddart, ed), Royal Society of Chemistry, London, 1994, i–xiii, 1–227.
3. L. Bachmann, W. Dasch, and P. Kutter, Ber. Bunsenges. Phys. Chem. *85*:883 (1981).
4. J. L. Burns, Y. Cohen, and Y. Talmon, J. Phys. Chem. *94*:5308 (1990).
5. J.-H. Fuhrhop and M. Baccouche, Liebigs Ann. Chem. 2058 (1976).
6. J. S. Nowick, J. S. Chen, and G. Noronha, J. Am. Chem. Soc. *115*:7636 (1993).
7. J.-H. Fuhrhop, H. H. David, J. Mathieu, U. Liman, H.-J. Winter, and E. Boekema, J. Am. Chem. Soc. *108*: 1785 (1986).
8. R. A. Walker, J. A. Gruetzmacher, and G. L. Richmond, J. Am. Chem. Soc. *120*:6991 (1998).
9. N. Kimuzuka, T. Kawasaki, K. Hirata, and T. Kunitake, J. Am. Chem. Soc. *120*:4094 (1998).
10. I. Sakurai, T. Karrabura, A. Dahurai, A. Kegami, and T. Setoi, Mol. Cryst. Liq. Cryst. *130*:203 (1985).
11. K. D. M. Harris, J. Solid State Chem. *106*:83 (1993).
12. S. Förster and M. Antonietti, Adv. Mater. *10*:195 (1998).
13. J. Tao, S. Stewart, G. Liu, and M. Yang, Macromolecules *30*:2738 (1997).
14. M. A. Hempenius, B. M. W. Langeveld-Voss, J. A. E. H. van Haare, R. A. J. Janssen, S. S. Sheiko, J. P. Spatz, M. Möller, and E. W. Meijer, J. Am. Chem. Soc. *120*: 2798 (1998).
15. L. Zhang and A. Eisenberg, J. Am. Chem. Soc. *118*: 3168 (1996).
16. Y. Yu and A. Eisenberg, J. Am. Chem. Soc. *119*:8383 (1997).
17. S. Klingelhöfer, W. Heitz, A. Greiner, S. Oestreich, S. K. Förster, and M. Antonietti, J. Am. Chem. Soc. *119*: 10116 (1997).
18. L. Balagh and D. A. Tomalia, J. Am. Chem. Soc. *120*: 7355 (1998).
19. B. M. Discher, Y. Y. Won, D. S. Ege, J. C. M. Lee, F. S. Bates, D. E. Discher, and D. A. Hammer, Science *284*:1143 (1999).
20. D. Voet, and J. G. Voet, *Biochemistry*, John Wiley & Sons, Inc., New York, 1990.
21. W. Hinrichs and W. Saenger, J. Am. Chem. Soc. *112*: 2789 (1990).
22. M. B. Bever, Encyclopedia of Materials Science and Engineering, Pergamon Press, Oxford, New York, Toronto, Sydney, Frankfurt, Vol. 6, p. 4083, 1986.
23. C. Draeger, C. Boettcher, C. Messerschmidt, L. Ruhlmann, J.-H. Fuhrhop, U. Siggel, and L. Hammarstroem, Langmuir *16*:2068 (2000).
24. H. Ohno, S. Takeoka, and E. Tsuchida, Polymer Bull. *14*:487 (1985).
25. H. Ringsdorf, B. Schlarb, and J. Venzmer, Angew. Chem. Int. Ed. Engl. *27*:113 (1988).
26. T. Komatsu, E. Tsuchida, C. Böttcher, D. Donner, C. Messerschmidt, U. Siggel, W. Stocker, J. P. Rabe, and J.-H. Fuhrhop, J. Am. Chem. Soc. *119*:11660 (1997).
27. J.-H. Fuhrhop, and S. Svenson, Stereochemistry and Separation Processes in Lipid Aggregates, in K. Kalyanasundaram and M. Grätzel (eds), Kinetics and Catalysis in Microheterogeneous Systems (Surfactant Science Series), M. Dekker Inc., New York, 273–302, 1991.
28. J. Köning, C. Böttcher, H. Winkler, E. Zeitler, Y. Talmon, and J.-H. Fuhrhop, J. Am. Chem. Soc. *115*:693 (1993).
29. J.-H. Fuhrhop and B. Rosengarten, Synlett 1015 (1997).
30. S. Svenson, B. Kirste, and J.-H. Fuhrhop, J. Am. Chem. Soc. *116*:11975 (1994).
31. S. Svenson, J. Köning, and J.-H. Fuhrhop, J. Phys. Chem. *98*:1022 (1994).

32. J.-H. Fuhrhop, C. Demoulin, C. Böttcher, J. Köning, and U. Siggel, J. Am. Chem. Soc. *114*:4159 (1992).
33. J.-H. Fuhrhop, U. Bindig, and U. Siggel, Chem. Comm. 1583 (1994).
34. J.-H. Fuhrhop, U. Bindig, and A. Schulz, New J. Chem. *19*:427 (1995).
35. A. H. Herz, Adv. Colloid Interface Sci. *8*:237 (1977).
36. J.-H. Fuhrhop, D. Spiroski, and C. Böttcher, J. Am. Chem. Soc. *115*:1600 (1993).
37. J.-H. Fuhrhop, P. Blumtritt, Lehmann, and P. Luger, J. Am. Chem. Soc. *113*:7437 (1991).
38. J.-H. Fuhrhop and C. Böttcher, J. Am. Chem. Soc. *112*: 1768 (1990).
39. J.-H. Fuhrhop, U. Bindig, and U. Siggel, J. Am. Chem. Soc. *115*:11036 (1993).
40. J.-H. Fuhrhop, L. Ruhlmann, C. Messerschmidt, W. Fudickar, J. Zimmermann, and B. Roeder, Pure & Appl. Chem. *70*:2385 (1998).
41. W. Fudickar, J. Zimmermann, L. Ruhlmann, J. Schneider, B. Roeder, U. Siggel, and J.-H. Fuhrhop, J. Am. Chem. Soc. *121*:9539 (1999).

35

Synthesis and Properties of Amphiphile-Based Gene Carriers

NILY DAN Drexel University, Philadelphia, Pennsylvania

I. INTRODUCTION

Gene therapy techniques aim to eliminate disease by correcting the cell genome, which requires introducing healthy genes that can replace damaged ones into the cell environment. Techniques for the identification and synthesis of disease-related genes are now common. However, gene therapy cannot be implemented without the development of a reliable and efficient mechanism for gene delivery into affected cells.

"Naked" genes cannot enter cells because the coil dimension of the rigid DNA is too large to allow transport through the intact cell membrane [1,2]. Moreover, cell membranes carry a net anionic charge that repels the highly charged, anionic DNA. Therefore, to transport genes into and through cells one must develop a mechanism that will both condense and invert the charge of the DNA coil [3–8].

Current techniques utilize recombinant viral vectors to deliver genes into cells. Although such carriers are highly efficient, they require extensive manipulation to eliminate toxic and immune responses. Their use in human therapy is also challenged by the need to prevent the generation of replication-competent viruses by the carrier, the limited size of genetic inserts, and their inability to target specific cell populations [3–8].

Synthetic gene carriers utilize complexes between DNA and either cationic polymers or cationic lipids. The cationic agent condenses the expanded DNA coil and reverses its anionic charge, thereby lowering the barriers for transmembrane transport. The structure and properties of synthetic gene carriers have been clearly linked to their performance in vivo [9–11]. Implementation of synthetic gene carriers requires, therefore, the ability to control their material properties during the synthesis stage.

In this chapter we examine the synthesis of *lipid*-based gene carriers, concentrating on the relationship between system parameters (e.g., lipid type, solution composition) and the properties of the DNA-lipid complex. Although this system is somewhat specialized, many aspects of its phase diagram are relevant to the synthesis of other amphiphile-based, multicomponent materials that are ordered on the nanoscale. Examples include surfactant-colloid complexes [12], surfactant-protein aggregates (O. Regev, private communication), and surfactant-polymer assemblies [13–15].

DNA-lipid complexes are formed by mixing DNA with cationic liposomes. The synthesis of such materials is based, therefore, on *self-assembly* between oppositely charged molecules as well as the inherent self-assembly tendencies of the amphiphilic lipids.

The liposomes used are typically unilamellar and contain a mixture of cationic and nonionic lipids. System parameters include, therefore, three compositional variables (DNA concentration, the ratio of DNA anionic charges to cationic lipid charges in solution, and the concentration ratio between the cationic and nonionic lipids). Other parameters include pH, salt concentration, and the valence of the solution ions.

The structures formed by DNA-lipid complexation are characterized by two length scales. One scale relates to internal ordering (generally of order 1–10 nm);

the other describes the globular or overall aggregate dimensions (0.1–1 μm). Various studies suggest that whereas the internal structure is reversible and reproducible, the globular structure is nonreversible and dependent on sample history [16–19]. Moreover, there are strong indications that carrier efficiency is linked to the internal rather than globular structure [13–21]. Therefore, we will focus the discussion in this chapter on the *internal* composition and geometry of DNA-lipid complexes.

What determines the internal structure of DNA-lipid complexes? One possibility is that each component (i.e., DNA coil and lipid liposomes) keeps its preferred form. Indeed, Felgner et al. [22] suggested that the liposomes may adhere, intact, to the DNA chain in a manner similar to that of surfactant micelles adhering to an oppositely charged polyelectrolyte [23]. Another possibility is that one component may impose its geometry on the other; this will lead to tubular DNA coated by a lipid bilayer [22] or to spherical liposomes coated by an adsorbed DNA layer. However, it is well known that mixing amphiphiles can lead to the formation of new phases (e.g., equilibrium vesicles [24]). Mixing lipids with DNA may result in similarly unexpected structures.

Several types of DNA-lipid complexes have been observed, as sketched in Fig. 1. The complex structures were shown to depend on the choice of the cationic and nonionic lipids as well as the different composition parameters [16,20,21,25–34]. In Section II we review the different aggregate types and their characteristics. In Section III we discuss the DNA-lipid phase diagram, and we conclude with a brief summary in Section IV.

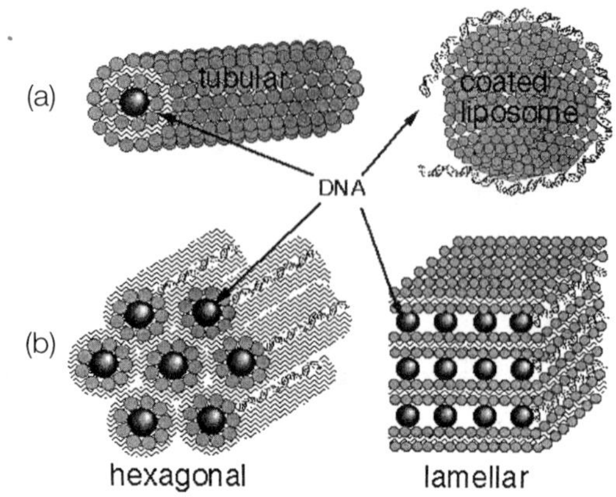

FIG. 1 The different types of complexes observed for DNA-liposome aggregates. (a) Nonequilibrium structures. Coated liposomes are formed when the anionic DNA adsorbs onto the exterior of the cationic liposome [18,19]. Tubular aggregates form when the DNA is encapsulated within a lipid bilayer [9,33,34]. (b) Equilibrium structures. Multilamellar structures obtain when ordered DNA layers alternate, in a stack conformation, with cationic bilayers. The DNA layers are characterized by a finite spacing between adjacent DNA units, which is, as a rule, somewhat larger than close packing (see Fig. 2) [16,25–28]. Inverted hexagonal structures are formed when DNA molecules, encapsulated by a lipid monolayer, order in a hexagonal array [20,21,29].

II. DNA-LIPID AGGREGATES: GEOMETRY AND CHARACTERISTICS

The transfection properties of lipid-based gene carriers have been investigated for a couple of decades [3–8]. Lately, attention has focused on systematic investigation of their phase diagram and the relationship between structure and performance [25–27].

Several experimental methods have been used to investigate the structure of DNA-lipid complexes. These include freeze-fracture and cryoelectron microscopy, which provide direct images on scales of 10 mm to 1 μm; high-resolution synchrotron x-ray diffraction measurements (SAXS), which can identify order on the 1–10 nm scale; and fluorescence and circular dichroism, which enable probing on the molecule scale.

As will be shown presently, the structure of DNA-lipid complexes depends not only on the system compositional parameters but also on the type of lipid mixture. Systems investigated include the cationic lipids *N*-(1-(2,3-dioleoyloxy)propyl)-*N,N,N*-trimethylammonium chloride (DOTAP), *N*-(1-(2,3-dioleyloxy)propyl)-*N,N,N*-trimethylammonium chloride (DOTMA), 3β-[N-(*N'*,*N'*-dimethylaminoethane)cabamoyl]-cholesterol (DC-Chol), *N,N*-dioctadecyl-lysineamine (lipid 43), *N'*,*N'*-dioctadecyl-1,2,6-triaminohexane (lipid 47), 1,2-dimyristoyloxypropyl-3-dimethyl-hydroxyethylammonium bromide (DMRIE), and 2'-(1'',2-dioleoyloxypropyldimethylammonium bromide)-*N*-ethyl-6-amidospermine tetratrifluoro-acetic acid (DOSPA), all of which are bilayer forming. The nonionic lipids used were 1,2-dioleoyl-*sn*-glycero-3-phosphatydylcholine (DOPC), which is bilayer forming, and 1,2-dioleoyl-*sn*-glycero-3-phosphatylethanolamine (DOPE), which forms inverted hexagonal phases. In some cases the liposomes also contained cholesterol (Chol), hexanol, monooleoyglycerol (MOG), or the zwitterionic 1,2-dimyristoyl-*sn*-glycero-3-phosphocholine (DMPC).

A. DNA Complexes with Unperturbed Liposomes

A common architecture of polyelectrolyte complexes with oppositely charged micelles is one where the micelles are attached, by electrostatic attraction, to the polyelectrolyte coil [23]. The similarities between this polyelectrolyte-micelle system and DNA-liposome systems have led to speculation that the latter may display such configurations [22]. To date, however, such arrays have not been observed.

Another type of assembly between relatively unperturbed liposomes and DNA is one where the DNA adsorbs and coats the exterior of the liposome, as sketched in Fig. 1. Such structures have been inferred from zeta potentials [17] or directly observed by electron microscopy [18,19] in several different lipid systems, such as liposomes containing DMRIE, lipid 43, lipid 47 or DOSPA [17], DOTAP-DOPE and DOTAP-Chol mixtures [18], or DMPC–DC-Chol mixtures [19].

B. Tubular Lipid-DNA Structures

Tubular DNA-lipid structures are structures in which the DNA is coated by a lipid bilayer, as sketched in Fig. 1. Such structures, characterized by a 7 nm diameter, were first identified in freeze-fracture electron micrographs by Sternberg et al. [33]. This diameter corresponds exactly to the DNA diameter plus the width of a bilayer. The liposomes used contained mixtures of DC-Chol and DOPE, and the DNA-to-lipid charge ratio was varied over a relatively large range. In systems that were allowed to incubate for long periods of time or in which the DNA concentration was high, globular objects of order 100 nm were found to attach to the tubular DNA complexes. The authors assumed that the globules were unperturbed cationic liposomes attached to the coated DNA strand either by electrostatic interactions or by fusion [33].

Tubular DNA aggregates were found in other systems as well. Hui et al. [9] observed, using freeze-fracture electron microscopy, tubular-fused globule arrays in liposomes containing either DOPC or DOPE. Xu et al. [34] observed tubelike aggregates whose diameter was 10 nm, partially adsorbed or extending from the surface of fused globules, in systems containing DOTAP-MOG, pure DOTAP, DOTAP-DOPC, or DOTAP-DOPE. However, the tubular aggregates were extremely short except in the case of DOTAP-DOPE complexes.

C. (Inverse) Hexagonal Arrays

Other experiments identified hexagonal (H^{C}_{ii}) DNA-lipid phases [20,21]. In this type of aggregate (see Fig. 1) DNA strands, coated by a lipid *monolayer*, are packed into a dense inverted hexagonal array. The hexagonal phases were observed in two types of systems: (1) liposomes containing a mixture of DOTAP and DOPE [20,21,29] and (2) liposomes containing a high ratio of hexanol to DOTAP-DOPC mixtures [20,21].

Circular dichroism studies conducted by Zuidan et al. [29] for DNA complexes with DOTAP-DOPE liposomes showed that DNA molecules in these complexes display both secondary and tertiary structure transitions, interpreted as the formation of tightly packed phases characterized by long-range chiral order. They [29] concluded that in these aggregates most of the DNA is packed into an inverted hexagonal lipid phase, which imposes long-range order and spatial organization.

D. Lamellar Lipid-DNA Phases

One of the most common phases obtained by the synthesis of DNA-lipid aggregates is that of lamellar aggregates, where two-dimensional liquid crystalline arrays of DNA are sandwiched between cationic bilayers (see Fig. 1). This type of aggregate may seem the most obvious way of packing lamellar sheets with rodlike objects while allowing each component to keep its original geometry. However, it was not clearly identified until recently [25–28].

To date, lamellar phases have been observed in DNA mixtures with cationic liposomes containing pure DOTAP [25–28,34] or mixtures of DOTAP and DOPC at all ratios at which the two lipids mix [20,21]. Using DOTAP-DOPE liposomes led to the formation of lamellar arrays when the DOPE concentration in the liposome was low or moderate [20,21,31]. Multilamellar aggregates were also observed using cryomicroscopy in systems of DMPC–DC-Chol at low DNA-to-lipid charge ratios [19].

What is the configuration of DNA in the multilamellar aggregates? SAXS experiments [26–28] found that the spacing between adjacent bilayers was equivalent to the diameter of a DNA molecule, suggesting that the DNA is flatly adsorbed to the lamellar surfaces. This is in agreement with experiments of D. Hirsch-Lerner and Y. Barenholz (private communication) that show strong dehydration upon complex formation.

DNA was found to order in a two-dimensional, liquid crystalline array between the multilamellar bilayers (see Fig. 1). The DNA array was characterized by a finite spacing limited to a relatively narrow range, between close packing (2.5 nm) and slightly larger values (6 nm), regardless of system parameters [19,25–28]. Similar configurations with similar DNA spacing were found in DNA adsorption on a single, supported cationic bilayer [35].

The DNA spacing in multilamellar aggregates was found to vary with the ratio of anionic DNA charges and cationic lipid charges in the solution, as sketched in Fig. 2 [21,26–28]. The DNA packing was found to decrease rapidly with the DNA-to-lipid charge ratio in a narrow region around the isoelectric point (where the number of DNA charges in solution equals that of the cationic lipids). On either side of this transition region, the DNA spacing remained relatively constant [21,26–28]. The width of the transition region was found to vary as a function of the salt concentration and fraction of nonionic lipid [21]. Only complexes obtained from solutions at exactly the isoelectric point were charge neutral. Complexes obtained at any other solution composition were shown to be either undercharged or overcharged [21,26–28]. In systems where the cationic lipids were in excess, fluorescence data showed the presence of many lamellar defects, indicating boundaries between DNA-rich regions and bare bilayer one [31].

The invariance of the DNA spacing as a function of DNA-to-lipid charge ratio in the limit of high DNA content can be interpreted as a saturation process, although one might have expected that the saturated structures would be charge neutral rather than overcharged. However, the constant DNA spacing in the limit of low DNA content and the coexistence between bare bilayer regions and regions containing complexed DNA indicate that, in this limit, there is a preferred DNA spacing that is set by the system parameters.

The effect of (monovalent) salt concentration on the DNA was also examined [21]. In isoelectric systems

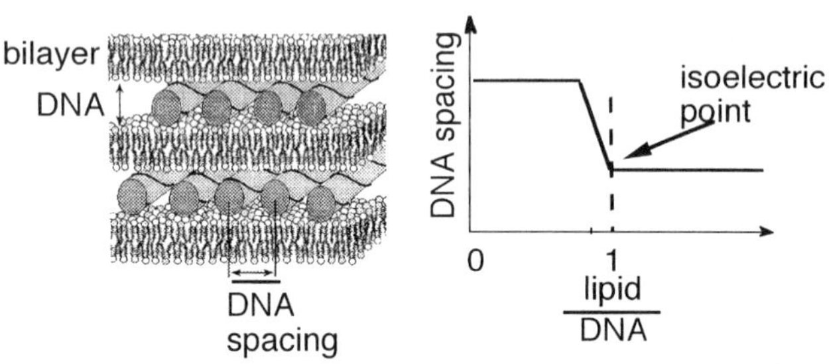

FIG. 2 The effect of the DNA-to-cationic lipid charge ratio on the spacing of DNA in multilamellar aggregates [21,25–28]. The spacing is defined as center to center and is in most cases slightly larger than close packing (2.5 nm). At a low DNA-to-lipid charge ratio (namely, when the liposomes are in excess), the DNA spacing is usually of order 6 nm and does not decrease when more DNA is added to the solution. Around the isoelectric point, where the DNA-to-lipid charge ratio is unity, a sharp decrease is observed. The transition region ends somewhat above the isoelectric point. Beyond this region, the DNA spacing does not decrease any further. The lower value of the DNA spacing is equal to close packing only when the lipid bilayer is purely cationic. In mixed cationic-nonionic bilayers, this spacing is usually of order 4 nm.

with a high cationic-to-nonionic lipid ratio, the DNA spacing was found to increase strongly with salt concentration. This increase was associated with release of DNA from the complex into solution [21]. However, quite unexpectedly, in an isoelectric system in which the nonionic lipid content was high, the DNA spacing was found to vary *nonmonotonically* with salinity, slightly increasing and then decreasing with salt [21].

III. DNA-LIPOSOME PHASE DIAGRAM

DNA-lipid complexes are multicomponent systems. Experiments show that their structure and properties depend on the type of lipids used as well as three compositional parameters: the ratio of anionic DNA to cationic lipid charges, the concentration ratio of cationic to nonionic lipid, and the salt concentration. Several theoretical models examined different aspects of the DNA-lipid phase diagram and complex properties [36–43]. The analysis is complicated by the large number of variables and the dominant effect of the electrostatic interactions, some of which are, to date, not well understood. We therefore focus here only on the main features in qualitative terms.

A. DNA-Lipid Interactions

The details of the DNA-lipid phase diagram cannot be understood without examining the forces driving the complex formation. Generally, salts form when the Coulomb attraction between oppositely charged ions overcomes the entropy of the individual ions. In small molecules dissolved in water, entropy usually wins and complete dissociation ensues [44]. However, the electrostatic attraction between highly charged macroions (such as polyelectrolytes or colloidal particles) and their counterions can overcome, to some degree, the counterion entropy. As a result, a fraction of the counterions condenses, namely remains in the vicinity of the macroion [44,45]. Complexation between two oppositely charged macroions releases some of their condensed counterions, resulting in a gain in system entropy. This gain is especially high for DNA because approximately three fourths of its counterions are condensed [45].

This qualitative description of macroion association (or precipitation) seems to indicate that such complexes would always prefer to be charge neutral. Yet, the phenomenon of charge inversion is well known in adsorbed polyelectrolyte layers [46,47] and polyanion-polycation complexes. Similarly, DNA-lipid complexes are rarely charge neutral [20,21,25–28].

Why are charge-neutral complexes unstable? This instability is due to the fact that the degree of entropy loss of the condensed ions increases with increasing macroion charge density. For example, the thickness of the Gouy-Chapman layer near flat surfaces, which contains the condensed ions, decreases with increasing surface charge density [44]. Similarly, the fraction of counterions condensed on a rigid rod increases with the line charge density [45]. As a result, the system prefers not to form charge-neutral complexes coexisting with highly charged macroions where the condensed counterions are closely trapped. Instead, it forms aggregates with some excess charge that can redistribute over a large area, thereby reducing the overall charge density and counterion condensation penalty [36,37]. Although this analysis is obviously oversimplified, it explains qualitatively why equilibrium charge-neutral DNA-lipid complexes are unstable and may be obtained only in isoelectric solutions in which the lipid-to-DNA charge ratio is unity [21,26–28].

B. DNA-Lipid Phases

Four types of DNA-lipid structures have been observed, as sketched in Fig. 1. The experiments suggest that two of these are nonequilibrium, namely the coated liposomes [18,19] and the tubular ones [9,33,34]. Equilibrium phases include the inverted hexagonal and multilamellar arrays.

DNA-coated liposomes seem to be the precursor, or initial stage, in the formation of dense phases. Electron micrographs show that the DNA either induces invagination or leads to coalescence of liposomes. In either case, the formation of multilayer complexes ensues [18,19], as can be seen in Fig. 3.

The role of tubular aggregates is somewhat less clear. It is possible that they are an equilibrium structure that is difficult to identify, given the techniques commonly used to examine these systems. However, it is more likely that the tubular aggregates are a metastable state accommodating portions of DNA strands that either "dangle" from multilamellar aggregates or connect two such aggregates. Indeed, tubular aggregates are always observed near dense, spherical objects of order 100 nm to 0.1 μm [9,33,34], as can be seen in Fig. 3. Sternberg et al. [33] speculated that these objects are intact liposomes. However, the multilamellar aggregates observed by Radler et al. [26] are also spherical objects of the same order of magnitude and are connected by strands whose diameter is much smaller than the sphere dimensions. The speculation that tubular aggregates are metastable is supported by theoretical models that predict that these are unstable when compared with either lamellar [40] or hexagonal [41] arrays but should be preferred to mixtures of naked DNA and bare lipid bilayers.

The two types of equilibrium DNA-lipid complexes identified to date are inverted hexagonal and lamellar

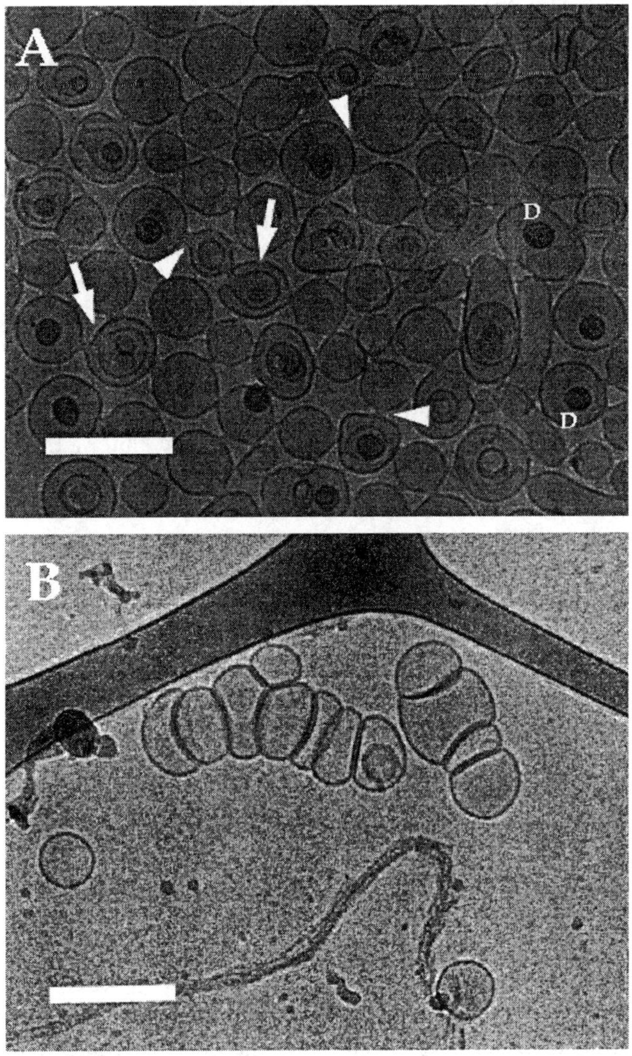

FIG. 3 Cryoelectron micrographs of DNA-lipid complexes (D. Danino and Y. Talmon, unpublished). The liposomes are composed of a mixture of 1:1 mole ratio DOTAP to DOPE. The bar represents 0.2 μm. (A) DNA-to-cationic lipid charge ratio 1:10. The DNA, which is seen here as darker, thicker lines, adsorbed on the liposome exterior, thereby inducing invagination, which may lead, with time, to the formation of intra-multilayered structures. (B) DNA-to-cationic lipid charge ratio 1.5. The DNA adsorbs onto the exterior of the liposomes, thereby leading to adhesion between neighboring liposomes. This type of aggregate may evolve, with time, to an inter-multilayer structure.

complexes. In the hexagonal aggregates, the DNA is encapsulated within an array of lipid monolayers [20,21,29]. In the multilamellar aggregates, ordered DNA layers alternate with bilayers [16,25–28].

What determines the type of equilibrium aggregate? The experimental evidence clearly shows that the type of equilibrium DNA-lipid complex formed is mostly dominated by the lipid properties [9,16–34]. Yet, how does the lipid type affect the equilibrium structure?

Comparing the packing of DNA in the hexagonal versus lamellar arrays (Fig. 1), we see that in the former the DNA can be in much closer contact with the cationic lipids than in the latter. This close contact is expected to optimize the electrostatic interactions, which are very sensitive to the distance between opposing charges [44]. However, wrapping the DNA around in the hexagonal phase may be associated with a high bending penalty that may become prohibitive in the case of lipids that form lamellae.

As a result, one would expect that hexagonal phases would be favored when the lipids used tend to form hexagonal phases. Indeed, such phases were observed when the lipid mixture contained a high fraction of a nonionic lipid known to form, in its pure phase, hexagonal arrays [20,21,29]. Complexes between DNA and lipid mixtures that contain a large fraction of lamella-forming cationic lipids or in which both cationic and nonionic components form lamellae are multilamellar [25–34].

Hexagonal phases are also likely to form in systems where the (monolayer) bending energy is relatively low, so that the energetic gain associated with the close lipid-DNA contact overcomes the penalty for monolayer bending [43]. The experiments of Koltover et al. [20,21] show that, as expected, reducing the lipid monolayer bending modulus leads to the formation of hexagonal phases, even though the preferred lipid phase is lamellar.

C. DNA Packing in Lamellar Aggregates

The density or spacing of DNA in lamellar aggregates is closely related to aggregate stability and gene transfer activity [3–11,20,21,25–28]. As a result, understanding the parameters controlling the DNA packing in multilamellar aggregates is not only of scientific interest but also of technological importance.

What sets the DNA spacing? In charge-neutral complexes it is obviously controlled by the fraction of cationic lipid in the complex. Indeed, Radler et al. [26,28] and Koltover et al. [21] found that the DNA spacing in charge-neutral complexes that are formed from isoelectric solutions consistently increased with the fraction of nonionic lipid in the mixture.

As discussed earlier, charge-neutral aggregates are unstable except in isoelectric solutions [36,37]. Indeed, the experiments show that around the isoelectric point the DNA spacing in the multilamellar complex varies sharply as a function of DNA-to-lipid charge ratio [21,26–28]. However, on both sides of this transition region the DNA spacing was found to be mostly independent of the DNA content, regardless of the type of lipid mix or salinity. This behavior seems to indicate that the system's energy is optimized when the complexes are at a specific degree of undercharging or overcharging, regardless of the amount of excess DNA or cationic lipid in the system.

Using a Poisson-Boltzmann mean field model for the system electrostatics, Bruinsma [36] and Bruinsma and Mashl [37] showed that the charge-neutral complex is unstable to the addition of either excess DNA or excess lipid. This instability is due to the complexation mechanism discussed earlier, namely that the overall energy of a system containing charge-neutral complexes and highly charged (excess) DNA or bilayers is higher than that of an overcharged or undercharged complex with a lower charge density. The Bruinsma model [36,37] predicts a spacing versus composition plot qualitatively similar to those observed experimentally. It should be noted, however, that although the model is based on the assumption that the interrod spacing is much larger than the rod diameter, the experiments [25–28] found that the interrod spacing never exceeded three times the DNA diameter.

The dependence of DNA spacing on the ionic strength was also investigated experimentally [21]. In general, increasing the salt concentration should increase the screening in the system, thereby reducing the penalty associated with counterion condensation and, thus, the driving force for complex formation. On the other hand, added salt also screens the inherently repulsive DNA-DNA interactions, which should allow the rods to pack more closely [50,51]. Koltover et al. [21] found that increasing the salt concentration generally leads to a sharp decrease in the DNA spacing, indicating that the DNA spacing is affected by the repulsive DNA-DNA interactions. More surprising, though, is their observation that at low salt concentrations the DNA spacing in systems in which the DNA content is low *increases* with salt [21]. A qualitatively similar trend was observed by Fang and Yang [35], who found that the spacing of DNA adsorbed on a single purely cationic bilayer increased with salt.

The Bruinsma [36,37] model cannot explain the dependence of the DNA spacing on the salt concentration because it was developed for the limit where no salt is present. Using a simple mean field model that can incorporate the effects of salinity, Dan [38,39] suggested that the DNA spacing in the lamellar aggregates is set, in the limit of low DNA concentration, by a balance between an attractive and a repulsive DNA-DNA interaction. The repulsive interaction is due to the electrostatic repulsion between similarly charged rods [50,51]. The attractive interaction was assumed to arise from the DNA-induced perturbation of the equilibrium bilayer packing. This assumption has been verified, to some degree, by the experiments of Hirsch-Lerner and Barenholz [31], who found that complexation with DNA perturbs the equilibrium packing of the lipids.

The Dan model predicts that there is an optimal spacing (in the limit of low DNA-to-lipid charge ratio) that is set by the balance between these two interactions [38,39]. This optimal spacing would remain constant upon increasing the DNA content until the entire lipid bilayer is occupied. Above this point, additional DNA will continue to adsorb due to the high entropy gain associated with complexation, thereby reducing the spacing and leading to an instability near the isoelectric point. Although the Dan model cannot explain overcharging above the isoelectric point, it does predict that in the limit of low DNA concentrations the DNA spacing should first increase and then sharply decrease with added salt [39], as indeed observed by Koltover et al. [21]. This counterintuitive result indicates that, although electrostatics dominate lamellar complexes in systems where the DNA charge concentration is similar to or exceeds that of the cationic lipids, membrane perturbation is important in the opposite limit where the DNA content is low.

IV. SUMMARY

The study of DNA-lipid complexes is of interest not only because of their technological application as gene carriers but also because they provide a model system for studying complex formation in charged, multicomponent, self-assembling systems. Understanding the forces dominating such assemblies will facilitate developing methods for the synthesis of novel materials characterized by ordered domains on the nanoscale.

Analysis of the DNA-lipid system shows that the driving force for complexation is the release of counterions that are condensed on the highly charged DNA [45] or near the bilayer surface [42,43]. As a result, charge-neutral complexes are *unstable* and, in fact, may be obtained only from isoelectric mixtures where the ratio of the DNA anionic charges to lipid cationic charges in solution is unity [21,26–28]. This observation has significant practical implications because to target a specific cell population the net charge of the gene carrier must be small [49]. The instability of the charge-neutral complexes means that it is unlikely that lipid-based gene complexes may be used in systems where targeting is required.

The counterion release mechanism for the formation of aggregates between oppositely charged macroions is not limited to the DNA-liposome system. It has been found to lead to blisters in membrane adhesion to surfaces [52] and the formation of colloidal rafts on oppositely charged vesicles [12].

Two types of equilibrium complexes have been identified to date: inverted hexagonal and multilamellar (Fig. 1). The type of equilibrium structure is set by the lipid properties. Therefore, complexes formed with a moderate to high content of hexagonal phase forming lipids are hexagonal, and those composed of bilayer-forming lipids are lamellar [21]. Hexagonal complexes may also be obtained in systems where the bilayer bending modulus has been greatly reduced, for example, by introducing a cosurfactant such as cholesterol [21].

Two other types of complexes have also been observed: tubular aggregates and coated liposomes (Fig. 1). Tubular aggregates are a metastable state in which dangling DNA ends that cannot be incorporated into lamellar complexes are coated by a lipid bilayer rather than remaining naked (tubular). Coated liposomes, with DNA adsorbed to the exterior of liposomes, are an intermediate state that leads, with time, to the formation of lamellar or hexagonal phases (coated), as shown in Fig. 3.

The DNA spacing in lamellar complexes determines the net complex charge as well as the size or number of genes that a complex may carry. Experiments show that the DNA spacing is fixed for either large or low DNA content in the system except for a narrow transition region near the solution isoelectric point [26–28]. DNA spacing increases, for any solution composition, with increasing fraction of nonionic lipids [21]. However, increasing the salinity led to nonmonotonic behavior in the low DNA concentration limit. Analysis indicates that in solutions near and above the isoelectric point, DNA spacing is set by the system electrostatics [36,37], and in the limit of low DNA content membrane perturbation energy may play a role [38,39].

ACKNOWLEDGMENTS

Thanks to D. Danino for helpful discussions and for sharing her unpublished data (with Y. Talmon). The support of NSF-BES 0096004 is acknowledged.

REFERENCES

1. G. Maret and G. Weill, Biopolymers *22*:2727–2744 (1983).
2. S. Smith and A. J. Bendich, Biopolymers *29*:1167–1173 (1990).
3. V. A. Bloomfield, Curr. Opin. Struct. Biol. *96*:334–340 (1996).
4. P. L. Flegner, ed., *Artificial Self-Assembly Systems for Gene Delivery*, ACS Conference Proceedings Series, American Chemical Society, Washington, D.C., 1996.
5. E. Marshall, Science *269*:1050–1055 (1995).
6. C. P. Hodgson, G. Xu, F. Solaiman, and M. A. Zink, J. Mol. Med. *75*:249–258 (1997).
7. D. D. Lasic, *Liposomes in Gene Delivery*, CRC Press, Boca Raton, FL, 1997.
8. J. P. Behr, Bioconj. Chem. *5*:382–389 (1994).
9. S. W. Hui, M. Langner, Y. L. Zhao, P. Ross, E. Hurley, and K. Chan, Biophys. J. *71*:590–599 (1996).
10. B. Strenberg, K. Hong, W. Zheng, and D. Papahadjopoulos, BBA (Biochimica et Biophysica Acta) *1375*: 23–35 (1998).
11. M. I. Angelova, N. Hristova, and I. Tsoneva, Eur. Biophys. J. *28*:142–150 (1999).
12. H. Aranda-Espinoza, Y. Chen, N. Dan, T. C. Lubensky, P. Nelson, L. Ramos, and D. A. Weitz, Science *285*: 394–397 (1999).
13. M. Swanson Vethamuthu, P. L. Dubin, M. Almgren, and Y. J. Li, J. Colloid Interface Sci. *186*:414–419 (1997).
14. M. Macdonald, Colloid Surf. A Phys. Eng. *147*:115–131 (1999).
15. E. Sokolov, F. Yeh, A. Khokhlov, V. Y. Grinberg, et al., J. Phys. Chem. B *102*:7091–7098 (1998).
16. N. J. Zuidan, D. Hirsch-Lerner, S. Margulies, and Y. Barenholz, BBA *1419*:207–220 (1999).
17. S. J. Eastman, C. Siegel, J. Tousignant, A. E. Smith, S. H. Cheng, and R. K. Sheule, BBA *1325*:41–62 (1997).
18. N. Smyth-Templeton, D. D. Lasic, P. M. Frederik, H. H. Strey, D. D. Roberts, and G. N. Pavlakis, Nat. Biotech. *15*:647–652 (1997).
19. S. Huebner, B. J. Battersby, R. Grimm, and G. Cevc, Biophys. J. *76*:3158–3166 (1999).
20. I. Koltover, T. Salditt, J. O. Radler, and C. R. Safinya, Science *281*:7881 (1998).
21. I. Koltover, T. Salditt, J. O. Radler, and C. R. Safinya, Biophys. J. *77*:915–924 (1999).
22. P. L. Felgner, Y. J. Tsai, L. Sukhu, C. J. Wheeler, M. Manthorpe, J. Marshall, and S. H. Cheng, Ann. N. Y. Acad. Sci. 126–139 (1995).
23. B. Cabane and R. Duplessix, J. Phys. Fr. *48*:651–657 (1987).
24. E. W. Kaler, K. L. Herrington, A. K. Murthy, and J. A. N. Zasadzinski, J. Phys. Chem. *96*:6698–6707 (1992).
25. D. D. Lasic, H. Strey, M. C. A. Stuart, R. Podgornik, and P. M. Frederik, J. Am. Chem. Soc. *119*:832–834 (1997).
26. J. O. Radler, I. Koltover, T. Salditt, and C. R. Safinya, Science *275*:810–814 (1997).
27. C. R. Safinya, I. Koltover, and J. Raedler, Curr. Opin. Colloid Interface Sci. *3*:69–77 (1998).
28. J. O. Radler, I. Koltover, A. Jamieson, T. Salditt, and C. R. Safinya, Langmuir *14*:4272–4283 (1998).
29. N. J. Zuidan, Y. Barenholz, and A. Minsky, FEBS Lett. *457*:419–422 (1999).
30. N. J. Zuidam and Y. Barenholz, Int. J. Pharmacol. *183*: 43–46 (1999).
31. D. Hirsch-Lemer and Y. Barenholz, BBA *1370*:17–30 (1998) (probing).
32. N. J. Zuidam and Y. Barenholz, BBA *1368*:115–128 (1998) (electron and structural).
33. B. Sternberg, F. L. Sorgi, and L. Huang, FEBS Lett. *365*:361–366 (1994).
34. Y. Xu, S. W. Hui, P. Frederik, and F. C. Szoka, Biophys. J. *77*:341–353 (1999).
35. Y. Fang and J. Yang, J. Phys. Chem. *101*:441–449 (1997).
36. R. Bruinsma, Eur. Phys. J. B *4*:75–88 (1998).
37. R. Bruinsma and J. Mashl, Europhys. Lett. *41*:165–170 (1998).
38. N. Dan, Biophys. J. *71*:1267–1272 (1996).
39. N. Dan, Biophys. J. *73*:1842–1846 (1997).
40. N. Dan, BBA *1369*:34–38 (1998).
41. D. Harries, S. May, W. M. Gelbart, and A. BenShaul, Biophys. J. *75*:159–173 (1998).
42. S. May and A. BenShaul, Biophys. J. *73*:2427–2440 (1997).
43. K. Wagner, D. Harries, S. May, V. Kahl, J. O. Radler, and A. Ben-Shaul, Biophys. J., in press.
44. S. A. Safran, *Statistical Thermodynamics of Surfaces, Interfaces and Membranes*, Addison-Wesley, Reading, MA, 1994.
45. G. Manning, Q. Rev. Biophys. *11*:179–246 (1978).
46. J. F. Joanny, Eur. Phys. J. B *9*:117–122 (1999).
47. S. A. Sukhishvili and S. Granick, J. Chem. Phys. *109*: 6861–6868 (1998).
48. A. V. Kabanov, T. K. Bronich, V. A. Kabanov, K. Yu, et al. Macromolecules *29*:6797–6802 (1996).
49. D. V. Schaffer and D. A. Lauffenburger, J. Biol. Chem. *273*:28004–28009 (1998).
50. R. Podgornik, D. C. Rau, and V. A. Parsegian, Biophys. J. *66*:962–997 (1994).
51. H. H. Strey, V. A. Parsegian, and R. Podgornik, PRL *78*:895–898 (1997).
52. J. Nardi, R. Bruinsma, and E. Sackmann, Phys. Rev. E *58*:6340–6347 (1998).

36

Synthesis of Microporous Materials from Reverse Micelles

RAMSHARAN SINGH, MARIO CASTAGNOLA, and PRABIR K. DUTTA The Ohio State University, Columbus, Ohio

I. INTRODUCTION

This chapter discusses the research progress that has been made on synthesis of microporous materials from reverse micelles. Although materials synthesis, especially the synthesis of controlled-size nanoparticles from reverse micelles, has been extensively studied, the synthesis of microporous materials with controlled porosity from reverse micelles is a relatively new effort. The synthesis of microporous materials is very sensitive to the reactant composition and hence presents a departure from the commonly practiced precipitation reactions carried out in reverse micelles. As reported here, the influence of the reverse micellar environment on synthesis of microporous materials is found to be quite profound. These effects include control of crystal growth pathways, inability to crystallize open-pore frameworks in certain reverse micelles, morphology control, and seeded crystal growth.

There are five sections in this chapter. In Section I, we discuss basic features of microporous materials, with particular emphasis on their synthesis. The nature of reverse micelles is briefly described and typical syntheses done in this medium are described. Section II focuses on the synthesis of zincophosphate sodalite from AOT reverse micelles. Minor variations in reactant composition lead to significant alterations of crystal growth pathways, and a description of these results forms the major thrust of this section. Section III discusses the reasons why open-pore structures such as zincophosphate X cannot be synthesized from AOT reverse micelles. Section IV discusses the synthesis of the zincophosphate X framework from cationic reverse micelles, in particular using the two-tailed surfactant dioctyldimethylammonium chloride (DODMAC). The use of DODMAC-based reverse micelles has opened up the range of microporous frameworks that can be synthesized. Small zincophosphate clusters (~15 nm) formed in DODMAC reverse micelles also provide excellent seed crystals for zincophosphate growth. Section V summarizes the current status of the field and points out potential new developments in this area.

A. Microporous Materials

Microporous materials include a large group of solids of varying chemical composition as well as porosity. These materials are characterized by channels and cavities of molecular dimensions. The framework structure is made up of interconnecting T—O—T′ bonds, where T and T′ can be Si, Al, P, Ga, Fe, Co, Zn, B, and a host of other elements [1]. Materials with Si—O—Al bonding in the framework are called zeolites and are extensively used in many applications [2]. The presence of Al in the zeolite framework leads to ion-exchange capabilities. In cases in which the ion-exchange sites are satisfied by hydrogen ions, the zeolites show remarkably strong superacid-type behavior [3]. Figure 1 shows the framework structures of some of the most extensively studied zeolites. The two frameworks that are of most relevance to this study are zeolite X and sodalite. Table 1 lists characteristic features of some zeolites, including their void volume and

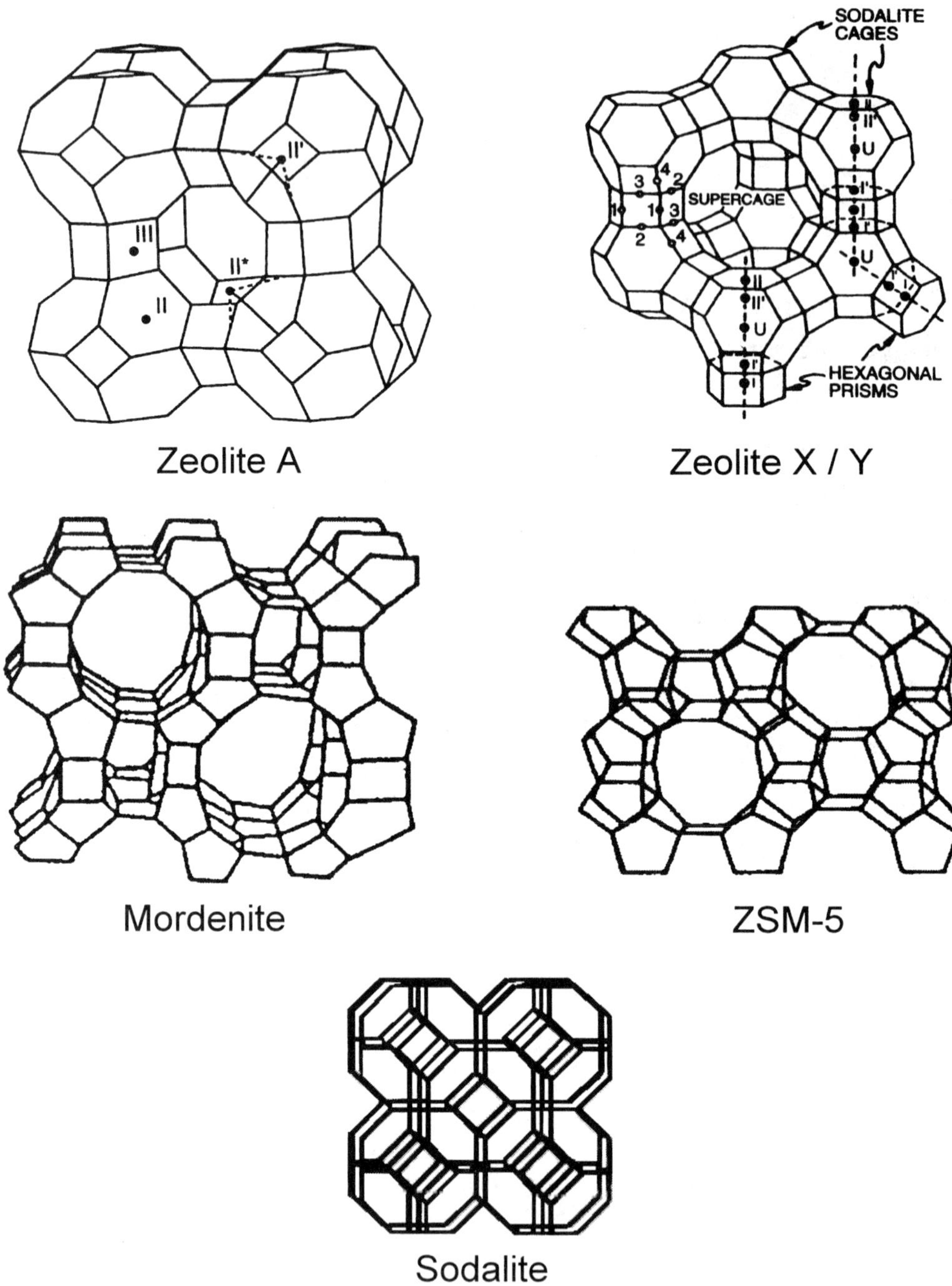

FIG. 1 Framework structures of some extensively studied zeolites (of importance to this chapter are the zeolite X framework and sodalite).

the kinetic diameter of molecules that can enter the porous structure. Table 2 shows examples of how zeolites are used commercially. As is evident from Table 2, the microporous nature and high surface area of zeolites are used in adsorption and separation applications [4]. Ion-exchange properties of these materials are exploited in the consumer and environmental industries [5]. Chemical and petroleum industries use zeolites as catalysts in hydrocarbon transformations [6]. Synthesis of new microporous frameworks has led to the development of new technologies, and thus considerable effort worldwide is expended in their discovery [2].

TABLE 1 Physical and Chemical Properties of Some Commercially Important Zeolites

Type	Unit-cell compound (typical, fully hydrated)	Pore structure	Typical void volume Si/Al	(cm^3/cm^3)	Kinetic diameter (nm)
A	$Na_{12}[(AlO_2)_{12}(SiO_2)_{12}]\cdot 27H_2O$	3D, 4.1 Å	1.0	0.47	0.39
X	$Na_{86}[(AlO_2)_{86}(SiO_2)_{106}]\cdot 264H_2O$	3D, 7.4 Å	1.0–1.5	0.50	0.81
Y	$Na_{56}[(AlO_2)_{56}(SiO_2)_{136}]\cdot 250H_2O$	3D, 7.4 Å	>1.5–3	0.48	0.81
L	$K_9[(AlO_2)_9(SiO_2)_{27}]\cdot 22H_2O$	1D, 7.1 Å	2.6–3.5	0.32	0.81
Mordenite	$Na_8[(AlO_2)_8(SiO_2)_{40}]\cdot 24H_2O$	1D, 6.5 × 7.0 Å	4.7–5	0.28	0.63
ZSM-5	$(TPA,Na)[(AlO_2)(SiO_2)_{30}]\cdot 10H_2O$	3D, 5.4 × 5.6 Å 5.1 × 5.5 Å	10–100	0.34	0.60

Because this chapter discusses the synthesis of microporous materials in reverse micelles, we will present in some detail information about the synthesis process itself. Microporous materials are typically synthesized by a hydrothermal process, which involves mixing appropriate reactants in an aqueous medium followed by heat treatment [7]. Even though the process is relatively simple, the development and control of porosity, which determines the ultimate practical use of these materials, are not easy to predict. This is primarily because the crystal growth of these materials is a very complicated chemical process [8]. For example, in zeolite formation, the silicon- and aluminum-containing reactants dissolve in the presence of base to produce many soluble species [9]. Speciation is strongly influenced by the pH, temperature, cations, and structure-directing agents [10]. Insoluble aluminosilicates (commonly referred to as gel) are rapidly formed by reaction of the solubilized species. Thus, this system is typically in a state of supersaturation for many of the aluminosilicate species. After an induction period that can extend from hours to weeks, crystals are formed. Nuclei formation that precedes crystal growth can occur by solid-state restructuring of the gel or precipitation from the supersaturated solution. The nuclei grow using nutrients from the solution or the dissolving gel to form crystals. Figure 2 is a schematic description of the crystallization process of microporous materials [11]. The complexity of the process is evident from the numerous chemical processes that occur during crystallization.

Typically, the porous structures that are formed are kinetic intermediates and transform to more condensed structures with time. Because of this kinetic stabilization, profound effects are observed on changing the synthesis conditions. The interplay between the inorganic reactants and organic additives also influences the crystal growth process. Crystal morphologies are controlled by preferential growth of crystal faces and are strongly influenced by changes in the synthesis conditions. Thus, zeolite synthesis is an exciting area of research, with major discoveries of frameworks being reported on a regular basis. However, the critical discoveries of new frameworks usually occur by trial and error. Thus, developing a comprehensive molecular-level understanding of the synthesis process that may lead to designed synthesis is of considerable interest.

Spectroscopic probes that have found considerable use in the past decade include nuclear magnetic reso-

TABLE 2 Examples of Industrial Applications of Zeolites

Zeolite	Process	Products
Faujasite (zeolite Y)	Cracking	Gasoline, fuel oil
Faujasite (zeolite Y)	Hydrocracking	Kerosene, jet fuel, benzene, toluene, xylene
Mordenite	Hydroisomerization	iC_6, C_7
Mordenite	Dewaxing	Low-pour-point lubes
Z5M-5	Xylene isomerization	*p*-Xylene
Z5M-5	Benzene alkylation	Ethylbenzene (styrene)
Zeolite A	Ion exchange	Detergents
Zeolite X	Adsorption	Separation of gases
Clinoptilolite	Ion exchange	Radioactive waste cleaup

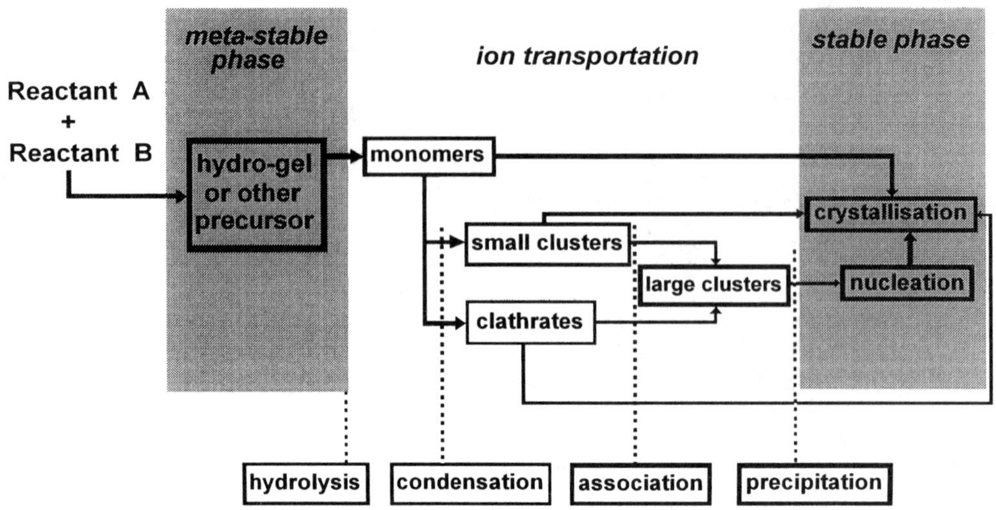

FIG. 2 Schematic representation of the crystallization process of microporous materials (reactants A and B can be sources of silicate and aluminate, respectively, or zinc and phosphate, respectively). (Adapted from Ref. 11.)

nance (NMR) spectroscopy, which has provided information about the nucleation process [12,13]. Other techniques that have provided information on the early stages of zeolite nucleation include small-angle and wide-angle x-ray and neutron scattering [14].

Although considerable research has been done on analysis of the structural aspects of species that are present during zeolite nucleation, much less is known about how these species assemble into crystals. This issue is critical for several reasons. The competitive growth of nuclei into crystals determines the final crystalline product. Even though nuclei of a certain zeolitic framework may be formed readily, the rate of crystal growth may be limiting. Growth of large crystals and seeded crystal growth are also dependent on the type of crystal growth. Finally, the morphology of the crystals depends on the crystal growth process.

We have reported several studies related to zeolite nucleation and crystal growth, primarily using Raman spectroscopy [15]. Two of these studies are relevant to how we got started in using reverse micelles for microporous material synthesis. The first study dealt with using amorphous material from a zeolite X synthesis, ion-exchanging it with a series of monovalent cations, and then continuing the synthesis by putting the solid back into a basic solution of the corresponding cation [16]. The products were analyzed by x-ray diffraction. Table 3 shows the results. Because the aluminosilicate and the hydroxide ion concentration were kept identical in all cases, the change in the framework structure must arise from the influence of the cation. Raman spectroscopy had shown that the amorphous gel is primarily composed of four-membered aluminosilicate rings. The model shown in Fig. 3 assigns specific building blocks for each structure. These intermediate structures, derived from four-membered rings, were hypothetical and derived on the basis of their specificity for a particular zeolite as well as their ready convertibility by rearranging a few T—O—T′ bonds. It was proposed that the formation of these units was controlled by the presence of specific cations in the reaction medium. Why and how do the cations direct the formation of such structures? We reasoned that the different electrostatic fields around the cation were responsible for stabilizing the different structures. Because reverse micelles provide novel cation-water environments, they appeared to be attractive in examining microporous material synthesis.

TABLE 3 Zeolitic Frameworks That Are Formed from the Same Aluminosilicate Composition in the Presence of Different Monovalent Cations[a]

Cations	Zeolite
Na^+	Zeolite A
K^+	Zeolite R (chabazite)
Cs^+	Zeolite Cs-D (edingtonite)
$N(CH_3)_4^+$	Zeolite ZK-4
$N(C_2H_5)_4^+$	Zeolite P (gismondine)[b]
$N(C_3H_7)_4^+$	Zeolite X[b]
$N(C_4H_9)_4^+$	Zeolite P

[a] Major constituent, other phases also present.
[b] Adapted from Ref. 16.

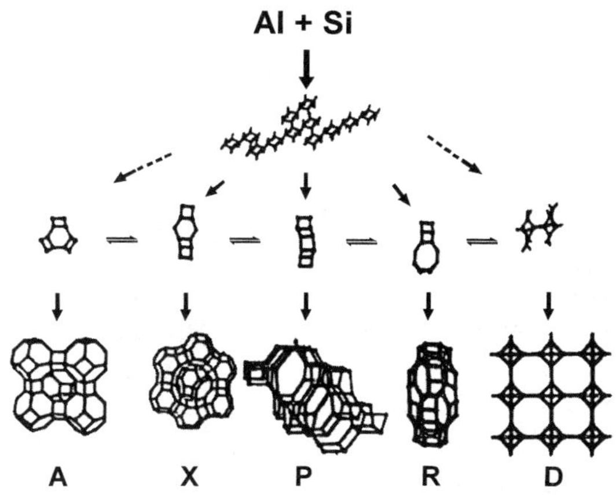

FIG. 3 Proposed intermediate building block structure for zeolitic structures made in the presence of different monovalent cations (see also Table 3). (Adapted from Ref. 16.)

A second study along these lines dealt with the synthesis of zeolites of low Si/Al ratio in the presence of ethanol [17]. At lower levels of alcohol (<40%), open-pore structures such as A, X, and Y were formed, whereas, at higher levels, condensed structures such as sodalite and cancrinite were formed. Figure 4 shows the powder diffraction patterns of the solids obtained as a function of ethanol content and indicates a change from zeolite A to X to cancrinite with increasing alcohol. Structural aspects of water-alcohol mixtures have been studied extensively [18]. It is well recognized that at low alcohol concentrations the viscosity, reciprocal self-diffusion coefficient, dielectric relaxation time, and NMR relaxation times of the water molecules are greater than in pure water. These observations have been interpreted as the alcohol stabilizing the structure of water, which is commonly referred to as hydrophobic hydration. At levels of alcohol at which the water structure is most rigid (20–40%), nucleation of large-pore zeolites, such as zeolite X, is facilitated. At higher levels of alcohol, the water structure disintegrates, leading to nucleation of more condensed species, such as sodalite and cancrinite. This model is supported by the fact that condensed zeolitic frameworks are formed by raising temperatures of synthesis and adding large amounts of salts, both of which are known to disrupt the structure of water. Because the structure of water in reverse micelles can be different from that of bulk water, it provides a novel medium for examining the crystallization of microporous materials in reverse micelles.

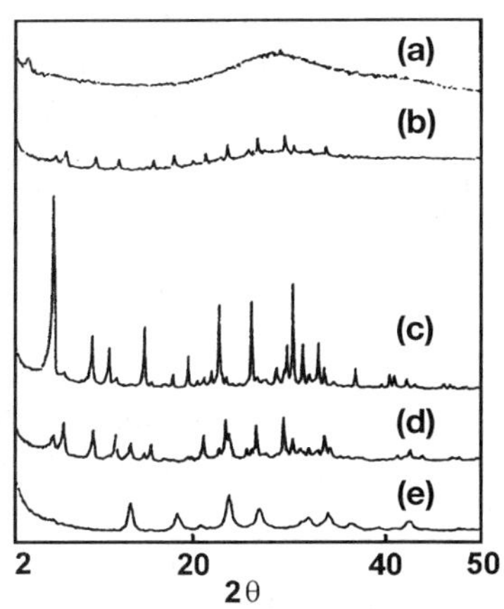

FIG. 4 X-ray powder diffraction patterns of solids obtained from a fixed aluminosilicate composition with varying amounts of ethanol: (a) 0%; (b) 10%; (c) 25%; (d) 50%; (e) 75% in volume percent ethanol. (Adapted from Ref. 17.)

B. Reverse Micelles

Certain surfactant molecules, dissolved in organic solvents, are capable of solubilizing water in the polar core, and these entities are called reverse micelles [19]. Reverse micelles belong to a class of thermodynamically stable dispersions of two phases stabilized by a surface-active agent. Examples include both water-in-oil and oil-in-water microemulsions. The dispersed phase is usually present as nanometer-size droplets. Like zeolites, microemulsions are important technologically, being used in food, cosmetic, and pharmaceutical industries as well as in enhanced oil recovery [20].

This chapter focuses on water-in-oil reverse microemulsions. These can be made with a larger number of surfactants, three of which that are relevant to this chapter are shown in Fig. 5. These are an anionic molecule, sodium bis(2-ethylhexyl)sulfosuccinate (AOT); a neutral molecule polyoxyethylenesorbitan trioleate (Tween 85); and a cationic molecule, dioctyldimethylammonium chloride (DODMAC). A schematic representation of a reverse micelle is presented in Fig. 6. The most extensively studied reverse micelles are those made with AOT. In the absence of water, various physicochemical measurements indicate that AOT in hydrocarbon consists of 15 Å aggregates [21]. Water can be readily incorporated into the core of the micelle, with the radius of the micelle increasing as the [water]/[AOT] ratio ($=\omega_0$) increases [22]. The radius (r) of the polar core is related to ω_0 as r (Å) = 1.8 ω_0. At low ω_0 values of 2–8, the water is held tightly by the polar sulfosuccinate groups and the Na^+ ions and has distinct spectroscopic features [22]. At larger ω_0, the water in the core region of the micelle resembles bulk water.

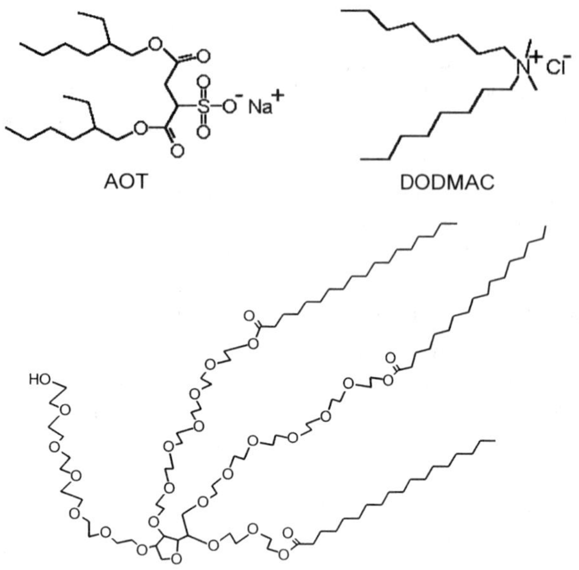

FIG. 5 Molecular structure of surfactants used for making reverse micelles for microporous material synthesis.

Thus, three types of water as shown in Fig. 6 can be distinguished: tightly bound to the headgroup, bulklike water in the middle of the reverse micelle, and an intermediate type of water bridging the two. The size of these water-swollen reverse micelles can vary from 5 to 50 nm.

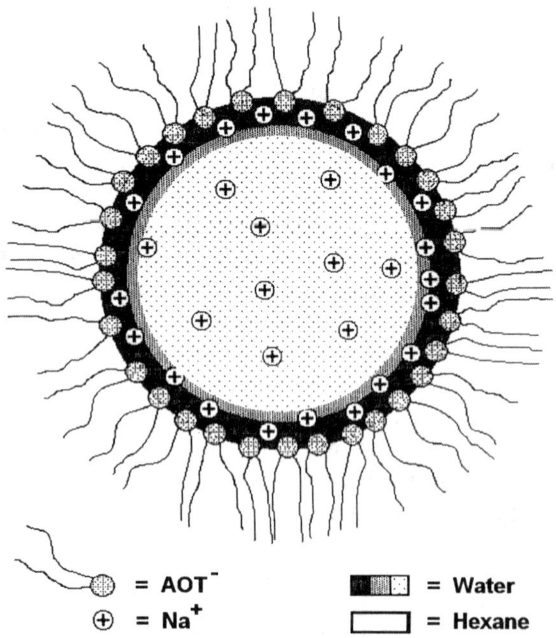

FIG. 6 Schematic representation of a reverse micelle.

Reverse micelles are dynamic entities. They move by Brownian diffusive motion and collide with rate constants of $\sim 10^6-10^8$ L mol^{-1} s^{-1} [23]. During these collisions, brief fusions of micelles occur lasting for microseconds. This provides enough time for exchange of water pools and thereby the reactants enclosed in them [24].

Extensive studies have been done with reverse micelles in both biological systems and materials science. Reverse micelles have the ability to solubilize macromolecules and enzymes, and enzymatic chemistry in these systems is dependent on the water content of the micelles [25]. Polymerization chemistry is also possible in reverse micelles, including control of gel formation and/or flocculation. Synthesis of inorganic particles with particular focus on size control has been extensively done with reverse micelles [26–29]. Semiconductors, metals, oxides, and carbonates have been prepared. A connection between the micellar environment and the final morphology of the material, including the state of aggregation, has been found to exist. The materials syntheses reported in the literature using reverse micelles involve reduction, hydrolysis, or precipitation chemistry. A schematic representation of the precipitation chemistry route to particles is shown in Fig. 7 [30]. Two reverse micelles containing reactants can undergo collisions and exchange reactants, which then combine to form a precipitate in the core of the reverse micelle.

C. Choice of Zincophosphates as the Microporous Materials for Growth in Reverse Micelles

Our group has developed procedures for the use of reverse micelles as reactants for synthesis of microporous materials. It needs to be made clear that, unlike the situation in most syntheses reported with reverse micelles, zeolitic-type materials are not the direct product of the reaction between two components. This is apparent from a comparison of Figs. 2 and 7. In both cases, the first step is the formation of a precipitate, but in microporous material synthesis, this is just the beginning of the process rather than the end. Nucleation and crystal growth need to occur. Considering the sensitivity of the crystal growth of microporous materials to the aqueous environment, reverse micelles provide a novel host for examining nucleation and crystal growth of these important materials.

Our first attempts to grow aluminosilicate zeolites at >60°C were unsuccessful because of instability of reverse micelles at high temperatures [31]. Intermicellar

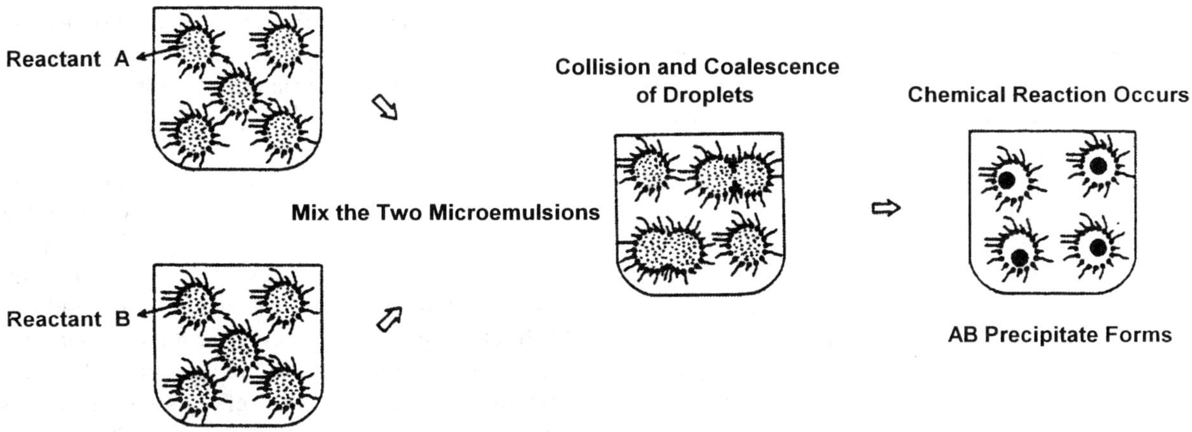

FIG. 7 Schematic representation of particle formation in reverse micelle via precipitation reaction between two reactants initially in two separate reverse micelles. (Adapted from Ref. 30.)

attractive forces increase upon raising the temperature, resulting in phase separation. Thus, in order to study the crystal growth characteristics of microporous materials in reverse micelles, we had to limit ourselves to frameworks that can be made under ambient conditions and chose to work with zincophosphates. The advantage of microporous zincophosphates over their aluminosilicate analogues relies on their low temperature and mild condition synthesis. For instance, whereas zeolite X is typically synthesized from a highly caustic gel between 70 and 100°C, zincophosphate X (ZnPO-X) with the same topology (Fig. 1) is prepared around pH 8 at room temperature. Microporous zincophosphate materials were first synthesized in the early 1990s by Stucky and coworkers [32,33]. The first examples of these type of compounds were the analogue structures of zeolite X, sodalite and zeolite Li-A(BW).

II. SYNTHESIS OF ZINCOPHOSPHATE SODALITE FROM AOT REVERSE MICELLES

A. Synthesis Procedures with AOT Reverse Micelles

For synthesis of zincophosphates from AOT reverse micelles, two reverse micelle solutions have typically been used, one containing zinc ion (identified as Zn) and the other containing phosphate, sodium hydroxide, and tetramethylammonium hydroxide (TMAOH) solution (identified as P) [34,35]. Tetramethylammonium ions were necessary for the uptake of phosphate ions into the micelle. Table 4 shows the typical characteristics of each micelle system. These micelles were made by equilibrating an AOT solution in hexane with the aqueous solutions and used for the zincophosphate synthesis experiments. Compositional changes were brought about in two ways. First, aging of the reverse micelles altered the intramicellar pH of the P micelles, making them less basic because of the reaction of the hydroxide ion with the ester functionality of the AOT headgroup. Zinc micelles, on the other hand, were acidic because of the hydrolysis of the zinc ions. Second, different volume ratios of the Zn and P micelles were mixed to vary the composition of the synthesis medium.

A typical experiment begins with mixing the Zn and P micelle solutions. The solutions are clear upon mixing the micelles. Then at various times, a white product settles out. After completion of the settling process, the product is removed, washed, and analyzed by powder x-ray diffraction, the primary method for identification of the frameworks.

Influence of aging of the reverse micelle preparations on the formation of zincophosphates has also

TABLE 4 Composition of AOT-Based Zinc and Phosphate Micellar Solutions

Zn micelle	P micelle
$[Zn^{2+}] = 0.0075$ M	$[P] = 0.0125$ M
$[Na^{+}] = 0.0525$ M	$[Na^{+}] = 0.55$ M
$[NO_3^{-}] = 0.00507$ M	$[TMA^{+}] = 0.346$ M
[AOT] = 0.065 M	[AOT] = 0.065 M
$[AOT]/[H_2O] = 13$	$[AOT]/[H_2O] = 21$
Micelle size = 8.5 ± 1 nm	Micelle size = 15 ± 1 nm

Source: Adapted from Ref. 35.

been examined [35]. Equal volumes of Zn and P micelles are mixed, and both solutions are aged for periods of time varying from 0 to 13 days. The products recovered with the solutions aged for 2 days of reaction are a mixture of hexagonal sodium zinc phosphate and sodalite. Pure sodalite is formed when the micelles are aged for more than 6 days. In experiments in which the volume of Zn micelle is three times that of the P micelle (both samples aged for 8 days), hopeite (zinc phosphate) is formed. This is consistent with the conventional synthesis of zincophosphates, in which hexagonal zinc phosphate is reported to form at pH 10, sodalite at neutral pH, and hopeite at acidic pH values [33]. Overall, the micellar chemistry and aqueous chemistry appear to follow similar pathways.

Because among these zincophosphate frameworks, only sodalite can be considered to belong to the family of microporous materials (framework structure in Fig. 1), this system was examined in more detail.

B. Formation of Zincophosphate Sodalite from AOT Reverse Micelles

Sodalite is a member of the microporous family of frameworks and has been extensively studied in the aluminosilicate [36] as well as in silica systems [37]. Solutions of P and Zn micelles aged for 8 days were used as starting materials for sodalite synthesis. The solids formed upon mixing these two micelle solutions in different volume ratios were examined. In particular, three compositions with relative ratios of the Zn and P micelle solutions of 0.8, 1, and 1.2 were found to have interesting crystal growth dynamics [35]. These were identified as compositions A, B, and C, respectively. The overall zinc and phosphate concentrations in these compositions are shown in Table 5. Profound differences were noted in the rate of appearance of product formation in these three compositions, although they all ended up forming zincophosphate sodalite. Next, we discuss in detail the reported characteristics of these three reverse micelle–led reactions and the implications for microporous material growth.

TABLE 5 Compositions of AOT-Based Compositions for Forming Zincophosphate Sodalite

Composition	Volume (mL)		Moles	
	Zn micelle	P micelle	Zinc	Phosphate
A	40	50	3×10^{-4}	6.25×10^{-4}
B	40	40	3×10^{-4}	5×10^{-4}
C	50	40	3.8×10^{-4}	5×10^{-4}

Source: Adapted from Ref. 35.

1. Composition A

Upon mixing the Zn- and P-containing micelles for composition A, the solution remained clear and evidence of reaction was provided by solids appearing at the bottom of the reaction vessel for the first time after 2 days. The peaks due to sodalite were evident in the diffraction patterns obtained from this sample. With time, the crystals increased in amount and were always the sodalite framework. Light scattering studies indicated continuous growth in size as shown in Fig. 8. A scanning electron micrograph of the crystals obtained after settling (4 days) and the powder diffraction pattern are shown in Fig. 9. The sizes of these crystals were between 500 and 600 nm. The morphologies were cubic crystals or pyramids (half-cubes). Centrifugation of the reaction mixture after 18 h of reaction afforded a small quantity of solid, which was examined by transmission electron microscopy (TEM) as shown in Fig. 10. The morphology of the crystals is similar to that observed after 4 days, although the crystals appeared to be smaller (100–500 nm) and peaked between 200 and 220 nm. Selected area diffraction of the crystal showed the material to be crystalline. Thus, it appears that at all observable stages of growth, the morphology of the sodalite crystals remains similar, with size and yield increasing in time. These observations indicate that sodalite is being formed while suspended, without any detectable intermediate amorphous phase. The morphology of the crystals suggests that they grow by deposition along specific crystal planes proceeding toward a cubic structure.

2. Composition B

For composition B, the visual observations were quite distinct from those for composition A. The reactor became cloudy within the first 12 h, followed by settling out of particles over the next 24 h. Light scattering data, as shown in Fig. 11, suggested that rapid growth of particles occurred after a size of 75 nm. The diffraction patterns of the solid recovered by centrifugation as soon as the solution became cloudy (12 h) as well as samples recovered after settling for 24 h were characteristic of the sodalite framework. Figure 12 shows a scanning electron microscopy (SEM) picture of the particles obtained for composition B prior to settling. The suspended particles were crystallites of sizes less than 600 nm. The crystals recovered after settling are similar in particle size and morphologies to

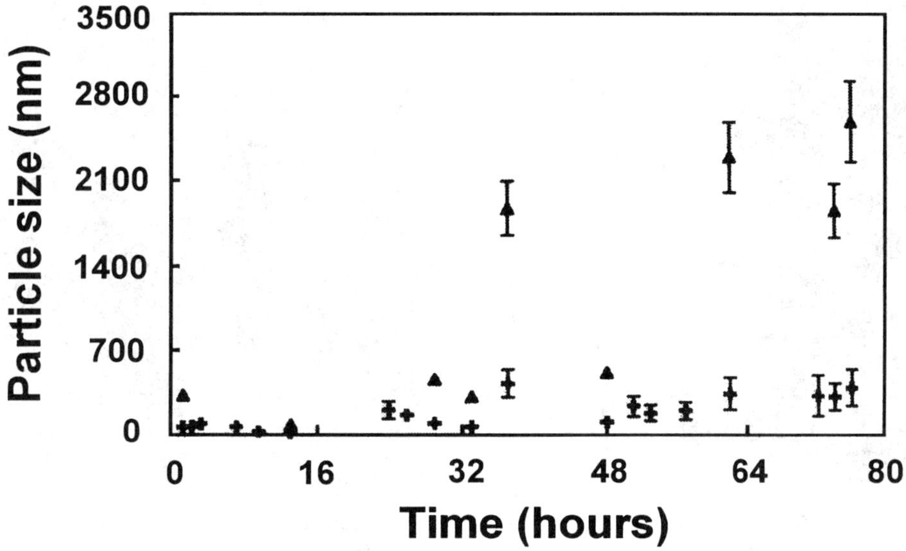

FIG. 8 Change of particle size with time for composition A, as determined by light scattering (+ and ▲ represent the smaller and larger sizes, respectively, obtained by fitting the data to the exponential sampling method). (Adapted from Ref. 35.)

(a)

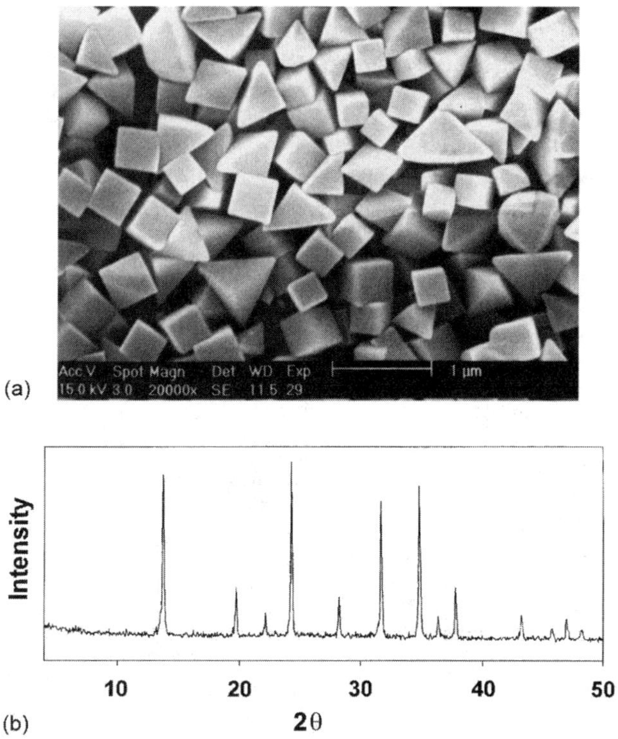

(b)

FIG. 9 (a) SEM pictures of sodalite crystals grown via composition A (4 days). (b) X-ray powder diffraction pattern, typical of the sodalite structure, grown from composition A (4 days). (Adapted from Ref. 35.)

the suspended particles, except that they appear agglomerated and 2–3 μm in size. Solids recovered after 6 h of reaction by centrifugation (prior to the appearance of any cloudiness) were examined by TEM and showed aggregates of small particles. Selected area diffraction showed that the material was crystalline. Thus, for composition B, crystals also appear without any apparent intermediate amorphous phase at the earliest stages examined.

3. Composition C

For composition C, the pathway was marked by the rapidity of the initial reaction to form a white solid. The reactant mixture turned turbid in 1 h, and complete settling of the solid was seen in 4 h. The growth pattern as measured by light scattering for composition C (Fig. 13) was consistent with these observations. The particles formed immediately after the appearance of turbidity and settling were found to be amorphous. These were discrete particles of approximately 5 μm and they agglomerated and formed a contiguous solid with time. Sodalite crystals grew out of this settled solid phase over a period of 4 days with the morphology shown in Fig. 14a.

In order to keep the particles suspended during formation of sodalite from composition C, the experiment was repeated in a rotating cell. Particles in a fluid that

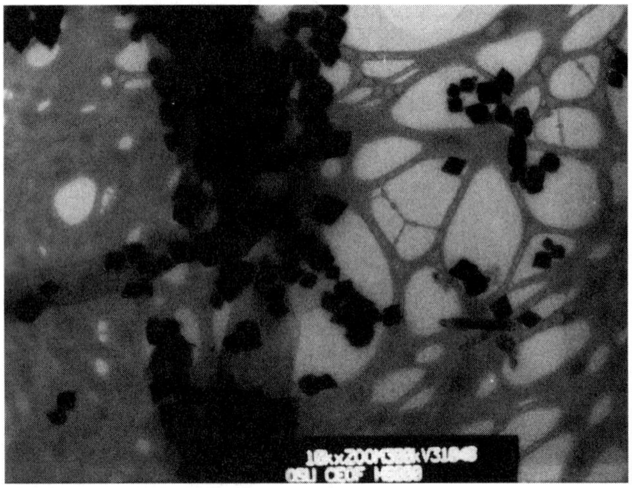

FIG. 10 TEM picture of crystals obtained at the early stages (18 h) of composition A. (Adapted from Ref. 35.)

FIG. 12 SEM picture of crystals obtained from composition B prior to settling. (Adapted from Ref. 35.)

will sediment due to gravitational forces (Stokes' law) can be kept suspended by rotation of the reaction chamber [38]. The reactor was rotated around its central axis at speeds up to 7 to 45 rpm. The orbits assumed by the particles were approximate circles around a center displaced horizontally from the axis of rotation. A minimum rotation speed was required to ensure that this center lies within the dimensions of the reactor. At very high rotation rates, particles with densities higher than the fluid (as in this case) spiraled out and hit the wall of the reactor. The choice of the rotation speed thus had to be optimized depending on the particle-fluid system.

Formation of sodalite crystals by pathway C could be completed in the rotating cell at 11 rpm with minimal sedimentation on the walls of the reactor over a period of 10 days. From the individual amorphous particles, sodalite crystals were seen to grow, as shown in Fig. 14b.

C. Role of the Intramicellar pH in Influencing Crystal Growth Pathways

Variation of the Zn-to-P micellar volume ratios over a narrow range (0.8–1.2) has a profound influence on the rate and mechanism of sodalite crystal growth as well as the final crystal morphology. Because such a minor change in composition has no influence in the conventional synthesis of sodalite, the micellar environment

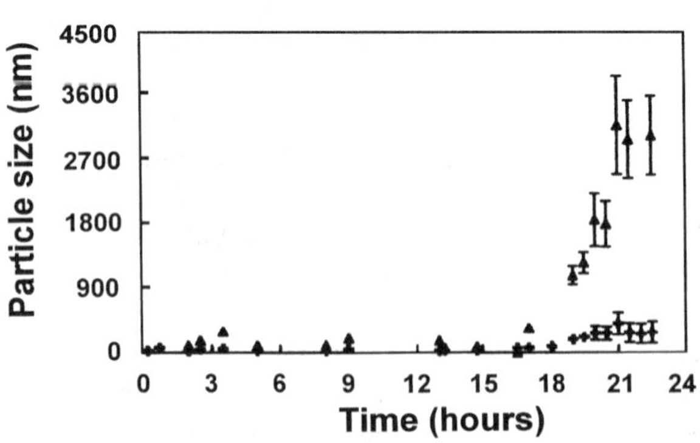

FIG. 11 Particle growth characteristics for composition B, as measured by light scattering (+ and ▲ represent the smaller and larger sizes, respectively, obtained by fitting the data to the exponential sampling method). (Adapted from Ref. 35.)

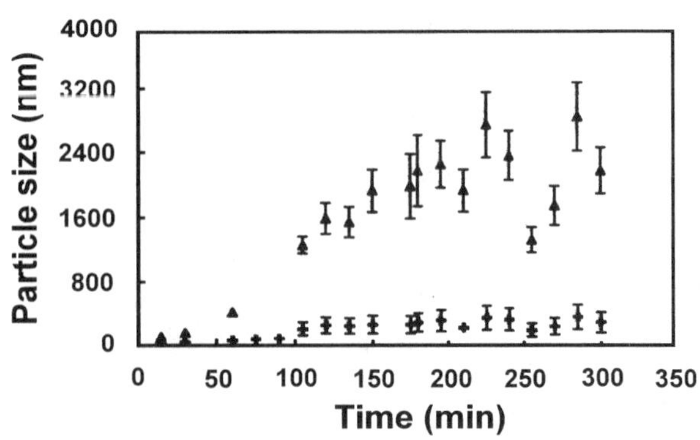

FIG. 13 Particle growth characteristics for composition C, as measured by light scattering (+ and ▲ represent the smaller and larger sizes, respectively, obtained by fitting the data to the exponential sampling method). (Adapted from Ref. 35.)

(a)
(b)

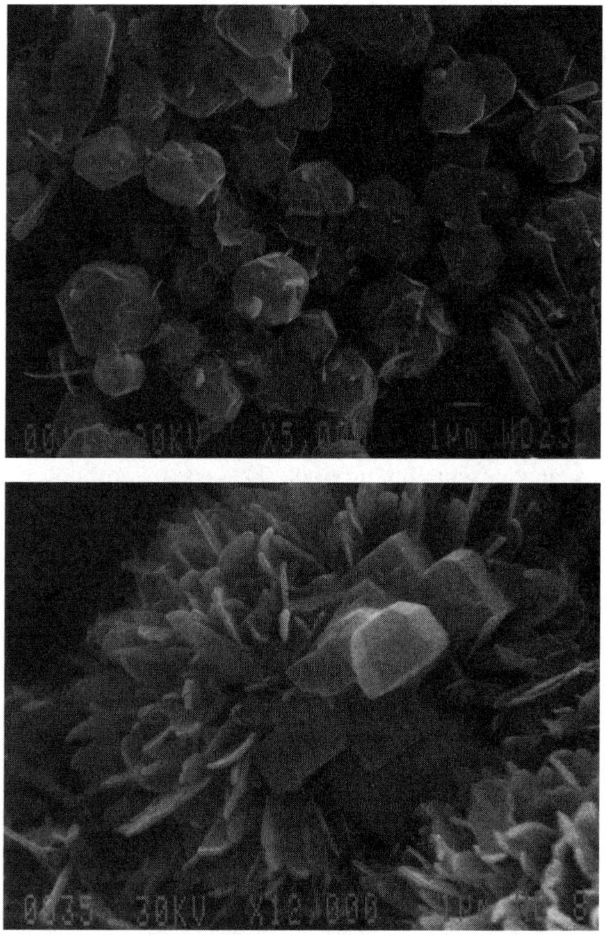

FIG. 14 (a) SEM pictures of sodalite crystals obtained from the amorphous phase of composition C. (b) SEM of sodalite crystals growing from suspended amorphous particles in a rotating cell from composition C. (Adapted from Ref. 35.)

obviously plays a major role. The major differences in the compositions A–C are in the relative ratios of zinc and phosphate ions present. The phosphate ions are always in excess, with the phosphate-to-zinc molar ratio decreasing from 2.1 in composition A to 1.3 in composition C (Table 5). Increasing the content of the Zn micelle makes the overall composition more acidic. From the pH-NMR calibration data [39], the pH values in compositions A, B, and C were found to be 7.6, 7.2, and 6.8, respectively, reflecting the acidity of the Zn micelle [35].

Crystallization has been extensively studied over many decades [40,41], and we provide some instances of systems in which pH has had a major effect on the crystallization. A good example is the crystallization of calcium hydrogen phosphates [42]. The hydroxide and phosphate ion concentrations appeared in the ion product $[Ca^{2+}]^{10}[PO_4^{3-}]^6[OH^-]^2$ as exponents, and a slight increase of pH considerably increased the supersaturation. Thus, a change in pH from 7.4 to 7.8 resulted in a considerable increase of crystal growth rates. Indeed, above a pH of 7.8, it was difficult to prepare a supersaturated solution without spontaneous precipitation of calcium phosphate. For $Cr(OH)_3$, an amorphous precipitate was formed at pH below 10, whereas a crystalline material was formed at pH above 10 [43]. Lazic [44] has reported a reduction of the induction period in hydroxyapatite formation from amorphous calcium phosphate as a function of pH. Beyond pH 10.2, the decrease in induction period was correlated with deprotonation of HPO_4^{2-} and the increase in concentration of $CaPO_4^-$. The effect of pH on crystal growth can also arise from surface charge on the particle. For example, with both $CaHPO_4 \cdot 2H_2O$ and hydroxyapatite, crystals grow more slowly at pH values less than the pH corresponding to the point of zero charge [45].

The structure of the nucleating species in sodalite growth is not known. As a matter of fact, even for simple precipitation reactions, the nature of the nucleating species is not obvious. For example, in the precipitation of Ag_2WO_4, it is not WO_4^{2-} but rather intermediate species such as $W_6O_{21}^{2-}$, $W_{12}O_{41}^{10-}$, and $HW_6O_{21}^{5-}$ [46]. The same is true for formation of ferric hydroxide, in which the important reaction is between $Fe_9(OH)_{20}^{7+}$ and $Fe_3(OH)_4^{4+}$ rather than between Fe^{3+} and $3OH^-$ [47]. Complicated reactions involving polymeric nucleating intermediates have been proposed for formation of TiO_2, $Cr(OH)_3$, and $Mg(OH)_2$ [41]. Hydrolysis of the Zn—O—P bond is essential in the development of nuclei of sodalite. Hydrolysis reactions have been studied in reverse micelles. Base-catalyzed hydrolysis of esters [48], as well as the formation of ferric oxyhydroxide species by hydrolysis of ferric ammonium sulfate, has been reported [49]. Hydrolyses of alkoxides in micelles have shown that the concentration of the reactants and water determine supersaturation levels [50].

Walton [51] has summarized the effects of initial supersaturation on crystal morphology. It was noted that the morphology varies from compact well-defined crystals to poorly developed crystals and finally amorphous aggregates as the initial supersaturation changes. An example is $ZnC_2O_4 \cdot 2H_2O$, whose crystals have different habits in different supersaturation ranges [52]. In the case of $CaHPO_4$, with increasing supersaturation, rhombohedral, intergrown, and twinned crystals of $CaHPO_4 \cdot 2H_2O$ and ultimately spherical agglomerates of $CaHPO_4$ are formed [53].

D. Reasons for Crystallization Pathway Control with AOT Reverse Micelles

After the micelles equilibrate, there is a distribution of zinc, phosphate, and hydroxide ion occupancies in the different micelles that exist in the medium. Depending on the intramicellar concentrations and pH, there is a fraction of these micelles in which saturation conditions are exceeded. This fraction was proposed to increase as the composition changed from A to C. For composition A, where few nuclei were formed in the micelles, the growth had to occur by acceptance of solution species present in other micelles. For composition C, on the other extreme, the particles grew rapidly by aggregation. Increased aggregation with supersaturation has been noted for silver clusters in AOT micelles [54]. Composition B represents an intermediate situation in which more particles are formed but there is enough time prior to aggregation to form sodalite nuclei. Figure 15 depicts this distribution schematically, assuming a Gaussian distribution of components (for illustration purposes), and shows that the fraction of micelles in which supersaturation has been exceeded increases from A to C.

In conventional aqueous systems, in which there exists uniformity of compositions throughout the system, it is impossible to control the supersaturation levels by minor changes. Indeed, the compartmentalization of reactants in reverse micelles and the nature of the exchange of reactants upon collision lead to the unique behavior of such micellar systems.

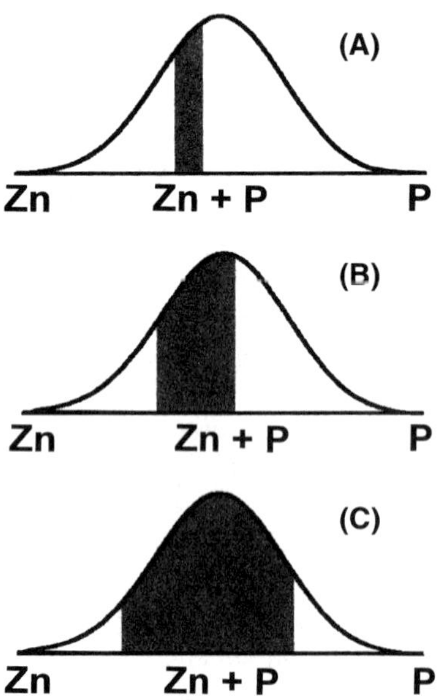

FIG. 15 Schematic representation of the differences in the fraction of micelles in which supersaturation has been exceeded for compositions A–C.

In composition A, where the supersaturation was lowest, crystal growth proceeded slowly, controlled by surface attachment kinetics. The surfaces of the crystals were examined by atomic force microscopy (AFM), which indicated layers 10 Å thick, typical of a sodalite cage [35]. Crystals of compact shapes are known to form at low supersaturation because the minimum overall energy of the crystal surface is reached under very slow growth, equilibrium-like conditions [55], which is consistent with the morphology observed for composition A.

The growth process in pathway B has been analyzed as an aggregation process. The early morphology of these crystals appeared cubic, suggesting that the initial growth process may be similar to that of composition A. The crystals aggregated via diffusion and convection. Such diffusion-controlled micellar collisions have been proposed for growth of silica and carbonate particles [56,57].

In pathway C, the intramicellar conditions resulted in the highest supersaturation. This led to rapid nucleation, and because the induction time for crystal formation was longer, amorphous particles were formed. The morphology of the particles formed initially in composition C supports the high supersaturation hypothesis. If the rate of particle growth is very high, the heat of precipitation cannot be transferred efficiently into solution. This leads to convection, and the particle is surrounded by depleted regions. The particle extends its surface highly anisotropically. This leads to structures as shown in Fig. 14b.

The formation of an amorphous phase in systems with high supersaturation has been noted in the crystallization of $CaCO_3$ [58], hydroxyapatite, $Mg(OH)_2$ [59], and $Al(OH)_3$ [60]. In all cases, the amorphous phase finally transforms to crystals. The explanation for this phenomenon is that the nuclei formed from highly supersaturated solutions do not have an exactly defined structure and hence crystals are not formed. Crystals are formed from these amorphous materials by dissolution into the mother liquor and a solution-mediated transformation. However, in the reverse micelles, nutrients cannot dissolve in the organic medium. There are two possibilities for crystal growth. The first is direct transformation in the solid state, as reported for the transformation of vaterite to calcite. In this case, the morphology remained unchanged upon transformation [58], which is not consistent with the observations for composition C. The second possibility is based on precipitation of metal hydroxides, in which nonstructural water that is retained in the precipitate can be liberated during the growth process [61]. Thus, after settling of the amorphous zincophosphate, with time it is surrounded by a thin water layer that can transport nutrients. Such a mechanism is supported by the rotating cell experiment, in which the suspended amorphous particles transformed directly to sodalite crystals (Fig. 14b) with the gel particles appearing to act as sites for growth.

E. Attempts at Crystal Growth of Zincophosphate X from AOT Reverse Micelles

The sodalite framework, as seen from Fig. 1, has limited porosity. Attempts were made to grow open-pore framework zincophosphate X (ZnPO-X) from AOT reverse micelles but without success [62]. In the following we outline the attempts at growth of ZnPO-X from AOT reverse micelles and the proposed reasons for failure.

Several procedures for synthesis of ZnPO-X using AOT reverse micelles were attempted [62]. The most straightforward procedure was similar to that for sodalite using two reverse micelle (AOT/hexane) solutions, each made by equilibration with aqueous zinc and phosphate solutions. The compositions of the aqueous solutions were chosen such that in the absence of the micelle, they would have resulted in formation of ZnPO-X. In a relatively fast reaction (24 h), sodalite was produced.

Another approach taken was to consider the intramicellar composition that was previously optimized for sodalite formation [35] and to alter it to a composition more appropriate for ZnPO-X. In the reverse micelles for sodalite formation, the aqueous Zn^{2+} solution used was 0.2 M and the Zn/P ratio was found to be 0.6. The optimal Zn/P ratio for aqueous synthesis of ZnPO-X is 0.94. To keep this ratio in the reverse micelles, the concentration of Zn^{2+} in the aqueous solution was increased to 0.32 M. For the phosphate solution, the [TMAOH] was increased by 34% to a final pH of 12.7. To keep the intramicellar pH to a maximum, the AOT/hexane solutions were equilibrated with the aqueous zinc and phosphate solutions for 3 h and the solutions mixed without any aging. After 4 weeks of reaction, no product was recovered.

Thus, the zincophosphate framework with the sodalite structure appeared to form preferentially in the AOT reverse micelle system. Under no composition conditions was it possible to form ZnPO-X.

Typically, in the synthesis of inorganic materials from reverse micelles, the chemistry in the reverse micelle is similar to that observed in the bulk solution. Clearly, in the case of microporous material synthesis, the reverse micelle environment has an influence on the product, even though from a composition point of view, ZnPO-X should be formed with AOT reverse micelles. The possible reasons why AOT reverse micelles preferentially direct the formation of sodalite were examined.

III. EVALUATION OF WHY ZnPO-X IS NOT FORMED WITH AOT REVERSE MICELLES

In the conventional synthesis for preparing ZnPO-X, the starting Na/Zn ratio was 0.5 [32]. If NaCl was added to this reaction system, it resulted in the appearance of sodalite [62]. On the other hand, with the composition typical for sodalite, addition of Na^+ complexing agents, such as 18-crown-6-ether or Kryptofix, in increasing amounts led to a decrease in the amount of sodalite and zincophosphate X appeared as the dominant product. The following relationship appears to exist between the two zincophosphate frameworks:

$$\text{ZnPO-X} \underset{-Na^+}{\overset{Na^+}{\longleftrightarrow}} \text{sodalite}$$

Each AOT detergent molecule contributes a Na^+ to the water pool of the reverse micelle, which leads to $[Na^+]$ ~4 M inside the water pool of the reverse micelle. Up to 28% of the sodium counterions in a reverse micelle can be in the bulk water region of the micelle [63]. This high concentration of Na^+ is proposed to be responsible for the nucleation and growth of sodalite in AOT reverse micelles under varying compositional conditions. This hypothesis would predict that complexation of the sodium ions in the reverse micelle should lead to formation of ZnPO-X. In order to convert from sodalite to ZnPO-X, the Na/Zn ratio in the reactant composition had to be lowered from 1.08 to 0.5. For comparable Na/Zn ratios in the reverse micelles, the free sodium concentration had to be lowered by 75%. The amount of crown ether necessary to bring the sodium ion to these low levels was estimated from the reported formation constants [64]. Addition of these amounts of crown ether to the phosphate reverse micelle led to phase separation.

The water structure in the reverse micelles may also play a role in destabilizing ZnPO-X. Figure 16a com-

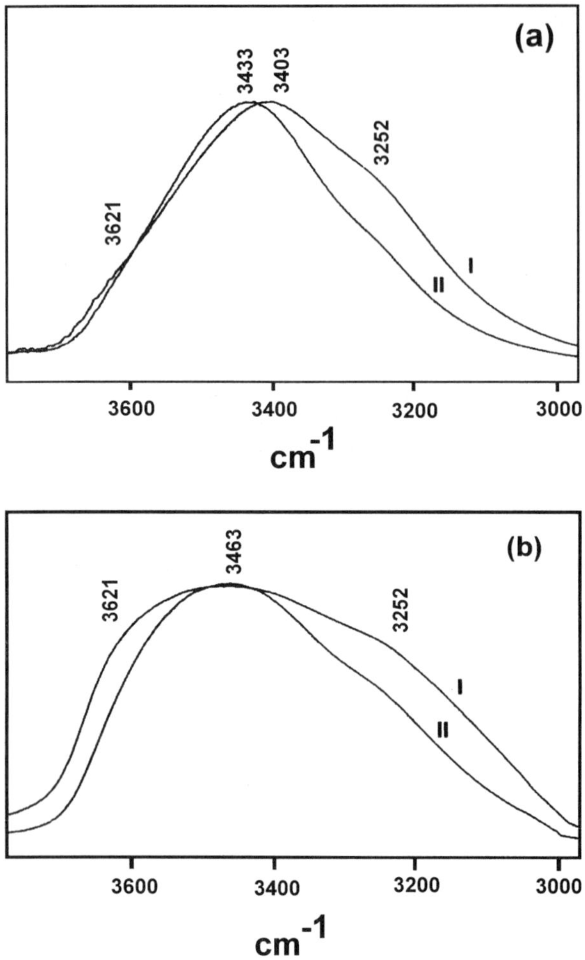

FIG. 16 Infrared spectra in the O—H stretching region for (a) H_2O (I), aqueous NaCl solution (1 M) (II) and (b) reverse micelle system with H_2O (I) and both Zn and P (II) (reverse micelles made with AOT). (Adapted from Ref. 75.)

pares the infrared (IR) spectra of the O—H stretching region in pure water with aqueous saturated NaCl. Three types of water have been proposed to exist around the added ions with different peak O—H stretching frequencies [65,66]: an innermost region of weakly hydrogen bonded, ion-immobilized water (type I, 3600 cm^{-1}), an intermediate structure broken region (type II, 3450 cm^{-1}), and an outer region with the normal liquid water structure (type III, 3350 cm^{-1}). At high NaCl concentrations, there is a decrease in intensity of the type I and III structures relative to type II structure. This would suggest a disordered, structure-broken form of water at high concentrations of salt [67].

Figure 16b compares the IR spectra of the ν O—H stretching in water-swollen AOT reverse micelles and the mixture of the Zn and P AOT reverse micelles [62]. Three types of water were also reported in reverse micelles [68,69]. These were bulklike water (3300 cm^{-1}), bound water separating the interface between bulk and the micelle interface (3490 cm^{-1}), and water trapped at the interface (3600 cm^{-1}). In the case of the reverse micelles that were used in making zincophosphates, there was a decrease in relative intensity of the bulklike and trapped water as compared with interfacial water even though the water-swollen micelles contained a large concentration of Na^+. This effect arose primarily from the Zn micelles. Why Zn^{2+} causes this enhanced disordering of water is not quite clear. Nevertheless, there is a parallel observation between NaCl aqueous solutions and Zn-containing AOT reverse micelles, i.e., an increase in disordered water. Because the water structure plays an important role in nucleation of open-pore frameworks, the intramicellar environment is not appropriate for crystallization of ZnPO-X.

This led to the study of nonionic surfactants, in particular, the polyoxyethylene surfactants [70]. Zincophosphate sodalite was readily synthesized from reverse micelle using Tween 85 in the presence of isopropanol in hexane solvent. However, attempts at preparing ZnPO-X from Tween 85 were unsuccessful.

IV. SYNTHESIS OF ZnPO-X FROM DODMAC REVERSE MICELLES

A. Nature of Zinc and Phosphate Reverse Micelles

Cationic micelles generally solubilize less water than their anionic counterparts. However, Vera and coworkers [71–74] reported that reverse micelles of the two-tailed cationic surfactant dioctyldimethylammonium chloride (DODMAC) had a high water uptake capacity. Most of the applications of DODMAC reverse micelles have been in extraction of biological molecules from aqueous media. Several reports detailing the water uptake by DODMAC reverse micelles, influence of the counterion on the reverse micelle, and the role of alcohols as cosurfactants have been published.

The surfactant DODMAC, commercialized as Bardac-80, can be obtained from Lonza. Bardac-80 contains 80% DODMAC, 10% water, and 10% ethanol. The ethanol and a fraction of the water can be evaporated under vacuum. DODMAC (0.16 M) and the cosurfactant 1-decanol (0.225 M) dissolved in isooctane were used as the medium for making reverse micelles for ZnPO-X formation [75,76].

TABLE 6 Composition of the Zinc and Phosphate Aqueous Solutions (before and after Winsor II Equilibration) and of the Aqueous Core Inside the Reverse Micelles (Surfactant, DODMAC)

Solution	Original solution		Remaining behind		In micelle (M)		Analytical method
Phosphate	P	0.66 M	P	0.63 M	P	0.69	ICP–OES
	Na	0.32 M	Na	0.38 M	Na	0.25	ICP–OES
	TMA^+	1.38 M	TMA^+	1.90 M	TMA^+	0.79	Raman
	pH	12.2	pH	12.3			
Zinc	Zn	0.366 M	Zn	0.11 M	Zn	0.70	ICP–OES

ICP–OES: Inductively Coupled Plasma–Optical Emission Spectroscopy.
Source: Adapted from Ref. 75.

Because there has been no previous report of using DODMAC-based reverse micelles for synthesis of materials, we provide some details of the nature of the zinc, phosphate, and water reverse micelles formed with this system [75]. In the preparation of the Zn and P reverse micelle solutions, it was noted that a turbid mixture was formed after the aqueous and organic phases were mixed. After a few minutes, two phases (an organic phase at the top and an aqueous phase at the bottom) were distinguishable, characteristic of a Winsor II system. The turbidity of the organic phase, due to water droplets suspended in it, disappeared during the first 2 days. In the case of the H_2O reverse micelle solution, mixing the organic and a limited volume of distilled water produced a slightly cloudy mixture. After additional shaking, the mixture turned into a completely clear single-phase solution characteristic of a Winsor IV system.

Table 6 details the aqueous solutions used for preparing the micelles, analysis of the aqueous solutions after equilibration with the surfactant solution, and the composition of the reverse micelle. The tetramethylammonium (TMA) ion in the phosphate reverse micelle was used as a structure-directing agent for ZnPO-X.

Table 7 shows the water uptake (ϕ_w) of the reverse micelles, the micelle size, the polydispersity, and the conductivity. The aqueous solution uptake of the Zn and P reverse micelles was calculated from the excess volume of aqueous phase recovered from each Winsor II system. Approximately 1.4 mL of solution A (phosphate solution) and 1.3 mL of solution B (zinc solution) went into the 50 mL of reverse micellar solutions. The P micelles solubilize slightly more water than the Zn micelles. This is not surprising because the P micelles contain Cl^-, OH^-, and PO_4^{3-} as counterions whereas the Zn micelles contain Cl^- and NO_3^-. Studies of DODMAC reverse micelles have found that, relative to chloride, polyvalent anions and hydroxide increased the water uptake in these micelles and nitrate decreased the water uptake. No significant differences in the water uptake between different cations have been reported for DODMAC micelles [77]. Note that some of these concentrations are higher in the remaining solutions because of preferential water uptake.

Figure 17a and b show the change in particle size and conductivity of the different reverse micelles as water uptake (ω_0) increases. In the case of the Zn and water micelles, in the initial stages of water uptake, there is an increase in size as a function of water incorporation, followed by a decrease. This could be due to change in shape of the particle from a cylindrical to a more spherical shape. The Zn and P conductivity pro-

TABLE 7 Results of Light Scattering and Conductivity Measurements Performed on Water, Phosphate, and Zinc Reverse Micelle Solution[a]

Reverse micelles	ϕ_w (% v/v)	Micelle size (nm)	Polydispersity index	Conductivity (μS/cm)
H_2O	3.1	23.2 ± 0.1	0.077	7.8
P	3.1	8.9 ± 0.1	0.005	0.65
Zn	2.9	20.7 ± 0.2	0.081	2.0

[a] ϕ_w represents the aqueous volume uptake per volume unit of reverse micelle solution (surfactant, DODMAC).
Source: Adapted from Ref. 75.

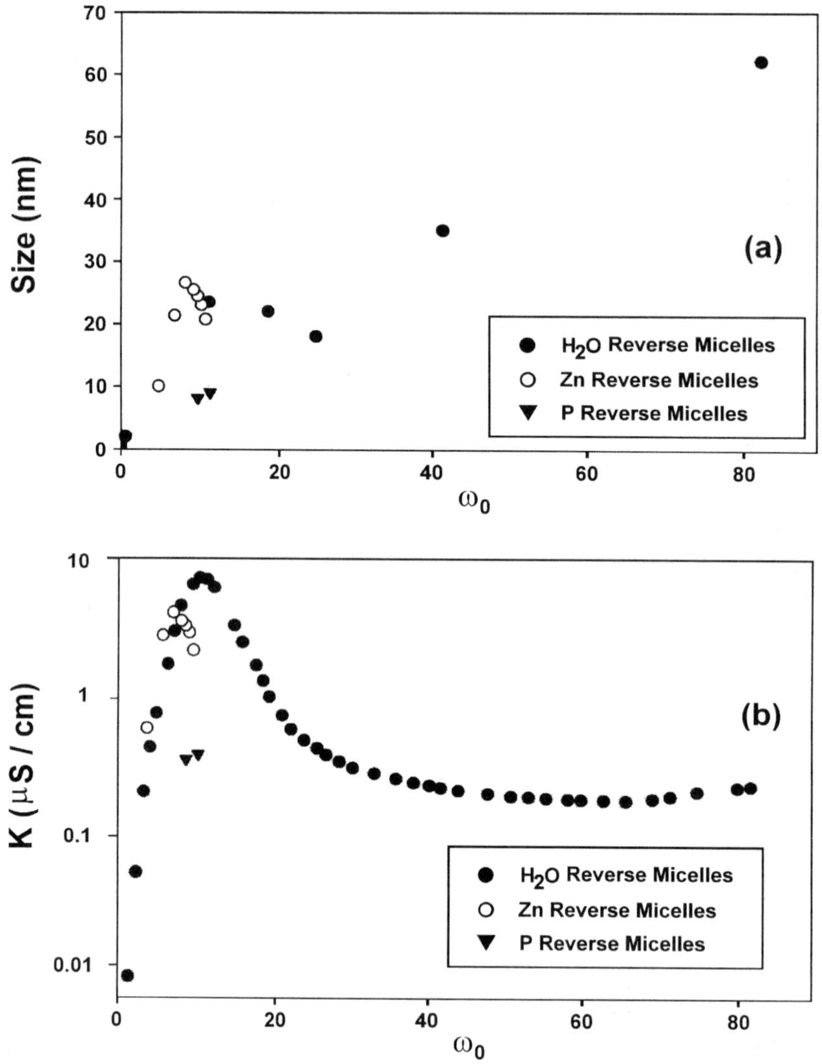

FIG. 17 (a) Size of H_2O, Zn, and phosphate reverse micelles (DODMAC) as a function of water content. (b) Conductivity of H_2O, Zn and phosphate reverse micelles (DODMAC) as a function of water content. (Adapted from Ref. 75.)

files appear at lower values than that of H_2O. This is consistent with the literature, the conductivity values for pure water reverse micelles being generally larger than for reverse micelles containing electrolytes [78]. It is interesting to observe that, in the case of H_2O reverse micelles, a percolation phenomenon is not observed at any water uptake. This behavior has also been observed in other cationic reverse micelle systems [79].

Infrared spectroscopy of the water in the AOT and DODMAC reverse micellar core also provided information on the differences between these two reverse micelles. Figure 18 compares the IR spectra of bulk water, water inside reverse micelles of AOT, and water inside reverse micelles of DODMAC. As discussed earlier, the high-, medium-, and low-frequency parts of the OH stretching band region are due to nonhydrogen, stressed-hydrogen, and hydrogen-bonded O—H. These three types of O—H observed in water and AOT are also present in DODMAC reverse micelles. In the case of both surfactants, the low-frequency region that describes the hydrogen-bonded water seems substantially less intense than that found in bulk water. This is probably not surprising considering the ions that are present in both of these micelles. Differences are observed in the interfacial water. Because of strong interactions with the polar headgroup and sodium counterion, AOT reverse micelles contain larger distributions of non-hydrogen-bonded water molecules at the interface. In

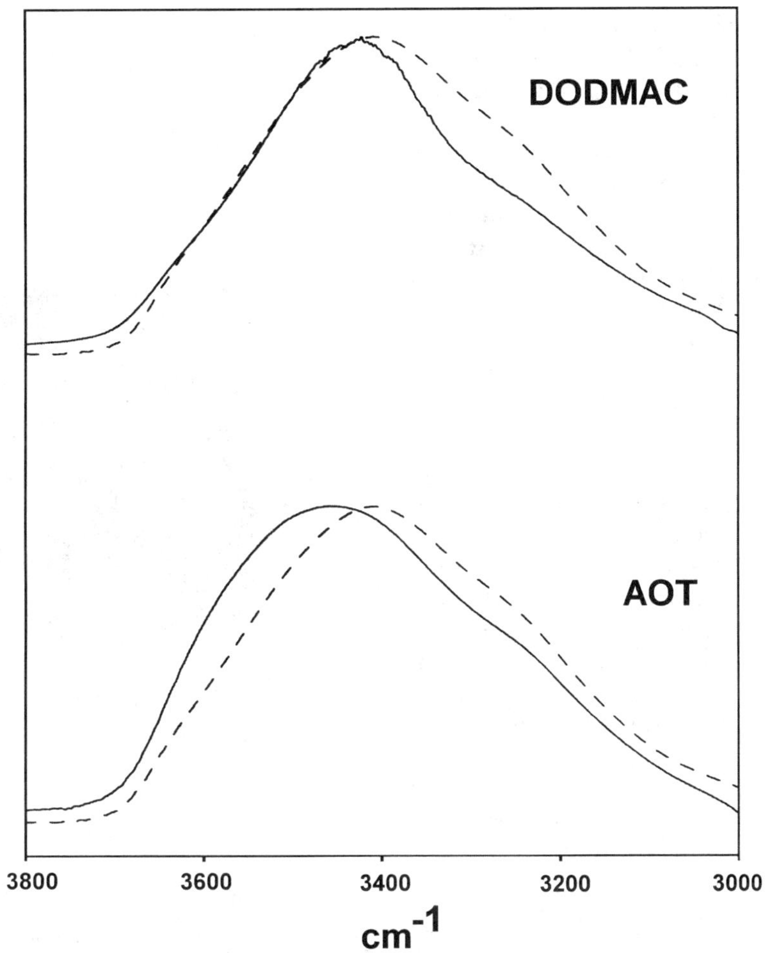

FIG. 18 Infrared spectra in the O—H stretching region for AOT and DODMAC reverse micelles (dashed line represents pure water). (Adapted from Ref. 76.)

contrast, DODMAC does not disrupt the water structure at the interface because quaternary ammonium salts have a "structure-making" influence.

B. Synthesis of ZnPO-X Using DODMAC Reverse Micelles

The primary goal of using DODMAC reverse micelles was to synthesize porous zincophosphate frameworks, in particular ZnPO-X. Two templating agents have been studied, tetramethylammonum ion and 1,4-diazabicyclo[2,2,2]octane (DABCO), whose molecular structures are shown in Fig. 19. We discuss these results separately.

1. TMA as a Templating Agent

Stucky and coworkers reported the synthesis of ZnPO-X at 4°C, and the procedure needed to be modified to carry it out at room temperature (25°C) [32]. The changes required were to make the solution dilute and more basic and increase the Zn/Na ratio [62]. The optimized procedure for making ZnPO-X involved three DODMAC reverse micelles made via the equilibration method [79]. An aqueous solution P was prepared by

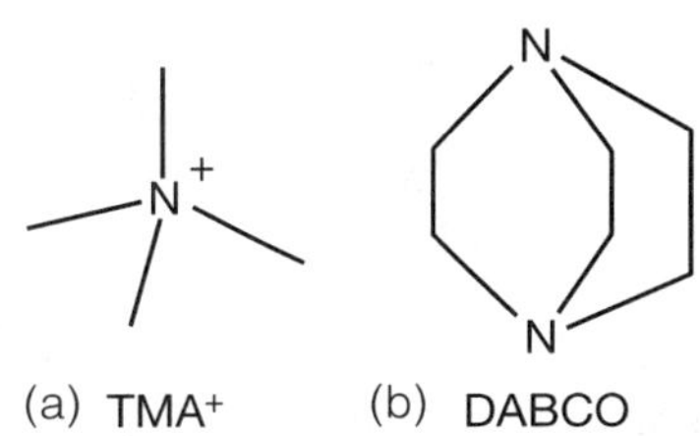

FIG. 19 Structures of the templating agents used for synthesis of ZnPO-X: (a) tetramethyl ammonium ion (TMA^+); (b) 1,4-diaazabicyclo[2,2,2]octane (DABCO).

combining 15.1 mL of 1.5 M H_3PO_4, 0.453 g of NaOH, 17 mL of 25% (w/w in water) TMAOH, and 2.15 mL of H_2O. Solution Zn contained $Zn(NO_3)_2$ (0.366 M). Solution TMA contained tetramethylammonium (TMA^+) bromide (1.0 M). Reverse micelles were made by equilibration of 3 mL each of P and Zn and 1 mL of TMA separately with 50 mL of the isooctane solution (containing the surfactant). The aqueous part that was not incorporated into the reverse micelle was removed from each solution. The three reverse micellar solutions were mixed in the Zn/P/TMA volume ratio of 6:10:5.

Particle growth in the solution was monitored by dynamic light scattering and showed an initial size of 12 nm for the reverse micelles. For the first 40 min, the size remained relatively constant, followed by accelerated growth of the particle size. Visual observations indicated that the solutions turned cloudy during the first 12 h, followed by the appearance of particles at the bottom of the reactor. The amount of settled particles increased with time. Upon ultracentrifugation of the mother liquor before any cloudiness was visually evident, a small amount of solid was recovered. These suspended crystals are small, on the order of a few hundred nm. Enough sample could not be recovered for diffraction analysis, but micro-Raman spectroscopy showed bands at 765, 983, 1014, and 1119 cm^{-1}, characteristic of faujasitic zincophosphate (ZnPO-X) [62]. The settled product was isolated by filtration, and Fig. 20a shows the x-ray powder diffraction pattern, which can be unambiguously assigned to ZnPO-X. Figure 20b shows the SEM picture of the sedimented crystallites, with sizes between 1 and 2 μm and the octahedral morphology expected of ZnPO-X.

Tetramethylammonium ion is used in the synthesis of ZnPO-X because it acts to template the faujasitic structure. Whether the tetraalkylammonium unit of the headgroup of DODMAC also has a preferential influence on the nucleation of ZnPO-X is of interest. Raman spectroscopy indicated entrapment of the TMA^+ in ZnPO-X made from DODMAC reverse micelles, but there was no indication that the surfactant was entrapped in the framework.

The dynamics of the crystal growth could be controlled to some degree by changing the Zn/P ratio. For example, if the Zn/P ratio was altered from 0.6 to 0.66, the reaction was accelerated by a factor of 4. This observation is consistent with studies in conventional hydrothermal synthesis of zincophosphates [33] and in the sodalite synthesis in AOT reverse micelles [35]. By increasing the Zn^{2+} concentration, the acidity of the reaction composition is increased and the rate of crystallization increases. Sodalite, hopeite, and the hexagonal phase $P6_1$ compete with ZnPO-X in the conventional zincophosphate system [33]. By altering the composition, it was possible to crystallize these phases from DODMAC reverse micelles; e.g., hopeite was formed at Zn/P ratios >0.8.

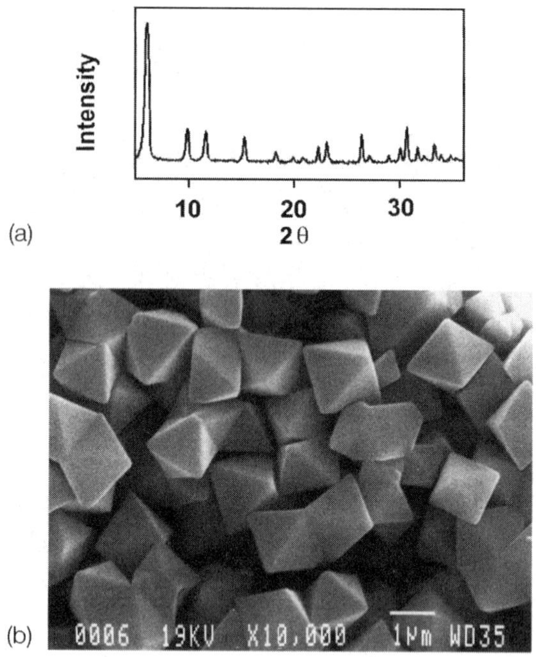

FIG. 20 (a) Powder x-ray diffraction pattern and (b) SEM of crystal obtained with DODMAC reverse micelles. (Adapted from Ref. 76.)

2. DABCO as a Templating Agent

With this templating agent, the DODMAC reverse micelles were made by an injection method. This method involves injecting a known volume of aqueous solution of the reactants into the surfactant solution, which completely goes into the reverse micelle formation and the system remains a single phase. The surfactant solution used was 0.2 M DODMAC and 0.33 M 1-decanol in isooctane. For the optimal synthesis of ZnPO-X, the following reactant micelle composition (abbreviated A) was developed. Two solutions were prepared: first, 0.16 M $Zn(NO_3)_2 \cdot 6H_2O$ aqueous solution; and second, 0.11 M NaOH, 0.58 M DABCO, and 0.27 M H_3PO_4 solution as another aqueous solution. In a typical reverse micelle reactant preparation, 100 mL of surfactant solution was placed in a bottle, 2 mL of an aqueous solution of a reactant was added, and the bottle was vigorously shaken for about 1 min and then equilibrated at room temperature (25°C). After 24 h of equilibration, the reverse micelle reactant solutions were used for synthesis. These reactants were stable over several weeks as no phase separation occurred at any time. Light scattering experiments showed that the diameters of the Zn- and P-containing reverse micelles were 8 and 6 nm, respectively.

Reaction was carried out by mixing the Zn and P micelles with volume ratio of 1:1. The reaction mixture was clear at the beginning of the reaction. Particles of the products started settling after about 8 h of reaction. This settling process became complete in about 3–4 days. The products were separated by centrifugation, washed with ethanol and water, and dried at room temperature under reduced pressure. Powder diffraction patterns as well as the octahedral morphology confirm the formation of the ZnPO-X structure. The yields of zincophosphate X from composition A were of the order of 15–20%, indicating that large fractions of the zinc and phosphate species were still present in the micellar medium. Light scattering indicated that the size of the clusters remaining in solution was of the order of 15 nm.

Table 8 compares the conventional composition for ZnPO-X with that used in reverse micelles. The ratio for the reactants is the same as for the conventional synthesis. However, the aqueous solution used for the reverse micelle preparation is about four times diluted. In the reverse micelle formation, some amount of water is used up for the hydration of the headgroup of the surfactant. Therefore, the effective composition and the concentration of the reactants in the water pool are probably similar to those under the normal solution synthesis conditions. Interestingly, when diluted aqueous solutions of the reactants were used for the normal solution synthesis keeping the same reactant ratio, no ZnPO-X was crystallized.

3. Seeding Experiments

Reverse micellar systems also provided a novel medium for studying seeding phenomena in the growth of microporous materials. The addition of seed crystals to speed up the crystallization process has been practiced for microporous material synthesis for four decades. Seeding can take various forms: addition of macroscopic seed crystals obtained from a previous synthesis, aging of reactants at a lower temperature to form seeds in situ, and addition of a "seed solution" created by a brief hydrothermal process to a reactant composition [80–82]. The mechanism for rate enhancement is eventually related to the small seeds. Macroscopic seeds promote nucleation by providing nuclei that exist on their surfaces (secondary process), and small seeds, by virtue of their high surface area, consume reactants and grow rapidly into crystals.

Two approaches to seeding have been examined in reverse micellar systems. The first dealt with taking well-washed ZnPO-X crystals prepared from composition A and added back to the reverse micelles of composition A. In order to keep the seed crystals suspended, the reaction was carried out in a rotating cell. Figure 21 shows the SEM pictures obtained from the synthesis. It appears that the seed crystals do not grow further and that a new crop of small crystals of reasonably uniform size grows on the seed crystals. The more interesting experiment dealt with taking the mother liquor from a composition A synthesis after the crystallization was completed and using this solution as a seed solution. As mentioned before, light scattering indicated that the size of the clusters present in the mother liquor was of the order of 15 nm.

To use the mother liquor as an effective seed solution, a second composition B involving three micellar solutions was prepared. This composition does not produce ZnPO-X and it takes about 2 days for a solid to first appear, thus providing a good source of nutrients. The three solutions in composition B were: 0.2 M $Zn(NO_3)_2 \cdot 6H_2O$, 0.11 M NaOH and 0.27 M H_3PO_4, and 0.58 M DABCO solution. Reaction was carried out by mixing zinc, phosphate, and DABCO reverse micelle solution with the volume ratio of 1:1:1. Upon adding the mother liquor to composition B, a large

TABLE 8 Composition of Reactants Used in Synthesis of ZnPO-X with DABCO as Structure-Directing Agent

Conventional synthesis	Aqueous solution used to prepare micellar solution	
	Composition A	Composition B
$[Zn^{2+}]$ = 0.8 M	$[Zn^{2+}]$ = 0.16 M	$[Zn^{2+}]$ = 0.20 M
$[Na^{+}]$ = 0.43 M	$[Na^{+}]$ = 0.11 M	$[Na^{+}]$ = 0.11 M
$[PO_4^{3-}]$ = 1.13 M	$[PO_4^{3-}]$ = 0.27 M	$[PO_4^{3-}]$ = 0.27 M
[DABCO] = 2.33 M	[DABCO] = 0.58 M	[DABCO] = 0.58 M

FIG. 21 SEM pictures of products from seeding experiment. (a) Dried ZnPO-X crystals as seeds. (b) Mother liquor from a completed ZnPO-X synthesis used as seed solution.

number of very uniform ZnPO-X crystals are produced as seen in Fig. 21b. The rodlike crystals are the product from composition B and are also found in the absence of the seed solution.

The seeding experiment involving addition of mother liquor to composition B was repeated in columns of different heights to examine the influence of longer suspension on crystal size. Figure 22 shows the SEM data for crystals obtained from columns of height 0.71, 1.78, and 2.62 m. With increasing column length, the crystal sizes increased and average ZnPO-X crystal sizes of 3, 6, and 15 μm were observed. The increase in size of crystals as a function of suspension time indicates that the crystals grow from the seed nuclei by incorporating nutrients, indicative of pathway A in sodalite growth from AOT reverse micelles.

FIG. 22 SEM pictures of ZnPO-X crystals obtained from columns of different heights: (a) 0.71, (b) 1.78, and (c) 2.62 m.

V. CONCLUSIONS

Reverse micelles have been demonstrated over several decades to be a unique reaction medium for synthesis of nanoparticles of a wide class of materials. This chapter demonstrates that microporous materials can also be synthesized under appropriate conditions from reverse micellar reactants. The metastable nature of microporous materials requires a well-controlled compositional environment for synthesis. This is in contrast to most syntheses that are carried out in the reverse micellar medium, which involve direct precipitation chemistry. Thus, the open-pore framework ZnPO-X could not be synthesized with AOT reverse micelles because the high levels of Na^+ in the water pool thwarted the nucleation of ZnPO-X and promoted the nucleation of sodalite. Understanding the reasons for failure of AOT reverse micelles in growing ZnPO-X led to the examination of cationic reverse micelles, especially the two-tailed surfactant DODMAC. ZnPO-X was successfully synthesized from DODMAC reverse micelles with two different structure-directing agents, tetramethylammonium ion and 1,4-diazabicyclo-[2,2,2]octane. Reverse micelles could be made by equilibration with excess aqueous phase containing the zinc and phosphate reactants (Winsor II) or by injection of the correct amount of aqueous phase into the surfactant-cosurfactant-isooctane medium (Winsor IV).

Several other unique aspects of the reverse micellar medium were discovered as compared with conventional synthesis. A particularly interesting feature was the control over crystallization pathways by minor changes in the reactant composition. This was well demonstrated in the crystallization of sodalite from AOT reverse micelles. Depending on the intramicellar pH, the fraction of the reverse micelles in which supersaturation conditions were achieved could be varied. For compositions in which the fraction was low, the number of nuclei formed was small and these nuclei grew slowly by incorporating nutrients from nonnucleated reverse micelles. On the other hand, for compositions where the fraction of reverse micelles in which supersaturation was exceeded was large, rapid precipitation of an amoprhous zincophosphate occurred from which sodalite crystals appeared, much like the observations in conventional synthesis. The reason that reverse micelles provide this degree of control is that they provide compartmentalized reaction centers that can exist in the same system but with different compositions.

A second discovery related to the DODMAC reverse micelles was the use of this medium as a seed solution. Typical synthesis yields of ZnPO-X from DODMAC reverse micelles were of the order of 15–20%, indicating that at the end of the synthesis reaction, a large fraction of the reactants are still present in the organic medium. They were found to be present as clusters of size 15 nm and would grow rapidly into ZnPO-X crystals if provided with a source of zinc and phosphate nutrients.

Because reverse micelles provide a novel reaction environment, it is expected that in the future new framework structures can be synthesized. Extension of reverse micellar systems for synthesis of mesoporous materials is also of interest. Reverse micelles that are more stable at higher temperatures can be used for the synthesis of zeolites (aluminosilicates).

ACKNOWLEDGMENTS

We acknowledge funding from NASA.

REFERENCES

1. R. Szostak, *Handbook of Molecular Sieves*, Van Nostrand, New York, 1992.
2. J. Weitkamp, H. G. Karge, H. Pfeifer, W. Holderich, eds., *Zeolites and Related Microporous Materials: State of the Art 1994*, Elsevier, Amsterdam, 1994.
3. A. Corma and H. Garcia, Catal. Today *38*:257–308 (1997).
4. E. R. Geus, M. J. den Exter, and H. van Bekkum, J. Chem. Soc. Faraday Trans. *88*:3101–3109 (1992).
5. H. G. Smolka and M. Schwuger, Colloid Polym. Sci. *256*:270–277 (1978).
6. J. W. Ward, Appl. Ind. Catal. *3*:271–392 (1984).
7. D. W. Breck, *Zeolite Molecular Sieves*, Wiley, New York, 1974.
8. M. E. Davis and R. F. Lobo, Chem. Mater. *4*:756–768 (1992).
9. C. T. G. Knight, Zeolites *10*:140–144 (1990).
10. R. M. Barrer, *Hydrothermal Chemistry of Zeolites*, Academic Press, London, 1982.
11. E. N. Coker, J. C. Jasen, J. A. Martens, P. A. Jacobs, F. DiRenzo, F. Fajula, and A. Sacco, Jr., Microporous Mesoporous Mater. *23*:119–136 (1998).
12. E. Engelhardt and H. van Koningsveld, Zeolites *10*: 650–656 (1990).
13. S. L. Burkett and M. E. Davis, Chem. Mater. *7*:920–928 (1995).
14. P.-P. E. A. de Moor, T. P. M. Beelen, and R. A. van Santen, J. Phys. Chem. B *103*:1639–1650 (1999).
15. P. K. Dutta, J. Inclusion Phenom. Mol. Recognit. Chem. *21*:215–237 (1995).
16. P. K. Dutta, M. Puri, and D. C. Shieh, Mater. Res. Soc. *111*:101–106 (1988).

17. P. K. Dutta, M. Puri, and C. Bowers, ACS Symp. Ser. *398*:98–109 (1989).
18. H. G. Hertz, Angew. Chem. Int. Ed. Engl. *9*:124–138 (1970).
19. M. Bourrel and R. S. Schechter, *Microemulsion and Related Systems: Formation, Solvency and Physical Properties*, Surfactant Science Series, Vol. 30, Marcel Dekker, New York, 1988.
20. B. H. Robinson, Nature, *320*:309 (1986).
21. M. Kotlarchyk, J. S. Huang, and S. H. Chen, J. Phys. Chem. *89*:4382–4346 (1985).
22. J. Eastoe, B. H. Robinson, D. C. Steytler, and D. Thorn-Leeson, Adv. Colloid Interface Sci. *36*:1–31 (1991).
23. P. D. I. Fletcher, A. M. Howe, and B. H. Robinson, J. Chem. Soc. Faraday Trans. 1 *83*:985–1006 (1987).
24. C. Oldfield, J. Chem. Soc. Faraday Trans. *87*:2607–2612 (1991).
25. F. M. Menger and K. Yamada, J. Am. Chem. Soc. *101*: 6731–6734 (1979).
26. M. P. Pileni, J. Phys. Chem. *97*:6961–6973 (1993).
27. J. H. Fendler, Chem. Rev. *87*:877–899 (1987).
28. M. L. Steigerwald, A. P. Alivisatos, J. M. Gibson, T. D. Harris, R. Kortan, A. J. Muller, A. M. Thayer, T. M. Duncan, D. C. Douglas, and L. E. Brus, J. Am. Chem. Soc. *110*:3046–3050 (1988).
29. W. F. C. Sager, Curr. Opin. Colloid Interface Sci. *3*: 276–283 (1998).
30. V. Pillai, P. Kumar, M. J. Hou, P. Ayub, and D. O. Shah, Adv. Colloid Interface Sci. *55*:241–269 (1995).
31. P. K. Dutta and D. Robins, Langmuir *7*:1048–1050 (1991).
32. T. E. Gier and G. D. Stucky, Nature *349*:508–510 (1991).
33. T. E. Gier, W. T. A. Harrison, T. M. Nenoff, and G. D. Stucky, *Molecular Sieves*, Van Nostrand Reinhold, New York, 1992, pp. 407–426.
34. P. K. Dutta, K. S. N. Reddy, L. M. Salvati, and M. Jakupca, Nature *374*:44–46 (1995).
35. K. S. N. Reddy, L. M. Salvati, P. K. Dutta, P. B. Abel, K. I. Suh, and R. R. Ansari, J. Phys. Chem. *100*:9870–9880 (1996).
36. J. Felsche, S. Luger, and C. Baerlocher, Zeolites *6*:367–372 (1986).
37. D. M. Bibby and M. P. Dale, Nature *317*:157–158 (1985).
38. G. O. Roberts, D. M. Kornfeld, and W. W. Fowlis, J. Fluid Mech. *229*:555–567 (1991).
39. R. E. Smith and P. L. Luisi, Helv. Chim. Acta *63*:2302–2311 (1980).
40. J. W. Mullin, *Crystallization*, Butterworth-Heinemann, Oxford, 1993.
41. O. Sohnel and J. Garside, *Precipitation—Basic Principles and Industrial Application*, Butterworth-Heinemann, London, 1992.
42. J. L. Meyer and G. H. Nancollas, Arch. Oral. Biol. *17*: 1623–1627 (1972).
43. M. Siddiq, T. Gilanyi, and A. C. Zettlemoyer, J. Colloid Interface Sci. *64*:192–193 (1978).
44. S. Lazic, J. Cryst. Growth *147*:147–154 (1995).
45. H. Hohl, P. G. Koutsoukos, and G. H. Nancollas, J. Cryst. Growth *57*:325–335 (1982).
46. J. B. Jensen and J. Lou, Acta Chem. Scand. *A37*:617–629 (1983).
47. J. H. A. van der Woude and P. L. deBruyn, Colloids Surf. *8*:55–78 (1983).
48. F. M. Menger, J. A. Donohue, and R. F. Williams, J. Am. Chem. Soc. *95*:286–288 (1973).
49. J. G. Darab, D. M. Pfund, J. L. Fulton, J. C. Linehan, M. Capel, and Y. Ma, Langmuir *10*:135–141 (1994).
50. T. Hirai, H. Sato, and I. Komasawa, Ind. Eng. Chem. Res. *32*:3014–3019 (1993); *33*:3262–3266 (1994).
51. A. G. Walton, in *Dispersion of Powder in Liquids*, 2nd ed. (G. D. Parfitt, ed.), Applied Science Publishers, London, 1973, pp. 175–220.
52. E. Kirkova and M. Djarova, *Industrial Crystallization*, Proc. Symp. 7th, Meeting Date 1978, 1979, pp. 81–89.
53. H. Furedi-Milhofer, E. Oljica-Zabcic, B. Purgaric, B. Kosar-Grasic, and N. Parkovic, J. Inorg. Nucl. Chem. *37*:2047–2051 (1975).
54. C. Petit, P. Lixon, and M.-P. Pileni, J. Phys. Chem. *97*: 12974–12983 (1993).
55. S. D. Loades, S. W. Carr, D. H. Gay, and A. L. Rohl, J. Chem. Soc. Chem. Commun. 1369–1370 (1994).
56. K. Osseo-Asare and I. J. Arriagada, Colloid Surf. *50*: 321–339 (1990).
57. K. Kandori, K. Kon-No, and A. Kitahara, J. Dispersion Sci. Technol. *9*:61–73 (1988).
58. O. Sohnel and J. W. Mullin, J. Cryst. Growth *60*:239–250 (1982).
59. J. L. Meyer and C. C. Weatherall, J. Colloid Interface Sci. *89*:257–267 (1982).
60. W. Kibalczyc, I. V. Melikhov, and V. F. Komarov, *Industrial Crystallization*, Proc. Symp. 8th, Meeting Date 1981, 1982, pp. 291–292.
61. O. Sohnel and J. Garside, *Precipitation—Basic Principles and Industrial Application*, Butterworth-Heinemann, London, 1992, chapter 4.
62. M. J. Castagnola and P. K. Dutta, Microporous Mesoporous Mater. *20*:149–159 (1998).
63. M. Wong, J. K. Thomas, and T. Nowak, J. Am. Chem. Soc. *99*:4730–4736 (1977).
64. E. T. Roach, P. R. Handy, and A. I. Popov, Inorg. Nucl. Chem. Lett. *9*:359–363 (1973).
65. H. R. Wyss and M. Falk, Can. J. Chem. *48*:607–614 (1970).
66. W. A. P. Luck, in *Water, A Comprehensive Treatise, Water in Crystalline Hydrates: Aqueous Solutions of Simple Non-Electrolytes* (F. Franks, ed.), Vol. 2, Plenum, New York, 1973, chapter 4.
67. R. E. Verrall, in *Water, A Comprehensive Treatise,*

Aqueous Solutions of Simple Electrolytes (F. Franks, ed.), Vol. 3, Plenum, New York, 1973, pp. 211–264.

68. G. Onori and A. Santucci, J. Phys. Chem. *97*:5430–5434 (1993).
69. T. K. Jain, M. Varshney, and A. Maitra, J. Phys. Chem. *93*:7409–7416 (1989).
70. C. F. Komives, D. E. Osborne, and A. J. Russel, J. Phys. Chem. *98*:369–376 (1994).
71. W. Wang, M. E. Weber, and J. H. Vera, J. Colloid Interface Sci. *168*:422–427 (1994).
72. M. S. Zabalay and J. H. Vera, J. Chem. Eng. Data *41*: 1499–1504 (1996).
73. W. Wang, M. E. Weber, and J. H. Vera, Ind. Eng. Chem. Res. *34*:599–606 (1995).
74. W. Wang, M. E. Weber, and J. H. Vera, Biotechnol. Bioeng. *46*:343–350 (1995).
75. M. J. Castagnola, Thesis dissertation, Ohio State University, 1999.
76. M. J. Castagnola and P. K. Dutta, Microporous Mesoporous Mater. *34*:61–64 (2000).
77. H. R. Rabie, M. E. Weber, and J. H. Vera, J. Colloid Interface Sci. *174*:1–9 (1995).
78. A. Derouiche and C. Tondre, J. Dispersion Sci. Technol. *12*:517–530 (1991).
79. A. Jada, J. Lang, R. Zana, R. Makhloufi, E. Hirsch, and S. J. Candau, J. Phys. Chem. *94*:387–395 (1990).
80. E. Narita, J. Cryst. Growth *78*:1–8 (1986).
81. J. Warzywoda and R. W. Thompson, Zeolites *11*:577–582 (1991).
82. L. V. Hou and R. W. Thompson, Zeolites *9*:526–530 (1989).

37

Mesoscopic Films at Interfaces

SRINIVAS MANNE, R. K. WORKMAN, and J. L. WOLGEMUTH University of Arizona, Tucson, Arizona

I. INTRODUCTION

The rapid growth of mesoscopic materials in the last decade or so confirms the adage that a gradual scientific understanding usually *follows* rapid technological development rather than the other way around. The synthesis of surfactant-templated mesoporous silica by scientists at Mobil research laboratories [1] triggered an explosive technological development in mesoscopic materials that continues apace; the original paper of Beck et al. has, in the past 8 years, garnered well over a thousand references. The basic idea of surfactant templating has since been extended to the synthesis of a large variety of oxides, sulfides, and metals and to a variety of material geometries—powders, crystals, films, and fibers [2]. Although great strides have been made in reaction optimization and especially morphological control of the end product, we have only recently begun to understand how intermolecular and surface forces among surfactant molecules, interfaces, and soluble precursor ions actually determine the ultimate morphology.

Surfactant-templated synthesis relies on inorganic polymerization (or, less often, heterogeneous nucleation) at the interfacial region between a surfactant aggregate and a solution in which the inorganic precursors are initially dispersed (Fig. 1). If the surfactant is soluble (which is most often the case), the surfactant-solution "interface" is dispersed in the form of micelles, and interfacial reactions yield a surfactant-inorganic nanocomposite in the form of a lyotropic liquid crystalline phase. The specific nanocomposite phase depends critically on the interactions between surfactant molecules and the solubilized inorganic ions in ways that are not yet fully understood. Controlled heating of the precipitate condenses and rigidifies the inorganic phase while pyrolyzing the surfactant phase, resulting finally in a mesoporous material with voids in the size range of micelles. By using both conventional surfactants and amphiphilic block copolymers, mesoporous oxides with void periodicities of 3–30 nm have been produced.

Early work on mesoscopic materials focused primarily on bulk synthesis, which yielded fine powders with typically submicrometer particle sizes [1]. These were primarily useful for investigating the pore structure and organization, via powder diffractometry, adsorption isotherms, and so on. The arrangement of void spaces has been referred to as the material's "primary structure" [3], corresponding to its mesoscopic lattice symmetry or space group. However, future applications of these materials hinge on controlling not only the primary structure but also the crystallographic coherence and spatial orientation of the pore lattice over macroscopic length scales. Thus, much of the recent literature on mesoscopic materials has focused on controlling these so-called secondary and tertiary structures. Secondary structures refers to the persistence length or coherence length (domain size) of a given lattice orientation, and tertiary structure refers to the orientation of the domains with respect to each other and with respect to the "lab frame" over macroscopic distances [3].

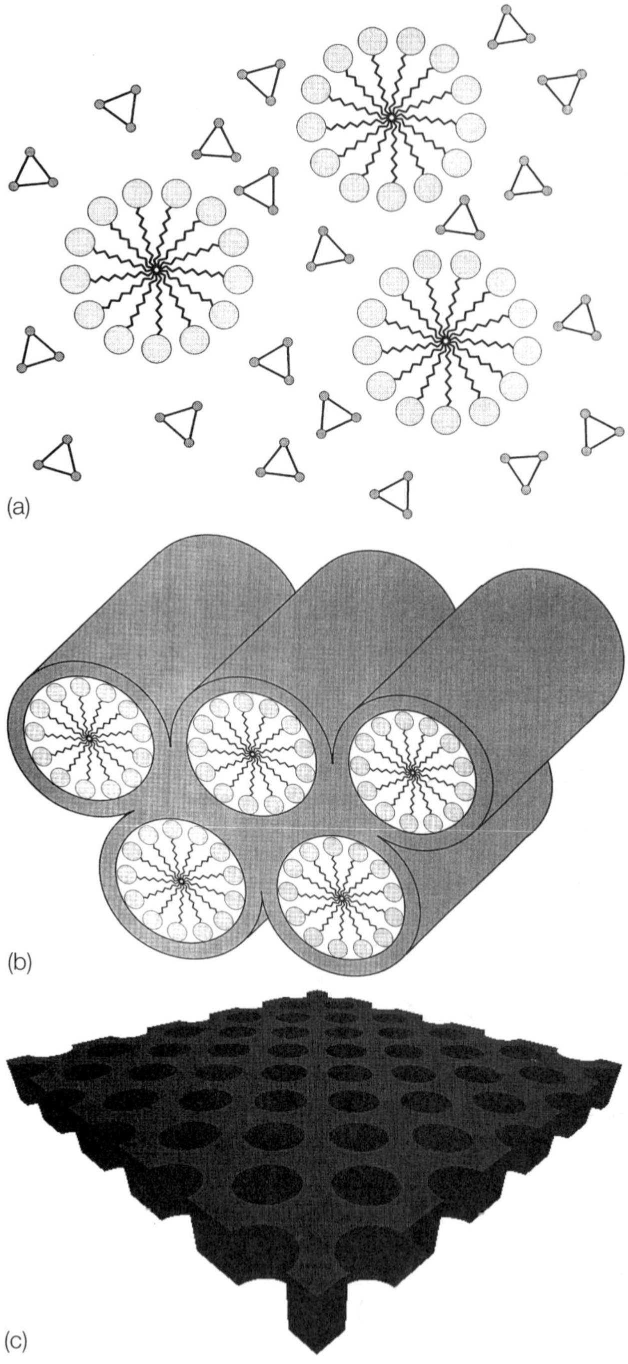

FIG. 1 Formation of a hypothetical mesoporous membrane (schematic). (a) Surfactant micelles initially coexist in solution with solubilized silicate species (triangles). (b) Surfactants and silicates begin to coassemble in solution to form surfactant-silicate mesophases. The silicates oligomerize and cross-link around the micelles. (c) Pyrolysis and calcination vaporize the surfactant and complete the silicate condensation, resulting in a membrane with regularly spaced holes of mesoscopic dimensions. An idealized membrane is shown with holes running perpendicular to the film plane; such a configuration has not yet been achieved.

In one way or another, nearly all attempts to achieve control over such secondary and tertiary structures have used *interfaces*, and the reaction products have been in the form of *mesoscopic films*. Typically, these films have submicrometer thicknesses but macroscopic lengths and widths and are composed of mesoscopic subunits with unit cells in the size range of micelles. In this chapter we review both the technological development of these films and our current state of scientific understanding of their self-assembly. This is intended not to be an exhaustive literature review but rather an attempt to find order and coherence in a rapidly expanding field—an attempt to identify the overall scientific and technological goals, achievements, and challenges ahead.

Technologically, the goal of surfactant-templated film synthesis can be identified as a robust membrane of *finite and controllable thickness* and *arbitrary lateral size*, whose pores are *individually addressable* and arranged in a lattice that is *crystallographically coherent* over the entire membrane surface. The simplest example is a thin film containing columnar pores normal to the film plane and arranged in a lattice (Fig. 1c). Such a membrane would find immediate use as a molecular filter, perhaps in separating polymers by molecular weight or separating branched from linear hydrocarbons. Membranes with uniformly functionalized pore walls can serve as catalytic supports, and the confined reaction space can be used to catalyze the synthesis of nanoparticles or nanowires. On the other hand, membranes with locally functionalized pores, where the chemical functionality varies in a known way across the membrane surface, can be used as chemical or biological sensors and separators. Although the ideal end product of Fig. 1c has yet to be realized by surfactant-based templating, several innovations *have* produced membranes of macroscopic size and with far greater control over secondary and tertiary structure than was previously available.

Scientifically, the ultimate goal is to successfully explain and predict the primary, secondary, and tertiary structures given the surfactant geometry, inorganic precursor species, interface properties, and reaction conditions. The chief difficulty is the number of competing interactions. Self-assembly in pure surfactant solutions involves a competition between attractive (hydrophobic) interactions and repulsive (hydrophilic) interactions, whose interplay gives rise to the characteristic micelle curvature and lyotropic phase behavior [4]. Whereas surfactant self-assembly in pure solution is fairly well understood, the effects of interfaces and large multivalent ions have started becoming clear only in the last few years. As a result of this work, much of the *primary* structure of mesoscopic films can now be understood in the context of competing surfactant-surfactant, surfactant-precursor, and surfactant-interface interactions. However, secondary and tertiary structures still remain a challenge in terms of both technological control and theoretical understanding. Because this review in part concerns our current state of understanding, it is natural to begin by considering how interfaces and solubilized precursor ions affect the self-assembly morphology of surfactant molecules in solution.

II. SELF-ASSEMBLY: EFFECTS OF INTERFACES AND PRECURSOR SPECIES

The self-assembly of soluble surfactants in free aqueous solution has been experimentally investigated and theoretically modeled for several decades, and it is worth briefly recalling the salient features [4]. At very low concentrations, surfactants dissolve completely and exist as solubilized monomers in water. But above the critical micelle concentration (cmc), surfactant molecules begin to self-assemble into finite liquid crystalline aggregates (micelles). This aggregation is *driven* by intertailgroup attractions (hydrophobic and van der Waals forces) but is *limited* by interheadgroup repulsions (electrostatic and steric/entropic forces). The balance between these intermolecular forces determines the final shape of the micelle. To first order, micelle shapes are described by the "dimensionless packing parameter" [5] $g \equiv v/(a_0 l)$, where v and l are the volume and extended length, respectively, of the hydrocarbon chain, and a_0 is the optimal headgroup area per molecule in the micellar environment. Values of $g < 1/3$ lead to spherical micelles, whereas $1/3 < g < 1/2$ favors cylindrical micelles and $g > 1/2$ favors flat bilayers [5]. For a saturated, unbranched alkane, the chain volume v (in nm^3) and chain length l (in nm) are given in terms of the number of carbon atoms n by Tanford's formulas [6], based on known bond lengths and angles for alkanes:

$$v \approx 27.4 + 26.9 \times n$$

$$l \approx 0.154 + 0.127 \times n$$

For typical surfactants, $n > 10$ and the second term dominates the expressions for v and l. The packing parameter therefore becomes independent of n and is approximately given by $g \approx (0.212\ \text{nm}^2)/a_0$. Thus, the micelle shape is determined almost exclusively by the optimal headgroup area for typical single-tail surfactants. The most commonly encountered micelle shapes are approximately spherical and are composed of order ~100 molecules.

Above the cmc, the monomer concentration remains roughly constant, with the newly added molecules going exclusively toward micelle formation. Some surfactants at moderate concentrations express a *second* cmc, which is often identified with the onset of micelle shape change, e.g., from spherical to cylindrical [7]. At higher concentrations, surfactant micelles pack into regular arrays and begin to form lyotropic liquid crystalline phases. For conventional surfactants (single head, single tail), the first lyotropic phase formed is usually one of three types. Surfactants with the largest value of a_0, whose micelles remain spherical up to high concentrations, form a *discontinuous cubic phase*; this phase is not completely understood but is thought to consist of discrete micelles in some kind of cubic lattice [8]. On the other hand, surfactants with smaller values of a_0, which undergo a sphere → cylinder micelle shape transition at the second cmc, typically form a *hexagonal phase* consisting of cylindrical micelles arranged in a hexagonal pattern. Finally, surfactants with the smallest headgroup areas, which form micelles in the shape of flat bilayers, form *lamellar phases* consisting of stacked, uniformly spaced bilayers at high concentration. The first two phases—discontinuous cubic and hexagonal—are the most relevant for mesoporous film synthesis because the pyrolytic removal of surfactants (the discrete phase) leaves a stable continuous phase of silicates. In contrast, lamellar surfactant-silicate mesophases typically collapse upon pyrolysis because the silicate phase is not continuous in this case.

The presence of third components in the surfactant solution—such as foreign counterions, polymers, and interfaces—can dramatically alter the surfactant self-assembly and lyotropic phase progression. This is because foreign species modify intersurfactant interactions and introduce additional competing interactions with surfactant or solvent molecules. We now consider these effects in detail.

A. Self-Assembly at Planar Interfaces

Surfactants are, of course, generally attracted to interfaces. All hydrophobic interfaces attract surfactant tailgroups via hydrophobic and van der Waals interactions, and many hydrophilic interfaces attract surfactant headgroups via screened electrostatic and hydrogen-bonding interactions. These interactions can lead to effective interfacial concentrations that are orders of magnitude greater than bulk surfactant concentrations. In 1955 it was first proposed that this enrichment gave rise to distinct interfacial aggregates analogous to bulk micelles [9]. In the last few years, atomic force microscopy (AFM) confirmed this view and obtained the first direct images of interfacial micellar phases [10,11]. AFM studies have since rapidly elucidated the aggregate structures of a variety of surfactants (ionic, nonionic, and zwitterionic) at a variety of solid surfaces (hydrophilic and hydrophobic) in contact with aqueous micellar solutions [10–27]. Most investigations have used atomically flat model surfaces, often the ordered cleavage planes of layered solids such as mica and graphite. These surfaces have proved ideal both for self-assembly studies and for mesoscopic film synthesis (see later). We now review the salient results for interfacial self-assembly morphologies, with special emphasis on mica and graphite; representative AFM images are shown in Fig. 2.

1. Hydrophobic Solid Surfaces

On the cleavage plane of highly oriented pyrolitic graphite (HOPG), early results [10,11] showed that interfacial self-assembly was determined almost entirely by the crystalline anisotropy of the substrate. AFM images revealed aggregates in the form of parallel stripes oriented perpendicular to an underlying symmetry axis and spaced apart by roughly twice the length of a surfactant molecule (Fig. 2b). Upon comparison with adsorption isotherm data, the AFM images were interpreted as half-cylindrical micelles, where the bottom row of molecules is oriented horizontally along a symmetry axis. Similar parallel stripes were also found at the same relative orientation for the crystalline cleavage plane of MoS_2 [11]. The parallel half-cylindrical morphology has since been observed on graphite for a variety of charged [11,14,18] and uncharged surfactants [16,19,22], over a surprisingly wide range of surfactant geometries. In general, surfactants capable of cylindrical curvature (i.e., those exhibiting a bulk hexagonal phase) have self-assembled into oriented half-cylinders on crystalline hydrophobic surfaces. This high degree of surface control has been attributed to the anisotropy of interaction between the horizontal alkane tail and the crystalline surface [10]. As long as the tail exceeds a certain minimum length (found to be ~10 carbon atoms [22,25]), the linear contact area between the tail and surface is thought to enhance the direction sensitivity of the interaction, leading to adsorption directions along the symmetry axes.

When the anisotropy is removed—i.e., when the hydrophobic substrate is amorphous—interfacial aggregates appear to revert to half-spheres (Fig. 2d) for a variety of charged and uncharged surfactants [28]. This may be considered the "natural" curvature for surfactants that form spherical micelles in solution but are subjected to a hydrophobic boundary condition. (These results are, however, more equivocal than those for graphite, and observed morphologies on silanated silica are quite sensitive to surface flatness and hydrophobizing technique [22,28].)

2. Hydrophilic Solid Surfaces

The most widely studied systems in this class have been ionic surfactants on oppositely charged surfaces—primarily cationic surfactants on flat, anionic surfaces of mica and silica. Here surfactant headgroups interact electrostatically with the surface, leading to "full" aggregates above the cmc, with headgroups facing the solution as well as the surface. Aggregate morphologies depend on both the surfactant geometry and the density of oppositely charged surface sites. On amorphous silica, AFM studies with single-tail cationic surfactants have revealed globular interfacial aggregates that resemble pinned micelles [11]; the arrangement of these globular aggregates generally resembles a glassy phase, having a well-defined nearest neighbor distance (comparable to a micelle diameter) but lacking long-range order over the surface. Single-tail cationics on the cleavage plane of mica have shown both spherical and cylindrical aggregates, depending on the surfactant geometry. Surfactants with large or unusually repulsive headgroups (e.g., triethylammonium or divalent headgroups) give rise to spherical aggregates arranged in a hexagonal lattice (Fig. 2a) [14,17]; the symmetry axes of this "micelle lattice" have been observed to align with the underlying mica lattice, indicating interfacial epitaxy [14]. Cationics with comparatively smaller or less repulsive headgroups have revealed parallel, meandering stripes (Fig. 2c), consistent with close-packed but flexible cylindrical aggregates lying horizontally on the mica surface [11,14,17]. Here also the cylinders can be partially aligned by the surface lattice, although to a lesser degree than on graphite.

These results can be qualitatively understood by considering each oppositely charged surface site as a potential adsorption site for a surfactant molecule and therefore a potential nucleation site for an interfacial aggregate above the cmc. When the substrate charge density is smaller than that of the micelle surface, interfacial self-assembly is determined primarily by intermolecular interactions, and the natural curvature is only minimally perturbed; thus, we expect globular aggregates resembling bulk micelles at surfaces with low charge density. This is exactly what is observed on silica, whose surface potential is typically small com-

FIG. 2 AFM images (200 × 200 nm) of interfacial surfactant micelles obtained in situ in surfactant solutions at twice the critical micelle concentration (cmc). (a) The divalent gemini surfactant $C_{18}H_{37}N^+(CH_3)_2(CH_2)_3N^+(CH_2)_3 \cdot 2Br^-$ (or $C_{18\text{-}3\text{-}1}$ for short) self-assembles into hexagonally close-packed spherical micelles on the anionic cleavage plane of mica. (b) Sodium dodecyl sulfate (SDS) on hydrophobic graphite, showing rigid stripes consistent with parallel half-cylindrical micelles. (c) Tetradecyltrimethylammonium bromide (TTAB) on mica, showing flexible stripes consistent with full cylindrical micelles. (d) The nonionic surfactant octaethyleneglycol monododecyl ether (or $C_{12}E_8$ for short) on an amorphous hydrophobic surface (alkylated silica) showing hemispherical micelles.

pared with that of a micelle (~100 mV). On the other hand, if the effective charge density of the substrate exceeds that of micelles, adsorbed headgroups are forced closer together than would be the case in a micellar environment, and the effective packing parameter g increases. This can cause a transition to lower aggregate curvature for surfactants that are already close to a transitional value of g but not for surfactants whose g is far from a transitional value [14]. This is exactly what is observed on mica, whose *effective* charge density is very large ($\sim 2e^-/\text{nm}^2$) due to exchangeable surface cations. Thus alkyltrimethylammonium bromides, which have g values close to 1/3 and undergo sphere $\rightarrow$ cylinder transitions at moderate bulk concentrations, self-assemble as close-packed cylinders on mica (Fig. 2c). On the other hand, divalent surfactants, which have larger headgroup areas and lower g values, favor high curvature and self-assemble as close-packed spheres on mica (Fig. 2a). These hexagonally close-packed spheres can be induced to switch to parallel flexible cylinders by the addition of strongly binding counterions such as salicylate [14], which effectively lowers the micelle surface charge and increases the packing parameter.

For nonionic surfactants based on oligomeric ethylene oxide (EO) headgroups, a_0 and g are determined simply by the physical headgroup size or the number n of EO units; a large n leads to spherical micelles and discontinuous cubic phases in bulk solution, whereas a small n leads to bilayers and lamellar phases [29]. Experiments with nonionic surfactants on silica show a similar trend at this interface [22]. Surfactants that form spherical micelles in bulk also reveal globular interfacial aggregtaes (with no long-range order), and surfactants that form bilayers in solution also from bilayers at the silica surface. Here the headgroup adsorption, which serves a foundation for interfacial self-assembly, is driven not by electrostatics but by hydrogen bonding between the EO oxygens and the surface SiOH units. That the interfacial self-assembly is only minimally perturbed from bulk self-assembly indicates that interfacial headgroup areas are comparable to a_0 in bulk solution, which in turn implies that the density of H-bonding sites on a silica surface falls short of the density of H-bonding sites (i.e., oxygen atoms) at a micelle-solution interface.

3. Air-Liquid Interface

Although the air-water interface is amorphous, hydrophobic, and relatively flat (owing to the large surface tension of water), it differs in a crucial and obvious way from, say, hydrophobized silica: Surfactant molecules can *penetrate* the air-water interface, giving rise to a diffuse density profile rather than a sharp boundary. Neutron reflectometry studies have indicated that carbon atoms nearest to the headgroup are oriented normal to the interface, whereas outer carbon atoms are tilted progressively further from the normal, suggesting the presence of interfacial micelles [30]. This is also in agreement with molecular dynamics simulations [31]. So far it has not been possible to determine the micelle shape directly by AFM at the air-liquid interface.

4. Secondary and Tertiary Structures of Interfacial Surfactant Aggregates

A glance at Fig. 1 readily shows that interfaces, in addition to determining the curvature of *individual* aggregates, often influence how the aggregates are arranged relative to each other and relative to the surface. These can be regarded as the secondary and tertiary structures of the interfacial aggregate layer and are of key importance in the nucleation of mesoscopic films. The long-range order is most pronounced for crystalline substrates such as mica and especially graphite, where epitaxy between surfactants and surface enhances both the interaggregate order (secondary structure) and the orientation relative to the substrate symmetry axes (tertiary structure). However, it is important to note that epitaxy can also *interfere* with tertiary structure, because surface lattices typically have two or more symmetry directions that can serve as equivalent orientation axes for interfacial aggregates. For instance, half-cylindrical or cylindrical aggregates, instead of orienting along a single direction over the entire surface, typically show a "patchwork" domain structure with two or three equivalent orientations relative to the underlying lattice (see Fig. 2b). Although there has been no systematic study of domain sizes to date, they are usually in the $1-10\ \mu\text{m}$ range and may be sensitive to local defects in the underlying substrate. These domain boundaries present a significant impediment to the nucleation of oriented, continuous mesoscopic films.

III. SURFACTANT MICELLES: EFFECTS OF INORGANIC PRECURSORS

Solubilized inorganic species can completely alter the micelle shape and the lyotropic phase progression of surfactants. For example, highly hydrated anions (such as chloride) bind relatively weakly to alkyltrimethylammonium ions, and the consequent electrostatic repulsion between headgroups favors highly curved micelles and (often) discontinuous cubic phases. In contrast, weakly hydrated anions (such as bromide) bind more energetically to alkyltrimethylammonium headgroups, favoring sphere-to-rod transitions at moderate salt concentrations and (usually) hexagonal mesophases [32]. Similarly, during the synthesis of mesoscopic materials, inorganic precursor species do not simply "coat" the surface of existing surfactant micelles by polymerization; rather, inorganic species actively *alter* the self-assembly by binding to surfactant headgroups via electrostatic forces or hydrogen bonds.

The model of simultaneous self-organization of surfactants and polymerizable inorganics, discussed in the following, has alternatively been termed "cooperative assembly" or "cooperative templating."

Even in the first work reported by Mobil researchers [1], conditions that favored *spherical* micelles in pure surfactant solutions nevertheless yielded a hexagonal-phase nanocomposite composed of *cylindrical* micelles in the presence of dissolved silicates. The mechanisms by which inorganic ions change the preferred micelle curvature and induce mesophase formation has since been a central preoccupation in the science of mesoscopic materials. The structure of the final material itself does not give definitive insight into this problem because the effects of intervening processing steps (condensation, dehydration, pyrolysis) after solution-phase self-assembly are also unknown. This problem has been addressed by probing the structure of solution mesophases *before* the formation of a solid phase, and this work has led to valuable insights into the effects of precursor ions on self-assembly [33–36]. These results will be discussed in detail.

Inorganic-headgroup interactions depend on the state of the solubilized inorganic species, which in turn depends on the synthesis route. For the case of mesoscopic silica—by far the most common mesoscopic material synthesized—two general routes have been in common use: acidic and basic synthesis. In each case the starting material is usually an oily alkoxide precursor, typically tetraethoxysilane (TEOS) or, less frequently, tetramethoxysilane (TMOS). In aqueous solution these compounds quickly hydrolyze into monomeric and small oligomeric units that are soluble to varying degrees. In highly basic conditions, monomer units are fully deprotonated SiO_4^{4-} ions; linear or cyclic oligomers, composed of a few (typically <10) monomer units, are likewise anionic in basic conditions. In highly acidic conditions, monomers are fully protonated $Si(OH)_4$ groups; for $pH < pI_{silica} \approx 2$, oligomers are neutral or weakly cationic. The distribution and charge of monomers and oligomers are highly sensitive functions of pH [37]. Consequently, the nature of silicate-surfactant interactions also varies with pH.

A. Basic Synthesis Conditions

Basic synthesis was historically the first route to mesoscopic silica [1], and it is the more extensively studied and better understood of the two synthesis routes. Conceptually, anionic silicate monomers and oligomers are electrostatically attracted to cationic surfactant headgroups (typically trimethylammonium or TMA^+), eventually leading to high silicate concentrations and condensation reactions around surfactant aggregates. Thorough and beautiful studies [33–36] have revealed the detailed mechanism by examining the formation of the surfactant-silicate lyotropic liquid crystal in solution, independently of the subsequent polymerization of the inorganic phase. They reached similar conclusions, essentially confirming the cooperative templating models proposed earlier from the structure of the final product materials [38]. The results of these investigations can be summarized as follows.

Dynamic light scattering and nuclear magnetic resonance (NMR) data show that micelles of alkyltrimethylammonium halides, which are spherical in pure surfactant solution, become cylindrical at all concentrations in the presence of silicate anions [33,35]. Thus, the highly charged silicate ions evidently displace the halide ions, and their energetic binding to the cationic headgroups induces a sphere-to-rod transition. However, this is not, by itself, enough to cause surfactant-silicate mesophase formation; indeed, for $pH > 13.5$, where silicate ions primarily exist as monomers and are prevented from condensing, no mesophase formation is observed [33]. This behavior is exactly what would be expected of a "classical" surfactant solution whose concentration exceeds the second cmc but falls short of its hexagonal phase. In this regime, entropy and intermicellar electrostatic repulsions prevent further ordering among cylindrical aggregates. Thus, the original alkyltrimethylammonium halide has been transformed by ion exchange into an alkyltrimethylammonium silicate; this still behaves like a surfactant solution in its isotropic phase, albeit with a different micelle geometry.

Mesophases are observed to form only at $pH < 13.5$, as the silicate species begin to condense and oligomerize [34]. This is also a compelling result, because silicate condensation is accompanied by charge reduction, thereby reducing interaggregate repulsion, presumably to the point where attractive van der Waals interactions start to dominate [34]. Under the influence of net attractive interactions, cylindrical aggregates orient parallel to one another in a hexagonal pattern to minimize their free energy. As the aggregates line up adjacent to one another, silicate groups that were initially associated with a single micelle become free to cross-oligomerize with those of adjacent micelles, finally creating a hexagonal mesophase with partial cross-linking.

At this point the surfactant-silicate nanocomposite begins to phase-separate from the solution; however, it is still not a "solid" precipitate but rather a weakly cross-linked lyotropic liquid crystal. Interestingly, the degree of silicate oligomerization in this mesophase is found to be extremely well defined, markedly different from that of the water-rich phase, and highly specific to the headgroup type [34]. For example, when quaternary ammonium surfactants are used as templates, silicates in the mesophase are found to consist almost entirely of octamers (double-four rings) if the headgroup is trimethylammonium or hexamers (double-three rings) if the headgroup is tri*ethyl*ammonium, while the water-rich phase is dominated by monomers and dimers [34]. Thus, the intermediate oligomers formed *en route* to condensation are evidently quite sensitive to the headgroup geometry. A change in temperature at this point can shift the oligomer distribution away from its optimal value for the headgroup, and the resulting modification of electrostatic forces can in some cases cause a global transformation to a new surfactant-silicate mesophase [34,36]. Such phase transitions are a further indication that the dense phase is not yet solid at this point in the reaction.

Extensive silicate condensation and cross-linking occur gradually in the reaction vessel over a period of several days (and are completed by drying and heating the precipitate). This reaction step can be accelerated by destabilizing the solution via pH or temperature; however, this sometimes disrupts the delicate liquid crystal phase and gives poor yield qualities (G. D. Stucky, personal communication). In general, best results are obtained when the cooperative assembly of surfactants and silicates is encouraged as far as possible before the condensation step. This is corroborated by work showing that dilute micellar solutions and low silicate/surfactant ratios are important for ordered pore structures [36].

Interesting parallels exist between mesophase formation under basic conditions and interfacial surfactant aggregation on highly charged surfaces such as mica. In both cases, electrostatic attractions between headgroups and multivalent species create local regions of high surfactant concentrations that resemble lyotropic mesophases. Of course, the binding geometry is significantly different in the two systems; charge neutralization by silicates occurs symmetrically over the entire micelle surface, whereas charge neutralization by the mica plane occurs only on the bottom surface of the micelle. Nevertheless, when three-dimensional mesophases in basic surfactant-silicate synthesis are compared with two-dimensional mesophases on mica, we find that the latter morphologies are two-dimensional slices of the former for a wide variety of surfactant geometries [14]. Thus, similarities between headgroup-counterion and headgroup-"countersurface" interactions give rise to interfacial morphologies that can serve as faithful templates for three-dimensional mesophases. This is an important factor in successful mesoscopic film synthesis on mica.

B. Acidic Synthesis Conditions

Despite the early importance and extensive studies of basic synthesis, most mesoscopic materials today are synthesized under acidic conditions (typical pH < 2.5). The chief advantage of acidic synthesis is that it has been found to promote secondary and tertiary structure [39]. Under these conditions, solution-phase silicates are neutral or weakly cationic, and the nature and strength of their interactions with cationic surfactants are expected to be very different from those under basic synthesis conditions. It is therefore surprising—even confounding—that the two synthesis routes often yield the *same* mesoscopic product morphology. Attempts to model the self-assembly under acidic conditions have stressed the putative role of mediating counterions. In symbolic terms, self-assembly under basic conditions has been described as involving S^+I^- interactions (S = surfactant and I = inorganic), whereas under extremely acidic conditions the putative interaction has been denoted $S^+C^-I^+$ (C = surfactant counterion) [39]. Unfortunately, no solution-phase studies of surfactant-silicate liquid crystals have yet been reported for acidic conditions to confirm this model. It is worth noting that simple coions added to surfactant solutions do not, in general, significantly affect self-assembly or cause mesophase separation. If the $S^+C^-I^+$ model is correct, the I^+ species may be oligomeric and multivalent (partially condensed), thereby binding several headgroups simultaneously. However, even in this case, it is worth noting that copolyelectrolytes in solution are not known to bind significantly to ionic surfactants [40].

Acidic synthesis near the isoelectric point has also been used with *neutral* surfactants as the structure-directing agents, the so-called S^0I^0 pathway [41,42]. These surfactants have included alkylamines [41], oligoethyleneoxides [42], and, more recently, polyethyleneoxide-polypropyleneoxide (PEO-PPO) block copolymers [42,43]. Here also, the addition of silicate precursor to a dilute micellar solution of amphiphile leads to mesophase formation, phase separation, and condensation to form surfactant-silicate nanocomposites. Although mesoscopic silicates produced by this pathway are similar in most respects to those produced using ionic surfactants, some differences have been noted. Mesosilicates templated by nonionic surfactants typically have thicker silicate walls (~2–7 nm thick) than those templated by ionic surfactants (~1–2 nm thick), making them more robust and thermally stable [41–43]. Whereas early results using nonionic surfactants in moderately acidic conditions gave materials with poor secondary structure [41,42], later work using PEO-PPO copolymers in highly acidic conditions yielded products with domain sizes comparable to those of ionic surfactants [43].

The initial stages of self-assembly in the S^0I^0 pathway are thought to be driven by H-bond formation and/or electrostatic interactions between headgroup sites and silicate species. For example, attractive interactions between the electronegative oxygen atoms (of oligoethyleneoxide groups) and silicate species can take the form of (1) H bonds with SiOH groups, for neutral monomers around the isoelectric point [42], and/or (2) electrostatic interactions with cationic silicate oligomers in strongly acidic solutions [43]. H-bond formation between headgroups and silicates is consistent with the observed thickness of silicate walls, as the headgroup region of PEO-based surfactants is considerably more diffuse than for ionic surfactants; wall thickness roughly scales with PEO chain length [43], further corroborating this model. Electrostatic interactions have been thought to promote secondary structure [41], and this is consistent with the greater degree of order observed in highly acidic conditions.

In a recent breakthrough, the S^0I^0 pathway has been extended to synthesize a host of other mesoscopic metal oxides, such as TiO_2, ZrO_2, and Al_2O_3, using PEO-PPO block copolymers as the templating agents [44]. Key to this synthesis was the use of ethanol as the reaction medium, which limits the inorganic polymerization rate (while still permitting mesophase formation) so that condensation and crystallization proceed slowly while self-assembly is being established. As with aqueous S^0I^0 synthesis, the inorganic oxide walls are found to be relatively thick and even partially crystalline. Here, in addition to H bonds and electrostatic interactions, the complexation of metal ions by ethylene oxide groups is proposed as a possible driving mechanism for cooperative assembly [44].

While electrostatic interactions, H bonding, hydrophobic interactions, and metal ion complexation are all *possible* mechanisms of mesophase formation under acidic conditions, the central question is whether the interaction energy gained by these mechanisms is enough to offset the significant entropy loss associated with mesophase formation. This is difficult to address from first principles. However, headgroup-polysilicate interactions are implicated in cooperative assembly, and it is instructive to compare this system with the closest well-defined and well-studied analogue: polymer-surfactant complexes [45]. The analogue for the S^+I^- pathway is cationic surfactants in the presence of anionic polyelectrolytes; such a combination is known to lead to micellar aggregation and mesophase formation in dilute solutions, exactly as observed in basic surfactant-silicate mesophases. However, in the remaining synthesis pathways, the analogy does not fare as well. The analogue for the $S^+C^-I^+$ pathway is cationic polyelectrolytes in the presence of cationic surfactants—a system that is not known to significantly affect self-assembly or lead to mesophase separation. The analogue for S^0I^0—nonionic (or weakly cationic) polyelectrolytes with nonionic surfactants—likewise does not lead to qualitatively different phase behavior. Thus, the mechanism for micellar shape transitions and mesophase formation in acidic conditions remains something of a mystery.

C. Liquid Crystal Templating

All of the pathways just discussed rely on spontaneous, cooperative assembly of surfactant molecules and silicate species in relatively dilute solutions, with the resulting mesophase depending critically on the nature and strength of interactions between silicate species and surfactant headgroups. A new, conceptually simpler approach uses silicate polymerization to solidify lyotropic liquid crystalline phases that are *performed* in highly concentrated solutions [46,47]. This approach effectively seeks to decouple the physics from the chemistry of mesoscopic materials; i.e., intermolecular forces build the templating framework, which is then solidified by hydrolysis/condensation reactions. The potential advantage of this technique is that it introduces a certain perdictability to the final structure; the surfactant phase progression reveals the range of available meoscopic morphologies. However, successful application depends on the degree to which silicate polymerization in the aqueous channels destabilizes the existing lyotropic template. Potential causes of disruption include silicate-headgroup binding, hydrolysis products such as methanol, and temperature changes due to heats of reaction, all of which can alter the optimal surfactant packing geometry. Indeed, methanol produced by the hydrolysis of TMOS has been found to destroy the liquid crystalline template [46], although the template reformed when the methanol was removed by a gentle vacuum. Hexagonal, bicontinuous cubic, and lamellar phases have been "cast" into mesoscopic silicate materials in this fashion.

IV. MESOSCOPIC FILM NUCLEATION AT INTERFACES

Although mesoscopic single crystals of ~1 μm size have been grown from bulk solution [1], coherent domains are more typically in the ~100 nm size range [41] and often even smaller. Recent work has centered on ways to enhance secondary and tertiary structure to macroscopic length scales, to facilitate applications in catalysis, separation science, chemical and biosensing, and so on. In practice, structural coherence of the final product can be achieved only by controlling the size and alignment of the labile liquid crystalline domains before condensation and solidification have taken place. This necessitates a reasonable grasp of the interactions among neighboring aggregates as well as those between aggregates and any external orienting fields. One report uses strong external magnetic fields (12 tesla) to orient a surfactant-silicate hexagonal phase in bulk solution via small differences in magnetic susceptibility between the liquid crystal and the surrounding solution [48]. Because the liquid crystalline phase is highly viscoelastic, the orientational order is maintained after removal from the magnetic field. The silicate oligomer phase is then destabilized by reducing pH, thereby effecting condensation. An estimated 72% of the final calcined material is found to be oriented along the original field axis [48]. In addition to magnetic fields, another report finds evidence for enhancement of secondary structure with the use of fluoride in acidic synthesis [49].

With the exceptions noted, all other attempts to enhance higher order structure have involved interfacial confinement in some fashion. When interfaces are involved, surface forces become the dominant orienting factor, and these are often much stronger than the body forces that can be exerted by external electric, magnetic, gravitational, and flow fields. Of course, surface forces are also short ranged, which might seem to limit the degree of higher order structure that can be achieved. However, this can be largely mitigated by interaggregate interactions. A single interfacial aggregate layer, which is highly ordered by strong surface forces, can "communicate" its structure to subsequent overlayers by interaggregate interactions, thereby serving as a template for mesophase nucleation. The interfacial morphology, and the degree to which it can be manipulated for applications, depends on *both* surfactant-surface and surfactant–inorganic precursor interactions and the competition (if any) between them.

A. Nucleation at Crystalline Surfaces

Mesoscopic films nucleated at crystalline surfaces such as graphite and mica [50–52] have revealed a significant enhancement of secondary and tertiary structure; this is not too surprising given the high orientational order observed for surfactant aggregates at these interfaces. Almost all published examples have used acidic synthesis conditions similar to those used for bulk mesosilicate synthesis, with the nucleating interface (either mica or graphite) immersed in the solution. When the reaction is complete, the nucleant surface is rinsed, dried, and calcined in the usual way. The end product is a supported mesoscopic film of order 0.1 to 1 μm thick with pore spacings in the size range of micelles (Fig. 3). Typical domain sizes are in the 10 to 100 μm range (two to three orders of magnitude greater than for bulk product), and domains are strongly aligned with respect to the substrate symmetry axes.

Only two types of mesoscopic film morphologies have been synthesized to date, both of which are consistent with the interfacial morphology expected with the particular surfactant-surface combinations used. Silicate syntheses in the presence of alkyltrimethylammonium surfactants and a graphite surface have yielded hexagonal-phase films in which pore channels are oriented parallel to the film plane and perpendicular to underlying symmetry axes [51]. This is consistent with the picture of hemicylindrical interfacial aggregates serving as a template for the heterogeneous nucleation of a surfactant-silicate hexagonal phase. Two-dimensional domains are in the form of elongated strips or tapes, with the long axis parallel to the cylinder axis. A very similar morphology is also observed when mica is used as the nucleant in the presence of alkyltrimethylammonium surfactants [50,51]. Here the templating interfacial layer consists of full rather than half-cylindrical aggregates, but the end result is still a hexagonal phase with pore channels oriented in the film plane (Fig. 3). Although AFM images of the surfactant-silicate aggregates in the reaction solution do not show the kind of strong *local* alignment observed on graphite [51], the long-range structure (Fig. 3a) clearly reveals epitaxial alignment by the substrate, with domains highly elongated along the pore channel axes. Despite the enhanced higher order structures, these films suffer from one technological drawback: The pores are aligned strictly parallel to the film plane, making them difficult to access for applications.

A very different film morphology is obtained on mica when conventional surfactants are replaced with

FIG. 3 Images of hexagonal-phase mesoscopic silica films grown at a mica-solution interface in acidic reaction conditions. (a) Low-magnification SEM image showing oriented strips or tapes (scale bar = 10 μm). (b and c) TEM images of the calcined film showing the cross section and transverse views, respectively. The tubules run parallel to the film plane. (From Ref. 51.)

divalent surfactants such as $C_{18\text{-}3\text{-}1}$. These have low packing parameters and therefore favor close-packed spherical micelles (analogous to a discontinuous cubic phase) on the mica surface [14] and in the bulk synthesis of mesosilicates [53]. Not surprisingly, mesosilicate films nucleated on mica are consistent with both of these morphologies. The hexagonally ordered interfacial micelles evidently serve as a template for the nucleation of a mesophase in which silicates condense on the surface of, and in the interstices between, hexagonally close-packed discrete micelles [52]. Consistent with this primary structure, the secondary structure shows isotropic disklike islands (rather than elongated domains), sometimes with distinct hexagonal faceting. It might be imagined that surfactant removal by calcination would give rise to a silica framework consisting of discrete, unconnected, roughly spherical voids. However, this appears not to be the case. Gas adsorption isotherms reveal an effective surface area and a lack of hysteresis that are consistent with complete connectivity throughout the mesoscopic film [52]. This implies the presence of small "necks" that connect neighboring void spaces. Although not much is known about this neck distribution, a couple of details are apparent. First, each void must be connected to at least two neighboring voids for complete connectivity to be established. Second, because the film lattice symmetry is unchanged by calcination, the neck distribution must either be random (the more likely case) or limited primarily to the direction normal to the film plane [52]. The void structure and extensive connectivity make this type of mesoscopic film inherently more accessible than the horizontal cylindrical pores discussed before; however, if the connectivity is random, the pores are still not *individually* addressable.

Film uniformity on crystalline surfaces is limited by two factors: multiple surface nucleation sites and equivalent surface symmetry directions. The former causes thickness nonuniformities and plays an important role in all surface nucleation processes. The latter is unique to crystalline surfaces and gives rise to twin boundaries in the mesoscopic film (see Fig. 3a). For films with horizontal cylindrical pores, twin boundaries manifest themselves as abrupt changes in the elongation direction of the striped domains [50]. The role of lattice defects in the substrate merits further investigation, as such defects could play a role in the establishment of a twin boundary or a new nucleation site.

B. Nucleation at Isotropic Surfaces

Whereas crystalline substrates can actively orient interfacial mesophases along specific in-plane axes, isotropic surfaces can at best sterically limit one of the mesophase symmetry axes to lie normal to the substrate plane. For example, an amorphous surface can restrict an interfacial hexagonal phase so that its cylindrical micelles lie parallel to the substrate plane, but it cannot orient the micelles along specific directions in the plane. This is consistent with observations of mesosilicate films nucleated at a silica surface [51,54,55] or at a polymer surface [56] using alkyltrimethylammonium surfactants. Experiments have revealed a primary structure consisting of hexagonally packed cylindrical pores but a very complex higher order structure in which the cylindrical micelles bend and meander over the substrate plane and spiral outward normal to the substrate plane [51,55]. An inexact but helpful analogy is a bowl of cooked noodles upturned onto a flat surface. The noodles immediately adjacent to the surface predominantly adopt a horizontal (but otherwise isotropic) conformation, whereas those further from the surface can adopt orientations with significant normal components. The silica substrates used have been either cover glass or the native oxide on a silicon water; in the latter case, the crystalline nature of the underlying Si surface has usually been disregarded because the oxide overgrowth has been assumed to be amorphous. This may not always be a valid assumption; mesosilicates nucleated on Si(100) and Si(111) wafers show isotropic domains, whereas those grown on Si(110) wafers show elongated domains aligned along the [001] direction [55]. This effect has been attributed to a strong anistropy in the arrangement of Si atoms on the (110) surface, which is thought to be preserved by the thin surface oxide layer.

Mesosilicate nucleation using alkyltrimethylammonium surfactants on silica differs from that on mica and graphite in one interesting respect. The favored surfactant-silicate morphology in bulk solution (hexagonal phase) is consistent with the interfacial micellar layer for mica and graphite (cylinders or half-cylinders), whereas it is not for silica (spheres). Thus, the silica interface does not serve a templating function as such but may simply act as a nonspecific boundary.

An important special case of isotropic surfaces is the air-liquid interface, as films grown at this interface are freestanding and can be more easily shaped and manipulated for applications. Alkyltrimethylammonium surfactants have been used to template hexagonal-phase mesosilicate films [57], with channels running parallel to the film plane, and the divalent surfactant $C_{18\text{-}3\text{-}1}$ has been used to nucleate a film morphology consisting of spherical close-packed voids [52]. Although the *c* axis of both films is oriented normal to the film plane, there is little evidence of long-range order in the film plane; x-ray diffractometry of the films shows very weak secondary peaks. This is consistent with the air-water interface serving as a nonspecific boundary, in the same fashion as the silica surface already discussed.

Electron microscopy of films grown at air-water interfaces shows that the air-facing side of the film is extremely flat, whereas the water-facing side is comparatively rougher [52,57]. This roughness is observed to have two distinct length scales. A smaller micelle-sizes roughness is thought to constitute a "snapshot" of silicified micelles in the process of attaching themselves to the interfacial mesophase [57]. A larger micrometer-scale roughness is also observed, consisting of dendritic bumps that have nucleated off the film and grown down into the solution [52,57]. These are similar to the hillocks observed on silica surfaces. They may likewise be caused by an "entropic strain relief" into the solution or alternatively by homogeneous nucleation occurring simultaneously in the solution.

C. Nucleation at Uniaxially Textured and Patterned Surfaces

A surface that is anisotropic only along one axis is potentially ideal for mesoscopic film nucleation because this axis can provide a strong interfacial orienting field while also preventing twin boundaries. Such a surface must consist of linear molecules in some kind of nematic order; polymers are therefore ideal for this purpose. Two reports show some success with this promising method [58,59]. They also serve to illustrate some of the subtle and interesting differences between nucleation mechanisms involving passive nonspecific confinement of those involving active templating.

In the first case [58], uniaxial anisotropy is achieved by unidirectional rubbing of a thin polymer film (5-nm-thick polyimide) spin-coated onto a glass substrate. Mesosilicate films nucleated on this rubbed surface, templated by alkyltrimethylammonium surfactants, show elongated domains of order 30 μm long consisting of cylindrical pores parallel to the rubbing direction. The tertiary structure is observed to be very uniform and without twin boundaries. Rubbed polymer films have long been known to align thermotropic liquid crystalline phases; similarly, in this case, they align the lyotropic surfactant-silicate mesophase before (or as) it condenses. The alignment is preserved upon calcination, which pyrolyzes both the surfactant and polymer film, leaving the aligned mesosilicate film attached to the glass substrate.

In the second case [59], uniaxial anisotropy is achieved by the Langmuir-Blodgett (LB) deposition of hydrophobic polymer layers (polyimide) on a silica substrate. Controlled withdrawal of the substrate through the Langmuir monolayer aligns the macromolecules parallel to one another along the withdrawal direction. When alkyltrimethylammonium surfactants are used as templating agents, this oriented polymer film serves to nucleate a uniaxially aligned hexagonal mesophase, resulting in cylindrical pores parallel to the film plane. Again, the tertiary structure is maintained over macroscopic distances, and twin boundaries are absent. However, unlike the rubbed polymer film just discussed, in this case the pores are found to align *perpendicular* to the polymer molecules, i.e., perpendicular to the substrate withdrawal direction. This is attributed to a difference in the alignment mechanism [59]. On the polymer LB film, surfactant tails are thought to adsorb horizontally and align themselves parallel to the hydrophobic polymer; the tail-to-tail horizontal layer of surfactant molecules serves as a template for half-cylindrical micelles, whose axes are therefore oriented perpendicular to the substrate axis (as on graphite) [10]. On the other hand, in the spin-coated polymer film just discussed, polymer chains are *not* aligned parallel to one another, and the rubbing process does not appreciably align these chains. Rather, rubbing creates mesoscopic grooves that partially expose the substrate; the large aspect ratio of these grooves, combined with reaction conditions that favor nucleation on silica over that on polyimide, is thought to be responsible for alignment of the mesosilicate film parallel to the rubbing direction. Thus, the parallel film morphology on rubbed polymer surfaces is putatively due to steric confinement of whole cylindrical aggregates or domains in long grooves, whereas the perpendicular film morphology on LB-deposited polymer surfaces is due to active alignment or templating of the liquid crystalline phase by direct surfactant-surface interactions.

D. Order Enhancement Using External Flow and Electric Fields

In all of the preceding examples, the orienting field acting on the surfactant-silicate mesophase originates from substrate anisotropies. Others have combined heterogeneous nucleation with external orienting fields such as fluid flow and applied electric fields [60–62]. Films grown in this manner have all been on amorphous silica; this is an ideal substrate for this application because no surface anisotropies are present to compete with the orienting effects of external fields.

Shear flows are known to enhance secondary and tertiary order by aligning cylindrical micelles and hexagonal phases in the direction of flow [63]. Similarly, shear flow of acidic reaction solution past a silica surface has been used to form mesosilicate domains elongated along the flow direction, with their cylindrical axes presumably oriented along this direction [60]. (Control films grown in quiescent solutions show disk-like, isotropic domains.) Calcination preserves the elongation orientation, although it also induces cracks in the film. Tungsten oxide–cetyltrimethylammonium bromide (CTAB) composite films have also been oriented successfully with shear flow, resulting in a film with enhanced tertiary structure and distinct optical anisotropy [61]. However, these films did not survive calcination.

Fluid flow has also been used in conjunction with confinement and electric field alignment to grow oriented mesosilicate bundles on surfaces [62]. Here an elastomeric stamp with ~1-μm-square grooves is pressed down onto a silica surface, and the surfactant-silicate reaction solution is wicked in from one end of the stamp; electro-osmotic flow is maintained by applying an electric field parallel to the grooves. As the reaction proceeds, mesosilicates form from the groove surfaces inward, eventually (after a few hours) closing off the grooves completely. The stamp is then lifted off, revealing long bundles of mesosilicates on the silica surface. This textured film was observed to consist of predominantly cylindrical pores (consistent with the alkyltrimethylammonium surfactants used) oriented parallel to the groove axis. These bundles have a square cross section composed of order 10^5 pores, reflecting the size and shape of the elastomer grooves.

At least three factors are thought to be responsible for the enhanced secondary and tertiary structure—confinement, fluid flow, and electric fields; however, the relative importance of each is not well known. It is instructive to compare these results with a similar, separate report in which confinement is the only important factor [64]. Here an elastomeric stamp is used to "ink" 3–10 μm lines of alkanethiols onto a gold surface, and this substrate is exposed to a quiescent reaction solution that favors growth only on the thiolated regions. Hexagonal-phase mesosilicate films are observed as before; however, although the pore axes are restricted to the film plane, they are more or less isotropic in this plane and *not* significantly oriented along the thiolated lines. Thus, fluid flow and electric fields appear to play a more dominant role than simple confinement in inducing uniaxial orientation.

Mesoscopic films grown inside the channels of an elastomeric stamp can be regarded as *hierarchically ordered* films; self-assembly imparts order at 10 nm length scales and the stamp at 1000 nm length scales. In one report [65], colloidal latex spheres are employed to pattern surfaces also in the middle length scale of ~100 nm. The colloidal suspension is first wicked through the grooves of an elastomeric stamp pressed onto a silica substrate. As the solvent dries, capillary forces drive the self-assembly of colloidal particles into ordered arrangements within the confines of the grooves. The silicate reaction solution is then wicked through the grooves packed with latex spheres, and the solution is allowed to condense over several hours. The stamp is then removed and the film calcined, pyrolyzing the surfactant as well as the latex particles. The end result is a silicate surface macroscopically textured by the stamp, wherein the silicate contains macropores from the latex particles and micropores from the surfactant self-assembly. Because PEO-PPO block copolymers are used as the structure-directing agents, both cubic and hexagonal phase primary structures can be obtained, depending on the lengths of the blocks [65].

E. Nucleation at the Solid-Liquid-Vapor Contact Line

In the preceding examples, mesoscopic films form by heterogeneous nucleation, with the interfacial phase near chemical equilibrium with the bulk solution. Although this near-equilibrium approach favors uniform films with high secondary and tertiary order, it limits film morphologies to those consistent with equilibrium interfacial surfactant structures, and it can require long reaction times—typically several days. An alternative and promising approach uses dip-coating, or controlled withdrawal of a substrate out of a reaction solution, to both generate and orient mesoscopic films [66,67]. The nucleation of these films occurs at the solid-liquid-vapor contact line, where solvent evaporation creates steep concentration gradients, giving rise to reaction conditions far from equilibrium.

In a unique departure from other approaches, dip-coating uses predominantly *ethanolic* reaction solutions, with ethanol/water molar ratios of order 5:1. Because ethanol is itself somewhat amphiphilic, it tends to interfere with surfactant self-assembly in solution. Micelle formation can require far higher surfactant concentrations in ethanol-water mixtures than in purely aqueous solutions [68]. Similarly, ethanol interferes with intermicellar ordering and can delay (or suppress altogether) the onset of lyotropic liquid crystalline phases [69]. Dip-coated solutions can therefore support moderate surfactant concentrations (~0.1 M) without forming surfactant-silicate mesophases in solution; instead, mesophases form *only* at the contact line, where rapid preferential evaporation of ethanol supersaturates the remaining (mostly aqueous) reaction solution. This process can be visualized as a trajectory in the ternary water-ethanol-surfactant phase diagram, beginning near the ethanol corner and ending near the water-surfactant line [67]. Ethanol has a significantly higher vapor pressure than water and evaporates rapidly at the contact line, allowing moderate dip-coating velocities (~1 mm/s) and processing times on the order of minutes. A flat, amorphous substrate such as silica is used in order to avoid competition with surface templating effects. After dip-coating, the supported film is calcinated in the usual way.

This technique has the advantage that films with a wide variety of primary structures can be produced with the *same* structure-directing agent; for instance, surfactant-silicate films of hexagonal, bicontinuous cubic, and lamellar phases have all been produced using alkyltrimethylammonium surfactants (although the lamellar phase, not surprisingly, did not survive calcination) [66]. The initial surfactant amount can be chosen so that the final concentration in the reaction layer, once the ethanol has evaporated, falls in the desired aqueous phase region. Stated another way, the trajectory in the ternary phase diagram need not end at the most dilute aqueous mesophase. (This would probably be the case if dip-coating were performed in aqueous solutions because solvent evaporation would slow considerably when a mesophase is established.) Despite the flexibility of this approach, the details of the ternary phase trajectories are not yet well understood, and the final morphology may be difficult to predict. In particular, it may not always be clear whether a given surfactant-silicate film represents an equilibrium or a kinetically stabilized mesophase encountered along the phase trajectory. Mesophases have been modified by thermal cycling, providing evidence for such kinetic traps [66].

The secondary and tertiary structures of dip-coated films are influenced, not surprisingly, by both the air-liquid and solid-liquid boundary conditions. Hexagonal-phase films, dip-coated from reaction solutions with comparatively low surfactant concentrations, reveal three distinct zones in cross section [66]. The regions of the film closest to the air-liquid and silica-liquid interfaces *both* show tubules aligned in the film plane but not oriented in any specific direction in this plane; this morphology is perfectly consistent with films heterogeneously nucleated at these interfaces, as discussed earlier [51,57]. However, between these two regions of planar alignment (each $\sim$50–100 nm thick) lies an interior region where the cylindrical tubules are completely disordered, denoting a disordered bulk mesophase. In contrast, when the reaction solution is prepared with a higher surfactant concentration, so that the phase trajectory ends up in the bicontinuous cubic region of the phase diagram, the resulting bicontinuous cubic film is ordered throughout its thickness [66].

Whereas the preceding results were obtained with alkyltrimethylammonium surfactants as templates, the dip-coating technique has been extended to nonionic surfactants and PEO-PPO-PEO triblock copolymers [43]. Here also, ethanol evaporation destabilizes the solution during dip-coating; the resulting nanocomposite layer, after calcination, yields a mesoporous silicate film that is extremely well ordered. A variety of cubic and hexagonal primary structures have been produced, and the primary structure can be controlled by the initial amphiphile concentration (as before) as well as by the ratio of hydrophilic to hydrophobic (EO/PO) block lengths. Amphiphiles with large polar blocks typically give rise to phases with discrete close-packed micelles (leading to a mesosilicate film with a cage structure), whereas those with smaller polar blocks lead to hexagonal-phase films (Fig. 4) [43]. A high degree of secondary structure is evidenced by x-ray diffraction patterns showing reflections of several orders. Tertiary order is also greatly enhanced; unlike the dip-coated films discussed before, these are observed to be structurally uniform throughout their thickness, and hexagonal-phase films are observed to be aligned with tubules parallel to the coating direction. These enhancements may be due to the somewhat higher coating speeds used in this work or to the much larger micelles associated with block copolymers.

Mesoscopic oxide films *other* than silicates have also been prepared by dip-coating, this time from *purely* ethanolic reaction media, using PEO-PPO block copolymers as templating agents [44]. As discussed in Section III.B, ethanolic solutions were necessary to slow the hydrolysis/condensation of alkoxide precursors; metal coordination complexes between the EO headgroups and condensed oligomers were proposed as a coassembly mechanism. Interestingly, the final product was observed to contain some nanocrystalline domains of the standard bulk oxide phase within the amorphous walls of the mesoscopic material [44]. In most cases the films were hexagonal phases with tubules running parallel to the film plane.

V. CONCLUSION

Interfacial confinement, often in combination with externally applied fields, has led to important advances in mesoscopic materials. These advances include primary structures that cannot be accessed in bulk solutions, enhanced secondary and tertiary structures, and hierarchically ordered morphologies. Although the idealized membrane of uniform and individually addressable pores (Fig. 1) has not yet been achieved, this has not prevented mesoscopic films from being used for some device applications, for instance, as catalytic supports [70] or as mesoscopic waveguides for mirrorless

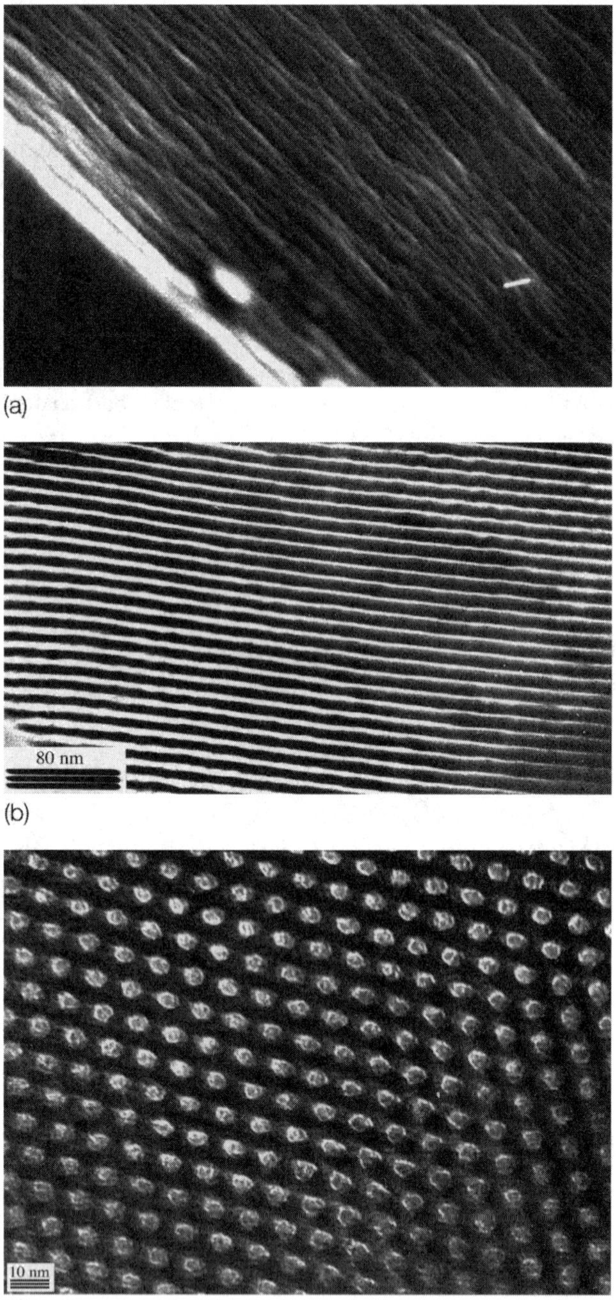

FIG. 4 Images of a hexagonal-phase mesoscopic silicate films deposited by dip-coating from a reaction solution onto an amorphous silica substrate. The structure-directing agent was the amphiphilic triblock copolymer $EO_{20}PO_{70}EO_{20}$. (a) Low-magnification SEM image of the as-deposited film showing striations parallel to the tubule axis and parallel to the coating direction. (b and c) TEM images of the calcined film showing the transverse and cross section views, respectively. The tubules run parallel to the film plane. (From Ref. 43.)

lasing [71]. Notwithstanding these advances, a membrane of uniform columnar pores remains a desirable goal. The major hurdle to columnar self-assembly is the energy cost associated with the end surfaces of the cylinders; whether these are flat terminations (exposing the interior hydrophobic tails) or hemispherical caps, they involve a local curvature at odds with the global cylindrical curvature of the micelle. Because unfavorable end surfaces are minimized by cylinders lying parallel to the film plane, this will in normal circumstances remain the preferred morphology. Whether interfaces can be suitably prepared to reduce the energy cost of the end surfaces, and thereby coax the surfactants into columnar self-assembly, remains to be seen.

ACKNOWLEDGMENTS

We thank G. D. Stucky for useful discussions. We gratefully acknowledge support from Procter & Gamble Corp. and from the University of Arizona while writing this review.

REFERENCES

1. (a) C. T. Kresge, M. E. Leonowicz, W. J. Roth, J. C. Vartuli, and J. S. Beck, Nature *359*:710–712 (1992); (b) J. S. Beck, J. C. Vartuli, W. J. Roth, M. E. Leonowicz, C. T. Kresge, K. D. Schmitt, C. T. W. Chu, D. H. Olson, E. W. Sheppard, S. B. McCullen, J. B. Higgins, and J. L. Schlenker, J. Am. Chem. Soc. *114*:10834–10843 (1992).
2. Reviews: (a) F. Schüth, U. Ciesla, S. Schacht, M. Thieme, Q. Huo, and G. Stucky, Mater. Res. Bull. *34*: 483–494 (1999); (b) P. T. Tanev, J. R. Butruille, and T. J. Pinnavaia, in *Chemistry of Advanced Materials* (L. V. Interrante and M. J. Hampden-Smith, eds.), Wiley-VCH, New York, 1998, pp. 329–388; (c) C. J. Brinker, Curr. Opin. Solid State Mater. Sci. *1*:798–805 (1996); (d) S. Manne and I. A. Aksay, Curr. Opin. Solid State Mater. Sci. *2*:358–364 (1997).
3. J. E. Martin, M. T. Anderson, J. Odinek, and P. Newcomer, Langmuir *13*:4133–4141 (1997).
4. Reviews: (a) D. F. Evans and H. Wennerström, *The Colloidal Domain*, Wiley-VCH, New York, 1994, chapter 4; (b) J. N. Israelachvili, *Intermolecular and Surface Forces*, 2nd ed., Academic Press, London, 1992, chapters 16 and 17; (c) R. G. Laughlin, *The Aqueous Phase Behavior of Surfactants*, Academic Press, London, 1994.
5. J. N. Israelachvili, D. J. Mitchell, and B. W. Ninham, J. Chem. Soc. Faraday Trans. I *72*:1525–1568 (1976).
6. C. Tanford, *The Hydrophobic Effect*, Wiley, New York, 1980.

7. (a) R. Zana, in *Cationic Surfactants: Physical Chemistry* (D. N. Rubingh and P. M. Holland, eds.), Marcel Dekker, New York, 1991, pp. 41–85; (b) J. G. Weers and D. R. Scheuing, ACS Symp. Ser. *447*:87–122 (1991).
8. (a) R. R. Balmbra, J. S. Clunie, and J. F. Goodman, Nature *222*:1159–1160 (1969); (b) K. Fontell, Colloid Polym. Sci. *268*:264–285 (1990).
9. A. M. Gaudin and D. W. Fuerstenau, Trans. AIME *202*: 958–962 (1955).
10. S. Manne, J. P. Cleveland, H. E. Gaub, G. D. Stucky, and P. K. Hansma, Langmuir *10*:4409–4413 (1994).
11. S. Manne and H. E. Gaub, Science *270*:1480–1482 (1995).
12. Review: S. Manne and G. G. Warr, in *Supramolecular Structure in Confined Geometries* (S. Manne and G. G. Warr, eds.), ACS Press, Washington, DC, 1999, chapter 1.
13. M. Jaschke, H. J. Butt, H. E. Gaub, and S. Manne, Langmuir *13*:1381–1384 (1997).
14. S. Manne, T. E. Schäffer, Q. Huo, P. K. Hansma, D. E. Morse, G. D. Stucky, and I. A. Aksay, Langmuir *13*: 6382–6387 (1997).
15. S. Manne, Prog. Colloid Polym. Sci. *103*:226–233 (1997).
16. H. N. Patrick, G. G. Warr, S. Manne, and I. A. Aksay, Langmuir *13*:4349–4356 (1997).
17. H. N. Patrick, G. G. Warr, S. Manne, and I. A. Aksay, Langmuir *15*:1685–1692 (1999).
18. E. J. Wanless and W. A. Ducker, J. Phys. Chem. *100*: 3207–3214 (1996).
19. W. A. Ducker and L. M. Grant, J. Phys. Chem. *100*: 11507–11511 (1996).
20. L. M. Grant and W. A. Ducker, J. Phys. Chem. B *101*: 5337–5345 (1997).
21. R. Lamont and W. A. Ducker, J. Colloid Interface Sci. *191*:303–311 (1997).
22. L. M. Grant, F. Tiberg, and W. A. Ducker, J. Phys. Chem. B *102*:4288–4294 (1998).
23. R. E. Lamont and W. A. Ducker, J. Am. Chem. Soc. *120*:7602–7607 (1998).
24. W. A. Ducker and E. J. Wanless, Langmuir *15*:160–168 (1999).
25. N. B. Holland, M. Ruegsegger, and R. E. Marchant, Langmuir *14*:2790–2795 (1998).
26. J. F. Wall and C. F. Zukoski, Langmuir *15*:7432–7437 (1999).
27. I. Burgess, C. A. Jeffrey, X. Cai, G. Szymanski, Z. Galus, and J. Lipowski, Langmuir *15*:2607–2616 (1999).
28. J. L. Wolgemuth, R. K. Workman, and S. Manne, Langmuir *16*:3077–3081 (2000).
29. D. J. Mitchell, G. J. T. Tiddy, L. Waring, T. Bostock, and M. P. McDonald, J. Chem. Soc. Faraday Trans. 2 *72*:1525–1568 (1976).
30. J. R. Lu, Z. X. Li, J. Smallwood, R. K. Thomas, and J. Penfold, J. Phys. Chem. *99*:8233–8243 (1995).
31. J. Bocker, M. Schlenkrich, P. Bopp, and J. Brickmann, J. Phys. Chem. *96*:9915–9922 (1992).
32. R. G. Laughlin, in *Cationic Surfactants: Physical Chemistry* (D. N. Rubingh and P. M. Holland, eds.), Marcel Dekker, New York, 1991, pp. 1–40.
33. A. Firouzi, D. Kumar, L. M. Bull, T. Besier, P. Sieger, Q. Huo, S. A. Walker, J. A. Zasadzinski, C. Glinka, J. Nicol, D. I. Margolese, G. D. Stucky, and B. F. Chmelka, Science *267*:1138–1143 (1995).
34. A. Firouzi, F. Atef, A. G. Oertli, G. D. Stucky, and B. F. Chmelka, J. Am. Chem. Soc. *119*:3596–3610 (1997).
35. Y. S. Lee, D. Surjadi, and J. F. Rathman, Langmuir *12*: 6202–6210 (1996).
36. Y. S. Lee, D. Surjadi, and J. F. Rahtman, Langmuir *16*: 195–202 (2000).
37. R. K. Iler, *The Chemistry of Silica*, Wiley, New York, 1979.
38. (a) Q. Huo, D. I. Margolese, and G. D. Stucky, Chem. Mater. *8*:1147–1160 (1996); (b) Q. Huo, D. I. Margolese, U. Ciesla, D. G. Dermuth, P. Feng, T. E. Gier, P. Sieger, B. F. Chmelka, F. Schüth, and G. D. Stucky, Chem. Mater. *6*:1176–1191 (1994); (c) A. Monnier, F. Schüth, Q. Huo, D. Kumar, D. Margolese, R. S. Maxwell, G. D. Stucky, M. Krishnamurty, P. Petroff, A. Firouzi, M. Janicke, and B. F. Chmelka, Science *261*: 1299–1302 (1993).
39. Q. Huo, D. I. Margolese, U. Ciesla, P. Feng, T. E. Gier, P. Sieger, R. Leon, P. M. Petroff, F. Schüth, and G. D. Stucky, Nature *368*:317–320 (1994).
40. V. Shubin, P. Petrov, and B. Lindman, Colloid Polym. Sci. *272*:1590–1601 (1994).
41. P. T. Tanev and T. J. Pinnavaia, Science *267*:865–867 (1995).
42. S. A. Bagshaw, E. Prouzet, and T. J. Pinnavaia, Science *269*:1242–1244 (1995).
43. (a) D. Zhao, J. Feng, Q. Huo, N. Melosh, G. H. Fredrickson, B. F. Chmelka, and G. D. Stucky, Science *279*: 548–552 (1998); (b) D. Zhao, P. Yang, N. Melosh, J. Feng, B. F. Chmelka, and G. D. Stucky, Adv. Mater. *10*:1380–1385 (1998).
44. (a) P. Yang, D. Zhao, D. I. Margolese, B. F. Chmelka, and G. D. Stucky, Nature *396*:152–155 (1998); (b) P. Yang, D. Zhao, D. I. Margolese, B. F. Chmelka, and G. D. Stucky, Chem. Mater. *11*:2813–2826 (1999).
45. Reviews: (a) P. Hansson and B. Lindman, Curr. Opin. Colloid Interface Sci. *1*:604–613 (1996); (b) Y. J. Li and P. L. Dubin, ACS Symp. Ser. *578*:320–336 (1994).
46. G. S. Attard, J. C. Glyde, and C. G. Göltner, Nature *378*:366–368 (1995).
47. C. G. Göltner, S. Henke, M. C. Weissenberger, and M. Antonietti, Angew. Chem. Int. Ed. *37*:613–616 (1998).
48. S. H. Tolbet, A. Firouzi, G. D. Stucky, and B. F. Chmelka, Science *278*:264–268 (1997).
49. P. Schmidt-Winkel, P. Yang, D. I. Margolese, B. F. Chmelka, and G. D. Stucky, Adv. Mater. *11*:303–307 (1999).

50. H. Yang, A. Kuperman, N. Coombs, S. Mamiche-Afara, and G. A. Ozin, Nature *379*:703–705 (1996).
51. I. A. Aksay, M. Trau, S. Manne, I. Honma, N. Yao, L. Zhou, P. Fenter, P. M. Eisenberger, and S. M. Gruner, Science *273*:892–898 (1996).
52. S. H. Tolbert, T. E. Schäffer, J. Feng, P. K. Hansma, and G. D. Stucky, Chem. Mater. *9*:1962–1967 (1997).
53. Q. Huo, R. Leon, P. M. Petroff, and G. D. Stucky, Science *268*:1324–1328 (1995).
54. M. Ogawa, Chem. Commun. 1149–1150 (1996).
55. H. Miyata and K. Kuroda, J. Am. Chem. Soc. *121*: 7618–7624 (1999).
56. M. Ogawa and T. Kikuchi, Adv. Mater. *10*:1077–1080 (1998).
57. H. Yang, N. Coombs, I. Sokolov, and G. A. Ozin, Nature *381*:589–592 (1996).
58. H. Miyata and K. Kuroda, Chem. Mater. *11*:1609–1614 (1999).
59. H. Miyata and K. Kuroda, Adv. Mater. *11*:1448–1452 (1999).
60. H. W. Hillhouse, T. Okubo, J. W. van Egmond, and M. Tsapatsis, Chem. Mater. *9*:1505–1507 (1997).
61. H. W. Hillhouse, J. W. van Egmond, and M. Tsapatsis, Langmuir *15*:4544–4550 (1999).
62. M. Trau, N. Yao, E. Kim, Y. Xia, G. M. Whitesides, and I. A. Aksay, Nature *390*:674–676 (1997).
63. C. A. Herb and R. K. Prud'homme, *Structure and Flow in Surfactant Solutions*, American Chemical Society, Washington, D.C., 1994.
64. H. Yang, N. Coombs, and G. A. Ozin, Adv. Mater. *9*: 811–814 (1997).
65. P. Yang, T. Deng, D. Zhao, P. Feng, D. Pine, B. F. Chmelka, G. M. Whitesides, and G. D. Stucky, Science *282*:2244–2246 (1998).
66. Y. Lu, R. Ganguli, C. A. Drewien, M. T. Anderson, C. J. Brinker, W. Gong, Y. Guo, H. Soyez, B. Dunn, M. H. Huang, and J. I. Zink, Nature *389*:364–368 (1997).
67. C. J. Brinker, Y. Lu, A. Sellinger, and H. Fan, Adv. Mater. *11*:579–585 (1999).
68. (a) R. Zana, in *Cationic Surfactants: Physical Chemistry* (D. N. Rubingh and P. M. Holland, eds.), Marcel Dekker, New York, 1991, pp. 41–85; (b) R. Zana, S. Yiv, C. Strazielle, and P. Lianos, J. Colloid Interface Sci. *80*:208–223 (1981).
69. K. Fontell, A. Khan, B. Lindström, D. Maciejewska, and S. Puang-Ngern, Colloid Polym. Sci. *269*:727–742 (1991).
70. A. Corma, Chem. Rev. *97*:2373–2419 (1997).
71. P. Yang, G. Wirnsberger, H. C. Huang, S. R. Cordero, M. D. McGehee, B. Scott, T. Deng, G. M. Whitesides, B. F. Chmelka, S. K. Buratto, and G. D. Stucky, Science *287*:465–467 (2000).

38

Synthesis of Mesoscopic Silica Films at Fluid-Fluid Interfaces

YOON SEOB LEE and JAMES F. RATHMAN The Ohio State University, Columbus, Ohio

I. BACKGROUND

This chapter provides an overview of efforts to synthesize structured mesoporous films at fluid-fluid interfaces and presents a detailed description of a novel method in which silica films are synthesized at a liquid-liquid interface. Potential advantages of making films by fluid-fluid techniques over conventional dip-coating and sol-gel processes are discussed.

A. Surfactant Self-Assembly on Solid Surfaces

The adsorption of soluble surfactants from aqueous solution onto gas, liquid, and solid interfaces has been extensively studied. The configuration adopted by an adsorbed surfactant molecule at a liquid-liquid interface may be much different from that observed at gas-liquid or liquid-solid interfaces. For example, at an oil-water interface the strong affinity of the surfactant tails for the oil phase results in the tails being oriented perpendicularly to the interface [1], even at very low coverage, where the surfactant tails lie down on the surface in gas-liquid or liquid-solid systems. As the amount of adsorption increases, surfactant-surface and surfactant-surfactant interactions result in the formation of self-assembled structures at surfactant concentrations well below the critical micelle concentration, the concentration at which self-assembly in solution is first observed.

Adsorption to a solid is further complicated by the chemical and physical heterogeneities of the solid, which strongly influence the adsorption and aggregation of surfactants [2,3]. Solid surfaces are often classified as hydrophobic (Mylar, polyethylene, Teflon) or hydrophilic (borosilicate glass, silica gel). Charged sites on the surface may increase or decrease the amount of ionic surfactant adsorbed. Ionic surfactants interact with hydrophobic surfaces primarily through dispersion forces, whereas the adsorption of ionic surfactants to hydrophilic surfaces is governed predominantly by electrostatic or charge-dipole interaction. The formation of hemimicelles or admicelles induces a sharp increase in adsorption as a function of solution concentration [4,5].

Surfactant hemimicelles have been imaged by atomic force microscopy (AFM) for the adsorption of the cationic surfactant hexadecyltrimethylammonium bromide ($C_{16}TAB$) on a hydrophobic graphite surface [6]. The observed hemimicelle structure is consistent with the preferential horizontal configuration of surfactant tails on the surface. The same structure has been observed for other quaternary ammonium surfactants of varying alkyl chain lengths [7] and having different geometries (e.g., gemini surfactants) [8] on hydrophobic surfaces such as graphite and MoS_2. On hydrophilic surfaces such as silica and mica, adsorption of cationic surfactants from aqueous solution results in admicellar structures; the interaction is primarily between the surface and surfactant headgroups. Spherical admicelles have been observed on highly charged silica surfaces, a process that is highly dependent on pH. Aggregation of the same surfactant on a mica surface resulted in cylindrical admicelles because of ion exchange between the K^+ ions on the surface and the charged headgroups, resulting in a higher adsorption density [7,8].

B. Synthesis of Mesoporous Silica Films

The discovery of M41S-type mesoporous materials in 1992 led to a tremendous level of research activity [9,10]. Their ordered, uniform pores and remarkably high internal surface area make these materials potential candidates for many applications where size selectivity and/or sorption capacity is important. Recent publications have reported various new synthetic schemes [11–18], development of new methods for improved characterization [19–22], and demonstrations of the application of these materials as sorbents [23,24] and catalysts [25]. A wide variety of reactions are possible. One of the most common involves the use of tetraethylorthosilicate (TEOS) as the silicate source:

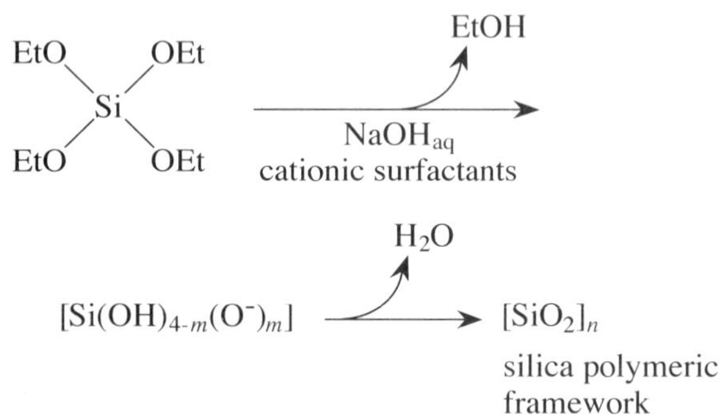

Although nearly all work to date has focused on the synthesis of mesoporous materials in particulate (powder) forms, other forms such as monolithic solids [26], membranes [27], hard spheres [28], long hollow fibers [29], and thin films can be synthesized. Films have unique potential applications as membranes, membrane supports, surface coatings to improve catalyst selectivity and prevent poisoning, and size-specific coatings for sensors, electrodes, and other microelectronic devices [30]. For example, a surface acoustic wave (SAW) sensor coated with microporous zeolite films has been developed for the selective sorption of gas molecules [31,32]. Small pore sizes limit zeolite films to molecules smaller than 1.5 nm; mesoporous films will increase this selectivity to molecules up to roughly 10 nm in size [33].

As in the synthesis of bulk mesoporous materials, the cooperative organization of cationic surfactant molecules with silicate species is the key factor in the synthesis of mesoporous films. The difference is that the reaction is now performed at an interface, and confinement of the reaction to the interfacial region provides an additional level of control over the structural evolution that occurs during the reaction.

Mesoporous films have been synthesized at air-solid and liquid-solid interfaces. For the syntheses at air-solid interfaces, conventional sol-gel methods have been widely applied. Much effort has been given to the formulation of precursor silicate gels containing cationic quaternary ammonium surfactants and tetramethylorthosilicate (TMOS) or tetraethylorthosilicate (TEOS) as the silicate source, usually under acidic conditions in order to inhibit the polymerization reaction prior to depositing the gel on the solid. Dip coating and spin coating are two commonly used methods. The rapid evaporation of solvent before the formation of silica by silicate condensation is important to prevent the formation of amorphous product. Ogawa synthesized layered [34,35] and hexagonal [36] silica-surfactant nanocomposites using both single and dichain quaternary ammonium surfactant on glass surfaces. Martin et al. [37] used ammonia gas as a catalyst instead of rapid solvent evaporation to increase the rate of silicate condensation after coating. This method, referred to as gas-catalyzed thin-film synthesis (GCTFS), has been used to prepare hexagonal mesoporous silica films on silicon and glass surfaces. Lu et al. [38] reported the synthesis of mesoporous films on silicon by dip coating. They monitored the evolution of lamellar, cubic, and hexagonal structures using a fluorescence depolarization method. The rapid formation of continuous nanolaminated films that mimic nacre of abalone shell on silicone surfaces by dip coating was also reported [39]. This result is one of the pioneering works in the synthesis of biomimetic nanocomposite assemblies. Their technique relies on the rapid evaporation of solvent in the early stage of reaction to ensure the formation of the desired surfactant-silicate mesophase after coating. Both the final mesophase structure and the film thickness of the ordered region are highly dependent on the process time scale (coating rate) and on the chemical properties and process used for pretreatment of the substrate. In another study, an acidic silicate gel was used in the microscopic patterning of orientated hexagonal silica film on solid surfaces [40]. The sol-gel evaporation method has also been performed in a microcapillary mold using an electric field guidance technique [41]. Patterned porous silica, nobia, and titania mesoporous films were synthesized. These techniques are potentially important in the development of low-cost lithographic techniques.

Mesoporous silica films at liquid-solid interfaces are synthesized in much the same way, again using acidic silicate precursor gels. Instead of applying a thin layer of gel directly on the solid surface, the substrate is immersed into the bulk gel, allowing the formation of surfactant-silicate self-assemblies on the solid surface. The final film is obtained by removing the solid from the liquid gel and drying in air. Mica, graphite, and glass surfaces have been widely used as solid surfaces for these films. Aksay et al. [42] synthesized well-ordered mesoporous films on mica and graphite, but the product on glass transformed to a spiral-like disordered structure soon after the drying process. A hexagonal mesoporous silica film has also been synthesized on mica [43]. Attard et al. [44] coated mesoporous platinum films on a gold electrode using the same approach.

The main drawback of the gas-solid and liquid-solid methods is that the film formation reaction takes place in contact with a solid substrate, so the substrate may strongly influence the resulting film structure. Syntheses under identical solution conditions but on different solid substrates often yield different film products. The influence of a solid surface on the self-assembly of surfactant molecules near the surface is well known and can be significant; indeed, it is often the most important factor. The ordering of molecules in the first layer next to a surface influences subsequent ordering in layers progressively farther away from the surface, and this effect can be propagated for thousands of layers into the liquid phase. Several researchers have suggested that this effect can be exploited to make highly sensitive sensors by taking advantage of liquid crystalline phase changes induced by minor chemical or physical properties of a surface. For film synthesis, substrate effects are a severe problem because they limit the type of film that can be produced on a given surface; for example, it turns out to be very easy to produce hexagonal-structured films on hydrophilic surfaces such as glass but much more difficult to synthesize cubic films using these methods. Liquid phase compositions that yield cubic-structured bulk (powder) material are generally observed to produce hexagonal films if the reaction is performed at a hydrophilic solid surface.

These results suggest that synthesizing films at fluid-fluid interfaces (gas-liquid or liquid-liquid) may offer a way to eliminate undesired substrate effects on film structure. Yang et al. [45] synthesized surfactant-silicate mesophases at air-liquid interfaces and observed the formation of thin hexagonal films. A cooperative relation between surfactant self-assembly at the interface and micelles in solution with the silicate ions (dual templating) was suggested. Another group reported the synthesis of mesoporous silica films of hexagonal and cubic structures on air–aqueous solution interfaces [46–49]. They suggested two discrete stages in the mechanism: induction and transition/growth. The long *c* axis in hexagonal films ran parallel to the surface, and the cubic film detected in the early induction period was transformed during reaction into an hexagonal structure in the final product [48]. Transitions such as this that occur in film structure during reaction are common.

The remainder of this chapter describes a new method for synthesizing mesoporous silica films. A surfactant-silicate precursor gel is first spread at a liquid-liquid interface and allowed to react for a prescribed period of time, after which the partially formed wet film is transferred to a solid substrate. This method eliminates structural reorganization within the film caused by the solid surface and so provides a way to synthesize selectively films of a desired mesostructure (e.g., lamellar, hexagonal, or cubic) on a wide variety of different solids.

II. EXPERIMENTAL

A. Materials

Cationic surfactants dodecyl-, tetradecyl-, and hexdecyltrimethylammonium chloride, denoted C_{12}TAC, C_{14}TAC, and C_{16}TAC, respectively, were from Fluka Chemical Co. Also, 25 wt% aqueous solutions of C_{16}TAC from Aldrich Chemical Co. were used. Alcohols with alkyl chains of *n*-octyl (C_8OH, Aldrich), *n*-decyl (C_{10}OH, Sigma), and *n*-dodecyl (C_{12}OH, Aldrich) were used as cosurfactants. Tetraethylorthosilicate (TEOS) from Aldrich (98 wt% aqueous solution) was used as the silicate source in all reactions. Benzene and sodium hydroxide were from Fisher and Jenneile, respectively. All chemicals were used as received.

Three solid surfaces were used as supporting solid substrates for films: borosilicate glass, indium-tin oxide (ITO)–coated glass, and Teflon. Precleaned microscope cover glasses were from Fisher Scientific and Surgipath Medical Industries, Inc. and were used as received. Teflon sheets (1/32 inch) from Scientific Instrument Services, Inc. were soaked in 0.1 M HCl solution for 1 day, washed with distilled water, and finally dried and stored in a desiccator at room temperature. ITO-coated glass samples were provided by the liquid crystal display (LCD) division of Hyundai Electronics, Korea. Two types of ITO glasses were used: 40 and 120 nm ITO layer thickness with 65 and 20 ohms resistance, respectively, both coated on 0.7-mm glass. They were pretreated using conventional cleaning procedures [50,51].

B. Methods

1. Precursor Silicate Gel

An aqueous solution containing 25.0 wt% cationic surfactant C_nTAC was first prepared and the pH adjusted to 12.5 by addition of solid sodium hydroxide. An alkyl alcohol was then added to give the desired alcohol/surfactant molar ratio. C_mOH/C_nTAC molar ratios from 0.01 to 1.50 were investigated; at higher alcohol content these mixtures were viscous gels. The alkaline alcohol–surfactant solution was stirred for 20 min at high speed. Benzene was then added in excess amount (6 mL benzene per mL aqueous solution) and the resulting two-phase mixture was stirred for 2 h. TEOS was then added, again with continuous stirring.

Following addition of the silicate species, a white precipitate began to appear within 5–15 min and the mixture became noticeably more viscous. This solution was stirred for 2 h, during which time the reaction flask was ventilated to allow evaporation of ethanol produced during the hydrolysis of TEOS. Molar silicate/surfactant (Si/C_nTAC) ratios ranging from 0.5 to 4.5 were studied. The final reaction mixture was stored without stirring at 25°C for 2 days, during which time it separated into two clear liquid phases. The time required for phase separation was dependent on the length of the surfactant chain; $C_{12}TAC$ systems separated relatively quickly in 4–5 h, whereas $C_{16}TAC$ systems required 5 days. Based on chemical analysis of the two phases by Fourier transform infrared (FTIR) spectroscopy, the top phase was essentially pure benzene and all other components were contained in the bottom phase. The bottom phase was thus saturated with benzene; the benzene content was approximately 8 wt%, with benzene molecules residing predominantly in the surfactant micelles. The final molar ratio of benzene to C_nTAC in this phase ranged from 1.0 to 1.5, depending on the specific surfactant system. The aqueous bottom phase was separated and stored in a sealed glass bottle; this solution is referred to hereafter as the precursor silicate gel.

2. Films

The precursor silicate gel just described was used for the synthesis of all mesoporous silica films at liquid-liquid and air-solid interfaces in this study. For the liquid-liquid technique, the two phases were benzene and water. The role of benzene, both as a component of the precursor gel and as the bulk organic liquid phase, is discussed in the next section. A known volume of silicate gel was carefully delivered to the interface using a micropipet positioned at the benzene-water interface, and the gel spread spontaneously across the entire available interfacial area. The reaction begins immediately because exposure of the gel to water induces the desired condensation polymerization of the silicate monomers and oligomers. The reaction was allowed to proceed for a specific length of time and the film was then transferred to a solid substrate by simply drawing the solid slowly through the interface in a single pass. Reaction times at the liquid-liquid interface varied from 30 min to 120 h. The transferred film was placed in a ventilated container and dried at 25°C for 2 days. Unlike conventional sol-gel techniques in which rapid drying is required [52–55], the much slower drying process used here was important for obtaining high-quality products. Finally, films were calcined to remove all organic material. Dried films were placed in a furnace at room temperature and the temperature was increased at a rate of 2–5°C/min to 550°C. This temperature was maintained for 10 h with continuous flowing air supplied. The furnace was then shut off and the samples were slowly cooled to room temperature. For characterization purposes, samples of wet, dried, and calcined films were collected at regular intervals throughout the process.

For the synthesis of films directly on a solid, two methods were used. Using the conventional dip-coating method, the solid substrate was immersed into the gel and then withdrawn. The other method was simply to deposit the gel on the solid surface using a pipette. In both cases, the film was then dried at 25°C for 48 h and calcined as described before.

C. Characterization

X-ray powder diffraction (XRD) patterns were obtained on a Scintag PAD-V diffractometer with Cu $K\alpha$ radiation of 1.54060 Å wavelength at 20 mA, 45 kV at room temperature. Each film was analyzed in two ways. A portion of film was peeled off the solid substrate, crushed, and mounted on the silicon holder as a powder, using vacuum grease. For the second analysis, the substrate coated with the intact film was directly mounted on the sample holder. These two experiments provide information about the structure of the film and the alignment of mesostructures with respect to the substrate surface. In all experiments, samples were scanned at 1.5–30° (or 10°) 2θ with 0.01–0.005° step size and 1.2 s per step. Both the standard (4 and 2 mm for x-ray source, 0.5 and 0.3 mm for detector) and the low-intensity (2 and 1 mm for x-ray source, 0.3 and 0.1 mm for detector) slit setups were selected depending on the intensity of the first highest peak and sample conditions. The angles (2θ, ω) were recalibrated for every measurement.

^{29}Si magic angle spinning (MAS) solid-state nuclear magnetic resonance (NMR) spectra were recorded on a Bruker AM 500 MHz spectrometer using a delrin rotor. The measurement conditions for high-quality spectra were 99.364 MHz resonance frequency, 45° radiofrequency pulse width of 3 μs, 36.0 kHz spinning speed, and 6000 scans with 10 s interval time. External tetramethylsiliane (TMS) was used as the standard. ^{14}N solution NMR spectra were also recorded on a Bruker AM 500 MHz spectrometer with external aqueous ammonium chloride solution as standard. All spectra were obtained at 25°C, and standard deconvolution techniques were used to determine chemical shifts and peak areas.

SEM images were obtained on a Philips XL-30 FEG scanning electron microscope. Both dried and calcined film surfaces were coated with gold for better conductivity. For side view observations, the film was fractured and mounted vertically on the sample holder using epoxy.

Freeze-fracture transmission electron microscope (FF-TEM) images were obtained on a Philips CM12 or Philips 300 TEM. The films were fixed vertically or horizontally in 2-mm round holes of a gold sample holder using glycerol or molten agar and frozen quickly in liquid ethene. The sample was then transferred and stored in liquid nitrogen for 2 h. Frozen films were fractured in a vacuum chamber using a metal knife at −170°C and coated with platinum and carbon. High-quality film replicas were obtained by dissolving the films in KOH-NaOH solution for 1 h. This replica was then transferred on a 5-mm copper grid and observed under the TEM.

A Nicolet 550 Fourier transform infrared spectrometer was used in the analysis of the silicate gel and the upper benzene phase at room temperature.

III. RESULTS

A. Precursor Silicate Gel

The strategy behind incorporating C_mOH and benzene into the gel with a C_nTAC cationic surfactant was that the concentrations of these three components could be varied in order to "tune" the micellar packing parameter (V/aL), where L (effective length of hydrocarbon surfactant tails) is directly related to the surfactant and alcohol chain lengths (m and n). Addition of the nonionic alkyl alcohol cosurfactant decreases a (optimal headgroup area) by reducing electrostatic repulsion between cationic surfactant headgroups, and V (effective molar volume of the hydrocarbon tails) is increased by solubilization of benzene into the interior of the surfactant aggregates. Many organic solubilizates and nonionic cosurfactants other than benzene and n-alkanols could be used to achieve essentially the same result. The alcohol and benzene also have an important influence on two additional properties of the gel. First, the inclusion of these organic compounds in the aqueous gel results in the gel having a density (0.85 g/cm^3) intermediate between those of pure water and pure benzene, so delivering the gel to a benzene-water interface is not problematic. Second, the water-gel and benzene-gel interfacial tensions are such that the gel spreads spontaneously at the liquid-liquid interface, making it easy to obtain highly uniform film thickness across the entire sample.

The use of gels for producing films is common in the literature. Silicate gels containing short-chain alcohols such as methanol, ethanol, and propanol as cosolvents [37] have been prepared under acidic conditions for use as precursors for film coating on solid surfaces. Despite their low pH, intended to inhibit the silicate polymerization reaction, these gels still exhibit poor chemical stability, as polymerization is observed within minutes after preparing the gel. In contrast, the precursor gel prepared here is extremely stable. FT-IR analysis showed no significant changes in composition in a gel sample stored for 1 year in a sealed container at room temperature. No phase separation was observed either, so this gel is remarkably stable both chemically and physically. Because the gel is also alkaline, simply exposing it to excess water as it spreads at the liquid-liquid interface is sufficient to initiate the reaction to form the silica film product.

Both ^{29}Si and ^{14}N NMR spectroscopy can be used to determine the degree of silicate polymerization and also whether the surfactant micelles are homogeneous [56–60]. These spectra indicate that the gel is isotropic and contains very little silicate in the form of monomers. Nearly all of the silicate is present as oligomers, so that the reaction in fact did proceed to a limited extent but was then inhibited.

B. Effect of Surfactant Type

Films were synthesized at a benzene-water interface and then transferred to borosilicate glass surfaces. Three silicate gels were prepared having identical compositions but with different cationic surfactants $C_{12}TAC$, $C_{14}TAC$, and $C_{16}TAC$.

1. Lamellar Mesoporous Silica Films

Syntheses using the $C_{16}TAC$ silicate precursor gel resulted in two different types of lamellar mesoporous silica films having *d*(100) spacings of 29–33 Å. Figure 1 shows the XRD pattern of lamellar films prepared for various reaction times at the benzene-water interface. Six well-resolved peaks were observed in 2θ between 2 and 10. These peaks do not match any known single structure; however, they are perfectly consistent with a lamellar structure having two distinct layer spacings. The three peaks designated as w:100, w:200, and w:300 match the lamellar *h00* indexing and the *d*(100) spacing value (29.1 Å) is typical of bulk lamellar products synthesized in aqueous $C_{16}TAC$ solution [61]. The other three peaks designated as b:100, b:200, and b:300 also match the lamellar *h00* indexing but with a higher *d*(100) spacing of 32.3 Å. The lamellar film synthesized at the liquid-liquid interface therefore exhibits two distinct distances between the sheets. Freeze-fracture TEM images confirmed this observation. Figure 2a shows the TEM image of this film fractured along the film surface; the lamellar structure is clearly evident and well developed. The surface roughness of this sample was a result of the freeze-fracture process used to prepare the sample for TEM. TEM images of the film edges (side view) are shown in Fig. 2b for the edge near the surface that was originally in contact with the benzene phase and in Fig. 2c for the edge near the water side. The average periodicities between layers estimated from these images are 34 and 30 Å, respectively, which are consistent with the *d*(100) spacing values from XRD.

The reason for the spacing between layers being larger on the benzene-contacted side of the film than on the water-contacted side is most likely benzene solubilization from the bulk phase into the reacting gel phase during the initial stages of the reaction. The swelling effect of organic solubilizates such as benzene is well known not only for lamellar surfactant aggregates but also for hexagonal, cubic, and simple micellar structures and has been extensively studied [62–64]. This effect has also been observed in the synthesis of bulk mesoporous materials, where the increased micelle size due to the incorporation of organic molecules inside the micelle resulted in an increase in the unit cell sizes of various mesoporous materials [65,66]. In this case, the silicate gel was saturated with benzene before being delivered to the benzene-water interface. Because the silicate polymerization reaction is initiated by exposure to water, film formation most likely begins at the water-gel interface and then proceeds relatively slowly across the gel phase. As the reaction zone moves toward the benzene-gel interface, the advancing water phase expels excess benzene from the gel.

The XRD results indicate that there is not a gradual increase in layer spacing as one moves from the water side to the benzene side; rather, the film exhibits two distinct spacings. The relative amounts of these two fractions in a film can be estimated from the XRD spectra by calculating the areas of the w:100 and b:100 peaks, *A*(w) and *A*(b), respectively. Figure 3 illustrates how the *A*(w)/*A*(b) ratio varies as a function of reaction time at the benzene-water interface for films having a thickness of 5 μm. For short reaction times, low values of *A*(w)/*A*(b) indicate that a considerable amount of benzene remains in the film, throughout most of the thickness. As reaction time increases, the *A*(w)/*A*(b) ratio increases and then levels off. These results are con-

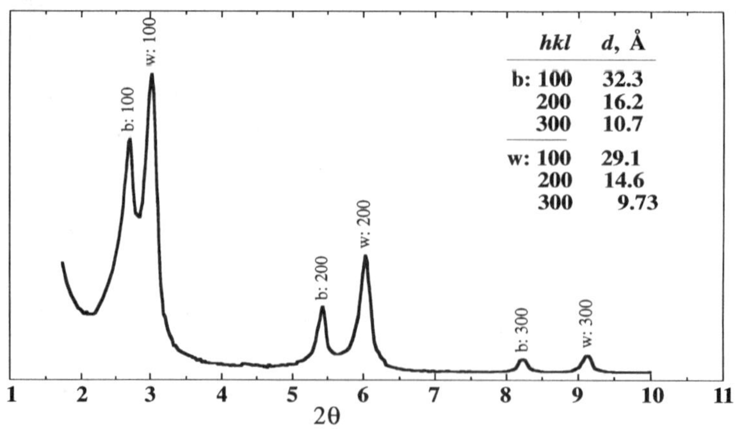

FIG. 1 Typical powder x-ray diffraction (XRD) pattern of lamellar mesoporous silica film having two different sets of *d* spacings.

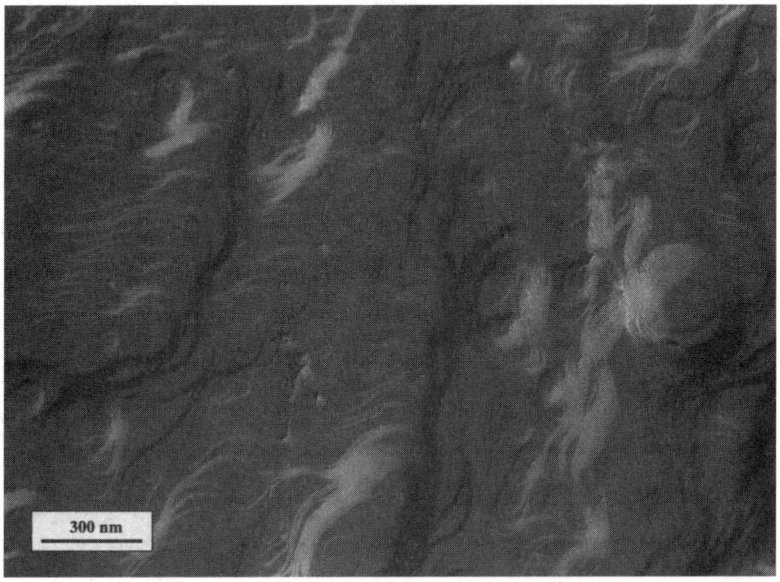

(a)

(b)

FIG. 2 Freeze-fracture transmission electron microscopy (FF-TEM) images of platinum-carbon replicas of the lamellar mesoporous silica film shown in Fig. 1: (a) the image taken from the surface fractured along the glass surface and the images of the side views taken near (b) the benzene-contacted surface and (c) the water-contacted surface.

sistent with the expulsion of some benzene from the film as a water-rich front advances through the film during reaction. Figure 4 is a schematic representation of the lamellar layers adjacent to the two bulk liquid phases.

2. Cubic Mesoporous Silica Film

Syntheses using the C_{14}TAC silicate precursor gel gave mesoporous silica films having cubic structures. Figure 5 shows a typical XRD pattern. Fifteen peaks were resolved and indexed as a cubic mesophase with *Pm*3*n* symmetry having a unit cell size of $a = 115.4$ Å. Unlike the case of the lamellar products obtained from C_{16}TAC silicate gel, the liquid-liquid reaction time had no effect on the unit cell dimension—all cubic films synthesized over a range of reaction times from 30 min to 5 days had $a \approx 115$ Å. Although analogues of all three surfactant bicontinuous cubic mesophases (*Pm*3*n*, *Ia*3*d*,

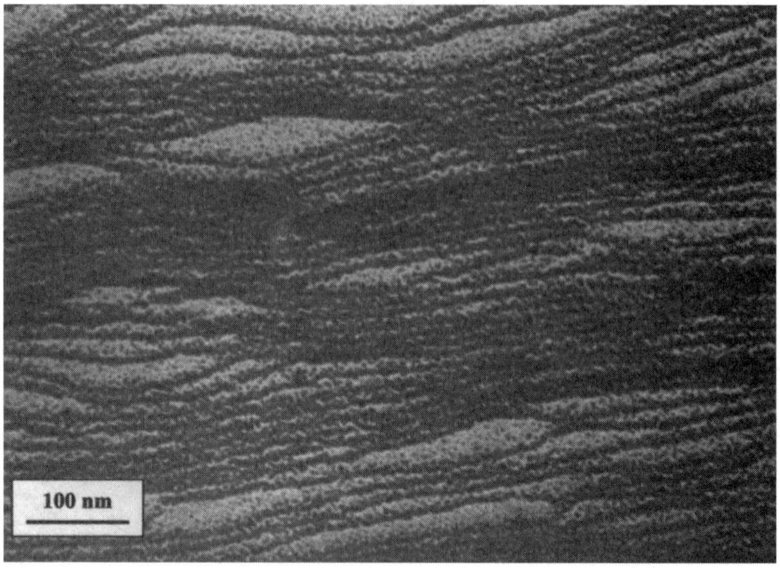

(c)

FIG. 2 Continued.

and *Im*3*m*) have been successfully generated in the syntheses of bulk cubic mesoporous materials [9,67,68], the *Pm*3*n* product reported here is the only cubic film observed to date. The unit cell sizes for the cubic films are significantly larger than those reported for bulk cubic materials, which are on the order of 105 to 110 Å [67,68]. This is somewhat surprising because the bulk materials were synthesized using C_{16}TAC, whereas the films were synthesized in this work using C_{14}TAC. This result may be a consequence of the multiconnected water and oil channels of the *Pm*3*n* cubic structure [69–73], which may prevent expulsion of benzene molecules from the structure during film formation.

3. Hexagonal Mesoporous Silica Film

The C_{12}TAC silicate precursor gel was used to prepare mesoporous silica films with hexagonal structure. A

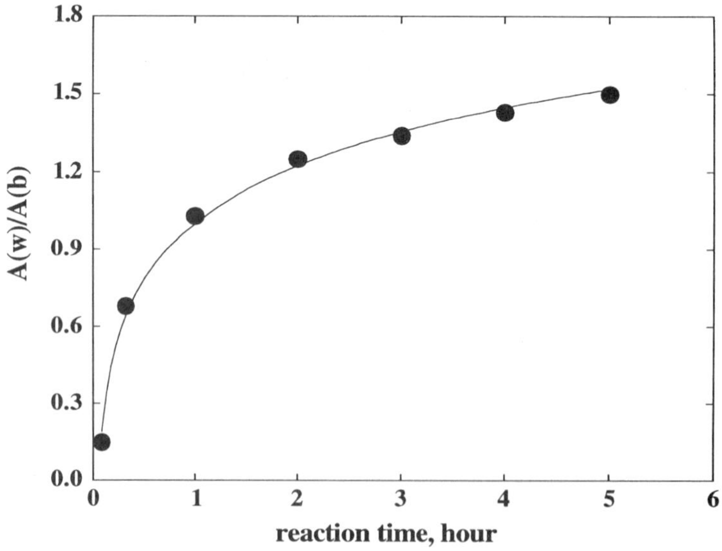

FIG. 3 Change of the concentration ratio of the small *d* spacing lamellar to the large *d* spacing one, A(w)/A(b), as a function of the reaction time at the benzene-water interface.

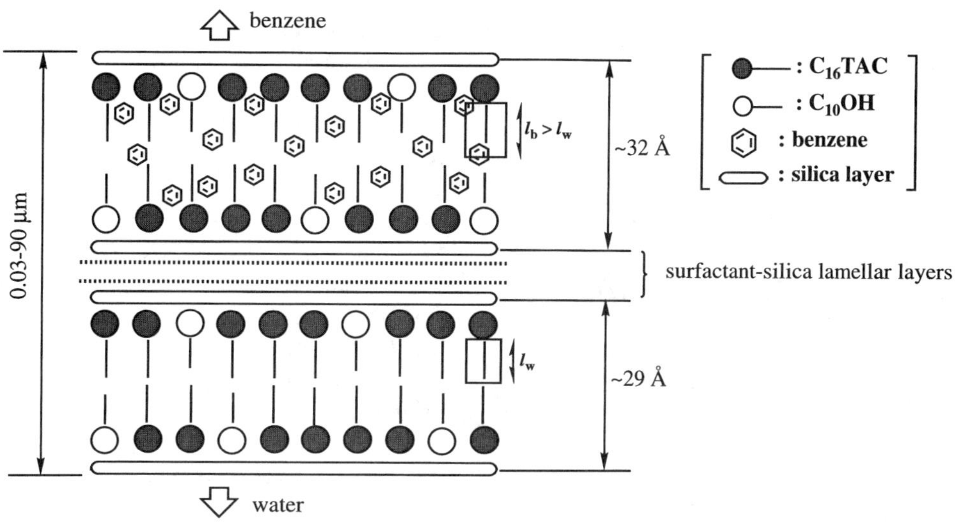

FIG. 4 Schematic representation of the cross section of a lamellar mesoporous silica film having two different sets of *d* spacings.

typical XRD spectrum for relatively thick films (5–100 μm) is shown in Fig. 6a. The four most clearly resolved peaks match the pattern indexed for hexagonal *hk0* symmetry, which is essentially identical to the pattern observed for hexagonal bulk mesoporous materials such as MCM-41. The larger *d*(100) spacing value (31.9 Å) reflects a swelling effect of benzene. The presence of relatively weak high-angle peaks at 110 and 210 indicates that although some of the pore channels are aligned otherwise, most are aligned along the *c* axis of the unit cell, parallel to the film surface.

Figure 6b shows the XRD pattern observed here for thin films with thicknesses less than 3 μm. All reaction conditions were the same as before; the only difference was the amount of gel delivered to the benzene-water interface at the beginning of the reaction. In similar fashion to what was observed for the lamellar films, the XRD peaks for thin hexagonal films cannot be in-

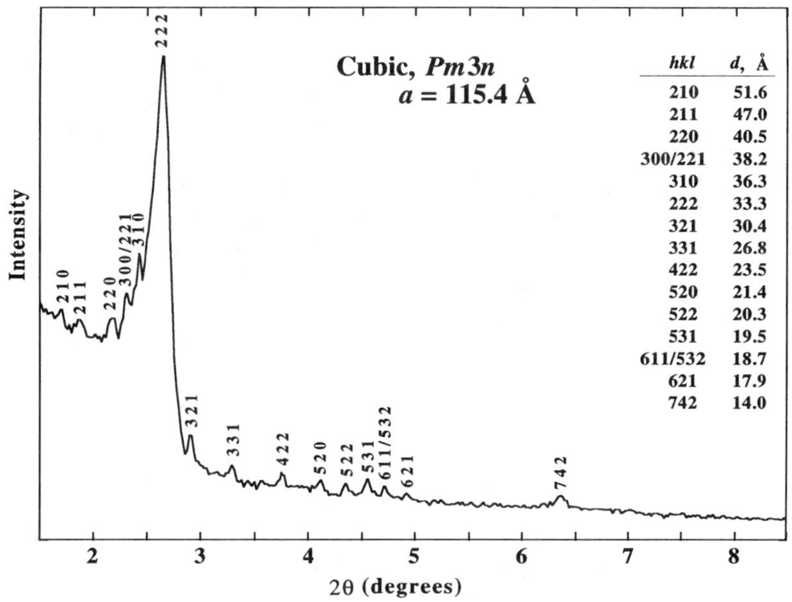

hkl	*d*, Å
210	51.6
211	47.0
220	40.5
300/221	38.2
310	36.3
222	33.3
321	30.4
331	26.8
422	23.5
520	21.4
522	20.3
531	19.5
611/532	18.7
621	17.9
742	14.0

FIG. 5 Typical XRD pattern of cubic mesoporous silica film.

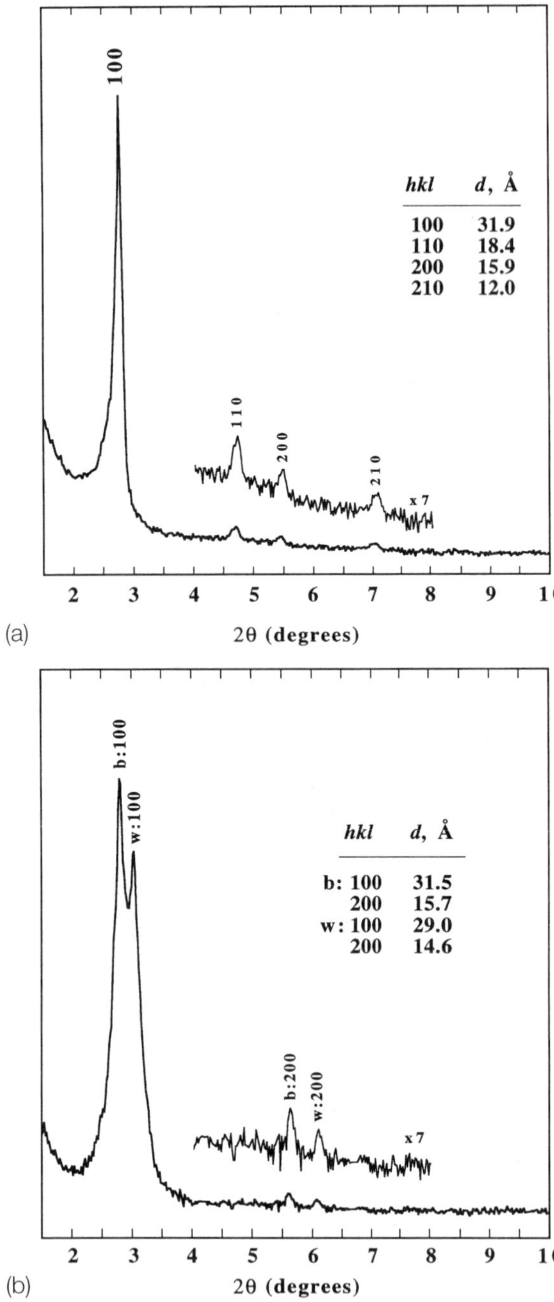

FIG. 6 Typical XRD patterns of hexagonal mesoporous silica films: (a) film with a thickness range of 5–100 μm and (b) film with a smaller thickness below 3 μm.

dexed as any single mesophase but they do match the pattern expected for a hexagonal structure containing pore channels of two distinct sizes. The average pore diameters on the benzene and water sides were 31.5 and 29.0 Å, respectively. Absence of (110) and (210) peaks indicates that the long *c* axis of each unit cell is oriented parallel to the film surface; i.e., the cylindrical pore channels run parallel to the surface and are not transverse. Synthesis at an air-solid interface of a hexagonal mesoporous silica film having similar structural features using acidic C_{16}TAC solution has been reported [43]. The 29 Å *d*(100) spacing value was the same as that observed here for the water side of the film and the same as those reported for the bulk hexagonal particles synthesized in aqueous C_{12}TAC solution. The higher *d*(100) spacing (31.5 Å) determined for the benzene side again is due to the swelling effect of benzene molecules solubilized into the surfactant aggregate during reaction.

Unlike that of the lamellar films described earlier, the proportional distribution of small and large pores [as determined by the *A*(w)/*A*(b) ratio] in hexagonal films did not exhibit any dependence on the time allowed for reaction at the liquid-liquid interface. The reason for this is not clear, but again it seems to be related to the benzene present in the gel during synthesis. Unlike that in a lamellar structure, benzene solubilized inside hexagonal micellar aggregates is much less mobile and therefore the expulsion of benzene from a hexaganol film during formation would be expected to be much less than for a lamellar film.

C. Deposition of Films on Solid Substrates

Mesoporous silica films on borosilicate glass, Teflon, and ITO glass surfaces were prepared by two different methods as described earlier. In one method, the precursor gel was spread at a benzene-water interface and allowed to react for a specified length of time, after which the film was transferred to the solid substrate, where the reaction was allowed to go to completion. The second method involved application of a conventional dip-coating technique in which the film was synthesized directly on the solid substrate. Films were then dried, calcined, and characterized by XRD to determine the mesostructure.

Table 1 summarizes the results of these experiments. The most important conclusion is that the liquid-liquid method allows one to synthesize selectively lamellar, cubic, and hexagonal films on any of the three surfaces. The major disadvantage of the dip-coating method is that the solid surface is the controlling factor in determining film structure; for example, regardless of the type of precursor gel (C_{12}-, C_{14}-, or C_{16}TAC), reaction on glass surfaces always resulted in hexagonal films, only lamellar films were obtained on Teflon, and none of the attempts to make cubic films by dip coating were

TABLE 1 Comparison of the Structure Obtained for Mesoporous Silica Films on Solid Substrates Using Two Different Methods[a]

Method	Solid surface	Surfactant in precursor gel[b] $C_{16}TAC$	$C_{14}TAC$	$C_{12}TAC$
Transfer from water-benzene interface	Glass	L	Q	H
	Teflon	L	Q	H
	ITO glass	L	Q	H
Dip coating from solution (sol-gel)	Glass	H	H	H
	Teflon	L	L	L
	ITO glass	H	H	H

[a]The method in which a partially formed film at a liquid-liquid interface is transferred to the solid allows the mesostructure to be selectively controlled by varying the gel composition (e.g., type of surfactant). In the conventional dip-coating method, the mesostructure is essentially determined by the solid surface.
[b]L, lamellar; Q, cubic; H, hexagonal.

successful. In contrast, the initial reaction period at the liquid-liquid interface allows the mesostructures to become sufficiently well developed that the final film structure is determined. When this film is subsequently transferred to a solid surface, specific interactions between the film and surface occur, effectively bonding the film to the surface, but any structural changes resulting from these interactions are not propagated into the film layer. The structure formed while at the liquid-liquid interface is preserved after transfer.

D. The Control of Film Properties

1. Film Thickness

Using the liquid-liquid interfacial reaction technique, the thickness of a film can be simply and very accurately controlled by adjusting the amount of silicate gel initially spread at a benzene-water interface. Figure 7 shows the variation in thickness for dried lamellar films as a function of the amount of gel delivered to the interface. Optically transparent films with thickness from 0.03 to 90 μm were synthesized. Much thicker films were also successfully prepared, although these were only semitransparent. XRD, SEM, and TEM results showed uniform well-developed structures even for films 300 μm thick.

One useful quantitative measure of film quality is the surface roughness, defined as the ratio of the average distance between high and low points on the surface to the total thickness of the film. Figure 8 shows SEM images of lamellar film surfaces. Both benzene-contacted and water-contacted sides have smooth surfaces, with an approximate surface roughness of 5%. Figures 9 and 10 show SEM images of cubic and hexagonal film surfaces, respectively. In both films, surface roughness was less than 5%.

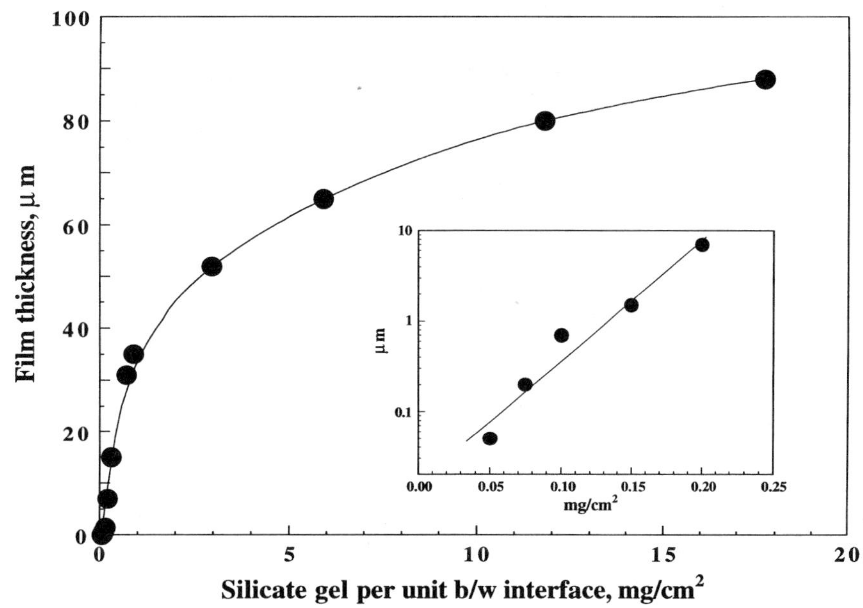

FIG. 7 Variation of the thickness of lamellar mesoporous silica films as a function of the silicate gel amount delivered at the benzene-water interface.

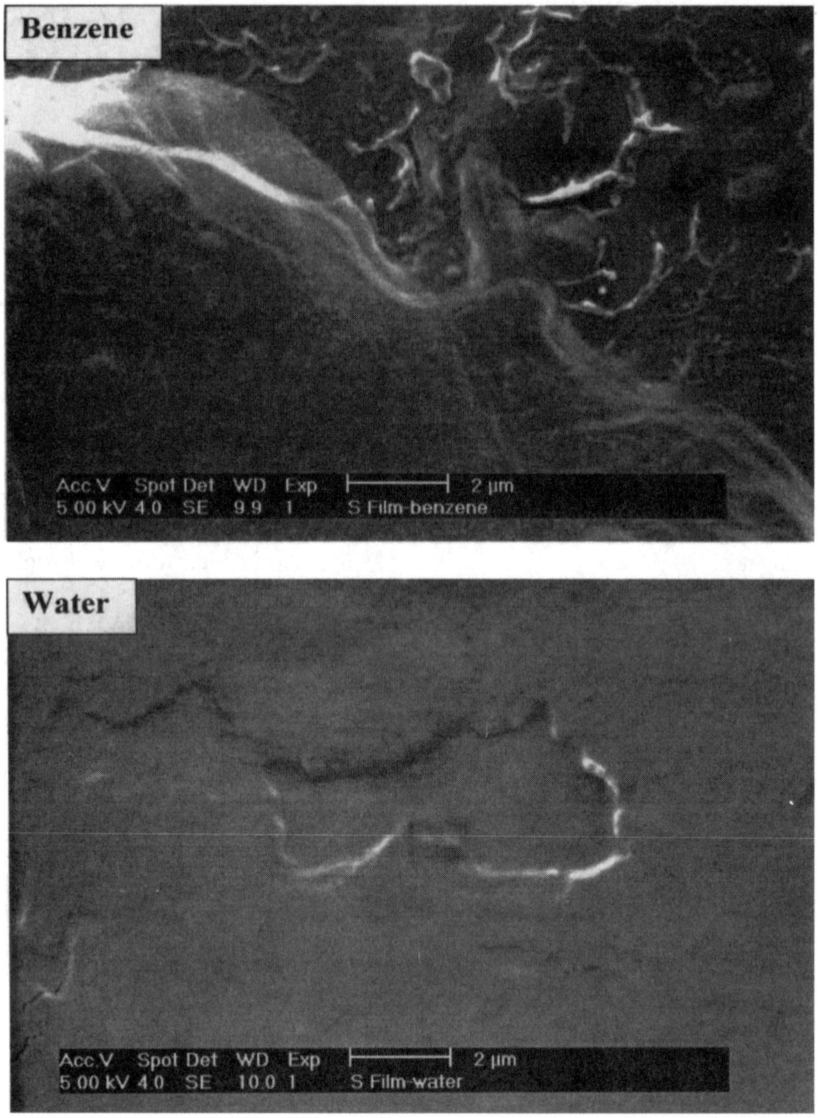

FIG. 8 Scanning electron microscopy (SEM) images of the surfaces of lamellar mesoporous silica film. Benzene and water on the images represent the benzene-contacted and the water-contacted surface, respectively.

SEM images of the side edges of lamellar, cubic, and hexagonal films are shown in Fig. 11. The lamellar film exhibits perfectly developed silica sheets; this uniform structure was observed across the entire thickness of the film, even for very thick films. The SEM image of the cubic film edge clearly shows the interconnected "spongelike" morphology characteristic of the *Pm*3*n* structure. The SEM image of the hexagonal film is characteristic of hexagonal structures viewed along the major axis along which the rod-shaped micelles are aligned [9]. These results demonstrate that the structural properties of a product can be selectively controlled by exploiting both the self-assembly of surfactant molecules in solution and the additional structure-directing effects associated with confining the reaction to an interfacial region. Combined, these strategies provide a means for obtaining not only uniform mesoscopic features (pore size and alignment) but also longer range uniformity so that the macroscopic morphology is also controlled.

These results also demonstrate that the liquid-liquid technique provides a route to making improved bulk mesoporous materials because thick films (thickness 1 μm or greater) can easily be made. Conventional processes for synthesizing bulk M41S-type materials yield particulate powders that have well-defined mesostruc-

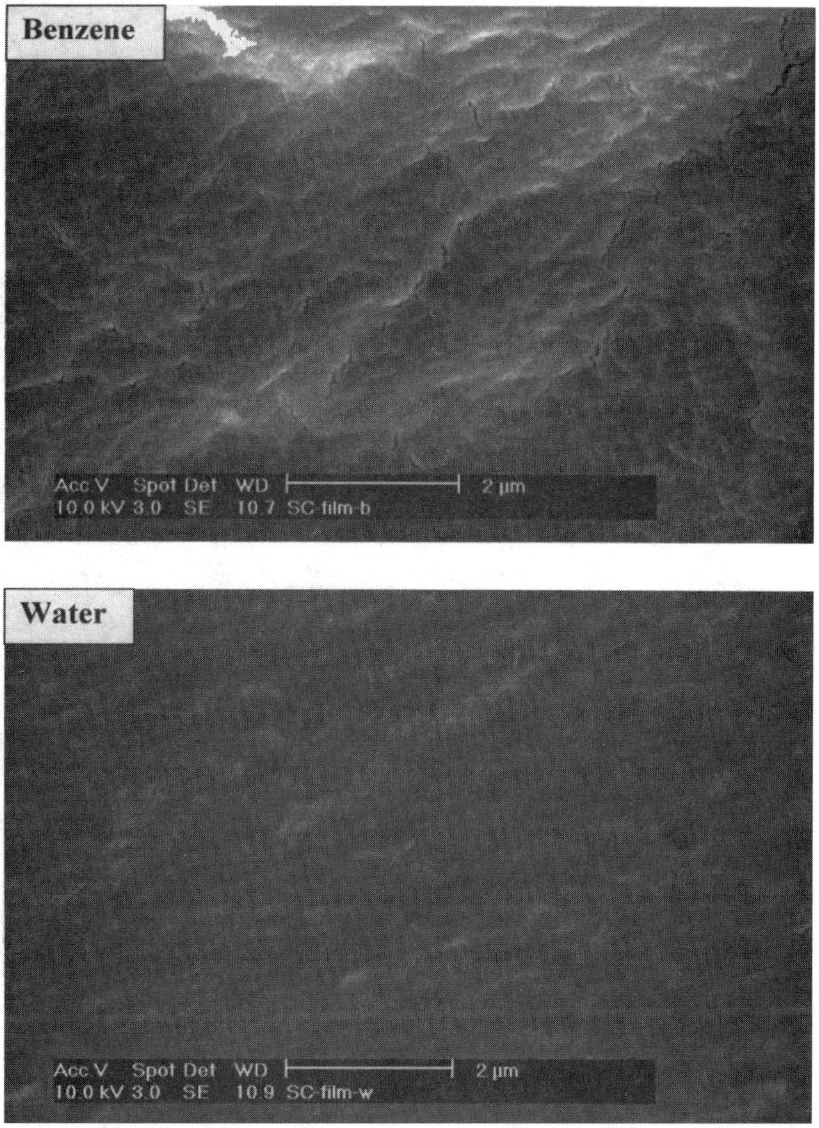

FIG. 9 SEM images of the surfaces of cubic lamellar mesoporous silica film. Benzene and water on the images have the same meaning as in Fig. 8.

ture, but this structure is relatively short range. Lamellar and hexagonal mesoporous silica particles, for example, are composed of randomly oriented microdomains, as illustrated for a lamellar product in Fig. 12. In comparison, thick films synthesized at a liquid-liquid interface exhibit much longer range ordering. Thus, highly ordered particles can be prepared by synthesizing the film and then simply crushing it to a desired size. Crushing the films in Fig. 11 can produce micrometer-sized particles that have uniform structure throughout; i.e., the long axes of unit cells within a given particle would all be aligned.

2. Other Strategies for Controlling Film Mesostructures

Results from a broad range of experiments have shown that the water content in the gel, the molar silicate/surfactant ratio, and the amount of cosurfactant are the main factors in controlling the properties of films synthesized from a precursor gel. The effects of organic solvents and short-chain alcohols such as methanol and ethanol have been extensively studied in the synthesis of bulk mesoporous material [65,74,75]. Crystallinity, unit cell size, and mesophase type are greatly affected

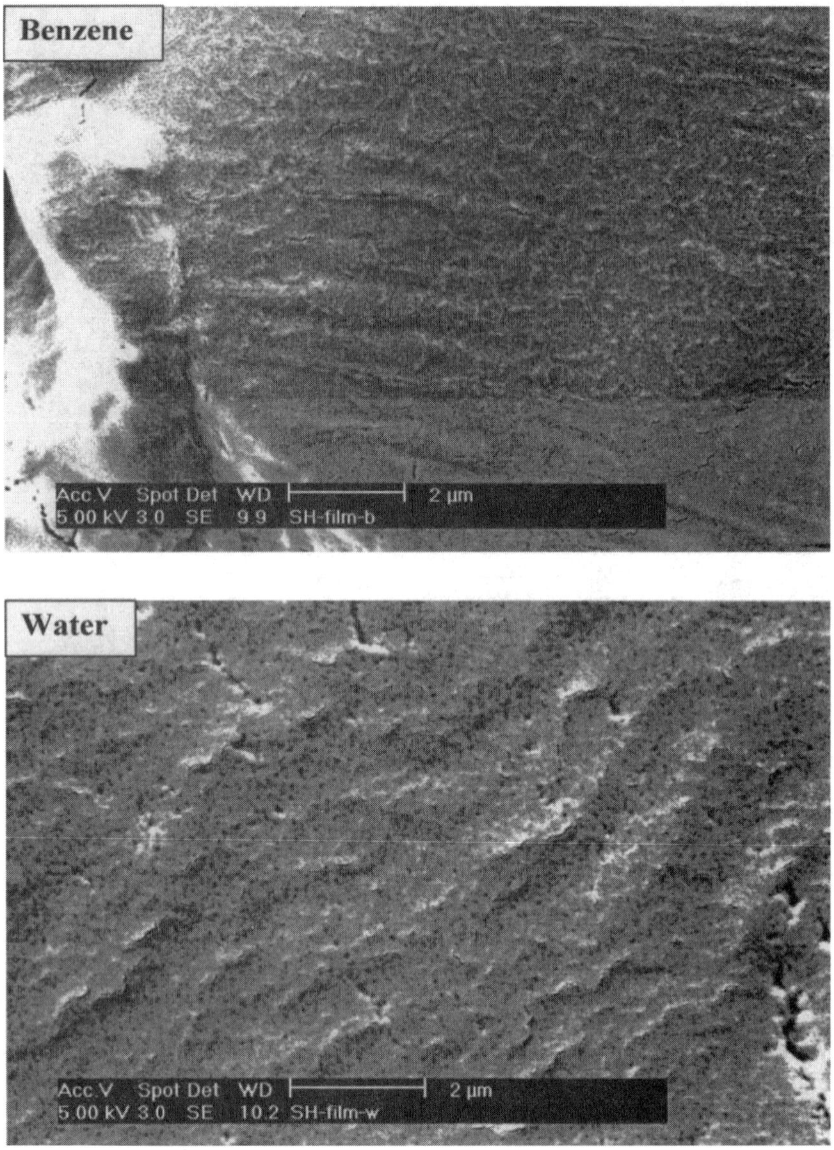

FIG. 10 SEM images of the surfaces of hexagonal lamellar mesoporous silica film. Benzene and water on the images have the same meaning as in Fig. 8.

by these simple additives. The elimination of the *gauche* defects in the surfactant chains has been observed in mixed surfactant systems of cationic surfactant and long-chain alcohol because of the strong chain-chain interaction [76,77]. Mixed micelle formation of a long-chain alcohol and cationic surfactant is thermodynamically favorable because the alcohol reduces electrostatic repulsion between the headgroups of the ionic surfactant. This reduces the effective area per headgroup and so may result in a transition in micelle structure. Long-chain alcohols can thus greatly influence both the mesophase type [68] and density of mesoporous materials synthesized in the presence of surfactant aggregates.

Mesoporous silica films have many potential applications. One currently being investigated is their use as a protective desiccant layer on ITO-coated glass. The ITO surface acts as an anode in multilayer organic electroluminescent liquid crystal display units [78–80]. The luminescent organic layer tends to absorb moisture from the air, resulting in a loss of display efficiency. A thin coating of mesoporous silica film on the ITO will retain moisture because of the mesoporous silica's extremely large surface area (500–1000 m^2/g) and thus

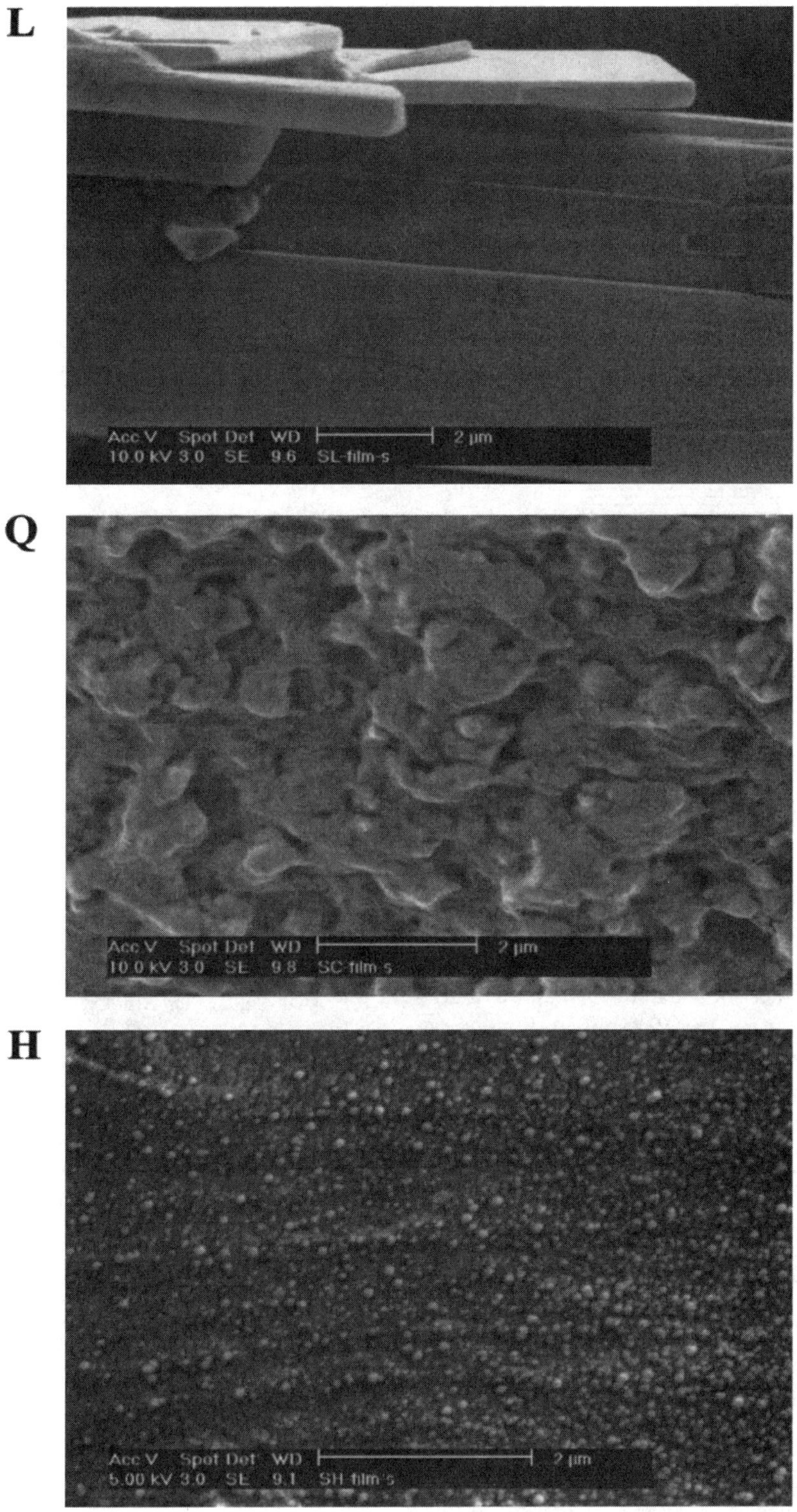

FIG. 11 SEM images of the side view of lamellar (L), cubic (Q), and hexagonal (H) mesoporous silica films.

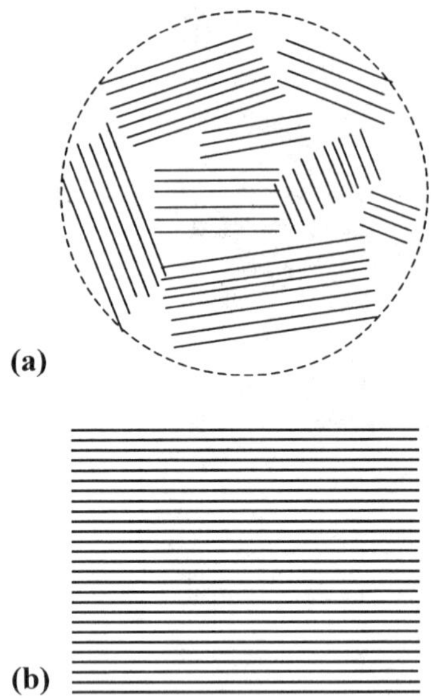

FIG. 12 Comparison of particles composed of lamellar silica synthesized by (a) a conventional sol-gel process and (b) a process in which the reaction is performed at a liquid-liquid interface. Synthesis of thick films provides a route toward preparing bulk particles having extremely long-range uniform structure.

prevent water from reaching the organic layer. The silica film does not decrease the conductivity of the metal oxide coating, so a substantial extension of the useful life of the display unit can be achieved without compromising any performance characteristics. Other possible applications for mesoporous films include their use in surface acoustic wave (SAW) sensors for larger molecules, selective membrane separations, quantum optics, and other electronic devices [81]. Naturally occurring laminated nanocomposites are observed in many biological systems and are responsible for the tensile strength, hardness, and toughness of many natural materials. The production of synthetic lamellar films is similar to biomineralization processes [39,42], and so these systems are ideal for biomimetic studies.

IV. SUMMARY

The synthesis of mesoporous silica films at a liquid-liquid interface using surfactant-silicate precursor gels offers several unique advantages over conventional methods. The preorganization and initial reaction at a liquid-liquid interface before transfer to a solid substrate eliminate effects of the surface on film structure. Lamellar, cubic, and hexagonal mesoporous silica films were selectively prepared on borosilicate glass, Teflon, and ITO glass surfaces. The mesostructures of the films were well developed and long range, and uniform structure was observed throughout the films. Films of a wide range of thicknesses were synthesized, from 0.03 to 300 μm. Lamellar and hexagonal films having two different d spacings within a single film were synthesized.

ACKNOWLEDGMENT

The authors thank Dr. Sang-Eon Park of KRICT (Korea Research Institute of Chemical Technology) and Dr. Woo Young Kim (Hyundai Electronics, Korea) for providing ITO-coated glasses. Yani Angsani and Janine Lawrence made substantial experimental contributions to this work.

REFERENCES

1. J. N. Israelachvili, *Intermolecular & Surface Forces*, Academic Press, New York, 1992.
2. A. W. Adamson, *Physical Chemistry of Surfaces*, 5th ed., John wiley & Sons, New York, 1990.
3. J. H. Clint, *Surfactant Aggregation*, Blackie, 1992.
4. T. Gu and Z. Huang, Colloids Surf. *40*:71 (1992).
5. R. G. Laughlin, *Cationic Surfactants: Physical Chemistry*, Marcel Dekker, New York, 1991.
6. S. Manne, J. P. Cleveland, H. E. Gaub, G. D. Stucky, and P. K. Hansma, Langmuir *10*:4409 (1994).
7. S. Manne and H. E. Gaub, Science *270*:1480 (1995).
8. S. Manne, T. E. Schäffer, Q. Huo, P. K. Hansma, D. E. Morse, G. D. Stucky, and I. A. Aksay, Langmuir *13*: 6382 (1997).
9. C. T. Kresge, M. E. Leonowicz, W. J. Roth, J. C. Vartuli, and J. S. Beck, Nature *359*:710 (1992).
10. J. S. Beck, J. C. Vartuli, W. J. Roth, M. E. Leonowicz, C. T. Kresge, K. D. Schmitt, C. T.-W. Chu, D. H. Olson, E. W. Sheppard, S. B. McCullen, J. B. Higgins, and J. L. Schlenker, J. Am. Chem. Soc. *114*:10834 (1992).
11. M. T. Janicke, C. C. Landry, S. C. Christiansen, S. Birtalan, G. D. Stucky, and B. F. Chmelka, Chem. Mater. *11*:1342 (1999).
12. M. J. Kim and R. Ryoo, Chem. Mater. *11*:487 (1999).
13. S. S. Kim, W. Zhang, and T. J. Pinnavaia, Science *282*: 1302 (1998).
14. F. Caruso, R. A. Carsuo, and H. Möhwald, Science *282*: 1111 (1998).
15. B. T. Holland, C. F. Blandord, and A. Stein, Science *281*:538 (1998).
16. A. Imhof and D. J. Pine, Nature *389*:948 (1997).
17. T. Sun and J. Y. Ying, Nature *389*:704 (1997).
18. W. Zhou, J. M. Thomas, D. S. Shephard, B. F. G. Johnson, D. Ozkaya, T. Maschmeyer, R. G. Bell, and Q. Ge, Science *280*:705 (1998).

19. C. G. Sonwane and S. K. Bhatia, Langmuir *15*:2809 (1999).
20. M. Kruk, M. Jaroniec, and A. Sayari, Chem. Mater. *11*: 492 (1999).
21. S. K. Bhatia and C. G. Sonwane, Langmuir *14*:1521 (1998).
22. S. Inoue, Y. Hanzawa, and K. Kaneko, Langmuir *14*: 3079 (1998).
23. H. Hata, S. Saeki, T. Kimura, Y. Sugahara, and K. Kuroda, Chem. Mater. *11*:1110 (1999).
24. X. Feng, G. E. Fryxell, L.-Q. Wang, A. Y. Kim, J. Liu, and K. M. Kemner, Science *276*:923 (1997).
25. T. Maschmeyer, F. Rey, G. Sankar, and J. M. Thomas, Nature *378*:159 (1995).
26. S. H. Tolbert, A. Firouzi, G. D. Stucky, and B. F. Chmelka, Science *278*:264 (1997).
27. D. Zhao, P. Yang, B. F. Chmelka, and G. D. Stucky, Chem. Mater. *11*:1174 (1999).
28. Q. Huo, J. Feng, F. Schüth, and G. D. Stucky, Chem. Mater. *9*:14 (1997).
29. P. Yang, D. Zhao, B. F. Chmelka, and G. D. Stucky, Chem. Mater. *10*:2033 (1998).
30. M. M. Collinson, C. G. Rausch, and A. Voigt, Langmuir *13*:7245 (1997).
31. T. Bein and K. Brown, J. Am. Chem. Soc. *111*:7640 (1989).
32. A. J. Ricco, G. C. Frye, and S. J. Martin, Langmuir *5*: 273 (1989).
33. D. H. Olson, G. D. Stucky, and J. C. Vartuli, U.S. Patent 5,364,797 (1994).
34. M. Ogawa, J. Am. Chem. Soc. *116*:7941 (1994).
35. M. Ogawa, Langmuir *13*:1853 (1997).
36. M. Ogawa, Chem. Commun. 1149 (1996).
37. J. E. Martin, M. T. Anderson, J. Odinek, and P. Newcomer, Langmuir *13*:4133 (1997).
38. Y. Lu, R. Gangull, C. A. Drewlen, M. T. Anderson, C. J. Brinker, W. Gong, Y. Guo, H. Soyez, B. Dunn, M. H. Huang, and J. I. Zink, Nature *389*:364 (1997).
39. A. Sellinger, P. M. Weiss, A. Nguyen, Y. Lu, R. A. Assink, W. Gong, and C. J. Brinker, Nature *394*:256 (1998).
40. M. Trau, N. Yao, E. Kim, Y. Xia, G. M. Whitesides, and I. A. Aksay, Nature *390*:674 (1997).
41. P. Yang, T. Deng, D. Zhao, P. Feng, D. Pine, B. F. Chmelka, G. M. Whitesides, and G. D. Stucky, Science *282*:2244 (1998).
42. I. A. Aksay, M. Trau, S. Manne, I. Honma, N. Yao, L. Zhou, P. Fenter, P. M. Eisenberger, and S. M. Gruner, Science *273*:892 (1996).
43. H. Yang, A. Kuperman, N. Coombs, S. Mamiche-Afara, and G. A. Ozin, Nature *379*:703 (1996).
44. G. S. Attard, P. N. Bartlett, N. R. B. Coleman, J. M. Elliott, J. R. Owen, and J. H. Wang, Science *278*:838 (1997).
45. H. Yang, N. Coombs, I. Sokolov, and G. A. Ozin, Nature *381*:589 (1996).
46. A. S. Brown, S. A. Holt, T. Dam, M. Trau, and J. W. White, Langmuir *13*:6363 (1997).
47. A. S. Brown, S. A. Holt, P. A. Reynolds, J. Penfold, and J. W. White, Langmuir *14*:5532 (1998).
48. J. L. Ruggles, S. A. Holt, P. A. Reynolds, A. S. Brown, D. C. Creagh, and J. W. White, Phys. Chem. Chem. Phys. *1*:323 (1999).
49. S. A. Holt, G. J. Foran, and J. W. White, Langmuir *15*: 2540 (1999).
50. I. Prieto, A. J. Fernández, E. Muñoz, M. T. Martín, and L. Camacho, Thin Solid Films *284*/*285*:162 (1996).
51. A. J. Fernández, M. T. Martín, J. J. Ruiz, E. Muñoz, and L. Camacho, J. Phys. Chem. B *102*:6799 (1998).
52. J. Sanchez and A. V. McCormick, J. Non-Crystalline Solids *167*:289 (1994).
53. J. Kim, J. L. Plawsky, E. V. Wagenen, and G. M. Korenowski, Chem. Mater. *5*:1118 (1993).
54. A. E. McHale, Phase Diagrams and Ceramic Processes, Chapman & Hall, London, 1998.
55. C. K. Narula, *Ceramic Precursor Technology and Its Applications*, Marcel Dekker, New York, 1995.
56. S. E. Rankin, C. W. Macosko, and A. V. McCormick, J. Polym. Sci. A Polym. Chem. *35*:1293 (1997).
57. J. Sanchez, S. E. Rankin, and A. V. McCormick, Ind. Eng. Chem. Res. *35*:117 (1996).
58. L. V. Ng, P. Thompson, J. Sanchez, C. W. Macosko, and A. V. McCormick, Macromolecules *28*:6471 (1995).
59. C. T. G. Knight, J. Chem. Soc. Dalton Trans. 1457 (1988).
60. A. Steel, S. W. Carr, and M. W. Anderson, J. Chem. Soc. Chem. Commun. 1571 (1994).
61. Y. S. Lee and K. W. Woo, Bull. Korean Chem. Soc. *14*: 599 (1993).
62. Y. S. Lee and K. W. Woo, J. Colloid Interface Sci. *169*: 34 (1995).
63. Y. S. Lee and K. W. Woo, Bull. Korean Chem. Soc. *14*: 392 (1993).
64. Q. Huo, R. Leon, P. M. Petroff, and G. D. Stucky, Science *268*:1324 (1995).
65. Q. Huo, D. I. Margolese, U. Ciesla, D. G. Demuth, P. Feng, T. E. Gier, P. Sieger, A. Firouzi, B. F. Chmelka, F. Schüth, and G. D. Stucky, Chem. Mater. *6*:1176 (1994).
66. J. M. Seddon and R. H. Templer, Philos. Trans. R. Soc. Lond. A *344*:377 (1993).
67. M. E. Cates, Philos. Trans. R. Soc. Lond. A *344*:339 (1993).
68. S. T. Hyde, Pure Appl. Chem. *64*:1617 (1992).
69. U. Henriksson, E. S. Blackmore, G. J. T. Tiddy, and O. Soderman, J. Phys. Chem. *96*:3894 (1992).
70. E. S. Blackmore and G. J. T. Tiddy, J. Chem. Soc. Faraday Trans. 2 *84*:1115 (1988).
71. M. T. Anderson, J. E. Martin, J. G. Odinek, and P. P. Newcomer, Chem. Mater. *10*:311 (1998).
72. M. T. Anderson, J. E. Martin, J. G. Odinek, and P. P. Newcomer, Chem. Mater. *10*:1490 (1998).

73. P. B. Miranda, V. Pflumio, H. Saijo, and Y. R. Shen, J. Am. Chem. Soc. *120*:12092 (1998).
74. R. H. Yoon and S. A. Ravishankar, J. Colloid Interface Sci. *179*:391 (1996).
75. Y. Ma, H. Y. Chao, Y. Wu, S. T. Lee, W. Y. Yu, and C. M. Che, Chem. Commun. 2491 (1998).
76. A. Kraft, A. C. Grimsdale, and A. B. Holmes, Angew. Chem. Int. Ed. Engl. *37*:402 (1998).
77. H. J. Brouwer, V. V. Krasnikov, A. Hilberer, and G. Hadziioannou, Adv. Mater. *8*:935 (1996).
78. M. T. Anderson, J. E. Martin, J. G. Odinek, and P. Newcomer, Microporous and macroporous materials, in *Materials Research Society Symposium Proceedings* (R. F. Lobo, J. S. Beck, S. L. Suib, D. R. Corbin, M. E. Davis, L. E. Iton, and S. I. Zones, eds.), Vol. 431, Materials Research Society, 1996.

39

Inorganic Nanostructure Design with Amphiphilic Block Copolymers

CHRISTINE GÖLTNER* Max-Planck-Institute of Colloids and Interfaces, Golm, Germany

I. INTRODUCTION

While searching for new materials with improved properties, scientists are increasingly focusing on the generation of composite materials instead of the synthesis of new molecular species. The main objective of these efforts is the combination of macroscopically immiscible components (e.g., metals and polymers or organic and inorganic compounds in general) on a mesoscopic length scale. Mesoscopically structured hybrid materials are fascinating because they often exhibit the typical characteristics of the components they comprise, such as the properties of a metal along with the solubility of an organic compound.

Chemistry on the dimension of a few nanometers ("nanochemistry") [1] is particularly intriguing because materials can be obtained that possess entirely new features, while their chemical composition is well known. Nanochemistry demands the application of specific nanochemical techniques and instruments [2] because the composition of a material occurs on a length scale between the dimension of a molecule and that of a macroscopic solid. Methods of classical synthetic chemistry applied to the generation of mesoscopic matter ("bottom-up") are tedious and time consuming. The opposite approach, "top-down" synthesis, which utilizes microlithographic techniques (as impressively demonstrated by Whitesides et al. [3]), is broadly applicable. However, the sensitivity of this route to defects increases with decreasing size of the object to be generated.

A more facile and less time-consuming route toward the construction of mesostructured hybrid materials would, therefore, involve a synthetic procedure during which the desired structure forms without external manipulation. This can be achieved only through the mechanisms of spontaneous, supramolecular self-assembly [4]. Supramolecular self-assembly or "self-structuring" is a principle often observed in nature and commonly takes place on exactly the length scale that is desired for the synthesis of nanochemical structures. Self-structuring is a consequence of competing interactions (e.g., Coulomb interactions, hydrogen bonding, hydrophobic interactions) and often produces structures of surprising regularity.

A. Mediation Between Incompatible Materials

The generation of a hybrid material can be achieved only by combining molecularly immiscible (incompatible) components; otherwise, some kind of alloy or homogeneous mixture (i.e., solution) would be obtained. The inherent problem of generating a mesostructured composite is that extraordinarily large interfacial areas are created, which introduce an interfacial energy of many kT into the system. In this case, macroscopic separation of the components is therefore energetically favored, and such hybrids macroscopically demix or, which is more commonly the case, do not form to begin with. However, this problem can be addressed by in-

**Current affiliation*: University of Bristol, Bristol, England.

troducing amphiphilic mediators into the system, which decrease the interfacial energy and therefore compatibilize the immiscible components. These amphiphiles consist of at least two covalently linked parts, of which one is miscible with one component and the other with the other component of the desired hybrid. The best-known and most widespread amphiphiles are surfactants, which allow the mixing of oil and water, a classical application of detergents.

In addressing other incompatibility problems, amphiphilic block copolymers (ABCs) have proved to be versatile [5]. These polymers, which consist of at least two immiscible polymer blocks, are produced through methods of modern polymer chemistry. ABCs are able to stabilize almost all imaginable interface areas, hence allowing the word "amphiphilicity" to be used in its most general sense, namely "loving both." For example, this general amphiphilicity is the foundation for the creation of composite materials made from a metal and a typical polymeric species. In the absence of a mediator, both components are mutually incompatible.

B. Amphiphilic Block Copolymers Versus Surfactants

Amphiphilic block copolymers are functional polymers [5,6] whose structure can be tailored so that the interfacial energies between materials of very different chemical natures, polarities, and cohesion energies—in short, incompatible materials—can be lowered much more than would be possible with common low-molecular-weight surfactants. They are therefore technologically relevant emulsifiers, dispersants, foaming agents, thickeners, and compatibilizers. For many applications, ABCs are able to substitute low-molecular-weight surfactants or extend their utilization.

By analogy with low-molecular-weight surfactants, amphiphilic block copolymers form aggregation structures in so-called selective solvents that are compatible with only one of the blocks. The structurally simplest aggregate is the spherical micelle, which consists of an insoluble core and a soluble corona. When the more polar block of the polymer points outward, that is, forms the shell, the micellar structure is called regular. In the opposite case, namely where the less polar block forms the shell, the micelle is called reverse. This classification, known from classical surfactants, is also applied to more complex aggregate structures such as lyotropic liquid crystal phases.

In dilute solutions of selective solvents, most amphiphilic block copolymers form spherical micelles [7–12], although cylindrical (wormlike) micelles have also been observed [13–15]. The synthesis of amphiphilic block copolymers is not covered here. Therefore, the interested reader may want to refer to excellent reviews [16–18]. At higher concentration, lyotropic mesomorphism is observed for many ABCs, which often comprises a wide variety of morphologies, many of which are known from lyotropic surfactant phases [19–23].

Because of their ability to stabilize interfaces between incompatible compounds, amphiphilic block copolymers are particularly suited to direct the structure of inorganic-organic composites. Their characteristic feature, namely amphiphilicity, gives rise to their spontaneous self-assembly into microphase-separated aggregation structures in the presence of solvents or even as neat polymers. This self-assembly can be utilized for a wide number of applications, of which a particularly interesting research area, namely the design of nanostructured inorganic-organic hybrid materials, is reviewed here.

This chapter is dedicated to a representative selection of principles and mechanisms that demonstrate the versatile nature of amphiphilic block copolymers as structure-directing agents for nanostructure design. At first sight, the aspects covered in this section originate from rather different approaches and may appear haphazardly chosen. They range from biomimetic crystallization to the synthesis of porous ceramic oxides, from low-concentration block copolymer solutions to bulk polymers, and from precipitates to nanostructured macroscopic objects. However, all of these research areas have in common that at some stage during synthesis, amphiphilic block copolymers play a structure-directing role.

II. AMPHIPHILIC BLOCK COPOLYMER MICELLES AS NANOREACTORS

One aspect of the self-assembly or aggregation of amphiphilic block copolymers is their ability to undergo micellization in selective solvents, hence compartmentalizing one microphase in the other [7–15]. The resulting compartments can be utilized for the synthesis of nanosized inorganic particles, which are separated and protected from the "outside world" by the micellar corona. In contrast to the micelles formed by low-molecular-weight surfactants in water, those formed by amphiphilic block copolymers are kinetically more stable, and molecular exchange dynamics are considerably slower. Therefore these simplest aggregate structures of amphiphilic block copolymers can be used as nanosized vessels in which chemical reactions can be conducted [24]. This procedure is called *exo*-templating

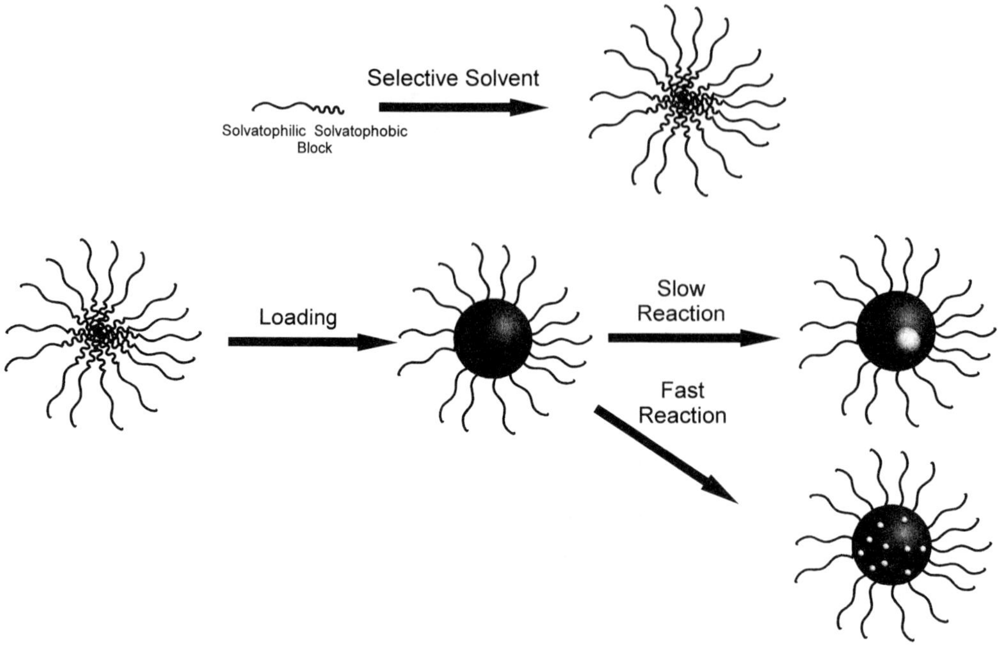

FIG. 1 Schematic representation of the ABC nanoreactor. The amphiphilic block copolymer undergoes micellization in a selective solvent. The micelles are loaded with an inorganic precursor. The chemical reaction is confined to the cores of the micelles, hence affording colloidal particles. (From Ref. 6.)

(see Fig. 1) because the structure-directing casting mold is wrapped around the inorganic object. ABC micelles, in which chemical reactions are performed, are referred to as *nanoreactors*.

The size [25], shape, and properties of these nanoreactors can be tailor-made by means of modern polymer chemistry [16–18]. For successful *exo*-templating, the polymer block, which forms the micellar core, must selectively interact with one or more of the starting materials, while the corona-forming block provides sufficient solubility of the nanoreactor in the surrounding solvent: Only then can the chemical reaction be trapped within the mesoscopic confinement of the nanoreactor.

One method of loading block copolymer micelles is to form a complex of a polymerizable ligand with the inorganic precursor [26], but the number of metal-complexed monomers that are suitable for living polymerization is very limited. It is therefore more common to bind the inorganic precursor to the ready-made block copolymer from homogeneous, nonaggregated solution. For example, it is possible to bind Zn^{2+} [27] selectively to the hydrophilic blocks of poly(styrene-*b*-methacrylic acid) (PS-PMAc) from THF solution. Complexation of the Zn ions by PMAc causes this polymer block to become insoluble in THF; consequently, THF is turned into a selective solvent in which micellization occurs.

Most commonly, however, the confinement of an inorganic precursor within block copolymer micelles occurs via a metal-ligand interaction in which the polymer block acts as a ligand (e.g., for $Pd(AcO)_2$ [27]) or via "ion exchange"; i.e., the inorganic precursor species represent the counterions for an ionic polymer block [28]. Metal colloids, for example, can be made by reducing an inorganic precursor, usually a metal salt, within the micellar core. The usually hydrophilic precursor can be solubilized even in nonpolar organic solvents if these solvents contain the appropriate amphiphilic block copolymer. This nanochemical methodology even allows the dissolution of substantial amounts of sodium chloride in toluene. One representative example of the utilization of block copolymer micelles as nanosized reaction vessels is the synthesis of gold colloids in the presence of micellar poly(butadiene)-*b*-poly(2-vinylpyridine) (PB-P2VP, see Scheme 1).

While the poly(butadiene) block is soluble in toluene (the selective solvent), the P2VP block, which shows high affinity for metal salts, is insoluble and induces micellization. Thus, a gold precursor (e.g., tetrachloroauric acid) is enriched within the micellar cores. The introduction of the inorganic species into "prefabricated" micelles is the most commonly applied method of loading block copolymer micelles because

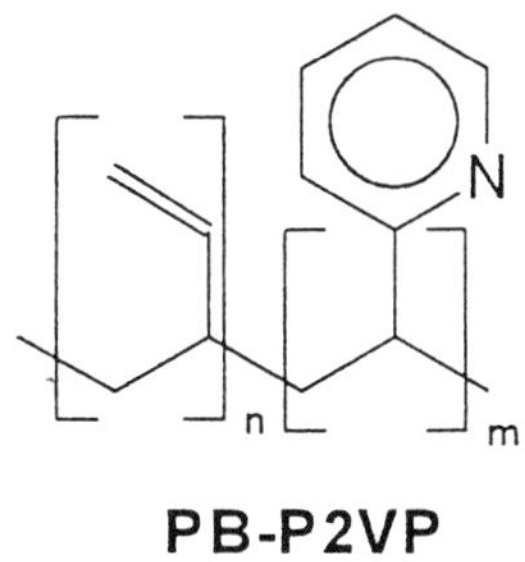

SCHEME 1 Chemical structure of poly(butadiene-*b*-2-vinyl pyridine).

it allows precise control of the amount of salt introduced, is simple to conduct, and is suitable for many solvent/block copolymer/precursor systems [28–31].

Within these loaded micelles, gold colloids are produced by chemical reduction, which is performed by adding a reducing agent to the micellar solution. The reducing agent reaches the micellar core via diffusion and is likewise incorporated. The stability and size of the gold nanoparticles are mainly determined by the micellar reactors (e.g., aggregation number, molecular weight, and the chemical nature of the block copolymer) in which they are generated as well as the amount of precursor introduced into the micelles. In addition, the features of the nanosized products can be influenced externally by choosing the appropriate reduction conditions: Fast reduction causes fast nucleation, so that many small particles are formed within the core of each micelle, whereas slow nucleation affords one colloidal gold particle per micelle (Fig. 2).

This image of a mesoscopic "nanoreactor" applies well to amphiphilic-block copolymer micelles because a nanometer-sized object is filled with starting material, and a second starting material or even a catalyst is fed at a determined rate (or the reaction can be conducted photochemically [32]). Finally, the reaction is restricted to the dimensions of the vessel; i.e., it neither explodes nor boils over! As with a bulk reactor, reaction kinetics can be controlled externally. With the help of a nanoreactor, stable noble metal particles are obtained that are ideal candidates for catalytic processes because of their extraordinarily large surface areas [33–37]. Conducting reactions in different-sized nanoreactors or block copolymer micelles (determined by the block length) generates colloids whose size corresponds directly to that of the primary nanostructure as well as the degree of loading. This represents an additional tool to fine-tune the size of colloidal particles, which is very important whenever particular optical or magnetic [38] properties are required.

The concept of the nanoreactor can also be applied to the production of colloidal metal oxides and sulfides. Quantum-size cadmium sulfide or zinc oxide can be made, safely wrapped in a stabilizing polymer shell. Most of the preparations of sulfide semiconductor nanoparticles involve the purging of solutions containing precursor-loaded ABC micelles with hydrogen sulfide [27] or causing precipitation of an oxide via external manipulation of the pH of the system.

The nanoreactor need not be constructed by engineers, maintained, or cleaned. The supramolecular self-construction of amphiphilic block copolymers in selective solvents occurs spontaneously. The latter can be predetermined and influenced by the methodology required of a nanochemical approach, in this case polymer chemistry combined with the intuition necessary

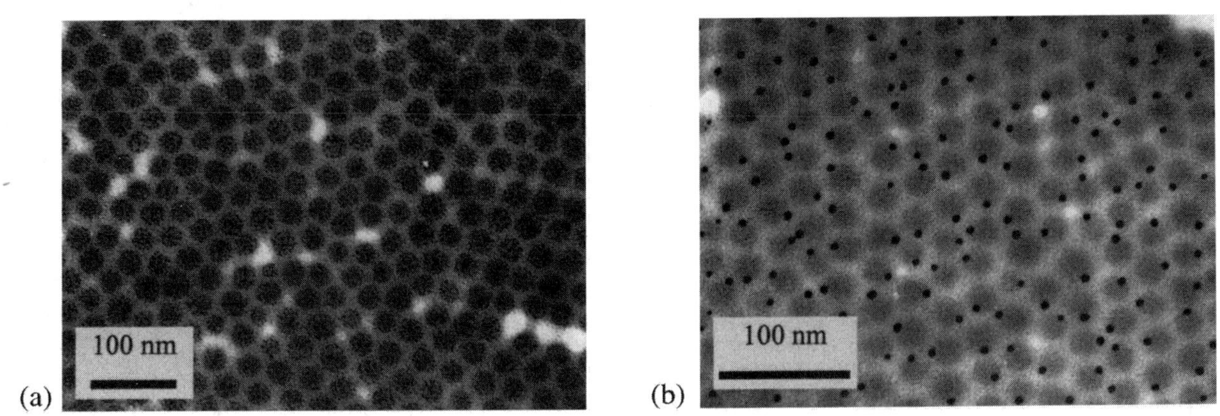

FIG. 2 TEM images of two different gold colloids produced in ABC nanoreactor micelles. (a) Fast reaction causes the generation of many small colloids in each micelle; (b) slow reaction gives rise to a single colloid per micelle. (From Ref. 6.)

to estimate which inorganic species can be compartmentalized in which ABC micelles.

III. DESIGN OF POROUS CERAMICS

Micellar nanoreactors represent *exo*-templates, because the product is enveloped by the structure-directing amphiphilic block copolymer, which determines the size and shape of the resulting colloidal object. The opposite approach to nanostructure design of inorganic materials is called *endo*-templating, in which the structure-directing medium is included within the inorganic material during the synthesis. The result is a solid, regularly structured inorganic-organic composite from which the organic matter can be removed. The remaining residue represents a highly porous ceramic material.

This approach allows the creation of pore systems in otherwise dense inorganic materials, thereby providing them with completely new properties. These properties are again due to the increase of the interfacial area or, in this case, of the surface area. Large specific surface areas in inorganic solids usually expose a vast number of surface sites where sorption processes or even chemical reactions, such as catalytic conversion, can take place. In combination with a pore system that causes an entrapped species to follow particular diffusion pathways (i.e., defining directionality and kinetics of diffusion), these materials are expected to be unique catalyst supports or stationary phases for chromatographic separation. Why the pore systems obtained by using amphiphilic block copolymers as structure-directing agents are superior to those obtained by other methods is explained in the following.

A. Porosity Through the Displacement of Porogens

The creation of inorganic pore systems is commonly achieved by growing an inorganic polymer (ceramic oxide) within or around an organic (sometimes even an inorganic) medium, the porogen [39,40]. Once the chemical preparation of the inorganic pore system is accomplished, the porogen is removed, usually by evaporation, calcination (i.e., firing), or extraction. Depending on the nature of the porogen and its compatibility with the inorganic product, materials with different pore structures, pore size distributions, and pore sizes can be manufactured.

For simplicity, the mechanisms underlying pore generation can be divided into two classes, namely the undirected displacement of a porogen and templating. If the porogen shows no specific interaction with the inorganic solid or is even miscible with the inorganic material at all times during the synthesis, the pore system arises simply from physical volume exclusion or molecular displacement out of homogeneous solution. In this case, the characteristic length of the structure that arises will depend on the steric demands of the molecularly dissolved porogen. If demixing occurs at any stage during the preparation, the volume excluded by the porogen (i.e., the later pores) will depend on the degree of repulsion, the amount of possible stabilizer, as well as the time at which the macroscopic phase separation occurs. One way to tailor the pore size distribution of macroporous ceramic oxides is to generate the inorganic solid in an emulsion. The oily emulsion droplets displace the inorganic precursor from their volume, thus creating macropores [41,42]. One example of the controlled demixing of two inorganic components is the manufacturing of *Vycor* glass or *controlled-pore* glass (CPG) [43–45].

Because no driving force pushes the system into a particularly ordered state in either case, the structures of the resulting porous materials represent a snapshot of a statistical situation; therefore, broad pore size distributions are usually obtained. In the following, this strategy will be referred to as *displacement of porogen*.

B. Micropores Through Molecular Templating

The undoubtedly more intriguing pathway toward the design of inorganic pore systems is *templating*, in which a more or less specific interaction of the inorganic with the porogen causes the creation of substantially more regular pore systems because molecular recognition and supramolecular organization are the structure-directing principles. One example of very specific templating, where the template creates a precisely defined pore system in silicates, is the synthesis of crystalline zeolites (see Chapter 40 for an exhaustive review).

In this case, quaternary ammonium salts are included in the cagelike voids of a crystal, thus generating a microporous inorganic material. The size and shape of the pore system in zeolites are defined by the templating species, usually one molecule or ion. As zeolites are regularly structured crystalline materials, their pore size distributions are peak shaped. Although zeolites with various structures and pore connectivities are available, their crystallinity does not permit the synthesis of pore systems with diameters larger than 1.3 nm [46,47]. The reason is that the walls that surround the pores in a zeolite are represented essentially by one chemical bond and are therefore too fragile to support larger pore diameters. The latter, however, are mandatory whenever a species to be catalytically converted is too bulky or too hydrophobic to enter a zeolitic pore system.

To extend the limits of classical zeolites, synthetic methods have been developed that favor the production of thicker walls surrounding larger voids. The synthesis of defined mesopores [48] is a templating procedure in which the specific molecular templates known from zeolite synthesis are replaced by supramolecularly aggregated amphiphiles. Templating can therefore be divided into two subgroups, namely *specific molecular* and *supramolecular* templating.

C. Mesopores Through Supramolecular Templating

In many respects, supramolecular templating bridges the gap between specific molecular templating and the undirected displacement of a porogen. On the one hand, it allows the synthesis of materials whose pore diameter is not limited by the structure of the product as is the case for zeolites. On the other hand, the pore size distributions, albeit narrow, can never be peak shaped. This is because of one of the fundamental principles of supramolecular chemistry. In a supramolecularly assembled, dynamic system, there always exists a statistical probability of defect sites, such as domain boundaries, which are continuously repaired but also introduced simultaneously at the same rate. A defect-free supramolecular system is therefore a contradiction in terms [2,4]. In spite of this, the structures of supramolecularly templated porous ceramic oxides are astonishingly regular. The pore size distributions that are obtained from different templating mechanisms are represented in Fig. 3.

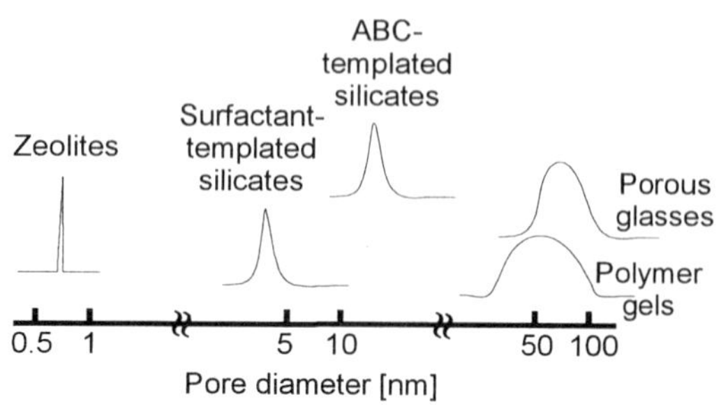

FIG. 3 Pore size distributions attainable with different methods. (From Ref. 50.)

Although this is still a field that receives great attention, it is not the aim of this chapter to dwell on the subject of low-molecular-weight surfactant templates, as review articles [49–55] and Chapter 40 in this volume provide more detailed insight for the interested reader. However, some general principles are outlined in the following to emphasize the striking similarities that can be found between low-molecular-weight surfactants and amphiphilic block copolymers and to demonstrate in what respect ABCs are often the superior choice.

1. Low-Molecular-Weight Surfactant Templates

During the past 8 years, surfactants have received increasing interest as porogens for the generation of inorganic nanostructures [49–55]. The microphase separation of a surfactant solution in water into aqueous and hydrophobic domains is the foundation of supramolecular templating. The principle is as simple as this: The inorganic precursor (hydrophilic) is expelled from the areas where the hydrophobic surfactant chains are located. The result of combining a soluble inorganic precursor with a surfactant solution is the formation of a regular hybrid material with characteristic lengths between 2 and 5 nm. Removal of the organic matter leaves behind a regularly structured, mesoporous molecular sieve.

The first mesoporous siliceous nanostructure was initially "overlooked" as curiously discovered much later: Another property attributed to this material, namely its low density, was patented without the regular nanostructure being detected at all [56,57].

The mesoscopic structure of similarly prepared porous ceramic oxides was finally discovered and systematically investigated by researchers at Mobil Research and Development Corporation. This new class of molecular sieves subsequently became known under the general name M41S [58–63].

M41S-type materials form as a consequence of cooperative interaction between micellar aqueous solutions of ionic surfactants and charged inorganic precursor species. Ion matching, sometimes via a mediating counterion, causes a surfactant-rich gel phase to precipitate from aqueous solution. This precipitate consists of a regularly structured assembly of surfactant aggregates "filled" with the inorganic solid. The structures observed resemble those of lyotropic liquid crystals, for example, hexagonally packed rodlike aggregates and gyroid or lamellar phases. Nonionic surfactants used in this precipitation procedure produce less regular materials because the attraction between the aggregating species, which in the absence of charges is due only to hydrogen bonding and hydrophobic interaction, is weaker. Nevertheless, the materials obtained exhibit narrow pore size distributions [64–66]. The precipitation procedures can be conducted hydrothermally or via sol-gel processing.

A related approach utilizes the regular structures of performed lyotropic liquid crystal phases of nonionic surfactants as structure-directing media [67–70]. In a sol-gel process, the hydrophilic inorganic silica precursor undergoes solidification confined within the aqueous domains of the liquid crystal, hence producing a cast or replica of the lyotropic system without changing the supramolecular structure. The materials obtained are similar to M41S-type mesoporous molecular sieves. However, because they are prepared in the presence of nonionic templates, which do not demand electrostatic compensation on a specific length scale, the wall thickness can be individually adjusted via the precursor content. Furthermore, the nanostructure derived from liquid crystalline bulk phases is a result of a homogeneous mixture solidifying as opposed to a precipitation process. Therefore the dimensions of the inorganic particles are far larger. All of these mesoporous ceramic oxides are amorphous on the atomic level but show periodicities on the nanometer length scale and a narrow pore size distribution (Fig. 3).

Based on the knowledge collected from utilizing surfactants as templates for mesoporous inorganics, amphiphilic block copolymers have been established as structure-directing media for porous nanostructure design [71,72]. They assist in overcoming the limitations inherent in their low-molecular-weight analogues, the classical detergents, in many aspects. The most commonly applied route toward ABC-templated inorganic nanostructures is the technologically widespread sol-gel process, the underlying chemistry of which is explained briefly using the synthesis of silica as a representative example.

2. Sol-Gel Processing of Silica

The sol-gel process for the production of silica is an industrially widely applied procedure. As it occurs in homogeneous solution, it is possible to manufacture clear macroscopic silica objects, such as fibers or lenses. Furthermore, this process can be conducted at quiescent temperatures, and no increased pressure is necessary. Therefore, although the precursors for silica synthesis are usually more expensive than those required for other methods, such as hydrothermal processing, it has found wide application. The starting material for sol-gel processing of silica is an orthoester of the general structure $Si(OR)_4$, which is hydrolyzed in order to formally yield silicic acid. The latter undergoes polycondensation into a three-dimensional network of silicon dioxide whose surface is saturated by silanol (Si—OH) groups. The various reactions occurring simultaneously in a sol-gel mixture (see Scheme 2) represent equilibria that can be influenced by parameters, such as the pH of the system, temperature, or solvent content, and that all have a substantial effect on the structure of the resulting silica gel. Whereas base-catalyzed saponification of the tetraalkylorthosilicate usually produces highly condensed particulate materials, acid-catalyzed hydrolysis and polycondensation afford a weakly branched polymeric sol. The most common alkoxide precursors are tetraethylorthosilicate (TEOS) and tetramethylorthosilicate (TMOS) because of their relatively low cost, easy handling, and relatively fast hydrolysis.

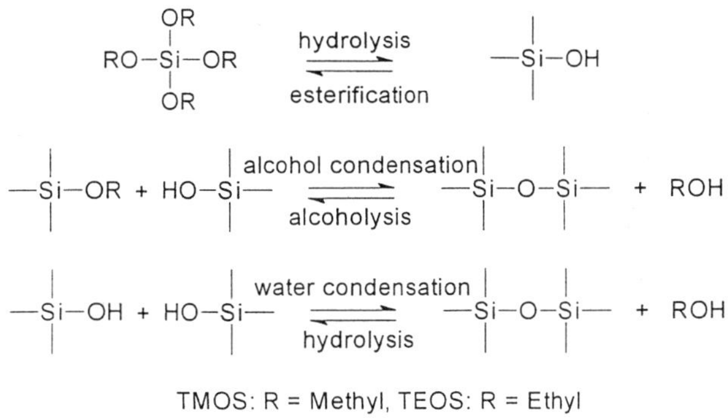

SCHEME 2 Reaction pathways during the sol-gel synthesis of silica.

The reader interested in deeper insight into the physics and chemistry of sol-gel processing may want to refer to a monograph by Brinker and Scherer [73], which discusses the subject in greater detail and provides an invaluable bibliography.

The sol-gel processing of silica is advantageous for the design of inorganic-ABC nanocomposites because the kinetics of the hydrolysis as well as the polycondensation of silicic acid can be fine tuned via the pH, which is of considerable importance for some of the procedures described in the following.

3. Generating Porosity with Amphiphilic Block Copolymers

The creation of ceramic nanostructures with controlled structure is a rapidly emerging field that greatly profits from the self-assembly of amphiphilic block copolymers as well as the variety of ABCs available. The ordering characteristics of amphiphilic block copolymers can be almost continuously tuned by varying the chemical nature of the blocks, solvent contents, molecular weight, block length ratios, and copolymer architecture or even by varying external parameters that influence the aggregation behavior, such as temperature or the addition of salt.

Simplistically, amphiphilic block copolymers can be assumed to act as "large surfactants" in that they allow self-organization into larger aggregate structures than their low-molecular-weight analogues. Like the micellar systems used as nanoreactors, the more complex aggregation structures of amphiphilic block copolymers exhibit decreased exchange dynamics, which go along with higher kinetic stability. Furthermore, amphiphilic block copolymers extend the synthetic methods for mesoporous ceramic nanostructures over the inherent limits of low-molecular-weight surfactant templates. The latter are available only up to certain alkyl chain lengths (usually 22 carbon atoms at the most). Accordingly, the pore diameters available for mesoporous inorganic nanostructures are limited to a maximum of 4.5 nm unless inert organic auxiliaries are introduced into the system during synthesis.

In contrast, amphiphilic block copolymers can be made (or even purchased) with considerably higher molecular weight, so that the synthesis of materials with larger pores is possible. In particular, modern polymer chemistry provides the tools to make these block copolymers in a vast variety of shapes and sizes, giving rise to the expectation that an equally rich variety of nanostructures could be produced.

The objective of this section is to discuss different methods of exploiting amphiphilic block copolymer templates for the generation of sol-gel–derived inorganic-organic hybrid structures with periodicities larger than 4 nm. All approaches to such materials have in common that the growth of the inorganic matter is displaced from the hydrophobic domains of an amphiphilic block copolymer aggregate structure. The compatibilizing amphiphilic block copolymer, therefore, not only stabilizes the interfacial area of a nanoscopic structure but also acts as a porogen, leaving nanometer-sized voids behind after its removal. Amphiphilic block copolymer templates replace the classical low-molecular-weight surfactants that were used previously, introducing mechanical stability and access to a wider range of pore diameters.

Three different approaches can be distinguished, depending on what causes the final inorganic-organic hybrid structure to form and on the amount of template present during the synthesis. The following section gives an overview of the different routes to porous ceramic nanostructures using amphiphilic block copolymers as structure-directing media.

4. Precipitation of Inorganic-ABC Hybrid Structures

By analogy with the preparation of M41S materials, the combination of an alkoxide silica precursor and a solution of amphiphilic block copolymers in a water-alcohol mixture causes the formation of a regularly structured gel phase, which precipitates from aqueous solution [74–76]. The synthetic procedure, which is conducted under acidic conditions using TEOS as the precursor utilizes mainly nonionic Pluronic-type triblock copolymers or nonionic star diblock copolymers as structure-directing agents* because of their good solubility in water, commercial availability, low cost, and biodegradability. The solidification of the silica network occurs within this polymer-rich precipitate, while the supramolecular structure is preserved. The amphiphilic block copolymer template can be completely removed from the hybrid material by solvent extraction or calcination, both of which leave behind the purely inorganic nanoporous ceramic oxide.

The structure of the final product is a result of a macroscopic separation of a microphase-separated, ordered phase from water as a solvent. The solidification of the sol-gel mixture takes place in the ordered environment of this microphase-separated system. The synthesis is conducted at relatively low polymer concentrations, above the critical micelle concentration, which are far lower than those necessary to form an ordered amphiphilic block copolymer structure, i.e., a lyotropic mesophase. The system tolerates the presence of considerable amounts of inert swelling agents, usually mesitylene, so that despite the relatively low molecular weight of the hydrophobic templating block, large pores (up to 30 nm) can be created. It is also possible to fine-tune the final ceramic structures via other synthesis parameters: The hydrophilicity of poly(ethylene oxide) is temperature dependent [77]; therefore raising the temperature during the sol-gel process formally increases the hydrophobic volume of the template, which is an elegant method of finely adjusting the pore size [74].

The pore diameter of the calcined samples depends on the molecular weight of the template (Table 1), mainly that of the hydrophobic block. In contrast, the

*Amphiphilic triblock copolymers of the general structure poly(ethylene oxide)-*b*-(propylene oxide)-*b*-(ethylene oxide) (PEP-PPO-PEO, Pluronics) are a trademark product of BASF, Mt. Olive, NJ.

TABLE 1 Pore Diameters and Specific Surface Areas of Pluronic-Derived Silicas

Pluronic-type template	Pore diameter (nm)	BET surface area (m^2/g)
$EO_5PO_{70}EO_5$	117	630
$EO_{20}PO_{70}EO_{20}$	96	690
$EO_{20}PO_{70}EO_{20}$	98	780
$EO_{20}PO_{70}EO_{20}$	100	820
$EO_{20}PO_{70}EO_{20}$	105	920
$EO_{20}PO_{70}EO_{20}$	104	850
$EO_{17}PO_{55}EO_{17}$	81	770
$EO_{20}PO_{30}EO_{20}$	78	1000
$EO_{26}PO_{39}EO_{26}$	88	960
$EO_{13}PO_{70}EO_{13}$	80	950
$EO_{19}PO_{33}EO_{19}$	71	1040

Source: Ref. 74.

specific surface areas, which are obtained from nitrogen sorption experiments, appear to be substantially affected by the size of the hydrophilic block.

In fact, the specific surface areas are far larger than can be expected from simple geometrical considerations (S. Förster, unpublished results): As the interfacial area in the microphase-separated system roughly represents the specific surface area of the later purely ceramic oxide, the calculation of exactly this area should allow an estimation of the surface area contribution arising from the mesopore system. The following equation outlines the dependence of the volume fraction ϕ for a "binary" system consisting of hexagonally packed cylinders:

$$\phi = \frac{8\pi}{3} \frac{R_{cyl}^2}{a^2} \tag{1}$$

with R being the radius of the cylinders (or pores), a the distance between cylinders, and d the dimensionality of the system (in this case 2). The interfacial area A/V per unit volume between hydrophilic and hydrophobic domains is given by

$$\frac{A}{V} = d \frac{\phi}{R_{cyl}} \tag{2}$$

The pore radius R can be obtained from transmission electron microscopy (TEM) analysis, the distance between cylinders from small-angle x-ray scattering experiments. The density ρ of the silica matrix can be either determined experimentally or assumed to be constant for different amorphous silicas (about 2.1 g/cm^3).

For a defect-free domain the specific surface area is obtained from Eq. (3):

$$A_{spec} = \frac{d\phi}{(1 - \phi)R_{cyl}\rho_{SiO_2}} \tag{3}$$

Even taking into account structural defects, domain boundaries, and external particle surface, it is evident that the specific surface areas of most mesoporous ceramic oxides obtained by different experimental approaches show a significant contribution arising from structures with smaller length scales, which are mainly due to microporosity. Suitable experiments have been devised to determine the mesopore surface area selectively by filling the micropores (B. Berton et al., unpublished results; B. Smarsly et al., unpublished results). The phenomenon of extraordinarily large specific surface areas gives rise to the assumption that the hydrophilic poly(ethylene oxide) chains of the template are "dissolved" or anchored in the ceramic material during and after solidification. The substructure of the resulting nanostructured hybrid material is therefore expected to be a two-phase system in which one microphase consists of an inorganic-poly(ethylene oxide) hybrid material. After removal of the organic structure-directing copolymer, the hydrophilic blocks leave behind smaller (micro)pores. Therefore, the poly(ethylene oxide) groups additionally act as porogens in their own right. By displacing the silica growth, they give rise to substantial microporosity and possibly small-scale surface roughness (B. Berton et al., unpublished results). This finding underlines the fact that in a mixture of incompatible components, each molecule of the compatibilizer contributes to the interfacial area.

As the amphiphilic block copolymer templates are nonionic, the wall thickness of the resulting mesoporous silicate is substantially higher than that of MCM-41–type materials, whose structure formation occurs via a complex interplay of ion matching. Therefore the mechanical and hydrolytic stabilities of the products are greatly improved. As typical polymeric species, amphiphilic block copolymers introduce considerable mechanical stability into the inorganic-organic hybrid material, which would be brittle in the absence of the template. Therefore, it is possible to cast homogeneous, smooth films by dip coating a silicon wafer into the sol-gel mixture [76] and to manufacture membranes whose structure and pore connectivity can be individually adjusted. This route toward ABC-templated mesoporous ceramic oxides also makes it possible to influence the shape of the particles via experimental parameters such as stirring. The improved mechanical performance of the hybrid product is a typical example of favorable effects that are observed when the hybridization of matter is conducted on the nanometer length scale.

One of the most impressive and pioneering uses seems to be its combination with other methods of "shaping matter" [78]. The simultaneous application of ABC templating and micromolding (pathway A in Fig. 3) as well as optionally using previously established polymer latex spheres [79] as additional templates (Fig. 4, pathway B) affords patterned materials with an exceptional micrometer-scale morphology, which show additional order on the mesoscopic length scale. Hierarchical ordering over several discrete and tunable length scales (tiers) ranging from 10 nm to several micrometers is the result.

Compared with the complicated multitier structure of these materials, the overall synthetic strategy is simple: In a "top-down" approach, the template with the largest length scale is prepared first by micromolding a poly(dimethylsiloxane) (PDMS) matrix, which is used as a stamp to imprint the micrometer-scale morphology [80–82]. This stamp is then pressed into a liquid sol-gel mixture consisting of templating Pluronic-type ABC, tetraethylorthosilicate (or other sol-gel precursors, e.g., for the synthesis of niobia), and water at a suitable pH. The sol-gel process is confined to the spaces where the micromold prevents it from wetting the substrate (usually a silicon wafer). Introducing one more step in the hierarchical order is achieved by additionally using latex templates, which were previously shown to be useful porogens in porous-silica synthesis and have been established as templates in a process called *rigid-colloid templating* [83]. The latex dispersion is introduced into the voids of the PDMS micromold, where it forms a colloidal-crystalline array (Fig. 4, pathway B). One example of a hierarchical ceramic oxide structure is shown in Fig. 5.

The precipitation of regularly structured hybrid materials from solution is a result of cooperative supramolecular aggregation. As the phase structure forms during the synthesis, that is, upon combination of the organic structure-directing agent and inorganic precursor, the material is self-structured, but the structure on the nanometer length scale is *induced*. Although a given template always produces the same product un-

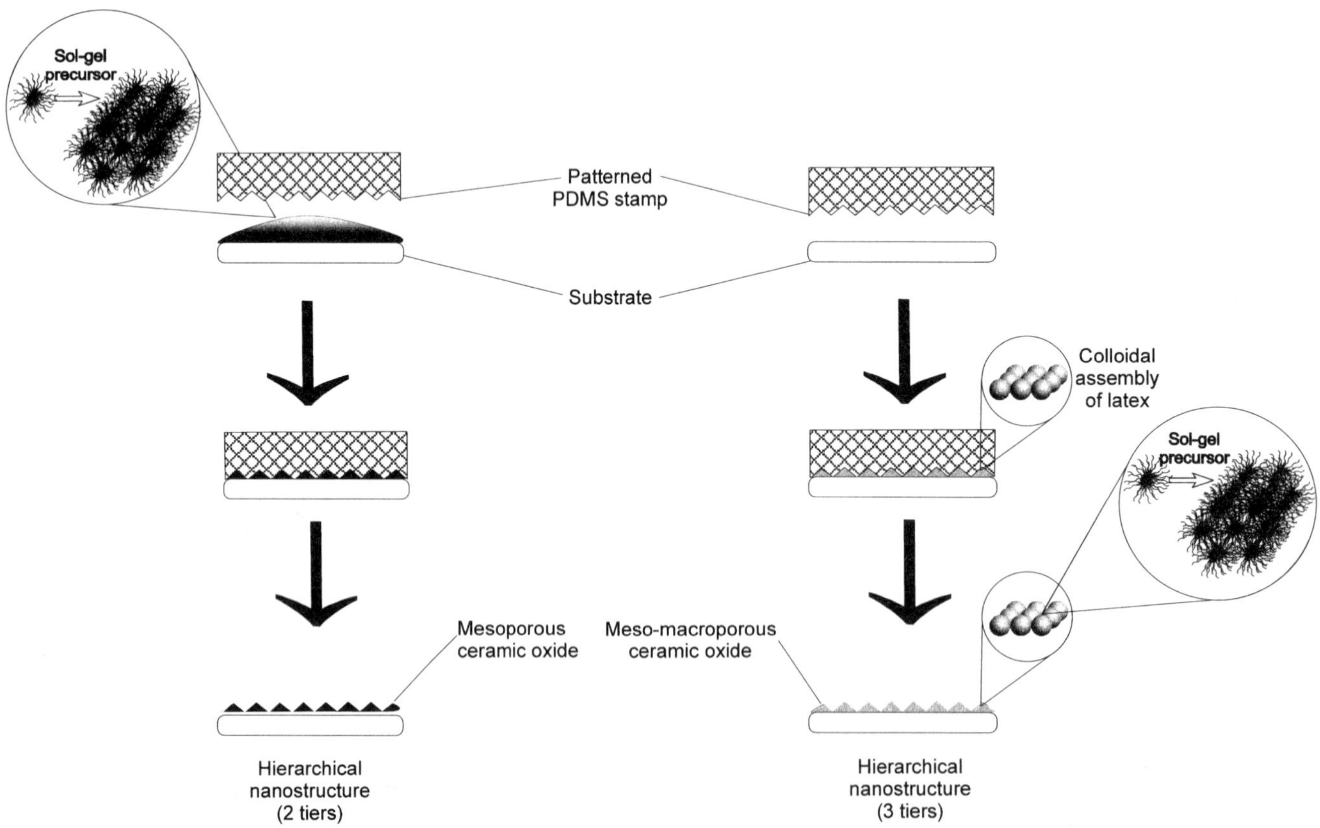

FIG. 4 Schematic representation of the generation of hierarchically structured ceramic oxides. (From Ref. 78.)

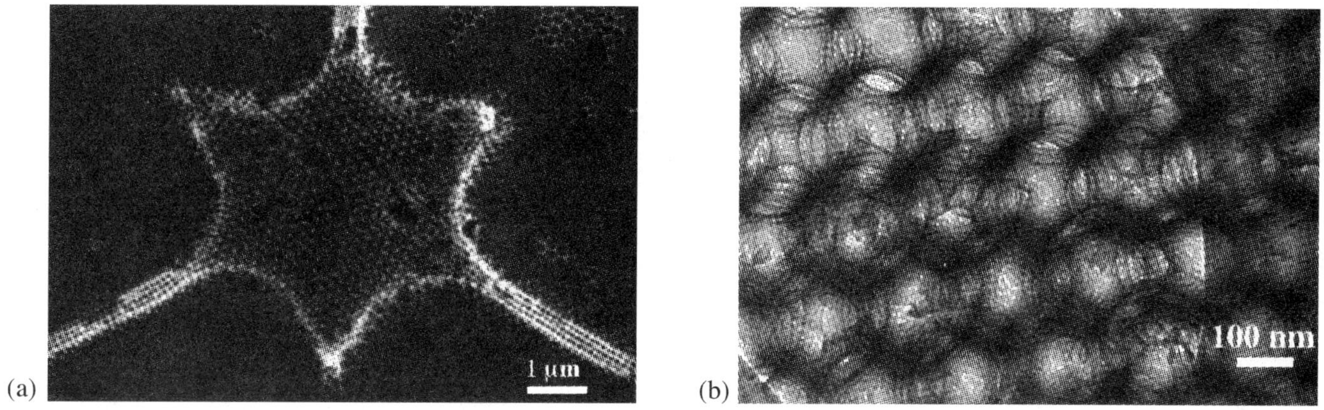

FIG. 5 Hierarchically structured (three tier) silica. (a) SEM picture of the micromolded (tier 1) structure; (b) TEM image of the latex-templated (tier 2) mesoporous substructure (tier 3). (From Ref. 78.)

der the same reaction conditions, the final structure can be determined only after the synthesis, which has the disadvantage that for a new template various experiments are necessary to map the "phase diagram" of the nanostructured precipitates.

The structures induced resemble those of the complex aggregates formed by many amphiphilic species regardless of molecular weight: The structure of lyotropic liquid crystalline phases, which have only recently been proved to be versatile templates for inorganic nanostructure design in a process called "nanocasting."

5. Lyotropic ABC Phases for "Nanocasting"

Precipitation of inorganic ceramic oxides in the presence of amphiphilic block copolymers is only one method of preparing large-pore mesoporous materials that show a high degree of order. A more predictable route toward inorganic-organic nanostructured hybrid materials is the direct utilization of lyotropic amphiphilic block copolymer phases. This approach can be understood as the vitrification of a microphase-separated medium, during which no changes of the phase structure occur.

The sol-gel synthesis of silica is conducted within the aqueous domains of the microphase-separated medium comprising amphiphilic block copolymer and water. Simplistically, the polycondensation of silicic acid is confined to exactly those aqueous domains, during which a replica of the lyotropic phase structure is formed (Fig. 6). Therefore, this procedure was termed *nanocasting*, and in many ways it resembles a nanoscale analogue of the lost-wax casting technique, which is still applied for manufacturing bronze statues and church bells. Nanocasting is the method that produces an exact cast of the lyotropic block copolymer phase structure.

Using the binary phase diagram of an amphiphilic block copolymer with water as a guideline, the structure of the final product can be predicted a priori: The *nanocast*, an ordered hybrid material consisting of silica and the templating block copolymer, is the result of adding a component to a binary lyotropic liquid crystal phase without ultimately changing the phase structure. This method is as applicable for nonionic [71,72,84,85] as for ionic [86] amphiphilic block copolymers, which is a strong indication that supramolecular templating of the lyotropic mesophase governs the mechanism of nanocasting.

Lyotropic lipid crystalline phases of nonionic surfactants in water have previously proved to be versatile structure-directing media [67,68] for the synthesis of regular mesoporous ceramic oxides. This approach to controlled inorganic pore systems (the "true liquid crystal approach") has been extended into the field of amphiphilic block copolymers [71].

The nonionic ABC templates used for nanocasting consist of a hydrophobic soft block (T_g below or around room temperature in order to warrant sufficient solubility at room temperature), such as poly(butadiene) [87], poly(ethylene-*co*-butylene) (Kraton Liquid*) or relatively short poly(styrene),* and a poly(ethylene oxide) block as the hydrophilic moiety. During nanocast-

*Poly(styrene-*b*-ethylene oxide) (SEs) (for SE10/10 the average molecular weight is 1 kg/mol; for SE30/30, it is 3 kg/mol) of various block lengths and low-dispersity poly[(ethylene-*co*-butylene)-*b*-(ethylene oxide)]s (KLEs) are products of Th. Goldschmidt AG, Essen, Germany.

FIG. 6 Schematic representation of nanocasting. The aqueous domains of a lyotropic mesophase (here isolated cylindrical micelles) are swollen with the silica precursor (TMOS), upon which a siliceous mesophase is formed. The system solidifies in the bulk, and after calcination an exact replica of the colloidal nanostructure is obtained.

ing, as is the case for the nanoreactor principle as well as the precipitation of regularly structured hybrids, the templating amphiphilic block copolymers are again assumed to be "big surfactants" in that their aggregation behavior is expected to be similar to that of low-molecular-weight surfactants, which is indeed the case.

The binary phase diagrams of aqueous amphiphilic block copolymer solutions follow the general phase sequence shown by surfactants. While at low block copolymer concentration the formation of micelles is observed, high concentrations give rise to more complex aggregates, such as cylindrical micelles or lyotropic liquid crystal phases of different structures. A rich lyotropic polymorphism is observed that can be utilized directly for porous nanostructure design. The phase diagrams correspond to those of low-molecular-weight surfactants in that they are more defined the narrower the molecular weight distribution. While heterodisperse block copolymers often show a tendency to form ill-defined phases, the phase diagrams of monodisperse amphiphilic block copolymers exhibit wide regions of high order. In general, the lyotropic liquid crystal phases of amphiphilic block copolymers are thermally more stable than those of classical nonionic surfactants. In most cases no clearing point is observed even at temperatures as high as 95°C. The temperature dependence of the phase structure, which is very pronounced for nonionic surfactants [88], is also less noticeable for the polymeric amphiphiles.

Consequently, heterodisperse SE templates afford nanostructures that appear to be affected by high defect densities where the defect sites determine the overall structure [71]. In contrast, nanocasts of low-polydispersity amphiphilic block copolymers are generally more ordered. The structure elucidation of the latter by imaging or scattering methods is, therefore, considerably simplified.

To a certain extent, the lyotropic phase behavior of amphiphilic block copolymers can be predicted as a function of the ABC's chemical nature, the overall molecular weight, and the block length ratio. These phase diagrams represent a valuable guideline for the structure design. Therefore, it is especially useful to map them for commercially available templates or for those that are likely to be used for more than one future experiment. The analytical tool for the structure elucidation of lyotropic liquid crystal phases is usually polarized-light optical microscopy or small-angle X-ray scattering. After mapping the phase diagram with respect to composition and temperature, the variation of one experimental parameter during the synthesis, namely the template concentration, allows the precise tailoring of pore systems with respect to their shape, density, and connectivity. Figure 7 shows TEM images of three samples, all obtained with the same template (KLE 3729) but at varying concentration [84].

Isolated, bent cylindrical pores result from a system at relatively low template contents, and a hexagonally packed cylindrical pore array is achieved at higher concentration. Increasing the template contents even further gives rise to an entirely different structure, which is again in accordance with the lyotropic phase diagram of the polymer. Multilamellar vesicles are found in coexistence with isolated fragments of sheetlike structures. Calcination, that is, removal of the organic scaffolding from a lamellar system drives the structure to fall apart.

Nitrogen adsorption-desorption isotherms again provide valuable information about the structure of the materials and allow one to gain insight into the pore structure. They support the conclusions drawn from the TEM micrographs, but in addition they illustrate the fact that each template molecule delivers its individual contribution to the overall interface area. Within the

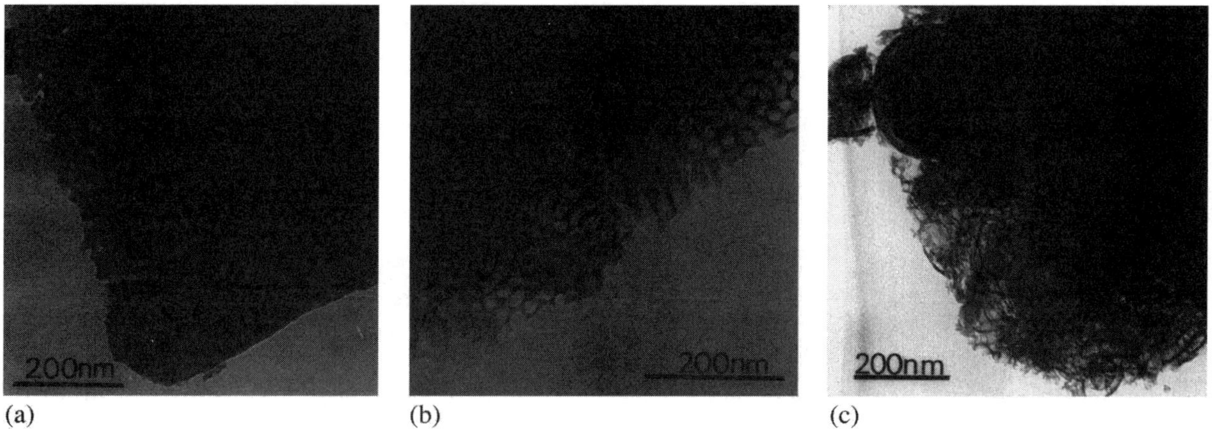

FIG. 7 TEM images of three samples obtained via nanocasting of KLE 3729 after calcination. (a) At 30 wt% polymer in water, isolated cylinders are obtained; (b) 50 wt%, cylindrical pores are assembled on an hexagonal lattice; (c) 70 wt% polymer give rise to a lamellar structure, which collapses upon calcination. (From Ref. 84.)

regime of one particular pore structure, the specific surface area increases significantly with growing template concentration (see Table 2). They also demonstrate that the same double templating action can be observed for amphiphilic block copolymers used in nanocasting as well as during the precipitation of mesostructured hybrid materials discussed before: The resulting nanoporous ceramic oxides possess an undeniable microporosity, which is due to the hydrophilic poly(ethylene oxide), which is initially homogeneously distributed in the siliceous microphase, displacing silica growth from its own molecular volume (B. Berton et al., unpublished results; B. Smarsly et al., unpublished results).

Some phase diagrams of low-polydispersity amphiphilic block copolymers exhibit areas of coexistence over a relatively wide range of compositions (see Fig. 8) [85]. This is probably due to kinetic inertia or to the fact that at the borderline between two thermodynamically stable phases the energetic differences between two structures are marginal. Swelling these coexisting phases with a siliceous precursor affords a microphase-separated siliceous phase, which has the same structure as the binary mixture consisting of water and amphiphilic block copolymer. As a result of the inorganic precursor undergoing polycondensation, the bulk phase solidifies without altering its superstructure.

TABLE 2 Relationship Between Template Contents and Specific Surface Area for PB-PEO–Templated Silica

Weight ratio template/water (%)	BET surface area (m^2/g)
30	770
50	820
70	1160

Source: Ref. 85.

The fact that the structural integrity of the binary lyotropic phase does not seem to be harmed by the nanocasting procedure suggests a reversal of the principle. If the nanostructure of the silica can be predicted a priori, the a posteriori analysis of a silica cast should provide valuable information on the structural composition of an unknown binary ABC phase. Assuming the noninvasive character of nanocasting, this method can help to elucidate more complex hyotropic phase struc-

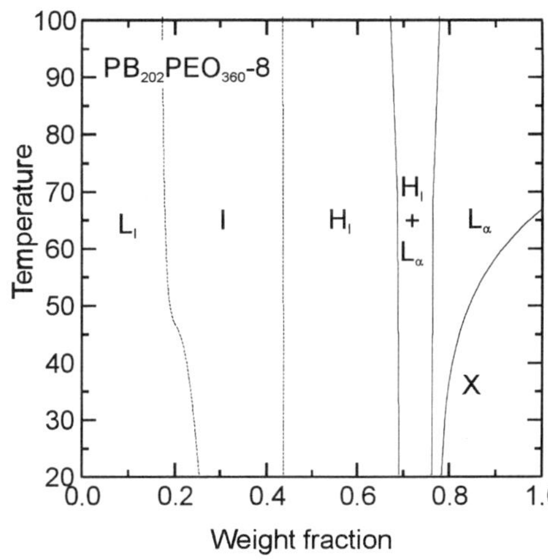

FIG. 8 Binary phase diagram of $PB_{202}PEO_{360}$ in water. L_I = micellar, H_I = regular hexagonal, L_α = lamellar, X = crystalline.

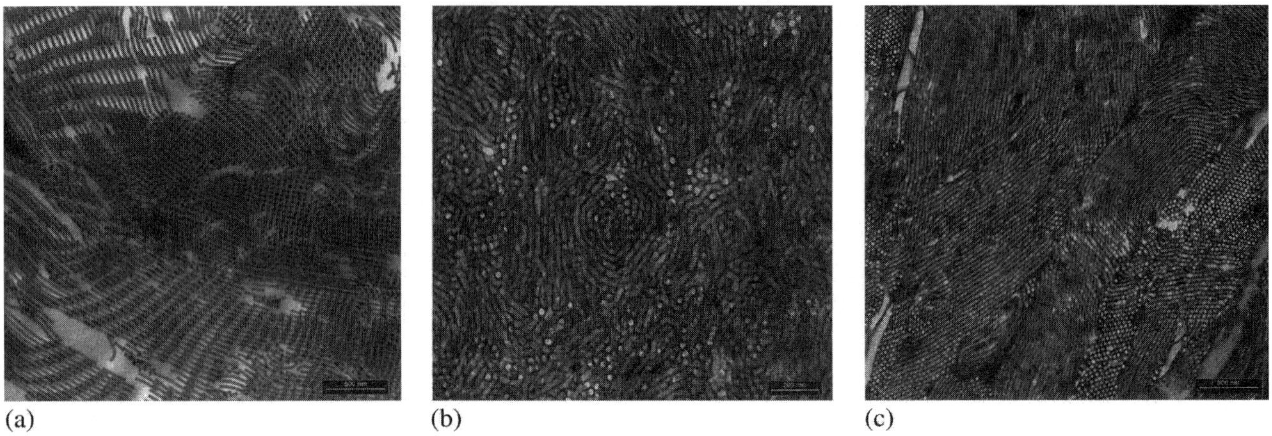

FIG. 9 TEM images of (a) thin-sectioned polymer gel, obtained by cross-linking a hexagonal lyotropic ABC phase, (b) silica nanocast of the lyotropic phase formed at the same polymer concentration, and (c) reference sample obtained from casting the polymer gel. (From Ref. 89.)

tures of amphiphilic block copolymers as a complementary technique to diffraction methods and may help to avoid time-consuming preparations (e.g., cryo-TEM, freeze etching). The phase structure simply has to be "frozen" into a solid silica cast, which is perfectly stable in a high vacuum under the electron beam.

The applicability of nanocasting as an analytical tool has been demonstrated [89] by comparing the silica structures obtained from a lyotropic phase, which has been cross-linked using γ-rays in order to provide sufficient mechanical stability to allow thin sectioning, with those of a silica nanocast obtained from a lyotropic phase of the same composition (Fig. 9). The similarity between the structures is striking. A reference sample was prepared by filling the pore system of the cross-linked polymer gel with silica and subsequent calcination. The pictures prove without doubt that the sol-gel process indeed does not have any structurally disrupting effect on the liquid crystalline phase [89].

In contrast to precipitation procedures, nanocasting allows the fabrication of objects (monoliths) that are macroscopically devoid of cracks and defects (Fig. 10) [71]. The porosity of these monoliths is as high as 85%. Moreover, diffusion pathways can be individually designed by templating a particular phase structure. Above all, the pore system of a macroscopic object is exclusively determined by the pore system, whereas particulate powders show a significant contribution to the surface area caused by the nonstructured particle surface.

Nanocasting can also be used to generate hierarchical pore systems. The synthesis of silica is performed in the presence of a polymer latex dispersion as the porogen, giving rise to spherical pores in the size and size distribution corresponding to those of the templating latex dispersion (see Fig. 11a). By adding an additional amphiphilic block copolymer as a second template, materials can be prepared with bimodal pore size distributions (see Fig. 11b) [79] that as monolithic species would be ideal candidates as stationary phases for chromatographic separation.

Nanocasting of lyotropic ABC phases allows to design predictably the structure, size, connectivity, and shape of nanoscopic pore systems in sol-gel–derived silicates. The generation of defined diffusion pathways, combined with the possibility of shaping macroscopic objects, leads to the highest expectations regarding the

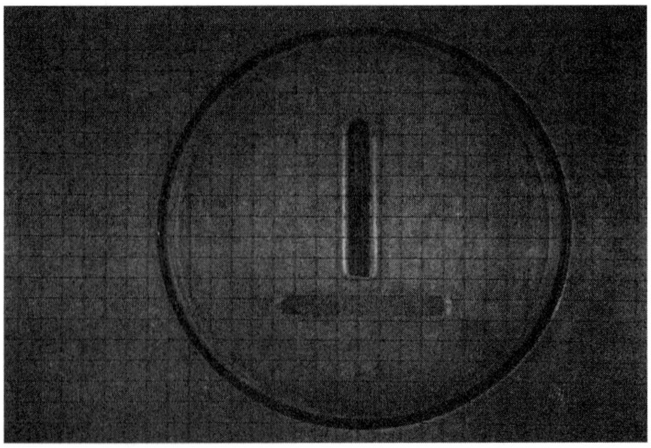

FIG. 10 Optically transparent silica monolith containing SE 10/10 as the template. The liquid crystalline mixture was pressed into a cylindrical mold. (From Ref. 71.)

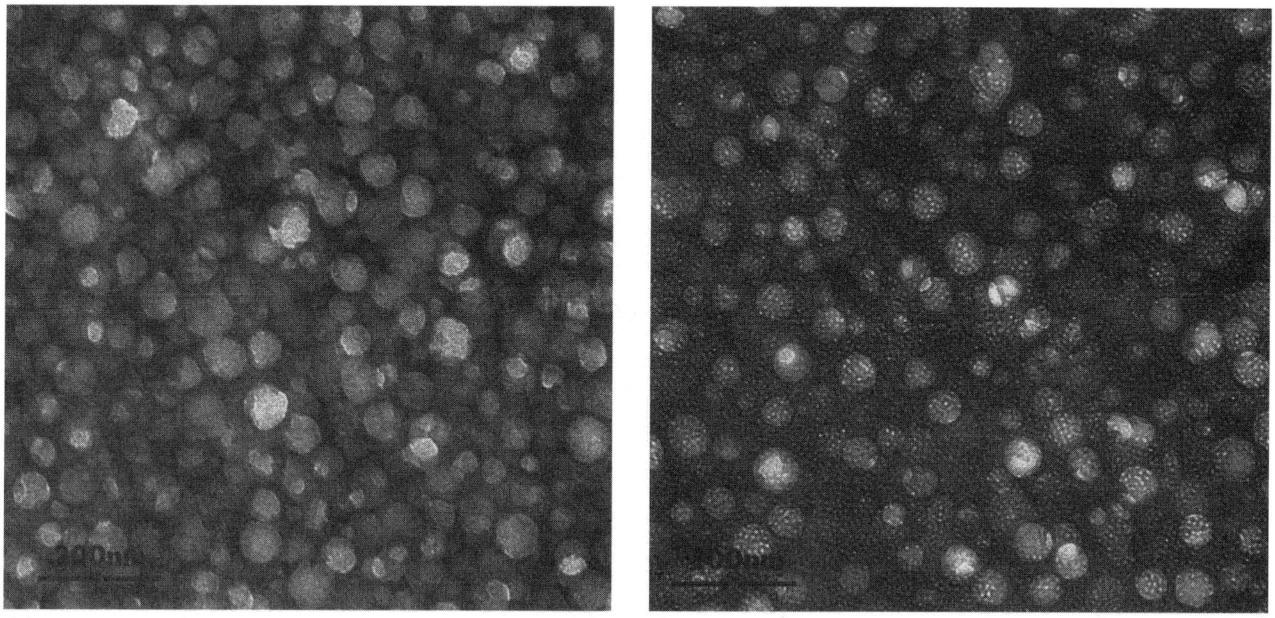

(a) (b)

FIG. 11 TEM images: (a) the silica sample templated with a poly(styrene) latex dispersion is purely macroporous; (b) the material obtained in the presence of additional ABC stabilizer clearly shows a hierarchical macro-mesophorous structure. (From Ref. 79.)

application in separation processes. Increasing the concentration of amphiphilic block copolymer present during the production of inorganic-organic nanostructured hybrid systems from the structure induction in micellar solutions to the utilization of prefabricated mesoscopic casting molds (nanocasting) leads to the ultimate question of whether amphiphilic block copolymer bulk phases can be used as structure-directing media for the synthesis of mesoscopic hybrid materials. The following section is focused on this aspect of nanostructure design with ABCs.

6. Sol-Gel Processing in ABC Bulk Phases

In one respect, amphiphilic block copolymers differ significantly from their low-molecular-weight analogues. Whereas a surfactant either decomposes or undergoes melting into an isotropic liquid state, the melts of amphiphilic block copolymers of sufficient molecular weights are generally microphase separated. The variety of polymorphism observed for block copolymer bulk phases is as at least as wide as that of their lyotropic phases and mainly depends on the volume fraction of each block. This bulk phase behavior is also temperature dependent. The phase behavior of amphiphilic block copolymers in the bulk is the focus of widespread research interest because it is not only the chemical composition of the block copolymer but also the structure of the microphase-separated phase it forms that defines the macroscopic mechanical performance of such materials.

Another approach to the synthesis of amphiphilic block copolymer–templated inorganic nanostructures is the solidification of a prehydrolyzed mixture of aluminum and silicon alkoxides in amphiphilic block copolymer bulk phases and subsequent formation of an ordered nanostructured alumosilicate [90]. Again, the formation of a solid, nanostructured inorganic-organic hybrid material is a consequence of the strict microphase separation between a hydrophobic (or better "silicatophobic") poly(isoprene) block and the "silicatophilic" poly(ethylene oxide) interacting with the inorganic sol-gel precursor.

The novelty of this process lies in the inorganic sol acting as a swelling agent, whose volume fraction determines the overall microphase structure of the hybrid material. As the microphase-separated structure that forms depends on the volume fraction of each block (or microphase), the amount of inorganic "microphase" added to the amphiphilic block copolymer determines the structure of the resulting inorganic-organic hybrid material. Interestingly, the chemical nature of the inorganic phase appears to be independent of the inorganic precursor/template ratio, hence manifesting the main difference from the ABC templating processes discussed earlier. In ABC bulk phase templating, it is the amount of inorganic that directs the structure of the hybrid material rather than the amphiphilic block copolymer. As a consequence, the whole phase diagram of this bulk block copolymer with respect to block length ratio can be traversed without having to synthesize amphiphilic block copolymers with different block length ratios. Depending on the amount of inorganic sol present in the block copolymer, a whole variety of differently structured hybrid materials can be produced (see Fig. 12).

Nuclear magnetic resonance (NMR) spectroscopic analysis finally proved the assumption made for the previous structure-directing methods, namely that the poly(ethylene oxide) blocks are firmly anchored in the inorganic phase rather than being located at the interphase adjacent to the hydrophobic domains. Solid-state NMR spectroscopy revealed that this anchoring leads to a substantial hindrance of the conformational mobility in the poly(ethylene oxide) chains compared with the relatively mobile hydrophobic poly(isoprene) [91].

Two possible scenarios can be envisaged for the structure of the hybrid material (see Fig. 13). The first is that the poly(ethylene oxide) block, albeit strongly interacting and partially penetrating, forms a pure PEO layer at the interface with the hydrophobic poly(isoprene) (Fig. 13, left-hand sketch) ("three-phase" system). The other possibility is the complete "dissolution" of the PEO chains in the aluminosilicate, which results in the "two-phase" system depicted in the right-hand sketch of Fig. 12b. Spin-diffusion NMR experiments showed that there appears to be no dynamic heterogeneity in the poly(ethylene oxide) chains, as would be expected for a three-phase system, giving rise to the conclusion that the hydrophilic domains constitute one homogeneous hybrid phase consisting of poly(ethylene oxide) and the inorganic.

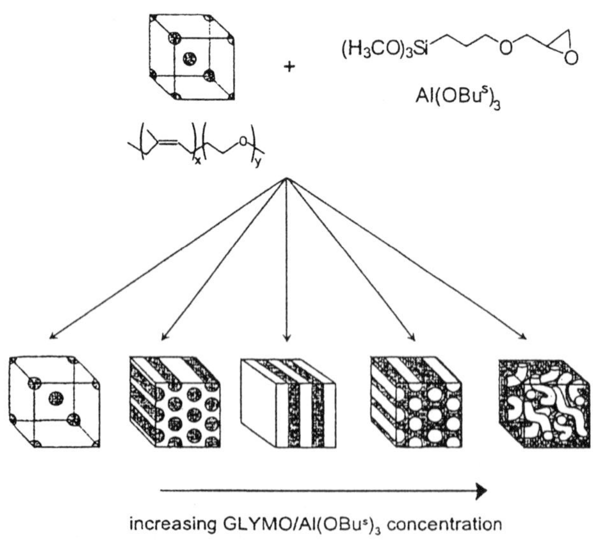

FIG. 12 Schematic representation of the phase structures created by swelling a bulk block copolymer with a prehydrolyzed sol-gel mixture. (From Ref. 91.)

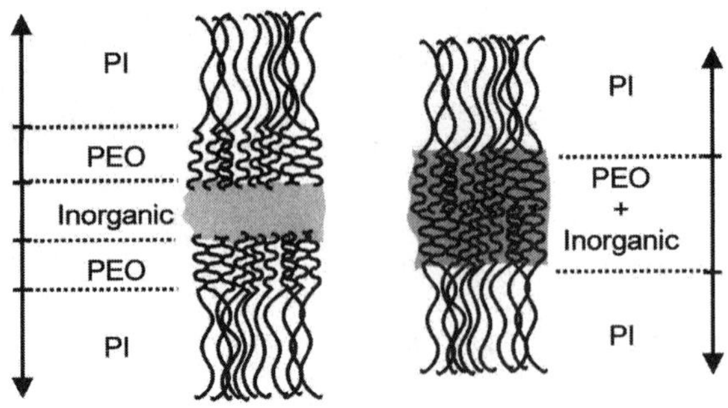

FIG. 13 Schematic representation of possible hybrid structures. Left: The poly(ethylene oxide) blocks of the template form a separate microphase at the interface to the hydrophobic block (three-phase system). Right: The poly(ethylene oxide) blocks are homogeneously distributed in the inorganic phase. (From Ref. 91.)

This approach to nanostructured inorganic-organic hybrid materials is the first one to allow the synthesis of inverse-topology systems, in which the hydrophobic polymer blocks represent the outside of the microphase-separated structure. After solidification of the inorganic sol, the hydrophobic phase can be swollen with organic solvents: This procedure allows the isolation from one another of colloidal objects, such as spheres or ceramic rods (see Fig. 14), which are sterically stabilized because the hydrophilic block is firmly anchored in the ceramic material [92]. Formally, this process represents a conversion of the initial *exo*-templating into *endo*-templating.

These bulk hybrid materials cannot be calcined without seriously affecting their structure. The reason is that the extremely large interface/surface area generated in such small objects makes a ceramic colloid unstable in the absence of a sterically stabilizing polymer/solvent layer. However, the steric stabilizer need not be removed in order to envisage applications of such ceramic nano-objects. The bulk preparation of their colloidal dispersions would open a variety of physicochemical aspects. For example, nanoscopic rods, when sufficiently concentrated, should give rise to colloidal nematic phase behavior in solution. Aligning such phases and incorporating the aggregated rods

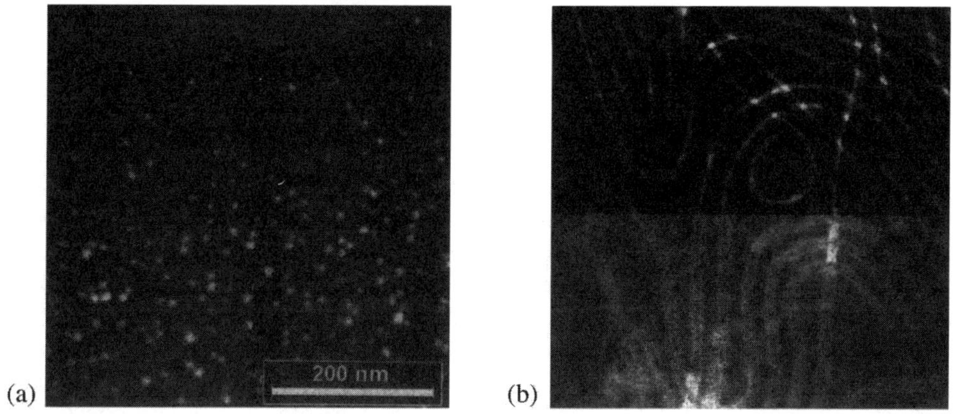

FIG. 14 TEM images of ceramic nano-objects: (a) spherical alumosilicate colloids; (b) rodlike particles obtained from a hexagonal phase. (From Ref. 90.)

into a solid polymer matrix should create unique anisotropic mechanical properties (nano-reinforced hybrid materials).

IV. BIOMIMETIC MINERALIZATION WITH ABCs

There are numerous examples of mesostructured inorganic-organic hybrid materials occurring in nature. For example, the hard outside and the iridescent inside of a shell chemically consist of the same material, namely calcium carbonate, which exists in two different crystal modifications. Mammals are able to fabricate the substance of bones and teeth, both high-performance materials, from a single mineral, hydroxyapatite. The mechanical properties of these have not yet been accomplished in any man-made ceramic.

Nature completes the controlled crystallization of such minerals in the presence of certain structure-directing proteins, which are able to affect not only the crystal structure but also morphology (e.g., disks, needles, cubes) and the orientation and spatial relationship of the crystals with each other. Excellent reviews have been published, which are highly recommended to the interested reader [93–95]. Whenever a chemist attempts to precipitate calcium carbonate or hydroxyapatite in a beaker, both minerals occur in the shape of macroscopic brittle crystals of one, mostly the thermodynamically most stable, modification. Because of packing or structural defects, the resulting precipitate is mechanically useless because crystallization occurs without any directing auxiliaries. Controlled crystallization, however, as demonstrated by nature, can be mimicked [96–99]. For example, structure-directing proteins can be isolated from seashells and used to influence the crystallization of minerals. It is possible to grow calcite and aragonite, two modifications of calcium carbonate, on one substrate alternatingly by adding first one, then another protein to the supernatant calcium salt solution. This procedure, however successful it may be, is not applicable to industrial or even laboratory scale experiments because the isolation and purification of the biological structure-directing entities are tedious and time consuming. However, the chemical and physical mechanism from which the ultimate structure control arises can be mimicked with simple methods of polymer chemistry. In a biomimetic approach, amphiphilic block copolymers replace the complex protein as the structure-directing compound and have the effect of avoiding unspecific crystallization processes. These auxiliaries fulfill two tasks: On the one hand, they stabilize the primary seed and avoid agglomeration and macroscopic precipitation [100,101]. On the other hand, they selectively bind to certain crystal planes, hence directing the mineralization process, which allows control of crystal structure as well as morphology.

During biomimetic crystallization, the structure-directing block copolymer acts as an amphiphile in the most general sense. One part of the molecule has strong affinity for the crystal, while the other shows only weak interaction with starting materials as well as product mineral. The task of the latter is solely stabilization of the dispersion, solubilization, and the prevention of macroscopic precipitation. As water is the most common reaction medium, both polymer blocks have to be hydrophilic. Therefore, these structure-directing block copolymers are called "double-hydrophilic" block copolymers. Only one block (the handle) interacts specifically with the crystal, while the second one (the head of the tool) adopts the role of a stabilizer, i.e., avoids macroscopic precipitation [100,101]. For example, appropriately choosing the chemical nature of double-hydrophilic block copolymers facilitates the control of particle size and morphology as well as the crystal modification of calcium carbonate. In a series of experiments presented here, the stabilizing block was poly(ethylene oxide). The specific interaction with the crystal was accomplished by either poly(ethylene imine tetraacetate) (PEDTA), poly(aspargic acid) (PAsp, well known from the biomineralization of seashells), or a synthetic poly(phosphate) (Pphos). The three polymer structures are shown in Scheme 3.

Despite their apparent similarity, the three polymers have vastly different effects on the growth of calcium carbonate, hence demonstrating the sensitivity of the precipitation to the chemical nature of the "head" of the structure-directing tool. Control experiments in the absence of any additives afford calcium carbonate in the shape of submillimeter-sized rhombohedral crystals (see Fig. 15a). In the presence of the phosphate block, monodisperse calcium carbonate spheres are obtained (Fig. 15b), which are approximately 3 μm in diameter. Hollow spheres with a radius of 10 μm are formed in the presence of PEO-PEDTA (Fig. 15c), and the addition of PEO-*b*-PMMA-Asp gives rise to the formation of twins, which apparently originate from rods or platelet-shaped seeds (Fig. 15d).

One of the most important features of biomimetic crystallization is the fact that the resulting product morphology is formed as a consequence of combining the mineral solution with the structure-directing block copolymer. It therefore reflects an individual interaction between organic and inorganic components, which serves to minimize the interfacial energy. Biomimetic mineralization is therefore another aspect of *induced* nanostructure. Another example of the induction of structure is the controlled, nanoscopic mineralization of hydroxyapatite, the main constituent of bones and teeth. In the presence of an amphiphilic block copolymer, micrometer-sized colloids with a fibrous structure are obtained, whose individual "whiskers" are as thin as 3 nm (see Fig. 16) [102]. Spinning and weaving of such fibers could lay the foundation for a biocompatible bone substitute.

This delicate structure can be detected neither in classical hydroxyapatite nor as an aggregate of the pure double-hydrophilic block copolymer in aqueous solution. Furthermore, this morphology appears to be specific for this particular combination of inorganic and double hydrophilic block copolymer and has not been detected for any other mineral, which confirms the as-

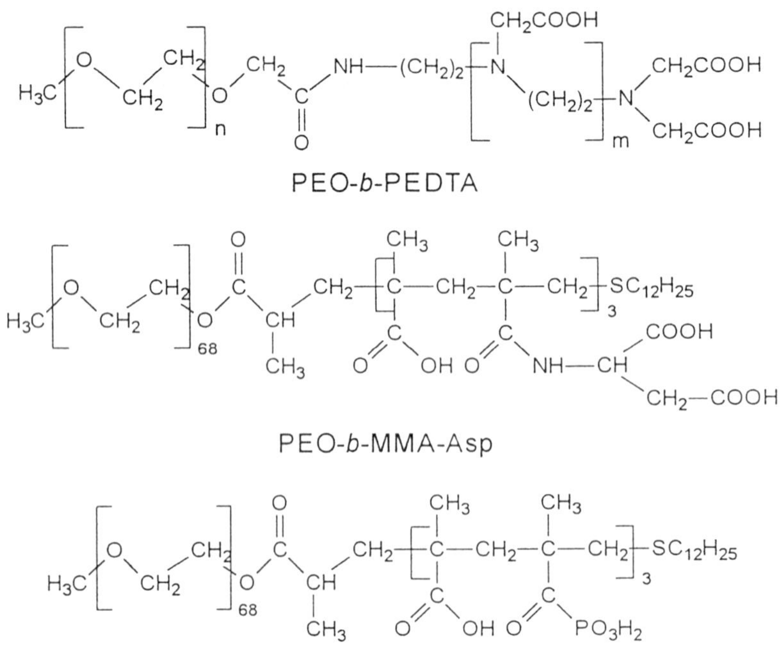

SCHEME 3 Double-hydrophilic block copolymers used as structure-directing agents for biomimetic mineralization.

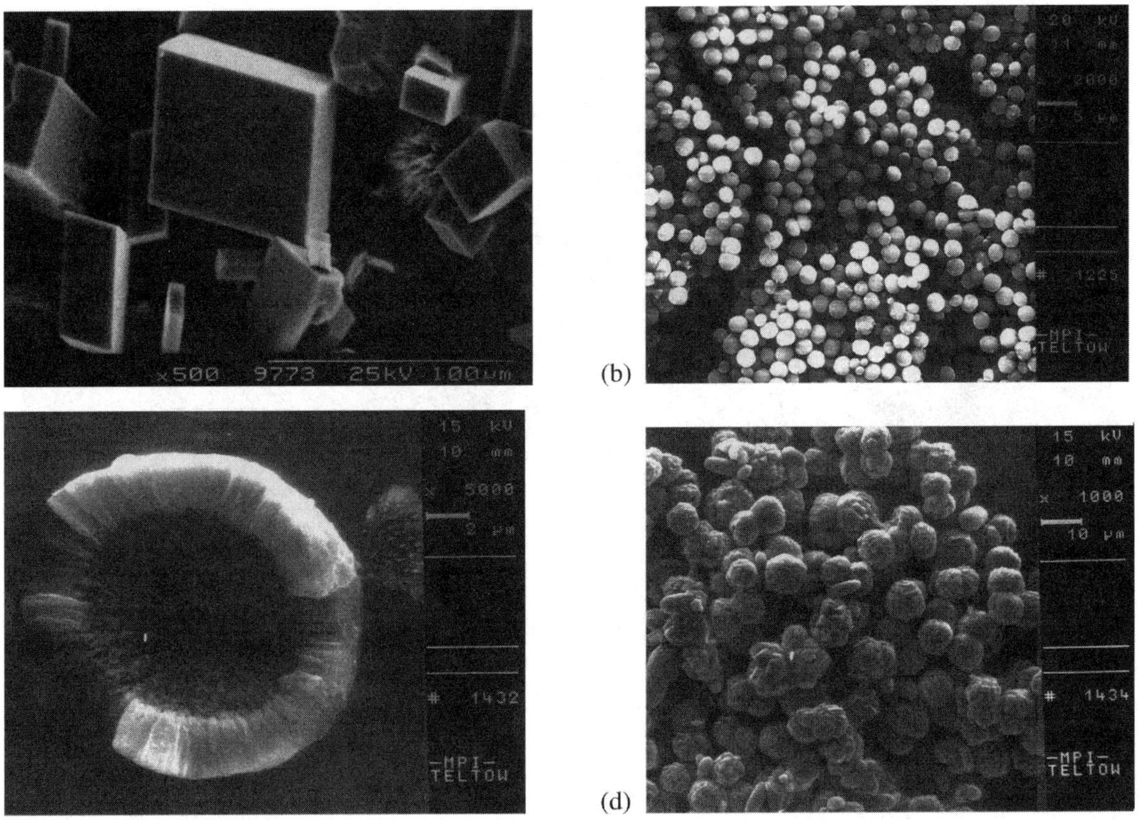

FIG. 15 SEM images of calcium carbonate samples: In the absence of structure-directing polymers (a) rhombohedral crystals are formed. (b) Monodisperse spherical particles; (c) hollow spheres; (d) twins. (From Ref. 101.)

sumption that an individual, specific interaction between the inorganic and the template is reflected and the process is truly biomimetic in this respect. The same extreme specificity and individuality are often observed for biomineralization, which raises the question of whether the naturally occurring processes are actually as complex as assumed or the process of biomimetic mineralization with block copolymers is as simple as it appears.

V. SUMMARY AND OUTLOOK

Five representative methodologies were chosen to demonstrate how the principles of self-structuring can be applied to the preparation of nanostructured inorganic-organic composites. The control of the resulting interface opens facile access to highly organized hybrid materials, which show typical structural elements on the length scale of a few nanometers. Amphiphilic block copolymer adopt the role of molecular "assembly teams" that take care of organizing and fitting of the components.

Amphiphilic block copolymers are versatile auxiliary agents for the synthesis of mesoscopically structured hybrid materials. Relatively little time and effort are necessary to tailor the chemical composition of these structure-directing media, and almost any compatibility problem can be addressed by simply choosing the right structure, size, and chemical nature of the compatibilizer, the ABC. The word "amphiphile" can indeed be used in its most general sense whenever there is a demand for interface stabilization on the nanometer length scale.

Whether the structure of the resulting hybrid material is induced upon combining inorganic and organic components, as in biomimetic mineralization or supramolecular templating of sol-gel processes with micellar solutions, or whether the hybrid is generated in a "preassembled" supramolecular casting mold, whether *endo*-templating produces nanometer-size particles or *exo*-templating yields porous monolithic objects, the flexibility of ABCs as structure-directing media is in the hands of modern chemists and materials scientists. A short while ago the problem of mediating between

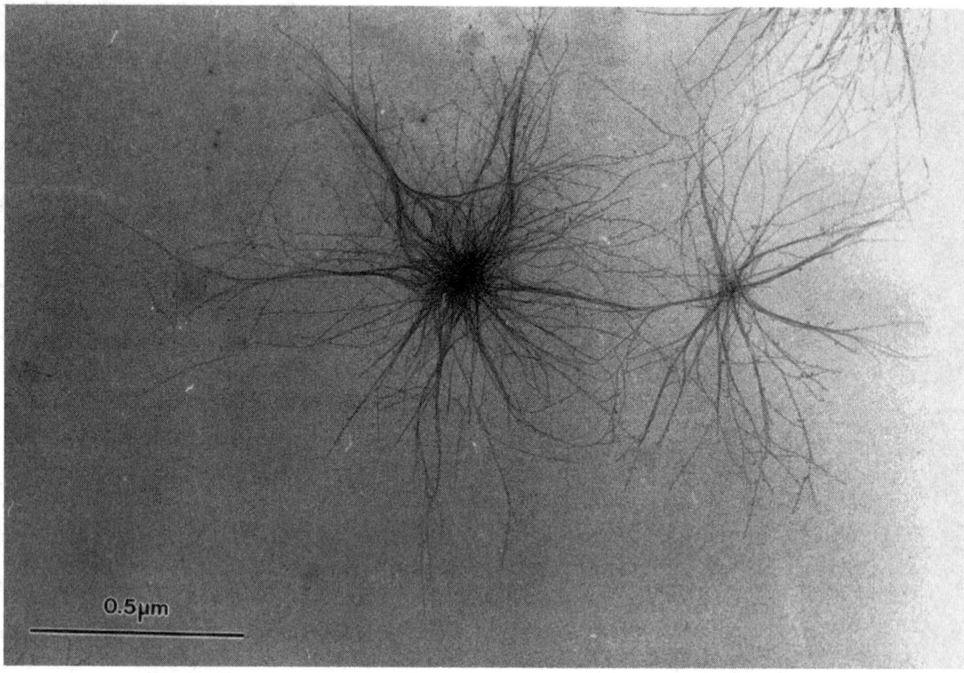

FIG. 16 TEM image of a hydroxyapatite morphology obtained from biomimetic crystallization in the presence of a structure-directing double-hydrophilic block copolymer. (From Ref. 102.)

two incompatible materials prevented the synthesis of nanoscale hybrid materials.

With the new trends and developments in amphiphilic block copolymer synthesis, access is opened to the generation of new technologies whose utilization depends on the imagination, creativity, and skills of nanochemists and, of course, on nanomarket demands.

ACKNOWLEDGMENTS

C. G. G. would like to thank M. Antonietti for his support and helpful discussions, M. P. Hase for inspiration, and the Max-Planck-Society for financial support.

REFERENCES

1. G. A. Ozin, Adv. Mater. *4*:612–649 (1992).
2. M. Antonietti and C. Göltner, Angew. Chem. Int. Ed. Engl. *36*:910–928 (1997).
3. Y. N. Xia, J. A. Rogers, K. E. Paul, and G. M. Whitesides, Chem. Rev. *99*:1823–1848 (1999).
4. J. M. Lehn, *Supramolecular Chemistry*, VCH, Weinheim, 1995.
5. S. Förster and M. Antonietti, Adv. Mater. *10*:195–217 (1998).
6. S. Förster, Ber. Bunsen Phys. Chem. *101*:1671–1678 (1997).
7. Z. Tuzar and P. Kratochvil, in *Surface and Colloid Science* (E. Matijevic, ed.), Plenum, New York.
8. A. Gast, NATO ASI Ser. E 303:311, 1988.
9. B. Chu, Langmuir *11*:414 (1995).
10. P. Alexandridis, Curr. Opin. Colloid Interface Sci. *1*: 490–501 (1996).
11. M. Moffit, K. Kougas, and A. Eisenberg, Acc. Chem. Res. *29*:95 (1996).
12. J. Selb and Y. Gallot, in *Developments in Block Copolymers* (I. Goodman, ed.), Vol. 2, Elsevier, Amsterdam, 1985, p. 27.
13. P. C. Conheim, T. V. Laxly, C. Price, and R. V. Stubbersfield, J. Chem. Soc. Faraday Discuss. I *76*:1857 (1980).
14. C. Price, E. K. M. Chan, A. L. Hudd, and R. B. Stubbersfield, Polym. Commun. *27*:196–198 (1986).
15. K. Schillen, W. Brown, and R. M. Johnsen, Macromolecules *27*:4825–4832 (1994).
16. M. Moffitt, H. Vali, and A. Eisenberg, Chem. Mater. *10*:1021–1028 (1998).
17. J. P. Spatz, A. Roescher, and M. Möller, Adv. Mater. *8*:337–340 (1996).
18. R. S. Kane, R. E. Cohen, and R. Silbey, Chem. Mater. *8*:1919–1924 (1996).
19. M. Svensson, P. Alexandridis, and P. Linse, Macromolecules *32*:637–645 (1999).
20. P. Alexandridis, Macromolecules *31*:6935–6942 (1998).

21. D. L. Zhuo, P. Alexandridis, and A. Khan, J. Colloid Interface Sci. *183*:339–350 (1996).
22. P. Alexandridis, D. L. Zhou, and A. Khan, Langmuir *12*:2690–2700 (1996).
23. G. Wanka, H. Hoffmann, and W. Ulbricht, Macromolecules *27*:4145–4149 (1994).
24. M. Möller and J. P. Spatz, Curr. Opin. Colloid Interface Sci. *2*:177–187 (1997).
25. S. Förster, M. Zisenis, E. Wenz, and M. Antonietti, J. Chem. Phys. *104*:9956–9970 (1996).
26. Y. N. C. Chan, G. S. W. Craig, R. R. Schrock, and R. E. Cohen, Chem. Mater. *4*:885–894 (1992).
27. T. Schneider, M. Haase, A. Kornowski, S. Naused, H. Weller, S. Förster, and M. Antonietti, Ber. Bunsges. Phys. Chem. *101*:1654–1656 (1997).
28. S. Förster and M. Antonietti, Adv. Mater. *10*:195–217 (1998).
29. M. Antonietti, E. Wenz, L. Bronstein, and M. Seregina, Adv. Mater. *7*:1000 (1995).
30. M. Antonietti, S. Förster, J. Hartmann, and S. Oestreich, Macromolecules *29*:3800–3806 (1996).
31. M. Antonietti, S. Förster, J. Hartmann, S. Oestreich, and E. Wenz, Nachr. Chem. Tech. Lab. *44*:579 (1996).
32. L. Bronstein, D. Chernyshov, P. Valetsky, N. Tkachenko, H. Lemmetyinen, J. Hartmann, and S. Förster, Langmuir *15*:83–91 (1999).
33. S. Klingelhöfer, W. Heitz, A. Greiner, S. Oestreich, S. Förster, and M. Antonietti, J. Am. Chem. Soc. *119*: 10116–10120 (1997).
34. E. Sulman, Y. Bodrova, V. Mateeva, N. Semagina, L. Cerveny, V. Kurtc, L. Bronstein, O. Platonova, and P. Valetsky, Appl. Catal. A Gen. *176*:75–81 (1999).
35. L. M. Bronstein, M. V. Seregina, O. A. Platonova, Y. A. Kabachii, D. M. Chernyshov, M. G. Ezernitskaya, L. V. Dubrovina, T. P. Bragina, and P. M. Valetsky, Macromol. Chem. Phys. *7*:1357–1363 (1998).
36. O. A. Platonova, L. M. Bronstein, S. P. Solodovnikov, I. M. Yanovskaya, E. S. Obolonkova, P. M. Valetsky, E. Wenz, and M. Antonietti, J. Colloid Polym. Sci. *5*: 426–431 (1997).
37. E. Sulman, V. Mateera, A. Usanov, G. Demidenko, L. Bronstein, D. Chernyshov, and P. Valetsky, J. Mol. Cat. A Chem. *146*:265–269 (1999).
38. M. Antonietti, S. Förster, and S. Oestreich, Macromol. Symp. *121*:75–88 (1997).
39. S. Wallace and L. L. Hench, in *Better Ceramics Through Chemistry* (C. J. Brinker, D. E. Clark, and D. R. Ulrich, eds.), Elsevier, North-Holland, New York, 1984, pp. 47–52.
40. H. Böttcher, K. H. Kallies, H. Haufe, and J. Seidel, Adv. Mater. *11*:138–140 (1999).
41. A. Imhof and D. J. Pine, Adv. Mater. *10*:697–700 (1998).
42. A. Imhof and D. J. Pine, Nature *389*:948–951 (1997).
43. T. H. Elmer, in *ASM Engineered Materials Handbook* (S. J. Schneider, Jr., ed.), Materials Park, OH, 1991.
44. W. Haller, Nature *206*:693 (1965).
45. W. Haller, J. Chem. Phys. *42*:686 (1965).
46. M. E. Davies, C. Saldarriaga, C. Montes, J. Garces, and C. Crowder, Nature *331*:698–699 (1988).
47. L. B. McCusker, C. Baerlocher, E. Jahn, and M. Bülow, Zeolites *11*:308–313 (1991).
48. K. S. W. Sing, D. H. Everett, R. A. W. Haul, L. Moscou, R. A. Pierotti, J. Rouquerol, and T. Siemieniewska, Pure Appl. Chem. *57*:603–619 (1985).
49. D. Antonelli and J. Ying, Curr. Opin. Colloid Interface Sci. *1*:523–529 (1997).
50. P. Behrens, Adv. Mater. *5*:127–132 (1993).
51. P. Behrens and G. D. Stucky, Angew. Chem. Int. Ed. Engl. *32*:696–699 (1993).
52. C. G. Göltner and M. Antonietti, Adv. Mater. *9*:431–436 (1997).
53. J. Liu, A. Y. Kim, L. Q. Wang, B. J. Palmer, Y. L. Chen, P. Bruinsma, B. C. Bunker, G. J. Exarhos, G. L. Graff, P. C. Rieke, G. E. Fryxell, J. W. Virden, B. J. Tarasevich, and L. A. Chick, Adv. Colloid Interface Sci. *69*:131–180 (1996).
54. D. Zhao, P. Yang, Q. Huo, B. F. Chmelka, and G. D. Stucky, Curr. Opin. Solid State Mater. Sci. *3*:111–121 (1998).
55. W. Meier, Curr. Opin. Colloid Interface Sci. *4*:6–14 (1999).
56. V. Chiola, J. E. Ritsko, and C. D. Vanderpool, U.S. Patent 3,556,725 (1971).
57. F. DiRenzo, H. Cambon, and R. Dutartre, Microporous Mater. *10*:283–286 (1997).
58. C. T. Kresge, M. E. Leonowicz, W. J. Roth, J. C. Vartuli, and J. S. Beck, Nature *359*:710–712 (1992); J. C. Vartuli, C. T. Kresge, M. E. Leonowicz, A. S. Chu, S. B. McCullen, I. D. Johnson, and E. W. Sheppard, Chem. Mater. *6*:2070–2077 (1994).
59. J. S. Beck, J. C. Vartuli, G. J. Kennedy, C. T. Kresge, W. J. Roth, and S. E. Schramm, Chem. Mater. *6*:1816–1821 (1994).
60. J. S. Beck, J. C. Vartuli, W. J. Roth, M. E. Leonowicz, C. T. Kresge, K. D. Schmitt, C. T. W. Chu, D. H. Olson, E. W. Sheppard, S. B. McCullen, J. B. Higgins, and J. L. Schlenker, J. Am. Chem. Soc. *114*:10834–10843 (1992).
61. A. Monnier, F. Schüth, Q. Huo, D. Kumar, D. Margolese, R. S. Maxwell, G. D. Stucky, M. Krishnamurty, P. Petroff, A. Firouzi, M. Janicke, and B. F. Chmelka, Science *261*:1299–1303 (1993).
62. Q. Huo, D. I. Margolese, U. Ciesla, D. G. Demuth, P. Feng, T. E. G. Fier, P. Sieger, A. Firouzi, B. F. Chmelka, F. Schüth, and G. D. Stucky, Chem. Mater. *6*:1176–1191 (1994).
63. Q. Huo, D. I. Margolese, U. Ciesla, P. Feng, T. E. Gier, P. Sieger, R. Leon, P. M. Petroff, F. Schüth, and G. D. Stucky, Nature *368*:317–320 (1994).
64. P. T. Tanev, M. Chibwe, and T. J. Pinnavaia, Nature *368*:321–323 (1994).

65. S. A. Bagshaw, E. Prouzet, and T. J. Pinnavaia, Science *267*:865–867 (1995).
66. S. A. Bagshaw and T. J. Pinnavaia, Angew. Chem. Int. Ed. Engl. *108*:1102–1105 (1996).
67. G. S. Attard, J. C. Glyde, and C. G. Göltner, Nature *378*:366–368 (1995).
68. G. S. Attard, M. Edgar, and C. G. Göltner, Acta Mater. *46*:751–758 (1998).
69. G. S. Attard, M. Edgar, J. W. Emsley, and C. G. Göltner, MRS Conf. Proc. *425*:179–184 (1996).
70. G. S. Attard, M. Edgar, J. W. Emsley, and C. G. Göltner, MRS Conf. Proc. *425*:185–189 (1996).
71. C. G. Göltner, S. Henke, M. C. Weißenberger, and M. Antonietti, Angew. Chem. Int. Ed. Engl. *37*:316 (1998).
72. M. C. Weißenberger, C. G. Göltner, and M. Antonietti, Ber. Bunsenges. Phys. Chem. *101*:1679 (1997).
73. C. J. Brinker and G. W. Scherer, *Sol-Gel Science: The Physics and Chemistry of Sol-Gel Processing*, Academic Press, San Diego, 1990.
74. D. Zhao, J. Feng, Q. Huo, N. Melosh, G. H. Fredrickson, B. F. Chmelka, and G. D. Stucky, Science *279*:548–552 (1998).
75. D. Zhao, Q. Huo, J. Feng, B. F. Chmelka, and G. D. Stucky, J. Am. Chem. Soc. *120*:6036–6042 (1998).
76. D. Zhao, P. Yang, N. Melosh, J. Feng, B. F. Chmelka, and G. D. Stucky, Adv. Mater. *10*:1380–1385 (1998).
77. R. Zana, Colloids Surf. A Physicochem. Eng. Aspects *123*:27 (1997).
78. P. Yang, T. Deng, D. Zhao, P. Feng, D. Pine, B. F. Chmelka, G. M. Whitesides, and G. D. Stucky, Science *282*:2244–2246 (1998).
79. M. Antonietti, B. Berton, C. Göltner, and H. P. Hentze, Adv. Mater. *10*:154 (1998).
80. Y. Xia and G. M. Whitesides, Angew. Chem. Int. Ed. Engl. *37*:550 (1996).
81. E. Kim, Y. Xia, and G. M. Whitesides, Adv. Mater. *8*: 245 (1996).
82. C. Marzolin, S. P. Smith, M. Pretiss, and G. M. Whitesides, Adv. Mater. *8*:571 (1998).
83. C. G. Göltner, Angew. Chem., in press.
84. C. G. Göltner, B. Berton, E. Krämer, and M. Antonietti, Chem. Commun. 2287–2288 (1998).
85. C. G. Göltner, B. Berton, E. Krämer, and M. Antonietti, Adv. Mater. *11*:395–398 (1999).
86. E. Krämer, S. Förster, C. Göltner, and M. Antonietti, Langmuir *14*:2027 (1998).
87. S. Förster and E. Krämer, Macromolecules *32*:2783–2785 (1999).
88. D. J. Mitchell, G. J. Tiddy, L. Waring, T. Bostock, and M. P. McDonald, J. Chem. Soc. Faraday Trans. *79*: 975–1000 (1983).
89. H. P. Hentze, E. Krämer, B. Berton, S. Förster, M. Antonietti, and M. Dreja, Macromolecules *32*:5803–5809 (1999).
90. M. Templin, A. Franck, A. Du Chesne, H. Leist, Y. Zhang, R. Ulrich, V. Schädler, and U. Wiesner, Science *278*:1795–1799 (1997).
91. S. M. De Paul, J. W. Zwanziger, R. Ulrich, U. Wiesner, and H. W. Spiess, J. Am. Chem. Soc. *121*:5727–5736 (1999).
92. R. Ulrich, A. Du Chesne, M. Templin, and U. Wiesner, Adv. Mater. *11*:141–146 (1999).
93. S. Weiner and L. Addadi, J. Mater. Chem. *7*:689–702 (1997).
94. P. Calvert and P. Rieke, Chem. Mater. *8*:1715–1727 (1996).
95. S. Mann, J. Webb, and R. J. P. Williams, eds., *Biomineralizaton: Chemical and Biochemical Perspectives*, VCH, Weinheim, 1989.
96. S. Mann, ed., *Biomimetic Materials Chemistry*, VCH, Weinheim, 1996.
97. R. Brace and E. Matijevic, J. Inorg. Nucl. Chem. *35*: 3691 (1973).
98. A. Chittofrati and E. Matijevic, Colloids Surf. *48*:65–78 (1990).
99. R. S. Sapieszko and E. Matijevic, J. Colloid Interface Sci. *74*:405 (1980).
100. M. Sedlak, H. Cölfen, and M. Antonietti, Macromol. Chem. Phys. *199*:247 (1998).
101. H. Cölfen and M. Antonietti, Langmuir *14*:582–589 (1998).
102. M. Antonietti, M. Breulmann, C. G. Göltner, H. Cölfen, K. K. W. Wong, D. Walsh, and S. Mann, Chem. Eur. J. *4*:2493–4500 (1998).

40

The Role of Surfactants and Amphiphiles in the Synthesis of Porous Inorganic Solids

ANDREAS STEIN and BRIAN J. MELDE University of Minnesota, Minneapolis, Minnesota

I. INTRODUCTION

For several decades now, organic molecules have been used to direct the structure of porous inorganic solids. For example, Barrer and Kerr (Mobil) pioneered the use of organic cations, such as tetramethylammonium ions as ingredients for zeolite synthesis [1,2]. The application of such structure directors has been extended to a large number of zeolite syntheses [3] leading to over a hundred zeolite structures [4]. The pore openings in these crystalline materials have spanned the range from about 3 to 13 Å. Because zeolites have shown tremendous commercial success in applications including ion exchange, size- and shape-selective catalysis, and separation [5], much research effort has focused on tailoring material compositions and pore sizes to optimize materials properties and to accommodate larger guest molecules. In the search for larger pore materials, a breakthrough was made in the early 1990s when Mobil announced the synthesis of a new family of mesoporous molecular sieves (M41S) that were prepared in surfactant solutions [6–9]. The new class of silicates and aluminosilicates possess ordered arrays of uniform channels that are tens of angstroms in diameter. In the surfactant-templated syntheses, aggregates of surfactant molecules rather than individual molecules fill the space that eventually forms the channels in the porous solids. Since the original publications, hundreds of scientific papers as well as several review articles have appeared describing research on ordered mesoporous sieves and related mesostructured materials. The compositional range for mesoporous materials has been greatly expanded, and new structures and morphologies have been developed by tuning the synthesis procedures. This chapter focuses on the role that surfactants and other amphiphiles play in controlling and modifying the structure of porous inorganic solids.

II. TEMPLATING OF SILICATES AND ALUMINOSILICATES WITH CATIONIC SURFACTANTS UNDER BASIC CONDITIONS

During their studies of surfactant-clay intercalation compounds, researchers at Mobil discovered structures with unexpected hexagonal pore arrays. They soon determined that similar structures could be obtained with a variety of silica and alumina sources under alkaline hydrothermal conditions when a cationic alkylammonium surfactant was present. Depending on reaction conditions, they observed silicate or aluminosilicate structures with hexagonal arrays of uniform channels (MCM-41), cubic structures (MCM-48), or lamellar structures (MCM-50). The new class of materials was labelled M41S. Their structures are shown in Fig. 1. MCM-41 consists of amorphous silicate walls (approximately 8–9 Å thick) [10] that surround cylindrical surfactant micelles. On the basis of transmission electron microscopy (TEM), infrared (IR), nuclear magnetic resonance (NMR), and Raman data, Davis et al. [11] concluded that the walls of MCM-41 exhibit a broad range of T—O—T bond angles, resembling amorphous silica or aluminosilicates rather than crystalline zeolites in

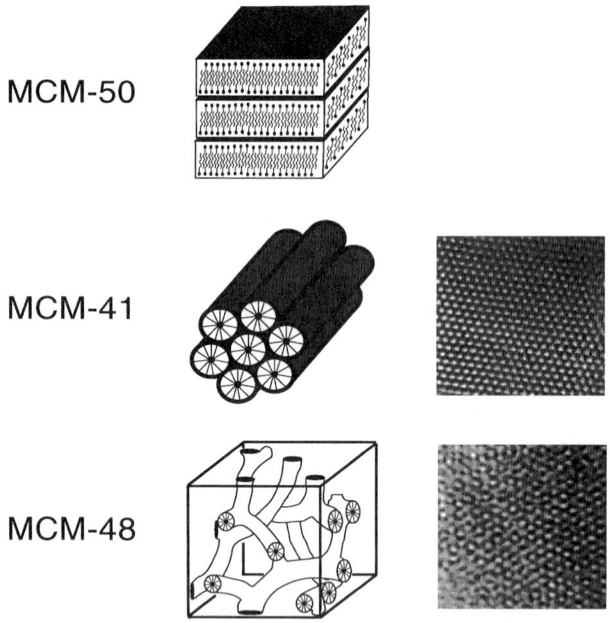

FIG. 1 Schematic structures of lamellar MCM-50, hexagonal MCM-41, cubic MCM-48, and TEM images of MCM-41 and MCM-48. (The TEM images were obtained by C. F. Blanford.)

terms of the local structure and bonding. A subsequent SiK XANES study by Fröba et al. [12] showed that the inorganic components in mesoporous silica are somewhat more ordered than "truly" amorphous silica phases.

The hexagonal wall structure is maintained after removal of the surfactant by calcination at about 360–550°C or by extraction with nonaqueous acid (such as HCl-methanol mixtures). The surfactant-free products have very high surface areas exceeding 1000 m^2/g (numbers reaching 2000 m^2/g have been reported), pore volumes in the range 0.7–1.2 cm^3/g, and high hydrocarbon sorption capacities for cyclohexane, benzene, or other nonpolar molecules (>0.7 cm^3/g). The pore sizes are controlled by the length of the cationic surfactant, ranging from 18 to 37 Å for $C_nH_{2n+1}(CH_3)_3N^+$ ($n = 8$–16) in silicates [7]. Larger pores can be obtained by modifications of the synthesis (see later).

Mesostructured materials were obtained under a wide range of conditions. Sources of silicon include sodium silicate, tetramethylammonium silicate, colloidal silica, silica gel, tetraalkylorthosilicates, and oligomeric silica clusters. Aluminum sources include alumina, sodium aluminate, and aluminum sulfate. In syntheses of M41S materials, hydrothermal reactions are carried out with cationic alkylammonium surfactants using hydroxide ions as mineralizers. Under these conditions silica forms oligomeric anionic species, often with multiple charges. Mesostructured products can be formed between room temperature and 150°C in reactions lasting from a few hours to days. In addition to the temperature [7,10,13,14], factors that influence the M41S structure include the reagent concentrations [10,15,16], the ratio of surfactant to inorganic species [17], alkalinity [15,18,19], hydrophobicity of the alkoxide precursors, the surfactant chain length, and the type of surfactant [7,15,18,20]. These parameters also influence the kinetics of polymerization. If extraneous amorphous impurity phases are present, they can be removed by washing the as-synthesized mesoporous sieve in mild base (e.g., 0.25 M Na_2CO_3 solution), which leads to preferential dissolution of the impurity [21].

A. Proposed Mechanisms for Formation of MCM-41

Before discussing possible formation mechanisms of the surfactant-inorganic composite materials, we need to define the terms "mesostructured" and "mesoporous." "Mesostructured" phases possess periodicity on a 20–500 Å scale. As in pure surfactant chemistry, hexagonal, lamellar, and gyroid cubic phases may be found in surfactant-inorganic composites, and transformations are possible, especially if the inorganic component is not yet highly condensed. If the surfactant template can be removed to obtain accessible pores with periodicity on the 20–500 Å scale, one obtains a "mesoporous" material.

The earliest proposed mechanism for the formation of MCM-41 was a liquid crystal templating (LCT) mechanism (Fig. 2a), in which a hexagonal array of rodlike micelles formed initially, before the anionic inorganic precursors assembled in the water region between cylindrical micelles (with positive surface charge) and condensed to form silica walls around the liquid crystal template [6]. In this mechanism the liquid crystal phase is intact before silicate species are added. Davis and coworkers [22,23] followed the synthesis of MCM-41 by in situ ^{14}N NMR spectroscopy, a technique that can detect the presence of the hexagonal phase. They did not detect this phase, thus disproving the LCT mechanism, at least for common reaction conditions [22,23]. True liquid crystalline properties may, however, exist at pH ≥ 12 and temperatures below 100°C, when anionic silicate species remain unpolymerized in aqueous solutions [24]. For silicate-surfactant mesophases to form, the surfactant concentration in the precursor organic solution must be above the critical mi-

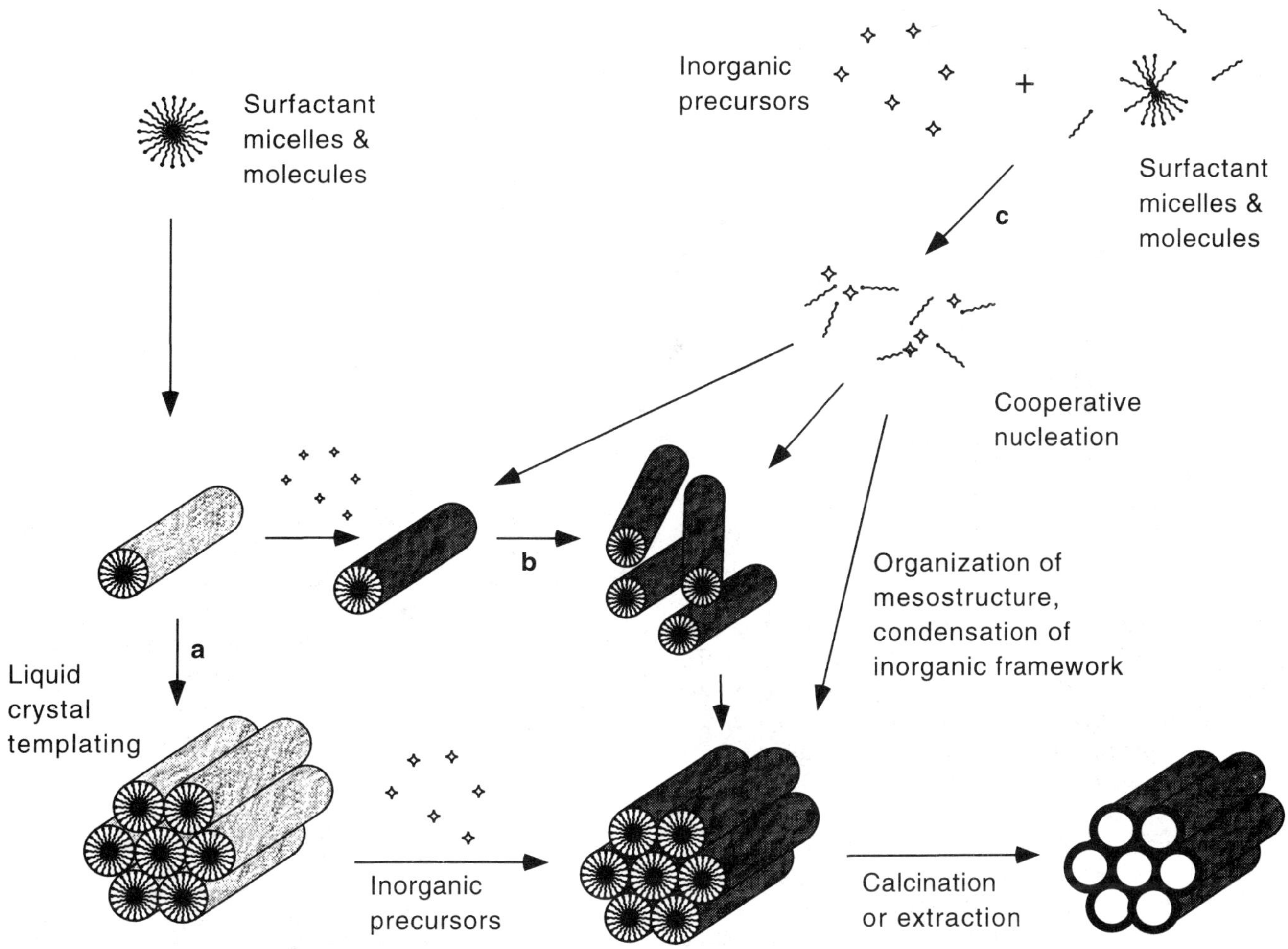

FIG. 2 Proposed mechanisms for the formation of MCM-41 [7,22,31]. (a) Liquid crystal templating. (b) Assembly of silica-coated cylindrical micelles. (c) Cooperative nucleation and mesostructure formation. In each mechanism, condensation of the inorganic framework can continue after the surfactant-inorganic mesophase has formed.

celle concentration (cmc, the temperature and concentration at which finite organized ordered arrays can first be detected). Silicate species can then be induced to polymerize by lowering the pH or increasing the temperature [25]. Under these conditions, one observes a surfactant-poor isotropic phase and a surfactant-rich liquid crystal phase. It should be noted that the shapes of the surfactant micelles are not directly related to the shapes of the inorganic-organic composite because of the additional interaction forces introduced by the inorganic component. In spite of its limited applicability, the LCT scheme has been useful to illustrate the synthetic and structural concepts involved in the M41S system.

An alternative mechanism was also proposed by the Mobil group, suggesting that the presence of the inorganic species mediates the ordering of the surfactant or of silicate-encased micelles. Once an ordered array was established, subsequent processing permitted removal of the surfactant and further stabilization of the walls [7]. In support of this mechanism, Davis et al. suggested the following formation pathway, based on ^{14}N NMR spectra of gels obtained in situ during the synthesis of MCM-41 and MCM-48. In those measurements a single isotropic line was observed and assigned to the presence of rapidly rotating rodlike micelles. These randomly ordered micelles are coated with two or three monolayers of charge-balancing silicate species (Fig. 2b). Partial condensation of the silicate oligomers leads to spontaneous organization of the cylinders into hexagonal arrangements. Unlike the formation of binary surfactant-water systems, this is a kinetically controlled process. The silicate precursors cannot condense completely as charged SiO^- species are required to balance the charge on the surfactant molecules. However, further condensation occurs during calcination [11,22,23].

Klinowski and coworkers [26] investigated the role of surfactant micelles as templates and as catalysts in the hydrolysis and polymerization of tetraethylorthosilicate. They determined that MCM-41 can be synthesized only when the cmc is met or exceeded (cmc = 0.0013 M for CTACl in water [27]). Their results also support a model in which individual micelles are initially formed and coated with silicate anions; the silica-coated micelles then self-assemble into the solid mesophase [28].

Under certain conditions a layer mechanism was observed, involving a lamellar-to-hexagonal phase transformation [10]. In this proposed mechanism, silicate oligomers initially function as multidentate ligands with a high charge density that effects a lamellar organization of the surfactant. As polymerization of the silica progresses, the reduced charge density of the growing silica polyanions increases the average headgroup area of the surfactant assembly, inducing a lamellar-to-hexagonal phase transformation. However, a layered intermediate is not always observed [29]. It was later shown by in situ x-ray diffraction (XRD) measurements that the lamellar-hexagonal phase transition depends on the pH of the reaction mixture and that it proceeds via dissolution of the lamellar phase [30]. The lamellar phase was observed at high pH (~13). Upon acidification, the charge density on the silicate oligomers was reduced and the lamellar phase dissolved. At a pH close to 11.6 the hexagonal phase appeared. The hexagonal phase became most ordered at pH 10.7.

The formation mechanism may be different under different synthesis conditions. Even though extensive surfactant-water phase diagrams have been developed, it was determined early on that surfactant-inorganic mesostructures could form under conditions where the surfactant alone would not form a liquid crystal phase (but above the cmc, i.e., when micelles are present) or silicate alone would not condense [10]. Stucky and coworkers pointed out that the electrostatic interaction and the matching of charge density at the surfactant-inorganic interfaces should govern the cooperative assembly of the mesostructure [10,25,31,32]. Tailoring of the structure (for example, adjustment of curvature) is possible by adjusting the electrostatic interactions, other bonding interactions, and charge density matching at interfaces. The mechanism formulated by Stucky's group [16,31,32b] and confirmed by others [33] involves the following processes (Fig. 2c):

1. Multidentate binding of oligomeric silica polyanions (three to eight Si atoms) to the cationic surfactant results in strong surfactant-silica interactions. An example of such a multiply charged anion is the double four-ring (D4R) oligomer. The polyanions replace monovalent anions (OH^-, Cl^-, Br^-) by ion exchange. The new ion pairs can reorganize themselves into mesophases with liquid crystal structures, the nature of which depends on the composition of the mixture, temperature, pH, and reaction time [16]. Because different silicate oligomers can have different charges, they interact differently with the surfactant headgroups.
2. Preferential silicate polymerization at the interface region due to the high concentration of silicate in this region, assisted by partial screening of negative charges by the cationic surfactant.
3. Charge density matching between the surfactant and the silicate during the condensation process. Before condensation, the surfactant-silica composite has the characteristics of a salt; it can be redissolved when placed in distilled water [31]. The condensation process leads to a reduction of the framework charge and to the formation of a rigid structure: $\equiv Si—O^- + HO—Si\equiv \rightarrow \equiv Si—O—Si\equiv + OH^-$. As the charge density on inorganic precursors decreases during condensation, some surfactant molecules have to leave the composite structure.

The formation of MCM-41 has been studied by several in situ methods. In a combined small-angle x-ray scattering (SAXS) and cryo-TEM study, Regev [34] observed a transition of spherical to elongated micelles as TEOS was added to a CTACl-NaOH-water mixture. He observed the formation of 50-nm-diameter ordered clusters of elongated micelles, which were "wrapped" with a silica film. During silica polymerization, as the formation of MCM-41 proceeded, the silica layer penetrated the cluster arrays and reduced the correlation length between micelles until eventually a hexagonal phase with ordered arrays of 5-nm pores was obtained. In an in situ ATR-FTIR study carried out with colloidal silica, sodium aluminate, and CTAOH above the critical micelle concentration and temperature, the investigators observed a gradual and irreversible loss of the colloidal silica starting material as an intermediate silicate phase was produced [35]. The silicate mesophase was formed when a temperature of 120°C was reached. The surfactant tails exhibited less order with increasing temperature, whereas the headgroup configurations became more ordered. The increased headgroup ordering was attributed to loss of hydration water in the headgroup region, decreased average headgroup spacing upon exchange of hydrated OH^- ions with silicate anions, and headgroup-counterion interactions. The IR study also revealed that at room temperature the silicate starting material is largely the double four-ring octamer ($Si_8O_{20}^{8-}$), which decomposes to low-molecular-weight oligomers upon heating. The silicate polyanions present at pH > 12 reduce repulsions between cationic headgroups by forming multidentate interactions with the headgroups. As a result, the average headgroup area becomes smaller.

Klinowski and coworkers [26] pointed out the catalytic action of the surfactant. In the absence of surfactant, TEOS and water are initially immiscible. As TEOS is hydrolyzed a homogeneous solution is formed within 1 h, but precipitation of solid amorphous silica products can require days. When TEOS hydrolysis is instead carried out in a surfactant solution, hydrolysis and precipitation of a surfactant-silica mesostructure can occur in a few minutes. Klinowski et al. [26] determined that in the presence of a cetyltrimethylammonium chloride or hydroxide surfactant the rate of silicate polymerization was 2000 times faster than without the surfactant. They proposed that the cationic surfactant can concentrate hydroxide ions at the surfactant-silica interface via halide ↔ hydroxide exchange, thus promoting the hydrolysis and condensation rates.

B. Molecular Templating Versus Supramolecular Templating

Beck and coworkers [13] explored in detail the ability of $C_nH_{(2n+1)}(CH_3)_3N^+$ (C_n, $n = 6$, 8, 10, 12, 14, and 16) surfactants to serve as structure-directing agents or templates for the formation of mesoporous as well as microporous molecular sieves. They carried out hydrothermal reactions with surfactant halides and sodium silicate at constant surfactant/silica mole ratios of 0.5, 11 wt% surfactant concentration, and pH 10. They investigated varying alkyl chain lengths and temperatures and, depending on these parameters, obtained either amorphous, zeolitic, or mesoporous structures. Their results are summarized in Table 1.

It is interesting to note that for the C_{12} series, the surfactant was present and intact in both the mesoporous and the zeolitic phases; solid-state NMR evidence indicated that the surfactant molecules constituting micellar arrays within the mesoporous structures were more mobile than the isolated surfactant molecules confined in the smaller zeolite pores. Although multiple zeolite phases were sometimes present, zeolitic and mesoporous phases were never observed together, suggesting different formation pathways for these phases. Based on their experiments, Beck and coworkers proposed the mechanistic pathways for the synthesis of micro- and mesoporous materials illustrated in Fig. 3.

C. The Cubic MCM-48 Phase

The cubic MCM-48 phase consists of two independent, interwoven networks of mesoporous channels (space group $Ia\bar{3}d$) [10,36]. The bicontinuous structure has been described as consisting of a single sheet that winds through space following a gyroid surface [37]. This phase has been studied less than MCM-41 because the synthesis appears to be less reliable [38]. Although MCM-48 can be formed over a wide range of surfactant/silicon ratios (from 0.12 to 0.8), successful product formation depends critically on the source of silica em-

TABLE 1 Products Obtained from Surfactant Templated Syntheses for Varying Surfactant Chain Lengths and Temperatures[a]

$C_nH_{2n+1}(CH_3)_3N^+$	Temperature		
n	100°C	150°C	200°C
6	Amorphous	ZSM-5	ZSM-5
8	Mesoporous	ZSM-5	ZSM-5, ZSM-48, dense phase
10	Mesoporous	Mesoporous	ZSM-5, ZSM-48, dense phase
12	MCM-41	Mesoporous	ZSM-5, ZSM-48, dense phase
14	MCM-41	MCM-41	Amorphous and zeolitic
16	MCM-41	MCM-41	Amorphous

[a]In this table, "mesoporous" refers to structures that showed one or two broad diffraction peaks in the X-ray diffraction pattern, but were less ordered than true MCM-41.
Source: Adapted from Ref. 13.

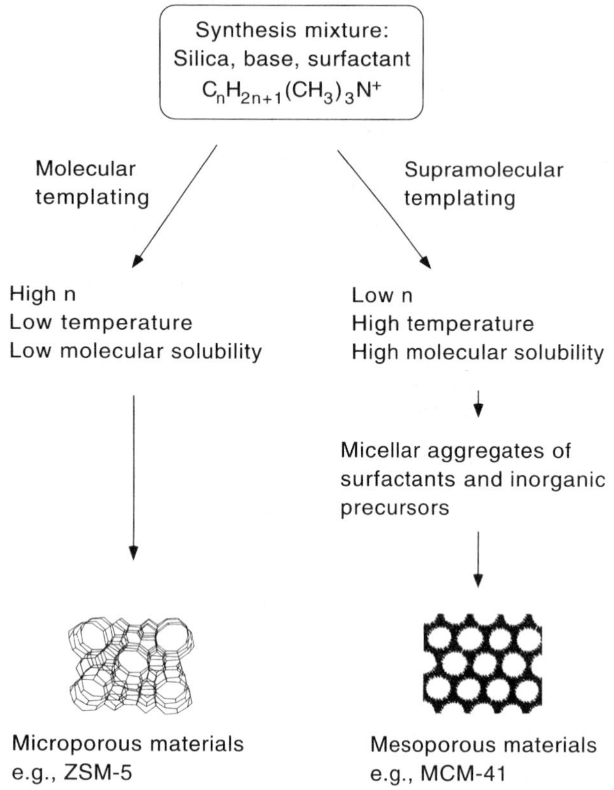

FIG. 3 Proposed mechanistic pathways for the formation of microporous and mesoporous materials. (Adapted from Ref. 13.)

ployed, the initial pH, temperature, and synthesis time [7,10,17,33,38–44]. Cubic phases can be favored by using surfactants with larger headgroups to increase curvature [18]. During the synthesis, several phase transitions can occur. For example, in a synthesis based on TEOS as the silicon source, with CTAB at pH 11.8 and 373 K (1 TEOS: 0.23 Na_2O: 0.55 CTAB: 112 H_2O), first a disordered tubular mesophase is formed, which transforms to a layered phase, then slowly to a cubic MCM-48 mesophase, and finally to another layered phase [38]. It is therefore important to time the synthesis of MCM-48 correctly to avoid production of the second layered phase.

In a TEOS-based synthesis with a sufficient amount of surfactant, MCM-48 was obtained in dilute base solution, whereas a lamellar MCM-50 phase was produced at high base concentration ($Na_2O/H_2O = 0.049$) [45]. It was proposed that the cubic phase is favored by interaction of the surfactant with monomeric silicate precursors, whereas more condensed silicate oligomers produce the layered phase. ^{29}Si MAS NMR showed a significantly higher number of Q^3 species for as-synthesized MCM-48 than for MCM-41, indicating a larger number of surface hydroxyl groups in the cubic phase [46].

A synthesis of micrometer-sized MCM-48 single crystals (truncated rhombic dodecahedra) was described by Ryoo and coworkers [47]. In this preparation, alcohol (MeOH, EtOH) was employed as an additive for mesophase control in the sodium silicate/CTAB solution. The mixture was heated to 100–130°C in a sealed autoclave to prevent loss of alcohol. If alcohol was allowed to evaporate during the synthesis, a disordered phase was obtained, indicating that the alcohol affects the structure of the surfactant-inorganic composite during condensation. A lamellar phase was obtained with excess alcohol.

D. Phase Transitions in Systems with Strong Interface Interactions (S^+I^-)

When considering the cooperative self-organization of organic surfactant molecules and inorganic solution species, one has to take into account multiple bonding interactions (electrostatic, hydrogen bonding, covalent bonding, van der Waals forces) and multiple interfaces (inorganic-inorganic, organic-organic, organic-inorganic, precursor-solvent) [40]. The self-assembly process is dominated by surfactant behavior if condensation of the inorganic component is negligible. On the other hand, if the inorganic component has strong interactions (e.g., fast polymerization/condensation), the structure is mostly determined by this component. The interface charge matching of amphiphilic surfactants with inorganic species controls the assembly; these interactions must be balanced to control the structure.

Even though the phase diagrams of surfactant-inorganic-solvent systems differ from those of pure surfactant-solvent systems, a comparison has proved valuable [40]. For classical surfactant-solvent systems the liquid crystal phases are determined by the effective surfactant ion pair packing parameter, defined as $g = V/a_0l$ (V = total effective molecular volume of hydrophobic surfactant chains plus any cosolvent species present, a_0 = effective surface area of surfactant headgroup at the hydrophilic-hydrophobic micelle aggregate interface, l = kinetic surfactant tail length) [48]. The value of g depends on a variety of factors, including the number of carbon atoms in the surfactant tail, the degree of chain saturation, the charge of the polar headgroup, the size of the headgroup, the ionic strength of the solution, pH, and temperature. The relationship is strictly valid only for dilute solutions where interactions between aggregates can be neglected. Nevertheless, an investigation of a large number of surfactants with varying tail lengths, headgroup sizes, and charges indicated that the molecular packing parameter model can be used to a first approximation to predict inorganic-surfactant composite structures [49]. Micellar shapes for cationic surfactants in the dilute micellar concentration regime were compared by Manne et al. [50]. They are summarized in Table 2.

The phase (hexagonal or lamellar) of the mesostructure is dependent on the pH and on the surfactant/silicate ratio [10,17]. Lamellar structures are favored when headgroups can pack tightly, for example, with double chain surfactants. Mixtures of charged and neutral surfactants with similar chain lengths also favor lamellar structures [41]. Curvature increases when polar headgroups occupy a larger surface area, and when g is less than 1/3, globular aggregates are preferred [24,51,52]. The headgroup area can be increased, for example, by addition of water, which decreases the charge density of the inorganic region. However, with neutral amine surfactants the phase responds differently to the water content in the reaction. At a higher water content (1 hexadecylamine: 1 TEOS: 200 H_2O) a lamellar product is obtained, whereas at a lower water content (1 hexadecylamine: 1 TEOS: 60 H_2O) the product is hexagonal [53].

In addition to dilution, the mesophase can be controlled by the surfactant-to-silicon ratio. Beck et al. [7] observed hexagonal MCM-41 with surfactant/Si ratios less than 1. The cubic form (MCM-48, $Ia\bar{3}d$) was obtained with surfactant/Si ratios greater than 1. With even greater surfactant/Si ratios the lamellar phase (MCM-50) was produced, which contained slightly thicker walls (10–11 Å) than MCM-41 (8–9 Å) [10].

If the structure remains flexible enough during polymerization, the reduction in framework charge density may lead to phase transitions. Phase transitions that have been observed with different surfactants include lamellar to MCM-41, MCM-41 to lamellar, MCM-41 to MCM-48, and MCM-48 to lamellar. These phase transitions are believed to occur in the solid phase rather than in solution [54]. A phase transformation from hexagonal to cubic was exploited in a rapid MCM-48 synthesis [42]. Using TEOS as the silicate precursor with cationic surfactants under basic conditions, the alkoxide was permitted to hydrolyze under rapid stirring for 1–3 h at 35–40°C, resulting in a hexagonal intermediate phase. The mixture was subsequently heated at 150°C for 3–5 h in a closed autoclave, producing well-ordered MCM-48 phases that were stable to calcination. It was shown that ethanol (hydrolysis product of TEOS or added cosolvent) was necessary to permit the phase transformation. The proposed role of ethanol as cosolvent was to increase the surfactant packing parameter g by increasing the effective surfactant volume.

Cosolvents can be added to reaction mixtures to influence the product phase. Nonpolar cosolvents, which associate most strongly with hydrophobic surfactant tails, result in swelling of micelles. Their use in controlling the pore dimensions of MCM-41 is discussed later. Polar cosolvents (e.g., ethanol or methanol) interact more strongly with the headgroups or with tail sections closer to the headgroup, leading to increases in the total volume V. Anderson et al. [55] studied the effect of the MeOH concentration on CTAB micellization and on the formation of mesoporous silica. They observed that addition of MeOH led to an increase in cmc for CTAB from 1.3×10^{-3} M for 0 wt% MeOH to 5.5×10^{-2} M for 60 wt% MeOH in 0.22 M NaOH solution. However, the long-range order of the mesoporous silica product decreased as the MeOH concentration increased. Ordered channel arrays formed between 0 and 60% MeOH, in which range the concentration of CTAB exceeded the cmc. At higher methanol content the cmc concentration was not ex-

TABLE 2 Morphologies of Surfactants and Surfactant-Silicate Aggregates in Dilute Aqueous Solutions[a]

Cationic surfactant mesophase	g	Micelle shape in aqueous solution	Micelle shape in silicate
Asymmetric gemini	<1/3	Spheres	Spheres (3D hexagonal)
Conventional alkyltrimethylammonium	≈1/3	Spheres	Cylinders (hexagonal)
Symmetric gemini, $s \geq 4$	≈1/3	Spheres/spheroids	Cylinders (hexagonal)
Symmetric gemini, $s = 2$	$1/3 < g < 1/2$	Cylinders	Bilayers (lamellar)
Conventional dialkyldimethylammonium	≈0.62	Bilayers/vesicles	Bilayers (lamellar)

[a]Here, s refers to the length of the spacer unit between two ionic headgroups in a surfactant of the type $C_{n\text{-}s\text{-}m}$. The values of n and m refer to the length of the two surfactant tails.
Source: Adapted from Refs. 40 and 50.

ceeded and wormlike structures resulted. In the methanol-rich solvent with its lower dielectric constant the surfactant no longer acted as an effective chemical dipole, thereby lowering the tendency for cooperative assembly into an ordered surfactant-silica phase.

As the cosolvents or auxiliary organics interact with the micelles, they cause rearrangement of the surfactant structure. This interaction is temperature dependent, and at higher temperatures the organic additives become less soluble in the hydrophobic regions and are expelled into the water-rich phase. This can lead to an increase in curvature at the surfactant-silica interface, resulting in a phase transition from lamellar to hexagonal [24]. The structural arrangement and interaction with solvent/cosolvent/auxiliary organics can be altered by modifying the surfactants with appropriate functional groups. When no cosolvents were used, Chen et al. [14] observed a transition from a hexagonal mesostructure to a lamellar phase to microporous silicalite as the hydrothermal temperature was raised from 100 to above 165°C. Davis and coworkers [56] calculated that the energy differences between alternative mesoscopic configurations are very small, thus permitting the formation of a number of different structures that have been observed. The enthalpy of calcined MCM-41 is only 15 kJ/mol higher than that of quartz for pores ranging from 28 to 50 Å in diameter.

III. WEAKER INTERFACE INTERACTIONS

M41S syntheses are based on direct electrostatic interactions between cationic surfactants and negatively charged silicate precursors in an alkaline environment (S^+I^-). These strong interactions often involve multidentate binding. Under these conditions (and also for S^-I^+ combinations) [31] the product structure is independent of the liquid crystal structure of the organic component alone [40]. Especially in dilute surfactant solutions the structural organization is influenced by van der Waals interactions involving the organic tail and the shape or size of organic groups (head, tail) that are present. However, usually both Coulombic and van der Waals components must be taken into consideration. Counterions present in solution contribute to the Coulombic forces and influence the structure of cationic surfactant micelles.

Stucky and coworkers pointed out that participation of counterions can be exploited to create mesostructures with weak interfacial interactions [$S^+X^-I^+$, $S^-M^+I^-$, $(S^0H^+)(X^-I^+)$] [31,40]. The latter combination involves hydrogen bonding in acidic media when the pH is below 2 for silica mesophases, i.e., below the isoelectric point of silica. Under these conditions, the product structure is more closely related to the structure of the pure organic phase (amphiphilic surfactants, nonionic surfactants, polymers). Counterion-mediated syntheses can be used to control the product morphology, as well as the structure on a molecular level. Other systems involving weak interfacial interactions are listed in Table 3.

For silica systems, the charge of the precursor is controlled by the pH. Assembly through weak hydrogen-bonding interactions is achieved when neutral (S^0) or nonionic (N^0) surfactants are used in neutral media [20] or when counterions (e.g., halide ions) are invoked under acidic conditions. Acid syntheses carried out at pH < 0 have produced an array of phases similar to those produced by base syntheses. Many of the early acid-based preparations were developed by the group of Stucky and coworkers. The product phases were denoted SBA-1 (cubic, $Pm\bar{3}n$), SBA-2 (3D hexagonal, $P6_3/mmc$, consisting of hexagonally close-packed spheres, ellipsoids, or interconnected ellipsoids), and SBA-3 (hexagonal, *p6mm*) [54]. A lamellar phase has also been observed under acidic conditions.

In sol-gel chemistry it is known that silica hydrolysis at a pH below the isoelectric point of silica leads to chainlike polymerization, whereas base hydrolysis results in denser clusterlike agglomerations [57]. One can therefore expect a different wall structure in acid than in base synthesis. It has been observed that the walls in acid-synthesized materials are thicker after calcination than in MCM-41 materials obtained from base synthesis. Because the cationic charge of the template is balanced by an anion, no counterion is required during surfactant extraction, and the template can be removed by extraction with ethanol [18]. It is notable that auxiliary organics do not affect the pore size strongly under acidic conditions, but they affect the product morphology [32b]. One can direct the morphology of the mesoporous structure by applying hydrodynamic forces (e.g., stirring at different speeds) to shape the emulsion phase (see later).

IV. EXTERNAL FIELDS

If assembly forces are weak, it is possible to orient the mesostructure by application of external fields, such as magnetic, electric, or shear fields. Chmelka and coworkers [24] used a strong magnetic field to align and characterize unpolymerized silicate-surfactant mesophases with the intention of controlling the structuring of the mesoporous solids. They followed the effects by ^2H NMR spectroscopy and showed that the initially

TABLE 3 Interfacial Interactions Between the Organic and Inorganic Phases[a]

Strong interfacial interactions	Covalent bonds	S—I	
	Covalent bonds	S—I	
	Ionic interactions	S^+I^-	
	Ionic interactions	S^-I^+	
Weaker interfacial interactions	Ionic interactions	$S^+X^-I^+$	
	Ionic interactions	$S^-M^+I^-$	
	Hydrogen bonding	$S^0H^+X^-I^+$	
	Hydrogen bonding	S^0I^0	
	Hydrogen bonding	N^0I^0	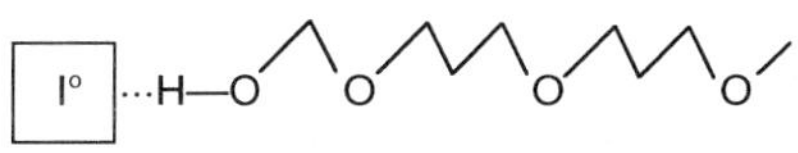

[a] "S" refers to the cationic, anionic or neutral surfactant, "N" refers to a nonionic surfactant, and "I" refers to the inorganic component.

randomly oriented domains in the lyotropic liquid crystal could be oriented over sample-size length scales. To facilitate the alignment process and overcome viscous and elastic forces, the sample was heated above the anisotropic-isotropic transition temperature and then slowly cooled in a high magnetic field. At room temperature the aligned liquid crystal domains remained stable for months without losing their orientational order outside the magnetic field. The prealigned structure could be preserved, keeping macroscopic orientational order even after polymerization of the silicate species by acid treatment followed by calcination [58].

V. METHODS OF CONTROLLING PORE DIMENSIONS IN MCM-41

The channel diameters in MCM-41 can be controlled in a number of ways. The obvious method of changing the alkyl chain length of surfactant molecules $[C_nH_{2n+1}(CH_3)_3N^+]$ is successful within the limits of $n = 8$ to about 18. With shorter surfactant molecules one obtains zeolitic phases [13]. With $n \geq 20$ the lamellar phase is favored over MCM-41. The low water solubility of surfactants with alkyl chains longer than $n = 18$ makes these less practical during the hydrothermal synthesis, although with mixed surfactants (gemini surfactant added) it is still possible to obtain MCM-41 with $n = 22$ [40,49].

Even in their early work, the Mobil group demonstrated that the channel diameters can be increased significantly by adding auxiliary organic chemicals (e.g., mesitylene, benzene, toluene, hexane, hexanol) to the reaction mixture [9]. For cetyltrimethylammonium ions as templates under basic conditions, the *d* spacing of the mesoporous silica product gradually increased from 37 to 65 Å as the mesitylene/SiO_2 ratio was increased from 0 to 1. A large variety of auxiliary organic substances have been found effective in increasing the pore size diameters of MCM-41, including many aromatic and aliphatic hydrocarbons and amines. Depending on their relative polarities and dielectric properties, the auxiliary organics can solvate various components of the surfactant array, such as the headgroup or the tails. Nonpolar organic cosolvents interact most strongly with the hydrophobic tails of the surfactants and lead to swelling of the micelles. This in turn produces MCM-41 channels with greater diameters. Within limits, the increase in d_{100} spacing is approximately proportional to the amount of organic swelling agent added. For example, a unit cell of 75 Å was obtained with 20 wt% mesitylene in the gel [9]. However, under acidic conditions, auxiliary organics were found to have little effect on the pore size [32b].

Analogous to cosolvents, mixed surfactant assemblies permit control of the charge density at headgroups and linear control of d spacings on a subangstrom scale, at least over small ranges. Ozin and coworkers [59] employed varying ratios of cetylpyridinium chloride and CTACl in the synthesis of MCM-41-type materials. The two surfactants were present as mixed micelles. No evidence for the presence of two distinct micelle types was seen. By varying surfactant ratios, d_{100} spacings from 41 to 43 Å were obtained.

Much larger variations in d_{100} spacings were achieved by a postsynthesis restructuring method also described by Ozin's group [60]. By hydrothermally treating the synthesis gel and aging it, pore expansions and increases in lattice parameter with d_{100} spacings up to 62 Å were observed with CTABr as the surfactant. The restructuring process appears to involve random expansion of domains of mesopores.

Changes in pore dimensions have also been related to adjustments in reaction conditions [6]. Various degrees of pore swelling during the synthesis of MCM-41 were attained in the absence of auxiliary organics by varying the CTA^+/SiO_2 ratio, the crystallization temperature, and the crystallization time in a CTAB/TMA^+-based synthesis [61]. For highly crystalline MCM-41 products with large pores, the optimal CTA^+/SiO_2 was found to be between 0.09 and 0.15. Swelling was also observed when TEA^+ or Na^+ was used to replace TMA^+, but less stable products were formed. A tentative proposed mechanism of swelling involves replacement of CTA^+ species by tetraalkylammonium ions at the interface between the micelles and the silica surface. In a synthesis of pure silica MCM-41 from fumed silica, CTAB, and TMAOH, the d_{100} spacing increased from 35 to 55 Å and the channel diameters from 27 to 36.5 Å as the synthesis temperature was increased from 70 to 165°C in a 48-h reaction [62]. Similarly, at a constant reaction temperature of 165°C the d_{100} spacing increased from 37 to 55 Å and the channel diameters from 26 to 36.5 Å with increasing reaction time between 1 and 48 h. The samples prepared at higher temperature contained thicker walls and exhibited higher thermal stability. The proposed mechanism for the pore size enlargement involves swelling of the surfactant micelles as some $C_{16}H_{33}(CH_3)_3N^+$ cations decompose at 165°C to form $C_{16}H_{33}(CH_3)_2N$ molecules. These electrically neutral molecules accumulate at the center of the micelle, thus increasing its diameter up to a certain limit.

For MSU-X materials synthesized by the N^0I^0 pathway (see later), the pore size is influenced by the assembly temperature [63]. With the commercial PEO surfactant Tergitol 15-S-12 [$C_{11-15}H_{23-31}O(CH_2CH_2O)_{12}H$], the average pore diameter can be increased from 2.1 nm (synthesis at 25°C) to 4.5 nm (synthesis at 65°C). Surfactants with a high cloud point are best suited for controlling the pore sizes of MSU-X mesostructures by adjusting the reaction temperature.

VI. TYPES OF SURFACTANTS USED IN THE SYNTHESIS OF MESOPOROUS MATERIALS

In addition to alkylammonium surfactants, many other types of surfactants have been employed as structure directors or space fillers in inorganic mesostructures to permit better control of the product architecture. Surfactant types that have been used include cationic, anionic, neutral, nonionic, zwitterionic, gemini, fluorinated, and chiral surfactants, as well as block copolymers. Some examples are listed in Table 4. The surfactant structure and surfactant composition influence Coulombic interactions between charged headgroups and the inorganic building blocks, van der Waals interactions between surfactant molecules, solvation energies, and other interfacial interactions.

Great flexibility in the packing parameter is possible with surfactants containing multiple headgroup charges and/or multiple tails (e.g., gemini surfactants) [40,49, 50,54]. Gemini surfactants of the type $C_nH_{2n+1}N^+$-$(CH_3)_2(CH_2)_sN^+(CH_3)_2C_mH_{2m+1}$ are composed of two single-tail surfactants that are joined at their ionic headgroups through a spacer unit. The size and properties of the spacer have a strong influence on the shape of micelles formed by the gemini surfactant. The structural variation permits adjustment of the g parameter; the spacer length can be used to adjust the effective headgroup size, a_0. It has been observed that gemini surfactants with short spacers favor the lamellar phase, those with longer spacers the hexagonal phase [54].

Headgroup functionalization is another approach to modifying interfacial interactions. For example, in the case of hydroxy-functionalized surfactants of the type $C_nH_{2n+1}N^+(CH_3)_2(CH_2)_mOH$, the less hydrophobic headgroup remains in contact with water or with the inorganic phase [40]. This leads to a decrease in the effective cationic headgroup area a_0; therefore lamellar phases are favored. When nonionic surfactants are used in neutral media, assembly occurs by hydrogen-bonding interactions. The resulting mesoporous silica or alumina products contain wormlike disordered channels [20,64]. These materials are discussed in more detail later.

TABLE 4 Some Surfactant Types Used in the Synthesis of Inorganic Mesostructures

Cationic	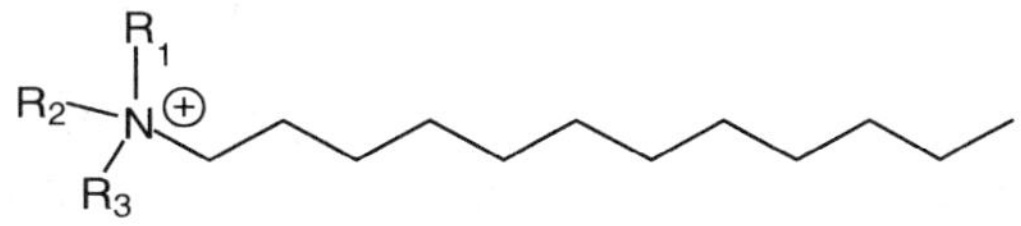
Anionic	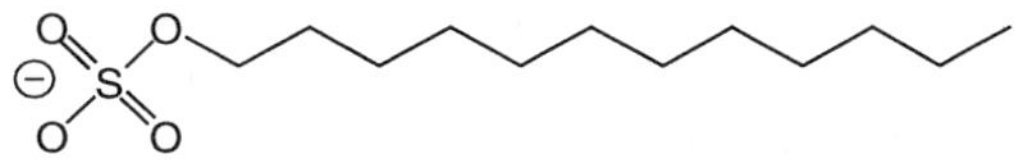
Neutral	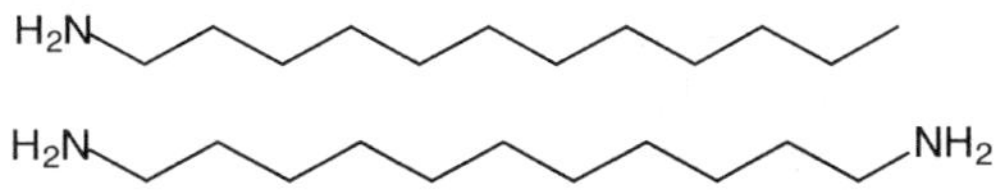
Nonionic	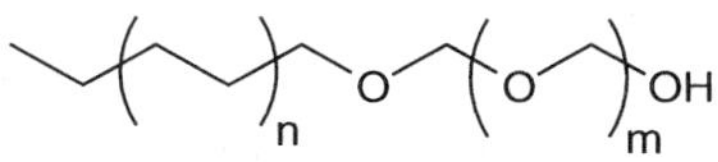
Zwitterionic	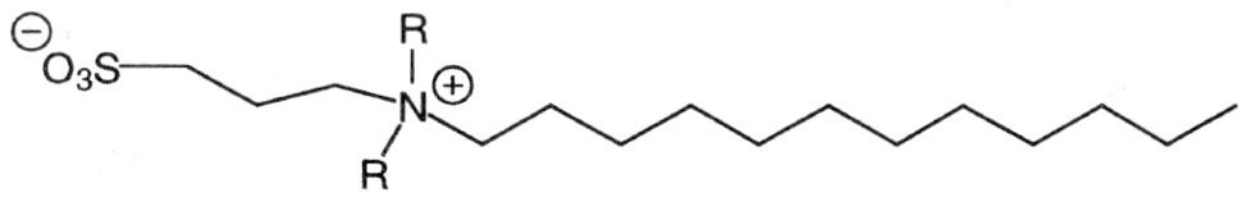
Gemini	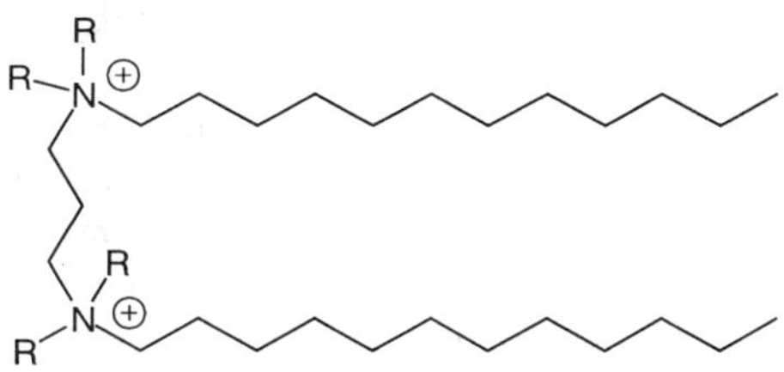
Double-headed	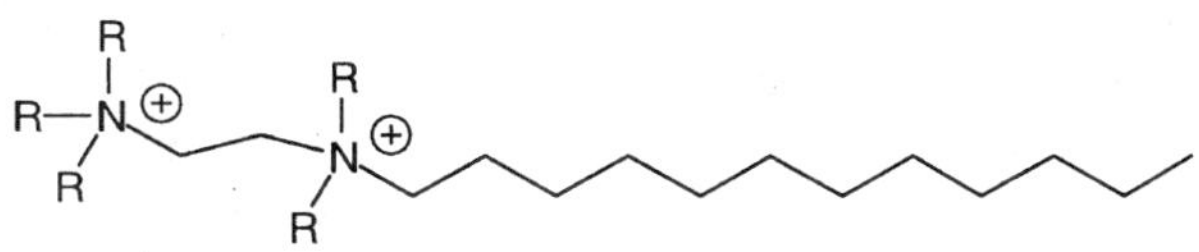
Organosilane	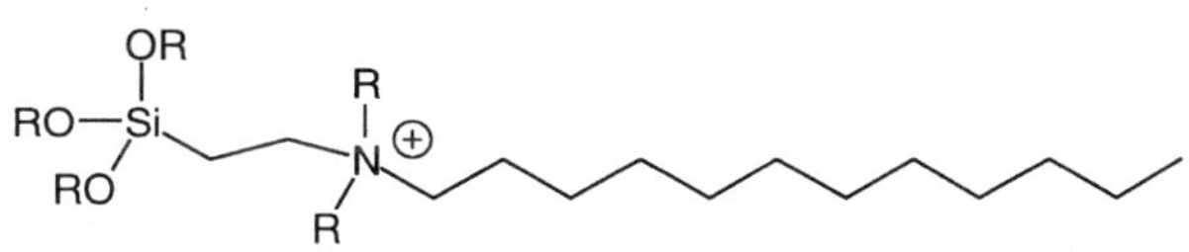
Functionalized	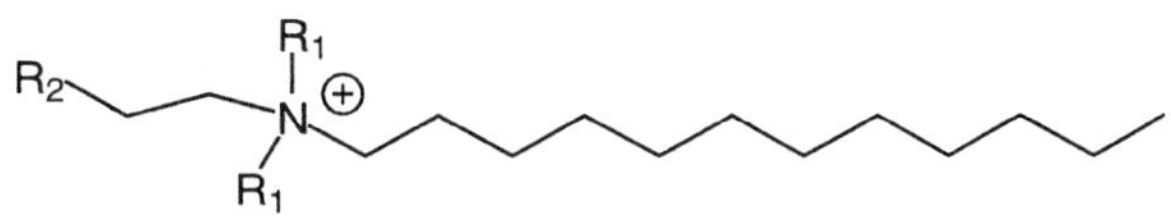

Mesoporous silicates with larger mesopores and thicker walls than M41S materials have been synthesized by employing nonionic polyethylene oxide (PEO) oligomeric surfactants or poly(alkylene oxide) block copolymers in acid media [65,66]. Silica structures were obtained with cubic $Im\bar{3}m$ (SBA-16), cubic $Pm\bar{3}m$ (SBA-11), 3D hexagonal ($P6_3/mmc$) (SBA-12), 2D hexagonal ($p6mm$) (SBA-15), and lamellar ($L\alpha$) symmetries. Large d_{100} spacings ranging from 104 to 320 Å were observed in the hexagonal form synthesized in the presence of triblock copolymers (PEO-PPO-PEO) [67]. Silica wall thicknesses ranged from 31 to 64 Å. The pore size and wall thickness can be adjusted by varying the heating temperature and reaction time. With their thicker walls, these porous solids exhibited higher hydrothermal stability in boiling water than MCM-41. In syntheses employing amphiphilic block copolymers consisting of an apolar polystyrene block and a polar polyethylene oxide block, the channels in the mesoporous product were not highly ordered [68].

In block copolymer syntheses, mesoporous silica formed in acid only (HCl, HBr, HI, NHO_3, H_2SO_4, H_3PO_4) but not in base. The pH must be $\ll 1$ to obtain mesoporous structures. At higher pH values, either no precipitation occurs (pH 2–6), disordered mesopores are formed (pH 7), or amorphous silica or silica gels precipitate (basic conditions). The reaction rate depends on the radius or charge of the anion and on the strength of the acid. The proposed pathway involves the $(S^0H^+)(X^-I^+)$ mechanism, where charge-associated EO units and the cationic silica species are assembled together by combination of electrostatic, H bonding, and van der Waals interactions [65,66]. When surfactants with short EO segments are employed, lamellar products are obtained. Because of their weaker interactions with the inorganic framework, certain block copolymers require lower calcination temperatures than low-molecular-weight cationic surfactants (~145°C vs. ~360°C) [65].

The block copolymer method has been extended to other metal oxide compositions [69], foams [70], fibers [51] and surface patterns [71]. Foams of mesostructured silica were produced using block copolymers $(EO_{20}PO_{70}EO_{20})$ with mesitylene as cosolvent in an aqueous acid synthesis. The foams are constructed of spherical cells up to 36 nm in diameter, which are interconnected through windows. The ratio of cosolvent to block copolymer affects the cell size [70]. In fiber syntheses, two-dimensional and three-dimensional hexagonal mesopores up to 63 Å were obtained, depending on the type of block copolymer, and fiber diameters ranged from submicrometer dimensions to several hundred micrometers. The channels in the fibers were uniaxially aligned [51].

A. Syntheses with Neutral or Nonionic Surfactants

Surfactant templating of mesoporous materials is often driven by charge matching between ionic species. Pinnavaia and coworkers have studied a pathway based on hydrogen bonding and self-assembly between electrostatically neutral primary amine surfactants (S^0) and neutral inorganic precursors (I^0). The effects of such neutral templating, when compared with similar ionic methods, may include thicker pore walls, loss of long-range structural order, smaller crystallite size, and increased textural mesoporosity.

Primary amines were used to template hexagonal mesoporous silica (HMS), the amines having a carbon chain length of C_8, C_{10}, C_{12}, C_{14}, C_{16}, or C_{18} [72]. TEOS was combined with the amine in an acidic ethanol-water solution, forming a gel that was aged for 18 h at ambient temperature. The template could be removed by calcination or by stirring in hot ethanol. Products yielded a single low-angle XRD peak and diffuse scattering suggestive of a broadened d_{hk0} reflection. Strong electrostatic interactions and charge matching seem to be necessary to obtain long-range order. A single XRD peak and diffuse scattering imply weaker nonionic interactions. Alcohol extraction preserved crystallinity better as a more intense XRD reflection was obtained. N_2 adsorption-desorption experiments produced type IV isotherms with hysteresis loops in the low relative pressure region characteristic of framework confined mesoporosity. Hysteresis loops also appeared in the high pressure range (>0.8), indicative of textural, or interparticle, mesoporosity. The textural mesoporosity was a consequence of the fine particle morphology (<40 nm) observed by electron microscopy. N_2 adsorption also revealed high surface areas (1000 m^2/g or greater) and pore walls that were considerably thicker than those templated by ionic surfactants of similar chain lengths.

Physical properties of HMS products were thoroughly studied and compared with those of S^+I^- and $S^+X^-I^+$ mesoporous silicas [73]. Transmission electron microscopy showed that HMS had hexagonal-like order, but less defined and not over as long a range as an ionically templated structure. Scanning electron microscopy produced images of nonuniform aggregates of small HMS particles, as opposed to the larger monolithic particles of S^+I^- and $S^+X^-I^+$ silicas. HMS textural mesoporosity was not found to occur at the expense of framework mesoporosity; it was suggested that fine particle size and interparticle porosity might facilitate diffusion. ^{29}Si MAS NMR studies yielded significantly higher Q^4/Q^3 ratios for HMS compared with S^+I^- and $S^+X^-I^+$ silicas, a result of more extensive cross-linking. This probably contributed to the greater thermal stability of HMS. It survived calcination for 4 h at 900°C, unlike the ionically templated silicas, which collapsed or suffered severe damage.

A formation mechanism was proposed in which surfactant molecules form rodlike micelles [72]. ^{14}N NMR experiments proved that the amines are electrostatically neutral, as spectra of amine solutions and wet HMS did not show a tetrahedral resonance from protonated amines. $Si(OC_2H_5)_{4-x}(OH)_x$ species (hydrolyzed TEOS) hydrogen bond to surfactant headgroups. Further hydrolysis and condensation result in the short-range ordered framework. The thicker pore walls can form because of the lack of strong electrostatic interactions [73]. Wall growth for an ionically templated material should terminate when charge compensation is achieved, a limiting effect not present during S^0I^0 templating. The amount of textural mesoporosity was found to depend on the relative amounts of water and ethanol in the cosolvent mixtures [74]. More water resulted in more rapid nucleation, smaller particle size, and greater textural mesoporosity. Addition of mesitylene was found to increase the pore size in "water-rich" (90:10 v/v H_2O/ethanol) mixtures and to decrease the pore size in "ethanol-rich" (35:65 v/v H_2O/ethanol) mixtures.

Mesostructures designated MSU-X templated by nonionic polyethylene oxide surfactants (N^0) have also been studied, where X is a numeral referring to a specific material templated by a particular family of surfactants [20]. PEOs have the advantage of forming spherical to flexible rodlike or wormlike micelles at critical concentrations approximately one hundredth those necessary for ionic surfactants. PEO surfactants used in syntheses included Tergitol 15-S-12 [$C_{11-15}H_{23-31}$-$O(CH_2CH_2)_{12}H$, Union Carbide], Triton X-100 [$(CH_3)_3$-$CCH_2C(CH_3)_2C_6H_4(CH_2CH_2O)_{10}H$, Union Carbide], and Igepal RC-760 [$CH_3(CH_2)_{11}C_6H_4(CH_2CH_2O)_{18}H$] (Rhône-Poulenc). TEOS was added to a surfactant solution, and the resulting product was aged for 12–16 h with moderate stirring and air dried. Surfactant was removed by calcination or an ethanol wash. As-synthesized MSU-X silica yielded a broad XRD shoulder. The peak greatly increased in intensity after calcination, coinciding with the appearance of a broad secondary reflection. No evidence of textural mesoporosity was seen from N_2 adsorption-desorption experiments. TEM showed that the channels were regular in diameter but lacked long-range order, suggestive of templating by wormlike micelles. Addition of NaF to a PEO-based synthesis improved XRD intensity as the fluoride ions catalyzed hydrolysis and improved silicate cross-linking [63]. Increasing the synthesis temperature up to 65°C (for Tergitol 15-S-12) could increase pore size by 2.4 nm, accompanied by a decrease in pore wall thickness. This was believed to be due to weaker hydrogen bonding between water and ethylene oxide segments as the temperature approached the surfactant's cloud point. All amphiphilic behavior is lost at the cloud point temperature and phase separation occurs. During higher temperature syntheses, a conformational change from coil to rod in ethylene oxide segments caused both the hydrophilicity and overall size of the EO headgroups to decrease. Micelle curvature decreased as the size of the hydrophobic core increased, leading to thinner pore walls and greater pore diameter. As synthetic temperatures approached the cloud point of a PEO surfactant, XRD peaks decreased in intensity and broadened.

Porous lamellar silicas were templated by vesicular assemblies of neutral diamine bola-amphiphilic surfactants [75]. TEOS was combined with a diamine (carbon chain length of C_8, C_{10}, or C_{12}) in water-ethanol, the product of which was stirred vigorously for 18 h and then air dried. Templates were removed by calcination or ethanol extraction and calcination. The resulting silica structures, designated MSU-V, showed (001) and (002) reflections in their XRD patterns. Electron diffraction patterns were typical of lamellar structures, but the calcined products retained their crystallinity. TEM showed the existence of elliptical multilamellar vesicles with diameters in the range 300–800 nm. N_2 adsorption experiments yielded type I isotherms with type H4 hysteresis loops resembling those of pillared clays. Pore size distribution maxima were located in the micropore region, 0.6–1.2 nm. The materials had high surface areas comparable to those of MCM-41 from their framework microporosity and textural mesoporosity.

The formation mechanism was postulated to be similar to natural biomineralization processes. Electrostatically neutral diamine surfactants assemble into multilamellar vesicles, the multilamellar regions composed of closely packed layers of surfactant separated by water layers. $Si(OC_2H_5)_{4-x}(OH)_x$ species penetrate the vesicular interfaces, diffuse into the multilamellar regions, and participate in hydrogen-bonding interactions with lone electron pairs on the surfactant headgroups. As this is a neutral process, neither electrostatic repulsions between silica oligomer species nor charge matching between surfactants and inorganic species needs to be invoked. The close proximity of the layers and further polymerization of adjacent silicate species result in the growth of parallel silica layers and intergallery pillars. Interpenetration of vesicles provides structural defects that may facilitate the removal of surfactant and water, thus avoiding framework steaming and decomposition during calcination.

The morphologies of MSU-V silicas could be altered by using bola-amphiphiles with C_{16}, C_{18}, or C_{22} chains [76]. Temperatures up to 55°C were required to make the longer surfactants soluble enough to template ordered products. Lamellar order was observed by XRD in as-synthesized products. Materials templated by diamines with C_{16} and C_{18} chains survived calcination, but the d_{001} reflection of the C_{22}-templated structure was only a shoulder. Transmission and scanning electron microscopies displayed diverse morphologies. C_{16}-templated products had a mixture of multilamellar vesicles, elongated multilamellar vesicles, platelike, and spiral ribbon morphologies. It was suggested that vesicles initially formed, decreasing the viscosity of the reaction mixture. Agitation rates increased and produced shear forces that could elongate the vesicles. The elongated vesicles could in turn split to form plates and spiral ribbon structures. C_{18}-templated materials consisted of partially interpenetrating multilamellar vesicles of varying diameters (250–800 nm). C_{22}-templated products were composed of large biocell-like, nonspherical hollow aggregates. Multilamellar regions were present only near the surfactant-water interfaces at the "hollow" cores; dense silica existed at the periphery. N_2 adsorption-desorption experiments indicated that the longer chain-templated materials were mesoporous, very little of it from textural mesoporosity. Longer surfactants templated larger pores.

Structures consisting of randomly ordered cylinders with uniform pore size distributions are observed in many syntheses of mesoporous materials [22]. Both noodlelike disordered and branched network–like disordered structures have been observed. Materials related to MSU-1 have been described by Ryoo et al. [77]. These structures, labelled KIT-1, contain channels with regular diameters, but the channel arrangement is a disordered three-dimensional channel network, interconnected through branches. They were obtained by adding $EDTANa_4$ to the synthesis mixture (sodium silicate, CTACl).

B. Ligand-Assisted Templating of Porous Transition Metal Oxides

Ionic templating such as that used in the synthesis of M41S materials is not easily applied to the synthesis of transition metal oxide mesostructures. The large ionic radii of early transition metals favor the formation of insoluble oxide oligomers that do not react sufficiently with surfactant to form a mesostructure. Ying et al. developed an alternative approach, ligand-assisted templating (LAT), in which the surfactant headgroup is ligated to the metal precursor. The covalent bond between surfactant headgroup and metal ensures a strong interaction before hydrolysis and is maintained throughout condensation and aging processes.

The LAT approach has been successful in the synthesis of various mesostructures of niobium oxides [78–81], titanium oxides [82], tantalum oxides [83], and zirconium oxides [84]. The effects of different headgroups were studied by melting dodecyl aliphatic surfactants in their neutral protic forms with niobium(V) ethoxide in a 0.5:1 ratio, adding excess water to hydrolyze overnight, and aging at 100°C for 5 days [78]. Amino and phosphate headgroups were successful in producing discernible hexagonal structures, aliphatic amines being preferred as they were more easily removed by solution techniques. The researchers determined that the best materials were obtained from the most basic headgroups and those with greater hapticity. Use of dodecylamine yielded materials displaying far greater order than those made with 4-dodecylaniline. Substitution at the amine headgroup with one or two alkyl groups produced a steric effect, weakening the Nb—N bond and the order of the product.

Different niobium oxide mesostructures were obtained from aqueous mixtures of $Nb(OEt)_5$ and dodecylamine by varying the surfactant-to-metal ratio. Dodecylamine/Nb ratios of 0.3–1.0 yielded materials with hexagonal XRD patterns, the intensity and sharpness of the reflections improving with increasing surfactant/Nb ratios. This material type was labeled Nb-TMS1 (niobium transition metal oxide molecular sieve). Nb-TMS2 materials were synthesized using a surfactant/metal ratio of 1.5:1. Nb-TMS2 was a hexagonal phase that could be indexed to a unit cell of the space group $P6_3/mmc$. Surfactant/metal ratios of 1.0–1.25 yielded either a mixed phase or a single phase not as well crystallized as Nb-TMS2. Nb-TMS4, a layered phase analogous to MCM-50, was obtained for a surfactant/Nb ratio of 2.0.

The effects of dodecylamine/niobium(V) ethoxide ratios were studied in ethanolic mixtures to which water was added as a reactant. Atmospheric moisture was allowed to diffuse into an ethanolic mixture of metal alkoxide and dodecylamine surfactant at a set temperature and aged further at 80°C. Varying the surfactant/metal ratios from 0.3 to 2.0 yielded only hexagonal phases. XRD peak intensity and sharpness increased with increasing proportion of surfactant. The observation of ordered products from precursors diluted in ethanol suggests that the mechanistic route to Nb-TMS1 is different from that proposed for MCM-41. Because performed micelles are not known to exist under high dilution in ethanol, it was proposed that Nb-TMS1 forms by self-assembly through condensation as opposed to condensation around preformed micelles. Evidence for a covalent interaction between Nb and N was provided by solution and ^{15}N solid-state MAS NMR studies [79]. Amine nitrogen appears to remain bonded to Nb throughout the course of the synthesis.

Crystallinity was found to be dependent on the rate of vapor diffusion into the ethanolic mixture and the temperature of hydrolysis [78]. Product crystallinity viewed by XRD improved with increasing temperature in the range −78 to 50°C and began to deteriorate at 78°C. Micelle formation is favored at lower temperatures. However, as hydrolysis of the metal alkoxide proceeded at low temperatures there was a tendency for incompletely hydrolyzed gel to separate from the ethanol solution. It solidified completely only upon warming to room temperature and adding more water. The deterioration of crystallinity at 78°C may be explained as higher temperature disruption of the self-assembly process. Slow diffusion of water into the ethanolic $Nb(OEt)_5$-dodecylamine mixtures allowed the researchers to grow monoliths up to several millimeters in length.

The effects of longer surfactant chain lengths and the addition of mesitylene in the vapor-diffused ethanol mixtures were examined. Octadecylamine mixed with $Nb(OEt)_5$ in a surfactant/metal ratio of 0.75 resulted in a hexagonally ordered product; a ratio of 1.0 produced a cubic phase indexed to space group $Pm\bar{3}n$ (Nb-TMS3). A layered phase was obtained with a surfactant/Nb ratio of 1.25. Only hexagonal phases were observed using dodecylamine with the same surfactant/metal ratios. Nb-TMS1 pore sizes determined by N_2 adsorption were 22 Å when templated with dodecylamine and 33 Å when templated with octadecylamine. Addition of mesitylene to the $Nb(OEt)_5$-octadecylamine mixture could increase the pore size to a limit of 39 Å, indicating that the nanotubes have a saturation point for absorption of the swelling agent.

Amine surfactant was typically removed from the materials by protonation in an organic solvent. The most effective method was the addition of one molar equivalent of triflic acid in dimethoxyethane at −78°C. Excess amounts could not be added because Nb-TMS1 materials are acid sensitive. Pyrolytic sublimation and calcination caused structural collapse. Rupture of the Nb—N bond results in open coordination sites that lead to structural rearrangement. These open coordination sites can be blocked by water or solvent in the solvent extraction methods.

Microporous niobium oxide molecular sieves (Nb-TMS5) with hexagonal packing order were prepared by templating with small molecules [80]. $Nb(OEt)_5$ was hydrolyzed in water, forming a suspension of loosely bound niobium-oxyalkoxide precipitates, followed by addition of hexylamine. The reaction mixture was treated hydrothermally at a temperature in the range 25–180°C for 1 to 2 days. Nb-TMS5 could also be templated with butylamine, amylamine, and heptylamine, varying the d_{100} spacing between 17.7 and 25.1 Å. Similarly to mesoporous Nb-TMS1, Nb-TMS5 could also be synthesized in an ethanolic solution to which water was added as a reactant to hydrolyze. Microporous niobia may also be templated by bifunctional molecules, such as 1,12-diaminododecane [81].

Surfactants with phosphate headgroups were used successfully with titanium isopropoxide in the synthesis of hexagonally packed mesoporous titania [82]. Acetylacetone was added to the titanium precursor to lower the rate of hydrolysis and to stabilize the gel. Only amorphous titania or anatase was obtained from reactions without acetylacetone. KCl was present in the aging medium of Ti-TMS1; otherwise a poorly crystalline material developed. As-synthesized Ti-TMS1 had a low hydrothermal stability, degrading at temperatures above 120°C.

The ligand-assisted templating approach was applied to the synthesis of hexagonally packed mesoporous tantalum oxide, Ta-TMS1 [83]. Only primary amines templated materials with ordered mesoporosity. Unlike the case of Nb-TMS materials, varying the surfactant/metal ratio did not produce different phases, nor did it noticeably alter XRD patterns. Using primary amine surfactants to template Nb-TMS and Ta-TMS1 materials requires that the pH be neutral or higher to avoid protonation and ensure a covalent interaction. Ta-TMS1 was synthesized at neutral pH as the material becomes soluble in basic conditions.

VII. MORPHOLOGY OF MESOPOROUS SOLIDS

In typical syntheses of M41S-type materials, fine powders with particle sizes between 0.05 and 2 μm are obtained [7,10]. Several procedures have been developed to control the product morphology and to produce larger uniform particles for chromatographic packing materials, shaped particles, fibers, and films for membranes or sensor applications. With few exceptions these syntheses involve acidic conditions (pH < 2, the isoelectric point of silica) in which interfacial interactions are weaker and the product structure is more closely related to the structure of the pure organic phase.

Mesoporous silica spheres can now be synthesized selectively over a wide range of sizes by surfactant templating methods: 0.4–1 μm (stirred, basic conditions) [85], 2–6 μm (quiescent, acidic conditions, mixed cationic-nonionic surfactants) [86], 1–10 μm (quiescent, acidic conditions) [87], 1–100 μm (hollow; prepared by stirred emulsion templating, acidic conditions [88], 0.1–2 mm transparent hard spheres (stirred, basic conditions cationic surfactant) [89]. Under static conditions the acidity influences the product morphology. At higher acidity, gyroids form by fast and local polymerization of a growing silicate liquid crystal seed; at lower acidity multigranular spheres result due to slower polymerization [87]. Under stirred conditions, the stir rate can affect the size of spheres, as the stirring affects the size of emulsion droplets [89]. Particle shapes can also be influenced by the stir rate, i.e., by shear fluid flow. Static conditions can be amenable for the growth of films. Fibers may form with slow stirring; as the stir rate is increased, spheres are produced, which become hollow at even higher stirring speeds [32b,88,89]. Even under quiescent conditions, shaped mesoporous silica particles can be obtained in dilute, acidic, homogeneous solutions [90]. Shapes that have been observed include fibers, ropes and arcs (with mesoporous channels running parallel to the fiber axis), discoids, gyroids, and spirals (where mesoporous channels are coiled around the particle body). Their particle sizes range from 1 to 70 μm. The synthesis of other inorganic materials with complex shapes derived from monolayer sheets, vesicle spheres, lipid tubes, and surfactant rods has been discussed, for example, in a review article by Mann and Ozin [91]. The appearance of particular mesoporous silica forms has been associated with the existence of dislocation or disclination defects, which initiates the growth and determines the shape of the resulting silicate mesophase [92].

Mesoporous powders have been prepared by spray drying of alkoxide-surfactant solutions [93]. The solutions were atomized into droplets in a heated air stream. Depending on the solution composition, powders either were mesoporous throughout or consisted of hollow spherical particles with mesoporous shells.

Typical fibers produced under acidic conditions have ranged from 1 to 15 μm in diameter and from 100 μm up to 5 cm in length [32b]. Fibers with helix morphology have been observed at a hexane-water interface at low stirring speeds. The mesoporous fibers may be used as optical fibers, as they are optically transparent in the visible region, and channels in fibers are parallel to the length of the fiber [94]. Polymers can be added to the synthesis mixture to increase viscosity during spinning [93]. These polymer fibers are incorporated into the framework.

Mesoporous silica films have been prepared at air-water interfaces, at oil-water interfaces [88], and on solid supports. They are usually formed under dilute, acidic conditions and they can be transparent, in some cases even after calcination [95]. Continuous thin films of ordered mesostructured silica formed at an air-water interface possessed a root-mean-square surface roughness of about 3 Å [96–98]. The channel structure in these films was retained to at least 650°C. It was proposed that the mesoporous silica film formed by silicification of a surface lyotropic silicate mesophase. Consistent with a cooperative assembly mechanism, the resulting films possessed a liquid crystalline texture with channels oriented parallel to the film surface. The proposed formation mechanism involves the collective interaction of a "hemimicellar" surfactant overstructure at the air-water interface and micellar aggregates in solution with polymerizable silicate oligomers. Within the film various channel designs were observed, including concentric circles, herringbones, fingerprints, and hairpin patterns. As in the 3D particles, the patterns are thought to arise from defects, such as disclinations and dislocations. The film thickness can be controlled by adjustment of the pH ($\leq$0.5 μm at low pH, up to 70 μm under less acidic conditions in room temperature reactions). Channels perpendicular to the normal of the film (*P6$_3$/mmc* structure) were synthesized from cationic gemini quaternary surfactants under acidic conditions at an air-water interface or on mica [99].

Various solid supports have been used for the growth of mesoporous films, including silicon [100], quartz [100–102], glass [230], mica [99,103], and pyrolitic graphite [104]. Some epitaxial growth and preferred orientation of the one-dimensional channel structure with pore channels oriented parallel to the substrate surface have been observed [103,105]. For example, for mesoporous silica films grown on the hydrophobic surface of freshly cleaved pyrolitic graphite evidence was given for registry between the film channels and the surface structure of the underlying graphite [104]. Dip-coating or spin-coating methods have been employed [100–104,106,230]. For example, hexagonal mesoporous silica films were prepared by spin coating an aqueous solution containing partially hydrolyzed TMOS and CTACl onto a glass substrate [102]. Lu et al. synthesized both cubic and hexagonal mesoporous silica thin films on silicon by sol-gel dip coating [100]. The starting solution (TEOS, EtOH, water, HCl) contained surfactant (CTAB) at a concentration below the cmc. This concentration increased during solvent evaporation, while the surfactant/silica ratio remained constant. A progression from lamellar to cubic to hexagonal structure was observed during aging or heating as the siloxane framework continued to condense. The resulting films were pinhole free and had accessible pores. Large-headgroup cationic surfactants (CTEABr) or gemini surfactants also lead to three-dimensional accessible pore structures, such as a 3D cubic (*Pm$\bar{3}$n*) or 3D hexagonal (*P6$_3$/mmc*) mesostructure [230].

Martin et al. [107] described a synthesis of mesoporous silica thin films consisting of aggregated submicrometer particles (150–500 nm). The synthesis involved diffusing ammonia into a homogeneous micellar coating solution on a nonporous substrate. Continuous film coverage was accomplished by controlling the formation kinetics by employing TMOS as a silica source, using highly concentrated, homogeneous coating solutions, rapidly diffusing ammonia into the micellar coating solution, and letting the solution evaporate under controlled conditions. The ammonia acts as a catalyst for hydrolysis and condensation. Because of the random orientation of the mesoporous particles, the hexagonally ordered pores were accessible.

VIII. STABILITY OF MESOPOROUS SILICATES

Whereas MCM-41 exhibits good thermal stability, its hydrothermal stability to hot water or steam is poor. The structural loss in boiling water is due to silicate hydrolysis, particularly at low degrees of condensation. Small-pore calcined materials are more hydrolytically stable than products with large pores. For example, MCM-41 calcined at 500°C from a $C_{12}TMA^+$ synthesis system gives a good XRD pattern after 3 h of heating in water at 100°C, whereas a large-pore (~55 Å) material loses its structure under the same conditions. High-temperature calcination can increase the hydrothermal stability of these materials. A large-pore (~55 Å) sample calcined at 800°C shows five or more peaks in the XRD pattern after 2 h of heating in water at 100°C [49].

Both the thermal and hydrothermal stability can be improved by contacting the as-synthesized mesostructure with TEOS or other metal alkoxides that render the surface more hydrophobic and eliminate hydroxide groups at terminal silicate units, the structural points most susceptible to attack by hydrolysis [40,108,109]. However, modification of the direct hydrothermal synthesis has permitted partial stabilization of MCM-41 products. An increase in the wall thickness of aluminosilicate MCM-41 was achieved by decreasing the OH^-/SiO_2 ratio in the synthesis gel, suggesting that the surfactant micelles were surrounded by a thicker silica-alumina coating under these conditions. The thicker walled materials exhibited greater thermal and hydrothermal stability [19]. As mentioned earlier, thicker walls (31–64 Å) have also been obtained in syntheses of hexagonal silica structures templated from PEO-PPO-PEO triblock copolymers, and the thicker walls have resulted in improved hydrothermal stability in boiling water [67]. Syntheses of MSU-G mesostructured silica vesicles (30 Å thick walls and 27–40 Å pores), which are based on neutral double-headed surfactants of the type $C_nH_{2n+1}NH(CH_2)_2NH_2$ (n = 10, 12, 14) with TEOS in ethanol-water mixtures, are stable in boiling water for >150 h [110]. Hydrogen bonding is believed to be the primary interaction force between the surfactant molecules and the inorganic framework walls during the assembly pathway. The stability of the bicontinuous pore structures was attributed to the high degree of cross-linking in the silicate framework (Q^4/Q^3 ratios were 6.2:1) and the thickness of the walls.

Salts (NaCl, KCl, sodium acetate, EDTA sodium salt) added to the synthesis mixture (base synthesis) can also improve the hydrothermal stability of MCM-41 in boiling water, although the hydrothermal stability may also decrease, depending on the type and relative amount of salt added to the reaction mixture [111]. For example, with 3 mol NaCl per mol CTACl, the resulting MCM-41 product is hydrothermally stable in boiling water for 12 h, whereas MCM-41 synthesized without added salt completely loses its structure with this treatment. Increased stability was also observed during calcination in oxygen with no structural loss at 960°C. It is believed that the salts affect the structure of water and therefore the local structure of the surfactant-silicate composites during hydrothermal synthesis. The greater stability results in part from a higher degree of condensation (greater Q^4/Q^3 ratio in the product) when salt was added to the mixture.

IX. MESOPOROUS MATERIALS DERIVED FROM KANEMITE AND OTHER LAYERED PRECURSORS

A different strategy of synthesizing mesoporous silicates was first reported by Yanagisawa et al. in 1990 [112] and has been more fully developed by Inagaki and coworkers [113–120]. The precursor in this synthesis was kanemite ($NaHSi_2O_5 \cdot 3H_2O$), a layered polysilicate composed of ~4 Å thick, single-layered sheets of SiO_4 tetrahedra. Yanagisawa and coworkers prepared a mesoporous structure by intercalation and subsequent condensation reactions. Surfactant-kanemite complexes were formed by cation exchange, where long-chain alkyltrimethylammonium ions replaced interlayer sodium ions, a process that had been previously demonstrated by Beneke and Lagaly [121] for a number of organic cations. During this process, the silicate components rearranged themselves around the surfactant, forming what appeared to be a wormlike channel structure. Surfactant removal by calcination between 700 and 1000°C resulted in the formation of mesoporous solids with surface areas of ~900 m^2/g. As in the synthess of MCM-41, the pore sizes were controllable by altering the chain length of the surfactants. The calcined materials were named FSM-*n*, where *n* refers to the carbon chain length of the organic species used in the synthesis; for example, FSM-16 refers to the product obtained by calcination of a hexadecyltrimethylammonium-kanemite complex [122].

These mesoporous structures were less ordered and had significantly smaller pore volumes than MCM-41 [123]. However, by raising the pH during the initial surfactant ion exchange from 8.5 to 11.5 or 12.3, products with greater hexagonal regularity, narrow pore size distributions, and pore volumes comparable to those of MCM-41 were obtained [114–116]. Although the wall thickness of FSM-16 appeared to be the same as that of MCM-41 (~8 Å after calcination or after extraction of the surfactant with acid) [124], FSM-16 exhibited higher thermal and hydrothermal stability than MCM-41. FSM-16 was thermally stable to at least 900°C. It has been suggested that the improved stability arose from a higher degree of condensation in the silica walls; the Q_4/Q_3 ratio based on ^{29}Si NMR was 88:12 in FSM-16 compared with 45:55 in a typical MCM-41 sample. Based on water vapor adsorption isotherms, the surface of FSM-16 was hydrophobic during the first exposure to water vapor [118].

The mechanism of FSM-16 formation appears to be different from that of MCM-41 [29,116,123–125]. Since the first report of the kanemite-based synthesis, the mechanism has been studied in detail by XRD, ^{29}Si MAS-NMR spectroscopy, HREM, SEM, adsorption, studies and rheology measurements. It is shown schematically in Fig. 4. In the first step of the synthesis, the cationic surfactant is intercalated between kanemite single-layer sheets, displacing interlayer sodium ions. As the previously charged surfactant headgroups interact with the anionic silicate layers, the kanemite layers swell. The pH of the suspension depends on the kanemite-to-surfactant solution ratio (K:S). Initial pH values of 9.0 and 10.8 have been measured for K:S ratios of 1:150 and 1:20 (w/w), respectively [124]. These increase by about 0.5 pH units during the exchange reaction. The extent of surfactant incorporation increases with pH. At this stage of the synthesis (FSM-A), the XRD and ^{29}Si NMR data indicate a transitional state from kanemite to a hexagonal phase. During this early period, in situ XRD also reveals an intermediate lamellar phase [29]. In the optimized FSM-16 synthesis, the kanemite-surfactant complex is then redispersed in water and the pH adjusted to a value of 8.5 (FSM-B). ^{29}Si NMR measurements indicate that condensation between silicate units occurs during this stage. Whereas the kanemite precursor exhibits only a sharp Q_3 resonance, the intercalated complex shows an additional Q_4 peak, which grows in intensity with further heat treatment. The intensity ratio Q_4/Q_3 increases with decreasing pH, indicating that condensation between SiO_4 sheets of kanemite occurs during lower pH treatment. Based on XRD and HREM

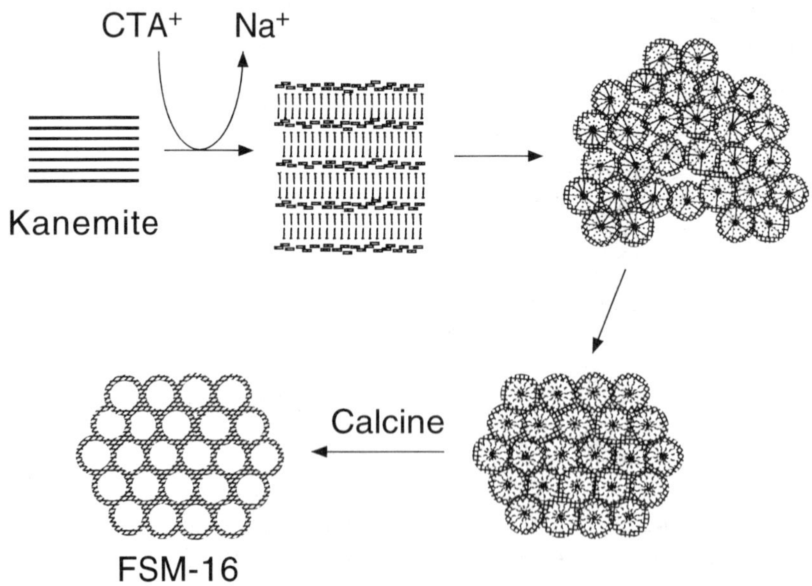

FIG. 4 Proposed mechanism for the transformation of kanemite to the hexagonal FSM-16 phase [115,116,124]. See text for explanation.

information, the layered structure is transformed into a more three-dimensional porous structure with pores that are initially wormlike and varying in length.

Whereas originally a sheet-folding mechanism was proposed, more recent studies indicate that the sheets are most likely broken down into smaller units (especially at high pH) and rearrange locally around the micellelike surfactant aggregates, forming the pore structure. The morphology of the samples does not change from that of the starting kanemite in the preparations carried out at lower pH, but it does change at high pH, indicating that partial dissolution and rearrangement of silicate species occur at higher pH. To prepare pure FSM-16, it is necessary to remove the fraction of the silica precursor that dissolves completely at higher pH values [116]. Condensation of the walls progresses when the sample is heated to 70°C for 3 h (FSM-C) and longer (FSM-D). The extent of conversion from layered structures to more three-dimensional structures increases with exchange time. During the heating process small hexagonal domains are observed; these domains merge with further condensation [125]. After subsequent calcination to remove the surfactant, a highly ordered hexagonal structure is formed (FSM-16). The wall thickness appears to increase from 4 Å (a single silicate layer) to 8 Å (a double layer) upon calcination. Even at the earlier stages, calcination of the kanemite-surfactant complex leads to further wall condensation (an increase in the Q_4/Q_3 ratio) and to the formation of a mesoporous product. The order of this product increases in the series FSM-A through FSM-D. After calcination, even the layered structures or mixed layered-mesoporous phases such as FSM-A, are converted into structures with a single low-angle diffraction line at 47 Å [124]. The layered structures disappear most likely from structural collapse during the calcination, while the mesopores remain intact.

The role of the surfactant in the FSM-16 synthesis is notable. The intercalation reaction is carried out at a much lower surfactant concentration than that used in the original MCM-41 synthesis: 3.2 wt% compared to 26 wt%. This concentration is still higher than the cmc of CTACl in aqueous solution at 25°C (0.05 wt%, 1.4 mM). However, no solution structure has been detected by rheological measurements at the surfactant concentrations used in the layered silicate system [123]. The exchange most likely involves free cation chains rather than micellar aggregates. Surfactant molecules remain intact after intercalation [123]. It has been suggested that the intercalated surfactant molecules reorganize from a bilayer structure to cylindrical micellelike aggregates as they interact with silicate sheet fragments [124]. This hypothesis is supported by the observation that syntheses employing small molecules (e.g., tetramethylammonium chloride) that do not form micelles do not result in mesoporous products [120]. For the most ordered structures, a minimum amount of surfactant must be present in the kanemite-surfactant complex. The degree of cation exchange and thus the surfactant/silica ratios in the complex increase with increasing pH. Hence the poorer hexagonal structure of complexes prepared at pH 8.5 might be attributable to an insufficient exchange ratio of surfactant ions. Removal of the surfactant is possible not only by calcination but also by extraction with acid, where the lost positive charges are compensated by protons. Inagaki et al. [120] examined readsorption of surfactants in FSM-16. They found that FSM-16 adsorbed many types of cationic, anionic, and nonionic surfactants, especially alkyltrimethylammonium (maximum amount 60% of the original template). The amount of alkyltrimethylammonium adsorbed increased steeply at the cmc of the surfactant. It appeared that micellelike surfactants were adsorbed more readily than molecular ones.

Acidic mesoporous structures were obtained by incorporating aluminum during different stages of the FSM-16 synthesis. In zeolites, surface acidity is attributable to isomorphous substitution of Al^{3+} for Si^{4+}. Treatment of the hexadecyltrimethylammonium-kanemite complex with aqueous aluminum trichloride followed by calcination resulted in substitution of some tetrahedral aluminum into the framework, but extra-framework octahedral aluminum was also present. As in the case of aluminosilicate MCM-41, the structural order was lower than in pure silica samples. With a surface area of 600 m^2/g, the aluminum-substituted FSM-16 had an acidity comparable to that of amorphous silica-alumina but only half that of ZSM-5 with the same Al_2O_3 contents [113]. The acidity could be increased to ~0.7 times that of ZSM-5 by adding aluminum in the form of $Al(NO_3)_3 \cdot 9H_2O$ or $NaAlO_2$ during the synthesis of the kanemite precursor [119]. The resulting crystalline layered sodium silicates with Si/Al ratios between 2.5 and 100 were then transformed into mesoporous aluminosilicates by surfactant intercalation and calcination. These products contained only tetrahedral aluminum and exhibited structural regularities higher than those of MCM-41 materials with the same Si/Al ratio. With $Al(NO_3)_3 \cdot 9H_2O$, cristobalite was formed as a side product, but not with $NaAlO_2$.

The intercalation strategy has also been examined with other layered precursors, such as magadiite (multilayered silica) [124] and kenyaite (12 Å thick layers)

[123]. However, mesoporous structures were not obtained in those instances. It appeared important that the precursor layers were easily deformable or easily broken up into smaller units, and hence single layers were preferred. Other clay minerals that have been studied as precursors for mesoporous silicates include sepiolite, hectorite, vermiculite, kaolinite, and chrysotile [120]. High-surface-area mesoporous products with varying degrees of pore regularity were formed from these precursors.

The understanding of the synthesis mechanism, although not complete, has reached a level where product optimization for different applications has become possible. A number of applications for FSM-16 have been studied. These include oxygen adsorption by a porphyrin/FSM-16 complex [126] and the catalytic and magnetic properties of nanostructured Pt clusters in the mesopores of FSM-16 [127]. The favorable thermal stability and the possibility of forming mesoporous aluminosilicates with greater acidity and order than similar MCM-41 materials make FSM-16 a promising material for applications where these characteristics are critical, including catalytic processes.

X. COMPOSITIONS CONTAINING NONSILICON HETEROATOMS

Mobil's original work described mesoporous sieves with silicate and aluminosilicate compositions [7]. Incorporation of aluminum atoms in the walls of aluminosilicate MCM-41 was confirmed by two-dimensional solid-state NMR spectroscopy [128]. Early patents suggested that other elements could also be incorporated in the material [8]. Over the years, many elements have been included in mesoporous materials by doping, grafting, and direct syntheses.

Doping of mesoporous silica with heteroatoms other than aluminum is possible but often permits only very small loadings. Various metals, including Al [7,128–138], Ti [139–143], V [144,145], B [143,146,147], Mn [39,148], Ga [146,149], Sn [150], Y [150], La [150,151], Ce [151], and Fe [146,150,152], have been incorporated in mesoporous silicates during the hydrothermal synthesis. Except for aluminum, the heteroatom content is typically much lower than the silicon content. This topic has been covered in a review [153]. Grafting of metals to surface hydroxyl groups of mesoporous sieves has been carried out using silane-coupling agents or reactive ligands. These methods have also been reviewed [154].

Using direct syntheses, Stucky and coworkers prepared a large number of mesostructured metal oxides, many of which formed lamellar phases [18,31]. These include a zinc phosphate mesophase (lamellar), an alumina mesophase (lamellar), and mesophases of Pb^{2+}, Fe^{2+}, Fe^{3+}, Mg^{2+}, Mn^{2+}, Co^{2+}, Ni^{2+}, Zn^{2+}, Al^{3+}, and Ga^{3+}, obtained with various anionic surfactants (all lamellar, although with Pb^{2+} a hexagonal mesophase was obtained). Many of these structures are probably salt-like and conversion to condensed inorganic structures requires special treatment (see, for example, the section on cluster-surfactant mesostructures later). When anionic sulfonate or phosphate surfactants are employed, they compete with oxide/hydroxide groups for coordination. Thus their anionic polar headgroups can be incorporated into the inorganic framework [18,31].

Some other mesostructured metal oxides that have been reported include oxides of W [155–157], Mo [156,157], Sn [158,159], Zr [32b,160–164], V [156,165,166], Hf [167], Al [20,64,168,169], Mn [170], Nb [78,79], Ti [82], Ta [83], Sb [18,31], and Fe [171]. Examples of mesostructured metallophosphates include Ti [82], Zr [84,160,163,172,173], Al [174–179], Ga [174], and V [166,180,181]. In many of these compositions the surfactant could not be removed without loss of the mesostructure. Some notable exceptions are oxides of Zr, Al, and Ni and phosphates of Ti, Zr, and Al. In a mesostructured titania sample the surfactant could be removed by calcination after stabilization with phosphoric acid [32b]. The pore size distribution remained narrow; however, the pores were disordered. Syntheses employing block copolymers as templates are also amenable to a variety of mesoporous metal oxides with pores up to 140 Å and relatively thick amorphous framework walls [69]. TiO_2, ZrO_2, Al_2O_3, Nb_2O_5, Ta_2O_5, WO_3, HfO_2, SnO_2, and mixed oxides have been synthesized using amphiphilic poly(alkylene oxide) block copolymers ($EO_{20}PO_{70}EO_{20}$) as structure-directing agents together with inorganic salt precursors in nonaqueous solution. The block copolymers are believed to complex to the inorganic metal oxide species by a chelating process. Both two-dimensional hexagonal and three-dimensional cubic ($Im\bar{3}m$) mesostructures were obtained by this method.

Although lamellar phases are often observed for aluminophosphate-surfactant mesostructures (synthesized from pseudoboehmite alumina, phosphoric acid, and neutral amine surfactants) [177], the layers need not be planar. Chenite et al. [176] observed coaxial cylindrical bilayers of inorganic-surfactant phases. These structures resemble vesicles, such as those found in aqueous solutions of phospholipids. Ozin and coworkers [175,182,183] observed hierarchichally structured aluminophosphates in syntheses using nonaqueous glycol-based solvents or glycol-water cosolvent systems. The lamellar aluminophosphates with alkylammonium surfactant bilayers form millimeter-sized spheroidal particles or particles with other —frequently regular [90]—shapes, often with intricate surface patterns. Bilayer and vesicle templating is believed to control the surface architecture in these materials. The glycol cosolvent is proposed to be responsible for inducing bilayer curvature. Some of the product shapes have been compared with radiolarian and diatom microskeletons. Hollow multilayer aluminophosphates were formed from phase-separated water–decylammonium dihydrogenphosphate (DDP)–tetraethylene glycol (TEG) or air-DDP-TEG liquid crystal microemulsions [90]. A modification of this synthesis has been applied to the preparation of a composite of hydroxyapatite and calcium dodecylphosphate lamellar phase as a bone mimetic system [184].

Because surfactant removal from lamellar phases normally does not result in mesoporous products, few applications of lamellar phases have been pursued. However, these phases may be useful to generate inorganic-organic composite mesostructures. These are now starting to be investigated. One example is an iron oxide–surfactant composite that is of interest for magnetic materials [171]. This phase consists of surfactant bilayers separated by iron oxide layers. The material is produced by controlled precipitation and hydrolysis of Fe^{II} or Fe^{III} cations in aqueous solution. The iron oxide layer thickness can be varied between 3 and 20 Å (one to six layers of iron–iron oxide) by adjusting the hydrolysis and solubility of iron oxide via redox chemistry and pH control. When only one or two layers are formed, the structures are surfactant salts; with three to six layers they appear to consist of cross-linked iron oxide layers separated by surfactant bilayers.

XI. CLUSTER-SURFACTANT MESOSTRUCTURES

During investigations of a variety of transition metal oxide precursors for the surfactant-based synthesis of mesostructures, Stein et al. found that tungsten, vanadium, niobium, and molybdenum oxides formed cluster-surfactant salts rather than connected open frameworks [155,185]. For example, a hydrothermal reaction of sodium metatungstate with cetyltrimethylammonium hydroxide produced a TEM image and powder x-ray diffraction patterns that were superficially similar to those of mesoporous silicates. However, closer examination revealed that in this system, the precursors formed thermodynamically stable anionic Keggin ions, leading to the salt $(C_{19}H_{42}N)_6(H_2W_{12}O_{40})$. These are very stable species and show little tendency to condense in the surfactant-salt structures. Hence, any attempt at removing the organic "template" resulted in dense WO_{3-x} phases.

Other salts of anionic metal oxide clusters with cationic surfactants that have been studied include $[C_{12}H_{25}N(CH_3)_3]_6(H_2W_{12}O_{40})\cdot xH_2O$ (a monoclinic structure), as well as salts composed of alternating metal oxide and surfactant layers, $[C_{12}H_{25}N(CH_3)_3]_{0.5}(MoO_{3.25})$ and $[C_{12}H_{25}N(CH_3)_3]_{2/3}V_2O_{5.33}\cdot H_2O$ [156]. The surfactant templates could not be removed in any of these samples without destroying the mesostructures. A single-crystal study of a lamellar decavanadate cluster-surfactant composite $(DTA_4H_2V_{10}O_{28}\cdot 8H_2O)$ confirmed that this material consisted of discrete $V_{10}O_{28}^{6-}$ clusters separated by long-chain cationic surfactant molecules (dodecyltrimethylammonium, DTA) [186]. The surfactants were tilted with respect to the normal of the plane and antiparallel to each other. The clusters were not directly bonded to each other but joined via hydrogen bonding from water molecules to form inorganic layers.

Cationic clusters of the type $MO_4Al_{12}(OH)_{24}(H_2O)_{12}^{7+}$ (M = Al or Ga) (denoted as Al_{13} or $GaAl_{12}$) readily precipitate as ordered salts with anionic surfactants, including sodium dodecyl sulfate (SDS), cetylphosphate, and dodecyl benzenesulfonic acid (DBSA) [174]. All of these salts are layered. The cluster-surfactant superstructure is controlled by electrostatic interactions between the clusters and the polar headgroups of the surfactant molecules and by hydrophobic interactions among the surfactant tails. The high charge density on the inorganic clusters is partially screened by the surfactant charges, resulting in a large enough ratio of chain volume to headgroup area for layering. In the case of Al_{13}-SDS salts, the SDS molecules appear to form interpenetrating bilayers.

As discussed in an earlier section, for M41S materials the long-range order of the oxide arrangement depends in part on the surfactant concentration. Low surfactant concentrations favor a hexagonal phase, and high concentrations lead to lamellar phases. The situation is different for cluster-surfactant salts. Whereas the dimensions of the Keggin salt depend on the hydrocarbon chain length [155,156], a wide range of surfactant concentrations can lead to the observed structure for a given surfactant. In the Keggin-surfactant gel, the cooperative forces between cations and anions become even more significant than in the silicate gels because of the high charge on a Keggin ion and the strong association of the surfactant molecules with a specific Keggin ion. Charge effects may also be responsible for the difference in structures between the tungsten Keggin salt and niobotungstate salts. For example, $(H_2W_{12}O_{40})^{6-}$ clusters form pseudohexagonal salts with cetyltrimethylammonium (CTA) ions, and similar Keggin ions $(PW_{12}O_{40})^{3-}$ or $(H_4PW_{11}O_{39})^{3-}$ form layered phases with CTA. This structural change is not surprising because only half as many surfactant molecules are associated with the −3 cluster as with the −6 cluster [185]. (The size difference of these anions is relatively small; diameters range from 8 to 11 Å.)

In order to investigate the feasibility of converting a cluster-surfactant salt into mesoporous transition-metal oxide networks, Stein et al. developed a "salt-gel" reaction to connect clusters within the salt [155,185]. The principle of the salt-gel reaction is to transform a crystalline array of molecular or low-dimensional building blocks, which are not connected, into a fully connected three-dimensional array by adding a component that forms covalent links between them. The process, shown in Fig. 5, resembles the sol-gel method except that the precursor phase is already ordered via ionic and/or noncovalent interactions. This method was demonstrated for niobotungstate clusters linked with silicate groups. Similar to the vanadium oxide clusters, $Nb_xW_{6-x}O_{19}^{(2+x)-}$ clusters ($x = 2, 3, 4$) formed layered salts with cetyltrimethylammonium cations. The lamellar salt was reacted with TEOS. The TEOS molecules were absorbed, presumably into the hydrophobic portion of the structure, and could be hydrolyzed to form silica within the salt. It was shown by infrared and solid-state NMR double resonance spectra that the silica was anchored to the clusters via covalent Nb—O—Si linkages. After removal of the surfactant by extraction with HCl-ethanol, high-surface-area mesoporous materials were created, although the final network displayed only weak long-range order.

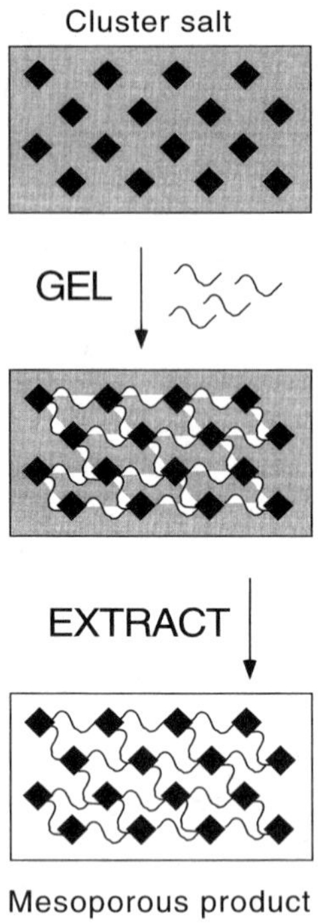

FIG. 5 Principle of the salt-gel method. The diamonds (♦) refer to charged clusters that form a salt with oppositely charged surfactant molecules (gray area). Linking molecules (~), such as TEOS, penetrate the surfactant-cluster salt and form a network with the clusters. Extraction of the surfactant results in a mesoporous product.

Al_{13} or $GaAl_{12}$ cluster-SDS salts could be converted into mesoporous solids by reaction with TEOS to form aluminosilicates [187] or with a buffered phosphate solution followed by acid treatment to synthesize aluminophosphates [174]. In both syntheses it did not appear that the precursor salt was dissolved during the reaction but appeared that the linking groups penetrated the salt structure. During the reaction with phosphate the surfactant/aluminum ratio was reduced from about 0.55 to 0.43. An increase in the ratio of octahedral to tetrahedral aluminum indicated that the clusters disintegrated as an extended aluminophosphate network was formed. Most of the linkages occurred between tetrahedral phosphate groups and octahedral aluminum atoms. The layered material was transformed into a nonlayered mesostructure with a pseudohexagonal array of channels. This transformation was related to two effects. First, upon acid addition some of the clusters broke up into smaller fragments that carried a lower positive charge per fragment than the original Al_{13} clusters. Second, the negative phosphate groups "neutralized" some of the positive charges on the polyoxocations. Thus, they decreased the charge density of the inorganic aluminophosphate region as well as the ratio of chain volume to headgroup area of the surfactant assembly. This resulted in an increased curvature of the inorganic-surfactant interface and the concomitant transformation from a lamellar to a hexagonal phase. The surfactant could be removed by extraction, resulting in open channels with a narrow pore size distribution centered around 17 Å and a BET surface area of 630 m^2/g. This product was demonstrated to be a good anion-exchange material with anion-exchange capacities for chromate and several monoanionic and dianionic organic dyes in the range from 1.3 to 1.6 mEq/g [174b].

In reactions of Al_{13}/SDS salts with TEOS, the alkoxide diffused through the cluster-surfactant salt, reacted with the clusters, and transformed the layered precursor into a nonlamellar mesostructured material with a surface area of 431 m^2/g [187]. Although the resulting products were less ordered than hydrothermally prepared mesoporous aluminosilicates, use of the salt-gel method permitted high aluminum content in the products with Si/Al ratios between 1.97 and 0.86. Similarly to the reaction with phosphate, an increase in the ratio of observable tetrahedral to octahedral aluminum suggested that the clusters disintegrated during the linking process, probably into octahedral units of lower nuclearity. The aluminum was then incorporated into an aluminosilicate framework, forming tetrahedral building blocks. The weakly ordered mesostructure was maintained after removal of the surfactant by calcination.

More ordered mesoporous silicates and aluminosilicates were obtained by a similar two-step method based on silicate and aluminosilicate polyanions as building blocks [41,188]. Fyfe and coworkers investigated double four-ring (D4R) silicate polyanions, $Si_8O_{20}^{8-}$ (Si_8) or the D4R aluminosilicate anion $Al_4Si_4(OH)_8O_{12}^{4-}$ (Al_4Si_4) as precursors, and suggested $AlSi_7$, Al_2Si_6, and Al_3Si_5 as additional feasible cluster precursors. These polyanions possess cubelike structures. Through electrostatic interactions with cationic surfactants they precipitate as mesoscopic salts. The salts were treated with acidic vapors, leading to the condensation of clusters. The Si/Al ratio in the condensed product was controllable by the choice of the precursor, although calcination resulted in partial dealumination and conversion of tetrahedrally coordinated Al into octahedral Al. By increasing the duration of the acidification treatment it was possible to effect structural transformations from lamellar to cubic and hexagonal mesostructures. Based on evidence from FTIR spectra, it was concluded that intact D4R clusters were present in the condensed products (before calcination). Fyfe and Fu rationalized that the structural transformations occurred as the charge densities on the clusters were partially reduced by protonation during the acid vapor treatment. This resulted in larger head-to-chain volume ratios of the surfactant assembly and in an increase in curvature of the interface between the surfactant and the inorganic wall.

The charge density balance at the cluster-surfactant interface could be varied by adjusting the charge of the cluster or that of the surfactant [188]. Although Si_8 and Al_4Si_4 were both cubic molecules, the difference in cluster charge resulted in different initial structure organizations. Notably, the charge-structure relationship was opposite to that observed for the acid Keggin or phosphorus Keggin structures mentioned earlier. The Al_4Si_4 cube with the lower charge formed a rod-based phase with curved interfaces as a surfactant composite under pH conditions where a layered phase predominated for the more highly charged Si_8 precursor. Another reason for the product differences may have been the different terminal groups (O^- vs. OH) present in Si_8 and Al_4Si_4 clusters. The latter condensed much more readily even in the absence of acid during the vapor-phase treatment. The effect of the surfactant charge on the product structure was studied by substituting quaternary ammonium surfactants with primary ammonium ions. CTA/Al_4Si_4 systems were hexagonal, cetylammonium (CAM)/Al_4Si_4 structures remained lamellar. This was interpreted by considering the partial deprotonation of CAM in the basic (pH 11) reaction mixture. As a result, both charged ammonium ions and neutral amine groups could act as structure-directing agents. With reduced headgroup repulsions, the head-to-chain volume decreased, lowering the curvature and leading to layered mesostructures. In addition, the headgroup of each individual CAM ion was smaller than that of a CTA ion.

XII. SURFACTANT-ASSISTED GALLERY FORMATION IN CLAYS

Hybrid materials consisting of layered silicates intercalated with alkylammonium surfactants have been employed in a wide variety of industrial and scientific applications [189]. Using a method related to the salt-gel synthesis, Pinnavaia and coworkers [190] synthesized mesoporous solids, so-called porous clay heterostructures (PCHs), by intercalating cosurfactant mixtures into layered clays and forming silicate structures within the interlayer species of clays. A variety of clays (rectorite, fluorohectorite, magadiite, vermiculite) were intercalated with quaternary ammonium surfactants by an ion-exchange reaction. The expanded products were then stirred in mixtures of TEOS and a long-chain neutral amine. The neutral amine acted as a cosurfactant, leading to further swelling of the clay, and providing accessible space for TEOS molecules. In addition, the amine directed the hydrolysis and condensation of the silicate precursor, which then formed columns between the layers. It is worth noting that both surfactant components were necessary for the formation of a gallery structure. An open pore system was obtained by calcination, and the product was active as a solid acid catalyst, for example, in the selective dehydration of 2-

methylbut-3-yn-2-ol to 2-methylbut-3-yn-1-ene [191]. TEOS/amine ratios ≥7.5 were required to form a uniform phase. With less TEOS two phases were observed: a surfactant-templated PCH and a silica-intercalated derivative with a basal spacing of 12.6 Å. Pore widths in the range 14–22 Å were obtained. The gallery heights depended on the chain lengths of the neutral amine as well as those of the cationic amine but were independent of the type of clay used. Figure 6 shows the proposed mechanism for this synthesis.

Additional selectivity can be imparted on a swelled porous material by employing a layered zeolitic precursor, such as the precursor to the lamellar molecular sieve, MCM-22 [192]. This material consists of aluminosilicate layers connected together by a layer of organic material, such as hexamethyleneimine. The ~25 Å inorganic layers are believed to consist of hexagonal arrays of pockets (rings of 12 ≡Si—O—Si≡ units) on each face, with ~7 Å window openings and depths of ~7 Å [193]. Within the layers, rings of 10 ≡Si—O—Si≡ units provide additional pores. When the MCM-22 precursor is reacted with a hexadecyltrimethylammonium surfactant and base and subsequently pillared with TEOS, a significantly more open structure is created [194]. The product, called MCM-36 [195], contains mesopores in the 30–35 Å range between inorganic layers in addition to the 6–7 Å micropores within the layers. The larger pores provide increased accessibility to the micropores in catalytic reactions. Similar results can be obtained by calcining the surfactant-intercalated MCM-22 precursor directly without any additional TEOS [193]. The more open product exhibited improved performance in the catalytic cracking of heavy gasoil fractions compared with the nondelaminated precursor.

XIII. SURFACE-MODIFIED MESOPOROUS SIEVES

Inclusion chemistry in mesoporous host materials, such as MCM-41 or MCM-48, has been reviewed by Moller and Bein [154] and by Ozin et al. [196]. Modifications of mesoporous solids include ion exchange, complexation, inclusion of molecules, clusters, polymers and enzymes, surface attachment of ligands and functional groups, and cocondensation reactions for the synthesis of hybrid materials. Here we will emphasize examples of functionalized mesoporous sieves where the surfactant molecules play an important role in the surface modification or where the additional surface groups influence the overall structure of the material through interactions with the surfactants.

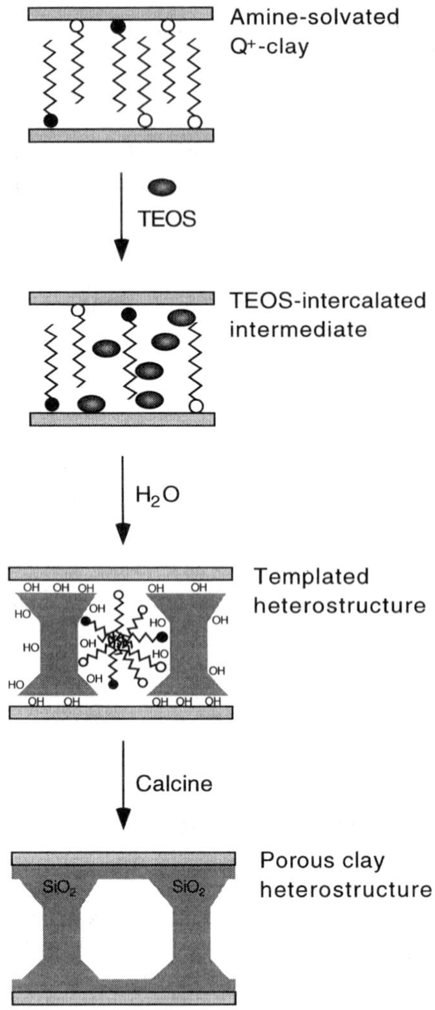

FIG. 6 Proposed mechanism for the formation of a porous clay heterostructure by gallery-templated synthesis. The Q^+-clay is prepared by ion-exchanging Li^+-fluorohectorite (clay layers indicated in gray) with a quaternary ammonium surfactant (chains with filled headgroups). The layers are further swollen with neutral amine molecules (open headgroups). TEOS displaces some of the amine molecules. Rodlike micellar arrays form between the layers and are surrounded by hydrated silica structures. A porous clay heterostructure forms after calcination. (Adapted from Ref. 190.)

A. Surfactants as Space Fillers

Instead of acting as templates or structure-directing agents, surfactant arrays can play the role of space fillers during the synthesis of surface-modified mesoporous sieves. Isolated metal centers have been grafted to mesoporous silica using, for example, the organometallic precursors titanocene dichloride [197,198], manganese decarcarbonyl [199], and vanadium tetrakisisopropoxide [200]. For each of these systems, the organometallic precursor was introduced into the open channels of calcined MCM-41. With certain reactive precursors, such as $TiCl_4$, it has been found that the tendency of $TiCl_4$ to polymerize uncontrollably led to the formation of large titania agglomerates and clogging of channels. The space-filling properties of the surfactant were exploited in the solution-phase grafting of titania onto the pore surface of MCM-41, a method that avoids pore clogging (Fig. 7) [201]. As-synthesized MCM-41, which still contained the surfactant, was dried under vacuum at a temperature low enough to avoid decomposition of the surfactant. The sample was then stirred for up to 24 h in a hexane solution of $TiCl_4$ under nitrogen. During this room temperature reaction, $TiCl_4$ entered the channels, probably by diffusing through the hydrophobic region of the micelles, which should be considered as dynamic rather than static entities. When $TiCl_4$ moved to the polar head of the surfactant, it formed a bright yellow chloride-amine complex. It was hydrolyzed and condensed as it encountered surface hydroxyl groups or trapped water molecules. The physical constraint of the pore walls and the chemical constraint of the surfactant acted as barriers to the formation of large titania agglomerates. After the reaction, the samples were washed with excess hexane under nitrogen. Further condensation occurred during subsequent exposure to air and calcination to remove the surfactant. Extensive characterization indicated that the mesoporous product contained a significant amount of titania (~15 wt%), which in careful preparations was present as isolated $(TiO_2)_n$ clusters ($n \sim 30-70$) that were attached to the silica walls and were uniformly distributed throughout the channels. The products were active catalysts in photooxidation and thermal oxidation reactions. This technique of constraining precursors for surface groups and minimizing uncontrolled polymerization may be amendable to other compositions as long as the precursor molecules can migrate through the micellar array.

B. Hybrid Organic-Inorganic Mesoporous Materials

Organic functionalization by grafting trimethylsilyl derivatives onto mesoporous silicates was first demonstrated in 1990 for materials derived from kanemite [202] and in 1992 for MCM-41 [7,203]. Addition of chlorotrimethylsilane to surface silanol groups allowed reduction of the pore sizes. Other surface attachment methods using silane-coupling agents or reactive ligands have been reviewed [154].

Mann and coworkers [204,205] prepared phenyl- and alkyl-functionalized mesoporous silicas by the co-condensation of the corresponding organosiloxanes with tetraethoxysilane in the presence of the $C_{16}TMABr$ surfactant. Subsequently, mesoporous sieves with reactive functional groups were prepared, including aminopropyl [206,207] and cyanoethyl [207], vinyl [208,209], mercaptopropyl [207,210,211], as well as allyl-, imidazole- and epoxy-functional [207] groups. The ordered mesostructure was significantly disrupted for the last three compositions and was also lower than

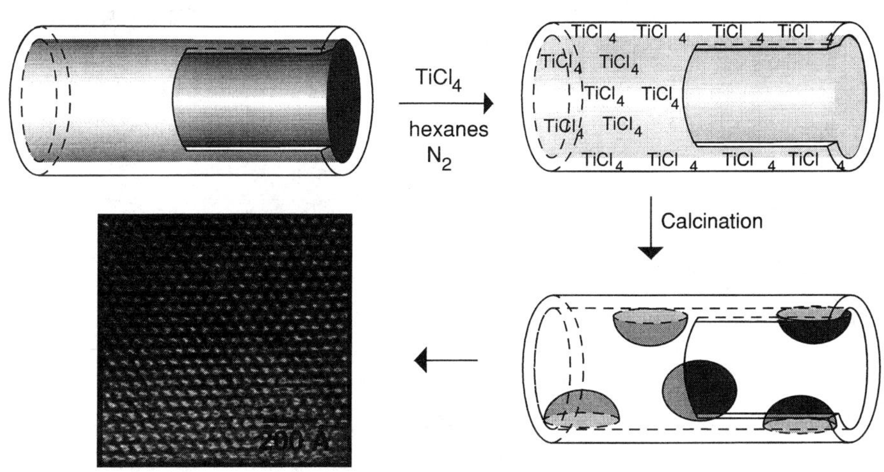

FIG. 7 Schematic of the titania grafting procedure in surfactant-containing MCM-41 [201]. A single pore is shown. The white region represents the silica framework, the gray gradient represents the surfactant, and the flattened hemispheres the grafted titania agglomerates. The TEM image shows the ordered channel structure of the grafted product.

typically observed for MCM-41 for aminopropyl materials. The difference in order may be attributed to different hydrolysis/condensation rates of tetraalkoxysilanes (TAOSs) and organotrialkoxysilanes (OTOSs) as well as to the effect of organic groups on the interaction with the surfactant.

Stein and coworkers synthesized mesoporous silicates with reactive organic functional groups, such as vinyl-MCM-41 [208,208b,212] and thiol-MCM-41 [210], and carried out further functionalizations within the mesoporous channels. Vinyl groups, for example, could be brominated in solution or gas-phase reactions. The intrachannel reactions provided proof for the location of the organic functional groups mainly within the mesopores. Further confirmation of the location of these groups was obtained from nitrogen adsorption measurements and small-angle neutron scanning experiments, using contrast matching techniques [212]. Moller and Bein [213] carried out a similar cocondensation of TMOS and 3-(trimethoxysilyl)propyl methacrylate in the presence of CTACl and also showed that methacrylate groups could be brominated. Thiol-MCM-41 was oxidized to produce a high-surface-area mesoporous support with sulfonic acid groups. This material was active as an acid catalyst for the dehydration and protection of alcohols [210]. Jacobs et al. [211] showed that sulfonic acid–MCM-41 is an efficient catalyst for the formation of polyol esters and bisfurylalkanes.

The cocondensation reactions can be carried out at room temperature. However, elevated temperatures up to about 95°C have led to more ordered products with more highly condensed walls. The surfactant templates were typically removed by acid extraction without breaking the Si—C bond through hydrolysis. In all these cases the organic functional groups were incorporated in the wall structure via covalent linkage to framework silicon atoms. The organic groups appeared to be uniformly distributed [208,208b,212].

Figure 8 is a plot of d_{100} spacings, pore diameters and wall thicknesses for hybrid MCM-41 samples obtained by cocondensation of PTES/TEOS [204] or VTAS/TAOS [208b] with varying concentrations of organosiloxane in the synthesis mixture. As the concentration of organosiloxane increases, both the d_{100} spacings and the pore sizes of the channels are significantly reduced, even when the same surfactant is used (acetyltrimethylammonium surfactant). At the same time, the apparent wall thickness increases, partly because organic groups extend into the channels. One possible reason for the shrinkage in cell dimensions may be a stronger interaction between the phenyl or vinyl groups and the less polar tails of the surfactant molecules, which draw the organic precursors further into the mi-

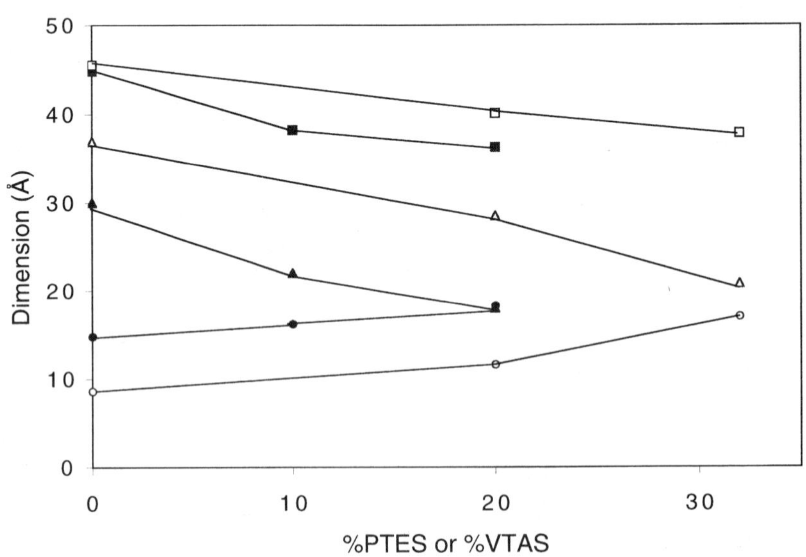

FIG. 8 Effect of organosilane concentration on pore dimensions of organically functionalized MCM-41 prepared by cocondensation of PTES with TEOS, VTMS with TMOS, or VTES with TEOS. PTES/TEOS system: (■) unit cell dimension, ao; (▲) HK pore size; (●) wall thickness. VTMS/TMOS or VTES/TEOS system: (□) unit cell dimension, ao; (△) HK pore size; (○) wall thickness.

celles. Similar pore shrinkage was observed when other organosiloxanes, such as mercaptopropylsiloxane, were included in a direct cocondensation reaction. At too high an organosiloxane concentration the products no longer exhibit hexagonal order.

XIV. TEMPLATING WITH INTERCALATED MICELLES

Functional organic molecules of various degrees of complexity can be encapsulated in inorganic glasses by sol-gel methods [214]. Normally, dense packed, amorphous structures are obtained by these methods. If, on the other hand, the organic molecules are intercalated in surfactant micelles during a typical synthesis of mesoporous silicates, they can be embedded in accessible channels. For example, phthalocyanine complexes [215,216] and porphyrins [217] can be incorporated in mesostructured hosts directly during the hydrothermal synthesis. In these cases the organic complexes are included in the surfactant micelles as the molecular sieve is formed. More ordered mesostructures are typically obtained at high ratios of surfactant to additional guest, where the additional guest provides only a minor perturbation in the structure. In case of an MCM-41 material containing encapsulated cationic porphyrin guests the surfactant molecules could be extracted without removing the porphyrin molecules [217]. The subsequently metallated porphyrin rings were accessible to other guest molecules and were active catalysts in the oxidation of an azo dye. The porphyrin molecules were stabilized to autoxidation by entrapment in the mesoporous sieve.

Enzymes (cytochrome *c*, papain, trypsin) have also been immobilized in MCM-41 to stabilize the enzymes and exploit their catalytic activity [218]. The encapsulated enzymes were catalytically active in the hydrolysis of *N*-α-benzoyl-DL-arginine-4-nitroanilide but to a lesser extent than with other immobilization techniques. Stein, Lim and Schroden (unpublished work) incorporated herring sperm DNA in surfactant-containing synthesis mixtures for mesoporous silicates.* Surfactants are known to form aggregates with DNA and stabilize the DNA over a wider pH and temperature range than in the absence of the surfactants. In room temperature reactions at an initial pH 9, mesostructured products were obtained whose unit cell dimensions were up to 20 Å larger than in products synthesized without DNA. The presence of DNA in the product was confirmed by ultraviolet (UV)-visible spectroscopy (absorption at 255 nm). Subsequent template removal was possible by extraction with methanolic HCl solution.

*The synthesis mixture contained 1 DNA, 10 C_nTMA^+, 75 TEOS, 12.5 NaOH, 10,300 H_2O; n = 12, 14, 16; DNA was in its acid form, $(C_{39}H_{51}O_{25}N_{15}P)_n$, and was not highly polymerized.

XV. CONCLUSIONS

During this decade much progress has been made in controlling the architecture of porous inorganic solids by using organic molecules or molecular aggregates as structure directors, space fillers, or templates. Functional porous nanostructures can now be designed with a high degree of complexity by combining processes from a number of fields, including sol-gel chemistry, zeolite chemistry, surfactant chemistry, colloid chemistry, and polymer chemistry. Beautiful hierarchical structures that were synthesized in the laboratory but resemble skeletons of natural organisms have already been described [175,182,183]. Surface patterning of mesoporous structures is also possible. For example, Ozin and coworkers deposited mesostructured surfactant-silicate assemblies on a gold substrate patterned with self-assembled monolayers (SAMs) of alkanethiolates, using a surfactant-alkanethiolate heterobilayer as a "buffer layer" [90,219].

Other recent methods of creating solids with pores on multiple length scales have involved combining surfactant templating with latex sphere [220–226] or emulsion templating of inorganic oxides [227,228]. The resulting structures contain organized arrays of pores in the submicrometer range, surrounded by walls that are themselves mesoporous [220,222] or even microporous [229]. Combination with stamping techniques permits patterning on an additional length scale [71].

The flexibility in choosing organic, inorganic, or hybrid building blocks and combinations of templates allows one to control the materials properties and to optimize them for each desired application. In the next few years one can expect new developments, not only in the nanometer architecture of porous materials but also in the design of macroscopic shape, morphology, and interfaces with other materials.

ACKNOWLEDGMENTS

Portions of some of the work described here were funded by 3M, Dupont, the David & Lucile Packard Foundation, the McKnight Foundation, the Petroleum Research Fund, the NSF (DMR-9701507 and the MRSEC Program of the NSF under award number

DMR-9809364), and the Office of the Vice President for Research and Dean of the Graduate School of the University of Minnesota.

ABBREVIATIONS

2D: two dimensional
3D: three-dimensional
ATR: attenuated total reflectance
CAM: cetyl ammonium
cmc: critical micelle concentration
C_nTMA . . .: alkyl ammonium group with three methyl functionalities and one linear alkyl chain containing *n* carbon atoms
CTA, CTA^+: cetyltrimethylammonium (ion)
CTACl: cetyltrimethylammonium chloride
CTAB or CTABr: cetyltrimethylammonium bromide
CTAOH: cetyltrimethylammonium hydroxide
CTEABr: cetyltriethylammonium bromide
D4R: double four-ring
EO: ethylene oxide
HMS: hexagonal mesoporous silica
HREM: high-resolution electron microscopy
FTIR: Fourier transform infrared spectroscopy
LAT: ligand-assisted templating
LCT: liquid crystal templating
MAS-NMR spectroscopy: magic angle spinning–nuclear magnetic resonance spectroscopy
M41S: designation by Mobil for the class of materials including MCM-41, MCM-48, MCM-50
MCM-41: hexagonal mesoporous structure
MCM-48: cubic mesoporous structure
MCM-50: lamellar mesoporous structure
PCH: porous clay heterostructure
PEO: polyethylene oxide
PPO: polypropylene oxide
PTES: phenyltriethyoxy silane
Q_3: a Q_3 resonance corresponds to SiO_4 tetrahedra with one terminal hydroxyl group [HO—Si(—OSi≡$)_3$]
Q_4: a Q_4 resonance corresponds to fully connected SiO_4 tetrahedra [Si(—OSi≡$)_4$]
SAM: self-assembled monolayer
SAXS: small-angle x-ray scattering
SDS: sodium dodecyl sulfate
SEM: scanning electron microscopy
T: tetrahedral framework atom such as Si, Al, or P
TAOS: tetraalkoxy silane
TEA^+: tetraethylammonium ion
TEM: transmission electron microscopy
TEOS: tetraethyl orthosilicate, tetraethoxy silane
TMA^+: tetramethylammonium ion
TMAOH: tetramethylammonium hydroxide
TMOS: tetramethyl orthosilicate, tetramethoxy silane
VTAS: vinyl trialkoxy silane
VTES: vinyl triethoxy silane
VTMS: vinyl trimethoxy silane
XANES: X-ray absorption near edge structure
XRD: X-ray diffraction (in this chapter refers to powder x-ray diffraction)

REFERENCES

1. B. M. Lok, T. R. Cannan, and C. A. Messina. Zeolites *3*:282–291 (1983).
2. R. Szostak, *Molecular Sieves*, Van Nostrand Reinhold, New York, 1989.
3. H. Robson and K. P. Lillerud, Microporous Mesoporous Mater. *22*:511–663 (1998).
4. W. M. Meier, D. H. Olson, and C. Baerlocher, Zeolites *17*:1–230 (1996).
5. J. M. Newsam, in *Solid State Chemistry, Compounds* (A. K. Cheetham and P. Day, eds.), Clarendon Press, Oxford, 1992, pp. 234–280.
6. C. T. Kresge, M. E. Leonowicz, W. J. Roth, J. C. Vartuli, and J. S. Beck, Nature *359*:710–712 (1992).
7. J. S. Beck, J. C. Vartuli, W. J. Roth, M. E. Leonowicz, C. T. Kresge, K. D. Schmitt, C. T.-W. Chu, D. H. Olson, E. W. Sheppard, S. B. McCullen, J. B. Higgins, and J. L. Schlenker, J. Am. Chem. Soc. *114*:10834–10843 (1992).
8. C. T. Kresge, M. E. Leonowicz, W. J. Roth, and J. C. Vartuli, U.S. Patent 5,098,684, Mobil Oil Corp. (1992).
9. J. S. Beck, U.S. Patent 5,057,296, Mobil Oil Corp. (1991).
10. A. Monnier, F. Schüth, Q. Huo, D. Kumar, D. Margolese, R. S. Maxwell, G. D. Stucky, M. Krishnamurty, P. Petroff, A. Firouzi, M. Janicke, and B. F. Chmelka, Science *261*:1299–1303 (1993).
11. C.-Y. Chen, H.-X. Li, and M. E. Davis, Microporous Mater. *2*:17–26 (1993).
12. M. Fröba, P. Behrens, J. Wong, G. Engelhardt, C. Haggenmüller, G. van de Goor, M. Rowen, T. Tanaka, and W. Schwieger, Mater. Res. Soc. Symp. Proc. *371*:99–104 (1995).
13. J. S. Beck, J. C. Vartuli, G. J. Kennedy, C. T. Kresge, W. J. Roth, and S. E. Schramm, Chem. Mater. *6*:1816–1821 (1994).
14. X. Chen, L. Huang, and Q. Li, J. Phys. Chem. B *101*: 8460–8467 (1997).
15. G. S. Attard, J. C. Glyde, and C. G. Goltner, Nature *378*:366 (1995).
16. A. Firouzi, D. Kumar, L. M. Bull, T. Besier, P. Sieger, Q. Huo, S. A. Walker, J. A. Zasadzinski, C. Glinka, J. Nicol, D. Margolese, G. D. Stucky, and B. F. Chmelka, Science *267*:1138–1143 (1995).
17. J. C. Vartuli, K. D. Schmitt, C. T. Kresge, W. J. Roth, M. E. Leonowicz, S. B. McCullen, S. D. Hellring, J.

S. Beck, J. L. Schlenker, D. H. Olson, and E. W. Sheppard, Chem. Mater. *6*:2317–2326 (1994).
18. Q. Huo, D. I. Margolese, U. Ciesla, P. Feng, T. E. Gier, P. Sieger, R. Leon, P. M. Petroff, F. Schüth, and G. D. Stucky, Nature *368*:317–321 (1994).
19. N. Coustel, F. Di Renzo, and F. Fajula, J. Chem. Soc. Chem. Commun. 967–968 (1994).
20. S. A. Bagshaw, E. Prouzet, and T. J. Pinnavaia, Science *269*:1242–1244 (1995).
21. C. T. W. Chu and C. T. Kresge, U.S. Patent 5,104,515, Mobil Oil Corp. (1992).
22. M. E. Davis, C. Y. Chen, S. L. Burkett, and R. F. Lobo, Mater. Res. Soc. Symp. Proc. *346*:831–842 (1994).
23. C.-Y. Chen, S. L. Burkett, H.-X. Li, and M. E. Davis, Microporous Mater. *2*:27–34 (1993).
24. A. Firouzi, D. J. Schaefer, S. H. Tolbert, G. D. Stucky, and B. F. Chmelka, J. Am. Chem. Soc. *119*:9466–9477 (1997).
25. A. Firouzi, F. Atef, A. G. Oertli, G. D. Stucky, and B. F. Chmelka, J. Am. Chem. Soc. *119*:3596–3610 (1997).
26. C.-F. Cheng, Z. Luan, and J. Klinowski, Langmuir *11*: 2815–2819 (1995).
27. A. W. Ralston, D. N. Eggengerger, H. J. Harwood, and P. L. DuBrow, J. Am. Chem. Soc. *69*:883 (1947).
28. C. F. Cheng, H. He, W. Zhou, and J. Klinowski, Chem. Phys. Lett. *244*:117–120 (1995).
29. S. O'Brien, R. J. Francis, S. J. Price, D. O'Hare, S. M. Clark, N. Okazaki, and K. Kuroda, J. Chem. Soc. Chem. Commun. 2423–2424 (1995).
30. A. Matijasic, A. C. Voegtlin, J. Patarin, J. L. Guth, and L. Huve, Chem. Commun. 1123–1124 (1996).
31. Q. Huo, D. I. Margolese, U. Ciesla, D. G. Demuth, P. Feng, T. E. Gier, P. Sieger, A. Firouzi, B. F. Chmelka, F. Schüth, and G. D. Stucky, Chem. Mater. *6*:1176–1191 (1994).
32. (a) G. D. Stucky, A. Monnier, F. Schüth, Q. Huo, D. Margolese, D. Kumar, M. Krishnamurty, P. Petroff, A. Firouzi, M. Janicke, and B. F. Chmelka, Mol. Cryst. Liq. Cryst. *1994*:187–200 (1994). (b) F. Schüth, U. Ciesla, S. Schacht, M. Thieme, Q. Huo, G. D. Stucky, Materials Research Bulletin *34*:483–494 (1999).
33. A. A. Romero, M. D. Alba, W. Zhou, and J. Klinowski, J. Phys. Chem. B *101*:5294–5300 (1997).
34. O. Regev, Langmuir *12*:4940–4944 (1996).
35. D. C. Calabro, E. W. Valyocsik, and F. X. Ryan, Microporous Mater. *7*:243–259 (1996).
36. R. Schmidt, M. Stöcker, D. Akporiaye, E. H. Tørstad, and A. Olsen, Microporous Mater. *5*:1–7 (1995).
37. M. W. Anderson, Zeolites *19*:220–227 (1997).
38. J. Xu, Z. Luan, H. He, W. Zhou, and L. Kevan, Chem. Mater. *10*:3690–3698 (1998).
39. D. Zhao and D. Goldfarb, J. Chem. Soc. Chem. Commun. 875–876 (1995).
40. Q. Huo, D. I. Margolese, and G. D. Stucky, Chem. Mater. *8*:1147–1160 (1996).
41. C. A. Fyfe and G. Fu, J. Am. Chem. Soc. *117*:9709–9714 (1995).
42. K. W. Gallis and C. C. Landry, Chem. Mater. *9*:2035–2038 (1997).
43. W. Z. Zhang and T. J. Pinnavaia, Catal. Lett. *38*:261 (1996).
44. R. Schmidt, H. Junggreen, and M. Stöcker, Chem. Commun. 875 (1996).
45. J. Lujano, Y. Romero, and J. Carrazza, in *Zeolites: A Refined Tool for Designing Catalytic Sites* (L. Bonneviot and S. Kaliaguine, eds.), Elsevier Science, Amsterdam, 1995, pp. 489–494.
46. R. Schmidt, M. Stöcker, and O. H. Ellestad, in *Zeolites: A Refined Tool for Designing Catalytic Sites* (L. Bonneviot and S. Kaliaguine, eds.), Elsevier Science, Amsterdam, 1995, pp. 149–156.
47. J. M. Kim, S. K. Kim, and R. Ryoo, Chem. Commun. 259–260 (1998).
48. J. N. Israelachvili, D. J. Mitchell, and B. W. Ninham, J. Chem. Soc. Faraday Trans. 2 *72*:1525–1568 (1976).
49. G. D. Stucky, Q. Huo, A. Firouzi, B. F. Chmelka, S. Schacht, I. G. Voigt-Martin, and F. Schüth. Proceedings, 11th International Zeolite Association Conference, Seoul, Korea, 1996.
50. S. Manne, T. E. Schäffer, Q. Huo, P. K. Hansma, D. E. Morse, G. D. Stucky, and I. A. Aksay. Langmuir *13*:6382–6387 (1997).
51. P. Yang, D. Zhao, B. F. Chmelka, and G. D. Stucky, Chem. Mater. *10*:2033–2036 (1998).
52. U. Henriksson, E. S. Blackmore, G. J. T. Tiddy, and O. Soderman, J. Phys. Chem. *96*:3894 (1992).
53. N. Ulagappan, Neeraj, B. V. N. Raju, and C. N. R. Rao, Chem. Commun. 2243–2244 (1996).
54. Q. Huo, R. Leon, P. M. Petroff, and G. D. Stucky, Science *268*:1324–1327 (1995).
55. M. T. Anderson, J. E. Martin, J. G. Odinek, and P. P. Newcomer, Chem. Mater. *1998*:1490–1500 (1998).
56. A. Navrotsky, I. Petrovic, Y. Hu, C. Y. Chen, and M. E. Davis, Microporous Mater. *4*:95–98 (1995).
57. C. J. Brinker and G. W. Scherer, *Sol-Gel Science: The Physics and Chemistry of Sol-Gel Processing*, Academic Press, San Diego, 1990.
58. S. H. Tolbert, A. Firouzi, G. D. Stucky, and B. F. Chmelka, Science *278*:264–268 (1997).
59. D. Khushalani, A. Kuperman, N. Coombs, and G. A. Ozin, Chem. Mater. *8*:2188–2193 (1996).
60. D. Khushalani, A. Kuperman, G. A. Ozin, K. Tanaka, J. Garcés, M. M. Olken, and N. Coombs, Adv. Mater. *7*:842–846 (1995).
61. A. Corma, Q. Kan, M. Navarro, J. Pérez-Pariente, and F. Rey, Chem. Mater. *9*:2123–2126 (1997).
62. C. F. Cheng, W. Zhou, and J. Klinowski, Chem. Phys. Lett. *263*:247–252 (1996).
63. E. Prouzet and T. J. Pinnavaia, Angew. Chem. Int. Ed. Engl. *36*:516–518 (1997).
64. S. A. Bagshaw and T. J. Pinnavaia, Angew. Chem. Int. Ed. Engl. *35*:1102–1105 (1996).

65. D. Zhao, J. Feng, Q. Huo, N. Melosh, G. H. Fredrickson, B. F. Chmelka, and G. D. Stucky, Science *279*:548–552 (1998).
66. D. Zhao, Q. Huo, J. Feng, B. F. Chmelka, and G. D. Stucky, J. Am. Chem. Soc. *120*:6024–6036 (1998).
67. D. Zhao, P. Yang, Q. Huo, B. F. Chmelka, and G. D. Stucky, Curr. Opin. Solid State Mater. Sci. *3*:111–121 (1998).
68. C. G. Göltner, S. Henke, M. C. Weissberger, and M. Antonietti, Angew. Chem. *110*:633–636 (1998).
69. P. Yang, D. Zhao, D. I. Margolese, B. F. Chmelka, and G. D. Stucky, Nature *396*:152–155 (1998).
70. P. Schmidt-Winkel, W. W. Lukens, D. Zhao, P. Yang, B. F. Chmelka, and G. D. Stucky, J. Am. Chem. Soc. *121*:254–255 (1999).
71. P. Yang, T. Deng, D. Zhao, P. Feng, D. Pine, B. F. Chmelka, G. M. Whitesides, and G. D. Stucky, Science *282*:2244–2246 (1998).
72. P. T. Tanev and T. J. Pinnavaia, Science *267*:865–867 (1995).
73. P. T. Tanev and T. J. Pinnavaia, Chem. Mater. *8*:2068–2079 (1996).
74. W. Zhang, T. R. Pauly, and T. J. Pinnavaia, Chem. Mater. *9*:2491–2498 (1997).
75. P. T. Tanev and T. J. Pinnavaia, Science *271*:1267–1269 (1996).
76. P. T. Tanev, Y. Liang, and T. J. Pinnavaia, J. Am. Chem. Soc. *119*:8616–8624 (1997).
77. R. Ryoo, J. M. Kim, C. H. Ko, and C. H. Shin, J. Phys. Chem. *100*:17718–17721 (1996).
78. D. M. Antonelli, A. Nakahira, and J. Y. Ying, Inorg. Chem. *35*:3126–3136 (1996).
79. D. M. Antonelli and J. Y. Ying, Angew. Chem. Int. Ed. Engl. *35*:426–430 (1996).
80. T. Sun and J. Y. Ying, Nature *389*:704–706 (1997).
81. T. Sun and J. Y. Ying, Angew. Chem. Int. Ed. *37*:664–667 (1998).
82. D. M. Antonelli and J. Y. Ying, Angew. Chem. Int. Ed. Engl. *34*:2014–2017 (1995).
83. D. M. Antonelli and J. Y. Ying, Chem. Mater. *8*:874–881 (1996).
84. M. S. Wong and J. Y. Ying, Chem. Mater. *10*:2067–2077 (1998).
85. M. Grun, I. Lauer, and K. K. Unger, Adv. Mater. *9*:254 (1997).
86. L. Qi, J. Ma, H. Cheng, and Z. Zhao, Chem. Mater. *10*:1623–1626 (1998).
87. H. Yang, G. Vovk, N. Coombs, I. Sokolov, and G. A. Ozin, J. Mater. Chem. *8*:743–750 (1998).
88. S. Schacht, Q. Huo, I. G. Voigt-Martin, G. D. Stucky, and F. Schüth, Science *273*:768–771 (1996).
89. Q. Huo, J. Feng, F. Schüth, and G. D. Stucky, Chem. Mater. *9*:14–17 (1997).
90. N. Coombs, D. Khushalani, S. Oliver, G. A. Ozin, G. C. Shen, I. Sokolov, and H. Yang, J. Chem. Soc. Dalton Trans. 3941–3952 (1997).
91. S. Mann and G. A. Ozin, Nature *382*:313–318 (1996).
92. H. Yang, G. A. Ozin, and C. T. Kresge, Adv. Mater. *10*:883–887 (1998).
93. P. J. Bruinsma, A. Y. Kim, J. Liu, and S. Baskaran, Chem. Mater. *9*:2507–2512 (1997).
94. Q. Huo, D. Zhao, J. Feng, K. Weston, S. K. Buratto, G. D. Stucky, S. Schacht, and F. Schüuth, Adv. Mater. *9*:974–978 (1997).
95. R. Ryoo, C. H. Ko, S. J. Cho, and J. M. Kim, J. Phys. Chem. B *101*:10610–10613 (1997).
96. H. Yang, N. Coombs, and G. A. Ozin, J. Mater. Chem. *8*:1205–1211 (1998).
97. H. Yang, N. Coombs, Ö. Dag, I. Sokolov, and G. A. Ozin, J. Mater. Chem. *7*:1755–1761 (1997).
98. H. Yang, N. Coombs, I. Sokolov, and G. A. Ozin, Nature *381*:589–592 (1996).
99. S. H. Tolbert, T. E. Schäffer, J. Feng, P. K. Hansma, and G. D. Stucky, Chem. Mater. *9*:1962–1967 (1997).
100. Y. Lu, R. Ganguli, C. A. Drewien, M. T. Anderson, C. J. Brinker, W. Gong, Y. Guo, H. Soyez, B. Dunn, M. H. Huang, and J. I. Zink, Nature *389*:364–368 (1997).
101. M. Ogawa, J. Am. Chem. Soc. *116*:7941 (1994).
102. M. Ogawa, Chem. Commun. 1149–1150 (1996).
103. H. Yang, A. Kuperman, N. Coombs, S. Mamiche-Afara, and G. A. Ozin, Nature *379*:703–705 (1996).
104. H. Yang, N. Coombs, I. Sokolov, and G. A. Ozin, J. Mater. Chem. *7*:1285–1290 (1997).
105. I. A. Aksay, M. Trau, S. Manne, I. Honma, N. Yao, L. Zhou, P. Fenter, P. M. Eisenberger, and S. M. Gruner, Science *273*:892–898 (1996).
106. D. Zhao, P. Yang, N. Melosh, J. Feng, B. F. Chmelka, and G. D. Stucky, Adv. Mater. *10*:1380 (1998).
107. J. E. Martin, M. T. Anderson, J. Odinek, and P. Newcomer, Langmuir *13*:4133–4141 (1997).
108. S. B. McCullen and J. C. Vartuli, U.S. Patent 5,156,829, Mobil Oil Corp. (1992).
109. T. Tatsumi, K. A. Koyano, Y. Tanaka, and S. Nakata, Stud. Surf. Sci. Catal. *117*:143–150 (1998).
110. S. S. Kim, W. Zhang, and T. J. Pinnavaia, Science *282*:1302–1305 (1998).
111. R. Ryoo and S. Jun, J. Phys. Chem. B *101*:317–320 (1997).
112. T. Yanagisawa, T. Shimizu, K. Kuroda, and C. Kato, Bull. Chem. Soc. Jpn. *63*:988–992 (1990).
113. S. Inagaki, Y. Fukushima, A. Okada, T. Kurauchi, K. Kuroda, and C. Kato, in *Proceedings of the 9th International Zeolite Conference* (R. von Ballmoos, ed.), Butterworth-Heinemann, Montreal, 1992, pp. 305–311.
114. S. Inagaki, Y. Fukushima, and K. Kuroda, J. Chem. Soc. Chem. Commun. 680–682 (1993).
115. S. Inagaki, Y. Fukushima, and K. Kuroda, Stud. Surf. Sci. Catal. *84*:125–132 (1994).
116. S. Inagaki, A. Koiwai, N. Suzuki, Y. Fukushima, and K. Kuroda, Bull. Chem. Soc. Jpn. *69*:1449–1457 (1996).

117. S. Inagaki, Y. Sakamoto, Y. Fukushima, and O. Terasaki, Chem. Mater. 2089–2095 (1996).
118. S. Inagaki, Y. Fukushima, K. Kuroda, and K. Kuroda, J. Colloid Interface Sci. *180*:623–624 (1996).
119. S. Inagaki, Y. Yamada, and Y. Fukushima, Stud. Surf. Sci. Catal. *105*:109–116 (1997).
120. S. Inagaki, S. Ogata, Y. Goto, and Y. Fukushima, Stud. Surf. Sci. Catal. *117*:65–76 (1998).
121. K. Beneke and G. Lagaly, Am. Mineral. *62*:763–771 (1977).
122. P. J. Branton, K. Kaneko, N. Setoyama, K. S. W. Sing, S. Inagaki, and Y. Fukushima, Langmuir *12*:599–600 (1996).
123. J. C. Vartuli, C. T. Kresge, M. E. Leonowicz, A. S. Chu, S. B. McCullen, I. D. Johnson, and E. W. Sheppard, Chem. Mater. *6*:2070–2077 (1994).
124. C.-Y. Chen, S.-Q. Xiao, and M. E. Davis, Microporous Mater. *4*:1–20 (1995).
125. Y. Sakamoto, S. Inagaki, T. Ohsuna, N. Ohnishi, Y. Fukushima, Y. Nozue, and O. Terasaki, Microporous Mesoporous Mater. *21*:589–596 (1998).
126. K. Moller, T. Bein, and R. X. Fischer, Chem. Mater. *11*:665–673 (1999).
127. M. Sasaki, M. Osada, N. Sugimoto, S. Inagaki, Y. Fukushima, A. Fukuoka, and M. Ichikawa, Microporous Mesoporous Mater. *21*:597–606 (1998).
128. M. T. Janicke, C. C. Landry, S. C. Christiansen, D. Kumar, G. D. Stucky, and B. F. Chmelka, J. Am. Chem. Soc. *120*:6940–6951 (1998).
129. K. R. Kloetstra, H. W. Zandbergen, and H. van Bekkum, Catal. Lett. *33*:157–163 (1995).
130. R. B. Borade and A. Clearfield, Catal. Lett. *31*:267–272 (1995).
131. Z. Luan, C.-F. Cheng, W. Zhou, and J. Klinowski, J. Phys. Chem. *99*:1018–1024 (1995).
132. Z. H. Luan, H. Y. He, W. Z. Zhou, C. F. Cheng, and J. Klinowski, J. Chem. Soc. Faraday Trans. *91*:2955–2959 (1995).
133. Z. Luan, C.-F. Cheng, H. He, and J. Klinowski, J. Phys. Chem. *99*:10590–10593 (1995).
134. A. Corma, V. Fornes, M. T. Navarro, and J. Pérez-Pariente, J. Catal. *148*:569–574 (1994).
135. A. Stein and B. Holland, J. Porous Mater. *3*:83–92 (1996).
136. M. Janicke, D. Kumar, G. D. Stucky, and B. F. Chmelka, in *Zeolites and Related Microporous Materials: State of the Art 1994* (J. Weitkamp et al., eds.), Elsevier Science, Amsterdam, 1994, pp. 243–250.
137. R. Schmidt, D. Akporiaye, M. Stöcker, and O. H. Ellestad, J. Chem. Soc. Chem. Commun. 1493–1494 (1994).
138. R. Schmidt, D. Akporiaye, M. Stöcker, and O. H. Ellestad, in *Zeolites and Related Microporous Materials: State of the Art 1994* (J. Weitkamp et al., eds.), Elsevier Science, Amsterdam, 1994, pp. 61–68.
139. A. Corma, M. T. Navarro, and J. P. Pariente, J. Chem. Soc. Chem. Commun. 147–148 (1994).
140. T. Blasco, A. Corma, M. T. Navarro, and J. P. Pariente, J. Catal. *156*:65–74 (1995).
141. W. Zhang, M. Fröba, J. Wang, P. T. Tanev, J. Wong, and T. J. Pinnavaia, J. Am. Chem. Soc. *118*:9164–9171 (1996).
142. M. Morey, A. Davidson, and G. Stucky, Microporous Mater. *6*:99–104 (1996).
143. P. T. Tanev, M. Chibwe, and T. J. Pinnavaia, Nature *368*:321–323 (1994).
144. K. M. Reddy, I. Moudrakovski, and A. Sayari, J. Chem. Soc. Chem. Commun. 1059–1060 (1994).
145. J. S. Reddy and A. Sayari, J. Chem. Soc. Chem. Commun. 2231–2232 (1995).
146. A. Tuel and S. Gontier, Chem. Mater. *8*:114–122 (1996).
147. A. Sayari, I. Moudrakovski, C. Danumah, C. I. Ratcliffe, J. A. Ripmeester, and K. F. Preston, J. Phys. Chem. *99*:16373–16379 (1995).
148. M. Yonemitsu, Y. Tanaka, and M. Iwamoto, Chem. Mater. *9*:2679–2681 (1997).
149. Z. Gabelica, J. M. Clacens, R. Sobry, and G. Van den Bossche, in *Zeolites: A Refined Tool for Designing Catalytic Sites* (L. Bonneviot and S. Kaliaguine, eds.), Elsevier Science, Amsterdam, 1995, pp. 143–147.
150. T. M. Abdel-Fattah and T. J. Pinnavaia, Chem. Commun. 665–666 (1996).
151. W. Zhang and T. J. Pinnavaia, Chem. Commun. 1185–1186 (1998).
152. B. Echchadhed, A. Moen, D. Nicholson, and L. Bonneviot, Chem. Mater. *9*:1716–1719 (1997).
153. A. Corma, Chem. Rev. *97*:2373–2419 (1997).
154. K. Moller and T. Bein, Chem. Mater. *10*:2950–2963 (1998).
155. A. Stein, M. Fendorf, T. P. Jarvie, K. T. Mueller, A. J. Benesi, and T. E. Mallouk, Chem. Mater. *7*:304–313 (1995).
156. G. G. Janauer, A. Dobley, J. Guo, P. Zavalij, and M. S. Whittingham, Chem. Mater. *8*:2096–2101 (1996).
157. U. Ciesla, D. Demuth, R. Leon, P. Petroff, G. Stucky, K. Unger, and F. Schüth, J. Chem. Soc. Chem. Commun. 1387–1388 (1994).
158. N. Ulagappan and C. N. R. Rao, Chem. Commun. 1047–1048 (1996).
159. K. G. Severin, T. M. Abdel-Fattah, and T. J. Pinnavaia, Chem. Commun. 1471–1472 (1998).
160. U. Ciesla, S. Schacht, G. D. Stucky, K. K. Unger, and F. Schüth, Angew. Chem. Int. Ed. Engl. *35*:541–543 (1996).
161. A. Y. Kim, P. J. Bruinsma, Y. L. Shen, and J. Liu, Mater. Res. Soc. Symp. Proc. *435*:131 (1996).
162. J. S. Reddy and A. Sayari, Catal. Lett. *38*:219 (1996).
163. A. Kim, P. Bruinsma, Y. Chen, L. Q. Wang, and J. Liu, Chem. Commun. 161–162 (1997).
164. J. A. Knowles and M. J. Hudson, J. Chem. Soc. Chem. Commun. 2083–2084 (1995).
165. V. Luca, D. J. MacLachlan, J. M. Hook, and R. Withers, Chem. Mater. *7*:2220–2223 (1995).

166. P. Liu, I. L. Moudrakovski, J. Liu, and A. Sayari, Chem. Mater. *9*:2513–2520 (1997).
167. P. Liu, J. Liu, and A. Sayari, Chem. Commun. 577 (1997).
168. F. Vaudry, S. Khodabandeh, and M. E. Davis, Chem. Mater. *8*:1451–1464 (1996).
169. M. Yada, M. Machida, and T. Kijima, Chem. Commun. 769–770 (1996).
170. J. Luo and S. L. Suib, Chem. Commun. 1031–1032 (1997).
171. S. H. Tolbert, P. Sieger, G. D. Stucky, S. M. J. Aubin, C.-C. Wu, and D. N. Hendrickson, J. Am. Chem. Soc. *119*:8652–8661 (1997).
172. P. Liu, J. S. Reddy, and A. Sayari, Mater. Res. Soc. Symp. Proc. *431*:101 (1996).
173. M. S. Wong, D. M. Antonelli, and J. Y. Ying, Nanostructured Mater. *9*:165–168 (1997).
174. (a) B. T. Holland, P. K. Isbester, C. F. Blanford, E. J. Munson, and A. Stein, J. Am. Chem. Soc. *119*:6796–6803 (1997). (b) D. A. Kron, B. T. Holland, R. Wipson, C. Maleke, A. Stein, Langmuir *15*:8300–8308 (1999).
175. S. Oliver, N. Coombs, and G. A. Ozin, Adv. Mater. *7*: 931–935 (1995).
176. A. Chenite, Y. L. Page, V. R. Karra, and A. Sayari, Chem. Commun. 413–414 (1996).
177. A. Sayari, I. Moudrakovski, J. S. Reddy, C. I. Ratcliffe, J. A. Ripmeester, and K. F. Preston, Chem. Mater. *8*:2080–2088 (1996).
178. D. Zhao, Z. Luan, and L. Kevan, Chem. Commun. 1009–1010 (1997).
179. P. Feng, Y. Xia, J. Feng, X. Bu, and G. D. Stucky, Chem. Commun. 949–950 (1997).
180. T. Abe, A. Taguchi, and M. Iwamoto, Chem. Mater. *7*:1429–1430 (1995).
181. T. Doi and T. Miyake, Chem. Commun. 1635 (1996).
182. G. A. Ozin and S. Oliver, Adv. Mater. *7*:943–947 (1995).
183. S. Oliver, A. Kuperman, N. Coombs, A. Lough, and G. A. Ozin, Nature *378*:47–50 (1995).
184. G. A. Ozin, N. Varaksa, N. Coombs, J. E. Davies, D. D. Perovic, and M. Ziliox, J. Mater. Chem. *7*:1601–1607 (1997).
185. A. Stein, M. Fendorf, T. P. Jarvie, K. T. Mueller, M. E. Garcia, and T. E. Mallouk, in *Advances in Porous Materials* (S. Komarneni, D. M. Smith, and J. S. Beck, eds.), Materials Research Society, Pittsburgh, 1995, pp. 69–79.
186. G. G. Janauer, A. D. Dobley, P. Y. Zavalij, and M. S. Whittingham, Chem. Mater. *9*:647–649 (1997).
187. B. T. Holland, P. K. Isbester, E. J. Munson, and A. Stein, Mater. Res. Bull. *34*:471–482 (1999).
188. G. Fu, C. A. Fyfe, W. Schwieger, and G. T. Kokotailo, Angew. Chem. Int. Ed. Engl. *34*:1499–1502 (1995).
189. R. A. Vaia, R. K. Teukolsky, and E. P. Giannelis, Chem. Mater. *6*:1017–1022 (1994) and references therein.
190. A. Galarneau, A. Barodawalla, and T. J. Pinnavaia, Nature *374*:529–531 (1995).
191. A. Galarneau, A. Barodawalla, and T. J. Pinnavaia, Chem. Commun. 1661–1662 (1997).
192. M. E. Leonowicz, J. A. Lawton, S. L. Lawton, and M. K. Rubin, Science *264*:1910 (1994).
193. A. Corma, V. Fornes, S. B. Pergher, T. L. M. Maesen, and J. G. Buglass, Nature *396*:353–356 (1998).
194. W. J. Roth, C. T. Kresge, J. C. Vartuli, M. E. Leonowicz, A. S. Fung, and S. B. McCullen, Stud. Surf. Sci. Catal. *94*:301–308 (1995).
195. C. T. Kresge, W. J. Roth, K. G. Simmons, and J. C. Vartuli, U.S. Patent 5,229,341, Mobil Oil Corp. (1993).
196. G. A. Ozin, E. Chomski, D. Khushalani, and M. J. MacLachlan, Curr. Opin. Colloid Surf. Sci. *3*:181–193 (1998).
197. T. Maschmeyer, F. Rey, G. Sankar, and J. M. Thomas, Nature *378*:159–162 (1995).
198. J. V. Walker, M. Morey, H. Carlsson, A. Davidson, G. D. Stucky, and A. Butler, J. Am. Chem. Soc. *119*: 6921–6922 (1997).
199. R. Burch, N. Cruise, D. Gleeson, and S. C. Tsang, Chem. Commun. 951 (1996).
200. M. Morey, A. Davidson, H. Eckert, and G. Stucky, Chem. Mater. *8*:486–492 (1996).
201. B. J. Aronson, C. F. Blanford, and A. Stein, Chem. Mater. *9*:2842–2851 (1997).
202. T. Yanagisawa, T. Shimizu, K. Kuroda, and C. Kato, Bull. Chem. Soc. Jpn. *63*:1535–1537 (1990).
203. J. S. Beck, D. C. Calabro, S. B. McCullen, B. P. Pelrine, K. D. Schmitt, and J. C. Vartuli, U.S. Patent 5,145,816, Mobil Oil Corp. (1992).
204. S. D. Sims, S. L. Burkett, and S. Mann, Mater. Res. Soc. Symp. Proc. *431*:77–82 (1996).
205. S. L. Burkett, S. D. Sims, and S. Mann, Chem. Commun. 1367–1368 (1996).
206. D. J. Macquarrie, Chem. Commun. 1961–1962 (1996).
207. C. E. Fowler, S. L. Burkett, and S. Mann, Chem. Commun. 1769–1770 (1997).
208. (a) M. H. Lim, C. F. Blanford, and A. Stein, J. Am. Chem. Soc. *119*:4090–4091 (1997). (b) M. H. Lim, A. Stein, Chem. Mater. *11*:3302–3308 (1999).
209. F. Babonneau, L. Leite, and S. Fontlupt, J. Mater. Chem. *9*:175–178 (1999).
210. M. H. Lim, C. F. Blanford, and A. Stein, Chem. Mater. *10*:467–470 (1998).
211. W. M. Van Rhijn, D. E. De Vos, B. F. Sels, W. D. Bosaert, and P. A. Jacobs, Chem. Commun. 317–318 (1998).
212. M. H. Lim and A. Stein, Mater. Res. Soc. Symp. Proc. *519*:89–94 (1998).
213. K. Moller and T. Bein, Stud. Surf. Sci. Catal. *117*:53–64 (1998).
214. B. C. Dave, B. Dunn, J. S. Valentine, and J. I. Zink, Anal. Chem. *66*:1120A–1127A (1994).

215. I. Honma and H. S. Zhou, Chem. Mater. *10*:103–108 (1998).
216. R. Hoppe, A. Ortlam, J. Rathousky, G. Schulz-Ekloff, and A. Zukal, Microporous Mater. *8*:267–273 (1997).
217. B. T. Holland, C. Walkup, and A. Stein, J. Phys. Chem. B *102*:4301–4309 (1998).
218. J. F. Diaz and K. J. Balkus Jr., J. Mol. Catal. B Enzymatic *2*:115 (1996).
219. H. Yang, N. Coombs, and G. A. Ozin, Adv. Mater. *9*: 811–814 (1997).
220. M. Antonietti, B. Berton, C. Göltner, and H. P. Hentze, Adv. Mater. *10*:154–159 (1998).
221. B. T. Holland, C. F. Blanford, and A. Stein, Science *281*:538–540 (1998).
222. B. T. Holland, C. F. Blanford, T. Do, and A. Stein, Chem. Mater. *11*:795–805 (1999).
223. C. F. Blanford, T. N. Do, B. T. Holland, and A. Stein, in Mater. Res. Soc. Symp. Proc. (P. W. Lednor, D. A. Nagaki, and L. T. Thompson, eds.), MRS, Pittsburgh, *549*:61–66 (1999).
224. J. E. G. J. Wijnhoven and W. L. Vos, Science *281*: 802–804 (1998).
225. O. D. Velev, T. A. Jede, R. F. Lobo, and A. M. Lenhoff, Nature *389*:447–448 (1997).
226. O. D. Velev, T. A. Jede, R. F. Lobo, and A. M. Lenhoff, Chem. Mater. *10*:3597–3602 (1998).
227. A. Imhof and D. J. Pine, Nature *389*:948–951 (1997).
228. A. Imhof and D. J. Pine, Adv. Mater. *10*:697–700 (1998).
229. B. T. Holland, L. Abrams, and A. Stein, J. Am. Chem. Soc. *121*:4308–4309 (1999).
230. D. Zhao, P. Yang, D. I. Margolese, B. F. Chmelka, G. D. Stucky, Chem. Commun. 2499–2500 (1998).

Index